Florida Ethnobotany

Florida Ethnobotany

Daniel F. Austin

Fairchild Tropical Garden, Coral Gables, Florida
Arizona-Sonora Desert Museum, Tucson, Arizona

with more than 500 species illustrated by

Penelope N. Honychurch, Priscilla Fawcett,
Vivian Frazier, Regina O. Hughes, and others

THE SOCIETY FOR ECONOMIC BOTANY EB FOUNDED 1959

CRC
CRC PRESS
Boca Raton London New York Washington, D.C.

FLORIDA NATIVE PLANT SOCIETY serenoa repens

COVER ART
Front Cover: *Passiflora incarnata*

Back Cover, Top to Bottom: *Ipomoea alba* —visited by a hawkmoth (*Manduca sexta*); *Agave decipiens*; *Cephalanthus occidenalis*; *Piscidia piscipula*

Senior Editor: John Sulzycki
Project Editor: Christine Andreasen
Developmental Editor: Donna Coggshall
Production Coordinator: Patricia Roberson
Assistant Project Editor: Katy Smith
Typesetting Services: Datapage, Dublin, Ireland
Indexing Services: Techni-Dexes
Cover Designer: Elise Weinger
Marketing Manager: Carolyn Spence

Library of Congress Cataloging-in-Publication Data

Austin, Daniel F.
Florida ethnobotany / by Daniel F. Austin.
p. cm.
Includes bibliographical references and index.
ISBN 0-8493-2332-0 (alk. paper)
1. Indians of North America—Florida—Ethnobotany. 2. Ethnobotany—Florida. I. Title.

E78.F6A87 2004
581.6′3′09759—dc22 2004049378

No claim to original U.S. Government works
International Standard Book Number 0-8493-2332-0
Library of Congress Card Number 2004049378

Disclaimer/Warning/Caveat

Do not consult these pages unless you agree not to hold the author or publisher liable for any errors or omissions. The uses are given as they are found in the literature, and from translations of the common names. The information **is not recommended** for use. These data are compiled to provide an idea of what people have used the plants for, and some (but not all) of the potential impacts on human physiology.

Data were gathered from the literature on botanical, economic, and medical plants, none of which will ever be completely known phytochemically. These data were compiled by a human being from published and personally communicated sources by human beings. Therefore, the originals and the following material **will** contain errors.

For any serious studies, values need to be rechecked and personally obtained for chemical data. Self-diagnosis and self-medication are **not** recommended. Serious scientific studies of plant uses are recommended, but **self-testing is not considered an option**. All plants, like all animals, contain toxins and carcinogens. Even commonly ingested food plants (like bell peppers, peanuts, and strawberries) can cause fatal allergies in sensitive people.

To my wife, Sandra Kay Roderick Austin,
for putting up with my innumerable idiosyncrasies over the years,
and for unfailing support

Foreword

James A. Duke
Economic Botanist, USDA (ret.)

How pleased I was to see the final draft of this marvelous resource, *Florida Ethnobotany*, the culmination of almost a half century of ethnobotanical investigations. The career of Dan Austin is well documented in the introduction, which is fascinating reading. Dan and I are both "graduates" of the Missouri Botanical Garden, so our professional acquaintance goes back about 40 years. We're both first taxonomists (later ethnobotanists) whose prior training was augmented and enhanced by our horizon-broadening tours of the great Missouri Botanical Garden. I was there, off and on, from 1960 to 1963, benefiting, as did Dan, from exposures to Edgar Anderson, Henry Andrews, Hugh Cutler, Calaway Dodson, Robert Dressler, Fritz Went, and R.E. Woodson. I don't know about Dan, but I went to Missouri, a very local "yokel," i.e., a southeastern botanist, trained and well versed in the flora of the Carolinas. But, I came out of Missouri 4 years later, with a much bigger neotropical outlook, better versed in the pantropical flora. In retrospect, I was a post-doc at Missouri before I knew the word postdoctoral.

That tropical experience has rather outweighed my southeastern experience to this day. I now have more than 6 treasured years spent in various parts of tropical America and I became a tropical ethnobotanist (Panama, 1961, 1965–1968) before I myself knew the word *ethnobotany.* That's why, as an ethnotaxonomist myself, I am so pleased to write this foreword to a book written by an ethnotaxonomist. By agreeing to indulge myself this pleasure, I granted myself yet another indulgence, the ability to get an earlier look at the draft of this long-awaited treasure.

I settled in, searching it immediately, as I was working on the draft of a *Frutal Medicinal Tropical,* my vision of a fruit orchard for Latin America, comprising species that have edible fruits as well as medicinal uses. Dan and I had exchanged views on the importance of colloquial names in Latin America. We both share the view that scientific names are the medium of communication between international ethnotaxonomists, but colloquial names are the means of communication (in addition to sign language) between ethnotaxonomists and their ethnic subjects.

Dan's text is appropriately based on the premise that "If people in other places use the plants, then people in Florida used them" (Dan Austin). Let me add a corollary that is equally useful and that is the premise of my *Frutal Medicinal Tropical*—"If poor or rich people in Florida can use them, then poorer people, elsewhere in the Caribbean, e.g., Antigua, Belize, Cuba, Dominique, Haiti, and Puerto Rico, can also use them similarly, to good advantage" (Jim Duke). Using his idea, Dan proceeded, and commenced and successfully concluded a comparative study of Florida and the surrounding regions, compiling ethnobotanical and utilization data on the plants that occur, naturally or naturalized (but not strict cultivars) in Florida. In the process, he compiled his exciting, fascinating, interesting, and utilitarian *Florida Ethnobotany.* His ethnoflora covers almost 900 species with common names, interesting historical accounts, and various uses, from building materials, dyes, foods, fuels, medicines, ornaments, mystical, and ritual. Dan Austin's ethnoflora mentions more than 400 species not covered in Dan Moerman's monumental book (1998), the most recent and complete compilation of plants used by indigenous people in North America (north of Mexico). So Austin nearly doubles the number of utilitarian species reported for Floridian species as reported by Moerman from Florida or elsewhere in North America. Not all of the covered species are restricted to Florida and the Caribbean. Many also occur in other southeastern states. Some are circum-Caribbean or even pantropical.

I'm delighted to report that Dan's text stresses common names (with translations into English when possible) to place the species in their cultural contexts. Dan's emphasis is on people living in Florida, but he makes comparisons with other indigenous people from other areas around that region. I can strongly recommend this book to all biologists, botanists, ethnologists, linguists, tropiphiles (those addicted to the tropics like I am), and yes, even health-care professionals, be they allopathic, alternative, herbal, homeopath, naturopath, nurse, nurse practitioner, . . . and yes, even psychopaths. Great reading!

Preface

Sir Ghillean T. Prance
School of Plant Sciences, University of Reading, Reading, U.K.

Given the diversity of peoples from many different backgrounds and the varied history of Florida it is surprising that a comprehensive study of their ethnobotany has not already been written. Ethnobotany is at present a growing and dynamic science so this volume is most timely. It is also good that the author has taken a broad approach to his definition of ethnobotany that includes much more than solely the botany of the indigenous Native American peoples. The result is a comprehensive treatment where comparison can be made between the names and uses of plants by the different peoples of Florida and of neighboring countries. This is a most thoroughly researched work, the contents and the use of which will extend far beyond the boundaries of Florida. It reflects the multiracial composition of the people of Florida where each group has its own ethnobotanical knowledge that has contributed to the vast amount of information that is included here. It also reflects the long history of interaction between Florida and her neighbors. There is a large amount of information about Mexico, the Caribbean, and other parts of the southern United States. As a nonspecialist on the flora of Florida, I not only learned a lot about the botany of the state through reading this fascinating text, but also a great deal about its intriguing history. A work of this magnitude could only have been achieved through the author's lifelong gathering of information and interaction with the flora and the people of Florida. The information given about each genus frequently does not start in Florida. Details of the early history of the nomenclature of species often takes us back to the work of European botanists such as Linnaeus or even earlier in the history of botany. This book combines a wealth of locally gathered information with a comprehensive study of the early literature about each species included.

When a detailed, well-researched ethnobotanical study is made, it highlights the extent to which the population uses a local flora. Details are given here of 814 of the 2654 species of flowering plants believed to be native to Florida and a further 74 have less well-documented uses. Use for 33% of a flora is high and reflects the number of different peoples and traditions that have searched the plant diversity of this tropical and subtropical vegetation to provide for their needs. In the accounts of the genera of useful plants we are taken on a journey from *Abutilon*, the original source of marshmallow, to *Zornia*, a legume that has many medicinal uses especially as a laxative and a diuretic. In between, the variety of local uses for the plants is amazing. I particularly like all the information given about the derivation of names and the large number of different local names that exist for each species. That the local nomenclature is so varied shows the importance of the scientific names as standard fixed references to the species. However, when a list of local names over a wide region is provided, it reveals much about the ethnobotany of the species.

The book is further enhanced by the various personal experiences, interactions, and reflections of the author that are scattered throughout the text. Dan Austin has personally interacted with many of the current ethnobotanists of the world and he has made good use of the information that comes from these discussions. He has also ensured accuracy of the information given here by sending out parts of the text to the world experts on each plant family. This thoroughly researched and well-verified text will be of use to all people interested in plant uses for many years to come.

Acknowledgments

It is hard to know when I began gathering the data for this book. Probably it started before my recollection of things when my parents and grandparents began teaching me about plants. My collecting had been going on for some time when a friend of the family found me gathering and pressing plants for Gordon Hunter, one of my favorite undergraduate professors. Each time I would bring out a plant to press, she would ask, "But, what good is it?" Mostly, I did not have an answer for her, but I began to learn.

Certainly, this book began forming the day David M. McJunkin walked into my office in 1970. I had just moved to Florida after working in Belém, Brazil and St. Louis, Missouri. It was difficult adjusting to southern Florida, as it was a life between the real tropics of Brazil and the temperate zone of St. Louis. David was an undergraduate in the Geography Department at Florida Atlantic University, who introduced himself by handing me an interlibrary loan article that I had requested some weeks before. If I remember correctly, I asked him who in thunder he was, and what in blazes he was doing with *my* material. However, I doubt I used those exact words in asking. Whatever his reply, we began our acquaintance by comparing our shared interest in people and plants. He had an amateur's knowledge of the technical aspects of plants, but he was an expert at finding and recognizing different kinds in the field. After some years of working together, with him sharing his knowledge and experience with geography and anthropology, and I mine on plants, we published "An Ethnoflora of Chokoloskee Island, Collier County, Florida" (Austin and McJunkin 1978). So far as I am aware, we coined the word "ethnoflora." I most recently have seen "our word" in a chapter title in David Yetman's book *The Guarjíos of the Sierra Madre* (2002).

About 1970, my graduate major professor, Walter H. Lewis, had become interested in ethnobotany—a topic I had incorporated into my dissertation. He and his wife, Memory Elvin-Lewis, published their *Medical Botany: Plants Affecting Man's Health* (Lewis and Elvin-Lewis 1977). Also, while I was at the Missouri Botanical Garden, Edgar Anderson and Hugh Cutler had shown great interest in those topics, and I always gleaned everything I could from them long before the topic came in vogue.

Before I began teaching plant taxonomy in the spring of 1970, I realized that the way I was taught was not going to work. Therefore, I began incorporating people and their relationships with plants into my course. In 31 years of teaching, I had only a handful of individuals who came into the course with an avowed interest in plants. Yet, most of them left my class proclaiming a new interest in the green world around them.

All of these interactions were enhanced by studies of the morning glories, the Convolvulaceae. I credit Walter Lewis for directing me into examining that group, while admitting my reluctance. At the time Walter suggested morning glories, I was fascinated with the mints. Walter's suggestion was, as always, excellent.

It is impossible to study the morning glories and not become entangled in their interactions with people. Temperate people, knowing only the bindweeds (*Convolvulus arvensis*), usually vehemently hate the plants. Tropical people, who know morning glories under different conditions, dislike some, but love others. It was through studies of the genus *Ipomoea* that I worked in the tropics and gathered a more complete appreciation of how people outside Florida lived with plants. *Camotes* (sweet potatoes or yams to many, *Ipomoea batatas*) surely has the longest and most intimate association with people.

The experience working with David McJunkin made me realize how little was recorded on Florida ethnobotany. Later, Mike Duever, then of the Audubon

Society, asked me to talk about the "Environmental Effects of the Glades Indians on Southern Florida." The talk was at the March 1980 meeting of the Association of Southeastern Biologists and the Ecological Society of America. Gathering information for that meeting in Tampa prompted me to formalize much of the information I had gathered on pre-European Glades people. The talk was not only well received at that meeting, but a version of it was so popular that it was given at least annually for the next 20 years.

Somewhere along the line, I met Cameron Donaldson, editor of *The Palmetto*, the newsletter of the Florida Native Plant Society. As I had been contributing to that publication since its beginning, Cammie and I got along well. Then, at a symposium organized for the meeting of the Florida Native Plant Society in Miami in 2000, Brad Bennett asked me to give a talk on uses by indigenous people. Specifically, he wanted me to talk about the Glades people since he was to talk about the Seminoles. Standing in the wings listening was Cammie Donaldson, and afterward she asked me to write on the topic for *The Palmetto*. I selected from the many possible plants, one of the least known and rarest in the southern end of the state. We published "Chiggery Grape. Discovering Florida's Ethnobotany. The People and Plant Interactions Series" (*The Palmetto* 20:6–9, 2000). It was an article that turned into a series. As I write this, the ninth piece is due off the Florida Native Plant Society presses.

It was with due prodding and cajoling from both Cammie and Kathy Craddock Burks (Florida Department of Environmental Protection) that I began thinking about turning all the data I had collected into a longer collection. As I thumbed through Dick Wunderlin's *Guide to the Vascular Plants of Florida* (1998) to make the list of species in Florida that had been used, I almost decided against it. The list seemed endless.

Still, when I compared my list of species that had been used with Dan Moerman's for North America (Moerman 1998), I realized the enormous gap remaining. Therefore, I began working on the manuscript soon after leaving Florida in August of 2001.

From 1970 until now hundreds of people have helped, educated, corrected, and aided me in countless ways. Prominent among them, since 2000, are the following individuals. I thank all of you for what you taught me!

Mary Adair (University of Kansas, Lawrence)—*Chenopodium* and archaeology
Jonathan D. Amith (Yale University, New Haven, CT)—Náhuatl words
Henrik Balslev (University of Aarhus, Hinnerup, Denmark)—Danish names
Mardie Banks (Florida Atlantic University, Boca Raton, FL)—Graphics, and genius combined with boundless energy in helping with computer-enhanced images
Bradley C. Bennett (Florida International University, Tamiami Campus, Miami)—Many hours of discussion about all sorts of topics relating to Florida plants and their relationships to people there and elsewhere
Joanne L. Birch (The University of Texas, Austin)—For introducing me to the work of Dr. Gideon Lincecum and his studies of the southeastern tribes
Anthony R. Brach (Missouri Botanical Garden and Harvard University Herbaria, Cambridge, MA)—Saururaceae
Keith A. Bradley (The Institute for Regional Conservation, Miami, FL)—Rare southern Florida plants
George Aaron Broadwell (University at Albany, SUNY, Albany, NY)—Choctaw
Rick Brusca (Arizona-Sonera Desert Museum, Tucson, AZ)—Support by the museum
Alberto Burquez Montijo (Universidad Nacional Autónoma de México, Hermosillo, Sonora)—Mexican plant names
Steve Cafferty (Natural History Museum, London)—Linnaean names

Michael Chamberland (Cheekwood Botanical Garden, Nashville, TN)—Cactaceae

Shih-Huei Chen (National Hualien Teachers College, Hualien, Taiwan)—*Stachytarpheta*

Fernando Chiang (Universidad Nacional Autónoma de México, D.F.)—*Lycium*

Deborah J. Clifton (University of Louisiana, Lafayette)—Southeastern languages

Mireya Correa (University of Panama and Smithsonian Tropical Research Institute, Canal Zone)—Panamanian names

Walter Courtenay (Professor Emeritus, Gainesville, FL)—Fishes

Melissa Darby (Portland, OR)—Archaeology and etymology of languages

Anne Cox (Lake Wales Ridge State Forest, Frostproof, FL)—Annonaceae, particularly *Asimina*

Kathy Craddock-Burks (Florida Department of Environmental Protection, Tallahassee)—Plants of northern Florida

Gerrit Davidse (Missouri Botanical Garden, St. Louis, MO)—Mesoamerican botany

George Diggs (Austin College, Sherman, TX)—Data on species shared with Florida

Mark A. Dimmitt (Arizona-Sonora Desert Museum, Tucson, AZ)—Succulents and support at the museum

Cameron Donaldson (Just Cause Media, Melbourne, FL)—Getting me to begin this study and Herculean editorial aid with the articles for *The Palmetto*

Adam Edwards (Florida International University, Tamiami Campus, Miami)—*Ilex*

Alana M. Edwards (Riverwoods Field Lab, Riverwoods Circle, Lorida, FL)—Lepidoptera

Eckart Eich (Freie Universität, Berlin, Germany)—Chemical aspects regarding pharmacological uses

Mary W. Eubanks (Duke University, Raleigh, NC)—*Tripascum* and other grasses

Laurence Fahrney (Arizona-Sonora Desert Museum, Tucson, AZ)—Technical support with computers

Paul R. Fantz (North Carolina State University, Raleigh)—*Centrosema* and *Clitoria*

Richard C. Felger (Drylands Institute, Tucson, AZ)—Northwestern Mexico's people

Jack B. Fisher (Fairchild Tropical Garden, Coral Gables, FL)—Illustrations and Florida botany

Sandi Frank (The New York Botanical Garden Press, The Bronx)—Literary and other help

Paul Fryxell (University of Texas, Austin)—Malvaceae

Kristen Gremillion (The Ohio State University, Columbus, OH)—*Chenopodium* and Caddo

Gerald F. Guala (National Science Foundation, Arlington, VA)—Grasses

Bruce F. Hansen (University of South Florida, Tampa)—Apocynaceae

Wendy C. Hodgson (Desert Botanical Garden, Phoenix, AZ)—Edible plants, especially *Agave*

Thomas Hollowell (Smithsonian Institution, Washington, D.C.)—Guyanese names

Penny N. Honychurch-Billingham (Ed Stone and Associates, Fort Lauderdale, FL)—Caribbean names in three languages

Maggie Hurchalla (former County Commissioner, Stuart, FL)—Martin County plants

Charles Jarvis (Natural History Museum, London)—Linnaean names

Philip Jenkins (University of Arizona, Tucson)—Solanaceae
Jackie Kallunki (New York Botanical Garden, The Bronx)—Specimens at New York
Lawrence Kaplan (Boston University, retired)—*Phaseolus*
Joe Kirkbride (USDA Agricultural Research Service, Beltsville, MD)—Help obtaining the *Ipomoea alba* image
Sven Koeltz (Koeltz Scientific Books, Koenigstein, Germany)—Illustrations
Suzanne Koptur (Florida International University, Miami)—Insects and plants
Leslie R. Landrum (Arizona State University, Tempe)—Myrtaceae
François Laroche (South Florida Water Management District, West Palm Beach)—Aid with Creole French
Matthew Lavin (Montana State University, Bozeman)—Fabaceae
Blanca León (University of Texas at Austin)—Peruvian names
Walter H. Lewis (Washington University, St. Louis, MO)—Medical botany
Memory Elvin-Lewis (Washington University, St. Louis, MO)—Medical botany
Jason Lindsay (Boca Raton, FL)—Evolution of musical instruments
Barney Lipscomb (Botanical Research Institute of Texas, Fort Worth)—Botany, taxonomy, illustrations, contacts, and general support
K.S. Manilal (St. Joseph's College, Bangalore, India)—Indian botany, Malayalam names
Jack B. Martin (College of William and Mary, Williamsburg, VA)—Creek and Mikasuki
Charles T. Mason, Jr. (University of Arizona, Tucson)—Gentianaceae
David J. Middleton (Harvard University Herbaria, Cambridge, MA)—Dutch
Regis B. Miller (USDA Forest Service, Madison, WI)—Common name and traits of trees
Marc Minno (St. John's River Water Management District, Palatka, FL)—Lepidoptera
Thomas Mione (Central Connecticut State University, New Britain)—Solanaceae
Daniel E. Moerman (University of Michigan–Dearborn)—Documentation
Francisco Molina-Freaner (Instituto de Ecologia Universidad Nacional Autónoma de México, Hermosillo, Sonora)—Sonoran vines
Elizabeth Morton (Arizona-Sonora Desert Museum, Tucson)—Untold hours creating electronic copies of drawings
Richard Moyroud (Mesozoic Landscapes, Boynton Beach, FL)—Horticulture and French names, although sometimes he wished he had never heard of Creole designations
Lytton John Musselman (Old Dominion University, Norfolk, VA)—Parasitic plants and Middle Eastern uses and names
Gil Nelson (Tallahassee, FL)—Florida trees
Dan H. Nicolson (Smithsonian Institution, Washington, D.C.)—Comparisons of Asian and Caribbean botany, especially historical documents and nomenclature
Inger Nordal (University of Norway, Oslo)—Norwegian names
Delena Norris-Tull (University of Alaska Fairbanks, Fairbanks)—*Chenopodium*
Robert Pemberton (U.S. Department of Agriculture, Fort Lauderdale, FL)—Insects
Ching-I Peng (Institute of Botany, Academia Sinica, Taipei, Taiwan)—Chinese plants
Barbara Pickersgill (University of Reading, Whiteknights, Reading, U.K.)—*Capsicum*
John Pipoly (Fairchild Tropical Garden, Coral Gables, FL)—Myrsinaceae
Ghillean T. Prance (The Old Vicarage, Silver Street, Lyme Regis, Dorset, U.K.)—Chrysobalanaceae

John Rashford (College of Charleston, Charleston, SC)—Jamaican views and words

Amadeo M. Rea (San Diego, CA)—Pima ethnobotany

Laura Redish (Native Languages of the Americas, URL: http://www.native-languages.org)—Information on relationships between indigenous languages

John and Charlotte Reeder (University of Arizona, Tucson)—Grasses and grass artists in Hitchcock and Chase (1950)

Ana Lilia Reina-Guerrero (Arizona-Sonora Desert Museum, Tucson)—Sonoran ethnobotany

Kenneth R. Robertson (Illinois Natural History Survey, Champaign, IL)—Rosaceae

Rhonda Ruff (Seminole Tribe of Florida, Fort Lauderdale)—Florida Seminoles

Thomas Schimming (Freie Universität, Berlin, Germany)—Pharmaceutical uses of plants in Germany and translations of names

John C. Semple (University of Waterloo, Waterloo, Ontario, Canada)—Asteraceae

Beryl Simpson (University of Texas, Austin)—*Krameria* and other groups

Bruce D. Smith (National Museum of Natural History, Smithsonian Institution, Washington, D.C.)—*Chenopodium* and cultivated American plants

Alice Micco Snow (Seminole Tribe of Florida, Brighton)—Creek and Mikusuki

James C. Solomon (Missouri Botanical Garden, St. Louis, MO)—Distributions; vouchers

Marie S. Standifer (Louisiana State University, Baton Rouge)—Plants in archaeological context

Susan E. Stans (Gulf Coast University, Fort Myers, FL)—Creek and Mikusuki names

Mary Stiffler (Missouri Botanical Garden Library, St. Louis)—Aid above and beyond the call of duty with obscure literary references

José Luis Tapia-Muñoz, Silvia Hernández Aguilar, Lilia Can Itza, and Filogonio May Pat (Centro de Investigacion Cientifica de Yucatan A. C. [CICY], Mérida, Yucatán, Mexico)—Mayan names

Fabienne Taylor (Fairchild Tropical Garden, Coral Gables, FL)—Creole French

Walter Taylor (University Central Florida, Orlando)—Information about André Michaux and other Florida historical data

Nancy J. Turner (University of Victoria, Victoria, British Columbia, Canada)—Plants, languages, and people in the Northwest

Carmen Ulloa (Missouri Botanical Garden, St. Louis)—Ecuadorian names

Thomas Van Devender (Arizona-Sonora Desert Museum, Tucson)—Sonoran data on many topics

John C. Volin (Florida Atlantic University, Davie)—Lively discussions and stimulating intellectual challenges

Valeria Volin (Division of Natural Resource Protection, Broward County, FL)—Linguistics and translations, especially Italian

Yoav Waisel (George S. Wise Faculty of Life-Sciences, Tel Aviv University, Tel Aviv, Israel)—Hebrew and Arabic words

Deena Deckert-Walters (Montgomery Botanical Center, Miami, FL)—Cucurbitaceae

Terrence Walters (Montgomery Botanical Center, Miami, FL)—Cucurbitaceae

Daniel B. Ward (University of Florida, Gainesville, retired)—Decades of stimulating conversation and exchange of views on Florida plants, especially *Sisyrinchium*

Grady L. Webster (University of California, Davis)—*Phyllanthus*

Jonathan West (University of Newcastle, Newcastle-upon-Tyne, U.K.)—Germanic etymology

Hugh D. Wilson (Texas A&M University, College Station)—*Chenopodium*

The late Roy Woodbury (Stuart, FL)—Miccosukee plants and names, locations, and identifications of plants, not only for me, but also for William Sturtevant's dissertation; were it not for Roy's identifications, many of the identifications in Sturtevant's study would have been suspect

Richard P. Wunderlin (University of South Florida, Tampa)—Plant nomenclature and distribution

David Yetman (University of Arizona, Tucson)—Mayo and Guarajío plant uses

Carole J. Young (The New York Botanical Garden Press, The Bronx)—Illustrations, and literature published at the garden

Ray Young (University of Wisconsin, Madison)—Properties of woods

Scott Zona (Fairchild Tropical Garden, Coral Gables, FL)—Arecaceae

I thank the Florida Native Plant Society for permission to use my articles. Although they have been revised and updated since they first appeared, accounts are based on material first published in the "Discovering Florida's Ethnobotany. The People and Plant Interactions Series" (Austin 2000a,b, 2001a,b, 2002b,c,d,e, 2003). I also have drawn heavily from other papers I published in *The Palmetto* (cf. References).

Illustrations have been taken from several sources, but Penelope Honychurch Billingham (signed PNH) gave permission to use the largest number (103). In addition, line art has been used from the following publications. The 89 illustrations by Priscilla Fawcett were taken from Correll and Correll (1982). Permission has been given by both Sven Koeltz, of Koeltz Scientific Books, Koenigstein, Germany, and Fairchild Tropical Garden to use Priscilla's work. Three other drawings from Correll and Correll (1982) by Gordon W. Dillon (orchids) and Jane W. Roller (*Thelypteris*) are used courtesy of Fairchild Tropical Garden.

Other drawings are used from Bucholtz (1968) with permission of the College of Agricultural, Consumer and Environmental Sciences, University of Illinois. *Tournefortia hisutissima* was used with permission of the Missouri Botanical Garden Press, originally appearing in Nowicke (1969). Illustrations of *Viola* are from Russell (1965), courtesy of the Botanical Research Institute of Texas, Fort Worth. John C. Semple provided his original drawings of *Chrysopsis* and *Pityopsis.* Aquatic plant drawings are used with permission of the University of Florida Center for Aquatic Plants (Gainesville). University of Arizona Press gave permission to use L.B. Hamilton's drawing of *Trianthema.*

Numerous drawings are from Britton and Brown (1896–1898), Sargent (1905), and Hitchcock and Chase (1950). Drawings by Vivian Frazier are from Correll and Correll (1972), and those by Regina O. Hughes are from Reed (1971). Beryl Simpson kindly gave permission to reproduce the drawing of *Krameria* by M.C. Ogorzaly. Shih-Huei Chen permitted the use of his drawing of *Stachytarpheta.* The illustration of *Iva annua* is from Tournefort (1823). Two drawings are pre-Columbian: *Zea* was skillfully adapted by Mardie Banks from Fuchs (1542); *Zephyranthes* is from Morrison (1699).

All of the photographs except three were taken by the author. *Alvaradoa* was provided by Thomas R. Van Devender. *Cucurbita okeechobeensis* was contributed by Terrance Walters and Deena Deckert-Walters. David N. Neubauer kindly permitted use of the photograph of *Torreya.*

Last, but not least, I thank four outstanding individuals at CRC Press for their kind assistance and help preparing the manuscript for the printer: Christine Andreasen, Donna Coggshall, Pat Roberson, and John Sulzycki.

Daniel F. Austin

The Author

Daniel F. Austin, Ph.D., is a Research Associate of the Fairchild Tropical Garden, Coral Gables, Florida, and the Arizona-Sonora Desert Museum, Tucson, Arizona. He is also an Adjunct Professor, Department of Plant Sciences, University of Arizona, Tucson; a Research Associate in the Department of Biological Sciences, Florida International University, Miami; a Research Associate at the Desert Botanical Garden, Phoenix, Arizona; a Research Associate at the Drylands Institute, Tucson; and an Emeritus Professor at Florida Atlantic University, Boca Raton.

Dr. Austin attended graduate school at Washington University and Missouri Botanical Garden, St. Louis, Missouri, where he worked with Walter H. Lewis. He received his M.A. in 1969 and his Ph.D. in 1970, and then spent 31 years teaching in the Florida State University System. While in Florida, he received three Green Palmetto Awards and the Mentor Award from the Florida Native Plant Society, the Outstanding Member Award from the Florida Exotic Pest Plant Council, and was named Environmental Champion by NatureScaping at the Earth Day Celebration in 2001. He also serves as Technical Advisor to The Nature Conservancy, Office of Endangered Species, Tallahassee; he is on the Advisory Board of The Wildflower Society, Florida and Caribbean Region, Delray Beach; he is part of the South Florida Ecosystem Recovery Strategy Team and Recovery Team Member for *Cereus eriophorus*, U.S. Fish and Wildlife Service, Vero Beach; and he is on the Sweetpotato Crop Germplasm Committee on Sweet Potato, U.S. Department of Agriculture, Beltsville, Maryland.

Dr. Austin is a Fellow of the Linnaean Society of London and a member of the International Association of Plant Taxonomists and the Society for Economic Botany. He has served the Society for Economic Botany as Book Review Editor (1992–present), as a Council Member (1993–1996), a member of the Publications (1992–present) and Publicity Committees, and in other capacities. He has published 7 books, 45 book chapters, 120 research papers, and 67 popular articles, plus numerous book reviews and abstracts. After taking an early retirement, he moved to Arizona to continue his studies. His current research focuses on the ethnobotany of plant species shared between the Caribbean and northwestern Mexico and adjacent United States, and on the systematics and evolution of the Convolvulaceae.

The Author

[illegible]

Contents

People and Plants

Introduction

Florida was the first mainland region north of Mexico to be explored and settled by Europeans. Although there have been individual studies of plants used by the inhabitants of Florida (Austin 1980, 1997, Larson 1980, Hann 1986), no synthesis has ever been made of this pivotal region in the history of New World exploration. That fact is especially remarkable because Florida supports one of the richest floras in North America, combining temperate and tropical elements. Perhaps part of the reason Florida has been neglected botanically is because the Spanish explorers failed to find large native populations that might be exploited for mining and food production. The Spaniards also failed to find people who had large quantities of precious metals (Hudson et al. 1989). Another factor is doubtless that the early records were sequestered in Spain for centuries.

Juan Ponce de León set foot on the peninsula on 2 April 1513 and named it *La Florída* because it was the *Pascua de flores* or *Pascua florída*, the Easter season (Bloodworth and Morris 1978, García-Pelayo y Gross 1980). Although officially discovered in 1513, there are tantalizing bits of historical evidence that the peninsula was known well before then (e.g., Williams [1837] 1962, Quinn 1955, Hudson 1975). For example, the "Cantino" map of 1502 shows a decidedly Florida-like peninsula extending toward the Antilles (Jane and Skelton 1960). Probably because the Spanish were so jealous of their newly acquired lands, little became known about them during the period between 1513 and 1763 outside Spain's inner circle of the Crown, the Church, and the economic elite.

The first expedition to explore *La Florída* was led by Pánfilo de Narváez in 1528. That venture was followed by the more famous one led by Hernándo de Soto in 1539. Spanish chronicles of those events remained in Madrid for centuries, and the limited information they contained about people and plants have comparatively recently become available (e.g., Swanton 1939). One of the few available early commentaries on Florida plants was by Nicholas Monardes in 1569, translated by John Frampton in 1577.

The first British survey was by German-born William G. DeBrahm, who became surveyor general of East Florida in 1764 (DeVorsey in DeBrahm [1775] 1971). DeBrahm began the job on 11 February 1765, and it lasted until October 1770. In 1771, he was recalled to England to stand charges levied against him by East Florida's governor James Grant (DeVorsey in DeBrahm [1775] 1971). DeBrahm and Grant had argued over essentially everything from the time they met in St. Augustine. DeBrahm started negotiating to publish his work as early as 1769, and it appeared in two books—*The Atlantic Pilot* ([1772] 1974), and his *Report of the General Survey in the Southern District of North America* ([1775] 1971).

Bernard Romans, born in Holland, but partly educated in England, also surveyed Florida for the British (Romans [1775] 1961). He was sent to America in 1755 as a civil engineer. Between 1760 and 1770 he was the deputy surveyor of Georgia, and then he went to East Florida as a surveyor of Lord Egmont's estates on Amelia Island and the St. Johns River. From 1769–1770 he was appointed as chief deputy surveyor for the Southern District and the first commander of vessels in service, at first working for DeBrahm. This position allowed him to go (at his own expense) to the Bahamas and the west coast of Florida as far as Pensacola, where he arrived in 1771. When Peter Chester, governor of West Florida, discovered that Romans was proficient in botany, he became the king's botanist in the province (Richardson in Romans [1775] 1961).

There were also three significant botanical surveys in the northern part of the state during the British period. The first of these was by John Bartram who, with his son William, explored the St. Johns River in the fall and winter of 1765–1766. However, that account was eclipsed by the journey made by William Bartram between 1773 and 1777, when he traveled alone over much of the same Florida territory he and his father had covered. His *Travels* became a major literary item in Europe and overshadowed much of his father's contributions (Harper 1958). Last, but certainly not least, was the exploration made by the Frenchman André Michaux. Not only did Michaux cover much of the same ground that the Bartrams had in northeastern Florida, but he also found plants neither of them had reported (Taylor and Norman 2002).

The Adams-Onís Treaty of 1821 ceded Florida to the Americans, and General Andrew Jackson was appointed governor. Jackson formalized the old East and West Florida into two counties, Escambia (west of the Suwanee River) and St. Johns (east of the Suwanee River). Shortly afterward, there was a flurry of books promoting this region (e.g., Forbes [1821] 1964, Simmons [1822] 1973, Vignoles [1823] 1977). Captain John LeConte made the first Florida survey in 1822 (Adicks 1978). LeConte propagated the myth that

Lake Mayaco, which Fontaneda ([1575] 1944) called Lake Mayaimi (now Lake Okeechobee), did not exist.

The botanical and geographical void of southern Florida is exemplified by people believing that Lake Okeechobee was a myth well past the time when the region became a territory of the United States and on into the mid-1800s. Both DeBrahm and Romans had left some botanical information about the coastal regions and the Florida Keys (Austin 1980), but the interior was a blank. The word "everglades," now famous as the vast saw-grass marsh that formerly covered much of southern Florida, did not come into existence until 1823. These sparse data, coupled with the fact that the indigenous people of the peninsula had essentially disappeared by the 1760s, allowed little to be recorded of the human use of plants in the region.

Methods

This compilation is a cross-cultural study of plants shared by different floras and peoples. I have used the common names and historical sequence to weave together the more esoteric information about the individual species. Most published studies that include common names do not translate them. That absence makes names mostly or completely inaccessible to many readers. That tendency is certainly understandable, because obtaining translations is time-consuming and difficult. However, omitting meanings of names precludes fantastic glimpses into the ways people's minds work. Sometimes names also show how the individuals fit in the society where they live or give a notion of how those societies function. To me, that is at least as important as any particular use a plant has.

Any study of ethnobotany is destined to be nothing more than a snapshot of the information gathered by a few individuals. No matter how cooperative those involved, most of what is known is not passed on to the person or persons recording the information. I know from my own childhood that my father and grandfather constantly amazed me with the intricate details they knew of certain plants. I also know that they did not pass all that information on to me. All of our recorded information from any person or people is destined to be incomplete.

With that in mind, I began working in the early 1970s on what the inhabitants of Florida may have been using among the state's plants. The information given here is little more than a glimpse of what those indigenous people, and those who interacted with them, learned and practiced. The following discussion is flawed and incomplete, but it gives what I believe is a better idea of how people used Florida plants than anything currently existing.

Typically, because of a northern European bias, past studies in the United States have examined humans and plants outside of Florida. The newest compilation of human botany in North America by Moerman (1998), as detailed and thick as the volume is, barely touches on what is known about Florida plants. The following discussion picks up where Moerman left off and treats the Florida plants, particularly those of tropical origin.

Since 1967, when I first went to Mexico with Bruce MacBryde and Royce Oliver, I have talked with people in much of the region discussed here. There are only three of the United States—all in New England—that I have not visited in gathering information on plants and people. I have visited Mexico numerous times, with most of the trips concentrating on lands of the Gulf coast, and from Mexico City south. In the Caribbean, I have made excursions to the Bahama Islands, Dominican Republic, Jamaica, Puerto Rico, and Trinidad. Similarly, plants and people were studied in Brazil, Colombia, Ecuador, Guyana, Peru, Suriname, and Venezuela. In Central America I visited only Costa Rica, but I made comparative field studies for an Old World comparison in Australia, Ceylon, Pakistan, Taiwan, and Thailand.

The Species

The species and the scientific names included here were compiled from the flora of Florida (Wunderlin 1998). We do not agree in all cases about the nativity of those species, but mostly we are in accord. Species included in this discussion are those that are probably native to Florida (Table 1 and Table 2). Three groups of species are excluded: (1) those related to species of known Florida usage (Table 3); (2) those related to species known to be used outside Florida (Table 4); and (3) weedy species whose provenance is uncertain, at least to me (Table 5). Those species would surely boost the total to more than the 888 that are recorded, but that inclusion might reflect too many historical changes in usage. Because the story is already complex, adding those other species would make for an even more unwieldy array.

There are three exceptions to the "native" rule. Those deviations are the cultivated *Nicotiana, Phaseolus,* and *Zea.* All of these domesticated plants were pivotal in the indigenous cultures, and were firmly established long before Europeans arrived. Any discussion of indigenous use of plants would be incomplete if these three were omitted.

Common Names

The next step was to compile the common names of the species, and the starting point for many plants was comparison with Morton (1981). Because Morton did not give the country of origin of the names she listed, it was necessary to compare them with other sources. Where no country of usage is given for a common name, it usually comes from Morton (1981) and has not been found elsewhere. For trees, there are also some missing provenances from Miller and Wiedenhoeft (2000). Many of the names in Morton (1981) turn out to be incorrectly spelled, and none has the diacritical marks. Some of the names included here may suffer from that same defect, but I have tried to make those corrections.

Primary among the sources for common names were flora and plant lists. Those sources include Chapman (1897), Britton and Brown (1896–1898), Johnston (1909), Standley (1920–1926, 1937), Pittier (1926, 1957), Small (1933), Standley and Record (1936), Fernald (1950), Martínez (1969, 1979), Adams (1972), Boldingh (1909, 1914), Britton and Millspaugh (1920), Leon and Alain (1946–1953, 1957–1963), Goodding et al. (1965), Correll and Johnston (1970), Liogier (1962, 1974), Fournet (1978), Correll and Correll (1982), Liogier and Martorell (1982), Proctor (1984), Kindscher (1987, 1992), Brako et al. (1995), Acevedo-Rodríguez et al. (1996), Diggs et al. (1999), Felger (2000), Felger et al. (2001), Yetman and Van Devender (2001), and Yetman (2002). English names were also sought in Britten and Holland (1886), Clute (1923), and Vickery (1995), although no attempt has been made to include all of the names for any species. In addition to these, searches were made in other literature cited in the text.

Even many of the individuals involved with ethnobotanical research are unaware of the impact that plants have had on our daily speech, regardless of which language we speak. As primarily European culture derivatives, we know most about the English, Spanish, Portuguese, French, and Dutch used in the Americas. History of use, etymology, and sometimes other information are available for several languages in the OED (1971), the Spanish dictionaries by Santamaría and García (1974) and Neves (1975), the dictionary of Portuguese (Ferreira ca. 1978), and French sources by Larousse (1987), Bloch and von Wartburg (1994), and Picoche (1994). Other language resources used were Dwelly (1933), De Vries (1959), Van Goor (1979), and Ragazzini and Biagi (1986). European common names also are listed by Linnaeus ([1749] 1979), Polunin (1969), Pignatti (1982), Polunin and Smythies (1988), Scott (1995), and Mossberg and Stenberg (1994), and Meyer et al. (1999).

Indigenous language names are more problematical. For those, I have used Siméon ([1885] 1981) for Náhuatl, and asked Jonathan Amith to check many of them. Mayan names were translated and put in context by my friend José Luis Tapia-Muñoz and his colleagues. Taino names were taken largely from Coll y Toste (1972). Tupí names were found in Sampaio (1970) and Ferreira (ca. 1978). A few Carib names are from Aublet (1775), but most came from Hodge and Taylor (1957) and Klooster et al. (2003).

Names for plants in indigenous languages of the United States were taken primarily from Hunter ([1823] 1973), Gatschet (1877, 1878, 1880), Mooney (1885–1886), Dorsey and Swanton (1912), Loughridge and Hodge (1914), Gilmore (1919), Gatschet and Swanton (1932), La Flesche (1932), Speck (1934), Vestal and Schultes (1939), Haas (1940), Geary (1955), Mahr (1955a,b), Sturtevant (1955), Feeling (1975), Siebert (1975), Crawford (1978), Daniels-Sakim (1983), Howard (1984), Sylestine et al. (1993), Kimball (1994), Munro and Willmond (1994), Drechsel (1997), Martin and Mauldin (2000), Snow and Stans (2001), and Rudes and Costa (2003). A large problem exists with publications by nonbotanists because most made no great effort to identify plants. English common names are usually all that are given for identifications, and these are often impossible to combine with the proper species. Many indigenous names of plants were omitted because I could not identify them.

No attempt has been made to bring the Muskogean words into line with the alphabet that has been adopted (e.g., Daniels-Sakim 1983, Martin and Mauldin 2000, Snow and Stans 2001). Instead, it seems more instructive to reproduce the spellings as they appear in the original sources. In fact, Snow and Stans (2001) have done exactly that by using the Creek alphabet for Creek names, but the English alphabet for those in Mikasuki. Spellings for Muskogee are those given in Loughridge and Hodge (1914), Howard (1984), and Martin and Mauldin (2000). Similarly, names from other languages are presented as they were found.

Common names are wonderfully mercurial. The words change shape and pronunciation depending not only on who says them, but also on who writes them. I have attempted to combine names that seem to be the same by including the variants within brackets. For example, *cihuapatli* [*ciguapazle, cihuapastle, ciguapaztle*] is the Náhuatl name and some variants derived from it for *Chromolaena odorata.* In most cases, I have tried to list a "correct" spelling first, and include the variants later. With many words adopted into American Spanish, they have been accompanied by a variant spelling or at least pronunciation. That is certainly true for Náhuatl words, and surely is the same for the hundreds of other languages that have come into contact with Spanish. The same must be true for French and Dutch.

English names come in an amazing variety of forms because our language is as eclectic in adopting and incorporating foreign words as the Aztecs were with religions (Soustelle 1962). Enriching that complexity are the wonderful additions that people of African heritage have made to what seems to be English. Sometimes the "English" spoken in the Caribbean is called British West Indian (or simply BWI) because of the syntactical and rhythmic alterations that they have used to replace those known to native English speakers, whether of British or American stock.

Apparently, the same kinds of changes have been made in Creole French and Dutch. Those who wrote down the names surely wrote what they heard, but the spellings are so beautifully imaginative that people with European languages as their first language sometimes have no idea what they are. My main consultant for French names, Richard Moyroud, fluent in continental French, often simply said he had no idea, or that there was no such word. When he could not tell me what the names meant, often Penny Honychurch could. Her years of experience with Creole French in Dominica enabled her to "interpret" many words for me. When they failed, Fabienne Taylor often translated them.

At times, I was able to decipher Creole French and Dutch names, not because I know French or Dutch, but because I do not. I simply pronounced them to myself and looked them up in a dictionary. Sometimes that process worked; sometimes it did not. For example, it took me some time to realize that the names for *Capsicum annuum* given as *piment zouézeau* or *piment zouézo* were based on *oiseau* for "bird," as in "bird pepper."

Comparatively little is known about the indigenous languages, and only a small percentage of ethnographic studies with botanical components bothered to learn the meanings of local language names. Outstanding among those that *have* recorded these instructive data are those by Alcorn (1984) and Felger and Moser (1985). Frustrating in its fragmentary translations of common names is the otherwise fine study by Messer (1978).

Although many of the species have long lists of common names, those given here are but a sample of those in use for most species. A number of other sources give other names, including Britten and Holland (1886), Bergen (1892–1898), and Clute (1923). Van Wijk (1909–1916) has devoted three enormous volumes to listing common names in four languages. Similarly, de Cleene and Lejeune (2002) give many names in Dutch, English, French, and German.

Spelling and Pronunciation

Each language has its own rules governing pronunciation, and there is insufficient space to summarize them here. Instead, a few of the conventions used for special sounds in select languages are provided.

Cherokee (Mooney 1885–1886, Feeling 1976, Moondove 1997)

u or **v** as in English *gun*, nasalized to ***unh***; ***do:-tsu*** (*Fraxinus*) and ***ujilvʔi*** (*Gossypium*).

' indicates accent on the syllable that it follows; when following a consonant it means that the vowel has been omitted; ***a'-tsi-na'*** (*Juniperus*).

ʔ within a word or at the end is a glottal stop, as the sound in English *oh'oh*; ***wa-neʔ-i*** (*Carya*) and ***ujilvʔi*** (*Gossypium*).

: indicates that the vowel is long and takes twice as long to say; ***so:hi*** (hickory nut).

Letters in parentheses are sometimes omitted from everyday speech; ***sa'-l(i)*** (*Diospyros*).

Chickasaw (Munro and Willmond 1994)

' indicates a glottal stop, as the sound in English *oh'oh;* ***ampohko'li'*** (*Campsis radicans*).

ł has no equivalent in English, but is near *thl* as in English *fifthly*; ***piłchi'***, thistle; also in Alabama, Choctaw, Creek, Koasati, and Mikasuki.

Chinantec, Paya, Tarahumara (Lipp 1971, Bye 1986, Lentz 1993)

ʔ within a word or at the end is a glottal stop, as the sound in English *oh'oh*; Paya, ***àʔù*** (*Zea mays*), Tarahumara, ***wum kaʔ*** (*Capsicum annuum*).

Superscript numbers indicate tones, as in Chinantec ***hyni***^{3} (*Phaseolus vulgaris*).

θ is a voiceless fricative, resembling "th" in English; Chinantec, ***ku θow*** (*Zea mays*).

Choctaw (Haag and Willis 2001)

o̲ indicates nasalization; ***hato̲ falaha*** (*Allium*).
ɬ has no equivalent in English, but is near *thl* as in English *fif<u>thl</u>y*; ***ɬali***, "to pour"; also in Alabama, Chickasaw, Creek, Koasati, and Mikasuki.

Mikasuki (Sturtevant 1955, Martin and Mauldin 2000)

A phonetic transcription was created by Mary R. Haas in the 1930s and 1940s specifically for use with the Muskogean languages. Sturtevant (1955) used his own modification of this method. Neither Bennett (1997) nor Snow and Stans (2001) used this transcription.
asaykho:mi:ci (*Acer rubrum*). Written by Snow and Stans as ***ashak homeche***.
~ indicates nasalization; ***cafã:mcó:bî*** (*Cicuta maculata*).
ɬ has no equivalent in English, but is near *thl* as in English *fif<u>thl</u>y*; ***oɬã:nî*** /*orã:nî*/ "cane"; also in Alabama, Chickasaw, Choctaw, Creek, Koasati.

Muskogee/Creek (Martin and Mauldin 2000)

In these Muskogee examples, the phonetic transcription follows the word but is included within //. It should be pointed out that Sturtevant was inconsistent in spelling. For example, he spelled "medicine" as both *hilíswa* and *hiliswâ*, even on the same page, e.g., 514. I have followed his rendition as nearly as possible.
v as in English *<u>a</u>go*; **<u>v</u>ce** /ací/ "corn"
e as in English *<u>i</u>f*; **<u>e</u>cke** /ícki/ "his/her mother"
u as in English *p<u>u</u>t*; **f<u>u</u>swv** /fóswa/ "bird"
a as in English *f<u>a</u>ther*; **<u>a</u>fke** /á:fki/ "hominy grits"
o as in English *r<u>o</u>de*; **<u>o</u>fv** /ó:fa/ "inside"
c as in English *in<u>ch</u>*; **<u>c</u>esse** /císsi/ "mouse"
ɬ has no equivalent in English, but is near *thl* as in English *fif<u>thl</u>y*; **ɬvɬo** /ɬaɬo/ "fish"; also in Alabama, Chickasaw, Choctaw, Koasati, and Mikasuki.

Format

Each genus or group of species within a genus forms an entry. Rarely are there two entries for a genus (e.g., Milkweed and *Asclepias*). When this occurs, the information in one did not fit comfortably with the others. Within the discussion, there is an introduction to the group, with some history, including etymology of generic and specific names. This is followed by uses, beginning with general uses such as wood, fuel, weaving materials, food, and then medicine. In many cases, some of the chemical and bioactive properties are given. The uses are given first by indigenous tribes when endemic to the New World. Where genera or species are shared, they are first introduced as the Europeans would have known them about the time of discovery. Indigenous uses then follow. To bring the information to the present, a few key discussions are presented from as near New World discovery as possible, and then more toward the present.

Entries are in two formats. Usually tropical species are introduced with a common name, and discussed briefly. In the discussion, the other common names that reveal similar views and concepts are grouped and discussed together. The common names for most of the temperate species are listed in alphabetical order, and the discussion follows. These two types of entries emphasize the tropical vs. temperate affinities, but they are not mutually exclusive. Although some species are tropical, so little information about them is available that they are handled like the temperate taxa. Available information was the overriding factor.

Where possible, the origins of the common names are given in historical perspective. Since little of that information is available for most languages, there is a dominance of English, French, and German chronologies of forms and varied meanings.

Rarely, synonyms follow the accepted binomial. These are included only in cases where it seems likely that there may be confusion, and to eliminate the problem of having to look up the names in Wunderlin (1998) or in Wunderlin and Hansen (2002), which is updated as changes become known. Because neither of these sources contains historical information on genera, those data were taken from TROPICOS (www.mobot.mobot.org). Type information, when discussed, was cross-checked with Wunderlin and Hansen (2002) and with the Linnaean Herbarium and Linnaean Typification Projects Web sites. Both Linnaean databases are at www.nhm.ac.uk.

Historically important individuals are cited with their full names and dates of birth and death in the first entry. Afterward, some of those who are well known are abbreviated to their last names. Some commentary on their nationalities and where these scientists worked is provided for a more complete view of the international effort that produced the current understanding of these plants. Although Linnaeus named the great majority of plants now growing within the political boundaries of Florida, scientists from around the world have been involved in their collection, naming, and current arrangement into taxonomic categories. This international effort makes for a complex ethnobotany layer superimposed on the strata already in place among indigenous people when the New World was rediscovered by the Old in 1492.

Definitions

Florida—Then and Now

The geographic delimitation of "Florida" used here is the current political unit, and not the limits used in the past. Over the centuries, the delimitation of "Florida" has varied considerably.

From the first known Spanish maps of the 1500s, the entire eastern coast of North America was labeled *La Florida.* For example, the Kerrigan translation of Barrientos's ([1567] 1965) original Spanish says that, "Florida has a promontory which extends a hundred leagues into the sea and narrows to a cape called Head of the Martyres. ... On the east is the Land of Cod and Newfoundland; to the west, New Spain; to the north, China and Tartary." The region called *los Martires* we now know as the Florida Keys. "*Nueva España*" was the viceroyalty (from 1521–1821) including land now in the southwestern United States, Mexico, and Central America to Panama. Thus, *la Florida* extended from Mexico to Newfoundland in Canada.

The region is labeled to conform with this idea on the Gerónimo de Chaves's map believed to date from the second half of the 16th century (Milanich and Milbrath 1989). Some maps gave added detail, as in Diego Rivero's 1529 map, which shows the *Tierra de Ayllón* in what is now coastal Georgia and the Carolinas. That region was named for Santo Domingo's royal judge Lucas Vásquez de Ayllón's attempt at colonization in 1526 (Milanich and Milbrath 1989).

However, by the time the Spanish relinquished the region to the English in 1763, the British colonies of Georgia and the Carolinas had been separated from the southern region. The crown also created the colony of East Florida and appointed James Grant, a hero of the Cherokee War, as its royal governor. Thus, by 1763, Florida was divided into "West Florida" (what is now called the Panhandle) and "East Florida" (the peninsula), with the two divided by the Suwanee River.

British rule of Florida lasted until 1783, when again Spain became the official governing country. However, during the 20-year period when Britain held the land, two major surveys were commissioned (DeBrahm [1775] 1971, Romans [1775] 1961). Spain formally ceded Florida to the United States in 1821. Then, in 1845, Florida was admitted to the Union (Bloodworth and Morris 1978).

Ethnobotany

The term "ethnobotany" has been defined and discussed many times since it was created by John Harshberger in 1895. I find unacceptable even the most recent discussions of the term, such as those published by Schultes and Reis (1995), Balick and Cox (1996), Cotton (1996), and Wickens (2000). All of those definitions take a "Eurocentric" view that I find prejudiced by assigning ethnobotany to the study of "traditional" people. I submit that we are all "traditional" in one way or another.

Although those authors using that definition tried to avoid the Caucasian bias expressed from Harshberger in 1895 to Schultes and Reis (1995), they have simply substituted another word for "primitive." Instead, my view of the discipline of ethnobotany is the complete and total study of people and plants. I taught a course for almost 30 years using that definition. However, when the class was team-taught at Arizona State University in Tempe with Wendy Hodgson and Gary Nabhan there was no consensus on the definition.

Florida's Lost Ethnobotany

People everywhere in the world use some percentage of the flora. A problem in discussing this usage is how to determine the extent to which plants are used. Some have suggested complicated ways of analyzing the data statistically (Moerman 1991). Others think that a more straightforward method is simply to look at the percentage of species used in each area (Yarnell 1964, Griffin 1967). However crude the percentage method, it provides one estimate of how plants were used. For many reasons, estimates are always below actual utilization.

Griffin (1967) thought that at least North American cultures used about 20% of the available species. That was the percentage used by natives of the Upper Great Lakes and by the earliest inhabitants of one island in southwestern Florida (Yarnell 1964, Austin and McJunkin 1978). Because Austin and McJunkin (1978) found a percentage that matched the one reported by Yarnell (1964) and Griffin (1967), one aspect of this investigation will be to "test" the 20% theory against a more rigorous data set. Early examination showed that the data for indigenous groups contained markedly fewer than the predicted numbers (Austin 1980, 1997, Hann 1991). Those reports also made it clear that the data for indigenous use of Florida plants were unusually fragmentary. Therefore, examination of a broader sample of people was needed.

The Florida People

The Florida peninsula north of Lake Okeechobee was populated by two large cultural groups when Europeans arrived. In the region around what is now Tallahassee (*tulwa*, town, *ahassee*, old, Creek) were the Apalachee (*apelachi*, allies, Choctaw). These were Muskogean people, from whom the Appalachian Mountains received their name. To their west were Choctaws, such as the Pensacola (*pan-sha*, hair, *okla*, people) (Swanton 1946, Simpson 1956).

Occupying most of the Florida peninsula were the people called the Timucua, from *atimoqua* (lord or ruler), in their language (Swanton 1946). Typically, the Timucua are divided into the Eastern and the Western Timucuan. The Western group included the *Acuera* (*acu*, moon or all, *ero*, year, Timucua) of the modern Ocala region south of Gainesville, the *Ocale* [*Cale, Ocali*] (meaning uncertain; *cale* is fruit in Timucua, *kali* is spring in Choctaw), the *Tocobago* (maybe "place where gourds are produced," Choctaw) around Tampa Bay, and the *Utina* (powerful, Timucua), between the Santa Fe and Suwanee Rivers. Eastern groups included the *Saturiwa* [*Saturibe*] on the upper St. Johns River, and the *Surruque* [*Horruque, Serropé, Surreche*] in the Cape Canaveral region (Swanton 1946). In total, there were 14 "Provinces," interpreted as tribes (Milanich 1975).

Some include the *Ais* of Indian River as part of the Timucuans, while others consider them related to the Glades people (cf. Milanich 1975). Most of what has been recorded of the Timucua language was published by Gatschet (1877–1880). Their language was, according to most, a distinct linguistic group. Swanton (1946), as usual, considered them "undoubtedly a branch of the great Muskogean stock," as he did almost everyone in the southeast. Reflecting modern understanding, Drechsel (1997) and Martin and Mauldin (2000) considered Timucua linguistically isolated. These people were virtually extinct by the 1740s (Swanton 1946, Simpson 1956, Tebeau 1968, Hudson 1976).

Archaeologists have called the first residents of southern Florida the "Glades Indians" (Goggin and Sturtevant 1964), in spite of the fact that they lived along the coasts and not in the Everglades (cf. maps in Swanton 1946, Milanich 1975). The Glades people met the Spanish when they arrived in 1513, and they preceded the Seminoles in the region. It was a Glades arrow that killed Ponce de León when he landed and tried to capture them for slaves (Douglas 1947, Hudson 1976).

The Glades people consisted of tribes or villages usually called *Ais* ("deer"), *Jaega* [*Jobé, Hobe*], *Tekesta* [*Chequescha, Tequesta*], *Calusas* ("fierce people"), and *Miamis* ("big lake") (Goggin and Sturtevant 1964). Fontaneda ([1575] 1944) recorded that dialects of those in the south and southwest he could understand, but the Ais and Jaega speech was linguistically distinct and he could not understand it. In spite of Fontaneda's comments, Swanton (1946) believed the five major groups spoke essentially one language. Only 11 words are known from their languages (Milanich and Proctor 1978, Austin 1997), so the point is moot.

Native populations declined rapidly after European contact, partly because of their enslavement by the English and Spanish, but mostly because of introduced diseases (Crosby 1972, Hann 1991). By 1763, Glades people were culturally extinct (Romans [1775] 1961, Hudson 1976, Sturtevant 1978). Because the Glades and Timucua cultures disappeared so quickly, little was recorded about their use of plants. We have only sketches from Le Moyne (1564), Fontaneda ([1575] 1944), Laudonnière ([1586] 1928), Dickinson in 1699 (Andrews and Andrews 1945), and Spanish priests (Hann 1986, 1991) as first-hand visitors, along with the archaeological record (e.g., Gilliland 1975, Hogan 1978, Milanich and Fairbanks 1980, Sears 1982). Much of what is known about utilization of plants has been summarized recently (Austin 1980, 1997).

During the 1600s and 1700s, the Glades people regularly moved in canoes between the Keys and Cuba (Barcia 1951, Sturtevant 1978). These southern Florida natives had ample opportunity to learn Spanish when they were in contact with priests and their accompanying escorts, during the time they spent in Havana, and when they visited *rancherias* along Florida coasts. Knowledge of that language led to their also being called "Spanish Indians" (for summary, see Austin and McJunkin 1978). Contact with Cubans for so long surely led to exchange of information regarding plants.

When Florida changed from rule by the Spanish to the British in 1763, the region had largely been depopulated by disease, slaving, and transplanting of people to Cuba (Swanton 1946). Into the unpopulated regions left by the demise of the Timucua and Glades peoples, groups of Muskogeans began emigrating under pressure from the north. People now called the Miccosukee ("Mikasuki" now applies only to the language) were among the first to move down the peninsula and populate empty southern Florida after the Second Seminole War (1835–1842). Ancestors of the Miccosukee were the the Hitichiti of southern Georgia (Swanton 1946, Simpson 1956, Tebeau 1968, Hudson 1976). Martin and Mauldin (2000) called their language the Hitichiti-Mikasuki. These people were followed by Creeks who spoke variants of Muskogee, with that spoken on the Brighton Reservation on the north end of Lake Okeechobee formerly being called "Cow Creek" (Sturtevant 1955). This I have called "Creek" following the terminology of Snow and Stans (2001). Their relatives in Oklahoma still speak Muskogee, and the two are similar enough that they can mostly understand each other (Snow and Stans 2001). DeBrahm ([1775] 1971), Romans ([1775] 1961), and both Bartrams, but especially William Bartram ([1791] 1958), left some information on the early Muskogeans of northern Florida.

However, well into the 1800s people dubbed the "Spanish Indians" remained in southern Florida. Goggin (1940), Swanton (1946), and Sturtevant (1953, 1978) considered these Spanish-speaking people as remnant Glades tribes. Older literature calls them "Calusas." One of the "Spanish Indians" was *Chakaika* [*Chekika*]. He and his followers lived and fought during the Second Seminole War (1835–1842) with Miccosukee leaders *Holatta-Micco* (*holahta*, village headman, *miko*, chief), who was called Billy Bowlegs by the whites, and *Hospetarke* (Mahon 1985). From that association, the Seminoles surely learned new plant uses, and this may be reflected in the data collected by Sturtevant (1955) among the Mikasuki-speaking Seminoles.

Neighboring People

Historical records confirm that southern Florida people traded with Cuba (Fontaneda [1575] 1944, Barcia 1951). Other historical accounts deal with trading within the eastern United States, in Mexico, and elsewhere. For example, Smith (1933) commented on the Potawatomi trading plant use with the Ojibwa and Menomini. Similarly, Latorre and Latorre (1977) noted how the Kickapoo learned from their neighbors when removed to Coahuila, Mexico. The similarity of uses of many plants within and outside Florida is so great among people of different linguistic stocks that exchange of information seems assured. Indeed, Howard (1968), Walker (1981), and Jackson (2003) suggested that all people in the southeast had "common understandings of cosmology ... [and] roughly standard medical practices and ethnobotany." Therefore, a synoptic introduction is provided to establish where geographically some of the important people were who are mentioned in the species accounts.

Comments on where these people lived at the time of contact are largely taken from Swanton (1946), Grosvenor et al. (1972), Billard (1974), Crawford (1978), Bennett (1982), Garrett et al. (1982), and Drechsel (1997). As Swanton (1946) made clear, the location of virtually all Southeastern tribes changed as they fought with each other and were driven from their native lands by European intrusion. Therefore, no attempt has been made to summarize the locations where these people have lived.

Southeastern Tribes

These are people with whom the Timucua certainly were in contact and surely traded, either directly or indirectly. One example of the trade network in plants is detailed under Osage orange (see *Maclura*). Although *M. pomifera* is native to Texas, it was traded as far east as Florida and probably reached North Carolina.

Although the groups periodically made war on each other, they continued trading. Arrival of the Europeans finally gave them a common enemy, and some established the Creek Confederacy, also called the Five Civilized Tribes. That political alliance included four Muskogean speakers (the Chickasaw, Choctaw, Creek, Seminole), plus the linguistically distinct Cherokee (Billard 1974, Martin and Mauldin 2000). Bartram ([1791] 1958) recorded the polyglot nature of the region, and Thomas (1954), Callendar (1978), Walker (1981), and Jackson (2003) have shown that some towns in the 1700s included various mixes of Catawba (Siouan-Catawaban), Cherokee (Iroquoian), Creek (Muskogean), Koasati (Muskogean), Natchez (language isolate), Nottoway [Nattaway] (Iroquoian), Shawnee (Algonquian), Tuskegee (Muskogean), and Yuchi (language isolate). Single towns speaking different Algonquian, Iroquoian, Muskogean, and isolated languages existed.

There were many other tribes in the region. In addition to those mentioned below, Swanton (1946) listed most of the others, their fates, population estimates, and other pieces of their history. Billard (1974) gave some of the same data updated. Booker et al. (1992) and Jackson (2003) have most recently provided a glimpse of the complex mixture of people and languages living in the region. Of the hundreds of names given to various villages and tribes in the southeastern United States, only the dominant or best known are listed here. These groups were selected largely because the extant literature on them includes information about plants.

Alabama [*Alabamu, Albama, Alebamon, Alibama, Alibamou, Alibamon, Allibamou*] (*alba-amo*, "medicine gatherers" or "thicket clearers," in Choctaw; from *alba*, vegetation, *amo*, clear, collect, gather; Muskogean language family) The tribal legend is that their name came from *alibamu*, "here we rest." That popular belief stems from a folk etymology given wide currency in the 1850s through the writings of

Alexander Beauford Meek. However, the first known use of this derivation appeared earlier in an unsigned article in a July 27, 1842 issue of the *Jacksonville Republican*. There is no linguistic evidence within Muskogean to support such an interpretation (Sylestine et al. 1993). De Soto found these people in 1540 in the northern part of Mississippi, west or northwest of the Chickasaw. However, by 1702, they were living on the Alabama River just below the junction of the rivers Coosa (Elmore County, AL) and Tallapoosa (Haralson County, GA) in what is now central-western Alabama. Most of them now live near Livingston in eastern Texas, and the others live in Louisiana (Sylestine et al. 1993).

Apalachee (people on the other side [of a river], or maybe Choctaw *apalichi*, rulers or allies; Muskogean language family) The first mention of these people was when Narváez and his company were attacked by them in 1528, and the Apalachee were one of the reasons for the Spanish misfortunes. Later, warriors from the Apalachee repelled de Soto and his men from what is now the Tallahassee area in the Florida Panhandle in 1539. These people constantly harassed de Soto's band from 1 October 1539 until 3 March 1540. The Apalachee were extinct by the late 18th century (Billard 1974), although some families were absorbed into the Creeks (Swanton 1946). Their language was closest to Alabama, Hitichiti-Mikasuki, and Koasati (Kimball 1994).

Apalachicola [*Palachicola*] (people on the other side [of a river], or people of the ruling place; Muskogean language family) These were Hitichiti-speaking villagers who lived on the lower part of the Chattahoochee [*Cvto-Hocce Hvcce*] (*cvto*, rock, *hocce*, marked, *hvcce*, stream) River. The early usage is vague, but by 1675 and 1686, the name was applied to villages in the Florida Panhandle and nearby southwestern Georgia and southeastern Alabama (Swanton 1946, Billard 1974). Most of them joined the Creek Confederation, with some families joining the Seminoles, and others being removed to Okmulgee, Oklahoma.

Caddo (contraction of *Kadohadacho*, true chiefs; Caddoan language family) The Caddoan word *taysa* (friend) was used to name the state we now call Texas. Confusion has it that the state was named with the Spanish word *tejas*, tiles, originally spelled, *texas*, but that is a convergence of sounds in distinct languages (Billard 1974). De Soto encountered one of the tribes, the *Tula*, in southwestern Arkansas in 1541 near the modern Caddo Gap. Others within this linguistic family included the *Hasinai* and the *Natchitoches* (maybe from *nashitosh*, pawpaws), among others. The principal range of the Caddo was in the southwestern corner of Arkansas, and adjacent southeastern Oklahoma, northeastern Texas, and northwestern Louisiana. There were almost 1000 of them living on the Brazos River in 1937.

Carolina Algonquians (Algonquian language family) Harriot ([1590] 1972) left most of the little that is known about these people in historic times. Among the groups that he named there, the principal ones are the *Chawanook*, *Secotan*, and *Weapemoc*. In his illustrations he concentrated on people in the towns of *Pomeiooc*, *Pommeioocke*, *Secota*, and *Roanoac*. Swanton (1946) gave a sketch only of the *Weapemoc*. Grosvenor et al. (1972) showed, in addition, the *Croatoan*, *Pamlico*, *Roanoke*, *Secotan*, and some other small groups along the coast. These people remain as the Lumbees (see www.lumbeetribe.com). This tribe includes remnants of several of the former groups, and has official recognition by North Carolina, but is not recognized federally.

Catawba (from *yi̱ kátapu*, "people of the fork" [of the river, cf. Young and Siebert 2003]; Siouan-Catawban language family, cf. Oliverio and Rankin 2003) The largest population called themselves *yi̱ í·suwa̱* [*Nieye, Esaw*] (*yi̱ í*, people [of the], *suwa̱*, river). The second dialect was spoken by the *yi̱ sá·wa̱* [*Saraw, Saura, Sauro, Cheraw*] (*yi̱ í*, people [of the], *sá·wa̱*, peninsula). Their home at the time of contact was in north-central South Carolina and near Augusta, Georgia. Spanish Captain Juan Pardo first encountered them in 1566 under the name *Ysa* [*Yssa*] in Georgia. The next year, he found the *Usi* living in Carolina. The Catawba now live in Oklahoma and South Carolina.

Cherokee (*tciloki*, people of a different speech, Muskogee; Iroquoian language family) In their own syllabary, created by *Sikwa'yi* [Sequoyah] (ca. 1760–1843), whom the whites called George Gist, Cherokee is rendered *Tsa'lagi* or *Tsa'ragi*, depending on dialect. Mooney (1885–1886) listed about 50 different spellings of "Cherokee" at various times in history. They call themselves the *Yun'wiya* (*yunwi*, people, *ya*, real) (Mooney 1885–1886) also glossed *aniyvwiya'i* (Walker 1981). The Cherokee historically occupied the eastern portion of the Appalachian Mountains in southwestern West Virginia, western North Carolina, and eastern Tennessee (*tanasi*, the Cherokee name of the Little Tennessee River). De Soto may have encountered these people in 1540 near Murphy, North Carolina; he called their principal town the *Guasili* [*Guasule*; *Cauchi* in Booker et al. 1992] (*savanna*, Catabwa). There are now Cherokee in North Carolina and Oklahoma.

Chickasaw (*chikki*, old, *asha*, to reside, or old residence, Choctaw. Muskogean language family) While most consider theirs a distinct language (Munro and Willmond 1994), others view it as a dialect of

Choctaw (Billard 1974). The area from the Jackson Purchase of western Kentucky (planted field, in Iroquoian) south along the Mississippi (big river, in Illinois) River in Tennessee and into northern Alabama and Mississippi was the contact-era region claimed by the Chickasaw. De Soto found them in northeastern Mississippi between the heads of the Tombigbee and Tallahatchie (*etvlwa*, town, *hvcce*, river, Creek) Rivers in 1540. Romans ([1775] 1961) gave a detailed early record of the tribe. Most Chickasaw now live in Oklahoma (cf. *Zea* for treaty and removal).

Choctaw [*Chata, Chactaw*] (they call themselves *Chahta*, river people; Muskogean language family) The largest of the Muskogean-speaking tribes in the southeastern United States, the Choctaw historically occupied parts of Mississippi, Alabama, and Louisiana. De Soto found these people in 1540 but recorded them under the name *Mabilia* (Halbert in Byington et al. 1915 suggested it was from *moeli*, "to paddle," Choctaw; Drechsel 1997 was doubtful; see *Arundinaria*). Also recorded by the de Soto chronicles are the names *Apafalaya* and *Pafallaya*, both variants of the Choctaw *pans-falaya* for "long hair," the Pensacola. The name *Pensacola* distinguishes Choctaws from the tribes who had the hair cut short on both sides, leaving only a "roach" or "mohawk" down the middle of the head (Swanton 1931, 1946). Romans ([1775] 1961) commented that the Choctaw were called "Flatheads" because of their modifying skull shape in infancy. The Choctaw, accomplished farmers, now live from Oklahoma east into parts of their former homeland (Swanton 1931, Billard 1974).

Creek (people called "Creeks" originally were named after the Ochesee Creek where they lived; shortened by 1725; that creek, now the Ocmulgee River [from *oke*, water, *mulgee*, place, Hitichiti], is usually translated "where the water boils up"; also called the Muskogee) The Mųskogee were the "core" Creeks. It is difficult to estimate the first contact with Creeks because they were so scattered into villages or subdivisions with so many names (see Booker et al. 1992). Perhaps the first were those recorded as *Kasihta* or *Coweta*, who were descended from the *Coosa* (maybe from *koskosa*, the cry of a bird, Creek). All three were reported by de Soto in 1540 as the *Coosa* [*Cosa, Coça*]. Various branches of the Creeks occupied most of Georgia, parts of Alabama, and the Panhandle of Florida. Most of their descendants are now in Oklahoma (*okla*, people, *homma*, red, in Choctaw).

Hitichiti (possibly from *Atcki-hata*, or *cekhicetv*, to pile up, a name given to the Hitichiti-speaking people by the Lower Creeks; said to refer to a "heap of white ashes piled close to the ceremonial ground"; Muskogean language family) The Hitichiti were originally in south-central Georgia, where they lived on the Altamaha River (*akh*-, Choctaw passive, *tamaha*, town), and northern Florida. De Soto found them on the Ocmulgee River and called them the *Ocute* [*Achuse, Achusi, Ochus*] (*ochisi*, a person speaking a different language). Part of them were removed to Oklahoma with the Creeks and settled near Hitichiti Station and Okmulgee. Those remaining in the east joined the Seminoles, and were partly resettled in the northern part of their territory in Oklahoma. Dreschsel (1997) considered them one of the many linguistically isolated groups in the southeast; others say they are Muskogean (Kimball 1994, Martin and Mauldin 2000).

Houma (*hooma*, red; probably shortened from Choctaw *shaktei*, crawfish, *homma*, red; linguistically isolated) These were neighbors of the Natchez and lived to their southeast in Louisiana and Mississippi. While de Soto was among the Choctaw in the winter of 1541, he sent a force to attack a nearby tribe they called the *Sacchuma* [*Saquechuma*]. The *Sacchuma* had abandoned and burned their village when the Spanish arrived, so the Europeans actually never saw them. These people were first visited in 1682 by the French priest La Salle, who called them the *Chakchiuma* (*shaktei homma*). In 1682, La Salle's lieutenant Henri de Tonti made an alliance with them. Pierre Le Moyne Iberville, the founder of Louisiana, renewed the agreement in 1699. Today more than 500 members of the United Houma Nation live in Texas with the Alabama (Alabama-Coushatta Tribe of Texas 2003). Dreschsel (1997) makes a convincing argument that these were one of the many linguistically isolated groups in the southeast.

Koasati [*Acoste, Acosta, Coste, Costehe, Cussatees, Cushatta, Coushatta*] (perhaps related to Choctaw *konshak*, cane, particularly since they think of themselves as the "White Cane" people; Muskogean language family) De Soto first mentioned these people under the names *Acosta* [*Acoste*], *Coste*, and *Costehe*. They were then living on what is probably Pine Island in the Tennessee River; it has been called "River of the Cussatees." After years of displacement, they finally settled in Louisiana. Most of them still live in Louisiana; others live with the Alabama in Texas (Sylestine et al. 1993, Kimball 1994).

Miccosukee (*mikko*, chief, *sukee*, pig, but maybe originally from *niko*, eater, *suki*, pig, because of their taste for introduced Old World swine, *Sus scrofa*; linguistically isolated or Muskogean) After living for many years on the edge of Lake Miccosukee in the Panhandle, these people were driven south by a variety of treaties and battles with the Americans. Today their principal home is the Miccosukee Reservation in southwestern Florida, although there are families scattered up the peninsula. Drechsel (1997) pointed out that Miccosukee is a synonym for Hitichiti, but both terms are retained for clarification.

Muskogee (probably from Shawnee *muskiegui*, lake or pond; related to Creek *muskeg*, swamp; Ojibwa *maskig*, swamp; Muskogean language family) See also Creek.

Natchez [*Nah'-Chee, Nauche, W'Nahk'-Chee, Nvce*] ("fast warrior(s)," in their own language) The Natchez (French spelling) language was formerly considered distantly related to Muskogean. Dreschsel (1997) and Martin and Mauldin (2000) considered Natchez "a language lacking close relatives." De Soto first learned about these people in 1542 when he reached the Mississippi River. They were probably the warriors who attacked his company after he died, under the leader called "Quiguattam" [*Quigualtanqui*]. The next encounter with them was in 1682 by La Salle, whose treatment was totally different; the French were welcomed into one of their towns. Survivors of a 1729 French attack joined the Catawba, Cherokee, and Creeks (Young and Siebert 2003), and their descendants live in Oklahoma and elsewhere. Officially recognized as the Natchez Nation, they consider themselves direct descendants of the Mound Builders (Fields 2003). The Natchez at contact lived in southwestern Mississippi and east-central Louisiana (Swanton 1946, Grosvenor et al. 1972). Fields (2003), the current *Cunv Uvse* (Great Sun) of the Nation, records towns in Oklahoma, Georgia, and South Carolina.

Powhatan ("falls in a current of water"; Algonquian language family) The Powhatan were in the tidewater of Virginia on both sides of Chesapeake Bay in 1607 when Virginia was established. The fabled Pocahontas, a real individual, was the daughter of the ruling leader at the time of European contact. Although the British called this leader Powhatan, his name was "Wahunsonacock." Several of the allied group names live on in localities, including Chickahominy, Potomac, Powhatan, and Rappahannock.

Seminole (either "people of a distant fire" or "wild ones," which is about the same; Muskogean language family) Simpson (1956) thought "Seminole" was derived from Creek *isti semoli*, people of a distant fire. The Seminole tribe of Florida publishes that their name came from Spanish *cimarrón,* wild or untamed. Fairbanks (1978) found historical data that the name "Seminole" came into existence about the 1760s, when they began being called "Seminoles" or "Siminoles" in English and *Cimarrones* in Spanish. The name was applied by the Spanish because these new immigrants in Florida "had left their settled towns and had established themselves in wild, vacant lands." The Muskogean tongues have no sound corresponding to the "r" in English or Spanish, and they first pronounced the word "Simalones" or "Simanoles." At first it seems incongruous that the Seminoles, who rarely spoke any Spanish, would have their name derived from a Hispanic word. However, there are at least two possible reasons for this, and for the presence in their languages of other loan-words from Spanish. People who became the Seminoles mingled with members of the Yemassees (*yi̱ musí*, ancient people, Catawba) and the Glades people. Both the Yemassees and the Glades people had been under Spanish influence for a long time and knew some of the language (cf. Simmons [1822] 1973, Swanton 1946).

Timucua (from *atimoqua*, lord or ruler; Timucuan language family) Timucua was the original name of a group later called *Utina* (powerful). Ponce de León first encountered the Fresh Water Timucua in 1513. Several others visited the area between then and 1528 when the Narváez expedition passed through their territory. However, it was the de Soto expedition in 1539 that gave the clearest description of the Western Timucua, and the French groups under Jean Ribault in 1562 and Ludonnière and Le Moyne in 1564–1565 that described the Eastern Timucua. At the time of contact, the Timucua dominated the Florida peninsula north of Lake Okeechobee. They were extinct by 1740, but Swanton (1946) thought that the Seminole town of *Etaine* was the same as *Utina* and that it might contain remnants of the people.

Yuchi [*Euchee, Uchi*] ("far away people" in their language, or perhaps a form of the Hitichiti word *ochesee*, "people of another language"; an isolated language) Their language has been considered Muskogean (Speck 1909, Swanton 1946) or Siouan (Billard 1974), but most agree they speak a unique language (Martin and Mauldin 2000, Jackson 2003). When de Soto passed near their region, the Yuchi were living west of the Cherokee in the hilly country of eastern Tennessee. Later there were scattered groups of Yuchi, one on the western parts of the Appalachians or the plateau lands beyond, and others throughout the southeast from Alabama to Georgia. The village in Alabama was visited first in 1701. For more information, see Swanton (1953). Some of the Yuchi went to Oklahoma and settled on the western part of the Creek territory. Others went to Florida and joined the Seminoles (Speck 1909, Jackson 2003).

Caribbean People

The Caribbean was populated by two basic groups, the Arawakan and Cariban tribes. The distinction between these two was not always clear, as there was intermarriage between them in the Lesser Antilles.

Taino The northern Caribbean, from the Bahamas and Cuba to Puerto Rico, was populated by Arawakan people typically called the Taino, although that designation oversimplifies reality (Sauer 1969). Columbus provided the first glimpses of these agricultural people when he arrived in the Bahamas and later in Cuba, Hispaniola, and Jamaica (Jane and Skelton 1960). The number of Taino words incorporated into Spanish is remarkable (Coll y Toste 1972), and many of the plant names used as far away as western Mexico and Peru are Taino words. Some familiar Taino words in English are Cuba, maguey, maize, and tobacco.

Carib Although perhaps the only remnant of Cariban culture remaining in the Antilles is on Dominica, the Carib have relatives in South America (Hodge and Taylor 1957, Taylor 1977). One of the unique aspects of the Antillean Caribs is that they stole their women from Arawakan villages. Men ostensibly continued speaking their language, while the women spoke their native Arawakan tongues. In reality, Taylor (1977) indicated that the hybrid languages were more Arawakan than Cariban.

Columbus's understanding of the Taino also added the words "Carib" (akin to *Galibi*, literally "strong men") and "cannibal" (from *caniba*, Arawakan) to English. For example, on 23 November 1492, his journal contains an entry that the Taino whom they captured in Cuba said that to the east was an island Columbus misunderstood as being called *Bohio* (house, Taino). That island we now know as Hispaniola, and the captive indicated that "this land was very extensive and that in it were people who had one eye in the forehead, and others whom they called 'Cannibals.' Of the last, they showed great fear, and when they saw that this course was being taken, they were speechless."

Mexico's People

At the time of contact with Europe, Mexico was home to hundreds of languages. Although plant names are given from several of these languages, the majority are taken from three dominant groups, the Aztecs, the Maya, and the Zapotec.

Aztecs Focused around the area that is now Mexico City were the people who called themselves *Mexicas* or *Aztecas.* Their language was Náhuatl, and an unusually complete vocabulary was compiled by Spanish priests (Siméon [1885] 1981). Much of the Creole Spanish spoken in central Mexico retains words from their language. Farther from that region, other languages may prevail, but there is still a heavy influence from former Aztec domination.

Maya In the Yucatán region of Mexico and south into much of Central America, the dominant language is Mayan, with multiple dialects (actually considered languages by many) such as Huastec. Few publications on plants have translated words from Mayan, and outstanding among these is the work by Alcorn (1984).

Zapotec A third linguistic group in southern Mexico of considerable historic importance is the Zapotec. Reko (1945) has given an exemplary account of their knowledge and beliefs regarding plants.

The trade systems of all three of these groups are staggering. According to Hernándo Cortés the central square of *Tenochtitlan* (*Tenoch*, chief of the Aztecs who founded the city in A.D. 1325, *tlan*, near) was surpassed by the one at *Tlateloloc* (Soustelle 1962). Cortés wrote that in *Tlateloloc* (*tlatelli*, height or elevation, *tlatia*, burned or hidden), there were more than 60,000 people trading every day and that "there are all kinds of merchandise from all the provinces, whether it is provisions, victuals or jewels of gold and silver."

The Aztecs in the Valley of Mexico kept regular contact with the Gulf and Pacific coasts. Their leaders, the Moctezumas, are said to have dined on fresh fish taken in the mornings on seacoasts. Given the lists of items in markets by Bernal Díaz del Castillo between 1517 and 1521 (Díaz 1956) and those recorded for the Moctezumas' meals by Fray Bernardino de Sahagún in the 1500s, that is believable (Soustelle 1962).

There is even evidence that people in Mexico City traded far to the north. Comparisons between the images of *Kokopelli* (maybe from *koko*, god, *pelli*, a humpbacked robber fly, Hopi) in the Pueblo regions of New Mexico and Arizona and the *Xochipilli* (*xochitl*, flower, *pilli*, lord) of Mexico City suggest that there were traders moving between those areas also (Schultes and Hofmann 1979, Young 1990, Bob Bye, personal communication, 1993). Even the names *Kokopelli* and *Xochipilli* may be cognates. Some think that these names and images represent the actual traders moving between regions.

By the time the La Venta center of Olmec culture on coastal Tabasco, Mexico, was flourishing from 800 to 400 B.C., they engaged in trade with the highland Maya area (Schlesinger 2002). These north–south routes brought them, among other items, jadeite, obsidian, and surely also plants. The Maya also had trade routes connecting them with *Teotihuacan* (*teotl*, god, *iua*, sent, *can*, place) in the central valley of Mexico. Hay et al. (1940) had already discovered trade routes that connected the region of Guatemala City with San José in Costa Rica. Among the many items traded by the Maya from the Pacific coast were chocolate seeds (*Theobroma cacao*). Thus, at a minimum, the Maya were trading from Mexico City to San José, Costa Rica, and with both the Gulf and Pacific coasts.

In his second voyage of 1526–1527, Francisco Pizarro's pilot Bartolomé Ruiz was the first to report rafts off the coast of Panama traveling on the open sea like the Spanish ships (Prescott 1892). Those rafts were actually Peruvian, and belonged to traders from that Incan empire. To the Spanish, these rafts were *almadías* or *balsas*, while the Incas called them *huampu*, and gave names to at least 13 different kinds. Inhabitants of Darien, Panama, called them *pucro*. Urteaga (1978) has given one of the most detailed of recent discussions regarding transport of trading cargo aboard balsas.

There is historical and archaeological evidence that Peruvians used these *balsas* to trade with the Zapotecs in western Mexico (Eubanks 1999). Similarly, Madsen et al. (2001) were convinced that people from Puná Island in Ecuador were also involved in that trade system.

Northern South American People

Along coastal Venezuela and in the Guianas, there was a mosaic of Arawakan- and Cariban-speaking people. Because these areas were among the first places visited by the Spanish, considerable plant information was exchanged. Sometimes it is possible to find distinctions between the continental Arawakan or Cariban names of plants and those used on the islands. Frenchman Jean Baptiste Christophore Fuseé Aublet (1720–1778) visited the Cayenne region in the 1770s and gave many indigenous names, including working them into scientific names (Plotkin et al. 1991). That usage scandalized some Europeans like Linnaeus who thought scientific names should be based on Greek and Latin, but Aublet (1775) produced an important ethnobotanical commentary.

Surprisingly, there are even a few Tupí names from Brazil in plant names used in the Caribbean. Tupí and its relative language Guaraní were dominant along the eastern coast of South America at the time of European contact, and countless Tupí words have become elements of Creole Portuguese (Sampaio 1970, Ferreira 1978). Some of those words have even moved into English, e.g., genip, ipecac, or ipecacuanha.

Old World Newcomers

European Settlers

The early European settlers along the eastern coast of North America included the Dutch, English, French, German, Spanish, and Swedish. Among others who visited the region, Kalm ([1753–1761] 1972) noted the Dutch, German, and Swedish people in the New Jersey region and eastern Canada. It was, however, the English, French, and Spanish who are best known in the regions farther south. The southern Europeans brought with them Gaelic views merged with those of later people. All the Europeans were heavily influenced by Greek and Roman thinking, especially when it came to plants (Arber 1986, Baumann 1993, Meyer et al. 1999).

The first Europeans to interact with indigenous Floridians were the Spanish, officially beginning in 1513. However, as Williams ([1837] 1962) and others have noted, the region may have first been seen by Sebastian Cabot in 1497 when, under the flag of England, he sailed along the eastern shore. Beaufort, South Carolina, claims to be the second landing site of the Spanish in the New World, where Captain Pedro de Salazar landed in 1514 (Anonymous 2003). One of the first interactions between the Spanish and people in what is now the Carolinas occurred when a slaving raid was sent from Hispaniola in 1518. That excursion set the pattern for Spanish attempts to obtain workers for their mines (Williams [1837] 1962). Lucas Vásquez de Allyón made two exploratory voyages to the Atlantic Coast in 1520 and 1523 (Milanich and Milbrath 1989). On the first reconnaissance, Allyón found a cluster of islands off the coast of what is now South Carolina and named them *Santa Elena* (Stormer 1993). Then, in 1526, Allyón brought settlers to *La Florida* and tried to establish a colony at *San Miguel de Gualdape*, probably on the Georgia coast. They lasted only 3 months.

Better known is that there were French colonies on the *Río de San Juan* (St. Johns River) and at *Santa Elena* in South Carolina (Barrientos [1567] 1965, Bloodworth and Morris 1978). Jean Ribault, leader of the Huguenots, and his men built Charlesforte in 1562 on present-day Parris Island off Beaufort, South Carolina. In Florida, they built Fort Caroline in 1564. This intrusion of Protestants into the Spanish Catholic realm was not tolerated, and they were quickly eliminated (see *Sassafras* for more details). In spite of their short tenure in Florida, the drawings made by their artist Jacques Le Moyne provide much of what is known from the time.

Pedro Menéndez reestablished Spanish dominance at *Santa Elena* in 1566, after expulsion of the French in 1564. The name *Santa Elena* came to represent an area on the southeastern coast of North America including most of present-day South Carolina, North Carolina, and parts of Georgia. Later it was the name of the first permanent settlement by the Spanish on present-day Parris Island, South Carolina, and the first capital of Spanish holdings in North America. Nicholas Monardes (Frampton [1577] 1925) gave a first glimpse of Florida and medicinal plants in 1569.

The Spanish capital was moved to St. Augustine in 1565 (Bloodworth and Morris 1978). The full name of this settlement originally was *San Augustín de las Florídas, el siempre fiel Ciudad de San Augustín* (Saint Augustine of the Floridas, the ever faithful city of St. Augustine). Saint Augustine remained the Florida "capital" until Spain ceded the lands to the British in the 1763. Nearby was *San Mateo*, a new name given to Fort Caroline by Menéndez when he captured it on Saint Matthew's Day.

Although not successful in Florida, the French were able to settle in what is now eastern Canada, the Mississippi River valley, and Louisiana. There they interacted with indigenous people and even engaged them in raids against other tribes and settlers from other countries. For example, when Henri de Tonti was among the Illinois in 1682, he told about his instigation of intertribal conflict. He wrote: "The savages there [Illinois] are active and brave ... in war, when they think nothing of seeking their enemies at a distance of 500 or 600 leagues from their own country. This constantly occurs in the country of the Iroquois whom, at my instigation, they continually harrass" (Tonti in Borland 1965).

By the time Speck (1941) worked with the Houma in Louisiana, they had lost their language and spoke only Creole French. Swanton (1946) made repeated reference to indigenous southeastern people being driven from English-dominated regions to French-held Louisiana.

The early British did not try to compete with the Spanish in Florida, but settled farther north up the coast. Their most famous early attempts at settlements include Roanoke Island in 1584 (see Harriot [1590] 1972, Hulton 1972), Jamestown, Virginia in 1607, the Plymouth Colony, Massachusetts in 1620, and Charleston (as Charles Towne), South Carolina in 1670. The Roanoke colony was a failure, but there are still people who claim to be descended from mixing of those colonists with indigenous tribes (see www.lumbeetribe.com). Simmons ([1822] 1973), Vignoles ([1823] 1977), and Williams ([1837] 1962) all provided insight into the early years of Florida becoming a territory.

Europeans had completely different views on the human's place in the world than the indigenous people. The philosophy of these light-skinned races from the Old World was Judeo-Christian. Essentially all of these newcomers believed the biblical comment that they should "conquer the land." Kalm ([1753–1761] 1972) was one of many who complained about their disregard for the future by devastating their resources.

Africans

Probably the first Africans to be brought to the southeastern region were left in 1526 when the Spanish abandoned the abortive colony at *San Miguel de Gualdape* (Milanich and Milbrath 1989). These Africans were members of the *Mende* [*Minde*] tribe from what is now called Sierra Leone in West Africa. Allyón had brought these people specifically because they were experienced in growing rice (Stormer 1993).

Subsequently, other African slaves were brought to the area because of their knowledge of rice (Carney 2001). Among these people were the *Mende, Kisi, Malinke*, and *Bantu* (Stormer 1993). Because they could not speak the same language, the people created a new language and culture, melded from Elizabethan English and their own individual dialects and languages. Although outsiders call them *Gullah*, these people formerly called themselves the *Geech* or *GeeChee*. According to the *Encyclopedia Britannica* (2003), both *Gullah* and *Geechee* are West African tribal names, although one source says *Gullah* was derived from *Angola*. Their descendants still occupy coastal regions in Florida to South Carolina (see www.coax.net/people/lwf/gg_coal.htm).

Some of the Africans brought to the New World would not tolerate being slaves and escaped. As the ancestors of the Seminoles were forced farther south, some of these blacks joined them and fought by their sides during their wars with the British and then the Americans. Fairbanks (1978) recognized two main patterns of interactions between blacks and Seminoles.

Williams ([1837] 1962) said Seminoles held slaves, but Fairbanks (1978) called these individual client relationships. The blacks were advisors, particularly in matters involving whites. Both cite the same example of *Wealusta* (Black Creek) and a black individual named Abraham, who served as interpreter when the Seminoles sent a deputation to Washington, D.C. in the early 1800s.

The second pattern occurred where the Africans maintained farming villages beside those of the Seminoles. This arrangement hastened acculturation of the Seminoles but rarely had an impact on their gene pool; some say never, but Sturtevant (1955) said Billy Bowlegs was half black. These Africans were allies of the Seminoles and fought beside them. Williams ([1837] 1962), for example, recorded a group of about 50 blacks who fought with the Seminoles against Major Dade at the battle at the Big Withlacoochee (*we*, water, *łako*, big, *chee*, little, or "little river"), later known as the Dade Massacre (Simpson 1956, Bloodworth and Morris 1978).

The attitude of the Seminoles toward the Africans was markedly distinct from the general feeling among the Creeks and other Muskogean people (Fairbanks 1978). These tribes sometimes held blacks in slavery. On other occasions, as in the battles at Pensacola in 1816, the Choctaw were enlisted to help the English capture and squelch a "Negro" fort (Williams [1837] 1962).

Grimé (1976) attributed the African use of American plants to independent discovery. However, comparison with use by indigenous people suggests that they learned much, if not all, of their uses from them. On the other hand, Africans surely taught the indigenous tribes uses of Old World plants.

Africans brought with them a worldview more like indigenous people than the Europeans (Iwu 1993). They believe that everything, living or inanimate, has a "life force that can only be properly harnessed and utilized by the knowledgeable and the initiated." This has incorrectly been called animism. The spirits are not worshiped, but instead consulted to achieve a goal. Daily life is considered inseparable from religion.

The Ethnoflora

First European Impressions

Although they were typically sparse, many of the first records of the New World contained comments on plants and indigenous uses. Often, these reports of the New World in Europe sounded like modern advertising propaganda. Columbus started it all.

Columbus wrote of his first landing in the *Lucayos* [Bahamas] on 11 October: "When they had landed, they saw very green trees and much water and fruit of various kinds" (Jane and Skelton 1960). He was impressed with the trees, and on 13 October he wrote that the people who lived there "came to the ship in boats, which are made of a tree-trunk like a long boat and all of one piece. They are very wonderfully carved . . . and large, so that some forty or forty-five men came." The Taino called these boats *canoas,* which came into English as "canoes."

On 16 October Columbus wrote, "I saw many trees very unlike ours, and many of them had many branches of different kinds, and all coming from one root; one branch is of one kind and one of another, and they are so unlike each other that it is the greatest wonder in the world. How great is the difference between one and another! For example: one branch has leaves like those of a cane and another leaves like those of a mastic tree." That was probably the first European literary record of strangler figs (*Ficus*) growing on palms.

A month later, on the island of Cuba, Columbus was at a loss to explain what he had seen. His scribes wrote for him: "He praises all this very highly to the Sovereigns, and declares that he felt inestimable pleasure and delight at seeing it, and especially the pines, because here could be built as many ships as might be wished, bringing out the materials, except timber and pitch, of which there would be an abundance here. And he affirms that he has not praised it to the hundredth part of that which might be said, and that it pleased Our Lord always to show him something better, and that continually in his discoveries up to then he had gone from good to better, as well in the matter of the land and trees, herbs, fruit and flowers, as in that of the people, and always in a different manner, and as this was the case in one place, so it was in the next."

After being in Florida in 1562, French promoter of the Protestant Huguenot colony Jean Ribaut (ca. 1520–1565), wrote to his commander Admiral Gaspard de Châtillon, comte de Coligny (1519–1572), about finding "merueylous commodities and treasures of the country" in an English version that was published in London in 1563. Farther along he wrote: "we sailed and veewed the coast all along with an unspeakable pleasure of thodiferous smell and bewtye of the same. . . . The Indians received us verry gentelly with great humanytie, putting us of these fruites, even in our boates, as mulberries, respices and such other frutes as they found redely by the waye. We enterd and vewed the cuntry therabowte, which is the fairest, frutefullest and pleasantest of all the world, habounding in honey, veneson, wildfoule, forrestes, woodes of all sorts. the hiest, greatest and fairist vynes in all the wourld with grapes accordingly. There growith also many walnut-trees, hazeltrees and small cherytrees very fine and great, and generally we have sene there of the same symples and herbes that we have in Fraunce and of like goodness savour and tast."

That Ribaut was not above exaggeration is shown by his mention of honey. Bees (*Apis melifera*) were not yet introduced into North America, and there could have been no honey there yet. Moreover, it is likely that he was describing the region in what is now South Carolina because hazel trees (*Corylus*) do not grow in Florida.

Spanish biographer Barrientos ([1567] 1965) was similarly exuberant in writing about the exploits of Pedro Menéndez de Aviles. The *Adelantado* could not have known of many of these items from personal experience, so Barrientos must have used documents from the de Soto or some other expedition in producing his report. Kerrigan has rendered the archaic Spanish into modern English easier to understand, but he has retained the enthusiasm of the original. Barrientos wrote: "Although the land is flat, far inland there are some mountain ranges. Everywhere are extensive woods, undergrowth and trees.

Some of the trees are marvelously tall; among them are walnut [*nogales*], laurel [*laureles*], liquidambar [*liquidanbares*], savine [*savinas*, junipers], wild olive [*acebuches*], live oak [*encinas*], oak [*robles*], pine [*pinos*], and plum [*ciruelos*]. There are also wild vineyards [*parras silvestres*], cereals [*servales*, but now *cereales*], and palmettos [*palmitos*] like those of Andalucia. Mulberry trees [*morales*], suitable for silk-making, are plentiful. Large lakes are numerous, and many of them are deep. Maize fields [*mayzales*] are on every hand." A bit farther along he continued that he found among the "produce of the earth," beans [*frisoles*], maize [*maize*], squash [*calabazas*], walnuts [*nueces*], chestnuts [*castañas*], and filberts [*avellanas*].

Of the "countless grapes," Barrientos wrote that "muscatel grapes [*ubas moscateles*] abound, and they are as good as the Spanish variety. Extensive oak groves yield an abundance of acorns [*vellotas*]." Showing that he knew of prior expedition reports he wrote that "enclosed gardens are to be found in Coza," a Muskogean village in what is now Georgia. At *Cofitachiqui*, another Muskogee village in Georgia, "mulberry groves are so thick that it is necessary to cut them down in order to make plantings. From the mulberries the Indians make raisins [*pasas*], which are good to eat."

Indigenous and Other Utilization

Little information exists about plant use by the first Florida people. Examination of the extant documents reveals that we have about 3% of the flora recorded as used by Glades and Timucuan people (Austin 1980, 1997). Elsewhere in North America, we know that people used ~20% of flora (Yarnell 1964, Austin and McJunkin 1978). The only comparable American figure for a nearby area is Mexico, where Bye (1993) estimated a 23% ethnoflora (5,000/30,000 species). However, only 3352 species (~11%) have documented use in Mexico. A report by Manandkhar (2002) that about 20% of the flora of Nepal is used by people is also tantalizing.

We have about half the number of records (10%) for native plant usage by Mikasuki and Creek speakers of southern Florida (Sturtevant 1955, Snow and Stans 2001) as for other American groups (Bushnell 1909, Swanton 1928a, Hamel and Chiltoskey 1975, Bye 1993). Some of that knowledge has been preserved in modern traditions remaining among those people, and in the literature (Swanton 1928a, Sturtevant 1955, Howard 1984, Snow and Stans 2001). However, no previous study has compared the potential biodiversity with the species that have documented uses.

It is my experience that people are inquisitive and inventive with plants; comparison of the literature suggests the same. People in the New World have experimented with introduced plants to the point that they have incorporated hundreds of alien species into their ethnofloras (see Morton 1981, Arvigo and Balick 1993, Moerman 1998, Bennett and Prance 2000). The Old World peach (*Prunus persica*) is one example where indigenous Americans quickly adopted and spread a new species. As with the swine and opossums (see *Sagittaria*), old words were substituted for these new fruit trees. The Powhatan called plums *poskam*. Later, cognates in related Algonquian languages were applied to the peach, as *poʔkama* in Shawnee, and *pokkitons* in Ojibwa. A similar substitution took place in the Muskogean languages. Muskogee for the *Prunus americana* is *pvkánvhe* (plum tree). The Creeks and Seminoles call peaches *pvkanv*, while their relatives say *tohã:nî* (Mikasuki), *takkolá* (Chickasaw, Choctaw, Koasati), and *takkola* (Alabama). The peach was so widespread among indigenous people that biologists thought for decades it might be native to the New World.

Humans not only experiment with plants to see what impacts they have on people, but they exchange information with their neighbors. That does not mean that each person within even the same village or town will use a single species the same way. Instead, they often have different applications for the same species, demonstrating personal bias and experiences (cf. Snow and Stans 2001). Therefore, this compilation is based on a simple premise: *If people in other places use the plants, then people in Florida used them.*

Archaeological and historical records make it clear that the people in the southeastern United States were fervent travelers and traders. Mississippian Complex people in the period that preceded historical contact had a trade network that extended from the Gulf of Mexico to the Great Lakes, and from the Atlantic Coast to the Rocky Mountains (Brown et al. 1990, Booker et al. 1992). Drechsel (1997) makes the convincing argument that the Mississippian Complex peoples, with their many and varied unintelligible languages, developed the Mobilian Jargon, at least in part, so that they could carry out trade across the region. He wrote that the jargon was "the language of voyages, in which Southeastern and other Indians engaged enthusiastically and over great distances." There are documented records of native people traveling by both land trails and natural waterways. These travelers often went for several hundred miles and occasionally even farther than a thousand miles (Drechsel 1997). The use of the Mobilian Jargon in historic times reached at least to, but not including, the Cherokee in Tennessee and the Catabwa of the Carolinas, north to the Osage or Oto of the Missouri River, and west through about the eastern third of

Texas to the Caddo and Natchitoches. The language extended beyond the Muskogean region and incorporated a few words from their Algonquian neighbors on their north (cf. pecan in *Carya*, persimmon in *Diospyros*, and *sagamité* or hominy within *Zea*).

Glades people and Timucuans had ample opportunity to exchange information with their neighbors. We have historical records that Glades people interacted with the pre-European people in Cuba (Fontaneda [1575] 1944), with the Spanish once they arrived on that island and on mainland Florida, and with the Timucua and other people living to their north in the peninsula (Fontaneda [1575] 1944, Barcia 1951, Hann 1991). Therefore, comparison with nearby areas (Mexico, North, Central, and South America, the Bahamas, the Caribbean) should provide a Florida Ethnoflora. That comparison is made in Table 1, which lists all of the Florida native plants that are known to have been used by people surrounding Florida.

Comparative information has been taken from Alcorn (1984), Arvigo and Balick (1993), Ayensu (1981), Bartram ([1791] 1958), Densmore (1928), Fernald et al. (1958), Foster and Duke (1990), Gilmore (1977), Honychurch (1987), Krochmal and Krochmal (1973), Liogier (1974), Little and Wadsworth (1964), Morton (1974, 1981), Roig (1945), Small (1933), Standley (1924), Standley and Record (1936), Taylor (1940), and Balick et al. (2000). Other primary references containing pertinent information are Arnason et al. (1980), Coe and Anderson (1996), Lewis (1992), Millspaugh (1892), Schetky (1964), and Vogel (1970). Numerous additional publications are cited within the text dealing with individual species.

The species accounts treat 813 of the 886 Florida species with documented uses. Summary information on the remaining species is given in Table 2. Elsewhere, Moerman (1998) has given the most complete synopsis available. However, even that large compilation is inadequate for southern Florida, as he himself noted (Moerman 1991). An attempt has been made to include all kinds of uses, ranging from utilitarian (wood, posts, dyes, fibers for ropes, etc.) to food, medicinal, and in religion.

Wunderlin (1998) calculated that Florida has 2654 native species. Although that figure might be debated, it represents a working number for comparison. The 888 species (Table 1) in the Florida Ethnoflora represent 33.5% of the native, pre-Columbian flora. That figure is considerably above the percentage documented as used by indigenous Florida people ($\sim$10%) and by people in other areas (20%) as recorded by Yarnell (1964), Griffin (1967), Austin and McJunkin (1978), Bye (1993), and Manandkhar (2002). My compilation suggests that Florida's lost ethnoflora consists of at least 13.5% of the native plants, or the difference between the 20% ethnoflora in other areas and the one reconstructed for Florida (33.5%).

However, data in Table 3 and Table 4 indicate that even those figures probably are too low. If, indeed, those species in these tables were also used, then the Florida ethnoflora would be 888 + [Table 3 =] 59+ [Table 4 =] 36 = 983 species. That information might indicate that 37% of the flora was used by indigenous people.

I am convinced that even 37% is too low. There are many trees (especially genera like *Carya* and *Quercus*) that provide useful wood and fruits that are not accounted for in this discussion. The more I study indigenous attitudes, languages, and my own cultural background, the more I am convinced that we have greatly underestimated the number of plants used by our ancestors. I suspect that indigenous people used 100% of the woody species, and perhaps more than 50% of the other life-forms. Recent discussions with Brad Bennett suggest that he may be gathering data on people in Panama to support that view. Future students of the ethnofloras of the United States and elsewhere should be able test my hypothesis.

TABLE 1
Florida's Ethnoflora

Scientific Name	Comments	Family
Abutilon permolle (Willdenow) Sweet		Malvaceae
Acacia angustissima (Miller) Kuntze		Fabaceae
Acacia farnesiana (L.) Willdenow		Fabaceae
Acacia tortuosa (L.) Willdenow		Fabaceae
Acalypha rhomboidea Rafinesque	Table 2	Euphorbiaceae
Acanthocereus tetragonus (L.) Hummelinck	Table 2	Cactaceae
Acer negundo L.		Sapindaceae
Acer rubrum L.	Gp	Sapindaceae
Acer saccharinum L.		Sapindaceae
Acer saccharum Marshall		Sapindaceae
Achillea millifolium L.		Asteraceae
Acmella oppositifolia (Lamarck) R.K. Jansen		Asteraceae
Acoelorrhaphe wrightii (Grisebach & H. Wendland) H. Wendland ex Beccari		Arecaceae
Acrostichum aureum L.		Pteridaceae
Acrostichum danaeifolium Langsdorff & Fischer		Pteridaceae
Actaea pachypoda Elliott		Ranunculaceae
Adiantum capillus-veneris L.	T	Pteridaceae
Adiantum melanoleucum Willdenow		Pteridaceae
Adiantum tenerum Swartz		Pteridaceae
Aesculus pavia L.		Sapindaceae
Agalinis tenuifolia (Vahl) Rafinesque		Orobanchaceae
Agave decipiens Baker		Agavaceae
Agave neglecta Small		Agavaceae
Ageratina altissima (L.) R.M. King & H. Robinson		Asteraceae
Ageratum littorale A. Gray		Asteraceae
Agrimonia incisa Torrey & A. Gray		Rosaceae
Agrimonia microcarpa Wallroth		Rosaceae
Agrimonia rostellata Wallroth		Rosaceae
Aletris aurea Walter		Nartheciaceae
Aletris farinosa L.		Nartheciaceae
Allium canadense L.		Alliaceae
Alnus serrulata (Aiton) Willdenow		Betulaceae
Alternanthera flavescens Kunth		Amaranthaceae
Alvaradoa amorphoides Liebmann		Picramnaceae
Amaranthus australis (A. Gray) J.D. Sauer		Amaranthaceae
Amaranthus hybridus L.	Gp	Amaranthaceae
Ambrosia artemisiifolia L.		Asteraceae
Ambrosia hispida Pursh		Asteraceae
Ambrosia trifida L.		Asteraceae
Amelanchier arborea (F. Michaux) Fernald		Rosaceae
Amianthium muscaetoxicum (Walter) A. Gray		Melanthiaceae
Amorpha fruticosa L.		Fabaceae

TABLE 1 *(Continued)*

Scientific Name	Comments	Family
Amorpha herbacea Walter		Fabaceae
Ampelopsis arborea (L.) Koehne		Vitaceae
Amphicarpaea bracteata (L.) Fernald		Fabaceae
Amphitecna latifolia (Miller) A.H. Gentry	Table 2	Bignoniaceae
Amsonia tabernaemontana Walter	Table 2	Apocynaceae
Amyris balsamifera L.		Rutaceae
Amyris elemifera L.		Rutaceae
Andropogon floridanus Scribner		Poaceae
Andropogon gerardii Vitman		Poaceae
Andropogon glomeratus (Walter) Britton et al.		Poaceae
Andropogon virginicus L.		Poaceae
Angadenia berteri (A. de Candolle) Miers		Apocynaceae
Angelica venenosa (Greenway) Fernald		Apiaceae
Annona glabra L.		Annonaceae
Antennaria plantaginifolia (L.) Richardson		Asteraceae
Apios americana Medikus	G, Gp	Fabaceae
Apocynum cannabinum L.		Apocynaceae
Aquilegia canadensis L.		Ranunculaceae
Aralia spinosa L.		Araliaceae
Ardisia escallonioides Schiede & Deppe ex Schlechtendal & Chamisso		Myrsinaceae
Argusia gnaphaloides (L.) Heine		Boraginaceae
Arisaema dracontium (L.) Schott		Araceae
Arisaema triphyllum (L.) Schott		Araceae
Aristida spp.		Poaceae
Aristolochia serpentaria L.		Aristolochiaceae
Arnica acaulis (Walter) Britton et al.		Asteraceae
Arnoglossum atriplicifolia (L.) H. Robinson		Asteraceae
Artemisia campestris L.		Asteraceae
Arundinaria gigantea (Walter) Walter ex Muhlenberg		Poaceae
Asarum arifolia Michaux		Aristolochiaceae
Asarum virginianum L.		Aristolochiaceae
Asclepias humistrata Walter		Apocynaceae
Asclepias incarnata L.		Apocynaceae
Asclepias tuberosa L.		Apocynaceae
Asclepias viridiflora Rafinesque		Apocynaceae
Asclepias viridis Walter		Apocynaceae
Asimina incana (W. Bartram) Exell		Annonaceae
Asimina pygmaea (W. Bartram) Dunal		Annonaceae
Asimina reticulata Shuttleworth ex Chapman		Annonaceae
Asimina triloba (L.) Dunal		Annonaceae
Asplenium erosum L. (= *A. auritum* Swartz)		Aspleniaceae
Asplenium serratum L.		Aspleniaceae

TABLE 1 (*Continued*)

Scientific Name	Comments	Family
Asplenium trichomanes L.		Aspleniaceae
Aster carolinianus Walter		Asteraceae
Aster dumosus L.		Asteraceae
Aster lanceolatus Willdenow		Asteraceae
Aster lateriflorus (L.) Britton		Asteraceae
Aster linarifolius L.		Asteraceae
Aster praealtus Poiret		Asteraceae
Athyrium felix-foemina (L.) Roth ex Mertens		Aspleniaceae
Atriplex pentandra (Jacquin) Standley		Amaranthaceae
Aureolaria flava (L.) Farwell		Orobanchaceae
Aureolaria pedicularia (L.) Rafinesque		Orobanchaceae
Aureolaria virginiana (L.) Pennell		Orobanchaceae
Avicennia germinans (L.) L.		Avicenniaceae
Axonopus compressus (Swartz) Beauvois	Table 2	Poaceae
Azolla caroliniana Willdenow	Table 2	Azollaceae
Baccharis halimifolia L.		Asteraceae
Bacopa caroliniana (Walter) B.L. Robinson		Veronicaceae
Bacopa monnieri (L.) Pennell		Veronicaceae
Baptisia alba (L.) Ventenat		Fabaceae
Baptisia lecontii Torrey & A. Gray		Fabaceae
Batis maritima L.		Bataceae
Bejaria racemosa Ventenat	Table 2	Ericaceae
Berchemia scandens (Hill) K. Koch		Rhamnaceae
Betula nigra L.		Betulaceae
Bidens bipinnata L.		Asteraceae
Bidens frondosa L.		Asteraceae
Bidens laevis (L.) Britton et al.		Asteraceae
Bidens mitis (Michaux) Sherff		Asteraceae
Bignonia capreolata L.		Bignoniaceae
Blechnum occidentale L.		Blechnaceae
Blechnum serrulatum Richard		Blechnaceae
Bletia purpurea (Lamarck) de Candolle		Orchidaceae
Blutaparon vermiculare (L.) Mears		Amaranthaceae
Boehmeria cylindrica (L.) Swartz		Urticaceae
Borrichia arborescens (L.) de Candolle		Asteraceae
Borrichia frutescens (L.) de Candolle		Asteraceae
Botrychium virginicum (L.) Swartz		Ophioglossaceae
Bourreria succulenta Jacquin		Boraginaceae
Brasenia schreberi J.F. Gemlin	Table 2	Cabombaceae
Brickellia eupatorioides (L.) Shinners	Table 2	Asteraceae
Buchnera americana L.	Table 2	Orobanchaceae
Bulbostylis capillaris (L.) C.B. Clarke	Table 2	Cyperaceae

TABLE 1 (*Continued*)

Scientific Name	Comments	Family
Bursera simaruba (L.) Sargent	G	Burseraceae
Byrsonima lucida (Miller) de Candolle		Malphigiaceae
Caesalpinia bonduc (L.) Rydberg		Fabaceae
Caesalpinia major (Medikus) Dandy & Exell		Fabaceae
Cakile edentula (Bigelow) Hooker		Brassicaceae
Cakile lanceolata (Willdenow) O.E. Schulz		Brassicaceae
Callicarpa americana L.		Lamiaceae
Calycanthus floridus L.		Calycanthaceae
Calyptranthes pallens Grisebach		Myrtaceae
Campsis radicans (L.) Seemann ex Bureau		Bignoniaceae
Canavalia rosea (Swartz) de Candolle		Fabaceae
Canella winteriana (L.) Gaertner		Canellaceae
Canna flaccida Salisbury		Cannaceae
Capparis cynophallophora L.		Brassicaceae
Capparis flexuosa L.		Brassicaceae
Capraria biflora L.		Veronicaceae
Capsicum annuum L. var. *glabriusculum* (Dunal) Heiser & Pickersgill		Solanaceae
Cardamine concatenata (Michaux) O.E. Schulz		Brassicaceae
Cardiospermum halicacabum L.		Sapindaceae
Cardiospermum microcarpum Kunth		Sapindaceae
Carphephorus corymbosus (Nuttall) Torrey & A. Gray		Asteraceae
Carphephorus odoratissima (J.F. Gmelin) H. Hebert		Asteraceae
Carpinus caroliniana Walter		Betulaceae
Carya alba (L.) Nuttall		Juglandaceae
Carya aquatica (F. Michaux) Nuttall		Juglandaceae
Carya cordiformis (Wangenheim) K. Koch		Juglandaceae
Carya floridana Sargent		Juglandaceae
Carya glabra (Miller) Sweet		Juglandaceae
Carya illinoiensis (Wangenheim) K. Koch		Juglandaceae
Carya pallida (Ashe) Engler & Graebner		Juglandaceae
Cassytha filiformis L.		Lauraceae
Castanea dentata (Marshall) Borkhausen		Fagaceae
Castanea pumila (L.) Miller		Fagaceae
Catalpa bignonioides Walter		Bignoniaceae
Catopsis berteroniana (Schultes f.) Mez		Bromeliaceae
Catopsis floribunda (Brogniart) L.B. Smith		Bromeliaceae
Catopsis nutans (Swartz) Grisebach		Bromeliaceae
Cayaponia americana (Lamarck) Cogniaux		Cucurbitaceae
Cayaponia quinqueloba (Rafinesque) Shinners		Cucurbitaceae
Ceanothus americanus L.		Rhamnaceae
Celtis iguanaea (Jacquin) Sargent		Celtidaceae
Celtis laevigata Willdenow	Gp	Celtidaceae

TABLE 1 (*Continued*)

Scientific Name	Comments	Family
Celtis occidentalis L.		Celtidaceae
Celtis pallida Torrey		Celtidaceae
Cenchrus echinatus L.	Table 2	Poaceae
Cenchrus incertus M.A. Curtis (=*C. spinifex* Cavanilles)	Table 2	Poaceae
Centrosema virginianum (L.) Bentham		Fabaceae
Cephalanthus occidentalis L.	Gp	Rubiaceae
Ceratiola ericoides Michaux	Table 2	Ericaceae
Ceratophyllum demersum L.	Table 2	Ceratophyllaceae
Cercis canadensis L.		Fabaceae
Chaerophyllum procumbens (L.) Crantz	Table 2	Apiaceae
Chamaecrista fasciculata (Michaux) Greene		Fabaceae
Chamaecrista nicitans (L.) Moench	Gp	Fabaceae
Chamaecyparis thyoides (L.) Britton et al.		Cupressaceae
Chamaelirium luteum (L.) A. Gray		Melanthiaceae
Chamaesyce mesembryanthemifolia (Jacquin) Dugand		Euphorbiaceae
Chaptalia tomentosa Ventenat		Asteraceae
Chenopodium album L.		Amaranthaceae
Chenopodium berlandieri Moquin-Tandon		Amaranthaceae
Chimaphila maculata (L.) Pursh		Ericaceae
Chiococca alba (L.) Hitchcock		Rubiaceae
Chiococca parvifolia Wullschlagel ex Grisebach		Rubiaceae
Chionanthus virginicus L.		Oleaceae
Chromolaena odorata (L.) R.M. King & H. Robinson		Asteraceae
Chrysobalanus icaco L.	G	Chrysobalanaceae
Chrysophyllum oliviforme L.		Sapotaceae
Chrysopsis mariana (L.) Elliott		Asteraceae
Cicuta maculata L.		Apiaceae
Cirsium horridulum Michaux		Asteraceae
Cissampelos pareira L.		Menispermaceae
Cissus verticillata (L.) Nicolson & C.E. Jarvis		Vitaceae
Citharexylum fruticosum L.		Verbenaceae
Cladium jamaicense Crantz		Cyperaceae
Clematis baldwinii Torrey & A. Gray		Ranunculaceae
Clematis crispa L.		Ranunculaceae
Clematis virginiana L.		Ranunculaceae
Clethera alnifolia L.		Clethraceae
Cnidoscolus stimulosus (Michaux) Engelmann		Euphorbiaceae
Coccoloba diversifolia Jacquin	G	Polygonaceae
Coccoloba uvifera (L.) L.	G	Polygonaceae
Coccothrinax argentata (Jacquin) L.H. Bailey	G	Arecaceae
Cocculus carolinus (L.) de Candolle		Menispermaceae
Collinsonia canadensis L.		Lamiaceae

TABLE 1 *(Continued)*

Scientific Name	Comments	Family
Collinsonia serotina Walter		Lamiaceae
Colubrina arborescens (Miller) Sargent		Rhamnaceae
Colubrina elliptica (Swartz) Brizicky & W.L. Stern		Rhamnaceae
Commelina erecta L.		Commelinaceae
Commelina virginica L.		Commelinaceae
Conocarpus erectus L.		Combretaceae
Conoclinium coelestinum (L.) de Candolle		Asteraceae
Conopholis americana (L.) Wallroth		Orobanchaceae
Conyza canadense (L.) Cronquist		Asteraceae
Corallorhiza odontorhiza (Willdenow) Poiret		Orchidaceae
Cordia globosa (Jacquin) Kunth		Boraginaceae
Cordia sebestena L.		Boraginaceae
Coreopsis leavenworthii Torrey & A. Gray		Asteraceae
Coreopsis major Walter	Table 2	Asteraceae
Coreopsis tripteris L.	Table 2	Asteraceae
Cornus alterniflora L.f.		Cornaceae
Cornus amomum Miller		Cornaceae
Cornus asperifolia Michaux		Cornaceae
Cornus florida L.		Cornaceae
Cornus foemina Miller		Cornaceae
Crataegus aestivalis (Walter) Torrey & A. Gray		Rosaceae
Crataegus crusgalli L.		Rosaceae
Crataegus flava Aiton		Rosaceae
Crataegus marshallii Eggleston		Rosaceae
Crataegus spathulata Michaux		Rosaceae
Crataegus uniflora Münchhausen		Rosaceae
Crinum americanum L.		Amaryllidaceae
Crossopetalum rhacoma Crantz		Celastraceae
Crotalaria pumila Ortega		Fabaceae
Croton linearis Jacquin		Euphorbiaceae
Ctenium aromaticum (Walter) A.M. Wood		Poaceae
Cucurbita okeechobeensis (Small) L.H. Bailey		Cucurbitaceae
Cucurbita pepo L.	G	Cucurbitaceae
Cuscuta americana L.		Convolvulaceae
Cuscuta compacta Jusseau ex Choisy		Convolvulaceae
Cuscuta gronovii Willdenow ex Schultes		Convolvulaceae
Cuscuta umbellata Kunth		Convolvulaceae
Cynoglossum virginicum L.		Boraginaceae
Cyperus haspan L.		Cyperaceae
Cyrilla racemiflora L.		Cyrillaceae
Cyrtopodium punctatum (L.) Lindley		Orchidaceae
Dalbergia ecastophyllum (L.) Taubert		Fabaceae

TABLE 1 (*Continued*)

Scientific Name	Comments	Family
Daucus pusillus Michaux		Apiaceae
Descurania pinnata (Walter) Britton		Brassicaceae
Desmanthus illinoiensis (Michaux) MacMillan ex B.L. Robinson & Fernald		Fabaceae
Desmodium incanum de Candolle		Fabaceae
Desmodium paniculatum (L.) de Candolle		Fabaceae
Desmodium rotundifolium de Candolle		Fabaceae
Diclipta sexangularis (L.) Jusseau		Acanthaceae
Diodia teres Walter		Rubiaceae
Dioscorea villosa L.		Dioscoreaceae
Diospyros virginiana L.	T	Ebenaceae
Dirca palustris L.		Thymeliaceae
Distichlis spicata (L.) Greene		Poaceae
Dodonaea viscosa (L.) Jacquin		Sapindaceae
Drosera brevifolia Pursh		Droseraceae
Drosera capillaris Poiret		Droseraceae
Drosera filiformis Rafinesque		Droseraceae
Drosera intermedia Hayne		Droseraceae
Drosera tracyi (Diels) T. Macfarlane		Droseraceae
Drypetes lateriflora (Swartz) Krug & Urban		Euphorbiaceae
Echinacea purpurea (L.) Moench		Asteraceae
Echites umbellata Jacquin		Apocynaceae
Eleocharis equisetoides (Elliott) Torrey		Cyperaceae
Eleocharis geniculata (L.) Roemer & Schultes		Cyperaceae
Eleocharis interstincta (Vahl) Roemer & Schultes		Cyperaceae
Eleocharis rostellata (Torrey) Torrey		Cyperaceae
Elephantopus elatus Bertoloni		Asteraceae
Encyclia cocleata (L.) Dressler (=*Prosthechea cocheata* (L.) W.E. Higgins)		Orchidaceae
Encyclia tampensis (Lindley) Small	Table 2	Orchidaceae
Epifagus virginiana (L.) Barton		Orobanchaceae
Epigaea repens L.		Ericaceae
Equisetum hyemale L.		Equisetaceae
Eragrostis elliottii S. Watson		Poaceae
Erechtites hieraciifolia (L.) Rafinesque ex de Candolle		Asteraceae
Erigeron annuus (L.) Persoon		Asteraceae
Erigeron philadelphicus L.		Asteraceae
Erigeron pulchellus Michaux		Asteraceae
Eriogonum longifolium Nuttall		Polygonaceae
Eriogonum tomentosum Michaux	Table 2	Polygonaceae
Erithalis fruticosa L.		Rubiaceae
Ernodea littoralis Swartz		Rubiaceae
Eryngium aquaticum L.		Apiaceae
Eryngium yuccifolium Michaux		Apiaceae

TABLE 1 (*Continued*)

Scientific Name	Comments	Family
Erythrina herbacea L.		Fabaceae
Erythronium umbellatum C.R. Parks & Hardin		Liliaceae
Eugenia axillaris (Swartz) Willdenow		Myrtaceae
Eugenia confusa de Candolle		Myrtaceae
Eugenia foetida Persoon		Myrtaceae
Eugenia rhombea Krug & Urban ex Urban		Myrtaceae
Eupatorium capillifolium (Lamarck) Small		Asteraceae
Eupatorium compositifolium Walter		Asteraceae
Eupatorium cuneifolium Willdenow		Asteraceae
Eupatorium fistulosum Barrat		Asteraceae
Eupatorium hyssopifolium L.		Asteraceae
Eupatorium perfoliatum L.		Asteraceae
Eupatorium pilosum Walter		Asteraceae
Eupatorium purpureum L.		Asteraceae
Eupatorium rotundifolium L.		Asteraceae
Eupatorium semiserratum de Candolle		Asteraceae
Eupatorium serotinum Michaux		Asteraceae
Eustachys glauca Chapman	Table 2	Poaceae
Eustoma exaltatum (L.) Salisbury ex G. Don		Gentianaceae
Euthamia graminifolia (L.) Nuttall		Asteraceae
Evolvulus alsinoides (L.) L.		Convolvulaceae
Evolvulus convolvuloides (Willdenow) Stearn		Convolvulaceae
Evolvulus sericeus Swartz		Convolvulaceae
Evonymus americanus L.		Celastraceae
Evonymus atropurpureus Jacquin		Celastraceae
Exostemma caribaeum (Jacquin) Schultes		Rubiaceae
Exothea paniculata (Jusseau) Radlkofer ex T. Durand		Sapindaceae
Fagus grandiflora Ehrhart		Fagaceae
Ficus aurea Nuttall		Moraceae
Ficus citrifolia Miller		Moraceae
Fimbristylis spadicea (L.) Vahl	Table 2	Cyperaceae
Flaveria bidentis (L.) Kuntze		Asteraceae
Flaveria trinervia (Sprengel) C. Mohr		Asteraceae
Fleischmannia incarnata (Walter) R.M. King & H. Robinson	Table 2	Asteraceae
Forestiera segregata (Jacquin) Krug & Urban		Oleaceae
Fragaria virginiana Duchesne		Rosaceae
Fraxinus americana L.		Oleaceae
Fraxinus caroliniana Miller		Oleaceae
Fraxinus pennsylvanica Marshall		Oleaceae
Froelichia floridana (Nuttall) Moquin-Tandon	Table 2	Amaranthaceae
Galactia volubilis (L.) Britton		Fabaceae
Galium aparine L.		Rubiaceae

TABLE 1 (*Continued*)

Scientific Name	Comments	Family
Galium circaezans Michaux		Rubiaceae
Galium tinctorium L.		Rubiaceae
Galium triflorum Michaux		Rubiaceae
Galium uniflorum Michaux		Rubiaceae
Gaylussacia dumosa (Andrews) Torrey & A. Gray		Ericaceae
Gaylussacia frondosa (L.) Torrey & A. Gray ex Torrey		Ericaceae
Gaylussacia mosieri Small		Ericaceae
Gelsemium sempervirens (L.) W.T. Aiton		Gelsemiaceae
Genipa clusiifolia (Jacquin) Grisebach		Rubiaceae
Gentiana catesbaei Walter		Gentianaceae
Gentiana saponaria L.		Gentianaceae
Gentiana villosa L.		Gentianaceae
Geranium carolinianum L.		Geraniaceae
Gleditsia triacanthos L.		Fabaceae
Gnaphalium obtusifolium L. (=*Pseudognaphalium obtusifolium* (L.) Hilliard and B.L. Burtt)		Asteraceae
Gnaphalium purpureum L. (=*Gamochaeta purpurea* (L.) Cabera)		Asteraceae
Goodyera pubescens (Willdenow) R. Brown		Orchidaceae
Gordonia lasianthus (L.) J. Ellis		Theaceae
Gossypium hirsutum L.		Malvaceae
Gouania lupuloides (L.) Urban		Rhamnaceae
Gratiola aurea Pursh	Table 2	Veronicaceae
Guajacum sanctum L.		Zygophyllaceae
Guapira discolor (Sprengel) Little		Nyctaginaceae
Guettarda elliptica Swartz		Rubiaceae
Guettarda scabra (L.) Ventenat		Rubiaceae
Guzmania monostachya (L.) Rusby ex Mez		Bromeliaceae
Gymnanthes lucida Swartz		Euphorbiaceae
Habenaria floribunda Lindley		Orchidaceae
Hamamelis virginiana L.		Hamamelidaceae
Hamelia patens Jacquin		Rubiaceae
Harrisia aboriginum Small ex Britton & Rose	Table 2	Cactaceae
Harrisia fragrans Small ex Britton & Rose	Table 2	Cactaceae
Harrisia simpsonii Small ex Britton & Rose	Table 2	Cactaceae
Hedeoma hispidum Pursh		Lamiaceae
Helenium amarum (Rafinesque) H. Rock		Asteraceae
Helenium autumnale L.		Asteraceae
Helianthemum corymbosum Michaux	Table 2	Cistaceae
Helianthus annuus L.		Asteraceae
Helianthus occidentalis Riddell		Asteraceae
Helianthus strumosus L.		Asteraceae
Heliopsis helianthoides (L.) Sweet		Asteraceae
Heliotropium angiospermum Murray		Boraginaceae

TABLE 1 *(Continued)*

Scientific Name	Comments	Family
Heliotropium curassavicum L.		Boraginaceae
Heliotropium polyphyllum Lehmann		Boraginaceae
Hepatica nobilis Miller (= *Anemone ameriana* (de Candolle) H. Hara)		Ranunculaceae
Heterotheca subaxillaris (Lamarck) Britton & Rusby		Asteraceae
Hibiscus moschatus L.		Malvaceae
Hieracium gronovii L.		Asteraceae
Hippocratea volubilis L.		Celastraceae
Hippomane mancinella L.		Euphorbiaceae
Hordeum pusillum Nuttall		Poaceae
Hydrangea arborescens L.		Hydrangeaceae
Hydrocotyle umbellata Michaux	Gp	Araliaceae
Hymenocallis latifolia (Miller) M. Roemer		Amaryllidaceae
Hypelete trifoliata Swartz	Table 2	Sapindaceae
Hypericum fasciculatum Lamarck		Clusiaceae
Hypericum hypericoides (L.) Crantz		Clusiaceae
Ilex cassine L.	G, Gp, T	Aquifoliaceae
Ilex decidua Walter	Table 2	Aquifoliaceae
Ilex glabra (L.) A. Gray		Aquifoliaceae
Ilex opaca Aiton	Table 2	Aquifoliaceae
Ilex verticillata (L.) A. Gray	Table 2	Aquifoliaceae
Ilex vomitoria Aiton		Aquifoliaceae
Illicium floridanum J. Ellis	Table 2	Illiciaceae
Illicium parviflorum Michaux ex Ventenat	Table 2	Illiciaceae
Impatiens capensis Meerburgh		Balsaminaceae
Ipomoea alba L.		Convolvulaceae
Ipomoea imperati (Vahl) Grisebach		Convolvulaceae
Ipomoea indica (Burmann) Merrill		Convolvulaceae
Ipomoea macrorhiza Michaux	Table 2	Convolvulaceae
Ipomoea pandurata (L.) G. Meyer		Convolvulaceae
Ipomoea pes-caprae (L.) R. Brown		Convolvulaceae
Ipomoea sagittata Poiret		Convolvulaceae
Ipomoea violacea L.		Convolvulaceae
Iresine diffusa Humboldt & Bonpland ex Willdenow		Amaranthaceae
Iris hexagona Walter	Gp	Iridacaeae
Iris verna L.		Iridacaeae
Iris virginica L.		Iridacaeae
Itea virginica L.	Table 2	Iteaceae
Iva annua L.		Asteraceae
Iva frutescens L.		Asteraceae
Jacquemontia curtisii Peter ex Hallier f.	Table 2	Convolvulaceae
Jacquinia keyensis Mez		Theophrastaceae
Juglans nigra L.		Juglandaceae

TABLE 1 (*Continued*)

Scientific Name	Comments	Family
Juncus bufonius L.		Juncaceae
Juncus effusus L.		Juncaceae
Juniperus virginiana L.		Cupressaceae
Justicia angusta (Chapman) Small		Acanthaceae
Kalmia latiflora L.		Ericaceae
Kalstroemia maxima (L.) Hooker & Arnott		Zygophyllaceae
Koanophyllon villosum (Smith) R.M. King & H. Robinson		Asteraceae
Kosteleztkya virginica (L.) C. Presl ex A. Gray		Malvaceae
Krameria lanceolata Torrey		Krameriaceae
Krigia virginica (L.) Willdenow		Asteraceae
Krugiodendron ferreum (Vahl) Urban		Rhamnaceae
Lachnanthes caroliniana (Lamarck) Dandy		Haemodoraceae
Lactuca canadensis L.		Asteraceae
Lactuca floridana (L.) Gaertner		Asteraceae
Lactuca graminifolia Michaux		Asteraceae
Lagenaria siceraria (Molina) Standley	G	Cucurbitaceae
Laguncularia racemosa (L.) C.F. Gaertner		Combretaceae
Lantana depressa Small		Verbenaceae
Lantana involucrata L.		Verbenaceae
Laportea canadensis (L.) Weddell		Urticaceae
Lasiacis divaricata (L.) Hitchcock		Poaceae
Lechea minor L.	Table 2	Cistaceae
Leersia hexandra Swartz	Table 2	Poaceae
Leitneria floridana Chapman		Simaroubaceae
Lepidium virginicum L.		Brassicaceae
Lespedeza capitata Michaux		Fabaceae
Leucothoë axillaris (Lamarck) D. Don		Ericaceae
Liatris chapmanii Torrey & Gray		Asteraceae
Liatris gracilis Pursh		Asteraceae
Liatris graminifolia Willdenow		Asteraceae
Liatris tenuifolia Nuttall		Asteraceae
Licania michauxii Prance	G, Gp	Chrysobalanaceae
Licaria triandra (Swartz) Kostermans		Lauraceae
Lilium catesbaei Walter		Liliaceae
Lilium iridollae M.K. Henry		Liliaceae
Limonium carolinianum (Walter) Britton		Plumbaginaceae
Linaria canadensis (L.) Chazelles	Table 2	Veronicaceae
Linderna benzoin (L.) Blume		Lauraceae
Liquidambar styraciflua L.		Altingiaceae
Liriodendron tulipifera L.		Magnoliaceae
Lithospermum carolinense (J.F. Gmelin) MacMillan		Boraginaceae
Lithospermum incisum Lehmann		Boraginaceae

TABLE 1 *(Continued)*

Scientific Name	Comments	Family
Lobelia cardinalis L.		Campanulaceae
Lonicera sempervirens L.		Caprifoliaceae
Ludwigia palustris (L.) Elliott		Onagraceae
Ludwigia virgata Michaux		Onagraceae
Lupinus perennis L.		Fabaceae
Lycium carolinianum Walter		Solanaceae
Lycopodiella cernua (L.) Pichi-Sermolli		Lycopodiaceae
Lycopus virginicus L.		Lamiaceae
Lyonia fruticosa (Michaux) G.S. Torrey		Ericaceae
Lysiloma latisiliquum Bentham		Fabaceae
Lythrum alatum Pursh		Lythraceae
Maclura pomifera (Rafinesque) C.K. Schneider		Moraceae
Magnolia acuminata (L.) L.		Magnoliaceae
Magnolia grandiflora L.		Magnoliaceae
Magnolia macrophylla Michaux		Magnoliaceae
Magnolia pyramidata W. Bartram		Magnoliaceae
Magnolia tripetala (L.) L.		Magnoliaceae
Magnolia virginiana L.		Magnoliaceae
Maianthemum racemosum (L.) Link		Ruscaceae
Malus angustifolia (Aiton) Michaux		Rosaceae
Manfreda virginiana (L.) Rose		Agavaceae
Manilkara jaimiqui (C. Wright ex Grisebach) Dubard		Sapotaceae
Marshallia obovata (Walter) Beadle & F.E. Boynton		Asteraceae
Maytenus phyllanthoides Bentham		Celastraceae
Medeola virginiana L.		Liliaceae
Melanthera nivea (L.) Small		Asteraceae
Melanthium virginicum L. (= *Veratrum virginicum* (L.) Aiton f.)	Table 2	Melanthiaceae
Melochia tomentosa L.		Malvaceae
Melothria pendula L.		Cucurbitaceae
Menispermum canadense L.		Menispermaceae
Metopium toxiferum (L.) Krug & Urban		Anacardiaceae
Mikania scandens (L.) Willdenow		Asteraceae
Mitchella repens L.		Rubiaceae
Modiola caroliniana (L.) G. Don		Malvaceae
Monarda punctata L.		Lamiaceae
Monotropa uniflora L.		Ericaceae
Morinda royoc L.		Rubiaceae
Morus rubra L.		Moraceae
Mosiera longipes (O. Berg) Small	*	Myrtaceae
Mosses	Table 2	Multiple families
Muhlenbergia capillaris (L.) Trinius		Poaceae
Myrcianthes fragrans (Swartz) McVaugh var. *simpsonii* (Small) R.W. Long		Myrtacaeae

TABLE 1 (*Continued*)

Scientific Name	Comments	Family
Myrica cerifera L.		Myricaceae
Myrica heterophylla Rafinesque		Myricaceae
Myrica inodora W. Bartram		Myricaceae
Nelumbo lutea Willdenow		Nelumbonaceae
Nicotiana rustica L.		Solanaceae
Nicotiana tabacum L.		Solanaceae
Nothoscordum bivalve (L.) Britton	Table 2	Alliaceae
Nuphar advena (Aiton) W.T. Aiton	Gp	Nymphaeaceae
Nymphaea odorata Solander	Gp	Nymphaeaceae
Nymphoides aquatica (J.F. Gmelin) Kuntze		Menyanthacee
Nyssa aquatica L.		Cornaceae
Nyssa ogeche W. Bartram ex Marshall		Cornaceae
Nyssa sylvatica Marshall		Cornaceae
Obolaria virginica L.		Gentianaceae
Ocotea coriacea (Swartz) Britton		Lauraceae
Oenothera biennis L.		Onagraceae
Oenothera fruticosa L.		Onagraceae
Okenia hypogaea Schlectendahl and Chamisso	Table 2	Nyctaginaceae
Onoclea sensibilis L.	Table 2	Woodsiaceae
Onosmodium virginianum (L.) de Candolle	Table 2	Boraginaceae
Opuntia humifusa (Rafinesque) Rafinesque		Cactaceae
Opuntia stricta var. *dillennii* (Haworth) Haworth	G	Cactaceae
Orbexilium pedunculatum (Miller) Rydberg		Fabaceae
Orobanche uniflora L.		Orobanchaceae
Orontium aquaticum L.		Araceae
Osmanthus americanus (L.) A. Gray		Oleacee
Osmunda cinnamomea L.		Osmundaceae
Osmunda regalis L.	Gp	Osmundaceae
Ostrya virginiana (Miller) K. Koch		Betulacee
Oxydendrum arboreum (L.) de Candolle		Ericaceae
Oxypolis filiformis (Walter) Britton		Apiaceae
Packera anonyma (A.W. Wood) W.A. Weber & A. Löve (= *Senecio anonymus* A.W. Wood)		Asteraceae
Packera aurea (L.) A. Löve & D. Löve (= *Senecio aureus* L.)		Asteraceae
Panicum hemitomon Schultes	Gp	Poaceae
Panicum laxiflorum (Lamarck) Gould (= *Dichanthelium laxiflorum* Lamarck)		Poaceae
Panicum strigosum Muhlenberg ex Elliott (= *Dichanthelium strigosum* (Muhlenberg ex Elliott) Freckmann)		Poaceae
Parietaria floridana Nuttall		Urticaceae
Parthenocissus quinquefolia (L.) Planchon		Vitaceae
Paspalidium geminatum (Forsskål) Stapf		Poaceae
Paspalum conjugatum Bergius		Poaceae
Paspalum floridanum Michaux	Table 2	Poaceae

TABLE 1 (*Continued*)

Scientific Name	Comments	Family
Passiflora incarnata L.		Passifloraceae
Passiflora sexflora Jussieu		Passifloraceae
Passiflora suberosa L.		Passifloraceae
Pavonia paludicola Nicolson ex Fryxell		Malvaceae
Pectis prostrata Cavanilles		Asteraceae
Pedicularis canadensis L.		Orobanchaceae
Pediomelum canescens (Michaux) Rydberg		Fabaceae
Peltandra virginica (L.) Schott & Endlicher		Araceae
Penstemon laevigatus Solander		Veronicaceae
Pentalinon luteum (L.) B.F. Hansen & Wunderlin		Apocynaceae
Penthorum sedoides L.		Penthoraceae
Peperomia obtusifolia (L.) A. Dietrich		Piperaceae
Persea borbonia (L.) Sprengel	Gp	Lauraceae
Phalaris caroliniana Walter		Poaceae
Phasaeolus polystachios (L.) Britton et al.		Fabaceae
Phasaeolus vulgaris L.		Fabaceae
Phegopteris hexagonaptera (Michaux) Fée	Table 2	Thelypteridaceae
Phlebodium aureum (L.) J. Smith		Polypodiaceae
Phoradendron leucocarpum (Rafinesque) Reveal & M.C. Johnston		Viscaceae
Phragmites australis (Cavanilles) Trinius ex Steudel		Poaceae
Phryma leptostachya L.		Phrymaceae
Phyla lanceolata (Michaux) Greene		Verbenaceae
Phyla nodiflora (L.) Greene		Verbenaceae
Phyla stoechadifolia (L.) Small		Verbenaceae
Phyllanthus caroliniensis (Swartz) Swartz		Euphorbiaceae
Physalis walteri Nuttall		Solanaceae
Physocarpus opulifolium (L.) Maximowicz		Rosaceae
Phytolacca americana L.	Gp	Phytolaccaceae
Picramnia pentandra Swartz		Picramnaceae
Pilea microphylla (L.) Liebmann		Urticaceae
Piloblephis rigida (W. Bartram ex Bentham) Rafinesque		Lamiaceae
Pilocereus robinii (Lemaire) Byles & G.D. Rowley (= *Pilocereus polygonus* (Lemaire) Byles & G.D. Rowley)	Table 2	Cactaceae
Pinckneya bracteata (W. Bartram) Rafinesque		Rubiaceae
Pinguicula lutea Walter		Lentibulariaceae
Pinguicula pumila Michaux		Lentibulariaceae
Pinus clausa (Chapman ex Engelmann) Vasey ex Sargent		Pinaceae
Pinus echinata Miller		Pinaceae
Pinus elliottii Engelmann	G, Gp	Pinaceae
Pinus glabra Walter		Pinaceae
Pinus palustris Miller		Pinaceae
Pinus serotina Michaux		Pinaceae
Pinus taeda L.		Pinaceae

TABLE 1 (*Continued*)

Scientific Name	Comments	Family
Piscidia piscipula (L.) Sargent		Fabaceae
Pisonia aculeata L.		Nyctaginaceae
Pithecellobium keyense Britton ex Britton & Rose		Fabaceae
Pithecellobium unguis-cati (L.) Bentham		Fabaceae
Pityopsis graminifolia (Michaux) Nuttall		Asteraceae
Planera aquatica J.F. Gmelin	Table 2	Ulmaceae
Plantago aristata Michaux		Plantaginaceae
Plantago rugelii Decaisne		Plantaginaceae
Plantago virginica L.		Plantaginaceae
Platanthera ciliaris (L.) Lindley		Orchidaceae
Platanus occidentalis L.		Platanaceae
Pleopeltis polypodioides (Liebmann) E. Fournier		Polypodiaceae
Pluchea camphorata (L.) de Candolle		Asteraceae
Pluchea foetida (L.) de Candolle		Asteraceae
Pluchea odorata (L.) Cassini		Asteraceae
Pluchea rosea R.K. Godfrey		Asteraceae
Plumbago scandens L.		Plumbaginaceae
Podophyllum peltatum L.		Berberidaceae
Polygala balduinii Nuttall	Gp	Polygalaceae
Polygala boykinii Nuttall		Polygalaceae
Polygala cruciata L.		Polygalaceae
Polygala lutea L.		Polygalaceae
Polygala nana (Michaux) de Candolle		Polygalaceae
Polygala polygama Walter		Polygalaceae
Polygala rugelii Shuttleworth ex Chapman		Polygalaceae
Polygonatum biflorum (Walter) Elliott		Ruscaceae
Polygonum densiflorum Meisner		Polygonaceae
Polygonum hydropiperoides Michaux	Gp	Polygonaceae
Polygonum punctatum Elliott		Polygonaceae
Polymnia uvedalia (L.) L. (=*Smallanthus uvedalia* (L.) Mackenzie ex Small)	Table 2	Asteraceae
Polyporus sp.		Polyporaceae
Polypremum procumbens L.		Tetrachondraceae
Polystichum acrostichoides (Michaux) Schott		Dryopteridaceae
Pontederia cordata L.		Pontederiaceae
Populus deltoides W. Bartram ex Marshall		Salicaceae
Populus heterophyllus L.		Salicaceae
Prenanthes serpentaria Pursh		Asteraceae
Prunus americana Marshall		Rosaceae
Prunus angustifolia Marshall		Rosaceae
Prunus caroliniana (Miller) Aiton		Rosaceae
Prunus geniculata R.M. Harper		Rosaceae
Prunus myrtifolia (L.) Urban		Rosaceae

TABLE 1 (*Continued*)

Scientific Name	Comments	Family
Prunus serotina Ehrhart		Rosaceae
Prunus umbellata Elliott		Rosaceae
Pseudophoenix sargentii H. Wendland ex Sargent		Arecaceae
Psychotria nervosa Swartz		Rubiaceae
Psychotria sulzneri Small		Rubiaceae
Ptelea trifoliata L.		Rutaceae
Pteridium aquilinum (L.) Kuhn		Dennstaedtiaceae
Pterocaulon pycnostachyum (Michaux) Elliott		Asteraceae
Pycnanthemum albescens Torrey & A. Gray ex A. Gray		Lamiaceae
Pycnanthemum flexuosum (Walter) Britton et al.		Lamiaceae
Pyrrhopappus carolinianus (Walter) de Candolle		Asteraceae
Quercus alba L.		Fagaceae
Quercus chapmanii Sargent		Fagaceae
Quercus elliottii Wilbur (= *Q. pumila* Walter)	T	Fagaceae
Quercus falcata Michaux		Fagaceae
Quercus geminiata Small		Fagaceae
Quercus incana W. Bartram		Fagaceae
Quercus inopina Ashe		Fagaceae
Quercus laevis Walter		Fagaceae
Quercus laurifolia Michaux		Fagaceae
Quercus margaretta Ashe ex Small		Fagaceae
Quercus marilandica Müenchhausen		Fagaceae
Quercus michauxii Nuttall		Fagaceae
Quercus minima (Sargent) Small		Fagaceae
Quercus muehlenbergii Engelmann		Fagaceae
Quercus myrtifolia Willdenow	T	Fagaceae
Quercus nigra L.		Fagaceae
Quercus pagoda Rafinesque		Fagaceae
Quercus phellos L.		Fagaceae
Quercus shumardii Buckley		Fagaceae
Quercus sinuata Walter		Fagaceae
Quercus stellata Wangenheim		Fagaceae
Quercus velutina Lamarck		Fagaceae
Quercus virginiana Miller	G, Gp, T	Fagaceae
Randia aculeata L.		Rubiaceae
Ranunculus abortivus L.		Ranunculaceae
Rapanea punctata (Lamarck) Lundell		Myrsinaceae
Ratibida pinnata (Ventenat) Barnhart	Table 2	Asteraceae
Reynosia septentrionalis Urban		Rhamnaceae
Rhabdadenia biflora (Jacquin) Müller Argoviensis		Apocynaceae
Rhamnus caroliniana Walter		Rhamnaceae
Rhexia virginica L.		Melastomataceae

TABLE 1 (*Continued*)

Scientific Name	Comments	Family
Rhizophora mangle L.	G	Rhizophoraceae
Rhus aromatica Aiton		Anacardiaceae
Rhus copallina L.		Anacardiaceae
Rhus glabra L.		Anacardiaceae
Rhynchosia minima (L.) de Candolle	Table 2	Fabaceae
Rhynchosia reniformis de Candolle	Table 2	Fabaceae
Ribes echinellum (Coville) Rehder		Grossulariaceae
Rivina humilis L.		Petiveriaceae
Rorippa palustris (L.) Besser	Table 2	Brassicaceae
Rosa carolina L.		Rosaceae
Rosa palustris Marshall		Rosaceae
Roystonea regia (Kunth) O.F. Cook		Arecaceae
Rubus argutus Link		Rosaceae
Rubus cuneifolius Pursh		Rosaceae
Rubus flagellaris Willdenow		Rosaceae
Rubus trivialis Michaux	Gp, T	Rosaceae
Rudbeckia hirta L.		Asteraceae
Ruellia caroliniensis (J.F. Gmelin) Steudel		Acanthaceae
Rumex verticillatus L.		Polygonaceae
Sabal etonia Swingle ex Nash		Arecaceae
Sabal minor (Jacquin) Persoon		Arecaceae
Sabal palmetto (Walter) Loddiges ex Schultes & Schultes f.	G	Arecaceae
Sabatia bartramii Wilbur		Gentianaceae
Sabatia brevifolia Rafinesque		Gentianaceae
Sabatia campanulata (L.) Torrey		Gentianaceae
Sabatia stellaris Pursh		Gentianaceae
Sagittaria lancifolia L.	Gp	Alismataceae
Sagittaria latifolia Willdenow		Alismataceae
Salicornia bigelovii Torrey		Amaranthaceae
Salicornia perennis Miller		Amaranthaceae
Salix caroliniana Michaux	G, Gp	Salicaceae
Salix eriocephala Michaux		Salicaceae
Salix humilis Marshall		Salicaceae
Salix nigra Marshall		Salicaceae
Salvia lyrata L.		Lamiaceae
Sambucus nigra L. ssp. *canadensis* (L.) R. Bolli	Gp	Adoxaceae
Samolus ebracteatus Kuntze	Gp	Primulaceae
Samolus valerandi L.		Primulaceae
Sanguinaria canadensis L.		Papaveraceae
Sanicula canadensis L.		Apiaceae
Sanicula marilandica L.		Apiaceae
Sanicula odorata (Rafinesque) Pryer & Phillippe		Apiaceae

TABLE 1 *(Continued)*

Scientific Name	Comments	Family
Sanicula smallii E.P. Bicknell		Apiaceae
Sapindus saponaria L.		Sapindaceae
Sarracenia flava L.		Sarraceniaceae
Sarracenia minor Walter		Sarraceniaceae
Sarracenia purpurea L.		Sarraceniaceae
Sassafras albidum (Nuttall) Nees	T	Lauraceae
Saururus cernuus L.		Saururaceae
Schaefferia frutescens Jacquin		Celastraceae
Schizachyrium scoparium (Michaux) Nash		Poaceae
Scirpus americanus Persoon		Cyperaceae
Scirpus tabernaemontani C.C. Gmelin		Cyperaceae
Scoparia dulcis L.		Veronicaceae
Scrophularia marilandica L.		Scrophulariaceae
Scutellaria havanensis Jacquin		Lamiaceae
Sebastiania fruticosa (W. Bartram) Fernald	Table 2	Euphorbiaceae
Senna ligustrina (L.) H.S. Irwin & Barneby		Fabaceae
Serenoa repens (W. Bartram) Small	G, Gp, T	Arecaceae
Sesbania drummondii (Rydberg) Cory	Table 2	Fabaceae
Sesbania vesicaria (Jacquin) Elliott	Table 2	Fabaceae
Sesuvium portulacastrum (L.) L.		Aizoaceae
Sideroxylon foetidissimum Jacquin		Sapotaceae
Sideroxylon lanuginosum Michaux	Table 2	Sapotaceae
Sideroxylon salicifolium (L.) Lamarck		Sapotaceae
Silene virginica L.	Table 2	Caryophyllaceae
Silphium compositum Michaux		Asteraceae
Simarouba glauca de Candolle		Simaroubaceae
Sisyrinchium angustifolium Miller		Iridaceae
Sisyrinchium nashii E.P. Bicknell		Iridaceae
Sium suave Walter		Apiaceae
Smilax auriculata Walter	G, T	Smilacaceae
Smilax bona-nox L.	G, T	Smilacaceae
Smilax glauca Walter		Smilacaceae
Smilax havanensis Jacquin	G	Smilacaceae
Smilax laurifolia L.	G, Gp, T	Smilacaceae
Smilax pumila Walter		Smilacaceae
Smilax smallii Morong		Smilacaceae
Smilax tamnoides L.		Smilacaceae
Smilax waleri L.		Smilacaceae
Solanum bahamense L.		Solanaceae
Solanum carolinense L.	Table 2	Solanaceae
Solanum donianum Walpers		Solanaceae
Solanum erianthum D. Don		Solanaceae

TABLE 1 (*Continued*)

Scientific Name	Comments	Family
Solidago canadensis L.		Asteraceae
Solidago fistulosa Miller		Asteraceae
Solidago gigantea Aiton (=*Solidago leavenworthii* Torrey and A. Gray)		Asteraceae
Solidago nemoralis Aiton		Asteraceae
Solidago odora Aiton		Asteraceae
Solidago rugosa Miller		Asteraceae
Solidago sempervirens L.		Asteraceae
Solidago stricta Aiton		Asteraceae
Solidago ulmifolia Muhlenberg ex Willdenow		Asteraceae
Sophora tomentosa L.		Fabaceae
Sparganium americanum Nuttall		Typhaceae
Spartina alterniflora Loiseleur		Poaceae
Spartina bakeri Merrill	G	Poaceae
Spartina patens (Aiton) Muhlenberg		Poaceae
Spermacoce assurgens Ruiz and Pavón	Table 2	Rubiaceae
Spermolepis divaricata (Walter) Rafinesque	Table 2	Apiaceae
Spigelia anthelmia L.		Strychnaceae
Spigelia marylandica (L.) L.		Strychnaceae
Spiranthes sp.		Orchidaceae
Sporobolus pyramidatus (Lamarck) Hitchcock		Poaceae
Sporobolus virginicus (L.) Kunth		Poaceae
Stachys floridana Shuttleworth ex Bentham		Lamiaceae
Stachys tenuifolia Willdenow		Lamiaceae
Stachytarpheta jamaicensis (L.) Vahl		Verbenaceae
Staphyllea trifolia L.		Staphyleaceae
Stenandrium dulce (Cavanilles) Nees		Acanthaceae
Stillingia aquatica Chapman		Euphorbiaceae
Stillingia sylvatica L.		Euphorbiaceae
Strumpfia maritima Jacquin		Rubiaceae
Stylosanthes hamata (L.) Taubert		Fabaceae
Styrax americana Lamarck	Table 2	Styracaceae
Suriana maritima L.		Surianaceae
Swietenia mahagoni (L.) Jacquin		Meliaceae
Symphoricarpos orbiculatus Moench		Caprifoliaceae
Symplocos tinctoria (L.) L'Héritier		Symplocaceae
Taxodium distichum (L.) Richard	G	Cupressaceae
Taxus floridana Nuttall ex Chapman		Taxaceae
Tephrosia florida (F. Dietrich) C.E. Wood		Fabaceae
Tephrosia hispidula (Michaux) Persoon		Fabaceae
Tephrosia virginiana (L.) Persoon		Fabaceae
Tetragonotheca helianthoides L.	Table 2	Asteraceae
Tetrazygia bicolor (Miller) Cogniaux		Melastomataceae

TABLE 1 (*Continued*)

Scientific Name	Comments	Family
Teucrium canadense L.		Lamiaceae
Thalassia testudinum Banks & Solander ex J. König		Hydrocharitaceae
Thalia geniculata L.		Marantaceae
Thalictrum cooleyi Ahles		Ranunculaceae
Thalictrum macrostylum Small & A. Heller		Ranunculaceae
Thalictrum revolutum DC.		Ranunculaceae
Thalictrum thalictroides (L.) A.J. Eames & B. Boivin		Ranunculaceae
Thaspium trifoliatum (L.) A. Gray	Table 2	Apiaceae
Thelypteris kunthii (Desvaux) C.V. Morton		Thelypteridaceae
Thelypteris palustris Schott		Thelypteridaceae
Thrinax morrissii H. Wendland	G	Arecaceae
Thrinax radiata Loddiges ex Schultes & Schultes f.	G	Arecaceae
Tilia americana L.		Malvaceae
Tillandsia fasciculata Swartz		Bromeliaceae
Tillandsia recurvata (L.) L.		Bromeliaceae
Tillandsia usneoides (L.) L.	Gp	Bromeliaceae
Tillandsia utriculata L.		Bromeliaceae
Torreya taxifolia Arnott		Taxaceae
Tournefortia hirsutissima L.		Boraginaceae
Tournefortia volubilis L.		Boraginaceae
Toxicodendron radicans (L.) Kuntze		Anacardiaceae
Tragia saxicola Small		Euphorbiaceae
Tragia urticaefolia Michaux	Table 2	Euphorbiaceae
Trema lamarckianum (Schultes) Blume		Celtidaceae
Trema micrantha (L.) Blume		Celtidaceae
Triadenum virginicum (L.) Rafinesque		Clusiaceae
Trianthema portulacastrum L.		Aizoaceae
Trichostigma octandrum (L.) H. Walter		Petiveriaceae
Triodanis perfoliata (L.) Nieuwland		Campanulaceae
Tripsacum dactyloides (L.) L.		Poaceae
Tripsacum floridanum Porter ex Vasey		Poaceae
Typha domingensis Persoon	Gp	Typhaceae
Typha latifolia L.	Gp	Typhaceae
Ulmus alata Michaux		Ulmaceae
Ulmus americana L.		Ulmaceae
Ulmus rubra Muhlenberg		Ulmaceae
Uniola paniculata L.	Table 2	Poaceae
Uvularia perfoliata L.		Colchicaceae
Uvularia sessiliflora L.		Colchicaceae
Vaccinium arboreum Marshall		Ericaceae
Vaccinium corymbosum L.		Ericaceae
Vaccinium darrowii Camp		Ericaceae

TABLE 1 (*Continued*)

Scientific Name	Comments	Family
Vaccinium myrsinites Lamarck	Gp	Ericaceae
Vaccinium stamineum L.		Ericaceae
Valeriana scandens L.	Table 2	Caprifoliaceae
Vanilla barbellata Reichenbach f.		Orchidaceae
Vanilla phaeantha Reichenbach f.		Orchidaceae
Verbesina virginica L.		Asteraceae
Vernonia angustifolia Michaux	Table 2	Asteraceae
Vernonia missurica Rafinesque	Table 2	Asteraceae
Veronica peregrina L.	Table 2	Veronicaceae
Veronicastrum virginicum (L.) Farwell		Veronicaceae
Viburnum acerifolium L.		Adoxaceae
Viburnum dentatum L.		Adoxaceae
Viburnum nudum L.		Adoxaceae
Viburnum rufidulum Rafinesque		Adoxaceae
Vicia caroliniana Walter		Fabaceae
Vigna luteola (Jacquin) Bentham		Fabaceae
Viola bicolor Pursh		Violaceae
Viola conspersa Reichenbach		Violaceae
Viola sororia Willdenow		Violaceae
Vitis aestivalis Michaux	Gp, T	Vitaceae
Vitis rotundifolia Michaux	G, Gp, T	Vitaceae
Vitis shuttleworthii House	G, Gp	Vitaceae
Vittaria lineata (L.) Smith		Vittariaceae
Voyria parasitica Schlechtendal & Chamisso (=*Leiphiamos parasitica* (Schlechtendal and Chamisso) Ruyters and Maas)		Gentianaceae
Xanthorhiza simplissima Marshall		Ranunculaceae
Ximenia americana L.		Olacaceae
Xyris ambigua Beyrich ex Kunth		Xyridaceae
Xyris baldwiniana Schultes		Xyridaceae
Xyris brevifolia Michaux		Xyridaceae
Xyris caroliniana Walter		Xyridaceae
Xyris chapmanii E.L. Bridges & Orzell		Xyridaceae
Xyris difformis Chapman		Xyridaceae
Xyris elliottii Chapman		Xyridaceae
Yucca aloifolia L.	G	Agavaceae
Yucca filamentosa L.		Agavaceae
Yucca gloriosa L.		Agavaceae
Zamia integrifolia L. f.	G	Zamiaceae
Zanthoxylum americanum Miller		Rutaceae
Zanthoxylum clava-herculis L.		Rutaceae
Zanthoxylum fagara (L.) Sargent		Rutaceae
Zanthoxylum flavum Vahl		Rutaceae
Zea mays L.		Poaceae

TABLE 1 *(Continued)*

Scientific Name	Comments	Family
Zephyranthes atamasco (L.) Herbert		Amaryllidaceae
Zephyranthes simpsonii Chapman		Amaryllidaceae
Zephyranthes treatiae S. Watson		Amaryllidaceae
Zigadenus densus (Desrousseaux) Fernald	Table 2	Melanthiaceae
Zizania aquatica L.		Poaceae
Zizia aurea (L.) W.D.J. Koch		Apiaceae
Zornia bracteata J.F. Gmelin		Fabaceae

Note: Not in Moerman (1998) = 422. Percent of Florida ethnoflora added = 422/888 = 47.5%.
Abbreviations: Glades (G), Timucua (T), and pollen from coprolites at Fort Center (Gp). Family names have been brought in line with Wunderlin and Hansen (2003).
*Discussed and illustrated under *Eugenia.*

TABLE 2
Other Species Used

Scientific Name	Common Names	Uses	Group(s) or Area Using	Chemistry
Acalypha rhomboidea	copperleaf	Reputedly medical (Hocking 1997); several others are used in Middle America (Morton 1981)	Eastern USA	
Acanthocereus tetragonus	dildoe cactus, triangle-cactus	Fruits edible (Morton 1968c)	Florida; Caribbean	
Amphitecna latifolia	black calabash (Florida, English Antilles), *maigüira* (*güira,* the Taino name for *Crescentia,* Cuba), *güira de olor* (*de olor,* fragrant, Cuba), *higüero* [*jigüero*] *galion* (*higüero*, calabash, *galion*, resembling a Rubiaceae used to cure leather, Dominican Republic), *güira cimarrona* (wild calabash, Dominican Republic), *higüerillo* [*cimarrona*] ([wild] calabash, Puerto Rico), *calebasier vénéneux* (poisonous calabash, French Antilles), *calabasse zombi* (zombi calabash, Guadeloupe, Martinique)	Wood is used for yokes and plows (Liogier 1974); fruit has the same properties as the *higüero* (*Crescentia cujete*) and is used as an aperitive and to treat chest problems, sores, headache, dysentery, and other problems (Roig 1945, Liogier 1974, Morton 1981, Austin 1992, 1995)	Florida; Caribbean	
Amsonia tabernaemontana	eastern bluestar	Central nervous system stimulant (Hocking 1997)	USA	Indole alkaloids
Axonopus compressus	tropical [broad-leaf] carpet-grass, *pahí hiná:ba* (*pahí*, grass, *hiná:ba*, hard, Koasati)	Named and perhaps used (Kimball 1994); flowers chewed and used as face poultice in Samoa (Hocking 1997)	Louisiana	
Azolla caroliniana	mosquito fern, mosquito plant, *ifa kika* (Arawak, Suriname)	Believed to drive mosquitoes away (Hocking 1997)	Tropical America	
Bejaria racemosa	tarflower	Shortly after I arrived in Florida I was told that the plant was used as fly-paper before screens were available; no published source has been found to verify that	Florida	
Brasenia schreberi	watershield	Meskwaki used the plant to treat stomach ailments and during menses (King 1984); recent use includes treatment of dysentery and tuberculosis (Hocking 1997)	Meskwaki	
Brickellia eupatorioides	prairie false boneset, false boneset	Bitter tonic, diaphoretic; old injury, coughs (Hocking 1997, Moerman 1998)	Florida; Navajo	
Buchnera americana	blue-hearts	Seminoles used this in 1919 as an emetic for headache, dizziness (von Reis and Lipp 1982)	Florida; Caribbean	
Bulbostylis capillaris	*partillo chico* (little ?)	Mouthwash for toothache (von Reis and Lipp 1982)	Paraguay	
Cenchrus echinatus	bur-grass, West Indian bur-grass	Fruits eaten and used to make alcoholic beverage (Hocking 1997)	Tropical America, southern USA	

Cenchrus incertus (=*C. pauciflorus*, *C. spinifex*)	coast sandbur, *oktvfonv* (Muskogee), *roseta* (rosette, New Mexico), *sacate cochonillo* (little pig grass, San Luis Potosí)	A decoction of the plant used in making blue cornmeal mush, and also given to nursing mothers to increase lactation (Curtin 1947, von Reis and Lipp 1982)	Mexico, New Mexico	
Ceratiola ericoides	rosemary	Domestic medicine (Hocking 1997)	Florida	
Ceratophyllum demersum	coontail, hornwort	To treat biliousness, scorpion stings (Hocking 1997)	Southeastern USA	
Chaerophyllum procumbens	*cerofoglio selvatico* (wild elegant leaf, Italian), *cerfeuil savage* (wild chervil, French), chervil (from Latin *chaeraphylla*, through Old English *caerfille*; akin to Old High German *kervala*, and Middle Dutch *kervele*), spreading chervil	Campbell (1951) reported that the Chickasaws considered *C. procumbens* poisonous and used it as an emetic; subsequently, the specimens in the Gideon Lincecum (1793–1874) collection at Texas have now been determined as *Osmorhiza longistyla* (Torrey) de Candolle, and the report was incorrect (Joanne Birch, personal communication, June 2003)	Southeastern USA	
Coreopsis major	starry coreopsis	Leaves used to dye eggs (Hocking 1997)	Southern USA	
Coreopsis tripteris	tall tickseed	Decoction of stems for pain and internal bleeding (Moerman 1998)	Meskwaki	
Encyclia tampensis	butterfly orchid, *tha-tha or ta-ta-he* (minnow plant, Mikasuki), *suhatkochi inhombitat* (squirrel's food, Creek)	Seminole children carve pseudobulbs into animals (Bennett 1997)	Florida Seminoles	
Eriogonum tomentosum	deer tongue, wild buckwheat	Leaves put in pillow to relieve asthma (Murphee 1965)	Florida Panhandle	
Eustachys glauca	star grass, saltmarsh finger grass	Tea made from the roots are laxative (Murphee 1965)	Florida Panhandle	
Fimbristylis spadicea	*cabezón* (Venezuela), *esparrago* (Veracruz?), *esparto*, *kuna-biaya* (Guajiro?)	In northwestern Venezuela around Río Negro, Perija, this sedge used to heal umbilical cord in infants; paper has been made from the stems in Veracruz (Morton 1981)	Venezuela	
Fleischmannia incarnata (=*Eupatorium incarnatum*)	*mata* (herb, Mexico), pink boneset, pink eupatorium,	The drying plant is vanilla-scented (Fernald 1950, Steyermark 1963); formerly added to pipe smoking tobacco in Mexico; one of the bioactive compounds in the herbs is coumarin (Hocking 1997); in Belize, *F. pratensis* is listed as a medicinal herb (Balick et al. 2000)	Missouri; Mexico, Belize	Coumarin
Froelichia floridana	cotton weed [cottonweed], [field, plains] snake cotton (Florida, Texas), wild cotton (Alabama)	It was apparently a personal observation that Hocking (1997) recorded the plant being used in domestic medicine in Redbay, Florida in 1940; unfortunately, he gave no details	Florida	
Gratiola aurea	hedge hyssop	Roots as drastic and anthelmintic and for gout (Hocking 1997)	Southeastern USA	
Harrisia aboriginum	prickly apple-cactus	Fruits edible (Morton 1968c)	Florida	
Harrisia fragrans	Caribbean apple-cactus	Fruits edible (Morton 1968c)	Florida	
Harrisia simpsonii	Simpson's apple-cactus	Fruits edible (Morton 1968c)	Florida	

TABLE 2 (*Continued*)

Scientific Name	Common Names	Uses	Group(s) or Area Using	Chemistry
Helianthemum corymbosum	rock rose, frostweed, ice plant	Alterative, tonic, to treat wounds, diarrhea, dysentery (Hocking 1997)	Eastern USA	
Hypelete trifoliata	white-ironwood (Bahamas), inkwood (Florida), *hueso de costa* (coastal bone, Cuba), *chandelle marron* (wild candle, Haiti)	Judging from the common name, used for ink; the wood used in the Caribbean (Little et al. 1974)	Caribbean	
Ilex decidua	possum-haw, *solopí ahissi* (*solopí,* ghost, *ahissi,* medicine, Koasati)	Named and perhaps used (Folsom-Dickerson 1965)	Louisiana	
Ilex opaca	American holly, *ittohissi halokpá* [*thishalokpá*] (*ittó*, tree, *hissí*, leaf, *halokpá*, sharp, Koasati); *iti hishi halupa* (*iti*, tree, *hishi*, leaf, *halupa*, sharp, Choctaw)	Bark boiled in water and infusion rubbed on itching skin by Koasati (Taylor 1940); Alabama, Catawba, and Choctaw used the infusion to wash sore eyes (Taylor 1940); Cherokee chewed berries for upset stomach (Hamel and Chiltoskey 1975); leaves and bark contain neutral bitter principle used as bitter, and to treat fevers in malaria, and rheumatism by Mississippi blacks (Hocking 1997)	Louisiana, Mississippi	
Ilex verticillata	yaupon	Yanovsky (1936) said the leaves were use to make tea in the northeastern states; Vogel (1970), King (1984), and Moerman (1998) found records of the Iroquois, Ojibwa, and Meskwaki treating diarrhea and sores with the plants	Iowa to Canada	
Illicium floridanum	stink bush (Florida, South Carolina), purple anise, tree anise	Flowers and crushed leaves smell like dead fish; bark aromatic tonic; leaves and fruits toxic (Hocking 1997)	Southeastern USA	
Illicium parviflorum	star anise, yellow anise tree	Bark source of anise oil; roots used like sassafras (Hocking 1997)	Southeastern USA	
Ipomoea macrorhiza	largeroot morning glory (Florida), *liane à minguet* (French)	A rare plant in Alabama, Florida, Georgia, Mississippi and the Carolinas (Beaufort and Georgetown Counties SC; Brunswick County, NC); the species is mostly associated with Indian mounds and perhaps that is why Hocking (1997) says "used medicinally"		Probably contains resins in the roots
Itea virginica	sweet-spire, tassle-white, Virginia tea, Virginia willow	A common name alludes to a use as a tea; the Old World *I. chinensis* is used in a fruit decoction to aid digestion (Hocking 1997)	Southeastern USA	
Jacquemontia curtisii	white jacquemontia	Seminoles rub the whole plant decoction on ailing legs (von Reis and Lipp 1982)	Southern Florida	
Lechea minor	[thyme-leaf] pinweed, *bakloski:kiyikcî* (diarrhea medicine, Mikasuki), *narkaya: kitahiliswâ* (diarrhea medicine, Creek), *yehot nuwi'hare* (tie-leaf, Catawba)	Used to treat a "Bird Sickness," diarrhea, and fever (Sturtevant 1955); Catawba used some *Lechea* to treat sores (Speck 1934)		

Leersia hexandra	*herbe rasoir* (rasor herb, French)	Used in tuberculosis in Senegal (Hocking 1997)	Africa; also Americas?	
Linaria canadensis	[wild] toad flax	Used as diuretic, laxative, and ointment on hemorrhoids (Hocking 1997)	Eastern USA	
Melanthium virginicum (= *Veratrum virginicum*)	bunchflower, black flower	Decoction of roots toxic; used to treat scabies (Hocking 1997)		
Mosses	*tikbá icokhiskó* (beard of the fire, Koasati), *iithicooski* (fire whiskers, Mikasuki)	Seminoles and other people named and used several kinds of mosses; Kimball (1994) only recorded that the Koasati used the moss to hide where a fire had been; however, Seminoles used it to treat "Fire Sickness" or anything involving too much "heat" (Sturtevant 1955, Bennett 1997); Sturtevant (1955) found several genera of mosses called by the single Mikasuki name	Southeastern USA	
Nothoscordum bivalve	crow-poison, false-garlic, *hakhintowá* (*hakhá*, raven, *im*, its, *towá*, onion, Koasati)	Maybe given a name simply to make sure people knew it differed from *Allium*	Louisiana	
Okenia hypogaea	beach peanut (Florida), *mata cancer* (cancer stopper, Mexico)	Hinton simply wrote on the herbarium label, "uses: medicinal"	Mexico	
Onoclea sensibilis	sensitive fern	Roots eaten by the Iroquois (Yanovsky 1936) and used as medicine by the Ojibwa (King 1984)	Canada	Contains filicin, filmarone, and oleoresin; very poisonous (Swerdlow 2000)
Onosmodium virginianum	false gromwell, wild Job's tears, gravel weed	Roots and seeds diuretic and expel kidney stones (Hocking 1997)		
Paspalum floridanum	blue grass, Florida paspalum	A tea made with as much grass as can be circled with the forefinger and thumb is the right amount (Murphee 1965)	Florida Panhandle	
Phegopteris (= *Thelypteris*) *hexagonaptera*	broad beech-fern	Reported to have hypotensive activity (Hocking 1997)		Contains no phloroglucides or filicin
Pilocereus robinii (= *Pilocereus polygonus*)	key tree cactus, tree cactus	Fruits edible (Morton 1968c)	Florida	
Planera aquatica	American planetree, planertree, plene, sycamore, water elm [waterelm], water elmaha, *ittofoló* (*ittó*, tree, *foló*, spoon, Koasati)	Wood used by fishermen as floats (Harrar and Harrar 1946); wood is light, soft, not strong, close-grained, light brown, with a nearly white sapwood (Sargent 1905); apparently, Koasati used the wood for spoons	Eastern USA	
Polymnia uvedalia (= *Smallanthus uvedalia*)	bearsfoot, yellow leafcup	Used as a tonic and stimulant, in rheumatism, arthritis, gastritis, hepatic enlargement, cervix-uteri hypertrophy (Hocking 1997); *Polymnia canadensis* used by Houma (Moerman 1998)	Southeastern USA	Root contains resins, tannins, reducing sugars

TABLE 2 (*Continued*)

Scientific Name	Common Names	Uses	Group(s) or Area Using	Chemistry
Ratibida pinnata	cone-flower	Toothache (Hocking 1997); Moerman (1998) reported the Cheyenne, Dakota, Lakota, and Navajo using *R. columnifera* for medicine	Meskwaki	
Rhynchosia minima	*frijolillo* (Puerto Rico), least snout bean [snoutbean], *mehenibbech* (*mehen*, short cycle or life span, *ib* = little bean or white bean, *bech*, quail, Maya)	In Curaçao people make a leaf decoction for an eyewash to relieve inflammation; a stronger brew is used to treat venereal disease and as an abortifacient; considered diuretic and used for heart troubles (Morton 1981)	Morton (1981) says the species is native to Asia; Wunderlin (1998) says native to Florida	
Rhynchosia reniformis	dollar weed	Diaphoretic for kidney stones (Hocking 1997)		
Rorippa palustris	hispid yellow-cress, marsh [water] cress	Ceremonial eyewash (Moerman 1998); antiscorbutic, tonic, stimulant, diuretic; in salads (Hocking 1997)	Navajo	
Sebastiania fruticosa	gulf sebastiana	Alabama chewed root as a laxative (Hocking 1997, Moerman 1998)	Alabama	
Sesbania drummondii	rattlebox, coffee-bean, rattle bush, senna bean	Seeds as coffee substitute, but toxic (Hocking 1997)	Southeastern USA	Saponins
Sesbania vesicaria	coffee-bean [weed], bagpod, bladderpod	Seeds apparently edible after cooking (Hocking 1997)		Seeds with fatty oils and toxic principles
Sideroxylon lanuginosa	*afoló intató (afoló,* screech owl, *in,* its, *tató,* honey-locust, Koasati); *coyyí notá okwo <t>lí* (*coyyí,* pine, *notá,* underside, *okwo <t>lí,* dewberry, Koasati)	Yanovsky (1936) said the fruits were eaten in the southern states	Southeastern USA	
Silene virginica	fire pink	Vermifuge and nervine (Hocking 1997); Foster and Duke (1990) suggest that there may have been confusion between use of *Spigelia marilandica* and this as both are called "pinks"	Core (1967) found no evidence of the plants being used	Saponins (Hocking 1997)
Solanum carolinense	Carolina nightshade, [Florida] horse-nettle, *wasó i <t>lakí* (*wasó,* wolf, *im,* its, *<t> lakí,* arrow, Koasati)	The Cherokee took an infusion to expel worms, put it on sores and poison ivy, gargled it to relieve sore throat and goiter, and made a preparation of it to treat mange in dogs; they also put a string of roots around babies' necks to relieve teething problems, and put crushed leaves in milk (after cows were imported into the New World) to kill flies; the Koasati made a tea of it to wash out a sore mouth (Folsom-Dickerson 1965); Hocking (1997) found that people in Alabama had used it to treat thrush	Cherokee, Koasati	Fruits are rich in alkaloids, including solanidine, soladine, solasonine, solamargine and related glycoalkaloids, and alkamines (Perkins and Payne 1978, Hocking 1997)

Spermacoce assurgens	*cuaci-iche* (maybe *cuachtli*, snake, *ichcatl*, cotton, Veracruz), *l'herbe acouette* (extra herb, Dominica), woodland false buttonweed (Florida)	People in Veracruz use the herb in birth, and to treat snakebite (Vásquez and Jácome 1997); in Dominica it is made into a tea to treat colds (Von Reis 1973)	Dominica, Veracruz	
Spermolepis divaricata	fool's parsley, forked seedscale, rough-fruited spermolepis	Plant poisonous, nauseating (Hocking 1997)		
Styrax americana	[American] snowbell, mock orange, spring orange, storax	Used as an expectorant, aromatic, vulnerary, for chest disease (Hocking 1997)		
Tetragonotheca helianthoides	Florida gensing, pine-land ginseng	In domestic medicine, presumably as a ginseng substitute (Hocking 1997)	Florida?	
Thaspium trifoliatum	[purple] meadow-parsnip, round heart [plant], [golden] alexander	Vulnerary; in uterine disorders, chorea (Hocking 1997)		
Tragia urticaefolia	nettle-leaf noseburn, *pahí lambí* (*pahí*, grass, *lambi*, sting, Koasati)	Named and perhaps used (Kimball 1994); at least a warning to be avoided	Louisiana	
Uniola paniculata	sea oats (southeastern U.S.), *araña* (spider, Cuba), *espiga del mar* (seaside spike, Texas), *millo de mar* (seaside millet, Cuba), wild rice (Bahamas)	Indigenous people in the U.S. may have used the seeds as food as people in Central America use them today (Austin et al. 1991)	Southeastern USA, Caribbean	
Valeriana scandens	*caoirin-leana* (*caoirin*, little berry, *leana*, meadow, for *V. dioica*, Gaelic)	Starchy roots eaten (Hocking 1997); Moerman (1998) lists eight other species used across North America; Menominee and Meskwaki used *V. edulis* on cuts and wounds, and to stop bleeding; Menominee used *V. uliginosa* in several medicines	Menominee and Meskwaki	
Vernonia angustifolia	ironweed	Roots used as a bitter tonic, an alterative, an aperient, and in female disorders (Hocking 1997)		Sesquiterpene lactones glaucolide-beta and baldvernin
Vernonia missurica	Missouri ironweed, Drummond's ironweed, *a-'kai'kawdl'a, ee-pa-la* (Matress plant, Kiowa)	Kiowa chewed blossoms, made pad for overnight bedding, used decoction to treat dandruff, and purple dye (Vestal and Schultes 1939); aqueous decoction used to treat dandruff; flowers give purple dye (Hocking 1997)	Kiowa	
Veronica peregrina	purseland speedwell, neckweed	Domestic medicine to treat scrofula (Hocking 1997)		
Zigadenus densus	*enú·su* (Catawba), black snakeroot, crow poison [crowpoison], Osceola's plume	Planted by the Natchez, whom the Catawba called *yi enú·su* (people of crow poison), because they planted it with their corn (Young and Siebert 2003); presumably, they did that to protect the corn from pests	Natchez	Other species known to contain colchicine-like alkaloids, zygadenine (plus hypotensive esters of zygadine), and germine

There is a single source for these species being used. Not enough details are given to provide more information on these species, but they are also part of Table 1.

TABLE 3
Species in Florida for Which No Documentation of Use Has Been Found, but That Are Related to Others Known To Be Used

Species	Comments
Acrostichum aureum	Not in Moerman (1998); *Acrostichum danaeifolium* used, see text
Aletris bracteata	Not in Moerman (1998); other species used, see text
Aletris obovata	Not in Moerman (1998); other species used, see text
Allium cuthbertii	Not in Moerman (1998); other species used, see text
Aristolochia tomentosa	Not in Moerman (1998); other species used, see text
Astragalus michauxii	Other spp. in Moerman (1998)
Astragalus obcordatus	Other spp. in Moerman (1998)
Astragalus villosus	Other spp. in Moerman (1998)
Cardamine bulbosa	Not in Moerman (1998); other species used, see text
Cardamine parviflora	Not in Moerman (1998); other species used, see text
Cardamine pennsylvanica	Not in Moerman (1998); other species used, see text
Ceanothus microphyllus	Not in Moerman (1998); other species used, see text
Cliftonia monophylla	Honey source (Hocking 1997); not in Moerman (1998)
Corydalis flavula	Poison fide Lewis and Lewis (1977); other spp. in Moerman (1998)
Corydalis micrantha	Other spp. in Moerman (1998)
Crataegus michauxii	Not in Moerman (1998); other species used, see text
Crataegus phaenopyrum	Not in Moerman (1998); other species used, see text
Crataegus viridis	Not in Moerman (1998); other species used, see text
Croton punctatus	Not in Moerman (1998); other species used, see text
Ctenium floridanum	Roots of *C. aromaticum* used, see text; not in Moerman (1998)
Dalea candida	Root chewed or made into tea by Nebraska tribes (King 1984); see also Moerman (1998:192)
Dalea purpurea	Root chewed or made into tea by Nebraska tribes (King 1984); see also Moerman (1998:193)
Dioscorea quaternata	Hocking (1997) considered it a synonym of *D. villosa*, so perhaps same uses; also considered synonymous in *Flora of North America* (Committee 2002); not in Moerman (1998)
Eryngium aromaticum	Not in Moerman (1998); other species used, see text
Gleditsia aquatica	*Gleditsia triacanthos* used, see text
Hymenocallis choctawensis	Not in Moerman (1998); other species used, see text
Hymenocallis crassifolia	Not in Moerman (1998); other species used, see text
Hymenocallis duvalensis	Not in Moerman (1998); other species used, see text
Hymenocallis godfreyi	Not in Moerman (1998); other species used, see text
Hymenocallis occidentalis	Not in Moerman (1998); other species used, see text
Hymenocallis palmeri	Not in Moerman (1998); other species used, see text
Hymenocallis puntagordensis	Not in Moerman (1998); other species used, see text
Hymenocallis rotata	Not in Moerman (1998); other species used, see text
Hymenocallis tridentata	Not in Moerman (1998); other species used, see text
Iris brevicaulis	Not in Moerman (1998); other species used, see text
Iris fulva	Not in Moerman (1998); other species used, see text
Iris tridentata	Not in Moerman (1998); other species used, see text
Litsea aestivalis	Other spp. used (Hocking 1997); not in Moerman (1998)
Lupinus diffusus	None of FL spp. in Moerman (1998); other species used, see text
Lupinus villosus	None of FL spp. in Moerman (1998); other species used, see text
Lupinus westianus	None of FL spp. in Moerman (1998); other species used, see text

TABLE 3 (*Continued*)

Species	Comments
Mimosa quadrivalvis	Not in Moerman (1998); other species used (Hocking 1997)
Mimosa strigillosa	Not in Moerman (1998); other species used (Hocking 1997)
Neptunia pubescens	Not in Moerman (1998); other species used (Hocking 1997)
Orobanche minor	Not in Moerman (1998); other species used, see text
Pachysandra procumbens	Ornamental ground cover (Hocking 1997); not in Moerman (1998)
Passiflora pallens	Not in Moerman (1998); other species used (Hocking 1997)
Passiflora sexflora	Not in Moerman (1998); other species used (Hocking 1997)
Photinia pyrifolia	Not in Moerman (1998); other species used (Hocking 1997)
Phyllanthus abnormis	Not in Moerman (1998); other species used, see text
Phyllanthus liebmannianus	Other species used, see text
Phyllanthus pentaphyllus	Other species used, see text
Salix floridana	Not in Moerman (1998); other species used, see text
Smilax ecirrhata	Not in Moerman (1998); other species used, see text
Smilax lasioneuron	Not in Moerman (1998); other species used, see text
Smilax walteri	Not in Moerman (1998); other species used, see text
Spigelia gentianoides	Not in Moerman (1998); other species used, see text
Tradescantia ohiensis	Slender shoots of *T. occidentalis* eaten (Yanovsky (1936); Meskwaki made a medicine with it (King 1984); see also Moerman (1998)
Urtica chamaedryoides	Not in Moerman (1998); other species used (Hocking 1997)

TABLE 4
Genera within Florida That Have Related Species Known To Be Used in Nearby Areas

Genera	Sources
Aeschenomene	Morton (1981:273); Hocking (1997:19)
Anemone	Hocking (1997:50); Moerman (1998:72)
Argythamnia/Ditaxis	Hocking (1997:68)
Berlandiera	Hocking (1997:109); Moerman (1998:121)
Bigelowia	Hocking (1997:112)
Calamintha	Hocking (1997:139)
Callirhoë	Hocking (1997:142); Moerman (1998:131)
Campanula	Hocking (1997:145); Moerman (1998:135)
Carex	Moerman (1998:137); Moerman (1998:137)
Cupania	Morton (1981:487)
Dalea	Morton (1981:307); Moerman (1998:192)
Delphinium	Morton (1981:215); Hocking (1997:248); Moerman (1998:196)
Dyschoriste	Hocking (1997:267)
Elytraria	Morton (1981:840); Hocking (1997:276)
Gaillardia	Hocking (1997:323); Moerman (1998:241)
Hedyotis	Hocking (1997:364)
Hygrophila	Morton (1981:841); Hocking (1997:385)
Hymenopappus	Hocking (1997:386); Moerman (1998:271)
Hyptis	Morton (1981:753, 758); Hocking (1997:388); Moerman (1998:273)
Lygodesmia	Moerman (1998:324)
Macranthera	Moerman (1998:327)
Mimulus	Hocking (1997:500); Moerman (1998:343)
Nolina	Hocking (1997:530); Moerman (1998:357)
Ocimum	Morton (1981:228, 773); Hocking (1997:536)
Oncidium	Hocking (1997:544)
Palafoxia	Moerman (1998:376)
Philadelphus	Hocking (1997:586); Moerman (1998:391)
Physostegia	Hocking (1997:592); Moerman (1998:397)
Setaria	Lentz (2000)
Spermacoce	Morton (1981:853, 878); Hocking (1997:743)
Suaeda	Hocking (1997:765); Moerman (1998:546)
Trillium	Hocking (1997:817); Moerman (1998:568)
Valerianella	Hocking (1997:837); Moerman (1998:587)
Veratrum	Hocking (1997:841); Moerman (1998:588)
Zigadenus	Hocking (1997:884); Moerman (1998:612)
Zizyphus	Morton (1981:502); Hocking (1997:885); Moerman (1998:614)

TABLE 5
Species Excluded from the Florida Ethnoflora Because of No Firm Evidence That They Are Native to the State

Species	Morton	Hocking	Moerman	Comments
Argemone albiflora	—	(1997:68)	—	Blue-stem prickly poppy (Florida); sap used to treat sore eyes (Hocking 1997)
Argemone mexicana	(1981:241, 397)	(1997:68)	—	Prickly poppy (Florida), *cardo santo* (holy thistle, Cuba); used to treat skin problems, as an emetic, and to kill head lice (Morton 1981, Foster and Hobbs 2002); see also Vásquez and Jácome (1997)
Bidens pilosa	(1981: 913)	(1997:111)	—	Spanish needles (Florida), *romerillo* (little rosemary, Cuba); smoked as a tobacco substitute, as a vulnerary, to treat dysentery, and other problems (Morton 1981); see also Vásquez and Jácome (1997)
Boerhavia diffusa	(1981:191)	(1997:118)	—	Red spiderlings (Florida); used to treat a wide variety of ailments in the Caribbean (Morton 1981)
Centella asiatica	(1981:643)	(1997:169)	—	Pony food, spade-leaf (Florida); most of the uses are in Asia where it is known as *gotu kola*, among other names (Austin 1998c)
Chamaesyce hirta	(1981:425)	(1997:175)	—	Poll-pod sandmat (Florida), *lechera* (milky one, Cuba); considered antiasthmatic, antispasmodic, febrifuge, and diuretic, among other applications (Morton 1981, Hocking 1997, Vásquez and Jácome 1997)
Chamaesyce hypericifolia	(1981:427)	(1997:175)	(1998:153)	Graceful sandmat (Florida), *yerba de la niña* (little girl's herb, Cuba); Cubans consider it diuretic (Roig 1945)
Chamaesyce maculata	—	(1997:176)	(1998:153)	Spotted sandmat (Florida), *hierba de la golondrina* (swallow herb, Mexico); used for eye problems (Hocking 1997)
Chamaesyce nutans	—	—	(1998:153)	Eye-bane (Florida); Houma rubbed the latex on itchy skin, eczema, sores, and a decoction of the plant was given to infants (Speck 1941)
Chamaesyce prostrata	(1981:428)	(1997:175)	—	Prostrate sandmat (Florida), *hierba de la niña* (little girl's herb, Cuba), *golondrina* (swallow, Cuba)
Chamaesyce thymifolia	(1981:429)	—	—	Gulf sandmat (Florida); the latex is a drastic irritant purgative, used on skin problems; an essential oil that repels flies, fleas from dogs; it has a number of medical uses (Morton 1981)
Chenopodium ambrosioides	(1981:176)	(1997:178)	(1998:155)	Mexican tea (Florida), *apasote* (from Náhuatl, *epatl*, skunk, *tzotl*, filth, Cuba); widely used to expel intestinal worms
Cyperus articulatus	(1981:47)	(1997:239)	—	Jointed flat-sedge (Florida), *enea* (cattail, Cuba), *junco* (rush, Cuba); used to treat toothache, but ingestion said to cause drunkenness (Hocking 1997)
Cyperus erythrorhizos	—	—	(1998:190)	Red-root flat-sedge (Florida); seeds were cooked for mush by Kamia in Washington State (Hocking 1997)
Cyperus odoratus	—	(1997:239)	(1998:191)	Fragrant flat-sedge (Florida), *chintule* (Mexico); tea used to treat colic, cause drunkenness, and give luster and fragrance to the hair (Hocking 1997)
Cyperus squarrosus	—	—	(1998:191)	Bearded flat-sedge (Florida); acoma, Keres and Laguna tribes ate the roots (Moerman 1981)

TABLE 5 (*Continued*)

Species	Morton	Hocking	Moerman	Comments
Datura stramonium	(1981:794)	(1997:245)	(1998:194)	Jimpson weed (Florida), *chamico* (Cuba); a source of belladonna alkaloids well known around the world (Hocking 1997)
Drymaria cordata	(1981:212)	(1997:265)	—	West Indian chickweed (Florida); leaves eaten as salad; also used to treat headache, earache, liver, stomach, and kidney problems (Hocking 1997, Vásquez and Jácome 1997)
Eclipta prostrata	(1981:754, 924)	(1997:271)	—	False daisy (Florida); roots eaten, and leaves used to treat coughs and other maladies in Trinidad (Hocking 1997)
Elusine indica	(1981:30)	—	—	Goose-grass (Florida), *pata gallina* (hen's foot, Cuba), *grama de caballo* (horse grass, Cuba); seeds eaten, and the plants used to treat diarrhea, dysentery, and convulsions (Morton 1981)
Erechtites hieracifolia	(1981:927)	(1997:283)	—	Fire-weed (Florida), *achicoria de cabra* (goat's chickory, Cuba), *borraja* (borrage, Mexico); used as emollient, astringent, emetic, and to treat cough and hemorrhoids (Hocking 1997)
Hippobroma longiflora (=*Isotoma longiflora*)	(1981:897)	(1997:375)	—	Madam-fate (Florida), *revienta caballos* (horse breaker, Cuba); although it is a skin irritant and the fruit is toxic, it has been used to treat various diseases (Hocking 1997)
Micromeria brownei (=*Satureja brownei*)	(1981:784)	(1997:497)	—	Savory (Florida), *poleo* (pennyroyal, Cuba); used to treat stomach and lung problems and other maladies (Hocking 1997, Vásquez and Jácome 1997)
Oxalis corniculata	(1981:361)	(1997:554)	(1998:373)	Wood-sorrel, sour-grass (Florida), *vinagrera* [*vinagrillo*] (sour one, Cuba); leaves antiscorbutic, antipyretic, and used to treat bruises, corns, and for cleansing wounds (Hocking 1997, Vásquez and Jácome 1997)
Oxalis violacea	—	(1997:554)	(1998:374)	Wood-sorrel, sour-grass (Florida), *vinagrillo* (sour one, Cuba); considered antiscorbutic, diuretic; eaten in salads (Hocking 1997)
Pedilanthus tithymaloides	(1981:456)	(1997:572)	—	Devil's backbone (Florida), *itamo real* (royal *itamo*, Cuba), *gallito* (little rooster, Cuba), *zapatitos* (little shoes, Cuba); Millspaugh (1913) described the Florida plants as a distinct subspecies; Dressler (1957) wrote: "The plants were probably carried from Cuba at an early date by Spanish ships"; plants are caustic, irritating, emetic, toxic, and have been used in several medicines (Hocking 1997)
Peteveria alliaca	(1981:189, 366, 459, 514, 780)	(1997:582)	—	Guinea hen weed (Florida), *anamú* (specific Taino name, Cuba); considered antispasmodic, to stop convulsions, a counterirritant for toothache, and useful for many other maladies (Hocking 1997, Vásquez and Jácome 1997)
Poinsettia heterophylla	(1981:463	(1997:297)	—	Wild poinsettia (Florida), *hierba lechosa* (milkweed, Cuba), *corazón de María* (Mary's heart, Cuba); used to treat pulmonary problems, malarial chills, erysipelas, and as a substitute for castor oil (Hocking 1997, Vásquez and Jácome 1997)
Portulaca oleracea	(1981:206)	(1997:625)	(1998:434)	*Lobuna* (Choctaw), purslane (Florida), *verdolaga* (Cuba); eaten as greens; medicinal (Morton 1981, Hocking 1997, Moerman 1998)
Priva lappulacea	(1981:750)	(1997:629)	—	Velvet-bur (Florida), *farolito* (little lantern, Cuba), *amor seco* (dry love, Cuba); used to treat wounds, headache, whooping cough, and other problems (Hocking 1997)

Rhipsalis baccifera	—	(1997:665)	—	Mistletoe cactus (Florida), *disciplinilla* (Cuba); used to tie up fractures (Hocking 1997)
Sida acuta	(1981:533)	(1997:722)	—	Wire-weed (Florida), *malva de caballo* (horse mallow, Cuba), *chak'misbil* (red broom, Maya, Yucatán), *chichipe* (Yucatan); branches are made into brooms; smoked as a stimulant and adulterant for marijuana; leaves contain ephedrine (Martínez 1969, Schultes and Hofmann 1979)
Sida rhombifolia	(1981:535)	(1997:722)	—	Indian hemp, Indian tea (Florida), *malva de cochino* (Chinese mallow, Cuba); same use as *S. acuta*
Waltheria indica	(1981:558)	(1997:862)	(1998:600)	*Malva blanca* (white mallow, Cuba), *hierba del pasmo* (Sinaloa); used from Tarahumara of Chihuahua (Ford 1975) to blacks of Guyana (Austin and Bourne 1992)

"Caswalóhlik ilpá óɬaha, hiná:yap naksahká sowi:sowík." (One hears that it is time to eat the green corn, for now the cicada is calling) (Koasati; Kimball 1994).

"Nu̲'ti· hatkulʔha kai̲'ʔhade tci'pse ki̲·ye nuti·mi'ha. Wi·ti·'-kuseʔ kuse'pi'tʔkoda kai'ʔhode. Iwi̲·'ti·ki̲·ye yamu ke'hade atce̲'haʔ mo'no-tuke'hade duke'hade waha' matciri'k howe pa'i̲ʔhowe mo̲tu'hawe tcapako'ʔkutci·we. Du'g mi· howe buru'kdugro're mo'notuke'yu'we." (Cut the top of the medicine on the side of the sunset. Cut the medicine vine behind the side of the sunrise. Put the root in water, and put a little back in the ground, in order that it may not vanish and there will be no more. It will come back again when you put it back in the ground) (Catawba; Speck 1934).

"I think it would be interesting to compare how ethnobotanists, anthropologists, and linguists do field work on plants some time: it's a little like the story of the three blind men who feel different parts of an elephant" (Jack B. Martin, personal communication, 23 Oct. 2003).

The Ethnobotany

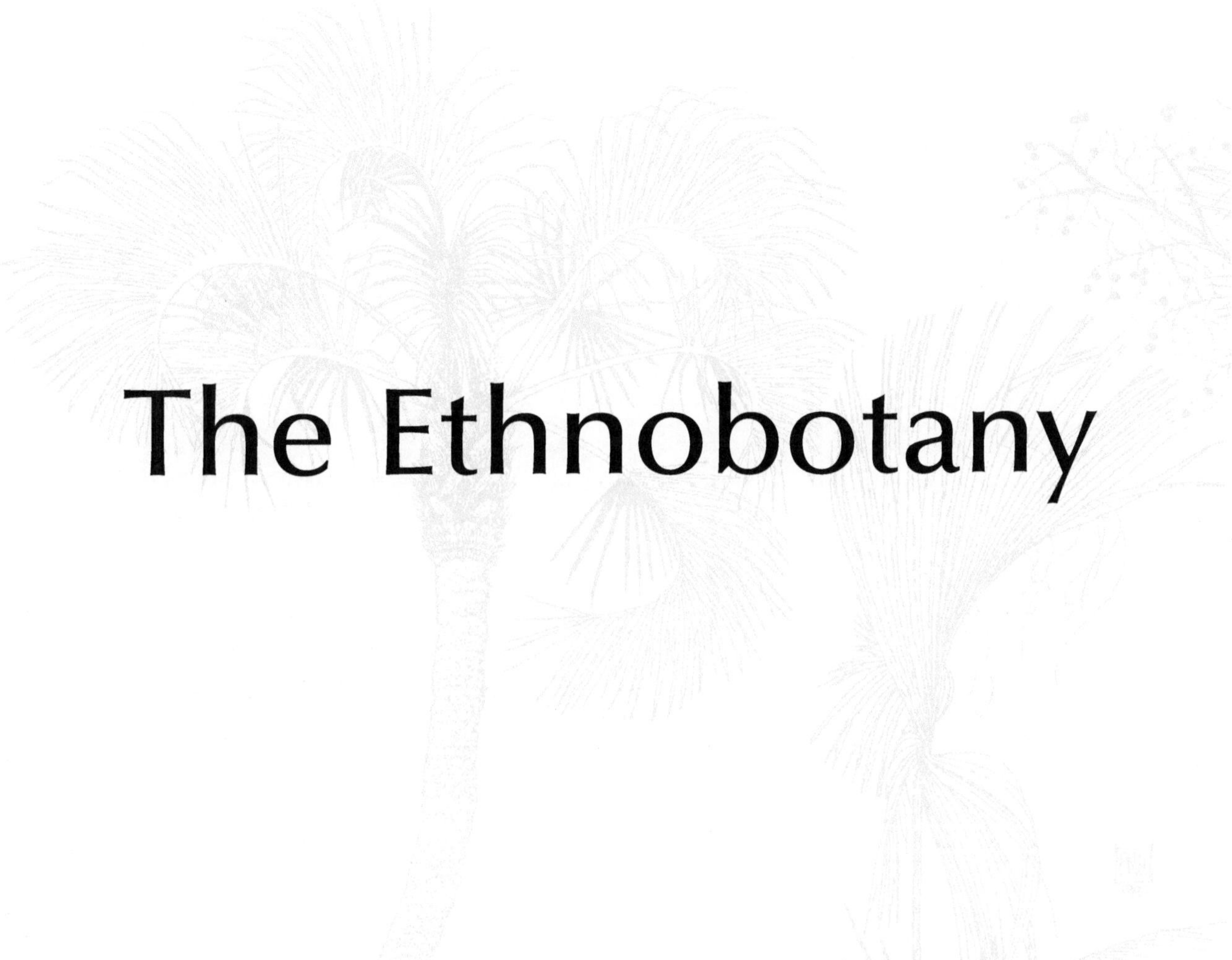

A

Abutilon

(British gardener Philip Miller adapted the Arabic name for the plant, *aututilun,* used by Arabic physician Ibn Sina [Avicenna], who died in A.D. 1037)

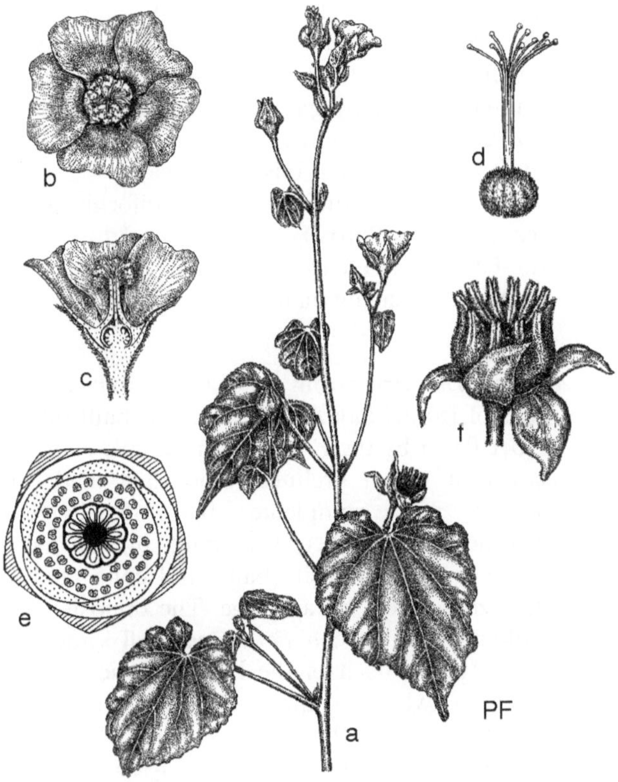

Abutilon permolle. a. Tip of flowering and fruiting branch. b. Flower, from above. c. Flower longitudinally dissected. d. Pistil. e. Floral diagram. f. Fruit. *Drawn by Priscilla Fawcett.* From Correll and Correll 1982.

Abutilon permolle (very softly pubescent)
arepa (the Taino word for a round cake of maize, typically served with butter and surrounding a filling of meat; the application to this plant is obscure, Belize)
butter print (from the fruit being used to make impressions on the surface of butter; a name shared with *A. theophrasti*, a common field weed in the midwestern United States)
butter weed (Bahamas)
daddy Joe (Bahamas)
[coastal] Indian mallow (Florida, Caribbean)
sacmizbil (*sak*, white, and *misbil*, broom, Maya, Yucatán; the related *Sida acuta* is *chak'misbil,* red broom)
velvet leaf [common] (Turks and Caicos)
velvety abutilon (Bahamas, a book name)
zacxiu (white herb, Maya, Yucatán)

Abutilon theophrasti was the first species of this genus that Europeans saw. These fiber plants were introduced through the Arabian trade route from Asia, and became known as Chinese jute, Chinese hemp, and Manchurian jute (Mabberley 1997). After the New World was discovered, Europeans found related species in the tropics. Probably, explorers were struck by the yellow flowers, or simply brought them back to Europe because the family (Malvaceae) was noted for producing medicinal species.

The original specimens and seeds of *A. permolle* were collected by Alexander von Humboldt in *America meridianalis* (Middle America). The plants were first named *Sida permolle* by Carl Ludwig von Willdenow (1765–1812), the director of the Berlin Botanical Garden. That name was created in his *Enumeratio Plantarum Horti Botanici Berolinensis* (List of Plants in the Berlin Botanical Garden), published in 1809. It was not until 1826 that the British botanist Robert Sweet moved the species into *Abutilon*; he did that in the *Hortus Britannicus* (Cultivated Plants of Britain).

The genus is also notable for mucilage that has been extracted from many species. Several species were exploited, and the extract was used as a poultice for tumors (Morton 1981). The mucilage serves as an emollient, and many use a decoction of it to clean wounds. In the Bahamas and Yucatán, the heated leaves are used to draw boils, and the decoction used to bathe sores. That usage is a continuation of medical application of the family that goes back hundreds, perhaps thousands, of years in Europe and the Americas. In that use, it is also like *Althea officinalis* (*Althea*, Greek, to heal; *officinalis*, of medicine), the mallow of Europe that provided the sweet originally called marsh mallow.

Emperor Charlemagne (742–814) thought so much of marshmallow that he ordered its cultivation. Doubtless, he had it used for both medicine and sweets. That original sweetmeat product was markedly different from our modern version—now the only plant extract marshmallows contain is the sucrose, and that sweetener does not come from mallows. Originally, the mucilage was extracted from marshmallow roots by boiling. It was used medicinally largely to soothe the skin and mucous membranes. That is essentially the use that *Abutilon* still has among some people.

The published information about the species does not record its use as a broom. However, the Maya common name clearly demonstrates that usage. In addition, in Yucatán, the fibers from the stems are made into twine.

Acacia

(Greek, *akakia*, tip, thorn, sharp point, used by Theophrastus, 372–287 B.C., and Dioscorides, fl. A.D. 40–80)

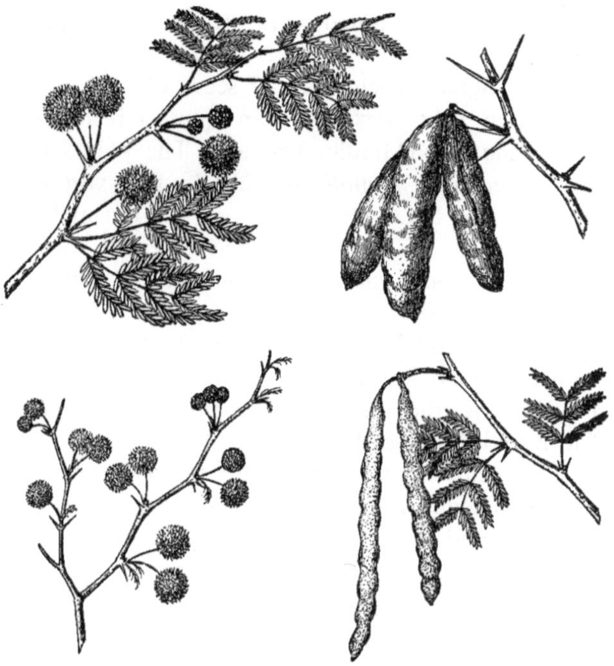

Acacia. *Acacia farnesiana* (above). *Acacia tortuosa* (below). Both from Sargent 1905.

acácia (Portuguese); *acacia* (English, Dutch, French, Spanish); *achachia* (Greek); *Akazie* (German); *akakia* (the juice, Coptic); *âqâqîya* (the juice, Arabic); *gaggia* (Italian)
mimosa (although applied to plants by at least 1619 in other languages, not appearing in English until ca. 1704); *mimosa* (from Latin *mimus*, Greek *mimos*, an actor or mime, with the feminine suffix *-osa*, resembling; said to allude to the mimicry of conscious life, French, Portuguese, Spanish); *Mimosa* (German)
wattle (origin uncertain, but appearing in Old English as *watul* ca. A.D. 700; possibly from Old Teutonic *wadlo-z*, to weave; originally meaning branches and twigs interwoven to form fences, walls, and roofs of buildings; also for the branches themselves, and then about 1810 to Australian legumes providing the branches)

Acacia angustissima (very narrow or slender)
cantemó (*kan*, yellow, *che'*, tree, *moc*, naked, Maya, Tabasco)
fern [prairie] acacia (Florida, Texas)
guaje [*huaj*] (*huaxin* [*hoatzin, hoaxin, uaxim, waxim*], simple term, often used for *Leucana*, Náhuatl, Yucatán); *guajillo* (little *guaje*, Sinaloa)
kuká (Guarajío, Sonora)
palo de pulque (*pulque* bush, Oaxaca)
timbe [*timbre*] (meritorius, Baja California, Oaxaca, San Luis Potosí)
white-ball acacia (Arizona, Texas)
xaax [*xa'ax*] (having many scales, Maya, Yucatán)

Acacia farnesiana (from the Farnese Palace Gardens of Rome. Palazzo Farnese was built on the Tiber River by the Farnese family, who commissioned it to an architect who then died; the project was completed by Michelangelo. Another palace then was built by the same family on the opposite bank and connected to Palazzo Farnese by a bridge. The opposite side building is called *La Farnesina*, well known to every Italian as it is the home of the Foreign Ministry, which is more often referred to as *Farnesina*, rather than *Ministero degli Esteri.*)
aroma [*aromo*] (fragrance, Puerto Rico, Yucatán, Nicaragua, Costa Rica, Colombia, Peru)
bihi (Zapotec, Oaxaca)
binorama [*vinorama, vignorama*] (Yaqui and Tarahumara, Baja California, Sonora, Sinaloa, Chihuahua)
casha [*cashaw*] (variant of *Acacia*, Puerto Rico, Belize)
cu'uca (Mayo, Sonora)
cuntich (*cun*, bewitch, *tich*, abscess on the eyelids like a grain of corn, Maya, Belize)
guacamaya francesa (French parrot, Cuba)
huisache [*guisache, huichin, huizache, huixachin, uisatsin*] (Texas, San Luis Potosí, Nuevo León, Tamaulipas, Coahuila, south to Veracruz and Chiapas, from *huit-axin*, *huit* from *uitqui*, to strike, *axin*, unguent, Náhuatl, Mexico)
opoponax (healing juice, from Greek, southern USA)
sweet acacia (Florida)
zubin-ché [*zubin*] (*che'*, tree, Maya, Yucatán)

Acacia tortuosa (twisted, referring to the pod)

casia [*akasee*] (variant of *Acacia*, Puerto Rico)
huisache (Durango, Aguascalientes)
poponax [wild] (healing juice, from Greek, Florida, Caribbean)
twisted acacia (Puerto Rico)

Europeans living around the Mediterranean were familiar with several species of *Acacia* before the New World was discovered (Flora Europaea 2003). They had names for them that they inherited from the Greeks and Romans (de Cleene and Lejeune 2002). When they arrived in the Americas, Europeans found even more species. Soon they were moving plants from continent to continent, and *A. farnesiana* was introduced from the Americas into Australia before there were European settlers there (Mabberley 1997). Not only were the flowers of several species pleasantly fragrant, but Mediterranean people knew the genus as a source of medicines.

The bark of *A. angustissima* is used to induce fermenting in *pulque* (see *Agave*), and in Yucatán as a remedy for enlarged spleen (Morton 1981). The species is high in tannin and is used in curing skins (Standley 1920–1926).

The wood of *A. farnesiana* is used. Bark and fruit provide materials for tanning and dyeing, and the fruit is used to make ink (Standley 1920–1926, Ford 1975). Gum from the trunk is used like gum-arabic and to cement broken china. Medicinally, the plant is used to treat dysentery, headache, dyspepsia, and bites from poisonous animals, particularly scorpions (Ford 1975, Morton 1981, Vásquez and Jácome 1997). Perfume is made from the flowers. The name *aromo* is said to be given because, when roots are chewed, they cause the spit to smell like excrement (Santamaría 1959).

A decoction of *A. tortuosa* leaves is applied for skin problems (Morton 1981). Root decoctions are also taken for gonorrhea and by women after childbirth.

Acer

(The classical name of the maple tree, Latin *acer*, *aceris*, literally meaning sharp, perhaps from the early use of the wood for making spikes and lances; akin to Akkadian *arku*, long-tailed, tall)

acero (Italian); *Ahorn* (German); *arce* (Spanish); *érable* (French); *asarot* (French)
awohátqua (Onondaga)
ayan̲' to'hayudi' (*ayan̲*, tree, *tohi*, blue, *ayudi*, wood; the bark made a blue dye, Biloxi)
héno (Muskogee)
lønne [*lønn*] (wages or pay, Norwegian)

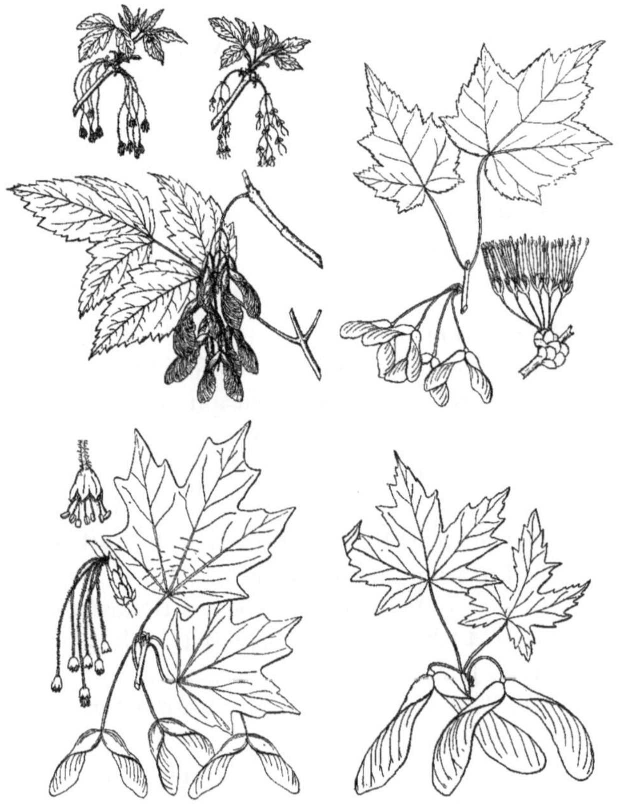

Acer. *Acer negundo* (top left). *Acer rubrum* (top right). *Acer saccharum* (lower left). *Acer saccharinum* (lower right). All from Sargent 1905.

maple (from Old English *mapel* or *mapul*, originally used in the compound *mapeltréow*, maple tree, akin to Old Saxon *mapuldur*)
schiechi kiminschi (*kiminschi*, plant, Delaware)

Acer negundo (a name from Hindi, *nirgandí*, because of the similarity of its leaves to *Vitex negundo*)
acero (Italian); *arce* (Spanish, Texas); *érable à giguère* (haunch maple, Quebec)
acezintle [*acecincle, acecinte*] (from *acezintli*, literally bad herb, Náhuatl, Chiapas)
ash-leaved maple
[California] box elder [boxelder, box-alder, box-tree] (in use by 1787, presumably because the wood was used for making boxes, and the leaf's similarity to the elder, *Sambucus*)
fresno de guajuco (ash *de guajuco*, Texas)
kaw-señ-añ-daw [*kaw-seì-aì-daw*] (*daw*, tree, Kiowa)
Manitoba maple
mistikosokaw (tree sugar, Cree); *sokawahtik* (sugar tree, Cree)
nahosh (Winnebago)

ósako (Pawnee)
palo de venagre (vinegar tree, Mexico)
raxoch (Guatemala)
sool (Navajo)
tashkadan (Dakota)
zhaba-ta-zhon (*zhaba*, beaver, *zhon*, wood, Omaha-Ponca); *zha'-be-ta-zhon* (*zha'-be*, beaver, *ta*, his, *zhon*, wood, Osage)

Acer rubrum (red)
ashak homeche [*asaykhô:mî:cî*] (*ashak*, acorn, *home*, bitter, *che*, little; an allusion to a pestle made from red maple altering the taste of acorn meal, Mikasuki)
chukchu (Choctaw. Swanton 1931 wrote that mortars from them gave food a taste "sufficiently bad to ruin one's stomach")
cicig'me'wc (Ojibwa); *cicigimé-wic* (Potawatomi)
héno (Muskogee); *hinô* [*haino*] (Creek, cf. Simmons [1822] 1973)
ipłimithatká (*ipło*, squirrel, *im*, its, *ittó*, tree, *hatká*, white, Koasati)
plaine (plane tree, from Old French *plasne* from the 1300s, itself derived from *Platanus* because of the broad leaves, Quebec); *plaine rouge* (red plane tree, Quebec)
red [scarlet, soft, swamp] maple
she-she-gum-maw-wis (sap flows fast, Ojibwa)
yap sake'here (red tree, Catawba)

Acer saccharum (sugar)
achsüñamihschi (*achsüña*, sugar, *mihschi*, tree, Delaware); *a'nina'tig* [*adjabubi'mi*, *inenatig*, *ininâtig*, *nin-au-tik*] (*migit*, tree, *nin*, our, Ojibwa); *inina'tig* (our tree, Potawatomi); *kisinamic* [*kisinamîc*] (cold tree, Potawatomi); *senamishi* (cold timber, Meskwaki)
can-n' (*can*, tree, *n'*, sap, Osage); *jan-ni* (Kansa, Omaha-Ponca); *nanni* (Iowa)
chan-ha san (*chan-ha*, bark, *san*, pale or whitish, Dakota); *nan-sank* (*nan*, wood, *sank*, real, genuine, Winnebago)
chukchu chito (*chukchu*, maple, *chito*, big, Choctaw); *chukchu hapi champuli* (*chukchu*, maple, *hapi*, salt, *champuli*, sweet, Choctaw)
érable à sucre (sugar maple, Quebec)
[Caddo, chalk, Florida, hard, rock, striped, sugar, northern sugar] maple
o-loñ-da'ke'li' (tree sap, Oneida); *u-ren'-na'-kri'* (tree sap, Tuscarora)
ta'niju'·ra (water wood, Winnebago)
thenomeysee (Shawnee, fide Edgar 1891; *mihschi*, tree, is the Delaware cognate for the second element)
waronawoenta girhit (*waronawoenta*, sugar, *girhit*, tree, Onondaga)
Zucherbaum (sugar tree, German); *Zukerholz* (sugar [maple] woods, German)

Europeans had lived with maples their entire history, and none of their native species has ever been used to produce sugar (Cheatham et al. 1995). However, when Europeans arrived in the New World, they found indigenous people harvesting sap from maples and other genera to produce a sweet drink and to use in cooking (Henshaw 1890). Although it may be difficult to comprehend from the perspective of this century, with its immediate and cheap sources of sugar, maples were an important source of sweets to indigenous Americans. History notes that, to the indigenous people in Quebec, *Acer saccharum* "affords great quantities of a cooling and wholesome liquor from which they make a sort of sweet" (Jefferys 1760 in Hedrick 1919). The month of March to the Menomini was *sho-bo-maw-kun-ka-zho* (sugar moon).

In 1870, the Winnebago and Ojibwa (Chippewa) sold the Northwest Fur Company 15,000 pounds of sugar a year made from this species (Hedrick 1919). During 1880 the United States produced 36 million pounds of maple sugar (Henshaw 1890). Considering that it took 30 to 40 gallons of sap to produce 1 gallon of syrup, the harvest was considerable. Some estimated that it took 4 gallons of syrup to yield 1 pound of sugar (Cheatham et al. 1995). In spite of the importance of these trees, they were commercially overshadowed by other sugar sources, perhaps because they could not produce enough to meet the demand.

Of the three maples in Florida, we have records of all being used for sugar (Smith 1933, Yanovsky 1936, Cheatham et al. 1995, Moerman 1998). *Acer negundo* was used for sugar by at least the Chiricahua (Arizona), Cheyenne (Montana and Oklahoma), Cree (Alberta, Saskatchewan to Montana), Dakota (Montana, Dakotas, Nebraska, Minnesota), Ojibwa (upper Midwest and southern Ontario), Mescalero (New Mexico to Texas), Montana (Montana), Omaha (Nebraska), Pawnee (Missouri River region), Ponca (Nebraska, South Dakota), and Winnebago (Wisconsin). *Acer rubrum* was used by the Abenaki (Quebec), Algonquin (New England, Quebec), Iroquois, including Cayuga and Onondaga (New York and Quebec). *Acer saccharum* was a source for the Algonquin including the Miami, Dakota, Iroquois, Malecite (New Brunswick), Menomini (Wisconsin), Meskwaki (Iowa), Micmac (Nova Scotia), Mohegan (Connecticut), Ojibwa, and Potawatomi (Wisconsin).

Yanovsky (1936) found records that the inner bark of *A. rubrum* was used to make bread.

The Koasati (southern Tennessee, northern Alabama) drank a medicinal infusion of red maple bark, mixed with *Ulmus americana* and *Nyssa sylvatica*, and washed wounds with it (Taylor 1940). Seminoles used red maple to treat "Ballgame Sickness" (sores, back or limb pains, hemorrhoids) (Sturtevant 1955). Red maple bark (collected from injured areas on the tree) is mixed with pine, live oak, laurel oak, pop ash, rabbit tobacco, wild persimmon, strap fern, gum bark (strangler fig), mulberry bark, and elderberry to treat postpartum problems, back pain, and postmenopausal ailments. The medicine, known as *ahecha keeke* (caretaking), is taken orally and used in a vaginal wash. Men are forbidden to touch this medicine (Bennett 1997).

Box-elder wood was used in ceremonies, to cook meat, for prayer sticks, and pipe stems. The Cherokee used red maple wood to make baskets, for carvings, bowls, lumber, and furniture (Hamel and Chiltoskey 1975). The Seminoles also used red maple wood to make spoons, arrowheads, and ox yokes (Bennett 1997). Sugar maple wood was more important than most other species (Moerman 1998). Box-elder wood was used to make charcoal for ceremonial painting and tattooing (Gilmore 1919).

Kalm ([1753–1761] 1972) noted that the European settlers of New England were making a dark blue dye of *A. rubrum*. That was probably a variation of a technique learned from indigenous people. However, use of the bark to make black ink was a decidedly European twist.

The Cherokee (western Carolinas) took an infusion of red maple and box-elder bark for cramps, dysentery, "female troubles," measles, and hives, and they used it to bathe sore eyes (Hamel and Chiltoskey 1975). Farther north, the Potawatomi and Mohegan made medicines from red maple (Smith 1933, Cheatham et al. 1995). Sugar maple had several medical uses, including as a blood purifier, against itch, and as an eye medicine, cough remedy, and expectorant (Cheatham et al. 1995, Moerman 1998). The Meskwaki and Ojibwa took an infusion of box-elder inner bark as an emetic (Moerman 1998).

Achillea

(Named in honor of the hero of the *Iliad*, Achilles, in Greek *Akhilleus*, son of Peleus and Thetis, who learned from Chiron the Centaur the healing properties of the plant)

cruaidh lus (*cruaidh*, sword, *lus*, herb, Gaelic)

Achillea millefolium. a. Habit. b. Stem segment showing detail of leaves. c. Flower head. d. Female and male flowers. e. Seeds. *Drawn by Regina O. Hughes*. From Reed 1971.

èarr thalmhuinn (*èarr*, noble, *thalmhuinn*, powerful, Gaelic)
herba militaris (military herb, name said to have been given the plants after the Greek Trojan Wars because of the use in staunching the wounds of soldiers)
lus chosgadh na fola (*lus*, herb, *chosgadh*, staunching, *na*, of, *fola*, blood, Gaelic)
milfoil (from Latin *mile, mille*, thousand, *folium*, leaf, appearing in English about A.D. 1000); *milefólio* (Portuguese); *milefolio* (Spanish); *milenrama* (Spanish); *millefeuille* (French)
Schafgarbe (sheep-yarrow, German)

Achillea millefolium (thousand-leaf)
a'djidamo'wano (squirrel tail, Ojibwa)
alcanfor (camphor, Veracruz, Yucatán)
alhucema (a name usually applied to *Lavandula*, Central America)

astaweskotawan (to put out a campfire, i.e., relieves burning pain, Cree); *kîshkatoa'soanûk* (flying squirrel tail, Potawatomi); *nokwe'sîkûn* (perfume reviver, Potawatomi)
athair-thalmhainn [*cathair-thalmhainn*] (*athair,* father, *thallmhainn,* powerful, Gaelic)
bloodwort
ciento en rama (hundred branches, Mexico)
cola de ardilla (squirrel tail, Central America)
common milfoil; *milenrama* [*milen rana*] (Spain, Central America); *mil-folhas* (thousand-leaf, Brazil); *milhojas* (thousand-leaf, Central America); thousand-leaf [-seal]
fani hasambish holba (*fani,* squirrel, *hasambish,* tail, *holba,* like, Choctaw)
flora de la pluma (feather flower, Spanish)
hank-sintsh (*hank,* woodchuck, *sintsh,* tail, Winnebago)
herbe à dindes (turkey herb, Quebec)
hierba del soldado (soldier's herb, Central America)
ihiseeyo (cough medicine, Cheyenne)
miskigonimaskigiah (head medicine, Cree); *ongunimasgigah* (bone medicine, Cree)
nose-bleed (USA, England)
plumajillo (little feather, Coahuila, Hidalgo, Puebla, Valley of Mexico, Veracruz, and others)
sanguinary (for the blood)
sereno de invierno (winter serenity, Dominican Republic)
taopi pexuta (wound medicine, Lakota); *xante canxlogan* (cedar weed, Lakota)
tlanquequetzal (remain kneeling, recorded 1615 by Ximénez for Aztecs, Náhuatl, Mexico)
we'-ts'a-çin̲-dse e-gon̲ [*wetsaoindse egon*] (like a rattlesnake's tail; *we'-ts'a,* snake, *çin̲-dse,* tail, *e-gon̲,* like, Osage)
yarrow [common, fragrant, western] (etymology uncertain, dating from ca. A.D. 700; cognate with Old English *gearwe,* Dutch *garwe,* German *garbe*)

According to legend, Achilles stopped the flow of blood from the wounds of his Myrmidon warriors with this plant (Baumann 1993, Quattrocchi 1999). Although the name and fame of this herb dates from the classical Greeks, its association with people in the Old World long predates that period. Yarrow was found in a Neanderthal grave dating back 60,000 years (Swerdlow 2000). Therefore, Europeans were well acquainted with the plants when they arrived in the New World, and there is still debate as to whether or not it is native here (Mabberley 1997, Diggs et al. 1999). Given yarrow's widespread use by indigenous people, there were surely some native forms and the species was likely circumboreal.

Use by the Illinois and Miami tribes on cuts was first reported in 1724 (Vogel 1970). The Cherokee used milfoil to treat spitting blood, bloody diarrhea, fever, bloody urine, and to aid sleep (Hamel and Chiltoskey 1975). The Chickasaws used it in a remedy for cramps in the neck (Swanton 1928b). The Creeks used the plant as a toothache remedy (Swanton 1928a). Additionally, the plant was used medicinally by at least the Abenaki, Algonquin, Bella Coola, Blackfoot, Carrier, Chehalis, Chipewyan, Ojibwa, Clallam, Cowlitz, Cree, Crow, Delaware, Flathead, Gitksan, Gosiute, Haisla, Hesquiat, Karok, Kallam, Kutenai, Kwakiutl, Lakota, Lummi, Mahuna, Makah, Malecite, Mendocino, Menomini, Micmac, Mohegan, Montagnais, Nitinaht, Ojibwa, Okanagan-Coville, Okanagon, Paiute, Potawotomi, Quileute, Quinault, Saanich, Salish, Shuswap, Skagit, Slave, Snohomish, Squaxin, Swinomish, Thompson, Winnebago, and Yuki (Gilmore 1919, King 1984, Moerman 1998, Marles et al. 2000). Others used it to make tea, as incense, and as a smudge to keep mosquitoes away (Moerman 1998).

Yarrow contains chamazulene, among other components of its essential volatile oils. That compound acts as an anti-inflammatory, antiallergenic, and helps stop spasms (Swerdlow 2000).

Acmella

(Diminutive of Greek *akme,* a point, edge, blossom, in reference to the taste of the foliage)

Acmella oppositifolia (having leaves opposite each other on the stem) (=*Spilanthes americana*)

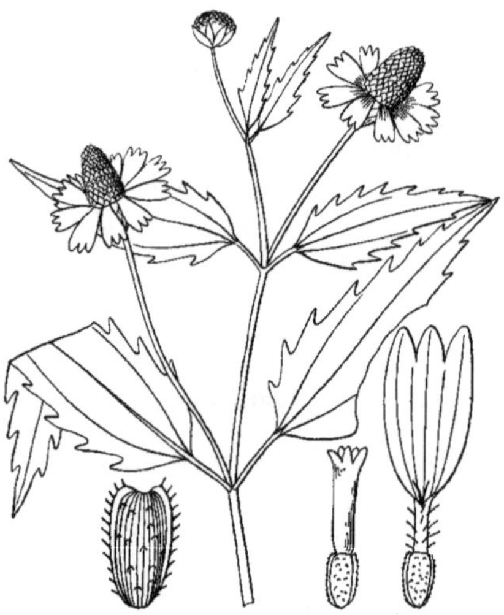

Acmella oppositifolia. From Britton and Brown 1898.

abecedária (alphabet plant; named this because it is locally believed that the plant will facilitate the tongues of babies and make them talk more easily, Brazil)
agrião bravo [*do Brasil, Pará*] (wild and bitter [from Brasil, Pará], Brazil)
botão de ouro (gold button, Brazil); *botón de oro* (gold button, Spanish); *botoncillo* (little button, Darien, Panama, Colombia, Ecuador)
calabaza (little head)
chisaca [*risacá*] (Chocó, Panama and Colombia)
cresson de Pará (cress from Pará, French)
flor de María (Mary's flower)
grana de oro (golden grain)
guaca (Colombia)
jambu (from *ya'mbu*, Tupí, Brazil); *jambu* [*mambu*] *açu* (big jambu, Tupí, Brazil)
ni (Huastec)
pimenta d'agua (water pepper, Brazil)
quemadera (it burns, Colombia)
rem [*rem'q en*]
spot-flower [spotflower] (Florida)
tripa de gallo (chicken guts)
xux (wasp, Maya, Yucatán)
yoyo quemado (burned *yoyo*, Colombia)

For many years *Acmella* was considered either a synonym or subdivision of *Spilanthes*, but that changed in the mid-1980s when it was reinstated as a separate genus (Mabberley 1997). There are about 30 tropical species in *Acmella*, but none is native to Europe. Spot-flower was first described by Thomas Walter in 1788. As a result of complications, however, Walter's species name cannot be used, and the correct binomial is *A. oppositifolia*.

Several species in the genus have become pantropical weeds, perhaps because people carried them purposefully. In South America the Brazilian or Pará cress, *A. oleracea*, is a medicinal and edible plant (LeCointe 1947). However, most people find its spilanthol-induced tingling of the tongue unpleasant (Mors et al. 2000).

The Tupisa Chocó eat the leaves, raw or cooked, and they are used to relieve liver trouble (Duke 1968). In Colombia the herbs are medicine for sore throat and mouth sores, and are used as cooked greens (Morton 1981).

Botoncillo contains the same spilanthol that *S. oleracea* does, and produces a similar tingling. Brazilians use the plants in a variety of medicines for mouth and throat ailments, but still consider it an "obligatory condiment in some regional dishes" (Mors et al. 2000).

Acoelorrhaphe

(Greek, *acoelo*, without the hollow, *rhaphe*, seam, because the seeds lacks the impressed raphe of many palms)

Acoelorrhaphe wrightii (named for Charles Wright, 1811–1885, a botanical collector in Cuba and elsewhere)

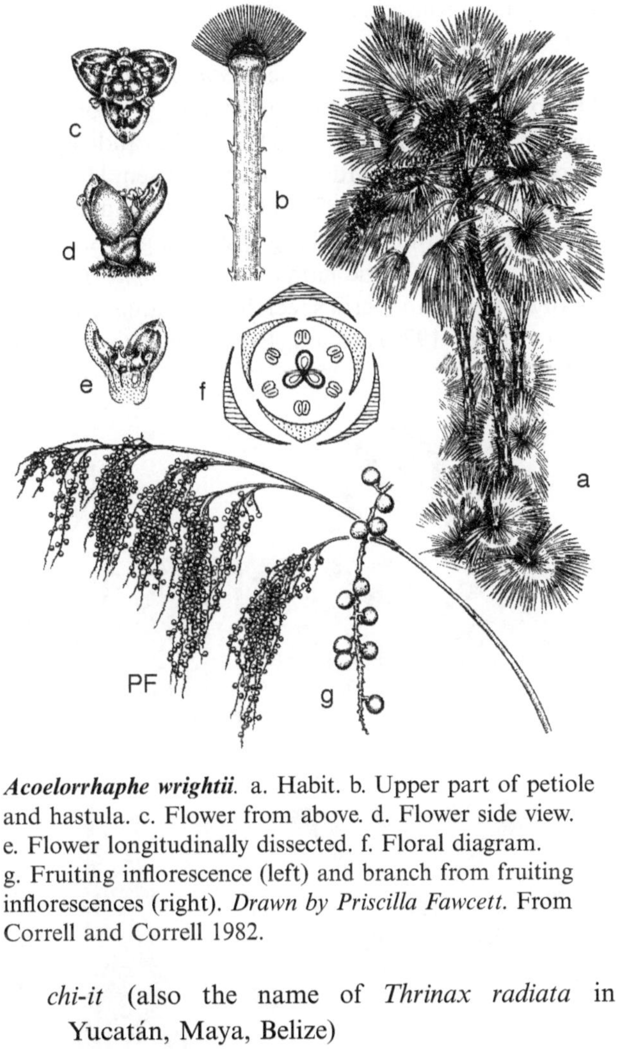

Acoelorrhaphe wrightii. a. Habit. b. Upper part of petiole and hastula. c. Flower from above. d. Flower side view. e. Flower longitudinally dissected. f. Floral diagram. g. Fruiting inflorescence (left) and branch from fruiting inflorescences (right). *Drawn by Priscilla Fawcett*. From Correll and Correll 1982.

chi-it (also the name of *Thrinax radiata* in Yucatán, Maya, Belize)
Everglades palm (Bahamas)
guano prieto (*guano* is a Taino name for a tree, presumably originally a *Ceiba*; *prieto*, black, meaning false, Cuba)
hairy tom [palmetto] (Belize)
harádan (Garífuna, Nicaragua)
palma (Belize)
palmetto (Belize)
papta (corruption of "palmetto," Belize)
[Honduras] pimenta [*piementa pim-iént, pim-int, prementa, primenta*] [palm] ("pimenta" and

other variants are corruptions of "palmetto," Belize)

tasiste [*taciste, ta-sis-te*] (*ta*, front, *sis*, something fresh, *te'*, to collect, Maya, Belize, Yucatán)

The Everglades palm is monotypic, and its single species grows from southern Florida, through the Bahama Islands, in Cuba, and from Chiapas, Tabasco, Yucatán, and Quintana Roo in Mexico, south to Nicaragua. From there it is apparently disjunct to northern Colombia.

The species was first named *Copernicia wrightii* in 1866 by German professor of botany at Göttingen, August H.R. Grisebach (1814–1879), and by the director of the royal garden in Herenhausen, Hermann A. Wendland (1825–1903). Then Wendland realized that the species did not belong with that genus, and he gave it the new generic name *Acoelorrhaphe* in 1879. Finally, the Italian world specialist on palms Odoardo Beccari (1843–1920) created *Acoelorrhaphe wrightii* in 1908.

Leaves are used for thatch and rope; fruits are edible (Morton 1968b). In Belize, the plants are used for food, medicine, construction, and in making products (Balick et al. 2000). Scott Zona (personal communication 2003) informs me that in Cuba, nurserymen now use trunks of this palm as supports for climbing aroids.

Acrostichum

(Greek *akrosticon, akros*, terminal; *stichos*, a row or line, a reference to the fertile apical portion of the leaf)

Acrostichum aureum (golden)
helecho (fern, Belize)
payuri (Carib, Suriname)
tiger bush (Belize)

Acrostichum danaeifolium (*danae*, the fern genus *Danaea*, *folium*, leaf, meaning that the leaves resemble those in the fern *Danaea*)

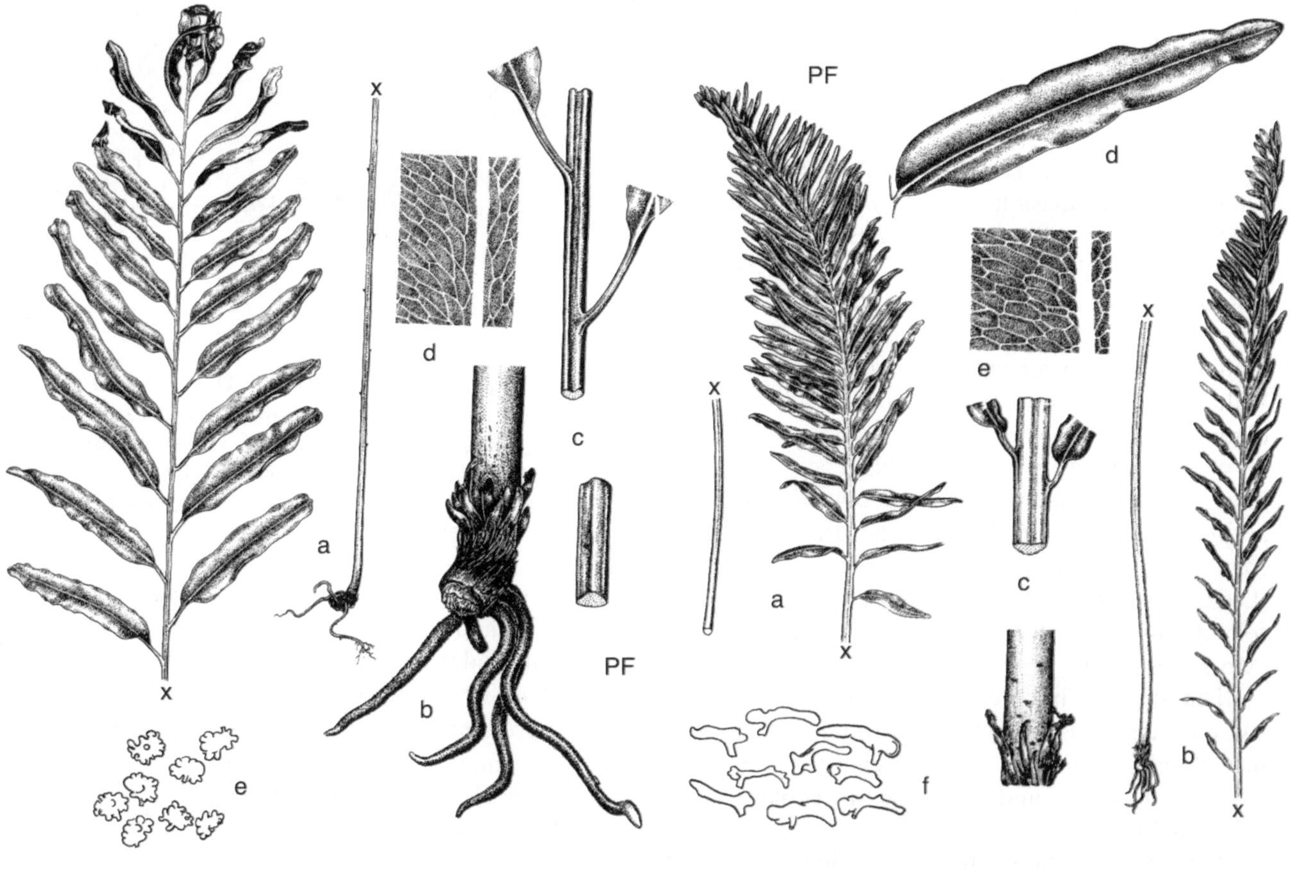

Acrostichum. *Acrostichum aureum* (left). a. Blade on left, stipe on right. b. Base of stipe with adventitious roots and scales. c. Above, pinnule attachment; below, section of stipe showing marginal spine. d. Areoles in midrib region of pinna. e. Paraphyses. *Acrostichum danaeifolium* (right). a. Sterile blade on right, section of sterile stipe on left. b. Fertile blade on right, stipe of fertile blade on left. c. Base of stipe with scales and scale scars below; middle of blade showing pinnae attachment above. d. Single pinna. e. Areoles in the midrib region of pinna. f. Paraphyses. *Both drawn by Priscilla Fawcett.* From Correll and Correll 1982.

avenção (based on *avenca*, fern, so "big fern," Brazil)
giant fern (Bahamas)
tapintcó:bi [*tapintcobi*] (big fern, Mikasuki)
toca:ɬakko [*toca:rakko*] (*ɬakko*, big, Creek)

The genus *Acrostichum* and the species *A. aureum* were both created by Linnaeus in *Genera Plantarum* in 1754. It was not until 1810 when *A. danaeifolium* was described from the state of Santa Catarina in southern Brazil where it had been collected by Georg Heinrich von Langsdorff (1774–1852).

Acrostichum aureum is used in Belize as food and medicine (Balick et al. 2000). Young leaves from this pantropical species are also used as food in Malesia, and the old leaves are used to thatch houses in Vietnam (Mabberley 1997).

Seminoles use *A. danaeifolium* as a febrifuge; an infusion of plant taken and rubbed on the body to reduce high fevers (Sturtevant 1955). Fiddleheads (unfurled fronds) are edible and eaten like asparagus (Morton 1968b).

Actaea

(From Latin *actaea*, Greek *aktea*, the name of the elder, Sambucus, used by Pliny, A.D. 23–79; the name was transferred to these plants by Linnaeus, perhaps because of the similarity of the leaves)

baneberry ("bane" as used here means "poison," and was combined with "berry" by 1755)
barba di capra (goat's beard, Italian)
Christophskraut (Saint Christopher's herb, German); Herb-Christopher (the name of European *A. spicata*, which Turner called Christophoriana or "Grapewurt, because it hath many blacke beries in the toppes lyke grapes" in 1548); *herbe de Saint Christophe* (French); *hierba de San Cristobal* (Spanish)
trollbaer (troll berry, Norwegian)
wasgobîdji'bîkeok (sweet root plant, Potawatomi); *wi'cosidji'bik* (drawing root, as in drawing out poisons, for *A. rubra*, Ojibwa)

Actaea pachypoda (with a thick flower-stalk [pedicel])
American herb-Christopher
baneberry
cohosh [blue, white, cohush] ("cohush" was recorded by 1796 for an *Actaea*, although the species is in doubt; authors in the mid-1800s said that it was from an "Indian word," although neither region nor tribe is indicated)
China-berry [chinaberry] (usually applied to *Melia azedrach*, Meliaceae)
doll's eyes ("doll" has been used in English since ca. 1560 as a pet form of the name *Dorothy*; apparently, it was not until ca. 1700 when the word was applied to an image of a human, particularly one used by girls as a plaything; Henry David Thoreau wrote in 1852 that *Actaea* fruits "have just begun to be white, as if they contained a pearly venom—wax white with a black spot ... imp-eyed"; in other words, like a doll's white china eyes)
false-aloe (maybe because both are laxative?)
masqua-mitruin (bear's root, because bears are believed to eat the roots, Plains Cree)
necklace-weed [necklaceweed] (alluding to the spikes with fruits on them like beads on a necklace)
pearl berry
racine d'ours (bear's root, Quebec)
rattlesnake herb; rattlesnake master (a name usually given to *Eryngium*)
richweed (Virginia)
[button]-snakeroot (usually applied to *Eryngium*, especially *E. yuccifolium*)
toadroot (maybe because both toads, *Bufo*, and plants are poisonous)
white baneberry
white-beads [-berry]

When Linnaeus created *Actaea* in 1753, he was familiar with the European species *A. spicata*. It was not until 1821 when Steven Elliott (1771–1830) described *A. pachypoda* in his flora of the Carolinas. There are now eight species known in the genus, with two in Europe, and two species and their hybrid in the New World (Mabberley 1997).

The Blackfoot took a decoction of the roots to treat colds and coughs (Moerman 1998). The Cherokee made an infusion to treat itch, as a gargle, for toothache, and they and the Meskwaki used it to rally patients nearing death (Hamel and Chiltoskey 1975). The Iroquois used the plant to treat men's urinary problems. The Menomini also used it to relieve headache caused by eye strain (Foster and Duke 1990). The Ojibwa gave root decoctions to children and adults with convulsions (King 1984).

People in Canada, presumably the Cree, made a decoction of roots with spruce fir (*Abies*) to treat gastric distress. Others have used the plants as an emetic and purgative (Hocking 1997).

Constantine Samuel Rafinesque (1830) was among the first in America to note the medicinal properties of

Actaea. He wrote that the roots were repelleant, nervine, and used to treat debility in Canada. Millspaugh (1892) was not as impressed with the American species, and instead recommended the European *A. spicata.* He considered *Actaea* useful for "reflex uterine headache," and some types of rheumatism. Later it was used for coughs, menstrual irregularities, colds, and constipation. It is thought that *A. pachypoda* may benefit circulation (Foster and Duke 1990), but ingestion of the plant may cause severe gastrointestinal inflammation and contact with the skin may result in skin blisters.

Adiantum

(Pliny, A.D. 23–79, used the Latin *adiantum* for the maiden-hair fern; Greek *adiantos,* unwetted or "wet-proof," not capable of being wetted, referring to the leaves that shed water or raindrops and remain dry)

Adiantum capillus-veneris (Venus's hair)
adiante (adiantum, French, from 1575); *adianto bianco* (white adiantum, Italian, from 1551)
avenca (fern, Brazil); *avenca comum* (common fern, Brazil)
black maiden-hair [maidenhair] ("maidenhair" was in use by ca. A.D. 1450 for *A. capellus-veneris*); common maiden-hair; southern maiden-hair
cabello de Venus (Venus's hair, Brazil); *capelevenere* [*capil uenere*] (Venus's hair, Italian, from 1551); *capillaire* (hair, Quebec); *cheueux de Venus* (Venus's hair, French, from 1550); Venus's hair [fern, from 1568] (Venus heyre)
chib (Maya?)
cuamaquistli (variant of *quauhmochitl,* snake jaws?, Náhuatl, Mexico)
culantrillo (little coriander, San Luis Potosí, Nuevo Leon, Central America); *culantrillo del pozo* (little coriander of wells, Spain)
dudder grass [dodder grass] (from *dother* or *dither,* trembling, shutter, or shiver, from 1658, now dialectic)
failtean flonn (*failtean,* head-band, *flonn,* white, Gaelic)
Frawenhar [*Frauen haar*] (maiden hair, German, from 1542)
iunkfrawen hare (young maiden's hair, German, from Turner [1548] 1965); *Junckfrawenhar* (young maiden's hair, German, from 1542); maiden-hair [maidenhair, true maiden haire] [fern] (from 1597)
ochquewi miecheken (*ochquèu,* woman, *milach,* hair, Delaware)
steenruyte (stone rue, Dutch, from 1549)
weewilsquee (Shawnee, fide Edgar 1891; probably cognate with the Delaware)

Adiantum melanoleucum (black and white)
fragrant maidenhair

Adiantum tenerum (tender, soft)
black stick (Belize)
[brittle] maidenhair [fern] (USA, Belize)
palo negro (black stick, Belize)
roc-che-cwan (*roc,* rock, *che',* plant, *kwa',* pain; mixture of English and Maya, Belize)

There are comparatively few ferns used in medicines. Perhaps one of the oldest known from Europe is *Adiantum capellus-veneris.* That Latin epithet was first applied to it by Pseudo-Apleius about the 4th century A.D. (Meyer et al. 1999). Herbalists since the mid-1500s have praised this fern, and many of its names are derived from that old Latin. Culpeper (1653), for example, recommended it to treat "coughs, asthmas, pleurisy, etc., and on account of its being a gentle diuretic, also, in the jaundice, gravel [kidney stones], and other impurities of the kidneys."

Off the Galway coast of Britain, people collected the fronds, dried them, and used them as a tea substitute (Hedrick 1919). However, many considered that drink expectorant, astringent, and a tonic for coughs, throat problems, and bronchitis (Foster and Duke 1990). Kalm ([1753–1761] 1972) comments that *herba capillaris* (*A. pedatum*) was exported to Europe in the 1740s when he was in Quebec, being "reckoned preferable in surgery to that which we have in Europe," which he identified as *A. capellus-veneris.* Locally, the French were using it for a tea to treat "consumptions, coughs, and all kinds of pectoral diseases." He continues, "This they have learnt from the Indians."

Among the few records of *A. capellus-veneris* use in North America is by the Mahuna of California and Navajo of northeastern Arizona. The Californians used it to treat rheumatism, and in Arizona, they used it to treat bee or centipede stings and mental problems (Moerman 1998). The Mexicans boiled the fern into tea used to treat retarded menstrual periods (Ford 1975). Hann (1986) listed *Adiantum* sp. as being used by the Apalachee; the only species known in their range is *A. capellus-veneris.* Surely, the other two native Florida maiden-hair ferns were also used.

We also know that the related *A. pedatum* was widely used in the southeastern United States and elsewhere. Cherokees called it *ka'ga-asguntagi* (crow's shin, Mooney 1885–1886). Among the eastern people using this species were the Cherokee, Menomini,

Meskwaki, Micmac, and Potawatomi (Smith 1933, Moerman 1998). A number of groups in western North America from British Columbia to California also used this species. Uses of this species are the same as for *A. capellus-veneris* (Foster and Duke 1990, Moerman 1998).

Although nothing has been found about Old World people washing the hair with ferns, that was a practice with both *A. capellus-veneris* and *A. pedatum* in the New World (Foster and Duke 1990). Indigenous Americans considered the maiden-hair ferns especially good for making their hair shiny (Foster and Duke 1990). One has to wonder if that was not the original reason the Old World plants were compared to Venus's hair.

Aesculus

(The Latin name for a tree, probably originally for the Italian oak; akin to Greek *askra*, Akkadian *isu, esu*, Summerian *gis*, tree)

Aesculus pavia (for Peter Paaw, 1564–1617, Dutch botanist)
atai (mostly given to *A. flava*, Choctaw); *vlv* (mostly given to *A. flava*, Muskogee)
[flame, scarlet, red, woolly] buckeye ("buckeye" is said to have been applied by the 1840s, "from the hilum of the fruit [seed] having the appearance of a stag's eye")
firecracker plant (an allusion to the red flowers)
fish poisonbush
ta-shka'-hi (Osage, probably for *A. flava*)

The first Europeans who arrived in the New World were completely unfamiliar with *Aesculus.* The Old World species, *A. hippocastanum*, native to Albania, Greece, Bulgaria, and Yugoslavia, was introduced into Europe in 1550 (Linnaeus [1753] 1957, Millspaugh 1892, Clapham et al. 1987). Called the horse chestnut, the tree reached England in the 1600s and was first cultivated there by John Tradescant in 1633 (Millspaugh 1892).

Not long afterward, in 1696, Leonard Plukenet recorded a new species, which he called "*Saamouna pisonis seu siliquifera brasiliensis arbor, digitatis foliis serratis, floribus teucrii purpureis* ("Saamouna" for fish or Brazilian tree with siliques, palmate leaves with serrate leaflets, purple flowers resembling teucrium). Because of confusion over the origin of his specimen, he thought the plants were from Brazil. Even Linnaeus, when he thankfully changed that long name to *Aesculus pavia*, listed the species from the Carolinas and Brazil. Linnaeus took the species name from Hermann Boerhaave who had created *Pavia* as a genus in 1720. Linnaeus had not only seen the Plukenet plate, he had also studied live plants at the *Hortus Cliffortianus.*

Thus, the first record of *A. pavia* in Europe noted that it had been used as a fish poison. Steyermark (1963), Uphof (1968), and Hocking (1997) also noted that the powdered seeds and crushed branches have been used to catch fish. Although no mention is made of that use for this species, Moerman (1998) does note the application by indigenous people for both *A. glabra* and *A. californica.* The roots contain a mixture of saponins called aescin, which have been used for soap (Steyermark 1963, Foster and Duke 1990), and those chemicals are surely why buckeyes are effective in killing fish.

Relatively few uses are recorded for *A. pavia* among indigenous people. The Cherokee made a poultice of the nuts for tumors and sores, and used a preparation to relieve colic and dyspepsia (Hamel and Chiltoskey 1975). They mixed scarlet buckeye with other plants used during childbirth to stop bleeding afterward, for swellings and sprains, and to prevent fainting (Hamel and Chiltoskey 1975).

Apparently, indigenous Americans also started the belief still prevalent in whites and blacks that carrying the seeds would help rheumatism, bring good luck, and relieve piles (Hamel and Chiltoskey 1975). Seeds from any species are considered equally effective in achieving those results (Moerman 1998).

Foster and Duke (1990) suggest that applications of all *Aesculus* in the Americas are essentially equal. Thus, the Delaware used *A. glabra* to treat earache, and the Iroquois and Mohegan used *A. hippocastanum* to relieve chest pains (Moerman 1998).

Scarlet buckeye is so small that its wood would probably not have been used as was yellow buckeye, *A. flava.* The Cherokee at least used wood of that related species for lumber, to make baby cradles, and in masks and dough trays (Hamel and Chiltoskey 1975).

Early colonists learned to be wary of the New World species because they were more toxic than the Old World plant (Millspaugh 1892). Still, the active chemicals are not systematically absorbed, and they continue to be used outside the United States to reduce swelling and seal capillaries (Foster and Duke 1990). All of the species irritate the mucous membranes, causing upset stomach if taken internally, but no fatalities have been recorded in the United States (Foster and Duke 1990).

Agalinis

(From Greek *aga*, wonder, and Latin *linum*, flax, because C.F. Rafinesque, 1783–1840, thought some species superficially resembled flax)

Agalinis tenuifolia (narrow-leaved)
kidney weed
slender [slender-leaved] gerardia (*Gerardeia*, formerly a genus name, commemorating the celebrated herbalist John Gerarde, 1545–1611)
slender-leaf false foxglove [slenderleaf false foxglove] ("foxglove" compares this herb with *Digitalis*)

Agalinis is an American genus with about 40 species (Mabberley 1997). It was defined previously as containing more species, but those species have since been placed in other genera.

The Meskwaki of Iowa used an infusion to treat diarrhea (King 1984). Morton (1974) found the people of South Carolina using the old stems as a diuretic. They also used the tea as a "nice drink anytime." Hocking (1997) considered the plant toxic to livestock.

Agave

(Greek *agavos*, meaning admirable, noble, splendid)

a'ud [*a'od*] (Akimel O'odham, Hai-ced O'odham, Tohono O'odham, Arizona, Sonora)
American aloes (Europeans)
bamboo (Bahamas)
bitoba [*toba*] (Zapotec, Oaxaca)
cactus (Americans)
century plant (Americans)
haamxö (Seri, Sonora)
maguey (from Taino, Spanish)
metl (Náhuatl, Mexico)
nootah (Navajo, New Mexico)

Agave decipiens (ensnare or deceive)
pasalátkicó:bî (big yucca, Mikasuki)
false sisal (Florida)

Agave neglecta (neglected, overlooked)
wild century plant (Florida)

Most English-speaking Americans call these spiny, drought-resistant, succulent herbs either "century plant" or "cactus." Europeans call the genus "American aloes." Bahamians say "bamboo." Indeed, none of these terms is biologically correct. The plants do not take a century to flower (10 to ~50 years, depending on the species), nor are they a "cactus" except in the original meaning of the word—"spiny" or "prickly." Because the Europeans were familiar with aloes (*Aloe*), they compared the American plants to them—they are not only similar in appearance, but both also provide medicines. Why Bahamians call them "bamboo" is a curiosity. To me there is no resemblance.

The heartland of the agaves is Mexico, where they are famous for producing intoxicating beverages called *tequilla, mescal,* and *pulque* (De Barrios 1971, Hutson and Marshall 1995). In addition, the plants were formerly a more important source of food, fibers that were used for clothing, ropes, and rigging for sailing ships, paper, soap, and medicines. These plants were "the corner grocery" for inhabitants of pre-European Mexico (Standley 1920–1926, Morton 1968b, 1981, Gentry 1982). Strangely, there were two native species in Florida—*Agave decipiens* and *A. neglecta.* The only non-European name we have for either of these prickly plants is from the Mikasuki—*pasalátkicó: bî* (big yucca). These people, originally from southern Georgia and the vicinity, were familiar with *Yucca aloifolia* that grew along the coasts in that area and farther north into the Carolinas and west into Mississippi (Sturtevant 1955). Since the Mikasuki speakers were relatively late arrivals in central and southern Florida, we have little idea of what the earlier inhabitants may have called the plants or how they used them.

Both of Florida's *Agave* have been anomalies since they were discovered and described. *Agave decipiens* was named in 1892 by the keeper of the herbarium at Kew, John Gilbert Baker (1834–1920). Although *A. neglecta* was officially named in 1903 by John Kunkle Small (1869–1938), the plants had been recorded (misidentified as *A. vivipera*) at the Cape Canaveral area of Florida by William Bartram after he lived there in December 1766. Indeed, Small named the species based on a Bartram drawing (Harper 1958) and cultivated material sent to him from Eustis (Lake County).

There are no species native to the area between Texas and southern Florida where the *Agave* are found. Moreover, *A. decipiens* belongs with what the late *Agave* guru Howard Scott Gentry (1982) called group *Rigidae*, and its nearest relatives are Mexican. Similarly, *A. neglecta* is in group *Sisalanae*, also Mexican. The nine species native to the Bahamas, and those in the other Caribbean islands, are more distantly allied to the Florida plants. Moerman (1998) reported the Apache using *A. decipiens*, but it does not occur in their range.

One of *A. decipiens'* probable closest relatives is *A. angustifolia*, itself famous for being the most widespread species in the genus. The Mexican *A. angustifolia* has been cultivated by perhaps every indigenous group in the New World from northwestern Sonora and northern Tamaulipas south to Costa Rica. Because of this cultivation, and the incredible array of habitats in which the plants grow, from sea level to 1500 m, there is a vast morphological spectrum in traits (Colunga-García Marin et al. 1996, 1997, 1999). Also related are *A. fourcroydes*, famous for fiber production (called *henequén* in Spanish, although originally Taino), and *A. tequilana*, from which the alcoholic beverage is obtained. Surely, with relatives like these, the Florida *A. decipiens* had similar uses among the indigenous people.

Perhaps the nearest relative of *A. neglecta* is *A. weberi* of the arid Mexican highland or the horticultural *A. desmettiana*, which is disturbingly variable. All three are related to *A. sisalana*. That Mexican species, the source of *sisal* fibers, was introduced into Florida in the 1800s by physician Henry Perrine (1797–1840). Perrine actually gave the species its scientific name before his untimely demise on Key Biscayne during a Seminole raid. Florida native *A. decipiens* is sometimes called "false sisal" to distinguish it from the cultivated Mexican introduction. *Agave neglecta* is so rare that no one knows much about it, and the species should be considered endangered.

The most widespread common name for *Agave* among Spanish-speaking Americans is *maguey*. This is the name usually encountered in Mexico, although it is a Taino word from Haiti. *Maguey* originally probably referred to the *Agave* look-alike, *Furcraea*, and it appeared in the literature by 1555. That original source (Eden 1555 in OED 1971) said, "*palmarum puta Maguieorum, que est herba*" (of the palm called Maguey, which is an herb)." Richard Eden went on to compare the plants to European orpine (*Sedum telephium*, Crassulaceae) because both have fleshy leaves. That may have been the earliest commentary on convergent evolution in print.

By 1745, in his book *A true and impartial journal of a voyage to the South-Seas and round the globe*, Englishman Pascoe Thomas wrote: "They call it a Maguey tree, and from it they get Wine, Vinegar, Honey, Thread, Needles, Stuffs for clothing, or Sails for Canoes and small Boats, and timber for building." Before him, the first three people who mentioned the plants made their first reference to the "wine" it produced.

By the time Linnaeus got around to naming the genus *Agave*, he decided to use a word that had been important to the Greeks—*agavos*, admirable, noble, splendid. Doubtless, he also had in mind the Greek mythological person Agave, who was a daughter of Cadmus and Harmonia (Quattrocchi 2000). Surely, Linnaeus intended the classical Old World name for New World plants to emphasize the important roles they played in the lives of the American natives.

While the Caribbean word overshadowed the Mexican name for the plants, many in that country still have names for the plants in their own languages. The Aztecs called the genus *metl*. Moreover, they continue to use a binomial nomenclature that predated that "innovation" by European scientist Linnaeus. Gentry recorded species in Mexico called *metometl* (worm agave), *pizometl* (small agave), and *papalometl* (talkative agave).

People living north of the Aztecs, but still speaking related Uto-Aztecan languages, called agaves by very different names. Three of these groups lived in Arizona (Akimel O'odham, Hai-ced O'odham, Tohono O'odham), and they say the plants are *a'ud* or *a'od*. The Seri, who also speak a related language, sometimes use the name *haamxö* as a generic for *Agave*. However, they also have names for each kind, such as *hamoc* or *coptoj* (different varieties of *A. angustifolia*), and *xica istj catitic* (soft-leaved thing) for *A. cerulata*.

Huastec Mayan speakers who live in the states of Veracruz and San Luis Potosí call different species *weey*, *wiich*, and *tsi'lim*. The Zapotec who live south of the Aztecs in Oaxaca and vicinity have a "generic" *Agave* name—*bitoba*, sometimes shortened to *toba*.

The genus *Agave* is a virtual chemical cauldron at different stages of its life. Perhaps the most important stage to it and people is just before it flowers. At that time the plant has stored up many carbohydrates to fuel the last phase of its life since the plants are monocarpic, flower once, and then die. People throughout the range of the genus used this time to either harvest the stem (the *piña*, because it resembles a pineapple after the leaves are removed), roast it, and consume the sugar-rich products, or they tapped the budding flower stalk and drained off the juices (*aguamiel*, honey water in Spanish). Once this *aguamiel* had fermented, it became *pulque*. The plants from which this was taken were associated with deities; the principal one was *Mayahuel*, the Aztec goddess of *pulque*.

If the plants were harvested just before the flower stalk began growing, they could then be roasted in pits to produce a sugar-rich food. This was done by cutting off the leaves from the *piña* or *cabeza* for roasting. The cooked product was called *mescal* (from *mexcali*, Náhuatl). This process was developed to the point that certain Mexican cultures, and those in what is now Arizona and New Mexico, depended on *Agave* for food during certain times of the year. We even have the practice "immortalized" in the Athabascan group called the "Mescalero Apaches." Other people in the southwestern United States also depended on this food source, but those called Mescalero Apaches were formerly a particularly warlike group. Only in the years since about 1990 have we learned how important this *Agave* food also was to the people who originally lived in the valley of Phoenix, Arizona and nearby (Dobyns 1988, Adams and Adams 1998).

Leaves were taken from the plants, roasted, and used by the Aztecs to produce paper, especially the leaves of the apical growing part of the plants. Other plants served as paper sources more often (cf. *Ficus*: Strangler fig), but there are still extant early post-Columbian manuscripts on paper made of agave fibers.

Distilled alcoholic beverages were unknown to the natives of the Americas before the Europeans arrived. However, indigenous Americans soon adopted distillation and began producing the Mexican equivalent of U.S. "moonshine." Until recently, these were comparatively small operations, but now large companies have begun producing quality beverages. Some of these companies with a reputation for fine products are Cuervo, Sauza, and Robert Denton.

Leaves contain sapogenins, sterols, terpenes and vitamins (Dewidar and el-Munajjed 1970, Blunden et al. 1974, 1980, 1986, Kintja et al. 1975, Cripps and Blunden 1978, Reichelt and Stransky 1978, Bedour et al. 1979, Xu and Zhou 1982, Sati et al. 1985, Camp et al. 1988, Ding et al. 1993, Yokosuka et al. 2000, Abdel-Khalik et al. 2002, da Silva et al. 2002). However, all the compounds that cause dermatitis are not known. Nor do we know what the pain-causing exudate is on the spines, but recently calcium oxalate raphides have been implicated (Salinas et al. 2001). Many people use the leaves as poultices for itch, sores, bruises, and wounds. Some also use them as a source of fish poisons, and one Mexican species (*A. lechuguilla*) has been used to poison arrows (Gentry 1982). A by-product of fiber operations in Mexico and East Africa is an extract that is pivotal in the manufacture of steroid drugs, including cortisone and birth control pills. Laboratory evidence supports many of the uses (Bianchi and Cole 1969, Davidson and Ortiz de Montellano 1983, Segura et al. 1991, Uniyal et al. 1991, Verastegui et al. 1996, Peana et al. 1997, Garcia et al. 2000). The chemistry of the plants is so impressive that even cultures where they are not native have incorporated them into medical practices (Duncan et al. 1999, Kassu et al. 1999).

"There is good and evil in everything," goes an old adage. The rise of popularity of *Agave* products that parallels the population growth in Mexico and the United States has its downside. In some areas, *Agaves* are thought to be overharvested (Burwell 1995). In other regions, those that remain after harvests from the wild are not producing fruits. That lack of reproduction is said by some to be directly related to the decline in pollinators, including nectar-feeding bats and hummingbirds (Arizaga et al. 2000a,b, Slauson 2000). So, the next time you drink a toast with an *Agave* product, remember to thank the bats and birds that made it possible!

Ageratina

(A diminutive form of the genus *Ageratum*, from Greek *ageraton*, not growing old, an allusion to the long-lasting flowers; perhaps originally applied by Pliny, A.D. 23–79, to what is now *Achillea*)

Ageratina altissima (tallest) (=*Eupatorium ageratoides, Eupatorium rugosum*)
deerwort-boneset ("boneset," a name applied by 1670 to comfrey, *Symphytum officinale*; applied to these American plants later)
hemp-agrimony (comparing it with *Agrimonia*, which see)
milk-sickness plant (from the technical physician's name *morbeo lacteo*)
noota ikheesh (*noti'*, tooth, *ikhsnsh*, medicine, Choctaw, Chickasaw)
richweed
[Indian, white] sanicle (comparing it with *Sanicula*, which see)
squaw-weed (North Carolina)
stevia (after the genus *Stevia*?, Wisconsin)
thoroughwort
white snakeroot (one of many plants used by indigenous people to treat snakebite)
white-top

Ageratina is a genus of perhaps 290 species in the eastern United States, Central America, and western South America (Mabberley 1997). Although the French botanist Édouard Spach (1801–1879) named the genus in 1841, the species have been mostly placed in *Eupatorium* until recently. Robert M. King and Harold Robinson emphasized the distinctiveness of *Ageratina* from *Eupatorium* in 1970, and most have accepted the two since then.

The Cherokee used the herb to treat ague (from Latin *acutus*, acute; malaria, see Desowitz 1991), diarrhea, fever, kidney stones, and urinary diseases, and as a diuretic, stimulant, and tonic (Hamel and Chiltoskey 1975). The Chickasaw chewed the root or held it in the mouth to stop toothache (Campbell 1951). The Choctaw used it against toothache like the

Chickasaw, and also as a stimulant and tonic (Campbell 1951). The Iroquois used the herb to treat venereal disease, as a physic, in sweat baths, and to treat a prolapsed or inflamed womb. The Meskwaki also used it in steam baths and to revive an unconscious patient (Moerman 1998). The Shawnee used it to treat snakebite (Coffey 1993).

Morbeo lacteo (milk sickness), also called the "trembles," was a mysterious and feared disease that attacked farm families and their livestock in the 1800s (Coffey 1993). The malady was more common in the South and Midwest, particularly in North Carolina, Ohio, Indiana, and Illinois. In 1818, Abraham Lincoln's mother, Nancy Hanks Lincoln, died of this disease in Little Pigeon Creek, Indiana. At least partly because of that, Thomas Lincoln and his new wife Sarah and young Abraham moved to Illinois. Although the cause of the problem was determined in the 1830s, county officials in Kentucky complained in 1852 that the disease was so prevalent that it deterred settlers.

It was a Shawnee woman who showed Illinois physician Anna Pierce the poisonous plant causing the disease (Coffey 1993). That fugitive from the forced relocations of local tribes, known only as Aunt Shawnee, explained the relationship between the white snakeroot and both milk sickness and trembles. However, it was not until 1917 that physicians were convinced that the plant caused the problems.

Tremetol, a toxic alcohol, is the chemical causing "milk-sick," or milk sickness in humans or trembling in other animals. The plant causes ketosis (excess ketone bodies; overly acidic body chemistry) and is transmitted through cow's milk to people. The breath takes on a fruity odor as acetone diffuses from the lungs. The symptoms are progressive and include lassitude, nausea, vomiting, stomach pains, intense thirst, prostration, coma, and death.

Later people considered the roots diuretic, diaphoretic, antispasmodic, aromatic, and vulnerary (Millspaugh 1892, Hocking 1997).

Ageratum

(From Greek *ageraton*, not growing old, an allusion to the long-lasting flowers; a name probably first applied by Pliny, A.D. 23–79, to *Achillea ageraton*)

Ageratum littorale (of sea coasts)
Cape Sable whiteweed
hauayche (*hauay*, expanded, *che'*, plant, Maya, Yucatán)
tup-lan-xix (*tup*, extinguish fire, *lan*, immerse, *xix*, excess, Maya, Belize; see also *Melanthera*)

Ageratum was named by Linnaeus in 1753, based on plants he knew from the Americas. Subsequently, discoveries have expanded the genus to 44 tropical American species. The rarest of these, at least in Florida, is *A. littorale*. The species was first described as part of the genus *Coelestina* by physician John Torrey (1796–1873) and Harvard professor Asa Gray (1810–1888) in 1841 from a specimen collected in Key West. Then, Gray realized that it was best put in *Ageratum*, and he made the change to *A. littorale* in 1881. These plants are so rare they are considered endangered in Florida (Coile 2000).

Martínez (1969) recorded the Mayan name in Yucatán, and said that the plant was used to treat *culebrilla* (ringworm). Balick et al. (2000) indicated that it was also considered medicinal in Belize but gave no details. The herb grows in Key West, Quintana Roo, Belize, Honduras, and Guatemala.

Agrimonia

(A plant mentioned by Pliny, A.D. 23–79, usually considered to be a corruption of *Argemone*)

Agrimonia. *Agrimonia microcarpa* (left). *Agrimonia rostellata* (right). Both from Britton and Brown 1987.

agrimony [agrimonie, egremoine, egremounde, ergrimonie] (English, variant spelling from 1578–1597); *agrimonia* (Italian); *agrimonia* (Spanish from 1557); *aigremoine* (French from 1550); *Odermenig* (German pronunciation of "agrimony," from 1542)
Bruchwurtz (fracture herb, German from 1542)
eupatorio (Italian from 1551; from *Eupatorium*, which see)
geurach (sharp, Gaelic)
mirean (Gaelic)
muir dhroighinn (*muir*, sea, *dhroighinn*, bramble, Gaelic)

Agrimonia incisa (cut into, alluding to the compound leaves)
incised agrimony

Agrimonia microcarpa (small-fruited)
slender groovebur
small-fruit agrimony

Agrimonia rostellata (with a small beak)
woodland agrimony [groovebur]

A single species of agrimony was well known to Europeans by the time they reached the New World. Linnaeus gave it the modern name of *A. eupatoria* in 1753. The common names listed under the genus were current from the time of Leonard Fuchs.

That Old World species was also known as church steeples, cocklebur, philanthropos, and stickwort, and was considered a panacea through the ages from the classical Greek and Roman times to the late 19th century (Dobelis 1986). The ancient Greeks used the plant to treat eye problems, diarrhea, gall bladder, liver and kidney ailments; Anglo-Saxons extended it to treat wounds, and in the 1800s it was prescribed for skin diseases, asthma, coughs, sore throat, and gynecological complaints (Dobelis 1986).

No records have been found of the Florida species that were used. However, it is known that the Cherokee, Iroquois, Meskwaki, Ojibwa, and Potawatomi used native American species (Smith 1933, Moerman 1998). Since these American species were used in many of the same ways as the Old World *A. eupatoria*, it seems likely that they contain many of the same compounds.

Aletris

(From Greek *altho, aleo,* to grind; Aletris was a legendary female slave who ground grain; apparently, it was the mealy texture on the outside of the corolla that caught Linnaeus's attention and led him to use this name)

Aletris aurea (golden)
golden colic-root [golden colicroot]

Aletris farinosa (mealy)
ague grass [root] (used to treat ague or malaria)
aletris (Brazil); *alétris farineaux* (mealy aletris, French); *mehlige Aletris* (mealy aletris, German)
aloe [star] root [wort]; [yellow] star-grass (North Carolina in 1737)
bitter grass; bettie grass; bitter plant

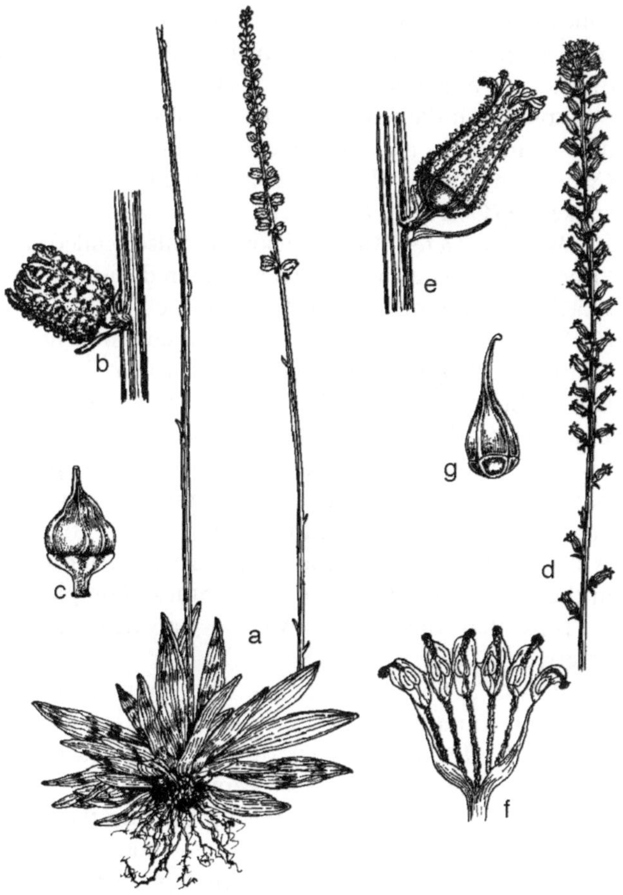

Aletris. a–c. *Aletris aurea.* a. Habit. b. Flower. c. Capsule. d–g. *Aletris farinosa.* d. Upper part of plant. e. Flower. f. Corolla split longitudinally. g. Capsule. *Drawn by Vivian Frazier.* From Correll and Correll 1972.

black root (usually *Pterocaulon pychnostachyum*)
blazing star (usually applied to *Liatris* spp.)
cadenaco (horn plant, Brazil)
colic [cole] root (to treat colic); *raiz das cólicas* (colic root, Brazil); yellow colic root
cometo-orinito (urine maker, Brazil)
crow corn
Devil's bit (usually applied to *Chamaelirium luteum*)
erva-da-febres (fever herb, Brazil); *erva-estrela* (star herb, Brazil)
[false] unicorn [plant, root]
wa̲'sa wi̲·'ti (cane medicine, Catawba); [*wa̲' sahawane*] (cane leaves, Catawba)
wisakon [*wissakan*] (it is bitter, Powhatan, Virginia; Capt. John Smith on Roanoke Island, North Carolina, in 1585 wrote: "the herbe which the savages call *Wysauke* werewith they cure their wounds which they receive by the poisoned arrows of their enemies"; Strachey wrote of *wighsacan* in 1612, "a small root which they

bruise and apply to wounds ... so this cureth their hurts" and "taken for purgative in spring"); cognates include *we?sakan* (Menomini); *wi?θa-kanwi* (Shawnee); *wisakan* (Cree)

Some of the early physicians and botanists who came to the eastern coastal regions of America found indigenous people using *Aletris* to combat colic and dysentery. As both of those were common problems at the time, a remedy was important. At first, it was thought that only one species was effective. However, now it is known that most, if not all, of the American species have the same medical attributes (Millspaugh 1892, Ellingwood and Lloyd 1898, Lewis and Elvin-Lewis 1977).

The first record after the failed colonization attempt on Roanoke Island by Sir Walter Raleigh seems to have been by John Brickell. This resident of North Carolina wrote in 1737 that "stargrass" was "used with good Success in most Fevers in this Country." About the same time, John Bartram wrote European physician Cadwallader Colden that "ye star root is chiefly used by several for ye pain of ye stomach thay boil ye root in water & drink ye decoction after ye method of our Indians from whom they learned ye use of it" (Vogel 1970). Both Brickell and Bartram had the Catawba as their sources.

Rafinesque, who also worked in the same region during the 1820s, wrote that *Aletris* was used by country physicians and Indian doctors and was highly valued by them for its tonic and stomachic properties. Some even considered it slightly narcotic. The same view was echoed by Dr. Jacob Bigelow in *American Medicinal Botany*, published between 1817 and 1820. Although several uses were attributed to the plants, most considered them tonic and best used against stomach problems, especially colic (Millspaugh 1892, Vogel 1970). Hocking (1997) still considered *A. aurea* tonic. In spite of this reputation, Yanovsky (1936) said the "bitter bulbs" were eaten in Louisiana.

By the 1930s, studies among the Catawba recorded that they still used the plants by soaking the leaves in water to produce an antidiarrheal tea, and the root for any type of colic or stomach problem (Speck 1937, 1944). Although some sources say that the roots are a remedy, especially against colic in infants, careful reading of the literature of the 1800s and before suggests that digestive upsets in adults as well as infants were termed "colic." By the mid-1950s, it was known that *Aletris* roots contained saponin and a bitter principle. At that time, the genus was still being used as a digestive aid, but also as a "uterine tonic" and a diuretic. The diuretic properties made the genus useful against dropsy (congestive heart problems). Since the roots are laxative, they were used against that—although they were also used to treat diarrhea. Some have also recommended the medicine for coughs and colds (Ellingwood and Lloyd 1898, Morton 1974).

For some time the use and properties of *Aletris* fell out of popularity. This may well have been caused by Harvey Wickes Felter, M.D. In 1922, Felter, in his version of *The Eclectic Materia Medica*, proclaimed that all the attributes touted for *Aletris* were because of confusion with *Chamaelirium* (Felter 1922). He went further to say that *Aletris* was worthless and should not be considered medicinal. Many subsequent books on medical plants have done what the authors of *World Economic Plants* did in 1999 (Wiersema and León 1999). Under the heading "ALETRIS" they wrote "Medic. (folklore)." If it is *just* folklore, then people are making millions of dollars in the current herbal remedy industry selling the roots of U.S. *Aletris*.

Allium

(Classical Latin name for garlic, *Allium sativum*; it has been suggested that the name is based on the Celtic word *all*, pungent, cf. Kindscher 1987; that word does not occur in Dwelly's 1933 Gaelic dictionary, but *alla* means fierce, wild, which may be the same)

aglio (from *allium*, Italian); *ail* (French); *ajo* (Spanish); *ajuin* (Dutch); *alho* (Portuguese)
ahkohtew'vslpse (Plains Cree)
cacomite (from *cacomitl*, Náhuatl, Mexico)
garlic (a compound word derived from Old English *gárléac*, containing *gár*, a spear or javelin, *léac*, leek, the culinary herb; used in English by ca. A.D. 1000); *geirlauk-r* (Old Norse); *gairleag* (Gaelic)
hato̱ falaha [*hato̱ṉ falaha*] (*hato̱*, winter, *falaha*, onion, Choctaw)
hvitløk (*hvit*, white, *løk*, onion, Norwegian)
Knoblauch (German); *knobloch* (Dutch)
kosáoma [*kosooma*] (an onion or garlic, but also something that stinks, Alabama)
onochschà (Onondaga)
schordon (Greek)
shumim (Hebrew); *sumu* (Akkadian)
svgi (Cherokee)
tôm (Arabic)
uleepen (Delaware)
winúnschi (Delaware)

Allium canadense (of Canada)
ail du Canada (Canada garlic, Quebec)
a̱nwa̱n'fka (Ofo)

Allium canadense. From Institute of Food and Agricultural Sciences.

cîgaga'wûn (skunk plant, Potawatomi); *jigâgamanj* (Ojibwa)
hashtola' atofalla'a' (*hashtola'*, winter, *atofalla'a'*, onion, Chickasaw); *kosáoma hissi okchákkoya ilpa* [*kosáoma hissi okchákkoya*] (*kosáoma*, onion, *hissi*, hair, *okchakko*, blue-green, *ilpa*, eat, Alabama)
kha-a'-mot-ot-ke'wat (skunk testes, Cheyenne); *kha-ohk-tsi-me-is'tse-hi* (skunk, it smells, Cheyenne)
meadow [wild] garlic
mon-zhon-xe [*mon-zonxe*] (*mon-zhon*, earth, *wa-xe*, to bury; akin to *mon-shon'-dse*, a hole in the ground, Osage)
psin (Lakota)
tafvmpe vhake [*ta-fvm'pe, tafv'mp-háke*] (*fvmpe*, stink, *vhake*, resembling, Muskogee); *tafvmpuce* (*fvmpe*, stink, *oce*, little, Muskogee)
towá bosóhka (*towá*, onion, *bosóhka*, bad smelling, Koasati)
ûñktcinsayi (Biloxi)
wild onion ("onion" from Latin *union*, a rustic Roman name for a single onion bulb, otherwise a pearl)

Members of the genus *Allium* were so important to Old World people that each language had a name for each of the species. In English, for example, there are chives (*A. schoenoprasum*), garlic (*A. sativum*), leeks (*A. porrum*), onions (*A. cepa*), ramsons (*A. ursinum*), scallions (*A. cepa*), and shallots (*A. ascalonicum*). A similar list could be made in the Romance and other Germanic languages. Gaelic, the most distinct of the western European languages, called them *saidse* (chives), *gairleag* (garlic), *leigis* (leeks), *uinnean* (onion), *gairgean* (ramsons), and *siobaid* (scallions). If they have a name for shallots, it has not been located.

In 1674, the Franciscan priest Jacques Marquette and his party traveled from Green Bay to the present site of Chicago eating almost nothing but *A. canadense* and *A. tricoccum* (Hedrick 1919, Fernald et al. 1958). The name of the city Chicago is corrupted from the Menomini name *shika'ko* (skunk place) because of the quantity of onions there and their dependence on them. Steyermark (1963) wrote that the bulbs "have a rather sweet flavor and are edible, while the bulblets of the inflorescence can be pickled and eaten as such." He further recorded that the entire plant, before flowering, can be boiled and made into soup and that the young fresh leaves were used in salads or as a garnish for flavoring meats, cheese, and other dishes.

Prehistoric people used onions, probably also as food. Gilmore (1931) found bulbs in the Ozark Bluff-Dweller site in Arkansas. Historic people from New England (Iroquois), Wisconsin (Menomini, Potawatomi, Winnebago), the Dakotas (Lakota), Nebraska (Omaha, Pawnee, Ponca), Iowa (Meskwaki), Illinois (Sauk-Fox), and the Carolinas (Cherokee) used these onions. These people ate them in soups, dried the bulbs for seasoning, and used them fresh as a relish (Smith 1933, Yanovsky 1936, Moerman 1998). Related species were similarly used. Oddly, no records seem to exist of southeastern people using the plants, but at least the Alabama, Chickasaw, and Muskogee have names for them. Maybe they were like the Powhatan, of whom Strachey ([1612] 1953) said, "the Inhabitans cannot abyde to eate of them." Moerman (1998) said the species was used by the Mahuna (California), but Hickman (1993) does not list the species.

In spite of this wide usage, it is said that in large quantity wild onion has killed cattle, and gastroenteritis has been reported in children (Lampe and McCann 1985, Diggs et al. 1999).

Alnus

(From the classical Latin name for the alder; Akkadian *alanu*, oak, acorn)

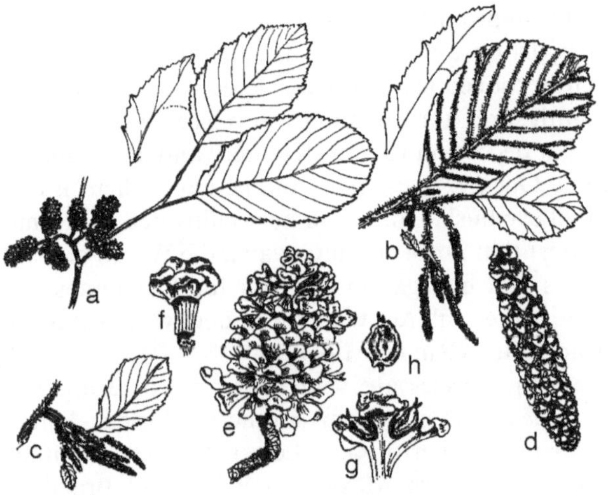

Alnus serrulata. a. Branch with pistillate aments. b. Branch with staminate aments. c. Young staminate aments. d. Staminate ament. e. Pistillate ament. f. Bract of pistillate ament. g. Bracts and nutlets. h. Nutlet. *Drawn by Vivian Frazier*. From Correll and Correll 1972.

alder (from Old English *alor, aler*, in use since ca. A.D. 700, related to Old Norse *elrir*, Old High German *elira, eirla*, and modern German *Erle, Eller*); aller (by Turner [1548] 1965; he wrote: "Alnus is called in greke Clethra, in Englishe an alder tree or an aller tree, in duche erlinbaum"); cognates include *aile* [*ilite*] (Spanish); *aune* [*aulne*] (French); *erlin* [*els, elst, elzenboom*] (Dutch)
atob (bitter, Potawatomi); *tópi* (*takw*, healing, *pi*, material, Delaware); *wadôb* (Ojibwa)
fèarn (Gaelic; fourth letter in the old *Ogham* alphabet)
gatogèhha (Onondaga)
okowwaykací (*okí*, water, *wáyka*, hanging over the edge, Koasati)
ontano (Italian)
uhl (Swedish)
verne (French)
wa'p?tu (Catawba); *yap si· wi saka're* (red blossom tree, Catawba)

Alnus serrulata (saw-toothed, in reference to the leaf margins)
alder [American, common, green, hazel, red, smooth, speckled, tag]

The enthusiastic and energetic British gardener Philip Miller (1691–1771) created the genus *Alnus* in the fourth edition of his *The Gardeners Dictionary* in 1754. Although this book went through four subsequent editions, his rendition of the Roman name for the alder tree stuck (de Cleene and Lejeune 2002). As exploration of the New World continued, Royal Gardener at Kew William Townsend Aiton (1731–1793) described plants from Pennsylvania as *Betula serrulata* in 1789. It took the director of the Berlin Botanical Garden, Carl Ludwig Willdenow, to realize that these were really *Alnus* and to create the modern name in 1805.

Moerman (1998) lists most of the American species of *Alnus* as having been used by indigenous people. In the southeastern United States, *A. serrulata* is recorded as used by only the Cherokee (Hamel and Chiltoskey 1975) as an analgesic, as a blood tonic, to treat coughs and skin and kidney problems, and as an emetic and purgative. The virtues of the plant were taught to the whites, and Porcher (1863) praised it highly for treating skin problems and internal or external bleeding. The bark contains tannic acid, which was formerly used in treatment of intermittent fever, as an alterative and astringent (Vines 1977, King 1984).

Alternanthera

(Latin *alterno*, to change or alter; and *anthera*, anther; the stamens alternate between fertile and sterile)

Alternanthera flavescens (yellowish)
alegría (happiness, Spanish)
amor seco del monte (wild dry love, Mexico)
chaff-flower ("chaff" is an old Teutonic-based word for the husks left from thrashing grain; the word was derived from Old English *ceaf*, which is cognate with Dutch *kaf*; "chaff" was in English by ca. A.D. 1000; "chaff-weed" was applied by Turner [1548] 1965 to a *Gnaphalium*, and the word "flower" was combined later for members of the Amaranthaceae)
immortele (brought into English from French *fleur immortelle* by 1832; meaning "live-forever")
jiquilete (*xiuitl*, herb, *quilitl*, edible herb, Náhuatl, Mexico)
joyweed (maybe simply a translation of *alegría*)

live-forever (in use by Gerarde's *Herball* of 1597 for a number of plants with long lives)
perpétua (perpetual or live-forever, Spanish); *perpétua-da-mata* (live-forever of the forest [wild], Brazil)
straw-flower ("straw" is from Old English *stréaw*, from a Teutonic origin with cognates *stroo* in Dutch, *Stroh* in German, *strå* in Swedish, and *straa* in Danish; by ca. A.D. 1000, it meant specifically the stalks of grasses, but subsequently has been applied to dried stems of many plants)
zac-muul [*sac-muul*] (*sak*, white, *muul*, sandbur, Maya, Yucatán)

Alternanthera maritima (by the sea)
beach alternanthera
seaside joyweed

When the Spanish *conquistadores* entered what is now Mexico City, they found storehouses of seeds in Moctezuma's granaries. The Aztecs called the seeds *tzoalli*, and showed the Spaniards the fields of red-topped plants they came from. When the Europeans discovered the importance of the seeds in indigenous religions, they declared it a crime to grow them.

To the Aztecs, the red of these plants was a symbol of immortality. Although the Spanish were horrified by the celebrations using *tzoalli*, they understood the symbolism. Red for them too was the color of royalty. In southern Europe, the plants became *alegría* (happiness or joy) or *perpétua* (perpetual or live-forever). Immortele became a common name in English, although it was originally used for another plant in an entirely different family (*Xeranthemum*, Asteraceae).

The seeds called *tzoalli* were *Amaranthus* (which see for details). Yet, in local usage, the common name *perpétua* or "live-forever" came to be applied to one of its relatives. The related genus is *Alternanthera*, and it was named by Linnaeus's Danish student Pehr Forsskål (1732–1763). However, his *Flora aegyptiaco-arabico* did not appear until 1775, after he had died of exposure in the deserts of the Arabian Peninsula. There now are about 200 species in the genus, most of which are scattered throughout tropical America (Mabberley 1997).

The species most commonly seen in Florida is *A. flavescens*, long known as *A. ramosissima*. This herb is called by several names in various parts of its range. In English, the plants are known as "chaff-flower" or "straw-flower," because of the papery scales between the small, almost insignificant flowers. They are also called "live-forever," an allusion to, or perhaps even a translation of the Spanish *perpétua*. In Brazilian Portuguese, this became *perpétua-da-mata* (wild live-forever) to distinguish them from those people cultivate (simply perpétua).

Those species for which we have recorded uses are largely applied to the same suite of problems (Morton 1981, Arvigo and Balick 1993, Mors et al. 2000). Decoctions of the plant are used to relieve coughs, or an infusion of the plant is considered sudorific. Internally, they are remedies for flu, colds, urinary problems, tiredness, and internal infections (especially of the reproductive organs), and they are used as a postpartum tonic. Externally, the juice of mashed or lightly cooked leaves is considered a good poultice for mouth sores, thrush, and to stop bleeding. Barbadians take a decoction to halt dysentery and as an abortifacient. One (*A. brasiliana*), and perhaps others, is called "Evil eye bush," and is an herb cultivated as a remedy against that problem. In Guyana, the Warao boil some leaves in a half gallon of water for 15 minutes. Then they bathe children with the water twice a day from head to toe (Rinders 1993).

Some species are wide ranging. For example, *A. flavescens* extends from southern Florida, Jamaica, Hispaniola, Puerto Rico, Lesser Antilles to Trinidad, and southeastern Mexico to Belize and Brazil. The other native species in Florida, *A. maritima*, is found on beaches in southern Florida, the Bahamas, the Caribbean, and South America. It also occurs in western Africa, a pattern fairly common among ocean-dispersed plants (see also *Annona*). Although no specific uses for "beach alternanthera" were located, perhaps the widespread uses apply to it as in other species.

In Belize, *A. flavogrisea* is called *golondrina* [*colondrino*] (dove) by Spanish speakers, and *xix-can-lol* (*xix*, bread crumbs, *ka'an*, yellow, *lol*, flower, so "yellow flower that looks like bread crumbs") in Maya. In the French Antilles, it is *herbe à albumine* (albumin herb). The species has all of the uses listed earlier, and others that are not listed (Morton 1981, Fournet and Hammerton 1991).

Alternanthera brasiliana has been introduced into Florida, but it is native through much of the Caribbean and south into Brazil. It has common names including *caaponga* (Tupí for what is usually called in Portuguese *beldroega-pequena* or *Portulaca oleracea*), *ervaço* (usually applied to *grão de bico, Cicer arietinum*), *infalível* (infallable), *perpétua-da-mata* (wild live-forever), *perpétua-do-brasil* (live-forever of Brazil), *bouton blanc* (white button, French Antilles), and *herbacé par le haut* (herb for heat, French Antilles). In all the areas where it is native, *A. brasiliana* has medicinal uses, and the name *herbacé par le haut* suggests that it might have been used to treat fever or maybe even hot flashes.

Alternanthera repens occurs throughout Mexico and Central America, south to Peru and Argentina, and in western Africa. Names recorded for this species include *ardosa jardin* (ardent garden), broad path (rendered *herbe à bordure* in French Antilles), *cabal-xtez* (prostrate plant that resembles a spiny bush, from *kabal*, prostrate plant, *x-tes*, spiny bush, Mayan), *caroca* (mistranscription of "carioca"?), *hierba de toro* (bull's herb), *sacachiuquim* (*saca*, remove, *chiuquim* =?), *sanguinaria* (bloody, applied to rocks, plants, etc. with a red color), *tianguiz-pepétua* (store-[bought] live-forever), washer-woman, and *yerba del moro* (Moor's herb). The species does not appear to be in the Caribbean, although older literature records it there.

One species in Brazil (*A. achyrantha*) has the intriguing name *erva-de-pinta*. Until the book by Desowitz entitled *Who Gave Pinta to the Santa Maria?*, most of us would not have known that *pinta* was a form of venereal disease. Brazilians use this herb to treat that and several other diseases (Mors et al. 2000). Indeed, several of the species in the genus are listed as being used to treat syphilis and other venereal diseases, but that typically is because they are diuretic and provide temporary relief from symptoms. Many plants in various parts of the world were (and probably still are) used to treat venereal diseases because, at least in part, they relieve symptoms. They do not alleviate the problem.

Alternanthera sessilis in the French Antilles is "rabbit meat," which may mean it is used as food for domestic rabbits, to "make meat." Or, it simply might mean that it was recognized as something that a rabbit might eat. This species is also called *maglwa* or *magloire* (my glory, my fame; a surname used by many in Dominica) in the French Antilles. In the Dominican Republic, the name *sanguinaria* is applied generically to several species of *Alternanthera*, but is used particularly for *A. sessilis*, also known as *santoma cimarrona* (wild Saint Thomas, or wild *Gomphrena globosa*). Both are used against urticaria and either cultivated or considered weeds in cultivated plots. Panamanians call it *sanguinaria* and eat the leaves as a famine food (Duke 1968). This same species in Taiwan is known as *han lien tsao*, and is substituted for *Eclipta prostata* in herbal remedies to treat the liver and kidneys. Both are considered antibacterial, astringent, and anti-inflammatory (Hsu 1986). *Alternanthera pungens*, also medicinal, is called *yerba de pollo* (chicken herb) in the Dominican Republic.

Comparatively little has been found on the chemistry of *Alternanthera*. Some species contain coumarin analogues (Yang et al. 1989), flavonoids (Zhou et al. 1988), ionone derivatives (Ragasa et al. 2002), sterols (Patterson et al. 1991), and triterpene saponins (Sanoko et al. 1999). Laboratory studies suggest that some species are anticarcinogenic (Aruna and Sivaramakrishnan 1992), antidiarrheal (Zavala et al. 1998), antinociceptive, at least in mice (Macedo et al. 1999), antiviral (Zhang et al. 1988, Unag et al. 1989, Qo et al. 1993), and hepatoprotective (Lin et al. 1994).

The most aberrant reference was for *A. lehmanii*, called *borrachera* or *chicha* in Colombia (Schultes and Raffauf 1990). According to them, this plant was occasionally added to *yagé* brews in Putumayo, Colombia, reputedly making a stronger beverage. A Web source (Gnostic Garden 2001) says, "Chemically obscure, needs further investigation." While that was intended for the species, it speaks for the whole genus.

Alvaradoa

(Named after Pedro de Alvarado, companion of Cortés and conqueror of Guatemala and Salvador)

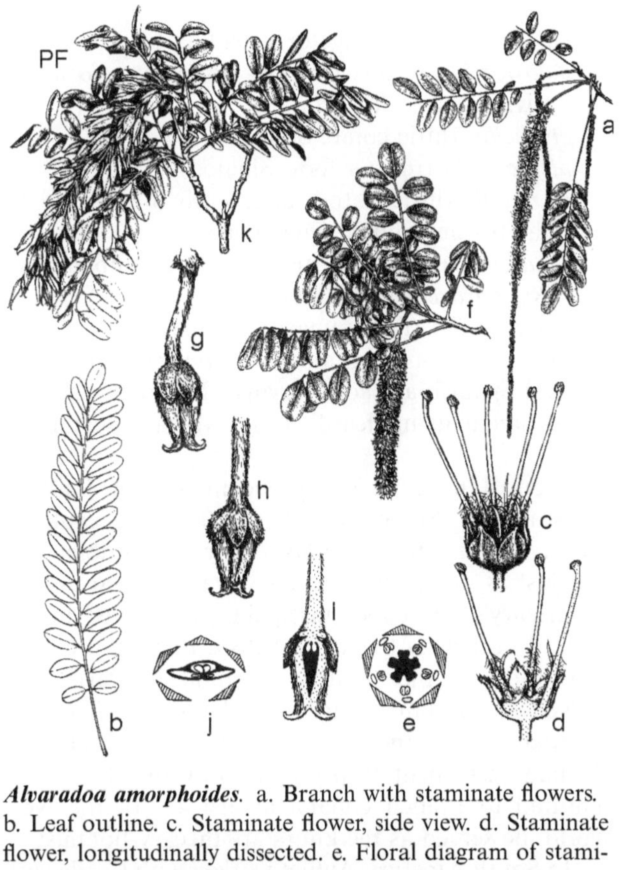

Alvaradoa amorphoides. a. Branch with staminate flowers. b. Leaf outline. c. Staminate flower, side view. d. Staminate flower, longitudinally dissected. e. Floral diagram of staminate flower. f. Branch with pistillate flowers. g. Pistillate flower, from the outside. h. Pistillate flower from the inside. i. Pistillate flower longitudinally dissected. j. Floral diagram of pistillate flower. k. Branch with fruits. *Drawn by Priscilla Fawcett*. From Correll and Correll 1982.

Alvaradoa amorphoides (resembling *Amorpha*)
aretillo (little earring, Sonora)
aroma blanca (white perfume, Cuba)
bel-cini-ché [*beezinik-ché, belsinik-ché, besinick-ché, besinik-ché,* and *xbesinic-ché*] (tree that is full of ants, Maya, Yucatán); *palo de hormigas* (ants' tree, Yucatán); *suetsinic-ché* (tree covered with ants, Maya, Yucatán)
camarón (shrimp, because of the flower/fruit color, Chiapas)
charagallo (*gallo,* rooster, Chiapas)
cola [*rabo*] *de ardilla* (squirrel tail, apparently in reference to the fruiting spikes, Mexico, Central America)
cortacuero (leather cutter, because of use in tanning, Belize)
gavilán (hawk)
guaji (Mayo, Sonora)
guetzé-guetzé [*gueguetzí, cuetzé, huezequí*] (Zoque, Chiapas)
huichipil [*huachipil, guacipil*] (maybe Náhuatl, Guerrero)
ma'acita (Guarijío, Sonora, Chihuahua)
paenepa (Mayo, Sonora)
palo de sobo (massage or kneading tree)
palo torcido (crooked wood, Sonora); *palo torsal* (twisted tree, Sonora)
peinecillo (little comb, Sonora)
pié de gallo (rooster foot, Sinaloa)
plumajillo (little feather, apparently in reference to the fruiting spikes, Guatemala)
ruda cimarrona (wild rue, notes a resemblance to *Ruta graveolens,* Quintana Roo)
tamarandillo (little tamarind, Cuba)
tarajay (resembling tamarisk, *Tamariscus*)
wild Spanish armada (Cayman Islands)
zo (language unidentified, not Maya, Yucatán)

From 1762 to 1763, Frenchman Jean Baptiste Christophore Fuseé Aublet (1720–1778) explored French Guiana for the king of France. Aublet was not just a botanist, but also was trained as an apothecary's apprentice in Spain, had served in the medical branch of the army, helped prepare prescription drugs at the Charity Hospital in Paris, and worked as "Botanist and First Apothecary" of the Company of the Indies at *Ile de France* (Mauritius). He had just spent 9 years in Mauritius before he ventured to French Guiana at age 42. In the New World, he served as the king's apothecary-botanist.

In South America, Aublet worked with Caribs and Galibis, coastal tribes of Cariban stock who remain there today. In addition, he worked with people called Coursari, Garipon, Noirague, and Creole Negroes. Thus, his contacts spanned languages as distinct as Carib, Tupí, and Creole French. He described medical uses for many of the plants he found, and he has been called the "founding father" of neotropical ethnobotany (Plotkin et al. 1991).

Among the 124 plants Aublet described from 56 different families was *Picramnia* (formerly Simaroubaceae, now Picramniaceae). Not surprisingly, several members of that family are effective medicinals, and Aublet described two of the most important genera—*Simaba* and *Simarouba.* Linnaeus had described *Quassia,* perhaps the most famous genus, while Aublet was still in French Guiana. *Quassia* and *Simaba* are used against fever, malaria, and dysentery (Hocking 1997).

Danish explorer Frederick Michael Liebmann (1813–1856), director of the botanical garden in Copenhagen, described *Alvaradoa.* The genus now contains five American species (Mabberley 1997).

In Yucatán, a decoction of *Alvaradoa* bark is drunk as a tonic for the digestive tract, and leaf decoctions are taken for urinary complaints (Morton 1981). Leaves mixed with honey and corn silk are used to stop bleeding. Leaves are put in warm baths for patients with rheumatism. The wood is strong and used in carpentry. Wood is also valued as fuel because it burns slowly and for a long time (Sousa 1940).

Although there seem to be comparatively few records in the literature on the uses of *Alvaradoa,* it has an impressive list of common names. Generally, when plants have multiple names across their ranges, they are valued parts of the local economies.

As usual, in the United States where people rarely retain the uses of tropical plants, there is no common name except the genus—alvaradoa. The average citizen will have no idea what you are talking about if that is used as a genus or common name. However, the deeper you go into the tropics, the more names for the shrubs and small trees you will find. In Cuba, it is called *tamarandillo* because of its similarity to the native *tamarindo* (*Lysiloma bahamensis*) and the exotic (*Tamarindus indicus*). The related Cuban endemic *A. arborescens* is *carmín* (scarlet). In Hispaniola, the related *A. haitiensis* is known as *abbé marron* (brown abbey). Jamaica has two endemic species so rare that no names have been recorded for them. The genus is absent from the Virgin Islands, Puerto Rico, and the Lesser Antilles, but it reappears in Yucatán where the Maya call it *bel-cini-ché* (tree that is full of ants). The name presumably is given because ants attend the floral (and extrafloral?) nectaries.

In the United States *Alvaradoa* occurs only in southern Florida, and there only in Miami-Dade County. The literature records them for Monroe County, but that seems to be an error (Coile 2000).

These Florida shrubs grow in the edges of rockland hammocks, mostly now in Everglades National Park. Formerly, they were scattered throughout other rockland hammocks in the area, but urbanization has eliminated most of them. Sometimes *Alvaradoa* grows with *Picramnia pentandra*, another member of the family.

The bush sulfur butterfly (*Eurema dina helios*) feeds exclusively on this species. It has perhaps the most restricted range in Florida as it is confined to the region of Miami-Dade County where the plants grow (Gerberg and Arnett 1989).

It is easy enough to imagine how most members of this family arrived in Florida. Fruits on *Simarouba* and *Picramia* are colorful and are eaten by birds. Presumably, avians are the primary dispersal agents within Florida, and perhaps inadvertently they brought them in their digestive tracts from nearby areas. However, it is more difficult to see how *Alvaradoa* fruits might attract birds. The fruits are neither fleshy nor showy, although they do turn pink at one stage. Moreover, they have wings that have prompted people to describe them as "samaras." Could these be plants that were brought by people traveling between the Caribbean Islands and peninsular Florida in pre-European times? Their uses outside Florida suggest that possibility.

Amaranthus

(Greek *amarantos*, unfading, an allusion to the lasting quality of the flowers, persistent bracts and sepals)

achita (Quechua, Peru); *kiwicha* (Cusco, Quechua, Peru)
alegría (happiness, Mexico)

Amaranthus hybridus. From Britton and Brown 1896.

amaranth (English); *amarante* (French); *Amarant(h)* (German); *amarant* (Dutch); *amaranto* (Italian, Portuguese, Spanish)
bledo (Spanish); blite (in English by 1420); *bredo* (Portuguese)
chithal tooro (bull *A. leucocarpus*, Huastec, San Luis Potosí)
chuhugia [*cuhukkia*] (Hia-ced O'odham, Arizona); *ki'aki* [*giád*] (Akimel O'odham, Tohono O'odham, northwestern Sonora and southwestern United States)
huautli [*huanhthi*] (Náhuatl, Mexico); *huisquiliti* [huisquilite] (for *A. spinosus*, Náhuatl, Mexico); *quelite* (edible greens, from *quilitl*, Náhuatl, Mexico); *quintonil espinosa* (spiny *quilite*, *A. spinosus*, Mexico)
koyo (Aymara, Peru)
nocuana guezetao naxiñaa (*nocuana*, edible herb, fruit, or remedy, *guezatao* means Moorish, but that often also indicates foreign, *naxiñaa*, red, Zapotec); *nocuana laza naxiñaa* (*nocuana*, edible herb, fruit, or remedy, *laza*, amaranth, *naxiñaa*, red, Zapotec)
pigweed (in use by 1844)

Amaranthus australis (southern)
careless (Florida)
cikiláfálî (*ciki*, house, *láfálî*, burner, Mikasuki)

Amaranthus hybridus (hybrid, because it was similar to two other species; not a hybrid)
ixtez [*xtez*] (Maya, Yucatán); *quiec tes* (Quiché, Guatemala)
quelite (Veracruz); *quelite morado* (purple *quelite*, New Mexico)

As a child, I learned never to walk barefoot in a barnyard or pig lot. There are two dangers. One is that neither animal has indoor plumbing; the other is that there may be spiny plants. I grew up calling one kind of spiny herbs "pigweeds," even though they grew in cattle pastures and not in pig lots. Later I learned that they were *Amaranthus*. The one that was particularly bothersome to tender bare feet was *A. spinosus* (spiny).

The genus *Amaranthus* remains an important food source for many people in the world. In Spanish, the generic name is *bledo*, which is rendered *bredo* in Portuguese. These terms originated with Greek *blíton* that was rendered *blitu* in Latin. Ultimately, for English speakers, this was changed into "blite." This type of name, which cannot be broken down further, is called by linguists a monomorphic lexeme, and is taken as a marker for a long usage, regardless of the language. Historical data on *bledo* or *blite* support that view, so the plants and the name have been with Old World people for thousands of years.

Although the compilers of the *Oxford English Dictionary* were in doubt regarding the relationship, blite probably evolved into a word of derision, "blight." That suggestion is made and then rejected by the OED (1971), but the historical sequence and origin of blight with farmers and gardeners convince me they are connected.

Blite made its way into English in the first recorded document in 1420. It was not until 1669 that the rendition "blight" appeared. Perhaps the changing attitudes about useful vs. nonuseful plants in the Old World, connected with novel views from the New World, led to the change.

When the first Europeans arrived in the New World, amaranths were among the plants they encountered being used by indigenous people. One of the first records was left by the Gentleman of Elvas about Hernando de Soto's Spanish troops in the river or swamp of *Cale* [*Ocale*], near what is now the west bank of the Withlacoochee River in Citrus County. This was southwest of the region now known as Ocala. These soldiers were "experiencing great hardship from hunger and bad roads as the land was very poor in maize, low, and very wet, swampy, and covered with dense forests, and the provisions brought from the port were finished. Whenever any village was found there were some *blites* and he who came first gathered them and having stewed them with water and salt, ate them without anything else" (Swanton 1946).

At first, they were not surprised because the plants were well known in their homelands, even if they were not the same species. However, their countrymen soon discovered that some Americans were using their plants in ways never dreamed of in Europe.

A Mexican Spanish name for *Amaranthus* is *quelite*, rendered *kelite* in Belize. *Quelites* served and still serve as the daily food for many people (Linares and Aguirre 1992). An example is the Tarascans of conquest times for whose daily food: "*Maíz, pescados secos, charales y quelites forman la base*" (Corn, dry fish, *charales* [a small bony fish], and *quelites* form the base) (Santamaría 1959). Perhaps the earliest record of Mexicans using *Amaranthus* in the 1500s is from Fray Bernardino de Sahagún, who wrote: "*Otra yerba se come, cocida llámase quintonilli, tiene las hojas anchelas: cuando es pequeña esta yerba es comestible, y cuando ya es grande llámase petzicatl, cuécese con salitre (es decir tequixquitl) esprímese del agua para comerla*" (Another herb is eaten, cooked it is called *quintonilli*; it has narrow leaves. When it is small, this herb is eaten, and when it is big, they call it *petzicatl*, cooked with salt (that is *tequixquitl*) and removing the water to eat it) (Santamaría 1959). *Quintonil* is from *quilitl*, greens, *tonilli*, cooked (Náhuatl). This same plant is the amaranth whose seeds are used to make breads called *alegría* (happiness).

Actually, *alegría* is not only a name for the plant from which *quelites* are taken to make *quintoniles*, but the seeds and the bread are all *alegría*. From the seeds, Mexicans made atole, little tamales called *huauquiltamales*, and a dough. The bread, made by mixing the toasted seeds with sugar and boiling, was commonly sold in markets (Santamaría 1959). Both hemispheres used the leaves as greens, and the seeds were eaten as food. So far so good.

However, at first the priests, and then the remaining devout Catholics, discovered that Aztecs mixed the seeds of *Amaranthus* with blood (sometimes human), and sculpted the resultant dough into "idols" that they offered to their deities. Both practices outraged the pious religious leaders. Catholic priests started a campaign to suppress the practice, and much of it was stamped out, or at least driven underground.

The celebrations that started the campaign against *Amaranthus* revolved around *A. leucocarpus*, plants now called "prince's feather." That species was the one that had been grown in Mexico since ca. 5000 B.C., and the Aztec emperor Moctezuma is said to have received an annual tribute equivalent to 200,000 bushels of seeds from this cultigen (Purseglove 1968). Each seed is usually less than 1 mm in diameter, so one must imagine how many plants and people it would take to gather that many bushels.

Dough made from the seeds (*tzoale, tzouali, zoale, soale*, from Náhuatl *tzoalli*) was made from a plant Aztecs called *uautli* or *huautli*. Historical records show that *tzoalli* seeds were used to make a statue of *Uitzilopochtli* (*uitzilin*, hummingbird, *opochtli*, left-handed) to celebrate his festival during the month of *toxcatl*, which occurred from the last part of April and the first part of May. The first day of the celebration was dedicated to *Tezcatlipoca* (*tezcatl*, mirror, *poca*, brilliant), and culminated in the sacrifice of a human to this deity. Eight days later they began the celebration to *Uitzilopochtli*. From the *tzoale* seeds, they created the idols, and also made sweets and candies that were offered to their deities (Siméon [1885] 1981).

It is not too far from being disgusted with a practice to having the whole organism tainted with that emotion. European farmers would not have liked the *Amaranthus* competing with the foods they were trying to grow in the first place. Adding religious bias to that negative feeling made *Amaranthus* even more undesirable. Maybe that is why Europeans began calling the *Amaranthus* "pigweed," because they were not fit for people, but were served to their barnyard animals.

Regardless, Americans and many other people used *Amaranthus* as food and medicine. Spanish speakers distinguished the individual species with modifiers, e.g., *bledo espinoso* for *A. spinosus.* Aztecs called the cooked leaves *quilitl.* That became Hispanized *quelite,* and the name has become generic in Mexican Spanish; it is now applied to many edible greens (Linares and Aguirre 1992). *Huautli* is now applied to fish or insect eggs, and not to a plant at all (Ebeling 1986).

A relative of *Amaranthus* is spinach, so it should not be surprising that the French call both *épinard* (French Antilles). In Creole English in the Dutch Antilles, Belize, and Panama, *Amaranthus* is called *calaloo* [*calaloe, pison calaloo, callalu*] (a Caribbean name, etymology uncertain). Technically, *calaloo* is the greens from leaves of *Xanthosoma,* but Asian *Colocasia,* also in the Araceae, has been substituted. So has *Amaranthus.*

In the United States, various people used *Amaranthus* as food. Moerman (1998) included 12 species, eaten by the Acoma, Apache (Chiricahua, Mescalero, White Mountain), Cahuila, Cocopa, Gosiute, Havasupai, Hopi, Iroquois, Isleta, Jemes, Keres, Keresan, Klamath, Laguna, Mendocino, Mohegan, Mojave, Montana, Navajo, Papago (Tohono O'odham), Pima (Akimel O'odham), Tewa, and Zuni. Felger and Moser (1985) add the Seri, and Hodgson (2001) added the Cochimí, Guarajío, Northern Tepehuan, Quecha, Tarahumara, Yaqui, and Yavapais. It would not be surprising to learn that all people in the New World used *Amaranthus* leaves and/or seeds.

Not only are *Amaranthus* eaten, but many people have discovered that they may also be used medicinally. The Cherokee used *A. hybridus* and *A. retroflexus* in "Green Corn" medicine, and to stop profuse menstruation (Hamel and Chiltoskey 1975). The Keres used an *Amaranthus* infusion for stomach problems, the Iroquois used it in witchcraft, and the Mohegans treat hoarseness with an infusion (Moerman 1998).

A leaf decoction has been used in Veracruz, Yucatán, and Costa Rica for stomach pain and liver trouble, and a root decoction serves as a diuretic and remedy for bladder problems (Morton 1981, Vásquez and Jácome 1997). Several Central American groups have learned that bathing in a plant decoction relieves fever, and the plant may be applied to external sores and wounds. Crushed or lightly cooked leaves cure sores, especially those that are slow to heal. Roots are also used to wash boils to help them heal (Morton 1981).

In Florida the most common species is *A. australis,* historically called "careless" because it does not care where it grows. This may have been among the plants called *cikiláfálî* (*ciki,* house, *láfálî,* burner) by the Miccosukee. They claimed that, when the dead herbs blow like tumbleweeds across a camp, sometimes they get in the cooking fires (Sturtevant 1955). When *cikiláfálî* did that, there was a good chance that they would set the house on fire. Perhaps that is one reason why the Seminoles kept a cookhouse *chikee* separate from where they slept.

Ambrosia

(From Greek *ambrosios,* delicious, divine, immortal, derived from *a,* not, *mbros,* mortal; in Greek mythology, the food and drink of the gods)

Ambrosia trifida. a. Upper portion of a flowering branch. b. Portion of a flowering raceme. c. Achene. *Drawn by Regina O. Hughes*. From Reed 1971.

ambrosia (used in English by ca. 1590); *ambrosia* (Italian); *ambrosie* (French); *Ambrosie* (German)
iubar-sléibhe (*iubar,* yew, *sléibhe,* heath, Gaelic)

Ambrosia artemisiifolia (with leaves like *Artemisia*, sagebrush or wormwood)
l'ache à dinde (turkey celery, Quebec)
alcanfor (camphor, Cuba)
altamisa (transcription of ambrosia, Mexico, Cuba)
beskambrosia (bitter ambrosia, Norwegian)
bitter weed
canhlo'gan onzi-pakinte (herb to wipe the rear, Lakota); *canhlo'gan was'te'mna* (sweet-smelling herb, Lakota)
hay-fever weed
herbe à poux (louse herb, Quebec)
hogbrake [hog-weed, hog-wort] ("hog" is probably derived from Cornish *hoch* or Welsh *hwch*, swine or perhaps yearling; "brake" here meaning thicket; "hot-weed" must be a mistake for "hog-weed"; cf. Small 1933)
kcicka (Biloxi)
mon̲-hin̲-pa (*mon̲-hin̲'*, grass, *pa*, bitter, Osage); *pexhuta pa* (bitter medicine, Dakota)
poipiye (to doctor swellings with, Lakota)
[annual, common, short] ragweed ("ragweed" from "ragwort," where "rag" denotes the incised margins of the leaves)
Roman wormwood ("wormwood" from the use as a vermifuge)
sarriette (from Latin *Satureja*, the aromatic old world mint, which must have been confused with New World aromatic composites, Quebec)
stammer-wort [stammerwort]
vtakrv homv (*vtakrv*, plant, *homv*, bitter, Muskogee)

Ambrosia hispida (hairy, with stiff trichomes)
altamisa de playa (beach ragweed, Puerto Rico)
bay tansy ("tansy" is taken from Medieval Latin *athanasia* and Greek *athanasia*, immortality; akin to "everlasting" or *immortelle* because of the long-lasting flowers; now *Tanacetum vulgare*, Asteraceae; these are coastal plants, hence "bay," Puerto Rico, Dominica); bay vine (Bahamas)
carquesa (an oven that is used to temper glass, Spanish)
coastal ragweed
sept Villes (seven towns, Dominica)
tapis vert (green carpet, Dominica)
wormwood (Dominica)

Ambrosia trifida (three-parted)
bloody weed (translation of Kiowa by Vestal and Schultes 1939; children broke off a branch to watch the "bloody" liquid exude; revered by the Kiowa)
buffalo-weed [buffaloweed]
canhlo'gan was'te'mna (sweet-smelling herb, Lakota)
giant [great] ragweed ("ragweed" was first applied to *Senecio*, then to *Ambrosia*)
hesterambrosia (horse ambrosia, Norwegian)
horse-cane
horse-weed (this is the name that I grew up using in western Kentucky; no one explained it, but we always thought the word "horse" referred to the large size)
Taubenkraut (dove herb, German)
yamnu'mnuga iye'ceca (it is like making noise crunching with teeth, Lakota)

Anyone with an allergy to the pollen of *Ambrosia* will wonder why Linnaeus used the Greek word for immortality to name the genus. He kept quiet about that topic, but he was following the lead of previous authors including Gaspar Bauhin (1560–1624) and Rembert Dodoens (1516–1585) who had used the name for *A. maritima*, a seaside plant growing around the Mediterranean. By the time Linnaeus described *A. artemisiifolia* and *A. trifida*, both were well-established weeds in Europe (Linnaeus [1753] 1957, Clapham et al. 1987).

Giant ragweed was first found by Melvin Gilmore in archaeological deposits of the Ozark Bluff-Dwellers. Gilmore (1931) wrote: "There is evidence that the ancient Ozark Bluff-Dwellers also had certain other species of plants not cultivated at the present time. The ground for this statement lies in the fact that supplies of the seeds of these other species of plants were carefully put away together with the selected seed of corn, beans, sunflowers, squash and pumpkins." Among the seeds "carefully put away" were those of *A. trifida*. Gilmore argued that the size of the seeds he found were larger than wild plants and therefore ragweed had been cultivated. Payne and Jones (1961) found that the variability in wild *A. trifida* included the sizes found by Gilmore, and concluded that the plants were not cultivated. They suggested that the seeds were probably used for medicine and not food.

A subsequent report of *A. trifida* from an archaeological site near Mobridge, South Dakota, also concluded a medicinal use (Kindscher 1987). However, a third site in west-central Illinois has been studied by Asch and Asch (1982), who concluded that this Woodland period (ca. 1500 B.C.) culture was cultivating *A. trifida*. Their conclusion was based on both the fact that the seed coat was missing in most of the samples (presumably because they had already been processed), and that it was second in abundance to the known cultigen of marsh elder (*Iva annua*).

When Gilmore (1931) talked with an Arikara man about *A. trifida*, he was told, "I do not know how this plant may have served those ancient people, but now I will tell you a use which our people made of it. If you crush the flower heads in your hand you will see that it makes a red stain. . . . Our people formerly employed the blossoms of this plant to make a red dye." An Omaha also told him that "our people used to make beautiful red stain from its flowering tops." That use is not recorded in Gilmore's (1919) study of the Missouri River tribes, nor is it mentioned by Moerman (1998), although Vestal and Schultes (1939) alluded to it when they reported that the Kiowa revered this plants. Tull (1999) also noted the Omaha use, and that a green dye can be made from the leaves, and that fiber can be extracted from the stem.

There is other documentation that historic tribes used *A. trifida* as a medicine. The Cherokee used *A. trifida* in "Green Corn" medicine, rubbed leaves on insect stings and hives, put juice from leaves on infected toes, and took a leaf infusion to lower fever and to treat pneumonia (Hamel and Chiltoskey 1975). The Iroquois used a decoction to stop diarrhea with bleeding; the Lakota used the seeds medicinally; and the Meskwaki chewed the roots to relieve fear (Moerman 1998). Hocking (1997) recorded use of the plant as an astringent and detergent, and the root was chewed as a sedative among "Wiscon[sin] tribes" (Payne and Jones 1961).

The Cherokee used *A. artemisiifolia* in "Green Corn" medicine, rubbed leaves on insect stings and hives, put juice from leaves on infected toes, and took an infusion of leaves to lower fever (Hamel and Chiltoskey 1975). The Houma took a decoction of roots for menstrual troubles (Speck 1941).

Farther north, the plants were used by the Dakota, Delaware, Iroquois, Lakota, and Oto (Moerman 1998). It has been reported as used by the Luiseño and Mahuna of California (Moerman 1998), but it is a recently introduced weed there (Hickman 1993). In a rare comment on a "forbidden topic," the Lakota used the leaves for toilet paper (Kindscher 1987). In Mexico, the entire plant was used in an infusion to aid digestion, as a febrifuge, and to expel worms (Martínez 1969). In Cuba, it is used in aromatic baths against rheumatism, and in decoctions to treat fever (Roig 1945). *Artemisia hispida* is used essentially for the same problems as *A. artemisiifolia*, especially by Caribs (Honychurch 1987).

Millspaugh (1892) considered ragweed useful in treating poison ivy rash, possibly because of its astringency. For that same reason it was applied to other problems of the mucous membranes, including leukorrhea, gonorrhea, diarrhea, dysentery, and enteritis. Plants contain essential oils, including *p*-cymene, limonene, alpha-pinene, and alicyclic ketone (Kindscher 1987).

Amelanchier

(From Provençal *amelanquier*, which in turn is based on *amelanco*, from the Gaulois language)

Amelanchier arborea. From Britton and Brown 1897.

amélanchier (French)
blåhegg (*blå*, blue, *hegg*, hedge, Norwegian)
Felsenbirne (rock pear, German)
guillomo (Spanish)
Felsenmispel (rock medlar, German); mespilus [mespile] (first used in English for the medlar, *Mespilus germanica*, about 1398; later expanded to include *Crataegus* and *Amelanchier*)
pero corvino (crow pear, Italian); *petites poires* (little pears, Quebec); *poirier* (pear tree, Quebec)

Amelanchier arborea (like a tree)
June-berry [juneberry]
poires (pears, Quebec)
sa-bas-sa (Osage)
saskatoon (sometimes applied generically, and other times to *A. alnifolia*, from Cree *misaaskwatoomin*; akin to *guzigwa'kominaga'wunk*, thorny wood, Ojibwa, and *bosîkwa'komînûn*, plum berry, Potawatomi)
[common, downy] service-berry [common service-berry, downy serviceberry, sarvice berry, sarvice-

berry, sarviss tree, sarvas] ("service" derived from *Sorbus*)
shad-blow [shad-blossom, shad-bush] ("shad" is perhaps from Gaelic *sgadan*, herring; "blow" from Old English *blówen*, from the Old Teutonic root *bló*, cognate to Latin *flos*, flower or blossom; some say the name was applied because of the plants being in bloom when shad, *Alosa sapidissima*, migrate back upstream)
sugar plum (Missouri)

There is a single species of *Amelanchier* native to Europe (Mabberley 1997). At first it seems odd that Linnaeus did not name the species; instead it was named by Alfred Rehder (1863–1949) in 1920, while he was curator of the herbarium at Harvard's Arnold Arboretum. However, Linnaeus called the European plants *Mespilus amelanchier.* Now Linnaeus's genus *Mespilus* has had most of the species placed in other genera such as *Pyracantha, Cotoneaster,* and *Amelanchier.* The Florida species was not named until 1812 when André Michaux described it, also as *Melispus.* Finally, Merritt Lyndon Fernald (1873–1950), another Harvard botanist, gave us *Amelanchier arborea* in 1941.

Across the colder parts of North America, people mixed dried, pounded meat with berries and fat to create what the Cree called *pimekan* (from *pimiy*, grease). That word came into English as "pemmican" by 1801, although European settlers had been using the mixture far longer. This mixture of protein-rich meat, berries to add flavor and serve as a preservative, and fat was essential to surviving the long winters. The food also served as rations for long journeys (Davidson 1999). This mixture was used all across the cold winter belt, from the Iroquois of New York to the Ojibwa of Ontario, and into the Great Plains (Gilmore 1919, Densmore 1928, Yanovsky 1936, King 1984). It was used at least as far south as the Osage and Omaha, who called it *ta'-pshe* (La Flesche 1932).

Steyermark (1963) recorded that the indigenous people of Missouri mixed the dried fruits of *Amelanchier* with cornmeal to make bread. That mixture is more akin to the preparations made by most southeastern people than those farther north. No record was found of people in the southeast using anything like pemmican. Instead, southern people had dishes often based on corn, beans, and squashes, and they dried meat, fruits, and berries for use during the winter months (Hudson 1976). Stews were more a part of the diet in the southeastern people, and they mixed in a variety of plant products (Hudson 1976).

To some people fruits of *Amelanchier* are not that good when taken directly from the plants, but they are said to be improved by drying and cooking (Fernald et al. 1958). Others consider the berries good (Hedrick 1919). Oddly, the small round fruits were compared with pears in French (*poires*), German (*Felsenbirne*), and English where they sometimes were called Maine sweet pear and grape-pears (Hedrick 1919). Fruits of *Amelanchier* were used by people from Alaska and New Brunswick to California and at least as far southeast as the Cherokee (Gilmore 1919, Moerman 1998). Given the culinary similarity between the Cherokee and their Creek neighbors (Hudson 1976), there is every reason to believe that all southeastern people used the fruits.

Cherokee also used *A. arborea* to expel worms, and as a spring tonic (Hamel and Chiltoskey 1975). The Iroquois used a bark decoction against gonorrhea and as a blood remedy (Moerman 1998). The same people used a fruit infusion to stop the after pains and bleeding from childbirth. The other eastern species have a wider range of recorded uses that perhaps apply to the Florida species.

Amianthium

(Asa Gray named these herbs with Greek *amiantos*, unspotted, pure, alluding to the lack of glands on the perianth)

Amianthium muscaetoxicum (fly-poison)
fly-poison

In 1788, Thomas Walter described herbs from South Carolina he called *Melanthium muscaetoxicum.* When Asa Gray worked on the genus he realized that this species was more closely related to other North American species. Gray created *Amianthium muscaetoxicum* in 1837, based on the species name Walter proposed. As late as Mabberley (1997) the species was put in a third genus, *Zygadenus* [*Zigadenus*] (Greek *zygos*, a yoke, and *aden*, a gland, because of paired glands).

Leaves and bulbs are poisonous, especially to livestock (Steyermark 1963). Although the species grows from Pennsylvania and West Virginia to Missouri and Oklahoma and south to Florida, there seem to be records of people using it only in the Carolinas. The Cherokee used it to treat itch, and to poison unwanted animals (Hamel and Chiltoskey 1975). All plant parts, but especially the bulbs, contain the toxic compounds zygadenine and zygacine (Foster and Caras 1994).

Amorpha

(Greek, *amorphos*, deformed)

Amorpha fruticosa. a. Upper part of a plant with flowers. b. Branch with fruits. c. Young leaflets. d. Flower. e. Pistil. f. Fruit cluster. *Drawn by Vivian Frazier.* From Correll and Correll 1972.

Amorpha fruticosa (shrubby)
ayikchô:mî (bitter medicine, Mikasuki); *tohó:mî* [*kado ho-mi*] (*eto,* wood, *home,* bitter, Creek, Muskogee)
bastard indigo (from its use as a source of blue dye); [desert] indigo bush; [dull-leaf, false] indigo
kitsuhast (*kitsu,* water, *hatsu,* strings, Pawnee)
lead plant (believed to indicate the presence of lead ore; in use by 1864, probably before)
polecat tree (maybe related to the retting process in preparing dyes)
tse'-biu-k'a-mon̲-ta hi (*tse'-biu-k'a,* frog, *mon̲'-ça hi,* arrow shaft stalk, Osage)

By the early 1700s, another American plant from the Carolinas had caught the attention of the European horticultural trade. A London book on cultivated plants listed for sale *Barba jovis americana, pseudo-acaciae foliis, flosculus purpureus minimis* (American Jobe's beard, with leaves resembling acacia, flowers very small and purple) (*Angliclici hortulani* 1730). This species appeared in two books in the 1730s and two in the 1740s, and then Linnaeus finalized its name in 1753 as *Amorpha fruticosa.* Somewhere along the line, it became known in English as "lead plant," with the belief that lead could be found wherever it grew (OED 1971).

Linnaeus called these shrubs *Amorpha* because the flower has only a banner wrapped about the staminal column. The other four petals that are normally present in legumes are entirely missing. Subsequently, exploration of the Americas has disclosed more kinds, and we now know that there are 15 species in North America (Mabberley 1997). *Amorpha* ranges from Canada through the United States, and barely reaches into Mexico in Baja California and Sonora.

In the eastern United States, there are six *Amorpha* species; those in Florida are *A. fruticosa* and *A. herbacea.* Both species range down most of the peninsula, with only a few individuals of either species known in the southernmost five counties.

Although this is a temperate genus (Wilbur 1975), there seem to be comparatively few records of how it was used by people in the past. Now, the attitude has evolved to the point that the USDA has listed one, *A. fruticosa,* as an invasive species, even though it is rarely encountered even within the heart of its range. Clearly, views of "lead plant" have evolved considerably since Europeans first arrived in the New World.

Native Americans had more regard for the plants. In Florida we know that *A. fruticosa* was called *ayikchó:mî* (Mikasuki). Related Creek names for the same plant were *kado ho-mi* (Muskogee) and *tohó:mî* (Creek).

Florida Seminoles used *ayikchó:mî* against what they termed the "Moving Sickness," a general malaise with a pain that moved from one area of the body to another (Sturtevant 1955). At least part of that illness seems to have stemmed from gastric distress. Regardless, roots of *Amorpha* were mixed with *Sisyrinchium* and *Andropogon* to treat the malady. A different recipe is given by Oklahoma Seminoles (Howard 1984). They boiled the leaves and stems, and the resultant tea was left to stand overnight. The following day the liquid was reheated and then was ready to use. The "tea" was taken four times a day, before meals and at bedtime, as a general tonic.

For rheumatism pains the doctor chews the *Amorpha* and applies it to the area that hurts (Sturtevant 1955). The Seminoles employed a mixture of medicines for chronic sickness, and the longer an individual was sick, the more plants were included. Sometimes this medicine consisted of 15 plants, and was strong enough potentially to kill the patient. One of the ingredients was *ayikchó:mî,* with buds and small bits of the stem employed. The same plant was given in medicine at birth ceremonies (Sturtevant 1955).

The Omaha of Nebraska used the leaves of *A. canescens* in some of the same ways as the Seminoles (Gilmore 1919). In addition to the shared uses, the

Omaha had others, including blowing powdered, dried leaves into cuts and open wounds to help them heal. They also considered *Amorpha* a counter-irritant; one end of a stick was attached to skin with saliva, then burned down from the opposite end until the fire touched the skin (Gilmore 1919). The Meskwaki also used *A. fruticosa* as medicine (King 1984).

In their Lakota-related language, the Omaha called *A. canescens* shrubs *te-hunton-hi* (buffalo belly plant). This name was derived from the fact that blooming is synchronous with the buffalo rutting season (Gilmore 1919). Williamson (1987) summarized their worldview when he wrote: "Native Americans lived very close to the land. Whether they were hunters, gatherers, or agriculturists, all groups depended heavily on the grace of nature and their own wits to survive. They had to learn weather patterns, the cycles of plant growth, and the movement of game in order to live." Sometimes, their common names reflect that diverse knowledge.

Several other people left other applications of *A. fruticosa*. Among the Pawnee, the plant was called *kitsuhast* (water-strings). Presumably, that reference is like the one to *A. canescens*, which was called in English "shoestrings," and referred to long tough roots. Among the Kiowa, the long stems were used as a foundation for bedding material (Vestal and Schultes 1939). Moreover, stems were made into mats, rugs, and bedding. The Lakota, like the Seminoles, used the stems to make arrows (Sturtevant 1955). The Pawnee also spread *Amorpha* on the ground to receive meat while butchering to keep it clean (Gilmore, 1919).

There are other records of two related species being used in several other ways that probably apply to Florida plants. *Amorpha canescens*, which is essentially the Great Plains species, ranging from Michigan to Montana and south to New Mexico and Texas, was called "prairie shoestrings," or just "shoestrings" (Coffey 1993). It was also called "wild-tea" because at least the Oglala Sioux made a drink of the leaves. The Ojibwa took a decoction of roots for stomach pain, and the Meskwaki of Iowa used an infusion of leaves to kill pinworms or other intestinal worms. Infusions also were applied to eczema. The Oglala also crushed dried leaves as finely as possible, mixed them with fat, and smoked the mixture. That particular preparation must have been medicinal, although it is not specifically stated as such (Moerman 1998).

Finally, *A. nana*, which grows from Minnesota and the Dakotas south to Oklahoma and New Mexico, was used by the Navajo (Elmore 1944). These Athabascan people call the plants *ni'pís-tjah* (plant that is lying on the ground). The Navajos use it as a *tc'iic'azee* (cold medicine), by making a snuff, which they used against catarrh. English names for the plant are largely lackluster, including small-leaved bird's foot, small-leaved deer weed, false indigo, shoestrings, and indigo bush. The latest floras on the region call it simply "dwarf indigo," translating the Latin name.

The medical uses have a chemical basis, as there is evidence that the plants contain alkaloids (Diggs et al. 1999). Moreover, livestock browsing the shrubs are poisoned, but poisons in the correct dosages are medicines. Other bioactive chemicals are also known, and *Amorpha* contains cannabinoid-like compounds, essential oils, rotenoids, flavanones, and isoflavones (Kemal et al. 1979, Ognyanov and Somleva 1980, Rozsa et al. 1982, Somleva and Ognyanov 1985, Li et al. 1993, Crombie and Whiting 1998, Ohyama et al. 1998, Lis and Gora 2001). Several of these compounds inhibit skin tumors and microbes (Mitscher et al. 1981, 1985, Petkov et al. 1983, Konoshima et al. 1993, Li et al. 1993).

English names for *Amorpha* are not nearly as interesting, probably because the plants had less importance in European-derived cultures. The most widespread known use for *Amorpha* is as a dye source. Like the highly valued blue from true indigo (*Indigofera suffruticosa*), at least *A. fruticosa* produced a blue dye, although some say it was inferior to true indigo. That usage resulted in the plants being known as "bastard indigo," "dull-leaf indigo," "false indigo," "indigo bush," and "desert indigo bush." While it may not be immediately obvious, the common name "polecat tree," recorded for northern Florida in the 1930s (Baker 1938), probably also reflects usage as a dye plant. Some accounts of retting and then mixing plants with mordants in preparing plant materials to use as dyes mention a particularly offensive odor.

That there was widespread use of *Amorpha* among indigenous people as a blue dye should not be surprising. Blue was an important color to many tribes; it is the color of the sky, and to some the color of a cardinal point. To the Hopi, blue was the color of the southwest. Zuni considered blue the color of the west (Williamson [1837] 1962). Plants that provided dyes colored like the cardinal points were regarded with reverence. Colors symbolized order in the world. Order is a lesson that would aid modern chaotic worldviews bent on "using up" the world's resources.

Ampelopsis

(Greek *ampelos*, a vine, *kissos*, ivy; referring to foliage and similarity to *Cissus*; created by André Michaux, cf. Taylor and Norman 2002)

Ampelopsis arborea. From Britton and Brown 1897.

Ampelopsis arborea (tree; why Linnaeus chose to call these twiners the "tree" is one of those unexplained quirks of his sometimes bizarre sense of humor)
ha'fáli (Mikasuki, fide Stirling 1941; however, Sturtevant 1955 found *ha'fáli* and *hili:háhli* being used for *Toxicodendron radicans*; see also *Parthenocissus* and Poison-ivy)
ifá imittó (*ifá*, dog, *im*, its, *ittó*, tree, Koasati; same name for *Parthenocissus*)
parrita de playa (little beach vine, Tamaulipas)
pepper-vine (USA)
Schreinrebe (basket vine, German)
vigne vierge (maiden vine, French)

Neither J.K. Small nor anyone else gives us a clue why pepper is associated with this climber. Probably, the name came into existence because the oxalic acid in the leaves and stems made it "peppery," but it does not seem to have been recorded. We know, at least, that several members of the Vitaceae have vegetative parts loaded with oxalic acid. It is so bad in some that people who work in vineyards develop painful rashes from contact with the plants. Tull (1999) warns that people have died from eating Virginia creeper (*Parthenocissus quinquefolia*) fruits. Others, with milder poisoning, have their kidneys damaged for life. Considering that, at one time, Virginia creeper was considered an *Ampelopsis*, is reason for being careful.

As now delimited, *Ampelopsis* consists of 25 species, occurring in temperate and subtropical America and Asia (Mabberley 1997). Most of the species are Asian, and only three seem to be in North America. Those American species are the pepper vine *A. arborea*, the partly sympatric *A. cordata* that grows in the United States and Mexico (Veracruz, Chiapas, Yucatán), and *A. mexicana* that grows from Sonora to Chiapas.

Of these species, both *A. arborea* and *A. cordata* have been used as a dye. According to Tull (1999), they are both good for dyeing gold and brown, if the vine tips and leaves are used with alum. This is a solar dye, where the plant material is added to water in a gallon jar, and left for a week or so. When the water is colored, wool is put in the dyebath. A gold color results, which Tull considers outstanding. If the alum is not used, the color is brown, which can be darkened through various shades by using copper or iron mordants.

The late Julian Steyermark (1963) wrote of *Ampelopsis cordata:* "In August the fruit, about the size of a pea, begins to ripen, turning from green to changing shades of orange, rose, purple, and finally turquoise blue. It is very ornamental in this multicolored display, and is commonly eaten by songbirds, such as brown thrashers, wood thrush, and others." He could just as well have been talking about *A. arborea.*

Maybe the fruits of *A. arborea* are blacker than "turquoise blue," but they are just as easily seen by foraging birds. Indeed, the plants spring up wherever there is a perch for birds to sit. Once established in open areas, they may sprawl long distances if there is not a nearby support. These mats of pepper vine may be a meter or more in thickness and effectively shade out other plants, native and alien. One study done in southern peninsular Florida found *A. arborea* in every habitat type (Austin et al. 1987b).

Given that birds, and probably other vertebrates, actively move the plants from place to place, why is neither of the two eastern U.S. species in the Bahamas?

Amphicarpaea

(From Greek *amphi*, of both kinds, and *carpos*, fruit, the allusion to two kinds of fruits, aboveground and belowground; also spelled *Amphicarpa*)

Amphicarpaea bracteata (with leaf-like structures [bracts] on the flower-cluster)
American licorice
aʔnapa·hkow (he conceals food in the ground, Menomini); *assanciko* (Ojibwa); *astahcikow* (Cree); *assantamin* (Strachey in [1612] 1953 wrote the word as *assentammens, ossantamens,* and *ottassantamens*; the same year Capt. John Smith wrote *assentamens*; all of these Powhatan words they translated as "peas"; the Powhatan has the same root as Cree */astaht-/*, Ojibwa

Amphicarpaea bracteata. From Britton and Brown 1897.

/*assant-*/ and means "concealed by the ground"; the final element is /**-min-i*/, berry, drupe, grain)
ati-kuraru (*atit*, beans, *ku*, its, *uraru*, earth, ground, Pawnee)
bûgwûdj'miskodi'simin (unusual reddish bean, Ojibwa)
goober (from Angolese *nguba*, mostly used for the peanut, *Arachis hypogaea*, South)
hinbthi-abe (*hinbthi-hi*, bean plant, Omaha-Ponca); *hon̲-bthin̲-çu* (Osage); *honink-boije* (Winnebago); *maka ta omnicha* [*onmnicha*] (*maka*, ground, *ta*, its, *onmnicha*, beans, Dakota)
hog-pea (North Carolina); [wild, wild-bean, pea] vine
[American] hog peanut; [ground, wild] peanut

In 1753, Linnaeus knew these plants only from Jan Gronovius's description in his *Flora Virginica* and the specimen collected by John Clayton. Neither Linnaeus nor the others seemed to realized that the plants had cleistogamous flowers that produced underground fruits larger than those aboveground. To Linnaeus's credit, he did realize that the plants were similar to what we now call *Apios*, and he put both in the genus *Glycine* (soybean).

The first record of these food plants was given by Strachey and Capt. John Smith in 1612. However, those obscure references to American "peas" used by people around the Chesapeake Bay were lost on most Europeans. Later it was rediscovered that indigenous people ate the subterranean fruits. The Cherokee used these seeds for food and sometimes made bean bread of them (Hamel and Chiltoskey 1975, Moerman 1998). The Dakota, Omaha, Pawnee, and Ponca thought the seeds too small to be used alone, but mixed them with other foods to add taste and nutritive valve (Gilmore 1919). The Meskwaki, like the other people in the Missouri River region, stole hoards of fruits from mice and voles (Moerman 1998). The Osage and Ojibwa treated them similarly (La Flesche 1932, Moerman 1998). At least the Omaha also ate the small roots (Gilmore 1919). Yanovsky (1936) had records of people eating *Amphicarpaea* throughout the eastern states.

Gilmore (1919) was one of the first to record this particular species being taken from the stores of voles, but the Shoshone woman Sacajawea had shown Lewis and Clark how to rob such stashes from a variety of animals in 1804–1805 (De Voto 1981). One of Gilmore's Pawnee informants wrote to him in 1913: "We call them *atikuraru* ... In winter time the women robbed rat's [*sic*] nests and got big piles of them. Nowadays when the old women see lima beans they say they look like *atikuraru* in Nebraska." Dakota women also ate the smaller aerial beans, which were about the size of lentils. These same women confided that they never robbed the rodent nests without leaving some food in exchange. Often, they left corn and replaced an equal quantity to what they took.

During the Civil War, Porcher (1863) wrote, "The subterranean pod is cultivated as a vegetable." Fernald et al. (1958) thought the beans good enough to dig, at least when camping.

Several tribes also considered the plants medicinal (Moerman 1998). The Cherokee used an infusion of the root to treat diarrhea, and blew it on snakebites to cure them (Hamel and Chiltoskey 1975). The Iroquois mixed the plants with others to treat a bad stomach caused by tuberculosis. The Lakota made a poultice of the leaves and used it to treat swellings. The Ojibwa mixed the root with others to make a physic (Densmore 1928).

Amyris

(Greek for much balsam)

Amyris balsamifera (balsam bearing)
balsam torchwood (Florida)
bois chandelle (candle tree, Hispaniola); candil (simple oil lamp, Venezuela)
bois des roses (rosewood, Puerto Rico)
cuaba [*blanca, de monte*] (Taino name for the tree [white, wild], Cuba)

Amyris elemifera. a. Flowering branch. b. Leaf outline. c. Flower triad, side view. d. Flower triad from above. e. Flower longitudinally dissected. f. Floral diagram. g. Fruiting branch. *Drawn by Priscilla Fawcett.* From Correll and Correll 1982.

guaconejo (Hispaniola)
limoncillo (little lemon, Mexico)
quigua [*tigua*] (Cariban?, Venezuela)
úsim yuera (kills little boys, Mayo, Sonora)
rosewood (Puerto Rico, Jamaica)

Amyris elemifera (bearing a fragrant gum)
amyris-wood (Jamaica)
bois chandelle (candle wood, French Islands); *bois chandelle blanc* (white candle wood, Guadeloupe); *bois flambeau* (flaming wood, Guadeloupe); candle wood (USA, Virgin Islands); *chandelle blanc* (white candle, Hispaniola); *chandelle marron* (wild candle, Hispaniola); white candlewood (Cayman Islands)
bois pini (pine wood, Guadeloupe); waika pine (Belize)
chilillo (dull, Honduras)
cuaba amarilla de costa (yellow coastal *cuaba*, Cuba); *cuaba de costa* (coastal *cuaba*, Cuba); *cuabilla* (little *cuaba*, Cuba, Puerto Rico)
flambouyant (St. John; name usually given to introduced *Delonix*)
guaconejo (big *cuaba*, Hispaniola)
melón (dull, El Salvador),
palo de tea (torch tree, Hispaniola); *tea* (derived from Latin *taeda*, torch, Spanish from Puerto Rico to Panama); *teílla* (little torch, Puerto Rico); torchwood (Bahamas, Jamaica, USA, Virgin Islands); white torch (Bahamas)
pimienta (spicy, Honduras)
roldán (twisted or talkative, El Salvador)
sea amyris (USA)
taray (tamarisk, Honduras)
trois paroles (three words or promises, Hispaniola)

Formerly, there was a forest along Florida's eastern coast. The continuous canopy ranged from Key West to a little north of Cape Canaveral. This narrow strip of trees was composed of virtually all tropical taxa in the south, and an increasing number of temperate species toward the north. Yet, this whole length, from latitude 24°55″ to 29° N, contained torchwood, *Amyris elemifera.* From Key West, the same species grew south through the Bahamas, Cuba, Jamaica, Hispaniola, Puerto Rico, Virgin Islands, the Lesser Antilles from St. Martin to St. Vincent, Grenadines, Grenada, and in Guatemala, Belize, Honduras, and El Salvador. Curiously, it has never been found in Mexico.

If you visit one of the remnants of this long coastal forest in Florida, you will find these shrubs or trees still scattered in isolated islands of vegetation that have survived urbanization. Walk up to one and pick a leaf, crush either the whole leaf or one of the usually three leaflets, and the aroma of citrus fills the air. Cultures throughout the world know that aromatic plants are useful, and certainly torchwood is among those.

Perhaps the most widespread use of "torchwood" is noted by its common name in English. In addition, it is the "white torch" (Bahamas) or "white candlewood" (Cayman Islands).

In Spanish use as a torch is most often indicated by calling it *tea* (torch). Although our English "torch" and the Latin *taeda* somewhat resemble each other, they are thought to have distinct origins. It appears that "torch" came to English from Latin *torquere*, to twist. The relationship is close, however, as often the end of a resinous stick, like pine, was lighted. Yet, a light was more easily obtained with a twisted bunch of grasses or straw. Either way, the result provided illumination to show the way in a pre-battery era. In British English, a flashlight is still called a "torch."

In the Americas, one of the dominant uses of torches from *Amyris*, and probably other woods also, was for night-fishing, especially for crawfish (spiny lobsters). The names give an idea of the geographic and cultural distribution of this practice. In English areas *Amyris* was also called "candle wood" (USA, Virgin Islands). French areas called it *bois chandelle* (French Islands), *bois chandelle blanc* (Guadeloupe), *chandelle marron* (Hispaniola), *chandelle blanc* (Hispaniola), and *bois flambeau* (Guadeloupe). "Flambouyant" (St. John) in English may be related to the French *bois flambeau,* but "flambouyant" is usually applied to the poinciana (*Delonix regia*) and its extravagant red flowers. French *bois pini* (Guadeloupe) may be given because the stems smell somewhat like pine when they are cut. In Hispaniola, and perhaps elsewhere, it is *palo de tea.* As with so many common names, Puerto Rico is different, with *teílla* a diminutive form of *tea.*

Some of the other names have no obvious relationship to torches. In English, some are artificial, as in "sea amyris" (USA), and probably "amyris-wood" (Jamaica). Users rarely are versed in Greek, and the probability of them using such names is almost nonexistent. Those names surely came from academics who had not bothered to talk with the people. The name "waika pine" (Belize) seems like an English name, but it is not clear what "waika" may be.

In the Greater Antilles, the plants are some variant of *cuaba.* That word is Taino, and originally referred to the related *A. balsamifera.* Perhaps at the time of European contact, the term meant any fragrant plant. In Cuba, *A. elemifera* is *cuaba de costa, cuaba amarilla de costa,* or *cuabilla* (Cuba, Puerto Rico). In Hispaniola, it is *guaconejo* (big cuaba). Some of the original meaning of *cuaba* may be revealed in that *Suriana* is called *cuabilla de la costa.* This plant ("bay cedar" in English) has leaves that, when crushed, give a fragrance similar to *Amyris.*

In Honduras, people call the trees *chilillo* and in El Salvador it is *melón.* At first, those seem to have nothing in common, but both turn out to have slang meanings of stupid or dull. The other Salvadorean name *roldán* may be related, or may simply refer to a twisted trunk. Surely, *pimienta* (Honduras) has a literal translation referring to the fragrance of the leaves, or even their taste. The Honduran name *taray* (tamarisk) seems to be totally unrelated.

Torches were not the only use that people had for *Amyris.* The wood is hard, heavy, strong, close-grained, extremely durable, and takes a brilliant polish (Little and Wadsworth 1964). Moreover, it is repellent to dry wood termites. Its high resin content allows green wood to burn, and green or dry it made for quick-burning cooking fires. Wood was also used as posts, for furniture and cabinet-making. However, in many areas within and outside Florida trees are so rare and small they would no longer serve those purposes.

In addition to the wood, other parts served medicinal purposes. Leaf decoctions or leafy branch tips were both used. Leaf decoctions in the Bahamas are used as a febrifuge, and applied to cuts and sores (Morton 1981). The decoction also helps reduce fever in cases of diarrhea (Hocking 1997). In Venezuela, the same decoction was used against venereal diseases (Roig 1945).

Also in the Bahamas, leaves were boiled with *Myrcianthes fragrans* and this mixed decoction was used to treat flu internally, and externally as a bath to speed recovery. This decoction mixed with *Sideroxylan americanum* was taken to ensure good recovery after childbirth (Morton 1981).

In the Caicos Islands, leafy branch tips are boiled and the resultant tea drunk to stop diarrhea. When the branch tips are boiled with lignum vitae (*Guajacum sanctum*) and *Myrcianthes fragrans* they yield a bath used to reduce fever.

In Cuba the root is macerated, mixed with sweet wine, and left in the sun for 9 days. The mixture then is used against shortness of breath (Roig 1945).

The most famous use is as a source of American or West Indian or Yucatán elemí. Gums of several different oleoresins of varying origins are called "elemí." This elemí is used in lacquers, and medicinally; it has been official in ten or more pharmacopoeias (Hocking 1997). The resins are related to one of the commercial compounds called "Balm of Gilead" (*A. gileadensis*).

Amyris elemifera is not the only source of elemí. Indeed, the other most important source is *A. balsamifera.* Both species are used to extract an essential oil containing caryophyllene, cadinene, and cadinol (Lawless 1995). These oils are used to make varnishes, in medicines, in perfumery, cosmetics, and soaps (Hocking 1997). The oils also are used in incense, where they are known as "rosewood," "rhodium wood," or "balsam tree bark."

In Florida, *A. balsamifera* has never been found anywhere except in one Miami hammock. However, its range is much greater than *A. elemifera,* going from Miami south to Venezuela and Colombia, and up along Mesoamerica through Mexico to Sonora. This species shares several names with *A. elemifera,* being called *cuaba, cuaba blanca, cuaba de monte* (Cuba), *guaconejo* or *bois chandelle* (Hispaniola), *candil* (Venezuela), balsam torchwood, and *bois des roses* (Liogier 1974, Hocking 1997).

On the mainland, *A. balsamifera* has other names. Throughout Mexico, the most common name is *limoncillo* (little lemon), which it does resemble in many respects. People in Sonora say it also smells lemony when burned. The Mayo in Sonora call the tree *úsim yuera* (kills little boys), but cannot explain that application (Yetman and Van Devender 2001).

Nor can they explain why they also call *Agonandra racemosa* (Opiliaceae) by the same Mayo name. In Venezuela, the tree is *quigua* or *tigua*, which may be Cariban words.

A number of *Amyris* species have been studied chemically. *Amyris elemifera* was found to have 41 chemical components, but beta-caryophyllene, caryophyllene oxide, and alpha-eudesmol are the major ones (Pino et al. 2000). Ragosti et al. (1998) also found texalin, an oxazole, in *A. elemifera.*

Within *Amyris*, the most common chemicals besides the essential oils are coumarins. These have been found in *A. balsamifera* (Burke and Parkins 1979), *A. diaptrya* (Laguna 1984), *A. elemifera* (Rastogi et al. 1998), *A. madrensis* (Dominguez et al. 1977), *A. texana* (Fuente et al. 1991), *A. lineata* (Laguna 1985, Laguna et al. 1987), and *A. simplicifolia* (Hernan et al. 1974). Related chemicals known from the genus are benzamides (Hasbum and Castro 1988), dihydrostyrilamides (Hasbum et al. 1988), and furoquinolines (Fuente et al. 1991).

Other species are known to have alkaloids (Laguna 1984), nicotinamides (Burke and Parkins 1978), and sesquiterpenes (Rohmer et al. 1977). Trees also contain a wide array of essential oil components (Beek et al. 1989), with *A. diaptrya* having more than 80 (Adams et al. 1998). In addition to being antibacterial and antifungal (Rastogi et al. 1998), Badawi et al. (1981) have suggested that extracts from *Amyris* may be useful against cancer.

Amyris consists of 40 species in tropical America (Mabberley 1997). Clearly, its potential medicinal and economic importance did not escape the early explorers of the region. *Amyris* was named by Patrick Browne (1720–1790) in his book *Civil and Natural History of Jamaica*, published in 1756. Not long afterward, Michel Adanson (1727–1806) renamed it *Elemi* in 1762, so both these men were aware of its medicinal potential.

In Haiti *trois paroles* is an intriguing name for *A. elemifera.* Technically, *trois paroles* means simply "three words." However, the word *parole* is idiomatically used as a promise. For example, *manquer à sa parole* (to pledge one's word) and *tenir parole* (to keep one's word). Could it be that *trois paroles* is an illusion to dishonesty? If so, what or who is the guilty party? The tree or the person? Might it be that since *Amyris* smells like a citrus, it "promises" to give fruit, but fails to produce?

Andropogon

(From Greek *andros*, man, *pogon*, a beard; the hairy spikelets resembled a man's beard to Linnaeus)

Andropogon*. Andropogon floridanus* (top left). From Hitchcock and Chase 1950. *Andropogon gerardii* (lower left). From Hitchcock and Chase 1950. *Andropogon glomeratus* (top right). From Institute of Food and Agricultural Sciences. *Andropogon virginicus* (lower right). *Drawn by P.N. Honychurch.*

barbón (bearded one, Quebec); *Bartgras* (beard grass, German)
bluestem (in use by 1864 for *A. gerardii*, because of the bluish, glaucous leaf sheathes along the stems; subsequently applied to several species of *Andropogon* and *Schizachyrium*)
Hühnerfuss (hen foot, German)
sanguinelle (little bloody one, Italian)

Andropogon floridanus (of Florida)
Florida bluestem
pahikitíscî [*pahce chate*] (grass red, Mikasuki); *pahatâ:fí* (*pahi*, grass, *tâ:fí*, broad, Mikasuki); *pahcî* (*páhce, pv'he*, grass, Creek, Muskogee)

Andropogon gerardii (named for French botanist Louis Gerard, 1733–1819) (=*Andropogon furcatus* of older literature)
barbón de Gerard (Gerard's beard, Canada)

big bluestem (named for its height, growing to 2 m tall)
hade-zhide (*hade*, hay, *zhide*, red, Omaha-Ponca)
mûckode'kanes (small prairie, Ojibwa)
pe'-çka zhu-dse (*pe'*, forehead, *çka*, white, *zhu-dse*, red, Osage; a sacred plant)
turkeyfoot

Andropogon glomeratus (clustered)
bushy beard grass [bushy beardgrass]; *yerba barbuda* (bearded herb, Puerto Rico)
bushy bluestem
ch'it-suuk (*ch'it*, simple term for *Thrinax* palm, *sak*, white, Maya, Yucatán)
matojillo (little grass, Puerto Rico)
matojo escoba de fogón (grass for the oven broom, Puerto Rico)
pajón (big grass, Dominican Republic)
panache (plume, Haiti)
wi·ti sara'k (root grass, Catawba)

Andropogon virginicus (of Virginia)
l'arbre à balai [or probably *l'herbe à balai*] (broom grass, Houma, Louisiana)
ashuck hooma (red grass, Chickasaw, cf. Romans [1775] 1961; *hashshok* (grass, *homma*, red, cf. Munro and Willmond 1994); *hashuk ba<u>n</u>si* (*hanshuk*, grass, *isht bashpoa*, a broom, Choctaw)
barba de velho (old man's beard, Ilha Marajó, Brazil); Virginia beard-grass
[Virginia, yellow, yellow-sedge] bluestem
broom sedge [sage] (early settlers employed the culms of this grass as brooms, hence the name, Steyermark 1963; only books call it "sedge"); broom sedge bluestem [broomsedge bluestem]; brown sedge (Puerto Rico)
capim membeca (*capim*, grass; *membeca* from Tupí *me'mbeka*, smooth, Ilha Marajó, Brazil)
escoba (broom, New Mexico)
fever grass
matojo de escoba (broom grass, Puerto Rico)
pahikitíscî (red grass, Mikasuki)
w<u>i</u>·ti· sara'k [*surá·k*] (root grass, Catawba)

In 1541, when de Soto's expedition reached southeastern Arkansas, a group was sent to the northwest to explore the countryside. The record they left was probably the first European account of tall-grass prairie. One of the leaders of the smaller group was Luiz Hernández Bidema, who wrote of coming to "some wide plains, on which grew a plant [*Andropogon*] so rank and high, that even on horseback we could not break our way through" (Swanton 1939). That plant would have been *A. gerardii.*

Apparently throughout its range, the stems of *Andropogon* have been used in thatching (Gilmore 1919, Jones 1936, LeCointe 1947, Pittier 1957). Within historic times, the Omaha and Ponca used *Andropogon* on poles to support earth coverings of lodges (Gilmore 1919). To demonstrate and test some of the old methods at the "Fort Ancient" site in Dayton, Ohio, several buildings have been made with *A. gerardii* thatch (Daumeyer 2002). Similar to use as thatch is the use in making cradles, as lining for cradles, graves, and cache pits, and as bedding materials (Dellinger 1932, Jones 1936, Daumeyer 2002, Standifer 2003). Webb and Funkhouser (1936) interpreted other remains as the remnants of grasses used for seating and floor coverings.

Houses were not the only things covered with the grass. Gilmore (1931) discovered that *A. gerardii* had been used to make breech-clouts. Not only could the belt be made of the cords, but a wisp of the longest basal leaves "was taken and a knot tied at the lower end . . . in putting it on the knot on the end of the wisp was tucked under the girdle at the back, just over the lumbar vertebrae. Then the wisp was passed between the thighs and up in front under the girdle and the distal end was draped over it in the front." He did not know what to make of the many discarded garments until a mummified body of a man was found with one of them in place.

Stems and leaves were widely used to make cords, ropes, and strings for weaving. Gilmore (1931) noted that, for this purpose, "a handful of the fallen basal blades lying at the base of the stalks in a loose mat all over the ground on the prairie was grasped and twisted, lifted slightly and twisted again, until a rope of the desired length was formed; this method is well known to every farmer who makes by hand a temporary rope from a straw pile or haystack." Gilmore (1931) identified ropes and cords at the Ozark Bluff-Dweller site in Arkansas made from big bluestem, and later Whitford (1941) recorded the same preparation in Ohio.

Those same cords were used to weave net bags (Gilmore 1931, Whitford 1941). There are prehistoric and historical records of those cords being used in footwear; sandals for warm weather, and insoles for warm padding in winter moccasins (Gilmore 1931, Webb and Funkhouser 1936, Whitford 1941, Standifer et al. 2003). The twisted stems were even woven into "overshoes" at the Ozark Bluff-Dweller site (Gilmore 1931).

Young boys of the Omaha and Ponca of Nebraska used *A. gerardii* stems, termed *peska*, to make arrows (Gilmore 1919). Arikara, Hidatsa, and Mandan of

North and South Dakota also used them, inserting a thorn from *Crataegus* for the arrow point. The arrows were used primarily to hunt frogs in training for larger game.

At least the Cherokee made a yellow dye of *A. virginicus* (Hamel and Chiltoskey 1975). Tull (1999) recorded a similar use in Texas.

The most widespread application of the genus seems to have been as medicine. The Ojibwa made a root decoction of *A. gerardii* to relieve stomach pain and as a diuretic (Densmore 1928). Gilmore (1919) learned from an Oto medicine man that decoctions of *A. gerardii* lower leaf blades were made to treat general debility where the cause was unknown. The remedy was thought to reduce fever.

The Cherokee used *A. virginicus* to stop diarrhea, in "Green Corn" medicine, to treat piles, poison ivy rash, frostbite, and to help cure sores (Hamel and Chiltoskey 1975). In the Bahama Out Islands, the sweetened plant decoction is drunk as a febrifuge (Morton 1981). The Brazilians take a decoction to relieve urinary problems (LeCointe 1947). The grass is used in domestic medicine in the United States as an alterative, tonic, and diuretic to treat erysipelas (Hocking 1997). The Houma used some *Andropogon* to strengthen mother and baby in pregnancy (Speck 1941). The Catawba of the coastal Carolinas used *A. glomeratus* roots for backaches (Moerman 1998). Several other species of *Andropogon* are used to treat fevers and considered diuretic (LeCointe 1947, Duke and Vásquez 1994, Mors et al. 2000).

Perhaps the greatest number of medical uses is recorded for *A. floridanus* (Sturtevant 1955). One of the common uses of this grass was as an analgesic. For that purpose it was mixed with other plants or used alone. An infusion was taken for "Wolf Sickness" (vomiting, stomach pain, diarrhea, frequent urination), "Moving Sickness" (moving pain in waist), and both drunk and used as a bath for "Gopher-Tortoise Sickness" (cough, dry throat, noisy chest).

Angadenia

(From Greek *ang*, vessel, box, case, and *aden*, gland, probably alluding to the stigma being enclosed)

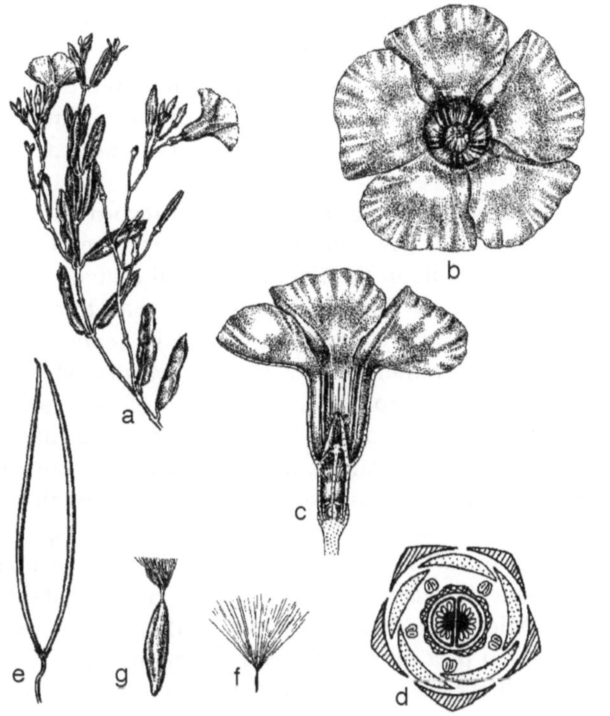

Angadenia berteri. a. Flowering branch. b. Flower from above. c. Flower longitudinally dissected. d. Floral diagram. e. Fruits. f. Seed with coma. g. Seed. *Drawn by Priscilla Fawcett.* From Correll and Correll 1982.

Angadenia berteroi (named for Carlo Luigi Guiseppe Bertero, 1789–1831, an Italian explorer of the American tropics. Because of a curious situation where an original author erroneously described two supposed species that are really one, the Corrells applied another name to the same plant in the Bahamas. They called them *Angadenia sagraei*, for Ramon de la Sagra, 1798–1871, Spanish botanist who explored Cuba. Not only was it strange that Alphonse DeCandolle proposed both *Echites berteri* and *E. sagraei*, but Miers transferred them both to *Angadenia* without realizing that they were the same plants.)

golden trumpet (Florida)

ismo:kha:kâ:ki:hiliswâ (*enokketv*, sickness, *heleswv*, medicine, Creek)

lice-root (Bahamas)

sanahahcayikcî (chronic sickness medicine, Mikasuki)

tumba gente (person tumbler, Dominican Republic)

One of the new plants that the Seminoles encountered when they came to southern Florida was the geographically restricted *Angadenia berteri*. These plants live only in the pine rocklands of Miami-Dade and Monroe Counties, and their restricted range and habitat have led to them being called "pineland golden trumpet." The species is threatened in Florida (Coile 2000).

Among the Seminoles, the "golden trumpet" was called *sanahahcayikcî* (Mikasuki) or *ismo:kha:kâ:ki:hiliswâ* (Creek). For the cure of the intercourse taboo, the doctor took small sections of the roots. These were boiled in water for 2 to 3 hours. The medicine was blessed, and then given to the patient. In cases of chronic illness, the roots were also boiled to produce a decoction. That medicine was either drunk for internal problems or applied directly to persistent external sores.

No other people in Florida have a record of using the plants, not even the newer settlers from Cuba. However, people in the Bahamas call the small shrub "lice-root," and that name suggests that they employed it. Similarly, in Hispaniola it is *tumba gente.* That name is reminiscent of *tumba-vaqueros*, which is used in Mexico for many members of the family Convolvulaceae. Liogier (1974) does not indicate in Hispaniola if the plant is used as a laxative, but the similarity of names to those in Mexico suggest that.

This is a small genus, with only two species (Mabberley 1997). The second is endemic to Cuba. Although *Angadenia* was described by British botanist John Miers (1789–1879) in 1878, the species were considered part of *Rhabdadenia* for many years. John K. Small even described the Florida population as a distinct species he called *Rhabdadenia corallicola.* However, as discerning as he was, he rarely examined floras of nearby regions to see if the plants he published as new really were different from existing species.

Angelica

(From Greek *angelikos*, a messenger or angel, referring to the cordial and medicinal properties of some species; akin to Akkadian *kallu*, messenger)

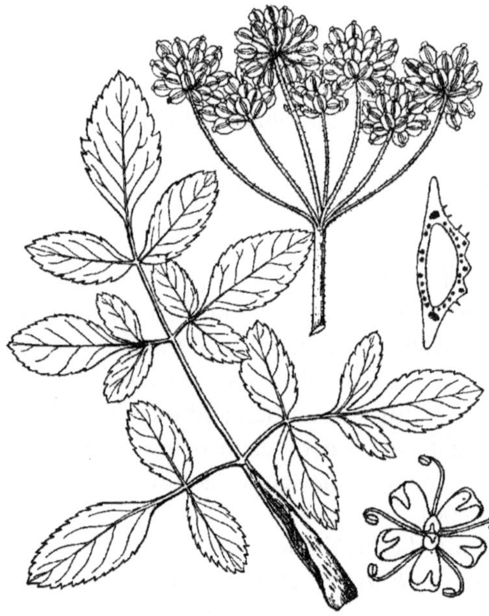

Angelica venenosa. Britton and Brown 1896.

angelica (Italian, Spanish); *Angelick* (German in 1542); *angelick* (Dutch in 1549); *angélique* (French); *Engelwurz* (angel herb, German)
ground ash (probably comparing the leaves with those of *Fraxinus*)
heptaphyton (from Greek *hepta*, seven, and *phyton*, plant, probably alluding to the compound leaves; used by Theophrastus, 372–287 B.C.)
Holy Ghost (translation of Latin *herba angelica*)
lus nam buadh (herb of virtues, Gaelic)
sløke (simple term, Norwegian)

Angelica venenosa (very poisonous)
angelico; hairy angelica
notossv [*notoss*] (Muskogee)

Europeans had long been familiar with *Angelica* when they arrived in the New World. Greeks and Romans knew the plant, and spread its fame through Europe. In Medieval Latin it was *herba angelica*, the "angelic herb," or "Herb of the Holy Ghost." When and by whom it was named is unknown, but Otho Brunfels (1488–1534) was the first to record the English name in 1530. It has been suggested that the plants were considered "holy" because they flower on the feast of St. Michael the Archangel (Coffey 1993).

Leonard Fuchs wrote in 1542: "It is especially effective against poisons. It wards off the contagion of raging, devastating pestilence. It protects the body from the wasting of the pestilence, they say, if only held in the mouth" (Meyer et al. 1999). Similarly, the English translation of Dodoens's 1554 herbal said, "The rootes of *Angelica* are contrarie to all poyson."

Not only were the plants considered medicinal, but they were cultivated in England from 1568 and used to make a confection called "Candied Angellica." Roots have been candied in England since at least 1629 when John Parkinson wrote in his *Paradisi in Sole* (Paradise in the Sun, a play on his name) that the candied stem and roots were used "to comfort and warme a colde and weake stomach."

When the species in Florida was discovered, the person who named it in 1793 thought it was another species of *Cicuta.* Because that is a notably poisonous plant, he called it *Cicuta venenosa.* It was not until 1943, when M.L. Fernald (1873–1950) realized that the plants were actually a distinct *Angelica.* However, by the rules of International Nomenclature, the first name given a species must be retained, even if it is erroneous. In this case, a basically nonpoisonous plant is called "venomous" (*venenosa*).

The Iroquois made a poultice of the plant to treat strained muscles and twisted joints. Moerman (1998) also records that the Iroquois ate the roots to commit suicide. However, the species is not listed as poisonous in any source I have checked (Hardin and Arena 1974, Perkins and Payne 1978, Turner and Szczawinski 1991). The plants contain furanocumarins, as do most members of the Apiaceae including celery (*Apium graveolens*) and parsnip (*Pastinaca sativa*), but there is no indication that they are likely to be fatal (Duke et al. 2001). Surely there was a misidentification of the plants in the report that Moerman used.

Small (1933) reported that the leaves of hairy angelica were used to discourage the tobacco habit. There is little other reference to *A. venenosa*, with most of the comments on *Angelica* concentrating on *A. atropurpurea* (Millspaugh 1892, Fernald et al. 1958, Foster and Duke 1990, Coffey 1993) or related species (Turner and Szczawinski 1991, Tyler 1993, Bremness 1994, Duke et al. 2001). Those other species are used in flavoring gin and liqueurs such as Benedictine and Chartreuse (Tyler 1993, Bremness 1994).

Annona: Pond Apples

(From *anón*, a Hispaniolan Taino word for the fruit)

Annona glabra. *Drawn by P.N. Honychurch.*

In 1514, when Gonzalo Fernández de Oviedo y Valdés arrived in Darién, Panama, he began writing what became one of the first notices of edible fruits from the Americas. His book had a chapter devoted to praise of the *guanabaná* (Oviedo [1526] 1969). This fruit was also called *anón* by the indigenous Taino of Hispaniola, and that name eventually became the genus *Annona*.

During the next 200 or so years, other species of *Annona* became known to Europeans. The *anón* was discovered by the British when physician Hans Sloane (1660–1753) arrived in Jamaica in 1687 as personal physician to the duke of Albemarle. Sloane's *Catalogus plantarum,* published in 1696, included the fruit as the sour sop. Pond apple was introduced to Europeans in 1734 by Mark Catesby. Linnaeus gave both the names used today, with sour sop becoming *A. muricata* and pond apple *A. glabra.*

Since those discoveries, it has been found that *Annona* is widespread in the Americas, and that it includes several species grown for their edible fruits. *Annona glabra* is not one of those grown for food. The fruits can be eaten, but most think they are not particularly tasty. According to one description, the fruit "has a strong smell, resembling ether, a special flavor, giving a hint of menthol" (LeCointe 1947). Be suspicious of anything described as "special."

In Florida we know the tree and its fruit as "pond apple." Other names also allude to a similarity with the apple, and supposedly to the animals that eat the fruits, e.g., alligator apple, *cachiman cochon* (pig *cachiman,* while *A. reticulata* is simply *cachiman,* Guadeloupe, Martinique), monkey apple, and dog apple. That origin is dubious. The names more likely mean that people were distinguishing between cultivated and wild "apples."

Pond apple is used by people throughout its range. The soft wood has been used in rafts, as floats on fishing lines, as corks for bottles, and to hone razors (LeCointe 1947). Hence its names corkwood in English; in Spanish *corcho* (cork), or *arbol* [*palo*] *del corcho* (cork tree); in Portuguese *areticum cortiça* (cork annona, Brazil); and in French *bois flot* (cork wood, Guadeloupe, Martinique). The Maya used the same idea with *mag* [*mak', maak, xmaac*] (cork, Yucatán). Others call it bobwood (Belize). Some reserve the roots for making substitute corks. Amazonian Brazilians considered the wood good for carpentry, making boxes and small products (LeCointe 1947). The Seminoles call these trees *etotakwe* [*totáka*] (Creek) or *totakwî* (Mikasuki). Those names literally mean "hollow tree," but may also allude to the spongy wood.

Other immediate and practical uses are also widespread (Morton 1981). As the seeds contain especially potent chemicals, they have been used to poison fish. However, people have been blinded by getting powder from seeds in their eyes. Similarly, seeds and leaves are insecticidal. Leaves placed in hen nests kill lice on the fowl. The bark causes dermatitis in some people. However, pond apple's greatest use is medicinal.

Medicinal traits are reflected in some common names the tree has in its range from southern Florida to Argentina. In Cuba and Hispaniola it is the *palo bobo* (crazy tree), and in Brazil *araticum do brejo* [*araticu du brejo*] (wizard's *Annona*, from the Tupí name for the genus, *arati'ku*, Brazil). Some of the other names have obvious meanings, and others do not: *at'p'* (Mixe, Oaxaca), *bagá* (Cuba), *catiguire* (Venezuela), *cayur* [*cayures*] (from *caya*, Taino, Dominican Republic), *cajuda* (Aruba, Bonaire, Curaçao), *corazón cimarron* (wild heart, Puerto Rico), *corossol sauvage* (wild annona, French Guiana), *couassol* (from *couer*, heart, French, Dominica), *guanabaná cimarrona* (wild *guanabaná*, Puerto Rico), *anonillo* (little *anón*), *mamin* (breast, French Antilles), *mamon de perro* (dog's breast).

Seeds, bark, and leaves of pond apple and its relatives contain many chemicals, including acetogenins, apomorphine alkaloids, cyclopeptides, diterpenoids, phenylpropanoid, reticuline, and (−)-*N*-methyl-lacto-daphnine (Yang et al. 1971, Haggag et al. 1981, Gupta 1995a, Chang et al. 1998, Gallardo et al. 1998, Li et al. 1998, 1999a,b, Liu et al. 1998, 1999a, b,c, Chen et al. 2000). Each compound alone has a marked impact on human physiology, and together they provide a "witches' brew" (sorry Shakespeare!) of poisons (Chávez et al. 1997). As medicines are synonymous with poisons, people have used and continue using *Annona* medicinally. Tests on these chemicals reveal tumor inhibition and other bioactivity (Warthen et al. 1969, Padmaja et al. 1995, Chávez et al. 1997, Chang et al. 1998, Liu et al. 1999a).

The acetogenins are especially potent because they inhibit the electron transport chain. In the wrong concentrations, that would inhibit oxygen availability in the body. No oxygen, no life. Theoretically, in smaller quantities, these chemicals are medicinal. Apomorphine alkaloids are largely emetics. So, if you think you have eaten something that is poisonous, you can "clean" the body with pond apple tea.

By far the most common use of pond apple is for lung and chest problems (Roig 1945). Syrup from boiled ripe fruits is used for the coughing associated with tuberculosis. A diffusion of the leaves is drunk as a beverage, often with the morning meal in Curaçao and elsewhere (Morton 1981). A stronger brew provides adults with medicine against abdominal cramps, colic, diarrhea, and dysentery (Ayensu 1981). The tea is a vermifuge (especially against hookworms), and has been used in Brazil against rheumatism (Roig 1945). Weaker draughts have been used for infant diarrhea. A decoction of fruit rind has been used for pulmonary problems in Yucatán. Leaves and flowers boiled together are used to treat liver trouble, especially jaundice, from Mexico to Argentina. However, in spite of the widespread usage, there is evidence that teas from annonas can cause symptoms resembling Parkinson's disease (Raloff 1992, Weiss 1999). So, caution is in order.

In Florida, the pond apple ranges across the state from southern St. Lucie to Manatee Counties, and south through the Keys. Elsewhere, *A. glabra* is a coastal mangrove tree; that is the reason for the names *areticum do mangue* (Brazil) and its English equivalent "mangrove annona." From Florida the trees range down the Atlantic Coast of Mexico, Middle America, and South America, and on the Pacific coast down to northern Peru. From the Americas, the species is disjunct to western Africa, where it is found in Senegal, Ghana, and southern Nigeria. Trees in Africa also have common names (Dalziel 1937); however, translations were found for only the Yóruba *àfe* (a fishing float). That name was carried by Africans taken as slaves to the Bahia region of eastern Brazil, where it is pronounced *apa* (Voeks 1997).

People also have carried the species to other parts of the world. Long ago, pond apple was introduced into Australia. Now, it has escaped there and become a pest in wetlands. So many thousands of acres of Australian wetlands are overrun by pond apple that those responsible for managing them are on the verge of despair. Pond apple, like American rabbits and prickly pears, has become a plague on that island continent. Yet, there may be hope for Australians. Some entrepreneurs think that the medicinal use may be turned to profit. They sell shampoos, soaps, and other bath items that proudly sport labels proclaiming them to contain "pond apple" extracts.

The impact of these products on human vitality is ambiguous in the language used on the labels. However, they use spin-doctoring logic—since they are "natural," they **must** be healthy. These advertising agencies and the consumers who accept their hype forget, or never knew, that strychnine, aflatoxin (from mold on peanuts; considered the most poisonous compound known), and hemlock (à la Socrates) are also perfectly "natural." So are hurricanes, tornadoes, and earthquakes. Natural does not mean good for living organisms. Nevertheless, maybe the soap companies will save Australia from invading American alien pond apples.

Antennaria

(Joseph Gaertner, 1732–1791, used the Latin *antenna*, a feeler, referring to the pappus)

antennaire (French)
bambagia selvatica (wild cotton-wool, Italian)
cat's foot (England); *kattefot* (cat's paw, Norwegian); *Katzenpfötchen* (cat's paw, German)
pied de chat (cat foot, French)

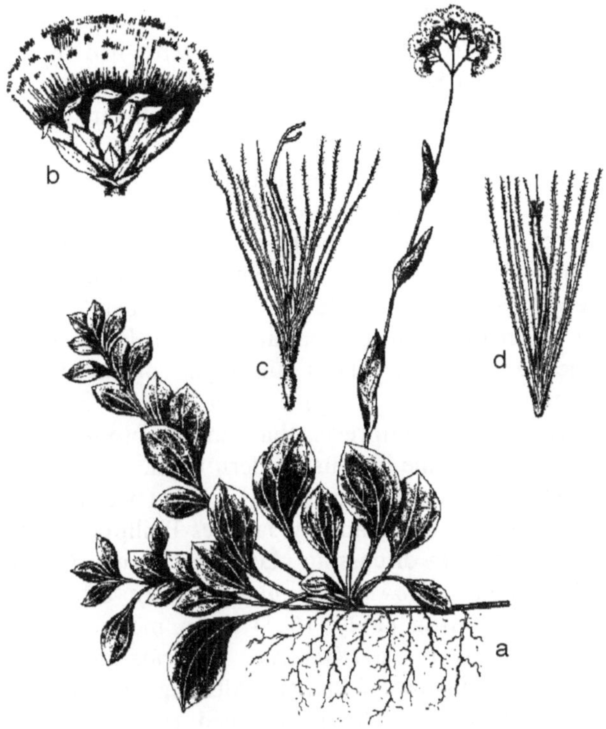

Antennaria plantaginifolia. a. Portion of plant showing rooting of creeping stem. b. Detail of staminate head. c. Female flower. d. Male flower. From Buchholtz 1968.

wi̲'ti hasami' (medicine leaves flat or stink terrapin, Catawba for *A. neglecta*)

Antennaria plantaginifolia (with leaves like plantain)
immortelle (immortal, alluding to the long-lasting dried flowers, Quebec)
lady's tobacco; woman's tobacco
pussy's toes [pussytoes]

Europeans were long familiar with mountain everlasting (*Antennaria dioica*) when they arrived in the New World. That medical species had a venerable history of use in Eurasia as an aromatic, bitter, vulnerary, diuretic, expectorant used to treat bronchitis, and in teas to improve milk flow in nursing women, to treat diarrhea, and for gallbladder problems (Hocking 1997). The species and its uses were discussed by Rembert Dodoens in 1616, Gaspar Bauhin in 1623, J.G. Gmelin between 1747 and 1749, and several times by Linnaeus between 1737 and 1753.

When the Europeans arrived in the New World they found similar plants that became known as *A. plantaginifolia*. They found indigenous people using that species and its relatives. The Iroquois in New England used woman's tobacco to treat leukorrhea, and as a mouthwash for toothache (Moerman 1998). The Meskwaki in Iowa used the plants to prevent sickness in childbirth (Moerman 1998). The Cherokee gave an infusion of the plant to children for bowel problems, and to regulate menstruation in women (Hamel and Chiltoskey 1975). The Natchez used the warm infusion of the tops and roots for colds and to stop coughs (Swanton 1928a, Taylor 1940). Due to similar uses, perhaps the American species has volatile oils, resins, and phytosterols like the Eurasian plants (Hocking 1997).

Apios

(Greek *apios*, a pear, from the shape of the tuberous roots)

Apios americana. From Britton and Brown 1897.

Apios americana (of America)
ahá caká:wa (*ahá*, tuber, *caká:wa*, rough, Koasati); *ahe kamassa* [*ahkamassa*] (*ahe*, potato, *kamassa*, firm, Choctaw; maybe this belongs here); *ahe-lo'sa* (*ahe*, tuber, *losa*, black, Choctaw; maybe this belongs here, but the roots are not typically black; Bushnell 1909 wrote: "The roots are first thoroughly boiled, then mashed, and served as food"); *akkalv* (*aklowvhe*, mud, Creek); *kan-shak ahe* (*kun̲shak*, reed, *ahe*, tuber, Choctaw); *ocká:hî* (water tuber, Mikasuki); *vh-aklowahe* [*ahakli:wá:ha*, *ahakliwahî*] (*vhv*, tuber, *aklowvhe*, mud, Creek)
blo (Teton); *mdo* [*modo*] (Dakota); *nu* (Omaha-Ponca); *tdo* (Winnebago)
ʔc'skame'li (*ʔc's'ka*, root, *me'li*, black, from French *liane noire*, Tunica)
chapelet (Quebec)
ground-nut [groundnut]
Indian potato; potato-bean [potatobean]
its (Pawnee)
ko'nan a'yip [*ko'nen a'yip*] (*ko'nan*, potato, *a'yip*, swamp, Atakapa)

mac-mac (fide Core 1967)
nu'na igatehi (*nuna*, potato, *igatehi*, swamp dwelling, Cherokee)
ochnenáta (Onondaga)
ohpenak [*ohpen*, singular] (Strachey wrote the name in [1612] 1953 as *ouhpunnauk*, Powhatan, Virginia); *okeepenauk* (Carolina Algonquin, North Carolina); cognates are *ahpenya* (Fox); *oʔpenya* [*peneeakee*] (my potatoes, Shawnee); *ophen* (Menomini); *oppin* (my potatoes, Ojibwa); *openauk* (Carolina Algonquians); *ppan* (my groundnuts, Penobscot); akin to *mûkwo'pîni* (bear potato, Potawatomi), *hópenis* [*hapniss*, *hoppenis*, *hóbbenis*, *hóbbenac*] (my potatoes, Delaware), *sagabon* (Metoac Algonquin, New York); *penninaugh* (the string of tubers, Powhatan, Virginia)
patates en chapelet (potatoes in rosary, Quebec)
pénacs (from the Algonquian name, Quebec)
pomme de terre (earth apple, the name now applied to the potato, *Solanum tuberosum*, Quebec)
ulí (Cherokee)
wild bean

Along the eastern coast of Long Island in New York is a locality called Sag Harbor. Odd as it may seem, that place owes its name to a small twining legume (*Apios americana*). The harbor got its name from the Metoac Algonquin people who formerly lived there and called these legumes *sagabon* (Brooklin online 2002). The English who settled there sometimes shortened the name to *pen, pin, pon,* or *bun*, and nearby in Massachusetts they were simply *pen*. The Algonquian speakers in New England named one of the rivers *Penecook* (sing., *pen*, pl., *penak*), again because of the abundance of this vine. The word, as *pénacs*, is still occasionally used for the plants by French-speaking Canadians.

Harriot ([1590] 1972) gave us another version of the *openauk* with Algonquin groups in North Carolina. He wrote: "Openavk are a kind of roots of round forme, some of the bigness of walnuts, some far greater, which are found in moist & marshy grounds growing many together one by another in ropes, or as thogh they were fastened with a string. Being boiled or sodden they are very good meat."

In [1612] 1953 Strachey recorded the Powhatan name as *ouhpunnauk*, and Siebert (1975) has it phoenetically as *ohpen*, with plural as *ohpenak*. Related words are *hópenis* (Delaware), *ahpenya* (Fox), *ophen* (Menomini), *oppin* (Ojibwa), and *ppan* (Penobscot). Edgar (1891) gave the Shawnee name as *peneeakee*, but Siebert (1975) wrote *oʔpenya*.

Core (1967) gave yet another story. He says that the plant is *mac-mac* but declines to give his source or mention the tribe. Probably that is the same as the tribe Micmac, Algonquian people who live in and around Nova Scotia. There is a triploid race growing in the area that these people surely cultivated (Bruneau and Anderson 1988).

In English, the plant usually is called Indian potato, although some call it potatobean, wild bean, or even groundnut. The name groundnut has confused people from the early settlement times until the present because of confusion with another "ground nut" more often called "peanut" (*Arachis hypogaea*) in American English. The Latin name of the Indian potato groundnut is *Apios americana*, a name it received in 1759. However, that binomial took almost a century to come into its modern form.

The first scientific mention of what we now call Indian potato was made by Jacques Philippe Cornut (1626–1651). In discussing plants in Canada, Cornut called them *Apios americana.* Shortly afterward, Robert Morrison (1620–1683) again mentioned the plants in his book *Plantarum historiae universalis* (Universal History of Plants). Although Cornut had named the vines, Morrison called these American vines *Astragalus perennis spicatis americanus, scandens caulibus, radice tuberosa* (the perennial American *Astragalus* with a spike, climbing stem, and tuberous root). When Linnaeus was studying the plants of the world, he examined the plants and specimens collected by Clayton in Virginia. Linnaeus decided that they were better called *Glycine apios*, combining names from several of the people who had studied the plants before him. He and Cornut had used the Greek word *apios.* German botanist Friedrick Casimir Medikus (1736–1808) disagreed that this species was a *Glycine.* He thought that the species should be in its own genus called *Apios*, and in 1787 he proposed the name we now use—*Apios americana.*

Not only did the Latin name vary, but several linguistic groups in North America also had their own names for the plants. The second oldest recorded American name for the plant seems to be *hopniss* or *hapniss.* That word was in use in New Jersey by the Delaware and Swedes when Linnaeus's student Peter Kalm visited in the 1700s (Kalm [1753–1761] 1972). The Delaware's neighbors, the Onondaga, called the plant *ochnenáta* (Zeisberger and Horsford 1887). After the introduction of the turnip (*Brassica rapa*), it became known as *hóbbenis* and *otechra*, in a new application of old names.

The Cherokee in the Carolinas called it *ulí,* a simple generic word for tubers (Hamel and Chiltoskey 1975) or *nu'na igatehi,* swamp dwelling potato (Mooney 1885–1886). Even though the Cherokee and Seminoles speak linguistically distant languages, an apparently cognate word for tuber was also applied by the Seminoles. Creek speakers call the vines *ahakli: wa:ha*, and their relatives call it *ocká:hî* (Mikasuki). That the plant was important to the Seminoles is recorded in the name of a village, originally *ahapopka* (root-eating place, Creek), now Lake Apopka, that once stood near the head of the Oklawaha River (Orange County). Their relatives the Choctaw say *kanshak ahe.* People farther north and west called the plants *blo* (Teton), *its* (Pawnee), *mdo* [*modo*] (Dakota), *nu* (Omaha-Ponca), or *tdo* (Winnebago) (Gilmore 1919). These simple words probably simply mean "tuber" or "root."

At first, French speakers, especially those in Canada, called this plant *pomme de terre*, the name now applied to the potato (*Solanum tuberosum*). About the same time they said *patates en chapelet*, or sometimes simply *chapelet*, because of the string of tubers on a root. French explorer Le Jeune said that Indians eat "a root ... that our French people call 'rosary' [*chapelet*], because it is distinguished by tubers in the form of beads" (Vogel 1970). At Port Royal, Nova Scotia, the French troups with Charles Biencourt de Saint-Just (1591–ca. 1623) scattered about the woods and shores digging these roots for food in 1613 (Hedrick 1919).

An indication of the importance of this wild food is the extent to which it was recorded in the literature on indigenous Americans (Smith 1933, Yanovsky 1936, Moerman 1998). Among the people either boiling or roasting the tubers, or substituting the seeds for pinto beans, were the Cherokee (western North Carolina and northeastern Georgia), Cheyenne (Montana), Dakota (Montana, North and South Dakota, Nebraska, Minnesota), Delaware (New Jersey and vicinity), Huron (southern Ontario and Michigan), Menomini (Wisconsin), Meskwaki (Iowa), Mohegan (Connecticut), Ojibwa (upper Midwest and southern Ontario), Omaha (Nebraska), Pawnee (upper Missouri River region of Nebraska), Ponca (Nebraska and South Dakota), Potawatomi (Wisconsin), Creek/Seminole (Georgia and Florida), and Winnebago (Wisconsin). John Bartram told Peter Kalm ([1753–1761] 1972) that the indigenous people who lived more inland not only ate the roots, "but likewise take the pease which lie in the pods of this plant, and prepare them like common pease."

The plant itself ranges even more widely than the people for whom we have records of uses. We know that *Apios* grows from Quebec to Minnesota and North Dakota to Montana, and south to north central Colorado down to Texas and east to Florida. Thus, people living in much of its range appear to have left no record of having used it. One example of absence of historical records is Florida. There, the pollen of *Apios* is known from Glades Indian sites that predate Europeans (Hogan 1978). So, we do have evidence that the original Florida people used the plants.

Some consider Indian potato to be held in highest esteem as food by Americans. Even many Europeans wrote of it with enthusiasm. Capt. John Smith, speaking of Virginia in 1626, wrote of "Ground nuts as big as Egges, and as good as Potatoes, and 40 on a string not two inches under ground." There was supposed to be a 1654 town law among the Pilgrims that ordered an indigenous person to be put in stocks if discovered digging groundnuts on English land. A second offense was to be punished by whipping (Fernald et al. 1958). There is even evidence that the species was cultivated outside its normal range. Indeed, the northern plants are sterile triploids that would not survive if people did not propagate them (Bruneau and Anderson 1988). Requiring cultivation speaks well for their former reputation as a starch source in that region. Williams ([1837] 1962) wrote that the "Seminoles raised great quantities for food."

Although many natives and the first European settlers in North America used the roots for food, not everyone was complimentary about it. Even some indigenous tribes considered it famine food. According to a study done in 1954, the Huron ate the roots with acorns during famines. Many groups saved the roots to use during winter. Raw roots have been described as "tough, with a very viscid, milky juice, but of a pleasant sweetish, turnip-like taste" (Fernald et al. 1958). Reportedly, the young tubers eaten raw leave a rubber-like coating on the teeth. Most of the recorded uses overcame these problems by roasting or boiling the tubers and then drying them. Some ground them for later use like flour.

Typical of the American paranoic approach to the world, today one group of researchers is trying to reintroduce the Indian potato as a crop (Blackmon and Reynolds 1986, Reynolds et al. 1990), while another is looking for chemicals to destroy it as a weed (Devlin and Deubert 1982). Still other investigators are finding that this wild plant has healthy chemicals that could be incorporated into modern diets (Wilson et al. 1987, Krishnan 1998, Mazur et al. 1998). Among other chemicals, the roots contain genistein, an anticarcinogenic compound (Krishnan 1998), isoflavoinoids and lignans (Mazur et al. 1998), and amino acids (Wilson et al. 1987).

These vines are not that common in southern Florida, but they can be found growing on stream banks and in other swampy areas. When they are found, a lucky person will find them in flower. The blossoms appear in the summer, they are an unusual purple color, and the clusters are often tucked back under the foliage. By carefully watching, a person can find the pollinators—flies (Westerkamp and Paul

1993). That may not help their reputation as a food plant, but fly-flowers are comparatively uncommon in Florida.

Apocynum

(From Greek *apokynon*, composed of *apo*, away, and *kyon*, a dog, for a plant supposedly poisonous to dogs)

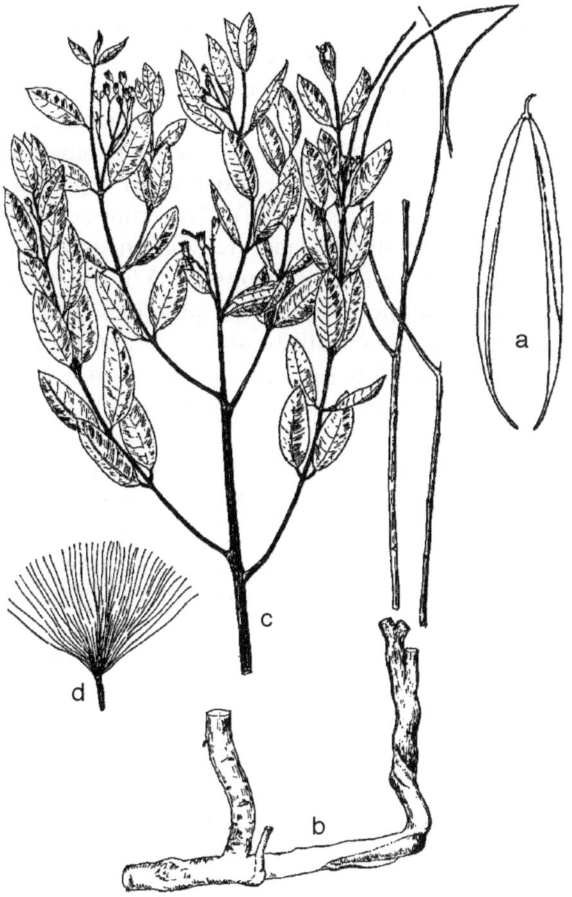

Apocynum cannabinum. a. Seed pods. b. Section of stems and rootstock. c. Upper part of stem with leaves and flowers. d. Seed with attached coma. From Buchholtz 1968.

Apocynum cannabinum (hemp, because of its use for fibers)
American [-Indian] hemp [American Indian hemp]
American ipecacuanha ("ipecacuanha" from Tupí-Guaraní *ipe-kaa-guéne*, creeping plant causing vomiting, 1682, originally *Cephaelis*)
beba'mokodjibika'gisin (bear root, found here and there, Ojibwa)
bowman's root (from use of fibers to make bowstrings)
Choctaw root [Choctawroot] (the Choctaw were people living in Mississippi, Alabama, and Louisiana; originally spelled *Chacta,* in 1675, and later *Chahta*)
[hairy, hemp, prairie, smooth, willow] dogbane [dog's bane] (a translation of the Greek); *Hundsgift* (dog's poison, German)
dropsy root (from use to treat dropsy, by ca. A.D. 1290 and Middle English *ydropsy*, now "hydropsy," where liquid leaks from the venous system into the surrounding tissues)
gatuñ'lati (wild hemp, Cherokee; there is a locality in Fannin County, Georgia called *Gatun'iti'yi*, "Hemp place")
'gho-la [*ghola*] (Kiowa)
[blind, Canada, Canadian, Indian] hemp (from Old English *haenep*, which was derived from Old Teutonic *hanapiz*; that word was cognate with Greek *kannabis*, and Latin *cannabis*; originally used ca. A.D. 1000 for what is now *Cannabis sativa*, although later used for New World plants); *canadische Henf* (Canadian hemp, German); *chanvre du Canada* (Canadian hemp, French); General Marion's hemp (so called because it was a favorite remedy of General Frances Marion, 1732?–1795, the "Swamp Fox" of the Revolutionary War, South Carolina); *wilskt hampa* (Indian hemp, Swedish)
herbe à la puce (flea herb, particularly applied to *A. androsaemifolium*, Quebec)
houatte (probably a variation of *ouate*, fiber used in stuffing or quilting, Quebec)
Indian physic (from Middle English *fisike*, in English since A.D. 1297, in turn from Latin *physica*, the art of healing with plant products)
kendyr (English?)
lechuguilla (little lettuce, New Mexico)
mukosaka'sakuk (sharp podded weed, Meskwaki); *mukoseki' ashikiki* (like a milk weed, Meskwaki)
nape'oi'lekiyapi (to burn it in the hand, referring to the fact that the fluffy seeds can be placed in the hand, lit, and burned without hurting the hand, Lakota)
nuxa-pist (little blanket, Blackfeet)
old amy root (South Carolina)
pimenaw (given by Strachey [1612] 1953 as *pemenaw*, and in Spanish as *pemmenáu*; Siebert 1975 interprets it as "hemp dogbane fiber or plant"; Willoughby 1907 said Byrd [1728] 1980 called it "silk grass" and that it had "stems about as large as the little finger"; see also *Yucca*)
rheumatism weed
snakes' milk
wormroot (supposedly translated from Penobscot name)

Linnaeus coined the species name in 1753, and thus confirmed the plants' use as a fiber source. He knew the plants from both Canada and Virginia, and could have learned about it from either place because people through its range extracted the fibers. One of the first reports of the plants being used in historic times was by David P. de Vries in 1642, but it is unlikely that Linnaeus saw that record (Vogel 1970). Peter Kalm also reported it was used by indigenous people and Swedish settlers in 1750 in Pennsylvania and New Jersey.

Kalm ([1753–1761] 1972) also published in 1753 that: "The people prepared the stalks of this plant, in the same manner as we prepare those of hemp or flax. It was spun, and several kinds of stuffs were woven from it. The savages are said to have had the art of making bags, fishing-nets, and the like, for many centuries together, before the arrival of the *Europeans.*" Later, he wrote: "On my journey through the country of the Iroquese [Iroquois], I saw the women employed in manufacturing this hemp. They made use neither of spinning-wheels nor distaffs, but rolled the filaments on their bare thighs, and made thread and strings of them, which they dyed red, yellow, black, &c. and afterwards worked them into stuffs, with a great deal of ingenuity." Later he wrote of a hempen rope 14 yards long that the Swedes purchased from the Iroquois for a piece of bread.

This is the same use that Strachey recorded in [1612] 1953 with *peminak* (rope), *pemunakqweraneind* (twined thread), *peyminakoó* (thread), *pemanakaon* (cord or small line or a thread), and *pemunkuan* (rope or cord) (Harrington 1955, Siebert 1975). The words, and doubtless the plants, occur in cognates for thread, including *piminahkwan* (Cree), *pimenahkwan* (Menomini), *piminakkwan* (Ojibwa), and *piminʔkwa* (Shawnee).

Gilmore (1931) reported that fibers of *Apocynum cannabinum* had been used by the Ozark Bluff-Dwellers in Arkansas to make twine and cords. Later, Jones (1936) identified the fibers in textiles and cords at the rock shelter at Menifee County, Kentucky. Whitford (1941) found both Adena and Hopwellian woven fabrics made from Indian hemp. He also found its fibers in Arkansas cords, Fox and Sauk bags, a fish net used by the Nanticoke, and an Iroquois burden strap.

As a medicine, apparently one of the oldest names for these plants in the southeastern United States is "Choctaw root." These people constituted the largest group in the southeast, with the exception of the Creek Confederacy (Swanton 1946). They had first been encountered by the Spanish expeditions through the area, including de Soto (Swanton 1946), and previously were known as the *Pansfalallya* [*Apafalaya,* hence Pensacola], long hair, because the men let their hair grow full length, unlike the surrounding people. Constantine Rafinesque reported in 1828: "This is a very active plant, highly valued by the Southern Indians. It is tonic, emetic, alterative and antisyphilitic. The root is the most powerful part; but it must be used fresh, since time diminishes or destroys its power . . . it acts as an emetic, equal to Ipecacuana; in smaller doses, it is a tonic, useful in dyspepsias and fevers. The Chicksaws and Choctaw Nations employ it in syphilis and considered it as specific."

Perhaps it was plants from the southeast about which Porcher (1863) wrote: "This plant has been proved by Prof. Thouin, of Paris, to possess a stronger fibre than that of hemp; and it is used by the American Indians for making cordage, fishing nets, and coarse cloth."

People from the Iroquois, Menomini, and Potawatomi of the northeast, to the Okanagan-Coville of the Canada, Washington border, and south to Cahuilla of southern California and adjacent Baja California Norte used the plants for medicine (Smith 1933, Moerman 1998). Oddly, there seem to be no records south of the Cherokee and Kiowa (Moerman 1998). There is a similar gap for the related *A. androsaemifolium* that covers much the same range. Surely, this reflects absence of data and not lack of use. As some of the names indicate, the plants have been used to treat dropsy, rheumatism, worms, constipation, pox, uterine obstructions, Bright's disease, asthma, wounds, diarrhea, biliousness, sore mother's breasts, and stomachache (Moerman 1998).

The root bark is an emetic and cardiac stimulant (Mabberley 1997). However, it is poisonous to animals such as cats, dogs, and livestock due to the presence of resins and cardiac glycosides, including cymarin, apocannoside, cyanocannoside, apocynamarin, and others (Turner and Szczawinski 1991). As little as one-half ounce of the plant ingested by livestock may kill a large farm animal such as a cow (Foster and Caras 1994). While sickness and death have been reported among humans who used it in medicine (Diggs et al. 1999), Foster and Caras (1994), and Turner and Szczawinski (1991) say there are no records of humans having died after using it.

The plant is also used as a brown or black dye (Porcher 1863).

Aquilegia

(Derived by some from *aquila,* the eagle, from resemblance of the spurs to the claws; by others from *aqua,* water, *legere,* to collect, from the nectar at the base of the hollow spur)

Akelei (Dutch, German); *akeleie* (Norwegian, dedicated to the goddess Frigg)
amor nascosto (hidden love, Italian)
ancolie [*anquelie, angorie*] (in use by 1325; borrowed from medieval Latin *aquilea* derived from Latin *aquilegus,* "that which collects water," because of the small cavities in the flower; later, the name was brought closer to Latin *aquila.* The initial nasal vowel is probably influenced by *melancolie,* justified by the position of the flowers; since the 15th century, *ancolie* has been the symbol of sorrow, French); *aquileña* (Spanish)
columbine (from French *colombine,* based on Latin *columbina,* dove-like, from the appearance of the five flower spurs resembling a cluster of doves)
erva pombinha (little dove herb, Portuguese); *lus a'chalmain* (dove herb, Gaelic)

Aquilegia canadensis (of Canada)
akeley (surely from French *ancolie*); *ancolie* (Quebec)
bells (Ohio); [red, rock]-bells
cluckies (Nova Scotia)
[red, wild] columbine
culverwort
gants de Notre-Dame (Mary's gloves, Quebec)
honey-horns; honeysuckle (a name usually applied to *Lonicera,* but used for this in Illinois, Massachusetts, New Hampshire, Wisconsin)
inubthon-kithe-sabe-hi (black perfume plant, *inubthon,* fragrant, *kithe,* to make, *sabe,* black, *hi,* plant, Omaha-Ponca); *mon̲'-bi-xon̲ ça-be* (*mon̲'-bi-xon̲,* perfume, *ça-be,* black, to distinguish it from other *mon̲'-bi-xon̲,* or other aromatic seeds used for perfume, Osage)
jacket-and-breeches; Jack-in-trowsers (Massachusetts)
meeting houses (New England)
rock-lily (New Hampshire)
skalikatit [*skarikatit*] (*skali,* seed, *katit,* black, Pawnee)

Linnaeus knew only this single American species and only two European kinds, although there are 19 species native to Europe (Mabberley 1997). Gaspar Bauhin was one of the first Europeans to use the name *Aquilegia* in 1623. Neither he nor Linnaeus explained its etymology, and the meaning is still debated. Judging from the Gaelic, Portuguese, and Old Latin names, the plants were probably considered more like doves than eagles. Frenchman Jacques Philippe Cornut (1626–1651) seems to have been the first European to discuss this American *A. canadensis* in his *Canadensium plantarum* (Canadian plants) of 1635.

The Cherokee used an infusion of the plant for diarrhea and heart trouble (Hamel and Chiltoskey 1975). Otherwise, records are from farther north. The Iroquois made a compound infusion of the plants with others to treat poison ivy and itch and for kidney problems (Moerman 1998). Similarly, the Meskwaki, Ojibwa, Omaha, Pawnee, and Ponca used the plant in a variety of ways (Gilmore 1919, Moerman 1998).

Aralia

(From French-Canadian *aralie,* the original specimens having been sent by the Quebec physician Michel Sarrasin de l'Étang, 1659–1734, to J.P. Tournefort, 1656–1708, under that name)

Aralia spinosa. From Sargent 1905.

Aralia spinosa (with thorns or spines)
American angelica tree
devil's walking stick [devil's walkingstick] (because of the spiny trunk)
Hercules' club
ittohalokpá (*ittó,* tree, *halókpa,* sharp, Koasati)
pick-tree
pigeon-tree (because birds eat the fruits)
prickly ash (comparison of the leaves with *Fraxinus*); prickly elder (comparison of the leaves with *Sambucus*)
toothache-tree (from analgesic compounds)

Leonard Plukenet (1641–1706) published on *Aralia spinosa* in 1696, and a year later Jan Commelin (1629–1692) wrote about them. Commelin had them in cultivation in Amsterdam, and presumably Plukenet had them growing in London. Linnaeus knew this species, plus one from China and two others from the New World. There are now about 36 species in the bicentric genus, with species in North America and east Asia (Mabberley 1997).

The Creeks used the root to stop internal bleeding; the Choctaw applied a poultice to swollen leg veins; and the Koasati put a cold infusion of the roots on sore eyes (Taylor 1940). The Choctaw also used *A. racemosa* root (*tally apola*, maybe *tali*, the end, and *apoli*, to pass through slowly) as a stimulant and expectorant (Campbell 1951).

Neighbors of the Muskogean people, the Cherokee, used *A. spinosa* to treat rheumatism and flatulence, in a salve for persistent sores, as a diaphoretic, a strong emetic, a tonic, to treat toothache, and against venereal disease (Hamel and Chiltoskey 1975). The Rappahannock of Virginia also made a salve to treat boils and sores, and used it as a remedy for fever (Moerman 1998).

Ardisia: Marlberry

(Name based on Greek, *ardis*, a point or arrow point, referring to the anthers)

Ardisia escallonioides. Drawn by P.N. Honychurch.

Hammocks (West Indian hardwood forests) along southern Florida coasts have shrubs in them that people call marlberry. Botanists call the plants *Ardisia escallionoides,* a name applied to it in 1831.

In Florida and in the Bahamas, this shrub is sometimes also called dogberry. Names that have "dog" attachedto them really have nothing to do with canines, but simply mean "wild." In most cases, the wild plant reminds people of something that is "tame" and they need a way to distinguish one from the other.

Fruits on these shrubs often remind people of blackberries (*morita*, Oaxaca), cherries, or other fruits. More people seem to think of it as a cherry, and formerly *Ardisia* was called "cherry" in Florida. That name is also applied by people throughout much of Mexico and the Caribbean. For example, our species is called *capulín* (cherry, San Luis Potosí), *capulín agrio* (sour cherry, San Luis Potosí), *capulín de pajaro* (bird cherry, Morelos), or *yagalancito* (*yagalan*, cherry or *Prunus*, *cito*, little, Zapotec, Oaxaca). In Panama, any species is likely to be called *iskarmas* (Cuna) or in Spanish *uvito* (little grape). In Martinique, they also compare it with grapes, and other species are called *bois raisin* (raisin tree). The Asian exotic *A. elliptica* (often called *A. solanacea*) is called "blackberry" in Jamaica. Some of the Puerto Rican species are called *mameyuelo* (little mamey). The *mamey* is the Taino name of *Mammea americana* (Sapotaceae), a fruit prized throughout the Greater Antilles.

As all these names suggest, the fruit is edible. Some think it good, or they would not compare it with blackberries, grapes, and raisins. Others think it is "unappealing" (Morton 1968b). Perhaps how good or bad it tastes depends on how hungry a person is. In Panama the Bayano Cuna use the plants medicinally, and stain their bodies with the berries (Duke 1972). Most literature mentions only the edible fruits and useful wood, but people in Panama and Mexico have medical usage for these plants. Other species also have medical applications (Hocking 1997).

Among the species used are American and Asian plants. More information is available on the Old World species than the American. For example, *A. crenata*, a Malesian species cultivated in Florida, is mixed with other plants in a tea to treat bruises, broken bones, and sprains. Application is both external and internal. From India through China, southeastern Asia to Japan, *A. crispa* roots are used to treat fevers, excessive salivation, to heal injuries, and as a diuretic. In Indochina, *A. isoraefolia* fruits are boiled as part of a vapor inhaled to stop pain in decayed teeth. Although several others are used, the most often noted is *A. japonica*, whose root decoction is used to treat influenza, to serve as a carminative, and as a poison antidote. The leafy shoots are made into a decoction used to stop coughs and uterine bleeding. Perhaps our species should be investigated further.

Several species are known to have a number of chemicals including anthocyanins, ardisicrenosides, glycocides, quinones, resorcinol, saponins, and triterpenoid saponins. Some of these compounds have been named ardisin, ardisinol-I, ardisinol-II, 2-hydroxy-5-methoxy-3-pentadecenyl-benzoquinone, bergenin, embelin, ilexol, myricitrin, quercitrin, and rapanone and (Podimuang et al. 1971, Ahmad et al. 1977, Jansakul et al. 1987, Tian et al. 1987, Wang et al. 1992, Jia et al. 1994, Baldini et al. 1995, Nguyen et al. 1996, Piacente et al. 1996, Horgen et al. 1997, Huang et al. 2000).

Laboratory experiments show that some of the extracts have antiviral activity, including human immunodeficiency virus and certain kinds of cancer.

A number of people have used the leaves as tobacco substitutes, or at least in mixtures. The Miccosukee called the plants *akcomakáhka:plô:cî* (black tobacco seasoning tree). Their Seminole relatives the Creeks call it *hici:apa:kalásti* (*hece*, tobacco, *vpakv*, seasoning, *lastv*, taste). Although Sturtevant (1955) recorded them as mixing the "leaves with tobacco to extend it," the common name suggests a more complicated purpose. Several tribes throughout the Americas mixed other plants with their tobacco for different reasons, including changing the flavor, but particularly for weakening a strong mixture. The Seminoles probably did that and simply did not reveal the information to Sturtevant. According to some individuals now on the reservation, they withheld many of their customs from Sturtevant. In Martinique, another species is called *bois chique* (tobacco tree). Perhaps smoking of the leaves was more widespread in the past.

Throughout their ranges, several species provide useful wood. The Miccosukee used the plants to make arrows, and as sticks for roasting meat. They also ate the berries as did other people.

The name *pejté* (Huastec, San Luis Potosí) has been reported for this species, although Alcorn (1984) did not find it being used. Instead, she found her informants calling the plants *pelat puchun,* meaning dove legs. Children ate the fruits, but adults mixed them with *atole* for food. The root and shoots were boiled with some other plants (*Malvaviscus arboreus, Paullinia tomentosa*, and *Cirsium* sp.), and used against diarrhea, stomach trouble, and vomiting.

The Maya in Yucatán have several names for *A. escallonioides*, but none of them suggests medical usage. Their application seems to be more immediately practical. They call the plants *ch'onche'* (*ch'on*, vulture, *che'*, tree, or tree where vultures sit), *sak'hoktub* [*sak-hok-tub, zachoclub*] (*sak*, white, *hok*, pull out, *tub*, around, probably meaning white tree that grows on the margins of planted fields [*maíz*]), and *sak'loobche'* (*sak'*, white, *loob*, to make storage bins, *che'*, tree). That last name is associated with use in creating storage bins where they keep their seeds for planting next season.

These same people also call the plants *xook* (chair, Maya, Yucatán). Presumably, they use the wood for making seats, although that too may be an allusion to the vultures sitting on them. A version that remains unclear is *xooknum* [*xook gum* by error]. The word *nuum* means spines that serve as pins, incapable, slow, or difficult. None of these makes sense to José Luis Tapia or me.

Some other names for marlberry have not been translated. These are *chaquis* (maybe based on *chaqué*, a kind of frock coat, Morelos), *guitumbillo* [*huitumbillo, huitumbio*] (Chiapas), *hullaba* (Belize), *residan* (Belize), and *mantákihui* (Totonac, Veracruz).

Flowers on these plants fool some people when they first see them into thinking that they are members of the Solanaceae. However, closer examination reveals that they are in the Myrsinaceae. The anthers are distinctive in this genus, and led Olaf Peter Swartz (1760–1818) to coin the name *Ardisia* in 1788. The flowers are fragrant and attract a number of pollinating insects (Pascarella 1997). There is also a close association between the species and a particular group of moths (Pascarella 1996).

When I first arrived in Florida, there was a popular commercial on billboards and elsewhere advertising Marlboro cigarettes. There was a tough-looking cowboy with a cigarette dangling from his mouth and the slogan, "This is Marlboro country." One of my colleagues decided to use that to his advantage to help people remember the name of the marlberry. He would take students into hammocks, point out the plants, and remark that "This is marlberry country."

That name "marlberry" always struck me as odd, because I did not find the plants growing on marl, a particular soil type where clay is mixed with calcium carbonate. Some places in the southern end of the state have large deposits of marl that were created when the sea levels were higher during the geologic past. Other places have accumulations that are more recent because the process is ongoing, marl being deposited by accretion of calcium carbonate crystals on cyanobacteria (blue-green algae-like bacteria). The process continues in the wet summers when water gathers in lower areas.

Few organisms can grow in marl, as it is nutrient poor and a harsh environment, either waterlogged or brick-hard when dry. It is only when marl becomes mixed with organic detritus that plants can grow in it. So, I wondered, why do people call *Ardisia* marlberry? Finally, I came across the probable answer. The word "marl" not only refers to the soil type, but it has evolved as a colloquialism. It seems that "marl" has become a contraction of "marvel." Historically, this abbreviation appeared in the literature in the early 1600s. The earliest example was written by Ben Johnson in 1609. In a sentence dealing with the collar and cuffs on a shirt, he noted: "'Tis mar'l you ha 'hem on now" (It is a marvel that you have them on now).

That is probably also the origin of the name "marlberry" as a contraction of "marvel berry." Perhaps the early settlers in the Bahamas and Florida were pleased with the berries and thought them a "marvel," but more likely they thought they looked like "marvels" (marbles). I grew up with people who pronounced "marble" as "marvel." So, our marlberry is really a "marble berry."

Argusia: Sea Lavender

(From Greek *Argos*, the hundred-eyed keeper of Io)

***Argusia gnaphaloides**. Drawn by P.N. Honychurch.*

Sea lavender (*Argusia gnaphaloides*) is a gray shrub that grows on the beaches of most of the New World tropics, through southern Florida, Bermuda, the Bahamas, West Indies, Yucatán, Mesoamerica, and Venezuela. Indeed, the species name *gnaphaloides* means that it resembles the white-gray plants in the genus *Gnaphalium* (Asteraceae). The plant was common on beaches throughout its range before people began developing the coastal areas for homes, businesses, and recreation. Moreover, it has white flowers, not lavender.

Before understanding why people call Florida's gray plant "lavender," it must be understood that the flower, leaf, and stem color was not the immediate source of the common name. "Lavender" is a comparison with an Old World mint. To Europeans, the "lavender" plant is *Lavandula*, a fragrant mint native to the Mediterranean, the Middle East, and India, and used for hundreds, if not thousands, of years. The word came to English through Middle English (*lavandre*), which in turn came from Middle Latin (*lavandula*). These words for the plants have all been used since at least the 1200s. They are perhaps all based on Latin *lavare* (to wash or bathe). Yet, it is possibly a corruption of the original word Latin *lividus*, blue or bluish, which was mispronounced. Perhaps it is also possible confusion or comparison with "bluing," a colloidal solution based on ferric hexacyanoferrate that is still sometimes used to make washed clothes brighter. Either way, since the Middle Ages, *lavandula* has indicated specific blue-flowered mints in Latin, Italian (*lavanda*), Spanish, and French (*lavande*).

Portuguese speakers say *alfazema*, which is surely cognate with *alhucema*, the alternate Spanish name for the plants. These seemingly anomalous words for the same plants are actually based on Arabic *al-khuzamâ* (the pleasant-scented one).

Still, since at least the Middle Ages the common name "lavender" has indicated the color of the flowers, and not the herbage. The word has been used so long in American English for a color that now few think of the application to the whole plant. Yet, the European aromatic herb was in the minds of newcomers to the New World where they encountered a similar plant growing beside the sea. So, what could be more natural than to call it "sea lavender" (Bermuda through Florida to Jamaica and Virgin Islands), "bay lavender" (Bahamas, Puerto Rico, Virgin Islands), or "white lavendel" (Dutch Antilles) in English, and *alhucema de costa* (coastal lavender) in Spanish. Others thought it better to call this aromatic new plant after some other mint. So, it became *salvia marina* (marine sage, Cuba), *romarin* (rosemary, Lesser Antilles), *romarin blanc bord-de-mer* (white seaside rosemary, Lesser Antilles), or *incienso de costa* (coastal incense, Cuba), *incienso de marino* (marine incense), *incienso de playa* (beach incense), or even *balsamillo* (little balsam, Cuba).

Plants frequently have dead leaves clinging to the lower branches, and these apparently looked like tobacco to some of the people examining them. Since the tobacco (*Nicotiana*, Solanaceae) was not a plant familiar to people in the Old World before arriving in the Caribbean to see the Caribs and Arawaks using it, perhaps names alluding to this American plant came from indigenous people. One of the first places Europeans settled was the island of Hispaniola, and there the sea lavender is called *herbe à chiques* (tobacco herb, Dominican Republic), *liane à chiques* (tobacco vine, Dominican Republic), *liane chiques* (tobacco vine, Dominican Republic), *racine chiques* (tobacco root, Dominican Republic). Nearby, and reaching to the Netherland Antilles off Venezuela, it is *tabaako die piskadoo* or *tabaco de pesadores* (fisherman's tobacco), *tabacón* (big tobacco), *tabacón piscador* (fisherman's tobacco), *tabacu di pescado* (fisherman's tobacco), *vesserstabak* (fisherman's tobacco, Dutch Antilles). If people actually used these plants as a tobacco substitute, I have found no evidence of it. Indeed, with the other uses that they have, it seems doubtful that they were smoked or chewed.

Medicinal uses are those most often recorded for sea lavender, and a variety of common names supports those observations. In the Dominican Republic they are called *herbe à malingres* (herb for the sickly), or *té marino* (marine tea). Elsewhere they are *té del mar* (sea tea, Puerto Rico), *temporana* (the early one, Puerto Rico), or *bois de lance noir* (black lance bush, Lesser Antilles), also suggesting medicinal uses. The name iodine bush (Morton 1981) is either a misapplication or mistaken identity, because that is a name usually applied to *Allenrolfea occidentalis* (Amaranthaceae). Either way, it suggests medical use.

Like its relative chiggery grapes (*Tournefortia hirsutissima*), this seaside plant has been used against those bothersome little spiders. It is called *nigua* (chigger), or *nigua de playa* (beach chigger) in both Puerto Rico and the Dominican Republic.

Several of the names associate the plants with animals. Some of these associations are obvious, and others are not. For example, the name *crête coq* (rooster's comb, Dominican Republic) is a common theme in the family (Boraginaceae). The comparison is made because of the curving inflorescence that reminds people of the comb on the rooster's head. A little more obtuse is *yerba cotorra* (parakeet's herb, Puerto Rico), but it probably means that these small Psittacidae eat the fruits or drink nectar from the flowers. Since both occur on beaches, the name crab bush (Virgin Islands) seems straightforward. Totally incomprehensible to me is *loup marine* (sea wolf, Lesser Antilles).

The Mayan name *ts'its ik maay* [*sicimay, siki-may*] (*ts'its ik*, strong, *maay*, perfume, Yucatán) is given because early in the morning, before the sun gets hot, the plant is at first pleasantly fragrant. Later the odor becomes unpleasant. Unfortunately, we do not know the meaning of the name *kokorobana*, used in Aruba, Bonaire, and Curaçao, but it resembles indigenous words in northern South America.

The scientific name in most modern literature for sea lavender is *A. gnaphaloides*, but it varies in different sources. For a short while it was called either *Mallotonia gnaphaloides* or *Messerschmidia gnaphalodes*, but the name most commonly used is *Tournefortia gnaphaloides*. This variation is distressing to everyone, but results from varying opinions about how different groups of species must be from others to be recognized at generic rank. The genus *Tournefortia* comprises largely twining or climbing plants, and that caused some people to exclude these erect shrubs. Other traits for the genus are fruits that are corky outside and fleshy leaves, but the habit is most easily noted. The genus *Argusia* was proposed for these species in 1760 by Georg Rudolf Boehmer, a German botanist in Leipzig. The species serving as the focal point for this genus concept is *A. sibirica*, originally described by Linnaeus as *T. sibirica*. Therefore, plants on opposite sides of the world have been united into the genus as discussed here. Since the fruits of most *Tournefortia* are fleshy and adapted for animal dispersal, some think that the corky fruit of *Argusia*, adapted for floating and water dispersal, is simply a convergent adaptation in distant species within a single genus.

Sea lavender, like its namesake *Lavendula*, has aromatic foliage and blossoms. Honeybees visit the flowers and produce honey in areas where the plants still thrive. Yet, while people scent their clothes and foods with *Lavandula*, no such use has been made of sea lavender, nor is such a use likely. Indeed, Morton (1968) tells us that when leaves are burned in the house it drives out fleas. Petersen (1974) confirms at least that the plant is used as an insecticide, but gives no further details. Both these authors and others indicate that the stems and leaves are used in "bush baths" to clean wounds and ripen boils and to treat other skin problems. As a footbath it is used to treat stiff legs. Leaves also are washed and held in the mouth to alleviate oral inflammations. Some localities in the Caribbean have prominent places among the herb vendors for sea lavender branches. In Curaçao, some individuals take a weak infusion of the leafy twigs each morning, with sugar, as a tonic and stomachic. Supposedly, this beverage also expels gravel (kidney stones) from the bladder and kidneys. One of the common uses among men in at least the Dutch Antilles is against venereal diseases. The leafy branches are mixed with *Stemodia maritima* and Epsom salts and drunk to relieve those problems. As the plants are listed among those that are diuretic, they may give temporary relief. It might even be that the Puerto Rican name *temporana* is an allusion to that temporary "fix." A decoction of leafy twigs is used as an antipyretic (Morton 1981, Hocking 1991), and several sources indicate that the plants have been used as an emmenagogue. Some believe that a decoction of the plants is an antidote for fish poisoning. Cubans even use a decoction against catarrh.

Perhaps the spotty literature on the species is due to the reluctance by some people to admit that they use the plants. Morton (1981) found women in the Dutch Antilles either reluctantly confessing that they use the plants, or at least crossing themselves before admitting it. That reluctance was because the plants are also used to induce abortion. Those people to whom Morton spoke were Catholics and in admitting use of the plant they ran the risk that they would be suspected of that practice. However, the same decoction in weaker doses is said to ease delivery for pregnant women (Morton 1968a).

Arisaema: Jack-in-the-Pulpit

(From Greek *aris*, a kind of arum, and *haima*, blood, from the spotted leaves in some species)

The only time I ever saw *Arisaema dracontium* in southeastern Florida was in a moist swamp forest in Martin County. These small herbs are related to more tropical plants, but their range in North America is temperate. Our two species (*A. dracontium, A. triphyllum*) are known by a variety of names. Americans call them Canada turnip, dragon root, green dragon, Indian turnip, and wild turnip. We also call at least *A. triphyllum* memory root—bite into a raw root, and you never forget the taste!

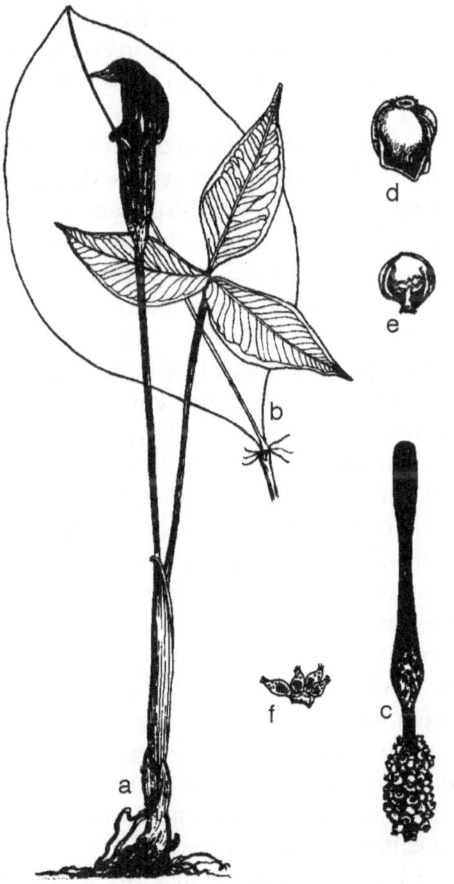

Arisaema triphyllum. a. Habit. b. Outline of one leaflet. c. Spadix. d and e. Berries. f. Seeds. *Drawn by Vivian Frazier*. From Correll and Correll 1972.

Europeans say *Drachenwurzel grüner* (green dragon-herb, German), *dreiblättiger Aron* (three-leaved Arum, German), *Feuerkolben* (fire herb, German), and *gouet à trois feuilles* (gouge of three leaves, referring to the pointed leaflets, French).

However, the best-known name now is Jack-in-the-pulpit. That name has always seemed strange to me. Who was Jack? Why was he in the pulpit? Why was there not a preacher there?

The name "Jack" is a "pet-name" (or by-name) that dates from at least the 1300s. It was used then much as its synonym "John" is now used for a frequenter of houses of prostitution. This allusion to someone without using his or her real name has been a widespread practice in Western societies throughout time. The allusion also occurs in most other societies, but often for different reasons. Regardless of the reason for usage, one may hear allusionary names such as a Jack (knave), Jack-tar (sailor), Cheap-Jack or Steeple-Jack (laborer), Jack-of-the-clock (the figure of a man who struck the gong on wind-up timepieces), Jack-of-all-trades (a person who can do about anything, with or without the added expression "master-of-none"), male animals (Jack or male donkey), and Jack-o-lantern (the will-o-the-wisp, but also a person who carried around a lantern at night and called out the time). The word is also used in names for plants. Jack-in-the-pulpit falls in that category, with facetious overtones. First, if "Jack" were a bona fide minister of the gospel, a derogatory term probably would not be used for him. Instead, the "person" in the pulpit (in these plants, the spathe wrapped around the reproductive structures) is a layperson and not a minister.

At first I thought this symbolism and allusion came about because the American plants were not part of the accepted medicinal pharmacopoeia of Europe, but instead used by the "savage" (literally, living in the forest, from Latin *silvaticus*, but historically a condescending term) indigenous people. That is not the case.

Since ca. 1450 the British have used derogatory names for their plants. At first they called their related *Arum maculatum* "cuckoo pindle," and then shortened that to "cuckoo pint." Both versions refer to the spadix with words meaning "penis." Even the British name "Lords and Ladies" emits phallic overtones (Vickery 1995). One has to wonder if missionary George Loskiel knew what he was saying when he called the American species "cuckow pint" in 1794. Jack-in-the-pulpit seems to have been an American extension (applied by 1884) of the old British allusions.

Indigenous names for these plants have been found from only two southeastern groups. *Arisaema dracontium* was *takko* to the Creeks, and *A. triphyllum* was *hichi* to the Choctaw. It is not clear which species the Choctaw called *kifash*, whose root was *hichuk*. However, among the Missouri River people, Gilmore (1919) found for *A. triphyllum* the names *mikasi-makan* (coyote medicine, Omaha-Ponca), and *nikso kororik kahtsu nitawau* (*nikso*, corn, *kororik*, resembles, *kahtsu*, medicine, *nitawau*, that resembles, Pawnee). The Plains Cree said *mistusko semina*. An apt name for *A. triphyllum* among the Osage was *e-haw-sho-ga* (bite the mouth; *i-ha'*, mouth, *tha-xta-ge*, bite). They also said *shon̲-ge i'-t'se-the* (wolf killer; *shon̲-ge*, wolf, *i'-t'se-the*, with which to kill). The Potawatomi called them *mûkwodji'bî* (bear's root), and the Menomini say *owässäutci'pa*. The Delaware said *maksáweek* [*maashk shaweek*] (*maashk*, defecation, *shawe*, in a hurry, *eek*, that which causes).

There are records of one or both species being used by at least the Cherokee, Choctaw, Delaware, Iroquois, Malechite, Menomini, Meskwaki, Micmac, Mohegan, Ojibwa, Pawnee, Penobscot, Potawatomi, and Rappahannock. Whether the Creeks of Oklahoma used *A. dracontium* before being relocated there or whether they learned of its use from other relatives is unknown (Howard 1984). Choctaws, for example, boiled *Arisaema* roots in water "to make blood" (Bushnell 1909, Taylor 1940, Vogel 1970). Since the Creek's neighbors in the southeast used it, they likely took the knowledge with them. The Florida Seminoles have apparently lost the knowledge of these plants.

Both species have medicinal uses (Moerman 1998). They have analgesic, antidiarrheal, carminative, diaphoretic, febrifugal, and sedative qualities. They are used as blood and cough medicines, as a cold remedy, to relieve cramps and lameness, as a dermatological aid and expectorant, as eye and throat medicines, for respiratory, skin, and female disorders, as a nose medicine (to induce sneezing), and to treat swelling, tuberculosis, and snakebite.

The seeds were put in gourds to make rattles for children, and the plants also were used in sacred bundles (Menomini) and gave the power of supernatural dreams to the owner (Plowman 1969). The Menomini called *A. dracontium* "owl's foot." The plant was used in witchcraft (pulverized root used in lip incision to counteract "witchery" to face), and divination (seeds as diagnostic medicine to predict recovery or death). Moreover, *Arisaema* served as a poison during war (finely chopped roots put in meat for enemies; root cooked with meat in abandoned vessels).

All parts contain acrid compounds, in addition to calcium oxalate, and are strongly irritant and poisonous if eaten raw. Both North American species have been part of the official pharmacopoeia and are listed as either *Radix Aronis dracontii* or *Radix Aronis triphylli* (Hocking 1997). *Arisaema* was recommended by the Eclectics for respiratory disorders, fever, and throat problems (Felter 1922). *Arisaema* was considered especially good for treating chronic laryngitis associated with singing or public speaking. A strong tincture of the fresh root was recommended for application as 1 to 5 drops every half-hour or hour. Additionally, a throat wash of the tincture was taken in a glass of water (Felter 1922). *Caveats* are appropriate.

The roots of both species were highly prized as food by all North Americans (Yanovsky 1936). *Arisaema* was prepared in several ways, but a common method was heating it to expel the poisonous compounds. Some suggest that the roots were sliced thin and roasted "for three days" to get rid of poisons. Millspaugh (1892) had eaten the roasted or boiled roots when young. He declared them good, saying, "I have roasted them myself, when a boy, and enjoyed a repast very pleasing to an empty stomach." Still, Foster and Duke (1990), Turner and Szczawinski (1991), and Duke et al. (2002) recommend against eating the roots because of the residual and irritant oxalic acid crystals.

Early experiments suggested that the poisons were expressed as a volatile gas that was slightly explosive when mixed with air (Millspaugh 1892). In his summary of folk uses of the Araceae, Plowman (1969) found that those "very labile compounds" had not been identified. The chemicals known from the family include starches, steroidal saponin-like compounds, polyphenolics (including tannins, anthocyanins, and flavones), cyanogenic substances unlike either amygdalin- or linamarin-types, and oxalic acid raphides. *Arisaema* contains the calcium oxalate, mucilage, maybe alkaloids, and probably many of the other chemicals (Plowman 1969, Vogel 1970, Foster and Duke 1990, Foster and Caras 1994). The Old World *A. vulgare* contains a volatile, coniine-like base (Plowman 1969).

American interest always was more in the residual starches until Europeans introduced better sources. Millspaugh (1892) compared the *Arisaema* starches with "arrow root" of the tropics (usually from *Maranta arundinacea*, but also from other species). Those same starches are now used in cosmetics, where they are called "Cupress powder."

When I was a professor in Florida, a mother and her young son came to me for "advice" about a science project that the boy was doing. He had apparently discovered independently that juice from the live plants applied to wounds stopped the pain. Since he was still in grade school, it seems unlikely that he had scoured the old literature and learned that the natives of North America used the sap in the same manner. Regardless, he was doing an experiment that involved getting as many volunteers as possible to prick their fingers with a needle and then apply juice directly from the plants. I too became one of his subjects with plants that they had imported from New England. We dutifully cleaned the instruments, drew blood from the end of my finger with a needle, and then applied the juice. The pain stopped immediately upon contact with the liquid. They told me that each person they had tested had exactly the same reaction.

Since these plants are no longer common because of habitat loss, I certainly do not recommend having pharmaceutical companies run out and harvest them wholesale to create a new pain-fighting product for the market. However, if you do happen to be near a wild region where the herbs grow, and you wound yourself, keep the young man's experiment in mind. You just might stop the pain before you get back to civilization and apply a more expensive remedy.

Aristida

(From Latin *arista*, the awn or beard of an ear of grain)

Aristida stricta*. Drawn by P.N. Honychurch.*

three-awn (a notation of the number of awns on the "seed")

Linnaeus knew a single species of *Aristida* in 1753. That plant, *A. adscensionis,* had been found in Jamaica and brought to Europe by Hans Sloane before 1696. Now there are about 330 species known in the genus, with one native to Europe (Mabberley 1997).

Gilliland (1975) suggested that the plant was used in thatch by the Calusa. Modern Seminoles on the Big Cypress Reservation still use this genus in weaving baskets. Western species of *Aristida* were utilized in medicines by the Keres, Navajo, and Hopi, but no records have been found of eastern species being used (Moerman 1998).

Aristolochia

(From Greek *aristos*, best, and *lochia*, delivery, from its value in aiding childbirth)

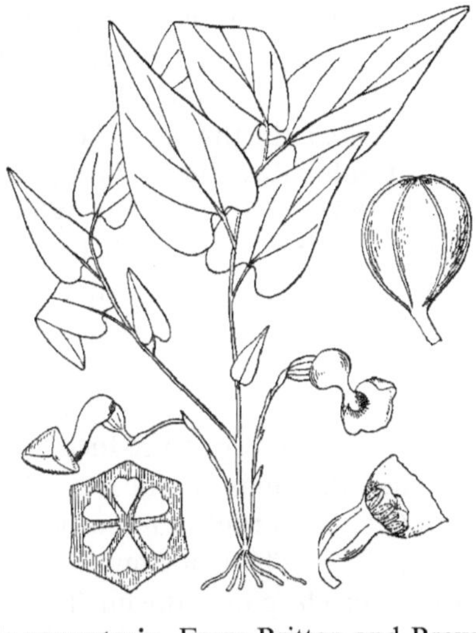

Aristolochia serpentaria. From Britton and Brown 1896.

aristoloche (French); *aristolochia* (Italian); *aristoloquia* (Portuguese); *Osterluzei* (derived from *Aristolochia*, through *Ostirlucie* in the 13th century, with variations in different locations like *Osterlizen* in Thuringia, *Osterlazei* in Lower Franconia, and *Oschelezei* in the Rhineland. Thus, it has nothing to do with East [*Ost*] or Easter [*Ostern*], and *luzei* has no meaning)

birthwort

culurain (Gaelic)

legeholurt (*lege*, doctor, *hol*, hollow, *urt*, herb, Norwegian)

Aristolochia serpentaria (referring to the medicinal use to treat snakebite)
cintó ahissí (*cintó*, snake, *ahissí*, medicine, Koasati)
[Virginia] Dutchman's pipe (comparison with the flower shape and the similarly curved pipe like the one used by Sherlock Holmes)
p'lew'óteek (that which has [grown] from where it started, Delaware)
pa-us-ka-ton-ga (like horse hairs; *pa-xin̲*, hair, *ka'-wa*, horse, *e-gon̲*, like, Osage)
tiak shua [*tiaóksês shuwa*] (*tiak*, pine, *shua*, smell, Choctaw)
tocopate (*toctli*, green cane, *patli*, medicine, Náhuatl, Texas)
unaste'tstiyú (very small root, Cherokee)
Virginia [white] snakeroot ("snakeroot" has been used as a name for medicinal plants since at least 1635, but originally applied to either *Polygala senega* or *Aristolochia serpentaria*, or both)

Linnaeus described this species in 1753, and talked about 12 other species. Seven of those species were native to Europe, although 19 are now considered native to the region around the Mediterranean (Mabberley 1997). Linnaeus gave by far the most space to discussing *Aristolochia serpentaria*. Leonard Plukenet had called the American plants *Aristolochia ... seu Serpentaria virginiana* (birthwort ... or Virginia snakewort), as had John Ray (1628–1705) in 1704 and Mark Catesby (1682–1749) in 1731. While the European species were known as medicine used during childbirth, it was these new plants from North America that became famous as a treatment for snakebite.

The Natchez of the lower Mississippi Valley took a warm decoction of snakeroot to reduce fevers (Swanton 1928a). The Alabamas mashed the leaves for a snakebite poultice, and drank a tea of leaves to assist the cure (Vogel 1970). The Choctaw took an infusion of roots for stomach pain (Bushnell 1909, Taylor 1940). Their neighbors the Cherokees in the inland Carolinas took an infusion as an abortifacient, as an anodyne against sharp pains in the breast, as a remedy for headache, and to mediate rheumatism. The roots were chewed as a remedy for colds. The plant was also used to treat coughs, fever, sore throat, dyspepsia, ague, pleurisy, and an infusion as an antiseptic. A plant poultice was applied to treat a nose sore from constant blowing, snakebite, and toothache (Hamel and Chiltoskey 1975). The Rappahannock in Virginia put a poultice of crushed roots on spider and snakebites. They also took an infusion of leaves to treat chills (Moerman 1998). People to the north, including the Delaware (eastern New Jersey and Pennsylvania), Micmac (Nova Scotia), Mohegan (Connecticut), and Penobscot (Maine and maritime eastern Canada), also used the plants in medicines (Moerman 1998).

Snakeroot was long used by early settlers for a variety of medicinal applications including snakebite (e.g., Catesby 1731, Porcher 1863). *Aristolochia* is popular in herbal medicine, although it can irritate the gastrointestinal tract and kidneys, potentially causing coma and death due to respiratory failure (Duke 1985, Foster and Duke 1990). Murphee (1965) also found it used to "build up blood." The vines contain a number of physiologically active compounds including aristolochine and aristolochic acid. Aristolochic acid is carcinogenic and nephrotoxic (Diggs et al. 1999).

Arnica

(Fernald said the origin of the name is disputed; Gledhill says from Greek *arnakis*, lamb's-skin, from the leaf texture; Ferreira says from Greek *ptarmiké*, a plant whose odor provokes sneezing, through Latin *ptarmica*, to Medievial Latin *armica*)

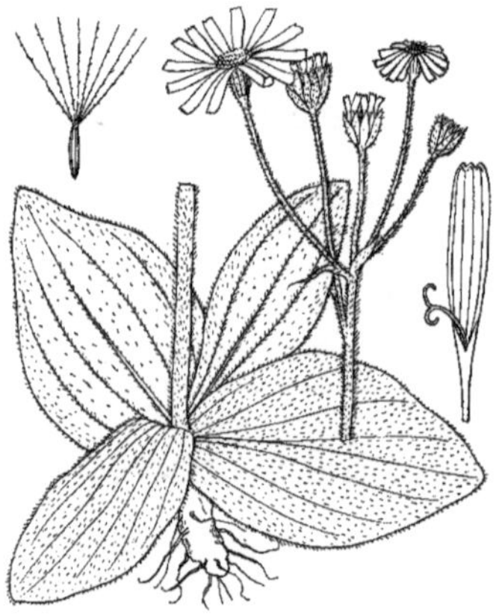

Arnica acaulis. From Britton and Brown 1898.

arnica (in use by 1753 when Ephriam Chambers wrote the supplement to his *Cyclopedia; or, an universal dictionary of arts and sciences*; he penned: "*Arnica* ... a species of doronicum, with plantane leaves"); cognates are *arnica* (French, Italian, Portuguese); *árnica* (Spanish); *Arnika* (German)
leopard's bane (Europe and Americas)
mountain tobacco (Europe)
solblom (sunflower, Norwegian)

Arnica acaulis (stemless)
leopard's bane (Maryland)
wi̲·ti· ya'mubahe̲'? (root in the water, Catawba)

Linnaeus created *Arnica* in 1753 without explanation, although most of the authors before him had called the plants *Doronicum*. He reserved that name, another of unexplained etymology, for a different group of Asteraceae. The most famous species in the genus is the European *A. montana*. Although some sources say that the plant has been used medicinally since the 16th century, it is surely older.

Not only is the herb famous as a medicine, but it probably killed Swiss herbalist Konrad von Gesner (1516–1565). Gesner decided to test arnica by taking a dose himself. Soon afterward, he wrote to a friend that there had been no ill effect. However, less than an hour later, he was dead—ostensibly of arnica poisoning (Dobelis 1986).

In spite of the Gesner incident, the plants remain a popular remedy in many parts of Europe. Not only may internal use of extracts be painful, but the sesquiterpene lactones increase blood pressure. The famous poet, philosopher, dramatist, and botanist Johann Wolfgang von Goethe (1749–1832) took arnica in his old age to relieve his angina (Bown 1995, Swerdlow 2000). Duke et al. (2002) say *A. montana* should not be used internally, although there is good evidence that external applications are effective. The species is still used internally in Germany, but is restricted to external use in the United Kingdom and the United States (Bown 1995).

When Europeans familiar with arnica arrived in the New World, they found indigenous people using other species. *Arnica fulgens* (Saskatchewan to North Dakota, west to British Columbia and California) was used on wounds and sore muscles by local tribes (Dobelis 1986). Vogel (1970) also lists *A. acaulis* (as *A. nudicaulis*), *A. cordifolia* (Arkansas to Arizona, east to South Dakota), *A. sororia* (Wyoming to British Columbia, Utah, and California), and *A. mollis* (Gaspé Peninsula and Quebec to Maine and New York west to Colorado and California) with medical properties that were used by native people. Moerman (1998) found records confirming that *A. cordifolia* and *A. latifolia* were used by the Okanagan-Coville, Shuswap, and Thompson tribes.

The only species in the southeastern United States is *A. acaulis*. This species ranges from Pennsylvania and Delaware south through the Carolinas, Georgia, and northern Florida. The only record found of local use was by the Catawba, who made *A. acaulis* into medicine that they drank for back pain (Vogel 1970, Hocking 1997). Hetz and Sosa (1988) have established that this species contains sesquiterpene lactones like many others in *Arnica*, and Merfort (1994) confirmed that it also contains flavonoid glycosides.

Although various species of arnica have been used for a variety of ailments, the only use now sanctioned by most is external (Bown 1995, Duke et al. 2002). Externally, the plant is considered bactericidal and vesicant, and is used to treat bruises, sprains, abrasions, and slight wounds. During World War II, the American species were so widely used that they became considered equal in healing power to the European *A. montana* (Vogel 1970). Mabberley (1997) considered *A. fulgens* more effective than *A. montana*.

Arnoglossum

(From Greek *arnos*, lamb, and *glossa*, tongue, referring to the leaves, and reapplied by Rafinesque to these plants; *Arnoglossa* is the ancient Greek name for plantain, *Plantago major*, and *P. lagopus*)

Arnoglossum atriplicifolia. From Britton and Brown 1898.

Arnoglossum atriplicifolia (having leaves like *Atriplex*) (= *Cacalia atriplicifolia*)
da'yewû (it sews itself up, because the leaves were thought to grow together again when torn, Cherokee)
pale Indian plantain (perhaps because the indigenous people used it as Europeans did *Plantago*)
tassle flower

Arnoglossum is another of the myriad new names created by Rafinesque. Many of the names he proposed are untraceable, but this one is. He proposed *Arnoglossum* in his *Florula Ludoviciana* (Louisiana Flora) of 1817. However, it was Linnaeus who described the species in 1753 as *Cacalia atriplicifolia*. Indeed, it was not until Harold Robinson studied the two genera in 1974 that they were separated. As now understood, *Arnoglossum* consists of seven species in the eastern United States (Mabberley 1997).

Core (1967) indicated that the leaves were eaten as a pot-herb. The Cherokee sprinkled the powdered leaf over food like salt (Mooney 1885–1886) and also used *da'yewû* as a poultice to remove blood or poisonous matter from cuts, bruises, and "cancers." The bruised leaf was bound over the spot and frequently removed and replaced (Hamel and Chiltoskey 1975).

Artemisia

(Named for *Artemisia,* the Greek Diana, goddess of chastity, as the plant was thought to bring on early puberty; Pliny, A.D. 23–79, used the name for the mugwort, *A. vulgaris*; he said that the name was in honor of Artemisia, wife of Mausolus, king of Caria)

amarella (little bitter one, Italian)
armoise (from *artemisia,* French); *artemisia* (Spanish); *artemísia verdadeira* (real artemisia, Portuguese)
Beifuss (close to your foot, German)
bollan bane (worn on 5 July, midsummer day on the Old Calendar, to repel witches on the Isle of Man); *bollan feailleoin* (synonym for *bollan bane*)
burmaid [*burmaill*] (for *A. absinthium,* Gaelic)
burot (cage root, Norwegian)
hierba de San Juan (Saint John's herb, Spanish)
laanah (for *A. herba-alba,* Hebrew)
liathlus (*liath,* sweet milk or feast, *lus,* herb, Gaelic, for *A. vulgaris*)
malurt (cognate with English mugwort, for *A. absinthium,* Norwegian); motherwort (a name applied to plants used to treat diseases of the womb, formerly *A. vulgaris,* now *Leonurus*); *muggart* (Scottish); mugwort (first recorded ca. A.D. 1000, of Teutonic origin from West Germanic *muggiwurti*)
muse'odji'bik (worm root, for *A. absinthium,* Ojibwa)
pe'-zhe-xu-a (*pe'-zhe,* herb, *xu-a,* sacred, Osage)
sage-brush [sagebrush, -bush] (apparently first used by Elliott Coues in 1872 in his discussion of North American birds; the allusion is to the odor of the plants resembling *Salvia*)
wormwood (originally confined to *A. absinthium*; first used ca. A.D. 1400; there are two versions of the origin of the name: one says it is a reference to its use in expelling intestinal worms, and also "worms" such as moth larvae and fleas that infested clothes and bedding; the other version is that it is based on Teutonic *wer,* man, and *mod,* courage; that version alludes to aphrodisiac and healing traits); *Wurmut* ("preserver of the mind," for *A. absinthium,* German)

Artemisia campestris (of fields or plains)
armoise rouge (red wormwood, Quebec)
aruone sauvage (wild wormwood, Quebec)
Déné k'áze 'eya ha nadié (person's throat hurts for this medicine, Chipewyan, Canada); *kotzezi naydi* (Slave, Canada)
[thread-leaf, western] sagewort
[Canadian, plains, sea] wormwood

Linnaeus ([1753] 1957) knew 19 species of *Artemisia.* According to Mabberley (1997), 55 of the ~350 *Artemisia* species are native to Europe. Given its history as medicine and culinary flavoring, no doubt explorers in the New World were pleased to see these familiar aromatic herbs (de Cleene and Lejeune 2002).

Like several of the other *Artemisia* species, this one has been broken up into a number of subspecies (Kartesz 1994). Moerman (1998) lists four subspecies of *A. campestris* that have been used by North American tribes.

The Blackfoot used the herbs to help tan hides, and the Lakota used them as body perfume. The Navajo used the seeds in a food.

In the eastern United States, the Menomini used this herb in a medicine to control menstruation. They also used it with *Angelica* for a poultice to relieve pain (Hocking 1997). The Meskwaki made a medicine of it for burns. Farther west on the plains, the Lakota made medicines that were diuretic, laxative, and sedative. The Blackfoot used it for abortions in difficult pregnancies, applied it to rheumatic parts, to remove scalp infections, for eczema and sore eyes, and for stomach trouble and coughs. Other people using the plants included the Chipewyan, Havasupai, Navajo, Okanagon, Shuswap, Slave, Tewa, and Thompson.

The genus is noted for chemicals that are toxic and under legal restrictions in some countries (Bown 1995, Swerdlow 2000). Among those compounds are thujone, cineole, and santoxin.

Arundinaria

(A genus created by French botanist André Michaux from Latin *arundo,* a reed, plus the suffix *-inis*)

Arundinaria gigantea (very large or giant)
ʔa'la (Tunica); *afho'ti* (cane-brake, Ofo)
axoki' (Biloxi; related terms are *kiduni,* canebrake; *axo'k dutca'ti,* split cane; *axo'k dutca'ti natcon̲'*

Arundinaria gigantea. *Drawn by Leta Hughey.* From Hitchcock and Chase 1950.

nkon̲ nkon̲ nda'sk nkon̲—I make baskets and mats out of split cane)

bamboo cane ("bamboo" perhaps from Malay, Sudanese, and Javanese *bambu*; maybe from Canarese *banbu* or *banwu*; first appearing in English by 1586)

cane [giant, large, river, southern, switch] ("cane" is from Middle English *canne*, appearing in English by 1398); *canne* (Houma, French)

iɬani (Alabama); *iɬaní* (Koasati); *ɬáwv* [*ráwv*] (Muskogee); *oɬã:nî* [*orã:nî*] (Mikasuki)

koça [*konshak*] (Choctaw); *kóhv* [*kohá, kóha, koha*] (Creek)

nissakan (Strachey [1612] 1953 wrote the Powhatan name *nissakan, nisake,* and *nehsaakan*, translating it both as "reed" and "cane"; Siebert 1975 found the root /**ni·?T-*/ in Menomini as *ne·?n-*, Ojibwa as *ni·ss-*, and *nehse-* in Penobscot)

oski [*oske, houski*] (Mobilian, cf. Drechsel 1997, from Choctaw *oski* [*uske, uski*]); *oski'* (Chickasaw)

pahiló:cî (*pahi*, grass, *ló:cî*, black, Mikasuki; the name of a dance fide Sturtevant 1955)

Rohr [*gras*] (German)

tu (Atakapa)

wa̲sa' (Catawba)

Although I have no idea how old I was, I still remember my first encounter with a cane-brake. My grandparents lived just above the floodplain of the Tennessee River in a remote part of west-central Kentucky. Beside their home was a small creek that we had to cross to visit them, walking when the water level was low, in a boat when the water level was high. Just above floodline on one bank, we always passed a large stand of canes. When I was a fairly young child, I ventured to explore that thicket alone and found a mysterious green wonderland. Some of the cane stems were 5 cm (2 inches) in diameter and reached 4.5 to 6 m (15 to 20 feet) tall. At the time, I had no idea of their famous reputation as sleeping places for black bears (*Ursus americanus*) or they would have been even more exciting. Looking back, I have to wonder if part of the mystery was the same feeling that indigenous people in the southeast had about this arguably most useful of American grasses.

Creeks used an expression *kóha-v'lkv* (cane-brake, Muskogee) to describe these thickets. Choctaw said they were *uski pata* (*uski*, cane, *pata*, broad or spread). In Cleburne County, Alabama, there is a stream called Cahulga Creek, meaning "canebrake creek," from Muskogee *kohv*, cane, *vlke*, grove. Perhaps that was the original source of the name "cane-brake" being used in English by 1769. The Choctaw called what is now the State of Kentucky *uski anun̲ka* (among the canes).

Bartram ([1791] 1958) described observing what he thought might be "the most extensive Cane-brake that is to be seen on the face of the whole earth" near the Creek village of Talahaschote (*talwa*, town, *ahassee*, old, *chee*, little) on the Suwanee River in Levy County, Florida. As explanation, he added that in the Carolinas they are called "cane meadows."

Extensive "cane-brakes" of *Arundinaria* grew in the southeastern states until about the early 1950s, particularly along larger waterways. Virtually all of those "brakes" were destroyed by grazing cattle, for agriculture, and housing to the point that the species is rare throughout its former range (Platt and Brantley 1997).

The word cane applies to several different plants, but two, sometimes three, native Florida grasses have that designation. One is *Arundinaria,* used for fishing poles and for constructing the most expensive fly-fishing rods; the other, *Lasiacis*, currently is not used by people in Florida. Sometimes people also call the giant reed (see *Phragmites*) a cane. People elsewhere in the Americas use all three genera and distinguish between them with common names, while at the same time noting their similarity.

Moreover, the words *reed* (see *Phragmites*), *cane*, and *bamboo* are intermixed in current and historical usage. All originally had somewhat different meanings and origins; yet, all came into English because of grasses.

"Cane" came to modern English from Middle English *canne*, through Old Frisian *cane*. These are based on Latin *canna*, and Greek *kanna*, reed. According to Zohary (1982) the Greek and Latin words were originally taken from the Semitic words in Hebrew *kaneh* or *qáneh*, and Arabic *qanáh*. In Latin the sense was extended from "(hollow) reed or cane" to "tube or pipe." The word "cane" appeared in English in 1398 in the sentence: "A noyse as it were wyth a canne other a great reyd."

"Bamboo" is the word most recently incorporated into English, although the origin is unclear. Some think it came from Malay, Sudanese, and Javanese *bambu*, but others think the word was introduced there from elsewhere. The people who think that it is a loan-word in those languages suggest that the original word was Canarese *banbu* or *banwu*. Regardless, the word was used first in Portuguese by Garcia de Orta, who had worked as a physician in Asia. In his 1563 book *Simples e Drogues*, he wrote "*Aquellas canas daquella arvore chamam os Indios, onde nasce, mambu*" (The canes of that tree the Indians where it grows call *mambu*). From that point, "bamboo" became incorporated into English within about two decades.

Obviously, the concepts of "reed," "cane," "bamboo," and "pipe" are all interrelated and applied to numerous plants. The first requirement is that the grasses are comparatively large, and have either a woody stem or one that is somewhat lignified. The most useful part of the grasses is the stem, and that structure has served people throughout the world in multiple ways.

Probably the first account of American canes occurs in the records of the de Soto expedition. On 13 October 1540, the group entered the province of *Mabilia* (maybe "to paddle," Choctaw) in what is now Clarke County, Alabama. Although the modern city of Mobile took its name from that region, it occupies a different location. There were hostile encounters and the Spaniards were forced to cross the Alabama River on cane rafts near the modern city of Claireborne (Swanton 1946). The chronicler known as the Gentleman of Elvas wrote that the Spaniards made the crossing on "rafts of canes and dry wood" (Swanton 1939).

Another early record was left by Henri de Tonti in his account of meeting the Taencas in 1692 in what is now northeastern Louisiana. He and La Salle had just left the *Arkancas* [*Arkansea*] (Arkansas, their name came from what the Illinois called them) near the mouth of where the river that bears their name met the Mississippi, and traveled south. The *Arkancas* lived in "cabins made with the bark of cedar." However, after entering the village of the Taencas, he wrote: "I was surprised to find their cabins made of mud and covered with cane mats" (Tonti in Borland 1965). Tonti and La Salle had finally entered the cultures of the southeast where the canes were dominant.

The first scientific record of giant cane in Europe was when Thomas Walter published his *Flora of the Carolinas* in 1788. He called the species *Arundo gigantea*, thinking they were related to that famous Mediterranean grass *A. donax*. André Michaux realized that the American plants were distinctive, and created the genus *Arundinaria* in 1813. Other people also found these giants interesting, and there are no fewer than 21 other synonymous names that have been given to them. The southeastern species has even been put in the Old World genus *Bambusa*.

American bamboos grow in the river systems connected to the Mississippi and Coastal Plain. The species is known in southern Missouri, Illinois, Indiana, Ohio, south into Florida, eastern Texas, and southeastern Oklahoma (Hitchcock and Chase 1950). There have been $\sim$50 other species put in *Arundinaria*, but some have been moved to other genera recently. Other species grow in South America and in Asia (McClure 1973, 1993, Mabberley 1997).

To most, if not all, southeastern people, *Arundinaria* canes were items used on a daily basis. Canes were used for arrow shafts, blowguns, blowgun darts, blowing tubes for medicines (*istoopofilka*; *ist-oo*, blowing, *pofilka*, pipe, Alabama), bows (*eccv kotakse*; *eccv*, gun, *kotakse*, crooked, Creek), building material for wattle and daub, fiber for weaving moccasins (Whitford 1941), flutes (*fehpv*, Creek), fuel, knives, lighting (candles), mats (*pachalh*, Choctaw; *kohv topv*, scaffold cane, and *kohv tvrke*, woven cane, Creek), medicine, and war spears (Sylestine et al. 1993, Munro and Willmond 1994, Moerman 1998, Martin and Mauldin 2000).

William Bartram ([1791] 1958) provided some of the most detailed information about these grasses in Alabama, Florida, Georgia, Louisiana, and North Carolina. He too recounts making a raft from the stems. At Otassee, Alabama, near where the de Soto army made their rafts, he wanted to cross a branch of the Alabama River that was flooded from rains. He and his companions decided they had to ferry their goods across, and he wrote: "I undertook to collect dry Canes, and my companions dry timber and logs and [grape] vines to bind them together: having gathered the necessary materials, and laid them in order [we] ... set about building our raft." They put the tree trunks together to form the base, and then gathered the canes, "in bundles, each near as thick as a man's body," to make the upper layer.

The Creeks made a unique arrangement of canes into a spiral fire to illuminate meetings in buildings. Bartram ([1791] 1958) described the fire when he was in the town of Attassee [Otassee] (properly *Atasi*, an Upper Creek town on the south side of Tallapoosa River, just below the mouth of Calebee Creek, Macon County, Alabama, Harper 1958). He wrote: "As their vigils and manner of conducting their vespers and mystical fire in the rotunda, [*sic*] is extremely singular, and altogether different from the customs and usages of any other people, I shall proceed to describe it. In the first place, the governor or officer who has the management of this business, with his servants attending, orders the black drink to be brewed, which is a decoction or infusion of the leaves and tender shoots of the Cassine: this is done under an open shed or pavillion, at twenty or thirty yards distance, directly opposite the door of the council-house. Next he orders bundles of dry Canes to be brought in; these are previously split and broke in pieces to about the length of two feet, and then placed obliquely crossways upon one another on the floor, forming a spiral circle round about the great centre pillar, rising to a foot or eighteen inches in height from the ground; and this circle spreading as it proceeds round and round, often repeated from right to left, every revolution encreases its diameter, and at length extends to a distance of ten or twelve feet from the centre, more or less, according to the length of time the assembly or meeting is to continue. By the time these preparations are accomplished it is night, and the assembly have taken their seats in order. The exterior extremity or outer end of the spiral circle takes fire and immediately rises into a bright flame ... which gradually and slowly creeps round the centre pillar, with the course of the sun, feeding on the dry Canes, and affords a cheerful, gentle and sufficient light until the circle is consumed, when the council breaks up."

Bartram ([1791] 1958) was among the first to describe canes being used to make hairpins, and of the Creeks spearing "salmon trout" (*Micropterus salmoides*, now called largemouth bass) with a cane "harpoon."

When talking of Creek dress, Bartram ([1791] 1958) wrote: "The men shave their head, leaving only a narrow crest or comb, beginning at the crown of the head, where it is about two inches broad and about the same height, and stands frizzed upright; but this crest tending backwards, gradually widens, covering the hinder part of the head and back of the neck; this lank hair behind is ornamented with ... [silver quills and silver plates], and usually the middle fascicle of hair, which being by far the longest, is wrapped in a large quill of silver or the joint of a small reed, curiously sculptured and painted, the hair continuing through it terminates in a tail or tassel."

Neither Romans ([1775] 1961) nor Bartram ([1791] 1958) was complimentary about southeastern music. Romans wrote that Chickasaw men send their women to work in the fields and then "they play on a aukward kind of flute made of a cane." Bartram described Creek music as "both vocal and instrumental; but of the latter they have scarcely any thing worth the name, the tambour, rattle-gourd, and a kind of flute, made of the joint of a reed." He thought that with the flute they "performed badly, and at best it is rather a hideous melancholy discord."

Of catching fish with cane spears, Bartram ([1791] 1958) wrote: "One of our Indian young men, this evening, caught a very large salmon trout, weighing about fifteen pounds, which he presented to the Col. [Barnet; but Harper (1958) thought that was a misprint for Col. James Barlett] who ordered it to be served up for supper. The Indian struck this fish, with a reed harpoon, pointed very sharp, barbed, and hardened by the fire. The fish lay close under the steep bank, which the Indian discovered and struck with his reed; instantly the fish darted off with it, whilst the Indian pursued, without extracting the harpoon, and with repeated thrusts drowned it, and then dragged it ashore."

Among some villages in the Creek confederacy, cane also played a role in marriages. Bartram ([1791] 1958) described a marriage: "When a young man has fixed his affections, and is determined to marry, he takes a Cane or Reed, such as they stick down at the hills of their Bean vines for their support: with this ... he repairs to the habitation of his beloved, attended by his friends and associates, and in the presence of the wedding guests, he sticks his Reed down, upright in the ground, when soon after his sweet-heart comes forth with another Reed, which she sticks down by the side of his, when they are married; then they exchange Reeds, which are laid by as evidences or certificates of the marriage."

However, the most famous use of canes by the Catabwa, Cherokees, Chitimacha, Choctaws, Creeks, Houma, Seminoles, and Yuchi, and perhaps others in its range in the southeastern United States was as blowguns. The "gun" or tube was called *oski'* (cane, Chickasaw), *kohv motke* (*kohv*, cane, *motke*, cut off short, Creek), *wókko* (Muskogee) by the Taskigi Creeks, *kaklu'mpa* or *uski lumpa* by the Choctaw (Bushnell 1909). To the Alabama, they were *iɬanibihi* (*iɬani*, cane, *bihi*, bow). These tubes were often 2 m (7 feet) long, but sometimes among the Cherokee reached 3 m (10 feet). Choctaw darts were made of cane (*uske*) or pine. Creek darts were *takfvlvlvke*; to the Choctaw, they were *shumatti* [*shuma'nte*] (from *shumo*, thistle down, *holutti* =?). The darts were "feathered" with thistle down (*Cirsium*) or fireweed down (*Epilobium* or *Erechtites*). Depending on the length of the tube and the skill of the user, they were accurate from 7.6 to 18 m (25 to 60 feet) (Swanton 1946), although Romans ([1775] 1961) says up to 20 yards. The southeastern region was the only one in North America that used this largely South American hunting instrument. While most tribes long ago ceased using the blowgun, demonstrations of former use can be seen in the replica of an old village in Cherokee, North Carolina.

In 1761, Lieutenant Henry Timberlake observed that Cherokee children, "at eight or ten years old, are very expert at killing with a *sarcaban*, or hollow cane, through which they blow a small dart ... and they seldom miss" (Swanton 1946).

Jean Bernard Bossu, writing in 1768, noted that the blowguns used by the Choctaw were about 7 feet long, and that, "when they see something [which they want to hit] they blow into it, and they often kill small birds" (Swanton 1946). Bernard Romans ([1775] 1961) visited the Choctaw not long afterward and recorded that they "use a very strait cane, eight or nine feet long, cleared of its inward divisions of the joints; in this they put a small arrow, whose one end is covered one third of the whole length with cotton or something similar to it; this they hold nearest their mouth and blow it so expertly as seldom to miss a mark fifteen or twenty yards off and that so violently as to kill squirrels and birds therewith."

Although few people now realize it, American cane was formerly an important food for people within its range. Until recently, two species were recognized. *Arundinaria gigantea* is the larger of the forms, reaching 8 m (25 feet) tall and fruiting erratically, perhaps every 40 to 50 years. Smaller plants, growing to 1 to 4 m tall (3 to 12 feet), were called *A. tecta* (switch-cane, reed, small-cane), and fruited every 3 to 4 years (Fernald et al. 1958). Even the Choctaw recognized the large cane as *oskonush* (*oski*, cane, *onush*, wheat, probably originally for *Arundinaria* and later shifted to *Triticum*), and the small one as *oskish* or *oski shifilli*. Now, the smaller plants are considered a subspecies of *A. gigantea* (McClure 1973).

The smaller subspecies *tecta* flowers and fruits regularly and people throughout the southern states harvested the seeds and used them as we do wheat (*Triticum*) flour (Hedrick 1919). In 1828, Timothy Flint recorded in his book *A Condensed Geography and History of the Western States, or the Mississippi Valley* that *Arundinaria* "produces an abundant crop of seeds with heads very much like those of broom corn. The seeds are farinaceous and are said to be not much inferior to wheat, for which the Indians and occasionally the first settlers substituted it."

The parent plant dies after fruiting, and the remaining seeds sprout to produce new plants. The young shoots on those were similarly gathered and eaten as potherbs, much as we now use oriental bamboo shoots (Fernald et al. 1958). Although the larger subspecies *gigantea* fruited irregularly, it provided many acres of gigantic fruiting stems when the event did occur. At those times, the grasses were not only important food sources for humans, but also for many seed-eating birds and other animals.

At least the Choctaw, Houma, and Seminoles used *Arundinaria* as medicine. The Choctaw made a root decoction to relieve chest pains (Taylor 1940). The Houma made a decoction by boiling the root. The resultant liquid was "drunk by persons past forty-five or fifty to renew strength, and specifically to stimulate the kidneys" (Speck 1941). The liquid was to be "drunk as hot tea every day for nine days." The Seminoles treated *layksínkoslihcî* (constipation, or "feces stopped-up" sickness, Mikasuki) by boiling the root of *oɬá:nî* and drinking the mixture (Sturtevant 1955). As with several other grasses, the seeds are sometimes consumed by a toxic fungus and become dangerous (Foster and Duke 1990).

Bartram ([1791] 1958) described visiting a Seminole village near a white settlement called Charlotia, on the St. Johns River between what is now East Palatka and San Mateo in Putnam County. There he was greeted in 1774 by the people and ushered into a "grand, airy pavillion in the center of the village. It was four square; a range of pillars or posts on each side supporting a canopy composed of Palmetto leaves, woven or thatched together, which shaded a level platform in the center ... all covered with carpets or matts, curiously woven of split canes dyed of various colours; here being seated or reclining ourselves, after smoaking tobacco, baskets of choisest fruits were brought and set before us."

Cane splints from the stems historically were used to make mats and baskets by tribes in the Creek Confederacy, including the Cherokees. The Alabama called these baskets *iɬanikolbi* (*iɬani*, cane, *kolbi*, basket). The Alabama boiled *ittokchakkosi* (wild cherry tree) bark to make a red dye that was used to color cane in the baskets (Sylestine et al. 1993).

This art is not lost; it can still be found on the Cherokee reservation in North Carolina. However, the grasses are now so rare, and the skill so demanding, that the items command prices of at least three figures.

Asarum

(Linnaeus continued using the Greek name for the plants, *asaron*, used by Dioscorides, fl. A.D. 40–80)

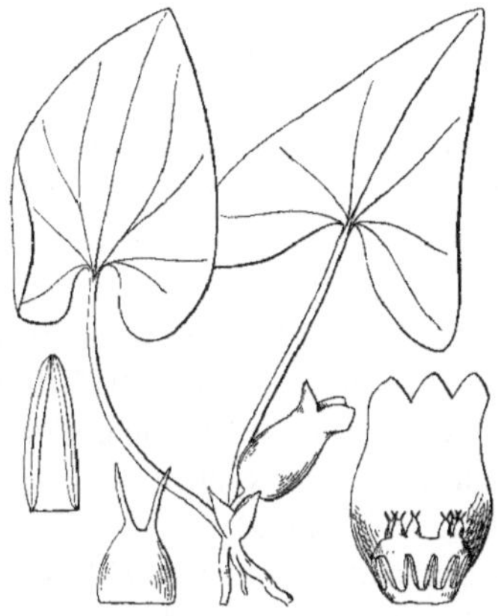

Asarum arifolium. From Britton and Brown 1896.

asair (Gaelic); *asaret* (French); *asaro* (Italian)
aschquoerìsta (probably *A. canadense*, Onondaga)
azarabacca [asarabacca] (apparently from *asarum*, the plant, and *bacca*, fruit, Turner [1548] 1965)
e-haw-nes-ni (water comes in the mouth; *i-ha'*, mouth, *ni*, water, for *A. canadense*, Osage)
fole-foot (used by Turner [1548] 1965; alludes to the shape of the leaf resembling a young horse's [foal] hoof, cf. Britten and Holland [1886] 1965)
gingembre sauvage (wild ginger, for *A. canadense*, Quebec)
Haselwurz (hazel wort, German)
name'pin (sturgeon plant, for *A. canadense*, Ojibwa)

Asarum arifolia (with leaves like *Arum*)
ha' mot' pa'tki ye (*mot' pa'tki ye*, flat, *ha'*, leaf, Catawba)
heartleaf (so called by A.W. Chapman because of the cordate leaf shape, Florida in the 1800s)
little brown-jug (an allusion to the shape of the flowers)
wild ginger ("ginger" from Old English *gingiber*, from the Latin *zingiberi*, akin to Sanskrit *crngavera*, perhaps compound from *crnga*, horn, *vera*, body, from the similarity of the rhizome to a stag's horn)

Asarum virginianum (of Virginia)
Virginia heartleaf

Linnaeus ([1753] 1957) knew four species of *Asarum*, with two of them native to Spain, Portugal, and other parts of Europe. The most famous of the Old World species, *A. europaeum* (asarabacca or hazelwort), was well known as a medicine. Not only was it used to treat alcoholic excess (emetic), but it is considered antiasthmatic and an immune stimulant (Bremness 1994, Mabberley 1997). After the 1520s, it became fashionable to use snuff, and hazelwort was sometimes used alone or added to tobacco.

The most widespread species in the eastern United States is *A. canadense*, and that is the one with the most documented uses (Densmore 1928, Yanovsky 1936, Moerman 1998). That species has been used throughout its range for medicine, food, snuff, and perfume (Moerman 1998). Canadian people used the root for many complaints, including heart problems (Vogel 1970). The Menomini, Meskwaki, Ojibwa, and the Potawatomi thought the root made food palatable to sick people, regardless of the malady they suffered (Vogel 1970, Moerman 1998). Several groups considered the root good for treating throat trouble, earache, lung troubles, and stomach cramps. Virtually throughout its range, people used the roots to season food. Although they do not list them, Fernald et al. (1958) note that other species may be used like *A. canadense*. Bown (1995) also listed a number of species that have been similarly used for seasoning and medicine. These and other data suggest that other species were substituted for *A. canadense* outside its range. For example, the Catawba are known to have used *A. arifolia* for heart problems (Speck 1932, Vogel 1970).

Asclepias

(Named for legendary Greek physician and deity Asklepios or Aesculapius)

Asclepias tuberosa (with tubers)
archangel (Rhode Island)
asclepiade tubereuse (tuberous milkweed, French)
butterfly's bed (Oklahoma); butterfly-[flower, milkweed, root, weed] ("butterfly-weed," in use by 1830); *tvffo* [*em*] *ohlikitv* [*ohliketv*] (*tvffo*, butterfly, [*em*, its], *ohlikitv*, chair, Muskogee); *yepananẹ'? wi·ti(re)* (butterfly medicine, Catawba)
Canada-root
chigger-flower [chiggerweed] ("chigger" was in English by 1769, having been derived from an African word akin to Wolof *jiga*, insect, Missouri)
colic root [colicroot] (a name more often given to *Aletris*)

env'ce-enok'ke-heléswv (*env'ce-enok'ke*, pleurisy, *heléswv*, medicine, Muskogee)
flux root [fluxroot]; [pleurisy, rubber, tuber, white]-root; wind [weed, root] [windweed, windroot]
fly catcher [flycatcher] (because of the "trap" that catches the legs of flower visitors)
grasshopper's bed (Oklahoma)
hatapushik okhinsh (*hatapushik*, butterfly, *ikhinsh*, medicine, Choctaw)
Indian-[bokay, paint, plume, posy, nosy] (Michigan, Texas)
ishkobo' homma (*ishkobo'*, head, *homma*, red, Chickasaw)
kiu makan (wound medicine, Omaha-Ponca); *makan saka* (*saka*, raw [as fresh wounds], *makan*, medicine, Omaha-Ponca); *makan saka thata i* (medicine they eat raw, Omaha-Ponca)
knollige Schwalbenwurz (tuberous swallow herb; the modifier distinguishes it from dogbane, which see, German)
kuku' (Cherokee; the word gave rise to the modern localities of Coco Creek, of Hiwassee River, and Coker post office, in Monroe County, Tennessee)
orange [apocynum, milkweed, swallow-wort, swallowwort, root]
silk-grass (the name was in use by 1620 for a variety of plants; it was first applied to a kind of "flag," then later to other fiber-producing species; in use by 1750 for *Asclepias*)
silkweed (in use by 1837: "This beautiful plant has already, by the French nation, been cultivated to advantage. The pappus is spun with raw silk for gloves, the juice collected for opium, and the leaf useful in dying," Williams [1837] 1962; as *A. phytolacchoides*)
swallow-wort [swallowwort]
yellow milkweed (Massachusetts)
zaip-ya-daw (*daw*, tree, Kiowa)

Europeans were well acquainted with *Asclepias* when they arrived in the New World. However, they were not prepared for the orange flowers of this herb and quickly took them back to the Old World. By 1687, the plants were already in cultivation in the Netherlands, although they were surely imported through England or Spain. The first records of *A. tuberosa* being grown in London are from Dillenius's *Hortus Elthamensis* of 1732, although it probably arrived there even earlier. By 1738, Linnaeus had studied the plants at the *Hortus Cliffortianus*.

Because of its showy flowers and habit of growing on moist roadsides, *A. tuberosa* is the best-known species in the eastern United States. These were plants of multiple uses, including dyes, food, fiber, and medicine. The Meskwaki used *A. tuberosa* as a dye. Millspaugh (1892) recorded: "The Western Indians boiled the tubers for food; prepare a crude sugar from the flowers, and eat the young seed-pods after boiling them with buffalo meat. Some of the Canadian tribes use the young shoots as a pot-herb after the manner of asparagus." The Dakota and Delaware cooked and ate them as they did *A. incarnata* (Yanovsky 1936). Still, Fernald et al. (1958) were not enthusiastic about the species as food. The Cherokee wove the stem fibers into belts (Hamel and Chiltoskey 1975).

Its reputation as a medicine was slower to develop. One of the first records of medical use is from 1750 (Vogel 1970), when Jane Colden wrote in a manuscript that remained unpublished until 1963 that "Silkgrass" root was "an excellent cure for the colick, about half a spoonfull at a time. This cure was learn'd from a Canada Indian, & is called in New England Canada Root." John Bartram wrote in 1751 that for pleurisy "the root must be powdered and given in a spoonful of rum, or rather as the Indians give it, bruise the root and boil it in water and drink the decoction. Peter Kalm saith it is excellent for hysteric passion."

Colden, Bartram, Kalm, and others learned uses from indigenous people. We now have records that pleurisy root was used by the Catawba, Cherokee, Delaware, Iroquois, Menomini, Meskwaki, Mohegan, Natchez, Navajo, Omaha, Ponca, Potawatomi, Rappahannock, and Seminoles (Smith 1933, Vogel 1970, Howard 1984, Moerman 1998). Undoubtedly, more tribes applied it.

The Catawba made a decoction of the roots to stop dysentery (Vogel 1970). The Cherokee used it as an analgesic, to stop diarrhea and bloody flux, as an expectorant, in treating heart trouble, and as a laxative (Hamel and Chiltoskey 1975). The Delaware treated pleurisy and rheumatism, and used it to aid childbirth (Moerman 1998). The Iroquois used it, probably among other purposes, for added strength in sports. Huron Smith called it "one of the most important Menomini medicines" (Vogel 1970). This tribe used it for bruises, cuts, lameness, swellings, and wounds; they also took it as a tonic. They pulverized the root and used it for cuts, wounds, and bruises, as well as mixing it with other roots for other remedies. The Natchez thought it the best remedy for pneumonia and "winter fever." They boiled the roots and took a cupful at a time (Vogel 1970). The Mohegans treated pleurisy with the dried root. The Omaha and Ponca ate the raw root to treat bronchial and pulmonary troubles (Gilmore 1919). They also chewed the root and put it on wounds and sores that were slow to heal. The plants also played an important part in the Omaha Shell Society and its rituals. The Potawatomi used *A. tuberosa* as a poison antidote (Smith 1933, Vogel 1970). The Rappahannock put the bruised leaves on snakebite (Moerman 1998). Howard (1984) was told by the Oklahoma Seminole Willie Lena that he boiled the plant and told his patients to drink the infusion; baths of it also serve as a general tonic. The Seminoles used it in 1919 to treat snakebite (von Reis and Lipp 1982).

Millspaugh (1892) found that early reports of using *A. tuberosa* as medicine were left by J.D. Schopf in 1787, Benjamin Barton in 1810, John Eberle in 1825, A. Nathaniel Chapman in 1827, and Benjamin Parker in 1841. By the time of the Civil War, Porcher (1863) gave a thorough discussion of the species. He wrote: "This plant is actively diaphoretic and expectorant, without being stimulant. . . . In large doses, it is purgative. It has been advantageously used in rheumatism, in most pectoral affections, catarrh, subacute pneumonia, and in phthisis, as a palliative. It has also been favorably employed in dysentery. . . . The powdered root has been employed as an escharotic, for restraining the growth of fungous flesh in ulcers. When the diaphoretic effect is desired the decoction of one ounce of the root to one quart of water is best, given in doses of a teacupful every two hours. Dose of powdered root, twenty grains to one drachm several times a day.

"In the neighborhood of Camden, South Carolina, the root of silk-weed (pleurisy root) is much relied on in rheumatism. The root is macerated in brandy. It is believed by many that it has a marked influence in promoting the excretion of bile, and the tincture is said by those who use it to have a laxative effect. It is used as a substitute for calomel. . . .

"The milky juice exuding from *Asclepias, Leontodon, Lactuca*, and the Euphorbiaceae yield caoutchouc. I would suppose that the queen's delight (*Stillingia*), which is abundant, would also furnish it. It might be procured from those which give a large exudation of milk when cut. I have collected and dried the juice of *Asclepias*.

"The downy substance attached to the seed of the silkweeds may be used for many purposes—for stuffing beds, cushions, etc." (See *Asclepias*: Milkweed below for fiber uses.)

Millspaugh (1892) wrote that, up to his time, "The pleurisy-root . . . has been regarded . . . as subtonic, diaphoretic, alterative, expectorant, diuretic, laxative, escharotic, carminative, anti-spasmodic, anti-pleuritic, stomachic, astringent, anti-rheumatic, anti-syphilitic. . . . It has been recommended in low typhoid states, pneumonia, catarrh, bronchitis, pleurisy, dyspepsia, indigestion, dysentery, helminthiasis, and obstinate eczemas."

Asclepias tuberosa was official in the U.S. Pharmacopoeia from 1820 to 1905 and in the National Formulary from 1916 to 1936. Roots of this and other members of the family are considered diaphoretic and expectorant, and in large doses emetic and purgative (Vogel 1970).

As with several other species, *A. tuberosa* contains asclepain, which has been used to remove warts (Bremness 1994). Milkweeds also contain resinoids, notably galitoxin, several cardiac glycosides, and alkaloids (Turner and Szczawinski 1991).

Any application of the genus is considered dangerous (Foster and Duke 1990, Turner and Szczawinski 1991, Foster and Caras 1994). Cooking apparently destroys the toxins except in highly sensitive individuals (Turner and Szczawinski 1991). In spite of the poisonous chemicals, Bown (1995) and Duke et al. (2003) note that the plants continue to be used medicinally. Duke et al. (2003) give the species a lukewarm recommendation, a single plus.

Asclepias: Milkweed

(See preceding entry for etymology)

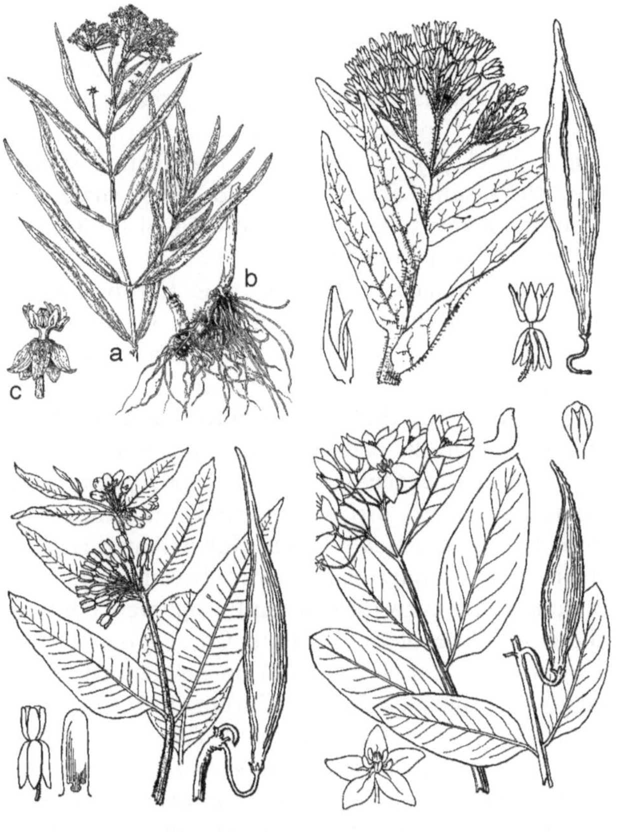

Asclepias. *Asclepias incarnata* (top left). a. Habit. b. Basal part of plant showing roots. c. Flower. *Drawn by Vivian Frazier*. From Correll and Correll 1972. *Asclepias tuberosa* (top right). From Britton and Brown 1898. *Asclepias viridiflora* (lower left). From Britton and Brown 1898. *Asclepias viridis* (lower right). From Britton and Brown 1898.

According to Greek legend, the renowned physician and deity Asklepios or Aesculapius was the son of Apollo. He was born in a forested area of Pelion (Pilion) in Thessaly and grew up there under the watchful eye of the wise centaur Chiron who taught him the art of medicine. He thus learned about the properties and uses of herbs, and became a healer. At the summit of his art he tried to extend the limits of human existence and to restore the dead to life. This made Hades angry and afraid that Asklepios would depopulate his kingdom of the dead. Therefore, Hades had Asklepios struck down with a thunderbolt. Yet, the legacy that Asklepios left behind was such that sanctuaries were built in his honor. The shrines were built on carefully chosen sites near streams, near springs, or in the shade of forests—areas of the source of his medicinal plants. There people seeking cures came to pray or to offer sacrifices. Miraculous cures are said to have taken place in these temples to Asklepios. Seven surviving accounts from the 4th century B.C. record numerous cures of the sick at Epidaurus in Peloponnese, the central of those sanctuaries.

Eventually, the disciples of Asklepios founded a temple on Cos and, with Hippocrates, their great master of popular medicine, founded the first school of medicine. Hippocrates and the others at this school continued use of the snake-staff that further symbolized the healing arts of Asklepios. That snake-staff symbol remains today as the caduceus of the medical profession (Baumann 1993).

When Linnaeus named plants in Europe and the other parts of the then-known world, he found that several were widely used in medicines. The most widely used and respected of these he named after the Greek deity, modifying the spelling into Latin *Asclepias.* Oddly enough, the one that is now taken as the focal point to define the genus (the type species) that Linnaeus called *Asclepias syriaca* (the milkweed from Syria), is actually from North America. That error in geography came about because the early botanists who knew the plant in Europe were confused about its place of origin. Both Charles de l'Escluse [Carolus Clusius] (1525–1609), in his history of rare plants and Jacques Phillipe Cornut, in his book on Canadian plants, thought the herbs originated in Syria. Clusius called them *Apocynum syriacum* and Cornut said *Apocynum majus syriacum rectum.* Such was his respect for these authors, and so bad was Linnaeus's geographic ability, that he called them *Ascelpias syriaca.*

Regardless, milkweeds enjoy a long reputation among all people for medicinal application. Among those chemicals in the latex are nitrogen- and sulfur-containing cardenolides with thiasoline/thiazolidine rings. These compounds include voruscharin, uscharidin, and calotropagenin. Leaves contain cardenolides, uscharidin, uscharin, calotropin, calacti, calotropagenin, and calotoxin (Morton 1981, Nellis 1997). Roots contain cardenolide aglycones azarigenin, corotoxigenin, and coroglaucigenin. The cardenolides have especially dramatic impact on the heart, and can either aid its function or stop it dead in its tracks.

Florida does not have a large number of *Asclepias,* but five have recorded uses—*A. humistrata, A. incarnata, A. tuberosa, A. viridis,* and *A. viridiflora.* Because some of these are uncommon, they are not well known. Surely, the best known is *A. tuberosa,* because they grow on roadsides and in open fields (handled separately in the previous entry). The others grow in more specialized habitats (*A. incarnata* in swamps, *A. viridis* and *A. viridiflora* in calcareous ravines or woods), or have less showy flowers.

Of several species known to have been used by prehistoric and historic North American people, only Florida's *A. incarnata* and *A. tuberosa* occur in the record as fiber sources. Gilmore (1931) simply said that *Asclepias* spp. fibers were identified in the Ozark Bluff site in Arkansas. Whitford (1941) found only four examples of *A. incarnata* in Hopewellian and rock-shelter sites as rope and fabric. Whitford (1941) identified *A. tuberosa* in 12 places "from all parts of its geographical range." The species was used by Hopewellian people to weave textiles in Ohio. Bluff-Dwellers in Arkansas left behind a bundle of knotted fibers. Historic people in Massachusetts used the fiber in woven textiles about the time of European contact, and the Iroquois used it to make wampum belts.

Asclepias humistrata was in use by people in the Panhandle when Murphee (1965) talked with them. They took the "milk" from milkweed and applied it to warts to remove them.

Asclepias incarnata has no recorded uses in Florida, but it was used by the Iroquois (New York), Lakota (Dakotas), Menomini (Michigan), Meskwaki (Iowa), and Ojibwa (Canada and USA) (Moerman 1998). The Lakota called it *wahinheya ipi'ye'* (medicine used to doctor gopher or swollen glands). Since the species ranges from Quebec to Manitoba and south to Florida and Texas and New Mexico, other people surely also used the plants. We have records of these tribes using the plants in an infusion of roots as a strengthening bath for children and adults, a cold infusion of roots applied to heal a baby's navel, a diuretic decoction (or to treat for too much urine), to treat lame backs, to expel tapeworms, and as a carminative and a laxative. Dried stems, with their strong fibers, were made into cords for tooth extraction. Moreover, flower heads were added fresh to soup or cornmeal mush or dried for winter use (Yanovsky

1936). In addition to normal strengthening, a decoction was taken to increase one's strength to fight witches.

There is little information about use of two other species. *Asclepias viridis* barely reaches Florida and is mostly found from Tennessee to eastern Texas and Nebraska. In Oklahoma (Howard 1984), the Creeks call it *hvtke kafkv* [*hv'tke-kafkv, hutki kathka*] (*hvtke,* white, *kafkv,* milk-exuding) and *tuji h ʃlʃswa* [*tuce heléswv*] (kidney medicine). There was apparently a generic term for all milkweeds—*nochi'* (Chickasaw), *nuchi* (Choctaw), and *noce* [*nóce*] (Muskogee). In English, it is antelope horn (Texas).

Creeks boiled roots to make a tea for kidney problems. Kidney stones were believed to be caused by rabbits; therefore, this is a "Rabbit Medicine." Because they brought this belief of rabbit medicine with them from the southeastern United States, they undoubtedly also took the remedy to Oklahoma (Howard 1984). Florida Seminoles retain the "Rabbit Sickness," but use other plants (Sturtevant 1955).

Although *A. viridiflora* grows from Florida to Texas and New Mexico and north to Ontario to Manitoba, the Creeks also use it. They call it *hutki kafka mabijadi* (*hvtke,* white, *kafkv,* milk-exuding, *mvbe-cate,* red-one), and use it to reduce soreness in an inflamed navel. Plants are boiled in water, and then a towel is soaked in liquid and used to apply the medicine (Howard 1984).

As with other plants, humans are the newcomers in using these herbs. Long ago they were discovered by insects, and evolved a close relationship with butterflies. Not only do many species of Lepidoptera sip nectar from the flowers, but the larvae of the genus *Danaus* feed exclusively on their foliage. That genus includes monarchs (*D. plexippus*), queens (*D. gilippus*), and soldiers (*D. eresimus*). Adult females seek out the members of the genus *Asclepias* and lay their eggs exclusively on them. When the larvae hatch, they feed on the plants and store the poisonous chemicals from their food. When these caterpillars finally hatch into adults, they remain equipped with those same poisonous compounds. Any bird or other animal that is foolish enough to catch and attempt to eat these colorful butterflies quickly learns that they taste horrible. Experiments with uninitiated birds have shown that only a single contact with these foul-tasting insects is enough to teach their potential predators to leave them alone.

There is a totally unrelated butterfly that looks so much like the monarchs that the two are distinguished with difficulty by people. That animal is the viceroy (*Limenitis archippus*). For many years, it was thought that these mimics simply copied the colors and patterns of the monarchs to avoid predators without using bad-tasting chemicals. However, recently in Florida it has been found that there are at least some distasteful viceroys also. By using the mimicry of monarchs and by having some individuals that taste bad, the species achieves added protection from would-be predators because the animals learn quickly to let them all alone (Glassberg 2001).

Asimina: Papaw, Pawpaw, or Payaya?

(From the Virginia Powhatan name *assimin*)

Asimina reticulata*. Drawn by P.N. Honychurch.*

Although I was only about 10 when I first saw fruits on pawpaw, I still remember them. On our trips to visit my grandparents in the wilds along the Tennessee River, we had to cross through a creek bottom that supported a wild profusion of plants. At one point beside the trail, there was a shaded grove of small trees scaled down to a youngster's size. Their crowns were only 10 or 15 feet above the ground. Each time I passed them, I felt that the spot was made for me. Then one trip, I found a large fruit, about 5 inches long and 2 inches wide. It was mostly green, but was turning yellow and was spotted with black. When I picked it up, the air was filled with a scent similar to overripe bananas. My parents told me the fruit was a pawpaw, and my dad recited a short verse.

> Pickin' up pawpaws, puttin' em in my pocket.
> Way down yonder in the pawpaw patch!

We took the fruit to my grandparents' home and kept it until it ripened. Then, we opened it and ate the sweet, fragrant pulp from around the large chocolate-colored seeds. Years afterward I learned that the plants were *Asimina triloba.*

More than a decade passed before I tasted the first custard-apple or cherimoya fruit (*Annona cherimola*) in Brazil. The same overpowering fragrance and a similar custard-like texture brought back memories of that first pawpaw.

Davidson (1999) wrote about "papaws" that the name was "sometimes spelled 'PAWPAW,' a corrupted name which is, confusingly, often also given to the completely different PAPAYA." If he had looked beyond the edibility of the fruits, he would have found a rich intermingling of Caribbean and eastern North American history, culture, and words.

The original "pawpaw" that Davidson was talking about is *Carica papaya.* By 1598, English-speaking people in the Caribbean were calling these plants "pawpaws" or "papaws." Then, in 1624, Capt. John Smith wrote of the "most delicate Pine-apples, Plantans, Papawes" from Bermuda. The names "papaw-bush," "papaw-thicket," and "papaw-tree" were in use by 1704 for plants being sold in Europe.

However, "pawpaw" had been adopted into English from Spanish *papaya,* which in turn was taken from the Taino *papaya,* a name Oviedo first recorded in 1535 (Coll y Toste 1972). *Papaya* is related to Carib *ababai,* which is cognate with *papaia,* the word used by the Venezuelan Caribs called the Ottomac. By the time Capt. Smith talked about them, this species had been taken around the world to India where its fruits were soon popular.

People settling in eastern North America had either found *Carica* for sale in their homelands or in the Caribbean. When they got to the temperate Americas, they found another tree with a similarly aromatic, sweet fruit. It reminded them of the "papaya," which had already become "papaw," and so that is what they called these different plants.

Although the books say that "papaw" is the preferred name for the North American *Asimina,* and "pawpaw" for *Carica,* I have found just the opposite to be true. Jamaicans, for example, use "papaw" and may not understand you if you say "pawpaw." Similarly, Adams (1971) recorded only the spelling "papaw." From Illinois to Florida all the people that I have talked with used the name "pawpaw" for *Asimina.* If you say "papaw," they think you do not know what you are talking about.

It was later that the scientific community discovered the North American species. After traveling in the Carolinas, Florida, and the Bahamas between 1722 and 1726, English naturalist Mark Catesby published a two-volume folio book containing his paintings and comments. One of the shrubs that he found in the Carolinas he called *Anona fructu lutescente levi scrotum Arietis referente* (the Annona with a smooth yellowish fruit that resembles a ram's scrotum). Indeed, the fruit he drew resembles exactly what he called it. Independently, Frenchman André Michaux thought the same, because in 1803 he called the plant *Orchidocarpum* (testicle fruit).

While these perhaps overstimulated European explorers saw one analogy, people also gave the plants other names. Kalm ([1753–1761] 1972) and his mentor agreed with Catesby's generic placement, and Linnaeus ([1753] 1957) dubbed them *Annona triloba.* In 1817 Michel Félix Dunal (1789–1856), a professor at Montpellier, France, moved the species to create *A. triloba.*

By 1760, the names "papaw" and "pawpaw" were being applied to *A. triloba.* By 1796, it was called not only papaw, but also "custard-apple" in Ohio, and later all the way to Texas. Other people call them "fetid-bush."

These temperate North American plants have edible fruits like their tropical namesakes, but not everyone likes them. Catesby said that the fruit "contains a yellow Pulp, of a sweet luscious taste." However, he added, "All parts of the Tree have a rank, if not a foetid Smell; nor is the Fruit relished but by very few." Timothy Flint, another early American, said the fruit pulp "resembles egg-custard in the consistency and appearance. It has the same creamy feeling in the mouth and unites the taste of eggs, cream, sugar, and spice. It is the natural custard too luscious for the relish of most people" (Fernald et al. 1958). Charles Sprague Sargent (1905) wrote: "The sweet and luscious wholesome fruit is sold in large quantities in the cities and towns in those parts of the country where the tree grows naturally." More recently, the taste of the fruit has been described as "cloyingly sweet" and "nauseating" (Vines 1977). The fruit, which usually falls green, must be stored until ripe before it can be used. People sometimes eat them raw with cream as breakfast food. Others use the fruits baked or as a filling for pies, or combined with eggs, cornstarch, and gelatin for a dessert.

Not only do people eat *A. triloba* fruits, but they have used them in different ways. French botanist André Michaux, who lived in the Americas for several years, wrote: "At Pittsburgh some persons have succeeded in making from it a spiritous liquor; but, notwithstanding this experiment, very feeble hopes can be entertained of cultivating the tree with profit for this purpose."

This difference of opinion is based at least in part on individual likes and dislikes, but there is some biological basis for it. Taste and texture vary a great deal among plants; some are mild, others strong. Recent study indicates marked genetic variation between populations (Huang et al. 1997, 1998, 2000). Fruits are often compared to bananas in taste and texture, although they are filled with large seeds and are much more aromatic than that tropical fruit. I find pawpaw fruits tolerable in small quantities, but they are not something that I would want often.

Animals, on the other hand, find them delectable. It is hard to find ripening fruits before some wildling has nibbled on them. Birds, gray foxes, opossums, raccoons, mice, gopher tortoises, and squirrels are among the animals recorded as feeding on *Asimina* fruits (Vines 1977).

The more northern *A. triloba* is not the only species in the genus. Florida has eight species, and some extend west to Texas. There is a fossil species known from the Eocene in London, England (Mabberley 1997); however, the genus is now restricted to eastern North America (Kral 1960).

The first record of *A. triloba* came from Virginia where its name in Powhatan was *assimin* (Siebert 1975). Strachey wrote in [1612] 1953 that these people called it *assessemin*, and translated that "a wheat plumb" (Harrington 1955). Siebert (1975) indicated that the spelling was the result of an accidental reduplication during copying from his original notes, and that *assimin* was closer to the original. The word is cognate with Fox *asimini*, Ojibwa *assimin*, and Shawnee *aʔsimi*. The Shawnee even include in their calendar a *aʔsi·mini-ki·sTwa*, "pawpaw month" or September (Siebert 1975). Farther west, the Caddo called it *nashitosh*.

The first record of the species in the Mississippi Valley was by Sieur de La Salle, but he did not give its Algonquian name (cf. *Maclura* for more biographical information). His companion Louis Joliet was the first who attempted the name in his reference to "*des assons, qui sont des petits fruits qui ne sont point en Europe*" (some *assons*, which are little fruits which do not occur at all in Europe) (Crawford 1978).

French explorer Julien Binneteau later wrote in 1699 that the Illinois called a fruit *asimines* (Crawford 1978), properly spelled *rassimina* (Drechsel 1997). Pierre François Xavier de Charlevoix also described the *acimine* that he saw on his voyage down the Mississippi River in 1721. Kalm ([1753–1761] 1972) said the French in Canada called them *acimine.* With the early French contact, it should be no surprise that it was Frenchman Michel Adanson who named the genus *Asimina* in 1763.

Three of the species are endemic to Florida, *A. obovata* (big-flower pawpaw), *A. reticulata*, and *A. tetramera* (four-petal pawpaw). Of these, *A. tetramera* has the smallest range, found in restricted parts of two counties, Martin and Palm Beach. As a result of habitat loss, the species is considered globally endangered (Coile 2000). The second most widespread species is *A. reticulata*, known as dog apple or Seminole tea in Florida. The other, *A. incana* (flag-pawpaw or gopher plant), also grows in nearby Georgia. Tea is made from *A. incana* roots to treat rheumatism (Murphee 1965).

Curiously, no record has been found of any species having been used by the Seminoles in spite of its common name "Seminole tea." However, the Seminoles and other Creeks surely used *Asimina* because of the place name Alcovy in Newton and Walton Counties, Georgia. "Alcovy" was derived from "*Ulco-fau*" in the Creek name *Ulco-fau-hatchee* found on an 1818 map. The original Creek meant "pawpaw thicket river" (*ołko* [*orkv, orko*], pawpaw, *ofv*, location, *hvcce*, stream). The Choctaw said *umbi*, and the Koasati *ombó*, while their Siouan neighbors in Louisiana said *ephu* (Ofo). The Atakapa, linguistic isolates, said *ol'* (sweet).

Indeed, there is also evidence that Algonquian, Siouan, and Iroquoian speakers throughout the east valued *A. triloba.* The Cherokee ate fruits (Hamel and Chiltoskey 1975), probably in ways similar to their Iroquoian relatives. The Iroquois tribes mashed the fruits into small cakes and dried them for future use. Those cakes were later soaked in warm water and cooked as a sauce or mixed with corn bread (Moerman 1998). Osage called the fruit *to'-zhon̲-ge.* The tribes through the southeast and central states ate the fruits (Yanovsky 1936).

No records have been found of indigenous people using fruits from other species, but Bartram ([1791] 1958) ate them from *A. pygmaea* when he was traveling in Florida in the 1760s. At Cuscowilla (near present Micanopy, Alachua County) he wrote of "the fruit of the size and form of a small cucumber, the skin or exterior surface somewhat rimose or scabrous, containing a yellow pulp of the consistence of a hard custard, and very delicious, wholesome food." Presumably, all species have edible fruits.

Few other uses have been recorded, but Sargent (1905) wrote that the fibrous inner bark, "stripped from the branches in early spring is used by fishermen of western rivers for stringing fish." One must recall that Sargent was a Bostonian, and to him everything west of New York City was "the west." Hamel and Chiltoskey (1975) later noted that it was the Cherokee, probably among numerous other tribes, who used bark to make strings and ropes.

Those were not simply isolated reports of using the fibers from the bark. The first to record indigenous use of *A. triloba* fibers at the Salts Cave seems to have been B.H. Young's book *The Prehistoric Men of Kentucky* published in 1910 (Jones 1936). Jones said that Gilmore found fiber being used at the Ozark Bluff-Dweller site in Arkansas, but only the seed is listed in his paper (Gilmore 1931). Jones (1936) identified the fibers in both textiles and cords at the Newt Kash Hollow shelter in Menifee County, Kentucky. Whitford (1941) added Bushwick Cave site in Ohio, and identified *A. triloba* fibers in historic Menomini and Potawatomi bags.

As the name Seminole tea suggests, the genus contains a variety of alkaloids and other chemicals. These compounds are known to cause rashes in some people, simply from handling the fruits. Seeds contain the alkaloid asiminine, which is emetic. Bark has the alkaloid analobine, and was once used as a medicine (Millspaugh 1892, Vines 1977). In addition, acetogenins have been found in some species (Woo et al. 1995, He et al. 1996, McLaughlin et al. 1996, Gu and Gordon 1999, Kim et al. 2000). Fruits also contain carbohydrates and lipids (Wood and Peterson 1999).

Another group of organisms learned of the poisonous, but useful, chemicals in *Asimina* long before humans. Those animals are the Lepidoptera, the butterflies. There is a single species of butterfly in Florida that has adapted to use the poisons to its advantage. That butterfly, the zebra swallowtail (*Eurytides marcellus*), uses the chemicals to repel potential predators (Minno and Minno 1999).

Glassberg (1999) recorded both *A. triloba* and *A. parviflora* as the "major foodplants" of larvae of this species, but in southern Florida they may feed exclusively on *A. reticulata.* When that was the only *Asimina* present, and in abundance, the butterfly was equally common. Sometimes there were three or perhaps four broods in warmer areas. Increasing urbanization eliminated the *Asimina* and its butterfly from the Miami area by about the 1950s (Klots 1951). They disappeared from the Boca Raton area in the early 1970s while I was living there. Ranges of the plant and the insect continue decreasing as more and more habitat is changed into parking lots, homes, and buildings. Soon Seminole tea and the zebra swallowtail will be known from southern Florida only as museum specimens.

Asplenium

(From Latin *splenion* or Greek *asplenon,* spleen; Dioscorides, fl. A.D. 40–80, considered the plants useful in treating maladies of the spleen)

dubh chasach [*dubhchasach*] (*dubh,* black, *chasach,* foot, Gaelic)
spleenwort (in use by 1578, although most people of the time preferred the name "miltwast[e]")

Asplenium erosum (gnawed away, in reference to basal lobes on leaflets) (= *Asplenium auritum,* eared)
culantrillo (little coriander, Yucatán)
eared spleenwort (Florida)

Asplenium serratum
birdnest fern; New World [wild] bird's nest [birdnest] fern
culantrillo (little coriander)
doradilla de la tierra (little golden one of the land); *doradilla de las Antillas* (little golden one of the Antilles); *doradille en scie* (saw-toothed little golden one)
feto macho do Pará (male fern [*Aspidium felix-mas*] of Pará, Brazil)
lengua de vaca (cow's tongue, Peru)
oreja de burro (donkey's ear, Oaxaca)
tooth-leaved spleen-wort

Asplenium trichomanes (hairy)
common spleenwort
doradille chevelue (hairy little golden one, Quebec)
English maidenhair (in use by 1562); maidenhair spleenwort (by ca. A.D. 1400 the name "maidenhair" was being applied to these ferns; in a publication dated between 1400 and 1450, it was called "Maydenheer or watirwourt, *capellis virginis*"; by 1837, the name "maidenhair spleenwort" appeared in books, combining names from common usage to distinguish genera)
polytrichon (many-haired; in use by 1562 in English, but based on an old Greek name)
svartburkne (*svart,* black, *burkne,* fern, Norwegian)
water-wort ("watirwourt" was in use between 1400 and 1450; used by Gerarde in 1597)

Originally this plant was used to treat the spleen, at least in Europe. One of the first records we have of spleenworts was left by Leonard Fuchs (for whom the cultivated plant *Fuschia* is named). He recorded names for these plants in his 1542 herbal—German

Hirszungen, Dutch *hertstonge*, French *langue der cerf*, Italian *hemionite*, and Spanish *lengua de ciervo*. They all translate to the same meaning as the English name he listed—"Hartis tunge" (Hart's tongue). He wrote that Pliny (A.D. 23–79) called the plant *splenion* (spleen, Latin) and Dioscorides (fl. A.D. 40–80) said *scolopendrio* (a centipede). We now call that plant *A. scolopendrium*.

The next usage of the name was in Henry Lyte's 1578 translation of herbalist Rembert Dodoens's *Cruydeboek* when he wrote of "spleenwort or miltwast." However, he and Fuchs were simply formalizing a usage of many centuries standing.

"Wort" is an English word, albeit of archaic usage and now mostly hidden in common plant names. It simply means herb or plant. Yet, the dominant name in English from Dodoens's time until the 1700s was "miltwaste" (again English, milt = spleen, waste = refuse, or refuse from the spleen). Dodoens himself said: "This herb is called ... in English ... Scaleferne, Finger ferne, and Myltewaste." Finally, those names dwindled in popularity, and spleenwort became the appellation of choice. In Italian *Asplenium* became *asplenio*, in Portuguese *asplénio*, and *asplénie* in French, little changed from the Latin source. In modern German the genus *Asplenium* became *Streifenfarn* (striped fern). *Hartzungen* now is applied to totally different plants.

Several species in the genus grow in Florida, but two with records of use are *A. erosum* and *A. serratum*. Both are considered endangered in the state (Coile 2000), but extend into the tropics where they are still frequent in a few areas. However, even those areas are being affected by human population growth, and are in danger of being destroyed.

Asplenium erosum is used along with several others for baths to overcome fevers and as a cure-all in northwestern Venezuela (Pittier 1926, Morton 1981).

However, the species with the most recorded uses is *A. serratum*. In Haiti, the leaves and rhizomes are powdered and made into a decoction or sweetened syrup that is taken to relieve problems of the liver and spleen (Liogier 1974). The same medicine is also used against diarrhea and spasms of hysteria. Some say it is used to banish nervous tics.

In Cuba, a decoction of *A. serratum* is reputed to dissolve calcifications in the liver, and is recommended for pulmonary inflammation and bronchitis (Roig 1945). People in northwestern Venezuela employ an infusion of various species, including *A. serratum* and *A. erosum*, among others, for bathing to overcome fever. Duke and Vásquez (1994) report that in the upper Amazon *A. serratum* is called *lengua de vaca* and used to treat liver diseases.

Some other American species have recorded uses. These include *A. pumilum*, from Yucatán, where it is called *culantrillo* (little coriander) or *sisil-ch'en* [*sisiltunich, zizalchen*] (Mayan, Yucatán). In North America, the Cherokee of western North Carolina and northeastern Georgia used at least two species. *Asplenium rhizophyllum* (walking fern) was used to create a compound to treat swollen breasts, and a decoction of the plant was taken as an emetic (Hamel and Chiltoskey 1975, Moerman 1998). This species, as with all those recorded, is considered astringent (Hocking 1991). Similarly, a decoction of maidenhair spleenwort (*A. trichomanes*) was taken to regulate menses, an infusion was used to treat breast diseases and "acrid humors," to relieve coughs, and for liver problems (Hamel and Chiltoskey 1975).

Aster

(Latin and Greek *aster*, a star, referring to the shape of the radiate heads)

aster (in use by 1603 in English; cognates include *aster* (French), *Aster* (German); *asters* (Norwegian); *astro* (Italian)

cod-wort (scrotum plant, in use by 1706; the Teutonic word "cod" was simultaneously applied to the scrotum, a long bag, and the outer covering of any fruit or seed ca. A.D. 1000)

daisy (Chaucer spelled the word "daysyes" ca. 1385; this is what we now call *Bellis perennis*; later the word was applied as a generic term for all members of the Asteraceae)

etoille (star-shaped, in use by 1549, French)

fall rose (Ohio)

frost-flower [frostweed] (Maine, New Brunswick, New Hampshire)

goodbye-summer (North Carolina)

inguinaria (from Latin *in Inguine*, because of use on sores of the flank, in use by 1557, Spanish)

it-brings-the-fall (translation of name used by "New York Indians," perhaps Iroquoian)

malmequer (you-love-me-badly, Portuguese)

Michaelmas daisies (this festival celebrates the Archangel Michael; originally, the name was applied to a British species; it was given to the American plants when the Gregorian calendar was introduced because it brought the feast forward by 11 days to 29 September, the time of flowering)

neòinean (strictly applied, *Bellis perennis*; otherwise generic for *Aster*, Gaelic)

sharewort [sharewurte, shareworts] ("share" is an old name for flank, as this name was used by

Henry Lyte in his 1578 translation of Dodoens's *Cruydeboek* of 1554, and by John Gerarde in his herbal of 1597)
starwort [starwoorts, steerewort] (star-herb, used by H. Lyte in his 1578 translation of Dodoens's *Cruydeboek* of 1554; later by John Gerarde in his herbal of 1597)
Sternkraut (star herb, used by Fuchs in 1542, German); *sterrecruyt* (star herb, in use by 1549, Dutch); *stjerne* (star, Norwegian)

Aster carolinianus (of Carolina)
climbing aster
wahó:ɬ íntayhî [*wahoɬe entayhe, waaho(th)e entahe*] (water moccasin's bed, Mikasuki)

Aster dumosus (bushy)
[bushy, rice-button]-aster
frost-flower
starwort

Aster lanceolatus (lanceolate, or with leaves resembling lance points)
[panicled, tall white, white panicle] aster

Aster lateriflorus (with one-sided flower-clusters)
[calico, California, cascade, hairy-stemmed, side-flower, starved, white woodland] aster
devil-weed (West Virginia)
farewell-summer (West Virginia)
nail-rod
old-field sweet (West Virginia)
old-Virginia stickweed (West Virginia)
rosemary (a confusion or comparison with *Rosmarinus*, Maryland)
Tradescant's aster (translation of an synonymous name, *Aster tradescantii*)
white-devil (West Virginia)
wire-weed (West Virginia)

Aster linarifolius (with leaves resembling *Linaria*, toadflax)
[flax-leaf, flaxleaf aster, savory-leaved, stiff-leaved] aster
ionactis (violet ray or beam, Greek)
[pine, sand-paper]-starwort

Aster praealtus (very tall)
[tall, will, willow-leaf, willowleaf] aster

Linnaeus ([1753] 1957) recognized 30 species of *Aster*. Since that time, *Aster* has grown to a genus of about 250 species even with the recent removal of several species groups (Mabberley 1997, Diggs et al. 1999). Neson (1994) redefined the genus such that *Aster* is a group of ca. 180 species confined to the Old World, and he put the New World species in 13 genera. Because a consensus has not yet been reached on generic limits, many authors continue to use the old definition of *Aster* (e.g., Wunderlin 1998, Diggs et al. 1999).

John Gerarde ([1597] 1975) said that the asters were called *inguinalis*, because a poultice of the plants was used on "botches, imposthumes, and vererous bubones (which for the most part happen *in Inguine*, that is the flanke or share) doth mightily maturate and suppurate them." Culpeper (1653) agreed, including two species of *Aster*. For what Culpeper called *A. hortense* (the cultivated aster), he said: "The leaves are only used, which are commended against buboes and swellings in the groin." The sea-starwort (*A. tripolium*) he recommended as "cooling, and good for burns, scalds, and inflammations, in any part. The seed is narcotic and soporiferous, and rarely used."

Aster caroliniensis was used by the Seminoles to treat "Snake Sickness" (itchy skin) (Sturtevant 1955). Leaves of this aster along with a whole *Paspalidium* plant, and *Ludwigia* roots were boiled in water (see also *Paspalidium, Ludwigia*). Patients drank the mixture and steamed their bodies to achieve the cure.

The only records found of *A. dumosus* being used are among the Tewa of New Mexico. They ate the fruits (Moerman 1998).

Aster lanceolatus was considered a febrifuge among the Iroquois (Moerman 1998). Matilda C. Stevenson's *Ethnobotany of the Zuñi* recorded: "The entire plant is ground between stones in the fraternity chamber of the *Shu'maakwe*, at noon during the elaborate ceremony of the preparation for the fraternity of medicine in August, by the *A'wantsi'a* ('great mother of the fraternity'). . . . This medicine is in the exclusive possession of the *a'kwamosi* (director of medicine), and is used only on the faces of the personators of the *Shu'maikoli*, patron god of this fraternity" (Coffey 1993). She added that the plant was also used to treat arrow or bullet wounds. For nosebleeds, the blossoms were crushed and sprinkled on live coals, and the smoke inhaled.

The Meskwaki boiled the leaves of *A. lateriflorus* as greens. They also used the entire plant as a smoke or steam bath to cure mental problems (Hocking 1997, Moerman 1998).

The Cherokee made a tea of *A. linarifolius* to relieve fever; they also drank it to stop diarrhea (Hamel and Chiltowsky 1975). When they had pains of various kinds, they made a poultice of the roots and applied them to the area that was hurting. The Cherokee also sniffed the liquid from the roots to stop catarrh (Hamel and Chiltowsky 1975).

The Meskwaki used *A. praealtus* to revive an unconscious patient (King 1984). The Navajo used another variety to treat snakebite, as an eyewash, to stop stomachache, and to cure internal injury (Moerman 1998).

Athyrium

(From the Greek *athyros*, doorless, the growth of the sporangia only tardily forcing back the outer margin of the indusium)

Athyrium felix-femina (lady fern)
a'sawan (Ojibwa)
fougère femelle (lady fern, Quebec); lady fern
nonagonawusk (milk weed, Potawatomi)

Athyrium is a genus of about 180 species throughout the world (Kato 1993). North America has two species, the western *A. alpestris*, and *A. felix-femina*. Although the species was considered exotic by Fernald (1950), it is now thought to be circumboreal (Kato 1993).

Moerman (1998) found documentation of use among the Iroquois, Meskwaki, Ojibwa, and Potawatomi in the eastern part of North America. In the west, the ferns are also used by the Bella-Coola, Cowlitz, Hesuiat, Karok, Kwakiutl, Makah, Nitinaht, Okanagan-Coville, Quileute, Quinault, Suswap, and Thompson.

The Iroquois mixed it with aster (*Aster nova-anglae*) for intestinal fevers, with other ingredients to keep a woman's water from breaking prematurely before giving birth, and men used it to treat venereal diseases. The Meskwaki made a medicine to treat breast pains associated with childbirth. The Ojibwa used it as a diuretic, a healing powder on sores, and to induce breast milk flow in women. The Potawatomi also used it for breast problems (Smith 1933). The Potawatomi name for the fern reflects that usage (Core 1967).

Hocking (1997) considered lady fern anthelmintic, a remedy for intestinal fevers, of help in preventing premature births, and venereal disease. Foster and Duke (1990) note that an antigout potential has been identified in the related *A. mersosorum*.

Atriplex

(Ancient Latin name used by Pliny, A.D. 23–79, for the orach, the kitchen vegetable, *A. hortensis*)

armoles (Portuguese); *armuelle* (Spanish)
atraphaxi (*atra*, black, *phaxys*, braid, Greek, Fuchs 1542); *atriplice* (from Latin *atriplicem*, Italian); orach [*orache, orage, orech, orege*] (from French *arroche*; in use in English by ca. 1430); *arasche* (Anglo-French by ca. 1265; this gave rise to *arche* in English, cf. Turner 1568); *arroche* (French dialect designating plants of salty soils, from Latin *atrapica* or *atripica*, from *atriplex*, altered from Greek, *atraphaxys*)
belle dame (pretty woman, French); *bonne dames* [*bonnedame*] (good woman, French, 1549)
chamiso (New Mexico)
melde (Dutch from 1549); *Melde* (German; formerly *Milten* and *Molten*, cf. Fuchs 1542 and Turner [1548] 1965); *melde* (used for family, Norwegian); *melden* (Dutch, cf. Harriot [1590] 1972)
praiseach-mhin chladaich (smooth broth of the shore, Gaelic)
salt-bush (an allusion to the plants growing in areas high in salt, whether near the ocean or on inland salines; in use by 1863)
shad-scale (for shad; see *Amelanchier*; presumably, the enlarged bract-like sepals reminded people of the scales on shad, *Alosa sapidissima*)

Atriplex pentandra (with five anthers)
crested salt-bush [crested saltbush]
sea-beach orach [seabeach orach]

The European orach, *Atriplex hortensis*, grows all around the Mediterranean coast, and as far east as Siberia. These herbs, which reach 7 feet (2.13 m) with arrowhead-shaped leaves, were cultivated as a green vegetable as we now use spinach (*Spinacia oleracea*) from the time the Mediterranean civilizations began (Davidson 1999). The plants were mentioned by Dioscorides (fl. A.D. 40–80), Galen (A.D. ca. 129–ca. 199), Pliny (A.D. 23–79), and Simeon Seth (fl. A.D. 1070–1080). Orach was used in England in the 1500s, but became less popular as the availability of spinach increased.

Europeans certainly would have recognized the several species of *Atriplex* that grow in the eastern United States as being like the Old World orach (but see *Chenopodium* for different taxonomic views). Fernald et al. (1958) comment on two species eaten in New England and eastern Canada. However, no records have been found of eastern tribes eating *Atriplex*. All of the records seem to be of western tribes. Moerman (1998) listed 18 species that were eaten by people from the Gosiute of the Great Basin west to California.

Those same tribes used several species for medicine and dyes (Curtin 1947, Moerman 1998).

Aureolaria

(From diminutive of Latin *aurea, aureola*, golden, probably with Greek *aria*, oak; maybe Rafinesque was alluding to the color of the corollas, and the fact that the plants were hemiparasitic on oak roots)

false foxglove (false to distinguish this from the true "foxglove," *Digitalis*; foxglove is derived from Old English *foxes glofa*, first recorded ca. A. D. 1000; related to *revbjelde*, fox bell; that the tubular corollas are similar to thimbles or fingerstalls is obvious; unexplained is how they became associated with foxes)
Klapperkopf (rattle head, German)
oak-leech [oakleech] (the name alludes to the tapping of *Quercus* roots by *Aureolaria* to obtain liquids and dissolved nutrients)

Aureolaria flava (yellow, the flowers)
[downy, smooth, wild, yellow] false foxglove
golden oak (a literal translation of the genus; or was this where Rafinesque got the genus?)

Aureolaria pedicularia (resembling *Pedicularis*)
fernleaf yellow false foxglove; yellow foxglove
fever weed (alluding to the use in lowering fevers)
lousewort (because it resembles *Pedicularis*)

Aureolaria virginica (of Virginia)
downy yellow false foxglove

Fernald (1950) included this genus with *Gerardia*, perhaps at least partly because Rafinesque named it. Now it is thought that there are ten species of *Aureolaria*, with all growing in the eastern United States, and one in Mexico (Mabberley 1997). Most of them are hemiparasitic on the Fagaceae, although they also grow on Ericaceae.

Probably the first record of this plant being associated with people was given by Gilmore (1931). He identified some species in the Ozark Bluff-Dweller site in Arkansas and wrote: "It is not known whether it had any use." These prehistoric people probably used the genus as did historic tribes.

The Cherokee used *A. flava* and *A. pedicularia* to halt dysentery and in treating stroke (apoplexy) (Hamel and Chiltoskey 1975). The Chickasaw considered *A. pedicularia* emetic but also used it as an antiscorbutic (Campbell 1951). The Cherokee also took a decoction of *A. virginiana* to treat dysentery (Taylor 1940).

Porcher (1863) wrote of *A. flava*: "This plant, it is said, will prevent the attacks of yellow and other flies upon horses; probably owing to its great viscidity. ... It is pubescent and highly viscous. It has very little taste, unless chewed for some time. Upon a subsequent examination (1862) ... I find that each hair with which the plant is covered secretes from a gland at its summit a tenacious, gummy substance, to which insects may adhere. Under the microscope it is an interesting object."

Subsequent practitioners considered *A. flava* a stimulant, diaphoretic, and sedative (Hocking 1997). *Aureolaria pedicularia* was considered antiscorbutic, diaphoretic, sedative, and emetic, and was used to treat fever (Hocking 1997).

Aureolaria virginiana was considered an antidote against venomous snakes (Hocking 1997). However, the first record of using this genus against poisonous snakes probably came from Stephen W. Williams in 1849, when he listed *A. flava* (under the synonym *Gerardia quercifolia*) as a Sioux remedy (Vogel 1970). Because Hocking (1997) did not give his sources, the use of *A. virginiana* remains uncertain.

Members of *Aureolaria* are said to be poisonous (Diggs et al. 1999). Presumably, that comes from their iridoid (aucubin) compounds. Those chemicals are sequestered by the Baltimore checkerspot (*Euphydryas phaeton*) butterflies (Mabberley 1997). Since these orange and black butterflies also feed on *Chelone glabra*, that is probably correct because they, too, are loaded with toxic chemicals (Glassberg 1999).

Avicennia: Black Mangrove

(Named for Iranian philosopher and physician Abu Ali Al-Husayn Ibn 'Abd Allah Ibn Sina, 980–1037, alias Avicenna)

Throughout the world, trees growing near the ocean margin in salty water are called mangroves. Regardless of species and language spoken where these trees grow, people have assigned colors to different kinds. The colors are not always the same among different language groups, but they use colors as modifiers to distinguish among the species.

The origin of "mangrove" in English is obscure. Some attribute it to Spanish *mangle* (thicket), first applied by Gonzalo Fernández Oviedo in 1535. Cognates, appearing somewhat later in the 16th century, occur in Portuguese as *mangue*, and in French as *manglé*. Yet others attribute all of these to the now extinct Malay term *mangi-mangi* denoting these

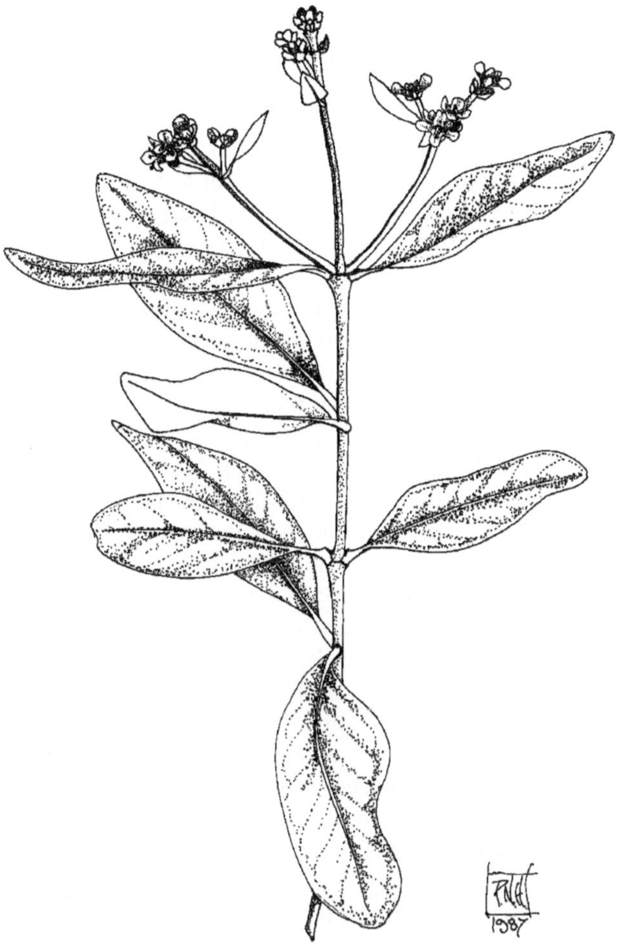

Avicennia germinans. *Drawn by P.N. Honychurch.*

plants. Regardless, the mangle (derived from French by ca. 1361) of branches on many of these plants caught the imagination of people who saw them. Perhaps they were just appalled at the prospect of climbing through the jungle gym of branches and roots to get ashore.

Black mangrove (*Avicennia germinans*) is the name most commonly applied to this American species. However, other species in the genus in the Old World also have the dark bark and foliage that, especially when viewed from a distance, apparently inspired this name. Among the names referring to the black color are *ahilo:clo:cî* (black tree, Mikasuki), black bush (British Antilles), black mangrove (Florida to Panama), blackwood bush (Bahamas), *itolastilásti* (*eto*, tree, *lvste*, black, Creek), *mangle negro* (black mangrove, Cuba, Puerto Rico, Panama, Sonora, Yucatán), *mangle prieto* (black mangrove, Cuba, Colombia, Venezuela), and *manglier noir* (black mangrove, Haiti, French Guiana). Some do not think they are black, but gray, as in *mangle gris* (gray mangrove, French Antilles) and *palétuvier gris* (gray mangrove, French Antilles). Even in some of the same regions, people attribute other colors to the trees. Hence, they are *mangel blanc* (white mangrove, French Antilles), *mangel blancu* (white mangrove, Dutch Antilles), *mangle blanco* (white mangrove, Veracruz, Oaxaca, Tabasco, Yucatán, Puerto Rico, Cuba), *mangue branco* (white mangrove, Brazil), *palétuvier blanc* (white mangrove, Haiti, French Guiana)—or, even *mangle amarillo* (yellow mangrove, Spanish), and Portuguese *mangue amarelo* (yellow mangrove, Brazil). The Mayo of Sonora simply say they are *ciali* (green).

No one can look at the trees long without noticing the accumulations of salt, both on the bottom of the leaves and on the rhizophores (root extensions like fingers below the trees). The salty aspect is recorded in *árbol de sal* (El Salvador), *madre de sal* (mother of salt, Chiapas), *palo de sal* (salt tree, Costa Rica, Nicaragua), *salado* (salty, Puerto Rico), salt pond tree (Dutch Antilles), and saltbush (USA).

Other people note associations with animals and the trees. Some of the associations are obvious, and others are not. For example, there is *chifle de vaca* (cow whistle or song, Puerto Rico). Who ever heard a cow sing or whistle? Perhaps this name, like others for some plants, simply indicates a nonsensical sense of humor. Other names are more easily understood across cultures. Because the trees grow on the margins of the ocean, they are in or near places where sea turtles nest, hence green turtle bough. Along coastal northwestern South America, the few remnants of mangroves remaining often have dozens of fat, lazy iguanas lying on the branches or nibbling young growing leaves and buds. There they are called *iguanero* (place for iguanas, Colombia) or *mangle iguanero* (iguana mangrove, Ecuador). I would bet that *siete cueros* (seven skins, Puerto Rico) refers to the number of animal skins that one might tan with the tannin extracted from a black mangrove. Finally, in Brazil they are called *siriuba* [*seriba, ciriuba*] (crab-tree in Tupí). There are almost always crabs of two or three species burrowing near the inland regions occupied by these trees, and this name shows that observation by inhabitants in that region.

In addition, there are names that fall into several categories. These are mostly indigenous names that have not been translated, although some interesting analogies are also included. No meanings have been found for *aili* (Cuna, Colombia), *apari'ë* [*apario, apalioe, apariyu*] (Carib, Suriname), *ishtatén* [*istatén*] (Mayan?, El Salvador), *koroda* (Arawak, Suriname), *parwa*, and *puyeque* (Sinaloa). Some of those making different observations are *algarrobo* (usually applied to mesquite, *Prosopis*, Fabaceae); this might be a mistake. Perhaps *bois de mèche* (wick tree, French

Antilles) indicates a usage that no one has recorded otherwise. The large size of old trees may be hidden in *calumate* [*columnate*] (variant of *columna*, column?, Costa Rica). In Ecuador, I have seen dugout canoes large enough for four or five people made from the trunks, and their forests are almost gone now. It is known that old trees grow to remarkable sizes. Someone from a temperate zone must have given the name *chêne* (oak, Guadeloupe). The name *courida*, also given as *cruda,* possibly comes from *couro* (skin in Portuguese, in reference to bark, Guyana). This may be another reference to tanning properties. The name *guapirá* comes from Tupí, meaning "bitter to eat." Finally, some call them limewood (Trinidad).

Before Linnaeus used this name, authors in India called them *Anacardium* (a name we now use for the cashew tree, an American native) and *Oepata* (an indigenous name on the Malabar coast of India). *Avicennia* contains about eight species, with half of them in the Old World tropics. In addition to our *A. germinans*, the American tropics also have *A. bicolor* and *A. schaueriana.*

Throughout its range, black mangrove is visited by honeybees. Several sources indicate that it is the major contributor to "mangrove honey," although Tomlinson (1986) was somewhat dubious about that. Surely, the most common use of all species in the past has been as a source of wood and fuel, especially charcoal. The wood of some species is prized, but ours tends to crack easily upon drying. Still, it has been used for posts and flooring. The bark not only is used in tanning animal skins, but also provides dyes. Black mangrove gives a black dye, and some of the Asian species yield a brown dye prized for batiks. Medicinally, *A. germinans* is considered a powerful astringent and, given the tannin content, that is surely true. That same tannin would account for the antihemorhagic and antidiarrhetic properties claimed for the medicines.

An obvious trait of the trees is the salt on them. This salt has been utilized by people throughout its range where the leaves are gathered as the source. Salt gathers on other parts of the plants, largely roots, but is less convenient to harvest. While the Seminoles had names for the trees, they claimed no use for it. However, given their nomenclature of this and white (*Laguncularia racemosa*) and red mangroves (*Rhizophora mangle*), that seems dubious.

There has long been disagreement about whether or not black mangroves should be put in the Verbenaceae, where they were initially placed, or removed into their own family, the Avicenniaceae. Tomlinson (1986) and others who have studied the group are convinced that they should be in a distinct family. Supporting that view are distinctive wood anatomy (secondary thickening), leaf anatomy, placentation, incipient vivipary, and seedling morphology. However, the TROPICOS Web site (Solomon 2003) lists the genus in the Verbenaceae. Since the Verbenaceae itself has been undergoing reorganization by the molecular genetic students, the situation, like so many in biology, remains debated and debatable. Unlike the "hard" sciences of chemistry and mathematics, biology remains open to interpretation and distinctive views. There is no single answer to many biological problems, in part because of human bias, differences in interpreting the same data, and because "Mother Nature" has solved problems in a jury-rig manner. That uncertainty is what makes the whole topic of biology an anathema to some, and a delight to others.

B

Baccharis: Saltbush

(Perhaps from *Bacchos*, the Greek god of revelry, but see discussion for alternatives)

Baccharis halimifolia. From Britton and Brown 1898.

When I first moved to Florida in 1970, I began learning about the shrub people called saltbush. As the summer finally changed to fall, I began seeing white-topped bushes everywhere. It soon became clear that these shrubs thrived best where the soil was moist but not wet. So, I would often find them in the edges of marshes, swamps, prairies, and other wetlands. As more land was cleared and drainage continued, the shrubs began appearing in more places that ceased having standing water. When I finally was able to fly over southern Florida in a small plane at low elevation in the fall, I was astounded to see the quantity of white-topped saltbushes that occupied the inner ponds on cypress domes, marshlands, wet prairies, and swamp edges. Then, when I got back to the plants on foot, I discovered the almost infinite number of seeds that each produced. It was no wonder that they were a dominant feature of the landscape. With that many seeds it would be more remarkable if they were not everywhere.

Now that I live in Arizona, I still see saltbushes everywhere. The species here are not the same as in the Caribbean wetland borders, but they are still in moist places. They come up beside sprinkler outlets near my plants, beside my barrel cacti (*Ferocactus wislizeni*), by the wall in my back yard, in desert washes and arroyos, and on periodically moist roadsides. The impression that they produce phenomenal numbers of seeds is enhanced by the plumed seeds floating in the air when I walk outside, and when I sweep my front porch to remove the drifted cottony masses.

Saltbushes belong to the genus *Baccharis*. This genus originally was proposed by Linnaeus for a dominant species in Florida, *B. halimifolia*. There are two opinions about the source Linnaeus used for these plants, which is exactly what he delighted in—multiple meanings. The standard answer is that the genus came from *Bacchos*, the Greek god of revelry. This is the interpretation given by botanists Fernald (1950), Munz (1973), and Correll and Correll (1982). Yet, what the Greek god of wine has to do with a fragrant American plant where wine was unknown before Europeans arrived is never addressed.

Another view has recently been published. This interpretation is offered by physician-historian Umberto Cuattrocci (1998). He says the name comes from "Greek *bakkaris*, an unguent made from 'asaron'; *bakcharis*, an ancient Greek name used by Dioscorides (fl. A.D. 40–80) for sowbread," which is *Cyclamen hederaefolium* (Primulaceae). Everyone knew that the name was applied to some other plant before Linnaeus, but Cuattrocci does not even mention a connection with *Bacchos*. Perhaps they are both correct interpretations, because the original *bakkaris* was a plant with a fragrant root that was used to add another flavor to wine. *Bacchos* (and Linnaeus) would have liked that.

There is little to connect saltbush with wine in North America, but the common names point to formerly widespread use as a medicine. In English *B. halimifolia* is called consumption weed (used to treat consumption or tuberculosis). The second most common name is groundsel bush or groundsel tree. That name is more complicated, but still may reflect medical usage because the etymology of the term "groundsel" is unclear. The second element seems to be based on Old English *swelgan*, to absorb, swallow. It appears that the earliest reference to groundsel was in the 10th century as *grundaeswyligiae*, which apparently meant "pus-swallower." This is thought to be an allusion to the medicinal uses in Europe of the related genus *Senecio*. That genus contains numerous alkaloids and continues to be used as an herbal medicine today. Groundsel bush then means a bush that might be used medicinally like the groundsel. Similarly, sea sage, silver sage, and silverling (sometimes applied only to *B. glomerulifora*, other times also to *B. halimifolia*) suggest medical plants.

Baccharis is also lowbush merkle, sea muckle, and sea myrtle. Those are names based on "myrtle," the common name for *Myrtus communis* (Myrtaceae) a common medicinal shrub in southern Europe. Use of myrtle as a medicine goes back in English literature to a handbook for surgeons written about 1400, and even older in Latin and Greek. So, "lowbush" and "sea" are compound names that distinguished these New World plants from familiar Old World medicines.

There also are indigenous names for these plants. In Florida, the Miccosukee called them *í:fâyhô:mî* [*í:frâyhô:mî*] (like dog shot, or *ifilayko:mi*, like dog feces?, Mikasuki). No use has been recorded, but the fact that they even had a name for them suggests that they probably were used. In Latin America, the shrubs are called *bajaquilla* (probably an indigenous word), *chilca*[*chilec*] (indigenous name, Colombia), and *ciro* (probably wax, *cirio*, Colombia).

In the southeastern United States, root decoctions are taken for colds and stomach pain. *Baccharis halimifolia* contains baccharine, and is used in domestic medicines against respiratory diseases and for fever. Sometimes, leafy branches are boiled with cudweed (*Gnaphalium obtusifolium*) and used as medicine for colds. Branches boiled with the tops of buttonbush (*Cephalanthus occidentalis*) also are given for colds and whooping cough (Morton 1974). Several species contain glycosidal saponin and are poisonous to farm animals (Dayton 1937). Lewis and Elvin-Lewis (1977) note that almost all members of the Asteraceae contain sesquiterpene lactones, and many of these are cytotoxic. Among those known to contain these lactones is *B. pteronioides* (*yerba-de-pasmo*). It is likely that these poisons are also among the active compounds in the medicinal mixtures.

Morton (1974) noted that the species previously was used more often. Physician Porcher (1869) wrote: "This plant is of undoubted value, and of very general use in popular practice in South Carolina, as a palliative and demulcent in consumption and cough. I have seen it used with advantage, and have often heard those employing it confess a benefit from it. A strong decoction of the root may be drunk several times a day. It is slightly bitter and mucilaginous to the taste." Few people now are even aware that it was ever used as medicine.

One more recent application is by beekeepers. The genus as a whole seems to be prized by alien honeybees (*Apis mellifera*), native bees, and butterflies. Honey from these plants has been sold seasonally from at least South Carolina to Florida. One need only sit beside the plants for a few minutes to see that hundreds of individuals of dozens of species of insects visit the flowers.

Yet another use of these plants that has disappeared from the United States is bundling the branches for brooms. Before brooms were commercially available, people all over the world made their own from whatever plants were handy. Those with abundant small, stiff branches were best, and *Baccharis* is particularly good. This usage goes all the way back to the Aztecs of Mexico who called one of their species (*B. ramulosa*) *tepopotl* (*popotl*, brush, Náhuatl). Other species in Mexico are called *escoba del monte* (countryside broom) and *escobilla* (little broom). One Mexican species is even called *limpiatuna* (prickly pear cleaner). Prickly pears are commonly eaten in Mexico, and it is not unusual to be served peeled fruits, green or ripe, in restaurants for breakfast. Since these fruits have small, barbed spines on them (called glochids), they have to be removed before eating. Usually, it is a good idea to remove the spines even before trying to peel the fruits. Some people have used brushes of *Baccharis*, but many "modern" rural people now sometimes use old pieces of cloth that they drag over the fruits before attempting to remove them from the plants.

Unfortunately, common names are not like scientific names, and apply to only one kind of plant. There are actually other species of plants in Florida (and elsewhere in the Americas) called "saltbush." Mostly, that name applies to the genus *Atriplex* (Amaranthaceae), and there are two species that grow along the Florida coasts. A totally separate derivation of the name has resulted in both *Atriplex* and *Baccharis* being named "saltbush." That route involved the tendency to grow in saline or salty areas, particularly near the ocean. However, elsewhere either genus may grow in areas with internal drainage that forces the water to evaporate and leave behind dissolved salts. Therefore, in Florida one may find *B. angustifolia* growing along with *Atriplex* on the flats near the ocean. Both are saltbushes, and they are distantly related and "united" only by ecology and common names.

What seems to be an "invented" name (although it was used by John K. Small in 1933) for *B. angustifolia* is "false-willow." Newer books call it "saltwater falsewillow," presumably to distinguish it from other species. The ecology is correct, but no Florida country person will have any idea what you are talking about if you ask for that plant.

Bacopa Water-Hyssop

(From *bacopa*, a French Guiana Carib name)

***Bacopa caroliniana*.** From Institute of Food and Agricultural Sciences.

To understand some of the complexities of how water-hyssop (*Bacopa monnieri*) came to get its common name, I must begin with a digression. Unless you are thoroughly familiar with the paraphernalia of Catholic Church rituals, or conversant in Hebrew, you will not understand the contorted history that led to an American plant being called "water-hyssop," an Old World name. Even if you are completely familiar with the Catholic Church or Hebrew, you still may not know the entire story.

First, *hyssop* is not derived from Greek, Latin, or even Germanic languages, but from Hebrew. The term was used in the Bible, and the Hebrew word is *esov*. This word in the Bible is translated as "hyssop," but earlier variants show the evolution from the original to the final form. Greeks, about A.D. 1000, first adopted the word and transcribed it as *ussupos*. Beginning much later, it was spelled in Old English as *ysopo, ysope, ysop, isop*, to finally "hyssop" in modern English. There has been debate about the identity of the original *ezov*, but that has been resolved to my satisfaction by biblical scholar botanist Michael Zohary (1982). He argues that, since the plants known botanically as *Hyssopus officinalis* do not grow, or only rarely do, in the region of the Bible, then they were unlikely to have been the important species used by Jews in the ancient purification rituals. He argues convincingly that the common plant *Origanum syriacum* must have been the *ezov* of the Bible, particularly since that remains its modern Hebrew name. Both plants are mints, but Linnaeus, who named both in 1753, could hardly have realized that what he called *Hyssopus* was not common in the Holy Land and *Origanum* was. Indeed, it is well known that Linnaeus was a poor geographer. For example, Linnaeus could not seem to understand the differences in the New World among Canada, Carolina, and Florida (Austin 1993).

In biblical times and now, bundles of *ezov* are used to anoint and purify alters and other items. When the comparatively young Catholic Church came into existence, it retained the purification ritual, but changed the details. Instead of bundles of *ezov*, Catholic priests use a holy-water sprinkler to bless items and people. They call that sprinkler a "hyssop," probably without knowing that it was named for an original bundle of mints used by Jewish rabbis.

Once the word *hyssop* was established in English, it became incorporated into common names of a series of plants that reminded people of the original mint. By the early 1500s, the literati of northern Europe knew how to distinguish hyssop from other similar medicinal herbs, but the knowledge of common people never keeps pace with scholars. Or, maybe it is the reverse, with scholars always being behind. For example, physician Leonard Fuchs's great herbal of 1542 lists *hyssopus hortensis* [the hyssop of horticulture] as medicinal, and quotes Dioscorides (fl. A.D. 40–80) and Pliny (A.D. 23–79) as earlier users of this plant (Meyer et al. 1999). At the time, the common names of the plant were "gardin hyssop" in English, *Garten Ispen* in German, *ysopé* in Dutch, and *isopo* in Spanish.

Nonphysicians, however, had their own nomenclature for similar species. Among the oldest names recorded are anise hyssop (*Agastache foeniculum*), and bastard hyssop (*Teucrium pseudo-hyssopus*), both true mints (Lamiaceae). There was also giant hyssop (*Lophanthus* sp.). After a while, plants other than mints also became known as hyssops. From Miller's (1730) *Dictionary*, we also find hedge hyssop (*Gratiola officinalis*, Veronicaceae), wild hyssop (*Verbena hastata*, Verbenaceae), and water hyssop (*Bacopa monnieri*, Veronicaceae). That same water hyssop is one that now grows in Florida, where it retains that old name.

It is uncertain if water hyssop (*B. monnieri*) is native to Florida. Wunderlin (1998) thought it was, as do many others. However, the species was present in India so long that it has Sanskrit names, *brahmi* and *mandúki*. Having Sanskrit names is sometimes taken as evidence that species are native to India. Still, there are numerous cases of plants having pantropical distributions, and being in India does not rule out its also being native in the New World.

Another old American name for *B. monnieri* is herb-of-grace. That name was also applied to *Verbena* in the 1540s, and shows the pervasive importance of several of these plants. Modern names for *Bacopa* show the continued confusion between these two genera with *véronique* and *petite véronique* (French Antilles). Indeed, the name herbe-of-grace was given to rue (*Ruta graveolens*) in the 1730s. In Puerto Rico, *Bacopa* is called *yerba de culebra* (snake herb) and in Cuba it is also *verdolaga de costa* (coastal purslane).

Bacopa monnieri has also been called *graciola* (derived from Latin *gratia*, because of medicinal value). That confusion is long-standing because Linnaeus put the species in *Gratiola* in 1753. The plants also have been known botanically as *Herpestis monnieri* and *Bramia monnieri*, so botanists and others have confused them. Roig (1945) compared *Bacopa* with the European *graciola* and said it was almost as good a medicine. He attributed purgative and hydragogue properties to both, and noted also that it was used against rheumatism, as a diuretic, against mouth infections, and to treat *escorbuto* (a lack of vitamin C) and catarrh of the lungs. Either Cubans had discovered it independently or they had brought with them a knowledge of the use against rheumatism, because that was one of the oldest known uses for the plants in India. The plants are also known diuretics (Lewis and Elvin-Lewis 1977).

The herbs have a number of chemicals including brahmine and bacco-side A (Hocking 1997). These compounds have been used as nerve tonics in mental disorders, against psychosis, and for epilepsy (Hindus). In the Old World, they also were used to treat thread-worm.

It is certain that *B. caroliniana* is native. Moreover, this herb has many of the same uses as *B. monnieri.* Yet, because of several distinctive traits, these two are always distinguished from each other by common names. The blue flowers of *B. caroliniana* cause it to be called blue hyssop or blue water-hyssop. When the leaves and stems are crushed, they give off a strong lemon scent, and this resulted in its being called lemon hyssop. To the Seminoles these were *wi:katcahilíswa* (*owe*, water, *kaccv*, puma, *heleswv*, medicine, Creek), and *okikó:wayikcî* (water puma medicine, Mikasuki). They use the plants as cough medicine and as a sedative against what they termed "Turtle Sickness," trembling, shortness of breath, cough. The species ranges from Virginia south to Florida and west to eastern Texas, so it is endemic to the southeastern United States.

There are other species in the genus that have caught the attention of people near them. *Bacopa* was named by Aublet in 1775, and he took the genus from a Carib name for the French Guiana species. That species was *B. aquatica*, and the French settlers in that country called it *herbe-aux-brulures* (burn herb) and "claim that once applied, it cures burns in a short period of time." Since that time, botanists have found some 56 species in the genus, most of which are native to the Americas (Mabberley 1997).

At least to the Maya of Yucatán, Mexico, *B. monnieri* has long been an important plant. They call it *ya'ax-kach* (*ya'ax*, green, *kach*, place). They know that where *ya'ax-kach* grows, there always is abundant water. And to the Mayas, where there is water, there is life.

Baptisia

(Etienne Pierre Ventenat, 1757–1808, created the genus from Greek *bapto*, to dip in dye)

baptisie (French)
cintó okcáyya ahissí (*cintó*, snake, *okcáyya*, living being, *ahissí*, medicine, Koasati; given to *B. australis* that they used to make yellow dye)
Färberhülse (dyer's pod, German)
ke-u'-gthe hi (Osage)
poafachi (this may belong here; Byington et al. 1915 wrote, "a weed that dyes black"; Choctaw)
wild indigo

Baptisia alba (white)
false indigo; white wild indigo
yvhv em vlikv (*yvhv*, wolf, *em*, its, *vlike*, clan, Creek)

Baptisia lecontei (named for Lewis Leconte, also sometimes listed as "Le Conte" (1782–1838), a naturalist from Liberty County, Georgia, who was a brother of Capt. John Eatton Leconte who searched for the headwaters of the St. Johns River) (= *Baptisia tinctoria* misapplied)
horsefly-weed
rattleweed
[pineland] wild indigo

At least the Cherokee, Koasati, and Ojibwa used several *Baptisia* species to make a blue dye (Hamel and Chiltoskey 1975, Moerman 1998). That information on usage as a dye was apparently transmitted to European botanists after Linnaeus described *Crotalaria alba* in 1753, which is now *B. tinctoria. Baptisia* was distinguished from other legumes in the eastern United States in 1808 because it was considered a substitute for *Indigofera*, the true indigo. It is now known that this North American genus contains 17 species endemic to the eastern United States (Mabberley 1997). Because the pods of *Crotalaria* and *Baptisia* are similar, and both genera are poisonous, the mistake by Linnaeus is understandable (Lampe and McCann 1985, Foster and Caras 1994). The entire *Baptisia* plant contains numerous alkaloids, but especially cytisine (Lampe and McCann 1985). Since *Baptisia* is used in modern herbal preparations in Europe, that is where most recent poisonings have taken place (Foster and Caras 1994).

Although *B. tinctoria* does not occur in Florida, a narrow endemic, *B. lecontii*, does. The two are so similar that substitution of the Florida species for the more widespread *B. tinctoria* is likely. Not only was *B. tinctoria* used for dyeing, but it was also an important component in indigenous medicines. People from the Micmac of Nova Scotia in Canada south to the Cherokees of the Carolinas used it in medicine. Infusions were made to stop vomiting, as an emetic, purgative, toothache remedy, to clean cuts and ulcers, as a douche, and to treat rheumatism, kidney trouble, cramps, sprains, and venereal diseases (Moerman 1998).

Baptisia alba was used by the Choctaw to treat swelling, and by the Koasati against rheumatism (Taylor 1940); thus, their neighbors the Creeks surely also applied it. Farther west, the Meskwaki of Iowa used a root mixture on wounds and sores, and to treat piles, catarrh, and snakebite (Moerman 1998). They also put it into a compound infusion for dropsy. The Osage used some species for an eyewash (La Flesche 1932).

Porcher (1863) and other physicians of the southeast at the time used the plants, considering them to act "violently as an emetic, cathartic, and subastringent antiseptic." Porcher (1863) says that a "Dr. Thacher speaks highly of its efficacy as an external application to obstinate and painful ulcers." That use reflects a common theme among indigenous people.

Batis: Saltwort

(From Greek *batis*, an ancient name applied to some other seacoast plant in Greece)

If you go to Everglades National Park and drive nearly to land's end at Flamingo, there is a trail that leads southwest along the coast toward Cape Sable. They call this the "Prairie Trail." Be sure to make the walk in the wintertime, preferably just after a really cold night. Otherwise, the mosquitoes will remove your entire blood supply before you have walked a half-mile. On a cool morning, it is usually possible to walk a couple miles and back with loss of only a few pints of blood.

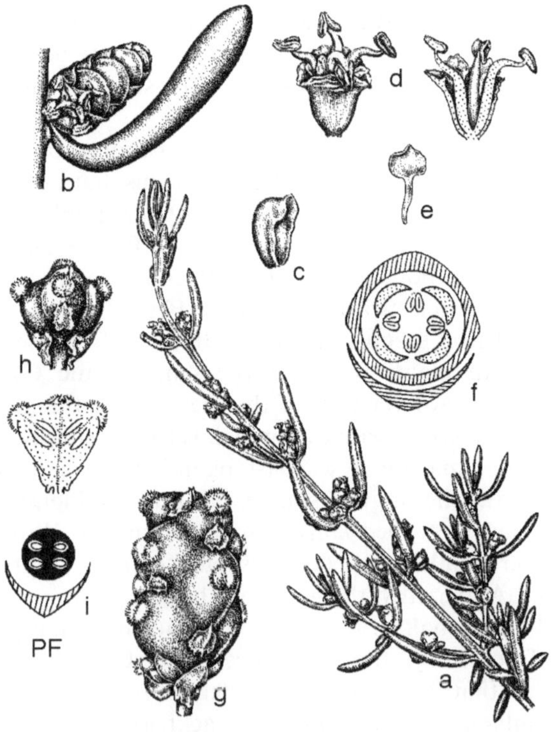

Batis maritima. a. Branch. b. Staminate inflorescence. c. Staminate flower bud. d. Complete staminate flower (left) and staminate flower dissected (right). e. Petal. f. Floral diagram of staminate flower. g. Pistillate inflorescence. h. Complete pistillate flower (above) and pistillate flower dissected (below). i. Floral diagram of pistillate flower. *Drawn by Priscilla Fawcett.* From Correll and Correll 1982.

Along the trail, there are large open "flats" that are covered with marl and little vegetation. These places led the first British surveyors mapping along the Florida coast to believe that they were old Indian gardens (DeBrahm [1775] 1971, Romans [1775] 1961). Yet, they also commented on the saline vegetation along that coast. Between the open areas are expanses of a southern Florida version of salt marsh, which is actually a tropical variant that barely reaches north to the United States. Dominant among these halophytic plants are glassworts (*Salicornia*), oxeye daisies (*Borrichia*), Christmas berry (*Lycium carolinianum*), and saltwort (*Batis maritima*). If the walk occurs during the mass migration of great southern white butterfly (*Ascia monuste*), the reward for blood loss is amply compensated. This butterfly is distinguished by its unique phosphorescent blue antennal tips, and saltwort is its favorite food. These butterflies flock to saline shores and deposit their eggs on *Batis* to assure their next generation will have ample food. This strategy has worked well for them for untold thousands of years because there have always been flat regions on tropical shores where the saltwort and the butterflies thrive.

Batis came to be recognized in Europe comparatively late. It was described by Linnaeus, but not until the tenth edition of his *Systema Naturae,* published in 1759. He took the generic name from Patrick Browne who had proposed it in 1756. Both continued use of an ancient Greek word, *batis,* that Pliny (A.D. 23–79) applied to a seacoast plant in Greece. Of course, the American plants were not the same as the Greek. There are two versions of the meaning of the genus. The usual conservative approach is to say, simply, "an ancient name for a seacoast plant." Some add that Pliny was the original source, but that the identity of the plant he called *batis* remains unknown. Without further detail, Borror (1960) wrote, "*bati, -s* . . . the ray fish." Apparently, the name has been used for a plant and an animal, so that really does not help much. To confuse things further, Correll and Correll (1982) wrote that the genus was taken "from a fancied resemblance to the fruit of a blackberry." What they did not add was that *batos* in Greek is the blackberry, and *batus* in Latin is a bramble, or blackberry. Yet, these two words, *batis* and *batos,* are not the same because Borror (1960) listed them both in distinct entries. So, let's leave it at the vague "a name used by Pliny for some coastal plant."

Whatever language they speak, people throughout the Americas have names for these plants. Since *B. maritima* ranges from North Carolina to Florida, Bermuda, the Bahamas, the West Indies, and from southern California to Peru and Venezuela, several languages are involved. Curiously, *Batis* also occurs in Hawaii, so there is yet another language.

In English *B. maritima* is called beachwort, pickleweed (USA), salt plant (Virgin Islands), sea fennel (Jamaica), and turtle-weed (Bahamas), in addition to being known as saltwort. Some even call them samphire, which is an English loan-word. Originally samphire was a French common name, but it evolved into an almost unrecognizable form as English speakers mispronounced it. The first record in French seems to have been 1545, when the name *herbe de Saint Pierre* (St. Peter's herb) appeared in Middle French, but it must be older. Even Gaelic speakers used the name *saimbhir.* In the Old World where the name originated, it was applied to both *Crithmum maritimum* (Apiaceae) and *Salicornia* (Amaranthaceae). People living in the Mediterranean and in the Caribbean heard the French name and rendered it *sampiere* or *sampere,* among other variations. Eventually, it became essentially standardized as samphire, and remains so in Jamaica and several other places. In those same places, the name "glasswort" seems to have replaced "samphire" for *Salicornia* (which see).

In Spanish-speaking areas, saltwort is known as *barilla* (Puerto Rico to Panama), *barilla de las antillas* (Antillian *barilla*), *chamís* (maybe a form of *chamiso,* usually used for *Atriplex,* Amaranthaceae, or *Chamissoa* in Amaranthaceae, Sinaloa), *lechuga de mar* (sea lettuce, Nicaragua), *planta de sal* (salt plant, Puerto Rico), *saladilla* (little salty one, Tamaulipas), *vidrillo* (little glassy one, Sinaloa), or *vidrio* (glass). Usually, the last two are applied more to *Salicornia,* but this emphasizes the confusion, or alternate classification, by people living in the range of the plants.

In French-speaking areas, the small shrubs are *crisse marine* (marine cress, Hispaniola) or *herbe-a-crabes* (crab herb, French Antilles). Cress is the English name usually given to several members of the Brassicaceae, and some of these mustard relatives grow beside *Batis.* The inhabitants of Aruba, Bonaire, and Curaçao call it wild banana, *banana di rif* [*bananiref*] ("reef banana"), or *rifbanantje* (reef banana tea). There is nothing that resembles a banana to most of us. Perhaps these people have something else in mind, as they use the same name for a *Heliconia.* Hawaiians call it *'akulikuli kai* (*'akulikuli,* succulent plant, usually applied to *Sesuvium portulacastrum, kai,* toward the sea).

People have good reason to have names for these plants. First, the leaves are edible; food was likely the first use. Duke (1972) has called them a "presalted salad." *Batis* has also been used as a potherb, a puree, and a pickle. Pickled in vinegar, they are used in Cuba as an antiscorbutic (Roig 1945). Other medicinal uses include applying a plant decoction internally and externally for syphilis, skin diseases including psoriasis and ulcers, for menstrual pains, against gonorrhea, and even asthma (Morton 1981). Some consider them aperitif and diuretic. A few people consider a daily salad of the plant a remedy for constipation, but they also take it for rheumatism and gout.

Many of the older publications point out that the plants are rich in iodine and bromine. These are important micronutrients that coastal people rarely lack, but inland groups sometimes need them, particularly the iodine, to avoid goiter. The plants have also been used as a source of soda (sodium bicarbonate). Perhaps most surprising, the Seri in Sonora, Mexico, use the crushed roots to sweeten their coffee (Felger and Moser, 1985), apparently a unique usage.

One of the most important other uses was to make soap. The plants were burned, and the ashes mixed with animal fat in the old tried-and-true method. Because of the high potash (potassium carbonate) content, the resultant soap might take off as much skin as grime and dirt, but it filled a need that preceded modern commercial products. This kind of soap was often referred to as "lye soap," and was highly alkaline.

Berchemia: Rattan Vine

(The name honors 17th- or 18th-century Dutch or French naturalist Jacob Pierre Berthoud van Berchem)

Berchemia scandens. From Britton and Brown 1897.

Perhaps the rarest woody vine in southern Florida is rattan vine, *Berchemia scandens.* However, the farther north one goes up the peninsula, the more common this temperate species becomes. The range of the genus is almost entirely within the United States, going from Virginia to Florida, and from Missouri across southeastern Oklahoma to Texas. There are herbarium specimens of the species in Chiapas and Guatemala, but little else was found about them in those regions. If they grow anywhere, they should be in Tamaulipas. So, perhaps the species is not really endemic to the United States.

There are not many common names for this species, and some of them reflect a complicated, confusing history. The dominant name for the species seems to be "rattan vine," with vine rattany, supplejack (or Alabama supplejack), and blackjack (Louisiana) less common.

As a common name, the elements of "rattan vine" date to the 1600s. The first usage of the word "rattan" in English was in a report written by a British visitor to Sri Lanka (Ceylon) in 1681. "Rattan" is another loan-word in English. This one is not borrowed from the European languages, but from East Indies Malay. The original spelling was *ratan,* and that is based on *raut,* to trip, peel, or strip. Those words applied originally to the palm genus *Calamus.* While the "true" rattan still comes from palms, it is not confined to that genus, and may be taken from several genera. The long, flexible strips of palm stem are the items originally meant by "rattan." People in the range of *Calamus* long ago learned to use the flexible wands for weaving a variety of items even though production is now mostly confined to chairs and lounges.

"Vine," of course, was originally a word specifically to mean the grape vine (*Vitis vinifera*). Western cultures then broadened the word to mean any twining, climbing, or creeping life-form (Austin 1994). Indeed, in places using older versions of English, such as Sri Lanka, "vine" still refers to only the grape vine, *Vitis.*

The name "vine rattany" also has a contorted history. It seems that the second element, rattany, was derived from a South American word, apparently Quechua of the Andes. The original was *ratânia* [or *ractania, rataña*], and referred specifically to the genus *Krameria* (which see). The confusion between *Berchemia* and *Krameria* apparently dates from the late 1700s and early 1800s when people used more of a *Krameria* extract as medicine and to adulterate port wine. Rattany, for *Krameria,* in that context first appeared in 1808. Even sea grape (*Coccoloba uvifera*) has been called false rattany; so there were perhaps several plants involved with similar uses.

"Supplejack" is a compound word that was first applied by British physician Hans Sloane in 1725 to plants in Jamaica. He complained about the stems of the plants being so thick that they barred his progress in the countryside. In its original sense, it was restricted to plants in the Sapindaceae, including *Cardiospermum* (which see) and *Paullinia.* Later, it came to be used specifically for *Berchemia,* apparently about the time that Frederick Pursh was publishing his *Flora Americana Septentrionalis* (Flora of North America) in 1814. Pursh was the first to apply the common name to *B. scandens* in the Dismal Swamp near Suffolk, Virginia. Williams ([1837] 1962) called the vines "supple jack" and wrote "walking canes of this vine are much admired."

Other than transcription of the generic name into other languages (e.g., *Berchemie,* German, and *berchémie,* French), there seem to be common names in only four indigenous languages. Creek said *istinokwana:ya* (*este,* person, *nokwv,* neck, *vyofetv,* to heal) and Mikasuki *cokaslaknî* (*cokasî,* grapevine particularly *Vitis shuttleworthii, lakni,* yellow). Koasati simply said *apihcí okcakkó* (*apihcí,* stem, *okcakkó,* green). Siouan speakers said *he'sanme'li* (*he'same,* vine, *me'li,* black, Tunica), and the Biloxi *to'hu sû'pka* (*tohu'di,* rattan, *sû'pka,* black). The Seminoles used a complex infusion of stems for chronic ailments, and in childbirth. However, even though their names are not recorded, other Muskogean people also used the plants (Taylor 1940, Speck 1941). The Choctaw used the vines for blood problems. The Houma called it *liane noire* (black vine) and used decoctions of leaves and bark for "impotency" in either males or females. Presumably, that meant being barren, although some claim that it was used by "American Indians to restore youthful vigor and sexual vitality!" (Coyote Creek 2003). A cold infusion of the burned stems was used by the Koasati as a cough medicine (Taylor 1940). Hocking (1997) indicated that the root has been used against syphilis, but gave no details. Although there seem to be no reports of chemicals in the American species, several glucosides and phenols have been found in its Japanese relative *B. racemosa* (Inoshiri et al. 1988, Inoue et al. 1990).

Other than medicines, native inhabitants of North America used the stems to fasten dugout canoes to shore and for other fastening purposes. Also, the woody stems were used to weave baskets and chair seats. At least in Mississippi the vines are also valued as important pollen sources for honey.

Berchemia was named in 1825 by A.P. de Candolle in his *Prodromus*, a flora of the then-known world. The genus contains perhaps 12 species (Mabberley 1997). These plants are found in southeastern Asia, Africa, Madagascar, and one single species in North America. Outside the Americas, there is evidence that at least one other species has been used; *B. discolor* in tropical and southern Africa has edible fruits utilized in Uganda. Rattan vine, however, should be treated with caution because it is considered mildly poisonous (Russel et al. 1997).

Rattan vine is associated with wetlands and disappears with habitat loss. That may be one of the reasons it is considered endangered in the State of Illinois (Illinois Natural History Survey 2003). Considering its apparent rarity, perhaps it should be considered endangered in Mesoamerica also. There are herbarium collections at Missouri Botanical Garden from Chiapas (by D.E. Breedlove in 1976) and Guatemala (by A.J. Sharp in 1945), but neither of the floras of these areas lists the genus.

Betula

(*Betula*, the Latin name for birch; akin to Hebrew *betula*, maiden bride, Hebrew *bat*, and Akkadian *bîntu*, girl, daughter)

beith (Gaelic)
birch (from Old English *berc* or *beorc*, from Latin *betula*); cognates include *abedul* (Spanish); *betulla* (Italian); *birck* [*berk*] (Dutch); *birk* (Danish); *Birke* (German); *bjork* (Norwegian); *björk*

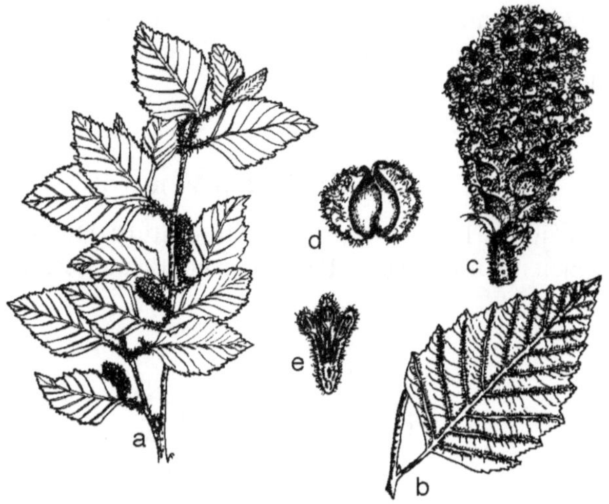

Betula nigra. a. Tip of branch. b. Leaf. c. Pistillate catkin. f. Nutlet with bract. e. Bract. *Drawn by Vivian Frazier.* From Correll and Correll 1972.

(Swedish); *bouleau* (Quebec, and Europe, cf. Turner [1548] 1965)
mîanoo's (Potawatomi)
onagúchscha (Onondaga)
semida (Greek, cf. Turner [1548] 1965)
vidoeiro (from Latin *betula*, through an archaic *vidoo* or *vido*, plus the suffix *-eiro*, tree, Portuguese)
wigwass (Ojibwa); *wínachk* [*wíhinachk*] (Delaware); *wuskwiy* (Plains Cree)

Betula nigra (black)
akcelelas'kv [*ak-cecvláskv*] (*oke*, water, *cele*, birch?, *lvste*, black, Muskogee)
[black, red, river, water] birch (from Old English *berc* or *beorc*); sugar birch (Quebec, recorded by Kalm [1753–1761] 1972)
lokapí (Koasati); *opahaksun* (Choctaw)
onàget (Onondaga)
yap koko'ha (tree breaks brittle, Catawba)

Birch trees were important to all European cultures, having names in Latin, Greek Hebrew, and Gaelic. It was *beith*, the white birch (*Betula alba*) that formed the first letter of the old Gaelic alphabet (Dwelly 1933). The dwarf birch, *B. nana*, was known as *beith bhog*, and formed the letter "P" in the Gaelic alphabet. Gaelic expressions included, "*Beith, luis, nuin*," the equivalent of "A, B, C" in English, as a euphemism for alphabet. The expression *'Sa bheith chubhraidh* means "in the fragrant birch," and the Buchanan clan had the tree as their badge. Four-time winner of the Pulitzer prize for poetry Robert Frost was a newcomer to admiration of *Betula* when he published a collection of poetry including the poem "Birches" in 1915 (Frost 1992).

The Creeks used *B. nigra* to treat pulmonary tuberculosis (Swanton 1928a). The Catawbas boiled the buds for a syrup and added sulfur to make a salve to treat ringworm and sores (Speck 1937). The Alabama boiled the bark to treat sore hooves in horses (Vogel 1970). The Cherokee chewed the leaves to stop dysentery or made an infusion of them as a remedy for dysentery, colds, and urinary problems (Hamel and Chiltoskey 1975). Farther north, the Ojibwa took a decoction for stomach pains (King 1984).

The essential oil methyl salicylate in *Betula* now is produced synthetically, although it formerly was an economic item distilled in Appalachia (Foster and Duke 1990). Although it was once widely used to alleviate rheumatism, gout, scrofula, bladder infections, neuralgia, and as an anti-inflammatory and analgesic, the oil is toxic and easily absorbed through the skin. Fatalities have been reported following its use.

Wood has been used for furniture, woodenware, wagon hubs, and fuel (Vines 1977). Sugar was obtained from the sap in the spring (Hedrick 1919). People in Scotland made a fawn dye from the bark of *Betula*, and perhaps the American species also served the same purpose (Dwelly 1933).

Bidens

(From Latin *bis*, twice, *dens*, tooth, because the fruit [cypsela, *sensu* Spjut 1994] has two "teeth" or barbs)

bident (from Latin *bidentis*, Quebec), double-tooth (a notation of the two-parted pappus resembling teeth); *Zweizahn* (two tooth, German)
brønsle (simple term, Norwegian)
bur[burr]-marigold (the proper name Mary, presumably alluding to the Virgin Mary, plus gold, their color; in use by about A.D. 1300, the name of several yellow-flowered Asteraceae, originally *Calendula*)
cuckold (*cuckoo*, the bird, from French *coucou*, plus with the pejorative suffix *-auld*. The word was in English about A.D. 1250 for the husband of an unfaithful wife; perhaps applied because the fruits were careless about where they attached themselves)
fourchette (Quebec)
harvest lice

Bidens laevis. From Institute of Food and Agricultural Sciences.

Spanish-needles (first used about 1845 for *B. bipinnata*)
sticktights

Bidens bipinnata (twice-pinnate or lobed)
cuckolds
high-brighties (southern Appalachia)
[soapbush, Spanish] needles

Bidens frondosa (leafy, referring to the outer involucre)
beggar's-lice (in use about 1880 for several plants with sticky fruits, including *Galium*, several Boraginaceae, and Asteraceae, Maine); beggars' tick(s) (Thoreau apparently wrote the word first in *Walden* in 1854); devil's beggar-ticks
boot-jacks ("boot-jack" is a forked instrument to remove boots; in use for that tool since 1841; apparently applied here because of the similarly forked fruit, Connecticut, Pennsylvania)
common bur-marigold
cow lice
cuckles [cockles] (a variant of cuckold, Massachusetts, New York)
devil's pitchfork (Massachusetts, Vermont)
old-ladies' clothes-pins (Massachusetts)
rayless marigold
stick-seed [stickseed, sticktight(s)]

takfun lvste (*takfun*, an archaic name, probably specifically for *Bidens* plants, *lvste*, black, Muskogee)

Bidens laevis (smooth)
smooth beggartick
wild-goldenglow

Bidens mitis (ripe; the original specimens had mature fruits)
ha:sâ:bî (sun replica, Mikasuki); *hasi aha:ka* (*hvse*, sun, *vhaketv*, resembling, Creek)
small-fruit beggar-ticks [smallfruit beggarticks]

My students learned about a plant known locally as "Fakahatchee Gold" soon after we began working on the vegetation of Fakahatchee Strand State Preserve in Collier County (Austin et al. 1986, 1990a,b). The plant name sounded like one for marijuana (*Cannabis*), but actually consisted of the dried leaves of *Bidens alba* used as a tobacco substitute. A smoker among them said they were not too bad in a pinch.

While Linnaeus created *Bidens* as a genus, it had been used before him by J.P. de Tournefort in his *Institutiones rei herbariae* (Institution of the King's Herbarium) of 1700. Linnaeus knew 11 species in 1753; there are now about 200 (Mabberley 1997). Several of these are widespread, but *B. alba* is pantropical. There are between 40 and 50 native species within the United States (Kartesz 1994).

Bidens bipinnata was used by the Cherokee to treat intestinal worms and the leaves were chewed to relieve sore throat (Hamel and Chiltoskey 1975). Hocking (1997) recorded that *B. frondosa* has been used as an emmenagogue, but does not state if by indigenous people or settlers. Paiute in the Great Basin used *B. laevis* for food, presumably the leaves cooked as greens (Moerman 1998).

The only record found within Florida is among the Seminoles. Both Sturtevant (1955) and Moerman (1998) recorded the plants as *B. coronata,* but Wunderlin (1998) says that plant is not in the state. Apparently, the plants used were *B. mitis.* Seminoles used an infusion for "Sun Sickness" (eye disease, headache, high fever, diarrhea), "Fire Sickness" (fever, body aches), and "Mist Sickness" (eye problems, fever, chills).

The only species that made it into most books is *B. bipinnata.* That species was discussed by several Europeans before being named by Linnaeus in 1753. Robert Morrison knew the plants by 1680 from cultivation in London; Paul Hermann had cultivated the herbs by 1687 in Leiden; and Leonard Plukenet had it by the 1690s. Linnaeus's former student Peter Kalm ([1753–1761] 1972) complained about their seeds sticking to his clothing in Pennsylvania in 1748.

In 1955, the U.S. Dispensatory reported, "The roots and seeds were popularly used as emmenagogues, and by the eclectics in laryngeal and bronchial diseases." Hocking (1997) added that the plants are used, apparently in Argentina, as ear drops, to treat conjunctivitis, and as a hemostyptic. Foster and Duke (1990) warn that the chemicals in this plant may be irritant. Duke et al. (2001) discuss only *B. pilosa* and *B. tripartita*; neither is native to Florida.

Long after my students tried "Fakahatchee Gold," I learned from my biochemist friend Kelsey Downum at Florida International University that *B. alba* contained so many potent compounds that he had never seen insects eating the leaves. After a little searching, I found that only the dainty sulphur butterfly (*Nathalis iole*) ate the leaves. Hsu (1986) and Foster and Duke (1990) say that *B. pilosa*, often confused with *B. alba*, has phytosterin B, a compound that acts as a central nervous system depressant and lowers blood-sugar level. No wonder a single butterfly can eat the leaves.

Bignonia

(With this name, Linnaeus honored Abbé Jean-Paul Bignon, 1662–1743, court librarian to King Louis XIV in Paris)

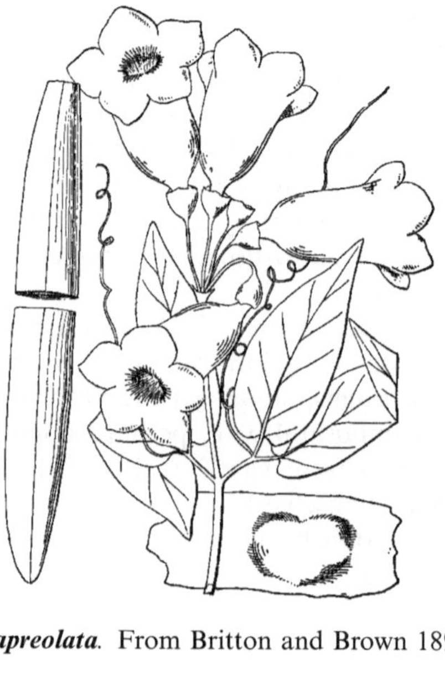

Bignonia capreolata. From Britton and Brown 1898.

Bignonia capreolata (twining)
batałó:li (Koasati)
chukfi imaseta (*chuki*, rabbit, *in*, its, *aseta*, string, Choctaw)
crossvine (alluding to the cross that is formed by the vascular bundle in a cross section of small stem)
quartervine
Trompetenblume (trumpet flower, German)

The first person to send crossvine to Europe has not been found, but it was possibly John Tradescant the Younger. He collected plants first on the James River of Virginia in 1637, and made other voyages as late as 1654. The vine appeared as an illustration published in 1674 by Italian botanist Paolo Boccone (1633–1703). Later, Linnaeus studied the climbers at the *Hortus Cliffortianus*.

Creeks used the plant as medicine, probably as did their neighbors (Swanton 1928a). The Choctaw made a decoction of the mashed bark to treat dropsy (Taylor 1940); the Houma made a root infusion to gargle for diphtheria (Speck 1941); and the Koasati used a decoction of the bark as a bath and drank it to relieve headaches and rheumatism (Taylor 1940). In the Carolinas, the Cherokee used an infusion of the leaves to purify the blood (Hamel and Chiltoskey 1975). Byington et al. (1915) said Choctaw used *chukfi imaseta* to treat dysentery, and simply that it was "the vine which grows and adheres to the bark of trees, having a trumpet-shaped blossom." Perhaps they meant *Campsis*, but Bignonia is more likely. Porcher (1863) records that, among South Carolina tribes, "a strong decoction of the bark cured cases of putrid sore throat. It is useful also in pleurisy."

Blechnum

(Pliny, A.D. 23–79, modified the Greek name *blechnon* into Latin *blechnum* for the male fern *Aspidium felix-mas*; Linnaeus reapplied it to these relatives)

Blechnum occidentale (western)
Abendlaendischer Rippenfarm (western saw-fern, German)
doradillo (golden one)
hammock fern
helecho (fern); *helecho de ladera* (slope fern, Veracruz)
New World midsorus fern (a book name, Wherry 1964)
palma boba (crazy palm)

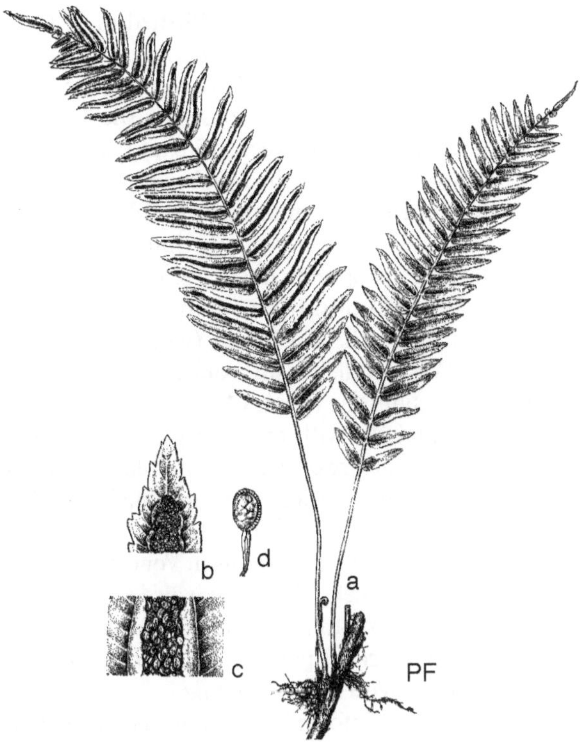

Blechnum serrulatum. a. Fertile plant. b. Tip of fertile pinna. c. Central portion of pinna to show sporangia. d. Sporangium. *Drawn by Priscilla Fawcett.* From Correll and Correll 1982.

paso de negro (black walk, Cuba)
uap-hia

Blechnum serrulatum (toothed)
asa-jike (Arawak, Suriname)
blechne (French)
Rippenfarn (saw fern, German); saw fern
swamp fern (Florida)
tapenykafa'bi (*tapeny*, fern, *kafa'bi*, maybe a mistranscription of *cvfahe*, lacking pattern or *kvfake*, scrambled, Creek); *tapintkafâ:bî* (*tapínti*, fern, Mikasuki)
toothed midsorus fern (a book name, Wherry 1964)

Linnaeus ([1753] 1957) named both *Blechnum orientale* and *B. occidentale*. Unfortunately, he mixed them up and called them by reverse names. He listed *B. orientale* as occurring in middle America, and *B. occidentale* as occurring in China. That only caused problems for a short while because he corrected his mistake in 1763.

Linnaeus named *B. occidentale* based on *Lonchitis justa nervum pulverulenta* (*Lonchitis* with dusty veins) of James Petivier's illustration, published in 1712. All Linnaeus said of origin was "*in America meridionale,*" but James Petiver (ca. 1663–1718) knew the plants from Jamaica and that is the basis of the Linnaean name. There are now between 150 and 200 species around the world included in *Blechnum*, with most of them in the Southern Hemisphere (Mabberley 1997).

In Florida, *B. occidentale* is the hammock fern or sinkhole fern. This species has a range similar to *B. serrulatum*, being known from southern Louisiana, in a few places in northern Florida (where it is endangered), the West Indies, and south to Mexico and Argentina. To date, the species has been located once in Collier County. As with most ferns, there is little in the literature about uses. Vásquez and Jácome (1997) recorded its use in treating bronchitis and coughs in Veracruz. Morton (1981) found uses only among the indigenous people of northwestern Venezuela and in Argentina. In Venezuela, the plant is used in medicinal baths and as a cure-all. People in Argentina treat nervous hysteria, constipation, bladder stones, and kidney problems with *B. occidentale.*

That paucity of information on human application of *B. occidentale* contrasts with the Asian *B. orientale* (de Winter and Amoroso 2003). In the Philippines and Papua New Guinea it is eaten. Flour is made from the rhizomes in Irian Jaya, Indonesia, and Australia. Plants are eaten after proper processing, because it also is used to induce sterility in women in Papua New Guinea. The liquid extracts from the rhizomes are also believed to stimulate the uterus. In the Philippines, plants are considered a diaphoretic, an aperitive, and are applied to cure boils and swellings. Elsewhere they are used to treat worms and bladder problems.

Given its abundance in modern Florida, it seems curious that recognition of *B. serrulatum* was comparatively late in Europe. Any wetland, particularly swamps and marshes, sometimes has almost monocultures of these ferns extending for hundreds of yards. Particularly in areas with seasonal fluctuations, little else can be found. Still, *B. serrulatum* was not described until 1792, when Louis Claude Richard (1754–1821) named it based on plants from French Guiana.

One of the most famous members of the genus is *B. indicum*, originally described in 1768 from India by Nicolaas Laurens Burmann (1734–1793). Formerly, the rhizome of that species was traditional foodstuff in northern and northeastern Australia. This species has been confused with *B. serrulatum*, and the uses of both are intermingled in the literature. As an example, Ayensu (1981) attributed medicinal uses to *B. indicum* in the West Indies, when those plants are properly *B. serrulatum*. Indeed, *B. serrulatum* grows from Florida, through the Bahamas, Cuba, Hispaniola, Puerto Rico, Mexico (Tabasco), Belize, Costa Rica, Guatemala, Honduras, Nicaragua, Bolivia, Paraguay, Peru, Ecuador, Venezuela, to southern Brazil.

Swamp fern has been used to treat wounds in the West Indies (Ayensu 1981). This usage is similar to that of many other species in the genus. However, the common names of the fern do not reveal any such uses. The only indigenous American name found for the species is *tapintkafa:bi* (*tapinti*, fern, Mikasuki), rendered *tapenykafa'bi* in Creek (Sterling in Sturtevant 1955).

The plants also are known as saw fern in English, one of the oldest names for them. That reference to the toothed pinnule margin is reflected in German *Rippenfarn* (saw fern), although the French *blechne* is a transcription of the old Greek name for the related plants. Swamp fern and saw fern both seem to have been created by people outside the academic realm, and they convey ecological and morphologically useful information. There are also some "invented" common names that seem to date from Edgar Wherry's (1964) treatment. He called them "toothed midsorus fern," certainly nothing that a non-academician would understand.

Although there is a paucity of information on the use of *B. serrulatum*, there is more documentation for other species. Some species have long histories of usage (Pizarro 1959). Three other tropical American species are used: *B. chilense,* called *palmilla* (little palm, Chile), is applied to wounds in the umbilical cord; *B. fraxineum* (tropical America) is given in an infusion as a cure-all; and *B. hastatum* (*palmilla*, Chile) is considered an emmenagogue and abortifacient, whose roots are anthelmintic.

From northern California to British Columbia, *B. spicant* (deer fern) is used as medicine, food, and fiber (Moerman 1998). The Hesquiat (British Columbia) chewed leaflets as a treatment for internal cancer. They also used the fronds in medicine for skin sores, and said they learned the practice from watching deer that rub their broken antler stumps on the plant. The Kwakiutl (northern Vancouver Island and southern British Columbia) took a decoction of root, a compound mixture, or held the decoction in the mouth to treat diarrhea. The Makah (Olympic Peninsula, Washington) chewed green leaves to relieve stomach distress and lung trouble. The Quinault (Olympic Peninsula, Washington) chewed leaves for colic. The Haisla (British Columbia), Hanaksiala (British Columbia), and the Nitinaht (Vancouver Island, British Columbia) peeled young stalks and ate the center portion as famine food. The Makah (northwest tip of Olympic Peninsula) used fronds to add flavor by placing them under items to be cooked.

The Nitinaht and Quinault (both Vancouver Island, British Columbia) placed fronds above and below food being steamed, particularly camash roots (*Camassia quamash*). The Yurok (northern California) used the leaves for bedding.

Blechnum occidentale contains flavonoid glycosides and four lignans (de Winter and Amoroso 2003). Water extracts from that species showed activity against *Bacillus subtilis, Candida utilis, Escherichia coli, Micrococcus luteus, Pseudomonas aeruginosa,* and *Staphylococcus aureus.* The rhizomes are considered antiviral, and they have strong inhibitory effects on influenza virus.

There are several known chemicals in different species of *Blechnum* that may be active in some of the remedies. Most common among the species are tannins and essential oils. However, a number of other compounds have been found in different species, including flavonoids, glycosides, and ecdysones (deoxycrustecdysone, deoxyecdysone, and alpha-ecdysone) (Chong et al. 1970, Jizba and Herout 1974, Miraglia and Padua 1985, Mesquita and Gottlieb 1985, Suksamrarn et al. 1986, Yusuf 1994).

Bletia: Pine Pink

(Named by Hipólito Ruiz López, 1754–1815, and José Antonio Pavón, 1754–1844, in honor of the Spanish botanist Luis Blet, an apothecary and owner of a botanical garden in Algeciras, Spain; this pharmacist accompanied Ruiz and Pavón on their expedition to the New World in the late 1700s, and they "thanked" him by creating this genus bearing his name)

> ***Bletia purpurea*** (purple, the flower)
> pine pink (Florida)
> *yie le* (Zapotec, Oaxaca)

In spite of its common name in English, the pine pink is not a "pink," i.e., a member of the Caryophyllaceae. Instead, it is a pink-purple orchid that grows in pinelands and upland regions near hammocks and swamps. In Florida the plants are restricted to the southernmost counties (Broward, Collier, Lee, Miami-Dade, Monroe, including the Keys, and Palm Beach), and the species is listed by the state as threatened with extinction (Coile 2000). The species is also listed as endangered in Colombia. Throughout its range, this endangerment has been brought about by drainage, changes in fire frequency, and increased human intervention in habitats. Essentially, people are now occupying the space that formerly was used by the plants. As Luer (1972) found recorded in the early 1970s, "mile after mile of virgin pineland is succumbing to the advancing tomato patches" in Miami-Dade County. Not only pineland, but other habitats also continue being developed in many parts of southern Florida, although some areas have been set aside as preserves. For example, these terrestrial orchids are still reasonably common in the Corkscrew Swamp, Fakahatchee Strand State Preserve, Big Cypress National Preserve, and Everglades National Park.

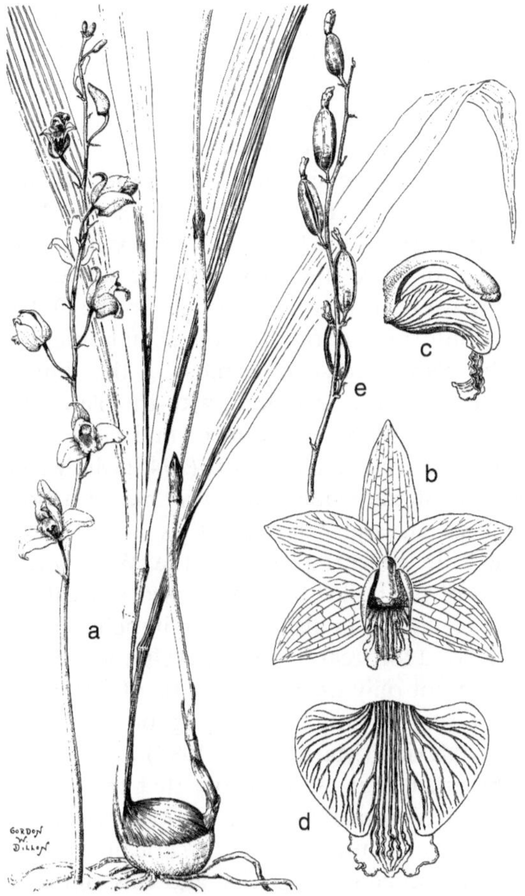

Bletia purpurea. a. Flowering plant. b. Flower, front view spread open. c. Lip and column, side view. d. Lip, spread out. e. Fruiting inflorescence. *Drawn by Gordon W. Dillon.* From Correll and Correll 1982.

John K. Small (1933) was one of the first people to study the pine pink in Florida, and he noted it as "An epiphyte form growing on the bases of cypress trunks, stumps and 'knees' in the Big Cypress Swamp . . . has apparently cleistogamous flowers with the lip nearly similar to the two lateral petals." Luer (1972) confirmed Small's observations on the orchid and noted, "The flowers of some plants do not open completely; those of others, not at all." Because this orchid is on the northern fringe of its range in southern Florida, it has "outrun" its pollinating bees. In Colombia those are Euglossine bees, with *Euglossa hemichlora* the primary visitor. Even though swallowtail butterflies (*Papilio* sp.) pollinate the orchids in Colombia, there is no evidence of them doing so in Florida.

There may have been more than mere appreciation for Blet's aid being shown by Ruiz and Pavón when they named the orchid after him. Luer (1972) noted: "Perhaps he [Blet] used the pseudobulbs in concocting potions, a practice still carried out by native people." It may have been Blet who brought to the Old World the first information on these medicinal orchids. Now that the plants are so rare, most of that knowledge seems to have been lost along with the plants. One of the few other places where medicinal use is recorded for this terrestrial orchid is Roig (1945) who wrote that a handful of the whole plant was put in a bottle of boiling water, and "then used as an effective stomachic." In several places in the West Indies, an infusion of the boiled bulbs was used to treat poisoning by ciguatera in fish. The split bulbs were considered good for treating wounds (Uphof 1968). Dried bulbs are not only used as the basis of a tonic, surely the application mentioned by Roig, but also as glue in Sonora, Mexico (Martin et al. 1998).

This medicinal use is so old that the plants have also acquired something of a mystical association in the Catholic Church in the American tropics. The orchids resemble candlesticks, and this similarity led to their Spanish common name. They are *candelaria* (candlestick), and have been incorporated into the Feast of Purification on February 2, also known as Candlemas Day. This old Jewish religious tradition dates from long before the Bible, and it is recorded in Luke 2:22–39. The celebration originated with the Jewish tradition of women being required to go into seclusion for 40 days after giving birth. According to custom, Mary and Joseph brought the baby Jesus to the temple on February 2, at the end of her seclusion, to "redeem" their son as a first-born male with an offering, probably of two turtledoves. The story goes that they were greeted by the aged Simeon who burst into song on first seeing the infant. His song, called *Nunc Dimittis*, has a long and prominent role in Christian worship. Because of a phrase in Zephaniah 1:12 that reads "I will search Jerusalem with candles" this day is also known as Candlemas, and sometimes observed with a candle-lit procession.

Blutaparon

(Perhaps derived from Latin *blitum,* historically meaning *Amaranthus*, and Greek *paron*, *para*, beside or near; Rafinesque may have been noting that the genus was near or related to *Amaranthus*)

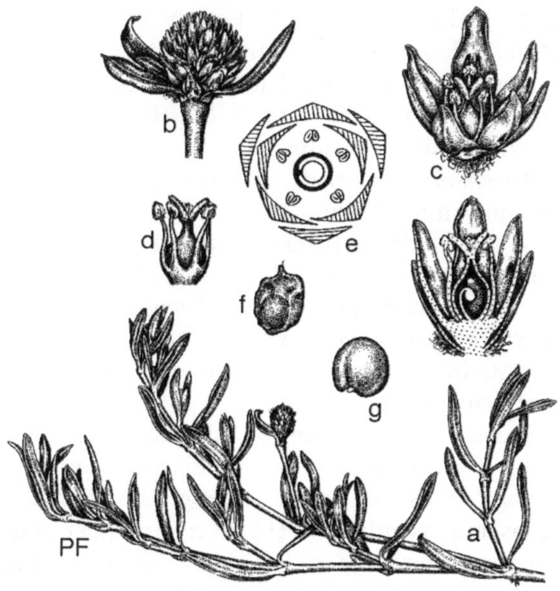

Blutaparon vermiculare. a. Branch with inflorescence. b. Flower head. c. Complete flower (above) and flower longitudinally dissected (below). d. Flower with perianth removed. e. Floral diagram. f. Fruit. g. Seed. *Drawn by Priscilla Fawcett.* Correll and Correll 1982.

Blutaparon vermiculare (resembling worms)
amaranthe bord de mer (amaranth at the sea edge, French Antilles)
beach carpet
bledo carbonero (charcoal saltwort)
hierba de sal (salt herb, Panama)
k'uk'uk [*kuk'uk*] (young part of a plant, Maya, Yucatán)
perejil de costa (coastal parsley)
samphire (applied to several medicinal and edible herbs; see also *Batis*, which is one of the other plants called this)
silverhead (of unknown origin, but it apparently refers to some morphological trait; applied by Wunderlin 1998)

The problematic biologist Rafinesque published a book in 1838 called the *New Flora and Botany of North America.* In the fourth volume of that work, he proposed the generic name *Blutaparon.* As with most of his erratic publications, the scientific establishment ignored this generic name until the 1980s. Then, James A. Mears resurrected *Blutaparon* and separated it from *Philoxerus,* into which it had previously been submerged. As now understood, *Philoxerus* is an Australasian genus that was originally described from Australia by Robert Brown in 1810 (Mears 1982a,b, Kanis 1984). As now circumscribed, *Blutaparon* contains four species in the Ryukyu Islands, the

Americas, and West Africa. One species is endemic to the Galapagos Islands (*B. rigidum*) and another to the far Asian orient, particularly the Ryukyu Islands (*B. wrightii*). The only species in the Americas is *B. vermiculare.* The species grows from the coasts of Florida and Texas, to Mexico and Central America, the Antilles, and south to Ecuador and Brazil. It also grows in west tropical Africa, and the Macronesian islands.

These plants are also like *Amaranthus* in that the stems and foliage are edible. *Blutaparon* too provides an already salted potherb or greens for meals. Some of their common names indicate that people know that trait and make use of it.

No chemical studies have been found of *B. vermiculare.* However, there have been investigations of related species. *Blutaparon portulacoides* contains an acyl steryl glycoside, flavonol (3,5,3′-trihydroxy-4′-methoxy-6,7-methylenedioxyflavone), glycosylated methylene bisflavonoid, isoflavone irisone B, and the steroids stigmasterol, sitosterol, and campesterol (Ferreira and Davis 2000, Salvador et al. 2000, De Oliveira et al. 2003). Some of these compounds have been found to inhibit *Trypanosoma cruzi, Leishmania amazonensis,* and Gram-positive and Gram-negative bacteria, along with the yeasts *Candida albicans* and *C. tropicalis* (Salvador et al. 2000).

Boehmeria

(N.J.B. von Jacquin, 1727–1817, named the genus to commemorate Georg Rudolph Boehmer, 1732–1803, a professor at Wittenberg)

Boehmeria cylindrica (cylindric, probably an allusion to the stems)
bog-hemp
[small-spiked] false nettle
mora de piedras (rock blackberry, Cuba)

Linnaeus ([1753] 1957) called this herb *Urtica cylindrica.* He based the name on information from Jamaica by Hans Sloane, and from Virginia by Gronovius. All three thought it an *Urtica.* It was not until Olaf Swartz studied the Caribbean plants that he put the species in *Boehmeria* in 1788. There are now about 80 species in *Boehmeria,* mostly in the lower latitudes (Mabberley 1997). The most famous of them is *B. nivea,* a commercial fiber plant usually known as "ramie."

Boehmeria cylindrica is found from Florida to Maine, southern Quebec, southern Ontario, west to Minnesota and Texas (Fernald 1950). It also occurs in Bermuda, Cuba, Puerto Rico, Jamaica, Trinidad, and tropical America (Leon and Alain 1946–1953, Adams 1972, Liogier and Martorell 1982).

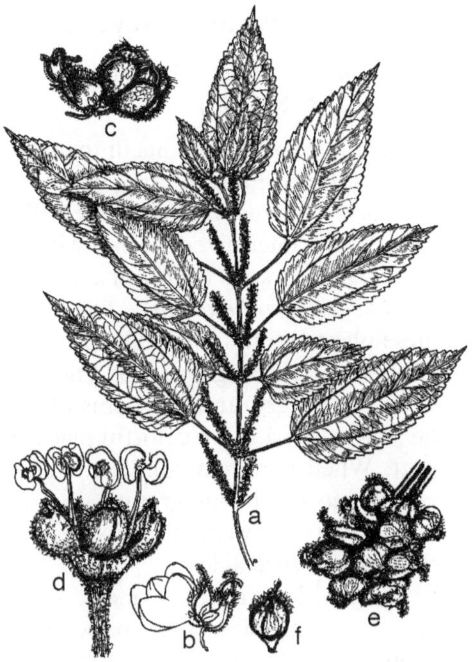

Boehmeria cylindrica. a. Top of plant. b. Staminate buds and cluster of pistillate flowers. c and e. Clusters of pistillate flowers in different stages. d. Staminate flower. f. Fruit. *Drawn by Vivian Frazier.* From Correll and Correll 1972.

Moerman (1998) did not include this species, but there are archaeological and historical data that it was important to indigenous people. Whitford (1941) had material from Bushwick Cave in Arkansas, and said that the plant was used for fiber "by practically all the tribes" covered in his survey; he gave examples from only five. The Cherokee used it to make a string for a feather charm. The Delaware used the fiber in stringing wampum, and there were two wampum belts from the Iroquois using cord from *Boehmeria.* The Ojibwa made a bowstring of it. The Micmac fastened the handle of a knife to the blade with cords of the fiber and used the plant to make a cord on one of their fish spears. In his later paper, Whitford (1943) thought that this genus was used less often than either *Laportea* or *Urtica* for fibers. Still, all were "much prized for fine work."

Given the absence of records of indigenous use, the chemical studies on the species are unusual. Beginning in the 1960s, research found chemicals in the herbs that show potential antiviral activity (Farnsworth et al. 1969, Krmpotic et al. 1972). Apparently, no work has been done on that possibility since Al-Shamma et al. (1982).

Borrichia: Sea Oxeye

(Named for Danish botanist Ole Borch, Latinized as Olaus Borrichius, 1628–1690)

Among the most colorful plants in the tropical salt marshes in southern Florida are the sea oxeyes (*Borrichia arborescens* and *B. frutescens*). These two species often grow intermixed along with saltwort (*Batis maritima*), glasswort (*Salicornia*), and Christmas berry (*Lycium carolinianum*). The polychrome green of these plants is often highlighted by the varied colors of *Borrichia* leaves (gray in *B. frutescens*), red fruits of Christmas berry, and red stems of glasswort, making the salt marsh a more colorful place than one might expect. When the great southern white butterflies (*Ascia monuste*) are migrating in mass and drinking nectar on the flowers of sea oxeye, it becomes a lively and colorful spectacle.

Many botanical keys have technical, and difficult to see, ways of telling these two species apart. Yet, there are comparatively easy ways to separate them. First, the pubescence of the leaves usually works—*B. arborescens* has green leaves (in the Florida Keys exclusively so, but plants with gray, green, and mixed gray and green leaves occur in the West Indies), and *B. frutescens* is gray with pubescence. If there are still doubts, simply pinch the base of the flowers. If your fingertips come away with small punctures, then the plant is *B. frutescens*—the involucral bracts are spine tipped. Perhaps that is a painful way to separate species, but it works better than the technical traits in some publications. Probably the people who use them as medicines do not distinguish them anyhow because their common names are not distinctive. Both are compared with either an "ox-eye" or a "daisy," or both as "sea oxeye daisy."

"Ox-eye" seems straightforward, and it is—people were reminded of the large eyes of that work animal. What is surprising is that so many things have been compared to ox eyes. Maybe because most of us have never even seen an ox up close, we think it odd that so many comparisons should have been made. Perhaps more surprising is that the name is not really applicable to species of *Borrichia* because the centers of the flowering heads are not dark, but yellow like the rays. This error is likely a result of the history of the naming of the species and its original grouping with a European genus. My friend and colleague John Semple suggests that the alternative would be to assume that some plants are named for extremely jaundiced oxen with yellow eyes. Well, maybe ...

The first comparison made in medical literature was published ca. 1400, where the plants were referred to as *Butathalmon vel butalmos, oculis bovis* (ox-eyes, spelled three different ways). In 1551, William Turner's herbal said, "Buphthalmus or oxey ... hath leaves lyke

Borrichia. *Borrichia arborescens* (left). a. Flowering branch. b. Flower head, side view. c. Ray flower on left, side view. d. Floral diagram of ray flower. e. Disk flower on left, side view; longitudinally dissected on right. f. Floral diagram of disk flower. g. Achene. *Drawn by Priscilla Fawcett*. From Correll and Correll 1982. *Borrichia frutescens* (right). a and b. Top of two different plants showing leaf variation. c. Phyllary. d. Disk corolla. e. Anthers. f. Achene. *Drawn by Vivian Frazier.* From Correll and Correll 1972.

fenel and a yellowe floure greater than Camomil, lyke unto an eye, wherupon it hath the name." Therefore, the original "ox-eye" plant was the Old World genus *Buphthalmon* (*bos,* ox, *phthalm,* eye), the cultivated ox-eye. Indeed, when Linnaeus named the New World plants in 1759, he called them *Buphthalmum frutescens.* Apparently, no one recognized *Borrichia arborescens* until they were named by Alphonse de Candolle in 1836. At the same time, he also moved Linnaeus's *Buphthalmum frutescens* into the genus Michel Adanson called *Borrichia.*

After the 1450s, many things were named for ox-eyes. Among them are sore eyes in humans, at least nine bird species, what we recognize now as seven genera of Asteraceae, two genera of fish, a drinking cup, a small glass model of an eye, and a small concave mirror. By 1866, two West Indian composites had been called ox-eyes—the creeping ox-eye or West Indian marigold (*Wedelia*), and the seaside ox-eye (*Borrichia*).

The name "daisy" is a similarly complicated story, but it is much older. "Daisy" seems to have been applied to European plants about A.D. 1000 in a list of medical plants. Part of the list consisted of "gearwe, fifleafe, daezesege, and synnfuelle." By Chaucer's time, ca. 1385, daisy was more recognizable as "daysyes." This is what we now call *Bellis perennis.* This species was a favorite of poets and others and figures prominently in the literature following Chaucer, including notables like Spenser and Shakespeare. "Daisy" has subsequently been applied to at least ten genera of composites, a sea-anemone, a term of endearment, a slang expression for something considered first-rate, and a number of other things.

Borrichia frutescens has comparatively few common names, and they all overlap with those applied to *B. arborescens*—ox-eye daisy, sea ox-eye daisy, and sea ox-eye. These are applied to both throughout English-speaking regions. In the Bahamas, they add to this seabush and samphire (see discussion of saltwort, *Batis*).

Otherwise, the plants are compared to a number of others, including bay candlewood (allusion to *Amyris balsamifera,* Caymans), bay marigold (a comparison with *Tagetes*), *clavelón de playa* (beach carnation, Puerto Rico), *romero* (rosemary, Cuba), *romero de costa* (coastal rosemary, Cuba), *salvia marina* (marine sage, Cuba), *té marino* (marine tea, Hispaniola), *té de playa* (beach tea, Hispaniola), and *verdolaga del mar* (sea purslane, Puerto Rico). All the plants compared are similarly medicinal or otherwise useful. Many of these same names are used for some *Croton* (Euphorbiaceae) species that have similar uses.

The uses include boiling the leafy branches and taking the decoction for colds and coughs. This "tea" is also used against whooping cough, back pain, and colds (Morton 1981). Juice from the leaves is given to babies with colds and coughs in the Bahamas. Decoctions are also taken for chest complaints, asthma, and malaria in Cuba (Roig 1945). Most surprisingly, an infusion of leaves is widely considered an antidote for eating poisoned fish (ciguaterra) in Barbuda (Von Reis 1973).

Properties listed for the plants include being antiscorbutic and stimulant. Maybe it is the bitter taste that is stimulant, because the leaves have a balsam-like bitter flavor, and are eaten with vinegar in salads (Roig 1945).

A third species of *Borrichia* was described in 1920 by Britton and Blake from a Cuban collection. Semple (1977, 1978) demonstrated that this was really just the hybrid between *B. arborescens* and *B. frutescens*; it is intermediate in leaf color between the two parents and has soft spines on the heads. This hybrid, *B.* ×*cubana* can be found at many locations in the Florida Keys from Miami to Key West in places where both parents grow.

Botrychium

(From diminutive of Greek *botrys,* a cluster of grapes, because of the appearance of the fertile fronds)

moonwort (translation of Medieval Latin *lunaria,* for the European fern formerly called *Lunaria minor,* now *Botyrichium lunaria*; first used in English in Lyte's 1578 translation of Dodoens's herbal, with "The herbe is now called in Latine *Lunaria* ... in English Lunarie or Moonewort"); cognates are *maankruid* (moon herb, Dutch); *Mondkraut* (moon herb, German)

Botrychium virginicum (of Virginia)
botryche de Virginie (Virginia botrychium, Quebec)
[common, Virginia] grape fern
hoeta hocksish (*hoeta,* vomit, *hocksish,* to plant, Chickasaw)
pubic weed
rattlesnake fern

Botrychium is a genus of ~50 species spread in a circumpolar distribution, and disjunct to tropical mountains (Mabberley 1997). There are records of *B. virginicum* being used among the Abnaki, Cherokee, Chickasaw, Iroquois, Ojibwa, and Potawatomi (Smith 1933, Moerman 1998).

The Abnaki used the plants as a demulcent and gave it to sick children. The Cherokee made a decoction of the roots and gave it as an emetic; they also rubbed the liquid on snakebites (Hamel and Chiltowsky 1975). The Chickasaw used the fern as an emetic as did the Cherokee, but also considered it diaphoretic and expectorant (Campbell 1951). The Iroquois treated coughs with the plant, especially those associated with tuberculosis. The Ojibwa considered the fresh root a cure for snakebite, but also thought it was repellent to the ophidians (Densmore 1928). They made a poultice they applied to cuts and used the fern to treat tuberculosis. The Potawatomi also used it medicinally (Smith 1933).

Bourreria: Strong Back

(Named for J.A. Beurer, a Nuremberg apothecary)

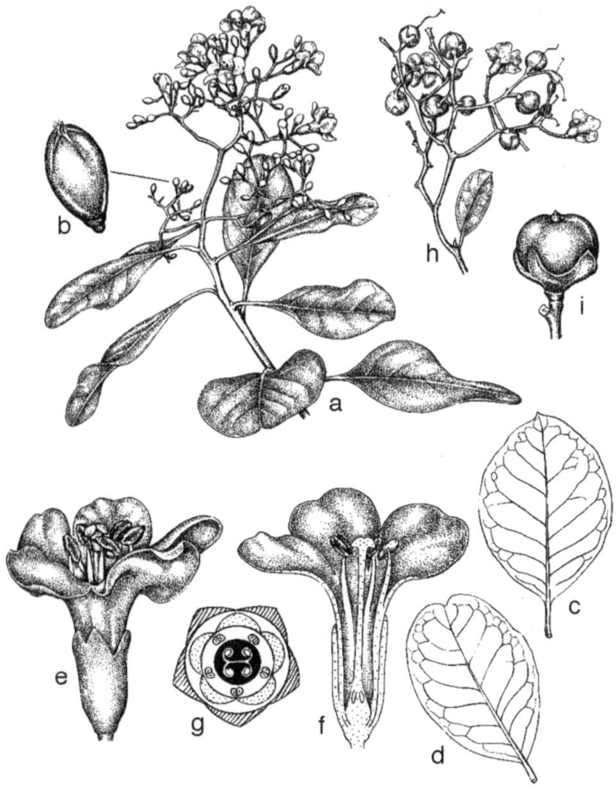

Bourreria succulenta. a. Flowering branch. b. Flower bud. c and d. Leaf outlines to show some variation. e. Flower, side view. f. Flower longitudinally dissected. g. Floral diagram. h. Fruiting and flowering branch. i. Fruit. *Drawn by Priscilla Fawcett.* From Correll and Correll 1982.

From the time I arrived in Florida in 1970 and began working at Fairchild Tropical Garden, I heard about the "strongback" of the Bahamas. Nothing was ever overtly said about it, but the innuendo spoke loud and clear—the name had sexual overtones. Somewhere along the line I learned, probably from a student, that it allowed one to make love longer, with a good, strong back. Yet, nothing of that has appeared in the books, at least the floras of Florida or the Bahamas. Helen Correll, coauthor of the most recent flora of the Bahamas, would not allow Don to include common names that were vulgar. Having known Don's earthy sense of humor, I cannot help wondering what he left out.

Botanically, strong back [strongback] is *Bourreria succulenta,* usually kept separate from *B. ovata* in the older literature. The species was named by Nicolaus Jacquin in 1760, and placed in a genus created by Patrick Browne earlier. There are at least three spellings of the generic name, perhaps because of its derivation as a patronym. However, the spelling used here is retained because of international agreement. This species ranges from southern Florida, the Bahamas, the Greater Antilles, and the Lesser Antilles south to Trinidad, then along the coastal parts of South America in northern Venezuela and the Dutch Antilles to Panama.

One of the most common remedies made from these shrubs is used to cure thrush and other oral inflammations (Roig 1945). This usage is reflected in the common name *curaboca* (cures the mouth, Cuba). In the Dutch Antilles, a leaf decoction or an infusion of the leaves in rum is considered an aphrodisiac, "for more energy," verifying my student's earlier interpretation about the Bahamas. This usage and the others for "weak back" are indicated in names such as strongbark (Bahamas), *doncella* (little lady, Puerto Rico), *Madame Jeanne* (Haiti), *palo bobo* (crazy tree, Dominican Republic), hall bark (Bahamas), and maybe *cateicito* (little tester?, Cuba) and *goeaana* (good Ann?, Dutch Antilles).

Fruits are eaten throughout the range of this and other species in the genus. This observation was noted in names including *acomat côtelette* (coastal acomat, *Homalium racemosum,* Flacourtiaceae, Guadeloupe), *ateje de costa* (coastal ateje, *Cordia collococca,* Cuba), *bois caprit bâtard* (false [bastard] goatwood, probably because of similarity of fruits to *Aegiphila martinicensis,* Guadeloupe), *bon-bon rouge* (red candy, French Antilles), *café marron* (wild coffee, Haiti), *cerecillo* (little waxy one, Puerto Rico), *mapou gris* (gray mapou, *Ceiba pentandra,* Haiti), *cotlette* (variant of *côtelette*?, Dominica), pigeon wood (Virgin Islands), and pigeon-berry (Puerto Rico). Flowers are used for nectar by honeybees, and the wood is burned as fuel. Application as fuel or other uses of the wood is probably implied in chinkswood (Antigua), and its probable variant, white-chank (Dutch Antilles), *roble*

[*de*] *guayo* (*guayo* oak, Cuba, Puerto Rico), *roble negro* (black oak, Cuba), and spoon tree (English Antilles). "Cutlass" (Grenadines) may be related as a cutting instrument needed to harvest the wood. The Caribs use the stems for torches (Hodge and Taylor 1957).

A broad array of other plants are mixed with *B. succulenta* in the Bahamas to treat various maladies (Ayensu 1981). For example, leafy twigs are boiled with five-finger (*Tabebuia heterophylla*) or fiddlewood (*Citharexylum fruticosum*) to treat pain in the back or waist. The plant is also used to quiet the nerves, and to strengthen "weak-backed" children. Here a weak back may refer to kidney problems. A young plant is boiled with goatweed (*Capraria biflora*) and ram-goat dash-along (*Turnera ulmifolia*) to treat diarrhea. Fever, headache, and stiffness of the limbs are treated by boiling a branch with lignum vitae (*Guaiacum sanctum*); that mixture also is used in bush baths to treat the same maladies. A decoction of three-finger (*Thouinia discolor*) and fiddlewood (*Citharexylum fruticosum*) is applied externally to treat running sores in infants. Steeped inner bark with mahogany (*Swietenia mahagoni*) and gumbo limbo (*Bursera simaruba*) is used to treat low blood pressure. The Caribs in Dominica use the plant alone by taking the emollient fluid from the inner bark, which is used to soothe inflamed eyes (Hodge and Taylor 1957).

Since it is compared to *Exostema* in the common name *frutica de carey* (shrub carey, *Exostema*, Cuba), it may be because of the antiseptic use of both. Probably, the names *palo de vaca* (cow tree, Puerto Rico, Haiti), *raspa lengua* (rough-tongue, Cuba), and *raspa lengua de costa* (coastal rough-tongue, Cuba), are allied because of a bad taste. No translations have been found for *taiwai* [*taway, tauai*] (from *iauái*, Carib, Dominica), and *watakeeri* [*watakeeri, watakeli*] (probably Carib and related to "wai" in *taiwai*, Aruba, Bonaire, Curaçao).

Bursera: Gumbo Limbo

(From Latin *bursa*, sacks, of aromatic oils within the plant tissues or for the German physician Joaquim Burser, or both)

Linnaeus first named the species in 1753, but he called it *Pistacia simaruba*, thinking it was related to the pistachio nut–producing genus. He had seen the specimen from Jamaica and a plate published in 1705–1725 by British physician Hans Sloane. Linnaeus also had seen the text and plate by Mark Catesby published in 1731–1732. In the interval between these two early works and his *Species Plantarum* of 1753,

Bursera simaruba. *Drawn by P.N. Honychurch.*

Linnaeus had discussed the plant in his book, *Materia Medica*, published in 1749. There he simply used the name *Simaruba*.

At the time the distinction between plants called *simaruba* (or *simarouba*, a Carib word) was not clear. Linnaeus did not know those trees now called *Simarouba* because that genus was not named until 1775 by Jean Fuseé Aublet. Still, plant material called *simaruba* had been known in Europe since about 1713, when it was brought back to Europe as medicinal bark from French Guiana. Some time between his *Materia Medica* and the *Species Plantarum* of 1753, Linnaeus decided that the specimen collected by Sloane in Jamaica and the one illustrated by Catesby must be an American relative of the European pistachio. So, he listed it along with the real pistachio (*Pistacia vera*), the mastic (*P. lentiscus,* which flavors the Greek wine *ouzo*), and some others.

By 1762, Linnaeus realized he had made a mistake in 1753, and adopted a name proposed but not published by Jacquin, when he created the genus *Bursera* in the second edition of *Species Plantarum* (1762). There are two stories about how this name originated. One has it that Jacquin noticed that there were *bursa* (sacks) of aromatic oils within the tissues of the plants and thus created *Bursera.* However, it is more likely that he and Linnaeus had in mind the German physician Joaquim Burser (Latinized Joachimus Burserus, 1583–1639), professor of botany and medicine in Denmark from 1625–1639, traveler, plant collector, and friend of Gaspar Bauhin. Linnaeus also created a new name for the species he had named in 1753, and now called it *B. gummifera.* By modern international agreement, that is not permitted, but it remained until 1890 for Charles Sprague Sargent to transfer Linnaeus's *Pistacia sumaruba* from the incorrect genus into *Bursera*, creating the name *B. simaruba* we use today.

Catesby and others of the time used an old European medical term as the common name for these plants, gum-elemí (English) and *gomme elemi* (French). That designation was in use in the Bahamas where Catesby drew the plants, and probably through most of the remaining European settlements in the Americas. However, the term was comparatively new in Europe, first noted in a Latin grammar published in Rome in 1517. In the interim it spread to Italian as *elemi*, French as *élémi*, and Portuguese as *gumílemi*. Some think that the word was brought into Europe via the Silk Road with Arabic traders as *lami*, and later had the English "gum" added. Regardless, it was a long-held tradition to treat certain ailments with plasters and ointments made of this aromatic tree gum, regardless of which species it came from. Perhaps the first gum-elemí came from *Canarium* (Philippines). Later, gum-elemí came from the Caribbean (*B. simaruba*) or Brazil (*Protium*). All three belong to the Burseraceae.

In the New World people created an array of variations on this Latin-based English name. Among the variations on the name are simply elemi, American elemi or West Indian elemí, gum elemi, gum-alamí, gummi elemi, gomali, gomalimi, gum elemy, gum tree (Bahamas, Dutch Antilles), lime tree (Dutch Antilles), and even aerial yam. Most of the designations are now archaic and have been replaced by other names, but they were widely used by different languages for many years. Now, they have been replaced by a name created by African slaves. That name is "gumbo limbo" [gumbo-limbo, gumbolimbo] (Florida to Panama), which may have been taken from *ngombo ulimbo* (slave's birdlime) (Austin 1978). Sometimes there are color modifiers such as *gumbolimbo blanco* (white *Bursera*, Belize), red gumbo limbo (Belize), or white gumbolimbo (Belize).

French speakers preferred to emphasize the gum part of the old designation, and use variations on that. They call it *gomme mombin* (gum tree, Granada), *gommier* (gum tree, Guadeloupe), *gommier barrière* (fence gum tree, alluding to its use for living fences, Guadeloupe), *gommier blanc* (white gum tree, Haiti), *gommier maudit* (wretched gum tree, St. Lucia), *gommier rouge* (red gum tree, Guadeloupe, Martinique), a mixture of English and French with peeling-bark *gommier* (Trinidad), and dryland *gommier* (Trinidad).

Judging from the number of names and the worldwide use of related plants, *B. simaruba* must be one of the most widely used plants in the American tropics. However, because it is one of the most aromatic, it is reasonable that it was used so much. The species has more than 100 common names from essentially all people within its range. *Bursera simaruba* ranges from Sonora in northwestern and Tamaulipas in northeastern Mexico, southern Florida south through the Bahamas, West Indies, the remaining tropical states in Mexico through Central America to Colombia and Venezuela and Guyana. Among the maladies treated with the plants are ache (stomach), bite (snake), calculus, cancer (stomach), cystitis, debility, diarrhea, dropsy, dysentery, enterorrhagia, fever, gangrene, hernia, impotence, intestine problems, kneecap problems, lactogogue (veterinary), nephritis, obesity, rash, renitis, rheumatism, sores, stomach problems, strain, swelling, tumors, hunger, venereal problems, wounds, and yellow fever (Austin 1978, Morton 1981). It has also been used as a depilatory, diaphoretic, diuretic, insect repellent, purgative, and vulnerary. The preparation was considered antiseptic, aphrodisiac, and expectorant.

Following the similarity to pistachio and mastic, the plants became in Spanish *almácigo* (mastic, Puerto Rico, Costa Rica, Dominican Republic, Panama, Colombia), *almácigo blanco* (white mastic), *almácigo colorado* (red mastic, Cuba), *almácigo encarnado* (flesh-colored mastic, Puerto Rico). Even in English some called them mastic trees, a name usually applied to *Sideroxylon*.

Because of the aromatic gum, the natives of the Americas had been using the trees long before the Europeans arrived. A common use was as incense. So, the plants became incense tree in English, rendered in French *bois d'encens*, and in Spanish *palo de incenso*. That use also gave rise to common names including balsam tree, turpentine tree, and torchwood. Two words for the aromatic resin were introduced into Europe from the Aztecs of Mexico. One of these was *tacamahaca* [*takahamaca, tacamahac, tamacahack, taccamahac, tacamacha, tacka mahacca, thecomahaca, tacamaque*]. This apparently came from the Náhuatl word *tecomahiyac* (*tecomatl*, clay vase + *xicaltl*, white). Nicholas Monardes in 1569 and the English translation by John Frampton of 1577 were perhaps the first to introduce this term to Europeans. The English translation says, "Thei do brying out of the new Spain an other kinde of Gumme or Rosine, that the Indians doeth call tacamahaca ... it is a Rosine taken out by incision of a tree ... and is of a verie sweete smell." The Mexicans used the gum to treat swellings of any body part, and to treat colds, flatulence, toothache, stomachache, women's problems, headache, sciatica, aching joints, and the complexion. Much later the name *tacamahaca* was confused or misapplied to the temperate *Populus balsamifera* to the confusion of many (e.g., Vogel 1970).

Oddly, neither Monardes nor Frampton seem to realize *tacamahaca* was the same as *copal*. They say, "The Copall is a Rosine verie white, and of muche brightnesse, it is brought in certaine greate peeces, which are ... verie cleare." Later, "With this Copall the Indians did make perfumes in their sacrifices, and so the use thereof was frequented in the Temples, for their priestes." Monardes goes on to describe how the incense is used against head colds and other respiratory problems. That indigenous word for incense stuck in all European languages as *copal* [*copalillo*] (*copalli*, incense, Náhuatl, Mexico). The source of the incense was the *copalcuahuitl* (incense tree, Náhuatl). Catholic priests were also quick to realize the value of this incense and adapted it to their religious services.

Each linguistic group had its own name for the trees, some referring to incense, and others to other traits. In other languages this tree was known as *suchicopal* (Tabasco), *yalaguito* (*yala*, perfume, *quito*, from *quie*, flower, and *ito*, small, Zapotec), *cachibú* (probably Carib *kasíbu*, sharp-pointed, Cuba), *caraña* (name given to Clusiaceae, Costa Rica), *carate* (synonym of *jagua* [tree], Panama), *caratero* (tree, Colombia, Panama), *kal-no* (choked or clogged up, perhaps to indicate the medicinal use, Maya, Yucatán), *limsi* (Miskito, Nicaragua), *lon-sha-la-ec* (Chontal, Oaxaca), *mara* [*maro, marero*] (from Latin *marum*, Mediterranean *Teucrium marum*), *percha* (perhaps because the sap resembles gutta percha, or a mispronunciation of *piocha*), *pom*, *quiote* (Sinaloa), *resbalamono* (Colombia, Panama), *tasun* [*taxun*] (Tepehua, Veracruz), *ta'sun* [*tasuni, tasun, tsun*] (Totonac, Veracruz, Puebla), *teuc, tsic* (Otomí, Veracruz), *tsok* (Oaxaca), and *tzaca* (Huastec, San Luis Potosí). Other variations on the medicinal theme are conveyed by variations on *jiote* [*jicote, jinote*] (from *xiuitl*, herb, Náhuatl, Sinaloa, Guatemala, El Salvador, Honduras, Costa Rica), *jiñocauaba* [*jiñocaube. jiñocuave, jiñote, jiniquite, jino*] (Nicaragua, Costa Rica, Panama), *jiote colorado* (red leaf, Sinaloa), *palo chino* (curly tree, Yucatán, Guatemala, Honduras, Belize), or simply *chino* (curly, Guatemala), and *palo jiote* (Veracruz, Tabasco, Chiapas, Guatemala, Honduras, El Salvador, Belize). Similarly, medical uses are implied in *chicchica* (maybe from Náhuatl *chichica*, bitter or sour), and *chicohuiste* [*choco-huite, chocohuite, chocogüite, chohuite, cohuite*] (*chichica*, bitter + *cuahuitl*, tree, Náhuatl, Chiapas), and *hupuk* [*huk'up*] (something one introduces [as medicine], Maya, Yucatán).

Still others were impressed by the peeling bark of the trees. People from temperate areas compared it with *Betula*, and called it birch gum, birch, birchwood, red birch, and West Indian birch. Natives also noticed the trait and call it *pellejudo* (with much skin, Guerrero), *chinacahuite* (*chinamitl*, separate, as in the peeling bark, *cuahuitl*, tree, Náhuatl, Guatemala), *chacah* [*chaca, chachah, chakah, chakag, chak chakaj, chakan, cajha, ca-ch, ca-cah, cha-ca, xa-ca, xaka*] (*chacah*, cooked, boiled, because of the appearance of the exfoliating bark on the trunk that makes it appear to have been cooked, Maya, Tamaulipas, Veracruz, Tabasco, Yucatán, Quintana Roo, Belize), *chacaj* (Tojolobal, Chiapas), *chaca piocha* (Tamaulipas), *chacah colorado* (red chacah, Maya, Yucatán), *chacai* (perhaps a loan-word from Mayan to Náhuatl, Puebla), *chacajiota* (chacah leaf, Náhuatl, Puebla, Veracruz), *sak chaka* [*sac-chacah*] (white chacah, Maya, Yucatán), and *tzaca*. Perhaps the names *ginicuite* (*xini*, peeling, *cuahutil*, tree, Náhuatl) and *jenequite*, are related. In Florida, the Seminoles called it *ahiciáhki* (peeling bark, Mikasuki).

Europeans also compared the trees with the red skin of the American natives. They call it naked Indian (Panama), naked boy, *indien nue* (naked Indian, French), *indio desnudo* (naked Indian, Belize, Panama), *indio en cueros* (naked Indian, literally, in skin), *indio peludo* (shaggy Indian, Belize), *pellejo de indio* (Indian's skin), *palo colorado* (red tree, Sinaloa), *palo retinto* (chocolate-colored tree, Sinaloa to Tamaulipas, Veracruz, Yucatán, Chiapas), *palo mulato* (mulato tree, Sonora, Nayarit, Tepic, Chiapas, Tabasco, Oaxaca, Veracruz, Yucatán), *palumulat* (mulato tree, Pima, Sonora), *mulato* (Oaxaca), and *to'oro mulato* (mulato gumbo limbo, Mayo, Sonora).

Other common names either confuse the trees with other species or make comparisons with them. These include *guacimo* (a name usually given to *Guazuma ulmifolia*, Malvaceae), *jobo* (Spanish, a name usually given to *Spondias mombin*), *gomme mombin* (mombin gum tree, Granada), and *jobo pelón* (*Spondias* with peeling bark, Spanish). In Costa Rica it is simply called *aceitero* (olive), but everyone knows the difference between it and the cultivated olive (*Olea europea*). In the Dutch Antilles it is *paaloe sieja doesji* (tree seed sweet), *paalkoe sieja maatsjoe* (tree seed forest), *palu di sia corra* (copper seed tree), *palu di sia machu* (male seed tree), *sia* (seed), and *sia blancu* [*sieja blanko*] (white seed).

Some names refer to uses, as in *copón* (big cup?, Honduras), *azucarero* (sugar giver), However, the Creek speakers among the Seminoles have a direct name for the tree they learned about from the Spanish—*itotiná:span* (*eto*, tree, *espane*, Spanish).

Byrsonima: Locust Berry

(Named by Louis C.M. Richard, 1754–1821, and Karl S. Kunth, 1788–1850, from Greek *byrsa*, hide and *nema*, thread, referring to the bark of some species)

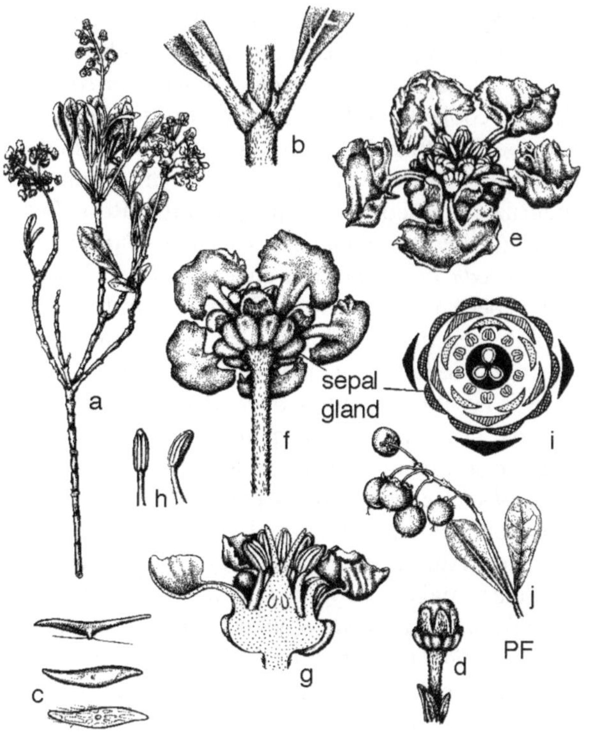

Byrsonima lucida. a. Flowering branch. b. Node showing clasping leaf bases. c. T-shaped trichomes from leaf surface, side and above, greatly magnified. d. Flower bud with basal bract and pair of bracteoles and glands on sepals. e. Flower from above. f. Flower from below. g. Flower longitudinally dissected. h. Stamens. i. Floral diagram. j. Branch with fruits. *Drawn by Priscilla Fawcett.* From Correll and Correll 1982.

In the Everglades where periodic fires maintain the pine flatwoods, the locust berry (*Byrsonima lucida*) is a small shrub, rarely over a meter (3 feet) or so tall. The farther south one goes in the Florida Keys, especially in the older hammock forests, the larger the plants. On the ecotone between hammock and salt marsh or even marl flats in some of the southern Keys the species reaches tree size. When I first read that they reached 7.6 m (25 feet) tall, and had trunks 15 cm (6 inches) in diameter, I did not believe it. But then, I had only seen the Everglades National Park plants. Later, I saw them on Cudjoe Key, and they matched the book descriptions—maybe exceeded them.

In the spring and summer these shrubs and trees are covered with small whitish to pale pink flowers that turn darker pink when they have been pollinated. For years, I thought that they were pollinated by bees and that they must provide nectar and pollen for them. I was right on the first point, but wrong on the second. One of the unique traits of the Malphigiaceae are the glands on the outside of the sepals. Those glands secrete an oil that is highly visible in some family members, but obscure in others. In either case, that oil is gathered by bees to construct their nests (Sazima and Sazima 1989, Steiner and Whitehead 1991).

People too have long used the genus *Byrsonima.* The genus is mostly rich in tannins that have been used to cure hides across the range. *Byrsonima lucida* is among those used as the name *bois tan* (tanner's tree, French Antilles) attests. Those same tannins have made *Byrsonima* useful for other purposes. Fruits and other parts are astringent and have been used against dysentery. Roots and bark have been used to combat that common problem in tropical areas and also have been used as a fungicide, against fevers and catarrh, to treat snakebite, and as a pectoral (Morton 1981). Some of the common names suggest long association with humans as medicines and for other uses. Medicinal names include *bois dysentérique* (dysentery tree, French Antilles), *doncella* (the Virgin, Dominican Republic), *sangre de doncilla* (blood of the Virgin, Cuba, Puerto Rico), *carne de doncilla* (Virgin's flesh, Cuba). Others probably alluding to uses of the wood or bark include candle-berry (Bahamas), *mar-icao* (Puerto Rico), *peralejo de pinares* (*Byrsonima* of the pines, Cuba), *piragua* (Taino, Cuba), and *sabica de costa* (coastal *sabicú*; *sabicú* is *Lysiloma sabicu*, Cuba).

Byrsonima lucida probably was used for dyes as related plants have been. Some species make brown-red dyes that prolong the life of the products colored because of the tannins and other chemicals. For example, *B. crassifolia* furnished one of the most important dyes in Amazonia. Because of the tannins and associated compounds, the plant was used by *caboclos* (river people) along the Amazon to make a brown-red dye that was used for fishing lines, sails, and clothes (LeCointe 1947: 319). Apparently, *B. lucida* has not been studied, but several species in the genus contain beta-amalin in the bark that makes it active against bacteria (Mors et al. 2000). In many places, the wood is used for carving small objects and for furniture (Little and Wadsworth 1964, Morton 1981).

Fruits of locust berry are edible, although that name does not seem to imply that they are. Some compare them positively to cranberries, while others say they have a "soapy" taste. Many of their common names allude to edible fruits, including *aceituna* (olive, Puerto Rico), gooseberry (Anegada), guana-berry (Bahamas), *merusier doré* (golden cherry, French Antilles).

The plants in Florida and the Bahamas are in the northern part of their range. Otherwise, they grow in Cuba, Hispaniola, Puerto Rico, the Virgin Islands, and from Anguilla and Barbuda to Barbados. Yet, this is one of the least used species in the genus. In Central and South America, there are others more famous for human utilization.

Perhaps the most famous of the group are several species called in Brazil *murucí* (from *muri'si*, Tupí). These include *B. crassifolia* and *B. verbascifolia*, among others. These species are called by several local names including *murici-do-campo* (wild murici), *murici-açu* (big murici), *murici-da-praia* (beach murici), *murici-da-mata* (forest murici), *murici-miúdo* (small murici), *murici-rasteiro* (prostrate murici), *murici-pe-nima* (murici the size of a penny), *chaparro* (thicket), *orelha-de-burro* (donkey's ear), *matega* (tea tree), *caá-açu* (big plant, Tupí), *pau-de-curtume* (tanning tree), *pessegueiro-de-curtume* (tanning prune). The best-tasting and most widely used for food is *B. crassifolia.* In Brazilian markets, they are almost always just called *murici.* They may be found fresh in markets, or made into juices, wines, or even ice cream. The taste is delightful and even temperate palates quickly appreciate these tropical flavors. Vacationers in Brazil even pack the fruits in water with sugar, and they are said to keep for as long as 2 weeks (Cavalcante 1976).

Europeans were first introduced to these delectables by Georg Marggraf [Marcgrav, Marcgrave] in the middle 1600s when he wrote, that *Murici* has "*bagos da figura e tamanho dos frutos da roseira braba, que são comidos*" (berries of the shape and size of the fruits of the wild rose, that are eaten). Now, *B. crassifolia* is so popular, for ornament and fruits, that it is grown up the Pacific coast of Mexico and even into southern California. In these areas the berries are eaten fresh, with salt, in soups, or as flavoring on meat.

Many of these same species are used medicinally in Amazonia (LeCointe 1947, Mors et al. 2000). The bark is made into a cough medicine and used to treat tuberculosis and other bronchial inflammations. In all the species, the berries are astringent and antidysenteric. In some the bark and berries are mildly laxative. Some are considered diuretic and emetic and have been used against syphilis.

Many other people use *B. crassifolia,* and it has common names throughout much of its range. In Yucatán, the Maya call it *chi.* The Zapotec of Oaxaca call it *mami-hña.* However, the most widely used name is a variant on the Aztec name *nantzincuahuitl* [*nanzinxócotl*] (*nantzintli* or *nantli,* mother, *cahuitl,* tree). This is typically shortened to a variant of the first word and becomes *nance* [*nanche, nanchi, na-nantze, nancis, nandzin*], or *nance agrio* (sour nance), *nanche amarillo* (yellow nance), *nance dulce* (sweet nance), and *nanche de perro* (dog's nance) throughout southern Mexico.

Florida's *Byrsonima* is "locust-berry." That sounds like a derogatory name. Surely, it means something only a foraging swarm of locust grass-hoppers would eat. However, the name turns out to be highly complimentary. It is true that the first meaning now in use in English of the word is for the rapacious insects. However, a historically important meaning—arguably more important—was a name of leguminous trees, particularly the locust tree or carob (*Ceratonia siliqua*). These trees were important food in the eastern Mediterranean, and reputedly provided the main food for John the Baptist of the Bible.

So, comparing the bushes and trees in the American tropics with that religious personage and his food surely originally was meant as a compliment. This interpretation is further supported by evoking the Virgin Mary in several of the common names (e.g., *sangre de doncilla, carne de doncilla*). Plants important enough to be associated with these holy biblical icons have been considered important regardless of how unassuming they now appear.

Moreover, there are several other "locust" trees in the New World, including temperate honey locust (*Gleditsia* spp.) and black locust (*Robinia pesudoacacia*). Within the tropics, the most important locust is the West Indian locust (*Hymenaea coubaril*). The West Indian locust not only figures prominently in medicines, but its ancestors have also left behind much of the fossilized resin we call amber.

C

Caesalpinia: Gray Nicker

(Named for Andrea Caesalpino)

Caesalpinia bonduc. *Drawn by P.N. Honychurch.*

Andrea Caesalpino (Andreas Caesalpinus, 1519–1603) was an Italian who studied at the University in Pisa with botanist Luc Ghini (1490–1556) and anatomist Realdo Colombo (ca. 1510–1559). Later, Caesalpino was the prefect of the first botanical garden in Pisa, and in 1563 he founded the second garden. That was surely why Linnaeus named the genus *Caesalpinia* in his honor in 1753. In continuing that name Linnaeus followed the lead of Hans Sloane, John Ray, and Mark Catesby who had studied the plants before him.

The gray nicker (*Caesalpinia bonduc*) grows in coastal regions on beaches or near mangroves. Every part of the plant except the flowers is armed with recurved spines that discourage close examination by ripping at the skin and clothes. Once the yellow, comparatively inconspicuous flowers are gone, they are replaced by spiny brown pods that contain the gray, marble-like seeds. Those seeds gave rise to the primary English name, "nicker." This ancient word has an obscure origin, as does its base, "nick." Probably, the name was first an allusion to the mythical water demon also known as a *kelpie* (Gaelic, *cailpech*). As such, it dates back to the time of Beowulf (6th century B.C.). It may date back to the Sanskrit word *nij*, to wash. Regardless of its exact derivation, its association with the sea resulted in it being applied to these plants. The word took on many meanings over the years, including "one who cheats at games" (from the 1660s), "one who fits objects nicely" (from the 1670s), and a game (from the 1670s). From the game, it came to mean a marble. In the Caribbean, it has most recently meant to compare the round seeds of these vines to marbles. However, that usage is recent, dating from Hans Sloane's 1707 mention of nicker-tree in Jamaica.

Variations on the nicker theme include ash-colored nicker (English Antilles), *canique* (nicker, Hispaniola), *canique grise* (gray nicker, Guadeloupe, Martinique), gray nicker (Florida, Bahamas, Barbados), horse nicker (Barbados), and nicker nut (Bahamas to Panama).

Comparison to other legumes include *haba de San Antonio* (San Antonio's fava bean, Veracruz), *jabilla* (little fava, Veracruz), *habilla de mar* (little sea fava, Sinaloa), *juquiri-rana* (from Tupí, *yuki'ri*, the name of *Machaerium ferox* + *rana*, false, Brazil), *taray* (a name usually given to *Eysendhartia polystachya* in Veracruz and to *Salix chilensis* in the central part of the country), and simply seabean (Panama).

Several places have confused *C. bonduc* with the Brazil wood (*C. echinata*) that was so much in demand from the Brazilian coastal forests that it was almost harvested to extinction in the 1800s. These are *bois de Brésil* (Brazil wood, French Antilles), *brasil* (coals, also the original comparison with Brazil wood because of the red color, Oaxaca), *brasilete colorado* (red Brazil wood, Cuba), and *palo fernambuco* (Pernambuco tree, Cuba).

Some of the names refer to the downright nasty, spiny nature of the plants. These are brier bush (English Antilles), brier (English Antilles), *guacolote* [*cascalote, nascalotl, huiscolote*] (perhaps derived from *huitzcocólotl*, spiny branch, Náhuatl; Cuba, Veracruz), *guacolote prieto* (black spiny branch, Cuba), *inimbó* (from Tupí, *inimicu*, hostile, Brazil), *inimboja* (very hostile, Brazil), *mato* (thicket, Panama), *mato azul* (blue bush, Puerto Rico), *silva da praia* (beach tree, Brazil), and *villa de mar* (beach villa, Spanish Antilles).

Other names have compared the seeds to plants, animals, and other objects. Some of these are likely to have dual meanings and also refer to medicinal uses. Among these are *arriózes* (originally the name given the ball from a muzzle-loading gun [blunderbuss], now applied to the seeds of this legume, Brazil), *bonduc* (derived from the Arabic *bunduk*, a little ball, Jamaica), *cojónes de gato* (cat testicles, Yucatán), *garrapata de playa* (beach tick, Venezuela), *mato de playa* (beach bush, Puerto Rico), *oeil de chat* (cat's eye, French Antilles), *ojo de águila* (eagle's eye, Mexico), *ojo de venado* (deer's eye, Sinaloa), *olho de gato* (cat's eye, Brazil), *piedra del águila* (eagle's stone, Oaxaca), *quaritt-oeil-de-chat* (cat's fourth eye, French Antilles), warree seed (Belize), and *yeau de chat* [*z'yeau à chatte*] (cat eye, Guadeloupe).

However, one of the most important categories of names includes the many references to medicinal application. These often allude to the use of the seeds to combat malarial fevers. As the seeds have been used around the world for this purpose, they have been called the "quinine of the poor." Names related to this use are *calentura* (heat or fire, Panama), fever nut (English Antilles), *graines quinine* (quinine seed, Hispaniola), *mate de costa* (coastal tea, Hispaniola), *mate* (tea, Hispaniola), *quasha* (probably a variant of *quassia*, West Indies), *quinine* (Hispaniola), and *quinine jaune* (yellow quinine, Hispaniola).

When names are old enough not to have a translation, they usually have been in the language so long that they designate specific ideas or objects. It is likely that *tikó:lî* (Mikasuki) and *i:kofkapó:yî* (Creek) belong in this category. It appears that the Creek name is based on *ekofkv*, breechclout, and *po:y-i*, completed, but that makes no obvious sense, unless it means that the spines might tear off even that garment. While Sturtevant (1955) found no Seminoles who knew of uses for the plants, he noted that there was a necklace of its seeds in a museum collection. Most likely, the informants who worked with Sturtevant simply had lost the old medicinal uses.

Indeed, the use of the oils in the seeds (up to 23% by volume) in medicines is founded on a variety of potent chemicals, especially the bitter bonducin, which is tonic and antiperiodic (Hocking 1997). Also present is a group of furanoditerpenes designated as caesalpins (Pascoe et al. 1986). These chemicals surely account for the seeds being used around the world since the plants were adopted by Americans and Europeans when they were brought back from the tropics. The seeds have also been used to combat colic, convulsions, dropsy, fever, hydrocele, leprosy, palsy, to remove pimples, against snakebite, to soften skin, as a tonic, against toothache and venereal diseases, and as a vesicant (Morton 1981, Mors et al. 2000).

The seeds are used in necklaces and in maracas. They are carried by natives as talismans or amulets, and they are still used as marbles.

A second species has been continually confused with this one, even though it has yellow seeds. This is *C. major,* yellow nicker, *guacolote amarillo, mato de playa.* The seeds of that species are so similar to the fruits of the mastic tree (*Sideroxylon foetidissimum*) that I have mistaken them for those. I only realized my error when I noticed that there was no mastic present. Like the mastic, these yellow seeds are almost certainly dispersed in Florida by the white-crowned pigeon (*Columba leucocephala*). That dispersal method is so different from the gray nicker that it makes the yellow species highly local throughout its range. Seeds of the gray nicker are dispersed by ocean currents. They travel remarkable distances, and the species is pantropical. Almost unbelievably, the seeds wash ashore in Scotland where they historically were known as Molucca beans. The Scots, however, had another name for all sea beans. They called them *arna Moire* (Mary's kidney, Gaelic), and they prized them.

Cakile

(From the old Arabic words *kakeleh, qaqila,* or *qaqulla*, the name for cardamom, *Elettaria cardamomum*)

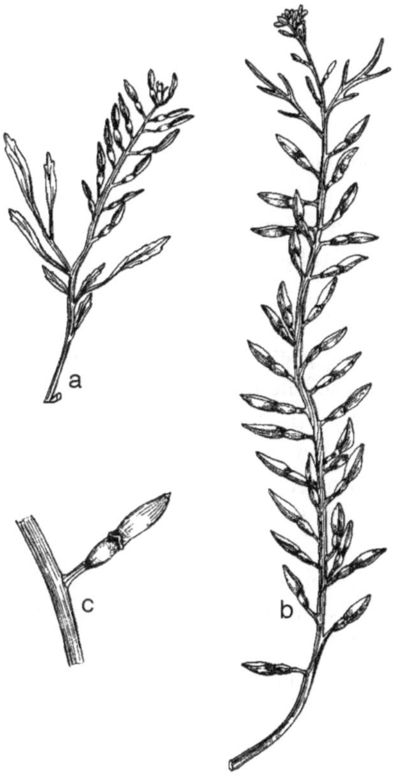

Cakile lanceolata. a. Tip of branch with fruit and leaves. b. Tip of branch with mature fruit. c. Fruit. *Drawn by Vivian Frazier.* From Correll and Correll 1972.

eruca marina (ocean arrugula, Portuguese); *oruga marina* (ocean arugula, Spanish); *roquette de mer* (sea rocket, French); sea rocket ("rocket" is from Middle French *roquette,* which came from

Old Italian *rochette*, a diminutive form of *ruca*, the garden rocket or arugula, *Eruca*; first in English by ca. 1530)
Meersenf (sea mustard, German)
radicetta marina (sea radish, Italian); *strandreddik* (seashore radish, Norwegian)

Cakile edentula (edible)
caquillier (sea-rocket plant, Quebec)
[American] sea-rocket (Florida)

Cakile lanceolata (with leaves shaped like lance points)
berro de costa (coastal water cress, Cuba)
gardena (misapplication of *Gardenia*?, Bahamas)
mostacilla del mar (sea mustard, Puerto Rico)
pork bush (Bahamas)
[southern, coastal] sea-rocket (Florida, Bahamas, Puerto Rico)

There are two species of *Cakile* on the shores of Europe and five in North America. The others are scattered around the globe (Kartesz 1994, Mabberley 1997). A total of seven species are now recognized, although Philip Miller knew only one when he described the genus in 1754.

People in southern Labrador have long cooked the plants as greens, and the common names suggest that other people have done the same. Fernald et al. (1958) were not overly enthusiastic about them. Kalm ([1753–1761] 1972) noted that people in the province of Quebec at Cap aux Oyes pounded the root, "mixed with flour, and eaten here, when there is a scarcity of bread." Tussac (1808–1827) recorded that blacks in the Antilles boiled the leaves and put them on persistent sores as a detersive.

Callicarpa: Beauty Berry

(Greek, *kalli*, beautiful, *karpos*, fruit)

To me the generic name of these plants is one of Linnaeus's strokes of genius—apparently without borrowing from anyone this time. Four authors had discussed this plant before him, and the names they used were lackluster. The nearest to creating something of interest was John Mitchell's *Sphondylococcos* (*spondyl*, vertebra, *coccus*, fruit) of 1748. Plukenet in 1696, Catesby in 1734, and Gronovius in 1739 used long but unimaginative phrase names. Linnaeus had actually seen these plants growing in the Uppsala gardens, and he called them *Callicarpa*, or beauty berry in English.

If you are walking in a pinelands in the southeastern United States, the clusters of small, purple, beautiful berries are sure to catch your attention. These small shrubs grow to about head high, and there may be a half-dozen rings (verticils) of fruit clusters scattered up the stems. In the late summer and fall, the fruits also catch the eyes of birds. These feathered dispersers spread beauty berry seeds around with a load of nutrients that helps the young plants get started in life. When the shrubs are in flower, they are not that showy, but fruit clusters are a completely different situation.

This is one of the American species that was taken into cultivation in Europe and consequently admired for its beauty and little else. The memory of its utility among native people and colonists faded and has almost disappeared. Views were different 230 years ago.

When William Bartram was traveling through southern Georgia in late 1773 or early 1774, he commented that the people of the area had long grown the plants. Around old Indian settlements on the Little River, he wrote, "I observed, in the antient [*sic*] cultivated fields, 1. Diospyros, 2. Gleditsia triacanthos, 3. Prunus Chicasaw, 4. Callicarpa, 5. Morus rubra, 6. Juglans exaltata, 7. Juglans nigra, which inform us, that these trees were cultivated by the ancients, on account of their fruit, as being wholesome and nourishing food" (Bartram [1791] 1958). While some might argue that the fruits of *Callicarpa* are not that good or nourishing, his observation presaged the worldwide cultivation of this species. In March of 1774, he was in St. Augustine, Florida, where he found a large earthworks and other signs of long occupation. He wrote, "There grew in the old fields on these heights great quantities of Callicarpa" (Bartram [1791] 1958). Bartram mentions *Callicarpa* at least 15 times, and in 6 of these comments associates it with being cultivated by people of the Creek Nation.

We might question the reasons for the plants' presence in the oldfields. Perhaps they were there simply because the Creeks permitted them to remain and did not cultivate them, or perhaps they grew there because birds spread their seeds into cultivated lands. However, I am inclined to believe Bartram because he talked with the local people and learned as much about their customs as he could. He was greatly interested in their cultivated and medicinal plants.

This interest in the plants by the Creeks and their relatives is reasonable because we have common names for the plants in Mikasuki (*kalá:cí:rá:pî*, tree having shells, perhaps because they note that snakes lay eggs among the roots), Creek (*ca:tapholó:ta* also transcribed as *catv-folotv*, red worm), and Choctaw (*shoklapa*, maybe *shokulbi*, the inside corner of a field,

Callicarpa americana*. Drawn by P.N. Honychurch.*

where the plants maybe grew). Koasati said *cosimilpá* (*cosí*, wren, *im*, its, *ilpá*, food). This is probably the plant the Chickasaw called *nita' lobahli'* (*nita'*, bear, *lobahli'*, raisin). Perhaps the Alabama name gives more indication of their views—*haakhimbałka* or *hak-himbałka* (*haakha*, pelican, *im*, its, *bałka*, raisin).

We do know that the Alabama used the plants in sweat baths against edema and malaria (Taylor 1940). The Choctaw used them against colic (Bushnell 1909). The Seminoles too had a number of uses, but those that have been recorded include remedies for what they call "Snake Sickness" and "Urine Retention Sickness" (Swanton 1928a). For Snake Sickness, the medicine is applied to the outside of the body as a steam, but the patient also drank some of the liquid. In Urine Retention Sickness, a decoction effects a diuresis that cleans the kidneys and bladder. However, that may not be all that is going on because the remedy is also thought to have been used for a number of urinary problems, including gallstones and bladder disorders. There is even some indication that it may have been used against gonorrhea. Chemical and laboratory studies have shown that root decoctions of the plants have antiviral activity against measles and polio pathogens (Hocking 1997). Perhaps there is also activity against other microbes. Although details are not given, there is one record of the plants being involved in a remedy for skin cancer (Duke 2003).

Because the foliage and other parts of these shrubs are aromatic, they have been used in medicinal teas in North America and elsewhere. In Cuba the species is called *filigrana morada* (purple filigree), or simply *filigrana* (filigree, but particularly a name given to this plant). That name links the medicine with *Lantana camara*, which Cubans also called *filigrana*. The name also links *Callicarpa* with *L. involucrata* because it is called *filigrana cimarrona*. All three are in the Verbenaceae and have similar odors.

In Cuba, *Callicarpa* is also called *filigrana de mazorca* (young corn filigree), although its relationship to corn is obscure. Users also understand the plant's relationship with habitats because they also call it *filigrana de piña* (filigree of the pine) and *filigrana de piñar* (filigree of the pinewoods). On that island, a medicinal tea of the leaves and small branches is particularly popular against chest inflammations (Roig 1945).

The other names for these plants revolve mostly around two themes. One is the fruit, and includes American beautyberry (Europe), bunchberry (USA), foxberry (USA), French mulberry (Europe), purple beautyberry (USA), *Schönfrucht amerikanische* (American beautyberry, German), and turkeyberry (USA). Perhaps sour-bush (USA) also refers to the fruits, but it might just as well apply to the tea that is made from the plants.

The other theme among names, oddly enough, compares *Callicarpa* with the mulberry (*Morus* spp.). These names include Bermuda mulberry (USA), dwarf mulberry (USA), and Spanish mulberry (USA). There is nothing about the vegetative parts that looks much like a mulberry; perhaps the color of the fruit reminded someone of that tree.

This shrub is one of those that is basically North American. Beauty berry grows from Florida west to Texas and north to Maryland and Oklahoma, However, *C. americana* also occurs outside that region in places where it seems odd that it should grow including Cuba, Coahuila, and Bermuda. That it would grow in Coahuila, Mexico, is not too surprising because Texas is just next door. Yet, how did the plants get to Bermuda and Cuba?

Several of the species in the genus have been used as fish poison, and *C. americana* is among them. Apparently, the only species that has been studied chemically in this regard is *C. candidans* (southern Asia, Indonesia to Australia), and it contains callicarpone. That chemical is said to have a potency equal to rotenone (derived from the legume *Derris*), which makes it powerful indeed (Hocking 1997).

One of the species apparently closely related to *C. americana* is *C. acuminata*, which ranges from Mexico to Colombia and Bolivia. This species is considerably larger than the Florida plants and may grow to 6 m. However, it too has a strong tendency to coppice, and produces clumps of many stems like *C. americana*. The predominant common name for *C. acuminata* is the Mayan *sak-puy'in* [*sak-pukin, zacpucin*] (*sak,* white), but it is also known simply as *puk'in*. The simple name is ancient enough that no translation exists—that is simply the name of the species. Other Mexican names include *elete'* [*elté*] (spring tree, Huastec, Hidalgo, San Luis Potosí), *alahuete* (perhaps an incorrect transcription of *elete'*, Huastec, San Luis Potosí), *granadilla* (a name usually given to *Passiflora*, Chiapas), *hierba de la chachalaca* (Chachalaca herb, the bird *Ortalis*, Mexico), *patsumacáshil* (Totonac, Veracruz), and *uvilla* (little grape, Tamaulipas).

This species grows all along the Mexican east coast and well into the interior. Like its U.S. counterpart, this species has a variety of medicinal uses. The Huastec people of central eastern Mexico have good documentation of usage (Alcorn 1984). They use the plants in root decoctions against internal problems, including hemorrhage associated with childbirth. For that problem the woman is asked to drink one small glassful. The Huastec people also mix the roots with other plants to make a decoction that treats constipation and inability to urinate. However, they are not limited to medicinal application of the plants. Leaves are used as dish scrubbers. Stems are burned as one of the most commonly preferred fuels and also used as supports for bean vines and as one type of rafter in house construction.

Among the uses by others are as a diaphoretic, a purgative, and, oddly enough, a decoction is given to halt dysentery and diarrhea. That dual usage is common in indigenous medicines, and the concentrations often vary—concentrated doses cause evacuation of the bowels, and weaker doses halt that problem. However, usually that dual usage philosophy is not easily separated from the idea that, if you have a problem, a good laxative will clean it out and get rid of it.

Calycanthus

(Linnaeus created the genus from Greek *kalyx*, a calyx, and *anthos*, flower, because the calyx and petals are the same color)

Calycanthus floridus. From Britton and Brown 1897.

Calycanthus floridus (flowering)

bubby-blossoms [bubby-shrub] ("bubby" is an English word meaning breast, in use by ca. 1686; cognates are German *Bübbi*, and French *poupe*; any connection between breasts and these flowers is obscure unless it might be the color)

Carolina allspice ("allspice" compares *Calycanthus* to *Pimienta dioica*, a species that is thought to combine the flavors of cinnamon, *Cinnamomum verum*, nutmeg, *Myristica fragrans*, and cloves, *Syzygium aromaticum*; the comparison dates from about 1601; most places in the world call the fruits of *P. dioica* either "pimento" or some variation on that Spanish word)

Gewürzstrauch (spice bush, German)

pineapple-shrub (presumably because of the fragrance of the crushed flowers) strawberry bush (applied by 1847 because the fruits resemble the strawberry in their red color) [common, South Carolina] sweet-shrub [sweet shrub]

As might be anticipated by the name "allspice," the bark has been used as a substitute for cinnamon (Hedrick 1919, Yanovsky 1936). The Cherokee put the bark on children's sores and made an infusion to relieve hives and to instill as eye drops. They also used the roots to create a strong emetic and a medicine for bladder problems (Hamel and Chiltoskey 1975, Moerman 1998). Fragrance from the flowers was used as perfume (Uphof 1968).

Hocking (1997) says that the seeds were used as a uterine tonic by indigenous people, but no other publication has been found verifying that. Sometimes now, a decoction of root bark or seeds is used to ease muscle cramps. That practice is not without danger because the seeds are poisonous and have been used by the Cherokees to destroy animals (Moerman 1998). Seeds contain calycanthine.

Porcher (1863), however, noted that it was used "as an antispasmodic tonic, in the cure of chronic agues. A strong decoction of the seed or bark of the root is given. The wood is strongly camphorated, especially the root, and Mr. Nuttall thinks [it] will probably produce this drug as abundantly as the *Laurus camphora.* Seeds seldom mature."

The bark is cinnamon-clove scented, and the wood smells like camphor (Bremness 1994). The plants have been used as an aromatic stimulant and to treat malaria (Hocking 1997). In Alabama, the bark heated with lard and elder bark (*Sambucus*) forms "fire grease," for use in treating burns (Hocking 1997).

Calyptranthes

(Greek *kalyptra*, veil, lid, *anthos*, a flower, referring to the operculum of the flowers)

Calyptranthes pallens (pale in color)
arrayán (well-rooted, Dominican Republic)
lemon tree (Belize); *lemoncillo* (little lemon, Dominican Republic)
[pale] lid-flower [lidflower] (Florida, Puerto Rico)
spice-wood (Bahamas)
tapón blanco (white lid or stopper, Puerto Rico)
white stopper (Bahamas, Puerto Rico)

The Swede Olaf Swartz, a professor in Stockholm, named this genus in 1788 based on plants he found in Jamaica. Now there are 130 species known in the American tropics (Mabberley 1997). Some of the species along the Amazon are fish dispersed, and the fruits of all are edible. Leaves of some have been used as spices.

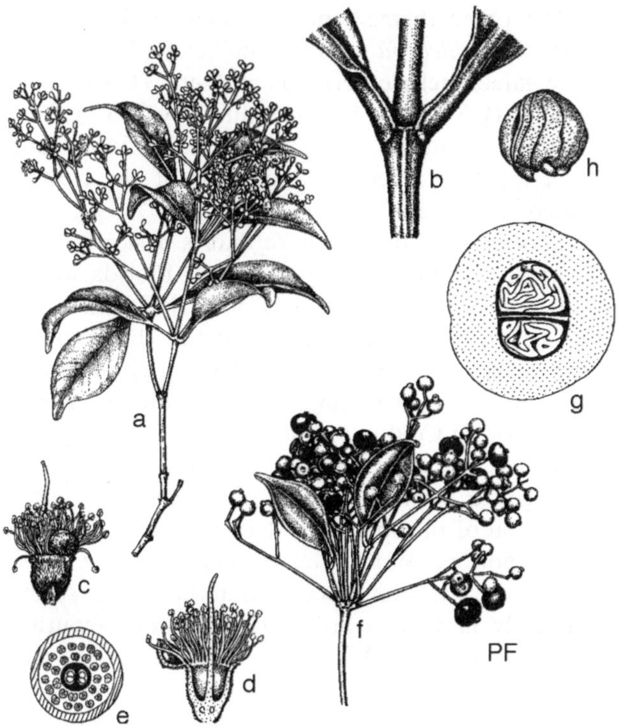

Calyptranthes pallens. a. Branch with flowers. b. Node and internodal ridge. c. Flower side view, with calyx lid still attached. d. Flower longitudinally dissected. e. Floral diagram. f. Fruiting branch. g. Fruit in transverse section. h. Embryo. *Drawn by Priscilla Fawcett.* Correll and Correll 1982.

Fruits are edible but not very good (Liogier 1974). The tree is considered medicinal in Belize (Balick et al. 2000).

Campsis

(From Greek *campsis*, curvature, an allusion to the curved stamens)

Campsis radicans (rooting)
ampohko'li' (Chickasaw)
cow-itch [cow-vine, vine] (according to Small 1933, the plants are mistaken for poison ivy, *Toxicodendron radicans*, and that gave rise to the name—although cows do not get poison ivy; could it be that the word "cow" does not refer to the bovid, but is dialectic, first recorded in the 1500s, for "bundle of twigs"?)
trumpet-[creeper, vine, flower, honeysuckle] ("trumpet" came into English by the 1300s from French *trumpette*, a musical wind-instrument; the word has been used for plants with

flowers resembling trumpets since 1705; Asa Gray, in his *First Lessons in Botany* published in 1857, is the first record found combining trumpet with creeper)

Campsis radicans is a unique member of the New World Bignoniaceae. There are only two species in the genus, with the other, *C. grandiflora,* native only to China (Mabberley 1997). The genus was created by Portuguese botanist João de Loureiro (1717–1791) in his *Flora Cochinchinensis* in 1790 for *C. grandiflora.*

Because the plants are so unique, it took people a long time to determine their relationships. Linnaeus called the New World plants *Tecoma radicans.* John K. Small (1933) too was unclear on the relationships, but he knew the vines were not part of *Tecoma.* He put this species in *Bignonia,* calling it *B. radicans.* In doing that, he mirrored what C.P. Thunberg had done with the Chinese species in 1784 when he called it *B. grandiflora.*

The Old World *Campsis* has been famous as a medicine since ancient times. It was mentioned in the *Shen nung pen tsao ching* (Shen Nung's Herbal) of ca. 25 B.C. as a high-grade drug (Hsu 1986). Traditional use of the Chinese species was to remove postpartum obstructions and reduce bleeding.

As famous as the Chinese species is as medicine, it seems odd that so little has been recorded about the American plants. Steyermark (1963) considered *C. radicans* narcotic. Hocking (1997) found that the root was used as a diaphoretic and vulnerary in domestic medicine. The species is not mentioned by Porcher (1863), Millspaugh (1892), Uphof (1968), Vogel (1970), Foster and Duke (1990), Moerman (1998), or Duke et al. (2002).

Canavalia

(Derived from the Malabar common name in India, *kavavali,* from *kanam,* forest, *valli,* climber)

jack-bean (this name was not in use until ca. 1885; perhaps the name originated with Malayalam *chakka,* the name of *Artocarpus,* which was rendered *jaca* in Portuguese by Gardia de Orta in 1563; as both were eaten in Asia, there may have been a transfer of names or equally possible is that it originated in English with the sense of "dwarfed or undersized," as people began using the combination with several other plants in the 1800s)

sword-bean (from the large, flat shape of the legumes; apparently first applied to *Canavalia* in 1883)

Canavalia rosea*. Drawn by P.N. Honychurch.*

Canavalia rosea (rose-color, of the flowers)

bay bean (Bahamas, Jamaica, Virgin Islands); *frijol de mar* (ocean bean, Honduras); *frijol de playa* (beach bean, Costa Rica); *haba de playa* (beach bean, Puerto Rico); *habichuela playera* (beach bean, Puerto Rico); sea [seaside] bean (Barbados, Jamaica, Cayman Islands)

horse bean (a name usually given to *C. ensiformis,* Bahamas)

June bean

kumatara [*kumatará*] (Carib, Suriname)

mate de costa (coastal plant, Cuba, Dominican Republic); *mato de playa* (beach plant, Puerto Rico)

ouari [*graines*] (maybe from West African slaves; the Katsina of Nigeria call one species *ro'bari,* Haiti)

pois bord de mer (seaside bean, Guadeloupe, Martinique); *pois cabrit* (goat bean, Guadeloupe, Martinique)

pois liane (bean vine, Haiti)

pois Makendal (Makendal's bean; François Makendal was a mystical figure who played an important role in the slave revolt in Haiti. He had only one arm and was well versed in voodoo. In the 1750s he organized a revolt against the French by going from plantation to plantation and talking to the slaves secretly. But the slaves were so afraid of retaliation that the movement failed. He was captured and tied to a post in *Cap Haitien* (*Cap Français* at the time). The French thought that by setting him on fire in front of all the slaves it would show them what happens when a slave revolts. But the flames burned the ropes that bound Makendal to the post and, according to the legend, he ran and disappeared before they could catch him. The slaves did not believe that he had died and the rebellion that he had started continued with Boukman; Guadeloupe, Martinique.)

pois maldioc lan mer (damned seaside bean, Haiti)

pois zombi (zombi bean, "*zombi*" is of Niger-Congo origin, akin to Kongo *nzambi*, god, Haiti)

vonvon (probably from the verb *aller*, as in *ils vont*, they go redoubled or "they go go," as it is a running beach plant, Guadeloupe, Martinique)

Ignorance, they say, is bliss. So, my students and I were certainly blissful for years. At one time I had an anthropology student who was interested in plants used by people. She convinced me that the seeds of bay bean were edible by gathering a handful in the field, cracking the seed coats with her teeth, and eating the contents. I tried them, and they were tasty—nutty and sweet. Then, I read about the alkaloids and proteins found in the seeds, and discovered that they were considered by many to be toxic. I quit eating them and stopped showing students how to eat them.

However, there is a long history of people eating many of the species in *Canavalia*. For example, *C. ensiformis* (horse bean, overlook bean, sword bean) is grown as a vegetable in tropical areas (Sauer 1964, Purseglove 1968, Herklots 1972). The young pods are sliced and eaten like French beans (Goodding et al. 1965). Alain and Leon (1946–1953) recognized six *Canavalia* in Cuba. *Canavalia ensiformis* is used for forage, and its fruits and seeds are edible; *C. gladiata* seeds are edible, and its foliage is used for fodder.

Mabberley (1997) says young pods of *Canavalia* are edible, although unripe seeds are considered toxic, and that all species contain alkaloids. Burkill (1966), on the other hand, says that the Malays eat *C. rosea* flowers as a flavoring, and that they are also eaten in Malaysia. In that region, some people consider the young seeds a good "pease porridge, not uncommonly eaten." Burkill even said that chemical studies showed no alkaloids or cyanogenic glycosides.

More recent studies of chemicals have established that, as Mabberley (1997) said, *Canavalia* contains several toxic chemicals. Among those are lectins, cyanides, complex proteins, and alkaloids (Bressani et al. 1987, Okolie and Ugochukwu 1989, Rosenthal and Berge 1989, Pérez et al. 1991, Abe 1993, Carlini and Udedibie 1997). Yet, other studies have confirmed what humans discovered thousands of years ago—processing of the seeds renders them edible (Apata 1994, Akpapunam and Sefa-Dedeh 1997, Vadivel 2001, Chel-Guerrero et al. 2002).

No records of *C. rosea* being used for food have been found in the New World. Only in Asia has it been recorded as food, although there are many records of other species being eaten in the New World.

There are numerous records of people feeding the foliage to livestock (e.g., Alain and Leon 1946–1953, Liogier 1974). In the Caribbean, the seeds are considered toxic and corrosive (Liogier 1974), but the leaves are used in medicines. Alain and Leon (1946–1953) noted, "*Dicen que ahuyenta las bibijaguas*" (they say it expels intestinal worms). In Hispaniola, the leaves are used in bitter tonics, and the roots are considered antiblennorrhagic. The plant is also used to treat asthma, although its toxic properties make that unadvisable (Liogier 1974).

Canella: Cinnamon Bark

(From Latin *canna*, and Greek *kanna*, a reed, referring to the rolled bark used as a spice)

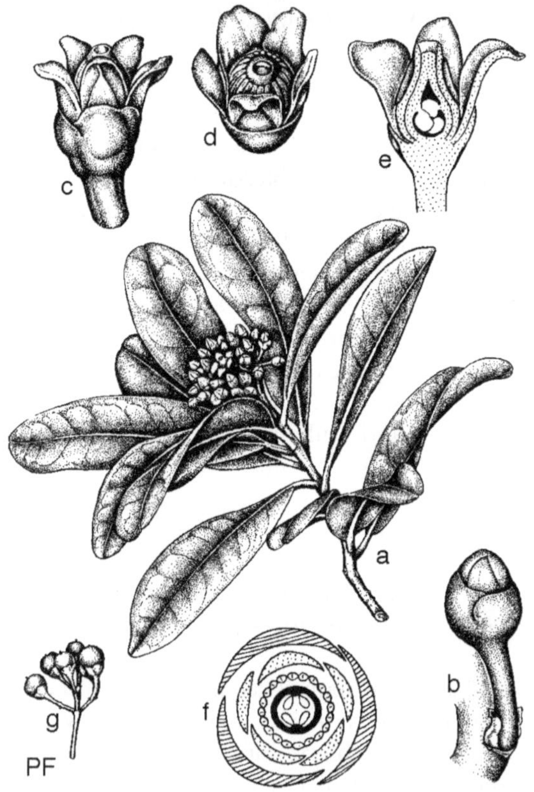

Canella winterana. a. Flowering branch. b. Flower bud. c. Flower, side view. d. Flower from above. e. Flower longitudinally dissected. f. Floral diagram. g. Fruits. *Drawn by Priscilla Fawcett.* Correll and Correll 1982.

In the 1970s, I was visiting Fairchild Tropical Garden to consult its living and herbarium collection. A prominent Harvard botanist was also visiting. As we passed in the hall we began talking, and I noticed he had a branch in his hand. I asked about it, and he told me that it was *Canella winterana*, known in some areas as cinnamon bark. He informed me that the leaves had a wonderful taste, and I knew what was coming. He had the reputation of being a practical joker, and he had decided that I was to be his next victim.

Although I had not been in Florida long, I had heard about cinnamon bark. Fresh leaves are fiery, and they have been used to season foods throughout its range. Indeed, in Cuba they have the common name *pica-pica* (it bites, referring to the stinging taste). Although I had never tasted them, I knew they would be pungent. He offered me a leaf to taste, and I took an ample bite from its tip. Then, after waiting a few seconds, I commented with a deadpan expression, "It has a nice spicy taste." He said nothing, frowned, and stalked off to his work area without another word. Sometimes, knowing about plants allows us to save face, even if our lips and tongue are burning like crazy.

These trees grow from southern Florida to the Bahamas, Cuba, Jamaica, Puerto Rico, Virgin Islands, the Lesser Antilles from St. Martin and Barbuda to St. Lucia and Barbados. They are naturalized in Brazil and cultivated in Venezuela. The species had been known to Europeans since at least 1623 when it was called *Cinnamomum s[ive] canella peruana* (Peruvian cinnamon or canella) by herbalist Gaspar Bauhin. Although Bauhin thought it came from Peru, Linnaeus listed it from Jamaica, Barbados, and "Carolina." As usual, his biology was good, but his geography was bad. The "Carolina" he thought of as the place for the plants was actually extreme southern Florida.

Linnaeus first wrote about the plants in cultivation in his *Hortus Cliffortianus* published in 1738, where he called them simply *Winteranus.* Then in 1753, he decided to call them *Laurus winterana*, but noted, "*Flores ipse non vidi*" (I have not seen the flowers of these). If he had, he would not have put them with the genus *Laurus* and other members of the group later called Lauraceae in 1789 by Antoine Laurent de Jussieu (1784–1836). There are many differences between the flowers in Lauraceae and these plants. *Canella* does not have poricidal dehiscence in the anthers and their arrangement is not the 3-merous whorls of the Lauraceae. Instead, *Canella* has stamens that are connate by their filaments into a tube that almost or completely envelops the pistil. Those traits would have caused Linnaeus to put the genus elsewhere. However, where he may have put it is open to question. It was not realized how unique the plants were until 1759 when it was separated as the Winteranaceae. As a result of a technical problem, that name has been rejected and the species is now placed in the Canellaceae described by Martius in 1832.

Indeed, all of the people from Bauhin to Linnaeus who had studied these plants knew them only as trees with pungent spicy leaves. Bauhin called them a *Cinnamomum*, and Plukenet said it was a "Cassia," which is a name used for another species of *Cinnamomum.* Sloane said that it was an *Arbor baccifera laurifolia aromatica* (tree with fleshy fruits and aromatic laurel-like leaves).

It was 1756 when Patrick Browne segregated out these plants as the genus *Canella.* In 1788, Joseph Gaertner put Linnaeus's *Laurus winterana* into *Canella* to produce the modern name *C. winterana.*

From the beginning, *Canella* was known in Europe as a spice tree. Moreover, there was confusion between it and the true cinnamon sources of the Orient. Those plants are cinnamon (*Cinnamomum verum* = *C. zeylanicum*) and cassia (*C. aromaticum* = *C. cassia*). Because the part used is a rolled "stick" or "quill" of bark, kinds were distinguished largely by taste. Few could tell them apart by morphological traits. Yet, before modern times few inhabitants of the Old or New Worlds could afford to taste Asian cinnamon or cassia. For example, in 13th-century Britain, cinnamon sold for 10 shillings a pound and sugar went for 12 shillings. The prices remained proportional for centuries. Few rich people in the Americas could afford real cinnamon or cassia, and no poor people could use it. Rich and poor were forced to use the local equivalent, *Canella.* Olaf Swartz wrote in 1791: "This bark, together with the fruit of *Capsicum*, were formerly common ingredients in the food and drink of the Caraibs [Caribs]" (Grimé 1976).

Most of the common names in the Americas also reflect spice usage. It is known as Bahama whitewood [bark] (English Antilles), *bois canelle* (cinnamon tree, Guadeloupe), *caneel* (variant of *canela*, Virgin Islands), *canela* (cinnamon, Puerto Rico), *canela blanca* (white cinnamon, Cuba), *canela de la tierra* (cinnamon of the land, Dominican Republic), *canelilla* (little cinnamon, Dominican Republic), *canella* (cinnamon, Puerto Rico, USA, English Antilles), canella bark (USA, Barbados), *canelle poivré* (cinnamon pepper, Haiti), *canellier bâtard* (bastard cinnamon tree, Guadeloupe), *canellier blanc* (white cinnamon tree, Guadeloupe), cilliment bush (variant of cinnamon bush, Virgin Islands), cinnamon bark (Bahamas), cinnamon canella (USA), *corteza de winter falsa* (bark of false winter), *cúrbana* (maybe Taino, Cuba), false winter's bark, Jamaica cinnamon (Jamaica), pepper cinnamon (Virgin Islands, Dutch Antilles), whitewood bark (English Antilles), and wild cinnamon (Virgin Islands). Williams ([1837] 1962) wrote, "The bark is strongly aromatic, and flavored much like *Laurus cinnamon* [*sic*], but is more pungent."

Most use the bark to season food, but the leaves (Virgin Islands) and berries (Puerto Rico) are also used. Supposedly, the berries are hot like black pepper when gathered green and dried. Recently, my niece Stephanie Hicks Lindsay took a cooking course in New York City. In that course, she was taught to use "Mexican cinnamon," which they also called "canella." She raved about the superior flavor of this condiment, so I went looking for it in the local Tucson store. Perhaps because I shopped in a market that is mostly frequented by Hispanics, I found the "canella cinnamon" with no trouble. The checkout clerk, however, was not Hispanic. I had to tell her what it was so she could look up the price. My wife used the *Canella* instead of the Old World cinnamon on an apple cobbler. The taste is good, but I experienced an aftertaste that was bitter and unpleasant. Not everyone has that problem, and so the product is experiencing a revival in some areas.

A decoction of the bitter bark of *Canella* is used as a febrifuge and gargle to relieve sore throat in Brazil. It is also a febrifuge in Cuba where it is called *malambo* (Taino or African?). Similar uses include the infusion as a tonic (Puerto Rico), to relieve indigestion and fever in new mothers (Cuba), and to treat "female weakness" in the Bahamas (Roig 1945, Morton 1981). Not only is it a remedy for new mothers, but in Cuba it has been used as an abortifacient (Roig 1945).

The bark is macerated in alcohol or rum and rubbed on rheumatic and other pains in Cuba and Jamaica (Roig 1945). Leaf decoctions are also used in baths for rheumatism in the Bahamas, or the leaves and bark combined with other "bush" (medicinal plants) and made into a bush bath in the Virgin Islands (Petersen 1974, Morton 1981). The bark mixed with lime leaves and ginger helps stimulate appetite after illness in the Virgin Islands (Petersen 1974). Similarly, a leaf decoction is used to alleviate headache. The decoction of the wood is applied to the head or the chipped wood is smoked in a pipe in the Bahamas. Rum containing the bark is drunk for stomach pain in Jamaica (Morton 1981). The plant is considered so useful by some that it has been called *paratudo* (for everything, Brazil).

There are potent chemicals in the plant and they have been shown to have insecticidal activity (Udino 1994). Both leaves and stems are toxic to poultry. Those same poisons make it useful as a fish poison called *barbasco* [*marbasco*] (American variant of *verasco*, general common name for *Pterocaulon* and *Verbascum,* but also used for several fish poison plants, Puerto Rico). The bark formerly was exported to Europe and the United States for pharmaceutical use. Bark yields 0.75 to 1.25% volatile oil containing l-*a*-pineol, cineole, caryophyllene, ~8% resin, and 8% manitol. There is also drimane sesquiterpene present (Kioy 1989, Kioy et al. 1989).

One point about the plants has bothered and intrigued me since I first learned about them in the 1970s. Catesby in 1731–1732 said that the pharmacies in London sold the remedy as "Winter's Bark." Linnaeus first called them *Winteranus* (*Hortus cliffortianus* 1738) and some of the common names contain the word "winter." My confusion is this—does "winter" refer to a person or a season?

Canna

(Greek *kanna,* a reed, Latin *canna,* a reed or cane; also "reed" in Akkadian, *qanu,* and Hebrew, *qane*)

Canna flaccida (flabby)
bandana-of-the-Everglades (Florida)
bird-shot (Florida, cf. Williams [1837] 1962); Indian-shot (Texas)
sawakmalî:tî (rattle's contents, Mikasuki); *sáwko matihíta* (*skokv,* rattle, *em,* its, *vtehkv,* container, Creek)

Linnaeus began *Species Plantarum* with "*Classis I. Monandria. Monogynia.*" The first species on page 1 was *Canna indica.* He had three species, but there are now eight to ten known from the American tropics. Starch from *C. edulis* is considered easily digestible food for invalids, but most of the species are not eaten. Seeds may be viable for about 600 years, and they have been used as beads (Mabberley 1997).

The hard, black seeds were put inside a turtle shell or other item to make a rattle used in Seminole ceremonies (Sturtevant 1955). Judging from the common names "Indian-shot" and "bird-shot" the seeds have also been used as bullets in guns when no other material was available.

Capparis: Capers

(From Greek *káppari* and Latin *capparis*; from Arabic *kabar,* head, because the buds resemble heads, cf. Marafioti 1970)

The first time I saw Florida caper fruits (*Capparis flexuosa*), I thought they looked like dentures. The pods are long and red, and embedded within are white seeds about the size and spacing of teeth. Admittedly, the red pod looks like a bad case of pyorrhea, but that was my first impression. I had read about and eaten the Old World capers, but the one in the coastal hammock was certainly different.

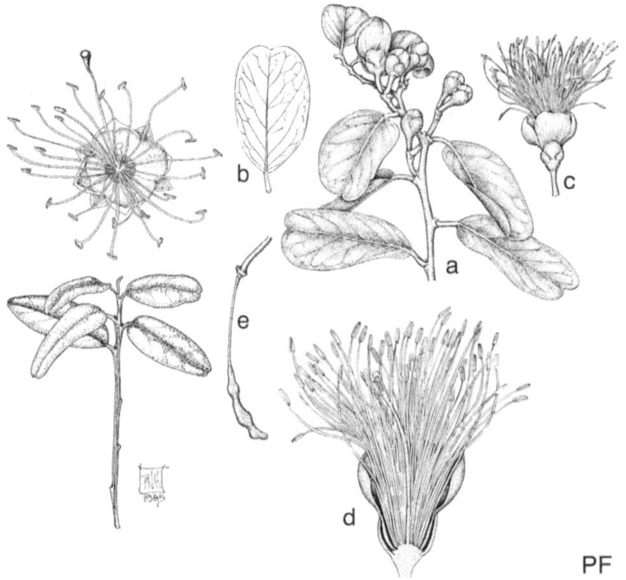

Capparis. *Capparis cynophallophora* (left). Above, flower, top view. Below, sterile branch. *Drawn by P.N. Honychurch. Capparis flexuosa* (right). a. Branch with flower buds. b. Leaf. c. Flower, side view. d. Flower longitudinally dissected. e. Fruit. *Drawn by Priscilla Fawcett.* From Correll and Correll 1982.

Linnaeus also realized that the American plants were different from those around the Mediterranean when he named both in 1753. He knew both species as living plants from cultivation, because he mentioned them in his *Hortus Cliffortianus* published in 1738. Moreover, he seems to have been impressed above all else by the long, slender, red fruits in the American plants. Certainly, those fruits were different from the short, fat pods of the European species. The Old World plants Linnaeus called *C. spinosa* because of their thorny nature. This and many other plants growing in the chapparal or *maccia* vegetation around the Mediterranean have armament of some type. The best-known American plants he called *C. cynophallophora.* This species name is one of the most famous examples of Linnaeus's uses of ribald names. *Capparis cynophallophora* means "the caper that bears a dog's penis." No other person or people in the world seemed to have noted that similarity until Linnaeus pointed it out. His seems to have been a uniquely vulgar mind.

The cultivated caper, *C. spinosa,* is commonly used as a food condiment in the Mediterranean countries. The unopened buds of these shrubby plants are used by people as diverse as the Saudis (*shafallah*), Greeks (*káppari*), Italians (*cappero*), Spanish (*alcaparra*), and Portuguese (*alcaparra*). Moreover, the usage has long been important in the cuisine of the region. Pickled, the buds add a unique flavor to particular dishes that are characteristic of southern Europe.

Whether the caper was spread from the Mediterranean through the Old World or introduced to there from the Orient is unresolved. The wide use of cognates for the Greek *káppari* argues for the first explanation, and many scholars support that view (Watt 1898, Hedrick 1919). In the major languages on the Indian subcontinent, names are related to the Greek phoneme. People in Bombay say *kabar,* and in widespread Hindi it is *kabra.* Across the Orient from east to west it is *kabra* (Bengal), *kabarra* (Punjab), *kabar* (West Pakistan), *kabarra* (Afghanistan), *kebír* (Persia), *kabár* (Syria), *kabar* (Arabic), and *kabarish* (Turkey) (Watt 1898). Those names apparently are not cognate to the Sanskrit *himsra,* or, if they are, they diverged from that language long ago.

The classical Eurasian literature had numerous records of the caper plants. For example, they were mentioned by authors like Antiphanes (4th century B.C.), Theophrastus (disciple of Plato and Aristotle, 372–287 B.C.), Zenon (ca. 260s–240s B.C.), Pliny (A.D. 23–79), Dioscorides (fl. A.D. 40–80), and Athenaeus of Naucratis (A.D. 170–230), author of *Deiphnosophistai* (The Learned Banquet). Even the Greek prostitute Phryne (400–330 B.C.) was a dealer in capers (Hedrick 1919). According to Jean de la Ruelle (Ruellius), 1474–1537, capers were cultivated in France before 1536 (Hedrick 1919). Scholars seem to be divided over why the ancients compared them to goats (*kapra*). There is nothing much about the plants that resembles a goat, except perhaps the odor. Both the plants and the bovid have strong, fetid, and to Westerners at least, markedly unpleasant odors. In other words, they stink.

Yet, to the people of the Mediterranean, the goat was historically an essential part of daily and religious life. Goats provided food, wool, and skins that were important for people of all classes. In addition, the animals provided offerings to the deities in the form of sacrifice. Perhaps from their point of view, a comparison of a food additive to a goat was one of the greatest compliments that could have been paid.

Another critical consideration in using the plants for flavoring foods is that people from Portugal to India prefer strong-tasting food. Many Americans want bland food. As I used to tell my classes, if Americans can taste the food, they spit it out. That statement is an obvious exaggeration, but it is not that far from reality.

This cultural difference between these parts of the world is difficult to understand until one realizes that it has been only the past few decades that the chile has become popular among some groups in the United States. Even now, there are regions within the United

States that consider *Capsicum* inedible unless it is the bell pepper, without capsaicin.

When the Europeans arrived in the New World, they found plants that they recognized as relatives to those of the Mediterranean *alcaparra*. These new plants were the American capers, but they were small trees, much larger than the small shrubs previously known. Moreover, the taste of these in foods was not the same as those from Spain and Portugal.

Indigenous people knew these American plants, and used them in seasoning, but they were not the important condiment that the genus was in the Old World. The Americans also used the plants in ways that were not familiar to Europeans. The larger plants allowed them to be used as wood for construction, tools, or even fuel (Little and Wadsworth 1964). Indigenous people also compared the fruits with others they knew and used names that showed that they had watched birds eating the fruits. Only the name *chile de perro* (dog [wild] chile, Chiapas) hints at a local use as a condiment.

Common names for these two species are not applied consistently, and there is considerable overlap in application. Both have similar or the same uses. An infusion or decoction of the leaves or roots is used as a diuretic and emmenagogue (Morton 1981). A decoction of the leaves is applied to various skin diseases, and the fruits are considered both sedative and antispasmodic. Not surprisingly, perhaps because of their long, slender, red fruits, both have been used to treat venereal diseases (Roig 1945). In the Cayman Islands, *C. cynophallophora* is called headache bush, suggesting another medical application.

Although Old World species of *Capparis* have been studied chemically more thoroughly, the New World species have received little attention. To date, it has been found that American *Capparis* have volatile oils, glucosinolates, and flavonoid aglycones (Kjaer and Schuster 1971, Gramosa et al. 1997, Pelotto et al. 1998).

Among the indigenous names for *C. cynophallophora* that do not seem to have been influenced by introduced languages are *arete* (Tabasco), *boxche'* (tree with black wood, Maya, Yucatán), *carbonero* (charcoal tree, Cuba), *ciguaraya* (bay tree, usually applied to a species of Meliaceae, Taino, Cuba), *endurece maiz* (it hardens corn, Nicaragua), *frijol* (bean, Dominican Republic), *frijolito* (little bean, Dominican Republic), *hoerihoeri* (probably Arawak, Aruba, Bonaire, Curaçao), *paaloe die loora* (parrot tree, Papaimento, Aruba, Bonaire, Curaçao), *raaba* (Papaimento, Aruba, Bonaire, Curaçao), *stokki* (Papaimento, Aruba, Bonaire, Curaçao), and *zic* (Guatemala).

Some of the local names are simply based on the Old World word "caper." These include caper tree (British Antilles, USA) and Jamaica caper (USA). Others compare the plant uses to another Old World condiment, mustard (*Brassica nigra*). These include mustard tree (English Antilles, Dutch Antilles), *jeerba mostard* (herb mustard, Aruba, Bonaire, Curaçao), *mostacilla* (little mustard, Cuba), and simply Spanish for mustard (*mostaza*). These names are applied because the roots of the capers are ground and used much like mustard or horseradish (Little and Wadsworth 1964, Little et al. 1974).

Other names compare the trees to other plants. Some of these are black wattle and black willow (Bahamas, Jamaica). In these, the modifier "black" simply means "wild." A "wattle" in Australia, for example, is the genus *Acacia*. However, other people apply the name to any shrub or tree that could be used in this type of construction. "Wattle and daub" houses are a type used by people around the world where the "wattle" refers to the walls woven of various plant parts, and the "daub" is a mud covering to keep out hot and cold winds.

A number of other places make comparisons that are not obvious. Several describe the plants with the color black. These include *bois noir* (black bush, Guadeloupe), *burro* (Puerto Rico), *burro prieto* (Spanish Antilles), *palo de burro* (burro bush), and *palo de burro prieto* (black burro bush). The reason for calling them black is obscure. Some of the other names are complimentary and others are not. Among these are *bejuco Ingles* (English vine, Puerto Rico), *bois caca* (feces tree, Haiti), *bois d'argent* (silver bush, Haiti), *bois Sénegal* (Senegal bush, Haiti), man of war bush (Dutch Antilles), *oliba* [*olivo*] (olive, Dominican Republic), *palo diablo* (devil's bush, Cuba), *pois mabouye* (from Taino *maboya*, evil spirit bean, Puerto Rico), *sapo* (toad, Puerto Rico), wild orange (Bahamas), and zebrawood (USA).

Names that may be applied more uniquely to *C. flexuosa* include bay leaved caper (USA), flexible caper (Florida), *bois rave* (enraptured tree, Haiti), *bokanche'* (*bo*, skin, *kan*, viper, *che'*, tree, or tree that looks like a viper's skin, Maya, Yucatán), bottle wiss (wiss = vine, Tortola), *cansa caballo* (tires horses, Venezuela), *jos* (San Luis Potosí), *mimbre de monte* (wild *Chilopsis linearis*, called desert willow in English, Sinaloa), *naranjuelo* (little orange, Colombia), *palanguán* (big plate, Puerto Rico), *palo barba de indio* (Indian's beard bush, Cuba), *pan y agua* (bread and water, Venezuela), *potal* (anchor?, Guatemala), *tabarjeca* (Mayo), *tablelojeca* (Sinaloa), *thiuh* (Huastac, San Luis Potosí), *vela de muerto* (dead man's candle), and *xbayuamak* [*xbayumak, xpayumak*] (climber that destroys, Maya, Yucatán).

Both species are also used as a chewstick or toothbrush (Roig 1945, Morton 1981). The ends of small branches are chewed until they are frayed and only the fibers remain. These fibers are effectively used to clean the teeth and work at least as well as dental floss. This is not the only plant used in this way, and some also are known to contain fluorine. Maybe the American capers also fluorinate the teeth and help prevent dental caries. No one seems to have examined that possibility. However, no one knowing the scientific name *Capparis cynophallophora* would be able to use it without remembering Linnaeus's euphemism.

Capraria: Goatweed

(From Latin *caprarius*, pertaining to goats, akin to Greek *capra*, goat)

Capraria biflora. *Drawn by P.N. Honychurch.*

The association with goats and these herbs came comparatively late to the scientific community. Two of the earlier Europeans who were familiar with the species had no real idea how unique the species was and compared it with *Lysimachia* (Plukenet in 1696), or with *Gratiola* (Commelin in 1701). Of those two, Commelin was by far the better botanist. It fell to Paul Hermann in 1698 to coin the name *Capraria*. Presumably, that name was applied because goats grazed on the plants when no other farmyard animals would touch them. At least that is true today and the animals have not changed that much in the intervening centuries. The plants contain so many potent chemicals that even few insects graze on the foliage. The only butterfly capable of dealing with the poisonous compounds is the common buckeye (*Junonia coenia*). Although these Lepidoptera feed also on other members of the Veronicaceae, they must have cast-iron digestive systems if they feed on *Capraria* and *Scoparia*.

Linnaeus accepted Herman's suggestion, basing the genus on the Greek word *kapros* (goats), and gave us the modern name *Capraria biflora*. This is apparently a case where the scientific name led to the common name "goatweed." Often it is the common name or use that leads to the scientific name. Linnaeus knew the plants from Herman's report on the island of Curaçao, and also from cultivation in Uppsala, Sweden.

These plants are basically American, known from Bermuda, the Bahamas, Florida, the West Indies, southern Texas, and Mexico, to Peru and Brazil. *Capraria* also has been introduced into a few parts of the Old World tropics.

Many of the common names given to these plants revolve around their use as a tea. Among these are *chá da américa* (American tea, Brazil), *chá das Antilhas* (Antilles tea, Brazil), *chá de Lima* (Lima tea, Brazil), *chá de Marajó* (tea from Marajó, Brazil), *chá do Mexico* (Mexican tea, Brazil), *chá preto* (black tea, Brazil), *hierba [de] te* (tea herb, Panama), Jamaica tea, *tan-chi [tanchi, tantje]* (something to put in the mouth, Maya, Belize), *tantsji* (Aruba, Bonaire, Curaçao), *tasajo* (Chiapas), *té* (tea), *te de las Antillas* (Antilles tea), *té de monte* (wild tea), *té de salud* (health tea), *té de Santa María* (Saint Mary's tea), *té del país* (country tea), *té nacional* (national tea), *té silvestre* (wild tea), *thé d'Amerique* (American tea, French Antilles), *thé de Antilles* (Antilles tea, French Antilles), *thé de Guadeloupe* (Guadeloupe tea), *thé de sante* (saint's tea, French Antilles), *thé des Anglais* (English tea, French Antilles), *thé du pays [du thé pays, dite payi]* (country tea, French Antilles), *thé muraille* (mulberry tea, French Antilles), *thé pays* (country tea, French Antilles), *thé suisse* (Swiss tea, French Antilles), West Indian tea, and tea (Dutch Antilles).

First and foremost, this is a substitute beverage for ordinary tea (*Camellia sinensis*) from Belize and the Bahamas south to Venezuela. In addition, this "tea" is an almost universal panacea in some places, or at least considered a highly effective tonic. The beverage is used against colds, colic, coughs, diabetes, diarrhea, flatulence, fevers, gonorrhea, hemorrhoids, indigestion especially in pregnant women, influenza, kidney or bladder problems, leukorrhea, measles, menstrual cramps in young girls, *pasmo* (blood stagnation or congestion), sore eyes, rheumatic pains, and tiredness (Morton 1981, Austin and Bourne 1992, Mors et al.

2000). The decoction is astringent, tonic, and stops vomiting. Juice drops also are put in the ears to combat earache. Not only is the decoction used as a tea, but also as bush baths. The common name that reflects this wide medicinal usage is *pericón* (useful for everything, or heal-all). The Bahamaian names stow wort and stow-weed probably mean essentially the same. An archaic meaning of "stow" is to "stop."

Several uses are associated with childbirth. A decoction combined with *Heliotropium angiospermum* is given to all week-old babies when they are "changing skin" in Curaçao (Morton 1981). In the Bahamas, it is given to expectant mothers to expedite labor. In Trinidad, it is drunk after childbirth. The same mixture is used to alleviate painful menstruation. In at least Yucatán and Cuba, it is a vaginal douche and likely effective against *Candida* infections.

Some of the common names reflect the medicinal uses without adding the word for tea. Among these names are earache bush, grannybush, worry bush, *viuda* (widow), *pasmo-wa-xi-uil* [*pasmoxiu*] (herb in the shape of a tortilla [*wah*, tortilla, *xi-uil-xiw*, herb] used to treat *pasmo*, Maya, Belize, Yucatán). Sometimes that is simply shortened to *pasmo*. Probably the name *chokuilxiw* [*chakuil-xiu, chocuilxiu*] (herb of the rat god, Maya, Yucatán) also reflects medicinal uses.

Other common names reflect a more frequent and utilitarian use as a broom. Most obvious of these are *escoba* (broom) and *escobilla* (little broom). However, it is likely that *claudiosa*, *claviosa*, *esclaviosa*, and *escabiosa* are mispronunciations of those names or misunderstandings by people who recorded them. These names refer to such disparate things (a person named Caludio, a heart, and the plant called "scabios" in English) that they must be errors of one kind or another. *Feregosa* probably belongs in this list, too. Surely, *cola de gallo* (rooster tail, Panama) is also associated with use as a broom.

In Oaxaca, Mexico, the plants are called *lengua de gallina* (chicken tongue). That must be an allusion to the long slender leaves as there is nothing else about the herbs that resembles a tongue. In Veracruz, it is *peludillo* (little hairy one), an inappropriate name because the plants are essentially glabrous. Could this be like the English expression: "It will grow hair on your chest"?

Some names either misidentify the plants or have a nomenclature system different from that used by the scientific community. These include *malvavisco* (usually confined to *Malvaviscus*, of the Malvaceae, Oaxaca) and *verbena* (verbena, Verbenaceae). Names given by Morton (1981) that have not been located elsewhere are *magüiro*, *majuito*, and *makjuito*. These resemble indigenous words but their provenance and meaning have not been determined.

Water extracts of dried leaves have hypoglycemic activity in mice (Pérez et al. 1984), and root extracts are cytotoxic against certain kinds of cancer (Nascimento et al. 1985). The antibiotic alkaloid biflorine was isolated from roots (Gonçalves et al. 1953, 1958). In large doses, biflorine causes dizziness, intoxication, and stupification. One study showed that leaves from goatweed were a central nervous system depressant. That depression probably leads people in the Netherland Antilles to chew the plant as a tobacco substitute.

Capsicum: Bird Peppers

(From Greek, *kapto*, to bite, or from Latin, *capsa*, a box, a reference to the fruit's shape)

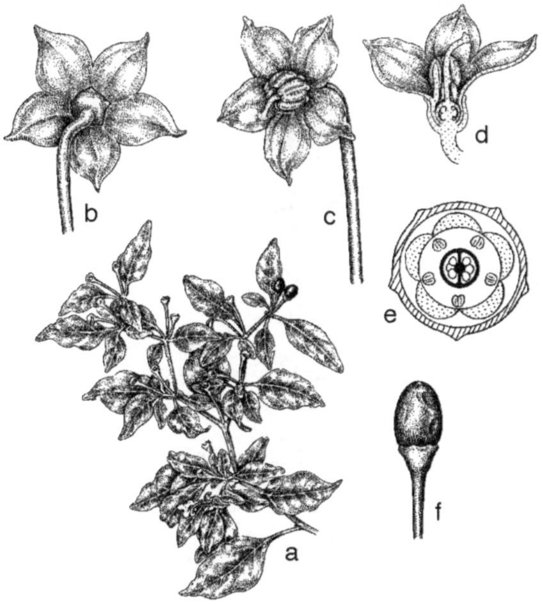

***Capsicum annuum* var. *glabriusculum*.** a. Branch of flowering and fruiting plant. b. Flower, back view. c. Flower, front view. d. Flower, longitudinally dissected. e. Floral diagram. f. Fruit. *Drawn by Priscilla Fawcett.* Correll and Correll 1982.

From Florida south into the tropics are herbs about 1.5 m (4.9 feet) tall that have small, inconspicuous, white flowers. Fruits smaller than the end of your little finger follow and then turn red when ripe. Birds love these little fruits and greedily defend the area where the bushes grow. Humans, too, love the fruits and compete with the birds for the harvest. This competition has been going on since there were humans and birds in the Americas. English-speaking people usually call the fruits bird peppers (Florida to Belize). Several others associate them with birds in their common names, including some that are simple translations, and others that give different ideas. For example, Haitians say *piment z'oiseaux* [*piment zoué-zeau, piment zouézo*] (bird pepper), while it is *ají pajarito* (little bird pepper) to Colombians, and the Huastec of Mexico say the same with *ts'itsin its.* However, the Huastec have noticed another aspect of this bird/plant association and also call them *taa' ts'itsin* (bird excrement). Birds do "plant" peppers regularly.

These herbs are *Capsicum annuum* (annual). Some call them *C. annuum* var. *aviculare,* and others use *C. annuum* var. *glabriusculum* (completely without hair). The correct name is the second, although it is too bad we cannot retain the one that nicely translates the common name—*aviculare* = of birds. Linnaeus called it *Capsicum* in 1753. Surely, this is another of Linnaeus's dual meaning genera intended to cause us to wonder, or perhaps, even to convey both meanings (Austin 1993).

The first Spanish in the Caribbean asked the natives what they called those blazingly hot little fruits that they added to their food. The Taino people of Cuba, related to the Arawaks of northern South America, replied that they were *ají.* This word, variously written as *axí, ajé, agí,* and *así,* came to mean *Capsicum* from Cuba to Peru. The Dominican Caribs still call them *áthi.* Their Central American Island Carib relatives say *áti,* Arawaks say *háthi,* and Guajiro *hasi* (Taylor 1977). The Garífuna of Nicaragua say *áti.* Mexican people using similar if not cognate names to *ají* are the Mayan who say *i'k* [*ik, ic*], the Huastec with *its* [*itz*], and perhaps *corí* [*coriquí*] (Tarahumara, Chihuahua), *ng-i* [*fúguñi*] (Otomí, Hidalgo), *pi'n* [*stac'apin, lactzu-pi'n*] (Totonac, Puebla), *pitzaj* (San Luis Potosí), and *ya'a* (Mixtec, Oaxaca). More distant, but still perhaps related, are *uchu* (Quechua, Peru) and *huayca* (Aymara, Peru).

Many other people have names for the bird peppers that apparently mean simply *Capsicum.* These include *chicnigüi* (Chiapas), *có'cori* [*cócorit, ci'ocori*] (Mayo, Sonora), *conguito* (Colombia), *tichushi* (Oaxaca), *gapuchín* (Veracruz), *guina* [*guiña, quiña*] (Zapotec, Oaxaca), *jonguñi* (Otomí, Hidalgo), and *jonnigüi* [*jonnihui*] (Zoque, Chiapas). The Tarahumara recognize three varieties, *wum pei, wum ganario',* and *wum ka?* (Chihuahua). The Jicaque of Honduras say *cela* (Lentz 1986).

Throughout the Greater Antilles, people still use the basic Taino word for the fruits, although now it is Spanish. Variations include *ají bobito* (crazy little chile, Hispaniola), *ají caballero* (gentleman pepper, Puerto Rico), *ají caribe* (Carib chile, Hispaniola), *ají chirel* (insipid chile, Colombia, Venezuela), *ají chivato* (goat pepper, Colombia), *ají guaguao* [*huarahuao*] (wild chile, Taino, Cuba), *ají montesino* (little wild chile, Hispaniola), *ají picante* (spicy pepper, Puerto Rico), and *ají tití* (monkey chile, Hispaniola).

The name *ají* was carried by the Spanish from the Caribbean to Peru and surrounding countries, where it was adopted into the locally spoken Creole. However, this did not occur in Mexico and other parts of North America, or in Brazil.

The dominant people in Mexico were the Aztecs. In their Náhuatl language, they called these spicy fruits *chilli.* Variations on that in Spanish became *chile* or *chil* from California through Arizona to Texas south to Venezuela. There are dozens of varieties and variations of chiles, but the one called bird pepper was the *chiltecpin* (literally "flea pepper," because it bites, Náhuatl). This too has variants including *chillipiquin, chiltepín,* or *chile piquín* (Sonora, Hidalgo, Puebla), *chiltepe, tepín,* and *pequin* (Mexico to Costa Rica). The biting aspect of the chile is also reflected in the names *chile diente de tlacuache* (opossum's tooth chile, Tamaulipas), *cuachile* [*guachile*] and *chile cuachile* (chile that bites, Náhuatl, Oaxaca), and *tempechile* [*tenpechile*] (*tentli,* lips, apparently an allusion to the burning of the lips, Náhuatl, Chiapas).

In addition, chile has other modifiers that are applied, either locally or to note particular variants and/or uses. Among these are *chile amash* (papery? chile, Tabasco), *chile bravo* (wild pepper), *chile de monte* (wild pepper, Sonora, Yucatán), *chile de zope* (vulture's pepper), *chile quimiche* (*quimchin,* rat, Náhuatl, Oaxaca), *chile siete caldos* (chile enough for six stews, Chiapas), and *chililllo* (little chile, Yucatán).

Other people have variations on the biting taste of these hot little fruits. In Yucatán, the Maya call bird peppers *max'ik* [*max, a'h'max-ik*] (*max,* bumblebee, plus chile, or chile that stings like a bee). In Costa Rica, they say *quiebra-muelas,* a chile so hot it is a "tooth breaker." Colombians simply say *pimiento morrón* (pepper that hits). In French areas, it is *piment caribe* (Carib chile, a reference to a formerly cannibalistic tribe, Dominica), or *piment enragé* (mad chile, French Guiana).

Some make other comparisons. As examples, they are little chiles to the Seri of Sonora (*coquée-quizil* [*quishil*]). The Huastec of San Luis Potosí also call them little chile, *tsakam its* [*tzacam-its, chakam ich*], but they recognize an almost indistinguishable type as *ts'aah its* (vine chile). The Tohono O'odham (Papago) of Arizona call all chiles *ko'okol* [*o'olas*] (from *s-ko'ok,* to hurt), but their relatives, the Akimel O'odham (Pima), call bird chiles *o'olas ko'okol* (spherical chile). English speakers have variants too, with goat pepper, heart-pea, heartseeds, red pepper, and wild pepper. The southeastern people had no name for these new plants, and said simply *hishi homi humma* (pungent red leaf, Choctaw).

Brazilians and other South Americans also use variations of the original word that Columbus called them in his journal—*pimiento.* This name is akin to Portuguese *pimento* and French *piment.* All are derived from Medieval Latin *pigmenta,* a spiced drink, hence also spice or pepper. The same Latin *pigmenta* meant drugs or juice of plants, and color.

We use the word "pimento" in English, but now usually to apply to one bland cultivar. In addition to those already listed are *pimenta* (Brazil), *pimientones* (Colombia), and *pepre* (pepper, Suriname). Brazilians have a variety like one of the longer cultivars of *chiltepín,* but they call it *pimenta malagueta.* The word *malagueta* may have come from Málaga, Spain, which was a center of Portuguese trade in spice (Ferreira ca. 1978). Or it may have come from Melle, in Niger, as Fuchs (1542) used the name *melegetae,* which appeared in European literature by A.D. 1214. *Melegetae* originally was *Aframomum melegueta* (Zingiberaceae), which was used as a substitute for black pepper (*Piper nigrum*) before the New World was discovered.

Our word "pepper" had a different origin. That came into English by the 1100s and was *peper* in Middle English and *pipor* in Old English. All three are cognate with German *Pfeffer,* Dutch *peber,* Norwegian *pipar,* and Swedish *peppar.* They are based on Latin *piper* and Greek *peperi* and were adopted from those Mediterranean languages before the 4th century A.D.

When Cristobal Colón (Christopher Columbus) first arrived in the New World, he found the natives of the Caribbean eating deliciously spicy condiments with their food. Since the most valuable resource of the time was pepper, Columbus was delighted. He wanted gold, but surely he could get plenty of that with these fiery fruits. He wrote: "Finally, to compress into a few words the entire summary of my voyage and speedy return, and of the advantages derivable therefrom, I promise, that with a little assistance afforded me by our most invincible sovereigns, I will procure them as much gold as they need, as great a quantity of spices."

Columbus promised gold and spices in equal quantities, but most historians do not even mention the plants. In fact, spices provided greater wealth to Europeans than those metals. It is difficult to believe the value Old World people put on pepper (*P. nigrum*). For example, in A.D. 410, Alaric the Goth captured Rome and demanded as tribute 30,000 pounds of silver, 5000 pounds of gold, and 3000 pounds of pepper (Howard 1996). In the Middle Ages rents, dowries, and taxes sometimes were paid with peppers. Landlords wanted rental in peppercorns rather than money. Trade in spices, predominantly pepper, made the famous centers of commerce of Alexandria, Genoa, and Venice the richest parts of Europe up until the 1500s. These cities had trade monopolies with the Arabic merchants and then sold the produce at inflated prices to the other Europeans (Rosengarten 1969).

It was largely an attempt to break that domination by these centers that initiated exploration of new routes to the tropics to obtain spices. Expeditions of both Columbus and Vasco de Gama were prompted by Portugal and Spain's desire to find cheaper spices. Indeed, the royalty expected to earn more from the spice trade than from gold and silver. Columbus found what he thought was the "orient," but brought home a red spice called *ají* and not the black pepper that Europe knew. It took time for merchants to realize how useful that condiment was, and they were still hungry for the familiar black pepper. Vasco de Gama left Portugal searching for a sea route to the spice-producing Malabar Coast of India in 1498. His success broke the trade monopoly and eventually led to the economic decline of Alexandria, Genoa, and Venice (Rosengarten 1969).

Although black pepper is still the most important spice in the world, chiles run a close second. Chiles are even better because they provide benefits unachievable with black pepper. Chiles are healthy food and provide chemicals needed by the human body that are not in black pepper. Chiles have many compounds besides the hot stuff (capsaicin), including vitamins A and C. In addition to capsaicin, compounds in chiles include solanidine, solanine, solasodine, and scopoletin. Duke and duCellier (1993) list 63 chemicals in *Capsicum* suspected of having cancer-preventative or antitumor activity.

All those chemicals make it easier to understand why people have used chiles in remedies to treat so many maladies, including asthma, backache, cancers, cough, gonorrhea, hemorrhoids, liver congestion, rheumatism, sore throat, swollen feet, toothache, tumors, and varicose veins. The Aztecs used chiles to punish children (Ross 1978), and others have used them as a calmant and to stop thumb-sucking and nail-biting. In addition, there is evidence that they stabilize blood pressure and reduce excessive bleeding. Although it is difficult to believe the mothers could stand the cure, some indigenous people in the southwestern United States have used chiles to wean infants from breast-feeding.

So, the next time you bite into a chile-laced dish, and the tears start streaming down your cheeks, remember that you are eating healthy food. As Mexicans say, "*Chiles dan sabor a la comida*" (Chiles give taste to food).

Cardamine

(From the Greek *kardamis*, a kind of cress)

billeri (bilious, Italian)
biolair ghriagain (*biolair*, cress, *ghriagain*, deer, Gaelic)
cardamine (from Greek *kardamis,* or the carding tool for wool, French)
cuckoo flower
karse (cress, Norwegian); *Schaumkress* (foam cress, German)
mastuerzo de prado (meadow mustard, Spanish)

Cardamine concatenata (with a chain)
cutleaf toothcup
bittercress ("cress," *Lepidium sativum*, is either derived from a Old High German *chresan*, to creep or creeper, or from Latin *crescere*, to grow)
cardamine (French)

Linnaeus included six species in *Cardamine* in 1754. All of them were from the Old World. Now there are 200 temperate species known, with 31 of them native to Europe (Mabberley 1997). Several of the European plants have been put in *Nasturtium*, and that suggests that they may have been eaten.

Hedrick (1919) listed eight species of *Cardamine* used from the Straits of Magellen to New Caledonia, so the comment by Fernald et al. (1958) that species, other than the two they listed, "doubtless have similar qualities," is surely correct. Most who comment on the genus suggest that it has "just the taste of the English water-cress" (Hedrick 1919). The comparison in the name to "cress" records that long-standing impression.

The Iroquois used *C. concatenata* as food, eaten raw or with salt and boiled (Moerman 1998). Several other species are known to have been used by a variety of indigenous people as food or flavoring in food (Moerman 1998).

Additionally, this species was used as a medicine. The Iroquois mashed the roots and applied them to the head to relieve headaches, to treat colds, to stimulate the appetite and regulate the stomach, to aleviate heart problems, and to treat injuries (Moerman 1998). The herbs were also considered somewhat hallucinogenic, and were used in devination, as a hunting medicine, and as a "love medicine" (Moerman 1998).

Cardiospermum: Balloon Vine

(From Greek *kardia*, heart, and *sperma*, a seed; refers to the shape of the seed or to the heart-shaped spot on the seed)

In southern Florida, you may find a bank of vines covering much of the other plants on a fence, or sprawling over shrubs on the edge of a disturbed site. These vines are mostly in Monroe and Miami-Dade Counties, but they may occur as far north as Seminole and Polk Counties. The fruits are eye-catching because they look like little balloons—hence the name balloon vine used from Florida and the Bahamas south to at least Panama. The plants are probably *Cardiospermum halicacabum*, although they might be *C. microcarpum*; the former is the more common of the two.

The flowers are small and greenish, but the foliage and fruits are a magnet for three butterfly species. These are the amethyst hairstreak (*Chlorostrymon maesites*), the silver-banded hairstreak (*Chlorostrymon simaethis*), and the Miami blue (*Hemiargus thomasi*). The amethyst hairstreak also uses a number of other plants to feed its larvae, but the silver-banded hairstreak prefers to put its eggs on the *Cardiospermum.* If you open the fruits, you find the caterpillars inside—or at least the frass from where they have been feeding. Since the plants are comparatively rare, these butterflies too are uncommon. In the early 1970s, it was hard to find a fruit without some sign of the silver-banded hairstreak having used it.

These plants occur around the world. They have attracted the attention of people in many areas and even made it into cultivation for the ornamental "fern-like" foliage and swollen seed pods. Most scholars think that *C. halicacabum* is native to the American tropics, including the southern United States. This may be one of the many medicinal plants that were carried around the world by the early Europeans. However, equally probable is that it has been spread around the world by nonhuman means. At least Smith (1985) was of the opinion that it is easily dispersed and probably indigenous throughout its range.

Cardiospermum halicacabum is found in the United States from Florida to Texas, and north to Kansas and Delaware. Small (1933) thought that it had been introduced from tropical America, and that is a possibility. Others have treated it as native, at least to Florida, and we make that assumption here. Outside the United States, the species is found throughout much of Mexico, Mesoamerica, the Caribbean, and in South America to Peru and Argentina. It is cultivated in Brazil.

Balloon vine is also widespread in the Old World. Indeed, in 1753 Linnaeus knew it from cultivation in Europe and from the wild only in Sri Lanka. The species name *halicacabum* was taken from some plant the Greeks called *halikakabos.* That was also a plant Plinius called *vesicaria.* Linnaeus adapted the specific epithet from the classical usage.

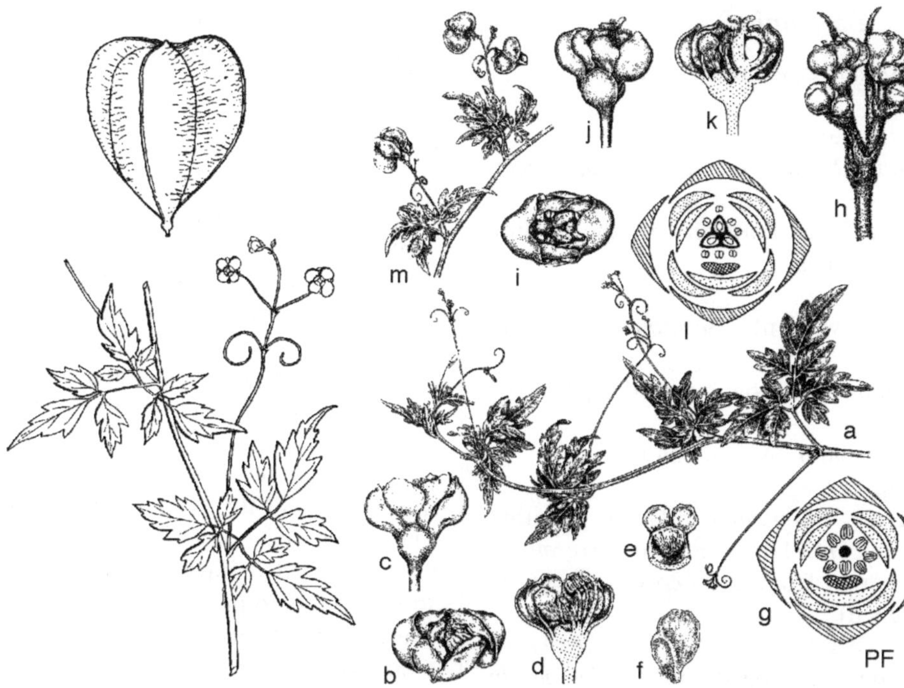

Cardiospermum. *Cardiospermum halicacabum* (left). Flowering branch (below left) and fruit (above left, enlarged). From Britton and Brown 1897. *Cardiospermum microcarpum* (right). a. Portion of flowering stem. b. Staminate flower from above. c. Staminate flower from side. d. Staminate flower longitudinally dissected. e. Glandular staminode-like tissue. f. Petal. g. Floral diagram of staminate flower. h. Inflorescence. i. Pistillate flower from above. j. Pistillate flower from side. k. Pistillate flower longitudinally dissected. l. Floral diagram of pistillate flower. m. Fruiting section of branch. *Drawn by Priscilla Fawcett.* From Correll and Correll 1982.

The species has been in India for so long that it has the Sanskrit names *indravalli* (Indra's vine; Indra is the Hindu god of thunder and storms), *Jyautishmati,* and *káraví.* Having a name in an extinct language is taken by most people as good evidence of nativity to a particular region, and balloon vine qualifies. Moreover, there are common names in Bengali, Gujarati, Hindi, Kannada, Malayalam, Marathi, Tamil, and Telegu. The fact that these names all differ from one another, and from American names, further suggests a pre-Columbian history in the Indian region.

American common names, in addition to balloon vine, are varied. Many of the other names also refer to the fruits, including heartpea (USA, Jamaica), heartseed (USA), love-in-a-puff (horticulture), and *bonnet carré* (square hat, Hispaniola). A variant is a bit more vulgar in Mexico where they are called *huevo de gato* (cat eggs [testicles], Durango). Apparently, the shape of the fruits also led to names like *bombilla* (electric lightbulb, Puerto Rico), *bombilla menor* (little electric lightbulb, Puerto Rico), *farolito* (little lantern, Cuba, Puerto Rico, Panama, Colombia), and *farolito de la Virgen* (the Virgin's little lantern, Venezuela).

Other names refer to other aspects. In India, and perhaps other areas, the plant is known as the blister creeper, a reference to irritating chemicals it contains. Trinidadians call it bich grass or *zeb bich* ("bich" is an intestinal malady). Mexican goats eat the twiners, and people call them *hierba de chivato* (goat's herb, Tamaulipas). There are even comparisons to other plants. To some the finely lobed and toothed leaves resemble parsley (*Petroselinum crispum*), with names like *bâtard persil* (false parsley, Dominica) or *persil bâtard* (Hispaniola). Other people think the vines look more like the sweet pea (*Lathyrus odoratus*), and call them *guisante maravila* (marvelous sweet pea, Puerto Rico) or *pois [de] merveille* (marvelous sweet pea, Hispaniola). They are also called wild supple jack (Jamaica) and winter cherry (USA, India). Some people simply refer to the fact that the plants are cultivated and say garden wine (wine = vine, Turks and Caicos).

There are also names that refer to medical uses such as sprainbush wine (wine = vine, Dutch Antilles). Several names make obscure references. These uncertain names include *muditos* (little changling, Oaxaca), *revienta caballos* (horse breaker, Cuba), and *toffe-toffe* (a *touffe* is a clump of grass, Hispaniola). Presumably, the first two allude to medicinal uses. Since the plants are considered laxative, the name *revienta caballos* may be an allusion to that. Perhaps the Hispaniolan name alludes to the massed clusters in which the plants grow. Biloxi in Louisiana said *po'poxtyi'* (*popoxi+tyi*, swelling medicine) and used the root in a remedy for dropsy (Dorsey and Swanton 1912).

Americans use the plants in a variety of ways. The most prominent medicinal treatment is for kidney problems. In Mexico, Cuba, Hispaniola, and the Turks and Caicos Islands the plants are considered diuretic (Standley 1920–1926, Roig 1945, Liogier 1974, Morton 1977, Ayensu 1981). In Mexico the roots are used, in Cuba the leaves, and in the Turks and Caicos the seeds. Each area has its own variant of medicinal preparations, including decoctions of roots and leaves. Others use the seeds, which are parched, ground like coffee, and boiled (Ayensu 1981). The preparation in each area, however, is prescribed for swelling of the legs and feet associated with kidney trouble. This same decoction is used to treat rheumatism and arthritis in Argentina.

The same or similar preparations are given in Mexico as sudorifics, emetics, purgatives, diaphoretics, and rubifacients in rheum (Standley 1920–1926). Cubans also use it as a sudorific (Roig 1945). In Hispaniola, the same basic decoction, either boiled or not, is also considered a refreshing drink (Liogier 1974). Similarly, in Dominica, the leaves are crushed in water by Caribs to make a refreshing and cooling drink (Hodge and Taylor 1957). People in Hispaniola also use the leaves and stems in emollient poultices for abscesses and sores. In Trinidad, leaves are used as an infusion for "bich" (lower intestine disease) and indigestion (Morton 1981).

According to Parrotta (2001), extracts from the plant have sedative effects on the central nervous system, and act as an analgesic, anti-inflammatory, vasodepressant, and antispasmodic. The plant contains at least saponins that would make it effective as a diuretic. The same saponins would be responsible for a slimy consistency that would be effective in poultices. A related species has a sulfur-containing cyanogenic glucoside (Hübel and Nahrstedt 1979). Probably *C. halicacabum* has the same or similar glucosides. In addition to medicinal uses, the saponin foams when the plants are rubbed in water, and the solution has been used as a soap substitute (Morton 1981). Caution is indicated because the chemicals in the plant irritate the skin in some people.

People also have uses for the plants other than medicinal. In the coastal Carolinas, the young foliage is cooked and eaten as a potherb (Morton 1968b). That usage may have been introduced by Africans because it is known in Africa and Asia. In Colombia, the tough stems are used mostly as a ready-made twine to tie up rustic constructions. In many places, the seeds are used in necklaces and other novelties. In some parts of Brazil, the plants are cultivated for the seeds that are made into small bands worn by men. They believe that these bands will ward off snakebite (Mabberley 1997). Outside the Americas some also believe that the plants are aphrodisiac.

The plant was first mentioned by German herbalist Leonard Fuchs (1542). He noted it as a novelty in his country that none of the ancient authorities apparently knew. Soon afterward, another European herbalist deduced that the plants should be used medicinally for heart disease. That conclusion was not based on that use elsewhere, but rather on the similarity of the shape of the fruit to the heart.

Fuchs (1542) also recorded common names that are no longer used. He put the plants immediately after what he called *Halicacabum vulgare* (now *Physalis alkakengi*) and said that the new plant was *H. peregrinus* because it had recently "traveled" to Germany. The English of his time called the plants "blackewinter cherie." That surely gave rise to the modern name "winter cherry." In German it was *welsch Schlutten* (foreign ground cherry), and in Dutch *roomsche crieken van over zee* (Roman [Catholic] cherry from across the sea). In French it was *halicacabon estrange* (foreign *halicacabum*), and in Italian *halicacabo salvatico* (wild *halicacabum*). The Spanish called it *yerba mora desta suerte estangera* (foreign bramble herb). Fuchs gave the plant no medicinal value in spite of writing a book on that topic, but he suggested that this new plant would be "suitable for planting on arbors." That statement presaged history.

Carphephorus

(Greek *carpho*, dried up, withered, *karphe*, hay, *phorus*, bearing)

Carphephorus odoratissima. From Britton and Brown 1898.

Carphephorus corymbosus (having flowers in flat-topped clusters)
coastal plain chaff-head [coastalplain chaffhead]
paint-brush [paintbrush] (given to the instrument with bristles to apply paint by 1837; later to plants)

Carphephorus odoratissima (very fragrant)
deer['s] tongue ("deer" from Old English *díor*; entymologically related to *dier* in Dutch and *Tier* in German, both with the more general meaning of "animal"); combined with tongue to allude to leaf shape; in use by 1883, southeastern USA)
dog tongue (southeastern USA); hound's tongue (comparison of the leaves with *Cynoglossum*, which see, southeastern USA)
vanilla leaf [vanillaleaf] (apparently in use after 1856, Florida); vanilla plant ("vanilla" came into English by 1662 from Spanish *vainilla*, diminutive of *vaina*, sheath, referring to the pods of the orchid *Vanilla planifolia*; the first record of it found associated with this Asteraceous herb is by Asa Gray in his *Manual of Botany* from 1856; he wrote of the "Vanilla-plant ... Leaves exhaling the odor of Vanilla when bruised," southeastern USA)

In 1816, French specialist in the family Asteraceae, Alexandre-Henri Gabriel Comte de Cassini (1781–1832), created the genus *Carphephorus* for an American plant. It is now known that the genus contains four species, all endemic to the southeastern United States (Correa and Wilbur 1969, King and Robinson 1987).

Murphee (1965) found residents of the Florida Panhandle putting these herbs in pillows to relieve asthma. *Carphephorus odoratissima* leaves contain coumarin and have been used as a flavoring agent, particularly in smoking tobacco, perfumes, and to imitate vanilla extract (Small 1933, Christensen 1946, Lewis and Elvin-Lewis 1977, Hocking 1997). It also has been used as a aromatic tonic, diaphoretic, stimulant, and diuretic (Hocking 1997). *Carphephorus corymbosus* is a urinary antiseptic and is sold under the proprietary name Rasapen.

Carpinus

(The classical Latin name of the hornbeam, said to be from *carpentum*, a Roman horse-drawn vehicle with wheels made from its hard wood; akin to Akkadian *karru*, mountain range, and Hebrew *pinna*, pinnacle)

agn-bøk (*agn*, husk, *bøk*, beech, Norwegian)

Carpinus caroliniana. From Sargent 1905.

carpino (from *carpinus*, Italian); *charme* (French)
leamhan-bog (marsh elmwood, Gaelic)

Carpinus caroliniana (of Carolina)
amerikanische Weib-Buche (American female-beech, German)
amerikansk avenbok (American night-beech, Canada)
blue-beech [bluebeech]
bois de fer (ironwood, Quebec)
broom-wood [broomwood] (presumable the branches were used to make brooms)
carpino [*carpe*] *americano* (American *Carpinus*, Guatemala, Honduras)
charme [*d'Amerique*] ([American] hornbeam, Quebec); *charme de Caroline* (Carolina hornbeam, French)
Hainbuche (thicket beech, German)
[American, hop, hophornbeam] hornbean ("hornbeam," the name of a European species from the 1570s or before, because of its hard, tough, close-grained wood; formerly called "hardbeam," USA, Canada)
icimilapí (Koasati)
ironwood (name given, often locally, to various trees with extremely hard wood; dating from at least as early as the 1650s)
ittopankachi (*itto*, tree, *panayka*, curled, *chi*, repetitive suffix; probably akin to "muscle-tree," Alabama); *ittosankachi* (*itto*, tree, *sanayka*, twisted, *chi*, repetitive suffix, Alabama)
lean-tree [leantree]; *otantahrteweh* [*o-tan-tahr-te-weh*] (lean tree, Iroquoian)
lechillo (little bed, Texas, Michoacán)
mora de la sierra (mountain mulberry, Sinaloa)
muscle-tree [musclewood] (named for the undulations on the trunk that resemble muscles)
palo barranco (ravine tree, Michoacán)
palo liso (smooth tree, Mexico)
smoothbark ironwood
tzutcamay (Zoque, Chiapas)
water-beech

Linnaeus created *Carpinus* in 1753, including two species, *C. betulus* and *C. ostrya.* In an unusual confusion, he attributed both to the Americas and to Europe. *Carpinus betulus* is native to Europe and extends from southern Norway south into the Mediterranean mountains. The plants he called *C. ostrya* ultimately became known as *Ostrya virginiana* in the Americas. It was not until 1788 when Thomas Walter gave the American *Carpinus* its name in his *Flora Caroliniana.*

Europeans were well acquainted with the useful wood of *Carpinus*, and they quickly understood the benefit of having a related species in the Americas (Edlin and Mitchell 1985). We now know of 26 northern temperate species, with 2 in Europe (Furlow 1990, Mabberley 1997).

Logs were used as the main supporting post for the ridge pole of Ojibwa lodges (King 1984). Colonial settlers in America used *Carpinus* to make bowls and dishes because it rarely split or cracked. Hornbeam has a thick, nearly white sapwood and a heartwood, which is pale yellow to tan. It has no characteristic odor or taste. The wood is hard and heavy, with a dry specific gravity of 0.58 (Miller and Wiedenhoeft 2000). The wood is difficult to work but has been used to make tool handles, levers, parts for farm machinery, and fuel wood. The wood is said to be an irritant to some people (Miller and Wiedenhoeft 2000), although it is not listed by Hardin and Arena (1974), Lampe and McCann (1985), Blackwell (1990), or Turner and Szczawinski (1991).

These trees were used through much of the Eastern Deciduous Forest. Among the people making medicines of them are the Delaware and Iroquois of New England, and the Cherokee of the Carolinas (Moerman 1998). The most widespread use was a mixture of the bark and other plants to treat diarrhea (Hamel and Chiltoskey 1975, Moerman 1998). Other uses for the tree included treatment of urinary problems, women's diseases, "general debility," itch, and consumption, and to facilitate childbirth.

Carya

(From Greek *karya*, a walnut tree)

anaskomins (Powhatan, Virginia); *osámener* (yellow nuts, Carolina Algonquins); *msim* (the nuts, Delaware)
asonòka (hickory nuts, Onondaga); *wane?i* [*wa-ne?-i*] (Cherokee)
hickory (shortened from *pohickery, pakahochory, powcohicoria,* or *pawcohiccora*; originally meaning the milk-like emulsion of the nuts; from *pakähikâré*, it is pounded [brayed], Virginia Algonquian; first used by Strachey in 1612)

Carya. *Carya alba* (upper left). From Sargent 1905. *Carya aquatica* (right). a. Branch with fruits. b. Fruits. c. Fruit, side view. *Drawn by Vivian Frazier.* From Correll and Correll 1972. *Carya glabra* (left middle). From Sargent 1905. *Carya illinoiensis* (lower left). From Sargent 1905.

kapko ⌊*kapun, yuhe*⌋ (nuts of shagbark or *C. ovata*, Choctaw)
ocá (Koasati); *o'ce* (Muskogee); *ocha* (Alabama); *oksak* [*uksak*] (the nuts, Choctaw); *osak* [*oksak*] (the nuts, Chickasaw); *osak tali'* (the nuts, *osak*, hickory, *tali'*, rock, Chickasaw); *otche* (Creek, cf. Simmons [1822] 1973)
pakan'ahtik (Plains Cree), *pummockóner* (nuts that lie in heaps [storage], Carolina Algonquians)
wanaku'?ka (Catawba)
ya' (hickory nuts, Yuchi)
yahu'lu [*yahu'li*] (Cherokee)

Carya alba (white) (= *C. tomentosa*)
assenamins (Strachey wrote this Powhatan word as *assunnomeindg* in [1612] 1953. He also spelled it *assinenans*, and identified both as "walnuts." Because of a linguistic comparison, Siebert 1975 included the name here. Related words in Fox, Ojibwa, and Delaware refer to *Acer saccharum*, although that species does not grow near Jamestown where Strachey recorded the names. The root of these words is /**a?senyi*/, rock or stone, and that is not applicable to the maple. The final /*-*eminsy*/ means drupe-bearing bush or tree. Siebert found the Munsee dialect of Delaware using the name *asená·minsi* for these trees, but the Ontario Delaware has *ptukwíim.*)
astichk (Onondaga)
hard-bark hickory; white [white-heart] hickory
hickory grubs (used by Williams [1837] 1962)
mi'tigwabak' (*mi'tigwab*, bow, *ak'*, wood, Ojibwa)

mockernut [hickory] (a name applied to these trees by the 1840s, presumably because they resembled a more desirable species)
pin̲hudi' (*pin̲*, hickory, *hudi'*, wood, Biloxi)
skipeimeysee (Shawnee, fide Edgar 1891; /*-*eminsy*-/ means drupe-bearing bush or tree)
uksak hata (*uksak*, hickory, *hata*, white, Choctaw)

Carya aquatica (growing in or near water)
bitter pecan
penossv (Muskogee)
swamp hickory [swamp-hickory]
water hickory [pignut]

Carya cordiformis (in the form of a heart)
bitter-nut [bitternut hickory]
[red, white, swamp]-hickory
noyer dur (hard walnut, Quebec)
o:cáktíkbî (astringent nut, Mikasuki); *oce-tofkv* [*tóktv, oci:tó:kfan*] (*oce*, hickory, *tofkv*, bitter and sour, Muskogee); *ochatakba* [*ochaatakba*] (*ocha*, hickory, *takba*, a taste between bitter and sweet, but not sour, Alabama)
[pig-hickory] pignut [hognut] (originally applied by Shakespeare in 1610 to *Bunium* in the Apiaceae; the allusion is problematical regarding whether it meant nuts eaten by pigs or part of their anatomy; the literary figure *Don Quixote* used the term for acorns in 1711, but by 1760 it was being applied to the fruits of *Carya*)
pin'hu paxka' (*pin̲*, hickory, *'hu*, wood, *paxka'*, sour, Biloxi)

Carya floridana (of Florida)
scrub hickory

Carya glabra (without pubescence, smooth)
[broom, black, brown, red, switch, switchbud, white] hickory
pignut [hickory] (see notation under *C. cordiformis*)

Carya illinoiensis (of Illinois)
aph'fo'tka (Ofo)
damzá (Otomí, Hidalgo)
nogal (Texas); *nogal de Cuilapan* (Culiapan walnut, Oaxaca); *nogal de nuez chica* [*de nuez lisa, liso, morada, pecanero*] (small [smooth nut, smooth, purple, pecan] nut walnut, Texas, Mexico)
nuez encarcelada (nut in jail, Texas)
o:cákî (hickory nut, Mikasuki); *o:cáimaknî* (nut meat, Mikasuki); *o:cáká:pî* (*á:pî*, tree, Mikasuki)
oce [*ocî, otche*] (the nut, Creek); *oce entvkacketv* (*oce*, hickory, *cvpko*, long, Creek); *ocE mahhe* [*o:cimahhî, ocí:máhh-i*] (*oce*, nut, *mahhe*, real, Creek); *ocí:yápî* (*oce*, nut, *apí*, tree, Creek); *oce-cvpkce* [*oce-cápkuce, oce-tvkác-kv, oce-cvpko*] (*oce*, hickory, *cvpko*, long, Muskogee); *oksak fula* (*oksak*, nut, *falaia*, long, Choctaw); *osak falha'* (*osak*, nut, *falaa'*, long, Chickasaw), *fala* (Choctaw; cognate with Chickasaw)
on-ku-a [*do'nä'i, don'ai*] (oily tree, Kiowa)
pacana (Spanish); *pacane* (French); pecan [pecaunes, paccan, pekan] (dating from the 1770s in English; from the native name for "nuts" in various Algonquian languages, such as Abenaki *pagann*, nut, Abenaki *pagon*, walnut; Cree and Menomini *paka:n* [*pekan*], large nut; Fox *paga:ni*, large nut; Miami *pakani*, nut; Miami-Illinois *pakaani*, nut, pecan); Ojibwa *pagan*; Shawnee *paka:ni* [*paka:na*], nut; Atakapa *pa'kan nec*; the Unami-speaking Delaware called the pecan *ká·se·m*; the cognate Shawnee was *ka·Te·mi*; it is not clear if this is cognate with "pecan"
pin̲txo'gon̲ udi' (pecan tree, Biloxi; *pin̲txo'gon̲ni'*, the nut)
ru (Tunica; this may belong here, or it may be generic)
so:-hi a-ni-nv-hi:-da' (long nut, Cherokee)
wa-da'-çto-çta [*hi*] (Osage)
yaga-sola (*yaga*, tree, Zapotec, Oaxaca)

Carya pallida (light in color)
[pale, sand] hickory

The first Europeans to see hickories in the New World were on the de Soto expedition in 1539, when they recorded them northwest of Gainesville. Soon after leaving *Ocale* (modern Ocala) the group "passed an uninhabited region ten or twelve leagues in extent which lies between the two provinces [*Ocale* and *Potano*], in which there are large forests of walnuts, pines, and other trees unknown in Spain" (Garcilasco in Swanton 1939). No walnuts grow in that area, so he must have meant hickories (Wunderlin and Hansen 2002).

This genus emphasizes how spotty and idiocyncratic ethnohistoric records of indigenous people can be. It is well known that people throughout the southeastern states valued the fruits of both hickories and pecans, as witnessed by the Powhatan name of hickory. Harriot ([1590] 1972) was among the first people to record the importance of *Carya* in the New World, although the term "walnut" at the time included both *Carya* and *Juglans*. Under the heading "Oyle" [oil], he wrote: "There are two sorts of *Walnuttes* both holding oyle, but the one farre more plentifull than the other. When there are milles &

other deuises [devices] for the purpose, a commodity of them may be raised [gathered]."

Moerman (1998) lists only the Choctaw of the Muskogean people using them; however, there are numerous other records of *Carya*. One of the oldest associations of indigenous people with pecans was from the prehistoric remains at the Ozark Bluff-Dweller site in Arkansas. There Gilmore (1931) found pecan hulls. Leonard Blake, the renowned student of plant remains who worked with Hugh C. Cutler (1912–1998) at the Missouri Botanical Garden, wrote Marie Standifer in 1988. In that letter he clarified that the pecan "does not grow in the area [of Ozark Bluff], carried in from lowlands of the Arkansas valley."

Perhaps the first historic account of Muskogean people using *Carya* came from the de Soto expedition. Historical accounts of that time noted the Chickasaw, in what is now New Madrid on the Mississippi River in Missouri, using pecans (Hedrick 1919). *Carya* use was further elaborated by surveyor Bernard Romans ([1775] 1961). He wrote that the Chickasaws and Choctaws went "gathering nuts and making their milk." While he did not record it, the Choctaw name of hickory milk is *uksak* [*oksak*] *ulhkomo*. Of the Chickasaws, he noted, "they likewise use hickory nuts in plenty, and make a milkey liquor of them, which they call milk of nuts; the process is at bottom the same as what we use to make milk of almonds; this milk they are very fond of, and eat it with sweet potatoes in it." He recorded that the Choctaws "make many kinds of bread . . . with the help of water, eggs, or hickory milk; they boil the esculent convolvulus [*Ipomoea batatas*]."

In 1773, Bartram ([1791] 1958) recorded that *C. ovata* was important to the Creeks of Georgia. He wrote: "The Creeks store up the . . . [shellbark nuts] in their towns. I have seen above an hundred bushels of these nuts belonging to one family. They pound them to pieces, and then cast them into boiling water, which, after passing through fine strainers, preserves the most oily part of the liquid; this they call by a name which signifies Hiccory milk; it is sweet and rich as fresh cream, and is an ingredient in most of their cookery, especially homony and corn cakes." This practice extended at least as far north as New Jersey where the Delaware also extracted hickory milk (Kalm [1753–1761] 1972).

Alabama-speaking people have a dish they call *ochahołka* (*ocha*, hickory, *hołka*, sofkey). Creeks call it *oce nehv osafke* (*oce*, hickory, *nehv*, fat, *osafke*, sofkey). This is a dish made by adding hickory nut meat to corn after it is cooked (Sylestine et al. 1993). Sofkey itself, a word taken from the Creek *osafke* [*sáfki*] cognate with Alabama *hołka*, was served hot or cold and generally without other seasoning. It is drunk or eaten with a spoon, depending on the consistency, which varies between a thin gruel and watery porridge.

The Choctaw called a dish made of hickory nut meat and parched corn beaten together *oksak bahpo* (*oksak*, nuts, *bahpo*, ground up maize). No mention has been found among other people of securing nuts from insects by "the smoke-drying process and then, in small quantities encased in airtight mud cells in the same manner that the mud daubers (*lukchuk chanuschik*) preserved their spiders" (Campbell 1959).

Pecans were eaten by people in Louisiana, surely both the Choctaw and Natchez, and they valued *C. illinoiensis* oil for seasoning their food (Hedrick 1919). The Seminoles also consumed the nuts of *Carya*, although their name is generic for both pecans and hickory nuts (Sturtevant 1955). Hedrick (1919), Gilmore (1919), and Moerman (1998) give ample records of other indigenous groups using *Carya*.

Carya alba and other species were used by the Cherokee as food (Hamel and Chiltoskey 1975). Similarly, the Iroquois ate *C. cordiformis*, the Omaha ate *C. glabra*, and the Comanche and Osage ate fresh and stored pecans (Hunter [1823] 1973, Moerman 1998).

Indigenous people used wood from *Carya* to make baskets, bows and arrows, corn beaters, blowgun darts, and in tool handles and other items (Gilmore 1919, Sturtevant 1955, Hamel and Chiltoskey 1975). During the 1800s, the wood was considered prime for barrel hoops (Porcher 1863). More recently, *C. cordiformis* wood was used for wheel stock, handles, and fuel, the trees being first cultivated in 1689 (Vines 1977). *Carya glabra* branches were made into brooms because of their tough, flexible stems, and used for fuel, tool handles, wagons, and agricultural implements (Vines 1977). Pecan wood has been used for furniture, flooring, agricultural implements, and fuel (Vines 1977). Pecans have been cultivated by descendants of Europeans in America since 1766 (Vines 1977). Bark was made into rope, called *baluhchi* (Choctaw) or *ochahobaski* [*ochaobaski*, *ochaabaski*] (*ocha*, hickory, *ho-*, third person plural prefix, *baksi*, rope, Alabama).

Several species of *Carya* were used to make dyes. Kalm ([1753–1761] 1972) wrote that the people in Pennsylvania used it: "Woollen and linen cloth is dyed yellow with the bark of hiccory." Porcher (1863) wrote, "A dye for woollens used on the plantation is made from that of most of the species." He added, "To color yellow.—Take three-fourths of hickory bark, with the outside shaved off, and one-fourth of black oak bark done in the same manner; boil them well together in a bell metal kettle until the color is deep; then add alum sufficient to make it foam when stirred up, then put the yarn in and let it simmer a little while; take it out and air it two or three times, having a pole over the kettle to hang it on, so that it may drain in the kettle; when dry rinse it in cold water."

The Choctaw spread leaves of *C. alba* to discourage fleas (Bushnell 1909). The Iroquois mixed hickory nut meat with bear oil as mosquito repellent (Moerman 1998).

The Cherokees made remedies from *Carya.* They used them for female problems, against colds, excessive bile, as a diaphoretic and emetic, and the bark was applied to cuts and chewed for sore mouth (Hamel and Chiltoskey 1975). The Delaware used the genus for female problems and as a tonic; the Iroquois and Meskwaki used at least *C. cordiformis* as medicine (Moerman 1998). Pecans were medicinal as far west as the Comanche and Kiowa (Moerman 1998).

Porcher (1863) noted of *Carya* ashes: "A cement for cisterns, as hard as marble, and impenetrable by water forever, is made of wood ashes two parts, clay three parts, sand one part, mixed with oil." Ashes were also used to make lye, which was used to process corn into hominy.

Cassytha: Love Vine

(From Aramaic *kesatha*)

When Linnaeus was writing his *Species Plantarum,* he was faced with two completely different genera of parasitic vines. Authors before him had put these vines in the same genus, *Cuscuta.* Linnaeus realized that two genera, in two distinct groups that are now called families, were involved, and he began searching for a second name. Being inventive and proficient in languages, he borrowed from cognate sources and came up with both names. One of these he retained as *Cuscuta* (Convolvulaceae). The other he called *Cassytha* (Lauraceae).

Several scholars have said that *Cassytha* is from Greek *kassyo, kassyein,* to sew or patch; in Latin *cassis* means web, hunting net, spider's web; in Akkadian *kasitu,* binding. They also record that *Cuscuta* was derived from Arabic *kusuta, kshuta, keskhut,* or *kusu.* What they have not generally said is what is meant by the Arabic and its cognate Aramaic word *kesatha.* Many years back, I looked into the situation and discovered that *Cassytha* is based on Arabic and/or Aramaic words meaning "a tangled wisp of hair" (Austin 1979). That aptly describes the masses of matted stems that cover the host plants.

Since Linnaeus gave us the genus *Cassytha,* many more species have been discovered. We now know of 20 species in the genus, most of them confined to the Old World, and with 13 endemic to Australia (Mabberley 1997). Usually a single species, *C. filiformis,* is considered indigenous to the Americas.

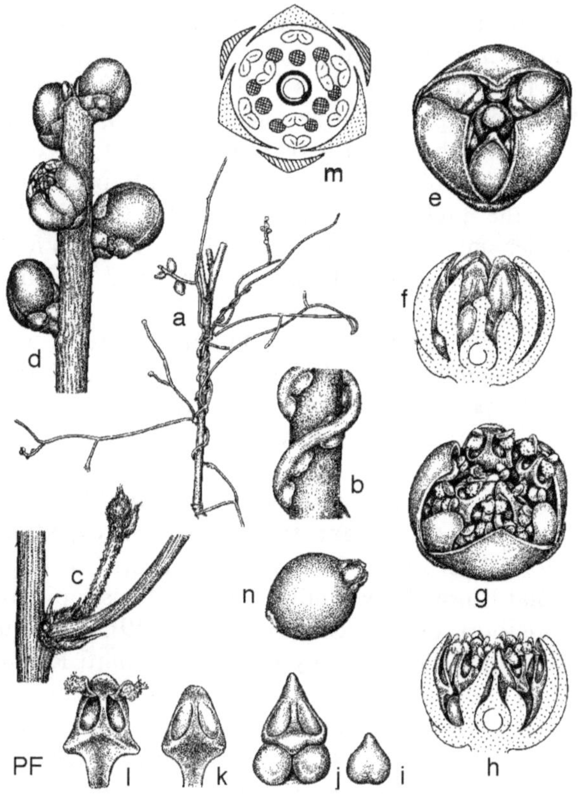

Cassytha filiformis. a. Stem parasitizing another plant. b. Stem on host with haustoria. c. Detail of branching. d. Inflorescence. e. Flower in first (or female) stage. f. Longitudinal section of first stage. g. Flower in second (or male) stage. h. Longitudinal section of second stage. i. Staminodal nectary. j. Stamen and nectaries. k. Stamen before dehiscence. l. Stamen dehisced. m. Floral diagram. n. Fruit. *Drawn by Priscilla Fawcett.* Correll and Correll 1982.

Cassytha is a member of Lauraceae, the same family as avocado (*Persea americana*), red bay (*P. borbonia*), laurel (*Laurus nobilis*), lancewood (*Ocotea coriacea*), sassafras (*Sassafras albidum*), and other familiar plants. Yet, all these familiar species are trees. Many wonder how it is that a parasitic vine is included in the same family. All the external habit differences disappear if one examines only the flowers and fruits. Flowers in *Cassytha* are built exactly the same as in the other genera, with whorls of anthers in threes and valvately dehiscent anthers. Fruits in *Cassytha* are drupes like all the others. One thing that is missing, or at least present in small quantities, is the aromatic oils so prominent in the other genera. However, most would conclude that those antiherbivore chemicals were probably unnecessary when the life-form was reduced to the parasitic mode.

Parasitism has arisen in numerous plant families in Florida including the Convolvulaceae (*Cuscuta*), Ericaceae in the broad sense (*Monotropa*), and Gentianaceae (*Voyria*). Many of these like *Cassytha* have been proposed as monotypic families with only the parasites included. However, all modern students have concluded that each is simply a highly specialized member of a larger group.

Cassytha filiformis is the only pantropical member in the genus. It has many names in indigenous languages in the Old World, especially in the Indian peninsula region. The first common name recorded for the Old World was from the Malabar Coast of India where Dutch botanist Hedrik van Rheede recorded the name *acatlia-valli* in the late 1600s. That name is most similar to the Sanskrit *akásvalli* (sky vine), but it is also related to *akáshvel* (sky vine, Bengali), *akáshwel* (sky vine, Marathi), *acatsjabulli* (maybe also sky vine, Malayalam), and *amarbeli* (live-forever vine, Hindi).

Similarly, there are many indigenous names in the Malay Peninsula. Those for which there are translations are Javanese *rambut puteri* (princess's hair) and *tali puteri* (princess's thread). In Sudanese, it is *sangga langit* and in Bangka, it is called *akar pengalasan.*

Cassytha filiformis is considered a native of the New World although no indigenous language names for it have been found. That particular fact makes me wonder if the recorded uses are autochthonous or imported with people from the Old World. At least some of them probably are native to the New World because they are mixtures with other plants known only from that region. So, maybe the absence of evidence among common names is insufficient to exclude native usage. On the other hand, perhaps the uses with native American plants have been invented since the Old World uses were introduced. We have insufficient data to draw a conclusion at this point.

Many of the common names in the Americas refer to love, as in love vine, *liane amitié* (love vine, Hispaniola, Guadeloupe, Martinique), love (Montserrat), and old man berry (Cayman Islands). That concept is related to the tenacity with which the vines cling to their hosts. Surely, it has given rise to the belief that the plants are aphrodisiac in the Bahamas and Montserrat. There complex directions are given to make someone fall in love, including cooking the vine with gumbo limbo (*Bursera simaruba*) and drinking the tea with or without cream (Ayensu 1981, Morton 1981). Perhaps related is the use of *Cassytha*, either by itself or combined with *Pluchea carolinensis,* then drunk or used in baths for 9 days after giving birth (Morton 1981). Love and birth are related events.

The tangled, lush growth of the vines also has been proposed as giving rise to the belief that it promotes growth of hair or stops its loss. That view is held in the Old World (Malay Peninsula, India) and in the New World (Bahamas, Jamaica, Guyana). In all these areas, decoctions are used for washing the head and stopping falling hair (Burkill 1966). The sap is also extracted from the crushed plant and used to rid the head of lice (Morton 1981). Some think the abundant muscilage is the principal aid in hair medicines.

Yet, the "Doctrine of Signatures" concept (a plant that looks like something is useful to treat it) may not be all that is involved with these and other uses of the plants. There are potent chemicals in *Cassytha* that may even be fatal to humans in elevated doses, so surely they would inhibit or kill certain disease-causing organisms. The plants are known to contain the alkaloids nantenine (Merchant and Desai 1973), laurotetanine (Abraham 1912), the aporphine and oxoaporphine bases (Cava et al. 1968), and dulcitol (Fuzikawa et al. 1940). Of these, laurotetanine will cause cramps in small doses and may be fatal if ingested in large amounts.

In spite of those chemicals, the stems are edible in small quantities. In Jamaica and Guyana, the crushed stems are considered a vermifuge (Morton 1981). Dried and powdered plants are taken with nutmeg to relieve indigestion and stomachache. Mixed with butter and ginger they are applied to tumors. Mixed with sugar, *Cassytha* is a remedy for sore eyes and headache. In Montserrat, a tea from the entire plant is drunk to treat colds and as a purificant. *Bidens alba* leaves are mixed with love vine and boiled, and the tea is used as a treatment for high blood pressure. In the Bahamas, a tea of the plant is used to treat waist pain. In Brazil, small branches are made into an "agreeable drink" for improving the blood and stopping hemorrhage (Mors et al. 2000).

Many of the common names in the Americas compare the plants to other things. Strings and pasta are common themes. Other objects include *alcanfora del Japón* (Japanese camphor), *bejuco dorado* (golden vine), *cabellos de angel* (angel hair; also the name for cotton candy), *corde à violon* (violin string, Guadeloupe, Martinique), *duivelsnaaigaren* (Devil's hair-net, Suriname), *fideos* (noodles, Cuba, Hispaniola), *fidello* (little noodles), *liane ficelle* (string vine, Guadeloupe, Martinique), *tente en aire* (tent in the air), *vermicelle* (Guadeloupe, Martinique), and yellow dodder (Montserrat). Woe-vine (USA, Bahamas) surely refers to the problems these plants cause when they parasitize plants that people wish to grow. Jaundice tie-tie (Belize) notes both medical use and the climbing habit.

Another name often used for the plants in English-speaking regions is dodder. That name and love vine are shared with *Cuscuta,* so scientists and nonscientists had and still have trouble distinguishing between these parasites. "Dodder" is a curious name for these plants. Since the 1300s, the meaning in English has been "to tremble or shake from weakness or age." A more recent meaning, dating from about the late 1500s or early 1600s, is to move (totter is cognate) or sway, as in the wind. There is a European grass (*Briza media*) that is called both dodder grass and quaking grass. The same concept probably applies to both *Cuscuta* and *Cassytha*—they tremble, at least their branch tips do, in the slightest wind. Can this also be why people in the Cayman Islands call it old man berry?

Dodder has been applied to *Cuscuta* since about 1265, so it is reasonably certain that use of that name for *Cassytha* is more recent. *Cuscuta* is native to Europe; *Cassytha* is not.

Some of the references in common names are not at all obvious. For example, in Brazil *Cassytha* is called *cipó-[de-]chumbo* (lead vine) or *erva-chumbo* (lead herb). I can think of only two possible reasons these names might have been applied. Either the tea made from the plants tastes or smells like lead, or drinking it induces trembling symptoms like lead poisoning. In Suriname, the Arawaks call it *karoba körö* [*karobakoro*].

My personal favorite among the common names is *liane sans fin* (vine without end, Guadeloupe, Martinique). That is reminiscent of the name given to *Cuscuta* in Sri Lanka (Austin 1979). There the Sinhalese call that parasite *aga-mula-neti-wel* (plant without beginning or end). How much more accurately poetic can one be?

Castanea

(From Latin *kastanea*, Greek *kastanon*, the chestnut)

Castanea dentata. From Sargent 1905.

castagno (from Latin *kastanea*, Italian); *castan* (Gaelic); *castanheiro* (the tree, Portuguese); *castaño* (Spanish); *châtaignier* (the tree, French); *Kastanie* [*Edelkastanie*] (German); *kastanje* (Dutch); *kesten* (German, cf. Turner [1548] 1965)
chestnut (from *chesten* or *chestine* nut, applied ca. 1519; presumably *chesten* was based on *castanea*, although the Gaelic names make it likely that it was "chaste")
cnò gheanmnuidh (*cnò*, nut, *gheanmnuidh*, chaste, Gaelic)
geanm chnò (*geanm*, chaste, *chnò*, nut, Gaelic)

Castanea dentata (toothed)
afata (Timucua)
apimin [*opemens*] (first recorded as *opemens* by Strachey in [1612] 1953, Powhatan, Virginia; Siebert 1975 said the word *wápimin* occurs in all eastern Algonquian dialects, but to the Fox and Menomini it meant "maize" and to some Ojibwa it was "apple"); *wapim* [*wabim, woapim*] (Delaware)
[American, sweet] chestnut
ohehhahtah [*o-heh-hah-tah*] (prickly bur, Iroquoian); *ohelliáta* (Onondaga); *unige'nia* [*u-ni-ge'-n(i)(a)*] (Cherokee)
oti [*uti*] (Choctaw); *oto* (Creek); *otowóske* (Muskogee)
prickly bur
sagatémener (roasting nuts, Carolina Algonquians)
stsaù (Atakapa)
yarapi'· (Catawba)

Castanea pumila (dwarf)
[Allegheny] chinquapin [*checkinakamin,* chincapin, chincopine, chicopin, chinkapin] (Powhatan, Virginia; Strachey [1612] 1953 wrote it *chechinquamins*; Vogel 1970 thought it meant "great seed," but Siebert (1975) said it was from /**ci·ʔt-*/, jab, prick, /**-i·nkwe·-*/, eye, and /**-min-i*/, grain; cognates occur in Cree, Fox, Menomini, Ojibwa, and Penobscot)
hachofakti [*chefuncte, kefonctei*] (*hacho* means "mad" in Creek and is an honorable war title; any possible connection is obscure, Choctaw); *hachiäfakto* [*hachiifakto, achiifakto*] (*hachi*, tail, Alabama); *hotiyyá:ya* (maybe from *hó:ti*, sores, *yá:ya*, elder sibling, Koasati)
hi̱tu tu' (seed small, Catawba)
na'u o'xox (*na'u*, bristle, *o'xox*, sharp, meaning the husk, Atakapa)
sapúmmener (little-stony-fruit tree, Carolina Algonquians)
uhnagi:ni'a [*u-hna-gi:-n(i')(a)*] (Cherokee)

It is impossible now to comprehend the changes in the North American landscape that began in 1930 when chestnut blight (*Cryphonectria parasitica* or *Endothea parasitica*) was accidentally introduced from Asia. Before the disease attacked the trees, chestnuts were common in the Eastern Deciduous Forest (Nixon 1997). *Castanea dentata* reached 120 feet tall and 3 to 4 feet in diameter (Sargent 1905, Harrar and Harrar 1946), but there are records of one tree whose branches did not begin until 70 feet above the ground and that had a trunk 22 feet in diameter (Millspaugh 1892). Chestnuts were the dominant emergent trees in the forests, and their loss totally altered the dynamics of the region in ways that will never be completely known.

Chestnuts were important food not only for animals like the now-extinct Carolina parakeet (*Conuropsis carolinensis*), passenger pigeons (*Ectopistes migratorius*), and myriad vertebrates still extant, but also for people. Chestnuts were eaten as food throughout their range (Fernald et al. 1958, Moerman 1998, Davidson 1999).

The first record of New World chestnuts came in 1539 from Rodrigo Ranjel, secretary and chronicler of de Soto's excursion. He described *C. dentata* as "chestnuts in the land ... which are like those of Spain, and grow on as tall chestnut trees" (Swanton 1939). Near the Santa Fe River in northern Florida, de Soto's company entered a Timucua village called *Cholupaha* (*paha*, house) along what is now the Santa Fe River in Alachua County, where "They found much food and many small chestnuts, but the trees that bear them are only two palms high and they grow in prickly burrs." That small chestnut was *C. pumila.*

Harriot ([1590] 1972) recorded use of chestnuts by the Algonquians in North Carolina with "Chestnvts, there are in diuers places great store: some they vse to eate rawe, some they stamp and boil to make spoonemeate, and with some being sodden they make such a manner of downbread as they vse of their beanes."

The usual interpretation is that the "acorn" later mentioned by Harriot ([1590] 1972) with the name *sagatémener* also was a chestnut. Why he used two entries with distinct names is unexplained unless the second entry was *C. pumila.* That Capt. John Smith and William Strachey also called *C. pumila* an "acorn" supports that view.

In 1612, Capt. John Smith wrote that the Virginia Powhatans, after boiling the nuts for 4 hours, made "both broath and bread for their chiefe men or at least their greatest feasts" (Rosengarten 1984). That same year Strachey wrote the name *chichiquamins* and a "kind of grain to eat." He also said *chechiquamins*, a "nut like a small acorn, good meat." Capt. John Smith recorded in 1624 that the Virginians "haue a small fruit growing on little trees, husked like a Chestnut, but the fruit most like a very small Acorne. This they call *Chechinquamins*, which they esteeme a great daintie." In 1682, Thomas Ash recorded that the people in the Carolinas used the chinquapin, and "Of the Kernel is made *Chocolate*, not much inferior to that made of Cacao" (Fernald et al. 1958).

Romans ([1775] 1961) wrote of the Creeks that for food "they dry peaches and persimmons, chestnuts and the fruit of *chamaerops* [*Serenoa*]." Bartram ([1791] 1958) only compared the taste of acorns to chestnuts, and noted chestnuts in the forests. Swanton (1946) wrote that both chestnuts and chinquapins were used as food by the Creek and Choctaw. Without mentioning who used them, Fernald et al. (1958) recorded: "The nuts were cooked in their corn-bread or, when roasted, were used as coffee." Apparently, they were referring to the Cherokee, who boiled the nuts, pounded them with corn, and kneaded and wrapped the mixture in a green corn husk to be boiled and eaten; they also used the seeds as a coffee substitute (Hamel and Chiltoskey 1975). The Iroquois made a similar bread mixture, and the preparation seems to have been common through the east (Moerman 1998). Yanovsky (1936) said seeds of *C. pumila* were eaten by people from Pennsylvania to Texas.

Of the wood, Porcher (1863) wrote: "The wood of the chestnut, though brittle, is very durable in weather. I am informed that fence-rails made of it will last over twenty years." It was later prized for posts, furniture, hop-poles, hoops, and all kinds of elastic props and handles, railroad cross-ties, and fuel (Vines 1977).

The Cherokee made a brown dye from the bark (Hamel and Chiltoskey 1975). As usual, little information is available on dyes. Perhaps the chemicals are much like those in the related *Quercus*, being largely tannins related to quercetin and quercetagetin (Cannon and Cannon 2003).

The Cherokee and Iroquois used *C. dentata* for more purposes, including as a cough syrup, to treat sores, for the stomach, to stop bleeding after childbirth, for heart trouble, to relive itch, as baby powder, to oil the hair, and to treat colds, rheumatism, and whooping cough (Vogel 1970, Hamel and Chiltoskey 1975, Moerman 1998). The Kosasti used *C. pumila* roots to relieve stomach trouble (Taylor 1940). The Cherokee used it for other problems, including headache, fever blisters, chills, cold sweats, and fevers (Hamel and Chiltoskey 1975).

Porcher (1863) wrote: "The roots contain an astringent principle; that of the chinquapin boiled in milk is much used in the diarrhoea of teething children. I would advise a tea made of this to be used extemporaneously in diarrhoea by our soldiers in camp. The bark of both trees contains tannin, and may be used in tanning leather." Later he wrote that a local physician had "used the decoction of the root and bark of the chinquapin frequently as a substitute for quinine in intermittent and remittent fever, and with decidedly satisfactory results ... Hot water is poured over the root and bark, and a large quantity taken during the twenty-four hours." Millspaugh (1892) noted that the leaves were part of the U.S. Pharmacopoeia.

Catalpa

(Named from *catapla* [*kathulpa*], the language of the Catawbas of the Carolinas where Mark Catesby discovered the tree in 1726)

Catalpa bignonioides. From Sargent 1905.

Catalpa bignonioides (resembling *Bignonia*, which see)
bean-tree; Indian-bean
candle-tree
catalpa (Sargent 1905 thought that the name was from Cherokee; Martin and Mauldin 2000 said from Muskogee *kvtvłv*); *catalpe* (French); catalphy [patalphy] (Kentucky)
Catawba-[tree] (taken from the name of the people by that name; their original home was from north-central South Carolina to near Augusta, Georgia)
Indian-cigar [-tree]; smoking-bean
ittotassikáya (*itto*, tree, *tassikáya*, catalpa, Alabama)
katałpá (obsolete: in wider use is *wá:ka impalaná*, cow's bean, Koasati); *kvtvłv* [*kvtvrv*] (Muskogee)
southern catalpa
Trompetenbaum (trumpet tree, German)

In the 1720s, Mark Catesby was traveling in the interior of South Carolina and discovered a beautiful tree. He not only painted the tree for his book, but he also brought its seeds back to the coast. He wrote, "This tree was unknown to the inhabited Parts of Carolina, till I brought the Seeds from the interior Parts of the Country." Seeds were distributed among the settlers, and sent to England. Catesby (1731–1732) called these trees *Bignonia Urucu foliis flore sordida albo, intus maculis purpureis & luteis asperso, siliqua longissima & angustissima* (*Bignonia* with leaves like Urucu [*Bixa*], flowers dirty white and marked with purple and yellow patches scattered within, silique long and slender) or the "Catalpa-tree."

Apparently, the seeds collected by Catesby thrived in the Carolinas and Europe because Linnaeus had studied living plants in the *Hortus Cliffortianus* he published in 1738. Linnaeus followed Catesby's lead and called the trees *Bignonia catalpa.* Later it became known that *Bignonia* consisted of twining species and these trees were distinctive. By the 1770s, Italian botanist Giovanni Antonio Scopoli (1723–1788) realized that the trees should be in a distinct genus and published *Catalpa* in 1777. Thomas Walter agreed, and in his *Flora Caroliniana*, published in 1788, he named the plants Catesby had introduced into Europe *Catalpa bignonioides.*

Both William Bartram and André Michaux found *C. bignonioides* farther south than had Catesby. Bartram ([1791] 1958) found the trees in Liberty County, Georgia. He wrote that in April of 1773, he was traveling down the high road "towards Fort Barrington, on the Alatamaha, passing through a level country, well watered by large streams, branches of the Medway and Newport rivers. ... The road is ... bordered on each side with ... tall and spreading trees, as the Magnolia grandiflora, Liquid Amber, Liriodendron, Catalpa" and numerous other forest trees.

Although Bartram ([1791] 1958) does not indicate locating *Catalpa* in Florida, he did find it there. Research by Taylor and Norman (2002) created a list of Florida plants documented by both John and William Bartram and André Michaux; the Bartrams found *Catalpa* along the St. Johns River above Lake Dexter. Although the date is not given, the plants would have been discovered between November 1765 and September 1774.

Wood is grayish brown to lavender tinged, with the sapwood lighter in color, straight grained, and durable. It weighs about 28 pounds per cubic foot and shrinks little. It is soft and moderately shock resistant but bends poorly. Uphof (1968), Vines (1977), and Hocking (1997) recorded that the wood of southern catalpa is used for posts, poles, rails, cross-ties, interior finish, and cabinet work.

No information has been found on the Catawba or any other tribe using these trees. However, *Catalpa* figures prominently in the pharmacopoeia of the Appalachians, and to some extent the coastal plain. Surely, the settlers learned to use the plants from indigenous people.

Bark from *Catalpa* has been used to expel intestinal worms, and a decoction is also mildly laxative. Not only are the seeds used to treat intestinal worms, but they also induce vomiting. Leaves have been applied as a poultice to skin wounds and abrasions (Krochmal 1968, Krochmal and Krochmal 1973).

Porcher (1863) was heavily influenced regarding this native *Catalpa* by the reports of medical use of another species in Asia. He wrote that *Catalpa* "[g]rows in the upper and lower country of South Carolina; collected in St. John's. . . . The physicians at Naples, after the favorable report of [Carl Pehr] Thunberg and [Engelbert] Kæmpher, as well as those of Brera, have given incontestable evidence of the advantages resulting from its use in asthma. The decoction of the fruit is also employed. . . . Kæmpher says he also applied the leaves to the painful part. Poultry are said to thrive on and to be fond of the seeds. The timber makes durable posts. I find no notice of the plant in the American works. The honey collected from the flowers is somewhat poisonous—resembling, though less active, that collected from the yellow jessamine [*Gelseminum sempervirens*]."

Porcher's reference to both C.P. Thunberg (1743–1828) and E. Kaempfer (1651–1716) alludes to their work in Japan. Kaempfer was a German physician with the Dutch East Indies Company who arrived in Japan in 1690. He produced the first list of plants known from that island, published after his death. Thunberg was a Swedish physician also attached to the Dutch East Indies Company. Sir Joseph Banks at Kew Botanical Gardens promoted his work in Japan, and Thunberg published his *Flora Japonica* in 1784. Both made comments on local medical uses of *C. ovata.*

Small (1933) found that local children in the southern states smoked the ripe pods. Judging from the common names, they may have learned that from indigenous people. Simply stealing a forbidden smoke among children may not be the only reason for Small's observation. Perkins and Payne (1978) record that the pods are smoked to relieve asthma, as originally indicated by Porcher (1863). Pods are also thought to have cardiotonic properties (Foster and Duke 1990).

Sargent (1905) wrote that *C. bignonioides* contains "a bitter principle and is a tonic and diuretic." Later these compounds were identified as catapol, an iridoid glycoside (Mabberley 1997). Perkins and Payne (1978) noted that the bark and pods include the glycoside catalpin, a bitter principle, sugars, and tannin. Presumably, the catalposide and allied compounds found in *C. ovata* of Asia are related chemicals (Hsu 1986).

Moreover, at least two species of *Catalpa* in Asia are used as medicines (Duke 2003). Those plants include *C. bungei* and *C. ovata* (= *C. kaempferi*, cf. Wu and Raven 1998). Nearer to Florida, *C. longissima* is used to treat bleeding, blennorrhagia, dysentery, fever, and sore throats in Hispaniola (Liogier 1978, Duke 2003).

The catalpa sphinx moth (*Ceratomia catalpae*) is able to sequester the poisonous compounds in the leaves and feed with impunity on them (Kimball 1965). When it does so, the leaves produce more extrafloral nectaries that attract insects that attack or remove the eggs or larvae of the moths (Mabberley 1997).

Catopsis

(Probably from Greek *katopsis*, sight, *katoposios*, visible)

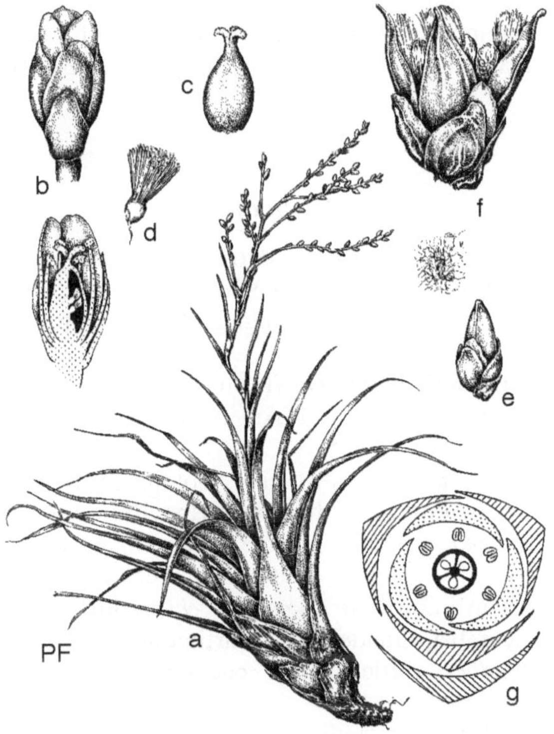

Catopsis floribunda. a. Flowering habit. b. Complete flower (above) and longitudinally dissected (below). c. Ovary. d. Ovule. e. Capsule, closed. f. Capsule dehisced (above) and seed (below), enlarged. g. Floral diagram. *Drawn by Priscilla Fawcett.* Correll and Correll 1982.

asõmco:bî (big Spanish moss, Mikasuki); *assonłakko* [*assonrakkô*] (*assúnwv*, Spanish moss, *łakko*, big, Creek)

wild pine ("pine" is short for "pineapple," Florida)

Catopsis berteroniana (for Carlo Giuseppe Bertero, 1789–1831, Italian physician, botanist and explorer)

mealy wild pine (Bahamas)

pata de gallo [*pié de gallo*] (chicken foot, Chiapas)
powdery strap airplant (Florida)

Catopsis floribunda (many-flowered)
ananas sauvage (wild pineapple, Guadeloupe, Martinique)
Florida strap airplant (Florida)

Catopsis nutans (nodding)
air-plants
wild-pine

August Grisebach created the genus *Catopsis* in 1864, when he named both *C. nitida* and *C. nutans.* In the years since, enough new kinds have been found so that there are now 19 tropical American species (Smith and Downs 1977).

Seminoles used the water trapped in the base of the plants for cooking during the dry season (Sturtevant 1955).

Cayaponia

(Brazilian António-Luiz Patricio da Silva Manso, 1788–1818, named the genus for the Cayapó, a group of Gé-speaking people of highland eastern Brazil)

Cayaponia americana (from America)
American melon-leaf [American melonleaf] (Florida)
bejuco de finca (farm vine, Dominican Republic); *bejuco de torero* (bullfight vine, Puerto Rico)
concombre [*bâtard, diable, hallier, marron*] ([false, devil's, thicket, wild] cucumber, Guadeloupe, Martinique); *ti concombre chien* (little dog cucumber, Guadeloupe, Martinique); *ti concombre hallier* (little thicket cucumber, Guadeloupe, Martinique)
torchon (dish cloth, Haiti)

Cayaponia quinqueloba (with five lobes on the leaves)
five-lobe melon-leaf [fivelobe melonleaf] (USA)

The two Florida species are rare, and I never saw them in 31 years of fieldwork. *Cayaponia americana* is known only from Miami-Dade County (Small 1933, Wunderlin 1998). This species also occurs in the Bahamas, and south through the Caribbean to the Lesser Antilles, but it is absent from Jamaica (Gooding et al. 1965, Adams 1970, Correll and Correll 1980, Acevedo-Rodríguez et al. 1996).

Cayaponia quinqueloba is known only from Alachua County (Small 1933, Wunderlin 1998). The northern Florida species is endemic to the southeastern United States, ranging from east-central Texas, to Georgia, South Carolina, and northern Florida (Radford et al. 1968, Correll and Johnston 1970, Diggs et al. 1999).

Most of the species seem to be Brazilian, and that is where the greatest use has been (Hocking 1997). Quattrocchi (1999) lists over 30 common names for Brazilian species. People on Ilha Marajó at the mouth of the Amazon River in Brazil call *C. triangularis* the *purga de gentio* (Indian's purge, literally "gentile's purge"). In the 1940s, it was a "popular medicine" as *purgativo enérgetico* (drastic purgative) (LeCointe 1947). All species appear to have cucurbitacins, which are somewhat toxic triterpenes with a bitter taste (Mors et al. 2001). Commercial preparations in Brazil are made from several species and even other genera (Mors et al. 2001).

Cayaponia kathematophora is grown as an ornament, and jewelry is made from its seeds (Robinson and Decker-Walters 1997). *Cayaponia ophthalmica* is a treatment for conjunctivitis (Mabberley 1997). The crushed leaves and stems of *C. glandulosa* of Peru and Colombia are used for liver problems and as an insect repellent by the Tikunas of Río Loretoyacu (Schultes and Raffof 1990). Seven species are used in northwestern South America by the Kofán, Kubiyari, Makuna, Matapié, Puinave, Taiwano, and Tikunas (Schultes and Raffof 1990). *Cayaponia racemosa,* called *camara* (room?, Mexico, Salvador, Panama), and *xta'keej* (deer feces; *xta',* feces, *keeh,* deer, because of the fruit, Yucatán), is sold in Mexican pharmacies as a tonic, especially for cattle (Hocking 1997). There are no recorded uses in the Caribbean, but the name *brionia* (noting the similarity to *Bryonia,* Cuba) compares it to a notoriously laxative European cucurbit.

Many places have names for their species, including *ak'il k'aak* [*akilkax*] (chicken vine, Maya, Yucatán) for *C. alata.* People in Honduras say *sandía de monte* (wild watermelon) or *granadilla de monte* (wild passiflora) for *C. microdonta.* In El Salvador, *C. attenuata* is *taranta* (craziness), *hierba de tamagaz* (snake herb, the *tamagás* is the viper *Bothrops*), *retámara* (resembling *retama* or *Spartium*), and *bola de ratón* (mouse testicle, Sinaloa). Salvadorians have used the stems to clean clothes (Standley 1931).

Ceanothus

(From Greek *keanonos* or *keanothos,* applied by Theophrastus, 372–287 B.C., to a corn-thistle, *Carduus arvensis*)

Ceanothus americanus. From Britton and Brown 1897.

Ceanothus americanus (of America)
kikuki manito (spotted snake spirit, Meskwaki, Menomini, Potawatomi, Ojibwa)
red-root [redroot]
Säckelblume (purse flower, German)
tabe-hi (*hi*, plant, Omaha-Ponca)
tauchima hobok (*hoba*, similar to, Choctaw)
[Indian, New Jersey, Jersey]-tea

Linnaeus described the first species of *Ceanothus* in 1753, calling it *C. americanus.* Leonard Plukenet had commented on the species in London in 1696 and Jan Commelin had discussed the species from the plants grown before 1697 at the garden in Amsterdam. It is not obvious that either of them knew of the medical traits of this New World plant. There are now 55 species in this North American genus (Mabberley 1997).

Missouri River people (Lakota, Omaha, Pawnee, Ponca) made a tea from the leaves, and burned the wood when other kinds were not available (Gilmore 1919). The Menomini and Winnebago of Wisconsin and Meskwaki of Iowa also used the leaves as tea (Yanovsky 1936). Although during the Revolutionary War *Ceanothus* was used as a substitute for tea (Vogel 1970, Krochmal and Krochmal 1973), there seem to be no records of indigenous people in the northeastern states using it that way.

People in the eastern parts of North America mostly used *Ceanothus* as medicine. For example, the Alabama made a decoction of the roots as a remedy for injured legs and feet (Swanton 1928a, Taylor 1940). The Cherokee took a root infusion for bowel problems and put it on aching teeth (Hamel and Chiltoskey 1975). The Ojibwa used a root infusion as a laxative and for pulmonary troubles (Moerman 1998). The Iroquois used the roots as an abortifacient, against diarrhea, for colds, and venereal diseases; they used the bark to treat sores and sore mouth (Moerman 1998). The Menomini considered a root infusion good for all stomach problems, and the Meskwaki mostly used it for bowel troubles (Moerman 1998). Probably this is the species used by the Choctaw for lung problems (Taylor 1940).

The roots are strongly astringent, with 8% tannin content, and they contain an alkaloid that is mildly hypotensive (Foster and Duke 1990).

Celtis: Hackberry

(Greek name used by Pliny, A.D. 23–79, for a lotus with sweet berries; described by Herodotus, fl. 484 B.C., Dioscorides, fl. A.D. 40–80, Theophrastus, 372–287 B.C., and Homer, 9th–8th? century B.C.)

Hackberry, our name for *Celtis*, is a corruption of Scandanavian names. Why the name was applied first to northern European plants (*Prunus*) and then to an unrelated species from the Mediterranean (*Celtis*) is uncertain. However, it may have been because of edible fruits in both genera.

In the high Scandanavian latitudes, there are two wild bird-cherries called "hagberry" or "hegberry." Those names appeared in English with John Gerarde's herbal in 1597, but they are probably based on Old Norse, *heggr.* The names are still given to *Prunus padus* and *P. avium*, as Danish *haegge-baer*, Norwegian *hegge-bär*, and Swiss *hägg-bär.* Modern Danish speakers think that "hedge" was the original reference with *hæg*, derived from *haek.* Still, an alternative meaning, at least originally, was the female personification of evil we call a "hag." The allusion to evil still lingers because some older people in the British Isles will not use the wood of *P. padus*—they consider it a witch's tree (Vickery 1995). At least the fruits are not regarded as dangerous because they have been prized for jams since ancient times. Gaspar Bauhin, in the 1600s, was an exception—he considered them inedible. Besides jams, both species have been used in liqueurs, and *P. avium* is still the flavoring in *Kirschwasser.*

Linnaeus gave us *Celtis* in 1753 when he adopted the name given to the plants by the 1st century A.D. Roman author Pliny the Elder. Gaspar Bauhin had also discussed the plants and called them *Lotus fructu cerasi* (the lotus with a cherry fruit). Within that seemingly simple phrase is hidden a complicated history.

Early Greeks wrote about the *lotus*, which was supposedly the food of the Lotus-eaters (*Lotophagi*). Some say the *lotus* was "the food of the Lotophagi, which Heroditus, Dioscorides, and Theophrastus

Celtis. *Celtis iguanea* (left). a. Flowering branch. b. Close-up of section of flowering branch. c. Staminate flower with dehiscing anthers. d. Staminate flower longitudinally dissected. e. Floral diagram of staminate flower. f. Young pistillate flower. g. Pistillate flower with perianth removed to show pistil and staminodes. h. Pistillate flower longitudinally dissected. i. Fruit. j. Diagram of pistillate flower. *Drawn by Priscilla Fawcett.* From Correll and Correll 1982. *Celtis laevigata* (top right). From Sargent 1905. *Celtis occidentalis* (bottom right). From Sargent 1905.

describe as sweet, pleasant and wholesome, and which Homer says was so delicious as to make those who ate it forget their native country" (Johns in Fernald et al. 1958). Many, like Joachim Camerarius (1534–1598), equated *lotus* with *celtis* in 1586. However, not everyone agrees on that. Some state that the plants eaten by those *Lotophagi* were *Ziziphus lotus*, and others think the delicacy may have been *Diospyros lotus*.

William Turner ([1548] 1965) was one of the first to discuss what we now call *Celtis*. He said, "It hath a leafe lyke a Nettel, therefore it may be called in englishe Nettel tree or Lote tree." Another comparison, made by John Martyn (1741 in OED 1971), was about "the nettle tree, the fruit of which is far from that delicacy which is ascribed to the Lotus of the ancients."

Nettle-tree was originally applied to the European species, *C. australis*, and later extended to those in North America. John H. Balfour (1849, *Manual of Botany*) compared the names with "*Celtis*, the Nettle-tree, or Sugarberry, has a sweet drupaceous fruit." The American plants were still being called nettle-tree by Chapman (1897), but that name was abandoned by Sargent (1905) in favor of hackberry and sugarberry.

Florida has four species of *Celtis*, three of which have restricted ranges. Both *C. laevigata* and *C. occidentalis* are temperate species, while *C. pallida* and *C. iguanaea* are tropicals on the northern edge of their distribution within Florida. All American species have been used by people wherever they grow.

Celtis laevigata (smooth) is called sugarberry in the United States. That name is akin to honeyberry, a name used by the Greeks for *C. australis*, and both refer to the sweet fruits. Similarly, Catawba said *yap nyu'stciwe?* (tree sweet). Yanovsky (1936) had records of tribes eating the berries in Minnesota, Nebraska, New York, North Dakota, South Dakota, and Wisconsin. Southern tribes also ate them.

The Seminoles ate the fruits and called Florida plants *istapõ:cki:kî* (to crunch, Mikasuki) or *kapapó:ckô* (cruncher, Creek). Similarly, the Muskogee-speaking people called these trees *kvpopockv* [*kapvpóckv, kvpvpockv*] (from *kvopocketv*, to crunch), which is cognate. The Alabama called the trees *kapko*, which is cognate to the first part of the other Muskogean names. Perhaps *kapko* is a simple term, although Martin and Mauldin (2000) thought the name *kvpopockv* was an imitation of people crunching the fruits.

People all across the southern United States used the plants for food or medicine. Among these were the Creek relatives the Houma, who used a decoction of bark for sore throat or the bark mixed with other items to treat venereal disease (Speck 1941). The Comanche beat the fruits into a pulp, mixed it with fat, rolled that into balls, and roasted them over a fire (Moerman 1998). Several groups in the southwestern United States, including the Eudeve, Akimel O'odham (Pima), Tohono O'odham (Papago), and Yavapai ground the seeds for food, either cooked or raw (Hodgson 2001).

Celtis occidentalis (western) is the upland species that is more common in the northeastern United States and adjacent Canada. The species reaches its northern limit in Quebec (Houle and Bouchard 1990) where it is called *sucré baie* (sugar berry) or *micocoulier* (from 1547, an unattributed quote, "*Lotos* is a tree named *micacoulier* in Provence"). The current spelling is derived from Provençal, although it originally came from Greek, *mikrokoukouli* (from *mikros*, small, and *kuklos*, round) (Boisacq 1923). The French name *bois inconnu* (unknown tree) hints at yet another early mystery. It is not now clear why French colonists began calling the trees that, but it may have been from La Salle's first contact with the Natchez people of Louisiana in 1682 (Swanton 1946). The Natchez venerated *Celtis* and burned its branches in a sacred fire (Speck 1941). By the late 1930s, the Houma, who by then spoke only Creole French, had altered the original *bois inconnu* for *Celtis* to *bois connu* (known tree).

Within the Missouri River drainage region, *C. occidentalis* was used for food and wood. Names through this region include *gube* (Omaha-Ponca), *kaapsit* (Pawnee), and *wake-warutsh* (raccoon food, Winnebago). The Dakota call it *yamnumnugapi* (to crunch, because animals crunch its berries), and used it to season meat. When the Dakota first saw black peppercorns (*Piper nigrum*), they thought they looked like *Celtis* berries, and called them *yamnumnugapi washichun* (white man's *yamnumnugapi*) (Gilmore (1919). To the south, the Kiowa say it is *ya-ai-gaw* [*ya-aip-hàp*], and the Atakapa say *kon nec* (*kon,* heavy, *nec,* wood).

The Houma used *C. occidentalis* as they did *C. laevigata* (Speck 1941). The Iroquois used a bark decoction to regulate menses and as an abortifacient. The Western Keres, Meskwaki, Omaha, Pawnee, and Kiowa ate the fruits. Many of them ground the berries to make mush or mixed them with fat and parched corn to make a gruel (Moerman 1998).

This species continues down into Mexico where it is called *palo blanco* (white wood, Coahuila, Tamaulipas). In the United States and Mexico, the wood is used as fuel, in furniture, flooring, crating, cooperage, and posts. In addition, the bark is considered an anodyne and a refrigerant (Hocking 1997).

The two tropical species are better adapted to dry habitats than the usual moist regions in Florida. Perhaps that is why they have such restricted ranges in the peninsula, or maybe they grow where they do because humans moved them there.

Celtis pallida (pale) has been found only in Lee County on old Indian middens (Wunderlin 1998). Outside Florida, it reappears in Texas, New Mexico, and, Arizona, thence south through Tamaulipas to Argentina, Bolivia, and Paraguay (Correll and Johnston 1970). It is called desert hackberry (Arizona, Florida), *granjeno* [*grangeno, granjén*] (little seeds, Arizona, Chihuahua, Coahuila, Durango, Nuevo Leon, Texas), and *granjeno huasteco* (Huastec seeds, Tamaulipas).

Many other people refer to its armament and call it spiny hackberry (Sonora, Arizona, Florida), *garabato* (iron hook, Sonora, Sinaloa), *garambullo* (spiny plant, Mayo, Sonora), *gec cehd* (*gec*, thorny, Zapotec, Oaxaca), *palo de águila* (eagle's tree, Sonora), *rompecapa* (cape tearer, Sonora, Oaxaca), or *vainora* [*bainora*] (probably because of similarity with *baiguo, Pisonia capitata*, Sonora).

Probably the most common allusion is to the fruits, with the names *acebuche* [*acibuche*] (wild olive, Sonora), *cumbro* [*gumbro*] (Mayo, Sonora), *cúmero* (either Mayo is the basis for the Spanish or vice versa, Sonora), *cungo* (Yaqui, Sonora). The name *capul* [*capuí*] (Sonora, Durango, Texas) is a comparison between *Celtis* and *Prunus*, but this refers to *Prunus serotina*, usually named *capulí*.

Presumably, the other names for the plants in western Mexico are simple terms. These include *koom* (O'odham, Sonora), *kuwavul* (Pima Bajo, Sonora), and *ptaacal* (Seri, Sonora).

The Mayo eat the fruits, make the roots into a laxative tea, and formerly used the branches for hunting bows. As with all species, the wood is used for fires and household items (Yetman and Van Devender 2002). Many other people in the southwestern United States and nearby Mexico ate the fruits, although some comment that they are astringent. I find them a little waxy, but otherwise good. The Seri preferred the wood for bows and also made cradles from it (Felger and Moser 1985). Some have used the leaves to relieve head and stomach pains (Hocking 1997).

Judging from the number of common names and the literature on it, *C. iguanaea* (of iguanas) is the most used species in the Americas. The trees range from Texas through Mexico, Mesoamerica to Peru, Paraguay, and Uruguay. In Florida, it is known from two disjunct areas on Indian middens in Lee and Collier Counties.

This species shares the names *garabato, cumbro, palo blanco*, and *vainora* with *C. pallida.* Many of its other names refer to the nasty, curved thorns on its branches, including *berraco* (screamer, because it tears at skin, Colombia), cat-claw (Virgin Islands), *clavo verde* (green nail), cockspur (Puerto Rico, Virgin Islands), *croc à chien* (dog's tooth, Guadeloupe, Martinique), *gallito* (little rooster, Dominican

Republic), *garabato blanco* (white iron hook, Sinaloa), *guachar-agüero* (bad luck), *liane pagne* (pain?, vine, Haiti), *manca montero* (cowboy crippler, Cuba), *rompe caite* (sandal breaker), *tala gateador* (cat's claws), *thak loh* (white *loh* or *Pisonia aculeata*, Huastec, San Luis Potosí), *uña de gato* (cat claw, Cuba, El Salvador), *zarza* (blackberry, Cuba), *zarza blanca* (white blackberry), and *zarzaparilla cimarrona* (wild sasparilla, Cuba). Several others allude to the edible fruits—*azufaifo* (sulfury one, Puerto Rico), *cagalera comestible* (edible *cagalera*, Nicaragua), *granejo* [*granjeno*] (little seed, Veracruz and elsewhere), *grão de galo* (rooster grain, Brazil), iguana hackberry (Florida), tropical hackberry (Sonora), *jujubier des iguanes* (iguana's jujube [*Ziziphus*] tree, French Antilles), and *meloncito blanco* (white little melon). Sometimes it is called *zitsmuc* (*sits'* means snack or *hors d'oeuvre*, probably in reference to the edible fruits, Maya, Yucatán). In Belize it is wild cherry.

The other names have mixed meanings, or they have not been translated. These are *baboyana*, *barimiso*, *cagalero* [*cagalera*, *cacatera*] (El Salvador, Panama), *cagalera blanco* (El Salvador), *carocá'a* (Guarajío, Sonora), *chaparro blanco* (white thicket, Sonora), *chêne blanc* (white oak, Haiti), *coca* (cocaine, Hispaniola), *irurapia* (Brazil), *joáo miúdo* (little John, Brazil), *liane crabe* (crab vine, Haiti), *marimiso*, *palo de arco* (bow wood, Sonora), *piscucuy*, and *muc* (cover, because the leaves are used to line and cover the pits where some traditional dishes are cooked, Maya, Yucatán).

There are many uses recorded in addition to eating the fruits and using the wood. The Mayas in Yucatán apply sap to sore eyes, but an excess is irritating and may cause blindness. They also use a leaf decoction bath to lower fever. The Huastecs use the shoot to bring boils to a head (Alcorn 1984). In Hispaniola, fruits and bark are considered an emollient and a pectoral. Wood from the trunks is used to make *canastas*. In Brazil, a leaf infusion is used in vaginal injections for treating leukorrhea (Mors et al. 2000). Boiled, mashed fruits are considered a specific remedy for dysentery and intestinal catarrh.

Beyond studies involving allelopathy (Lodhi and Rice 1971), few studies have been made of chemicals in the genus. It is known that *C. iguanaea* contains the sugar alcohol quebrachitol (Plouvier 1958). Moretenol has also been isolated (Santa-Cruz et al. 1975). More recently, extracts have been found cytotoxic on carcinoma cells (Ruffa et al. 2002).

In addition to all these uses, two others are seldom mentioned. All *Celtis* species may be used as a tan dye, and at least the nonthorny kinds may be used in weaving baskets (Tull 1999). The slender, flexible stems were favorites among some weavers.

Many of the fruits eaten during earlier centuries tended to be small and unassuming. The positive side of those small fruits was that they often had more flavor than those now widely available. Regardless, the fruits of the hackberries seem to be among those that have been essentially discarded by modern societies.

There remains yet another mystery about *Celtis* that to me overshadows the others. No one seems to have ever explained if iguanas really do eat that spiny *C. iguanaea.*

Centrosema: Butterfly Pea

(From Greek *kentron*, a spur, prickle, sharp point, the center, and *sema*, a signal, referring to the spurred standard petal)

Centrosema virginianum. From Britton and Brown 1897.

We now call these plants *Centrosema virginianum*, although they were not always called that. Indeed, until 1837 when George Bentham transferred the species to this genus, it was placed in *Clitoria*. Later, there was a debate whether or not *Centrosema* or *Bradburya* was the correct genus.

Linnaeus, in his underappreciated role as dirty-old-man, put this species in the genus *Clitoria*. He and several of his predecessors were impressed with the similarity of the flower and that female structure. Gronovius had previously called the plants *Clitoria foliis ternatis, calycibus campanulatis geminis* in 1739–1743. Even earlier, it was *Clitorius trifolius, flore menore caeruleo* to Dillenius in 1732. They and Linnaeus knew the plants grew in Virginia, the state named for the Virgin Queen, making the names even more suggestive.

The similarity was not limited to botanist-physicians of Europe because several of the indigenous names for these plants in the Americas also make the same or related allusions. These names show a more euphemistic approach than Linnaeus's direct comments. Some of those names are *colincito* (little hill, Hispaniola), *conchita* (little shell, Puerto Rico), *conchita de Virginia* (little shell from Virginia, Puerto Rico), *consuelo de los caminantes* (traveler's consolation), *crica de negra* (negresses' cat, Cuba), *cucha de negra* (negresses' spoon, Puerto Rico), *diversión de los caminantes* (traveler's diversion, Hispaniola), *divierte caminante* (traveler's delight, Cuba), *flor de conchitas* (little shell flower, Puerto Rico), *frijol marrullero* (wily bean, Cuba), *liane fifi* (Fifi's vine, Hispaniola), *marrullero* (wily one, Cuba), *papito* (little potato, Cuba, Hispaniola), *papito de la reina* (queen's little potato, Cuba), *papo de la reina* (queen's potato, Hispaniola), *sonajera azul* (blue tamborine, Puerto Rico), *totico* (blackbird, Hispaniola), *vaillant garçon* (brave waiter, Hispaniola), *yerba de bulla* (uproar herb, Puerto Rico), and *zapatito de la reina* (queen's little shoe, Oaxaca).

Whether or not Florida's common name for these vines has been bowdlerized or simply evolved in a different direction is unknown. Regardless, we typically call the plants butterfly pea or coastal butterfly pea. In Cuba, it is simply *azulada* (the little blue one). Mayan people in Yucatán call it *cantsin* [*k'antsin*]. Others simply refer to the twining habit by saying *corde violón* (viola string, Hispaniola), *liane corde violón* (viola string twiner, Hispaniola), or *liane violón* (string twiner, Hispaniola). The comparison with violin strings seems curious, because the same basic names are given to the parasitic *Cassytha*. Others prefer to make a comparison between *Centrosema* and the other beans by calling it *pois marron savane* (wild bean of the savanna, Guadeloupe, Martinique), *pois sauvage* (wild bean, Guadeloupe, Martinique), *pois-pois* (bean-bean, Guadeloupe, Martinique), *pois-pois vrai* (true bean-bean, Guadeloupe, Martinique), or *ti-pois* (small bean, Guadeloupe, Martinique). Only one name has been found that refers to medicinal use, *grippez* (for influenza, Hispaniola).

As might be expected from the common names, the plants have mostly been used to treat sexual problems. In most places, the plants are considered an aphrodisiac, doubtless because of the resemblance to the human anatomy. They also have been used directly in medicines.

To make the situation more complicated than many, few professional botanists and virtually all nonprofessionals do not distinguish among all the species recognized in either *Clitoria* or *Centrosema*. Moreover, they do not always distinguish between those two genera. Because of that complication, here more than in some other cases, the expert on this group, Paul Fantz, feels that many of the names cited under *C. virginianum* are actually applied to *Clitoria* species (personal communication, Jan. 2002). His experience suggests, for example, that *conchita* is a name applied exclusively to *Clitoria* and not *Centrosema*. The same may be said of other names in the list above. However, as most of us cannot distinguish these, the plants themselves are likely to be used without regard to species. That may cause problems for those being treated.

In Cuba, a decoction of roots alone or roots and flowers is considered emmenagogue. This mixture is made by placing a handful of cleaned and macerated roots in a bottle of water. A glass taken in the evening is said to promote menstruation and induce uterine contractions and to aid in *el flujo loquial*. A stronger dose of the same liquid is used as a vaginal douche. An infusion of the flowers is used in the same ways. By combining a handful each of flowers and roots in a bottle of good wine one makes a *medicamento magnífico* (magnificent medicine) that is taken in one cup a day to treat *clorosis* (a malady of adolescents involving "impoverishment of the blood," probably anemia) and for liver and intestinal problems. Seeds are said to be laxative, vermifugal, and slightly emetic. Similarly, in Hispaniola the species is used against anemia and general debility.

Within the Americas, this species ranges from Florida to Texas, and from New Jersey to Kentucky and Arkansas, then in Mexico (Sonora and Tamaulipas south), the Bahamas, Cuba, Dominican Republic, Haiti, Jamaica, Puerto Rico, Turks and Caicos Islands, several Virgin and Leeward Islands, and in South America to Paraguay and Argentina. Roig (1945) reported *C. virginianum* as being in Africa, but that turns out to be a misidentification of *C. molle* (Fantz 1996) as *C. pubescens*. The true *C. pubescens* is found at higher elevations in southern Mexico and otherwise confined to Central America (Fantz and Predeep 1992).

The other species in the Caribbean and Mexico are reported to have the same medicinal traits. Within Cuba, there are two other species, *C. plumeri* and *C. pubescens*, that are used (Leon and Alain 1946–1953). Those species are known by the same names as *C. virginianum*. However, in Puerto Rico, *C. pubescens* is also called *flor de pito* (whistle flower), having the same suggestive overtones as the other names. In Mexico, it is known as *bejuco de patito* (little duck's vine, Oaxaca). Also in Mexico, *C. plumeri* is called *bejuco de palomita* (little dove's vine, Oaxaca), *frijolillo* (little bean, Tabasco), *mariposa* (butterfly, Campeche), and *patitos* (little ducks, Tabasco). Yet other species in

Mexico are called *talmohuite* (*C. sagittatum*, Hidalgo) and *patabuil* (*C. hastatum*, Sinaloa). So, throughout the range several species have attracted attention.

Pittier (1926) reported for Venezuela what Roig (1945) called *C. violacea,* known there as *generala* (woman with the rank of general). That plant has a reputation of being effective in curing venereal diseases and leprosy. However, it turns out that the name was completely incorrect in Roig (1945), and the plants are *Rhynchosia violacea.*

The genus *Centrosema* has 35 species in the warmer parts of the Americas (Mabberley 1997). Some are used as "green manures," especially under plantations of rubber (*Hevea brasiliensis*) and coconut (*Cocos nucifera*). In the past four decades, agronomists have discovered that species of *Centrosema* make excellent forage crops, providing more crops per season than other plants, and with key nutrients utilized better by animals. Germ plasm is being investigated at stations in both Old and New World tropics. One of the last species described, *C. carajasense,* is used as a fish poison in Amazonian Brazil.

One of the distinctive traits of both *Centrosema* and *Clitoria* is that the flowers are rotated 180° and the banner points downward. In that position, it serves as a landing platform for pollinations. In Costa Rica, carpenter bees (*Xylocopa*) are a major pollinator of at least *Centrosema* (Janzen 1983). Bumble bees (*Bombus*), as long-distance fliers, are also important in mixing the gene pool. Also important there is Euglossine Merian's orchid bee (*Eulaema meriana*). The Euglossine bees are known mostly as orchid pollinators, but this one collects nectar from *Centrosema.* Ranges of several species of *Clitoria* are similar to some in *Centrosema,* and apparently some of the same large bees visit its flowers.

Some years ago, I was leading a class field trip near the beach in southern Florida. As we walked along, we passed over a carpet of butterfly pea in full flower. I no longer remember why I did not point it out. Still, this time I passed over the plant without comment. From behind me, one of the students yelled, "Hey, what is this plant?" Without looking back, I answered, "It is *Centrosema,* called butterfly pea. It made a much better story when it was in the genus *Clitoria.*" The student retorted, "Well, hell, look at the plant next time before answering!"

Cephalanthus: Button-Bush

(Greek *kephale,* head, *anthos,* flower)

Since about the year 1340, people who speak English have been using buttons, although the literature of the time spelled it "boutounz." Yet, as far as I am aware, people in the Americas never invented that seemingly simple fastener for clothes and other items. Some time after the Europeans invented the closing device, they began to apply the word to other structures. Among the things that resembled buttons to some people are plants like button-wood and button-bush. Younger people who saw the plants, and particularly the fruits, would not have given them the same names. The fruit clusters do not resemble modern buttons in the slightest. However, the fruit clusters do look like old buttons that were globular.

Cephalanthus occidentalis. From Institute of Food and Agricultural Sciences.

The first record we have of these plants being known in Europe was published by Leonard Plukenet in 1696. He called them *Scabiosa dendroides americana* (the American tree *Scabiosa*). The only similarity they have with *Scabiosa* (Dipsacaceae) is the globelike flower cluster, and Linnaeus noted the difference. In 1738, Linnaeus proposed the new name *Cephalanthus* for these trees. That name was picked up by his contemporaries and used in 1739 by Jan Gronovius and in 1740 by Adrian van Royen (1705–1779). In 1753 Linnaeus gave us the name we use today—*Cephalanthus occidentalis.*

This species is almost confined to mainland North America, but it also ranges into Central America and the Caribbean. The trees in the United States and Canada are known from New Brunswick and Ontario, south to Florida, and west to Texas, Arizona, and California. In Mexico, they grow from Chihuahua to Veracruz, Guerrero, and Tabasco, and south to Honduras. Button-bush is also in Cuba, but in none of the other Caribbean islands—not even the Bahamas. Given that the fruits are eaten by wildfowl, including at least seven species of ducks (Martin et al. 1951), one might reasonably expect the small trees somewhere in the Bahamas, but they are not there.

Now there are six species known in the genus; most are tropical, and *C. occidentalis* is the only one in North America. With a name like *C. occidentalis*, there should also be a *C. orientalis*, and there is. Linnaeus described that species at the same time in 1753, and he knew the species from India and Africa. There are also species in China and Taiwan, and others in the Americas, ranging as far south as Peru and Argentina.

Cephalanthus occidentalis has a long history of medicinal use, but apparently none of the other species has similar usage. The Chinese, who are famous for finding the medicinal aspects of virtually everything, do not seem to have used the genus either.

In Florida, the Miccosukee call these shrubs or small trees *halpáti:hosô:tî* (alligator shader); their relatives say *sakco 'meto* [*sakcomitto*] (*sakco*, crawfish, *em,* its, *eto,* tree, Creek) or *sakcó imittó* (*sakcó,* crawfish, *im,* its, *ittó,* tree, Koasati). The Choctaw called them *notêm pisa* (*noti,* tooth, *im,* its, *pisa,* eyes or vision) or *akkasoli,* and the Chickasaw know it as *shakchimitti'* (*shakchi,* crawfish, *im,* its, *itti',* tree). Sometimes the Chickasaw say *shakchikoyyo* (*shakchi,* crawfish, *ikoyyo,* a word always combined). Kiowa said *ee-toñ-bà-a.* North in Canada, they are called *bois de plomb* (lead flower), apparently because of the similarity of the fruit cluster to a bullet or lead sinker used in fishing.

A number of names have been recorded in Mexico. In the northern part of its range, it is known as *mimbre* (Sinaloa) or as *mimbro* (Mayo, Sonora). That is a name given to several genera in Mexico, including *Cephalanthus*, *Chilopsis* (Bignoniaceae), *Cornus* (Cornaceae), and *Forestiera* (Oleaceae). The only similarity between them is the leaves, so the reason this name is applied to them and also to the genus *Monstera* (Araceae) is obscure. *Cephalanthus* is also called *rosa de San Juan* (St. John's rose, Texas, Durango), although association with that saint is obscure unless there is a medicinal connection. Some people think they smell like jessmine (*Jasminum*, Oleaceae), and say they are *jasmín* (Michoacán, Guerrero). Both have nocturally fragrant flowers, so the analogy is reasonable. No meaning was found for *cahuapate* (Náhuatl, Oaxaca), but the Zapotec simply say *yo yado* [*ya yado*] (flower, Oaxaca). The southernmost names located are *uvero* [*umbero*] (grape tree, Tabasco). The rounded fruit cluster is reminiscent of grapes. In Cuba, the species is known as *aroma de ciénaga* (marsh perfume, Cuba), *aroma de laguna* (bay perfume, Cuba), and *cavellina de ciénaga* (marsh carnation, Cuba). All of those names refer to the fragrance of flowers similar to *jasmín* in Mexico, but add the habitat as a "binomial" to distinguish taxa.

In the United States, the most frequently used name is button-bush, but there are others that are (or at least have been) used. These names include crane-willow, globe-flower, honey-balls, pincushion flower, pond dogwood, and snowball. Some have objected to several of these names because they have also been used to refer to other plants. For example, the name button-bush historically was also applied to *Platanus occidentalis* (Platanaceae). We also call *Platanus* (which see) the American plane tree. Snowball is also applied to *Symphoricarpos* (Caprifoliaceae), and globe flower is used for *Trollius* (Ranunculaceae). However, that is part of the beauty of common names—they describe something that has caught the fancy of the people who apply them. As common terminology is not bound by international rules, there will be overlap. Instead of causing distress, that should cause delight in the ingenuity of the human intellect.

Even more than the indigenous names show, this was an important plant in indigenous people's pharmacopoeias. *Cephalanthus* was involved in at least five remedies by the Seminoles and was "a favorite remedy" among the Choctaw (Bushnell 1909). We have records of its use also among the Chickasaw (Campbell 1951), Kiowa (Vestal and Schultes 1939), Koasati (Taylor 1940), Meskwaki (Smith 1928), and Comanche (Moerman 1998). These represent four major language groups (Algonquian, Aztec-Tanoan, Muskogean, Shoshonean), and any plant affecting that many diverse people was important indeed.

These indigenous people used button-bush as a remedy for dysentery (Choctaw, Seminoles), as a diuretic (Seminoles), as an emetic (Meskwaki), as a febrifuge (Choctaw), for eye trouble (Chickasaw, Choctaw), and to stop hemorrhages (Kiowa). Moreover, it was an all-round tonic and analgesic, used to treat headaches (Seminoles), rheumatism (Koasati), stomachaches (Seminoles), and toothaches (Choctaw). The pollen was found in human coprolites in the Fort Center archaeological site, a Glades construction (Hogan 1978). There is no reason to assume that the pollen might be ingested for other than medicinal application by those southern Florida people who preceded the Seminoles.

Several parts of the tree were used, mostly in decoctions but also in poultices. A root poultice was applied to sore eyes. Decoctions of bark were used for dysentery and for eye problems. Root bark and trunk bark were febrifuges. Bark was also used to stop toothache. Decoction of leaves relieved rheumatism.

The Comanche even used the wood to make game sticks.

Europeans and Americans in many places used the plants much as had the indigenous people. Root tea was gargled and swallowed to relieve throat irritations, severe colds, and to bring up phlegm. Mixed with other plants, it was considered a prime treatment for colds and even whooping cough. At one time, a mixture of root bark and comfrey (*Symphytum officinale*) was even considered a cure for consumption (Porcher 1863). Decoctions were also used treat skin infections. Decoctions of the bark were once considered effective for treating malarial fevers and syphilis.

Yet, by the late 1800s, the plant was considered "not important" in medicine. Although a formula for preparing the plant as a medicine was given by Millspaugh (1892), he was not impressed. He said, "The tincture ... has a light, clear, orange-brown color, but transmitted light, a bitter, astringent taste, and an acid reaction." He too attributed tonic, laxative, and diuretic properties to the plants. Murphee (1965) found people in the Panhandle putting roots in whiskey to treat asthma and dropsy.

Then people began realizing that there were problems with using the plants and some of those treated developed side effects worse than the malady being treated. Some people developed violent vomiting, convulsions, and paralysis (Perkins and Payne 1978). The chemistry of the plants was examined and found to contain the toxic glycosides cephalanthin and cephalin. Both of these are hemolytic (destroys blood corpuscles) and even lead to death (Foster and Duke 1990). Also present are substantial amounts of tannins and volatile oils. Although the leaves are toxic to horses, they are browsed with impunity by deer. Therefore, it seems that the chemicals react differently in different organisms.

Cercis

(From Greek *kerkis*, the ancient name used by Theophrastus, 372–287 B.C., for the redbud or Judas tree)

arbol del amor (tree of love, Spanish)
Judas' tree; *alberi du guida* (Italian); *arbre de Judée* (French); *arvore de Juda* (Portuguese); *Judasbaum* (German)
olaia (Portuguese; this Middle Eastern tree does not grow in Portugal, but was introduced and cultivated for ornamental purposes)

Cercis canadensis (of America)
ga-lo-we:-d(i) (Cherokee)
gaînier (girdle-maker, but also a name for *Cercis*, whose pod resembles a girdle, French)
kee-à-gu-la (Kiowa)
łotakkvpe [*rvtákvpe, rotakkvpe*] (*łotak*, maybe a simple term for the plants, although Martin and Mauldin 2000 suggest it imitates the crack of a whip when used in compounds, *vpe*, stem, Muskogee); *taka'sta'* [*takastaka', takastak, takasta'*] (Chickasaw); *takaská:ci* (akin to *takáhkan*, crack in the earth or mud, Koasati)
[eastern] redbud
yap pete̲'? (flat tree seed, Catawba)
zhon̲'-sha be hi [*zhon̲'-sha be the hi*] (*zhon̲*, wood, *sha-be*, dark, *hi*, tree, Osage)

Cherokee children are known to have eaten the blossoms (Moerman 1998). Hedrick (1919) recorded that French Canadians "use the flowers in salads and pickles." At least the Canadians seem to have been influenced by the edibility of the Old World species, *Cercis siliquastrum*, which has edible flowers and fruits (Hedrick 1919). In Texas, the buds, flowers, and young pods are eaten raw or sautéed (Vines 1977, Cox and Leslie 1991). However, others say the bud is poisonous because of saponins (Diggs et al. 1999).

Indigenous Americans used the tree mostly as medicine. The Alabama used an infusion of roots and inner bark for fever and congestion (Swanton 1928a, Taylor 1940). The Cherokee used a bark infusion to treat whooping cough (Hamel and Chiltoskey 1975). The Delaware used an infusion of the bark to treat fever and vomiting (Moerman 1998).

The Osage used charcoal from the wood to paint the face during battle. Only those so painted were awarded honors (La Flesche 1932).

Chamaecrista

(Conrad Moench, 1744–1805, composed this name from Greek *chamai*, on the ground or dwarf, and Latin *crista*, a crest)

Chamaecrista fasciculata (clustered)
bala okchi (Choctaw)
beeflower
dormidera (little sleeper, Cuba, Venezuela)
golden cassia
kowwàykinchastoki (*kowwayki*, quail, *im-*, its, *chastoki*, beans, Alabama); *kowwaykí impalaná* (*kowwaykí*, quail, *im*, its, *palaná*, bean, Koasati)
ñajú (Venezuela)
ołkofimpátâ:kî [*o(th)kofempaataake*] (from *ołkofi*, persimmon, *im*, its, *pataki*, bed, Mikasuki)
[showy] partridge pea

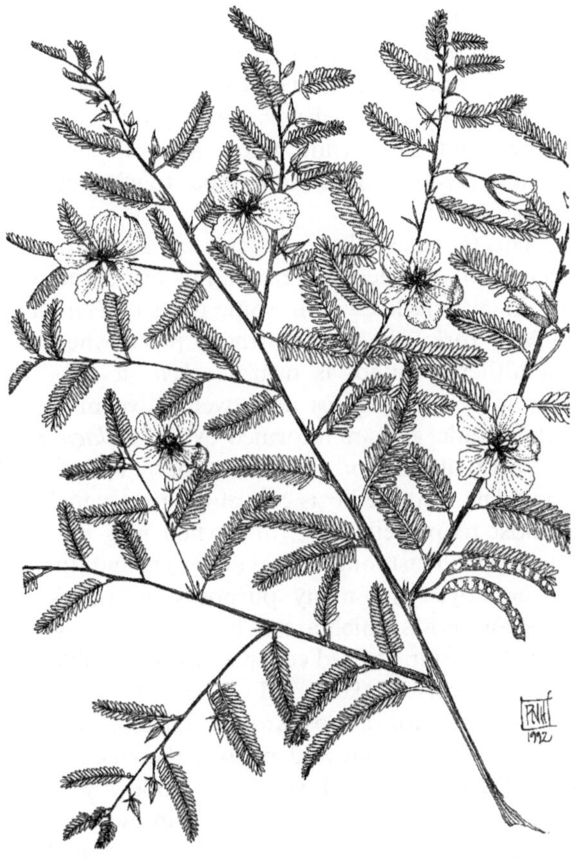

Chamaecrista fasciculata. *Drawn by P.N. Honychurch.*

[American, prairie, wild] senna (Carolinas)
sleeping-plant [sleepingplant]
unnagéi (black, Cherokee)

Chamaecrista nicitans (blinking, moving)
hassihasàaya (*hassi*, grass, *hasaaya*, to sulk or play possum, Alabama)
ołkofimpátâ:kî [*orkofimpátâ:kî*] (persimmon's bed, Mikasuki); *satá impatá:ka* (*svtv*, persimmon, *em*, its, *espvtakv*, bed, Creek)
[sensitive] partridge pea
sensitive pea (this and the specific name alludes to the "sleep" movement where the leaflets move against the rachis in the evening)
sim-sim (*sim*, nasal mucous or semen, Maya, Belize)
tohkil (Choctaw)
wild sensitive plant

Linnaeus called the sensitive pea *Cassia nicitans* in 1753. Before that date Plukenet had called it *Senna spuria virginiana, mimosaea foliis, floribus parvis nicantibus* (false *Senna* from Virginia, with leaves like mimosa, and small flowers that sleep) in 1696. Linnaeus had also studied the living plants at both the *Hortus Cliffortianus* and *Hortus Upsaliensis.* When Conrad Moench (1744–1820), professor at Marburg, Germany, put the species in *Chamaecrista* in 1788, most people thought he was a devout "splitter," just creating names where they were not needed. Subsequent studies have indicated that plants called *Cassia* were actually from more than one lineage that had converged on a similar morphology. André Michaux described *Cassia fasciculata* in the early 1800s, and it was moved to *Chamaecrista* by Edward Lee Greene (1843–1915), another devout splitter who created many "microspecies."

The earliest record of *C. fasciculata* in Florida was pollen associated with pre-Columbian Glades people at the Fort Center site on Lake Okeechobee (Hogan 1978). Small (1933) recorded the Seminoles using *C. fasciculata* "in cases of nausea." Sturtevant (1955) said the same species was used as medicine to treat stomachache. The Seminoles also treat urinary tract infections with a medicine made by boiling the entire plant, including roots (Bennett 1997). The Houma boiled the roots with wild beans to make a tea for treating typhoid fever (Speck 1941).

Mooney (1885–1886) found that the Cherokee drank a decoction of *C. fasciculata* for fever and an unidentified disease with symptoms including black spots and partial paralysis. The Cherokee also bruised moistened leaves with water for dressing sores (Mooney 1885–1886) and used a medicine made from roots of *C. fasciculata* or *C. nicitans* to keep ballplayers from tiring and to treat fainting spells (Hamel and Chiltoskey 1975). The Meskwaki soaked the seeds of *C. fasciculata* in water until they became mucilaginous and then ate them to relieve sore throat (Vogel 1970).

Porcher (1863) said of *Cassia fasciculata*: "It is said to be as safe and as certain in its operation as the imported senna [*Senna italica*], but more apt to gripe; this may be corrected by infusing fennel seed or some other aromatic with the leaves. It is prepared in large quantities by the Shakers, and is generally collected after the seeds ripen; one ounce of the leaves is added to one pint of hot water, of which the dose is one to three ounces, repeated." Under the synonym *C. chamaecrista,* he wrote, "The leaves are said to be purgative."

The Mikasuki and Creek names for the plants allude to their use in ripening persimmons. They do that by "placing them in a pit lined with partridge pea plants. Layers of the plant are laid over each layer of persimmons until the pit is full" (Bennett 1997).

Chamaecyparis

(From Greek *chamai*, on the ground or dwarf, and *kyparissos*, cypress, used originally by Pliny, A.D. 23–79)

Chamaecyparis thyoides (like or resembling *Thya* or *Thuja*)
achena (Muskogee); *a'tsina'* [*a'-tsi-na', a-tsi:-n'*] (a loan-word from Muskogean, Cherokee)
amerikansk vit-ceder (American white cedar, Dutch)
cedre blanc d'Amerique (American white cedar, Quebec)
cedro [*cipresso*] *bianco* [*americano*] ([American] white cedar, Italian); *cedro blanco americano* (American white cedar, Spanish)
cowahlá (Kosasti)
cypress (for etymology, see *Taxodium*)
kogelcypres (ball cypress, Dutch)
ochsochráta (Onondaga)
rakíock (soft wood, Carolina Algonquian, North Carolina)
Scheinzypress (shiny cypress, German)
vit-cypress (white cedar, Dutch)
[Atlantic, coast, southern, swamp] white cedar (see *Juniperus* for etymology of "cedar")
witte Amerikaanse ceder (white American cedar, Dutch?)

Linnaeus recognized three species of *Cupressus* in 1753—*C. sempervirens* in Europe, *C. distica* in Virginia and South Carolina, and *C. thyoides* in Canada. He knew the first two from living plants at the *Hortus Cliffortianus.* The third he knew only from Plukenet's drawing and from his student Peter Kalm. The first species remains in *Cupressus*, and *C. distica* was moved to *Taxodium* by A. Richard in 1810. *Cupressus thyoides* was not moved to the genus *Chamaecyparis*, created by Edouard Spach in 1841, until 1888 when N.L. Britton (1859–1934), E.M. Sterns (1846–1926), and J.F. Poggenberg (1840–1893) made that change.

The basis of Linnaeus's comment on the plants was information brought back by his student Peter Kalm. On 5 May 1749, Kalm was in New Jersey where he made a special collecting trip. He wrote: "A tree which grows in the swamps here, and in other parts of *America*, goes by the name of *White Juniper-tree.* Its stem indeed looks like one of our old tall and strait juniper-trees in *Sweden*: but the leaves are different, and the wood is white. The *English* call it *White Cedar,* because the boards which are made of the wood are like those made of cedar. But neither of these names are just, for the tree is of the cypress kind. It always grows in wet ground or swamps: it is therefore difficult to come to them, because the ground between the little hillocks is full of water. The trees stand both on the hillocks and in the water: they grow very close together, and have strait, thick, and tall stems; but they were greatly reduced in number to what they have been before. . . .

"The marshes where these trees grow are called *Cedar Swamps.* These cedar swamps are numerous in *New Jersey*, and likewise in some parts of *Pensylvania* and *New York.* The most northerly place, where it has been hitherto found, is near *Goshen* in *New York*, under forty-one degrees and twenty-five minutes of north latitude, as I am informed by Dr. *Colden.* For to the North of *Goshen*, it has not been found in the woods. The white cedar is one of the trees which resist the most to putrefication; and when it is put above ground, it will last longer than under ground: therefore it is employed for many purposes; it makes good fences, and posts which are to be put into the ground; but in this point, the red cedar is still preferable to the white; it likewise makes good canoes. The young trees are employed for hoops around barrells, tuns, etc. because they are thin and pliable; the thick and tall trees afford timber, and wood for cooper's work. The houses which are built of it, surpass in duration those which are built of *American* oak. Many of the houses in Rapaapo were made of this white cedar wood; but the chief thing which the white cedar affords in the best kind of shingles" (Kalm [1753–1761] 1972).

He goes on to note that the shingles are more durable than other woods, except for red cedar (*Juniperus virginiana*). Shingles are light, so there is no need for strong beams to support them, which makes it unnecessary to build thick walls. When fire breaks out, shingles do no harm when they fall, and they soak up water to prevent fire spreading. If they catch fire, they do not easily set new fires elsewhere. Apparently, there had long been a trade in white cedar shingles because he comments at length on the young shoots remaining. Kalm then criticizes the short-sighted approach the locals had taken and how it would be hundreds of years before the cedar trees could grow back. He notes, "They are here (and in many other places) in regard to wood, bent upon their own present advantage, utterly regardless of posterity." Kalm was far in advance of the ecological movement in preaching sustained yields.

When Bartram traveled in the southeast in the 1770s, he still used the name *Cupressus thyoides.* He was not sure it was the same as that species and, while on the Tombigbee River in Alabama, said, "I observed a species of Cypress; it differs a little from the white Cedar of New Jersey and Pennsylvania (*Cupressus thyoides*) the trunk is short and the limbs spreading horizontally" (Bartram [1791] 1958). That slight variation Bartram saw is part of what A.J. Rehder (1863–1949) divided into 30 varieties and forms in 1949 (Michener 1993).

It is remarkable how little information exists on the early uses of white cedar apart from that given by Kalm. The only note found is that the Ojibwa of Minnesota in 1891 used a decoction of leaves in steam for headaches and backaches and a poultice of crushed leaves and bark for headache (Moerman 1998).

Morton (1974) found the people of South Carolina using an infusion of cones in whiskey to treat asthma. People there still chew the twigs for the flavor. Morton (1974) found records of the branches being stimulant, aromatic, diaphoretic, and stomachic.

Chamaelirium

(From Greek *chamai*, on the ground, dwarf, and *leiron*, a lily; the genus was established from an under-developed specimen)

Chamaelirium luteum. From Britton and Brown 1896.

Chamaelirium luteum (yellow)
angel's-wand
blazing star (used by John Bartram in 1751; usually given to *Liatris*)
colic root [colicwort] (usually given to *Aletris*)
devil's bit (translation of Medieval Latin *morsus diabili*, devil's bite, in German *Teufels-abbisz*; originally applied to *Scabiosa pratensis*, and in English ca. A.D. 1540; transferred to *Chamaelirium* by 1751)
drooping starwort
fairy wand
grub root [grubroot]
heloinas (from Greek *helos*, marsh, as some species frequent bogs; formerly this species was included in this genus, as *H. dioica* and *H. sabiosa*, although as now delimited, *Helonias* is distinct)
rattlesnake root
squirrel-tail
star root [starwort]
[false, true] unicorn [root] (allusion to *Aletris*)
unicorn plant

Although Millspaugh (1892) attributes the name "devil's bit" to "an aboriginal legend," the name was in use in the Old World during the Middle Ages (A.D. 500–ca. 1500), well before the New World was discovered. However, Millspaugh was alluding to the legend recounted in Jonathan Carver's book *Travels through the Interior Parts of North America in the Years 1766, 1767, and 1768* (Vogel 1970). The concept of "devil" did not exist among indigenous people; that is a European belief.

The name "devil's bit" was explained in 1616 as being "so called, because it sheweth as though the middle, or the heart of the root, were gnawed or bitten by some Diuell ... as though the Diuell did enuie the good which it bringeth vnto men by the incredible vertues that are therein."

As Vogel (1970) explained, the name "devil's bit" has been applied to three native American plants (*Aletris farinosa, Chamaelirium luteum*, and *Liatris scariosa*), and at least one other from Europe. Hunter ([1823] 1973) called a plant by that name, and editor Richard Drinnon interpreted it as *C. luteum.* However, the Osage and Kansas with whom Hunter spent his youth lived in an area where that plant does not grow (Steyermark 1963, Barkley 1986). The Osage name *sho-ma-cas-sa-es-ra-rah* means "it kills wolves" (*shon̲-ge*, wolf) so it was a poisonous plant. That is further evidence that Hunter was talking about some other species, because *Chamaelirium* is not that poisonous. Probably, Hunter was talking about *Melanthium woodii*, which does grow in the eastern two thirds of Missouri and is notoriously poisonous to mammals (Steyermark 1963, as *Veratrum woodii*; Wiersema and León 1999). However, he may have mistaken the plant for *Arisaema triphyllym,* which is *shon̲-ge i'-t'se-the* (wolf killer, Osage).

As now delimited *Chamaelirium* contains the single species, which is endemic to eastern North America (Mabberley 1997). The plants grow from Massachusetts, New York, and southern Ontario west to Michigan and south through Illinois, to Arkansas and Florida (Fernald 1950, Browne and Athey 1992, Smith 1994). The first record of the species was the specimen sent to Gronovius by John Clayton in the 1730s. That sheet and a record from Canada formed the basis of Linnaeus's description in 1753.

John Bartram was perhaps the first to record the medical qualities of these lilies (Coffey 1993). He wrote in 1751, in the appendix to the *Medicina Britannica*, printed by Benjamin Franklin, of the "[b]lazing Star, as it is called by the back [country] inhabitants, by others, Devil's bit. ... This precious root is a great resister of fermenting poisons, and the previous pains of bowels, taken in powder, or the root bruised and steeped in rum."

Moerman (1998) does not mention these plants, although Vogel (1970) devoted almost a page to them. As they also were recorded by Jonathan Carver and George Henry Loskiel, both of whom visited the northeastern United States, the tribes were probably in New England. Those indigenous people [unspecified] chewed the root to relieve coughs and stop miscarriage (Millspaugh 1892). However, the most common use of these plants appears to have been against colic, worms, and fevers.

The first record of the use of the herbs where a scientific name was given was by physician Benjamin Barton who, in 1793, noted that they were called "blazing star or devil's bit" (Vogel 1970). He recorded that a water infusion of its roots was used as a tonic and anthelmintic. Later physicians also used the plants for a variety of maladies, most of them associated with "male and female reproductive organs" (Foster and Duke 1990). The plants were listed in the National Formulary under the names "Helionas, Blazing Star, or False Unicorn" between 1916 and 1947. They were considered diuretic and a uterine tonic (Vogel 1970).

Core (1967) did no more than quote Millspaugh although he established that the species was used in the southern Appalachians. In that region, powdered roots were used to treat uterine disorders, pain in the head, side, or loins, poor appetite, depression, and colic. The rootstock also was used to treat intestinal worms, vitamin C deficiency, and to induce flow of urine and saliva. *Chamaelirium* is also emetic (Krochmal and Krochmal 1968). Hocking (1997) considered the plants a uterine tonic, emmenagogue, diuretic, anthelmintic, and alterative. Duke et al. (2001) concur, but provide a warning that it should be avoided during pregnancy and that Canadian regulations prohibit its use. According to Millspaugh (1892), the plants contain chamaelirin, and Hocking (1997) added only that it also had starch.

Chamaesyce

(Name used by Dioscorides, fl. A.D. 40–80, from Greek *chamai*, on the ground, *sykon*, a fig; plants are prostrate and have figlike fruits, at least to S.F. Gray who named them)

Chamaesyce mesembryanthemifolia*. Drawn by P.N. Honychurch.*

Chamaesyce mesembryanthemifolia (with leaves like *Mesembryanthemum*, Aizoaceae)
box-leaved spurge ("box" alludes to *Buxus*, Florida)
coastal beach sandmat (Florida); sandmat (Florida)
chicken weed (Belize)
revienta caballo (horse breaker, Cuba)
yerba mala (bad herb, weed, Cuba)

British naturalist Samuel Frederick Gray (1766–1828) created the genus *Chamaesyce,* removing the species from the larger *Euphorbia.* Indeed, when Jacquin named this species in 1760 he called it *E. mesembryanthemifolia.* Those who recognize *Chamaesyce* as distinct from *Euphorbia* think that it contains about 300 species spread around the world in temperate and tropical zones (Diggs et al. 1999).

On the coast north of Havana, sea bathers use the latex from these plants to cure the wounds made by urchins (Roig 1945). Elsewhere along the coast the horses who eat the plants are said to die, hence the name *revienta caballo.* The herb is also used against gonorrhea and as a laxative. It is used medicinally in Belize (Balick et al. 2000).

Chaptalia

(Named by Etienne Pierre Ventenat for the French chemist Jean Antoine Claude Chaptal Count de Chanteloup, 1756–1832)

Chaptalia tomentosa (pubescent)
i:chakcobî (deer's ear, Mikasuki); *icohackó:ma:* [*cohácko*] (*eco*, deer, *hvcko*, ear, Creek)
sunbonnets (Florida)

The Seminoles used these herbs as medicine to treat "Deer Sickness" (arms and/or legs become useless and painful; common dreams of deer, dead and alive) and urine retention (Sturtevant 1955). Several of the other usually narrowly endemic species in the genus have been used as medicines. For example, *Chaptalia texana* is used by the Mountain Pima (Laferrière 1994), and *C. nutans* has a name, *motitas* (Tabasco), comparing it to *mote* (*Erythrina glauca*). In Haiti, *C. nutans* is called *fwa pa nan kont* (literally, "liver is not struggling"), *pinga nèg* (the man's substance), and *toro tig* (tiger bull). All three of the Haitian names allude to cleansing the blood and digestive system as part of an aphrodisiac. *Chaptalia nutans* is used in medical preparations (Beauvoir et al. 2001). Moreover, *Chaptalia* has been shown to inhibit *Bacillus subtilis* and *Staphylococcus aureus* (Beauvoir et al. 2001).

Chenopodium

(From Greek *chen*, a goose, and *pous*, foot, an allusion to the shape of the leaves; probably the plant called *chenopus* in Greek by Pliny, A.D. 23–79)

Chenopodium album (white, alluding to the color of the leaves)
Ackermolten (field atriplex, Fuchs 1542, German)
ansérine (of geese, Quebec)
armuelles silvestre (wild atriplex, from a 1557 herbal, Spanish); *arroche sauvage* [*arroche salvuaige*] (wild orach, from a 1575 herbal, French); *atriplice salvatica* (wild atriplex, from a 1551 herbal, Italian)
bàtl-sai-añ [*bàtl-sai-ya-doñ*] (stink weed, Kiowa)
blue weed
chou gras (fat cabbage, Quebec)
dirty dick (England)
dungweed (England)
eto^n ditli (leaf is blue, Slave)
fat hen (in use by 1795); *poulette grasse* (fat hen, Quebec)
Gänserfuss weisser (white goosefoot, German); [white] goosefoot (in use by 1548)
gensz fusz (in use by 1548, Dutch)

Chenopodium. *Chenopodium album.* a. Habit, small plant. b. Flowering branch. c. Flowers, with sepals and petals erect (above) and reflexed (below). d. Utricle. e. Seed. *Drawn by Regina O. Hughes.* From Reed 1971. *Chenopodium berlandieri* (inset in upper right corner). From Britton and Brown 1896.

[common] lamb's quarter(s) (in use by 1773 in England, USA, Bahamas)
muckweed (England)
[common] pigweed (in use by 1844, USA, Bahamas)
shokhimpa (*shokha'*, hog, *impa'*, food, Chickasaw)
wild spinach
wilde areche (used by William Turner in 1558)
wilde melde (wild atriplex, from a 1549 herbal, Dutch)
withiniwpakwatik (Cree)

Chenopodium berlandieri (named for its discoverer Jean Louis Berlandier, 1805–1851)
ashecani (with a name like "ash-can" one has to wonder if Martínez 1979 was not kidding us!)
belle dame sauvage (wild *Chenopodium*, Du Pratz in Louisiana in 1758)
choupichoul (Natchez name given by Du Pratz in 1758)
chuale [*chual, cual*] (from Náhuatl, *tzohualli*, seeds used to make tamales called *chual*, Mexico)

cotasula (Guarijío, Chihuahua)
huauzontle [*huauzontl, huauzontli, huanzoncle*] (*quauhtli*, eagle, *tzontli*, hair, Náhuatl, Mexico)
kovi (Akimel O'odham [Pima], Arizona)
lamb's quarter(s)
melden (Dutch, cf. Harriot 1580)
pit-seed goosefoot
quelite cenizo (ashy greens, México); *quelite salado* (salty greens, México)
taani' (Chickasaw); *tahwv* (Muskogee)

When Linnaeus named *Chenopodium* in 1753, he knew 22 species; 19 were from Europe and nearby Asia. From the Americas he knew *C. ambrosioides, C. anthelminticum*, and *C. virginicum*. In adopting the genus he continued a name that had been given by Swiss herbalist Gaspar Bauhin in 1623, and much earlier by Pliny in southern Europe. Today the genus contains ~100 species (Mabberley 1997).

Earlier Europeans lumped what is now called *Chenopodium* with other plants that we now put in *Amaranthus* and *Atriplex*. Fuchs (1542) included *Atriplex hortulensis* with *Chenopodium album* under the heading *De Atraphaxi* (*atra*, black, *phaxys*, braid). This name presumably alludes to the "braid" of the fruiting cluster with its black fruits, called by most people the "seeds."

The first comment on these plants in English was by William Turner ([1548] 1965) who wrote of them, "*Pes anserinus* is called in duch gensz fusz and it may be called in Englishe Goose-fote." Later, in 1568, he referred to both *Chenopodium* and *Atriplex* as *areche* (we now spell it "orach").

Chenopodium album is native to the Old World (Flora Europaea 2003). Archaeological studies in Canada that identified the species in old Blackfoot sites dated to A.D. 1500 (Tull 1999) cast doubt on whether the species was native or introduced. Unfortunately, Tull did not give a specific reference to her information, and she was no longer able to find her source in her notes (Delena Norris-Tull, personal communication 2003).

There are ample instances of *C. berlandieri* being identified as *C. album* (cf. Hodgson 2001). The question of nativity remains unresolved, but Mary Adair (2003, personal communication 2003) suspects that plants used by the people in the region of the Blackfoot were *C. leptophyllum*. Similarly, King (1984) said that there was no evidence that *C. album* was used by indigenous people. *Chenopodium berlandieri*, on the other hand, is unquestionably native in North America, and was cultivated by indigenous tribes well before Europeans arrived (Smith 1992).

In the Americas, the first mention of what may have been *Chenopodium* was in Virginia by Harriot ([1590] 1972). He wrote in the 1580s: "There is an herbe which in Dutch is called *Melden*. Some of those that I describe it vnto, take it to be a kinde of Orage; it groweth about foure or fiue foot high: of the seede thereof they make a thicke broth, and pottage of a very good taste: of the stalke by burning into ashes they make a kinde of salt earth, wherewithall many vse sometimes to season their brothes; other salte they know not. Wee our selues, vsed the leaues also for potherbes." Later, in speaking of interplanting between hills of maize, he wrote: "By this meanes there is a yard spare ground betwene euery hole [of maize]: where according to discretion here and there, they set as many Beanes and Peaze: in diuers places also the seedes of *Macócqwer* [*Cucurbita*], *Melden* and *Planta Solis* [*Helianthus annuus*]." Harriot ([1590] 1972) also recorded that the seeds were eaten in gruel by the Carolina Algonquians.

Frenchman Antoine Simon Le Page Du Pratz visited Louisiana in the 1720s and witnessed what Gilmore (1931) thought was *Chenopodium* being cultivated. Heiser (1985) rejected that view. Smith (1992) reexamined the account in detail, and came to the same conclusion as Gilmore.

Smith (1992) went to the original French published by Du Pratz in 1758, and discovered that some previous translations were misleading. One of the key segments of Du Pratz's text said, "*J'ai vû les Natchez & autres Naturels sémer une graine qu'ils nommoient* Choupichoul, *sur les battures; ce sable n'étoit nullement cultivé, & les femmes & les enfans avel leur pieds couvroient telement quellement cette graines sans y regarder de près. Après cette sémaille, & cette espéce de culture, ils attendoient l'Automne, & recueilloient pour lors une grande quanttité de cette graine: ils la préparoient comme du millet, & elle étoit très-bonne à manger. Cette plante est ce qui l'on nomme* Belle Dame sauvage, *qui vient en tout pays, mais il lui faut une bonne terre; & quelque bonne qualité qu'ait une terre en Europe, elle ne vient que d'un ped & demi de haut; & sur ce sable de Fleuve, sans culture elle s'éléve jusqu'à trois peds & demi & quatre peds*" (Du Pratz in Smith 1992). In 1911, John R. Swanton translated this segment. He wrote, "I have seen the Natchez and other natives sow a grain which they call *choupichoul* on these sand banks. This sand is never cultivated and the women and children cover the grain, with a great deal of indifference, with their feet, almost without looking at it. After this sowing and this kind of cultivation they wait until autumn and then gather a great quantity of this grain. They prepare it like millet and it is very good eating. This plant is that which is called "beautiful savage lady" [he footnoted the French *Belle dame saluvage*] and which grows in all countries, but it needs a good soil, and however good is the quality of any European soil it there reaches a height of only 1.5 feet, while on this river sand without cultivation it reaches a height of 3.5 to 4 feet" (Swanton in Smith 1992).

Smith (1992) compared etymology, detailed plant and seed morphology, and ecology of today's wild members of *Chenopodium* near the area where Du Pratz made his observations. He concluded that *C. berlandieri* was the species that Du Pratz described. Thus, cultivation of *C. berlandieri* extended into the 1720s and possibly longer into the historical period.

Using "archaeological sites and regions that provide information on agricultural origins," Smith (1992) mapped the range of pre-European distribution of cultivated *C. berlandieri*, and the other Hopwellian cultivars. The seed crop was grown from the Appalachians west through the Ohio Valley and Mississippi Valley west to the Great Plains between 34° and 40° latitude. This region he dubbed the "interior mid-latitude zone of domestication of native North American seed crops at 2000 to 1000 B.C." Outside this region, along the Atlantic and Gulf coastal area and across the northern latitudes, Smith (1992) concluded that forager economies were based largely on wild species of animals and plants (except for *Cucurbita* cultivation) until the A.D. 800 to 1100 shift to maize-centered agriculture. Fritz (2000a) has recently given additional details on the agricultural systems of the Mississippi Valley during that period.

Smith (1992) concluded, "Le Page Du Pratz witnessed and described the closing chapter of an agricultural economy that had existed in eastern North America across 4,000 years." Since Smith's (1992) publication, an archaeological study by Gardner (1994) has found *C. berlandieri* among historical Caddo in Boissier Parish, Louisiana. The localities in northwestern Louisiana were occupied by these neighbors of the Natchez in the 1770s, and include both cultivated and wild types of this chenopod. This is the latest historic record of the herbs being cultivated in the southeast for their edible seeds.

Chenopodium berlandieri was cultivated longer in the southwestern United States, and it is still cultivated in Mexico and South America (Smith 1992). Hodgson (2001) found that Cahuilla and Louiseño gathered *Chenopodium* seeds and ground them into flour. The River Pimas (Akimel O'odham) parched and ground the seeds, then ate the flour in *pinole* or combined it with other meal before it was eaten. They stopped growing *kovi* for food in the 1870s when the drought years hit (Rea 1997). The Yavapais collected the infructescences in burden baskets, spread them on a flat surface, and beat them with a stick to separate the seeds. Winnowed seeds were then parched with coals in a basket, ground, boiled, and eaten. The same pattern holds from California, Arizona, and New Mexico south into Mexico.

Southwestern tribes including the Apache, Baja Californians, Cocopa, Mohave, Paiute, Pima, Tarahumara, Tipais, Tohono O'odham, and Yavapais ate the leaves and young shoots. The usual method was to boil and then fry them in grease (Curtin 1949, Ebeling 1987, Hodgson 2001). Rarely did these same people eat the leaves and young stems raw.

People in the southeast did not eat raw plants (Hudson 1976). An exception was *Allium*, but that was an unusual case. Even *Allium* was usually cooked. In the southeast, the few green plant parts that were eaten historically were put in soups and stews (Hudson 1976). More recently, the Cherokee and Muskogean people fried onions, ramps, and other greens before they ate them. Whether they did this before contact with Europeans is not clear.

My own family, going back in my memory to grandparents on both sides, had a marked aversion to raw vegetables. When plant vegetative parts were eaten, it was common practice to fry them, as did the indigenous people of both the southeast and southwest. There was only a short time in the spring when some few vegetables like new lettuce were eaten raw, and even then the lettuce was preferably spread with hot bacon grease ("wilted") before being consumed.

Europeans discarded *Chenopodium* after spinach (*Spinaca oleracea*) was introduced from Asia about A.D. 1000, although the process took until the mid-1500s (Hedrick 1919, Fernald et al. 1958, Mabberley 1997, Davidson 1999). Some people retained a taste for it, but *Chenopodium* eventually became regarded as a weed. The chronology of the names tracks the attitude change; fat hen and goosefoot dominated from the time of Pliny to the 1790s. Then, *Chenopodium* still had a "favorable" name of lamb's quarters from the 1770s to the early 1800s. After the mid-1800s, it became known as dungweed, pigweed, and muckweed.

The same change occurred in the Americas when maize was brought in from Mexico, and *Chenopodium* of the old Eastern Agricultural Complex declined (Hudson 1976, Smith 1992). Apparently, the records by Harriot ([1590] 1972) and Le Page Du Pratz in the 1720s were the last remnants of generalized *Chenopodium* cultivation witnessed by Europeans in eastern North America.

Chimaphila

(From Greek *cheima*, winter weather, cold, and *philos*, loving, an allusion to the evergreen habit)

Chimaphila maculata. From Britton and Brown 1897.

amiskwathowipak (Cree)
bittergrønn (bitter greens, Norwegian)
ga'gige'bug (everlasting leaf, for *C. umbellata*, Ojibwa)
i·pi· sara'k [*surá·k*] (fire grass, Catawba)
ne-was-char-la-go-ne (good for colds or coughs, for *C. umbellata*, Osage)
Winterlieb (winter loving, German)

Chimaphila maculata (mottled or purpled)
herbe à clé [*herbe a clef*] (key herb, Quebec)
hierba de higado (liver herb, Mexico)
mountain tea
pipsissewa [pipsisseway] (from Cree, *pipisissikweu*, in use by 1789; later recorded in 1814 by Pursh as *sip-si-sewa*)
prince's pine [striped prince's pine, prince's flower] (in use by 1891, also applied to *Pinus banksiana*; the allusion is obscure)
rats-bane [ratsbane]
rheumatism-root (New England)
waxflower
wintergreen [spotted] (a name applied by the 1540s to plants remaining green all winter, primarily Ericaceae)

The genus *Chimaphila* was created in 1814 by Frederick T. Pursh (1774–1820), a traveler, gardener, and plant collector born in Saxony who settled in Pennsylvania. He published the name in his *Flora Americae Septentrionalis* (North American Flora). Since then, other species have been found and the genus now contains four to five species in Eurasia, and North and tropical America (Mabberley 1997). Northern Europe has a single species, the circumboreal *C. umbellata*, which has been used to treat bladder problems.

The names wintergreen and pipsissewa have been used for a variety of species in *Chimaphila, Gaultheria*, and *Pyrola*. For example, "wintergreen" was mentioned in 1542 by Fuchs (*Pyrola rotundifolia*), Turner ([1548] 1965), Parkinson (1640), and in 1653 by Culpepper (*Pyrola minor*). A use for one is likely to be the same as the other.

The Cherokees used *C. maculata* to treat colds, fevers, milk sickness, pain, rheumatism, scrofula, tetter and ringworm, ulcers and "cancers," and urinary problems. It was also used as an emetic for babies (Hamel and Chiltoskey 1975). The Nanticoke in Delaware used this species against ague (Moerman 1998).

There are many more uses by indigenous groups for *C. umbellata*, and it is likely that *C. maculata* was used in the same ways (Moerman 1998). Hocking (1997) found records of *C. maculata* being used as a diuretic, a urinary antiseptic, to treat rheumatic gout and weak kidneys, as a tonic and alterative, and in cough syrups. He also noted that boiled leaves were a kidney remedy on the Gaspe Peninsula of Quebec.

Physician B.S. Barton recorded in 1810 that he used a strong infusion of wintergreen (*Gaultheria procumbens*), which "evidently possessed of a stimulant and anodyne quality." He also wrote that it was used by "Indians" (not specified) to treat asthma (Vogel 1970).

Physician F.P. Porcher recorded in 1849 that *C. maculata* was used by the Indians of South Carolina for kidney, scrofulous, and rheumatic problems (Vogel 1970). *Chimaphila maculata* also was recommended for ulcers and skin problems. Hocking (1997) reported that people in Charleston, West Virginia, still use the leaves to treat kidney problems.

Chimaphila umbellata contains eriocolin, arbutin, chimaphilin, and tannin (Uphof 1968, Hocking 1997). Presumably, *C. maculata* contains the same compounds.

Chiococca: Snowberry

(Patrick Browne immortalized the fruits in Greek with *chion*, snow, *kokkos*, berry)

Throughout the West Indian hardwood forests called hammocks in southern Florida, there are scrambling, woody climbers reaching to the canopy. The leaves on these lianas are distressingly similar to a number of other plants, and there is not much distinctive about them. This is snowberry (Florida) or West Indian snowberry (Bahamas, Puerto Rico) (*Chiococca alba*), and typically, this is the last

Chiococca alba*. Drawn by P.N. Honychurch.*

hammock plant that students learn to recognize. I always tried to help them by telling them in practice sessions that it was a "nondescript, scrambling vine."

Even Linnaeus and his predecessors had problems with these plants. They did not really know where to put them, and most of the early names put them closer to honeysuckles (Caprifoliaceae) than Rubiaceae, where they are now. Like those before him, Linnaeus first called these *Lonicera* in 1753. Soon afterward, he realized his mistake, and renamed them *C. racemosa* in 1759. However, by international rules, one cannot simply change a name because a mistake was made. So the modern name is the one Linnaeus first gave to the species (*alba*). That species finally was put in the genus *Chiococca* (created by Patrick Browne in 1756) by A.S. Hitchcock in 1893.

Chiococca alba is widespread, ranging from southern Florida and southern Texas to the Bahamas, through much of the West Indies, and south from Baja California, Sonora, and Tamaulipas, to Peru, Brazil, and Argentina.

Several of the common names of *C. alba* refer to animals, although for different reasons. For example, they are associated with pigs as in *bejuco de berac* (pig vine, Puerto Rico), *bejuco de berraco* [*verraco*] (pig vine, Cuba, Puerto Rico), and *raiz de verraco* (pig root, Cuba). Those common names are probably associated with the odor of the stem when cut and peeled. Others think the odor is more like that of a skunk. The Maya call the scrambling shrubs *pay-che'* [*pax-che*] (skunk tree, Belize). Because of the Mayan word construction, the name may also be given because the plant's dark leaves and white fruits reminded them of the mammal, or both reasons. Spanish speakers say it is a *zorillo* (little skunk, Belize), and English speakers in the same area say skunk root.

In addition, the species is compared to a mouse as in *croc souris* (mouse-colored hook, Hispaniola) and *oreja de ratón* (mouse's ear, Sonora, Yucatán), or a rat with "man rat root" (Bahamas), and "rat-bush" (Bahamas). It is even compared to a bat with *raíz de murciélago* (bat root, Colombia). Perhaps these too refer to the foul odor of the cut stem, as each of these mammals has a distinctive smell.

One of the old names for the vines that is rarely heard now in Florida is snakeroot (Bahamas, Florida). That, and the Brazilian translation, *raiz de serpentária,* made no sense until I finally discovered the etymology of the most widespread name applied to these plants. That name is *canica* (Texas, Cuba, Mexico). Variants of this name occur as far south as Brazil, with *caimica* (Columbia), *cainana* (Brazil), *caínca* [*cainca*] (Brazil), *cainica* (Columbia), *casinga* [*caringa*] (Brazil), the *casinga cheirosa* (fragrant, *casinga* is the medicinal *Laetia suaveolens,* Flacourtiaceae), *quimaque* (Hispaniola), and finally *caninana* (Mexico, Brazil). These made sense only after determining that *caninana* is based on the Tupí word *ñakani'nã* (head by your foot), and refers to the snake *Spilotes pullatus,* a colubrid. Probably, without knowing the meaning, people carried that name throughout the New World wherever these plants grew. The variant *canica* resembles both Náhuatl (*canin*) and Otomí words (*caní,* herb), but presumably is unrelated to either. Morton records the name *falsa canica* (false canica) but does not give the region of its use. Although the spelling is markedly different, I suspect that *timaque* [*timacle*] (Hispaniola) and *bejuco timaque* [*timacle*] (timaque vine, Hispaniola) are cognates to *canica.*

Other common names make other comparisons. In Barbados they are briny roots, presumably in reference to taste. Brazilians think the somewhat square stems resemble a cross and call them *cipó cruz* (cross vine) or *cruzeirinha* (little cross). Martín Sessé and José Moçino, two Spanish explorers in the 18th century, called the vines *perlilla* (little pearl). Either that name was misunderstood or it has changed, because they are now called *perilla* (pear-shaped lobe, Nicaragua, Peru, Mexico). Either way, the names refer to the white fruits. Also alluding to the fruits is the name *acetillo* (little sour one, El Salvador).

The only reasons that seem to exist for comparing these vines to heaths (Ericaceae) is that both are medicinal, and both have somewhat tubular flowers. Yet, the comparison is made with *bois branda* (heather bush, Guadeloupe, Martinique) and *petit branda* [*ti branda*] (little heather, Guadeloupe).

Comparisons to *Jasminum* (Oleaceae) indicate that the flowers are fragrant at night for moth pollination. Those names include *jasmin bâtard* (false jasmine), and *jasmin bois* (jasmine bush, Guadeloupe, Martinique). In the Creole French of the islands, *bois* (literally tree) means "herb," with a medical connotation. Another suggestion of nocturnal fragrance is *dama de notche de pine ridge* (Pine Ridge lady of the night, Belize).

Several of the names make specific reference to the roots, one of the preferred medicinal parts. These are *purga preta* (black purge, Brazil), *quina de raiz preta* (black-rooted quinine, Brazil), *racine noire* (black root, French Antilles), and *raiz preta* (black root, Brazil).

Virtually all of the other common names refer directly or indirectly to medical usage. David's milkberry (Texas), David's root (Commerce), and milkberry (Texas) allude to previous commercial use of the herbal medicine in the United States and Europe. In those pharmacopoeias the root bark was also known as *Radix Caincae.* Other names that refer to use as a medicine include *cuanecatashich* (*cuana*, remedy, Zapotec, Hidalgo), *dambrê* (apparently a malady, Brazil), *lágrimas de María* (Mary's tears, Panama), *lágrimas de San Pedro* (Saint Peter's tears, El Salvador), *liane de sorciers* (sorcerer's vine, French Antilles), *madreselva* (mother of the forest, Nicaragua), *madreselva de las Antillas* (Antillean mother of the forest), *pegajoso* (easily caught, Sinaloa), pissabed (Bahamas), *poaia* (from *pu'aya,* ipecac, an emetic, Tupí, Brazil), *raiz de frade* (brother's [eclesiatical] root, Brazil), *raiz de quina* (quinine root, Brazil), *suelda con suelda* (money-with-money, Yucatán, Cuba), tim-tom bush (Barbados), *tori naca* (Mayo, Sonora), and *x-kanchak-che'* [*xcanchac-che', canchacche', kanchak-che'*] (*ka'an,* yellow and *chak*, red, *che'*, tree, in reference to the wood, Yucatán, Maya).

The medicine from different parts of the snowberry is considered acrid, tonic, astringent, vomitive, and pectoral. It is said to be anti-asthmatic, antihydropic, diuretic, emmenogogue, emollient, febrifuge, hydragogue, and a purgative. It has been used against ailments including alcoholism, colitis, constipation, delayed menstruation, dementia, depression, dropsy, dysentery, edema, endometriosis, lymphatic system disorders, nervousness, and pain. It is used to expel gas, ease menstrual cramps, as an aphrodisiac, to cure sores, rashes, and ulcers, for snakebite, to dispel witchcraft and evil eye, to treat ulcers of the stomach or intestines, and against venereal diseases (Roig 1945, Morton 1981, Arvigo and Balick 1993, Vásquez and Jácome 1997, Mors et al. 2000).

Most users are quick to point out that this is a strong remedy and not to be taken internally by the elderly and very weak (Arvigo and Balick 1993). Preparations vary depending on the problem being treated, but roots are most often used internally (Mors et al. 2000). Boiling the roots produces a liquid that is ingested alone, or it may be prepared with other plants added. For example, the roots are boiled with *Picramnia pentandra* to treat gas and relieve menstrual cramps (Morton 1981). Flowers are emollient and are used to treat skin diseases.

One of the most interesting remedies is to inhibit alcoholics from drinking (Arvigo and Balick 1993). The method is to place one handful of chopped root in 1 quart of rum, vodka, or gin. This mixture is soaked in the sun for 5 days and strained. People seeking the cure should take one shot of this daily until the mixture is gone, or finish it all at once. Apparently, they will vomit violently after each drinking session, and after that the smell of alcohol will make them nauseous for years.

The plants are loaded with chemicals. Compounds identified include coumarins, alkanes, carbohydrates, lignans (El-Hafiz et al. 1991), cardiotonic glycosides, saponins, tannins, flavonoids (Schapoval et al. 1983), alkaloids (El Abbadi et al. 1989), and a triterpene (Bhattacharyya and Cunha 1992). An alcohol extract of dried leaves showed *in vitro* cytotoxic activity against cancer (65% inhibition). Stem extracts showed 77% inhibition, while roots gave 69.6% (Nascimento et al. 1990).

If you move into the pinewoods, there is another plant that resembles snowberry. This one, however, has smaller fruits and slightly darker flowers. That plant is *Chiococca parviflora*, which Wunderlin (1998) lists as synonymy of *C. alba.* Although opinions have varied over the years, elsewhere in the tropics others retain them as distinct. Roots of this species have been used to treat tuberculosis and bed-wetting in children. Much less is known about this species, but medicinally it may be similar to *C. alba.* Both of these species have been used medicinally, and the information on one is perhaps applicable to both. Some people, like some taxonomists, may not separate them.

Chionanthus

(From Greek *chion*, snow, and *anthos*, flower; alluding to the snow-white flower clusters)

Chionanthus virginicus (of Virginia)
flowering ash; poison ash (a name used for *Rhus vernix* by 1760; the result of confusion with *Chionanthus,* as by Porcher in 1863); white ash (the name usually given to *Fraxinus americana*)

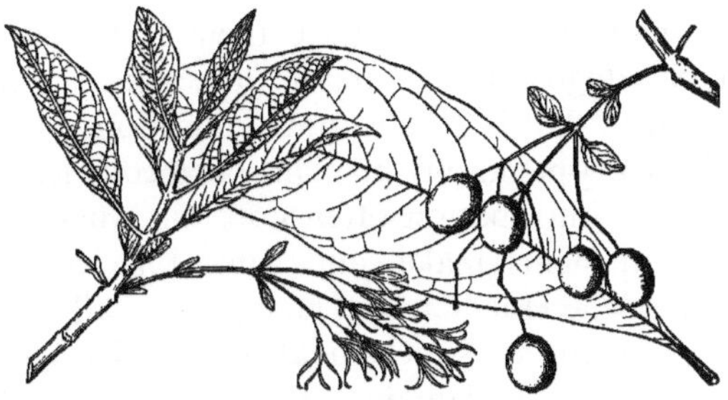

Chionanthus virginicus. From Sargent 1905.

[American, white] fringe [tree, fringe-tree] (probably first used by Mark Catesby in 1731)
fringe-flower (in use by 1882, when John Smith wrote the name in his *Dictionary of Popular Plant Names*); fringe-tree (in use by 1730); white fringe tree (used by Williams [1837] 1962)
Grancy graybeard; granddaddy's beard; graybeard [tree] (first applied for the lichen *Usnea barbata* by 1599; later extended to the tree)
hataks pone nepakwibe (*hatak*, man, Choctaw)
nita imìlipa (*nita*, bear, *im,* its, *ilpa*, food, Alabama); *nitá inłakakcó* (*nitá*, bear, *im*, its, *łakakcó*, swamp privet, *Viburnum rufidulum*, Kosasti)
old man's beard (recorded for *Clematis* in 1542; later for *Tillandsia usneoides* by Patrick Brown for Jamaican plants in 1756; later applied to a number of plants including *Saxifraga* and *Chionanthus*)
privet tree (misapplication of the name for *Ligustrum* by John Brickell in 1737, although they are in the same family)
shavings (application obscure unless possibly the stems were used for handles of shaving brushes)
snow-drop tree [snow-drop-tree, snowdrop tree] (by 1731 when Philip Miller used it in his *Gardeners Dictionary*)
snow-flower(s) [snowflower(s), -tree] (snow-flower was used by 1664, probably for *Galanthus* as a commentary on medieval earring fashions, not weather; by 1862 applied to *Hepatica*; finally, in 1862 it was being used for *Chionanthus*, with or without the suffix -tree)

Probably the first record of *Chionanthus* in Europe was the appendix written by London apothecary James Petiver in John Ray's *Historia plantarum* (History of Plants) in 1704. Petiver was an avid collector of plants from around the world, but he kept virtually no records of where they came from or when he got them. Therefore, the specimen that is now in the Linnaean herbarium has no provenance information.

Mark Catesby illustrated the plants in 1731, and Linnaeus studied them in cultivation at the *Hortus Cliffortianus* in 1738. Only a year later, Adrian van Royen listed the species being grown in Leiden. There is small wonder that the plants were spread so quickly as they are breathtaking when in flower. Catesby used the name "fringe tree" in 1731.

The Choctaw boiled the bark and used the extracts to bathe wounds. They also used the mashed bark as a poultice on cuts and bruises (Bushnell 1908). The Alabama and Koasati used the bark to treat toothache, and the Koasati used it on cuts and bruises like their relatives the Choctaw (Taylor 1940).

Rafinesque (1830) thought the roots made a cataplasm that would heal wounds without infection. Porcher (1863) wrote: "An infusion of the roots is given in long standing intermittents. It is tonic and febrifugal, with some acro-narcotic properties; used in the form of cataplasm as an application to wounds and ulcers." Millspaugh (1892) too was enthusiastic about the medical properties of the plants. He wrote: "The previous use of the bark of this shrub as an astringent vulnerary, and the bark of the root as a tonic after long and exhaustive diseases, is one that has a great merit. The bark in infusion is a remedy that was too long often neglected for foreign drugs."

Krochmal (1968) and Krochmal and Krochmal (1973) noted that the Appalachian pioneers applied a poultice to cuts and bruises. Leaves and flowers were used to treat inflammations and sores, ulcers in the mouth and throat, and diarrhea. Hocking (1997) considered the plant an alterative, bitter tonic, for use against liver problems, as a stimulant, and in controlling fevers. However, Foster and Duke (1990) warn that overdoses cause vomiting, frontal headaches, and slow pulse.

The best-known active chemicals in the plants are chionanthin and tannins (Hocking 1997). Duke et al. (2001) are not as enthusiastic as was Millspaugh (1892). They give the remedy a single plus (not as safe as a cup of coffee).

Chromolaena: Bitter-bush

(From Greek *chroma*, color, and *chlaena*, cloak or blanket, perhaps referring to the bracts)

Along roads and trails, especially in the Florida Keys, there is often a mass of green sprawling over other plants. Near Christmas, the white to pale pink or blue flowers of this herb paint the landscape. These flowers are visited by a number of bees and flies and make the area buzz with life. Some people call the herb bitter-bush (Florida, Jamaica, Trinidad, Puerto Rico).

Many of those who have immigrated from the Caribbean call it Christmas bush (Jamaica, Virgin Islands, Trinidad), Christmas rose, *fleurit-Noël* (Christmas flower, Guadeloupe, Martinique), or *flewi nwel* (Christmas flower, Trinidad). A variety of different Florida plant species blooming near the end of the year are called Christmas flowers.

Opinion is divided about what the scientific name should be for this sprawling herb, as it has been since Linnaeus described *Eupatorium odoratum* in 1759. Some retain it in that genus, while others say that grouping is artificial and segregate this and 164 other species into another genus. They call *Chromolaena odorata* bitter-bush. Indeed, *Chromolaena* was created in 1836 by Swiss botanist Augustin-Pyramus de Candolle (1778–1841), but the specialists in the Asteraceae have not agreed about generic delimitation. Still, with added information views change, and the trend now is to accept the isolation of *Chromolaena* (King and Robinson 1987).

The name *Chromolaena* is often interpreted as referring to the bracts. Some think that instead the name alludes to the tendency of the herbs to smother other vegetation, and many of the common names support that view. The old name *Eupatorium* was in remembrance of Mithridates Eupator, the ancient king of Pontus, who used plants of this genus for preparing medicines. Among those who use the plants as cures, *Eupatorium* would have been a more appropriate term. The name *cihuapatli* [*ciguapazle, cihuapastle, ciguapaztle*] (*cihua*, woman, *patli*, medicine, Náhuatl, San Luis Potosí) notes its use by women. That name is also applied to *Montanoa tomentosa*, another medicinal composite. The specific name *odoratum* refers to the fragrance given off by the leaves. That fragrance almost guarantees that people will have uses for the plant. Perhaps the common names paraffinweed, parafinbush, turpentine weed, *varejón de caballo* (horse's [wild] sage, Colombia), and vanilla (Jamaica) also note the fragrance.

Bitter-bush was restricted mostly to the Florida Keys for many years and then, as the area was developed, it spread through more of the peninsula. *Chromolaena odorata* also is known from the Bahamas, the West Indies, Mexico (almost all states), and south to Argentina and Paraguay (Solomon 2002). Comparatively recently, it was introduced into the Old World tropics where it is a weed of croplands and forests (Cruttwell-McFadyen 1989, Rouw 1991, Peng and Yang 1998). Presumably, confusion about the origin of the plants led to the common name Siam weed.

Although these large herbs were in Florida before Europeans settled there, comparatively few residents have names for them. That is not true outside Florida, where many people recognize them. There are two primary categories of common names for the plants, with a third that is an uncertain collection.

Many people know the plants as medicines. Those having this information often note the cross that the leaves and stems make, and some take this as an emblem of the Holy Cross. These people call the herbs *crucetillo* (little cross, Oaxaca), *crucita* [*crucito, cruzeta*] (little cross, Jamaica, Texas, San Luis Potosí, Trinidad), *crucita oleroso* (fragrant little cross, Nicaragua), *cruz-de-campo* (wild cross, Brazil), *cruz-quem* (whose cross, Brazil), *cruzeiro* (cross, Brazil), *curarina* (little curer), and *curarina de monte* (wild little curer). Other names making holy references are *Santa María* (Saint Mary, Puerto Rico) and *caraiquillo Santa María* (Saint Mary's *caraiquillo,* Puerto Rico). The reference to *caraiquillo* is to *Lantana camara*, and it is probably the same as in the name *alba-haquilla* (white *haquilla*, Cuba). Perhaps it is based on some Arawak word. Archangel (Jamaica, Trinidad) is surely another of these holy references. Perhaps kingsweed also belongs in this grouping. In Veracruz, it is called *krus tok'te* (cross *tok'te* which compares it to *E. albicaule*, Huastec).

The basis for some of the medicinal names is obvious, while for others it is obscure. For example, *langue à chat* [*langue chatte, langa chata, langi-chatte*] (cat tongue, Hispaniola, French Antilles) refers to the rough leaves. More obvious is *mejorana* (makes it better). The Cuban *filagrana* [*filagrama*] *de sabana* (savanna filigree) is an allusion to *Lantana camara*, a notorious medicinal plant. The more widespread *rompe-zaragüey* (shoe breaker, the same name given to *Vernonia menthaefolia*, Cuba, Hispaniola) and *zaragüey* (shoe, Hispaniola) presumably denote the curing of medical problems. The name hemp agrimony (Jamaica, Trinidad) compares it with agrimony (*Agrimonia*, which see), a well-known herbal medicine. In Panama, it is *paleca*, which may be a reference to *palco*, a mouth sore found in children.

In Cuba and Venezuela, some call it *chimuyo* [*chimojo*], a reference to *chimó*, a mixture of tobacco and lime that is chewed. Whether or not the mixture is considered medicinal depends on the viewpoint. If one is a field hand working in the hot sun with little food, it might seem like medicine against pain.

The remaining references are unknown: Armstrong's weed, Jack-in-the-bush [Jackney-bush, and Jack-in-bush] (Belize, Jamaica, Trinidad), and *zanca de grullo* (crane's stilts, Cuba).

As a medicine, the plant has been used in various ways (Standley 1920–1926, Roig 1945, Martínez 1969, Liogier 1974, Petersen 1974, Morton 1981, Arvigo and Balick 1993). In several places, flowers are made into a tea used against coughs and diabetes. Leaf decoctions are considered a remedy for colds and coughs in at least Barbados, Belize, Panama, and Trinidad. In Jamaica the plant is used the same way, but mixed with other plants.

Internally, the medicine from bitter-bush has been used by the Maya of Yucatán against stomachache and kidney trouble. Guatemalans consider the bitter root tonic and use it against stomach problems. Other people in Mexico once used it against malaria, as an emmenagogue, and to treat gonorrhea. In Belize and Hispaniola, the tea is considered good for calming the nerves and to alleviate laryngitis, insomnia, and depression. In Jamaica, Cuba, the Virgin Islands, Barbados, and Belize it is used externally to treat fever, and in baths for skin diseases. Crushed leaves are even put on cuts to help them heal. There is a known anticoagulant property to the plants (Mabberley 1997).

The plants contain a number of toxic compounds, which is common for all medicines. Some of these compounds are so potent that the plant has been used as a fish poison. The species contains ceryl alcohol, ß-sitosterol, aromatic acids, the flavone isosakuranetin, chalcone (odoratin), sesquiterpene alcohol (eupatol), trihydric alcohol, anisic acid, terpene alcohols (lupeol, ß-amyrin), and flavone (salvigenin) (Baruah et al. 1978, 1993, El-Sayed et al. 1988, ller et al. 1994, Irobi 1997, Morton 1981, Nguyen et al. 1992, Wollenweber et al. 1995, Wollenweber and Roitman 1996).

Antibacterial activity of dried leaves and stems has been demonstrated against *Bacillus subtilis.* Dried leaves have a coagulant activity in rats and decreased prothrombin time. Fresh leaf extracts have antifungal activity against *Rhizopus* and *Ustilago maydis.* Leaf essential oils are antibacterial against *B. subtilis, Escherichia coli, Klebsiella acrogenes, Staphylococcus aureus* (Arvigo and Balick 1993), and a number of other microorganisms (Gupta 1995a).

The only reference to the plants being used as food may be from Venezuela (Pittier 1926). There they call it *pesebrito.* A *pesebre* is a manger for feeding animals. Perhaps the people there use the rampant growth of greens as fodder for animals, although that seems doubtful. The plants are known to be toxic to some types of livestock (Mabberley 1997, Apori et al. 2000).

Those who hate the plants have a number of pejoratives for it. One of the derogatory things one might do is compare it with a blood-sucking ectoparasite, which is done by calling it a *garrapata* (tick, Nicaragua). People who have pulled it from their cultivated fields call it *guérit-tout* (conquers all, Hispaniola, Guadeloupe, Martinique), *guérit-trop-vite* (conquers all life, Guadeloupe, Martinique), *guérit-vite* (conquers life, Hispaniola), *mata finca* (farm killer, Hispaniola), *rey del todo* (king of everything), *suplicio* (torture, an allusion to covering everything), and *tokabal* [*tocaban, x-tokabal, xtok'aban*] (plumb thief, Maya, Yucatán). That Mayan name is given because this is a frequent weed in groves of *ciruela* (*Spondias monbin,* Anacardiaceae). When allowed to grow below the trees, there is a marked decrease in the number of fruits produced. The Maya knew the allelopathic effects on maize long before they were discovered by others (Oguntimein and Elakovich 1991).

Meanings or references for the following names are obscure. The herbs are called *kluset* (Creole French, Trinidad), *niquibey* (probably Taino, Hispaniola), *sich* (Tzeltal Maya, Chiapas), tonka bean (Jamaica), triffid weed, *tsitsaque-teutzushu* (Totonac, Veracruz), and *yaxhatz* [*hotz*] (Belize).

In a deliciously ironic twist, these medicinal and poisonous plants are now providing another useful service in areas where they are not native (Holm 1991). Someone had the idea that they might provide good "green manure" under crops of black pepper (*Piper nigrum*). The side benefit is that they also deter a roundworm pest (*Heteroderma marioni*) of black pepper and the root-rot fungi (*Pythium complectans* or *P. splendens*).

Chrysobalanus: Coco-Plum

(Based on the Greek, *chrysos,* golden, and *balanos,* acorn or glans penis)

Chrysobalanus icaco. *Drawn by P.N. Honychurch.*

In areas where the beach is still intact in Florida, one occasionally finds a small shrub with dark green, egg-shaped leaves. There is a definite rosaceous look about the plants although technically they are not in that family. If you visit during the right season, you find drupes about the size of a quarter. The fruits are distinctive of the coco-plum, and come in several colors: white, yellow, red, or purple. If you go inland from the beach, you find the same kinds of plants in hammocks, on the edges of wetlands, and scattered throughout the Everglades. As the water table has been lowered by urbanization and "flood control" the shrubs have become more abundant than they were formerly.

Although the name coco-plum sounds like there should be some resemblance to the coconut (*Cocos nucifera*), there is no relationship between the names. The resemblance is a case of convergence in totally different languages to words that are spelled and pronounced the same (homonyms).

While this species was recorded by Fernández de Oviedo y Valdes Gonzalo (1478–1557) in the 1500s, it first became known to most Europeans when Hans Sloane published his study of Jamaican plants in 1696. Sloane called them *Myrobalanus minor, folio fraxini alato, fructo purpureo* (small perfumed acorn, with leaves winged like the ash, and purple fruits), but he pointed out that there were yellow-colored variations (*... fructo luteo*).

It was not until 1703 when Charles Plumier (1646–1704) told Europeans that the indigenous name for these plants in Haiti was *icaco*. However, Spanish explorers had long known the name because Barrientos (1567) reported that Menéndez and his men were forced to gather and eat "*cacos*" during their short war against the French Huguenots in 1565. That common name was so similar in languages from Florida to Venezuela that eventually it too became part of English. Linnaeus used Plumier's information to give it its specific name. This shrub or small tree's scientific name is *Chrysobalanus icaco*, applied by Linnaeus in 1753.

Most texts translate *Chrysobalanus* as the golden acorn. However, true to his personality, Linnaeus had selected a Greek word with a dual meaning. *Balanos* means an acorn, but it also refers to the glans penis of the male reproductive organ. It was not an accident that Linnaeus also applied the same word to the barnacle, *Balanus*.

The specific name taken from Plumier's report on the island of Hispaniola included the indigenous name for the plants there in Taino. *Hicaco* [*icaco, jicaco*] was the word for the plants in Hispaniola, and there are variants in the languages spoken from Florida to Venezuela and Colombia. The Seminoles of Florida call the plants *hekako* [*hiká:ko*] (Creek) and *hiká:kî* (Mikasuki). Probably, this is the *tococo* (eat new fruit) of the Timucua. In Mexico and Central America there are variations including *caco* (Chiapas), *gicaco, hicaco, higágu* (Garífuna, Nicaragua), *icaco* (Cuná, Panama), *icaque, icaquier, jicaco, jicaquillo, xicaco* (Oaxaca), and *zicaque* (from Tamaulipas and Guerrero, Mexico, to Honduras). Surely *ka-ka-tà* (Belize) is a variation of the same name.

"Coco-plum" is used in the Bahamas as well as Florida. In the Caribbean it is *ecacs* or *ekaakes* (Aruba, Bonaire, Curaçao), *hicaco* [*jicaco*] plum (Belize), *icaco de costa* (coastal coco plum), *icaco de playa* (beach coco plum), *icaco dulce* (sweet coco plum), kocho-rhum (Belize), *prunier icaque* (coco plum tree), *zikak* (Trinidad), or *z'icaque* (Dominica). It may be that *uichup* (Cuná, Panama) is cognate to these names.

There are other names in different languages. To the Zapotecs of Oaxaca, Mexico, it is *nocuana bebebe* [*nocuana pebebe, nocuana be-bee*] (*nocuana*, edible herb, *bebebe*, fruit), *pe-pepe niza-tao-pani* (seaside fruit), or simply *pepe* [*pe-pepe*] (fruit). Other people think it is more like a plum or prune, and it is *ciruelo de algodón* (cotton plum), *prune-coton* (cottony plum), *ciruelo de paloma* (dove plum, Yucatán), *prune de guyane* (Guyana plum, French Guiana), *prune de l'anse* (cove plum, French Guiana), or simply *pruim* (plum, Suriname). Although the name is usually given to *Coccoloba*, it is also known as pigeon plum (Guyana). My favorites are "fat pork" (Bahamas to Guyana) and "pork-fat apple" (Bahamas). The name fat pork is not pronounced as we would in the United States, but rendered "fat poke."

With the shift of dominant indigenous languages in Brazil, the name becomes *abajeru* (from Tupí *waye' rá*, apparently formerly *Chrysophyllum*, Sapotaceae, Amazonas, Pará). Variations on this name include *ariu, ajurú, abagerú, guajuru, guagerú*, and *uajurú*. In Guyana and Suriname, the name is *kurimiru* [*caramio, koelimiro, kulimiro*], which is Arawak (Klooster et al. 2003).

The most widespread use of this plant is as food. The common name typically refers to the fruit. Fruits are consumed in several ways. Perhaps the most common form is fresh, directly from the plant. Fruits may also be canned or made into jam or jelly. According to some, the fruits are improved by soaking overnight and then peeling and stewing with sugar. Some even pierce them through the kernel before doing that because the seed is edible along with the pulp. Moreover, they were formerly the basis of businesses in Cuba, Brazil, and Costa Rica where they were canned and exported (Morton 1968b).

Students in my classes over the years were not always overly impressed with the flavor of the fruits. They gave a number of comparisons to the slightly sweet pulp of the fruit, but the one I favored was that it tasted like "sweetened cotton."

Most who are familiar with the fruits know that eating the pulp from around the seed is only the first step. Breaking open the seed coat exposes the kernel within. This is edible raw or toasted, and it tastes somewhat like almonds. In Brazil, the oil from the kernel has been substituted for almond oil in medical ointments (Morton 1981).

Less known in the United States is that the seeds contain enough oil that they may be strung on sticks and burned like candles (Morton 1981). They were the original torches of the regions where they grew. Not only do the seeds produce light, but the oils also have been used to make candles, soap, axle grease, and other similar products. These kernels, along with the leaves, create a black dye. Along the Amazon, the kernels are used to dye fishing nets to make them resistant to decay (LeCointe 1947).

Oil in the seeds may be 20 to 22% by volume. The kernel contains palmatic acid, stearic acid, archidic acid, conjugated octadecadienoic acid, alpha-eleostearic acid, alpha-parinaric acid, alpha-licanic acid, and 4-oxooctadeca-*cis*-0, *trans*-11, *trans*-13, *cis*-15-tetraenoic acid.

Among the Seminoles, the plant is called *hika:ka: pi* (hikako, *apí*, tree, Mikasuki). The fruit is *hika:kî* [*hekaake, hikaaki*] (Mikasuki) or *hika:ko* (Creek). They considered them a source of food (fruits), arrows, and medicine (Sturtevant 1955, Bennett 1997). Indeed, the Seminoles used coco-plum in a medicine to remake a person through *aknoksahkî*. In situations where a person gossips too much, or whose presence makes other people no longer want that person around, a "cleansing" is needed. This cleaning is not just washing the outside of the body, but using a treatment that cleans the whole person. If the job is done correctly, everyone will like the person better. Logs are cut a specific size from live oak (*Quercus virginiana*), willow oak (*Q. laurifolia*), mastic (*Sideroxylon foetidissimum*), satinleaf (*Chrysophyllum oliviforme*), coco-plum (*Chrysobalanus icaco*), and wax myrtle (*Myrica cerifera*). The logs are burned and the ashes boiled. After a particular song is sung over the medicine to bless it and make it potent, some is put on the patient's tongue. Then he or she bathes in cold water. The patient may not eat in the morning after taking the cure, but may either eat at noon or fast all day. Another bath is taken with specially prepared mud from a crawfish chimney, and then the patient returns to camp. The treatment is finished and the person should behave differently and be viewed differently by those around him or her.

Throughout its range, *C. icaco* is considered medicinal by all who know it (LeCointe 1947, Morton 1981, Mors et al. 2000). The bark and leaves are high in tannin and therefore astringent. One of the primary uses is to treat dysentery. For example, in Trinidad, a decoction of bark and roots is used to treat that problem. In El Salvador and Brazil, an infusion of fruits, leaves, and bark or roots in boiling water is taken to treat diarrhea.

In Brazil and elsewhere, the bark is considered antidiabetic. Roots, bark, and flowers are used against chronic diarrhea, blennorrhagia, leukorrhea, and bladder problems. The mixture of boiled leaves and bark or roots is also used to treat hemorrhages. In Costa Rica, a bark infusion is drunk to relieve kidney problems.

Tea from the leaves produces hypoglycemia in mice and protects against lethal doses of alloxane. The tea also inhibits intestinal absorption of glucose and controls blood sugar levels in type II diabetes.

Chrysophyllum: Satinleaf

(Greek, *chrysos*, gold, *phyllon*, leaf)

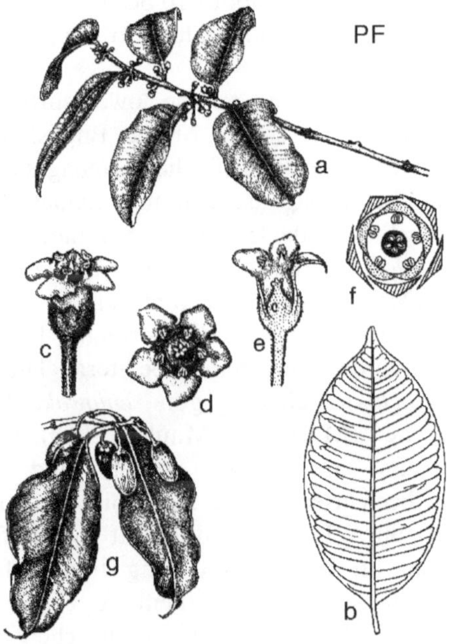

Chrysophyllum oliviforme. a. Flowering branch. b. Leaf outline showing venation. c. Flower, side view. d. Flower, from above. e. Flower, longitudinally dissected. f. Floral diagram. g. Branch with fruits. *Drawn by Priscilla Fawcett.* From Correll and Correll 1982.

Before urbanization, there was an almost continuous West Indian forest along the southeastern Florida coast extending from at least Miami to West Palm Beach. Scattered through this forest were small trees with leathery leaves that are dark green above and satiny brown below. When the wind blew across these forests, they glistened like the gold that the Spanish *conquistadores* sought because of the exposed undersides of the leaves. The trees are stainleaf (Florida, the Bahamas, Puerto Rico), *Chrysophyllum oliviforme.*

By the time I arrived in Florida in 1970, most of the forests had been replaced by housing, and the few remaining fragment hammocks were devoid of the satinleaf in Martin, Palm Beach, and Broward Counties. In Miami-Dade County, there are still places where the trees persist. Moreover, on many of the tree islands in the Everglades and Big Cypress Swamp region, the trees remain. Sometimes their leaves can be seen from a distance as you approach those hammocks, and it is always an adrenalin boost to see this comparatively rare tropical plant.

Satinleaf is essentially circum-Caribbean, and is found in Florida, the Bahamas, Cuba, Hispaniola, Puerto Rico, Jamaica, Chiapas, San Luis Potosí, Oaxaca, Yucatán, Belize, Honduras, and Panama. It belongs in the genus *Chrysophyllum* with about 40 other species (Mabberley 1997). Some 15 or so species are confined to Africa, about 10 occur in Madagascar, and 2 or 3 grow from Indomalesia to Australia. All the others are American.

The satiny covering on the lower surface of the leaves is surely the source of the English common name. However, speakers of other languages and those deeper in the tropics use common names that have nothing to do with the leaves. The Seminoles call them *hilókwa inlokcí:yaca:kitâ* (Creek) or *hacalo:pí:há: yickí:kî* (Mikasuki). Sturtevant (1955) translated both words to mean "to chew strangler fig berries." However, that does not seem to be literal. The Creek is based on *helokwv,* chewing gum, *enlowake,* bud, and *yvcaketv*, chew (Martin and Mauldin 2000). The small, black fruits are chewy, sweet, and gummy, and have been used by the Seminoles and others as "chewing gum." They are edible, but you are forced to chew them an inordinately long time before you feel comfortable swallowing them. A related name is the Mexican *chicle de monte* (wild chewing gum, Chiapas).

A large variety of birds and mammals consume the fruits and spread them. It is largely because of animal dispersal that the seedlings appear in distant tree islands across southern Florida. The same occurs outside Florida where they are eaten by more tropical animals. In the Yucatán peninsula of Mexico, the Maya call the trees *chi'ik che'* [*chikeh*] (coatí tree). There the *tejon* or *coatimundi* (*Nasua narica yucatanica*) not only eat the fruits, but also make their temporary nests in the tree branches.

The leaves and the fruits are similar to a related species from deeper in the tropics that has long been a favorite. That is the *caimito* (*C. cainito*). Although the word *caimito* was adapted by the Spanish from the Taino word in Hispaniola, it is now used and understood throughout Spanish-speaking Latin America.

The *caimito* was the first species that became known to Europeans. That name appeared first in 1633 in a book by Jan de Laët entitled *Novus orbis, seu Descriptionis Indiae occidentalis* ... (The New World, or Description of the West Indies ...). At first, people in Europe did not know what to do with this strange fruit. Hans Sloane thought it was more like a pond apple and called it *Annona foliis subtus ferrugineis, fructu rotundo majore laevi purpureo* (Annona with leaves rusty below, and having a large somewhat purple round fruit). Linnaeus capitalized on the leaves being reddish below and coined the genus *Chrysophyllum* in 1753. He did not learn about the satinleaf until later and named it in 1759 when he published *Systema Naturae.*

Many of the common names for the satinleaf have references to its relative *caimito,* usually called star apple in English. For example, *caimitillo* [*cayumito*] (little caimito, Chiapas, Cuba, Dominican Republic), *caimito* (from Haitian Taino, Caribbean, Oaxaca, Honduras), *caimito cimarrón* [*cayumito silvestre*] (wild caimito, Chiapas, Cuba), *caimito de perro* (dog [wild] caimito, Hispaniola), wild star apple (Belize, Jamaica), *zapote caimito* (caimito zapote, Oaxaca), and *zapotillo* (little zapote, Mexico). In those names referring to *zapote,* the speakers are using a Náhuatl word, *zapotli,* meaning "sweet fruit."

Two instances of apparent misidentification are recorded with *canela* (cinnamon, Oaxaca), and *palo de canela* (cinnamon tree, Oaxaca). If misidentification is not involved, the application of *canela* is broader than has been found elsewhere.

Two of the common names compare the fruits to others. In Belize *C. oliviforme* is called damsel, and elsewhere damson-plum. That name compares it with the "true" damson plum (*Prunus* × *domestica* ssp. *institia*). The comparison with damson is apparently because both are small acid purple fruits. The tastes of the two surely cannot be compared. In the Bahamas the people call it saffron-tree. The comparison with saffron (*Crocus sativus*) is uncertain unless it refers to the color of the leaf undersides.

Some names make other allusions. The comparison between the fruit and a mammary gland is clear in *teta de burro* (burro's tit, Puerto Rico). Exactly what is meant by *quebracoyol* (*quebra,* break, *coyol,* a nut-bearing tree, Náhuatl, Chiapas) is not obvious. However, this may be another allusion to testes. Finally, *ocatlán* (place of *ocatl,* Náhuatl, San Luis Potosí) has the typical Náhuatl ending meaning place, *–tlán*. Unfortunately, no translation has been found of the first part of the word unless it might be transcribed as *ocotl*, which means pine resin. Any connection between pines or resins and *Chrysophyllum* is abstruse. *Thijul* [*thituy*] has been recorded as the name used by the Huastec of San Luis Potosí. That seems to be a variant of their name *thiiw*, for *C. mexicanum*, which is not translatable. Either the names are the same for both species, or there is some slight orthographic variation.

The remaining names all appear to be untranslatable and are perhaps simply the indigenous phonemes for the species. These include *chijilté* (Chiapas), *chuni* (Chiapas), *isi* (San Luis Potosí), *nensarrajo* (Choco, Panama), *pacuschumí* [*pascuschuni*] (Chiapas), *pipa* (Cuná, Panama), and *tuko* (Choco, Panama).

As with many other members of *Chrysophyllum*, the fruits are not large, and they are especially favored by children. The wood of all is valued. According to Liogier (1974), the wood of satinleaf is hard and difficult to work. Still, it is used for posts and rafters, and it is made into charcoal. Leaf decoctions are used to heal infected contusions. This is also one of the species that the Miccosukee use in a medicine to "cleanse" a person (see *Chrysobalanus*: Coco-Plum, for details). Comparatively little seems to be known about this species, but saponins, coumarins, and an alkaloid have been reported for the genus (Schultes and Rafauf 1990). Furthermore, cyanogenesis has been found in one species. Myrcetin glucosides are known to be in *C. cainito*.

Certainly, the best-known species in the genus *Chrysophyllum*, and the one to which satinleaf is compared by most people, is *C. cainito*. That species apparently grows nowhere in the wild but is cultivated at least in Hispaniola and Cuba to Mexico, Central America, and northern South America. Some consider it to be native in the Caribbean and think that it subsequently was taken to the mainland in cultivation. From what is known, the uses of satinleaf and those of *C. cainito* are largely comparable.

As a comparison for common names *C. cainito* is called *abiu do Pará* (*abiu* or *Pouteria caimito* from Pará, Brazil), *caïmite* (Hispaniola, Trinidad), *caïmite des jardins* (garden caimito, Hispaniola), *caïmite franche* (free caimito, Hispaniola), *caimitero* (caimito tree, Puerto Rico), *caïmitier* (caimito tree, Hispaniola), *caimito* (Puerto Rico, Cuba, Colombia), cainit (English Antilles), *cainitier* (caimito tree, French Antilles), *cainito, cayumito* (Mexico), goldenleaf, *grande caimite* (big caimito, Hispaniola), *kaimit* (Trinidad, Dutch Antilles), *kê-sê-weé-ree* (Yukuna, Arawakan), *so-da-deé* (Andoke, Witototan language group), star apple (Florida to Guyana), star plum, *sterappel* (star apple, Dutch Antilles), and *ta-uh* (Tikuna, Arawakan?).

The wood of this species is used in the same way that satinleaf is used. However, the fruit of *C. cainito* is large enough and sweet enough to provide a real treat. Many prize the fruits, while others dislike the gummy aftertaste. The fresh fruit is high in niacin (0.8 to 1.05 mg/100 g). Supposedly, the green fruit is astringent and becomes laxative when mature. Apparently, ample quantities must be eaten because a normal portion has no obvious impact on most people. To make the claim more confusing, it is also said that "overindulgence in fruit induces constipation and gastralgia" (Morton 1981). Leaves of this species too are used in medicine, and all parts of the tree are high in tannin. The latex has been used to adulterate rubber, and the Yakuna use latex for fungal infections of the crotch. The Tikunas use the latex on infected gums (Schultes and Raffauf 1990).

Chrysopsis

(From Greek *chrysos*, gold, and *opsis*, aspect or resembling; from the golden inflorescences)

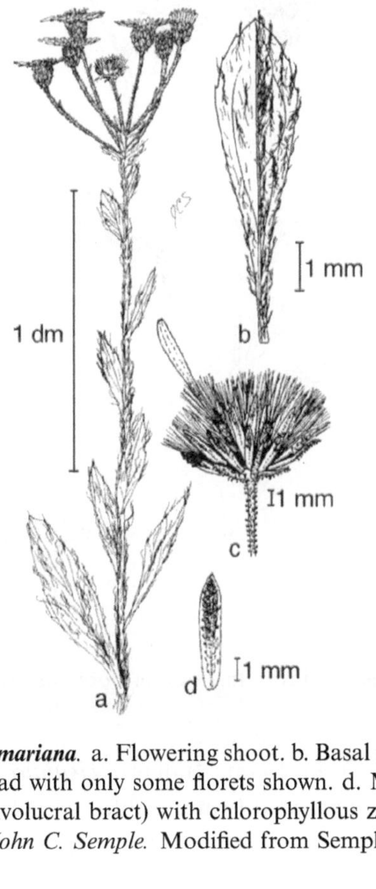

Chrysopsis mariana. a. Flowering shoot. b. Basal rosette leaf. c. Dried head with only some florets shown. d. Mid series phyllary (involucral bract) with chlorophyllous zone dark. *Drawn by John C. Semple.* Modified from Semple 1981.

Chrysopsis mariana (of Maryland)
[Maryland] golden-[aster, star] ("golden," in reference to color has been combined with various plants since at least the 1570s, and probably before; the OED 1971 lists 33 combinations; "aster" and "star" are not among them)

This genus was long part of the confusion involving what are now known as *Heterotheca* and *Pityopsis*. John Semple (1977, 1980, 1996, Semple et al. 1980) resolved those problems. More discussion is given under *Heterotheca*.

The Delaware used an infusion of roots to calm sick infants, and as a tonic for them (Moerman 1998).

Cicuta

(Latin for the poison hemlock, *C. virosa*; the Greek philosopher Socrates, charged by the government with impiety and corruption of youth, drank hemlock to commit suicide in the spring of 399 B.C.)

Cicuta maculata. From Institute of Food and Agricultural Sciences.

cicuta (from Latin, Italian); *cicutaire* (French); *ciguë aquatique* (water cicuta, French)
cowbane
fealladh-bog (*fealladh*, false, *bog*, marsh, Gaelic)
selsnepe (*selleri*, celery, *nepe*, turnip, Norwegian)
Wasserschierling (*Wasser*, water, *Schierling*, hemlock, German)

Cicuta maculata (spotted, noting colored spots particularly on stems) (= *C. curtissii*)
beaver poison (Texas)
cafã:mcó:bî (*cafã:m* = ?, *có:bî*, big, Mikasuki); *ka: kîmba* (Creek)
carrotte de Moreau (Moreau's carrot, from a Frenchman who died from eating them; see Fernald et al. 1958)
[spotted] cowbane (USA)
machiskatask [*maciskatask*] (Cree); *manito-skatask* (*manito*, spirit, Plains Cree)
muskrat-weed (USA); musquat-root (variant of muskrat, USA)
[common, spotted] water hemlock (USA)
yagodi (Slave, Canada)
zhon̲'-xa-shton̲-ga (*zhon̲'-xa*, a counting stick, *shton̲*, habitually, *ga*, by blows, Osage)

That the Greek scholar Socrates (ca. 470–399 B.C.) used *Cicuta* to commit suicide indicates the antiquity of European familiarity with this genus, particularly *C. virosa*, the Old World species. However, there are seven other species in the genus, including the American plants. Mabberley (1997) thought that possibly *Cicuta* contains the most violent poisons of all northern temperate plants.

Although *Cicuta* contains the virulent poison cicutoxin, a resinoid that affects the nervous system, people have used it as a medicine. A single bite is said to be enough to kill a human (Diggs et al. 1999). Although it is a drastic poison, the Seminoles used it to treat fever (Sturtevant 1955). The Osage used the root to treat bruises and swellings and to reduce inflammation (La Flesche 1932). Farther north in Canada, the Ojibwa also used it and even "smoked the root" to attract deer (King 1984).

Cirsium: Thistles

(From Greek *kirsion*, some kind of thistle)

We all know a thistle, even if we do not know much about plants. A thistle is a nasty prickly herb reaching head-high that sticks you when you pass. Essentially, the word "thistle" has been a generic name for spiny, armed plants since about A.D. 725. Not long after that date it was restricted to plants in the Asteraceae usually with lavender or purple flowers. We now put many of these in the genus *Cirsium*, and

Cirsium horridulum. *Drawn by P.N. Honychurch.*

most of the others go in the related genera *Carduus, Cnicus,* and *Onopordum.*

As common as the plants are in temperate regions, there is a mystery about the origin of their English common name, thistle [*thistil, thistel, thystel*]. The best that can be said is that it is Germanic. There are variants in German and Dutch (*distel*), Old Norse (*thistell* or *thistill*), Swedish (*distel*), and Danish (*tidsel*). From about 725 until the 1500s, much of the literature equated the common name with *Carduus.* Then, with its creation in Philip Miller's *The Gardeners Dictionary* (4th edition, 1754), there was a shift to equating *Cirsium* with the common name "thistle." Since then, there has been a shuffling of species from *Cirsium* to *Carduus* with few agreeing exactly where one ends and the other begins. As it is now defined, *Cirsium* comprises about 250 species, with 50 in Europe and about 50 in temperate North America (Kartesz 1994, Mabberley 1997). Florida has seven species, only one of which is alien (Wunderlin 1998). At least six of the Florida species have been called *Carduus* at one time or another. Only *C. horridulum* and *C. nuttallii* are common species. Some consider Florida lucky to be free of the worst weedy species (Holm et al. 1991).

Plants called *Cirsium* have a venerable history. There was a kind of thistle called *kirsion* in Greek by Dioscorides (fl. A.D. 40–80), who used it to treat varicocele and other problems of the veins. Greek legend also records that Earth made the thistle in a moment of grief for the loss of Daphnis, shepherd and musician, poet and hunter. Although the Greek lamented the loss of Daphnis, the Norse esteemed the plants. Norse mythology records that Thor, the thunderer, protected the plant, along with all those who wore it. The Norse called them the "lightning plant" (Lym and Christianson 1996). Pliny (A.D. 23–79) Latinized the Greek name to *Cirsion,* and that use was continued by Miller in the 1750s with the spelling changed to *Cirsium.*

There is a strong association with Scotland and the thistle, including an image of one on its heraldic emblem. According to a notice published in 1488, that emblem is a "couering of ... purpir tartar browdin with thissillis and a vnicorne" (covering of ... purple tartan bordered with thistle and a unicorn).

According to legend, the Scottish people venerate the thistle because it saved them during a war. The event took place when Norsemen invaded Scotland during the reign of Malcolm I (A.D. 938–958). These Norsemen were trying to subdue Staines Castle. One night the invaders took off their footgear to wade through the moat surrounding the castle. Instead of finding water in the dark they found the moat dry and filled with cotton thistles (*Onopordon acanthium*). Their painful yells and curses aroused the garrison and the Norse were defeated. In memory of this victory the thistle became the floral emblem of Scotland, and the "cotton thistle" was renamed the "Scottish thistle." In 1540 King James V established the *Most Noble and Most Ancient Order of the Thistle.* King James VII revived it, knighting eight noblemen in 1687. In 1703 Queen Anne increased the number knighted to 12.

The thistles are such important plants in Scotland that there are two generic words for them in Gaelic. Some species are designated *fòthannan* (as in *fòthannan achaidh,* field thistle, *Cirsium arvense*; or *fòthannan Baltach,* Baltic thistle, *Carduus crispus*), and others are called *cluaran* (as in *cluaran deilgneach,* prickly thistle, *Cirsium vulgare*; or *cluaran Lèana,* English thistle, *Cirsium palustre*).

People have uses for various species throughout the range of the genus *Cirsium.* By far, the most noted of the two common species in Florida is *C. horridulum* (the horrible thistle). That Latin name was given to these plants by Frenchman André Michaux in 1803 in his book *Flora Boreali-Americana* (Flora of North America). The continent seemed much smaller at that time because the Lewis and Clark expedition had not yet taken place. That expedition's famous trip up the Missouri River system to the Pacific did not start until 1804. There is another connection between Michaux and that famous journey—that Frenchman almost became the botanist on the trip (Taylor and Norman

2002). If President Jefferson had not been fighting with the French over Louisiana and with the Spanish over the lands near California, it might have become the "Michaux Expedition."

For some odd reason, the "accepted" common name of *C. horridulum* is yellow thistle in the United States. Since the flowers are purple in Florida (at least in the southern part), that name seems wrong. Reports say that flowers may be yellow elsewhere. Other people may call *C. horridulum* the bull thistle (Texas to the Bahamas), or simply say it is a thistle. This plant is almost confined to the southeastern United States coastal plain, from Texas to Florida to Maine and the Bahamas. There is also a record of it in Chiapas, and Correll and Correll (1982) indicate it has been found elsewhere in Mexico.

The Miccosukee know the plants as *tokifáskî* (sharp thistle), their relatives the Creeks say *aká:co* [*aga'djo*], and in Muskogee the plants are called *vkaco* [*vkácv, vkacv, kotacv*]. Their relatives, the Chickasaws, call their plants *oktaak pilhchi'* (*oktaak*, prairie, *piłchi'*, thistle). Choctaw say *shumatti* (blowgun dart) or *shumo* (down), which is the same meaning as Koasati *łohpó*.

The only use attributed to these plants by the Seminoles who talked with Sturtevant in the 1950s was to make blowgun darts. Presumably, they used the pappus from the fruit clusters as "feathering" on the base of their darts so that they would catch air in the gun. People in South America use a similar material from kapok trees (*Ceiba pentandra*) for their darts. Relatives of the Seminoles, the Houma of Louisiana, have more uses for thistle, which they call *chardron*, a local form of the French word (Speck 1941). They use an infusion of leaves and roots in whiskey and find it a strong astringent. Further, an infusion of leaf and root will clear phlegm from the lungs and throat. The tender white hearts are eaten raw. Probably, the Seminoles knew of those uses in the past. Even by the 1950s many were forgetting old herbal uses or dying without training apprentices. Alice Snow no longer uses *Cirsium* (Snow and Stans 2001), and Bennett (1997) did not find it in use. Other people also found that an infusion of *C. horridulum* could be used against colds, dyspepsia, and diarrhea (Hocking 1997).

People in the Bahamas also use the plants. As with the Houma, they eat the succulent stems. To do that they chop them up and eat them fresh, or they put them in salads. Sometimes they pickle them in vinegar (Correll and Correll 1982).

Raw thistle achenes are bitter but may be roasted and eaten. The young leaves of several species with the spines removed can be served as a salad or cooked like spinach (Fernald et al. 1958). The inner stem pith can be steamed and served like asparagus. The heads can be used as rennet to curdle milk for cheese making (Lym and Christianson 1996).

Both the Spanish and French depart in their common names from the *Cirsium* root words and call these armed herbs *cardo* (Spanish), *cardillo* (New Mexico), and *chardón* (French). Those words are based on Latin *Carduus*. This was the name used by classical authors Virgil (70–19 B.C.) and Pliny (A.D. 23–79), and by Linnaeus ([1753] 1957) for a variety of bristly plants related to *Cirsium*.

There are reports of several groups using *C. mexicanum*. The Zapotec call it *guia dana* [*guie tana*] (*guia*, flower, *tana*, dry), and the Maya say *omil*. Other people in Mexico say it is the *cardo santo* (holy thistle) or *cardo santo macho* (male holy thistle), and use it to treat colds as with other species. In the French Antilles the species is known as *chardón* (Guadeloupe, Martinique) or *chardón-amourette* (bitter thistle, Guadeloupe, Martinique). Other species are called *cardo* in Mexico and Central America, and sometimes called *cardón*. The cacti *Cereus aragoni* (Costa Rica) and *Pachycereus pringlei* (Baja California, Sonora) are both called *cardón*. Calling a cactus *cardón* never made any sense to me until I realized it was a comparison to other spiny plants in the same sense as using the word *cactus*. Both *cardón* and *cactus* are generic terms for prickly plants.

From the way people treat the names "thistle" and *cardo*, they seem to regard the species in *Cirsium* as being all the same. There are records of at least 29 species from Europe and the Americas being used. Most of the literature on the genus consists of attempts to control some weedy species, but a few discuss medically active compounds. Within the genus, people have identified acetates, alkynes, flavonoids, glucosides, flavone glycosides, and volatile compounds (Takano and Kawaminami 1988, Takaishi et al. 1990, Binder and French 1994, Martínez-Vazquez et al. 1998). These substances have been reported in Japanese, Mexican, and European species. The chemicals have been shown to reduce pain and inflammation, and control blood sugar, at least in the laboratory (Martínez-Vazquez et al. 1998, Iwashina et al. 1999, Pérez et al. 2001). As those are maladies the plants are used to treat, there is support for their medical effectiveness.

Several groups and uses are reported with only the generic name, which may represent several species. Even if that is correct, the uses reflect what was probably done historically in Florida. The Huastec call thistles *cholich* [*tzolich*] (San Luis Potosí, Mexico), and they are used against dysentery, aches and pains, and coughs. These same people consider them a *milpa* (corn field) weed that is also a forage readily eaten by horses (Alcorn 1984). Within the United States and

Canada, some species of *Cirsium* has been used, in addition to those named above, by the Abenaki, Apache, Atsugewi, Blackfoot, Cahuilla, Cherokee, Cheyenne, Chiricahua and Mescalero, Comanche, Costanoan, Cowichan, Cree, Delaware, Flathead, Gosiute, Hopi, Iroquois, Kawaiisu, Kiowa, Kwakiutl, Meskwaki, Montagnais, Navajo, Nitinaht, Ojibwa, Okanagon, Paiute, Potawatomi, Quileute, Ramah and Kayenta, Spokan, Sushwap, Thompson, Tubatulabal, and Western Keres (Smith 1933, Moerman 1998). That pretty well covers all four corners of North America and Mexico.

Cissampelos

(From Greek *kissos*, ivy, and *ampelos*, a vine; referring to foliage and similarity to *Cissus*)

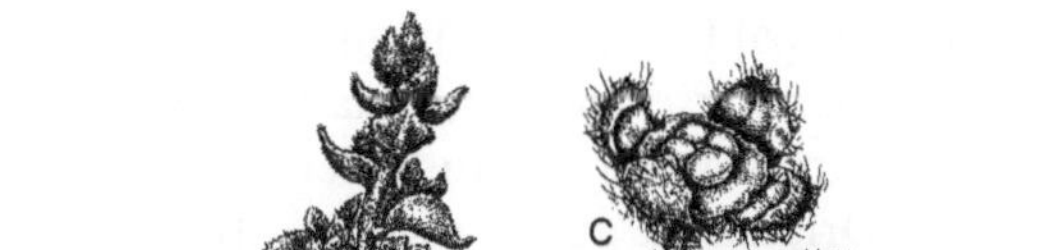

Cissampelos pareira. a. Lower part of the stem. b. Part of the pistillate inflorescence. c. Segment of staminate inflorescence with open flower. d. Staminate flower longitudinally dissected. e. Floral diagram of staminate flower. f. Pistillate flower. g. Pistillate flower longitudinally dissected. h. Floral diagram of pistillate flower. i. Fruit. *Drawn by Priscilla Fawcett.* Correll and Correll 1982.

Cissampelos pareira (from Portuguese *parreira*, vine, alluding to a resemblance to the grape vine, *Vitis*)

abútua (twining plant in the family Menispermaceae, Tupí, Brazil)
alcotán [*de la tierra*] [*cotán*] (*alcotán* is usually a name given to a type of hawk; thus, *alcotán de la tierra* would be "hawk of the earth," Nicaragua, Guatemala, El Salvador)
aristoloche lobée (lobed *Aristolochia*, Haiti)
barbasco (Peru)
bejuco azul (blue vine, Costa Rica); *bejuco de alcotán* (hawk vine, El Salvador); *bejuco de mono* (monkey vine, Puerto Rico); *bejuco de pitilla* (little flower vine, Cuba); *bejuco de preñada* (pregnant woman's vine, Guatemala); *bejuco de ratón* (mouse vine, Dominican Republic); *bejuco de sabana* (savanna vine, Cuba); *bejuco de terciopelo* ([snake] *Bothrops* vine, Cuba); *bejuco ñame* (yam vine, Dominican Republic)
butua (Colima, Guerrero, Veracruz)
caapeba (*caa*, plant, *peba* from *pewa*, large boat, Tupí, Brazil)
cipó de cabras (goat vine, Brazil); *cipó de cobra* (snake vine, Brazil); *cipó de gota* (drop vine, Brazil)
curarina [*de monte*] ([wild] little healer, Chiapas, Guatemala)
cuxbá [*cuxogui*] (Quecchí, Guatemala)
erva de nossa senhora (Mary's [our Lady's] herb, Brazil)
estrella de la preñada (pregnant woman's star, Guatemala)
false pareira root (a root confused with the "real pareira," *Chondodendron tomentosum*)
feuille coeur (heart leaf, Haiti)
guaco [*huaco*] (Veracruz, Guatemala; for more on *guaco*, see *Manfreda*)
hierba de la vibora (snake herb, San Luis Potosí)
hierba ratón (mouse herb, Venezuela)
ice vine
imchich masha (Quechua, Peru)
ischochichac (Totonac, Veracruz); *ixcatú-can* (*kan*, yellow, Maya, Guatemala); *k'on k'ach* (Huastec, San Luis Potosí)
iztac-coanenepilli [*istaconenepilli*] (*iztac*, white, *coatl*, serpent, *nenepilli*, tongue, an herb used as an antidote, Náhuatl, Mexico)
liane [*a*] *serpent* (snake vine, Haiti, Guadeloupe, Martinique); *liane à coeur* (heart vine, Haiti, Guadeloupe, Martinique); *liane amère* (bitter vine, Guadeloupe, Martinique); *liane corde* (cord vine, Guadeloupe, Martinique); *liane gourde* (gourd vine; probably a corruption of *liane corde*, Haiti); *liane molle* (soft vine, Guadeloupe, Martinique); *liane patte cheval* (horse hoof vine, Haiti); *liane quinze jours* (15 day vine, Guadeloupe, Martinique)

oreja de ratón (mouse's ear, Guerrero, Michoacán, Guatemala, Puerto Rico)
orelha de onça (jaguar's ear, Brazil)
palikur (probably from *palikar*; derived from Greek *palikaris*, valiant, Peru)
pareira (used in English by 1715, from Portuguese *parreira*, vine); *pareira brava* (wild vine, Oaxaca, Veracruz, Jamaica, Barbados); *parreira brava* (wild vine, Brazil); *parreira do mato* (wild [forest] vine, Brazil)
pataka [pathaca] (variant of *patate*, potato, Virgin Islands); *patacón* (big potato, Trinidad)
patte cheval [*pat chawal*] (horse hoof, Haiti)
pegamo (Colima)
peteltún (Maya, Yucatán)
picamano (hand stinger, Nicaragua)
quacori (Cayman Islands)
tamagás [*tamasás*] (the poisonous viper *Bothrops*, Guatemala, Nicaragua)
tomatillo de sabana (savanna tomato, Cuba)
tortilla de los sapos (toad's tortilla, Veracruz)
ts'its'uk-ak' [*tsutsuc*] (*ak*, vine, Maya, Yucatán)
uxum ichich (woman medicine, Huastec, San Luis Potosí)
velvet-leaf (Jamaica, Virgin Islands, Barbados)
venadera (poisonous one, Costa Rica)
walik ts'ohool (evil eye plant, Huastec, San Luis Potosí)

Cissampelos pareira is one of those nondescript climbers that people rarely look at closely, even if they are supposed to be critical. This climber occurs in the American tropics, through the Old World tropics, and in Australia (Mabberley 1997). The vine has 101 scientific names that have been applied to it (synonyms) because of its variability and the fact that people did not examine it closely.

Pareira brava became known in Florida when it was collected in 1929 by E.C. Marquad in what is now Matheson Hammock Park, and then listed in the *Manual of the Southeastern Flora* (Small 1933). In Florida, the plants were confined to a small area in Miami-Dade County. Indeed, the plants have always been so rare in southern Florida that they almost did not exist. There are no specimens in the database maintained by Fairchild Tropical Garden, and I never saw the species in 31 years of fieldwork there. The Florida Department of Agriculture considered the vines endangered in the state (Coile 2000) and wondered if they had been extirpated. So far as known, the last plants were accidentally killed around 1993 at Matheson Hammock State Park when a restoration crew mistook the plant for an exotic that was meant to be destroyed (Gann et al. 2002).

The species is so widely and commonly used by people that it might even qualify as a floristic element introduced by humans. However, the bright red, fleshy fruits are relished by birds, so the human-introduced conclusion is at least uncertain.

The first European reference to these plants was made in 1648 by German pioneer in tropical medicine Georg Marggraf, who found it being used in eastern Brazil. He called them *Caapeba brasiliensibus.* From that point until Linnaeus studied them for his *Materia Medica* of 1749, confusion about their identity reigned. John Ray in 1686–1688 considered them a *Convolvulus.* Hans Sloane in 1696 thought the vines should be a *Clematis.* Leonard Plukenet in 1700 thought they were a *Hedera.* Charles Plumier thought he found another species in the Caribbean but continued calling them *Caapeba* in 1703.

In 1715, the British journal *Philosophical Transactions* of London published an article that included the following: "The *Pareira Brava* is a Root which comes to us from Brazil by way of Lisbon." The OED (1971) says that "pareira" is a "drug made of the root of a Brazilian shrub, used in disorders of the urinary passages. Originally understood to be the root of *Cissampelos pareira* or 'Velvet leaf,' the *pareira brava* of the Portuguese, whence the name: now said to be that of a different shrub, *Chondodendron tomentosum*; the 'Velvet leaf' being distinguished by some as the *Spurious pareira.*" The same source confirmed that the "real pareira" is *Chondodendron tomentosum;* so, *Cissampelos pareira* is relegated to being the false pareira. Both twiners are in the same family, the Menispermaceae, and they have been used for the same purposes.

People throughout the range of *pareira brava* have uses for the plants. It was those uses in Brazil that captured Georg Marggraf's (1610–1644) attention when he and Dutch physician Willem Piso (ca. 1611–1678) went there in 1637. In Mexico, where the Aztecs knew the plants under the name *iztaccoanenepilli*, one of the important uses was to treat bites of poisonous animals, particularly snakes (Martínez 1969, Vásquez and Jácome 1997). Standley and Steyermark (1946) wrote: "Throughout much of tropical America they have a high reputation as a remedy for bites of snakes and other poisonous animals." Uphof (1968) also noted that the vines were used by Chinantecs and others for snakebites. Others noting that use include Liogier (1974), Duke and Vásquez (1994), and Hocking (1997). The snakebite aspect was even noted by Rudyard Kipling's 'Kim' (Hocking 1997).

The species was treated among indigenous remedies in the journal *Economic Botany* in nine articles between 1947 and 1996 (Jain 1965, Duke 1970, Jain and Tarfder 1970, Lipp 1971, Wong 1976, Arenas and Moreno 1977, Bedi 1978, Manandhar 1991, Coe and Anderson 1996). Five of those were American uses, and the others were from India and Nepal. Medical use is also listed by Roig (1945), Uphof (1968), Liogier (1974), Morton (1981), von Reis and Lipp (1982), Duke and Vásquez (1994), Hocking (1997), Mors et al. (2000), and numerous other American publications. Duke et al. (2002) summarized the uses, listing its activities as alexeteric, analgesic, antiabortive, anti-inflammatory, antipyretic, astringent, bitter, curare, depurative, diaphoretic, diuretic, emmenagogue, expectorant, hemostat, litholytic, myorelaxant, piscicide, poison, sedative, stomachic, tonic, and vulnerary. They continue with uses that include treatment for acne, amenorrhea, asthma, bleeding, calculus, catarrh, childbirth, cough, cystosis, dermatosis, diarrhea, dropsy, dysentery, dyspepsia, dysuria, enterosis, fever, gastrosis, gravel, hemicrania, hemorrhoids, inflammation, insomnia, jaundice, nephrosis, nervousness, pain, prolapse, rash, sinus infection, snakebite, sore, sting, stomachache, stone, tuberculosis, ulcer, urethrosis, urogenitosis, uterosis, venereal disease, water retention, worms, and wounds. It would seem to be a panacea, but they give it a single "plus," which means they think it is not as safe as drinking a cup of coffee (and Jim Duke noted that he often drinks more than three cups of coffee a day).

Surely because of the worldwide use by indigenous people, and because the vines belong to the same family as some of the most famous sources of curare, they have been subjected to more chemical analysis than many others. Uphof (1968), Wong (1976), Morton (1981), Duke and Vásquez (1994), and Mors et al. (2000) note that the roots contains alkaloids including bebeerine (pelosine), cissamine, cissamparine, dehydrodicentriine, curine, cyclenine, dicentrine, hyatidine, hyatine, hyatinine, insularine, isochondodendrine, menismine, pareirine, and tetrandine. Of these, cissamparine has shown antitumor activity (Morton 1981), tetrandine is analgesic, anti-inflammatory, and febrifuge (Duke and Vásquez 1994), and chemicals of the bisbenzylisoquinoline type exhibit high cytotoxic activity (Mors et al. 2000). Hyatine has muscle relaxant and hypotensive effects (Wong 1976).

Cissus: Possum Grape

(From the Greek *kissos*, ivy, referring to the climbing habit)

Although the plants had been known in Florida for many decades before David Fairchild introduced new genetic strains into the state, he worried about them. Fairchild quipped that that those he introduced might lead to him being remembered as a "Damnable introducer of weeds" (Fairchild 1947). Anyone who has tried to pass through a tangle of these plants would understand his comment.

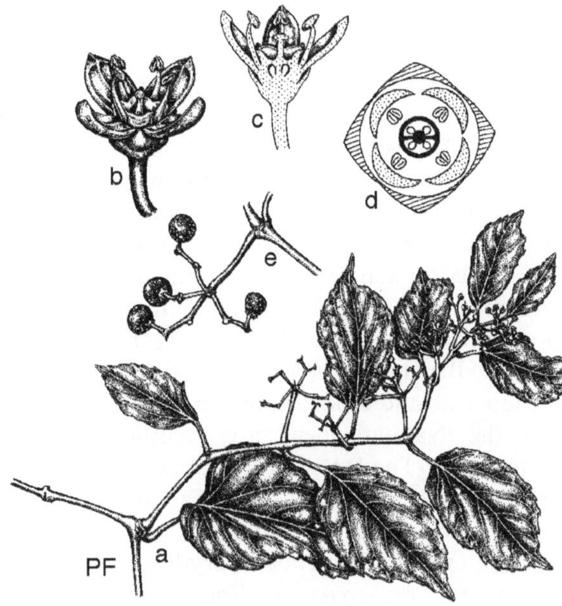

Cissus verticillata. a. Branch. b. Flower, side view. c. Flower longitudinally dissected. d. Floral diagram. e. Fruits. *Drawn by Priscilla Fawcett.* From Correll and Correll 1982.

In Florida the species is called possum grape, and for many years it was *Cissus sicyoides*, a name Linnaeus gave it in 1759. However, in one of his uncommon lapses, Linnaeus had actually given these plants a name in 1753. The earlier name is the one that must be used. It appears that Linnaeus was misled by the citation from Hans Sloane published in 1696. In his treatise on Jamaican plants, Sloane had written that one was a *Viscum ramulis & folis longis dentissimis straitis & radiatis* (mistletoe with branches and long leaves toothed, ridged and rounded), and that it was a parasite. So, Linnaeus called Sloane's plant *Viscum verticillatum.* When he encountered the same species in 1759, he did not recognize it and called it *Cissus sicyoides.* This vine was known as *C. sicyoides* until Dan Nicolson (Smithsonian Institution) and C.E. Jarvis (Natural History Museum, London) discovered the mistake and corrected it in 1984. This pair gave us the correct modern name of *C. verticillata.*

Although Sloane's was the first record of these American plants known to Linnaeus and most other Europeans, his was not the first report of them. Earlier, Francisco Hernández (1651) listed the climbers and recorded that the Aztecs of Mexico called it *tlacamazatcazqui y papan* (*tlacatl,* human, *maçatlozqui,* medicinal plant, *papan,* upon, apparently meaning "medicinal plant for use on humans"). More recently, it has been recorded in Náhuatl as *molonqui* (sprawler, Valley of Mexico), and *tepemecatl* (*tepetl*, in the mountains, *mecatl*, rope, so it is wild rope). It was supposedly also listed by Hernández as *temecatl* and considered a laxative herb. Its Spanish name *tumba-vaqueros* (cowboy tumbler, Valley of Mexico, Morelos, Hidalgo) has the same connotation. *Tumba-vaqueros* is also applied to many species of Convolvulaceae that are notably laxative. That interpretation is further supported by the name bastard bryony (Jamaica, Trinidad). Bryony is *Bryonia*, a member of the Cucurbitaceae, and a notorious purgative.

All the Náhuatl ideas are recorded in other modern names. The fact that it "sprawls" over everything nearby is consistent with Fairchild's worries. Some of the names that convey the notion of scrambling plants covering everything are *achite* (synonym for *anil-trepador*, Brazil), *anil trepador* (blue climber, Brazil), *bejuco caro* (expensive vine, Puerto Rico, Hispaniola), *bejuco Castro* (Castro's vine, Colombia), *bejuco chirriador* or simply *chirriador* (*chirriar* is slang for going on a spree, plus vine, Colombia), *bejuco comemano* (hand-eating vine, Guatemala, Honduras, Central America), *bejuco de caro* (expensive vine, Puerto Rico, Dominican Republic, Venezuela), *bejuco loco* (crazy vine, Tabasco, Panama), *bejuco loco rojo* (red crazy vine, Tabasco), *caro* (expensive, Cuba), and *picamano* (hand biter).

Surely, a derogatory connotation is also implied by several other names. These include *bejuco de gallina* (chicken vine), *tripa de Judas* (Judas's guts, Valley of Mexico, Morelos, Hidalgo, Oaxaca), *tripa de vaca* (cow guts, Guanajuato), and *tripa de zopilote* (vulture guts, Sinaloa).

Names showing that the plants were used to create strings or rope are similarly varied. In Florida, the Creek-speaking Seminoles called it *chu:los:sho:a:kee* (*culoswuce*, possum grape, *shoakee* = ?) or *bakslô:cî* (black strings). Their relatives the Mikasuki speakers have the same name in *baksô:cî* (black strings). Another Seminole name is *afoslasti*, based on Muskogee *vfuswv*, strings, and *lvste*, black. Both use the vines to make strings and ropes, and medicinally (Sturtevant 1955). Historically, Seminoles used the fibers from the stems to make bowstrings.

The Maya of Yucatán also use the stems for ropes. They say it is *tab-kanil* (*tab*, string, *kanil*, secondary root). In other words, it is a plant used to make cords that are used to carry things on the shoulders. The Maya also call it *tab-kaan* [*ta-kan*] (*tab*, cord, *kaan*, yellow, Yucatán, Belize) or *ya'ax-tabkanil* (yellow string rope). In Panama it is rock rope. Although the names do not say so directly, the stems are also used in making baskets (Mexico, Jamaica). Maybe *cortina de pobre* (curtain of the poor, Brazil) indicates fabrication of curtains from the stems.

Many of the names indicate medicinal uses. In Peru, two names associate it with souls or spirits. These are *ampato huasca* (defending spirit, Quechua) and *zapo-huasca* (toad's spirit, Quechua). In Brazil, the plants also are called by two different names. They are either *pucá* or *cipó-pucá* [*cipó-pucí*] (*pucá* vine, from Tupí *pi'sá*, fruit of *Rauvolfia bahiensis*) to compare them with another medicinal plant. Alternatively, they may be called *caavurana de cunha* (*caá*, herb, *caavurana* is medicinal *Solanum caavurana*, *cunha*, woman, Tupí). The most common usage of *Cissus* is for skin diseases, and this Tupí name compares it with one of the plants that they use to treat leprosy. Perhaps *caavurana* is a man's medicine, and *caavurana de cunha* is for women.

Other names that suggest remedies for skin problems include blisterbush (Jamaica, Trinidad), *cuacua* (a creole word, Nicaragua), *herbe à ulcères* (herb for ulcers, Hispaniola), *pico de mano* (hand sticker, Nicaragua), scratch wys (scratch vine, Jamaica, Trinidad), and yaws bush (Jamaica, Trinidad). Some specific examples of uses include crushed leaves with an oily carrier used to treat itchy skin in Curaçao (Morton 1981). In Mexico, the bark is crushed in alcohol and used to treat skin problems (Martínez 1969). In Jamaica the stems are mixed with fat or castor oil and used on boils and bruises. In a number of places the leaves are used as poultices on tumors, boils, hemorrhoids, abscesses, and sores. The sap is also used on sunburn. In Guatemala the sap is used to treat gangrene.

Surely, the most inclusive common name is *sánalo todo* (cure-all, Veracruz). Sometimes the vines are compared to others, such as wild yam (yam is *Dioscorea*, in Jamaica, Trinidad) or *yedra* (ivy). Perhaps those associating the plant with snakes are alluding to its medicinal properties. Those names include *paja de la culebra* (snake straw), snake bush (Jamaica, Trinidad), and snake wys/withe (snake vine, Jamaica). Perhaps poison wyth (English Antilles) has the same meaning. Mary bush (Jamaica, Trinidad) carries an unspecified application. Probably the names *iasu* or *bejuco iasu* (*iasu* vine, Costa Rica) and *liane a minguet* (*minguet* vine, Hispaniola) indicate uses that are obscure.

Additionally, the leaf decoction taken internally or in baths is used to treat rheumatism in Mexico, Cuba, and Hispaniola (Roig 1945, Martínez 1969, Liogier 1974). Gargles of the decoction of the leaves are used against throat infections and flu (Hispaniola). In Brazil, tea from leaves is drunk as a treatment for heart problems, tachycardia, hydropsy, high blood pressure, and anemia (Mors et al. 2000).

As with the common name possum grape, many others make comparisons with other Vitaceae. The vines are called *bejuco de parra* (grape vine, Hispaniola), *bejuco ubí* (grape vine, Cuba), *parra de monte* (wild grape, Mexico), pudding bush (Jamaica, Trinidad), pudding vine (English Antilles), pudding wys/withe (Jamaica), *ubí* (grape, Cuba), *uva* (grape, Oaxaca), *uva brava* (wild grape, Panama, Brazil), *uva cimarrona* (wild grape), *uva silvestre* (wild grape, Mexico), *uvilla* (little grape, Nicaragua), *uvilla silvestre* (little wild grape), and *vid silvestre* (wild vine, Hidalgo, Veracruz). Fruits are consumed by many animals; however, that should not be taken to imply that the fruits are edible for humans.

Although the sap contains enough saponin to make it lather like soap and be used like soap, it is also consumed. The names *bejuco de agua* (water vine, Colombia), *yucohuira* (water vine, Mayo, Sonora), and *liane a eau* (water vine, French Antilles) suggest that it may be used as a source of water. It is said to have a slightly acidic flavor, and it is used simply as a water source in Colombia and Sonora (Pérez-Arbelaez 1978, Yetman and Van Devender 2001). However, most people use the "water" medicinally. In Hispaniola a tea from the plant is used for colon inflammation to reduce fever, and it becomes diuretic when mixed with cactus fruits, maize silk, and sugar cane (Liogier 1974). In Amazonian Brazil, the sap is used against epilepsy, and preliminary data suggest possible anticonvulsive traits (Mors et al. 2001). Moreover, in addition to saponins, the species contains anthocyanins (Toledo et al. 1983). Presumably, it also contains stilbenes, flavonols, quercetin, kaempferol, resveratrol, piceatannol, pallidol, and parthenocissine a, like its relative *C. quadrangularis* (Adesanya et al. 1999).

In Brazil, *C. verticillata* is called *tinta dos gentíos* (gentile's dye). There and elsewhere it provides dyes (Morton 1981, Mors et al. 2000). In Colombia, Mexico, and Jamaica, the fruits are used to prepare a blue dye.

There are several names with obscure allusions. These include *boeng abie ma* [*boen atti mama*] (tree heart mother, Sranan, Suriname), *cabuja di tranke* [*kaboeja di trankera*] (*kaba*, end, *di*, for, *tranga*, strong, Sranan, Dutch Antilles, Suriname), *fei mush* [*feuille mouche*] (fly leaf, Jamaica, Trinidad), *feuille cotaire* (numbered leaves, Hispaniola), godmort (St. Lucia), lambrall (English Antilles), *omhein-ingtouw*, *paramaru* (Carib, Suriname), and pinakoop (English Antilles). *Prenada segunda* (Belize) seems to mean a "second light"; perhaps the stems were used for candles.

My favorite name is *coronilla* (Belize), the little crown. Although trying to get through a tangle of these climbers will elicit four-letter Anglo-Saxon words, they do crown their host admirably.

Citharexylum: Fiddlewood

(From Greek *kithara*, an ancient lyre-like instrument, and *xylon*, wood)

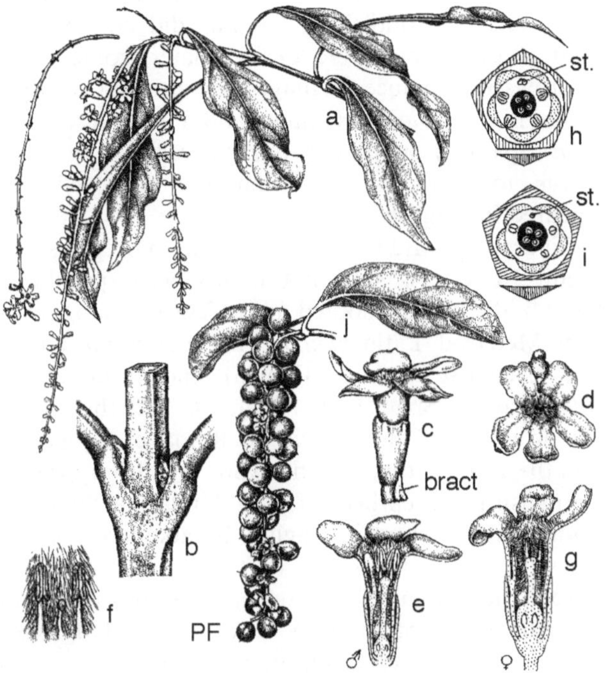

Citharexylum fruticosum. a. Branch with flowers, on left, old male spike with most flowers fallen. b. Node, with axillary bud enclosed by leaf base. c. Flower, side view. d. Flower, from above. e. Staminate flower longitudinally dissected. f. Staminode and two stamens in staminate flower. g. Pistillate flower longitudinally dissected. h and i. Floral diagram of staminate and pistillate flowers, respectively. j. Fruiting branch. st. = staminode. *Drawn by Priscilla Fawcett.* From Correll and Correll 1982.

There are two stories about how fiddlewood (*Citharexylum fruticosum,* Florida to Panama and Trinidad) got its name. One is that wood from it was used to make fiddles (violins for those of highbrow tastes). The other is that the tree was known as *bois fidèle* (faithful wood) to the French, and that the English misunderstood and mispronounced the name. Take your pick.

Either way, the wood is involved, and its use in musical instruments is recorded in the name *palo de guitarra* (guitar tree, Puerto Rico). We are left to presume that Linnaeus learned of this usage in making musical instruments and selected the generic name to commemorate that fact. Linnaeus based the generic name *Citharexylum* on Greek *kithara*, an ancient lyre-like instrument with a triangular body and 7 to 11 strings. Indeed, wood from the trees has been used to make fiddles, violins, guitars, and other musical instruments (Little and Wadsworth 1964).

To us in the 21st century, it seems curious that Linnaeus would choose the Greek musical instrument *kithara* in naming this genus. However, he tried to create generic names based on classical language words, and that was the one available to him. Modern words such as "fiddle" or "violin" did not exist in classical times, but were derived later, some from Latin-based Romance languages. The original versions of violin were *víolina* (Italian and Portuguese) and *violin* (Spanish), supposedly diminutive forms of *viola*. It was not until the 1570s that the word "violin" appeared in English.

The word "fiddle" has an even more obscure origin. No one is sure, but it has been speculated that "fiddle" was adopted into Teutonic languages from Medieval Latin *vidula*. That Latin word may have given rise to *viole* (French), and ultimately to *viola* (Italian, Spanish, Portuguese). Yet, the words *viol* or *viola* did not appear in Romance languages until the 15th century, and the word "fiddle" appeared in English much earlier than did "viola" or "violin." Fiddle was recorded in English by about A.D. 1205. Since the Germanic usage of these words predates those in the Romance languages, a Teutonic origin and subsequent adoption into Latin-based languages seems equally or perhaps more plausible.

Some books record our plants (*C. fruticosum*) as Florida fiddlewood, but people in the field will not know what you mean by that name. That is probably a name applied in the lumber trade. Supposedly, there is reason to call it white fiddlewood in the Caribbean and Central America where there is also a yellow fiddlewood (*C. caudatum*). However, *C. fruticosum* is called yellow fiddlewood in Jamaica.

Several other names refer to the quality of the wood. It is so hard that it is called *bois de fer blanc* (white ironwood, Guadeloupe, Martinique). In the same region, it is called *bois carré* (turtle wood), presumably because of its dense wood, and perhaps because it grows near the coast.

Perhaps because they have a different taxonomic scheme from ours, some people compare fiddlewood to coffee. They know the difference, but still see a similarity. These people call it *café cimarrón* [*marron*] (wild coffee, Dominican Republic), or wild coffee (Bahamas). Others think it looks more like pomegranate (*Punica granatum*) and say it is *grenade* (pomegranate, Haiti) or *grenade marron* (wild pomegranate, Haiti). Yet others say it is like cinnamon (*Canella winterana*) and call it *canilla de venado* (deer's cinnamon, Cuba). In Panama, it looks to some like a *manglillo* (little mangrove). Cubans use an old Taino word and call it *guayo blanco* (white *guayo*).

Some think the drooping fruit clusters resemble necklaces and say it is a *collarette* (little necklace, Cuba). People in the Greater Antilles sometimes are more direct and simply say the flower and fruit clusters hang down with *penda* (dangling, Cuba, Dominican Republic), *péndula* (drooping, Puerto Rico, Dominican Republic), and *pindoula* (drooping, Haiti). Bahamians simply think it is a dirty tree and call it trash-a-berry.

However, many have discovered medicinal properties in the plants and given them names that reflect those properties. Balsams are universally recognized as medicinal, and *Citharexylum* is *bálsamo* (balsam) in Puerto Rico. The reference is obscure, but medical usage is indicated by the names long Tom (Bahamas) and *madame Claude* (Haiti). The most widespread name after fiddlewood is old woman's bitters (Puerto Rico, Virgin Islands, Panama), a clear indication of herbal usage.

The strangest name found, *susanaleche* (Virgin Islands), may also reflect medicinal use. The name may be derived from St. Susana who was beheaded in Rome in A.D. 295. The Catholic festival celebrating that event is held on 11 August. So, in keeping with their earthy and (to outsiders) often obtuse humor, the people in the islands refer to the tree as giving milk from Susan, the virgin saint.

Citharexylum fruticosum grows from the southern part of Florida, in the Bahamas, the West Indies from Cuba to St. Vincent, Grenada, and Trinidad, and from Venezuela to Suriname. Much of that range it shares with several of the ± 70 species in the genus (Mabberley 1997). Other species grow as far south as Argentina. One of the species with which it shares its range was also named by Linnaeus. Indeed, it was the first species he named in the genus, *C. spinosum*. Both kinds have been recognized as distinct by all who studied them until recently when a revision was published that considered them the same. If indeed they are the same, the Florida plants will have to be called *C. spinosum*, as Wunderlin (1998) did. However, pending more information, the old familiar name is continued here.

Various parts of fiddlewood have been used by people throughout its range. Probably the most commonly used part is the wood. Not only is the wood used for fuel and fence posts, but it is also prized for cabinetwork and musical instruments in several places (Little and Wadsworth 1964). There is disagreement about the color of the wood, since Little and Wadsworth (1964) say "reddish," Morton (1981) describes it as "red," and Liogier (1974) calls it "ash-colored." As they do not specify, they could be talking about different parts, the heart and sapwoods, which are usually different colored. Because I have not seen the mature wood, I cannot determine which may be which. Both Morton and Liogier agree that the wood is hard, heavy, and compact.

Fruits are somewhat like cherries, and they may be eaten. Most agree that only children like them, and Morton (1968b) says that the "ripe fruit is edible but not good." Like many things, it probably depends on how hungry you may be.

As the common name "old woman's bitters" suggests, there are several medical applications of fiddlewood. When a plant has a bitter taste, it is almost certain that it has been used in medicines; likewise, when it is aromatic. In both cases, people have experimented and used them to treat some malady.

The whole view that bitter is medicinal was enforced when quinine bark (*Cinchona* spp.) was introduced as a cure for malaria in the late 1600s. Because the remedy came from the Catholics, and was called Jesuit's powder and then Jesuit's bark, it was shunned by the Protestant world until 1820. Once established in the Catholic and Protestant realms, bitters became even more popular.

In the Bahamas, and judging from the common names many other places, the plants are medicine (Morton 1981). In the Turks and Caicos, the leafy twigs are boiled and the decoction taken to banish pain from exposure to chilly wind. When that same decoction is mixed with strongback (*Bourreria succulenta*) and spoonbush (*Thouinia discolor*), the decoction is administered internally and externally as a remedy for "bright" (running sores on a child's body).

Fiddlewood boiled with mahogany (*Swietenia mahagoni*), lignum vitae (*Guajacum sanctum*), roots of doctor's club (*Zanthoxylum coriaceum*), and snowberry (*Chiococca alba*), with latex of papaya (*Carica papaya*) added, makes a tea to dispel indigestion.

Fiddlewood is probably cultivated more than anything else for the fragrant white flower. Those small flowers perfume the air, especially at night, and attract a wide variety of insects. During the daytime they are visited by bees, wasps, and butterflies, and at night by moths. The plant is a virtual grocery store for insects, and those who grow it find both the insects and fragrance an addition to an already attractive tree.

Cladium

(Greek *kladion*, a branchlet; named by Patrick Browne in 1766, possibly referring to the inflorescence)

Cladium jamaicense (from Jamaica)

***Cladium jamaicense*.** From Institute of Food and Agricultural Sciences.

caña amarga (bitter cane, Dominican Republic)
cortadera (the cutting one, Dominican Republic); *cortadera de ciéniga* (marsh cutting one, Puerto Rico); *cortadera dos filo* (cutting leaves, Brazil)
cumaní (Taino?, Dominican Republic)
cutting grass (Belize)
herbe [*z'herbe*] *rasoir* (razor herb, Haiti)
holche [*holsache*] (*hol*, hollow, *che'*, herb, Maya, Chiapas; unlike most sedges, the stems [culms] are hollow)
pahikóslî (cutting grass, Mikasuki); *pahikosnâ* (*pvhe*, grass, *kaske*, scrape, Creek)
paint brush grass (Belize)
prickly sedge
saibal (Lacandon, Chiapas)
[Jamaica, swamp] saw-grass [sawgrass] (a comment on the cutting leaf margins; in use by 1822, Florida to Panama)
serrucho (little saw, Puerto Rico, Panama)
siba (Yucatán)
sticky grass (Bahamas)

When Patrick Browne (1720–1790) named this species from Jamaica in 1766, he created a genus that now has three species. Two are found in North

America, and the other, *Cladium mariscus*, is almost cosmopolitan. The Old World species was known as "elk sedge" in Anglo-Saxon and was formerly used in thatching houses in Great Britain (Mabberley 1997).

The growing heart (apical meristem) is edible (Morton 1968b). It is a bit tricky to extract the heart without being cut by the leaves, but it provides a moist bite when successful. There is a slight resemblance to palm heart, but sawgrass hearts are not usually as sweet. Sturtevant (1955) found the Seminoles using the hollow stems as a *'sakpofketv* (medicine blowing tube). This refers to the Creek tradition of "blessing" medicinal plants. Part of that ritual included the *heles hayv* (traditional doctor) bubbling his breath through the liquid using a tube (Howard 1984).

Juice of the grated fruit is used to treat high blood pressure, and a plant decoction soothes chicken-pox eruptions in the Bahamas (Ayensu 1981, Morton 1981). Leaves have been used to make baskets and cheap paper (Eldridge 1975, Morton 1981). Balick et al. (2000) simply record that the sedge is used in making products (including paint brushes according to its name), as an irritant, in medicine, and for fibers in Belize.

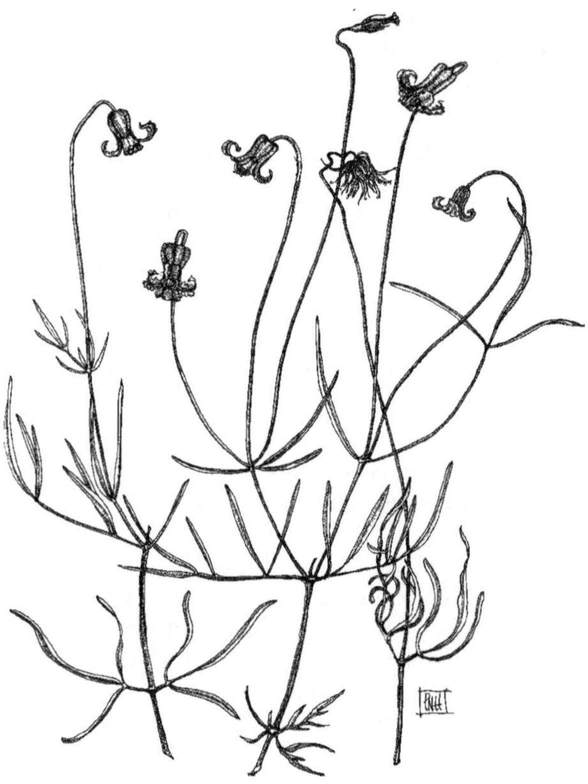

Clematis baldwinii. *Drawn by P.N. Honychurch.*

Clematis

(Greek *klematis*, a shoot or vine-branch or twig; named by Linnaeus, but originally used by Dioscorides, fl. A.D. 40–80, for a climbing plant with long and lithe branches)

Altermannshart (hard old man [the devil], German)
ampelomelaena (*ameplos*, a vine, *melaina*, black, was used by Dioscorides for *Clematis vitalba*, Greek)
clématite (from Greek, French); *clematite* (Portuguese)
hierba de los pordioseros (beggar's herb, Spanish)
old man's beard (the reference here to "old man" is probably another allusion to the devil)
skogranke (*skog*, forest, *ranke*, climber, Norwegian)
traveler's joy (John Gerarde claimed to have coined the name in 1597 when he wrote, "Decking and adorning waies and hedges, where people trauell, and thereupon, I haue named it 'Traueiler's Iioe'"; the name was used later by Williams [1837] 1962)
tysk klematis (German clematis; introduced, Norwegian)
Virgin's bower (applied to the European climber, *Clematis vitalba* by John Gerarde in 1597; later extended to American species)
Waldrebe (woods vine, German); *woudtwijngaert* (wood twiner, Dutch by 1549)

Clematis baldwinii (commemorates the Pennsylvania physician and botanist William Baldwin who died on an expedition to South America in 1819)
[dwarf, marsh] clematis (these have the earmarks of book names; plants do not grow in marshes, and few non-academics call them clematis)
leather flower
pine-hyacinth ("pine" denotes the typical habitat, in pinelands; "hyacinth," from Greek *uakinthos*, rendered *hyacinthus* in Latin, originally a precious blue stone or simply the color; thus, the name alludes to the pineland habitat and the blue flowers)
virgin bower [virgin's bower]

Clematis crispa (crinkled, referring to the sepals)
blue-bells
blue-jasmine [jessamine]
curl-flower; curly-clematis
[swamp] leather-flower (in use by 1866, and originally referring to *Clematis viorna*, which has thick leathery purple sepals)
marsh-clematis

Clematis virginiana

devil's [darning] needles (in used by 1854, but usually applied to insects also called dragonflies); devil's hair (alluding to the tangled stems, Virginia); devil's thread (alluding to the tangled stems)

gander vine ("gander" has meant a male goose since about A.D. 1000; the allusion seems to be that the vines wander aimlessly like a gander, South)

herbe aux gueux (beggar or rascal's herb, Quebec)

ladies bower ("ladies" is a form of "Our Lady's," and was in use by the time John Gerarde wrote his herbal in 1597)

leather flower

love vine (a name used for virtually any plant that twines about others)

old man's beard

pipe-stem (perhaps the stems were used in making pipes)

traveler's ivy; traveler's joy (New Hampshire)

Virgin's bower [virginsbower] (applied to this species by 1668)

wild hops (comparing it to *Humulus* because both twine, Maine, New Hampshire)

woodbine (originally a European term for *Convolvulus* and *Hedera*, dating to about A.D. 875, and alluding to the tendency of the climbers to wrap around others, Maine)

Europeans and Americans have had mixed feelings about these twining plants. Some obviously consider them sent by the devil himself, as in the names "devil's hair," "devil's thread," and "old man's beard." Presumably, that is because they climb over other plants, but it might be because of their irritant chemicals. Others are enchanted with the elongated, plumose stigmas, and colorful flowers. That view is indicated in names like "traveler's joy" and "Virgin's bower." Many have found the plants medicinal, dating from the 16th century herbalists including Leonard Fuchs (1542), William Turner in 1568, and John Gerarde ([1597] 1975).

Clematis baldwinii, endemic to the southeastern United States, was used by Seminoles to treat sunstroke (Small 1933). Cherokee used *C. virginiana* in "Green Corn" medicine, for stomach and kidney problems, and mixed with other plants for backache (Hamel and Chiltoskey 1975). Iroquois used it for venereal sores, kidney problems, and as a wash to induce dreams (Moerman 1998).

In addition to *C. virginiana*, Moerman (1998) listed eight other North American species that have been used from coast to coast. Rafinesque (1830) was among the first Europeans to record American uses for these plants. He recommended an oily liniment for curing the itch, and considered the plant, in minute doses, good for chronic rheumatism, palsy, and ulcers.

Porcher (1863) wrote of *C. crispa:* "This plant is ... employed in secondary syphilis, ulcers, porrigo, etc.; given internally with the powdered leaves applied to the sore. It acts also as a diaphoretic and diuretic ... it possesses the properties of the *C. vitalba*, which is a dangerous vegetable caustic, used as a substitute for cantharides, and applied to rheumatic limbs, and in paralysis and gout. The decoction of the root is alterative and purgative; and is also said to be valuable in washing sores and ulcers, in order to change the mode of their vitality, and to make them cicatrize. Shecut remarks that 'the Spanish or blistering flies are very fond of the *Clematis crispa*, and it would be well for medical gentlemen in the country to propagate the plant about their residences, in order to secure a constant succession of these valuable insects.' ... The American species are deserving of particular attention, and we would invite further investigation of them."

The reference to "cantharides" and "Spanish flies" are to the soldier beetles (Cantharidae), considered a counterirritant and aphrodisiac since the 14th century (Borror and Delong 1954, Gillott 1980). Alcock (1985) discusses the chemical in association with beetles and their use of it as an antipredator compound.

Porcher (1863) said of *C. virginiana* plants that they had "also a caustic property, and are employed internally as diuretics and sudorifics in chronic rheumatism; and externally, in the treatment of eruptions, and as vesicants ... a yellow dye may be extracted from both leaves and branches; the latter are sufficiently tough to make withs and fagots. The fibrous shoots may be converted into paper, and the wood is yellow, compact, and odoriferous, furnishing an excellent material for veneering." Hocking (1997) added that the species was diuretic, diaphoretic, and used to treat rheumatism and indolent ulcers.

Clematis contains acrid, toxic compounds, including anemonin, a dilactone derived from protoanemonin, that are highly irritating to skin and mucous membranes (Lampe and McCann 1985, Diggs et al. 1999). Rafinesque (1830) noted: "Bark and blossoms acrid, raising blisters on the skin; a corrosive poison internally, loses the virulence by cooking." The French (*herbe aux gueux*) and Spanish (*hierba de los pordioseros*) names referring to beggars are a reference to their past use to irritate the skin to simulate sores and induce sympathy from potential donors (Bown 1995). Ingestion of any part of extracts may cause vomiting of blood, severe diarrhea, and convulsions (Foster and Duke 1990).

Clethra

(Linnaeus adapted the classical Greek name *klethra* for an alder and applied it to this group)

Clethera alnifolia. a. Habit. b. Inner surface of sepal. c. Inner surface of petal with stamens. d. Young style and stigmas. e. Capsule with part of calyx removed. *Drawn by Vivian Frazier.* From Correll and Correll 1972.

Clethra alnifolia (alder-leaved)
[spiked, white]-alder (Carolinas)
puppy-tail [puppytail] (Carolinas)
Scheineller (shining elder, German)
soap-bush (Carolinas)
summer-sweet
[coastal] sweet-pepper-bush [coastal sweetpepper-bush]
white-bush

Linnaeus created the genus for these plants in 1753, although they had been discussed by Plukenet in 1696 and Catesby in 1731. Both Plukenet and Catesby called the trees *Alnifolia americana floribus pentapetalis albis in spicam dispositis* (The American alder-leaved plant, bearing five-petaled white flowers in spikes). So, Linnaeus translated the name "alder" into Greek. There are now 64 species in the genus, most of them in tropical America, but also found in Asia to Malesia (Mabberley 1997).

Moerman (1998) found no records of *Clethra alnifolia* being used by indigenous people. However, the Cherokee are known to have utilized *C. acuminata* as an emetic (specifically for bile). The bark was mixed with bark from wild cherry (*Prunus serotina*) to lower high fever, and a bark infusion of *Clethra* alone was used to treat bowel problems (Hamel and Chiltoskey 1975, Moerman 1998).

Root decoctions are used as a foot bath for "ground itch"; leaves and flowers are diaphoretic, stimulant, and antipyretic, and applied for pulmonary disorders. The leaves are used to treat inflammations and swellings (Morton 1974, Hocking 1997). Leaves contain considerable amounts of tannin (Porcher 1863).

Cnidoscolus

(Greek *knide*, nettle, *skolos*, thorn or prickle)

Cnidoscolus stimulosus. *Drawn by P.N. Honychurch.*

Cnidoscolus stimulosus (stimulating, an understatement!)
bull nettle ("bull" implying potent)
finger-rot (an exaggerated comment on the impact of the stinging trichomes, Carolinas to Florida)
fly-fly (Carolinas)
hashtapola [*hashtapotla, hatapusha*] (Choctaw)
spurge nettle ("spurge" is from Old French *espurge*, in turn based on *espurgier*, and Latin *expurgare*, to cleanse or purify from impurity;

the verb has been in English since about 1303 A.D.; the noun was in use by 1387)
stinging nettle (in use in English by 1525, although it was spelled "styngynge nettylles"; originally applied to European *Urtica*, later to American plants)
tread-softly (an admonishment to be careful of where one walks)

After the Austrian botanist Johann Baptist Emanuel Pohl (1782–1834) traveled and collected plants in Brazil, he returned home and published a book on his discoveries in 1827. That book, *Plantarum Brasiliae Icones et Descriptiones* (Descriptions and Drawings of Brazilian Plants), included one of his more stimulating finds, *Cnidoscolus.* Earlier, in 1813, André Michaux had named the southeastern United States species *Jatropha stimulosus.* It was not until 1845 that St. Louis physician George Engelmann (1809–1884) and Harvard professor Asa Gray realized that the Michaux species belonged in *Cnidoscolus.* There are now approximately 75 species in this American genus, with 4 in the United States (Kartesz 1994, Mabberley 1997).

Seeds and roots are edible (Morton 1968b) but barely worth the effort to obtain them. Murphee (1965) found people in the Panhandle making a tea to "give a man courage" (a sex aid). In the Carolinas, milky sap of the root was mixed with *Smilax laurifolia* and steeped in whiskey or gin to increase male potency (Morton 1974). Porcher (1863) wrote, "It might be employed like the nettle (*Urtica*), as a counter-irritant in epilepsies, and diseases requiring stimulating applications."

Coccoloba: Seagrape and Pigeon Plum

(From Greek *kokkos,* a berry, *lobos,* lobe or pod)

Coccoloba. *Coccoloba diversifolia* (left). *Coccoloba uvifera* (right). *Both drawn by P.N. Honychurch.*

The name kino sounds like (and is) a card game Mark Twain played in the 1800s while making notes to write *Following the Equator* (Twain [1897] 1992). However, kino is also the plant known in Florida as seagrape (*Coccoloba uvifera*). Apparently, the first usage of the word "kino" dates from 1757, when it was recorded as a medicine by the prominent patron of botanical exploration physician John Fothergill (1712–1780) in his *Medical Observations.*

The first kino used in Europe was "Mandingo kino" or "Gambia kino" from western Africa. Then, in 1774, it was recorded as *gummi kino.* Later, kino came to be a commercial name, applied to several tannin-rich resin-like products, from several species. 'Kino' was described as a substance of brittle consistency and dark reddish-brown color. The material was widely used in medicines, for tanning, as an astringent, and for dyeing cotton. That name was even translated into Spanish as *quino* (Colombia).

Some time later, materials called West Indian kino, or Jamaica kino, or American kino became known. That variety supposedly was taken from what we call seagrape (Florida to Panama). However, it probably was also taken from the pigeon plum (*C. diversifolia*) because few distinguish between those two species.

A more obtuse reference to the same astringent extracts began about 1839, when the material was called false rhatany. The name has also been applied to rattan vine (*Berchemia,* more details are there and under *Krameria*). False rhatany was taken from both *C. uvifera* and *C. diversifolia.*

The first tendency is, however, to compare the fruits of *C. uvifera* with the cultivated grape. Like true grapes (*Vitis*), fruits of *Coccoloba* are purple when ripe and hang in clusters. For example, in the Virgin Islands and Belize *C. uvifera* is simply "grape." In other English-speaking areas, it is seaside grape (Bahamas, Jamaica, Trinidad, Guyana), wild grape, or bay grape (British West Indies).

The comparison with grapes had been made when Captain Gonzalo Fernández de Oviedo y Valdés provided the first record of the seagrape in the early 1500s. He wrote: "The Christians give the name *uvero* to the tree the Indians call *quiabara.*" Subsequently, that indigenous name has been rendered *guiabara* and *buibasa.* Oviedo probably was using a Taino name because people still use *guayabón* (big *guiabara,* Dominican Republic) and *guaya-canejo* (Cuba) to refer to *C. diversifolia.* Many of the names for seagrape are also applied to pigeon plum, although there is little doubt that the people using the names know the difference between the two species.

Other people call the trees or their fruits *agraz* (unripe grape, Colombia), *papaturro* (sea grape, Honduras, Costa Rica), *papaturro extranjero* (foreign sea grape, Nicaragua), *uva* (grape, Yucatán, Veracruz, Dominican Republic), *uva caleta* [*uva de galeta*] (coastal grape, Cuba, Dominican Republic), *uva de* [*la*] *mar* (sea grape, Tamaulipas, Yucatán, Oaxaca, Puerto Rico, Dominican Republic), *uva de playa* (beach grape, Tamaulipas, Veracruz, Panama, Venezuela, Puerto Rico), *uvero* (grape tree, Tamaulipas, Panama, Colombia, Cuba, Puerto Rico), *uvero de playa* (beach grape tree, Tamaulipas, Veracruz, Panama, Costa Rica, Venezuela), *uva de la costa* (coastal grape), *uvero macho* (male grape), *uvilla* (little grape, Dominican Republic), or *uvita de playa* (little beach grape, Panama, Costa Rica). Pigeon plum is called *uva cimarrona* (wild grape, Dominican Republic), *uverillo* (little grape tree, Cuba, Puerto Rico), *uvilla de sierra* (wild [forest] little grape, Dominican Republic), and *uvilla* (little grape, Cuba, Dominican Republic).

French Caribbean areas also compare the sea grape with the grape. They call the seagrape *squinom raisin bord-de-mer* (wild sea grape) or *raisin* [*raisinier*] *du bord de mer* (edge-of-sea raisin tree, Dominica, Guadeloupe, French Guiana), and the fruits *raisin la mer* (sea raisin, Haiti) or *wezen* (raisin, Trinidad). Pigeon plum is called *raisin marron* (wild raisin, Haiti).

Dutch speakers also think the fruits look like grapes. These people say *druif* [*dreifi, druifi*] (grape, Dutch West Indies, Suriname), *dreifi de la mar* (sea grape, Suriname), *dreifi die lamán* (sea grape, Aruba, Bonaire, Curaçao), *zeedriefi* [*zeedruif*] (sea grape, Dutch West Indies, Suriname), and *zusterdruif* [*schuster-druif*] (sister grape, Suriname). These same people say that *C. diversifolia* is *dreifi sjimarón* (wild grape, Aruba, Bonaire, Curaçao) or *paaloe die lamán* (seaside tree, Aruba, Bonaire, Curaçao). No meaning has been found for *bessji die kalakoena* (Aruba, Bonaire, Curaçao).

Some names are unique to *C. diversifolia.* For example, *escobillo de monte alto* (tall, wild little broom, San Luis Potosí, Veracruz, Quintana Roo) is a reference to using the branches for brooms. Mountain grape distinguishes the more inland *C. diversifolia* from the coastal *C. uvifera.* However, unless *palo bobo* (crazy tree) refers to the sometimes complex branching pattern, it is difficult to understand what it may mean.

When I first arrived in Florida I found some seagrape jelly for sale and bought it. On the first taste, I decided that I had been suckered into spending money on apple jelly falsely labeled as seagrape. Only much later did I learn that the two are so similar in taste that few can distinguish them. Some names for seagrape confirming that similarity are *manzano* (apple tree, Sinaloa) and *manzana extranjero* (foreign apple).

Other people make different comparisons, some obvious and others obscure. These names include *aragueque* (from Aragua, Venezuela), *baya de praga* [*boga de praja*] (beach berry), *cumare blanco* (white big bed), hopwood (British West Indies), *Juan garrote* (John thick tree, Colombia), *kiiche* [*kii-che, niiche', niiche*] (*nii*, good flavor, *che'*, tree, Maya, Yucatán, Belize), *matora* [*matula*] (Arawak, Suriname), *micongo* (Colombia), *murta* (myrtle, Colombia), *nula* (Cuna), *papalón* [*papatón*] (big potato, Central America), and pigeonwood (British West Indies).

The name pigeonwood crosses over into *C. diversifolia*, which is also called doveplum (Florida, Puerto Rico), pigeon-plum (Florida), and plum-bush (Bahamas). In Spanish, that is rendered *fruta de paloma* (dove fruit, Cuba) or *uva de paloma* (dove grape, Cuba).

Surprisingly, there are comparisons with mangroves. Artist Mark Catesby in 1732 called it the "Mangrove Grape Tree." In Spanish areas seagrape may be *mangle de falda* (mangrove of the edge), while Dutch regions say *manggel die sabbana* (savanna mangrove, Bonaire, Curaçao).

Some names refer to the wood instead of the fruits. These people are impressed with the reddish color and say *C. uvifera* is *carnero* [*camere*] (meat-colored, Mexico), *corralero* (little coral tree, Colombia), horsewood (British West Indies), or *palo mulato* (mulato tree, Panama). People often do not distinguish *C. diversifolia* and call it red-wood (Dutch Antilles) or *gateado* (color of a cat, a reference to the streaked wood). Oviedo compared the wood to *madroño* (*Arbutus*, Ericaceae) because both use red, but said *C. diversifola* was stronger. In the early 1500s, people used the wood for houses, butcher blocks, and shoe lasts.

The Glades people used these plants because the pollen has been found in pre-Columbian coprolites from Fort Center on the western side of Lake Okeechobee (Hogan 1978). Moreover, fruits were among those of several plant species offered to Jonathan Dickinson and his party when they were shipwrecked off the Jupiter Inlet in 1699. So, the earliest known inhabitants of Florida used the fruits and most likely the wood.

As much as other people valued the fruits and other parts of the seagrape, there is confusion about whether the Seminoles of Florida used the species. They did, however, use the pigeon plum (*C. diversifolia*). Different individuals used the same names for both species, and it is unclear whether or not they distinguished between them (Sturtevant 1955). The name they apply is *bihcô:bî* (Mikasuki) or *ki:hołákko* [*ki:horákko*] (Creek), both meaning "big mulberry." Because the mulberry (*Morus rubra*) was a tree familiar to them in their southern Georgia and northern Florida homelands, their designation of *Coccoloba* suggests that they were simply comparing it with an old familiar plant.

Presumably, the Seminoles used the fruits of both species in the same way, although those of *C. diversifolia* need more treatment to be edible. Another common name for the pigeon plum is tie-tongue, an allusion to the abundant tannin in the fruits. That astringency is diminished by drying and rehydrating, and the Seminoles are said to have used that method (Morton 1968b). Fruits of both seagrape and pigeon plum have not only been eaten and made into jelly and syrups, but some have also been made into potent alcoholic drinks.

Wood from both species has been used as fuel and in building cabinets (Little and Wadsworth 1964). Both yield a red dye, colored largely by the tannins. That tannin content has made them useful in medicines, whether called "kino" or not. The roots also are rich in tannins and have been used to treat dysentery. Bark from at least *C. uvifera* has been used as a febrifuge. Seagrape leaves have served as plates or were pinned together with thorns and used as hats.

One of the first Spanish uses was as impromptu paper. Oviedo, writing in the early 1500s, recorded: "At the time the wars were going on in Hispaniola and the other islands and on Terra-Firme, as the Christians did not carry with them paper and ink, they used these leaves like paper. The leaves are thick and green ... and with a pin or sharp point one can write anything on them, from one end to the other ... Thus written upon, the leaves were sent by an Indian wherever the Spaniards wished them to go" (translation from Standley 1920–1926).

Although Oviedo told Spaniards in Europe about the seagrape, few other Europeans knew of his report. The oldest records Linnaeus had available when he named the species were two published in 1696 by Hans Sloane and Leonard Plukenet. Sloane called the plants *Prunus maritima racemosa, folio rotundo glabro, fructo minore purpureo* (racemose marine plum, with round glabrous leaves, fruits small and purple). Perpetuating Sloane's error, Mark Catesby also called the plants "plums," but Plukenet and Linnaeus knew better. Unfortunately, they really did not know what to make of the trees either. Plukenet called them *Uvifera littorea, foliis amplioribus fere orbiculatis crassis, americana* (seaside American grape, with leaves broad to almost orbicular and thick). Linnaeus in 1753 called seagrape *Polygonum uvifera* and wrote, "*Ex habitu hue retuli (flores non vidi), vel enim hujus generis vel promimi erit*" [From the habit, I retain the species in *Polygonum* (I have not seen flowers), because this genus has prominent nodes].

The generic name of seagrape and its relatives was first created by Patrick Browne in 1756. In his book, *The Civil and Natural History of Jamaica*, Browne introduced Europeans to many plants that were previously unknown. Browne created the genus *Coccoloba*, so Linnaeus simply transferred his own species to Browne's genus to create *C. uvifera*. Pigeon plum did not become known in Europe until 1769, when it was named by Swiss botanist Nicolaus Jacquin. Now *Coccoloba* contains 120 species, all American (Mabberley 1997).

Humans are certainly not the only ones using the fruits from seagrape and pigeon plum. The common name pigeon plum itself indicates bird usage, and avians also consume seagrape fruits. One of the first years I was in Florida, I was walking along the remains of a tropical hammock beside the ocean in Palm Beach County. I heard a rustling in the branches in front of me and paused to look. There, sitting on a seagrape branch, was a mockingbird (*Mimus polyglottos*). Something about the bird seemed odd; it was actually swaying unsteadily back and forth on the branch. Then, the gray and white bird reached out and ate an overripe seagrape fruit, and I realized that I had met a drunken Florida native.

Coccothrinax

(Named by Charles Sprague Sargent in 1899 with Greek, *kokkos*, berry, *thrinax*, a fan)

Coccothrinax argentata (silvery)
guano (palm, Taino, Dominican Republic); *palma de guano* (Dominican Republic)
latanier [*latangnier*] (*latania* is an indigenous word from Mauritius for a palm there; apparently, this native plant became confused with the one from Africa, Haiti); *latanier* [*balai, d'Afrique*] ([broom, from Africa] palm, Haiti)
miraguano (*mira* probably corruption of *lana*, ocean, *guano*, palm, Taino, Cuba)
palma coyau (palm heart, Haiti)
silver [palm, thatch] (Florida)
yarey (*yerén* [*lerén*], fruit, Taino, Haiti); *yuraguana* (*yerén* [*lerén*], fruit, *guano*, palm, Taino, Cuba)

After collecting plants in the Caribbean, Austrian professor Nicolas Joseph von Jacquin returned to Europe and published his book *Fragmenta Botanica* (Botanical Fragments) in 1801. In that publication, he described *Palma argentata*. It was the early 1900s when Cornell professor and horticulturist Liberty Hyde Bailey (1858–1954) moved Jacquin's species to

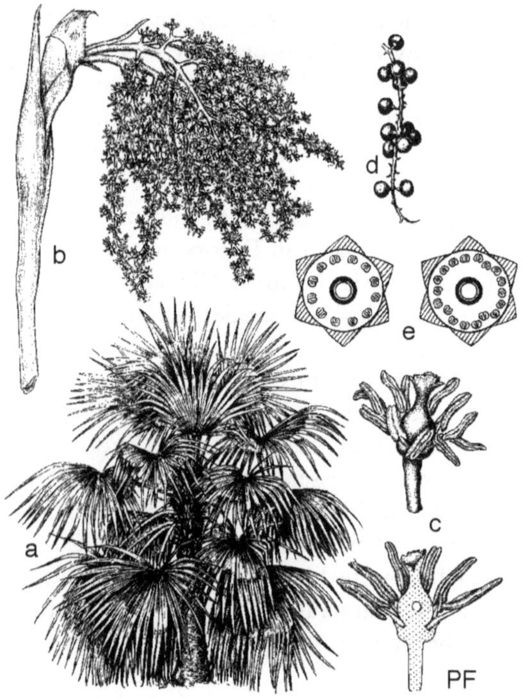

Coccothrinax argentata. a. Habit. b. Flowering inflorescence. c. Complete flower (above), flower longitudinally dissected (below). d. Branchlet with fruit. e. Floral diagram of two types of flowers. *Drawn by Priscilla Fawcett.* From Correll and Correll 1982.

Coccothrinax. There are now 47 Caribbean species in the genus (Mabberley 1997).

As with most palms, the heart is edible, as are the fruits (Morton 1968b). Oil from the seeds is used in Haiti to "*reavivar el olfato*" (renew the sense of smell) (Liogier 1974). Leaves are used to make brooms. Stems are hard and are used to make pilings in salt water and for fences; leaves are used to thatch houses and to make baskets, ropes, twines, and hats. Stems were used by the early Europeans, and perhaps by indigenous people, for turtle *craals* or corrals (Austin and McJunkin 1978). Scott Zona (personal communication 2003) says that the most important modern application is in making brooms.

Cocculus

(Diminutive of Greek *koccus* or Latin *coccum*, a scarlet berry, and *–ulus*, the diminutive; an old name, thought to be a source of dye, which turned out to be from cochineal insects on the berries rather than the fruits themselves)

Cocculus carolinus (of Carolina)
coral-[beads, berry, vine] [coralbead, coralberry]
hierba de ojo (eye herb, Coahuila, Tamaulipas)
margil (English?)
[Carolina, red-berry, red] moonseed (according to the OED 1971, the "moonseed" is a translation of the Latin and was first used by Philip Miller in the *Gardeners Dictionary* of 1739)
[Carolina] snailseed (compared with a snail because of the coiled seed)

The first European mention of what we now call *Cocculus* was in 1753. That year Linnaeus published *Species Plantarum* and wrote, "*Menispermum carolinum, foliis cordatis subtus villosus. Habitat in Carolina*" (*Menispermum* from Carolina, with cordate leaves villose below. Grows in Carolina). That is all most people know about this twining species.

It was not until 1817 when Augustin-Pyramus de Candolle created the genus *Cocculus.* Both genera belong to the Menispermaceae, a family notorious for species used in curare, fish poisons, sweeteners, and contraceptives (Mabberley 1997). The family is famous for containing sesquiterpenoids that act as poisons. However, little seems to be known about the North American members and Hardin and Arena (1974) simply say "may be poisonous." Perkins and Payne (1978) did not add much, but they make it clear that the related species *C. laurifolius* in southeastern Asia contains alkaloids cocculidine, coclalurine, coclamine, coclifoline, coclifoline, laurafoline, and trilobine. Of these, coclalurine is a curare-like compound. Hocking (1997) adds that it contains benzylisoquinoline derivatives. It seems likely that the species in the eastern United States contains many of the same chemicals, but no cases of poisoning have been recorded.

These vines grow in the heart of Muskogean territory in the southeastern United States. Plants range from North Carolina to northern peninsular Florida, west to Kentucky, southern Missouri, eastern Oklahoma, and Texas. Rhodes (1997) says they also grow in Mexico, maybe Coahuila or Chihuahua, but he does not say which state.

Given the poisonous nature of most members of the family, it would be best to heed Hardin and Arena's (1974) warning that they might be poisonous. Standley (1920–1926) said that the fruits were edible, but Correll and Johnston (1970) made it clear that they "had not tried them." Because the Choctaw used the berries to poison fish (Swanton 1946), eating them would not be advisable. Swanton's informant Jackson Lewis also pointed out that in catching fish, they poisoned a pool with berries but cut down bushes and piled them about it to keep livestock away and cautioned their people not to drink it. Eventually, when the Choctaw began raising European-introduced cattle and chickens, they stopped using the berries for beads because they were afraid they would kill those domestic animals (Swanton 1946).

The only other notation found is among the Houma who used the root to make a drink taken as a blood clarifier (Speck 1941). However, in the southwestern United States, *C. diversifolius* is used identically with *C. carolinus* (Hocking 1997). Brazilians use *C. filopendula* to promote menstruation and to treat liver and bladder afflictions and snakebite (Lewis and Elvin-Lewis 1977, Mors et al. 2000).

Collinsonia

(Named for Peter Collinson, 1694–1768, a London cloth merchant and naturalist who used his overseas agents to provide plants for his and his friend's gardens)

Collinsonia canadensis. From Britton and Brown 1897.

Collinsonia canadensis (of Canada)
archangel (named for St. Michael the Archangel, who held the highest rank among angels; the name has been applied to a number of mints since about A.D. 1000, presumably for their medical uses)
citronella (alluding to the strong lemon fragrance, resembling citron, *Citrus medica*)
Collinson's flower
[broadleaf, Canada] collinsonia (book names)
gravel-root (applied to *Eupatorium purpureum* by 1866; the name reflects treatment of kidney or bladder stones; "gravel" has been applied to urinary crystals since ca. 1400)
hardtack (figuratively "hard-fare"; a name applied to a ship-biscuit by 1841; probably derived from "hard-bread," which began to be used as a biscuit in the 1400s)
heal-all (anything considered a universal remedy, a panacea; in use by 1577 as a term for plants and other medicines)
horse balm (see horse-weed); horse-weed [horseweed] (explained by John Bartram in 1751 as "not only because horses are very greedy of it, but it also is good for sore gall'd backs"; Kalm added in 1753 that it had that name because "horses eat it in spring, before any other plant comes up")
knob grass [knobgrass, knob root, knobroot, knob weed, knobweed] (all names alluding to the knotty roots)
mountain-balm (akin to the name mountain-balsam or *Abies*, but alluding to use in external medicines, Virginia)
ox balm (see horse-weed)
rattlesnake-root (recorded by John Bartram in the 1740s, West Virginia)
richleaf; richweed (said to have been given because of the strong lemon fragrance; in use by 1830)
stone root [stoneroot] (two versions exist for the application of this name; one is that it was applied because of the stone-hard rhizome; the other is that it came to reflect use in treatment of kidney or bladder stones)

Collinsonia serotina (late-flowering)
citronelle (because of lemon scent, French)
horse balm
richroot
stone root

Collinsonia is an endemic North American genus of four species (Kartesz 1994, Mabberley 1997). The original specimen of *C. canadensis* was sent to Peter Collinson by John Bartram in 1735. Collinson sent the specimen to Gronovius and then it was used by Linnaeus in his *Hortus Cliffortianus* of 1738 where he coined the genus *Collinsonia.* The surveyor-general of New York, Cadwallader Colden (1688–1776), used the name *Collinsonia* in his *Plantae Coldenghamiae in provincia Noveboracensi Americes sponte crescentes* (Plants Collected by Colden in the Provinces Where They Grow in North America), published in 1743.

The Cherokee used *C. canadensis* on swollen breasts and as an emetic, and they used mashed flowers and leaves as a deodorant (Hamel and Chiltoskey 1975, Moerman 1998). The Iroquois applied leaves to stop headache, to treat diarrhea, rheumatism, the blood, boils, heart trouble, and kidney problems, and as a bath to strengthen children (Moerman 1998). John Bartram recorded that Conrad Weiser, his interpreter when he visited the Onondaga in 1743, had been told that the Onondaga used the plants to counteract rattlesnake bites (Berkeley and Berkeley 1982, Coffey 1993).

Rafinesque (1830) also commented on *Collinsonia.* He called it "richweed," and said indigenous people used it in a tincture to heal wounds. People in Appalachia make a tea from the plant and use it to treat headache, cramps, and indigestion (Krochmal and Krochmal 1968). Hocking (1997) considered it diuretic, diaphoretic, astringent, a tonic, and an expectorant, and useful in treating hemorrhoids.

Perhaps no distinction is generally made between species because *C. serotina* has some of the same names as *C. canadensis.* Hocking (1997) wrote that it was used as tea ("French tea"), and was medicinal (Hocking 1997).

These medicines are not without potential problems. Peter Kalm ([1753–1761] 1972) said that *C. canadensis* "has a peculiar scent, which is agreeable, but very strong. It always gave me a pretty violent head-ach [*sic*] whenever I passed by a place where it stood in plenty, and especially when it was in flower." Foster and Duke (1990) warn that minute doses of fresh leaves may cause vomiting. Hocking (1997) attributes those reactions to volatile oils, resins, and mucilates. Still, Duke et al. (2001) gave *C. canadensis* two plus marks, indicating it is as safe as a cup of coffee.

Colubrina: Snakewood

(From Latin *colubrinus,* like a serpent, alluding to the stems or stamens)

My former student, now a working anthropologist, had surveyed one region of the Big Cypress for archaeological sites used by indigenous people. In one hammock, he had found a tree that he did not recognize. The area was covered with historical and pre-European materials and had been used by more than one ethnic group, including the Glades and Seminole people. He suggested we visit the hammock to check on the unknown tree. As we walked across the wet prairie, mostly dry in this season, we talked about the area we were going to visit and some of the surrounding region.

Climbing up a slight rise into the hammock we had to dodge piles of cow dung and bones, because a herd of Seminole cows was roaming the fields and forests. We stopped at the tree my friend said had stumped him. At first, I too was at a loss as to its

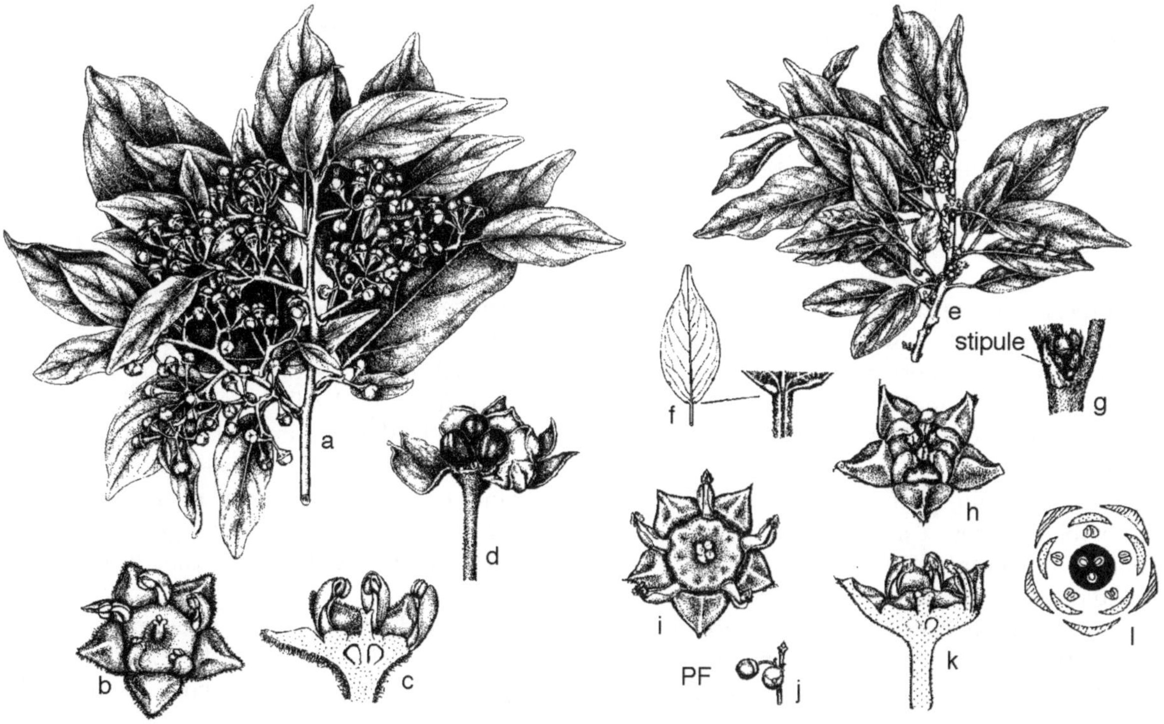

Colubrina. *Colubrina arborescens* (left). a. Fruiting branch. b. Open flower. c. Flower longitudinally dissected. d. Open capsule. *Colubrina elliptica* (right). e. Branch with flowers. f. Leaf outline and inset detail of marginal glands. g. Node, with persistent stipule and lateral flower cluster. h. Flower, in early stage. i. Flower from above, in late stage. j. Fruit. k. Flower in middle stage, longitudinally dissected. l. Floral diagram. *Both drawn by Priscilla Fawcett.* From Correll and Correll 1982.

identity. The branches were limber and the egg-shaped leaves were coated with red pubescence on the lower surface. I kept running through my mind the possibilities from this area southwest of Lake Okeechobee in Hendry County. Then, I realized what the plant was. I said to him, "Well, if I were in Everglades National Park, I would call this *Colubrina arboresens* or snakewood."

The other areas where I had seen this species were far to the south of this spot. According to the books, this species did not grow this far north (Little 1978). Obviously, this tree had not read the books because it was growing about 50 miles north of its known range. The species is tropical, and only within the temperature-moderating hammock forest where it grew could it withstand the freezes of this region.

Typically, this snakewood was known from southern Miami-Dade County on the mainland, the Florida Keys, the Bahamas, and the West Indies as far south as Barbados. Plants are also found in Central America, but they are said to have been introduced and primarily grown for shade from Yucatán through El Salvador and Honduras to Guatemala (Morton 1981). There are herbarium records of the species also extending south into Colombia, Ecuador, and Peru.

Several other species of *Colubrina* are in southern Florida, all on the northern fringe of their tropical ranges. Some of the species are so rare in the United States that they are considered endangered. *Colubrina arborescens*, *C. cubensis*, and *C. elliptica* are among those considered endangered within the United States (Coile 2000). Miami-Dade, Collier, and Monroe Counties are the regions where those endangered plants were formerly known. My student and I were looking at one of the endangered trees far north of those counties.

The name snakewood has a complicated history, and it is not clear whether or not the common names gave rise to the scientific, or vice versa. However, I suspect the name snakewood is much older. Apparently, the first recorded association with snakes was in 1760 with Nicolaus Jacquin. In his book *Enumeratio Systematica Plantarum* (Systematic Enumeration of Plants) he described plants now called *Colubrina ferruginea*. However, Jacquin did not call them that. Instead, he thought they belonged in another genus and called them *Rhamnus colubrinus*. The temperate genus *Rhamnus* was known to this Swiss botanist, and he did not realize that those he studied in the Caribbean were generically distinct. It fell to two French botanists, Louis C.M. Richard and Adolphe T. Brongniart (1801–1851), to create a new generic name for this red-pubescent plant and its relatives. This pair called these plants *Colubrina*, basing the name on the *Rhamnus colubrinus* of Jacquin.

No one is sure whether Jacquin meant to note the snakelike sprawling stems of the plants or their stamens (Quattrocchi 1999). There is nothing about the stamens that appears snakelike to me, but the stems do twist and turn in a way that might cause someone to think of a snake. Charles S. Sargent (1905) suggested that the name probably was given "on account of the peculiar twisting of the deep furrows on the stems of some of the species." Tomlinson (1980) agreed with Sargent and added that "the name was Latinized as *Arbor colubrina* (serpent tree) by early botanists."

Even today, the plants are called common snakebark (Bahamas), snake-bark (Puerto Rico), or snake wood (USA, Jamaica). The related species *C. elliptica* is called smooth snakebark (USA, Puerto Rico) or snake root (Virgin Islands). Curiously, no names in any other language make reference to snakes.

Some of the names compare snakewood with other known plants. For example, in Puerto Rico it is called *abeyuelo* [*abeuello, abuejuelo*]. That suggests that it resembles the *abey*, a tree known scientifically as *Jacaranda bahamensis* (Bignoniaceae). Both are favored herbal medicine sources. In Cuba, snakewood is sometimes called *achiotillo* (little achiote or *Bixa*) or *bijáguara* [*birijagua*] (*bijá* is *Bixa* in Taino). Both names compare the plants to *Bixa orellana* (Bixaceae). *Bixa* is the source of a medicinal red dye that is now used throughout the tropics. Snakewood has also been called wild ebony (Jamaica) and mountain ebony (Jamaica).

Several of the names indicate substitution for other plants, such as coffee colubrina (USA, Puerto Rico) and wild coffee (USA). As the plants have been used as soap substitutes, it is not surprising to find them being called scrubbing bush (British Antilles), soap bush (Bahamas), soap tree (Puerto Rico), and soapwood (Bahamas). Uses in treating digestive problems are surely shown by names like bitters (Bahamas) and *palo amargo* (bitter wood, Dominican Republic). Several species are said to be bitter, antiscorbutic, stomachic, tonic, laxative, a febrifuge, and a vermifuge (Roig 1945, Morton 1981). The bark contains saponins, particularly colubrinoside, and alkaloids (Colard et al. 1976).

Another curious name for the genus is nakedwood (USA). I have seen the plants in all seasons and even in droughts. Never have I seen them lose their leaves, so they could not be called "naked" because of that. The application dates to at least the late 1800s. Perhaps the genus was named because of plain wood. Sargent (1905) described the wood of *C. elliptica* as "heavy, hard, very strong, dark brown tinged with yellow, with thin light yellow sapwood." Almost the same description is given for *C. arborescens* by Little and Wadsworth (1964) and they commented that both species are used for posts. Formerly, *C. arborescens* was used as pilings because it was resistant to decay in water. Little and Wadsworth (1964) add that the specific gravity of *C. arborescens* is 0.7 and of *C. elliptica* is 0.8. That density probably explains why the plants were called *bois de fer* (iron wood, Haiti) and *bois fer blanc* (white iron wood, Haiti).

Other names for *C. arborescens* are black velvet (Jamaica), *corazón de paloma* (dove heart, Dominican Republic), *cuerno de buey* (ox's neck, Dominican Republic), *fuego* (fire, Cuba), and greenheart (Jamaica, Puerto Rico, Barbados). Of all the names, surely the most curious is *trejo* (Dominican Republic). This name seems to apply to a table shot in *trucos*, a game resembling billiards. Perhaps the balls are made from the wood of this tree. Black bead tree (Barbados) is the only name found that refers to making necklaces and other ornaments from the seeds.

Colubrina is a genus with rampant taxonomic uncertainty. Some sources (e.g., Standley 1922–1926) say that *C. arborescens* is not present in Central America and Mexico, but that the species there is actually *C. greggii*. These two names represent species so similar that they are not easily distinguished, at least in herbarium specimens. For example, the Missouri Botanical Garden database (TROPICOS) and the herbarium in Mérida, Yucatán, both have specimens listed as both species.

The situation is complicated by the fact that the "real" *C. arborescens* was brought into southern Mexico and Central America for cultivation long ago (Morton 1981). Whether wild or cultivated, there is a series of local common names for this species. Surely related to the cultivated plants is the name *chaquira* [*chaquirio*] (El Salvador), which alludes to an ancient vessel (*abalorlos*) that brought the Spanish to the Americas to trade with indigenous people. Perhaps the cultivation of the trees in Central America dates from the time of that ship.

In Chiapas, Mexico, the trees are called *cascalote* [*cascalata*] and *cascarillo*. Those names allude to the medicinal bark and also perhaps to its use to adulterate *cascara sagrada* (*Rhamnus purshiana*), a famous laxative from western North America. Since *cascara sagrada* is loaded with anthraquinones, as are several other members of the genus, perhaps *Colubrina* also contains those chemicals. This species is in turn related to *Rhamnus frangula* of Eurasia and the Mediterranean, which itself has a long history of purging inhabitants of the Old World (see *Rhamnus*). Probably related to *cascarillo* is the name *churumai* (Campeche); *churumo* is juice from something, alluding to a medical use.

Other names suggesting medical use include *guayabillo* (little guava, Guatemala), *pimiento-che'* (pimienta tree, Yucatán), and *tzecui* (Zoque, Chiapas). Both guava and canella remain useful for treating several maladies.

Two names recorded from Guatemala seem to be Mayan—*costex* and *coxté*. Because dialects of that language vary considerably the meaning cannot be certain, but an apparent cognate appears in the Maya of Yucatán. The infinitive of that word is *koxok* (to tie with cord). To *koxtik* is to tie something, and *k'oche'* is a bed or platform on which ill or deceased individuals are carried. It seems the flexibility and resistance of the wood make it useful in constructing these beds and also in constructing the framework for roofs on buildings covered with palm leaf thatch. Those uses do not seem to be recorded elsewhere for the genus, and they illustrate the ethnobotany often hidden in common names.

Colubrina: Soldier-Wood or Mabí

(See preceding entry for etymology)

Of all the Caribbean region members of *Colubrina*, the most famous is *C. elliptica*. That species is usually called *mabí* [*maabee, mabie*] in Puerto Rico, Haiti, and the Dominican Republic. There are variants of that name in several places, including *mabee tree, mambee* (Guadeloupe), *maubee* (Virgin Islands), *mawbie* (Lesser Antilles), *palo* [*de*] *mabí* (Puerto Rico), *mabit* (Haiti), and *bois mabí* (*mabí* tree, Haiti, Dominican Republic, Guadeloupe). Everything we know about that common name indicates a Caribbean origin, and it is a Taino word. A comparison with other Arawakan words is instructive. For example, "remedy or medicine" in Arawak is *ibihi*, and in Karina, the Cariban language spoken in Suriname, it is *e:pit*. In Island-Carib, medicine is *ibíe*. All those words are close enough to *mabí* that they could be cognates.

Anyone who has visited the island of Puerto Rico has seen stands along streets where *mabí* is sold. The bark is steeped in water to make a cooling drink, *mabí* or *mabí champan*. Alternative common names for this drink are *amargo* (bitter, Dominican Republic) or *palo amargo* (bitter wood, Dominican Republic).

Inhabitants of the Caribbean considered this beverage a tonic and antidote for indigestion and dysentery. Coll y Toste (1972) comments, "*Su hoja sirve en Puerto Rico para hacer una bebida refrescante, tónica, de uso generalizado. Viene a ser la cerveza criolla*" (Its leaf serves in Puerto Rico to make a refreshing, tonic beverage of general use. It is becoming the Creole beer). The plants have many other uses.

Because the bark is resinous, astringent, and bitter, a decoction is applied in Mexico to pimples and itching skin (Morton 1981). In Cuba, it is taken as a febrifuge and a remedy for liver and pulmonary diseases, dysentery, gonorrhea, and syphilis (Roig 1945). The leaf decoction is taken in Cuba as a treatment for fever, diarrhea, and dysentery. That preparation also serves as a vermifuge and an emmenagogue, and it is applied to ulcers of the gums. Names associated with the treatment of diarrhea include seaside buckthorn and wild guava (Cayman Islands).

A factory on Curaçao made *mabí* beverage, and Morton (1981) gave details of its manufacture. Bark is shipped to the island from farther north. This bark is brewed with cloves and brown sugar, and the froth is skimmed from the surface. A "starter" (small portion) from a previous batch is added to the new preparation to encourage fermentation. The decoction is left standing for 1 day, skimmed again, and then bottled but not capped. If closed, the bottle will blow off the cap because of the accumulated fermentation products. The product is much appreciated locally as a refreshing drink and a bitter tonic, as it is farther north in the Caribbean.

Few chemicals in the plant have been studied, but it is known that the leaves are high in tannin and the bark is high in saponins. A major component is the saponin mabioside (Seaforth et al. 1992, Oulad-Ali et al. 1994). The bark also contains a bitter chemical called colubrinin. Surely, the species contains colubrinol and colubrinoside, which have been found in related species (Wani et al. 1973, Colard et al. 1976a, b).

The name soldier-wood (USA) is perplexing. The combination of the word "soldier" with plants and animals has been made since at least the first part of the 1800s, but the other cases recorded are plants that have some red or purple coloring. The oldest instance with animals is soldier-insect, which dates from 1699. Other names applied to animals are from the 1800s. For example, there is soldier-beetle (1855), soldier-ant (1857), soldier-bird (1857), soldier-fish (1880), and soldier-moth (1880). The first records of soldier-wood were from the 1820s and applied to the legume *Calliandra purpurea.*

Sometime between the 1820s and the early 1900s, the name soldier-wood was applied to *Colubrina elliptica.* The fruit is the only thing about this plant that is red or purple and that would cause people to use the term soldier in the name. Perhaps that is the reason for the name. On the other hand, could it be a reference to medical use by soldiers?

There seem to be three basic uses of *mabí* in addition to the drink—soap, wood, and charcoal. In Mexico, the plants are called *amole* (soap, Náhuatl, San Luis Potosí). There and throughout the range of the species, from southern Florida, the Bahamas, West Indies, Veracruz to Yucatán, Mesoamerica, and northern Venezuela, the leaves are a soap substitute (Morton 1981).

Several common names refer to the wood. These include *bois costiere* (coastal tree), *bois de fer* (iron wood, Haiti), *palo de arco* (bow tree), *palo de hierro* (iron wood), *madera de hierro* (iron wood), and *guitarán* (big guitar, Puerto Rico). The references to iron allude to the hardness of the wood, which is also heavy, with a specific gravity of 0.8. The wood has the reputation of being durable and has been used for posts, among other things (Little and Wadsworth 1964).

Another name that is somewhat obscure is naked-wood (USA, Puerto Rico). Tomlinson (1980) thought this name was given because the bark of older trees peels off in strips. However, the name may allude to the wood color since there are similar references in other languages. In Puerto Rico, the plants sometimes are called *catire.* That word is a name for a pale-skinned person resulting from a cross between a white and mulatto. In Cuba, it is *jayajabito* [*jayajabico*]. Although the etymology is uncertain, it may be a combination of Taino *yaya* and *jabiya* (*Hura crepitans*). *Yaya* is the Taino common name of annonaceous *Oxandra lanceolata* and also colloquial Cuban for "become pallid." Thus, it would be the pale *Hura.*

The Maya of Yucatán make the most vivid allusion with *sacna-che'* [*sak-nak-che'*] (*sak,* white, *nak,* intestines, *che',* tree). That name is given because the *Colubrina* climbs through neighboring plants and the stem is the color of intestines.

Names such as *carbonero* (charcoal tree, Cuba), *carbonero de costa* (coastal charcoal tree, Cuba), and *palo carbonero* (charcoal tree) indicate the hard wood is made into charcoal. Charcoal is often in demand as fuel in many parts of the Caribbean, Mexico, and Central America.

To avoid confusion, the correct scientific name of the *mabí* is *Colubrina elliptica.* That name is based on *Rhamnus elliptica* given to these plants by Swedish professor Olaf Peter Swartz in 1788. Unfortunately, another name (*Ceanothus reclinatus*) was given to the plants that same year by the celebrated French magistrate and botanist Charles-Louis L'Héritier de Brutelle (1746–1800). Because of confusion or misinterpretation, L'Héritier's name was thought to be correct. The name used in the older literature is *Colubrina reclinata.* In some sources, the two are considered distinct species. It was only in 1958 that George K. Brizicky (1901–1968) and William L. Stern realized this complication and made the name *C. reclinata* synonymous with *C. elliptica.*

This *Colubrina* has the most restricted range of the species in Florida, being confined to small areas of hammocks in southern Miami-Dade County and nearby Monroe County. To see the species it is best to go to preserved areas on upper Key Largo. There, or on one of the wilder sites like Adams Key in the Biscayne National Monument, it is possible to find these trees reaching 50 feet in height. Their dark bark flakes off in large patches, and the inconspicuous yellow flowers turn into orange-red fruits. Those fruits are easily noticed against the green of the canopy by a variety of birds.

Given its restricted range in Florida, and the fruits that are probably dispersed by birds, we must wonder how the trees came to be there. Perhaps the seeds were carried by birds since they flower and fruit with little regard to season. However, knowing the value the Caribbean people place on the plant as a medicine and beverage, one has to wonder. There was a long history of Cuban and Bahamian plantations along the coasts of southern Florida during its early history. Could people on one of those plantations have brought the plants for local usage?

Commelina

(Commemorating three Dutch botanists of the name Commelin or Commelijn; Linnaeus said that the two blue petals of the original species reminded him of Jan, 1629–1692, and his nephew Kaspar [Caspar, Casparus], 1667–1732; the small white petal reminded him of a third relative who died young)

comméline (French)
Himmelsauge (heaven vessel, German)

Commelina erecta (upright)
bluebird (a creole name, Nicaragua)
canutillo (widespread Spanish name for the genus, maybe "little cane," from *caña*)
[white-mouth] day flower [whitemouth dayflower] (alluding to the duration of each flower, USA); dew-flower
espuelitas (little spurs, Coahuila)
hierba de pollo (chicken herb, Texas)
little bamboo (Belize)
mataliste [*mataliz, matalís*] (maybe from *mata*, plant, *lisa*, smooth, used for several members of the Commelinaceae, Tabasco, El Salvador)
okí ahissí [*oybá ahissí*] (water [*oybá,* literally "rain"] medicine, Koasati)
pah-tsá (*pah*, something bitter, *tsá*, to mix in, perhaps because the *Commelina* is mixed in with maize dough to aid fermentation in the preparation of *pozol*, Maya, Yucatán)
tamakusi (Carib, Suriname)
utek' (Huastec, San Luis Potosí)
x-habul-ha (*ha*, water, *bul*, submerge, Maya, Yucatán)
ya'ax-ha-xiu (*ya'ax,* green, *ha*, water, *xiw*, herb, Maya, Yucatán)

Commelina erecta. *Drawn by P.N. Honychurch.*

Commelina virginica (of Virginia)
[Virginia] day-flower [dayflower] (R. Holme wrote in 1688, "The Virginian Spiderwort ... may be called the Day Flower, for it opens in the day and closes in the night")
dew-flowers (maybe a comment on the nectar within the bracts)

When Linnaeus created the genus *Commelina* in 1753, he included nine species, although not all of those are now retained in the genus. None of those he knew is native to Europe; they come instead from the Americas, Africa, and Asia. The genus now has about 170 species growing in the tropical and warm parts of the world (Mabberley 1997).

P.P. Sheehan in 1912 (in von Reis and Lipp 1982) and J.K. Small (1933) were told that the Seminoles use the sap of *C. erecta* to soothe irritations even though it contains oxalic acid crystals that may be highly irritating to mucous membranes (Lampe and McCann 1985). Sturtevant (1955), Bennett (1999), and Snow and Stans (2001) were unable to find them still using the plants. The Huastecs of Mexico use the water trapped in the inflorescence bracts for eye problems (Alcorn 1984), and the plant is used in a similar way in Bolivia (von Reis 1973) and Belize (Balick et al. 2000). Kimball (1994) thought the Koasati formerly used it in tanning leather.

There has been so much confusion between *C. elegans* and *C. virginica* that it is likely both have been used similarly. A decoction is considered diuretic and is used for intestinal problems and to staunch the bleeding of surface wounds (Hocking 1997).

Conocarpus: Buttonwood

(From Greek *konos*, a cone, and *karpos*, fruit)

Conocarpus erectus*. Drawn by P.N. Honychurch.*

When I first arrived in Florida from the midwestern city of St. Louis, Missouri, I had read about "driftwood" and seen it for sale in shops. Those gnarled pieces of wood sold for high prices in the Midwest, and it had been something of a mystery to me where the wood originated. Not long after arriving in the state, I made a pilgrimage to the Florida Keys, where the answer to that old mystery was revealed. In several places along the road south, there were shops with stacks of "driftwood." I even found men in pickup trucks harvesting the wood from the edges of mangroves—driftwood and buttonwood were the same *Conocarpus erectus.* On one occasion, my class and I helped a man extract his truck from the mangrove mud where he had been stuck for some hours. The bed of his truck was loaded with several hundred pounds of buttonwood root systems and stumps. We never learned exactly what the harvesters were paid for the products, but it was markedly less than the selling price in St. Louis.

In the Romance languages, the name of a flower bud is a *botón* (Spanish), *botão* (Portuguese), *bottone* (Italian), or *bouton* (French). These are supposedly based on Latin *bottare,* meaning "to thrust or put forth." That sense is much older than in English where it came to mean that about A.D. 1400. Older in English is the meaning of a "fastener." That idea has been traced to about A.D. 1340. In English, linking the word "button" with a plant name has been traced to about 1598.

Still, when someone now says "buttonwood," are they talking about the fastener or the globose cluster of flower buds? Few would even be aware of the connection between buds and buttons.

In the past, buttons were not always flat objects. Button shapes have varied over the centuries, with the older kinds tending to be globose; those of more recent derivation tend to be flattened. Just when the tropical trees now called buttonwood received that name is unknown, but the usage surely must go back to the early 1600s. Yet, looking at the names given to the plants before we began calling them *C. erectus* in 1753, one might not suspect the antiquity of the common name.

Probably the first reports of the trees in the New World were published in 1696. That year Leonard Plukenet called them *Alnus maritima myrtifolia coriariorum* (the oceanside alders with leathery myrtle-like leaves). The same year, Hans Sloane, in his book on Jamaica plants, also called it an alder, but Linnaeus realized that the plants were not that similar to *Alnus.* To make a unique place for them, Linnaeus created a new generic name, *Conocarpus.* The fruit cluster is globose, but Linnaeus dissected out the individual parts and saw that each flower in the group had produced a cone-shaped mature ovary. Presumably, he was talking about the upright stance of the trees when he named the species *erectus.* By modern international rules of nomenclature, all scientific names must be based on a "type" (permanent reference point for identification). Sloane's drawing (plate 161, figure 2) published in 1725 in his second book, *A Voyage to the Islands Madera ... and Jamaica,* is the type for buttonwood.

People who use common names are often not aware of scientific appellations. Most people either allude to the buttonlike nature of the flower and fruit clusters, or to the fact that the trees are mangroves, or both. Some of the names noting the buttons are black buttonwood, *botoncahui* (button tree, mixture of Spanish, *botón*, and Náhuatl, *cahuitl*, Sinaloa), *botoncillo* (little button tree, Puerto Rico, Dominican Republic, Veracruz, Chiapas, Yucatán, Belize, El Salvador, Venezuela), button alder, button bush (Belize), button mangrove (USA), button tree, buttonwood (Florida to Virgin Islands, Panama), *palo botón* (button tree), silver buttonwood (Florida), and white buttonwood (USA). Those comparing them to buttons and mangroves say *mangle [de] botón* (button mangrove, Puerto Rico, Cuba, Panama), *mangle botoncillo* (little button mangrove, Campeche, Venezuela, Puerto Rico), *mangle botón* (button mangrove, Puerto Rico), or *mangue de botão* (button mangrove, Brazil). Surely, *mangle garbancillo* (little bean [garbanzo] mangrove, Colombia) also belongs in this group.

Those comparing the trees with the other mangroves often use colors. Some think the trees should be white and say *estacahuite* [*estachahuite*] (*iztac-cahuitl*, white tree, Náhuatl, Oaxaca), *manggel blanko* [*mangel blancu, manguel blancu*] (white mangrove, Aruba, Bonaire, Curaçao), *mangue branco* (white mangrove, Brazil), white mangrove (Belize), and *witt mangel* (white mangrove, Dutch Antilles). Others think the trees should be gray and call them *grize manguel* (gray mangrove), *grijze manggel* (gray mangrove, Aruba, Bonaire, Curaçao), *mangle cenizo* (ash mangrove, Oaxaca), *mangle gris* (gray mangrove, Guadeloupe, Martinique), *manglier gris* (gray mangrove, Guadeloupe, Martinique), and *paletuvier gris* (gray mangrove, Guadeloupe, Martinique). Still others think they should be black and say *mangle negro* (black mangrove, Costa Rica, Oaxaca) or *mangle prieto* (black mangrove, Dominican Republic, Tabasco).

Some make no reference to colors but simply say *mangel* (mangrove, Dutch Antilles), *mangle* (mangrove, Sonora, Campeche, Guerrero, Dominican Republic), *manglier* (mangrove, St. Lucia), *mangue* (mangrove, Brazil), *manguel* (mangrove, Dutch Antilles), or *paletuvier* (mangrove, Haiti).

Other names note that the trees are mangroves but indicate some association other than color. There is *mangle jelí* or simply *jele* (Ecuador), *mangle lloroso* (crying mangrove, Venezuela), and *mangle torcido* (twisted mangrove, Panama). The most complex of the group is *mangle marequito* or just *marequito* [*mariquito*] (Costa Rica). The word *mariquito* has several meanings. Three of the meanings seem to have nothing to do with these trees. Those are a black and red hemipteran, locally called *vaca de San Antón*; a climbing parrot; and in Cuba, *miel con almíbar con queso fresco* (honey with carmel with fresh cheese). Finally, it is simply translated as "mangrove."

The Seminoles of Florida claim uses for only the buttonwood, although they have named all the other mangroves (Sturtevant 1955). The Miccosukee call this tree *ahilo:chiskoposki*, and their relatives the Creeks say *tolastisilopócki*. Both names mean "small leaved black tree" (*eto*, tree, *esse*, leaf, *lopocke*, small, *lvste*, black). The Seminoles used the plants to dye buckskin and the wood in fires to barbecue meat. All who know the plants consider the wood of high quality for fuel.

Down on the Yucatán peninsula and Belize, the Maya also have names for the trees. Instead of making the color comparisons other people use, they sometimes say trees are the *xkan-che'* [*canche, kanche, k'ank-che', k'ank-ank-che', k'an-inche', x-kanche'*] (*ka'an*, yellow, *che'*, tree). An equally common name they use is *tabche'* [*taabche', x-tabchém taabche, tabche*] (*taab*, fixed or steady, *che'*, tree). The Mayans give the tree this name because they find the trees well established by their roots in muddy soils of the mangroves.

In Mexico, some say the trees are the *madre de sal* (mother of salt, Oaxaca, Chiapas). That name apparently alludes to the deposition of salt on the bases by seawater evaporation. In Panama and Colombia, buttonwood is called *saragoza* [*zaragosa*] (*Zaragosa*, the capital of Aragon, Spain). No reason for comparing the trees with that city has been found. Finally, in Cuba buttonwoods are called by their Taino name, *yana*.

Buttonwood has an extensive range. It is found in Florida, Bermuda, the Bahamas, West Indies, on both coasts of northern Mexico, south to Panama, and also on the coasts of South America. This is one of the comparatively few species that also occurs in West Tropical Africa. Pond apple (*Annona glabra*) is another of those species (See *Annona:* Pond Apple), and many think that the distribution of both species was by seawater dispersal.

The wood has been used for rafters, boats, and cabinets. Posts made from buttonwood are durable in the soil, whether wet or dry. Leaves and bark have been used throughout the range for tanning leather. In the Virgin Islands, buttonwood has been used as a fish poison (Petersen 1974). In Mexico, the bitter bark is used as an astringent and tonic medicine; it is applied to prickly heat in the Bahamas (Martínez 1969, Morton 1981). Brazilians take the fragmented bark and leaves and boil them together, strain them, and give them sweetened to weak, anemic, and convalescing individuals (Morton 1981). Cubans have given the dried powdered bark in a more palatable beverage to reduce fever (Roig 1945).

Charcoal making has been the most common use for buttonwood throughout its range in the Americas. Although few people have seen them, there were formerly great mounds in southern Florida where people had made charcoal from the wood of buttonwood (Tebeau 1955, 1966, 1968). The last charcoal mounds I found were in the Flamingo area of Everglades National Park and on Chokoloskee Island in Collier County. Those mounds in Collier County have now been replaced by trailer courts and houses.

Conoclinium

(From Greek *konos*, a cone, and *klinion*, a little bed, referring to the receptacle)

Conoclinium coelestinum. a. Habit. b. Flowering head. c. Receptacle with one flower. d. Anthers. *Drawn by Vivian Frazier.* Correll and Correll 1972.

Conoclinium coelestinum (sky blue, the flowers)
(= *Ageratum conyzoides*)
blue mistflower
bouton blanc (white bud, Dominica)
celestina [*azul, blanca*] ([blue, white] heavenly, Cuba)
herbe a bouc [*a femme, à la vierge, à pisser, aux sorciers*] (billy goat [woman's, Virgin's, urination, sorcerer's] herb, Dominique, Martinique)
mastruço (from Latin *masturtiu*, the crucifer; comparison obtuse, Brazil); *mondastro* [*mentastro*] (Puerto Rico)
petit pain doux (little sweet bread, Dominica, Dominique, Martinique)
rompezaragüey (shoe breaker, same name given to *Vernonia menthaefolia*, Dominican Republic)
yerba de cabro [*cabrio, chivo*] (goat herb, Puerto Rico, Dominican Republic)
zeb a fam (woman's herb, French Antilles)

Because of extensive parallelism and convergence, generic definition in the Asteraceae has long been problematical. Linnaeus put this species in *Eupatorium* in 1753, as had Plukenet, Dillenius, and Gronovius before him. De Candolle put it in *Conoclinium* by in 1836, and even then, some did not accept the disposition. However, as now understood, *Conoclinium* is a genus of three species endemic to the eastern United States and nearby Mexico (Mabberley 1997). Like many plants in *Eupatorium*, this one has a long history of medicinal use.

Bouton blanc is used in Dominica as a tea for colds or simply as a cooling tea (Honychurch 1987). Formerly the herb was widely used as a diuretic, against coughs and colds and dysmenorrhea, to induce abortion, and as a bath for skin eruptions (Roig 1945, Ayensu 1981, Honychurch 1987). Plants contain alkaloids, cyanogenic glycosides, and volatile oils such as eugenol, coumarin, and decoumarol (Ayensu 1981).

Conopholis

(From Greek *conos*, cone, and *pholis*, scale, a reference to the bracts on the flower and fruit clusters making them resemble pine-cones)

Conopholis americana. From Britton and Brown 1898.

Conopholis americana (of America)
cancer-root (by 1546, there were references to herbs called "cancer" or "herbe cancer"; John Gerarde's *Herball* of 1597 mentioned "cancerwort," which apparently referred to *Linaria* or

Veronica; however, "cancer-root" in association with plants seems to have appeared in 1714, and was clearly associated with this species by 1856)

squaw-root [squawroot] (apparently first published in association with "a medicinal plant put up by the Shakers," in 1848. By 1856, Asa Gray's *Manual of Botany* identified both "squaw-root" and "cancer-root" with *C. americana*)

These plants belong to the family Orobanchaceae, partly defined because many members are saprophytic or parasitic. *Conopholis americana* grows on the roots of trees, much like its relatives *Epiphagus* and *Orobanche.* There are only two species of *Conopholis*, and both grow in the southeastern United States.

Although Europeans had never seen *Conopholis* when they arrived in the New World, they were familiar with *Orobanche* (which see), as there are 45 species in Europe (Mabberley 1997). In fact, Linnaeus first named the Florida species in 1767 and called it *O. americana.* It was not until 1825 when Carl Friedrich Wilhelm Wallroth (1792–1857) created *Conopholis americana.* Some 19 years later, F.M. Liebmann described the second species, *C. alpina*, from Puebla, Mexico. Now, *Conopholis alpina* (= *C. mexicana*, = *C. alpina* var. *mexicana*) is known to grow from Texas, New Mexico, Arizona, and Mexico south to Panama (Correll and Johnston 1970, Mabberley 1997).

Although no records have been found of *C. americana* being used by indigenous people, there are records that *C. alpina* was used. Moerman (1998) recorded that the Keres near Albuquerque, New Mexico, used it to strengthen weakened tuberculosis patients. Their relatives in the same area, the Jemez, used it in magic to make runners faster.

Both Hocking (1997) and Mabberley (1997) note that *C. americana* is medicinal, and that it has been used as an astringent. The common name "squaw-root" perhaps indicates that the settlers learned the use of these herbs from women in eastern tribes. As the Shakers are known to have learned much of their herbal lore from local tribes, that interpretation becomes more likely. Whether or not the name "cancer-root" suggests use in treating malignant sores is uncertain.

Conyza

(Latin *conyza*, and Greek *konyza*, a flea, a name originally used by Aristotle, 372–287 B.C., Nicaner, Theophrastus, 372–287 B.C., and Dioscorides, fl. A.D. 40–80, for two kinds of European composites; Christian Freidrich Lessing, 1809–1862, reapplied the name to these plants)

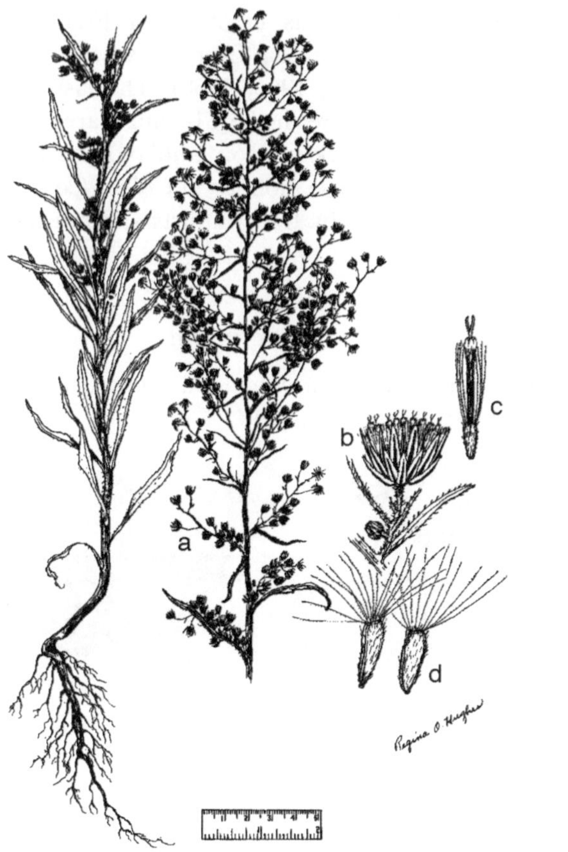

Conyza canadense. a. Habit. b. Flower head. c. Disk flower. d. Achenes. *Drawn by Regina O. Hughes.* From Reed 1971.

Conyza canadense (from Canada) (= *Erigeron canadensis*)

atakło:lasti [*atakro:lásti*] (*takłî*, bush, *lvste*, black, Creek); *takłô:cî* [*takrô:cî*] (black bush, Mikasuki); *vtakłv lvste* [*vtakrv lvste*] (*vtakłv*, plant, *lvste*, black, Muskogee)

atsil-sun'ti (fire-making material; *ajila*, fire, Cherokee)

bitterweed (New Mexico); butterweed (corruption of "bitterweed"?, USA)

canhlo'gan was'te'mna iye'cece (resembling sweet-smelling herb [*Ambrosia*, which see], Lakota)

colt's tail (New Mexico); horsetail (USA); [smooth] horseweed (USA)

fleabane (repels fleas, New Mexico)

fox tail (Dutch Antilles)

gababi'kwuna'tig (knotted tree, Ojibwa)

ha'mo u'teawe (leaf ball flowers, Zuni)

hierba de burro (donkey herb, Sinaloa)

no'sowini (sweat, because it is used in the sweat bath, Meskwaki)

pazotillo (little skunk feces or little *Chenopodium ambrosioides*; from Náhuatl *epatl*, skunk, *tzotl*, filth, New Mexico)

pride weed (New Mexico)

German physician Lessing created the genus *Conyza* in his *Synopsis Generum Compositarum* (Synopsis of Genera of Compositae) published in 1832. Most people continued putting this species in *Erigeron* until Arthur J. Cronquist (1910–1992), an influential student of the family Asteraceae who worked at New York Botanical Garden, insisted that it was distinct in 1943. Now the genus contains about 60 species that grow in temperate and warm regions around the world. Europe had no native species, but this one and another are naturalized there (Mabberley 1997).

People throughout the range of the species use it for medicine to treat a variety of ailments (King 1984, Kindscher 1992, Vásquez and Jácome 1997, Moerman 1998). Curtin (1947) recorded a number of uses for the herbs in New Mexico. The Seminoles told Sturtevant (1955) they used the plant to treat coughs and colds. They also told Bennett (1997) that it was still in use but would not explain the application. The Houma used it to treat leukorrhea (Speck 1941).

Corallorhiza

(From Greek *corallion*, coral, and *rhiza*, root; the name was applied to these orchids in 1755 by Abraham Gagnebin, 1707–1800)

Corallorhiza odontorhiza. From Britton and Brown 1896.

coral-root (translated from the Greek name, apparently not in English until ca. 1845)
freumh corail (coral root, Gaelic)
Korallenwurz (coral root, German); *korallrot* (coral root, Norwegian)
racine de corail (coral root, French)

Corallorhiza odontorhiza (toothed-root, from their shape)
chicken-toes
[autumn, late] coral-root [coralroot]
crawley root
dragon's claw [dragon-claw] (used for the orchid by 1832)
fever-root (in use by 1884 for the root-parasitic *Pterospora andromeda*, Ericaceae, and for *Triosteum perfoliatum*, Caprifoliaceae; use for *Corallorhiza* is either a confusion or comparison with *Pterospora*); dever-root [deverroot] (a corruption of "fever-root"?)
turkey claw

Willdenow named these terrestrial orchids *Cymbidium odontorhiza* in 1805. Shortly afterward, French botanist Jean Louis Marie Poiret (1755–1834) moved the species to *Corallorhiza*. Currently, the genus contains 15 species that grow in northern temperate regions around the world. One species is native to Europe (Mabberley 1997).

These orchids have no chlorophyll (Mabberley 1997). Instead, they rely on their relationship with the fungi on their roots for survival (mycotrophs). The plants are typically a reddish color, sometimes marked with darker blotches.

No records have been found of indigenous people using *C. odontorhiza*, but farther north and west *C. maculata* was used by the Iroquois, Navajo, Paiute, and Shoshone (Correll 1950, Moerman 1998). Correll (1950) speculated that it was used to make a tea to build up blood because of its red coloration.

Other indigenous uses of *C. maculata* are as a diaphoretic, febrifuge, and sedative. Those are the same as for non-indigenous application of *C. corallorhiza*, and it seems likely that both species were used similarly. Ellingwood and Lloyd (1898) and Felter (1922), prominent Eclectic physicians in their time, used the Florida species as a diaphoretic, for fever, and as a sedative. Ellingwood and Lloyd (1898) considered it a diaphoretic *par excellence*.

These are endangered plants in Florida (Coile 2000). Because of their special habitat requirements, they seem to be declining throughout their range.

Cordia: Blood-berries and Dog Tits

(Named by Linnaeus in honor of the German botanist and pharmacist Valerius Cordus, 1514?–1544, one of the fathers of pharmacognosy)

In the early 1970s, blood-berry (*Cordia globosa*) was one of the rarest shrubs in the pinelands of Miami-Dade and Monroe Counties. Then, the butter-

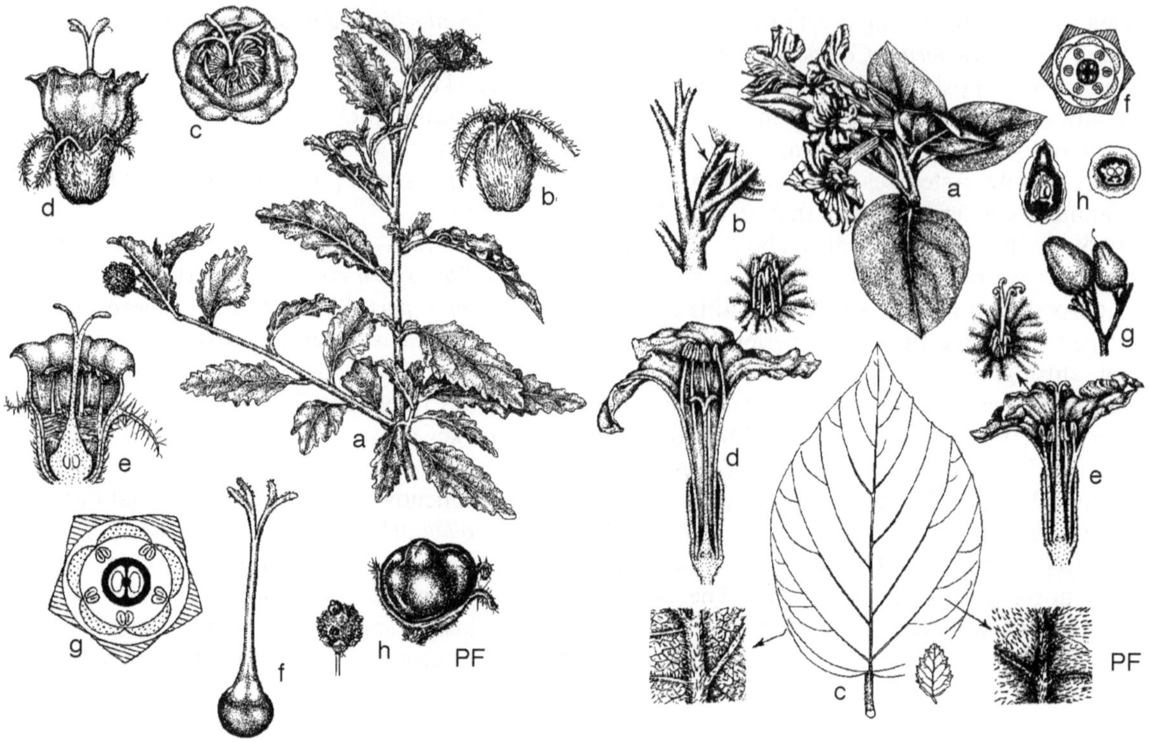

***Cordia**. Cordia globosa* (left). a. Flowering branch. b. Flower bud. c. Flower from above. d. Flower, side view. e. Flower longitudinally dissected. f. Pistil. g. Floral diagram. h. Fruit in cluster on left, enlarged fruit on right. *Cordia sebestena* (right). a. Flowering branch. b. Details of base of inflorescence (left), arrow showing sympodial renewal shoot from leaf axil. c. Leaf with details of lower (left) and upper surfaces (right); inset is of *C. globosa* for comparison of size. d. Long-stamened flower, longitudinally dissected, with detail of throat (arrow). e. Short-stamened flower, with detail of throat (arrow). f. Flora diagram. g. Fruits. h. Fruit in longitudinal section (left) and transverse section (right). *Both drawn by Priscilla Fawcett.* From Correll and Correll 1982.

fly enthusiasts discovered the plants and *C. globosa* appeared everywhere in cultivation. These devotees may well have invented the name blood-berry, because it is not in any of the older literature. The same people, or their horticultural friends, surely created the more recent name, butterfly sage.

Blood-berry is perhaps based on *yerba de la sangre* (blood herb, Cuba) or maybe *sanguinaria* (for the blood, Cuba). The red fruits following the small clusters of white flowers make the names seem appropriate. However, the Spanish versions were based on medical uses. In Cuba and Hispaniola, a decoction of the plants is highly esteemed as an astringent and hemostat in instances of pulmonary hemorrhage. Throughout the range of the species, the main application is for lung problems (Roig 1945, Liogier 1974, Morton 1981). That use holds through the range of *C. globosa* in southern Florida, the Bahamas, Cuba, Hispaniola, and Puerto Rico to Aruba, Bonaire, Curaçao, and Margarita, thence onto mainland Venezuela, Colombia, and in Central America through Belize, Guatemala, El Salvador, Honduras, Nicaragua, Costa Rica, and north to Mexico (Yucatán, Chiapas, Campeche, Morelos, Quintana Roo). Jamaicans use a tea to treat colds and tightness of the chest. In the Grenadines the plant is used as a remedy for colds, chest congestion, and for menstrual cramps (Morton 1981).

The name butterfly sage is also related to black sage (Caymans, Jamaica), man black sage (Jamaica), and wild sage. Those are simply English variants of the Spanish *salvilla cimarrona* [*savilla cimarrona*] (wild sage). People everywhere have learned that *Salvia* (Lamiaceae) is a useful medicine, and any plant that resembles it may have "sage" applied. People use the name sage because the leaves of the blood-berry are somewhat fragrant.

Many of the other common names make more or less obvious reference to medical applications. Several plants in the French Antilles are called *bonbon* (candy), with different modifiers, and some are medicinal. *Cordia globosa* is *bonbon rond* (round candy, Guadeloupe, Martinique). The Jamaican gout tea may be a valid name (and use), or it may be a misunderstanding of the French *guérit-tout* (cure all, Guadeloupe, Martinique, Trinidad). Several Caribbean medicinal plants are named after people, perhaps because they were renowned healers. For blood-berry, it is John Charles [John Charley] (Jamaica) or *Juan prieto* (black John, Dominican Republic).

There may be more than one message in *rompe camisa hembra* (female torn shirt, Cuba). That is especially true since the antonym is *rompe camisa macho* (*Turnera diffusa*). *Turnera,* known in the herbal industry as *damiana,* is touted as an aphrodisiac. Maybe *C. globosa* is also used that way.

Several other names are probably allusions to herbal medicines. Comparison with the avocado (*Persea americana*) with the name *aguacatico* (little avocado, Venezuela) surely denotes medicinal use. Some names comparing the shrubs with Verbenaceae include *caraiquito negro* (black *Lantana,* Venezuela) and *cariaquito de sabana* (savana *Lantana,* Venezuela). Maybe the inclusion of indigenous words *saraguaso prieto* (black *saraguaso,* maybe Taino, Puerto Rico) complicates this imperfectly known name. No translation has been found for *ateje* (Taino?, Cuba), but it is used for more than one *Cordia.*

Both *copillo* (little cup, Puerto Rico) and Curaçao bush (British Antilles) allude to medicines. However, it is not clear what is meant by *cuajatinta* (coagulates ink [or coffee], El Salvador), unless it is an allusion to a terrible taste. Other names with obtuse allusions are *herbe à boue* (filth herb, Guadeloupe, Martinique), *muñeco* (doll, Dominican Republic), *palo negro* (black stick [tree]), and *papita* (little potato, Cuba).

A decoction of blood-berry with fever grass (*Andropogon citratus*) and bamboo (*Bambusa vulgaris*) is used against fever (Ayensu 1981). A simple decoction of blood-berry has been used in Cuba and elsewhere as a depurative for troublesome skin eruptions (Roig 1945). This mixture is used internally and in baths. In Jamaica the plant decoction is used as a beverage. People in Yucatán employ the leaves as seasoning in cooking armadillo (*Dasypus novemcinctus*) to improve its flavor.

It is hard to see how *mierda de gallina* (chicken feces, Cuba) or *zompopo* (simpleton or a species of ant, El Salvador) might indicate any human use, but perhaps they do. *Cenigal* (*ceniza,* ash, with the suffix *-al,* an ash-colored thicket, Dominican Republic) may also suggest a use or simply be descriptive. The names *diente de chucho* (dog's tooth) and *diente de perro* (dog's tooth) surely refer to the distinctive leaves. In Maya, the name is simply a mention of a pest, *hauche* (*hau* [*haw*], open the road, *che',* tree, Maya). This name is given because the trees grow frequently in paths and need to be removed.

No further information was found on *achechive* [*achechibe*]; it seems to be an indigenous name. I would not be surprised if the name had not been derived from *anacahuite,* a Mexican name originally given to *C. boissieri* and then to *C. sebestena.*

Cordia globosa is not unique among species in the genus in having people use it. Linnaeus put three species in *Cordia* in 1753, and two of them had reputations as medicines. The one best known in Europe was the Old World *C. myxa* (Grigson 1986), which Linnaeus had grown at the botanical garden in Uppsala. He had previously reported on it in his book *Materia Medica* in 1749.

A first-hand report of American species was made by Hans Sloane from the time he spent in Jamaica. Although Linnaeus visited Sloane in Chelsea in 1736, he was dismayed that the herbarium was bound up in books and could not easily be studied. The names Linnaeus created based on Sloane came from his publications and not from the herbarium.

Because of the difficulty and time needed to browse through the 265 volumes of Sloane's specimens, perhaps it is understandable that Linnaeus did not mention what is now called *C. globosa* being among them. On the other hand, perhaps he did not know what to do with it. In those cases, he simply kept quiet.

Sloane's illustration of *Caryophyllus spurius inodorus, folio subrotundo scabro* (false odorless carnation, with almost round scabrous leaves) served as the basis for what Linnaeus called *C. sebestena.* We have to wonder if Linnaeus knew the meaning of the Arabic name *sebestena* (from *seg pistan,* dog tits, Grigson 1986), but I would bet he did.

The other reference Linnaeus used for *C. sebestena* was Mark Catesby. In his text, Catesby (1734–1747) reported that the wood of that tree "contains much Gum, in Smell and Appearance resembling Aloes, and is by the Inhabitants of the Bahama Islands (where it grows) called *Lignum* Aloes." Even today, aloe-wood remains a common name for the species in the Bahamas.

Because of the tree's similarity to the Mexican species *C. boissieri* it is known by the Náhuatl name for that species, *anacahuite* [*anacahuita, anacagüita, anaconda*] (paper-tree, Yucatán, Cuba, Puerto Rico). The Maya of Yucatán call it *kopté* [*copté, sak-k'opté, zak-k'opté, zac-copté*] (*sak,* white, *k'opte',* with a hard center). The Spanish *siricote* [*ciricote, ziricote*] (Yucatán, Guatemala, Belize) or *siricote blanco* (white *ciricote,* Yucatán) is surely a corruption of the Mayan. These names are used for the tree whose wood is hard, and whose fruits are used for food (Rico-G. et al. 1990). Fruits are made into preserves or jam, and they have a good flavor. The fresh fruits directly from the tree are considered edible, but the taste is not good. Apparently, sweetening and preserving fruits improve the taste.

Cubans call it *vomitel* (vomit) or *vomitel amarillo* (yellow vomit), and use it for stomach problems (Roig

1945). In the Lesser Antilles *C. sebestena* is called *bois râpe* (rough tree, because of the sandpapery leaf surface, Guadeloupe, Martinique), *manhage* (Aruba, Bonaire, Curaçao), scarlet accordia (Dutch Antilles), scarlet flower (Dutch Antilles), *sébastier* (Guadeloupe, Martinique), and *ti-soleil* (little sun, French Antilles). It is also called *caujaro Español* [*cawara spaño, karawaara spanjool*] (Spanish *caujaro*, Aruba, Bonaire, Curaçao). *Cordia alba* in Venezuela was the original *caujaro*, perhaps a name of Arawakan language origin.

Cordia sebestena is also in Florida, but it is questionable that it is native. The legend says that John James Audubon gave it the name "Geiger tree" because of the specimens that 19th-century ship captain John Geiger planted at Key West (Little 1979). According to Acevedo-Rodriguez (1996), the species is only native to the Greater Antilles, the Bahamas, and the Virgin Islands. He considers it introduced elsewhere as an ornamental. Indeed, it has been cultivated widely through the Americas and elsewhere for its orange-red flowers and white fruits. Some of the plants may also have been grown for their edible fruits and medicinal traits. Certainly, the species has a venerable reputation for both.

While no chemical studies seem to have been made on either of these Florida plants, several other members of *Cordia* have been examined. There are multiple biologically active compounds in those species, including arabinoglucan, essential oils, terpenes, meroterpenoid naphthoquinones, and taxifolin (Srivastava and Srivastava 1979, Velde et al. 1982, Basu et al. 1986, Bieber et al. 1990, 1994, Fun et al. 1990, Nakamura et al. 1997). Other studies have shown these and other *Cordia* extracts to be antifungal, anti-inflammatory, antiulcer, antiviral, hepatotoxic, and larvicidal (Tiwari and Srivastava 1979, Marston et al. 1988, Hayashi et al. 1990, Sertie et al. 1990a,b, Rapisarda et al. 1993, Akhtar and Ahmed 1995, Ioset et al. 1998).

A Lesser Antillean name summarizes the utility of *Cordia* although it refers directly to *C. sebestena*. *Mapou rouge* (red *mapou*, Guadeloupe, Martinique), contains a Cariban element (*mápuia*, bush-spirit, Hodge and Taylor 1957). Maybe the *Cordia* species house the essence of the healing force.

Coreopsis

(From Greek *koris*, a bug, and *opsis*, resembling, an allusion to the seeds)

Coreopsis leavenworthii (honors M.C. Leavenworth, 1796–1862, a student of southern U.S. botany)
tick seed (southeastern USA)

Coreopsis leavenworthii. *Drawn by P.N. Honychurch.*

Linnaeus was familiar with eight species of *Coreopsis* when he described the genus in 1753. All but one of those he knew was from North America. Later it was found that the genus is endemic to the Americas, and now it contains 50 species (Mabberley 1997). Several have been used in medicines.

The Seminoles used an infusion of *C. leavenworthii* to treat heat prostration (Small 1933). However, Sturtevant (1955), Bennett (1997), and Snow and Stans (2001) have not subsequently found the Seminoles using it.

Cornus

(The Latin name of the cornelian cherry, *Cornus mas*, from Greek *keras*, and Latin *cornum*, horn; akin to Akkadian *qarnu*, and Hebrew *qeren*)

chápacor (Carolina Algonquians); *vtchappoc* (Powhatan, Virginia) (both forms were interpreted by Geary 1955 as a *Cornus* whose roots were used to make a red dye)
còrn (Gaelic); cornel (from Latin *cornum*, horn, perhaps from the color; akin to Greek *kerasos*, cherry; first applied in the 1550s by William Turner and other herbalists); *cornouiller* (cornel tree, although originally referring to the fruit, French; related German names are *Cornell-baum*, cornel tree, and *Kornel-berre*, cornel berry)
dogwood (derived from "dog-tree," which in turn was taken from "dog-berries," also called

Cornus. *Cornus alternifolia* (top left). From Sargent 1905. *Cornus amomum* (top right). From Britton and Brown 1897. *Cornus asperifolia* (lower left). From Sargent 1905. *Cornus florida* (lower right). From Sargent 1905.

hound's tree or hound's berry, from at least 1550 and Turner's *Herbal*; the leaves were incorrectly considered a cure for hydrophobia)
hakchupilhko [*hakchipilhko, akchupilhko, hakchupilhkash*] (*hakchuma*, tobacco, *pilhko* =?, Choctaw)
Hartriegel (heart rail, German)
ka-na'-si-ta' (Cherokee)
nan'taxpa'yudi (Biloxi)
shen-don-shu-gah (bitter red berry, Osage)
vtvphv [*atvp'ha, atapha*] (Creek, Muskogee); *cofi* (Hitichiti)

Cornus alternifolia (alternate leaved)
alternate-leaved dogwood [alternateleaf dogwood, alternate-leaf dogwood]
[blue, purple] dogwood
green osier [green-osier] ("osier" from Latin *ausaria* or *osaria,* willow-bed, because the word was originally restricted to willows, *Salix*)
gwai-gee-àp-aip (Kiowa)
mowo'samic (moose wood, Potawatomi); *muj'omij'* (moose plant, Ojibwa)
pagoda [dogwood]; pagoda-cornel
pigeonberry
umbrella-tree
zaikh-'koñ-a [*sek'a'n, sai'kañ*] (arrow wood, Kiowa)

Cornus amomum (a Latin name of some shrub)
chan-shasha (*chan*, wood, *shasha*, red, Dakota); *ruhi-shutsh* (*shutsh*, red, Winnebago)
kinnikinnick [*killegenico, killickinnick, kinnakinnec*] (mixture, Algonquin)
ninigahi (*nini*, pipe, *igahi*, to mix, because it was mixed with tobacco for the pipe, Omaha-Ponca)
rapahat (*ra*, stick, *pahat*, red, Pawnee)
red dogwood; red willow
silky cornel; silky dogwood

Cornus asperifolia (rough-leaved)
mansa-hte-hi (*mansa*, arrow, *hte*, real, *hi*, plant, or "real arrow tree," Omaha-Ponca); *mansi-hotsh* (Winnebago); *mon̲'-ça hi* (arrow shaft stalk, Osage)
nakipistatu (*nahaapi*, tree, *kipis*, arrow, *tatu*, real, Pawnee)
rough [rough-leaved, tough-leaved, toughleaf] dogwood

Cornus florida (flowering)
American box; [New England] boxwood; cornel box; dog box; false box [false boxwood]
[Indian] arrow-wood; *bois fleche* (arrow-wood, French)
blomsterkornell (flowering cornel, Dutch); [flowering, white] cornel; Cornelian-wood; cornelius-tree; *corniolo americano* (American cornel, Italian); *cornouiller de la floride* (Florida cornel, Quebec)
cornus americano (American dogwood, Spanish)
country-timber
[common white, eastern flowering, Florida, flowering] dogwood; *bois de chein* (dogwood, French)
hattawanominschi [*hat-ta-wa-no-min-schi*] (*minschi*, plant, Delaware); *monhacanniminschi* [*mon-ha-can-ni-min-schi*] (*moon-hak-an*, digging stick, *hee*, thing indicated, *minschi*, plant, Delaware)
kanunsi'ta (Cherokee)
nahiskilá (Koasati); *naskila* (Alabama)
nature's mistake (some think the bracts are aberrant petals)
red [osier, willow]
ta̲'si yapare (dog tree/wood, Catawba)

Cornus foemina (female; so considered by Philip Miller, 1691–1771, as opposed to the "male," *C. florida*)
[blue-fruit, stiff cornel, stiff, swamp] dogwood
stiff-cornel [stiffcornel]

Europe has four native species of *Cornus*, although Linnaeus ([1753] 1957) included only three of them; the other two he described were American *C. canadensis* and *C. florida.* There are now about 65 species, most of them in the north temperate zones (Mabberley 1997). The common names in diverse European and American languages give some indication of how important the genus has been to people.

Two widespread uses of branches and leaves of *Cornus* were for making arrows and in smoking. At least the Comanche, Omaha, Pawnee, Ponca, and Winnebago used *C. asperifolia* stems to make arrow shafts (Gilmore 1919, Moerman 1998). The Comanche also used *C. florida* branches for arrows (Tull 1999). One of the widespread names for *Cornus*, especially *C. amomum*, was *kinnikinnick*. That designation was used for several plants, but often implied that they were smoked alone or added to tobacco. The Lakota, Menomini, Omaha, Pawnee, Ponca, and Winnebago added dried *Cornus* inner bark to tobacco (Gilmore 1919). The Micmac used dried *C. foemina* bark to mix with their tobacco (Moerman 1998). Probably among all these people, but at least with the Winnebago, *Cornus* was used as a ceremonial smoke (Moerman 1998).

The Cherokee used *C. alternifolia* wood for carvings and loom shuttles; the Ojibwa used the wood for awl handles, mauls, and war clubs, and the twigs for thatch and other materials (Moerman 1998). The Anglos used the wood for small articles, tool handles, wheel hubs, and pulleys (Vines 1977). Kalm ([1753–1761] 1972) added that settlers made weaver's spools, joiner's planes, and wedges. Unspecified tribes used the roots to make a dye (Vines 1977). The dye probably was like that made from *C. florida*, either rusty brown or bright yellow, depending on the mordant. Tull (1999) added that numerous tribes used several *Cornus* species to color baskets and porcupine quills.

Documentation was found for only two groups of Muskogean people using dogwood. The Houma made a decoction of *C. florida* and *C. foemina* roots or bark to alleviate fever, including malaria (Speck 1941). The Alabamas used the bark of *C. florida* to relieve diarrhea (Swanton 1928a). However, the Creek name *vtvphv* [*atapha*] and the Mikasuki name *atópa:bi* (dogwood replica; see *Psychotria:* Wild Coffee) suggest a wider application among Muskogean people. Considering the wide use of *Cornus* by indigenous groups and subsequent immigrants, it seems likely that all species were used by everyone who lived near them.

The tannins in *Cornus* are well known, and are at least some of the active compounds involved in their use as remedies (Yazaki and Okuda 1989, Hatano et al. 1990, Wedge et al. 1995). Felter (1922) recorded that *C. florida* contained 3% tannin, presumably in the bark. He also noted that it contained a bitter chemical called cornine [cornin]. Millspaugh (1892) was of the opinion that the bitter compound was an alkaloid; however, a more recent study did not find alkaloids in *C. florida* (Barbosa et al. 1990). Instead, cornin is an iridoid glycoside, and it seems to occur in many species of *Cornus* (Stermitz 1998, Tanaka et al. 2001).

Old World species are also known to contain medically active glycosides (Hatano et al. 1990). *Cornus officinalis* includes triterpenoids, ursolic acid, saponins, cornin, morroniside, loganin, and sweroside (Hsu 1986). As Millspaugh (1892) reported several of those compounds for *C. florida*, it would appear that there is a shared variety of chemicals in *Cornus*.

The tannins at least would explain the widespread use for skin problems and piles. The Cherokee, Iroquois, Menomini, and Winnebago used *C. alternifolia* and *C. amomum* to treat skin problems (blisters, cancer, navel, ulcers, venereal sores) (Hamel and Chiltoskey 1975, Moerman 1998). At least the Menomini used a bark infusion of *C. alternifolia* to treat piles (Moerman 1998). The Cherokee also employed a bark infusion of *C. alternifolia* and *C. florida* for measles (Hamel and Chiltoskey 1975).

One of the prime applications of *Cornus* was to treat chills, coughs, laryngitis, and fever associated with colds and flu. For this purpose decoctions or infusions of stems, bark, or roots from *C. alternifolia, C. florida,* and *C. foemina,* were used by the Cherokee, Ojibwa, Iroquois, and Potawatomi (Smith 1933, Hamel and Chiltoskey 1975, Moerman 1998). The Iroquois also prescribed a bark infusion of *C. amomum* for chest congestion and one of *C. alternifolia* for tuberculosis (Moerman 1998).

The Cherokee treated colic with both a root bark and flower infusion of either *C. alternifolia* or *C. florida* (Hamel and Chiltoskey 1975). The Iroquois and Winnebago made a remedy for indigestion with an infusion of the inner bark of *C. amomum* (Moerman 1998). The Cherokee, Menomini, Rappahannock, Winnebago (Hamel and Chiltoskey 1975, Moerman 1998) treated diarrhea with an infusion of root bark from *C. alternifolia, C. amomum,* or *C. florida.* Worms, especially in children, were expelled with a bark infusion of either *C. alternifolia* or *C. florida* by the Cherokee (Hamel and Chiltoskey 1975). The Iroquois used bark infusions made from either *C. alternifolia* or *C. amomum* (Moerman 1998) internally for urinary problems and venereal disease.

Cornus was considered both emetic and laxative. A bark infusion of either *C. alternifolia* or *C. amomum* was used by Iroquois and Ojibwa as an emetic, and an infusion of *C. amomum* was used as a laxative by the Iroquois (Moerman 1998). Root bark infusions or decoctions of *C. alternifolia* and *C. florida* were considered stimulant by the Cherokee (Hamel and Chiltoskey 1975); the Delaware and Rappahannock considered *C. florida* a tonic (Moerman 1998).

People from the Iroquois of New England to the Cherokee of the Carolinas considered *Cornus* analgesic. The Iroquois applied a bark infusion of *C. alternifolia* to swollen areas (Moerman 1998). The Cherokee used a bark infusion of *C. alternifolia* and *C. florida* to treat headache and women's backache (Hamel and Chiltoskey 1975). The Iroquois used a bark infusion of *C. amomum* as a bath to help babies sleep (Moerman 1998). Tannins and some of the other compounds in *Cornus* are known to be antibacterial, which would explain why the Cherokee, Ojibwa, Iroquois, and Potawatomi used a root bark infusion of *C. alternifolia* and *C. florida* to treat sore eyes (Smith 1933, Hamel and Chiltoskey 1975, Moerman 1998).

There was a widespread belief among indigenous people that the blood could become "tainted" and thus cause sickness. That belief persists in some people in the United States and among many Bahamians today. The Cherokee and Rappahannock believed a root bark infusion of *C. alternifolia* and *C. florida* improved the blood, and thus made a sick person well (Hamel and Chiltoskey 1975, Moerman 1998). Perhaps the Iroquois were also aware that goiter was caused by a diet deficiency, although they surely did not know it was iodine. In any event, they took a bark infusion of *C. amomum* for that malady (Moerman 1998).

Fruits of *C. florida* are typically considered somewhat poisonous, although they have been soaked in brandy as a "bitter digestive tonic and for acid stomach" (Foster and Duke 1990). Rafinesque said the people in Louisiana ate *C. amomum* fruits (Hedrick 1919).

A widespread historical application for *Cornus* stems was to make "chew sticks" for cleaning the teeth (Millspaugh 1892, Uphof 1968, Coon 1974, Lewis and Elvin-Lewis 1977). This practice was unknown among indigenous people although they used other materials to clean their teeth (Lewis and Elvin-Lewis 1977). The fact that they adopted use of dogwood stems to clean their teeth from the black slaves confirms they were close observers (Coon 1974). Both *C. amomum* and *C. florida* were preferred, but all were likely used. Coon (1974) noted "the astringent properties of the bark being valuable for hardening the gums." Millspaugh (1892) and Lewis and Elvin-Lewis (1977) similarly comment how effective *C. florida* stems were in removing plaque, leaving the gums pink and firm, and the teeth white and decay-free. However, Foster and Duke (1990) warn that the hard fibers tend to erode the teeth much as a hard-bristle toothbrush does.

An important ethnobotanical aspect of *C. florida* that is often overlooked was to keep track of seasons. Several groups planted corn when the dogwood bloomed because that was considered a sure sign that winter was over (Millspaugh 1892, Coon 1974).

Crataegus

(From Greek *krataigos*, an ancient name for a kind of flowering thorn used by Theophrastus, 372–287 B.C.; composed of *kratus*, strong, *akakia*, *akis*, thorn)

Crataegus. *Crataegus aestivalis* (top left). From Sargent 1905. *Crataegus crusgalli* (top right). From Sargent 1905. *Crataegus flava* (lower left). From Britton and Brown 1897. *Crataegus spathulata* (lower right). From Sargent 1905.

a'ho (Ofo)
aubépine (from popular Latin *spina alba* of Pliny, A.D. 23–79; coexisted as *aubepin,* masculine form, because *spinus* is shrub, while *spina* is spine, French); *biancospino* (white thorn, Italian); *epine blanche* (white spine, France); *Weissdorn* (white thorn, German); *wittedoorn* (white thorn, Dutch)
cenellier (Quebec)
chanafila [*shanafila*] (black haw, Choctaw); *selvwv* [*cate*] ([red] haw, Muskogee)
[red] haw [hawthorn] (from Old English *haga*, hedge; akin to Dutch *haag*, German *Hage*, Danish *have*, garden; see *Celtis:* Hackberry for more on etymology)
machkígen [*machkigenamúnschi*] (*kiminschi,* plant, Delaware); *mine'saga'wunj* (having fruit and spines, Ojibwa)
oki'neyak (Plains Cree)
pommettes (little apples, Quebec)
ta-çpon̲' (Osage)
tschigachrèchtes (Onondaga)
uath (Gaelic)

Crataegus aestivalis (summer-flowering)
a'stan̲tka' (Biloxi)
chommilámhi [*chommilámmi*] (*chommi,* anus, *lomhi*, hidden; the name refers to the dimple on one end of the fruit, Alabama)
[apple, May, May hawthorn] haw

Crataegus crusgalli (spur of a cock)
squimenuckee (Shawnee, fide Edgar 1891)
[cockspur, pin]-thorn [hawthorn]; thorn-[apple, bush]

Crataegus flava (yellow)
summer haw (used by Williams [1837] 1962); yellowleaf hawthorn

Crataegus marshallii (named for dendrologist Humphrey Marshall, 1722–1801, who first described it)
cittiłhómma cobá (*chi*, you, *ittiłi*, eye, *hommá*, red, *cobá*, big, Koasati)
parsley hawthorn (superficially, the leaves resemble those of parsley, *Petroselinum crispum*)

Crataegus spathulata (wedge-shaped, referring to the leaves)
chittihámma [*chittiomma, chittihomma, chittihomma sàwwa*] (*chi-*, you, *ittiłi*, eye, *homma*, red, *sawwa*, small, Alabama); *citiłhómma* [*tiłhómma*] (*cíl*, you, *ittiłi*, eye, *hommá*, red, Koasati)
little-hip hawthorn [littlehip hawthorn]; [pasture, small-fruit] hawthorn
spathulate haw (a book name); summer haw (used by Williams [1837] 1962)

Crataegus uniflora (one-flowered)
dwarf hawthorn; winter haw (used by Williams [1837] 1962)

Fernald (1950) recognized 103 species of *Crataegus* in the northeastern United States. Many people believe that there are fewer species than that. To illustrate the chaotic condition of the taxonomy of the genus, a story made the rounds when I was in graduate school at Washington University in St. Louis, Missouri, in the late 1960s. The prevalent expert on *Crataegus* was suspected of naming every minor variant as a new species. So, a cunning colleague went into the field, made multiple collections from a single tree, and gave them all fictional numbers and localities. These samples were then sent to the expert for identification. Specimens from the single tree were returned with more than a dozen species names. Mabberley (1997), who usually puts numbers of species with no comment, wrote for *Crataegus* that there were 186, plus maybe 78, species.

Europeans and indigenous Americans are known for eating the fruits of hawthorns—particularly when there is nothing better available. There are records of fruits having been eaten from *C. aestivalis, C. crusgalli, C. flava, C. marshallii, C. spathulata*, and *C. uniflora* (Yanovsky 1936, Hocking 1997). Moerman (1998) lists 12 other species whose fruits were eaten across North America.

The wood from all *Crataegus* is hard and heavy, and it was used by Europeans and Americans. Harrar and Harrar (1946) say that the famous Irish shillelaghs (cudgels) were made from the wood of black hawthorn. In the Americas, at least the wood of *C. crusgalli* has been used for tool handles (Uphof 1968).

The Cherokee considered an infusion of the bark of *C. spathulata* good for the circulation and to prevent spasms. Berries were eaten to improve the appetite, particularly when ill (Hamel and Chiltoskey 1975). They also drank a bark tea or bathed in it to "ward off tacklers" during ballgames (Hamel and Chiltoskey 1975).

There are three other *Crataegus* in Florida that surely were used. Foster and Duke (1990) consider all the species in the genus and comment that *Crataegus* is a famous herbal heart tonic used throughout the world. For example, Culpeper (1653) wrote: "The seeds in the berries beaten to powder being drunk in wine, are good against stone and dropsy." Lewis and Elvin-Lewis (1977), Foster and Duke (1990), Swerdlow (2000), and Duke et al. (2002) record that modern research has confirmed its use for hypertension, weak heart, angina pectoralis, and artherosclerosis.

Some of the other uses for species in Florida are in other medicines. Roots of *C. crusgalli* were aperatives, deobstructants, and diuretics in Mexico (Hocking 1997). *Crataegus marshallii* is considered a substitute for *C. oxyacantha* as a hypotensive (Hocking 1997). Unripe fruits of *C. uniflora* were used as an astringent in bladder disorders and as a cardiotonic among the Meskwaki (Vogel 1970, Hocking 1997).

Crinum

(From Greek *krinon*, a lily; Pliny, A.D. 23–79, used the name *crinon* for the red lily and Linnaeus reapplied it to the American plants)

Crinum americanum (from America)
halpatcolasî [*halpattcolaçi*] (alligator tongue, Mikasuki)
lirio de San Pedro (Saint Peter's lily, Cuba)
[Florida, southern] swamp lily

Linnaeus ([1753] 1957) knew four species of *Crinum* when he created the genus, one from the Americas, two from India, and the fourth from Africa. *Crinum americanum* he knew from cultivation in Uppsala and from the literature. There are now perhaps 120 species in the genus, all from tropical and warm temperate zones (Mabberley 1997).

Bulbs are used to heal wounds and flowers are pectoral (Roig 1945). Several species in the genus are used in the Caribbean for a variety of ailments (Morton 1981).

Bulbs, and presumably other plant parts, contain lycorine, crinidine, crinamine, and other alkaloids (Perkins and Payne 1978). The Lubber grasshopper (*Romalea microptera*), ranging from North Carolina to Texas, is adapted for using these poisons, and advertises its toxicity in its aposematic coloration (Capinera 2001).

Crossopetalum: Maiden-Berry

(Based on Greek *krossos*, fringe, and *petalon*, petal, an allusion to their ornamented margins)

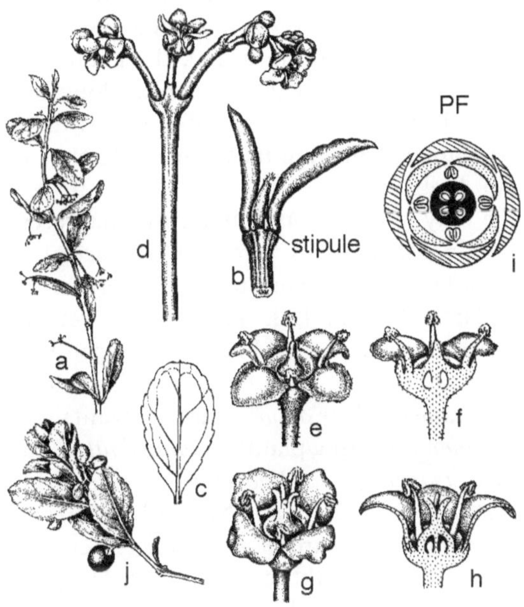

Crossopetalum rhacoma. a. Flowering branch. b. Terminal bud. c. Leaf outline. d. Flower clusters. e. Flower, front-side view, at early (male) stage. f. Flower at male stage, longitudinally dissected. g. Flower, front-side view, at later (female) stage. h. Flower at female stage, longitudinally dissected. i. Floral diagram. j. Fruiting branch. *Drawn by Priscilla Fawcett.* From Correll and Correll 1982.

Scattered through the pinelands of Miami-Dade and Monroe Counties are small shrubs some call Christmas-berry or maiden-berry. People apply these two names to different species of *Crossopetalum*, the first belonging to *C. ilicifolium* and the second to *C. rhacoma*.

We might expect the name "Christmas-berry" to be young, because that celebration is not that old (Austin 1998a). The word "Christmas" did not exist until 11 centuries after the time of Christ. Use of the name Christmas-berry goes back to at least the early 1900s, although maiden-berry seems to be of more recent application. However, the combination of "maiden" with other words in plant names is a venerable practice, dating to 1589 or earlier. The meanings of "maiden" are varied in English and in the plant names, and why the word has been applied to Florida plants is uncertain. We might speculate that the red fruits were compared with the blushing cheeks of a young woman, but it could just as well be one of the other meanings. For example, two other ancient interpretations are a plant grown from seed and a plant that has not been budded, lopped, pruned, or transplanted.

Older names for *C. rhacoma* in Florida and in other English-speaking areas include poison cherry (USA, Bahamas, Jamaica, Puerto Rico), snake berry (Cayman Islands), tobacco berry (Cayman Islands), wild cherry (USA, Puerto Rico), and wild tobacco (Cayman Islands). In these names "poison," "snake," and "tobacco" are modifiers applied to any unknown fruit that resembles a known and usually cultivated type. In other words, those modifiers are equivalent to "wild."

The comparison with fruits is also common in other parts of the Caribbean. For example, they are *palo de paloma* (dove tree, Puerto Rico), *pico de paloma* (dove beak, Puerto Rico), *palu di pushi* (*pushi* tree, Dutch Antilles), and *petit merisier* (little wild cherry, French Antilles). Comparisons with other species include *lemoncillo* (little lime, Cuba), *limonejo* (lime tree, Cuba for *C. latifolia*), *limonejo de costa* (coastal lime tree, Cuba), and *membrillo* (quince, *Cydonia oblonga*, Rosaceae).

Names even compare the fruits with candy, perhaps because of their bright red color. These include *bonbon rouge* (red candy, French Antilles), *coral* (Spanish), *ti bonbon* (little candy, Guadeloupe, Martinique), and *ti bonbon rouge* (little red candy, French Antilles).

There also are a number of obscure references. In Cuba the plants are called *jinca pata* (*pata*, foot), although I have been unable to find native speakers of Spanish who understand the word *jinca*. In the Dutch Antilles, *Crossopetalum rhacoma* is *kuragoga*, surely because of its similarity to the related species *C. uragoga*. One species is named *mala mujer* (bad girl, Puerto Rico for *C. latifolia*), although we do not know why she was bad. Elsewhere, *mala mujer* is used for thorny and (usually) stinging plants.

Also in Puerto Rico, *C. rhacoma* is *manto* (mantle), a name most often given to members of the Convolvulaceae that blanket others with their growth. Finally, there is *maravedi* (Cuba, Puerto Rico), which turns out to be another Arabic word in Spanish. Not only is that name applied to *C. rhacoma*, but in Cuba it is also applied to *C. uragoga* (see below). For some reason the word was taken from the name of an old gold coin and applied to these plants. Perhaps it was the similarity of the coin and the shape of the leaves.

Both of Florida's *Crossopetalum* species have extremely limited ranges, and both are considered endangered by the state (Coile 2000). Essentially, both species are confined to Miami-Dade and Monroe Counties, although historically *C. rhacoma* was also found in Broward, Palm Beach, Collier, Lee, and Sarasota Counties. Outside Florida, *C. rhacoma* ranges through the Bahamas, from Cuba to St. Lucia, Aruba, Bonaire, and Curaçao, and southern Mexico to Colombia and Venezuela. *Crossopetalum ilicifolium* has a similar range.

Both species of *Crossopetalum* have edible fruits, although each is small and not very filling. These small drupes are produced sporadically at least through the warm months and soon disappear because they are greedily consumed by birds and other fructivores. The most widespread and common use made of these rare plants is for treating kidney problems (Morton 1981). Throughout the Caribbean, a root decoction is used to expel kidney stones, as a diuretic, and against bladder and kidney infections. In Cuba, a decoction of leafy branches and roots is taken hot or cold as a diuretic and to counteract kidney inflammation (Roig 1945).

Perhaps because the plants are uncommon and small (rarely exceeding 1 m tall, although they can grow to 5 m), little has been written about their uses. That should not diminish the favor they have been given in regions where they occur, at least in the Caribbean.

The Celastraceae, to which *Crossopetalum* belongs, is rife with taxonomic problems and confusion. Historically, what is now one genus was thought for decades to be three—*Crossopetalum, Rhacoma*, and *Myginda*.

Crossopetalum was created by Patrick Browne in his *The Civil and Natural History of Jamaica* published in 1756. Subsequently, Linnaeus rejected the genus in his *Systema Naturae* published in 1759, and created *Rhacoma* (Latin *rhacoma* or *rhecoma*, root, rhubarb from Pliny, A.D. 23–79; Greek, *rhakoma*, rags, *rhakos*, tattered garment). He applied the name *R. crossopetalum* to the species Browne previously had named in 1756. That action is considered illegal, as there was an available genus that Linnaeus should have used. Therefore, his *Rhacoma* becomes a synonym of the correct name, *Crossopetalum*. That legal aspect was not corrected until 1766 when Heinrich J.N. von Crantz (1722–1797), professor of medicine in Vienna, made the change. Crantz should have been followed, but for decades, people followed Linnaeus and used the genus *Rhacoma*.

The latest interpretation is that *Crossopetalum* and *Myginda* also are the same genus even though the *Flora de Cuba* and other publications recognized them as distinct. Apparently, the authors of the *Flora de Cuba* kept them apart based solely on having four locules in *Crossopetalum* and two in *Myginda*, leaving a single species in the latter genus.

Myginda was created in 1760 by Nicolaus K. von Jacquin to commemorate his friend, the Austrian botanist Franz Mygind (1710–1789). Jacquin's *M. uragoga* is now correctly *C. uragoga*. That species grows in Cuba, Puerto Rico, Colombia, and in Mexico from Veracruz and Oaxaca south, and apparently disjunct, to Nicaragua. As indicated by the species name *uragoga* (promoting urine), the plant had been famous as a diuretic in the Caribbean long before it was described in Europe. In Cuba, this is the *hierba* [*yerba*] *maravedí*, making that allusion to the old coin with an Arabic word.

The only information found on human uses of *C. uragoga* in Mexico was provided by Alcorn (1984). She recorded that the Huastec had several names for these plants. In San Luis Potosí they were usually called *ts'amuts' uxkwe'* (old lady's earrings). However, they also call them *tsakam itsal koox* (koox-large bird chile), *ts'amuxlaab ts'ohool* (earring plant), *ts'amuts'laab ts'ohool* (earring plant), *ts'amuts' uthu'* (monkey earrings), *tsatsa' ilaal* (*Hylocereus undatus* medicine), and *ts'amuts' an thakchaam* (Náhuatl earrings). In the Veracruz dialect they are *ch'amuxlaab te'* (earring tree) or *chakam granadillo* (little granadillo). The plants are used for dermatological problems including blisters, calluses, and corns. The bark and root are mixed with other plants (*Ehretia anacua*, *Lantana camara*) and used to stop hemorrhage associated with obstetrical problems. Roots and bark also are boiled and used in medicinal baths to stop pain associated with dysentery and burning eyes. All of the applications are topical and not internal, differing markedly from the Caribbean use as a diuretic.

The family is noted for containing a variety of cardenolides, sesquiterpenes, and terpenoids. Several of those compounds have recently been found in the genus *Crossopetalum*. However, little is known about the chemistry of the genus. It is clear that the sesquiterpenes serve as deterrents to insect pests in the Panamanian species (Tincusi et al. 1998), and that cardenolides in the Mexican species are cytotoxic (Ankli et al. 2000). If similar chemicals occur in the Florida species, those traits would make them useful for several of the maladies they have been used to treat.

This group of medicinal plants is another whose application has almost been lost. Few people in the Caribbean and apparently in Mexico now seem to remember that there are curative traits associated with these small shrubs. Maybe we humans should blush like the maidens of maiden-berry's namesake at endangering useful organisms before we even know their potential.

Crotalaria: Rattlebox

(From Greek *krotalon*, a rattle or castanet)

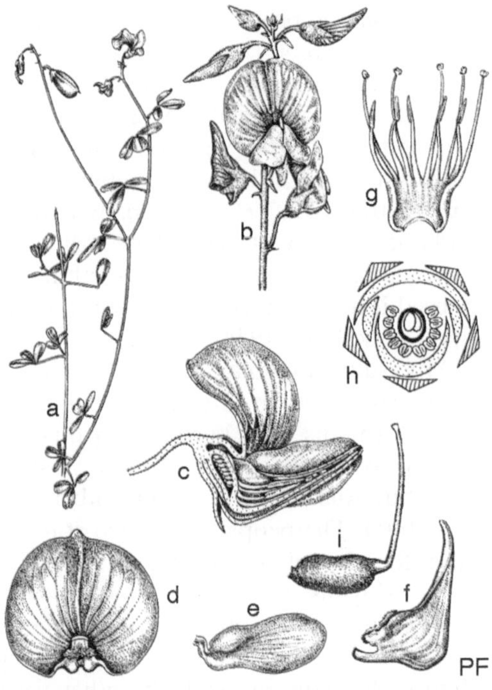

Crotalaria pumila. a. Branch of flowering and fruiting plant. b. Inflorescence. c. Flower longitudinally dissected. d. Standard (banner). e. Wing. f. Keel. g. Stamens united. h. Floral diagram. i. Pistil. *Drawn by Priscilla Fawcett.* From Correll and Correll 1982.

Europeans were fascinated by plants from other parts of the world that produced fruits like rattles. Beginning in the 1600s, herbalist-physicians like Leonard Plukenet and Hans Sloane recorded these *Crotalaria* from the Americas. Others, including Hendrik van Rheede and Paul Hermann, noted them in India, elsewhere in Asia, and in Africa. Some say that these authors used a translation of the Sanskrit *ghantár avamu* to establish their Greek name *Crotalaria* (Watt [1889] 1972). Linnaeus continued using the established name for the genus when he published his *Species Plantarum* in 1753, providing the foundation of modern usage. He knew of 13 species, and that list has now grown to about 600 (Mabberley 1997).

At almost the same time that the Europeans were learning about rattlebox legumes, they were encountering another kind of rattling American. The first to record that one in English was Capt. John Smith, who wrote in 1630 about "the danger of the rattell snake." It was not until 1759 that Linnaeus named that snake *Crotalus*, using the same Greek base word as *Crotalaria*. We now know that there are two genera of rattlesnakes (*Crotalus* and *Sistrurus*), but they are both poisonous pit vipers. The genus *Crotalaria* is also poisonous, and using the same base word for their genera was a good way of emphasizing that potential danger, although that was not Linnaeus's primary intent.

Obviously, the plants and the saurian relatives are poisonous in different ways, and they have distinctive toxins. Most, if not all, members of *Crotalaria* contain pyrrolizidine alkaloids (Williams and Molyneux 1987, Deyo and Kerkvliet 1990, Nellis 1997). Some species, like *C. incana*, have retrorsine in the whole plant, which causes liver and kidney neoplasms, plus the alkaloid integerrimine in the seeds, which causes liver and lung lesions. A related plant, *C. retusa*, has both retrorsine and microcrotaline (Nellis 1997). Since several species are used as bush teas in the Caribbean, they can cause major health problems. Those problems are not confined to adults, but can be passed to infants through mother's milk.

The snake, on the other hand, is highly evolved to deliver a mixture of poisons through hollow fangs serving much like hypodermic needles. Its poisons consist of enzymes, glycoproteins, polypeptides, and other chemicals that inhibit blood clotting and actually begin to digest tissues.

Additionally, the snake's "rattles" too are derived in totally different ways. The rattle in the ophidian is partly the result of incomplete shedding of the skin and accumulations of horny materials at the base of the tail. When snakes are agitated, they vibrate their tails and they "rattle," perhaps incidentally warning off organisms that possibly might otherwise step on the snake and harm it. The plants have legumes that become somewhat inflated when ripe, and their numerous small seeds detach from the placenta. When the plants are bumped or shaken by the wind, the seeds rattle against the walls of the dry fruits.

Shortly after I arrived in Florida, I had a student who decided to frighten the new professor. He knocked on my office door, stuck in his head, and said, "I have a present for you." With his hand behind his back and out of view, he shook the dry fruits of *Crotalaria*. Because I knew that he was full of mischief, and because I was fascinated by snakes of all kinds, I replied, "Fine! Bring it in and let's have a look." The smile faded from his face and he entered to display the dried legumes. The sounds of the fruit and the rattlesnake are remarkably similar, and I could not distinguish them then or now. I have been startled more than once in the field when I bumped into dry branches of *Crotalaria* and momentarily mistook them for *Crotalus.*

There are several *Crotalaria* introduced into Florida and nearby regions, including *C. incana, C. lanceolata, C. retusa,* and *C. spectabilis.* Burkill (1966) and others think that *C. incana* is native to the Americas, although a more exact provenance is not indicated. *Crotalaria lanceolata* is from South Africa; *C. retusa* is from the Old World tropics; and *C. spectabilis* was introduced by David Fairchild from the Old World when he worked for the U.S. Department of Agriculture. Throughout the native ranges of these species, they are used for a variety of purposes, with high-quality stem fibers used in ropes, strings, and fabrics. Those fibers have been substituted in some areas for the famous sunn hemp (*C. juncea*) of India. The second most widespread use of plants in the genus has been to provide a green-manure crop for fertilizing lands. Indeed, *C. spectabilis* was introduced by the U.S. Department of Agriculture in 1921 and touted as a soil conditioner and forage crop. Subsequently, it was discovered that the seeds were toxic, and even then some encouraged people to grow it to deter nematodes and other garden pests. According to them, if planted around garden species, the toxins in the leaves and stems discouraged aboveground pests and the poisons in the roots stopped nematodes from attacking below the ground.

Throughout the range of the genus *Crotalaria* in the Americas, the species often have common names alluding to the rattling dry seed pods. Perhaps the most famous to English-speaking people is shake-shake, mostly applied to *C. incana.* However, that species and others are known as *tcha-tcha* (onomatopoetic for the rattle sound) or *sonnette* (little bell) in the French Antilles. People in Belize call it wild-piss-a-bed. *Crotalaria pumila* is called *pois-savanne* (savanna bean), *pois-sonnette* (little bell bean), and *tcha-tcha* at least in Martinique and Guadeloupe. In Spanish areas the various species tend to be called *cascabelillo* (little bell), *matraca* (a type of musical instrument), or *sonajuelas* (little bells).

The American native species *C. incana* perhaps encouraged others to be used in curing rituals also. In several places in the Caribbean and in South America where the plants are called shake-shake, they are used to treat mute children. The reasoning seems to be that they will help "because the plant speaks" (Nellis 1997). Discussions with herbalists in the Guyanas indicate that they do not always distinguish between *C. incana* and several other species (Austin and Bourne 1992). Also, the widespread application of the "rattler" common names may denote lumping of similar species. Root teas from several species have been used in the Americas and the Old World to treat skin problems. The name *chipil* (Sonora) is probably related to treating skin diseases with *C. pumila.* That name seems to be based on *chipeliui* (Náhuatl), which refers to curing lesions caused by a skin disease.

Both *C. incana* and *C. retusa* are now in Florida, but they are not included in the native ethnoflora because they are aliens introduced within historic times. We do have one native that has been used. *Crotalaria pumila* appears in the vocabulary of the Seminoles as *casapó:cî* [*sacasapkó:cî, satcasapkotci*] (*svcvsakkv*, rattle, *o:ci*, small, Mikasuki). Those names are close if not identical to *shachachapkooche* (*svcvsakkv*, rattle, *oce*, little, Mikasuki) and *kvsvkakuce* (*kvsvkicetv*, rattle, *uce*, little, Creek) reported by Snow and Stans (2001) for *C. rotundifolia.* It may be that more than one species was used. The Catawba called it *sḛ́'sereʔha* (rattle) or *tapasi̱'sare* (something rattles), but they used it to treat mental problems and teething in infants (Speck 1934, 1944).

The Mikasuki name came from an unpublished 1941 manuscript used by Sturtevant (1955), but he found no informants who knew the plants. Perhaps any use previously known had been lost by his consultant, but Alice Snow now remembers it (Snow and Stans 2001). The people of Mexico still use it.

Names for the plants appear in Mexican languages. The Mayans of Yucatán know *C. pumila* as *sakpet* (*sak,* white, *pet,* small portion). The Mayan name is said to be used because the plants grow in small clumps that contrast with the other vegetation because of its whitish-green color. The plant is used in "small portions" as a condiment in some areas of Cozumel and perhaps elsewhere in Mexico (Valdez et al. 1989).

There are older records of people using the plants elsewhere in Mexico and Central America. A collection of *C. pumila* from Temascaltepec in the state of Mexico (*George B. Hinton 1636* NY) made in 1932 notes, "edible, boiled with ripeseed" (von Reis and Lipp 1982, Hodgson 2001). The Zapotecs in Oaxaca gather *C. pumila* from maize fields and call it *sij* in their language or *chipil* in Spanish (Messer 1978). To distinguish it from *C. incana,* they call that wild plant *sij gihs* (wild *sij*) or *chipil de campo.* The young foliage is eaten, often made into *tamales de chipil* with dough from *maize de elote* (green corn). Similar use has been made in Guatemala (Macbride 1945 in Messer 1978).

Farther north in Sonora, the plants are *cascabel de víbora* (snake rattle), *buli de Dios* (God's rattles,

Guarijío), *huevitos del toro* (little bull "eggs," usually a euphemism for testes), and *jojolino. Cascabel de víbora* notes the old similarity of the rattles in the plants and snakes. The Guarajío name alludes to a similar sound made by rattles during dances. *Jojolino*, however, seems to be a variant of *ajonjolí*, or sesame (*Sesamum indicum*). That common name, *ajonjolí*, is applied to *C. incana* in the Sonoran region (Hodgson 2001).

Seeds of *Crotalaria* and *Sesamum* are about the same size, and bear a slight resemblance to each other. Perhaps the common names comparing both these species to *Sesamum* are relicts from past widespread utilization.

Knowledge of human usage in Sonora has almost ceased to exist, except for a few records. Gentry (1942, 1963) observed the Guarijío eating the seeds of *C. incana*, and also grinding the seeds of an unidentified species before eating them as *pinole*. Modern Guarijío recall that the plants were eaten but have never tried them (Yetman, Feb. 2002, personal communication, Yetman 2002).

The common names plus the records in Mexico suggest that the practice of eating the plants as a condiment or the seeds as *pinole* was previously more widespread. Perhaps the Seminoles learned the usage from their predecessors in Florida, the Glades people.

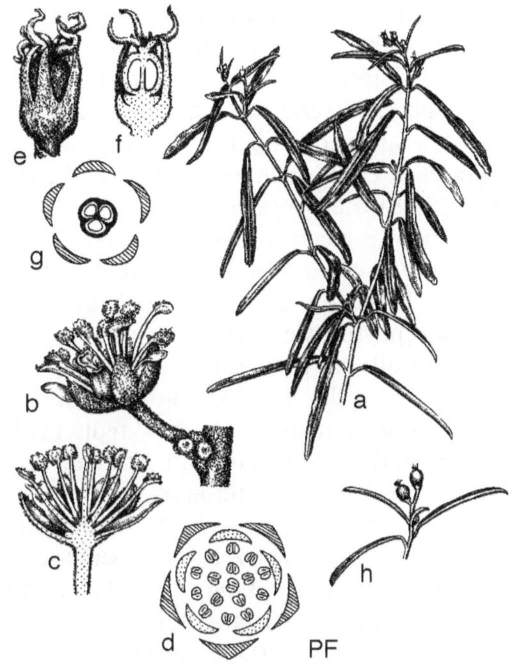

Croton linearis. a. Fertile branch. b. Staminate flower, side view. c. Staminate flower, longitudinally dissected. d. Floral diagram of staminate flower. e. Pistillate flower, side view. f. Pistillate flower, longitudinally dissected. g. Floral diagram of pistillate flower. h. Fruiting branch. *Drawn by Priscilla Fawcett.* From Correll and Correll 1982.

Croton: Grannybush

(From Greek *kroton*, the name of "ticks," noting the similarity of the seeds to those blood-engorged ectoparasites)

Start walking in the pinelands in the lower Florida Keys and you will find small herbs known locally as grannybush (Bahamas, Florida) or pineland croton (Florida). You find the same plants if you continue up the coast to Miami-Dade County. However, by the time you reach Broward County, the plants are gone. Just as you think they have disappeared completely, they reappear in northern Palm Beach and Martin Counties. These pineland herbs have a disjunct distribution in southeastern Florida.

Technically, the herbs are *Croton linearis*, and they usually grow up to about 1 m (3 feet) tall. Wherever they grow, they are the unique host for the Goatweed butterfly (*Anaea andria*), so its colonies are similarly disjunct.

In the neighboring Bahama Islands, the herbs may be called bay wormwood, marigold, or muckle (mispronunciation of "myrtle"). The plants grow throughout the islands and are not always associated with pinelands. There, as in Jamaica, they tend to be on limestone substrates. Maybe it is the overburden of acid to neutral sands that prevents them from being in Broward and most of Palm Beach Counties.

In classical Greek literature, the word *kroton* referred to plants and to the ectoparasitic arachnid we call a tick (Anglo-Saxon *tica*; also Greek, *zecke*; Germanic about A.D. 800 and equated with *ricinus*). When Linnaeus came to study the plants, he discovered the old confusion of naming plants and the parasitic animal with the same word.

There were other classical words available. For example, Romans such as Pliny (A.D. 23–79) also used *ricinus* (diminutive of Latin *rica*, little veil) to indicate what we now call the castor bean. Linnaeus continued that application with *Ricinus communis.*

Most of the earlier botanist-herbalists had used *Ricinoides* (resembling *Ricinus*) for the other group of plants. Linnaeus needed a better word to distinguish between these two different kinds of plants, and so he adopted the old Greek *kroton*. Both names now designate distinct members of the spurge family, Euphorbiaceae. To end the confusion with animals, he called the most common tick genera *Dermacentor* (Greek, skin-piercer) and *Ixodes* (Greek, *ixod*, birdlime, because they stick to the skin).

Anyone who has ever tasted, or even seen anyone taste, castor oil knows that there are incredibly foul chemicals in the liquid. The same may be said for extracts of *Croton*, and seeds in both genera contain oils that may be lethal if ingested in the wrong quantities. In spite of their inestimably poisonous nature, people have used extracts from seeds of both plant genera as medicines for thousands of years. Some species of *Croton* contain oil that can be lethal if as few as 20 drops are ingested.

One famous medicinal species is Asian *C. tiglium*, which has up to 30 to 50% oil. In the oils are compounds that have been called "crotin," but they are actually two phytotoxins (Hocking 1997). Properly diluted, these oils have been used as laxatives longer than history records. The oil and its constituents (including resins containing phorbol, tiglic acid, and other compounds) are drastic cathartics in humans, and expel worms in domestic animals. That oil has been used as an epispastic, and by malingerers to simulate skin diseases because it is a vesicular counterirritant and corrosive (Hocking 1997). The croton oil is also found in leaves and stems of many species but in lesser quantities. The compound itself is not carcinogenic but may augment carcinogenesis initially induced by aromatic hydrocarbons, at least on the skins of mice (Lewis and Elvin-Lewis 1977). The compounds in the seeds also have tumor-inhibitory properties.

From this extreme species, the other end of the spectrum of chemical constituents and human uses is *C. eluteria* (cascarilla bark, sweet bark, sweetwood bark, Eluthera bark, Bahamas). That plant has been used to flavor liqueurs (quinine substitute), mixed as an aromatic with tobacco, and as a fumigant (Morton 1981). Until about the 1920s, it was part of the U.S. Pharmacopoeia and U.S. National Formulary. These plants contain the volatile oils eugenol, limenone, vanillin, cascarillin, and they are bitter, tonic, and spicy.

Croton linearis was not named until 1760, after being discovered by Nicolaus von Jacquin. People living where it grows find it closer in the chemical spectrum to *C. eluteria* than *C. tiglium*. In fact, grannybush has been used as an adulterant for cascarilla bark.

John K. Small (1933) recorded that people in Florida used grannybush as a tea. That must have close to the end of the time it was used there because no usage has been found in the state since. However, people continue to use it in the Bahamas, Cuba, and Jamaica. They consider the plant decoction a remedy for colds, fevers, rheumatism, stomach trouble, menstrual cramps, and insect bites. To make the brew for insect bites, the plant is boiled longer to make it particularly strong (Morton 1981).

This is an important medicine for women's problems. Old grandmothers are its primary prescribers, and that is how it got the name "grannybush." For those purposes, the plant is tied in a bundle and boiled with rum. The resultant mixture is given as a refreshing, restorative drink to new mothers for 9 days after they give birth. To reduce fever, patients are bathed with a combined decoction of the plant and *Pluchea carolinensis*. Jamaicans use a hot bath of *C. linearis*, which they call rosemary (Caymans, Jamaica), to relieve colds. Formerly, the dried powdered leaves were prescribed in Jamaica as a remedy for colic. Jamaicans also use it, as they do true rosemary (*Rosmarinus officinalis*), in washing their hair.

The Jamaican common name is related to comparisons with rosemary elsewhere. For example, in the Dominican Republic it is *romarin* (false rosemary). Cubans say *romero de costa* (coastal rosemary). Others may say it is Spanish rosemary or wild rosemary (Bahamas, Jamaica). Those names, and *Croton* substitution for rosemary, are perhaps related to the Creole French name *trop faire pain mourri* (make too much bread, and you will die, Haiti).

Other names in the Caribbean are more obtuse and obscure. For example, in Haiti it is called *copahy*, a word whose meaning is unknown. Perhaps *copahy* was derived from Taino *cabs*, often rendered as *capá* (Coll y Toste 1972). Although it does not sound like it, the name *escupe culebra* (snake spit, Dominican Republic) alludes to a medicinal use. Surely, the appellation sassparilla does the same. For decades after its discovery in the Old World, all *Smilax* roots were called either China root or sassparilla (see *Smilax:* Green-Briar). There is no obvious similarity between *Smilax* and *C. linearis*, so the allusion must be to their uses as beverages or medicines.

The name *tremolino prieto* (black little trembler, Dominican Republic) may indicate the plants blowing in the breeze, but more likely that is another medicinal comment. That suggestion is made more probable when it is noted that several species of irritant *Croton* and *Oxalis* are called *tremolino* in Hispaniola. Moreover, the people there separate *tremolino prieto* from *tremolino blanco* (applied to *Corchorus hirsutus*, *Croton discolor*, and *Phoradendron anceps*).

Grannybush contains a series of alkaloids. Among those alkaloids are bases A (pronuciferine, *N,O*-dimethyl-crotonosine), B, and E. The plants also contain 8,14-dihydronorsalutaridine, 8,14-dihydrosalutaridine, linearisine, and pronuciferine. Plants are able to create crotonosine from linearisine (Willaman and Hui 1970). Additionally, proaporphine and aporphine alkaloids have been characterized (Casagrande et al. 1975), as has a diterpene with insecticidal properties (Alexander et al. 1991).

Of the 750 or so species in the genus *Croton* (Mabberley 1997), a surprisingly large number have been used by people. Most (possibly all) of the plants are apparently chemical factories for a complex of compounds. Extracts range from the most drastic laxative known (*C. tiglium*), to palatable and healthy beverages (*C. linearis*), flavor additives and enhancers (*C. eluteria*), host plants for lac insects (*C. laccifer*), hosts for butterflies (*C. linearis*), and as a source of timber (*C. megalocarpus*), and wound-healing chemicals (*C. lechleri*, Porras-Reyes et al. 1993). Hocking (1997) lists 36 species that have been used worldwide, and Moerman (1998) lists five for temperate North America.

Still, ingestion of *Croton* is a gamble that you have identified the species properly. One species is lethal, and another provides pleasing teas. Perhaps too many mistakes occurred and that is why grannybush is no longer used in Florida.

Ctenium

(From Greek *kteis* or *ktenos*, a comb, diminutive *ctenion*, a small comb, from the appearance of the flower spike)

***Ctenium aromaticum**. Drawn by Mary Wright Gill and Agnes Chase.* From Hitchcock and Chase 1950.

Ctenium aromaticum (fragrant)
tooth-ache grass [toothache] (first located in Williams [1837] 1962)

Ctenium was named in 1813 by German taxonomist Georg Wolfgang Franz Panzer (1755–1829), who called the grass *C. carolinianum*. Technically that species name is a synonym of *C. aromaticum*, because Walter had named it in another genus in 1788. However, Panzer's *C. carolinianum* was used until the principal at the Brooklyn Female Academy, Alphonse W. Wood (1810–1881), made the correction in 1861. As now known, *Ctenium* has 17 species confined to the Americas and Africa. Not only is the widespread North American *C. aromaticum* in Florida, but *C. floridanum* is endemic to the northeastern part of the state. Like wiregrass (*Aristida beyrichiana*) with which *Ctenium* grows, it flowers only after fires burn through the pinelands (Mabberley 1997).

No actual records of human use of *Ctenium* for toothache have been found. However, the common name suggests that its pungent taste has been used to relieve that problem. Moreover, Williams ([1837] 1962) wrote, "It affects the breath and milk of cows, who eat it when young and tender. The root is bitter, and affects the salivary glands." Hocking (1997) also indicated that the grass has been used as a sialagogue (to increase saliva flow). Fernald (1950) noted that the bruised roots are fragrant, perhaps containing some of the same chemicals as aboveground parts. Gamboa-Leon and Scott-Chilton (2000) identified isobutylamide as the numbing agent. Barkworth (2003) wrote, "The roots are spicy when freshly dug."

Cucurbita

(Classical Latin name for the bottle gourd, *Lagenaria*; reapplied by Linnaeus to the New World plants)

Cucurbita okeechobeensis (from Lake Okeechobee)
ciko:yi (Mikasuki)
Okeechobee gourd (Florida)

Cucurbita pepo (originally ripe or tender, but later applied to the melon as Latin *pepo* and Greek *pepon*)
Pumpkin (derived from Old World Greek *pepon* or Latin *pepo*, for a New World plant; *Pepon* evolved into French *pompon*, Italian *popone*, and Portuguese *abóbora* and finally reached English as "pumpkin" ca. 1654); *abóbora* (from *pepon*, Portuguese); *peapag* (from *pepon*, Gaelic)
calabaza (derived from Arabic *qar'ah yábisah*, dry gourd, a name originally for *Lagenaria*, Spanish, Mexico to Peru)
cocuzza (slang, originally for *Lagenaria*; probably locally applied to *Cucurbita*, Italian)
courge [*gourde*] (from *Cucurbita*, French)
giromons [*giraumon*] (*giraumont* was in use by 1734; then, Le Paige Du Pratz used it in Louisiana in 1758, mostly used for *C. maxima*

and *C. moschata*; from Tupí *yuru'mu*; French, Louisiana); *jerimum* [*jerimu, jerumu, jirimu, jirumum, jurumu, gerimum, gurumu, girimu, girumum*] (from Tupí, maybe first used by Piso and Marcgraf in 1658, Portuguese, Brazil)
Kürbis [*Kirbiz*] (German)
zucca (originally for *Lagenaria*, later extended to *Cucurbita*, Italian)
Squash (came into English from two directions: The older meaning, from ca. 1565, came from Latin *quassare*, to crush; the younger derivation, for *Cucurbita* fruits, came from Algonquian words appearing in print ca. 1634 as *isquoutersquashes*, with variant spellings as *squantersquash*, or *squoutersquashes*; Hedrick 1919)
abóbora (from *pepon*, Portuguese)
a͟nta aho͟nni ("pumpkin with rind bent," for crook-necked squash, Biloxi)
calabaza (originally used for *Lagenaria*, Spanish)
citrouille (in use as "*sitroules*" by Champlain in New England in 1605; resembling *Citrus*, French)
cushaw (winter squash with a curved neck; akin to Algonquian *askushaw* or *escushaw*, it is green, appearing by 1588; probably a word related to a root called *coscúshaw*; see *Peltandra*)
cymling [cimnel, simnel, simnel-gourd, simblin, simlin, symnels] (Hedrick 1919 thought this might be a native American word for white, scalloped squash, also called the pattypan; perhaps from Old French *simenel*, or *seminel*, maybe related to Latin *similia* and Greek *semidalis*; in use by 1648 at the mouth of the Susquehanna River; cymling in use by 1794, and Thomas Jefferson wrote cimnel in 1803)
Kürbis (from *Cucurbita*, German)
[vegetable] marrow (squash, from Old English *mearg*, of Teutonic origin; applied to *Cucurbita* by 1816)
zucca (Italian)
Gourd (from Middle French *gourde*, itself from *Cucurbita*; the diminutive of *gourde* or its variant *courge* is *courgette*, French)
calabash (from French); *calabasse* (French); *calabaza* (derived from Arabic *qar'ah yábisah*, dry gourd, a name originally for *Lagenaria*, Mexico to Peru, Spanish)
cetriolo (gerkin [and gourd?], Italian)
cocomero (watermelon [and gourd?], Italian)
Flaschenkürbis (bottle gourd [pumpkin], German)
Melon (from French *melon*, in use by A.D. 1387, when "melon" appeared in English); *melão* (Portuguese); *meloen* (Dutch); *melon* (from Latin *melopepo*, French, Spanish); *Melone* (German); *melone* (Italian)

North America Languages

askútasquash (eaten raw, Natick and Narragansett, Connecticut and Rhode Island)
atowetowe·w (cymbling or summer squash, Powhatan, Virginia; Strachey in [1612] 1953 recorded the name as *autowtaoh*, and translated it, "an ear of wheat"; Siebert 1975 found cognates in Penobscot as *atówatawan*, and *atówatawe*; at least the first element of the Narragansett *askútasquash* is cognate)
casila:nî (loan-word from the Spanish *castillano*, Creek)
choksi (squash, but especially winter squash, Alabama); *coksí* (Koasati, Mikasuki); *coksláknî* (*coksí*, squash, *lákni*, yellow, any yellow squash, Mikasuki); *cokslakno:cî* (little yellow pumpkin, Mikasuki); *cvse* [*cv'se, chasi, chassa*] (pumpkin, Creek, Muskogee); *kochi* (gourd, Alabama); *lokosh* (Chickasaw); *tvhóyv* [*tvhvyv*] (squash, Muskogee); *choksiáyyi* (*choksi*, squash, *tayyi*, woman, Alabama); *choksichàakka* [*choksichàka, choksichakka*] (*choksi*, squash, *chàakka* = ? *chàachi*, older brother; cushaw, a large, long-necked, striped squash, Alabama); *choksilaana* (any yellow squash, but especially pumpkin; *choksi*, squash, *laana*, yellow, Alabama); *choksiłahkachi* [*choksiłaakachi, choksiłahkachi*] (*choksi*, squash, *łahkachi*, becoming white-haired [striped squash, cushaw], Alabama); *isito* (Choctaw); *isito holba* (squash; *isito*, pumpkin, *holba*, resembling, Choctaw); *olbi'* (squash, Chickasaw); *ostahi* (squash, Chickasaw); *osto* (pumpkin, Chickasaw); *yátkitiscíncoksî* (Indian's pumpkin, Mikauski)
efépe, fépe (gourd, Muskogee)
gescundhàckall (Delaware)
i-ya [*iya*] (pumpkin, Cherokee)
macócqwer (included pumpkins, melons, and gourds, Carolina Algonquians); *máhkahkw* (pumpkin, Delaware; the word is cognate with other Algonquian languages where it means "tub, cask, box, keg" or "container"; those words include Cree *mahkahk*, Fox *mahkahkwi*, Menomini *mahkan*, Ojibwa *makkakk*, and Shawnee *mka?kwi*); *mahkahk* (Capt. John Smith and Strachey in [1612] 1953 also wrote it as *macock makokos*, *makawke*, and *mahcawk*, Powhatan, Virginia; Strachey's word *macaugscaunesmes* should be translated "pumpkin seeds")
mo'yum [*mo'yu*] (Atakapa)
osaw'puk (Plains Cree); *owgîssimaû'n* (tangled hairs, referring to the placenta, Potawatomi)
patnga (Hopi)

pemesk (Strachey wrote *pamyack* in [1612] 1953; Siebert 1975 suggests that the word is cognate with *wi·ne·mehkwan* in Menomini; this same tribe has transferred the word to the introduced watermelon, *Citrullus lanatus*, as *aski·-mehkwan*)
quaasiens (in 1642–1653, Van der Donck said, "a name derived from the aborigines," maybe from *askútasquash*, New York)
tocobago (has been translated as "place where gourds are produced," Timucua?)
wa'tap hany'? (pumpkin seed, Catawba)
wamnu (pumpkin and squash, Dakota); *wagamun* (pumpkin and squash, Teton); *watan* (pumpkin and squash, Omaha-Ponca); *wa-to<u>n</u>* (Osage); *watan hti* (real squash, Omaha-Ponca); *watan miha* (small, spherical, spotted black and green, Omaha-Ponca); *watan nide bazu* (large oval, pointed at both ends, greenish color, Omaha-Ponca); *watan kukuge* (speckled, Omaha-Ponca); *watan niha snede* (long *watan miha*, Omaha-Ponca); *watan miha ska* (white *watan miha*, Omaha-Ponca); *watan miha saba* (black *watan miha*, Omaha-Ponca); *watan miha zi* (yellow *watan miha*, Omaha-Ponca)

Middle American Languages

apí (Bribrí, Costa Rica)
ayotero (the plant, Náhuatl, Mexico through El Salvador and Guatemala)
ayotli [*ayote*] (pumpkin or squash fruit, Náhuatl, Mexico through El Salvador and Guatemala)
bát (Brunka, Costa Rica)
caay ixám (horse's gourd, Seri, Sonora)
calabaza (pumpkin, derived from Arabic *qar'ah yábisah*, dry gourd, a name originally for *Lagenaria*, Spanish, Mexico to Peru)
chichicayotli (*chichi*, bitter, *ayotli*, squash, Náhuatl, Mexico)
chiverres (Costa Rica)
gueto bichim (*gueto*, squash, *bichim*, deer, Zapotec, Oaxaca); *gueto* or *giht* (pumpkin or squash, Zapotec, Oaxaca)
ka (pumpkin and squashes, Maya, Yucatán)
ma she (gourd, Tarahumara, Chihuahua); *ma'she'* (pumpkin or squash plant, Chinantec, Oaxaca)
moe (Cuna, Panama)
nu (Jicaque, Honduras)
pís (Cabécara, Costa Rica)
póri-chla (Guatuso, Costa Rica)
xam (*Cucurbita*, Seri, Sonora); *xam cozalc* (ribbed squash, *C. pepo*, Seri, Sonora)

South American Languages

auyama [*ahuyama*] (pumpkin, first learned when Columbus arrived on Terra Firme, fide Sauer 1969, Arawak); *uauíama* (pumpkin, Carib)
zapallo (from *sapallu* or *zapallu*, mostly used for *C. moschata* and *C. maxima*, Quechua, Mexico to Panama)

There is a small park preserving the site of a historic indigenous village visited by the early Spanish on a peninsula west of Tampa Bay. The place was known as *Tocobago*, which been translated as "place where gourds are produced" (Fontaneda [1575] 1944). Maybe in part that translation is why John K. Small (1922) thought that southern Florida was the "Garden of Eden" for the pumpkin (*Cucurbita pepo*) and its relative the Okeechobee gourd (*C. okeechobeensis*).

The Okeechobee gourd is native to Florida, but its closest relative, *C. okeechobeensis* ssp. *martinezii*, grows in Mexico (Walters and Decker-Walters 1993). That means that the closest relative to the Okeechobee gourd is native to Chiapas, Oaxaca, San Luis Potosí, and Veracruz. If that were the only native gourd in Florida, it would be intriguing enough, but there is another species. That other species is *C. pepo*, the one that comprises the gourds, pumpkin, and squashes.

Columbus's arrival in the Caribbean in 1492 marked the Europeans' first encounter with the genus *Cucurbita*. This New World genus produced fruits similar to those of cucumbers and watermelons that the Spanish knew at home, but the explorers noted the differences. Columbus called *Crescentia*, *Cucurbita*, and *Lagenaria* by the only name he knew—*calabazas*, and he probably found all three in Cuba (Sauer 1969). That he did find *Cucurbita* there is indicated by his comment that these plants were "more delicate and lacking the little spines that the plant of the *calabaza* [*Lagenaria*] has."

The first historical record of *Cucurbita* in Florida was from the expedition of Pánfilo de Narváez in 1528 (Swanton 1939). After passing through *Apalachee* and what is now Tallahassee, the Spanish encountered squashes in the land of the *Aute* [*Ochete*] near what is now known as Wakulla Springs (Wakulla County) in the Florida Panhandle. This is probably the locality where Alvar Nuñez Cabeza de Vaca (1490–1564) recorded the following in his *Relaciónes* of 1555: "About their howses they have commonly square plotts of cleered grownd, which serve them for gardens, some one hundred, some two hundred foote square, wherein they sowe their tobacco, pumpons, and a fruit like unto a musk million [*Cucumis melo*], but less and worse, which they call macock gourds [*Cucurbita*], and such like, which fruiets increase exceedingly, and ripen in the beginning of July, and contynue until September; they plant also the field apple, the maracock [*Passiflora incarnata*], a wyld fruit like a kind of pomegranett, which increaseth infinitlye, and ripens in August, contynuing untill the end of October, when all the other fruits be gathered, but they sowe neither herb, flower, nor any other kynd of fruiet" (Gilmore 1919).

Many variations of *C. pepo* were developed by prehistoric people (Cutler and Whittaker 1961), and two lineages were brought into cultivation. The most famous lineage of pumpkins (*C. pepo* ssp. *pepo*) was domesticated in Mexico, but the other subspecies, which includes various squashes and ornamental gourds, was domesticated in the eastern United States (Decker 1988, Decker-Walters et al. 1993). *Cucurbita pepo* ssp. *ovifera* includes scallop and crookneck squashes and most ornamental gourds. The oldest record of this subspecies in Florida was as a wild plant at the end of the Pleistocene, and it has been in cultivation in the eastern United States since 5500 (± 1500) B.P. (Newsom et al. 1993, Fritz 2000b).

Squashes and pumpkins are cultivated lineages of *C. pepo*. There are summer squashes, winter crook-necked squashes, autumn squashes, and autumn pumpkins. To make it more confusing, the marrows or vegetable marrows are in the same group. The "Seminole pumpkin" (*C. moschata*) is an introduction of a plant domesticated in Mexico (Robinson and Decker-Walters 1997). Apparently, the ancestors of the Seminoles brought this variety into cultivation.

Investigations of two kinds of gourds unearthed in the late 1890s turned up an unexpected southern record for *C. pepo* ssp. *ovifera*. The other gourd was *Lagenaria siceraria*, the bottle gourd. Those remains of pre-Columbian gourds had been found by the flamboyant and sometimes controversial Frank Hamilton Cushing.

In May of 1895, Cushing became convinced that Colonel W.D. Collier's garden had been inhabited by prehistoric people. So, in February of 1896, he returned to excavate the site on Marco Island in Collier County. When he did, he uncovered some of the most spectacular botanical finds in North America (Cushing 1897).

While Marion S. Gilliland (1975) was preparing her study of Cushing's material in the 1970s, she called in renowned cucurbit expert Hugh C. Cutler of the Missouri Botanical Garden. Cutler quickly realized that the material was the first ever found of *C. pepo* south of Kentucky in the United States (Cutler 1975). *Cucurbita pepo* was one of two species of gourds used by the Calusas, who probably abandoned Marco Island in the late 1400s, just before the Spanish arrived.

The English names for these plants are themselves confusing. We use the word "pumpkin," derived from Old World Latin *pepon* or *pepo*, for a New World plant. The Latin *pepon* evolved into French *pompon*, Italian *popone*, and Portuguese *abóbora*. It finally reached English as "pumpkin" after being spelled *pepone* (1545), *pepon* (1578), *pompon* (1586), *pompion* (1588), and *pumpion* (1606). *Pepon* was even incorporated into Gaelic as *peapag*.

Originally, *pepo* simply meant "melon" and either referred to the "watermelon" (*Citrullus lanatus*) or "muskmelon" (*Cucumis melo*), or both. Those two genera are native to Africa and reached Europe long before the Americas were known. People in Europe knew the New World *Cucurbita* looked like the familiar melons, but at first they did not have words to distinguish them.

"Melon" itself is a word almost unrecognizable in the older literature. That is what Harriot ([1590] 1972) meant when he wrote "Mellions," and what Strachey meant in the 1600s with "million." Indeed, the literature from A.D. 1387, when "melon" appeared in English, also included the spellings *melone, millian, milon, myllon, milion, mellon, millon, million, mealon, meloune*, and *milleon*. In other European languages, they are *melon* (French, Spanish), *melão* (Portuguese), and *melone* (Italian).

The word "squash" came into English from two directions. The older meaning (ca. 1565) came from Latin *quassare*, to crush. The younger derivation, for *Cucurbita* fruits, came from New England. These Algonquian words appeared in print in 1634 as *isquouter-squashes*, with variant spellings as *squanter-squash*, or *squoutersquashes* (Hedrick 1919). Those names were from *askútasquash* (Natick and Narragansett, Connecticut and Rhode Island), which means "eaten raw" (Robinson and Decker-Walters 1997). Roger Williams, in 1643, was one of the first to record *askútasquash*, and he added, "their vine apples—which the English from them call squashes; about the biggness of apples of several colors." John Josselyn wrote in 1672 that these were "a kind of mellon or rather gourd, for they sometimes degenerate into gourds. Some of these are green, some yellow, some longish, like a gourd; others are round, like an apple; all of them are pleasant food boyled and buttered, and season'd with spice. But the yellow Squash, called the apple Squash, because like an apple, and about the bigness of a Pomewater [a large apple], is the best kind. They are much eaten by the Indians and the English."

Van der Donck, speaking of the fruits in "New Netherlands" in 1642–1653, added, "The natives have another species of this vegetable peculiar to themselves, called by our people *quaasiens*, a name derived from the aborignes, as the plant was not known to us before our intercourse with them. It is a delightful fruit, as well as to the eye on account of its fine variety of colors, as to the mouth, for its agreeable taste. . . . It is gathered in summer, and when it is planted in the middle of April, the fruit is fit for eating by the first of June. They do not wait for it to ripen before making use of the fruit, but only until it has attained a certain size. They gather the squashes, and immediately place them on the fire without any further trouble" (Hedrick 1919).

Historically, people used ripening time to distinguish between pumpkin and squash—squashes produce a summer vegetable; pumpkins ripen only in the fall. Gourds have bitter instead of sweet flesh, and a hard, somewhat woody rind. However, Robinson and Decker-Walters (1997) point out that this will not work. They define squash as all cultivated *Cucurbita.* These they divide into the summer (including marrow) and winter squashes depending on whether the fruit is used when immature or mature. The term "winter squash" refers to the ability of the fruit to be stored until the winter months. Summer squashes are typically *C. pepo*, but the winter squashes may be one of four species (*C. argyrosperma, C. maxima, C. moschata,* or *C. pepo*). Gourd is any cucurbit not used as food. The hard shell is not characteristic because many summer squash cultivars have a hard rind when mature; they are simply eaten before ripening.

If this confusion of names were not enough, the Spanish began calling pumpkins *calabaza* (derived from Arabic *qar'ah yábisah*, dry gourd, a name originally for *Lagenaria*, Mexico to Peru) or *zapallo* (from *sapallu* or *zapallu*, Quechua, Mexico to Panama; mostly used for *C. moschata* and *C. maxima*).

When Linnaeus tried to unravel all this babble of words, he settled on *Cucurbita* for the New World plants although it was the old Latin name for what we now call *Lagenaria.* From the word *Cucurbita,* we got the English "gourd" by way of Middle French *gourde.*

The uses for pumpkins, squashes, and gourds are many and varied. First and foremost, they are food. There are as many ways to eat them as there are people, and humans have selected a dazzling array of cultivars. Typically, the fall-fruiting, orange-fleshed types include pumpkins and acorn squashes. Those with lighter-colored flesh tend to be called squashes, including zucchini. Gourds come in many colors and shapes and now are mostly ornamental. The type called finger gourd, crown-of-thorns gourd, the gourd of the Ten Commandments, or the holy gourd is arguably the most unusual. There are also warted gourds, star gourds, and bicolored spoon gourds.

Fruits are boiled and baked as vegetables or used in soups, and they are also dried, ground into flour, and made into bread. Young shoots and leaves are cooked as greens. Flowers with the pistils removed are cooked and eaten or stuffed with meat and rice. Squashes and pumpkins also serve ceremonial purposes in many indigenous North American groups.

Cabeza de Vaca was already dead when Harriot ([1590] 1972) noted that the Carolina Algonquians called these plants "*Macócqwer*, according to their severall formes called by vs, *Pompions, Mellions*, and *Gourdes*, because they are of the like formes as those kindes in England. In Virginia such of several formes are of one taste and very good, and do also spring from one seed. There are of two sorts; one is ripe in the space of a month, and the other in two months."

According to Strachey, in the 1600s Virginia people "seeth a kind of *million* [melons], which they put into their walnut-milke, and so make a kynd of toothsome meat." Robert Beverley added in 1705 that people in the same area called the summer squash *macock,* which they "boil'd whole when the Apple is young, and the Shell tender, and dished with Cream or Butter, relish very well with all sorts of Butcher's Meat, either fresh or salt. And whereas the Pompion is never eaten till it be ripe, these are never eaten after they are ripe."

By 1758, Le Page Du Pratz may have been recording a Creole preparation of pumpkins (as *giromons*). He wrote, "they are cut into the shape of pears or other fruits and preserved thus with very little sugar, because they are naturally sweet. Those who are unacquainted with them are surprised to see entire fruits preserved without finding any seeds inside. The *giromons* are not only eaten preserved; they are also put into soups. Fritters [*bignets*] are made of them, they are fricasseed, they are cooked in the oven and under the embers, and in all ways they are good and pleasing."

Seeds alone are also used as medicine throughout much of the Americas. Typically, the seeds are given to children to rid them of intestinal worms. The active compound is cucurbitin, a water-soluble, non-proteineogenic amino acid (Dunhill and Fowden 1965). Besides being useful for those worms, the substance inhibits growth of immature liver flukes, *Schistosoma japonicum*, at least in culture (Fang et al. 1961). The bitter components are tetracyclic terpenes called cucurbitacins. These are antitumor compounds and are applied to ulcers that are difficult to heal. There is even evidence that the cucurbitacins may help with benign prostatic hyperplasia (BPH) (Swerdlow 2000, Duke et al. 2002).

Florida people had dry-fleshed gourd varieties that served as floats, rattles, and containers. The Calusa and probably other groups used the dry fruits as floats on nets they made from palm leaf fibers for fishing. In the 1560s, French artist Jacques Le Moyne found the Timucua using "two other rattle pumpkins filled with small stones or seeds" (Gilliland 1975). Jonathan Dickinson, the Quaker castaway on Florida's southeastern coast in 1699, recorded the Ais also using gourds as rattles and as containers for making black drink (*Ilex vomitoria*) (Andrews and Andrews 1945). All three tribes probably used gourds like those identified by Cutler (1975) from Marco Island.

The pumpkin is among the most famous of the gourd relatives because it is eaten and used at Halloween. Its use as a Jack-o'lantern at Halloween came into existence long after the Europeans came to the Americas. Celebrating Halloween is an Old World Gaelic event, and the Jack-o'lantern commemorates a man who bargained with the Devil. Since the Irishman kept outwitting the Devil and not delivering up his soul, the King of Darkness became angry. So, Mephistopheles damned the Irishman to eternal wandering. Before the man left, he dipped a turnip he had been eating into the coals of Hell. Now, he forever carries that torch before him at night.

Because turnips are poor lights, inventive Irish or Irish-Americans substituted the pumpkin for the turnip. A hollow lantern makes a better light. Now, the Old World legend and the New World plant are wedded in a custom that masks the origins of both elements.

Cuscuta

(From Arabic *kusuta, kshuta, keskhut,* or *kusu,* "a tangled wisp of hair"; a loan word from Aramaic *kesatha*; see Austin 1979)

cabellos de Venus (Venus's hair, Spanish); *cabelos* (hair, Portuguese); *cabelos de Jesus* (Jesus's hair, Brazil)
cetto em vpette (*cetto*, snake, *em*, its, *vpette*, shadow, Muskogee); *cetto entopv* (*cetto*, snake, *em*, its, *topv*, table [Oklahoma], bed [Florida], Creek)
lucv em pvtakv (*lucv*, turtle, *em*, its, *pvtakv*, pallet, Muskogee); *lucv em vpette* (*lucv*, turtle, *em*, its, *vpette*, shadow, Muskogee)
cuscuta (Italian, Portuguese); *cuscute* (French)
dodder (in use in English by A.D. 1265, and with cognates in German *Dotter*, Dutch and Danish *dodder*, and Swedish *dotra*; originally meaning yellow, as in the yoke of an egg; the word "dodder," as now used in English to mean "tremble," was not applied until 1627, and has a completely different etymology from "totter")
golden-thread
love vine (in use by 1885; however, "love-bind" was used by John Parkinson for *Clematis* in 1640)
Siede (German)
snyltetråd (*snylte*, parasite, *tråd,* thread, Norwegian)
strangle weed (originally applied to *Orobanche* by 1562; soon afterward used for *Cuscuta europaea*)
vamad giikoa [*vammat geekwa*] (snake headdress or crown, Akimel O'odham [Pima], Arizona); *vamad givud* (snake belt, Akimel O'odham, Arizona); *vamad vijin* (snake spider's web, Akimel O'odham, Arizona)
wegei washai (reddish grass, Tohono O'odham [Papago], Arizona)

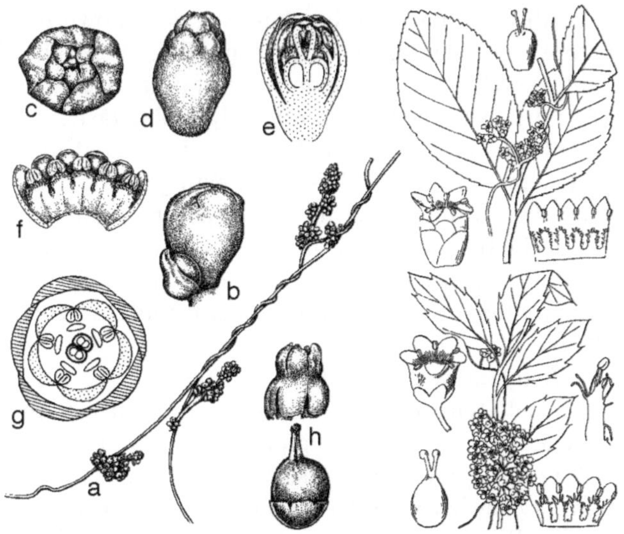

Cuscuta. *Cuscuta americana* (left). a. Section of flowering and fruiting stem. b. Flower bud. c. Flower from above. d. Flower, side view. e. Flower longitudinally dissected. f. Inner surface of corolla spread, showing the fringed scales and anthers. g. Floral diagram. h. Fruit below; detached corolla above. *Drawn by Priscilla Fawcett.* From Correll and Correll 1982. *Cuscuta compacta* (top right). From Britton and Brown 1898. *Cuscuta gronovii* (lower right). From Britton and Brown 1898.

Cuscuta americana (of America)
abrazos (hugs, Spanish)
aletria di mondi (*aletria* =?, *di mondi,* world's, Netherland Antilles)
amitié [*l'amitié*] (lover, Haiti, Martinique); *amor di neguer* (false love, Netherland Antilles); *amor falso* (false love); *amora falsoe* (false love, Netherland Antilles)
arhui-arhui

bass viola strings
bejuco amarillo (yellow vine, Dominican Republic); *bejuco de buey* (ox vine, Puerto Rico); *bejuco de mona* (monkey vine, Puerto Rico); *bejuco fideo* (noodle vine, Cuba); *bejucito de amor* (little love vine, Dominican Republic)
cipó chumbo (lead vine, Brazil); *cipó dourado* (golden vine, Belém, Brazil)
corde à violon (violin strings, Martinique); *cuerda de violín* (violin strings, Puerto Rico, Colombia); *cuerdilla* (little strings, Colombia)
cuscute d'Amerique (American cuscuta, Haiti)
devil's guts; *duivelsnaai-garen* (*duivel*, devil, *snaar*, string, *garen*, thread, Netherland Antilles); *hilu di diabel* (devil's string, Netherland Antilles)
dodder (Jamaica, Puerto Rico); yellow dodder (Virgin Islands)
enreda cotorra (parrot catcher, because the birds are easy to catch when they are tangled in the stems, Colombia)
fideillo (little noodles, Cuba); *fideítos* (little noodles, Dominican Republic); *fideo* [*fideos*] (noodles, Cuba, Dominican Republic, Puerto Rico, Colombia); *fios de ovos* (egg noodles, Belém, Brazil)
hell weed
herbe à amitié (lover's herb, Martinique); *herbe à amourette* (love vine, Martinique)
hierba mala (weed [literally bad herb])
hilo de oro (cord of gold, Venezuela)
kanlecay (*kan*, yellow, *le*, leaf, *kay*, fish; because the scales of the stems remind them of fish scales, Maya, Yucatán)
love bush [vine] (Jamaica, Virgin Islands, Barbados); love weed
planta de bruja (sorceress's plant)
planta sin pie (plant without a foot [base])
rejillo (Colombia)
strangle-weeds (South)
wild vermicelli
yellow love (Virgin Islands)

Cuscuta compacta (compact, referring to the dense inflorescences) (as *C. paradoxa*)
compact cuscuta (a book name; a translation of the Latin)
hakastahkata (*hakastah*, vine, *kata*, yellow, Pawnee)

Cuscuta gronovii (for Jan F. Gronovius, 1690–1762)
scaldweed (there are two meanings for "scald" and it is not clear which may be applicable here: perhaps the intent is to indicate that the plant can be used to treat "scalds" or burns from a hot liquid, a meaning in use since about A.D. 1300; the other meaning is "scaly," in use by the 1530s, and that is not obviously applicable)

Cuscuta umbellata (umbellate, the inflorescences)
bejuco fideo (noodle vine, Cuba)
cipó chumbo (lead vine, Brazil)

When I worked in Belém, Brazil, I was surprised to find people encouraging *Cuscuta* on their croton (*Codiaeum*) hedges in their front yard gardens. When we asked them about the plants, they told us they were *cabelos de Jesus* (Jesus's hair). They considered them not only ornamental, but also medicinal. The name seemed to be associated with the Doctrine of Signatures. However, when I looked more closely at the literature, I discovered that an older name was *cabellos de Venus* (Venus's hair). As both Venus and Jesus are associated with love, albeit different kinds, both are equally appropriate for the plants.

While people derived from Western European temperate regions uniformly regard *Cuscuta* as a pest, that is not true of other cultures. Among the indigenous people of North America, for example, there are records of the Cahuilla, Cherokee, Diegueño, Kawaiisu, Navajo, Paiute, and Pawnee using the plants for food, medicine, cleaning "pads," and dyes.

Only the Cherokee in the east were found to have used *Cuscuta*. They applied *C. gronovii* as a poultice for bruises (Hamel and Chiltoskey 1975). Farther west the Pawnee used *C. compacta* to predict whether or not their suitors were sincere, and made an orange dye from it (Gilmore 1919). Paiute knew the parasite as "woman without children," and ate it as a contraceptive (Moerman 1998). Wooton and Standley (1915) and Yanovsky (1936) found that the Navajo have eaten the parched seeds of *C. megalocarpa* for food, and Moerman (1998) said they have used it as an emetic in ceremonies. The Pima do not distinguish species but use generic terms for the plants. Historically they considered them poisonous, and believed that "if a snake sees them take the plant [on which it grows], the snake will get after them" (Curtin 1949). Now, they regard *Cuscuta* more as an agricultural pest (Rea 1997).

Cubans use *C. americana* to treat liver problems and as a laxative (Roig 1945). In the Bahamas, *C. americana* is considered an aphrodisiac and a bath for prickly heat and itching (Morton 1981). In Haiti it is used against jaundice. A decoction is used on sores as an antiseptic; it is considered antibilious, laxative, carminative, and antiparasitic (Liogier 1978). Hocking (1997) considered the plants tonic, astringent, and alterative. It is used with *Tecoma stans* to lower fever in the Virgin Islands. Jamaicans drink a tea for colds, colic, and marasmus in children (Morton 1981). They mix it with *Borreria laevis* and *Zebrina pendula* as an emmenagogue. In Trinidad, a *C. americana* infusion is taken for jaundice, and a decoction is a bath for marasmus. In Yucatán, *C. americana* is used to relieve dyspepsia. In Suriname it is used to stop dysentery and urinary problems.

To treat sores, Cubans use *C. umbellata* as an antiseptic, and they consider it diuretic, tonic, and good for the stomach (Roig 1945). LeCointe (1947) recorded that in the Lower Amazon in Brazil, a powder of the entire plant of *C. umbellata* was applied to cure sores and ulcers. He considered an infusion as astringent, stomachic, diuretic, and good to treat severe diarrhea. It was also gargled to relieve laryngitis. Hocking (1997) listed *C. umbellata* as vulnerary, and noted that the fresh juice had been used against angina and tuberculosis.

Mors et al. (2000) noted that most Brazilians used all *Cuscuta* similarly. They found that they were emollient, anti-inflammatory, stomachic, anticatarrhal, purgative, depurative, diuretic, and antiblennorrhagic. *Cuscuta* was used in treating throat infections, liver colics, bleeding diarrhea, tuberculosis, wounds, and sores.

Because of their yellow or orange colors, it is not surprising that some have been found to have beta-carotene, alpha-carotene-5, and they also have 6-epoxide, taraxanthin, and lutein (Hsu 1986). Ergoline alkaloids have been reported in some (Mandrile and Dongiorno de Pfirter 1990), but Eckert Eich (personal communication, May 2003) says they are incorrect. The species he and his students have examined do not have ergoline alkaloids. Most members of the Convolvulaceae have calystegines, but they too are absent from *Cuscuta* (Schimming et al. 1998).

Cynoglossum

(From Greek *kynos*, dog, and *glossa*, a tongue, referring to the leaves, which people thought resembled a hound's tongue)

äsûkitä'boe (stickers, for *C. officinale*, Potawatomi; compare with Cherokee below)
cinoglosso (hound's tongue, from Greek, Italian); *cynoglosa* (Spanish); *cynoglosse* (French); *hundetunge* (Norwegian); *Hundszunge* (German); *langue de chien* (French); *teang' a' choin* (Gaelic)

Cynoglossum virginicum (of Virginia)
dog-bur
hound's tongue
unistil'unisti (they stick on, Cherokee)
wild comfrey (comfrey is of obscure etymology; related to Old French *confirie*, and Medieval Latin *cumfiria*, perhaps related to *confirma*, to make well)

Europeans were well acquainted with hound's tongue when they arrived in the New World, although they had *Cynoglossum officinale* (medicinal). The American species is similar enough that they had no trouble distinguishing it as a new kind of hound's tongue. Culpeper (1653) wrote: "The root is cold, drying and binding, it is useful in catarrhous defluxions upon the lungs, and to temper the sharpness of the blood; and, by consequence, excellent for all kinds of fluxes and hemorrhages."

The Cherokee drank a decoction of root or tops for kidney troubles, and bruised the root with bear grease as an ointment to treat "cancer." They also drank a decoction of the plant to improve memory and used it in love charms (Mooney 1885–1886). The Cherokee also used it in "Green Corn" medicine, and against itch, particularly itching genitals (Hamel and Chiltoskey 1975, Foster and Duke 1990).

The Old World *C. officinale* contains the potentially cyanogenic alkaloids cynoglosine, consolidine, plus fatty oils, and choline. The alkaloids function as a central nervous system depressant (Foster and Duke 1990, Hocking 1997). Presumably, the American *C. virginicum* also contains those compounds.

Cyperus

(The ancient Greek *kyperos* or *kyperios*, the name of the galingale or sedge)

Cyperus haspan (an indigenous name in Sri Lanka for the sedge)
[haspan] flatsedge ("flat," not angled in this sense, is akin to Swedish *flat* and Dutch *flad*; possibly from Old Aryan *plat*, and related to Greek *platos*)
sedge (from Old English *seíg*, related to Low German *segge*, both probably from the Latin *secare*, to cut; akin to Old Celtic *seska*, Irish *seisg*, Welsh *hesg*, Breton *hesq*)
sokihatkhakcobâ:bî [*çokihathaktcobabi*] (opossum [white pig] ear replica, Mikasuki; cognate with *sokhv*, pig, *hvtke*, white, *hvcko*, ear, *vhake*, resembling, Muskogee)

The "galingale" originally was either one of the Zingiberaceae, maybe *Alpinia*, or a sedge, or both. The name "galingale" was imported into Europe by at least A.D. 1000, and probably earlier. The word is thought to have been derived from Middle Eastern and Asiatic languages (Arabic *khalanjan* or *khaulinjan*). Dodoens was one of the first Europeans to record it being a sedge, *Cyperus longus.* A 1578 translation of his Latin reads, "The roote of Cyperus or English galingal is hoote and dry in the third degree." Subsequently, the name has been applied to several species of *Cyperus.* Not only have people used *Cyperus* medicinally, but they also use the stems for weaving mats and filling pack saddles (Morton 1981).

Although not stated by Sturtevant (1955), the Mikasuki name suggests association with the "Opossum Sickness" (*sokihátkiinoka:cí:cihci*). While his collaborators had a name for the sedge, they revealed no uses.

Cyrilla

(Linnaeus named these shrubs and small trees in honor of Domenico Cirillo, 1734–1799, a professor of medicine at Naples)

Cyrilla racemiflora. From Sargent 1905.

Cyrilla racemiflora (having flowers in racemes) (= *C. antillana*)
- *barilla* (maybe from Taino, *baria,* the forest tree *Cordia gerascanthus,* Cuba)
- beet wood [beetwood] (because bark and heartwood are red as beets, Adams 1972, Jamaica)
- bloodwood (because bark and heartwood are red, Jamaica)
- *bois couché* (bed-wood, Guadeloupe)
- burn-wood (West Indies)
- *clavellina* (little carnation, Cuba)
- *colorado* (red one, Puerto Rico); *palo colorado* (red tree, Puerto Rico, Dominican Republic)
- [American, swamp] cyrilla (USA, Puerto Rico)
- *granado* (the same name is given to the pomegranate, *Punica granatum,* Dominican Republic)
- he-huckleberry (because these shrubs never produce fleshy berries, people have concluded they are "males"; the comparison to huckleberry is not as far-fetched as it may appear—they are related to *Vaccinium*)
- iron-wood (West Indies)
- [southern] leather-wood [leatherwood] ("leatherwood" in use since 1760, but originally applied to *Dirca,* which see)
- *llorona* (crying woman, Cuba)
- *olivier montagne* (mountain olive tree, Guadeloupe)
- *piojillo* [*piojito*] (little louse, Venezuela)
- *sabina macho* (male juniper, Dominican Republic)
- [black, red, white] titi [ti-ti] (etymology uncertain, maybe from "tie." In the United States, this name has been given to the buckwheat tree or black titi, *Cliftonia monophylla,* Cyrillaceae, or to leatherwood, red titit, or white titi, *Cyrilla racemiflora,* both applied by Chapman's *Flora of the Southeastern United States* of 1860; also to sourwood, *Oxydendron arboreum* by Small's *Flora of the Southeastern United States* of 1903. The ultimate origin is unknown in the United States. However, the word has been derived from Tupí, as the name of monkeys in the genus *Callithrix* from 1832; and from Maori, the name in New Zealand for the diving petrel, *Pelicanoides* or *Halodroma urinatrix.* Probably the southeastern U.S. application is yet another derivation, from the use of the strong inner bark of some species for tying, or from the fact that the tree "ties" up everything in its thickets. The local use in northern Florida implies that it came about because of the difficulty in passing through the vegetation consisting of these plants, hence local names such as "Titi Hell"; applied to members of Cyrillaceae by Williams [1837] 1962. See also *Dalbergia.*)
- *warimiri* (Arawak, Guyana, Suriname)
- *yanilla* (maybe from Taino, *baria,* the forest tree *Cordia gerascanthus*; or perhaps diminutive of Taino *yana,* the buttonwood, *Conocarpus erectus,* Cuba)

In the 1970s, I had a grant to study vegetation in the Fakahatchee Strand State Preserve in Collier County. The first person I hired to work for me was an unlikely field botanist, Julie Jones. She had terrible vision ("I am legally blind"), had spent her college years working in a laboratory, and knew nothing about wild plants. However, she was incredibly enthusiastic and turned out to be one of the best field assistants I ever worked with. She went on to become one of the first two women to be hired by the Florida Game and Fresh Water Fish Commission, and attributes passing the field exam to working for me.

The final exam for the commission recruits was to navigate through wild country they had never seen in the Florida Panhandle at night. They were given a compass, a small flashlight, and a topographic map. Those who made it to the rendezvous point the next morning passed the exam. Julie said it was a breeze after working with me in the thickets and unsurveyed wilds of southern Florida. There was one problem—she got diverted by a titi swamp, or as they knew it in the Tallahassee area, "Titi Hell." In spite of the "Titi Hell," Julie became the first female lieutenant colonel in Florida's history, and continues to be promoted.

The "Titi Hell" that temporarily stumped Julie was composed of both kinds of titi, *Cyrilla* and *Cliftonia.* No doubt, Alexander Garden, who collected the plants Linnaeus used to name *Cyrilla,* had fewer problems gathering them.

Alexander Garden (1730–1791) was a physician and amateur botanist who lived in Charleston, South Carolina. Linnaeus named the cultivated *Gardenia* in his honor, making his name famous in botanical and horticultural circles, although most of us had no idea who he was.

Between the publication of Linnaeus's *Species Plantarum* in 1753 and the appearance of his *Mantissa Plantarum* in 1767, Garden had collected and sent him a new kind of plant from the Carolinas. According to the label in Linnaeus's herbarium, he collected the small tree "*in Carolinae pinetis humentibus*" (in moist Carolina pinewoods). In 1767, Linnaeus called the plants *C. racemiflora*; at the same time, he created a new genus and species.

After 74 years passed, the Austrian director of the botanical garden in Vienna, Stephen F.L. Endlicher (1804–1849), published his book called *Enchiridion botanicum* in 1841. In that volume, he created the family Cyrillaceae, based on the genus *Cyrilla.* Although there have been changes since, the family is now considered to contain three genera—(*Cyrilla* (monotypic), *Cliftonia* (monotypic, southeastern USA), *Purdiaea* (12 spp. in tropical America, especially Cuba)—and 14 species (Brummitt 1992, Stevens et al. 2001). Those families are part of "core Ericales" along with Actinidiaceae, Clethraceae, Ericaceae, and Sarraceniaceae (Judd et al. 2002).

In Puerto Rico, *Cyrilla* wood is seldom used except for fuel, because it warps badly during seasoning. In Cuba and Hispaniola, the handsomely colored wood has been made into furniture (Little and Wadsworth 1964, Liogier 1978). The spongy bark at the base of the trunk is absorbent, pliable, and astringent, and has been used as a styptic to stop bleeding in several parts of its range (Liogier 1978). Porcher (1863) wrote: "It has been used with advantage as a substitute for agaric and other styptics. I learn that it is much confided in for this purpose by those living in Darlington district, South Carolina. When rubbed on the hand, it produces a sensation similar to that produced by the application of an astringent fluid. It has also been applied to ulcers when the indication is to cicatrize them. This plant merits further attention."

One of the active compounds in its function as a styptic is tannin. Cronquist (1981) indicated that the plants are high in tannins, and that is surely one of the compounds involved in their red color. Although no records have been found, surely indigenous people and settlers used these tannins in curing animal skins. That application is suggested by the name "leatherwood," as this tree does not have bark resembling that of *Dirca* (which see).

Thomas (1960) investigated variation in the species and concluded that clonal reproduction is correlated with a high incidence of seedless parthenocarpy in populations where self-pollination is high. Because sexual reproduction is low, these factors have combined to create a pattern in which morphologically distinct populations occupy different habitats.

Cyrtopodium

(Greek *kyrtos,* curved, convex, and *podium,* a small foot, in reference to the curved foot of the column of the flower)

Cyrtopodium punctatum (spotted, referring to the flowers)
abejitas (little bees, Dominican Republic)
cañuela (little cane, Cuba, Hispaniola, Puerto Rico); *canniette* (Haiti); *canuyele* (Haiti)
cebolleta grande (big onion, Venezuela)
cow-horn orchid (the pseudobulbs resemble horns to some)
orchidea de elote (*elote* is an unripe ear of corn, Sonora)
piñuela (little pine, Puerto Rico)

Linnaeus described these epiphytic orchids in 1759 as *Epidendrum punctatum.* Then, in 1813, when Robert Browne studied plants cultivated at Kew Gardens near London, he created *Cyrtopodium.* In 1833, John Lindley (1799–1865), also in London, moved the Linnaean species into Browne's genus to create the modern name. There are now 30 species in this American genus (Mabberley 1997).

A decoction of the pseudobulbs is used as a pectoral, against cough, bronchitis, and asthma (Roig 1945). The decoction is also used on skin infections, broken bones, and bruises in Cuba and Hispaniola (Liogier 1974). A mucilaginous pulp from the bulbs was used as glue by the indigenous people and early settlers (Luer 1972).

D

Dalbergia

(Named by Linnaeus f. for Swedish planter Carl Gustav Dahlberg (fl. 1753–1775), who collected for Linnaeus and who was a mercenary in Suriname; also for his brother Nils E. Dahlberg, 1730–1820, who was a Swedish court botanist-physician)

Dalbergia ecastophyllum*. Drawn by P.N. Honychurch.*

Dalbergia ecastaphyllum (Greek *ek*, out, *aster*, star, *phyllum*, leaf, in reference to the unifoliolate leaves)
akarerowoi [*akareroai*] (Carib, Suriname)
bejuco de peseta (coin vine, Dominican Republic); coin-vine (Florida)
bejuco serná (maybe *serná* is a variant of *sarna*, a contagious skin disease, Cuba)
chapero (from *chapa*, leaf or sheet, Belize)
dogwood (Belize)
hoop wood (used for barrel hoops, Barbados)
liane à barriques (barrel vine, Guadeloupe, Martinique); *liane à barriques bord-de-mer* (seaside barrel vine, Guadeloupe, Martinique); *liane à clous* (boil vine, *clou*, nail or boil, Haiti)
maraimaray [*marai-marai, maray-maray*] (Puerto Rico, Venezuela); *maramare blanco* (*mare* is a type of reed used for flutes, but this name probably came from an indigenous word; maybe Arawakan, Venezuela)
palo de pollo (chicken tree, Puerto Rico)
péndola (dangler, Cuba)
ti-ti (Bahamas; cf. *Cyrilla* for etymology)
z'herbe à clous (boil herb, *clou*, nail or boil, Haiti)

Dalbergia was named in 1782, not by Linnaeus, but by his son. In his *Supplementum Plantarum*, an addendum to his father's *Species Plantarum*, the younger Linnaeus created the name for an Old World species. In 1894, German botanist Paul Hermann Taubert (1862–1897) moved a species the elder Linnaeus had named to create *Dalbergia ecastophyllum.* Now there are 100 tropical trees and lianas in the genus, in both the Americas and the Old World (Mabberley 1997). Mostly, the genus is known for providing beautiful and useful wood, but there are other applications.

A decoction of coin-vine is taken as a diuretic, and the sap extracted from the young leaves and branches taken a spoonful per day is an emetic-cathartic (Roig 1945). The bark from the trunk and the seeds are vermicidal and considered poisonous by some.

Daucus

(From Greek *daukos, daukon,* sweet, carrot; Latin *daucus*; one of the classical names)

carota (Italian); *carotte* (French); carrot (in use by 1533; from French *carotte,* based on Latin *carota,* an adaptation of Greek *karoton;* Dioscorides, A.D. 40–80, called *carota* the *pastinaca silvestris*)
cenoura (from Arabic, Portuguese); *zanahoria* (spelled *çanahoria* in 1557, from Arabic *isfanaria* or *sannaria,* Spanish)
curran fiadhain (wild carrot, Gaelic)
gallinacci (literally little yellow one, but probably derived from Latin *Gallicam,* from France, Italian)
Gelbe Rübe (yellow carrot; rendered *Geelrüben* in 1542, German); *gulrot* (*gul,* yellow, *rot,* root, Norwegian)
Möhre (German)
pastinaca (from Latin *pastino,* dig ground in preparation for planting, and *-aceus,* pertaining to; used by Pliny, A.D. 23–79, for the wild carrot; in use by Fuchs 1542)
peen (in use by 1549, Dutch)
Queen Anne's lace (Queen Anne was the ruler of Britain from 1702 to 1712; why this common name appeared in the 1890s is unexplained)
staphylinos (classical Greek referring to a carrot or parsnip; in use by Fuchs 1542)

Daucus pusillus (very small)
American [wild] carrot

rattlesnake weed
zanahoria silvestre (wild carrot, Sonora)

Classical Mediterranean taxonomy was not the same as that used today. The Greeks called the root of the wild carrot *staphylinos*, and they cooked and ate them although they were rarely larger than a finger (Baumann 1993). Dioscorides (fl. A.D. 40–80) recorded that people also mashed the leaves with honey to clean ulcers. In Spain and Portugal, the Arabic influence was so strong that the plants became *cenoura* (Portuguese) and *zanahoria* (Spanish). The Romans called the same plants either *carota* or *pastinaca*, apparently reserving the latter name for those cultivated.

As late as the 1500s, the carrot was known in French as *carota* (by 1536) and *pastenade* (by 1561), and in Italian as *carota* (by 1561) and *pastinaca* (by 1597). By contrast, the parsnip (in English by 1398, from *pastinaca*) was either *siser* or *pastinaca* (Latin), *cheruy* (French), *rape gialle* (French turnip, Italian), or *chiriuia* (Spanish). Leonard Fuchs called the parsnip *sisaro* in 1542.

That variety of names resulted in those used today in the regions formerly under the rule of the Roman Empire. For the carrot, Charles de l'Ecluse in 1576 and Joseph Pitton de Tournefort in 1700 preferred *Daucus*, while Gaspar Bauhin (1623) used *Pastinaca*. Finally, in 1753, Linnaeus used *Daucus* for the technical name of the carrot, and the parsnip became *Pastinaca*.

When Europeans arrived in the New World, they found a small, annual, wild carrot, *D. pusillus*, already there. That species occurs naturally from Florida to California, and north to South Carolina, Missouri, Kansas, and British Columbia (Fernald 1950, Steyermark 1963, Barkley 1986).

Unfortunately, for unraveling the uses of native plants, the Old World carrot (*D. carota*) was introduced soon after Europeans arrived. It escaped almost immediately, and became known throughout the eastern states as Queen Anne's lace.

The problem with compiling a list of uses of carrots is separating the native from the alien species. As an example, editor Richard Drinnon identified the "Beaver root" in Hunter ([1823] 1973) as *D. carota*. However, there is no real evidence that the species was wild in the region of western Missouri and adjacent states occupied by the Osage and Kansas at that time. Carrots were in Virginia between 1609 and 1648, in Massachusetts by 1629, and in New York by 1779 (Hedrick 1919).

Hunter said the Osage name was *sha-ba-wa-nembra* (beavers eat it; *zha'-be*, beaver, *wa-non-bthe*, to eat), and that it "very much resembles the common garden carrot, in size and appearance, but in taste it is disagreeably bitter." Steyermark (1963), however, indicated that the roots of wild carrots are sweet and good. Fernald et al. (1958) concurred and thought that the roots raised from the seeds of wild plants were "remarkably sweet." The identity of the plant discussed by Hunter cannot be confirmed, but it might have been *D. pusillus*.

Only among the western tribes is there evidence that the native carrot was used. Yanovsky (1936) said the Nez Perce and Navajo ate the roots, boiled or raw. Moerman (1998) added the Costanoan, Clallam, Cowichan, Mendocino, Miwok, Navajo, Saanich, and Salish. These people used it to clean the blood, to stop itching, to treat fevers and snakebite, and as a good-luck charm. Hocking (1997) indicated that the native *D. pusillus* was used to treat infections and rattlesnake bites, and to promote wound healing.

Moerman (1998) also recorded that the roots were eaten. How much there would be to eat on an annual plant is problematical, and there was confusion with the introduced Old World species. It is pertinent that Hodgson (2001) does not include *D. pusillus* among the edible plants of the Sonoran Desert.

The record on *D. pusillus* looks like one of a native species being supplanted by one introduced from the Old World, which was a more easily obtained food plant and, perhaps, a more effective medicine. Because of the lack of properly documented records, that view remains open to interpretation.

Descurainia

(Named for the French pharmacist François Déscourain, 1658–1740, a friend of the French botanists Antoine and Bernard de Jussieu)

Besenrauke (*Besen*, broom, *Rauke*, mustard, German; *Rauke* is from Latin *eruca*, now the genus *Eruca*)
flixweed (variation of "flux-weed," alluding to a use to control diarrhea, English)
hundesennep (hound's late turnip, Norwegian)
sagesse (French)
sisymbre (based on similarity to *Sisyrimbium*, another mustard, French)
Sophienkraut (sophie herb; *Sophia* was an old generic name for some mustard, now *Descurainia sophia*, German)
tansy mustard ("tansy" is taken from Medieval Latin *athanasia* and Greek *athanasia*, immortality; akin to "everlasting" or *immortelle* because of the long-lasting flowers; now *Tanacetum*

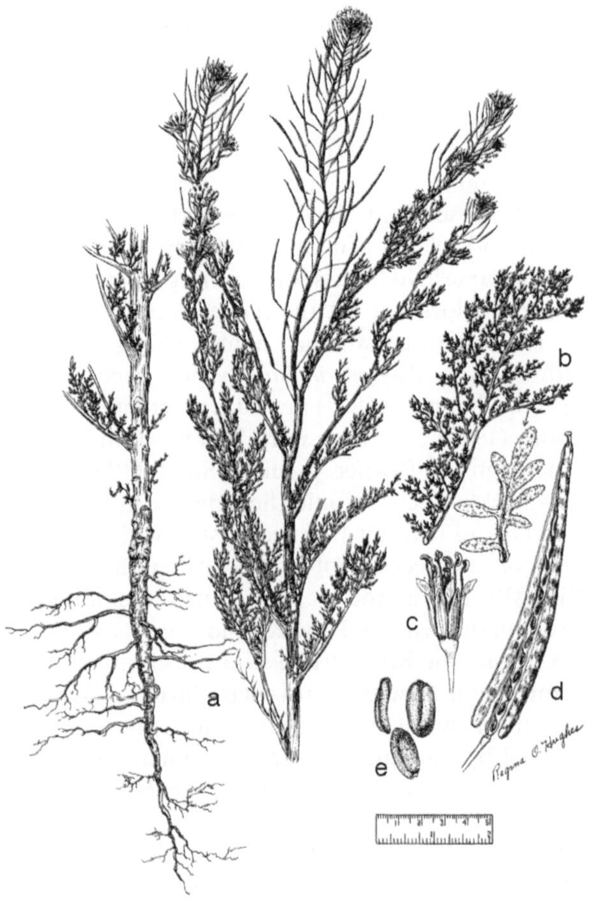

Descurania pinnata. a. Habit. b. Enlarged cauline leaf. c. Flower. d. Silique. e. Seeds. *Drawn by Regina O. Hughes.* From Reed 1971.

vulgare, Asteraceae; tansy-mustard was in use by 1856 when Asa Gray applied it to *Sisyrimbium*)

Descurainia pinnata (pinnate, the leaf lobing)
shuu'uvad [*rú-u-what, show-ou-wat*] (maybe *such'iavik,* recorded by Curtin 1949, is a faulty transcription, Pima, Arizona); *shuu'uwad* (Tohono O'odham, Arizona); *shu'awat* (Pima Bajo, Sonora)
[pinnate, western, yellow] tansy mustard

I spent the first 29 years of my life living around *D. pinnata* without knowing what it was. Then, when I went to Florida, I was south of tansy-mustard's range. It was not until I came to Arizona in 2001 that I learned to identify the plants. In my yard, and the surrounding desert, the plants came up in thick patches on disturbed sites after the winter rains in 2002. I could easily understand why people, at least those in the West, considered it a major seasonal food source.

Tansy mustard is a species that grows from Florida to California, and from Quebec to British Columbia (Fernald 1950, Barkley 1986, Welsh 1987). It also grows in Chihuahua, Sinaloa, and Sonora, Mexico (Wiggins 1980, Felger 2000, Hodgson 2001). People throughout the western part of its range ate the seeds, and probably those in the eastern states did during the old Eastern Agricultural Complex (Hudson 1976). As many of the western tribes remained gatherers, they may have retained more of the uses for small seeds that were lost in the East when maize, beans, and squash were introduced.

Descurainia was used by the Hohokam of Arizona. Indeed, some think it was possibly cultivated, and seeds have been recovered from pottery in two archaeological sites (Hodgson 2001). Among the historic people, the plants have been recorded as used among the Atsugewi, Cahuilla, Cocopa, Gosiute, Hopi, Kawaiisu, Keres, Maricopa, Mojave, Navajo, Paiute, Tohono O'odham (Papago), Akimel O'odham (Pima), and Quechan (Moerman 1998, Hodgson 2001).

Maricopa and Quechan baked the young growth as greens in a pit lined with hot stones. They piled alternating layers of greens and hot stones, covering the top with earth. After about 30 minutes, they ate the greens or dried and stored them (Hodgson 2001). Possibly other people ate them similarly, but no records of this have been found.

Mostly, people ate the seeds. Atsugewi, Cahuilla, Cocopa, Gosiute, Hopi, Kawaiisu, Keres, Maricopa, Mojave, Navajo, Paiute, Tohono O'odham, and Akimel O'odham are among the people known to have consumed them. This was still the most commonly used seed crop by some Tohono O'odham in the 1930s (Hodgson 2001). However, soon afterward, they were lost from their diet, probably sometime before 1945 (Rea 1997). Rea (1997) speculated that people stopped eating them because they objected to their mucilaginous nature.

There were different preparations, but the one among the Akimel O'odham and Tohono O'odham is perhaps typical of southern tribes. The plants were picked when the fruits were ripe. Then the minute seeds were shaken into baskets. The seeds were then roasted, mixed with water, and eaten like *atole* (Curtin 1949). Historically the Akimel O'odham also made a cold drink of raw seeds. Seed pods were gathered in a basket, the seeds extracted, winnowed, and then ground. The flour was added to cold water with sugar for a cold drink. "The Papago use it much in the summer, saying it cools them off" (Rea 1997).

The Pima also used the tiny seeds to extract foreign objects from the eye (Rea 1997). One seed is put in the eye and it brings out the foreign object. Seeds were sold in great quantity near the villages of

Altar and Guaymas, Sonora in 1900 (Hodgson 2001). More recent people living in the Guasabas-Granados area of northeastern Sonora used *pamita cimarrón* (unknown *Descurainia* sp.) as a stomach medicine. Similarly, the Paipai in Baja California used *pamita* (*D. pinnata*) as a remedy for *empacho* (indigestion), stomachache, and vomiting (Ford 1975). Mucilage around seeds may coat the digestive system and slow digestion, thereby helping control diabetes (Hodgson 2001). Seeds were used for stomach ailments by the Cahuilla and for sores by the Pima (Moerman 1998). A poultice of the herbage was used for toothache by the Navajo, and the Utes used the plants in an unspecified medicine. Hocking (1997) noted that the seeds were put in sugared water for liver diseases, but does not say which people did so.

The Hopi used the plant mixed with dark iron pigments as a dye. They used the pigment to decorate pottery (Moerman 1998).

Kearney and Peebles (1942) wrote that the herbs, "although avoided while the plants are green, are relished, especially by horses, when the buds are ripe." In spite of that, Diggs et al. (1999) reported that the herbs were poisonous to livestock, producing symptoms similar to selenium poisoning; animals become blind, wander aimlessly, and lose the ability to swallow; the sickness is called the "blind staggers" or "paralyzed tongue."

Desmanthus

(From Greek *desme*, a bundle, and *anthos*, flower, presumably from the flowers clustered in heads)

Desmanthus illinoiensis (from Illinois)
atikatsatsiks [*ati(t)ka tsatsiks*] (*atit*, bean, *tsatsiks*, spider, *ka*, inside, Pawnee)
[Illinois, prairie] bundle-flower [bundleflower] (bundle-flower is a translation of the genus)
kitsitsaris (*kits*, plant, *tsitsaris*, bad, Pawnee)
pezhe gasatho (*pezhe*, plant, *gasatho*, rattle, Omaha-Ponca)
prairie-mimosa (André Michaux, who originally described the species, put it in the genus *Mimosa*)
prickle-weed (from confusion with *Mimosa*?)
spider bean

Desmanthus is a small genus of 24 species mostly confined to the warm parts of the Americas (Mabberley 1997). The plants look markedly like *Mimosa*, but *Desmanthus* is always unarmed and has sessile fruits. *Mimosa* is often armed and has stalked legumes. *Desmanthus* is usually confined to prairie regions, occurring in the Great Plains, in the desert grasslands of southern Arizona, and similar areas. *Desmanthus illinoiensis* is centered in the southern Great Plains, with extensions to the east into Florida and west into Texas (Diggs et al. 1999).

The common name "prickle-weed" poses a conundrum. These plants lack the prickles or spines of *Mimosa* and related genera. So why do they have the common name? No help is forthcoming from the OED (1971) as it does not list the name. Asa Gray, sometimes the first to record such names, calls them "desmanthus" in his *Lessons in Botany* published in 1875. As usual, Chapman (1897) provides no common name. Gilmore (1919) said they were "spider-bean," taking his cue from the Pawnee. John K. Small (1933) called them "prickle-weed." Could the name be as late as the 1930s? Or did it result from confusion with *Mimosa*?

All of the records of people using these plants medicinally are in the western states, including the Keres, Omaha, Paiute, Pawnee, and Ponca (Moerman 1998). Along the Missouri River area, Gilmore (1919) was among the first to record their place. The Omaha-Ponca and Pawnee used the leaves to treat itch. Seeds were used by the Paiute to treat conjunctivitis (Moerman 1998). The Keres had a name for the plants, but no use was given.

Omaha-Ponca boys made toy rattles from pods (Gilmore 1919). That use is reflected in their name for the plants. The Meskwaki and Potawatomi used the root juice to lure animals into traps (Smith 1933, King 1984).

Seeds are reportedly high in protein (Foster and Duke 1990), but there seem to be records of only wild animals eating them. The Mayo of Sonora noted that their quail are fond of the seeds and they make use of that to hunt these birds (Yetman and Van Devender 2002). Although Martin et al. (1951) did not list the genus as bird food in the United States, it seems probable that the seeds are used in Florida.

Desmodium

(Desvaux used the Greek word *desmos*, a bond or chain, in reference to the jointed fruit pod, a loment)

Desmodium incanum (ash-colored with pubescence) (= *D. supinum, D. canum*)
amor de campo (wild love, Cuba?); *amor seco* (dry love, Cuba, Dominican Republic, Colombia)
bee bur [bush]
beggar's ticks (USA)
'bísbìská ? (Paya, Honduras)
bissalun̲ko (*bissa*, blackberry, *lun̲ko*, grunt, like a hog, Choctaw)

cadillo (usually an umbel, Colombia)
chausiya [*ischautzia*] (Totonac, Veracruz)
chickweed (Cayman Islands)
collant (sticky, Guadeloupe, Martinique)
common tick-trefoil (Bahamas); *grand trèfle* (big clover, Guadeloupe, Martinique), tick-[clover, trefoil] (USA); *trèfle savane* (savanna clover, Guadeloupe, Martinique)
cousin [kuzen] (Bahamas); *cousin grand-savane* (big-savanna cousin, Guadeloupe, Martinique); *cousin trois sous* (cousin three times, Haiti)
empanadilla (little turn-over, referring to the resemblance of the shape of the loments to the bakery treat, the *empanada* or turn-over, Cuba)
fever weed (Bahamas, Trinidad)
iron vine (Barbados)
kowike entvlako (*kowike*, quail, *em*, its, *tvlako*, beans, Muskogee)
man back (Belize?); strong back (Belize)
mozote (maybe from *moztla*, tomorrow, Central America)
peanut (Bahamas)
pega pega (sticker-sticker, Mexico, Costa Rica, Colombia); *pega ropa* (clothes sticker, Mexico)
sá-rá [*s-r*] (maybe a combination of Maya and Spanish or English; *sa*, cornmeal drink or *atole*, *ra*, there are no words beginning with "r" in Maya, Belize)
stick-tights (USA)
'stvlokpuce (*vlokpicetv*, to stick on, Creek)
sweetheart (Barbados)
tofó:ma (Creek); *tofoome* [*tofõ:mî*] (Mikasuki)
tortilla de ratón (mouse's bread, Mexico)
wild ground-nut [wild granite] (Bahamas)
wild pinder
zarzabacoa (a variant of *guasabacoa*, a *Desmodium*, Taino, Puerto Rico)

Desmodium paniculatum (tufted, the flower clusters)
l'herbe à collet rouge (sticky red herb, Houma)
panicled-leaf tick-trefoil (a book name)

Desmodium rotundifolium (round-leaved)
dollar leaf (USA)
flux vine (USA)
prostrate tick-trefoil [prostrate ticktrefoil]

The French professor in Angers, Nicaise Auguste Desvaux (1784–1856), created the genus *Desmodium* in 1813. Now, the genus contains 450 species, most of them in tropical regions (Mabberley 1997). In life-form they range from small herbs with pesky fruits (loments) that cling to trousers, socks, and other articles of clothing to trees that produce timber. Several are used in medicines.

As the name "flux vine" suggests, *D. rotundifolium* is known for stopping dysentery. It has been used for medicine near Tallahassee and in nearby Mississippi (Hocking 1997). Perhaps this is the same species that the Alabama used in that region for bad colds and lung problems (Taylor 1940). In Louisiana, the Houma used *D. paniculatum* for "one who suffers from weakness and cramps," by soaking the broken roots in whiskey and having the sick person drink it (Speck 1941).

Although Snow and Stans (2001) suggested their plant might be *D. lineatum* (sand tick-trefoil [sand ticktrefoil]), the photo they included is of *D. incanum*. Presumably, they were dealing with only a single species.

The many names applied to *D. incanum* reflect not only use by humans, but also their irritation at its apparent constant presence. The sticky loments adhere to virtually everything, making it an easily dispersed herb. However, those who raise livestock acknowledge its value as fodder and as food for wild animals. Howard's (1984) Seminole informant recognized that quail feed on the fruits.

Alice Snow knew that *'stvlokpuce* was used "for lots of different things" (Snow and Stans 2001). *'Stvlokpuce* is used for treating alcoholics and for pregnant women near delivery. The Seminoles also still use it when "you don't feel like eating and throw up" ("Snake Sickness") (Snow and Stans 2001).

Leon and Alain (1974) record that a decoction of the root is good for dysentery and liver problems; also diuretic and good to settle the stomach. Cubans consider it an excellent hemostat, and it was used in the wars for independence in the hospitals to heal wounds. It also has been used on that island to stop internal bleeding (Roig 1945).

In the Bahamas and Trinidad, herbal teas are taken for heat, urinary problems, and fever. The herb is boiled to make a tea for excessive or painful menstruation or with salt to lessen pain after parturition (Ayensu 1981). The Jamaicans drink a decoction to alleviate colds (Morton 1981). An infusion has been used in Guatemala and Honduras to treat gonorrhea (Hocking 1997). It is a medicinal plant in Belize (Balick et al. 2000).

Dicliptera

(Named by A.L. Jussieu from Greek *diklis*, double-folding, and *pteron*, wing, referring to the two-winged capsule)

Diclipta sexangularis. a. Leafy branch imposed on part of inflorescence. b. Flower, side view. c. Flower longitudinally dissected. d. Ovary and lower part of style. e. Floral diagram. f. Fruit. *Drawn by Priscilla Fawcett.* From Correll and Correll 1982.

Dicliptera sexangularis (six-angled) (= *D. assurgens*)
amor seco (dry love, the name may reflect confusion with *Desmodium*, which see)
foldwing (Florida)
nimiz (*nii*, nose, *miis*, cat, Maya, Yucatán)
penasmiento (the thinker)
tinta montañes (wild ink)
uxkwe' ts'ohool (old lady herb, Huastec, San Luis Potosí)
wits paya' (flower *paya'*, Huastec, San Luis Potosí)
yerba fluxion (flux [diarrhea] herb)

In 1696, Leonard Plukenet called a plant he found in Jamaica *Euphrasia, affines majori folio* (similar to *Euphrasia* but with larger leaves). Linnaeus ([1753] 1957) called it *Justicia sexangularis.* Both of those genera are other members of the family, but these herbs did not actually belong to either one. Then, in 1807, Jussieu moved Linnaeus's species to create *Dicliptera sexangularis.* Today, there are 150 species in the genus (Mabberley 1997).

In Yucatán, the plant is a remedy for asthma (Morton 1981). The Huastec of San Luis Potosí use the herb to stop hemorrhage, and to quiet restless babies (Alcorn 1984); it is used to treat eye problems in Veracruz (Vásquez and Jácome 1997). They also bundle the branches to make brooms.

Diodia

(From Greek *diodos,* a thoroughfare, because it often grows by waysides; Quattrocchi thought possibly from *dis,* twice, and *edos, edios,* form or kind, referring to the calyx)

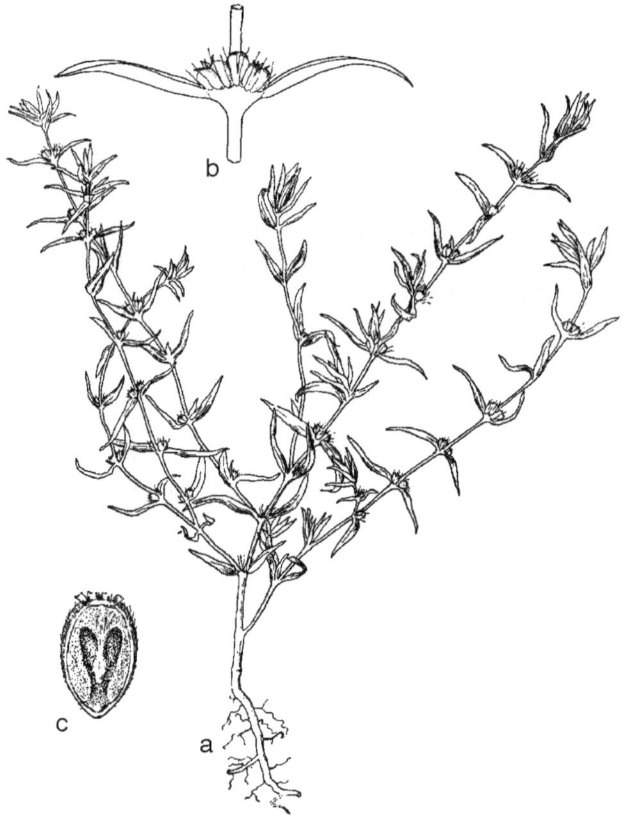

Diodia teres. a. Habit. b. Section of stem with leaves and young fruits. c. Seed. From Buchholtz 1968.

Diodia teres (round)
buttonweed [rough] (in use by 1878, perhaps first for a European Composite)
poor joe [pó jó] (the South, from Arizona to the Carolinas)
poor weed; poor-land weed (considered an indicator of almost sterile soils)

Diodia is a genus of about 30 species in tropical and warm parts of the Americas and Africa (Mabberley 1997). Linnaeus ([1753] 1957) knew the single species *Diodia virginiana* from Jan Gronovius's *Flora Virginica* and from cultivation at the *Hortus Cliffor-*

tianus. Diodia teres was not described until 1788 when Thomas Walter named it in his *Flora Caroliniana.*

This variable species ranges from Florida to Arizona and southern California, and north to New York, Pennsylvania, Ohio, Illinois, and Missouri. In Mexico it grows in Baja California Sur, Sinaloa, and Sonora. Hocking (1997) says that it has been used in domestic medicine, but gives no further details.

Dioscorea

(Linnaeus dedicated the genus to the Greek naturalist Dioscorides, fl. A.D. 40–80, a native of Anazarba in Sicily)

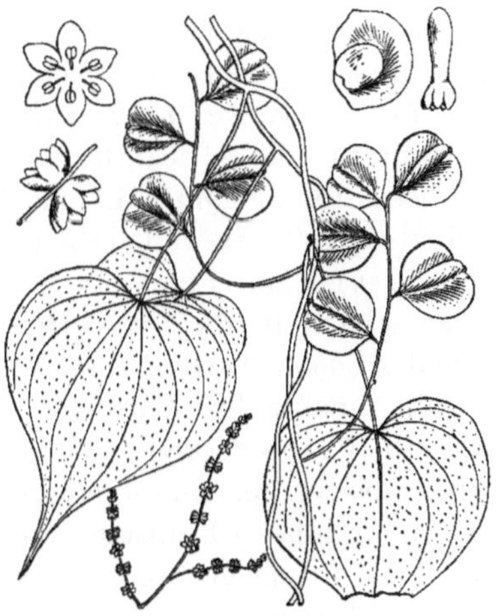

Dioscorea villosa. From Britton and Brown 1896.

Dioscorea villosa (soft-hairy)
China root (possible confusion with *Smilax*, which see)
colic-root (a name also given to *Aletris*, which see)
devil's bones
rheumatism root
[Atlantic, wild] yam [root] ("yam" from *ñamé*, an African word; the OED 1971 says that the word appeared in Spanish as *igname* in 1557, but Columbus spelled it *niames* with other variants in his journal of 1492; see Jane and Skelton 1960, Sauer 1969; rendered in Portuguese *inhame* by 1567, in French *igname* by 1575, and in English as "yam" by 1589)

Linnaeus ([1753] 1957) recognized eight species of *Dioscorea* when he accepted the genus as proposed in 1703 by Charles Plumier. All that Linnaeus had seen were from India except *D. villosa*, which came from Virginia and Florida. Now the genus contains about 850 species, mostly in the tropics, although one reaches Europe (Mabberley 1997).

The wild yam was not listed officially in the National Formulary as medicinal until 1916, and it remained there until 1942 with recommendations for use as a diaphoretic and expectorant (Vogel 1970). Long before that "official" acceptance, the Meskwaki of Iowa used it to relieve labor pains (Vogel 1970). While there is considerable history of the Maya and Aztecs using their species of *Dioscorea*, there is scant mention of the North American plants among indigenous groups.

Black slaves in the southeastern United States possibly learned to use these plants from the indigenous people, but they surely also knew uses of their own African species and made adjustments. They used *Dioscorea* to treat colic and to relieve rheumatism (Vogel 1970). During the Civil War, Porcher (1863) wrote: "The decoction of the root ... is eminently beneficial in bilious colic: one ounce is added to one pint of water, and half of this is taken at a dose ... acts with great promptitude, and ... Dr. Neville places much reliance on the tincture as an expectorant; it is likewise diaphoretic, and in large doses emetic." In South Carolina, Morton (1974) found people mixing it with *Asclepias tuberosa* as a remedy for flatulence and indigestion and with *Aletris farinosa* to treat a prolapsed uterus.

Tubers of *D. villosa*, like others in the genus, contain steroid saponins and resins (Hocking 1997). The most famous steroidal saponin is diosgenin, used in the manufacture of progesterone and other steroid drugs (Lewis and Elvin-Lewis 1977, Foster and Duke 1990). Although originally extracted from plants, diosgenin is now completely synthetic, but continues to be used in oral contraceptives, cortisone, progesterone, and other steroids (Swerdlow 2000). *Dioscorea villosa* is also a source of synthetic DHEA, a hormone that inhibits the production of excess fatty acids and cholesterol (Swerdlow 2000).

Diospyros

(Named by Linnaeus for the Greek deity Zeus, *dios*, divine or god, *pyros*, grain; *diospyron* was the ancient name used by Theophrastus, 372–287 B.C., for the fruit of the nettle tree, *Celtis australis*)

Diospyros virginiana (from Virginia)
axka' (Biloxi)
biłko (Alabama); *biłkó* [*biłkó cobá*, domesticated; *biłkó fihna*, wild, literally "real"], Koasati)
çta-iṉ'-ge (Osage)
date-plum (USA); *Dattelpflaume* (date-plum,

Diospyros virginiana. From Sargent 1905.

German); winter-plum (Texas)
edre'· (Catawba)
guajacana (Catesby 1734–1747 followed Plukenet from 1696 in using an old Italian name for *Diospyros lotus*)
Jove's-fruit (Texas)
kafi (Choctaw; compare with *ołkofi* and *ukov*)
mespila [Latin *mespilorum*] (LeMoyne [1564–1565] in Florida and then Harriot [[1590] 1972] at Roanoake, Virginia, compared them with European medlars, *Mespilus*)
ołkofi [*orkofi, o(th)koofe, olkofi*] (Mikasuki); *ongouffé* [*ougoufle*] (French spelling of Choctaw name *ukov*, from 1777); *onkof* [*inkof*] (Chickasaw); *ukov* [*unkof*] (Choctaw)
passi:min [*pessimin, pichamins, pushemins, putchamins*] (Strachey wrote the Powhatan words *pessemmins, pichamins,* and *pusheminas* in [1612] 1953; he added, "plum very delicious when it is ripe"; Capt. John Smith wrote that it was a "fruit like a medlar ... red when it is ripe ... delicious as an apricot." Powhatan, Virginia); persimmon [putchamins piakamine, pessimmon, 'simmons] (akin to *pasiminan*, dried seed, Cree, *pakimin*, broken seed, crushed fruit, Nipissing, *pakiminasigan*, preserves, Nipissing, *pashkiminassigan*, preserves, sweetmeats, *pasimen*, Delaware, Ojibwa, *piakemen*, cranberry, Menomini, *piakimina*, Illinois name recorded by La Salle in 1682, *piakimini*, Miami, *pokmlnesln*, cranberry, Potawatomi); *phishimons* (Catesby 1734–1747); piaquiminia (recorded by Henri Joutel in 1687 among the "Arkansas" [Quapaw]. Presumably, he used the Illinois name); *pishamia* (Plukenet's 1696 Latin rendering of the Algonquian name, Virginia. Siebert 1975 considered the root word to be */*pehs-/*, a peel or husk, and noted the cognates */piss-/* in Ojibwa, */pe?s-/* in Shawnee, and */pehs-/* in Penobscot); *plaquemine* (European French); *pliaquemine* (colonial version of indigenous name in use by 1751; retained in French-speaking regions of Louisiana and Missouri)
possum apple (a Cracker name, Florida); possumwood (Texas)
sah lee [*sa'-l(i), sali, sa'-thl(i)*] (Cherokee)
satá (simple term, Creek); *svtv* [*svt'tv*] (Muskogee; there is a creek called Cetaehlustee, from *svtvluste*, black persimmon, in Chambers County, Alabama)

Diospyros is a genus of ~475 species, with 80 in the Americas, 94 in Africa, about 100 in Madagascar, and 200 in Asia (Mabberley 1997). The genus is best known as the source of ebony wood, and it was favored by ancient Egyptians for making furniture. When Linnaeus ([1753] 1957) gave us the modern scientific name, the southeastern species had been known to Europeans for more than 200 years. However, within scientific circles, Caspar Bauhin first noted the trees in Europe in 1623 when he wrote that they were *Loti africanae similis indica* (Indian lotus which is similar to the African lotus) (cf. *Celtis:* Hackberry for discussion of *Lotus*). Leonard Plukenet recorded the plants in 1696. He called them *Guajacana loto arbori affinis virginiana Pishamia dicta* (Guajacana is the tree related to the lotus, from Virginia where it is called *Pishamia*). *Guajacana* seems to be the Tuscani Italian name for *Diospyros lotus*, also called *l'Ebano nero* (Nero's ebony).

Much earlier, however, persimmons were recorded in accounts from Virginia to Louisiana and north to Illinois. The de Soto chroniclers recorded persimmons in the late 1530s and early 1540s as *ameixas* [plums] (Swanton 1946), while others usually called them medlars or *Mespilus.*

In 1542, Hernando de Soto found that people on the Mississippi River made loaves of *Diospyros.* According to Hedrick (1919), the loaves were "like unto brickes, also plummes of the making and bigness of nuts and have three or four stones in them." Presumably, this site was at the Natchez village of Guachoya, where Swanton (1939) said that de Soto's men got corn, beans, and dried persimmons from local storage huts. That site is thought to have been near modern Ferriday, Louisiana (Swanton 1939).

Harriot ([1590] 1972) said that they were "Medlars a kind of verie good fruit, so called by vs chieflie for their respects: first in that they are not good vntill they be rotten: then in that they open at the head as our medlars, and are about the same bigness: otherwise in taste and colour they are farre different: for they are as red as cheries and very sweet: but whereas the cherie is sharpe sweet, they are lushious sweet."

Capt. John Smith also wrote about them in 1612 in Virginia. He said, "The fruit like medlers; they call *Putchamins*, they cast them vppon [upon] hurdles on a mat, and preserve them as Pruines." He further said, "if it be not ripe, it will drawe a mans mouth awrie with much torment; but when it is ripe, it is delicious as an Apricock" (Davidson 1999). In the same year, Strachey wrote that the *pichamins* were an "excellent plumb" (Harrington 1955).

La Salle wrote in a letter ca. 1682 that *"c'est à pimiteoui et au-dessous qu'on commence à trover d'un fruict que les Sauvages appelent piakimina, tres-ordinaire à tout la Louisiane ... C'est peut-etre ce que Jolliet apelloit des greandes"* (it is at Pimiteoui and below that one begins to find a fruit that the Indians call *Piakimina*, very common to all of Louisiana.... It is perhaps this that Jolliet called pomegranates) (Crawford 1978).

Henri de Tonti gave an account of a meeting between La Salle and a Taensa [Natchez] leader in Louisiana between 1678 and 1683. In that he recorded having asked to have the women *"faire des farines de bled d'Inde et des pastes d'un certain fruit qu'ils appelent Paquimina, lequel est fort bon"* (make flour from Indian corn and paste from a certain fruit that they call *Paquimina*, which is very good) (Crawford 1978).

Henri Joutel wrote in July 1687 that the "Arkansas" [Quapaw] have *"encore d'un espece de fruit qu'ils appelent piaquiminia"* (also a kind of fruit which they call *piaquiminia*) (Crawford 1978). That site was near where the Arkansas River joins the Mississippi in southeastern Arkansas. As the Quapaw spoke a Siouan language, Joutel was using the Illinois name.

Shortly afterward, Joutel recorded persimmon as one of the fruit trees growing on the Illinois River near the junction with the Mississippi. That junction of rivers is now in the suburb of St. Charles in the northwestern part of modern St. Louis, Missouri.

In 1748, Le Page Du Pratz found the Natchez using the persimmon. He wrote: "When [the persimmon] is well ripened the natives make bread of it, which keeps from one year to another, and the virtue of this bread, greater than that of fruit, is such that there is no diarrhea or dysentery which it does not arrest, but one ought to use it with prudence and only after being purged. In order to make this bread the natives scrape the fruit in very open sieves to separate the flesh from the skin and seeds. From this flesh, which is like thick porridge, and from the pulp they make loaves of bread 1½ feet long, 1 foot broad, and the thickness of the finger, which they put to dry in the oven, on a grill or, indeed, in the sun. In this latter fashion the bread preserves more of its taste. It is one of the merchandises which they sell to the French" (Swanton 1946).

Bernard Romans ([1775] 1961) said that the Choctaw in the 1760s still used persimmon, at least when their crops failed. He wrote, "They make bread of the different kinds of *Fagus*, of the *Diospyros*," and several other plants.

William Bartram wrote to physician Benjamin S. Barton in 1789 that the Cherokee and Creek used persimmon. He noted that they "have in use a vast variety of wild or native vegetables, both fruits and roots, viz.: *diospyros, morus rubra, gleditsia multiloba s. triacanthus*," and many others (Swanton 1946).

Swanton (1946) found that the Tunica who lived along the Mississippi in southwestern Mississippi mostly used corn in their diet. However, historically, they were reported to "have lived one entire month every year on the persimmons besides putting up a great quantity of persimmon bread." South of the Tunica, the Chitimacha who lived on the Mississippi delta in Louisiana also ate the fruit of the persimmon (Swanton 1946). There is now a place near where they lived that is called Plaquimine Bayou [persimmon creek; bayou from Choctaw *bayuk*].

We have documentation of *Diospyros* use by the Alabama, Apalachee, Carolina Algonquians, Catawba, Cherokee, Chitimacha, Choctaw, Comanche, Creek, Illinois [*Illiniwek*], Natchez, Quapaw, Rappahannock, Timucua, Tunica, and Seminoles, essentially people throughout its range (Harriot [1590] 1972, Swanton 1946, Howard 1985, Hann 1986, Sturtevant 1955, Crawford 1978, Drechsel 1997). Surely all of these people ate the fruits.

Fruits have been used to make beer and distilled liquor, and the seeds were roasted, ground, and used as a coffee substitute (Standley 1920–1926). Kalm ([1753–1761] 1972) provided a detailed recipe English and Swedes in New Jersey followed to make cakes, beer, and brandy of ripe fruits, and Porcher (1863) praised alcoholic beverages made from it. Tull (1999) also gave recipes for jam and bread. Swanton (1946) noted that the persimmon root sometimes was carved and Tull (1999) added tool handles and golf clubs to bring its use into the modern world.

The bark is rich in tannin and has been used as a substitute for oak in tanning leather (Porcher 1863). Black dye may be made from the bark also, but Tull (1999) preferred *D. texana* over *D. virginiana*.

Four tribes have been documented as using persimmon medicinally. The Alabama boiled roots for a tea used in "bowel flux" (Bennett 1997). The Catawba treated thrush in babies with a bark infusion (Bennett 1997). The Cherokee took medicine made from it for bowel problems, sore throat, and mouth, heartburn, bile, and liver problems, piles, thrush, toothache, and venereal disease (Hamel and Chiltoskey 1975). The Rappahannock took an infusion of the bark to treat thrush and sore throat (Moerman 1998).

Green fruits contain tannic acid and have been used against diarrhea, chronic dysentery, and uterine hemorrhage (Porcher 1863). Morton (1974) found people in the coastal region near Charleston, South Carolina using the green fruit to treat thrush in infants, and on the gums of teething children. Bark is astringent and bitter. Tea made from dried or fresh leaves is said to be rich in vitamin C (Fernald et al. 1958), but Tull (1999) did not find it that appealing.

Dirca

(Named for Dirce, wife of Lycus, who, after her brutal murder, changed into the fabulous fountain Dirce northwest of Thebes in Boeotia [*Voiotia*], a district in east central Greece northwest of Attica)

Dirca palustris. From Britton and Brown 1897.

Dirca palustris (of swamps)
bois de plomb (lead tree, Quebec)
cîbägob (dead man's [spirit?] bark, Potawatomi); *djibe'bûb* (spirit bark, Ojibwa)
drago (dragon, Mexico)
eastern leatherwood (in use by 1760)
moose-wood [moosewood] (by 1792 the name was applied to *Acer pennsylvanicum*, but by 1860 Asa Gray was using it for *Dirca*; the *Acer* was given the name because moose, *Alces alces*, fed on its branches; perhaps that is the reason the name was extended to *Dirca*)
rope-bark (used in a medical dictionary published in 1851)
wicopy (first recorded in [1612] 1953 by Strachey, who wrote the Virginia Powhatan word *weihkippeis*, and said it meant "hemp"; Siebert (1975) has associated this with *wikepi*, base or inner bark of linden or leatherwood; akin to Cree *wikupiy*; first connected with *Dirca* in 1778, when Jonathan Carver wrote in his *Travels through the interior parts of North America in 1776*, "The Wickopick or Suckwick appears to be a species of white wood"; related to a name applied to *Tilia*, which see)

John Clayton sent a specimen of a new shrub to European Jan Gronovius in the 1730s. When Gronovius published the *Flora Virginica* in 1739–1743, he called the plant *Thymelaea floribus albis primo vere erumpentibus* (A *Thymelaea* with white flowers, truly the first to flower). Linnaeus also studied the specimen, now deposited in the Linnaean herbarium in London, and decided it was not *Thymelaea*. So, he created the genus *Dirca* in 1751 in a thesis defended by his student Leonhard J. Chenon at Uppsala. In *Species Plantarum*, Linnaeus ([1753] 1957) wrote, "*Habitat in Virginiae paludosis*" (Grows in Virginia swamps).

In Florida, these are rare shrubs of bluff forests and river banks in Gadsden, Liberty, and Jackson Counties. The species also grows from New Brunswick and Quebec to Minnesota, and south to Arkansas, Oklahoma, Alabama, and Florida (Fernald 1950, Gleason and Cronquist 1963).

Dirca has three species and is now known to be endemic to North America, Kartesz (1994) said the only other species was *D. occidentalis*, but *D. mexicana* was named in 1995 from Tamaulipas, Mexico. The plants belong to the Thymeliaceae, a family with 50 to 53 genera, and about 750 species. The family is widespread, but especially abundant in Australia and tropical Africa. Much of the family consists of toxic plants, which produce glycosides and accumulate daphin (a coumarin) or related substances (Mabberley 1997).

In the northern part of their U.S. range, where these shrubs are more common, indigenous people made wide use of them. The most common application was for cords and fibers. There are records of use as cords among the Menomini, Ojibwa, Potawatomi, and Winnebago (Smith 1933, King 1984, Moerman 1998). Fernald (1950) wrote that the wood was white, soft, and brittle, but the fibrous bark was remarkably tough and was used by indigenous people to make thongs. Gilmore (1931) identified *Dirca* at the Ozark Bluff-Dweller site in Arkansas. Then Jones (1936) found them in the Newt Kash Hollow Shelter in Kentucky. Whitford (1941) found the species used from Arkansas to Wisconsin. Mats made of the fiber came from the Bushwick Cave in Arkansas, and cloth came from the Adena Culture in Ohio. Whitford (1941) also found unprocessed fibers simply twisted into cord. The Potawatomi in Wisconsin had used the plant as the weft for a bag, and the Winnebago had made a bicolored bag of *Dirca* and *Tilia.*

Perhaps as common, although not mentioned by Moerman (1998), was the incorporation of the flexible shoots into baskets (Mabberley 1997). Steyermark (1963) wrote: "The twigs are very supple and can be tied into knots without breaking. On account of their toughness, the bark and branches were used by Indians for making baskets and for tying purposes."

In spite of sometimes causing severe dermatitis, with redness, blistering, and sores (Foster and Duke 1990), the plants were used as medicines. Steyermark (1963) recorded that some individuals are sensitive to the bark, and their skin becomes irritated or blistered after contact with the fresh bark. Also he noted that the bark, if taken internally, induces vomiting and has purgative properties, and that the fruit was reputed to be narcotic. Minute doses cause burning of the tongue and salivation, but it was used in remedies for toothache, facial neuralgia, and paralysis of the tongue (Foster and Duke 1990).

At least the Algonquin, Iroquois, Menomini, Ojibwa, and Potawatomi used *Dirca* as a medicine (Smith 1933, Moerman 1998). The Algonquin used inner bark to make a laxative tea. The Ojibwa made green stalks into a physic and a diuretic remedy for urinary infections and lung troubles, and used roots to wash the hair. The Iroquois made a medicine of bark and roots for back pains, to purify the blood, to use as a purgative, and they put a poultice of branches on swellings and neck sores. As an emetic, it was used during pregnancy, for kidney problems, consumption, and venereal disease. The Menomini and Potawatomi used it as a diuretic (Smith 1933, King 1984).

Five phenolic glycosides were found in *D. palustris.* These chemicals showed variable activity as enzyme inhibitors and have potential as antioxidants and anti-inflammatory agents (Ramsewak et al. 1999).

Five insecticidal compounds have been isolated from the seeds (Ramsewak et al. 2001). The triglycerides were effective against the yellow fever mosquito (*Aedes aegyptii*) larvae and exhibited potent feeding-deterrent activity against larvae of corn earworm moth (*Helicoverpa zea*), gypsy moth (*Lymantria dispar*), lackey moth (*Malacosoma disstria*), and tussock moth (*Orgyia leucostigma*).

Dirca occidentalis yielded two antileukemic daphnane esters, Pimelea factor P2 (I) and dircin (II), plus three lignans, (−)-medioresinol (III), (+)-syringaresinol (IV), and (−)-lariciresinol (V), as well as the coumarin daphnoretin (VI). All were found to be cytotoxic constituents (Badawi et al. 1983). Genkwanin (I), (±)-lariciresinol (II), and sitoindoside II (IV) also were active against cancer (Suh et al. 1995).

Distichlis

(Rafinesque named this grass from Greek *distichos,* two-ranked, alluding to the leaves in two rows)

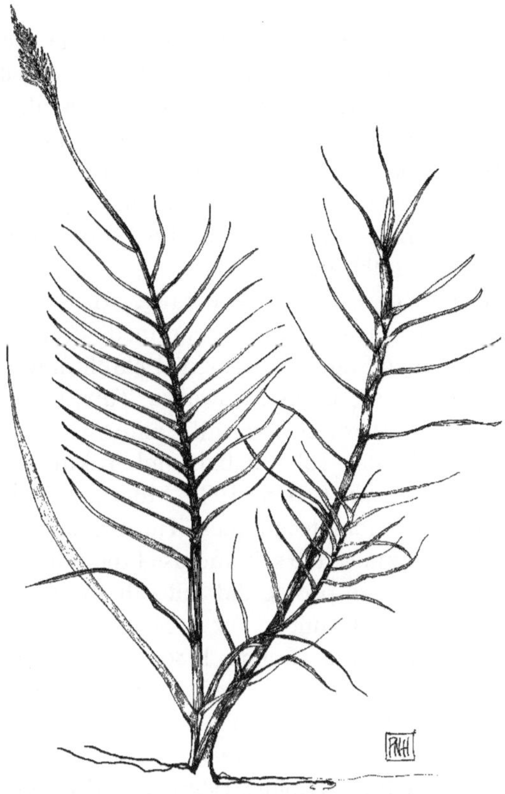

Distichlis spicata. *Drawn by P.N. Honychurch.*

Distichlis spicata (spiked, the inflorescence)
alkali-grass
chépica (Chile)
grama [*brama*] (grass, Chile); *grama de costa* (coastal grass, Cuba)
hapí pahí (*hapí,* salt, *pahí,* grass, Koasati; this may belong here as they once burned it and leached the ashes for salt)
mu'umk vashai (sharp grass, Akimel O'odham [Pima], Arizona)
'onk wasai [*onk vashai*] (salt grass, Tohono O'odham, Arizona, Sonora)
[desert, inland, seashore] salt grass; *pasto salado* (salt grass, Chile); *zacate salado* (salt grass, Sonora)

spike-grass
yerba del burro (burro herb, New Mexico)

Distichlis is a genus of four or five species, confined to saline and alkaline habitats in the Americas and Australia (Felger 2000). *Distichlis spicata* has an unusually large range. It is found from the Gulf of St. Laurence, Cape Briton, Prince Edward Island, and New Brunswick in Canada, along the eastern U.S. coast to Mexico, and south to Belize (Fernald 1950, Balick et al. 2000). In the West Indies, it appears to be confined to the Bahamas, Cuba, and the Cayman Islands (Leon and Alain 1946–1953, Correll and Correll 1982, Proctor 1984). It also grows along the Pacific Coast from Washington, south along both coasts of Baja California and down to the Guaymas region of Sonora, and in Sinaloa, thence into South America (Brako and Zarucchi 1993, Jorgensen and León-Yánez 1999, Felger 2000).

In spite of this extensive range, the records of people using the grasses are confined to the southwestern United States. Moerman (1998) found it being used among the Cahuilla, Kawaiisu, Tubatulabal, and Yokut, all southern and central California tribes.

The Cahuilla, Tubatulabal, and Yokut obtained salt from *Distichlis* (Ebling 1986, Moerman 1998), but they used different techniques. The Cahuilla burned the grass to remove the salt and used the salty ash as a condiment. These people did not salt the food when they cooked, but simply put a pinch of salt in the mouth while eating (Ebling 1986). The Tubatulabal pulled up the grass, stacked it in a basket, and thrashed off the salt much as people handle grain. They were able to gather several hundred pounds of salt in this manner (Ebling 1986). The Kawaiisu used both dried and green grass in water to make a beverage. The Tubatulabal used the leaves and stems as food. The Cahuilla also used the leaves for brushes, especially to remove cactus thorns.

There were also medicinal applications. The Kawaiisu took a decoction of the plants for insect bites, to slow a heart beating too fast, as a laxative, and to treat gonorrhea. The Yokut used a salt extract to treat bad colds and to improve a lost appetite.

Rea (1997) found the Pima still remembered the plant and its name, but they used it only for cattle food. Nonetheless, if the plants on the Colorado River delta are considered the same species, they and others ate the seeds. Some consider those plants to be *D. spicata* ssp. *stricta*, while most now call them *D. palmeri* (Felger 2000, Hodgson 2001). These plants were harvested in great quantities by the Cocopa prior to the devastation of the ecosystem by construction of upriver dams. Those people ate the seeds in *atole.*

Dodonaea: Varnish-Leaf

(Named in honor of Dutchman Rembert Dodoens)

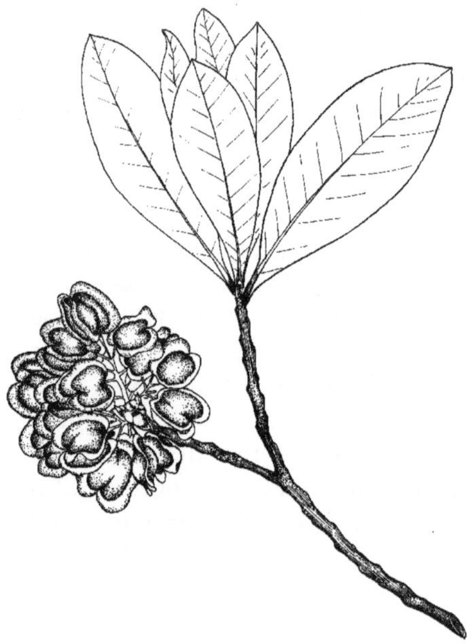

Dodonaea viscosa. From Little and Wadsworth 1974.

A particular pinelands on the Big Cypress Seminole Reservation has an abundance of varnish-leaf (*Dodonaea viscosa*) in the shrub layer. One of our students had difficulty remembering the plant, and was quizzed each time we went there. Finally, he answered, "varnish-leaf" without even looking when he was asked about any plant in that area.

Many people have a hard time recognizing this shrub because there is nothing very distinctive about it except the shiny leaf surface. The leaf certainly looks like it has been varnished, and that must have been the reason for originating the common name.

Still, that seems like a highbrow name for a plant since most poor people (who typically give such names) could not afford that expensive covering. The word "varnish" itself has been in English since at least 1341 and had taken on a variety of meanings by the 1500s. So, even if poor people could not afford the product, they might have heard about or seen items with that coating. Making me more suspicious that the name is a recent innovation is that neither Chapman (1897) nor Small (1913) cited the common name. It is only with Small's (1933) flora of the southeastern United States that the name appears.

Names more likely to have originated early include candlewood (Bahamas), hop-bush (Panama), hop shrub (Barbados), native hops, and sticky hop bush.

Those names convey practical uses for the plants, either as torches or candle substitutes, or in making beer. One Spanish name also mirrors the English candlewood—*ocotillo* (little torch, from Náhuatl, *ocotl*, Guanajuato, Hidalgo). Others note the sticky leaf surface, including *chilca* [*chirca*] (mostly applied to a resinous composite, Costa Rica), *chrica de monte* (wild *chilca*, Costa Rica), and *gui laga citi* [*quí-laga-cijti*] (*gui*, cane, *laga*, flower, *citi*, sticky, Zapotec, Oaxaca). The Zapotec name has also been rendered as *yak cij* (*yak*, tree, *cij*, sticky, Oaxaca). Closely related to those names is *incienso* (incense).

As the names candlewood and *ocotillo* suggest, the wood is easily ignited and that, along with the crushed leafy twigs, is used for torches in the Bahamas, Mexico, and Panama (Duke 1967, 1972, Morton 1981). Since the plant is known to some as "hops," it would be correct to conclude that they use it in place of *Humulus lupulus* to make beer (Panama, Barbados, USA).

The Miccosukee were comparatively late arrivals in peninsular Florida, and their only name for the plants is *takłi* [*takri, takthi*] (bush). There is no indication of what their predecessors within the range of the plants might have called this bush. In Florida, the species grows from St. Johns County down the east coast to Key West and up the west coast to Hernando County. Outside of Florida, the plant is widespread, growing in the Bahamas, West Indies, southern Arizona, and through Mexico and Central America to Chile and Argentina. This same species is widespread in the Old World tropics.

As there are records that the plants are used by people almost everywhere they grow (Prendergast and Pearman 2001), people in Florida probably used them. Uses range from purely utilitarian to medicinal. Because the plants typically have small, multiple stems, they are easily turned into brooms. *Dodonaea* has been used for brooms at least in Brazil, where they are called *faxina vermelha* (yellow broom [cleaner]), *vassoura de campa* (wild broom), and *vassoura vermelha* (yellow broom).

Although the stems typically do not grow very large, they do produce enough wood for some uses. In Mexico, a common reference is to *jaras* (arrows, the Spanish from Hebrew *khara*, to cast, Oaxaca), and *jarilla* [*jaril*] (little arrow, Sonora, Morelos, Oaxaca). Not only do those names reflect past usage for arrow-making, but *jarita* also has been used for *Salix* (Curtin 1947). When large enough, the close-grained, brown hard wood is used for engraving, turning, tool-handles, and walking sticks. People in Hispaniola use the wood in similar ways.

There is a variety of common names that either confuse or compare *Dodonaea* with other species. Some use the name *castaño* (chestnut), surely because of the edible seeds. Mayans compare the plants with *Talisia olivaeformis* (Sapindaceae) with the name *huayun ak* (*huaya* is *Talisia*, *ak*, vine). Both have edible seeds.

In Puerto Rico, *Dodonaea* is sometimes called dogwood, probably because both are bitter. Also because of the bitter flavor, it is compared with sorrel (*Oxalis*, Oxalidaceae) in the names switch-sorrel (Jamaica, Arizona) and *mangle-oseille* (mangrove sorrel, Guadeloupe, Martinique). The French name also makes a comparison with mangroves, as do *manglier petites* (little mangrove, Haiti) and *pativier* (variant of *paltuvier*, mangrove, French Antilles). That comparison is harder to understand because varnish-leaf is an upland plant. To some they are *granadina* [*granadillo, grenadina*] (little passionflower, Dominican Republic, Baja California), an allusion to species that do not seem similar to me.

Since the time of José Celestino Mutis (1732–1808), Colombians have called the species *hayo* (coca) or *hayuelo* [*ayuelo*] (fruit of the coca). Some resemblance of the habit and leaves is apparent, and Mutis is said to have explained to disappointed users that this was not the real coca (Pérez-Arbelaez 1978).

Presumably, the following indigenous words are simple—*chilim* (Morton 1981), *sen* (Morton 1981), *tarachiqui* (Opata, Sonora), and *pirimu* [*jirimu*] (Tarascan, Michoacán).

The most common use throughout the range of the species is as medicine. Some names that allude to medicines are *bois guillaume* (William's tree, French Antilles), *chamiso* (also applied to a medicinal grass and to *Ademostoma*, Rosaceae, Puerto Rico, Argentina, Uruguay), *chamana* [*chanamo*] (*chan* is *chia* or *Salvia*, Colombia, Peru), *chapuliztle,* [*chapul-ixctli, chapuliztoli*] (*chapul,* grasshopper, *ichtli*, fiber, because grasshoppers eat the leaves, Náhuatl), *chulita* (pretty little one), *cucaracha* (cockroach, Dominican Republic), *cuerno de cabra* (goat's neck, Oaxaca), *erva de veado* (deer herb, Brazil), *guachomó* (little orphan, Guarijío), *guatacán* (big ear), *guitarán* [*quitarán*] (big guitar, Puerto Rico), *hierba de la cucaracha* (cockroach herb, Durango, Patoni), *hierba del campo* (country herb), *munditos* (little worlds, Hidalgo), *palo de reina* (queen's herb), *palo de rey* (king's herb, Dominican Republic), and *pinchon* (big spine, Trinidad). The name *plomito* may be an allusion to lead (*plomo*), which would imply medical use, or to doves (*paloma*), which would not.

The leaves and their resinous exudate are aromatic, bitter, astringent, sudorific, purgative, and a febrifuge (Morton 1981, Hocking 1997). Both leaves and/or wood decoctions have been used against fevers (Mexico, Panama, Brazil). The leaf decoctions are also used against rheumatism and venereal disease (Mexico, Brazil). Bark extracts have been used in astringent baths and fomentations (Mexico, Brazil) to provide relief from several problems, including rheumatism (Amabeoku et al. 2001). Although people in some areas eat the seeds (Mexico), they and the entire plant contain enough saponin that people have also used them as *barbasco* to poison fish (Panama, Colombia). Leaves are chewed as a stimulant—maybe that is why Colombians confuse it with coca—or to treat toothache (Panama). The sap is applied to clear tumors (Brazil). Sap with other plant parts is used in cataplasms in the treatment of flatulence (Brazil), gout, and venereal diseases (Mexico, Brazil). Leaf infusions are hemostatic for people and animals (Colombia, Panama).

Some of the chemicals known to be in the plants include flavonoids and other polyphenols in the leaves and their exudates (Rao 1962, Paris and Nothis 1970, Sachdev and Kulschreshtha 1983, van Heerden et al. 2000). There are also diterpenoid acids based on ent-labdane and ent-clerodane skeletons (Sastry and Nayudamma 1966, Jefferies and Payne 1967, Hsü et al. 1971, Abdel-Mogib 2001, Ortega et al. 2001), and triterpenoids (Azam 1993). The bark contains tannins (Sastry and Nayudamma 1966). Fruits, seeds, and stems contain saponin esters, called dodonin and dodonosides A and B, with anti-exudative, phagocytosis-enhancing, and molluscicidal properties (Wagner et al. 1987). There are also prosaogenins (Han et al 1995). In addition, the plants contain coumarins and the lignocoumarins cleomiscosin A and B (Rojas et al. 1992, 1996, Gonzalez et al. 1993, Hemlata and Kalidhar 1994, Ghisalberti 1998, Siddiqui 1998). Laboratory studies suggest that crude methanol extracts are effective in inhibiting *Candida albicans, Bacillus subtilis, Escherichia coli, Pseudomonas aeruginosa,* and *Staphylococcus aureus.* Ethanol extracts showed similar inhibition of other bacteria.

Attempts have been made to divide the variability in the pantropical plants into several species and varieties. Most people still follow West (1984) in recognizing *D. viscosa* as a single species, although there are 68 species in the genus. Some 59 of those species are endemic to Australia (West 1984, Mabberley 1997).

This is one of comparatively few plants known since the 1700s that does not have a name created by Linnaeus. *Dodonaea* was created by Philip Miller in 1754 to honor Dutchman Rembert Dodoens (1517?–1585), one of the foremost physicians and herbalists of his day. Indeed, Linnaeus did not know what to do with these plants. In his *Hortus Cliffortianus* published in 1738, and in his *Flora Zeylanica* from 1747, Linnaeus called the plants *Dodonaea.* Then, in 1753, he changed his mind and named them *Ptelea viscosa.* To show how confused Linnaeus was, *Ptelea* is in the Rutaceae, and *Dodonaea* belongs in the Sapindaceae. It was not until 1760 that Jacquin created the binomial we now use.

Drosera: Sundews

(Greek *droseros,* dewy)

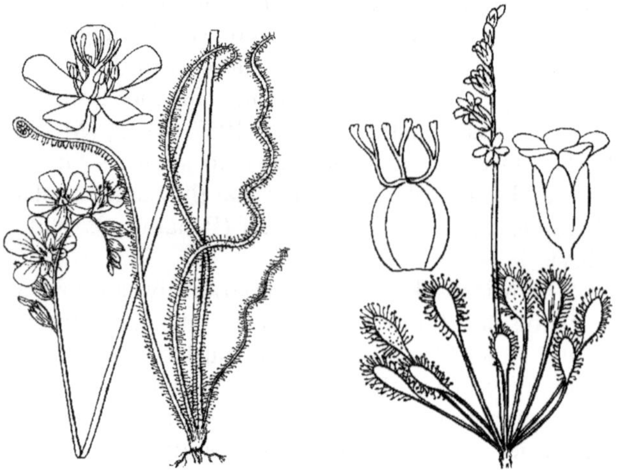

Drosera. *Drosera filiformis* (left). *Drosera intermedia* (right). From Britton and Brown 1897.

"This herbe is of a very strange nature and maruellous: for although that the Sunne doe shine hot, and a long time thereon, yet you shall finde it alwaies moist and be-dewed, and the small haires thereof alwaies full of little drops of water: and the hotter the Sun shineth upon this herbe, so much the moystier it is, and the more be-dewed, and for that cause it was called Ros Solis in Latin, which is to say in English, the dewe of the Sun, or Sundewe." Those words were written by Henry Lyte in 1578, as a translation of Rembert Dodoens's herbal of 1554. Throughout history, everyone has been fascinated with sundews.

When I took classes where sundews grow, I challenged the students to walk barefoot over these carnivorous plants to "test" the idea that they digest their prey. No one ever took me up on that suggestion. Even if they had, they would have no problem. The chemicals that dissolve their insect prey have no impact on human feet—I tried it!

We now think of sundews as the stereotypical plants that have turned the tables on insects, and "eat" them instead of being eaten. However, that is a relatively new view of the herbs. The early Latin speakers called them *ros solis* or *rosa solis* (sun rose) because of a fancied resemblance to the sun. That name carried over into Spanish (*roscío del sol*), Portuguese (*rorela*), and French (*rossolis*), but the sun comparison is not confined to the languages related to Latin. Norwegian *soldogg*, German *Sonnentau*, and Dutch *sundauw* all mean "sundew."

"Sundew" may not, however, have been the original sense. Britten and Holland ([1886] 1965) note an old German name *sindau*, which meant "ever-dewy." They said that name was from Anglo Saxon, Old Saxon, and Frissian *sin*, "ever," rather than from "sun." Another form of the word, as *Singreen* and *Syndaw*, is applied to *Alchemilla* (Rosaceae). About the latter form, Parkinson wrote in 1640 that it was "from the Germane name *Sinnaw*, because the hollow crumplings and the edges also of the leaves, will containe the dew in droppes like pearles, that falleth in the night."

In English, the first appearance of the name sundew seems to have been in the British physician William Turner's *Herbal*, published in 1568 (Coffey 1993). However, Gaelic speakers of western Europe had been using names for these small plants of bogs and marshes far longer. Speakers of the Celtic languages sometimes called these wetland plants *lus na fearnaich* (plants of the bloody-flux). That name came about because the farmers who grazed their cows and sheep in wet meadows thought the plants caused a plague in their animals. Indeed, from one view it seemed logical. The plants were red; animals that grazed where the plants grew were defecating bloody-red stools. That looks like cause and effect. Actually, the *fearnaich* was caused by something totally different and not the plants at all, but it shows the danger of jumping to conclusions based on limited information.

Other people noticed that cattle will not eat sundews, probably because they taste bitter and caustic (Watt [1889] 1972). This taste may be due to the acidic polysaccharide they contain. That acid (up to 4% by volume) is in the sticky droplets that catch insects (Mabberley 1997). This material not only helps trap, but it also kills insects by clogging their spiracles (breathing tubes).

Another Gaelic name for sundews is *lus na ghadmainn* (plant of the insects), showing the old observation of insect-catching (Vickery 1995). Sundews were also known as *lus an-t-soluis* (herbs of the sun), *lus na greih* (sun plant), *lus na eiyrts* (plant of strength), and *lus y ghruiaghtys* (plant with a sun-face). These last two names are associated with a belief that they were a potent aphrodisiac (Vickery 1997). The plants "clung" to anything that touched them, and the leaves were used to transfer that tendency to a person with whom someone wanted to fall in love. It probably worked as well as mistletoe (see Austin 1998a).

From the beginning, these small reddish rosettes were also used as medicines. That was the context of their first mention by Turner, and subsequently by other herbalist-physicians like Joachim Camerarius in 1588, John Gerarde in 1597, Gaspar Bauhin in 1623, and Nicholas Culpeper in 1653. These authors used phrase names for sundews, such as *Salsirora sive ros folis* (sprinkled with dew or rose leaf) and *Ros folis folio rotundo* (rose leaf with round leaves). Linnaeus eliminated the phrase name in 1753 and placed the five sundew species known to him in the genus *Drosera*.

Traditionally, the sundew was made into a tea or tincture and used to treat dry, spasmodic coughs, asthma, arteriosclerosis, chronic bronchitis, and as an aphrodisiac. For example, Gerarde ([1597] 1975) thought the plants a "remedy for consumption." Their limited effectiveness against bronchial problems is due to the quinones in the leaves (Mabberley 1997). William Turner wrote in 1568: "Our Englishmen nowadays set very much faith by it, and hold it good for consumptions and swooning, and faintness of the harte." Gerarde ([1597] 1975) even said that there was a "liquor made thereof, which the common people call Rose solis" (Britten and Holland [1886] 1965). That liqueur was popular in Britain, France, and Germany during the 17th century for its reputedly fortifying and aphrodisiac properties (Bown 1995).

Proteolytic enzymes are the basis of sundew use to treat warts and corns. Culpeper (1653) said that the sundew flowered in June, when the leaves were "fittest to be gathered." He believed: "Sun rules it, and it is under the sign of Cancer. The leaves, bruised and applied to the skin, erode it, and bring out such inflammations as are not easily removed. The juice destroys warts and corns." Culpeper's astrological references were spurned by the medical profession of his time. This mumbo-jumbo was adapted, partly out of context, from the Arabic magicians and alchemists, but has remained surprisingly popular then and now.

Plumbagin in sundew is active against Gram-negative bacteria (Foster and Duke 1990). Among the many Gram-negative bacteria are those that cause typhoid fever (*Salmonella typhi*), food poisoning (*Salmonella* spp., *Shigella* spp.), and Legionnaire's disease (*Legionella*).

Not suprisingly, people in various parts of the globe have learned to use different species in similar ways. Sundews are used in Chinese medicines against lung diseases, dysentery, sore throat, and ear infections. The Miccosukees (Sturtevant 1955) used the plants to treat ringworm, and called them *oláyíkcî* (ringworm medicine). Their Creek-speaking relatives call sundew *hilápa hilíswa* (*helvpa*, ringworm, *heleswa*, medicine), as do the Chickasaw with *halambo tish* (*halambo*, ringworm, *ittish*, medicine). The Kwakiutl of British Columbia, Canada, treat warts, corns, and bunions as did the Gaelic speakers of Europe (Moerman 1998). These Canadian Americans also used sundew as an aphrodisiac, saying that it is a "medicine to make women love-crazy."

People in India have a variant of the topical remedy, where the mashed leaves are applied to skin problems (Watt [1889] 1972). This technique raises a blister as a "counterirritant." This treatment seems akin to hitting yourself on the thumb with a hammer to make a headache disappear. Yet, strange as this may seem, the Indian logic sometimes works. The same counterirritant concept applies to the modern use of capsaicin (the hot stuff) from chile peppers (*Capsicum* spp., Solanaceae) in over-the-counter remedies for arthritis and shingles. Western medical researchers finally found that the chemicals in the treatments "confuse" the nerve endings reporting the problems (pain, itching, or whatever) to the brain, and temporarily stop the symptoms. The method is not a cure, but it provides relief while the body fights off the malady.

Sundews belong to the family Droseraceae, which contains four genera (Mabberley 1997). Remarkably, these plants are related to carnations and their relatives (Williams et al. 1994). The four genera are *Aldrovanda vesiculosa* (central Europe, Asia to northeastern Australia), *Dionaea muscipula* (endemic to Carolinas), *Drosera* (~110 spp., worldwide), and *Droserophyllum lusitanicum* (Portugal, South Africa, Morocco). All are carnivorous and grow on nitrogen-poor acidic soils or rocks. Most are in wetlands such as marshes and bogs, although *Aldrovanda* is aquatic, and *Droserophyllum* grows in dry areas. Their adaptation of carnivory allows them to get enough nitrogen to carry out their life cycles.

Five species of *Drosera* are in Florida; the others are *D. brevifolia, D. filiformis, D. intermedia,* and *D. tracyi* (Schnell 1976, Wunderlin 1998). *Drosera capillaris* is the most common and widespread, also occurring in Cuba, Hispaniola, Jamaica, Puerto Rico, Trinidad, Guyana, and Central America.

Discovery of carnivorous plants in the 18th century led writers and sensationalists to imagine "man-eating" plants in the "jungles" of unexplored lands. That myth persists and is rejuvenated periodically in Hollywood productions, like *The Little Shop of Horrors,* where a monster plant demands, "Feed me!"

Some years ago an undergraduate asked me, in all seriousness, "Where do the man-eating plants grow?" I dutifully explained that no such organism existed, but he really did not seem to buy the story. So, I dreamed up a hypothetical case of two kinds of people-eating plants living near each other. The students were asked to use the logic techniques we had studied, and the information I provided, to determine if the plants on opposite sides of a river were different species. I am not sure they appreciated the irony of my example, but for a change, they did laugh during that exam.

Drypetes

(Martin Hendriksen Vahl, 1749–1804, named these from Greek *drypetes*, fruit ready to fall from the tree, fully ripe; from *drys*, tree, and *petein*, to fall)

Drypetes lateriflora. *Drawn by P.N. Honychurch.*

Drypetes lateriflora (flowers produced laterally in leaf axils)
bois côtelette (cutlet tree, Haiti)
cirulela de Guyana (Guiana plum, Puerto Rico); Guiana plum (Florida, Bahamas, Jamaica); Guinea plum (Puerto Rico)
cueriduro (leathery and hard, Cuba, Puerto Rico)
hueso [*de monte, blanco*] ([wild, white] bone, Cuba)
mula (mule, El Salvador)
pae Manuel (father Manuel, Dominican Republic)
ramón blanco (white tree, Cuba); whitewood (Jamaica)

In 1788, Olaf Swartz named trees he found in the Caribbean *Schaefferia lateriflora*, a genus now included in the Celastraceae. It was not until 1892 when C.W.L. Krug (1833–1898) and I. Urban (1848–1931) realized that the plants were actually members of the Euphorbiaceae and moved the species to *Drypetes.* This is now a genus containing about 200 tropical species in the Americas, Africa, and Asia (Mabberley 1997). Many of them provide timber, but others are used in medicines and as chew-sticks.

Wood is used in construction (Liogier 1974); fruits are eaten (von Reis and Lipp 1982).

E

Echinacea

(From the Greek *echinos*, a hedgehog, referring to the prickly receptacle)

Echinacea purpurea (purple)
coneflower (applied to *Rudbeckia* by 1817, but later denoting a number of different Asteraceae)
[eastern] purple coneflower (*Echinacea* was meant, with or without the prefix "purple," since at least the time of Asa Gray in 1857)
Igelkopf (hedge crown, German)

Linnaeus called these plants *Rudbeckia purpurea* in 1753. Then, in 1794, Conrad Moench created the genus *Echinacea* to accommodate species in this group. Nine species are now known, all native to eastern North America (Mabberley 1997). Perhaps the Muskogean names were related to Koasati *innatho: pahissí* [*innotho:pahissi*] (*innathó:pan*, toothache, *ahissí*, medicine) for *E. pallida.*

The Choctaw chewed the root of *E. purpurea*, and swallowed the saliva or made a tincture of the root to relieve coughing (Campbell 1951). They used it in a similar way to relieve indigestion. The Delaware combined the roots with those of staghorn sumac (*Rhus typhina*) to treat venereal disease (Moerman 1998).

Considering the wide use of *E. angustifolia*, it is surprising how little seems to be recorded of indigenous people using *E. purpurea.* Foster and Duke (1990) note that both species are now used for identical purposes in the herbal trade and imply that historically that was also true. The people of the Great Plains, from at least the Cheyenne to the Lakota, considered *E. angustifolia* a panacea and magical herb (Vogel 1970, Swerdlow 2000).

Gilmore (1919) said Missouri River tribes used *E. angustifolia* more than any other plant. The Dakota called it *ichalipe-hu* (*ichalipe*, whip, *hu*, plant), and the Omaha and Ponca said *hika-hi* (*mika*, comb, *hi*, plant), *ikagahai* (to comb), or *inshtogalite-hi* (*inshta*, eye, *galite*, comb, *hi*, plant). The Pawnee said it was *ksapitahako* (*iksa*, hand, *pitahako*, to whirl), because of a game children played with it. Another name they used was *saparidu kahts* (*saparidu*, mushroom, *kahts*, medicine) because the head resembles a mushroom. Moerman (1998) listed this species as important among the Blackfoot, Cheyenne, Dakota, Kiowa, Meskwaki, Montana, Omaha, Pawnee, Ponca, Teton, and Winnebago. *Echinacea pallida*, sharing much of its range, was used similarly. These two were used to treat snakebites, spider bites, cancers, toothaches, burns, sores, wounds, flu, and colds.

Modern research shows that *Echinacea* has cortisone-like activity, and that it is antiviral, bactericidal, insecticidal, and an immunostimulant (Foster and Duke 1990, Swerdlow 2000, Duke et al. 2002).

Echites

(Patrick Browne named this from Greek *echis*, a viper, referring to the poisonous nature and twining habit of the plant; Pliny, A.D. 23–79, used the word *echites* for a kind of stone)

Echites umbellata. *Drawn by P.N. Honychurch.*

Echites umbellata (having partial or secondary umbellate flower clusters)
bejuco de leche (milk vine, Central America)
Danish (Bahamas)
Devil's potato-root (Bahamas); wild potato (Bahamas)
maroon weed (possibly alluding to the "Maroons" of the Cockpit country, Jamaica)
rubber-vine (Florida, Bahamas)
savannah flower (Jamaica)

Patrick Browne created this genus in 1756 when he published *The Civil and Natural History of Jamaica*, but provided no valid species epithet. Jacquin then formalized the concept by naming this species in 1760. The genus contains six Caribbean species (Mabberley 1997).

Used on a sore leg, it supposedly will cause vomiting if the wound is poisoned (Ayensu 1981).

Several of the common names note the poisonous nature of the plants. Barham, in his *Hortus Americanus* published in Jamaica in 1794, wrote that it was a widely known poison (Grimé 1976). Perkins and Payne (1978) simply noted that the root and its milky sap have not been studied chemically. Perhaps *E. umbellata* has chemical constitutents similar to plants reported as *E. hirsuta*, although they are correctly in the genus *Mandevilla*. That Old World species has the flavonoids naringenin, aromadendrin (dihydrokaempferol), and kaempferol; the coumarin fraxetin; the triterpene ursolic acid; and the sterol glycoside sitosteryl glucoside (Chien et al. 1979).

Eleocharis

(Robert Brown named this "beauty of the marsh"; from Greek *heleos*, a marsh, *charis*, grace, beauty)

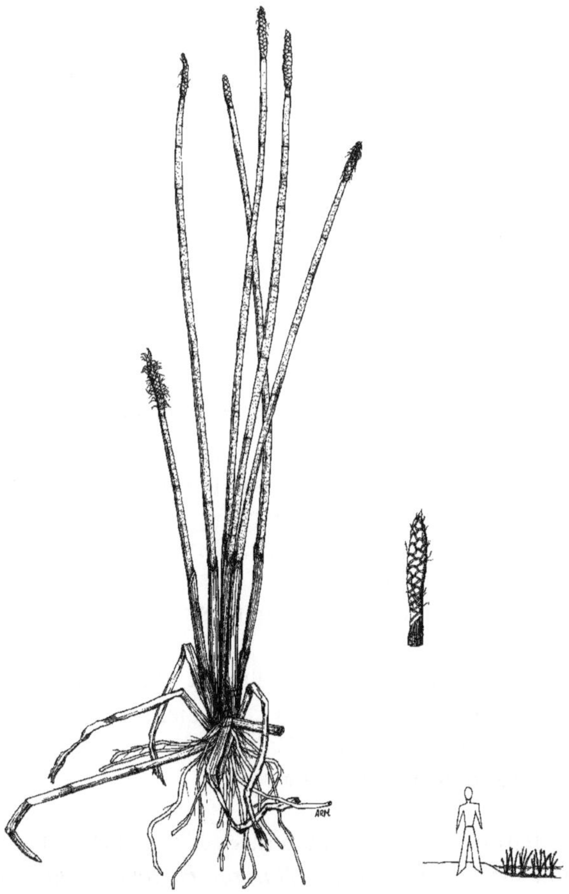

Eleocharis equisetoides. From Institute of Food and Agricultural Sciences.

giunco tondo (round juncus, Italian)
héléocharis (from *Eleocharis*, French)
jonc (Haiti) and *junco* (Cuba, Dominican Republic, Puerto Rico; common names applied widely to this genus and others based on the Latin *iuncus*, for a rush, derived from *iugere*, to tie or bind, from the use of stems for tying); *junquillo* (little juncus, Dominican Republic, Puerto Rico)
sivaks (*siv*, rush or reed, Norwegian)
spike-rush ("spike" has been used for a plant inflorescence since about 1711)
Sumpfried (*Sumpf*, swamp, *Ried*, reed, German)
tulillo (little tule, from *tollin* or *tullin*, Náhuatl; *tule* is applied to both *Juncus* and *Typha*, Sonora)
Wasserpest (water pest, German)

Eleocharis equisetoides (resembling *Equisetum*)
jointed spike-rush
jointed spike-sedge
pahifamí:câ (*pv'he*, grass, *fvmece*, sweet [smelling], Creek); *pahifotóhlî* (sweet [smelling] grass, Mikasuki)

Eleocharis geniculata (bending like a knee-joint) (= *E. caribaea*)
bioro (Arawak, Suriname)
Canada spike-rush
i:kana cokhíssi (*ekvnv*, ground, *kokhesse*, whiskers, Creek); *yaknicô:skî* (ground whiskers, Mikasuki)
kiririma (Carib, Suriname)
road-grass (Florida)
[Canada] spike-sedge [spike sedge] (Florida)

Eleocharis interstincta (spotted, referring to the septa of the culms or stems)
bioro (Arawak, Suriname)
ça (Osage; mats called *ça-zhin̲-ga* [*zhin̲-ga'*, small] made from it; a sacred plant used to make the shrine of the *wa-xo'be* or Sacred Hawk, a deity)
knotted spike-rush

Eleocharis rostellata (little-beaked, the seed)
[beaked, traveling] spike-rush

Eleocharis, a sedge genus with about 120 species, is widespread; Mabberley (1997) says it is cosmopolitan. The common name spike-rush came about because of the slender spikelike inflorescence on a rushlike plant. At least the larger of the species have been used around the world in weaving and making baskets, bedding, mats, and similar items. For example, in Hispaniola, *E. interstincta* stems are used in households to make seats, chairs, and mats, and as stuffing in pillows and similar items (Liogier 1978).

Oddly, no weaving uses have been recorded among the Seminoles, although they call it by the same name as the grass they use in weaving baskets, *pahifami:câ* (*pvhe*, grass, *fvmece*, sweet). Instead, *E. equisetoides* was used to make beads. No explanation is given by Sturtevant (1955), but presumably the jointed stems were strung. *Eleocharis geniculata* was used in a medicine by the Seminoles (in a mixture with several other plants) to treat "Fire Sickness," "Thunder Sickness," and urine retention (Sturtevant 1955). In Hispaniola, *E. interstincta* is used as a diuretic, against stomachache (Liogier 1978). *Eleocharis rostellata* was used as an emetic by the Raman Navajo (Hocking 1997, Moerman 1998).

Elephantopus

(From Greek *elephas*, elephant, *pous*, foot, said to be a translation of an aboriginal name in India)

Elephantopus elatus (high, lofty) (as *E. tomentosus*)
devil's grandmother (New England)
[woolly] elephant-foot (USA)
lengua de vaca (cow's tongue, Puerto Rico)
sikohákcobî (pig ear, Mikasuki); *sokkaha:ckô* (*sokhv*, pig, *hvcko*, ear, Creek)
tobacco-weed

Elephantopus was named by Linnaeus ([1753] 1957), when he described Indian and American species. An Italian professor of botany in Bologna, Antonio Bertoloni (1774–1850), named this American species in 1850, although numerous subsequent students of the flora considered it the same as *E. tomentosus.* There are about 30 species in tropical regions of the world (Mabberley 1997).

Several species in the genus are used medicinally, with *E. scaber* used in Colombia and Brazil as a tonic, febrifuge, and diaphoretic against cough, bronchitis, and asthma (Roig 1945). Although Sturtevant's (1955) informants considered it "useless," former reports of names suggest it may have been used much as *Chaptalia* (which see).

Encyclia

(From Greek *enkyklios, encyklia*, round or circular; Latin *encyclios*, belonging to a circle, the labellum encloses or encircles the column)

Encyclia cochleata (shell-shaped, an allusion to the labellum, which resembles a clam shell) (=*Epidendrum cochleatum, Prosthechea cocheata*)
cañelo (little cane, Cuba, Puerto Rico)
cangrejitos (little crabs, Dominican Republic)

Encyclia cocleata. *Drawn by P.N. Honychurch.*

[Florida] clam-shell orchid [clamshell orchid] (Florida); cockle-shell orchid (Puerto Rico); shell-orchid (Bahamas); *conchitas* (little shells, Dominican Republic)
flor de San Juan (St. John's flower; presumably St. John because it flowers near this saint's birthday on 24 June, Venezuela)
guiguitte [*quiquitte*] (Haiti)
javitas (Dominican Republic)
orquídea negra (black orchid, Puerto Rico)
ts'aak (glue, Huastec, San Luis Pososí)

Sir William Jackson Hooker (1785–1865) named this genus in 1828. In 1955, Albert Marie Victor Lemée (1892–1900) moved Linnaeus's *Epidendrum cochleatum* into *Encyclia* in his *Flore de la Guayane Française.* Some 235 American orchids are now put in *Encyclia* (Mabberley 1997).

"Root sap" (pseudobulbs?) is used as glue to make musical instruments and in medicinal baths to treat weight loss (tuberculosis) (Alcorn 1984). Pseudobulbs are a source of mucilage (Hocking 1997).

Epifagus

(Thomas Nuttall named this parasitic herb with Greek *epi*, upon, *phagus*, the beech)

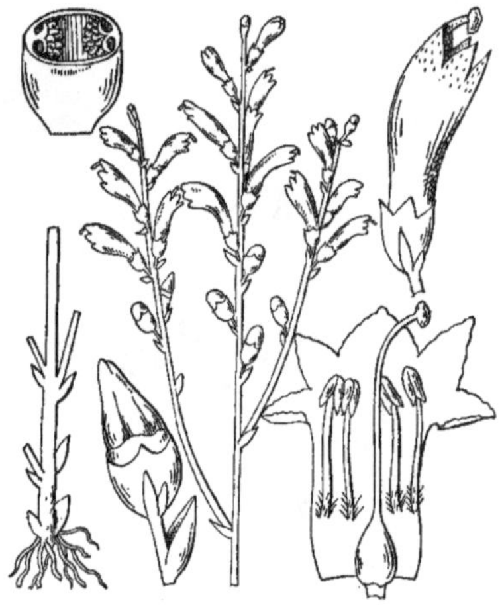

Epifagus virginiana. From Britton and Brown 1898.

Epifagus virginiana (of Virginia)
beech drops [beechdrops] (applied since at least 1876 to this parasitic plant and equated with "cancer-root")
cancer-root (first used in 1714 for this plant added to a mixture to cure "cancer," a term not always applied in the sense of a malignant growth; John Gerarde in 1597 referred to cancerwort, but that meant either *Linaria* or *Veronica*)

Although *Fagus* is circumboreal, there is a single species of *Epifagus*, which is parasitic on the roots of North American *F. grandifolia* (Mabberley 1997). Perhaps this American endemic is a remnant of differential extinction during the Pleistocene.

The Iroquois used an infusion of the plant to treat diarrhea (Moerman 1998). Homeopathic physicians used the root as a vulnerary, especially on sores and cankers (Hocking 1997).

Epigaea

(From Greek *epi*, upon, and *gaia*, the earth, alluding to the creeping habit)

Epigaea repens (running on the ground)
Bodenlorbeer (ground laurel berry, German); ground-laurel
fleur de mai (May flower, Quebec)
gravel weed [plant] (used by Shakers to treat kidney stones)
ground hog's forehead (North Carolina)
Mayflower (not because it blooms in May, but according to legend, the Pilgrims saw this flower on landing at Plymouth and named it after their ship, cf. Coffey 1993)
moss beauty

Epigaea repens. From Britton and Brown 1897.

mountain [winter] pink
rough-leaf
terrapin's foot ("terrapin" is the box turtle, *Terrapene* spp. Apparently both words were derived from the Algonqian word *torepe* and used in English by 1612. Related words are Abenaki *turepé* or *tourepé*, and Delaware *tulpe*. Breaking one of his own "rules," Linnaeus adapted an indigenous word into one that only resembles Latin as *terra*, the earth, and *pene*, nearly; used in North Carolina.)
trailing arbutus (compared with the "strawberry tree" of Europe, *Arbutus unedo*)
wa'bî'gon (white flower, Potawatomi)

Plukenet described and illustrated a small plant he called *Pyrola affinis repens fruticosa* (Prostrate woody shrub similar to *Pyrola*) in 1696. Linnaeus knew the plants from both Canada and Virginia, and realized they were unlike anything in Europe. He called them *Epigaea* because of their prostrate growth (Linnaeus [1753] 1957). This genus has a single species in eastern North America, a second one in the Caucasus and Asia Minor, and a third one in Japan (Mabberley 1997).

The Potawatomi of Wisconsin considered these their tribal flowers, and thought they were sent directly by their deity, called *kîti'manîtowiwin* (Smith 1933). The plants grow intermixed with mosses, flower in May, and bear fruit in July. Apparently, the Japanese do not revere their species, but they eat the fruits (Ohwi 1965).

The Algonquin in Quebec used an infusion of the leaves to treat kidney problems (Moerman 1998). The Iroquois made a compound mixture of plants including these to treat labor pains and rheumatism; a decoction of the whole plant for the kidneys; or a simple decoction of leaves to aid digestion (Moerman 1998). The Cherokee made a decoction to induce vomiting, for use as an infusion to stop diarrhea in children and to treat kidney and chest problems (Hamel and Chiltoskey 1975).

Eclectic physicians used the plant to treat bladder problems, particularly kidney stones; it contains the glucosides arbutin, urson, and ericolin, plus tannins (Millspaugh 1892, Felter 1922). Although arbutin is effective as a urinary antiseptic, it hydrolyzes to hydroquinone, which is toxic (Foster and Duke 1990).

Equisetum

(From Latin *equus*, horse, and *seta*, bristle)

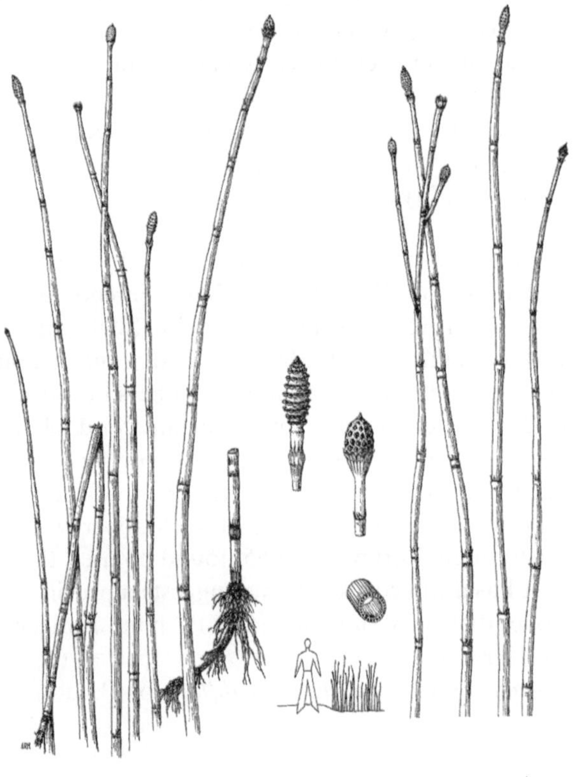

Equisetum hyemale. From Institute of Food and Agricultural Sciences.

Equisetum hyemale (of winter, because evergreen)
biorag (tusk, Gaelic)
cañuela (little cane, Texas, Arizona); *cañutillo* [*del Llano*] (little cane [of the Plains], New Mexico)
cuilg struth eich (horse's water bristles, Gaelic)
Dutch rush (apparently because it was used by the Dutch)
gijib'inuskon' (it is round, Ojibwa)
[common, greater, rough, winter] horsetail (appearing in the 1200s, from the resemblance of the branched plants to the animal's appendage)
joint rush
Kannenkraut (can herb, German; formerly used to clean milk cans)
mande idhe shnaha (*mande*, bow, *idhe*, use, *shnaha*, to smooth, or to-make-a-bow-smooth, Omaha-Ponca); *mi'-he-ga* (Osage)
nebratah (said by Rafinesque to be the name among Missouri tribes; similar and perhaps confused with *nibdhathka*, flat river, Omaha, or maybe both were so called)
pakarut (Pawnee)
pewter-wort ("Women scowre their pewter and wooden things of the kitchen therewith, which the Germaine women call *Kannenkraut*, and thereupon some of our huswives do call it *Pewter-woort*"; cf. Britten and Holland [1886] 1965)
prêle (rush, used since 1538, from *asprele*, in 1200s, based on Latin *asperella*, from the adjective *asper*, rough, from the stem texture, Quebec); *prêle des tourneurs* (scouring rush, Quebec)
sa-po-tun-a-kio-tio-yis (jointed water grass, Blackfeet)
Schachtelhalm (hollow stem, German)
[American, common, great, scrubbing tall] scouring-rush [scouring rush] (applied about 1815, particularly for this species; the reference is to the use of the silicon-containing stems in cleaning or polishing)
shave-grass (akin to scouring-rush)
snakegrass

This widespread species was used throughout the Northern Hemisphere (Foster and Duke 1990). The most common application through the *Equisetum* range is for scouring or sanding because of the silica in the stems.

Leonard Fuchs (1542) mentioned two species known in his time, *E. arvense* and *E. palustre*. These were mentioned by Dioscorides (fl. A.D. 40–80), Galen (A.D. 129–?200), and Pliny (A.D. 23–79) as medical plants, so their tradition in the Old World is venerable. Fuchs listed names for them as English, shavegrasse, horsetayle, water taile; German, *Kannenkraut*, *Katzenwedel* (cat's brush, in the sense of a "voluminous tail"), *Pferdtschwantz* (horse muzzle), *Roszchwantz* (horse muzzle), *Roszwadel* (horse muzzle); Dutch, *peersteert* (horse tail, Dutch); French, *queue de cheual* (horse tail), *asprelle* or *presle* (rush); Italian *coda di cauello* (horse tail), *poligono* (many joints), *sanguinaria femina* (for female blood); Spanish, *cien ñudillos yerua* (herb of a hundred little joints). The one name that needs further explanation is *Kannenkraut*. That refers to tin cans used as containers for milk and other items. The cans were cleaned in Germany with *E. arvense* stems due to their high content of silicic acid, the crystals of which show a mechanical cleaning effect (E. Eich and T. Schimming, personal communication, 2003).

Culpeper (1653) made no distinction between the species in Europe, saying that *Equisetum* was "very powerful to stop bleeding either inward or outward ... it also stays laxes or fluxes in man or woman, and heals the inward ulcers ... it solders together the tops of green wounds." He also considered it diuretic and good for curing kidney stones, coughs, and inflammation of the genitals.

The species is remarkable in that no use is mentioned among the people in the southeastern United States. Maybe Porcher (1863) gave some idea of why they are not mentioned when he wrote of *E. laevigatum:* "They contain a great deal of silica. The dried stems of *E. hiemale* and *E. arvense* are imported from Holland for cleaning wooden utensils and polishing cabinet work, turnery, and metallic wares. This plant might be profitably cultivated for the use of turners, cabinet-makers, and other artificers." Indeed, Gray (1875) says *E. hymale* is from Europe, but later authors say that the European plants are only varieties of the widespread species (Gleason and Cronquist 1963). Plants in the southeastern United States were known as *E. affine* in the 1800s.

The closest neighbors of the Muskogean people recorded as using *Equisetum* were the Cherokee. They used infusions for the kidneys and constipation (Hamel and Chiltoskey 1975). Otherwise, the Bella-Coola, Blackfoot, Carrier, Cheyenne, Costanoan, Cowlitz, Cree, Flathead, Gosiute, Havasupai, Hoh, Iroquois, Karok, Klamath, Lakota, Mahuna, Makah, Menomini, Meskwaki, Ojibwa, Okanagan-Coville, Quileute, Quinault, Sanpoil, and Thompson used it for medicine, food, and fibers (King 1984, Moerman 1998). Several of the tribes used the mashed stems to kill vermin and weeds. Winnebago children sometimes made whistles of the stems, but were warned that in blowing them they might call in snakes (Gilmore 1919).

It is now known that the plants contain alkaloids, including nicotine, but their toxicity is due to thiaminase breaking down the vitamin thiamine (Foster and Duke 1990, Mabberley 1997).

Eragrostis

(Greek *eros*, love, *agrostis*, field or country, apparently because it grows widely in cultivated and open areas)

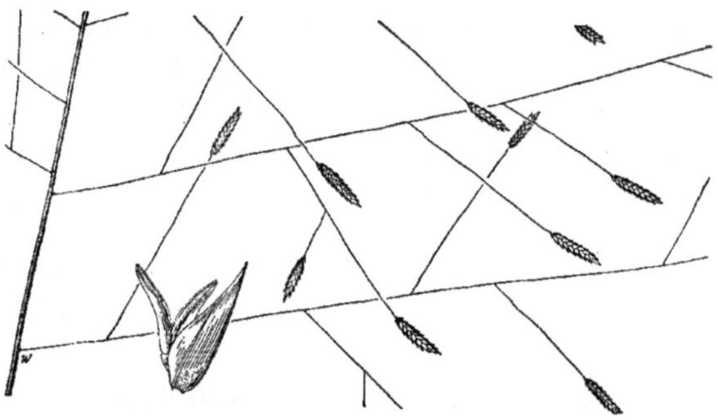

Eragrostis elliottii. *Drawn by Edna May Whitehorn.* From Hitchcock and Chase 1950.

Eragrostis elliottii (named for American botanist Steven Elliott, 1771–1830)
Elliott's lovegrass (a book name)
love grass [lovegrass] (in use by 1702, when London apothecary James Petiver wrote, "What is peculiar in this Love-grass is its having just under each spike, its stalk clammy," Europe, USA)
tickle grass (because the infructescences "crawl" up inside trouser legs, Kentucky)
yatlaytí:cî [*yatleititci*] (Mikasuki)

Eragrostis is a grass named in 1776 by Nathanael Matthaeus von Wolf (1724–1784). Sereno Watson (1826–1892) named this species in 1890, and the genus contains about 300 temperate and tropical plants. Two of the most famous members of the genus are *E. tef* (t'ef or teff), an important food plant in Ethiopia, and *E. pilosa*, which also produces edible seeds but is best known as being a part of the bricks used to build pyramids in 3359 B.C. (Mabberley 1997).

Although Sturtevant (1955) found no uses, the fact that it has a Miccosukee name suggests a previous use or one that was not shared with him. Indigenous Americans ate the seeds of some species (e.g., *E. mexicana* by Cocopa, *E. secundiflora* by Paiute).

Erechtites

(Greek, from *erechtho*, to rend, break, possibly referring to the dissected leaves; first used by Dioscorides, fl. A.D. 40–80, for some species of groundsel, *Senecio*; another suggestion is that it commemorates Erechtheus, a fabled king of Athens)

Erechtites hieraciifolia (having leaves like *Hieracium*, hawkweed)
achicoria de cabra (goat's chicory, Cuba, Puerto Rico)

American burn-weed [burnweed] (the herbs proliferate after disturbance, USA); fire-weed [fireweed] (a name also used for *Epilobium*, USA)
äsûkitä'boe (stickers, Potawatomi)
côte soleil (sun's rib, Haiti)
dog-weed (Bahamas)
lechuguilla (lettuce, Dominican Republic)
letron (big letter, Haiti)

Paul Hermann (1646–1695) had these plants in cultivation in the Netherlands in 1698. Later, Linnaeus saw them growing at both the *Hortus Cliffortianus* and the *Hortus Upsaliensis*. This herb was called *Senecio hieracifolius* by Linnaeus in 1753. Rafinesque did not think they should be in *Senecio* and created *Erechtites* in 1817. The genus includes 15 American species (Belcher 1956, Vuilleumier 1969, Barkley and Cronquist 1978).

The plant is used as a tonic, an astringent, an alterative, and to treat dysentery (Roig 1945, Liogier 1974).

Erigeron

(From Greek *eri*, early, and *geron*, old man; an ancient name of a Mediterranean plant said by Dioscorides, fl. A.D. 40–80, to be used "because the flowers in spring turn gray like hair"; reapplied by Linnaeus in 1753)

bakkestjern (*bakke*, hill, *stjern*, star, Norwegian)
Berufskraut (profession weed, German)
farewell-to-summer (Newfoundland, Labrador)
Feinstrahl (delicate line, German)
fleabane ("fleabane" first appeared in print in English when Turner published his book *Names of Herbes* in 1548; they were plants to repel fleas)
impia (Italian)
vergerette (bundle of rods or clothes brush, Quebec); *vergerolle* (French)

Erigeron annuus (living a single year)
asterana (from *aster*, star, Italian)
cespola (from *cespo*, tuft, Italian)
[little-, tall] daisies (Maine)
[annual, daisy] fleabane ("daisy" was first used about A.D. 1000; however, by Chaucer's time ca. 1385, it was more recognizable as "daysyes," now called *Bellis perennis*; subsequently applied to a number of Asteraceae)
lace-buttons
muletail-weed (Oklahoma)
[sweet-]scabious (confusion or comparison with *Knautia arvensis*, Dypsacaceae)

Erigeron annuus. a. Habit. b. Flower head. c. Disk flower. d. Ray flower. e. Achenes. *Drawn by Regina O. Hughes.* From Reed 1971.

white-top [weed] (Ohio)

Erigeron philadelphicus (of Philadelphia)
cocash (English?)
daisy (North Dakota, Ohio)
[daisy, Philadelphia]-fleabane
frost-root [frostroot] (variant of "frost-weed," used by Porcher in 1863; see *Verbesina*)
mourning-bride (same as "mourning-widow," North Carolina); mourning-widow ("mourning-widow" refers to dark-colored flowers, North Carolina)
scabish [skevish] (comparison or confusion with *Knautia arvensis*, Dypsacaceae, Pennsylvania); sweet-scabious (comparison or confusion with *Knautia arvensis*, Dypsacaceae)
squaw-weed

Erigeron pulchellus (handsome)
blue spring-daisy
plantain-fleabane (plantain refers to broad leaves)

poor Robin's plantain ("poor Robin" is a dialectic term in use since the time of Chaucer in 1374); Robin's fleabane (a variant of "poor Robin")
rattlesnake-plantain (alluding to use in treating snakebite)
Robert's-plantain (probably first used by physician Johann D. Schopf after visiting North America in 1783–1784; the OED 1971 found it only from 1846 and after)
rose-betty

Although William Turner was the first to use the word fleabane for European plants, there is no way to be sure what he was talking about. He wrote, "Coniza maye be called in englishe Flebayne." Now the name *Conyza* is applied to the American plant *C. canadense* (which see). That plant was introduced into Europe early and was praised by John Parkinson in 1640. At the same time, Parkinson called some plant *C. palustris major*, the "greater Marsh or water Fleabane." Not long afterward, Culpeper (1653) mentioned two "Marsh fleabanes"—*Erigeron viscosum* and *Senecio vulgaris*.

The practice of using plants to discourage fleas was a long-standing and ongoing practice well before Parkinson. For example, a book on gardening published in 1563 stated, "The Gnats ... also be chased away with the decoction of the herbe named Flebane, sprinkled on the bed." John Gerarde ([1597] 1975) wrote, "Conyza hath from time to time been called English fleabane." Culpeper (1653) wrote about five kinds of fleabanes—Canada fleabane (*C. canadense*, as *E. canadense*), marsh fleabane (*S. vulgaris*), mountain fleabane (*S. jacobaea*), small fleabane (*P. dysenterica*), and flea-wort (*E. viscosum*), also called marsh fleabane. Europeans had major problems with fleas and used whatever was available to try to rid themselves of the pests. As late as 1801, the New Hampshire American Samuel Stearns said, "the chief use of the *flea banes* is for destroying fleas and gnats, by burning the herbs so as to waste away in smoke" (Vogel 1970).

Moerman (1988) found no indigenous use of *E. annuus*. However, John Bartram wrote of it in 1751, "*Erigeron*, used by some for the bite of a snake: it bears a white flower in the spring, something like a large Daisy, about a foot high, the roots run ... of a hot taste; the Indians pound this root, and apply it to cold tumours to dissolve them" (Coffey 1993). Presumably, Bartram was referring to the Delaware near where he lived. Later, the plants were considered diuretic, diaphoretic, and stomachic in edema. They were the source of the pharmaceutical known as *Herba Asteri* (Hocking 1997). During the Civil War, Porcher (1863) said that it was "a favorite diuretic in the dysuria of children—used by Physick and Dewees in painful micturition dependent on nephritis. This also yields a styptic oil similar to that afforded by the *E. Canadense* [*Conyza canadensis*]." *Erigeron annuus* contains tannins and volatile oils (Hocking 1997).

Surely the best known of the fleabanes was *E. philadelphicus*. In the east, it was used by the Cherokee, Houma, Iroquois, Meskwaki, and Ojibwa (Vogel 1970, King 1984, Moerman 1998). Farther west, it was favored by the Blackfoot and Okanagan-Coville.

The Cherokee used *E. philadelphicus* to suppress menstruation, headache, and bleeding, and to promote sweating. It was used for epilepsy, colds, coughs, sores, weak eyes, and kidney problems (Hamel and Chiltoskey 1975). The Houma used it to treat menstruation problems (Speck 1941). The Iroquois made a wash to stop the itch of poison ivy and to cure running sores. They also used it to stop congestion of the lungs (Moerman 1998). The Meskwaki snuffed the powdered flowers to relieve headache, colds, and catarrh by sneezing (Vogel 1970). The Ojibwa not only burned the flowers and inhaled the smoke to make the patient sneeze to relieve head colds, but they also used an infusion to break fevers (Vogel 1970). The Blackfoot made a medicine to stop diarrhea and hemorrhage in childbirth. The Okanagan-Coville used it to relieve headaches.

Physician Benjamin Barton thought in 1804 that *E. philadelphicus* was good "for gouty and gravelly complaints.... It operates powerfully as a diuretic and also as a sudorific." Rafinesque (1830) found that these herbs "act in a mode peculiar to themselves, by means of their acrid quality. Their oil is so powerful that two or three drops dissolved in alcohol have arrested suddenly uterine hemorrhages."

Porcher (1863) wrote: "It is diuretic, without being offensive to the stomach.... In great repute as a remedy in calculus and in nephritic diseases. It was a favorite prescription in Philadelphia in dropsy, and Dr. Wistar recommends it in hydrothorax complicated with gout.... One ounce of the plant to be administered in infusion or decoction of one pint in twenty-four hours."

The Cherokee used *E. pulchellus* to suppress menstruation, headache, and bleeding, to promote sweating, and for epilepsy, colds, coughs, sores, weak eyes, and kidney problems (Hamel and Chiltoskey 1975). The Iroquois used a decoction to treat colds, coughs, and tuberculosis (Moerman 1998).

The German physician Johann Schopf wrote in 1788 that "the Robert's Plantain" was good to treat snakebite because, according to local doctors, it had "many times been of excellent use where signs of the poison taken up into the blood were already plainly manifest." He added, "it is raised there foresightfully in gardens, so as to be found in the night if occasion arises" (Coffey 1993). Hocking (1997) considered *E. pulchellus* to be diuretic, diaphoretic, and a tonic.

Eriogonum

(From Greek *erion*, wool, and *gonu*, knee, from the woolly stems and leaves and swollen joints of the stems)

Eriogonum longifolium (long-leaved)
Indian turnip
long-leaved [wild] buckwheat ("buckwheat" came from the Latin *fagus*, beech, and Greek *pyrum*, fruit or seed, comparing a plant introduced from Asia with one that was native and familiar. The word was recorded in Turner's *Names of Herbes* in 1548. Turner wrote, "Elatine [*Polygonum esculentum*] is lyke wythwynde, but it hath seedes and floures lyke Buckwheate [*Fagopyrum*]. Buckwheat is cognate with Dutch *boekweit* and German *Buchweize*.)
long-leaved eriogonum (a translation of the Latin)
umbrella-plant ("umbrella" appeared in English about 1611 for the portable shade; it is based on Latin *umbra*, shade; the ribs of the umbrella were first compared to branching in a palm in 1798, and dozens of plants have been associated since)

This genus, related to buckwheat (*Fagopyrum esculentum*) and rhubarb (*Rheum rhabarbarum*), contains about 240 species (Mabberley 1997). *Eriogonum* is endemic to western and southern North America, and some of the species are cultivated for their edible leaves or roots (Diggs et al. 1999). Prospectors believed *Eriogonum ovalifolium* indicated the presence of silver.

The only indications of use of this species are among the Comanche of northwestern Texas and the Kiowa who lived near the Arkansas and Red Rivers of Oklahoma (Moerman 1998). Kiowas ate the roots, and the Comanche took an infusion of them as a remedy for stomach trouble.

In spite of those few records, Moerman (1998) records another 36 species that have been used as foods, chewing gum, dyes, drinking tubes, tobacco pipe stems, brooms, medicines, and toys.

Erithalis

(Named by Patrick Browne, who knew it in Jamaica, from Greek *erithales, erithallos*, growing luxuriantly, flourishing)

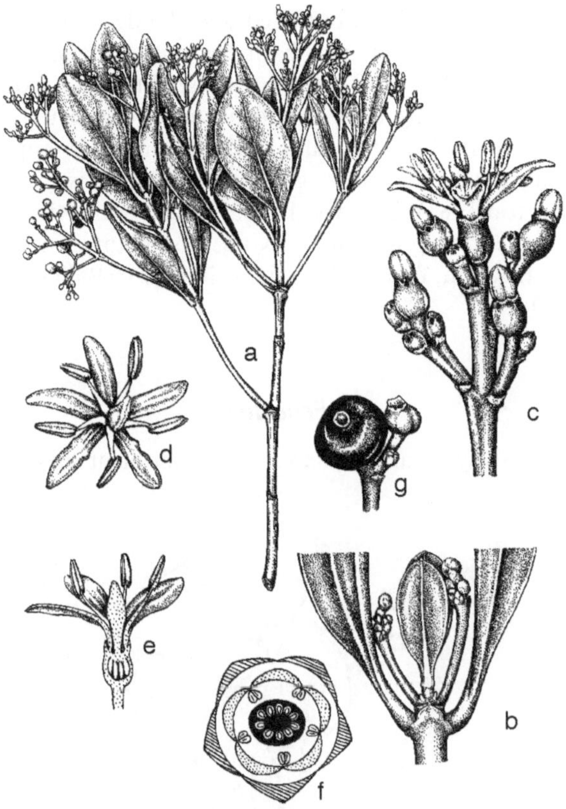

Erithalis fruticosa. a. Flowering and fruiting branch. b. Shoot apex with flower-bearing branches. c. Details of flowering branch. d. Flower, from above. e. Flower, longitudinally dissected. f. Floral diagram. g. Fruit. *Drawn by Priscilla Fawcett.* From Correll and Correll 1982.

Erithalis fruticosa (shrubby)
black-torch (Florida, Bahamas, Puerto Rico, Dominica); *bois chandelle* [*noir*] ([black] torch tree, Guadeloupe, Martinique, Dominica); [*bois*] *flambeau* (torch [wood], Dutch Antilles, Dominica); *lumbra blancu* (white torch, Dutch Antilles); *tea* (torch, Puerto Rico)
botoncillo (little button, Belize)
cuaba prieta (black *cuaba,* which may be from *cua*, place, *bana*, large, Taino, Cuba)
jayajabico [*jayabico*] (probably Taino, as their name for an unidentified tree is *jayabacaná*, Cuba, Puerto Rico)
parrot-apple (Tobago)
pigeon-berry (Bahamas)
rompe machete (machete breaker, Cuba)
tarro de chivo (goat horn, Mexico)
víbona (Cuba)

Linnaeus described *Erithalis fruticosa* in 1759, putting the species in a genus that Patrick Browne had created in 1756. Ten species are now known in the Caribbean (Mabberley 1977).

Leafy branches are boiled and the decoction used to treat piles, and in a bath for measles and sores. Parched and ground leaves are applied to dry a wound or stop bleeding of a newborn's navel (Ayensu 1981, Morton 1981). The name "black torch" is applied because of the smoke given off when it is burned (Morton 1981). The wood is hard, heavy, and durable, and used for posts and carving. The plant is used in construction, fuel, and medicine in Belize (Balick et al. 2000).

Ernodea

(Named by Swartz, Greek, *erno*, a shoot or offshoot)

Ernodea littoralis. *Drawn by P.N. Honychurch.*

Ernodea littoralis (of the seashore)
cough bush (Bahamas)
ernodea de playa (beach ernodea)
guana berry (from iguana?, Cayman Islands)
wild cherry (Belize)
yellow jugs (Belize)

Olaf Swartz named the genus and species in his *Nova Genera et Species Plantarum* (New Genera and Species of Plants) of 1788. Nine species are now included in this small genus endemic to the southeastern United States and the Caribbean (Mabberley 1997).

In the Turks and Caicos and Out Islands the leafy branch tips are made into a tea to treat cough (Morton 1981). Plants are used in Belize for food, for making small items, and as forage (Balick et al. 2000).

Eryngium: Kissing Comfits to Rattlesnake Master

(From *erygein,* diminutive of *eryngos,* to belch, in reference to its carminative properties; a name of a plant, and a goat's beard)

Eryngium yuccifolium. From Britton and Brown 1897.

Europeans were well acquainted with the genus *Eryngium* when they arrived in the New World. In the Old World *E. maritimum* (sea-holly) had been famous since the time of the Greeks, who called it *erungion.* That name was taken into Latin as *eryngion,* and then into Italian and Spanish as *eringio.* Eventually, it made its way into English as "eringos" about the time of Shakespeare.

Dioscorides (fl. A.D. 40–80) prescribed *erungion* for swellings and other people throughout its range collected and used it to the point that it became rare. Now it is prized as a tourist souvenir and threatened with extinction. The species grows throughout all but the coldest parts of Europe near the coasts, and people everywhere knew and used the plants for medicine and food. Candied roots were known in English as either eringos or kissing comfits. By both names, they also had the reputation of being aphrodisiac. In Shakespeare's *The Merry Wives of Windsor* from 1598, the aging Falstaff has a midnight rendezvous in the park:

> *Falstaff*: Who comes here? my doe?
> *Mrs. Ford*: Sir John? Art thou there, my deer? my male deer?
> *Falstaff*: My doe with the black scut? Let the sky rain potatoes; let it thunder to the tune of Greensleeves; hail kissing comfits and snow eringoes; let there come a tempest of provocation.

Kissing comfits, perfumed and preserved roots of *Eryngium*, usually dipped repeatedly in sugar (Grigson 1986), sweetened the breath—surely needed in a time when toothbrushes were unknown and bathing was considered unhealthy. The word "comfit" is alien to modern English speakers, but it was used from the 1480s through the 1700s, borrowed from Old French *comfiter.* Typically, it meant a preserve made with sugar.

Not only was the Old World *Eryngium* considered to be food, medicine, and an aphrodisiac, but it also was regarded as an alexiteric (antipoison). To the Gaels, *E. maritimum* growing near the ocean was *cuilionn tragha* (*cuilionn*, holly, *tragha*, seashore). With a reputation like that, it is small wonder that Europeans recognized the American *Eryngium* species when they arrived. Two American species were most similar to the Old World plants, *E. aquaticum* and *E. yuccifolium.* Both were used by indigenous people throughout their ranges (Moerman 1998).

Among the most famous uses for these plants was to treat snakebite, particularly from rattlesnakes. While modern people dismiss that application as mere superstition, it must be remembered that Europeans came from a background where members of royalty were regularly dispatched by poisons. Europeans had developed a conviction that certain plants, or other materials, would neutralize poisons. Europeans brought that worldview when they arrived in the Americas. When Europeans found indigenous people believing that certain plants cured bites from poisonous rattlesnakes, they were quick to adopt that same view.

The earliest record of a plant called a snakeroot is from 1635. That was a reference to the Seneca snakeroot (*Polygala senega*), but there were others in which some people had total belief. In 1679, a Mr. Moreau commented that a colleague should "make bold to send a small quantity of snake root, the best sudorific ... and the best counter poison that nature ... can afford." To give an indication of the breadth of that view, there are plants called American snakeroot, black snakeroot, button snakeroot, Canada or Canadian snakeroot, Red River snakeroot, Sampson's snakeroot, Seneca snakeroot, Texan snakeroot, white snakeroot, wild snakeroot, and Virginia snakeroot.

Immediately, these two new American species of *Eryngium* became known as button-snakeroot, corn snakeroot, and rattlesnake master. Those names apparently were not carried back to the Old World because the plants are known by the French as *panicant d'eau* (water grass [panic or *Panicum*]), and as *Wassermannstreu* (water *Eryngium*) by the Germans. Moreover, even within the United States, distinctive names have evolved. Academics call *E. yuccifolium* the yucca-leaf eryngo. In Louisiana, general knowledge of the plants was lost until a woman named Mary Curby reintroduced the medical applications. From then onward, locals called the plants *merrye-curvye*, their version of the woman's name (Touchstone 1983).

The American snakeroot names for at least *E. yuccifolium* surely were reinforced, if not prompted, by its name among the Cherokee, who called it *saikwa'* [*silkwa'yi, salikwa'yi*] (green snake). Although the Natchez were geographically far away and spoke languages in distant families, they had a name for the plants that seems to be a cognate, *awe'lwaih.* Since the Cherokee and Natchez lived together for some time, they perhaps traded ideas and words occasionally. Former neighbors of the Chereokee, the Koasati, called *Eryngium* either *kowabí* or *híssi tabátka* (*híssi,* hair, *tabátka*, to catch). Those names were of plants questionably identified as *E. aquaticum*, but they came from a period when the differences were not understood (Taylor 1940, Folsom-Dickerson 1965). Yuchi call button-snakeroot *f?âde',* and still take it as medicinal tea during their Green Corn Ceremony (Jackson 2003).

Among the historic Creeks of Georgia, this important plant was *pas'sv* (Muskogee). That became *pasa* [*pa:ssâ*] among the Creek-speaking Seminoles and *pasî* among the Mikasuki-speaking Seminoles. Snow and Stans (2001) wrote the words *passv* (Creek) and *pashe* (Mikasuki).

Gideon Lincecum recorded the Choctaw name for *E. yuccifolium* (incorrectly reported as *E. aquaticum*) as *pis-shok*, and *chul-la* (Campbell 1951), comparing the first word to a plant that yielded a red dye. The dye plant is correctly *pishuk* (redroot or *Sanguinaria*), and *chula* means fox. Byington et al. (1915) rendered this *bisakchula.* There is no obvious similarity between the plants so there must be some unrecorded connection. I have to wonder if the correct prefix was not cognate with Creek *passv* and Mikasuki *pashe*, and unrelated to *bissa* (blackberry) or *pishuk* (*Sanguinaria*).

As the ancestors of the Seminoles were pushed farther down the Florida peninsula, they became separated from many of the plants they had used traditionally. *Eryngium aquaticum* was one of those. However, they still had *E. yuccifolium* extending to the tip of the peninsula (*ekan yuksa*, Land's end in Creek). Either that species filled the gap and became even more important to them or was always preferred. Given the importance of *E. yuccifolium* among the people of the southeastern United States, it is tempting to identify the prominent *niye* or *nie* of the Timucua (Gatschet 1877, 1878, 1880) as the same. However, there is insufficient information to be sure. The last syllables of all the indigenous names resemble one another, and may or may not be related.

Millspaugh (1892), Krochmal and Krochmal (1973), and Moerman (1998) all record uses for rattlesnake master. In addition to using it to treat rattlesnake bite, the indigenous people and later American settlers used it against liver problems, as a diuretic, and as an emetic. Chewing the root increases saliva flow. Liquid from roots mashed in cold water was drunk to relieve muscular pains. Roots were used to treat rheumatism, respiratory ailments, kidney trouble, and bladder problems, and to increase perspiration. We know that the plant was used at least among the Cherokee, Creek, Seminole, Meskwaki, and Natchez, and we might assume that it had wider use. This species ranges from New Jersey and Connecticut to Ohio, Wisconsin, Michigan, and Minnesota, south to Kansas, from Texas to Florida and north through the Carolinas. Although not mentioned at all by Gilmore (1919) and not included by Moerman (1998), the fibers from *E. yuccifolium* have been identified as used to make bags found in Arkansas (Swanton 1946) and slippers in Kentucky (Gordon and Keating 2001).

Uses across the indigenous groups were reasonably consistent. For example, the Cherokee used the plants to treat whooping cough, and they rinsed their mouths with an infusion to stop toothache (Hamel and Chiltoskey 1975). The Creeks also used it for respiratory problems and pain (Swanton 1928a, Taylor 1940). They took a cold infusion of roots for neuralgia, kidney trouble, and rheumatism. As it is diuretic, they used it against "the clap." *Eryngium* was also used to treat diseases of the spleen. It was supposed to produce a feeling of peace and tranquillity, and to clean the blood. That may be why it was also used as a "war physic." Several of the early visitors to the Creeks mention the use of this physic. Although he does not mention it by name, Bernard Romans ([1775] 1961) talks at length about the uses of "physics" among the Creeks and their relatives the Choctaw. They used these plant extracts to cleanse the body physically and ritually through their emetic action. One Choctaw who had worked for Romans grew so fond of him that the Choctaw was preparing to go to war against the Creeks when it was thought that they had killed Romans. Romans ([1775] 1961) wrote: "when I was afterwards missing, and it was thought the Creeks had destroyed us in coming from the Chicasaw nation, this savage armed to avenge my death, and was actually taking the war physick as they term it, when news was brought to the nation of my arrival in Mobile."

The Meskwaki of Iowa not only considered *Eryngium* an antidote for poison, but they also took it for bladder trouble (Moerman 1998). They used leaves and fruit in their Rattlesnake Medicine song and dance. The Natchez parched the leaves and used them to treat dysentery. They also chewed the leaves and stems to stop nosebleeds (Swanton 1928a).

The Seminoles too had a "Snake Sickness," which the Miccosukee call *cíntico:binoka:cí:cihcî:* "diamondback rattlesnake [*Crotalus adamanteus*] caused sickness" (Sturtevant 1955). While the name implicates the diamondback, a water moccasin (*Agkistrodon piscivorus*) may also cause the malady. A decoction of *pasî* was applied to the skin that became itchy during that disease. They used the same extract to treat any persistent external sore, or they made a bath of it to treat rheumatism. In addition the Seminoles used *pasî* to treat the "Cow Sickness" (decoction of roots; lower chest pain, digestive disturbances and diarrhea), the "Dead People's Sickness," "Menstruation Sickness" (stomachache, headache, body soreness), and "Otter Sickness" (infusion of plants; severe diarrhea, bloody stools, severe stomachache). At least the first three of these maladies required purification and the ceremonial use of this emetic functioned well for that purpose.

Eryngium aquaticum is also called rattlesnake master, and it was a popular remedy among the Alabama, Cherokee, Delaware, and Koasati. An infusion was used as an emetic among the Alabama, Cherokee, and Koasati (Taylor 1940). The Delaware used it to expel intestinal worms (Moerman 1998).

Eryngrium yuccifolium was perhaps the most important plant in the Muskogean pharmacopoeia (Swanton 1928a, Sturtevant 1955, Howard 1984). Alice Snow wrote her observations in her instructions for gathering it: "Look for *passv* in a dry open ground with palmettos around it. … *Passv* blossoms arise on a little stem about two to three feet tall. … To collect the plant, dig it out of the ground to get the roots. Only one root is needed. Sometimes the medicine man asks for a root from a plant that does not have multiple roots, so you have to look carefully. When you look on top, the one with multiple roots will have a lot of heads. The single root will have just one head. The single one is hard to find. Prepare the root according to the doctor's instructions. The root can be used whole, cut in four pieces, or crushed. I think this is strong medicine because it is the one used for the Green Corn Dance" (Snow and Stans 2001).

Erythrina: Coral Bean

(Greek *erythros*, red; flowers and seeds are red in most species)

When Linnaeus created the genus *Erythrina* in 1753, he knew three American species. Now there are 112 species known, with 12 species in Asia, 32 in Africa, and almost 70 in the Americas (Mabberley 1997). Neither Linnaeus nor his predecessors, who had studied these plants in the American wilds, had any idea that they were as important in American cultures as they were. Nor could they have guessed the research

Erythrina herbacea. *Drawn by P.N. Honychurch.*

that would go into studying the chemicals in them from the 1930s into the early 1970s (*Erythrina* Symposium 1974, Hargraves et al. 1974).

The research push during World War II began with studies done by the irascible Boris A. Krukoff (1898–1983), one of the few botanists in history to become wealthy. During his study of curare, Krukoff found that *Erythrina* was one of its sources. Then it was discovered that the seeds of some *Erythrina* contain alkaloids with a curare-like action (Dominique and Altamirano 1877 in Symposium 1974, Folkers and Koniuszy 1940 in Symposium 1974). Since curare was an important muscle relaxant during surgery, another source was considered potentially useful. Finally, it was learned that chemicals in *Erythrina* (erythroidine, erysodine) have tertiary bases, not quaternary bases like curare alkaloids, and that they were not as useful as some had hoped.

Further research revealed that *Erythrina* contains many potent chemicals, including more than 30 different alkaloids. The Florida species, *E. herbacea*, contains erysopine, erysopine, erysothiopine, erysothiovine, erysovine, erythrinine, erythroresin, the emetic coralin, erythric acid, and hypaphorine (Standley 1920–1926, Morton 1971, Hargraves et al. 1974, Schmutz and Hamilton 1979).

In the southeastern United States, *E. herbacea* is known as cardinal spear, Cherokee bean, coral bean [plant, tree], and red-cardinal flower. Several of the other names applied to it elsewhere reflect the red color, including *alabarda de cardenal* (cardinal's saddle, Spain), *colorín* (color of life, Tamaulipas to Oaxaca), *corolillo* (little red one), *imortelle* (immortal), and *Korallenstrauch* (coral bush, German). Louisiana is the exception in the United States, where the plants are called cry baby tree and *mamou* (mamma) because the nectar is so abundant the trees "weep."

Erythrina herbacea is sometimes called the eastern coral bean to distinguish it from *E. flabelliformis*, the western coral bean. The southeastern species ranges from the coastal Carolinas to Florida, west to Texas and Arkansas, then in Tamaulipas, San Luis Potosí, Hidalgo, Puebla, Veracruz, Oaxaca, and Chiapas. *Erythrina flabelliformis* grows from Arizona and New Mexico south through Baja California, Sonora, Chihuahua, Sinaloa, Durango, Zacatecas, Jalisco, Michoacán, San Luis Potosí, Pueblo, and Morelos. *Erythrina americana*, a third species in section *Erythrina* (Krukoff and Barneby 1974), ranges in Mexico from Veracruz to Hidalgo, Mexico, Distrito Federal, Pubela, Chiapas, and Yucatán (Garcia-Mateos et al. 2001). Common names of the three are so intertwined that people may not distinguish among them.

In Mexico, these species have long been intimately tied to daily and religious life. The young flowers are eaten cooked like string beans. I have been told from Chihuahua to Oaxaca that flowers are most commonly cooked and eaten with eggs for breakfast, but they are also recorded as being cooked as a vegetable or used fresh in salads (Morton 1968b). The young leaves can also be cooked and eaten. The flowers are considered pectoral, and the roots sudorific and diaphoretic. Leaves are emmenagogue, and a decoction of the flowers is used to treat pulmonary disorders (Martínez 1969). Juice from the stem is used to treat scorpion stings. The wood is soft and light, and is used for corks, carvings of saints, masks, and similar items. The bark, in addition to giving a yellow dye, is astringent, and it has been used to reduce fever, relieve dysentery, and treat eye inflammation.

In addition, the seeds are used in necklaces, rosaries, and novelties. However, the seeds are poisonous because of the alkaloids they contain, and they have been used for poisoning rats and dogs. Bark and stems have similar properties, and both have been used to poison fish. The beans have been used internally to treat cancer. However, the seeds contain erythroidine, a central nervous system depressant. That depression of the nervous system is surely why they have been used as a hypnotic agent.

Mexicans call the seeds either *colorín*, *patol* (from Náhuatl *patolli*, banner), or sometimes *patol colorín* (red banner, Tamaulipas). Seeds are, or at least were, used in a game resembling dice called *patol*. The Maya name for the seeds was *tze-te* (poison), and the plants are discussed in the Popul Vuh, a text that recounts the origins of those people (Tedlock 1985). The name *tze-te* is usually equated with the Spanish name *árbol de pito.* The Zapotec called these seeds *bi-chilla* [*pi-chijlla*] (*bi, pi,* fruit, *chilla, chijlla,* luck, oracle) or *bzaa tutze* (*bzaa,* bean, *tutze,* poison). Their red color was synonymous with toxicity among the ancient and modern Zapotecs (Reko 1945). It was these red beans and the game *patol* that the Zapotecs and presumably the Maya and Aztecs used to predict the future. *Pi-chijlla* was also *Cipatli* (the Shark, or the first day of the month) in the Aztec *Tonalamatl* (*tonalli,* ray or sun or sign of nativity, *amatl,* paper), which is the calendar based on divination. To the Zapotec, this deity was *peyoo pichijllo,* the monster of the earth and the symbol of malevolence and of procreation. At the same time, he was the supreme deity (*chijlla-tao*) of the Zapotecs. Therefore, in Mexico, the playing of *patol* was originally no simple game.

Presumably, the names *chijol* and *chilicote* are based on the Zapotec word *chilla* (luck or oracle), and not on *chilli* from Náhuatl. The word *pitos* applied to the flowers literally translated in Spanish means whistle (the instrument). However, the usage looks suspiciously like the word has been derived from some other language, although which is unclear. The word *pito* seems to have been applied to the flowers of *Erythrina* because youngsters make (or at least made) whistles of them.

The Aztecs typically called *Erythrina* species *tzonpantli* [*zumpantle, tzompantli, tzampantle, tzompantle, zampantle, zumpantle*] (hair-banner, Náhuatl) or *tzon-pan-cuahuitl* [*tzan-pan-cuohuitl, tzinacanquahuitl*] (hair-banner tree, Náhuatl). The other Náhuatl name *macayxtli* (*maçati,* become an animal, *ixtli,* powerful) must be associated with the "hypnotic" effect of the seeds. Those seeds were used in communicating with the gods from the Tarahumara of Chihuahua to Guatemala (Schultes and Hoffman 1979). Extracts of one species have been compared to diazapam in calming effects (Garín-Aguilar et al. 2000). Moreover, an array of bioactive chemicals has been extracted from the circum-Caribbean species, including alkaloids and lectins (Hargreaves et al. 1974, Ortega 1990, Naidoo et al. 1997, García-Mateos et al. 1998, 2001, Garín-Aguilar et al. 2000).

In an unsupported statement about *E. herbacea,* Hocking (1997) noted that in the southeastern United States the seeds were "once an item of commerce among Indians." That statement is all the more fascinating because of the records of medicinal uses among Muskogean people. We have the Choctaw names *chilo'píntobét* (*shilop,* spirit, *in,* its, *tobi,* bean) and *nita inchastoki* (*nita,* bear, *im+chastoki,* its bean). However, other Muskogeans said *itco intcastuge* (*icho,* deer, *im+chastoki,* its bean, Alabama), *cointalá:ko* [*ichinchastoki, icho inchastoki*] (*eco,* deer, *em,* its, *tvlako,* , bean, Creek), *icó impalaná* (*icó,* deer, *im,* its, *palaná,* bean, Koasati), and *i:cí:salâ:lî* (deer's bean) from the Mikasuki (Bushnell 1909, Swanton 1928a, Taylor 1940, Sturtevant 1955, Sylestine et al. 1993, Kimball 1994). Three of the five have the same translations, but, if Bushnell (1909) gave a literal translation of the Choctaw, they are not cognates to the others. Sturtevant (1955) recorded use of the seeds in the "Horse Sickness," "Deer Sickness," and "Dog Sickness." He added, without elaboration, "mythological food." Sheehan was told in 1919 that the Seminoles gave it to dogs as an emetic (von Reis and Lipp 1982).

As the horse was a historical addition to the animals in the Americas, the other two illnesses must be more ancient. Moreover, the word for horse in Creek is *coakko* [*corakko*] (big deer), so Horse Sickness must have been derived from the older symptoms and treatment. One of the ingredients in the remedy for Horse Sickness was ground *Erythrina* seeds, either dried or green. At times, scrapings from the roots were dried and powdered as a substitute. This powder was added to some material from other plants, made into a decoction, and ingested. Variations were made for Deer and Dog Sicknesses. Especially because of the statement that *i:cí:salâ:lî* was "mythological food," these surely reflect ancient practices reminiscent of the Mexican uses.

Because *E. herbacea* is not native to Cuba, it is listed as a cultivated substitute for Cuba's native, *E. corallodendron.* Both species are known as *piñon* (big pine). Still, that name may be an indication of contact with Mexico because *E. corallodendron* is called *piñon de pito.* The word *pito* may have been brought from Mexico to Cuba by the early Spanish.

In spite of Alice Micco Snow's obvious extensive knowledge of plants used by the Seminoles (Snow and Stans 2001), *Erythrina* is not among those she listed. That absence may reflect shifting knowledge like the comments made about the Seri by Felger and Moser (1985) and about the Mayo by Yetman and Van Devender (2001). *Erythrina* were important plants in a religious context in the past, but they have now become either unimportant or relegated to food or some other secular application.

Erythronium

(The Greek name for the purple-flowered European species, from *erythros,* red)

Erythronium umbilicatum (navel-like) (= *Erythronium americanum* misapplied)

adder's tongue (used in the 1580s for the fern *Ophioglossum* in Europe, then applied to other plants)
amber-bell
dimpled trout-lily [troutlily] ("trout-lily" dating from ca. 1898, presumably in reference to the spotted leaves that resemble the fish)
dog's tooth [dog-tooth, dogtooth] violet (used in the 1580s to designate *Erythronium dens-canis*, being a literal translation of the Latin name); *Hundszahn* (hound's tooth, German)
fawn-lily

The Old World *Erythronium dens-canis* corms are a source of starch used for vermicelli and cakes in Japan (Mabberley 1997). Early travelers recorded that they are also eaten with reindeer or cow milk in Mongolia and Siberia, and that the leaves are boiled and eaten (Hedrick 1919). The Dutch herbalist Rembert Dodoens in 1578 and the Englishman John Parkinson in 1609 both listed the plants as ornamental and medicinal in Europe.

The only records found are for *E. americanum*, but given the similarity between these two species and the long confusion, both were probably used similarly. The Cherokee used *E. americanum* only as medicine. They warmed leaves and crushed the juice onto wounds that would not heal, gave an infusion of the root to reduce fever, and made a compound mixture to prevent fainting (Hamel and Chiltoskey 1975). Young Iroquois women took the raw plant, minus the roots, as a contraceptive; they made a poultice of roots for swellings and removing splinters (Moerman 1998).

Winnebago children ate the bulbs of *E. mesochoreum* in the spring (Gilmore 1919). The Blackfoot, Montanan, Okangan-Coville, Okanagon, Shuswap, and Thompson tribes ate *E. grandiflorum* bulbs (Moerman 1998). Kwakiutl ate both *E. oregonum* and *E. revolutum* (Moerman 1998).

Although the bulbs have been eaten fresh and cooked, physicians Jacob Bigelow in 1822 and A. Clapp in 1852 considered them emetic and a substitute for colchicum in treating gout (Millspaugh 1892). Porcher (1863) wrote: "The recent bulbs are emetic when powdered and given in doses of twenty to forty grains. When dried or cooked they become eatable. The berries are said to be more active and certain in their operation than the root." The roots were official in the U.S. Pharmacopoeia from 1820 to 1863, and they were used to treat gout (Vogel 1970). Recent experiments have found that water extracts from *E. americanum* are active against Gram-positive and Gram-negative bacteria; the related *E. grandiflorum* has slight antimutagenic activity (Foster and Duke 1990).

Eugenia: Stoppers

(Named for Prince Eugene of Savoy [Eugenio di Savoia, Françoise Eugène de Saavoie-Carignan], 1663–1736)

When I first arrived in Florida, one of the first groups of trees that I studied was the stoppers (*Eugenia*). I was told that there were three possible sources for the common name, which apparently began with Charles S. Sargent in his treatment of North American trees of 1894. As Sargent did not tell how the name stopper came into existence, we have a choice of possible scenarios.

First, the species are hard to pass through where they grow in dense thickets near the coasts. They "stop" a person's passage, hence the name.

More plausible to some is that the fruits are loaded with tannins and have long served as a remedy for diarrhea. Since they "stop" the problem, they are stoppers.

Unlikely as it is, my favorite explanation was offered by a math professor where I once worked. He was an amateur naturalist and delighted in his ability to tell everyone the names of common trees and shrubs. Since he could rarely tell the stoppers from each other, he was "stopped."

The genus *Eugenia* was created by Linnaeus in 1753. He accepted the suggestion of John Michell from 1748 who had proposed naming some Brazilian plants after Price Eugene of Savoy. That prince was, among other things, a teacher of Frederick the Great. However, the plants were named for him because he built the Belvidere Palace near Vienna at the beginning of the 18th century and made a collection of rare plants in its garden. The species Mitchell and Linnaeus used to create the genus name was *E. uniflora*, known as the *pitanga* in Brazil, the *cerezo de Cayena* in Cuba, and the *guinda* in El Salvador. We call it Suriname cherry. This Brazilian species was already widely grown in the 1750s for its edible fruits.

The name stopper is applied in Florida and the Bahamas, with variations being stopper berry (Bahamas), stopper bush (Bahamas), and white stopper (Florida, Bahamas) for *E. axillaris*. The other common species, *E. foetida*, is usually called Spanish stopper (Florida, Bahamas), although it formerly had another common name. In the late 1800s, *E. foetida* was sometimes called "gurgeon stopper." That modifier comes from French *grugeons*, lumps of crystalline sugar in brown sugar (dating from 1480s), and formerly also meant the smallest or most "writhen" fruit on the tree (first recorded in 1611) and connected with *gruger*, to crunch. "Grugeon" is no longer used by most English-speaking people, and few even know of the name's existence.

There are other species of *Eugenia* native to Florida, including redberry stopper (*E. confusa*) and red stopper (*E. rhombea*). These two are rare in the southern end of the state, but they have common names there and elsewhere. Redberry stopper has edible fruits and the wood, which is hard, strong, heavy, close grained, and red-brown, is used for cabinets (Little et al. 1974). In Puerto Rico, it is called *caracolillo* (little snail), *cieneguillo* (little swampy one), or simply ironwood (because of the hard wood). In Hispaniola it is *yayao* [*jayao, ayay, hayao, hayay, jayayo*], a name also used for *Daphnopsis crassifolia*, Thymeliaceae. *Daphnopsis* has bark with resistant fibers that are used to make strings, but, beyond the common names, there is no indication that *E. confusa* has such bark. The red stopper, *E. rhombea,* is called by the generic name *mije* in Cuba. That may be another Taino word, or perhaps it is a mispronunciation of *myrta.* In Hispaniola *E. rhombea* is called *arrayán* [*arraiján*] (well-rooted), *bois myrte* (myrtle tree), *escobón blanco* (white broom), *myrte* (myrtle, *Myrtus*), or *tu fais* (you make it). Wood in this species is chestnut colored, durable, hard, strong, and heavy, and is used for cabinets, posts, and stakes, and as fuel (Liogier 1974). Sometimes the flexible stems are used for hanging and drying tobacco.

Many of the common names for *E. axillaris* compare it with cherries, because both have black, edible fruits. Among names making that comparison are black cherry (Jamaica), *merisier pays* (wild cherry, Martinique), *merisier petite feuilles* (small-leaved cherry, Martinique), *merisier* (cherry, St. Barts, Martinique), and *merisier noir* (black cherry, Guadeloupe, Martinique). The name choaky-berry (Dutch Antilles) gives some idea of how unfavorably the fruits of stoppers compare with cherries.

Several other names refer to the edible fruits. The names *grajo* (Cuba, Puerto Rico) and *guairaje* (Cuba) may be words derived from Taino. In Cuba *E. axillaris* is also called *guairaje colorado* (red *guairaje*), *guairaje macho* (male *guairaje*), *guairaje preto* (black *guairaje*), and *guairajón* (big *guairaje*). Elsewhere the fruits are compared with others as in *guayabillo* (little guava, *Psidium guajava*, Mexico, Guatemala), pigeon berry (Dutch Antilles) and strawberry (Caimans). Maybe the name krum berry (Puerto Rico) is related to gurgeon as both refer to dregs.

The wood of *E. axillaris* is hard, heavy, fine textured, dark-red, and strong, and has been used for a variety of items (Small 1933). The Seminoles call the plants *ahikólkî* (heartwood? tree, Mikasuki) and the Creek name *hikolwâ* may be a loan-word with the same meaning. Sturtevant (1955) recorded that Seminoles historically used the stems for bows, and they surely used them for other items.

Presumably, the stems have been used as rods because they are called rodwood (Jamaica) or brownleaf rodwood (Jamaica). Indeed, rodwood is the generic name for *Eugenia* in Jamaica. What use the rods are put to is unstated, but the old saying "spare the rod and spoil the child" comes to mind. Perhaps they also were used historically as ramrods for old muzzle-loading guns. The Seminoles used muzzle-loading guns during the Third Seminole War, and it is likely that they too used the branches of these and other trees as ramrods.

The plants branch freely when the apical bud is damaged and produce a series of small, straight branches. Bundles of those branches have been used for brooms, as noted in the names *escobo* (brush, El Salvador), *escobón colorado* (red broom, Hispaniola), and *escobón de vara* (broom of sticks, Hispaniola). Although it is a totally different usage, the name wattle (Bahamas, Puerto Rico) indicates another use in building with the long branches.

The hardness of the wood is recorded in names like *guayacán negro* (black lignum vitae, based on the Taino word *guayacán,* El Salvador), *palo de hueso* (bone tree, Hispaniola), and *vaina de espada* (sword's sheath). The wood is valued for "turnery" in Guatemala (Morton 1981), and the young slim trunks are made into rafters to support thatch for roofs. The smaller branches are also used for making fish traps and animal enclosures.

Eugenia axillaris, like several others, has been compared with the *chamiso* (also applied, in other areas, to a medicinal grass and to *Ademostoma,* Rosaceae, El Salvador), and the *murta* (myrtle, usually the genus *Myrtus,* Puerto Rico). At least *Eugenia* and *Myrtus* are in the same family and there is a marked resemblance of leaves and flowers. The sweet, fragrant flowers are recorded in the name *oloroso de San Pedro* (St. Peter's perfume, Cuba), and probably in white bush (Bahamas).

Eugenia foetida shares many of these names. It is called *bois petite feuilles* (little leaf tree, Haiti), *escobón* (big broom, Hispaniola), *guairaje* (Cuba), *guairaje blanco* (white *guairaje,* Cuba), *guairaje colorado* (red *guairaje,* Cuba), *guayabillo* (little guava, Cuba), *merisier* (cherry tree, Haiti), *petite feuilles* (little leaf, Haiti), rodwood (Jamaica), and stopper-bush (Bahamas). However, *E. foetida* is called black wattle (Bahamas) to distinguish it from *E. axillaris. Eugenia foetida* too has fragrant flowers, and that is surely why it is called *bálsamo* (balsam, Cuba). Because of its hard wood, it is called ironwood (Bahamas). Still, the most anomalous name it has is *malaguette* (Hispaniola). That word is applied to a fiery small chile in Brazil, and there is nothing about this tree that resembles the chile. If the name is based on the locality of Málaga, Spain, it remains an oddity because the tree is native to the Caribbean (see also *Capsicum*: Bird Peppers).

Several stoppers have been used medicinally. Cubans and Bahamians use a decoction of *E. axillaris* to treat colds (Roig 1945, Morton 1981). The plant is boiled alone for tea to treat diarrhea, or with *Bourreria ovata* for building blood and energy. The most widespread use is as a sexual enhancer. According to Ayensu (1981), it is used for "building up men's energy and body." The root is boiled to increase sexual potency, and used in a bath for women, especially after childbirth. With that common use, it seems odd that no common names imply that application.

Spanish stopper is also used in baths, probably much as white stopper. Formerly, the leaves of Spanish stopper were sold fraudulently under the name *Uva ursi* in Havana (Roig 1945).

There is yet another "stopper" in southern Florida that has been the source of confusion from the time of its discovery in the mid-1800s. The plants have been placed in *Eugenia*, *Psidium*, and *Mosiera*. The species first became known in 1854 when it was described as *E. longipes* by Otto Carl Berg (1815–1866). John K. Small moved it to *Mosiera* in 1933, when he created a new, unique genus for the species. In 1973, Rogers McVaugh moved the species to the genus *Psidium* where Wunderlin (1998) put it, although he was uncomfortable with it being there. Everyone has agreed that the plants did not really "belong" in either *Eugenia* or *Psidium*, but there did not seem to be a good reason to segregate them. Finally, Les R. Landrum (1992) compared them with some Central American species. It then became clear that the Florida plants known as *Mosiera longipes* were members of a Mesoamerican group consisting of three species. Because the plants were disjuncts on the northern fringe of the genus range, they previously had not been connected with their relatives. These plants are called Bahama stopper (Bahamas), mangrove-berry (Florida), and wild guava (Bahamas).

Hedrick (1919) said the fruits of *Mosiera* have "the flavor of cranberries." I consider them to be the best of those plants called stoppers.

Eugenia. *Eugenia axillaris* (top left). *Drawn by P.N. Honychurch. Eugenia confusa* (top right). From Little and Wadsworth 1974. *Eugenia foetida* (lower left). *Drawn by P.N. Honychurch. Eugenia rhombea* (lower right). From Little and Wadsworth 1964.

Eupatorium

(Named for Mithridates Eupator, 120–63 B.C., king of Parthia [modern northeastern Iran], who used a species as a medicine or as an antidote for poison; he was one of Rome's most formidable opponents)

Eupatorium serotinum. a. Lower portion of a plant with roots. b. Upper portion of a plant in flower. c. Flower head. d. Seed. From Buchholtz 1968.

Eupatorium capillifolium (hair-leaved, because of the finely divided leaves)
copal (because of a fragrance thought to be similar to the real *copal* or *Bursera,* Cuba)
cypress-weed [cypressweed] (this and *E. compositifolium* commonly appear in prairires around cypress stands when the water is low)
dog fennel [eupatorium] (Florida, Bahamas)
hinojo de sabana (savanna fennel; *hinojo* from Latin *feniculum,* Cuba)
pahí wohoksí (*pahí,* grass, *wohoksí,* hairy, Koasati)
taki [*takri*] (bush, Mikasuki)
vtako-luste (*vtakv,* weed, *lvste,* black, Creek; the name was historically applied to an unidentified medicinal shrub with black berries, cf. Martin and Mauldin 2000; and there also may be confusion with *Conyza canadense*)

Eupatorium compositifolium (compound-leaved, alluding to the pinnately divided leaves)
dog fennel
yankee-weed [yankeeweed]

Eupatorium cuneifolium (with wedge-shaped leaves)
waxy thoroughwort (a comparison with *E. perfoliatum*)

Eupatorium fistulosum (tubular, alluding to the stems)
Joe Pye weed (the name "Joe Pye's weed" was in use by 1818, first for *E. purpureum* and then for other species)
queen of the meadow

Eupatorium hyssopifolium (having leaves like hyssop, *Hyssopus officinalis*)
hemp-weed [hempweed]
horehound (compared with the *Marrubium vulgare,* Lamiaceae)
hyssop-leaf thorough wort [hyssopleaf thoroughwort, hyssop-leaved thoroughwort] (having leaves like hyssop; see *Bacopa:* Water-Hyssop; because the stem appears to grow "thorough" [through] the leaves)
Justice's weed (so named because they were used by John Justice of South Carolina in 1800 for curing bites of snakes and other poisonous animals; he received a premium for disclosing his remedy)

Eupatorium perfoliatum (the stem grows through the leaf)
ague-weed [agueweed] (according to physician B.S. Barton in 1810, this is a translation of the plant's name among northeastern tribes)
[common] boneset (a name applied to comfrey, *Symphytum officinale,* by 1670; applied to these American plants by 1830); wood boneset
cross-wort [crosswort] (Virginia)
fever-wort [feverwort] (North Carolina)
howe chache (maybe *hobechechi,* from *hobechi,* steaming, Choctaw; used in sweat baths as an emetic to eliminate "cold and bile")
Indian-sage (so called because northeastern tribes used it so much to treat ague); wild-sage
Joe-Pye [Joepye] (North Carolina)
niya'wibûkûk' (Ojibwa)
shauashko (Choctaw)
sweating plant [weed]
teagel [teasel] (confusion or comparison with teasel, *Dipascus sativus*)
thorough grow [thoroughgrow] (Pennsylvania); thorough wort [thoroughwort, common thoroughwort] (originally "through-wort," because the stem appears to grow "thorough" [through] the leaves); thorough-stem [thoroughstem, throughstem]; thorough-wax [thoroughwax]; thorough-wort [thoroughwort]
vegetable-agrimony
wild-Isaac (maybe named for by Isaac Bierfield, of Newberry, South Carolina, who used "dog fennel" in tanning leather in 1861, cf. Porcher 1863, North Carolina)

Eupatorium pilosum (hairy)
rough boneset
wild horehound (comparison with *Marrubium vulgare,* the European horehound)

Eupatorium purpureum (purple, the flowers)
amadita'ti (water dipper, because water can be sucked through the hollow stem, Cherokee)
caca'bagsit (piercing leaf stem, Potawatomi)
[Indian] gravel root [gravelroot] (used to expel kidney stones); gravel weed [gravelweed]
haiyu<u>n</u>kpulo (also applied to any plant without a particular name, Choctaw)
hemp weed; hemp-agrimony
Joe-Pye weed [jopi weed, green-stemmed, sweet, sweet-scented] (named for Joe-Pye); jopi-root (variant of Joe-Pye, North Carolina)
kidney root (because of diuretic activity); kidney wort [kidneywort]
king of the meadow [prairie]; pride of the meadow; queen of the meadow (Indiana, Maine, Massachusetts); queen of the prairie
knot-root [knotroot]

motherwort (Massachusetts)
purple [tall] boneset
purpurfärbener Wasserhanf (purple-colored water hemp, German)
quillwort (were the stems used as ink pens?, West Virginia)
skunk-weed [skunkweed] (from the strong odor); stink-weed [stinkweed]
[purple] thorough-wort [thoroughwort]
trumpet weed [root] (from the hollow stems); turnip-weed (a corruption of trumpet)
yi'mohkuna'mitʔe (*yi'mohku*, herb, *na'mi*, stinking, Tunica)

Eupatorium rotundifolium (round-leaved)
false horehound
round-leaf thoroughwort [roundleaf thoroughwort]

Eupatorium semiserratum (partly toothed)
small-flower thorough-wort [smallflower thoroughwort]

Eupatorium serotinum (late-flowering)
late eupatorium (a book name)
late-flowering thoroughwort
silver rod

Eupatorium is a genus of about 45 species, most of them endemic to the eastern United States (Mabberley 1997). However, there is one famous species native to Europe—*E. cannabinum.* That plant was well known by people in all European countries when the New World was discovered. The herb was an important medicine, particularly in treating "ague" (malaria). Some of the names for it in Europe are *cainb-uisge* (water hemp, Gaelic), *canapa selvatica* (wild hemp, Italian), *eupatoire chanvrine* (hemp eupatorium, French), *eupatório de avecena* (Avicenna's eupatorium, Portuguese), hemp-agrimony (compares it with *Cannabis* and *Agrimonia*), *hjorterøst* (*hjort*, deer, *røst*, touched, Norwegian), *trevo cervino* (deer's three-leaves, Portuguese), and *Wasserdost* (water oregano, German).

When explorers and settlers arrived in the New World, they found a bewildering variety of *Eupatorium* species. However, they soon learned from the indigenous people that these new plants were as important in medicine as the one they knew at home.

There is documentation of use of several species among indigenous tribes. *Eupatorium perfoliatum* was used by the Abnaki, Cherokee, Creek, Delaware, Iroquois, Koasati, Menomini, Meskwaki, Micmac, Mohegan, Nanticoke, Ojibwa, Penobscot, Rappahannock, Seminole, and Shinnecock (Vogel 1970, King 1984, Moerman 1998).

Millspaugh (1892) and Core (1967) speculated that use of *E. perfoliatum* in treating dengue fever (bone-break fever) resulted in its name boneset. However, that may not be correct. The Abnaki used the herb to "mend bones," and the Iroquois put a poultice on broken bones to help them heal. Williams ([1837] 1962) noted that the Seminoles used it as a gentle emetic and as a sudorific in fevers.

The Cherokee use *E. perfoliatum* as a diuretic, emetic, purgative, sudorific, and tonic, against malaria and sore throat (Hamel and Chiltoskey 1975). John Williams (1827) recorded that the Seminoles made a preparation to treat fevers. Creek women with hip pain were treated with steam from boiling *E. perfoliatum*; those with epilepsy were treated with steam from its roots (Vogel 1970). The Delaware used an infusion of roots and leaves to treat chills, fevers, and stomach problems. The Iroquois used the plants to treat colds, fevers, headaches, kidney problems, pleurisy, pneumonia, stomach problems, syphilitic sores, and to stop the liquor habit. They also considered it laxative. The Koasati and Seminoles used *E. perfoliatum* as an emetic (Taylor 1940, Sturtevant 1955). The Menomini used it in a tea to reduce fever. The Meskwaki gave it to expel intestinal worms. The Micmac recommended it for kidney trouble, gonorrhea, and tuberculosis. The Mohegan took it for relief from colds, especially when mixed with *Prunus virginiana*, to treat fevers and stomach trouble, and as a general tonic. The Nanticoke made a medicine for chills and fever. The Ojibwa used it to correct irregular menses, applied a poultice to help rheumatism pains, and put the chewed plant on rattlesnake bites. The Penobscot used *E. perfoliatum* for gonorrhea, kidney trouble, and tuberculosis. The Penobscot and Rappahannock considered it a tonic. The Shinnecock took it for colds and fevers.

Later people used *E. perfoliatum* much as did the indigenous tribes. Visiting German physician David Schopf called it emetic, purgative, and diaphoretic, and advocated it for arthritis, gout, malaria, and rheumatism in 1788 (Vogel 1970). Settlers used it to combat the yellow fever epidemic of Philadelphia in 1793 (Vogel 1970). Porcher (1863) recommended: "Thoroughwort, drank hot during the cold stage of fever, and cold as a tonic and antiperiodic, is thought by many physicians to be even superior to the Dogwood, Willow, or Poplar, as a substitute for quinine. It is quite sufficient in the management of many of the malarial fevers that will prevail among our troops during the summer; and if it does not supply entirely the place of quinine, will certainly lessen the need for its use." Hocking (1997) listed the herbs as stimulant, tonic, diaphoretic, emetic, and bitter; to be used as a laxative, and to treat fever, malaria, typhoid, yellow fever, dyspepsia, pneumonia, and rheumatism. *Eupatorium perfoliatum* was in the U.S. Pharmacopoeia from 1820 to 1916 and in the National Formulary from 1926 to 1950 (Vogel 1970).

Eupatorium pilosum was used by the Cherokee to treat breast problems, tuberculosis, and colds. They considered it diuretic, laxative, and tonic (Hamel and Chiltoskey 1975).

The Cherokee used a hollow stem section of *E. purpureum* to blow medicine onto surfaces. It was also used to treat rheumatism and female problems, as a diuretic and tonic, and to remove sickness from "being near a dead person" (Mooney 1885–1886, Hamel and Chiltoskey 1975). Moerman (1998) said the Mahuna took the roots as a laxative, but Hickman (1993) does not list the species in California. The Meskwaki considered it a breath freshener and men kept it in the mouth when wooing women. The Menomini and Potawatomi used it to stop bleeding after childbirth; the Potawatomi put a poultice on burns; and the Rappahannock made a medicine for the blood (Smith 1933, Moerman 1998). The Ojibwa inhaled the vapors from an infusion made from the tops of the plants to alleviate colds.

Stories vary, but the name Joe Pye weed came from a "Joe Pye," who was either an 18th-century Massachusetts medicine man or a 19th-century Caucasian "Indian theme promoter." He used the root to induce sweating in the treatment of typhus fever (Vogel 1970, Coffey 1993). The rhizome and root (known as purple boneset, gravel root, and queen of the meadow) were listed in the U.S. Pharmacopoeia from 1820 to 1842. These plants are probably most famous as a diuretic (Vogel 1970).

The Alabama took *E. serotinum* to relieve stomach pains (Vogel 1970). The Houma made a decoction of flowers to treat typhoid fever (Speck 1941).

Several other species are known to have been used medicinally in Cuba (Roig 1945). Two common weedy species, *E. capillifolium* and *E. compositifolium*, are notoriously fragrant. Fresh juice from both has been put on insect bites and their herbage has been strewn on floors as an insecticide. Leaves from both have been used to treat arthritis (Hocking 1997). In Cuba, a decoction of *E. capillifolium* is used to stop dysentery and other intestinal problems (Roig 1945). Cubans also consider it hemostatic. *Eupatorium compositifolium* tea made from the roots was considered a remedy for fever (Murphee 1965).

When Thomas Nuttall was ill with "intermittent fever" in Arkansas in 1819, he took *E. cuneifolium*, as *E. perfoliatum* was not available. "This dose, though very nauseous, did not prove sufficient to operate as an emetic, but acted as a diaphoretic and gentle laxative, and prevented the proximate return of the disease" (Nuttall in Vogel 1970).

Eupatorium fistulosum has been used for kidney problems in Mississippi. A tea from the plants is said to reduce polyps (Hocking 1997).

Morton (1974) found people in South Carolina using *E. hyssopifolium* to relieve colds and fevers. Local people made a mixture containing these herbs, longleaf pine (*Pinus palustris*), and life everlasting (*Gnaphalium obtusifolium*), which they drank and used in baths. Leaves of *E. hyssopifolium* have been used to treat snakebite in Texas (Hocking 1997).

Porcher (1863) had heard that *E. rotundifolium* had been used to treat consumption. He had no personal experience with the species.

The chemistry of *Eupatorium* is complex and incompletely known. However, several species contain volatile oils. Among these are *E. capillifolium*, *E. serotinum*, and presumably *E. compositifolium* (Hocking 1997). *Eupatorium capillifolium* has phellandrene among those oils (Hocking 1997). Other compounds and bioactivity have been documented. *Eupatorium cuneifolium* contains eupacunin (germacranolide), which has significant antileukemic and tumor-inhibiting properties (Hocking 1997). *Eupatorium hyssopifolium* contains quercetin 3-glucoside (Wagner et al. 1972). *Eupatorium purpureum* contains euparin and coumarin (Hocking 1997). *Eupatorium rotundifolium* contains glycosides, some of which are tumor-inhibiting sesquiterpene lactones, including eupachlorin and eupatorin acetate (Hocking 1997). *Eupatorium semiserratum* contains antileukemic sesquiterpene lactones, including eupaserrin, in significant amounts. Those lactones inhibit human carcinoma (Hocking 1997).

Eustoma

(Greek *eu*, true, well, *stoma*, mouth, in reference to the showy flowers)

Eustoma exaltatum (raised high, lofty)
centaurée maritime (seaside centaury, an allusion to *Centaurium*, also in the Gentianaceae, French Antilles)
genciana de la tierra [*del país*] (country gentian, Cuba)
[beach, catchfly, marsh, seaside] gentian (Bahamas, Florida)
mountain-bob (Bahamas)

British botanist Richard Anthony Salisbury (1761–1829) named *Eustoma* from plants growing in *The Paradisus Londinensis* of 1806. Later he and

Eustoma exaltatum. a. Habit. b. Calyx. c. Petal. d. Stamen. e. Ovary. *Drawn by Vivian Frazier.* From Correll and Correll 1972.

George Don (1798–1856) moved Linnaeus's *Gentiana exaltata* to create *Eustoma exaltatum.* There are still only three species in the genus, all native to the Americas (Mabberley 1997).

Leaves are steeped in hot water as an eyewash (Ayensu 1981, Morton 1981). The plant, including the roots, is used as a stomachic and febrifuge (Roig 1945).

Euthamia

(Thomas Nuttall said the name was from Greek *eu*, well, and *thamees, thameo*, crowded, referring to the flower clusters, cf. Sieren 1981; Diggs et al. 1999 suggested perhaps from *euthemon*, neat or pretty)

> ***Euthamia graminifolia*** (leaves like grass)
> flat-top goldenrod
> *owesa'wenokûk* (yellow top, Potawatomi)

Euthamia is a genus of Asteraceae with eight species confined to North America (Sieren 1981). In the past, some of the species have been placed in *Solidago*, and the common name "goldenrod" indicates the similarity between the genera.

The Ojibwa made an infusion of the roots or flowers to alleviate chest pains associated with lung problems (Moerman 1998). The Potawatomi used blossoms to treat fever (Smith 1933).

Additionally, the plants were used as a hunting lure. The Ojibwa smoked the flowers, believing that the odor simulated that of a deer's hoof.

The plant was considered as a possible rubber source (Mabberley 1997).

Evolvulus: Ojitos Azules

(From Latin *e*, lacking, without, and *volvi, volutum*, twining)

One of several names for *Evolvulus alsinoides* in Yucatán is *ojitos azules* (blue eyes). That may not be a Florida name for them, but it certainly beats the "invented" name slender dwarf morning-glory (Wunderlin 1998). Besides, there are so many Spanish-speaking people in Florida that it could be a common name there.

Historically, these diminutive plants had no common name in Florida. The first names for them appeared when floristic biologists felt pressured by amateurs to provide common names. In many parts of the world, *Evolvulus* has common names, and several of the individual species also have them because people use them. However, in Florida, the only member of the genus that has enjoyed a well-known name in recent years is blue daze, *E. glomeratus.* That species was introduced by a nursery in Miami, and the horticultural name invented—much like slender dwarf morning-glory—was created for *E. alsinoides.* These southern Brazilian plants were introduced into horticulture in only about 1979 in Miami, but by the early 1980s they were cultivated throughout the tropics of the world (Austin 1984).

Although there are several members of *Evolvulus* that rival blue daze for beauty of their vegetation or flowers, the group is not well known to temperate residents. Linnaeus even started off misunderstanding them. The species name *alsinoides* (resembling the sandwort, *Alsine*, Caryophyllaceae) was put in *Convolvulus* by Linnaeus in 1753. Linnaeus was prompted to use this species epithet by the last part of *Convolvuli minimi species bisnagarica hirsuta, alsines folio* used by Leonard Plukenet in 1696. Other authors before him used different comparisons. Johannes Burmann, in his study of plants in Ceylon (modern Sri Lanka), compared the herbs with *Anagallis* (Primulaceae). Dutchman Hendrik A. van Rheede, in his *Flora Malabarica*, used the local Indian name *vistnu-clandi.*

That Malayalam language name is a variant of the one still used throughout India and nearby Sri Lanka, *vishnu-kranti* (Vishnu's halo, Tamil). The name is gravid with religious and secular meaning and allusion, and it has been used widely throughout the Indian subcontinent for thousands of years (Austin 1980).

The fact that all of the names familiar to Linnaeus in the 1750s were based on Old World collections is one of the incompletely resolved mysteries of biogeography. We have presumed that, because 99 of the 100 or so species in *Evolvulus* are native to the Americas, that all are. One of the two Old World species, *E. nummularius*, was introduced into that hemisphere almost certainly during World War II. In spite of that, the story of *E. alsinoides* is not as clear cut.

According to all studies to date, there is a Sanskrit name for *E. alsinoides*. That name is written *vishnu-kranta* or *vishnugandi*. Regardless of spelling, both are exactly the same words as *vishnu-kranti* (Watt [1889] 1972, van Valkenburg and Bunyapraphatsara 2001, Parrotta 2001). It is widely agreed that plants with names in Sanskrit are native to the Indian subcontinent. So, how did an American plant come to have a name in Sanskrit, a classical language (relegated to holy texts) that effectively has long been dead?

The conservative answer is that there were some plants used medicinally by pre-European people who spoke Sanskrit, and that the name was switched to the American plants after they were introduced. Such trading of plants and switching of names is well known in the Convolvulaceae (Austin 2000), and surely occurred in other families. Presumably, *E. alsinoides* was introduced into the Old World by the Spanish when they arrived in India about 1505. It is more firmly established that the Spanish introduced sweet potatoes (*Ipomoea batatas*), moonvines (*I. alba, I. turbinata*), quamoclits (*I. hederifolia, I. quamoclit*), wood rose (*Merremia tuberosa*), and other New World laxative members of the family at about the same time (cf. Austin 1980a,b, 1998). Another blue-flowered morning glory in Asia with American relatives makes the simple answer seem inadequate (Austin et al. 2001), but there is no better hypothesis at the moment.

Regardless of exactly the dates of origins of plants and names in Asia, in the Americas there are multiple names for *E. alsinoides*. In Yucatán, other than being called *ojitos azules*, the plants are known as *sia-siu* [*xia-xiu, xiatiu*] (*sian*, enchantment, *xiw*, herb, Maya). This name strongly implies that the herbs had and still have an ancient and religious significance for the advanced Mayan people of Mesoamerica. The Maya also call them *x-havay* [*x-haway*] (leprosy and other contagious skin diseases) and *tsoots ts'ul* (Spaniard's hair). The fact that other countries that are occupied by the descendants of the Maya have common names for them further suggests a long history. In El Salvador, they are *cenicito* (little ashy one) or *oreja de ratón* (mouse's ear). In Guatemala, the plants are *quiebra-cajete* (box-breaker), an allusion to medical treatment of bowel problems. In Honduras, people say *pata de paloma* [*pate paloma*] (dove's foot). Farther north in Mexico and adjacent Texas they are *ojo de víbora* (snake's eye), so maybe people in northern Mexico used them like the Maya.

Throughout the Caribbean, there are also names for these pineland herbs. They are chickweed evolvulus (Bahamas) and *ilusión haitiana* (Haitian dream, Hispaniola; the name also applied to introduced and cultivated *E. tenis*). The *ilusión haitiana* suggests an application like *sia-siu* or enchantment herb of the Maya. Some compare them with flax (*Linum usatissimum*, Linaceae) by calling them *lin* (flax, Guadeloupe, Martinique). In Cuba, they are *tebenque* (also applied in Oriente Province for *E. arbuscula*).

In South America, *E. alsinoides* is *corre-corre* (run-run, Brazil). Could that name be associated with treating bowel problems? Otherwise, they are *oretama de sabana* [*retama de sabana*] (*retama* of the savana, Venezuela). The name *retama* refers to the European legume *Lygos*, called white broom in English. Some Venezuelans say *Santa Lucía* (St. Lucie). That name also refers to *flor de Santa María*, which is *Chrysanthemum parthenium*, a popular medicine used for eye pains. Alternatively, they are simply a *yerba de sabana* (savana herb, Venezuela).

Indeed, all of these names are reminiscent of the reputation the species has in Asia as a "brain tonic." As such, *E. alsinoides* has been used to treat nervous disability and memory loss (Austin 1980a, Parrotta 2001). Cultured tissues of the plant reputedly accumulate ergot alkaloids, amides of the indole derivative of D-lysergic acid, which is biosynthetically derived from the amino acid tryptophan (Nambiar and Mehta 1981, van Valkenburg and Bunyapraphatsara 2001). However, studies by Eckart Eich (personal communication) failed to find ergoline alkaloids in the species. The German study did, however, find tropane alkaloids (tropan-3β-ol, formerly called tropine), tropan-3β-ol (= pseudotropine), tropan-3-one (= tropinone), merresectine C (an ester of 3Δ-tropanol), and other different similar esters of tropanol and nortropanol, respectively, and several pyrrolidine alkaloids (hygrine, norhygrine, 5-(2′-oxopropyl)-hygrine, 5-(2′-hydroxypropyl)-hygrine, 5-(2′-hydroxypropyl)-hygroline, 2′,3-*N*-methylpyrrolidinylhygrine, 2′,4-*N*-methylpyrrolidinylhygrine, cuscohygrine) (E. Eich, personal communication, September 2003).

While it is dubious that this species produces ergoline compounds, people have independently discovered its impact on human behavior throughout the tropics.

There are a number of other applications of *E. alsinoides* across its range, and several bioactive chemicals have been isolated from it. Other major applications are for gastric distress, duodenal ulcers, and fever, especially in children. Externally, it is applied to promote healing of ulcers (Dey and Das 1988) in Asia, and the Maya too use it to treat skin diseases. *Evolvulus alsinoides* has also been found to contain flavonoids, saponins, triacontane, and sitosterol (Metha and Shah 1958). Laboratory experiments show that alcohol extracts are antiulceric, anticatatonic, and a central nervous system depressant in rats (van Valkenburg and Bunyapraphatsara 2001). In Cuba, a decoction of the whole plant has been used against fevers and chronic diseases of the intestines (Roig 1945). Everyone considers it a bitter tonic and antipyretic.

Linnaeus corrected his mistake of 1753 when he moved the species *alsinoides* from *Convolvulus* to *Evolvulus* in 1762. Indeed, he created the genus for that species and a few others known at the time. Since the 1760s, there has been only one worldwide study of the genus—the doctoral dissertation by Simon J. van Ooststroom (1934). Only a few new species have been discovered more recently. All of the species are native to the Americas, and range from Canada to Argentina and Chile, with the vast majority of them found in Brazil.

The genus is not well known for medicinal application, although several species contain calystegines (Schimming et al. 1998). The North American endemic *E. nuttallianus* was used by at least the Kakyenta Navajo (Moerman 1998). In addition, the Florida species *E. convolvuloides* and *E. sericeus* are still used in Latin America. Villagers in Salango, coastal Ecuador, told me in 1999 that they make a medicinal tea from *E. convolvuloides* although I have not found it recorded anywhere in the literature. Known as *havay ak* (leprosy vine, Maya), *E. sericeus* is used in Yucatán to treat all kinds of burns (Tapia and Contréras, in press). *Evolvulus arbuscula* grows in the Bahamas, Cuba, Haiti, Jamaica and Little Cayman, and is used by people in all those places (Roig 1945, Morton 1981). A decoction of the entire plant is a popular cure for catarrh and neurasthenia in Cuba. Jamaicans make a tea considered a treatment for heart problems, flatulence, and bellyache. The plants are sometimes sold as a substitute for pencil flower, *Stylosanthes viscosa*. The related Bahamian *E. squamosus* is used in much the same manner, especially against fever or jaundice.

Evonymus

(Sometimes spelled *Euonymus*; ancient Greek name, from *eu*, good, *onoma*, name; Euonymus was the mother of the Furies, an appropriate allusion, as the plants had a bad reputation for poisoning cattle)

Evonymus atropurpureus. From Sargent 1905.

berrétta da prète (priest's cap, from the four-lobed capsule, which resembles the shape of the priests' hats in the first half of the 19th century, Italian)
bonetero (little hat, Spanish)
burning-bush (a common name applied by the 1880s to a number of plants having red flowers or fruits because they reminded people of the biblical story in Exodus III)
corallini (little red one, Italian)
feòras (Gaelic)
fusaggine (little spindle, derived from Latin *fusus*, Italian); *fusain* (from Latin *fusaginem*, through *fusago*, derived from Latin *fusus*, a spindle, for spinning thread, because of the hard wood used in the manufacture of spindles, French); *fusaria comune* (common spindle, Italian)
oir (Gaelic)
prick-timber ("The butchers doe make skewers of it, because it doth not taint the meate as other wood doe," Aubrey 1626–1697 in Vickery 1995)
Spindelstrauch (spindle shrub, German); *spolebusk* (spool bush, Norway)

Evonymus americanus (of America)
burning-bush
bursting-heart; hearts-bursting-with-love
[American] strawberry bush [American strawberrybush] (applied since 1856 or before because the fruits resemble those of strawberry, but only in their red color)

Evonymus atropurpureus (dark purple)
arrowwood; Indian-arrow [root, wood]
bitter-oak
bleeding-heart

bonnet de pretre (priest's hat, Quebec)
burning-bush
fusain (spindle, Quebec)
ni-ni'-ba zhon (*ni-ni'-ba,* pipe stem, *zhon,* wood, Osage)
spindle-tree
strawberry-tree
[eastern] wahoo [wa-a-hoo, wahoon] (Dakota *wahú,* literally, arrowwood, first used ca. 1810; however, see also *Ulmus*)
wananha-i-monthin (ghost walking-stick, Omaha-Ponca)

Although physician E.P. Claus said in 1861 that the indigenous Americans introduced the early settlers to *Evonymus* as a medicine, many of them also knew of its use from the Old World (Linnaeus [1749] 1979, Millspaugh 1892). Old World people formerly used the wood to make gunpowder, skewers, spindles, and toothpicks (Dwelly 1933). They also made purple and yellow dyes from the seeds and other parts. Mabberley (1997) said that the wood was used to make "violin bows (pegwood)," but bows have no pegs.

The two eastern species were used similarly, but not identically by different people. *Evonymus americanus* was used by the Cherokee in an infusion to relieve stomachache, spitting blood, and a prolapsed uterus, and in a mixture to treat urinary problems and venereal disease. They also put an astringent infusion as an antiseptic on swellings and other skin problems, and sniffed bark infusions for sinus trouble (Hamel and Chiltoskey 1975). The Iroquois used a decoction to either suppress or stimulate menstruation and to treat difficult urination (Moerman 1998).

The Meskwaki put a decoction of *E. atropurpureus* bark on sores that were slow to heal and used a bark infusion to treat sore eyes (Vogel 1970). The Mohegans used a leaf infusion as a physic (Vogel 1970). Winnebago women drank a decoction of the inner bark of *E. atropurpureus* to cure uterine trouble (Gilmore 1919).

Porcher (1863) wrote that both species were "[e]metic, discutient, and antisyphilitic. It is also thought to be narcotic. The seeds are said to be nauseous, purgative, and emetic, and are used in some places to destroy vermin in the hair. Leaves are poisonous to cattle."

Lampe and McCann (1985) say that only *E. europaeus* has been recorded as containing the poisons, although Foster and Duke (1990) consider the American species as probably dangerous. That Old World species contains evomonoside, a digitalis-like cardioactive glycoside, several alkaloids but mostly evonine, and a compound that inhibits protein synthesis (Foster and Caras 1994).

Evonymus atropurpurea contains a resin called euonymin, the alkaloid evonynol, evonysterol, and evatromonoside (which is a cardiac glycoside digitoxigenin and arabinose) (Hocking 1997, Diggs 1999). Wahoo was in the U.S. Pharmacopoeia from 1863 to 1916 and in the National Formulary from 1916 to 1947 (Vogel 1970). The root is a bitter tonic, laxative, increases bile flow, and stimulates the liver (Hocking 1997).

Exostema: Princewood

(Greek *exo,* outside, *stemma,* crown or wreath, referring to the exserted stamens)

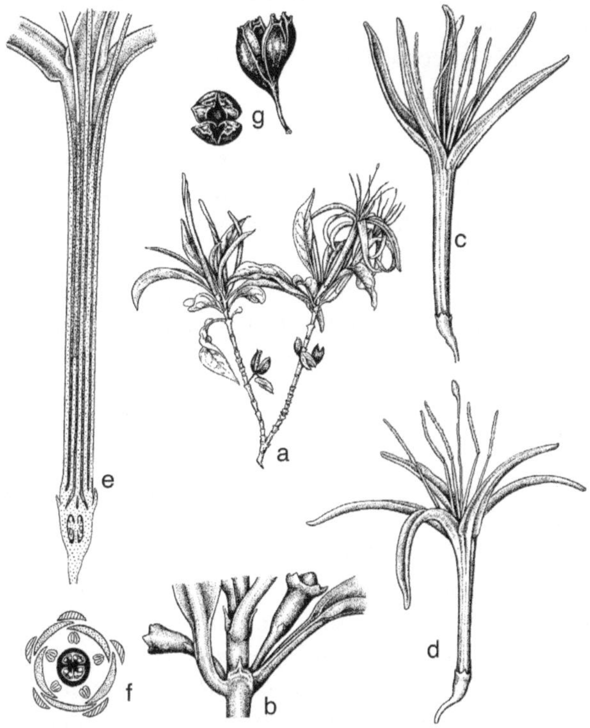

Exostemma caribaeum. a. Branch with flowers and fruits. b. Node, showing stipule and insertion of leaves and flower buds. c. Young flower, side view. d. Flower at anthesis, side view. e. Flower tube longitudinally dissected. f. Floral diagram. g. Fruits dehisced. *Drawn by Priscilla Fawcett.* From Correll and Correll 1982.

Forests in southern Florida, especially Miami-Dade and Monroe Counties, are composed largely of West Indian plants. Locally these forests are called hammocks, and they approach having 100% tropical species composition. Visit these forests in the peak mosquito season, and you may have an opportunity to see flowers of one of the state's rarest trees—if you survive the blood loss.

Arrive early in the morning and you will smell the fragrant vanilla odor and see the white flowers on princewood (Florida, Bahamas, Puerto Rico). Scientifically these trees are *Exostema caribaeum*, and they were first introduced to Europeans in 1760. That year Nicolaus Jacquin published his book *Enumeratio Systematica Plantarum* (Systematic Enumeration of Plants) where he called the tree *Cinchona caribaea* (the Carib's quinine). His reason for thinking it was a quinine was supported by two lines of evidence—the plants resemble *Cinchona*, and local people throughout the Caribbean call them quinine in their languages, and use them to treat malaria. It was not until 1819 when the Swiss Johann J. Roemer (1763–1819) and the Austrian Josef A. Schultes (1773–1831) revised the eighth edition of Linnaeus's *Systema Vegetabilium* that it was realized the species belonged to another genus in the family Rubiaceae. They moved the species to a genus created by Aimé J.A. Bonpland (1755–1858), the exploring companion of Alexander von Humboldt (1769–1859). In 1807, Bonpland created *Exostema*. In case you have seen more than one spelling or author citation of that genus be aware that its synonyms are *Exostemma* Augustin-Pyramus de Candolle and *Exostemon* George Edward Post and Carl E. Otto Kuntze (Lorence 1999).

We now know that *Exostema* is more closely related to *Chiococca* than to *Cinchona* (Bremer and Jansen 1991). However, there are compounds in *Exostema* that show *in vitro* activity against the organisms that cause malaria (Noster and Kraus 1990, Kohler et al. 2001). So, while they did not become famous for it, the indigenous people of the Caribbean discovered effective medicines against this killer disease as did those of South America who showed Europeans the source of quinine. History has made much of the multiple ways that humans combated malaria in the Old World (Desowitz 1987), but it has given short shrift to similar activity in the Americas.

Many common names still reflect the use to treat one of the most deadly diseases of the tropics. Among those names are Caribee bark tree (Jamaica), Caribbean princewood (Florida), *falsa quina* (fake quinine, Mexico), Jamaica Jesuit bark (Jamaica), *palo de quina* (quinine tree, Puerto Rico), *piñi-piñi* (maybe a variant of *quinquina*, Dominican Republic), princewood (Florida, Bahamas, Puerto Rico), *quina* (quinine, Hispaniola, Puerto Rico, Dominica), *quina caribea* (Carib's quinine, Cuba), *quina criolla* (creole quinine, Dominican Republic), *quina de las Antillas* (Antillean quinine, Cuba), *quinina* (quinine, Hispaniola), *quinina des Antilles* (Antillean quinine, Hispaniola), *quina de Michoacán* (Michoacán quinine, Mexico), *quinquina bord de mer* (seaside quinine, Guadeloupe, Martinique), *quinquina des caraïbes* [*quenquine caraïbe, quinquina caraïbe*] (Carib quinine, Haiti, Guadeloupe, Martinique), *quinquina pays* (wild [country] quinine, Haiti), *quinquina piton* (wild [mountain peak] quinine, Martinique), and West Indian quinine bark (Puerto Rico). Note that some names compare the trees to "Jesuit's bark," one of the old names for the quinine source. We do not know what the Caribs called this species, but they said *utáuaho* for the related *E. sanctae-luciae*. A similar word probably was used for *E. caribaeum*.

Treating malaria is not the only medicinal application for the plants. The bark is also emetic (Mexico) and, combined with leaves, has been used to treat colds and improve the appetite (Virgin Islands), and as an antipyretic (Morton 1981). Moreover, *Exostema* has been used as a fish poison (Virgin Islands). Wood with the bark removed is steeped or boiled and taken for stomachache, diarrhea, and piles, and given to children as a vermifuge (Bahamas). To treat these problems, the bark may be combined with *Picramnia pentandra*. Otherwise, the bark may be boiled with mahogany (*Swietenia mahagoni*) to make a tea believed to prevent anemia.

The plants contain a number of chemicals that may be effective against some or all of those complaints. *Exostema* contains coumarins, 4-phenylcoumarin, antineoplastic agents, and cucurbitacins (Mata et al. 1987, 1988, 1990, Noster and Kraus 1990, Bose and Banerji 1991, Calera et al. 1996, Kohler et al. 2001). Pharmacological assays of the bark and roots show antibiotic activity and depression of the central nervous system.

Most of the other common names applied to these trees allude to their use as torches. Probably the most obvious of these are black torch (Virgin Islands), *bois chandelle* (candle tree, Guadeloupe, Martinique), *bois flambeau* (flame tree, Guadeloupe, Martinique), *chandelle Anglais* (English candle, Haiti), princetorch (Bahamas), and yellow torch (Puerto Rico, Virgin Islands). Less obvious in use for light sources are the names *cera amarillo* (yellow wax, Cuba), *cerillo* (little wax tree, Cuba), *cerillo de costa* (coastal little wax tree), *cerillo de loma* (hill wax tree), and *tendre en gomme* (soft gum, Guadeloupe, Martinique).

Surely associated with the light source and a pleasant smell are *copalche* (literally, copal tree, Chiapas, Oaxaca), *copalchi de Jojutla* (copal tree from Jojutla, Morelos), *copalchi de Michoacán* (copal tree from Michoacán, Mexico), *jocotillo de cerro* (mountain plum, Chiapas), *chaktsiis* [*chactsus*] (*chak*, red, *ts'ix'*, resin, Maya, Yucatán), and *sabac-ché* [*sabakche*] (*sabak*, black spot, *ché*, tree, because the resin, when exposed to the air, turns black, Maya, Yucatán). However, several of these names suggest use of the resins for other purposes. The comparison with *copalche* alludes to the previous mixture of three species that were exported to Europe under that name. The other two species were *Coultarea hexandra* (Rubiaceae) and *Croton niveus* (Euphorbiaceae). All had bark with a similar odor and flavor, and all were used to treat intermittent fevers of malaria (Uphof 1968).

The other names seem to be comparisons; some are obvious and some are not. Names like *albarillo* (little apricot, Puerto Rico) and *poirier de montagne* (mountain pear, French Antilles) must refer to the general plant or leaf aspect because the fruits are dry capsules.

This species is called *carey* (Cuba) or *carey de costa* (coastal *carey*, Cuba). Other references using the word *carey* are *frutica de carey* (*Bourreria succulenta*) and *carey de costa* (*Krugiodendron ferreum*). Technically, the Taino word *carey* refers to the sea turtle (*Caretta caretta*) and the comparison between these plants and that animal is not immediately apparent. However, the polished wood of all of these is the saffron-like color that resembles the turtle shell, especially if polished. If there is another allusion, it is not obvious. Ironwood (Antigua) certainly notes that the wood is hard, heavy (specific gravity of 1.0), and strong. Maybe *macagua de costa* (coastal *macagua*, Cuba) is based on another Taino word.

If these names allude to the wood, it is because the wood has been used by many people. Modern use is predominantly for fence posts, but that surely was not the only pre-European application (Liogier 1974). Throughout its range the wood is used in cabinets, canes, and inlaid pieces (Standley 1920–1926, Liogier 1974). The wood must have been used in buildings and for other purposes, especially because we know that the Caribs used the related species *E. sanctae-luciae* to make paddles (Nicolson 1991).

Several of the common names refer to the aromatic white flowers. The blossoms are nocturnal, and that is further suggested by the comparison with jasmine (*Jasminum*, Oleaceae) in the name *palo de jazmin* (jasmine tree, Puerto Rico). The names *lechesillo* (little milky one, Puerto Rico), *lirio bobo* (fool's lily, Dominican Republic), and *lirio santana* (Santa Ana's lily, Cuba) also seem to refer to the milky white, lily-like flowers.

The comparisons with *hesito* (hesitation?, Nicaragua) and *cuero de sapo* (toad skin, Puerto Rico) completely elude me.

Exothea

(From Greek, *exoteros*, outside, expel, because the genus was removed from *Melicoccus*)

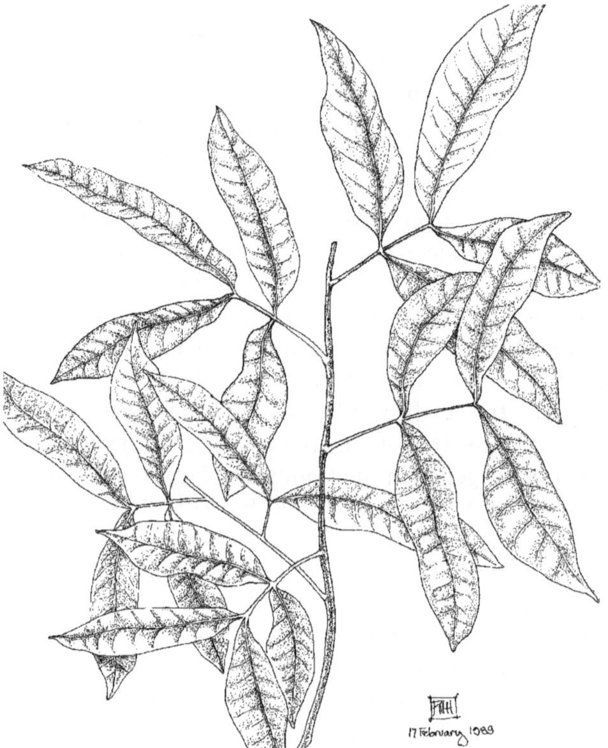

Exothea paniculata. *Drawn by P.N. Honychurch.*

Exothea paniculata (having paniculate inflorescences)
bitter bough (Bahamas); butter bough (a corruption?, Florida, Bahamas)
bois mûlet (mullet tree, Haiti)
cuerno de buey (ox neck, Dominican Republic)
gaita (Puerto Rico)
guacarán (*guacara*, lie, Taino, Puerto Rico)
guamacá (probably Taino, Cuba)
guénépier [*quénépier*] *marron* (wild genip tree, Haiti); wild genip ("genip" arose through confusion with *Genipa americana*, which is *jenipapo* in Portuguese from Tupí *ñandi'pap*; later applied to *Melicoccus bijugatus*, Jamaica; leaves on *Exothea* somewhat resemble those of *Melicoccus*)
inkwood (Florida)
nisperillo (little *nispero* [*Manilkara zapoda*], Dominican Republic)
white ironwood (Puerto Rico)
yaicuaje (probably Taino, Cuba)

Scottish botanist James Macfadyen (1798 [1800?]–1850) described *Exothea* in *The Flora of Jamaica* of

1837. In 1888, German botanist Ludwig Adolph Timotheus Radlkofer (1829–1927) moved Jussieu's *Melicocca paniculata* to create *Exothea paniculata.* Two other species have been added to the genus from the Caribbean and Central America (Mabberley 1997).

The wood is reddish, hard, compact, and used for posts, railroad ties, marine piling, boat building, cabinets, and tool handles (Liogier 1974, Little et al. 1974). Historically, ink was made from fruits, hence the name "inkwood."

F

Fagus

(*Fagus*, classical Latin name of the beech tree; Greek *phagos*, to eat, an allusion to the edible nuts; akin to Akkadian *paglu*, strong)

Fagus grandiflora. From Sargent 1905.

beech [bóece, béce, beche, bech, beetch, beeche] (derived from Old Teutonic *bokjon*, about A.D. 800, originally meaning a tree with an edible fruit; Old Teutonic *boka* was cognate to Latin *fagus*, and Greek *phagos*); *bech* (by Turner [1548] 1965); *beuk* (Dutch); boc (Old English); *boek* (Flemish); *bøk* (Norwegian); *bók* (Old Norse); *Buche* (from Old Teutonic *boka*, German; cognate to "book," because the earliest writing tablets were made from beech bark and boards)
faggio (from Latin *fagus*, Italian); *faidhbhile* (Gaelic); *fayard* (from 1371, used only in the dialect of Lyon, derived from *fagus*, French); *feagha* (Gaelic); *haya* (Spanish)
hêtre (from Frankish *haistr*, from the radical *haisi* bush, or thicket, French)
mast-beech (Culpeper 1653)

Fagus grandifolia (large leaved)
ajawe'mic (Potawatomi); *ajawêmij* (Ojibwa); *popsquawsewaymeysee* (Shawnee, cf. Edgar 1891); *schauweminschi* (*minschi*, plant, Delaware)
[American, red, ridge, stone, white, winter] beech
beechnut (said to be first used in 1739)
hatombalaha [*hatomalaha*] (Choctaw); *toomalàaha* (Alabama)
haya (Hidalgo)
hêtre [*americain, rouge*] ([American, red] beech, Quebec)
jozgóera (Onondaga)
nofó (Koasati)
tanicani (Delaware)
totocal (Hidalgo)
yap tukse'? (beech tree, Catawba)

Europeans knew *Fagus sylvatica* when they arrived in the New World. They used the straight-grained wood for kitchen utensils, turnery, and furniture (Mabberley 1997, de Cleene and Lejeune 2002). The wood contains toxic saponins and formerly was a source of creosote. Moreover, since prehistoric times the edible seeds ("nuts") had been used for human food. Their oldest name in Greek simply means "to eat." People later began calling them "beechnuts" or simply "mast," a term that includes acorns (Davidson 1999). According to Sargent (1905), the sweet seeds "are a favorite food of swine, and yield a valuable oil." He further recorded that, for *F. grandifolia*, "the sweet nuts are gathered and sold in the markets of Canada and of some of the western and middle states."

In France, the term *fou* (or *fayard*, or *foyard*) was used to designate large trees left for reproduction (and beech nuts?), while *hêtre* was used for young trunks repeatedly cut and regrown with multiple stems, as in a copse (Bloch and von Wartburg 1994). Later, as forest management changed, *hêtre* was used for the adult tree as well as the coppice growth.

The American beech produces prized wood, although so few stands remain that mature trees are uncommon. Along with birch (*Betula*) and maple (*Acer*) it has been used in the hardwood distillation industry for charcoal, wood alcohol, and acetate of lime (Harrar and Harrar 1946). Wood has been sold commercially in the United States for making brush backs, chairs, clothespins, cooperage, crates, flooring, fuel, shoe lasts, spools, spoons, tool handles, toys, and veneers (Harrar and Harrar 1946, Vines 1977). As a fuel, it has few superiors, making an excellent cooking fire when thoroughly dry (Harrar and Harrar 1946). Seeds have been sold in the northern states and Canada to make vegetable oil and as swine food (Vines 1977).

The Cherokee used the wood for lumber and buttons (Hamel and Chiltoskey 1975), while the Micmac made snowshoe frames from it (Moerman 1998). The Potawatomi carved it into bowls (Smith 1933, King 1984). Surely all tribes used it as fuel.

Indigenous people in the New World also made use of the edible fruits. The Algonquin, Iroquois, Menomini, Ojibwa, and Potawatomi are known to have eaten the seeds, but probably all the people within the range of the species ate them (Smith 1933,

Yanovsky 1939, Moerman 1998). Sometimes the nuts were eaten mainly by people working in the bush. Others gathered and stored them. They sought out caches left by chipmunks (*Tamias straitus*) or deer mice (*Peromyscus*) because it was known that they stored only sound seeds and it saved a great deal of work. Sometimes the seeds were crushed and boiled and the liquid used as a nourishing drink. At other times, they mixed the nuts with cornmeal and beans or berries to make into bread as they did with chestnuts, hickory nuts, and acorns. Oil was rendered from them as from chestnuts, hickory nuts, and acorns. The Iroquois mixed the oil with bear grease as a mosquito repellent and used it for a hair tonic (Moerman 1998).

There were also medicinal applications of *Fagus.* The Cherokee chewed the nuts to expel worms (Hamel and Chiltoskey 1975). The Iroquois used the bark when women were having difficulties with abortions (Moerman 1998). The Iroquois also used the trees in a "blood purifying" mixture; decoctions were used to treat burns or scalds, for yellow skin and gall, and to treat tuberculosis (Moerman 1998). Similarly, the Malecite used the leaves on sores. The Menomini mixed the bark and root in medicines, and the Micmac treated skin problems with leaves. The Ojibwa made a medicine for pulmonary problems. The Potawatomi treated burns and scalds, and restored frostbitten extremities with leaves (Smith 1933). The Rappahannock made a bark infusion to treat poison ivy (Vogel 1970). Extracts from the trees contain guaiacol (Hocking 1997).

Ficus: Strangler Fig

(*Ficus,* the old Latin name of the tree or fruit; akin to Akkadian *piqu* or *siqu,* narrow, *piaqum, siaqum,* to be narrow; Greek *sykon*)

In 1895, W.D. Collier, who lived on Key Marco in southwestern Florida, was digging muck from shell ridges on his property for his garden. Buried within the muck he found pieces of wood, cords, and shell implements. After preliminary visits by an amateur archaeologist, and by personnel from the Bureau of American Ethnology in Washington, D.C., an expedition was arranged to excavate the site. The project was headed by Frank Hamilton Cushing of the bureau, and resulted in what were the most remarkable plant artifacts yet discovered in Florida. Nets made of palm leaf fibers, carvings of animals, household utensils, and ceremonial masks had been preserved in the anaerobic conditions of the marine peat and muck deposits.

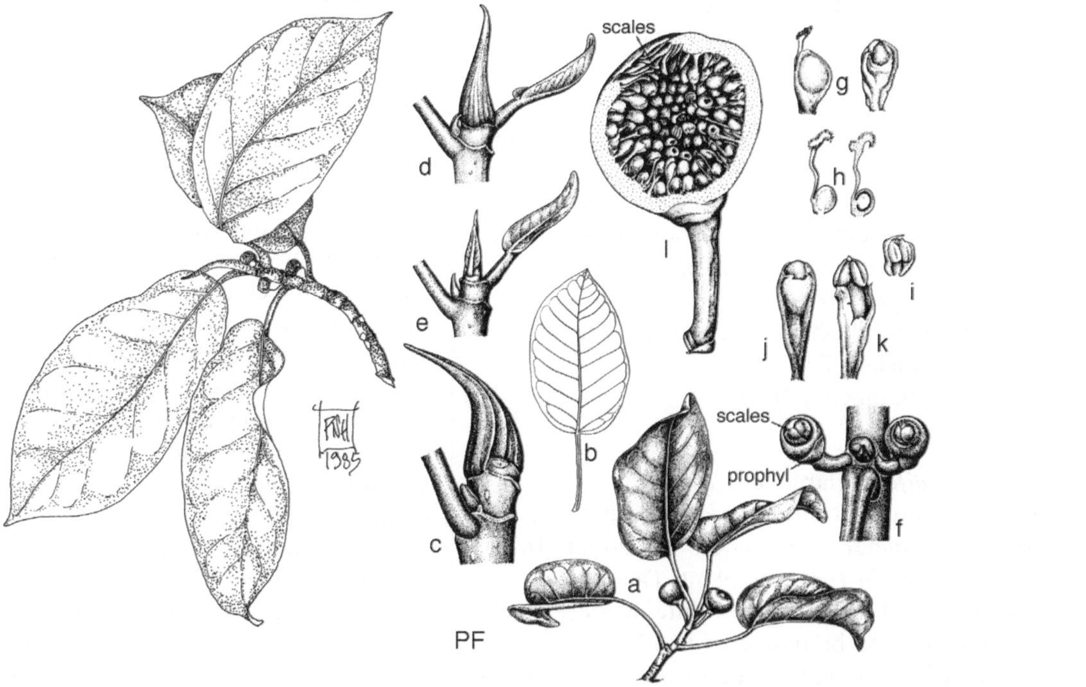

Ficus. *Ficus aurea* (left). Note sessile figs. *Drawn by P.N. Honychurch. Ficus citrifolia* (right). a. Leafy branch with figs. Note stalked figs. b. Leaf outline. c to e. Development of growing branch tip; c. Resting bud, d. Growing bud. e. Growing bud with stipule removed. f. Young fig pair. g. Young pistillate flowers, one on left with expanded style. h. Long-styled pistillate flower, one on right longitudinally dissected. i to k. Staminate flowers. i. Same age as pistillate in g and h, j, ripe, k, in anthesis. l. Fig in vertical section with wasp grubs hatching, seeds ripe. *Drawn by Priscilla Fawcett.* From Correll and Correll 1982.

Among the artifacts recovered was a bivalve molluscan shell (genus *Macrocallista*) with the figure of a male masked dancer drawn on the inner surface. That image is the only known self-portrait of the Calusa people, who had inhabited the site until about A.D. 1500. Studies done at the Smithsonian Institution Bureau of American Ethnology concluded that the image was created with the latex of a native fig (Gilliland 1975). Exactly how white latex became the black image of the dancer has not been explained, but presumably, the latex was used as an adhesive to retain black pigments (probably charcoal) on the shell surface. We do not know how much more the pre-European Glades people used figs. However, judging from the uses by the Seminoles and other people within the range of the species, their utilization was considerably more diverse than just artwork.

By the time Sturtevant (1955) worked with the Seminoles, he found that they used at least *Ficus aurea*. To the Miccosukee this tree was *hacaló:pî* [*hachelope*], and their relatives the Creeks said *hilokwapî* [*hilokhaga*]. Those terms, based on *vlokpicetv*, mean "sticks to you," although Sturtevant (1955) considered them simple. Both people ate the fruits, used the stems for arrows, made bowstrings of the bark on the roots, and used the same kinds of strings (*hapalch aske*, Mikasuki) for lashing house parts together, stringing up meat, and fish lines.

They used the latex as a chewing gum, and the plant figures prominently in their medicine to heal wounds and cuts. Martin and Mauldin (2000) called the tree *helokhakv*, and said the bark was beaten to extract the gum for chewing. The bark was then mashed and applied to wounds. The Creeks were also recorded as boiling bark of the root in water, and wrapping the sore with a bandage soaked in the medicine. Yet another record is of the chewed berries used as a "bandage" for the wound.

Since the Seminoles ate the fruits of *F. citrifolia* and used its latex as chewing gum, they probably used it in the same ways. At least the Cuna of Panama used *F. citrifolia* as a bandage to help wounds heal (Duke 1968), so the practice using both species was widespread. Indeed, Hernández, writing in 1651, noted that *F. petiolaris* was similarly used in Mexico. He recorded that "its milk cures sores of the lips and chronic ulcers" (Standley 1920–1926).

Figs of various kinds have been important in the lives of people wherever they occur. Europeans were well acquainted with one kind of fig when they arrived in the New World. That species was *F. carica*, the cultivated fig called *higo* (Spanish), *figue* (French), *fico* (Italian), *figo* (Portuguese), and *Feige* (German). Those names are derived from the Latin *ficus*, while the Greeks say *sykon*. Even Gaelic *fige* was based on the Latin root. The origin of the cultivated European species is disputed, with some contending that it came from southwestern Asian (Mabberley 1997) and others maintaining that it evolved in the Caspian region of northwestern Turkey (Zohary 1982). Regardless, it was in the Gezer site, on the western slopes of the Judaean mountains, by 5000 B.C. (Zohary 1982), and the fruit was known to the Hebrews as *teenim*. However, when Europeans arrived in the American tropics, they found trees that awed them. Oviedo ([1526] 1969) was so taken with the size of a fig that he not only spent a page talking about it, but also included a drawing.

The Spanish were not always complimentary about the New World figs. The general term for most of those was *matapalo* (tree killer). The name not only is used in Spanish-speaking areas, but also remains in English-speaking areas such as Trinidad. Another name that conveys the contempt that people feel for these trees is Scotch attorney (Trinidad). Usually, however, in English we simply say strangler fig. To distinguish species, modifiers are added, like Florida strangler fig (Florida), golden fig (Florida), or golden wild fig (Florida, Bahamas). These names refer to *F. aurea*, and emphasize its red fruits. Sometimes English speakers simply say rubber tree (Florida) or wild fig (Florida to Panama).

Elsewhere the plants are known as *figuier* (fig tree, Haiti, Guadeloupe, Martinique), *figuier blanc* (white fig, Haiti, Puerto Rico, French Antilles), *higo* (Panama), *higo cimarrón* (wild fig, Dominican Republic), and *jagüey hembra* (female fig, Cuba). The word *jagüey* is another of those Caribbean terms whose origin may be Taino.

Within Florida, trees of *F. aurea* range about halfway up the peninsula. They reach the Cape Canaveral area on the east coast and the Tampa Bay region on the west. When there are periodic freezes, they are sometimes killed to the ground, but rarely are the roots destroyed.

It may be this species that Williams ([1837] 1962) called "howey." That name has not been found elsewhere, but his description seems to fit. He wrote: "This tree bears a multitude of little figs, not bigger than a grain of corn. They are of a dark brown colour when ripe. In formation and taste, they are perfect figs."

The more tropical of the two native species in Florida is *F. citrifolia*. This species is typically restricted to the southern four counties (Broward, Collier, Monroe, Miami-Dade). During warm periods, it extends its range northward, as it has into southern Palm Beach County since the 1970s. Cold periods in the past have driven it back into the tropical parts of the peninsula. *Ficus citrifolia* also has the larger range of the two, being found from Belize, Yucatán, Florida, the Bahamas, Cuba, Hispaniola, Jamaica, and the Lesser Antilles to Paraguay. On the island of Barbados, it is known as the bearded fig. It has been suggested that Barbados (bearded) is derived from the earlier frequency of these fig trees (Goodding et al. 1965). This species too is basically called fig, with names based on that word including *figuier blanc* (white fig, Guadeloupe, Martinique), *figuier maudit* (accursed fig, Guadeloupe, Martinique), *higo* (fig, Dominican Republic), *higuerón* (big fig tree, Colombia, Darien), *higuillo* (little fig, Hispaniola), and short-leaved wild fig (Bahamas).

To distinguish it from *F. aurea*, it is sometimes called *jagüey macho* (Puerto Rico). But, in Puerto Rico, it is also *jagüey* or *jigüerillo*. Names alluding to the latex include *bois laglu* (the glue tree, Puerto Rico), *caucho menudito* (little rubber tree, Colombia), and *lechecillo* (little milky one, Puerto Rico). In Panama, it is either *saguagua* (Cuna) or *tugua* (Choco), but usually to the Cuna, it is *suu* [*tuu*]. In Guadeloupe and Martinique it is the *aralie-cerise* (joining cherry), presumably an allusion to the fused adventitious roots.

In the Bahamas and some other parts of the Caribbean, fig latex dissolved in water is taken for heartburn, palpitations, and chest pains (Morton 1981). In Hispaniola, the latex is used on hernias and for chest diseases (Liogier 1974). The latex contains ficin, a proteolytic enzyme, but what impact that would have on those problems in unclear. However, studies have shown that extracts from other *Ficus* species are antibacterial, anti-inflammatory, analgesic, antipyretic, and antidysenteric (Atique et al. 1985, Dey and Das 1988, Macfoy and Cline 1990, Mousa et al. 1994, Forestieri et al. 1996). The latex is chewed throughout the range of figs, and used at least by children as a birdlime to catch birds. Like the Seminoles, people in Hispaniola use the adventitious roots as cords.

Figs were more important to Mexican cultures than to those in the Caribbean. One of the widely used names for fig trees is *amate* (*amatl*, paper, Náhuatl). This may be applied to the tree, or some variant of *macahuite* (*ama-cuahuitl*, fig-tree) may be used. Not only was that done by the Aztecs, but the same also applied to the Zapotecs. In their language, the fig is *guiichi be-yo* [*quichi pe-yo*] (*guiichi*, paper, *be-yo*, fig tree, Oaxaca). As in Náhuatl, paper has the same name as the tree. Even the Tarascan name *siranda* (paper) is applied to fig trees. Figs for making paper were so important that there are places named after them, including *Amatepec* (hill of figs) and *Amecameca* (place where they wear fig shirts). However, the Aztecs also called them *coamichin* (*coatl*, snake + *michin*, fish, fish-snake or eel), probably because of adventitious roots. From that is derived the common name *camichín*.

Paper made from figs held special importance to the Aztecs and other people in Mexico. Bark preparation into paper was associated with curing and religion. Therefore, not long after the arrival of the Catholics, this practice mostly went underground. However, before it disappeared, a few of the texts that were in Aztec libraries were preserved. Not only did Mexicans use the paper "books" to record histories, but they were also used in correspondence among the elite. A common use that survives involves paper for *brujería* (witchcraft), where images of people, horses, and other animals are fashioned from the sheets. The images (*muñecos*) are employed in curing diseases, or buried in front of a house or corral for malevolent purposes. The use continues among the Otomí people of Puebla who cut out figures of the *nahuales*, their devils and spirits (Standley 1920–1926). *Amate* paper also is still being produced in Guerrero where it is brightly decorated with scenes of everyday life (Sandved et al. 1993). These bright sheets of paper not only depict modern village life, but also provide a glimpse into history that extends back for thousands of years of papermaking.

Flaveria

(Frenchman A.L. de Jussieu named the genus in 1789, with Latin *flavus*, yellow, referring to a yellow dye obtained from *Flaveria chilensis*)

Flaveria bidentis (having two teeth)
contra erva do Peru (Peruvian antipoison, Brazil); *contrayerba* (antipoison, Chile, Peru?)
dasdaqui (Chile); *daudá* (Chile)
matagusanos (worm-killer, Chile)
smelter's bush; *smelterbossie* (smelter's bush, South Africa)

Flaveria trinervia (three-veins, meaning the three large veins of the leaves)
contrayerba (antipoison, Cuba)
yerba de la vieja (old woman's herb, Cuba)

This small genus of 21 species is confined to the Americas except for one endemic to Australia and two that are pantropical (Powell 1978, Mabberley 1997). Presumably, *Flaveria bidentis* is one of the widespread species because it has names in both Afrikaans and English in South Africa (Quattrocchi 1999).

In Argentina, Brazil, Chile, and Peru, an infusion of *F. bidentis* has been used to treat indigestion, as an aromatic, stimulant, and vermifuge. Externally the herb has been used as a cataplasm for bites of poisonous animals and washing wounds (Hocking 1997).

Roig (1945) found that people in Cuba, especially in the Province of Havana, used *F. trinervia* to treat intestinal problems including gastroenteritis. Others considered *contrayerba* effective in treating the kidneys or diarrhea. The plants are also used as a medicine in Belize (Balick et al. 2000).

Forestiera

(Named by Jean Louis Marie Poiret for 19th-century French physician Charles Leforestier)

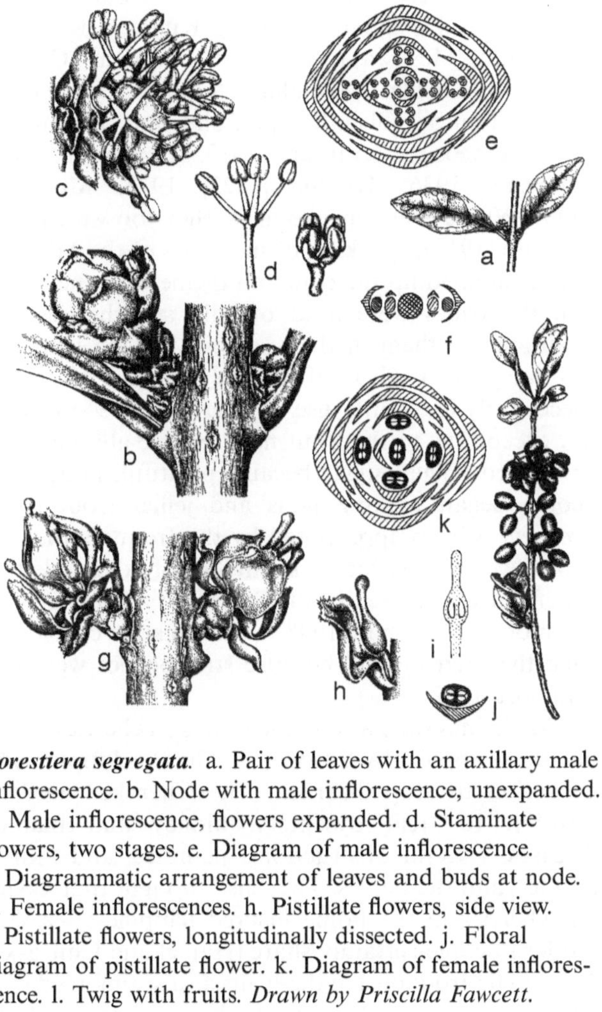

Forestiera segregata. a. Pair of leaves with an axillary male inflorescence. b. Node with male inflorescence, unexpanded. c. Male inflorescence, flowers expanded. d. Staminate flowers, two stages. e. Diagram of male inflorescence. f. Diagrammatic arrangement of leaves and buds at node. g. Female inflorescences. h. Pistillate flowers, side view. i. Pistillate flowers, longitudinally dissected. j. Floral diagram of pistillate flower. k. Diagram of female inflorescence. l. Twig with fruits. *Drawn by Priscilla Fawcett.* From Correll and Correll 1982.

Forestiera segregata (kept apart)
blueberry (Bahamas)
Florida privet (Florida, Bahamas)
ink-bush (Bahamas, Virgin Islands)
okfi:lî (Mikasuki); *oyakɬó:ska* [*oyakró:ska*] (Creek)
yanilla blanca (white little *yana*, comparing it to *Conocarpus erectus*, Taino, Cuba)

Jacquin named plants he found in the Caribbean *Myrica segregata* in 1788. Then, in 1810, Poiret described *Forestiera*, and in 1893 Jacquin's species was moved there. This North American genus now contains 15 species (Mabberley 1997).

The straight stems of *okfi:lî* were formerly used for arrows (Sturtevant 1955). The name "blueberry" probably simply notes the color of the fruits and does not make a comparison with *Vaccinium*. "Ink-bush" suggests that ink may have been made from the fruits in the past.

Fragaria

(From the Latin *fragans*, sweet-smelling, from the fragrance of the fruit; classical name of the strawberry)

Fragaria virginiana. From Britton and Brown 1897.

eerdthbesycruyt (earth berry herb, Dutch in 1542); *aardbei* (earth berry, Dutch); *eertbesien* (earth berry, Belgian); *Erdberre* [*Erdberr*] (earth berry, German)
fragola (written *fragharia* in 1542, then as *fraghe, fragole* in 1571, reaching modern form by 1696, Italian); *fraisier* [*fraysier*] (first written as *fresas* by Ruellius in 1536, then as *frayses*, but the modern form did not appear as *fraise* until 1542, Quebec, France); *freizen* (Dutch); *fresa* (Spanish)
jordbaer (field berry, Norwegian)
morango (from Latin *morus*, the mulberry, said to have been derived from a popular Latin *moranicu*, which may have come from *terrestria mora*, earth mulberry, from the 1400s, Portuguese)

strawberry [strawberrie, strayberry] (Anglo-Saxon *streowberie*, a name applied with variable spelling since about A.D. 1000; but in modern fashion since Turner in 1538; the element "straw" has been thought to come from a particle or straw or mote, alluding to the achenes on the surface of the fruit; another view is that it denotes the runners)

sùbh thalmhainn (*sùbh*, strawberry, *thalmhainn*, of the earth, Gaelic); *sùbhag làir* [*sùbh làir*] (*sùbhag*, strawberry, *làir*, mare, Gaelic)

tlachd sneisd (*tlachd*, beauty, *sneisd*, little, Gaelic)

Fragaria virginiana (of Virginia)

aʔni (Cherokee)

aparu-huradu (*aparu*, berry, *huradu*, ground, Pawnee)

a'stepa'x kotka' (*asi*, berry, Biloxi)

ba-stse'-ga (Osage); *bashte* (Omaha-Ponca); *haz-shchek* (*haz*, fruit, Winnebago); *wazhushtecha* (*wazhushtecha sha wi*; from *sha*, red, *wi*, moon [June], when strawberries are ripe; *wazhushtecha-hu*, the plant, Dakota)

biun̲ko (Choctaw); *biyyo'ka* (Chickasaw); *biyyokhá* [*obi:yokhá*] (Koasati)

cokake'nict li'l (soft blackberry, Atakapa)

ídzíaze (little heart, Chipewyan, Canada); *í*^{n}*dzheah* (Slave, Canada)

kepalv [*kepálv*] (Muskogee)

maskihkiwimin (recorded by Strachey in [1612] 1953 as *muskeskiwimins*, Powhatan, Virginia); *mskíhkawimin* (Penobscot, Maine; derived from Penobscot *mskìhko*, which is cognate with Cree *maskihkiy*, Fox *maskihkiwi*, Menomini *maskihkiw*, Ojibwa, *maskikki*, all meaning herb)

ode'iminidji'bik (*ode'i*, heart, *mini*, berry, *dji'bik*, root, Ojibwa); *otahimin* (heart berry, Cree)

otquechtarocu òchia (Onondaga)

scarlet Virginian strawberry (Porcher 1863); [mountain, scarlet, Virginia, wild] strawberry

wtèhim (Delaware); *wuttahimneash* (New England tribes fide Hedrick 1919, surely Algonquian)

In 1542, Fuchs found several names in use for the common European strawberry, *F. vesca* (Meyer et al. 1999). Most of them carried over into the names applied now.

Fuchs wrote that the juice of the plants with honey was good for the spleen. Added to white pepper it helped breathing difficulties. The fruit quenches thirst and is good for the stomach, especially if it is bilious. Juice pressed from the plant is good for face sores, red eyes, and excessive tears. Root decoctions soothe liver problems, settle the stomach, and staunch menstrual flow, while it firms the gums, soothes mouth ulcers, and makes the breath sweet (Meyer et al. 1999).

Fruits on the European plants were good, but not as good as those discovered in the New World. *Fragaria virginiana* is one of the two species (the other is *F. chiloensis*) involved in the hybrid now cultivated around the world. In 1643, Roger Williams (1603?–1683), who founded the Rhode Island colony, wrote that he heard a physician say, "God could have made, but God never did make, a better berry" (Williams in Hedrick 1919).

Fruits of *F. virginiana* were food to at least the Abenaki, Algonquin, Bella-Coola, Blackfoot, Cherokee, Cheyenne, Clallam, Cree, Dakota, Haisla, Hanaksiala, Hesquiat, Iroquois, Kitasoo, Klamath, Menomini, Meskwaki, Nespelem, Nitinaht, Ojibwa, Okanagan-Coville, Omaha, Osage, Oweekeno, Pawnee, Ponca, Salish, Sanipol, Shuswap, Thompson, and Winnebago (Gilmore 1919, Densmore 1928, Yanovsky 1936, Hunter [1823] 1973, Moerman 1998). Where other species grew, they too were eaten. Gilmore (1919), speaking of the tribes of the Missouri River, summed up the common theme when he wrote, "All the tribes were fond of wild strawberries and luxuriated in them in their season, but the fruit was too juicy to lend itself to the process of drying successfully for winter use." Moerman (1998) listed a few exceptions to that, but most people ate them raw or cooked. When sugar became plentiful, indigenous people began making jams and jellies from them. *Fragaria* was so appreciated by the Iroquois that the fruits were considered symbolic of their deities' beneficence and used in the Strawberry Thanksgiving ceremony (Moerman 1998). The Dakota considered June the "red moon" because strawberries were ripe (Gilmore 1919).

Medicines were also made from strawberries, much like those applied by Europeans. The Blackfoot used a decoction to stop diarrhea. The Cherokee used an infusion to treat dysentery, kidney and bladder diseases, visceral obstructions, jaundice, and scurvy, and to calm the nerves. They also thought that having strawberries in the home insured happiness—a view modern people certainly share (Hamel and Chiltoskey 1975). The Iroquois ate the fruits as a spring medicine. The Malecite and Micmac took an infusion of the plants to regulate menstruation. The Ojibwa used them to stop diarrhea in children and to treat stomachache. To heal sores, the Okanagan-Coville mixed deer fat with powdered leaves and put the powder into babies' mouths to cure thrush. The Thompson people also used them to stop diarrhea (Moerman 1998).

Porcher (1863) wrote, of *F. vesca*: "[Conrad] Gesner [in 1587] speaks of the good effects of the fruit in calculous disorders, and Linnaeus extols its efficacy in gout, having, he says, prevented paroxysms of it in himself by partaking of this fruit very freely. They are also supposed to possess vermifuge properties, and to be useful in phthisis [tuberculosis]. The leaves are astringent and are recommended for bowel complaints; and the roots are much used in Europe as diuretics; frequently given in dysuria, in infusion, made with an ounce to the pint of water. ... Lallemand ... states that strawberries are quite serviceable in relieving irritable conditions of the bladder and urethra."

Of *F. virginiana*, Porcher (1863) said: "It was introduced into England in 1629, and possessed a fame equal to the hautbois. The pulp has a fine flavor. ... This plant is well known, and its economical value and application require no description. The use of the fruit often acts beneficially upon dyspeptics, who are benefited by acids. The celebrated [French painter, Henri-Julien-Félix] Rousseau [1844–1910] was always relieved of a calculous affection by eating this fruit. ... The old Carolina strawberry is a well known and much esteemed variety. The pulp is colored and juicy, and has a fine vinous flavor." The first record of the new strawberry in England was by John Parkinson, and that was only 8 years after Edward Winslow found them in Massachusetts (Hedrick 1919).

Fraxinus

(The classical Latin name for the tree and a spear or javelin made from it; akin to Akkadian *burasu*, Hebrew *beros*)

Fraxinus caroliniana. *Drawn by Vivian Frazier.* From Correll and Correll 1972.

a'gimak' (snowshoe wood, Ojibwa); *pachgámmak* (Delaware)
ash [aesc, asse, aycha, assche, asche, ashe, aish, esche, ach] (from about A.D. 700, with the alternate spellings in chronological order; in the 1440s the German and English spellings were identical); *ask* (Norwegian); *ask-oz* (Old Teutonic); *askr* (Old Norse); *es* [*esc, esk, eisce, essce*] (Dutch); *Esche* (German)
do:-tsu (Cherokee); *gahuwéja* (Onondaga)
frassino (Italian); *freixo* (Portuguese); *fresno* (Spanish); *frêne* (French)
nuin [*nion*] (Gaelic, also the 12th letter of their alphabet)
wat (Atakapa)

Fraxinus americana (of America)
[American, Biltmore, Biltmore white, Canadian, cane, small-seed white, white] ash
eto-hv'tke [*'to hvtkv*] (*eto*, tree, *hv'tke*, white, Muskogee)
franc frêne ("*frêne*" from Latin *Fraxinus*, after having gone through *fraisne*, then *fresne*, the circumflex often indicates a lost "s"; anything with *franc* in Canada refers to an imported European species, since they are so commonly planted in France, so "imported ash," Quebec); *fresno* (Texas)
itcika' wanat (*itcika'*, bow, *wanat* =?, Catawba); *yap tukse'* (tree river bank, Catawba); *wactu'* (Catawba)
shinap (Choctaw); *sinapo* (always applied to this species, and sometimes to other *Fraxinus*, Alabama); *sinapó* (Koasati)

Fraxinus caroliniana (of Carolina)
[Carolina, swamp, water] ash
eto hvtke [*etchatka, tohatka*] (*eto*, tree, *hvtke*, white, Creek, Muskogee); *hikaape* (*hik*, white?, *aape*, stem or trunk, Mikasuki)
pop-ash [poppy-ash]

Fraxinus pennsylvanica (of Pennsylvania)
[green, pumpkin, red, river, swamp, water] ash
emkwansûk (spoon wood, Potawatomi); *mealawqua* (Shawnee, fide Edgar 1891)
frêne rouge (red ash, Quebec)
kiditako (Pawnee)
psehtin (Dakota); *tashnánga-hi* (*hi*, plant, Omaha-Ponca)

According to Norse legend, the world tree, an evergreen ash called *Yggdrasill,* overshadows the whole universe. Its roots, trunk, and branches bind together Heaven, Earth, and the Netherworld. Rooted in the primordial abyss Hel, the subterranean source of matter, it bears three stems. The center stem runs up through Midgard, the earth, which it supports. It issues out of the mountain Asgard where the gods assemble at the base of Valhalla. This heaven can only be reached by Bifrost, the bridge of the rainbow. The stem spreads its branches over the entire sky, the leaves are the clouds, and the fruits the stars. Four stags, Dain, Dvali, Duneyr, and Durathor, symbolizing the cardinal winds, live in these branches, feeding on the flower buds and dripping dew from their antlers to earth. The second stem springs up in Muspellsheim, the warm South, where Urth, the Past, Verdandi, Present, and Skuld, the Future, dwell, and the gods sit in judgment. The third stem rises in Nifleheim, the cold North, where all knowledge of mankind flows from the fountain of the Frost-giant, Mimir, the personification of Wisdom. The tree is the Nordic Tree of Life, symbol of strength and vigor, because the first Norseman, Ask, sprang from the ash tree (Lehner and Lehner 1960, de Cleene and Lejeune 2002).

The green ash (*Fraxinus pennsylvanica*) is also one of the two sacred trees to the Omaha; the other is the cedar (*Juniperus virginiana*). The *tashnánga-hi* is connected with the beneficent natural world, and part of the sacred pole of the Omaha and Ponca is made of it; the other part is cottonwood (*Populus*).

White ash (*F. americana*) wood was used to make handles for fishing spears by the Ojibwa, and the Micmac used it to make ax and knife handles (Moerman 1998). The Abenaki used it for snowshoe frames; the Iroquois for baskets and chair backs; and the Malecite for boat frames and snowshoes. The Meskwaki also used it to weave baskets, and the Ojibwa to frame snowshoes (Moerman 1998).

Fraxinus americana has long been an important timber tree; it has constituted 40% of the "ash" wood sold in the United States (Vines 1977). The wood has been used for butter tubs, cabinets, doors, farm utensils, frame parts for airplanes, fuel, millwork, musical instruments, railroad cars, railroad cross ties, refrigerators, sashes, silos, tanks, toys, vehicle parts, veneer, and woodenware (Vines 1977).

People from Maine and nearby Canada south to the Cherokee (Moerman 1998) made medicines with it. The Abenaki used an infusion of the bark to induce menses. The Iroquois used it as a physic, especially to purify the blood chewed bark to induce vomiting, or drank an infusion to stop stomach cramps. They used it as a laxative, and a compound mixture was used to become pregnant. Externally, the Iroquois used a bark infusion to treat syphilitic and neck sores (Moerman 1998). The Meskwaki put an infusion of bark on sores, itch, and vermin in the scalp. The Micmac used leaves to clean after childbirth, as did the Penobscot. The Ojibwa made a bark medicine for several problems. White ash bark was official in the National Formulary from 1916 to 1926 for use as a tonic and astringent (Vogel 1970).

Both the Iroquois and Meskwaki considered the leaves of the tree effective against snakebite. The Iroquois used a bark infusion to treat snakebite. The Meskwaki also used a decoction of flowers to treat snakebite.

Perhaps southeastern tribes believed the same because of the comments by Porcher (1863). He wrote: "The leaves of the F. Americana are said to be so highly offensive to the rattlesnake that that formidable reptile is never found on land where it grows; and it is the practice of hunters and others having occasion to traverse the woods in the summer months to stuff their boots or shoes with white ash leaves as a preventive of the bite of the rattlesnake." However, the belief that ash repels snakes is ancient in Europe (de Cleece and Lejeune 2002).

Popash (*F. caroliniana*) is not mentioned by Moerman (1998). However, Bennett (1997) found the Miccosukee using the stems to make pestles, *sofkee* spoons, and for firewood. Historically, they used the branches to make bows and arrows. The bark can be cut where the plant has been damaged and used to make "women's medicine."

Green ash (*F. pennsylvanica*) wood was used by the Cherokee, Cheyenne, Dakota, Havasupai, Lakota, Ojibwa, Omaha, Pawnee, Ponca, Potawatomi, and Winnebago. From it they made ball bats, baskets, butter paddles, cradle board, firewood, handles for hoes, axes, and other utensils, lumber, poles—especially for the Cheyenne Sundance Lodge—racks for drying meat, snowshoe frames, sleds, spoons, tent poles, pegs, and pins. It was most widely prized for making bows and arrows and for pipe stems (Moerman 1998). Whites use the wood for tool handles, furniture, interior finishing, cooperage, and wagons (Vines 1977).

There is less on medicine regarding this species, with records from the Algonquin, Ojibwa, and Omaha. The inner bark was made into an infusion to treat depression. It served as a tonic and was used in a number of rituals.

The Ojibwa even cooked and ate the inner bark of *F. pennsylvanica* (Yanovsky 1936). Supposedly, it tasted like eggs (Moerman 1998).

G

Galactia

(Named by Patrick Browne, from Greek *galaktos*, milk, presumably because some species contain milky sap)

***Galactia volubilis**. Drawn by P.N. Honychurch.*

Galactia volubilis (twining)
ayikchi: łî (good medicine, Mikasuki); *hilishiłî* (maybe from *heleswa*, medicine, *hełe* [*here*], good, Creek)
milk-pea

Linnaeus ([1753] 1957) seems to have had considerable problems with the twining legumes. This species he called *Hedysarum volubile.* Patrick Browne studied the plants in Jamaica, and named the genus *Galactia* in 1756. It was not until 1894 that New York botanist Nathaniel Lord Britton (1854–1934) moved the Linnaean species into *Galactia.* Since then the genus has grown to 140 species found mostly in the Americas and Australia (Mabberley 1997).

The Seminoles use the roots of milk-pea in "Baby Sickness," "Cow Creek Sickness," and in childbirth, against appetite loss, fever, headache, and diarrhea (Sturtevant 1955).

Galium

(From Greek *gala*, milk, alluding to the European *G. verum*, whose flowers were used to curdle milk for cheese making or to color cheese; early Christian tradition had it that *Galium* filled the manger in Bethlehem)

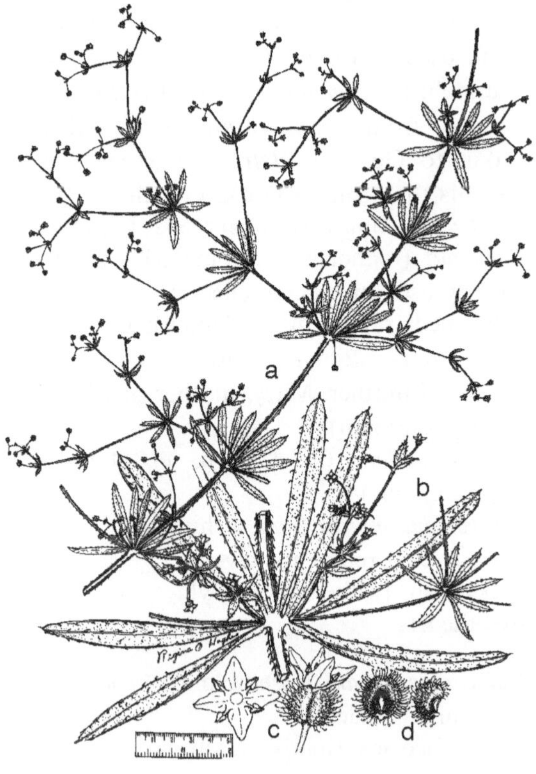

***Galium aparine**.* a. Habit. b. Enlarged node with leaf whorl and lateral inflorescences. c. Flowers, top view (left) and side view (right). d. Fruits. *Drawn by Regina O. Hughes.* From Reed 1971.

Galium aparine (an old generic name, interpreted as meaning to catch, cling, or scratch)
amor del ortelano (garden love, Spanish from 1542)
aparine (clinging, Italian, from 1542)
aspergula [*asperula*] (rough, 1542 pharmaceutical name); *aspérule* (French)
caille-lait [*gaillet*] (curdle milk, Quebec)
catchweed bedstraw ("bedstraw," since ca. 1527 applied to these plants because they resembled straw that was formerly used to stuff mattresses;

originally applied to one species, *Galium verum*, Our Lady's Bedstraw, and later as a generic name)
cheese-rennet (English in 1653)
[spring] cleavers [clyver, clevers] (from Old English *clife*, Old High German *chliba*, from root of *clifian*, to stick, adhere); *cleefcruyt* (cling herb, Dutch, from 1542)
cu-chullain (*cu*, dog, *chullain*, reed, Gaelic)
garbh-lus [*lus garbh*] (*garbh*, rough, *lus*, herb, Gaelic)
goosegrass [goose grasse, goosegrass cleavers] (since at least 1538 applied to this species and others because branches were used to feed geese)
gooshareth (English, from 1542)
grateron [*gratteron*] (scratcher, first used 1314, is still used in the west, influenced by *gratter*, to scratch; from *gleteron*, derived from old French *gleton*, from *cleton*, originally Frankish *kletto*; akin to German *Klette*, the name is used for plants that grab or hook, French)
Klebkraut (cling herb, from 1542, German)
Labkraut (rennet herb, from 1542, German; formerly used in making cheese)
lady's bedstraw; *bedstro* (Dutch); *Liebfrauenstroh* (*Liebfrau* refers to "*unsere liebe Frau*," our beloved mother, Mary, mother of Christ, *Stroh*, straw, German)
luibh-an-ladhair (*luibh*, herb, *an*, of the, *ladhair*, claw, Gaelic)
nan isht hummachi (*nan*, a thing, *isht*, for, *hummachi*, dyeing red, Choctaw)
sticky bobs [willy, stickywilly]
sweethearts

Galium circaezans (resembling *Circaea*, "enchanter's nightshade")
[forest, licorice, woods] bedstraw
cross-clover
wild licorice

Galium tinctorium (used for dyeing)
dye [stiff, marsh] bedstraw

Galium triflorum (three-flowered)
fragrant [sweet-scented] bedstraw
lady's bouquet
wau-inu-makan (woman's medicine, Omaha-Ponca)
wau-pezhe (woman's herb; *wau*, woman, Omaha-Ponca)

Galium uniflorum (single-flowered)
oneflower bedstraw (a book name)

Europeans certainly knew *Galium aparine* when they arrived in the New World as the species is native to all parts of Europe. Many consider it not native to the New World, but in either case its history sets the stage for the others that are native here for certain.

Lady's bedstraw contains coumarin, at least in the dried form, and has been used to treat a variety of maladies, including bladder and kidney inflammations, dropsy, kidney stones, and fevers (Foster and Duke 1990). The juice contains citric acid, which may have antitumor activity. In addition, some think the plant lowers blood pressure. It is known to contain asperuloside, an anti-inflammatory compound (Foster and Duke 1990). Juice is used to prevent scurvy, but may cause contact dermatitis. Culpeper (1653) praised it highly.

Moreover, lady's bedstraw was the source of important dyes (Cannon and Cannon 2003). It was used to color Scottish tartans, and in several countries, to color cheeses. Yellow dyes come from the flower tops, while the roots yield red. The tops with different mordants can yield a range of colors from yellow and green to orange and brown. Colors from roots are coral-reds to salmon-pinks, and they withstand light and washing without fading. Roots contain several anthraquinones including alizarin, purpuroxanthin, rubiadin, purpurin, and lucidin. According to Cannon and Cannon (2003), most species in *Galium* contain these pigments.

Apparently the only records of Muskogean people using *Galium* are the Choctaw who used *G. triflorum* (reported as *G. asperellum*), *G. pilosum* (reported as *G. boreale*) (neither in Florida), and *G. uniflorum*. They took both *G. triflorum* and *G. uniflorum* as an astringent, to treat the flu, and "in all cases requiring diuretic, and diaphoretic" (Campbell 1951). *Galium pilosum* was used to prevent pregnancy.

The Cherokee used *G. circaezans* to treat coughs, hoarseness, asthma, and as an expectorant (Hamel and Chiltoskey 1975). The Ojibwa used *G. tinctorium* for respiratory problems, and the Micmac made a red dye from the roots for porcupine quills (Moerman 1998).

Galium triflorum was the most popular, or at least the most recorded, species among indigenous people. The Cherokee took an infusion for gallstones (Hamel and Chiltoskey 1975). The Iroquois made a poultice to stop backache in babies and used a poultice to relieve swollen testicles or ruptures (Moerman 1998). The Iroquois, Karok, and Quileute used it as a love medicine, which may not be different from the use as perfume among the Makah, Omaha, and Ponca, or in washing hair by the Nitinaht (Gilmore 1919, Moerman 1998).

Gaylussacia

(Named for Louis Joseph Gay-Lussac, 1778–1850, the French chemist who discovered the law of combining volumes of gases)

Gaylussacia dumosa. *Drawn by P.N. Honychurch.*

Gaylussacia dumosa (dwarf)
Buckelberre (knot berry, German)
gopher-berry (New England)
[bushy, dwarf] huckleberry (see *Vaccinium* for derivation of huckleberry)
whortleberry (Williams [1837] 1962)

Gaylussacia frondosa (leafy, from the bracteate racemes)
blue-tangle; dangle-berry [dangleberry, tangle-berry]
highberry
[blue, dwarf] huckleberry
kuwaya [*ku-wa-ya'*] (Cherokee)

Gaylussacia mosieri (named for Charles A. Mosier, 1871–1936, the first superintendent of Royal Palm State Park, Florida's first state park, a friend and collaborator of J.K. Small)
woolly-berry

"You don't know about me without you have read a book by the name of *The Adventures of Tom Sawyer*; but that ain't no matter. That book was made by Mr. Mark Twain, and he told the truth, mainly. There was things which he stretched, but mainly he told the truth." Those words came from one of the most famous literary characters of all times—Huckleberry Finn. I do not recall if Twain ever had Huckleberry and Tom eating huckleberries, but surely he did.

Hedrick (1919) obviously was not familiar with *Gaylussacia frondosa* when he wrote: "The fruit is large, bluish, rather acid and is used in puddings. The fruit is sweet and edible according to Gray." According to Fernald and Kinsey (1943), the fruits will "make one of the most luscious of desserts, being remarkably juicy and with a rich, spicy and sweet flavor." Fernald et al. (1958) also spoke from experience when they wrote of *G. dumosa*, "On account of its habitat, in sphagnum bogs, the Dwarf Huckleberry is not generally known in the North; but its fruit is juicy and deliciously spicy. Southward, where it takes to dry sands, the fruit is readily accessible." Yanovsky (1936) found records of both *G. dumosa* and *G. frondosa* being eaten by tribes in the eastern and southern states.

The Chickasaw are the only Muskogean people found using *Gaylussacia*. Roots from an undetermined species were used to treat delirium (Taylor 1940). The Rappahannock used an infusion of dried or fresh roots to treat stomach problems (Moerman 1998). From the location of both tribes, the plants could have been either *G. dumosa* or *G. frondosa*. The more widespread northern *G. baccata* was used as food by the Cherokee, Iroquois, and Ojibwa. Both the Cherokee and Iroquois used that species to relieve dysentery and kidney problems, and to generally promote good health (Hamel and Chiltoskey 1975, Moerman 1998).

There is supposed to be a simple way to distinguish between huckleberries (*Gaylussacia*, berrylike drupe with ten hard nutlets) and blueberries (*Vaccinium*, fruit a many-seeded berry). Most people can distinguish between them, but, once they taste the delicious fruits, their minds wander in ecstasy, and they intermix the names.

Gelsemium

(A.L. de Jussieu, 1748–1803, Latinized the Italian *gelsomino*, jasmine, to create this genus in 1789)

Gelsemium sempervirens (always green, or living forever)
aatyammichi (*aati*, Indian, *yammi-chi*, to get someone intoxicated, Alabama)
Carolina jasmine (jasmine or jessamine, is derived from Persian *yasmin, yasman*; plants of the "real" jasmine are *Jasminum* in the Oleaceae)
cow-itch (may cause contact dermatitis; the name also used for *Campsis*, which see)

Gelsemium sempervirens. From Britton and Brown 1897.

evening trumpet flower
gelsemio (jasmine, Mexico); gelsemium (in use as an English word by 1875)
[false, wild] jasmine [jessamine] (yellow jasmine [jessamine] (used by Williams [1837] 1962); yellow jasmine root
jazmín amarillo [*de Carolina, silvestre*] (yellow [Carolina, wild] jasmine, Mexico)
madreselva (mother of the forest, Mexico)
[wild] woodbine [wood vine] (originally a European term for *Convolvulus* and *Hedera*, dating to about A.D. 875, and alluding to climbing; now more often used for *Parthenocissus*)

William Bartram ([1791] 1958) wrote, "what seems remarkable, the yellow Jessamine, . . . which is killed by a very slight frost in the open air in Pennsylvania, here on the summits of the Cherokee mountains [Oconee Station Mountain, South Carolina] associates with the Canadian vegetables, and appears roving with them in perfect blossom and gaiety." He saw the plants again in several places, but never commented on their toxicity. Perhaps that was because he thought they belonged in the genus *Bignonia.* Catesby (1731–1932) had seen them earlier and called them *Gelseminum, sive Jasminum luteum odoratum Virginianum scandens, semper virens* (*Gelseminum* or yellow, fragrant, climbing jasmine from Virginia, always green). Apparently, Jussieu took his clue from Catesby in naming the genus. What British botanist Catesby was doing using an Italian plant name is not explained. Catesby wrote, "Altho' Mr. Parkinson calls it *semper virens,* I have always found it lost its Leaves in Winter." He referred to John Parkinson's books on plants published in 1629 and 1640, so the climbers had been introduced into Europe long before.

Gelsemium has two species, with *G. sempervirens* ranging from the eastern United States to Guatemala. The other, *G. elegans,* grows in southeastern Asia to western Malesia (Mabberley 1997). The Old World species contains poisonous alkaloids like the American and has been used to commit murder and suicide. Formerly, the plants were included in the Loganiaceae, but now this genus and *Mostuea* are separated into the Gelsemiaceae (Cronquist 1981, Judd et al. 2002).

The Delaware in New England used *Gelsemium* roots as a blood purifier and in a healing salve (Moerman 1998). John Brickell praised the virtues of *Gelsemium* in 1737 and most likely learned of it from local tribes.

Apparently, the knowledge of using these vines for medicine was lost until it was again accidentally rediscovered in 1821. Francis Peyre Porcher (1825–1895) wrote in 1849 about a planter who was accidentally given a medicine from *Gelsemium* instead of the intended herb. Then, a proprietary nostrum called "Electric Febrifuge" began being sold, the primary ingredient of which was *Gelsemium* (Porcher 1863, Millspaugh 1892). This sparked interest in these plants that lasted for years. *Gelsemium* was listed in the U.S. Pharmacopoeia from 1863 to 1926 and in the National Formulary from 1926 to 1955. The plants are considered sedative, antispasmodic, and nervine, and are recommended as a treatment for asthma, cramps, fevers, hysteria, infantile paralysis, migraine, mydriasis, and tetanus (Hocking 1997).

However, extreme caution with these plants is advised. Ingestion of a single flower has caused death (Foster and Duke 1990). There are several alkaloids in the plants (Hocking 1997). The U.S. Dispensatory says those chemicals "in toxic doses [produce] almost complete paralysis" (Coffey 1993).

The most abundant toxins are gelsemine, gelsevirine, gelseminine, gelsemicine, gelsemoidine, gelsenicine, and koumidine (Lampe and McCann 1985, Nellis 1997). These chemicals act much like strychnine in blocking muscle activity and can mimic symptoms of tetanus. Even honey made from the nectar is suspected of causing human deaths (Nellis 1997). On the positive side, gelsemine is analgesic, slows the heartbeat, and reduces blood pressure. The *British Pharmaceutical Codex* suggests use of the powdered roots to treat trigeminal neuralgia and migraine headaches. However, dosage is critical.

Genipa

(From Tupí *ñandi'pab,* dark color on the chest, now rendered *jenipapo* in Brazilian Portuguese; formerly *Casasia,* named for Luis de las Casas, Captain General of Cuba)

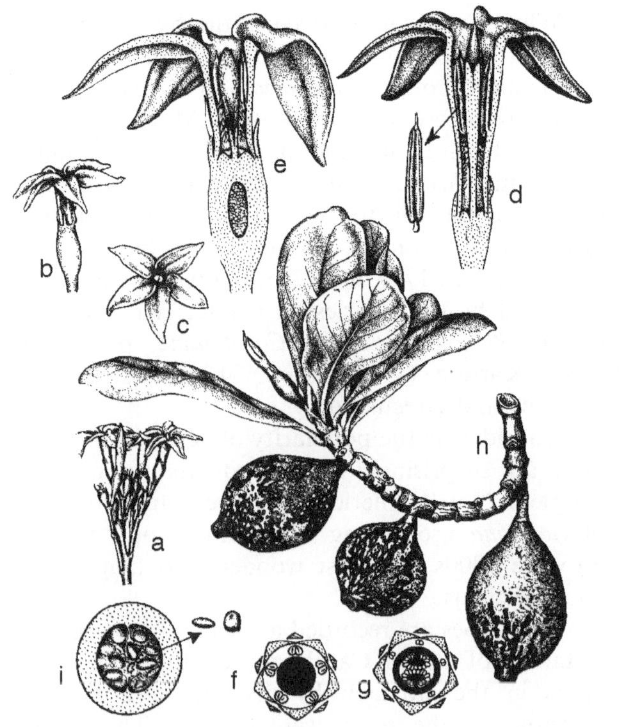

Genipa clusiifolia. a. Cluster of staminate flowers. b. Flower, side view. c. Flower, from above. d. Staminate flower, longitudinally dissected, with insert of stamen (arrow). e. Pistillate flower, longitudinally dissected. f. Floral diagram of staminate flower. g. Floral diagram of pistillate flower. h. Branch with fruits. i. Fruit in transverse section, with seeds as insert (arrow). *Drawn by Priscilla Fawcett.* Correll and Correll 1982.

Genipa clusiifolia (leaves resembling those of *Clusia*) (= *Casasia clusiifolia*)
jagua de costa (coastal *jagua*, Taino for Rubiaceae); *jagüilla* (little *jagua*, Cuba)
seven-year apple (Florida, Bahamas)

The best-known species in this genus is the one Linnaeus described in *Systema Naturae* in 1759, *Genipa americana.* However, Jacquin named this Caribbean species in 1796, and it was moved to *Genipa* in 1861. There are only seven species in the genus, all of them tropical American (Mabberley 1997).

Roig (1945) said *G. clusiifolia* has the same properties as *G. americana.* The most famous use of *jenipapo* is as a body dye among the indigenous people of South America. The iridiod genipin turns black when it combines with proteins in the skin, and it is widely used as body coloring. When applied to garments, it produces a blue dye. Fruit pulp also is used to treat dysentery and syphilis, but the seeds are emetic (Mors et al. 2000). Contents of the ripe fruit of *G. clusiifolia* are edible (Morton 1968b), but not good. I consider them disgusting.

Gentiana

(Pliny, A.D. 23–79, and Dioscorides, fl. A.D. 40–80, said that Gentius, last king of Illyria, ca. 180–168 B.C., discovered the medicinal virtue of the plants and they named the genus for him)

Gentiana saponaria. From Britton and Brown 1897.

Bitterwurz (bitter herb, German, used by Fuchs 1542)
creamh (nibble, Gaelic)
currac-an easbuig (*currac*, cap, *an*, of, *easbuig*, Bishop, Gaelic)
felworts ("fell" in this sense is from Latin *fell*, gall or bitter; wort, plant, England)
genciana (Spanish, Portuguese); gentian (first recorded in English about A.D. 1000); *Enzian* (German, used by Fuchs 1542); *gentiane* (French); *genziana* (Italian)
lus a' chrùbain (the crouching plant, Gaelic)
søte (sweet, Norwegian)
ton-ga-shin-ga (it gives strength to a child; *on-k'i*, to give, *zhin-ga'*, child, Osage)

Gentiana catesbaei (named for its discoverer, Mark Catesby, 1679–1749)
bitter-root [bitterroot]
blue-bells [bluebells] (a name now used for *Campanula*, Campanulaceae, *Mertensia*, Boraginaceae, or *Lupinus*, Fabaceae)
Catesby's snakeroot
[American, blue, southern] gentian
nipi lapushkichi (*nipi*, meat, *lapushkichi*, little soft, Choctaw)

Sampson's snake-root (South Carolina, Porcher 1863; common name also applied to *Gentiana villosa* and *Psoralea psoraloides*)

Gentiana saponaria (from resemblance of foliage to that of *Saponaria*, Caryophyllaceae, usually called "soapwort")
harvest-bells [harvestbells]
makan chahiwi-cho (*makan*, medicine, *chahiwi*, blossom, *cho*, blue, Winnebago)
pezhuta-zi (*pezhuta*, medicine, *zi*, yellow, a reference to the color of the roots, Dakota)
soapwort gentian

Gentiana villosa (long-haired)
blue-bells [bluebells] (a name usually given to *Campanula*, Campanulaceae, *Mertensia*, Boraginaceae, or *Lupinus*, Fabaceae)
Calathian-violet (originally used by John Gerarde in 1597 for another species; a calthus, from Greek *kalathos*, is a vase-shaped basket, an allusion to the corollas)
[blue, marsh, rough, straw-colored, striped] gentian
harvest-bells
Sampson's snakeroot

The writings of Pliny and Dioscorides affirm that people were using a Mediterranean gentian long before the New World was discovered. Indeed, the use of gentians as medicine is recorded on a papyrus found in an Egyptian tomb at Thebes and dated to ca. 1180 B.C. (Dobelis 1986).

The most famous of the gentians was what Linnaeus called *Gentiana lutea* in 1753. His was one of many publications mentioning the herbs as medicine, and it had been discussed by Fuchs (1542), Joachim Camerarius (1534–1598) in 1588, John Gerarde in 1597, Paul de Reneaulme (ca. 1560–1624) in 1611, Gaspar Bauhin in 1623, and others.

Gentiana lutea is a mountain plant of southern Europe, growing in much of the Mediterranean. Linnaeus ([1753] 1957) wrote, "*Habitat in alpibus Norvegicis, Helveticis, Apenninis, Pyrenaeis, Tridentinis*" (Grows in the mountains of Norway, Switzerland, the Apennines [of Italy], the Pyrenees [of France and Spain], and the Tridentine [Alps of northern Italy]). The underground stem of *G. lutea* is used to make gentian bitters and brandy (to aid digestion and stimulate the appetite). It was a nerve tonic used by the ancient Greeks as a poison antidote (Bremness 1994). The plant is also an ingredient in vermouth and Angostura bitters.

When Europeans arrived in the New World, they found a variety of new gentians. Linnaeus ([1753] 1957) named both *G. saponaria* and *G. villosa*, the first discussed before him by Robert Morrison's list of cultivated plants in London, Mark Catesby, and Jan Gronovius. *Gentiana villosa* was known to Linnaeus only from the Clayton collection from Virginia and Gronovius's comments in his *Flora Virginica* published between 1739 and 1743. *Gentiana catesbaei* was described by the Charleston planter, Thomas Walter (1740–1789), in his *Flora Caroliniana,* published in 1788. Kartesz (1994) listed 34 species in North America and Greenland.

Considering the popularity of gentians in Europe, there are surprisingly few records among indigenous people in North America. With the sparsity of records of *Gentiana* use among indigenous people until well into the 1800s, we must wonder who taught whom about their use.

Two tribes are recorded as using *G. catesbaei.* The Catawba of South Carolina used the herb to stop pains in the stomach by drinking a hot or cold infusion, or the roots were chewed. The Catawba also applied it to cure backache. The Choctaw treated snakebite with it (Swanton 1931). Smoky Mountain settlers used it against colic (Krochmal 1968, Vogel 1970).

Although they did not mention American tribes, two early books were surely influenced by local uses. Jacob Bigelow (1786–1879), who published *American Medical Botany* between 1817 and 1821, discussed use of *G. catesbaei.* Another early record was by John Woods, an Englishman who founded a colony of expatriates at Albion, Illinois, in 1821. He wrote that they adopted as medicine "snake-root, gentian, gensing, Columbia-root, sumach, and sassafras tree" (Kindscher 1992). Perhaps part of the lack of documentation is that people were not that careful to distinguish between *Gentiana* species, considering them all useful as "bitters." Although neither Bigelow nor Woods specifically mentioned indigenous uses, it has been assumed that they learned from them. Use was clearly widespread because *G. catesbaei* was official in the U.S. Pharmacopoeia between 1820 and 1882, as a bitter tonic and substitute for *G. lutea* (Vogel 1970). The U.S. Dispensatory in 1955 said the species was "reputed to be but little inferior to the official species [*G. lutea*], the most popular of all bitters in the treatment of atonic dyspepsia, anorexia, and similar complaints" (Coffey 1993).

Porcher (1863) praised "Sampson's snake-root" highly as bitters. He further indicated, "Gentiana purpurea, rubra, and lutea are used in England as substitutes for hops. No doubt our species would serve the same purpose; at any rate, they will give a bitter tonic property when used in the manufacture of ale, beer, etc." Millspaugh (1892) simply said Catesby's snakeroot was a substitute for *G. lutea.* Hocking (1997) wrote that *G. catesbaei* was used like *G. lutea*, as a bitter, particularly by German manufacturers as *Enzianschnapps*, and that it was a source of "official Gentian." Newsom (1989) provided an updated view of the species in medicine.

The Dakota and Winnebago used a simple or compound decoction of *G. saponaria* roots as a tonic (Gilmore 1919). Although no records of indigenous tribes using *G. villosa* have been found, it is famous in the Appalachians. Krochmal and Krochmal (1973) wrote: "Root tea is drunk as tonic, and a piece of the rhizome sometimes is carried or worn in the belief that it will increase one's physical powers. The rhizome has been used to treat indigestion, gout, and rheumatism, and to induce vomiting. It has also been used as a tonic, as a digestive aid, for nervous distress, and gout."

Wunderlin (1998) also lists *G. pennelliana* for Florida. It would bc surprising if people there had not also used it, but no records have been found.

Duke et al. (2002) list many attributes for *G. lutea* that are considered applicable to *Gentiana* generally (Moore 1979). The bitter compounds are typically glycosides (Simonetti 1990), and roots of gentians contain gentisin, gentianin, gentiamarin, and aromatics. Gentiopicrin is the compound contributing to the use as a bitter tonic and, when the roots are chewed, aids in breaking the alcohol habit (Hocking 1997). Regarding malaria, Porcher (1863) wrote: "It is believed by many that the use of bitters in spring and autumn will counteract the action of malaria. They certainly prevent debility, and increase the digestive and nutritive powers, and thus indirectly act as prophylactics, even when they possess no positive virtue as antiperiodic agents." That seems to be one of the few cases where people in the 19th century were conservative about treating that disease, because Moore (1979) noted that gentiopicrin is still used in pharmaceuticals to treat malaria. Gentisic acid is used now principally for rheumatic inflammations (Moore 1979).

Among the most recent chemical analyses of other *Gentiana* are several that report new types of chemicals and reconfirmation of effectiveness. Kakuda et al. (2001, 2002) found several new triterpenoids and secoiridoid glycosides along with those already known. Rojas et al. (1999, 2000) found in *G. spathacea* muscle relaxing (spasmolytic) properties that compared with those of *Datura.* Research on plants used in Tuscany, Italy by Uncini et al. (2000) found extracts from *G. kokiana* were antihypertensive and supported the herb's local use to treat hypertension. There may also be free radical–scavenging action in *G. decumbens*, an herb long used as medicine in Mongolia (Maygmar and Aniya 2000).

Moore (1979) considered a gentian: "The best possible remedy for a person with chronic indigestion and dyspepsia in the evening." He was not alone in that opinion. In 1983, at age 84, E.B. White, the author of *Charlotte's Web*, wrote in praise of gentians (Coffey 1993): "I can still buy Moxie in a tiny supermarket six miles away. Moxie contains gentian root, which is the path to the good life."

Geranium

(An old Greek name, *geranion* is from *geranos*, a crane; the long beak-bearing fruit was thought to resemble the bill of that bird)

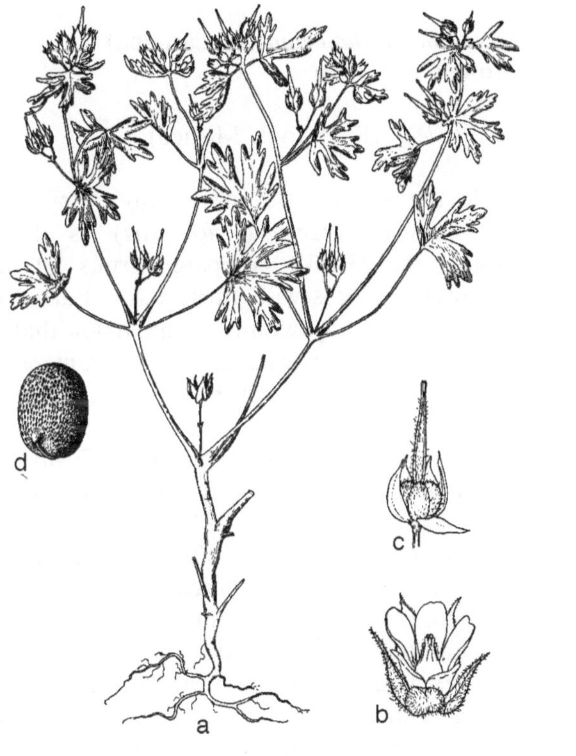

Geranium carolinianum. a. Habit of plant. b. Flower. c. Seed pod enclosed by calyx. d. Seed. From Buchholtz 1968.

anda'nkalagiski (it removes things from the gums, for *G. maculatum*, Cherokee)
be'cigodji'bigûk (one foot, for *G. maculatum*, Ojibwa)
bec a gru (crane's bill, in use by 1549, French); cranes bill [cranesbill] (in use by 1548 when Turner spelled it "Cranes byl")

erba roberta (Robert's herb, Italian); *erva roberta* (Portuguese); herb Robert [herb Roberte] (for *G. robertianum*, named for St. Robert; used in Henry Lyte's 1578 translation of Dodoens's *Cruydeboek*); *herbe à Robert* (in use by 1550, French); *hierba de San Roberto* (Spanish); *robrechtscruyt* (in use by 1549, Dutch); *Ruprechtskraut* (in use by 1542, German)
l'esquinancie (from medical Latin *cynanche*, Greek *kynankhe*, dog-collar, because of the strangling sensation of the ailment, which is called "quinsey" in English; *G. robertianum* was used to treat the problem, Quebec)
geranio (Spanish)
lus an ròis (plant of the rose, a disease; the same as quinsey?; for *G. robertianum*, Gaelic)
ruideal (probably "red-haired," for *G. robertianum*, Gaelic)
stankstorkenebb (stinking stork's bill, Norwegian)

Geranium carolinianum (of Carolina)
Carolina geranium
crane's bill
pata de léon (lion's foot, Mexico)

While the similarity is not always easy to see without knowing cranes (*Grus grus*) as did the Europeans, the Greeks compared fruits of these bright-flowering herbs with those birds. Early botanists were so captivated with the comparison that they continued the analogy with *Erodium* (from *erodios*, heron) and *Pelargonium* (from *pelargos*, stork) for related genera in the Geraniaceae.

Although there are more species of *Geranium* in Europe, Fuchs (1542) listed five (*G. dissectum, G. pratense, G. rotundifolium, G. robertianum, G. sanguineum*). The most famous of these is herb Robert. Culpeper (1653) and other herbalists recommended this plant for a variety of ailments, but mostly to stop bleeding and help wounds heal.

When they arrived in the New World, Europeans found that indigenous people were also using *Geranium* as medicine. Moerman (1998) listed eight species that are known to have been used. In the east, the most famous of these is *G. maculatum.* That species was used by the Cherokee, Choctaw, Iroquois, Menomini, Meskwaki, and Ojibwa. The major uses of this species were to stop bleeding, diarrhea, and dysentery, and to help cure sores. Because it has 10 to 20% tannin in the roots, it is highly astringent and effective for those purposes (Foster and Duke 1990).

Geranium carolinianum is a miniature version of *G. maculatum.* This diminutive herb grows from Massachusetts, Connecticut, and New York west to Michigan and southern British Columbia, and in the south from Florida to California (Fernald 1950, Steyermark 1963, Correll and Johnston 1970, Barkley 1986, Hickman 1993, Diggs et al. 1999). The herbs also continue south to San Luis Potosí in Mexico.

Hartwell (1967–1971) recorded that *G. carolinianum* had been used against cancer. In Japan, where the herbs are exotic, they are considered antidiarrheic and stomachic (Duke 2003). Hocking (1997) unfortunately neither lists his source nor tells who he found using the herbs. However, he records that the plant has been used to treat fever, as a diuretic, as a purge for children (not adults), and in a decoction for rash in children. Leaves and roots are crushed and boiled to make a syrup to treat dysentery.

Gleditsia

(Linnaeus named this tree for his contemporary German Johann Gottlieb Gleditsch, 1714–1786, director of the Berlin Botanic Garden)

Gleditsia triacanthos. From Sargent 1905.

Gleditsia triacanthos (three-thorned)
acacia (used by Plukenet in 1700); *acacia de tres espinas* (three-spined acacia, Spanish); *acacia negra* (black acacia, Argentina); *acacia tres puas* (three-spined acacia, Spain); three-thorned acacia
algarrobo de miel (honey mesquite, a name also given to *Prosopis*)
black locust (confusion with *Robinia*)
cawwimeysa (Shawnee, fide Edgar 1891); *pitelawe minschi* (*minschi*, plant, Delaware)
Christusdoorn [*christus-dorn, korstorne*] (Christ's thorn, Dutch); *Christusdorn* (Christ's thorn, German); *valsa Christusdoorn* (false Christ-thorn, Dutch)
Confederate pintree (USA)
Dornbaum (thorn tree, German); *Schotendorn* (pod thorn, Germany)

espinheiro da Virginia (Virginia spine tree, Portuguese)
févier à trois épines (three-spined *Gleditsia*, France)
honey [shuck, honeyshuck, shucks honey locust] ("honey" is a Teutonic word derived from Old English and Old Frisian *hunig*, dating from ca. A.D. 825; said to have been originally applied to the product of honeybee, *Apis mellifera,* and surely first used for other sweet products of nature as those animals were brought into northern climes from the Mediterranean)
[common] honey locust (in use by 1812)
kati (Choctaw); *kati'* (Chickasaw); *kvtókwv* (Muskogee); *tato* (Alabama); *tató* (Koasati)
kulse'tsi [*kulsétsi*] (Cherokee; *Kulsetsi'yi,* abbreviated *Kulse'tsi,* "Honey-locust place," was the name of several former settlements in the old Cherokee country. One was on Keowee River, near the present Fall Creek, in Oconee County, South Carolina; another was on Sugartown or Cullasagee or *Kulse'si* Creek, near the present Franklin, in Macon County, North Carolina; a third was on Sugartown Creek, near the present Morgantown, in Fannin County, Georgia. The local name was rendered Sugartown by the traders.)
locust (from Latin *locusta,* through Old French *locuste*; originally applied to the insects in the Acridiidae; in English about A.D. 1200; the same Latin word was also applied to legume fruits, first the carob, *Ceratonia siliqua,* in the Levant, and then to other species)
nit (Atakapa)
piquant amourette (prickly love affair, French)
spino di Giuda (Judas's spine, Italian)
squeak-bean; sweet-bean
sweet locust (in use by Porcher 1863)
thorn(y) [locust, tree, acacia] (USA)
ya'pat e' (flat [pod] tree, Catawba)
zinhawarãtschi (Onondaga)

Gleditsia is a legume genus of 12 species, with 2 or 3 in eastern North America, 1 in South America, 1 in the Caspian, and the rest from India to Japan and New Guinea (Mabberley 1997). The common names are often used interchangeably, but the "preferred" nomenclature is to call this genus "honey locust" to distinguish it from "black locust" (*Robinia*).

Trees were imported to Europe by the latter 1600s. Leonard Plukenet included them in his *Almagesti botanici mantissa* of 1700. He thought they were like those known from the Mediterranean and called them *Acacia americana, abruae folio, triacanthos* (American acacia with delicate leaves, three-spined). Mark Catesby reported them next and used Plukenet's name. However, the latest interpretations of his plate (#43) have identified his drawing as *G. aquatica* (Howard and Staples 1983, Wilbur 1990). That species was not formally named until Humphrey Marshall (1722–1801) described it in 1785.

By 1753, Linnaeus had studied living *G. triacanthos* in both the *Hortus Cliffortianus* published in 1738 and the *Hortus Uppsaliensis* of 1748. Between those books he decided to name it after Johann Gleditsch and coined *Gleditsia* in 1748.

The sapwood of honey locust is yellowish, while the heartwood is light red to reddish brown; it has no characteristic odor or taste. The wood is heavy, hard, tough, strong, and has a high luster (Sargent 1905, Harrar and Harrar 1946, Alden 1994a). Posts buried in the ground practically never rot (Coon 1974). The wood has been compared with black locust (*Robinia pseudoacacia*) in terms of strength, second only to that legume. The Cherokee used the wood, at least in later years for fence posts (Hamel and Chiltoskey 1975). Because they used posts in some of their buildings earlier, they were likely used then also. The Cherokee also made bows from the resilient wood (Steyermark 1963).

Porcher (1863) wrote: "The pores of the wood are very open. When perfectly seasoned, the wood is extremely hard. It is far inferior to the black walnut or wild cherry for cabinet-making. Hedges of it are rendered impenetrable by its long thorns." The wood has also been used to make posts, cross ties, wheels, and other items (Sargent 1905, Harrar and Harrar 1946). Hocking (1997) found that the bark had been used in tanning leather; Duke (2003) recorded ample tannin present.

Steyermark (1963) recorded uses that are probably typical of those elsewhere in its range. In Missouri, he found country people tying the large thorns to poles for use as "gigs" for bullfrogs. Other people used the thorns as large needles or pins for closing woolen sacks, and for carding wool. These uses are reflected in the name "Confederate pintree."

William Bartram gave the first record of Creek and Cherokee uses of honey locust in response to Benjamin Barton's questions in 1789. He wrote, "They have in use a vast variety of native or wild vegetables, both fruits and roots" (Bartram in Swanton 1946). Primary among those listed were persimmon, mulberries, and honey locust. Because of the comments by Mooney that the Cherokee "extracted their only saccharine from the pod of the honey locust, using the powdered pods to sweeten parched corn and to make a sweet drink" (Swanton 1946), there is corroboration for Bartram's comments. If either tribe used the roots of honey locust as Bartram implied, no other record has been found. However, the Catawba did eat the pods and mix them with bread and drink (Speck 1934).

Hocking (1997) wrote that the pods were used for sweetener, and mixed with sweet acorns to make edible cakes. He noted that legumes were used to make beer by slaves, but gave no source. Perhaps he was referring to Porcher (1863), who recorded, "Beer is sometimes made by fermenting the sweet pods while fresh." Fernald et al. (1958) found the pulp, "a well known pleasant nibble." Steyermark (1963) wrote from personal experience that the "thin jellylike pulp surrounding the seeds is sweet and provides a somewhat pleasant diversion to taste when one is tramping outdoors."

The people of the eastern United States also employed these trees for medicine. Among the tribes who use them are the Cherokee, Creek, Delaware, Meskwaki, and Rappahannock (Moerman 1998). The Cherokee used fruits to sweeten worm medicine and to treat dyspepsia, measles, and whooping cough (Hamel and Chiltoskey 1975). The Creeks made medicine to treat smallpox, and for children's "complaints" (diarrhea?) (Swanton 1928, Taylor 1940). The Delaware mixed the bark with that of prickly ash (*Zanthoxylum americanum*), wild cherry (*Prunus virginiana*), and sassafras (*Sassafras albidum*) to make a tonic to "purify the blood" and to relieve coughs. The Meskwaki made an infusion of the twigs to treat colds. They used the bark to treat fevers, measles, smallpox, and to help an ill person regain strength. The Rappahannock made an infusion of bark and roots to help recover from colds and coughs.

The species is oddly absent from both Millspaugh (1892) and Vogel (1970). Foster and Duke (1990) found that the pods were formerly made into tea to treat indigestion, measles, and catarrh of the lungs. An inner bark tea with sycamore bark (see *Platanus*) was once used to relieve hoarseness and sore throat. Juice of the legumes is antiseptic, and compounds from the leaves retard certain types of cancer.

Several chemicals have been isolated from *Gleditsia.* Duke (2003) found that the fruits contain (+)-catechin, (−)-epicatechin-3-glucoside, carbohydrates, fats, fiber, glucose, leucocyanidin, and protein. Seeds are known to have 7,3′,4′,5′-tetrahydroxyisoflavanol, acetic acid, calcium, carbohydrates, dihydroxystearic acid, fat, fiber, galactose, mannogalactan, mannose, phlobaphene, phosphorus, protein, sativic acid, tetrahydroxystearic acid, and tocopherols. There are a number of other compounds known in other plant parts.

Foster and Duke (1990) also warn that "All plant parts . . . contain potentially toxic compounds." However, Duke et al. (2002) record that the plants have analgesic, antidote, antitussive, cytotoxic, depurative, mydriatic, narcotic, oxytocic, poisonous, and toxic qualities, and are used as a vasodilator. They give the herbal remedy from *Gleditsia* two plus marks (as safe as coffee).

Gnaphalium

(Linnaeus based this on Greek *glaphallion, knaphallon,* soft down, woolly, referring to the pubescence of the plants)

Gnaphalium obtusifolium. a. Upper part of a flowering plant. b. Seed. From Buchholtz 1968.

cat-luibh [*luibh-a-chait*] (cat-herb, Gaelic)
cud-weed (English)
gnaphale (French)
impia (Italian)
Ruhrkraut (dysentery herb, German)

Gnaphalium obtusifolium (uneven leaves) (= *Pseudognaphalium obtusifolium*)
ahalla'ta' (Chickasaw)
ahá lvbvkca [*ahá labakca, aha-lopv'kcv, ahalobaktsi, halobaktei*] (*vhv,* root, *lvpvtke,* wild, Creek, Muskogee)

ahisi laksa (*ahisi*, tobacco, *laksa*, bitter, Alabama)
bashûchak (Choctaw)
cat-foot (USA)
chafe-weed (USA)
cudweed (chewed as cud, like tobacco, USA); fragrant cudweed [everlasting] (Texas)
[life, sweet] everlasting (for long-lasting dried flowers, USA); life alasses (Gullah, South Carolina)
Indian tobacco (USA); rabbit tobacco (Florida)
iskafotó stokhátka (*iskafotó*, mountain mint, *stokhátka*, gray, Koasati)
kasd'uta (simulating ashes, Cherokee)
napopokpoke (*nam pakanli*, a flower, Choctaw)
old field balsam (USA)

Gnaphalium purpureum (purple) (= *Gamochaeta purpurea*)
patte de chat (cat foot, Houma, Louisiana)

Linnaeus created the genus *Gnaphalium* in 1753, and included 36 species. Both *G. obtusifolium* and *G. purpureum* were on the list. He had seen neither of those alive, but knew them from herbarium specimens and literature, beginning with Robert Morrison's (1620–1683) account of both in 1680. However, Linnaeus did know living plants of the single European *G. dioicum*, and had discussed it in his *Materia Medica* of 1749. He had seen others cultivated at both the *Hortus Cliffortianus* and *Hortus Upsaliensis*. Mabberley (1997) included about 50 species in *Gnaphalium*, and he called the genus "cosmopolitan."

Virtually all who live near *Gnaphalium* have used it. The most pervasive common name in English is "rabbit tobacco," which indirectly records the indigenous practice of mixing the leaves with tobacco (*Nicotiana* spp.) for smoking. That combination was typically to avert sickness or witchcraft (Swanton 1928a, Taylor 1940, Howard 1984, King 1984). An old use still practiced is in treating colds and lung pains (Bushnell 1909, Swanton 1928a, Bennett 1997). The plants are either used alone or mixed with others to treat mumps, upset stomach, nervousness, and insomnia (Howard 1984). These two species were used by at least the Alabama, Cherokee, Chickasaw, Choctaw, Creek, Koasati, Menomini, Meskwaki, Montagnais, Potawatomi, and Rappahannock. Murphee (1965) found people in the Panhandle putting dried leaves in pillows to relieve asthma. Other people, and probably these too, used other species. Several other species are still used today (e.g., Laferrière 1994, Yetman and Van Devender 2001).

Goodyera

(Named for British botanist John Goodyer, 1592–1664, of Mapledurham, Hampshire, who is probably most famous for his translation of Dioscorides's *Materia Medica*, which was not discovered and published until 1934, some 250 years after it was written)

creeping lady's tresses (comparison with *Spiranthes*, which see)
goodyère (this is a recent name; the genus was not named until 1810, French)
kêdîgi'bagûk (spotted leaf, Potawatomi)
knerot (knee root, Norwegian)
mogairlean ealaidheach (creeping orchid, Gaelic)
Netzblad (net-leaf, German)

Goodyera pubescens (hairy)
adder's-tongue [cactus, violet] (Maine)
canker-root [cancer-weed] ("canker" is a variant of "cancer," but also perhaps a reference to thrush or *Candida* infection)
lattice-leaf
net-wort [networt]
[spotted, net-leaf]-plantain
ratsbane (North Carolina)
rattlesnake-[leaf, orchid, root, tongue, violet], (New Jersey, New York)
[downy] rattlesnake plantain (in use since 1846; from the indigenous people chewing the leaves and applying them to rattlesnake bites, cf. Core 1967; but a different explanation was offered by Frère Marie-Victorin in his *Flore laurentienne* of 1935; he wrote, "these plants superficially resemble the Plantains; on the other hand, the leaf patterns, which resembles a snake skin, suggested the idea [doctrine of signatures] to the first American settlers to use the Goodyers as antidotes to snake venom")
scrofula weed (scrofula is a tubercular disease)

Many early botanists in Europe commented on their species, *Goodyera repens*. They called these ground orchids by various names. Joachim Camerarius, in his *Hortus medicus* of 1588, and Christian Mentzel, in his *Pinax* of 1682, thought the herbs should be called *Orchis* (testicle). Gaspar Bauhin called it *Pseudoorchis* (false-testicle) in his *Pinax* of 1623. Peter Johannes Loesel, in the *Flora Prussia* of 1703, put it in *Pyrola* (little pear, from the leaf shape, Ericaceae). Linnaeus was consistent in naming it *Satyrium* (the Greek *satyr*) from his *Flora Lapponica* of 1737 to *Species Plantarum* ([1753] 1957). Thomas F.

Dalibard, in his *Flora Parisiensis,* published in 1749, agreed. *Flora Sibirica,* published between 1747 and 1749 and authored by Johann G. Gmelin (1748–1804), called it *Epipactis* (*epi*, upon, *paktos*, hellbore or *Helleborus*). Finally, in 1810, British botanist Robert Brown (1773–1858) created the genus *Goodyera* for these circumpolar orchids.

At first, it seems odd that there are only French names for these orchids in southern Europe. However, a comparison of its range reveals that they are absent from most of southern Europe, with the exception of the Pyrenees Mountains in France (Polunin 1969).

The Cherokee used *G. pubescens* as a blood tonic, for burns and colds, as an anemetic, to improve the appetite, and to treat sore eyes, kidney problems, and toothache (Hamel and Chiltoskey 1975). The Delaware and Mohegan used it for pleurisy, rheumatism, sores, and thrush (*Candida*), and to treat women after childbirth (Moerman 1998).

Frederick Pursh said in 1814 that the orchids had a reputation as an infallible cure for hydrophobia. Near the same time in New England, they were regarded as a remedy for skin diseases (Coffey 1993). Thoreau did not care much about the orchids' reputation as a medicine. He wrote in 1851: "[I]t may not be easy to convince a quibbler and proser of its [common name, rattlesnake-plantain] fitness. We want some name to express the mystic wildness of its rich leaves. Such work as men imitate in their embroidery, unaccountably agreeable to the eye, as if it answered its end only when it met the eye of man; a reticulated leaf, visible only on one side; little things which make one pause in the wood, take captive the eye" (Coffey 1993).

Porcher (1863) does not mention these plants, and Millspaugh (1892) simply says, "a reputed American remedy for scrofula." Vogel (1970) makes no mention of these orchids, nor do Duke et al. (2002). Perhaps the rarity of these orchids in Europe and the Americas led to them being overlooked by many. They were not mentioned by Europeans Fuchs (1542) or Culpeper (1653) either.

Beginning in 1918, Meyer and Meyer (1986) remembered: "The fresh leaves are steeped in milk and applied as a poultice to bruises, bites of insects, skin irritations. The leaves also may be used whole if preferred." Foster and Duke (1990) added: "Physicians once used fresh leaves steeped in milk as poultice for tuberculosis swelling of lymph nodes (scrofula). Fresh leaves were applied every 3 hours, while the patient drank a tea of the leaves at the same time." Then they quietly explain most of the absence from the herbal literature with "Of historical interest only. Too scarce to harvest."

Gordonia

(Named by John Ellis for the English nurseryman James Gordon, 1708–1781, gardener and correspondent of Linnaeus)

Gordonia lasianthus. From Sargent 1905.

Gordonia lasianthus (*lasi*, shaggy, hairy, *anthos*, flower)
black laurel
ittomico (*eto*, tree, *mekko*, chief, Creek; given by Simmons [1822] 1973; the name was used earlier by William Bartram for *Persea borbonia*, which see)
loblolly bay (wetland bay; see Austin 1991 for etymology of "loblolly")
red bay (Bartram [1791] 1958; a name usually given to *Persea*, which see)
to:li (Mikasuki)

Before I noticed that Linnaeus called loblolly bay *Hypericum lasianthus* in 1753, it never occurred to me that the two families were that similar. Both have many stamens, three to five carpels, and both may be woody. However, *Gordonia* lacks the pellucid dots that are so characteristic of *Hypericum.* Regardless, in 1771, John Ellis, an Irish-born merchant in London, (1710[1711?]–1776) put Linnaeus's species in *Gordonia*, a genus he created for that species. Today there are 70 species in *Gordonia*, with most native to southeastern Asia (Mabberley 1997). *Gordonia lasianthus* is the only species in North America (Kartesz 1994).

The Seminoles formerly distinguished between what we call *Persea borbonia* and *Gordonia lasianthus*, although now either is likely to be called *to:li* (Mikasuki) or *to:la* (Creek) (see *Persea*: Red bay). Bartram ([1791] 1958) called *G. lasianthus* both red bay and loblolly bay. Perhaps the two have been used similarly or even interchangeably. At least the Koasati used *Gordonia* medicinally (Taylor 1940).

Gossypium: Cotton

(From the Latin *gosspion,* used for the cotton plant by Pliny, A.D. 23–79)

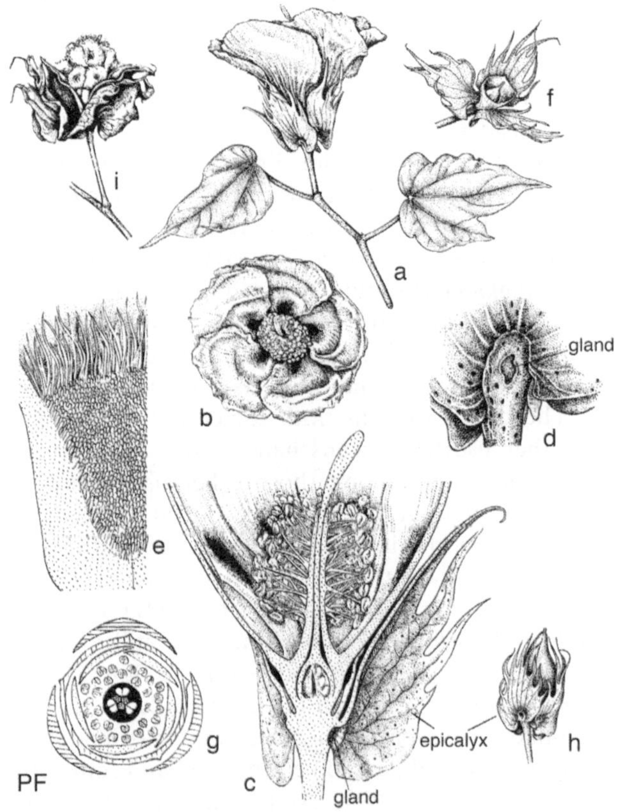

Gossypium hirsutum. a. Branch with flower. b. Flower, from above. c. Base of flower, longitudinally dissected. d. Detail of gland at base of calyx lobe. e. Calyx, from within. f. Calyx with young fruit. g. Floral diagram. h. Undehisced fruit. i. Dehisced fruit with cottony seeds. *Drawn by Priscilla Fawcett.* From Correll and Correll 1982.

Europeans were so accustomed to cotton that, when they found it in the Caribbean, they were none too impressed (Sauer 1969). Everywhere the earliest Spanish went in the Caribbean, they found two kinds of cotton (*Gossypium barbadense* and *G. hirsutum*), both different from those they knew at home (*G. herbaceum*).

Cotton was a comparatively new crop in Europe, having arrived in the Near and Middle East during Hellenistic and Roman times (Zohary and Hopf 1993). Indeed, the word "cotton" in English and the Romance languages was derived from Arabic (*qutun, qatn*) because traders brought the word and plant to Europe via Spain. That origin, instead of from *gossypion,* is one of the few instances in the Romance languages where the Latin word was superseded by another. Italian, usually closest to the Latin base word, uses *cotone.* Similarly, in French it is *coton,* and the Portuguese say *cotão.* The Spanish, misunderstanding the articles of Arabic, call it *algodón,* derived from *al-qutun* [*al-qoton*] (the cotton). The English began using words "*cotoun*" and "*coton*" about A.D. 1300, and even in Gaelic it is *cotan.* Germans say *Baumwolle* (tree wool), and the creole Dutch say *katuna* or *katoen* (Dutch Antilles). However, the people in the Dutch Antilles also call it *watten-struik* (wadding bush).

The Spanish were astonished at the quality of cloth produced from Cuba and Hispaniola to Mexico, and later from what became Arizona in the southwestern United States. Each of the peoples they met seemed to outdo the others in the quality of the cotton fabrics they created. One of the reasons that indigenous Americans produced such outstanding cloth, beyond their skill as weavers, was the lint or fibers from the cotton. American plants produced longer and higher-quality lint than Old World species. Now, in one form or another, American species constitute ~95% of the world production (Simpson and Ogorzaly 1995).

Florida's cotton is *G. hirsutum,* and it is usually just called wild cotton (Florida, Puerto Rico) or native cotton (Puerto Rico). The remarkable thing about that species in Florida is that it extends from there through Central and part of South America and up Mexico to the Pima of Sonora and modern Arizona. It is not an accident that modern production is dominated by "Pima cotton," although that name applies to three different entities (Rea 1997). What is grown now (*G. barbadense*) is not the same that was originally grown by the Pima (*G. hirsutum*), but the area still produces quality cotton.

Although recent genetic studies show that they are not viable taxa (Fryxell 1988, Wendel and Albert 1992, Wendel et al. 1992), it is still convenient to recognize three morphological groups within *G. hirsutum.* The Caribbean tree cotton (var. *marie-galante*) was the West Indian type, growing through the Greater Antilles, from Panama to Trinidad and in South America to northern Brazil. Variety *latifolium* was endemic to southern Mexico and Guatemala. This variant is known from about 3500 B.C. in Mexican archaeological sites (Smith 1968). Variety *punctatum* grew from the Bahamas and Florida around the coasts of the Gulf of Mexico to Yucatán. The modern forms called "Upland" cotton were derived from green-fuzzed seed types and named after their introduction into the southeastern United States in the 1700s to distinguish them from the black-seeded Sea Island cotton (*G. barbadense*). It was only in the 1800s that the gray- and white-fuzzed seeds were introduced to become the modern "Upland" cottons (Purseglove 1968).

Mostly, *G. hirsutum* is called tree cotton, as in *algodón arbórea* (Puerto Rico), *algodonero* (cotton tree), cotton bush (Virgin Islands), cotton tree (Belize), *cotonnier* (cotton tree, Guadeloupe, Martinique), *palo de algodón* (tree cotton, Morton), and tree cotton (Puerto Rico). There is such variation with *G. hirsutum* that people named the types in different ways. Some distinguish the wild from the cultivated with *algodón criollo* (creole cotton, Cuba), *algodón del país* (country cotton, Puerto Rico), *algodón de altura* (upland cotton, Venezuela), *algodón silvestre* (wild cotton, Puerto Rico), *coton pays* (wild [country] cotton, Guadeloupe, Martinique), *coton sauvage* (wild cotton, Haiti), and wild cotton (Belize). Others separate them by the color of the lint. While people refer to the lint as violet or red, the actual colors are not the tones these terms typically indicate. Colors in cotton range from white through various shades of tan to brown to a reddish-brown or chestnut. Still, people call them *algodón morado* (purple cotton, Dominican Republic), *coton blanc* (white cotton, Dominica), *koton wuge* [*coton rouge*] (red cotton, Trinidad), *coton violet* (purple cotton, Haiti), and red cotton (Trinidad). That is also done in indigenous languages, as in Huastec where they recognize *thak kwinim* [*cuinim*] (white cotton, San Luis Potosí) and *tsokoy* (brown cotton, San Luis Potosí).

Cotton has been important to indigenous people for so long that many groups have simple terms for the plants. In the West Indies, the Arawak called it *ikálotopue*, and the Caribs say *maulu* [*mauru*] (Dominica, Suriname). The Carib word is considered cognate with the South American Tupí name *amandiyu-b* (Hodge and Taylor 1957). Mainland names include *ampi* (Morton 1981), *dehti* (Otomí, Veracruz), *maʔhua* (Tarahumara, Chihuahua), *mojofono* (Choco, Panama), *odigé* (Hispaniola), *pishm* [*pishten*] (Mixe, Mexico; *pishten-kiup*, the plant; *pishten-puih*, the flower), *tsiin* (Morton 1981), *tüdy* (Otomí, Veracruz), *uchto* (Morton 1981), *upsana* (Cuna, Panama), and *utju* (Morton 1981). However, a few languages have more interesting variants on their names. The Seri say *mooj* (cotton's fruit, Sonora). The Maya say either *xchup* (stuff, as in pillows, Yucatán) or *taman* (Yucatán). After the Spanish arrived, the Maya began calling sheep and lambs *taman*, the name they had long used for cotton.

Tribes in the United States were also familiar with cotton. They called it *naafalli'* (Chickasaw), *o'ftati* (Ofo), *pakaalihatka* (*pakaali*, cotton, *hatka*, white, Alabama), *paykahatká* (*payka*, cotton string, *hatká*, white, Koasati), *pakpvke-hvtke* (*pakpvke*, flower, *hvtke*, white, Creek), *pokpo* (Choctaw), *ptçato* (Biloxi), *ro'ptini* [*ro'puhtini*] (Tunica), *ta-çka' hin* (*ta*, his, *çka*, white, *hin*, hair, Osage), *toki* (Pima), and *ujilvʔi* (Cherokee). The historical record suggests that the plants were more important in the southwest than in the southeast. Indeed, archaeological data suggest that the genus was used comparatively late by southeastern tribes. No remains of cotton were found in early archaeological sites (Gilmore 1931, Whitford 1941). While Webb and Funkhouser (1936) did find cotton, they considered it intrusive.

The Aztecs were more direct with their naming. They said *ichcatl* (cotton) or *ichcaxihuitl* (thread herb, Náhuatl). Indeed, the fibers were treated by Hernández in 1651 as *ychcaxihuitl seu gossypio* (thread herb or *Gossypium*). It should be noted that the word *ichtli*, which is the root of *ichcatl*, was derived from the Náhuatl word for fibers from *Agave*, their most important religious and utilitarian plant (see *Agave*).

The use as fiber in cloth sometimes overshadows the other uses Americans had for the cotton plants. Most important among those other uses was food. People throughout the Americas cooked the seeds and ate them. People in the Caribbean boiled the seeds before eating them (Duke 1968). The Pima roasted the seeds and ate them like peanuts (Rea 1997). In the late 1800s and early 1900s, the Pima were still pounding the seeds with mesquite pods to make flour (Russell 1908). Castetter and Bell (1942) recorded a "sort of cake-like tortilla baked in ashes" that was made from finely ground and parched cotton seed and some other meal source. In 1899, American promoter David Wesson learned that the meal and oil from cotton seed was edible (when properly processed) and began selling it (as Wesson Oil) instead of throwing the "waste" from gins away (Davidson 1999). Wesson thus started the modern vegetable oil industry.

There are also numerous medical applications for cotton. The roots were formerly in the U.S. Pharmacopoeia as an emmenagogue, antihemorrhagic, and abortifacient. Gossypol is among the many compounds in the roots and seeds (Morton 1981, Hsu 1986). Although that chemical is mildly poisonous, and bitter, some consider the raw seeds aphrodisiac, expectorant, laxative, and nervine. Seeds are also prescribed to relieve headache (Duke 1968). Flowers, seeds, and leaves are used in Haiti whenever an emollient is needed, and the seeds are a remedy for dysentery (Beauvoir et al. 2001). A leaf decoction is given for flu, fever, cough, consumption, and chest colds. Flowers are used in tea for flu and colds (Bahamas). Leaves and flowers contain glycosides, quercimeritrin, isoquercitrin, and salicylic acid that have bacteriostatic, analgesic, and antipyretic actions (Ayensu 1981). So multiple are the uses of different parts of the cotton plant that we might almost conclude that the plant will cure virtually any malady. Various preparations have been used to treat abdominal complaints, bat bite, constriction of the ovaries, dermatoses, earache, eye problems, fever, flatulence, gastralgia, hemorrhoids, inflammations of the ganglia, parturition, rheumatism, sores, tachycardia, and urinary problems; it is also a "cacatrizant."

Cotton is now considered an endangered species in Florida (Coile 2000). Among the causes of endangerment are habitat loss, and short-sighted, misdirected efforts by the U.S. Department of Agriculture (USDA).

When I first arrived in Florida in 1970, there were still people around who personally knew the "cotton-pickers" the USDA had sent into the Florida Keys and southern coastal regions to "eliminate" cotton. Apparently, the USDA considered the wild cotton the "intermediate" source for the boll weevil that had for so long plagued the crop (Burke et al. 1986). What the USDA had not considered, among other things, was the distance between the wild cotton and that cultivated farther north. Elimination of one single cotton host so far from the centers of cultivation would have had little impact on the weevil problems for growers.

However, the cotton-pickers themselves were smarter than the bureaucrats in Washington, D.C. who hired them. Realizing that total elimination of the species from the state would also mean termination of their jobs, they always left some plants behind. They were even accused of destroying individual plants, and then scattering seeds from them as they left the site. Now, that is real planning for the future.

Gouania: Chew-Stick

(In honor of French botanist Antoine Gouan, 1733–1821, professor of botany at Montpellier)

The first time I ever saw chew-stick (*Gouania lupuloides*) was in the Florida Keys. Back then, the only members of the Rhamnaceae that I knew were the temperate genera *Berchemia* and *Ceanothus.* These scrambling shrubs somewhat resembled *Berchemia,* but its fruits, with their usually three-winged margins, are markedly distinct. Worse, the "watchspring" tendrils baffled me. Eventually, I figured out the identity of the plants and their common name. Then I learned that Linnaeus did not know what to do with the species either, and so I felt a little better.

First, Linnaeus put the plants in *Paullinia* (Sapindaceae). Then 5 years later, in 1753, he called the chew-stick *Banisteria lupuloides.* Along with *B. lupuloides,* he described six others as *Banisteria.* Now we know that only *B. lupuloides* belongs to the Rhamnaceae, and the other six species are in the Malphigiaceae (in genera *Stigmaphyllon, Heteropterys,* and *Hiptage*).

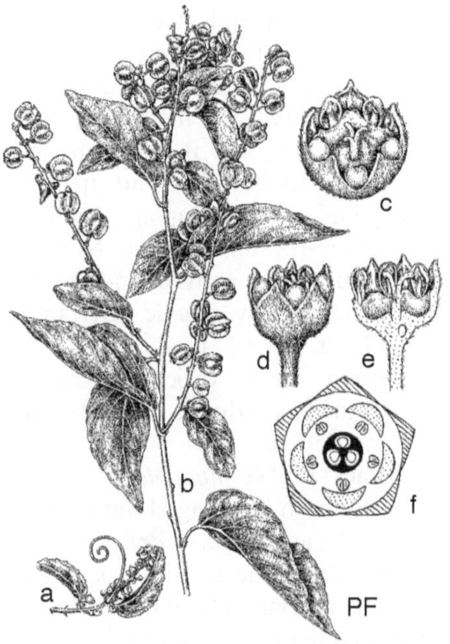

Gouania lupuloides. a. Flowering branch. b. Fruiting end of branch. c. Flower, from above. d. Flower, side view. e. Flower, longitudinally dissected. f. Floral diagram. *Drawn by Priscilla Fawcett.* From Correll and Correll 1982.

Studying independently, Nicolaus Jacquin proposed the genus *Gouania* in 1763. Finally in 1910, Linnaeus's species *lupuloides* (resembling the hop plant) was transferred to its proper place by Ignatz Urban, the tireless German student of West Indian botany.

Gouania lupuloides is a widespread member of the family, known from southern Florida, the Bahamas, the West Indies south to Grenada and from Mexico (Tamaulipas, Veracruz, Tabasco, Yucatán, Quintana Roo, Chiapas, Sinaloa, Colima, Nayarit, Guerrero) to Belize, El Salvador, Guatemala, Honduras, Nicaragua, Costa Rica, Panama, Colombia, Brazil, Ecuador, Peru, and Bolivia. Our species is one of the 50 to 70 in a genus spread around the world in the tropics (Mabberley 1997). Most of the species are loaded with saponins and some are used as shampoos.

The species in Florida is best known as a toothbrush, as implied by the names chew-stick, chaw-stick, cho-bush, chow-bush, chew-wis (Florida, Puerto Rico, Jamaica, Barbados, Trinidad, Belize), and toothbrush tree (Trinidad). That the usage is old is suggested by its recording as chaw-stick in Florida in the 1800s (Chapman 1897). Surely, by that date, the application, if not the name, had been in place for several hundred years. Other languages say the same with *brosse à dents pays* (wild toothbrush, Haiti) and *limpia-dientes* (tooth-cleaner, Belize).

Although younger people may not be familiar with chew-sticks, they were the norm for the world before the Chinese toothbrush was introduced to Westerners after 1498 (Panati 1987). Chew-sticks to clean the teeth were recorded by the Babylonians in 5000 B.C. (Lewis and Elvin-Lewis 1977), and are still used by many poor people in developing countries. Chew-sticks were apparently unknown in North American groups before Europeans arrived, and some think that they were introduced by Creole Africans from the Caribbean. Plants to be used as chew-sticks are carefully selected for a variety of characteristics, including foaminess, hardness, and bitterness (Lewis and Elvin-Lewis 1977).

Other names for *G. lupuloides* reveal different views. Some allude to learning about its use from the indigenous people, as in *bejuco [de] Indio* (Indian vine, Dominican Republic, Puerto Rico), *bejuco de la India* (Indian [woman's] vine, Dominican Republic), and *botán de indio* (Indian bud, Belize).

Due to the saponins, the plant has a soapy taste. Names reflecting that quality include *jaboncillo* (little soapy one, Colombia), *jaboncillo bejuco* (little soapy vine, Cuba), *liane-savon* (soap vine, Dominican Republic, Guadeloupe, Martinique), and soap-stick or soap vine (Puerto Rico). Presumably, *G. lupuloides* has been used for washing other items as the name soap-stick implies, and as other species have. However, Cubans sometimes just think it tastes bad, as in *bejuco amargo* (bitter vine).

Because of clean teeth, the breath is refreshed, and people note that with *bejuco de sopla* (breath vine, Puerto Rico), *mascapelo* (mask it, Puerto Rico), *yerba buena* (good herb, Dominican Republic), and *bon bois* (good tree, French Antilles). At least in some cases, good breath improves interpersonal relationships, because it is also called *bejuco cundeamor* (count-of-love vine, Dominican Republic; name given to several vines, including *Momordica charantia* and *Ipomoea quamoclit*). Presumably, enhanced interaction also applies to *bejuco cuaresma* (abstinence vine, Dominican Republic), although the allusion is not clear. *Cuaresma* for Catholics is between Ash Wednesday and Easter, so maybe they "give it up" for Lent.

Some names simply refer to the appearance of the plant. Both *rabo de mono* (monkey tail, Nicaragua, Colombia) and white root (St. Thomas, Florida) are straightforward. On the other hand, what the plant has to do with cloth is not obvious in *bejuco leñatero* (linen vine, Cuba). The Maya think that the leaves resemble those of *Persea*, as in *somoak* [*om-ak, xomoak, xomak, xmak*] (*om*, avocado, *P. ameriana*, *ak*, twiner, Yucatán), and *x-pahua-ak* (*pahua*, common name for *P. schiedeana*, *ak*, climber, Yucatán). As avocado leaves are sometimes used in cooking, perhaps there is more to these names than a physical appearance. In Belize, it is *baatch-pám* (*baat*, ax, *pám*, the toucan, *Aularcorhynchus prasinus prasinus*).

Quechua speakers also note the spiraled "watch-spring" tendrils and the twining stems. They say *chirapa-sacha* (*chirapa*, curly, *sacha*, wild plant, Peru).

The *Oxford English Dictionary* (1971) indicates that the first use of "chew-stick" for *G. lupuloides* was made by Peter Simmonds (1858). That may well be correct, but the name "chaw-stick" was well established before that. For example, Macfadyen (1837) praised the plant as follows: "In powder, it forms an excellent dentifrice; its aromatic bitter producing a healthy state of the gums, and the mucilage it contains working up by the brush into a kind of soap-like froth. A tincture also is prepared from it and much recommended, diluted with water, as a wash or gargle, in cases of salivation or disease of the gums. The principal use to which the Chaw stick is applied, however, is as a substitute for the tooth-brush itself. For this purpose, the extremity of a piece of branch, which is usually about the thickness of the little finger, is softened by chewing, and then rubbed against the teeth. In this manner, a tooth-brush and, with it, a powder are obtained equal, if not superior, to any in use in Europe."

In Mexico, Central America, and the Caribbean, this species has long been a popular tooth cleaner. The stems are both aromatic and bitter, and those are traits preferred by many. According to Standley (1920–1926) and others, dried and powdered stems were not only used as dentifrices, but were exported from the West Indies and Central America to Europe and the United States. There is still a Jamaican mouthwash known as "Chew-Dent" that contains *G. lupuloides.*

Leaves have also been used to treat stomach problems (Mexico, Caribbean). Ash from the stems is powdered and placed on sores to speed healing. That ash and the leaves are said to be diuretic in Hispaniola (Liogier 1974), and therefore useful in treating dropsy. The stem powder is also thought to strengthen those who have been ill (Hocking 1997).

There are triterpenoid saponins in the plants that are at least partly responsible for its uses (Kennelly et al. 1993). Other species in the genus have other compounds (Nair and Madhusudana 1993). These chemicals are doubtless why several species other than *G. lupuloides* also are used. In the Americas, *G. polygama* has similar uses to *G. lupuloides.* That species is known in Venezuela as *rema* (oar), *bejuco de rema* (oar vine), and *reuma* (rheumatism). Surely, *rema* is an error for *reuma.* There is nothing about the plants that would suggest an oar unless it is an allusion to the winged fruits.

In the Old World, a poultice of *G. javanica* roots is applied to sores in southeastern Asia (Lewis and Elvin-Lewis 1977). Similarly, the pulped stems, roots, and leaves of *G. leptostachys* are used to treat skin diseases in Java (Lewis and Elvin-Lewis 1977). In addition, the bark is added to water used to wash the hair and kill parasites in Indonesia (Hocking 1997). In Fiji, *G. richii* leaves are pounded in water and drunk for stomachache, a use mirroring that of *G. lupuloides* in the Americas (Hocking 1997). According to the *Protocol Internationale*, the roots of *G. tiliaefolia* are a soap substitute and the leaves are recommended as a wash for ulcers in Indonesia (Hocking 1997).

Jamaicans not only use the stems to brush their teeth and as medicine, but they also use them in brewing beer, as a substitute for hops (*Humulus lupulus*). Indeed, the reason Linnaeus said it resembled hops (*lupuloides*) is because Leonard Plukenet in 1696 called the vines *Lupulus sylvestris americana* (wild American hops). Surely, he was reminded of hops because the locals used the plants to brew beer. Clean teeth are important, but drinking beer takes priority.

Perhaps their use of chew-stick in brewing is why some Jamaican beers have such a distinctive taste. I wonder if the beer cleans the teeth while it pixilates?

Guajacum: Lignum Vitae

(From *guayacán,* a Hispaniolan Taino name for the tree)

Almost halfway down the Florida Keys there is a small roadside marker that points to an island on the bay side and notes that it is the Lignum Vitae National Botanical Site. When I first started going there, not only was there no sign, there were no Park Service boats available to take visitors across the sometimes rough ocean. We either canoed over or rented a powerboat.

Regardless of how you arrive, getting to the island is stepping into a time machine and arriving in the early 1900s. The headquarters is a Key Largo limestone blockhouse literally bolted onto its foundation by steel cables. Using sane building methods, the first level of the house is open to allow hurricane waves to break underneath the occupants. Modern buildings in the Florida Keys show no such ecological awareness and foresight in planning.

Near the house on a trail through the forest is the former state champion lignum vitae (*Guajacum sanctum*). When I first saw this tree in the early 1970s, I was impressed with its size and main trunk about 9 inches in diameter. Then I saw the current champion on Totten Key—it has "small" branches considerably larger than the trunk of the Lignum Vitae National Botanical Site "giant."

Guajacum sanctum. a. Flowering branch. b. Leaf outline. c. Flower, side view. d. Flower, longitudinally dissected. e. Pistil. f. Petal. g. Floral diagram. h. Fruit, closed (center) and open from side and above. *Drawn by Priscilla Fawcett.* From Correll and Correll 1982.

All things considered, it is a minor miracle that there are any lignum vitae trees remaining in Florida. As soon as the Spanish discovered the island of *Española* (modern Hispaniola), they contracted syphilis and took it back to Spain. By 1508, they had also taken back the local Taino remedy for the malady, tea from lignum vitae (Standley 1920–1926, Crosby 1972, Desowitz 1997). *Guajacum* was praised by Oviedo ([1526] 1969) and Nicholas Monardes in 1569 (Frampton [1577] 1925). Although few biologists know about it, there was already a medical book written on the syphilis remedy by 1540 (von Hutten 1540).

Whether or not the remedy helped the disease, the Fuggers of Augsburg, also noted as the most important bankers of the time, became the chief importers of the plants (Crosby 1972). Trees disappeared in the New World from much of their former range as the Fuggers commissioned shiploads to be brought to Europe to satisfy the market largely created by them. By the time the English surveyed the Florida coast, they commented that almost all the trees had been cut (Romans [1775] 1961).

Guayacán (Caribbean, Mexico, Guatemala, Nicaragua) is a Taino name taken from Hispaniola (Oviedo [1526] 1969); Oviedo also wrote it as *guayacá* and *guayacan* (without accent). Monardes (Frampton [1577] 1925) wrote *guaiacan,* but etymologists assert that the aboriginal Caribbean name was *guaiac* or *guayaco*. The name also has been spelled *Guaiacum* in much of the scientific literature, although Linnaeus wrote *Guajacum*.

The common names now used are *gaïac bâtard* (false guaiac, Haiti), *gaïac blanc* (white guaiac, Haiti), *gaïac cardasse* (comb guaiac, Haiti), *gaïac femelle* (female guaiac, Haiti), *gayac* (Haiti), *guayacabillo* (little guayac), *guayacán* (Taino, Caribbean, Mexico, Guatemala, Nicaragua), *guayacán bastardo* (false guaiac, Dominican Republic), *guayacán blanco* (white guaiac, Cuba, Puerto Rico), *guayacán de vera* (true guaiac, Puerto Rico), *guayacáncillo* (little guayac, Dominican Republic, Puerto Rico), *pyé gayac* (country guaiac, Haiti), *wajakka* (Abaco, Bonaire, Curaçao), *wajakaa maatsjoe* (forest guaiac, Dutch Antilles), *wayaca* (guaiac, Dutch Antilles), and *wayaca shimarón* (wild guaiac, Dutch Antilles). *Vera* [*bera, beera*] (Dominican Republic, Dutch Antilles) seems to be a shorthand form of *guayacán de vera*.

Other than the Taino name, the first Spanish reports called the trees *palo santo* (holy wood, Mexico, Puerto Rico). The name holy wood was applied early and is used as the Spanish name by both Oviedo ([1526] 1969) and Monardes (Frampton [1577] 1925). As late as 1651 Hernández devoted a chapter to *Hoaxacan, seu Ligno Sancto* (guaiacan or holy wood) (Standley 1920–1926). The French use the same name and call it *bois saint* (holy wood, Haiti), and the Portuguese say *pão santo* (holy wood, Brazil). Those names were used because of belief in the Doctrine of Signatures, which maintains that when God sent a pestilence to humans, He also sent a remedy from the same region (Crosby 1972). All but the best physicians of the time thought that *palo santo* was that remedy. It turned out that the merchants were touting a false remedy and people continued to die of the disease (Crosby 1972). Then as now, *caveat emptor*!

The first record in English of the common name *lignum vitae* (wood of life, Latin) appeared in 1594 for the wood. Then, in John Gerarde's *Herball* of 1597, the name was applied to the tree itself. These names are dominant in English, but some of the variants are almost unrecognizable. For example, in the Virgin Islands they may be called *lingy whity, lingy vitey,* or *niglum witey*. Those are attempts by people of African origin to pronounce the Latin words. Some still use a translation of the Latin and call it tree of life (Haiti).

When syphilis first appeared in Europe, physicians did not recognize the malady. Because syphilis began with ulcers and sores, Europeans compared it with a disease they knew, the "pox" (from the 1300s, applied to chicken pox). The French called it the Italian pox, and the Italians called it French pox. Others said it was the pox of Naples, and the medical literature called it the Great Pox. Indeed, people should have followed Oviedo's ([1526] 1969) recommendation to call it the "Pox of the Indies," because each nationality thought (probably correctly) it contracted the malady from the other. These old names remain in *Pockholz* (pox tree, German) and rarely in English as pockwood tree.

Monardes (Frampton [1577] 1925) had already noted that there was one species (*G. officinale*) on Hispaniola, and another on St. John near Puerto Rico (*G. sanctum*). We now know *G. sanctum* from southern Florida, the Bahamas, Cuba, Hispaniola, Puerto Rico, Yucatán, maybe Veracruz and Tabasco, Guatemala, Honduras, Nicaragua, and Panama. Subsequently, another species was found on the Mexican mainland. The Mexican species is *G. coulteri*, and grows from Sonora to Oaxaca. That tree also quickly acquired the Taino name *guayacán*. It too is called by the original Spanish names *árbol santo* (Puebla) and *palo santo* (Oaxaca, Jalisco). Indigenous people also had names for it. Those names include *matlaquahuitl* (maguey or *Agave* + tree, Náhuatl), *yaga-na* (*yaga, tree*, Zapotec, Oaxaca), *yutnu-tandaa* (Mixtec, Oaxaca), *yúyaguo* (Mayo), *moeni* (Seri), and *huha'awo*[*hyuawo*] (Yaqui).

In addition to being considered antisyphilitic for so long, resin from the tree was considered an effective treatment against gout, rheumatism, scrofula, and cutaneous diseases. Many consider the resin stimulant, alterative, diaphoretic, and purgative in large doses. Indeed, *Guajacum* was in the U.S. Pharmacopoeia until the 1920s. Although the bark and wood contain alpha-guaiaconic, guaiaretic, and other acids, enedione lignan, furanoid lignan, prosapogenin, sapogenin, saponins, and vanillin (Kratochvil 1971, Majumder and Bhattacharyya 1974, 1975, Ahmad 1984a,b, 1985, 1986a,b, 1989b,c, 1990a,b, Almeida et al. 1996, Hocking 1997), the medical application is no longer accepted in the United States or Europe. Lignum vitae is subject to legal restrictions in some countries (Swerdlow 2000).

Throughout the Caribbean *Guajacum* is still used against venereal diseases and considered stimulant and sudorific, traits attributed to the resin by the pharmaceutical trade (Hocking 1997). Tea of the wood is used against skin diseases, rheumatism, gout, asthma, scrofula, and chronic lung problems (Liogier 1974, Beauvoir 2001). The resin (guayacol is red-brown, changing to blue or green when acidified) is used to strengthen gums and is applied to relieve toothache. The resin is also applied externally to skin diseases, and taken orally for arteriosclerosis, rheumatism, blood pressure regulation, and gout. Bark infusions are used against chills, grippe, rheumatism, and scrofula.

Resin (*guaiaci resina*) is still one of the principal reagents in tests for occult blood (Hocking 1997). Moreover, certain extracts show promise as a moluscicide (Almeida et al. 1996), and the boiled bark is a fish poison (Petersen 1974).

Wood from all species is strong, hard, heavy, resinous, and lubricant. The wood is used for rollers, maces, pulley sheaves, deadeyes, bandsaw guides, awning rollers, furniture casters, mallets, bowling balls, turned items made on lathes (Liogier 1974), and even railroad ties (Standley 1920–1926). It is famous for making self-lubricating, sea-worm resistant bearings and bushing blocks on ships. The heavy wood (specific gravity 1.2 to 1.3) sinks in seawater.

After surviving the harvesting of the Fuggers of Augsburg, the trees in Florida are now threatened by both habitat loss and a scale insect. Although the scale has been attacking the trees since at least the late 1960s, the species doing the damage was only identified in the early 1990s (Williams 1993). An endemic insect that apparently lived as a minor pest of the trees until recently now threatens to destroy plants that gave the name to a Florida botanical site.

Guapira

(J.F. Aublet named the genus from an indigenous word, perhaps from Tupí, *wa'bi*, to eat, *rob*, bitter)

Guapira discolor. *Drawn by P.N. Honychurch.*

Guapira discolor (two surfaces of the leaves unlike in color)

barrehorno (*barre*, from *barrer*, to clean off vegetation, *horno*, oven, but also "at the most auspicious time," Cuba, Puerto Rico)
beef-[tree, wood] (Bahamas, Florida)
blolly (shortened form of "loblolly," Florida); narrow-leaf blolly (Bahamas)
cabbage-tree (Cayman Islands; could the leaves be edible?)
pigeon-berry (Bahamas)

Professor of medicine and botany at Halle, Germany, Kurt Sprengel (1766–1833) called this Caribbean species *Pisonia discolor* in 1825. It was not until 1968 when Elbert Little moved the species to *Guapira*, a genus Aublet had created in 1775. Now there are about 70 species in the genus (Mabberley 1997).

Berries are edible and the wood is useful (Little and Wadsworth 1964).

Guettarda: Velvet Seed

(To commemorate the French physician Jean Étienne Guettard, 1715–1786)

Both species of velvet seed (*Guettarda elliptica, G. scabra*) historically grew from about Fort Lauderdale (central Broward County) south along the eastern Florida coast almost to Key West (Little 1978). In addition to that, there is one small disjunct population in southern Martin County. This colony lies 40 to 50 miles north of the historical range of the others, although urbanization has now made the gap even larger. Even stranger for this shrub, which is normally a pine rockland species farther south, it grows with gumbo limbo (*Bursera simaruba*) in Martin County. This association also has other disjunct and unusual species for the region and former County Commissioner Maggy Hurchalla was finally able to have the unique site made into the Rocky Point Hammock Park.

Outside of the Miami-Dade County parks, the Everglades National Park, and the pine rocklands in the lower Florida Keys, few people in Florida know these plants. Where they occur, they are often abundant shrubs. Deeper into their range, through the Bahamas, the West Indies from Cuba to Trinidad and some of the other islands off northern South America, people know the plants well. There they not only know them from their red fruits, but they also compare them with other plants. They compare *G. scabra* with the guava (*Psidium guajava*) by calling it *bois goyavier* (guava bush, Guadeloupe, Martinique), *goyavier bâtard* (false guava, Guadeloupe, Martinique), *guayabillo blanco* (little white guava tree), and wild guave (Dutch Antilles). They distinguish *G. elliptica* with

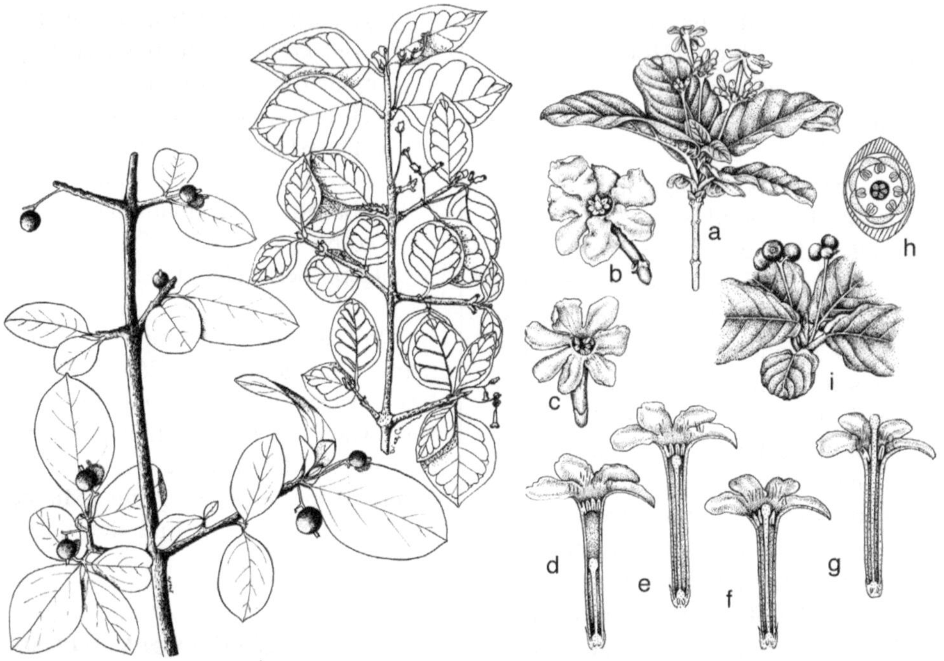

Guettarda. *Guettarda elliptica* (left). From Little and Wadsworth 1964. *Guettarda scabra* (right). a. Branch with flowers. b. Flower from above, stigma apparent. c. Flower from above, stigma not apparent. d to g. Flowers longitudinally dissected to show variation in style length, from short (d) to long (g). h. Floral diagram. i. Fruiting branch. *Drawn by Priscilla Fawcett.* From Correll and Correll 1982.

guayabillo prieto (black little guava, Mexico). For some reason they compare *G. scabra* with *chicharron de monte* (wild [mountain] peas, Cuba) and juniper (Grenada).

These two species sometimes are also compared by other people. In English, *G. elliptica* is usually called velvet seed (Florida, Puerto Rico), while *G. scabra* is rough velvet seed (Florida), roughleaf velvet seed (Florida, Puerto Rico), or velvet berry (Bahamas, Puerto Rico).

Inexplicably, in the Virgin Islands *G. scabra* is called greenheart. That common name elsewhere is given to *Chlorocardium rodiaei* (also called *Nectandra rodiaei* and *Ocotea rodiaei*, Lauraceae) of Guyana, sometimes called Demerara greenheart (Gérard et al. 1996). That species is a large rainforest giant that provides one of the main timbers exported from that country. It is a heavy timber, resistant to termites and borers. It is a mystery how the people in the Virgin Islands came to use the same name for *G. scabra*, which is a small tree of moist coastal woodlands on limestone, reaching a height of only 10 to 15 feet and only about 4 inches in diameter. Maybe it also has green heartwood?

Wood of these small trees is light brown, hard, heavy (specific gravity ~0.8), and fine textured (Little and Wadsworth 1964). The stems are used for stakes and fuel, and for other items like some of its related species.

Both *G. elliptica* and *G. scabra* are called *cucubano*, as are other species in Puerto Rico. *Guettarda scabra* is just *cucubano* or *palo de cucubano* (*cucubano* wood), while *G. elliptica* is *cucubano liso* (smooth *cucubano*). This is a Taino word, referring to both a lightning bug and an owl. *Cucubá* still is the Puerto Rican screech owl (*Otus nudipes*) (Raffaele 1983). This is the screech owl on that island, and its counterpart on Cuba is the bare-legged owl (*Gymnoglaux lawrencii*). On both islands these birds are also called "Cuckoo Birds" because their "love call is the mellow, sonorous 'coo-coo'" (Grossman and Hamlet 1964). The name *cucubá* may be an onomatopoeic imitation of the call.

What relationship the plant might have with a bird is obscure. The only apparent possible relationship is that the birds are active at night, and the plants flower then. In the evening the 3/4-inch-long flowers perfume the air with their heavy scent and attract half-blind sphinx moths (*Perigonia lusca*). That moth was considered rare when Holland (1965) studied the insects of the state. The moth is still known only from Palm Beach and Miami-Dade Counties (Struttmann 2002), and has been found feeding as a larva on many members of the Rubiaceae deeper within its range. Larvae have also been raised on *Ilex krugiana* (Aquifoliaceae), an unrelated species.

Could it be that someone thought the moth visitors were really birds? Is that why both are called *cucubano*?

Names for the two Florida species diverge at this point. Fewer names are known for *G. elliptica*. Several of its names refer to a cross, including *crucecilla* (little cross, Mexico), *cruceto negro* (black little cross, Venezuela), and *crucillo* (little cross, Mexico). That may be an allusion to medical use following the Doctrine of Signatures. There are other species in other families that have the cross allusion in their names, and all are used for some remedy.

Some of the other names for this species are particularly curious. In Venezuela, it is called *cabrito* (kid or young goat). All that is recorded about it there is that the fruits are edible. However, it makes one wonder if "kids" (humans and goats) are among the only ones eating those fruits. Also in Belize it is prickle wood, and in Venezuela it is called *punteral* (thicket of "stickers"). If you go walking in a patch of it, the stiff branches certainly would make those names appropriate. Pickle-wood (Belize) must be an error for "prickle-wood."

Oddest of names for *G. elliptica* are *cuero de sabana* (savanna skin, Cuba). No tannin content has been recorded or any use for tanning, so the "skin" in the common name must be the thick leaves. If you really work at it, they do resemble leather. *Cagüilla* (Cuba) must be another of the Taino names whose meaning, beyond the identity of the plant, has been lost.

The rough leaves of *G. scabra* were the source of its species name (*scabra*, rough), and the vernacular *serrasuela* [*serrazuela*] (little saw, Puerto Rico). That name is reminiscent of roughleaf velvet seed in English. Perhaps the name *carapacho* (carapace of crustaceans, Cuba) also alludes to the abrasive surface as in the carapace of the spiny lobster (*Panulirus argus*).

In the Bahamas, *G. scabra* is called gnat wood (Bahamas). Since there are often gnats around in the summer when the plants flower, that is a reasonable name. The species is called *palo de dajao* (*dajao* is a river fish) in Puerto Rico, so maybe it has been used as a fishing pole.

There do not seem to be many common names alluding to medicinal uses. Since the plants have been used widely as aphrodisiacs, that is somewhat surprising. Surely the names bat-leaf (Bahamas), blue copper (Tobago), *bois madame* (lady tree, Guadeloupe, Martinique, St. Lucia), *bois noire* (black tree, St. Lucia), candlewood (Antigua, Dutch Antilles), craw-wood (Barbuda), and four-man's strength (Bahamas) refer to giving "energy" or "restoring strength" for love-making. That application is apparently the most common use of any of the American species of *Guettarda*. Sometimes *G. scabra* is mixed with banana (*Musa* ×*paradisica*), or the leafy stems are tied in bundles with bark of *Bourreria ovata* and *Tabebuia heterophylla* and boiled to produce an aphrodisiac decoction. Beyond that, there seems to be little in the way of medicinal use for the genus in the Americas.

Related species contain a variety of chemicals, including essential oils (*G. poasana*, Costa Rica), glycosides (*G. platypoda, G. trimera*, New Caledonia), quinoline alkaloids (*G. ovalifolia*, Puerto Rico; *G. noumena*, New Caledonia), terpenoids (*G. angelica*, Brazil), and indole alkaloids (*G. acreana*, Ecuador; *G. trimera*, New Caledonia). Among the laboratory results exhibited by these compounds are activity as antiviral agents, antimicrobial agents, inhibitors of yeast, and as antispasmodics (Sousa et al. 1984, Kan-Fan et al. 1985, Matos et al. 1986, Aquino et al. 1988, 1989a,b, Cotterill and Fergusson 1993, Jiang et al. 1994, Montagnac et al. 1997, Capasso et al. 1998). There seem to be no studies of either of the Florida species, but presumably they contain related chemicals.

Linnaeus created *Guettarda* in 1753. The only species he knew was *G. speciosa*, from Java; it also occurs elsewhere in Asia and in Madagascar. That species is a noted source of "zebra wood," although wood sold under that name also comes from several other genera (Uphof 1968). The genus *Guettarda* is now known to contain about 80 species spread around the tropical coasts of the world (Mabberley 1997). Several of them have edible fruits, and maybe that is as good a human use as any—especially on a hot, muggy day in the rockland flatwoods where these shrubs grow. However, Floridians might introduce the trend of use of *G. speciosa* from the Pacific Islands. There, girls put them in their hair for perfume, and Fijians and Samoans make them into necklaces or leis.

Guzmania

(Named by Hipólito Ruiz and José Antonio Pavón for Anastasio Guzman, a Spanish naturalist of the 1700s)

Guzmania monostachya (one-spiked inflorescence) air plant

asõmco:bî (big Spanish moss, Mikasuki); *asso-nakko* [*assonrakkô*] (*assúnwv*, Spanish moss, *łakko*, big, Creek)

curujey bonito (pretty airplant, *curujey* is the Taino word for epiphytes, Cuba)

piña de agua (water pineapple, Dominican Republic)

tinaja [*tinajita*] ([little] well, Dominican Republic)

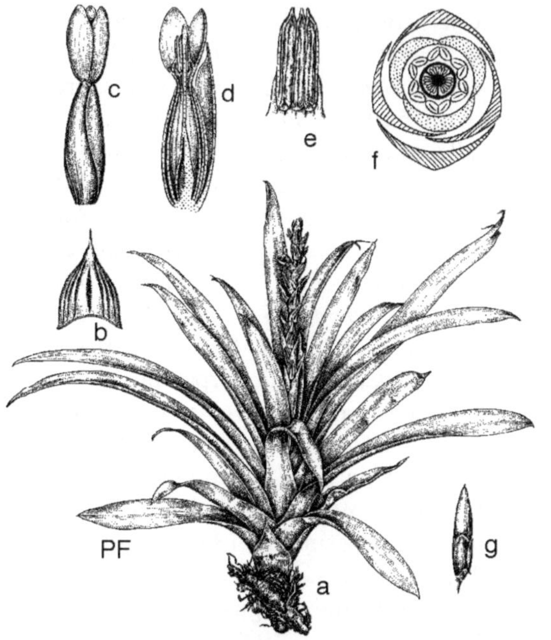

Guzmania monostachya. a. Flowering and fruiting habit. b. Floral bract. c. Flower and bract, side view. d. Flower and bract, longitudinally dissected. e. Fused anthers. f. Floral diagram. g. Fruit. *Drawn by Priscilla Fawcett.* From Correll and Correll 1982.

Linnaeus described *Renealmia monostachya* in 1753. That genus belongs in the Zingiberaceae. When people later looked at the plants, they realized that the species belonged in the Bromeliaceae. So, New York Botanical Garden's Henry Hurd Rusby (1855–1940) and professor at Königsberg Carl Christian Mez (1866–1944) moved the species to *Guzmania,* which had been described in Peru in 1802. Smith and Downs (1977) included 131 species in *Guzmania,* all endemic to the American tropics.

Seminoles used the water trapped in the base of the plants for cooking during the dry season (Sturtevant 1955). The common names suggest that other people used the water in the same way.

Gymnanthes: Crabwood

(From Greek *gymnos,* naked, and *anthos,* flower)

In South America, the early explorers of the French Guiana region found Cariban people calling one tree *carapa,* and another *caraipé.* It was probably inevitable that English-speaking people misunderstood these indigenous names, and thought they said "crab." However, that is what appears to have happened with one or both of these words. The first variant seems to have been *carapa,* the indigenous name that also gave us the Latin name of a tree related

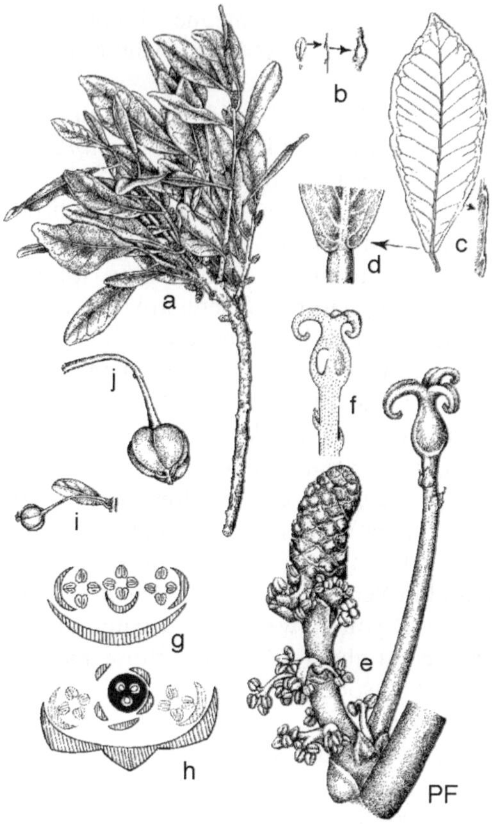

Gymnanthes lucida. a. Branch with catkins. b. Leaf primordium, with inset of margin, and marginal gland, highly magnified. c. Mature leaf, with inset of margin, and marginal notch, highly magnified. d. Leaf base. e. Single flowering branch, with lower flowers expanded. f. Pistillate flower, longitudinally dissected. g. Floral diagram of staminate flower. h. Floral diagram of pistillate flower. i. Fruit on twig. j. Fruit. *Drawn by Priscilla Fawcett.* From Correll and Correll 1982.

to mahogany and China-berry, *Carapa guianensis* (Meliaceae). These are large trees in the Guianas and their fruits wash out into the ocean and eventually drift onto the beaches of Florida. They only flower about once every 5 years, and trunks are harvested for timber (Gérard et al. 1996). The seeds serve as fish bait, as a source of *andiroba* (from *adi'roba,* bitter oil, Tupí), which is an oil that is used in lamps, soap-making and candle-making, as a medicine against arthritis and throat infections, and as an insect repellent, and the oil was formerly used by head hunters in shrinking heads (Hocking 1997). The common names for the wood includes bastard mahogany, *andiroba, tallicona,* and—you guessed it—crabwood.

Florida's crabwood is not the same plant at all, but a small tree in the Euphorbiaceae known scientifically as *Gymnanthes lucidus* (*lucidus,* shining). The genus and species were created by Olaf Swartz in his *Nova*

Genera et Species Plantarum (New Genera and Species of Plants) published in 1788. Another version is that crabs eat the bark and berries in the Bahamas, and that gave it the name (Von Reis 1973).

This is one of those confusing cases where the morphological similarities cause problems in interpretation of generic limits, and books keep changing names back and forth. Britton and Millspaugh (1920), Small (1933), and Leon and Alain (1957–1963) used the name *Gymnanthes lucidus.* Then, because of studies done on Jamaican and Bahamian plants (Adams 1972, Gillis 1974), the name was changed to *Ateramnus lucidus.* Subsequently, Tomlinson (1980) and Correll and Correll (1982) used *A. lucidus* for the trees. Finally, Webster (1983) examined the situation and discovered that *Ateramnus* was actually a synonym for *Sapium.* That revelation caused *Gymnanthes lucidus* to be restored.

Regardless, the word *Ateramnus* proposed by Patrick Browne in 1756 during his sojourn in Jamaica gave an apt commentary on these plants. Browne based the name on the Greek word *ateramnos,* meaning hard or stubborn. Almost certainly, he was referring to the dense (specific gravity 1.1), fine-textured trunk with its white sapwood and olive to dark brown heartwood (Little and Wadsworth 1964).

Florida's *G. lucidus* ranges through the Greater Antilles, the Lesser Antilles to Guadeloupe, and is also in Yucatán, Guatemala, and Belize (Leon and Alain 1957–1963, Adams 1972, Correll and Correll 1982, Balick et al. 2000). There are 15 species in the genus, with 5 in Jamaica, 2 in Nicaragua, and the others throughout Mesoamerica (Mabberley 1997).

Three variants of the common names seem to have arisen from confusion or comparison with the Guyanese *carapa.* These are "crab bush" and "man crab bush" (British Antilles) and "crabwood" (Bahamas, Jamaica, USA, Guyana). Even these may be references to the similarity of the wood to a formerly better-known plant. The following common names are comparisons or descriptions of the wood. These are *bois marbre* (marbled tree, Haiti), *bois madre* (maybe a corruption of *bois marbre,* otherwise mother-wood, West Indies), *bois vert* (green wood, Lesser Antilles), *branquilho* (little white one, Brazil), goat wood (St. John), false lignum vitae [lignum-vitae] (Belize), *Juan prieto* (black John, Dominican Republic), *palo de hueso* (bone wood, Dominican Republic), oysterwood (USA), Cuban oysterwood (Bahamas), and shiny oysterwood (USA). In Puerto Rico and elsewhere the wood is used for fence posts, canes, handles, backs of brushes, mirrors, and ornamental articles made by woodturning (Little and Wadsworth 1964).

There are even some names that compare these small trees with tobacco, as *palo de tabaco* (tobacco wood, Dominican Republic), *tabaco* (tobacco, Puerto Rico), and *tabacón* (big tobacco, Dominican Republic). While it seems unlikely, the leaves may have been used as tobacco substitutes. If that is the case, the people calling them by those names and using them were totally unaware of a long-term error in botanical literature. Beginning at least in the 1920s, perhaps with Paul Standley (1920–1926), there are notes that the trees contain milky sap that is extremely poisonous. Either that was the source of the common name poisonwood (USA) or the plants were confused with *Metopium.* That statement from Standley, who spent thousands of hours in the field in Middle America, is almost impossible to believe. He was an extremely astute botanist who listened to and critically examined the information provided by helpers and consultants from nearby villages. Yet, his *Trees and Shrubs of Mexico* is the first place that I have found this mistake. Instead, as Adams (1972) and others have pointed out, the sap in these plants is not milky, but clear. It never caused problems for me, and presumably did not for others, because there is no credible account of it in the literature. Instead, the erroneous comments from the 1920s were repeated by Standley and Record (1936) and Little and Wadsworth (1964).

Some of the other names refer to other aspects of the plants. These include *pij* (probably a simple Maya term, Guatemala), *ramón* (much-branched, Puerto Rico), and *haite* [*aité, aiite, aite, aiti, yaití, yaibi*] (*haytí,* high land, Taino, Cuba, Puerto Rico, Mexico), and *yaiti macho* (male *yaití,* Cuba). *Capixava* (from *kapi'xawa,* planted place, Tupí, Mexico, West Indies) is attributed to *G. lucidus,* but in Brazil, it is the name of *Sebastiana brasiliensis.*

Other names refer to other aspects of the plants, particularly fruits. These are *aceitillo* (little oily one, Cuba), *granadillo* (from *granada,* the pomegranate, because of the red color), and what seems to be simply a spelling variant, *greadillo* (little pomegranate, Dominican Republic), *granadilla boho* (maybe *boho* is a corruption of *bohio,* house, Taino, Dominican Republic), *huevo de chivo* (goat testicle [egg], Dominican Republic), and *jabacón* (perhaps a transcription error for *tabacón,* Dominican Republic).

In the Bahamas, including the Turks and Caicos Islands, leaves are chewed or branches boiled and the decoction taken for stomachache and "griping" (Von Reis 1973, Morton 1981). Whole small plants are boiled, root and all, and the liquid used to stop diarrhea or relieve other stomach problems. To cure piles, leafy twigs are boiled with *Bourreria succulenta* leaves, root of *Zanthoxylum coriaceum,* root of *Chiococca alba,* leaves of *Cassia bahamensis,* and aboveground parts of *Turnera ulmifolia* and *Capraria biflora.* All are bundled together and tied. The decoction is said to be an "unfailing remedy." In Cuba, a decoction of the bark is used against toothache (Roig 1945).

Taken individually, the common names are similar to those of many other Caribbean species. However, some may suggest a use not recorded in the literature. Given the small fruits (~10 mm across), the name *huevo de chivo* seems to come from an exaggerated imagination, or a snide comment. However, when it is compared with goat wood, oysterwood, and shiny oysterwood you begin to wonder if the trees were not used as aphrodisiacs.

H

Habenaria

(Carl L. von Willdenow named the genus from Latin *habena*, a strap, holder, rein; the lip is usually trilobed and always spurred)

Habenaria floribunda (many-flowered) (= *H. strictissima* var. *odontopetala*)
clavo cojincillo (*clavo*, club, *cojincillo*, pad, Veracruz)
false rein-orchid
hihkitî:nayikcî [*hihkitiinayikci*] (horned-owl medicine, Mikasuki); *istikinihiliswâ* (*estekene*, horned-owl, *heleswa*, medicine, Creek)
long-horned orchid (the "horn" is the spur)
mignonette orchid ("mignonette" from feminine of *mignonet*, dainty, French, and Middle French *mignon*, darling; in English by 1752)
oek ahyetche (water medicine, Mikasuki)
tooth-petal orchid

Habenaria was segregated from the other terrestrial orchids by Willdenow in 1805. Then John Lindley described *H. floribunda* in 1835. There are now about 600 pantropical and subtropical species in the genus (Mabberley 1997).

The Seminoles use the plant in funeral ceremonies to send off spirits and in a medicine to strengthen medicine men (Sturtevant 1955, Bennett 1997). The Zoque-Popoluca in Veracruz use it to stop vaginal bleeding (Vásquez and Jácome 1997).

Hamamelis

(Ancient Greek name of the medlar, *Mespilus*, or some similar tree; reapplied to the American plants)

Hamamelis virginiana. From Sargent 1905.

Hamamelis virginiana (of Virginia)
[spotted, striped] alder
bwaoke'ît (no heart wood, all white, Potawatomi)
café du diable (devil's coffee, Quebec)
iyanabi (*iyan*, from English iron?, *abi*, tree, Choctaw)
pistacio (first used as the genus by Plukenet in 1696, and then as a common name by C.S. Rafinesque in his *Medical Flora*, 1828)
tobacco-wood
winter-bloom [winterbloom]
[American, common, southern] witch-hazel [witch-hazel] ("witch," or "wych" applied by 1540s, in this case means having pliant stems; see *Ulmus*; "hazel" is a name applied to the fruits of *Corylus* since about A.D. 700; derived from Old Teutonic *hasaloz*, and cognate with Latin *corylus*, helmet, and perhaps Gaelic *coll*); snappy [snapping] hazel
Zaubernuss (magic nut, German)

"Witch-hazel" began as a common name for European plants. The first appearance in print was in Shakespeare's *Henry VIII* where he wrote, "That everie bowyer ... for every bowe that he maketh of Ewe, make fower other bowes ... of Elme, wyche hasill, ashe, or other Wood." Gerarde ([1597] 1975) equated the term "Witch hasell" with the hornbean (*Corylus*), and it too is famous for its pliable branches (Edlin and Mitchell 1985, Vickery 1995). It is not clear which species Harriot was talking about in what is now North Carolina. When he wrote in 1588 that the Carolina Algonquians had "wich-hazel, whereof the inhabitants vse to make their bowes" (Harriot [1590] 1972). The first equating of *Hamamelis* with witch-hazel was in 1760 (OED 1971).

In the Americas, *Hamamelis* branches are preferred as "witching-rods," "dowsing-rods," or "*virgula divina*." The translation of divining rods [staffs, sticks, wands] (appeared by 1656, translated from Latin) indicates the use in locating water or mineral deposits. However, according to Rosengarten (1984), the preferred plant in England has been the hazel, *Corylus*, since the Dark Ages (A.D. 476–1000). Panati (1987) places the date in the 7th century A.D.

Divining with hazel rods probably orginates with Greek mythology when Apollo and Mercury, the two sons of Jupiter, exchanged gifts. Apollo received a tortoise-shell lyre; he used its music to release the artistic spirit in humans. Mercury got a winged wand made of hazel-wood whose touch helped people express their thoughts through words. A winged hazel rod entwined with snakes became the symbol of communication. Romans also revered *Corylus*, and hazel torches were burned on wedding nights to ensure a happy marriage (Rosengarten 1984). The use of a divining rod made from a hazel branch probably can be traced back to mythology.

Some still practice this "art" today in both America and the Old World. Those who use this technique, called dowsers (etymology unknown), hold a forked branch in both hands. They say that the branch turns and points to the earth at the place where buried treasure, veins of precious metals, and streams of subterranean water are to be found. The Mohegans, apparently learning the technique from the Europeans, used this method to try to find water and buried treasures (Moerman 1998).

Rafinesque called *Hamamelis* "pistacio." The learned Englishman John Lindley said the seeds were edible (Hedrick 1919). *Hamamelis* seeds approach 10 mm in length, and have been eaten by some tribes (Yanovsky 1936). They were considered sacred beads by at least the Menomini (King 1984).

Overshadowing the dowsing potential of the shrubs is their medicinal use (Moerman 1998). The Cherokee took an infusion to treat fever, colds, sore throat, and tuberculosis. They also used it to wash sores and to alleviate other skin problems (Hamel and Chiltoskey 1975). The Iroquois made extensive use of the shrubs. They used a twig infusion to stop bloody dysentery and cholera and a poultice of branches for asthma, colds, coughs, consumption, and heaves, or drank a decoction. Decoctions or infusions were used to treat arthritis, kidney problems, and toothache, and to purify the blood. A decoction of *Hamamelis* and other plants was used to prevent hemorrhage at childbirth and as an astringent applied to external wounds and bruises. Although it was said to stimulate appetite, a bark decoction was emetic. The Ojibwa used an infusion for skin troubles and to wash sore eyes, and they took it as an emetic in cases of poisoning. The Menomini made a decoction and rubbed it on their legs to keep them supple during sports and to relieve back problems. The Potawatomi put twigs in sweat baths to relieve sore muscles (Smith 1933).

Medicinal extracts, including lotions and salves, still are made from leaves, twigs, and bark, now by distillation (Vines 1977, Foster and Duke 1990, Swerdlow 2000). The products are used to treat bleeding, lessen inflammation, and decrease mucous secretions. Tannins in the leaves and bark are thought to be responsible for the astringent and hemostatic properties (Foster and Duke 1990, Duke et al. 2002).

Hamelia: Firebush

(Named for the French forester, agronomist, and *Inspecteur générale de la Marine*, Henri Louis du Hamel du Monceau, 1700–1782).

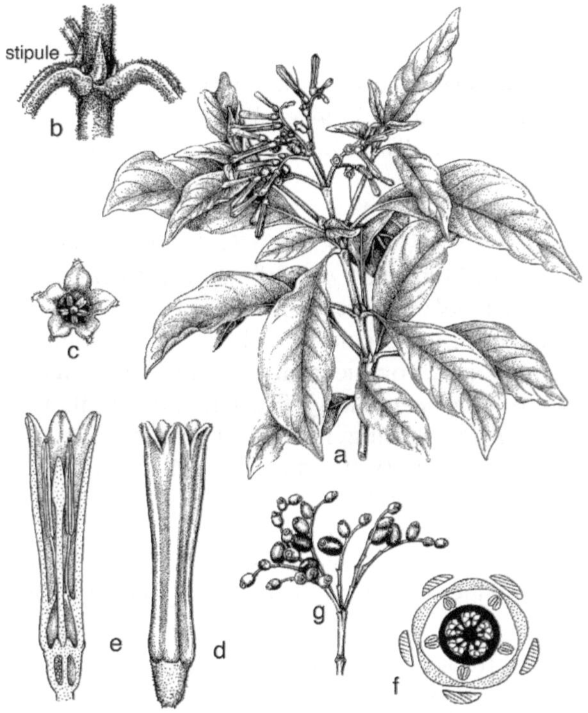

Hamelia patens. a. Branch with flowers. b. Node showing stipules. c. Flower, from above. d. Flower, side view. e. Flower, longitudinally dissected. f. Floral diagram. g. Fruits. *Drawn by Priscilla Fawcett.* From Correll and Correll 1982.

An old adage says that you remember something longer if you have to figure it out for yourself. What usually is not included is that you remember it better if you make a fool of yourself in the process.

Some time after moving to Florida, I decided that I wanted a small pollination garden outside one of my windows. I planted both firebush (*Hamelia patens*) and necklace pod (*Sophora tomentosa*)—nice complementary red (bird) and yellow (bee) flowers. It seemed that those colors would attract any pollinators around, whether vertebrate or invertebrate.

I watered the plants faithfully for a few weeks. Then, I noticed that the more I watered, the more the *Hamelia* improved and the *Sophora* declined. So, changing tactics, I watered less often. As the soil dried out, the *Sophora* brightened up and the *Hamelia* began to wilt. There was obviously an error in my strategy. Finally, it occurred to me that *Sophora* always grows in dry beach areas and *Hamelia* in moist-to-wet hammocks and swamps. Together, they were not going to work, so I quit watering. The *Sophora* thrived, and of course, the *Hamelia* died.

There are benefits to and costs for everything. Firebush in the major part of its range (the Bahamas, West Indies, from southern Mexico to Peru and Paraguay) depends largely on hummingbirds for pollination. Southern Florida has no resident hummingbirds, and only some of the firebush plants can be pollinated by the overwintering winged jewels, the ruby-throat (*Archilochus colubris*). Instead, the plants "use" butterflies as alternate pollen vectors. I suspect that many of the plants are also facultatively self-pollinating (autogamous), but I have seen no studies on that topic.

Within the areas where hummingbirds are the pollinators, even they have costs and benefits. Mites living in the flowers are transported from plant to plant by the visiting birds (Colwell 1995, Paciorek et al. 1995). Not only do the hummingbirds obtain the reward of nectar and the energy from it, but the mites are moved around and the plants are pollinated. There are negative impacts on both the plants and the birds to having the parasitic mites involved, but they are not so great that they impede reproduction.

To the birds and to humans, the most eye-catching aspect of *Hamelia* is the red color—in leaves, flowers, and fruits. Indeed, that color is the reason for the common name firebush. Some allusion to the color is found in about half the vernacular names for the species. Some compare *Hamelia* to other red plants, like *achiotillo colorado* (little red achiote, Honduras). *Achiote,* also called *annato,* is *Bixa orellana* (Bixaceae), perhaps the most commonly used red pigment in the Americas. Others think the flower cluster resembles a coral necklace and say *bois corail* (coral tree, French Antilles), *corail* (red one, Haiti), *coral* (red one, Cuba, Honduras), *coralillo* (Cuba, El Salvador, Costa Rica, Colombia), *coralito* (little red one, Venezuela), *fleur corail* (red flower, Haiti, Guadeloupe, Martinique), and *palo de coral* [*palo coral*] (red tree, Cuba).

Still other names make comparisons to a variety of items. In different places it is *cuetillo* [*cohetillo*] (little rocket, probably fireworks), *clavito* (little heart, Guatemala), *coloradillo* (little red one, Honduras), *cresta de gallo* (rooster comb, Colombia), *palo camarón* (shrimp tree, Costa Rica), Polly red head or red head (Belize), red berry (Panama), *sancocho* (parboiled meat, El Salvador), or just scarlet bush (Puerto Rico, Panama).

For some reason, there also is an association with the color blue, and people say *añileto* (little indigo [*Indigofera*], Costa Rica) or *azulillo* (little blue one, Costa Rica). Alluding to a medical use, it is called *bálsamo* (balsam, Puerto Rico), *bálsamo cimarron* (wild balsam), or *bálsamo colorado* (red balsam, Puerto Rico).

Other common names that allude to medical uses are *flor de baño* (bath flower, El Salvador), *herba de erisepala* (erysipelas herb, Guatemala), *hierba de cancer* (cancer [any sore] herb, Guatemala), *kunabiaya* (water-medicine in Kunana), *zorillo* (skunk, Costa Rica), and *zorillo real* (royal skunk, Costa Rica). The references to the skunk are probably because the plants resemble *Chiococca alba,* which is also called *zorillo.*

As with several of their names, the Maya have a distinctive viewpoint and say *k'anan* [*sax-canaan, canaan, ix-canan, kanan, neanan, ix-kanan, xcanal, xcanan, xkané*] (*ya'ax,* green, *k'anan,* necessary, Belize, Yucatán,) or *ix-canan amarillo* (yellow necessary, Guatemala). These people consider the plant "necessary" because it is important in religious ceremonies, and it is a traditional medicine for skin diseases. According to Arvigo and Balick (1993) *ix-canan* means "guardian of the forest," although no basis for that has been found except by allusion. The Maya also call it *chactoc* [*chadoco, chacloco*] (*chac,* red, *toc,* remove, Belize, Tamaulipas) because it "removes" the skin problems. In Belize, the plants are also *klaush-pàm* (*klaush* does not seem to be in Maya, *pám,* toucán, *Aularcorhynchus prasinus prasinus,* Maya) and *sac-le-much* (*sak,* white, *le',* leaf, *much,* frog, Maya).

Some names clearly allude to the edible fruits, including *recadito* (little gift, Columbia) and *uvero* (grape tree, Panama). The name *pata de pajaro* (bird foot, Puerto Rico) probably is an indication that the birds also have been seen eating the fruits.

Some names have obscure references. The allusions to the *canilla de venado* (deer's long bone, Nicaragua) and to a *canudo* [*canuto*] (tube, Guatemala) might be references to use in an indigenous musical instrument, because *Hamelia* is also called by the instrument's name *zambumbia* (El Salvador). The instrument is usually known as *zambomba,* and consists of a hollow tube, closed on one end by a skin, with a cane in the middle, and played by rubbing the cane with a moist hand. Maybe the reference to *canilla* and *canudo* means that *Hamelia* stems have been used for that "cane."

The meanings of some of the names are not readily apparent although we can translate them. What, for example, does *chichipin* [*chichipince, chichipú*] (to rain drop by drop, Náhuatl, Guatemala, Honduras, El Salvador) have to do with the plants? Other names that are not obvious in meaning are *cacauapaxtle* [*cacanapazue*] (chocolate moss, Náhuatl, Veracruz), *doncilla* (virgin, El Salvador), *estirnina* (little spoon, Nicaragua), *flor de cangrejo* (crab flower, Guatemala), *herva de rato* (mouse's herb), *hierba de Cuba* (Cuban herb, Yucatán), *leoncito* (little lion, Colombia), *pañete* (little cloth, Veracruz), and *xuchit paltimatía* (*palta* is avocado, *Persea americana,* El Salvador). Maybe spoons are made of the stems to give *estirnina*, and the leaves do look a bit like those of *palta.* Does *cacahuaxochitl* (chocolate, flower, Náhuatl, Veracruz) refer to the red flowers?

Simple and untranslated names include [*palo de*] *masamora* (Nicaragua), *pilí-tso* (Guatuso, Costa Rica), *pisí* (Costa Rica), *ponasí* [*bonasí*] (Cuba), *sibunkhen* (Kekchi, Yucatán), and *sisipinse* (Guatemala). Apparently, there is some relationship between the following names, but it is far from certain what the original may have been—*desyerba conuco* (remove weeds from a garden, Dominican Republic), *bencenuco* (maybe = *vence conuco*, or garden conqueror, Colombia), *cerva conuco* (deer garden, Dominican Republic), *buzunaco* [*buzunuco*] (Dominican Republic), and *buzunaco coral* (Cuba). If Spanish was not the original language, these all might have been derived from some Taino word resembling *buzunuco.*

Throughout its range, the fruits are utilized by people fresh from the plant or fermented into a drink (Standley 1920–1926). In addition, an antiscorbutic syrup is sometimes made from the fruits as a remedy for dysentery. The wood, which is light or dark brown and hard, has been used for various items.

However, the most common use seems to be for skin problems such as sores, rashes, burns, itching, cuts, fungus, and insect bites and stings. In Mexico, Belize, and the Caribbean, the crushed leaves are applied to cuts and bruises, and baths made with leaves and fruits are used for skin diseases (Standley 1920–1926, Roig 1945, Liogier 1974, Vásquez and Jácome 1997). Similar preparations are used for measles and rheumatism. Fruits and leaves are macerated to make a medicine used against sarcoptic mange. Leaves are used in vaginal washes and to treat migraine and other headaches.

The plants contain several bioactive chemicals, including ephedrine (Chaudhuri and Thakur 1991), flavanone glycoside (Aquino et al. 1990), and oxindolic alkaloids (Borges et al. 1977, 1979). Ethanol extracts in the laboratory act against *Staphylococcus aureus*, and water extracts combat *Escherichia coli, Salmonella typhosa, Sarcinia lutea, Serratia marcescens*, and *Shigella flexneri* (Jiménez et al. 1979b). Analgesic effects (Esposito-A. 1985) and antifungal activity against *Neurospora crassa* (López-A. et al. 1981) also have been shown. All of these extracts are astringent and antidiarrheal (Pérez et al. 1996).

Still, the best-known chemical in the plants is tannin. The high tannin content has also prompted people to use both leaves and stems in tanning leather (Standley 1920–1926, Liogier 1974). Many people just plant it as an ornamental because both fruits and flowers attract birds.

Only rarely did ruby-throated hummingbirds stop at my plants. Even when they did, they were rarely there more than a day or two. In their place, I was able to watch zebra longwing butterflies (*Heliconius charitonius*) all year long.

Hedeoma

(Christiaan H. Persoon altered this name from *hedyosmon*, an ancient name of mint, from the Greek *hedys*, sweet, and *osme*, scent)

Hedeoma hispidum (with short stiff hairs)

maka chiaka (*maka*, probably skunk, but also meaning ground, Dakota)

[mock, rough] pennyroyal ("pennyroyal" where "royal" is sometimes rendered *ryal* or *rial*; derived from Anglo-French *pulyole ryale*, and from Latin *pulegiol*, little thyme, plus *real*, royal; originally applied to European *Mentha pulegium*); rough false-pennyroyal [rough falsepennyroyal]

pe-zhe-tu-hu (*pe'-zhe*, herb, *to'-ho*, green, Osage for *H. pulegioides*)

rough hedeoma (a translation of the scientific name)

su'wa ki· wi·ti· (smell-root, Catawba; perhaps for *H. pulegioides*)

Frederick Pursh introduced *H. hispidum* in his *Flora Americana Septentrionalis* (Flora of North America) of 1814. That was only 8 years after Persoon (1761–1836) created the genus. This endemic American genus contains 38 species, ranging from the southern United States into South America (Mabberley 1997).

Hedeoma hispidum is one of the least-used members of the genus according to Moerman (1998). The Dakota used the leaves to treat colds and add flavor to food to make it more appealing particularly to those who were sick (Gilmore 1919).

The species most used in the eastern United States, *H. pulegioides*, was included in the pharmacopoeia of the Catawba, Cherokee, Chickasaw, Delaware, Iroquois, Mohegan, Nanticoke, Osage, Ojibwa, Rappahannock, and Shinnecock (Moerman 1998). Since the Chickasaw used a cold infusion of the roots for itching eyes (Taylor 1940), they and their relatives may have used *H. hispidum.*

Most, if not all, members of *Hedeoma* contain a variety of essential oils that provide the fragrance, particularly pelegone (Lawless 1995). These oils can be lethal if ingested (Millspaugh 1982, Foster and Duke 1990).

Helenium

(The Greek name of some plant, said by Linnaeus to be named after Helena or Helen of Troy, wife of King Menelaus of Sparta)

Helenium autumnale. a. Upper part of plant showing habit. b. Lower stem and roots. From Buchholtz 1968.

Helenium amarum (bitter)
bitter-weed [bitterweed]
pahí hó:mi (*pahí*, grass, *hó:mi*, bitter, Koasati)
sneeze-weed [sneezeweed]
Sonnenbraut (sun bride, German)
Spanish daisy
yellow-dicks [yellowdicks]

Helenium autumnale (flowering in the fall)
autumn [common] sneeze-weed [sneezeweed]
cabezona (big head, Valley of México)
cacamosiksan (it makes you sneeze, Cree)
chapuz (Michoacán)
fever-grass
Herbst-Sonnenbraut (autumn sun bride, German)
manzanilla (chamomile, Mexico); *manzanilla de Yucatán* (Yucatán chamomile, Mexico)
ox-eye
plumajillo (little feather, New Mexico)
staggerwort (from the symptoms of cattle after grazing on the plants)
[false, swamp] sunflower
tcatcamo'sikani (inhalant, Meskwaki)
yellow dog-fennel
yellow-star

According to Mish (1988), the first recorded use of the common name "sneezeweed" was in 1837, when it was used for *Helenium autumnale.* That may well be true, but sneezing had been associated with plant names for far longer. At least as long ago as Leonard Fuchs's *Herbal* of 1542, a European composite was called "sneezewoort." That plant was also called *Ptarmica* (Greek, to cause sneezing), and is now *Achillaea ptarmica* (Linnaeus [1753] 1957, Brako et al. 1995). Thus, knowing that plants caused sneezing is much older than the 16th century. One might say that "sneezeweed" dates at least to Galen, who lived between A.D. 129 and 200. He wrote, "The flowers of *Ptarmica* have the property of causing sneezing" (Galen in Meyer et al. 1999).

Today most cultures treat sneezing casually. However, sneezing was not always considered a trivial event (Panati 1987). In the 4th century B.C., both Aristotle (372–287) and Hippocrates explained sneezing as the body's attempt to expel some foreign substance. They further observed that sneezing associated with particular illnesses sometimes foretold death. It become a popular custom to offer benedictions such as "Long may you live!" or "May you enjoy good health!"

About 100 years later, the Romans held that sneezing was the body's attempt to expel a sinister spirit that would later cause illness. To suppress a sneeze was to court illness or even death. Therefore, the custom became one of offering "Congratulations!" when someone sneezed.

The Christian expression "God bless you!" was actually a popal fiat in the 6th century A.D. That was a time of a plague in Italy with one of the foreboding symptoms being severe sneezing. People often died shortly after beginning the sneezing bouts. Pope Gregory the Great asked the healthy to pray for the sick. He also asked that the current phrase of "May you enjoy good health!" be replaced and suggested the invocation "God bless you!" If no one was around to offer the blessing, the sneezer was advised to say "God bless me!"

Today, we have the residual of those 6th century events in multiple languages, and religions other than Catholicism. When someone sneezes, we say:

Gesundheit! (German)
Felicità! (Italian)
Salud! (Spanish)

One of the many useless bits of information that I still remember from high school Spanish classes was the following bit of doggerel. If someone sneezed and no one offered *Salud*!, we were to say—*Qué díos me bendiga hasta llegue a donde haya gente*! (May God bless me until I get where there might be some people!). It did not make sense at the time, but in historical context, it fits perfectly.

At least some indigenous Americans had their own version of the association between sneezing and maladies. The Cherokee used dry, powdered *Helenium* leaves to induce sneezing (Hamel and Chiltoskey 1975). The Menomini made a powder of flowers to cause a person to sneeze, and thus eliminate a headache and relieve a cold (Moerman 1998). The Meskwaki also inhaled the powder of the dried florets to relieve colds and catarrh (King 1984).

The Cherokees prescribed an infusion of the roots of *H. autumnale* to prevent menstruation after childbirth (Hamel and Chiltoskey 1975). The Comanches made an infusion of stems as a wash to lower fever. The Menomini also used the plant as an alterative (Moerman 1998).

The only records of Muskogean people using *Helenium* seem to be the Koasati. They used *H. amarum* in a sweat bath to reduce swelling, particularly that associated with dropsy (Taylor 1940).

Helenalin, a lactone apparently associated with sesquiterpenes, is known from all species of *Helenium* (Foster and Duke 1990, Hocking 1997). Helenalin kills fishes, worms, and insects (Foster and Duke 1990). The plants are poisonous to cattle and may cause contact dermatitis in some people. Both species are used identically in South Carolina, where people make a tea to reduce fever (Morton 1974). Morton (1974) records tenulin, a pseudoguaianolide, as the fever-lowering compound that is also responsible for poisoning grazing animals. However, Lewis and Elvin-Lewis (1977) attribute the poisoning of cattle to the glycoside dugaldin. Also present are aromaticin and amaralin (Morton 1974). Hocking (1997) lists the bitter glycosides tenulin and isotenulin in *H. amarum.*

Although what we now call *Helenium* was unknown to the classical Greeks, an association of Helen of Troy's name with healing plants by Linnaeus was appropriate (Baumann 1993). Helen gave a pain-soothing drink *nepenthes* to Telemachus and his companions to make them forget their warriors fallen in battle. Furthermore, a plant named *helenion* or *helenium* by the ancients was thought to have sprung from Helena's tears on the death of the seafarer Canapos who had taken her and Menelaus to Egypt after the fall of Troy. No one is sure what the ancients were calling *helenion*, but now a group of American plants commemorates this Greek heroine.

Helianthus

(Linnaeus used the Greek *helios*, the sun, and *anthos*, a flower)

Helianthus annuus. a. Habit. b. Ray flower. c. Disk flower. d. Achenes, two views. *Drawn by Regina O. Hughes*. From Reed 1971.

Helianthus annuus (annual)
áñil (blue, from use of seeds as dye, New Mexico)
auti cákhire (Catawba)
castill (probably through the mistaken idea that they were from Spain, La Bajada, New Mexico, Curtin 1947)
chimal-acatl [*chimalati*] (*chimalli*, shield, *acatl*, cane, Náhuatl, Mexico)

flor de sol (sun flower, Mexico); *flùr-na-greine* (*flùr*, flower, *na*, of the, *greine*, sun, Gaelic); *plùr-na-greine* (*plùr*, flower, *na*, of the *greine*, sun, Gaelic)
gigantón (big one, Mexico)
girasol (turns with the sun, Italian, Mexican Spanish); *mirasol* (looks at the sun, Spanish)
haash abé (*haash*, sun, *abé*, image, Mikasuki); *hvse-vháke* (*hvse*, sun, *vháke*, image, Muskogee); *hashi* or *hashi napakali* (*hashi*, sun, *napakali*, flower, Choctaw); *hashi' pakali'* (*hashi'*, sun, *pakali'*, flower, Chickasaw); *hvse hecv* (*hvse*, sun, *hecv*, literally "having sight," but surely meaning "seeing," Muskogee)
hivai (Akimel O'odham or Pima, Arizona)
ho-soñ-a (looking at you, Kiowa)
i'la akiktce'hi (*i'la*, sun, *akiktce'hi*, flower, Ofo)
kirik-tara-kata (*kirik*, eyes, *tara*, having, *kata*, yellow, Pawnee)
lampote (big lamp, Mexico)
maíz de tejas [*teja, Texas*] (literally Texas or "tile" corn, but also possibly based on the word *natasi* or *yatasi*, the name one tribe of Caddo called themselves, Mexico)
mapi'-i'-ti'a (*mapi'*, sunflower, *i'ti'a*, big, Hidatsa)
mi'-to-o-xthe (always looks to the sun; *mi'*, sun, *ta*, in that direction, Osage)
quisnaniquitonale (*ixneneuilia*, resembling, *quilitl*, greens, *tonilli*, cooked, Náhuatl, Guerrero)
soleil (little sun, Quebec); *tourne soleil* (turn with the sun, Quebec)
Sonnenblume (sunflower, German)
sunflower [sonne flower] (first used as a translation of *flos solis*, originally applied to *Heliotropium* in the 1560s. Probably first applied to *Helianthus* by John Gerarde in 1597 as "*Flos solis maior*. The greater Sunne flower." However, the first record of the sunflower in Europe was a drawing by Rembert Dodoens in 1568. Then Harriot called it *Planta Solis* in [1590] 1972.)
te'xlk la'k (*te'xlk*, flower, *la'k*, sun, Atakapa)
wahcha-zizi (*wahcha*, flower, *zizi*, yellow, Dakota); *zha-zi* (*zha*, herb, *zi*, yellow, Omaha-Ponca)
xaricarmata (Tarascan, Michoacán)
yendri (Otomí, Hidalgo)

Helianthus occidentalis (western)
fewleaf sunflower

Helianthus strumosus (scrofulous, from the thick-based hairs)
pale-leaf woodland sunflower [paleleaf woodland sunflower]

For decades, we have believed that the sunflower (*Helianthus annuus*) was domesticated in North America. That view was largely the result of Charles Heiser's research into the genus (cf. Heiser 1951, 1976). Now, a seed taken from an archaeological site in Central America has shaken that idea by predating all North American sites (Lentz et al. 2001). Heiser is resisting the new interpretation (cf. Heiser 2001, Lentz 2001a,b).

Regardless, when Europeans arrived, sunflowers had been cultivated in western North America since about 4200 B.P. (Fritz 2000b). Historic southeasterners were largely continuing the cultivation of the plants of the old Eastern Agricultural Complex, including sunflowers (Hudson 1976, Yarnell 1986, Smith 1992), which is often dated as beginning in the early second millennium B.C. In addition, Hudson (1976) writes, "And considering the importance of the sun in their belief system, the Southeastern Indians would not have failed to notice that the sunflower turns to face the sun when it rises, and then follows it across the heavens to where it sets in the west."

Just when the sunflower was introduced to Europe is not known, but it was there within the first decade of Columbus's arrival in the New World (Reveal 1992). Both Heiser (1976) and Hawkes (1998) agree that the first published record of the species in Europe was with Rembert Dodoens in 1568. However, Hawkes (1998) points out that by that date the species was widely cultivated in Europe.

The record in the east is patchy regarding who cultivated sunflowers and who did not. Harriot ([1590] 1972) saw them being cultivated among the Carolina Algonquians in the 1580s. Harriot wrote of sunflowers in 1588 when he said: "There is also another great hearbe in the forme of a Marigolde, about sixe foote in height; the head with the floure is a spanne in breadth. Some take it to bee *Planta Solis*: of the seedes heereof they make both a kinde of bread and broth" (Harriot [1590] 1972). Champlain found them grown near Lake Simcoe, Ontario, in 1615, where, in addition to maize, the Huron "plant likewise squashes, and sunflowers, from the seeds of which they make oil, with which they annoint the head" (Champlain in Gilmore 1919).

Bernard Romans ([1775] 1961) found the Choctaw mixing sunflower meal with maize. He wrote, "They cultivate . . . the *Helianthus giganteus* [*H. annuus*]; with the seed of the last made into flour and mixed with flour of the *Zea* they make a very palatable bread." Although no records of the other southeastern people have been found growing and eating sunflowers, Muskogeans are famous for mixing various seeds in their bread flour (see *Carya, Castanea, Fagus, Quercus*). Moreover, the foods of most of the Muskogean people were similarly made (Hudson 1976). Oddly, sunflowers were not mentioned by Bartram ([1791] 1958), nor does Moerman (1998) list a single eastern tribe as using them as food. For more information on indigenous use, see Heiser (1951).

Gilmore (1919) did not find sunflowers cultivated among the Dakota, Omaha, Pawnee, or Ponca. He did not find the Arikara, Mandan, and Hidatsa [Gros Ventre, Minitari] growing them, but they did. There is a long, instructive quote in Heiser (1976) on Hidatsa use of the sunflower. Indeed, the Hidatsa called the month of April *mapi-o'ce-mi'di* (sunflower planting moon); see also Waheenee et al. (1987).

Medicinal use of *H. annuus* is entirely among the western tribes in Moerman (1998). The Ojibwa, however, used *H. occidentalis* as a poultice to help bruises and contusions heal. The Iroquois used *H. strumosus* roots to worm their children (Moerman 1998), and the Meskwaki used the plant to treat lung troubles (King 1984). Roots of *H. strumosus* are edible and are now recommended as food for people with diabetes (Hocking 1997).

Heliopsis

(From Greek *helios*, the sun, and *opsis*, resembling, referring to the flowers)

Heliopsis helianthoides. From Britton and Brown 1898.

Heliopsis helianthoides (resembling *Helianthus*, which see)
[smooth] ox-eye [oxeye]
gi'ziso'bûgons' (small sun leaf, Ojibwa)
sunflower heliopsis (a book name)

In an unusual move, Linnaeus ([1753] 1957) gave the same species four names—*Buphthalmum helianthoides, Silphium helianthoides, Silphium solidaginoides,* and *Rudbeckia oppositifolia* (Reveal 1983). When Linnaeus made mistakes, he did it right. It was not until 1807 that the species was put in *Heliopsis.*

Described as a "bizarre individual and brilliant mycologist, born in South Africa, studied in Germany, lived in Paris," Christiaan H. Persoon created *Heliopsis* in *Synopsis Plantarum* in 1807 (Correll and Johnston 1970). That is now a genus of 15 upland herbs endemic to the Americas (Mabberley 1997). *Heliopsis helianthoides* occurs over much of southern Canada and eastern United States west to the Dakotas, Nebraska, Colorado, and New Mexico (Correll and Johnston 1970). Nearby to the south is *H. parvifolia,* growing in the Trans-Pecos area of Texas, New Mexico, Arizona, and south to Durango, Nuevo León, and Tamaulipas. Also in Mexico is *H. longipes,* which grows in Guanajuato, Querétaro, and San Luis Potosí (Martinez 1967).

Records have been found for only two tribes using *H. helianthoides.* The Ojibwa made a root decoction or chewed fresh roots and applied them to stimulate limbs (Densmore 1928). The Meskwaki made a medicine of the root for lung problems (King 1984).

In Mexico, *H. longipes* is called *chilcuan, chilcuague, peritre* [*pelitre*], or *peritre del país* (country *peritre*). The first two names are derived from Náhuatl *chili*, chile or *Capsicum*, and *mecatl*, thread, referring to the filiform roots that have a spicy taste (Martínez 1969). These plants were first noted by Francisco Hernández (1651) when he wrote about the "*Chilmecatl* seu *Ychcha*." The uses Hernández gave were virtually the same as those found by Martínez (1969).

The bark of the roots, when chewed, is spicy, numbs the tongue, and induces abundant saliva. The plant is added to beans as a condiment in place of chiles, or to alcoholic drinks to give them a special flavor. Roots are also chewed to relieve toothache. Alcoholic extracts have been used as a stimulant and as an anesthetic.

Root of *H. helianthoides* contains scabrin (butyl amide of an 18-carbon unsaturated fatty acid) and heliopsin (Hocking 1997). Scabrin is considered a promising source of insecticide (Mabberley 1997). *Heliopsis longipes* contains *N*-isobutyl-2,6,8-decatrienamide and scabrin. Both have been used as insecticides, especially to get rid of house flies (Hocking 1997).

Heliotropium

(From *heliotropion*, Greek *helios*, the sun, *trepo*, turning; ancient writers believed that the plant turned toward the sun in flowering)

Heliotropium. *Heliotropium curassavicum* (left). a. Flowering and fruiting branch. b. Apical section of inflorescence. c. Flower, from above. d. Flower, longitudinally dissected. e. Stigma. f. Floral diagram. g. Fruit. *Drawn by Priscilla Fawcett*. From Correll and Correll 1982. *Heliotropium polyphyllum* (right). *Drawn by P.N. Honychurch*.

Heliotropium europaeum (the Old World species known to settlers)

eliotropio (from Greek *heliotropion*, Italian); *héliotrope* (French); heliotrope (in English by about A.D. 1000); *heliotropio* (Spanish)

erba porraia (warty herb, Italian); *hierba verruguera* (warty herb, Spanish); *verrucaria* (warty herb, Portuguese)

Sonnenwende (turns with the sun, German)

tornassol (turns with the sun, Portuguese)

Heliotropium curassavicum (from the island of Curaçao)

absinthe bord de mer (seaside absinthe or *Artemisia absinthium*, a completely different kind of plant in the Asteraceae, Guadeloupe, Martinique)

alacranillo [*de playa*] (little [beach] scorpion, Cuba, Hispaniola); *hierba del alacrán* [*yerba de alacrán*] (scorpion herb, Dominican Republic); *cola de alacrán* (scorpion tail, Veracruz); *rabo de alacran de playa* (beach scorpion tail, Cuba)

ba:bad 'i:wagi (*ba:bad,* toad/frog, *'i:wagi,* eaten greens, Tohono O'odham, Arizona, Sonora); *cama de sapo* (toad's bed); *hierba del sapo* (toad herb, Sonora, Nayarit)

Chinese-pusley (Texas, Arizona)

cola de mico (monkey tail, Texas, Veracruz); *rabo de mico* (monkey tail, Yucatán)

cotorrera de playa (beach parrot, Puerto Rico)

cresta de galo (rooster comb, Brazil)

hant otópl (what the land sticks to, Seri, Sonora)

heliotropio cimarrón (wild heliotrope, Colombia); [alkali, salt, seaside, wild] heliotrope (Sonora, Arizona, Texas, Bahamas, Puerto Rico)

hierba del torojo (Sonora)

hierba del vidrio (glass herb); *pasto vidrio* (glass grass)

hoskos [*hoshos*] (Cahuilla of California)

jaboncillo (little soap)

kaakaichu sisivoda (quail's topknot, Akimel O'odham, Arizona, Sonora)

kakachu e'es [*kaakaichu e'es*] (quail's plant, Akimel O'odham, Arizona, Sonora); *kakaichu 'i:wagi* (*kakaichu,* Gambel's quail, *Callipepia gambelii*, *'i:wagi,* eaten greens, Tohono O'odham, Arizona, Sonora)

kokolode shimaron [*cocolode*] (wild *kokolode*, Netherland Antilles)

[sea, small seaside, wild] lavender ("sea lavender" is a name usually reserved for *Argusia gnaphaloides*, in the same family, Barbados)

pata de gallo (chicken foot)

pond weed (Bahamas)

potács camoz ("what thinks it is an iodine bush," *Allenrolfea*, Seri, Sonora)
quail-plant [quailplant] (maybe quail eat the seeds, Texas)
quinaquina bord de mer (seaside quinine, Guadeloupe, Martinique)
romarin blanc (white rosemary, Guadeloupe, Martinique)
rosa morada (purple rose, Nayarit)
sailor's tea
verveine bord de mer (seaside verbena, Guadeloupe, Martinique)

Heliotropium polyphyllum (many leaves)
[pineland, yellow] heliotrope (Florida, Bahamas)

Heliotropium curassavicum was described by Linnaeus ([1753] 1957). Before him, the plants had been discussed by European authors starting with Robert Morrison's list of plants cultivated in London in 1669. Although the species is native to the Americas, it had become naturalized in France by the time Thomas François Dalibard (1703–1779) published his *Flora Parisiensis* in 1749. Paul Hermann had the species in cultivation in Holland in 1687. All the pre-Linnaean writers called the plants *Heliotropium*, following the classical usage for *H. europaeum*.

In its native range of seacoasts and borders of other saline areas from Delaware to Florida, west to the western states, the Caribbean, Mexico, Central and South America, *H. curassavicum* has a history of use by many different people. Perhaps because other species are more potent, there are few records of use in the Caribbean. There a decoction is used to treat leukorrhea in Curaçao, and Cubans use it as a substitute for *H. indicum* (Morton 1981).

There are more records of people using *H. curassavicum* in the western United States than elsewhere in North America. Perhaps records of it are abundant because there are relatively few other species found in the western United States.

Although Curtin (1949) did not note use of *Heliotropium* among the Pima, it was mentioned earlier by a study by Frank Russel in 1908. At that time, "The upper part of the light yellowish root is dried and ground in mortars, dried again, and ground very fine upon a metate, when it is ready to be applied to sores or wounds after they have been washed" (Rea 1997). In his recent study, Rea (1997) found only that the modern Pima use the roots in a tea for sore eyes.

The Paiutes and Shoshone made a medicine to treat diarrhea from *H. curassavicum*, and a decoction served as a diuretic, emetic, and gargle for sore throat (Moerman 1998). The Shoshone of the Great Basin also used a decoction of the plant tops to bring out measles (Moerman 1998). The Tubatulabal of California took a decoction of the entire plant to relieve dysentery, and ate the seeds (Ebeling 1986). The Cahuilla formerly used it and had a name for it, but the informant could no longer remember what it was used for (Ebeling 1986).

Heliotropium curassavicum is used in Nayarit to help control coughs (Martínez 1969), and in Veracruz to treat asthma, fever, anemia, and scorpion stings (Vásquez and Jácome 1997). People use the plants as food and medicine in Belize (Balick et al. 2000). *Cresta de galo* is considered a depurative in Brazil (Hocking 1997).

Indians (unspecified) made a purple dye from *H. curassavicum* (Hocking 1997). The flowers of *H. polyphyllum* were used to make a yellow dye (Hocking 1997).

Like *H. angiospermum*, there are pyrrolizidine alkaloids in *H. curassavicum* (Mors et al. 2000). These chemicals are known to be liver toxic and cause veno-occlusive disease of the liver (Smith and Culvenor 1981, Foster and Hobbs 2002).

Heliotropium: Scorpion-Tail

(Greek *heliotropion*, from *helios*, the sun, *trepo*, turning)

Along the margins of hammock forests in southern Florida you are likely to see a small herb about 0.6 m (2 feet) tall with a flower stalk curved around like a scorpion's tail. That comparison is exactly what some people make when they see it, and the common name scorpion-tail (Florida, Puerto Rico) has been used for many years. Indeed, throughout its range, people have associated scorpions with *Heliotropium*. In southern Florida, *H. angiospermum* is the most common species. Since scorpions were once common where the plants grow in Florida, the name makes sense. These stinging animals were so common in the past that they prompted a comment from Jonathan Dickinson, who was forced to lie among them in a Jaega hut in 1699 (Andrews and Andrews 1945).

Other names that associate the plants with scorpions include *alacrán* (scorpion), *alacrán chico* (little scorpion), *alacrancillo* (little scorpion, Cuba, Hispaniola, Veracruz), *cola de alacrán* (scorpion tail, Veracruz), *flor de alacrán* (scorpion flower, Guatemala, Belize), and *rabo de alacrán* (scorpion tail, Cuba, Puerto Rico, Dominican Republic), or scorpion tail (Belize). Obviously, in different places, people associate the plant with different species, but in Florida the scorpion most likely to be found with the herb is *Centruroides gracilis.* This scorpion is known as the Florida bark scorpion, the brown bark scorpion, or the slender brown scorpion. In Cuba, this animal is known as *alacrán prieto* (black scorpion) or *alacrán azul* (blue scorpion). Of course, there are other role models for the plants elsewhere, but this is a widespread animal.

This allusion to the animal must be a tropical view, especially since the plants were put in the genus *Heliotropium* by Linnaeus and many of his predecessors. The name *Heliotropium* goes back to the classical Greeks, with ancient writers believing that the plant turned toward the sun while flowering. The name has even been rendered "turnsole" in English (from 1300s, based on Latin *tornare,* to turn, and *sole,* sun).

In Florida, the Seminoles say *ho:malastocî* (*home,* bitter, *lasetv,* taste, *oce,* small, Creek) or *hõ:mó:cî* (small bitter, Mikasuki). They usually black pepper (*Piper nigrum*) by that name, but sometimes they distinguish between the two with the modifier *-a:bî* (replica or resembling) for the *Heliotropium.* Although it has a name, Sturtevant's (1955) informants claimed it was not used. A similar comparison to a "pepper" is made by the Maya who say *nemax* [*nemaax, remax*] (*ne,* tail, *maax,* wild chile, Yucatán). That name is give because the inflorescences resemble the pedicel of the chile, *Capsicum.*

Others are impressed by the curved inflorescence, but compare it to other things. Some think of other tails, and say it is a *cola de mico* (monkey tail), dog's tail, or *rabo de mico* [*rabamàco, rabomíco, rabo mico*] (monkey tail, Yucatán, Nicaragua). Others think it looks like their chickens and say *crête de coq* (rooster's comb) or rooster comb (Bahamas, Virgin Islands). Some think it resembles musical instruments, particularly the *cabeza de violin* (violin head or crozier, Sonora).

Some of the names either refer to the inflorescence or pubescence, and their application is not clear. Since these plants are normally glabrous, perhaps the names originally referred to related species. The names are cat tongue (Bahamas, Virgin Islands), *pittone velue* (shaggy ring), and *veloutte* (velvety).

Several of the names make allusions that are lost to those outside their region of use. One of these is *cotorrerilla* (little parrot, Puerto Rico), probably meaning that birds eat the flowers and/or fruits. Maybe *creeré cacao* (Haiti) means that the plant grows with cacao (*Theobroma cacao*). What horse bush, *mocos de pavo* (turkey snot [mucus], Dominican Republic) and *mont joli* (pretty mount, Haiti) allude to is uncertain. There is a hotel called *Mont Joli* in Cap-Haitien referring to a nearby mountain, but I would not be surprised if the plant name referred to a female body part.

Translations are not available for some names. These include *cocolode* [*kokolode*] (Bahamas, Curaçao, Virgin Islands), *laven co'va* (Mayo, Sonora), *soguilla* (Sonora), and *zimbaclot* (Haiti).

A number of names have medicinal references. Among these are *herbe à chiques* (tobacco herb, Haiti), *herbe à clous* [*z'herbe à clous*] (boil herb, Haiti), *herbe à malingres* (herb for the sickly, Haiti), and sorebush. There are also comparisons within the same family, as in *borraja* (borage, *Borago officinalis*), another noted European medicinal herb.

Although there are few direct records of *H. angiospermum* having been used as a remedy for eye problems, the common names suggest that it has. Indeed, Ayensu (1981) and Morton (1981) record that in Jamaica and the Caicos Islands the plant is boiled and the decoction used to treat eyes that burn.

One name given that records the use as eye medicine is eyebright (Bahamas, Virgin Islands) and the variant bright-eye bush (Bahamas, Virgin Islands). The name "eyebright" in Europe is usually given to *Euphrasia officinalis* and its relatives. The first record found of that name being used was 1533 when Sir Thomas Elyot wrote his list of "Thynges good for the eyes: Eyebryght ..." in his medical book, *The Castel of Helthe.* That plant had the reputation in Europe for being a cure for weak eyes between 400 and 500 years ago. A variant of "eyebright" used in the Bahamas and the Virgin Islands is "sunlight."

Along the same line, *H. angiospermum* has been considered a variant of clary (*Salvia sclarea,* Lamiaceae) with vernacular names including small wild clary, white clary, and wild clary. Since about A.D. 1000 variants of the word "clary" have existed and this aromatic mint of southern Europe was used to adulterate wines, as a condiment, and as a medicinal herb. In the 1500s, the name also was transformed into clear-eye, Godes-cie, and see-bright to emphasize its use for treating the eyes.

Although it is used for the eyes, primary among the recorded uses for *H. angiospermum* is as a remedy for skin problems. In Cuba, scorpion-tail is used on burns. People take the dried leaves, make them into a powder, wet the burned spot, and shake on the powder (Roig 1945).

Hispaniolans use the leaves in decoctions to wash sores and for general cleaning of the baby's skin at birth, or for peeling baby skin (Liogier 1974, Ayensu 1981). A decoction also is taken internally to cure skin sores, and the pounded leaves are mixed with petroleum jelly and applied to boils. Boiled plants are used to treat headsores, boils, fresh cuts, and wounds. Mexicans consider the plant a remedy for nosebleed, gum disease, dysentery, flu, and skin problems (Standley 1920–1926, Vásquez and Jácome 1997). A decoction is used as a bath to treat lightheadedness, jaundice, and rheumatism (Liogier 1974).

The few studies that have been made of this species of scorpion-tail record that *H. angiospermum* contains arginine, amino alcohols, pyrrolizidine alkaloids, and ornithine decarboxylase (Birecka et al. 1985, 1987). In general, the pyrrolizidine alkaloids are potent poisons, and children have been killed by overdoses of the plant decoction in Barbados (Morton 1981). The pyrrolizidine alkaloids are notorious in the genus and family (Dattagupta and Datta 1977, Mohanraj and Subramanian 1978, Birecka et al. 1980, Davicino et al. 1988, Asibal et al. 1989, Roeder et al. 1991, Constantinidis et al. 1993, Trigo et al. 1993, Farsam et al. 2000). Other studies suggest that some species have antitumor principles (Kugelman et al. 1976), insecticidal and antifungal properties (Villarroel et al. 2001), and wound-healing effects (Reddy et al. 2002).

Several other species in the genus are used in a similar way as medicines. In the Caribbean, the best known and most widespread is *H. curassavicum.* That species ranges from Delaware, south through Florida, the Bahamas, and the West Indies, and from Oregon south through Mexico and Central America, and in South America to Chile and Paraguay. *Heliotropium curassavicum* not only shares several common names with *H. angiospermum,* but both have essentially identical uses.

Probably even better known is *H. indicum,* as it was also in the Old World before Columbus's voyages. This species is the most widely used of all. It is known in Ayurvedic medicine by the Sanskrit names *hatisunadá, srihastiní,* and *bhúrúndí.* This species has been carried around the world for its medical applications.

Insects applied the chemicals in these plants long before humans. Both moths and butterflies sequester the pyrrolizidine alkaloids. Moreover, like humans, they have different uses for the chemicals. Larvae of the arctiid moth *Hyalurga syma* store the poisons in a predator-defense system (Trigo et al. 1993). The Danaid butterflies (the monarchs, *Danaus plexippus,* and relatives) must use them to narcotize females before copulation can occur (Pliske and Saltpeter 1971, Edgar et al. 1974, Amladi 1975, Pliske 1975). Humans are relative newcomers in the chemical usage game.

Hepatica

(Feminine form of *hepaticus,* pertaining to the liver, referring to the color and shape of the leaves, and to their curative properties)

Hepatica nobilis. From Britton and Brown 1897.

blåveis (*blå,* blue, *vei,* road, Norwegian)
dithean Abraoin (*dithean,* daisy, *Abraoin,* April, Gaelic)
Edellebere (noble berry, German)
erba trinita (Italian); herb trinity; *Herba Trinitalis* (Sir T. Browne wrote in 1646 that the plant "obtaineth that name onely from the figure of its leaves, and is one kind of liverworte or Hepatica"); *trinitaire* (of the Trinity, Quebec)
hepatikos (Greek); *hepar* (liver, Latin); hepatica (recorded in English in 1578 in the translation of Rembert Dodoens's *Cruydeboek* of 1554 where he wrote, "The leaves of *Hepatica* are broade, and diuided into three parts"); *hépatique* (French); *hierba del higado* (liver herb, Spanish)
Leberblümchen (little liver flower, German)
liver-leaf [liverleaf] (dating from 1851)
liverwort [liver-weed] (plant for the liver; "liverwort" first used in English about A.D. 1100; perhaps originally applied to *Marchantia,* but later including *Hepatica*)
trefoil (a name now usually given to *Desmodium,* Fabaceae); *trifoglio epatico* (three-lobed liver, Italian)
wapukwunes (Plains Cree)

Hepatica nobilis (well-known) (= *Anemone americana, H. acutiloba, H. americana, H. hepatica, H. triloba*)
asawûsk (yellow weed, Potawatomi)
blue anemone; noble agrimonie (1578 translation of Dodoens's *Cruydeboek* of 1554)
kidney-wort
[common] liverwort; nobel liver-wort [nobel liverwort, nobel lyverwurte] (in English in 1578 translation of Dodoens's *Cruydeboek* of 1554; "liver-wort" akin to *Leberkraut* in German and *leverkruid* in Dutch); round-lobed liver-leaf [roundlobed liverleaf]; three leafe lyverwurte

(1578 translation of Dodoens's *Cruydeboek* of 1554)
squirrel-cups

The history of these herbs dates from the classical Greeks and Romans. *Hepatica* is a classic example of the Doctrine of Signatures; people believed it would cure problems of the liver. Problems treated included cowardice, freckles, and indigestion (Dobelis 1986). Herbalists knew and recommended the plant. Dodoens (1578 translation) wrote, "The Hepatica or Noble Lyverwurte is a soveraigne medicine against the heate ... of the Lyver." Culpeper (1653) added, "It is a singular good herb for all diseases of the liver, both to cool and cleanse it, and helps inflammations in any part, and the yellow jaundice ... a good remedy to stay the spreading of tetters, ringworm, and other fretting and running sores and scabs."

When the Europeans arrived in the New World, they found these woodland herbs growing from northern Florida into Canada. Although botanists have divided them into at least four species over the centuries, the tendency today is to recognize them as varieties of a single Circumboreal species.

Several of the eastern tribes used the herbs for liver ailments, digestive upsets, laxatives, and to wash external injuries. The Cherokee made an infusion as an emetic for abdominal pains, as a laxative, for the liver, and made a compound medicine rubbed on swollen breasts and drunk for indigestion (Taylor 1940, Hamel and Chiltoskey 1975). The Iroquois used *Hepatica* during childbirth, to prevent conception, to treat diarrhea and stiff muscles, and for shortness of breath in runners. Roots were chewed (by women?) to bewitch men, and used to tell fortunes (Moerman 1998). The Menomini used liverwort to treat female problems, especially leukorrhea, and dysentery (Smith 1923). The Meskwaki used an infusion to treat eyes and mouth problems (King 1984). The Nanticoke chewed petals to prevent summer fevers (Moerman 1998). The Ojibwa made a medicine of the roots as an abortifacient, to stop convulsions in children, and for liver ailments, and they made a poultice for bruises and inflammations. The Potawatomi took an infusion of root and leaves to cure vertigo (Smith 1933).

Use of these herbs had boom and bust years through the 1800s. In 1830, there was a great demand for potions containing the plants during the flu epidemic in New York (Hocking 1997). Then, during the Civil War, Porcher (1863) wrote: "A tonic and astringent, supposed by some to possess deobstruent virtues. It has been used to a considerable extent in haemoptysis and chronic cough; but Wood says it has fallen into neglect." Then, there was a "liver tonic" boom that consumed 450,000 pounds of leaves in 1883 alone (Foster and Duke 1990). But, by the time Millspaugh (1892) wrote his book *American Medicinal Plants*, he said, "It is now falling into disuse on account of its mild properties, forming as it does simply a slightly astringent, mucilaginous infusion." During its periods of popularity, it was collected for the drug market and sold as *Folia Hepatica*.

During its years of popularity, the primary use was as a mild mucilaginous astringent, to treat coughs, fevers, and liver ailments (Meyer and Meyer 1986, Foster and Duke 1990). Duke et al. (2002) do not even mention the plant; however, Foster and Duke (1990) considered it mildly astringent, demulcent, and diuretic. A leaf tea was used as the diuretic. Those of us old enough remember an old patent medicine called *Sal Hepatica* [liver salt] were misled. That laxative had only borrowed the plant's name (Ward 1999).

Hocking (1997) noted that *Hepatica* contains tannins, mucilage, and the glycoside hepatrilobin. He also found that the plants had been used to treat bronchitis, phthisis (tuberculosis), and influenza.

A single reference has been found to using the herbs as a dye source. The Potawatomi used roots to make a dye for mats and baskets (Smith 1933).

Heterotheca

(From Greek *heteros*, different, and *theca*, case, alluding to the dissimilar ray and disk achenes)

Heterotheca subaxillaris. From Britton and Brown 1898.

Heterotheca subaxillaris (having flower clusters near the axil where the leaves join the stem) *árnica* (name used by Kikapoo displaced to Coahuila, Mexico; see Latorre and Latorre 1977)

camphor-daisy [-weed] ("camphor" was taken into English from French *camfre* or *camphre*, ultimately from Greek *caphura* and Arabic *kafur*, a vegetable oil with a bitter taste and characteristic smell; in Eurasia, the compound was famous as an antaphrodisiac; by at least 1570 it was combined with other words to indicate distinctive species)

golden [gold]-aster ("golden," in reference to color, has been combined with various plants since at least the 1570s, and probably before; the OED 1971 lists 33 combinations, and "aster" is not among them)

When I first arrived in Florida in 1970, I soon found three common species of golden-asters. No flora of the region had been published since Small's (1933) *Manual of the Southeastern Flora*, so I used that book to identify the plants. One of them was *Pityopsis graminifolia*, another was *Chrysopsis floridana*, and the third was *Heterotheca subaxillaris*. The following year, Long and Lakela (1971) published *A Flora of Tropical Florida*, and two of the three names changed. Instead of *Pityopsis graminifolia*, they called it *Heterotheca graminifolia*. *Chrysopsis floridana* too was placed in the same genus and became *Heterotheca floridana*. Not knowing any better, I began following Long and Lakela (1971).

Then my friend John Semple (1977, 1980, 1996, Semple et al. 1980) studied the groups and put all three back in the genera where Small (1933) had them. He did, however, redefine those three genera.

At first, I resisted John's changes, but I finally studied them in detail to see what he was talking about. *Heterotheca* contains 25 species and all are native to the southern parts of North America, but some extend to Argentina.

There are chemical differences, too, that Semple and his colleagues did not discuss. *Heterotheca subaxillaris* is the most fragrant of them, as witnessed by the common name comparing it with camphor. My students never had problems remembering it after smelling crushed leaves and hearing its common name.

Ajilvsgi (1984) wrote that livestock avoid camphorweed. That comes as no surprise although Semple had found herbarium specimens with notes that it "might be used as a forb locally." He and my students agree they would be "very hungry cows."

There is no record of indigenous or other people using camphor-weed in the United States, but Semple has seen herbarium labels indicating that Mexicans use them for medicine. It turns out that the "Mexicans" were displaced Kickapoo from southern Wisconsin who moved there in 1852 to avoid persecution in the United States. They used *H. subaxillaris* (reported as *H. latifolia*) to treat heat rash, burns, and itch, and to ease menstrual pain (Latorre and Latorre 1977).

Moerman (1998) found the Cheyenne, Hopi, Isleta, Luiseño, and Navajo using *H. grandiflora* and *H. villosa*, basically across the southwestern United States. It would be logical for a similar use to extend to the species through the southeastern states.

Martínez (1969) found *H. inuloides* was called *acahual* (from Náhuatl *acahualli*, dry herbs to ignite ovens), *false árnica* (false arnica), and *cuauteteco* [*akawtomitl*] (from Náhuatl *cuahuitl*, tree, and *teteco*, a passive form of the verb *teki*, to cut or to pick/pluck) in Mexico. In Aguascalientes, Chihuahua, Guanajuato, Hidalgo, Jalisco, Nayarit, Nuevo Leon, Oaxaca, San Luis Potosí, the Valley of Mexico, and Veracruz an alcoholic decoction of flower heads is used to treat bruises, toothache, stomach problems, and coughs (Vásquez and Jácome 1997). *Heterotheca inuloides* does not contain the alkaloids found in the European *Arnica montana* (Martínez 1969). Von Reis and Lipp (1982) found the same species recorded for the same use on a specimen collected by G.B. Hinton (847, GH?) in 1932. Altschul (1973) found that *H. leptoglossa* has been used to treat colds and flu in Mexico. Semple (personal communication April 2003) considers that taxon part of *H. inuloides*.

Both *H. subaxillaris* and *H. grandiflora* contain sesquiterpenes. Bohlmann and Zdero (1979) and Bohlmann et al. (1982, 1985) found cadinane derivatives in *H. subaxillaris*, and Bohlmann et al. (1979) and El-Dahmy et al. (1986) found cadinane and norcadinane derivatives in *H. grandiflora*. Wollenweber et al. (1985) found both flavonoid glycosides and resins in *H. grandiflora* and *H. subaxillaris* (as *H. psammophila*). Jakupovic et al. (1987) found humulene derivatives in *H. villosa*. A related species in the southeast, *H. camporum*, contains flavonoids with both insecticidal and antibacterial properties (Waage and Hedin 1985), although no records have been found that it was used as a medicine.

Hocking (1997) noted that *Heterotheca* contains "volatile alcohols, probably borneol." Lincoln and Lawrence (1984) and Duschatzky (1998) described the volatile constituents of the leaves of *H. subaxillaris*. Soon afterward, Dellacasa et al. (1999) found that those essential oils exhibited antimicrobial activity against *Bacillus subtillis* and *Escherichia coli*.

Most of the chemical literature is on *H. inuloides*, probably because it is the most famous herbal medicine. The plant has long been known to contain coumarin, galangin, kaempferol, quercetin, and rhamnetin (quercitin 3,0 rhamnoglucoside) (Kubo et al. 2000, Delgado et al. 2001). It also contains sesquiterpenoids, as well as flavonoid aglycones, flavonoid glycosides, and terpine derivatives (Bohlmann et al. 1976, Jerga et al. 1990a,b, Kubo et al. 1995, 1996, Haraguchi et al. 1997). A series of studies indicates that these chemicals are analgesic, anti-inflammatory, antimicrobial, and both cytotoxic and antioxidative (Kubo et al. 1994, 1996, Haraguchi et al. 1996, 1997, Gene et al. 1998, Segura et al. 2000, Delgado et al. 2001).

Hibiscus

(Dioscorides, fl. A.D. 40–80, recorded the Greek name *ibiskos*, the mallow; Pliny, A.D. 23–79, used *hibiscus* and *ebiscum*, the marsh-mallow; Linnaeus reapplied *Hibiscus* and retained *Althaea officinalis* for the marsh-mallow)

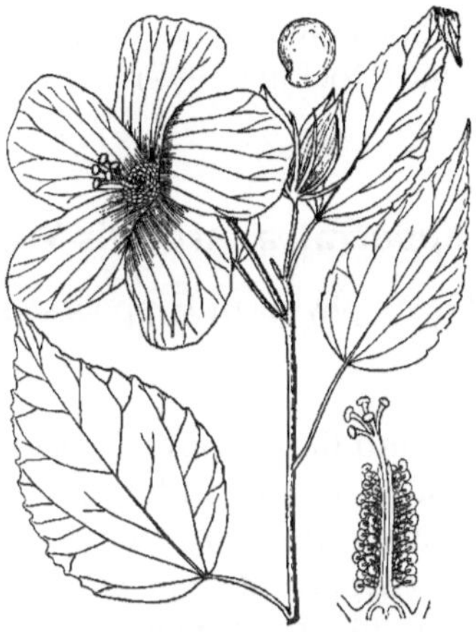

Hibiscus moschatus. From Britton and Brown 1897.

hibiscus (in use by 1706 when a book on European medicines included "*Hibiscum* or *Hibiscus*, the Herb Marshmallows, of known Vertue against the Stone and Gravel"); *ibisco* (Italian)
ketmie (from Arabic *khatmi*, the name of *Hibiscus esculentus*, French)
Roseneibisch (rose hibiscus, German)
Stundenblume (hour flower, German)

Hibiscus moscheutos (from its odor, suggesting the musk-rose)
breast-root ("In some parts of Virginia its root is used in the form of an emollient poultice in tumors of the breast, whence the name Breast-root," cf. Castiglioni 1790 in Coffey 1993)
crimson-eyed rose-mallow [crimsoneyed rose-mallow]; mallow rose (New York)
marsh hibiscus [mallow]; [swamp rose, water] mallow
musk [musk-plant] (Ohio)
[common] rose mallow (rose-mallow was in use by 1731 when Philip Miller included the name in his *Gardeners Dictionary* for *Althea*; then Linnaeus called it *Hibiscus moscheutos*, to make the hibiscus the musk-rose in 1753, but that was in Latin; rose-mallow was not applied to *Hibiscus* until Asa Gray used it in 1866)
wild cotton (a comment on the similarity of *Hibiscus* to *Gossypium*)

Europeans knew at least two species of *Hibiscus* when they arrived in the New World, although neither was native to most of Europe. *Hibiscus syriacus* (Syrian ketmia) and *H. trionum* (bladder ketmia) had both been introduced and cultivated. They were, however, quite familiar with the "mallows," having used *Althea* and *Malva* for medicines since at least the times of Pythagorus (fl. 6th century B.C.), Hippocrates (ca. 460–ca. 377 B.C.), and Theophrastus (372–287 B.C.) (Porcher 1863, Meyer et al. 1999).

According to Jacque P. Cornut in his book *Canadensium plantarum* of 1635, these American plants resembled the *Rosa moscheutos plinii* (the Musk-Rose of Pliny). That herb is now the cultivated *Rosa moschata*, native across southern Europe, northern Africa, and western Asia, but naturalized in North America (Fernald 1950, Hortorum 1976).

The only record of indigenous use seems to be among the Shinnecock, who lived on what is now Long Island, New York. They used an infusion of dried *H. moschatus* stems to treat bladder infections (Moerman 1998).

Settlers, however, made great use of members of the Malvaceae. Porcher (1863) wrote simply that *H. moscheutos* "is possessed of demulcent properties; a convenient substitute for" *Malva rotundifolia.* That species he recommended because it "possesses valuable demulcent properties. . . . It is very emollient, and is employed in catarrhal, dysenteric, and nephritic diseases, and wherever a mucilaginous fluid is required. It is administered in the shape of emollient enema, and it forms a good suppurative or relaxing cataplasm in external inflammations." He further noted that "*Pythagore regardait leur usage comme propre à favoriser l'exercise de la pensée*" (Pythagorus regarded its usage to help the exercise of thinking). And, he added, "Hippocrates employed it as we do, for gargles and collyriums, as an application to heated and inflamed parts, as a vehicle for pectoral and anodyne medicines, and for those administered in diseases of the urinary passages."

Coon (1974) wrote, "Readers should learn to know the values in *all* the mallows." He then recommended the infusions of the leaves, with one teaspoon of leaves to one cup of boiling water, "to loosen coughs and relieve sore throats, especially if honey is added to each dose." He continued that it was also good for irritations of the bowels, kidneys, and urinary organs. Externally, he thought the members of the family were good for poultices for inflammations and bruises and for reducing the pain and swelling of bee and wasp stings.

Foster and Duke (1990) wrote that *H. moscheutos* "abounds in mucilage." They noted that the leaves and roots, like those of related species and genera, were used as demulcents and emollients in treating dysentery, lung, and urinary ailments. Hocking (1997) listed the seeds as "cordial, stomachic, demulcent, emollient, nerve sedative; roots and leaves sedative."

Hieracium (including *Pilosella*)

(From Greek *heirax*, *heirakos*, hawk; Pliny, A.D. 23–79, believed that hawks ate the plants to sharpen their eyesight)

ɬawohokv [*rawohokv*, *tawohokv*, *rawvhokv*, *tawvhokv*] (Muskogee)
Habichtskraut (hawk weed, German); hawk-weed [hawkweed] (appeared in Anglo-Saxon about A.D. 1000; mentioned as medicinal in the herbals by William Turner in 1562 and John Gerarde in 1597)
hårsveve (*hårs*, hair, *veve*, weave, Norwegian)
pilosella (little hairy one, Italian); *piloselle* (French)
srubhan na muice (pig's snout, Gaelic)

Hieracium gronovii (in honor of Jan F. Gronovius, 1690–1762)
vein-leaf, [hairy] hawkweed

Norway has at least 24 "kinds" of hawkweed that are recognized (Mossberg and Stenberg 1994). A major problem in recognizing species in the genus is that there is so much apomixis that individual populations often seem distinctive (Fernald 1950, Mabberley 1997). Early experiments with genetics on this genus failed because it was not recognized that they were not outcrossing plants. Most of the plants are triploid or tetraploid apomicts and the modern trend is to call them "microspecies" (Mabberley 1997).

Hocking (1997) simply wrote that *Hieracium gronovii* was used medicinally. No other records have been found regarding this species, but Moerman (1998) lists seven others (including European *H. pilosella*) as having been given as a tonic, to treat diarrhea, and to help cure sores. These species were used by the Cherokee, Iroquois, Navajo, Ojibwa, Okanagan-Coville, and Rappahannock, people living from New England to Washington and British Columbia, and from the Carolinas to Arizona.

The wide distribution of the pharmacopoeias in which these *Hieracium* were used suggests independent discovery of their virtues. However, the European species were ancient remedies. John Gerarde wrote in 1597 of *H. aurantiacum*: "The decoction or the distilled water of this herbe taken inwardly, or outwardly applied, conduce much to the mundifying and healing of greene wounds; for some boyle the herb in wine, and so give it to the wounded patient, and also apply it outwardly." Culpeper (1653) considered *H. pilosella* a "cooling, somewhat drying and binding herb for a wide range of ailments."

Many species contain inulin. *Hieracium pilosella* contains antibiotic components, including umbelliferone and luteolin (Hocking 1997).

Hippocratea: Arthritis-Vine

(Named for Greek physician Hippocrates, ca. 460–ca. 377 B.C.)

Twining up into the canopy of trees in swamp and hammock forests in southern Florida is a plant known locally as arthritis-vine. That name is applied because the nodes of *Hippocratea volubilis* (climbing) are swollen and look like the similarly inflamed joints of arthritic humans. Although appropriate, and useful to new students learning to recognize the sometimes bewildering tropical species, that name is surely a recent application. Neither Chapman (1897) nor Small (1933) includes the vernacular name; they simply call it the "Hippocratea family" (Hippocrateaceae). More recently, Wunderlin (1998) dubbed the plants "medicine vine."

While there should be multiple medical uses for plants with those common names, and with the generic name it has, there are not many listed. Indeed, Morton (1981) does not include the species, but discusses only *H. celastroides* from Mexico and Central America.

Linnaeus's knowledge of the single species he described, *H. volubilis*, came completely from Charles Plumier and Pierre Barrère, explorers of the Caribbean. The choice to name these plants after that famous Greek physician followed the lead of Charles

Plumier, who wrote, "*Hippocrates Cous, seu in Insula Coo natus.*" (Hippocrates of Cous, or growing on the island of Coo). Several of the species contain references to Greek philosophers, including Theophrastus and Dioscorides in the following two entries. There is nothing in the comments by these authors that the climbers are used for medicines. Further, Linnaeus did not cite his own *Materia Medica* for the species, although he had seen it in cultivation at the *Hortus Cliffortianus.*

The genus *Hippocratea* was named by Linnaeus for the Greek doctor Hippocrates, the son of Heraclides who was the contemporary of Herodotus, Socrates, Plato, and Democritus. Many books consider Hippocrates the "Father of Medicine." Indeed, he had many "modern" ideas about medical practice, including advocating good diet, fresh air, clean water, and exercise. He separated medicine from philosophical speculation and mythology. He coordinated the available information on plant remedies and prescribed them when they appeared to be indicated by his diagnosis of the disease.

Not everyone agrees that the group should be split out into the distinct family Hippocrateaceae, which was created by Antoine Laurent de Jussieu in 1811. As a family, it contains more than 300 species, and is widespread in the tropics. Most species are traditionally put in either *Hippocratea* (100 to 120) or *Salacia* (200). Others divide *Hippocratea* into 12 genera and *Hippocratea sensu stricto* becomes monotypic (*H. volubilis*). Everyone agrees that those species are close to Celastraceae, and some include all the species within that family.

Regardless, Florida has the single species *H. volubilis,* which also grows in Cuba, Hispaniola, Puerto Rico, some of the Lesser Antilles, and from Mexico (Veracruz, Chiapas, Campeche, Yucatán) to northern Argentina. Oddly, the vines are absent from the Bahamas, Barbados, and Jamaica. Although not included in the recent *Flora de Nicaragua* (Stevens et al. 2001), the species does occur there. Perhaps the spotty distribution in the Caribbean is because the winged fruits are dispersed by wind. Some of the islands (e.g., Barbados) where the plants are not known are outside the "normal" reach of several wind-dispersed taxa.

Common names for the scramblers seem to focus on the fruits and colors. For example, several vernacular names compare the fruits to almonds (*Prunus dulcis,* Rosaceae). These comparisons include *bois z'amande* (almond tree, Guadeloupe, Martinique), *liane z'amande* (almond vine, Guadeloupe, Martinique), and *Waldmandelbaum* (forest almond tree, German). In Amazonian Brazil it is called *fava de arara* (macaw's bean). Those comparisons are made because the seeds are edible, although somewhat bitter. Only Roig (1945) makes the comparison to taste with almonds, and the seeds are used to make *horchata,* a mixture of ground seeds, sugar, and water that somewhat resembles a milkshake. The seeds contain 50% oil, which is odorless and colorless.

The Maya, because of fruit shape, say *tulubalam* (jaguar's ears, *t'ul,* long ears, *balam,* jaguar, Yucatán). Still, the Mayans exaggerate the resemblance to the feline's ears. The fruits are samaroid mericarps, usually three in a group, flattened, oblong or elliptic, and 4 to 6 or sometimes 8 cm long and about 3 cm broad. The fruits are not really much like the rounded ears of *Panthera onca,* but the Maya see some resemblance. Given the importance of this top predator in the Mayan pantheon, it would not be a surprise to learn that the name was an allusion to how well it could hear.

It is unclear whether or not the people who attribute colors to the climbers do so because of physical or medical properties, or some other reason. Regardless, in Puerto Rico and Brazil *H. volubilis* is considered a *bejuco prieto* (black vine) or *cipó preto* (black vine). Yet to the Haitians it is a *liane blanche* (white vine), while it is *liane rouge* (red vine) in Guadeloupe and Martinique. These colors are not obvious in cut stems, so they must come from some other source.

Two names are possibly medical allusions, although one of them is questionable. In Cuba, it is *amansaguapo* (man tamer), either a medical commentary or the result of someone trying to pass through a tangle of the stems. Perhaps, instead, that name is related to the Hispaniolan application of the slender stems as substitute ropes (Liogier 1974). However, the name *bejuco de vieja* (old woman's vine, Cuba) is surely identical in allusion to "arthritis vine."

No meanings have been found for three names. However, *jaiquimey* (Hispaniola) shares syllables, although not their sequence, with the common (*jaimiqui*) and specific names (*Manilkara jaimiqui*) applied to the wild sapodilla (Sapotaceae). It may be that all three words are based on the name of the island Jamaica. The Hispaniolan name *mayit* resembles a Taino word, while *zorain* might be "disorder" (*z,* of the, *orain,* alternate spelling of *orage,* or disorder, Haiti), in reference to the twining stems. Another reference to twining stems is barracuta tietie in Belize, where they make cords of them (Balick et al. 2000).

Although people use the plants as medicines, that apparently is not a common application. In Cuba, the seeds produce an emollient oil, and the flowers and leaves are pectoral, i.e., good for inflammation of the chest and larynx (Roig 1945). The *horchata* made from the seeds is considered tasty and pectoral. Comments by Standley (1920–1926) suggest that historically *H. volubilis* was more important as a pectoral, and as a remedy for snakebite. In Brazil, plasters of leaves are used to cure infected sores (LeCointe 1945), to control coughing, and as an expectorant (Roig 1945, Mors et al. 2001). According to Roig (1945), the flowers and bark are a febrifuge.

Other species in the genus are better known for different applications. Related species in Mexico are made into a tincture or paste used to kill parasites on the human body (Standley 1920–1926). Some examples include *H. acapulcensis,* which is called *hierba del piojo* (Michoacán, Guerrero), *cuanabichi* (louse-killer, Zapotec, Oaxaca), *mata-piojo* (louse-killer), *hierba del piojo* (louse herb), *bejuco de piojo* (louse vine), *pepitas de piojo* (louse eggs), and *semillas de piojos* (louse eggs). Similarly, people in Yucatán are recorded as taking a decoction of *H. celastroides* as a tranquilizer (Morton 1981), in addition to using it to kill lice on humans and animals. Maybe the name knot vine for the Australian species *H. barbata* also indicates medical use.

There are a number of bioactive compounds within the genus. *Hippocratea volubilis* contains triterpene alkaloids (Alvarenga and Ferro 2000), and other species have benzyl isothiocyanate, cyanogenic glycosides, essential oils, quinones, saponins, sesquiterpenes, triterpenoids, and volatile compounds (Palacios et al. 1989, Mata et al. 1990a,b, Calzada et al. 1991, Iwu et al. 1991, Mata 1993, Calzada and Mata 1995, Pérez et al. 1995, Jimenez-E. et al. 2000). These chemicals have antibacterial properties, and are antiinflammatory (Iwu et al. 1991, Pérez et al. 1995).

Considering all the applications that people have made of the plants in the Americas, it seems curious that the genus *Hippocratea* appeared in the journal *Economic Botany* only four times in the first 50 years of the journal's existence (Kaplan 2001). Of those times, there is only a single reference to American species (Palacios et al. 1989). The others are comments on Old World species (Irvine 1952, Jain and Trarafder 1970, Secoy and Smith 1983). At least two factors are involved in that paucity of mention. The first factor is the generally ready availability of more dependable and better substitutes for the nutritional and medical applications formerly given *Hippocratea*. The second factor must be the continued destruction of the forests where these vines grow. These two factors, although usually considered independently of each other, are working to erode the knowledge we humans formerly had of the living world around us. We are sterilizing our surroundings and putting limits on our imaginations.

Hippomane

(Theophrastus, 372–287 B.C., used the Greek words *hippo*, horse, and *mania*, madness because the fruits of the original plant, some say *Datura stramonium,* made horses crazy when they ate them; Linnaeus substituted the name for an American plant)

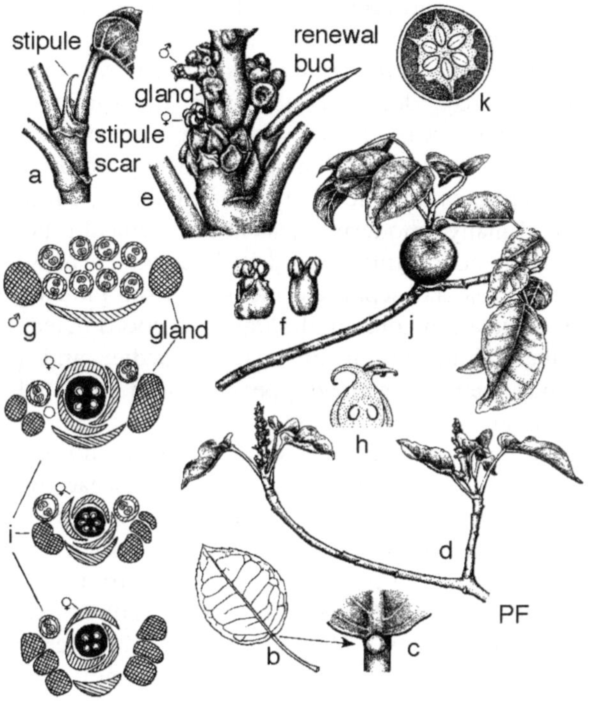

Hippomane mancinella. a. Terminal bud with stipule. b. Leaf outline. c. Base of leaf, with petiolar gland. d. Flowering branch. e. Detail of flowering spike, with renewal bud in axil of leaf to right. f. Staminate flowers, with and without perianth. g. Diagram of staminate flowering unit (glands cross-hatched). h. Pistillate flower, longitudinally dissected. i. Diagram of lower flowering unit with pistillate flowers, staminate flowers, and glands, in three different arrangements. j. Twig with fruit. k. Fruit in transverse section. *Drawn by Priscilla Fawcett.* From Correll and Correll 1982.

Hippomane mancinella (little apple)
arbol de la muerte (tree of the dead, Oaxaca)
hinchahuevos (swollen testes [literally, eggs], Mexico)
kinsah-uinik (*kinsah,* kill, *uinik,* man, Maya, Yucatán)
manchineel [*mançanila, manzanillo,* manchaneel] (little apple, English and Spanish variants, Oaxaca, Veracruz, Cuba, Puerto Rico, Dominican Republic, Belize, Panama, Colombia, Venezuela, Virgin Islands); *manzanillo de la playa* (little beach apple, Puerto Rico)
pinipiniche [*penipeniche*] (Cuba)

Oviedo gave perhaps the first account of these trees in the 1500s. He wrote that the milky juice was used by islanders to make poison arrows. The fruits and sap are poisonous internally, although some react to juice on the skin, while others are immune (Standley 1920–1926). Dry fruits are considered safe enough to use as a diuretic (Duke 1972). Smoke from burning wood causes inflammation of eyes; crabs that eat the fruits are poisonous if consumed by people (Duke 1972) but goats that eat the fruits are not. Seeds and bark are used as a vermifuge, but dangerous (Petersen 1974). Gum from the trunk is used against dropsy and venereal disease in Jamaica and elsewhere (Morton 1968, 1981, Petersen 1974). Writing in 1673, Richard Ligon noted: "The people who have lived long there [Barbados], say, 'tis not wholsom to be under the shade of this tree. The fellers, as the cut them down, are very careful of their eyes; and those that have Cipers, put it over their faces; for if any of the sap fly into their eyes, they become blind for a month. ... Yet, of this timber we make all, or the most part, of the Pots we cure our Sugar in; for, being sawed, and the boards dryed in the Sun, the poyson varpours out. ... The fruit is like an apple *John*, and 'tis said to be one of those poysons, wherewith the Indian Caniballs envenome their arrows" (Ligon in Standley 1920–1926).

Hordeum

(The ancient Latin name applied by Cato, 234–149 B.C., to Pliny, A.D. 23–79; akin to Latin *horreum*, a storeroom, Akkadian *qaritum, qiritu*, granary, storeroom, Hebrew, *qorah*, beam, roof)

barley (from Old English *baerlíc*, of unknown origin; in use by A.D. 966); barley-corn (in use by A.D. 1382, "corn" meaning grain)
barli-arisi [*barlhi arisi*] (Tamil); *barlibiyam* (Telugu)
bygg (Norwegian, for *H. vulgare*); *toradsbygg* (Norwegian, for *H. distichon*)
cebada [*sevada*] (San Luis Potosí); *cebada común* (Spain); *cevado* (from Latin *cibare*, to eat, feed, Portuguese)
chaîr (Tunisian)
dreche (French)
eòrna (Gaelic)
eiwt (Coptic)
Gerste (German)
juba [*jau, jeb, jóo*] (Bengali); *yuva* (Sanskrit)
orge (from Latin *oryza*, rice, Quebec); *orzo* (Italian)
seorah (pl. *seorim*) (Hebrew); *si'yr* (Modern Egyptian Arabic)

Hordeum pusillum (very small)
little barley

Both two-rowed barley (*Hordeum distichon*) and common barley (*Hordeum vulgare*) were essential grasses to Europeans at the time of contact with the New World. Barley was a critical source of malt used in manufacturing beer and whisky (Uphof 1968). The grains were also important elements in the diets of many Europeans. Old World people were also familiar with several other species, including the six-rowed barley (*H. hexastichon*), which has been found in the ancient tombs of Egypt (Manniche 1989, Hocking 1997). *Hordeum* is a genus of about 20 northern temperate species, with 8 in Europe (plus cultigens) (Mabberley 1997).

Hordeum pusillum was named by Thomas Nuttall in 1818. The species grows from Delaware to Washington, south to Florida, Louisiana, Texas, New Mexico, and Arizona (Fernald 1950, Kearney and Peebles 1951).

It has been known for some time that little barley was cultivated 2000 years ago by the Hohokam of Arizona (Gasser 1982, Asch and Asch 1985). Gayle Fritz also identified it in the Spiro site of Oklahoma (Yarnell in Asch and Asch 1985). However, it was not until somewhat later that it was realized that it also was cultivated in Illinois (Asch and Asch 1985).

Little barley may be one of the dominant "seeds" (caryopses) in Late Woodland or Mississippian archaeological sites, and dates to $2550–2770 \pm 70$ B.P. (600–800 B.C.). More recently, Fritz (2000b) concluded it was domesticated by 2000 B.P. However, little barley was one of several highly nutritious native foods in the Eastern Agricultural Complex that was abandoned when maize became the predominant American grain (see also *Chenopodium, Helianthus, Phalaris*). Moerman (1998) did not mention this species.

Hydrangea

(From Greek *hydor*, water, *angeion, aggeion*, a vessel, referring to the shape of the fruit)

Hydrangea arborescens (becoming tree-like)
hill-of-snow
[American, mountain, smooth, wild] hydrangea
seven bark[s] (the herbal drug from *H. arborescens*)

***Hydrangea arborescens*.** From Britton and Brown 1897.

In the 1730s, John Clayton visited an inland area and returned with a plant he had never before seen. He knew that he had found something different because he wrote on the specimen label, "*Novum genus pertinens a polyandria monogynia*" (New genus belonging with those of many stamens and one gynoecium).

Clayton sent the specimen to Europe where his colleague Jan Gronovius in the Netherlands studied it for the flora. Gronovius agreed with Clayton's assessment, and called the plants *Hydrangea* in his *Flora Virginica* published in 1739–1743. Linnaeus also concurred and called the shrubs *H. arborescens* in 1753.

In 1756, Dr. Alexander Garden of Charleston returned from a trip to the Cherokee country in the mountains. He brought back with him "the *Hydrangea* of Gronovius" (Vogel 1970). Clearly both of these early plant explorers had learned about the plants from the Cherokees because Mooney later recorded this tribe using *Hydrangea* roots in a decoction with other plants to treat women who "had strange dreams during their menstrual period" (Vogel 1970). Mooney got his information from the "Swimmer Manuscript," a series of prescriptions written by *A'yuninis'*, his Cherokee shaman informant.

That was not the only use to which the Cherokee put these shrubs. They also used *Hydrangea* to regulate menstrual periods, to stop vomiting, as an emetic, a purgative, and an antiseptic, to treat burns, tumors, ulcers, stomach trouble, high blood pressure, sprains, and swollen muscles (Hamel and Chiltoskey 1975). From another report, Moerman (1998) found that they also made tea from peeled twigs, and fried them like green beans. The Delaware used *H. arborescens* to treat gallstones (Moerman 1998).

That limited information is what has been found about indigenous people using a plant that has a wide range in the eastern states. *Hydrangea arborescens* grows from New York to Florida, and west to Oklahoma, Arkansas, Missouri, Indiana, and Ohio (Fernald 1950). It is mostly an Appalachian species and on the southern part of its range is most abundant in the northern half of Georgia. There are relict populations in Liberty and Walton Counties in the Panhandle that are considered endangered within Florida (Coile 2000). Although the plants are rare in some regions because of habitat destruction, they have been brought into cultivation. Brickell and Zuk (1997) list four formally named cultivars of the species.

People after physician Alexander Garden in the 1750s continued using these shrubs as medicine. Although neither Porcher (1863) nor Millspaugh (1892) mentions the plants, they were important to the medical community in the early 1900s. Culbreth (1910) recommended the plants as a "[d]iuretic, narcotic; renal and cystic calculi." The dried underground parts were official in the National Formulary between 1916 and 1926 for diuretic and diaphoretic applications (Vogel 1970). Felter (1922) said that *H. arborescens* was a "diuretic and sedative to cystic and urethra irritation, with passage of gravelly urine. It does not dissolve gravel, but is believed to be of value in preventing their formation, especially alkaline and phosphatic concretions." Krochmal and Krochmal (1973) give much of the same information, but add that the plants were used in colonial times to treat kidney stones.

Currently Bown (1995) suggests that this hydrangea serves as follows: "Internally for kidney and bladder stones, cystitis, urethritis, prostatitis, rheumatoid arthritis, gout, and edema. Works well with [Breakstone parsley] *Aphanes arvensis*, [Joe-Pye weed] *Eupatorium purpureum*, and [Bearberry] *Arctostaphylos uva-ursi*. Excess may cause dizziness and bronchial congestion." Hocking (1997) wrote that they were "[d]iuretic in kidney stones, cathartic, said to be narcotic."

However, these are not plants to be use without caution. As Foster and Duke (1990) reported, "In experiments, [*H. arborescens*] has caused bloody diarrhea, painful gastroenteritis, and cyanide-like poisoning." That result is not surprising as the plants contain hydrangin and neohydrangin (glycosides), volatile oils, and resins (Hocking 1997, Duke 2003).

Hydrocotyle

(Greek *hydro*, water, *kotyle*, a shallow cup, in reference to the peltate leaves of some species)

escudela da agua (water shield, Portuguese); *escuelle d'eau* (water shield, French)
lus na-peighinn (penny herb, Gaelic)

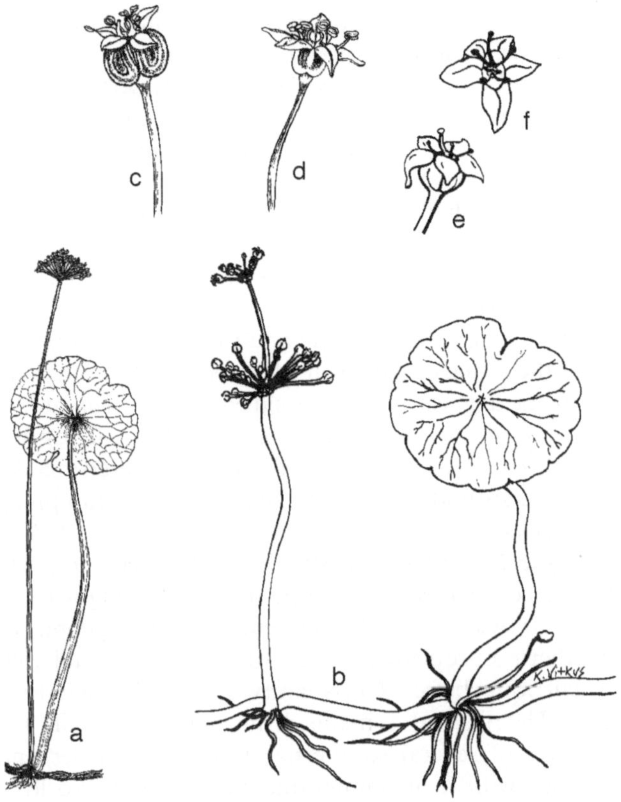

Hydrocotyle umbellata. a and b. Habit. c. Fruit. d and e. Flower, side view. f. Flower, top view. *a, c, and d drawn by Vivian Frazier.* From Correll and Correll 1972. *b, e, and f drawn by K. Vitkus.* From Institute of Food and Agricultural Sciences.

penny-wort [pennywort] (comparing the round leaf with a coin, this name has been used in English since ca. 1400 for *Umbelicus*, Crassulaceae. By contrast, *Hydrocotyle* was not named for the coin, but from its similarity to *Umbelicus*. Henry Lyte wrote in his 1578 translation of Dodoens's herbal that the name was used "[b]ycause of certayne similitude . . . that it hath with Penny-wurte of the wall [*Umbelicus*].")
skjoldblad (shield leaf, Norwegian)
soldinella aquatica (little water coin, Italian)
Wassernabel (water navel, German)

Hydrocotyle umbellata (having partial or secondary umbellate flower clusters)
acariçoba (*waka'ri*, with a shell, *çoba*, living, Tupí, Brazil)
berro macho (male cress, Venezuela)
erva de [*herva*] *capitão* (captain of herbs, Brazil)
locatokokocihiliswâ (*lucv-svoke*, box turtle, *heleswv*, medicine, Creek); *loksittish* (*loksi'*, turtle, *ittish*, medicine, Chickasaw); *yokcapolo:ckayikcî* (spherical turtle, i.e., box turtle medicine, Mikasuki)
nickels-and-dimes (Texas)
ombelico de Venus (Venus's navel, Texas, Puerto Rico)
[umbrella water, marsh, water] penny-wort (USA, Puerto Rico)
quitasolillo (little parasol, Cuba)
sombrerillo de agua (little water hat, Puerto Rico)
takpalule (Mikasuki, Bennett 1997)
yerba de cuatro (herb for four, Puerto Rico)

Linnaeus created *Hydrocotyle* in 1753, accepting a generic name proposed by Charles Plumer in 1703. At the same time, he named *H. umbellata*, which Marggraf and Piso had called *erva de capitão* and *acariçoba* in 1648. Linnaeus knew five kinds, although now *Hydrocotyle* is known to be a near-cosmopolitan genus containing perhaps 130 species (Mabberley 1997).

Pittier (1926) recorded use of the juice and root against leprosy in Venezuela. In Cuba the plant is considered aperitive and diuretic, but in higher doses emetic. *Hydrocotyle umbellata* is used against rheumatism and for liver and intestinal problems (Roig 1945). Seminoles use the plant to treat "Turtle Sickness" (Sturtevant 1955), and as a remedy for asthma and pneumonia (Bennett 1997).

Hymenocallis

(Greek *hymen*, membrane, *kallis*, beautiful, an allusion to the membrane uniting the stamens, the staminal cup)

Hymenocallis latifolia. *Drawn by P.N. Honychurch.*

Hymenocallis latifolia (broad-leaved)
chrysolite lily
liria sanjuanera (St. John's lily, Cuba)
lirio de playa (beach lily, Puerto Rico)
[Cayman Islands] spider lily (USA, Caymans)

Philip Miller called these plants *Pancratium latifolium* in 1768. It was not until 1812 that Salisbury realized that the tropical plants were distinct from the temperate *Pancratium.* Now there are 30 to 40 species of *Hymenocallis* known from warm regions around the world (Mabberley 1997).

Bulbs are used to heal wounds, and flowers are pectoral (Roig 1945). Several other species in the genus are used in the Caribbean for a variety of ailments (Morton 1981). Like their relatives *Crinum,* these species contain lycorine and other alkaloids; yet most species are used in the Caribbean as medicines (Perkins and Payne 1978, Lampe and McCann 1985).

Hypericum: The Saint's Weeds

(*Huper,* above, *eikon,* image, alluding to an idol or spirits [apparitions], or both; probably meaning that the plants had power over evil)

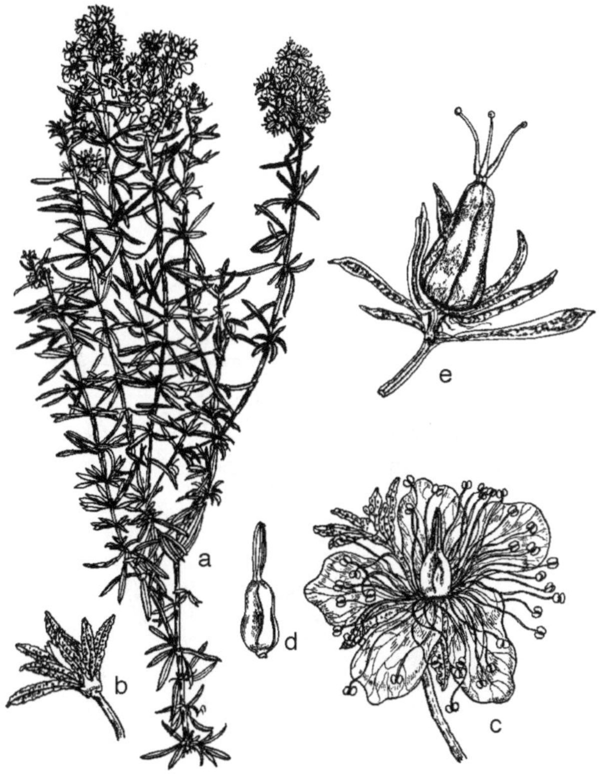

Hypericum. *Hypericum fasciculatum.* a. Upper part of plant. b. Bud. c. Flower. d. Capsule. *Hypericum galioides.* e. Capsule. *Drawn by Vivian Frazier.* From Correll and Correll 1972.

Once you learn some plant names in Florida you cannot help noticing how our wild areas have been affected by saints. Flatwoods has St. John's wort; scrub has other species. Different kinds of St. Andrew's cross are found in flatwoods and hammocks. Wetlands have St. Peter's wort. How saints' names became attached to Florida plants is a story reaching back to seasonal astronomical events and pre-Christian eras. My introduction to the saga began in the Amazon basin.

In June of 1969 my wife and I were studying plants near the Amazonian town of Belém, Brazil. We were living in a guest residence called *Hotel Açaí* at the *Instituto Agropecuaria do Norte.* As the end of the month drew near, dozens of people gathered on the institute grounds in a great open area and began piling up stacks of wood for bonfires. Having never before seen the Midsummer Festival, we attended the party that accompanied those large fires on the night of 24 June. It was a novel experience for two people of northern European heritage to witness a ritual that began long before Christianity spread through Europe. The festival remains firmly entrenched in the cultures of several southern European countries and traveled with their immigrants into the New World.

We noticed at the time that the date was near the Summer Solstice (21 or 22 June), and that was the ultimate beginning of the tradition. All over the world ancient people celebrated the Summer Solstice or Midsummer's Day because that is the point at which the sun stops its northern travel and returns south toward the equator. Indeed, several groups of people believed that, if the deities were not placated, the sun would continue north and the world would end.

Romans not only burned fires, but they also threw in plants that had just begun opening their yellow flowers, and were considered "sun-symbols." Romans and Greeks believed that the bright flowers were sent by the gods to provide a holy sacrifice to repel evil of all kinds, not just to compel the sun to return. Those plants were *hypericum* in Latin; the Greeks called them *hupereikon.* A different interpretation was offered by Walker (1976), who suggested that the old Greek name *hupereikon* was derived from *hypo,* approaching, and *ereike,* heather. Regardless, centuries later, those words would be transformed into the plant genus *Hypericum,* including the plants now also called St. John's wort, or St. Andrew's cross, or St. Peter's wort. Spanish speakers still sometimes call them *hipericón,* and Italians *iperico.* The Norwegians have retained part of the old Greek with *perikum.*

Romans also called the plants *fuga demonum* (demon chaser, Latin), and the French changed this to *chasse-diable* (chase the devil), the Dutch to *hexenkraut* (spell-herb), and the English to devil's scourge. Other names include amber (from their scent resembling ambergris), sweet amber, balm-of-warrior's wound, cammock (*camag*, for bent, curl, temple, Gaelic), Klamathweed (named for the Klamath River in California, where the European plants were first introduced), *lus Chaluim Chille* (St. Columba's herb, Gaelic), *lus na fala* (bloodwort, in reference to red extract, Gaelic), *lus an fhògraidth* (banishment herb, Gaelic), rosin-rose (Dioscorides wrote that the seed "is like in smell to the rosin of the pine"), and witch's weed (surely an allusion to use by women herbalists). The name *achlasan* [*seid, seun*] *Chaluim Chille* (St. Columba's oxterful [armpit package], Gaelic) alludes to St. Columba using the *Hypericum* in an ointment under the arms of a stressed shepherd to calm his nerves (Beith 1995). Gaelic speakers also call them *eala-bhuidhe* [*eala-bhuidhe*] (*eòlas*, knowledge of healing incantations, *bhuidhe*, yellow, Gaelic). That name became corrupted and led to the herbs being called cola-weed and eola-weed.

French speakers also call it *millepertuis* (1000 openings), *toute-sainte* (all saints), *tout-saine* (heal-all), or *tusane* (all wholesome). The French names are related to touch-and-heal and *tusan* of the English and Irish. The English say they use the name because the leaves are "touched" by their many dots. The name tipton-weed is related. Gaelic speakers also say one species is *beachnuadh boireann* (female St. John's wort); another is *beachnuadh firionn* (male St. John's wort). No one seems to be able to translate the first part of the name.

By the 1500s, Europeans had all been exposed to the Christian religion, and it had changed the names of many plants. By or before 1538, names for *Hypericum* were *saynt Iohans wurt* or *saynt Johannes worte* (St. John's plant, Old English). Cornish speakers had the same name but spelled it *losow sen Jowan*. It was also *saynt Iohns grys* (St. John's grass, Old English). The Germans and Dutch said it was *Iohns kraut* (John's plant), and later *Johanniskraut*. In Sweden it is *Johannis ort*. To the Spanish it is *hierba de San Juan*, and in Portuguese *erva de São João*. Names like God's-wonder-plant, penny-John, and speckled John are related.

All these references to saints were part of the *modus operandi* of the Christian priests working to convert people from other religions. When the priests discovered they could not stamp out the pagan festivals and ideas, they simply adopted and incorporated them, with the appropriate saints or festivals, into their own system. That amalgam satisfied both the converters and convertees. First, the priests used St. John because his birthday was on 24 June. That was close enough to the Midsummer Festival to work. Since the *Hypericum* blossoms begin near then, it was even better. Best of all, the plant was supposed to bleed on 29 August, the anniversary of St. John's beheading. All of the European names were originally applied to *H. perforatum*, but modifiers were tacked on for related species. For example, among others, the English recognize Irish, trailing, and marsh St. John's worts.

When the New World was discovered many species of *Hypericum* were added to the growing roster, including *H. fasciculatum* (with leaves in bundles). European St. John's worts have five petals; in the New World there are species with four petals, such as *H. hypericoides* (resembles *Hypericum* because it was previously placed in *Ascyrum*). These four-petal species were seen as plant versions of St. Andrew's cross because of their X-shape. So, around 1613, St. Andrew was added; then St. Peter was enlisted. St. Peter, the acknowledged gatekeeper of heaven, was included because of the St. Petertide festival on 29 June. This gatekeeper was depicted holding keys crossed into an X. So, both St. Andrew and St. Peter were originally tied to four-petaled flowers.

People occupying the Americas also had names for the plants. The Cherokees said *kowaya yust'* for *H. hypericoides.* Alabama speakers called the same plant *omagaga*, while the Koasati said *winihká ahissí* (*winihká*, thunder, *ahissí*, medicine). To the Creeks *H. fasciculatum* was *cissiwi:lano:ma:* (*cesse*, rat or mouse, *welvnwv*, feces), and their relatives the Miccosukee said *cisilaykô:mî*. Both have the same meaning, presumably in reference to the seeds. The Houma said *bois de marécage* (swamp bush).

Moreover, Americans and Europeans had similar uses for different species. Leaves of some species have been used in salads or to flavor liqueurs (Bremness 1994). Flowers have made appealing red and yellow dyes. Among the Old and New World people *Hypericum* was used in medicines. The most ancient remedy was to promote healing (e.g., balm-of-warrior's wound), and Greeks called one kind *androsaemon* (man's blood). Externally, the plants have been applied to heal cuts, wounds, burns, hemorrhoids, and varicose veins (Bown 1995). Extracts of flowers are antiviral, astringent, diuretic, and sedative, and are used to treat inflammation and diarrhea, and to improve blood flow. Research is ongoing to test for possible use in treating AIDS; PubMed (2003) returned 28 papers published since 1989 on that topic alone.

Probably the best-known current use is against nerves and depression. Most new books on medicinal plants recommend them, especially for women undergoing depression associated with menopause. However, Duke (1997) says that eating a bagel or biscuit for breakfast may be just as advantageous. There is a link between high-carbohydrate diets and higher levels of serotonin, and that is the chemical promoted by St. John's wort.

Not many of the recent books on herbal medicines mention bad side effects. One of the best-documented problems among those using St. John's wort is that it can cause dermatitis, or worse, photosensitization (Tyler 1993). The compound responsible for this is a reddish dianthrone pigment called hypericin. So, you have a big problem if you use *Hypericum* and then turn red like a boiled lobster after a few minutes in the sun. The reaction in some white sheep and cattle that accidentally graze on the herbs is so dramatic that they die. People have not died from using St. John's wort as medicine, but walking around looking like a boiled lobster could result in deepening the depression one had hoped to cure.

I

Ilex

(Latin *ilex, elex,* ancient name for European holm oak tree, *Quercus ilex*)

Ilex cassine (from Muskogee *esse,* leaf or *vsse,* tea)
ahihátkî [*ahe hatke, aheke*] (*ahe,* tree, *hatke,* white, Mikasuki)
dahoon (maybe from French; in use when Catesby visited the Carolinas, 1720s); *houx d'Ahon* (W.R. Gerard in Hale 1891 gave this French spelling; from *houx,* holly, *d'Ahon,* an attempt at the Catawba word *yaupon;* maybe the source of *dahoon*)
helok hakv (*helukwv,* black gum, *enhake,* resembling, Muskogee)
inlokci:ca:tî (*enlowake,* bud, *cate,* red, Creek)
yanilla blanca (white yanila, based on the Taino word *yana,* Cuba)

Ilex glabra (without pubescence)
gallberry (Florida)
ittohissishalokpa [*ittohissikalokpa, ittishalokpa, itichalokpa*] (*itto,* tree, *hissi,* hair, *ishalokpa,* sharpening tool, Alabama)

Ilex vomitoria (regurgitating; from the southeastern custom of using it as an emetic)
aatichoosanàyli (*aati,* Indian, *ichoo,* mouth, *sanay-li,* twist, Alabama)
Apalachine [*thé des Apalaches*] (Apalachee tea, early French visitor records, Florida Panhandle)
asi [*assi*] (Hitchiti-Mikasuki); *esse* [*assi, ussie*] (leaves, Creek, Muskogee; compare with *vsse,* tea); *cassina* [*cacina, cazina, cassine, cassena, cassiana, cusseena*] (1696 or before, derived from Creek *assi* or *assin*); *cassiné* (Timucua, fide Crawford 1988); *cossine* (Geary 1955 gives this as a person's name; a loan-word?); *usi* (based on *usinola,* the black drink singer, Apalachee); *vsse* [*vs'se*] (tea, Muskogee)
asilaknî (*asi,* leaf, *lakni,* yellow, Mikasuki; maybe this species)
assi lupub'ski (*assi,* leaves, *lopútke,* small, Creek; cf. Hale 1891)
awotaicita [*sawatcka*] (*vwoticetv,* to make vomit, Creek)
black drink (a name given by British traders, cf. Hale 1891; notably, the term "black drink" is not used by Catesby, Romans, or Bartram; perhaps first used by Williams in 1837; the name is not given in the OED 1971, but Hudson 1979 said it was in use by the 1600s)
canâ:fili (simple term, Mikasuki); *canâ:filápi* [*canâ:filá:pî*] (*canâ:fili,* cassine, *á:pi,* tree, Mikasuki)
chocholate del indio (Indian chocolate, Mexican visitors among Tawakoni Wichita, Texas)
cowwta oucoroge (Cherokee, North Carolina)
esta hoola (Bartram [1791] 1958 said this meant "the most beloved," or "beloved tree," Choctaw, Louisiana; Swanton 1931 glossed it *isht,* most, *ahollo,* beloved; Byington et al. 1915 translated the phrase as "magnificient, miracle" or "conjuration"; used to "make rain," as reported by Romans [1775] 1961)

Ilex. *Ilex cassine* (left). *Ilex vomitoria* (right). *Both drawn by P.N. Honychurch.*

foskey (Yemasee; cf. Hale 1891)
hishi' [*asi*] (leaves, Chickasaw; singer: *taloowa'*)
hoyaní:cî (from Creek *mikkohoyaní:cî*, chief physic; perhaps confusion with another species, cf. *Salix humilis*)
isu [*isota*] (herb, Timucua; Le Moyne recorded that the singer chanted *Ha, he, ya, ha*)
ittokchakko (*itto*, tree, *okchakko*, blue-green, Alabama)
ka'tchit (to drink, Chitimacha; questionable identification; cf. Hudson 1979)
katsgimilpa (catfish eat it; *katsg*, catfish, *im*, its, *ilpa*, food, Alabama)
no'ut (Chitimacha, Louisiana)
té del indio (Indian tea, early Mexican visitors among Tawakoni Wichita, Texas)
wai't'i (Chitimacha, Louisiana delta)
yaupon [*yopon, youpon, yupon, uupon*] (derived from a coastal Carolina Catabwa language, *yópon* or *yã'pã*, diminutive of *ʔyap* [*ye'p, yo'p, yup*], tree, 1709); related to Biloxi names for *I. opaca*, *apso'nd ayudi'* (*psûn̲ti*, sharp-pointed; so called because of thorns), and Catawba *yapʔ ha hitce* (tree leaves sharp)

Although Catesby (1734–1747) knew that *dahoon* and *cassine* were two species, Linnaeus created confusion in 1753 by including both under the name *Ilex cassine*. It was not until Daniel Carl Solander (1733–1782) and William Aiton studied them in 1789 that the original *cassine* became known, through a technical problem, as *I. vomitoria*. Hedrick (1919) and Yanovsky (1936) still had the commonly used species listed incorrectly under *I. cassine*. Native people knew all along which was which.

Muskogean people called *I. cassine*, *ahihátki* (Mikasuki) or *inlokci:ca:ti* (Creek). Perhaps it was the Catawba who dubbed it *dahoon* (1720s, or before), at least the "Carolinas by Col[onel] Bull's plantation" is where Catesby found the name used. Much later that plant became henderson-wood (Carolinas) or *yanilla blanca* (white little *yana*, comparing it to *Conocarpus erectus*, Taino, Cuba).

Sturtevant (1955) found the plant used as a soap substitute. The leaves are sometimes dried, roasted, and sold for use as tea along the South Atlantic Coast (Morton 1968b). In some they produce dizziness and have a laxative effect. Fruits are cathartic (Morton 1968, 1974).

Modern Seminoles use the stems as firewood, but they make a "smoky old woman's fire" (Bennett 1997). They still use it as a medicine, but are not willing to share that information. Otherwise, children mash the fruits, and blow them through tubes to make bubbles (Bennett 1997). Perhaps children are imitating a medicine man as he prepares other plants.

The original *cassina*, now known as *I. vomitoria*, was the basis of indigenous secular and ritual practices. People throughout southeastern North America knew this plant (Hudson 1979). To the Timucua, it was *isu, isota*, in spite of the fact that the *American Heritage Dictionary* (2000) says their word was *cassiné* [*kasine*]. Probably, the Timucuan word is a cognate or a loan-word for the Creek *asi* (Crawford 1988).

By at least the 1600s (Hudson 1979), a ceremonial beverage used throughout the range of the plants came to be called black drink. The Seminole leader, who became known as Osceola, was in fact called by his Hitichiti or Miccosukee title, *asi yahola* (black drink singer). In other Muskogean languages that title was based on *hishi' taloowa'* (*hishi'*, leaves, *taloowa*, singer, Chickasaw), *vsse yvhikv* (*vsse*, tea, *yvhikv*, singer, Muskogee), and *usinola* (*usi*, tea, *inola*, singer, Apalachee).

Some think that *I. glabra* may have been one of the hollies used in the black drink of the southeastern Indians. That is unlikely because, as the common name "gallberry" implies, the leaves and fruits are bitter and the whole plant is emetic (Morton 1974). Plants are now being used in xeriscape with success. The Alabama call it *ittohissishalokpa* (Swanton 1928a, Sylestine et al. 1993), but that is not likely to become a trade name.

Impatiens

(From Latin *impatiens*, impatient, alluding to the explosive release of seeds when the ripe capsule is touched)

Impatiens capensis (erroneously thought to be from the Cape of Good Hope, Africa)
[orange, wild] balsam [weed] ("balsam" came into English from Latin *balsamum*, about A.D. 1000, meaning an aromatic vegetable juice; the word was applied to *Impatiens* before 1542; "weed" was added later)
jewel-weed [jewelweed] ("jewel" is derived from Old French *joel*, a word with an uncertain etymology, maybe from Latin *jocalis*, game, but always meaning an article of great value; applied to these American plants by 1890 because of the "earring-like shape of the flowers, and the silvery sheen of the under surface of the leaf in water," OED 1971)
lady's earrings (from the "earring-like shape of the flowers")
quick-in-the-hand

Impatiens capensis. a. Habit, upper branch of plant. b. Basal part of plant, showing roots. c. Flower. *Drawn by Vivian Frazier.* From Correll and Correll 1972.

[spotted] snap-weed
[spotted] touch-me-not ("touch-me-not" was originally applied by John Gerarde in 1597 to the squirting cucumber, *Cucumis asininus;* the name was reapplied to *Impatiens noli-tantere* about 1659)
tapasi̱·'wi su're (flower wild, Catawba)
twatubigo'niak (Potawatomi)
wild [brook] celandine ("celandine" appeared in English in A.D. 1310; from Latin *chelidonia,* Greek *chelidonion,* the swallow; the name derived from a legend going back to the classical Greeks and Romans that the plant flowered at the times the swallows arrived in the spring, and withered when they migrated south in the fall; originally applied to *Chelidonium,* Papaveraceae; later applied to several genera)

Linnaeus and other Europeans knew their species, *Impatiens noli-tangere,* variously known as touch-me-not, yellow balsam, and quick-in-hand, long before finding another species in the New World. That species, ranging from Europe to the Kamtschatka peninsula of Russia, and south to Japan and China, has long been noted as a laxative and diuretic, and has been used to treat wounds (Uphof 1968).

Of all the uses by indigenous people, the most famous is to stop the rash caused by poison ivy (Vogel 1970). Foster and Duke (1990) noted that a 1957 study by a physician found extracts of jewel weed effective in 2 to 3 days in treating 108 of 115 patients. Unhappily, I would have fallen into the minority for whom it was not effective. When I was an undergraduate in Kentucky, there were many times when I applied it to my skin. It never kept me from developing a severe poison ivy rash or relieved the itching.

The Cherokee, Meskwaki, Omaha, and Potawatomi relieved the rash caused by wood-nettle (*Laportea canadensis*) with juice of jewel weed (Smith 1933, Vogel 1970, Hamel and Chiltoskey 1975, Moerman 1998). The Cherokee, Iroquois, Mohegan, Nanticoke, Ojibwa, Omaha, Penobscot, and Shinnecock treated burns, cuts, hives, liver spots, measles, raw or sore eyelids, and rashes with various preparations made from *Impatiens* (King 1984, Moerman 1998). The Cherokee included it as one of the plants in their "Green Corn" medicine, and used it to help ease childbirth pain (Hamel and Chiltoskey 1975). A related analgesic use by the Ojibwa was to relieve headache (Moerman 1998). Whole-plant infusions were used by the Cherokee and Potawatomi to treat chest colds and stomach problems (Hocking 1997). The Iroquois used it as a diuretic for dropsy and to treat fever (Hocking 1997, Moerman 1998).

Hocking (1997) said that this species is emetic, but he may have confused it with the related *I. pallida* (Porcher 1863). That species is not known in Florida, but it has a variety of uses, many similar to *I. capensis.*

The plant provided a yellow-orange dye for the Menomini, Ojibwa, and Potawatomi (Smith 1933, King 1984, Hocking 1997, Moerman 1998).

Ipomoea

(Greek *ips, ipos,* worm or bindweed, *homios, hoimios,* resembling, referring to the twining habit)

Ipomoea imperati (possibly from Latin *imperatus,* ruler or emperor)
batatilla (little potato, Dominican Republic)
beach morning glory (Florida)
bejuco de cadena (chain vine, Venezuela)
bejuco de costa (coast vine, Puerto Rico); *bejuco de puerco de costa* (coastal pig vine, Puerto Rico); *boniato de playa* (beach sweet potato, Cuba; *boniato* is a Taino word, perhaps originally applied as now to the white-fleshed, pink-skinned *I. batatas,* and extended to other members of the family for comparison)
bejuco de las tortugas (turtle vine, Guerrero)
liane manger cochon (hog-food vine, Haiti)

patate lan mer [*marron*] ([wild] beach potato, Haiti)

A strong decoction of the root has been used against dysentery, and an infusion of the stem has been taken for venereal diseases (Pittier 1926).

Ipomoea indica (from India)
aguinaldo morado [*rosado*] (purple [rosy] gift, Cuba)
bejuco de gloria (glory vine, Puerto Rico)
blue dawn flower (Florida)
bois patate marron (wild potato bush, Haiti); *patate marron* (wild potato, Haiti); *patate sauvage* (wild potato, Haiti)
campanilla (little bell, Honduras); *campanitas* (little bells, Dominican Republic)
estrella vespertina (morning star, Dominican Republic)
gloria de la mañana (Belize); [purple] morning glory (Florida, Bahamas)
liane manger cochon (hog-food vine, Haiti); *mange cochon* (hog food, Haiti)
liane purgative (purgative vine, Haiti); *liane purgative a Bauduit* (Bauduit's purgative vine, French Antilles)
quiebra-platos (plate breaker, Oaxaca, Veracruz)
quilamula (Quiché, Guatemala)
rue purgante (purgative rue, Haiti)
sayün (Quiché, Guatemala)
soyotquilit (*soyot* =?, *quilitl*, greens or edible herb, Náhuatl)

Roots of this species are purgative as is the milky sap (Liogier 1974, Morton 1981). A decoction of the leaves is used against sores and ulcers (Liogier 1974)

Ipomoea pandurata (fiddle or violin shaped, meaning the leaves)
High John the Conqueror (see also *Triadenum*)
man-root
man-of-the-earth
wild jalap (applied in the late 1700s although "jalap" is a Mexican species; more confusingly, "jalap" has been applied to *Podophyllum*, which see)
wild potato [wild-potato] [vine]

William Bartram wrote, "The dissolvent and diuretant powers of the root of the convolvulus panduratus [*Ipomoea pandurata*] so much esteemed as a remedy for nephritic complaints, were discovered by the Indians to the inhabitants of Carolina" (Bartram in Swanton 1928a). Missionary George Henry Loskiel recorded in 1794 that "convolvulus jalappa" grew in abundance in "Indian country, and is prescribed as a purgative. In the rheumatism of the legs they roast the roots, then slit and apply them to the soles of the feet as hot as the patient can bear" (Loskiel in Vogel 1970).

Yanovsky (1936) and a number of others said the roots were considered famine food by indigenous tribes. It is pertinent that neither Gilmore (1919) nor Hedrick (1919) listed the species. The resins ($\sim$5%) in the roots make them less than palatable and digestible. They were, however, used as a mild purgative, to expel gravel, and to treat malaria (Hocking 1997). Roots were used by Carolina tribes to treat kidney trouble, rheumatism, and purgative (Vogel 1970). Part of the confusion about edibility seems to have been misidentification of the "Potato Clan" among some eastern tribes with this plant. Their symbol was actually *Apios* (which see).

The mystical shops are selling what appears to be roots of this North American species under the name *I. jalapa* and High John the Conqueror (cf. Yronwode 1999). They tout the root as an aphrodisiac, mirroring its older names man-root and man-of-the-earth.

Ipomoea pes-caprae (goat's foot)
batatilla (little potato, Dominican Republic); *boniato de playa* (beach sweet potato, Cuba, Puerto Rico)
bay hops [winders] (Florida, Bahamas); *bejuco de mar* (sea vine, Veracruz); *bejuco de playa* (beach vine, Mesoamerica)
bejuco de la tortuga (turtle vine, Mesoamerica)
cowslip (Belize)
goat's foot (USA); *pata de cabra* (goat's foot, Veracruz)
liane manger cochon (hog-food vine, Haiti)
patate bord de la mer (seaside potato, French Antilles); *patate lan mer* [*marron*] ([wild] beach potato, Haiti)
railroad vine (USA)
riñonina (little kidney, Mesoamerica)

Used against arthritic pain and internal upset, also as a purgative and diuretic, and to lower blood pressure (Roig 1945, Vásquez and Jácome 1997). Elsewhere it has been used against "weakness in women," for bathing sores and wounds, as a febrifuge and emollient, and to treat animal stings and bites (Duke 1972, Morton 1981). Recent studies indicate that it contains compounds as effective as Benadryl® (Pongprayoon 1991, Pongprayoon et al. 1991, 1992). In Belize, the vines are used to make a dye (Balick et al. 2000)

Ipomoea sagittata (having arrowhead-shaped leaves)
glades [marsh, salt-marsh] morning glory (Florida)
l'herbe à congo (Congo herb, but *congo* here meaning the water moccasin, *Agkistrodon pisciverous*, Houma, Louisiana French)
snake grass
wahoɫe entayhe [*waaho(th)e entahe*] (*wahoɫe*, water moccasin, *entayhe*, bed, Mikasuki)

The Houma used it to purify the blood and as a remedy for snakebite (Speck 1941). The Seminoles use the same name for this, *Mikania scandens* and *Ludwigia*. The plant is used to treat Snake Sickness (Bennett 1997).

Ipomoea violacea (purple, a misnomer through confusion)
aguinaldo blanco de coasta (white coastal gift, Cuba)
beach [coast] moonvine [moon vine, moon-vine] (Florida, Puerto Rico, English Antilles)
bejuco de luna (moon vine, Puerto Rico); *flor de luna* (moon flower, Puerto Rico)
bejuco de vaca (cow vine, Puerto Rico)
ferrocarril (railroad)
jabilla (name usually used for *Hura crepitans*, Taino, Cuba)

The plant is used against skin and liver problems and considered emollient and vulnerary (Roig 1945).

Ipomoea: Moonvines

(See preceding entry for etymology)

Shortly after 1913, the Lake Okeechobee American Legion developed a cheer for a West Palm Beach parade:

Custard Apple, Moonvine, Catfish and Moonshine!
Ever-glades Post 20! Whoopee! Muck Rats!

That pretty well summed up the abundance of moonvine (*Ipomoea alba*) in southern Florida at the time (along with a few other prominent items). Moonvine formerly blanketed the canopy of a pond apple forest that stretched for 32,000 acres along the southern edge of Lake Okeechobee (Will 1964a). This forest, and the moonvines covering it, was totally destroyed and replaced by agriculture. A belt of trees and vines 50 miles long and 2 or more miles wide was eliminated in less than a decade (Small 1929). Although moonvines are not as common in Florida as they once were, they are still abundant in some places.

Ipomoea. *Ipomoea alba. Drawing by Regina O. Hughes.* Courtesy of the United States Department of Agriculture.

Moonvines, also called moonflowers in English and *flor de luna* (moon flower, Veracruz, Tabasco) in Spanish, have fascinated people since the first Europeans arrived in the New World. The flowers open in the evening, they are strongly scented, and they attract large nocturnal sphinx moths (Sphingidae). That combination of traits, plus some others, enticed the Spanish to carry the seeds back to Europe and then to the rest of the world within the first few years after 1492. The plants became pantropical so quickly that until recently there remained a cloud of confusion about their origin. Most people, from at least the 1750s to the present, thought that the first mention of the moonvine was by H.A. Rheede in his list of plants cultivated in Malabar, India, in the 1680s. However, it appears that the first record was made by an early Spanish historian. Shortly after his arrival in Panama in 1514, Gonzalo Fernández de Oviedo wrote about the *flor de la Y* that he found in Cuba (Oviedo [1526] 1969). All students of that island have equated that account with *I. alba*, and Oviedo's brief description fits.

The origin of the Taino word "*Y*" for these plants has been a mystery ever since. It may be a comparatively simple matter of related words (cognates) and partial misunderstanding on the part of the Spanish.

We know from Columbus's account that the Taino people of Hispaniola called *I. batatas* (sweet potato) *ajé* [also spelled *hage, age, aje, axe, ase*] (Jane and Skelton 1960, Sauer 1969). The Taino spoke a language related to the Arawakan people of northern South America. The Arawaks call the sweet potato *haliti*, and their relatives the Guajiro say *haisi*. Taino, Arawak, and Guajiro words are cognate with *á:hi* (Mikasuki), *ahe* (Choctaw), *vhv* [*aha*] (Creek), and even *nuna* (Cherokee). Through simple misunderstanding or poor transcription, the Taino word would be changed to "*é*" or "*y*" by Spanish-speaking listeners in the 16th century. So, the name Oviedo left behind for the moonflower, *flor de la Y*, may be translated "flower of the *ajé* or morning glory," or, more exactly, the "root with flowers."

Surely, the Cuban name *jabilla* is related to "*Y*." *Jabilla* (from *jabí*) too is probably based on the same root as Taino *hage* and Arawak *haliti*. This Cuban name is applied to *I. violacea*, a species that resembles *I. alba*. In both Spanish (*aguinaldo blanco de costa*, white coastal gift) and English (coast moonvine) the *jabilla* is considered the "coastal" moonvine because it is found on beaches and in mangroves. Could *mash* (Chiapas) be another cognate?

Other common names are easier to understand. In Hispaniola, moonvine is rendered *estrella vespertina* (evening star), and in Yucatán *oración* (prayer). Colombians say it is *galán de noche* (gallant at night). French Antilles people say *belle de nuit* (beauty of the night). Early settlers in Florida called it "nightbelle" (Williams [1837] 1962). This fascination with the evening flowers spilled over into Europe and caused Swiss botanist J.D. Choisy (1799–1859) to separate the species out into the genus *Calonyction* (beautiful at night, Greek) in 1833. Yet, the similarities in shape and time of flowering between moonvine and some other species in the family do not necessarily indicate relationships. More often, the flower shape reflects convergence onto a pollination type from several different lineages.

Hondurans prefer to call the moonvine *pañal de niño* (child's diaper), while people in El Salvador call it *garza* (egret). That Salvadoran name is probably a corruption of indigenous names for the plants, including *gamuza* (Chiapas), *huamol* (Chiapas), and *guamol* (Chiapas). Those words might be linguistically related to *amule* (variant of Náhuatl *amole*, soap, Oaxaca), or they may have some other relationship.

Moonvines are also called *bejuco de tabaco* (tobacco vine) in El Salvador, because stems were used to tie up tobacco leaves to dry (Standley 1920–1926). In Mayan-speaking areas from Chiapas to Yucatán, the climber is *nacta* (Lacandon), *naxh, huchuk* (hiccup), *huchuk ts'aan* (hiccup vine), *xpeten* (an island in water or of vegetation usually with a permanent pond within), *piroreta* (spiral), and *sutup* [*suput, xutu, xutub, zutub*] (something that turns or forms a spiral). Even though three of the translations suggest only descriptive terms regarding the twining or climbing life-form, their existence shows that people took note of the plants. Often when any people have a name for a plant, they have a use for it, whether or not they share that information. Other Maya call the plants *haapolin* (*ha*, water, *polin*, origin, or found near water, Yucatán). They know that where they see a vine, there will be a place to get water.

Venezuelans and Mexicans call these vines *nigua* (chigger). *Nigua* is another Taino word that was spread widely outside Hispaniola by the Spanish. Was moonvine really used against chiggers, as was chiggery grape (*Tournefortia hirsutissima*)? There are numerous uses for the plants, but nothing has been found that substantiates that use. Moreover, moonvine and several unrelated plants are called *nigua*. Perhaps *nigua* is a general name that refers to some physical trait resembling chiggers.

In Puerto Rico and Oaxaca, Mexico, the twiners are *bejuco de vaca* (cow vine), and *bejuco de puerco* (pig vine), because they are fed to those animals as fodder. There are reports of the immature seeds being eaten in India, and the plant has had the reputation there as a snakebite remedy since at least the 1600s (Watt 1898, Standley 1920–1926).

There are several other recorded applications, including use as a febrifuge, to stop hair loss and dandruff, as a laxative, and to expel gas (Liogier 1974). Some Mayans suggest that it "diverts hunger when tied around body." Oviedo ([1526] 1969) was the first to note the plant's laxative aspects, and that is likely why the Spanish took moonvine and many other American plants around the world. People in Oviedo's time were convinced that all human maladies came from being "poisoned" by material in their colons. Laxatives were the medical panaceas.

There are enough chemicals in the vine to have real impact on several of the problems it is used to treat. Moonvines contain calystegines, resins, indolizinic alkaloids, ipomine, and ipalbidine (hexahydroindolizine alkaloid) (Gourley 1969, Dawidar et al. 1977, Ikhiri et al. 1987, Schimming et al. 1998). The calystegines are poisons (= medicine, in small doses), the resins are laxative, and alkaloids have a number of effects on human physiology.

Milky juice from moonvine stems has a long pre-European history in Mesoamerica of being used to coagulate latex from the *arbol de caucho* (rubber tree, *Castilla,* Moraceae) (Standley 1920–1926). The vine is the *bejuco de cuajar hule* (vine for coagulating *Castilla,* Oaxaca), or simply *cuaja leche* (coagulate milk, Oaxaca). *Castilla* was an important rubber source for the Mayans who used it to make balls for their game *pelota,* among other things (Hosler et al. 1999). The Aztecs also adopted rubber use, and considered the sap the tree's "blood." Balls from this *caucho* were so valuable to the Aztecs that they paid their taxes with them.

Although humans have carried moonvine with them, it also travels well without them. The seeds are dispersed by water, and that is one of the reasons it was so common on Lake Okeechobee at the beginning of the 1900s. The seeds float so well that they arrive regularly on the coasts of Britain, Ireland, and presumably Norway (Gunn and Dennis 1976). Since the closest point of origin for those drift-seeds is the Cape Canaveral area of eastern Florida, they have a long seawater journey—at least 6759 km (4200 miles). Drift-seed enthusiasts in those areas report that they arrive literally by the thousands. One person reports that they come in "numbers beyond imagining." Incredibly, some small percentage is still alive when they arrive on the southwestern coast of England. A colleague there has been able to germinate them for illustrations for a book he wrote on the species found on British coasts (Nelson 2000).

One of the first things that I did when I first arrived in Florida was to survey the southeastern coastal area for morning glories. The most abundant species in wetlands was the moonvine. I found a spot on the edge of the Everglades where a vine sprawled over the shrubs and trees near a swamp. One evening I set up a camera with a flash on a tripod and focused on an unopened bud. Promptly at 7:00 P.M., the flower began to open and shed its musky fragrance into the mosquito-filled dusk. Within minutes after it opened a large tomato hornworm sphinx (*Manduca sexta*) arrived and probed the flower with its strawlike proboscis. I stepped over and snapped the picture. It seemed like the easiest thing in the world, and I anticipated many more photographs like it. But, 30 years later, I had added only one. Moths and moonflowers are unpredictable subjects.

Iresine: Bloodleaf

(From Greek from *eiresione, eiros,* wool)

My students often found a particular herb reaching to about a meter tall scattered through dry sites. When growing in those sites, the leaves and stems are often tinged with red, leading to the common name

Iresine diffusa. From Britton and Brown 1896.

bloodleaf (Florida). Other names that are related to the color of the plants include *camarón* (shrimp, Costa Rica) and red dandelion.

The Latin name for these members of the Amaranthaceae is *Iresine diffusa,* although some of the older literature uses the name *I. celosia.* The opposite leaves, swollen nodes, and red coloring made the plants comparatively easy for students to identify. However, sometimes when wading through swamps such as Fakahatchee Strand State Preserve, we would find larger herbs with more delicate leaves, longer internodes, and more open inflorescences. There was never a trace of red on those plants. Students rarely saw the similarities and usually tried to make it something more than an etiolated form of the more familiar variation of bloodleaf.

A more constant trait of these herbs is the flowering and fruiting clusters, which are always "cottony." Indeed, it was that trait that led to the name *Iresine.* That genus was named by Patrick Browne in his book, *The Civil and Natural History of Jamaica,* published in 1756. Some have proposed that the genus name came from Greek *iris,* a rainbow, and *isnos,* ribs of the leaves, an allusion to the color on the leaves, but that does not seem to be the case. If Linnaeus had named the genus, it would have been more likely that he used a dual meaning name, as he loved to make such commentaries.

The cottony aspect of the plants has been noted by people in the Caribbean and Central America. In those places, the herbs are known as *adorno de niño* (child's adornment, Guatemala), *algodoncillo* (little cotton, Central America), *barba de viejo* (old man's beard, Central America), and *herbe coton* (cotton herb, Guadeloupe, Martinique) to indicate that aspect. Probably the name *zacxiu* [*zactezxiu, zakyezxiu*] (*sak,* white, *xiw,* herb, Maya, Yucatán) makes the same suggestion. Since the related species *I. calea* of Mexico and Costa Rica is called *cola de cabra* or *cola de chivo* (goat's tail), perhaps *chivito* (little goat, Central America) applied to *I. diffusa* makes the same allusion. All the references to goats must be noting the white cottony appearance. Of course, the plants are also grazed by livestock, and that is probably also inherent in those common names.

The meaning is not clear, but the same comments may be indicated by some other names. These names include *hierba de la plata* (silver herb, Central America), *plumaria* (little feather, Colombia), and *velo de princessa* (princess's veil, Guatemala). They are also called *velo de novia del monte* (wild lady's veil, Puerto Rico). That seems to be a comparison with *velo de novia*, which is *Porana paniculata* (Convolvulaceae). Both the Amaranth and the morning glory have a profusion of small white flowers, and that may be the reason for these and the other common names.

When the wind blows through the plants, they dance. That "dancing" must be related to the common name juba bush [jubba bush, juba] (Jamaica). *Juba* is a dance that originally came with slaves from western Africa to the Caribbean in the 1500s, and to the United States in the 1600s (Major 1994). In their new homeland, these people continued their secular and religious "step dances," and even accompanied them with drums called *yubá* or *juba*. Originally, the American dance was a competitive event where opponents would try to outdo one another in feats of skill, even balancing a cup of water on their heads. As these African dances encountered Irish jig and clog dances, they incorporated elements from them. Eventually, the dances changed so much they became known as "tap dancing," and they were made famous by the early 1800s by people such as William Henry Lane (nicknamed "Master Juba"), born in 1825.

Surely, *felicidad* (happiness, Puerto Rico) also refers to the "dancing" plants. Perhaps similar movements lead to *taba de güegüëch* (*taba*, one who directs an Indian dance, Zapotec, Oaxaca, El Salvador). *Yerba de cabeko* (moving herb, Belize) is similar.

The name *taba de güegüëch* is odd, however, in that the word *taba* also refers to a foot bone. The bone reference seems to be related to the names *canillita de zanate* (grackle's long bones, Central America), *coyontura* (bone joint, El Salvador), *coyontura de zanate* (grackle's [bone] joint, Central America), and *coyontura de pollo* (chicken's [bone] joint, El Salvador). These five names may actually be comments on the swollen nodes of the stems, which somewhat resemble joints. Indeed, they may be names that indicate an application of the Doctrine of Signatures. That philosophy holds that, if a plant resembles a body part, it is good to treat maladies of that part. If that is the case, none of the medical uses I have found reflects that.

There are several medical applications of *I. diffusa*. People from the Catawba who live along the coastal Carolinas and Georgia to Colombia in South America have used the plants to obtain relief from coughs. The Catawba used a syrup of leaves and stems against whooping cough (Speck 1937). In Colombia, the plants are sold by vendors to use in a decoction to stop coughs (Pérez-Arbelaez 1978). That use is certainly reflected in the name *l'herbe à croupe* (whooping cough grass, Houma, Louisiana), and probably by bitter weed (USA). The plant is also used in Colombia to relieve pain in the side associated with violent coughing.

Related medical applications are to treat colds and edema in Jamaica. When a decoction is mixed with *Desmodium* and *Borreria laevis*, the tea serves as a diuretic and purgative used in Jamaica against dropsy and gonorrhea (Ayensu 1981, Morton 1981). *Hierba de orín* (urine herb, Veracruz) surely denotes the diuretic nature. A slight variant is used in Venezuela, where the plant is reduced to ashes, lemon or lime is added, and the mixture taken as a purgative. That same mixture might be used to treat bowel pain. In Cuba, a decoction is used for stomach ailments (Uphof 1968), including colic. This drink is given to women in childbirth as a tea mixed with *Piper nigrinodum*, salt, and a few drops of whiskey. In Guatemala, the juice has been used to treat erysipelas (Standley and Steyermark 1946).

Several of the common names may be related to medicinal uses. For example, the herb is *chancanil* (*chan*, small, *kanil*, roots, Queech', Guatemala), *mosquito* (Guatemala), *pié de paloma* (dove foot, Guatemala), *siete pellejos* (seven skins, El Salvador), *tabudo* (a forbidden [*tabú*] plant, Guatemala), and *tlancuaya* (*tlan*, near, *?cueitl*, skirt, Náhuatl, Veracruz). The *pié de paloma* may be a comment that doves consume and spread the seeds. Perhaps they are called *siete pellejos* because of their colors.

Although the genus *Amaranthus* is reasonably well studied, not many other genera are well known chemically. No chemical studies seem to have been made of *I. diffusa;* however, the related *I. herbstii* has been examined. That originally South American species, called beef plant, beefsteak plant, *chorcha de gallo* (rooster cork), and *gésier poule* (chicken gizzard), is known to have antibiotic activity (Wong 1976). Moreover, studies of the effects of aqueous extract from *I. herbstii* on the central nervous system indicate its possible use in catalepsy and nociception (DeFeo et al. 1996). Otherwise, the species shows promise as a source of betacyanins (Cai et al. 2001), of potential use in food coloring.

Iresine is a genus native to the tropical and temperate regions of the Americas and Australia with about 80 species (Mabberley 1997). Most species grow in dry, open areas in their native lands. *Iresine herbstii* and *I. lindenii* (blood leaf) are the most widely cultivated species and are grown for their red leaves.

Iresine diffusa is among the most common weeds in disturbed sites throughout its range from North Carolina to Louisiana and Texas, the Bahamas, the West Indies, from Sonora and Chihuahua in Mexico to Panama, and south to Peru and Argentina. Probably, that is why they are called new-burn weed (Bahamas). When a site is disturbed by fires, hurricanes, or something that moves the soils, these plants appear. The species is heliophylic and persists as long as there are openings and sunlight. They occur only where there are breaks in the canopy. There they can perch on cypress stumps, old logs, or other openings.

Also preferring open sunny sites dominated by grasses is the Juba Skipper (*Hesperia juba*), originally named by Samuel H. Scudder in the 1870s. Skippers are notorious for bouncing flight. No doubt Scudder saw the "dancing" butterfly and thought of the African dance then popular.

Iris

(Greek for rainbow and the ancient name of the flower; also the Greek goddess of the rainbow and messenger of the gods, daughter of Thauas by the oceanid Electra)

flag (of obscure origin, but perhaps from Gaelic where the word means bulrushes prepared for weaving; in English by A.D. 1387); *flaede* (Middle Danish); *flaeg* (Danish); *flag* (Dutch)
giaggiolo (Italian)
iris (from Greek; Iris was the goddess who was the messenger of the gods, and was held to display as her sign or appear as a rainbow; in English by A.D. 1490); *iris* (French)
lírio (lily, Portuguese, Spanish)
pakwiasko'ns (water weed, Potawatomi); *sênipa'spwûn* (Prairie Potawatomi)
Schwertlilie (sword lily, German); *sverdlilje* (sword lily, Norwegian)
seilistear (Gaelic, for *I. pseudoacorus*)

Iris hexagona (six angled)
Dixie iris ("Dixie" became a euphemistic name for the southern United States [the "South"] in 1859 after Daniel D. Emmett wrote the song *Dixie*, Florida)
pasí:nácákî (dwarf cat-tail, *pasí:nî*, cat-tail, Miksauki)
prairie iris (Florida)

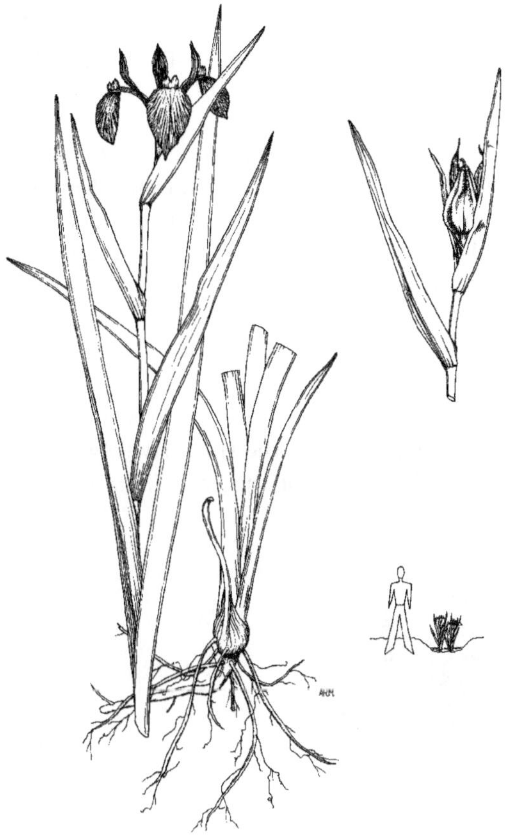

Iris virginica. From Institute of Food and Agricultural Sciences.

Iris verna (spring-flowering)
dwarf [dwarf violet, violet] iris

Iris virginica (of Virginia)
oko:wílli paká:li (*okí*, watery, *ilillá*, death, *paká:li*, flower, Koasati)
Virginia iris

Linnaeus ([1753] 1957) knew 18 species of *Iris*, with 12 of them growing in Europe. The genus was well known to most Europeans and had acquired a variety of common names in different languages. When they arrived in the New World, Europeans found species new to them, and the genus now contains about 210 species, with 30 in Europe (Mabberley 1997).

Hogan (1978) found *Iris* pollen in her samples from the Glades site at Fort Center on Lake Okeechobee. These people had been using the plants for something, although it is not clear what that may have been. In 1919, Sheehan and Small were told that *Iris* was one of the plants given to Seminoles bitten by alligators (Small 1933, von Reis and Lipp 1982). Sturtevant's (1955) informants were not familiar with that application. Taylor (194) also found the Koasati using the plants. Given the widespread application of *Iris* as medicine (Moerman 1998), doubtless several species had historical use.

The Cherokee used the roots of *I. verna* or *I. virginica* in a compound medicinal salve for ulcers and took an infusion for liver and urinary problems (Hamel and Chiltoskey 1975). Creek considered *I. verna* a powerful cathartic (Swanton 1928a).

Iva

(Said to be named after the similarly fragranced mint, *Ajuga iva*; also Latin for the yew, *Taxus*)

Iva. *Iva annua* (left). From Lamarck 1823. *Iva frutescens* (right). From Britton and Brown 1898.

Iva annua (living a single year; annual)
careless weed
[annual] marsh elder [marshelder] ("marsh elder" in use by 1755 to indicate the genus *Iva*; "marsh" is a tract of low-lying land that has standing water most of the year; in use since about A.D. 725, from Old English *mersc, merisc*; akin to Dutch *mersch*, German *Marsch*, Danish *marsk;* probably derived from Latin *mariscus*)
sump weed [sumpweed] ("sump" in use for a swamp, marsh, or morass since about A.D. 1425; related to Low German *sump, sumpf*, Middle Dutch *somp, sump*, Flemish *zompe*, and West Frisian *sompe*; akin to Greek *somphos*)

Iva frutescens (shrubby)
bigleaf sumpweed (Florida)
high-water shrub (Florida)
horse brush (Florida)
marsh elder (Florida)

Linnaeus ([1753] 1957) created the genus *Iva* with two species, *I. annua* and *I. frutescens.* The former was based on Adrian van Royen's list of plants cultivated in Leiden of 1740, his own *Hortus Uppsaliensis* of 1748, and his list of dissertations in *Nova plantarum genera* of 1751. He also cited a specimen given to him by A.L. de Jussieu. Both he and Royen had seen cultivated specimens of *I. frutescens,* and he had literature citations and herbarium specimens from other sources. There is nothing in the phrase names used before 1751 that would have been an inspiration for his using *Iva,* so he apparently took it from elsewhere.

All sources that I have seen waffle on the origin of the word *Iva.* Several authors say simply that it was the "old name of some medicinal plant" (e.g., Fernald 1950, Weber 1987). Gledhill (2002) added that it was used by Rufinus (Archbishop of Sorrento, 12th century) and "applied to various fragrant plants." Others say the genus is patterned after *Ajuga iva* (Munz 1973, Correll and Correll 1982, Hickman 1993, Diggs et al. 1999). No one has clarified why the mint has that name.

What no one seems to have pointed out is that Latin *ívus* is the basis of *íva* in Spanish, and that both are names for *Taxus.* It would appear that Linnaeus used *Iva* to compare this composite with a mint that grows under *Taxus.*

Iva, a genus endemic to North America and the Caribbean, comprises 15 species (Mabberley 1997). Florida has 5 species, with at least *I. asperifolia* introduced from Mexico, and there are several others in the United States (Fernald 1950, Correll and Johnston 1970, Barkley 1986, Hickman 1993).

Two of the species in Florida have been used by people. *Iva annua* was not only used by historic indigenous tribes, but a form was also cultivated by eastern people for their large edible seeds (Gilmore 1931, Jackson 1960, Black, 1963, Asch and Asch 1978, Yarnell 1978, 1986, Smith 1992). Archaeological sites yielding the large-seeded forms of sump weed have been found in Arkansas, Illinois, Iowa, Kentucky, Mississippi, Missouri, North Carolina, and Tennessee. More have been found near the Missouri and Mississippi River confluence than elsewhere, but the cultivar was widespread. Although sumpweed seeds are known from archaeological sites dated at ca. 3500 B.C., those are within the size of wild plants. Seeds increased in size from ca. 2000 B.C. to A.D. 1300, so most believe that people were cultivating them and selecting for that trait (Fritz 2000b).

Seeds were harvested by hand, and Asch and Asch (1978) have shown that a single person can obtain enough seeds in 1 hour to supply an adult's daily energy requirements. When taken from the plants the seeds are enclosed in a fibrous shell of the achene. Again, Asch and Asch (1978) suggest that these were boiled to weaken or split the achene. Once the fruits are dried, they can be winnowed to separate the shell from the seed. When they sampled the produce, they were pleasantly surprised at the nutty taste; all of the disagreeable odor of the fresh plant had been eliminated.

The seeds contain about 40% fat and 32% protein. While low in lysine, the seeds are high in thiamine, niacin, calcium, iron, and phosphorus. Calorie content is similar to the sunflower (*Helianthus*, which see). These extinct cultivars were displaced as food with the introduction of maize in the first millennium B.C. (Smith 1992). Only the wild forms remain.

Hocking (1997) said that sumpweeds were used by several tribes to treat stomachache or cramps, and claimed to be contraceptive. Neither Vogel (1970) nor Moerman (1998) included the species, and the source of that information has not been located. Crites and Terry (1984) and Zardini (1992) mention sump weed in relation to other cultivated seeds.

Hocking (1997) also said that the bark of *I. frutescens* was used to reduce fever. No corroborating source has been located, but Bohlmann and Zdero (1979) found two eudesman derivatives in *I. annua.*

J

Jacquinia

(Linnaeus dedicated the genus to Austrian Nicolas Joseph von Jacquin, 1727–1817, professor of botany and chemistry in the Medical Faculty of the University of Vienna, and later rector of the University)

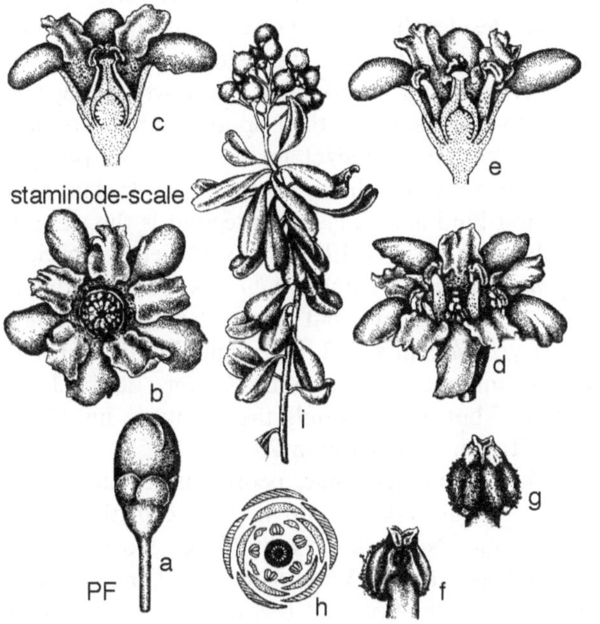

Jacquinia keyensis. a. Flower bud. b. Flower, from above at early phase. c. Flower, longitudinally dissected. d. Flower, from above at later phase. e. Flower, longitudinally dissected at later phase. f. Anther, from back. g. Anther, from front. h. Floral diagram. i. Fruiting branch. *Drawn by Priscilla Fawcett.* From Correll and Correll 1982.

Jacquinia keyensis (of the Florida Keys)
Joe-wood [Joe-bush] (Bahamas)
ironwood (Bahamas)

Jacquinia is a genus of 35 species in Central America and the Caribbean (Mabberley 1997). Fruits from several species are notoriously poisonous with rotenone and have been used to catch fish (Hocking 1997). Perhaps the most famous for that purpose was long called *J. barbasco,* although it is correctly *J. arborea.* The word *barbasco* (from Latin *verbasco,* mullein, *Verbascum,* Scrophulariaceae) now means "fish-poison."

The Florida species is the source of cudjoe wood (Mabberley 1997). Presumably, it too contains the poisons in the fruits.

Juglans

(Linnaeus created the name from a contraction of *Jovis glans,* the nut of Jupiter—although *glans* has an alternate meaning, which he no doubt intended)

Juglans nigra. From Sargent 1905.

cnò Fhrangach (*cnò,* nut, *Fhrangach,* French, Gaelic)
gall-chnò (foreign nut, Gaelic)
noce (Italian); *nogal* (Portuguese, Spanish; from Latin *nux,* through vulgar Latin *nucale*); *noyer* (French); *Walnussbaum* (walnut tree, German); *walnoot* (Dutch)

Juglans nigra (black)
assenamins (Strachey wrote this Powhatan word as *assunnomeindg* in [1612] 1953 and said that it was a walnut; however, because of comparisons with other Algonquian words Siebert (1975) thinks it belongs with *Carya alba,* which see; distinctions between "walnuts" and "hickory nuts" typically were not made until ca. 1810); *osamener* (Carolina Algonquians)
baganak (Potawatomi, for *J. cinerea*; akin to Menomini *paka:n,* see *Carya illinoiensis*); *paucauns* (Strachey used the word in [1612] 1953 under "walnuts"; see also "pecan" under *Carya illinoiensis*); *ptúcquim* [*pdukquim*] (Delaware); *tuk·wi·má·nsi* (walnut tree, Delaware)
chak (*chak-hu,* walnut tree, Winnebago); *chan-sapa* (*chan,* wood, *sapa,* black, Teton)
haahi (Alabama); *hahe* (Choctaw); *há:hi* (Koasati); *hayi* [*hayyi, hasyi*] (Chickasaw)
hma [*gma*] (Dakota)
noyer (Quebec)
ochatapaski (*ocha,* hickory, *tapaski,* fine textured, Alabama)
osõchqua (Onondaga)
poho'n-ä [*po-ho-na, poñ-hoñ-á-daw, poñ-hoñ-ai-gaw*] (the word in the name, *po-ho-na,* refers to a former tribal custom of the husband's right to

cut off a piece of his wife's nose if he finds she has been unfaithful, Kiowa)
sahtaku (Pawnee)
sedi [*se:-di*] (Cherokee)
ta'-ge [*sha-e*] (Osage); *tdage* (*tdage-hi*, walnut tree, Omaha-Ponca)
uksak [*okesok*] (Choctaw)
vhah'wv-enlókce (Muskogee)
[black, eastern black walnut] walnut (foreign nut, to distinguish it from the hazel nut, *Corylus*; derived from Anglo-Saxon *wealh*, Welsh, foreign, or alien, the second element is from Latin *nux*, nut)
waru' [*wararu'·*] (walnut, Catawba)
yegg (Atakapa)

My introduction to the black walnut occurred when I was a small boy. When my great-uncle learned that I was bothered with ringworm on my posterior, he vowed he could cure it very simply. All other remedies had failed my parents, and they agreed. I waited and watched with considerable apprehension while he walked across the clean-swept yard, hurrying the chickens out of the way, and picked up a green walnut fruit with the husk remaining. He broke the husk open, and ordered me to drop my pants. After I complied, he smeared the infected area with the dark juice from the walnut husk. Once it dried, he allowed me to put my pants back on. Within a week, the ringworm was gone. I gained new respect for my great-uncle and the black walnut. Many years later, I learned that he had applied a remedy that dated to the classical Greeks and Romans, perhaps earlier in the Americas (see also Murphee 1965).

Although the history of *Juglans* is much younger in Europe than in the Americas, Europeans were already familiar with walnuts when they arrived in the New World. *Juglans regia* was native as far west as the Balkan peninsula in Greece, Crete, Romania, Turkey, and Persia, but it had been introduced into Italy by 116 B.C. (Hedrick 1919). The Romans extended the cultivation of walnuts through ancient Gaul (France). By 1542, Leonard Fuchs said the walnut was cultivated throughout Germany, where it was called *Welchnusz* (Welsh nut), and the Dutch said *nootboom* (nut tree). The Italians were already saying *noce*, the French *noyer*, and the Spanish *nogal* (Meyer et al. 1999). When *Juglans* finally reached England, it first was called "Wealh nut" or "Gaul nut," and this evolved into "walnut" (Rosengarten 1984).

Records of hickories (*Carya*) and walnuts (*Juglans*) are difficult to sort out through the 1700s because they were lumped under *Juglans*. It was not until 1818 when Thomas Nuttall separated *Carya* that records became clear. Chroniclers of the de Soto expedition reported "walnuts" near modern Gainesville (Swanton 1939), but these were hickories (Wunderlin and Hanson 2002).

Harriot ([1590] 1972) wrote: "There are two kindes of Walnuts, and of them infinit store: In many places where very great woods for many miles together the third part of trees are walnuttrees. The one kind is of the same taste and forme or little differing from ours of England, but that they are harder and thicker shelled: The other is greater and hath a verie ragged and hard shell: but the kernell great, verie oylie and sweete. Besides their eating of them after our ordinarie maner, they breake them with stones and pound them in morters with water to make a milk which they vse to put into some sorts of their spoonmeate; also among their sodde wheat [maize], pease, beans and pompions which maketh them haue a farre more pleasant taste." The first kind is a *Carya*; the second is the walnut.

Bartram ([1791] 1958) recorded eight species of *Juglans*. The true walnut (*J. nigra*) was among them, as well as the pecan, but some he called "Juglans hickory." Michaux did the same in his notes and herbarium, recognizing both *J. nigra* and "Juglans hickory," but he listed only the walnut in his *Flora* of 1803 (Taylor and Norman 2002).

Throughout its range, people used black walnuts for food; Yanovsky (1936) listed use in the north—Minnesota, Nebraska, New York, North Dakota, South Dakota, and Wisconsin. The Cherokee mixed the nuts with hominy (*newada*) and beans into porridge much as did the Carolina Algonquians described by Harriot ([1590] 1972). The Cherokee also saved the dried nuts for later use (Hamel and Chiltoskey 1975). The Comanche ate fresh nuts and stored others for winter use (Moerman 1998). The Dakota, Omaha, Osage, Pawnee, Ponca, and Winnebago ate the seeds from the nuts directly, or made them into soup (Hunter [1823] 1973, Gilmore 1919). The *tdage* of the Omaha-Ponca is in their story of the "*Ishtinike* and the Four Creators," indicating the reverence in which they held the trees and their antiquity of use. The Meskwaki ate nuts, as did their neighbors (Moerman 1998).

The most-detailed records of food come from the Iroquois of New England. These people boiled the fresh nut meats and drank the liquid. Fresh seeds were mixed with cornmeal to make a bread, which might also include beans and berries. Similarly, seeds were included in corn pudding, corn soup, hominy, and with mashed potatoes after Europeans introduced that South American plant. Oil was extracted from the nuts by boiling and used as a condiment on cornbread, corn pudding, and mush by the False Face Society (Moerman 1998).

Walnut trees also provided other materials. The Cherokee pounded walnut bark and put it into small streams to catch fish (Mooney 1885–1886). The Iroquois mixed nut meat oil with bear grease as a mosquito repellent and used the inner bark fiber to weave burden straps (Whitford 1941, Moerman 1998).

Black or brown dyes were made from the roots, bark, and fruit husks. Leaves produced a green dye. These dyes were made by the Cherokee, Choctaw, Dakota, Kiowa, Meskwaki, Ojibwa, Omaha, Pawnee, Ponca, and Winnebago (Bushnell 1909, 1917, Gilmore 1919, Moerman 1998). Traditional Cherokee basket makers still use these dyes to stain their wares. One of these, a prized item I got in the Quala Coop store in Cherokee, North Carolina, in the 1970s, is made from hickory splints.

Walnut has been one of the most desirable woods in the eastern United States, to the point that "poachers" sometimes steal trees from yards. Rosengarten (1984) recorded that a single tree might be worth $30,000 for its wood. Walnut wood is used in making superior furniture, cabinets, veneers, caskets, coffins, posts, railroad ties, and fuel (Vines 1977). Walnut was prized for making gunstocks during the Civil War and World War I, and is still considered one of the best woods for that purpose (Rosengarten 1984).

Many people used walnuts as medicine. The only record among Muskogean people seems to have been with the Houma. They used an infusion of nutshells to relieve the itch, and a decoction of leaves to lower blood pressure (Speck 1941). Other groups used the inner bark in a tea as an emetic and laxative; the bark was chewed for toothache or made into an infusion to stop diarrhea and expel worms; and fruit husks treated ringworm, colic, and were poulticed on headaches, sores, and inflammations (Ford 1975, Foster and Duke 1990, Moerman 1998). Use is known among the Cherokee, Comanche, Delaware, Houma, Iroquois, Kiowa, Meskwaki, and Rappahannock (Moerman 1998).

Juglone is among the active compounds in walnuts. It is antiseptic, antifungal, and herbicidal (Lewis and Elvin-Lewis 1977, Foster and Duke 1990). Tannins are also abundant, and are effective antibiotics in addition to contributing to the dyes. Black walnut also contains lectins and ellagic acid, and thus is somewhat active against tumors (Lewis and Elvin-Lewis 1977).

The Delaware spread leaves to get rid of fleas (Moerman 1998). The Choctaws also scattered *okesok* leaves to drive fleas away (Bushnell 1909). Because of the smell of the leaves, and the tendency of fleas to congregate where people and their animals live, that probably worked.

Juncus

(From Latin *iuncus*, for a rush, derived from *iugere*, to tie or bind, from the use of stems for that purpose; akin to Akkadian *unqu, uqqu*, ring, Greek *onkos*, hook)

Juncus effusus. From Institute of Food and Agricultural Sciences.

Binse (German)
jonc (from Latin *juncus*, French); *junco* (Spanish)
rush (of uncertain etymology, but perhaps from Germanic words, from Old English *risc* [*rix, rise, rixe*], Middle Dutch *risch*, and Middle Low German *risch(e)* [*rysse, risk, ryse*] dating to about A.D. 725)
siv (Norwegian)

Juncus bufonius (of toads, from its occurrence in damp places)
[common, toad] rush

Juncus effusus (loosely spreading)
albatá incassí (*albatá*, alligator, *in*, its, *cassí*, corn, Koasati)
[common, rice] bullrush ("bullrush" is of uncertain etymology; from Middle English *bulrysche* or *bolroysche*, from about A.D. 1440; the term is also applied to *Scirpus* and *Typha*, which see)
cakekesqus (Strachey in [1612] 1953 said this was the Powhatan word for "rushes")
[common, Japanese mat, mat, soft, *tatami*] rush

Many of the Cyperaceae and Juncaceae have flexible, tough stems and have been called similar terms by people throughout Europe and North America. *Juncus* has been used in Egypt for matting since the Neolithic (Mabberley 1997). Leaves until recently were still used in Scotland for thatching roofs, and may still be seen in historical replicas of old houses. The genus contains about 300 species and is nearly cosmopolitan (Mabberley 1997). Europe has 53 species and Kartesz (1993) lists more than 100 for North America. In addition to these 2, Moerman (1998) listed 12 others that were used by North American tribes.

Perhaps it was one of the *Juncus* that was used by people on the Great Plains of Arkansas to cover their houses when they were visited by the de Soto expedition in 1541. Ruiz Hernández de Biedma then wrote: "[W]e came to some collections of huts, covered with rush sewed together. When the owner of one moves away, he will roll up the entire covering, and carry it, the wife taking the frame of poles over which it is stretched; these they take down and put up so readily, that though they should move anew every hour, they conveniently enough carry their houses on their backs" (Biedma in Swanton 1939). Biedma was critical of them because they had no interest in growing plants for food, but he failed to point out that there were so many buffalo ranging through the area that they would have been unable to protect their crops from them.

The Cherokee used *J. effusus* to bind oak leaves around dough before it was cooked (Hamel and Chiltoskey 1975). The Ojibwa used rushes for weaving mats, bags, and pouches (King 1984, Moerman 1998). All the other records are from western tribes, including the Cahuilla, Hesquiat, Karok, Mendocino, Pomo, Snuqualmie, Tolowa, and Yurok. All these groups used *Juncus* leaves for tying and weaving.

Porcher (1863) noted that *J. effusus* was "cultivated in Japan for making floor mats, chair bottoms, etc. It is sometimes employed in South Carolina for similar purposes. The pith, when dried and oiled, will serve as a wick." His reference to the Japanese using the rushes is for making *tatami* mats.

There are even records of the Mendocino, Okanagan-Coville, and Snuqualmie either eating the plants or feeding them to their animals (Moerman 1998). The introduced Old World horse and cow might eat the plants, but they hardly seem edible for humans.

Only the Cherokee and Iroquois are noted as having used *Juncus* for medicine in the eastern United States. The Iroquois used a decoction of *J. bufonius* as an emetic; runners drank it to prepare themselves to race (Moerman 1998). *Juncus effusus* was also considered an emetic among the Cherokee. Other uses included an infusion given to babies to prevent lameness and a wash to strengthen them. The Karok in California regarded the rush as sacred and their shaman burned it as an offering while praying.

Settlers and later Americans considered *J. bufonius* diuretic and cathartic (Hocking 1997). *Juncus effusus* was also used as a mild diuretic, to expel kidney stones (lithiasis), and for urination difficulties (strangury). It was also used as a cathartic (Hocking 1997). Porcher (1863) noted for *J. effusus* that "A decoction of the plant is said to be diuretic."

Hocking (1997) found that the roots of *J. effusus* contain tannin and a tripeptide.

Juniperus

(The classical name, from Latin *ieniperu, gigiperus,* names used by Virgil, 70–19 B.C., and Pliny, A.D. 23–79)

Juniperus communis (Latin *communis,* common; Circumboreal, but not in Florida)
aitionn [*aiteann*] (Gaelic; badge of the Highland clans Athole, Gunn, MacLeod, Murray, and Ross)
faile de dh'iubhar beinne (bath of the juice of the juniper, a favored treatment for headache, Gaelic)
ga'gawan'dagisid (deceptive, Ojibwa)
iubhar-chreige (*iubhar,* yew, *chreige,* craig or rock, Gaelic)
juniper (English by the 1400s) *enebro* (Spanish in 1542); *einer* (Norwegian); *geneure* [*ieneuve*] (French in 1542); *genever* (Dutch in 1542); *genève* [*genèvier*] (French); *ginepro* (Italian in 1542); *zimbro* (Portuguese)
kahka-kewatik (raven tree, Cree); *kahka-kiminiatik* (crow berry, Cree)
mbethi^n dzhi (owl berries, Slave)
Waldholder (woods elder, modern German; see also *Sambucus*); *Wacholder* (modern German; *Weckholder* in 1542; probably from *wachen,* guard, be awake, meaning evergreen; the Middle High German *queckholder* comes from *queck,* making alive; *Quêkis* [branches] were used in rituals to beat people and animals to ensure fertility)

Juniperus sabina (Latin *sabina,* juniper; Old World only)
'ar'ar (Arabic)
chedraia (Greek)
ciye (Coptic)

savin [saffern, saffron] (came into English about A.D. 1000 from Old French *savine*, from Latin *sabina*, juniper); *sabina* (Portuguese, Spanish by 1542); *samban* [*samhan*] (Gaelic, the plant is not native to northern Europe, and the Romans surely introduced it and its name; the Gaelic word has other meanings that predated this use); *sauine* (Italian in 1542, now *savina*); *sauinier* (*savine* tree, French in 1542); *savelboom* (savin tree, Dutch in 1542); *Sevenbaum* (savin tree, German in 1542)

Juniperus virginiana (from Virginia) (= *J. salicicola* fide Wunderlin 1998; *J. virginiana* var. *salicicola* fide Adams 1993)

acinî (Mikasuki); *vcénv* [*acinâ, achena, auchenau, auche-nau*] (Creek, Muskogee; in Levy County, Florida, there is an Atsena Otie Key, from *vcénv*, cedar, *ote*, island); *a'tsina'* [*a'-tsi-na', a-tsi:-n', ajina*] (loan-word from Creek, Cherokee)

atcuwe se'?; *se wyhukse'*; *aswetci'here* (Catawba; cognate or loan related to Creek?)

[Carolina, coastal red, southern red (the Florida form)] cedar ("cedar" came from Middle English and Old French *cedre*, taken from Latin *cedros* and Greek *kedros*, used in English by about A.D. 1000; the modifier "red" was applied to distinguish the American from the Old World species); *cèdre rouge* (red cedar, Quebec)

chowwaala [*chowaala, chowahla*] (Alabama); *chuala* [*chuatla*] (Choctaw); *chowaala'* [*chowa hla', chohwala', chomahala', chawala*] (Chickasaw)

ekawai:pv (Comanche)

genévrier (Quebec); juniper-bush; red juniper

hante [*hante sha*] (*hante*, eggs, *sha*, red, Dakota)

icuwése (Catawba)

khicuc [*khishoush*] (Atakapa)

'ko-kee-äd-la [*a-heeñ*] (conspicuous, Kiowa)

maazi (Omaha-Ponca)

mehokhócus (Delaware); *miskawa'wak* (red wood, Ojibwa)

moroke (Strachey identified this word as the "cedar" in [1612] 1953; it is not mentioned by Siebert 1975; cognate with Delaware and Ojibwa?)

noehntotakri (*ojenta*, wood, Onondaga)

pencil-wood (formerly popular for making lead pencils)

sabina [*colorado*] ([red] cedar, New Mexico); savin [red savin, savine]

sequaw (probably meaning red, Shawnee, fide Edgar 1891)

tar (Yuchi)

tawatsaako (Pawnee)

xon-dse hi (*zhu-dse*, red, *hi*, tree, Osage)

ya-'toñ-ba (wood for love flute, Kiowa)

Since the time of the classical Greeks and Romans, two species of *Juniperus* have been recognized in Europe, the juniper and the savin. Greek author Dioscorides (fl. A.D. 40–80) and Roman herbalists Galen (A.D. 129–?200) and Pliny (A.D. 23–79) discussed both species (Meyer et al. 1999). Because these two Old World species contributed heavily to the names used for American plants, they are listed above with scientific and common names.

During the Middle Ages, Europeans thought that juniper smoke gave protection against contagious diseases, particularly the plague and leprosy (Dobelis 1986). This belief prevailed even in the times of Rembert Dodoens, who wrote in 1616, "Iuniper or the berries thereof burned driueth away ... all infection and corruption of the ayre." Northern Europeans burned juniper during childbirth to prevent the *aillse* (fairies, Gaelic) from substituting a changeling for a newborn. Up until the early 1900s, many people in the British Isles believed "that he who cut down a juniper would die within a year" (Vickery 1995).

Those practices were likely holdovers from earlier times, probably going back to Gaelic and classical Greek and Roman cultures. Junipers have been considered wondrous by every culture living where the plants grow. Indigenous Americans consider them sacred.

Juniperus communis has given us two other familiar words. "Geneva" is an Anglicized version of the French *genève*, juniper, used for the city in Switzerland. The other word names a product that is said to have originated in Holland, where its Dutch name has been *genever*, juniper, since at least the 1500s. However, the product now is known as "gin," a highly contracted version of "Holland's geneva." This is an alcoholic drink made from grain and flavored with the crushed "berries" of *J. communis*. In addition, the fruits are used to flavor game, stuffings, marinades, and stews (Bremness 1994).

Harriot ([1590] 1972) may have been the first to record red cedar in the New World although he did not know the Algonquian name for it in North Carolina. He wrote, "Cedar, a very sweet wood & fine timber; whereof if nests of chests be there made, or timber thereof fitted for sweet & fine bedsteads, tables, deskes, lutes, virginales & many things else." Later he noted that others had reported white cedar (see *Chamaecyparis*), but he had not seen it.

Red cedar is prominent in the Cherokee story of "How the World Was Made." Their story relates that when the first plants and animals were made, they were told to stay awake for seven nights (seven, a holy number). One after the other, the animals fell asleep until finally only the owl and panther were awake. These two were given the power to see in the dark and prey on the others. Of the trees, only the cedar (*Juniperus*), pine (*Pinus*), spruce (*Picea*), holly (*Ilex*), and laurel (*Persea*) were awake on the seventh day. Therefore, "to them it was given to be always green and to be greatest in medicine" (Mooney 1885–1886).

A Yuchi story explained how the cedar got its red wood. A wizard was interfering with the movement of the sun across the sky. Two warriors were selected to dispatch the wizard, and when he appeared, they swung their clubs and knocked off his head. They returned to their people with the head, but it refused to die. The warriors were advised to tie the severed head to the tip of the cedar tree and only there did it die. Before dying the wizard's head bled through the cedar, staining its wood red. Ever since, cedar wood has been red and the tree a powerful medicine (Lankford 1987).

Cedar wood has long been one of the most prized in the Americas. The Cherokee carved the wood, made fence posts and furniture with it, and used it for mothproofing (Hamel and Chiltoskey 1975). The Kiowa used the red, aromatic heartwood to make "love flutes" (Vestal and Schultes 1939). The Ojibwa put split strips as fences and thatch on graves, made them into mats, and made cradle boards with the wood (Moerman 1998). The Potawatomi wove mats and bags from the bark (Smith 1933, King 1984). The Thompson people made a fuel of the wood for smoking skins to color them dark (Moerman 1998).

Wood has been used for novelties, posts, woodenware, millwork, paneling, closets, chests, and pencils (Vines 1977). The aromatic wood still is prized for chests because people think it repels insects. Porcher (1863) wrote: "The wood of this tree is well known. It is sometimes dug up in the mud of our swamps in a perfect state of preservation. It is aromatic, light, soft, bearing exposure to water and weather, and suitable for all kinds of cabinet work, in the construction of posts, staves, the inner work of houses, and particularly in the building of boats. Cedar boxes are not infested by insects, moths, etc., and are used for storing away woollens. The leaves also prevent the attacks of insects when spread over cloth."

Bark was used to build houses, wigwams, wickiups, and mats by the Ojibwa (Moerman 1998). They also used shredded bark as padding on cradle boards. The Potawatomi made bags from the fibers (Whitford 1941).

To the people in the Missouri River region, the tree was sacred to their legendary thunderbird. Therefore, the Lakota, Omaha, Pawnee, and Ponca put boughs on tipi poles to avert lightning (Gilmore 1919). Various versions of the "thunderbird" story existed throughout southeastern tribes, and perhaps they had a similar use of red cedar (Lankford 1987). Indeed, people as distinct as the Ojibwa and Navajo considered the trees capable of averting evil and burned branches as incense or carried wands in ceremonial dances (Moerman 1998).

Although Moerman (1998) listed all but three of the 13 species known in North America (Adams 1993), and three entries under *J. virginiana*, he does not include all the Muskogeans who used these plants.

The Chickasaws warmed the ends of cedar limbs in water with elder (*Sambucus*) and placed them on the head to alleviate severe headache (Swanton 1928b). The Creeks used cedar fumes to relieve cramps in neck muscles (Swanton 1928a). The Alabamas boiled sprigs for warm applications to rheumatic pains (Swanton 1928a). The Chickasaw's neighbors, the Choctaws and Natchez, used cedar for muscular pains, swollen legs, and mumps (Swanton 1928a).

More details are recorded among the Seminoles than any other southeastern group. They use the southern red cedar (*J. virginiana* var. *salicicola*) to treat cold symptoms (including cough, runny nose, stuffy head, and sore throat) and swollen joints (decoction as body rub and steam bath). Additionally, red cedar was used to treat "Eagle Sickness" (stiff neck and back), "Fawn Sickness" (swollen legs and face), "Ghost Sickness" (dizziness, staggering), "Hog Sickness" (unconsciousness), "Mist Sickness" (eye disease, fever, chills), "Opossum Sickness" and "Raccoon Sickness" (children dreaming about these animals), "Rainbow Sickness" (fever, stiff neck, backache), "Scalping Sickness" (fever, headache, backache), and "Thunder Sickness" (fever, dizziness, headache, diarrhea). A decoction was used as an emetic during religious ceremonies and to make a witchcraft medicine. Branches were kept with eagle tail feathers to prevent the powerful plumes from causing sickness (Sturtevant 1955).

The Ojibwa and Meskwaki also used cedar for headache and to season other medicines (Vogel 1970). Fruits to the Dakota were *hante itika* (cedar eggs), and they were boiled with leaves for a decoction used to treat coughs (Gilmore 1919). Smoke from burning twigs was inhaled to cure colds. Juniper was also used by the Omaha in ceremonial cleansing. Red cedar was used as medicine and incense by at least the Alabama, Cherokee, Chickasaw, Comanche, Cree, Creek,

Dakota, Delaware, Iroquois, Kiowa, Lakota, Meskwaki, Natchez, Ojibwa, Omaha, Pawnee, Ponca, Rappahannock, Seminoles, and Salish (Moerman 1998).

The chemistry of *Juniperus* has been comparatively well studied. To everyone's surprise, the genus contains the antibiotic podophyllotoxin, originally known from *Podophyllum* (which see) in the Berberidaceae (Lewis and Elvin-Lewis 1977). Otherwise, there is considerable variation in the other compounds present in different species.

The Old World *J. sabina* contains sabinol, sabinyl acetate, perpinene, pinene, sabinene, decyl aldehyde citronellol, geraniol, cadinene, and dihydrocuminyl alcohol. However, *J. virginiana* is so similar that it is sometimes substituted commercially for *J. savina* (Lawless 1995). These compounds are considered powerfully anthelmintic, diuretic, emmenagogue, rubefacient, stimulant, and a vermifuge, but they are so potent that they "should not be used in therapy, whether internally or externally" (Lawless 1995). Duke et al. (2002) admonish that extracts from *J. virginiana* may be fatal.

Justicia

(Named for the Scottish horticulturist James Justice, 1698–1763, who was elected a fellow of the Royal Society in 1730)

Justicia angusta (narrow, the leaves) (reported as *J. crassifolia*)
akciswantá:cí:kî (penis hardener, Mikasuki); *haswâ: isfa: pi:ckâ* (*hvswv*, penis, *enfapetv*, erection, Creek)
loose-flower water-willow [looseflower waterwillow]

Wunderlin (1998) lists seven species of *Justicia* in Florida, with three of these having been introduced from elsewhere. *Justicia angusta, J. cooleyi,* and *J. crassifolia* are endemic.

Miccosukee Josie Billie told Sturtevant (1955) about this plant that would "restore the virility to a man sixty to a hundred years old." A dried piece of the plant was kept by a man who knew the song that enabled the treatment. The song was considered "dangerous" and valuable. Billie said he would not refuse to treat any man who asked for the cure.

Sturtevant (1955) wrote of the plant, "probably *Justicia crassifolia.*" That determination seems unlikely. *Justicia crassifolia* is endemic to a small region in Hernando, Sumter, and Lake Counties. Therefore it seems unlikely that these would have been the plants used by the Seminoles. Much more likely is *J. angusta,* a plant widespread through the southern end of the peninsula. Doubt prevails because this is not one of the plants identified by Roy Woodbury for Sturtevant (1955).

Justicia pectoralis was introduced by the Cubans (and probably by other Latin Americans) as a medicinal plant from their homeland that they call *tilo.* That is the same word used in Spanish to designate *Tilia,* and there are similarities in the way these unrelated plants are used (see *Tilia*). This *Justicia* is famous for use throughout the Americas as a calming tea, among other uses (e.g., Schultes and Hofmann 1979).

K

Kallstroemia

(In honor of A. Kallstroem, an obscure contemporary of Johann Anton [Giovanni Antonio] Scopoli, 1723–1788, Austrian physician and professor of natural history at Pavia, who named the genus)

Kallstroemia maxima (largest)
abrojillo (little bur); *arbojo* [*terrestre*] ([land] bur, Cuba, Dominican Republic, Puerto Rico, Nicaragua, Colombia)
atarraya (something that detains or catches, Puerto Rico)
batatilla (little potato, Dominican Republic)
caldero (little cauldron, Puerto Rico)
caltrop [big, greater] ("caltrop" probably came from Latin *calcum*, heel, and *troppa*, trap or snare, and has been applied to a number of thorny or spiny plants. First applied to plants ca. A.D. 1000, and to *Tribulus* by 1511. Later applied to *Kallstroemia* and to an instrument of war ca. 1300. By the early 1500s, this trap was a ball with four sharp prongs or spikes so that one always stands upright regardless of how it falls on the ground; USA, Bahamas, Puerto Rico.)
câpre marron (wild caper, Haiti)
centipede root (Virgin Islands)
chax-chauxnuc (a name usually applied to *Tribulus cistoides*, but occasionally to *Kallstroemia* by confusion; *chak*, red, *xnuuc*, old; the name *"xnuuc"* or *"viejo"* is given to the ferruginosus pygmy-owl, *Glaucidium brasilianum ridgwayi*, and the spines on the fruits remind people of its feathered "ear" tufts, Maya, Quintana Roo)
cresson courant (twining [literally, running] cress, Guadeloupe, Martinique)
golondrina (dove, El Salvador)
guía [*hierba*] *de parra* (grape vine, El Salvador)
hierba [*yerba*] *de pasmo* (herb for *pasmo*, a disease with fever and aching bones, from Latin *spasmus*, Venezuela)
hierba de pollo (chicken herb, El Salvador, Panama, Colombia)
jepo
longlo (maybe from "long" and "low," describing the growth habit, Virgin Islands)
maconcherie (probably a mistake for *macoucheri*, as in the Macoucheri Estate, Dominica)
mata (wild plant or thicket, Costa Rica)
ni (Huastec, San Luis Potosí)
pale-flowered turkey blossom (Jamaica)
parsley (Cayman Islands)
patagon (big foot, Martinique)
perrito (little dog, Colombia)
police macca (*macca* means a thorn; so the name means that the thorns prevent a person from going to a particular place, in the same way the police would deter them, Jamaica)
pourpier bâtard (false purslane; *pourpier* is from Vulgate Latin *polpie*, in turn from *pulli pedem*, chicken foot, Guadeloupe, Martinique); *pourpier courant* [*rampant*] (twining [creeping] purslane, Martinique)
shanapmucui (Belize)
talcacao (maybe comparing these to the plant that Hernández called *tlalcacauatl*, the peanut, *Arachis hypogaea*, Náhuatl, Costa Rica)
taraya (this looks like an orthographic version of *tlalca* or *tlalza*, El Salvador)
tlalcacauatl [*tlalca-cauatl*] (probably from *tlalzahuatl*, where *zahuatl* is mange, because of the chiggers, *Eutrombicula*, that they harbor; in Sonora, the Mayo call *Kallstroemia* the *jímiri*, chiggers, because of that problem)
verdolaga (purslane or *Portulaca*, Yucatán, El Salvador, Costa Rica, Venezuela); *verdolaga blanca* [*de caballo, de monte*] (white [horse, wild] purslane, El Salvador); *verdolaga de abrojo* (bur purslane, Puerto Rico); *verdolaguita* (little purslane, El Salvador)
xichilak [*xichiak*] (*xich'il*, something that is hard or flexible and not easily broken, *ak*, vine, Maya, Yucatán, Quintana Roo)
yerba de gallina (chicken herb, Guerrero)

Two genera that have always given me trouble are *Kallstroemia* and *Tribulus*. When in fruit they are not that difficult to recognize. Fruits are prickly in *Tribulus* and tuberculate in *Kallstroemia*. Otherwise, one has to count the carpels—4 to 5 in *Tribulus*; 8 to 12 in *Kallstroemia*. Both genera have tropical affinities, and grow mostly in Florida's peninsula.

It turns out that Linnaeus ([1753] 1957) did not distinguish these two genera either. He called this species *Tribulus maximus* after studying plants cultivated at the *Hortus Cliffortianus*. Linnaeus had also studied the illustrations in Hans Sloane's book, *Catalogus plantarum que in insula Jamaica sponte proveniunt* (Catalog of Plants Growing Wild on the Island of Jamaica). Except for the cultivated plants, he knew the species only from Jamaica.

We now know that *Kallstroemia* is a genus of 17 species, mostly in the American tropics (Porter 1969). Florida has the single species scattered around the state that locally appears and disappears. Deeper in the tropics, the plants are more abundant and frequent.

People in El Salvador and Colombia sometimes cook young plants of *K. maxima* and eat them (Standley 1928, Pérez-Arbelaez 1978). Using them as food, however, is not a common practice. Moreover, it is typically done when nothing else is available.

A more common application is in medicines. Pittier (1957) wrote that Costa Ricans crush leaves and apply them to boils and other sores to bring them to a head. In Cuba, decoctions of the plant are applied to various skin problems, and either a decoction or an infusion is used to combat urticaria (Roig 1945, Leon and Alain 1946–1953, Liogier 1978). In Colombia, the herb is considered a diuretic and laxative (Pérez-Arbelaez 1978, Morton 1981). Costa Ricans and Venezuelans make a poultice to treat abscesses and tumors (Pittier 1957, Morton 1981).

Bremness (1994), Bown (1995), and Duke et al. (2002) do not mention these largely tropical plants. Hocking (1997), on the other hand, notes that decoctions of the plants have been used to treat skin diseases and that the leaves are mildly purgative.

It has been known for a long time that domestic animals are sometimes poisoned by *Kallstroemia* (Dollahite 1975, Wilson et al. 1982). Among the potent chemicals in the genus are saponins (Dominguez et al. 1985, Filho et al. 1989). In addition, the plant contains diosgenin and has been suggested as a commercial source of that compound (Khanna et al. 1981, Datta et al. 1983). That chemical is now commercially taken from *Dioscorea* and is a major element in building hormones (Lewis and Elvin-Lewis 1977).

Kalmia

(Named by Linnaeus for his student Pehr or Peter Kalm, 1716–1779, who explored northeastern North America; subsequently, Kalm was professor of economy and natural history at Åbo, modern Turku, Finland)

Kalmia latiflora. From Sargent 1905.

Kalmia latiflora (broad-leaved)

Berglorbeer (mountain laurel berry, German)

big-leaf ivy [big-ivy]; ivy[-bush] ("ivy" came from Old English *ifig*, and dates to about A.D. 800, originally applied to *Hedera helix*, and later to other evergreens); poison ivy (in use in the South during the Civil War by local farmers, fide Porcher 1863)

calico-bush ("calico" came into English about 1505, as a variant of the Indian city Calicut on the Malabar coast; it was applied to a cotton cloth ca. 1589; these plants had this name by 1847)

grand [*gross*] *Kalmie* (big kalmia, German)

lamb-kill (a note on its toxicity when browsed by domestic animals)

[American, mountain, rose-, round-leaved, sheep-] laurel ("laurel" derived from Latin *laurus*, originally applied to *Laurus nobilis*, and later to a number of similar plants; the modifiers distinguish this from the Old World species)

Lorberrose (rose laurel, German)

spoon-wood [spoonwood, spoon tree] ("spoon tree" dates from Kalm's book of 1753; by 1847 it became "spoon-wood" and was equated with "calico-bush" and "mountain laurel")

wicke (probably from an Algonquian word, Quebec)

According to Lewis and Elvin-Lewis (1977) and Foster and Duke (1990), the resins in *Kalmia* are among the most poisonous known. All plant parts contain andromedotoxin, a toxic diterpene causing slow pulse, lowering of blood pressure, lack of coordination, convulsions, progressive paralysis, and death. Arbutin, a glycoside of hydroquinone, is also present and poisonous (Lewis and Elvin-Lewis 1977). Surely Linnaeus was focusing on the beauty of the flowers when he named the genus after his student. We have every reason to believe that he considered the patronym a compliment (Reveal 1992a). Indeed, Kalm ([1753–1761] 1972) wrote, "Dr. *Linnaeus*, conformable to the peculiar friendship and goodness which he has always honoured me with, has been pleased to call this tree *Kalmia latifolia*."

Linnaeus had not seen living plants of *Kalmia* as he had so many of those in *Species Plantarum,* which appeared the same year as the first part of Peter Kalm's *En resa til Norra America* (Travels into North America). In Kalm's book he used the binomial given by his mentor and called the plants both "American dwarf laurel" and "spoon tree." Kalm noted that the Swedes of New Jersey called them spoon-tree because the indigenous people who formerly lived there "used to make their spoons and trowels of the wood of this tree." He brought a spoon back to Sweden with him. He gives no detail beyond "made by an *Indian*," but it was made on the site of Philadelphia, so the maker was probably Delaware.

Catesby (1734–1747) noted that deer fed on the leaves with impunity, while cattle and sheep died. Kalm himself wrote, "Their leaves are poisonous to some animals, and food for others." He added that deer ate the plants and the dogs became poisoned after eating the offal from butchered animals. People who ate the stag meat, however, had no problems.

Indigenous Americans knew of the poisonous nature of *Kalmia*, even though Europeans like Mark Catesby focused on the beauty of the flowers. The Cree, Delaware, and Mahuna considered the plants poisonous (Lewis and Elvin-Lewis 1977, Moerman 1998). However, the Cree still used the leaves as medicine (Moerman 1998). The Cherokee also used *Kalmia* as medicine (Hamel and Chiltoskey 1975). Leaves or an infusion of leaves were used to treat pain from rheumatism and brier scratches and cramps in ballplayers. They also made a salve from leaves, which was used "for healing." An infusion was assured to rid a person of vermin.

Porcher (1863) wrote: "A writer under the signature of 'Cunio' communicates the following to the 'Atlanta Commonwealth,' 1861 ... Mr. Charles Foster, long known as a wood engraver at Nashville, Tennessee, many years since, I can state that the wood of the maximum or mountain laurel, as well as its confrère, Kalmia latifolia, known by every farmer as poison ivy, are equalled only by the best boxwood." Farther on, Porcher recorded that physician Dr. Shoemaker "reports cases of obstinate diarrhoea cured by a decoction, thirty drops being taken four times a day. The leaves have been advantageously used in syphilis, and extensively applied in tinea, psora, and cutaneous affections. Dr. Barton states that nervous symptoms have resulted from the external use of the strong decoction, thirty drops taken internally six times a day producing vertigo. ... Dr. Shoemaker supposes it suitable to cases of hypertrophy of heart, and other diseases, when it is necessary to decrease the action of that organ; and from the tannin present that it is peculiarly fitted for cases of hemorrhage, dysentery, etc. He proposes that two ounces of the leaves be macerated in a pint of alcohol for a week, and then strained, the dose of which for an adult is thirty drops every two or three hours. If these observations are confirmed it will give the plant a high reputation as a sedative, and attention is invited to it." These applications in minute doses were often effective, but always dangerous (Foster and Duke 1990).

Koanophyllon

(Perhaps from Greek *choane*, funnel, and *phyllon*, leaf)

Koanophyllon villosum (shaggy pubescent or hairy) (= *Eupatorium villosum*)
albahaca de sabana (savanna basil, Cuba); *albahaquilla* (little basil, Cuba)
bitter bush [brush] (Jamaica); bitter sage (Bahamas)
filigrana (literally filigree, but also used as a common name for aromatic *Callicarpa* and *Lantana*, Cuba)
Florida Keys [velvety] thoroughwort (Florida)
jackmada (Bahamas)
rompezaragey de sabana (savanna shoe breaker, Cuba); *rompezaragüey* (shoe breaker; a name also used for *Conoclinium* and *Vernonia*, Cuba)
travesera (misbehaving one, Cuba)
trebolillo [*tribulillo*] (little clover, *trébolo* from Latin *trifolium*, Oriente, Cuba)
yerba amarga (bitter herb, Cuba)
zanca de grullo (crane's stilts, also used for *Chromolaena odorata*, cf. bitter bush, Cuba)

One of Linnaeus's students, Olaf Peter Swartz, found these plants in Jamaica and named them *Eupatorium villosum* in 1788. Later, this genus was named by Manoel Arruda de Cámara (1752–1810) in his *Travels in Brazil* published in 1816. The segregate genus was subsequently ignored by virtually all who studied the plants, and instead included as a synonym under *Eupatorium* (e.g., Chapman 1897, Small 1933, Alain and Leon 1946–1953). Robert M. King and Harold Robinson resurrected the genus in 1975. *Koanophyllon* now comprises about 120 species endemic to the American tropics (Mabberley 1997).

This is a shrub that grows in the Bahama Islands, Florida, Jamaica, and the Cayman Islands. It is particularly prone to growing in calcareous substrates, which perhaps explains why it is confined to the Florida Keys and Everglades Keys.

In Cuba, the entire plant is used in decoctions against diarrhea and other intestinal problems (Roig 1945). Other Cubans use it in baths to relieve the pain of rheumatism and as an astringent. In Jamaica it is a remedy for treating cholera and diarrhea; it serves as a diuretic, and the leaves or twigs are employed to reduce fever. Leaves are chewed to relieve flatulence in the Bahamas (Hocking 1997).

Chemical studies indicate that *K. villosum* has diterpenoid lactones (Manchand et al. 1979). Other species in the genus also have diterpenoid lactones and sesquiterpene lactones (Bohlmann et al. 1981, 1984, Jakupovic et al. 1987, Schuster et al. 1992).

Kosteleztkya

(Commemorates the Czech botanist Vincenz Franz Kosteletzky, 1801–1887, physician and professor of botany in Prague)

Kosteleztkya virginica. From Britton and Brown 1897.

Kosteleztkya virginica (from Virginia)
fen rose
marsh-mallow (originally *Althaea,* later extended to *Hibiscus,* which see, and *Kosteleztkya*)
oopaake embakshe (*oopaake,* owl, *em,* its, *bakshe,* threads, Mikasuki); *opv 'mvhoswv* (*opv,* owl, *em,* its, *vfuswv,* threads, Creek)
oopaake oshtaape shapoke (owl boots, Mikasuki); *opv enhvfvtehkv* (*opv,* owl, *em,* its, *sohtehkv,* boots, Creek)

Snow and Stans (2001) record that the Seminoles use *Kosteleztkya virginica* for cramps, heat, and to induce labor. The related species *K. pentasperma* is used in Jamaica and Nicaragua (Morton 1981, Coe and Anderson 1996).

Krameria

(Apparently named for the Austrian army physician, Johann Georg H. Kramer, d. 1742)

cadia de perro (dog bur, Curaçao)
cósahui (Yaqui)
crameria [*clameria, grameria*] (Mexico); krameria (introduced into England in 1806, presumably with the common name; soon afterward, blood-red extracts were used to adulterate port wine)
mesquitillo (little mesquite, Spanish)
raiz de culantrillo (parsley root, Mexico)

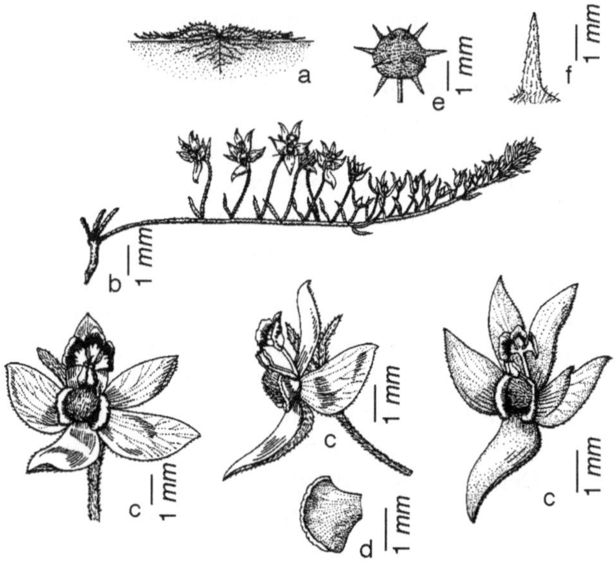

Krameria lanceolata. a. Habit. b. Flowering branch. c. Three views of the flower. d. Elaiophore. e. Fruit. f. Fruit spine. *Drawn by M.C. Ogorzaly.* From Simpson 1989.

ratany [rattany] (in English by 1808, from *ratânia* [or *ractania, rataña*], maybe from Quechua, meaning "ground-creeping" fide Simpson 1991)
wetahúpatci (Tarahumara)

Krameria lanceolata (with leaves shaped like lance-points)
clameria [*crameria*] (Mexico)
mezquitillo (little mesquite, Mexico)
mountain burr
prairie bur
raíz de cucillo (little cuckoo root, Mexico)
ratanhia du Texas (Texas rhatany, Paris)
sandspur (a name better known for *Cenchrus,* but both have spiny fruits, Florida)
Texas [trailing] ratany [root]
tícara (Chile)

Krameria is a small genus, with 18 species found from the southern United States to northern Argentina and Chile (Simpson and Salywon 1999, Simpson et al. 2004). There are two unusual traits about the family, in addition to their zygomorphic flowers and spiny fruits. Members of the genus are hemiparasitic, forming haustoria on the roots of a broad range of host plants (Musselman 1975, 1977). Flowers of *Krameria* are nectarless, and the rewards for visitors are fatty oils produced by modified external surfaces of the lower two petals (elaiphores). *Krameria* actually depends on oil-collecting bees and only *Centris* (Anthophoridae) has been reported visiting their flowers. A given *Centris* species or even individual, on the other hand, will visit other oil-producing plants, using the oils from the flowers for their larval food (Simpson and Salywon 1999).

The genus was named by Pehr Löfling (1729–1756), a Swedish student of Linnaeus. During his travels as one of Linnaeus's "Disciples," Löfling went to Venezuela, where he died. After his death, his book *Iter Hispanicum* was published in 1758, and in it he created the genus and described *K. ixine.* The genus is the only one in its family and has formerly been allied with the Fabaceae or the Polygalaceae. Recent molecular work (Chase et al. 1993, Gadek et al. 1996, Sheahan and Chase 1996, Savolainen et al. 2000, Soltis et al. 2000, Simpson et al. 2004) has shown that *Krameria* is most closely related to the Zygophyllaceae.

Linnaeus accepted Löfling's genus and species and changed the *ixine* to *ixina*, but *ixine* is the correct spelling (Simpson 1991). Now it is known that *K. ixine* grows in Puerto Rico, Hispaniola, the Lesser Antilles, Guatemala, Nicaragua, Honduras, Costa Rica, Colombia, Venezuela, and eastern Guyana. In Puerto Rico, it is known as *abrojo colorado* (red bur) and *ratania falsa* (false ratany). Elsewhere it is known as *cadia del perro* [*cadilla del perro*] (dog bur, Venezuela), *carreton* (roadside thing, Colombia), *estancadera* (resembles hips, Colombia), *pegasaja shimaron* (wild sticker, Curaçao), *ratanhia das Antilhas* (Antillean ratany, Brazil), *ratanhia de nova grande* (new big ratany, Brazil), and *ratanhia de sabanilla* (little savanna ratany, Brazil).

Hipólito Ruiz first found the South American plants being chewed by a Peruvian woman. She told him that they were useful to whiten and strengthen the teeth (Simpson 1991). At that time, no other uses by indigenous people were known, but Ruiz began experimenting with the plants. Later, he extolled *ratánia's* virtues and a variety of internal astringent uses were instigated in Peru and Spain. The medicine was introduced into England in 1806 and into Germany in 1818.

People in Curaçao use *Krameria* to ease intestinal irritations, as an emmenagogue, to promote easier childbirth, or even as an abortifacient (Morton 1981). In Venezuela, the decoction is used for liver and kidney problems. Brazilians use it to treat dysentery (Mors et al. 2000).

Four species of *Krameria* in addition to *K. lanceolata* grow in the southern United States (Correll and Johnston 1970, Simpson and Salywon 1999). *Krameria erecta* is known from California, Nevada, Utah, Arizona, Baja California (Norte and Sur), Chihuahua, Coahuila, Durango, San Luis Potosí, Sinaloa, Sonora, and Zacatecas. Rea (1997) has records of *K. erecta* being made into astringent medicines and dyes by the O'odham (Pima, Papago). The Mayo of Sonora know this species as *tajimsi* (sun beard), and others in the area say *tajuí* (Spanish) and range ratany. The Mayo also use the roots in astringent medicines and in dyeing (Yetman and Van Devender 2002). The Guarijío call it *tahué* [*tajué*] and use it for healing sores, much as do their neighbors the Mayo (Yetman 2002). The Seri call *K. erecta* the *haxz iztim* (dog's hipbone), but Felger and Moser (1985) record no use of this species. *Krameria erecta* is uncommon while *K. grayi* is common in the Seri region, and they do use it.

The similar *K. grayi* grows in much the same area as *K. erecta*. Both extend into Texas, with *K. erecta* occurring commonly in west Texas. *Krameria grayi* was used similarly among the Paiute (Great Basin) and Shoshoni (Nevada, Montana). The Pima also made a brown dye from it for baskets (Curtin 1949). Rea (1997) found the Pima calling the plants *eedho* [*oeto*], and many remembered its use as a red dye, but that practice has recently fallen into disuse. Some of the older people still remembered it being used for internal problems, particularly rheumatism. The Tohono O'odham (Papago) distinguish between *eedho* (*K. grayi*) and *ge'eged eedho* (large krameria, or *K. erecta*), and one of Rea's informants said the two were used interchangeably as medicines. Kay (1996) found *K. grayi* called *temitzo* (Ópata), *chacate* (Maricopa), *heepol* (Seri), *cósahui* [*cosagui*] (Yaqui, maybe a loanword from Náhuatl, where it means "yellow"), and white ratany. The Maricopa chew the roots for sore throat, take a tea of it for fever and coughs, and put it on the umbilical stump to prevent infection (Kay 1996). The Yaqui consider the plants good for the blood; The Tarahumara make a medicine for aching teeth (Kay 1996). The Seri historically used the roots of these plants to make a reddish brown dye for baskets (Felger and Moser 1985). The Seri also made a tea for upset stomach and diarrhea, and sprinkled powdered stems on skin sores that were slow to heal.

Krameria ramosissima is even more similar to *K. erecta.* Oddly enough, it is not the sister species of *K. erecta*, but basal to a clade containing *K. erecta* (Simpson et al. 2004). *Krameria grayi* is in a totally different clade; it is restricted to Texas (Brewster and Val Verde to Hidalgo Counties), and then south only into Coahuila, Nuevo Leon, and Tamaulipas. No uses were found for *K. ramosissima* (Simpson 1991).

Krameria lanceolata differs from the other U.S. species in that it is herbaceous and its branches sprawl on the ground. This is the most widespread species in the United States, known from Colorado and Kansas south to Arizona, New Mexico, and Texas, and east to

Georgia and Florida. It also grows in Chihuahua and Coahuila in Mexico. Martínez (1969) said that *K. lanceolata* grows as far south as the Valley of Mexico, but that seems to be confusion with *K. secundiflora*.

Although *K. lanceolata* has the widest range in the United States, it is not recorded as used among indigenous people there. However, other cultures record the roots as an astringent medicine (Hocking 1997). According to Simpson (1989), Burlage's 1968 *Index of Poisonous Plants of Texas* reported that the species was used to make a medicinal tea to combat diarrhea, hair and teeth loss, and cancer of the tongue.

Martínez (1969) said no distinction is made among the Mexican species; all are used as an astringent. They are employed to treat diarrhea and for skin and gum sores. Martínez (1969) noted that a Dr. L. Hernández Chávez had used them with some success in treating cancer of the tongue, stomach, and intestines.

Simpson (1991) provided convincing evidence that *Krameria* was not typically used as an internal medicine before 1843. She argues that the indigenous use of various species "followed the recognition of their being related to 'true' ratany and subsequent substitution for that species." Also because reports from South America postdate 1800, she was convinced that "modern" indigenous uses as a medicine were derived from contact with Old World people. Her argument is not as convincing regarding use for the teeth and sores.

Roots of *Krameria* contain sugars and *N*-methyl tyrosine (Simpson 1991). *Krameria* is also rich in catechin tannin (Nierenstein 1931, Morton 1981, Mors et al. 2000). More recent study has revealed procyanidin and complex polyphenols, which are formed from catechin and epicatechin (Scholz and Rimpler 1989). There are also neolignans and norneolignans in some species, including *K. lanceolata* (Achenbach et al. 1989).

Some of the species have been reported as being antiviral and inhibitory to *Bacillus cereus, B. subtilis, Escherichia coli,* and *Staphylococcus aureus*; others are hypotensive (Kay 1996). Duke et al. (2002) recommend *K. lappacea* as antihemorrhagic, anti-inflammatory, antiseptic, astringent, expectorant, fungicidal, and vulnerary, and most sources agree that many species are as effective. *Krameria* is used today in food additives and specialty toothpaste (Simpson 1991).

Krigia

(Named in 1791 for German physician David Krig or Krieg, 1670–1710, who was among the first to gather plants in Maryland; he collected there in 1698)

Krigia virginica (of Virginia)
dwarf dandelion
Virginia dwarf dandelion [Virginia dwarfdandelion]

Krigia is a genus with seven species endemic to North America (Ki-Joong and Turner 1992, Mabberley 1997). The small herbs were collected by John Clayton in the 1730s and sent to Jan Gronovius in the Netherlands. Gronovius called them *Hyoseris,* and Linnaeus ([1753] 1957) dubbed them *H. virginica.* It was not until 1803 when Carl Willdenow revised *Species Plantarum* that the species was placed in *Krigia.*

Clayton was not the only person to find these plants along the eastern coast of North America. Thomas Walter also found and named them in his *Flora Caroliniana* of 1788. They were named again by William P.C. Barton in his *Flora Philadelphia Prodromus* of 1815, and by Thomas Nuttall in the *Genera of North American Plants* of 1818.

Although these small herbs resemble dandelion (*Taraxacum officinale*), they were not mentioned by Fernald et al. (1958) as edible. Still, Hocking (1997) said that they were used for food like dandelion. He added, apparently from personal experience, that they were "much like that plant."

No other mention has been found of these herbs being used by people in North America until very recently. Some of the species are now considered attractive and are cultivated (Mabberley 1997). The *Hortus Third* (Hortorum 1976) lists three species in cultivation.

There is indirect evidence that indigenous people may have used these composites. Moerman (1998) found that the Menomini made a hunting lure from *K. biflora*. These people lived in Wisconsin and used the stem to imitate the cry of a fawn in distress. That noise would sometimes attract a doe.

Krugiodendron: Black Ironwood

(Dedicated to Carl Wilhelm Leopold Krug, 1833–1898, the German consul in Puerto Rico, businessman, botanist, and patron of the sciences, who studied the West Indian flora; Greek *dendron* makes a name that means "Krug's tree")

***Krugiodendron ferreum**. Drawn by P.N. Honychurch.*

Not long after moving to Florida, I discovered a wonderful fragment of tropical hardwood forest (hammock) along the coast in Boca Raton. Eventually, the parcel was purchased along with the beachfront and made into two preserves—the Patch Reef Park and Gumbo Limbo Nature Center. Within the hammock forest, I found trees lying on their sides and others with their tops broken out but still living. These trees were damaged by the hurricanes that hit the site in the mid-1940s (1946, 1947), yet they still dominated the forest. One tree that had not fared so well was a large black ironwood (Florida, Jamaica, Puerto Rico). On the ground beside the tree was the heart of a large trunk that had been snapped by one of the tropical storms. That heartwood was so hard that it remained essentially unchanged during the 31 years that I showed it to classes going through the site. Due to the kindness of my last undergraduate class and the nature center, that 50-odd-year-old branch now hangs outside my office where I watch it turning in the wind each day.

Black ironwood or ironwood (Puerto Rico) was named *Krugiodendron ferreum* by German botanist Ignatz Urban, Assistant Director of the Botanical Garden and Museum at Berlin-Dahlem. The species is *ferreum* (like iron), and the branch outside my office has lasted longer than iron would have in the salt spray of southern Florida's tropical climate.

The name "ironwood" has been used for trees with hard wood by people throughout the world. There are many unrelated species with wood so hard people think it is as "hard as iron." The first usage in English of the term was in 1657, almost without doubt in reference to *K. ferreum*. The publication stated, "Ironwood is called so, for the extreme hardness; ... 'Tis much used for Coggs to the Rollers." Among other species that have been called ironwood are *Bumelia lycoides* (North America, Sapotaceae), *Carpinus caroliniana* (North America, Betulaceae), *Copaifera mopane* (Africa, Fabaceae), *Diospyros virginiana* (North America, Ebenaceae), *Erythroxylum aureolatum* (Caribbean, Erythroxylacae), *Fagara lentiscifolia* (Caribbean, Rutaceae), *Hypelate trifoliata* (Florida, Caribbean, Sapindaceae), *Mesua ferrea* (Asia, Clusiaceae), *Metrosideros vera* (Asia, Myrtaceae), *Notelaea ligustrina* (Australia, Oleaceae), *Olea capensis* and *O. undulata* (Africa, Oleaceae), *Ostrya virginica* (North America, Betulaceae), *Reynosia sepentrionalis* (Florida, Caribbean, Rhamnaceae), *Sideroxylon foetidissimum* (Florida, Caribbean, Sapotaceae), *Sloanea jamaicensis* (West Indies, Malvaceae), *Stadmannia sideroxylon* (Asia, Sapindaceae), and *Xylia dolabriformis* (Asia, Fabaceae).

Other people in the range of *K. ferreum* also make comparisons to iron in their common names for it. They call it *acero* (steel, Cuba), *boafierro* [*guafierro*] (good iron, Dominican Republic), *bois de fer* (iron tree, Haiti, Virgin Islands, Guadeloupe), *bois fer* (iron tree, Haiti), ironberry (Dutch Antilles), ironwood (Florida, Puerto Rico, Virgin Islands), *palo de hierro* [*fierro*] (iron tree, Puerto Rico, Dominican Republic), and *guatafer* (*guata*, padding, *fer*, iron, Hispaniola, Virgin Islands).

Sometimes people compare the wood with other things. Along the lines of comparing it to iron are the names axemaster [exmaster] (Belize), *palo de diablo* (Devil's tree, Cuba), and *quiebra-hacha* [*quebracho*] (ax-breaker, Belize, Guatemala, Dominican Republic, Puerto Rico). The Maya too found the hard wood useful, but instead of cursing it, turned it to good use. They call it *chimtoc* [*chim-tok, chin-tok*] (*chim*, spear, *tok*, hard, Maya, Yucatán). Formerly, the plant was used to make weapons such as bows, arrows, and lances. Now the wood is used to make work tools. Because of the shine the wood takes when polished Cubans call it *carey de costa* (sea [coast] turtle) and *hueso de tortuga* (turtle bone). The wood color has led to the names *bariaco* (resembling the mahogany colored wood of the *baría* tree, *Calophyllum calaba*, Puerto Rico), *ciguamo* [*siguamo*] (laurel, Dominican Republic), *cocuyo* (also a name for a *Sideroxylon*, Cuba), *coronel* (dogwood, Cuba), and ebony [wood] (Virgin Islands).

Even though I knew the green stems would sink in ocean water, I could not understand the name leadwood (Florida) until I sawed a segment of the tree. The odor of the wood as it heated from being sawed was almost identical to that of molten lead. It is not clear why the tree is called *espejuelo* (little mirror, Puerto Rico) unless the wood can be polished to make mirrors. More likely, the wood has been used as frames for mirrors. The leaves are indeed narrow, eliciting the name *hoja ancha* (narrow leaf, Dominican Republic), but no more so than some of the others with which it grows. Why it would be called *manggel cora* (*cora* = ?, mangrove, Dutch Antilles) is not clear, as it grows on ground higher than mangroves.

The species grows in southern Florida, the Bahamas, the West Indies south to St. Vincent and Bequia, and in Bonaire and Curaçao, Yucatán to Guatemala, Belize, and Honduras. In those areas, it is consistently associated with other species in the facultatively deciduous tropical forests. Associated species in Florida include gumbo limbo (*Bursera simaruba*), Guiana plum (*Drypetes lateriflora*), poisonwood (*Metopium toxiferum*), and paradise tree (*Simarouba glauca*). Typically, these are all evergreen, but during years with little rainfall, they have the facility to drop their leaves and become deciduous. This type of forest dominates at least coastal sites throughout the range of black ironwood.

Krugiodendron is a monotypic genus containing only *K. ferreum*. Although Urban created the genus in 1902, the species had been known since 1793 when it was named as *Rhamnus ferreus* by Martin Vahl in Hans West's study of the plants of St. Croix in the Virgin Islands. Subsequently, other botanists dealing with the plants really did not know what to do with them. When Augustin-Pyramus de Candolle published the family in his *Prodromus Systematis Naturalis Regni Vegetabilis* in 1825, he decided the species belonged to the temperate North American genus *Ceanothus*. Only a year later, Adolphe T. Brogniart put the species in the tropical genus *Scutia*. In 1859, Grisebach called the trees *Condalia ferrea* in his *Flora of the British West Indian Islands*. Just before Urban's study, Charles S. Sargent called the trees *Rhamnidium ferreum* in 1891. While such shifting of species from one genus to the other frustrates professional biologists and amateurs by changing names, it is a normal part of the search for the "correct" answers. Because plants do not come with name-tags, people have to try to interpret their relationships with the best information on hand at any given time.

The most widespread use of the species is for its wood. That wood is one of the densest in the world, and this is the heaviest of native woods in Puerto Rico and the United States, with a specific gravity of 1.3 to 1.42 (Record 1926). The sapwood is light brown, and the heartwood is orange brown to dark brown, streaked, fine-textured, and resistant to decay and attack by dry-wood termites. Historically, there were surely larger trees throughout its range, as Sargent (1905) reported Florida plants reaching 30 feet (10 m) tall and 8 to 10 inches (20 to 25 cm) in diameter. Britton and Millspaugh (1962) record the stems to almost 20 inches (50 cm) in the Bahamas, and Adams (1971) indicates that Jamaican trees reach 55 feet (17 m) tall. Perhaps Florida trees were formerly equally as large.

Now throughout much of the range of the species most trees are small. Currently, they are used for posts, veneer, cross ties, and canes (Little and Wadsworth 1964). Schirarend (1987) has provided an anatomical study of ironwood.

In Yucatán, the Maya use the bark and root decoction as a mouthwash to relieve toothache and gum ailments (Steggerda 1943). A decoction of the roots is purgative. In the Turks and Caicos Islands in the southern Bahamas, people use the fruits to make jam and wine (Morton 1968b).

Standing on the top of the tower that penetrates the canopy in the forest at Gumbo Limbo Nature Center in Boca Raton, Florida, you can actually look down at the flowers and fruits on the trees. The small greenish flowers are visited by a variety of bees and wasps and soon produce fruits. Those small fruits are about 0.25 inches (7 mm) long and do not look tasty or large enough to produce much that might be edible. Perhaps the isolation of people in the Turks and Caicos Islands made everything begin to look edible.

L

Lachnanthes

(Named by Steven Elliott from Greek *lachne*, wool, *anthos*, flower, because of the pubescent or woolly flowers)

Lachnanthes caroliniana. From Institute of Food and Agricultural Sciences.

Lachnanthes caroliniana (of Carolina)

dye root; dyer's dilatris (*dilatris* old generic name, from *di*, two, *latris*, servants, an allusion to two small anthers and one large one, USA)

[Carolina] redroot; red weed

spirit-weed

tali'wa [*tale'wa*] (Creek)

wi̱·ti· ta'ktu [*tu'ktu, ta'ktuwi*] (red root, Catawba)

wool flowers (from the pubescence on their outside)

yellow tricoma (used by Williams [1837] 1962)

For many years the specific name of this herb, native from Massachusetts to Florida and Cuba, was *tinctoria* (Latin, for dyeing). That name, as did "dyer's dilatris" and "red root," arose from the herb's use as a dye. Porcher (1863) wrote, "The root ... [yields] a beautiful dye; hence the name." However, the story of this wetland herb as a dye plant among indigenous people goes back to William Bartram, who mentioned *tale'wa* as a dye in 1789 (Swanton 1946). Albert S. Gatschet in an unpublished Creek vocabulary from the late 1800s wrote that the name *tali'wa* or *tale'wa* in Creek meant an herb whose roots make red dye for the hair (Swanton 1946). This identity is confirmed by his noting that it was "a plant growing 1 to 2 feet high on sandy soil, with yellow flower. When the roots are fried in oil, the color of the oil changes to a beautiful, brilliant red of the claret hue and is used among other things to make hair oil."

James Adair wrote in 1775 that the Chicksaws also used the plant (Swanton 1946). Romans ([1775] 1961) made the same report of the color but could never locate the plant. One of Swanton's (1946) own informants, Jackson Lewis, who originally belonged to a Hitichiti town among the lower Creeks, knew a red dye source by the name *tale'wa*. Swanton (1946) thought it might be the "celandine poppy" (*Stylophorum diphyllum*), but that taxon is outside the range where he reported it, and Moerman (1998) does not list the species.

Millspaugh (1892) recorded: "The root was esteemed as an invigorating tonic by the Aborigines, especially the Seminoles, in whom it is said to cause brilliancy and fearless expression of the eye and countenance, a boldness and fluency of speech, and other symptoms of heroic bearing." The Cherokee considered the root astringent and used a decoction for sores, for bowel complaints, spitting blood, bloody piles, sore mouth and throat, and venereal disease (Hamel and Chiltoskey 1975). The Catawba used it similarly (Speck 1934). Red root was taken to England in 1812, and subsequently volunteered from seed (Millspaugh 1892). As far away from its native range as the Virgin Islands, Petersen (1974) recommended it be used for muscular strains.

Lactuca

(Linnaeus used the Latin *lactuca*, from *lacteus*, milky, full of milk, referring to the latex)

Lactuca floridana. From Britton and Brown 1898.

hass (Arabic)
lettuce (in English by about A.D. 1290; the origin is debated: some derive it from Latin *lactuca* or its adjectival derivative *lactucea*; akin to Middle English *letuse*, Old and Modern French *laitue*; those words were derived from *lac*, milk, referring to the latex of the plants); *laitue* (French); *Lattich* (German); *lattuga* (Italian); *lechuga* (Spanish); *liatas* (Gaelic); wild lettuce (USA)
thridax (Greek)
wy (Coptic)

Lactuca canadensis (of Canada)
agosdagisdi (*agisdi*, something to eat, *adosdi*, raw, Cherokee)
Canada [wild] lettuce
hissokcakkó (*híssi,* leaves, *okcakkó*, green, Koasati)
odjici'gomin (Ojibwa)
soló:pá: łíntókífáskî [*soló:pá:ríntókífáskî*] (ghosts' thistle, Mikasuki); *yafikcaka:câ* (*yvfekcv,* soul or ghost, *vkacv*, thistle, Creek)

Lactuca floridana (of Florida)
Florida lettuce; woodland lettuce

Lactuca graminifolia (with leaves like grass)
grassleaf lettuce (a book name)

Vigorous growing leafy stems before flowering and unexpanded inflorescences may be cooked as a potherb (Fernald et al. 1958). At least the Cherokee ate them historically, and probably others also. Morton (1968b) considered young plants and leaves fairly good when cooked as greens, but not good raw. However, Fernald et al. (1958) considered the greens good enough to use as a salad. Native people used *Lactuca* as a nerve tonic, sedative (Cherokee), pain reliever (Cherokee, Iroquois), and diuretic (Iroquois), to ease or promote lactation (Ojibwas, Meskwakis), and as a remedy for warts (Ojibwa), pimples, poison-ivy rash (Menomini), and other skin irritations, and to stop bleeding cuts (Iroquois) (Vogel 1970, King 1984, Foster and Duke 1990). External use is risky as the plant may cause dermatitis in some.

Early American medical practice considered the plants anodyne, diaphoretic, laxative, and diuretic (Millspaugh 1892). Pollen was in the samples examined by Hogan (1978) so the Glades people also used wild lettuce. It was long believed that the latex, called *lactucarium* (lettuce opium), was a narcotic, but it is only "slightly soporific" (Millspaugh 1892). Cultivated lettuce (*L. sativa*) lacks the bitter sesquiterpene lactones in the wild species (Diggs et al. 1999).

Lagenaria

(Named by Nicolas Charles Seringe, 1776–1858, from Greek *lagenos*, flask, bottle, or Latin *lagena*, a flask; alluding to the most common use of the fruits)

Lagenaria siceraria fruit being used as a water container by a Pomeioc woman in North Carolina in the 1580s. From Harriot (1590).

Lagenaria siceraria (Greek *sicera*, intoxicating, presumably derived from use of one of the six *Lageneria* species in making an intoxicating drink)
alahko (Koasati; they called sphinx moths *alahkobitlí*, gourd-dancer)
bagaña (Dominican Republic)

calabash (probably from Persian *kharbusa*, melon, or Arabic *qar'ah yabisah*, dry gourd, 1596, English); *calabas largu* (big calabash, Dutch Antilles); *calabaza* [*calabazo*] (Dominican Republic, Puerto Rico, Costa Rica); *calabasse* [*courant, franc, gros, longue, musquée*] (calabash [running, free, big, long, musk-smelling], Houma, Louisiana; Haiti)
camasa [*camaza*] (Venezuela)
candungo amargo (perhaps "bitter and useless," Dominican Republic)
carracho (Puerto Rico)
efepe [*eféèpe, fepé, hefepe, hefepe-nerkv, hefepe-neɬkv, hifipa*] (*efépe*, gourd, *heɬkv*, seed, Creek, Muskogee); *hifipî* [*hefepe-eele*] (*efépe*, gourd, *eele*, seed?, Mikasuki)
[bottle, dipper, etc.] gourd (from Latin *cucurbita* through French *gourde*, first recorded in A.D. 1300)
gourde des pèlerins [*trompette*] (pilgrim's [trumpet] gourd, Haiti)
güicharo (Puerto Rico)
güiro (gourd plant or fruit, Taino, Cuba, Haiti); *güiro cimarrón* (wild gourd plant, Cuba); *güiro guyaro* (*guyara* =?, *güiro*, gourd, Cuba)
i'-ni-i-zhe (Osage)
horoto (Arawak, Suriname)
ko (gourd; *akodi*, cup, Biloxi)
koro (Carib, Suriname)
nambiro [*marimbo*] (Chiapaneca, Costa Rica, Puerto Rico)
pehe (used for rattles, Omaha-Ponca)
shokshi' (Chickasaw); *shukshi okpulo* (*shukshi* now means watermelon; surely a transference from this older use, *okpulo*, bad or spoiled, Choctaw); *shukshubok* [*shukshihobok*] (*shukshi*, watermelon, *holba*, resembling, Choctaw); *kochi* (Alabama)
tä 'mbactu' (Yuchi)
tarro (horn, El Salvador)
tecomate (usually *Crescentia cujete*, from *tecomatl*, Náhuatl, Salvador); *tecomatillo* (little *tecomate*, El Salvador)
tol (gourd with narrow middle, El Salvador, Guatemala)
totuma (usually *Crescentia cujete*, El Salvador, Venezuela)
tula de mate (tea gourd, Panama)
wamnuha [*wakmu*] (used for rattles, Dakota)

Names of three genera are so badly intermingled in common usage that it is impossible to be sure which applies to *Lagenaria*. Several of the names given in this abbreviated list are of *Crescentia*, but also are often applied to fruits of *Lagenaria*. Even the English word "gourd" may have originally meant some other plant. Gourds were widespread in the New World at the time of European contact, although the confusion of species delayed their identification. De Candolle (1886), for example, did not think they were present when Europeans arrived in the New World.

Historically, there is evidence of gourds among the Carolina Algonquians in drawings labeled "A cheiff Ladye of Pomeiooc" and "Their manner of prainge [praying] with Rattles abowt te fyer [fire]" made by John White in the 1580s (Harriot [1590] 1972). In 1699, Jonathan Dickinson recorded the Jaega near modern Jupiter Inlet using rattles and mixing *casine* in gourds (Andrews and Andrews 1945). The Jesuit priest Jacques Graver found the Houma using *Lagenaria* in 1700 (Thwarts 1896–1901 in Vogel 1970). William Bartram found Seminoles near the Alachua Savanna (now Payne's Prairie) in Florida growing them in 1774 (Bartram [1791] 1958). It was not until Gilliland (1975) began studying the Marco Island artifacts that *Lagenaria* gourds were identified by Hugh Cutler among the materials left in a burned village about A.D. 1300 (Gilliland 1975). Robinson and Decker-Walters (1997) noted that there is archaeological evidence in Peru, Mexico, and Florida placing the vines there between 7,000 and 10,000 years ago. The Florida record, of probably wild plants, from the Windover Site in Brevard County, was dated at 7,290 years old (Doran et al. 1990).

It is now known that gourds were used by at least the North American tribes Akimel O'odham (Pima), Calusa, Cherokee, Cocopa, Congaree, Dakota, Havasupai, Hopi, Houma, Iroquois, Jaega, Keresan, Navajo, Ojibwa, Ponca, Carolina Algonquians, Seminoles, Tohono O'odham (Papago), and Yuma. The Seminoles and their Muskogean relatives the Houma made dishes, cups, water dippers, and water bottles of them and used them in medicine. Both Seminoles and Houma used them as an analgesic (Speck 1941, Sturtevant 1955). Seminoles call them *hefepe* or *hefepe nerkv* (Creek) or *hefepe eele* (Snow and Stans 2001) and still use the seeds to stop bad dreams. The Ojibwa ate the seeds (Yanovsky 1936). Residents of the Panhandle were still using them to cure "ground itch" when Murphee (1965) talked with them.

In the Caribbean, fruits were used for containers and cooking utensils, but the primary use was as a musical instrument (Liogier 1962). That instrument is called by its Taino name, the *guajey*. Seminoles in William Bartram's time ate the young fruits as do people elsewhere in the Americas (Liogier 1962), but people also use the fruit pulp and seeds as a laxative, and the root as a diuretic for dropsy (Liogier 1974). Syrup from the fruit pulp is a treatment for coughs and asthma (Liogier 1974). Heiser (1979) and Morton (1981) give more details.

Laguncularia: White Mangroves

(From Latin *laguncula*, a flask or bottle, in reference to the fruit shape)

Laguncularia racemosa*. Drawn by P.N. Honychurch.*

In the coastal fringes of southwestern Florida lies one of the largest remaining expanses of mangroves in the New World. Lands that are now in the Great White Heron National Wildlife Refuge, the Everglades National Park, and the Ten Thousand Islands region have areas where one may travel in mangroves for miles without seeing another person. Most of the time, traveling in those areas is perilous because of the number of salt-water mosquitoes that vie for a spot to drain your blood. Sometimes the air is so thick with mosquitoes that breathing without inhaling them is difficult.

If you do travel that region and stop occasionally on the drier spots, you might find a mound 2 to 3 m (6 to 9 feet) or even more across, now grown over with other vegetation. Some of those are remnants of where the old charcoal makers plied their trade. Making charcoal in the region was an important local endeavor through the first quarter of the 20th century, and it continued sporadically until the end of World War II. There were even businesses like the Manetta Company along Shark River and the Rutherford Company near Chatham River that were still limping along in the early 1950s (Tebeau 1968).

White mangrove (Florida to Barbados and Panama) is *Laguncularia racemosa*, the name given to it in 1807 by Carl Friedrich von Gaertner, a German physician and botanist in Stuttgart who specialized in seeds. He moved the specific name from *Conocarpus racemosus*, because Linnaeus thought it and *C. erectus* were close enough to be included in the same genus. Gaertner used *Laguncularia* to refer to the fruits. Indeed, the shape of the fruits is the same in both *Conocarpus* and *Laguncularia*, but there are numerous other differences. Linnaeus based his species name on specimens from the Lesser Antilles, but the trees range much farther. White mangrove is known from southern Florida and Mexico (Campeche and Sonora south), through Central America and south to northwestern Peru and northern Brazil. Unlike most American plants, it also occurs in tropical West Africa.

In Florida the Seminoles know these trees as *ahilo: chátkî* (white black tree, Mikasuki), *ahilo:csókpî* (spongy black tree, Mikasuki), or *tolastihatkî* (*eto*, tree, *lvste*, black, *hvtke*, white, Creek). While Sturtevant's (1955) informants said they did not use the plants, they did use *Conocarpus* and *Rhizophora*. Perhaps knowledge of these coastal plants was being lost even then by restriction of these people to inland reservations. Alice Micco Snow (Snow and Stans 2001), who has lived her entire life on the Brighton Reservation on the northern end of Lake Okeechobee, does not remember the plants at all.

People living along coasts still know these plants as mangroves, although their modifiers do not always match those in English. Some people just say "mangrove," with *mangel* (Virgin Islands), *mangle* (Sonora, Haiti), *mangue* (Brazil), or *palétuvier* (Guadeloupe, Martinique). To others, they are colored. They may be considered "white," as in *mangle blanc* (Guadeloupe, Martinique), *mangle blanco* (Tamaulipas and Sinaloa south to Panama, Puerto Rico, Dominican Republic), *manglier blanc* (Haiti, Guadeloupe), or *mangue branco* (Brazil). Others consider them "gray," as *mangle gris* (Guadeloupe, Martinique) and *palétuvier gris* (French Antilles). They are even said to be *mangle amarillo* (yellow mangrove, Cuba, Dominican Republic, Venezuela), *siari paseo* (*siari*, green, *paseo*, mangrove, Yaqui, Sonora), *mangle colorado* (red mangrove), or *mangle prieto* (black mangrove, Dominican Republic). There does not seem to be a single reason that different colors are applied, and it may depend on what part of the plant is being stressed. If they say "white" or "green" they are usually talking about the leaves in contrast to those of "black" mangroves (*Avicennia germinans*). Gray and yellow are more often comments on the bark. Red seems to be more a reference to the extracts of tannin made from the plants.

To other people they are false mangrove, *mangle bobo* (idiot mangrove, Puerto Rico), *mangle chaparro* (thicket mangrove), *mangle chino* (curly mangrove, Sinaloa), or *mangle marequita* (ocean mangrove, Costa Rica). I suspect that *bobo* and *chaparro* refer to the problems of navigating through them at high tide.

Like Linnaeus, other people compare them with buttonwoods (*Conocarpus*). To them, they are bastard buttonwood, buttonwood (Panama), or white buttonwood (Florida).

Sometimes people do not distinguish further and have simple names for them. To the Mayo of southern Sonora, they are *moyet*, and to the Seri, they are *pnaacoj*. However, the Seri also note that the white mangrove is *pnaacoj hacáaiz* (harpoon mangrove). In Cuba, it is *patabán* (Taino?), while in South America it is *akira'ë* [*akira*] (Carib, Suriname), *coil* (Guyana), or *furilia* (Arawak, Suriname).

In Mexico, the Aztecs called them *sincahuite* [*cincahuite*] (maybe *centli*, stem, *cuahuitl*, tree, Náhuatl). That word was adopted into modern Spanish and persists in several places (Mexico, El Salvador). In Yucatán and Belize, they are *sak-okom* [*tsakol-kon, zacalcom, zacolcom*] (*sak,* white, *okom*, posts that are the principal support for houses, Maya). People in Belize also call them *but-nuut* (*but*, develop, *nuut*, put something inside something else, Maya). That name too sounds like part of a building.

Costa Ricans consider them the *palo de sal* (salt tree), while in Brazil they yield *tinteira dos mangais* (mangrove dye). Bahamians consider the tree green turtle bough, presumably for making krawls (corrals) to retain these animals.

Like the early settlers in southern Florida, the most common use of the trees has been as a source of charcoal or tannin. The bark contains 10 to 14% tannin (Standley 1920–1926, Morton 1981), and all parts are used for tanning leather in much of coastal Mexico, the Caribbean, and south to Panama. The high tannin content also causes fishermen to dye their nets with the bark (Morton 1981).

Other uses of the wood are for lumber, planks, carts, gates, wooden fences, some construction and repair of farm buildings, flooring, work benches, machinery platforms, railroad ties, piling in areas without teredo worms, posts, tools, houseposts, and bridge lumber (Little and Wadsworth 1964, Uphof 1968, Duke 1972, Liogier 1974). The wood is hard, heavy, strong, attractive, dark greenish brown, with a moderately fine texture, lustrous, and fairly easy to season. However, it is difficult to saw, resistant to drywood termites, and durable in contact with soil.

The high tannin content in the bark also makes it astringent, and it is used as a tonic for fever, a vulnerary to cure skin problems including wounds and ulcers, and to treat scurvy and dysentery (Standley 1920–1926, Uphof 1968, Duke 1972, Liogier 1974, Morton 1981, Coe and Anderson, 1996, Hocking 1997). There is reported antitumor activity attributed to the tannin content (Hartwell and Abbott 1969). According to Hocking (1997), the leaves are boiled for greens in Scandinavia and Great Britain. If he is correct, they must be imported there because the climate will not allow *Laguncularia* to grow there.

In addition to the tannin, the plants contain gum polysaccharides, including galactose, arabinose, rhamnose, galacturonic acid, glucuronic acid, and its 4-*O*-methyl ether (Pinto et al. 1993). There are also several minerals in the leaves (Medina et al. 1995) and considerable variation in the foliar waxes, including triterpenoids and alkanes (Rafii et al. 1996).

For thousands of years humans lived beside mangroves and valued them as breeding grounds for fish, crabs, and other food they could easily harvest. As populations grew, values often changed, and mangroves were cleared so that people could live closer to the sea. Until recently, most countries had forgotten the valuable seafood that begins in these marine nurseries. Many countries have now passed laws intended to protect mangroves, but they continue destroying them for other enterprises. As the mangroves are removed or polluted so badly that the seafood either cannot begin growing there or is so poisonous humans cannot eat it, many still argue that mangroves are nothing but mosquito-breeding grounds.

There are so few old, large mangroves remaining in the world now that most of us cannot imagine how large the trees can grow. In 1999, I visited the mangrove preserve on the outskirts of Guayaquil, Ecuador, where shrimp farming is steadily encroaching on the old forest. While the remaining trees were large, none was nearly big enough to have been used to make the dugout canoes that lined the village bank. Some of those canoes were made from trees at least 1 m (3 feet) in diameter. Craighead (1971) recorded mangroves that size growing in southern Florida until 1960 when Hurricane Donna toppled trees with diameters of 1.2 m (4 feet). By the time I saw the path of that hurricane in the early 1970s, those fallen tree trunks mostly were covered over with mud, water, and herbs.

While wandering around in mangroves in southwestern Florida, I have often wondered what those old forests were like. Sometimes during those walks through the peaceful shade below mangrove canopies, mangrove skippers (*Phocides pigmalion*) would fly past. The ethereal iridescent turquoise blue stripes on their hind wings flickered like the ghosts of those old fallen trees, reminding me of ancient forests I would never see.

Lantana

(Virgil, 70–19 B.C., called a plant *lenta viburna*, flexible viburnum; now reflected in *V. lantana,* and reapplied by Linnaeus to these plants; see also *Lantana:* Button sage, next entry, and *Viburnum*)

> ***Lantana depressa*** (low, pressed down) (= *L. ovalifolia*)
> big sage (Bahamas)
> lantana (Florida)
> *svpeyv łakko* [*rakko*] (*svpeyv*, the mythical plant, *łakko*, big, Creek); *shapeye-choobe* (*choobe*, big, Mikasuki)

Linnaeus ([1753] 1957) included seven species in *Lantana*, with only one from Africa. Indeed, most of the 150 species in the genus are American (Mabberley 1997).

A decoction is used as a drink or bath for measles in the Bahamas (Ayensu 1981, Morton 1981). Since both *L. camara* and *L. involucrata* (see *Lantana:* Button Sage) are also called "big sage," uses of the three are likely interchanged. Indeed, Snow and Stans (2001) list *svpeye* or *shapeye* as *Polygala*, but say that *svpeyv rakko* is stronger for the same purpose. So, not only is more than one *Lantana* involved, but a second genus. See *Polygala* for further details.

Lantana camara contains triterpenoids (lantadense) and verbascoside, an inhibitor of protein kinase C, and can be poisonous (Nellis 1997). Distribution of these poisons in other species is poorly known.

Lantana: Button Sage

(See preceding entry for etymology)

An aromatic shrub called button sage (Florida, Puerto Rico) grows behind the dune crest from Cape Canaveral in Brevard County around the southern tip of the state and up to Pinellas County in the Tampa Bay region. Typical of common names, the plants are not sages (genus *Salvia*) but are actually a lantana (*Lantana involucrata*). To make matters more confusing, some people also call these fragrant bushes oregano (genus *Origanum*). This apparently lax use of names results from similarities with true sages and

Lantana involucrata. *Drawn by P.N. Honychurch.*

oreganos (e.g., zygomorphic flowers, aromatic foliage) and differing philosophies of classification.

Both "lantana" and "sage" are derived from ancient names dating back to the classical periods of Rome and Greece. "Oregano," on the other hand, appears to be from the 13th century, and had other names in older Greek.

English, Spanish, and French speakers consider these shrubs sages. In English, they are big sage (Bahamas), bitter sage (Caymans), black sage (Bahamas), dark sage (Bahamas), fine-leaf sage (Bahamas), rock sage (Barbados), roundleaf sage (Caymans), sage (Virgin Islands), teeny-weeny sage (Bahamas), white sage (Barbados), and wild sage (Bahamas, Puerto Rico). In Spanish lands they are *salvia santa* (holy sage, Guatemala). People speaking French say *baume blanc* (white sage, Guadeloupe, Martinique), *petit baume* (little sage, Guadeloupe, Martinique), and *sauge* (sage, Guadeloupe, Martinique).

The word "sage" in English traces its origin to the Latin *salvia.* From that language, *salvia* became *saluie* in Old English, *save* or *sauge* in Middle English (about the 4th century A.D.) and ultimately sage. Since English is a Germanic language, it is not surprising to find a similar alteration of the original *salvia* in them. From the Latin, the word became *salbeia* or *salveia* in Old High German, and then *salvie* or *selve* in Middle Low German. Today, it is *selie* in Dutch and *Salbie* in German. In Romance languages, the evolution followed a similar route, staying *salvia* in Italian and Spanish, but becoming *sauge* in French and *salva* in Portuguese. In spite of the 2000 or so years that these words have been evolving, the same exact meaning of *salvia* (to save or preserve, because of medical usage) is retained in English. Not only did the word have that meaning in the beginning, but we now know that many of the spices, including mints, are antibacterial (Chen et al. 1985, Sow et al. 1995, Tassou et al. 1995, Wen et al. 1995, Billing and Sherman 1998, Imai et al. 2001). However, the complete reliance on spices to protect from infection in foods will never become a reality because of evolving new virulent strains (D'Aoust 1994), and differential impact on microorganisms (Beddows et al. 2000, Ismail et al. 2001).

Those who think *L. involucrata* is like oregano say *orégano* (oregano, Tamaulipas) and *orégano de monte* (wild oregano, Nicaragua). The word "oregano" (from the Greek *oros*, mountain, plus *ganos*, brightness, joy, or pride), first was *oreiganos*, and later was used in English ca. A.D. 1265. Originally, the word applied to several species of mints in the genus *Origanum* (See *Bacopa:* Water-hyssop), including *O. vulgare* and *O. marjorum*. Before that *Origanum* was called *dictamnon* by the Greeks, or *artemidion* after Artemis (goddess of the chase). She could wound with her poisoned arrows and then cure the wound with *O. dictamnus* of Crete (Baumann 1993).

In spite of its common names, button sage is a *Lantana*. That word was the ancient Latin name for *Viburnum* and is now reflected in *V. lantana*. Between the 13th and 14th centuries, the spelling was *lentana*, but it changed before 1590 to *lantana*. In Sicilian, it remains *lantanu*, and the spoken Latin originally was *lentaginem*, from *lengus*, pliant (from flexible viburnum, *lenta viburna*, of Virgil, 70–19 B.C.). Linnaeus applied the name to American plants to create the modern application.

There are 150 species in *Lantana*, mostly native throughout tropical America, but with a few in southern Africa (Mabberley 1997). The best-known species is *L. camara*, one of the most widely used of the medicinal species, which is also an aggressive weed. The species is a weed in Florida and other parts of the United States, including Hawaii, where insects were introduced about 1902 for minimally successful biological control (Julien and Griffiths 1998), New Caledonia, South Africa (Cilliers and Neser 1991), and elsewhere. This is the most-studied species in the genus and contains toxic triterpenoids (the lantadenes) (Achhireddy et al. 1985, Sharma et al. 1989, 1990, 1991, Barre et al. 1997, Misra and Laatsch 2000), verbascoside (an inhibitor of protein kinase C; Herbert et al. 1991, Taoubi et al. 1997), and phytotoxins (Jain et al. 1989). This and other species also contain novel compounds (Singh et al. 1990, Abeygunawardena et al. 1991, Takeda et al. 1998, Muhayimana et al. 1998).

Button sage (*L. involucrata*) is native from southern Georgia and Florida, along the Gulf Coast states to Texas and in the Bahamas and the Caribbean from Cuba to Barbados. The species grows in Guatemala, Belize, southern Mexico (Campeche, Quintana Roo, Yucatán), and the Galapagos Islands. In those areas, it has yet other common names, and most of them make other references to medical uses. These names include *cariaquilla de Santa María* (Saint Mary's *caraiquilla*, Puerto Rico) and *cariaquito encarnado* (red *cariaquito*). In Venezuela, *cariaquito* is a generic name for *Lantana*, with *L. alba* being *cariaquito blanco*, *L. camara* is *cariaquito colorado*, and so on. There are also the names *chiligüe* (El Salvador), *cinco negritos* (five little black ones, El Salvador), *confite* (confidence, Sinaloa), *confituria blanca* (white confidence, Mexico), *cuasquito oloroso* (fragrant *cuasquito* Nicaragua), *doña sanica* (healthy lady, Dominican Republic), *doña Anita* [*Ana*] (lady Anita [Ana], Dominican Republic), *filigrana cimarrona* (wild filigree, Cuba, Puerto Rico), *hierba de la sangre* (blood herb, Cuba), *té de costa* (coastal tea, Cuba, Puerto Rico), wild balsam (Bahamas), and wild mint (Jamaica).

Other names given to the plants have obscure references. There seems to be nothing similar to peony (*Paeonia*, Paeoniaceae), but in Mexico it is called *peonía* (Tarahumara, Chihuahua) and *peonía colorado* (red peonia, Tamaulipas). In addition, in Mexico, *L. involucrata* is known as *tarepe* (Michoacán) and *sikilha'xiu* [*zicilhaxiu*] (*sikil*, *Cucurbita* seed, *ha'*, water, *xiw*, herb, Maya, Yucatán).

The most unexpected of all the common names of *L. involucrata* is *cerise* (cherry in French and Spanish, Hispaniola). There is nothing to me that resembles a cherry (*Prunus cerasus*, Rosaceae). To make it more intriguing, that is also the common name widely given to *Momordica charantia* in the Caribbean. In Haiti, this name may be spelled as *sorrosie* and *assorossie*, while in Jamaica and the Cayman Islands, it is written *serasee* or *cerasee*. My Jamaican students all pronounced it as "ser-see" and were astonished at the spelling in Caribbean herbal shops. The only possible connection that I can imagine between these plants is that they are medicinal. Perhaps the fact that the "medicine" as a tea also tastes terrible contributed to both being called by the same name.

Lantana involucrata has been used as a tea and bath for measles and chicken pox, to relieve itching skin (Ayensu 1981), to control high blood pressure (Honychurch 1987), against irregular menstruation (Morton 1981), and to aid parturition (Ford 1975). It is also considered effective if boiled with sugar and used as a drink to bring out the spots of measles and chicken pox (Ayensu 1981). In Mexico, pieces of leaves or stems have been put in the ears as a "cure" for deafness (Standley 1920–1926). Throughout much of its range people consider a leaf infusion sudorific and diuretic, and the pounded leaves in water as emetic. Boiled with *Serjania subdentata* (Sapindaceae) it "purifies blood" and reduces fever (Morton 1981).

Chemically, the species is similar to *L. camara* and contains lantanine, which has qualities similar to quinine (Liogier 1974, Ayensu 1981). Both are aromatic and *L. camara* is said to have 0.22% essential oils (Ayensu 1981). Other species have been studied more recently and have been shown to be antimicrobial (Saleh et al. 1999), anti-inflammatory, analgesic, and antipyretic (Forestieri et al. 1996), and to have antithrombin activity (O'Neill et al. 1998). *Lantana trifolia* also has been shown to be antimicrobial (Rwangabo et al. 1988). In spite of its poisonous chemicals, some cultures still eat the fruits (Herzog et al. 1994, Taoubi et al. 1997). If *L. involucrata* indeed has the same basic chemical spectrum as *L. camara*, many of the folk uses, while dangerous, may prove effective.

One beautifully ironic study has shown that the aqueous leachate of *L. camara* kills water hyacinth (*Eichhornia crassipes*) (Saxena 2000). If *L. involucrata* does the same, it would be a lovely bit of human justice to turn native and exotic Lantanas against the alien pest water hyacinth. *Eichhornia* alone has cost the taxpayers of Florida millions of dollars annually for decades. It would be true fair play—and a better use of taxes—to turn one alien plant against another.

Laportea

(Charles Gaudichaud-Beapré, 1789–1854, named the genus for François L. de Laporte, 1810–1880, Count of Castelnau, an entomologist who collected plants in Florida between 1843 and 1847)

Laportea canadensis (from Canada)
Brennpflanze (stinging plant, German)
Canadian wood-nettle [woodnettle, nettle of the woods] ("nettle" from Old English *netele, netel, netle*, based on an Old Teutonic name for the genus *Urtica*, applied by about A.D. 705); nettle patch; [Canada, stinging] nettle; woodland-nettle
ha'-do-ga (a sacred plant; fibers used to weave sacred rush mat case for the *wa-xo'be* or Sacred Hawk; *mon̲-hin̲ wa-xton̲ ha-do-ga* (*mon̲-hin̲*, grass, Osage)
hatka holhpa (*hatka*, man, but implying a Choctaw, *holhpa*, blister, Choctaw; also used for a stinging caterpiller)
masan (Cree, Menomini, Ojibwa); *masan* (Strachey wrote this *makuhsaan* in [1612] 1953, Powhatan, Virginia); *masana* (Fox, Shawnee)
stvtvlálv (Muskogee)

Although none of my sources lists the common name "stinging nettle," that is the one I was taught when growing up in western Kentucky. This is a plant of forested lowlands, often growing on floodplains. No one there had ever heard of "wood-nettle" or "Canada nettle." We also knew that there were two kinds of "stinging nettle," the other one the weedy upland *Urtica dioica*. We tried to avoid both, but sometimes failed. Once when leading a class field trip as a teaching assistant to an undergraduate professor, I accidentally walked deep into a patch of *Laportea* before they penetrated my Levis. Although I stopped everyone and we backed out, those wearing shorts had tears running down their cheeks as we left.

Lewis and Elvin-Lewis (1977) record that the stinging plant hairs of both nettles contain histamine-like substances. These compounds are held in the bulbous tips of the hairs that break upon puncturing the skin to release the irritant chemicals. Not surprisingly, indigenous people had remedies for the rash caused by nettles. The Potawatomi relieved the rash with the juice of jewel weed (*Impatiens capensis*) (Smith 1933, Vogel 1970). The Rappahannock treated nettle stings with urine or salt and water (Vogel 1970).

Use of *Laportea* fibers goes back to the Hopwellian Cultures where Whitford (1941) has identified fabrics woven of them. The fibers also appear as cords in prehistoric cave shelters in Arkansas, Kentucky, and Tennessee (Whitford 1941). Historically, at least the Abenaki, Fox, Menomini, Meskwaki, Sauk, and Ojibwa used fibers extracted from the stems to make baskets, twine, bags, cords, and sewing fibers (Whitford 1941, Uphof 1968, Moerman 1998). Those fibers are said to be 50 times as strong as the same diameter of cotton (Uphof 1968).

Among Muskogean people, only one tribe is noted as using the plants. According to Speck (1941), the Houma applied a decoction to lower fever. No other people is recorded as using the plant in this way and the genus does not appear in Porcher (1863) or Millspaugh (1892).

Several other eastern tribes used the plants for other purposes. The Iroquois used a decoction to counteract "love medicine," an infusion of roots to ease childbirth, and to treat tuberculosis (Moerman 1998). The Meskwaki used the root to treat incontinence and as a diuretic (Moerman 1998), seemingly diametrically opposed applications. The Ojibwa also used an infusion of roots to treat urinary problems (Hocking 1997, Moerman 1998).

Lasiacis

(Greek, *lasios*, shaggy, woolly, *akis*, tip, referring to the tuft of wool at the apex of the spikelet)

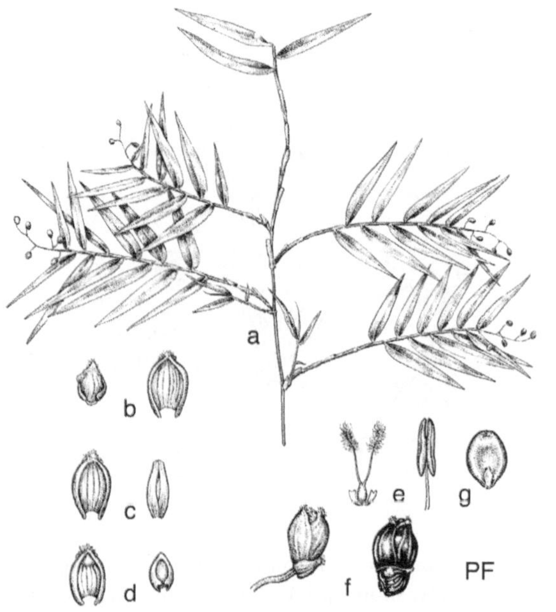

Lasiacis divaricata. a. Habit. b. First and second glume. c. Sterile lemma and palea. d. Fertile lemma and palea. e. Pistil and lodicules (on left) and stamen (on right). f. Young spikelet (on left) and mature spikelet (on right). g. Caryopsis. *Drawn by Priscilla Fawcett*. From Correll and Correll 1982.

Lasiacis divaricata (spread apart, alluding to the branches)
alcarrizo (reed, from Latin *carex*, a sedge, Hispaniola); *carrizo* (cane, Hispaniola, Honduras)
bambou marron (wild bamboo, Hispaniola); wild bamboo (Florida; see Small 1933); *petit bambou* [*calumet*] (little bamboo [pipe], Guadeloupe, Martinique)
calimete (variant of *calumet*, Hispaniola); *calumet* (pipe, French, Guadeloupe, Martinique; reed (from Latin *calamellus*, Greek *kalamos*, now "straw" or "pen," from quill or reed pens, in English, through *haulm* from Middle and Old English *healm*, Latin *culmus*, "stem")
cane-grass (Bahamas); wild cane (Bahamas)
mehensit (little reed, Maya, Yucatán)
pitillo de monte (wild little whistle); *pito de bejuco* (vine whistle, Cuba)
sit [*zit*] (cane or reed, usually used for *Arundo donax*, Maya, Yucatán)
small-cane [smallcane] (Florida)
tibisee (Florida, Bahamas); *tibisí* (name for reed, *Phragmites australis*, Taino, Cuba); *tibisí chico* (little reed, Cuba)

Wild bamboo (*Lasiacis divaricata*) is the smallest of the native bamboo-like grasses in southern Florida. *Lasiacis* was created in 1864 as *Panicum* sect. *Lasiacis* by German botanist August H.R. Grisebach. Later, in 1910, it was moved to generic rank by A.S. Hitchcock. The small plants (2 to 4 m; 6 to 12 feet) grow in southern Florida, the Bahamas, the West Indies from Cuba to Barbados, and from Baja California to Veracruz, and south to Peru and Argentina. In much of its range, uses of the grass overlap with uses of the other plants with the names "cane" and "reed" (see *Arundinaria, Phragmites*).

While it usually grows in hammocks, *Lasiacis* is not immune to water level and habitat loss changes. Although the species ranges from the southernmost Florida Keys to Cape Canaveral in Brevard County on the east and Fort Myers in Lee County on the west coast, now it is confined to tiny fragment sites. Most of the coastal regions where it grew are housing developments. We no longer use it, and we are about to lose it.

Leitneria

(Named for German physician Edward Frederick Leitner, 1812–1838, who was killed in Florida during the first Seminole War)

Leitneria floridana. From Sargent 1905.

Leitneria floridana (of Florida)
corkwood

This monotypic genus is endemic to the southeastern United States. There are Eocene fossils from London, England, that have been identified as this (Mabberley 1997), but Cronquist (1981) has questioned their identity. The single species was named by Alvin Wentworth Chapman (1809–1899), author of the *Flora of the Southern United States* in 1860. He described the genus and species in the first edition of that flora, and then went on to publish a second edition in 1883, and a third in 1897 (Correll and Johnston 1970).

Florally, the plants are so unique that George Bentham and Joseph D. Hooker (1817–1911) segregated them as the family Leitneriaceae in 1880 (Cronquist 1981). The modern view, however, based on molecular genetics, anatomy, and serological studies, is that they are members of the Simaroubaceae with flowers reduced for wind pollination (Judd et al. 2002).

These trees grow in isolated areas in the Bootheel area of southern Missouri (Butler, Ripley, Dunklin Cos.), Arkansas, Texas (Brazos River near Columbia), Florida (Appalachicola area), and southwestern Georgia (Sargent 1905, Fernald 1950, Steyermark 1963, Correll and Johnston 1970, Jones and Coile 1988).

Wood has been used for fishing corks (it is lighter than cork), floats for nets, and stoppers (Sargent 1905, Hocking 1997, Mabberley 1997). The plants are so rare they are considered threatened in Florida (Coile 2000), but Ward (1979) did not consider current harvesting for these purposes to be a threat.

Lepidium

(Greek classical name from *lepidion*, a little scale, an allusion to the flat fruit)

biolair-Fraing (French cress, for *L. sativum*, Gaelic)
Kresse (from Latin *cressa*, German)
lepidio (Spanish for *L. latifolium* from 1542); *lepidio maggiore* (big little-scale, Italian for *L. latifolium*, from 1551)
markkarse (field cress, Norwegian for *L. campestre*)
matkarse (food cress, Norwegian for *L. sativum*)
mostardina (little mustard, Italian)
passerage (cress, French for *L. latifolium*, from 1550)
pepercruyt (pepper herb, Dutch for *L. latifolium*, from 1549); *pepperwurt* [*pepperwoort, pepperwort*] (Dutch for *L. latifolium*, from 1500s); *Pfefferkraut* (pepper herb, German for *L. latifolium*, from 1542)
vollkarse (wall cress, Norwegian for *L. heterophyllum*)

Lepidium virginicum (of Virginia)
bird's pepper; poor man's pepper; Virginia pepperweed [Virginia pepperweed]

Lepidium virginicum. a. Habit of upper part of plant. b. Young plant, showing basal and cauline leaf forms. c. Flowers, top and side views. d. Silicles, mature (left) and immature (right). e. Seed. *Drawn by Regina O. Hughes.* From Reed 1971.

boiled-seed bush
cresor (cress, Houma, Louisiana); *cressón* (big cress, Puerto Rico); *cresson* [*a, de*] *savane* [*alénois, danois, savage*] (savanna [awl, Danish, wild] cress, Guadeloupe, Martinique, Haiti)
culantrillo (little coriander, El Salvador)
cupapayo (perhaps based on the Zoque name *cupapé*, or Mayan *k'opte'*, with a hard center, the name for *Cordia*, although any relationship between this herb and that tree is obscure, Yucatán)
hashshan [*haashan, hashan*] (Chickasaw); *tukáskénv* (Muskogee)
julipliegue (strong reprimand, presumably an allusion to the spicy taste, Guatemala)
lentejilla [*lentajilla, lentejuela, lentepilla, antejuela, antejuelilla*] (little lens, Texas, Chihuahua, Mexico, Guatemala, Puerto Rico)
[*hierba*] *mastuerzo* [*mastuerce, mastuerco*] (nasturtium, Yucatán, Belize, Cuba, Dominican Republic, El Salvador, Guatemala, Puerto Rico); *maltuerce* (Nicaragua)
mecheche (*che'*, plant, Maya, Yucatán); *mixixe* (Mayan?, Guerrero)
mostacilla (little mustard, New Mexico)

panalillo (a name usually given on the northwestern coast to *Thouinidium decandrum*, Sapindaceae, Mexico)
passerage de Virginie (Virginia cress, French Antilles)
[American, Virginia, wild] pepper-grass [pepper-bush] ("pepper-grass" applied ca. 1475 to *Lepidium sativum*, because of the spicy taste of the fruits, subsequently to other species; USA, Bahamas, Barbados, Puerto Rico)
perejil de la tierra (wild parsley)
put xiu (*put*, papaya [*Carica papaya*], *xiw*, herb, Maya, Belize, Yucatán)
putcan [*putkan*] (*put*, papaya, *kan*, yellow; the reference to papaya is because the odor of the mashed fruits resembles that plant, Maya, Yucatán)
sabelección (knows the lesson, Cuba)
sacabé [*sakbé*] (*sak*, white, *bé*, road, probably given because it grows along roads and pathways, Maya, Guatemala)
tongue-grass [tonguegrass] (northeastern USA)
Virginia cress (South Carolina, fide Porcher 1863); *Virginiakarse* (Virginia cress, Norwegian)
wild cress (Barbados)

Europeans were well acquainted with *Lepidium* by the time the New World was discovered. Indeed, classical authors Dioscorides (A.D. 40–80), Galen (A.D. 129–?200), and Pliny (A.D. 23–79) discussed *L. latifolium*, and Fuchs (Meyers et al. 1999) also detailed three other species. Culpeper (1653) wrote of *L. sativum*: "This is very successful for the sciatica, gout, and pains in the joints ... it also amends deformities or discolorings of the skin, and takes away scars, marks, scabs and burns. The juice is given in ale to procure women a speedy delivery in travail."

In addition to the medical applications listed by the classical authors, Fuchs added, "*Lepidium* has the power to thin thick and viscous fluids. Accordingly, it is a culinary herb and suitable for pickling" (Meyers et al. 1999). The Cherokee agreed and ate young *L. virginicum* plants either boiled or fried (Moerman 1998). Porcher (1863) added, "It is suitable to be used in winter and early spring salads, but is far less in request than some of the other cresses." Hedrick (1919) did not list this edible species, although he included six others.

Among the Muskogean people, the Houma took a mixture containing this and other herbs with whiskey to relieve tuberculosis (Speck 1941). Muskogean neighbors and their subsequent allies, the Cherokee, produced a poultice to help cure blisters and to ease croup, and made a remedy to give sick chickens to make them lay eggs (Hamel and Chiltoskey 1975). The Menomini used an infusion of the plant to wash the rash from poison ivy (Moerman 1998).

Foster and Duke (1990) record that the herbs were used to treat scurvy in addition to the applications noted among the indigenous tribes. Hocking (1997) says that the Mayas made a root decoction with *Aloe barbadensis* root and treated coughs and colds with it. The Maya have adapted the alien *Aloe* into their pharmacopoeia. Elsewhere in Mexico, a tea is made from *Lepidium* flowers and leaves to relieve digestive problems (Morton 1981). People from Mexico and Guatemala to Coro in Venezuela (Martínez 1969, Ford 1975, Morton 1981, Gupta 1995a) use the herbs in other applications. However, Foster and Duke (1990) warn that *Lepidium* may cause skin irritation and blisters.

Lespedeza

(Named after Vincente Manuel de Céspedes, the Spanish Governor of East Florida between 1784 and 1790, who sponsored Frenchman André Michaux's exploration of the region; Michaux intended for the name to be "*Cespedeza*" but it was misspelled in the printed book as *de Lespedez*, cf. Taylor and Norman 2002)

Lespedeza capitata (with flowers in heads)
[bush, dusty] clover
parus-as (*parus*, rabbit, *as*, foot, Pawnee)
roundhead lespedeza
te-hunton-hi nuga (*te*, buffalo, *hunton*, belly, *hi* plant *nuga*, male, Omaha-Ponca)

The species was named by Michaux in his *Flora Boreali-Americana* (Flora of North America) of 1803. *Lespedeza* is now a genus of 40 species, growing in temperate regions of North and South America, and also in Asia and Australia (Mabberley 1997).

Gilmore (1919) found the people of the Missouri River region had used these plants. Its name in Omaha-Ponca considers it a "male," while at the same time comparing it with *Amorpha* (see *Amorpha*: Lead Plant), which they called *te-hunton-hi miga* (female buffalo belly plant). These two tribes used the stems of *Lespedeza* as they did *Amorpha*—as a moxa, where one end was stuck to the skin and the other burned until it "cured" the ailment by scorching the patient. The "cure" was for rheumatism and neuralgia. The Meskwaki of Iowa considered the root an antidote for poison. The Comanche made a tea from the leaves (Moerman 1998).

In Mexico, *L. capitata* is used for kidney problems and anuria, including nephrosclerosis, oliguria, and uremia (Hocking 1997).

Leucothoe

(Roman writer Ovid, 43 B.C.–A.D. ?17, identified Leucothoë as the daughter of King Orchamus of Babylonia; she was one of the many loves of Apollo)

Leucothoe racemosa. From Britton and Brown 1897.

Leucothoe axillaris (in the "armpit," or having the flowers in the angle between the leaves and stems)
coastal dog-hobble [coastal doghobble]
drooping leucothoë (a book name)
fetter bush ("fetter" from Old English *feter*; with cognates in Dutch *velter*, Swedish *fjältrar*; in English since about A.D. 800. Then, as now, it means a chain or shackle of some kind around the feet of humans or animals. No indication of when it was applied to these bushes has been found, but the allusion is to the symptoms of animals being poisoned by eating the branches.)

Leucothoe axillaris was described by J.B. Lamarck in 1783 as an *Andromeda*. At almost the same time, Thomas Walter described the same plants under the name *Andromeda catesbaei* in his *Flora Caroliniana* of 1788. The name proposed by Lamarck is used because it preceded the name used by Thomas by some 5 years.

It was not until 1834 when David Don (1799/1800–1841) realized that these plants belonged in a new genus and created *Leucothoe*. Subsequently, other regions of the world were explored and this ericaceous genus now has eight species in North America, Japan, and the Himalayas (Mabberley 1997). There are five species endemic to North America, some of which are cultivated as ornamentals.

Considering the potent chemicals in the Ericaceae, it should not be surprising that these plants were used by indigenous people. This species contains andromedotoxin (Perkins and Payne 1978) and is as toxic as *Kalmia* (which see).

The Cherokee used an infusion to treat pain, rheumatism, and itch, as a stimulant, and to treat mange in dogs (Hamel and Chiltoskey 1975, Moerman 1998). That is the only tribe known to have used these shrubs, but Hocking (1997) said that they were "used as errhine" (increase nasal mucous) among "Indians."

No mention of the plants was made by Porcher (1863) or Millspaugh (1892).

Liatris

(Derivation of the name is unknown, and J.C.D. von Schreber, 1739–1810, left us no clue; could it be possible that he used the Gaelic *liatrus*, spoon-shaped, as an allusion to the sometimes bulbous roots?)

aontashe (Omaha-Ponca)
blazing star [blazingstar]
button-snakeroot
gay feather [gayfeather]
kahtsu-dawidu [*kahtsu-rawidu*] (*kahtsu*, medicine, *dawidu* or *rawidu*, round, Pawnee)
makan-sagi (hard medicine, Omaha-Ponca)

Liatris chapmanii (named for Alvin Wentworth Chapman, 1809–1899, who wrote the first flora of the southeastern United States)
blazing star
Chapman's gayfeather (a book name)
ecjomahi (*ici*, deer, *im*, its, *aahi*, potato, Mikasuki)

Liatris graminifolia (with leaves like grass)
backache-root (North Carolina)

Liatris gracilis (slender) (=*Liatris laxa, L. laevigata*)
button snakeroot (Florida in 1911); rattlesnake master (Florida)
'comvhv [*comahâ, ico imaha, i'djo maha, ijo maha, itimahi, itco imaha*] (*eco*, deer, *em*, its, *vhv*, potato, Creek, Muskogee); *i:címa:hî* [*eeche*

'mahe, iicimaahi] (*ici*, deer, *im*, its, *aahi*, potato, Mikasuki); *issim-ahi* (*issi*, deer, *im*, its, *ahi*, potato, Chickasaw)

Liatris tenuifolia (slender-leaved)
backache root
[slender-leaved] blazing star
deer bowl (Florida)

Liatris is a genus of 43 species endemic to eastern North America, including northeastern Mexico (Gaiser 1946, Mabberley 1997). As Ajilvsgi (1984) and Diggs et al. (1999) pointed out, one of the generic common names is "button-snakeroot," derived from the old use of the cormlike roots in treating snakebite. That name is also applied to other plants used to treat bites (see *Eryngium*). The application may date to prehistoric people like those at the Ozark Bluff-Dweller site in Arkansas. There Gilmore (1931) identified remains of two species and commented on *L. squarrosa* that "This was probably used medicinally, as it is still employed by Indians." Earlier (1919) he had learned of *Liatris* use among the Omaha-Ponca and Pawnee.

Swanton (1928) reported the Creeks using *Liatris* sp., and Taylor (1940) found their relatives the Koasati using *L. acidota.* Spoehr (1939) found the Creeks using *L. gracilis*, as did Sturtevant (1955) with the Miccosukee. Bennett (1997) found the people on the Big Cypress Reservation where Sturtevant worked using *L. chapmanii.* Given the similarity of several *Liatris* species, more than one may be used in any particular area. Several other species are used by various people, including the Blackfoot, Comanche, Kiowa, Koasati, Lakota, Meskwaki, Ojibwa, Omaha-Ponca, Pawnee, and Tewa (Moerman 1998).

The Seminoles use the roots for "Cow Sickness" (chest pain, digestive upset, diarrhea), "Deer Sickness" (pain in limbs, especially joints, swelling, bad dreams), and "Bird Sickness" (diarrhea, vomiting, appetite loss) (Sturtevant 1955). Part of their treatment is a decoction of this plant and several others used externally for rheumatic joints. *Liatris chapmannii* also is used to treat Deer Sickness, constipation, and asthma, and is used in other herbal mixtures (Bennett 1997). The Chickasaw use the root as a medicine to treat swelling (Munro and Willmond 1994).

Hocking (1997) noted that *L. chapmanii* contains chapliatrin, an antileukemic germacranolide compound, and that extracts show *in vitro* inhibition of human carcinoma (Kupchan et al. 1973). He also said that *L. gracilis* contains substances active against lung cancer, leukemia, and lymphocytic anemia. In his master's thesis, Hocking found that *L. tenuifolia* contains chapliatrin, spicatin (a sesquiterpene cytotoxic lactone), and ecuparin.

Licania

(Misspelled anagram of the Galibí name *caligni,* given to a French Guiana plant by J.F. Aublet; see Prance 1970)

Licania michauxii. *Drawn by P.N. Honychurch.*

Licania michauxii (named by Ghillean Prance to commemorate André Michaux, 1746–1802, French botanist who lived some years in the United States)
ayekche-chobee (*ayekche*, medicine, *chobee*, big, Mikasuki)
gopher-apple ("gopher," as *gaufre*, honeycomb, was applied to burrowing animals by French colonists because their dens are like "honeycombs"; "gopher" was the name already established for the tortoise that Linnaeus called *Gopherus polyphemus* in 1759; fruits are eaten by the gopher tortoise)
lucv-huehkv enhompetv (*lucv-huehkv,* gopher turtle, *em*, its, *hompetv*, food, Creek); *apaahoshe emempeeke* (gopher turtle food, Mikasuki)
maiden-blush [-rose] (in use for these plants by 1760; an allusion to the pink tinge on the white fruit)
obî:hosimîmpî [*abah'ha shemembe*] (wolf's food, Mikasuki)
yahami:láyka [*yamelikv*] (*yvhv*, wolf, *em*, its, *vlicv*, sickness brought on by, Creek)

Aublet created the genus *Licania* in 1775, and some of the Amazonian and Guianese species are large trees. Therefore, it was a surprise when Prance

(1970, 1972) concluded that *L. michauxii* was part of the genus. *Licania* contains 193 tropical species, with 1 in Africa, 2 to 3 in Asia, and the rest in the Americas (Mabberley 1997).

Hogan (1978) found pollen from *L. michauxii* at the pre-Columbian village site at Fort Center on Lake Okeechobee (Sears 1982). When Sturtevant (1955) worked on the Big Cypress Seminole Reservation, they still were using the plants. Fruits are edible (Morton 1968b), but to me taste like a new plastic shower curtain smells. Sturtevant (1955) found the species used in medicines to treat "Wolf Sickness" (vomiting, stomach pain, diarrhea, frequent urination), chronic sickness, insanity, and to ease childbirth. Bennett (1997) found similar applications to stomach pains and asthma. His informants and Snow and Stans (2001) indicated that it was most important in purification ceremonies following death.

No chemical studies have been made of this *L. michauxii*, but others are known to contain a variety of bioactive compounds. Among those chemicals are flavonoids, lupanes, terpenoids, and triterpenoids (Bilia et al. 1996, Braca et al. 1999a,b, 2000).

Licaria

(J.F. Aublet named the genus in 1775 with *licari,* a French Guiana Carib name)

Licaria triandra (with three-stamens)
bois fourmi (ant tree, Martinique)
cigua gorrita (little alien *cigua*, Dominican Republic); *cigua prieta* [*sigua prieta*] (black *cigua*; *cigua*, a laurel, Taino, Dominican Republic)
Gulf licaria (USA, Puerto Rico)
Gulf misanteca (misanteca, probably from Latin *minus*, fewer, Greek *anthus*, flowers, USA); [*palo*] *misanteco* [*misanteca*] (*misanteca* [tree], Puerto Rico)
laurel [*blanco, de Jamaica, de loma*] ([white, Jamaica, hill] laurel, Cuba); *laurier jaune* [*sassafras*] (yellow [sassafras] laurel, Haiti)
lebisa [*levisa*] (usually a name given a stingray, *Dasybatus torrei*, Taino?, Cuba, Puerto Rico)
[pepperleaf] sweet-wood [sweetwood] (Florida, Dominican Republic, Jamaica)

Olaf Swartz named these trees *Laurus triandra* in 1788. It was not until 1937 that the species was moved to *Licaria* by Dutch botanist André Joseph Guillaume Henri Kostermans (1907–1994), who lived in Asia. The genus contains 40 tropical American species, some producing oils and local medicines (Mabberley 1997).

Cubans have used a bark decoction for stomach ailments, and root bark and trunk extracts as stimulants and astringents (Roig 1945). A decoction of the bark has been applied to skin problems in Hispaniola, and the oil used to treat rheumatism (Liogier 1974). An undisclosed medicinal use is known from Belize (Balick et al. 2000). The wood is yellow-green, heavy (specific gravity 0.9), and has been used in posts, construction, and for making matches (Little and Wadsworth 1964, Liogier 1974).

Lilium

(Classical name from *lilium*, a lily, Greek *lirion*, Anglo-Saxon *lilie*)

Lilium catesbaei. From Institute of Food and Agricultural Sciences.

lily (from Latin *lilium* through Old English *lilie*, in English by A.D. 971); *giglio* (Italian); *lelie* (Dutch); *lili* (Gaelic); *Lilie* (German); *lirio* (Portuguese, Spanish); *lis* (Quebec)

Lilium catesbaei (named for its discoverer, Mark Catesby, 1679–1749)
Catesby's lily (a book name)
oksatlí ahissí (*oksatlí,* swelling, *ahissí,* medicine, Koasati)
pine lily (because it grows in pinelands, locally known as pine flatwoods)

Lilium iridollae (a rainbow)
Panhandle lily

Florida has four native species of *Lilium* (Wunderlin 1998). There are no records of *Lilium* having been used by indigenous Florida people, but many of the others in North America were used. Moreover, there is a long history of people using *Lilium.*

The Latin word, from Greek *leirion,* did not appear in English until about A.D. 971. The best known and useful of European lilies was *L. candidum,* the one mentioned by Shakespeare. According to an ancient Semitic legend, the lily sprang from the tears of Eve when she was expelled from the Garden of Eden, and it was the symbol of motherhood to Sumerians, Babylonians, Assyrians, and Egyptians. Lilies were the flowers of Hera, Greek goddess of the moon, and Roman Juno, the goddess of light (Lehner and Lehner 1960).

Fresh and dried flowers are used in medicines, the pollen is used against epilepsy, and the flowers are used to scent perfumes. Mediterranean use goes back at least 5000 years when *Lilium* was drawn on Minoan (Crete) frescos (Simpson and Conner-Ogorzaly 1995). This lily is probably the "Rose of Sharon" of the Bible, and has been in cultivation since 1500 B.C. With the spread of the Christian religion, the lily was associated with the Virgin Mary and became incorporated into many religious pictures. The species is now known as the "Annunciation lily" (based on the 25 March Catholic festival of the Virgin Mary), "Madonna lily," or "Bourbon lily" (de Cleene and Lejeune 2002).

Based on its name, the Koasati used *Lilium* as medicine. The Cherokee, the neighbors and sometime allies of the Muskogeans, used at least one species of lily. These people used *L. canadense* for food during famines, and made medicines of them to treat diarrhea and rheumatism (Hamel and Chiltoskey 1975). *Lilium iridollae* in the Florida Panhandle has been confused with *L. canadense*; both were probably used by the Creeks. *Lilium canadense* was also used by the Algonquin, Huron, Ojibwa, Malecite, and Micmac (Moerman 1998).

Limonium

(Philip Miller used the ancient Greek name for some marsh plant, *leimon,* a meadow or bog)

acelga silvestre (wild *acelga,* Spanish)
lavande de mer (sea lavender, French); sea lavender (comparison with the European mint called lavender, *Lavendula*)
marrisp (*mar,* sea, *risp,* scratch, Norwegian)
statice (from Latin *statice* and Greek *statikos,* to stand still, alluding to a use to stop bleeding, Italian)
Widerstoss (countershock, German)

Limonium carolinianum (from Carolina)
canker root (used to treat sores)
ink root (presumably used to make ink)
marsh mullein (comparison with European mullein, *Verbascum,* Scrophulariaceae)
marsh rosemary (comparison with European rosemary, *Rosmarinus*)
sea lavender

Philip Miller named *Limonium* in 1754. In 1778, Thomas Walter described plants from the Carolinas he called *Statice caroliniana.* N.L. Britton realized that these New World seaside herbs should be moved, and transferred the species to *Limonium* in 1894. Now the genus *Limonium* contains about 350 species in the Northern Hemisphere (Mabberley 1997).

Roots are gathered in winter, boiled, and the liquid drunk as a remedy for colds, fever, and especially diarrhea. Raw roots are also chewed to stop diarrhea (Morton 1974). The Micmac of Nova Scotia and vicinity used a decoction of roots to fight hemorrhage caused by consumption (Moerman 1998).

Lindera

(Named by Carl Peter Thunberg for Johann Linder, 1676–1723, a Swedish botanist)

Lindera benzoin (old name for some member of the Lauraceae, from Arabic *lo-benjuy* or *luban jawi,* frankincense of Java; taken into Spanish as *benjui, benjuy* by 1516, Portuguese as *beijoim* by 1498, and Venetian Italian as *benzoi* by 1461; originally applied to the resin from *Styrax benzoin*)
ascopo (although Harriot [1590] 1972 does not provide enough information to be sure *ascopo* is

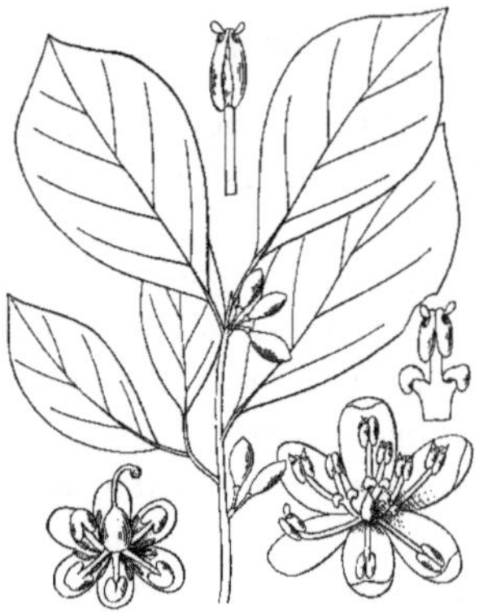

Linderna benzoin. From Britton and Brown 1896.

this species, his description makes it sound like that: "*Ascopo* a kinde of tree very like vnto Lawrell, the barke is hoat in tast and spicie"; Geary 1955 thought it might be *Magnolia virginiana*; Carolina Algonquians)
benjamin-bush (benjamin, a corruption of *benzoin*); *benjoin* (derived from Arabic, French)
fever-bush (in use by 1760s); *Fieberstrauch* (fever bush, German)
kaáchk'aachkeek (*kaachk'aachk*, dry, *eek*, that which is; in other words, brittle, Delaware)
kvpvpaskv [*kvpvpáskv*] (spice, Muskogee)
nohʔ-da-tli' [*na'tu'li*] (Cherokee)
pond-spice; [common, northern] spice-bush [spicebush, spice-wood] (applied to *Lindera* in New England before 1740 according to Kalm; OED says applied in English in 1845 because of the aromatic leaves; Oklahoma Yuchi still call it "spice-wood"); wild allspice
shábak [*tchápaak*] (*tchap*, medicine, *aak*, shrub, Delaware)
snap-bush

Early authors Leonard Plukenet in his *Almagestum botanicum* of 1696 and Jan Commelin in his *Hortus medici* of 1697 both noted that spice-bush smelled like the benzoin of Asia. So, Linnaeus ([1753] 1957) used "benzoin" as the species name. It is not known where Commelin obtained the specimens he cultivated. Regardless, indigenous American people had used the plants for decades.

Both the Cherokee and Ojibwa used the leaves or fruits for flavoring strong-tasting meats (Yanovsky 1936, King 1984). Opossums (*Didelphis virginiana*) and ground hogs (*Marmota monax*) were gamey meats that the Cherokees (Hamel and Chiltoskey 1975) noted were improved with this spice. That application is surely responsible for the common name "spice-bush." Physician W.P.C. Barton wrote that, during the Revolutionary War, the fruits were considered a substitute for allspice (*Pimenta dioica*, Myrtaceae) (Hedrick 1919). Ohio pioneers used young twigs and leaves as a substitute for tea and spice (Hedrick 1919).

These shrubs also had medical uses as noted by Commelin and Linnaeus. The Cherokee, Ojibwa, and Mohegan used leaves to make tea that was sometimes a simple beverage, but more often medicine. The Cherokee took the drink for blood and female problems, colds, croup, measles, and tuberculosis (Hamel and Chiltoskey 1975). The Iroquois also drank an infusion to relieve colds, fever, and measles, and the Mohegan took it to expel worms (Moerman 1998). We know only that the Ojibwa used it medicinally (Moerman 1998). The Rappahannock used an infusion for menstrual problems (Vogel 1970). In addition, the Cherokee used *Lindera* as a diaphoretic, and the Creeks used it as an emetic and in steam baths for aches (Swanton 1928a, Taylor 1940, Hamel and Chiltoskey 1975).

Settlers learned to use the bark as a febrifuge, a carminative to relieve flatulence and colic, and to treat rheumatism (Vines 1977, Foster and Duke 1990, Mabberley 1997). More recently, the plants have been recommended as an abortifacient, analgesic, candidacide, carminative, depurative, diaphoretic, emetic, emmenagogue, panacea, stimulant, tonic, and vermifuge (Duke et al. 2002). Most of those applications reflect uses by indigenous tribes.

Plants are known to contain laurotetanine, a potentially cytotoxic alkaloid. Moreover, extracts from the stem bark inhibited *Candida albicans* better than any of the other 53 plants tested (Duke et al. 2002). Yuchi still use it in a purification bath in March before entering the football game grounds (Jackson 2003).

Liquidambar

(A hybrid name composed of Latin *liquidus*, fluid, and Arabic *ambar*, amber, an allusion to the fragrant terebinthine juice or gum that the tree exudes)

Liquidambar styraciflua (an old generic name, from Latin *styrax*, gum, and *fluxus*, flowing)
Ambrabaum (amber tree, German)
American storax; *estoraque* (from *styrax*, the plants giving aromatic resins in Asia; Oaxaca, Chiapas, Guatemala)
a'yuxudi' (Biloxi; plus, *ayuxu' anaki'*, the fruit; *ayuxu' sinto̲n̲ni'*, the resin)

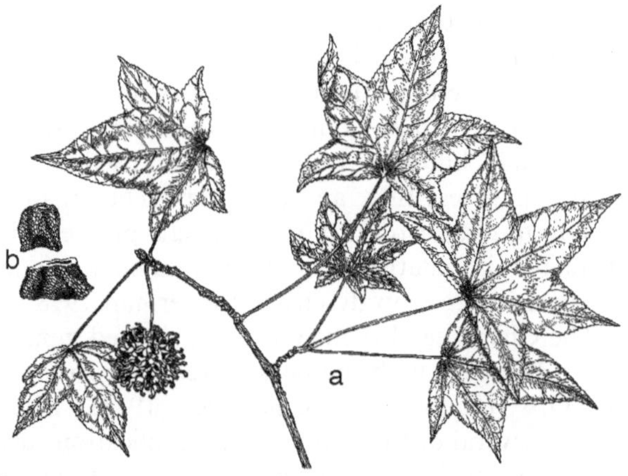

Liquidambar styraciflua. a. Branch with fruit. b. Seeds. *Drawn by Vivian Frazier*. From Correll and Correll 1972.

bálsamo copalme (copalme balsam; *copalme* based on *copalli*, resin, Náhuatl, Veracruz); *copalillo* (little copal, San Luis Potosí)
bilsted (presumably English; listed by Asa Gray 1875, USA)
buluchka'an (Mexico)
[alligator, American sweet, California red, red, star-leaf, white] gum [-wood]
hacaló:pî (Creek; compare with *helocoppi* by Simmons [1822] 1973, and *helocoppe* by Williams [1837] 1962)
helúkfvmécv (*helok*, black gum, Muskogee); *helokvpe* [*helúk wvpe*] (*helok*, black gum, *vpe*, tree, Muskogee); *helúkwv* [*helokwv, helukwa*] (Creek, Muskogee; compare with *Nyssa* and *Eugenia*); *hika* (Choctaw; the resin is called *hika nia* [*hekania*])
ien-gua-o (Oaxaca)
ikob [*icob*] (Huastec, San Luis Potosí); *itsché* (*its*, resin, *che'*, tree, Maya, Mexico); *ocop* [*occob, ocóm*] (Quecchí, Guatemala)
ingamo (maybe cognate with *ikob*, Oaxaca)
ko'ma [*ko'ma'liso*] (Totonac, Puebla)
kum hasana're (gum resin, Catawba)
lesquin (Central America)
liquid amber [liquid-amber] (in use in English by 1589; mentioned by Bartram [1791] 1958); *liquidámbar* [*diquidambo, qiramba*] (Oaxaca, Guatemala, Honduras, Nicaragua); *kirambo* (perhaps a corruption of *liquidambar*, Huastec, San Luis Potosí); oyl of amber (from 1657, England)
ma'lah² [*ma-la*] (Chinantec, Oaxaca)
maripenda (Tarascan, Michoacán)
nabá (Chol, Chiapas)
nijte-pijto [*nite-biito, nite biito*] (*pi, biito*, fruit, Zapotec, Oaxaca)
ocótzotl [*ocóxotl, ocozotl, ocozo, ocozote, ococote*] (*ocotl*, pine, *tzotl*, resin, Náhuatl, Veracruz, Oaxaca); *ocotzocuahuitl* (tree that gives pine resin, *ocotl*, pine, *tzotl*, resin, *cuahuitl*, tree, Náhuatl)
opossum-tree ("opossum" from *âpäsûm*, white beast, Virginia Algonquian)
pekiw (Strachey wrote this *pkekewh* and *pickuts*, compared it with "balsam," and said it came from a tree called "the Virginia maple." Those two clues suggest placement here. Siebert 1975 simply put the word under "gum," the meaning in all the related Algonquian tongues. Cognates include Cree *pikiw*, Fox *pekiwa*, Menomini *pekew*, Ojibwa *pikiw*, and Penobscot *peko*.)
somerio (San Luis Potosí)
suchete (Hidalgo); *toshcui* (Zoque, Chiapas); *tzoté* (*tzotl*, resin, Náhuatl, Guatemala; also Tzotzil, Chiapas)
sul'to'nko (Totonac, Puebla)
sweet gum [sweetgum] (dating back to at least the 1580s)
tìhtàmáʔa (Paya, Honduras)
trementina de pino (pine terebinth or resin, Mexico, Central America)
tsila'lu' (Cherokee; a former settlement on a small branch of Brasstown creek of the Hiwassee River, just within the line of Towns County, Georgia, was called *Tsila'lu'hi*, "sweet-gum place"); apparently cognate with *kafolá* [*kaholá*] (Koasati)
umwé (Jicaque, Honduras)
xochiocotzotl [*xochiocotzol*] (*xochitl*, flower, *ocotl*, pine, *tzotl*, resin, name for the gum, Náhuatl); *xochiocotzo quahuitl* [*xochiocotzocuahitl, xochicatscuáhuitl, xochiquahuitl*] (*xochitl*, flower, *ocotl*, pine, *tzotl*, resin, *cuahuitl*, tree, Náhuatl); *xochiatl quirámbaro* (*xochitl*, flower; the second word may be a corruption of *liquidámbar*, San Luis Potosí, Veracruz)
yaga bizigui [*bicigui*] (*yaga*, tree, *pi*, fruit, Zapotec, Oaxaca); *yaga huille* (*yaga*, tree, Zapotec, Oaxaca); *yaga pito* [*yaga bito, yaga pito, yaga vio, yaga vido, pijto, vito*] (*yaga*, tree, *pi*, fruit, Zapotec, Oaxaca)
yassaako [*yassàako*] (related to *yaska*, to chew, Alabama)

"Storax" came into English by A.D. 1382 from "*torax*" and in 1388 as "*storax*," a fragrant balsam from the bark of an Asian tree now known to be *Liquidambar orientalis*. Sometimes the resin is called the "Levant storax" because it came into Europe through the Holy Land. The gum is also the "balm of Gilead" of the Bible (Mabberley 1997).

There are two versions of the origin of the word "storax," although the argument that it is from Hebrew *tzori* is surely correct. Zohary (1982) noted that the name *tzori* appears six times in the Bible, three of those associated with Gilead. Its Hebrew synonym *kataf* or *nataf* also occurs. He agreed with an earlier suggestion that Greek *styrax* was derived from Hebrew *tzori*.

The other version is that "storax" came from the Greek and Latin *styrax*, a word now used for plants in the Styracaceae. Quattrocchi (1999) said that the Greek *styrax* or *styrakos* is a corruption of Arabic or Semitic *assthirak* for *Styrax officinalis*. Zohary (1982) did not accept that interpretation because the Hebrew name for *Styrax* in the Holy Land is *livneh*. Moreover, that name and the Arabic *libna* and *abhar* (white, bright) are cognates. More convincing, *S. officinalis* yields no gum. It is *S. benzoin* from Asia that produces the aromatic gum, and it was introduced later. Europeans may not have known storax plants by the time the Americas were discovered, but they certainly recognized that gum from the Holy Land.

One of the first references to sweet gum in the United States was in the 1580s by Thomas Harriot ([1590] 1972) although it is not clear that he was talking about *Liquidambar* alone. He wrote: "Sweete Gummes of diuers kindes and many other Apothecary drugges of which wee will make speciall mention, when wee shall receiue it from such men of skill in that kynd." Only shortly afterward, in 1589, the name *liquidambar* was used by John Florio in his book *A worlde of wordes*, which was a dictionary of Italian and English (OED 1971). In 1616, another author wrote, "*Liquid Amber*. A sweete Rosin brought from the West Indies, comfortable to the braine."

By the time Culpeper (1653) wrote his *Complete Herbal*, he had completely confused the American with the Old World storax. He called storax *L. styraciflua*, but gave its range as the southern part of Europe, Greece, Turkey, and Syria. Regardless, both species yield gums so similar they may not be distinguished without chemical analysis, and maybe not then (Duke et al. 2002).

The fragrant gum from *L. styraciflua*, also called copalm balm, American copalm, and storax, is obtained from the inner bark after wounding or gashing a tree. This gum has a long history of use by the Aztecs and people farther north (Standley 1920–1926, Martínez 1969). Romans ([1775] 1961) tells us that the Choctaw mixed dried, crumbled leaves of *Liquidambar* with their tobacco for smoking.

Three Muskogean tribes are known to have used *Liquidambar* gum as medicines, as did their neighbors the Natchez and Cherokee. The Choctaw made a poultice to help cure cuts and bruises (Bushnell 1909). The Houma similarly put a dressing on skin problems (Speck 1941). The Koasati took a decoction of the bark for "Night Sickness" (Taylor 1940). Du Pratz in 1758 recorded that the Natchez used the sweet gum, probably like their neighbors the Houma (Swanton 1946). The Cherokee used the rosin or inner bark to make in a salve for wounds, sores, ulcers, and itch (Hamel and Chiltoskey 1975). They also used the salve to treat diarrhea, flux, and dysentery. They made an infusion to treat certain types of female problems and considered it sedative. The Rappahannock made a bark infusion to stop dysentery (Moerman 1998).

The Cherokee also chewed the hardened gum and made a beverage with the bark mixed with *Evonymus* and *Vitis* (Hamel and Chiltoskey 1975). Settlers adapted several of these indigenous applications using it as chewing gum, antiseptic, and to treat skin diseases and dysentery (Standley 1920–1926). Sweetgum was popular with Confederate doctors (Mabberley 1997, Diggs et al. 1999). Porcher (1863) wrote: "The inner bark contains an astringent, gummy substance. If it is boiled in milk, or a tea made with water, its astringency is so great that it will easily check diarrhoea, and associated with the use of other remedies, dysentery also. The leaf of the gum when green I have also ascertained to be powerfully astringent, and to contain as large a proportion of tannin as that of any other tree. I believe that the Gum leaf and the leaf of the Myrtle and Blackberry can be used wherever an astringent is required; cold water takes it up. They can, I think, be also used for tanning leather, when green, in place of oak bark."

Apparently, the American species contains many of the same compounds as *L. occidentalis*. That species has cinnamic acid, cinnamein, styrol, vanillin, and resins (Hocking 1997). The oleoresins are used as the chewing gum. *Liquidambar formosana* has similar chemicals (Hsu 1986).

The first record of this species in Mexico was made by Bernal Díaz del Castillo between 1517 and 1521 (Díaz 1956). In describing the meals of the Mexican emperor Moctezuma, he says: "After he had dined, they presented to him three little canes highly ornamented, containing liquid amber, mixed with an herb they call tobacco, and when he had sufficiently viewed and heard the singers, dancers, and buffoons, he took a little of the smoke of one of these canes." The balsam and gum were used to flavor tobacco and as incense in houses and temples. Few people realize it today, but smokers burn a similar mixture; modern tobaccos have *Liquidambar* in them for added flavor (Meyer 1997).

Wood from *L. styraciflua* is known as "stain-walnut" in the trade. This wood has been used for flooring, furniture, veneers, woodenware, construction, boxes, cross ties, barrels, sewing machines, cabinets, molding, vehicle parts, conveyors, musical instruments, and tobacco boxes (Vines 1977). Foster and Duke (1990) also noted that sweetgum is an ingredient in "compound tincture of benzoin" still available in pharmacies.

Liriodendron

(From Greek *lirio,* lily or tulip, and *dendron,* tree, alluding to similarity with tulips, *Tulipa*)

Liriodendron tulipifera. From Sargent 1905.

Liriodendron tulipifera (bearing tulips; an old generic name for the plants)
basswood (comparing the bark with that of *Tilia*; see quote by Kalm below)
canary yellow-wood
canoe-tree (used by Swedes in Pennsylvania when Peter Kalm visited in 1748)
cucumber tree (some thought the "fruits" resembled cucumbers, *Cucumis sativus*)
es-pe-ton-ga (Osage)
old wife's shirt tree; old woman's smock (a name used by the English in Pennsylvania by 1748 because of the similarity of the leaf shape to a shirt)
[blue, hickory, soft, white, yellow] poplar ("poplar" is a term applied to a large variety of trees; see *Populus*; in use for these plants by 1705 in Virginia and by Englishmen in Pennsylvania by 1748; Kalm explained that people thought there were two "kinds" of trees; some yielding white wood, and others yellow)
popple (variant of poplar); sap poplar (maybe referring to extracting sap from the trees to make sugar); tulip-poplar [tulip poplar, tulip bearing poplar] (the word "tulip," first recorded in English about 1554; is derived from *tuliband,* a Turkish pronunciation of the Persian *dulband,* turban; the name was applied because of the resemblance of the expanded flower to the headgear); tulip tree [tuliptree, tulip wood] (although the first appearance in print in English was in 1705, the Latin version had been in use since at least the 1690s)
saddle-leaf [saddleleaf, saddletree] (alluding to the shape of the leaves)
whitewood (Plukenet may have used the word first in 1696 when referring to a tree from the West Indies, but he mentioned the tulip tree in the same sentence)
yap sake̲'? (tree red, Catawba)

Linnaeus gave us *Liriodendron tulipifera.* In naming it, he used a collection sent to Europe by John Clayton in the 1730s, but that was not the only material he had studied. The oldest record of this species in Europe was published by Leonard Plukenet in 1696. He called the plants *Tulipifera caroliniana, foliis productioribus magis angulosis* (Tulip-bearer from Carolina, producing large, angular leaves). However, Plukenet also had the trees in his book listed under the name *Tupilifera virginiana, tripartito aceris folio* (Tulip-bearer from Virginia, with three-parted leaves resembling maples). He thought that there were actually two different kinds of plants and he gave them distinct names. Linnaeus had also studied living plants at the *Hortus Cliffortianus* and *Hortus Upsaliensis.*

Moreover, by the time he published the name of these plants in 1753, Linnaeus had discussed them with his former student Peter Kalm ([1753–1761] 1972). Kalm knew firsthand about the size of old plants in New England and surely conveyed that to Linnaeus.

Kalm ([1753–1761] 1972) wrote, "canoes are boats made from one piece of wood ... for that purpose a very thick trunk is hollowed out." He listed as preferred woods both cedars (*Chamaecyparis, Juniperus*), chestnut (*Castanea*), white oak (*Quercus*), and tulip tree. His comment that the "Indians" made their canoes from them is surely one of the oldest on record. Moreover, he was talking about tribes that were not mentioned by Moerman (1998), probably the Delaware. Yet, it cannot be restricted to them because the trees were probably used by all in the region. People as far south as the Cherokees are recorded as making canoes 30 to 40 feet long from the trees (Hamel and Chiltoskey 1975).

These trees were so useful to the people of the region that Kalm ([1753–1761] 1972) spent considerable space commenting on them. He wrote: "Its wood is here made use of for canoes, boards, planks, bowls, dishes, spoons, door posts, and all sorts of joiner's work. I have seen a barn of considerable size, whose walls and roof were made of a single tree of this kind, split into boards. Some joiners reckoned this better than oak, because the latter frequently is warped, which the other never does, but works very easy; others again valued it very little. It is certain, that it contracts so much in hot weather, as to occasion great cracks in the boards, and in wet weather it swells so as to be near bursting, and the people hardly know of a wood in these parts which varies so much. ... The bark (like *Russia* glass) is divisible into very thin leaves, which are very tough like bast. ... The leaves, when crushed and applied to the forehead, are said to be a remedy against the head-ach. When horses are plagued with worms, the bark is pounded, and given to them quite dry. Many people believe its roots to be as efficacious against the fever as the Jesuit's bark."

George H. Loskiel in 1794 wrote that the fruit and bark were a powerful Indian remedy against agues (Vogel 1970). Unfortunately, no tribes are mentioned. The earliest direct record of indigenous people using *Liriodendron* was published by Hunter ([1823] 1973). He wrote that the Osage at least, and probably the Kansas, used the root bark to treat malaria (intermittent fevers). Fruits ("seed balls") were given to children to destroy worms.

Later studies found that the Cherokee gave an infusion of the bark to expel pinworms, to treat boils, cholera, coughs, dysentery, dyspepsia, fever, hysteria, rheumatism, snakebite, and wounds (Hamel and Chiltoskey 1975). The Rappahannock made a poultice for headache and chewed the bark as stimulant (sex vigorant) (Moerman 1998). Vogel (1970) noted that the modern Catawba were using the roots as a vermifuge.

Physicians and other scholars who praised the value of remedies from *Liriodendron* included Johann D. Schopf after visiting North America in 1783–1784, Benjamin Barton in his 1810 and 1812 books, Jacob Bigelow in *American Medicinal Botany*, published between 1817 and 1820, and Rafinesque (1830). They considered the plants effective against rheumatism and arthritis, fevers, snakebites, and as a diaphoretic (Vogel 1970). Porcher (1863) wrote, "Tulip Bearing Poplar (*Liriodendron*) and the Willow bark supply a remedy for the fevers met with in camp. Cold infusion given." Later he recommended "employing the *Liriodendron tulipifera* as a substitute for quinine during the stage of intermission of all mild cases of intermittent fever." Hocking (1997) summed it up with the statement that the bark was chewed as a tea substitute and that "poplar bark" was used to treat hysteria.

Liriodendron bark was official in the U.S. Pharmacopoeia from 1820 to 1882, as a bitter tonic, antiperiodic, and diuretic (Vogel 1970). It is now known that the plants contain liriodendrin (diglucoside of lirioresinol), a digitaloid substance, and tannin (Hocking 1997).

Lithospermum

(From Greek *lithos*, stone, and *sperma*, seed, referring to the hard nutlets)

Lithospermum carolinense. From Britton and Brown 1898.

erba-perla (pearl herb, Italian); *herbe aux perle* (pearl herb, French by 1550); pearl plant (so named because of the hard white pearl-like seeds, Britten and Holland [1886] 1965); *peerlencruyt* (pearl herb, Dutch)

granos de amor (grains of love, Spanish by 1557)

grémil (from Old French *gromil*, dating from the 1200s; etymology uncertain, although at one time thought to be from Latin *granum milii*, millet grains, Quebec); gromwell [graymile, grummell] (in use by ca. 1310, derived from Old French *gromil*)

lithospermo (stone seed, Italian by 1551); *steinfrø* (stone seed, Norwegian); *Steinsame* (stone seed, German)

migliarino (resembling *miglio*, millet, Italian)

mijo de sol (sun millet; *mijo* from Latin *milium*, Spanish by 1557)

painting root ("The girls in the North of Europe paint their faces with the juice of the root upon days of festivity"; Britten and Holland [1886] 1965)

puccoon (from the Powhatan word, Virginia; see *pakkan* under *L. carolinense*)

Lithospermum carolinense (of Carolina)

hoary [yellow] puccoon

iłaní slá:na (*iłaní*, cane, *lá:na*, yellow, Koasati)

odji'biknamuûn' (*biknam* may be cognate with *pakkan*, Ojibwa) *pakkan* (Strachey wrote in [1612] 1953 the Powhatan as *poughcone*, "red paint or dye," *pokcoons*, "a red dye," and *pocones*, "small root that grows in the mountains, which being dried and beat in powder turns red"; Siebert 1975 found cognates in Penobscot *pákahkan*, Mahican *pkáhkan*, and Nanticoke *pakahk*, all meaning "blood")

Lithospermum incisum (sharply cleft leaves)
narrow-leaf gromwell [narrowleaf gromwell]

Common names offer a glimpse into the thoughts and views of other people and cultures. However, those names must be used with caution. I learned *Lithospermum* as "yellow puccoon" when I was young. Another plant used in herbal medicines I learned as "goldseal" (*Hydrastis canadensis*). My mother's family talked about and gathered "yellow puccoon" when she was young, and I could never find that much of it growing around the area where they lived. There was plenty of goldseal, but *Lithospermum* was indeed rare. Sometimes I wondered if they had overcollected it.

It was only much later that I discovered that both *Hydrastis* and *Lithospermum* are known as "yellow puccoon." These plants have entirely different chemicals and properties and a mistake in using one for the other—for anything except coloring—might prove fatal.

Linnaeus ([1753] 1957) created the genus *Lithospermum* with six species, only one of them from the Americas. He did not invent the name but adapted it from Gaspar Bauhin's *Pinax* of 1623 and earlier authors. The best known of the species was what he called *L. officinale*, which had a long history of medical use. Leonard Fuchs (1542) called this *Lithospermon* (Greek), *Lithospermum* (Latin), and German *Meerhirsz* (sea millet) or *Steinsamen* (stone seed) (Meyer et al. 1999). Pliny (A.D. 23–79), Dioscorides (fl. A.D. 40–80), and Paul of Aegina (A.D. 625–690) also discussed the herb.

The only species Linnaeus knew from North America in 1753 was *L. virginianum*, a species that does not reach Florida. It was not until J.F. Gmelin revised *Systema Natura* in 1791 that one of the *Lithospermum* in Florida was named. Gmelin called the herbs *Batschia caroliniensis*, and it was not until 1892 that Conway MacMillan (1876–1929) created *Lithospermum carolinense.*

The Ojibwa used the dried root of *L. carolinense* to make a red dye. In making the dye, the women mixed dried and pounded roots with water and ochre. For coloring porcupine quills, they boiled the roots in water, added the quills that had previously been soaked in hot water, and let the mixture stand. If the color was not what they wanted, they repeated the process (Densmore 1928). The same tribe used the roots to make a red face paint (Moerman 1998).

Although neither Vogel (1970) nor Moerman (1998) gives the same information, Hocking (1997) said that the Ojibwa also used this species as a vermifuge. Otherwise, the only record found of medical use was among the Lakota who treated chest wounds with powdered roots (Moerman 1998).

The other species, *L. incisum,* appears to have no records of use in the east, but many in the western parts of North America. Records exist for use by the Blackfoot, Comanche, Hopi, Navajo, Okanagon, Shoshoni, Sioux, Thompson, and Zuni (King 1984, Moerman 1998). In Veracruz, the Zoque-Popoluca use it to treat snakebite (Vásquez and Jácome 1997).

The Shoshoni used the roots to make a blue dye. The Thompson people dipped the roots in hot grease and used them as red paint on their bodies and on tanned leather.

The Blackfoot used *L. incisum* roots to make a tea. They and the Shoshoni, Okanagon, and Thompson ate them boiled or roasted. The Blackfoot also burned the dried tops as incense. The Cheyenne put the dried plants on paralyzed body parts. They used an infusion or chewed the plant and put it on the patient to relieve irrational behavior and as a stimulant. The Hopi used *L. incisum* to stop hemorrhages and to build up the blood. The Navajo made a cold remedy by chewing the plant, took it as an oral contraceptive, to treat coughs, to treat sores of the navel, and as an eyewash. They considered it a panacea. The Sioux made a medicine of it to stop bleeding lungs. The Zuni considered it a remedy for swelling, skin abrasions, infections, stomachache, kidney problems, and sore throat.

Moerman (1998) found several other species in use including *L. canescens, L. multiflorum, L. ruderale*, and an unidentified species among the Pima and Keres. In addition, some of the names of other plants in the eastern states are instructive. Both *L. croceum* and *L. canescens* are "puccoon." *Lithospermum canescens*, one of the most famous eastern species, is sometimes called "Indian-paint," and the Omaha-Ponca dubbed it *bazu-hi* (*hi*, plant) (Gilmore 1919, Fernald 1950). Presumably, most of the species were used by various tribes to make dyes of one kind or another.

Lobelia

(Dedicated to Matthias de l'Obel, 1538–1616, a Flemish herbalist and physician to William Prince of Orange, who attended Lord Edward Zouche in his embassy to the court of Denmark and was physician to King James I of England)

Lobelia cardinalis. From Institute of Food and Agricultural Sciences.

Lobelia cardinalis (cardinal, from the color and shape of the corolla, like a miter, a pre-Linnaean generic name for *Lobelia*)
cardinal-flower (from French *cardinale rouge*, cardinal red, alluding to the red color of the flowers and cardinal's robes, from 1698 when James Petiver published the name)
héce-óme (*héce,* tobacco, *hóme*, bitter, Muskogee)
hierba de piojo (louse herb, Veracruz)
hog physic (fide Meyer and Meyer 1986)
inote'wi (said to mean "Indian tobacco," but maybe not literally, Meskwaki and Potawatomi)
lobélie éscarlate (scarlet lobelia, French); *Lobelie scharlachrote* (scarlet red lobelia, German)
red betty; red cardinal; red lobelia
totkv heleswv (*totkv*, fire, *heleswv*, medicine, archaic Muskogee)
totkv mocasv (*totkv*, fire, *mocasv*, new, Muskogee)
tsaliyu'sti (*tsálú*, tobacco, *iyu'slí*, like, resembling, Cherokee)

The name "lobelia" came into English in 1738 when Philip Miller published "Shrubby lobelia with a purslane leaf" in his *Gardeners Dictionary.* Miller simply translated the Latin phrase name *Lobelia frutescens, portulacae folio* used by Charles Plumier in 1703. Europeans knew *Lobelia* when they arrived in the New World. There are three species now recognized in *Flora Europaea* (2003), but only *L. urens* is widespread in the Mediterranean.

Not much notice was taken of *Lobelia* until the 1700s when Europeans in America discovered that the indigenous people considered some of the species cures for venereal diseases. Supposedly, the first account of use for this purpose was among the Iroquois, and Sir William Johnson, an official agent of the British government among the northern tribes, is said to have purchased the remedy from them in the early 1700s (Millspaugh 1892). Given the comment that the Iroquois "held this plant a secret in the cure of syphilis" and got Johnson to pay them for it, one has to wonder who was tricking whom.

Cadwallader Colden (1688–1776), a physician, botanist, and statesman in New York, was another person who worked among the Iroquois. He promoted among settlers use of the species Linnaeus called *L. syphilitica* in 1753. In 1758, Colden claimed that lobelia had cured soldiers of syphilis in less than a week (Vogel 1970). Not long after its introduction into Europe, people realized that *L. syphilitica* did not work and abandoned it (Millspaugh 1892).

Lobelia cardinalis is considered the sister species to *L. syphilitica*, and hybridizes with it (Diggs et al. 1999). Both of these were thought by indigenous people to be weaker than *L. inflata*, but employed for similar purposes. *Lobelia inflata* was so popular among New England people that it received the English common name "Indian tobacco." Cherokee applied their name, *tsaliyu'sti*, indiscriminately to more than one *Lobelia* and even to *Verbascum*, an introduced European weed (Mooney 1885–86). That name is appropriate for all, because *Lobelia* contains lobeline, which acts much like nicotine in poisoning the central nervous system (Diggs et al. 1999).

The first record of *Lobelia* probably was made by Francisco Hernández (1651) who called them *Rapuntium maximum, coccineo spicato flore* (the largest little turnip with scarlet flowers in a spike). Whether he was describing *L. cardinalis* or *L. laxiflora* is problematical. The Aztecs were well acquainted with the latter species, which they called *chiplanxochitl* (*chilpan*, medicinal plant, *xochitl*, flower, Náhuatl). They had long used that herb as an emetic, expectorant, and antispasmodic (Martínez 1969). Alternatively, it could just as easily have been *L. cardinalis,* as it goes south to Panama. People in Veracruz still use *L. cardinalis* to rid themselves of lice (Vásquez and Jácome 1997).

The Cherokee used a poultice of crushed *L. cardinalis* leaves to relieve headache and rheumatism. They also used an infusion to expel intestinal worms, relieve colds and croup, and cure sores, fever, stomach problems, nosebleed, and syphilis (Hamel and Chiltoskey 1975). The Delaware used a root infusion to treat typhoid (Moerman 1998). The Iroquois added *Lobelia* to many of their medicines to strengthen them. A poultice or infusion treated pain, sores, fevers, stomach problems, cramps, sore breasts, and menstruation problems (Moerman 1998).

The Iroquois, Meskwaki, and Pawnee employed cardinal-flower as a medicine to end quarrels between husbands and wives (Gilmore 1919, Moerman 1998). The Meskwaki also considered it a ceremonial "tobacco" that was not smoked but offered to the deities to ward off storms and spirits of the dead (Moerman 1998). Similarly, the Iroquois considered it a mystical plant that could ward off spells, and generally divert bad things (Moerman 1998).

Gilmore (1919) was convinced by the disjunct isolated populations of *L. cardinalis* that the Missouri River *wazathe* (healers, Omaha) had introduced them from farther east. He makes a convincing argument for the movement by indigenous healers of cardinal-flower, ginseng (*Panax quinquefolia*), sweet flag (*Acorus calamus*), columbine (*Aquilegia canadensis*), and some others.

The entire *Lobelia* plant is poisonous to humans and livestock, and can cause death. One view of toxicity was given in 1855, when Thomas C. Haliburton wrote in his *Nature and Human Nature,* "He foamed at the mouth like a hoss that has eat lobelia in his hay" (Haliburton in OED 1971). Porcher (1863) wrote, "This plant is used by the Indians as an anthelmintic—some say quite as efficient as the pink-root [*Spigelia marilandica;* see *Spigelia*]. Mérat says it is employed as a poison by the negroes at the Cape of Good Hope." Unless *L. cardinalis* had been introduced to southern Africa, it could not have been used as Porcher says. The species is native to the Americas.

Lobelia has a variety of pyridine alkaloids, with more than 14 having been isolated from *L. inflata* (Turner and Szczawinski 1991). The most abundant alkaloids are lobelamine and lobeline. Overdoses will cause vomiting, paralysis, depressed body temperature, coma, and death (Lewis and Elvin-Lewis 1977). Plants also contain the pungent volatile oil lobelanin.

Meyer and Meyer (1986) say the flowers were used by rustics as a red dye. Unfortunately, they did not say who the "rustics" were, or how to make the dye.

Lonicera

(Named to commemorate German herbalist Adam Lonitzer, 1528–1586; his name was rendered *Lonicerus* in Latin)

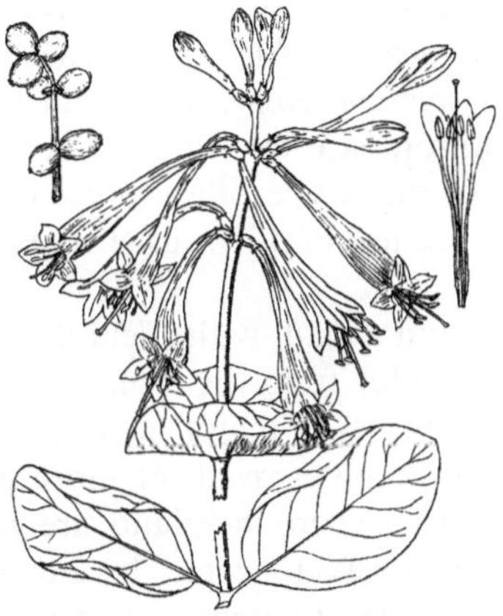

Lonicera sempervirens. From Britton and Brown 1898.

caprifoglio (goat leaf, Italian); *caprifolium* (goat leaf, Latin); *chèvrefeuille* (goat leaf, Quebec); *Geissblatt* (goat leaf, German)

féithlean (*feith,* sinew, *lean,* follow, Gaelic, for *L. periclymenum*)

honeysuckle (the name alludes to being able to suck nectar from the flowers; from Middle English *hunisuccle,* a name in use by ca. A.D. 1265, when applied to clover flowers; later applied to *Lonicera* by 1548)

ladh-shlat (twig that surrounds, Gaelic)

leddved (joint wood, Norwegian)

lus na meala (honey plant, Gaelic)

madre selva [*madreselva*] (mother of the forest, in use by 1557, Spanish); *madresilva* (mother of the forest, Portuguese); *matersylva* (mother of the forest, Latin)

mammekenscruyt (mother's herb, in use by 1549, Dutch)

okshulba (Choctaw)

owehkim ako'pukwa (Plains Cree)

pakaalakomahli (*paxkaali,* flower, *akomahli,* cantelope [*Cucumis melo*], a plant that was brought to the New World; surely a transfer from a native species, Alabama)

pa<u>n</u>hi<u>n</u> a'xiyehi' hupi'hi (vine with fragrant blossoms, Biloxi; this may be a name for introduced *L. japonica*)
papahitikwa (Plains Cree)
periclymenon (literally "famous twiner," Greek)
vincibosco (conquers forests, in use by 1551, Italian); *vincibosse* (conquers forests, in use by 1551, French)
volucrum maius (flying May, Latin)

Lonicera sempervirens (evergreen)
aatyammichi (*aati*, Indian, *yammi-chi*, to get someone intoxicated, Alabama; perhaps confused with *Gelsemium sempervirens*, which see)
coral [crimson, scarlet, trumpet] honeysuckle (referring to the red corollas; in use by 1753)
woodbine (in use by 1753; used by Porcher 1863)

The shift from a plant with a leaf that smells like a she-goat or the odor from the armpit (*caprifolium*) to honeysuckle reflects a major change of view. Or perhaps it was simply a result of more people living in cities and towns where the plants are grown for the flowers and their sweet nectar. In either case, Europeans had a variety of names and opinions regarding honeysuckles when they arrived in the New World.

Leonard Fuchs discussed *Lonicera periclymenum* in 1542 and used its pharmaceutical names *Caprifolium* and *Matersylva*. These plants were famous for having been discussed by Dioscorides (fl. A.D. 40–80), Galen (A.D. ca. 129–ca. 199), and Pliny (A.D. 23–79). At the time, that species was believed to heal ulcers, impetigo, and other skin problems, shrink the spleen, help breathing difficulties, speed childbirth, and clear facial spots (Meyer et al. 1999).

Linnaeus ([1753] 1957) put 15 species in *Lonicera*, a name he created in *Hortus Cliffortianus* in 1738. Of those, 5 were from the Americas. Before him the species he named *L. sempervirens* had been discussed by Paul Hermann in 1687 and August Rivinus in 1690, so the plants had been growing in Europe for more than a half century. That they were there is no wonder because of their arresting red flowers. These twining shrubs still command the attention of hummingbirds in the Americas and the attention of people with gardens throughout the temperate parts of the world.

Vogel (1970) found no records of *L. sempervirens* being used by indigenous people. Moerman (1998) did not find any either, but he discovered that 13 other species were used—including introduced Japanese honeysuckle (*L. japonica*). In the eastern parts of North America *L. canadensis* was used by the Iroquois, Menomini, Montagnais, and Potawatomi. *Lonicera dioica* was medicine for the Algonquin, Cree, Iroquois, Meskwaki, and Ojibwa. The Iroquois made remedies from *L. oblongifolia*.

During the Civil War, Porcher (1863) wrote of *L. sempervirens:* "The plant is not much used in medicine. The syrup made of the leaves is given in asthma, and in angina tonsillaris. The leaves and bark of the *L. caprifolium* of Linn. are styptic and acrid; the flowers diuretic; the latter in decoction calm the pain of colic (*coliques ou tranchées*) following childbirth." None of the Eclectics mentioned the genus (Culbreth 1910, Felter 1922), and Bown (1995) and Duke et al. (2002) discuss only Old World species. The Alabama name *aatyammichi* makes it seem that we have all missed something.

Ludwigia

(Named for German physician Christian Gottlieb Ludwig, 1709–1773, professor of medicine at Leipzig)

Ludwigia palustris. From Britton and Brown 1897.

Ludwigia palustris (of marshes)
akasî (Mikasuki; maybe from *ak*, water, and *assi*, leaf)
seedbox

Ludwigia virgata (like a twig)
hoktvlkolowv enlocowv vhakuce (Snow and Stans 2001 translate this as "old lady's paint pot," Creek; from *hoktvl*, old woman, *kolowv*, glue, *em*, her, *locowv*, clay jug, *vhake*, resembling, *uce*, little; see discussion below)
kolátbahcayikcî [*kolatbatcaiitci, kolotpahche ayekche*] (*kolotpahche*, wide-open [sore], *ayekche*, medicine, Mikasuki)
primrose willow

Decoctions of *Ludwigia virgata* root are used as a bath for "Snake Sickness" (itchy skin) (Sturtevant 1955). *Ludwigia palustris* also was part of the "Snake medicine" (Sturtevant 1955).

Although Snow and Stans (2001) list *L. virgata*, no use is given. The Creek name provides a great deal of information about their beliefs that is not obvious from the literal translation. The name *hoktvlkolowv* seems straightforward. However, it contains *Kolowa*, which was a feared demon, also called *vkolowetv*. The *vkolowetv* was a "being" that could pounce and hold with its feet and claws as does a cat. As Martin and Mauldin (2000) phrased it, the *Kolowa* was "a character much feared if it was heard halloing while a dance was being held—it could pounce like a panther." Clearly, there is much more in the name than the simple translations suggest.

Lupinus

(*Lupinus*, of the wolf, from Latin *lupus*, wolf, because of the erroneous belief that it consumed soil fertility; akin to Akkadian *luppu*, bean, Armenian *gail*, wolf, Greek *lykos*, wolf, Hebrew *lifnaj*, interior, Sanskrit, *vrkas*, wolf)

Lupinus perennis. From Britton and Brown 1897.

altramuz [*altramuzes*] (in use by 1557; maybe from Latin *altra*, other, and *mus*, playing card, Spanish)
fyg bene (fig bean, by Turner in 1568); *Feigbon* (fig bean, by Fuchs 1542, related to Old High German *figbôna*, Middle High German *vîge*, comparing the shape of the seeds to those of figs, German); *vijchboonen* (fig bean, Dutch by 1549)
garbancillo (little garbanzo, *Cicer arietinum*, New Mexico)
garden lupine (John Gerarde in 1597)
lupin (French by 1550); *lupin* (Norwegian); *Lupine* (German); *lupine* (in English by ca. 1420); *lupino* (Italian); *lupini bianchi* (white lupine, Italian by 1551)
patita de la paloma (dove's foot, New Mexico)
Quaker-bonnets
thermis (Greek, Fuchs 1542); *tremoço* (from Greek *thermos*, akin to Arabic *turmus*, the lupine, Portuguese)

Lupinus perennis (living longer than 2 years)
[blue, sundial, wild] lupine
monkey-faces (Ohio)
old-maid's-bonnets (Long Island, Massachusetts)
sandflat-sweet-pea (Texas)
wild pea (Massachusetts, Vermont)

We humans seem to have had an uneasy relationship with wolves (*Canis lupus*) from our beginnings. When our species lived closer to nature, people respected and even revered the wolf. The fact that we domesticated wolves to create dogs gives some idea of the early relationships between the ape and canid. As cultures began living in agricultural settlements, they took less note of the natural world. In the Old World, people even began to fear the wolf although no parallel to that existed in the New World.

That same duality extended to the lupines. From an ancient misunderstanding of how plants interacted with soils, the plant's name was applied because of the belief that it robbed the soil of nutrients. Fuchs (1542) indicated instead that the plants thrived in areas where the soils were poor for growing other plants, and surely others also knew.

Some ambivalence about these plants is reflected in old Greek tales. A story told about Zeno (ca. 333–262 B.C.) of Citium of Crete, the famous founder of the Stoic school, illustrates the view. Zeno had a surly but humorous disposition; after drinking enough wine, he compared himself to the lupine—which loses its bitterness when soaked in water (Baumann 1993).

Along the same lines, lupines were part of the special food for visitors to the Oracle of the Dead on the bank of the river Acheron, Epiros. These people had to be psychologically and physically prepared to communicate with the Underworld. They were fed alkaloid-containing lupine seeds, which prepared their perceptive states to make them believe they were talking with the dead. The species was also mentioned by Galen (A.D. 129–?200) and Pliny (A.D. 23–79) as medicinal.

It is not surprising that Dioscorides, fl. A.D. 40–80, recorded two kinds of lupines, one sweet (probably alkaloid free), and the other bitter. Both of these were probably *L. albus*, sometimes known as the Egyptian lupine or white lupine. This species was food for Egyptians, Greeks, and Romans who boiled or roasted the seeds to remove alkaloids and make them edible (Mabberley 1997). Still, Athenaeus of Naucratis (A.D. 170–230), author of *Deipnosophistai* (The Learned Banquet), said the lupine was a food in times of scarcity.

However, Athenaeus, a wealthy scribe, may have exaggerated the unimportance of lupines. Pliny, for example, said that no food is more wholesome or easier to digest than the lupine, and that they produced a cheerful countenance and a fresh complexion (Coffey 1993).

Halfway around the world, another group of people domesticated a second species of lupine. The Incas of the Andes brought into cultivation *L. mutabilis.* They call them *tarwi* [*harhui*]; their neighbors the Aimara say *tauri,* while in English they are pearl lupines. The Incas even developed a cultivar lacking toxic quinolizidine alkaloids (Mabberley 1997). These seeds are soaked in water and then roasted, and still are sold by the bagful by vendors in many Andean towns. They certainly lived up to Pliny's prediction and improved my disposition when I ate them.

Moerman (1998) found records of 23 or more species being used in North America outside of Mexico. Sometimes people ate the foliage, and later fed it to their horses, although both were done sparingly because of the poisons they contain. The Kwakiutl sometimes ate them raw, but only before going to bed because they became dizzy. Only the Costanoan seem to have eaten the seeds in *pinole*, with all others eating vegetative parts.

There are records for use for only one of the species in Florida. *Lupinus perennis* was made into a cold infusion and drunk by the Cherokees to stop hemorrhage and nausea (Hamel and Chiltoskey 1975). The Menomini fed it to horses to make them more spirited and lively (Moerman 1998), which is reminiscent of Pliny's comments about people's attitudes improving.

Lycium

(From Greek *lykion,* a name used by Dioscorides, fl. A.D. 40–80, and Pliny, A.D. 23–79, for a thorny shrub, dyer's buckthorn, perhaps a species of *Rhamnus*; said to have been used because the shrub was growing in *Lycia,* a southwest region of Asia Minor)

agutoli (Italian)
Bocksodorn (buck thorn, German); *bukketorn* (buck thorn, Norwegian); box-thorn [boxthorn] (probably derived from buck thorn; see *Pediomelum*)
cambronera (Spanish)
chico bush (New Mexico)
cilindrillo (little cylinder, Texas, Mexico)
desert-thorn; rabbit thorn (New Mexico)
Duke of Argyll's tea-plant (John Douglas Southerland Campbell, 1845–1914, was the ninth duke of Argyll, and the governor general of Canada between 1878 and 1883)
espinheiro de casca branca (white-barked thorn, Portuguese)
lyciet (from *Lycium,* French)
spina-cristi (Christ's thorn, Italian)
spina-santa (sacred thorn, Italian)
squaw-berry
tea tree (Europe)
tomatillo (little tomato, Texas, New Mexico, Mexico)
wolf-berry (first applied to *Symphoricarpos occidentalis* by 1834; apparently exended to *Lycium* later)

Lycium carolinianum (of Carolina)
Carolina wolf-berry (Texas)
Christmas berry (Florida)
ch'ilib tux (*ch'ilib,* shrub, *tux,* turkey; the name is applied because the ripe red fruits resemble the red spotting on the throat of the turkey, Maya, Yucatán)

It is not safe to assume that anything a bird or other mammal can eat is safe for humans (Hardin and Arena 1974). However, I wondered for years as I sat watching the mockingbirds (*Mimus polyglottos*) guarding fruiting bushes and gobbling down their fruits, if people could eat the berries of *L. carolinianum.* I now know the berries will not poison you, but they are not likely to become tomorrow's most popular fruit either. Fernando Chiang, the specialist on the North American members of the genus, told me, "Yes, the fruits are edible though (to *my* taste) not very palatable."

Lycium (Solanaceae) is a genus with 100 species that grows in the warm temperate regions of the world. Europe has three species and Australia one (Hitchcock 1932, Mabberley 1997). The duke of Argyll's tea-plant (*L. barbarum,* including *L. chinense* and *L. halmimifolium*) is naturalized in Great Britain and north to Norway (Mossberg and Stenberg 1994, Clement and Foster 1994, Mabberley 1997), although the species is native from southeastern Europe to China. Presumably, the shrub was imported as a substitute for Chinese tea (Standley 1920–1926). Young shoots of *L. europaeum* have been eaten as a vegetable in Spain and Italy.

Florida's single species is native from South Carolina to Florida and west to eastern Texas and Tamaulipas, Nuevo Leon, Oaxaca, San Luis Potosí, Veracruz, Yucatán, and Quintana Roo in eastern Mexico (Correll and Johnston 1970, Chiang-C. 1981). In an unusual distribution, the species reappears in Baja California, Sinaloa, Colima, Nayarit, and Michoacán (Standley 1920–1926, Wiggins 1980, Chiang-C. 1981). Other species begin in the Trans-Pecos region of Texas and extend through other western states and south through Mexico. Cuba has three, *L. acnistoides, L. carolinianum*, and *L. infaustum* (as *L. tweedianum*). The latter species, *L. infaustum,* also grows from the Bahamas south to Argentina (Correll and Correll 1982, Liogier and Martorell 1982, Bernardello 1986). In Hispaniola, *L. infaustum* is called *vidrio* (glass). Only *L. acnistoides* has a common name in Cuba, and it is called *belladona de la tierra* (wild belladonna) or *palo de gallina* (chicken tree). Cubans consider their endemic *L. acnistoides* poisonous and a substitute for the real belladonna (*Atropa belladonna*). Roig (1945) wrote that the leaves are more poisonous than the roots, which are more toxic than the fruits. Chickens are poisoned by the fruits.

Most species are listed as having edible berries, and Hardin and Arena (1974) say they are edible raw, cooked, or dried. They do not distinguish between species, and presumably included *L. carolinianum.* Fernald et al. (1958) do not mention the Florida species. Moerman (1998) lists five species, all of which have edible berries. Perhaps only *L. acnistoides* is the exception to the rule.

Other species are used medicinally. The Navajo use both *L. pallidum* and *L. torreyi* for toothache and other maladies. Both the Navajo and Zuni considered their *Lycium* sacred plants (Curtin 1947). According to Mayes and Lacy (1989), the Navajo call *L. pallidum* "god's food" (*haaschch'éé'dáá'*).

Lycopodiella

(Josef Ludwig Holub, 1930–, used the diminutive form of *Lycopodium*, from Greek *lykos*, wolf, *podion*, foot)

Lycopodiella cernua (nodding, drooping) (= *Lycopodium cernuum*)
adrangaman (Sranan, Suriname)
bejuco de amor (love vine, Dominican Republic)
bending club moss (Florida, English Antilles)
cabane de la vierge (the Virgin's house, Dominica)
cabello de sacerdote (priest's hair, Brazil)
club-moss (Florida)
enxofre vegetal (vegetable sulfur, Brazil); *soufre vegétal* (vegetable sulfur, French Antilles)
fern (Belize)
helecho (fern, Peru)
licopodio (lycopodium, Cuba, Peru)
lus-bhalgair (*lus*, herb, *bhalgair*, fox, Gaelic)
palma de São João (St. John's palm, Brazil)
samambaia (fern, from *ham a'bae*, that which twists in a spiral, Tupí, Brazil)
shapumba (Palikur, Peru)

Linnaeus knew these nonflowering plants as *Lycopodium cernuum* in 1753. Most of the world called them by that name until the 1990s. However, back in 1964, Holub created the genus *Lycopodiella.* There are about 40 species in that segregate genus, with 2 in Europe although most grow in the tropics (Wagner and Beitl 1993).

As several of the names suggest, there is widespread use of these herbs during various festivals. They are used in Christmas nativity displays (Duke and Vásquez 1994), but they are not confined to those occasions. The name *palma de São João* suggests that it was used as a substitute for palms in Christian festivities. No use by non-Christians is known.

Spores are put on cuts, used like talcum powder, as baby powder in diapers, and have been used to coat pills and lubricate condoms (Roig 1945, LeCointe 1947, Duke and Vásquez 1994). The plant has been used to treat bladder problems, diarrhea, and dysentery (Roig 1945, Duke and Vásquez 1994). Dominicans use the plant as a bitter tea to treat fever (Ayensu 1981). This is also a medicinal plant in Belize (Balick et al. 2000). Although it is not identified or named in the text, a drawing in Snow and Stans (2001 p. 68) is surely this plant.

Lycopus

(Based on Greek *lykos*, wolf, *pous*, resembling, because Linnaeus saw some similarity to a wolf's foot)

Lycopus virginicus (of Virginia)
aniwani'ski (talkers, Cherokee)
Virginia bugleweed [bugle-wort] ("bugle," in this sense, is from Latin *bugula*; the Latin *bugillo* was applied by Marcellus Empiricus about A.D. 400 for mints now called *Ajuga reptans*; akin to French *bugle*, Italian *bugola*, and Spanish

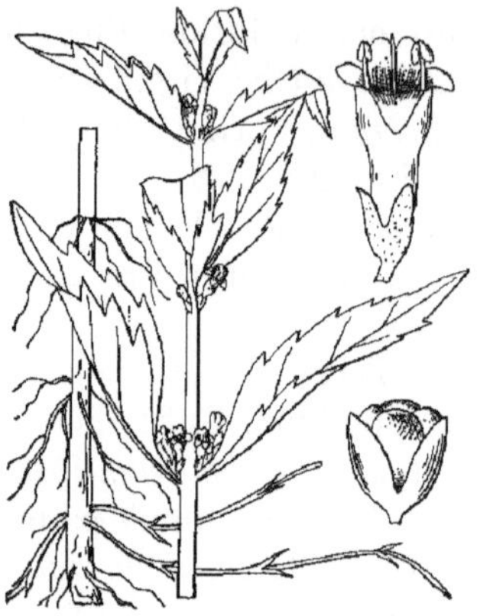

Lycopus virginicus. From Britton and Brown 1898.

bugula; in English by about A.D. 1265; later applied to *Lycopus* with modifiers)

Virginia water horehound ("horehound" was originally applied about A.D. 1000 to *Marrubium vulgare*. The name came from Old English *háre*, hoary, and *húne*, the name of a plant of uncertain origin, which became *hôrhowne* in Middle English; water horehound was in use by 1578 for the European *Lycopus europaeus*)

Linnaeus coined the genus *Lycopus* in 1738 for European mints, and continued its use for both *L. europaeus* and *L. virginicus* in 1753. What he saw that reminded him of a wolf is not apparent now. Today, the genus contains only four species. Two are native to Europe, and the others are native to America and Australia (Mabberley 1997).

The Cherokee took an infusion of this mint at the Green Corn Ceremony, and let their infants chew the root because it gave them "eloquence of speech" (Hamel and Chiltoskey 1975). It was also used as a snakebite remedy. The Iroquois considered it a poisonous plant (Moerman 1998). The similar *L. americanus* was used by the Meskwaki for stomach cramps (Smith 1928). Other species were used by the Blackfoot, Ojibwa, Okanagon, and Thompson tribes (Moerman 1998).

Lyonia

(Named for Scottish John Lyon, 1765–1814, a gardener in Philadelphia, Pennsylvania and Tennessee)

Lyonia fruticosa (shrubby)

aha:pó:cwântî (strong small stem tree, from *ah-*, tree, *a:p-*, stem, *-o:c-*, small, *want-i*, strong, Mikasuki)

có:skan (apparently comparing it to *coskv*, the post oak, *Quercus stellata*, Creek)

Lyonia fruticosa. *Drawn by P.N. Honychurch.*

staggerbush (cattle browsing on the plants are poisoned and stagger)

André Michaux called these shrubs *Andromeda ferruginea* var. *fruticosa* in 1803. They went under a number of other names following that, but George Safford Torrey (1891–1977) created *Lyonia fruticosa* in 1916. *Lyonia* now comprises 35 species in Asia, North America, and Mexico (Judd 1981).

The Seminoles used the wood for making pipe bowls (Sturtevant 1955). Now people harvest the gnarled stems and put plastic branches with plastic or silk leaves on them for indoor "planters."

Lysiloma: Wild Tamarind

(Greek, *lysis*, opening, *loma*, margin, the fruits)

Not long after I moved to Florida, I found a single wild tamarind (Florida, Bahamas) tree growing in a coastal hammock in the town of Ocean Ridge south of Boynton (Palm Beach County). According to the books I had available, the plants (*Lysiloma latisiliquum*) were native to the Florida Keys and southern Miami-Dade County. Because I was doing a study of that hammock for another purpose, we listed the species there with a note as a curious range extension (Austin and Weise 1972). It was only later that I learned, much to my embarrassment, that the tree had been brought from the Miami area and planted by the McGinty brothers. Those wealthy brothers, avid amateur malacologists, lived only a short distance

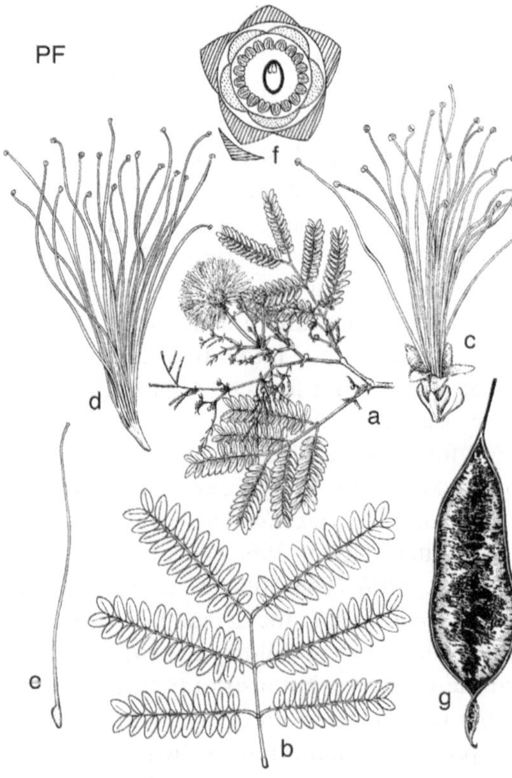

Lysiloma latisiliqum. a. Flowering branch. b. Leaf outline. c. Flower, with subtending bracts. d. Stamens. e. Pistil. f. Floral diagram. g. Fruit. *Drawn by Priscilla Fawcett.* From Correll and Correll 1982.

north of the hammock. When they collected tree snails (*Liguus*) farther south in the state, they occasionally released some of them in the tropical forest near their house. They thought that the snails needed another food source and knew that wild tamarind would serve admirably. So, later I quietly corrected my error (Kreiger and Austin 1975).

It has always struck me as curious why this tree came to be called wild tamarind. The true tamarind (*Tamarindus indica*) is native to the Old World, and belongs to either the subfamily Caesalpinoideae or the distinct family Caesalpinaceae, depending on the specialist consulted. *Lysiloma* belongs in the subfamily Mimosoideae (or family Mimosaceae). The gibbous flowers of the Caesalpinoideae differ from the radially symmetrical blossoms of the Mimosoideae. Their flowers are not the only difference. Fruits of *Tamarindus* are indehiscent and contain abundant sweet pulp around the seeds. The genus *Tamarindus* is Latinized from Arabic *tamar-hindi* (date of India), in reference to the wide ancient use of the fruit pulp by people in India and nearby countries. Fruits in *Lysiloma* are dehiscent and do not have pulp.

Since there are other species called "wild tamarind" beginning at earlier dates, as, for example, *Pithecellobium arboreum* in the 1830s and *P. dulce* in the 1860s, the name looks suspiciously like an extension of use from those plants. Perhaps slaves traded between the notorious importation and distribution centers of Jamaica and Trinidad brought with them a recollection of *P. arboreum* in those areas. Maybe they applied the name to the most similar trees they found elsewhere. Supporting that contention may be that both *Lysiloma* and *P. arboreum* also are called *tengüe* (Taino?, Cuba).

Finally, I discovered that the external traits that I had been using were not the similarities that led to several species being called "tamarind." The wood was the source of the names, as it is dark red, hard, heavy, and compact in all three genera. That same comparison of wood probably also led to the common name *bacona morada* (purple *bacona*, maybe Taino, Cuba). The tree called *bacona* is *Alchorea latifolia*, Euphorbiaceae. It is more certain that horseflesh (Bahamas, usually for *L. sabicu*), *sabecú* (Taino, Cuba), and *sabecú amarillo* (yellow *sabecú*, Cuba) refer to the wood. *Lysiloma* is also called *moruro de costa* (coastal *moruro*, Taino?, Cuba), and the similar name *moruro abey* (Cuba) is applied to *Peltophorum adnatum*, another legume with similar wood. The "model" for *abey* is either *Poeppigia procera* (Fabaceae) or *Ouratea littoralis* (Ochnaceae). It would not be a surprise to discover that the names *jigüe* (Taino?, Cuba), and *jigüe blanco* (white *jigüe,* Cuba) were comparisons of wood between *Lysiloma* and other species.

Three other names also seem to be indirect references to the wood. *Zapatero* (shoemaker) surely refers to using the wood as lasts or other shoe-making utensils. Perhaps *tabernan* [*tabernon, tavernon*] (innkeeper, Dominican Republic) and *taverneau* (variant of *tavernier*, innkeeper, Haiti) denote use of the wood in buildings. The wood is used in carpentry, for railroad ties, for making boats in the West Indies, and for bobbins, shuttles, and cabinetwork. Indeed, the wood has been exported to England for use in the textile industry (Morton 1981). Timber from the tree is considered good quality and durable in water.

The names *dormido* (sleeper, Cuba) and *soplillo* (delicate one, Cuba) have been given to *L. latisiliquum* (Barneby and Grimes 1996). Both refer to the leaves, with *dormido* a comment on the "sleeping" movements of the leaflets at night. That name is used for some other legumes with those movements. The *soplillo* must be a commentary on the fine leaflets. While those same authors attribute the common name "singing bean" to Florida, no others in the state have found that used in at least three decades.

Some of the other names are intriguingly obtuse. Why the trees are compared with candles in *candelón* (big candle, Hispaniola) or to snails with *caracolí* (snail, also for *Pithecellobium unguis-cati,* Hispaniola) and *caracolillo* (little snail, Dominican Republic) remains unclear. A romantic would suggest that the flowering trees resemble big candles. Maybe the curled fruits on the others led to the snail comparison. Or could it have been a reminder that the trees are favored hosts for tree snails (*Liguus*)? The Maya have three names for the species: they say *ts'aslam* [*tzalam, tzalán, salom*] (simple term, Yucatán, Belize), *tzukté* (*zuk,* horse's mane, Yucatán), or *xiaxek* (*ya'ax,* green, *ek,* star, a name typically given to *P. mangense,* Yucatán). Perhaps those Mayan names also allude to the wood.

There is a distressing confusion in the literature about the delimitation of *L. latisiliquum* and related species. The species was first introduced to Europeans by Charles Plumier in 1703 based on plants that he drew in Hispaniola. Then Linnaeus put the species in *Mimosa latisiliqua* in 1753 along with 38 others. Subsequently, it has been discovered that Linnaeus put multiple genera in his concept of *Mimosa.* Several of these species now have been segregated into *Acacia, Albizzia, Entada,* and *Pithecellobium.* When George Bentham, a long-time president of the Linnean Society of London, first studied these Caribbean trees he realized that they needed a distinct generic name. He created *Lysiloma* for them in 1844, and named those in the Bahamas *L. bahamensis.* Similar trees on Cuba he named *L. sabicu.* Later he moved Linnaeus's *Mimosa latisiliqua* to *L. latisiliquum.*

As recently as Liogier (1974), both *L. latisiliquum* and *L. bahamensis* were considered distinct. Subsequent study of the Bahama plants showed that these two were actually the same, and they were treated as one by Correll and Correll (1982). Indeed, in their treatment, they considered *L. latisiliquum* as growing only on mainland Mexico and Mesoamerica. The Corrells thought that *L. sabicu* was found only on Cuba, the Bahamas, and Hispaniola (Correll and Correll 1982). Later, Barneby and Grimes (1996) examined the problem and came to a different conclusion. This pair wrote that *L. latisiliquum* is native to peninsular Florida, the Florida Keys, and the Bahamas (but unknown from Andros), extends southeast from Grand Bahama to the Caicos Islands and the northwestern coast of Haiti and Cuba, and then reappears in Tabasco, Yucatán, and Quintana Roo, and south to Belize, Honduras, and Peten, Guatemala.

Barneby and Grimes (1996) also found that *L. sabicu* is native to Cuba, the Bahamas, Hispaniola, and Puerto Rico. They saw no Mexican specimens of this tree. This species is cultivated in southern Florida; it has been reported as locally naturalized (Barneby and Grimes 1996, Wunderlin 1998), but that is dubious.

Medicinally, *L. latisiliquum* is considered tonic, febrifuge, and hemostat. The leaves in enemas have been used to stop prolonged diarrhea (Liogier 1974). There must be abundant tannins in the plants because they have also been used as a hemostat after childbirth (Morton 1981). A mixture of *Lysiloma,* mahogany (*Swietenia mahagoni*), and pine (*Pinus caribaea*) is combined with the roots of Hercules's club (*Zanthoxylum coriacea*) to treat the bleeding. The bark is also used in a complex mixture to treat gonorrhea (Morton 1981). The bark is made into a beverage in the Caicos Islands as a substitute for tea (*Thea sinensis*). Others in those islands recommend a bark tea of wild tamarind for "inward piles."

In Mexico, *L. acapulcensis, L. candida, L. divaricata,* and *L. latisiliquum* have been used for tanning leather (Uphof 1968, Alcorn 1984, Barneby and Grimes 1996). The Huastec of San Luis Potosí also use *L. acapulcensis* medicinally. Mostly, it is used to stop bleeding during childbirth, as *L. latisiliquum* is used in the Bahamas, but also to treat sore gums, and pain from toothache, tooth decay, and tooth loss. For this purpose, it is made into a "tea" and used as a mouthwash, but it has the negative effect of darkening the teeth (Alcorn 1984). A decoction of the bark or root is also used to stop diarrhea. The same liquid is applied to external sores, rashes, and infected or incurable sores.

There are a number of gum exudates known for the genus (Anderson et al. 1990). Bark extracts from *Lysiloma* have antimicrobial activity (Pérez et al. 2001).

These trees are being used more in southern Florida as landscape plants, and the single tree that was planted near Palm Beach has been left standing beside a beachside condominium (ironically called "The Tamarind") that replaced the hammock there. Everywhere they are used they seem to thrive well north of their native range in the southernmost parts of the state. That ability to withstand cold makes me wonder if they are not comparatively recent arrivals to the flora of the peninsula. Perhaps current cultivation of the species is simply speeding up the natural migration north that would have taken place without humans.

Lythrum

(From Greek *lythron,* blood, gore, alluding to the color of the flowers, and probably to the use in stopping bleeding; a name used by Dioscorides for what is now called *Lythrum salicaria*)

Lythrum alatum. From Britton and Brown 1897.

Lythrum salicaria (an Old World species that influenced thinking about those in the New World)
bouquet violet (purple bouquet, Quebec)
kattehale (cat tail, Norwegian)
lisimaquia roja (red loosestrife, Spanish); lysimachia (from Greek *lysis*, a release from, and *mache*, strife, or from King Lysimachus of Thrace, the son of Agathocles; a name used by Pliny. According to tradition, the herb's virtues were discovered by Lysimachus who, when chased by a maddened bull, in desperation seized a plant of loosestrife and pacified the animal by waving the plant before him. The word was applied by Linnaeus to a totally distinct plant in the Primulaceae.)
[purple, spike] loosestrife (derived from *lysimachia*, ending strife)
lus na siochaint (plant of peace, Gaelic)
salcerella (little willow, Italian); *salicaire* (like a willow, French); *salicaria* (like a willow, Italian); *salqueirinha* (like a little willow, Portuguese); *Weiderich* (like a willow, German)

Lythrum alatum (winged, the stems)
caña (cane, Mexico)
lance-leaved loosestrife (a translation of *L. alataum* var. *lanceolatum*)
milk willow
winged loosestrife [lythrum] (book name)
yerba del cancer [*de México*] ([Mexican] cancer herb, Texas; the name is usually applied to *L. californicum*)

Lythrum salicaria began as a medicinal plant imported from Europe into the Americas to relieve the suffering of settlers. That it apparently did, because it is still considered a useful herbal remedy by Foster and Duke (1990), Bown (1995), and Duke et al. (2002). Even the Iroquois adopted it and made a remedy to banish fever and sickness caused by being near the dead (Moerman 1998). Unhappily, the plant escaped into the wild and is devastating native American wetlands in the northeastern states by crowding out native plants. This is a blessing that has become a scourge (see Diggs et al. 1999).

Although these purple-flowered herbs had been discussed by many authors before Linnaeus ([1753] 1957), it was he who established the genus *Lythrum* as it is now used. He knew 7 species, 3 of them from the Americas, and the genus now contains 36 (Mabberley 1997). There are 13 species in Europe.

It was not until 1814 that Frederick Pursh named *L. alatum* in his *Flora Americana Septentrionalis* (Flora of North America). He had the plants from a wetland in Georgia. *Lythrum alatum* grows from northern New York to British Columbia, and south to Florida and Texas. The variety *lanceolatum*, sometimes considered a distinct species, grows from Florida to Texas, and from Virginia to Arkansas and Oklahoma.

Although *Lythrum* was a popular remedy in Europe (Fuchs 1542 in Meyer et al. 1999, Gerarde [1597] 1975, Culpeper 1563), there are few records of the plants being used by indigenous Americans. The Cherokees took an infusion of *L. alatum* for the kidneys (Hamel and Chiltoskey 1975). Elsewhere the plants are considered astringent, and they have been used to treat women's disorders. Hocking (1997) says that they have been used to treat cancer in Mexico, but that may be confusion with *L. californicum.*

M

Maclura

(Thomas Nuttall named the genus for geologist William Maclure, 1763–1840; Maclure was born in Ayr, Scotland, and came to the United States in 1782; he visited Spanish Florida in 1817–1818 with ichthyologist Charles-Alexandre Lesueur, entomologist Thomas Say, lepidopterist Titian Ramsey Peale, and invertebrate zoologist George Ord)

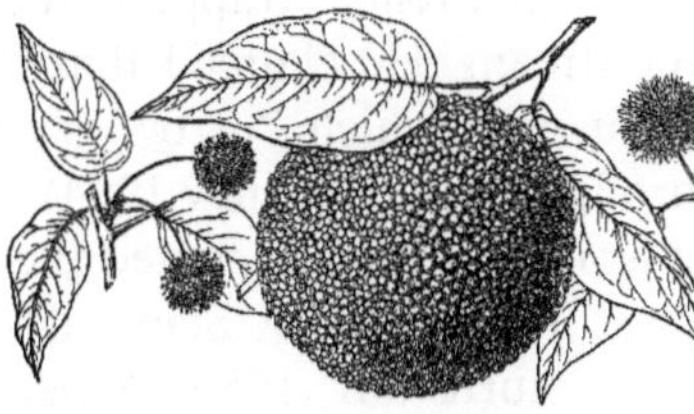

Maclura pomifera. From Sargent 1905.

Maclura pomifera (bearing pomes or apples)

ahilaknî (yellow tree, Mikasuki); *eto láne* [*'to lane, tola:nî, tulani*] (*eto*, tree, *láne*, yellow, Muskogee); *itti' lakna* [*itti' laknani'*] (*itti'*, tree, *lakna*, yellow, Chickasaw)

Bogenholz (bow wood, German); bow-wood (referring to this species, first found in the *Treasury of Botany*, published in 1866)

bois d'arc (bow wood, French); bodark [bodock, bodeck, bodare] (attempts to pronounce the French *bois d'arc*, Arkansas, Missouri)

bois jaune (yellow wood, from 1804 or before, cf. Jurney 1994); *geelhout* (yellow wood, Dutch); yellow-wood

hedge[-plant] (before the invention of barbed wire in the 1870s, these were the preferred hedge plants for fence rows; it met the farmer's criteria that a hedge must be "horse-high, bull-strong, and pig tight"; the legend is that the inventor of barbed wire took his inspiration from these plants and "put steel thorns" on twisted wire)

hedge-apple (originally applied by 1617 to crab apples and *Arbutus*; reapplied to *Maclura* later)

horse-apple [horse apple] (horses and cattle eat the ripe fruits, sometimes to their detriment; occasionally the large fruits stick in their throats and kill them); did the English give rise to the Koasati name *có:ba intakkolá* (*có:ba*, horse, *in*, its, *takkolá*, peach), or was it vice versa?; *issobintakolo* (*issoba*, horse, *im+takolo*, its peach, Chickasaw)

kamo (Alabama); *ittokamo* (*itto*, tree, *kamo*, the *bois d'arc*, Alabama)

mi<u>n</u>'-dse-shta-hi (*mi<u>n</u>'-dse*, bow, *shta*, smooth, *hi*, wood, Osage); *zho<u>n</u> -çi min-dse hi* (*zho<u>n</u>*, wood, *çi*, yellow, *mi<u>n</u>'-dse*, bow, *hi*, tree, Osage); *zhon-zi-zhu* (*zhon*, wood, *zi*, yellow, *zhu*, flesh, Omaha-Ponca)

mock orange [wild orange] (about the only similarity to an orange is that both are round; however, this folk-comparison persists)

nakitsku (Pawnee)

naranjo chino [*naranja China*] (Chinese orange, Texas, Mexico)

Osage (the Osage tribe spoke a Siouan language called *Dhegiha*, lived near the four-corners of southwest Missouri, southeast Kansas, northeast Oklahoma, and northwest Arkansas; these were the people with whom Hunter [1823] 1973 lived most of his early life); *Osage Dorn* (Osage thorn, German); Osage apple-tree; Osage-orange (the OED 1971 says in use since 1859, but Hunter was already using the name in 1823)

Ozark (this mispronunciation of *bois d'arc* finally gave the name to the Ozark Mountains, Missouri, cf. Steyermark 1963)

rootwood

zaip-gwawt-'ko-ya-daw (Kiowa)

Maclura grows in a few spots around Tallahassee (Leon County) and Marianna (Jackson County) (Kurz and Godfrey 1962). What is remarkable about those groves is that they are at least 700 miles east of what is considered the native range of the species (Little 1971, Jurney 1994). It is obvious that humans brought the plants to Florida. What is not clear is which group of people and in what time period.

Osage-orange is native to the homelands of the Caddo, Wichita, and Tonkawa, but not the Osage. Perhaps it was because of their bellicose nature that people associated *Maclura* with the Osage. However, the Osage and the Quapaw spoke the same language so perhaps they formerly were considered the same by settlers (Rankin 1988). The Caddo traded the wood with their neighbors, and the Quapaw took it as far east as the junction of the Arkansas and Mississippi Rivers in what is now southeastern Arkansas (Swanton 1946). To the west, this valuable resource was trafficked at least as far as the Tewa Pueblo in north-central New Mexico. However, the bow-wood was in demand even farther from its homelands.

The Arikara of South Dakota, the Blackfoot of northwestern Montana, and the Kiowa of north-central Oklahoma also prized the tree as bow-wood (Peattie 1953). In 1810, the Scottish traveler John

Bradbury found the Arikara using *bois d'arc* bows. They valued them so much that a bow of the wood was worth a horse and a blanket. Shortly afterward, Prince Maxmilian of Wied-Neuwied found the Blackfoot coveting the bows that they had aquired through barter. As late as 1853, after all individuals who could had switched to guns, the Kiowa met Lieutenant Whipple on the *Llano Estacado* of New Mexico. He wrote, "They carried superb bows of *bois d'arc*, ornamented with brass nails, silver plates, and wampum beads."

When they could get it, the Yuchi (along the Tennessee River in what is now eastern Tennessee) and the Taskigi or Tuskegee Creeks (on the Tennessee River in northeastern Alabama) used the wood for making bows (Swanton 1946). The Alabama also used the wood to make utensils and tools. Knowledge of *Maclura* wood spread throughout the southeastern United States.

The first European contact with the region where this wood originated was made by Jacques Marquette and Louis Joliet [Jolliet] between 1673 and 1675. The two traveled down the Mississippi River to its junction with the Arkansas River. There they met the Quapaw, who had for so long been instrumental in trading *Maclura* wood farther east. Following this expedition, the French sent settlers into the area, and René-Robert C. Sieur de La Salle created the first settlement near what is now Peoria on the Illinois River. Later, in 1699, Cahokia was founded east of modern St. Louis and named after a tribe it supplanted there. It was not until 1764 that Pierre Laclede Liguest established the beginnings of St. Louis. By that time, French trappers and other travelers had traversed much of the region learning about its natural resources, and the name *bois d'arc* had become established.

By the late 1700s, Americans were well aware of *bois d'arc* and its uses, but they had comparatively little information on the wild plants. At the time of the Louisiana Purchase in 1804, John Sibley and Merriwether Lewis reported to President Jefferson on *bois d'arc*. Based on the information they had, they concluded that the main source of trees was on the Red River in Oklahoma and Arkansas (Jurney 1994). At that time, the Caddo still maintained a lively trade among Plains and southeastern tribes for the wood. The Dunbar and Hunter expedition was sent in 1804–1805 to gather more data. They went along the Ouachita River in south-central Arkansas and found only trees that had been planted from more distant sources.

Peter Custis found a transplanted *bois d'arc* within 1 mile of Natchitoches, Louisiana, in 1806 (Jurney 1994). At the time, that was the known southeastern limit of cultivation of the species. The tree had been brought there from the Red River in southwestern Arkansas. Custis even found that there was a stream locally called "Bois d'Arc River" that fed into the Red River in southeastern Oklahoma. He thought that all occurrences of the tree east of that region were where it had been planted around old Caddo villages.

Anthony Glass, traveling overland from Natchitoches to the Wichita villages on the upper Red River in 1807, observed what he took to be native stands of *Maclura*. One site was in what is now Lamar County, and the other was in Fannin County, Texas. The second site was known as "Bois d'Arc Creek" and Glass was told by French trappers that it was a favorite beaver stream. He claimed that the Wichitas made bows of it with which "arrows were propelled completely through bison" (Jurney 1994). Other tribes who used the wood for bows included the Comanche, Omaha, Pawnee, Pima, Ponca, Seminoles, and Tewa (Gilmore 1919, Sturtevant 1955, Moerman 1998). Modern Cherokee in Oklahoma still prize the wood for bows, although no direct evidence has been found that they knew about the wood before they were removed from the Carolinas (Herrin 1999).

The sapwood of Osage-orange is narrow and light yellow, while the heartwood is golden to bright orange, and darkens upon exposure. The heartwood can also contain red streaks. The wood is hard, heavy, tough, resilient, and has a high luster when polished; it is ring porous and commonly confused with black locust (*Robinia pseudoacacia*) (Alden 1994b). In addition to bows, it has been used for fuel, fence posts, game calls, smoking pipes, artificial limbs, crutches, insulator pins, wheel rims and hubs of farm wagons, railroad ties, tree nails, and machinery parts. Wood from the trees is considered one of the woods most resistant to decay in North America (Diggs et al. 1999).

One of the driving forces for the Dunbar and Hunter expedition was the potential of the tree as a commercial dye. Hunter wrote in 1805 that *Maclura* "is more frequently called *Bois jaune* (Yellow Wood) [and] used by them [Caddo] and the inhabitants as a dye" (Jurney 1994). The Kiowa and Pima also used it for that purpose (Moerman 1998). Sargent (1905) identified the chemicals responsible for the dye as moric or moritannic acid. Cannon and Cannon (2003) say that the responsible chemicals are morin, a flavonol in the wood, while the fruits have isoflavones osajin and pomiferin. The dried, chipped wood is still available from dye suppliers in the United States and Europe. The bark has an orange cast, and it was used to make khaki dye during World War I; it is still used for that purpose today (Hocking 1997).

Much less known is that fiber from the bark has been used by indigenous tribes. Gilmore (1931) found cords made of *Maclura* at the Ozark Bluff-Dweller site in Arkansas.

Hunter ([1823] 1973), who had lived with the Osage most of his life, listed the species along with a series of edible fruits (plums, strawberries, gooseberries, whortleberries, blackberries, and grapes). That seems to imply that the Osage also ate the fruits, but no other record of edibility has been found.

The only record of medical use of these plants seems to be among the Comanche who made the roots into an eyewash (Vogel 1970). Any use is surprising because of the irritating compounds in the family, but especially in the latex. Some people develop dermatitis from contact with the latex, leaves, or fruits (Diggs et al. 1999). Fruits contain lupeol and lurenol mixed with some related chemicals; woods have morin and maclurin (Hocking 1997). George M. Toffel, of the University of Alabama, found in the 1940s that a single *Maclura* fruit in a room will drive away cockroaches. He attributed that to the cedar-like odor of the chemicals in these multiple fruits (Peattie 1953).

The only truly modern indigenous use appears to be among the Oklahoma Seminoles. There Howard (1984) found interesting views about the plants. He recorded that at least some individuals believed that a person contemplating a journey should get four spines from a *tulani.* Those spines are to be put under the hatband in front of the traveler's hat while facing the direction of intended travel. That action will cause dangerous snakes to move from the traveler's path. In the rocky hill country of Oklahoma where rattlesnakes are abundant, that might be an important piece of woods-lore.

Magnolia

(Linnaeus adopted the genus proposed by Charles Plumier for these plants in 1703; named after the Frenchman Pierre Magnol, 1638–1715, an innovator in classification and director of the botanical garden in Montpellier)

Magnolia acuminata (pointed, the leaves)
chawpesha (bitter, Osage)
cucumber tree [cowcumber tree] (a comparison of the fruits to those of *Cucumis sativus*; a name used in print by Thomas Jefferson in 1806; also used by Hunter [1823] 1973)
sha-ga-shingah (*sha'-ge,* fingers, *zhin̲-ga'*, little, Osage)

Magnolia grandiflora (large-flowered)
bull-bay [bull bay]

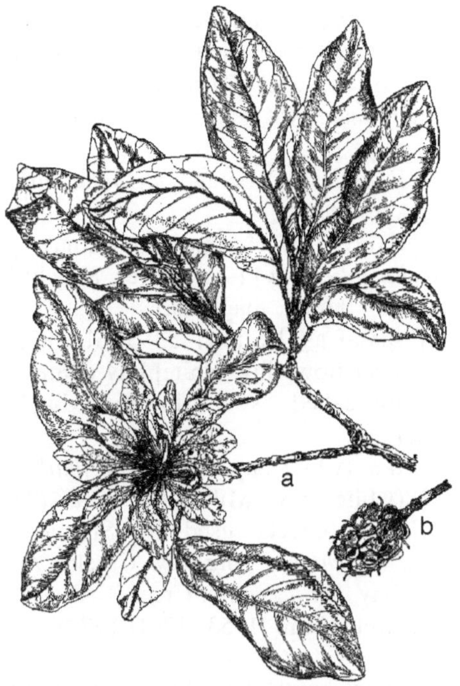

Magnolia virginiana. a. Branch with flower. b. Fruit. *Drawn by Vivian Frazier.* From Correll and Correll 1972.

great laurel magnolia ("magnolia" as a word in English dates from about 1748)
hi'lañ te nec (*hilan,* vulture, *te,* plural, *nec,* tree, Atakapa)
kaɬàaha [*kaɬɬàaha*] (Alabama); *kolaha* [*katlaha*] (Choctaw)
kokayudi [*okayudi*] (Biloxi)
loblolly [southern] magnolia ("loblolly" alludes to the trees growing in wet ponded areas and other swamps)
tolacobá (*tolá,* bay, *cobá,* large, Koasati)

Magnolia macrophylla (large-leaved)
[large-leaved] cucumber tree

Magnolia pyramidata (pyramid-shaped or triangular-leaved)
Bartram's magnolia (so called because Bartram described the species in [1791] 1958)
mountain [pyramid] magnolia
southern cucumber tree
wood oread (the oreads were mountain and hill nymphs in Greek mythology)

Magnolia tripetala (three-petaled)
elkwood
umbrella tree

Magnolia virginiana (from Virginia)
awûxû'xkudi' (Biloxi)

beaver-tree (recorded by Kalm in the 1740s; said to be called that "because the root of this tree is the dainty of beavers," *Castor canadensis*, and these animals taste best if they had eaten the bark)
Indian bark
kalaha [*katlaha*] (Choctaw); *kałàaha hatka* (white bay, Alabama)
laurier (laurel tree, Houma, Louisiana)
okeetosku [*oketoksu*] (sour water, Hitichiti)
swamp bay [laurel, magnolia]
swamp sassafras (Kalm [1753–1761] 1972)
sweet [sweet-flowering, white] bay (Catesby found it being called "sweet-flowering bay" in the 1720s)
to:lahá:tka (white bay, Creek); *tó:lhátkî* [*too-lat-kee*] (white bay, Mikasuki); *tola* (Timucua); *tolaliocko* (Creek, used by Simmons [1822] 1973); *tolochlucco* (big bay, Creek); *tolá hátka* (*tolá,* bay, *hátka*, white, Creek)
white laurel (Kalm [1753–1761] 1972)

William Bartram ([1791] 1958) wrote about finding seven species of *Magnolia* in his *Travels.* Those species were *M. acuminata, M. fraseri, M. grandiflora, M. macrophylla, M. pyramidata, M. tripetala,* and *M. virginiana* (Harper 1958). In other words, he missed only one species in the southeastern United States, *M. ashei* (Kartesz 1994). Perhaps it is not surprising that he gave more space to *M. virginiana* than any other because that species was favored by indigenous tribes.

The Cherokee used *M. acuminata* and *M. macrophylla* wood for furniture and building. They used the wood of *M. virginiana* for skewers for cooking meat, to add a sweetish flavor. Oddly, there seem to be no records of other tribes using the wood from these plants. Porcher (1863) wrote that *M. acuminata* "is soft, fine grained, and susceptible of a brilliant polish. It is sometimes sawed into boards, and used in the interior of wooden houses."

The most common application of the trees seems to have been medicinal. The Cherokees took an infusion of *M. acuminata* bark for stomachache or cramps, diarrhea, toothache, indigestion, and sinus trouble (Hamel and Chiltoskey 1975). The Iroquois, Kansas, and Osage used the fruit and bark to expel intestinal worms (Hunter [1823] 1973, Moerman 1998).

Bernard Romans ([1775] 1961) extolled the virtues of the root bark of "Magnolia major," presumably *M. acuminata,* mixed with other herbs in wine as a bitter infusion to treat malaria. Physician Johann D. Schopf, after visiting North America in 1783–1784, recommended both *M. acuminata* and *M. virginiana* to treat diarrhea, cough, tuberculosis, and fever (Vogel 1970). André Michaux wrote in 1802 that frontiersmen infused cones of *M. acuminata* in whiskey as a bitter drink to relieve malaria (Vogel 1970).

Porcher (1863) wrote that *M. acuminata* cones were "being employed in the form of a spirituous tincture in rheumatic affections. Used as a prophylactic in autumnal fevers. The flowers of most magnolias exhale a strong aromatic fragrance; the bark of all possesses a combination of bitter and hotly aromatic properties, without astringency, and that of many acts as a powerful medicine, in a similar way to Peruvian bark and Winter's bark." Hocking (1997) recorded that the bark of *M. acuminata* had been used as a bitter tonic, against fever, rheumatism, vermifuge, and indigestion.

The Choctaw washed with a decoction of *M. grandiflora* bark to relieve the itching of prickly heat, and made a drink or bath for dropsy (Bushnell 1909). The Koasati washed sores with bark (Taylor 1940).

Physician Benjamin Barton declared in 1793 that *M. grandiflora* mixed with "snakeroot" was a substitute in Florida for quinine in treating malaria (Vogel 1970).

Porcher (1863) wrote of *M. grandiflora*: "The medicinal and chemical properties of [*M. virginiana* and *M. grandiflora*] are supposed to be identical. Mr. Proctor ... found in this species volatile oil, resin, and a crystallizable principle analogous to the liriodendrine of Prof. Emmet, obtained from the *L. tulipifera* growing in this state (see *Liriodendron*) ... in Mexico the seeds are employed with success in paralysis." Bark of *M. grandiflora* was used to treat malaria, rheumatism, and bitter tonic. Seeds have fatty oils used in soap manufacture (Hocking 1997). People in Veracruz still use it as a tonic (Vásquez and Jácome 1997).

The Cherokee made an infusion of *M. macrophylla* bark for stomachache or cramps, diarrhea, toothache, indigestion, and sinus trouble (Hamel and Chiltoskey 1975). Regarding *M. macrophylla,* Bartram wrote to his friend Robert Barclay some time before his book was published in 1791: "I shall just observe that I discovered, in the Creek Nation & Wt. Of Georgia a Species of *Magnolia* (*auriculata*) very different from Mr. Frazers. The leaves of which were very large near 2 feet in length, the Flowers, white, Very large, & Fragrant & the Strobile or Seed Vessels 4.5 inches in length of a fine Crimson Color."

Porcher (1863) thought that *M. macrophylla* had the same properties as *M. virginiana.* All Hocking (1997) said was that it was used medicinally.

No records have been found of indigenous tribes using either *M. pyramidata* or *M. tripetala.* However, Hocking (1997) considered the bark of *M. pyramidata* tonic and astringent. Porcher (1863) thought that *M. tripetala* "acts so powerfully on the nerves as to induce sickness and headache." In spite of that view, or

Acoelorrhaphe wrightii — This is the "Everglades palm" (Florida), *chi-it* (Yucatán), or *guano prieto* (Cuba). The leaves are used for thatch and rope, and the stems are used in construction. (See p. 63.)

Agave decipiens — "False sisal" was probably used for fiber and medicine by the Glades people like its Mexican relative *A. angustifolia.* The Seminoles call it *pasalátkico:bî* (big yucca, Mikasuki). (See p. 68.)

Alvaradoa amorphoides — In Yucatan, this tree is the *bel-cini-ché* to the Maya and *palo de hormigas* to others. The wood is prized, and the bark and leaves are medicinal. Prehistoric people may have introduced the species into Florida. Photo by T.R. Van Devender. (See p. 77.)

Annona glabra — "Pond apple" fruit pulp is edible but not particularly good. The wood is made into corks, but seeds, bark, and leaves are medicinal. The Seminoles call the species *etotakwe* (hollow tree, Creek). (See p. 95.)

Asimina triloba — "Pawpaw" fruit pulp is edible, and the bark was an important fiber source. Two places in Georgia are called "Alcovy," which was derived from the 19th-century Creek name *Ulco-fau-hatchee* (pawpaw thicket river). (See p. 122.)

Asplenium erosum — "Eared spleenwort" or *culantrillo* was mixed with several other ferns and used in baths to reduce fever. (See p. 124.)

Aster carolinianus — The Miccosukees call the "climbing aster" the *wahó:tlíntayhî* (water moccasin's bed) and use it to treat Snake Sickness. (See p. 126.)

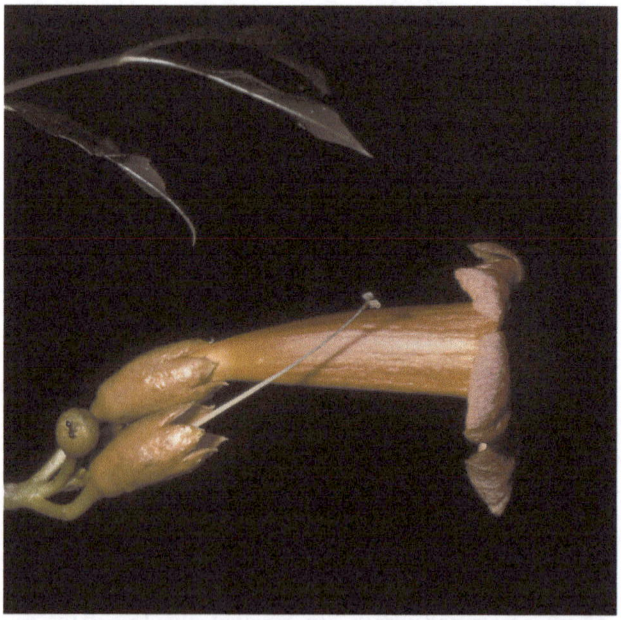

Campsis radicans — The Chickasaw call these vines *ampohko'li'*. A remedy made from the root helps reduce fever and heal wounds. Usually it is called "trumpet vine," but some know it as "cow itch." (See p. 160.)

Canna flaccida — This herb has hard, black seeds that the Seminoles call *sawakmalî:tî* (Mikasuki) and *sáwko matihíta* (Creek), both meaning "rattle's contents." The settlers called this "bird-shot" or "Indian shot" because seeds were substitute bullets in muzzle-loading guns. (See p. 164.)

Capparis flexuosa — The Maya call this shrub *bokanche'* or "tree that looks like a viper's skin." The Haitians say *bois rave* (enrapture tree). Both Florida species provide a mustard substitute. (See p. 164.)

Capsicum annuum — The Taino of Cuba called the fruits *ají*, and that became one Spanish word for red peppers as far away as Peru. Fruits were *chilli* to the Aztecs, and that too was adopted into Spanish and English as "chile." (See p. 168.)

Carphephorus corymbosus — "Paint-brush" is considered a urinary antiseptic. More famous is "deer tongue" or "vanilla leaf." The leaves have been used as medicine and to flavor tobacco and perfumes. (See p. 174.)

Carya floridana — *Pawcohiccora*, a name used by the Powhatan of coastal Virginia, became "hickory" in English. *Carya* was so important as a food and for wood that some species have distinct Creek names, as in *penossv*, *oce-tofkv,* and *oce-cvpkce.* (See p. 176.)

Cephalanthus occidenalis — The Miccosukee call these shrubs *halpáti:hosô:tî* (alligator shader); the Creeks say *sakco'meto* (crawfish's tree). Although the remedy may be lethal, this has been an important medicinal plant throughout its range. (See p. 190.)

Chiococca parvifolia — Medicine made from "snow berry" or "skunk root" has been used as a purgative and to treat asthma, expel gas, lower fever, and ease menstrual cramps. A preparation in Belize also helps alcoholics stop drinking. (See p. 201.)

Chromolaena odorata — "Bitter bush," "Christmas rose," and *fleurit-Noël* are all names for this herb. The plant is either a fish poison or medicine, depending on how the preparation is concocted. (See p. 204.)

Cordia sebestena — Artist Mark Catesby reported in the 1730s that the wood "contains much Gum, in Smell and Appearance resembling Aloes, and is by the Inhabitants of the Bahama Islands (where it grows) called *Lignum* Aloes." (See p. 239.)

Cornus foemina — Englishman Philip Miller considered this a "female" species as opposed to the "male" (*C. florida*). The Creeks call the shrubs *vtvphv,* pronounced *"atápha."* (See p. 241.)

Cucurbita okeechobeensis — The wild "Okeechobee gourd" is *ciko:yi* (Mikasuki) to the Seminoles. *Coksî* is their generic name for the cultivated *Cucurbita*, and that has cognates in Creek (*chasi*), Alabama (*choksi*), and Chickasaw (*lokosh*). Photo by Terrance Walters. (See p. 251.)

Drosera tracyi — Gaelic-speakers call their "sundews" the *lus na greih* (sun plant). The Miccosukees treat ringworm with the herbs and call them *oláyíkcî* (ringworm medicine). (See p. 276.)

Echinacea purpurea — Modern herb collectors make no distinction between the species of "purple coneflowers" and put them all in the same gathering. The Choctaw chewed the roots to relieve coughing. (See p. 279.)

Eupatorium capillifolium — The Cubans call this herb *copal* (incense), and many English speakers say "dog fennel." This species and *E. compositifolium* are notoriously fragrant, and their herbage strewn on floors is insecticidal. (See p. 296.)

Euthamia graminifolia — "Flat-top goldenrod" was used to treat chest pains and lower fever. The Ojibwa smoked the flowers as a hunting lure, saying that the odor simulated the smell of a deer's hoof. (See p. 299.)

Evolvulus sericeus — This species, *hauay ak* (leprosy vine) to the Maya, yields a medicine to treat burns. The Mayas call the related *E. alsinoides* the *sian xiw* (enchantment herb) and use it to treat mental problems. (See p. 301.)

Hypelete trifoliata — Known as *hueso de costa* (coastal bone), the wood is valued in the Caribbean. (See p. 42.)

Illicium floridanum — This plant is called "stink bush" or "purple anise." The flowers and crushed leaves smell like dead fish, but the bark is an aromatic tonic. (See p. 42.)

Ipomoea pes-caprae — "Goat's foot" or *riñonina* (little kidney) contains compounds as effective as Benadryl® in treating insect stings and bites. (See p. 366.)

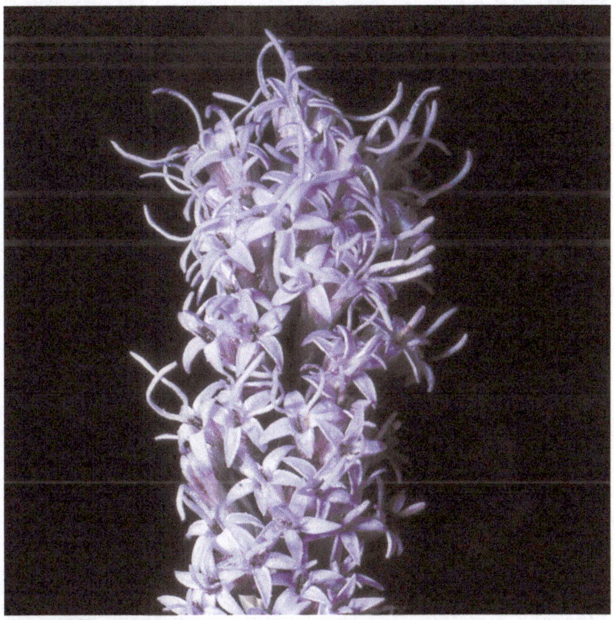

Liatris tenuifolia — The Creeks call several species *ico imaha* (deer's potato). The Creeks, Seminoles, and other southeastern people make medicine from the roots. (See p. 403.)

Lycium carolinianum — These are either a "wolf-berry" or "Christmas berry," depending on the speaker. Birds eat the fruits, and so can people, but *Lycium* specialist Fernando Chiang says they are not likely to replace cherry tomatoes. (See p. 416.)

Magnolia triapetala — The Timucua called the genus *tola*. The Seminoles used the same name, with modifiers added to distinguish species, as in *tó:lhátkî* (Mikasuki) and *to:lahá: tka* (Creek), both meaning "white bay," for *M. virginiana*. (See p. 425.)

Monarda punctata — The Creeks know these mints as *kofucka rakko* (strong fragrance). They and other tribes in the East made a number of medicines from them. (See p. 443.)

Morinda royoc — "Cheeseweed" has ripe fruits that smell like hot limberger cheese; the Jamaicans call it "duppy poison." Also called "yaw weed," the shrub was famous for treating that terrible malady by the 19th century. The plant is a source of yellow to red dye. (See p. 444.)

Nyssa ogeche — The species was first found by John and William Bartram in October 1765. William later wrote, "… they are called Ogeche limes, from their acid fruit being about the size of limes, and their being sometimes used in their stead." (See p. 460.)

Opuntia humifusa — John Harriot wrote in the 1580s that "prickly pears" to the Carolina Algonquians were *metaqvesvnnavk*. In some places, they were so abundant from indigenous use that William DeBrahm, who surveyed the Florida coast in the 1760s, thought that cochineal insects might be grown on them as a source of red dye. (See p. 465.)

Passiflora incarnata — Writing in 1612 on the James River of Virginia, William Strachey considered, "The *Maricock* apple, of the bigness of a green apple, and hath manie azurine or blew kernells, like as a pomegranate, a good sommer cooling fruit." (See p. 483.)

Peperomia obtusifolia — These wild peppers are used in medicines throughout much of their tropical range. (See p. 490.)

Phyla stoechadifolia — This "fog-fruit," endangered in Florida, is sacred to the Maya. They call it *kabalyaxmik*, a name referring to the herb's important place at the base of the *ya'ax che'* (green tree, *Ceiba pentandra*) that supports the world. (See p. 501.)

Piloblephis rigida — This is *hapo:sikâ:yî* or *haposhekaaye-choobe* (Mikasuki) and *kvfockv* (Creek). Seminole Alice Snow said: "My mother told us if you have a cold you could smell this. It was good smelling, so we would smell it all the time." (See p. 511.)

Pilocereus robinii — This tree cactus of the Florida Keys has edible fruits much like those of the more famous saguaro (*Carnegeia gigantea*) of northwestern Mexico and adjacent Arizona. (See p. 43.)

Pinguicula lutea — The Seminoles call this herb *la:nihiliswâ* (yellow medicine, Creek) or *talakcîhcayikcî* (raw medicine, Mikasuki). They use it to treat digestive problems. (See p. 512.)

Piscidia piscipula — The scientific name means "fish-poison for little fish." The tree is also "fish-fuddle" (USA, Jamaica), "fish-poison" (Florida to Panama), *morta* [*à*] *poissons* (Lesser Antilles), and *matapez* (Mexico, Colombia). The bark is medicinal. (See p. 514.)

Pisonia aculeata — Although it is now called "devil's claw," these spiny climbers were known as "blaspheme vine" in the early 20th century. The branches were made into barrel hoops and medicines. (See p. 516.)

Platanthera ciliaris — The Cherokee put roots of these orchids on hooks to make fish bite, but it only worked if they knew the proper song and spat on the hook. (See p. 522.)

Pluchea rosea — "Camphor-weed" is mixed with other plants in the Bahamas to reduce labor pain. The Choctaw thought *P. foetida* was a *hoshukkosona* (strong smell grass) and used it for treat fevers. (See p. 525.)

Plumbago scandens — There is an ambiguity toward these plants typified by the name *homme* [*à*] *deux faces* (man with two faces, Haiti). The herbs are pretty because of their flowers, but they are acrid, blistering, emetic, and medicinal. (See p. 527.)

Polygala boykinii — "Milkworts" are widely believed to induce milk flow in animals and humans. *Polygala*'s religious significance in Florida is indicated by the Miccosukee name *sápiyâ:bî* (resembling the mythical plant "*svpeyv*"). (See p. 532.)

Psychotria sulzneri — This hammock shrub is known to the Seminoles as *atópâ:bî* (dogwood replica, Mikasuki). (See p. 547.)

Quercus laevis — Acorns were so important to indigenous people that there is a river in central Florida the Timucua called *ajano hibita chirico* (river of little acorns). The Seminoles call one site *lokcha apopka* (place for eating acorns, Creek). (See p. 558.)

Rhizophora mangle — A European wrote in 1613 about a "tree called Mangrowes, they grow very strangely, & would make a man wonder to see the manner of their growing." The Seminoles call them *ahilo:ckitiscî* (Mikasuki) and *tolastica:tî* (Creek). (See p. 569.)

Sabal palmetto — These are *ta:larákko* (Creek) or *talcó:bî* (Mikasuki), both meaning "big pàlm." Fibers from the leaves were traded north to the Winnebago of southern Wisconsin and Iroquois of New York, 600–700 miles north of the closest "cabbage palm" stands. (See p. 585.)

Sabatia bartramii — "Rose-pinks" are *kococompahá:ka* (Creek) or *owa:ciká:bî* (Mikasuki) to the Seminoles. Both names mean "star replica." They use one species to treat *Ha:sthlayhi* (Sun Sickness), a malady that strikes in the hot months of May through August. (See p. 587.)

Sarracenia minor — Many indigenous tribes considered pitcher plants medicinal. "Hooded pitcher plant" is also called "smallpox plant" because it was used to treat smallpox and other skin rashes and eruptions. (See p. 603.)

Saururus cernuus — Indigenous American names include *ishuna ignone* (Choctaw), *oyihilíswa* (Creek), and *yahkakayíkcî* (Mikasuki). The Seminole names mean "widow or widower medicine" because it is used to help the bereaved get over their loss. (See p. 608.)

Spartina bakeri — Calusas living on Marco Island, Florida, thatched their houses with "cord grass" before the Europeans arrived. (See p. 640.)

Stachys floridana — "Florida betony," like other perennial species in *Stachys*, has edible tubers. John Lightfoot wrote of *S. palustris* in 1777 that they were "sweet, and in times of necessity... eaten by men, either boiled, or dry'd, and made into bread." (See p. 645.)

Stillingia sylvatica — Because the stems have milky juice, the Seminoles call this plant *owa:ctafânkî* (milky center, Mikasuki) or *pisi:ká:fkatî:ko* (pliable breast milking, Creek). A decoction is taken to treat diarrhea, vomiting, and appetite loss. (See p. 650.)

Strumpfia maritima — "Pride-of-Big-Pine" (Florida), also called *romero falso* (false rosemary, Cuba), is a stimulant and is used to treat bites of poisonous animals; a stronger decoction permanently sterilizes women. (See p. 652.)

Tournefortia hirsutissima — The Seminoles call the climber *cokashatkî* (white grapevine, Mikasuki) or *colo:fákâ* (grapevine cord, Creek). The Caribbean name "chiggery grapes" is used because some people treat itching chigger bites with the leaves. (See p. 675.)

Torreya taxifolia — Known as "stinking cedar" in the Panhandle, the wood was made into fence posts. Now the Florida endemic is globally endangered and too rare to harvest. Photo by David Neubauer. (See p. 674.)

Trema micrantha — To the Taino of Cuba, the "Nettle Tree" was *majagua*, and *cabuya* was their word for cords made from the bark. The Seminoles know the plants as *silî: tá:pî* (Mikasuki) or *lipá:pín* (Creek), slimy stem. People and birds eat the fruits. (See p. 679.)

Tripsacum floridanum — From at least the time of the pre-Columbian Ozark Bluff-Dwellers of Arkansas, seeds of *T. dactyloides* have been eaten raw or popped like popcorn. *Tripsacum floridanum* provides a local southern Florida alternative. (See p. 686.)

Vitis shuttleworthii — The Miccosukee call this grape *balbe* or *cokasî:*. The name in Creek is *pałko*. Grapes were an important food to indigenous people, and the stems were a source of emergency drinking water. It is hard to beat wildlife to these sweet fruits. (See p. 707.)

Zanthoxylum fagara — Creek-speaking Seminoles call the tree *tofvske*, and the Miccosukee say *caháhtî*. They used the wood for bows and arrows. Many people use various parts of the plant for food, medicines, and dyes. (See p. 725.)

Zephyranthes simpsonii — "Rain lilies" appear suddenly after spring and summer rains. That sudden emergence has led to names like *brujita* (little witch), and *duende* (little devil), "fairy-lilies," and "Zephyr-lilies." The Seminoles used *Zephyranthes* as an analgesic against toothache. (See p. 733.)

perhaps because of it, *M. tripetala* was formerly in the U.S. Pharmacopoeia (Hocking 1997).

Because *M. virginiana* grows from Massachusetts southward along the coast to Florida and west to Texas, it is associated with several other species used by indigenous people. The Rappahannock in Virginia have been reported to use *M. virginiana* as a stimulant or "mild dope" (Moerman 1998), although that seems unlikely. Regardless, the Choctaw and Houma used the leaves and twigs to "warm the blood," and the same decoction was used to treat colds (Bushnell 1909, Speck 1941, Vogel 1970). The Biloxi of Louisiana made a tea from bark and leaves and used it to promote perspiration (Dorsey and Swanton 1912). Probably the use of *Magnolia* is underreported because it shares the common name *tola* with *Persea* and *Ocotea*.

Tó:lhátkî is an important ingredient in the ceremony for the annual Seminole Busk festival (Sturtevant 1955). The Choctaw probably used it in the same way because they applied the same name (Bushnell 1909).

Magnolia virginiana became important to Americans who put the bark in brandy as a cough medicine, and used the liquid to treat fevers, diarrhea, dysentery, colds, rheumatism, and malaria, and for other uses similar to *M. acuminata* (Millspaugh 1892, Vogel 1970, Hocking 1997). Peter Kalm ([1753–1761] 1972) wrote, "The virtues of this remedy are universally extolled." He went on to note that the bark eased chest problems, internal pains, fever, dysentery, and colds. Externally, it was equally valuable. An elderly Swedish settler whom Kalm met had been afflicted with stubborn leg sores that were healed by a native (Delaware?) who used the wood of *M. virginiana*. The wood was burned down to charcoal, ground down to powder, mixed with fat, and rubbed on the open sores. Kalm reported that the remedy, "Dried up the holes, which before were continually open." Physician Benjamin Barton wrote in 1793 that *M. virginiana* was a favored remedy among the "Indians of the Kanawha River" (Tutelo tribe, Siouan language stock) of West Virginia where it was used to treat rheumatism and fevers (Vogel 1970). Barton's student Thomas D. Page wrote a dissertation on *M. virginiana* in 1802.

In Millspaugh's (1892) time, the bark was considered a bitter aromatic tonic, febrifuge, diaphoretic, antiperiodic, and gentle laxative. It was recommended for "acute coryzas, bronchial catarrhs, chronic rheumatism, dyspepsia, remittent and intermittent fevers and typhoid states."

Magnolia spp. were in the U.S. Pharmacopoeia from 1820 to 1894 (Vogel 1970). The recommended use was largely for intermittent fever (malaria) and rheumatism (Bown 1995). Duke et al. (2002) list only *M. virginiana*, which they said has been used as an anti-inflammatory, cardioactive, diaphoretic, stimulant, and tonic. They recommend it with two plusses (as safe as a cup of coffee) for chills, colds, dyspepsia, epilepsy, fever, gastrosis, inflammation, malaria, rheumatism, and typhoid.

Magnolia in the Americas contains salicifoline and magnoflorine (Hocking 1997), and Porcher (1863) said that it also contained liriodendrine. In Asia, *M. officinalis* contains magnocurarine, a chemical similar to the muscle relaxant curare (Bown 1995).

Maianthemum

(From Latin *maius*, May, and Greek *anthemon*, flower)

Maianthemum racemosum. From Britton and Brown 1896.

Maianthemum racemosum (having flowers in a raceme)
agoñgosi'mînaga'wic (chipmunk berries [with] little wood, Potawatomi)
false [wild] lily-of-the-valley ("lily-of-the-valley" is a literal translation of *lilium convallium*, which itself is a literal translation of the Hebrew phrase; applied in Europe by British herbalist Turner in 1538)
muguet (from *noix muguette*, in turn from *noix muscade*, nutmeg, although the relationship is obscure, Quebec)
Schattenblume (shade bloom, German)
scurvy-berries
[clustered, false, feather, feathery, large, two-leaved, zig-zag] Solomon's seal ("Solomon's seal" is translated from Latin *sigillum Solomonis*; used in English by 1526 in the *Grete Herball*, for *Polygonatum*); Solomon's plume [zigzag]
[American, wild, false] spikenard ("spikenard" is derived from Latin *spica*, spike, plus *nardi*, aromatic resin; an aromatic substance employed in Roman times in the preparation of a costly

ointment or oil, originally made from *Nardostachys grandiflora*, Valerianaceae or Caprifoliaceae; used in English by A.D. 1350)

treacle-berries ("treacle" combines the old pharmaceutical sense, an antidote against venomous bites, and sweet; the first dates to about A.D. 1200 in Old French, *triacle*, and 1340 in Middle English, *trya'cle*; OED 1971 says the sense of sweet dates from 1694, but earlier in Josselyn 1672, see below)

Europeans were well acquainted with Solomon's seal (see *Polygonatum*) when they arrived in the New World, and here they found different but similar new herbs. Scientists back in Europe did not know what to do with the plants. Linnaeus thought they were like lily-of-the-valley and called them *Convallaria racemosum*. Later Réné L. Desfontaines (1750–1833) decided they were in a related genus and called them *Smilacina racemosa*. That genus was a diminutive form of *Smilax*, indicating another similarity. Finally, in 1821, Johann H.F. Link (1767–1851) created the name now in use. The common names parallel the confusion among the scientists and compare the herbs to the same Old World plants.

American people used these herbs as medicines, but not like Solomon's seal had been used in Europe. Records have been found of many indigenous eastern tribes using American spikenard, but none for the Muskogean tribes. Their allies the Cherokee treated sore eyes with a cold infusion of roots (Hamel and Chiltoskey 1975).

Moerman (1998) recorded uses by other eastern tribes. The Abenaki used a decoction to stop bleeding in the lungs. The Algonquin used an infusion to wash back sores. The Delaware considered the herbs tonic. The Iroquois drank infusions with or without whiskey after miscarriages, to expel tapeworms, counteract poison, and relieve rheumatism. They also used it externally to soak sore feet, on swollen areas, and on snakebite. The Malecite and Micmac used an infusion to heal rashes and itching. The Menomini inhaled steam from it to relieve catarrh. The Meskwaki used it as a calmative and to loosen bowels. They also included it in several preparations to aid or divert magic. The Mohegan used it in cough medicine, for stomach problems, and as a spring tonic. The Ojibwa applied a root decoction or poultice to cuts, to relieve back pain, headache, and sore throat. They also considered it diuretic. The Potawatomi helped revive comatose patients with it (Smith 1933).

The Ojibwa also considered the roots edible, soaking them in lye to remove the bitter taste, parboiling, and then cooking them like potatoes (Fernald et al. 1958). Porcher (1863) noted that *Maianthemum* growing in the Confederate States "yield starch from their roots." Even when Fernald et al. (1958) were considering them as food plants, they cautioned that the herb was so uncommon it was "unwise to draw upon it when other vegetable food is available."

John Josselyn, writing of New England in 1672, recorded that berries were "called treacle-berries,—having the perfect taste of treacle when they are ripe. . . . Certainly a very wholesome berry, and medicinal." Fernald et al. (1958) added that the fruits are "bittersweet, suggesting bitter molasses, but they are cathartic and should be eaten with caution."

Malus

(From Latin *malus*, an apple, Greek *malea* or *melis*; akin to Akkadian *malum*, to be full, Hebrew *male*, full)

Malus angustifolia. From Sargent 1905.

abhal, ubhal (Irish); *abhall, ubhal* (Gaelic); *aepli* (Old Swedish); *afal* (Welsh); *apel* (Dutch); *Apfel* (apple, German); *aval* (French); apple (from Old English *appel*, the fruit, appearing about A.D. 885; akin to *aepplas*, of the eye, but it is not clear which is the earlier application, the fruit or the eye; suggestive is the Gaelic expression *clach-na-sula* [stone of the eye], which means the apple or the pupil of the eye; de Candolle suggested in 1886 that "apple" might have been derived from an Indo-European root of *ab, av, af,* or *op*); *elpe* (Norwegian); *iabloko* (Russian); *obolys* (Lithuanian)

craobb-ubhal-fhaidhain (*craobb,* crab, *ubhal,* apple, *fhaidhain*, prophet's, Gaelic)

fladh-abhal (deer-apple, Gaelic)

goirteag (sour, Gaelic)

maçã [*maciera, maçanzeiro*] (from Latin *malinus*, Portuguese); *manzano* (Spanish)

melo (from Latin *malus*, Italian)

na hollo intakkon (*na hollo,* white man, *in*, his, *takkon*, peach, Choctaw)

pommier (from Latin *pomum*, the apple, French)

sagara (Basque)

seb [*seo, sev*] (Afghanistani, Bengali, Hindi, from *seba* [*seva*], Sanskrit; akin to *sher,* Punjabi, *síb* [*séb, séf*], Persian, *súf,* Sindhi)
tappuah (Hebrew); *tuffah* [*tiffâh*] (Arabic)

Malus angustifolia (narrow-leaved) (=*Pyrus angustifolia,* not *Prunus angustifolia,* which see)
[southern, wild] crab [apple, crab-apple, crabapple] ("crab" is of uncertain origin, but maybe from Scandinavian *scrab, scrabbe,* the fruit, first used in English in A.D. 1420; if correct, then the combination with "apple," from 1712, is redundant)
éco-em-pvkánv [*eco em pvkanv*] (*éco,* deer, *em,* its, *pvkánv,* peach, Muskogee)
ki'wile o'l ne'c (*ki'wile o'l,* persimmon, *ne'c,* tree, Atakapa)
shakulap (Choctaw)
she' hi (apple tree, Osage)
simmenachkee (Shawnee, fide Edgar 1891)
svkta (Cherokee)
takkolchobasi (*takkola,* peach, *choba,* big, *-osi,* little, Alabama); *takkon masofa* (*takkon,* peach, said to be the "old form" by Haag and Willis 2001, *masunfa,* an apple, Choctaw); *takolo maso'fa'* (Chickasaw)

Europeans were acquainted with the apple when they arrived in the New World. Indeed, their languages contained a number of words to describe the basic apple and its dozens of cultivars and hybrids. Pallidus wrote in the 4th century A.D. that Rome grew 37 varieties (Bianchinni and Corbetta 1975); now between 7000 and 8000 cultivars are listed worldwide (Davidson 1999).

The latest studies suggest that the apple, historically called *Malus domestica, M. pumilus,* or *Pyrus malus,* may actually be a complex hybrid (Phipps et al. 1990, Manganaris Alston 1992, Mabberley 1997, Zhou and Li 2000, Robinson et al. 2001, Harris et al. 2002). The European "crab," often called *M. sylvaticus,* appears to have escaped from cultivated plants; it is not their ancestor. The first records of Old World apples being imported to North America was in 1639 when the seeds were listed in an inventory of items taken to the Massachusetts Company (Hedrick 1919). John Josselyn was treated to "pippins" in 1639 grown on Governor's Island in Boston Harbor, which was supposed to be the only place the trees were then grown in the New World (Hedrick 1919). "Pippins" has an obscure etymology (from Old French *pipin*), but it is a name applied to varieties of apples grown from seed since about A.D. 1430.

In the New World distinct kinds of apples were found, but they all were small-fruited trees that matched the European concept of the wild apple called a "crab." Several of these grow in North America; Kartesz (1994) listed 11 species and 5 named hybrids. Perhaps the first wild crab apple was recorded by Giovanni da Verazzano along coastal New England in 1524, probably *M. coronaria* (Hedrick 1919). In early records, there may also have been confusion between *M. coronaria* and *M. angustifolia* because Hedrick (1919) still was not convinced that they were distinct. Other American apples probably have been recorded under the name "medlar." For example, the journal of Lieutenant Diego Peña (1949) spoke of medlars among the Chickasaws in 1716.

The Cherokee ate the sun-dried fruits of both species in their part of the Carolina mountains (Moerman 1998). It seems likely that indigenous people mixed crab apples with their cornmeal cakes as they did other fruits.

Porcher (1863) wrote about what he called *M. coronaria,* although that species is rare in the geographic area he covered, saying: "It is not employed medicinally. The fruit is very acid to the taste, and is often made into preserves. The bark, with that of the white hickory, gives a yellow dye. Alum must be used as a mordant. The yarn should first be boiled with soap and water, then wrung out, and boiled in the preparation." Fruits are used in jellies, pickles, preserves, and cider; they contain abundant pectin (Fernald et al. 1958, Uphof 1968).

The Creeks used a crab apple, probably *M. angustifolia,* in an attempt to cure rabies (1928a). The other species have been applied to an array of maladies (Moerman 1998). Seeds and leaves, especially when wilted, are toxic. These are among the members of the Rosaceae containing the cyanogenic glycoside amygdalin, which releases prussic acid (hydrocyanic acid) upon hydrolysis (Hardin and Arena 1974). However, fruits are edible if the seeds are discarded.

Wood is hard, close grained, and light brown tinged with red. It has been used for tool handles and levers (Sargent 1905). In spite of widespread use, Harrar and Harrar (1946) called *M. angustifolia* a "weed tree."

Martin and Mauldin (2000) show that the name of the persimmon (*svtv*) was applied to the Old World apple when it arrived and eventually supplanted that native fruit. Eventually, *svtv* came to mean the apple and the native persimmon became the Seminole apple (*svtv semvnole*).

Similarly, the Alabama and Choctaw apply words that now mean "peach" (Sylestine et al. 1993, Haag and Willis 2001). Since the peach was introduced from the Old World, they have surely taken an old word and applied it to these new fruits. Maybe they have used the old word for the native plum, and reapplied it (see *Prunus*).

Manfreda

(Named for an ancient Italian writer; British naturalist Richard A. Salisbury, 1761–1829, dedicated the patronym in 1866 to "*MANFREDUS de Monte Imperiali, scriptor antiquus de Simplicibus*")

Manfreda virginiana. From Britton and Brown 1896.

Manfreda virginiana (of Virginia) (=*Agave virginica*, *Polianthes virginiana*)
false aloe
pa:ssa (Creek); *pasî* (Mikasuki) (these names are usually given to *Eryngium*, which see)
rattlesnake master (alluding to the old use to treat snakebite)
Virginia agave (a book name)

The belief was widespread in the Americas that certain plants could be used to counteract the poison of venomous snakes. That concept dominated from New England and nearby Canada at least as far south as Mexico (Santamaría 1959, Vogel 1970). Until recently, this was all considered nonsense. Now a study by Castro et al. (1999) has experimental evidence that plant chemicals partially neutralize the hemorrhagic effect of snake venom. Castro et al. (1999) wrote: "Chemical analysis of these extracts identified catequines, flavones, anthocyanines and condensated tannins, which may be responsible for the inhibitory effect observed, probably owing to the chelation of the zinc required for the catalytic activity of venom's hemorrhagic metalloproteinases." No one is ready to espouse use of plant extracts alone to treat snakebite, but perhaps this is another example of our predecessors knowing more that we do.

Linnaeus ([1753] 1957) was not alone in having a less-than-clear concept of the genus *Agave*. He described *A. virginica* based on a specimen sent by John Clayton in the 1730s. In the same book, he described the genus *Polianthes*. Subsequently, *Agave virginica* has been placed in both *Manfreda* and *Polianthes*. Wunderlin and Hansen (2002) now include the species in *Manfreda*. Molecular genetic and morphological data suggest that *Manfreda* and *Polianthes* are closely related (Cronquist 1981, Bogler and Simpson 1995, 1996).

Gentry (1972, 1982) considered these plants members of the distinct genus *Manfreda*, and many have subsequently agreed (Verhoek 1975, 1977, 1978, Bogler and Simpson 1995, 1996). As a separate genus, *Manfreda* comprises 22 species in the southeastern United States and Mexico (Mabberley 1997).

Campbell (1951) and Sturtevant (1955) found records that Creeks in Florida in the early 1800s used both *Manfreda* and *Eryngium* to treat snakebite. The Creeks even called them by the same name. In both reports a decoction of the roots was used internally or externally to treat the bite. Campbell (1951) also noted that the "root of this plant boiled in sweet milk and taken freely, or chewed and swallowed is a certain cure for the bite of a rattlesnake." That was surely a late variation after they adopted Old World cattle.

The Catawba and Cherokee also considered *M. virginiana* potent medicine. The Catawba applied the plant in the same way that the Creeks did for snakebite. They also drank an infusion of the roots to relieve dropsy and used it as soap (Taylor 1940). The Cherokee chewed the root to expel intestinal worms, to stop diarrhea, and for the liver (Hamel and Chiltoskey 1975). They considered it "strong medicine."

Porcher (1863) noted that the plants were "[c]alled by Negroes rattlesnake's master," and considered a "domestic remedy for flatulent colic; used in Charleston district for the bite of the rattlesnake." Neither Millspaugh (1892) nor Vogel (1970) mentions the species. Hocking (1997) recorded, "Roots bitter, used as carminative in colic, antispasmodic, stomachic."

Foster and Duke (1990) suggested that the plants might be a source of steroid synthesis, but warned that they may produce a strongly irritating latex. That comment is supported by the information that these plants and their Mexican relatives, like *Agave*, contain saponins. In Mexico, the plants are called *amole* (soap) or *huaco*.

The common name *huaco* or *guaco* complicates the story. Originally, *guaco* was a Taino name given to the genus *Mikania* in Cuba (Liogier 1962, Coll y Toste 1972). However, the word was taken to mainland

Mexico, where it was applied to an array of plants, including the genera *Aristolochia*, *Mikania*, and even *Manfreda* (Standley 1920–1926, Santamaría 1959, Liogier 1962, Verhoek 1978).

Santamaría (1959) quoted from Élisée Reclus who wrote about *guaco* in Mexico in 1903. Reclus wrote: "*Mordidio tres veces por serpientes, nunca experimentó ningún mal, porque desde su llegada al país tuvo el cuidado de inocularse el guaco (planta bien conocido, cujo jugo inoculado anticipadamente preserva con todo seguridad de la muerte, a los que son mordidos por serpientes venenosas)*" [Bitten three times by serpents, he never had a problem, because after his arrival in the country he took the precaution to inoculate himself with the *guaco* (a plant well known, whose sap taken in anticipation preserves completely from death, all of those that are bitten by venomous snakes)].

With time, the story grew more complex, and the name changed to *huaco*. Now *huaco* or *guaco* is translated "cow bird" but applied to two kinds of birds. One of them is also called *chicura*, *chileancho*, and *cuapaxtle* (*kwa:*, head, and *pa:xtli*, moss, Náhuatl), and is presumably the genus *Molothrus*, although which species is problematical. The other is the laughing falcon (*Herpetotheres cachinnans*). Legend has it that this bird was able to cure itself after a snakebite, and that gave rise to use of *Manfreda* to cure people (Verhoek 1978).

Manilkara: Wild Dilly

(Michel Adanson created the genus in 1763 from *manil-kara,* a Malayalam vernacular name cited by H.A. van Rheede in 1683 from the Malabar Coast of western India)

My first introduction to the genus *Manilkara* came in 1969 when I lived at the mouth of the Amazon River in Belém, Brazil. One of the indigenous *mateiros* (woodsmen) with whom I worked in the forest studying vines brought me a fruit about the size of a tennis ball. The outside of the berry was rough, brown, and scaly, and it bled white milky latex from the stem where it had been attached to the tree. He told me it was a *sapoti.* Although I had never seen the fruit or the plants, I recognized the name as a Portuguese version of English "sapote" (*M. zapota*). At the time, I had no idea that both words and plants had been borrowed from the Aztecs of Mexico.

Because the fruit was unripe and bled white sap, I was not anxious to sample it. My friend, however, assured me that it would ripen and be "*muito sabroso*" (very delicious). The fruit was set aside and temporarily forgotten.

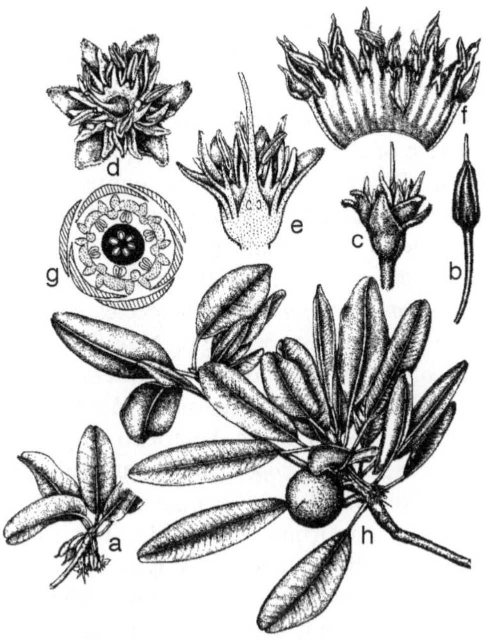

***Manilkara jaimiqui*.** a. Flowering branch. b to f. Flower: b. Early (female) stage; c. Side view; d. From above; e. Longitudinally dissected; f. Corolla open out, one lobe and its stamen folded forward to show scales. g. Floral diagram. h. Branch with fruit. *Drawn by Priscilla Fawcett.* From Correll and Correll 1982.

A few days later I rediscovered the fruit. It had shrunken somewhat and become wrinkled. Instead of bleeding white when it was broken, it showed a brown interior and had a wonderful smell. To me the taste was more like brown sugar than anything else.

When I moved to Florida in 1970, I was reintroduced to *M. zapota,* although this time the locals called tree and fruits "sapodilla" or "dilly." In the interim, I had learned more about the tree, partly from my friend Paulo Cavalcante (Cavalcante 1974, 1976, 1979).

These trees, originally from southern Mexico and Central America, were important to the indigenous people of the region long before the Spanish arrived. Trees not only provided edible fruits, but also medicines from the leaves, a rubbery sap the Spanish learned to call *chicle,* and hard, heavy, strong, and durable dark red wood. The Aztec name *tzapotl* for these trees was altered to *zapote* in Spanish, and later to "sapote" in English. *Manilkara zapota* was sometimes called *sapotilla* (little zapote) by the Spanish in an incorrect translation of one of the Aztec names. In Náhuatl, these trees were either *xicozapotl* (*xicotli,* large bee, *tzapotl,* sapote) or *chictzapotl* (chicle sapote) (Simeon 1981). That sounded like "*chico zapote*" (little sapote) to the Spanish, and became *sapodilla.* People everywhere have a tendency to abbreviate words, and *sapodilla* became "dilly" in several English-speaking places.

Not long after arriving in Florida, I found wild sapodillas in the Keys that were not exactly like the cultivated plants I knew from the mainland or Brazil. It did not take long to determine that this plant was a wild relative of the cultivated species, known locally as "wild dilly." At the time, the name in use was *Manilkara bahamensis* (Long and Lakela 1971), although John K. Small (1933) had called them *Mimusops emarginata.* Now the Florida and Bahama trees are *Manilkara jaimiqui* subspecies *emarginata.*

These Florida trees belong to *Manilkara,* a tropical genus of 65 species found in the Americas, Asia, the Pacific, Africa, and Madagascar (Mabberley 1997). All of the 20 species in the American tropics produce rubbery latex, but *M. zapota* was historically the most important to indigenous people, and has become commercially important worldwide. The first choice for use among the Aztecs and Mayas, as now, was for chewing gum. That may seem incomprehensible, but chewing is an ancient need of our species.

Different people chewed different materials. Northern Europeans chewed spruce gum, among other things, into the 1700s. The Spanish, however, learned much earlier of a better choice. That material became known as *chicle* (*tzictli* or *chictli,* Náhuatl). Bernardino de Sahagun, writing between 1519 and 1540, was perhaps the first to record use of the material. He wrote, "The people chew [*chicle*] because of the popping it caused on being mashed by the teeth" (Simeon 1981). Americans are addicted, as were the Aztecs. In the late 1980s we bought chewing gum at the rate of "ten million pounds a year" (Panati 1987).

The Aztecs thought so much of the tree producing chicle that they named localities after it, as in *Tzapotlan* (*tzapotl,* sapote trees, *tlan,* place of). They also had a *Tzapotlan teohuatzin,* which was a priest in charge of making arrangements for the deity *Tzapotlatenan* (*tzapotlan*+*nantli,* mother), the deity of the hot baths and of medicines. She was thought to have discovered *oxitl* (resin from pines), which was used to treat a variety of illnesses.

This favorite chew of the Aztecs and other people in Mexico did not immediately become popular with Europeans. It was 1707 before the words "chewing-balls" appeared, and they referred to spruce and tobacco. Children chewed spruce; men preferred tobacco.

In 1845, Antonio López de Santa Anna, defeated Mexican commander who led the destruction of the "Alamo" in Texas in the 1830s, settled on Staten Island, New York. Thomas Adams, a local New York inventor, learned of the chewable sap from Santa Anna, and imported a quantity gathered by *cichleros* in Mexico and Mesoamerica (Simpson and Conner-Ogorzaly 1995). Later, he sold the sweetened *chicle* formed into balls and marketed them as a substitute for the paraffin then being sold as a "chew" for youngsters. The first *chicle* balls were sold in 1870. Although the expression "chewing gum" was in common use in local speech, it took Mark Twain to give literature "chawing-gum" in 1871 (Panati 1987). "Chicle" did not appear in English until 1889.

Among the formerly isolated people of the Caribbean islands, *M. jaimiqui* had assumed an importance similar to *M. zapota* on the mainland. *Jaimiqui,* the species name of this tree, is taken from the Taino language, and it is still the name in Cuba, Hispaniola, and Puerto Rico. People in Hispaniola have shortened *jaimiquí* and say *jaiquí,* a name they also apply to *Sideroxylon salicifolium* (see *Sideroxylon:* Mastic and Bustic). Although *acaná* [*jácana, hácana*] (Taino, Puerto Rico) has been listed as a common name of *M. jaimiqui,* it is usually applied to *M. bidentata.* However, since indigenous people used both trees in similar manners, those names may also be "correctly" applied to *M. jaimiqui.*

Other people call *M. jaimiqui* by other names. *Acubá,* usually now rendered *ausuba,* is another Taino name also applied to *M. zapota* in Puerto Rico. Additional comparisons include *zapote de costa* (coastal sapote, Puerto Rico) and *mameyuelo* (little *mamey,* Puerto Rico). The Taino name *mamey,* which appeared in English in 1572 and became "mamee," was an alternative name for both *M. zapota* and *M. jaimiqui.* However, it usually applied then and now to another member of the Sapotaceae, *Mammea americana.*

Europeans in the Dominican Republic began calling these trees *nisperillo* (little *níspero*) and *nisperillo de hojas finas* (little *níspero* with fine leaves). Those are diminutive forms of *níspero,* which is *Manilkara zapota.* Hard as it is to believe, *níspero* is based on the Latin *mespilus,* which was the classical name for what in English is the medlar (*Mespilus germanica,* Rosaceae). The Spanish and English names have the same Latin root word, and they were originally applied to an Old World plant that largely was supplanted in tropical American diets by various *Manilkara* species.

Second to chewing the sap, eating the fruits was probably the most widespread use of wild dilly. If enough trees were available, they were also used for their wood. Trees still are used for timber in Hispaniola and Puerto Rico, and the wood is prized for its wine-red color. *Manilkara jaimiqui* wood is hard, strong, and used for roof supports, braces, and naval construction (Liogier 1974).

Like its more famous relative *M. zapota,* wild dilly is medicinal (Eldridge 1975). In the Bahamas, fruits are eaten in moderation because they make the bowels "hard to move" (Ayensu 1981). That difficulty with constipation after eating too much fruit is thought to be because of the latex (Morton 1981). Tea from leaves is used at least in the Bahamas to relieve flu and fever (Ayensu 1981, Morton 1981). In Cuba, the plants are famous as an astringent and antiseptic (Roig 1945).

People in Trinidad and Tobago still use the seeds of *M. zapota* and the related *Pouteria sapota* and *Mammea americana* to control ectoparasites on their dogs (Lans et al. 2000). At least in *Manilkara*, they work for that purpose because seeds contain poisonous saponins. According to a study by Singh et al. (1984), the potency is high, with the lethal dose $LD_{50} = 30$ to 50 mg/kg.

Wild dilly is rare in Florida and apparently elsewhere in its range (Liogier 1974, Liogier and Martorell 1982). In Florida, the species is considered threatened by the State of Florida (Administrative Code Rule Chapter 5B-40, Coile 2000). This resource, so prized by former natives of the New World, is now being bulldozed as coastal forests are destroyed to build expensive housing for the wealthy. Mostly, these native trees are being replaced in landscaping with imports from other continents. The values of some people changed as we became a mechanized society divorced from and largely ignorant of the living world around us. Even worse, people living in those homes will never know what they have denied future generations.

Marshallia

(Dedicated to American Moses Marshall, 1758–1813, nephew of dendrologist Humphrey Marshall, 1722–1801, and cousin of John Bartram)

Marshallia obovata (leaves are obovate, that is, like the inverse of an egg or spoon-shaped)
[spoon-shaped] Barbara's buttons

Once in the early 1970s I found a *Marshallia* growing in a pine flatwoods in Jonathan Dickinson State Park in Martin County. Each year for the next two decades I returned to the site. I was unable to relocate the plants although Wunderlin (1998) shows *M. tenuifolia* growing south into Palm Beach County. I cannot say how rare these Barbara's buttons are farther north, but they are certainly uncommon in the southern end of their range.

That was the only time I ever saw the genus, and that is not too surprising because two of the three species in Florida are on the state endangered list (Coile 2000). *Marshallia*, a genus endemic to North America, contains seven species (Channel 1957). The group was named in 1791. At the time Henry [formerly Gotthilf H.E.] Muhlenberg (1753–1815) was a Lutheran minister and pioneer botanist in Pennsylvania. He learned that his countryman Johann C.D. von Schreber, a professor in Erlangen, Germany, was revising the eighth edition of Linnaeus's *Genera Plantarum.* Muhlenberg requested that Schreber name this new genus of Asteraceae for Moses Marshall.

Marshallia obovata is endangered in Florida where it grows only in Jackson County (Coile 2000). Elsewhere the species is known from Virginia to Georgia, and in Alabama (Coile 2000). Fernald (1950) reported var. *platyphylla* also from Missouri, but Steyermark (1963) did not include the species. Regardless, Wunderlin (1998) included the variety within the species.

The species was named by Thomas Walter in 1788, but it was not until 1901 that it was placed in *Marshallia.* Apparently, the plants are not that common anywhere in their range, and little seems to have been recorded about them. Among indigenous people, we know only that the Catawba used this herb as a remedy for some disease (Moerman 1998).

Maytenus

(From the Araucan name *maiten, mayten,* or *mayton,* used in Chile for *M. boaria*)

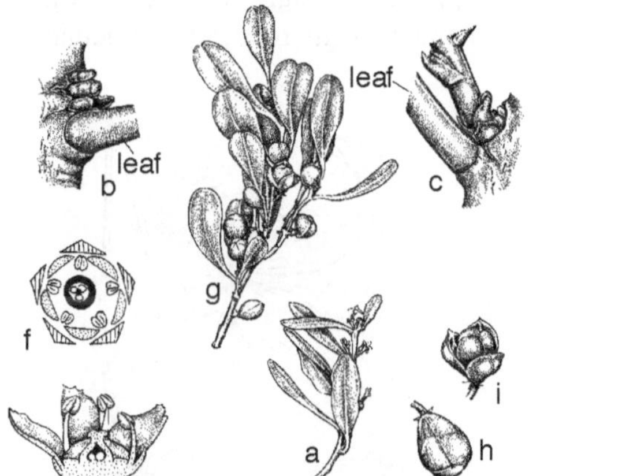

Maytenus phyllanthoides. a. Flowering branch. b. Leaf axil with three serial dormant buds. c. Leaf axil with lower bud developed as a branch. d. Flower, from above. e. Flower, longitudinally dissected. f. Floral diagram. g. Branch with fruit. h. Unripe fruit. i. Dehisced fruit. j. Fruit longitudinally dissected. *Drawn by Priscilla Fawcett.* From Correll and Correll 1982.

Maytenus phyllanthoides (resembling *Phyllanthus*, a genus of Euphorbiaceae, because of the fruits below the leaves in both)
Florida mayten (Florida)
gutta-percha mayten (Florida)
leather-leaf (Texas)
mangle aguabola (water trunk mangrove, Mexico)
mangle dulce (sweet mangrove, Texas)
verdolaga (a name usually given to *Portulaca oleracea*, Cuba)

Maytenus phyllanthoides was named by George Bentham in 1844. The genus contains about 200 species growing in the tropics and warm parts of the world (Mabberley 1997). Several species throughout the range of the genus furnish products used by humans.

Leaves of *M. phyllanthoides* produce a gutta percha, which is used as a remedy for toothache and scurvy in Mexico (Martínez 1969, Hocking 1997). The wood is used as fuel.

Medeola

(Named for the Greek princess, priestess, and witch, Greek *Medeia* or Latin *Medea*; daughter of the Colchian king Aeetes, Jason married her in exchange for her help, and brought her home with him to Hellas; so named for its reported medicinal virtues)

Medeola virginiana. From Britton and Brown 1896.

Medeola virginiana (of Virginia)
concombre sauvage (wild cucumber, apparently both French *concombre* and English cucumber were derived in the 1300s from Latin *cucumis*, Quebec)
[Indian] cucumber-root; *Gurke indische* (Indian cucumber, German)
jarnotte (Quebec)
Virginian medeola (a book name)

When Linnaeus adopted the genus *Medeola* in 1753, he accepted the name proposed by Jan F. Gronovius in his *Flora Virginica* published between 1739 and 1743. Leonard Plukenet had also described these New World plants in 1696, but he incorrectly put them in *Lilium*. The monotypic genus is endemic to eastern North America, and it is probably related to *Clintonia*, which has four species in North America and one in eastern Asia (Mabberley 1997, Utech 2002).

Manasseh Cutler is considered the first to record that the root was eaten by indigenous people in New England in 1785 (Fernald et al. 1958). Presumably, the people referred to by Cutler were the Iroquois (Moerman 1998), but it may have been several tribes because Yanovsky (1936) says in the "northeastern states." By 1814, Frederick T. Pursh in his *Flora Americae Septentrionalis* reconfirmed Cutler's observation (Hedrick 1919). Asa Gray said in 1868, apparently from personal experience, that the tuberous, white rootstock has a taste like the cucumber (Hedrick 1919).

Most references suggest that the first record of the plant's medicinal use among indigenous people came from the time of Barton. For example, Porcher (1863) wrote, "Dr. [B.S.] Barton [in 1810] thought it useful in dropsies. It enjoys some reputation as a hydragogue . . . it is esteemed a very active diuretic." However, the generic name applied by Gronovius and Linnaeus was apparently already based on knowledge that *Medeola* were used as remedies. Gray (1875) wrote, "Named after the sorceress *Medea*, from the wholly imaginary notion that it possesses great medicinal virtues." Fernald (1950), always a careful student of etymologies, similarly wrote, "Named after the sorceress *Medea*, for its imagined great medicinal virtues." Both her name and the genus were derived from Latin *medela*, or *medella*, meaning healing or remedy (Quattrocchi 1999).

However capricious Linnaeus may have been about using names with double or even triple meanings, he did not invent or use them without some basis. It is clear that the Iroquois used *Medeola* to cure maladies (Moerman 1998). That information would have been conveyed to Linnaeus, especially since it is a Peter Kalm specimen that typifies *M. virginiana* (Linnaeus herbarium No. 468.1, LINN). More certain is that Linnaeus continued use of Gronovius's name *Medeola*. Gronovius had written "An Ipecacuanah" on the sheet that John Clayton collected in 1734, thus comparing it to the famous South American emetic *Cephaelis ipecacauanha*, Rubiaceae. Clearly, both Gronovius and Linnaeus had information about medical application, probably in both Virginia and New England.

Moerman (1998) indicated that the Iroquois used an infusion of crushed dried *Medeola* berries to treat infants with convulsions and incorporated the plant into their "Little Water Medicine." Fernald et al. (1959) wrote, "In view of the attractive appearance of the root, it is surprising that it is not mentioned by other students of Indian foods." Although they were talking about foods, the statement is equally applicable to medicine.

Melanthera

(From Greek *melas*, black, and *anthos*, anther, noting how the anthers contrast with the white corollas)

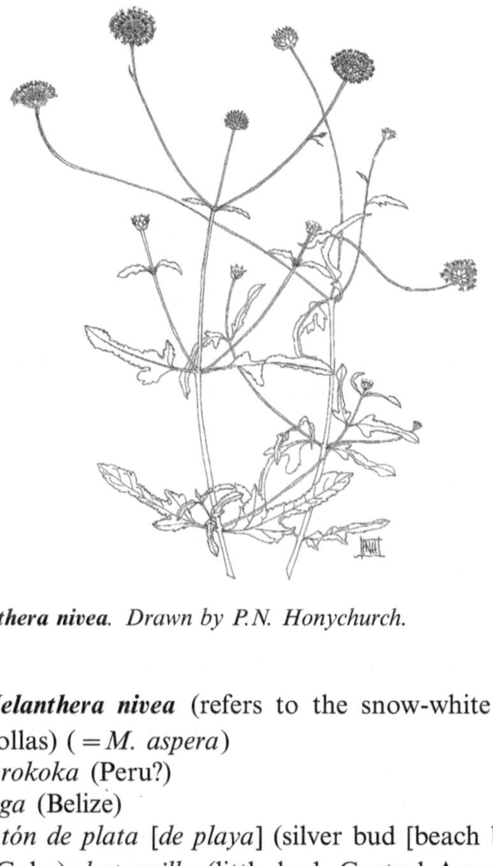

Melanthera nivea. Drawn by P.N. Honychurch.

Melanthera nivea (refers to the snow-white corollas) (= *M. aspera*)
aprokoka (Peru?)
bega (Belize)
botón de plata [*de playa*] (silver bud [beach bud], Cuba); *botoncillo* (little bud, Central America); *botón blanco* (white bud, Central America); *bouton blanc* (white bud, Guadeloupe, Martinique)
cabeza negra (black head, Central America)
caraquillo [*caraquito*] *blanco* (white little *caraqui*, Puerto Rico, Venezuela)
cat tongue (Bahamas)
la chinoise (the Chinese, Guadeloupe, Martinique)
clavel blanco (white carnation, Dominican Republic); *clavellina de monte* (little wild carnation)
cosagané (Central America)
ehtiil kelem (like *kelem* [*Bidens alba*], Huastec, San Luis Potosí)
flor de la vida (flower of life, Central America)
hierba de caballo (horse herb, Central America)
hotz (to spit, Maya, Belize)
julio (July, the month it flowers?)
margerite bord-de-mer (border of the sea margerite; margerite given to *Leucanthemum*, formerly *Chrysanthemum*, in Europe, Guadeloupe, Martinique); *marguerite blanche* (white marguerite, Guadeloupe, Martinique)
orozúz (recorded by Morton 1981, but given for *Conyza canadense* by Liogier and Martorell 1982)
ouache-ouache (padding?, Hispaniola)
paira (bow for shooting arrows in Taino, but surely from some other source, Costa Rica)
salaíllo (Puerto Rico)
sirvulaca (Peru?)
Spanish needle (Belize)
square-stem [squarestem] (Florida)
totalquelite (all-greens, Nicaragua)
tup-lan-xix (*tup*, extinguish fire, *lan*, immerse, *xix*, excess, Maya, Belize; see also *Ageratum*)
yerba de cabra (goat weed, Puerto Rico)

Linnaeus ([1753] 1957) named 11 species of *Bidens*. One of them, *B. nivea*, we now know as *Melanthera nivea*. It was not until 1792 that Julius Philip Benjamin von Rohr (1737–1793) created *Melanthera*, and not until 1903 that J.K. Small put the Linnaean species in it. Today there are at least 40 species, excluding those in the Far East (Mabberley 1997).

People in the Caicos Islands use a decoction, paste, or powder to cure sores (Morton 1981). Náhuatl speakers in Veracruz still treat headache with it (Vásquez and Jácome 1997). In Panama, people use the plant as forage for cattle (Standley 1928). People in northwestern Venezuela rub the plant on their faces before hunting, believing it will improve their marksmanship (Morton 1981).

Several other species of *Melanthera* are used (Parks 1973, Hocking 1997), and there appear to be several bioactive compounds involved (Rivera et al. 1980, Hanson et al. 1994, Penders and Delaude 1994).

Melochia

(Linnaeus named this from Greek *meli*, honey, *echo*, to hold; also possibly from *meli* and *locheia*, childbirth, ancient name for a plant supposed to ease parturition;

Arabic *melochich*, is the name for a *Corchorus* in the same family)

Melochia tomentosa (pubescent)
basora cora (weed broom); broom-weed [-wood] (Puerto Rico, Virgin Islands)
bretónica (little tail, Puerto Rico, Venezuela); *bretónica afelpada* (hairy little tail, Puerto Rico); *bretónica morada* (purple little tail, Venezuela)
cariaquito [*morado*] ([purple] usually applied to *Lantana*, Venezuela)
deshinchadera (swelling reliever)
escoba roja (red broom); *escobilla* (little broom, Dominican Republic)
malva (mallow); *malvavisco* (usually applied to *Malvaviscus*, Puerto Rico)
raichie (Jamaica)
red rope (Bahamas)
tea bush (Jamaica)
toton
varita de San José (St. Joseph's little broom, Nicaragua)
velvet-leaf (Cayman Islands)
zac-chichibe (*sac*, white, *chichibe*, something in the path that sticks, the name of *Stylosanthes hamata*, so "white" is added to distinguish *Melochia*, Mayan)

Linnaeus ([1753] 1957) named *Melochia*, and described *M. tomentosa* from Jamaica in 1759. Since then the genus has grown to 54 species, which are especially common in the American tropics (Mabberley 1997).

None of the literature mentions it, but the most widespread use of *Melochia* is as a broom—as noted in many of the common names. In addition, there are a number of medical applications. This is not surprising because of the novel chemicals in the genus (Tschesche and Reutel 1968, Palmer 1975, Kapadia et al. 1975, Breuer et al. 1982).

Bahamians boil roots into a decoction used to relieve back pain (Ayensu 1981, Morton 1981). In Curaçao a decoction is gargled for sore throat and used as a cold remedy in Jamaica (Ayensu 1981, Morton 1981). The decoction is also drunk for stomachache and recommended to stop diarrhea in Aruba. A decoction of powdered roots is used for an emmenagogue (Morton 1981). In Venezuela the plant is either taken internally or used in baths to reduce swelling of legs and feet; it is also a remedy for eye problems (Pittier 1926). Like *M. pyramidata* in Cuba (Roig 1945), *M. tomentosa* is used as an emollient in much of Latin America (Morton 1981, Vásquez and Jácome 1997).

Melothria

(Greek *melothron*, a "melon" or apple, an ancient name used by Theophrastus, 372–287 B.C., and Dioscorides, fl. A.D. 40–80, for some plant, perhaps *Bryonia*; Latin *melothron* or *melotrum* applied by Pliny, A.D. 23–79, to "*Vitis alba*")

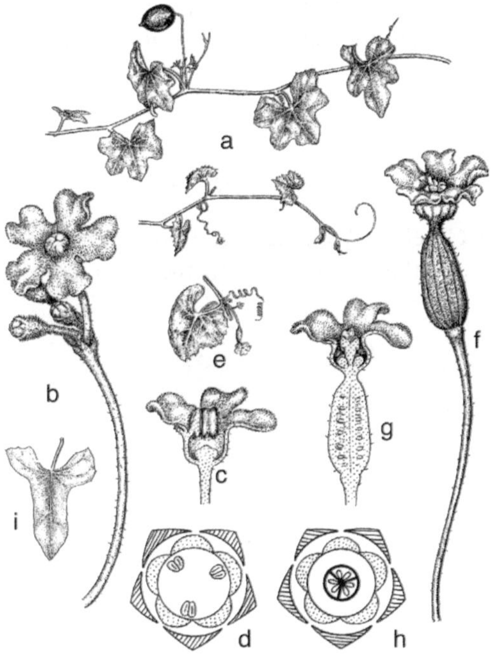

Melothria pendula. a. Section of fruiting stem (above) and tip showing tendril (below). b. Inflorescence of staminate flowers. c. Staminate flower, longitudinally dissected. d. Floral diagram of staminate flower. e. Branch with pistillate flower. f. Advanced pistillate flower, side view. g. Advanced pistillate flower, longitudinally dissected. h. Floral diagram of pistillate flower. i. Unusual leaf form. *Drawn by Priscilla Fawcett.* From Correll and Correll 1982.

Melothria pendula (hanging down, meaning the fruits)
bejuco sin madre (vine without a mother, Dominican Republic)
cereja purga (purging cherry, Brazil)
creeping cucumber (Florida, Bahamas); *pepinillo* [*cimarrón*] ([wild] little cucumber, Cuba, Puerto Rico); *pepinito* (little cucumber, Puerto Rico); *pepino cimarrón* (wild little cucumber, Cuba); *snekikomkomro* (snake cucumber, Sranan, Suriname)
l'herbe à congo (snake herb, Houma, Louisiana; see also *Ipomoea sagittata*)
melón de ratón (mouse's melon, Belize)
teta de perra (dog's tit, Cuba)

Sir Hans Sloane and Leonard Plukenet both recorded these vines from the Caribbean in 1696. Linnaeus had seen them cultivated at both the *Hortus Cliffortianus* and *Hortus Upsaliensis* before he named them *Melothria pendula* in 1753. Although both Sloane and Plukenet had called them a *Cucumis*, Linnaeus appropriated for them an old term for a melon. There are now ten species, all native to the New World; several have edible fruits and are used locally in medicines (Mabberley 1997).

Fruits are eaten in the Bahamas, while some consider them poisonous (Diggs et al. 1999). According to Roig (1945), four seeds are enough to purge a horse. In spite of that potency, they are used when purges are needed in Cuba. In Belize, the plants are used as medicine, food, and poison (Balick et al. 2000).

Many years ago I mentioned to a Bahamian woman that they were poisonous, and she replied that she had eaten them raw and pickled for years. Fruits are also eaten in Hispaniola (Liogier 1974). The Houma used pulverized leaves and gunpowder to treat a water moccasin bite (Speck 1941).

Menispermum

(From Greek *mene*, moon, *sperma*, seed; referring to the shape of the seed)

Menispermum canadense. From Britton and Brown 1897.

Menispermum canadensis (from Canada)
maple vine [vine-maple] (some think the leaves look like those of the maple, *Acer*, which see)
ménisperme du Canada (Canadian moonseed, Quebec)
moon-creeper
[Canada, Canadian, common, Virginian] moonseed (according to the OED 1971, the common name came from the Latin and was first used by Philip Miller in his *Gardeners Dictionary* of 1739; the modifiers "Canadian" and "Virginian" were added by 1822); *Mondsame* (moonseed, German)
raisin de couleuvre (snake grape, Quebec)
[American, Texas, yellow] sarsparilla (from Spanish *zarza*, briar, *parilla*, small grape vine; the name "saraparilla" has been applied to a number of plants, Indiana, Ohio; apparently, "safarilla" is a corruption of sarsparilla, Kentucky)
yellow parilla (the word "parilla" is apparently a shortened form of "sarsparilla"; although *parilla* is Spanish, a diminutive form of *parra*, grape vine, that seems to be incidental)

Linnaeus ([1753] 1957) named these plants based on literature and personal knowledge of cultivated specimens in Europe. He had studied living plants at the *Hortus Cliffortianus* and at the *Hortus Upsaliensis*. In publications on both places he followed Jan Gronovius's *Flora Virginica* from 1739–1743 and called the plants *Menispermum foliis peltatis subrotundis angulatis* (moon-seed with peltate, almost round and angular leaves). Linnaeus adopted the first two words of Frenchman Joseph Pitton de Tournefort's (1656–1708) polynomial from 1705 when he named the vines *Menispermum canadense*. Happily, he was not swayed by Leonard Plukenet's name of 1696, *Hedera monophylos virginiana, convolvuli foliis* (Virginia ivy with one leaf, leaves like *Convolvulus*).

We now know two (maybe four) species in *Menispermum*, and these are shared between eastern North America and eastern Asia (Mabberley 1997). The Japanese *M. daurica* is the ecological analogue if not the sister to *M. canadense* (Ohwi 1965). That Japanese plant has also been used in the same ways as the American (Foster and Duke 1990).

Only three tribes are known to have used *M. canadense*, but probably more did. The Cherokee took a remedy of the vines for weak stomach, diarrhea, as a laxative, for venereal disease, and to strengthen "weakly females" (Hamel and Chiltoskey 1975). The Delaware put the plants in a salve used on chronic sores (Moerman 1998). The Ojibwa used the vines for an undisclosed medicine (Vogel 1970). Rafinesque said indigenous people used the root in scrofulosis, but did not say which tribes (Millspaugh 1892).

Porcher (1863) wrote: "[Steven] Ell[iott] never saw it, but thinks that it grows in the mountains. Dr. [Asa] Gray determined a specimen sent from St. John's, Charleston district, by H.W. Ravenel, Esq., to be this. ... It is said to be much used in Virginia by physicians; and in domestic practice, as a substitute for sarsaparilla, in scrofulous and cutaneous affections ...

the roots are tonic, alterative, and diuretic. . . . It is also employed by the vegetable practitioners. . . . Said to be laxative and tonic, and used in debility and in giving tone to the stomach and nervous system."

Millspaugh (1892) recorded that early settlers found the root useful as a diuretic to relieve strangury in horses. He also noted that the root was used much like sarsparilla, i.e., treating syphilitic, scrofulous, and rheumatic diatheses; as a laxative; as a tonic for use against general debility, atonic dyspepsia, and "kindred problems"; and as a remedy for pleural adhesions and inflammations of the alimentive mucous membranes.

As late as the 1980s, the remedy was used in Appalachian Kentucky to "cleanse the blood," where "one span of sasfriller root is boiled in a quart of water until one-half pint of liquid remains. One half pint of whiskey and a heaping tablespoon of sulfur are added to this. One spoonful is taken before each meal until it is all used up" (Bolyard in Coffey 1993). With that much whiskey each day, it would at least relax patients enough so that they might heal.

Moonseed was official in the U.S. Pharmacopoeia between 1882 and 1905, and in the New York edition from 1831 to 1842 (Vogel 1970). Duke et al. (2002) list the vine's qualities as alterative, cyanogenic, diaphoretic, diuretic, laxative, nervine, poisonous, stomachic, and tonic. They state that it has been used to treat cancer, constipation, debility, dermatosis, enterosis, fever, gastrosis, gout, scrofula, sores, syphilis, venereal disease, and water retention. They rate it in terms of safety along with a single cup of coffee, but refrain from commenting on its probable utility.

There are certainly bioactive compounds in *M. canadense.* Among the chemicals known are the isoquinoline alkaloids, including dauricine, a compound with curare-like action. The grapelike fruits contain isoquinoline alkaloids and have been fatal to children eating them (Lewis and Elvin-Lewis 1977, Turner and Szczawinski 1991).

Metopium: Poisonwood

(Uncertain; the classical Greek name *metopion* is the gum of an African tree now unindentified; most sources translate *metropium* as forehead)

There are two native trees in southern Florida that cause dermatitis (or worse) in sensitive individuals. Probably the most famous is the manchineel (*Hippomane mancinella*); the other is the poisonwood (Bahamas, Florida). The manchineel is in the Euphorbiaceae, while the poisonwood is *Metopium toxiferum* in the Anacardiaceae. In other words, the poisonwood is related to poison ivy (*Toxicodendron radicans*), poison oak (*T. pubescens*), poison sumac

Metopium toxiferum*. Drawn by P.N. Honychurch.*

(*T. vernix*), cashews (*Anacardium occidentale*), and pistachios (*Pistacia vera*). It is a family famous for causing human misery and delight.

Although I was familiar with both poison ivy and cashews when I arrived in Florida, the poisonwood was new to me. It took me some years to learn to distinguish it from gumbo limbo (*Bursera simaruba*), although now I find them so distinct that I wonder why I ever had trouble. I quickly learned the reputation of poisonwood and carefully avoided contact with it. That was not always true of my students. Repeatedly over the years, different individuals brought in leaves or branches of poisonwood either totally puzzled about the identification or thinking that they were gumbo limbo. Few of us actually ever got dermatitis from the specimens, although a couple who were gathering ecological data for me came down with rashes in unexpected places after working alone in the field.

My only direct experience with the rash came from a "free" trip with Donovan Correll during the creation of his *Flora of the Bahama Archipelago* (Correll and Correll 1982). We made a collecting trip to Andros, and I followed him and our local guide through *Metopium* and *Clusia* forests much of the day. That would not have been so bad except that they kept breaking poisonwood branches in front of me and leaving them to drip resin as I passed. It was not until we were in the small plane on our way back to Miami that I discovered a large blister growing on one finger. That "free" trip cost me two visits to the dermatologist's office and two injections of cortisone. If I had only known of the Bahamian and Mexican remedies it might have saved me considerable expense and discomfort. In the Bahamas, they take leaves from the "maiden fern" (*Thelypteris kunthii*), parch and powder them, and sprinkle the dust on dermatitis caused by poisonwood. Mexicans use the leaves of gumbo limbo to relieve the dermatitis, and there seems to be some chemical basis for that (Peraza-Sánchez and Peña-Rodriguez 1992).

Catesby (1731–1732) introduced Europeans to these plants when he wrote, "From the trunk of this Tree distils a liquid black as Ink, which the Inhabitants say is Poison." He had found the plants on Providence, Eleuthera, and other islands of the Bahamas where I had my unpleasant contact with the sap and unwittingly verified the reports of the Bahamians who told him about the poison. Catesby called the plants *Toxicodendron foliis alatis fructu purpureo pyriformi sparso* (the poison wood with winged leaves and sparse purple pear-shaped fruits). That comment about the fruits is odd because they are yellow to orange.

The Catesby text and plate were the only source that Linnaeus had for the species when he named it in 1759. Perhaps that is why poisonwood was a puzzle to him. Linnaeus did not realize that the plants were related to those we now put in the Anacardiaceae. For some odd reason, Linnaeus called the trees *Amyris toxifera* in 1759, thereby putting them with *Amyris, Citrus,* and *Zanthoxylum* in the Rutaceae. Indeed, Linnaeus ignored that Patrick Browne had created a genus for these poisonous plants in 1756 when he published *Metopium* in *The Civil and Natural History of Jamaica* for what we now call *M. brownei.* Linnaeus instead called them *Rhus metopium,* and missed the connection with Florida's *M. toxiferum.*

Browne's name for the plants has puzzled people from the beginning. He applied to New World plants the classical Greek name *metopion* for the gum of an African tree now unidentified. Most sources translate *metopium* as "forehead," but it is unknown exactly what Browne meant. I suspect that he, or someone he knew, had brushed against resin on the trees and developed a rash on the forehead. Since he would have been familiar with the classical Greek, he saw the connection with the old word and called the plants *Metopium.* Another interpretation is that the Latin means "poison carrying," and that would also make sense (Nellis 1997).

The poisonwood in southeastern Florida also occurs in the Bahamas, Cuba, Hispaniola, Puerto Rico, and Anguilla. Within its range, it has several common names, most of which allude to its poisonous nature. Sometimes in Florida it is called the coral sumac, and in the Bahamas it is poisontree. Oddly, it is Florida poisontree in Puerto Rico although it is native to that island. Cubans usually call it *guao de costa* (coastal *guao,* Taino, Cuba). That reference is a comparison with the "true" *guao* (Puerto Rico), *Comocladia glabra* in the same family. Sometimes the *Metopium* also is called *guao* (Hispaniola).

There is almost complete overlap of common names in Hispaniola with *Hippomane.* There the *Metopium* is called various renditions of *manchenille* (little apple, Haiti) or *manzanillo* (little apple, Dominican Republic). Mostly, the trees are simply *manchenille* or *mancenillier* (little apple tree, Haiti). However, there are other names variously pronounced so that they actually have different meanings in French. These other names include *machanoise* (chewer, Haiti), *machandeuse* (chewer, Haiti), and *maximier* (maximizer, Haiti). Haitians also call the trees *bois mulâtre* (mulatto tree).

The first mention of this tree in Florida may have been by DeBrahm ([1775] 1971), who called it *papajos,* although his description of the plants reads like a mixture of satinleaf (*Chrysophyllum olivaeforme*) and something else (Austin 1980). The name *papayo* is used for *Metopium* in Puerto Rico, but elsewhere it means papaya tree (*Carica papaya,* Caricaceae). DeBrahm's book contains many curious comments, confusions, and allusions to contacts with the Caribbean, where he may have heard the name. In Puerto Rico, the trees also are known as *almendrón* (big almond) and *cedro prieto* (black cedar).

Poisonwood contains alkylcatechols (urushiols), as do its relatives in *Toxicodendron* (Rivero-Cruz et al. 1997). Urushiol is a mixture of compounds, each with a catechol ring having a 15-carbon side chain with one or two double bonds. In the presence of oxygen in the skin, the urushiol is converted into a quinone that binds to proteins (Nellis 1997). Not surprisingly, those compounds are allelopathic to higher plants and inhibitory to fungi (Rivero-Cruz et al. 1997, Anaya et al. 1999). The toxin is stable and remains in the leaves and wood as long as they remain intact. In spite of that, people have used the wood for hundreds of years, although they are often irritated by the sawdust. Once finished, the wood is harmless, takes a fine polish, and is used for furniture and other items.

As incredible to those of us who are sensitive to those poisons in the Anacardiaceae find it, *Metopium* has been used medicinally. The resin is known as hog gum or doctor gum (Uphof 1968). It is emetic, purgative, diuretic, sudorific, and sedative (Roig 1945). The wood has been used to adulterate *Quassia amara* (Uphof 1968).

Several other authors simply say that the resin has "medicinal properties." Those who give details on use of the resin indicate that it has been used to treat erysipelas and inflammatory rheumatism. Poisonwood also has been used as an abortifacient; however, the "remedy" is notorious for also killing the mother (Ayensu 1981, Morton 1981). That result should not be a surprise considering that the "cure" contains 13 packets of Epsom salts, and 1 quart of chlorox bleach in addition to *Metopium* and *Maytenus buxifolia.* Another widespread usage is to "remove" an aching tooth. Apparently, dabbing the "milk" (resin) on the offending tooth kills the nerve to the point that it can be removed without problems. I shudder to think of the impact the urushiol would have on the sensitive mucous membranes.

There are only three or four species in *Metopium.* In addition to *M. toxiferum,* the other widespread Caribbean plants are *M. brownei,* known from Veracruz, Chiapas, Campeche, Yucatán, and Quintana Roo in Mexico, and south through Belize and Guatemala to the islands off Colombia. Within the Caribbean, the species grows in Cuba, Jamaica, Hispaniola, and the islands of Aruba, Bonaire, and Curaçao. The tree is oddly missing from Puerto Rico. *Metopium venosa* is endemic to Cuba, and *M. gentlei,* which Mabberley (1997) does not recognize, was described from Guatemala.

Of the related species, *M. brownei* has been used in all of the same ways as *M. toxiferum.* The wood of this species is especially esteemed, and it is prized for making furniture in spite of the hazardous sawdust. The Maya of Yucatán do not seem to completely share that opinion of the trees. They call the trees *boxcheché* (*box,* black, *che',* tree, *chem,* canoe). They also call the trees *kabal chechem,* where *kabal* is a derogatory term meaning "low in stature." That's how I felt when I was contaminated with Bahamian poisonwood sap—low in morale and stature.

Mikania

(Named for Bohemian Joseph Gottfried Mikan, 1743–1814, professor at Prague)

Mikania scandens (climbing)
climbing hempweed [hemp vine] (Florida)
guaco (indigenous Taino name for the vines, Cuba); *guaco blanco* (white *guaco,* Colombia)
halosvkatv entihv [*a:ló:saká:ta intayhóci, halocsha kate entahe*] (moccasin's scaffold; *halosvkatv,* water moccasin, *entihv,* bed, Creek); *wahołe entayhe* [*wahó:rintayhî, waaho(th)e entahe*] (water moccasin scaffold; *waaho(th)e,* water moccasin, *entaahe,* bed, Mikasuki)

This was another of the species that Linnaeus included in his broad concept of *Eupatorium* in 1753. It was not until Willdenow studied them in 1800 that he created the genus *Mikania* for what Linnaeus called *E. scandens.* Since then, the genus has grown to about 430 species in the tropics of both hemispheres (Mabberley 1997). Several species are used as medicines.

Seminoles use the same name for *M. scandens, Aster caroliniensis, Ipomoea sagittata,* and *Ludwigia.* These vines are used to treat "Snake Sickness"

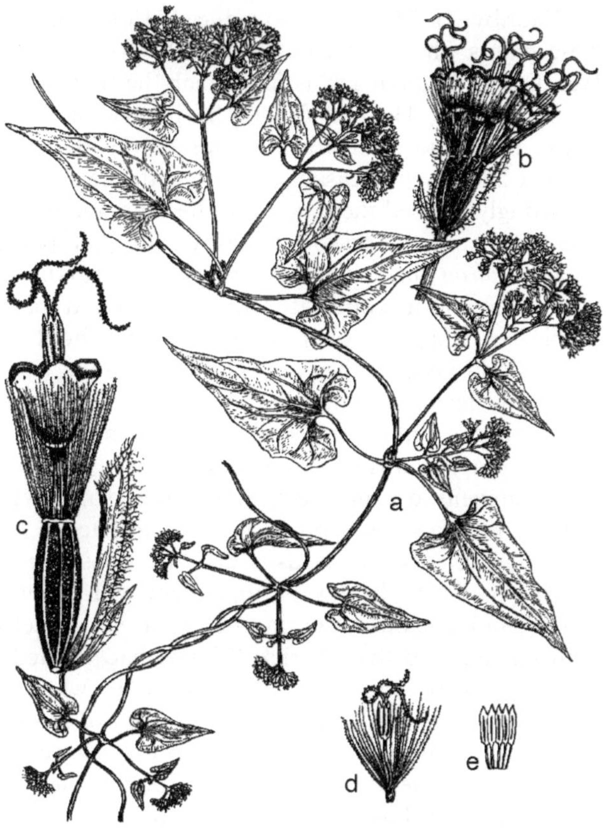

Mikania scandens. a. Section of stem. b. Head of four flowers. c. Single flower. d. Flower with corolla removed showing anthers united into a tube. e. Anthers spread out. *Drawn by Vivian Frazier.* From Correll and Correll 1972.

(Sturtevant 1955, Bennett 1997). A medicine to treat snake bite still is made from the stem and leaves of *M. scandens* mixed with other plants (Bennett 1997). This plant is used in Cuba for intestinal and stomach problems, including diarrhea (Roig 1945).

Cubans also treat intestinal ailments and rheumatism with the leaves of *M. cordifolia* (Roig 1945). In Brazil and the French West Indies, a decoction is used to treat snakebite. *Mikania guaco* is used as a febrifuge and snakebite remedy throughout the West Indies and Mesoamerica (Morton 1981).

Mitchella

(Commemorating physician John Mitchell, 1711–1768, a correspondent of Linnaeus who resided in Virginia; he was elected a fellow of the Royal Society in 1748)

Mitchella repens (creeping)
[box, chicken, cow, deer, fox, Jesuit, one, snake, squaw, tea, turkey]-berry (Alabama, Massachusetts, Michigan, New York)

Mitchella repens. From Britton and Brown 1898.

[creeping] checker-berry [creeping checkerberry] (the OED says these plants came to be called "checker-berry" by 1849 because of confusion with *Gaultheria procumbens*; perhaps the name refers to the fruit, which Thomas More said in 1723 has "a small red berry white in the middle, but, while not fully ripe, white and red prettily streaked like a Pomegranate," Massachusetts)
eye-bright (in Europe the name is usually given to *Euphrasia officinalis*; reapplied to a variety of plants in the Americas, Massachusetts)
fi:tó imilpá (*fi:tó*, turkey, *im*, its, *ilpá*, food, Koasati)
hive-vine (New York)
mountain-tea (New York)
pain de perdriz (partridge bread, the spelling *perdriz* is archaic; modern French uses *perdrix*; from the Latin *perdix,* Quebec); partridge-berry [partridgeberry, partridge-vine] (so named by 1714 because they were considered a food favored by the "partridge." That bird got the English name from the Latin *perdix*. In Europe, this bird was the grey partridge, *Perdix perdix*. New Englanders began calling the ruffed grouse, *Bonasa umbellus,* a partridge, and then people in Pennsylvania and Virginia added the quail, *Colinus virginianus.*)
[heath-hen, pigeon, pudding, snake, squaw]-plum (Maine, New York)
Rubhuhnberre (red hen berry, German)
running box (New York)
squaw vine (said to have been coined by colonists who saw indigenous women using the herb)
twin-berry (the most commonly used modern name for these plants, this refers to the two fruits resultant from paired flowers; sometimes the two fruits are partly attached to each other); two-eyed berry [two-eyed chequer-berry]

Linnaeus ([1753] 1957) gave the scientific name to these small, prostrate, evergreen herbs. He was not the first European to study these diminutive plants, but he was the first to realize that they were unlike any in Europe. The American plants were examined by Leonard Plukenet in 1696, who thought they were *Syringa* (lilacs). John Mitchell, in 1729, thought they were *Chamaedaphe* (Ericaceae). Jan Gronovius, using the John Clayton collection for his *Flora Virginica,* called them a *Lonicera* (honeysuckles). So, Linnaeus commemorated Mitchell's error by naming the plants after him.

Well toward the end of the 19th century people thought that the *Mitchella* growing in Japan was the same as that in North America (Millspaugh 1892). However, in 1846, P.F. von Siebold (1796–1866) and J.G. Zuccarini (1797–1848) realized they were different and named the Japanese plants *Mitchella undulata.* Even then, some considered the Japanese plants only a variety of *M. repens.*

We do not know the provenance of Plukenet's plants or the information exchanged about them. However, in 1723, Thomas More sent a package of plants from Boston to William Sherard in London. Included was *Mitchella,* which he claimed had leaves that make "the best tea in the world and is counted a great catholicon among us. Colonel Brown, a man of great figure here, told me his body was so big of a dropsie as a sack of malt, but by drinking this tea he soon recovered a very healthy state of body" (More in Coffey 1993).

In 1810, Manasseh Cutler opined: "The leaves are much celebrated by the common people as a diuretic and sweetener of the blood, but are of very little efficacy. It makes an ingredient in their diet drinks." Physician H.B. Skinner had a different opinion. In his book *The Family Doctor,* published in 1844, he claimed, "a tea of this plant gives much relief to women in labor" (Vogel 1970).

A "Canadian Indian doctor" gave Dr. Stephen W. Williams information in 1849 that he used *Mitchella* with *Spiraea* for the "cure of salt rheum." Williams valued *Mitchella* as a diuretic, used extensively for dropsy, and the astringent berries were used to stop diarrhea (Vogel 1970).

Later, Porcher (1863) wrote: "An infusion of the stems and leaves is used in dysuria, its diuretic powers, however, not being of any importance. The 'Cherokee Doctor' declares that the 'decoction taken freely is an excellent article to facilitate childbirth. It should be used daily for two or three weeks before that period!' "

Millspaugh (1892) disagreed with Porcher. He wrote: "*Mitchella* is one of the many plants used by the American Aborigines as a parturient, frequent doses of a decoction being taken during the few weeks just preceding confinement. It has also been found to be a valuable diuretic and astringent, and to have an especial affinity to various forms of uterine difficulties." In spite of the bickering about its effectiveness, *Mitchella* entered the National Formulary where it remained from 1926 to 1947 as an astringent, tonic, and diuretic (Vogel 1970).

All the uses were learned from indigenous people, with tribes through much of the range of *Mitchella* recorded as valuing the plants. Some shared their uses while others were more guarded. The Creeks used the herb to treat fevers according to William Bartram in material written in 1773–1774 (Sturtevant 1955). The Penobscots would only tell Speck that it was to be steeped for use as a medicine (Vogel 1970). Some Ojibwa agreed that they used the plant, giving no details, while others said they smoked it during ceremonies (Moerman 1998).

The Abenaki made a poultice to reduce swelling (Moerman 1998). The Cherokee took it as a diaphoretic and diuretic, and to cure dysentery. They also used it for women's problems, including menstruation pains, sore nipples, and to facilitate childbirth (Hamel and Chiltoskey 1975). The Delaware used it to treat female reproductive organs and sore muscles and stiff joints associated with rheumatism (Moerman 1998). The Iroquois used it as a diuretic, as an analgesic (for labor and other pains), and to stop vomiting, flatulence, fever, and various stomach problems. They also applied it to cuts and rashes to stop bleeding, to prevent rickets, swollen testicles or ruptures, and to cure venereal disease (Moerman 1998). The Menominis steeped the leaves for a drink to stop insomnia (Vogel 1970). The Montagnais made a remedy against fever (Moerman 1998). William Bartram noted in 1774 that the Seminoles extol the plants "as an infalable remedy in Nephritic complaints" (Bartram in Sturtevant 1955).

Vogel (1970) wrote that the plant contains a "bitter principle" and tannin; fruits contain a saponin. Plants also contain glycosides (Swerdlow 2000).

Fruits are edible and may have first been mentioned by Thomas More in 1723 when he wrote: "The berries are rather of an agreeable taste, and are sometimes eaten by children in milk." Yanovsky (1936) simply said the fruits were eaten in Texas and the eastern states. Porcher (1863) added, "The fruit is slightly acid, and is edible." Not everyone was as impressed with the flavor. Fernald et al. (1958) damned *Mitchella* with faint praise when they wrote, "The berries are dry and very seedy, but without disagreeable flavor."

Modiola

(From Latin *modiolus*, Greek *modiolos*, the nave of a wheel)

Modiola caroliniana (of Carolina)
[bristle-fruited, Carolina, Carolina bristly, red-flowered] mallow
pilapila [*pila pila*] (Chile)

In his *Species Plantarum*, Linnaeus ([1753] 1957) described 15 species that he considered members of the genus *Malva*. One of those he called *Malva caroliniana*, a name he based on studying living plants at the *Hortus Cliffortianus* and *Hortus Upsaliensis*, and on the publications by John Martyn from 1728 and Jan Dillenius in 1732 on plants grown in London. Both Martyn and Dillenius thought the plants were *Abutilon*, but Linnaeus disagreed.

Conrad Moench disagreed with all three of them. When he studied plants from the New World, he decided these mallows were distinct enough to be in a new genus. He called the species *Modiola multifida* in 1794. It was not until 1831 that George Don realized that the Linnaean name had priority, and he gave us *Modiola caroliniana*.

It is not surprising that there was disagreement about these plants, because we now know that this species represents a monotypic genus with the single species native to the Americas (Mabberley 1997). The species has as a diagnostic trait a transverse partition between the seeds in the fruit locules, but otherwise it looks like *Malva*.

Carolina mallow is considered native from Virginia to Florida and southern Texas. From there, it grows through eastern Mexico and tropical America to Argentina (Fernald 1950, Correll and Johnston 1970). Although reported from California as if native, it was considered naturalized by Munz (1973).

The only record found of indigenous people using the plants was among the Houma. These people of the lower Mississippi River valley made a gargle for sore throat, tonsillitis, and diphtheria (Speck 1941). Neither Porcher (1863) nor Millspaugh (1892) mentioned the species. Similarly, it is absent from the remedies of the Eclectics (Culbreth 1910, Felter 1922) and modern herbal lists (Bown 1995, Duke et al. 2002).

Hocking (1997) does include the plants. He lists *Modiola* as having been used as an emollient and sedative, to treat edema, and for throat problems in Chile. It was used in a cold water extract as a healing bath. Grimé (1976) found records of the plants being used among blacks to treat menstrual problems.

Wunderlin (1998) considers the species native to Florida. I have followed him, although the pattern of distribution and use is suspiciously like that of an introduced plant.

Monarda

(Commemorates the Spanish physician Nicolás Bautista Monardes, ca. 1493–1578 or 1588, best known for his 1574 book *Historia Medicinal de las cosas que se traen de nuestras Indias Occidentales que serven en Medicina*)

Monarda punctata (dotted)
a-shem-bra (to make sleep; *a-zhon*, I sleep, Osage)
horsemint; *witsagwai wi·ti* (horse medicine, Catawba)
kofucka łako [*kofutcka rakko, takko*] (*kofucka*, fragrance, *łakko*, big or strong, Creek)
rignum (probably derived from *Origanum*, the genus of oregano, New England)
shinuktileli [*shinuktitleli*] (*shinuk*, sand [probably originally *shilup*, ghost], *tileli*, to drive out, Choctaw); *sholop tilhi* [*shoptilhi'li*] (*sholop*, ghost, *tiłi*, to send away, Chickasaw)
sholop tilhi iskanno (*sholop*, ghost, *tiłi*, to send away, *iskanno*, smaller, Chickasaw; probably belongs here); *sholop tilhi losa'* (*sholop*, ghost, *tiłi*, to send away, *losa'*, black, Chickasaw; probably belongs here)

John Bannister of Virginia collected the first plants of *Monarda punctata* in 1680 and sent them to Oxford University in England. However, it was best known to Linnaeus from garden and herbarium specimens collected by John Clayton in the 1730s (Reveal 1992a).

Swanton (1928a) found the Creeks using horsemint, and the plant is still used in Oklahoma (Howard 1984). An infusion of the plant causes perspiration. Mixed with *ahá labakca* (*Gnaphalium obtusifolium*) it relieves a delirious patient (Swanton 1928a, Howard 1984). The Alabama, Chickasaw, Choctaw, and Creek used it to ward off rheumatism and to protect the living (Howard 1984). The Catawba used a leaf decoction to relieve headache (King 1984). The Osage used it as a sudorific and for pain relief from coughs, chest complaints, and fevers (Hunter [1823] 1973). It was also used as perfume (La Flesche 1932). The plant also was used by the Delaware, Menomini, Meskwaki, Mohegan, Nanticoke, Navajo, Ojibwa, and probably others (Vogel 1970, King 1984, Moerman 1998).

Monotropa

(From Latin *mono*, one, *tropos*, turn, referring to the summit of the flowering stem being turned to one side)

Monotropa uniflora. From Britton and Brown 1897.

Monotropa uniflora (one-flowered) (= *M. brittoniana*?)
bird's nest (New Hampshire); nest-plant [nest-root]
broom-rape [broomrape] (New Hampshire; usually used for *Orobanche*)
convulsion-root [convulsion-weed] (New Hampshire, New Jersey)
corpse-plant (New Jersey); ghost-flower [ghost-plant] (Labrador, Maine, New Brunswick, Newfoundland)
Dutchman's pipe (New Jersey); Indian pipe (in use by 1817); fairy-smoke (Maine); pipe-plant; tobacco-pipe
eyebright (Maine)
fit-root [fitsroot, fit-root-plant] (South Carolina, Maine, New Brunswick)
[American] ice-plant [iceplant] (New Hampshire)
mipitahmaskihkih (Cree); *weakosin'nk* (in a bunch, Potawatomi)
ova-ova (English?)
[one-flowered] wax-plant

Monotropa uniflora is a circumboreal saprophyte that extends into the tropics at higher elevations in Central America (Mabberley 1997). Linnaeus ([1753] 1957) knew it from Canada to Virginia, and accepted the name *Monotropa* as proposed by Gronovius. Those who preceded his *Flora Virginica* considered it an *Orobanche*. Apparently, there are only two species in the genus, this and *M. hypopitys*.

Millspaugh (1892) recorded that the Americans learned about medical applications of these plants from the indigenous people; his source was Rafinesque's *Medical Flora.* Rafinesque said that juice from *Monotropa* "mixt with water deemed specific lotion for sore eyes" by native people. Porcher (1863) wrote, "This is used by the steam practitioners." Others consider it good for spasms and fainting spells, which explains some of the common names (Coffey 1993). The Potawatomi used the root for "female troubles" (Smith 1933).

Morinda: Cheeseweed

(Ostensibly commemorates the French physician Louis Morin, 1636–1715, although Linnaeus probably composed the name from the Latin *morus+indicus,* Indian mulberry)

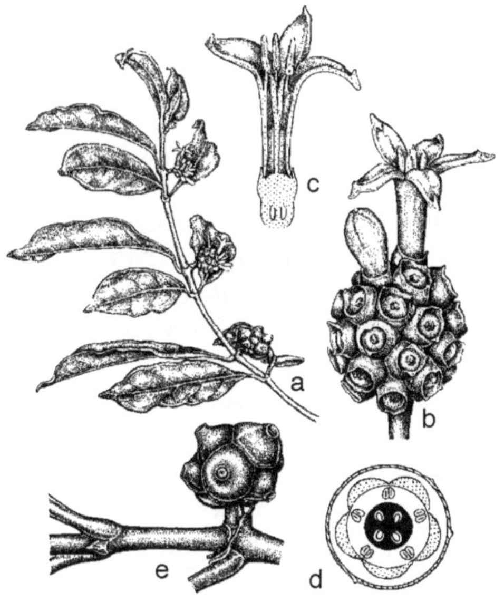

Morinda royoc. a. Flowering and fruiting branch. b. Syncarp, with young fruits and flowers. c. Flower, longitudinally dissected. d. Floral diagram. e. Branch with fruit. *Drawn by Priscilla Fawcett.* From Correll and Correll 1982.

Although it is also called Indian mulberry or red gal [redgal], most people who have been in Florida a long time call these shrubs cheeseweed. Technically, the plants are *Morinda royoc,* and belong in the Rubiaceae with coffee (*Coffea*), quinine (*Cinchona*), and gardenia (*Gardenia*).

I did not really understand why the shrubs were called cheeseweed until an "accident" by one of my students. She was gathering data for me on endangered species when she found a low shrub with small white flowers she did not know in one study plot. She put the fruiting branch in the back of her car at the end of a long day drove home and forgot about the branch. The next morning she drove to school early and parked as usual. That evening when she returned and opened the steamy hot car, an overpowering stench made her stagger backward. She described the odor in various ways, but the least damning comparison was to hot Limburger cheese. So, perhaps cheeseweed is as good a name as any.

Nearby Florida in the Bahama Islands, people call the species duppy poison, rhubarb, stiff cock, wild mulberry, yaw weed, and yellow ginger. Those names are related to several of the uses that people have for them there and elsewhere, as aphrodisiacs, medicines, and dyes.

The meanings of most of the words in these names are obvious, but "duppy" may not be. "Duppy" appeared in English literature for the first time about 1774. The word originated in the language of the Ashanti of Ghana, a matrilineal Akan people who brought the word to the Caribbean when they were sold there as slaves. The first statement published about their beliefs concerning duppies in Jamaica was, "They firmly believe in the apparition spectres. Those of deceased friends are *duppies*; others ... like our raw-head-and-bloody-bones are called *bugaboos.*" By the 1830s, the view had changed so that both duppies and bugaboos were feared. Nowadays that view has changed again, and duppies are mostly considered mischievous, as in the expression, "Is the dutty duppy man dweet" (The dirty ghost did it). One of the things duppies seem to enjoy is throwing stones (Williams 2002). The name "duppy poison" means not fit for human consumption, but it can also mean duppies use it for poison, or that it can be used to poison duppies.

A widespread use is as an aphrodisiac. Some of the names are uneasy allusions showing as much fear as belief. In addition to the names from the Bahamas, there are duivel's brood (devil-spawn), *loembra maatsjoe* (woman's bush?, Aruba, Bonaire, Curaçao), *palo garañon* (tree's semen), *pam di djabel* [*pam die diaabel*] (devil's spawn?, Aruba, Bonaire, Curaçao), *pepa de ratón* (rat's seed, Venezuela), and strong back (Jamaica). *Morinda* may be used alone or mixed with other plants; both *Chiococca alba* and *Smilax havanensis* have been added to the mixture (Ayensu 1981). Although some consider *Morinda* for men, others recommend it for women to regulate menstruation (Morton 1981).

Medicines of other kinds are also common. Some of the names relating to them are *curamagüey* (literally, cure's *Agave,* Cuba), painkiller (Jamaica), *piñipiñi* (Cuba), *raíz de indio* (Indian root), and probably *xcruz che'* (plant in the form of a cross, Maya, Yucatán). The Maya of Belize say *ax-ak* (*ax,* verrugose or lumpy, *ak,* vine). One of the common uses applied from Cuba to Curaçao is as a laxative or, depending on dosage, a purge. Names probably relating to that use are *bataata heel* (complete potato, Aruba, Bonaire, Curaçao), *estologe* (for the stomach, Hispaniola), *paja de ratón* (rat's straw), *piña kan* (snake's pineapple, Yucatán), *revienta caballo* (horse-splitter, a name generally applied to a poisonous member of Caribbean Campanulaceae, Cuba), rhubarb root (Cayman Islands), and *ruibarbo de la tierra* (wild rhubarb, Cuba). Wart vine (Belize) is unique.

Fruits are eaten by people and hogs. The latter use is reflected in the Jamaican name hog apple.

In Haiti, the Bahamas, and other places, *Morinda* is considered tonic, stomachic, and useful against diarrhea (Liogier 1974, Ayensu 1981, Morton 1981). One of the tonic remedies is to mix it with *Cissus verticillata*, *Smilax* spp., and *Desmodium incanum*. Fruits are used to make an antiscorbutic, depurative drink (Duke 1968, Liogier 1974). Young leaves have been eaten as a potherb, and the seeds of some species are roasted and eaten. The name sore throat (Bahamas) reflects that the leafy stems are boiled and the liquid gargled for that problem. The decoction is also used to ease teething pain in infants (Liogier 1974). Generally, the plant is analgesic and the wilted leaves may be pressed on painful parts of the body or used as poultices for headaches. Crushed leaves in lard or camphor are placed on the head for colds and neuralgia. In Haiti, it is considered a contra-venom against the *Tuatua* (Liogier 1974).

Morinda royoc is called yaw weed, and has been famous for treating that terrible malady since at least the 1800s. Although it is a disease unfamiliar to most people in temperate zones, it formerly was common in the Caribbean where it is thought to have been introduced by slaves from sub-Saharan Africa (Desowitz 1997). The cause is *Treponema pertenus*, a nonvenereal spirochete transmitted by contact other than sexual. Although the problem begins with an innocuous-appearing lesion of the skin, it later destroys not only the skin but also the underlying bone.

The genus is a noted dye plant. Names for *M. royoc* alluding to dyes are *paaloe zapateero* (shoemakers tree, Aruba, Bonaire, Curaçao), red gal (Jamaica), *safrán* (saffron, usually a name for *Cucurma longa*, Haiti), *safrán du pays* (country saffron, Haiti), *safrán du St. Domingue* (Santo Domingo saffron, Haiti), and yellow root (Cayman Islands). Colors vary with mordants and may be either saffron or red (Duke 1968, Morton 1981).

In English-speaking regions, *M. royoc* is sometimes called Indian mulberry. However, that name is used mostly for the related Asian plant, *M. citrifolia*. Linnaeus named both *M. royoc* and *M. citrifolia* in 1753. He based *M. citrifolia* on Rheede's *Flora Malabarica* where it was called *cada-pilava* (misspelled by Linnaeus as *coda-pilave*; modern spelling, *kadapilavu*; today the name of the breadfruit *Artocarpus altilis*) in the Malayalam language. The modern name of *M. citrifolia* in Malayalam is *kakapazham* (*kaka*, crow, *pazham*, fruit). Of course, the genus name is another of Linnaeus's dual meanings. The specific name *royoc* came from Charles Plumier's *Royoc humifusum, fructu cupressino* (the humble *royoc* with the fruit of the cypress), although no one knows where he got the word *royoc*. Perhaps it is based on *hoyoc*, the Mayan name for *M. yucatanensis*.

Roig (1945) recorded that the roots of *M. royoc* contain tannin, agalic acid, mucus, and calcium oxylate, and no subsequent chemical studies appear to have been made. Considering the chemicals found in other species, that seems a particularly short-sighted void. Many other species in the genus have been used by people, with 26 mentioned in the journal *Economic Botany* in the first 50 years of its existence (Kaplan 2001). In *M. citrifolia* alone, chemicals have been found that are analgesic, insecticidal, nematicidal, antineoplastic, and immunomodulative.

It should come as no surprise that *M. citrifolia* also has an impressive history of use because it is important to people from at least India to Hawaii (Watt [1889] 1972, Burkill 1966, Degener 1973, Kaaiakamanu and Akina 1976, Heinicke 1985, Aalbersberg et al. 1993, Dittmar 1993, Dornenburg and Knorr 1994a,b, Hagendoorn et al. 1994, Farine et al. 1996, Dixon et al. 1999, Hirazumi and Furusawa 1999).

Other species that have been used by people indicate analgesic, antifungal, antileukemic, antimicrobial, antinematodial, antineoplastic, antiviral, cytotoxic, and hypoglycemic activities (Durodola 1974, Chang et al. 1982, Younos et al. 1990, Kamanyi et al. 1994, Rath et al. 1995, Kwo and Craker 1996, Bankole 1997, Mackeen 1997, Ali et al. 2000), with possible control of fascioliasis and schistosomiasis, *Plasmodium falciparum*, trypanosomes, and ulcers (Adewunmi and Kayode Adesogan 1984, Asuzu and Chineme 1990, Koumaglo et al. 1992, Udem et al. 1993, Olajide 1998). Many of these effects are due to a large number of anthraquinones, iridoid glycosides, flavone glycosides, flavonoids, sterols, and nitrosatable amines (Singh and Tiwari 1976, Adesogan 1979, Vermes and Wagner 1980, Venkateshwar Rao and Rao 1983, Poulsen et al. 1991, Srivastava and Singh 1993, Dryas et al. 1994, Legal et al. 1994, Cimanga et al. 1995a,b, 1997, Leistner 1995, Uhegbu and Madkuagwu 1995, Ismail et al. 1997, Wang et al. 2000, Sang et al. 2001). Given the similar uses for *M. royoc*, it surely contains related chemicals.

Cheeseweed grows in southern Florida from Martin County around the southern tip to Lee County. Disjunct plants are known from as far north as Cape Canaveral in Brevard County on the east coast and Hillsborough County on the west coast. The shrubs also grow in the Bahamas, Cuba, Hispaniola, Jamaica, Grand Cayman, and southern Mexico to Ecuador and the northern coast of Venezuela. *Morinda royoc* is absent from Puerto Rico and the Lesser Antilles. There are now 80 species known in the genus, with all but about 22 species confined to the Old World tropics. The related *M. yucatanensis* is called *xhoyoc* [*joyoc, hoyok*] or *x-yen-cab* [*xhoyenkab*] (Maya) and *piñuela* (little pineapple, Yucatán). However, not all fruits called pineapples are as tasty as their namesake.

Morus: Mulberry

(From classical Latin *morum* and Greek *moron* for the mulberry and blackberry)

Morus rubra. From Sargent 1905.

Humans and other vertebrates near mulberry trees (*Morus rubra*) quickly become aware of the edible fruits when they are in season. If the people do not notice them immediately, the red-black stains left by the birds as they digest the sweet fruits are ample indication of their value. Although people throughout temperate Europe now prize mulberries, that is a comparatively recent phenomenon.

No mulberries are native to Europe or most of the Mediterranean region; only the black mulberry (*M. nigra*) grows wild in northern Persia on the shores of the Caspian Sea (Hedrick 1919, Mabberley 1997). Pliny (A.D. 23–79) noted that *M. nigra* was successfully reared in Italy by Lucius Vitellus after several failed attempts. At the time of Paladius and Athaneus (A.D. 170–230), the tree was uncommon. Silk culture under Justinian (A.D. 483–565) gave *Morus* new importance and it spread, reaching France in 1500 (Hedrick 1919). The white mulberry, *M. alba*, was brought from Asia into the Levant and then into Tuscany about 1434 (Hedrick 1919). Subsequently, *M. alba* replaced *M. nigra* in raising silkworms.

Although it hardly seems plausible, the word "mulberry" is based on the Latin *morus.* From *morus* or *morum* in Latin, the fruits became *mora* in Italian, Spanish, and Portuguese, and *mûre* in French. *Morus* plus berry was transcribed into *múlbere* in Old High German to become *Maulbeer* in modern German, *moerberie* in Dutch, and mulberry in English.

Mulberry is first mentioned in print in English in 1382: "The kyng saue in to Jerusalem ... cedres as long mulberries (Vulg. *cedros quasi sycamoros*)." The use of the word *sycamoros* is not an accidental comparison. Greeks like Theophrastus (372–287 B.C.) called the mulberry *sukamnos* or *sykamnos* [modern Greek = *sycaminus*]. The Greek words *sykamnos, sycamore,* and *sycomore* are variations of *sykomoros* (*sykon*, fig + *moros,* mulberry). To further complicate the situation, "sycamore" was first used in English in 1388 to refer to a fig (*Ficus*), then to a maple (*Acer*), and after the New World was discovered, yet another tree (*Platanus*). Cognate words occur in Hebrew as *shiqmah* and in Sumerian as *sukannu,* although in the Bible the tree is also *tut* in Macabees and *mesukan* in Isaiah (Zohary 1982). *Mesukan* is cognate to *messikanu* in Sumerian.

When Europeans arrived in the New World, they found another species, the red mulberry *M. rubra.* Linnaeus knew this tree from the publications of Leonard Plukenet who called it *Morus virginiensis arbor, loti arboris instar ramosa, foliis amplissimis* (mulberry tree from Virginia, a tree branched like lotus, leaves large).

Many early explorers of the New World commented on the native use of the trees. Hernando de Soto noted mulberries cultivated in what is now the Panhandle of Florida while en route to Apalachee in 1540 (Standley 1920–1926, Swanton 1939). Barrientos (1567) wrote that the de Soto expedition, upon arriving at *Cofitachiqui,* a Muskogee village in Georgia, found that "mulberry groves are so thick that it is necessary to cut them down in order to make plantings."

In 1682, Henri de Tonti was traveling with Sieur de la Salle down the Mississippi. In a Taencas (Natchez) village he "saw the chief seated on a camp bed ... surrounded by more than 60 old men, clothed in white cloaks, which were made by the women out of the bark of the mulberry tree" (Tonti in Borland 1965).

William Bartram, traveling near Wrightsborough, Georgia in the 1770s, also saw cultivated trees. Of old Creek villages Bartram says: "These are always on or near the banks of rivers, or great swamps, the artificial mounds and terraces elevating them above the surrounding groves. I observed, in the ancient cultivated fields ... *Morus rubra*" (Bartram [1791] 1958).

Red mulberry was used at least by the Alabama, Atakapa, Biloxi, Catawba, Cherokee, Chickasaw, Chitimacha, Choctaw, Comanche, Creek, Delaware, Iroquois, Koasati, Lakota, Meskwaki, Natchez, Ofo, Omaha, Onondaga, Powhatan, Rappahannock, Timucua, Seminoles, Shawnee, and Yuchi (Yanovsky 1936, Swanton 1946, Moerman 1998). The species is native from New England (western Massachusetts, Connecticut, Long Island), southern Ontario, west to Michigan, southern South Dakota and Texas, and southward to the tip of mainland Florida. In the southeast, at least the Choctaw, Creek, Koasati, and Yuchi had a "mulberry month." The Choctaw called it *hashe* [*hvsh*] *bihi* (*hvsh*, month, *bihi*, mulberry).

The Onondaga call the fruit *sachiissegóna òchia* (*òchia,* fruit). Algonquian speakers in New England, the Delaware, call the mulberry tree *oakhattim inschi.* Strachey ([1612] 1953) recorded the Powhatan name *muskimins.* The Shawnee, another Algonquian tribe, say *mtekwa·palwa.* Louisiana Siouxan-speakers said *an̲san̲kûdi'* (Biloxi) and *fafanaki* (Ofo). To their west the Atakapa said *cec.* Catawba said *tcutca'?.*

There are records of several cognate words for mulberry in the southeast. These include *bí:hî* [*biihi, behe*] (Mikasuki), *ke* [*kei, ki*] (Muskogee), *kí:han* (Creek), *bihi'* (Chickasaw), *bihi* (Choctaw), *bihi* [*bihala*] (Alabama), *bihalá* (Koasati), and *kuwa* [*ku:wa, koo wah'*] (Cherokee).

The importance of the species to these native people is apparent in names of towns (Swanton 1946). There was a town called *Kayomalgi* (mulberry place) by Gatschet, which Swanton (1946) also writes as *Cayomulge.* In both instances, the word for mulberry is *cayo,* equivalent to Creek *kei* or *ke.* The Cherokee have two or more settlements in South Carolina named *Kuwahi'yi,* abbreviated *Kuwahi',* "Mulberry-grove place."

Although the most frequently used names in English are simply mulberry or red mulberry, *M. rubra* is also called black mulberry, bulberry, Virginia mulberry, and Virginia mulberry tree. To the French it is *murier sauvage* (wild mulberry tree), while the Spanish say *moral* (mulberry grove), and the Germans *Maulbeerbaum* (mulberry tree).

The fruit was the part of the plant most used by people in North America, at least in its season. The earliest record is from the de Soto expedition in the Muskogee village *Cofitachiqui* (Barrientos 1567). The chroniclers recorded, "From the mulberries the Indians make raisins [*pasas*], which are good to eat." Various people employed it in basically the same ways, either eaten fresh or dried, and mashed and made into cakes for future use. Probably the early use was like the Iroquois who soaked the dried fruit cakes in warm water during the winter and then cooked them into a sauce or mixed them with corn bread.

The Cherokee historically mixed the berries with sugar and cornmeal to make dumplings (Hamel and Chiltoskey 1975). There are records of the Cherokee, Comanche, Iroquois, Omaha, and Seminoles eating the fruits, but it is likely that all people within the range of the species found *Morus* welcome food. Europeans and Americans in the New World quickly learned to make the fruits into jelly, jam, pies, and other sweets.

The Seminoles and probably others made bows from the branches (Sturtevant 1955). Americans used the wood for casks and tubs, fence posts, cross ties, furniture, and boats. Throughout the southern states, the trees were planted as food to fatten hogs and for poultry (Sargent 1905). Several groups used the inner bark to make cloth, including the Caddo and Tonkawa of Texas (Newcomb 1961) and the Natchez of Louisiana (Tonti in Borland 1965).

One of the first records of indigenous mulberry dye came from Frenchman Jacques Le Moyne during his stay in northeastern Florida (Swanton 1946). He recorded that the Timucua used the trees to dye materials in the mid-1560s. Colors derived from *Morus* vary depending on whether the leaves, stems, or fruits are used. Tans and yellow-greens are obtained from branches with an alum mordant, while gray-lavender to purple comes from berries with alum (Tull 1999). Cannon and Cannon (2002) noted that dyes from *Morus* result largely from the flavonol morin but rutin is also present. Bark and wood contain various flavones, including mulberrin, and these create golden yellows, bronze, olive, and browns with different mordants.

A mulberry caveat is that unripe fruits and sap from other parts can cause gastrointestinal distress and skin rash (Tull 1999). That toxicity also suggests that bioactive compounds are present. The mulberry was considered medicinal by the Alabama, Cherokee, Creek, Meskwaki, and Rappahannock (Moerman 1998). The Alabama and Creeks used a decoction of the roots to treat urinary problems (Swanton 1928a, Taylor 1940). The Cherokee took an infusion of bark to expel worms, to stop dysentery, and as a laxative (Hamel and Chiltoskey 1975). The Meskwaki considered the root bark a medicine for any illness. The Creeks also used the root as an emetic (Swanton 1928a). The Rappahannock rubbed on the sap to cure ringworm. As late as Millspaugh (1892), the bark was still being recommended as a cathartic and anthelmintic. Krochmal and Krochmal (1973) noted that the fruit could be made into a cooling drink to reduce fever, and it was also used as a mild laxative. A decoction of the outer bark was used to treat worms, while the inner bark was scraped and boiled to make a tea used as a children's laxative.

Little is known about the chemistry of *M. rubra* beyond the phenolics in the wood (Deshpande et al. 1974). However, a great deal is known about other species of *Morus*, and surely much of that applies to *M. rubra*. Various parts of *Morus* contain alkaloids, coumarins, cyanidin, ecdysterone, flavones, flavonoids, inokosterone, isoquercitrin, mucilage, pectin, phenolics, tannins, and triterpenoids (Kino et al. 1977, Hsu 1986, Nomura et al. 1988, Mihara et al. 1995, de Souza et al. 2000, Kim et al. 1999, 2000, Asano et al. 2001, Nomura 2001, Shi et al. 2001a,b, Kusano et al. 2002). The extracts are considered antibacterial, anti-inflammatory, antioxidant, antipyretic, diuretic, hypotensive, immunomodulating, and sedative (Chaterjee et al. 1983, Hsu 1986, Yen et al. 1996, Chen et al. 1998, Kim et al. 2000). In addition, *Morus* has shown potential for treating diabetes (Ionescu-Tirgoviste et al. 1989, Basnet et al. 1993, Chen et al. 1995a,b, Nojima et al. 1998, Limus et al. 1999, Andallu et al. 2001, Kubo et al. 2001, Petlevski et al. 2001) and hypolipidemia (Yassin et al. 2001).

Often when I see a mulberry tree I am reminded of the children's rhyme "Here we go round the mulberry bush." That rhyme always seemed odd to me because the mulberry is a tree, not a bush. It turns out that the use of the word "mulberry" in English is equivalent to Spanish where *mora* also means blackberry (*Rubus*). Vickery (1995) has traced the history of the word and found that the "mulberry bush" in the game was actually a *Rubus* and not a *Morus*, and the mixing of their names goes back to Horace (65–8 B.C.), Vergil (70–19 B.C.), and Ovid (43 B.C.–A.D. ?17). Indeed, a glossary published in the 11th century by R.P. Wülcker (Wright 1884) recorded that *mora* was used as a generic name for all berries.

One should be wary of identifications whether they are common names or Latin names.

Muhlenbergia

(J.C.D. von Schreber named this in honor of Gotthilf Heinrich [Henry] Muhlenberg, 1753–1815, a German-American Lutheran minister and amateur botanist)

Muhlenbergia capillaris (hairlike)
[long-awned] hair-grass [hairgrass] (Texas, Bahamas, Puerto Rico); *yerba de pelos* (hair grass, Puerto Rico)
[Gulf, hairy-awn, slender] muhly (Texas)
purple-grass (Bahamas, Puerto Rico)

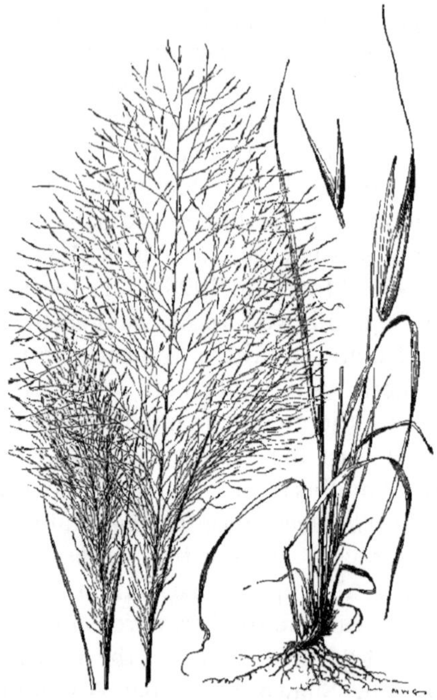

Muhlenbergia capillaris. *Drawn by Mary Wright Gill.* From Hitchcock and Chase 1950.

pvhe cvmpv (*pvhe*, grass, *cvmpv*, sweet, Muskogee); sweet-grass (English name used by Big Cypress Seminoles)
rabo de pavo real (peacock's tail, Puerto Rico)

Jean Baptiste Lamarck called this grass *Stipa capillaris* in 1791, only 2 years after Schreber proposed the genus *Muhlenbergia*. Subsequent exploration found other species, and the genus has grown to about 160 species in the world tropics (Mabberley 1997).

This species has a long history of use as fodder and a forage plant (Hocking 1997). Less well known is that the Seminoles now use, and probably historically used, it to make baskets. The weavers on the Big Cypress Seminole Reservation call the stems "sweet-grass," and books typically translate that as *Glyceria septentrionalis, Anthoxanthum odoratum,* or *Hierochloe odorata* (Fernald 1950, Radford et al. 1968). None of these genera is native to Florida, with all having ranges farther north.

Independent examination by Bennett (personal communication, 2000) and by me revealed that the Seminoles use several grass species. The individuals we talked with call them all "sweet-grass" and were aware that there were different kinds. Both *M. capillaris* and *Aristida* were used.

The baskets now offered by the Seminoles for sale in their shops are a design adapted from the blacks who lived with them in the past. The coiling techniques used to make the items resemble those employed by the Gullah (Geeche) of the coastal sea islands of the Carolinas, Georgia, and Florida. Pre-European baskets in the Southeast were woven and not sewn (Swanton 1946, Cooke and Sampley 1947).

Myrcianthes

(Named by Otto Carl Berg from *myrcia*, perfume + *anthos*, flower, for having flowers like *Myrica*)

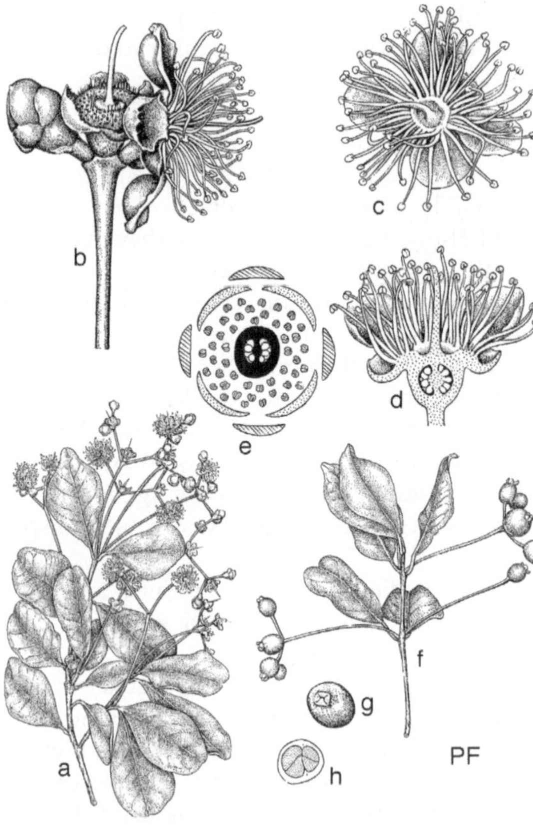

Myrcianthes fragrans. a. Branch with flowers. b. Distal part of flowering branch, with three-flowered dichasial unit; the bracteoles subtending lateral flowers have fallen. c. Flower, from above. d. Flower, longitudinally dissected. e. Floral diagram. f. Branch with fruit. g. Fruit. h. Fruit (two-seeded) in transverse section, showing fleshy cotyledons of embryo. *Drawn by Priscilla Fawcett.* From Correll and Correll 1982.

Myrcianthes fragrans (perfumed) **var. *simpsonii*** (for Charles Torrey Simpson, a naturalist and book author in Miami, well known in the early 1900s)

ahitiyã:ncó:bî (*ahi*, tree, *tiyán*, twisted, *có:bî*, big, Mikasuki); [*tokina:naɬákko*] *tokina:narákko* (*eto*, tree, *konhe*, crooked, bent, *ɬakko*, big, Creek)

arrayán (well rooted, Dominican Republic)

bois d'Inde (Indian bush, Haiti)

ciya:fá:põmí:kî (to make ax handles, Mikasuki)

goyavier bois (guava tree, Martinique); *guayabillo* (little guava, Tamaulipas); *guayabito* (little guava, Venezuela); *guayabacón* (*guayabacán*, Taino name for a *Myrica*, Puerto Rico)

merisier (cherry tree, Guadeloupe)

naked-wood [nakedwood] (Florida, Bahamas; also a name given to *Colubrina*, which see)

pale stopper (Bahamas); Simpson's stopper (Florida)

perfume bush (Bahamas)

pimienta (spicy one, Cuba); *pimientilla* (little spicy one, Tamaulipas)

Neither Morton (1981) nor Wunderlin (1998) recognizes varieties within the species, but var. *simpsonii* is readily distinguishable in the field. Although *Myrcianthes fragrans* var. *simpsonii* is endemic to Florida, the other variety is more widespread. In Mexico the species grows from Durango, San Luis Potosí, Tamaulipas, Veracruz, and Tepic, and throughout Central America and the Caribbean, to Colombia and Venezuela. Mexicans use the wood for building homes and other purposes (Standley 1920–1926) as do people in the Caribbean and elsewhere (Morton 1981). The Seminoles use the wood for pestles, food paddles, blowgun darts, and ax handles (Sturtevant 1955). Sometimes they called it *ciyá:fá:pomí:kî* (to make ax handles).

In the Caicos Islands a decoction of branch tips is a febrifuge, a remedy for aches and pains, and a perfume. Elsewhere in the Bahamas, the decoction is mixed with *Pluchea rosea* (camphor-weed) and used to reduce labor pain (Morton 1981). It is mixed with *Amyris elemifera* to drink or to bathe in to treat the flu (Ayensu 1981).

Myrica: Wax Myrtle

(Greek *myrike*, the name of the tamarisk, *Tamariscus*, or some other fragrant shrub; perhaps from *myrizein*, to perfume)

The view from my office windows when I first went to Florida was of a 90-plus acre triangle of land that retained much of the natural vegetation (Austin 1990). Scattered throughout the parcel were numerous wax myrtle bushes, scientifically *Myrica cerifera*. Each fall and winter, the tree swallows (*Tachycineta bicolor*) would come into the area in flocks of several thousand. The small birds, normally insect eaters during the summer, would spiral down to the wax myrtle bushes and snatch off berries, or land and then

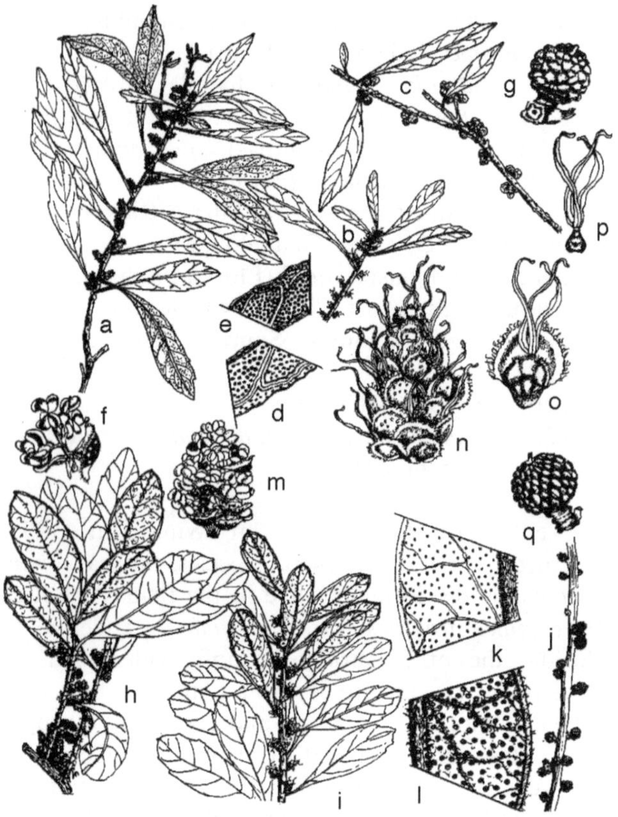

Myrica. a to g, *Myrica cerifera.* a. Branch with staminate flowers. b. Branch with pistillate flowers. c. Branch with fruit. d. Top of leaf. e. Underside of leaf. f. Staminate flowers. g. Fruit. h to q. *Myrica heterophylla*. h. Branch with staminate flowers. i. Branch with pistillate flowers. j. Branch with fruit. k. Top of leaf. l. Underside of leaf. m. Cluster of staminate flowers. n. Cluster of pistillate flowers. o. Pistillate flowers. p. Pistil. q. Fruit. *Drawn by Vivian Frazier.* From Correll and Correll 1972.

fly back up into the column. When I first saw this, there were so many birds in the group I thought it was a tornado. This process would continue until individual bushes were stripped of their berries, and then the funnel would move on to the next shrub.

I have been in the field when one of these flocks moved in, and the sound made me think there was a rain shower coming toward me. The sound resulted from a "shower" of wax myrtle seeds passing through bird guts and raining down around me. They are impressive displays of birds dispersing the seeds of their food plants.

Wax myrtle is more temperate than tropical but grows in both regions. The plants range from New Jersey to Florida, Bermuda, the Bahamas, the West Indies (Cuba, Hispaniola, Puerto Rico to Guadeloupe), west to Arkansas and Texas, from Veracruz and Chiapas to Yucatán, and Mesoamerica south to Costa Rica. *Myrica cerifera* became known to Europeans when it was illustrated by Mark Catesby (1731–1732). He called the plants *Myrtus brabanticae similis caroliniensis humilior, foliis latioribus & magis serratis* (similar to *Myrtus* from Belgium, but from Carolina and smaller, leaves wider and larger teeth) and put them in two different drawings (plates 13, 69).

Catesby was not the only person who thought the plants looked like myrtle (*Myrtus communis*). People from the southern United States to Belize call them mickleberry (Carolinas), muckle (Carolinas), muckleberry (Carolinas), myrtle (Belize), or sweet myrtle (Carolinas). Those names are variations of "myrtle."

Regarding use of the wax by people in the New World, Catesby wrote: "In November and December, at which time the Berries are mature, a Man with his Family will remove from his Home to some Island or Sandbanks near the Sea, where these Trees most abound, taking with him Kettles to boil the Berries in. He builds a Hut with Palmetto-Leaves, for the shelter of Himself and Family, while they dry, which is commonly Three or Four Weeks.

"The Man cuts down the Trees, while the Children strip off the Berries into a Porridge-Pot, and having put Water to 'em, they boil them 'till the Oil floats, which is skimmed off into another Vessel. This is repeated until there remains no more Oil. This, when cold, hardens to the Consistency of Wax, and is of a dirty green Colour. Then they boil it again, and clarify it in brass Kettles, which gives it a transparent Greenness.

"These candles burn a long time, and yield a grateful Smell. They usually add a fourth Part of Tallow, which makes them burn cleaner."

Use as a wax source for making smokeless candles was a long-established practice by the time Catesby recorded it, and many people note that in their plant names. They call the shrubs *arbol de cera* (wax tree, Hidalgo, Veracruz, Oaxaca, Dominican Republic), *arbre a suif* (tallow shrub, Canada), candleberry (USA), candleberry myrtle (Catesby 1731–1732), *cera vegetal* (vegetable wax, Central America), *cerero* (waxy one, Puerto Rico), *cerier* (wax tree, Houma, Louisiana), *cirier de la Louisiane* (Louisiana wax tree, USA), *palo de cera* (wax tree, Dominican Republic), southern wax myrtle, tallow berry (Carolinas), tallow shrub (USA), wax berry (Puerto Rico, Jamaica), southern wax myrtle (Florida), wax tree (Bartram [1791] 1958), and wax wood (Jamaica). The wax is harder and more brittle than beeswax, and it is still used to make and scent soap in North America and the Caribbean (Bown 1995).

The Houma also used *Myrica heterophylla* (evergreen bayberry) berries to make candle wax (Speck 1941). If that is any indication, they probably also used

M. inodora, as it was being used ca. 1800 by inhabitants of Alabama to make candles (Hocking 1997). Indeed, Bartram ([1791] 1958) found it on the Upper Tensaw River near Hall's Creek, Alabama, where it was used for "production of wax for candles, for which purpose it answers equally well with beeswax, or preferable, as it is harder and more lasting in burning."

Shortly after Catesby's book was published, the plants were growing in the garden maintained by wealthy Anglo-Dutch financier George Clifford at Hartekamp in Haarlem, Holland, where Linnaeus studied them for his book *Hortus Cliffortianus* of 1738. About 10 years later Linnaeus recorded the plants growing in the botanical garden at Uppsala, Sweden, where he was a professor. It was from Catesby's drawings and from personal knowledge of living plants that Linnaeus gave them the name in use today—*Myrica cerifera* (wax bearing).

The Latin name refers not only to the fruits as a source of wax for making candles, but also to the fragrance of the leaves. That fragrance also was enjoyed as a flavor in foods. Names that reflect that use are bayberry (USA, Puerto Rico), bayberry tallow (USA), southern bayberry, and sweet bay (Carolinas). The name "bayberry" is redundant because "bay" comes from the Latin *baca* (berry), through the Old French *baie*, to *beg* in Old English in the 11th century, and finally to *baye, baie,* or *bay* about 1398.

"Bayberry" appeared in English about 1576 when Henry Lyte translated the Dutch herbalist Rembert Dodoens's comments on the Old World species *Laurus nobilis* as, "Called in Latine *Lauri baccae*, in English Bay berries." Sweet bay is a name for *M. cerifera* and for *Magnolia virginica.* The name spicebush indicates to others that the leaves were used to season foods, but that name is typically applied to *Lindera benzoin* (which see).

Food was not the only item that people flavored with the leaves of *Myrica.* The Seminoles call the shrubs *cowã:nócâ:pî* [*tcowano'tcapi, cowaanocaapi, chowane emochaape, chuono chopp, chuone em ochopee*] (*chowane*, fly, *em*, its, *ochaape*, garden, Mikasuki) or *soli:capî* [*shule chopee*] (*sule*, buzzard, *cupof*, garden, Creek) and their neighbors say *hungwekilo* (*wiikih*, heavy, Choctaw). All three used the leaves to make their tobacco last longer or as a substitute for *Nicotiana* (Sturtevant 1955, Bennett 1997). That usage is clearly indicated in the name used by their relatives the Alabama, who called these shrubs *ito hakchomma* (*ito*, tree, *hakchomma*, tobacco) or *choyyiniyaachi* [*choyyinihaachi*] (*choyyiniha*, resin, *chi*, small). The Koasati simply said *ittoikíllo* (*ittó*, tree, *ikíllo*, undying).

Other names for the plants record different views. In the Caribbean, they are *arrayán* [*arraiján, arraigan*] (well rooted, also applied to *Eugenia, Myrcianthes, Myrciaria*, Cuba, Puerto Rico), *caca-ravet* (*caca*, feces, *ravet*, rat, Guadeloupe, Martinique), *perico* (a type of parrot, Hispaniola), *sause bastardo* (false willow, Belize), and *tiguapén* (*tigua* is an aquatic bird, Hispaniola). In Mexico, they are *chac-ol'ol* [*chac-olol*] (*chak*, red, *ol'ol,* renewal, in reference to young leaves, Maya, Oaxaca), *gua-ut*, or *huanacanalá* (Veracruz, Oaxaca).

Medicinal uses are reflected in the names tea-bark (Belize, Carolinas), tea-box (Belize, Carolinas), and wild tea (Carolinas). Millspaugh (1892) said the first record of using the plants as medicine was in Louisiana in 1722. That document noted that the water in which the wax had been boiled was a cure for the most violent cases of dysentery. Plants were used by the Choctaw (Bushnell 1909), Houma (Speck 1941), Koasati (Taylor 1940), and Seminoles (Sturtevant 1955), and the Houma and Koasati lived near the area where medical use was first recorded by Americans. Usually, European-derived people learned to use New World plants from natives whether or not history has recorded that exchange.

Other people who used *Myrica cerifera* as medicine were the Micmac of New Brunswick and Nova Scotia, although all the people in eastern North America used other species of *Myrica* as medicine, including at least the Bella Coola, Woodlands Cree, Delaware, Mohegan, Ojibwa, and Potawatomi (Smith 1933, Moerman 1998). The Micmac used *M. cerifera* as an analgesic, an exhilarating beverage, a diuretic, emetic, febrifuge, gargle for inflamed tonsils, to treat headache and external inflammation, as a vermifuge, and for stomachache. People in South Carolina, the Caribbean, and Central America used the plants in the same ways (Martínez 1969, Morton 1974, 1981). Eclectic physicians (Felter 1922) considered the wax ("myricin") useful for problems of the mucous membranes such as "stomatitis, nursing sore mouth, ulcerated fauces [throat], bowels and stomach" (Millspaugh 1892). Murphee (1965) found people in the Panhandle using it to speed up labor during childbirth. The plants are still considered medicinal by the Herb Society of America, with mostly the same uses as in the 1700s (Bown 1995).

Among the bioactive chemicals in *M. cerifera* are the flavonoid glycoside myricitrin, the triterpenes myricadiol, taraxerol, and taraxerone, and myrica acid, an oleanane triterpenic acid (Paul et al. 1974, Nagai et al. 2000). Also, the nonpeptide endothelin receptor antagonist triterpenoid myriceric acid A is known (Fujimoto et al. 1993, Mihara and Fujimoto 1993, Sakurawi et al. 1996). *Myrica cerifera* also contains palmitic acid and tannins (Kapadia et al. 1976).

In German *M. cerifera* is *Wachsbusch* (wax bush) or *Wachsgagle* (wax gale). The latter name compares it with the European and American *M. gale.* That shrub from Europe is called *roid* (Gaelic), *bwrle* (Welsh), and *pors* (Norwegian, Swedish). In northeastern North America these plants are sweet gale, meadow-fern, *piment royal* (royal pepper, Canada), or *bois sent-bon* (sweet-scent bush, Canada).

You may be more familiar with *Myrica* than you think because the leaves also are used to improve the foaming of beer. The next time you drink beer, especially European brands, remember that the head is probably due to *Myrica.*

N

Nelumbo

(Linnaeus first used *nelumbo* in 1753 for the species name of the sacred lotus, now *N. nucifera;* he derived *nelumbo* from the Sinhalese name, *nelun* [*nelum*]; in 1763, Michel Adanson reapplied the word to the genus)

Nelumbo lutea. a. Habit. b. Pod, longitudinal section. c. Seed. d. Stamens. *Drawn by Regina O. Hughes.* From Reed 1971. Fruit (inset, upper right). From Institute of Food and Agricultural Sciences.

Nelumbo lutea (yellow)
akkotorkv (Creek), *aco:batabí* [*aco:bata:bí, co:batabí*] (Koasati)
alligator buttons (South Carolina); alligator peas (South Carolina); sacred-bean (Louisiana)
American nelumbo (Mississippi)
[big]-bonnet (Mississippi, Tennessee)
can-dock (Louisiana)
duck-acorn (Louisiana)
graine à voler (stealing seed, maybe a reference to the tendency to "steal" habitat and crowd out other aquatics; or maybe a food worth stealing, Louisiana); *vole* (thief, Louisiana)
grand oval (big oval [leaf], Louisiana)
great yellow water-lily (New York); yellow pond-lily (Arkansas)
jacquinot (maybe based on Algonquian *yankapin,* Missouri)
knock-knocks (South Carolina)
kat'ho'pe (Atakapa)
[American] lotus (in use by 1584 in English for these aquatic plants, from Egyptian *lotos*; in modern Arabic *bashnyn abklyad,* white lotus or *nîlûfar,* for *N. nucifera*; see *Celtis*: Hackberry, for alternate etymology); *Lotosblume* (lotus flower, German); yellow lotus (Arkansas)
[lily, monaca, pond, rattle]-nut (Louisiana, Wisconsin)
tethawe (Omaha-Ponca); *tewape* (Dakota); *tse'-wa-the* (*tse'-wa,* bison, *the,* this, Osage); *tsherop* (Winnebago)
tukawiu (Pawnee)
wampain [*wankain, wonkapin, yankapin, yonkapin*] (related to *chinquapin,* the Powhatan name in Virginia, first recorded in 1624 by Capt. John Smith but applied to *Castanea*; Arkansas, Illinois, Indiana, Louisiana); *wagipîn* (crooked potato, Potawatomi); water chinquapin; *yonkapin-bonnet* (akin to *yankapin* and *wampain,* Algonquian, Tennessee); yawk-nut (akin to *yonkapin,* Louisiana)

The American lotus was discovered late by European scientists. Maybe the plants have always been uncommon in the northern part of North America, or perhaps they were rare because the indigenous people had eaten almost all of them. William Bartram actually collected one of the first plants in the 1770s; his collection forms the basis of the name *Nelumbo lutea.* That name was finally published in 1799 when C.L. von Willdenow created it as he revised Linnaeus's *Species Plantarum.*

However, it may be that visitors to the New World thought they were seeing a plant they had known since the 1600s or perhaps earlier. The Asian species had been known in Europe since at least the time of Rheede's *Hortus Indicus Malabaricus* in 1667–1703. Linnaeus had studied living plants for his *Hortus Cliffortianus* of 1738 and herbarium material for his *Flora Zeylanica* of 1747.

Bartram ([1791] 1958), for example, noted water lilies from Alabama, Florida, New Jersey, and North Carolina. He called them *Nymphaea nelumbo,* the name that Linnaeus had applied to the Indian water lotus in 1753.

Bartram ([1791] 1958) provided one of the first accounts of personal experience eating the seeds of these New World aquatic herbs. He wrote: "The seed vessel when ripe, is perforated, each cell containing an oval osseus gland or nut, of the size of a filbert; when they are fully grown, before they become quite hard, they are sweet and pleasant eating, and taste like chestnuts: I have fed freely on them without any injury, but found them laxative."

Rafinesque wrote in 1817 that the roots were also eaten by indigenous people, but said little more than that they had to be prepared first (Hedrick 1919). Thomas Nuttall gave perhaps the first details about indigenous people using *Nelumbo.* In 1821, he wrote of the Quapaw in Arkansas: "This morning I observed the wife of the chief, preparing for her family a breakfast from the nuts of the Cyamus (or Nelumbo). They are first steeped in water, and parched in sand, to extricate the kernels, which are afterwards mixed with fat, and made into a palatable soup. The tubers of the root, somewhat resembling patatas or sweet potatoes, when well-boiled, are but little inferior to a farinaceous potato" (Nuttall in Coffey 1993). In 1837, Nuttall wrote of the Osage that *N. lutea* "is everywhere made use of by the natives, who collect both the nuts and roots" (Gilmore 1919).

Most of the recorded uses are as food. Various parts were used by the tribes and required distinctive preparation. Seeds were used as food by at least the Comanche, Creek, Dakota, Meskwaki, Ojibwa, Omaha, Osage, Oto, Pawnee, Ponca, Potawatomi, Quapaw, Seminoles and Winnebago (Gilmore 1919, Smith 1933, Fernald et al. 1958, Moerman 1998). Roots were eaten by the Dakota, Huron, Ojibwa, Omaha, Osage, Oto, Pawnee, Ponca, Potawatomi, Quapaw, Seminoles, and Winnebago (Gilmore 1919, Smith 1933, Fernald et al. 1958, Kindscher 1987, Moerman 1998, Snow and Stans 2001). Young leafstalks and unrolling leaves were used by some as a potherb (Fernald et al. 1958). Perhaps that is what Moerman (1998) meant when he wrote about "terminal shoots cut crosswise, strung on string, and dried for winter use" by the Meskwaki. However, the more widespread use, as among the Ojibwa, was to cut the rhizomes and store them for winter (Moerman 1998). The Missouri River people peeled and then cut tubers into pieces about an inch long and dried them for winter (Gilmore 1919).

The Dakota, Omaha, Pawnee, Ponca, and Winnebago attributed mystical powers to *Nelumbo* (Gilmore 1919). The reason for that is not given, but they held that tubers collected by a tall person would be long, while those gathered by someone short would be short. Perhaps that same attitude led to people using *Nelumbo* as medicine. Huron Smith wrote in 1933 that the Potawatomi used the roots, either fresh or dried, as a poultice for many "inflammatory problems" (Smith 1933). The same or a different view might have been the case when physician Cadwallader Colden told Peter Kalm in 1751 that the "Indians [probably Iroquois] use the roots" as a preventative for scurvy (Vogel 1970). Although Colden called the plants "a *Nymphaea*," he may have meant the species. It could just as easily have been a *Nymphaea*, or even both genera, as either would serve the purpose.

Alice Snow had only heard that the green part of the seed (embryo?) had been made into a remedy to treat whooping cough (Snow and Stans 2001). She noted, however, that the seeds (*aktvpehv*, *ak*, in water, *tvph*, flat) were still eaten when they could be found. Snow observed, "There aren't many of these plants any more because the white people take them out by the trailer full and take them up north and sell them. They use them for flower arrangements" (Snow and Stans 2001).

Nicotiana

(Linnaeus named tobacco after the French diplomat Jean Nicot, 1530–1600, ambassador to Portugal in 1560, who promoted *N. tabacum* to the Portuguese and French courts; Nicot considered the plants of great medicinal value)

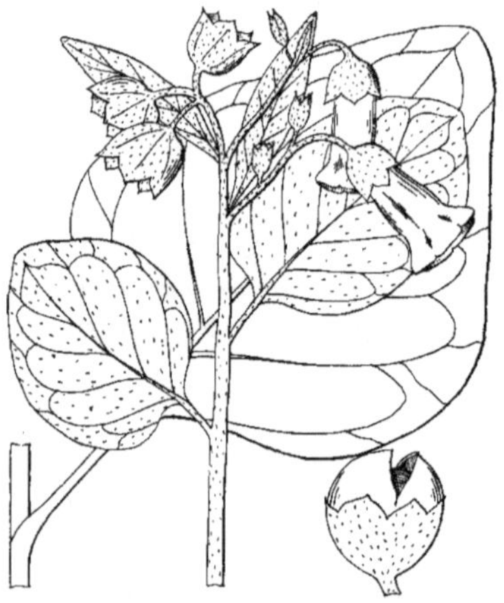

Nicotiana rustica. From Britton and Brown 1898.

Nicotiana rustica (of the country)
Nicotiana tabacum (Taino name for rolled leaves of the plant used for smoking)
aka (Sumu, Nicaragua)
akomî [*akahke*] (tobacco, Mikasuki); *hece* [*héce, hicî*] (tobacco, Muskogee); *hini* (Timucua; Crawford 1988 gave *epa* as "use tobacco," in

Timucua, but Gatschet 1877–1880 gave *hini*); *hisí* [*ahisi*] tobacco, Alabama); *hitci* [*ahecha, heche, hitche*] (said to mean "finding," because that was the warrior's name given to this powerful plant after it was discovered, cf. Swanton 1929; however, that derivation is doubtful, Creek); *hici pvkpvki* [*hitci pakpagi*] (*hici*, tobacco, *pvkpvki*, blossom, Creek; William Bartram was called *Puc Puggy* [*pvkpvki*, blossom] because he hunted flowers, cf. Bartram [1791] 1958; *Hici pvkpvki*, mentioned in the Creek "Emergence and Migration" story recounted in 1735 by *Chekilli*, cf. Lankford 1987; also in the Creek "Origin of Tobacco" story as "the greatest medicine there is," cf. Swanton 1929; probably *N. rustica* although Swanton thought it might be *Lobelia*); *hitci atculi pakpagi* [*itch-au-chu-le-puc-pug-gee, eschalapootche, hici lapucki*] (*hici*, tobacco, *vcule*, old man, *pvkpvke*, blossom; often "old man's tobacco" or "blossom of the ancient people's tobacco," *N. rustica*, Creek, Muskogee, cf. Howard 1984)

chandi [*tchondee*] (used for *N. quadrivalvis* by Gilmore 1919, but no distinction made when *N. tabacum* became available, Dakota, cf. Catlin [1844] 1973); *non-ni'-hi* [*no-ne-aw, no-ní-hi*] (*non-ni'*, tobacco, *hi*, plant, Osage; archaic for any smoking material)

charhû' [*tcarhu, jarhooh*] (fire to hold in the mouth, Tuscarora); *tsa'lu* [*tsalu, tsárú, tso·la, szólaw, jola*] (Cherokee)

chomak (Chickasaw); *hakchomma* (*hakcho*, tobacco, *homma*, red, Alabama); *hakchuma* [*achumon*] (Choctaw; probably from two words as in Alabama); *hakcommí* (Koasati)

i'pa (Catawba)

i'tci [*i*] (Yuchi)

iste'waw (Plains Cree); *se'ma* (Potawatomi); *theyamah* (Shawnee, fide Edgar 1891)

iúri (Garífuna, Nicaragua; cf. Coe and Anderson 1996)

kschátey (Delaware)

kuutz [*cutz, k'utz*] (Maya)

lapscon (Arikara, cf. Catlin [1844] 1973)

mannah sha (Mandan, cf. Catlin [1844] 1973)

ojenqua (Onondaga)

petume [*petum, petema, petima*] (from *pe'tim*, Tupí, Brazil)

pistacan (Blackfoot, cf. Catlin [1844] 1973)

punche [*punchi*] (New Mexico; cf. White 1941, 1942, Curtin 1947, Ford 1975, Cobos 1983, Howard 1984)

tac (Ofo)

tobacco [tobaco, tobago, topaco] (to the Taino of the Caribbean, *tabaco* meant what we today call a cigar; "tobacco" was applied after the plant was introduced to Europe); *pyé tabak* (country tobacco, Haiti); *tabac* [*tabak*] (France, Haiti); *tabac mannoque* (*manoque* is from *manoca*, a fistful, and specifically refers to a bundle of leaves tied together at the base, using one of the leafstems, Haiti); *tabacco* (Italian); *tabaco* (Puerto Rico); *tabaco de olor* (fragrant tobacco, Dominican Republic), *Tabak* (German); *tombaca* [*tobac*] (Gaelic)

tsit [*tsigg*] (Atakapa)

twahko [*twahku*] (Miskito, Nicaragua); *wì'rú* (Paya, Honduras)

uppówoc (they puff it, Carolina Algonquian, North Carolina); *vhpooc* [*ahpooc, apookan, apooke, pooc, uhpoocan, upooc*] (*N. rustica*; Powhatan, Virginia; Strachey 1612)

yani' (Biloxi)

yetl (Náhuatl, Mexico); *picietl* [*pecielt*] (from *piciliui*, reduce in size, *yetl*, tobacco; *N. rustica*, Náhuatl, Mexico)

zèb a la rèn (kidney herb, Haiti); *herbe à la reine*, queen's herb; a mistake for *zèb a la rèn*?)

My introduction to tobacco came early because my father and grandfather smoked, but my most vivid early recollection is about the plant as medicine. When I was still too young to be in school, I developed a terrible earache. We lived outside the city limits, and that was before there were telephones to summon a physician. I remember lying on the couch sobbing from the painful throbbing. My father lit up a cigarette, squatted beside me, and gently blew smoke into my ear. Almost immediately the pain lessened. A few more puffs, and the aching was gone. I was greatly relieved, and can easily understand why early Europeans like Jean Nicot considered the weed medicinal.

There has been confusion about the identity of tobacco and the Taino name for it since Columbus first recorded finding the plant in the Caribbean. Part of the problem is that he found Taino people carrying "a half burnt weed in their hands, the herbs for smoking which they are in the habit of using" (Jane and Skelton 1960) and elsewhere noted they called it *cojiba* [*cohiba*]. Coll y Toste (1972) emphasized that the *cojiba* was the plant, and the rolled leaves used for smoking were *tobaco*. In 1498, the Spanish wrote that the Taino called one island *tobaco*, although we now call it Tobago (Millspaugh 1892, Coll y Toste 1972). Spaniards were using tobacco in Yucatán by 1520, and Hernando de Toledo introduced it from there to Spain in 1559 (Millspaugh 1892). Columbus had probably already taken it back to Spain. It was being grown at Angoulême in France in 1556 (Hawkes 1998), before Nicot is supposed to have introduced it to the Portuguese and French courts.

It was Bartolome Las Casas in 1552 who, after Columbus, first gave the definition of *tobaco* as what we now call a cigar (from Spanish *cigarra*, the insect cicada, because of similarity of the rolled tobacco leaves to the insect's body; in use by 1730). At least Nicholas Monardes used the modern concept of *tobaco* in 1569 and illustrated *N. tabacum* (Frampton [1577] 1925). Rembert Dodoens had illustrated *N. rustica* in 1554, even though he called it "yellow henbane" (Goodspeed 1954). Sir Francis Drake and Sir Walter Raleigh found *N. rustica* being used by Canadian tribes and brought it to England in 1586 (Millspaugh 1983, Mabberley 1997).

There are 67 *Nicotiana* species in the Americas, the South Pacific, and Australia, with 7 or so used by indigenous North Americans (Goodspeed 1954, Mabberley 1997, Moerman 1998). We really do not know which tobacco Columbus found in use on Hispaniola (Sauer 1969), but the assumption is that it was *N. tabacum*. That species seems to have been confined to the Caribbean or nearby tropical mainland, and John Rolfe introduced *N. tabacum* from the West Indies into Virginia in 1610 or 1611 (Winter 2000). In doing so, Rolfe launched the basis of plantation life that still dominates the agricultural economy of the southeastern United States (Vogel 1970). *Nicotiana rustica* was cultivated from Peru to Canada, and some sources call it "Aztec tobacco" (e.g., Lewis and Elvin-Lewis 1977).

Because the de Soto narratives did not mention tobacco, Swanton (1946) concluded that it "had not attained the social significance it enjoyed in the seventeenth and eighteenth centuries." However, there is another interpretation. Since at least the 16th and 17th centuries it was a universal custom to smoke in friendly situations (Winter 2000), so there is no reason to expect the de Soto narratives to comment on tobacco. De Soto and his men ransacked every village they found, so the pipe certainly would not have been proffered.

Jacques Le Moyne, writing of the French Huguenot colony in Florida in 1564–1565 said, "They have a certain plant [*hini*], which the Brazilians call *petum* (*petun*), and the Spaniards *tapaco*. The leaves of this, carefully dried, they place in the wider part of a pipe; and setting them on fire, and putting the other end in their mouths, they inhale the smoke so strongly, that it comes out their mouths and noses, and operates powerfully to expel the humors" (Swanton 1946).

Barlowe reported tobacco growing along with corn in the fields of the Algonquians of North Carolina (Swanton 1946) in 1584. Similarly, Harriot ([1590] 1972) wrote at length about tobacco among the same people, recording the importance the plants already had throughout the southeast: "This *Uppówoc* is of so precious estimation amongest them, that they thinke their gods are maruelously delighted therewith: Whereupon sometimes they make hallowed fires & cast some of the pouder therein for a sacrifice: being in a storm vppon the waters, to pacifie theyr gods, they cast some vp into the aire and into the water: so a weare [weir] for fish being newly set vp, they cast some into the aire; also after an escape of danger, they cast some into the aire likewise: but all done with strange gestures, stamping, and sometimes dauncing, clapping of hands, holding vp of hands, & staring vp into the heavens, vttering therewithal and chattering strange words & noises."

In 1612, Strachey recorded the Powhatan name tobacco and several associated words. These Virginia people called the pipe *apokan* [*vppocane, vhpoocan*] (Harrington 1955), which has been transcribed phoenetically as *ohpokan* by Siebert (1975), who also gave cognates for the word "tobacco pipe." These related Algonquian words for "tobacco pipe," and hence tobacco, are Cree (*ospwakan*), Delaware (*hopékan*), Fox (*ahwakana*), Menomini (*ohpuakan*), Miami (*pwakana*), and Ojibwa (*oppwakan*). The Powhatan called the tobacco pouch or bag a *vttamahcoih*.

Le Page Du Pratz recorded in 1758 that the tribes of the lower Mississippi used the calumet "in their treaties of peace and in their embassages, the principal usage of which is that the deputies of the two nations smoke it [tobacco] together" (Swanton 1946). He further commented that the *machonchchi* (*basho'nkchi*, the sumac that bears the purple bud, Mobilian or Choctaw) was mixed with the tobacco, "because in smoking they do not care to have the tobacco so strong" (Du Pratz in Swanton 1946).

The Creeks continued using *N. rustica* in medicines, in busk preparations, and as an offering in post holes where new cabins were erected in a "Square Ground" (Swanton 1946). Similarly, the Chickasaw were using it for those purposes in 1775 (Adair in Swanton 1946). Romans ([1775] 1961), however, found the Choctaw raising "some tobacco, and [they] even sell some to the traders, but when they use it for smoking they mix it" with sumac (*Rhus glabra*) and sweetgum (*Liquidambar styraciflua*).

A Creek name that probably refers to the old tobacco, *hici pvkpvki*, was recorded in a 1735 legend (Lankford 1987). Indeed, *N. rustica* was incorporated into indigenous stories by the Cherokee, Creeks, Hitichiti, and Yuchi. The Creeks, Hitichiti, and Yuchi share a legend, although the Yuchi are linguistically isolated (Swanton 1929, Jackson 2003). Here is a synopsis of the Yuchi version (Lankford 1987):

> A man and woman went into the woods and had intercourse, and some semen fell on the ground. They went their own ways, but much later she again passed by the spot. A plant she had never seen before was growing there. Her lover did not know the plant, so she took her son to see it. It is implied that the son came from the earlier union, as he was "fatherless." He was not familiar with the plant either, but decided to name it. He called it "*i*," tobacco. He took some home, planted and grew it, experimented with it and showed the rest of his people how to grow and use it.

The legend among the Cherokee was considerably different. It concentrated on the beginning of the world when people and animals were the same (Mooney 1885–1886).

> There was only one *tsa'lu* plant in the beginning and all came to that plant for their tobacco. Then the *Dagul'ku* geese stole it and carried it far to the south. People were suffering without it, and there was one woman who grew so thin and weak everyone thought she would die soon without tobacco.
>
> Different animals tried to retrieve the plant. Both large and small were killed by the *Dagul'ku* before they could get the plant. Even the Mole tried, but he too was killed.
>
> Finally, the Hummingbird offered to try. They said he was too small and should stay home. To prove his worth, he told them to point out a plant in the field and see how he would steal tobacco. They pointed out a plant, and suddenly there he was sitting on it. The next moment he was back beside them. No one had seen him move because he was so swift. "That is the way I'll do it," he said. So, they let him try.
>
> He flew off and soon saw the tobacco. The *Dagul'ku* were all around it but he was so small and swift they did not see him. He darted down on the plant, snatched off the top with leaves and seeds, and was off before the *Dagul'ku* knew what had happened. Before he got home the old woman had fainted and they thought she was dead. But, he blew the smoke into her nostrils, and with a cry of "*Tsa'lu*!" [tobacco] she opened her eyes and was alive again.

Presumably, the tobacco of the ancients was abandoned, or at least relegated to religious purposes, when the Caribbean *N. tabacum* became available. Swanton (1946) concluded that little time was devoted to cultivating *N. rustica*, in spite of Barlowe's and Romans's comments to the contrary and the clear indications in the legends that it was grown. Typically, *N. rustica* was "encouraged" but not always cultivated in the sense of ground tilling as for maize and beans (Winter 2001).

Although Small (1933) included *N. rustica* in the Florida flora, Wunderlin (1998) did not. Perhaps inclusion or exclusion is a question of what constitutes part of the "native" flora, because Chapman (1897) did not even discuss the genus in the southeastern United States. Kartesz (1994), like Small, also allows *N. rustica* among the indigenous North American plants. Asch and Asch (1985) have dated *N. rustica* in Illinois at A.D. 500–600, and in Iowa at A.D. 450–650, and Winter (2001) gives more information about it. Although it is clear that this tropical American species was cultivated by 2000 B.P. at the latest and introduced by people into the eastern states (Fritz 2000b), it arrived long before the Europeans and therefore meets my criteria for inclusion.

Nuphar

(Named by James Edward Smith, 1759–1828, using an Arabic name, *nîlûfar*)

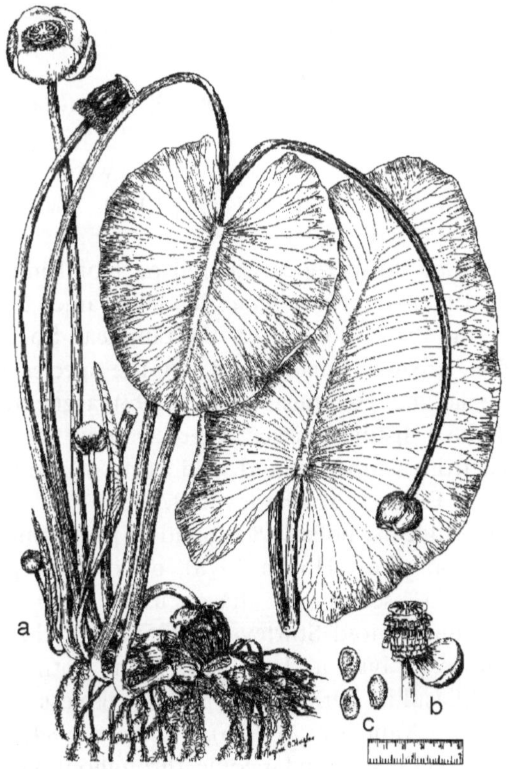

Nuphar lutea. a. Habit. b. Petal and stamen detail. c. Seeds. *Drawn by Regina O. Hughes.* From Reed 1971.

golfão amarelo (yellow gulf; *golfão* is the augmentative from Greek *kolpos* and Latin *colup* or *golfu*, a gulf or large bay, presumably applied because these plants grow there, Portuguese)
nenúfar amarillo (yellow nuphar, Spanish); *nénuphar jaune* (yellow nuphar, French)
ninfea gialla (yellow nymph, Italian)
Teichrose (pond rose, German)

Nuphar advena (newly arrived, adventive, even though it is native)
aco:batabí [*aco:bata:bí, co:batabí*] (Koasati); *aktapi:horákko* [*aktapi:hoɬákko*] (*okotafketv*, floating, *ɬákko,* big, Creek); *okpaloɬcáyhî* (tall water floater, Mikasuki)
çiṉ'-moṉ-noṉ-ta (Osage; used in ceremonies)
cyamus (from Greek *kyamos*, a bean; Williams [1837] 1962 wrote that the fruit had edible seeds and that "Hogs will swim in the water to obtain the fruit")
kîtewi (pine snake, name given because of the appearance of the roots when the water dries away and exposes them, Potawatomi)
malangueta (this odd name, resembles *malagueta* for chile)
mulefoot lily (Florida)
pwakumosikum (Cree)
spatterdock (USA)
teghaizé (Chipewyan, Canada); *teta* (Slave, Canada)
yellow pond-lily (Florida); yellow water chestnut (Florida, cf. Williams [1837] 1962)

This was formerly considered part of Linnaeus's ([1753] 1957) *Nymphaea lutea.* In 1809, James Edward Smith, the British founder of the Linnaean Society of London, created *Nuphar* and put the species there. Wiersema and Hellquist (1994, 1997) argue that it should be a distinct species. There are now 16 species known from temperate and cold parts of the world, with 2 in Europe (Mabberley 1997).

Although the pollen was found in pre-Columbian Glades coprolites (Hogan 1978), no modern people in Florida seems to have used the plants. The Seminoles told Sturtevant (1955) that the plant was useless, but it has names in both Creek and Mikasuki. Perhaps it is not useless to all of them. Yanovsky (1936), King (1984), and Moerman (1998) found people throughout virtually its entire range using the plants for a variety of purposes.

Nymphaea

(From Greek *nymphaia*, water nymphs)

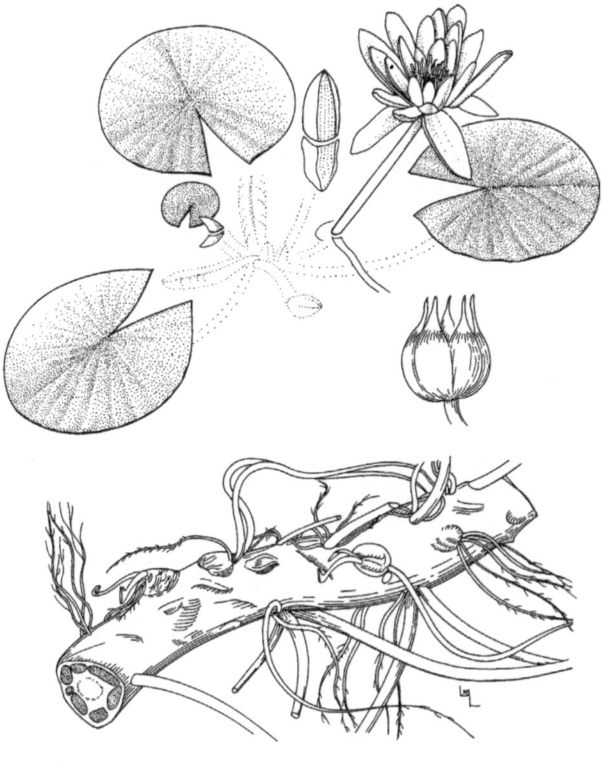

Nymphaea odorata. From Institute of Food and Agricultural Sciences.

bior-ròs (pointed rose, Gaelic)
duilleagh-bhàite bhàn (leaf from down [below], Gaelic); *lili-bhàn* (down [below] lily, Gaelic)
golfão branco (white gulf; *golfão* is the augmentative from Greek *kolpos* and Latin *colup* or *golfu*, a gulf or large bay, presumably applied because these plants grow there, Portuguese)
lotos (Greek)
nenúfar (yellow nuphar, Spanish); *nénuphar* (nuphar, French); *nîlûfar* (Arabic); *ninfea* (Spanish); *ninfea bianca* (white nymph, Italian)
Seerose (ocean rose, German)

Nymphaea odorata (fragrant, the flowers)
a'yip te'xlk ta'lkop (*a'yip*, marsh, *te'xlk*, flower, *ta'lkop*, blue, Atakapa)
aco:batabí [*aco:bata:bí, co:batabí*] (Koasati); *akkotórkv* (maybe from *okotafketv*, floating, Muskogee); *ayiipifo pakaali* (*ayiipi*, lake, *fa-o*, place, *pakaali*, flower, Alabama); *okifoopatka* (*oki*, water, *fa*, place, *oo-*, within, *pat(àa)ka*, to be put, Alabama); *oktaak okpalli* [*okpal*] (*oktaak*, prairie, *okpalli*, to float, Chickasaw);

ue-ak-tv'phe (from *uewv* or *oke*, water, *tv'phe*, broad, Muskogee)
alligator-blankets [-bonnets] (Louisiana, South Carolina)
beaver-root (Newfoundland)
chumbacua [*chumbagua, chumbahua*] (Tarascan, Michoacán)
cow-cabbage (Texas); water-cabbage [-nymph, -queen] (Texas)
fairy-boats (Texas)
flor de agua (water flower, Panama)
guia benne [*guia beñe*] (*guia*, flower, *benne*, mud, Zapotec, Oaxaca)
lampazo (spot-light, Nuevo León)
[pond, toad, water, white water]-lily [waterlily]
locasolkahilíswa (*loca*, turtle, *sulka*, many, *hilíswa*, medicine, Creek); *yokcacó:kayikcî* (many turtle medicine, Mikasuki)
nabagûck (Potawatomi); *pok-shicken-i-uck* (Ottawa tribe of Ojibwa); *wabi'nabagac* (white water lily, Potawatomi)
néufar blanc (white nuphar, Quebec)
ova de galleta (cracker's egg)
pagayeur (paddler, Louisiana, Quebec)
se'np'odal-e' [*saiñ-po-dal*] (Kiowa)
star lotus (South Carolina)
tapalule chobe (*chobe*, big, Mikasuki)

In 1791, Dr. Cadwallader Colden wrote to Peter Kalm that indigenous people did not get scurvy like the Europeans because they ate the roots of a *Nymphaea* (Vogel 1970). Possibly that was *N. odorata* because Huron Smith found the Ojibwa cooking the flowerbuds of that species for food (Yanovsky 1936). Millspaugh (1892) noted a number of medical uses learned from indigenous people, and King (1984) and Moerman (1998) record the Meskwaki, Micmac, Ojibwa, Okanagan-Coville, Penobscot, and Potawatomi using *Nymphaea* for medicines and food.

In Florida, Hogan (1978) found lilies among pre-Columbian Glades people. Williams ([1837] 1962) recorded that the Seminoles used the root to cure "felons" (a small abscess or boil; apparently based on Latin *ulceribus*; used for these sores by about A.D. 1116). Sturtevant (1955) was told that there were two kinds, but he was given a single name, *yokcacó: kayikcî,* for it. At the time they used them for "Turtle Sickness," "Chronic Sickness," and in childbirth. Bennett (1997) was told that a root infusion in winter was used to treat diarrhea.

No information has been found on similar uses among indigenous Americans, but *N. alba* in Scotland is *cairt-locha* (lake-bark, Gaelic) or *rabhagach* (joker, Gaelic), and juice from its roots was boiled as a mordant for dyeing wool (Dwelly 1933). Possibly New World species were used in a similar way.

Nymphoides

(Named by Jean François Séguier, 1703–1784, with Greek *nymphaia*, water nymphs, *-oides,* meaning resembling *Nymphaea*)

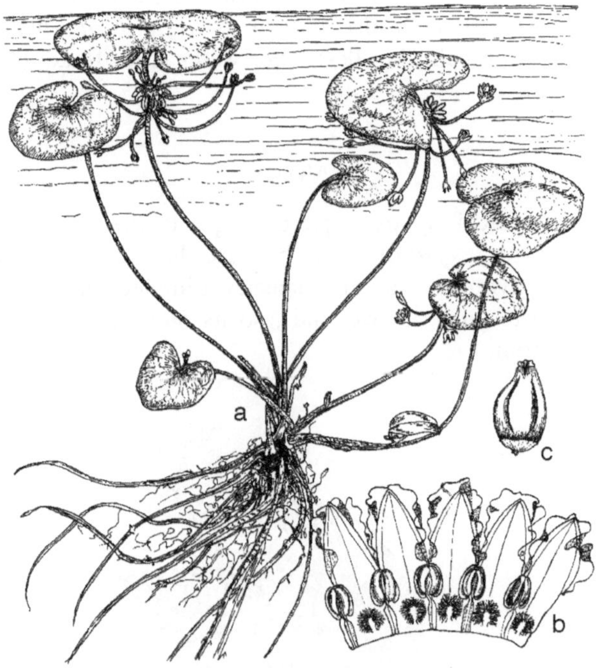

Nymphoides aquatica. a. Habit. b. Corolla spread open. c. Capsule. *Drawn by Vivian Frazier.* From Correll and Correll 1972.

Nymphoides aquatica (growing in water)
aktvpehv [*vktvpehv*] (water floater, from *okotafketv*, floating, Creek); *okpalóhlî* [*akpoloh-looche*] (water floater, Mikasuki)
[big] floating-heart (Florida)
icha:swalópi ahá:ka (*echaswv*, manatee, *lope*, liver, *vhvka*, roots, Creek); *posá:flô:pâ:bî* (manatee liver replica, Mikasuki); *posá:flô:pî* (manatee liver, Mikasuki)

Linnaeus put *Nymphoides* (as a species) in the genus *Menyanthes*, which was an improvement over the Gentianaceae and other families that later students used. However, it was only a year later, in 1754, when Jean François Séguier created the genus *Nymphoides.* There are 20 species in the genus, widespread in wetlands, with the only *N. peltata* in Europe (Mabberley 1997). Some have edible tubers, and others have medicinal seeds.

Sturtevant (1955) found the Big Cypress Seminoles using these herbs for "Turtle Sickness" (trembling, shortness of breath, cough), "Chronic Sickness," and in childbirth. Snow and Stans (2001) applied one of the same names but gave no remedies.

This species occurs in Virginia and the Carolinas to Florida. There are several other species nearby with uses. In Cuba and Puerto Rico, *N. indica* is the *trebol de agua* (water clover), and is used against stomach problems, as an antiseptic, a vermifuge, and for fevers (Roig 1945). It is also known as *tréfle d'eau* (water clover, French Antilles) and water snowflake (Jamaica).

Nyssa

(From Greek *nysso*, to pierce or prick; the name was taken from *Nysa* or *Nyssa* because she was one of the water nymphs; the only species Linnaeus knew, *N. aquatica*, grows in wetlands, so its name is a double mnemonic)

Nyssa aquatica. *Drawn by Vivian Frazier.* From Correll and Correll 1972.

Nyssa aquatica (growing in wetlands, aquatic)
bottle-arsed tupelo (so called by Romans [1775] 1961)
Chickasawatchie whitewood (*Chickasawatchie* or *Chickasaw-hatchee*, meaning "Chickasaw river")
cottonwood [bastard, gum] (could this comparison be made because their foul-tasting fruits turn the mouth to cotton?)
gum [black, bowl, cotton, gray, ladle, papaw, sap, sour, swamp, water, white, yellow] ("gum" first used by Chaucer in 1385 in the sense of a viscid plant secretion; differs from a resin in that it is soluble in both hot and cold water. Derived from Old French *gomme*, cognates include Portuguese and Spanish *goma*, Italian *gomma*, Latin *gumma*. First applied as an abbreviation of "gum-tree" in 1802, and then in 1817 it was being used to mean a hive for bees cut from a tree stem section. Perhaps that last name came into being because preference was given to hollow *Nyssa* and *Liquidambar* for making hives.)
hornbeam (classical name of *Carpinus*; reference here obscure)
hush apa ("to kill birds," so named because birds eat the berries and are more easily killed for food, Choctaw)
łobonó (Koasati)
okifa itto (*oke*, water, *fa*, place, *itto*, tree, Alabama)
[wild] olive-tree [olivetree] (an allusion to the resemblance of the fruit to an olive, *Olea europaea;* a comparison that was made by Mark Catesby in 1730)
pepperidge (derivation obscure, but from "pipperidge"; first used by 1538 in Europe for the barberry, *Berberis*; by the 1820s, in Suffolk, England and in New England the word was changed to "pepperidge" or "pepperage"; it retained the reference to barberry in the Old World, but was reapplied to *Nyssa* in the Americas)
poplar [bay, hickory, swamp] (allusion obscure, cf. *Populus*)
[big, gum, large, swamp, rootwood, water] tupelo [tupeloo, tupola, tupolo, tupelow] (from *eto*, tree, *opelwv*, swamp, Muskogee; used in print by 1730; one of the first to use it was Mark Catesby; "tupelo" is the "preferred" common name)
yassaakohatka [*yassàakohatka*] (*yassaako*, sweet gum, *hatka*, white, Alabama)

Nyssa ogeche (the *Ogeeche* were the Yuchi living on the Georgia and South Carolina border near the time of Bartram's visit in 1773; maybe from the Hitichiti word *Ochesee*, "people of another language")
gopher plum ("gopher," as *gaufre*, honeycomb, was applied to burrowing animals by the French because their dens are like "honeycombs." "Gopher" was the name already established for the tortoise that Linnaeus called *Gopherus polyphemus* in 1759. Perhaps applied because the trees are often hollowed, or because of some perceived connection to the tree of the Bible, Hebrew *gopher*, whose wood made the ark.)

limetree [wild] (name used by John Bartram in 1765)
Ogeechee [Ogeche, Ogechee, Ogeeche] lime (found by John and William Bartram in October 1765, but the first published record is Bartram [1791] 1958; fruits were used as substitutes for lime, *Citrus auriantiaca*, in 1765 in the region of southern Georgia and northern Florida where the species grows)
Ogeechee tupelo (the "preferred" common name)
[gum, lone, sour, white] tupelo (from *eto*, tree, *opelwv*, swamp, Muskogee)

Nyssa sylvatica (of the woods)
a͟ntuda'yudi' (Biloxi)
bee gum (bee-hives or "gums" made from stem sections; in use by 1817); beetle-bung (surely a corruption of "bee gum"); [plain, quartered] black gum [blackgum] (this is the "preferred" common name); bowl gum (stems used to make bowls); cotton-gum [cotton gum] (an obscure allusion to cottonwood, *Populus*); gum (see under *N. aquatica*); sour gum (already in use in the southeast by the time of William Bartram's visit, 1773–1774); yellow gum [yellow gumtree]
black tupelo; *tupelo negro* (black gum, Spanish); *tupelo noir* (black gum, French); tupelo (see also *N. aquatica*); moeras tupelo (modifier obscure); [*svart, zwarte*] tupelo (surely modifiers are German *schwarz*, or Dutch *zwart*, black)
cabo de luc (maybe "Cape Lucas," Mexico)
chiste (a sweet also known as *alfajor*, Mexico)
helukwv-vpe [*helocoppe*] (*helukwv*, black gum, *vpe*, stem, Muskogee; compare with *Liquidambar* and *Eugenia* names); *helúk-lv'ste* (*helukwv*, black gum, *lv'ste*, black, Muskogee); *hikahí* (Koasati)
hornbeam [hornbine, hornpine, hornpipe] (classical name of *Carpinus*; application here obscure, especially since the wood of *Nyssa* is soft)
iti ani [*itani*] (*iti*, tree, *ani*, berry, Choctaw)
kosapa [*kosapi, kosapo*] (Alabama)
leni tachkw'ákan [*lenni tachquohácan*] (grinding-device [mortar] making-tree, Delaware)
nyssa sylvestre (translation of scientific name into French, Quebec)
palo de papaxi (the name seems to compare Nyssa with *Randia*, the *papache*, Mexico)
pepperidge (see under *N. aquatica*)
petcui (Mexico)
snag-tree ("snag" probably of Scandinavian origin, cf. Norwegian dialectic *snag*, sharp point, stump; akin to Icelandic *snagi*, peg; used for "stump" in English by 1577; also in dialectic the sloe, *Prunus spinosa*, by 1587; perhaps an allusion to the fruits)
stinkwood (since 1300, "stink" has meant an offensive smell; application obscure)
tetzicohuitl (*tetzico*, restrains, *cuahuitl*, tree, Náhuatl, Mexico)
uda'lana (Cherokee)
wild pear-tree (an allusion to the fruits)
yap beʔbe'ʔ (tree immovable, Catawba)

John W. Swanton (1946) created the single most valuable resource on how indigenous people in the southeastern United States lived from the time of European contact until the early 1900s. However, that book is one of the most maddening in existence because of its inconsistencies and disorganization.

One of the problems with the book surfaces with consideration of *Nyssa*. Swanton purports to give historical synopses of tribes and translations of tribal names (pp. 81–219). However, nowhere in those lists does he mention the Ogeeche or any variant spellings of it. He discusses the Ogeeche elsewhere in the text, and shows where they lived on the maps, but there is no hint of them in pages 81–219.

Regardless, by making comparisons within the text it appears that *Ogeeche* may be the same as *Ochesee*. Earlier (Swanton (1939) said that *Ochesee* was a Hitichiti word with the same meaning as *celokke* [*chilokee*] in Muskogee, or "people of a different language" (Martin and Mauldin 2000). That tribal name is the source of Florida's third *Nyssa* species name and was applied by William Bartram during his excursion through the South in the 1770s.

Near Fort Barrington along the Altamaha River in Georgia, Bartram hired a Creek to take him in a boat to see the plants now called *N. ogeche*. He wrote: "They grow in the water, near the shore. There is no tree that exhibits a more desirable appearance than this, in the autumn, when their fruit is ripe, and the tree divested of its leaves; for then they look as red as scarlet, with their fruit, which is of that colour also. It is of the shape, but larger than the olive, containing an agreeable acid juice. The leaves are oblong lanceolate and entire, somewhat hoary underneath, their upper surfaces of a full green, and shining; the petioles short, pedunculis multifloris. The most northern settlement of this tree, yet known, is on Great Ogeeche, where they are called Ogeche limes, from their acid fruit being about the size of limes, and their being sometimes used in their stead." The place he refers to as the "Great Ogeche," the Ogeechee River, is one of the longest in Georgia and flows southeast between the Savannah and Altamaha Rivers.

My own impression of these trees was not much like Bartram's. Those I saw in Apalachicola National

Forest were scrawny little things compared to their two cousins in the eastern states. Admittedly, I saw them only in leaf and flower and not at their peak of fruit as did Bartram. Still, I am not the only one who was not as taken with them as he. In 1824, South Carolina professor Stephen Elliott wrote of "the pleasant acid of its fruit," but added, "its last flavour is austere" (Elliott in Fernald et al. 1958). On the other hand, Small (1933) considered a beverage of the fruits good.

Even Elliott's comment would be a compliment to the fruits of *N. aquatica.* Fernald et al. (1958) described them as having "disagreeably tasting fruits as large as a small olive." However, they say the fruits of the northern variations of *N. sylvatica* "are bluish stone-fruits with a very thin pulp, but sharply acid and pleasant to roll in the mouth in autumn and winter." Perhaps we humans simply do not appreciate what constitutes "good" fruits, as these are among the most desirable of snacks for wood ducks (*Aix sponsa*). I have fond memories of watching these dazzlingly colored waterfowl feed on *N. aquatica* fruits.

The primary use of *Nyssa* among indigenous people seems to have been as a dye or medicine. The Choctaw burned the bark and mixed in red oak ash to make a red dye (Bushnell 1909, 1917). Since the oak alone would have given a red color, perhaps the *Nyssa* was added as a mordant. The Alabama, Creek, Houma, and Koasati made medicines from *Nyssa,* at least *N. sylvatica,* and possibly other species. The Alabama used the plants to treat "cancer" (Hocking 1997). The Creeks made a preparation of the bark and chips to treat tuberculosis (Swanton 1928a, Taylor 1940). The Houma made a root or bark decoction to expel intestinal worms (Speck 1941). The Koasati applied a bark decoction to gunshot wounds and drank some to help the healing (Taylor 1940).

The Cherokee used *N. sylvatica* in most of the same ways as their neighbors did, plus some unique applications. They too expelled worms with it, but they also used medicine made from it to stop diarrhea, as an emetic, during childbirth, and for urinary infections (Taylor 1940, Hamel and Chiltoskey 1975). The Cherokees also dripped ooze from the roots into the eyes to help them heal (Hamel and Chiltoskey 1975).

Swanton (1946) recorded indigenous southeastern people using *Nyssa* wood in making drums (species not indicated, perhaps all). Farther north, the Ojibwa used the wood of *N. sylvatica* to make awl handles, mauls, and war clubs because it does not split or check (Moerman 1998). Wood has been used in more recent years as veneer, plywood, railroad cross ties, boxes, cooperage, pulp, woodenware, hubs, wharf piles, handles, and planing-mill products (Vines 1977). Twigs were made into chewsticks for cleaning the teeth and into snuff dipsticks for making the teeth dirty (Hocking 1997).

Settlers also found inventive ways to utilize *N. aquatica.* Wood from those trees was made into boxes, woodenware, and fruit crates; the light, spongy roots were used to make net floats and corks (Vines 1977). Uses hidden within common names, but rarely recorded elsewhere, are to make bowls and ladles. It is also called "sap gum," so perhaps sugar was extracted from it.

During the Civil War, Porcher (1863) recorded: "The roots are white, spongy, and light, and are sometimes used in the Confederate States as a substitute for cork. ... The genus exhibits a constant peculiarity of organization ('the fibres are united in bundles and interwoven like a braided cord'), hence the wood is extremely difficult to split, unless cut into billets — much used for hubs of wheels; also preferred for the sideboards of carts. ... Trays, bowls, dippers, mortars, and other utensils are manufactured from it. ... Mr. W. Gilmore Simms suggests to me the use of the tupelo, on account of its lightness, for making cartridge-boxes." Porcher also writes at length about the wood being made into *sabots,* wooden shoes worn by slaves.

One of the strangest applications for *Nyssa* wood is construction of a "tupelo-tent," a surgical "tent" made from the spongy wood of the roots. Perhaps that too came about during the Civil War. No explanation has been found, but these "tents" were perhaps like the towel covering used in modern surgery to keep areas surrounding the incision covered—and to soak up excess fluids. Hocking (1997) also compares the use to "Laminaria pegs," i.e., as "tents" and dilating bougies for distending body cavities during surgery. The pegs also have been used to induce abortion.

The first American use of the name "pepperidge" was in a literary and not a scientific context. None other than James Fenimore Cooper, in his 1826 book *The Last of the Mohicans,* referred to a "trencher, neatly carved from the knot of a pepperage." That was probably a passage that Cooper never expected anyone to remember, but it provided ethnobotanical information on a tribe that is not recorded elsewhere.

O

Obolaria

(From Greek *obolos*, Latin *obolus*, a small Greek coin, referring to the thick rounded leaves)

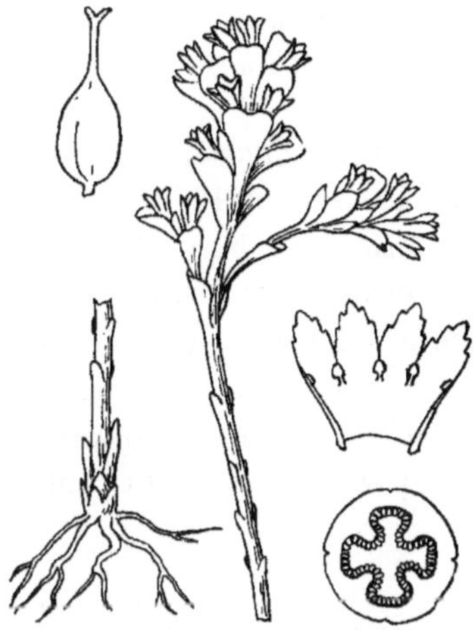

Obolaria virginica. From Britton and Brown 1897.

Obolaria virginica (of Virginia)
klotchoachokama (*achokma*, good, Choctaw)
[Virginia] pennywort

Obolaria is a small mycotrophic plant with the single species endemic to eastern North America (Mabberley 1997). The first record of this herb in Europe was published by Robert Morrison in 1680 about a cultivated plant in London. He, Leonard Plukenet, and John Ray called the plants *Orobanche virginiana, radice coralloide summo caule foliis subrotundis* (*Orobanche* from Virginia, with roots like coral and stem having almost round leaves). Gronovius in the *Flora Virginica* of 1739–1743 called it *Obolaria*, and Linnaeus finalized the name as *O. virginica* in 1753.

Although formerly placed with *Orobanche*, these are relatives of gentians. This is another of the saprophytic members of that family (see also *Voyria*). We now know that the species grows from Florida to Texas, north to New Jersey and Pennsylvania, West Virginia, Ohio, Indiana, and southern Illinois (Fernald 1950, Mohlenbrock and Voight 1959, Correll and Johnston 1970).

The Cherokee made a medicine of these plants for colds, coughs, and colic, and used it as a diaphoretic (Hamel and Chiltowsky 1975). The Choctaw used the water from the boiled roots, mixed with boiled sweetgum roots, on cuts and bruises (Bushnell 1909).

Ocotea

(J.F. Aublet named these trees after a Carib Garipon word from French Guiana; the species they call *ajouhou-ha*, he rendered as *Ocotea guianensis*)

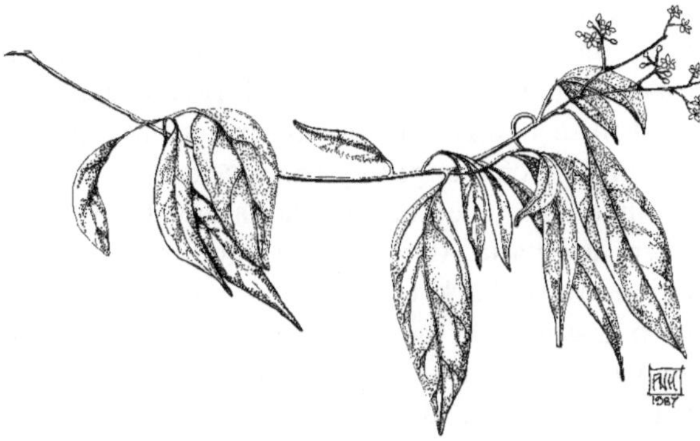

Ocotea coriacea. *Drawn by P.N. Honychurch.*

Ocotea coriacea (leathery)
aho:slákní [*ahilákno:cí*] ([small] yellow tree, Mikasuki)
avispillo (little wasp, Puerto Rico)
bastard [black] torch (Bahamas)
bois doux négresse (sweet black wood, Guadeloupe)
cabirma aromática (aromatic *cabirma*, Taino word for some tree, Dominican Republic)
Cap Henri (a locality, Haiti)
cap-berry [capberry] sweetwood (Jamaica); rock sweetwood (Panama)
cigua [*sigua*] (laurel, Taino, Cuba, Puerto Rico, Panama); *cigua blanca* (white laurel, Dominican Republic)
cip-che' (*kip*, smooth or slick, *che'*, tree, Maya, Campeche)
cochoch-che' [*hochokche, xhoch-che'*] (*koch*, something narrow, *hoch*, something hollow, *che'* tree, Maya, Quintana Roo, Yucatán); *hoc-che'* (*hoc*, something hollow, *che'*, tree, Maya, Yucatán); *jobon-che* (*hobon*, hollow, *che'*, tree, Maya, Quintana Roo)
lancewood (a reference to leaf shape, Florida)
laurel (laurel, Mexico, Belize, Puerto Rico); *laurel avispillo* [*cigua*] (wasp laurel [laurel laurel], Puerto Rico); *laurel perdiz* (quail laurel, Dominican Republic); *laurel verde* (green laurel, Yucatán); *laurier* [*blanc*] ([white] laurel tree, Haiti); *laurier marbré* (marble laurel tree, Grenada)
lebisa (usually a name given a fish, *Dasybatus torregi*, Taino?, Cuba)

palo de gas (gas tree, Yucatán)
pepper cillament (pepper cinnamon, Virgin Islands)
siguaraya (Taino?, Cuba)
sweet torchwood (Bahamas); sweetwood (Jamaica, Belize, Dutch Antilles)
tola:nocî (*tola*, red bay, *noce*, milkweed, Creek)

People are told that one of the traits of tropical forest trees is that leaves of many species are similar. Indeed, one of the last species in the West Indian forests along the southeastern Florida coast that I learned to recognize was the lancewood. My students were similarly challenged by the species.

Unlike many members of the family, the foliage is not very aromatic. It requires a combination of leaf venation traits, bark comparisons, and considerable luck to distinguish the lancewood from the Guiana plum when they are neither flowering nor fruiting. Perhaps that same difficulty is part of the reason botanists have had such a hard time deciding on the proper name of these short-lived trees.

Discounting the placement of the species in *Laurus*, the discussion is whether the trees should be in *Nectandra* or *Ocotea*. *Nectandra* (from Greek *nektar*, nectar, plus *andros*, stamen) was described by Daniel Rolander (1725–1793) and Christian Friis Rottboell (1727–1797) in 1778. Three years earlier, Aublet had named *Ocotea* and used it for a French Guiana tree that the Garipon Caribs applied to drain tumors and swollen glands (Aublet 1775).

Olaf Swartz first described this species in 1788 and called it *Laurus coriacea*. Subsequently, it has been considered either a *Nectandra* or *Ocotea* (Chapman 1897, Sargent 1905, Britton and Millspaugh 1920, Small 1933, Long and Lakela 1971, Correll and Correll 1982, Rohwer 1993).

Historically, the only recorded use in Florida is among the Seminoles. Their warriors used the wood for bows (Sturtevant 1955).

Wood, with light brown sapwood and dark brown heartwood, has been used in carpentry, cabinetwork, for poles, and to make charcoal (Small 1933, Little and Wadsworth 1964, Liogier 1974). In Mexico, the trees are used for wood in Yucatán. However, the common Mayan names make me wonder how much they value them. No other uses have been found for *N. coriacea* although *N. rodiaei* (Greenheart) of South America is well known as a source of wood (Standley 1920–1926). Bark of that tree contains the alkaloids bebeerine, sipirine, and nectandrine, which are tonic, astringent, and a febrifuge like cinchona. Perhaps some of the same compounds are in other members of the genus.

Oenothera

(Said to be based on *oinos*, wine, and *thera*, catcher; a name used by Theophrastus, 372–287 B.C., for some Onagraceae; however, Pliny, A.D. 23–79, wrote, "*onothera, sive onear, hilaritatem afferens in vino*" [ass-catcher, or ass-hunter, a plant whose juice in wine causes happiness; or, as translated in 1601, "Oenothera, otherwise Onuris, an herb good also in wine to make the heart merry"])

enagra (from Latin *onager*, Greek *onagros*, the wild ass, *Equinus onager*, Italian); *onagre* (French); *hierba del asno* (ass herb, Spanish)
evening primrose ("primrose" from Latin *prima rosa*, the earliest rose, applied to *Primula* since 1400s; combined with "evening" since the 1760s because of the nocturnal flowers)
Nachtkerze [*Nachtkerz*] (night candle, German)

Oenothera biennis (biennial, maturing in two seasons)
canhlo'gan hu'nla (rattle weed, Lakota)
coffee-plant (an infusion used as a drink in the harvest field, cf. Coffey 1993, North Carolina)
cure-all (used as a domestic medicine, cf. Coffey 1993, North Carolina)
evening [night] primrose [common, eveningprimrose] (because the flowers open in the evening and close in the morning; applied to these plants by the 1760s)
fever-plant (used as a diaphoretic in fevers, cf. Coffey 1993, Kansas)
[large] German rampion ("rampion" has been the name for bellflower, *Campanula rapunculoides*, since the 1760s; related to French *raiponce*, Italian *ramponzolo*, German *Rapunzel*, Kansas)
golden-candlestick (Kansas)
king's cure-all (South)
nattlys (night light, Norwegian)
night willow-herb ("willow-herb" has been applied to a variety of plants because their leaves resemble those of *Salix*; originally recorded in 1574 by Rembert Dodoens)
owesa'wanakuk (yellow top, Potawatomi)
scabbish [scabish, scabiose, scavey, scurvish] (Josselyn in 1638 thought it was confused with "scabious," or *Knautia*, Maine, New Hampshire, Ontario, South Carolina)
speckled-John (Kansas)
tree primrose (used since 1638)

Oenothera fruticosa (shrubby, a misnomer)
sundrops (used since the 1790s for any of the species whose flowers open when touched by the morning sunshine)
narrowleaf evening primrose [eveningprimrose]
wild beet (used as a potherb in West Virginia, Coffey 1993)

For some unexplained reason, Linnaeus told us that *Oenothera biennis* was introduced from Virginia into Europe in 1614. After that time, a number of herbalists and botanists commented on the plants. Gaspar Bauhin commented on them in his *Pinax* of 1623. Gronovius included them in his *Flora Virginica* published between 1739 and 1743.

The perplexing thing about Linnaeus commenting that the plants were introduced from Virginia is that now some contend that *O. biennis* is not an American native. Both Clapham et al. (1987) and Mabberley (1997) say that it is native to the Old World.

In his wonderful way, Linnaeus created a Gordonian knot when he used the name *Oenothera*. There simply is no single translation for that word. Most people either avoid the problem, as when Fernald (1950) wrote, "Name used by Theophrastus for a species of *Epilobium*," or simply "Greek, meaning wine-scented" (Munz 1973). However, as Coffey (1993) correctly points out, it can mean "wine-scenting," "wine-imbibing" (from *oinos*, wine, *thera*, imbibing), "donkey-chase" or "ass-catcher" (if based on *onos*, donkey, *thera*, hunt or chase). An enticing epilogue was added to the etymology in 1860 when Joseph E. Worcester's *A Dictionary of the English Language* recorded, "The roots of the species *Oenothera biennis* are eatable, and were formerly taken after dinner to flavor wine, as olives now are; hence the name *Oenothera*, or wine-trap." Porcher (1863) interpreted the history somewhat differently. He wrote, "The ancients thought the plant possessed the power of allaying intoxication and calming the most ferocious animals."

John Josselyn, writing in 1674, explained some of the common names in *An Account of Two Voyages to New-England Made During the Years 1638, 1663*. He wrote that it was the "yellow Lysimachus of Virginia," and that, because it has "the Flower yellow and like Primroses, [it is] therefore called Tree-Primrose." He went on to say that the English settlers in New England took it to be "scabious" (*Knautia arvensis*, Dypsacaceae) (Coffey 1993). Peter Kalm's observations made 75 years later support Josselyn's. While Kalm was in Montreal in 1749, he noted *O. biennis* in abundance in open woody hills and fallow fields. He wrote: "An old *Frenchman*, who accompanied me as I was collecting its seeds, could not sufficiently praise its properties of healing wounds. The leaves of the plant must be crushed, and then laid on the wounds" (Kalm [1753–1761] 1972). Perhaps Josselyn's and Kalm's informants did not know the proper scabious plant, and simply knew that *Knautia* was used to heal sores like the Tree-primrose.

During the 1800s, *O. biennis* was cultivated in English and German gardens for its edible roots (Hedrick 1919). However, the French looked askance at it as an edible and grew it only as a curiosity. During the 1860s, it was even sold by enterprising Americans to their gullible compatriots as "German Rampion."

Porcher (1863) wrote: "The root and herb have been employed in cutaneous diseases. Dr. Griffith has used it with success in tetter, applying the decoction to the affected part several times a day, and giving it internally at the same time. He has been successful with it in subsequent trials. The plant should be gathered about the flowering season. The young sprigs are mucilaginous, and can be eaten as salad. ... Its roots have a nutty flavor, somewhat similar to those of rampion, and are used in Germany and some parts of France, stewed and raw, in salads, with mustard, oil, salt, and pepper, like common celery."

The Cherokee cooked and ate the leaves as greens and boiled the roots like potatoes (Hamel and Chiltoskey 1975, Moerman 1998). Otherwise, the eastern tribes used the herbs as medicine. The Cherokee took an infusion to treat obesity and to relieve piles (Hamel and Chiltoskey 1975). The Iroquois too used the plant to treat piles and boils (Moerman 1998). The Ojibwa put a poultice on bruises (Moerman 1998). The Potawatomi used the seeds for "medicine" (Smith 1933).

Scotia Pharmaceutical, Inc. in England discovered that *O. biennis* was rich in GLA (gamma-linolenic acid) compounds, which are important in the production of fatty acids and prostaglandins. It has been selling these oils in health-food stores for years, and Mabberley (1997) records that the company did a £36 million business in 1993. The company suggests that its product can alleviate, if not cure, a variety of diseases. Many remain skeptical, and Mabberley (1997) noted that placebo tests were as effective as the GLA.

Opuntia: Prickly Pear

(Philip Miller gave us the generic name *Opuntia*, taking it from Latin *herba opuntia*, from *Opus*, *Oputis*, a town of Locris in Greece; in Greek *Opous* or

Opountos; the Latin name was used by Theophrastus, 372–287 B.C., for some unknown plant, and Miller applied it to the New World cacti)

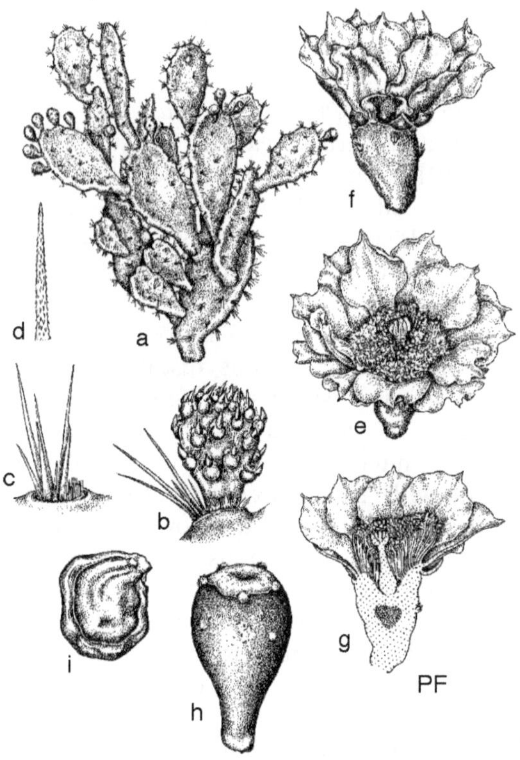

***Opuntia stricta* var. *dillennii*.** a. Habit. b. Young joint, with leaves remaining in areoles. c. Areole. d. Glochid from areole. e. Flower, from above. f. Flower, side view. g. Flower, longitudinally dissected. h. Fruit. i. Seed. *Drawn by Priscilla Fawcett.* From Correll and Correll 1982.

In 1989, Carol Lippencott found the cactus moth (*Cactoblastis cactorum*) on the endangered semaphore cactus (*Opuntia corallicola*) in the lower Florida Keys (Habek and Bennett 1990). In March of 1992, we found the moth on the prickly pear cacti (*O. stricta* var. *dillenii*) at the Florida Atlantic University campus in Boca Raton more than 100 miles north of the Keys site. Cacti that had been almost 2 m tall (6 feet) in Boca Raton were eaten to the ground within less than a year.

Now the insects are found throughout the Florida peninsula and are spreading through native *Opuntia* farther north. As the moths spread, cattle ranchers' hopes that they will eliminate cacti from the state escalate along with their visions of increased profits. Biologists are terrified that the insects will continue around the Gulf of Mexico to eradicate the many *Opuntia* species that populate the Chihuahuan and Sonoran Deserts of the United States and Mexico. If they spread as feared, millions of people will be impoverished.

Several species of prickly pears are an essential part of the lives of people in Mexico and the southwestern United States. *Opuntia* have been used as food and medicines for thousands of years. The plants were so important to the Aztecs that they named cities for them (e.g., *Tenochtitlan*, place where nopal grows on rocks), and *Opuntia* now appears on the Mexican national flag.

Spaniards found *Opuntia* cacti on their first trips to the New World. Soon afterward, the indigenous names *nopal* and *tuna* became elements in the Spanish language. *Nopal* (the plant, from *nopalli*, Náhuatl) is applied to all of the flat pad cacti (*Opuntia*), although sometimes *Nopalea* is considered a distinct genus. *Tuna* (the plant and fruit) comes from Caribbean Taino language although it is superficially similar to *atún* (Spanish) and *thunnus* (Latin), which refer to the fish in the family Scombroidae. The importance of *Opuntia* in Mexico is partly indicated by their designation in various languages. They are *biichi* [*bitzu*] (*bi*, fruit, *ichi*, spiny, Zapotec), *câhâ* (fruit, Otomí), *xâthâ* (plant, Otomí), *nochtli* (fruit, Náhuatl), *pak'ak'* [*pacac*] (Huastec), *pak'an* (Maya), *pare* (Tarascan), *taat* (Mixe), and, in Cochimí, the single vowel "*a*." The Seri and Mayo do not have a generic name for prickly pears, but they use distinct names for each species (Felger and Moser 1985, Yetman and Van Devender 2001).

Europeans had never seen plants like prickly pears before arriving in the New World because the family Cactaceae was endemic here. Oviedo ([1526] 1969) conveys something of the astonishment they felt when he wrote: "There are some wild plants . . . called *Tunas.* And they grow on some very spiny thistles [*cardos*] and produce the fruit that is called *tunas*, that is similar to young figs [*brebas*] or figs [*higos*], being about the same size. . . . And within are very red and have small seeds like figs."

Eating the fruits produces effects that are surprising to novices. When they are in fruit around my home in Tucson, the coyotes eat them greedily, and their scat becomes as red as blood. Knowing that, it was not a surprise to learn of Oveido's experience in the 1500s. Again, he wrote: "After one eats three or four of them (and better eating a larger number) if the one who has eaten them goes to urinate, the urine comes out exactly resembling true blood. And in this manner it happened to me the first time I ate them. And after an hour the urine began to flow (and the fruit greatly causes this) that when I saw the color of the urine I was astonished and frightened and greatly suspicious of my health."

Historically, Florida shared no species with Mexico, however, *O. stricta* var. *dillenii* is shared with the Caribbean. Bahamians call these plants scurgeon needle, and Puerto Ricans say *higo del mar* (sea fig). Elsewhere *O. stricta* is prickly pear (Florida, Jamaica, Virgin Islands, Barbados) or *tuna brava* (wild cactus), while it is bull sucker or miss blyden in the Virgin Islands. In Jamaica, it is seaside tuna. People in the French Antilles say *cactier en raquette* (cactus with pads), *mal rachette* (bad pad, Dominica), *raquette* (pads, Hispaniola), *raquette à fleurs jaunes* (pads with yellow flowers, Guadeloupe, Martinique), *raquette à piquanats* (spiny pads, French Antilles), or *raquette bord de mer* (pads by the sea, French Antilles). The other indigenous name from the Caribbean is *bata* [*pata*] (Carib, Dominica). Mayan speakers in Yucatán call it either simply *pakan* [*pak'an*] (prickly pear) or *ya'ak-pak'an* [*yaaxpakan, yax-pak'an*] (*ya'ax*, green, *pak'an*, nopal). The latter name is used to distinguish it from *O. cochinillifera.*

In the southeastern United States, Harriot ([1590] 1972) said, "a kind of pleasaunt fruite almost of the shape and bignes of English peares, but that they are of a perfect red colour as well within as without." Harriot noted that in the West Indies the dye called cochineal grew, but he was not sure if it could be obtained from prickly pears in Virginia. That dye is from the cochineal insect *Dactylopius coccus.* DeBrahm ([1775] 1971) knew the plants and the dye source, and noted *Opuntia* in East Florida (the peninsula). Similarly, he and Romans ([1775] 1961) recommended the area for cochineal production because of the number of *Opuntia.* The Carolina Algonquians called the plant *metaqvesvnnavk* [*metaquessunnauk*, from *mä tahkwi-senyä-ki*, what is eaten bare or raw], given to *O. humifusa* in North Carolina (Harriot [1590] 1972, Geary 1955). That species is probably what the Biloxi called *maxon̲'tkxo'hi a'naki.* The Osage said they were *tse'-wa-xa-ga ton̲-ga* (*tse'-wa-xa-ga*, bison bur; cocklebur, *ton̲-ga*, big).

In Virginia, Strachey wrote in [1612] 1953 that the Powhatan called the leaf of the "prickle pear" *matakuskc* (Harrington 1955). Siebert (1975) wrote this phonetically as *matakask*, and said it was composed of /**matw-*/, uneven, jagged, /**-ak-*/, capsule, nut, and /**-askw-i*/, plant. He added that Algonquian speakers classify all fruits into /*pak-*/, capsules, nuts, or /*min-*/, berry, grain, legume. The root word is /*pak-*/, shortened to /*–ak-*/.

The Seminoles and Creeks know prickly pears and the de Soto chronicles mention them (De La Vega [1605] 1962). To them, the plants are *tvlłakko* [*tvlrakko, tvráko, tvnłakko, tvlhvcko*] (*tvl* <*etvlwv*, tribe, *łakko*, large). Others have written the name *łakko* [*rakko*] (*łakko*, big, Muskogee), or *kanłakko* [*kanrakkô*] (*kan* <*ekvnv*, land, *łakko*, big, Creek). To the Miccosukee in Florida, it is *hacolkatô: nî* (Mikasuki). Other Muskogean names are *talaapàkha* [*tali aapakha*] (magic stone, Alabama) and the apparently cognate *talhpakha* (Choctaw), and *talipatká* (*talí*, stone, *pátka*, bed, Koasati).

Because of early references like Oveido, *Opuntia* is sometimes called cactus pear (Europe), Indian fig (USA, Europe), or even Barbary pear (Europe), although these names are usually applied to *O. ficus-indica,* the species best known in Europe. *Opuntia ficus-indica* became popular early in the Old World (Jain and Tarafder 1970, Barbera et al. 1992, Brutsch and Zimmerman 1993). *Opuntia* became an important medicinal plant in China between the *Pen tsao kang mu* (Compendium of Chinese Medicinal Herbs) of 1590 and 1765 when the *Pen tsao kang mu shih* (Supplement to the Pen tsao kang mu) appeared (Hsu 1986).

Recent laboratory studies of several *Opuntia* species support Oveido's observation that the fruits (and other parts) are diuretic (Galati et al. 2002). In Mexico and the Caribbean, *O. stricta* is used to treat urinary problems, as a poultice and emollient for cuts, tumors, splinters, and boils and for stomach ulcers. Pads are heated and made into poultices to treat pleurisy and rheumatic pain. Other species are used as diuretics, against skin and eye inflammations, to halt diarrhea and dysentery, as a beverage for chest problems and fever, as an analgesic, and to treat stomachache, high cholesterol, and diabetes (Martínez 1969, Morton 1981).

Laboratory experiments indicate that *Opuntia* remedies are effective as an analgesic (Loro et al. 1999), anti-inflammatory (Park et al. 1998, 2001, Loro et al. 1999), antiulcer (Galati et al. 2001, Lee et al. 2002), antifungal (Gupta and Banerjee 1972), and an antiviral agent (Ahmad et al. 1996). While *Opuntia* helps control diabetes (Meckes-Lozyoa et al. 1986, Frati-Munari et al. 1989a,b,c,d, 1991, 1992, Roman-Ramos et al. 1991, Trejo-Gonzalez et al. 1996, Bwititi et al. 2000, 2001, Shapiro and Gong 2002), some of the experiments also show toxicity to the kidney (Bwititi et al. 2000). Moreover, the plants help control cholesterol (Fernandez et al. 1990, 1992, 1994, Cardenas-Medellin et al. 1998) and act as antioxidants (Budinsky et al. 2001, Acuña et al. 2002), and monoamine oxidase B inhibitors (Han et al. 2001).

Most of the studies about *Opuntia* concentrated on action rather than chemistry. Still, the consensus is that the highly soluble fiber in the pectin decreases plasma low density lipoprotein concentration (Fernandez et al. 1990, 1992, Shapiro and Gong 2002), and is presumably responsible for several of the other actions. beta-Sitosterol is considered the anti-inflammatory component (Park et al. 2001), but the plants also contain triterpenes, malic acid, and succinic acid (Hsu 1986).

Chemical reports have concentrated mostly on nutritional value. The most abundant components of fruit pulp and skin are ethanol-soluble carbohydrates. Pulp contains glucose (35%) and fructose (29%) while the skin contains essentially glucose (21%). Protein content is 5.1% in the pulp, 8.3% in the skin, and 11.8% in the seeds. Starch is found in all parts of the fruit, with the pulp fibers rich in pectin (14.4%), and 29.1% cellulose in the skin and 45.1% in the seeds. The skin is remarkable for its content of calcium (2.09%) and potassium (3.4%) (El Kossori et al. 1998).

In addition to having the tasty and nutritious fruits mentioned by Oviedo and many others since (e.g., Saenz et al. 1998, El Kossori et al. 1998), the mucilaginous pads are edible raw or cooked. Young pads are usually cut into pieces and boiled, while mature pads may be roasted and eaten immediately or dried for future use. In Dominica and probably elsewhere the pads are used to wash the hair (Honychurch 1986). All aboveground parts of the plants are apparently effective medicines for a variety of problems. The plants host an insect that produces a dye that has been used for thousands of years by Mexicans and it was the source of "Royal Purple" for centuries (McJunkin 1991). Although cattle ranchers hate them, prickly pears are groceries and pharmacies hiding behind spines.

Orbexilium

(Rafinesque coined this genus without explaining what it meant; perhaps from Latin *orbi*, a circle, and *exili*, thin, referring to the fruits)

Orbexilium pedunculatum. From Britton and Brown 1897.

Orbexilium pedunculatum (with flowers on a stalk) (=*Psoralea psoralioides; Orbexilium gracile*)

Bob's-root [bobsroot]; Congo root

Sampson's snakeroot (Samson or Sampson is the legendary Hebrew hero of great strength whose exploits are recounted in the Bible, Judges 13–15, but the individual named here is some other person, perhaps a black herbalist); snakeroot (used as a name for medicinal plants, but originally applied to either *Polygala senega* or *Aristolochia serpentaria*; later given to other plants)

scurf-pea ("scurf" has been a disease of the skin since about A.D. 1000 where scales or flakes of the epidermis continually fall; presumably, the name reflects the scaly morphological condition of the plant)

wi̲·ti· ki̲· warupʔha (root round, Catawba)

The concept of a snakeroot, or the root of a plant that counteracts the poison of snakes, appears to be of American origin. Europeans, however, were well versed in poisons, and firmly believed in counterpoisons. In Spanish counterpoisons were *contrayerbas.* Europeans were also familiar with poisonous snakes, having several species of *Vipera* (Viperidae) spread from England and Germany south. However, these "adders" were not as large or as toxic as those encountered in the New World (Ditmars 1931, Darlington 1957).

According to the records, the Seneca tribe of western New York taught the settlers to use American or Seneca snakeroot (*Polygala senega*) for snakebite. John Gerarde was among the first to record that Virginia snakeroot (*Aristolochia serpentaria*) was a potent counterpoison in 1633. No fewer than 13 species in 10 genera have been considered snakeroots during eastern North American history. The first record of Sampson's snakeroot seems to be from 1892 when that name was applied to *O. pedunculatum.*

Both Carolina's Catawba of the coastal plains and the Cherokee mountain people used these plants as medicines. The Catawba used a poultice or salve made from the roots to treat boils, wounds, sores, and broken bones (Speck 1934, Moerman 1998). The Cherokee used an infusion to induce abortions, as a diaphoretic against colic and indigestion, and as a tonic (Hamel and Chiltoskey 1975). There is no suggestion that indigenous people used *O. pedunculatum* against snakebite, and perhaps Vogel (1970) was correct in suggesting that the original species called Sampson's snakeroot was *Gentiana catesbaei.* Indeed, it is under the latter scientific name that Porcher (1863) discussed the values of "Sampson's snakeroot" along with other *Gentiana.*

The use against snakebite must have come about because of African influence since the plants are commonly used in voodoo (Hocking 1997). That too must be the source of "Sampson" in the common name, and surely is for the name Congo root. Plant roots contain oleoresins that are used as an aromatic and bitter tonic and stimulant (McChesney and Adams 1985, Hocking 1997).

Orobanche

(Named from Greek *orobus*, vetch, and *anchein*, to strangle, because several species specialize in growing on legumes)

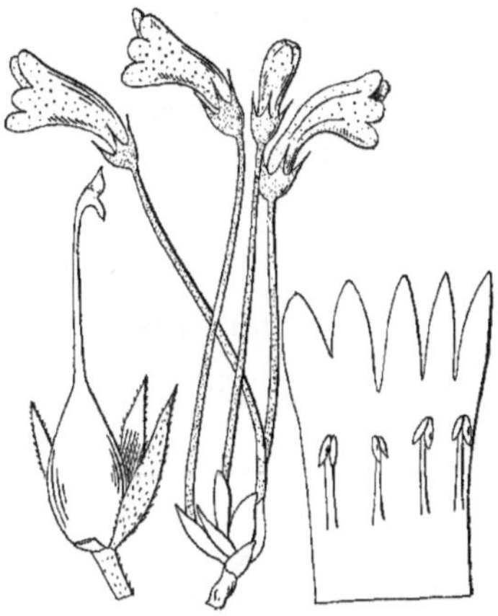

Orobanche uniflora. From Britton and Brown 1898.

brincalheta (from Latin *vinculu*, a confining band, alluding to the parasitic habit, Portuguese)
broom-rape (a translation of the Medieval Latin *rapum genistae*, meaning "a knob or tuber of *Genista*"; refers to the parasitic growths on the roots of broom or *Genista tinctoria;* the Latin was applied to *O. minor* by Rembert Dodoens in 1554 and translated into English in 1578)
erba lupa (wolf herb, Italian)
fiamma (flame, Italian)
orobanche (French)
penachos (from Italian *pennacchio*, plumed or beset with plumes, as feathers in hatbands, Portuguese)
snylterot (root parasite, Norwegian)
Sommerwurz (summer herb, German)
succiamele (from *succhione*, sucker, *miele*, honey, Italian)

Orobanche uniflora (one-flowered)
[one-flowered] cancer-root (by 1546 there were references to herbs called "cancer" or "herbe cancer"; John Gerarde's *Herball* of 1597 mentioned "cancerwort," which apparently referred to *Linaria* or *Veronica*; however, "cancer-root" in association with plants seems to have appeared by 1714; see also *Conopholis*)
naked [pale] broomrape
pipes (Maine)
squaw-drops [squawdrops, squawroot] (North Carolina)
squirrel's-grandfather (California)

The opinionated British Protestant clergyman William Turner wrote in 1548, "Orobanche ... may of his properties be called Chokeweed, because it destroyeth and choketh the herbes that it tyeth, and claspeth with his roote." That summed up European disgust for these parasitic herbs.

Europeans may have already known about medical uses of these plants when they arrived in the New World because a book on medicine published in 1671 said, "Broom-rape ... easeth pains in the Reins [kidneys]." Alternately, they may have learned uses from indigenous Americans.

Moerman (1998) found four or five species used for food and medicine among the Akimel O'odham (Pima), Blackfoot, Cahuilla, Gosiute, Keres, Montana, Navajo, Paiute, and Zuni. Among the medicinal uses were to treat wounds and sores, hemorrhoids, pneumonia and other lung troubles, cancer, and birth injuries.

No reference has been found that indigenous people in the southeastern United States used *Orobanche uniflora* or any other species. However, given the wide application by western tribes to external sores, wounds, and ulcers, southeastern use seems likely.

Porcher (1863) wrote that *O. uniflora* was "used as a remedy in carcinomatous affections, ulcers." He also indicated, "Dr. [Benjamin Smith] Barton [1810] thought [both *O. uniflora* and *Epifagus* were] ingredients of a secret remedy for cancer, known as 'Martin's Cancer Powder.'"

Saadoun and Hameed (1999) found an antibacterial activity in the extract from *O. cernua*. That plant was effective against *Staphylococcus aureus, Streptococcus, Bacillus cereus,* and *Escherichia coli.* They wrote, "the *O. cernua* extract displays remarkable activity against some Gram-positive bacteria such as *Streptococcus* and *Staphylococcus.*"

Orontium

(According to Galen, Greek *orontion* was a name for a water-plant and remedy for jaundice; also the Syrian river called *Orontes*; Latin *Oronteus*, poetic for Syrian; one of Aenea's companions in the *Aeneid* was "ardent" Orontes; reapplied by Linnaeus to this American endemic)

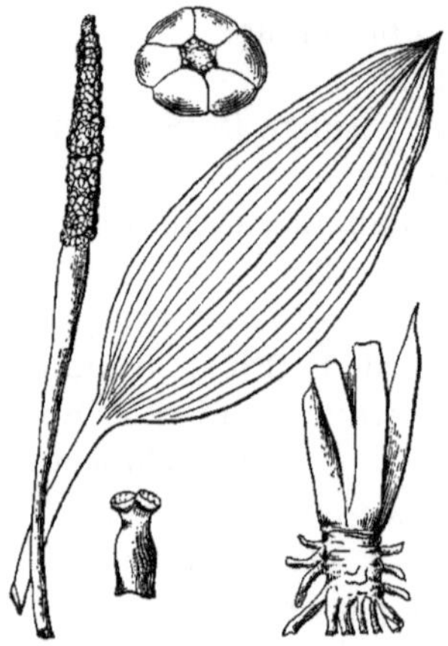

Orontium aquaticum. From Britton and Brown 1896.

Orontium aquaticum (of wetlands, aquatic)
bog-torch (Georgia)
bull-tongue (a comparison with the leaf shape, Georgia)
fireleaf
floating arum
golden-club [goldenclub] (refers to the golden-yellow inflorescences; in use by 1860, USA)
never-wet [never-wets] (because submerged leaves resurface dry, Georgia)
taw-kee [*tackvim, tawkee, tawkin*] (Algonquian, in Delaware, New Jersey, Pennsylvania)
tuckahoe [*tockwhogh, tockawhoughe*] (by confusion, as this was historically used for *Peltandra*, Powhatan, Virginia)
water-dock (a resemblance of the leaves to *Rumex*)

The confusion between *tuckahoe* (*Peltandra*) and *taw-kee* (*Orontium*) may be due to faulty history, or it may have resulted from misunderstanding by Capt. John Smith. Perhaps, however, Fernald et al. (1958) were wrong when they wrote that the indigenous people used the two and their names indiscriminately. Neither the Swedes who learned from the local tribes nor Peter Kalm had any trouble telling the two apart.

When Kalm ([1753–1761] 1972) was in New Jersey in 1749, he discussed these wetland herbs with the settlers. They told him: "The *Indians* pluck the seeds, and keep them for eating. They cannot be eaten fresh or raw, but must be dried. The *Indians* were forced to boil them repeatedly in water, before they were fit for use; and then they ate them like pease. When the Swedes gave them butter and milk, they boiled or broiled the seeds in it. Sometimes they employ these seeds instead of bread; and they taste like pease. Some of the Swedes likewise ate them; and the old men among them told me, they liked this food better than any of the other plants which the *Indians* formerly made use of." Core (1967) had the use correct, but he applied the wrong name.

Perhaps the indigenous people also ate the roots as they did *tuckahoe.* Yanovsky (1936) said that both roots and seeds were eaten by indigenous people in at least New York, Pennsylvania, and Virginia, but that report is suspect. Their roots were eaten by settlers (Plowman 1969). Porcher (1863) wrote from hearsay, "The root is acrid, but becomes eatable by roasting." He implied that he knew firsthand that "both the seeds and roots were eaten by the Indians."

Cherokee babies were bathed in a decoction every new moon (Vogel 1970).

Osmanthus

(From Greek *osme,* odor, and *anthos,* flower, because of the fragrant flowers)

Osmanthus americanus. From Sargent 1905.

Osmanthus americanus (American)
[American] devil-wood [devilwood] (a name in use by 1818; perhaps the name originated because of the hard wood)
[American, scrub wild, wild] olive (fruits resemble olives, *Olea europaea,* which are in the same family, Oleaceae)

Linnaeus named these trees *Olea americana* (American olive) in his *Mantissa Plantarum* of 1767. It was not until 1790 when Portuguese missionary João de Lourerio created the genus *Osmanthus* for Asian plants in the *Flora Cochinchinensis.* Almost a century passed before George Bentham and Joseph Hooker (Kew Gardens) realized that the American plants belonged in the Asian genus and created the modern name in 1878.

Osmanthus now contains 15 species, and grows in southern Asia and the southeastern United States. The genus has been incorrectly reported in Hawaii and

Polynesia (Sargent 1905), but all Asian species grow from Russia to Japan and Taiwan (Fernald 1950, Ohwi 1965, Czerepanov 1994). Among the Asian species, *O. fragrans* is native to China where it has long been cultivated for its fragrant flowers that are used to perfume tea and confectionary (Hedrick 1919, Mabberley 1997).

Osmanthus americanus grows from Florida to eastern Louisiana where it reaches the Mississippi River, and east along the Coastal Plain to Cape Henry, Virginia (Fernald 1950, Radford et al. 1968, Jones and Coile 1988). *Osmanthus megacarpus*, the scrub wild olive, grows in the central Florida peninsula (Wunderlin 1998).

No records of indigenous people using the plants have been found. Moreover, there are few indications of other cultures using either *Osmanthus* species.

Conflicting reports of fruit edibility exist. In his *Flora Americana Septentrionalis* (Flora of North America) of 1814, Frederick Pursh said the fruit was edible. George Vasey, however, in a USDA report of 1875, disagreed. He said fruits were of no value (Hedrick 1919); however, they are oleaginous (Hocking 1997). Presumably, like the olive (*Olea europaea*), preparation of the drupes is necessary to make them edible. Neither Fernald et al. (1958) nor Tull (1999) mentions the species.

Wood from *Osmanthus* is heavy, hard, strong, resistant, difficult to work, dark brown, with thick light brown or yellow sapwood (Sargent 1905, Hocking 1997). Trees are uncommon, which is why use is only local. Bark from *Osmanthus* has been used as a bitter astringent, laxative, and emollient (Hocking 1997). However, the genus is absent from Porcher (1863), Vogel (1970), Foster and Duke (1990), Bremness (1994), Bown (1995), and Moerman (1998).

No chemical studies have been found for the American species. Asian species, on the other hand, have been examined by a number of laboratories. Some of the Old World species contain monoterpene alkaloids, essential oils, furanosides, glycosides (iridoid, lignan, phenolic, phenylethanoid, phenylpropanoid, and secoiridoid), pyranosides, verbascoside, and other chemicals (Inouye et al. 1975, Kaiser and Lamparsky 1978a,b,c, Kaiser et al. 1978, Wen et al. 1983, Inagaki et al. 1991, Sugiyama and Kikuchi 1991, 1993, Sugiyama et al. 1993a,b, Benkrief et al. 1998, Omura et al. 2000).

Osmunda

(Origin uncertain, just as Linnaeus preferred; according to some, for *Osmunder*, the Saxon name for Norse Thor, the god of thunder, weather, and crops)

Osmunda. *Osmunda cinnamomea* (left). *Osmunda regalis* (right). Both from Britton and Brown 1896.

Osmunda cinnamomea (cinnamon-brown)

buckhorn brake [buckhorn fern] ("brake" perhaps of Scandinavian origin, akin to Old Swedish *braekne*, fern)

cinnamon fern (here and in the species name, the reference is to the reproductive fronds that are markedly different in color from the green sterile fronds)

fiddleheads (used for the ornamental carving at the bow of a vessel carved like the head of a violin; in use by 1599; applied to young fern leaves by 1882, perhaps earlier)

flowering fern (ferns are nonflowering plants, but the fertile fronds look like flower remnants to some)

Osmunda regalis (royal)

bog onion (local name in the British Isles, alluding to the similarity of the caudex to an onion bulb)

buckhorn brake

flowering fern (used since at least the time of Porcher 1863)

hamockolo:pintapintî [*hamótckolópintápinti*] (*hamockolo:p*, old paint woman, *im*, her, *tápinti*, fern, Mikasuki)

helecho de espigas (fern with spikes, Cuba)

helecho real (royal fern, Cuba); *raineach rìoghail* [*righ-raineach*] (royal fern, Gaelic); royal fern (first in English by 1600; however, reference to royalty is much older in other languages); *kongsbregne* (king's fern, Norwegian); *Königs-Rispenfarn* (King's panicle-fern, German)

hoktalkoló: intocâ (*hoktvl*, old woman, *kolowv*, glue, *em*, her, *locowv*, clay jug, Creek)

Osmund royal (used by 1548, when written in English "Osmende royall"; apparently from Old Swedish *osmunder*, dating to 1350); Osmund the

Waterman (from 1578, used by Rembert Dodoens in Dutch in 1554)
St. Christopher's herb (from 1578, used by Rembert Dodoens in Dutch in 1554)
water-fern (from 1578, used by Rembert Dodoens)
yatcheiwashaguen taapente [*yatahuwashat taapente*] (lost person's fern, Mikasuki)

Most sources simply say that *Osmunda* is probably derived from *Osmunder,* or *Thor* in Saxon. However, that oversimplifies what is known about the name. Cognates to "Osmund" go back to Medieval Latin *osmunda,* and French *osmunde* from the 12th century. In all of the English texts, the reference is to ferns, but not usually the plants now meant. Originally, the term "Osmund" indicated some of the "male-ferns," including what is now *Dryopteris felix-mas.* It was not until the late 1600s or early 1700s that "royal fern" became fixed with *O. regalis.* Frenchman Charles Plumier was one of the first to use that concept in 1705.

Other options exist for derivation of the name *Osmunda.* One is Osmundus, an 11th-century Scandinavian writer. Another likely model is Osmun, Bishop of Salisbury, who died in 1099, and Quattrocchi (1999) implies that there are other possibilities. Regardless, all of these seem to postdate the legendary Osmunder and must be derived from that Scandinavian base.

Royal fern has a wide range, including both the Old and New Worlds. The plants grow from Newfoundland to Saskatchewan and south to Florida and Texas. They are also in Bermuda, the Bahamas, the West Indies, Mexico, and south to Paraguay. Europeans knew them when they arrived in the New World because they are in wetlands in Europe (from southern Norway south) and in western Asia to India and Korea (Morton 1981, Mabberley 1997). At least in northern Europe, there is a history of medical application of royal fern.

According to Vickery (1995), *O. regalis* is called both "royal fern" and "bog onion" in the British Isles. Up until at least the 1880s, it was still used to treat sprains, rheumatism, sciatica, and bruises. The caudex was beaten, covered with cold water, and allowed to "sit" overnight. The thick starchy fluid that resulted was used to bathe the affected areas.

Vickery (1995) also found a story about the ferns that I encountered among people in western Kentucky when I was growing up. Some British and Kentuckians believed that the fern "flowered," but that it always did so at night. With daylight, the flowers disappeared, but we might know they had been there because they left their "seeds," i.e., the spores.

People in the New World also took advantage of the royal fern. Hogan (1978) found *Osmunda* spores in the pre-Columbian samples she examined from the Fort Center site. The density in the coprolite samples led her to conclude that the people were using the ferns and that they were not contaminated by windborne spores.

We do not know how the Glades people used the ferns. The Seminoles used an infusion of the plant to steam and bathe the bodies of people who were "insane," or disoriented (Sturtevant 1955). Modern Seminoles make a medicine from the roots to treat senility (Bennett 1997). A medicine was also made for use during childbirth and to treat chronically ill babies.

One application was to treat the malady called "Old Paint Woman's Sickness" (Sturtevant 1955). That illness was caused by a supernatural being called *hamockolo:pi* (the old paint woman, Mikasuki). Her name is part of the Mikasuki word for royal fern that was, in the 1950s, *hamockolo:pintapinti* (*tapinti,* fern). To inflict her problems, she "hugs" the patient; the symptoms are weakness of the limbs and neck. A mixture of her fern with others is the remedy.

At least the Iroquois and Menomini also used this fern (King 1984, Moerman 1998). The Iroquois mixed fronds of royal fern with wild ginger to expel worms from children. They also made a remedy for blood and menstrual problems, and kidney ailments. The Menomini used the roots in a medicine to promote lactation.

During the Civil War, Porcher (1853) wrote of royal fern: "It is sometimes employed in dropsy, as an astringent in injuries, and by Dr. Heidenreich in the radical cure of hernia ... it purges mildly in doses of two to four drachms of the powder. It acts upon the bile, augments digestion, and strengthens chylification. The extract has been thought peculiarly suited to cases of children affected with caries, mixed in milk or water, and continued for some time . . . has been employed successfully in doses of three drachms in the rickets. ... The leaves have been selected to make cradles for delicate children, from some supposed good effects derived from their use."

Osmunda cinnamomea is less common in the area around Lake Okeechobee where Hogan (1978) studied the Fort Center Glades site, but it becomes more common farther into the temperate zones. The Koasati in Louisiana used *O. cinnamomea* as a medicine (Taylor 1940). The Cherokee of the Carolinas used a decoction to treat rheumatism and chills (Hamel and Chiltoskey 1975). The Iroquois used a decoction to relieve headache and rheumatism pains, colds, and menstrual problems (Moerman 1998). The Menomini ate the young leaves and made a medicine to increase milk flow (Yanovsky 1936).

Several tribes knew that the young, incompletely unfurled leaves were edible (Moerman 1998). The Abenaki ate them as a snack, and the Menomini added them to soup. The Menomini also figured that eating the young leaves would make them smell like deer that had been browsing on the plants, and perhaps make it easier to approach the deer. Fernald et al. (1958) wrote that croziers were eaten by small boys but not appreciated by adults. They do tend to have a bitterish flavor that most people of European lineage do not like.

Ostrya

(The Greek name for a tree with hard wood; Theophrastus, 372–287 B.C., was probably talking about the southern European *O. carpinifolia*)

Ostrya virginiana. From Sargent 1905.

carpinella (Italian)
charme houblon (hop hornbeam, French)
Hopfenbuche (hop bush, German)

Ostrya virginiana (of Virginia)
American hornbeam ("hornbeam," the name of a European species from the 1570s or before, because of its hard, tough, close-grained wood; originally used for *Carpinus*)
bois de fer (ironwood, Quebec)
deerwood (probably in the sense of a "wood" or woodland where deer are likely to be found)
hardhock (maybe from "hock," derived from "hook"; a rod or stick with a hook at the end, used since 1570; perhaps related to the name "hardhack" used locally in New England for *Spiraea tomentosa*)
[eastern, woolly American] hop-hornbeam ("hop" used as modifier because of the similarity of the fruit cluster to that of the hop, *Humulus lupulus*; used by 1794)
ironwood (name given, often locally, to various trees with extremely hard wood; dating from at least as early as the 1650s)
ittowanha (*itto*, tree, *wanha*, hard, Alabama); *itukawiloha* [*itukahiloha*] (Choctaw)
leverwood ("lever" is a bar used to dislodge some heavy or firmly fixed object)
ma'nanous' (Ojibwa)

Southern Europeans knew hop-hornbeam (*Ostrya carpinifolia*) when they arrived in the New World. The new one they found was *O. virginiana.* Wood from the New World species was much like that in the Old World. It is strong, hard, tough, durable, light brown tinged with red or often nearly white, with thick pale sapwood (Sargent 1905). Both the Lakota and the Malecite used the wood to make bows (Moerman 1998). Probably other tribes used it similarly. The Ojibwa used the trunks to build dwellings (King 1984). Among people of European heritage the wood has been used for fence posts, golf clubs, tool handles, mallets, and woodenware (Vines 1977).

The Cherokee, Delaware, Iroquois, Ojibwa, and Potawatomi made medicines from *Ostrya* (Smith 1933, King 1984, Moerman 1998). Bark decoctions were used to build up the blood, soothe sore muscles, and relieve toothache by the Cherokees (Hamel and Chiltoskey 1975). The Delaware used a root mixture for female problems and as a tonic. The Iroquois made decoctions to treat cancer, coughs, swellings, and tuberculosis. The Ojibwa made medicines to treat coughs, lung hemorrhages, urinary problems, and rheumatism. The Potawatomi used the bark to stop diarrhea and hemorrhages.

Oxydendrum

(From the Greek *oxys*, sour, and *dendron*, tree, because of the acidic foliage)

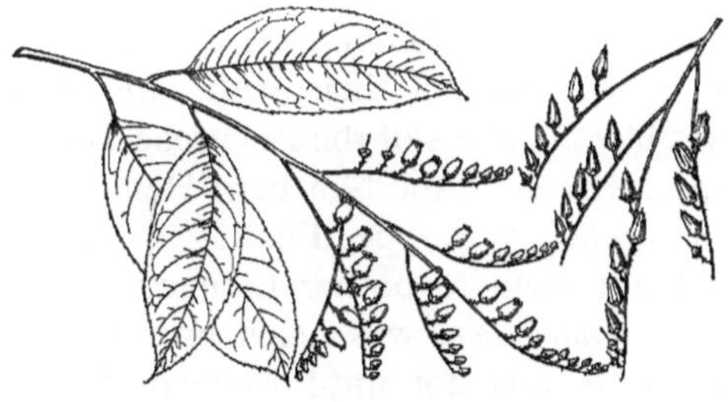

Oxydendrum arboreum. From Sargent 1905.

Oxydendrum arboreum (becoming a tree)
arbre oseille [*arbre l'oselle*] (sorrel tree, French); sorrel-tree ("sorrel" from *surelle*, diminutive of French *sur*; "sorrel-tree" used in 1730 by Mark Catesby; in 1856 Asa Gray equated it with "sourwood")
arrow-wood (a name widely applied to many genera because of the straight young stems that were made into arrow shafts)
elk-tree ("elk" is of uncertain history, but probably derived from Latin *alces* and Greek *alce*, both

referring to the European Cervid, *Alces alces*, the animal known in North America as "moose"; later applied also to wapiti or *Cervus canadensis*; presumably, the common name alludes to the leaves being favored as food)

iti kosoma (*iti*, tree, *kosoma*, fetid, Choctaw)

lily-of-the-valley-tree (an allusion to the small white flowers that are reminiscent of those on lily-of-the-valley, *Convallaria*)

pin-ne-se-ga (astringent taste, Osage)

Sauerbaum (sour tree, German); sourwood ("sour" derives from Old English *súr*, with cognates in French *sur*, German *sauer*, Dutch *zuur*; used by 1856 where equated with sorrel tree by Asa Gray); sour gum [sourgum-bush]

titi (etymology uncertain, maybe from "tie"; the local use in northern Florida implies that it came about because of the difficulty in passing through the vegetation consisting of these plants, hence local names such as "Titi Hell"; originally applied to members of Cyrillaceae by 1860; J.K. Small linked the name with *Oxydendrum* in 1903; or could it be from Choctaw *iti*, tree?)

ʔyápske [*yap hita'ʔare*] (tree sour, Catawba)

When Catesby, Gronovius, and Linnaeus were confronted with the trees now known as *Oxydendrum*, they did not know what to do with them. Catesby called them *Frutex foliis oblongis acuminatis* (shrub with oblong acuminate leaves). Gronovius did not do much better with *Andromeda arborea, foliis oblongo-ovatis integerrimis* (Tree *Andromeda* with oblong-ovate, entire leaves). Linnaeus used the Gronovius name and called them *Andromeda arborea.*

The plants they put them with, *Andromeda*, are small shrubby members of the Ericaceae. Both *Andromeda* species grow in the Northern Hemisphere with one widespread in Europe. Finding a tree form of these did not seem to bother Linnaeus because his concept of *Andromeda* was much broader than that used today. It was not until Augustin-Pyramus de Candolle studied the plants for his *Prodromus* of the world flora that the genus *Oxydendrum* was created in 1839. The single species is confined to the deciduous forests of eastern North America, ranging from Pennsylvania and West Virginia west to Ohio and Indiana, and south to Florida and Louisiana (Fernald 1950). *Oxydendrum* has been reported from Texas, but Correll and Johnston (1970) could not verify its presence there.

Although Hamel and Chiltoskey (1975) say that sourwood was used for firewood by the Cherokee, that was not always the case. Indeed, Mooney (1885–1886) said that his informants told him that wood from the tree was never burned. They believed that lye made from the ashes would "bring sickness to those who use it in preparing their food." They also maintained that, if a person slept beside a sourwood fire, the wood "will barbeque him."

However, these taboos did not prevent the Cherokee from using branches of sourwood to cook their meat before the fire. It was a preferred cooking wood because its acidic flavor gave the meat a better taste (Mooney 1885–1886). Spoons and combs were also carved from the wood. Other uses among the Cherokee were to make sled runners, arrowshafts, and pipe stems (Hamel and Chiltoskey 1975).

Yanovsky (1936) said that the young leaves were eaten in salads in the southeastern states. This is one of several places where he extrapolates too far. Southeastern tribes normally did not eat raw vegetables (Hudson 1976).

Both the Catawba and the Cherokee used sourwood in medicine (Moerman 1998). The Catawba used an infusion as a menstrual aid and for women

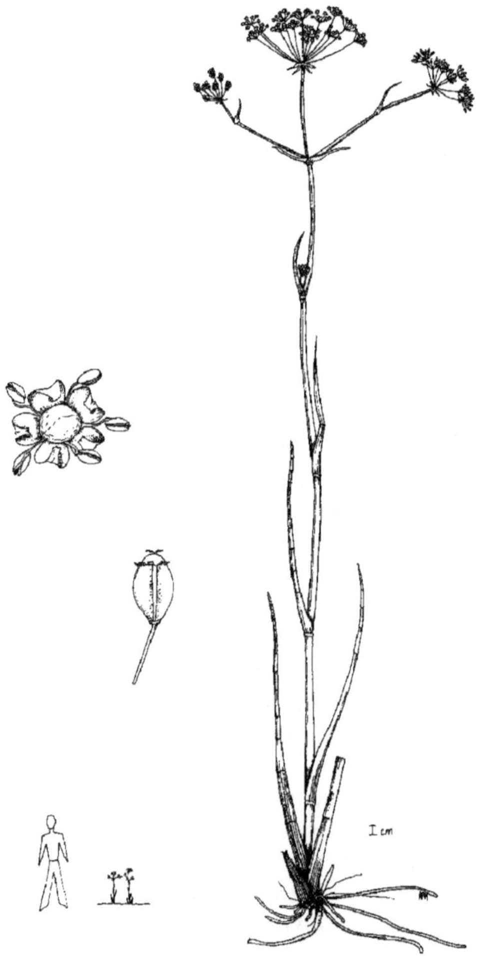

Oxypolis filiformis. From Institute of Food and Agricultural Sciences.

undergoing menopause. The Cherokee used the plant in an infusion to stop diarrhea and as a tonic for indigestion, nervousness, asthma, and spitting blood (Hamel and Chiltoskey 1975). Bark sap was used on itch, and the bark was chewed to heal mouth ulcers.

Settlers continued using the plants as medicine. Porcher (1863) wrote: "The leaves, when chewed, allay thirst. A decoction of the bark and leaves is also given as a tonic." Foster and Duke (1990) noted that "Leaf tea [is] a Kentucky folk remedy for kidney and bladder ailments (diuretic), fevers, diarrhea, and dysentery."

Oxypolis

(Named by Rafinesque in 1825, from Greek *oxys*, sharp and *polos*, axis, referring to the leaves)

Oxypolis filiformis (threadlike, another reference to the leaves)
akkolo:ɬka [*akkolo:rkâ*] (*akkolo:ɬka* is usually *Nelumbo*, Creek; a comparison or a misuse)
cafã:mâ:bî (*a:bi,* replica, Mikasuki)
water dropwort (Florida, Bahamas)

Thomas Walter described *Oenanthe filiformis* in 1788 based on specimens from South Carolina. It was not until 1894 that N.L. Britton moved the species to *Oxypolis*. There are now seven species in this endemic North American genus (Mabberley 1997).

There are records of two species having been used. The Cherokee used *O. rigidior*, stiff cowbane, roots for food (Hamel and Chiltoskey 1975). The Seminoles had names for *O. filiformis* but they remembered no uses (Sturtevant 1955).

P

Packera/Senecio

(*Packera*: Named in 1976 by Askell and Doris B.M. Löve for Canadian botanist John G. Packer; *Senecio*: From Latin *senex*, an old man, alluding to the white pubescence of many species, or the white pappus)

Packera aurea. a. Plant in flower. b. Ray flower. c. Disk flower. d. Longitudinal section. e. Seed. From Buchholtz 1968.

åkersvineblom (field-wine-bloom, *Senecio vulgaris*, Norwegian)
bualan [*am bualan*] (Gaelic)
cardo morto (dead thistle, *S. vulgaris*, Portuguese)
erba calderugia (bowl herb, maybe from the resemblance of the involucre to a bowl or volcanic caldera, *S. vulgaris*, Italian)
gemeines Kreuzkraut (common cross-herb, *S. vulgaris*, German)
Greiskraut (gray [aged] herb, German)
groundsel (derived about A.D. 700 from Anglo-Saxon *grundeswelgiae*, pus-absorber; perhaps, but less convincingly, from *grundeswylige*, ground-absorber, because of its rapid spread)
herba cana (white-haired herb, *S. vulgaris*, Spanish)
lus Phara léith (gray-headed Patrick's herb, Gaelic)
ragweeds (a name applied to *Senecio* since ca. 1658, and expanded to *Ambrosia* ca. 1866)
ragwort (probably applied because of the ragged leaves; used in English by A.D. 1450; applied at least by John Gerarde in 1597 to St. James wort, *S. jacobaea*)
sciopeti (Venetian dialect for "little exploders" because you can take the bracteal cup and hit the back of your hand and it will make a noise like a little explosion, Italian)
séneçon (old one, French); *suzón* (*S. vulgaris*, Spanish)
tasneirinha (diminutive of *tasneira* or *taneceto*, from *Tanacetum*, alluding to the resemblance between that member of the Asteraceae and *S. vulgaris*, Portuguese)

Packera anonyma (nameless) (=*Senecio anonymus*)
Small's ragwort (Nathaniel L. Britton tried to name this species *S. smallii*, for John K. Small, but he proposed that name in 1890, and Alphonse W. Wood had already named the plant in 1861)

Packera aurea (golden) (=*Senecio aureus*)
butter weed (unexplained by Williams [1837] 1962; perhaps for the butter-yellow flowers)
false valerian (the name alludes to *Valeriana*, a widely used medicinal herb)
female regulator (alluding to use in controlling fertility or use during childbirth?)
fireweed (usually given to Onagraceae, formerly *Epilobium*, and to Asteraceae, *Erechtites*)
golden ragwort
golden senecio (book name)
Goldenes Kreuzkraut (golden cross-herb, German)
hanatcuwi·'hare (make [child] come quick, Catawba)
life-root [liferoot]; life-wort [lifewort]
squaw-weed ("squaw" from the Narragansett *squaws*, Massachusetts *squa*, woman, with related forms in other Algonquian languages; combined with "weed" by 1847 by Darlington who wrote that it "had been denounced ... as being poisonous to sheep")

unkun (English? a name unexplained by Millspaugh 1892)

When Wunderlin (1998) published his *Guide to the Vascular Plants of Florida*, he used the long-established genus *Senecio*, based on the Latin name of a plant used by Pliny. The name alludes to the white pubescence of many species or to the white hairs of the pappus. Later, Wunderlin and Hanson (2002) accepted several changes in the genera of Asteraceae and began using the genus *Packera* for a group of 60 to 65 species, mostly in North America, with 16 in Mexico, and others in Siberia. Kartesz (1994) continued using *Senecio* in the broad sense, so both options are given here.

Regardless of the name, the species contain potent pyrrolizidine alkaloids (e.g., senecionine), which cause liver damage if ingested (Lewis and Elvin-Lewis 1977, Lampe and McCann 1985). That damage occurs through acute venous occlusions in the liver (Budd-Chiari syndrome), and can lead to cirrhosis and, in some cases, death (Lampe and McCann 1985, Foster and Duke 1990).

In spite of their toxic properties, some species were still being used externally in Europe in the 1970s as poultices for wounds and abscesses (OED 1971). Bown (1995) found that *P. aurea* still is grown in Belorussia, central Russia, and the Ukraine for the pharmaceutical industry.

Plants in this group of Asteraceae have been used by people for thousands of years in both hemispheres. Famous species in Europe are *S. cineraria, S. jacobaea,* and *S. vulgaris* (Polunin 1969). Vickery (1995) found people still applying *S. vulgaris* to cuts, treating ague with it, and using it as a laxative. *Senecio jabocaea* was associated with witches and fairies in the British Isles and is known in Gaelic as *buadhghallan buidhe* [*buaghallan, boholàun, boholàun buidhe*] (*buadh*, virtuous, *ghallan*, branch, *buidhe*, yellow).

In the Americas, the Catawba used *P. anonyma* to treat consumption (Moerman 1998). The Cherokee used *P. aurea* as a contraceptive and for heart trouble (Hamel and Chiltoskey 1975). The Iroquois used *P. aureus* for the blood, as a diaphoretic, to reduce fever in children, and to treat broken bones (Moerman 1998).

Porcher (1863) wrote, "It is said by [David] Schoepf to have been a favorite vulnerary with the Indians; the juice of the plant in honey, or the seeds in substance, are employed." Millspaugh (1892) knew that the plants had been used by indigenous people to stop bleeding, as an abortivant, and a vulnerary. It was recommended for controlling bleeding in the lungs, for uterine problems, as a diuretic, pectoral, diaphoretic, and tonic.

Bown (1995) still maintained that *P. aurea* is a bitter, astringent herb that is diuretic, stimulates the uterus, and controls bleeding. However, she added that it should be used by "qualified practitioners only," and that it is subject to legal restrictions in some countries. Duke et al. (2002), who typically give level-headed recommendations, list *P. aurea* as "XXX"—not to be used (think of each X "as a skull and crossbones"). Hocking (1997) noted that *P. anonyma* has been said to have antitumor properties, but gives no reference.

Moerman (1998) found ten other species of *Senecio* and *Packera* being used by different tribes within North America. Thus, it would be surprising if other species (*P. glabella, P. obovata,* and *P. paupercula*) had not been used by indigenous tribes. However, of these, *P. glabella* has an enormous range in North America and no records have been found of anyone using it.

Panicum

(From the classical Latin name of bread, *panis*, or millet, *panus*; related to Akkadian *panu*, Italian *pane*, bread)

cockspur (originally the spur on a cock or male chicken, used since at least the 1590s; awns on some grasses led to the comparison with the fowl)
Hirse (German)
millet (from Latin *millium*, having a thousand grains, French); *miglio* (Italian); *milho* (Portuguese)

Panicum hemitomon (halved, from the somewhat one-sided spikes)
cintha:câ:bî (snake tail replica, Mikasuki); *cintha:cî* (snake tail, Mikasuki)
maiden-cane [maidencane] (perhaps meaning "grass resembling harvest maiden," from the old tradition of forming the last handful of wheat into the shape of a woman; harvest maiden also known as kirn-baby and kirn-doll, from the 1770s but surely much older as "kirn" dates to the 1300s, USA)
pahitoɬpiɬî (*pahi*, grass, *toɬpiɬl*, knees, Mikasuki); [*pahitóɬpiɬó:cî*] *pahitórpiró:cî* (*pahi*, grass, *toɬpiɬl*, knees, *oci*, small, Mikasuki)

Panicum hemitomon. From Institute of Food and Agricultural Sciences.

Panicum laxiflorum (loosely flowered) (as *P. xalapense*) (= *Dichanthelium laxiflorum*)
open-flower witch-grass [openflower witchgrass] (Florida)
soft-tufted panic-grass [soft-tufted panicgrass]

Panicum strigosum (with appressed hairs) (as *P. polycaulon*) (= *Dichanthelium strigosum*)
cofimássi (*cufe*, rabbit, *em*, its, *vsse*, dried leaves, Creek); *cokfimasí* (*cokfi*, rabbit, *im*, its, *híssi*, leaves, Koasati); *cokfimpatâ:kî* [*tcokfwimpataki*] (rabbit's bed, Mikasuki)
cushion-tufted panic-grass [cushion-tufted panicgrass]
rough-hair witch-grass [roughhair witchgrass] (Florida)

Hogan (1978) found pollen of *Panicum hemitomon* in the coprolites of the Glades people that she studied. As they were living beside vast stands of the wind-pollinated plants, that is no clear indication that they used the grass. However, being in a pre-Columbian context and having historical documentation of use is provocative. The Seminoles too know these formerly abundant grasses (Sturtevant 1955), and at least use them to the extent of "reading" the landscape. Where there is *pahitółpiłó:cî*, they know that the water remains near a certain depth. That constitutes part of their knowledge of the landscape.

Because of doubtful identifications, Moerman (1998) placed all the grasses used by the Creeks, Natchez, and Seminoles under *Panicum* sp. However, the species recorded by Sturtevant (1955) are reliably known. Since Sturtevant (1955) indicated that a third species was similarly used, perhaps others were used by the Seminoles' relatives, the Creeks.

The Creeks and Natchez used a *Panicum* leaf infusion for fevers, especially malaria (Swanton 1928a, Taylor 1940). Symptoms of those diseases are close enough to the Seminole malady called "Gopher Tortoise Sickness" (cough, dry throat, noisy chest) to suggest that the Creeks and Natchez used the same grasses. The Seminoles also used *cokfimpatâ:kî* for "Rabbit Sickness" (muscle cramps) (Sturtevant 1955).

The Cherokee used some *Panicum*, possibly more than one species, for padding inside their moccasins (Hamel and Chiltoskey 1975).

Parietaria

(Linnaeus based this on Latin *parietis*, a wall, in reference to its frequent occurrence there)

bartram [bertram] (an English corruption of Greek *pyrethrum*, from *pyros*, fire; the name was originally given to *Anacyclus pyrethrum* or pellitory of Spain by at least 1578 with Henry Lyte's translation of Dodoens's *Cruydeboek* of 1554; later the name became secondarily applied to *Parietaria*, because both were called pellitory)
blidnesle (gentle nettle, Norwegian)
Glaskraut (glass herb, German)
lus a' bhalla (wall herb, Gaelic)
parietaria (from Latin *parietis*, Italian, Spanish); *parietária* (Portuguese); *pariétarie* (French); pellitory (from Latin *parietis*)

Parietaria floridana (from Florida)
herbe à murailles (wall herb, Haiti)
herbe gras (fat herb, Haiti)
paille à terre (country straw, Haiti)

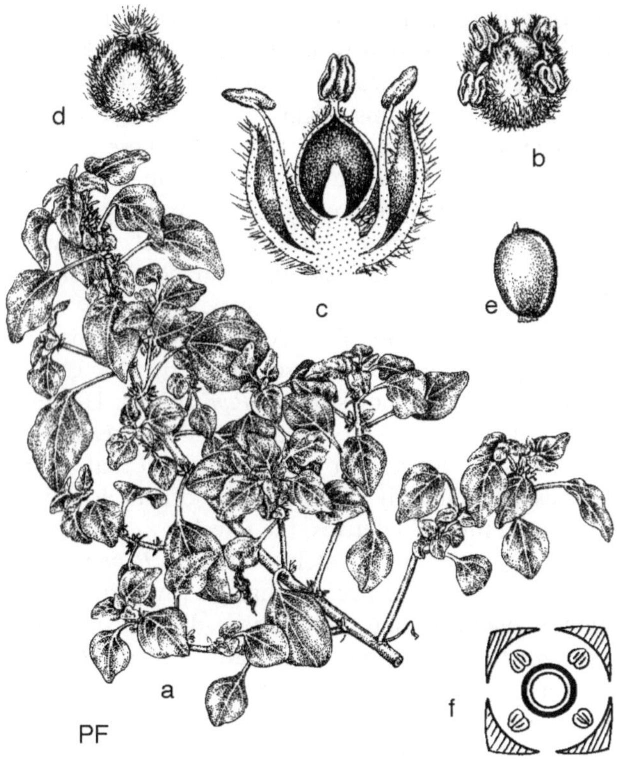

Parietaria floridana. a. Section of plant. b. Flower, side view. c. Flower, longitudinally dissected. d. Fruit. e. Achene. f. Floral diagram. *Drawn by Priscilla Fawcett.* From Correll and Correll 1982.

parietaria (pellitory, Dominican Republic); *pariétarie* (pellitory, Haiti); *pariétarie capitee* (pellitory with heads, Haiti); *pariétarie sauvage* (wild pellitory, Haiti); pellitory (Florida)
thé del malheureux (tea for the unfortunate, Haiti)

Newcomers to the New World were able to recognize this southeastern plant because they knew European "pellitory-of-the-wall" or "mind-your-own-business" (*Paritaria judaica* in western and southern Europe and *P. officinalis* in central and southern Europe). *Parietaria officinalis* was famous and was known in 16th-century English as parietorior or pellitoire of the wall, in German as *Tag und Nacht* (day and night), *Sanct Peterskraut* (St. Peter's herb), *Glaszkraut* (glass herb), or Dutch *glascruyt* (glass herb). Keeping closer to the Latin, the French said *parietaire* or *laparitoire*, the Italians and Spanish *parietaria.* Old World people took their use from classical authors Dioscorides (fl. A.D. 40–80), Galen (ca. A.D. 129–ca. 199), and Aëtius (A.D. 527–565), and used it for kidney stones, as a diuretic, and to treat hemorrhoids (Meyer et al. 1999). Finding a similar plant in the Americas, they used it largely for the same problems.

In Hispaniola, fresh *P. floridana* is diuretic and is also used for painful hemorrhoids (Liogier 1974). Haitians consider an infusion of the plant a diuretic for use in treating angina and gout (Morton 1981). *Parietaria* is also used for erysipelas and earache.

Parthenocissus

(Jules Emile Planchon, 1823–1888, combined Greek *parthenos,* virgin, and *kissos,* ivy, perhaps alluding to the unisexual flowers)

Parthenocissus quinquefolia. *Drawn by P.N. Honychurch.*

Jungfraurebe (virgin vine, German); *vigne-vierge* (virgin vine, French)
vite del Canadà (Canada grape, Italian)

Parthenocissus quinquefolia (with five leaflets)
American ivy
false grape; *parrita cimarróna* (little wild grape)
five-leaves; *l'herb à cinq feuilles* (five-leaved herb, Houma, Louisiana)

ifá imittó (*ifá*, dog, *im*, its, *ittó*, tree, Koasati)
ingtha hazi itai (ghost grapes; *hazi*, grapes, Omaha-Ponca)
manido'bima'kwud (*manido*, spirit, Ojibwa)
omakaski'bag (toad weed, Potawatomi)
sa-tai-al-go (paint berries, Kiowa)
vfala omat [*afala oma*] (*vfala*, poison ivy, *omat*, like, Creek)
vigne-vierge (virgin vine, Quebec); Virginia [Virginian] creeper ("creeper" is another word for climber or twiner)
woodbine [woodbind, wild wood-vine] (originally a European term for *Convolvulus* and *Hedera*, dating to about A.D. 875, and alluding to the tendency of the climbers to wrap around others, USA)

The first record of these vines in the New World is Jacques Philippe Cornutus's *Edera quinquefolia canadensis* (five-leaved Canadian ivy), published in his book *Canadensium plantarum ... historia* of 1635. Linnaeus knew this and several other sources. He had studied live plants at the *Hortus Cliffortianus.* Perhaps because he was influenced by Cornutus and others, he called the vines *Hedera quinquefolia.*

Probably the first report of interaction of the plant and people was left by Capt. John Smith in 1624. He wrote that, in Virginia, there was a "kind of woodbind ... which runnes vpon trees, twining it self like a Vine: the fruit eaten ... worketh ... in the nature of a purge." In spite of Smith's report, Yanovsky (1936) said that the fruit could be eaten raw, and the stalks were peeled and boiled for food. He reported those uses in Minnesota, Montana, and Wisconsin. One of them is wrong, and I doubt that it was Smith.

Native people in the New World had long been familiar with the twiners. The plants were used by the Cherokee, Creeks, Houma, Iroquois, Jemez, Keres, Kiowa, Meskwaki, Montana tribes, Navajo, and Ojibwa (Densmore 1928, Moerman 1998). Most of these people used the plant as medicine, but they also made dye from it, and some claim to have eaten the roots. In the southeast, the Cherokee used an infusion against jaundice (Hamel and Chiltoskey 1975). The Houma used a hot decoction of stems and leaves to reduce swelling, to treat wounds, and against lockjaw (Speck 1941). The Creeks used it against venereal disease (Taylor 1940). Snow and Stans (2001) published a picture of the vine (p. 80), but gave no common name, or included it in the text. Some Seminoles still use this medicine.

In Mexico, the bark has been used as an alterative, tonic, expectorant, and against dropsy; crushed leaves are a counterirritant, producing blisters when applied to the skin (Standley 1920–1926). Early American physicians also used a tincture, which was sometimes called *Decoctum ampelopsis* or *Infusum ampelopsis*, reflecting an old scientific name, *Ampelopsis quinquefolia* (Millspaugh 1892). Not much seemed to be known about its chemistry in the 1890s, and the same is true today. Foster and Duke (1990) cryptically wrote, "Berries reportedly toxic," although Capt. John Smith wrote the same in the 1620s. The leaves contain calcium oxylate and cause dermatitis in some people (Foster and Duke 1990, Foster and Caras 1994), compounding the difficulty people have distinguishing between this and poison ivy (*Toxicodendron radicans*).

Paspalidium

(Diminutive of *Paspalum*, Greek *paspalos*, for millet; the genus was separated from *Paspalum* by Otto Stapf, 1857–1933)

Paspalidium geminatum (twins or double) (=*P. paludivagum*)
akkotó:ɬka [*akkotó:rka*] (*akkotorkv*, usually *Nelumbo*, Creek; a comparison or a misuse)
ciktohacî (*cekto*, snake, *hvce*, tail, Creek)
Egyptian panicum (some consider the species native, like Allen 2003, while others think it was introduced from Egypt)
kissimmee grass ("kissimmee" was rendered as *Cacema* on the Moll Map of 1720 and *Casseeme* on William's Map of 1837; the locality was not mentioned by Swanton in either 1939 or 1946; language and meaning unknown)
pahitóɬpiɬî [*pahitórpirî*] (*pahi*, grass, *toɬpiɬ*, knees, because the stem is jointed, Mikasuki; this name is also used for *Panicum hemitomon*, which see)
water panic grass (Florida)

Pehr Forsskål called plants he found in Egypt *Panicum geminatum* in 1775. However, it was not until 1919 that the species was transferred to *Paspalidium* in the *Flora of Tropical Africa.* Godfrey and Wooton (1979) considered the species native, while Crins (1991), Wunderlin (1998), and Diggs et al. (1999) thought it introduced. The problem is considered unsolved by Gerald Guala (personal communication, Oct. 2003).

The Seminoles use a decoction of the plant to treat "Snake Sickness" (itchy skin) (Sturtevant 1955).

Paspalum

(From Greek *paspale* or *paspalos*, meal or millet)

Paspalum conjugatum. *Drawn by Mary Wright Gill*. From Hitchcock and Chase 1950.

Paspalum conjugatum (joined together)
ahan toom (elote-roasting ear grass, Huastec)
bed grass (Bahamas)
cañamazo amargo (big sour cane, Cuba); *cañamazo* [*hembra*] ([female] big cane, Dominican Republic)
capim de marreca (mallard hen grass, Marajó, Brazil); *capim gorda* (fat grass, Marajó, Brazil)
gengibrillo (little ginger)
grama (grass, Dominican Republic); *grama de antena* (antenna grass, Central America); *grama comun* (common grass)
herbe sûre [*z'herbe sûre*] (sour herb, Haiti); Jamaican sour grass (Jamaica); sour-grass; *yerba agria* (sour-herb)
horquilla (little hairpin, Central America); *horquetilla* [*blanca*] ([white] little hairpin, Puerto Rico); *pasto horqueta* (hairpin grass)
paja de panela (sugar grass)
sarataya (Siona, Ecuador)
sour paspalum (a book name)
tarurco [*torurco, toro-urco*]
trencilla (Central America)
turvará (indigenous name, Costa Rica, western Ecuador)

Around the edges of wetlands where it is neither wet enough to be marsh nor dry enough to be pinelands, there is an intermediate zone locally known as wet prairie. These areas dominated by graminoid plants are where *Paspalum conjugatum* most often is encountered. The flowering scapes stand above the mostly prostrate stems and leaves, and the two-branched inflorescences have reminded people of *horquillas* (hairpins). Occasionally in *Paspalum* we find a dark mass of mycelia that has replaced the seeds, and we know that an ergot fungus (*Claviceps*) has infected the grass. Those fungi contain LSD-like chemicals and can cause problems or be used to relieve maladies. I must wonder if these grasses were used for medicines because of this fungal infection.

Judging from the names of this species throughout its range, people have had strong feelings either for or against this grass. Some pastoral groups dislike the grass because it does not provide good forage for their animals—ostensibly because of its sour taste. Others praise it highly as food for their animals. For people, though, the grass is sometimes important in remedies.

In Cuba, *cañamazo amargo* is used as a bath for patients with malaria (Roig 1945). People in Trinidad use a decoction to relieve fever, flu, pleurisy, pneumonia, and fatigue (Wong 1976). The Bahamians make a medicine for tuberculosis with prickly pear (*Opuntia*) and wood ashes (Eldridge 1975). The grass is also medicinal in Belize (Balick et al. 2000).

Passiflora: Passion-Flower

(Linnaeus reversed the Latin *Flos passionis*; crucifixion flower, originally used by Nicolas Monardes)

Virtually every discussion about passion-flowers tells of the comparison between the passion of the crucifixion of Christ and the flowers. Jesuit priests made that analogy in their efforts to convert the New World people. Presumably, the "*leaf* symbolizes the spear. The five *petals* and five *sepals* the ten apostles (Peter who denied, and Judas who betrayed, being omitted). The five anthers, the five wounds. The *tendrils*, the scourges. The *column* on the ovary, the pillar of the cross. The *stamens*, the hammers. The three *stigmas*, the three nails. The *filaments* within

***Passiflora suberosa**. Drawn by P.N. Honychurch.*

the flower, the crown of thorns. The *calyx*, the glory or nimbus. The *white* tint, purity. The *blue* tint, HEAVEN" (Coffey 1993). What most accounts do not add is the historical sequence behind that story.

Nicolas Monardes was perhaps the first to use the Latin term *Flos passionis* (flower of the crucifixion) in 1582. Since that story and the plant were not part of John Frampton's ([1577] 1925) translation, English-speaking people were rarely aware of Monardes's role. Supposedly, the same species was brought to Europe by Jac Boccio in 1610, and that introduction may have been the source of cultivated plants in England, Holland, and Sweden (Svanidze et al. 1974).

The same passion-flower, later named *P. incarnata*, was mentioned by Strachey ([1612] 1953) on the James River of Virginia: "Here is a Fruict by the Naturalls called a *Maracock* this groweth generally lowe and creepeth in a manner amongest the Corne ... yt is of the bignes of a Queene-apple, and hath many azurine or blew kernells, like as a Pomegranett, and it bloometh a most ssweet and delicate flower, and yt is a good Sommer Cooling fruict, and in every field where the indigenous people plant their Corne be Cart-loades of them." His original notes on the Powhatan used *maracah* (Harrington 1955). That same year Capt. John Smith reported that the indigenous people planted "*Maracocks*, a wild fruit like a lemmon, which also increase in fruit" (Coffey 1993). Subsequently, herbalist Gaspar Bauhin recorded the species in 1623. Parkinson ([1629] 1976) wrote that the plant "Maybe called in Latine, *Clematis Virginiana*; in English, the Virgin or Virginia Climer; of the Virginians, *Maracoc*; of the Spanish in the West Indies, *Granadillo*, because the fruit ... is in some fashion like a small Pomegranate on the outside."

However, it was the name *Granadilla hispanis, flos passionis italis* (little Spanish pomegranate, Italian passion flower) published by Francisco Hernández in his book of 1651 that was the earliest firsthand record Linnaeus had as the basis for *Passiflora* in 1753. Not only did Linnaeus have the description and drawing from Hernández, but he also knew that the plants had been cultivated in England from the 1600s. Indeed, Linnaeus had studied the live plants at the *Hortus Cliffortianus* (Holland) and *Hortus Uppsaliensis* (Sweden). All these names had been applied to the species that Linnaeus called *P. incarnata* (flesh-colored). He was mistaken about the flower color because they are blue.

The Latin *Flos passionis* became *flor de la pasión* (Spanish), *fleur de la passion* (French), *flor da paixão* (Portuguese), and passion-flower as generic equivalents of the genus *Passiflora*. The apparent lone exception to these names is in Puerto Rico where the genus is called *parcha* (from *palcha*, Quechua). Today *parcha* is mostly associated with the introduced South American *P. edulis*. Probably the plant and its name were introduced at the same time from Peru where now *P. edulis* is called *maracuya*.

The names *maricock* and *maracocks* gave rise to maracoc, maycock, maypop (Alabama, North Carolina), mayapple (Alabama, North Carolina), Molly-pop (Alabama, North Carolina), pop-apple (North Carolina), apricot (North Carolina), and apricot-vine (Texas). All of these names are supposedly derived from *mahcawq* [*mäkak, mä'kâwk*] (Powhatan), akin to *machkak* (Menomini), *mäkäk* (Cree, Ojibwa), and *ma'ka'kwi̲* (Fox). Although similar, there seems to be no relation to Tupí *mboruku'ya* or *maraú-yá*, in Portuguese *maracujá*, and *maracuya* in Spanish, names for *P. edulis* (Gerard 1907).

Passiflora incarnata is also known as *granadilla* (little pomegranate, Texas, Florida fide Williams [1837] 1962), Holy-Trinity flower (Texas), *pasionaria* (of the crucifixion, Texas), passion-vine (North Carolina), and purple passionflower (Florida). *Opako* is the Alabama name, and it is almost identical to the Koasati *apakó*, Muskogee *opvkv* [*opv'kv*], and Miccosukee *opakî*. Probably belonging here is *làanasi* (*laana*, yellow, *osi*, suffix meaning extremely, Alabama). The plant designated by the Alabama name is described as having a "small sweet melon, smells like a honeydew, makes the mouth itchy, size of an orange; a vine with a fruit similar to passion fruit (if one eats too much of it, it will blister the tongue and mouth)" (Sylestine et al. 1993). The species ranges from Virginia to Missouri, south to Florida and Texas and Bermuda, and it is introduced farther north in the United States.

In addition to eating the fresh fruits (*uwa'ga*), the Cherokee made a social drink of them, mixing the juice with cornmeal as a thickener. That would be a drink akin to *horchata* in Latin America. Most surprising is that the Cherokee also parboiled and then fried the leaves in hot grease as a potherb (Hamel and Chiltoskey 1975).

The Cherokee and other indigenous people used *P. incarnata* as medicine. A compound infusion of the roots was used to treat boils, and roots were used to stop inflammation of wounds. Infants were given infusions of the roots to aid in weaning, and it also was used to relieve earache (Hamel and Chiltoskey 1975). The Houma called it *chassepareille incarnata* (flesh-colored saw brier), and took an infusion of the roots as a blood tonic (Speck 1941). Whole plants have traditionally been used in a tea as an antispasmodic and as a sedative for neuralgia, epilepsy, restlessness, painful menses, insomnia, and tension headaches.

Linnaeus ([1753] 1957) also published the Florida species *P. lutea* (yellow passionflower, Florida), *P. multiflora* (whiteflower passionflower, Florida; *fruta de perro*, dog fruit, Cuba; *pasionaria vainilla*, vanilla of the crucifixion, Cuba), and *P. pallens* (pineland passionflower, Florida). Similarly, Linnaeus named *P. sexflora* and *P. suberosa*, which are better known within the Caribbean.

Passiflora sexflora grows in Florida, the Greater Antilles to Puerto Rico, and from Mexico to Panama and Colombia. In those areas the plant is known as *ala de murcielago* or bat wing (Jamaica), duck foot (Jamaica), duppy pumpkin (Jamaica), goat foot (Jamaica, Florida), and *pasionaria de cerca* (fence passion vine, Cuba). Guatemalans make a sedative preparation of the flowers for nerves, insomnia, and diarrhea.

Even more noted is *P. suberosa*, native from southern Florida to southern Texas to northern South America and the West Indies, introduced into the Old World. That species like all the others has edible fruits, although they are smaller than most. The edible fruits gave rise to the names *baleeyail an 'its'aamal* (deer watermelon, Huastec, San Luis Potosí), *huevo de gallo* (rooster egg, Cuba), juniper berry (Bahamas), *meloncillo* (little melon, Cuba), *morita* (blackberry, Dominican Republic), *parcha yedra* (ivy passionfruit, Puerto Rico), *parchita de culebra* (snake's little passion-fruit, Venezuela), and wild pumpkin (Caymans). Fruits are also used to make black ink as noted by the names indigo berry (Virgin Islands), ink berry (Virgin Islands), and ink vine (Barbados).

There are always names that are at odds with others. *Passiflora suberosa* is called corky passionflower in Florida because some scholar simply translated the Latin name. In Yucatán, the Maya say *kansel-ak* (*kants'il*, like cotton, *ak*, vine). Supposedly, it is called that because it is covered with trichomes like cotton, but most Florida plants are glabrous. In Hispaniola it is *leontafia* (lion's *aguardiente* or whiskey) or *tidiane* (little Diane).

When I first experienced juice from *maracujá* in Belém, Brazil in 1969, I was told that it was good to drink with dinner because it aided digestion and calmed the nerves. At the time, I drank it because it tasted good, but I could not really notice the effects.

Subsequently, I discovered that several species are considered calmants or even intoxicants (Fellows and Smith 1938, Speroni and Minghetti. 1988, Medina et al. 1990, Solbakken et al. 1997, Dhawan et al. 2001a,b,c). Some experiments support the tranquilizing conclusion, while others do not (Coleta et al. 2001, Volz 2001). One other caveat is that the commercial extract is possibly toxic, at least in some individuals (Fisher et al. 2000). However, most results indicate that plant extracts are mildly sedative, reduce blood pressure slightly, and decrease motor activity.

In addition to the supposed sedative effects of *Passiflora*, several others are known, including as an adjuvant agent in the management of opiate withdrawal, and as an antibacterial, anticonvulsant, antifungal, and antioxidant (Nicolls et al. 1973, Medina et al. 1990, Akhondzadeh et al. 2001, Murcia et al. 2001, Taglioli et al. 2001). Perhaps the most surprising result was in neutralization of hemorrhage from the fer-de-lance (*Bothrops atrox*) venom (Otero et al. 2000), although bleeding decreased only 25%.

Chemicals have been identified from several species, including *P. edulis, P. foetida, P. incarnata,* and *P. quadrangularis*. Compounds found are the alkaloid passiflorine, benzylic beta-D-allopyranosides 1 and 2, which are representatives of a rare class of natural glycosides, flavonoids, cycloartane triterpenoids and six related saponins, harmane and harmine, oxygenated monoterpenoids, and passifloricins (polyketides alpha-pyrones) (Lutomski and Wrocinski 1960, Bennati 1971, Morton 1981, Osorio et al. 2000, Yoshikawa et al. 2000a,b, Christensen and Jaroszewski 2001, Echeverri et al. 2001). Harmaline and harmine have been used to treat Parkinson's disease (Swerdlow 2000).

In the 1960s, it was uncommon to find passion-fruit juice in groceries, even in mixtures. Then a television commercial made "passion-fruit" an everyday word because it was an ingredient (minor) in "Hawaiian punch." The ad showed one cartoon character "punching" another when they said they wanted the drink. Everyone laughed at the catchy phrase, although few realized what a passion-fruit was. Botanists knew it was *Passiflora*, and somehow we felt smug knowing that bit of trivia.

Pavonia

(Antonio José Cavanilles named this for José Antonio Pavón y Jimenez, 1754–1844, a Spanish explorer who toured Chile and Peru with Hipolito Ruiz López and Joseph Dombey)

Pavonia paludicola (swamp-loving) (= *P. spicata*)
cadillo de ciénaga (marsh sticker, Puerto Rico)
cotton (Belize); wild cotton (Belize)
gombo-mangle (mangrove okra, Guadeloupe, Martinique)
kayuwaballi (*kayuwa*, mahoe or *Hibiscus tiliaceus*, *balli*, resembling, Arawak, Suriname)
mahot mangle [*mare*] (mangrove [ocean] fiber tree, Taino and French, Guadeloupe, Martinique); *mahuat* (fiber tree, Taino, French Antilles); *majagüilla* (little fiber tree, Hispanized Taino, Cuba, Hispaniola); smaller mahoe (Jamaica)
mangrove mallow (Guadeloupe, Martinique)
sunabao (Guadeloupe, Martinique)
swamp bush (Bahamas, Puerto Rico)

Pavonia was created in 1786 by the director of the botanical garden in Madrid, Cavanilles (1745–1804). However, it was not until 1989 that Dan Nicolson and Paul Fryxell described *P. paludicola* from the Lesser Antilles. That species brought the genus to a total of 150 species found in tropical and warm regions of the world (Mabberley 1997).

In the French Antilles, the leaves are applied to inflammations, boils, and abscesses (Morton 1981). In Haiti, an infusion is gargled for tonsillitis. Taken regularly, it is laxative.

Pectis

(From Latin *pecten, pectinis*, a comb, referring to the bristles along the margins of the leaves or the papus)

Pectis prostrata (lying flat)
cominillo [*tomillo*] (little dwarf, Venezuela); *comino de piedra* [*de sabana, rústico*] (stone [savanna, wild] dwarf, Venezuela)
contra-yerba (herb against, typically meaning that it can be used to treat any malady)
hierba de gallina (chicken herb); *hierba de chinche* (bedbug herb)
romero macho (wild [male] rosemary, Puerto Rico)
tebenque [*tebenki, tebink, theebink*] (probably Taino, Cuba); *tebink moge* (probably Taino, Cuba?)
zacato-coche (car grass; probably because it is common on roadsides)

Pectis prostrata has been reported from a number of places in the Caribbean, but there are indications that those are misidentifications. For example, the *Flora of Cuba* reported the plants from Jamaica, but Adams (1972) could not verify that they had ever been there. Similarly, Morton (1981) recorded medical use in Puerto Rico, but Liogier and Martorell (1982) do not include the species. TROPICOS lists specimens from Texas, Mexico (Chiapas, Tabasco, Yucatán), Belize, Costa Rica, El Salvador, Guatemala, Honduras, Nicaragua, Panama, and Ecuador.

Maybe the true distribution is less important for ethnobotanical comments because people often do not distinguish species. In Venezuela, for example, both *P. ciliaris* and *P. prostrata* have the same names (Pittier 1926). Perhaps that is because the aromatic traits of both are similar.

Pectis prostrata has been taken to stop diarrhea, dispel flatulence, as an emmenagogue, and for venereal diseases in Venezuela (Pittier 1926). In Jamaica and Puerto Rico, it is taken for colds and tuberculosis (Morton 1981). The species is used as a medicine in Belize (Balick et al. 2000). Hocking (1997) reported that it had been used to treat colds and tuberculosis, to expel flatulence, and as an emmenagogue.

Pedicularis

(Named from Latin *pediculus*, a louse, because Europeans believed that cattle or sheep feeding where *P. palustris* grew became covered with lice; also *herba pedicularis*, lousewort, because it was used to kill lice)

kallgrås (*kall*, cold?, *grås*, grass, Swedish)
Läusekraut (louse herb, German)
myrklegg (*myr*, bog, *klegg*, gadfly, Norwegian)
pédiculaire (French); *pediculare* (Italian)
riabhach (gray or grizzled, Gaelic)

Pedicularis canadensis (of Canada)
beefsteak-plant (Long Island)
betony [betong, beton lousewort, head-betony] ("betony," from Latin *betonica*, which Pliny, A.D. 23–79, said was a Gaulish name; *betonica*, from *vettonica*, derived from Vettones, people of Lusitania, originally applied to *Stachys officinalis*, New York); wood-betony ("wood" meaning growing wild, as opposed to the cultivated betony, *Stachys officinalis*)
cagacka'ndawesoanûk (flying squirrel tail, Potawatomi)
chickens'-heads (Long Island)

Pedicularis canadensis. a. Habit. b. Flower. c. Stamen. d. Pistil. e. Fruit. *Drawn by Vivian Frazier.* From Correll and Correll 1972.

[Canadian, common, early, early fern-leaf] lousewort ("louse" from a Teutonic base-word, as Old English *lús*, with cognates in German *Laus*, Danish and Swedish *lus*; first applied to *Helleborus* in the 1540s by Leonard Fuchs, as *Laüszkraut*, and later to *Pedicularis* by John Gerarde in 1597); lousewort-foxglove

snaffles (presumably from "snuffles" or "sniffles," a nasal catarrh, used since the 1820s; a local name in England, usually applied to *Rhinanthus*, also in the Orobanchaceae, formerly Scrophulariaceae; cf. Coffey 1993)

Linnaeus included Sweden's *Pedicularis* in his *Flora Oeconomica* of [1749] 1979. When he published *Species Plantarum* in 1753, all of its 14 species were Old World plants. It was not until 1767 when Linnaeus published *Mantissa Plantarum* that he gave us *P. canadensis*, based on a collection by his student Peter Kalm in Canada.

People of the northeastern United States had known about this *Pedicularis* and other species for a long time by the 1760s. About 1750, Jane Colden wrote concerning what is now *P. canadensis:* "This pedicularis is call'd by the country people Betony. They make a Thee [tea] of the leaves, et use it for the fever & ague et for sickness of the stomach" (Colden in Coffey 1993). Settlers may have known about medicines from Old World species, or they may have learned from the indigenous tribes.

Moerman (1998) recorded use among eight tribes. The Catawba made an infusion of roots to treat stomach problems. The Cherokee treated dysentery, coughs, and stomachaches with it (Hamel and Chiltoskey 1975). They and the Iroquois also rubbed an infusion of roots on sores. The Iroquois treated women's menstrual problems, heart troubles, and bleeding tuberculosis with the plants. The Menomini used it as a love charm. The Meskwaki treated external sores and tumors, and also made a love medicine with it (King 1984). The Mohegans used an infusion of leaves to induce abortion. The Ojibwa used a root infusion to counteract anemia, to treat stomach ulcers, sore throats, and as a love potion. The Forest Potawatomi used the roots as a physic, while the Prairie Potawatomi used the roots to reduce both internal and external swelling (Smith 1933).

The Menomini and Potawatomi mixed lousewort with other plants to fatten their horses. Both the Cherokee and Iroquois ate the leaves and stems, sometimes cooking and seasoning them with salt, pepper, and butter (Yanovsky 1936).

In the 1750s some Europeans still believed that cattle or sheep feeding where *P. palustris* grew became covered with lice. By the 1900s, others had totally changed views. Vickery (1993) recorded that people in the Shetland Islands called *P. vulgaris* "bee-sookies" or "honey-sookies" because of its "nectar-filled flower-tubes," which children sucked for their sweet flavor.

Pediomelum

(Per Axel Rydberg, 1860–1931, segregated these plants from *Psoralea* with Greek *pedion*, field, and *melon*, an apple or fruit)

Pediomelum canescens (grayish-pubescent) (= *Psoralea canescens*)

buck-horn [buck-thorn] ("buck-thorn" is now applied to several genera, but was applied to *Rhamnus catharticus* by Dodoens in 1554; the

Latin *cervi spina* was applied to *Rhamnus* by Valerius Cordus, 1514?–1544)
buckroot (known by this name in 1765 when John Bartram visited the Carolinas, Berkeley and Berkeley 1982)
hoary scurfpea ("scurf" is dry, scaly skin, especially on the head; probably from Old English *scurf*; akin to Swedish *skorv*, Danish *skurv*, Dutch *schurft*, and German *Schorf*)
owá:lá:rî: ínsawá:kî (Sturtevant wrote "prophets' [plural; singular, *owá:lî*] coconut," Mikasuki; "*owá:lî*" is also translated as "wise-man," "magician," or in Creek, as "knower"); *owa:lâlki insawkô* (*owalv*, knower, *em*, his, *svokv*, rattle, Creek; the "coconut," *Cocos nucifera*, is *talasvokv*)

These herbs are restricted to parts of Virginia, Georgia, Florida, and Alabama (Radford et al. 1968). Because of their limited range, not much has been written about them, yet the Seminoles knew and used them as late as the 1950s (Sturtevant 1955). Moreover, they were familiar with the restricted range of the plants in Florida, noting that they did not grow south of Punta Gorda. They would make special trips into the area where the plants grew to obtain stocks of the roots to dry for use in medicine. Hedrick (1919) and Yanovsky (1936) say the roots have been eaten in the southern states.

According to Josie Billie, one of Sturtevant's collaborators, these legumes were analgesic when the warmed root was applied externally. To treat rheumatism, they took a root, dipped it in water, warmed it over the fire, and then pressed it against the sore spots. They considered it strong and expected the pain to be gone by the next morning. In addition, the roots were used in a medicine to treat colds and coughs. Sturtevant (1955) gave a lengthy account of his personal experience treating his own cold, and he was convinced it helped.

Neither Foster and Duke (1990) nor Duke (2002) even mention the genus *Pediomelum*. Under *Psoralea*, Hocking (1997) says that this species has been used to treat gastric distress.

Peltandra

(Rafinesque named this with Greek *pelte*, a shield and *andros*, stamens)

Peltandra virginica (from Virginia)
[green] arrow [arum] (USA)
ocfô (Creek); *okõ:nî* (Mikasuki)

***Peltandra virginica*.** a. Habit. b. Outline of leaf. c. Spadix. d. Berry (submersed). *Drawn by Vivian Frazier*. From Correll and Correll 1972.

Pfeilaronstab (arrow arum stick, German)
takwahahk (Capt. John Smith wrote the Powhatan name as *tockwhogh*, *tocknough*, and *tockawhoughe*. He said that it was the "chief root they have for food ... like a flag in low muddy freshes ... of the greatness and taste of potatoes ... raw it is not better than poison ... roasted ... in summer they use this ordinarily for bread." Strachey wrote in [1612] 1953 that it was a "bread made of a wort called *taccahoappoans*." Siebert 1975 considered the root to be /**takw-*/, to pound fine, beat to a powder. The word written by Strachey includes the element *appoans*, which became "pone" in English; see also *Zea*); cognates include *takáham* (Delaware); *takwaham* (Cree); *takwahamwa* (Miami); *takwham* (Nipmuck-Pocumtuck); *taw-ho* [*tawho, tawhim, tawhim, tuckah*] (Delaware, New Jersey); *tquogh* (Mohegan); *tukwhah* (Shawnee); *nitakhwa* ("I pound him for bread," Shawnee); *otakwaʔa·n* (Ojibwa)

tuckahaw (according to Romans [1775] 1961, there was a Chickasaw town named for this plant, and using the Algonquian word)
tuckahoe [*tockwhogh, tockawhoughe, taw-ho*] (from Powhatan, Virginia); *coscúshaw* (Carolina Algonquians; see also Geary 1955)
[Virginian] wake robin (New England)

Although there is considerable confusion about the original identify of *tuckahoe*, it has become associated with *Peltandra* since being discussed by Kalm ([1753–1761] 1972), who had personal knowledge of the plants and the people using the name. That Powhatan word is related to *petukqunneg* (cake of bread, from *petukqui*, bread, *pitikwah*, made round, Cree). The distribution of the words presumably corresponds to part of the range of usage. However, most of these terms were transferred to maize when it was introduced, and that has complicated the situation.

The first record of the indigenous use of these roots is in 1612 when Capt. John Smith wrote, "In Iune, Iulie and August they feede vpon the rootes of *Tockwough*, berries, fish and green wheat [maize]" (OED 1971). Strachey recorded the plant and name the same year. The Seminoles also use the plant for food (Sturtevant 1955). Tull (1999) indicated that long periods of drying (sometimes months) and baking are necessary to render the acrid roots palatable. Both roots and fruits were eaten after detoxification. Methods of preparing them are given by Fernald et al. (1958).

Harriot ([1590] 1972) had recorded another name and use among his list of fruits. He called them *sacqvenvmmener*, and not long afterward Strachey and Capt. John Smith called them *ocoughtanamins*. Harriot ([1590] 1972) wrote that *sácqvenvmmener* were "a kinde of berries almost like vnto capres [capers, *Capparis*] but somewhat greater which being grow together in clusters vpon a plant or herb that is found in shallow waters: being boiled eight or nine hours according to their kind are very good meate and holesome, otherwise if they be eaten they will make a man for the time franticke or extremely sicke." Smith thought they should be boiled half a day (Swanton 1946).

Yanovsky (1936) may have confused *Orontium* with *Peltandra*. Still, he listed *Peltandra* as having been eaten in all the southeastern states, and in New York, Pennsylvania, and Virginia, a distribution agreeing with its Algonquian names.

Moerman (1998) indicates that the Nanticoke of Delaware grated the roots in milk and gave it to babies for some unstated medical reason.

Penstemon

(From Greek *pente*, five, and *stemon*, stamen, referring to the four fertile stamens and one sterile staminode)

Penstemon laevigatus (smooth)
[eastern smooth, foxglove, hairy] beard-tongue

Europeans were familiar with *Digitalis* when they arrived in the New World, but here they found plants somewhat different from that old medicinal herb. When Casimir Christoph Schmidel (1718–1792), a German physician at Erlangen, was working with plants grown in Kew Gardens outside London, he decided that the American plants should have a distinct name. He called them *Penstemon* in 1763 because of their androecial arrangement. The genus now has grown to 250 species, with all but one confined to North America (Mabberley 1997).

One of the species discovered after Schmidel was *P. laevigatus*. This herb, also described from plants grown at Kew, was named by William Aiton in 1789, and the specimen he used to name the species is in the Fothergill collection at the British Museum of Natural History. Smooth beard-tongue ranges from New Jersey, Pennsylvania, and West Virginia south to Florida (Gadsden and Jackson Counties), and west into Alabama and Mississippi.

Not much information is available about this plant. The Cherokee used an infusion of *P. laevigatus* to stop cramps (Hamel and Chiltoskey 1975). This is probably the same species used by the Creeks and Natchez for colds, coughs, consumption, and whooping cough (Swanton 1928a).

Porcher (1863), Millspaugh (1892), Vogel (1970), and Foster and Duke (1990) do not mention the genus. However, other *Penstemon* species were used by the Iroquois, Kiowa, Lakota, Pawnee, and several tribes farther west. Moerman (1998) found records of 23 other species being used.

Some *Penstemon* contain iridoid glycosides, especially catapol, and both moths and butterflies have adapted to sequester and use those compounds. The inchworm and looper moths *Neoterpes graefiaria* and *Meris alticola* take catapol from *Penstemon*, and the Arachne checkerspot butterfly (*Polydryas arachne*) also uses it.

Pentalinon

(Friedrich S. Voigt, 1781–1850, named this with Greek *pente*, five, and *linon*, rope, a reference to the elongated anther appendages)

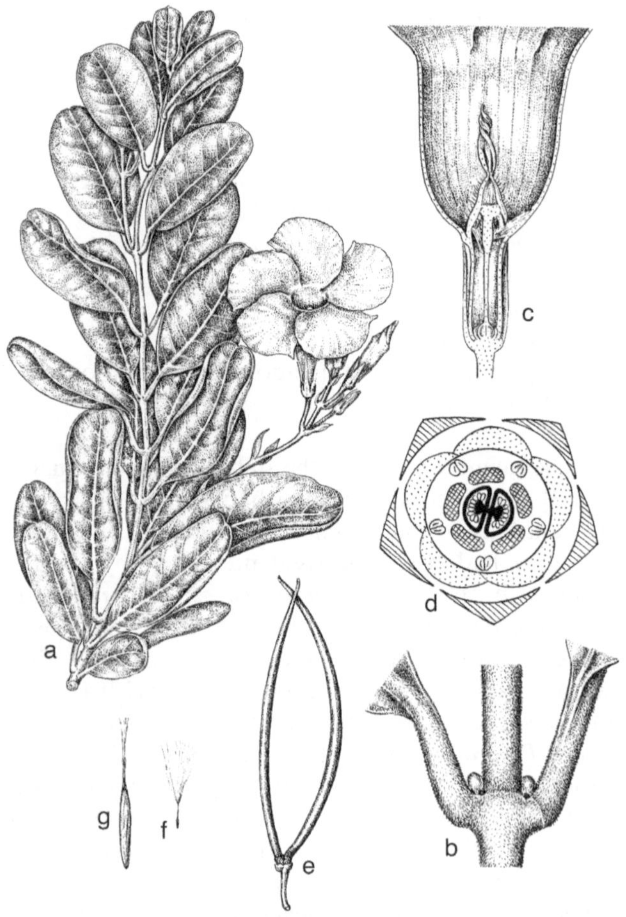

Pentalinon luteum. a. Flowering branch. b. Node. c. Flower tube, longitudinally dissected. d. Floral diagram. e. Fruits. f. Seed with coma. g. Seed. *Drawn by Priscilla Fawcett.* From Correll and Correll 1982.

Pentalinon luteum (yellow) (= *Urechites lutea*)
[*bejuco*] *ahoga vaca* (cow strangler [vine], Dominican Republic)
barbeiro amarillo (yellow beard, Puerto Rico)
bejuco marrullero (false? climber, Cuba)
Catesby's vine (Bahamas)
clavelitos (little carnation, Cuba)
corne cabrits (goat horn, Haiti)
curamagüey (Taino?, Hispaniola)
Dominican viper tail (Dominican Republic); hammock viper's tail [viperstail] (Florida)
Jamaica nightshade (Jamaica); yellow nightshade (Jamaica?); nightshade (Cayman Islands)
wild allamanda (Florida)
wild unction (unction = ointment, Bahamas)

Linnaeus ([1753] 1957) called these climbers *Vinca lutea.* Then, for many decades, they were called *Urechites lutea,* a genus established in 1860 by the Swiss botanist Johannes Müller of Aargau. However, Bruce Hansen realized that *Pentalinon,* described from plants grown in the Calcutta Botanical Garden in 1845, was an earlier and valid generic name (Hansen and Wunderlin 1986). *Pentalinon* now contains two species, both native to Florida, Central America, and the Caribbean.

In the Dominican Republic *Pentalinon* is used to treat heart disease (cardiotonic), edema, fever, and colic, and as a purgative (Hocking 1997). Plants are used to treat headache in Guatemala (Rosatti 1989). However, doing so is dangerous because the latex is poisonous, having been used to poison arrows in tropical countries (Rosatti 1989). It is poisonous to cattle; people powder the leaves to kill destructive insects and animals (ants, dogs) (Liogier 1974). Among the poisonous compounds are the cardenolides oleandrin, urechitin, and urechitoxin (Gibbs 1974).

Penthorum

(From Greek *pente,* five, and *horos,* a column or pillar, referring to the five-parted flowers)

Penthorum sedoides. a. Top of plant. b. Part of procumbent stem of plant with roots. c. Cluster of flowers and fruits. *Drawn by Vivian Frazier.* From Correll and Correll 1972.

Penthorum sedoides (like *Sedum*)
[ditch, Virginia] stonecrop (from Old English *stáncrop,* combining "stone," a rock, and

"crop," as gathered from the top of an herb for culinary or medical purposes; combined, the designator has been in use since about A.D. 1000, first as the common name for European *Sedum acre;* modifiers were added later for other plants)

Penthorum has been an oddity since it was discovered. Jan Gronovius first included it in his *Flora Virginica* of 1739–1743. Linnaeus described it in 1744, and in *Species Plantarum* ([1753] 1957) said only of it that it had a "Habitat in Virginia."

Depending on how it is interpreted, the single American species either belongs in the Saxifragaceae along with one to three others that grow in Asia, or in its own isolated family, the Penthoraceae (Cronquist 1981, Mabberley 1997). In some regards, the plants are transitional between the Saxifragaceae and Crassulaceae (Cronquist 1981).

Penthorum sedoides grows from New Brunswick, southwestern Quebec, southern Ontario, Michigan, Wisconsin, Minnesota, and Nebraska south to Florida and Texas. Within that large range, only two tribes have been recorded as using the plants.

In her 1975 master's thesis at the University of Tennessee, Knoxville, Myra Jean Perry recorded that the Cherokee used the leaves as a potherb (Moerman 1998). That is an odd use, indeed, because few to none of the other members of the family are considered edible. In Iowa, the Meskwaki made a cough syrup of the seeds (King 1984).

Although the plants are not mentioned by Porcher (1863), they are discussed by Millspaugh (1892). He wrote, "It has always held a place in domestic practice as an astringent in diarrhoea and dysentery." According to him, two physicians brought the plants to notice in 1875 as a remedy for irritation of the mucous membranes and to treat maladies like pharyngitis, vaginitis, and tonsillitis. In 1931, Maud Grieve's book *A Modern Herbal* included the report that "this plant has of late attracted much notice ... as a remedy for catarrh, catarrhal inflammation of the larynx, chronic bronchitis ... and affections of the stomach and bowels. It has also been employed with success in treatment of diarrhoea, haemorrhoids and infantile cholera" (Coffey 1993). Foster and Duke (1990) echo the same information, but the species is not mentioned by Bremness (1994), Bown (1995), or Duke et al. (2002).

Peperomia

(Ruíz and Pavón combined Greek *peperi*, pepper, and *homoimos*, resembling)

Peperomia obtusifolia (leaves blunt or round at apex)
agronemia (of cultivated places?, Hispaniola)
climbing pepper (Belize); wild pepper (Florida, Bahamas)
cupeycito (little cupey, *Clusia rosea*; Hispanized Taino, Dominican Republic)
lentejuela (applied to *Lepidium virginicum* in Morton 1981)
pàrahá (Paya, Honduras)
peperomia (Florida)
tep-pim (*tep*, something adorned, *pim*, fat or large, Maya, Belize)

The genus *Piper* was the only one that Linnaeus ([1753] 1957) recognized, and he included 17 species. He called these plants *Piper obtusifolium*, and noted that they had been discussed previously by Charles Plumier in 1693 as *Saururus repens, folio orbiculari nummulariae facie* (prostrate lizard's tail, with rounded leaves resembling coins). Then, after their exploration of Peru, Hipólito Ruiz López (1754–1815) and José Antonio Pavón (1754–1844) created the genus *Peperonia* in 1794 in their book *Flora Peruvianae, et Chilensis Prodromus* (Preliminary Flora of Peru and Chile). The genus now contains 1000 tropical species, mostly in Americas (Mabberley 1997).

Several of the *Peperomia* are widely used as medicines. *Peperomia pellucida* is the most famous (Liogier 1974, Ayensu 1981, Morton 1981), but *P. magnifoliifolia* is more similar morphologically to *P. obtusifolia*; it has even been recorded in botanical literature under the latter name. *Peperomia magnifoliifolia* is used in Barbados as a remedy for coughs and colds (Morton 1981). In Veracruz, the Zoque-Popoluca treat erysipelas with it (Vásquez and Jácome 1997).

Persea: Red Bay

(A classical name doubtfully from *Persica*; from Persian, or from Greek *persis*)

Americans and Bahamians call *Persea borbonia* the red bay. They have used that name since at least the time of Mark Catesby (1731). Calling trees "bays" is obvious because they belong to the same family of plants (Lauraceae) called by that name since the time of the Greeks and Romans. "Bay" is derived from Latin *baca.*

Until recently, it never occurred to me to ask why the plants were called "red" bays. When a colleague asked me why, I found that "red" in the name refers to the wood. Of the trees that he saw in Virginia and

Persea borbonia. a. Branch with flowers. b. Bud. c. Flower. d. Anther. e. Branch with fruit. *Drawn by Vivian Frazier.* From Correll and Correll 1972.

Carolina, Catesby said, "The wood is fine-grain'ed, and of excellent use for Cabinets, etc. I have seen some of the best of this Wood selected, that has resembled Water'd Sattin, and has exceeded in Beauty any other Kind of Wood I ever saw." Sargent (1905) wrote that the wood was "heavy, hard, very strong, rather brittle, close-grained, bright red." The wood has been used from at least the time of Catesby for cabinets, and also on the interior finish of houses. Formerly, it was used in ship- and boatbuilding. Comparing its wood to mahogany with "Florida mahogany" is high praise indeed.

Williams ([1837] 1962) also praised the wood. He wrote, "This tree produces timber inferior only to mahogany, which it closely resembles." He then added, "The young leaves are often used for tea, which is a most pleasant and healthful beverage."

The Flemish Charles de l'Ecluse, who became the king's botanist to James I of England, was the first to apply the name *Persea* to these plants. Philip Miller picked up the name and continued its use. That Greek name originally was used by Theophrastus and Hippocrates for an unknown Egyptian tree. Of that tree, Pliny wrote, "Persea ... is far different from the Peach-tree Persica and beareth fruit like vnto Sebesten, of colour red" (from a 1602 translation). The derivation of *Persea* was thought by Pliny to be the same as *Persica* (from Persia), but that is dubious. Apparently, Persia is derived from Greek *persis*, which in turn was probably taken from Arabic *fars*. Some have speculated that the Old World plants called *Persea* were *Cordia myxa* (Boraginaceae), but their identity remains uncertain.

Curiously, Linnaeus did not follow either l'Ecluse (alias Clusius) from 1601 or Gaspar Bauhin from 1623 in keeping these plants distinct from *Laurus*. Instead, Linnaeus called the plants *Laurus persea*, which we now know as *P. americana* (avocado). Linnaeus's reluctance to keep *Laurus* separate from *Persea* is reflected today in the chaotic status of genera in the family.

Other names for red bay and its variations (including *P. pubescens*) are laurel-tree, shore bay, swamp bay, swamp red-bay, sweet-bay, and tiss-wood. The last name, "tiss-wood," may be related to its use in tisanes for a beverage or medicine. The first record of it that I saw was by Vignoles ([1823] 1977) who spelled it "tiswood." Almost certainly, this plant and sassafras (*Sassafras albidum*) have been used interchangeably since Europeans encountered people using them. Leaves of both have served as the basis of gumbos (from Choctaw, *kumbo*), especially those including crabmeat.

The Miccosukee call these trees *tó:lî*, their relatives the Creeks say *tó:la*, and the Koasati say *tolá*. Surprisingly, because they were supposed to have a distinct language, the Timucua also said *tola*. Those are simple terms that cannot be translated. However, the Alabama call the tree *ittoissi kosáoma* (*itto*, tree, *hissi*, hair, *kosooma*, stinking). This may be what the Choctaw called *iti chinisa* (striped tree).

William Bartram ([1791] 1958) recorded that the trees were called *eto mico* [*itto micco, eto micco*] (*eto*, tree, *mekko*, tree, Creek). Simmons ([1822] 1973) noted that the Seminoles were still using the name. This tree is perhaps the most important plant among modern and historic Seminoles. The Seminoles used the leaves to make a beverage like tea. They also used the dried leaves in cooking like their relatives the Choctaw, and they made spoons from the wood. They were not the first to use the plants in Florida, as the pollen of *P. borbonia* has been found in a pre-Columbian site near Lake Okeechobee (Hogan 1978).

Red bay has been used by the Seminoles as an abortifacient, analgesic, antiemetic, diuretic, aphrodisiac, emetic, febrifuge, a laxative, a love medicine, a panacea, a psychological aid, in childbirth, to cure dreams, and to improve the appetite, as well as in a ceremonial context (Sturtevant 1955). The Creeks and their relatives the Seminoles diagnose diseases in ways different from Europeans and Americans. Each series of symptoms is identified with a name as in our system, but their nomenclature involves an animistic worldview. Among the diseases they recognize, *Persea* has been used to treat "Bear Sickness" (fever, headache, thirst, constipation, and blocked urination), "Bird Sickness" (diarrhea, vomiting, appetite loss), "Buzzard Sickness" (vomiting in children), "Cat Sickness" (nausea), "Dead People's Sickness" (grief, cough, appetite loss, vomiting; sometimes the same as "Ghost Sickness"), "Deer Sickness" (numb, painful limbs and joints), "Fire Sickness" (fever and body aches), "Ghost Sickness" (grief, cough, appetite loss, vomiting, dizziness, staggering, sometimes the same as "Dead People's Sickness"), "Hog Sickness" (unconsciousness), "Mist Sickness" (eye disease, fever, chills), "Opossum Sickness" (appetite loss and drooling in babies), "Otter Sickness" (diarrhea and vomiting), "Raccoon Sickness" (diarrhea in babies), "Rainbow Sickness" (fever, stiff neck, backache), "Scalping Sickness" (severe headache, backache, low fever), "Sun Sickness" (eye disease, headache, high fever, diarrhea), "Thunder Sickness" (fever, dizziness, headache, diarrhea), "Turkey Sickness" (dizziness or "craziness"), and "Wolf Sickness" (vomiting, stomach pain, diarrhea).

Additionally, red bay has been used to treat sickness caused by adultery, including headache, body pains, and "crossed fingers." As an aphrodisiac, the leaves are sung over to attain the love of a particular girl. If the aphrodisiac has worked, the leaves may be rubbed on the mother's body during protracted labor.

Sometimes red bay is even considered a panacea where the leaves are used for everything and can be added to any medicine. As a psychological aid, the leaves are used to cure fear in babies caused by dreams about raccoons or opossums. In addition, an infusion of leaves may be used to steam and bathe the body of an "insane" person. The plant may be burned to work the same cure.

There are also particular ceremonial applications. A decoction is taken as an emetic by doctors to strengthen their medicine. That may be the use at any time, or particularly when a death has occurred. The leaves are used as emetics during funeral ceremonies, carried by every member of the burial party, placed on top of casket, and burned to keep the soul of the recently departed from returning home. Leaves are also added to food after a recent death.

Some recognize a wetland form as a distinct species, calling it *P. palustris.* The Creeks have used the root of that form as a "hydrogogue" and alterant. The decoction is considered diaphoretic in "fevers of all descriptions" by the Choctaws (Bushnell 1909).

Persea is notorious for having poisonous compounds although it provides food and medicine for humans and other animals. The red bay and its variations contain an array of essential oils, including camphor, cineol, eucalyptol, and p-cymene (Tucker et al. 1997). Those, and probably other chemicals yet to be identified, are responsible for the tight evolutionary relationship between the spicebush swallowtail butterflies and members of the Lauraceae. The spicebush swallowtail (*Papilio troilus*) and the Palamedes swallowtail (*P. palamedes*) are among the few insects that can detoxify or sequester the poisons, and they specialize on *Persea* (cf. Lederhouse et al. 1992, Carter and Feeny 1999).

There are perhaps 200 tropical Asian and American species in *Persea* (Mabberley 1997). Many of them are part of local pharmacopoeias, but none is as famous as the avocado. Although most people know that species as the primary ingredient passed down from the Aztecs almost without change in the recipe for *guacamole* (Coe 1994), few realize that it is also medicinal. The leaves and other parts contain a variety of toxic compounds that are dangerous to vertebrates if consumed in quantity (Hargis et al. 1989, Grant et al. 1991, McKenzie and Brown 1991, Stadler et al. 1991, Burger et al. 1994, Oelrichs et al. 1995). In spite of the toxic chemicals in the leaves, Latin Americans regularly use them in small quantities as a spice. They give foods an anise flavor (Hearon 1993). Those same compounds show potential as insecticides (Oberlies et al. 1998) and give some protection against *Giardia* (Ponce-Macotela et al. 1994).

Avocado and other species in the genus show considerable dietary and medicinal potential (Zanobi et al. 1974, Meade et al. 1980, Mohan and Kekwick 1980, De-Oliveira et al. 1985, Ballot et al. 1987, Ma et al. 1989, Sheldon et al. 1990, Guevara et al. 1994, Eccleston and Harwood 1995, Kimura et al. 1995, Koua et al. 1998, Castro et al. 1999, Chiapella et al. 2000, Domergue et al. 2000, Kim et al. 2000, Kruthiventi and Krishnaswamy 2000, Caballero-George et al. 2001, Hashimura et al. 2001, Kawagishi et al. 2001, Kut-Lasserre et al. 2001, Schlemper et al. 2001, Stucker et al. 2001, Lequesne et al. 2002). Therefore, their congener *P. borbonia* is showing the same patterns as its relatives. In spite of its extreme importance among indigenous people in the southeastern United States, red bay has never appeared in the journal *Economic Botany* (Kaplan 2001).

Although modern southern culture owes as much to the Creeks as any group of people, their botanical heritage has been slighted (Hudson 1976). The Creeks and their relatives the Seminoles were aware of all the bays. Bartram ([1791] 1958) told us Creeks called *P. borbonia* the *eto miko,* and *Magnolia grandiflora* was *tolochlucco* (big bay). By the 1950s, the Seminoles had shortened those names to *to:li* and *to:lhátkí* (white bay, *M. virginica*). What will be remembered in the next 200 years?

Phalaris

(Greek *phalaris, phaleris*, used by Dioscorides, fl. A.D. 40–80, for some kind of grass; presumably from *phalaros*, having a patch of white or crest, alluding to the inflorescence)

alpiste (French)
canaria (Italian); canary-grass ("canary," referring to the Canary Islands or *Canariae Insula* [Isles of Dogs], a name used by Pliny, A.D. 23–79); *kanarigras* (*kanari*, canary, *gras*, grass, Norwegian)
Glanzgras (*glanz*, shiny, *gras*, grass, German)
phalaride (French)
reed-grass (a name most often used for *Phragmites*)
ror-flen (*ror*, reed, *flen* = ?, Swedish)
scagliola (Italian)
strandror (*strand*, seashore, *rør*, reed, Norwegian)

Phalaris caroliniana (of Carolina)
baabkam [*papkam*] (pl., *baahpakam*) ("it has a grandfather," Akimel O'odham [Pima]; Rea's informants could not give him a reason for the name; he speculated it must be an ancient poetic allusion)
Carolina canary-grass [canarygrass]
may-grass [maygrass] ("May-grass" was in use by 1830, but applied to *Panicum latifolium*; apparently applied to *Phalaris* ca. 1974, originally by archaeologists)

Europeans knew at least three species of *Phalaris* when they arrived in the New World. Reed canary-grass (*P. arundinacea*) and Harding grass (*P. aquatica* L.) were grown for hay, while seeds of canary grass (*P. canariensis*) were eaten by people. There are about 20 species in *Phalaris*, with 7 native to Europe (Mabberley 1997). *Phalaris caroliniana* was named by Walter in 1788, and now grows from Florida to California and Mexico, north to Maryland, Tennessee, Missouri, Kansas, Colorado, and Oregon.

May-grass is a starchy-seeded annual grass whose grains (caryopses) are dominant in Middle and Late Woodland archaeological sites (Crites and Terry 1984, Asch and Asch 1985). The grains have been associated with people in Illinois, dating 1700 ± 70 B.P. in one site and 1400 ± 70 B.P. at another site. It was already an important cultivated food in Kentucky in 1000 B.C. Fritz (2000b) concluded that it was a domesticated crop by 3000 B.P.

Moerman (1998) found the species being used only by the Gila River Pima, based on a report by Amadeo Rea. Subsequently, Rea (1997) provided more information on may-grass use. While reading through a 1908 report by Frank Russell, Rea noticed an unidentified grass the O'odham had called *papkam*. He realized that it was what modern people call *baabkam*. Russell wrote, "The heads are tied in bunches and dried in the sun. They are then shelled, screened, the seeds parched, ground on the metate, and eaten in pinole. They are 'not sweet.'" One of Rea's O'odham friends had told him the seeds were once eaten but did not taste good. These grasses are another member of the Eastern Agricultural Complex that was abandoned when maize agriculture became widespread.

Phaseolus

(From Greek *phaseolos*, a little boat, light vessel, referring to the similarity between the pod and the craft; used by Dioscorides, fl. A.D. 40–80, for what Agnes Arber identified as *Vigna unguiculata*, cf. Meyer et al. 1999; known as *faseolus* or *phaseolus* to Romans)

Phaseolus polystachios. From Britton and Brown 1897.

Phaseolus polystachios (many-spiked)
[thicket, wild] bean [vine]
sacsac (Delaware)

Phaseolus vulgaris (common)
acayotle (maybe from *acatl*, cane, *etl*, bean, Náhuatl, recorded by Humboldt fide Hedrick 1919)

aífi (Garífuna)
añkonaki' (Ofo)
assentemmens (Strachey [1612] 1953 compared these beans with Italian *fagioli*, Powhatan, Virginia)
a'teba'kwe (Abenaki, Maine)
bala (Choctaw); *bala'* (Chicksaw); *palaná* (Koasati), *salá:lî* (Mikasuki); *tvlako* [*tala:ko*] (Creek, Muskogee); *tobi* (Choctaw)
[common] bean ("bean" from Old English *béan*, perhaps cognate to Latin *faba*, but some think that doubtful); *Bohne* (German)
bisoloma (Arawak, Suriname)
cannellini (Italian name for a large white kidney bean)
chin (Jicaque, Honduras)
etl (bean, Náhuatl); *exotl* (green bean still in the pod, Náhuatl)
fagiolo [*fagiuolo*] (from Latin *faseolus*, Italian); *feijão* (from Latin *faseolus*, with a change in the suffix, Portuguese); *frijol* (Spanish); *phaséole* (French)
French-bean (called French or Roman beans in 1632, probably because they were imported from the Mediterranean area into northern Europe)
haricot (etymology uncertain, from French *haricot*, shortened in the 1600s from *hericoq de mouton*, of the 1300s, a stew usually of mutton, reapplied to the beans about 1653; in 1693, Joannes de la Quintyne used both *aricos* and *haricauts* in his book *The Compleat Gard'ner*)
ho<u>n</u>-bthi<u>n</u>-ge (Osage)
húce [*nuntce'*, *yúce*] (Catawba)
hynï³ (Chinantec, Oaxaca)
kidney-bean (a reference to the similarity of shape between the beans and human organ; used by Turner in [1548] 1965 as "kydney beane")
kima't ta'l (*kima'ti*, middle, *ta'l*, inside, Atakapa)
koje's (Potawatomi)
malachxil [*malachxitall*] (Delaware, New Jersey)
mîskodi'ssimi (Potawatomi); *mushaquissedes* (Pequod, Rhode Island, Connecticut)
navy bean (specifically meaning the dried, white haricot, in English by 1856)
ogaressa (Huron, Ontario)
okindgier (beans in pods, Carolina Algonquians); *okinsher* (Powhatan, Virginia)
ossahèta (Onondaga)
peccataas [*peketawes*] (recorded by Strachey in [1612] 1953, Powhatan, Virginia); *peswe'min* (Plains Cree)
pinto bean (apparently first applied in English to the spotted beans of the southwestern United States in 1916)
pònair Fhrangach (French beans, Gaelic)
purutu (Quechua, Peru)
sahe [*sahu*] (Micmac?, St. Laurence)
si'hpari (Tunica)
snap bean (in use since 1770 for the green pods that are broken into pieces and cooked as a vegetable while still young)
tsodi' (Yuchi)
ta<u>n</u>tka (Biloxi)
teppuhguam-ash (twiners, Algonquian)
tuya (Cherokee)
wickonzówer (pleasant when cooked, Carolina Algonquians)

When I was growing up, beans were a part of each noon and evening meal. My mother varied the dishes; sometimes they were white beans, and sometimes brown or pintos. Seasonally, we had green beans. I always liked the beans the second time a dish was served because they were then cooked to mush. Little did I know that Algonquian people of coastal Virginia preferred them that way too in the 1580s when Thomas Harriot was there.

Classical mention of "beans" and the early introduction of American plants into Europe caused confusion about their nativity for hundreds of years. Indeed, Columbus saw beans on his first voyage in 1492, and brought them back to Spain in 1493. Of the plants he saw in Cuba, Columbus wrote of "fields planted with *faxones* and *habas* very different from those of Spain" (Hedrick 1919). Jane and Skelton (1960) corrected the name *faxones* to "beans," and *habas* as "kidney beans" although Columbus thought the former was *Vigna unguiculata* and the latter *Vicia faba.* Their translation is what he saw, not what he thought he saw.

Phaseolus vulgaris reached England in 1594 (Hawkes 1998). Leonard Fuchs in 1542 thought what we now call *P. vulgaris* was the bean of classical Rome and Greece and called them *Faselen.* Similarly, Hieronymus Tragus in 1552 and Davidus Kyber in 1553 thought the same as Fuchs. That view held until the late 1800s, but it has subsequently been confirmed that *Phaseolus* is a New World crop (Smartt and Simmonds 1995). Archaeological studies indicate beans in Mexico by 6000–7000 B.P., and in Peru 7000–8000 B.P. (Smartt and Simmonds 1995, Kaplan and Lynch 1999, Lentz 2000). They reached eastern North America some time after those dates (Fritz 2000b), apparently between 700 and 1000 B.P.

Beans were widely cultivated among the northern Florida tribes before Europeans arrived (Hedrick 1919). Pánfilo de Narváez found beans in western Florida in 1528, as did Hernando de Soto in 1539 (Swanton 1936, 1946). Similarly, Jean Ribaut found beans cultivated by the Timucua in northeastern Florida in 1562.

Earlier explorers also found beans farther north (Hedrick 1919). Florentine navigator Giovanni da Verrazzano (alias John Verazanno) (1485–1528) found them among the indigenous people of NorumBega, Maine, in 1524. As the first European visitor to the New England coast, he had never before seen kidney beans. He wrote of the people, presumably the Pequod, "Their ordinarie food is of pulse, whereof they have great store, differing in colour and taste from ours, of good and pleasant taste." Later, Jacques Cartier found beans "of every color" among the Hurons at the mouth of the St. Lawrence River of Quebec in 1535.

In the same region, Samuel de Champlain (1567–1635) found the Abenaki of the Kennebec River on southern Maine eating multicolored beans in 1605, and Capt. John Smith (1580–1631) knew them among the New England tribes in 1614 when the Pilgrims first landed. Before 1670, John Josselyn wrote of "French beans: or rather, American beans. The herbalists call them kidney-beans from their shape and effects: for they strengthen the kidneys. They are variegated much, some being bigger, a great deal, than others; some white, black, red, yellow, blue, spotted" (Hedrick 1919).

Harriot ([1590] 1972) wrote of North Carolina that the *okindgier* was "called by vs beanes, because of in greatnesse and partly shape they are like to the Beanes in England, sauing that they are flatter, of more divers colours, and some pide. The leafe also of the stemme is much different. In taste they are altogether as good as our English peaze."

Beans are documented as grown for food among the Abenaki, Algonquin, Apache, Aztecs, Cherokee, Choctaw, Creek, Delaware, Havasupai, Huron, Iroquois, Menomin, Navajo, Ojibwa, Onondaga, Papago, Pequod, Potawatomi, Santee, Seminoles, Sia, Tewa, Tuscarora, and Zuni (Romans [1775] 1961, Bartram 1719, Hedrick 1919, Smith 1933, Swanton 1946, Berkeley and Berkeley 1982, Moerman 1998). All the agricultural tribes in the eastern United States surely made use of them.

Harriot ([1590] 1972) wrote that the people of Virginia cooked corn and beans together to "make them victuall either by boyling them all to pieces into a broth; or boiling them whole vntill they bee soft and beginne to breake as is vsed in England, eyther by themselues or mixtly together: Sometime they mingle of the wheate with them. Sometime also beeing whole sodden, they bruse or pound them in a morter, & therof make loaues of lumps of dowishe bread, which they vse to eat for varietie." Romans ([1775] 1961) found the Choctaw doing much the same by boiling corn and beans together, and calling it *holhponi.* We now call this mixture "succotash" (an Algonquian word akin to Narraganset *msekwatas*; in use by 1751). Capt. John Smith recorded for Virginia a dish of unripened corn, roasted in hot ashes, and eaten boiled with beans during the winter. He called it *pausarowmena* or *pausarawmena.* Le Page Du Pratz found the Natchez in 1758 cooking corn bread with beans, a dish they called *co oëdlou* (Swanton 1946).

The Caddo of southwestern Arkansas, northwestern Louisiana, and adjacent Texas had a unique bean dish. Henri de Joutel wrote about 1615 that they "do not make much mystery in preparation of them." These people cooked them in a big pot without any preparation and kept them covered with leaves until they were almost done. Then, they poured warm, salted water over them before serving. Those eating the pods were expected to eat strings, stems, and other parts or remove the pieces they did not want (Swanton 1946). This tribe was also unusual in salting their food.

By the time Bartram ([1791] 1958) was among the Seminoles of northern Florida in the 1770s, he found them growing "beans" (*Phaseolus*). They were also growing what he called "pease," the introduced cowpea, *Vigna unguiculata.* Indeed, along with native produce, Bartram recorded a number of other introduced plants among these villagers.

At Palatka (from Creek *pilotaikita*, crossing), in what is now Putnam County, Florida, Bartram visited a Seminole garden. The field was planted "chiefly with corn (*Zea*), Batatas [*Ipomoea batatas*], Beans [*Phaseolus vulgaris*], Pompions [*Cucurbita pepo*], Squashes (*Cucurbita verrucosa*) [*C. pepo*], Melons (*Cucurbita citrullus*) [*Citrullus lanatus*], Tobacco (*Nicotiana*) &c. are abundantly sufficient for the inhabitants of the village." At the same village he later wrote, "The fields surrounding the town and groves were plentifully stored with" exotic crops including peas (*Vigna unguiculata*), potatoes (*Solanum tuberosum*), peaches (*Prunus persica*), figs (*Ficus cairica*), and oranges (*Citrus sinensis*).

Later Bartram visited Cuscowilla (*taska*, warrior, *weli*, plunderer, Choctaw), near the northwestern corner of Tuscawilla Lake, and east of the present Micanopy, Alachua County, close to the Alachua savanna. There the Seminoles planted, "but little here about the town, only a small garden spot at each habitation, consisting of a little Corn, Beans, Tobacco, Citruls [*Citrullus lanatus*], &c." Instead, they had their major crops elsewhere.

Bartram also described an extensive maize field near Cowee (a Cherokee town formerly on the Little Tennessee River, Macon County, NC). He rode "near two miles through Indian plantations of Corn, which was well cultivated, kept clean of weeds and was well advanced, being near eighteen inches in height, and the Beans planted at the Corn-hills were above ground."

Sturtevant (1955) found the southern Florida Seminoles growing what the Miccosukee called *sala: lkitíscî* (red bean), *sala:llákní* (yellow bean), and *sala: llákno:cî* (small yellow bean). The beans were *Phaseolus vulgaris.*

Fernald et al. (1958) indicate that *P. polystachios* is edible, and Freytag and Debouck (2002) concur. Both books, however, point out that there are few seeds borne on the plants and that they are small. Both Uphof (1968) and Hocking (1997), on the other hand, write that the seeds are a popular food, being dried and cooked. Uphof (1968) wrote they were "highly prized by North American Indians." Since Moerman (1998) did not list the species I wonder where Uphof got his information. However, our expert on *Phaseolus,* Lawrence Kaplan, wrote me (Apr. 2003), "When I was trying crosses ... [i]t produced well in the greenhouse, certainly enough to warrant gathering."

Beans have been so important to Europeans and other settlers in the New World that expressions have grown up around them. Generally they have a good reputation, and that led to the saying that someone is "full of beans." That phrase means that the person is in an energetic, cheerful mood (Davidson 1999). The expression in Portuguese is *cheio de feijão.*

There is an impolite rhyme dealing with the difficulties humans have digesting the proteins in *Phaseolus.* This incomplete digestion results in an unusually large amount of methane, and subsequently flatulence. As an ancient saying goes:

Beans, beans, the wonderful fruit.
The more you eat, the more you toot.
The more you toot, the better you feel.
That's why I eat beans at every meal!

Phlebodium

(Greek *phlebodes,* full of veins; John Smith moved Robert Brown's section *Phlebodium* to generic level)

Phlebodium aureum (golden, in reference to the rhizome scales) (= *Polypodium aureum*)
avenca dourada (golden fern, Brazil)
cabbage palm fern (Florida)
calaguala [*calahuala*] (Cuba, Veracruz, Honduras, Panama)
istimá:ha imanâ (*este,* person, *im,* his, *aha,* tuber, *em,* its, *mahe,* height, Creek); *yá:tcáyhimá:hî* (*yá:tcáyh,* tall person, *im,* his, *á:hî,* tuber, Mikasuki); *yaatchayhen taapente* (tall person's fern, Mikasuki)
mano de léon (lion's paw, Totonac, Veracruz)
ni^{23} *za*^{3} (Chinantec, Oaxaca)
[golden] serpent fern (Florida, Bahamas)

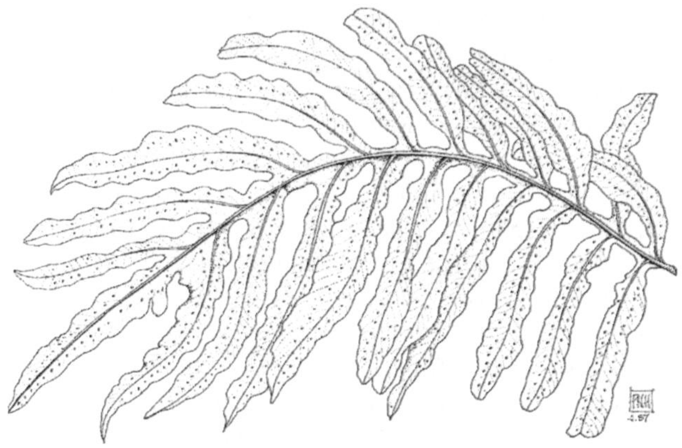

Phlebodium aureum. *Drawn by P.N. Honychurch.*

Linnaeus ([1753] 1957) called this fern *Polypodium aureum,* following Charles Plumier's *Polypodium majus aureum* (big red fern) of 1693. Many people still agree with him, but others segregate this and two to four other species into *Phelbodium* following John Smith (1798–1888) and Nauman (1993).

The name *calaguala* is widespread in Latin America for ferns famous as medicines (Duke and Vásquez 1994, Gupta 1995a). While that name subsumes several species, the most famous has been *Polypodium calaguala,* a South American species. However, people from Cuba through Veracruz, Mexico, and south through Guatemala, Honduras, and Panama mostly use *P. aureum* (Gupta 1995a). Cubans consider the rhizome of *P. aureum* good to treat wounds, falls, and external sores (Roig 1945). They also consider *P. aureum* anthelmintic and sudorific, and use it to treat rheumatism, heart disease, high blood pressure, asthma, and colic. Totonacs in Veracruz use *P. aureum* to treat coughs, internal bruises, and the kidneys, and in medicinal baths (Vásquez and Jácome 1997). In northern South America, Duke and Vásquez (1994) comment that *calaguala* is the first plant mentioned when medicinal plants, particularly treatments for cancer, are discussed. Florida is the northern limit of this fern.

A primary use of the fern among the Seminoles is for chronic sickness that had not responded to other treatments (Sturtevant 1955). Typically, the mixture for chronic sickness is a combination of many plants called *ayikctanahkó:cí* (small gathered medicine, Mikasuki) or *atilo:kocí* (*vetelokv,* gathered, *oce,* small, Creek). *Phlebodium aureum* is one of four ferns that might be included. Another mixture is given for "insanity," any kind of confusion or odd behavior (Sturtevant 1955). Still another use is during childbirth.

Gupta (1995a) reported that an ethanolic extact of cabbage palm fern showed activity against certain viruses. It has not been confirmed that the species has the same activity as others called *calaguala.*

Phoradendron

(Thomas Nuttall named this with Greek *phoros,* bearing, *dendron,* tree)

Phoradendron leucosperma. From Britton and Brown 1896.

Viscum album (although European mistletoe is placed in a genus distinct from the American plants, they differ in technical details; people familiar with the plants in Europe would have had no trouble recognizing the eastern North American species)
druidh lus (*druidh,* Druid, *lus,* herb, Gaelic); *gui* (French); *guis* (sticky, Irish)
mistletoe (a word from Old English in use by A.D. 1100; from *mistle,* dung, and *tan,* twig; akin to Old High German *mistil,* mistletoe, and *zein,* twig; the name came into existence because people believed that the mistle thrush, *Turdus viscivorous,* excreted the seeds on limbs; it actually scrapes them off its bill; see Austin 1998a); *Mistel* (German); *misteltein* (Norwegian); oak mistletoe (USA)
muérdango (Spanish)
uil'-ioc (all-heal, Gaelic)
vischio (from Latin *viscos,* sticky, Italian); *visco* (Portuguese)
Vogel-liem [*Vogellym*] (birdlime; cf. Turner [1548] 1965)

Phoradendron leucocarpum (white-fruited) (=*P. flavescens,* misapplied)
a:tilhiciksó imitto (*a:tilhiciksó,* elf, *im,* his, *ittó,* tree, Koasati)
antcka' nûta'wayi' (Biloxi)
fani shapha (*fani,* squirrel, *shapha,* flag, Choctaw); *fanishapha'* (*fani',* squirrel, *im,* its, *shapha',* flag, Chickasaw); *ipɫiäsapha* (squirrel's flag, from *ipɫo,* squirrel, *im+sapha,* its flag, Alabama)
fani' hasimbish (*fani',* squirrel, *hasimbish,* tail, probably is this plant, Chickasaw)
gui (mistletoe in French, Houma, Louisiana)
hinɫimásókcî [*hinrímásókcî, henle 'mashokche*] (squirrel tea, Mikasuki)
'to eleko [*ito-iliko, toiliko, tuhiligu, eto-eleko, tóh-elleko, tohiríkko, tohiɫikko, toheleko*] (*eto,* tree, *eleko,* feet, Creek, Muskogee)
uda'li (it is married, Cherokee)

Mistletoe has a long association with native people throughout the eastern United States, much as it has among the Europeans. The Cherokee, for cxample, used it to treat headache, epilepsy, in "medicine for pregnant women," to lower high blood pressure, and to cure vomiting (Hamel and Chiltoskey 1975). Among the Muskogean people, the Chickasaw made a remedy from it to treat "Red Squirrel Sickness" (toothache, swollen jaw, and sometimes nosebleed) (Swanton 1928a,b). The Houma made a decoction of the plant to aid debility and paralytic weakness, considering it something of a panacea (Speck 1941). The Yuchi use it to treat "Raccoon Sickness" (diarrhea, usually in babies) (Howard 1984). The Creeks made medicine from leaves and branches for lung troubles, including consumption (Swanton 1928a). The Florida Seminoles use it to treat "Deer Sickness" (numbness, painful limbs and joints), as an emetic during ceremonies, and as a medicine for chronically ill babies (Sturtevant 1955). The Oklahoma Seminoles make an infusion of leaves and berries for ringworm sores and to treat hemorrhoids, and give it to children for whooping cough (Howard 1984). Modern Florida Seminoles include it as part of the "Death Medicine" (Snow and Stans 2001).

Phragmites

(From Greek *phragma,* hedge, fence, or screen, plus *–ites,* resembling; the word was used by Pliny, A.D. 23–79, for a reed)

Phragmites australis (southern)
abo'djigun (something turned out or over, Ojibwa)
act (Atakapa)

Phragmites australis. *Drawn by Mary Wright Gill and Agnes Chase.* From Hitchcock and Chase 1950.

aqraban (Bahrain)
arrow [bamboo] grass
atask (Strachey wrote *attasqwas* in [1612] 1953, and Capt. John Smith penned *attasskuss* for the Powhatan word; both called them "weeds" but Smith added "grasses"; the root /**aTw-*/ occurs in their word *atons*, for "arrow"); cognates include *assa·kanask* (Ojibwa); *asta·kanask* (Cree)
bennels (from Hedrick 1919, source not traced)
caña de Indio (Indian cane, Panama, Puerto Rico); *caña de pantano* (marsh cane, Puerto Rico); cane grass [wild cane] (New Mexico); *cañoto* (wild cane)
carrizo (Arizona, New Mexico, Sonora to Honduras, Hispaniola)
chaume (more often applied to straw as for a thatch roof, from Latin *calamus*, French)
hajna (Saudi Arabia)
halal (cane like corn stalks, Maya, Yucatán)
i'hya (Cherokee)
kaneh [*qáneh*] (Hebrew)
koha:ha:kâ (*kohv*, cane, *vhake*, replica, Creek); *ołã: nâ:bî* [*orã:nâ:bî*] (cane replica, referring to *Arundinaria*, Mikasuki); *ku<u>n</u>shak* (Choctaw); *oskoba* (from *oski holba*, resembling cane, Choctaw; a local synonym for *ku<u>n</u>shak*)
lók'aa' (Navajo)
pull reed ("A long reed used for ceilings instead of laths"; cf. Britten and Holland [1886] 1965); [common] reed [grass] (Florida to New Mexico, Bahamas, Puerto Rico Panama); *riet* [*rietje*, *rif*] (reed, Dutch)
Rohr (from Gothic *Raus*, German); *takrør* (roof reed, Norwegian)
roseau (from Old French *ros*, related to German *Raus*, French); *roseau commun* (Quebec); *petit roseau* (little reed, Guadeloupe, Martinique)
saesgean (also an area dominated by moors or fens, Gaelic)
sak-halal [*zak-halal, zachalal*] (*sak*, white, *halal*, cane-like corn stalks, Maya, Yucatán)
soccos ("seed" cane, for planting, Spanish)
taa gui (*taa*, mat, matting, *gui*, cane, Zapotec)
tibisí (reed, Taino, Cuba, Hispaniola)
tlh'ogh elgha^n nachel ghi^{n}la (grass which is connected together, i.e., large nodes, Chipewyan, Canada)
wa:pk [*vaapk*] (O'odham, Arizona, Sonora)
xapij (Seri, Sonora)

The next time you listen to orchestra music, remember that you owe many of the sounds to plants. Stringed instruments and woodwinds (oboes, clarinets, bassoons, etc.) are made partly or completely of wood. However, the topic here is the music coming from the woodwinds, due to plant parts called reeds.

The word "reed" applies to several different plants, but three species are native to Florida. One is the musical reed source (*Phragmites*), one for fishing poles (wild cane; see *Arundinaria gigantea*), and the other is wild bamboo (see *Lasiacis divaricata*), which currently is not used by people in Florida. People elsewhere in the Americas use all three and distinguish between them with common names, while at the same time noting their similarities.

Moreover, the words "reed," "cane," and "bamboo" are intermixed in current and historic usage. All originally had somewhat different meanings and origins, yet those words and "calumet" all came into English because of plants (see also *Arundinaria*; *Lasiacis*).

The oldest of these terms is "reed." That word is related to Old English *hréod*, Old Frisian *(h)reid*, Old Saxon *hriad*, and Old Teutonic *hreundo*, all meaning tall straight stems of grasses. Originally, reed applied to either *Phragmites* or *Arundo*. "Reed" dates from ca. A.D. 725, being equated in a publication that year to *harundo* and *canna*. Although similar, *harundo* is not related to Latin *hirundo*, the swallow.

Obviously, the concepts of reed, cane, bamboo, and pipe are all interrelated and applied to numerous plants. The first requirement is that the grasses are comparatively large and have either a woody stem or one that is somewhat lignified. The most useful part of the grasses is the stem, and that has served people throughout the world in many ways.

Perhaps at least partly because *Phragmites* is most widely distributed, it is a well-known plant (e.g., Howard and Powell 1963, Kenk 1963, Baranov 1967, Core 1967, Johnston 1970, Morton 1970, Turner and Bell 1971, de la Cruz 1978, Heiser 1978, Bailey and Danin 1981, Wolverton 1982, Cunningham and Milton 1987, Timbrook 1990). The species grows from southern Canada to the central United States, to California, Louisiana, Florida, the West Indies, and from Mexico to Chile and Argentina. *Phragmites* also grows widely in Europe, Asia, Africa, and Australia. The genus *Phragmites* was created by French botanist Michel Adanson who called it that because the grass grows along streams and "fences" them from nearby areas. These are the plants called *kaneh* or *qáneh* in Hebrew versions of the Bible (Zohary 1982), and ancient Egyptians used the culms as "quill" pens. *Phragmites* was well known to the Europeans when they first arrived in the Americas.

In the New World, Europeans continued their old names for the plants. Dutch speakers use a word closest to English, when they say *riet* [*rietje, rif*]. The Germans call the reed *Rohr*. The French borrowed that term and say *roseau* (from Old French *ros*). Speakers of French also have a Latin derivative, *chaume* (from Latin *calamus*), but this is more often applied to straw as for a thatch roof. Romance languages continued using their Latin-based words, in Portuguese *cana*, Spanish *caña*, and Italian *canna*. Gaelic speakers called it *saesgean*, which also means an area dominated by moors or fens.

Europeans in the Americas also compared *Phragmites* with the Old World sugarcane (*Saccharum officinarum*), saying *caña de Indio* (Panama, Puerto Rico) and *soccos*. They use that comparison because both plants provide sweets. Fernald et al. (1958) described the process of *Phragmites* extraction. Stems are gathered before flowering, dried in the sun, and ground or beaten into flour. The finer parts are sifted out and moistened to make a gummy mass that is roasted by a fire until it swells and browns slightly. The material is eaten like taffy or marshmallows.

There is also a sweet edible gum exuded from damage caused by insects (the mealy plum aphid, *Hyalopterus arundinis*). Although Hodgson (2001) suggested that the use of the sweet extract from *Phragmites* called *manna* by the Europeans and *cadece* (juice of the reed) by the Cochimis was originally restricted to the California area, its use seems to have been more widespread. It may be the abundance of historical documents from the California region that gives the impression of a focus there. Many people, including Akimel and Tohono O'odham (Pima and Papago), Paiutes, Panamints, Yavapais, Cocopas, and others certainly showed the Europeans the sweet material. Europeans then left behind written accounts.

Other native American names compare *Phragmites* with cane (*Arundinaria gigantea*) or maize (*Zea mays*). The Seminoles say *koha:ha:ká* (Creek) and *ołá:ná:bí* (Mikasuki), referring to *Arundinaria*, which they call by the simple terms *kohá* (Creek), *kóha* (Muskogee), or *ołá:ni* (Mikasuki). In Yucatán, the Maya say it is *halal* (cane-like corn stalks), or *sak-halal* [*zak-halal, zachalal*] (*sak*, white, *halal*).

From British Columbia through the western United States to Panama, the young shoots are served as a potherb (Fernald et al. 1958, Duke 1968, 1972, Moser and Felger 1985, Hodgson 2001). In the 1870s the English botanist Mrs. Phoebe Lankester noted that the young shoots, especially where protected from light, "made an excellent pickle" (Fernald et al. 1958). As late as 1942, Mr. and Mrs. Whittrock commented that indigenous people in New Jersey boiled the rhizomes like potatoes, and in the early spring they cooked the young shoots like asparagus (Fernald et al. 1958). The rhizomes may be harvested during any season and eaten raw or ground into flour.

The seeds are hard to extract from the chaff, but they were cooked and eaten by people from New England to Oregon and south to the Sonoran Desert and probably elsewhere (Fernald et al. 1958, Mabberley 1997, Hodgson 2001). Sometimes the chaff was allowed to remain on the seeds. Again in New Jersey, the Whittrocks recorded that local indigenous people did not remove the hull, but cooked the whole grain into a reddish gruel, colored by the hull. They added that it was "wholesome as a food, though not too appetizing in appearance" (Fernald et al. 1958).

People from the Thompson of British Columbia and the Okanagon of the Washington and Canada border to the Zapotecs of Oaxaca use *Phragmites* to prepare mats (Reko 1945, Mabberley 1997). Surely, people in the eastern parts of North America did the same although they made more use of cane (see *Arundinaria*). Some of the mats were used to dry food, while others were for sleeping and other purposes (Rea 1997, Moerman 1998).

In the southwestern United States and probably elsewhere in its range, *Phragmites* stems have been used to make many items. One of the most frequent applications before indigenous people obtained guns was as shafts of arrows and atlatl darts. There are records of at least the Apache, Aztecs, Havasupai, Hopi, Hualapai, Kawaiisu, Klamath, Navajo, Paiute, and Tewa making arrows from them (Elmore 1944, Ebeling 1986, Mabberley 1997). Some other uses include making baskets, boats, containers, flutes, nets, pipes (or "cigarettes"), pipe stems, prayer sticks, roofing, screens, thatching, and weaving rods (Kearney and Peebles 1951, Felger and Moser 1985, Ebeling 1986, Mabberley 1997, Hodgson 2001). Both stems and leaves were used to make cordage for nets and snares (Ebeling 1986).

We can still purchase *Phragmites* reed "screens" to use as visual barriers along fences and other areas. In addition, the species is used commercially to make paper, cellophane, cardboard, synthetic textiles, fiberboard, fuel, alcohol, insulation, and fertilizer (Hocking 1997). The Seminoles probably also used reed much as they did cane (see *Arundinaria*).

In the southeast, only the Seminoles are recorded as using *Phragmites* as medicine. These people used the hollow stems as a tube to blow medicine into wounds or onto skin problems, and even as a medication for boils and carbuncles (Sturtevant 1955). Similarly, the Navajo use *lók'aa'* for stomach and skin problems (Mayes and Lacy 1989). The Apache made a medicine to treat diarrhea and other digestive problems (Moerman 1998). The Blackfoot made a decoction of the whole plant as an emetic. The Iroquois mixed *Phragmites* with bottle-brush grass (*Elymus hystrix*) to make "corn medicine," a mixture in which they soaked corn kernels before planting them. In the Bahamas and Cuba, a root decoction of *Phragmites* is taken as a diuretic and antiseptic. On North Caicos the root is boiled with *Chiococca alba* to relieve "pain in the back" (kidneys) (Roig 1945, Morton 1981).

Human alterations of landscapes have resulted in major changes in reed abundance. Formerly, when the land was wetter, people in the southwestern part of the United States and nearby Mexico named places because of the stands of *Phragmites*. Some examples include Carrizo, Carrizo Butte, Carrizo Mountains, Carrizo Ridge, and Carrizo Wash (Apache County, Arizona), Carrizo Creek (Catron County, New Mexico), Carrizozo (Lincoln County, New Mexico), *Xapij an Hax* (reedgrass inside water), and *Hax Cáail* or *Pozo Carrizo* (Kino Bay, Sonora). Now, drainage has resulted in near disappearance of the species. The alternate situation applies in the eastern states where the apomictic form has become a weed covering vast areas in monocultures.

"Reed" in woodwinds is a slightly different meaning from all the others. There, a segment of the stem is removed and thinned so that air moving over it in the confined space of the instrument will cause it to vibrate. Sounds from that vibration are amplified by the remainder of the instrument to produce the music. Occasionally, the reeds are made of *Arundo donax*, another reed; however, the preferred "reed" is from *Phragmites*. No other plant produces sounds as pleasing as it, and synthetic products are usually not acceptable to the musicians who demand excellence in sounds. An ancient grass used to make Greek panpipes is still the best.

Phryma

(Derivation unknown; possibly from Greek *phryna*, a toad)

Phryma leptostachya. From Britton and Brown 1898.

Phryma leptostachya (slender-spiked; an old generic name)

[American] lopseed ("lop," to hang loosely, was in English by 1578; "lopseed" was recorded in a common name by 1850)

Apparently, the first that Europeans learned of these herbs was from Jan Gronovius's *Flora Virginica*. In 1739, he called the plants *Verbena racemosa simplicissimo, flores sessilibus, calycibus fructus reflexis racemoque appressis* (A verbena with a simple raceme, flowers sessile, the fruiting calyx reflexed and appressed to the raceme). In 1740, Adrian van Royen used the same phrase to describe plants growing in the

Netherlands. Not long afterward, in 1751, Linnaeus coined the generic name *Phryma* and left everyone wondering on what he based the name. *Species Plantarum* (Linnaeus [1753] 1957) states simply, "*Habitat in America septentrionalis*" (Grows in North America).

That problematic etymology of the generic name set the stage for trying to understand the relationships of these plants. They have been put in their own family, the Phrymaceae, or the Verbenaceae. The most recent molecular genetic studies support *Phryma* being removed from the other families and recognized as a distinct lineage (Wagstaff and Olmstead 1997).

Although Linnaeus did not know it when he named the herbs, the genus is disjunct between Asia and North America. There has been a tendency to call the Old World plants a different species from the American, but the differences are minor (Hara 1969). Since the late 1960s, most have considered them a single, bicentric species in the monotypic *Phryma.*

In North America, the plants are considered native from New Brunswick to Manitoba, south to Florida, Alabama, Louisiana, and eastern Texas (Fernald 1950, Steyermark 1963, Barkley 1986, Jones and Coile 1988, Diggs et al. 1999). These plants are widespread and important medicines in Asia, but few records exist of their application in the Americas. The Ojibwa chewed roots or gargled a root decoction to relieve sore throat, and drank the decoction when they had rheumatism (Densmore 1928, Moerman 1998).

Hocking (1997) noted that the leaves and roots are insecticidal and have been used to kill houseflies in Japan and China. Elsewhere in Asia, the plants are used to treat fevers, ulcers, ringworms, scabies, boils, carbuncles, and cancers (Hsu 1986, Foster and Duke 1990). The herbs contain phryumarol, a sterol (Hocking 1997).

Phyla

(João de Louriero, 1717–1791, used Greek *phyle*, a tribe or clan, an allusion to the many flowers in tight heads)

fog fruit (in use for *P. nodiflora* in the Americas by 1886; the old meaning of "fog" was for "grasses" that sprang up in fields immediately after the hay was harvested; that sense dates from the 14th century A.D.; the concept of moisture in the air did not appear until the 1500s, although it now precludes the older use); frog fruit (corruption of "fog fruit")

Phyla lanceolata. *Drawn by Vivian Frazier.* From Correll and Correll 1972.

Phyla lanceolata
lance-leaf fog-fruit

Phyla stoechadifolia (leaves resembling those of the mint *Lavandula stoechas*) (=*Lippia stochadifolia*)
azulejo (little blue one)
epazotillo (little epazote, *Chenopodium ambrosioides*, Totonac, Veracruz)
guia huace [*quie huace*] (*guia*, flower, *huace*, smell of roses or sweet fruit, Zapotec, Oaxaca)
kabalyaxmik [*cabalyaxnic*] (*kabal*, small plant, *ya'ax*, green, *nik*, joined to, in other words, small plant that grows with the *ya'ax che', Ceiba pentandra*, the green tree that supports the world; Maya, Yucatán)
marsh phyla (a book name)
orozús (literally, "juicy-root," a name usually given to a legume, Belize)
poleo [*poley*] (a mint resembling *"yerbabuena"* or *Mentha*, Puerto Rico)
southern fog-fruit (Florida)

té [*cimarron, negro*] ([wild, black] tea); *té de Yucatán* [*del país*] (Yucatán [country] tea, Yucatán)

Linnaeus ([1753] 1957) had not seen living specimens of what he called *Verbena stoechadifolia.* Instead, he knew them from the reports of Charles Plumier from 1703, Sebastien Vailant in 1718 (1669–1722), and Adrian van Royen in 1740. Presumably all three earlier authors had studied the plants in cultivation in France and then in the Netherlands. It was 1803 when André Michaux discovered and named the second species as *Lippia lanceolata.* Both species have been moved from genus to genus as understanding of their relationships evolved. Most of us still do not know the differences between *Lippia* and *Phyla,* but the latter genus contains about 11 tropical and warm region species around the world (Mabberley 1997).

Little beyond the common names has been recorded for *Phyla.* The Mahuna of southern California used *P. lanceolata* to treat rheumatism (Moerman 1998). The Houma used *P. nodiflora* (*l'herbe à la tortue,* turtle herb) as a bath to make lazy babies walk (Speck 1941). Numerous other uses are recorded by Ayensu (1981) and Morton (1981) for that species. The reverence accorded *kabalyaxmik* by the Maya summarizes attitudes generally toward several species (cf. Arvigo and Balick 1993, Vásquez and Jácome 1997). *Phyla stoechadifolia* is endangered in Florida (Coile 2000).

One of the problems with attributing uses to *Phyla* is its confusion with related genera *Lippia* and *Aloysia.* Standley (1920–1926), for example, put them all in *Lippia.* Related *L. umbellata* is used as a remedy for colic. *Lippia dulcis* (*hierba dulce*), is a well-known sweetener, and *L. graveolens* (*orégano*), better known as *A. graveolens,* is a spice used like the mint that is its namesake (*Oreganum*).

Phyllanthus

(From Greek *phyllon,* leaf, and *anthos,* flower, because the flowers of some species are borne on leaflike dilated branches)

Blattblume (leaf flower, German)
cherek hantu (ghost's diarrhea plant; *P. frondosus,* Malay)
dukong anak (the child pick-a-back; *P. urinaria,* Malay)
graines en bas feuilles (seeds below leaves, French Antilles)
keezhu (below, a reference to the fruits being borne under the leaves, Malayalam)

Phyllanthus caroliniensis. From Britton and Brown 1897.

leaf-flower; *phyllanthe* (leaf flower, French)
quebra pedra (stone breaker, Portuguese)

Phyllanthus caroliniensis (of Carolina)
cababesinixte (a reference to a plant with small fruits that resemble ants; *cabal,* small, *bel,* clothes, *sinik,* ants, *te,* toward, *che',* plant, Maya, Yucatán)
canastilla (little basket, Veracruz)
Carolina leaf-flower (a book name)
gale-of-wind
[*graines*] *en bas feuilles* ([seeds] under the leaves, Martinique)
rock phyllanthus (a book name, Bahamas)
yerba de la muina (indisposition herb, Veracruz)

The first time I saw *Phyllanthus caroliniensis* in Florida was in the herb layer of a pinewoods that had recently burned. The plants were abundant for a year after the burn, and added a lacy green to the ground. Then, as other plants grew up, they disappeared. Soon afterward, the whole region became a housing development and those small herbs were gone. It all happened so fast that I never was able to learn much about these curious "leaf-flowers."

Phyllanthus is a genus of about 800 species growing in the tropical and warm parts of the world (Mabberley 1997, Webster 2002). The genus ranges in life-form from large trees (e.g., *P. acidus*) to small herbs like *P. caroliniensis.* Linnaeus created the genus in 1753, although people had been using the word *Phyllanthus* for these plants long before then. The name was actually coined by Jan Commelin in 1697, and used by Catesby (1731–1732) and Albert Seba in

1734–1735. Linnaeus had adopted *Phyllanthus* beginning with *Hortus Cliffortianus* of 1738.

In *Species Plantarum*, Linnaeus recorded the six species that he then knew: *P. grandifolia*, *P. emblica*, *P. epiphyllanthus*, *P. maderaspatensis*, *P. niruri*, and *P. urinaria*. Three species were based on specimens from India (*P. emblica*, *P. maderaspatensis*, *P. urinaria*), and the others from the Caribbean (*P. epiphyllanthus*, *P. grandifolia*, *P. niruri*).

It was not until Thomas Walter published his *Flora Caroliniana* in 1788 that *P. caroliniensis* became known. These small herbs have now been found from Pennsylvania to Missouri and Kansas south to Brazil, Bolivia, and Argentina (Fernald 1950, Hocking 1997). These herbs are like *P. niruri* and *P. urinaria* in aspect.

Vásquez and Jácome (1997) found people using *P. caroliniensis* to treat bronchitis, bruises, the heart, and skin problems. It is used as a diuretic in the French West Indies (Hocking 1997) and as a medicinal in Belize (Balick et al. 2000). The information available indicates that these herbs were used in the same ways as *P. niruri* and *P. urinaria*.

However, the story becomes more complicated once those other two species are considered. *Phyllanthus niruri* was described by Linnaeus in 1753. He thought it was from India and cited both Rheede's list of 1678–1703 containing plants used on the Malabar Coast of India and John Martyn's (1699–1768) list of plants in cultivation in London in 1728. Although Linnaeus's name continues in use, *P. niruri* has never been found in India (Webster 1955, 1956, 1970, 2002, personal communication 2003).

The native range of *P. niruri* is from southern Texas, through Mexico to Argentina (Stevens et al. 2001). The plants that are so famous for medicine in India are actually *P. fraternus*, not named until studied by Grady L. Webster (1955). Given the difficulty of identification of these small herbs, plus the prevalence of some widespread and weedy species, I suspect that the small native, and often rare, *P. caroliniensis*, has been relegated to a plant that is now neglected.

However, *P. caroliniensis* contains several bioactive compounds that would have made it useful. Phytosterols, quercetin, gallic acid ethyl ester, and geraniin have been identified in it by Cechinel Filho et al. (1996a,b). These authors wrote, "Pharmacological analysis also revealed that quercetin, gallic acid ethyl ester and a semi-purified fraction of flavonoids (1–100 mg kg-1, i.p.) exhibited graded and significant antinociception against acetic acid-induced abdominal constriction. The mean ID50 values (mg kg-1) for these effects were: 18.8, 34.7 and 5.3, respectively. It is concluded that quercetin, gallic acid ethyl ester and some as yet unidentified flavonoids might account for the antinociceptive action reported for the HE [hydroalcoholic extract] of *P. caroliniensis*." Moreover, Santos et al. (1999) and Narayana et al. (2001) found that the bioflavonoids in *P. caroliniensis* were not only antinociceptive, but also anti-inflammatory.

The plant of choice in most, if not all, of the Americas is now *P. niruri*. However, Caribs in Dominica made a tea of the Old World exotic *P. tenellus* to induce abortions, and it kills rabbits and guinea pigs if they eat it (Hodge and Taylor 1957, Honychurch 1987).

In Cuba, *P. niruri* is known as *yerba de la niña* (little girl's herb). Elsewhere it is called niruri (Florida), gale of wind (Florida, English Antilles), *peronilla del pasto* (grass pear, Puerto Rico), *viernes santo* (Holy Wednesday, Puerto Rico, Colombia), *erva* [*herva*] *pombinha* (literally little dove herb, but slang for female genitals, Brazil).

There are two main themes in the common names—use to break up kidney stones and use as a quinine substitute. For the kidney stone theme, it is called *arranca pedras* [*arrebenta pedras*] (stone breaker, Brazil), *malva pedra* (probably a lapse for bad stones, Brazil), and *quebra pedra* (stone breaker, Brazil). Names that allude to reducing fever are *feuilles la fièvre* (leaves for fever, Haiti), *quinina* (quinine, Haiti), *quinina créole* (Creole quinine, Haiti), *quinina criolla* (Creole quinine, Dominican Republic), *quininina* [*quinine du*] *pays* (country quinine, Haiti), *quininito* (little quinine, Haiti), *quinino de pobre* (poor person's quinine, Puerto Rico), and *sulfate pays* (country sulfur, Haiti). In Paraguay, the herbs are *paraparai mí* (from *parapa'rá*, Guaraní).

Phyllanthus nirun has names noting seeds below the leaves. Those include *des dos* [*de dou*] (from the back[s], Haiti), *derrière dos* (behind the back, Haiti), and *graines sur dos* (seeds on [the] back, Haiti).

In Cuba *P. niruri* is used to treat malarial fever, diabetes, liver problems, biliousness, and dysentery, and it is used as a diuretic. In Hispaniola, these bitter plants are used to treat malaria (Liogier 1978). In Brazil, the herbs are used against urinary disorders, diabetes, jaundice, and malaria (Mors et al. 2000). The fruit juice is considered antidiabetic.

Phyllanthus contain flavonoids, lignans, and glycosides (Mors et al. 2000). Given the confusion about the taxonomy, the identity of the species reported as *P. niruri* is dubious, but the genus is nonetheless chemically bioactive. Experiments show significant increase in diuresis and sodium and creatine excretion (Mors et al. 2000, Barros et al. 2003). In humans, persistent ingestion of tea from the roots expels kidney stones. *Phyllanthus* extracts show lipid-lowering activity (Khanna et al. 2002). There are also indications of antimalarial activity (Tona et al. 2001) and inhibition of HIV (Qian-Cutrone et al. 1996).

Over 30 years after I first saw *P. caroliniensis* in Florida, I learned that they were dispersed by mourning doves (*Zenaida macroura*). I had wondered for years how that region was able to support so many of these doves, and a partial answer came in a study by Tyler and Jenkins (1979). Doves are opportunistic and feed on an array of seeds—most of them small. The Oklahoma study found 17 species of plants in dove crops. So, the year of the burn in the pine flatwoods was also a good food year for the mourning doves. Now the few doves there are obliged to sustain themselves at feeders outside residents' homes.

Physalis: Ground Cherries

(From classical Greek *physa* for a bladder, meaning the inflated calyx)

Physalis walterii. *Drawn by P.N. Honychurch.*

After spending years in Mexico, Francisco Ximénez published his book *Naturaleza y Virtudes de las plantas* (Nature and Virtues of Plants) in 1615. In that book was an herb called *coztomatl* (yellow tomato, Náhuatl). The herb is *Physalis* but no one is sure which of the species. The characteristic enlarged and inflated calyces growing over the tomato-like berry are obvious in the Ximénez drawings. Today we know and eat fruits of a Mexican species as *tomatillo* in *salsas.* In Ximénez's time, people in Europe were already familiar with their native *P. alkekengi* and some others from elsewhere.

Classical Greeks called the Old World plants *halikakabon* (bladder) or *phusalis,* and the Romans simply translated the Greek into Latin, *vesicaria.* Later people in southwestern Europe took their names from the Greeks, and said *halicacabon comun* (common bladder, French 1575) or *halicacabo uolgare* (common bladder, Italian 1551). This was also rendered *Boberellen* (bladders, German, 1536). The French sometimes said *couile bobes* (*bobe*'s testicles, 1549).

By the mid-1500s, Europeans from Italy to Holland and Germany were familiar with plants now known as *P. alkekengi.* Each country had its own names for these plants. Many sources say the specific name, *alkekengi,* is an Arabic name. However, that does not appear to be correct. According to Pignatti (1982), "*alkekengi* corresponds probably to the *Halikakabos* of Dioscorides." This name was picked up because of the similar sound in the medieval Arabic *hab-kakeng e* and then, in the Italian dialects, it became *alchechengi, accatengi, arachengi, arcachenzi, chechingi, chichingi, chechingela.* Other popular names in Italy now are *coralli* (coral) and *palloncini* (baloons).

The name *alkekengi* was spread with the plants from southeastern Europe, and it also was spelled *alkakinge* (English, 1542), *akcakeng* (Turner [1548] 1965), *alquequanges* (French 1549), and *alkanges* (French 1551). Others compared the fruits to cherries, calling them winter chirrir (1568), red winter cherries (1597), *Judenkirsen* (Jew's cherry, German, 1542), and *Schlutten* (ground cherry, German, 1542).

Some of the European names suggest that the plants were introduced comparatively recently. These allusions include *otra especie de yerua mora* (another species of blackberry herb, Spanish 1557), *baguenaudier* (comparison with the bladder mallow, *Colutea arborescens,* French, 1550), and *criecken van over zee* (foreign *criecken,* Dutch, 1549).

The plants came to northern Europe from the Middle East. That route is suggested by the German names *Judendocken* (Jew's bundle, 1542), and *Judenhutlin* (variant of *Judenhütlein,* Jew's little hat, 1542). Those words are comparable with Jerusalem cherry still used in English. The name *rot Nachtschatten* (red night-shade or *Solanum,* 1542) may also indicate a comparison of a new plant with an old familiar one.

Nine species of *Physalis* are native to Florida (Wunderlin 1998). If indigenous Americans had as much trouble distinguishing the different types as botanists have, they probably considered them all the same. With little regard for species, plants are called by a number of generic common names. Those names include *alkekengi* [*alquequenje*] (USA, Puerto Rico), Barbados-gooseberry (USA, Europe), cherry tomato (New York), Chinese lantern (USA), ground-cherry (USA), husk-tomato (USA, Puerto Rico), Japanese lantern (USA), Jerusalem cherry (USA), strawberry tomato (USA in 1800s), *tomatillo* (Mexico), wild cherry (USA), and winter cherry (USA, Europe).

One of the most widespread species in Florida is *P. pubescens.* That plant was first reported in Brazil by Marggraf in 1648, with the indigenous name *camaru.* The species is still known there as *camapu* (from Tupí *kama'pu*). A synonym is *juá-poca* (from Tupí *yua'-poca, yua',* fruit). The only other South American name found for the species is *sousourouscorou* (Carib) or *pororu wokuru* (Carib, Suriname). However, in Yucatán, the Maya say *yooch ik bach* (*yooch,* animal food, *ik,* chile, *bach,* the chachalaca, *Ortalis vetula*).

Many names refer to the inflated fruits. These include *bonnets de grandmaman* (grandma's bonnet, Louisiana), *coque-molle* (flabby shell, Hispaniola), *herbe a cloques* (blister herb, Hispaniola, Guadeloupe, Martinique), *maman laman* [*maman lanman*] (*maman,* mother, Haiti), *vejiga de perro* (dog bladder, Cuba), and *poc* (from French *pouche,* pouch, Guadeloupe, Martinique).

Several names note the impact *Physalis* has when either eaten or taken as medicine. These are *bate-testa* (hits the stomach, Brazil), *sacabuche* [*peludo*] ([hairy] stomach emptier, Puerto Rico), *coqueret de la Barbade* (Barbados flirt, French Antilles), *pantomima* (pantomime, Cuba), and *revienta caballos* (horse breaker, Cuba). Other names denote edibility or inedibility, as in *erva moura do peru* (Peru blackberry, Brazil), hairy-ground-cherry (USA), *tomate freadilla* (damaging tomato, Sonora), *tomatillo* (little tomato, Belize), and *yerba de sapo* (toad herb, Venezuela).

Joachim Camerarius recorded the foreign species *P. angulata* in European medical gardens in 1588. This species was known to African slaves when they arrived in the New World, and some brought their names with them. These words persist as *topo-topo* [*topotopo*] (Guyana, Trinidad), *topatop* (Jamaica), *topeton* (Panama), *tope-tope,* and *tophe-tophe* (Hispaniola). Variants are derived from the African Gold Coast, where the species is *totototo* or *tutotuto* (Twi), *wotowoto* (Ewe) or *gbato-gbato* (Ewe). Those names represent the sounds of children popping the inflated pods in play (Hutchinson and Dalziel 1937) and are related to poppers (Bahamas) and pops (Barbados). The Barbadian name has been in use at least since 1750 (Hedrick 1919). In the Great Plains, Omaha-Ponca children called the fruits *pe igatush,* Pawnee children said *nikakitspak* (to pop, forehead), and both played a "popping" game similar to that of the Africans (Gilmore 1919).

Some names for *P. angulata* are allied with those for *P. pubescens.* Those referring to the fruit are *balãozinho* (little balloon, Brazil), *bucho-da-rã* (frog's mouth, Brazil), *juá-da-capote* (hooded fruit, Brazil), *tomate* (tomato, Sonora), *tomate de cascara* (husk tomato, Sonora), and winter cherry (Trinidad, Jamaica). Others voicing their like or dislike are *sacabuche* (stomach emptier, Puerto Rico) and *matafome* (hunger-killer, Brazil).

Sometimes the Aztecs called this and other *Physalis* simply *tomatl,* now in Spanish *tomate* (tomato, Oaxaca). The plant we now call "tomato" (*Lycopersicon esculentum*) the Aztecs called *xitomatl* [*jitomatl*] (red tomato, Náhuatl). Spanish speakers in the New World generally call wild *Physalis* either *tomatillo* (little tomato) or *tomate del campo* (wild tomato).

The only species with a Seminole name is *P. walteri,* formerly known as *P. viscosa* (Sullivan 1985). While the Miccosukee call that species *łalilsatoklicí:kî* [*ralilsatoklicí:kî, łaliçatoklitcigi*], Sturtevant (1955) made no comments. When Sheehan was among the Seminoles in 1919, he found them treating coughs and colds with the plants (von Reis and Lipp 1982). *Physalis walteri* was first recorded in Europe when Francisco Hernández published his *Rerum Medicarum Nova Hispanae Theasurus* in Rome in 1651. He called this herb *miltomate* (wild tomato, Náhuatl, Oaxaca), although people in Belize use the name for *P. pubescens.* In Spanish, this is rendered *tomatillo de monte* (wild tomatillo, San Luis Potosí). The Huastec say *tuthay ch'ohool* (tomato herb, Huastec, Veracruz), *tuthay ts'ohool* (tomato herb, Huastec, San Luis Potosí), *its an t'ot* (tomato herb, Huastec, San Luis Potosí), or *tuthaayil an t'ot* (buzzard tomato, Huastec, San Luis Potosí). To the Zapotecs, it is either *bathus* (Oaxaca) or *bithus gihs* (wild *Physalis,* Oaxaca).

Other North Americans who use different species of *Physalis* are the Akimel O'odham, Apache, Cherokee, Dakota, Diegueño, Hualapai, Iroquois, Isleta, Keres, Meskwaki, Mohave, Navajo, Omaha-Ponca, Pawnee, Seri, Tarahumaras, Tohono O'odham, Winnebago, Yuma, and Zuni (Rea 1997, Moerman 1998, Hodgson 2001). All of these people ate the fruits and used the herbs as medicines (Yanovsky 1936, Curtin 1947, Felger and Moser 1985, Rea 1997, Moerman 1998, Yetman and Van Devender 2001). Some of the names they used for different species are *hanpok-hischasu* (owl eyes, Winnebago), *kekel viipid* (old man's testicles, Akimel O'odham), *makan bashahon-shon* (crooked medicine, Omaha-Ponca), *tamaniolipe* (Dakota), *tombrisi* (Mayo), and *tulumisi* (Guarijío). The Seri recognize two they call *xtoozp* and *insáacaj.*

Fruits of most species are edible when ripe, either raw or cooked. They are fried, boiled, or stewed to make marmalade by cooking them with syrup. Leaves and green fruits are somewhat poisonous, as they are laxative and diuretic because of their solanine content (Morton 1981, Turner and Szczawinski 1991).

Species are often used indiscriminately as medicines (Lipp 1971, Hussey 1974, Wong 1976, Bye 1979, Coe and Anderson 1996). In the Bahamas, Cuba, Jamaica, Hispaniola, Colombia, Brazil, and Mexico, the fruits are considered febrifuge, and both leaves and fruits are used as diuretics. The leaves or sap are applied as emollient plasters, for rheumatic ailments, or as drops for sore eyes in children. Leaf teas are taken to dispel indigestion, cure nephritis, cystitis, otitis, jaundice, herpes, and bowel problems called bich. Leaf suppositories are also used for bich, and the roots are used to treat toothache.

Some consider the genus narcotic and stimulant, at least partly because of its alkaloid, tannin, potassium chloride, and potassium citrate content. Berries contain acetylcholine, calystegines, ascorbic acid, and vitamin A (Zennie and Ogzewalla 1977, Morton 1981, Mello and Afiatpour 1985, Asano et al. 1995, 1997). Several species contain physalins, and the alkaloid hygrine is known from the roots (Willaman and Hui-Lin-Li 1970, Kawai et al. 2001). Withanolide glycosides have been found in *Physalis peruviana* (Sakurai et al. 1976, Lee et al. 1991, Ahmad et al. 1999), and the alkaloid 3-alpha-tigloyloxytropane is known from *P. alkekengi* (Yamaguchi and Nishimoto 1965).

Aqueous extract of leaves is antimicrobial (Ayensu 1981, Cáceres et al. 1991). Other laboratory studies indicate possible use against gonorrhea (Cáceres et al. 1995), cancer (Chiang et al. 1992a,b), and diabetes (Roman-R. et al. 1992). Modulation of the pituitary and basomedial hypothalamic lysyl-aminopeptidase activities has been shown in *P. alkekengi* (Vessal et al. 1996). Nutritional and quality analyses have been made of *tomatillos* (*P. ixocarpa*), and other species are presumably similar (Bock et al. 1995). Fruits are high in niacin (Morton 1981).

Two recent studies have been made of the relationships of the genus, one based on morphology (Alelius 1996) and the other on molecular genetics (Mione et al. 1994). The results are that *Physalis* may be paraphyletic, and *P. alkekengi* may not be closely related to the other species. The American *Physalis* seems to be most closely related to the endemic genera *Chamaesaracha* and *Margaranthus*. If that is true, then French botanist Joseph Pitton de Tournefort was correct when he segregated *Alkekengi* as a distinct genus in 1742.

Physocarpus

(From the Greek *physa*, a pair of bellows, and *karpos*, fruit, a reference to the inflated carpels)

Physocarpus opulifolium. From Britton and Brown 1897.

Physocarpus opulifolius (powerful leaves, but probably a comparison with the leaves of *Viburnum opulus*)
Blasenspiere (bladder spit, German)
[common, eastern] ninebark (dating from 1796 or earlier when it was applied to *Spiraea*, and alluding to multiple bark layers)

Linnaeus called these shrubs *Spiraea opulifolia* in 1753. He had studied living plants at the *Hortus Cliffortianus* and specimens sent by John Clayton from Virginia. They did not seem all that different from the other ten species of *Spiraea* that he described in *Species Plantarum*, and he saw no reason to separate them. However, Carl Johann Maximowicz, 1827–1891, studying them over a century later, decided they were distinctive. The fruits were inflated, unlike *Spiraea*, so he named it *Physocarpus*. There are now ten species known in the genus, which grows in North America and Asia (Mabberley 1997).

The best-known species in *Physocarpus* is the one in Florida, and it also grows from Virginia and West Virginia to Tennessee south to Alabama and Georgia. Ninebark barely extends into Florida, known only from Calhoun and Jackson Counties near Tallahassee (Wunderlin and Hansen 2002).

Moerman (1998) listed three eastern tribes who used the plants. The Iroquois used ninebark as a gynecological aid. The Menomini made a bark drink to cleanse the woman's system and enhance fertility. The Ojibwa used the root as an emetic (King 1984). Tribes in the western states used it to treat gonorrhea and tuberculosis, and as a laxative (Foster and Duke 1990, Moerman 1998).

These plants are potentially toxic (Foster and Duke 1990). Presumably that toxicity is due to the cucurbitacins and triterpenes (Sarker et al. 1999, Kim et al. 2000).

Phytolacca

(Greek, *phyton*, a plant, and *lacca*, derived from Hindi, *lakh*, referring to a crimson dye)

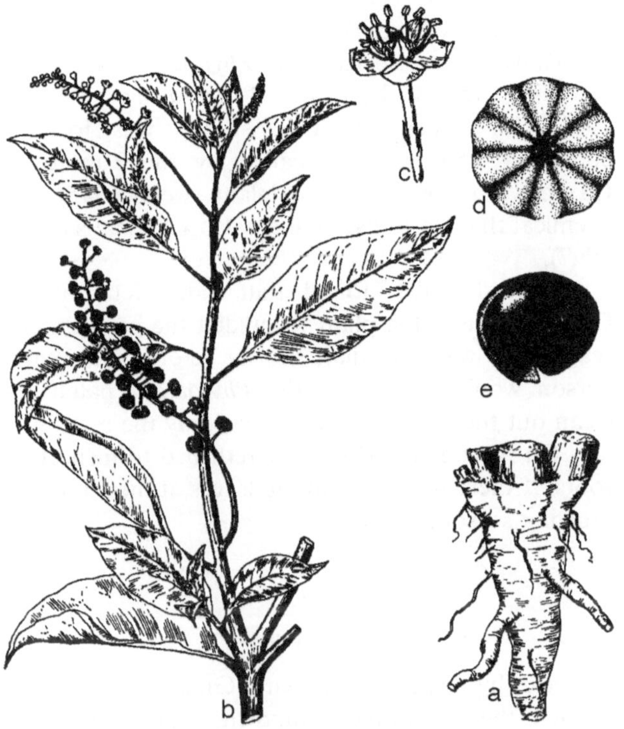

Phytolacca americana. a. Part of fleshy taproot. b. Upper branch with flowers and fruits. c. Flower. d. Upper surface of berry. e. Seed. From Buchholtz 1968.

Phytolacca americana (American)
amerikanische Scharlachberre (American scarlet berry, German)
bledo carbonero (charcoal saltwort; also used for *Blutaparon*, which see)
cancer root (one of many plants with this name, Carolinas)
chou-gras (fat cabbage, Louisiana, fide Rafinesque)
cokan (northern tribes, fide Millspaugh 1892; Algonquian and cognate with *pakkan*)
coskîlpâ [*oskílpá*] (Creek; in Oklahoma, the word applies to *Sambucus*)
crow-berry (USA)
dla:-ya-de (Cherokee)
garget (the throat, originally from Old French *gargette*; the association is not clear)
herbe de la laque (lacquer herb, Quebec)
ink-berry (Carolinas)
Kermesbeere (scarlet berry, based on kermes or cochineal insects, German)
koshe [*kó:sî, ko:si*] (Mikasuki); *koshiba* (Choctaw); *koshibba'* (Chickasaw); *kosiba* (Alabama); *kosabí* (Koasati); *os'a* [*osa, osá:*] (Creek, Muskogee)
kox ti̲npka (Biloxi)
morelle a gràppes (bunch blackberry, Quebec)
paok (by Swedes in Pennsylvania in 1740s, fide Kalm [1753–1761] 1972; from an Algonquian word; cognate with *pakkan*)
pigeon [poke] berry (Carolinas)
poke salat (although "salat" and "salad," are etymologically the same word, from Latin *salata*, they are not the same preparation; "salat" is made by boiled the young green leaves as a potherb)
[American] poke weed [pokeweed] (in use by 1751, USA); [Virginian] poke [skoke] (from Virginia Powhatan *pakkan*, based on Algonquian cognates meaning "blood"; in use by 1708; see also *Lithospermum*; OED 1971 incorrectly derived it from the Powhatan *uppówoc* or *apooke*, smoke, and cognate with the Narragansett *puck*, smoke)
tabosso (sometimes used for this plant, although also for sumac, *Rhus*, Alabama)
wild spinach (Carolinas)

Each spring when I was growing up, we went farther into the country to gather "poke" for greens. When just the first young leaves and stems came through the ground, we cut off the tops and took them home. After monotonous winter meals mostly lacking vegetables, those "greens" were a treat. We called the meal "poke salat."

Anyone growing up in the eastern United States is likely to have eaten this spring potherb. The dish was shared by the Cherokee, Malecite, Iroquois, and Mohegans (Moerman 1998). Interestingly, the modern Alabama have noticed that "white people eat the young leaves," but they do not (Sylestine et al. 1993). "Poke salat" is a dish of the young *Phytolacca* shoots and/or leaves eaten well cooked and with two changes of water. The potherb, however prepared, is somewhat laxative, and historically constituted an important addition of vitamins and minerals across the South (Morton 1968b, Nellis 1997).

The entire plant is poisonous when raw (Morton 1974, Foster and Duke 1990). Perhaps the first record of that was left by Mark Catesby (1731, 1:24). He wrote of the bird we now call the mourning dove (*Zenaida macroura*), "They feed much on the Berries of Poke, i.e., *Blitum Virginianum*, which are poison." Common names crow-berry and pigeon berry also note that birds feed with impunity on the fruits.

Strained juice of ripe fruits may, however, be used to color food, when carefully separated from the poisonous seeds (Morton 1968b). Several groups used the red-purple fruits as dyes (Moerman 1998), and use in coloring is how the herb got its name.

The association of "poke" with *Phytolacca* dates from the first English colonists sent in the late 1500s by Sir Walter Raleigh to the Virginia coast. There they found Algonquian-speaking people referring to a number of plants as *pakkan*. That word became "puccoon" for *Lachnanthes, Lithospermum*, and *Sanguinaria*. The fourth one, *Phytolacca*, became "poke," although based on the same root. The color is from the betacyan pigment caryophylline (Nellis 1997).

Juice of the berries was also used in coloring wines (Nellis 1997), but Porcher (1863) reported that practice was outlawed in Europe. He wrote, "The French and Portuguese mixed it with their wine, to give it color, and this was prohibited by royal ordinance of Louis XIV, 'on pain of death, as it injured the flavor!'" Juice was also made into ink. Porcher (1863) wrote, "With alum to fix the color, I have used the juice of the pokeberry as a red ink. The directions to the printer for this volume were written with this; before adding alum I found that the red color was fugitive. I consider it, prepared as above, an excellent substitute for carmine ink."

The first records of Florida Glades people using *Phytolacca* were the pre-Columbian pollen samples studied by Hogan (1978). How Glades people used the plant is unknown, but they probably used it in the same ways as others in the United States. Yanovsky (1936), Sylestine et al. (1993), and Moerman (1998) recorded that the Alabama, Cherokee, Delaware, Iroquois, Mahuna, Micmac, Mohegan, Rappahannock, and Seminoles used the plants as medicine. More indigenous groups probably shared the custom.

Sturtevant (1955) found the Seminoles using the berries as an analgesic, especially for rheumatism. The Cherokee and Delaware used the berries exactly like the Seminoles, but the Iroquois and Rappahannock treated that problem with different preparations of the herbs (Hamel and Chiltoskey 1975, Moerman 1998). The Alabama still make a decoction of the leaves as an emetic and formerly used the berries in making a medicinal whiskey (Sylestine et al. 1993). Other tribes and European settlers used various plant parts as emetics, expectorants, cathartics, and poultices on sores (Porcher 1863, Millspaugh 1892, Foster and Duke 1990). Root tea from the herb was still in use to cure "ground itch" in the Panhandle when Murphee (1965) talked with people there.

Ingestion of seeds, leaves, or other parts is especially dangerous because of the toxins, although small quantities are not obviously immediately poisonous for most people (Lewis and Elvin-Lewis 1977, Foster and Duke 1990, Nellis 1997). Among the toxic chemicals are the alkaloid phytolaccin and the triterpene saponins phytolaccagenin and phytolaccatoxin. The latter is similar to cicutoxin of water hemlock (see *Cicuta*). Among the poisons are proteinaceous mitogens that produce blood cell abnormalities when ingested or absorbed through the skin (Foster and Duke 1990, Nellis 1997).

Many other species of *Phytolacca* are used, including the widespread *P. icosandra* (Morton 1981), for which the U.S. species was formerly confused. That plant is called by the Zapotec of Oaxaca, *biaa* or *piaa*, soap plant (Reko 1945). The chemical that led to use as soap also kills snails (Nellis 1997).

Howard (1984) found that older Seminoles in Oklahoma used the plant to predict the next growing season. He was told that, shortly before New Year, a person would go out to the *Phytolacca* patch and clean out the dead growth so that only the roots were left. On New Year's Day, they returned to inspect the roots. If there were new young leaves, it meant a good growing season.

Picramnia

(Named by Olaf Swartz with Greek *pikros*, bitter, *amnion*, the membrane around the fetus, in reference to the bark)

Picramnia pentandra (five anthers)
aguedita (little ague, ultimately derived from Latin [*febris*] *acutus*, referring to malaria, Cuba, Dominican Republic)
bitter-bush [root] (Bahamas, Puerto Rico)
bois montagnes [*moudongue, Madame*] ([lady] mountain bush, Guadeloupe, Martinique)
bois poisson (fish bush, Haiti, Guadeloupe); *bois sardine* (sardine bush, Haiti)
café marrón (wild coffee, Dominican Republic)
fwenn (possibly from *foin*, hay, Haiti)
graines dorées (golden grains, Guadeloupe)
guarema (probably Taino, Puerto Rico)
hueso (bone, Puerto Rico)
ojo [*palo*] *de peje* [*pez*] (fish eye [tree], Dominican Republic)

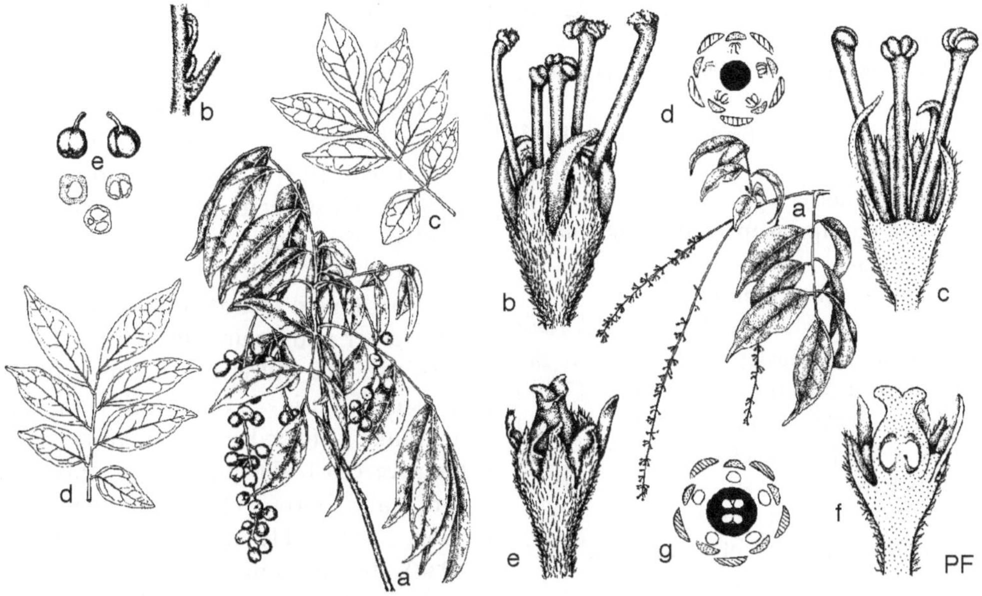

Picramnia pentandra. (Left). a. Branch with pistillate inflorescences. b. Staminate flower, side view. c. Staminate flower, longitudinally dissected. d. Floral diagram of staminate flower. e. Pistillate flower, side view. f. Pistillate flower, longitudinally dissected. g. Floral diagrams of pistillate flower. (Right). a. Branch with fruit. b. Serial buds in leaf axil. c. Leaf with opposite leaflets. d. Leaf with alternate leaflets. e. Fruits, complete and transversely dissected, showing variation in seed numbers. *Drawn by Priscilla Fawcett.* Correll and Correll 1982.

quina [*quinina*] *de la tierra* [*del país*] (wild quinine, Cuba)
snake-root [snake-stick, snake-wood] (Bahamas)
vaillant garçon (strong waiter, Haiti)

Olaf Swartz gave us both the genus and species names for these trees in the late 1700s. He founded the genus on plants from Jamaica, and *P. pentandra* he discovered on Montserrat in the Lesser Antilles. Today the genus contains 45 species, all confined to tropical America (Mabberley 1997).

As suggested by the references to ague and quinine in the common names, this is a venerable remedy for fevers, malarial and otherwise (Roig 1945, Liogier 1974, Ayensu 1981, Morton 1981, Beauvoir et al. 2001). Throughout the Caribbean, the tree is used as a tonic and febrifuge (leaves, roots, bark), and against diarrhea. Roots are boiled with *Chiococca alba* and the decoction used to alleviate gas and menstrual cramps. That tea is used against colds and tuberculosis, and to increase the appetite (Roig 1945, Ayensu 1981). In Hispaniola, the wood is rarely used, but a red dye is extracted from the flowers (Liogier 1974). Haitians consider a decoction made from leaves and bark effective against fever, indigestion, dysentery, intestinal worms, and anorexia (Beauvoir et al. 2001). Caribs on Dominica soak the wood chips to make a bitter drink to alleviate appetite loss. Sap from the bark is put on lesions from yaws (framboesia) (Hodge and Taylor 1957). Among the Warao of Guyana, it is used to "bitter the blood" (Reinders 1993).

Pilea

(John Lindley named this with Latin *pileus*, cap or hat; he thought the sepals of pistillate flowers covering the achene resembled the "felt caps" used by Romans)

Pilea microphylla (small-leaved)
alfombra (carpet, Venezuela)
artillery plant (from the way it "shoots" its seeds into nearby areas, Florida, Jamaica, Puerto Rico, Virgin Islands); *escopetilla* (little shotgun, Venezuela); gunpowder plant (was it really used in manufacturing powder? or is this another allusion to "shooting" its seeds?, England); kanonneer-plant (Dutch Antilles)
baby puzzle (Jamaica, Lesser Antilles); baby's lace [lace plant] (Bahamas, Jamaica, Lesser Antilles, Belize)
beldroega (derived from Arabic *burd(u)lagá*, usually applied now to *Portulaca*, Brazil)
botisuelo (ground cover, Hispaniola)
brilhantina (shiny one, Brazil)
dentelle [*petite dentelles, tidentelles*] (little tooth [teeth], Hispaniola)

doradilla (little golden one, in reference to resemblance to a medicinal fern, Venezuela)
erva gorda (fat herb, Brazil); *folha gorda* (fat leaf, Brazil)
erva mucuim (worm herb, from *mukoo'í*, Tupí, Amazonas, Brazil)
frescursa (unkept, Cuba)
gale of wind (a comparison with *Phyllanthus*, which see, Florida)
golondrina (swallow, Hispaniola)
ha'il tsan (snake water, Huastec, San Luis Potosí)
hierba de la viruela (measles weed); *hierba del tejado* (roof herb, from its growing there, Cuba)
madre perla [*madreperla*] (mother-of-pearl, Puerto Rico); *madre selva* [*madreselva*] (mother of the forest, Puerto Rico)
mañanita (little tomorrow, Cuba)
marposa (probably a variant of *"mariposa,"* butterfly)
palma del norte (northern palm, El Salvador); *palmilla* [*palmita*] (Hispaniola)
panchita (little belly, Hispaniola)
paragüita (little umbrella)
parietaria (wall weed)
[*petit, ti*] *teigne* [*tengn*] [*blanche, plan*] (little [full, white] ringworm, Lesser Antilles)
pitsits wal (scattered eyes, Huastec, San Luis Potosí)
rockweed (one of the most common places it grows is in rocky crevices)
samancito (little useless one, Venezuela)
schaam-teloos
señorita (lady, Cuba)
sereno de invierno (winter serenity, Hispaniola)
smoke plant
tumiin ts'ohool (money plant, Huastec, San Luis Potosí)
urtiga (comparing it to *Urtica*, Brazil)
verdolaguilla (little purslane, Puerto Rico)
wild thyme [wild tim]
yedra [*del tejado, de vidrio*] ([thatching, glass] ivy, Cuba, Puerto Rico)
yomha
zizal-xiu [*siisalxiw*] (*zizal* or *sisal*, *Agave sisalana*, *xiw*, herb, so called because it occurs frequently in the shade at the base of *Agave*, Maya, Yucatán)

Linnaeus called these plants *Parietaria microphylla* in 1759. At the time, European *Parietaria* was one of the few relatives that he knew, and he was inclined to include these tropical plants there rather than create a new genus. It was not until 1821 when John Lindley created the genus *Pilea*, and then in 1851 F.M. Liebmann put the Linnaean species in it to create *Pilea microphylla*. Subsequently, about 200 species, from tropical and warm regions around the world, have been discovered (Mabberley 1997).

In the northeastern United States, the Iroquois and Cherokee have used *P. pumila* as a medicine (Moerman 1998). The more tropical *P. microphylla* has a much wider range and many more people have used it, although no records were found within the United States.

Roig (1945) recorded that Cubans use the plant as a refresher and against liver problems. He also noted local use as a diuretic, often used to treat inflammations of the urinary tract, as it is throughout much of its range (Meléndez 1989, Mors et al. 2001). The Huastec crush the plant and apply it to sores, itchy sores, and burns. They also treat sore eyes, mouth sores, *susto*, and fever in children with measles (Alcorn 1984). Brazilians not only use it as a diuretic, but also as an antithermic; poultices hasten maturation of boils (Mors et al. 2001). Other people use it to expel worms (Asian) and to treat wounds, female problems, inflammation, and tuberculosis.

Piloblephis

(Rafinesque named these Florida near-endemics with Greek *pilos*, hair, and *blepharis*, an eyelid, so a "hairy eyelid," in reference to the sepal pubescence)

Piloblephis rigida*. Drawn by P.N. Honychurch.*

Piloblephis rigida (stiff, refers to the short leaves) (=*Satureja rigida*)
hapo:sikâ:yî [*hapóçigaiyi, hapoosikaayi*] (simple term, Mikasuki); *haposhekaaye-choobe* (big *haposhekaaye*, Mikasuki)
kvfockv [*kafócka*] (simple term, Creek)
pennyroyal [pennyrile, penyryall] (dating from A.D. 500–600, derived from Anglo-French *pulyole ryle*, Old French *puliol real*, and other variants, ultimately from Latin *pulegium real*, royal thyme, or *Mentha pulegium*, Florida, Bahamas)
wild pennyroyal (mentioned by Williams [1837] 1962, Florida)

William Bartram ([1791] 1958) was the first to find these shrubby mints in Florida. However, it was not until 1834 when George Bentham officially created the name *Satureja rigida*, based on a suggestion by Bartram. It was under that name that the plants were discussed for many years, even though Rafinesque had put them in *Piloblephis* in 1838. Only since the 1980s has Rafinesque's name for them been adopted, and this seems to be the only species in the genus.

It was thought that these mints were endemic to Florida until they were found in the Grand Bahama region by the Corrells (1982). Presumably, those Bahamian plants are native and not introduced, but with the reputation "pennyroyal" has as medicine and food seasoning, that is problematical.

Well before Sturtevant (1955), the Creeks were known to have used *kafócka*, particularly in what they call "Cow Creek Sickness." During his work with the Miccosukees on the Big Cypress Reservation, Sturtevant (1955) found them using this mint to treat "Hog Sickness" (periodic unconsciousness with shallow and slow breathing) and fever, and in medicines used at births, deaths, and during the busk (see *Pterocaulon*). An infusion also was used to treat sores and ulcers on the legs and feet. When used to treat these maladies, it was either mixed with several other plants or used alone, depending on the problem.

Alice Snow wrote: "My mother told us if you have a cold you could smell this. It was good smelling, so we would smell it all the time. *Kvfockv* is used in different treatments. Usually one or two sprigs of the leaves are needed" (Snow and Stans 2001). She still uses pennyroyal to treat congestion; boiling it and breathing the steam.

An important application has long been for flavoring food. *Hapo:sika:yi* is documented as having been added to soups (Sturtevant 1955), but probably to other dishes. Currently, the leaves are added to pots of turtle meat (Bennett 1997).

Morton (1968) recommended branches be used to make tea, but she did not mention that fresh is better than dried. Once the branches have dried, they lose some of their fragrance and taste.

Pinckneya

(André Michaux named this genus after General Cotesworth Pinckeney, 1746–1825, appointed as minister to France in 1796, but the revolutionary regime there refused him, cf. Taylor and Norman 2002)

Pinckneya bracteata. From Sargent 1905.

Pinckneya bracteata (having leaf-like structures in the inflorescence)
fever-tree [fevertree] (Florida)
Georgia bark tree (Georgia); bitter bark (Florida)

John and William Bartram found these plants near Fort Barrington in Georgia, and later William named the species *Bignonia bracteata* (Harper 1958). André Michaux did not realize that the plants he found on the St. Mary's River were the same, and published *Pinckneya pubens*. Rafinesque created the name now used in 1827. There are 17 species placed in the genus, with only 1 in the southeastern United States (Rogers 1987, Mabberley 1997).

Michaux originally recorded that the plant was "very useful in intermittent fever" in his *North American Sylva* of 1817–1818 (Porcher 1863, Taylor and Norman 2002). He learned that from either indigenous people or the Americans they had taught. The bitter bark, containing cinchonin, was used during the Civil War as a substitute for quinine in treating malaria (Millspaugh 1892, Mabberley 1997). Porcher (1863) recorded, "Dr. Law, of Georgia, cured six out of seven cases with it. It did not distress the stomach." Later he noted, "In Georgia a handful of the bark is boiled in a quart of water till the liquid is reduced to one-half; the infusion is given."

Pinguicula

(From Latin *pinguiculus*, somewhat fat, referring to the viscid appearance of the upper leaf surface it uses to catch insects)

butterwort (name used from at least the time of John Gerarde's *Herball* of 1597, who wrote, "called Pinguicula, of the fatnes or fulnes of the leafe, or of fatening; . . . it is called Butterwoorts, Butter roote, or white roote")
earning grass ("earning" is a North Country British Isles word for cheese rennet)
erba grassa (fat herb, Italian); *Fettkraut* (fat herb, German); *grasilla* (little fat one, Spanish); *grassette* (little fat one, French); *measgran* (buttery, Gaelic)
la:nihiliswâ (*lane*, yellow, *heleswa*, medicine, Creek)
mòthan (Gaelic)
talakcíhcayikcî (raw medicine, Mikasuki)
tåt-ort (*tåt*, to plug or stop up, *ort*, herb, Swedish); *tettegras* (*tette*, to plug or stop up, *gras*, herb, Norwegian)

Pinguicula lutea (yellow, the flowers)

Pinguicula pumila (dwarf)

Linnaeus had grown up with butterworts, which he knew as *tåt-ort* (Linnaeus [1749] 1979). A few *Pinguicula* are circumboreal in acid bogs, and some grow south into Portugal and Italy. Butterworts were used to curdle milk. They were important medicines and were believed to avert evil in the British Isles (Beith 1995, Vickery 1995). Linnaeus (1753) knew only 4 species, but there are 46 in the Americas and Europe (Mabberley 1997).

Both Florida species have been used to treat "Raw Meat Sickness" (Sturtevant 1955). That malady stems from the belief that undercooked meat not only remains in the body, but "grows," and causes abdominal pain. For mild problems the infusion is drunk cold, but if severe, the medicine is boiled.

Pinus

(Ancient Latin name for pines)

ʔe'ksha (Tunica)
chooye [*co:yî*] (Mikasuki); *choyyi* (Alabama); *coyyí* (Koasati), *cule* [*cúle, colî*] (Creek, Muskogee); *tiak* (Choctaw); *tiyak* [*ittiyak*] (pine tree, Chickasaw)
furu (Norwegian); *Föhre* (German)
Keifer (German); *keper* (Dutch)
nohji (Cherokee)
ocote (torch, from *ocotl*, Náhuatl)
pin (from Latin *pinus*, French); *pinho* (Portuguese); *pino* (Italian, Spanish)
wazi (Dakota)

Pinus clausa (closed, meaning the partly serotinous cones)

Pinus. *Pinus clausa* (left). *Pinus elliottii* (right). *Both drawn by P.N. Honychurch.*

co:yihiskopóskî [*tcoihiçkopoçki*] (Mikasuki; cognate to Creek *cule*, pine, *esse*, leaf, *lopockuce*, little)
pitch pine (called "*P. rigida*," which is not in Florida; the trees were "low, poor timbered tree, but produces turpentine and tar," which seems best to describe this species, cf. Williams [1837] 1962)
sand [scrub] pine

Pinus echinata (spiny)
ácuwe [*etcuwe'*, *ícuwe*] (Catawba)
choyyihissi hobaski (*choyyi*, pine, *hissi*, hair [needle], *hobaski*, short, Alabama); *coyyí acakí* (*coyyí*, pine, *acakí*, dwarf, Koasati)
[long-tag, shortleaf, short-leaf, yellow] pine
taklaha (*tiak*, pine, *lakna*, yellow, Choctaw)
teetpa hitcuwe' (chewed pine, the resin, Catawba; also applied to *P. palustris*)

Pinus elliottii (commemorates Stephen Elliott, 1771–1830, professor in Charleston who published a flora of South Carolina and Georgia in the 1820s)
chooye enlepaatooche, chooye ebeele (pine branch tips, Mikasuki)
chooye [*co:yí*] (Mikasuki); *cule* [*colí, chuli*] (Creek); *tcuyi* (Alabama)
cule emohlowakuce (*cule*, pine, *em*, its, *elvcce*, branch, Creek)
este lopockuce or *este-lopocke emeto* ("where the little people live"; from *este*, person, *lopockuce*, small, *em*, its, *eto*, tree, Creek)
nat'tsi (Cherokee)
slash pine (Florida)
yaat-hoboske emahe (where the little people live, Mikasuki)

Pinus glabra (smooth)
spruce pine (Florida)

Pinus palustris (of swamps, a misnomer)
atcuwe nu're (pine fat [rich], Catawba); *ansudi'* [*ansûdi'*] (generic for pine or *P. palustris*, Biloxi); *ansudi' nitan'xti* (Biloxi)
choyyihissi hobaski (*choyyi*, pine, *hissi*, hair [needle], *hobaski*, long, Alabama)
choyyinaani (*choyyi*, pine, *naani*, male, Alabama); *coyyí ná:ni* (*coyyí*, pine, *ná:ni*, male, Koasati)
longleaf pine (southeastern United States)
ne'c mañ ne'c (*ne'c*, tree, *mañ*, long, *ne'c*, tree, Atakapa)
tiak fanya (*tiak*, pine, *fani*, squirrel, Choctaw); *tiak hobak* (*tiak*, pine, *holba*, resembling, Choctaw)
yellow pine (once the "principal timber, used for plank and scantling in the southern states; and also produces turpentine and tar," Williams [1837] 1962; is this a translation of the Koasati *coyyilá:na, coyyí*, pine, *lá:na*, yellow?)

Pinus serotina (late, alluding to the cones typically not opening until burned)
many cored pine (Williams [1837] 1962)
[marsh, pond] pine

Pinus taeda (Latin for torch; ancient name for a resinous pine)
choyyihoba (*choyyi*, pine, *hoba*, castrated, Alabama)
coyyí ná:ni (*coyyí*, pine, *ná:ni*, male, Koasati)
loblolly pine ("has much sap," Williams [1837] 1962)
oldfield pine (the plants quickly re-seed and sprout in old agricultural areas)

Pines are among the most widely used of plants throughout their range (Standley 1920–1926, Moerman 1998). Many pines have been employed in North American medicines, e.g., *Pinus palustris* and *P. rigida* have stimulant, laxative, diuretic, pectoral, vermifuge, discutient, antiherpetic, detergent, baslasic, and vulnerary properties (Vogel 1970). Pine tar was in the U.S. Pharmacopoeia between 1820 and 1950 and has been in the National Formulary from 1950 to the present; it is considered an antibacterial, irritant, parasiticide, and expectorant.

The earliest records of pines being used in Florida were among the Glades people (Gilliland 1975, Hogan 1978, Austin 1980). Wood is the best-known product, although there are numerous other uses by people (Hamel and Chiltoskey 1975, Austin and Smith 1997). The inner bark and roasted seeds were eaten as famine food. Needles were cooked or eaten raw. Resin was an effective salve alone or as a medicine base, and teas from leaves were expectorants, antidiarrheics, and vermifuges. Lighter pine (the resinous heartwood) has long been used as fuel.

The Creeks used the wood for beams, ball poles, houses, torches, bighouse seats, baskets, and religious scarification (Swanton 1928a). The Seminoles continued those practices (Sturtevant 1955). In addition, the Seminoles formerly used the resin as arrow point glue and extracts of bark for tanning. When Sturtevant (1955) worked with the Miccosukee, they were using pines for "Ballgame Sickness" (caused by getting hit by a ball during the traditional single-pole game) and to treat rheumatism.

Snow and Stans (2001) note that the tips of branches are the best medicine. The Choctaw used them similarly (Taylor 1940). Four pieces, 4 inches long, are taken, crushed, and put in water with *vtakroluste* (dog fennel, *Eupatorium capillifolium*) and boiled. [Note: The multiple of four, an allusion to the cardinal points, is a reverential request for the aid of the spirits.] Sometimes the bark is used as medicine. The liquid is used to clean sores; for women's medicine, a forked stem is used. Snow uses that branch in a medicine to prevent miscarriage.

Piscidia: Fish-Poison

(From *piscis,* fish, and *caedo,* to kill or poison, Latin)

Piscidia piscipula. From Sargent 1905.

I vividly recall the first time I ever saw a fish-poison tree (*Piscidia piscipula*) in flower in the spring of 1970. Not only was the tree impressive for the multitude of white, pink, and yellow legume blossoms, but it was literally alive with insects and birds. Standing in the Florida Keys on the edge of a road segment that had been abandoned in 1935 after the Labor Day Hurricane, I was able to distinguish at least ten kinds of bees. Moreover, I watched as birds from four families sipped nectar. There were ruby-throated hummingbirds (Trochilidae), several species of warblers (Emberizidae), spot-breasted orioles (Icteridae), and doves (Columbidae).

Europeans learned of these trees when British physician Hans Sloane returned from a tenure in Jamaica that began in 1687. In his book on the plants of that island, Sloane wrote *Coral arbor polyphylla non spinosa, fraxinis folia, siliqua, alis foliaceis exstantibus, rotae molendinariae fluviatilis, vel seminum laserpitii instar aucta* (Unarmed red tree with many leaves, leaves resembling ash, fruit pods with winged leaves standing out like windmill blades along rivers, or having seeds resembling carrots). That may not sound much like the plants we now call *Piscidia piscipula,* but the specimen in the herbarium at the British Museum of Natural History in London confirms the identity. Furthermore, Sloane wrote, "The Indians and negros make use of this bark to take fish, especially in deep holes in inland rivers." Later, 57 years after Sloane's book was published, Linnaeus used its description and plate as the basis of the name *Erythrina piscipula* (little fish). Linnaeus was fooled by Sloane's reference to the flowers as "red," and he put the plants in the wrong genus.

Patrick Browne, however, knew the trees were distinctive and proposed the name *Ichthyomethia* (fish-wine, Greek) in 1756. Much later, Browne's name was rejected by international agreement and replaced by *Piscidia.* Charles Sprague Sargent gave us the modern name in 1891 when he made the combination *Piscidia piscipula,* and Rudd (1969) reinforced that placement. Unfortunately, some modern publications have not reached the 19th-century level; they retain the old synonym *Piscidia erythrina.* The incorrect name appears in many of the chemical records (Heller and Tamm 1975, Pietta and Zio 1983, Delle Monache et al. 1984, Redaelli and Santaniello 1984, Labbiento et al., 1986, Della Loggia et al. 1988, Tahara et al. 1991, 1992, 1993a,b, Moriyama et al. 1992, 1993). Surveyors of the literature must be aware of this and similar nomenclatural complexities that may cause confusion.

Many of the common names, extending from the time of Hans Sloane in the late 1600s until today, allude to use of these trees to catch fish. To do that the bark and leaves are mashed and put in water. Some people simply say *Piscidia* is *barbasco* (fish poison), a Spanish word applied to a variety of plants, ranging from other trees in the Fabaceae to herbs in the Asteraceae. Others say *Piscidia* is *barbasco amarillo* (yellow fish poison, Colombia, Venezuela), alluding to the wood color to distinguish it from others called "*barbasco.*" Wood in *P. piscipula* is heavy, hard, close-grained, and clear yellow-brown. It is more certain that the Huastec alternate name *k'anaw te'* (yellow maize tree, San Luis Potosí) is an allusion to the wood. There the wood is preferred for house posts; the heartwood is used to cure a variety of ailments, to divert evil eye in adults and even unborn children, and to ensure a long life (Alcorn 1984). There is a popular Mexican belief that the wood petrifies after many years, beginning with the heart. Perhaps that is why the heartwood is considered so potent in medicines. Other people note that the wood takes a fine polish and is durable in water, and therefore is used for boat building and piles of wharves. It has also been used as fuel, to make charcoal, and for cartwheels.

In the Lesser Antilles *Piscidia* is known as the *morta* [*à*] *poissons* (fish killer). Elsewhere it is the *colorin de peces* (red [flower] for fish, Guerrero, Michoacán), fish-fuddle (USA, Jamaica), fishpoison (Florida, Bahamas, Puerto Rico, Panama), or *mata-pez* [*mata-pescado, mata-peces*] (fish killer, Mexico, Colombia).

People used chemicals in the plants to stupefy fish and themselves. When a sedative was needed, extracts from *Piscidia* served the purpose, thus gaining it the names *borracho* (drunkard, Venezuela), *palo emborrachador* (intoxicating tree, Venezuela), and *bois* [*à*] *énivrer* [*énvirage*] (intoxicating tree, Martinique, Guadeloupe). The dry bark, especially from roots, has a strong odor of opium and tastes unpleasant (Standley 1920–1926). This bark produces burning in the mouth, and both odor and taste have resulted in the names *guamá* [*guaná*] *hedondida* (stinking *guamá*, a Taino word for several legumes, Cuba), and stinkwood [stink tree] (Virgin Islands, St. Eustatius, St. Martin). Mexicans said *cocuile* [*cocuite*] (*cococ*, pungent and *huitl*, herb, Náhuatl, Mexico) because of the taste. Maybe the name *cachimbo* (pipe, Costa Rica) is related. The names *palo de zope* [*palo-zope*] (vulture tree, Guatemala), and *zopilocuave* (*zopilote*, vulture, from *zopilotl* or *tzopilotl*, and *cuave* from *uaxin* or *huaxin*, Náhuatl, El Salvador) more clearly refer to the odor and medical uses. The Aztec word *uaxin* [*guaje, huaje, huaxin, guaxi*] refers to both *Leucana esculenta* (Fabaceae) and *Crescentia cujete* (Bignoniaceae).

Other people simply view the trees, perhaps as I did when they were in flower, as a *candelón* (large candle, Cuba), *guamá candelón* (candle *guamá*, Cuba), and *guamá* [*guaná*] *de costa* (coastal *guamá*, Cuba). Those who say *flor de papagallo* (parrot's flower, Mexico) were surely noting the presence of those nectar-loving avians.

Alternatively, people may compare these trees with others. People call *Piscidia* the black mahoe (Trinidad, Tobago), dog (Jamaica), dogwood (Belize, St. Eustatius, St. Martin, Jamaica, Puerto Rico), black dogwood (Jamaica), Jamaica dogwood (Florida, Bahamas), *Jamaika hundsholz* (Jamaica dogwood, Dutch Antilles), or *frijolillo* (little bean, Mexico). Surely the name *chijol* (*chi-xo-lli*, swollen pod, Náhuatl, Tamaulipas, Veracruz) also refers to the legume fruits. Apparently, this Aztec name occurs in the Veracruz dialect of Huastec as *chiihol*, and among San Luis Potosí speakers as *ts'ijol* [*tzijol*], although Alcorn (1984) did not comment on the relationship.

Some of the names are either simple or obscure. These include *chaperno* (Guatemala), *haabi* [*habi, haabim, haabin, habim, jabin, jabine, jamcui, jamguij, javin, jebe, jubi*] (*ha*, water, *bin*, imperfect future form, or "that which will be green after the rains," Maya, Yucatán), may bush (Belize), *cahuirica* [*cahuinga, cahuiricua*] (Michoacán), *cuchiván* (Venezuela), *tatzungo* [*taizungo, zatzumbo*] (Tarascan, Michoacán), *tiazab* [*tiaxib*] (Guatemala, Belize), and *tuncuy* (Mexico).

Several names clearly allude to medicines. Use against inflamed bowels is inherent in *colango naranja* (orange *colango* [from *col(e)*], derived from Greek *cholé*, bile+*angi(o)*, derived from Greek *aggeîon*, vessel, Brazil). That same sense is inherent in *ventura* [*ventiera*] (for the belly, Puerto Rico). Since the chemicals are toxic, it is easy to see why the names *bois de chiens* (dog bush, French Antilles), *matapoijo* (louse killer), *palo de gusano* (worm tree, Belize), and worm-wood [wormwood] (Belize) are used.

There are seven species in *Piscidia* distributed throughout Central America, the West Indies, and northern South America. There are differing opinions on how these species should be circumscribed (Rudd 1969, Liogier 1974, Liogier and Martorell 1982). As now delimited, *P. piscipula* grows in southern Florida, the Bahamas, through the West Indies to Trinidad and Tobago, southern Mexico (Veracruz, Chiapas, Campeche, Yucatán, Quintana Roo), and through Central America to Panama and perhaps Colombia and Venezuela.

The bark contains compounds that are narcotic and analgesic. It has been used as an anodyne in neuralgia, nervous insomnia, and whooping cough (Standley 1920–1926, Goodding et al. 1965, Duke 1972). In addition, the bark extract has been used against toothache. Other uses include as a uterine depressant with low toxicity, a hypnotic, antispasmodic, diuretic, diaphoretic, and expectorant. In addition to applications to humans, it has been used to cure mange in dogs and as an insecticide.

Plants contain isoflavones (erythbigenin, erythgenin, jamaicin, junipegenin A, lisetin, piscigenin), saponins, piscidiasaponin, resin, rotenone and rotenoids, glycosides, piscidic acid, and the Neutral Bitter Principle piscidin, which is actually two compounds (Heller and Tamm 1975, Pietta and Zio 1983, Redaelli and Santaniello 1984, Labbiento et al. 1986, Della Loggia et al. 1988, Tahara et al. 1991, 1992, 1993a,b, Moriyama et al. 1992, 1993, Hocking 1997). Laboratory studies indicate that the extracts are antifungal (Cáceres et al. 1991), antiviral against poliovirus II, herpes, influenza, and vaccinia (May and Willuhn 1978), a uterine relaxant (Pilcher 1916, Butler and Mullen 1955), and a molluscicide (Domínguez and Alcorn 1985). Previous use as a sedative or analgesic also is supported by experiments (Della Loggia et al.

1981, Pietta and Zio 1983). As poisonous as many of these compounds are, there are still insects that have been able to detoxify and consume them in *Piscidia* and the notoriously poisonous related genus *Lonchocarpus* (Cervantes 1999).

In a testament to Huastec attention to environmental interactions, they use the trees as predictors of the future. When the flowers dry up without setting fruit, farmers know that their *milpas* (corn fields) will fail that year (Alcorn 1984).

Pisonia: Devil's Claw

(Commemorates the Dutch physician and pharmacist Willem [Wilhelm] Piso [Guillaume Le Pois], ca. 1611–1678)

My friend Barbara Hiaasen, who works at a Fort Lauderdale animal rescue center, once brought in a small plastic bag full of bird feathers. She said that a cedar waxwing (*Bombycilla cedrorum*) had been rescued with its feathers so full of these fruits that it could not fly. I pulled the mass of breast feathers and sticky fruits from the bag and examined them with a hand lens. They were devil's claw (*Pisonia aculeata*) fruits. It has long been known that the fruits of this scrambling bush endanger birds and Paul C. Standley, in his book *Trees and Shrubs of Mexico*, published in the 1920s, said that the "fruits stick to bird feathers so much that they cannot fly." Standley was a focused botanist, and for him to comment on birds getting stuck required unusual circumstances. That trait is not confined to this species; in Malaysia there is another called the "bird-catching tree" (Dalziel 1937).

These armed climbers have been known to Europeans since at least the late 1690s, when they were mentioned by Leonard Plukenet (1696). Indeed, he provided the first common name for the plants when he listed them as being called *fringego* (Jamaica). That name remains today in the Bahamas as *fingrigo*, although no one seems to know its meaning or origin. Given the armed nature of the plants I cannot help wondering if that name is a version of *fin de gringo* (end of whites). The name more likely came from *fingerer* (thief), a word more commonly used in English in the 1500s and 1600s. Anyone who has ever been caught by these plants will understand that the claws on the stems steal bits of flesh, clothes, and other items from their victim.

Pisonia aculeata is known in southern Florida, the Bahamas, the West Indies, Mexico (Campeche, Chiapas, Oaxaca, Yucatán, Quintana Roo), Central America and South America through coastal Brazil to Paraguay and Bolivia. The same species occurs in Africa, Asia, and Australia, and that is the part of the range where it is perhaps not native. Morton (1981) attributed the devil's claw Asian and African range to human introduction, but that is uncertain.

Most of the common names for these nasty lianas refer to its armament. The names pull-and-haul-back, hold-back, pull-back, and pullback (Florida, Bahamas) date to the 1800s or earlier in Florida. Other names in English that note the curved, paired thorns on the stems are black thorn (Barbados), cat's claw (Florida, British Antilles), cock spur (Bahamas), devilsclaws (Florida), prickly mampoo (Puerto Rico, Virgin Islands), tiger nail embra (Belize), and wait-a-bit (Florida, Bahamas).

In Spanish, the plant becomes a *bueno amigo* (good friend, Colombia), *cruceta espinuda* (little spiny cross, Oaxaca), *cruz espino* (spiny cross, Chiapas), *goma de uña* [*coma de uña*] (sticky claw, Tamaulipas), *uña de diablo* (devil's claw, Michoacán, Guerrero), *uña de gato* (cat claw, Tamaulipas, Michoacán, Guerrero, Cuba, Dominican Republic, Puerto Rico), *uña de tigre* (jaguar claw), or simply *zarza* (spiny like *Rubus*, Cuba). To some people, these plants are a *caballero* (gentleman, El Salvador), probably derived from *cagalero* [*cargalera*] or *cagalero negro* (black *cagalero*). The word *cagalero* must be from an indigenous language because the similarly thorny *Celtis pallida* is *cagalero comestible* (edible *cagalero*, Nicaragua). Another comparison with that hackberry is *garabato* (iron hook, Durango, Sinaloa), and *garabato prieto* (black iron hook, Michoacán, Guerrero). They are also *tutum prieto* (black *tutum*, San Luis Potosí), maybe a comparison to *tutumushte* (El Salvador) or *tutumuste* (Guatemala), both references to tree *Ipomoea* (Standley 1922–1926).

French speakers say *P. aculeata* is an *amourette* (little lover, Guadeloupe, Martinique) or *croc à chien* [*croc de chien*] (dog's tooth, Haiti, Guadeloupe, Martinique). One of the names in Brazil suggests that it is the *espora de gallo* (rooster spur).

Indigenous languages convey the same ideas with *beeb* [*hbeeb*] (vine with spines, an interjection of surprise or pain upon encountering the spines, Maya, Yucatán), *huitzcocólotl* [*huiscalote, caltute*] (spiny, from *uitzcolotl*, Náhuatl, San Luis Potosí, Puebla; also applied to *Caesalpinia bonduc* in Cuba), *itsik' mitsu'* (cat claw, Huastec, San Luis Potosí), and *mitz-matlayahuale* (*mitz*, you, *matlaçautl*, pest, Náhuatl, San Luis Potosí). The last Náhuatl name apparently was taken to the Dominican Republic by the early Spanish where it became *mayacaule.* There are also simple and untranslated names like *istiján-uaiya* (Totonac, Veracruz) and *loh* [*loj, lo*] (simple term, Huastec, San Luis Potosí).

My personal favorite among all these names is blaspheme-vine (Florida). Some of my students have provided their own invectives after getting inextricably caught by these plants.

Other people see the plants differently. Although I find nothing in the botanical literature about it having tuberous roots, it is called *espino y camote* (spine and tuber, Oaxaca, Chiapas) and *guechi-gu* (*guechi*, spine, *gu*, potato, Zapotec, Oaxaca). Those people were using those roots as laxatives (Standley 1920–1926), the universal cure-all.

Individual plants may climb for tens of meters over the surface of others or into nearby trees. Where they get so large, the scramblers form tangles of barbed stems that led Watt ([1889] 1972) to comment that they form "most excellent impenetrable fences." Indigenous people also note those tendencies and call them *tapaciriba* (clothing the countryside, probably from *tapeçaria*, Brazil), *tampamolón* [*tampamaloh*] (place covered with *loh*, Teenek, San Luis Potosí), and *yaguapendá* (from Tupí, *ya'wara*, ground, *pi'nima*, painted, Argentina).

Another name in Brazil is *cipó mole* (soft vine), presumably an allusion to the soft wood. The related *P. excelsa* of the Andaman Islands in Asia has such soft wood that elephants have been known to eat the branches with zest (Burkill 1966).

Devil's claw belongs to the family Nyctaginaceae, including more familiar plants like cultivated *Bougainvillea*, four-o-clock (*Mirabilis*), and the devil's bouquet or scarlet musk-flower (*Nyctaginea capitata*). Many members of that family have nocturnal blossoms. For its night-time flowers *P. aculeata* is called *huele de noche* (fragrant at night, Oaxaca, Guatemala).

Linnaeus gave us the genus *Pisonia* in 1753. The basis for that genus is *Pisonia aculeata*, called the "type species," or the permanent reference by which other species may be judged as related or not. In coining the name, Linnaeus honored Willem Piso who served the Dutch settlement in Bahia, Brazil with Georg Marggraf [Georgius Marcgravius]. Together they wrote one of the earliest books on medicinal plants of Brazil, the *Historia naturalis Brasiliae: De Medicina Brasiliensi* (book 4).

People throughout the Americas have used *P. aculeata* in a variety of ways. The branches formerly were made into barrel hoops in Jamaica (Standley 1920–1926). In Mexico, the plants are burned in *milpa* (corn field) preparation, the ash serving as a fertilizer (Alcorn 1984). Mostly, however, the plants have the reputation as medicines. Throughout its range, the root of *P. aculeata* is considered purgative, and decoctions or infusions of bark and leaves are used externally or internally against rheumatism, joint inflammation, and venereal disease (Standley 1922–1926, Roig 1945, Martínez 1969, Liogier 1974, Grimé 1976, Mors 2001). The Mayas in Yucatán use it for urinary problems and to treat itch or mange (Martínez 1969, Morton 1981). The Huastecs in San Luis Potosí and Maya in Yucatán use the plants to treat boils and the ash mixed with lime and *nixtamal* water for other dermatological problems (Martínez 1969, Alcorn 1984). *Nixtamal* is the preparation of maize (*Zea mays*) kernels treated with the alkali in ashes to remove the outer coating prior to making the cornmeal for tortillas. Liquid from the raw stems and leaves is drunk to reduce fever.

No laboratory studies of *P. aculeata* have been found, but its Puerto Rican relative *P. borinquena* has shown 50% to 100% parasite suppression at 5 μg/ml of resistant malaria (*Plasmodium falciparum*), and 85% or more inhibition of microbial growth at 100 μg/ml of tuberculosis bacteria (*Mycobacterium tuberculosis*) (Antoun et al. 2001). Another species, *Pisonia umbellifera*, contains six saponins, three of which are oleanolic acid saponins, and two of which have an unusual seco-glycopyranosyl moiety (Lavaud et al. 1996). Since other members of the family are known for diuretic and wound-healing activity, perhaps similar chemicals are in devil's claw (Lewis and Elvin-Lewis 1977).

Surely, the most famous medical relative of devil's claw is the four-o-clock (*Mirabilis jalapa*). That species contains a number of alkaloids such as trigonelline, saponins that make it soapy, and probably resins that make it a drastic purgative (Morton 1981). The plants are considered poisonous by most, and are now grown only for ornament. In Florida, however, the best-known relative of devil's claw is the *Bougainvillea*, cultivated in many yards and anywhere a beautiful, but armed, barrier is desirable. Anyone who has trimmed a *Bougainvillea* plant will relate to the common name of *P. aculeata*, "blaspheme vine."

Pithecellobium: Black Bead

(From Greek *pithekos*, an ape or monkey, and *ellobion*, ear-ring, in reference to the unique coiled fruits)

British leader Oliver Cromwell's fleet took Jamaica from the Spanish in 1655, and 32 years later, in December 1687, Hans Sloane arrived on the island. Sloane, who was to become Britain's most prominent botanist of his century, lived and worked on Jamaica until his return to England in 1689. His herbarium specimen of plants we now call *Pithecellobium unguis-cati* provides documentation for the first common

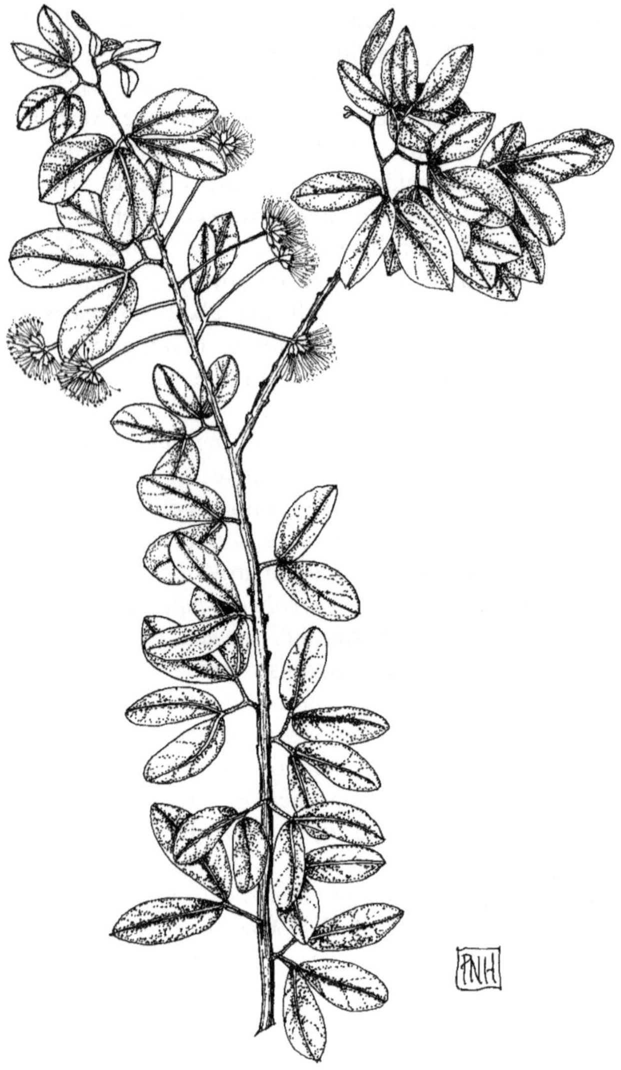

Pithecellobium keyense*. Drawn by P.N. Honychurch.*

name—a necklace of its black seeds. The name "black bead" was already in use in the 1680s.

Both Leonard Plukenet (1642–1706) and his sometime rival Sloane simultaneously introduced black bead to England in 1696, although they called the tree an *Acacia* in their books. It was not until 1837 that Karl Friedrich Philipp von Martius (1794–1868), German botanist and professor in Munich, created the genus *Pithecellobium.* The "ape" (*pithekos*) he saw eating the fruits was a New World monkey, probably a howler (*Alouatta* spp.).

Some names suggest other fanciful items. To people in Guadeloupe and Martinique the trees are *acacia à bracelets* (bracelet acacia) or *collier* [*à*] *diable* (devil's necklace). Other people note the leaves with *acacia à quatre feuilles* (four-leaf acacia, French Antilles), or the bark with *vainita de iguana* (iguana bark, Venezuela). People even think the curved fruits resemble horns and call the plant goatbush (Virgin Islands).

Several designations note the black seeds with their contrasting red arils. Among those names in English are beefsteak (Grenadines), black bead (Florida and Bahamas to Panama), black Jessie (Trinidad), bread-and-cheese (Virgin Islands, Barbados, Trinidad), mangrove beadtree (Barbados), and mangrove black bead (Barbados). In Spanish, they are *coralillo* (little red one), more of a reference to the aril than the seed. The Maya name *t'siu-che'* [*t'sin-che', tsim-che', tzin-che', otsuiché*] (*t'siu,* red-eyed bird, *che'*, tree, Yucatán), notes a bird of the region (*Molothrus aeneus* or red-eyed cowbird), whose eyes resemble the red in the fruits. In Belize, it is *xa-coy* (*xa,* a word to introduce a question, *coy,* tranquil or serene, Maya).

One French name in the Lesser Antilles is a curiosity. It is either *tendre à caillou* (tender [thing] with pebbles) or *tendre à caillou-rivière* (tender [thing] with pebbles that grows by the river (Dominique, Martinique). Perhaps that is a reference to the edible aril. A German name is *Antillen kieselholz* (Antillean pebble-tree), but its allusion too is unclear, as the "generic" name for *Pithecellobium* is *Spiralhülse* (spiral bean) (Nikolov 1996).

A number of common names have other allusions. *Aromo* (perfume, Puerto Rico) presumably comments on the odor of the flowers, although I have never noticed much fragrance. *Cinazo* (ash-colored, Dominican Republic) could be another reference to the pinkish flowers. In Costa Rica, *P. saman* is known as *cenizero* (ashy one), so the names are related (Janzen 1982). Some think the trees look like privet (Jamaica), the temperate *Ligustrum* (Oleaceae). People in northwestern Mexico say *guamuchililllo* (Sonora) because it is a small version of *guamúchil* (*Pithecellobium dulce*).

There are also several names apparently in indigenous languages that have not all been identified or translated. These names include *dinde* (Colombia), *güichere* (Venezuela), *orore* (Cumanagota, Cariban, Venezuela), *pidua* (Chocó), and *yacure* (Arawak?, Venezuela). In Barbados, the tree is moabite. However, any relationship between the plant and the people of Moab, a region that bordered on the territories of the trans-Jordanic Israelites in the 1380s or earlier, seems unlikely.

By far the most common allusions are to the plant's armament. The best-known English names mentioning the thorns are cat's claw [catclaw] (Florida, Puerto Rico), catclaw blackbead (Florida), crab prickle (Virgin Islands), and crab wood (Dutch Antilles). Other languages render these *uña de gato* (cat claw, Cuba, Dominican Republic, Puerto Rico, Nicaragua, Panama), *uña de gatu* (cat claw, Dutch

Antilles), *griffe-chatte* (cat's claw, Dominique, Martinique), or *bois crabbe* (crabwood, French Antilles).

People have some wonderfully obtuse ways of cursing these plants. They may be the *arranca pellejo* (skin puller, Venezuela), the *beshi di Juana* (Jane's kiss, Dutch Antilles), *bois traînant* (dragging tree, Dominique, Martinique), *manca montero* (wild impediment, Cuba), *rolón* (enveloper, Puerto Rico), or the *rolón escambrón* (scaly enveloper, Puerto Rico). To others, they are *avaramo* (Brazil) because they resemble the *avará* or *Arecastrum* palm, both notably spiny. They might be *escambrón colorado* (big red scale), *espino de mar* (sea spine), *espino de playa* (beach spine, Nicaragua), or the *espinuelo* (spiny one, Venezuela). Nevertheless, everyone's view after being caught in the claws and branches of this plant is that it is the *diaballe* (devil, Martinique).

Linnaeus ([1753] 1957) captured the bristly nature of the tree when he called it *Mimosa unguis-cati* (cat's claw mimosa). He knew the plants from the botanical garden in Uppsala, Sweden, and from Sloane's Jamaican herbarium specimen. However, the species ranges from southern Florida, through the Bahamas, the West Indies, and Mexico (Tamaulipas and Sinaloa) south through Central America to Bonaire and Curaçao, Venezuela, and Guyana. Throughout its range, everyone has used the plants. The wood is clear brown, hard, and heavy, and has been used by all who know it, especially in construction (Morton 1981). The fruit yields a yellow dye.

The Spaniards apparently thought that the black seeds resembled the kidneys and the immature white aril the fat surrounding them. Therefore, by the Doctrine of Signatures, they concluded that the plant was good to treat kidney problems. Some think the Spaniards started using the bark and fruits to treat kidney problems, but people who preceded them in the Americas had used the plants long before Europeans arrived. Since the bark and fruit are rich in tannin and astringent, they had been used to treat bronchitis, chronic diarrhea, and hemorrhaging for generations (Standley 1920–1926, Roig 1945, Mors et al. 2001). The bark was also considered tonic and diuretic and was used to treat fever, and to cure chronic sores. During early European history the plant in Jamaica was considered the "sovereign medicine for the stone and gravel," and also for the liver and spleen (Standley 1920–1926). In Curaçao the leaf decoction is taken to relieve colds (Morton 1981).

The related species, *P. keyense* (of the Keys), is similarly rich in tannins, and is used to stop bleeding. In the Bahama Islands, leafy twigs are chewed or made into a "tea" to stop bleeding 3 or 4 months into pregnancy (Morton 1981). That species, easily confused with *P. unguis-cati*, is also known as black bead (Florida, Bahamas), ram's horn (Bahamas), or aroma (Bahamas).

People consider the aril edible, but some are sweet and others are astringent. Perhaps if they are eaten while still white they would be tastier, but those I have tried were red, in fully open fruits, and tasted too much like alum. Seeds are still strung into necklaces and rosaries as they were in Sloane's time.

The most famous member of the genus is the American *P. dulce*. That species is most often known under some variation of the Náhuatl name *quauhmochitl* (snake jaws), including *huamúchil, cuamúchil, guamúchil, quamochitl, guamuche, camanchil, camonsil,* and *camachile*. While the plant was first recorded by Hernández in 1651, the connection between the name and the plants carried around the world by the Spanish remained unknown for centuries. Indeed, the vagaries of where a plant is first collected and reported as new to science are unpredictable. *Pithecellobium dulce* was described in 1795 from the Coast of Coromandel in India, and for decades people thought it was native to the Old World. Then it was discovered that it had been taken there by the Spanish, first to the Philippines, and then to mainland Asia (Standley 1920–1926).

Since no chemical studies have been found for *P. unguis-cati* or *P. keyense*, the best that can be done is to compare it with congeners. *Pithecellobium dulce* contains tannins (Steynberg and Hemingway 1994), triterpene glycosides (Yoshikawa et al. 1997, Nigam et al. 1997), triterpene saponins (Sahu and Mahato 1994), and vernolic, malvalic, and sterculic acids (Hosamani 1995). An Indonesian species contains flavan-3-ol gallates and proanthocyanidins (Lee et al. 1992). Triterpene saponins are anti-inflammatory (Sahu and Mahato 1994), triterpenoid glycosides are antifungal (Khan et al. 1997), and other compounds are antimicrobial (Ali et al. 2001). Related species *P. saman* and *P. mangense* yield gum exudates (Leon De Pinto et al. 1995), and *P. saman* contains saponins similar to those in *P. dulce* (Varshney et al. 1985).

Pithecellobium flexicaule produces seeds that are consumed in northeastern Mexico (Alanis Guzman et al. 1998) and that have been used in afforestation in Texas (Vora and Labus 1988). Related species have uses like *P. unguis-cati* and maybe similar chemicals.

Sometimes Florida common names are specific: *P. keyense* is blackbead and *P. unguis-cati* is catclaw. Nowhere else in their range are they distinguished, and botanists did not even separate the two until *P. keyense* was described in 1928. However, a small gall-wasp (*Tanaostigmodes pithecellobiae*) uses only *P. keyense* and ignores *P. unguis-cati* (Weekley

2000). Once again, insects are better biologists than humans.

Pityopsis

(Thomas Nuttall named this Greek *pitys*, pine, and *opsis*, resembling)

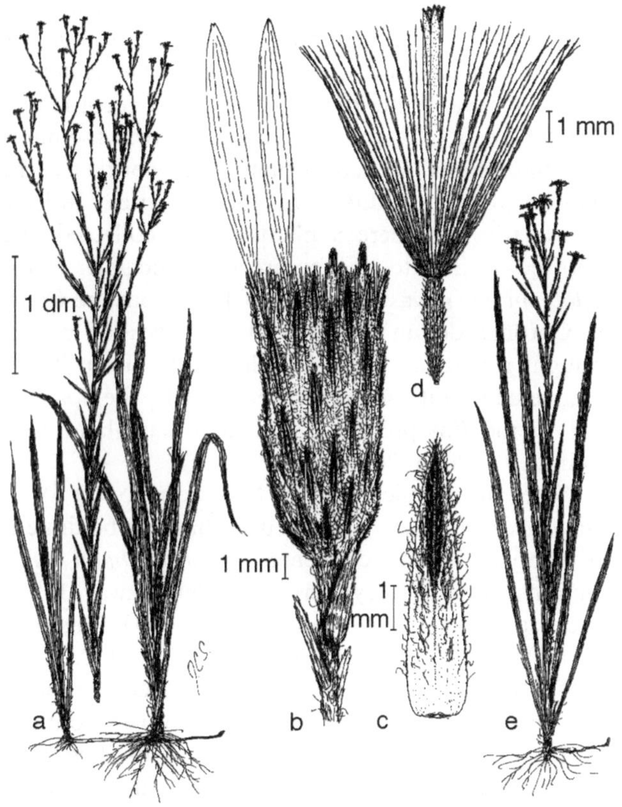

Pityopsis graminifolia. Polyploid varieties. a. *Pityopsis* var. *latifolia*; habit of a robust plant. b to e. *Pityopsis* var. *tracyi*. b. Habit. c. Head; only some florets drawn; the heads of var. *latifolia* are smaller. d. Mid-series phyllary; chlorophyllous zone dark. e. Mature cypsela with disk floret attached. *Drawn by John C. Semple*. Modified from Semple and Bowers 1985.

Pityopsis graminifolia (with leaves like grass)
fever grass (Florida Panhandle)
golden aster (Florida)
gopher grass (Florida Panhandle)
[silk, silver]-grass (reference to the white pubescence on the stems and leaves)
nashoba impisa (*nashoba*, wolf, *im*, its, *pisa*, eyes, Choctaw)
paháɬlo:cí (white-black grass, Mikasuki)
pahe hatkooche (small white grass, Mikasuki)
paho:catlî (Mikasuki)
pvhe hvtkuce [*pvhke hvtke*] (*pvhe*, grass, *hvtke*, white, *oce*, small, Creek)
solopí ahissi (*solopí*, ghost, *ahissi*, medicine, Koasati)
scurvy-grass (a use not recorded elsewhere)

The genus *Pityopsis* comprises eight species ranging from Canada to northern Mexico (Semple et al. 1980, Semple and Bowers 1987). In the southeastern United States there are records of use among several native groups. Indigenous people in the coastal Carolinas used the plant as a poultice for sprains (Morton 1974). Now people in the Carolinas make a tea of fresh leaves before the plants flower to treat kidney and bladder problems. The tea is considered a good diuretic to relieve dropsy (edema).

The Choctaw use burned plant ashes to treat mouth sores (Bushnell 1909). The Creeks used a mixture of plants including *Pityopsis* to reduce fever (Spoehr 1939 in Sturtevant 1955). The Seminoles on the Big Cypress Reservation were using an infusion of the plants for colds and fevers when Sturtevant (1955) worked with them. He also recorded that they used the plant for "Cow Creek Sickness," in childbirth and in the busk (see *Pterocaulon*). Modern Seminole Alice Snow recorded that the plant is used to treat headache but gave no further information (Snow and Stans 2001).

Sturtevant (1955) was told about a plant called *pahátlo:cî* (white-black grass, Mikasuki). He had no voucher and did not know its identity. Based on the uses and other information he gave about that plant, it might have been *P. graminifolia.* Murphee (1965) found residents of the Panhandle continuing the use of "gopher grass" to treat boils and infected cuts, fevers, gout, ground itch, and rheumatism.

Plantago

(Linnaeus continued the Latin *planta*, footprint or sole of the foot, *ago* from *agere*, to bear or resemble, following use by Pliny, A.D. 23–79)

cuach Phàdruig (*cuach*, drinking cup, *Phàdruig*, of Father Druid, for *P. major*, Gaelic)
groblad (healing leaf, for *P. major*, Norwegian)
kjempe (giant, Norwegian); *koempe* (translated "warrior" by Coffey 1993; supposedly called that because of a game in which children try to knock the "heads" off each other's "weapons," which are the flower stalks; generic for genus, Danish)

***Plantago*, comparing alien and native species.** 1. *Plantago lanceolata*. a. Flower. b. Capsule. c. Seeds. 2. *Plantago major*. a. Flower. b. Capsule. c. Seeds. 3. *Plantago rugelii*. a. Flower. b. Capsule. c. Seeds. *Drawn by Regina O. Hughes*. From Reed 1971.

llantén [*yanten*] (the broad one, from Latin *planos*, Spanish)

plantain (from Latin *plantagenim*, sole of foot, in reference to the large broad leaves; first used in English about A.D. 1255); *piantaggine* (Italian); *plantain* (French)

slàn lus (healthy herb, Gaelic)

waybread [waybred] (from Old English *waybráde*, meaning "broad-leaved plant growing beside the way"); *vejbred* (waybread, Danish); *weegbree* (Dutch, from Middle Low German, *wegebrede*); *Wegebreite* (German, from Old High German *wegbrieta*); *Wegerich* (of roads, German)

white-man's footprint ("New England Indians," tribe not specified, cf. Millspaugh 1892)

Plantago aristata (bearing bristles)

buckthorn [buckhorn plantain] (from Latin *cervi spina*; applied by Valerius Cordus, 1514?–1544 to *Rhamnus catharticus*; later applied to other genera)

[bottlebrush, bracted, large-bract, largebract] plantain (the modifiers refer to the inflorescence)

ribwort (an allusion to the prominent veins in the leaves; originally applied to the European *P. lanceolata* about A.D. 1440; later expanded to the American species)

Plantago rugelii (named for its discoverer Ferdinand Rugel, 1806–1879, German-born planter and explorer in the southeastern United States)

[blackseed, common, Rugel's] plantain

Plantago virginica (of Virginia)

bo-u-na (Kiowa)

[dwarf, hoary, pale-seed] plantain

It has always struck me as odd that temperate and tropical plants (*Musa, Plantago*) are both called "plantain." Apparently, I have not been alone in that reaction, as it was discussed by Father Acosta in 1590 and numerous authors since (de Candolle [1886] 1959, OED 1971). Current opinion suggests that the *Musa*, named "plantain" or *plátano* in Spanish, came from Latin *platanus*, referring to the broad leaves. The original comparison was not to the temperate deciduous tree, but simply a reference to wide leaves (see also *Platanus*). In Portuguese, *plátano* still means *Platanus*, even in Brazil where *Musa* is *banana*.

A dictionary of Galibí published in 1664 glosses *Musa* as *palatana*, and a Carib dictionary of 1665 says they were known as *'balátana*. In Tupí, they were *prátane* (OED 1971). De Candolle ([1886] 1959) suggested that those words were derived from *bala* or *palan* from Malabar and introduced by the early Portuguese. The Brazilian Tupí now is *pacoba* or *bacoba*, while in French Guiana it is *bacove* (both from Tupí *pa 'kowa*, leaf for wrapping). Regardless of how the name for *Musa* was derived, it is clear that the name for *Plantago* came from the Old Latin name for the temperate plants.

It probably is impossible now to separate what indigenous people did with *Plantago* from uses they learned from European settlers. The plantain had long been used in medicines in Europe, and it figured prominently in the writings of Pliny (A.D. 23–79), Dioscorides (A.D. 40–80), and Galen (A.D. 129–?200). Two species were discussed by Leonard Fuchs in 1542, William Turner in 1568, and John Gerarde in 1597, among others. During this time period, *P. major*, the best-known species, was called weybrede or plantayne by Turner, while in German it was *Wegrich* (orthographic variant of *Wegerich*), in Dutch *wechbree* (orthographic variant of *weegbree*), in Italian *piantagine*, and in Spanish *llantén*. Then, as when discussed by Millspaugh (1892), it was largely considered vulnerary and demulcent, and recommended for coughs and bronchial problems.

Plantago aristata was used by the Cherokee to treat headache, dysentery, poisonous bites and stings, burns, and sore eyes, and in a douche (Hamel and Chiltoskey 1987). The Kawaiisu and Mendocino also used it in the western United States (Moerman 1998). The Menomini used *P. rugelii* as a dressing on burns or to treat inflammation (King 1984).

Possibly a formerly more widespread view of these plants was recorded by Vestal and Schultes (1939). They found that the Kiowa regarded *P. virginica* as a symbolic plant. Old men tied garlands or wreaths around their heads during dances as a symbol of health.

Platanthera

(Louis Claude Marie Richard created the name from Greek *platys*, broad, *anthera*, anther)

Platanthera ciliaris (fringed, the petals)
orange-plume
Waldhyazinthe (forest hyacinth, German)
yellow finger orchid; yellow fringed orchid [orchis]

When I was growing up, I learned from watching to spit on the baited hook before beginning fishing. The practice always struck me as odd, but many of the older fishermen did it routinely. I never knew if it was part of my Gaelic or Cherokee ancestry.

Then I learned that the Cherokee used roots of the yellow-fringed orchid on hooks to make fish bite (Hamel and Chiltoskey 1975). Members of that tribe have not always used the orchid, but they formerly used the Venus fly-trap (*Dionaea muscipula*). Mooney's (1885–1886) paper included the following prayer for catching fish by his Cherokee informant *A'yuninis'*, whom he called Swimmer: "Listen! Now you settlements have drawn near to harken. Where you have gathered in the foam you are moving about as one. You Blue Cat and the others, I have come to offer you freely the white food. Let the paths from every direction recognize each other. Our spittle shall be in agreement. Let your and my spittle be together as we go about. The fish have become a prey and there shall be no loneliness. Your spittle has become agreeable. I am called *A'yuninis'*. *Yu*!"

The Cherokee took a cold infusion of the orchid to stop headache and diarrhea (Hamel and Chiltoskey 1975). The Seminoles used the roots in a remedy to treat snakebite (Sturtevant 1955).

Platanus

(The original Greek, *platanos*, broad, in reference to the leaves, a name for the plane-tree used by Pliny, A.D. 23–79; the Greek Theophrastus, 370–288 B.C., spelled it *platys*)

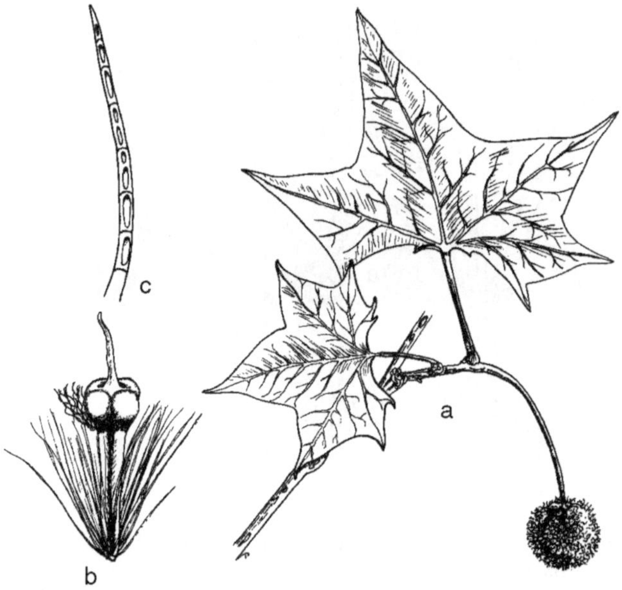

Platanus occidentalis. a. Branch with fruit. b. Achene. c. Trichome. *Drawn by Vivian Frazier*. From Correll and Correll 1972.

Platanus occidentalis (western, i.e., of the Western Hemisphere)
akhatka (*oke*, water, *hatka*, white, Alabama); *akhv'tkv* [*akhatkv*] (*owv*, water, *hvtke*, white, Muskogee)
álamo [*álama blanco*] (usually used for *Populus*, Tamaulipas, Nuevo León, San Luis Potosí, Veracruz)
American plane [plane-tree, planetree] (to distinguish it from the European plane tree, *P. orientalis*, Canada, USA); plane [plane-tree, planetree] (from Old French *plasne*, in turn from *planus*, broad, because of the broad leaves, used in English by 1382; perhaps originally applied to *Acer pseudoplatanus*); *Platane* (German); *platane* (Quebec, France); *platano* (Italian); *platano occidental* [*occidentale, de Virginia*] (western [Virginia] plane-tree, Canada, USA);

westerse plataan (western plane-tree, Dutch, Canada, USA, West Indies)
aya (Veracruz)
ayan̲' san̲han̲' udi' ("strong wood tree" or *san̲*, white?, Biloxi; compare with Omaha *jan̲ san̲*, white wood and Osage *san̲san̲*, white sycamore)
bois puant (stinking tree, Canada, France)
butterwood (a corruption of "button"?, North America); buttonball[-tree] (USA); buttonwood (surely an allusion to the round, buttonlike fruits; recorded in Swedish by Kalm in 1753; first published in English by Frederick Pursh in 1814 for this species, Canada, United Kingdom)
ciłó (Koasati); *sini* (Choctaw)
cotonier (cotton tree, French, USA)
keisewaquata (Shawnee, fide Edgar 1891)
kuwajunega (Cherokee)
oo-da-te-cha-wunnes (big stockings, Delaware?, USA)
[American] sycamore tree [sycomore] ("sycamore" was originally applied to *Ficus sycamorus*, a fig in Egypt, Syria, and nearby countries. The word is a combination of Greek *sykos*, fig, and *moros*, the mulberry, and came into English about A.D. 1300. By ca. 1588, it was applied to European *Acer pseudoplatanus*, and then to American *Platanus* by 1814 or perhaps before.)
Virginia maple (name in New England in the 1740s, fide Kalm 1753)
wasbok [*watenbok*] (water beech, used by Swedes in Pennsylvania in the 1740s, Kalm 1753); water beech [waterbeech] (published in Swedish by Peter Kalm in 1753; the OED says first in English by Frederick Pursh in 1814)
yap hi̱·tuwi·' (tree of burrs); *yap taktce' hi̱·tcuwi·'-hare* [*ya bwe*] (tree white [of] many burrs, Catawba)

The common names "plane-tree" and "sycamore" have been confused in English since the 1300s. That confusion reflects the application of the names to *Acer pseudoplatanus* and *Platanus orientalis*. For example, Vickery (1993) devotes almost two pages to the "sycamore" (*Acer*), yet that maple in Norway is the tanbark plane (*platanlønn*). In the United States, a "sycamore" is *Platanus*.

The name "plane" was in English before the "official" introduction of *Platanus*. In 1398, a John De Trevisa wrote, "The plane is a colde tre and a drye, and ye leaves therof helep in hoot eveles." Therefore, by that period, Europeans as far north as England were making medical use of a "plane" tree that must have been *Acer*.

William Turner's *Herball* recorded that *P. orientalis*, a native to Asia Minor, was introduced into Britain around 1562 (OED 1971). Then, 78 years later, John Parkinson told us that John Tradescant had *P. occidentalis* sent to him from Virginia and introduced it into England. Neither of these two species did well in the British Isles. In 1675, people noticed a different plane-tree. Although these hybrid trees (*P. ×acerifolia*) are now called the "London Plane," they probably originated in France or Spain (Edlin and Mitchell 1985). From that point, the "plane" became one of the most common street trees in England and much of the rest of Great Britain.

The Cherokee were one of the few tribes recorded as using the wood of *Platanus* (Hamel and Chiltoskey 1975). Settlers used the wood for crates, interior finishing, furniture, cooperage, rollers, butcher blocks, and tobacco boxes (Vines 1977). In spite of this usage, the technical sheet from the U.S. Forestry Service is not overly complimentary. It states that, although the wood is not durable, "It holds its shape well after steaming and machines, but requires high speed cutter heads to prevent chipping. It shrinks moderately in drying and is inclined to warp when flat sawn." The sapwood is white to light yellow, while the heartwood is light to dark brown. Sycamore is classified as moderate in weight, hardness, stiffness, shock resistance, strength in bending, endwise compression, and nail-holding ability. It has a close texture, glues well, and resists splitting due to interlocked grain. The Forestry Service noted modern application in furniture (especially drawer sides), containers, millwork, flooring, veneer, pallets, boxes, plywood, pulp wood, paper, and particle boards.

Although there seem to be no medical applications of *Platanus* in Europe, indigenous Americans used it throughout its range. The Cherokee considered the trees, but especially the bark, cathartic and emetic, and used it for menstrual and urinary problems and measles. They also treated dysentery with an inner bark infusion. Externally, the Cherokee and Iroquois used sycamore in a steam bath for rheumatism (Hamel and Chiltoskey 1975, Moerman 1998). The Creeks, Delaware, and Meskwaki treated colds, sore throat, hoarseness, and tuberculosis with the bark. The Cherokee, Iroquois, and Meskwaki used bark or root infusions to treat skin eruptions, scabs, eczema, and knife and ax wounds (Swanton 1928a, Taylor 1940, Hamel and Chiltoskey 1975, Moerman 1998). Moerman (1998) recorded Mahuna use of this species, but this tree does not grow there (Hickman 1993).

Foster and Duke (1990) found that the bark was once recommended by descendants of Europeans for rheumatism and scurvy. They found no evidence of efficacy. Lewis and Elvin-Lewis (1977) mention only that the airborne pollen causes allergies. Hocking

(1997) affirms that the leaves were formerly used in eye lotions, and that the plant yielded antipyretic tonics. He further notes that it, like *P. orientalis*, contains cyanogenic glycosides. All agree, however, that the greatest current use for sycamore is as an ornamental tree, particularly in parks and along streets.

Pleopeltis

(From Greek *pleos*, full or many, and *pelte*, a shield, in reference to the paraphyses)

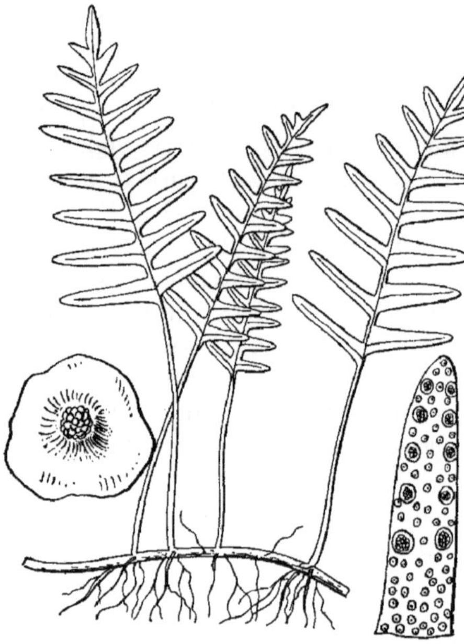

Pleopeltis polypodioides. From Britton and Brown 1896.

Pleopeltis polypodioides (resembling *Polypodium*) (=*Polypodium polypodioides*)
coladilla (little tailed one)
doradilla (little golden one, Cuba, Dominican Republic)
fougere grimpante (climbing fern, French Antilles)
gray polypody (a book name, Bahamas)
helecho que resucita (resurrection fern, Dominican Republic); resurrection fern (USA)
ihosi:cokhíssi (*ehose*, a mystical being who causes people to get lost, *kokhesse*, whiskers, Creek); *iwa:satkáhcicô:skî* (cognate to the Creek name, with the same meaning, Mikasuki)
istilibuski entaapente [*yaato posheke entaapente, yatapushkaale entaapente*] (*isti*, person, *libuski*, little, *entaapente*, his fern, Mikasuki)
koo' te' (tree caretaker, Huastec, San Luis Potosí)
kutsiil Boo'waat (*Boo'waat* dance baton, Huastec, San Luis Potosí)
lakcv-cvmpv sohhontat (little thing that sprouts on sweet acorn [live oak]; *lakcv*, acorn, *cvmpv*, sweet, *sohhontat*, from *s-*, with, *oh-*, on, *hont-* sprout, -*us-*, little, -*at*, the one that, Creek)
liane anolis (lizard [*Anolis*] fern, Haiti)
liane coresse (running fern, French Antilles)
okecheske entapente [*okiciskintapinti*] (*okecheske*, many small bases, *entaapente*, its fern, Mikasuki)
polypode (many foot, French Antilles)
tsooy ahaatik (lord rib, Huastec, San Luis Potosí)
weering grass (Creole, Nicaragua)

Pleopeltis is often included within *Polypodium*. As now circumscribed, it contains 11 to 50 species (Andrews and Windham 1993, Mabberley 1997).

Resurrection fern is one of the most widespread ferns in the New World, ranging through much of the eastern United States south through most of Latin America (Andrews and Windham 1993). The Houma made a cold infusion of fronds used to treat babies' sore mouths (thrush) and use a decoction of the fronds for headache, bleeding gums, and dizziness (Speck 1941). The Seminoles used the fern in a complex mixture to treat chronic sickness, certain kinds of mental problems, and in certain medicines used in childbirth (Sturtevant 1955). Bennett (1997) found that modern Seminoles mix *P. polypodioides* with shoestring fern (*Vittaria lineata*), boil it in water, and drink the decoction, or administer it in a steam bath to treat depression. They also told him that the plant was used in another medicine but would not share particulars with him. Alice Snow calls the plant *okecheske entapente* (Snow and Stans 2001).

In Cuba, *doradilla* is considered a prime remedy for liver problems (Roig 1945). Some there also think it good for heart problems. Whole plants, but mostly the leaves, are put in water containers and give a slight earthy flavor. In Roig's time, there was a patented remedy called *Elixir de Doradilla.*

The plants are considered a cure for liver problems in Hispaniola (Liogier 1974). Morton (1981) found the dry fern being sold by herb vendors in Puerto Rico as a remedy for high blood pressure, and Hocking (1997) found that the entire plant had been used for bronchitis. In Mexico, Alcorn (1984) found the Huastec used the fern to treat fever and dizziness, and to scrub the head. A tea is made from the fern to clean the stomach and lungs in Nicaragua (Barrett 1994).

Pluchea: Cure-for-All

(Named to commemorate the French abbot Noël-Antoine Pluche, 1688–1761, a seminary teacher and naturalist)

In 1817, the French botanist Alexandre Henri Gabriel de Cassini (1781–1832), who specialized in the Asteraceae, described plants in a new genus he called *Pluchea.* The reference species (type) of the genus is *Pluchea marilandica,* which another French botanist, André Michaux, previously had placed in the genus *Conyza.* Indeed, Michaux was following Linnaeus who placed in *Conyza* several species that today are segregated into other lineages.

Such changing of names is a normal part of scientific learning because evolution has selected among different phylogenies to the point that they are difficult to distinguish. Not only have scientists had trouble sorting out the genera and species in the Asteraceae, but nonscientists sometimes apply names in what appear to be random ways. However, there are two frequent and essentially tropical *Pluchea* that people use across their ranges. Of the five native species in Florida, this discussion focuses on *P. carolinensis* and *P. odorata.* The second of these also appears in the literature under the names *P. purpurascens* and *P. symphytifolia* (Gillis 1977, Khan and Jarvis 1989), so it is often difficult to sort out uses. Since both species have mostly the same applications, exact identity is not critical.

Only two indigenous common names were found in the southeast. The Choctaw call *P. foetida hoshukkosona* (*haskuk,* grass, *kosoma,* strong smell), and they boiled the leaves and used the water extract "during attacks of fever" (Bushnell 1909). The Koasati call *P. camphorata ittohapakó* (*ittó,* tree, *hapakó,* death), and also used it to treat fever (Taylor 1940). Given the importance of the plants to the remaining Americans, other linguistic groups probably also had names for them.

In Mexico, the Aztecs knew these plants and used them. Both *P. carolinensis* and *P. odorata* were known as *ciguapatle* [*ciguabate, cichapatle, cipatle, seguapeti, siguapete, siguapote*], a name derived from *cigua,* woman, *patl,* medicine (Náhuatl, Mexico to El Salvador). However, alternate names were applied.

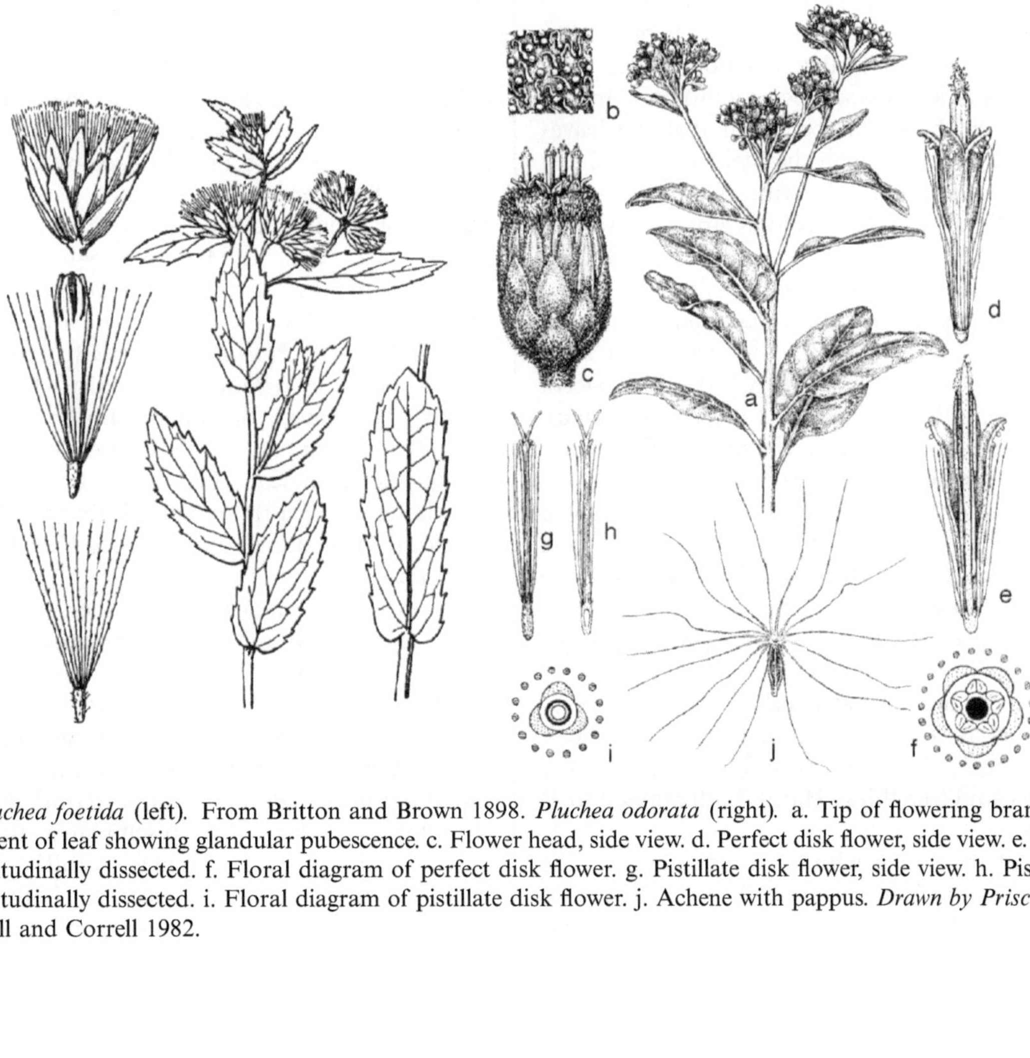

Pluchea. *Pluchea foetida* (left). From Britton and Brown 1898. *Pluchea odorata* (right). a. Tip of flowering branch. b. Enlargement of leaf showing glandular pubescence. c. Flower head, side view. d. Perfect disk flower, side view. e. Perfect disk flower, longitudinally dissected. f. Floral diagram of perfect disk flower. g. Pistillate disk flower, side view. h. Pistillate disk flower, longitudinally dissected. i. Floral diagram of pistillate disk flower. j. Achene with pappus. *Drawn by Priscilla Fawcett.* From Correll and Correll 1982.

Perhaps *comalpatli* (*comal*, cooking disk to prepare tortillas, *patl*, medicine) is simply another way of saying woman's medicine, because cooking is traditionally considered the female realm. Probably both *ahuapatli* [*alaa-patli*] and *nahuapate* [*nahuapote*] are variants of *nahuac*, of *anahuac* (Mexico, El Salvador), meaning a medicine growing near water. Both species are most frequent near marshes. Perhaps the modern *hoja de la playa* (beach leaf) comes from the Náhuatl source.

To the Mayas, both plants were *chal-ché* [*chalchay, chalche, chalché, ix chal-ché, chal-ch, caal-ca*] (wash quickly, Maya, Yucatán, Belize, Guatemala, Nicaragua). That name is an allusion to medical use before, during, and after childbirth, much as in the Náhuatl names.

Several apparently indigenous names applied in different areas have not been translated. These include *alinanche* (Sinaloa), *clina*, *suquinay* [*suquinayo*] (Guatemala, El Salvador), *teposa*, and *tewa zahn* (Zapotec, Oaxaca).

Perhaps the most widespread view of these plants is shown by the names cure-for-all [curforal] (Florida, Barbados), and *guérit-tout* [*gueritoute, gueri-tout, guerir tou, jewi tut*] (cure-all, Haiti, Guadeloupe, Martinique). Most of the common names clearly indicate medical applications. Both cough bush (Bahamas) and sour bush (Bahamas) note usage as a remedy for colds and associated symptoms (Morton 1981). Although the simple comment that the leaves are smoked like tobacco is in the literature, the many names comparing the leaves with *Nicotiana* surely suggest an adjunct to relieve cold symptoms. Among those names are bitter tobacco (Jamaica), Indian tobacco, riverside tobacco, *tabac à Jacot* [*tabac à jako*] (false tobacco, Haiti), *tabac diable* (devil's tobacco, Haiti, Guadeloupe, Martinique), *tabac marron* [*tabac mawon, tabac sauvage*] (wild tobacco, Haiti), *tobacco cimarron* (wild tobacco, Panama), *tabac zombie* (ghost tobacco, French Antilles), and wild tobacco (Bahamas, Jamaica).

Of these names, *tabac à Jacot* requires some explanation. The name *Jacot* [*Jacquot*] is used in a game English speakers call "Simon Says." In the game "*Jacques* [or *Jacquot*, the diminutive] *a dit*," one imitates as a parrot [*Jacquot*] imitates people. So, the plants "imitate" the real tobacco.

Some of the names that allude to medicinal uses are obtuse as in *feuilles la chose* (the thing's leaf, Haiti), fox-leaf, *la chause* [*la chose, la choille, la choise, la choige, la choy*] (the thing, Haiti), both applied to *P. carolinensis.*

Several of the names with religious references are notations of medical applications. These include *flor de ángel* (angel flower) and *flor de Guadalupe* (Guadalupe's flower), applied to only *P. odorata*, and *hierba de Santa María* (Saint Mary's herb, Mexico, Nuevo León), given to *P. carolinensis* and *P. odorata.* Others say *Santa María* (Saint Mary, Tamaulipas, Veracruz, Yucatán, Belize, Nicaragua) or *Santa María cimarron* (wild Saint Mary).

Several people compare these aromatic shrubs with *salvia* (Latin, to save or preserve) not only because both are aromatic, but also because both are medicines that "preserve" health. In French, both *P. carolinensis* and *P. odorata* are *grande sauge* (Haiti, Martinique, Guadeloupe), [*la*] *sauge* (sage, Haiti), or *sauge rouge* (red sage, Haiti, Martinique). In Spanish, they become *salvia* (sage, Puerto Rico, Dominican Republic, Nicaragua), *salvia colorada* (red sage), *salvia de remedios* (medicinal sage, Dominican Republic), *salvia blanca* (white sage, Dominican Republic), *salvia cimarrona* (wild sage, Cuba), *salvia de playa* (beach sage, Cuba), *salvia del país* (country sage, Cuba), *salvia real* (royal sage), or *salvia santa* (holy sage). Although it is not at first obvious, the name saab [saav, big-saab] (Bahamas) is probably a Creole version of *salvia.* Other names that note the fragrance are camphor-weed (*P. odorata*, Florida), *canela* [*canelón*] (cinnamon, Mexico, Baja California), or sweet scent (Puerto Rico, Virgin Islands).

Until the early 1800s, many species now considered *Pluchea* were placed in the genus *Conyza* (Greek, a flea), and the Greek gave rise to the common name *conyse odorante* (fragrant horsetail, Dominican Republic). *Conyza* was originally used by Dioscorides (fl. A.D. 40–80) for some kind of fleabane, so the Hispaniolan name follows a venerable usage although the original was a Mediterranean plant. Several members of the Asteraceae have been considered "fleabanes" and used to repel those annoying animals (see *Erigeron*). *Pluchea carolinensis* too is called bushy fleabane (Bahamas), hairy fleabane, and shrubby fleabane (Bahamas). Both *P. carolinensis* and *P. odorata* are called sweet-scented fleabane (Bahamas. Jamaica, Trinidad), but usually only the latter is [salt] marsh fleabane (Arizona to Florida). Their names suggest they have been used to repel fleas.

Some of the names denote other uses or comparisons. In Guadeloupe and Martinique, *P. odorata* is *bois liège* (cork bush), and the wood is presumably the part used. People in Belize say *pito sico.* A *pito* is a flower, and *sico* may be a variant of *seco*, dry, probably a name used like "straw-flower" in English, because of the dry flowers. Perhaps *corail* (coral, Haiti) notes the pinkish flowers. Any resemblance to a *z'orielle mouton* (sheep's ear, French Antilles) must also have to do with pubescence, as both are white and woolly. Residents of Mexico call *P. odorata* the *jara.* It is not clear if they are comparing the species with rockrose (*Helianthemum*, Sonora) because both have white pubescent leaves, or if they are calling it an "arrow" (the Spanish from Hebrew *khara*, to cast).

The common name cure-for-all is an exaggeration, but both species are used to treat many problems. Among the afflictions treated are asthma, bronchitis, cholera, colds, digestive and eye complaints, fever, gout, headache, hoarseness, hypertension, menstrual difficulties, nervousness, neuralgia, night sweats, rheumatism, sores, spasms, stomach disorders, swelling in any part of the body, toothache, twitching muscles, and wheezing. They are also used to treat new mothers and women in labor (Standley 1920–1926, Roig 1945, Martínez 1969, Morton 1981).

Laboratory studies have shown that *P. odorata* has antifungal, antimicrobial, antibacterial, and anthelmintic properties (Dominguez and Zamudio 1972, López-Abraham et al. 1979, Scholz et al. 1990, 1994). These traits apparently stem from terpenes. Most of the known compounds in *P. odorata* are sesquiterpenes (Dominguez et al. 1981, Arriaga-Giner et al. 1983, Arriaga and Borges-del-Castillo 1985, Ahmad et al. 1992a,b, Loayza et al. 1992, Ahmed et al. 1998). Sesquiterpenes also are among the most reported chemicals in other species (Ahmad and Fizza 1988a, b, Ahmad et al. 1989a, 1991, 1992a,b, Ando et al. 1994, Zdero and Bohlmann 1989, Uchiyama et al. 1991, Guilhon and Muller 1996, 1998, Mahmoud 1997, Shimoma et al. 1998). However, monoterpenes, triterpenes, and triterpenoids are known to have similar properties (Uchiyama et al. 1991, Kaith 1996, Pérez-Garcia et al. 1996). Also reported are anti-inflammatory (Srivastava et al. 1990, Sen and Nag Chaudhuri 1990, 1991, Sen et al. 1993a, Pérez et al. 1995), anti-ulcer (Pal and Chaudhuri 1989, Sen et al. 1993a), hepatoprotective (Sen et al. 1993b), and trypanocidal (Zani et al. 1994) properties.

Common names indicate an ancient Náhuatl and Mayan application for women, especially during and after childbirth, and suggest an original specialized medicine. With the influx of several European cultures and confusion with the Old World fleabanes, that specialization may have given way to a broader utilization. Indeed, considering the array of uses for *Pluchea*, it is easy to see why it became known as a cure-for-all or *guérit-tout.* Given the Old World use to treat poisonous snakebites, it is surprising that people in the New World did not include that among their pharmacopoeia. Experimental data for *Pluchea* actually indicate that it works for that problem (Alam et al. 1996).

Plumbago: Plumbago or Leadwort

(Latin *plumbum*, lead, *ago* from *agere*, to bear or resemble)

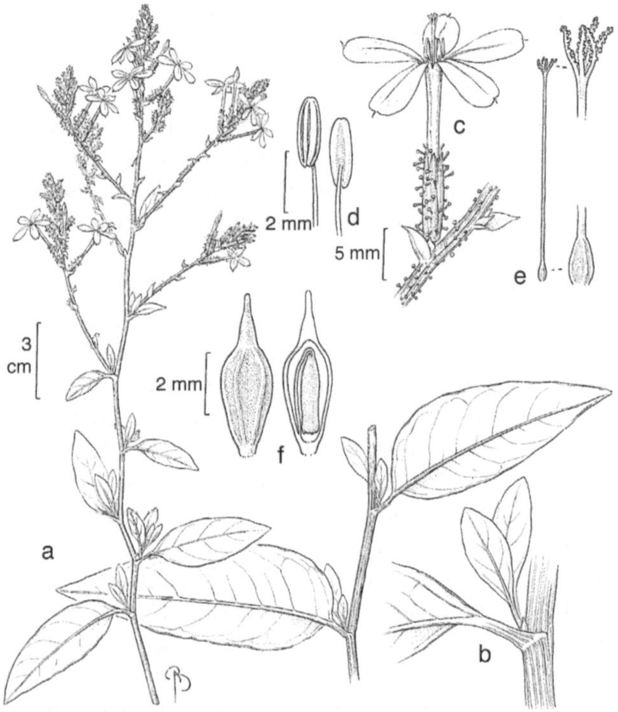

Plumbago scandens. a. Flowering branch. b. Detail of node. c. Flowering and subtending bract and bracteole. d. Stamens, front and side view. e. Gynoecium, with details of stigma and ovary. *Drawn by Bobbi Angell.* From Acevedo 2003.

The word "plumbago" appeared in English by 1784 although it has a venerable history in southern European languages. Most speakers of English call the plants leadwort. Both names refer to the genus *Plumbago* and the family Plumbaginaceae. The history of the words "plumbago" and "leadwort" is long and complicated, and sometimes the plant is inseparable from the metal except by inference.

Dioscorides (fl. A.D. 40–80) used two Greek words to designate both a metal and a plant. Pliny (A.D. 23–79) Latinized those words as *molybdaena* and *plumbago*, and used both for what is now *galena*, lead ore. However, for the plants, Pliny always used *plumbago.* In Philemon Holland's 1601 translation Pliny says, "There groweth commonly an herbe named in Greeke Molybdaena, that is to say in Latin, Plumbago, euen vpon euery corne land." Matthioli's *Commentary on Dioscorides* was translated into French in 1572, and subsequently most French, Italian, and German speakers reapplied "plumbago" to the metal. Linnaeus

(1753) formalized the use of *Plumbago* to the plant and it has remained that way.

The white plumbago (Bahamas, Puerto Rico) in southern Florida is *P. scandens*. That sprawling herb is also found in the Bahamas, the West Indies, Mexico (Sonora and Tamaulipas south) to Costa Rica and South America to Peru, Bolivia, and Argentina. The plant has multiple names throughout its range because it is notorious for several reasons..

The names for the American plants were complicated because Europeans already knew *P. europaea*. They called the Old World plants by two sets of names—one mentioning the metal lead, and the other teeth.

Mostly, *P. europaea* is associated with the metal in the names leadwort (English), *piombaggine* (lead plant, Italian), *plúmbago* (Sonora, Spanish), and *Bleiwurz* (lead herb, German). Those names are either based on an old belief that it could be used to cure lead poisoning (Bremness 1994), because of the "lead-colored" flowers (OED 1971), or because the roots impart a lead color to the hands when handled (Walker 1976). Few English speakers recognize the word plumbago, and they use its translated form, leadwort. According to Nathan Bailey's *A Universal Etymological English Dictionary*, "leadwort" was first used in English in 1727.

The other names for *P. europaea* are based on its use in treating toothache. Surely the oldest name of those referring to teeth was *Lepidium Dentillaria dictum* (little scale called dentillaria) published by Gaspard Bauhin in 1623. That name evolved into *dentelaire* (French) and *dentilária* (Portuguese). This knowledge of use was carried to the Americas where *P. scandens* became *dentelaire* [*dentelle*] (Haiti) and *dentelária* (Oaxaca, Cuba).

The European Spanish *hierba blesa* (pretty herb) and *velesa* (pretty one) are related, and these were applied to *P. scandens* as *belesa* [*beleza*] (Puerto Rico). These probably gave rise to *bella Emilia* (pretty Emily, Dominican Republic, Colombia, Argentina), *belleza enredadera* [*veleza enredadera*] (pretty twiner, Puerto Rico), and *embeles* [*embeleso*] (Yucatán, Cuba). There is either a linguistic relationship between these and Spanish *embelezar* (to enchant) because of the beauty of the flower or simply as a result of confusion with that word.

Throughout the Americas, there is an ambiguity of feelings toward these plants that is typified by *homme* [*à*] *deux faces* (man with two faces, Haiti). The herbs are pretty because of their flowers, but like their European cousin, they are "acrid, blistering, and emetic" (Polunin 1969). Many of the common names warn against those traits.

Some of the names are direct, as in blister leaf (Virgin Islands), *chilillo* (little pepper, Veracruz), *herbe brûlante* (burning herb, Guadeloupe, Martinique), *queimadeira* (burner, Brazil), *sarne vejiguilla* (blister herb, Peru), *yerba de vejigatorio* (irritant herb, Guyana), *tlalchinchinolli* (from *tlachichinacapololli*, cruelly tortured, Náhuatl, Querétaro), and *tlepatli* (fire medicine, Náhuatl, Querétaro). Others damn them by inference or association, as in *caataia* (*caa*, plant, *taia*, from *ta'yá* = *Zanthoxylum violaceum* in the Rutaceae, a family famous for volatile oils, Tupí, Brazil), *folhas de louco* [*louco*] (crazy leaf, Brazil), *malacara* (bad-face, Cuba), and *herbe bourrique* (she-ass herb, Haiti). Surely, the fiery juice led to the comparison with mustards and the names *moutard pays* (wild mustard, Guadeloupe, Martinique) and *sinapisme* (from Latin *sinapis*, mustard, Guadeloupe, Martinique).

Some people consider the plants the devil incarnate, with *erva do diabo* [*erva de diabo*] (devil's herb, Brazil), *herbe au diable* (devil's herb, Haiti), *hierba del diablo* [*yerba del diablo*] (devil's herb, Dominican Republic, Colombia, Argentina), and *mauvaise herbe* (damned herb, Haiti). At the opposite extreme, some people consider them an *erva divina* (divine herb, Brazil).

Other names are more obtuse allusions to medicines. These include doctor-bush [doctorbush] (Bahamas, Florida), *herbe à Madame Bihoret* [*zèbe Man Bihoret*] (Guadeloupe, Martinique), *folha de louro* (parrot leaf, Brazil), *hierba del negro* (negro's herb, Oaxaca), *hierba del pajaro* (bird's herb, Panama), *hierba de alacrán* (scorpion herb, Sinaloa, Jalisco, San Luis Potosí, Oaxaca, Guanajuato), *lagaña de aura* (vulture's eyelid, Cuba), *lagaña de perro* (dog's eyelid, Morelos, Cuba), and *muela de alacrán* (scorpion's molar, Mexico).

People also find the sticky fruits worthy of comment. Some of these names are positive, such as *collant* (sticky one, Guadeloupe, Martinique), *erva de amor* (love herb, Brazil), and *meladillo* [*melallillo, mielilla*] [*silvestre*] ([wild] honey-giver, Puerto Rico, Cuba, Mexico). Typically, *meladillo* is applied to *P. auriculata* in Puerto Rico. Other names are negative, as in *pegajoso* (sticky one, Tamaulipas, Sinaloa), *pega-pollo* (chicken-catcher, Dominican Republic), *pegoso* (sticky), *mata-pollo* (chicken-killer, Dominican Republic), and *visqueira* (the viscous one, Brazil).

Some people do not know of the medicinal uses. They may be the ones calling the plants *canutillo* (little tube, maybe the flower, Sinaloa), *guapote* [*guapito*] (from *guapo*, handsome, Venezuela, Peru), *estrenina* (little gift, Sonora), *Isabel* (Hispaniola), *jazmín azul* (blue jassmine, Yucatán), and *jazmim azul* (blue jassmine, Brazil).

Several indigenous names are descriptive or simple terms. Among those are *caajandiva* [*caajandivas*] (*caa*, plant, *jandiva*, from *ñandi'á*, catfish, Tupí, Brazil), *caaponga* [*caapononga*, *caponga*] (*caa*, plant, *ponga*, climber, Tupí, Brazil), *chabak* [*chab-ak, chabac, chapak*] (*chab*, bad odor, *ak*, vine, Maya, Yucatán), *guacochile* (if Náhuatl, maybe from *uactoc*, mature, *chilli*, chile, El Salvador), *guaicuru* (maybe Taino), and *turicua* [*tiricua, jiricua*] (Guanajuato, Tamaulipas). The Huastec call it *hurika*, and some people have spelled that *eureka* (San Luis Potosí). The second spelling has no linguistic relation to "eureka" in English.

Leaves and roots when applied to the skin produce almost instant reddening and in a short time blisters. An external decoction is used to treat erysipelas, felon, itch, mange, warts, external ulcers, leprosy, and similar problems (Bye 1986). Root juices have also been used to cauterize ulcers in horses (Mors et al. 2001). The same parts are poisonous if taken internally (Tokarnia and Dobereiner 1982, Medeiros et al. 2001, Mors et al. 2001). *Plumbago europea* has similar properties and is used as an emetic; when chewed, the plant increases saliva flow. Roots are used from both species to relieve pains in ears and teeth, for swollen joints, and as a purgative (Mors et al. 2001). Beggars have applied the leaves to raise sores on their body to arouse pity (Standley 1920–1926). Leaves and roots are macerated in alcohol and applied to the skin to treat rheumatism (Pittier 1926, Petersen 1974, Liogier 1974).

Roots and leaves contain plumbagin (Harborne 1967), a naphthoquinone derivative characteristic of most species in the genus (Mors et al. 2001). There are related compounds in *P. scandens* and other species (Sankaram et al. 1979, Bhattacharyya and Carvalho 1986, Dinda and Chel 1992, Dinda 1994, Gupta et al. 1999). Plumbagin is active against several fungi and bacteria (Gonçalves de Lima et al. 1968, Vijver and Lotter 1971, Bambode and Shukla 1974, Durga et al. 1990, Ahmad et al. 1998, Ali-Shtayeh and Abu Ghdeib 1999, Beg and Ahmad 2000). In addition, plumbagin and related compounds have shown promise in cancer treatment, for inhibiting inflammation, in controlling cholesterol, and improving cardiotonic action (Melo et al. 1974, Itoigawa et al. 1991, Sharma et al. 1991, Oyedapo 1996, Devi et al. 1999).

Several studies show that the chemicals in *Plumbago* inhibit feeding by insects (Kubo et al. 1980, Sharma 1984, Hassanali and Lwande 1989). In spite of the potpourri of toxins in the plants, one genus of North American butterfly uses *Plumbago* for its larvae. Those animals are butterflies in the genus *Leptotes*, but specifically *L. cassius* (Cassius blue) and *L. marina* (marine blue) use *Plumbago* (Brown 1990, Minno and Minno 1999). How the larvae of those tiny blue gems can ingest the scalding compounds with immunity is unknown. Yet, when they become adults, they can also sip nectar from the flowers.

Podophyllum

(From Greek *pous, podos*, a foot, and *phyllon*, a leaf; Linnaeus modified the name from Catesby's *Anapodophyllum*, duck foot leaf)

Podophyllum peltatum. From Britton and Brown 1897.

Podophyllum peltatum (shield-shaped)

behen (from Medieval Latin *behen*, corrupted from Arabic *bahman* [*behmen*], a kind of root; the word appeared in English in Henry Lyte's 1578 translation of Dodoens's *Cruydeboek* of 1554 as "Called ... of herboristes at this day Behen, or Been album")

che-sa-ne-pe-sha (it pains the bowels, Osage; also called *shon'-gthin-dse*)

citron (fruits resembling *Citrus medica*, among the French in 1700, Missouri); *citron sauvage* (wild citron, *Citrus medica*, Quebec)

devil's-apple; hog apple (Iowa); Indian apple; mug-apple

duck's foot (in use by the 1730s by Catesby); wild duckfoot

fala imisito (*fala*, crow, *im*, its, *isito*, pumpkin, Choctaw); *fala intanchi* (*fala*, crow, *in*, its, *tanchi*, corn, Choctaw); *falaanosi'* (*fala*, crow, *aanosi'*, bed, Chickasaw)

ipecacuana (from *ipega'kwai*, duck's penis, or *ipe-kaa-guéne*, creeping plant causing vomiting, Tupí, Carolinas, Catesby 1731–1732)

kvpotoyv [*'kvpotoyv, kvpolóye*] (*kvpoteyetv*, something to put on the head, Creek)
[ground, wild] lemon; yellow-berry
Maiapfel (May apple, German); May-apple [American, common, mayapple] (Catesby called the herbs May Apple in 1731, although the OED 1971 says the name dates from Philip Miller in 1733 as *Pomum Maiale*, May apple; either way Catesby found it in local use in the Carolinas by the 1720s); *pomme de mai* (May apple, Quebec)
[American, wild] mandrake[-pear] (pear-like fruits, compared with *Mandragora officinarum*, New Jersey)
mäskätámîn utcípa (*mäskätá*, defecation, *mîn*, berry, *utcípa*, root, Menomini); *maskiichtew* [*masgichteu*] (*maashk*, defecating, *kiich*, causes, Delaware)
maypop (a name usually applied to *Passiflora*, Virginia)
parasols (an allusion to the leaves, Ohio); umbrella plant [root] (Virginia)
podophylle (French)
Puck's foot ("Puck," from Old English *púca*, Old Norse, *púki*, Welsh, *pwca*, Irish *púca*, a mischievous demon; it is not clear if the word is of Teutonic or Celtic origin, but it was in English by about A.D. 1000; Middle English *the pouke* was identified with the devil; from the 16th century, as Puck, a tricky goblin or sprite; also called Robin Goodfellow, Hobgoblin; all three were used by Shakespeare in 1590)
raccoon berry (from *ärä'kun*, for *Procyon lotor*, Virginia Algonquian; in use by 1884)
vegetable-calomel ("calomel," from Greek *calos*, beautiful, *melos*, black or from Ethiopian *calos*, beautiful, *melos*, black, because white or pale bodies rubbed with it become black; mercurous chloride [Hg_2Cl_2] or *mercurius dubius*; used as a laxative in medicine from 1676)
vegetable-mercury (the same allusion as "calomel")
wild jalap ("jalap," from *xalli*, sand, *atl*, water, *pan*, upon, Náhuatl, used to identify the city as Xalapa and, in this context, *Ipomoea purga*, by 1675)

When I was still a child, I was told that there were "male" and "female" plants of May apples. They were easy to tell apart, because the "male" had a single leafstalk, and the "female" had a branched leafstalk with a leaf on each branch. Moreover, the "female" had flowers in the place where the stalks split into two. I was skeptical. I planted some near my home. Each rhizome node (joint) bore a single leaf—which was unbranched one year and branched the next. So much for male and female plants.

The first record Linnaeus had of these plants was in German physician Christian Mentzel's (1622–1701) book of 1682, the *Pinax*, where he called them *Aconitifolia humilis, flore albo unico campanulato, fructu cynosbati* (A weak herb with leaves like Monk's-hood, a single, bell-shaped, white flower, a fruit like a dog briar). Mark Catesby (1731–1732) called the plants *Anapodophyllum canandense* (duck foot leaf from Canada). Thankfully, Linnaeus shortened the name to *Podophyllum* in his *Hortus Cliffortianus* of 1738. Catesby added that, because the herb flowers in May, the local residents in South Carolina called it "May apple."

Podophyllum has two species (Mabberley 1997). *Podophyllum peltatum* is endemic to the eastern United States and ranges from Florida to Quebec and Ontario in Canada, and west to Minnesota and eastern Texas (Fernald 1950, Correll and Johnston 1970, Diggs et al. 1999). The other species, *P. hexandrum*, grows from the Himalayas to eastern Asia, including Afghanistan, Pakistan, northern India, and China (Hsu 1986, Hocking 1997).

Fernald et al. (1958) noted that the fruits are relished by most individuals when fresh, although Harvard guru Asa Gray (1810–1888) was the exception. Gray wrote that the fruit was "mawkish, eaten by pigs and boys." Although the taste is "peculiar," as Fernald et al. (1958) described it, the flavor is agreeable. Native people in the Americas concurred. At least the Cherokee, Delaware, Iroquois, Menomini, Meskwaki, Osage, Penobscot, and Ojibwa ate the fruits fresh, cooked, or mashed and made into small cakes that were dried for future use (Hunter [1823] 1973, Yanovsky 1936, Vogel 1970, King 1984, Moerman 1998). As needed, the cakes were soaked in warm water, cooked as a sauce, or mixed with corn bread.

There are accounts of indigenous people committing suicide by eating the root of May-apple (Vogel 1970). Because the roots are purgative, emetic, and irritating, that would be an excruciating way to die, although Benjamin S. Barton wrote in 1810 that the root was "possessed of some degree of an anodyne, or narcotic quality."

The Cherokee used *Podophyllum* to treat intestinal worms, rheumatism, ulcers, and sores, as a laxative, and dropped in the ears "to restore hearing" (Hamel and Chiltoskey 1975). The Choctaw treated biliousness with the root (Swanton 1931). The Delaware used the herb as a laxative and spring tonic (Moerman 1998). The Iroquois used it as a strong physic, to treat boils, as a laxative, and as a poison to soak corn seeds in before planting. The Menomini also used it to kill insects on cultivated plants (Mahr 1955a). The Meskwaki used the root to treat rheumatism, as an emetic, and as a physic (King 1984). The Osage used the powdered root as a cathartic, gave it as an antidote for poison, and used it to ward off fever (Hunter [1823] 1973). The Penobscot used May-apple to cure warts (Vogel 1970).

Tiny quantities of roots, leaves, or green fruits are poisonous, and powdered root and resin can cause skin and eye problems. *Podophyllum peltatum* contains more than 15 biologically active compounds, including podophyllin, and a bitter resin containing lignins and flavonols (Diggs et al. 1999). Podophyllin, concentrated in the root, is the drug of choice for treating genital warts although it is highly allergenic (Foster and Duke 1990). Etoposide, a semisynthetic derivative from the roots, is the first-choice drug for treating small-cell lung carcinoma and testicular cancer (Foster and Duke 1990, Coffey 1993, Hocking 1997, Mabberley 1997, Swerdlow 2000). Teniposide is used to treat brain tumors and childhood leukemia (Swerdlow 2000). *Podophyllum hexandrum* has larger rhizomes than the American species and contains more resins (12%) with about twice the content of podophyllotoxin than *P. peltatum* (Hocking 1997). *Podophyllum hexandrum* also contains podophyllotoxin, desoxypodophyllotoxin, isopicropodophyllone, astragalin, hyperin, quercetin, and kaempferol (Hsu 1986).

The major disperser of *P. peltatum* in Delaware forests is the box turtle (Mabberley 1997). Apparently, that is why the herbs can get away with "hiding" their fruits below the canopy of leaves where birds and many other animals have difficulty finding them.

Polygala: Milkworts

(From Greek *poly*, much, *gala*, milk)

Milkworts are everywhere in the pine flatwoods in late April. Sometimes thousands of plants belonging to nine species will be in view on ridges and swales. As I look at the plants, I can almost see the Greek physician and author Pedanios Dioscorides from the

Polygala. a. *Polygala cruciata*. b. *Polygala lutea*. c. *Polygala grandiflora*. d. *Polygala rugelii*. e through k. *Polygala nana*. e. Habit. f and h. Keel, two views. g. Flower. i. Filaments connected to petals. j. Mature seed. k. Pistil. *a through d drawn by P.N. Honychurch. e through k drawn by Vivian Frazier.* From Correll and Correll 1972.

1st century A.D. selecting samples. Bending over, he gently snaps off the stems, and thinks that the bitter plants will produce *polygalon* (much milk). They will be useful to the farmer who had recently come to him for help renewing the flow of milk in his cow. Dioscorides tucks the plants gingerly into a small bag he has draped over his shoulder and walks on.

Speakers of different Mediterranean languages called these plants their linguistic equivalents of milkwort (wort, plant) long before the Dutch herbalist Rembert Dodoens published the Latin name in 1554. Henry Lyte's translation of Dodoens's Dutch text on *Polygala* says that it "engendreth plentie of milk; therefore it is good to be used of nurses that lack milk." Others following this idea include Italian *poligala*, Spanish *hierba lechera* (milk giving herb), Portuguese *erva leiteira*, and French *latier* (milk giver). Gaels also call them *lus a'bhainne* (milk herbs). In spite of these names and beliefs, there is no experimental evidence that the plant extracts increase milk flow.

Europeans know the plants by other names. Norwegians call them *blåfjaer* (blue-flower). While not all their species have blue flowers, they retain the basic name and add modifiers, such as *bitterblåfjaer* (*P. amarella*) or *storrblåfjaer* (*storr*, big, *P. vulgaris*). In Guernsey *Polygala* is *herbe de paralysie*, and it is used to treat or prevent paralysis or strokes. Gaelic speakers in Scotland call the herb *saibann nam bansidh* (fairy women's soap). Indeed, many species contain saponins (soaplike compounds). The Germans call it *Kreuzblume* (cross-flower). One can see a resemblance (with considerable imagination) to a cross in the flowers.

Many of the Florida species have common names that reflect their colorful flowers. Bachelor's buttons, the least creative among these, was adopted from unrelated European plants (*Centaurea cyanus*, Asteraceae). Sometimes that name is applied to all the species, but there are more often modifiers, as in white bachelor's-button (*P. balduinii*), bog bachelor's-button (*P. lutea*), dwarf bachelor's-button (*P. nana*), and yellow bachelor's-button (*P. rugelii*).

Some of the more intriguing common names for milkworts are given to *P. cruciata* (drum-heads) and *P. pauciflora* (gay-wings, bird on the wing, baby's toes, baby's feet, baby's slippers, satin flower, Indian pink, maywing). However, the most curious names for Florida's *P. incarnata* are procession-flower or Rogation-flower, both of which it shares with European *P. vulgaris.* Herbalist John Gerarde in 1597 explained these names by saying that the plants "flourish in the Crosse or Gang weeke, or Rogation weeke; of which floures the maidens which use in the countries to walk the Procession doe make themselves garlands and nosegaies, in English we may call it Crosse-floure, Rogation floure, and Milkewort, of their virtues in procuring milk in the breasts of nurses." Rogation Sunday is the fifth Sunday after Easter, and it is followed by Rogation Week when church processions (or "gangs"), led by a person carrying a cross, bless crops. This may be another pagan ritual incorporated into the Christian religion, because there is a word in Gaelic for the procession—*liodan.*

Milkworts, also known as candyweeds or candyroots (because of a licorice taste to the roots of some), are small herbs in North America and northern Europe, but in drier climates and within the tropics they may be shrubs or even trees (Mabberley 1997). Some species produce dyes, and one from tropical Africa (*P. butyracea*) yields a fiber.

Many species around the world are used in medicines, although only one in North America has received much publicity. That northern species, *P. seneca* (snake-root, Seneca snakeroot), became famous as a snakebite remedy when the first Europeans arrived in the New World. Although its effectiveness in treating snakebite is doubtful, the species became popular for treating pleurisy, the most common ailment in colonial Virginia (Coffey 1993). Even William Byrd (1674–1744), one of the surveyors of the line between North Carolina and Virginia, used it to treat gout in a member of his party in early 1728 (Byrd [1728] 1980).

Snake-root contains the glucoside senegin (a saponin), polygalic acid, resin, methyl salicylate, and fatty oils (Hocking 1997). Several of these make the plant potentially effective in medicines as an emetic, expectorant, cathartic, diuretic, antispasmodic, and sweat inducer, to regulate menses, for colds, and against croup, pleurisy, rheumatism, heart troubles, convulsions, and coughs, and as a poultice against swelling. The presence of methyl salicylate also supports its use against some of these maladies. This chemical is more familiar under the name of "wintergreen" and it has long been used in medicine and flavorings. Florida's *P. boykinii* shares methyl salicylate with others, and has been used in Mexico in a cold water infusion to correct dizziness (Hocking 1997). Similarly, Seminoles told Sheehan in 1919 that it was a medicine for treating vertigo (von Reis and Lipp 1982).

Pollen of the genus has been found in the pre-Columbian deposits at the Glades site of Fort Center on the western side of Lake Okeechobee (Hogan 1978). The Seminoles have different names for *P. rugelii* and related species, but they seem to use each one as a generic term for several species. The species has had religious significance in Florida for a long time, and this relationship is indicated by the Miccosukee name *sápiyâ:bî* or *sápiyî* (resembling the mythical plant "*sápiyi*") (Sturtevant 1955). Snow and Stans (2001) used *svpeyv* as the Creek name.

A more detailed explanation of *svpeyv* is given by Martin and Mauldin (2000). They wrote that this is the name of "a plant whose root is used as a charm," basing their comment on some notes made by Mary R. Haas ca. 1940. For the second meaning, they added, "a charm (once used for hunting, but now used to attract a suitor)."

Sturtevant (1955) wrote that the Creeks called *Polygala*, regardless of the species, *hi:lamási* (*hele*, medicine, *em*, its, *vse*, tea, Creek). He was told that they used both *P. lutea* and *P. rugelii*, but that the former was better. His informants used *Polygala* for "Sapiyi Sickness" (heart palpitations, yellow skin, body swelling, shortness of breath), "Chronic Sickness," snakebite, birth medicine, and sorcery.

Snow and Stans (2001) added the Creek names *svpeyv hvlwat* (*svpeyv*, polygala, *hvlwv*, high, *-at*, the one that is), and *svpeyv lopockuce* (*svpeyv*, polygala, *lopocke*, small, *uce*, small), plus the Mikasuki names *wootaacheeke em oekekche* (properly *wootaacheeke em alekche*, emetic medicine; from *wootaach-*, make vomit, *-eeke*, thing that, i.e., emetic, *em*, its, *alekche*, medicine), *tofoome chayhe* (*chayhe*, tall, Mikasuki; the first word may be cognate with Creek *eto-home*, bitter wood), and *shapeye peshkooshka* (small polygala). Bennett (1997) was told the Mikasuki name *eeye mashe* (*eeye*, foot, *em*, its, *ashe*, tea).

In the "Plant Identification Chart for Creek Speakers," Snow and Stans (2001) list *P. grandiflora*, *P. lutea*, and *P. rugelii* only under *svpeyv* and *wootaacheeke em oekekche*. There is apparently more about identification and use of the genus than the simple version given by Sturtevant (1955) indicates. Organisms that have uses older than cultural memory always have long and complex associations with people.

Snow and Stans (2001) illustrate *P. grandiflora* in their Plate 24, and call it *svpeyv*, with the English names "small one" and "candy root." Snow wrote, "You find this plant in an open place on dry or damp land. Get four whole plants with the root attached. The thin stem is about a foot tall with small purple flowers, and the root is white. You use *svpeyv* for treatment to clean the body and to vomit. It is used in 'on the wagon medicine' as well." In keeping with Snow's comment that *svpeyv* cleans the body, large doses of *P. rugelii* are reported to act as a strong laxative.

The Choctaw used *P. lutea* (bog bachelor's-button, candyweed, wild bachelor's-button), calling it *kwonokashaipsa* [*kwonokasha ipsa*] (*kowaanakaasha*, little people, *impa*, eat it), as a poultice to treat swelling by infusing dried blossoms in hot water (Bushnell 1909). Florida's other species with a recorded use is *P. polygama*. As with several species, it has been used to treat coughs. Both probably contain the saponin wintergreen.

John K. Small (1869–1938), the botanist from New York Botanical Garden who explored Florida during his winter breaks, was the first to recognize one endemic species as distinct. He called that herb *P. arenicola* in 1905. In the 1970s, Robert R. Smith and Daniel B. Ward at the University of Florida realized that, because of a legal technicality, the plants needed a new name. They commemorated this New Yorker's keen observations by dubbing the plants *Polygala smallii*.

Urbanization in southern Florida has pushed *P. smallii* to the edge of extinction. The species was proposed for the Federal Endangered Plant List, and it became one of the first from the region to be listed in 1985. Because no one knew much about the plants, or why they were so restricted, several studies were initiated.

Pamela Krauss, while at Florida Atlantic University, discovered by the early 1980s that the species was restricted to small sandy spots in rockland pine flatwoods in Miami-Dade County. Apparently, the Broward County plants were extirpated by then. About the same time John Popenoe, then Director of Fairchild Tropical Garden, found the species in Martin County's Jonathan Dickinson State Park. In the late 1990s George Gann rediscovered the plants in Martin and subsequently in nearby Palm Beach and St. Lucie Counties.

Even these northern populations reproduce erratically, and studies for the Florida Native Plant Society by Christine Lockhart are producing population data that will help future management of this highly endangered Florida endemic. One of the aspects of its biology that is a contributor to its spotty distribution is its isolation in "pockets" of sand within pinelands. Another of those aspects may be dispersal by ants.

Ant colonies have limited ranges, and that restricts where the seeds can be carried. To accomplish their dispersal, *Polygala* seeds have special "food-bodies" (elaisomes). Those tiny structures attract these small foraging insects. Ants carry the seeds back to their nests, eat the food, and then discard the seeds outside. The seeds, having been put in a rich garbage heap with plenty of open space and reduced competition, germinate and provide new colonies. Perhaps our disruption of native ants with pesticides and alien introduced ants is showing us the "ripple-effect" of disturbing one small segment of the web of life.

We humans are incredibly arrogant and egocentric, and we tend to think that only big animals are important. Still, we can make fun of ourselves, and ants. There is a song *High Hopes* (1959) where, although everyone said it could not, an ant carries off a rubber-tree plant. So, never underestimate the importance of your neighbors.

Polygonatum

(From Greek *polys*, many, *gonu*, knees, alluding to the many joints of the rhizome)

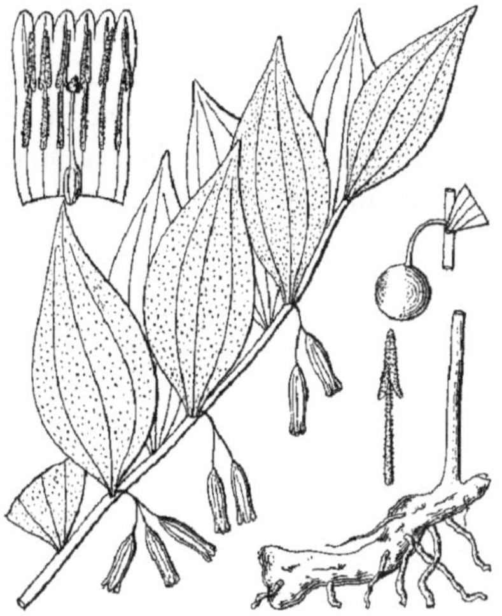

Polygonatum biflorum. From Britton and Brown 1896.

frassinella (little ash, Italian 1551)
Ladder to Heaven (John Parkinson wrote in 1640, "Wee in English [call it] Salomon's Seale most usually, but in some countries the peope call it Ladder to Heaven, according to the Latine *scala coeli*, which was anciently known to the Apotheccaries shoppes, from the forme of the stalke of leaves, one being set above another")
poligonia (derived from the Greek *polygonaton*, Spanish 1557)
Salomans seale (Gerarde [1597] 1975); Solomon's seal (translated from Latin *Sigillum solomonis*; used in English by 1526 in the *Grete Herball*); *Salomons segel* (Dutch 1549); *Salomonssiegel* (modern German); *sceau de Salomon* (French 1549); *sigillo di Salome* (modern Italian)
scala caeli (ladder to heaven, Turner said this was English in 1548 even though it is Latin)
Weiswurtz (white herb, German 1542)

Polygonatum biflorum (two-flowered)
seal-wort [sealwort]
[great, King] Solomon's seal; *sceau de Salomon* (Solomon's seal, Quebec)
utistugi' (Cherokee)

According to the OED (1971), people in the Middle Ages (A.D. 500–ca. 1500) began calling European plants *sigillium Solomonis* (Solomon's seal). That name was said to be based on an even older history, going back to the time of the biblical King Solomon. Greek physician Dioscorides (fl. A.D. 40–80) called the plants *polygonaton* and wrote, "Spreading on the root helps cure wounds. Indeed, it also removes and aids facial blotches." Roman doctor Galen (A.D. 129–?200) recorded, "The root of polygonatum is spread on wounds. It is with it that they clear birthmarks [moles] from the face." In medieval times the plants were used to "seal" or "mend" wounds, sores, and especially bruises. Fuchs wrote in 1542, "Women today still wash and color their faces with Solomon's-seal." (Meyer et al. 1999).

Fuchs noted in 1542 that the pharmaceutical name for the plants in Germany was *Sigillium solomonis* (Meyer et al. 1999). Passage of time led to confusion and there arose three versions of why the plants were named after Solomon. One of the stories said that transverse sections of the rootstocks resembled a seal used by the biblical king. Another held that it was not the section but the leaf scars that looked like his seal. The third view held that it was because the root was good "to seal and close up green wounds" (OED 1971).

The Cherokee, Menomini, Meskwaki, Ojibwa, and Rappahannock used the American plants as medicine (King 1984, Moerman 1998). The Cherokee treated dysentery, breast diseases, carbuncles and other skin problems, leukorrhea, stomach problems, and lung diseases with the herbs (Hamel and Chiltoskey 1975). They also considered the root a mild tonic. The Menomini used the roots as an analgesic. The Menomini and Meskwaki put the root in an incense to revive unconscious people. The Ojibwa used the root as a physic and cough remedy. The Rappahannock treated cuts, bruises, and sores with the roots in a salve.

The Cherokee used the dried roots to make flour for bread (Hamel and Chiltoskey 1975). Young stems were also eaten like asparagus. Yanovsky (1936) and Fernald et al. (1958) say that several northern tribes used them in the same ways. The Ojibwa burned the roots as incense before going to bed (Moerman 1998).

Polygonum

(Greek *poly,* many, *gonum,* knees or bends, in reference to the many swollen nodes on the stems)

arssmerte (for *P. hydropiper,* by Turner in 1548)
corriola bastarda (false morning glory, Portuguese)
glùineach (having large knees, Gaelic)
Knöterich (knotted, German)
knotgrass (from the knotted or jointed stem of *Polygonum aviculare,* in English by 1500)
lus an fhògair (banishment herb, Gaelic for *P. hydropiper*)
renouée (French)

Polygonum densiflorum (compactly flowered)
denseflower knotweed (USA)
guinea hen bush (Bahamas)
herbe pou poule (little chicken herb, Guadeloupe, Martinique)
John doctor (Bahamas)
lechuga (lettuce, Guatemala)
piment vache (cow pepper, Guadeloupe, Martinique)
snout smartweed (Texas)
yerba de hicotea (turtle herb, the *jicotea* is a freshwater turtle, Taino, Puerto Rico)

Polygonum hydropiperoides (resembling *P. hydropiper*)
American water pepper [water-pepper] (USA)
chillillo (little chile, Valley of Mexico)
flor de chajutal (hot flower, Guatemala)
swamp smartweed (Florida)
tasteless knotweed (a misnomer; the plants are spicy when chewed, USA)

Polygonum punctatum (dotted, meaning the tepals)
canilla de pava (strong turkey, Guatemala)
chile de perro (dog's chile, Costa Rica); [*chilillo*] *de perro* (little [dog's] chile, Guatemala); *chilillo rojo* (red chile, Veracruz); *piment vache* (cow pepper, Guadeloupe, Martinique)
curage (cleansing, French, Houma, Louisiana)
ojig'imin (fisher-berry, Ojibwa)
roríspwa (*korí,* chile or *Capsicum annuum,* Tarahumara, Chihuahua)
[American, dotted, water] smartweed (Florida, Texas, Bahamas, Puerto Rico)

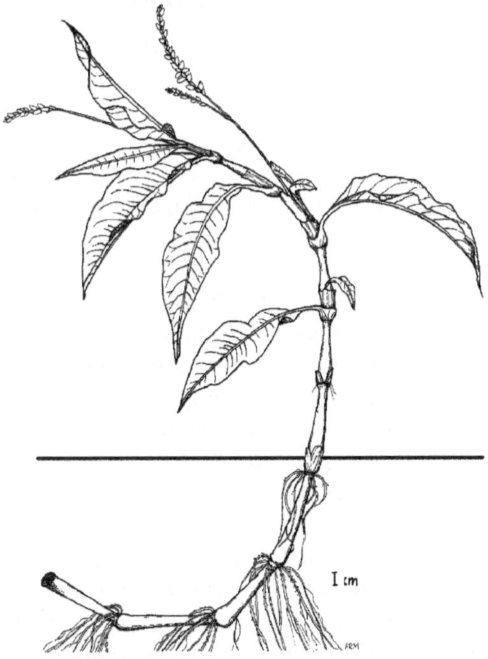

Polygonum. *Polygonum densiflorum* (left). *Polygonum hydropiperoides* (right). Both from Institute of Food and Agricultural Sciences.

yerba de burro (donkey's herb, Dominican Republic); *yerba de caimán* (caiman's herb); *yerba de hicotea* [*hycotea, jicotea*] (turtle herb, the *jicotea* is a freshwater turtle, Taino, Dominican Republic, Puerto Rico)

Most of *Polygonum* is referred to as "smartweed." Diggs et al. (1999) cite a comment by Kirkpatrick (1992) suggesting that the name was derived from the tendency of the sap to "smart" (burn) when it touched the skin. However, historical data support a slight modification of that view. "Smartweed" was applied to *P. hydropiper* by 1787. In that year, W.H. Marshall, writing on the rural economy of Norfolk, said, "*Smartweed*, biting and pale-flowered persicarias; arsmart." Indeed, the OED lists "arsesmart" as the common name for "smartweed." So, Kirkpatrick had the right idea, but the wrong part of the body.

Several species have a peppery taste, like *P. hydropiperoides* and its namesake *P. hydropiper* (water pepper). Indeed, Dutch herbalist Rembert Dodoens simply called the second species *Hydropiper* in 1554, surely reflecting its use in seasoning food. Fernald et al. (1958) noted that application for it and several relatives. The practice is problematical because some people develop contact dermatitis from the plants (Foster and Duke 1990).

There appear to be no records of people in the Americas using *P. densiflorum*, but Hawaiians use an infusion of the plants to purify the blood (Moerman 1998). Perhaps that species is used like some of the others in the New World although there are no records to support that. In the Bahamas, the root of *P. densiflorum* is crushed and inhaled to relieve headache (Higgs 1969).

The earliest association of *Polygonum* with Florida people was *P. hydropiperoides* in pre-Columbian Glades coprolites from Fort Center on Lake Okeechobee (Hogan 1978). *Polygonum hydropiperoides* contains tannins, rutin (3% in the leaves), quercitin, and kaempferol (Hocking 1997). It has been used to stop intestinal and uterine hemorrhage, for strangury (slow, painful urination), and as a stimulant. In Brazil, *P. hydropiperoides* is considered diuretic, and an emmenagogue, and used as a treatment for piles. It is used to promote conception in Mexico (Ford 1975, Hocking 1997, Vásquez and Jácome 1997).

Among the Houma, a decoction of *P. punctatum* roots was used to treat pains and swelling in the legs and joints (Speck 1941). Farther north, the Ojibwa took a decoction of leaves and flowers for stomach pain, and the Iroquois made a compound medicine for "loss of senses during menses" (Moerman 1998). The Tarahumara of northwestern Mexico used the plants as a fish poison (Pennington 1958). They bundled the plants, crushed them, and put them in large baskets, which they dipped into water until the water turned green. According to Pennington (1958), "fish rise to the surface almost immediately." Tarahumara also added young leaves to their corn dish *esquiate* for a spicy flavor.

In Hispaniola, *P. punctatum* and other plants in the genus are rubefacients when crushed; they are stimulant, astringent, diuretic, emmenagogue, anthelmintic, and antiarthritic. On that island, the species is used against urinary and gallbladder problems, and to treat hemorrhoids, intermittent fever, and malignant sores (Liogier 1974). Several of the common names mention dogs. Those are indications that the herb is wild, but perhaps also indicates that the plant has been used to treat mange (*jiote*, from *xiotl*, Náhuatl) in dogs (Morton 1981).

Europeans were familiar with *Polygonum* when they arrived in the New World. They had learned that the small seeds from weedy species like *P. aviculare* could be harvested and used, at least as a starvation diet (Fernald et al. 1958). Indigenous Americans also used *P. erectum* in the same way. Indeed, there is evidence they cultivated that species for food (Smith 1992). More recently, the seeds of at least *P. hydropiperoides* have been promoted as a protein source (Boyd 1968, Boyd and McGinty 1981).

Polyporus

(The genus was named by E.M. Fries, 1794–1878, from Greek *poly*, many, and Latin *porus*, small openings)

false truffle (USA)
okipen (earth tuber, Virginia Algonquian)
turma (truffle, name used by Fontaneda [1575] 1944)

When Hernando Descalante Fontaneda was rescued from his captivity among the Calusas of southern Florida, he returned to Spain and wrote an account of his time among those people (Fontaneda [1575] 1944). One of the few "plants" he mentioned was a root eaten by the Floridians. In speaking of people who lived on the *laguna de mayaimi* (Lake Okeechobee), he wrote, "*y sobre esta laguna que corre por en medio de la tierra dentro tiene muchos pueblos aunque son de treynta i quarenta vso. y otros tantos lugares tienen pan de rraises ques la comida ordinaria la mar parte del tienpo, aunque por caso de la laguna que crese mucho que no alcansan estas rraises por estorbo de la mucha Agua y ansi dejan de comer Algun tienpo este pan pescado mucho y muy bueno, otras Rayses a manera de turmas de las de Aca duses.*"

Buckingham Smith translated this as follows: "On this lake, which lies in the midst of the country, are many towns, of thirty or forty inhabitants each; and as many more places there are. They have bread of roots [*Smilax*], which is their common food the greater part of the time; and because of the lake, which rises in some seasons so high that the roots cannot be reached in consequence of the water, they are for some time without eating this bread. Fish is plenty and very good. There is another root, like the truffle over here, which is sweet."

In spite of Fontaneda's comparison of the second root to truffles, Smith interpreted the *turma* as being *Apios americana*. Smith had apparently never been in swamps where *Apios* grows, or he would not have selected that species. If the water was too high for their common food (*Smilax*), it was certainly too deep to gather *Apios*. Elsewhere I have made the argument that the "bread" noted by Fontaneda was the carpophore of *Polyporus* (Austin 1980). Gerard (1907) found the product also being eaten by the Virginia Algonquians.

Polypremum

(From Greek *polypremnos*, many-stemmed)

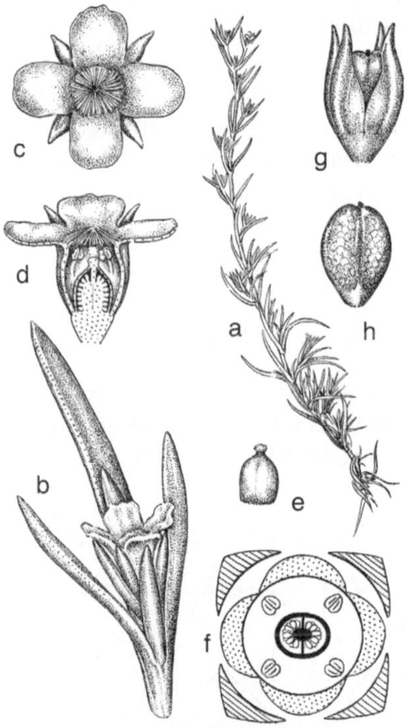

Polypremum procumbens. a. Flowering branch. b. Twig with flower. c. Flower, from above. d. Flower, longitudinally dissected. e. Pistil. f. Floral diagram. g. Fruit within calyx. h. Fruit. *Drawn by Priscilla Fawcett*. From Correll and Correll 1982.

Polypremum procumbens (prostrate or lying on the ground, meaning the stems)
rust-weed

Polypremum was named by Linnaeus in 1753. The genus contains a single species in the warm parts of the Americas. Although few disagree that it a distinct species, there is contention regarding its familial relationships (Cronquist 1981, Mabberley 1997).

The plant is used as a remedy for metritis (uterus inflammation) in El Salvador (Von Reis 1973).

Polystichum

(From Greek *polys*, many, and *stichos*, rows, alluding to the sori of some species being in ranks)

Polystichum acrostichoides (resembling *Acrostichum*, which see)
bear's bed (derived from the Cherokee name); *yana utseta* (the bear lies on it, Cherokee)
canker-brake ("canker" came from Old Northern French *cancre*, and appeared in English by about A.D. 1000; the word is cognate with "cancer"; historically, the disease was a malady of the mouth; "brake" is akin to Old Swedish *braekne*, fern)
Christmas fern
dagger fern (from the shape of the leaflets)
fougère à faucilles (sickle fern, Quebec)
shield fern
tapasi' moso'here (flower on [branch grows] rough, Catawba; not distinguished from *Pleopeltis*, which see)

No one had to explain the reason for calling the plants "Christmas fern" when I learned it on a college course field trip in the early 1960s. Around us the woodland was bleak from the winter cold, with bare trees, dead and fallen leaves on the ground, and ice covering the surface of the creek. Tucked in a hillside crevice of rich organic soil was this single green plant. From the name, I surmised that people had used this evergreen plant in decorations at Christmas time. I was correct, but the fern appeared in the literature under that name only in 1878. People of North America probably began using the name much earlier.

Polystichum is a genus with four European species (Mabberley 1997). While Kartesz (1994) recognized 24 species and 2 named hybrids in the United States and Canada, only 15 were included in the *Flora of North America* (Wagner 1993).

The German physician Albrecht W. Roth (1757–1834) created the genus *Polystichum* in 1799 for a species in his country. In doing that he moved a species Linnaeus called *Polypodium lonchitis* into his new genus. That species is now known to range from Europe to western North America, with isolated localities around the Great Lakes, Newfoundland, and Nova Scotia.

Later, André Michaux discovered the most common species in eastern North America in 1803, and they were given their modern name in 1834. Unlike some ferns, these have a comparatively rich history of uses.

Hamel and Chiltosky (1975) confirm that the Cherokee ate the fiddleheads. We question that because neither Hedrick (1919) nor Fernald et al. (1958) list the plants.

The more common application of the plants was as medicine. The Cherokee used the plants as an emetic, and to treat rheumatism, chills, fever, stomachache, bowel problems, and toothache (Hamel and Chiltoskey 1975). The Iroquois treated cramps, convulsions, diarrhea, fever, and rheumatism, used it as a blood purifier, and as an emetic for dyspepsia; it was taken before and after birth to clean the womb. The Malecite and Micmac chewed roots to relieve hoarseness (Moerman 1998).

Pontederia

(With this name, Linnaeus commemorated Italian physician Guilio Pontedera, 1688–1757, professor at Padua and also the prefect of the Botanical Garden of Padua from 1719 to 1757)

Pontederia cordata (heart-shaped, the leaves)
hicaknâ:blo:cî [*hitcakanablotci*] (*hicakna*, feral-taro, *a:bi*, replica, *lo:ci*, black, Mikasuki; see *Sagittaria* for more on etymology); *hikacha nabe* (*hicakna*, feral-taro, *a:bi*, replica, Mikasuki)
hishi shafuha (*hesse*, leaf, *cofokne*, pointed, Creek)
pickerel [pike]-weed (USA)
wampee (it is white, Shawnee); *wómpí* (Massachusetts)

Linnaeus ([1753] 1957) knew this species from publications by several previous authors before he studied the living plants at the *Hortus Cliffortianus*. He knew only one, but there are now five American species in *Pontederia* (Mabberley 1997).

Although Sturtevant (1955) was told this herb was useless, seeds may be eaten directly from the plant, boiled as cereal, or dried for future use (Fernald et al. 1958). Dried seeds were ground and made into flour for making bread (Morton 1968b).

Farther north, the Malechite, Micmac, and Montagnais people used the plants for medicine. The

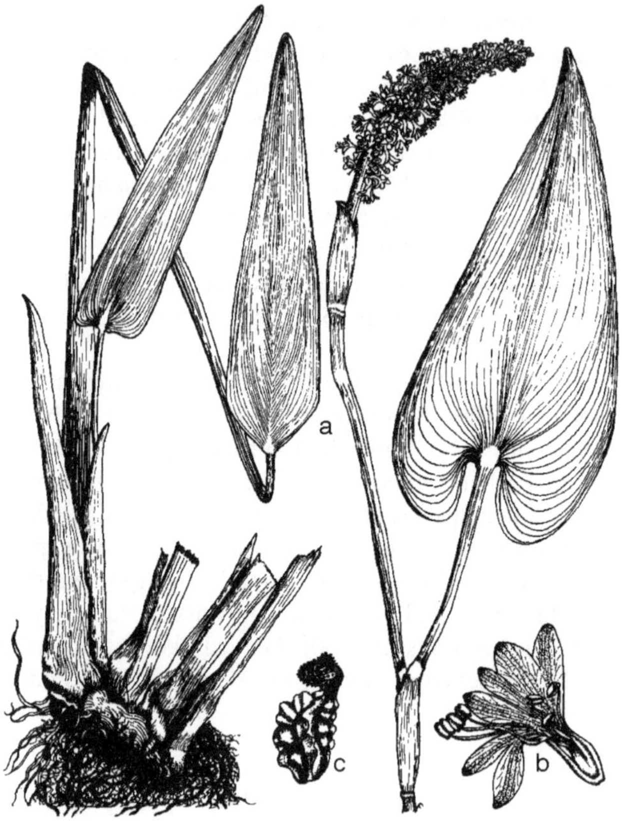

Pontederia cordata. a. Habit. b. Flower. c. Revolute-coiled perianth after flowering. *Drawn by Vivian Frazier*. From Correll and Correll 1972.

Malechite of New Brunswick and Micmac of Nova Scotia used *Pontederia* as a contraceptive (Moerman 1998). The Montagnais of eastern Quebec and Labrador made a treatment for "illness in general" from the herbs.

Populus

(Classical Latin name for the genus, maybe referring to people; or perhaps from Greek *paipaloo*, shutter, or *paipapalloomai*, tremble, vibrate)

Populus deltoides. From Sargent 1905.

álamo (based on *ala*, wing, in both Spanish and Portuguese; presumably from a resemblance of the fluttering leaves to wings on birds)
amocholhe (Delaware)
crintheann (trembling, Gaelic)
eadha (Gaelic)
poplar (from Latin *populus*, spelled "popler" by Turner [1548] 1965); *chopo* (from Latin *populus*, Spanish); *choupo* (Portuguese); *Pappel* (German); *pappel* [*popel*] (Dutch); *peuple* (French); *pioppo* (Italian); *poppel* (Norwegian, Swedish)
squejóna (Onondaga)

Populus deltoides (deltoid or triangular, from the outline of the leaf)
álamo (cottonwood, Texas); alamo cottonwood (a redundant name)
ashumbala [*shumbala*] (Choctaw); *hashoomala* [*hasho'mala, hashoomala', ashomala, ashoomala, ashoomala'*] (questionably from *hasha*, leaf, *im*, its, *ala* =?, Chickasaw)
ba'-k'a hi (a sacred tree, used in rites, Osage); *chan ya'hu* (*chan*, tree, *ya'hu*, peel off, in reference to their use of the bark as food for horses, Dakota); *maa zhon* (*maa*, cotton, *zhon*, tree, Omaha-Ponca); *wága chan* (*wága*, take off, *chan*, tree, Dakota)
[eastern, southern, yellow] cottonwood ("cotton," originally the wool surrounding the seeds of *Gossypium*, brought into English about A.D. 1300 from French *coton*, which in turn came from Arabic *al-qoton*, with the deletion of the article. Lewis and Clark called the trees "Cotton Timber" at their winter quarters with the Mandan in 1804. Although the OED 1971 says that "cotton" was combined with "wood" about 1823 to indicate the cottony fluff around the seeds of these trees, "cottonwood" was also used in the Lewis and Clark journal of 1804.)
ete hesha kaklahashe (noisy leaf tree; from *ete*, tree, *hissi*, leaf, *chashahachi*, rattle, Choctaw)
hecélwv [*hecelwv*] ("poplar," Muskogee; apparently not cognate with Creek *tvłtahkv*)
itti' tohbi' (*itti'*, tree, *tohbi'*, white, Chickasaw)
laird (from *lier*, to bind, related to Latin *ligare;* some books specifically mention fascicles or bundles, perhaps of wood, Quebec)
natakaaru (Pawnee)
peuplier (from Old French *peuple*, still in use by some country people, from Latin *populus*, dated ca. 1170, Quebec)
[Carolina-, necklace-, water-] poplar
pû'titu kayudi' (Ofo)
tai [*tay*] (Atakapa)
tvłtahkv [*tarrahkv, tartahkv, tvrtakhv, tałtahkv, tałtv'hkv, tałłahkv*] (Creek); *tałtahká* [*tałahká*] (Koasati)
ya-hee-hwai [*ä'hi'ñ, ä-heeñ*] (*ä-heeñ*, principal, *hwai*, tree, Kiowa)

Populus heterophylla (variable leaves)
[black, swamp] cottonwood
downy poplar

According to Greek legend, Hercules was wearing a crown of poplar twigs when he returned from Hades after vanquishing Cerberus, the guardian hell-hound (Baumann 1993). The Greek tree forming his crown was the white poplar (*Populus alba*), and the two-colored leaves symbolized respect for the Chthonian deities. The dark side of the leaf represented the Underworld, and the light side the living. This tree became the *arbor populi* (tree of the people) to the Romans, and it was used to decorate public places such as the *Piazza Poppolo* in Rome (Hocking 1997). Just how much that Greek association with the Underworld affected different cultures is partly indicated by the dominance of their word in languages throughout Western Europe. The species is now cultivated as far north as Norway and Sweden where it is known as *sølvpoppel* (silver poplar).

In a parallel view, people of the Missouri River region also held the cottonwood as sacred but surely for different reasons. The Sacred Pole of the Omaha was made from cottonwood, and the bark was used as fuel for roasting the clays to make paints for heraldic and symbolic paintings on their bodies (Gilmore 1919).

Many people throughout the range of *Populus* in North America have eaten the inner bark (Moerman 1998). The Cheyenne, Blackfoot, and Dakota ate the inner bark and sap, particularly in the spring (Gilmore 1919, Moerman 1998). Other people eating various parts of *P. deltoides* include the Coeur d'Alene, Flathead, Kutenai, Montana, Ojibwa, Omaha, Pawnee, Pima, and Ponca (King 1984, Moerman 1998). In the southwestern United States, many people ate *P. deltoides* ssp. *wislizeni* (Moerman 1998).

At least four dye colors were obtained from *P. deltoides*. The Cheyenne used the buds to make green, brown, and red dyes, although the methods are not given (Moerman 1998). Missouri River people also used the buds to make yellow dyes (Gilmore 1919). Tull (1999) indicates that yellow may be obtained with alum, tin, or chrome as mordants. Green results from copper or iron mordants.

Wood from *Populus* was an important resource throughout its range. Swanton (1946) found the southeastern tribes making their largest canoes of poplar. Southeastern people also used the wood to make stools, doors of houses, and fire sticks. *Populus deltoides* wood more recently has been used for paper pulp, cases and crates, tubs and pails, excelsior, veneer for plywood, musical instruments, dairy and poultry supplies, laundry appliances, and fuel (Vines 1977).

There are numerous records of "poplar" being used among indigenous people for a variety of medical treatments. Many of those references are impossible to pin to a species. Probably Duke et al. (2002) have made the best choice by listing them all under the name "poplar" (*Populus* spp.). Caution is advised, however, because reports of "popular" include both *Liriodendron* and *Nyssa*. Since the *Populus* species all contain salicin, it seems likely that they were used similarly in medicines.

The *Populus deltoides* was used by several southeastern tribes and by some in the northeast. The Catawba used an infusion of the bark, along with wild cherry and dogwood to treat expectant mothers (Vogel 1970). The Chickasaw boiled cottonwood and willow roots to make a drink to treat dysentery and fever (Swanton 1928a). The Choctaw boiled leaves and bark to treat wounds and made a combination of stems, leaves, and bark to cure snakebite (Bushnell 1909). The Creeks used a decoction of cottonwood to treat sprains and fractures and a decoction of roots as a remedy for dropsy (Swanton 1928a, Taylor 1940). The Delaware combined cottonwood with black haw and wild plum bark to make a woman's medicine (Moerman 1998). The Iroquois made a vermifuge of the plants (Hocking 1997). The Ojibwa used the cottonwood in two distinct ways. Buds were stewed with bear fat and used to treat earache, while the cotton from fruits was used as a absorbent on open sores (Vogel 1970).

Porcher (1863) wrote of *P. heterophylla*, "Upon examining the excrescences caused by an insect in large numbers on the leaves of the cotton-wood tree ... I find them possessed of great bitterness, and suggest an examination into their tonic properties." No other indication has been found of others following this lead. However, Porcher (1863) recommended using *P. deltoides* extract as a substitute for quinine in treating what was called at the time as "swamp fever," now known as malaria (Hocking 1997).

Early in its history another North American *Populus* became confused with tropical *Bursera*. Because of that confusion, *P. balsamifera* became known as *tacamahaca* (Vogel 1970, Bremness 1994). That Aztec name was originally applied to *Bursera* (see *Bursera:* Gumbo Limbo), but the literature on common names now largely associates it with *Populus*. In another oddity, the European *P. alba* became official in the U.S. Pharmacopoeia between 1895 and 1936 as a source of salicin, but native North American species were never listed (Vogel 1970). To fill that gap, Eclectic physicians recommended several native *Populus* (Culbreth 1910, Felter 1922).

Prenanthes

(From Greek *prenes*, drooping, and *anthe*, flower)

Prenanthes serpentaria. From Britton and Brown 1898.

Hasenlattiach (hare's lettuce, German)
lattuga montana (mountain lettuce, Italian)
prenanthé (French)

Prenanthes serpentaria (old name for various plants used to treat snakebite)
cancer weed
[white] canker weed [cankerweed, canker-root] ("canker" came from Old Northern French *cancre*, and appeared in English by about A.D. 1000; the word is cognate with "cancer"; historically, the disease was a malady of the mouth, perhaps a reference to thrush or *Candida* infection)
dado'cabodji'bik (milk root, for *P. alba*, Ojibwa)
DeWitt snakeroot; rattlesnake root; snake-gentian; snakeweed (names given alluding to use against snakebite)
drop flower (it is not clear if this English name gave rise to the genus created by Linnaeus in 1753, or vice versa)
earthgall [gall of the earth] (used since about A.D. 1000 for *Centaurium pulchellum*, alluding to it being bitter like bile; later applied to *Prenanthes*)

laitue blanc (white lettuce, Quebec); *weisser Lattich* (white lettuce, German); white lettuce
[ivy, joy] leaf
lion's foot (used for a *Helleborus*, Ranunculaceae, by Turner in 1538; and for *Alchemilla*, Rosaceae, by 1610; later applied to other plants); *ped d'leon* (lion's foot, French)
milkweed (usually applied to *Asclepias*, but also to others with white latex)

Two of the *Prenanthes* that Linnaeus ([1753] 1957) knew were from the New World, and the other five were from Europe and Siberia. However, it was not until 1814, when Frederick Pursh published his *Flora America Septentrionale*, that *P. serpentaria* was added to the list. Now there are 30 species known from the north temperate region, and the European plants have been reduced to a single species (Mabberley 1997).

This was one of the many plants used by the indigenous tribes to relieve the bites of snakes and other venomous animals. There are many more poisonous snake species in eastern North America than in Europe, and the settlers were quick to rely on the local remedies for these problems. Accounts are common in the literature of the time of a person being bitten by a reptile and recovering after chewing, ingesting, and/or applying a poultice of some plant.

Writing in 1728, William Byrd gave a typical account of the time. Not only were his observations dubious, but he also clearly had no regard for the safety of his dogs. He wrote: "The rattle-snake, has an utter antipathy to this plant, insomuch that if you smear your hands with they juice of it, you may handle the viper safely. Thus much I can say of my own experience, that once in July, when these snakes are in their greatest vigor, I besmear'd a dog's nose with the powder of this root, and made him trample on a large snake several times, which, however, was so far from biting him, that it perfectly sicken'd at the dog's approach, and turn'd its head away from him with the utmost aversion" (Coffey 1993).

James Adair listed several snakebite plants, including *P. serpentaria*, in a book about his trading among the Creeks along the Mississippi River in 1775. Prince Maximilian listed as a remedy a plant called "lion's heart" among the Delaware in the 1830s (Vogel 1970).

In the eastern states, several tribes used *Prenanthes* as a snakebite remedy. At least *P. alba, P. altissima, P. aspera, P. serpentaria*, and *P. trifoliata* were utilized (Moreman 1998). These all went under the name "snakeroot," "lion's foot," or "gall of the earth."

Northeastern tribes certainly applied *P. alba*, and there are records of both the Iroquois and Ojibwa using it on snakebite (Densmore 1928, Moerman 1998). The Iroquois also used *P. altissima* and *P. trifoliata* (Moerman 1998).

However, indigenous people did not confine the plant to those uses. In the south, the Choctaw used *P. virgata* (reported as *Nabalus asper*, and *P. aspera*, corrected by Joanne Birch) as an anodyne and as a diuretic, and for other problems (Campbell 1951). The Cherokee used both *P. serpentaria* and *P. trifoliata* roots in stomachache medicine (Hamel and Chiltoskey 1975). The Cherokee also cooked and ate the leaves of these two herbs.

Porcher (1863) did not say much about *Prenanthes*. He wrote simply of *P. alba*, "The root is excessively bitter; it is used in domestic practice in this state as a tonic. I would invite further examination."

Millspaugh (1892), on the other hand, had a lot to say about the plants. First, he complained, "This botanically difficult species ... includes in itself what were once considered to be 17 distinct species and varieties; and affords an interminable field of work for a botanist of Rafinesquian tendencies." He continued, "As Gall of the Earth, it has been known in domestic practice from an early date, and is said to be an excellent antidote to the bite of the rattlesnake and other poisonous serpents—one who searches through the domestic literature of medicinal plants, wonders why the bite of snakes ever has a chance to prove fatal." He recommended the species for relieving dysentery, anemic diarrhea, and as a stomach tonic. Hocking (1997) added that the plants were used as a bitter tonic, an astringent, and "to relieve hypertension in Negroes." The genus is not listed by Bremness (1994), Bown (1995), or Duke et al. (2002).

Millspaugh (1892) recorded that chemical studies of the plants had found resins, tannins, gums, and waxes. Hocking (1997) agreed that they contained tannin, and that it was known as *Herba Nabali serpentariae* among the homeopathics. The Latin pharmaceutical name was based on the old generic name *Nabalus*. There were no papers listed on the genus in either PubMed or AGRICOLA in January 2003.

Prunus

(Classical Latin name of the plum, *P. ×domestica*; Greek *proumne*)

Prunus americana (of America)
ciruela (Mexico)
kande (*kande-hi*, plum tree, Omaha-Ponca); *kante* (*kante-hu*, plum tree, Dakota); *ko<u>n</u>'-dse xo-dse* (*ko<u>n</u>'-dse*, plum, *xo-dse*, gray, Osage); *kantsh* (*kantsh-hu*, plum tree, Winnebago)

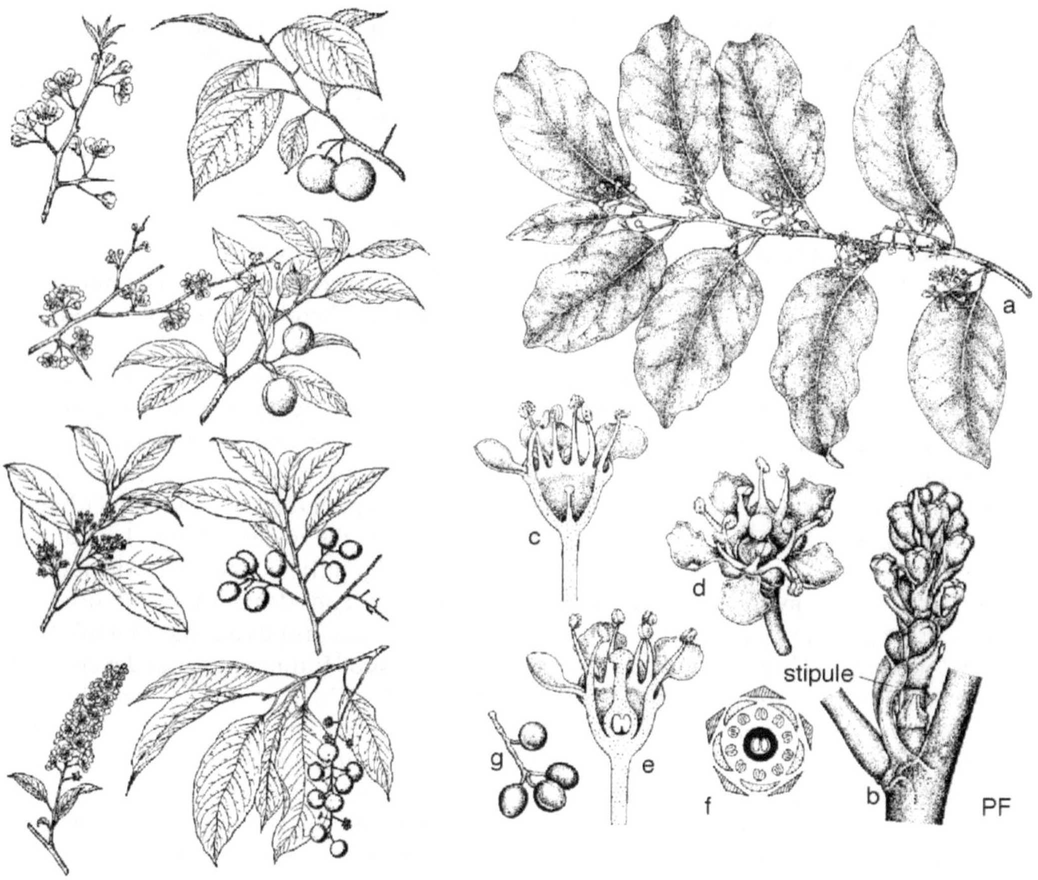

Prunus. *Prunus americana* (top left). *Prunus angustifolia* (upper middle left). *Prunus caroliniana* (lower middle left). *Prunus serotina* (bottom left). All from Sargent 1905. *Prunus myrtifolia* (right). a. Flowering branch. b. Lateral spike at early stage of expansion. c. Staminate flower, the pistillode longitudinally dissected. d. Perfect flower, front-side view. e. Perfect flower, longitudinally dissected. f. Floral diagram of perfect flower. g. Fruits. *Drawn by Priscilla Fawcett.* From Correll and Correll 1982.

kwanunsdiʔi (Cherokee)
niwaharit (*niwaharit-nahaapi*, plum tree, Pawnee)
pank-ai-da-lo (sour plum, Kiowa)
[August, Canadian, goose, hog, horse, native, river, thorn, wild] plum ("plum" from Old English *plumé*, which in turn was derived from Latin *prunus*); Potawatomi plum (a tribe in Michigan; the plum was introduced into Utah); red and yellow plumb (used by Williams [1837] 1962); [American, red, wild yellow] plum (Asa Gray 1875 distinguished red and yellow plums)
poskam (Strachey in [1612] 1953 wrote *poskamatk*; Siebert 1975 found cognates among several other Algonquian languages, some of which were transfers after contact with the imported European fruits; the Shawnee say *poʔkama* for the peach, *Prunus persica*, and used the diminutive *poʔkama Ta* for plum; the Ojibwa say *pokkitons*, but apply it to the pear, *Pyrus communis*); *bu'gesana'tig* (Ojibwa)
pvkánvhe (Muskogee)
sipuamantican (Delaware)
sloe (plum, from Old English *sláh*, used in English since about A.D. 725, related to Frisian *slé*, Dutch *slee*, German *Schlehe*)
tischo (Onondaga)
turi̲·'tcine' (fruit bitter, Catawba)

Prunus angustifolia (narrow-leaved)
akôn̲t atcû'ti (*akôn̲ti*, peach, *atcû'ti*, red, Ofo)
Chickasaw [Chicasa] plum ("Chickasaw," the name of a Muskogean tribe, appeared in English by 1674; used for the plum by André Michaux 1802)
contai (Quapaw)
eco empvkanv (*eco*, deer, *em*, its, *pvkanv*, peach, Creek)
icho intakkosáwwa [*ichintakkosàwwa*] (*icho*, deer, *im-*, its, *takkola*, plum, *sawwa*, small, Alabama); *isi intakkonlushi* [*isi takkonlushi*] (*isi*, deer, *in̲*, its, *takalo*, peach, *oshi*, small, Choctaw); *takkolsalbá* (*takkolá*, peach, *salbá*, maybe cognate with Alabama *sawwa*, small, Koasati)

iti alikchi [*italikchi*] (*iti*, tree, *alikchi*, doctor, Choctaw; an axiom of the tribe was never to kill a cherry tree; considered the best medicine for young girls)
kwah noon:' sdeʔee (Cherokee)
mountain cherry
[sand, sandhill, Indian] plum
stiiñki (Biloxi)
takkonlushi [*takkon lushi*] (*takkon*, originally plum, now peach, *oshi*, small, Choctaw); *takkoosàwwa* [*takkosàwwa*, *takkosáwwa*] (*takkola*, plum, *sawwa*, small, Chickasaw); *takoloshi'* (Chickasaw); *takoloshi' imilhlha'* (*takoloshi'*, plum, *imiłła'*, wild, Chickasaw)

Prunus caroliniana (from Carolina)
[American, Carolina] cherry or laurel-cherry [laurelcherry] ("laurel-cherry" was in use by 1789; the modifiers distinguish the New World from the Asian species)
cherry laurel (in English by 1664; the namesake, *Prunus laurocerasus* was introduced into Europe in 1576 from Turkey, according to Linnaeus; Old World plants were used to make "cherry-laurel water," a watery solution of the volatile oils from the plant containing prussic acid)
ittokchakkosi (*itto*, tree, *okchakko*, blue-green, *osi*, small, Alabama)
mock-[wild-]orange
wild-peach

Prunus geniculata (referring to nodes, sometimes resembling knees, the stems being zigzag)
[Harper's, scrub] plum

Prunus myrtifolia (with leaves like myrtle, *Myrtus communis*)
almendrillo [*almendrito*] (little almond, Cuba, Dominican Republic, Puerto Rico); *almendro* (almond, Venezuela); *amandier à petite feuilles* (little-leaf almond, Haiti); *amandier des bois* (Martinique)
ants-wood (Jamaica)
cassada-wood [wild cassada, wild cassava] (comparing it with "cassava" [*Manihot*], Jamaica)
cuajaní hembra (female *cuajaní*; a river in the Dominican Republic has the Taino name *cuaja*, Cuba); *cuajanincillo* (little *cuajaní*, Cuba)
durasnero de monte (wild peach, Brazil)
la mandit [*le mongier*] (the almond, Haiti)
marmelo bravo [*do matto*] (wild quince, *Cydonia oblonga*, Brazil); *membrillo* [*membrillito*] (little quince, Dominican Republic)
myrtle laurel-cherry (USA); West Indian [laurel] cherry (USA, Bahamas)
noyou (almond, Guadeloupe; a name also applied to *Merremia dissecta*)
palo de hacha (ax tree, Dominican Republic)
virarú (maybe from *virar*, leaning to one side, Brazil)
warimiaballi (*warimia*, is *Tapirira guianensis* in the Anacardiaceae, *balli*, resembling, Arawak, Suriname)
yaya boba (crazy *yaya*, the word *yaya* is surely Taino as it appears in several plant names, Dominican Republic)

Prunus serotina (late-ripening)
aguasique (Pima Bajo)
capulín [*capolín*, *capulí*] (Náhuatl, Texas, Sonora to Veracruz and Chiapas, Guatemala)
cereso [*cerezo*, *cereza*] (Chiapas, Guatemala); *cerises d'atomne* (autumn cherry, Quebec)
[cabinet, mountain, rum, southwestern, whisky, wild] cherry or black-cherry [blackcherry] ("cherry" was derived in the 1300s from Old English *ceris*, which was taken from Latin *cerasus*; cognates are Spanish *cereza*, Portuguese *cereja*, French *cerise*, German *Kirsche*, and Dutch *kers;* Greek *kerasos* is related)
[southwestern] choke-cherry [chokecherry, choke cherry] (a name applied to both *P. serotina* and *P. virginiana*, dating from about 1796 in the northeastern United States; the name refers to the astringent fruits)
detsé [*detzé*] (Otomí, Veracruz)
gthon' pa hi [*gthon-pa hiu*] (Osage)
ikwe'mic (Ojibwa); *okwe'mînûn* (grubworm berry, Potawatomi)
ittó fotóhka (*ittó*, tree, *fotóhka*, smells decayed, Koasati); *okòoła* [*okwaała*, *okwáoła*, *okooła*] (Alabama)
ittobaksa (*itto*, tree, *baksa*, thread, twine, Alabama; cf. also *Tilia*); *ittotalikcho* (*itto*, tree, *tałkcho*, rope, Alabama; identity doubtful)
jeco (Guarijío, Sonora)
kona ha'go (*konawv*, bead, *hayetv*, to make, Creek)
taunday (Zapotec, Oaxaca)
'to fvmpe [*tofv'mpe*, *tofompa*, *tofámbi*] (*eto*, tree, *fvmpe*, cherry, Creek, Muskogee; cf. Simmons [1822] 1973)
tup (Quiché, Guatemala)
usábi (Tarahumara, Chihuahua)
xeugua (Michoacán)

Prunus umbellata (inflorescence branches resembling the ribs of an umbrella)
[flatwoods, hog] plum
[black] sloe [of the South]

Like other youngsters in western Kentucky, I grew up nibbling the fruits of *Prunus serotina*. Although they are somewhat bitter, to a boy they were pleasant, and helped slake thirst during a hard day spent exploring the woods and fields. The more sinister side of the plants was completely unknown until a neighbor's horse decided to sample a wilted branch. Although I was not yet a teenager, the vivid image of that dying animal is still burned in my memory. Later, I learned that it was the release of hydrocyanic or prussic acid (cyanide) that killed the poor creature. My friends and I became more cautious of the fruits, even though our reading about them revealed that they were a significant food to indigenous Americans, and the bark was one of their important medicines.

Many years later I tried the fruits of *capulín* in Lima, Peru. All my agriculturist colleagues at the *Centro Internacional de la Papa* there could tell me was that the fruit was an important food in their country, and that it was *una ceresa peruana* (a Peruvian cherry). Eventually, I learned that the drupes I ate were simply a larger-fruited subspecies of what had been a part of my childhood diet.

Some 17 of the more than 200 species of *Prunus* grow in Europe (Mabberley 1997). Thus, Europeans brought a variety of common names into the languages they introduced into the New World. As an example, some names of Old World *Prunus* in Spanish are *albaricoque* (for *P. armeniaca*, the apricot), *cerezo de monte* (for *P. avium*, the wild-cherry), *ciruelo* (for *P. ×domestica*, the plum), *almendro* (for *P. dulcis*, the almond), and *melocoton* (for *P. persica*, the peach). A similar list could be given for other European languages.

When explorers and immigrants arrived in the New World, they found indigenous people using a number of native species. Seven of those grow in Florida; four native species are called "plums" and three are "cherries."

Originally, the name "plum" meant only *P. ×domestica* and dates in English to about A.D. 700. That word was derived from Old Low German *pluma*, which came from Latin *prunus*. The same Latin word or its alternate declension *prunum* also gave us the English word "prune" about A.D. 1345, and its cognate *prugna* in Italian. In the 1700s and 1800s, some considered plums to be genus *Cerasus*, while cherries were *Padus*. However, even that nomenclature is confused because the Greeks used the word *kerasos* for the Old World bird-cherry *Prunus avium*. Theophrastus (372–287 B.C.) used *Padus* (from Greek *pados* or *pedos*) for a cherry whose timber was used for axles.

According to historic usage, plums are usually larger than cherries, have a bloom or whitish wax on the skin, a somewhat flat pointed stone, and sweet pulp (OED 1971, Davidson 1999). Traditionally, a "cherry" has smaller fruits, no bloom on the skin, a rounded stone, and the pulp may be sweet, sour, or even bitter. The basis of the concept for "cherry" is *P. avium* or *P. cerasus*. However, in practice, usage breaks down, and *P. angustifolia* is called both a plum and a cherry. *Prunus angustifolia*, *P. americana*, *P. geniculata*, and *P. umbellata* are "plums"; alternately, *P. caroliniana*, *P. myrtifolia*, and *P. serotina* are "cherries."

Although Cortez's *conquistadores* had eaten *Prunus* in Mexico in 1519, it was the mid-1700s before the Florida species were named. Linnaeus knew only *P. virginiana* when he published *Species Plantarum* in 1753, and it had been discussed by Leonard Plukenet in 1696, Mark Catesby in 1731–1732, and J.F. Gronovius in 1739–1743.

However, *Prunus* was first recorded in what today is the United States by the de Soto expedition. The Spanish found the plants near what is now New Madrid, Missouri, in 1540 or 1541, and noted that the Chickasaw were using them for food (Hedrick 1919). Those plums were surely what Varner and Varner translated as "dried fruit" from Garcilasco De La Vega ([1605] 1962). André Michaux tried to commemorate that important association between that tribe and the fruits by calling the trees *P. chicasa* in 1802, but Humphery Marshall had already named them *P. angustifolia* in 1785. Not only did the people along the Mississippi eat the fruits, the Ozark Bluff-Dwellers of Arkansas used the seeds to make beads (Gilmore 1931).

Prunus angustifolia grows from Florida through the eastern two thirds of Texas, and north to Missouri, Kentucky, and east to New Jersey and Maryland. Fernald (1950) said that it occurred elsewhere, "as a relic of cultivation." Bartram ([1791] 1958) never saw them in the wild but always in deserted Creek villages. He speculated, "I suppose it to have been brought from the S.W. beyond the Mississippi, by the Chicasaws." Maybe that comment is why Harrar and Harrar (1946) and Little (1979) thought that the original range of *P. angustifolia* might have been central Texas and Oklahoma, but that the trees were early and widely spread by indigenous tribes.

The fruits were favored by the Chickasaws as noted by the common name. Diggs et al. (1999) record that "Native Americans" dried fruits on hot rocks so they could be stored for future use. Otherwise, Moerman (1998) listed only the Comanche as using the fruits as food, both fresh and dried. Yanovsky (1936) and Hudson (1976) wrote that all southeastern tribes ate plums of all kinds.

Prunus americana ranges from Florida to northern Arizona and New Mexico, north to Utah and Wyoming, and east through Iowa, Indiana, Wisconsin, Minnesota, New York, southern Ontario, Saskatchewan, and Manitoba (Fernald 1950, Kearney and Peebles 1951). Little (1979) said the American plum grows in the mountains of northern Mexico, but that appears to be a confusion with *P. mexicana* (Standley 1920–1926).

Fruits of *P. americana* were used as food by at least the Apache, Cherokee, Cheyenne, Creeks, Crow, Dakota, Iroquois, Isleta, Kiowa, Meskwaki, Ojibwa, Omaha, Osage, Pawnee, Ponca, and Winnebago (Gilmore 1919, Hunter [1823] 1973, Densmore 1928, Yanovsky 1936, Sturtevant 1955, Moerman 1998). Bundles of the flexible twigs were used as brooms throughout the Missouri River region, and the Omaha planted their beans, corn, and squash when *kande* came into bloom (Gilmore 1919). Isleta, Navajo, and Ojibwa made dyes from the bark, producing yellow or red, depending on the other plants mixed and mordants (Moerman 1998).

Several tribes used *P. americana* in medicine (Moerman 1998). The Cherokee made a cough syrup of the bark, and an infusion for kidney and bladder problems (Hamel and Chiltoskey 1975). The Cheyenne applied the mashed fruits to mouth diseases, and the Meskwaki used the root bark. The Mohegan made an infusion of twigs to treat asthma. The Omaha boiled root bark and applied it to skin abrasions (Gilmore 1919). The Ojibwa made a decoction of the roots for intestinal worms and put a decoction of the bark on cuts and wounds as a disinfectant (Densmore 1928); they also used the rootlets in a remedy for diarrhea. The Rappahannock used the red plum in an undisclosed medicine.

The last southeastern plum to be named was *P. geniculata*. Roland M. Harper (1878–1966) described that species from Lake County in 1911, and it is so rare that it has been considered federally endangered since 1987 (Ward 1979, Coile 2000, Chafin 2001). Little (1979) did not even discuss the name. This small plum has fruits to about 2 cm wide and grows in the heart of what was Timucua territory before the arrival of the Europeans.

Moerman (1998) does not list *P. myrtifolia* as used by native people, and with its limited distribution in Miami-Dade County, that is not surprising. Like other species in the genus, this one contains hydrocyanic glycosides. Bark, leaves, and twigs are used fresh for asthma and cough in Cuba (Roig 1945). In Hispaniola and Puerto Rico the trunks are used for electric line posts, in rural construction, and in cabinetwork (Liogier 1974, Little et al. 1974). The bark has a strong smell of almonds and is used to clean false teeth (Liogier 1974).

Stephen Elliott named *P. umbellata*, called the "sloe of the South" because of its black fruits. At the time he named the plum, Elliott (1821) recorded that people were using the fruits in preserves. When they are available, they are preferred because they are unusually rich in pectin (Harrar and Harrar 1946). I found no records that they were used as a substitute for the European sloe (*P. spinosa*), but it would be a safe bet that they were.

The laurel-cherry (*P. caroliniana*) has the second most restricted range of the species, growing on the coastal plain from southeastern North Carolina to central Florida, west to eastern Texas (Little 1979; Diggs et al. 1999). Alabama people boiled the inner bark to make a red dye used to color split cane woven into baskets. Porcher (1863) was enthusiastic about these plants during the Civil War but largely as an ornamental. He wrote: "This tree, the flowers of which are much frequented by bees, grows abundantly on the sea-coast of our states, and is certainly one of the most beautiful and manageable evergreens that we possess. It can be cut into any shape, and is of a most attractive green color. It forms an impervious hedge, and grows rapidly." About its potential as medicine, Porcher continued: "The black, oval berries contain an abundance of Prussic acid, as does the whole tree; but I do not know of any use to which it is applied. Dr. Thompson has found great use from Prussic acid, largely diluted, as a local application in impetigo." Hocking (1997) added that the leaves and bark contain prulaurasin, which hydrolyzes to racemic mandelonitrile and glucose and releases hydrogen cyanide. As with all members of the genus, laurel-cherry may be fatal to stock if browsed in quantity (Harrar and Harrar 1946).

Wild black-cherry (*P. serotina*) grows from Florida to Texas and southern Arizona, south through Mexico (Baja California Sur, Sonora east to Tamaulipas, south to Oaxaca and Chiapas), Guatemala, and north to North Dakota and Minnesota, Nova Scotia, New Brunswick, southern Quebec, and southern Ontario (Fernald 1950, Little 1979). The Mexican and tropical American plants are usually segregated into *P. serotina* ssp. *capuli* (Felger et al. 2001), formerly recognized as a distinct species. Fruits from this subspecies have long been made into alcoholic and non-alcoholic drinks, and they are eaten fresh and preserved throughout the Americas (Felger et al. 2001).

The cherry now called ssp. *capuli* was first noted in Mexico by Hernándo Cortez's men in 1519 when the fruits were an important food for Spanish soldiers during the siege of Mexico City (Standley and Steyermark 1946). Francisco Ximémez said in 1615 that the fruits "*no son nada inferiores a nuestras cerezas*" (are not inferior to our cherries). Among the tribes in the United States, black-cherry fruits were used as food by the Cherokee, Iroquois, Menomini, Ojibwa, and Potawatomi (Smith 1933, Moerman 1998). Fernald et al. (1958) noted that indigenous people first pounded dried fruits (including pits) and then leached out the poisonous chemicals before using them in foods. An alternate view is that, once pounded, the poisons volatilize and fruits are rendered harmless (Dunmire and Tierney 1997).

The Cherokee also used the wood for buildings, furniture, and carvings (Hamel and Chiltoskey 1975). Trees now are highly prized for the beautiful wood, which is used for furniture and cabinetmaking (Vines 1977, Diggs et al. 1999). Porcher (1863) was similarly complimentary, writing, "The wood of this tree is highly valuable, being compact, fine grained, and brilliant, and not liable to warp when perfectly seasoned. When chosen near the ramifications of the trunk, it rivals mahogany in the beauty of its curls." Others since have used the wood for furniture, cabinets, printer's blocks, veneer, patterns, panels, interior trim, handles, woodenware, toys, and scientific instruments (Vines 1977).

Ximénez recorded that the Mexicans made a decoction of the bark that was used to cure "*las cámaras de sangre*" (blood clots). Also, the powder of the bark was used to clarify the vision, heal inflammations, moisten the tongue when it is dry from the heat or fever, and was used to cure various external sores and lesions (Martínez 1969). Ximémez considered the bark useful against malaria, but later reports considered it simply antipyretic (Martínez 1969). Tarahumara used the leaves and bark to catch fish and in a tea to treat whooping cough (Pennington 1958). They also added young leaves to their corn dish *esquiate* to increase spiciness.

In the United States, the black-cherry is used as medicine by tribes throughout its range. Eastern tribes using it included the Catawba, Cherokee, Delaware, Illinois-Miami, Iroquois, Malecite, Micmac, Mohegan, Narragansett, Ojibwa, Osage, Penobscot, Potawatomi, Rappahannock, Shawnee, and Shinnecock (Smith 1933, Vogel 1970, Moerman 1998). People typically used the bark in tea or syrup for coughs, fever, cold, sore throats, diarrhea, lung ailments, bronchitis, pneumonia, and dyspepsia (Foster and Duke 1990, Moerman 1998, Duke et al. 2002). In addition to those uses, Vogel (1970) recorded the Cherokee using a bark decoction to treat measles, and the Missouri River tribes treating dropsy with it (Gilmore 1919). The Catawba mixed black-cherry bark with poplar and dogwood to treat expectant mothers (Vogel 1970). Like several other eastern tribes, the Illinois-Miami put root bark on infected gums (Vogel 1970). The Osage and Shawnee mixed black-cherry bark with *Arisaema* and snakewort to treat malaria (Vogel 1970).

Among the non-indigenous people, Porcher (1863) gave a typical view when he wrote, "This is undoubtedly one of the most valuable of our indigenous plants." He listed many maladies for which black-cherry had been used successfully, but like indigenous tribes considered it best against diarrhea, colds, and coughs. One of his favorite recipes was for "making 'Cherry' cordial by the Southern matrons in the lower country of South Carolina (Saint John's)—a most delectable drink at all times, but particularly valuable in the present emergency [Civil War]: Fill the vessel with cherries (not washed, if gathered clean). Cover with whisky. After several weeks pour off all the clear liquor and press the cherries through a sieve. Put into the juice thus pressed out five pints of brown sugar, and boil with syrup enough to sweeten the whole demijohn. Pour five pints of water on the thick part; boil and strain to make the syrup with the sugar." Murphee (1965) found residents of the Panhandle still using the gum to make a remedy to relieve colds and croup. The fruits are still a flavoring for rum and brandy (Uphof 1968, Diggs et al. 1999).

We partly owe the success of the Lewis and Clark expedition to choke-cherry bark (De Voto 1981). There were several occasions when the leaders were required to give remedies to individuals during the journey, and Captain Meriwether Lewis's mother had been a skilled herbalist or "yarb doctor" (Vogel 1970). On 11 June 1805, between Maria's River and the Great Falls on the Missouri River in Montana, Lewis became ill with abdominal cramps and fever. He had no medicine with him at the time so he decided to use choke-cherry. He brewed a strong decoction of small twigs and leaves and drank a pint at sunset. An hour later, he drank another pint. By 10 P.M., his cramps and fever were gone.

Lewis had used the other choke-cherry (*P. virginiana*), but his technique was based on what he learned for using the eastern *P. serotina*.

Pseudophoenix

(Charles Sprague Sargent adapted H.A. Wendlands's name *pseudo*, false, *phoenix*, genus of the date palm)

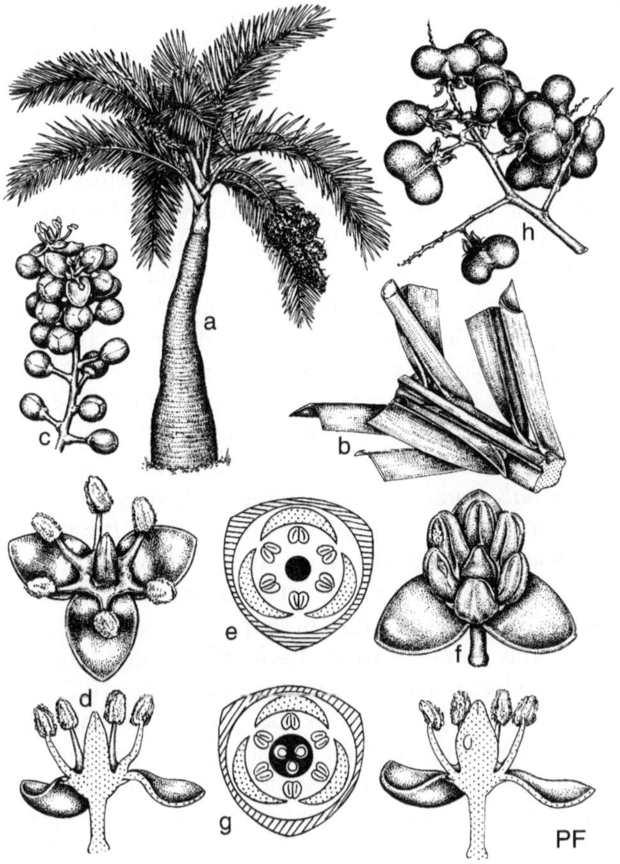

Pseudophoenix sargentii. a. Habit. b. Section of leaf midrib showing attachment of segments. c. Branchlet of inflorescence. d. Complete staminate flower (above) and longitudinally dissected staminate flower (below). e. Floral diagram of staminate flower. f. Complete perfect flower (above) and longitudinally dissected perfect flower (below). g. Floral diagram of perfect flower. h. Fruits. *Drawn by Priscilla Fawcett.* From Correll and Correll 1982.

Pseudophoenix sargentii (named for Charles Sprague Sargent, 1841–1927, director of the Arnold Arboretum of Harvard University)
buccaneer palm (Bahamas)
cathier (wine tree, Haiti); *chacha* [*cacheo, caicha, casei, catey*] (wine, Dominican Republic); *palma de cacheo* (wine palm, Dominican Republic)
[Florida, Sargent's] cherry palm (Florida, Puerto Rico)
hog cabbage-palm [palmetto] (Bahamas)
palma de Guinea (palm from Guinea, in the belief that the tree was from Africa, Cuba)
palme marron (wild palm, Haiti); *palmiste* [*à vin, mâle*] ([wine, male] palm, Haiti)

These palms were not even discovered until the late 1800s. Four species are known in the genus, all endemic to the Caribbean (Mabberley 1997).

In Hispaniola, people extract from the trunk a sweet liquid, which ferments into a drink called *cacheo* (Liogier 1974). In Dominica, Puerto Rico (Mona Island only), and Florida these trees are nearing extinction, perhaps due to overuse by humans (Coile 2000, James 2003). At one point the Hispaniolan *P. eckmanii* was thought to have been extinct since 1926 due to overharvesting for winemaking, but it has been relocated (Zona 2002).

Psychotria: Wild Coffee

(Based on Greek *psyche*, soul or life, plus *iatria*, medicine, or "to give life")

Psychotria nervosa. *Drawn by P.N. Honychurch.*

Imagine a world without coffee (*Coffea arabica*). There was no coffee in the Americas before the plant was introduced. Perhaps more surprising, there was no coffee in Europe when Columbus sailed in 1492.

German physician Leonard Rauwolf (1535–1596) was the first European to mention coffee after his trip to the Levant in 1573–1576. Still, it was 99 years after the New World was discovered before much attention was paid to the beverage by Prosper Alpinus in 1591. About 200 years after Europeans settled in the New World, the French introduced *Coffea arabica* into Martinique in 1717, and the Dutch took it to Suriname shortly before that. The plants reached Jamaica in 1728 (Hedrick 1919).

Before the comparison between cultivated coffee and wild plants could be made, *Coffea* had to be introduced into the New World. "Wild coffee" was first applied to its relative *Faramea occidentalis* by Philip Miller in 1730. That plant is also a member of the Rubiaceae, the family containing *Coffea*, along with what is now called wild coffee in Florida (*Psychotria nervosa, P. sulzneri*). When the name "wild coffee" was extended from *Faramea* to *Psychotria* is unknown. However, both genera are still known by that common name in English, French (*café marron*, Guadeloupe, Martinique), and Spanish in the Caribbean (*café cimarrón*, wild coffee, Dominican Republic), and Mexico (*kapee ts'ohool*, coffee herb, Huastec, San Luis Potosí). There is an indirect reference to the place of coffee origin in Africa with *palo moro* (Moorish tree, Puerto Rico).

Although *Psychotria* is compared with coffee, it is not used as a beverage. Several people have told me they tried using *Psychotria* seeds as a substitute "coffee," and got only bad taste and terrible headaches from the mixture. However, these plants are surely rich in tannins because mashed leaves are used as a hemostat in Cuba (Roig 1945) and Puerto Rico (Meléndez 1989). People in Barbados use a leaf decoction as a febrifuge, as a remedy for colds, and to treat stomach ailments. Leaves are combined with other plants as "bush tea" (Goodding 1940–1942), which is used to treat a number of maladies. In keeping with their multiple common names, the Huastecs of Mexico have the most varied uses (Alcorn 1984). Like Barbadians, the Huastecs use it to treat respiratory problems, especially asthma. Other uses include application to local swellings, such as swollen feet, and erysipelas, sores, boils, tumors, and skin fungus. Does *bois laitelle* (milk trees, Haiti) mean they use it with milk?

Before *Coffea* was introduced into the New World, people here used other names for the *Psychotria*. The only indigenous name from within Florida is for *P. sulzneri*. This hammock shrub is known by the Seminoles as *atópâ:bî* (dogwood replica, Mikasuki). According to Sturtevant's (1955) informants, the plant was "useless." However, that name was derived from Creek *vtvphv* [*atápha*] for dogwood (*Cornus*), and that comparison with another species suggests a loss of information (see *Cornus*). Every species of *Cornus* in North America was used by people of essentially every culture (Moerman 1998).

Florida's other "wild coffee" (Florida, Bahamas, Puerto Rico), *P. nervosa*, has a diversity of names outside the state. Among those names is *ipecacuana ondulada mayor* (big ipecacuanha with undulate leaves) in Cuba (Roig 1945). That name was given because the roots of *P. nervosa* were used as an emetic as were those in the real *ipecacuanha* (*P. ipecacuanha*). The comparison is instructive, because *ipecacauanha* has a long history.

Between 1570 and 1600, the Portuguese priest Manoel Tristão was in Brazil where he learned that the Tupí people used the roots of a plant for treating dysentery. When he returned to Europe, he told more people about the plant, and French physicians finally made it available in 1672 (Burkill 1966). We get our English name *ipecacuanha* from the Tupí-speakers in South America who called their medicinal plant *ipega'kwai* (duck's penis) or *ipe-kaa-guéne* (creeping plant causing vomiting). The Florida plants have no such imaginative names.

The only name other than "wild coffee" found in Florida is Seminole balsamo, given by Morton (1981). Since she gives no source for the name and it has not been found elsewhere, the origin remains a mystery. There is documentation for the name *bálsamo* in Puerto Rico (Liogier and Martorell 1982, Meléndez 1989). Literally translated, the word means "balsam." The probable sense of its use for "Seminole balsamo" is the now archaic meaning of "soothing." Etymologically, the word in Spanish and English has been equivalent to "balm," both having been derived from Latin *balsamum*. The very name itself suggests a long medical history—unless Morton invented the name.

Some of the other designations indicating use as a medicine are kidney bush (Caymans) and St. John's bush (Barbados). While perhaps not exactly "medical," the name strong back (Caymans) suggests enhanced stamina. Since the name implies two distinct applications of strength, it was perhaps an important plant. The interpretation most often given for such names is improved sexual stamina. However, there was an equally valuable application. Since our species became bipedal, we have suffered from innumerable back problems. In addition, as the name is endemic to the Virgin Islands where the population is largely derived from slaves, it takes on added meaning. Individuals who could not do their share of work during slaving times were not likely to survive long. Medicines to "soothe" aching backs would have been at a premium.

Other names show different perspectives. The names *penda* (dangler, Dominican Republic) and *tapa camino* (path coverer, Cuba) are surely related to the locations where these shrubs grow and are descriptive of their life-forms. *Plateado* (folded or plaited, Cuba) is a literal equivalent of the specific name of the technical synonym *Psychotria undata*. That name was proposed by Nicolaus von Jacquin in his book of 1798, *Plantarum Rariorum Horti Caesarei Schoenbrunnensis*, and he too was taken with the undulating pattern of the living leaves. For some reason not understood, the plant is called *huesito* [*husito*] (little bone) in Panama. If that were an isolated case, it would be one thing, but that word is applied to several Mexican species. None of the uses suggests a bone medicine, so perhaps the name refers to the wood.

Mayans call this shrub *ya'ax-k'anan* [*anal, anal xiv, sacxanal, canaan*] (*ya'ax*, green, *k'anan*, necessary, Yucatán), which is their same designation for *Hamelia patens*. Both plants are important in Mayan religious ceremonies, and other people also have names with reverent implications. Although the plants are not mentioned in the recent book on Haitian voodoo (Beauvoir et al. 2001), the names *cabra blanca* (white goat, Dominican Republic) and *cabra santa* (holy goat, Dominican Republic) suggest involvement with that religion. It is locally well known that the Santarías, a Cuban religious group related to voodoo (Voeks 1997), leave both goat and chicken offerings to their deities on the steps of the Miami-Dade County courthouse. So, *cabra santa* carries special meaning, and a *cabra blanca* is the best offering.

The most information available about the religious and medical importance of *P. nervosa* is among the Huastec (Alcorn 1984). There the shrub is *wats'ul*, a simple name denoting great age. *Wats'ul* also has a number of alternate names, including *tsakam wats'ul* (little *wats'ul*, Veracruz) and *wach'ul ch'ohool* (*wach'ul* herb, Veracruz), perhaps denoting a taxonomy that does not match ours. The names *baina ts'ohool* (sheath herb, San Luis Potosí), *itsal kw'a'* (toad chile, San Luis Potosí), and *tse'tsem ts'ohool* (woodpecker herb, San Luis Potosí) may also imply religious significance. The name *tsabalte' ts'ohool* (skin fungus herb, San Luis Potosí) points to a medicinal use, and the names *tse'tsem t'abat'* (little *Tabernaemontana alba*, San Luis Potosí), and *tsakam tsabalte'* (little *Cestrum dumetorum*, San Luis Potosí) compare it with other plants important among those people. People in Belize say it is *contra yierba* (herb against), which is the description often given to a strong plant used to counteract poisons, or at least some problem from an outside source.

The genus *Psychotria* was named by Linnaeus because of the reputed medical properties of some species. Which species he knew about being medicinal is more of a problem, as there are perhaps 1650 in the genus (Hamilton 1989a,b,c, Nepokroeff et al. 1999). However, discoveries in the late 1960s led to a renewed interest in *Psychotria*. Studies among people in the Amazon revealed that one or more members of the genus were added to the psychoactive plant mixture called *ayahuasca* (vine of the soul, Quechua), used for religious, medicinal, and social purposes (Pinkley 1969, Schultes and Raffauf 1992). Subsequently, several studies of the genus have revealed alkaloids, benzoquinones, cyclic peptides, and other chemicals that exert a variety of influences on human physiology (e.g., Beretz et al. 1985, Adjibade et al. 1991, Amador et al. 2000, 2001, Khan et al. 2001, Verotta et al. 2002). None of the studies examined the Florida species, and the large genus is considered paraphyletic (Nepokroeff et al. 1999). Inferences of what might be the potential chemical bases for uses of *P. nervosa* or *P. sulzneri* are impossible, particularly because not all species contain the same classes of active compounds (Leal and Elisabetsky 1996).

Still, the Huastec of Mexico use *wats'ul* as a love charm, and that is as good as any. You simply grind the seeds and carry them in a pocket. What could be easier?

Ptelea

(From *ptao*, to fly, the Greek name for the elm, reapplied by Linnaeus to this genus with a similar fruit; akin to Akkadian *petelu*, to wind, entwine)

Ptelea trifoliata. From Sargent 1905.

Ptelea trifoliata (three-leaved)

ague bark (from the use to treat malaria)

cola de zorillo (little skunk tail, Texas, Arizona, Sonora, Chihuahua, Tamaulipas, Veracruz); *vara de zorro* (fox bush, northern Mexico to Veracruz); *zorillo* (little skunk, Hidalgo; because plant and animals have a similar odor)

dreiblättrige Lederbaum (three-leaf leather-tree, German, fide Millspaugh 1893); *Lederstrauch* (leather bush, German)

[common, three-leaf, woolly] hop-tree [hoptree] ("hop-tree" dates from ca. 1877, with the substitution of the fruits of *Ptelea* for those of hops, *Humulus lupulus* for making malt liquor, and as a tonic and soporific)

orme de samaire à trois feuilles (samara elm with three leaves, French; fide Millspaugh 1892)

pickaway[-anise]

pinacatillo (comparing the smell with the pinacate beetle, Coahuila; from Náhuatl *pinacatl*, their name for the beetle *Elodes* spp., Tenebrionidae, a black insect that sprays an irritating, stinking chemical for defense)

potato chip tree [potatochip-tree] (alluding to the flat fruits)

prairie grub

psehtin (Lakota)

quinine-tree (a name comparing these shrubs with *Cinchona* spp., the source of the anti-malarial

compound quinine, an alkaloid first isolated in 1820 by French Pierre-Joseph Pelletier and Joseph-Bienaime Caventou; the word "quinine" appeared in English by 1826, although it had been used in Spanish as *quina* or *quinaquina*, taken from Quechua, since at least the 1640s)
sinaptahaspí (*sinapó*, ash, *tahaspí*, light [weight], Koasati)
skunk-bush [skunkbush] (the odor caused comparison with the skunk, whose name was taken from the indigenous Abenaki *seganku* or *segongw*, the weasel relative *Mephitis mephitis*)
stinking-ash; wafer ash ("wafer" refers to the flat, rounded fruits, like the thin, crispy cake that gave rise to the Old English *wafre*, and our word "waffle"; the word "ash" compares it with *Fraxinus* because of the leaves)
swamp dogwood (a misnomer or misidentification; this is an upland plant)
[shrubby, tree] trefoil (called "trefoil" from Latin *trifolio*, three leaves because of the compound, trifoliolate leaves)
wahoo (from *uhawhu*, a Creek word for the elm; however, the Dakota *wahú*, arrowwood, was applied to *Euonymus*)
wing-seed

The first living hop-trees to arrive in Europe were those sent to England by Reverend John Banister in 1704 (Millspaugh 1893). Those plants died, but Mark Catesby reintroduced the shrubs about 1724. Subsequently, hop-tree was cultivated in other places in Europe.

Both Leonard Plukenet in 1696 and Johann Jakob Dillenius in 1732 had studied specimens of these plants and called them *Frutex virginianus trifolius, ulmi samaris* (three-leaflet shrub from Virginia, with samara fruits like the elm). When Linnaeus examined the shrubs, he realized they were distinct from the elm, which he called *Ulmus* (which see). Linnaeus began calling the New World plants *Ptelea* in his *Hortus Cliffortianus* of 1738 and continued using the name in *Species Plantarum* ([1753] 1957).

Although Linnaeus knew *Ptelea* only from Virginia, William Bartram ([1791] 1958) recorded the shrubs in North and South Carolina, Alabama, Georgia, and northern Florida between 1773 and 1777. André Michaux, during his visit to Florida in 1788, found the plants in Alachua, Levy, Orange, and Putnam Counties (Taylor and Norman 2002). There are now about 11 species recognized in this North American endemic genus (Bailey 1962, Mabberley 1997). Numerous varieties of *P. trifoliata* have been named, and these intergrade in a confusing manner.

Perhaps the first record of medicinal use of the wafer-ash in the Americas was by the German physician Johann David Schoepf (1752–1800) who visited during the Revolutionary War in 1783–1784. He and others of the period considered it aromatic, bitter, stimulant, and an expectorant tonic useful against malaria (Millspaugh 1893). They surely learned about the plant directly or indirectly from indigenous people, but that does not seem to have been recorded.

Later, Rafinesque reported *Ptelea* in his *Medical Botany* of 1830. As usual, he obtained much of his information from indigenous people, but typically declined recording who and where. Rafinesque considered the plants vulnerary and vermifuge (Millspaugh 1893). Based on his report, the plants were most likely used by tribes where he explored; however, actual reports have been found for only two groups not visited by Rafinesque. The Menominis considered the root bark of *P. trifoliata*, a plant brought to Wisconsin from Kansas, as a sacred medicine and panacea to season and render other medicines potent (Vogel 1970). The Meskwakis used it similarly, and for lung problems, often in tea with other barks (Vogel 1970). Those reports are also the only ones found by Lewis and Elvin-Lewis (1977) and Moerman (1998) regarding indigenous people in the United States. However, in Mexico the plants are still used in remedies for dyspepsia, as a mild tonic, in a bath of an alcoholic infusion of leaves against chills, and to treat rheumatism (Martínez 1969).

Following Rafinesque, several American physicians promoted wafer-ash as a medicine. Eclectic physicians considered it second only to hydrastis (*Hydrastis canadensis*) as a tonic, and they thought a cold infusion of the plants especially good for debilitating fevers (Culbreth 1910, Felter 1922).

Martínez (1969) reported that the bark of the root contains the alkaloid berberine, but that has not been confirmed in more recent chemical studies. It does contain a variety of alkaloids, including dictamnine, pteleine, ptelecultinium, pteleatinium chloride, and several others. In addition to the alkaloids, the plants produce coumarins (Bailey et al. 1971, Szendrei et al. 1973, Mitscher et al. 1975a,b, Petit-Paly et al. 1987, 1989). At least the pteleatinium chloride is antibacterial, and some of the other compounds are antipyretic, antifungal (antiyeast), and effective against the tuberculum bacillus (Hocking 1997, Mabberley 1997). Decoctions from roots have been used as a stomach tonic and for oral and throat diseases (inflamed uvula). The leaves and shoots still are considered anthelmintic (Hocking 1997, Vásquez and Jácome 1997). Equal parts of fresh leaves and bark are used by German homeopathics (Hocking 1997). According to Diggs et al. (1999) the plants contain a poisonous saponin that causes photodermatitis.

Fruits have been substituted for hops in making beer since at least 1868 when Asa Gray recorded that use (Hedrick 1919, Vines 1977).

Pteridium

(J.A. Scopoli, an Austrian physician and professor of natural history in Pavia, named the fern with Greek *pteridion*, small fern, a diminutive of *pteris*)

Pteridium aquilinum*. Drawn by P.N. Honychurch.*

Pteridium aquilinum (of an eagle, from the wing-shaped fronds or leaves)

achshikimiinshi [*achsh'kíwmiinshi*] (*ach*, abundant, *sh'kíw*, urinate, *miinshi*, tree, Delaware)

Adlerfarn (eagle fern, German); eagle fern; *fougère d'aigle* [*fougère de aigles*] (eagle fern, Quebec); *helecho de aguilo* (eagle fern, Colombia)

[southern] bracken (perhaps from Scandinavia, akin to Old Swedish *braekne*, fern, USA, Bahamas)

[hog, pasture]-brake (from Old German, *brache*, "wasted" [i.e., wild] land)

fayti:yâ:bî [*faitiyabi*] (turkey leg replica, Mikasuki); *pinilio:mâ* (*penwv*, turkey, *ele*, leg, *ome*, resembling, Creek); *fi:tiyyí* (*fi:tó*, turkey, *iyyí*, foot, Koasati); *tapitapí* (an archaic name, Koasati)

fiddleheads [fiddlenecks] (the young fronds resemble the head of a violin, in use by 1599)

fourchette (table fork or breast bone, Haiti)

grande fougère (big fern, Quebec); *helecho* (fern, Panama, Belize)

poor man's soap (Alabama)

raineach-mhór [*rainich móire*] (Mary's fern, Gaelic)

Although Moerman (1998) and Balick et al. (2000) separate *Pteridium aquilinum* var. *caudatum* into *Pteridium caudatum*, Wunderlin (1998) retains both as a single species. Regardless, *P. aquilinum* is widespread and was used by people from Alaska to Mexico and Florida. Among the relatives of Florida people, the Koasati made a decoction of the roots for chest pain (Taylor 1940). The Seminoles used the plant for "Turkey Sickness" (permanently bent toes and fingers) (Sturtevant 1955). Murphee (1965) found people in the Panhandle making a tea from the plants to treat burns.

Farther north, the steamed mature fronds are used to make medicinal teas and inhalants for lung disorders and headaches (Moerman 1998). The young buds (fiddleheads) are edible. Older parts are poisonous (Morton 1968b), and King (1984) warns against using it for food at any stage. Yellow and green dyes are extracted from the roots (King 1984, Tull 1999).

Pterocaulon: Blackroot

(From Greek *pteron*, wing, plus *kaulos*, stem)

Within the United States, there are relatively few plants whose roots yield a black sap. The two in the southeast are called "blackroot." Scientifically, the herbs belong to the genus *Pterocaulon*. The green leaves with their white undersides merging into the stem and extending downward toward the ground, like wings, are as characteristic as their black roots. The first information Europeans had of these plants was when Linnaeus called them *Gnaphalium virgatum* in 1759. However, it was Stephen Elliott, an American professor in Charleston, South Carolina, who gave them the Greek name fitting their morphology in 1823.

For many years, it was believed that a single species grew in the United States, and it was called *P. virgatum*. Then, when the Frenchman André Michaux, who explored the Carolinas and Florida,

Pterocaulon pycnostachyum. *Drawn by P.N. Honychurch.*

looked more closely, he realized that there were two different species. One species grew in the southeastern states (North Carolina to Florida, west to Mississippi), and the other in Texas, eastern Mexico, and the West Indies. The species name *virgatum* had already been applied to the western species, so Michaux called the one in the southeast *Conyza pycnostachya* (*pycno*, thick, *stachyo*, spike, in reference to the flower cluster). Then, in 1823, Stephen Elliott moved the species to *Pterocaulon* to create the name used now, *Pterocaulon pycnostachyum.* Later, in 1836, Augustin-Pyramus de Candolle moved *virgatum* to create the name *Pterocaulon virgatum* for the western species.

Both species have long been called blackroot or Indian blackroot in the United States, but they sometimes also are called rabbit tobacco. The last name is more often applied to members of *Gnaphalium*, the genus where Linnaeus originally put these plants. Our word tobacco came into English from Spanish *tabaco*, first used in the 1500s but taken from a Taino word in Hispaniola. The modifier "rabbit" is equivalent to saying "wild" to distinguish it from the cultivated *Nicotiana.* Comparison with *Nicotiana* results from *Pterocaulon* leaves being smoked or chewed either as a medicine or simply because tobacco was not available.

The western species, *P. virgatum*, has a richer history of common names than the eastern plants, surely because that species is geographically spread from Texas to Argentina, and in the Caribbean. It too is called blackroot or Indian tobacco in the United States. However, Jamaicans call it golden cudweed. That Jamaican name is based on the Old English *cwidu* or *cudu*, which is sometimes also rendered *quid.* Those words date from about A.D. 1000 and refer to any plant substance that is held in the mouth and masticated, but not swallowed. Typically, after the introduction of *Nicotiana* into the Old World, the "cud" has been synonymous with that species. However, the practice of chewing a cud is much older in Europe than knowledge of the American *Nicotiana.*

Other people were more impressed with the looks of the aboveground parts of these plants. Due to the white pubescence, people call them *branqueja* (the white one, Brazil), *calça de velho* (old man's shoes, Brazil), or *oreille mouton* (sheep's ear, Martinique, Dominique). *Pterocaulon* is called *langue à vache femelle* (female cow's tongue, Martinique, Dominique) to distinguish it from *langue à vache*, which is *Elephantopus.* Because the plants retain their shape and color when dried, they are also called *siempreviva* (everlasting, Cuba). References to where one may find the plants are in some names—*alecrim das paredes* (rosemary of the walls, Brazil), *travesera de loma* (crosser of hills, Cuba), and *vela de sabana* (savana candle, Dominican Republic). Allusions to the potent chemicals contained in the herbs are *barbasco* [*verbasco*] (a fish poison, Brazil) and *cura nacío* (cures at birth, Dominican Republic).

The eastern *P. pycnostachyum* has names only in English and in two Seminole languages. Sturtevant (1955) found literature with various spellings of these names back to the 1930s, and then wrote that Mikasuki *picikcalah kayikcî* meant "blood saver medicine." That is the same sense as *pechekche alahke aayek* that the modern Mikasuki speakers translate more exactly (*pechekche*, blood, *alahké*, leftover, *ayekche*, medicine). "Blood medicine" is the Mikasuki name *pechekche emayekche* recorded by Snow and Stans (2001).

Williams ([1837] 1962) may have been the first person to record that this plant was used by the Seminoles. He wrote that it was "[t]he famous Indian remedy for pulmonary disorders." Sturtevant (1955) later recorded that the Seminoles used these plants widely in medicines where there was a real or perceived problem with the blood. The plant was used to treat chronic sickness, coughs and colds, "Cow Creek Sickness," fever, menstrual difficulties, "Otter Sickness," and was used in childbirth. Modern Seminoles who recall the old medicines still note similar uses (Snow and Stans 2001).

Chemical studies of several *Pterocaulon* species show that the plants contain a variety of coumarins (Debenedetti et al. 1981, 1991, 1992, 1994, 1996. 1997, 1998, 1999, Magalhaes et al. 1981, 1989, Boeykens et al. 1994, Vilegas et al. 1995, MacLeod and Rasmussen 1999, Palacios 1999, Vera et al. 2001). Coumarins have a long history of use as a blood anticoagulant (Lewis and Elvin-Lewis 1977). Coumarins are lactones, which are vitamin K antagonists, and they act as indirect anticoagulants with a delayed effect. Following oral ingestion, absorption, and metabolism in the liver, chemical action results from a reduction in synthesis of prothrombin, a plasma protein produced in the liver in the presence of vitamin K. Prothrombin is converted into thrombin during the clotting of blood. Alterations of these pathways change the ability of blood to clot, and noticing this change is surely partly what led people to use these plants. Clotting alteration would have been the obvious result when people used *Pterocaulon* as abortives and styptics, and to relieve menstrual difficulties (Sturtevant 1955, Uphof 1968, Morton 1974).

Coumarins and their derivatives are not the only bioactive chemicals in *Pterocaulon.* Other studies show that the plants contain caffeolyquinic acids (Martino et al. 1979), thiophene acetylenes and flavonols (Bohlmann et al. 1981), flavonoids (Semple et al. 1998, 1999), flavanones, and caryophyllene (MacLeod and Rasmussen 1999). These chemicals are antioxidant, prooxidant (Desmarchelier et al. 1997), and antiviral (Semple et al. 1998, 1999). Thus, the mixture of chemicals extracted into medicines made from *Pterocaulon* would be effective against colds and asthma.

Sturtevant (1955) also recorded the Creek name for these plants, and Snow and Stans (2001) confirm that it is still used. In Creek, the plants are *yvnvsv heleswv* [*yanasa hiliswâ*] (*yvnvsv,* buffalo, *heleswv,* medicine). The allusion to a buffalo may seem odd in the modern world largely devoid of them. However, *Bison bison* historically ranged into the woodlands of the eastern United States well within the area occupied by the Creeks and ancestors of the Miccosukees (cf. Caras 1967).

The Seminoles also sometimes used blackroot in the busk (Capron 1953), a term derived from their word *posketv* [*poskita*] (to fast, Creek). The busk, or Green Corn Ceremony, involves a 4-day festival, sometimes with 1 to 3 preliminary days added. The event occurs between late April and mid-July. Special busk grounds are prepared distant from both Seminole and white settlements. During the evening of the third day, fires are rekindled with flint and steel in the old manner, and the sacred medicine bundle is opened. The contents of the medicine bundle are usually kept secret, but at least three outsiders in recent history have been permitted to examine it. Spoehr (1939), Capron (1953), and Sturtevant (1954) were each permitted to make notes on the contents of different medicine bundles. Each of the three found similarities and differences in the contents. Capron (1953) is the only one who found *Pterocaulon* in the bundle. Regardless of the exact contents of the medicine bundle, a variety of its plant contents (up to 20 plants) was used to brew the *ayikctanahki* (gathered medicine) or *ayikctanahkco:bi* (big gathered medicine), a critical beverage in the cleansing of the old year and preparation for the new.

The only other record of use of *Pterocaulon* found from Florida was by Murphee (1965). She was told that people in the Panhandle used it to cure "teen age trouble" (illegitimate pregnancy).

The genus contains 18 species in the warm parts of the Americas and southeastern Asia, reaching to Australia and New Caledonia (Cabrera and Ragonese 1978). Within its range, medicines are known to be made from it in the United States, Argentina (Vera et al. 2001), Brazil (Magalhaes et al. 1989), and Australia (Cribb and Cribb 1981, Semple et al. 1998, MacLeod and Rasmussen 1999). Perhaps not surprisingly, the aborigines of Australia also use the leaves as a tobacco substitute (Cribb and Cribb 1981). Nothing has been found to corroborate the inference, but maybe that is why Uphof (1968) said blackroot was a "stimulant" and Morton (1974) simply wrote that it was "narcotic." Nothing about its chemistry suggests those impacts on users.

Possibly the leaves are poor substitutes for *Nicotiana.* That might be why it was called "rabbit" tobacco. Either way, the common name conjures up the image of some pensive rabbit sitting on its haunches puffing on a pipe.

Pycnanthemum

(André Michaux created this name with Greek *pycnos,* dense, and *anthemon,* a flower, from the compact inflorescences)

[American wild] basil (comparing the fragrance to the cultivated basil, *Ocimum basilicum*)

Dickblume (dense flowers, German)

mountain-mint (William Salmon included "mountain-mint" in 1671 in his *Synopsis Medicina* for what is now *Clinopodium,* with about 20 European species; the name later was shifted to the New World, and "mountain-mint" used for *Pycnanthemum* by 1861)

Pycnanthemum albescens. From Britton and Brown 1898.

Pycnanthemum albescens (whitish)
sak:fotó [*iskafotó*] (Koasati; probably cognate with Creek *kvfockv* [*kafócka*]; cf. *Piloblephis*)
shinuktiɬeli [*shinuktelele*] (*shinuk*, sand [probably originally *shilup*, ghost], *tiɬeli*, to drive out, Choctaw; this name is a variant of the one used for *Monarda punctata*, which see)
white [white-leaf] mountain-mint [whiteleaf mountainmint]

Pycnanthemum flexuosum (curved alternately in opposite directions)
Appalachian mountain-mint [mountainmint]
dysentery weed

My education about *Pycnanthemum* began when I was working with my undergraduate professor Gordon Hunter. He had become interested in the mints and hired me in the 1960s to collect and work with him on some of the problems in the western Kentucky species. I knew the plants in the Jackson Purchase region of Kentucky between the Tennessee and Mississippi Rivers, but he had me look more closely at them. During the time that I was gathering specimens for him, I did not know that they had a long and complicated history of association with people.

An early mention of what is now *P. virginianum*, published in 1697 by Paulo Boccone (1633–1703/04), an Italian monk, physician, and professor at Padua, alludes to its medical application. He called the plants *Serpentaria virginianus* (Virginia snakeroot), and plants with that name were considered remedies for snakebite. That common name has been applied to numerous plants in different genera over the past few centuries (cf. also *Aristolochia*).

In 1696, Leonard Plukenet discussed the same species, but he gave it the phrase name *Clinopodium, pulegii angusto rididoque foliio, virginianum* (*Clinopodium*, with narrow and rigid leaves like *pulegium* [pennyroyal], from Virginia). Both Jan Gronovius's *Flora Virginica* and Linnaeus's *Hortus Cliffortianus* followed Plukenet's lead in thinking the herbs were a *Clinopodium*, but another early student of the herbs thought differently. In his 1699 book, Robert Morrison emphasized that the Virginia plants should be a distinct species of *Pulegium*. After further study, Linnaeus changed his mind, and in 1753 he called the plants *Satureja virginiana*. Thus, in the first 57 years that single species had been known, it had been placed in four genera. People are still debating placement of many species in the Lamiaceae, as genera of mints remain problematical. However, most now agree that the species discussed from Plukenet to Linnaeus is *Pycnanthemum*, a genus created by Michaux in 1803. We now know of 17 species endemic to North America (Mabberley 1997).

Indigenous people in the eastern United States have long been familiar with *Pycnanthemum*. People from the Ojibwa of southern Canada to the Cherokee of the Carolinas are recorded as using the plants in seasoning food (Densmore 1928, Hamel and Chiltoskey 1975). As Fernald et al. (1958) noted, most of the aromatic mints have been used as food additives to improve flavors. The common name "basil" or "American wild basil" widely applied to *Pycnanthemum* alludes to a more widespread application in seasoning food than most literature suggests. It is known that *P. muticum* contains several volatile oils including pulegone, menthol, menthone, and limonene; presumably all species contain different mixtures of those aromatic compounds (Hocking 1997). Porcher (1862) was especially complimentary to the family when he wrote that it does "not contain a single unwholesome or even suspicious species; their tonic, cordial, and stomachic qualities are due ... to the presence of an aromatic, volatile oil, and a bitter principle."

Medicine and food blend in most cultures so that it is not always possible to distinguish one from the other. However, several tribes considered different species of *Pycnanthemum* a good remedy for colds (Moerman 1998). Foster and Duke (1990) say that all species were used interchangeably.

In the southeastern United States, the Choctaw used a leaf decoction to promote sweating in treatment of colds (Bushnell 1909, Taylor 1940). The Koasati put the whole *P. albescens* plant in water, then used it to stop nosebleed. Roots mixed with those of *Salix nigra* were boiled, and the decoction drunk to relieve headache (Taylor 1940). To relieve contagion of death (*onalfapó*) leaves were put in cold water and the resulting mixture drunk and used to bathe the body while facing east (Kimball 1994). The Choctaw, the

Koasati, and the Cherokee also used another species that does not grow in Florida, *P. incanum* (Hamel and Chiltoskey 1975, Taylor 1940). *Pycnanthemum incanum* was used similarly to the second Florida species, *P. flexuosum*. The Cherokee used Appalachian mountain-mint in a poultice to relieve headache, diarrhea, upset stomach, colds, and heart trouble, and to bathe inflammations, particularly those of the penis (Hamel and Chiltoskey 1975).

By the early 1860s, *Pycnanthemum* was used through much of the eastern United States, and it was also known in Europe. In 1861 the British botanists John Lindley and Thomas Moore published their *The Treasury of Botany*. They gave *Pycnanthemum* the common name "mountain-mint," a designation indicating medical use. As Lewis and Elvin-Lewis (1977) point out, *P. flexuosum* was famous during that time under the alternate name "dysentery weed," and it was widely used throughout the southeastern states for bowel problems. Curiously, neither Porcher (1862) nor Millspaugh (1893) mentions *Pycnanthemum*, although they discuss a number of other mints, largely European, that were in use as medicines. Perhaps they thought that these wild plants had no virtues that warranted their substitution for plants already in people's gardens.

Pyrrhopappus

(From Greek *pyrros*, fire or flame-colored, and *pappos*, fluff or downy appendage, an allusion to the pappus of the fruits)

Pyrrhopappus carolinianus (of Carolina)
boñ-yai (Kiowa)
Carolina desert-chicory [Carolina desertchickory]
[false, native]-dandelion
sunflower (Louisiana)
toñ-awdl-'kok-yai (Kiowa)

This genus of three species endemic to North America was named by de Candolle in 1838 (Kim and Turner 1990). Prior to that, Thomas Walter had named this species in his *Flora Caroliniana*, but he put it in another genus. All three species occur in Texas (Diggs et al. 1999), but only *P. carolinianus* ranges into the eastern states. In the east it grows from Texas to Florida, north to Delaware, southern Indiana, Missouri, and Kansas (Fernald 1950). The variation within the genus is complex because of apparent hybridization (Barber and Estes 1978, Petersen et al. 1990).

At least two—and perhaps all three—species were used by indigenous tribes. In the southeastern states, the Cherokee used an infusion of *P. carolinianus* to purify the blood (Hamel and Chiltoskey 1975). Farther west, in the southern Great Plains, the Kiowa ate the roots. They considered them good, but said they were much sweeter in the autumn (Vestal and Schultes 1939). In New Mexico and Arizona, the Navajo used the flower stalks of *P. pauciflorus* as an emetic, and the Keres knew the plant but professed no use of it (Moerman 1998).

Porcher (1863), Millspaugh (1892), Foster and Duke (1990), Bown (1995), and Duke et al. (2002) do not mention the plants.

Q

Quercus: Oaks, Live and Otherwise

(Classical Latin name of oaks)

Thomas Mann delighted in scavenging food. Tom was my friend and student in the early 1970s, but in another era, he would have been a "hunter-gatherer." In the 20th century, he only gathered.

From time to time, I provided Tom with information on plant edibility in exchange for a sampling of his vegetable products. His acorn bread is particularly memorable. I think I remember it because it was chocolate brown; he made it from acorns that had not been leached enough to remove the tannin. It was tasty, but left an astringent, puckering aftertaste.

Some of the American trees still producing acorns are astonishingly large, but they were larger when people from the Old World came to this continent. When Europeans from Spain, Portugal, and England arrived in North America, they found giant trees, the likes of which existed only in the memories of their older countrymen. As Europeans had long since cut all of the large trees that grew in those countries, no one reaching the New World had ever seen anything resembling the enormous sky-fingering lofty towers of the Americas. People who saw those trees were delighted because they knew the wood could be used for building boats, houses, and other items essential for daily living. Moreover, they knew that people remaining in their homelands would pay fortunes for those wood resources. Immediately, they began felling the trees and opening the land to remake it in the image of human-scarred, unforested, barren, cultivated Europe.

Europeans figured that they had arrived in a "desert" in the New World, and they were terrified of wild areas. Although it has another meaning now, our word "desert" originally referred to anywhere people did not live. Immediately the newcomers began trying to make the "desert" into a "town"—a gathering of people in an enclosure or fortress. They knew that deserts not only contained tangible, potentially dangerous people whom they called "Indians," but they also believed forests were inhabited by a plethora of other perils including trolls, goblins, spirits, ghosts, and similar malevolent entities. While "towns" were populated with many of those same shadowy hazards, other humans buffered the danger.

Quercus. *Quercus chapmanii* (upper left branch). *Quercus geminiata* (upper right branch). *Quercus minima* (lower left branch). *Quercus myrtifolia* (lower right branch). *Quercus virginiana* (tree on right). *All drawn by P.N. Honychurch.*

They were not aware of it, but their actions were the same as those of fish in schools and birds in flocks. Gatherings reduce the chances of individuals being singled out for predation.

Many of the large trees the Europeans found in the New World were in genera they knew, such as *Abies* (firs), *Acer* (maples), *Betula* (birches), *Castanea* (chestnuts), *Fagus* (beech), *Fraxinus* (ash), *Juglans* (walnuts), *Picea* (spruces), *Pinus* (pines), *Populus* (poplars), *Quercus* (oaks), and *Ulmus* (elms). The newcomers to the New World were so greedy for the wood from these trees that large specimens like those they first encountered are today difficult to imagine, much less find. A visit to the Joyce Kilmer segment of the Nantahala National Forest in western North Carolina will give an idea of the size of those old originals. Some of the remaining tulip trees (*Liriodendron tuplipifera*) there are so large that it takes three or more people linking hands to reach around their bases.

One of the most important genera to people then and now is *Quercus*, a Latin name applied since the days of the Roman Empire. Angel Oak on the Sea Islands near Charleston, South Carolina provides a glimpse of what those trees formerly were in the southeastern United States. Most of the other oak trees in North America are mere seedlings, no more than 200 or 300 years old.

Quercus in English is the "oak," a word that appeared by A.D. 749, although spelling varied to include *oke* as late as the 1500s. The English name is related to Old High German *eih*, and Middle High German *eich*, a word now rendered *Eiche* in modern German. In Dutch, it is *eycken* or *eyckenboom* (oak tree). In Spanish, the trees are *encinos* (evergreen or "live" species) or *robles* (deciduous species). In Portuguese, the generic term is *carvalho* (from the pre-Roman base word *carb* or *carv*, branched), and *roble* (from Latin *robore*, to be strong) is applied to only European *Q. robur*. To French speakers oaks are *chêne*, related to *chesne* as it was spelled by Fuchs in his *L'histoire des plantes* of 1550 (Meyer et al. 1999). Louisiana Houmas, who had lost their native language, said that *Q. virginiana* was *chéne vert* (green oak). Italians retain the Latin word only slightly modified into *quercia*.

In Hebrew oaks are *allon* or *elon*, both derived from the base-word *el* (god) (Zohary 1982). Oaks are called by that name because the trees are considered holy places where God speaks to mortals. Biblical legend has it that Abraham was visited by the angel of Jehovah under the branches of an oak (Lehner and Lehner 1960). Classical Greeks and Romans, including Pliny (A.D. 23–79) and Dioscorides (fl. A.D. 40–80), said *dryas*, and considered the oak the preferred tree of Zeus. Therefore, devotees of Zeus gathered under oak trees to worship, believing that, when their prayers were heard, the leaves rustled and birds began singing in the branches. That was how Zeus responded to the faithful (Lehner and Lehner 1960, Austin 1998a).

Celtic people also believed that spirits living in trees could be influenced by actions or ceremonies. The expression "knock on wood" (originally "knock-wood") is one of many Gaelic remnants in English. Knocking on wood, especially oaks, drew the attention of the spirits to the wish that had just been made (Panati 1987). One variant of the Celtic word for oak, *dair*, was the fourth letter of their alphabet, and an ancient holy site is the modern Irish city of Kildare (church in the oaks). In Gaulois, oak was *cassanus*, not Latin *quercus*, because as a sacred tree of the Druids, it kept its original name. Both *cassanus* and *quercus* are probably akin to Akkadian *kassu*, strong.

Many people throughout the Americas have generic names for oaks. Cherokees say *a-da:-ya*. In Náhuatl, they are *ahuatl* [*ahoatl, aoatl*]. Mexicans using that name also held the oaks in high regard, as witnessed by city names like *Ahuatepec* (oak hill), *Ahuachichilpa* (in the red oaks), and *Ahuatlán* (near the oaks). In Oaxaca, the Zapotecs say *yaga-yoo* [*yaga-xoo*] (*yaga*, tree, oak). Muskogean speakers, the Alabama, use the generic term *baya* [*bayo*]. Indeed, the genus is so important among indigenous people that Brown (2001) used it as an example of American nomenclature.

One measure of how important oaks were to southeastern tribes is reflected in the Muskogee names for them. Among the Seminoles in Oklahoma, there are ten species, each with distinctive names. Those are *cóskv* (post oak, *Q. stellata*), *kv'lv* (white oak, *Q. alba*), *kvlv'pe* (overcup oak, *Q. lyrata*), *lókcvpe* (red oak, *Q. rubra*), *lókcvpe-láne* (live oak, *Q. virginiana*), *lókcvpe-lv'ste* [*lakcvpe-lvste*] (black oak, *Q. velutina*), *méskol-vpe* (water tree oak, *Q. nigra*), *meskólwv* (chestnut oak, *Q. muehlenbergii*), *sécv* (blackjack oak, *Q. marilandica*), and *tokvmáhv* (Spanish oak, *Q. falcata*).

In Florida, the only recorded indigenous names for oaks come from Muskogean people. Some Seminoles call the live oak (*Q. virginiana*) *lakcv cvmpv* [*lakcacámpa, alatka chumpa, lakchachaba*] (*lakcv*, acorn, *cvmpe*, sweet, Creek), and others say *okiciskî* [*okecheshke*] (*oki*, water, *ciskî*, oak, Mikasuki). Muskgee speakers in Oklahoma say *lókcvpe-láne* (*lokcvpe*, acorn-tree, *lane*, yellow). The diminutive, but similar, *Q. elliottii* (= *Q. pumila*) is called *okiciskincó:cî* (*oki*, water, *ciskî*, oak, *inoci*, son, Mikasuki), while *Q. myrtifolia*, the myrtle oak, is *colokota:pî* (oak tree, Mikasuki) or *tohatka* (*eto*, tree, *hatka*, white, Creek).

The most common tree on the Big Cypress Seminole Reservation in southern Florida is *Q. laurifolia*, the laurel oak. That wetland species is the *asaykaapi* [*asayka:pî*] (*asayk*, acorn, *api*, tree, Mikasuki), *asaykhiskoposkia* [*ashak heskoposhke, oshe huhka puske*] (*asayk*, acorn, *hiski*, leaf, *oposkia*, screech owl, Mikasuki), *asykhoomi* (*asayk*, acorn, *hoomi*, bitter, Mikasuki), and *mishcolabi* [*miskotapî, meskolwv*] (*méskolwv*, water oak, *vpe*, tree, Creek).

Oaks are cherished by people throughout the world for their woods and fruits, although some are more coveted than others. At least 30 species have been used for food and oil by people within the United States (Hodgson 2001). Live oak (*Q. virginiana*) is one of the most prized of American trees, and surely the most valued in the southeastern states. The species grows from Mobjack Bay, Virginia (southwest end of Chesapeake Bay), along the coastal plain and islands to Florida and along the shores of the Gulf of Mexico to northeastern Mexico, spreading farther inland in Texas than in other states (Diggs et al. 1999). In Texas, *Q. virginiana* introgresses with *Q. fusiformis*, formerly a variant, but now separated (Simpson 1988, Nixon 1997). Live oak is disjunct from Florida to Cuba (Sargent 1905, Leon and Alain 1946–1954).

Live oak wood is heavy, hard, strong, and tough—traits many think were passed to its worshipers and users. Wood is also close grained, light brown or yellow, and retains thin nearly white sapwood. *Quercus virginiana* was so valuable during the time of sailing vessels that the U.S. Navy obtained large holdings of live oak forests for exclusive use in the government's shipyards (Harrar and Harrar 1946). The large, massive, and arching limbs were highly sought after for making ship "ribs and knees." The wood was also used for construction and in manufacturing hubs, wooden cogs, furniture, and flooring. This was one of the most important trees to early European settlers in the southeastern United States, and wood from live oaks still is considered the best quality available for many items.

Dyes and paints were made by indigenous people from oak bark. Bushnell (1909) found that the Choctaw were using oaks only to make a red paint. They mixed bark from three or more oaks, including *Q. laurifolia* and *Q. virginiana*, and boiled the mixture until the liquid thickened. The bark was then removed and the liquid boiled again until it was thicker. Just before removing the hot mixture from the fire, a piece of pine pitch (*P. palustris*) was added, melted, and thoroughly mixed with the other liquid. This paint was then ready for use. Houma made a similar preparation, but also used it to stain baskets (Speck 1941). Bark was used sparingly by settlers in the tanning industry (Harrar and Harrar 1946).

Fruits of oak trees are acorns (*ak*, oak, and *corn*, seed, Anglo-Saxon). In Spanish, they are *bellota*, in French *gland*, in Italian *glanda*, and in Portuguese *glande*. All but the Spanish name are derived from the Latin *gland* (acorn). Some oaks produce tasty acorns that are eaten either raw or roasted; others have tannin that must be leached out before being eaten.

The seed crop from oaks is called "mast," from Old English *maest*, akin to Old High German *mast*, and Old English *mete*, meaning food. The word is related to "meat" (from Middle English *mete*, and related to Italian *madére*, and Greek *madaros*, to be wet) in the sense of food for humans and swine. A scribe of the 17th century wrote: "Acorns ... (before the use of *Wheat-Corn* was found out) were heretofore the *Food* of *Men*, nay of *Jupiter* himself ... till their Luxurious *Palats* were debauched. ... And Men had indeed *Hearts* of *Oak*; I mean, not so *hard*, but *health*, and *strength*, and liv'd naturally, and with things easily *parable* and plain" (from Fernald et al. 1958).

Indigenous Americans also had names for acorns. The Mixé of Oaxaca and Chiapas call the acorns *kook* [*shokiup*], the Otomí in Hidalgo say *mêttza* [*ndezâ*], the Zoque in Chiapas say *camay-cuy*, and the Tarascan of Michoacan use *taraquen*. The Catawba said *hitumo̱'*. The Cherokees called acorns *gule*. The Siouan people in Louisiana said *aya̱n'yâ* (Biloxi) and *u'tu i'fhu* (oak seed, Ofo).

Muskogean tribes also had names for acorns; the Creek said *lakcv* [*lokcha, sockcha*] (Simmons [1822] 1973, Martin and Mauldin 2000), the Miccosukee *asayk* (Sturtevant 1955), the Chickasaw *nasi'* (Munro and Willmond 1994), the Choctaw *hane* (Crawford 1988) or *nusi* (Byington et al. 1915, Haag and Willis 2001), and the Alabama *bayaɬi* (*baya*, oak, *aɬi*, fruit, Sylestine et al. 1993). In Florida, the Timucua said *aha* [*aja*].

Indeed, there is a river in central Florida the Timucua called *ajano hibita chirico* (river of little acorns). The Seminoles also named places after the oaks, calling one *lokcha apopka* (*lokcha*, acorn, *apopka* place for eating, Creek). The first of these places we now call the Hillsboro River near Tampa, and the other is the Houston River in the Ten Thousand Islands of Collier County south of Naples. Live oak seeds were eaten by all tribes within its range (cf. Bartram 1791). Probably, it is live oak acorns in Timucua baskets that Frenchman Jacques Le Moyne drew from his visit to Florida in 1564–1565 (cf. Melanich and Milbrath 1989).

The first record of indigenous people using acorns to produce oil came from the de Soto expedition. The chroniclers reported people in the town of *Chiaha* (a Yemasee town) on the upper Tennessee River with "calabashes" (*Cucurbita*) filled with oil from nuts and acorns (Swanton 1939). Later, Capt. John Smith wrote: "the Acornes of one kind, whose barke is more white than the other, is somewhat sweetish; which being boyled halfe a day in severall waters, at last affords a sweete oyle, which they keep in goards to annoint their heads and joints. The fruit they eate, made in bread or otherwise."

Harriot ([1590] 1972) was the first to record oil being made from acorns by North Carolina Algonquians. He wrote that there are "severall kindes of *Berries* in the forme of Oke akornes, which also by their experience and vse of the inhabitantes, wee finde to yeelde very good and sweete oule." He selected one called *mangúmmenauk* [*mankwiminaki*, big nuts], "which beeing dried after the maner of the first sortes [chestnuts, walnuts, hickories], and afterwards watered they boyle them, & their servants or sometimes the chiefe themselues either for variety or for want of bread, doe eate them with their fish or flesh." Presumably, the Powhatan acorns came from the oak Strachey called *poawamindg* (Harrington 1955). The Powhatans called acorn oil *monohominy* (Strachey [1612] 1953).

John Lawson, writing in the early 1700s about the Carolinas, found much the same. He remembered: "The Indians beat them [acorns] into meal and thicken their venison broth with them, and oftentimes make a palatable soup. They are used instead of bread, boiling them till the oil swims on the top of the water, which they preserve for use, eating the acorns with flesh meat" (Swanton 1946). Later Lawson paid special attention to live oak acorns (*Q. virginiana*) in the coastal zone. He found: "The acorns thereof are as sweet as chestnuts, and the Indians draw an oil from them, as sweet as that from the olive, though of an amber color. . . . I knew two trees of this wood among the Indians, which were planted from the acorns, and grew in the freshies, and never saw anything more beautiful of that kind."

John Josselyn reported in 1672 that the Pilgrims in Massachusetts discovered baskets of *Q. alba* acorns hidden in the ground by the local Massachuset tribe in December of 1620. The New England settlers appropriated those seeds and prepared a palatable and nutritious dish of boiled sweet acorns. Rosengarten (1984) noted that acorns (*Q. alba*, which he emphasizes) contain 50.4% carbohydrates, 34.7% water, 4.7% fat, 4.4% protein, 4.2% crude fiber, and 1.6% ash. A pound of acorns is said to provide 1265 calories. Not only did native Americans eat acorns, they also used their oil for cooking and flavoring other foods such as hominy. We put butter on grits; they used acorn oil (see also *Carya, Juglans, Zea*).

Heavy crops of mast, or "mast years," are irregular, and this helps spread the species. There are so many seeds during those seasons that, by simple chance, more are able to escape insects, birds, squirrels, and other animals that consume them. Many animals eat acorns; acorn weevils (Curculionidae) specialize on them. Gall wasps (Cynipidae) deform leaves, stems, and flower clusters with imitation plant hormones to secure their eggs and larvae. For many seasons I watched blue jays (*Cyanocitta cristata*) and boat-tailed grackles (*Quiscalus major*) gorge themselves on live oak acorns in southern Florida.

In England, the rural economy of an area was formerly measured in terms of the number of swine that its oaks could support. One entry in the *Domesday Book*, compiled on the orders of William the Conqueror about A.D. 1086, says of a certain village, "There is wood for forty swine," and the village was taxed accordingly (Edlin and Mitchell 1985). In Europe, *Q. robur* was the species most prized for its acorns, and the American *Q. alba* is its New World equivalent (Fernald et al. 1958, Rosengarten 1984). The live oak runs a close second.

Oak bark and galls have been used to prepare permanent ink and medicines since antiquity (Meyer et al. 1999). Fuchs in 1542 recorded both uses and illustrated many galls on the leaves of *Q. robur*. Culpeper (1653) and other herbalists continued attributing medical properties to various parts of the oak. Florida and Oklahoma Seminoles came to the same conclusions (Sturtevant 1955, Howard 1984). Tannin is the known active chemical in oak medicines, and it has been shown experimentally to be antiviral, antiseptic, and antitumor, but also may be carcinogenic (see Simpson 1991 and Foster and Duke 1994 for divergent opinions).

Celtic and other people considered the oaks symbolic of strength. We still use that symbolism in the U.S. Army by presenting an Oak Leaf cluster for exceptional bravery. Our practice mimics that in the times of the Roman author Pliny (A.D. 23–79), when the oak symbolized bravery and a crown of oak leaves was a reward for outstanding military valor. Another European remnant is a Gaelic belief that oak leaves repel witches (Beith 1995, Vickery 1995). I wore a belt with oak leaves embossed on it for years and did not encounter a single witch. So, the charm obviously works.

R

Randia: White Indigo-Berry

(The genus honors the British apothecary and gardener Isaac Rand, d. 1743, who was the director of the Botanic Garden in Chelsea)

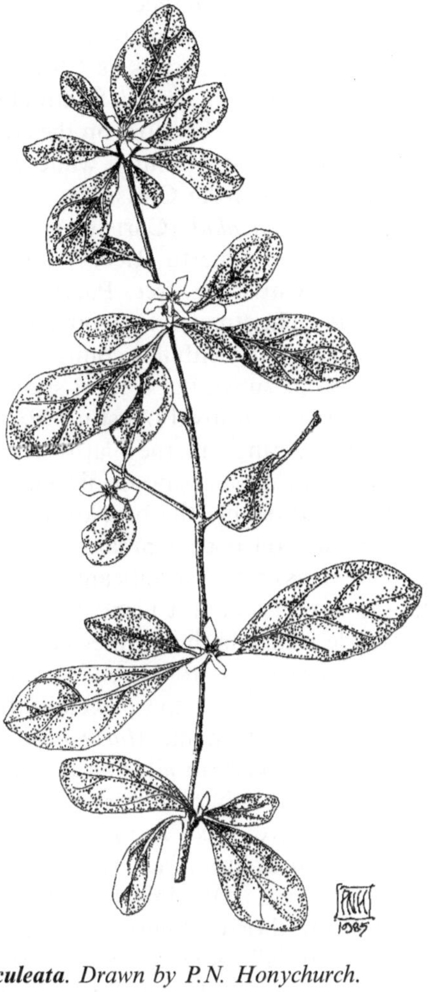

Randia aculeata*. Drawn by P.N. Honychurch.*

According to my students, there is never anything distinctive about white indigo-berry (Florida). Even when I pointed out the short internodes resulting in clustered leaves, the reverse lance-shaped (oblanceolate) leaves, larger on the distal end and tapering to a pointed base, or the white flower, they still said it was not enough. So, I advised them to identify it by process of elimination until they learn it better. "If it is nothing else we have studied, try *Randia aculeata*."

Linnaeus ([1753] 1957) named the plants *Randia*, but he did not originate the name. Instead, Linnaeus attributes *Randia* to Dutchman Adrian van Royen who took it from a De Jussieu, probably Bernard De Jussieu (1699–1776). Although the Linnaean Herbarium in London contains two specimens collected by Hans Sloane in Jamaica, Linnaeus gave no indication that he had seen them. Instead, the basis for *R. aculeata* is a specimen from Jamaica thought to have been collected by Patrick Browne. The species name *aculeata* (prickly, Latin) refers to the short branches that are effectively spines.

Linnaeus probably also knew the species from plants cultivated in Leyden and recorded by Royen. Those plants too were apparently originally from Jamaica, where the plant's common name now is box-briar (Bermuda, Bahamas, Jamaica, Panama, Puerto Rico). Perhaps the name was in use in Linnaeus's time since "briar" and "*aculeata*" both allude to spiny plants. Indeed, the prickly nature of these shrubs has led to names through much of its range from southern Florida, Bermuda, the Bahamas, West Indies, Tamaulipas, and Sinaloa south to Cozumel Island, Ecuador, and Venezuela.

Other similar names in English are prickly bush [prickle bush] (Bahamas, Jamaica) and sticky bush. In French, it is a *croc à chien* (dog fang, Haiti). Spanish speakers have a greater variety of names. They call it the *agalla de costa* (coastal gall, Cuba), *escambrón* [*cambrón*] (big scale, Puerto Rico), *palo de espinillo* (tree of little spines), *punta real* (royal point or spine), or *serrasuela* (saw's at shoe-soles). Although *Randia* is called *bois de lance* (lancewood, French Antilles) and lancewood (Cayman Islands), the reference might be to the "lances" of the spines. The stems often are too small to make effective lances for hunting or fighting, although some are recorded up to 10 m tall and 22.5 cm in diameter (Roig 1945, Morton 1981).

Unless it is based on some unrecognized indigenous term, the name *papachilla* (Sinaloa; other spp. called *papache* in Sonora, from *papachos*, to cherish or caress) also seems to allude to the "loving" nature of the thorns in grasping passers-by. The name *peetsch-kitam* [*pechcitam, xpe'ech kitam*] (*peetsch*, dwarf, *kitam*, javalina, *Tayassu tajacu*, Maya, Yucatán, Belize) is an allusion to the spines. Apparently, the name compares the canines of these pig look-alikes to the spines on the plants. Both are dangerous to the unwary.

Several of the names have not been translated or only partly so. Those from Cuba and Puerto Rico may be based on Taino words. For example, *Randia aculeata* is *cabai-nacte* [*cabai nagte*] (Puerto Rico), *dragale* (*draga*, dredge, Cuba), *yamaguey* (Taino, Cuba), and *yamaguey de costa* (coastal *Randia*, Cuba). The word *yamaguey* differs only slightly from

Camaguay, a province in Cuba, with both based on indigenous words. Or perhaps the province was named after these plants. The names *cracilla* (*crac*, crack, Mexico), *pitajoni bravo* (*pitajon*, big agave, *bravo*, wild, Cuba), and *pitajoni espinoso* (*pitajon*, big agave, *espinoso*, spiny, Cuba) all seem to be Spanish, but the allusions are uncertain. The onomatopoeic *crac* may refer to being stabbed by the spines. *Pitajon* probably refers to the spiny nature that resembles *Agave*. However, we are left wondering what kind of animal has a tail resembling some part of *Randia* in the name *raboe die kabasi* [*rabu di cabai, raboe die kabai*] (tail of *kabasi*, Dutch Antilles). The name *wakoera* [*wakura*] (Dutch Antilles) resembles an Arawak word as does *kabasi*, but nothing further has been found.

A number of names make understandable comparisons. To people in Cuba, these shrubs are like a *café cimarrón* (wild coffee). Some think *Randia* is more like a *Buxus* and say Florida boxwood, although others think that name should be confined to *Schaefferia frutescens* (Celastraceae). In Belize, they think the shrubs look like limes (*Citrus aurantiifolia*), and say bastard lime or wild lime; they also call them wild guava and wild okra. However, you might be able to make a fishing-rod (Antigua, Virgin Islands) from the stems. Surely those who say *maíz tostado* (toasted corn, Colombia) are talking about something not obvious to the rest of us. In the southern Dutch Antilles, people make food beaters from branches, which they call *leele* [*lele*] (Bonaire, Curaçao), and the tree is *palu di lele* (food beater tree).

The Florida name white indigo-berry is analagous to indigo berry (Florida, Jamaica, Virgin Islands, Barbados), ink berry (Puerto Rico, Jamaica, Barbados), *tintello* [*tantillo*] (ink bush, Puerto Rico), and *tintero* (ink bush, El Salvador, Colombia). All those names refer to the fleshy berries, which have been used as substitute ink. Similarly, the names *palo de cotorro* (dove tree, Puerto Rico) and *petit coco* (little fruit, St. Barthélemy, Guadeloupe) refer to the fruits. However, *corallero* (red one, Colombia) makes no sense because the fruits are white outside and blue-black inside.

People not only note the edible fruits on *R. aculeata*, but they also consume them from other species. However, edibility is variable and some think the fruit a last resort. The related species *R. echinocarpa* has been described as bitter and overly sweet (Yetman and Van Devender 2002), and that pretty much sums up *R. aculeata*. In much of Mexico, the green astringent fruit is used to treat dysentery, while the ripe fruit is eaten as food. Ripe fruit also gives a blue dye (Standley 1920–1926).

One series of names surprised me by referring to the cross of the crucifixion of Christ. For 31 years, neither my students nor I ever noticed a similarity, but we were not guided by the Doctrine of Signatures. Nonetheless, from Mexico to Colombia the shrubs are known as *crucecilla de la costa* (coastal little cross, Sinaloa), *crucesita* (little cross, Mexico), *cruceta negra* (little black cross, Venezuela), *cruceto* [*crucito, crucete, crucilla*] (little cross, Tamaulipas, Veracruz, El Salvador, Columbia, Venezuela), *espina cruz* (cross spine, Cuba, Veracruz), and *palo de cruz* (cross tree, Colombia). Even the name *Juan de la Cruz* (John of the Cross, Cuba) makes that reference.

It might be the comparison to the cross that led people to begin using the shrubs as Christmas trees. When I first saw them being used in Santo Domingo, I could not believe my eyes. Vendors on the streets were selling leafless stems, some painted white and others silver, to be used during the Christmas season. They call them *árbol de Navidad* (Christmas tree, Puerto Rico), Christmas tree (Puerto Rico, Virgin Islands), *palo de Navidad* (Christmas tree, Puerto Rico), and *ramo de Navidad* (Christmas tree, Dominican Republic). In a tropical country with few conifers available, it only makes sense to substitute a local species during that Druidic-Christian festival (Austin 1998a).

Throughout its range in the Caribbean, *R. aculeata* is a medicine. Curiously, no medical use has been found in Mexico. Probably the best-known and most widespread application is to combat fever. The name chill-bush (Bahamas) refers specifically to that use. In Cuba, the sap is considered a hemostat (Roig 1945). Leaves are made into a bath for infected sores in the Bahamas, and leaves or roots boiled with *Senna mexicana* are used to expel the placenta after childbirth (Morton 1981). The name *María angola* (Angola Mary, Colombia) is likely an allusion to those uses.

Not only is white indigo-berry used during childbirth, but some have used it in earlier stages of relationships, as an aphrodisiac. The names *resuelesuele* (return money, Dominican Republic), *sota caballo* (substitute horse, Puerto Rico), and stiff cock (Bahamas) are allusions to that (Morton 1981). Sometimes the plant is used alone for this concoction, but on other occasions, the root is boiled with *Chiococca alba* or both are boiled with *Smilax havanensis*.

No chemical studies seem to have been made of *R. aculeata*. However, other species of *Randia* include saponins (triterpene, triterpenoid) (Saharia and Seshadri 1980, Murty et al. 1989, Sati et al. 1989, Sotheeswaran et al. 1989, Dubois et al. 1990, Dubois et al. 1990, Lemmich et al. 1995, Jansakul et al. 1999, Sahpaz et al. 2000), iridoid glucosides (Varshney et al. 1978, Sainty et al. 1982, Uesato et al. 1982, Sati et al. 1986, Davioud and Bailleul 1988, Sahpaz et al. 2000), coumarins (Bashir et al. 1986, Mallick et al. 1998), triacontanoic acid esters (Mallick et al. 1998), and some that are uncertain (Quershi and Thakur 1977).

Within the laboratory these chemicals suggest potential use as analgesics (Reanmongkol et al. 1994) and as anti-implantation compounds (Prakash et al. 1985). The chemicals are also hemolytic (Sotheeswaran et al. 1989) and cicatrizing in ulcers (Pérez et al. 1993). In addition, saponins in the fruits of *R. siamensis* have been used to poison fish (Jansakul et al. 1999). The iridoid glucosides are precursors of indole alkaloids and those from a number of plants have been shown to be analgesic, antimicrobial, antiviral, anti-inflammatory, wound healing, antioxidant, and cancer chemopreventive.

In one study the anti-inflammatory properties of iridoids "showed a significant effect on TXB2-release from calcium ionophorestimulated human platelets, with inhibition percentages slightly lower than the reference drug ibuprofen" (Bermejo et al. 2000). So, if you do not have a commercial preparation with ibuprofen handy, you could always try eating some white indigo-berry. Who knows? It might work.

Ranunculus

(From Latin *rana*, frog, plus *–unculus*, small; applied to these plants by Pliny, A.D. 23–79)

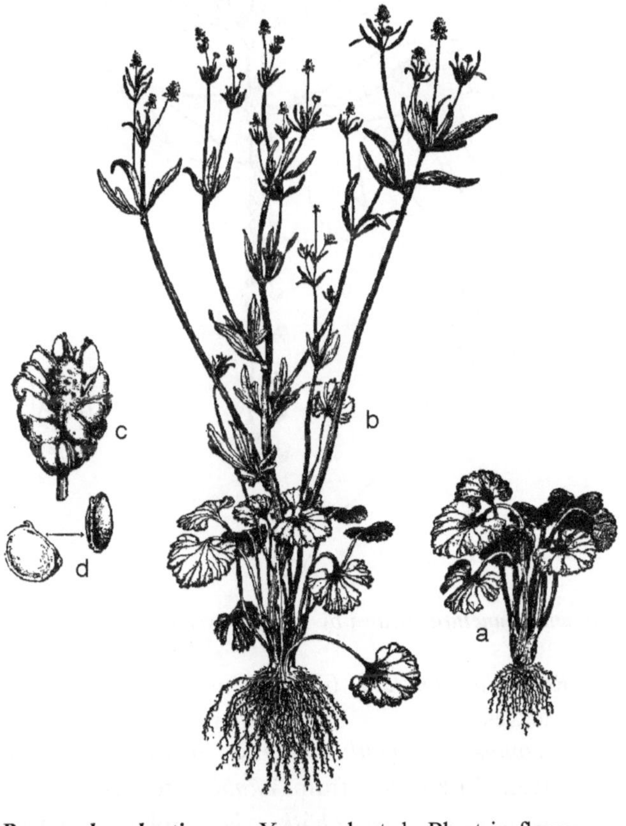

Ranunculus abortivus. a. Young plant. b. Plant in flower. c. Mature receptacle bearing seeds. d. Seeds. From Buchholtz 1968.

batrachion (frog, Greek)
buidheag (little yellow one, Gaelic)
buttercup (this is a comparatively late name for *Ranunculus*, having been in use by 1777; before that the plants were known as "butterflower," "king cups," and "gold cups")
butterflowers or gold cup (in use by 1578, when Henry Lyte translated Dodoens's 1554 herbal with "Some do also name it ... in Englishe Gold-cuppes, Gold-knoppes, and butter-flowers")
crowfoot (in use by about A.D. 1440 for *Ranunculus*, particularly those with divided leaves; subsequently extended to the entire genus; the comparison is with the foot of the carrion crow, *Corvus coronae*); *Hahnenfuss* (cocksfoot or crowfoot, German)
king cup (a name used by Turner in his *Herbal* of 1562 where he equated it with crowfoot)
llygad elrill (Welsh)
meadow-bloom
pilewort (Turner, in his *Names of Herbes* of 1548, called *R. ficiaria* the "fygwurt," where "fig" at that time was the name for piles)
ranoncule (from *ranunculus*, Quebec); *ranuncolo* (Italian); *ranúncolo* (Spanish)
Scharbockskraut (scurvy herb, German)
soleie (*sol*, sun, *eie*, owned, Norwegian)
spearwort (in use by 1387, when a book published in London said, "sperwort, *calidus est et siccus in quarto gradu*" [it is dry and hot in the fourth degree])
yellow-weed

Ranunculus abortivus (abortive, i.e., with reduced styles and petals)
[kidney-leaf, little-leaf] buttercup
chicken pepper
[small-flowered, smooth-leaved] crowfoot

Although we credit Linnaeus ([1753] 1957) with the creation of *Ranunculus*, the name had been in use since the time of Pliny (A.D. 23–79). That ancient application is still reflected in the Romance languages as *ranoncule* (French), *ranuncolo* (Italian), and *ranúncolo* (Spanish). Fuchs (1542) also used the Greek name *batrachion* (frog) for *Ranunculus*.

Europeans knew these plants well before they arrived in the New World. There were multiple uses and beliefs associated with the genus. Farmers in some areas were convinced that cattle eating these yellow-flowered plants in the spring passed the color along in their golden butter. Some cattlemen even rubbed the flowers on their cows' udders and hung the plants over the barn door to add color to the cream (Coffey 1993,

Vickery 1995). In the Inner Hebrides, *R. flammula* was used as a substitute for rennet in cheese making (Vickery 1995).

The belief that the gold of the flowers was passed to cream and butter was erroneous. Sap from the plants is acrid and grazing animals avoid them when possible. Indeed, the sap is so caustic that Fuchs wrote in 1542: "Healthy scoundrels and beggars abraid their legs and arms with this plant in order to extort alms shamelessly, and by this trick increase their gain" (Meyer et al. 1999).

Considering the acrid nature of the plants, it is difficult to believe that the Cherokee cooked and ate the leaves as greens (Hamel and Chiltoskey 1975). However, Fernald et al. (1958) had discovered edibility and experimented with *R. bulbosus.* They found that John Lightfoot, in his *Flora Scotica* of 1777, had experimented before them and written, "not withstanding this corrosive quality, the roots when boiled become so mild as to become eatable." Fernald et al. (1958) reported that the round bulbs, after having overwintered, were mild and sweet in the spring. When thoroughly dried they lost their acridity.

Vogel (1970) found that the Illinois-Miami treated gun or arrow wounds with the crushed root of a *Ranunculus.* No species was indicated; perhaps several were used. The crushed root was dissolved in warm water, and apparently was intended to prevent infection through its caustic or vesicant activity. Farther south, the Cherokee used small-flowered crowfoot in a poultice on abscesses, for "thrush" (*Candida*), as a sedative, and to relieve sore throat (Hamel and Chiltoskey 1975). The Iroquois took a decoction to treat epilepsy, blood disease, sore eyes, stomach trouble, stiff muscles, snakebite, and toothache. They also found it emetic, and used it to counteract poison and to dry up smallpox sores. The Meskwaki used the roots as a styptic to stop nosebleed (Vogel 1970). Several other species were used for similar purposes in the eastern forest region by the Abnaki, Montagnais, Ojibwa, and Potawatomi (Smith 1933, Vogel 1970, Moerman 1998).

In 1751, John Bartram wrote that a decoction of *R. abortivus* was considered a remedy for inveterate syphilis (Coffey 1993). However, it was the introduced European *R. bulbosus* that became official in the U.S. Pharmacopoeia between 1820 and 1882. That species was considered acrid, and it was applied externally as a counterirritant (Vogel 1970). Porcher (1863) recommended either *R. sceleratus* and *R. repens* as a rubefacient. Of *R. scleratus,* he added, "It has been administered with success in asthma, icterus, dysuria, rheumatism, pneumonia, and fixed pains. When it acts as a vesicant, it has not the disadvantage of producing strangury." Millspaugh (1892) gave much the same information, but noted that *R. acris* and *R. bulbosus* have similar properties.

Fernald et al. (1958) called the acrid principle in *Ranunculus* anemonol. However, Turner and Szczawinski (1991) explain that the acrid compound is protoanemonin, an enzymatic breakdown product of the glycoside ranunculin. When the plants are dried, the protoanemonin becomes the innocuous anemonin. Foster and Duke (1990) warn that the acrid chemicals are known for "causing intense pain and burning of the mouth, mucous membranes; blisters skin."

Rapanea

(Jean Baptiste Christophore Fuseé Aublet took the name from the French Guiana indigenous *rapana*)

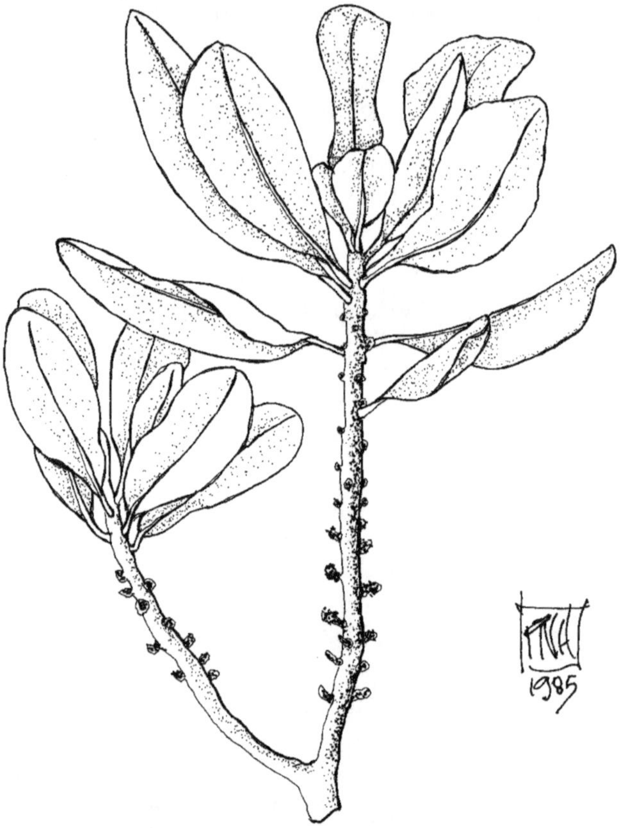

Rapanea punctata*. Drawn by P.N. Honychurch.*

Rapanea punctata (dotted, in reference to the leaves)
akcomakáhka:phatkî (white tobacco-seasoning tree, Mikasuki; from *akahke,* tobacco, *hatki,* white); *hici:apa:kahátki* (*hici,* tobacco, *vpakv,* seasoning, *hvtke,* white, Creek)
bádula (Puerto Rico)

caca-ravat (*caca,* feces, *ravet,* rat, the same name is used for *Myrica, Forestiera,* and *Trema,* Guadeloupe, Martinique)
camagüilla (from Taino *yamaguay*?, Cuba)
canelón (big cinnamon, Argentina); *fuelle canelle* (cinnamon leaf, Haiti)
chagualito (Colombia)
colicwood (Florida)
cucharo (spoon, Venezuela)
dakara (Arawak, for *R. guyanensis,* Guyana, Suriname)
fuelle (variant of *feuille,* leaf, Haiti)
kunaporan [*koonaparan*] (Carib, for *R. guyanensis,* Suriname)
mamayuelo (little mamey, Puerto Rico); *mameycillo* (little mamey, Venezuela)
manteco blanco (white butter, Venezuela)
myrsine (myrtle, Greek *Myrtus,* a related genus, Florida)
palo de sabana (savana tree, Dominican Republic)
poève (perhaps *poéle,* a pall at a funeral, Haiti)

Lamarck called this small tree *Sideroxylon punctatum* in 1794, thus placing it in the wrong family. It was not until 1969 that Cyrus Longworth Lundell (1907–1994) discovered the problem and moved it to *Rapanea.* The number of species in the genus is not clear as Mabberley (1997) considers it part of *Myrsine,* as have most other recent biologists.

As their names suggest, the Seminoles used this to extend their tobacco (Sturtevant 1955). No other uses are known, but the name "colicwood" indicates former use as a treatment for colic. Some separate the Greater Antilles plants from *M. guianensis* or *R. guianensis.* If that is done, the names from South America should be stricken.

Reynosia

(In 1866, Grisebach named this genus for Alvaro Reynoso, a Cuban chemist and agriculturist who died in 1888)

Reynosia septentrionalis (pertaining to the north)
Caicos [darling] plum [dollen plum] (Florida, Bahamas)
red ironwood (Bahamas)

It was not until 1899 when Urban described this species, with one of the specimens he cited having been collected in Florida. All of the other 17 species are all in the Caribbean except for 1 in Central America (Mabberley 1997). Fruits of many species are edible.

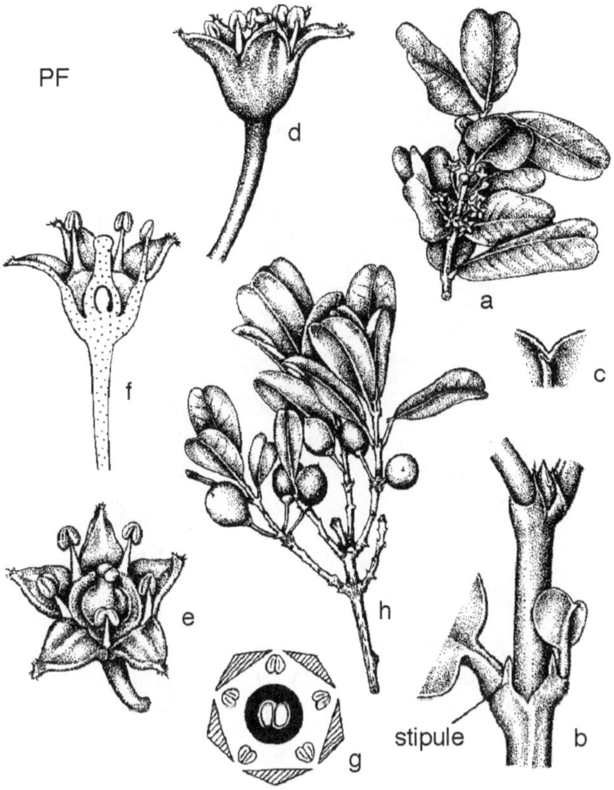

Reynosia septentrionalis. a. Branch with flowers. b. Detail of leaf axils and terminal bud. c. Leaf apex showing midrib extension. d. Flower, side view. e. Flower, from above. f. Flower, longitudinally dissected. g. Floral diagram. h. Branch with fruits. *Drawn by Priscilla Fawcett.* From Correll and Correll 1982.

In the Bahamas the trunk is scraped and boiled into a decoction used to stop diarrhea (Eldridge 1975, Morton 1981). The decoction is also used for "weakness." Fruits are edible, but make some people dizzy if there is nothing else in the stomach (Morton 1968, 1981).

Rhabdadenia

(Swiss botanist Johannes Müller Argoviensis, 1861–1920, named this with Greek *rhabdos,* a rod or stick, and *aden,* gland, possibly a reference to the slender fruit)

Rhabdadenia biflora (with two flowers)
bejuco de manglar [*de mangle*] (mangrove vine, Dominican Republic, Puerto Rico); mangrove [swamp] vine (Florida, Bahamas, Puerto Rico, Panama, Belize)
caoutchouc (rubber, alluding to the latex, Hispaniola); *palo de leche* (milk tree, Dominican Republic); rubbervine (Florida)

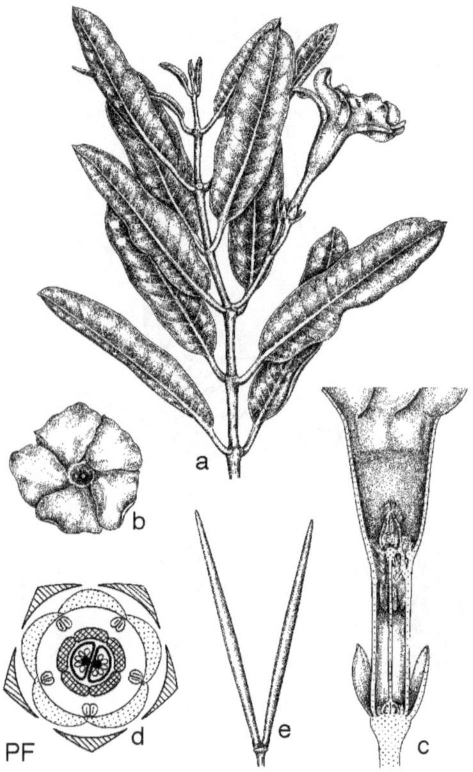

Rhabdadenia biflora. a. Apex of flowering stem, note that the inflorescence is not axillary. b. Flower, from above. c. Flower tube, longitudinally dissected. d. Floral diagram. e. Fruits. *Drawn by Priscilla Fawcett*. From Correll and Correll 1982.

clavelito (little carnation, Colombia); *clavelitos de manglar* [*de sabana*] (little mangrove [savanna] carnation, Cuba)
tataku (Miskito)

Jacquin named these climbers *Echites biflora* in 1760. Subsequently, the plants were studied for the *Flora Brasilensis*, and Swiss botanist Johannes Müller Argoviensis, who did the family for that book, moved the species to *Rhabdadenia*. Since then two other species have been found in tropical America (Rosatti 1989, Mabberley 1997).

The acrid juice is purgative and poisonous and produces blisters on the skin (Standley 1920–1926, Roig 1945). Taken internally the latex inflames the entire alimentary canal, although it is used to kill troublesome feral animals (Roig 1945). Mabberley (1997) indicated that it contains alkaloids.

Rhamnus

(Pliny, A.D. 23–79, used *rhamnus*, the ancient name "for a prickly plant," said to be based on Greek *rhamnos*; according to Millspaugh, the name is from Gaelic *ràmh*, branching; Quattrocchi thought the name was from Greek *rhabdos*, a stick; akin to Akkadian *rapasu*, to beat, Sumerian *rab*, stick, branch)

Rhamnus caroliniana. From Sargent 1905.

buckthorn ("buck" referred originally to either the male deer or goat, although now it is applied to a number of caprids; the word is perhaps of Germanic origin, from Old English *buc* and *bucca*, or maybe from Gaelic *boc*; combined with "thorn" dates from ca. 1578, and is a translation of Latin *cervi spina*); hart's-horn (cf. Culpeper 1653)
cervispina (from Valerius Cordus in 1561); *espino cerval* (deer's horn, Spanish); *spino cervino* (deer's horn, Italian)
Christes thorne (by Turner in 1548 and then in John Gerarde's *Herball* of 1597); *Kreuzdorn* (cross thorn, German)
frangula (from Latin *frangere*, to break, first used by Rembert Dodoens in 1554); *frangula* (from Latin *frangula*, Portuguese)
geitved (goat wood, Norwegian)
herb-ivy [herb-eve] (from Culpeper 1653)
nerprum [*nerprun*] (from Latin *niger prunus*, black plum; only in Gallo-Roman, French)
ohicta (Onondaga)
ram of Lybia (John Gerarde's *Herball* of 1597); ram-thorn (from 1633 in Thomas Johnson's enlarged version of Gerarde's *Herball*)
rhamn (called "rammyn or thefthorne" first in English about A.D. 1340; in 1388 glossed "rhamne" and translated "theve [thief] thorn")

Rhamnus caroliniana (of Carolina) (=*Frangula caroliniana sensu* Kartesz 1994, Moerman 1998, Diggs et al. 1999)
bog-birch
[alder-leaf, Carolina, southern, yellow] buckthorn
Indian-cherry (because the fruits were eaten by indigenous tribes, as medicine?)

polecat-tree ("polecat" from Middle English *polcat, pullcat,* where *pol* or *pull* probably came from Old French *poule*, chicken, and *cat*, the feline; first used A.D. 1320 for the animal, *Putorius foetidus* of Europe)
yellow wood [yellowwood]

When Europeans arrived in the New World, they were greatly enamored of plants called "buckthorn" or "*espino cerval.*" Those plants had been medicines since at least the 1st century A.D. when Pliny talked about them. The Greek physician Galen of the 2nd century A.D. also praised the plants as giving protection from witchcraft, demons, poisons, and headaches (Dobelis 1986). It is surely significant that the Temple of Nemesis is at the Rhamnus ruins on the coast northeast of Athens and Marathon in Greece (Baumann 1993).

The variety of common names used between the time of Pliny and the discovery of the New World is partly due to the rich history of mixing pagan with Christian views. The oldest of the names alludes to "branches," regardless of its linguistic stock. Languages as diverse as Gaelic, Greek, and Sumerian make that notation, surely because the bark was taken from branches for use in medicine. Even today, the connection with pagan spirits remains in the common name *trollhegg* (troll hedge or garden) for *R. frangula* in Norway (Mossberg and Stenberg 1994). With the rise of Christianity, its leaders incorporated this sacred pagan plant into their pharmacopoeia by giving it names like "Christes thorne" and "*Kreuzdorn.*" Europeans knew two important species now called *R. frangula* and *R. cathartica* (Culpeper 1653, Millspaugh 1893, Dobelis 1986).

Florida's buckthorn was described by Thomas Walter (1740–1789), a Charleston planter, in his *Flora Caroliniana* published in 1788. That same year, Frenchman Charles-Louis L'Héritier de Brutelle described *R. alnifolia* from eastern Canada. Those two species were described after Spanish priests in California dubbed a local plant the *cáscara sagrada* (holy bark) in 1777. Coon (1974) said that the priests gave the plant that name because of the high regard the local tribes had for it as a medicine. Clearly, it was highly regarded by numerous groups because Moerman (1998) listed 30 tribes that used it.

It was not until 1825 that botanists described the Pacific coast plants as *R. purshiana.* Although Europeans were so enthusiastic about their buckthorns that they introduced both species into the eastern United States where they became naturalized, they were reluctant to accept the American plants as medicines. It was not until 1877 that American and European physicians began accepting *cáscara sagrada* as an effective medicine (Doeblis 1989). Once physicians admitted *cáscara sagrada,* it became part of the U.S. Pharmacopoeia in 1894, and it is now the most commonly used laxative in the world (Heiser 1993). However, see also Bown (1995) and Duke et al. (2002).

I grew up hearing people talk about taking "cass-cáre-ah," and the older people considered it a remedy for whatever ailed them. It was long after I learned to speak Spanish that I connected "cass-cáre-ah" with *cáscara* (bark), and then with *cáscara sagrada* (holy bark). My parent's generation was dosed with that plant weekly as a general "preventative" for everything. Authors of most literature on the time and topic call *cáscara* a "tonic." However, those who actually were obliged to take it have other, more derogatory terms for *cáscara*—and its impact on the alimentary canal.

Lewis and Elvin-Lewis (1977) recognize three classes of purgatives. The first works by adding bulk, and *Plantago psyllium* is probably the best known of that category. Their second category includes lubricants, such as "mineral oil" (liquid paraffin). *Cáscara sagrada* is in their third category where the chemicals irritate the lining of the intestinal walls. The active compounds in *cáscara sagrada* and other *Rhamnus* are anthraquinone glycosides, particularly cascarin (Lewis and Elvin-Lewis 1977, Turner and Szczawinski 1991).

Indigenous Americans did not have the same preoccupation with their bowels as did Europeans, but they still used *Rhamnus* as medicine. One indication of the difference in views is that only two tribes are known to have used *R. caroliniana.* The Creeks took an infusion of the wood to relieve jaundice (Swanton 1928a). Oklahoma Delaware took a decoction of the bark to "remove bile from the intestines," and as an emetic and cathartic (Moerman 1998). Since the plants did not grow in the original homeland of the Delaware in New Jersey and Pennsylvania, uses must have been learned after their removal.

California tribes used *cáscara sagrada* as a dye (Moerman 1998). No records have been found for eastern tribes using *R. caroliniana* as a dye; however, Tull (1999) was assured that it provides dark brown and yellow colors. Cannon and Cannon (2003) record dyes obtained from several *Rhamnus* species, including yellows that Jews were forced to use to color their clothes to distinguish them from Christians in the Middle Ages. Presumably, the chemicals of the Old World species forming the dyes are the same as in American plants. Cannon and Cannon (2003) noted flavonols, including kaempferol, rhamnetin, rhamnocetrin, and quercetin, and the anthraquinones, alaterine, frangularoside, rhamnicoside, often with a dominance of emodin.

Some sources record the fruits of Carolina buckthorn as edible. For example, Tull (1999) noted that she had been assured that the fruits could be eaten, but after tasting them, she concluded that they were inedible. Certainly, nothing about the flavor is likely to induce people to eat many of them. While fruits are attractive reds or purples, the flesh is bittersweet. Fernald et al. (1958) noted that some *Rhamnus* fruits are "said to be palatable," while others are poisonous. Dodoens in the 1500s said that the berries "be not meete to be ministered but to young and lustie people." Fernald et al. (1958) have an even less favorable quote from New England where a young boy remarked, "Them is terrible things for the guts."

Rhexia

(From Greek *rhexis,* breaking or rupture, for "the breaking or bursting forth of the entrails of victims"; a name used by Pliny, A.D. 23–79, for some Boraginaceae reputed to be useful in curing ruptures)

deer-grass [deergrass] (OED 1971 says a name in use by 1866; however, Porcher 1863 had written earlier, "Deer grass ... The leaves of this plant have a sweetish, acid taste; and are eaten with impunity. Deer are said to be fond of them")
Maid-Marian (by 1525 the name of a female personage in the May-game and Morris-dance, held on the first of May; later the female companion of Robin Hood; presumably used for these plants because they begin flowering in May)
meadow-beauty [meadow beauty, meadowbeauty] (in use by 1866; unexplained unless it may be taken literally)
sorrel (used by Porcher 1863; this must be a comparison with *Rumex*, which is typically called sorrel because of the sour taste)

Rhexia virginica (from Virginia)
handsome-Harry [handsome Harry] (used by John K. Small in 1933)
[common] meadow-beauty
Robinhood (name used for the plant by Manasseh Cutler in 1785, maybe because of the alternate name Maid-Marian)
vinegar wood (an allusion to the sour taste)
Virginia soapwort (John Clayton wrote "soap wood" in the 1730s, but he compared the English to Latin *saponaria,* also called soapwort)

Linnaeus ([1753] 1957) had not seen living plants of *Rhexia* when he adopted the genus from Jan Gronovius. In his *Flora Virginica,* Gronovius called these herbs *Rhexia calycibus glabris* (Rhexia with a glabrous calyx). Linnaeus shortened that to *Rhexia virginica.* Either Linnaeus was impressed with the broken fruit on the specimen, or he had learned how easily the petals fall from the flowers. In the early 1970s, one of my students assumed that the name *Rhexia* was based on "wrecks-easy." She was almost correct. Although the flowers may be in their prime when stems are picked, the petals usually fall between that moment and closing the press to make herbarium specimens, no matter how short the interval.

This North American genus contains 13 species (James 1956, Kral and Bostick 1969). *Rhexia cubensis* extends into the Caribbean, being found in Cuba, Hispaniola, and Puerto Rico (Leon and Alain 1946–1953, Liogier and Martorell 1982).

Rhexia virginica grows from Nova Scotia to Ontario, and south to Georgia, Alabama, Tennessee, and Missouri. Although Correll and Johnston (1970) specifically excluded *R. virginica* from Florida, the plants do grow there. Indeed, Wunderlin (1998) says that it is "Frequent; northern cos." On his Web site, he shows the species from Escambia to Flagler across the whole tier of Florida counties.

Fernald et al. (1958) wrote that these small herbs were "too beautiful and local a plant for general use as a food, but in regions where it abounds it furnishes a pleasant salad, the leaves having a sweetish and slightly acid taste, while the tubers are pleasantly nutty in quality." They continue that, probably, "other species of the genus have similarly palatable qualities." Related to edibility is the use by the Montagnais of eastern Quebec and Labrador, who made the stems and leaves into a sour beverage. Micmac of New Brunswick and Nova Scotia and vicinity used it to clean the throat (Moerman 1998).

Porcher (1863) also considered the leaves edible. Unfortunately, there does not seem to be much other information on the edibility of *Rhexia.*

Rhizophora: Red Mangrove

(From Greek *rhiza,* root, and *phoros,* bearing, in reference to the aerial roots)

When Europeans first went from their temperate homes into the tropics, they found plants and animals that were incredible. Those lucky enough to return home brought fantastic tales to their homebound compatriots. One of those tales familiar to everyone is about mermaids. A story that has not been made so popular involves red mangroves—trees that seemed to be walking out into the edge of the sea. One of the first comments in English about these trees appeared in

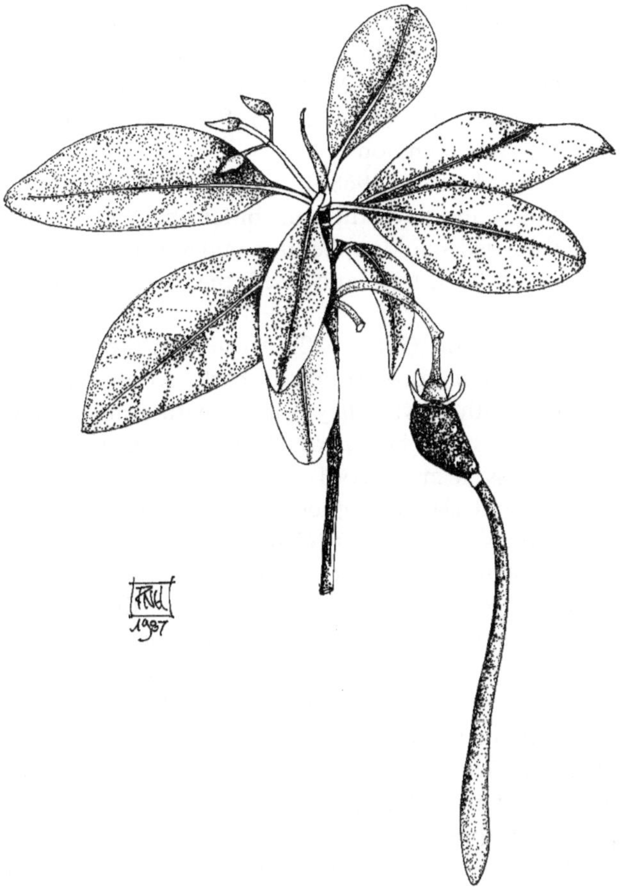

Rhizophora mangle. *Drawn by P.N. Honychurch.*

1613 with "Amongst all the rest there growes a kinde of tree called Mangrowes, they grow very strangely, & would make a man wonder to see the manner of their growing" (OED 1971).

Herbalist Johanne Bauhin wrote in 1650 that the tree was the *Mangle pyrifoliis* (thicket with pear-shaped leaves). However, Everardus Jacobus van Wachendorff (1702–1758), studying cultivated plants in Holland, coined the name *Rhizophora* in 1747. Linnaeus liked that, and adopted it in the name we currently use—*Rhizophora mangle*. The species name *mangle* means simply "thicket." Most common names still incorporate *mangle*, or translate it into a word meaning the same.

It seems odd, but people throughout the world, regardless of their languages, have a tendency to assign colors to the various kinds of mangroves. Inexplicably, they do not agree on what color a particular species is. In English, *R. mangle* is mostly red mangrove (Florida to Panama), although sometimes that becomes roode mangrove (Guyana) in Creole. The name "red mangrove" may have been first applied by part-time botanist and full-time pirate William Dampier in 1697. To others, it is simply mangrove (USA) or old lady mangrove. However, some think it should be black mangrove (Guyana), a name mostly given to *Avicennia germinans*.

In Spanish, *R. mangle* is simply *colorado* (red, Belize), *mangle colorado* (red mangrove, Tamaulipas, Veracruz, Tabasco, Oaxaca, Guerrero, Caribbean, and south to Panama), *mangle rojo* (red mangrove, Venezuela), or simply *mangle* (mangrove, Baja California, Sonora, Dominican Republic south). Others think of it differently, and say it is the *mangle caballero* (gentleman's mangrove, Costa Rica), *mangle dulce* (sweet mangrove, Baja California, Venezuela, Peru), *mangle de chifle* (whistle mangrove, Puerto Rico), *mangle gateador* (cat-colored mangrove, Costa Rica), or *mangle piñon* (pine mangrove). Still others note the adventitious roots and say it is a *mangle injerto* (grafted mangrove, Ecuador). Those with a more utilitarian view say *mangle salado* (salty mangrove, Panama, Venezuela), *mangle tinto* (dyer's mangrove, Veracruz), or *mangle zapatero* (shoemaker's mangrove, Puerto Rico). Even in Spanish, some say it is the *mangle prieto* (black mangrove).

French speakers are mostly agreed it is a *mangle rouge* (red mangrove, Guadeloupe), *manglier rouge* (red mangrove, Haiti, Guadeloupe), or *palétuvier rouge* (red mangrove, French commerce). Often that is shortened to *manglier* (mangrove, Haiti) or *palétuvier*. In French, too, it is *mangle noir* (black mangrove, Guadeloupe). Others think the propagules resemble candles, and say *manglier chandelle* (candle mangrove, Guadeloupe). Spanish speakers also say *candelón* (big candle, Veracruz, Colima, Sinaloa).

Brazilians say *mangue vermelho* (red mangrove), *mangue verdadeiro* (true mangrove) or just *mangue* (mangrove). They also describe the propagules as *mangue de pendão* (hanging mangrove, Brazil). Like people speaking Spanish, they note uses with *mangue de sapateiro* (shoemaker's mangrove), and *mangue-tan* (tanning mangrove). Sites where the trees are thick are a *mangarabeira* [*mangarobeira*] or a *managaixi* (both, place of mangroves).

In the old Dutch colonies, the trees are *romangel* (red mangrove), *mangro* (Sranan, Suriname), or *zwamp-mangro* (swamp mangrove, Suriname). People also allude to the extraction of tannin for curing animal hides with *mangel tan* (tan mangrove, Dutch Antilles), although *taan* is the continental spelling of the word. These same people have some wonderful deviations from the norm. They note that these coastal forests are a *duivenhout* (dove wood, Suriname), where many birds roost. They also allude to the many adventitious roots with *duizendbeen-boom* (thousand leg tree, Suriname) and *duizenhout* (thousand wood, Suriname).

Indigenous people have a multitude of names for these useful trees. Several may be simple terms with no translations. These include names from Sonora, Mexico, such as *canaro* (Mayo), *nahnawa'ara* (Yaqui), and *pnaaocj-xnazolcam* (Seri). In South America, they are *kakoetji-roe* (Guyana), *kakutiru* [*kakutira*] (Arawak, Guyana, Suriname), *kunapo* [*konapo*] (Carib, Guyana, Suriname), and *purgua* (Venezuela). To people on the Panama–Colombia border they are *aili kinnut* (Cuna).

The Maya are direct and say *tab-ché* [*tap-ché, xtabche*] (*tab*, salt, *che'*, tree, Maya, Yucatán, Belize). In Brazil, that is not enough. Brazilians say *apareiba* (place that receives waves?), *canaponga* (*cana*, cane or grass, *ponga*, put), *guapereiba* [*guaparaiba, guarapahy*] (edge of a depression, Tupí), and *ratimbo* (*rã*, frog, *timbo*, white or ash-color, Tupí).

To the Seminoles of southern Florida, the trees are *ahilo:ckitiscî* (*ahî*, tree, *kitisci*, red, *lo:ci*, black, Mikasuki), and *tolastica:tî* (*eto*, tree, *lvste*, black, *cvte*, red, Creek). Perhaps the trees were more important to them in the past when they clothed themselves with buckskins cured with the tannin from *Rhizophora*. However, by the time Sturtevant (1955) worked with them, they remembered no other uses.

Early visitors to the Caribbean noted the utility of these trees, the oysters that grew on their roots, and the fish that hid in their prop-roots during high tides. Wood was used to make oars, fuel, building wharfs and docks that resist attack by *Teredo* mollusks, huts, houses, doors, and window frames. Little and Wadsworth (1964) record that the sapwood is light brown, while the hardwood is reddish brown or dark brown. Wood is hard, heavy (specific gravity 0.9 to 1.2), and durable in the soil but susceptible to dry-wood termites. DeBrahm ([1775] 1971) said that the wood "is very close and fine grained, after cut and dryed grows as hard and near the Colour of Lignum sanctum [*Guajacum*], takes a very fine Polish." He also thought the coals of the wood "burn clear, and leave little or no Ashes."

However, most of all, the indigenous people and the Europeans found the tannins in the leaves and bark valuable for making leather. Natives and newcomers knew that adding salts of copper or iron to the bark would yield olive and brown dyes. Panamanians made a red dye from the young shoots, and DeBrahm ([1775] 1971) recorded that Morocco leather was made with red mangroves.

Oviedo noted that the native people ate the fruit. Supposedly, "It is tawny and within is a marrow or heart which the Indians eat when they can find no better fruit (for it is somewhat bitter), and they say it is wholesome" (translation from Standley 1920–1926). Far away in Sonora, Mexico, the Seri also ate the fruits (Felger and Moser 1985). Perhaps the practice formerly was common.

Julia Morton (1968b), who would eat almost anything, considered the fruit bitter and "not good." She also tried smoking the propagules as cigars, made tea from the leaves, and used the twigs as toothbrushes.

Hans Sloane (1707–1725) wrote that the blacks in Jamaica made cords from the bark. After removing the bark with knives, they beat it "with a mallet until the first gross one be separated from the second: of the more gross is made cords of other britches, for the negroes and slaves." Others recorded that they also made ropes of the fibers (Grimé 1976).

Medicinally, the bark is used as a febrifuge, to stop hemorrhage, and as a remedy for sore throat. *Curanderos* in Belize still use a decoction of the bark to treat sores, skin problems, leprosy, and swellings (Arvigo and Balick 1993). Morton (1981) included remedies for asthma, syphilis, elephantiasis, diarrhea, dysentery, leukorrhea, gonorrhea, bed-wetting in children, backache during pregnancy, and to cure fish poisoning (presumably ciguatera).

Laboratory experiments indicate that the bark extract is antibacterial (Rojas-H. and Coto-P. 1978, Melchor et al. 2001), helps heal ulcers (Perera et al. 2001), aids in controlling diabetes (Alarcon-A. et al. 1998), is antifungal (Cáceres et al. 1993), and helps enhance immunity (Rao et al. 1971). Moreover, the extract is insecticidal (Williams 1999).

When Frank Hamilton Cushing excavated the long-abandoned and buried Calusa village on Marco Island in 1896, he found that the houses had been built on pilings over the water in a mangrove swamp (Gilliland 1975). He was convinced these people were a semi-aquatic culture, like those in Venice, or those off the coast of Venezuela (meaning little Venice). He dubbed one part of the site "Court of the Pile Dwellers." The unique artifacts he recovered had been covered and then preserved in mangrove peat, an acid, anaerobic medium.

Memories of my first visit to the Marco Island region in the early 1970s are still vivid. As I stepped out of my car to examine bayside vegetation, my white trousers immediately turned black. Mangrove mosquitos were so abundant that my clothes and skin looked as though they were covered with black hair, and my skin burned from their bites. How did the Calusa thrive and practice their arts there without these tiny vampires draining their blood?

Rhus

(The ancient name used by the Greek philosopher Theophrastus, 372–287 B.C., probably based on Greek *rhodos*, red, from the tannin in the stems and the edible red fruits)

bati (Choctaw); *bati'* (Chickasaw)
sumac (from Arabic *summaq*); *sommacco* (Italian); *sumac* (French); *Sumach* (German); *sumagre* (Portuguese); *zumasque* (Spanish)

Rhus aromatica (fragrant)
hoyvnecv rakko (*honece*, wild, *łakko*, big; see Martin and Mauldin 2000; compare with *Salix humilis* or *mikko hoyvnijv*)
polecat-bush
[fragrant, lemon] sumac

Rhus copallina (from *copal*, because of a copal-like resin from the stems)
añil de pinar (pine-tree indigo)
aso:kóta:pî [*ashooket aape, asookotaapi, açokotapi, hekotaape*] (*ashooket*, berry, *aape*, tree, Mikasuki); also called *asó:kótî* or *asó:kotcó:bî* (*asó:kótî*, berry, *có:bî*, big)
[black, dwarf, flame-tree, mountain, shining, winged, wing-rib] sumac (USA); shoemake [shumaker] (Carolinas, Florida); *sumaque* (from Arabic *summaq*, Mexico)
tvwv [*tawâ, tawa, tv'wv, cúwv*] (Creek, Muskogee); *tabó* (Koasati); *tauwv lvste* [*tawa lasti*] (*tvwv*, sumac, *lvste*, black, Creek, Muskogee)

Rhus glabra (smooth)
bashukcha [*basho'nkchi, bushukcha, bushucha*] (said to mean "the sumac that bears the purple bud," Choctaw, Mobilian; maybe from *bashih*, to wither or fade, an allusion to the flowers?, *okchakki* [*okchakko*], blue-green, Choctaw); *bashanchik* [*homma', tohbi'*] ([red, white] sumac, Chickasaw)
chan-zi (*chan*, wood, *zi*, yellow, Dakota); *haz-ni-hu* (*haz*, fruit, *hu*, plant, Winnebago); *ni-ni'-gah hi* [*kin-ne-ne-kah*] (*ni*, live, *ni*, water, *hi*, plant, Osage; akin to *kinnikinnick*); *glikenícan* (Delaware; akin to *kinnikinnick*)
maw-hko-la (tobacco mixture, Kiowa)

Rhus. *Rhus copallina* (left). From Sargent 1905. *Rhus glabra* (right). a. Habit. b. Flowers; b′. detail of flower showing disk within stamens. c. Fruit. d. Seeds. *Drawn by Regina O. Hughes*. From Reed 1971.

minbdi-hi (*hi*, plant, Omaha-Ponca); *mon̲'-bi-dse* (Osage)
nuppikt (sour top, Pawnee)
otgòto (Onondaga)
smooth sumac (eastern United States)
tangomóckonomindge (little dye-bearing tree, Carolina Algonquin, North Carolina)
tvwv-łakko [*tvwv-rakko*] (*tvwv*, sumac, *łakko*, big, Muskogee); *tabó:so* (Koasati); *tabosso* (Alabama; see also *Phytolacca*)

On their arrival in the New World, Europeans were pleased and surprised to see sumac. They already knew *Rhus coriaria* of the Mediterranean, which had long been used for dyeing and tanning (Standley 1920–1926). They found Americans using the several native species in many of the same ways they knew. All of the southeastern tribes used the genus in myriad applications (Hedrick 1919, Yanovsky 1936, Moerman 1998).

The most widespread species in Florida is *R. copallina.* Among the Cherokee, the berries were eaten to stop vomiting and bed-wetting; an infusion of bark was poured over sunburn blisters and drunk to increase milk flow (Hamel and Chiltoskey 1975). The Choctaw used an infusion of its roots as a remedy for sore mouth (Bushnell, 1909, Swanton 1946). The Creeks used a decoction of the root to treat dysentery (Swanton 1928a). The Koasati made a decoction of leaves as a bath and gave it to babies to make them learn to walk (Taylor 1940). Sturtevant (1955) was told that the Seminoles used *tvwv* as a diuretic and for "Cow Creek Sickness." Snow and Stans (2001) recorded that the Seminoles collect and wash roots, peel off the skin, and put them in water. The decoction is used to induce vomitting in people being treated for alcoholism or to make widows throw up to remove the breath of their deceased spouses. *Tvwv* also was mixed with bay (*Persea*) and button snakeroot (*Eryngium*) as part of treatment for people who lose their wives or husbands. Bennett (1997) was told that the fruits were placed in cheesecloth, then steeped in water to make a tea. For emesis much as in Snow's preparation, a remedy is used for cleansing the body of "pollution" of any kind, including spirits and food poisoning.

In South Carolina's low country, a leaf decoction is used to bathe "painful places." The root is peeled and chewed as a remedy for diarrhea and dysentery. Medicines from the plant also are employed to treat leukorrhea, gonorrhea, and bladder trouble (Morton 1974). Murphee (1965) found that people in the Panhandle made a tea of the roots, mixed with cornmeal, to treat erysipilis and other skin problems.

Both *R. aromatica* and *R. glabra* have more restricted ranges in Florida, as they are more temperate. Natchez made a poultice of the roots of *R. aromatica* to treat boils (Swanton 1928a). Ojibwa used the roots in a medicine to treat diarrhea and the bark and berries in a medicinal ceremony (Moerman 1998).

There are records of more people using *R. glabra*, including the Apache, Catawba, Cherokee, Comanche, Creek, Dakota, Flathead, Gosiute, Iroquois, Kiowa, Kutenai, Lakota, Meskwaki, Micmac, Natchez, Ojibwa, Okanagan-Coville, Omaha, Osage, Pawnee, Ponca, Potawatomi, Sanpoil, Tewa, Thompson, and Winnebago (Hunter [1823] 1973, Smith 1933, Vogel 1970, Moerman 1998). Probably the most widespread use was mixing the leaves with tobacco for smoking. Dyes were also widely made (Cannon and Cannon 2003). Gilmore (1931) found empty inflorescences at the Ozark Bluff-Dweller site in Arkansas, and speculated that the fruits were used "as they were by Indians of historic times, in dyeing processes, or medicinally."

The Cherokee ate the berries to halt vomiting and bed-wetting, and they poured an infusion over sunburn blisters and other burns (Hamel and Chiltoskey 1975). The Creeks used a decoction of the roots to stop dysentery (Swanton 1928a).

Ribes

(Alphonse de Candolle thought *Ribes* was from the old Danish colloquial *ribs,* for *R. rubrum*, the red currant; others, including Leonard Fuchs in 1542, said that *ribes* was a name used by Moors and pharmacists; it may be from Arabic *ribas*, acid tasting, alluding to the fruit)

aelbesien (*besien*, possessing; in use for *R. rubrum* by 1549, Dutch)
casis [*cassis*] (French, Spanish; perhaps because liquors were made from the fruits of *R. nigrum* in the community of Cassis, near Marseille, France; see de Candolle 1886)
currant (a corruption of Greek *Korinthos*, Corinth, an ancient Greek city; that city also gave English "Raisins of Corinth," a small variety of grape)
gadellier (the plant, probably from Gaelic *gardis*, sour, Quebec)
gooseberry (having nothing to do with the bird called a "goose," but rather a corruption of the French *grosielle* or *grosell*, words that may have come from Latin *grossulus*, an unripe fig; used in English by 1532 when "Gose berrys" were equated with "*grosielles*")
gròsaid (Gaelic, for *R. grossularia*); *groeselier* (the plant, in use for *R. uva-crispa* by 1550; *groselier*, by 1575; French); *groseilles* (the fruit, French;

according to some, from Latin *grossulus*; an alternate explanation is that it came from Middle Dutch *croesel*, from *kroes*, crinkled); *groseillier* (the plant, Quebec); *groselha* (the fruit, Portuguese; from German *Kraüselbeere* through French *groseille*; cf. Ferreira de Holanda ca. 1978); *groselheira* (the plant, Portuguese); *grosella* (the fruit, Spanish; from Latin *grosullus*; see García-Pelayo y Gross 1980); *grosellero* (the plant, Spanish)

hiyo-lókcv (*hiyón*, thus, after this manner, *lókcv*, acorn, Muskogee)

Johannisbeere [*Johannes beer*] (John's berry, for *R. rubrum*; rendered *Johans beerlin* in 1542, German)

Krauselbeere (first given by Fuchs in 1542 as *Krüselbeer* for *R. uva-crispa*; from *kroes*, crinkled, plus *Beere*, berry, German)

pe'-çi [*pe'-çi-ga*] (Omaha)

raosar (Gaelic)

ribes (for *R. rubrum*, in use by 1568, English); *ribes* (for *R. rubrum*, Fuchs 1542, Arabic; in French by 1550; in Italian by 1551; in Spanish by 1557); *ribs* (Danish); *rips* (Norwegian); *risp* [*resp*] (Swedish)

stekelbesien (*stekel*, stickers, *besien*, possessing; in use for *R. uva-crispa* by 1549, Dutch); thorn grape (used by Henry Lyte in his 1578 translation of Dodoens's *Cruydeboek*); *uva spina* (modern Italian; see Ragazzini and Biagi 1986)

uva crispa (in use for *R. uva-crispa* by 1551, Italian)

Ribes echinellum (little hedgehog, a reference to prickly fruits)
[Florida, Miccosukee] gooseberry

Ribes was unknown to the Greeks and Romans, or at least it is not mentioned in their literature (de Candolle 1886). Gerarde ([1597] 1975) said the fruits were available in London markets, but were not cultivated (de Candolle 1886, Hedrick 1919). Although Hedrick (1919) said that neither plants nor fruits had a name in Gerarde's time, there are literary records from 1532 equating "Gose berrys" with "*grosielles*." De Candolle (1886) was convinced that the plants were foreign to the Mediterranean, and recorded that *R. rubrum* was still known as *groseillier d'outremer* (foreign current) in France. Similarly, in Geneva, they were *raisin de mare* (sea grape), and in the canton of Soleure *meertrübli* (*meer*, sea, *trübli*, grape).

De Candolle (1886) thought that the French and Breton names indicated cultivation of *R. rubrum* in western France before the Norman conquest. He thought that all the names in use were derived from Gaelic, suggesting *gardis* (sour) as their common ancestor. The examples he gave are *gardes* (in Rouen), *grades* (in Caen), *gradilles* (in Lower Normandy), and *castilles* (in Anjou). Dwelly (1933) confirmed the Gaelic word, although spelled *garg* or *gairge*, as tart or acrid.

Leonard Fuchs published one of the earliest herbals of the 16th century in 1542 (Meyer et al. 1999). He listed what are now the two European species, *R. rubrum* and *R. uva-crispa.* According to Fuchs, *R. rubrum* was cultivated in German gardens, although native in western Europe. He remarked under the "Properties from the Moors" (*Vires ex Mauritianis*): "The berries with their tart taste are pleasing. They cool an upset stomach, slake the thirst that torments both the feverish and others. They dispel aversion to food and arouse the appetite. They check foul effluvia of the belly caused by bile. They remove the gnawing and twitching. They soothe the heating of the blood, and tame the bitterness of bile. The Capuans [from Capua in Italy] preserve the juice of its fruit for year-round use, calling it *Rob de Ribe* [Essence of Currant]." That information agrees with the names in other herbals of the 16th century (Meyer et al. 1999).

Currants were well known in northern Europe at the time the New World was discovered. They were, however, still poorly known in the Mediterranean (de Candolle 1886). *Ribes* is a genus of 150 species growing in the temperate regions of the Northern Hemisphere and the Andes of South America (Fernald 1950, Diggs et al. 1999). There are 9 species in Europe (Mabberley 1997) and 57 (2 of hybrid origin) in North America (Kartesz 1994).

Probably the first record of *Ribes* in the New World was made by Capt. John Smith in 1609, who spoke of them in New England (Hedrick 1919). The species he found there was *R. oxyacanthoides.* In 1621, Edward Winslow mentioned the same species in Massachusetts. When Linnaeus ([1753] 1957) named this species, he had studied living plants at the *Hortus Upsaliensis* and specimens in the herbaria of Leonard Plukenet and Jan Dillenius. He knew only one other American species, *R. cynosbati.*

Florida has the single species, *R. echinellum*, and it is rare in Gadsden and Jefferson Counties in the Panhandle (Wunderlin 1998). The species was not even discovered until 1924, when it was described by F.W. Coville (1867–1937). In Florida, the plants grow in the region formerly occupied by the Muskogean-speaking Chatot (including the Apalachicola), who later merged with the Choctaws. After the Chatot migrated west into French territory (Swanton 1946), the ancestors of the Miccosukees settled there and nearby at Oscilla River and Tallahassee after 1818

(Williams [1837] 1962). Their principal village between about 1818 and 1836 was on the shores of the lake now called Miccosukee (Simpson 1956). The only other known population for the plants is in McCormick County, South Carolina (Radford et al. 1968, Kral 1983b, Catling et al. 1998).

Ribes echinellum is so rare that it is endangered in Florida and nationally threatened (Coile 2000). That rarity precludes much human use, and it is little wonder that there are few records of them. Still, fruits of many species of *Ribes* are eaten. Hocking (1997) wrote that *Ribes* fruits are small and sour, but are used in jellies, preserves, jams, pies, and juices. Most fruits are good sources of vitamin C. He made no mention of *R. echinellum*, but listed 23 others. Earlier, Hedrick (1919) had listed 24 species. Individual plants of *Ribes* vary in taste, and there are also differences in people's preferences (Hedrick 1919). It seems likely that most, if not all, species of the genus produce edible fruits.

Hocking (1997) wrote only that some species are used as brown dyes. Polunin (1969) indicated that, in Europe, leaves of both *R. rubrum* and *R. nigrum* yield a yellow dye. Their fruits provide black (*R. rubrum*) and blue or violet (*R. nigrum*) colors. No information is given by Moerman (1998), Tull (1999), or Cannon and Cannon (2003) on people using *Ribes* for dyeing.

Currants are well known for a number of chemicals that exhibit antibacterial (embelin), antioxidant, and antiviral activity (Moyer et al. 2002, Chitra et al. 2003, Vecera et al. 2003, Knox 2003). *Ribes* species also contain anthocyanins, phenolics, quercetin, and a variety of other compounds (Maatta et al. 2001, Frank et al. 2002, Erlund et al. 2003). Some Asian species are even considered an oral contraceptive for females (Shah 1971, Prakash 1985, Sharma et al. 2001). Most of the chemical analyses and bioactivity studies have been done on *R. nigrum*, but perhaps some of the other species have the same traits.

Rivina: Rouge-Plant

(Named for Augustus Quirinus Rivinus [Bachman], 1652–1723, a German physician in Leipzig)

Along the southeastern coast of peninsular Florida are still remnants of the formerly almost continuous tropical forests. Walk into one that has a mature canopy, and you will feel like you are in a cathedral. The trunks of the trees are widely spaced like columns, and they support the ceiling of the tree crowns, like a mural in the Vatican.

Your first impression will be that there are no understory herbs. Then, when you look more closely, you will find one single herbaceous species scattered about. That is rouge-plant (Florida to Arizona), *Rivina humilis*. This relative of pokeberry (*Phytolacca*) is the only one that grows in the low light of the tropical forest floors. Oddly enough, it also grows in sunny places. There are many odd things about these plants.

On many occasions, I have had students bring in these herbs and complain that they could not figure out what they were. According to them, none of the keys works, or they end up in the Solanaceae, and not the correct Phytolaccaceae. I pointed out to them that they are not alone in their confusion. Common names for the herbs include baby-peppers (Florida), *chile* (*Capsicum*), *chile de ratón* (rat's pepper), jumbee pepper bush (Virgin Islands), pepper bush, *tomatillo* (little tomato, Belize), and wild tomato. Those names came about because of the small, red berries that resemble chiles and tomatoes.

Indeed, most of the names given these plants in the Americas refer to the color of the fruits, or the berries themselves. Some people compare the fruits to seeds of *Bixa orellana* (also called *achiote, annato, roucou, urucu*), that famous tropical American substitute for the Old World saffron (*Crocus sativus*). Those names include *achiotillo* (little achiote, Honduras), rouge-plant (Arizona, Texas), and *kuxubcan* [*coxubcan, cusucan, cuxuban*] (*ku'xub*, achiote, *kan*, snake, Maya, Yucatán, Belize). Others think they are more like blood and say bloodberry (Jamaica, Cayman Islands, Barbados, Trinidad), cat's blood (Puerto Rico, Virgin Islands, Barbados), dog blood (British Antilles), dogberry (Jamaica), and *sangre de toro* (bull blood, Argentina, Uruguay).

To many, the fruits are just red, but they are inventive about saying that. Names meaning red include *carmín* (Costa Rica, Puerto Rico, Colombia), *coral* (Mexico), *coralillo* (Durango, Nicaragua, Belize), *coralillo carmin* (Nicaragua), *coralillo cimarrón* (Sonora), *coralito* (Texas, Durango, Cuba), and wild ruby. More imaginative allusions to the color are *arrebol* [*arrevol*] (red color of the sky, Puerto Rico), *cochinillo* (little cochineal, a cactus-feeding insect, *Dactylopius coccus*), and apparently to a blush in the face of a *demoiselle* (young lady, French Antilles). Indications that fruit was used as dye or ink are reflected in *pinta-pinta* (paint-paint, Colombia) and *saca-tinta* (ink maker, Nicaragua).

Huastec names are *taa' t'ele'* (baby excrement, San Luis Potosí), *tsakam taa' t'ele'* (little baby excrement, San Luis Potosí), and *akan t'ele'* (baby foot, San Luis Potosí). Those sound distressingly like a major digestive problem in the child. Because diarrhea has been perhaps the most common killer of infants throughout history the Huastec names allude to the use of these plants to treat that malady (Alcorn 1984). In the Virgin Islands and elsewhere diarrhea is one of the

Rivina humilis. *Drawn by P.N. Honychurch.*

prime sicknesses treated with *Rivina* (Petersen 1974, Ayensu 1981, Morton 1981). From the Virgin Islands to Curaçao, a "tea" is made of the entire plant to cure that malady.

Several of the names for the fruit include allusions to colors and medicinal uses and mention that birds feed on them. These are *bonbon codine* (codine candy, Haiti), *bonbon coq d'Inde* (rooster candy, Haiti), *coqueta* (flirt, Belize), fowl berry (Cayman Islands), *hierbe à charpentier* (woodpecker [carpenter's] herb, French Antilles), *hierba de los carpinteros* [*yerba de los carpinteros*] (woodpecker [carpenter's] herb, El Salvador), *hierba* [*herba, yierba*] *mora* (blackberry herb, Nicaragua), *mata gallina* (chicken killer, Dominican Republic), *ojo de ratón* (rat's eye, Cuba), and pigeonberry (Arizona, Texas, Sonora). A large variety of birds eat the fruits with impunity. Among the birds feeding on the berries in Florida and Texas are turkeys (*Meleagris gallopavo*), chachalacas (*Ortalis vetula*), mourning doves (*Zenaida macroura*), white-winged doves (*Zenaida asiatica*), mockingbirds (*Mimus polyglottos*), catbirds (*Dumetella carolinensis*), and robins (*Turdus migratorius*) (Texas A&M 2000).

It is not unusual for birds to eat with impunity plant parts that cause problems for humans. There is even disagreement on whether or not the fruits of these plants are edible. Standley (1920–1926) considered them edible, but Nellis (1997) disagreed. That same difference was the norm in my childhood regarding fruits of *Phytolacca.* We were admonished by adults, and by books we found in the library, that the berries were poisonous. One of my friends delighted in eating handfuls of pokeberries to "prove" that they were harmless. No one at the time ever considered quantity eaten, or variable sensitivity of individuals. Certainly none of us had ever heard of phytolacccatoxin, which is now known to be in *Phytolacca* (Lewis and Elvin-Lewis 1977). Worse, during that time no one even knew of mitogens in the plants and their insidious risks to health when absorbed through skin abrasions.

According to Nellis (1997), consumption of *Rivina* fruits causes numbness of the mouth within 2 hours, with a feeling of warmth in the throat and stomach. Those symptoms are followed by coughing, thirst, tiredness with yawning, vomiting, and diarrhea, sometimes bloody. Leaves and roots contain more toxins, and cattle eating them yield tainted milk, and stools that smell of *Rivina* (Burkill 1966).

The responsible toxins in *Rivina* have not been identified, but they may be similar to those in *Phytolacca* (Nellis 1997). It should be noted that *Rivina* possibly contains mitogens, but there is evidence that they are useful against microbes (Salvat et al. 2001). However, their effectiveness against bacteria is slight and use should be mediated by awareness of their potential dangers.

Surely Linnaeus did not know the poisonous nature of the herbs when he named them for Rivinus. On the other hand, perhaps he did. He was not above naming plants to insult as well as to compliment (Austin 1993). However, this time he seems to have simply been following the suggestion of Charles Plumier who called these plants *Rivina* before him.

Names that are surely references to poisoning or use as medicines are also common, probably because remedies for several maladies include these plants. *Rivina* has been put in medicines to treat colds, diarrhea, difficult urination, flatulence, gonorrhea, jaundice, and ovarian pain. The herbs treat *susto* (fright) in Puebla and Veracruz, Mexico and among the Huastecs of San Luis Potosí (Alcorn 1984, Vásquez and Jácome 1997). Names noting those applications probably include fleda bush (British Antilles), *hierbe blanche* (Guadeloupe, Dominique), *liane blanche* (Guadeloupe, Dominique), *quileto blanco* (white quelite), *flor blanca* (white flower, El Salvador), and *skwam butz snya* (*skwam*, herb, Zapotec, Oaxaca).

The name *yerba de cancer* (cancer herb) should not be interpreted literally as malignant tumors. In much of Spanish America, the word "cancer" simply means a dermal sore. Standley (1920–1926), Liogier (1974), and Morton (1981) are among those recording that the leaves are used to treat wounds, sores, and other problems that might be called *cancer* in Spanish.

Other names compare these plants with different species, usually those that are also medicinal. How any of these came to be compared to caimans (from Carib *cayman*, in use by 1577; including the genera *Caiman, Melanosuchus,* and *Paleosuchus*), the alligator relatives, is not clear. Still, plants are *caimancillo* (little caíman, Dominican Republic), *caimoní* (usually applied to *Wallenia laurifolia*, Myrsinaceae, Hispaniola), and *caimoní cimarrón* (wild *caimoní*, Hispaniola). Two other names that are comparisons to famous medicines are *hierba del zorrillo* (skunk herb, name also applied to *Petiveria alliacea*, Mexico) and *stack-mahack* [*stakamahatchi, stock ma hork, stopmahatje*] (possibly from *takahamaca,* a name for *Bursera*, Virgin Islands).

There are no obvious meanings for *kahold* (West Indies), *penzon mazoumba* (Haiti), *pimpín* (Colombia), *teyweso* (Guarijío, Sonora), and *ucusuiro* (Sonora). However, *laman la yé* (it cries for thee, Haiti), *maman la cri* (mama it cries, Haiti), and *bacot mútica* (snake's pillow, Mayo, Sonora) are ominous. In spite of these names, some tout the herbs as good goat fodder (Yetman and Van Devender 2001). Maybe the goat's reputation for being able to eat anything with impunity is its salvation.

These plants have a surprising tolerance for temperatures and moisture variations. The herbs grow from Florida to Texas, Arizona and Oklahoma, south through Mexico, the Bahamas, West Indies, and Mesoamerica to Peru and Argentina. It is almost incomprehensible that plants growing in the shade of a tropical West Indian forest can also grow in the desert canyons of Arizona and Mexico. Yet, when the rains are ample in the Sonoran and Chihuahuan Deserts, the herbs appear.

This *Solanum* look-alike is a monotypic genus, with the only species scattered through the Americas (Rogers 1985). The flowers are not particularly attractive to most people, but *Rivina* most likely has been brought into cultivation for its bright fruits. Maybe, like *Phytolacca,* it too will amaze those who examine its chemistry. Already, *Rivina* has surprised those who have examined the phylogeny of the family. Most recent data suggest that the family is polyphyletic, and strongly linked with Nyctaginaceae (Downie and Palmer 1994). Can this really be a highly modified *Bougainvillea*? Or, is *Bougainvillea* actually a drastically modified *Rivina*?

Rosa

(The ancient Latin name; related to Greek *rhodon*, Akkadian *russu*, red)

bakcó paká:li (*bakcó,* briar, *paká:li*, flower, Koasati)
goláp (Bengali); *gul* [*guli*] (Persian, Punjabi, Afghanistani); *guláb* (Hindi)
roas (Frisian); *rodea* [*rhodon*] (Greek); *roos* (Dutch); *ròs* (Gaelic); *rós* (Icelandic); *ros* (Swedish); *rosa* (Italian, Spanish, Portuguese); *rósa* (Old Norse); *rose* (Danish, French); *Rose* (German)
rose-hips [heps, hepps] (from Old English *héope*, related to Old Saxon *hiope*, Middle High German *hiefe*, Old Teutonic *heupón*, a thorn bush; equated to *Rosa canina* with "hip-bramble" in A.D. 1386)
vered (Hebrew)

Rosa carolina (from Carolina)
a'kiska (Ofo)

Rosa carolina. From Britton and Brown 1897.

Carolina [dog] rose
kati ancho (*kati*, thorn, *ancho*, mantle, Choctaw)
pasture rose
pussaqweinbun (Strachey said in [1612] 1953 that this Virginia Powhatan word meant a rose)
tsist-uni'gisti (rabbit food, Cherokee)

Rosa palustris (of marshes)
hvyo palecv [*hiyópalécv*] (*hvyo*, harvest, *palecv*, ten year, Muskogee)
swamp rose

Surely, the oldest use of roses is their edible fruits, called "hips" in English. Rose seeds have been found dated at 5000 B.P. in the Netherlands and at 3500 B.P. in England (Widrlechner 1980). In Britain, rose-hip seeds were found with the skeleton of a Neolithic woman from 2000 B.C. (Genders 1988). The discovery of that antiquity is one indication of the widespread and ancient use of this food. Hips are rich in vitamin C and contain malic acid. Malic acid is one compound that prevents gout, the excess accumulation of uric acid, so the tasty food is also medicine (Genders 1988).

Theophrastus, 372–287 B.C., distinguished between wild roses as *cynosbaton* (dog rose) and cultivated types he called *rhodon* (rose) (Widrlechner 1980). Pliny, A.D. 23–79, recorded one of the oldest stories for Eurasian roses, although there are even older tales related to Greek deities. He wrote that the Greeks called the plant *cynorrhodon* (*cyno*, dog, *rhodon*, rose), because a woman had a dream that it would cure the bite of the mad dog. She sent some to her son who was in the military and who had been bitten by a dog. Pliny wrote that the son was beginning to show a horror of water, "when a letter arrived from his mother, begging him to obey the heavenly warning. So his life was unexpectedly saved." The plants are still called dog rose, but Linnaeus put that name in Latin as *Rosa canina*.

An even older history regarding roses is known from the Middle East and India. Roses were grown in the physic gardens of western Asia and northeastern Africa 5000 years ago (Lehner and Lehner 1960, Simpson and Conner-Ogorzaly 1995). Roses are associated with many religious leaders, including Brahma, Buddha, Confucius, Jesus, Mohammed, Vishnu, and Zoroaster (Swerdlow 2000).

"Attar," the oil of the damask rose (*R.* ×*damascana*), Rosaceum, an ointment of rose oil and honey, and rosewater were the most lavishly used perfumes and cosmetics in Persia, Egypt, Greece, and Rome (Lehner and Lehner 1960). Being scented with attar is still the sign of devout Muslims and Hindus. However, use of rose oil is not confined to those religions. Rose oil is a worldwide component in 46% of men's perfumes and 98% of women's fragrances (Lawless 1995). Among the more than 300 components identified in rose oil are citronellel, farnesol, geraniol, nerol, phenyl ethanol, and stearopten (Lawless 1995).

According to an Indian story, rosewater was discovered in 1612 by Nur-i-Jehan Begun, on the occasion of her marriage to the emperor Jehangir. While they were walking in his palace gardens along the canals and fountains decked with rose petals in celebration of their wedding, she noticed an oily film on the water surface produced by the action of the sun on the rose petals. Fascinated by the heavy scent of this oil, she ordered it bottled for later use. It was named *Atar-Jehanghiri*, perfume of Jehangir. This *attar* (fragrance, *atir*, Persian, *'itr*, Arabic, *itr*, Hindi) was considered the most precious of all perfumes (Watt [1889] 1972). Sometimes it is still called by its full name in English and Persian, *atar-gul* (rose-perfume).

Europeans were fascinated with roses when the New World was discovered and imported them for both beauty and utility. The newcomers also found several species native to the eastern United States. The indigenous tribes did not share the Old World fascination with them, but several species were used (Gilmore 1919, Moerman 1998). The Cree and Iroquois used *R. acicularis* in food and medicine, particularly for treating coughs and sore eyes. The Ojibwa and Omaha used *R. arkansana* similarly. They also used it to heal wounds and as a tonic (Kindscher 1992). Both the Cherokee and Ojibwa used *R. virginiana* to expel intestinal worms from children, to treat sore eyes, and to heal wounds.

Records have been found for use of both native species in Florida. The Menomini ate the fruit skin of *R. carolina* for stomach trouble (Moerman 1998). The

Cherokee used an infusion of bark and roots of *R. palustris* to treat intestinal worms, and a decoction of roots to stop dysentery (Hamel and Chiltoskey 1975).

In spite of their names, the Chickasaw rose (*R. bracteata*) and Cherokee rose (*R. laevigata*) are not native species used by those tribes. These plants were imported from China and naturalized (Wunderlin 1998). The Chickasaw rose was imported in 1870, and it is now a pest in Louisiana, Mississippi, and Alabama (Hocking 1997). The Cherokee rose has a similar history, and it is a pest in the southern United States (Hocking 1997). So, a rose by any other name may smell as sweet, but it will smother more desirable natives in doing so.

Roystonea

(Orator Fuller Cook, 1867–1949, named the palm after General Roy Stone, 1836–1905, an American Army engineer in Puerto Rico)

Roystonea regia (royal)
ara'uwa (Carib, Suriname)
hiskiwisá:kî [*hiskiwisá:hî*] (*hiski*, leaves, *wisá:ki*, pointing-outward, Mikasuki)
jagua [*yagua*] (royal palm leaf, Taino, Mexico); *palma de yagua* [*cubana*] ([Cuban] palm giving leaves, Puerto Rico)
Konnings palm (Antilles)
manaca (Taino, Greater Antilles)
palma (palm, Dominican Republic); *palma cana* [*criolla*] (cane [Creole] palm, Cuba); *palma real* [*cubana*] ([Cuban] royal palm, Dominican Republic, Puerto Rico, Belize, Guatemala, Mexico); *palmiste* (palm, Haiti)
[Cuban] royal palm (Florida, Bahamas, Puerto Rico, Belize)
waa (Cuna)

Coll y Toste (1972) said that the first record of *Roystonea* came from Oviedo in the 1500s. Oviedo wrote, "*Entre las otras palmas hay un gnero dellas, que los indios llaman manaca*" (Among the other palms there is a genus of them that the Indians call *manaca*). That Taino name is no longer used for the royal palm, but it is used for *Calyptronoma* and *Prestoea* (Scott Zona, personal communication 2003). However, the Taino term *yagua* remains in common names for *Roystonea* and shows the importance of the fronds in thatching *bohios* (houses) of the *campesinos* (Coll y Toste 1972).

Europeans soon learned the utility of these large trees. The fruits were especially good for feeding swine (Austin 1980). Spanish ships leaving Cuba often were provided with not only pigs but also a store of royal palm fruits to feed them (Millspaugh 1907). When the Seminoles moved into Florida, they became particularly fond of pigs. Indeed, one translation of their tribal name Miccosukee is chief (*micco*) pig (*sukee*). The Gainesville area was a famous village for Seminoles and their fondness for pigs remains in "Hogtown Creek." Some still irreverently call the city of Gainesville "Hogtown."

Roystonea regia. From Little and Wadsworth 1974.

In the 1760s, when William Bartram was traveling in Florida, he found the royal palm in what is now Volusia and Lake Counties. He called the trees *palma elata*, and wrote that they had a "strait trunk ... sixty, eighty or ninety feet high, with a beautiful taper of a bright ash colour." There are no longer royal palms in the area, and Harper (Bartram [1791] 1958) speculated that the freeze of 1835 killed them. Zona (1996) discussed the northern Florida *Roystonea* in more detail.

Given the Seminole fondness for swine and the disjunct location of royal palms found by Bartram, it is tempting to believe that the Miccosukee planted them there. Whether or not that is the case, they still had a name for them when Sturtevant (1955) worked with them.

Cubans make the root into a diuretic medicine, and some think it good for diabetes, perhaps because it is emollient (Roig 1945). As with most palms, the terminal bud is edible, but harvesting kills the tree (Morton 1968b). The fruit pulp is also edible, as the Spanish and Miccosukee learned long ago.

Rubus

(Ancient Latin name of the blackberry)

baer (berry, Norwegian)
bakcó (Koasati); *bakhe* (Mikasuki); *bissa* (Choctaw); *bissa'* (Chickasaw); *kvco* (Muskogee)
blackberry (a name first applied to *Rubus fruticosus* in Europe about A.D. 1000; subsequently applied to all members of *Rubus*)
bramble (from Middle English *brambel* or *braembel*, first used in English about A.D. 1000); *braam* (Dutch); *Brombeere* (bramble berry, German)
cloudberry (originally applied to *Rubus chamaemorus* in Europe by at least the time of John Gerarde's *Herball* in 1597; the species name means "ground-berry," and the origin of "cloud" in the name is unknown)
co'kake'ñic (Atakapa)
dewberry (a berry with dew-drops on it; originally applied to *Rubus caesius* of Europe by the time of the English translation of Rembert Dodoens's book on herbs in 1578, and later applied to *R. canadensis* and other American species)
[*fir, feir*] *dhris* [*dreas*] (*fir*, chief, *dhris*, briar or bramble, Gaelic)
iya'fhu (maybe *iya*, raccoon, *i'fhu*, seed, Ofu)
kanugu'la [*kanugahli, kah noo gah hlee, kanugu'la, ka-nu'-ga-tli'*] (scratcher, Cherokee; there was a place called *Kanu'gulayi*, or *Kanu'gulun'yi*, "Brier place," a Cherokee settlement formerly on the Nantahala River, about the mouth of Briertown Creek, in Macon County, North Carolina)
kuskite'mina (Plains Cree)
mehowìmi [*mehowimínac, m'howimínac*] (Delaware)
nzúkii-lehlookihláashak (from *nzúkeew*, black, *lehlookíhlaash*, raspberry, Ontario Delaware, O'Meara 1996)
raspberry ("rasp" is of unknown etymology, but appears to be related to "raspis," from 1532, which was perhaps derived from Old French *raspeit* or *raspise*, and Italian *raspato*, a rose-colored wine, ultimately from Medieval Latin *raspatum* or *raspa*, a bunch of grapes. From ca. 1460 this was interpreted as being a wine made from raspberries. William Turner's *Names of Herbes* published in 1548 stated, "*Rubus ideus* [*sic*] is called ... in englishe raspeses or hyndberies." John Parkinson also mentioned "raspis" in *Paradisus Terrestris* in 1629. Strachey 1612 wrote of "raspices." However, by 1623 Europeans had begun applying "raspberry" to fruits of *Rubus idaeus*, and that trend continued with the eventual disappearance of "raspis.")
ronce (French); *rovo* (Italian)

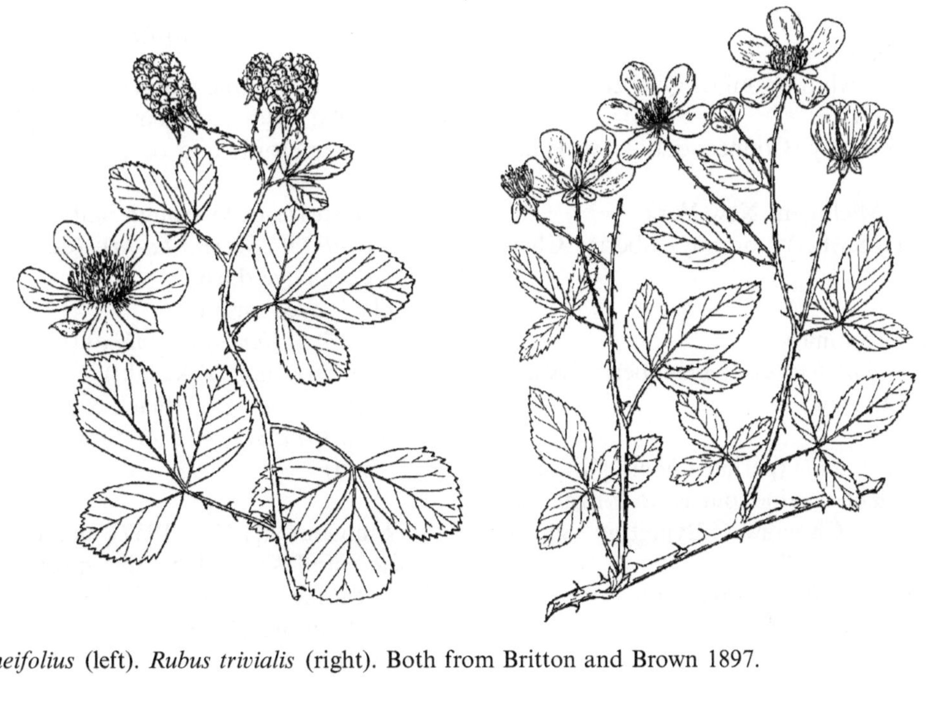

Rubus. *Rubus cuneifolius* (left). *Rubus trivialis* (right). Both from Britton and Brown 1897.

sájis óchia (*óchia*, berry, Onondaga)
silva (forest, from Latin, Portuguese)
smeur (blackberry; also means anoint or smear, presumably because it was used for dye, Gaelic)
thimble-berry (applied to the black raspberry by Thoreau in *Walden,* from 1854)
udlosinvda (Cherokee)
wa-çki'-the ça-be (*wa-çki'-the*, fruit, *ça-be*, black, Osage)
wi·ya̲'ro (blackberry, Catawba)
zarza (Spanish)

Rubus argutus (leaves sharply toothed or serrated)
okwołí (Koasati)
saw-tooth blackberry [sawtooth blackberry]

Rubus cuneifolius (cuneate-leaved)
bakcsalâ:lî [*bakhe shaalale*] (twining briar, Mikasuki)
brier-berry
kacohalkocî [*kv'co-hv'lkv, kv'co-palv'knv*] (*kvco*, briar, *hvlkvte*, climbing, Creek, Muskogee); *kvco huełv* [*huerv*] (*kvco*, briar, *huełv*, standing, Muskogee); *kvco-em-ette* (*ko'co*, briar, *emetv*, it gives [fruits], Muskogee)
knee-high [sand] blackberry

Rubus flagellaris (like a whiplash, alluding to the branches)
bumble-kites [bumblekites] (so called because "children eat so many that their kites or bellies rumble," according to Young 1945 in Coffey 1993)
cloudberry-root
finger-berry (New York)
northern dew-berry
ronce à flagelles (whip bramble, Quebec)
running blackberry
sow-teat [blackberry] (Connecticut, New Hampshire)
thimble-berry (Michigan, New York)
u-dlo'-si-nv-da (dewberry or boysenberry, Cherokee)

Rubus trivialis (ordinary)
biuntalali [*bissuntalali*] (*bissa'*, blackberry, *talaalih*, spreading, Choctaw)
blackberry
haiyantalali [*aiuntalali*] (perhaps *hayyi* [*hayyon*] is archaic for plant or vine, but is no longer used, *okpolo,* bad, Choctaw; Byington glossed *haiyunkpolo* as "weed")
iuntalali (from *talaalih*, to spread out, so "spreading vine," Choctaw)
southern dewberry

Rubus comprises 250 species, plus numerous apomictic lines (Mabberley 1997). European use of *Rubus* as food and medicine dates from at least as early as the Romans and Greeks, probably much earlier.

Hogan (1978) found pollen of *Rubus* at the Fort Center Glades village on Lake Okeechobee. Presumably, they were using the plants as have people throughout the range of the genus. It is known that the Timucua also used blackberries (Hann 1986). Fruits of all species were eaten as food (Hedrick 1919, Yanovsky 1936, Moerman 1998), and at least the Choctaw, Creek, Koasati, and Yuchi recognized a "blackberry month."

The Cherokee used *R. argutus*, *R. flagellaris*, and *R. trivialis* in an infusion of roots or leaves for diarrhea, rheumatism, as a diuretic, for venereal disease, and as a hemorrhoid remedy (Hamel and Chiltoskey 1975). They also chewed the root to cure a coated tongue. The Catawbas used *Rubus* to treat diarrhea (Speck 1937, 1944). The Alabamas used *Rubus* for a toothache remedy and in a poultice against pneumonia (Taylor 1940).

Small (1933) found the Seminoles using *R. trivialis* to treat stomachache. Sturtevant (1955) found the Miccosukee using *bakcsalâ:lî* in a mixture for chronic sickness, in birth ceremonies, and for food. Modern Seminoles on the Big Cypress Reservation eat the fruits and use the plant in medicine. The entire plant is placed in hot water and the decoction taken orally for "Ant Sickness," and to treat itchy skin; other plants may be added (Bennett 1997).

Rudbeckia

(In using this name, Linnaeus honored his professor Olaus Olai Rudbeck, 1660–1740, physician and professor of anatomy and botany at Uppsala)

Rudbeckia hirta (rough hairy)
a'wi-akta' [*a-wi-a-ka-ta*] (deer eye, Cherokee); deer's eye daisy
black-eyed susan
ha:słâyhî [*haash (th)aime*] (*haash*, sun, *łâyhî*, shot or shooting, Mikasuki)
memakate'nîngwe (black eyeballs, Potawatomi)
paká:li lá:na (*paká:li,* flower, *lá:na,* black, Koasati)

Linnaeus ([1753] 1957) named the genus and species, and since then, the species found in North America have brought the total to 15 (Mabberley 1997). Gilmore (1931) found some *Rudbeckia* at the Ozark Bluff-Dweller site in Arkansas, but offered no

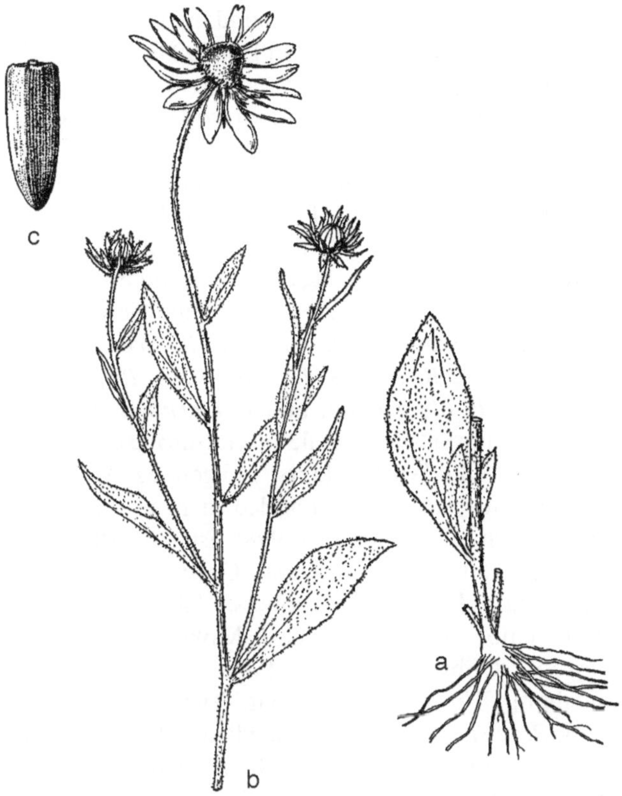

Rudbeckia hirta. a. Lower part of plant. b. Upper part of plant with flowers. c. Seed. From Buchholtz 1968.

suggestion as to its possible use. Given its popularity with historic tribes, it may have been medicinal.

Most people within the range of this species are recorded as having used it, including the Cherokee, Iroquois, Ojibwa, Potawatomi, Shuswap, and Seminoles. The Cherokee used the herbs to bathe sores, for diarrhea, and to wash snakebites (Hamel and Chiltoskey 1975). Both the Cherokee and Iroquois used it to treat worms (Moerman 1998). Small (1933) recorded that the Seminoles used a cold infusion to treat headache and fever, although Sturtevant (1955) found no common name or use for the plants. Bennett (1997) was told that modern Miccosukee use *haash (th)aime* to treat sunstrokes and headaches. That name was given (as *ha:słayhi*) by Sturtevant's (1955) informants to the "Sun Sickness" itself (see also *Sabatia*).

Leaves contain an oleoresin (Perkins and Payne 1978). Both Turner and Szczawinski (1991) and Mabberley (1997) noted that the plant was "toxic," without further comment.

Ruellia

(Named for the French herbalist Jean Ruelle, 1474–1537)

Ruellia caroliniensis (of Carolina)
East Tennessee pinkroot ("pinkroot" is a name usually used for *Spigelia*)
[false, Kentucky, wild] petunia

Ruellia is a genus of about 150 species found in the tropics and in temperate North America (Mabberley 1997). When he described it, Linnaeus ([1753] 1957) knew 8 species. Among those he described was *R. tuberosa* of the Caribbean, a plant of considerable medicinal reputation (e.g., Ayensu 1981, Morton 1981). Neither Vogel (1970) nor Moerman (1998) included the genus even though there are about 20 species within the geographic range covered by their research (Kartesz 1994).

Sturtevant (1955) found no Seminoles using the plants, but Hocking (1997) said that *R. caroliniensis* was used by them as an adulterant of pink root (see *Spigelia*). He also recorded that a decoction was used for stomach distress by the Seminoles.

Porcher (1863) included *R. strepens*, with the comment, "The leaves are said to be subacrid." Millspaugh (1892) did not include the genus, nor did Bremness (1994), Bown (1995), or Duke et al. (2002).

Rumex

(Classical Latin name for sorrel, *R. acetosella*; akin to Akkadian *ramaku*, to pour, Hebrew *romak*, spear or javelin)

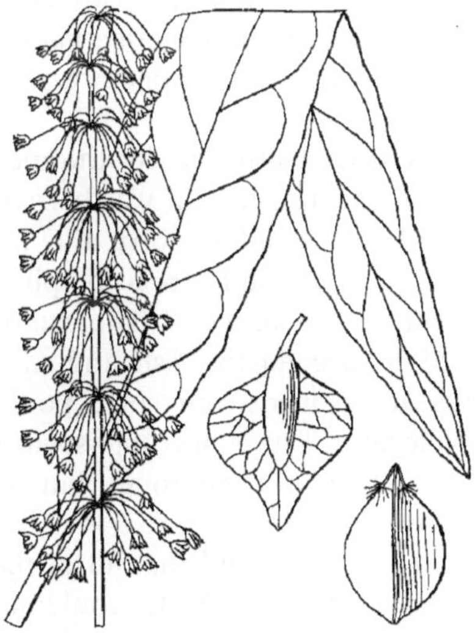

Rumex verticillatus. From Britton and Brown 1896.

acederas [*azederas*] (sorrel, Spanish for *R. acetosa* by 1557); *acetosa* (sorrel, Italian by 1551)
Ampfer [*Ampffer*] (dock or sorrel, German for *R. acetosa* by 1545)

bloodwort (because it "cleanseth the blood and strengthens the liver," Culpeper 1653)
cañaigre [*cañagre, cañaigría*] (from *caña*, cane, *agria*, sour; name for *R. hymenosepalus*, but sometimes used generically, Spanish)
dock [docken, singular] (from Old English *docce*, in use by the 1100s; related to Middle Dutch *docke*, Old Danish *adokke*, Old French *éadocce*; Gaelic *dogha*, cognate with English "dock," but designating "burdock," *Arctium lappa*)
geoirean (bitter or sharp [taste], Gaelic; they also call it *copag*, and *samh* for *R. acetosella*)
høymole (*høy*, hay or high, *mole* =?, Norwegian)
osielle (sorrel, French for *R. acetosa* by 1550)
patientie (for *R. alpinus* by 1549, Dutch)
sorrel (in English for *R. acetosa* by ca. 1400; William Turner in 1568 compared it to *Oxalis*; from the Old French *surele* from the 1100s, which is the diminutive of Germanic *sur*, sour; in modern German *sauer*)
sour-dock (used for *R. acetosa* or *R. acetosella*)
syre (sour, Norwegian for *R. acetosa* and allies)

Rumex verticillatus (having flowers in whorls around the stalk)
[swamp, water] dock
hoshukome (*hoshuk*, grass, *home*, bitter, Choctaw)

Europeans were already distinguishing between the "docks" and the "sorrels" by the time Old English was spoken in the 1100s. "Dock" typically meant the larger-leaved species of *Rumex*, with comparatively little sour taste. "Sorrel," on the other hand, meant the smaller-leaved species with a distinctly acid taste to the leaves.

Sorrel has been cultivated in Europe since the Middle Ages. For many years it was believed that the acid in the plants would dissolve the troublesome bones in fish, such as shad. To accomplish that, the English invented "greensauce." The mixture, which was intended to accompany fish and meat, got its name because it was made from sorrel pounded into a paste with vinegar or lemon juice and sugar (Davidson 1999). It did not remove the bones, but it added an interesting taste.

Fernald et al. (1958) said that all of the 15 or so species in the eastern United States could be eaten, but that care must be taken to gather only the young leaves as they become bitter with age. Two advantages to using *Rumex* greens is that they are wholesome, and that they lose little bulk in cooking so that a small bunch makes a larger meal than other greens. The first one or two waters in which they are boiled should be thrown away to remove the strong taste. At least the leaves of *R. crispus* contain abundant vitamin C and more vitamin A than carrots (Tull 1999).

Some Europeans used *Rumex* as food more than others. Culpeper, in 1653, had an explanation for that view and wrote, "Yet such is the nicety of our times (forsooth) that Women will not put it in the Pot because it makes the Pottage black; Pride and Ignorance (a couple of Monsters in the Creation) preferring Nicety before Health" (Fernald et al. 1958).

Soon after they arrived in the New World, Europeans introduced *R. acetosa, R. acetosella, R. conglomeratus, R. crispus, R. domesticus, R. patientia, R. pulcher, R. obtusifolius,* and *R. maritimus* (Fernald 1950). Within a short while after Europeans arrived with their Old World plants, indigenous Americans adopted those *Rumex* and added them to their foods and medicines. Moerman (1998) lists the Blackfoot, Cherokee, Cheyenne, Cocopa, Costanoan, Dakota, Havasupai, Delaware, Iroquois, Isleta, Kawaiisu, Mendocino, Micmac, Mohave, Mohegan, Montana, Navajo, Ojibwa, Paiute, Pima, Rappahannock, Shoshoni, Thompson, Yavapai, Yuri, and Zuni as using *R. crispus* for food and medicine. The Apalachee used *R. acetosa* in medicine (Hann 1986). That quick incorporation of European species is both a measure of indigenous Americans' adoption of alien plants, and their prior knowledge of native species.

The Cherokee ate *R. acetosa*; they and the Iroquois ate *R. acetosella* and *R. crispus* leaves (Moerman 1998). Fernald et al. (1958) indicated that indigenous tribes ate the seeds of various *Rumex*, and documentation has been found for the Cocopa, Costanoan, Kwasiisu, Mendocino, "Montana Indians," Navajo, Paiute, and Pima (Ebling 1986, Moerman 1998). Eastern tribes most likely also ate the seeds.

Dock leaves are eaten, but they are more famous as medicinal plants. In 1398, a book on medicines stated, "Al manere Dockys heele smyrtynge of Scorpions." Later, they were largely relegated to relieving the itching and burning cause by nettles (*Urtica dioica*). In a related use, Tull (1999) advocated using dock "tea" to treat poison-ivy rash.

Reverend Manasseh Cutler wrote in 1785 that the indigenous tribes around Boston used *R. verticillatus* to clean ulcers that would not otherwise heal (Vogel 1970). The Choctaw tried to ward off smallpox by bathing in a decoction of *R. verticillatus* leaves (Bushnell 1909, Taylor 1940). The Chickasaw used "common dock" as a remedy against jaundice (Swanton 1928b). The Houma used *R. salicifolius* to treat liver trouble, intestinal disorders, jaundice, and menstrual problems, and probably also used *R. verticillatus* (Speck 1941).

Indeed, indigenous tribes across North America used native and introduced *Rumex* species for a variety of maladies. Many of the uses concentrated on external problems like sores and cuts that needed help healing (Vogel 1970, Moerman 1998). Two alien species were incorporated into the American medicines; *R. obtusifolius* made the U.S. Pharmacopoeia between 1820 and 1905, and *R. crispus* between 1863 and 1905 (Vogel 1970). *Rumex* was first listed for treatment of skin diseases, as alteratives and depuratives, but it later was included among laxatives and tonics. All species are high in tannins and oxalic acid, and these in conjunction with the anthraquinones are among the chemicals responsible for their actions (Fernald et al. 1958, Vogel 1970). Anthraquinones are particularly notable for inhibiting fungi (Foster and Duke 1990). Herbal extracts are mildly antibacterial toward both Gram-positive and Gram-negative bacteria (Duke et al. 2002).

Several of the older references allude to the similarity between *Rumex* and rhubarb (*Rheum* ×*hybridum*), both in the Polygonaceae. Indeed, roots of both genera have long been applied as laxatives (Millspaugh 1893). Among the compounds found in *Rumex* are anthraquinones, particularly emodin and rumicin (Kay 1996). However, it should be noted that while the anthraquinones are laxative, the tannins are antidiarrheal (Foster and Duke 1990, Duke et al. 2002).

Gaelic people obtained a black dye from the roots mixed with copper (Dwelly 1933). Quick adoption of alien *Rumex* as a dye source suggests that other native species were formerly used. Tull (1999) indicated that tans, browns, green, gold, mauve, yellow, and orange can be made from different *Rumex*, and that the Navajo used them for those colors. The Cheyenne obtained a blue dye from some species that was used to color feathers and porcupine quills (Tull 1999), although Moerman (1998) listed only reds and yellows. The Choctaw made a yellow dye from the roots of what Bushnell (1909, 1917) identified as *R. crispus*, but the Houma used the native *R. salicifolius* roots to make a dye for cane and palmetto baskets (Speck 1941).

S

Sabal

(Michel Adanson used a vernacular American name for the generic name of these palms in 1763; could he have taken it from the Mayan word *sibul*, for a black bird?)

Sabal etonia (named because it grew in the "Eton Scrub" of northern Florida; ultimately, either from *Utina*, the name of a Timucuan people who lived in Putnam County after withdrawing from the Tomoka River, or from Creek *atan*, from, and *aia*, go, or "go elsewhere")
scrub palmetto (Florida)

Sabal minor (smaller)
[little] blue stem [bluestem, palm] (Alabama, Texas, Florida)
latania [*latanier*] (Louisiana)
[bush, dwarf, swamp] palmetto [palm] (Texas, Alabama, Florida, Georgia)
taala (Alabama)

Sabal palmetto (diminutive of Latin *palma*, the palm tree, probably through Spanish *palmito*)
big top (Bahamas)
[swamp] cabbage [palm, palmetto, tree] (Florida)
cani (the leaf, Timucua)
chaw-fo-ka-naw (language?, USA)
guana cana (*guana* cane; "*guana*" is a Taino name for some tree, perhaps this, Cuba); *guano rabo de cote* ("cote's" tail cane, Cuba)
latanier (*latania* is an indigenous word from Mauritius for a palm there; apparently, this native plant became confused with the one from Africa, Houma, Louisiana)
maxon̲tkayixki' (Biloxi)
palma [*cana*] ([cane] palm, Cuba)

Sabal. *Sabal etonia* (right, trunkless). The leaf and fruit cluster are from *S. etonia*. *Sabal palmetto* (left, with trunk). *Both drawn by P.N. Honychurch.*

[bank, Carolina, common, hat] palmetto[-tree] (perhaps first recorded as *palmito* by Barrientos 1567, who noted that Menéndez and his men had to resort to eating them, southeastern United States, Bahamas); *palamito* (Houma, Louisiana)
pond thatch (Bahamas)
tala (Choctaw); *tá:la* (Koasati); *taalachoba* (*taala*, palm, *choba*, big, Alabama); *ta:łakko* [*ta:larákko, talalocho, tallaloko, talv rakko*] (*talv* or *tala*, palm, *łákko*, big, Creek; maybe first recorded as *talalocho* by Simmons [1822] 1973); *talcó:bî* (*tala*, palm, *có:bî*, big, Mikasuki); *talimushi* (*tala*, palmetto, *imushi*, it's uncle, Choctaw)
[pond-, white-] top (Bahamas)

Of the three species of *Sabal* in Florida, the first discovered by European science was not the most obvious. People from at least the time of Menéndez in the 1560s mentioned cabbage palms (Barrientos 1597), but it was the trunkless *S. minor* that was first described by Jacquin in 1776. It was not long before *S. palmetto* was "discovered," and it was named by American Thomas Walter in 1788. Both put their species in *Corypha*. The first Florida palm to be called a *Sabal* was *S. etonia*, described in 1896. People had finally discovered Adanson's *Sabal* and realized that *Corypha* was an Old World genus. *Sabal* now has 15 species (Zona 1990).

Surely the oldest Florida use of *Sabal*, regardless of species, is of the leaves (Austin 1980). Fontaneda ([1575] 1944) recorded palm use by the Calusa. Of the men he wrote that they "*andan en cueros, sino solamente vnos brageros tejidos de palmaque*" (go naked, except only some breech-cloths woven of palm). Women, he wrote, wear "*un mantellin de vnas palmas Rajadas y tejidas*" (a shawl made of a kind of palm-leaf, split and woven). Jonathan Dickinson found the Jaega men wearing the same loincloth in 1699 (Andrews and Andrews 1945). Either of the three *Sabal* could have been used for these clothes, but *S. palmetto* was the most abundant.

It would appear that cabbage palm fibers were more often used within and outside their native range than is at first obvious. Whitford (1941) found that the materials had been traded with northern tribes and appeared among artifacts from the Winnebago of southern Wisconsin and the Iroquois of New York. Those two tribes were 600 to 700 miles north of the closest *S. palmetto* stands. He had identified palmetto fibers in cords used to make carrying straps for bags, although the Iroquois had covered them with softer fibers. The Cherokee were known at the time to make baskets with them (Whitford 1941).

The Timucua used palm leaves for thatch, and there still is a locality in Alachua County called "Kanapaha" (*cani*, palm, *paha*, house), based on their language (Simpson 1956). Leaves are still used for thatching traditional Seminole houses called *cuko* (Creek) or *ciki* (Mikasuki), in a style likely modified from the Glades people before them. In addition, the Seminoles use the leaves to make potato-drying mats, fish drags, and rope (Sturtevant 1955). The Alabama used strips of the leaves to hang meat for smoking (Sylestine et al. 1993). Vignoles ([1823] 1977) wrote of these palms: "There are a number of varieties of the palmetto, from whence the inhabitants and even the Indians form ropes: much of the same tackling of the fishing boats is made of this grass, which is rather the better for being kept wet, as it is apt to crack like rushes when dry: we also find in the country that palmetto, from whence the rigging, cordage and cables of most of the West India small craft are made."

From the trunks of *S. palmetto*, the Seminoles made house poles, food paddles, arrows, hunting dance staffs, heddles, ballsticks, and skin-drying frames (Sturtevant 1955). Wood was used for salt by the same process as potash, burning and using ashes (Morton 1968b).

According to Morton (1968b), one of the ways the Seminoles ate the fruits was to reduce dried fruits to meal, which they made into bread. They also ate raw fruits or made them into syrup.

Sturtevant (1955) recorded that the Seminoles used the plants for "fish poison"; no further details were given. They also mixed the fruits with an unidentified grass to treat "Grass Sickness" (low fever, headache, weight loss) (Sturtevant 1955). Bahamians have a similar use as a cure for fish poisoning by eating the palm heart or steeping it in gin for a medicinal drink (Eldridge 1975).

The best-known modern use of the palms is to cook and eat the terminal bud or "cabbage" (Morton 1968b). Both *S. etonia* and *S. minor* were used similarly (Yanovsky 1936). A lesser-known use was to extract the fresh pith of upper trunk, which was eaten as is or made into "pie."

The heart has been eaten since about the time of the first Spanish visitors to Florida. The first records known were of the Spanish troops under Pedro Menéndez de Avilés extracting palm hearts on their overland route to expel the French Huguenots at the south end of Anastasia Island (St. Johns County) in 1565 (Barrientos 1567). Apparently, the heart was not eaten before the Europeans brought metal tools into this region because the utensils available to the indigenous people were not adequate to extract the heart efficiently.

Sabatia

(Michel Adanson named this genus for Liberato Sabbati, b. 1714, an Italian surgeon in Rome who published in 1745)

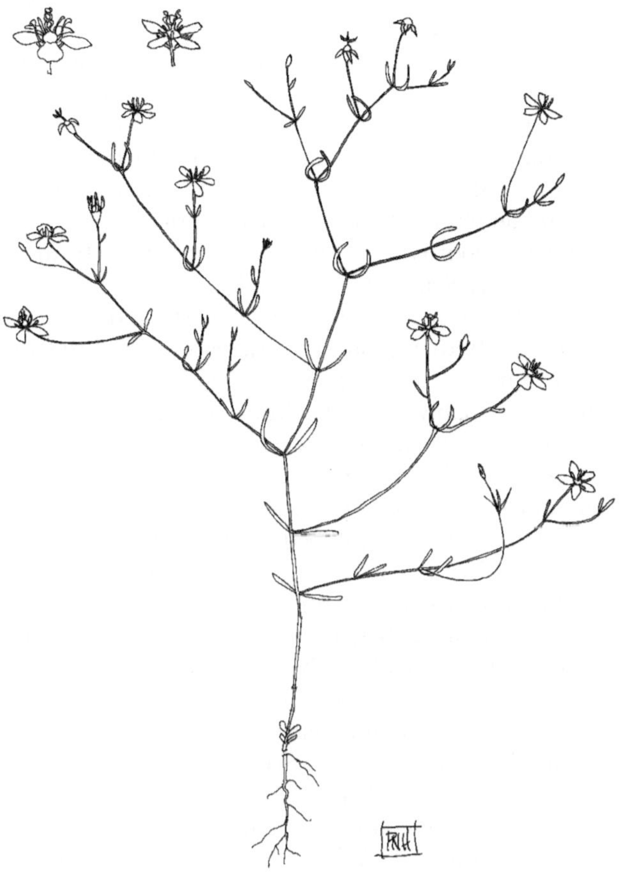

Sabatia brevifolia. *Drawn by P.N. Honychurch.*

Sabatia angularis (with angles on the stems)
Bitterwurzel (bitter herb, German)
[bitter-, pink-] bloom [-clover] (West Virginia)
[American, angular, red] centuary [centory, centry] (applied by 1831; "centuary" is the English rendition of *Centaurium*, a related genus described by London apothecary John Hill, 1716–1775)
eyebright (North Carolina; in Europe usually given to *Euphrasia officinalis*)
[square-stemmed] rose gentian ("gentian" is from the genus *Gentiana*, a related bitter medicine)
rose-pink [rosepink] (etymology of "pink" as a plant is obscure, but perhaps from "pink eye," small eye, related to French *oeillet*, the diminuitive of *oeil*, eye, from Medieval Latin *ocellus;* the name "pink" was originally given to *Dianthus*, Caryophyllaceae, in the 1500s; perhaps "rose-pink" simply refers to the color)
square-stemmed sabatia (a book name)
wild-succory ("succory," used since 1533; a corruption of chicory, *Cichorium*, Asteraceae, another bitter medicine)

Sabatia bartramii (named for the Bartrams, John, 1699–1777, and his son William, 1739–1823)
Bartram's rose gentian [rosegentian] (a book name)
marsh pink

Sabatia brevifolia (short-leaved)
quinine [flower, herb]
shortleaf rose-gentian [rosegentian]

Sabatia campanulata (bell-shaped)
marsh pink
slender rose-gentian [slender rosegentian]

Sabatia stellaris (starlike)
kococompahá:ka (*kocecvmpv*, archaic for star, *vhake*, resembling, Creek); *owa:cíkâ:bî* (star replica, Mikasuki)
marsh [sea] pink [sea-pink]
rose-of-Plymouth (Mrs. William Dana Starr wrote in 1900 that the inhabitants of Plymouth were convinced that the Pilgrims of 1620 named the plant *Sabatia* after the Sabbath; they maintained that was the holy day on which they first saw the flower, and that "strong objections are made if any other flowers are irreverently mingled with it in church decorations," Massachusetts)

According to Greek legend, the Kentauron (Centaur) Chiron was a physician famous for his knowledge of healing plants (Baumann 1993). Among his many feats, this half-man and half-horse was the master of Asclepius and teacher of Achilles. One version of his story has it that he used a *Centaurium* (Gentianaceae) to treat an arrow wound accidentally inflicted on him by Hercules. According to some, confusion by the 16th-century herbalists held that the plants used were *Centaurea* (Asteraceae), another genus named for the Centaur. Regardless, Chiron is believed to have discovered that *Centaurium* was a blood purifier, and it remains in use today to aid digestion (Baumann 1993, Duke et al. 2002).

Those stories of *Centaurium* were recorded by Theophrastus, 372–287 B.C., but the plant was still famous at the time Europeans arrived in the New World. Spanish speakers called the European medicinal plants *centaurea*, while in Portuguese it was *centáurea*. In French, the plants were *centurée*, in Italian *centauro*, and in English centuary. Germans also considered the plants valuable, but rendered them *Tausendgüldenkraut* (thousand-dollar herb).

When people arrived on the east coast of North America, they found plants similar to *Centaurium.* These herbs were promptly dubbed "centaury." Linnaeus described two of the Florida species in 1753, *Chironia angularis* and *C. campanulata,* based on specimens collected by his former student Peter Kalm. He too wanted to compare them to the Old World remedies, and chose to dedicate them to *Chiron.* It was only 1763, when Michel Adanson published his *Familles des Plantes,* that people realized Linnaeus had included two distinct genera in his *Chironia.* One of the genera remained *Chironia,* a group of about 15 species that grow in Africa and Madagascar. The American plants became *Sabatia,* with 17 species endemic to North America.

According to physician Benjamin S. Barton (1766–1815), *Sabatia angularis* is the plant, "which is called Centory or Centry in Philadelphia, &c., where it is so commonly employed both by physicians and as a domestic remedy in almost every family. ... I believe that no bitter has been more generally prescribed in the United States, in febrile and other affections, than this common American plant, especially since the memorable year 1793 [date of a yellow fever outbreak], when it was much employed in certain stages of yellow-fever; and in which I believe it was very often used with much benefit" (Coffey 1993). The plants were similarly praised by Jacob Bigelow (1787–1879) in his *American Medical Botany,* published in 1817–1820, and by John Eberle (1787–1838) in his *A Treatise of the Materia Medica and Therapeutics* from 1825 (Porcher 1863).

However, it was not until 1831 when the name "American centaury" appeared in the literature. That year, John Davies wrote in his *Manual of Materia Medica and Pharmacy Adapted to British Practice,* "American Centaury ... This plant is a pure bitter, justly held in estimation as a valuable tonic and febrifuge." Porcher (1863) recommended, "The cold infusion of one ounce of the herb to one pint of boiling water, taken in doses of a wineglassful every two hours ... or thirty grains to sixty grains of the powder, which also acts as a vermifuge. The decoction, extract, and tincture may be used." He considered *S. stellaris* as good a remedy as *S. angularis.*

The American centaury was included in the U.S. Pharmacopoeia between 1820 and 1882, and considered a "bitter stomachic, similar in its action to the other gentians" (Vogel 1970, Coffey 1993). This was also one of the *schwitzgegreider* (sweat herbs) the Pennsylvania Dutch mixed into whiskey (Vogel 1970).

The Cherokee are the only indigenous tribe found to have used *S. angularis.* They drank an infusion to treat malaria (Hamel and Chiltoskey 1975). However, Daniel Drake recorded use of *S. angularis* as a febrifuge in the Ohio Valley in 1850 (Vogel 1970). He did not reveal the people using it, but it was perhaps the Shawnee.

Sturtevant (1955) and Moerman (1998) recorded that the Seminoles use *S. campanulata.* That identification is probably not correct, as there was confusion regarding southern Florida plants dating from Small (1933) until Wunderlin (1998), even though Wilbur (1955) had clarified their distinctive ranges. The plants that were reported by Sturtevant (1955) were from the Big Cypress Seminole Reservation, where *S. campanulata* does not grow. Both *S. stellaris* and *S. grandiflora* are frequent on the reservation, but the one likely to be confused with *S. campanulata* is *S. stellaris.* That is probably what the Mikasuki call *owa:ciká:bi* and the Creeks call *kococompahá:ka.*

Owa:ciká:bi is part of a remedy for what the Seminoles call *ha:sɬayhi* (sun shot, Mikasuki), and Sturtevant (1955) dubbed the "Sun Sickness." *Ha:sɬayhi* is a malady that strikes in the hot months of May through August. In the early morning the eyes hurt, a headache begins, there is a high fever, and diarrhea. The patient can become blind and even die if not treated (Sturtevant 1955). The roots of *Sabatia* are mixed with *Bidens coronata* and *Persea* in cold water and used in the treatment. See also *Rudbeckia.*

Hocking (1997) listed two other Florida species. *Sabatia bartramii* was not named until it was described by Robert L. Wilbur (1955). Prior to that time, it had been confused with John K. Small's (1933) *S. dodecandra. Sabatia bartramii,* a multipetaled southern Florida herb, is used as a "bitter tonic" to aid indigestion by Seminoles and others (von Reis and Lipp 1982). Similarly, *S. brevifolia* has been used as a substitute for quinine.

Sabatia angularis contains the bitter principles erythrocentaurin and gentiopicrin, which promote appetite and digestion (Lewis and Elvin-Lewis 1977, Hocking 1997). Uphof (1968) and Hocking (1997) say that *S. brevifolia* and *S. campestris* also contain the same bitter compounds. *Sabatia campestris* has long been used by Mexicans and indigenous people of Texas as a "chill tonic" (Hocking 1997). Probably, all the species in *Sabatia* contain gentiopicrin and the related erythrocentaurin, as those chemicals also have been found in *Enicostema, Gentiana,* and *Swertia* (Ghosal et al. 1974, Sluis et al. 1983, El-Sedawy 1989a,b, Menkovic et al. 2000). Indeed, Cronquist (1981) considered the iridoid compounds, particularly gentiopicroside and related substances, characteristic of the Gentianaceae. Perhaps the bitter compounds are why Millspaugh (1893) remarked, "A peculiar property pervades the whole of this natural order—the species when fresh are all emetic and cathartic, and, when dry, tonic, and stomachic in varying degrees."

Sagittaria

(Linnaeus used the Latin *sagittarius*, belonging to an arrow, in reference to the leaf shape)

arrow-head [arrowhead] (in use for these plants by John Gerarde's *Herball* of 1597; the word for the points of arrows, in use by 1483)
cola de golondrina (swallow [bird] tail, Spanish)
erba saetta (arrow herb, Italian); *erva frecha* (arrow herb, Portuguese); *Pfeilkraut* (arrow herb, German); *fleche d'eau* (water arrow, French); *saeta de agua* (water arrow, Spanish); *sagittaire* (arrow, French)

Sagittaria lancifolia (lance-leaf, referring to leaves shaped like a lance point)

angelito (little angel, Dominican Republic)
[bulltongue, lance-leaved, scythe-fruit] arrow-head [arrowhead] (Florida to Texas)
arrow-leaf [arrowleaf] (Puerto Rico)
éneas (copper colored, Hispaniola)
flecha de agua (water arrow, Cuba, Puerto Rico); *saeta de agua* (water arrow, Puerto Rico, Panama); *flechera* (arrowhead, Cuba); *flechiere obtuse* (blunt arrowhead, Haiti); *sagitaria* (arrow, Cuba)
hicaknâ:bî (feral-taro replica, based on *hicaknî*, for *Colocasia esculenta*, an introduced plant, Mikasuki)
kibix (*ki*, henequen fiber, *bix*, uncooked beans, Maya, Belize)
lirio de sabana (savanna lily)
plécoire (Haiti)
rabadilla (little tailed one, Puerto Rico)
renoncule aquatique [*des savanes, sagittee*] (water [savanna, sagittate] ranunculus, Haiti)
swamp spaghetti (USA)

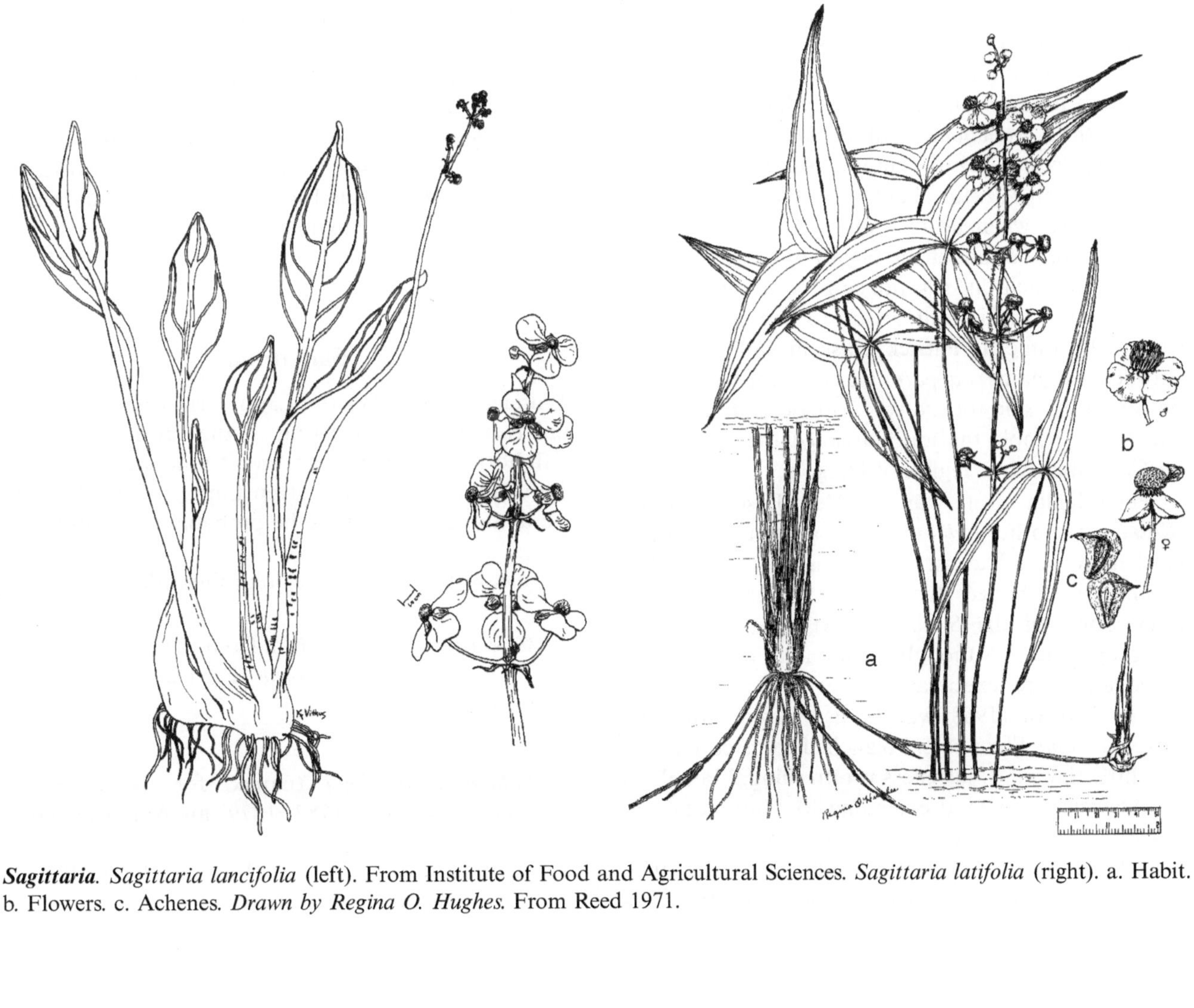

Sagittaria. *Sagittaria lancifolia* (left). From Institute of Food and Agricultural Sciences. *Sagittaria latifolia* (right). a. Habit. b. Flowers. c. Achenes. *Drawn by Regina O. Hughes.* From Reed 1971.

yautía [*yahutia, diahutia*] (first recorded by Bartolome Las Casas, whose copiest wrote *yahubía*, Taino name for *Zanthosoma sagittifolia*, later applied here, Hispaniola)

Sagittaria latifolia (broad-leaved)
[broadleaf, common] arrowhead (Florida)
çin̲ (Osage)
duck potato (USA)
kaishúepenauk (maybe "loose ground nuts," i.e., easily dug, Carolina Algonquians; see Geary 1955); *katniss* (Algonquian, probably Delaware, New Jersey); *wabasi'bînik* [*wabasi'-pînîk, kopînîak*] (white potato, Potawatomi)
wappato [*wappatoe, wappatoo, whápto, wapto, wapito, wapata*] (recorded in 1807 by Patrick Gass of the Lewis and Clark expedition in the Columbia River region where the word *wapato* is part of the Chinook jargon; probably a loan word adopted by the Cree traders during the fur trade era, who now call it *wapatowa*, white mushroom)

The Glades people living on Lake Okeechobee before the Europeans arrived were using *Sagittaria* (Hogan 1978). Given the rich history of the genus among North American people, they were likely using it for both food and medicine. They had available to them both *S. lancifolia* and *S. latifolia*, with the second species being the more useful of the two. Harriot ([1590] 1972) probably gave the earliest name of *S. latifolia* when he recorded, "*Kaishúepenauk*, a white kind of root about the bignes of hen egs & nere of that form: their tast was not so good to our seeming as the other." In his comparison to hen eggs, he did exactly the same as Patrick Gass in 1807 when he recorded the name *wappato*. Gass's record was from the opposite side of North America, among the Chinookan people the journal called the Skilloot in the Columbia River region of Oregon.

Peter Kalm wrote while in New Jersey in 1749: "*Katniss* is another *Indian* name of a plant, the root of which they were likewise accustomed to eat. ... The *Indians* either boiled this root or roasted it in hot ashes. Some of the *Swedes* likewise eat them with much appetite." Although Kalm and the Swedish Americans were exuberant about the roots of *S. latifolia*, Morton (1968) was not as pleased with those of *S. lancifolia*. She wrote, "blanched roots edible, but with little flavor; ... juiceness is refreshing." People in the Atlantic states, and extending to Minnesota and Wisconsin are known to have eaten the roots (Gilmore 1919, Hedrick 1919, Densmore 1928, Yanovsky 1936, King 1984). Similarly, Cubans, Chinese, and Panamanians ate the roots, either roasted or made into flour (Roig 1945, Duke 1972).

The Miccosukee name for *S. lancifolia* poses an interesting problem. According to Sturtevant (1955), *hicakna:bi* means "feral-taro replica," based on *hicaknî*, the name for *Colocasia esculenta*. The Creek name for *C. esculenta* is *hilá*. That plant was unknown to the people in North America until it was introduced from Asia (Smartt and Simmonds 1995). Perhaps this is another instance of the name of an old food being usurped by a new one. That replacement has been noted for the opossum (*Didelphis virginiana*) in both Cherokee and Mikasuki (Mooney 1885–1886). In Mikasuki the opossum is now *sokihatkhi* (white pig). Prior to the introduction of the Old World swine, the opossum was *soki*. That name was switched to the pig, and the "white" modifier attached to note the old familiar North American animal (Sturtevant 1955). Cherokee call the opossum *sikwa utse'tsti* (grinning pig); the hog is *sikwa*.

Regardless of its past uses, by the time Small (1933) was studying plants in Florida, the Seminoles had a single use for *S. lancifolia*. He learned that they used the plant to treat shock after an alligator bite, and his notation seems to have been the sole source for Sturtevant's (1955) entry.

Grimé (1976) found that blacks in Jamaica used the plant to treat wounds and yaws. In Cuba and Haiti, the roots are used as a rubifacient to stop toothache (Roig 1945, Morton 1981). Crushed leaves are poulticed on infected sores, chronic itch, and snake and insect bites in Panama (Duke 1972).

Salicornia

(Linnaeus used either the Latin derived from *sal*, salt, and *cornus*, horn, referring to the salt-loving plants with hornlike branches, or Arabic *sala-al-qarab*, giving rise to French *salicorne*; given Linnaeus's tendency to use dual-meaning names, he probably used both)

glasswort (in use since John Gerarde's *Herball* of 1597)
salicornia (probably from from the Arabic *sala-al-qarab*, Italian); *salicorne* (French)
sosa (Spanish)

Salicornia bigelovii (named for its discoverer, Jacob Bigelow, 1787–1879, an American botanist)
annual [dwarf] glasswort (Florida, Puerto Rico)
chifle (whistle, Puerto Rico)

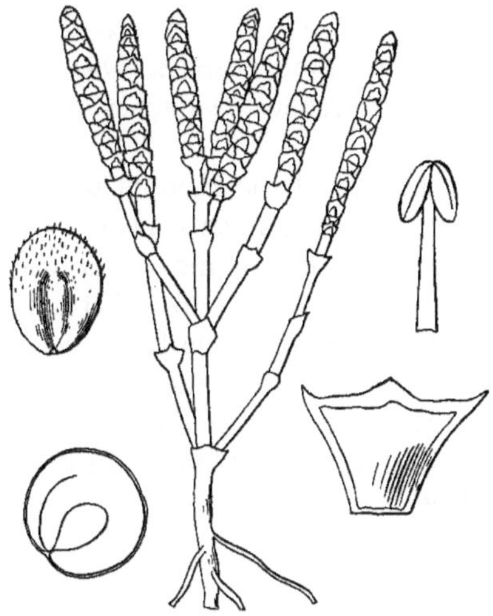

Salicornia bigelovii. From Britton and Brown 1896.

Salicornia perennis (perennial) (= *S. virginica*)
[American, perennial, Virginia, woody] glasswort (USA, Bahamas)
Guinea bead (Bahamas)
pickleweed (USA)
wild coral (Carolinas, Bahamas)
yerba de vidrio (glass herb, Spanish)

Salicornia is used by people all over the world. In Europe, *S. europaea* (samphire, chicken claws, pigeonfoot, glasswort) is pickled in vinegar. An ash called *barilla* was made of it to use in the manufacture of soap and glass, and it is considered diuretic and antiscorbutic (Hocking 1997). East Indians also use *S. brachiata* for carbonate of soda, in manufacturing (Morton 1974). In Australia, *S. australis* shoots are pickled and eaten (Uphof 1968).

Florida species are also cooked and eaten (Morton 1968b). In the Carolinas, a decoction is used to treat colds and for whooping cough (Morton 1974). Gosiute (Utah, Nevada), Heiltzuk (coastal British Columbia), and Salish (Vancouver Island) used them, but no records were found of indigenous people using them in the southeastern United States.

Salix

(Ancient name for the willow; possibly related to the Gaelic *suil* for the tree, and Akkadian *salihu*, sprinkler of water)

accaycay (Coptic)
ai-pee-á-'gaw (Kiowa); *señ-ä* [*señ-ya-daw*] (Kiowa)
itea (Greek)
ittokoowisa (*itto*, tree, *oki*, water, *homi*, bitter, Alabama); *okawisa* (Alabama); *okoowaykachi* (*oki*, water, *oo*-, within, *wihaachi* =?, Alabama)

Salix humilis. From Britton and Brown 1896.

jarita (to cast, as an arrow, from Hebrew *khara*, New Mexico); *xara* (arrow shaft, from Arabic *xara*, New Mexico)
k'aí (Chipewyan, Canada); *ka* (Slave, Canada)
kitapato (Pawnee)
nepise (Cree)
ozi'sigo'bimic (Ojibwa)
ruhi (Winnebago); *sin-des-nes-ni* (it grows by water; *ni*, water, Osage; also *thiu'-xe*)
safsaf (Arabic)
sallow (from Old English *sealh*, Anglican *salh*, cognate with Old Norse *selja*, Swedish *sålj* [*sälg*], *sålg* (Norwegian), and Danish *selje*; the word is also cognate in languages other than Germanic languages, with Latin *salix*, Irish *salieach*, and Welsh *helyg*; in English by about A.D. 700); saugh (the Old English nominative singular, derived from the Late Anglican *salg* or *salig*; in English by about A.D. 1000); *salgueiro* (from Latin *salix*, Portuguese); *salice* (Italian); *saúce* (Spanish); *saule* (French)
suil (the 16th letter of the Gaelic alphabet)
walipe-popa (Dakota)
willow (from Old English *welig*, dating to A.D. 750; akin to Middle High German *wilge*, and Greek *helike*); *walg* [*welg*] (Teutonic); *Weide* (German); *wilg* (Dutch); *wylch* or *willig* (Frisian)
yehiye̲' yap (water tree, Catawba)

Salix caroliniana (from Carolina)
ahwá:na: [*ahwánv, akawana, ahuane, awannah*] (*owv*, water, *wvnv*, string, Creek, Muskogee); *oke bakshe* [*okibaksî, okebakshe, bakshee*] (*oke*, water, *baksî,* string [rope, vine], Mikasuki)
di-l(i)-ga-li:-s-gi (Cherokee)
saule (French, Houma, Louisiana)
takoi̲n̲sha [*tikoi̲n̲sha, tako̲n̲wisha*] (maybe from *itt-ak-owisha* or *itt-ako-wisha*, tree within water, Choctaw)

[coastal plain, southern, swamp, ward] willow (Carolinas to Florida)

Salix eriocephala (cottony-headed, in reference to the fruits)
Missouri River willow

Salix humilis (low)
hoyaní:cî (Mikasuki; adapted from Creek); *hoyvnijv* (spreading, alluding to rhizomatous habit, Creek, Muskogee)
mikko hoyvnijv [*mikko hoyaní:câ, mikkohoyani:câ, micco hoyonee, micco hoyonvicha*] (*mikko,* chief, *honece,* wild, Muskogee; see Howard 1984)
small pussy-willow
tcáła [*tcára*] (Yuchi; still taken as medicinal tea during the Green Corn Ceremony; see Jackson 2003)
[dwarf, gray, prairie] willow

Salix nigra (black)
black willow
osí (Koasati)

It was not surprising that Hogan (1978) found willow pollen in the pre-Columbian Glades village on Lake Okeechobee. People around the world have long histories of using willows. In the 1500s, the Glades people were traveling between Key West and Havana in dugout canoes to trade birds and deer skins. The birds were kept in willow cages (Barcia 1565).

From day to day, the Seminoles used the wood to make toy bows for children, sofkee spoons, and stick ball rackets (Bennett 1997). They also constructed chairs and other items from the wood. Dye was extracted from wood and used to color deer hides (Bennett 1997). The Lakota made walking sticks of *Salix eriocephala,* and the Ute used it to make baskets (Moerman 1998). Whitford (1941) found that willow had been used by the Menomini, Winnebago, Ottawa, and Ojibwa to make bags, pouches, fishnets, and cords. He also found the fibers used to make a Micmac fishing spear.

Eryngium yuccifolium and *S. humilis* were the two most important plants in the Green Corn Ceremony of the Seminoles, Creeks, and Yuchi (Howard 1984). *Salix humilis* was also important to the Catawba, Cherokee, Delaware, Menomini, and Meskwaki (Hamel and Chiltoskey 1975, Moerman 1998). The plants were used to treat diarrhea, to reduce fever, to treat skin problems, and to stop bleeding.

The Cherokee, Houma, Iroquois, Koasati, and Micmac used *S. nigra* (Taylor 1940, Moerman 1998). The Cherokee used it to treat the same problems as with *S. humilis* (Hamel and Chiltoskey 1975). The Houma used it to reduce fever, and to strengthen a person with "thin blood" (Speck 1941). The Koasati took an infusion to relieve headaches and reduce fever (Taylor 1940).

We have records of the Houma using *S. caroliniana* for weak blood and fever (Speck 1941). As Alice Snow wrote, "It is used for most everything" among the Seminoles (Snow and Stans 2001). The classical use against fever as among the Houma is repeated in species after species (Moerman 1998). There are numerous other uses among the Seminoles of more religious and ritual significance (Sturtevant 1955, Moerman 1998).

Salvia

(Latin *salvus,* whole, well-preserved, safe; akin to Akkadian *salwu,* healthy, Persian *haruwa*)

Salvia lyrata. From Britton and Brown 1898.

athair liath (father gray-hair, Gaelic)
clary (from Old English *slaric,* changing to *claré* in the 1500s; still *sclarée* in French, from Latin *sclarea*; originally applied to *Salvia sclarea,* Lamiaceae; since about A.D. 1000 variants of the word "clary" have existed, and this aromatic mint of southern Europe was used to adultrate wines, as a condiment, and as a medicinal herb; in the 1500s, the name also was transformed into "clear-eye," "Godes-cie" [good-see], and "see-bright" to emphasize its use for treating eyes)
sàisde [*sàiste*] (from Latin *salvia,* Gaelic); *Salbei* (German); *salva* (Portuguese); *salvia* (Italian, Spanish); *salvie* (Norwegian); *sauge* (French); *savie* [*salve*] (Dutch)
torman (based on *tormach,* increase, Gaelic)

Salvia lyrata (lyre-shaped)
cancer-weed [cancerweed] (first used in 1546 for various plants to cure "cancer," not always applied in its sense of being a malignant growth; name used by Williams [1837] 1962; well established in South Carolina as a name by the time of Porcher 1863)
lyre-leaf sage [lyre-leaved sage]
meadow [wild]-sage

Europeans had a history of using *Salvia* as medicine dating to the classical Greeks and Romans, and probably earlier. Pliny (A.D. 23–79, Dioscorides (A.D. 40–80), Aëtius (A.D. 527–565), and Paul of Aegina (A.D. 625–690) discussed what we now know as *Salvia fruticosa, S. officinalis, S. pratensis,* and *S. sclarea* (Meyer et al. 1999). Fuchs in 1542 (Meyer et al. 1999) and later herbalists also commented on those European species. The very name "*Salvia*" reflects their importance (de Cleene and Lejeune 2002).

The most popular among the sages in early Europe was *S. officinalis.* Of that species, John Parkinson wrote in 1629, "Sage is much used of many in the month of May fasting, with butter and Parsley, and is held of most much to conduce to the health of a mans body." Not much later John Evelyn wrote in 1699 that "tis a plant endu'd with so many and wonderful properties, as that the assiduous use of it is said to render men Immortal."

Indigenous Americans also had a rich history of incorporation of *Salvia* into their lives. Moerman (1998) listed 7 species used in North America, but that is probably an underestimate. Historical records are also biased toward the western United States, perhaps partly because of the greater number of indigenous species there. For example, Fernald (1950) listed only 4 native species in the northeastern states, while Kearney and Peebles (1951) recognized 14 from Arizona alone.

The widespread *S. lyrata* is the most common species in the eastern states. This sage ranges from Pennsylvania to Florida and west to Texas, southeastern Kansas and Illinois. The Catawba made a salve from the roots to treat sores (King 1984). The Cherokee took an infusion to treat diarrhea, colds, nervous debility, and phelgmatic individuals (Hamel and Chiltoskey 1975); they used the same preparation to strengthen women and as a laxative. The Cherokee also mixed the leaves and flowers with honey and drank it to relieve asthma. Since they did not have honey before the introduction of the bees (*Apis melifera*), that was either a modification of an old remedy, or one they learned from Europeans. Porcher (1863) wrote, "The fresh radical leaves of the plant, when bruised and applied to warts, generally destroy them; continue the application for a day or two, and renew it every twelve hours." Seeds were also made into an ointment to cure wounds and sores (Coffey 1993).

Sambucus: Elders

(Derivation uncertain, may be from Greek *sambuce,* a musical instrument; more details in discussion)

Few people in the United States will understand if someone calls a plant an "elder," although they would if it is called an "elderberry" (USA to Virgin Islands). The word "elder" is applied to what scientifically is *Sambucus.* Both words have complicated histories.

"Elder" has been in English since at least A.D. 700, when it was spelled *ellaern* or *ellen* (Old English) or *ellern* (Anglo-Saxon). Related words in Middle Low German were spelled *elderne, alhorn,* or *elhorn.* Genders (1988) said those words were cognate to Saxon *aeld* (fire), because the hollow stems were used to kindle campfires, but that is dubious. Germanic Europeans were afraid to burn elder wood. De Cleene and Lejeune (2002) offer a more likely etymology.

Cognate name of the shrubs are *hyld* (Danish), and *hyll* (Swedish and Norwegian), and these are related to *Holle, Holunder, Holder,* or *Elhorn* (German). Those names and the Danish *Hyldemör* or *Hyllefrao* allude to Freda, who was the Norse goddess of fertility, and the most revered of all goddesses. She was called *Vrouw Arda, Vrouw Herta,* or *Vrouw Holle* in the Netherlands. This "Lady Elder" lived in an elder and became the symbol of reincarnation. She and the plant represent the European "Tree of Life."

Unhappily, the etymology of *Sambucus* is not as easily unraveled. *Sambucus* was once thought to be from Greek *sambuce,* an ancient flutelike musical instrument made from the readily removed tubes of bark (Fernald 1950). Recent suggestions are that *Sambucus* may have come from *sambuca* [*sambuke, sambuc*] (Latin *sambúca,* Greek *sambúka,* cognate to Aramaic *sabbeka*) "a triangular stringed instrument with a shrill tone" (OED 1971). Some say these musical instruments were made from the wood of elders (Walker 1976, Kowalchik and Hylton 1987). Quattrocchi (1999) was uncertain, saying Latin *sambucus* and *sabacus* were used by Pliny for an elder tree. The Portuguese name *sabugueiro* comes from *sabugo,* taken from Latin *sabacu.* Could all these words be related to Greek *sabakos,* rotten, because of the unpleasant smell of the foliage? Regardless, *Sambucus* has cognates in *sureau* [*siriyo, suyeau*] (French, Trinidad), *sambuco* (Italian), *saúco* (Spanish), and *sabugueiro* (Portuguese), although *senboqua* (Arabic) might be a loan-word.

***Sambucus nigra* ssp. *canadensis*.** From Institute of Food and Agricultural Sciences.

Other European languages also have names for elders, because there are three native species there. Before arriving in the New World, Europeans had long histories of interacting with *S. ebulus, S. nigra,* and *S. racemosa.* These have their own complicated name history.

The best known of the three is *S. nigra*, and it is called elder (English), boor [boer, bour] (akin to arbour and bower, British Isles), *sureau* (French), *sambuco* (Italian), *saúco* (Spanish), *sabugueiro* (Portuguese), *Holunder* (German), *vlier* (Dutch), and *svarthyll* (black elder, Norwegian). *Sambucus nigra* formed the 15th letter of the Gaelic alphabet as *ruis.* Hermann Boerhaave (1668–1738), the famous Dutch physician, is said to have held the *vlier* in such regard that he doffed his hat each time he passed it (Curtin 1947, Kowalchik and Hylton 1987). Supposedly, he did that not only because of his high regard for the plants, but also because European folklore required reverence in the presence of the elder. Vickery (1995) found *Sambucus* to be one of the "most enigmatic plants in British folk tradition." The plants are feared and associated with witches (it belongs to them and repels them), yet valued for its protective qualities as a fly repellent and in herbal remedies.

The "elder" in Florida is often called *S. canadensis.* It also goes under the English names American elder (USA), southern elderberry, elder-blow (USA, "blow" = blossom, related to German *Blume*), garden elder (USA), sweet elder (USA), West Indian elder (Barbados), and white elder.

However, determining what species are in North America turns out to be more complicated than is at first apparent. The most recent examination of the species is that by Bolli (1994) who made most American plants subspecies of *S. nigra.* Hence, Florida plants are *S. nigra* ssp. *canadensis.* Although Wunderlin (1998) uses *S. canadensis,* Stevens et al. (2001) adopt *S. nigra* ssp. *canadensis.* Plants in the southwestern United States and northern Mexico are *S. nigra* ssp. *cerulea* (Felger et al. 2001, Hodgson 2001). My discussion also follows Bolli (1994).

Elderberry, in this broad definition, ranges through the eastern woodland part of the United States and is disjunct across the Great Plains into western North America from Canada south through Mexico, Mesoamerica, and the West Indies, to Peru and Veneuela. In Spanish, this large shrub or small tree is *capulín silvestre* (wild cherry, New Mexico), *chacuaco* (maybe from *tzacualli,* shortcut, Náhuatl, New Mexico), *flor de sauz* (elder flower, New Mexico), *saúco* (Mexico to Panama), *saúco blanco* (white elder, Cuba), *saúco colorado* (red elder), *saúco extranjero* (foreign elder), or *tápiro* (Sonora and Arizona). French speakers say *sureau* [*blanc*] ([white] elder, Canada, France, French Antilles) or *fleur sureau* (elder flower, Haiti).

Numerous indigenous people in Mexico also name these plants. To the Aztecs, they were *xomatl* [*xumatl, xumetl, azumiatl*] (Náhuatl, Veracruz). That name is akin to *bixhumi* [*bahman, bajman*] (*bi*, fruit or leaf, *xhumi* from Náhautl *xomatl*, elderberry, Zapotec, Oaxaca). Elsewhere, they are *tsolos-che'* (*tsolos*, smooth, *che'*, tree, Maya). Tarahumara say *ñe ho* (Chihuahua). Although introduced in Yucatán, the Maya also call them *sak'ah-tsum* [*sacatsun*] (*sak'*, white, *ah-tsum*, male turkey). That name is given because they think the fleshy excrescence on the male turkey beak resembles the inflorescence. The remaining names appear to be simple terms, including *coyopa* (Zoque), *condumbo* [*cumdemba, cumdumba*] (Tarascan), *nttzirza* (Otomí), *shiiksh* (Mixe), and *yutnucate* (Mixtec).

We have records of the eastern ssp. *canadensis* being used by the Algonquin, Cherokee, Chickasaw, Choctaw, Creek, Dakota, Delaware, Houma, Illinois, Iroquois, Menomini, Meskwaki, Micmac, Mohegan, Ojibwa, Omaha-Ponca, Osage, Pawnee, Rappahannock, Sauk-Fox, and Seminoles. The western ssp. *cerulea* was used by the Cahuila, Chehalis, Clalla, Costanoan, Diegeño, Hiá-ced O'odham, Karok, Klallam, Klamath, Kawaiisu, Luiseño, Lummi, Maricopa, Mendocino, Mewuk, Miwok, Navajo, Okanagan-Coville, Paiute, River Pima, Pomo, Quinalt, Salish, Skagit, Skokomish, Squaxin, Thompson, Tohono O'odham, Yavapai, Yokut, Yuki, and Yurok (King 1984, Kay 1996, Moerman 1998, Hodgson 2001). The Dakota called elderberry *cahupta* or *chaputa-hu* (elder bush). To the Omaha-Ponca, they were *wagathahaskha* or *wagathahaskha-hi* (elder bush). *Skirariu* was the Pawnee name (Gilmore 1919). The Plains Cree called them *kehte'mina*. They are *hauku'usi* to the Mountain Pima, and *dahapdam* to the Akimel O'odham or Gila River Pima (Kay 1996, Rea 1997). To the Choctaw they are *bashan̲kchi*, and the Chickasaw say *basho̲'kchi'* [*basho̲'chi'*].

In Florida, these useful plants are known as *tá:bó:cî* (small elder, Mikasuki) and *coskîlpâ* [*echoskelepv, tcokiliba*] (*coskelepv*, Creek). Martin and Mauldin (2000) reported the names *łvło* [*rvro*] *em pvtake* (*łvło*, fish, *em pvtake*, its bed) and *hakkv hakv* (*hakkv*, spoon, *hakv*, resembling) for *Sambucus*. The first name referring to fish surely denotes an observation that fish like to lie in the shade of these streamside bushes. The second perhaps means that the wood was useful for carving spoons.

Sturtevant (1955) reported the elderberry being used by the Seminoles in a medicine to treat stomachaches, and as blowing tubes, toy blowguns, and scarcity food. The Iroquois treated fever with infusions made from elderberry bark or berries, and used them as a pain killer and anti-inflammatory (Vogel 1970). The Choctaw combined elderberry leaves with salt for a headache poultice (Bushnell 1909). The Creeks pounded roots, stirred them in hot water, and bound them on women who suffered from swollen breasts (Swanton 1928a). The Houma used them as an anti-inflammatory (Speck 1941), and bark tea was applied in difficult childbirth. The Menominis used dried flowers as an antipyretic (Smith 1928). Berries were considered official in the U.S. Pharmacopoeia from 1820 to 1831 and formed the base of a refrigerant and diuretic drink. Flowers were in the U.S. Pharmacopoeia from 1831 to 1905 and in the National Formulary from 1916 to 1947 (Millspaugh 1892, Kowalchik and Hylton 1987). Root decoctions were considered a cure-all in South Carolina (Morton 1974).

At least some of the active compounds leading to such widespread use of *Sambucus* as a medicine are the cyanogenic glycosides (Foster and Duke 1990, Buhrmester et al. 2000). They also contain astragalin, kaempferol, rutin, sambunigrin, and tannins (Kay 1996). Roots, bark, leaves, and unripe berries contain these toxins, and they are not safely consumed without preparation. In addition, the plants contain chemicals that are antiviral, at least to tobacco viruses (Chen et al. 2002). Other studies indicate the plants are not just poisonous, but truly medicinal (Chichon 2000, Seeram et al. 2001). Plants have anti-inflammatory, antibacterial, diuretic, diaphoretic, and laxative properties (Kay 1996). Duke et al. (2002) give it three plusses, safer than coffee.

Flowers are nontoxic, and eaten in pancakes, fritters, cooked in scrambled eggs, or made into teas and other beverages. Tea has been considered a mild stimulant, carminative, and diaphoretic, as well as good-tasting. Cooking detoxifies the glycosides and renders the cyanide harmless, so fruits are safely made into tarts, pies, jams, chutneys, preserves, and wine (Kowalchik and Hylton 1987). The Italian liquor *sambuca* is flavored with elder berries.

Additionally, the berries contain anthocyanins that have been used since at least the time of classical Greece for dyes (Johansen et al. 1991, Nakatani et al. 1995). Pan, the Greek god of shepherds, is reputed to have dyed his hair black with elderberries (Kowalchik and Hylton 1987). The berries also have been used since the Greeks and Romans to color wine (Baumann 1993). Adulterating wine got so bad in Portugal it became illegal to grow elderberries (Kowalchik and Hylton 1987). Leaves also have been included to impart a clear green tint to medicinal oils (Millspaugh 1892). The berries produce violet, blue, purple, lavender, blue-green, gray, brown, yellow-tan, and deep forest green shades for wool, baskets, or meat stamp dyes (Kowalchik and Hylton 1987, Tull 1999). Flowers have been added to perfumes (Millspaugh 1892).

Europeans and Americans formerly would have understood if you asked which tree was the "country medicine chest." Now, wholesale drainage and falling water tables in the United States have made the plants rare. Just at a time when herbal medicine is in a renaissance, some are realizing too late that it was worthwhile to have wild and wet areas with their animals and plants.

Samolus

(An ancient Latin name for plants used by Druids, said to refer to curative traits of this genus in diseases of cattle and swine; it may be Gaelic, from *sabh*, ointment or salve, or *samh*, fat, and *lus*, plant)

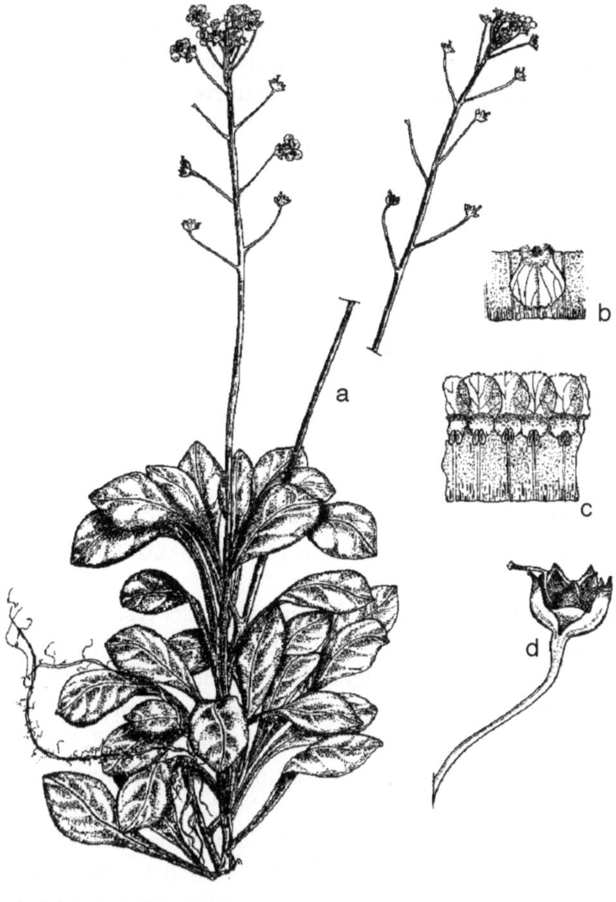

Samolus ebracteatus. a. Habit. b. Outer surface of flower to show reflexed petal. c. Corolla spread out, inner surface. d. Mature fruit with pistil attached to one segment. *Drawn by Vivian Frazier.* From Correll and Correll 1972.

brookweed (dates from ca. 1624, Europe)
Bunge (German)

Samolus ebracteatus (without bracts)
espuela de caballero (gentleman's spit, Cuba)
larger water-pimpernel [water pimpernel] (USA, Bahamas); pimpernel ("pimpernel" came from Medieval Latin *pipinella*, which perhaps came from *bipinella*, a diminuitive of *bipennis*, two-winged, apparently in reference to *Pimpinella*, Apiaceae)
limewater brookweed (Florida)

Samolus valerandi (presumably from Latin *valeo*, be strong, plus *-andi*, full of, or "full of strength")
alface dos rios (river lettuce, Portuguese)
[common] brookweed (Europe, USA); seaside brookweed
ehtiil ithim palats (like *Lobelia berlanderi*, Huastec, San Luis Potosí)
mouron-d'eau (water pimpernel, French)
palplina de agua (touching water, Spanish)
pineland pimpernel; water-pimpernel (Bahamas)
Salzbunge (salt pimpernel, German)
samolo (Italian); *strandsamel* (*strand*, seashore, *samel*, pimpernel, Norwegian)

This is another medical plant that Hogan (1978) found at the pre-Columbian Glades village at Fort Center. Hers, however, apparently is the only record of indigenous use of the plants in continental North America. Nearby in Cuba, *Samolus* is considered antiscorbutic and diaphoretic and is sometimes eaten in salads or as emergency food (Roig 1945). Although no use is given for *Samolus*, its Huastec name associates it with an edible plant, suggesting that it too was eaten (Alcorn 1984).

Sanguinaria

(Latin *sanguinarius*, bleeding, based on *sanguis*, blood, from the color of the juice)

Sanguinaria canadensis (of Canada)
achtuchwiminschi (*achtuch*, red, *minschi*, plant, Delaware)
bloodroot (originally applied by 1578 to a *Geranium* in Europe and then reapplied to this New World plant)
Blutwurz [*Blutwurzel*] (blood herb, German)
[red] Indian-paint
large-leaved sandwort
meskwaspen (Capt. John Smith and Strachey wrote in 1612 wrote of *musquaspenne*, "a root of the bigness of a finger, and as red as blood ... in drying, it will wither to almost nothing ... they use to paint their mats, targets, and such like"; Siebert 1975 reconstructed it to ancestral

Sanguinaria canadensis. From Britton and Brown 1897.

Proto-Algonquian *meçkwaxpenya*, "red root"); *mis'kodji'bik* (*mis'ko*, red, *dji'bik*, root, Ojibwa); *maskwa'we* [*mêskwa*] (red, Prairie Potawatomi); *mackwasdji'bîkûkûk* (*mackwas*, red, *dji'bîk*, root, *ûkûk*, plant, Potawatomi)

minigathe makan wau (*wau*, woman, *makan*, medicine, *minigathe*, seeking, Omaha-Ponca)

panson [pauson] (variant of "panse," a word from French *panser*, to take care of the sick or treat wounds)

peh-hishuji (*peh*, gourd, *hishuji*, to make red, probably referring to use for coloring gourds rattles, Winnebago); *shu-jee-hu* (*zhu-dse*, red, *hiu*, plant, Osage)

pishkak [*pishuk, pishaiyik*] (Choctaw; it is not clear that this name refers to a single species; may be *Sanguinaria, Lachnanthes*, and/or *Lithospermum*); *hahtok* (Chickasaw)

[red] puccoon [pochones, pochoon] (Strachey recorded the Powhatan in 1612 as *poughcone*, "red paint or dye," *pokcoons*, "a red dye," and *pocones*, "small root that grows in the mountains, which being dried and beat in powder turns red"; Siebert 1975 found cognates in Penobscot *pákahkan*, Mahican *pkáhkan*, and Nanticoke *pakahk*, all meaning "blood"); coonroot [coonroot, cornroot] ("coon" is an abbreviation of "puccoon;" and "corn" a corruption of that, West Virginia)

redroot (used by John Bartram, 1751, Maine)

sang-dragon (dragon blood, Quebec); *sanguinaire* (bleeding, Quebec)

snake-bite (New Hampshire)

sweet-slumber (Pennsylvania)

tetterwort ("tetter" from Old English *teter*, first used about A.D. 700; various skin diseases including ringworm, eczema, and herpes; "tetterwort" first applied to *Chelidonium majus* ca. 1400–1450; later to *Sanguinaria*, in the same family)

tumeric [tumerick] (used by John Bartram, 1751, Pennsylvania; the real tumeric is *Cucurma longa*, Zingiberaceae, and it too gives a red color)

white-puccoon (a compound name noting that this "puccoon" has white flowers, New York)

wi̱·ti· saka̱'ʔ [*siʔká̱, ske̱*] (*wi̱·ti*, root *saka̱'ʔ*, red, Catawba); *wayu'k̄ te'ʔ* (Catawba)

Probably the first published record of *Sanguinaria* was by Capt. John Smith in 1612, when he wrote that the Virginia Powhatan used it for dyes. Their red dye came from the "Pocones," which "is a small roote that groweth in the mountaines, which being dryed and beate in powder turneth red: and this they use for swellings, aches, annointing their joints, painting their heads and garments." Later people confused the origin of red dyes taken from *Sanguinaria* with *Lachnanthes*. Robert Beverley in 1705 wrote, "*Smith* talks of this *Puccoon*, as if it grew on the Mountains, whereas it is common to all the Plantations of the English" (Swanton 1946). Actually, Smith was writing about *Sanguinaria* while Beverley was describing *Lachnanthes* (which see).

William Strachey, also writing in [1612] 1953, said of the Virginia people: "They are generally of a cullour browne or rather tawny, which they cast themselves into with a kind of arsenick stone, like red patise or orphement, or rather red tempered oyntments of earth, and the juyce of certaine scrused rootes, when they came unto certaine yeares, and this they doe (keeping themselves still so smudged and besmeered) eyther for the custome of the countrye, or the better to defend them (since they goe most that naked) from the stinging of muskitoes, kinds of flies or biting gnatts . . . for which they daily anoint both face and bodyes all over . . . as can cast them into that stayne. . . . Their head and shoulders they paint oftenest, and those red, with the roote of pochone, brayed to powder, mixed with oyle of the walnutt, or bear's grease; this they hold in sommer doth check the heat, and in the winter armes them in some measure against the cold." Beverley added that the body painting also aided them against "all Lice, Fleas, and other troublesome Vermine." Various people used a mixture of roots of puccoon, wild angelica (*Angelica atropurpurea*), and bear oil to repel vermin and to retain skin moisture and reduce perspiration (Vogel 1970).

Perhaps the painting of their bodies led to the adoption of the term "redskins" that persists as the myth that the indigenous people of the New World had skins that color. Apparently, the first use of "redskins" dates from 1699 with "Ye firste Meetinge House was solid mayde to withstand ye wicked onslauts of ye Red Skins" (OED 1971). It is surely significant that forest tribes also used red face and body paint during war (Swanton 1946).

In Florida, the de Soto expedition found Timucua men near Tampa Bay using red body paint (Swanton 1946). Bartram ([1791] 1958) found the head, neck, and chest of the Creeks, Seminoles, and Cherokee painted with vermillion. James Adair found the Chickasaw similarly colored in 1775 and was told by the traders that the quantity of color was an estimate of a man's wealth (Swanton 1946).

Red face paint had special significance among the women. Capt. John Smith wrote, "at night where his lodging is appointed, they set a woman fresh painted red with Pocones and oile, to be his bedfellow" (Vogel 1970). The chaplain traveling with William Byrd's surveying party, drawing the line between North Carolina and Virginia in 1729, was bothered that these ladies, "colour'd with pochoon," visited the men (Vogel 1970). William Bartram ([1791] 1958) wrote that Creek women "never paint, except those of a particular class, when disposed to grant certain favors to the other sex." Among the Yuchi, only unmarried women who were willing to grant sexual favors painted their faces, and then they put "a circular spot in red, about one inch across, on each cheek" (Swanton 1946).

Puccoon, with various spellings, was most often identified as the plant source of the red coloring. While there were at least four "puccoons" (*Lachnanthes, Lithospermum, Phytolacca, Sanguinaria*), the one that has become most associated with the name is *Sanguinaria.*

The first scientific record of *Sanguinaria* was made by Frenchman Jacques Philippe Cornut (1626–1651) in his *Canadensium plantarum historia* (History of Canadian Plants) published in 1635. Cornut used the phrase name *Chelidonium majus canadense acaulon* (large stemless celandine from Canada). That name was subsequently repeated in lists of European cultivated plants including Robert Morrison's *Plantarum historia universalis Oxoniensis* (Universal History of Plants Cultivated at Oxon) published in 1669, and John Ray's *Historia plantarum* (History of Plants) appearing in 1686–1688. If either of these authors knew more about the plants than their curious blood-red roots, nothing was mentioned.

It is now known that *Sanguinaria* is a monotypic genus related to poppies (Papaveraceae) endemic to the deciduous forest region of the eastern United States (Mabberley 1997). The genus was created by Dillenius in 1732 in his *Hortus Elthamensis* and adopted by Linnaeus in 1753. Linnaeus simply emphasized the first locality he knew for the plants by adopting *canadensis* from Cornut's phrase name.

Indigenous people colored items other than their skins with *Sanguinaria.* The Natchez colored textiles red, and they and the Creeks colored baskets made of split cane (*Arundinaria*). Le Page Du Pratz noted red among the colors of cane mats among the Natchez in 1758. Mark Catesby (1731–1743) found people in Virginia and the Choctaw and Chickasaw making colored cane baskets. To the Chickasaw these were *oski' tannafo'* (*oski'*, cane, *tannafo'*, basket), while to the Alabama they were *iłanikolbi* (*iłani,* cane, *kolbi,* basket).

Le Moyne's illustrations of the Timucua show a variety of basket types, some containing red-stained gourds (Swanton 1946). Animal skins were colored red at Cofitachequi, and in Mississippi canoes were that color, among several others (Swanton 1946). Mark Catesby (1731–1743) said that the people of the Carolinas made garters and sashes of hair from buffalo (*Bison bison*) and raccoon (*Procyon lotor*), "which they dye black and red." Houma women on the lower Mississippi wore clothing colored red, yellow, and white (Swanton 1946).

As important as *Sanguinaria* was for coloring, it was more significant as medicine. There are records of the herbs being used medicinally among the Abenaki, Algonquin, Cherokee, Delaware, Iroquois, Malecite, Menomini, Meskwaki, Micmac, Mohegan, Ojibwa, Onondagas, Osages, Penobscot, Ponca, Potawatomi, Quapaws, Rappahannocks, Sauks, Shawnees, and Tuscaroras (Gilmore 1919, Densmore 1928, Smith 1933, Vogel 1970, Moerman 1998). These people used the root to treat asthma, bronchitis, burns, fevers, laryngitis, lung ailments, rheumatism, toothache, and warts, among other maladies (Foster and Duke 1990, Moerman 1998, Duke et al. 2002). They considered the extract an appetite stimulant, arterial sedative, and emetic.

Because the Virginia settlers were in infrequent contact with Europe, they relied on indigenous remedies, including bloodroot (Vogel 1970). Jonathan Carver traveling in the "Interior Parts of North America" between 1766 and 1768 called bloodroot, "a strong emetic, but a dangerous one." George H. Loskiel (in *History of the Mission of the United Brethern*) and physician Mannasah Cutler agreed. William Downey wrote a dissertation on puccoon or bloodroot to obtain his medical degree at the University of Pennsylvania; it was edited by physician B.S. Barton prior to 1808 (Vogel 1970). Self-proclaimed

"Indian" physician Robert D. Foster included bloodroot in his remedy for croup (Vogel 1970). John Goodale Briante who studied with the "St. Francis tribe of Indians, at Green Bay," with the Potawatamis, and other tribes used bloodroot in a medicine called *was-a-mos.* He claimed to cure cancer with the remedy in 1870.

Sanguinaria was official in the U.S. Pharmacopoeia between 1820 and 1926 and in the National Formulary from 1926 to 1965. It was considered an expectorant, emetic, tonic, and alterative (Vogel 1970). However, depending on the person, emesis is effected by doses from 30 mg to 1 g of root (Duke et al. 2002).

After intensive study in the 1970s and 1980s, the U.S. Food and Drug Administration declared *Sanguinaria* an unsafe herb (Duke 1985, Bennett et al. 1990, Kiger 1997). Duke et al. (2002) consider the herbs poisonous and recommend against their use.

In spite of these recommendations against *Sanguinaria,* the plant contains at least three chemicals with demonstrated bioactivity conforming to historical use by indigenous people (Lewis and Elvin-Lewis 1977). The isoquinoline alkaloid sanguinarine has demonstrated antibacterial activity that reduces caries. At least one modern brand of toothpaste contains sanguinarine. Historically, *Sanguinaria* was used in one of the most effective cancer remedies known (Lewis and Elvin-Lewis 1977). After learning of the use among indigenous people around Green Bay, Wisconsin, a cancer remedy was created by J.W. Fell in 1857. That treatment was tested in Middlesex Hospital in London. The Fell techniques have been associated with remission if not actual cures because two of the active compounds in the remedy are sanguinarine and chelerythrine. The third chemical, protopine, slows the rate of cardiac contraction (bradycardia) and has been applied for that purpose.

Because of the similarity of the sap to blood, there were also mystical beliefs associated with *Sanguinaria.* The Penobscots wore a necklace of bloodroot, believing that it prevented bleeding because of the red juice it contained (Vogel 1970). The Tuscaroras used *Sanguinaria* for divination (Vogel 1970). The Ponca put some of the root on their hands and contrived to shake hands with a woman they wanted. If successful, they believed that the woman would marry them within 5 or 6 days (Gilmore 1919).

Sitting on one of my bookshelves is a Cherokee basket dyed with walnut and redroot. The weavers used walnut (*Juglans nigra*) to produce chocolate brown and redroot (*Sanguinaria*) to create reds that have faded to rust. It is a pleasant reminder of colors of ancient usage and symbolically links me with my heritage of people using these dye plants.

Sanicula

(Name derived from Latin *sanare,* to heal, and *culus,* diminutive, alluding to the medical properties; the plants were formerly in great demand for healing wounds)

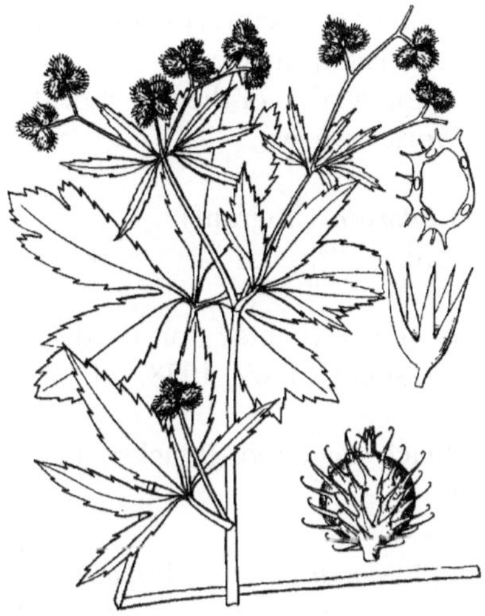

Sanicula canadensis. From Britton and Brown 1897.

bodan coille (*bodan,* penis, *coille,* woods, Gaelic)

erba frangolina (brittle herb, based on *frangere,* to break, an archaic form alluding to the brittle stems, Italian)

sanicle (from Latin *sanicula,* French); *sanicola* (Italian); *sanícula* (Portuguese, Spanish); *Sanikel* (German, Norwegian); *Saniket* (German); *senikel* (Dutch, in use by 1543)

self-heal (a name given to a variety of plants believed to have great healing properties, including *Prunella vulgaris,* Lamiaceae, *Sanicula europaea,* Apiaceae, and formerly to *Pimpinella saxifraga,* Apiaceae; the name began being used about 1387)

Sanicula canadensis (of Canada)

Canadian black snakeroot [blacksnakeroot] ("snakeroot" has been used as a name for medicinal plants since at least 1635, but originally applied to either *Polygala senega* or *Aristolochia serpentaria*; later applied to other plants)

mukkude'widji'bik (*mukkude'wi,* black, *dji'bik,* root, Ojibwa)

[Canada, Canadian] sanicle ("sanicle" from Old French *sanicle,* first used in the 12th or 13th centuries for *S. europaea*; the modifier distinguishes the New World species)

Sanicula marilandica (of Maryland)
[Maryland] black snakeroot [blacksnakeroot] (a name also given to *Cimicifuga racemosa*, Ranunculaceae)
pool-root [poolroot] (by analogy, a plant with a medicinal root growing beside a pool of water; first applied by 1890 to *Ageratina altissima* by John S. Billings in *The National Medical Dictionary*; later [or before?] applied to *Sanicula*)
[American, black] sanicle
self-heal (application of the Old World name to the New World species)

Sanicula odorata (fragrant)
cluster sanicle

Sanicula smallii (named for its discoverer, John Kunkel Small, 1869–1938, cf. Austin et al. 1987a)
Small's black snakeroot [blacksnakeroot] (a book name)

German physician Leonard Fuchs wrote in 1542 of what we now call *Sanicula europaea*: "It has been learned from experience that the juice of *Sanicula*, as a drink, remarkably benefits wounds. It is also agreed that the soaked herb, applied to tumors, stops them, or even the juice, rubbed on man or cattle. And, there is no better remedy when disease strikes throat or lungs. In short, *Sanicula* excels at everything *Symphytum* [Boraginaceae] does. It particularly helps the spitting of blood, dysentery, and kidney ailments" (Meyer et al. 1999). That was essentially the attitude of Europeans regarding *Sanicula* when they arrived in the New World.

In the Americas, the Europeans found numerous other kinds of *Sanicula*. Kartesz (1994) included 22 species for the United States, Canada, and Greenland. Fernald (1950) had 5 species in the northeastern states; 4 are in Florida (Wunderlin 1998).

There is a substantial problem in sorting out which of these plants might have been used by indigenous people. As Vogel (1970) pointed out in detail, a prevailing name for *Sanicula* is "snakeroot." That common name has been applied to so many genera in so many ways that we probably have information on but a small part of the historical accounts of *Sanicula*. For example, Romans ([1775] 1961) wrote that "snakeroots" were used in the South to treat malaria. Similarly, Greenlee (1944) wrote that Florida Seminoles boiled "snakeroots" in a pot and drank it for stomachache. *Sanicula marilandica* has been used to treat fever; *S. smallii* was used for stomach problems (Moerman 1998). Still, there is no way to know which of the half-dozen or so genera those "snakeroots" might be.

Moerman (1998) has sorted out the references that give enough information to be sure that they refer to *Sanicula*. All of the Florida species of *Sanicula* have been used by people within their range. The Ojibwa made a decoction of powdered *S. canadensis* root to regulate menses and to use after childbirth (Moerman 1998). The Houma made a root decoction for heart trouble (Speck 1940).

The Iroquois applied *S. marilandica* as an emetic to counteract poisons, to cure sore navels, for dropsy, as a laxative, and to treat venereal disease. The Malecite used an infusion of roots to correct irregular menstruation. They also thought *S. marilandica* was used by sorcerers for evil purposes. The Micmac used roots to regulate menses, to stop menstrual pain and rheumatism, to aid in parturition, for kidney trouble, and to treat snakebite. The Ojibwa treated fever with the roots and put pounded root on rattlesnake bite (Moerman 1998). Huron Smith wrote of the Ojibwa, "if this root be chewed it would cause eruptions on the epithelial lining of the mouth. They consider it a very potent remedy" (Coffey 1993).

Sanicula odorata was made into an infusion with spikenard (*Maianthemum racemosum*) to relieve back and side pain and kidney trouble in Malecite women. Menomini sorcerers used that species for evil. Meskwaki used it as an astringent and to stop nosebleed (Moerman 1998). The only known use of *S. smallii* was by the Cherokee who drank an infusion with pink lady's slipper (*Cypripedium acaule*) to treat stomach cramps and colic, and used the liquid as a liniment (Hamel and Chiltoskey 1975).

Several other allusions to indigenous uses are not confirmed in other sources. For example, George H. Loskiel, in *History of the Mission of the United Brethren* published in 1794, recorded that "Canadian sanicle" was used by "Indians" (Vogel 1970). André Michaux wrote on 18 August 1795 that a decoction of *S. marilandica* roots was a sovereign remedy for venereal diseases. Michaux did not state if he found indigenous people or American settlers using the plant (Vogel 1970). Porcher (1863) wrote of *S. marylandica*, "The Indians used it as we do sarsaparilla in syphilis, and also in diseases of the lungs." Millspaugh (1892) did not include the genus, and Duke et al. (2002) discussed only *S. europaea*.

Foster and Duke (1990) note that the leaves of *S. canadensis* contain allantoin as does *S. europaea*. That compound is apparently effective for treating bruises and inflammation (Hocking 1997). Presumably, that species and the others are likely to contain the chemicals known from the Old World plants. *Sanicula europaea* contains saponins, resins, and tannin-bitter substances (Hocking 1997).

Sapindus: Soap-Berry

(Latin *saponis*, soap, and *indicus*, from Indians)

Sapindus saponaria. From Sargent 1905.

Soap. We use it dozens of time each day, and never think about it. It is just there, along with other "essentials." It took the Old World lifestyle to invent the product. Soap requires alkaline bases, like wood ashes, and fat, originally from swine. Their mixture over heat and subsequent cooling allows them to harden into a lathering, cleaning material.

Soap has been a part of European lives for so long there is little variation in their words for it. Our first English records as *sápe* come from about A.D. 1000. Related Germanic languages like Dutch use *zeep*, German *Seife*, Norwegian *saapa*, and Finnish *saip(p)io*. The Romance languages are based on Latin *saponis* and *sapo* (first used by Pliny, A.D. 23–79). Soap is *sapone* in Italian, *savon* in French, *jabón* in Spanish, and *sabão* in Portuguese. Greek is *saponos*. One suggestion is that Europeans learned the words and soap-making process from the Tartars, and that may be correct. Still, all these words are related to Latin *sebum*, tallow. After all, there is no soap without oils and fats.

People in the New World had abundant wood ashes, but they were always short on fats. There were few native fatty animals in the Americas, and it was always a struggle to get enough in their diets to survive. Bears (*Ursus americanus*) and opossums (*Didelphis virginiana*) were among the few good sources of fats. To give an idea of the change in importance of fat sources, the Cherokee, before Europeans, called the opossum *si'-qua*. After the pig (*Sus scrofa*) was introduced, it became *si'-qua* and the opossum *si-quu:-tes-ts'* (smiling pig) (Mooney 1885–1886). A similar shift took place in Mikasuki, Choctaw (*shuhka*, pig, and *shukata*, from *shuhka*, opossum, and *hata*, white), and Creek (see also *Cyperus* and *Sagittaria*).

Indigenous Americans were aware of soapy plants. In Florida, certain trees were called soap-berry (Florida to Panama) by the Europeans. This tree, *Sapindus saponaria* (from Latin *saponis*, soap), was Indian soap. *Sapindus* grows from southern Florida through the Bahamas, the West Indies, and in Mexico (Sinaloa to Veracruz) south to Peru and Argentina. Depending on the definition used, there are either two other species or two variants of *S. saponaria*. A kind lacking winged rachises is *S. marginatus*; it grows in northern Florida and nearby adjacent states. The third is *S. drummondii*, or *S. saponaria* var. *drummondii*, which grows from Kansas and Louisiana to Arizona and northern Mexico. Ethnobotanically all may be considered together.

Gonzalo Fernández de Oviedo y Valdés was probably the first European to record these trees in 1535. He called the seeds *cuentas de xabón* (counting soap, Mexico). That name records two of the uses that Americans had for the seeds, counting, perhaps in beads (as necklaces and rosaries), and soap. The first idea persists today in *palo de cuentas* (counting tree, Oaxaca). Osage say it is *wa-no<u>n</u>-p'i<u>n</u>-hi* (*wa-no<u>n</u>-p'i<u>n</u>*, necklace, *hi*, tree). In Florida use in necklaces is recorded in *ahina:ka:sî* (bead tree, Mikasuki) and in Creek *tokona:wâ* (*eto*, tree, *konawv*, bead). The Biloxi of Louisiana said *aya'yiñk udi'* (*ayan*, tree, *yiñki'*, small, *udi'*, acorn).

The soapy trait was not confined to washing, but the saponins were also used to catch fish. Other people fishing with the seeds say *jequitinhaçu* [*jequitiguaçu*] (*ye'kei*, basket for fishing, *tinha*, false, *açu*, big, Tupí, Brazil). A Guaraní name *yequiti* (*ye'kei*, basket for fishing) is used in Argentina. Surely, *salta-martim* (jumping-kingfisher, Brazil) refers to catching fish. Is it possible that fishing is also the reference in *devanador* (*devanar*, to spool, wind on spool, Veracruz)?

However, people from the Aztecs to the Europeans noted the soapy traits for washing. To the Mexicans, seeds were *amolli* [*amole, yamolli, yamole*] (soap, Náhuatl). The Aztec word was so quickly adopted into Spanish that people were soon calling the fruits *amole de bolita* (soap balls, Mexico) or *amolillo* (little soapy one, Sonora). Any relationship between soap and the name *tehuistle* [*tehuztle, tehuixtle, tehuiztl, tehoizli*] (sharp rock, Náhuatl) is no longer apparent. The Zapotecs say *amole bibi* (*amole*, soap, Náhuatl, *bibi*, fruit, Oaxaca), *bibí* [*pibí, pipé, pipí, pipal, pepe, pepo*] (*bi*, fruit, Oaxaca), and *yaga bia* [*piaa*] (*yaga*, tree, *piaa*, fruit, Oaxaca). The Mixe use *pipe*, a loanword from their Zapotec neighbors.

While a variety of names are known for other people, translations are lacking. Many are perhaps simple terms. *Chorote* (Dominican Republic) may be from Taino. In Mexico and Central America trees are *güiril* [*guiril, gualulo, guayul, hualul, hualule, huaya, huayul, huiril*] (Oaxaca, Guatemala, Belize), *coyul* (Mexico), *cuyus* (Nicaragua), *choloco* [*cholulo*] (fruit), *jutuhui* (Guajíllo, Sonora), *ma-mu-jo* [*ma-mu-ho*] (Mexico), *morkowekuak* (Cuna), *pacón* [*pacun, pacum*] (Guatemala, El Salvador, Honduras), *snotpu'u* (Mexico), *tubchi* [*tupchi*] (Mayo, Sonora), *tzatzupu* (Mexico), and *ximbi'p* (Mexico). South Americans say *chumbimbo* [*chambimbe, chumbino*] (Colombia), *marikuru* (Arawak?), *michu* (Colombia), *jisotoúbo* (Bolivia), *para-para* [*paraparo*] (Venezuela), *quillay* (Mopori, Chile), and *sulluco* (Quechua, Peru).

Other Spanish words are more widely used. Most commonly used is *jaboncillo* [*jaboncelle, jaboncillal, jabonera*] (little soapy one, Sonora, Nuevo León, San Luis Potosí, Durango, Tamaulipas, Veracruz, Yucatán, Guatemala, Nicaragua, Costa Rica, Cuba, Puerto Rico, Panama, Colombia). In Argentina *Sapindus* is *palo jabón* (soap tree). Even the Maya mix Spanish and Maya as *jabón-ché* (soap tree, Mexico, Belize). Sometimes, people say *chirrión* (squeaky one, Mexico) or *luci* (*lucer*, to brighten, shine, Mexico), to note how clean things become.

In French colonies, people call the trees an *arbre savon* (soap tree, French Antilles), *bois de savon* (soap tree, French Antilles), *bois de savonette* (soap tree, French Antilles), *bois savonette* [*bwa savonette*] (soap tree, Haiti, French Antilles), *savonettier* (soap tree, Guadeloupe), or *savonier* (soap tree, Central America, USA). Sometimes, they use terms like *bois mausseux* (foaming tree, Guadeloupe), *savonette montagne* (wild soap, Guadeloupe), *savonette mousseuse* (foaming soap, Guadeloupe), *savonette pays* (wild soap, Haiti), or *savonnier des antilles* (Antilles soap, Lesser Antilles).

To Brazilians trees are *saboeiro* (soap tree), *saboneteiro* (soap tree, Pernambuco), or *sabãozinho* (little soap), and *fruta-de-sabão* are fruits. Sometimes they just say *sabonete* (soap, Ceará). Apparently, the Brazilian army had soldiers use the fruits, and that is why they are called *sabão de soldado* (soldier's soap) or *sabonete de soldado* (soldier's soap). Further indication of the wild source is *sabão de macaco* (monkey's soap).

In former Dutch colonies, the trees are *savonetapel* (table soap, Curaçao) or *sopo sirie* (soap seed, Sranan, Suriname). Germans call the tree *Seifenbaum* (soap-tree).

English names are surprisingly varied, probably indicating more uses than some other plants. The trees are Florida soapberry (USA), Mexican soapberry (Texas to Belize), mountain cherry (Belize), southern soapberry (USA), sumac-leaved soapberry (USA), winged [wing-leaf] soapberry (USA, Central America, South America), and West Indian soapberry (USA). A bit more imagination goes into the names soap-nuts (English) and soap-seed [soapseed, soap seed] tree (Belize), but not into soap-tree (Belize).

Some think they should be called false dogwood (USA), although the similarity escapes me. The comparison of seeds to marbles, in black nicker tree, makes more sense. Wild China-tree (USA) notes that the plants, fruits, and seeds resemble those in chinaberry, *Melia azedrach*. Several of the other comparisons are clear but some are obtuse. They are correctly *arbolio* (little tree, Sonora), and their seeds do resemble miniature balls for *boliche* (bowling, Sonora, Peru). Older mention of the black seeds is in *bois de merle* (blackbird tree, French Antilles) and *sibul* [*subul, zubul, sihom*] (*subul*, black bird, Maya, Yucatán, Belize).

Since the seeds are poisonous, one can see why they are *graine canique* (dog seed, Haiti) or maybe used as a *jurupe* (*jurar*, to swear, curse, Ecuador). Perhaps the same ideas are part of *mata de chivo* (goat bush, Dominican Republic), *mata muchacho* (boy killer, Sonora), *maté negro* [seeds] (black tea), and *palo amargo* (bitter tree, Dominican Republic). Surely, "seahone te" (Belize) is a corruption of *sabón té* (soap tea). Those last four names are surely associated with uses in teas as a febrifuge and tonic, and to treat kidney problems (Standley 1920–1926, Duke 1972, Morton 1981). Since the fruit contains up to 37% saponin, the names may also be connected with treating rheumatism, gout, tumors, and swellings (Takagi et al. 1980, Morton 1981). *Sapindus* has acetylated triterpene saponins (Kanchanapoom et al. 2001) that are known to be antimicrobial (Tamura et al. 2001) and that have even been used to detoxify snake venom (Castro et al. 1999). The compounds also inhibit tumor cells (Meyer-A. et al. 2001).

The bark is white, so that may have led to *palo blanco* (white tree, Chihuahua). Use of the wood may have given *zapotero* (shoe-maker). Comparisons no longer clear are *casita* (little house, Argentina, Paraguay), *chocho* [*choco*] (childish old man, Colombia), *limoncillo* (little lemon, Panama), and *palo de voladillo* (flying tree, Mexico). Maybe these names are associated with wood being used as fuel, treating wounds, or powdered seeds used as an insecticide (Standley 1920–1926, Duke 1972, Morton 1981, Dhar et al. 1996).

It is always surprising how we humans recycle ideas. Americans knew that *Sapindus* was a good soap for thousands of years. Still, in 2000, the idea was resurrected to win the Coca-Cola Award in

Ecoefficiency. A student at the Universidad Nacional Agraria La Molina, in Lima, Peru, took the prize with "*Obtencion de un detergente biodegradable y de un material de construccion a partir del boliche* (*Sapindus saponaria*)" (Extraction of a biodegradable detergent and construction material from *boliche*).

What is old is new.

Sarracenia

(Named by Linnaeus to commemorate Michel Sarrasin de l'Etang, 1659–1734, the Court Physician of Quebec)

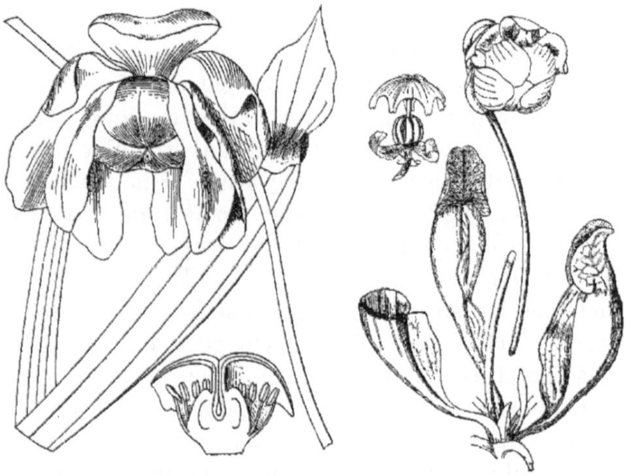

Sarracenia. *Sarracenia flava* (left). *Sarracenia purpurea* (right). Both from Britton and Brown 1897.

Sarracenia flava (yellow)
biscuits (probably a reference to the frisbee-shaped stigmas)
bonnet lily
[Eve's, water]-cup (tubular leaves hold water, and dead insects)
[umbrella-topped] flycatcher (used by Porcher 1863); flytrap
[trumpet, yellow-flowered] pitcher plant (first applied to *Nepenthes* of the Old World by 1835, and then to *Sarracenia* by 1857)
sidesaddle-flower (used by Porcher 1863)
[golden, yellow, southern] trumpet(s); trumpet-leaf [trumpet leaf] (in use by 1884); trumpet plant (apparently more recent than trumpet-leaf; not in the OED 1971); huntsman's horn
watches (see *S. purpurea* for derivation)

Sarracenia minor (smaller)
hooded pitcher plant
[spotted] side-saddle plant
smallpox plant
trumpet plant (Florida Panhandle)

Sarracenia purpurea (purple)
ayekitas [*ayikitas, ayikicas*] (little frog pants, Cree)
bleuets (little blue one, Quebec)
bog-bugle (South)
cochons de pelé (bald pigs, Quebec); *petits cochons* (little pigs, Quebec)
[Adam's, Eve's, fever, forefather's, huntsman's, Indian, meadow, water]-cup (Maine, Massachusetts, New Brunswick)
devil's-boot (South)
flytrap (South Carolina)
fox-glove [foxglove] (Maine)
frog-bonnet (South)
herbe-crapaud (toad herb, Quebec)
Indian jug [pipe, teakettle] (Newfoundland)
kokokoo'manasîn (owl's shoes, Potawatomi)
[Adam's, forefather's, Indian]-pitcher (Maine, Massachusetts, New Brunswick); [common, northern, purple] pitcher-plant
sabot ([horse] hoof, French)
sidesaddle-flower (Mark Catesby wrote in 1738 that "the under part of the flower is somewhat the seat of a side-saddle, from which in Virginia it has received its name")
skunk-cabbage (usually used for *Symplocarpous foetidus*, Araceae, Minnesota)
smallpox-plant (from medicinal use)
St. Jacob's-dipper (reported by Grace Greylock Niles in 1904, in the book *Bog-Trotting for Orchids*, Vermont)
ts'ëli tili (frog pail, Chipewyan, Canada)
[dumb] watches (reported by Grace Greylock Niles in 1904, "because children playing with the hard shells of the stigmas ... call them watches," New Jersey, Vermont)
whippoorwill's-boots [-shoes] (Maine, Pennsylvania)

The standard explanation for the dedication of the genus *Sarracenia* is that Michel Sarracin de l'Étang (1659–1734), physician at the Court of Quebec, sent the first specimens of at least *S. purpurea* to Europe (Fernald 1950, Quattrocchi 1999). However, Diggs et al. (1999) were not caught in that trap. They wrote simply that the name was dedicated to Sarracin. Indeed, a survey of the synonyms published by Linnaeus ([1753] 1957) makes it clear that both *S. flava* and *S. purpurea* were known in Europe before Sarracin was born. Plants now called *S. flava* were known to Matias de l'Obel before 1570, and those now called *S. purpurea* were discussed by Gaspar Bauhin in 1623.

Linnaeus ([1753] 1957) listed only those two species. Subsequently, more have been discovered to bring the total to eight, all endemic to eastern North

America (Mabberley 1997). The hybrids are difficult to recognize and people tend to confuse them with other species. Kartesz (1994) recognized 17 with formal names.

Although Hocking (1997) recorded use of three of the species in Florida, both Vogel (1970) and Moerman (1998) found only *S. purpurea* in the records of indigenous tribes. This sparse record perhaps reflects loss of information rather than lack of use. Regardless, *S. purpurea* was used by at least the Algonquin, Cherokee, Cree, Iroquois, Malecite, Menomini, Micmac, Montagnais, Ojibwa, Penobscot, and Potawatomi (Smith 1933, Core 1967, Vogel 1970, Moerman 1998).

Core (1967) noted that the people of the southern Appalachians, typically the Cherokee in his study, used the leaves as drinking cups when on trips. That same use was found among the Potawatomi, particularly when they were in the swamps where these plants grew (Smith 1933). Cree and Ojibwa children used the leaves as toy kettles.

The most widespread application of *S. purpurea* was as medicine. The Algonquin used it to ease the pain of childbirth and to treat urinary problems (diuretic). The Cree took it to remedy the sickness associated with absence of a menstrual period, to prevent sickness after childbirth, to help expel the afterbirth, to relieve lower back pain, and to treat venereal disease. The Iroquois used *S. purpurea* as "Basket Medicine," to alleviate thirst, chills, fever, liver problems and whooping cough, as a "love medicine," and as lacrosse medicine (presumably to make them play better). The Malecite, Micmac, and Penobscot took an infusion for tuberculosis and other pulmonary problems. The Menomini thought the plants were used by sorcerers. The Micmac used *S. purpurea* to relieve kidney problems and sore throat. The Montagnais treated smallpox with it. The Ojibwa and Potawatomi used the herb to ease the pain of childbirth (Smith 1933).

Although the indigenous people were well acquainted with the medical aspects of *S. purpurea*, settlers were not always privy to that information. John Josselyn wrote in 1672, "I wonder where the knowledge of this plant has slept all the while, i.e., above forty years" after he leaned about its properties. He continued, "It is excellent for all manner of fluxes."

Millspaugh (1892) was more guarded about pitcher-plants. He was involved in a debate about whether or not the plants were effective in treating smallpox. According to him, the people of northeastern Canada (Montagnais?) considered *S. purpurea* a specific in treating that malady. An incident is recounted where everyone was alarmed during an outbreak of smallpox in 1861. After the arrival of an old indigenous woman, "the alarm had greatly diminished," because she went among the ill and treated them with an infusion of *Sarracenia* roots.

A contrary opinion was proffered by another physician. He said that, "This medicine was thoroughly tested ... in the spring of 1864 at the Small-pox Hospital at Claremont, in Alexandria, Va., for a period of several weeks ... and proved to be without any curative powers in this disease." Offering an alternate view, Millspaugh (1892) quoted from another person who added that, not only did the plant save his brother's life, "but its use also appears to wholly obviate the unsightly pitting common to the disease." He then added, "if it is extracted and dried at the proper season."

Based on a study by Porcher of *S. flava* published in 1849, Millspaugh (1892) thought that *S. purpurea* caused "diuresis, moderate catharsis, and gastric excitation, as well an increased and irregular heart's action, and congestion of the head." His concluding remark was that an experiment in Europe "corroborates the aborigine's use of the drug, and adds one more proof to the many that are tending to reveal the fact that our American native practice was essentially correct." Hocking (1997) concluded that *S. purpurea* was useful as a "stimulant, tonic, laxative, in dysentery, diuretic, diaphoretic, smallpox, pain."

Porcher (1863) wrote of *S. flava*: "We are informed by several gentlemen of this state that these plants are used in dyspepsia with great service. The roots are undoubtedly possessed of bitter, tonic, and stomachic properties; and we are credibly assured of a number of cases in which relief has been experienced from them. The taste is disagreeable to those using them for the first time, but eventually it becomes pleasant, as we have ourselves experienced. An infusion might serve as a useful substitute for bitters."

He continues: "It is supposed by many to relieve most of the distressing symptoms of this affection [dyspepsia], among which may be cited: gastralgia, pyrosis, acidity, and the general feeling of malaise, so frequently attendant upon it. In some it induces considerable diuresis, and in others soreness of the mouth. In experiments made upon my own person, to ascertain its physiological effects upon a healthy individual, it exhibited a tonic, stimulating influence upon the digestive organs, producing some cerebral disturbance, when persisted in." Chemical examination by a colleague showed that "it contains besides lignin, coloring matter, and traces of a resinous body, an acid, or an acid salt, and also an astringent property, due neither to tannic nor gallic acid, and a salt of some alkaloid, related perhaps to cinchonia, which, should it prove new, may be called sarracenin. ... A bit of the fresh or dried root of either species may be chewed, and the juice swallowed, during the day before each meal; it may be given powdered in the form of pill, with a little rhubarb if necessary, or a tincture may be made by pouring a pint of brandy over several ounces of the root, of which half an ounce, diluted, may be taken three times a day. I have lately had cases reported to me, of its marked success in the relief of chronic diarrhoea and dysentery, and I am pleased to learn that it is now widely used in other portions of this state, and in Georgia, with very general approbation."

Hocking (1997) stated that *S. flava* had an astringent rhizome and that it had been used as a demulcent in dyspepsia, pyrosis, diarrhea, analgesic. A tincture is tonic, and stomachic in gastralgia. He found that it had been used as a food called "gator tator," and had been an ingredient in moonshine. Unfortunately, he did not give more details on the last two uses.

Hocking (1997) wrote that *S. minor* was formerly used to treat smallpox, skin rash, and eruptions. Because of its spots, he considered that an application of the Doctrine of Signatures. That philosophy certainly was not why people in the Panhandle used it as a remedy for coughs and indigestion, and to stop vomiting (Murphee 1965).

Modern chemical studies have confirmed that *Sarracenia* contains the alkaloid serracenine, plus serapine (anesthetic principle), anthocyanidins, beta-sitosterol and alpha-amyrin, the unusual enol diacetel monoterpene sarracenin, the triterpene betulinaldehyde, and volatiles (Miles et al. 1975, 1976, Bhattacharyya et al. 1976, Miles and Kokpol 1976, Hocking 1997, Newman et al. 2000, Sheridan and Griesbach 2001). Lupeol showed antitumor activity (Miles et al. 1974, Miles and Kokpol 1976). There is also the insect paralyzing agent coniine (Mody et al. 1976). Duke et al. (2002) wrote of *S. purpurea*, "If it really contains coniine ... it should be avoided."

Sassafras

(In naming this genus, C.G.D. Nees von Esenbeck, 1776–1858, is said to have applied an aboriginal name *sasafràs* used by the French in Florida)

Sassafras albidum (white)

ague tree (from its use in treating fevers, of malarial and other origins)

cahkanî [*cakani*] (Mikasuki); *chihkafo* [*ch‹hkafo*] (Alabama); *cihkafó* (Koasati)

cinnamon wood (Monardes first compared it to *Cinnamomum verum* and others in 1569)

Sassafras albidum. From Sargent 1905.

Fieberbaum (fever tree, German); *fiewerbaum* (Pennsylvania Dutch)

filé [*gumbo*] (powdered sassafras leaves used to thicken and season soups, stews, and gumbos; the past participle of French *filer*, to spin thread, from its effect when added to hot liquids, from Old French, Louisiana)

gehachgamápuek (Delaware); *pawquanemekysee* (Shawnee, cf. Edgar 1891)

hvsh (*akshih*, root, Choctaw; compare with *kombo ashish* and *kafe ashish* below)

kafe ashish (*kafe*, coffee, *akshih*, root, Choctaw, which incorporates the English loan-word; this is the name Bushnell 1909 gave for only the medicine with "The roots are boiled in water and the extract is drunk 'to thin the blood'")

kafi' (the same word is used for alien coffee, *Coffea*, although it is sometimes pronounced and spelled *kaafi'*, Chickasaw); *kafi* [*iti kafi*] (Choctaw; the same word, *kafi*, is used for *Diospyros*, which see)

kombo ashish (*kombo*, sassafras, *akshih*, root, Choctaw; the word "gumbo," as in French *filé gumbo*, the seasoning now attributed to the Acadians of Louisiana, is from Choctaw; this term was in use in 1909 when Bushnell worked among the people of Bayou Lacomb; however, Byington et al. 1915 did not include *kombo*; there is, however, the well-attested Chickasaw word *kombo'*)

ku'stapiʔ (Catawba)

kunstutsi' [*ka:-n'-s-ta-tsi', kúnstú'tsi, kansdaji*] (Cherokee; a gap in the Great Smoky range, about the head of Noland Creek, on the line between North Carolina and Sevier County, Tennessee was called *Kunstutsi'yi* "Sassafras place")

pauame (Timucua name given by Monardes in 1569)

salep [saloop] (archaically, a hot drink, sometimes used medicinally, made from salep, sassafras, or

similar aromatic herbs; derived from Ottoman Turkish *slep*, from Arabic *salab*, a kind of orchid)
sassafras [laurel] (Monardes in 1569 said the French Huguenot he met called the tree *sassafras*, and the Spanish from Florida corrupted that to *sassafragia*; maybe that is how the version originated that the name was from Latin *saxifragia*, a variant of *herba saxifraga*, a medicinal herb now applied to a completely different plant in the Saxifragaceae; however, even Monardes's story does not account for a totally distinct Timucua name, *pauame*, for the tree); sassafrax (fide Small 1933); saxifras (Strachey in 1612); white sassafras (this English name reflects an observation also made by the Seminoles; they recognized two "kinds" of sassafras; *cahkankitisci*, red sassafras, and *cahkansopatfi*, gray, or *cahkanhátki*, white; the root bark is red, and the stem bark gray, hence the distinction)
shi-kee (Osage; *she'-ke* is a rattlesnake, perhaps there is confusion)
smelling-stick
swanággerachs (it is not clear if this is an indigenous name, or an attempt by these people to pronounce "sassafras," Onondaga)
weso [*wéso, we:so, we-sho, wí:so, wi'su*] (Creek, Muskogee)
wìnachk [*wink'nachk*] (*wink'*, pleasant [taste], *wonáchk*, tree bark, Delaware); *winauk* (sticky tree, Powhatan, Virginia)

Sassafras is a genus of three species, one endemic to eastern North America, and two in China and Taiwan (Mabberley 1997, Peng 2000). *Sassafras* was discovered by the Spanish in Florida in 1528, although the OED (1971) gives no source for that information. Core (1967) was in error when he wrote that the tree was "discovered in 1602 by Bartholomew Gosnold." At least Monardes in 1569 (Frampton [1577] 1925), Harriot (1590), and Gerarde ([1597] 1975) preceded that 1602 date.

The first newcomers to the New World had no idea that this new tree would become such an economic resource for the Old World. Sassafras literally launched flotillas of ships to Florida, and made patrons and financiers in Europe incredibly wealthy. The demand was such in 1625 for the Virginia Company to "bestowe their moneyes ... upon two comodities onely, Tobacco and Sassafras matters of present profitt" (Vogel 1970).

As John Uri Lloyd wrote in the late 1800s, "this tree at one time created greater interest in the Old World than any other American product, not excepting tobacco" (Vogel 1970). As a medicine, sassafras eclipsed lignum vitae (*Guajacum sanctum*) and held that position for years. Harriot ([1590] 1972) reflected the opinion of the times when he wrote that sassafras was "farre better and of more uses than the wood which is called *Guaiacum*, or *Lignum Vitae*." The bark of sassafras was listed in the *Pharmacopoeia Londinensis* in 1618. Later, the dried root bark was in the U.S. Pharmacopoeia, 1820–1926, and the National Formulary, 1955–1965; sassafras oil was in the U.S. Pharmacopoeia, 1820–1955, and in the National Formulary, 1955–1965; and safrole was in the U.S. Pharmacopoeia 1905–1916 (Vogel 1970).

Authors of the Oxford English Dictionary (OED 1971) dismiss the story of sassafras discovery as medicine as told in 1569 by Nicholas Monardes (1493–1588). The OED says, "The Spanish writer Monardes ... regards the Sp[anish] name as adopted from Fr[ench], which seems unlikely." These bibliophiles were relying too much on etymology and too little on history. Vogel (1970) understood Monardes correctly. Monardes erred largely in believing that medicine from sassafras cured such a remarkable variety of maladies.

The history of a brief French presence in Florida is well known and documented (cf. Milanich and Milbrath 1989). Knowledge of French Huguenot René de Laudonnière's establishment at Fort Caroline on the River May (now the St. Johns River) reached Spain at the end of March 1565. That date was after Pedro Menéndez de Avilés's royal contract had been signed making him *adelantado* and requiring him to explore a Florida extending from the present Florida Panhandle around the Florida Keys thence north to Newfoundland. After establishing a settlement at St. Augustine in September 1565, he captured Fort Caroline. He then renamed that fort San Mateo because he captured it on St. Matthew's Day in 1565 (Bloodworth and Morris 1978). Next, he killed the French leader, Jean Ribaut, and many of his followers on the beaches of Matanzas, south of St. Augustine. However, some of the French were retained as captives.

Monardes's version follows that history particularly well. First, he told of a Frenchman teaching him about the tree and establishes the time. Monardes wrote in 1569: "It maie be three yeeres paste, that I had knowledge of this Tree, and a Frenche manne whiche had been in thos partes, shewed me a peece of it, and tolde me merveiles of his vertues. ... He tolde me that the Frenchemen, which had been in the Florida at that tyme, when thei came into those partes, thei had been sicke the moste of theim, of greevous and variable deseases, and that the Indians did shewe them this Tree, and the maner how thei should use it." That would make the year that Monardes met the Frenchman 1566, the year after Menéndez expelled the Huguenot colony.

Later Monardes continued, "After that the Frenche menne were destroied, our Spaniardes did beginne to waxe sicke, as the Frenche menne had dooen, and some whiche remaine of them, did shewe it [sassafras] to our Spaniardes." This is a highly synoptic version of the massacre that gave modern Florida the place name "Matanzas."

Still later Monardes noted that Pedro Menéndez (rendered almost unrecognizable as "Peter Mellendis" by Frampton [1577] 1925) "brought with hym in common this woode of the Sassafras." Monardes wrote that Menéndez saw many of his men healed with sassafras.

Monardes goes on to say, "The Tree groweth in some partes of the Florida, and doeth not grow in others." He specifically mentioned it growing at the port of St. Elen [San Helena, Leon County] and the port of St. Mathew [San Mateo, Duval County] "and not in others." Maps in Little (1978) and Wunderlin and Hanson (2002) confirm sassafras presence in both places. In Monardes's time, the interior of the peninsula was unknown and he had no way of knowing that the tree extended inland farther south.

According to Vogel (1970), sassafras "was used in some way by virtually all Indians living in its range." Moerman (1998) listed the following tribes as using sassafras: the Cherokee, Choctaw, Creek, Delaware, Houma, Iroquois, Koasati, Mohegan, Nanticoke, Rappahannock, Ojibwa, and Seminole. Other sources add Alabama, Catawba, Kickapoo, Osage, Powhatan, Quapaw, Carolina Algonquians, and Timucua (Frampton [1577] 1925, Harriot [1590] 1972, Vogel 1970, Howard 1984). Yanovsky (1936) said it was used by tribes in the eastern and southern states.

We know from Monardes that the Timucua "doeth use to put them [sassafras leaves] beaten or stamped upon bruses, or of any manne beaten with drie bloews, and beeying drie, thei are used in Medicinable thynges" (Frampton [1577] 1925). The Creeks boiled the roots with an unidentified grass (maybe *Andropogon*, which see) for patients with bowel and stomach pains accompanied by vomiting (Vogel 1970). Some of the Creeks also mixed sassafras with devil's shoestring (*Tephrosia virginiana*) to treat bladder problems and coughs (Swanton 1928a). The Seminoles visited by Greenlee in the 1940s used sassafras to treat coughing, gallstones, and bladder pain (Vogel 1970).

Sturtevant (1955) recorded the Miccosukees on the Big Cypress Reservation, well south of the natural range of sassafras, still using the trees. In part, their uses reflect their Creek heritage as some of the same maladies are recognized by their relatives (Swanton 1928a, Howard 1984). The Miccosukee used sassafras to treat "Cat Sickness" (nausea), "Cow Sickness" (chest pain, digestive upsets, diarrhea), "Dog Sickness" (appetite loss, drooling), "Horse Sickness" (nausea, constipation, blocked urination), "Monkey Sickness" (fever, itch, enlarged eyes), "Mythical Wolf Sickness," "Opossum Sickness" (appetite loss, drooling), "Otter Sickness" (diarrhea, vomiting), "Raccoon Sickness" (diarrhea), "Wolf Ghost Sickness" (diarrhea, painful defecation), and "Wolf Sickness" (vomiting, stomach pain, diarrhea, frequent urination). These are only the "major" uses and they had a number of other remedies involving sassafras (Sturtevant 1955).

Several other Muskogean tribes used sassafras. The Alabama made a tea of roots that was drunk to relieve stomach trouble or pain, used it against pneumonia, and considered it to be "good for the blood" (Howard 1984). Sometimes the Alabama made a gumbo soup to relieve the same problems (Howard 1984). Charles S. Sargent, in his *Manual of the Trees of North America* (1905), recorded that the Choctaw still were preparing powdered leaves to be sold for making *filé gumbo*. They were still adding those leaves to their soups when visited by Bushnell (1909). The Alabama boiled sassafras root with the head bones of a hog to make a dish they call *yakchi*.

The Houma used a decoction of fresh or dried roots to try curing measles and scarlet fever (Speck 1941). The Koasati made a poultice of mashed leaves to relieve bee stings (Taylor 1940).

The Choctaw made a tea "to thin the blood" (Bushnell 1909). A similar tea was used virtually throughout the range of the plants by indigenous tribes. In addition to the medicinal use, settlers of varying ethnic backgrounds quickly adopted the tea as a beverage or in more complex mixtures (Murphee 1965, Vogel 1970). Settlers added sassafras to pine, molasses, and spruce beer, and it was a critical component in root beer (Vogel 1970).

Eventually, researchers began looking at safrole, the best-known active ingredient in sassafras. Laboratory studies on mice showed that this compound, which is chemically related to myristicin and asarone, was both carcinogenic and hepatotoxic (Lewis and Elvin-Lewis 1977). Safrole was banned by the Federal Department of Agriculture (Foster and Duke 1990). Use of sassafras is prohibited by the European Union (Mabberley 1997). One report says that levels of 0.66 mg/kg may be hazardous to humans, and a cup of tea contains nearly five times that much (Newell et al. 1996). Like Duke et al. (2002), I prefer another perspective—sassafras tea is 1/13th as carcinogenic as a beer.

Saururus: Lizard's Tail

(From Greek *sauros*, lizard, *oura*, tail)

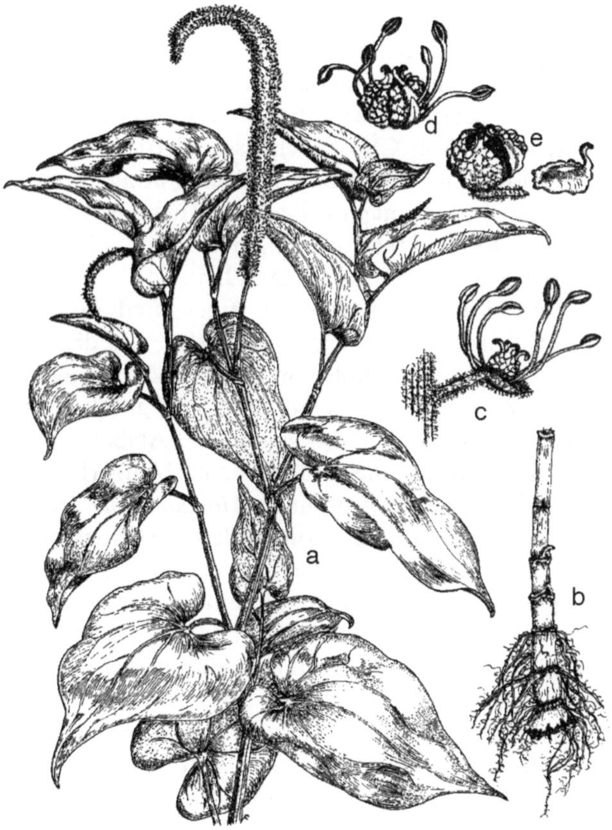

Saururus cernuus. a. Top portion of plant. b. Portion of lower stem showing adventitious roots. c. Flower, showing carpels united at the base only (no calyx or corolla). d. Fruit, with persistent stamens. e. Carpels breaking apart. *Drawn by Vivian Frazier*. From Correll and Correll 1972.

The odor is one of the things still fresh in my memory—apart from wondering if alligators really were dangerous—from the first time I waded thigh-deep into the Fakahatchee Strand in 1970. The large, green herbs in a pond gave off the scent of licorice and mint as I walked through them. All who have waded through swamps in Florida will have come across spots where there are acres of lizard's tail (*Saururus cernuus*). Even if they do not notice the rich green heart-shaped leaves, or the stems reaching waist high, most people, like me, note the plants. Once the stems and leaves are bruised by your passing, they release a pleasant spicy fragrance that demands attention. Yet, the species apparently remained unknown to Europeans until British botanist Leonard Plukenet published the first illustration in 1696 under the name *Serpentaria repens, floribus stamineis spicatis* (prostrate snakeroot, flowers with stamens in spikes). That was fully 200 years after Europeans arrived in the Americas.

Even then, professor of economic botany Peter Kalm did not mention the species from his visit to New England and nearby Canada in the 1740s (Kalm [1753–1761] 1972). Nor did that inveterate explorer of wetlands William Bartram remark on the plants in his book of travels through the southeast in the 1760s (Bartram [1791] 1958). Why this prominent wetland plant was so late in becoming known to Old World biologists remains something of a mystery. Given the importance of *Saururus* among the indigenous groups, it is tempting to speculate that these people withheld information on a sacred plant.

In 1739, Dutch botanist Jan Fredrik Gronovius's named the genus *Saururus*, based on a specimen that country clerk and amateur botanist John Clayton had sent from Virginia. If Gronovius had ever seen the living plants, he surely would have called the genus something else. Only dead, brown dried specimens are reminiscent of the namesake. Living plants have startling, if not spectacular, white flower stalks in the spring wetlands that contrast strongly with the dark green of the foliage. Linnaeus, who retained Gronovius's name for the plants, also gave *Saururus* the species name *cernuus* (nodding), to note a trait of the flower cluster.

There are only two species of *Saururus*, *S. cernuus* in North America, and *S. chinensis* in Asia (Mabberley 1997). The family (Saururaceae) contains only four genera, and like *Saururus* it is divided between southeastern Asia and southern North America. All are aromatic and medicinal.

In English the dominant common name has been lizard('s)-tail since about 1753. Other English names include water-dragon, swamp lily, black sarsparilla, and breastweed. There is some evidence that the last four are more recently applied. Spanish speakers call it *yerba mansa* (tame herb). The scientific name has simply been translated to create common names in German, *Molcheschwanz* (*Molch*, lizard, *schwach*, weak), and French, *saururé penché* (*saururé*, lizard, *penché*, nodding) or *queue de lézard*. However, there are indigenous American names for the plants, including *ishuna ignone* (*iskuna*, guts, *ik*, not, *nuna*, ripe, Choctaw), *oyihilíswa* (*owe*, widow, *heleswv*, medicine) in Creek, and *yahkakayikcî* in Mikasuki (widow or widower medicine).

Foliage and roots of *Saururus* were widely used by the Americans. For example, we know that people like the Cherokee of the Carolinas, the Choctaw of Louisiana, the Ojibwa of Minnesota, and the Seminoles of Florida all used the plants. On the other hand, it is not among the species enumerated as used by the people of the entire Missouri River region.

The Choctaw and other Americans used the plants in poultices for wounds, for inflamed breasts, and inflammations in general (Bushnell 1909). Treatment of skin problems was the most widespread type of use recorded, including treatment of spider bites. Tea was made of the whole plant to treat illness associated with fever and aching, rheumatism, and stomach ailments. Generally, the medicine from *Saururus* was considered good for pleurisy (inflammation of the plura or sacs around the lungs), rheumatism, as an emetic, an antispasmodic, an astringent, a sedative, an emollient, and as a mild tranquilizer.

In the 1790s French botanist André Michaux, living in Charleston, South Carolina, wrote that usage was common among residents. Furthermore, he recorded: "*Le Saururus cernuus est reconnu très bon remede por meurir les palyes qui viennent à suppuration et en diminuer l'inflammation. On fait bouillir les recines ou les broye, on y ajoute un peu de farine de froment pour en fair un cataplasme*" (*Saururus cernuus* is known as a very good remedy for sores and inflammations. The roots being ground or boiled, some wheat flour added, are then used as a poultice). That segment was quoted by Sturtevant (1955) from Michaux's journal of 1785 to 1796.

Not only is the Florida native medicinal, but members of this small family are used throughout the world. The Asian *S. chinensis* is also used to treat inflammation and as a sedative. Similarly, the related *Anemopsis* of the southwestern United States and adjacent Mexico has those same uses (Moore 1989, Kay 1996). Like the Florida plants, those in the west are called lizard's tail in English and *yerba mansa, yerba del mansa,* or *yerba el mansa* (tame herb or herb of the tame one) in Spanish. In addition to the similar uses, that species has a variety of names recorded by various people, including the Apache, Ópata, Pima, Mayo, Yaqui, Papai, and Seri.

It is probably safe to say about *Saururus,* as has been said about *Anemopsis,* that "everybody who has lived where *yerba mansa* grows has used it as a medicine" (Moore 1989). Widespread usage for poultices on external lesions and its emetic nature make the statement likely.

Possible infection from wounds was a constant problem before the discovery of modern antibiotics. For example, ignorance of ways to fight infections probably caused at least as many deaths during the Civil War as direct battles. Any plant that provided even marginal protection from this danger was sought by people throughout the world.

The aristololactam analogues cepharanone B and sauristolactam are not only antiseptic but also antitumor (Rao and Rao 1990). In addition, *S. cernuus* contains the antioxidant dehydrogeranylgeraniol and the sedative manassantin A (Rao et al. 1987, Rao and Reddy 1990, Rao et al. 1990, Rajbhandari et al. 2001). Each genus (*Anemopsis, Houttuynia, Saururus*) has its own characteristic essential oil, and these compounds inhibit several types of viruses, including herpes simplex, influenza, and human immunodeficiency (Childs and Cole 1965, Acharya and Chaubal 1968, Sanvordeker and Chaubal 1969, Tutupalli and Chaubal 1975, Tutupalli et al. 1975, Hayashi et al. 1995). The Asian *S. chinensis* also has been shown to fight infections, and provide liver protection because of its diastereomeric lignans, flavonoids, furanoditerpenes, saucernetilignans, and sesquilignans (Xu and Xu 1986, Sung et al. 2000, Sung and Kim 2000, Ahn et al. 2001, Sung et al. 2001, Ma and Ruan 2001, Hwang et al. 2002).

Emetics also were an essential part of the cultures of the people living in the southeastern United States, especially those whose languages were allied with Creek (Muskogean language group). There was a custom among all these groups of observing the *posketv* [*poskita*] (to fast, Creek). That event is best known as part of the festivities now called the Green Corn Ceremony. Anthropologist Charles Hudson (1976) has compared this ceremony to the combined festivals of Thanksgiving, New Year's, Yom Kippur, Lent, and Mardi Gras.

The Green Corn Ceremony was the single most important event of the year in all southeastern groups. Essential items for the ceremonies were herbal teas. Several of the teas consumed were mixtures, and included emetics. The black drink included *assí* (*Ilex vomitoria*), and it became the most famous. Often recorded as an additive to black drink was *hicí* or tobacco (*Nicotiana tabacum*). Among the other emetics used in the medicinal drinks were *pasa* or button snakeroot (*Eryngium yuccifolium*) and *oyihilíswa* (lizard's tail). One ritual and ceremonial medicinal tea called *ayikckanahki* (gathered medicine) contained up to 14 different plants.

Although the Muskogean people were accustomed to sharing their black drink and tobacco with visitors to the village square, the Green Corn Ceremony and its drinks were private affairs. Many secrets were kept about that event that they have not yet been shared with the non-Muskogean world. In surviving practitioners, medicines and their preparations were then, as now, important events that were passed only from teacher to apprentice. Surely, that accounts for at least some of the uneven reports of *Saururus* among eastern United States people.

The Green Corn Ceremony was the Muskogean time of renewal when old animosities were forgiven, criminals tried, individuals purged of evil, and new children named. Ceremonial beverages used by southeastern people during this time played important roles, and *Saururus* is prominent in several of them. In each case the beverage aided in cleansing and refreshing before starting again.

We in the United States borrowed part of that custom, reworked it, and use it in an almost unrecognizable modified form. We no longer use lizard's tail, but the Coca-Cola Company once advertised our replacement as the "pause that refreshes."

Schaefferia

(Jacquin named this for Jacob Christian Schaeffer, 1718–1790, a German mycologist and clergyman)

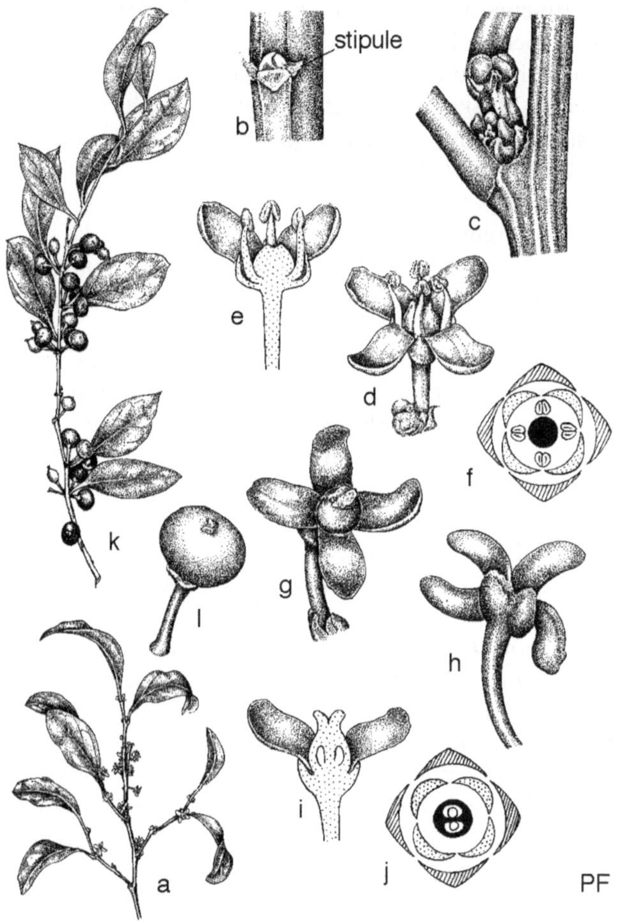

Schaefferia frutescens. a. Branch with pistillate flowers. b. Node, with leaf scar and bud with persistent lateral stipules. c. Node, with axillary branch cluster, stipules fallen. d. Staminate flower, front-side view. e. Staminate flower, longitudinally dissected. f. Floral diagram of staminate flower. g. Pistillate flower, from above. h. Pistillate flower, from below. i. Pistillate flower, longitudinally dissected. j. Floral diagram of pistillate flower. k. Branch with fruits. l. Fruits. *Drawn by Priscilla Fawcett.* From Correll and Correll 1982.

Schaefferia frutescens (shrubby)
amansa guapo (tamer of the wild, Cuba)
boi de Persia cimarrón (wild Persian tree, Cuba)
[Florida] boxwood (originally the Old World box-tree, *Buxus*, prized by wood-turners, wood-engravers, and used in the manufacture of mathematical and musical instruments, Florida, Puerto Rico, Dutch Antilles)
cabra [*cimarrona*] ([wild] goat, Dominican Republic)
cafecillo (little coffee, Cuba, Puerto Rico)
cambia voz (voice changer, Cuba)
[*bois*] *capable* (competent tree, Haiti)
fruta de paloma (dove fruit, Venezuela)
guairaie (the name resembles the Taino word *caguairin* for another Cuban tree)
jasmín (jasmine, Cuba)
jibá (also used for *Erithroxylum*, Taino, Puerto Rico)
limoncillo (little lemon, Colombia, Venezuela)
marguerite (a name usually applied to composites, such as *Melanthera*, Hispaniola)
[*petit*] *merisier* ([little] cherry, Guadeloupe, Martinique)
mije (an indigenous name, maybe Taino, Cuba)
palo de araña (spider tree, Dominican Republic)
petit bois blanc [*ti-bois blanc*] (little white tree, Haiti)
[*bois*] *petit garçon* (little waiter [tree], Haiti)
yellow-wood (Florida)

After collecting plants in the Caribbean, Jacquin described this genus and species in 1760. There are now 16 *Schaefferia* species known, all of them in tropical America (Mabberley 1997).

Roig (1945) found *Schaefferia* being sold in markets in Santiago de Cuba and Havana under the names *amansa guapo* and *cambia voz.* No one would tell him what it was being used for, but he speculated it was an aphrodisiac. Liogier (1974) confirmed Roig's suspicion. In addition, tea of the leaves is used in Haiti against flu and coughs (Liogier 1974).

The wood is used in construction (Liogier 1975). Little et al. (1974) described the wood as light brown to yellow, hard, moderately heavy, and fine textured. Under the names yellow-wood or boxwood it has served as a substitute for true boxwood (*Buxus*) in engraving.

Schizachyrium

(From Greek *schizo*, to split, and *achyron*, chaff or husk, alluding to the toothed lemma)

Schizachyrium scoparium. Drawn by Mary Wright Gill and Agnes Chase. From Hitchcock and Chase 1950.

Schizachyrium scoparium (broom-like)
[broom] beardgrass; broom grass [bluestem]; broom sedge [sage]
[little, pinehill] bluestem [little bluestem]
bunchgrass (this applies more to a growth form than a species)
prairie grass (indicators of prairie habitats)
wire grass (a name more often given to *Aristida*)

From at least 1753, one of the largest prairie grass genera was considered to be *Andropogon.* That genus received a number of species that were dominant plants in grasslands. Little bluestem remains one of the four most abundant and important grasses in the Great Plains, with *A. gerardii, Panicum virgatum,* and *Sorghastrum nutans* (Diggs et al. 1999). However, it was distinguished fairly late in history.

During the time André Michaux spent in South Carolina, he collected this grass. Subsequently, Michaux called it *A. scoparius* in his *Flora Boreali-Americana* (Flora of North America) published in 1803. Most authors retained it in *Andropogon,* but Christian G.D. Nees von Esenbeck (1776–1858), a German professor in Breslau, created a new genus for some of these grasses in 1829. He called the genus *Schizachyrium.* It was not until 1903, however, when John K. Small published the combination *S. scoparium* in his *Flora of the Southeastern United States.* Even then, most agrostologists continued using the older name (Hitchcock and Chase 1950).

As now understood, *Schizachyrium* is a genus of mostly tropical grasses with 60 species, particularly common in savannas. *Schizachyrium scoparium* grows from Quebec to Alberta and from Maine to Montana, south through Idaho, Arizona, and Florida; other varieties grow from Texas and Arizona to Coahuila, south to Nuevo León, and San Luis Potosí (Hitchcock and Chase 1950, Correll and Johnston 1970).

Comparatively few uses are recorded for grasses among indigenous American tribes. Most of those recorded have been used as thatch, and others are known as foods (e.g., *Zea, Zizania*) and medicines.

Schizachyrium scoparium was first reported among the Ozark Bluff-Dwellers of Arkansas by Gilmore (1931). Subsequently, Marie Standifer et al. (2003) found this grass among archaeological items in Arnold Research Cave, Missouri. People who used the Missouri area had padded their sandals with little bluestem. Presumably, they were to help them keep warm, just as they were by the historic people. The Lakota used the plants much like the prehistoric people in Missouri (Moerman 1998). These plains people rubbed the grasses into softness and used them as insulation in their moccasins during the winter. They also used *Andropogon gerardii* (big bluestem) and *Eryngium yuccifolium* for the same purpose (Standifer et al. 2003).

Other people had several uses for little bluestem. The Comanche made bundles of the stems for switches in the sweat lodge (Moerman 1998). The Comanche also burned the stems and applied the ashes to help cure syphilitic sores (Vogel 1970).

Scirpus

(Ancient Latin for the bulrush; the name used by Pliny, A.D. 23–79; a cognate word in Hebrew is *haber,* joined, united, as in weaving mats from the stems)

bull-rush (from Middle English *bulrysche* or *bolroysche,* from ca. 1440); club-rush
giunco (from Latin *juncus,* Italian); *jonc* (French); *junco* (Spanish)

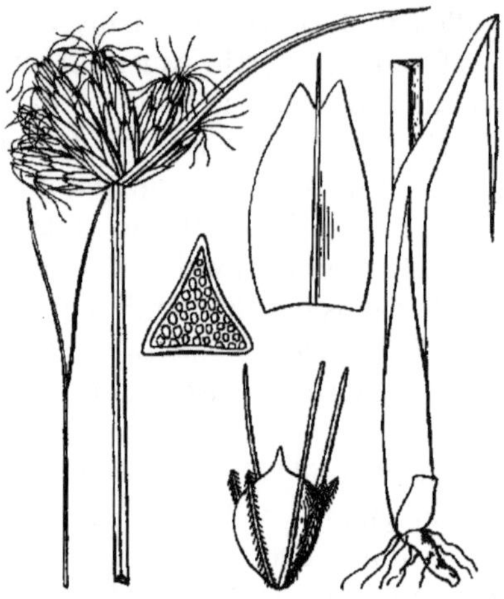

Scirpus americanus. From Britton and Brown 1896.

luachair (also means brightness, Gaelic)
meithean (Gaelic)
Simse (German)
souchet (refers to any sedge but mostly applied to *Cyperus* spp.; derived from *souche,* a stem or stock, itself from Gaulois *tsukka*, corresponding to German *Stock* (*st*- becomes *ts*- in Celtic; in use in French by 1375; cognates in Italian are *soc, socca*, and in Spanish *zoque*)

Scirpus americanus (American)
American bulrush

Scirpus tabernaemontani (named for German herbalist Jakob Theodor von Bergzaben, 1520–1590, who Latinized his name to Tabernaemontanus, "tavern in the mountain") (=*S. validus*)
ana'gûnûsk (mat weed, woven into mats, Potawatomi); *ana'kun* (Ojibwa)
[softstem] bulrush [bullrush] (USA)
chairmaker's rush; rush (perhaps from Germanic words, from Old English *risc* or *rix* and *rise* and *rixe*, Middle Dutch *risch*, and Middle Low German *risch(e), rysse, risk, ryse*, dating to about A.D. 725)
énea (copper colored, Spanish)
l'herbe a souris (mouse herb, Houma, Louisiana)
psa (Dakota); *sa-hi* (*hi*, plant, Omaha-Ponca)
rat grass (Louisiana)
sistat (Pawnee)
sword grass (descriptive of the culms)
tule (Spanish and English, from *tollin* or *tullin*, Náhuatl)

Siméon ([1885] 1981) wrote of the *tollin*, "*El* tollin *servía para hacer estreras o* petlatl, *lo que constituía, sobre todo en tierra de los toltecas, una industria importante*" (The rush served to make matting or *petlatl*, which constituted, through the whole land of the Toltecs, an important industry). Different species of these plants have been used for mats throughout the world (Hocking 1997). Moerman (1998) listed 12 species that have been used by indigenous people all over North America.

In North America perhaps the first report of prehistoric people using *Scirpus* was by Gilmore (1931). He identified *Scirpus tabernaemontani* that had been used to weave bags and mats at the Ozark Bluff-Dweller site in Arkansas. Later historic people continued weaving them.

Scirpus americanus was used by at least the northwestern groups including the Hanaksiala, Keres, Kwakiutl, Makah, Nitinaht, Pomo, and Salish (Moerman 1998). *Scirpus tabernaemontani* was used by the Cherokee, Cree, Dakota, Kawaiisu, Hopi, Iroquois, Lakota, Meskwaki, Menomini, Ojibwa, Omaha, Pawnee, Ponca, Potawatomi, and Ute (Gilmore 1919, Densmore 1928, Smith 1933, Yanovsky 1936, Moerman 1998). These people used both species for medicines, foods, and fibers. Yanovsky (1936) found records of *S. tabernaemontani* being used for food in California, Michigan, Minnesota, Missouri, Nebraska, New York, North Dakota, South Dakota, Montana, and Wisconsin. Roots were dried and pounded into flour that, mixed with pollen, was made into bread.

Seminole relatives the Koasati and Houma both used *Scirpus* (Taylor 1940, Speck 1941). Among the Houma *Scirpus* was used as a wash for weak legs, and given as a decoction to quiet nervous, fretful, and crying children (Speck 1941). They also made the dried plants into brooms.

Scoparia: Sweet-Broom

(Latin *scopa*, broom of twigs)

Sweet-broom, a widespread herb in the tropics, is commonly found as a weed near houses. It would go too far to say it was cultivated, but it is at least tolerated. The most common references to it are to use as a broom—stems are bundled and used to clean the area around and within homes. That this practice is old is reflected in the scientific as well as most of the common names.

This is *Scoparia dulcis,* the sweet broom. That name was applied by Linnaeus in 1753, but it was a shift from his earlier use. In his study of the private garden of George Clifford, published as *Hortus Cliffortianus* in 1738, Linnaeus called it *Capraria foliis ternis, corollis quadripartitis* (goatweed with leaves in [groups of] three and a four-parted corolla). He knew it from Jamaica and Curaçao, where its modern names

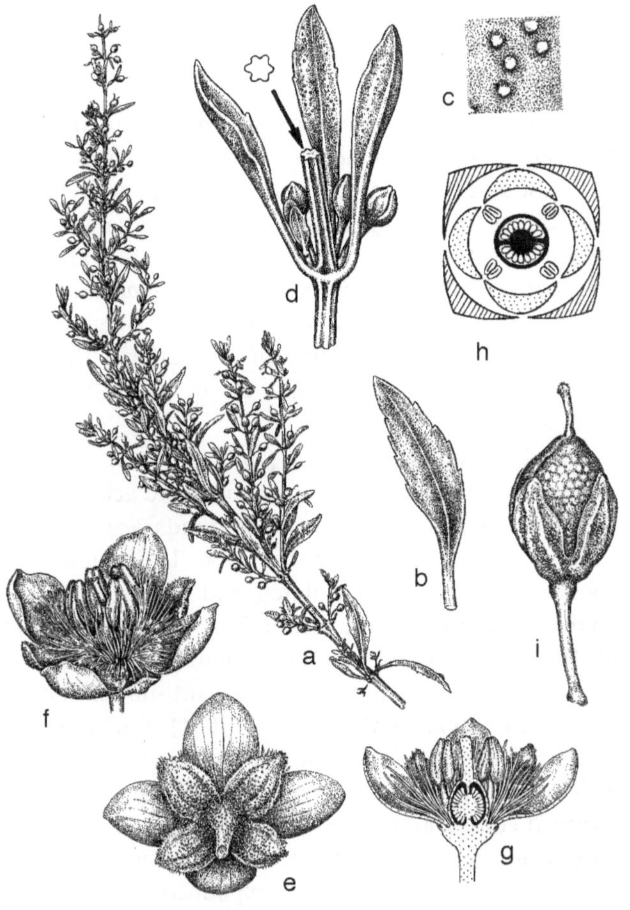

Scoparia dulcis. a. Fertile branch. b. Leaf. c. Section of leaf showing glands. d. Node. e. Flower, from below. f. Flower, front-side view. g. Flower, longitudinally dissected. h. Floral diagram. i. Fruit. *Drawn by Priscilla Fawcett*. From Correll and Correll 1982.

are sweet broom (Florida, Georgia, Louisiana, Texas, Jamaica), *bezemkruit* (broom weed, Dutch Antilles), and [*switi*] *sisibi-wiwiri* [*siebie wiwirie*] (*switi,* sweet, *sisibi,* sweep, *wiwiri,* leaf, Sranan, Suriname). In 1753, Linnaeus switched names and called another plant *Capraria,* and this one he based on a translation of its common name, sweet-broom.

Not everyone agrees that it should be called "sweet," but many agree on its use as a broom. In French, it is *balai doux* [*balié doux, balais doux, balé dou, balye du*] (sweet broom, Haiti, Dominica, Trinidad), *balai savane* (savanna broom, Guadeloupe, Martinique), *balié sauvaj* (wild broom, Haiti), or *herbe à balai* (broom herb, French Antilles). In Spanish, it becomes *escoba dulce* (sweet broom, Venezuela, Panama), *escobeta* (little broom, Central America), *escobilla* (little broom, Cuba, Honduras, Venezuela), *escobilla menudita* (tiny little broom Colombia), *escobo dulce* (sweet broom, Colombia), or *pichanga dulce* (sweet broom, Colombia). Even *miel de tierra* (earth honey, Argentina) is a reference to the sweetness, although not to the broom. In Nicaragua, it is *brum sirpi* (sweet broom, Miskito) and *brum tahplira* (sweet? broom, Miskito).

Those who do not think it sweet call it *escoba* [*escobita*] *amarga* (bitter [little] broom, Puerto Rico) or *escobilla amarga* (bitter little broom, Puerto Rico, Costa Rica, Panama, Venezuela). Sometimes it is *escoba de Castilla* (Spanish broom, Costa Rica), although it does not grow in Spain. People speaking Portuguese say *vassourinha* [*bassourinha*] (little broom, Amazonas, Brazil), *vassourinha de botão* (button broom, Brazil), or *vassourinha de varrer* (broom-branches, Minas Gerais, Brazil). Sometimes it is broomwort [sweet broomwort] (Texas) in English. All those names probably originally came from South America where it is *ñucñu-pichana* (sweet broom, Quechua, Peru), *piqui pichana* (sweet? broom, Peru), and *tupixaba* [*tupiçaba, tapixava, tapixapa, tapeiçaba, tupeiçaba, tupeicaba*] (broom, from Tupí *tui'xaba,* Minas Gerais, Brazil) (Pereira 1949).

Some of the local names are either untranslated or obscure. These include *bati matsoti* (Peru), *búñiga* [*buñiga*] (Hispaniola), *haraspata* (Miskito, Nicaragua), *kukibiu edan* (Arawak, Suriname), *mariquita* (also name for a beetle, Costa Rica), *paraguay* (presumably a word in Guaraní, Colombia), *shikiskiki tokon* (Carib, Suriname), *sirsaika* (Miskito, Nicaragua), and *sur'préjsì* (Paya, Nicaragua).

Two reported names are suspect (Pérez-Arbelaez 1978); both *azafran de la tierra* (wild saffron, Colombia) and *color* (roots used for coloring, Colombia) refer to use as a substitute for saffron (*Crocus sativus*). Those people who call the plants *chile de pajaro* (bird chile, Mexico) or *hierba de pajarito* (little bird's herb, Mexico) see something that is not obvious to the rest of us. The seeds are so tiny that they would barely be noticeable to all but the smallest seed-eating birds.

On the other hand, many of the comparisons are more easily understood, whether we agree or not. Because of its fragrance, one might compare it to anise (*Pimpinilla anisum*), as do people who call it aniseed [pine-ridge anise-seed] (Belize), anise-seed bush, or *anisillo* (Nicaragua). Others think it is more like coriander (*Coriandrum sativum* or *Eryngium foetidum*) and say *culantrillo* (little coriander, El Salvador, Belize), *culantro* (coriander, El Salvador), *culantro de pollo* (chicken's coriander, Central America), *culantro montés* (wild coriander, Central America), or *culantro-pim* (Belize). The same views apply to comparing it with *mastuèrzo* (nasturtium, *Tropaeolum majus,* Cuba, Hispaniola, Puerto Rico, Costa Rica).

The plants are small and herbaceous, but otherwise not much like *arrocillo* (little rice, Central America), rice weed (Nicaragua), or wild rice (Nicaragua). However, others make a similar comparison with *orozuz* (from Arawak, *oro çuç*, root of a legume, Puerto Rico) and *orozuz de pasto* (grass *orozuz*, Puerto Rico). Why *Scoparia* is compared to a *bruscón* (resembling *brusca*, medicinal herbaceous *Cassia*, Venezuela), *cancharagua* (*canchalagua*, a medicinal Gentianaceae, Hispaniola), or *verbena silvestre* (wild verbena) is obscure except that all are used as medicines.

Medicinal uses are common, and many of the names also carry that implication. In Mexico, it is *yerba del golpe* [*hierba del golfe*] (herb for bruises), and there is a widespread view that any bruise or other injury is cured by the plant. One of the common preparations is in teas. That product leads to the names licorice weed (USA, Puerto Rico), *té* (tea, Chiriguaná, Colombia), *thé savane* (savanna tea, Guadeloupe, Martinique), and *typychá kuratu* [*typycha caratu*] (probably Guaraní, in related Tupí, *tipi* is *Petivaria alliacea*; *chá*, tea, Paraguay). In "tea" it is applied as an abortifacient or contraceptive (with or without *Chenopodium ambrosioides*), antiemetic (with mother's milk for infants), antifilarial, aphrodisiac, in baths, to purify blood, for bronchitis, colic, coughs, diarrhea, earaches, as an emetic, to treat erysipelas (infected wounds), eye ailments, fevers, wounds, hemorrhoids, kidney disease, pain, in poultices for migraine, as a purgative, against snakebite in a plaster with salt, and to treat swelling, urine retention, and venereal diseases (Liogier 1974, Morton 1981, von Reis and Lipp 1982, Duke and Vásquez 1994, Gupta 1995a, Mors et al. 2001). According to the name teeth bush (British Antilles), the plant is used as a tooth cleaner. In Ecuador, it is named *tiatina* (Los Rios Province, Ecuador), presumably for Aunt Tina who may have been a *curandera*.

The plants contain alkaloids, salicylic acid, sisterol, hexacosanol, D-mannitol, tritriacontane, dulciol, scropanol, dulciolone, 6-methoxybenzoxazolione, betulinic acid, ifflaionic acid, scoparic acid A, scopadulcic acid A, scopadulcic acid B, and other compounds (Willaman and Schubert 1961, Chen and Chen 1976, Mahato et al. 1981, Morton 1981, Kawasaki et al. 1987, Hayashi et al. 1988c, 1990a, 1992). The whole plant has been shown to be a central nervous system depressant (Jiu 1966), so that may be why it has been used as a marihuana substitute (Duke and Vásquez 1994). Since the fresh or dried plants have been used to kill fleas and lice, and put in baths to kill vermin (Altschul 1973), smoking it is problematical and dangerous.

At one point it was thought that a compound called amellin from *Scoparia* was antidiabetic (Nath and Banerjee 1943, Whittaker 1948). However, a single subsequent study did not support that view and it was abandoned (Budde and Shriner 1950). Duke et al. (2002) suggested that the plants should be reexamined for that compound.

Extracts from *Scoparia* have even been used in Nicaragua to treat malaria. A study by Riel et al. (2002) showed that several compounds from the herbs have direct inhibitory activity against porcine gastric H(+), K(+)-ATPase. They further found that there was considerable activity against *Plasmodium falciparum* in the laboratory, and they suggested a mechanism for this inhibition of malaria through proton pump action (Asano et al. 1990, Hayashi et al. 1991).

Studies by Hayashi et al. (1988a, 1990b) and Riel et al. (2002) have shown that scopadulcic acid B from the herbs is antiviral. The compounds are effective against *Herpes simplex* virus type 1. In addition, compounds have been isolated and tested that were analgesic, diuretic, and anti-inflammatory (Freire et al. 1991, Ahmed et al. 2001). Scoparinol, a diterpene, showed significant analgesic and anti-inflammatory activity in animals (Ahmed et al. 2001). Furthermore, a sedative action of scoparinol was demonstrated with a significant effect on both onset and duration of sleep (Ahmed et al. 2001). There was also diuretic action by scoparinol. Freire et al. (1991) suggested that the analgesic activity might be explained by an anti-inflammatory activity of the triterpene glutinol.

Pereira-Martins et al. (1998) found a mutagenic effect of the flavone cirsitakaoside when it was tested in human blood cultures, but not when Swiss mice were treated orally. Cytotoxicity was also previously reported by Hayashi et al. (1988b). Riel et al. (2002) found antitumor activity in various human cell lines. Nishino et al. (1993) noted that the inhibition was due to scopadulcic acid B. Nishino et al. (1993) further indicated, "The potency ... proved to be stronger than that of other natural antitumor-promoting terpenoids, such as glycyrrhetinic acid."

Lans et al. (2000) found people in Trinidad and Tobago treating their dogs with herbs, including *Scoparia*. Leaves are used to control ectoparasites and to groom animals. Apparently, if an herb is good enough for people, it is all right to use on their dogs. Or, is it the other way around?

Many years ago, these herbs were taken to the Old World. Their common names *teh makao* (Macao tea) and *cha padang* (tea of open places) suggest that they were carried from China into the Malesian Peninsula. However, the Malay name for them distills sweet-broom's reputation in the New and Old World—*bunga baik salam* (good flower of peace).

Scrophularia

(Named this by Linnaeus because some species have fleshy knobs on the rhizomes; these were thought to cure "scrofula" and remove "fig-warts")

Braunwurz (brown herb, in use by 1542, German); *brunrot* (brown root, Norwegian)
escrofularia (from Latin *scrophularia,* Spanish, Portuguese); *scrofolaria* (Italian); *scrofularire* (French)
figwort (a name applied to a variety of plants; in use by the time Turner wrote his *Names of Herbes* in 1548; so called because it was thought to cure "fig-warts," a disease in humans resulting in a growth resembling figs or *Ficus*; applied both to piles and warts; the name of the disease was in use by the 1400s)
lus nan cnapan (knobbed plant, Gaelic)
water betony

Scrophularia marilandica (of Maryland)
brown-wort [brownwort] (in use as "broune wurte" by 1578)
bull-wort [bullwort] ("wort" is archaic English for "herb")
carpenter's square (because of the square stems); square stalk
[Maryland] figwort
heal-all (a name applied to several kinds of plants, including several mints)
Holme's weed
kernelwort
knotted root
Kropfwurz (swelling plant, German)
murrian grass
[great] pilewort (because of its used in treating piles or hemmorhoids; "pilewort" was applied to *Ranunculus ficaria* by the 1570s; later to *Scrophularia*)
scrofula plant; *Scrophelnpflanze* (scrofula plant, German)
stinking Christopher (an English dialectic name for *Scrophularia*; Millspaugh 1892 wrote, "All parts of the plant have a heavy, rank odor when bruised")

John Parkinson wrote in 1640 that a European species of *Scrophularia* was "effectual for the Kings Evil, or any other knots, kernels, bunches or wennes growing in the flesh wheresoever: it is of singular good use to be applied for the hemorrhoids or piles, when they grow painefull and fall down, and for other such knobbes or kernells as sometimes grow in and about the Fundament" (Coffey 1993). The name "Kings Evil" was a translation from Latin *regius morbus*, and from the 14th century A.D. it was thought in England and France to be curable by a king's touch. The same disease, called scrofula (from Latin *scrofularia*, a swelling or tumor), was tuberculosis of the lymph nodes, particularly in the neck.

Linnaeus described *S. marilandica* from plants John Clayton collected in Virginia, and those growing at the *Hortus Upsaliensis* at the university in Sweden. We now know that *S. marilandica* grows from southwestern Maine and southwestern Quebec to Minnesota, south to South Carolina, northern Georgia, Alabama, Louisiana, and Oklahoma (Fernald 1950). The species barely reaches Florida, being known only from Gadsden, Jackson, and Taylor Counties. Although the genus was popular with Europeans as a medicine, the indigenous people of the Americas apparently used it less. Only two tribes are recorded as employing the plants. The Iroquois gave an infusion to women with irregular menses. The Meskwaki used the plants in an undisclosed medicine (Moerman 1998). Elsewhere in North America, we know of only three other species used.

Porcher (1863) wrote tersely of *S. marilandica*, "It is vulnerary and soothing, when applied as a poultice to ulcers, burns, etc." On the other hand, Millspaugh (1892) thought that *S. nodosa* of Europe and the American *S. marilandica* were equal in effectiveness. He noted, "This herb is said to serve as a soothing poultice to inflamed tumors, suppurating mammae, ulcers, burns, hemorrhoids, etc.; it is also used alone and as a component of salves, for itch, various eruptions, and 'scabs' in swine, as well as a tonic and deobstructant in hepatic and glandular disorders."

More recently, Foster and Duke (1990) wrote that *S. marilandica* was applied in a poultice as a domestic cancer remedy, and used to allay restlessness, anxiety, and sleeplessness in pregnant women (Foster and Duke 1990). Hocking (1997) added that it was emmenagogue, alterative, and vulnerary (Hocking 1997).

Not much has been done on the chemistry of *Scrophularia.* Millspaugh (1892) reported scrophularin and scrophularosmin, a stearoptene in the plants. Lewis and Elvin-Lewis (1977) found that the genus contains cardiac glycosides, especially in *S. aquatica*, but poisoning is rare because of the unpleasant taste and odor. Bown (1995) found that *S. nodosa* contains aucubin, a mild laxative that increases excretion of uric acid and harpagide; it is recommended for skin problems. Bown (1995) and Duke et al. (2002) cautioned against use of the herbs by those with heart problems.

Scutellaria

(From Latin *scutella*, a small dish, in reference to the protuberance on the calyx)

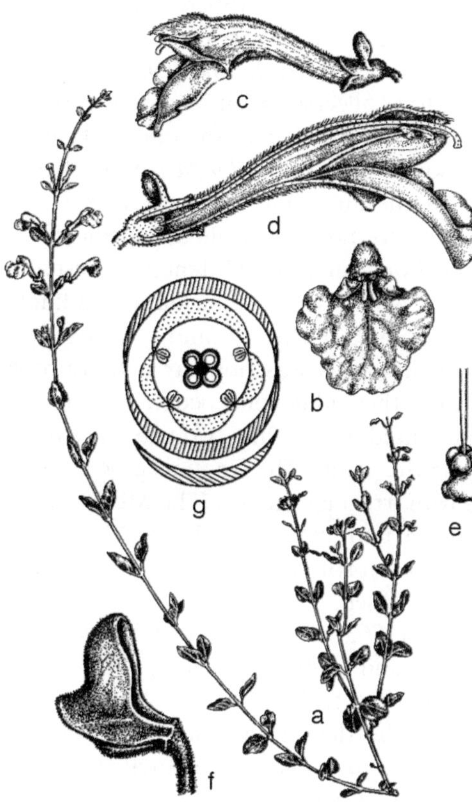

Scutellaria havanensis. a. Flowering stem. b. Flower, front view. c. Flower, side view. d. Flower, longitudinally dissected. e. Ovary. f. Fruit. g. Floral diagram. *Drawn by Priscilla Fawcett.* From Correll and Correll 1982.

Scutellaria havanensis (from Havana, Cuba)
escudo de la Habana (Havana shield, Cuba)
skullcap (in use by 1682 for any close-fitting cap; the helmetlike protuberance on the calyx prompted comparison by 1760, Florida, Bahamas)
toque de la Havane (Havana cap, Martinique)

Linnaeus ([1753] 1957) recognized 12 species in *Scutellaria* when he described the genus. He did not coin the generic name, but borrowed it from several authors before him. The genus has grown to about 350 species, and Mabberley (1997) considered it cosmopolitan. There are 13 species in Europe, and Kartesz (1993) admitted 44 for North America.

This herb is used in Cuba to treat spleen infections, edema, and *enasarca* (mange). Seeds are used against psoriasis, and sarcoptic mange (Roig 1945). Oddly, Morton (1981) did not include this species of *Scutellaria*; Ayensu (1981) did not even include the genus.

The Cherokees use three species of *Scutellaria* (Hamel and Chiltoskey 1975). Hairy scullcap (*S. elliptica*), hoary skullcap (*S. incana*), and blue skullcap (*S. lateriflora*) are considered abortifacient. They are used to relieve painful breasts and are mixed with other plants to expel the afterbirth. A mixture of other plants with roots of either species is also a kidney medicine. The Iroquois also use *S. lateriflora*, but for cleaning the throat (Moerman 1998). The Delaware use *S. incana*, and the Meskwaki *S. parvula*, so there was considerable application of the genus in the eastern United States.

In Jamaica, *S. viminea* has been boiled with ginger to treat colic (Morton 1981). Costa Rica's Bribri tribe use a decoction of *S. costaricana* as a febrifuge (Morton 1981). Presumably, the lack of data on Florida people using *Scutellaria* is an absence of information and not use.

Senna

(Philip Miller adapted the Arabic *sana* or *sanna* for the plants)

Senna ligustrina (resembling privet) (=*Cassia bahamensis*)
brusca (prompt, Hispaniola; surely a comment on use as a laxative)
canéfice bâtard (false cassia; *cassia* is an old Greek name, *kasia* or *kassia*, of Semitic origin, akin to Hebrew *qase'ah, qesi'ah*, or *quetsiah*, used for an aromatic and fragrant shrub, Guadeloupe, Martinique)
casse savane (savanna cassia, Guadeloupe, Martinique); *casse sene* (cassia senna, French Antilles)
Christmas flower (Bahamas)
dandelion (both have yellow flowers, Bahamas)
Gilead bush ("Balm of Gilead" [Jordan], from *Commifera*, is also medicinal, Bahamas)
piss-a-bed (Jamaica)
privet senna (Puerto Rico); privet wild sensitive plant (Florida)
sen [*de la tierra, del país*] (country senna, Cuba); *sénné zombi* (zombie senna, Guadeloupe, Martinique)
stink bush (Bahamas); stinking pea (Bahamas); *yerba hedionda* (stinking herb, Cuba)

Like many others in the genus, *Senna ligustrina* is a sometimes drastic purgative (Standley 1920–1926, Roig 1945). Surprisingly, this species is not among the poisonous species listed for Florida by Perkins and Payne (1978). Most, if not all, of the species owe their action to anthraquinones, although lectins and other compounds are known (Lewis and Elvin-Lewis 1977, Perkins and Payne 1978, Lampe and McCann 1985). The exact action on the body seems to be dose dependent.

In the Bahamas, leafy branches are boiled and the decoction taken for fever or stomach trouble. Branches are boiled with *Ambrosia hispida* and *Turnera ulmifolia* to treat menstrual cramps. Leaf juice is applied to hemorrhoids and used to draw out boils. Roots and leaves mixed with *Randia aculeata* help discharge afterbirth (Ayensu 1981). Cubans use the leaves as a purgative and to stimulate the bladder and uterus (Roig 1945). Haitians make remedies of the pods to treat chronic constipation (Morton 1981).

Several species occur in Florida, although most of them are introduced (Wunderlin 1998). The temperate species are covered in more detail by Moerman (1998).

Serenoa: Saw Palmetto

(Named by Sir Joseph Dalton Hooker for the Harvard botanist Sereno Watson, 1826–1892)

When I first saw the scientific name of the saw palmetto, *Serenoa repens*, I thought it unusually appropriate. The genus appeared to be based on "serene," and the species *repens* means reclining. How poetic it seemed for plants with their stems lying peacefully on the ground. Then, I learned that the genus was created to honor Sereno Watson, an introverted assistant to Asa Gray at Harvard University. So much for poetry.

The first European to mention saw palmetto was Spaniard Hernando Descalante Fontaneda ([1575] 1944). He called the shrubs both *palma* and *palmito*, and recorded how important they were to the people living in Florida. Quaker merchant Jonathan Dickinson called them a "shrubby palmetto" in the 1690s (Andrews and Andrews 1945). Historian Barcia (1565) and surveyor Romans ([1775] 1961) called the palm *Chamaerops*. They mistook *Serenoa* for a relative of the dwarf palm of the Mediterranean coast, *Chamaerops humilis*. It was William Bartram who coined the name "saw palmetto" that we use today (Bartram [1791] 1958).

Bartram also applied the Latin name *Corypha repens*, which became the basis for the modern name, *S. repens*. It was on St. Simon's Island, Glynn County, Georgia that Bartram first noted the plants, and he wrote of seeing "many curious vegetable productions, particularly Corypha palma (or great Cabbage Palm) [*Sabal palmetto*], Corypha pumila [*Sabal minor*], Corypha repens, frondibus expansis, flabelliformis, plicatis, stipit. spinosis (Dwarf Saw Palmetto)." In his *Travels*, Bartram ([1791] 1958) wrote that the berries were "delicious and nourishing food." Dickinson did not agree, at least on his first attempt, and wrote that the fruits were comparable to "nothing else, but rotten cheese steeped in tobacco" (Andrews and Andrews 1945). When Bartram was interviewed later about plants used by the Creeks, he admitted that the fruit "is a little bitterish and stinging on the palate, at first using it, but soon becomes familiar and desirable" (Bartram 1853).

These fruits were important to people throughout their range from North Carolina to Florida, west to southwestern Louisiana and southeastern Texas. Our first records of use in Florida come from a pre-Columbian Glades site on Lake Okeechobee (Hogan 1978). Hahn (1986) also found records of use among the Timucua and Apalachee. The Calusas with whom Fontaneda lived and the Jobé who met Dickinson ate them fresh. The Choctaw near the Mississippi River also dried the fruits for winter use (Romans [1775] 1961), as did their Siouan-speaking neighbors who called them *amashû'pka* (Ofo), *maxon̲tkxo'hi a'naki* (ancient palmetto fruit, Biloxi), or *ta'carahc* (Tunica). The Atakapa said *ma'kpel.*

Serenoa is *tala* (palm) to the Alabama and Choctaw. In related Muskogee it was *seyapho* [*seopho, shiopo*] (Simmons [1822] 1973, Bennett 1997). In a transfer resembling the one with the opossum and pig (see discussion in *Sagittaria*), the old name *seyapho* is now also applied to the date palm (*Phoenix dactylifera*) (Martin and Mauldin 2000).

Serenoa fruits are *siyá:pho intá:la* (*seyapho,* the fruits, *em+talv*, its palm, Creek) and *siyaphíntá:lî* [*sheopein taale, sheop'n taale, sheopo entaale*] (*sheope,* the fruits, *im,* its, *taale,* palm, Mikasuki). Older Seminoles still consume the fruits in a drink called *shiope sofkee* (palmetto gruel). Bennett and Hicklin (1998) consider this palm the most versatile plant in Florida, providing food, fiber, oil, medicine, wax, and roof thatch.

The Choctaws, Creeks, and Miccosukees used the leaves for making baskets (Bushnell 1909, Sturtevant 1955, Bennett and Hicklin 1998). Those baskets were used for a variety of purposes, including medicinal preparations. Leaves were also made into brushes, ropes, rattles, dance fans, fire fans, as punk for flint and steel fires, fish "drags," dolls, and tourist goods (Sturtevant 1955, Moerman 1998). Like most palms, the "heart" of saw palmetto is edible (Morton 1968b). There is, however, no documentation of native people in North America using palm hearts before Europeans arrived.

Fruits contain fatty oil, usually scarce in indigenous diets, with triglycerides including capric, caprylic, carioic, lauric, palmitic, and oleic acid (Bennett and Hicklin 1998). Berries also have phytosterols, their

Serenoa repens. Inflorescence, branchlet with fruits, and leaf (above). Habit (below). *Drawn by P.N. Honychurch.*

esters and glucosides, tannins, resins, and up to 28% sugars (Hocking 1997). American tribes used it is as a diuretic, sedative, and anti-inflammatory, for asthma, colds, coughs, chronic bronchitis, diarrhea, prostate problems, and migraine (Krochmal and Krochmal 1973, Foster and Duke 1990, Tyler 1993, Hocking 1997, Bennett and Hicklin 1998).

The species was listed in the U.S. National Formulary until 1952 (Bennett and Hicklin 1998). Then, in the 1960s, Europeans discovered that the fruits of *Serenoa* showed estrogenic activity, which was attributed to the high content of beta-sitosterol (e.g., Kloss 1966, Elghamry and Haensel 1969).

Later, it was rediscovered that benign prostatic hyperplasia (BPH) could be treated with phytosterols, particularly beta-sitosterol (e.g., Wagner and Flachsbarth 1981, Tarayre et al. 1983, Hiermann 1989). This discovery started a debate that remains unresolved.

In the past two or so decades literally dozens of papers have been written about "tests" of the efficacy of saw palmetto extracts. A small sample includes papers by Flamm et al. (1979), Carbin et al. (1990), Breu et al. (1992), Romics et al. (1993), Vahlensieck et al. (1993, 1996), Schneider et al. (1995), Gutierrez et al. (1996a,b), Kondas et al. (1996, 1997), Weisser et al. (1996, 1997), Bartsch et al. (1998), Ondrizek et al. (1999), Schilcher (1999), Wagner (1999), Medvedev et al. 2000), Sokeland (2000), Koch (2001), Lowe (2001), Dreikorn (2002), Ernst (2002), and Lowe and Fagelman (2002).

A recurrent theme among those against the use of saw palmetto is that there have been "no double-blind studies" (e.g., Lowe 2001, Ernst 2002, Lowe and Fagelman 2002). However, there are at least four studies using that approach that were not mentioned by these critics (Carbin et al. 1990, Weisser et al. 1997, Sokeland 2000, Prager et al. 2002). Another complaint is that past studies have not been long enough. Some argue that tests should be at least 6 months long to "show" if the phytomedicine is effective (Lowe 2001). That argument should have been silenced years ago, because Romics et al. (1993), Kondas et al. (1996, 1997), and Sokeland and Albrecht (1997) did studies longer than 6 months in duration. Moreover, Sokeland and Albrecht (1997) not only did a 1-year study, but it was also double blind.

Germans have pioneered both chemical and clinical research on saw palmetto and BPH. The impression gained by reading the literature on this topic is that some physicians are not willing to believe research done outside their own country. Surely, there is abundant healthy skepticism in the slowness of physicians to accept the possibility of the effectiveness of *Serenoa* extracts, but there is also an aura of sovereignty in their criticisms. Could it also be a remnant belief that only compounds synthesized by humans can be effective?

A major complication to accepting the idea that a plant extract can be effective is that no one understands just what happens in BPH. Everyone knows that there is tissue swelling, and that anti-inflammatory compounds help, but it is not that simple. There is no correlation between the amount of swelling and the magnitude of urinary retention. Several scenarios have been suggested for the mechanism and for treatment, but no one is yet sure which is correct. Worse, they may all be involved.

The standard approach by U.S. urologists is invasive surgery. Some doctors will attempt one of the three routes of medication, using the "party standards" alfuzosin and terazosin, while others will attempt the alternative synthetic drug finasteride (Wilde and Goa 1999). These drugs have different mechanisms of action, with the former an alpha-adrenoceptor blocker, and the latter inhibiting 5-alpha-(testosteron)-reductase (Schilcher 1998). Because phytomedicines are cheaper and as effective, and have none of the adverse side effects, many German physicians prescribe them (Schilcher 1998).

The debate over whether or not saw palmetto phytosterols are effective is not likely to end soon. Still, I have to point out a datum that no one has yet mentioned. Over two centuries ago Bartram ([1791] 1958) noted that bears (*Ursus americanus*) also ate saw palmetto berries. Has anyone ever heard of a bear having BPH?

What is not apparent to the physicians, or to their patients, is that the environment in the range of *Serenoa repens* is being ravaged in the public demand for this phytomedicine (Bennett and Hicklin 1998). Each year in Florida both state and national park officers are stopping more trespassers who are poaching saw palmetto fruits. No one knows the impact of continued harvesting of the fruits on the population dynamics of this formerly common plant. In some areas, the palmettos are now uncommon. At first, cattle farmers were delighted that there might be a reduction of these "weeds" in their pastures. Then, in years when prices are high, some have even excluded harvesters and gathered fruits from their own property.

In the United States, a country with an aging population, individuals rarely know or care about debates over effectiveness, much less habitat devastation. About 25% of men between 40 and 50 are afflicted with BPH, and nearly 80% older than 70 have it (Schilcher 1998). Most men suffering from the problem simply want relief. If saw palmetto works, and is cheaper than prescription drugs, they will use it.

If saw palmetto works for bears, why not people? Some of my friends say I am an "old bear," and it works for me.

Sesuvium

(Perhaps derived from Latin *sesuvii*, a Gallic tribe west of the Seine)

Sesuvium portulacastrum (with flowers resembling *Portulaca*)

azucena de mar (sea *azucena*; usually *azucena* is the tuberose, *Polianthes tuberosa*, Agavaceae, Puerto Rico)

banana di rif (reed banana, Aruba, Bonaire, Curaçao); *platanillo* (little banana, Dominican Republic)

bay flowers (Virgin Islands)

Sesuvium portulacastrum*. Drawn by P.N. Honychurch.*

beldroega de praia (beach purslane, Brazil); *pourpier bord de mer* [*la mer*] (seaside purslane, Haiti, Guadeloupe, Martinique); sea purslane [sea pusley, seaside purslane] (Florida to Panama); *verdolaga de la playa* (beach purslane, Cuba, Puerto Rico, Guatemala); *verdolaga rosado* (pink purslane, Puerto Rico, Costa Rica); *verdolaguilla* (little purslane, Dominican Republic)
camburito (little banana, Venezuela)
ceniceinta [*cenicilla*] (little ash-colored one)
conejera (mother of rabbits); *herba* [*yerba*] *de conejo* (rabbit herb, Venezuela)
coupe pieds (literally "foot-cutter," unexplained, French Antilles); *coupier* ("foot-cutter," French Antilles)
seaside samphire (Barbados)
tsaycan (viper's fang, because of the leaf shape, Maya, Guatemala)
vidrillo (little glass); *vidrio* (glass, Venezuela); *yerba* [*hierba*] *de vidrio* (glass herb, Puerto Rico, Venezuela)

Linnaeus at first called these plants *Portulaca portulacastrum* in 1753. Then, realizing his error, he created the name *Sesuvium* in 1759. There are now 12 species in warm parts of the world (Mabberley 1997).

Although these beach plants were apparently used more commonly in the past, few now seem to apply them. In Cuba, the stems and leaves are considered emollient, resolutive, refreshing, and were used against scurvy. On both Cuba and Hispaniola, the seeds are said to be a vermifuge (Roig 1945, Liogier 1974), although Morton (1968b) said they were a famine food. Plant stems and leaves are edible raw or pickled, are antiscorbutic, and salty, although in Haiti, the juice also is given as a vermifuge (Morton 1981). In Hispaniola, the plant is edible, and considered emollient and fortifying (Liogier 1974). *Sesuvium* is cultivated as a vegetable in Asia (Morton 1968, Hocking 1997). The common names suggest that sea purslane has been used as rabbit food.

In Argentina, a decoction is given for bladder, liver, and kidney problems. Hot poultices are used for chest congestion, and cold poultices plus a plant decoction are put on burns (Morton 1981). *Sesuvium* is one of richest sources of ecdysterone; it also contains alpha-ecdysone, a rare insect-molting hormone, and sesuvin (flavonol glycoside) (Morton 1981). In Asia, sea purslane is considered hemostatic, and used for wounds, especially for the stings of toxic fishes (Hocking 1997).

Sideroxylon: Mastic and Bustic

(From Greek *sidero*, iron, and *xylon*, wood, a reference to the hard wood)

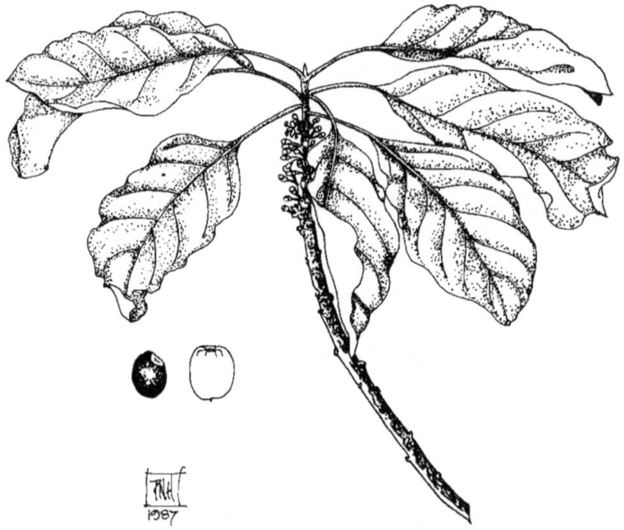

Sideroxylon foetidissimum*. Drawn by P.N. Honychurch.*

Spaniards arriving in the New World brought with them the word *almástiga* (from Arabic *al+mastika*). That became "mastic" in English, and both referred to the Old World tree *Pistacia lentiscus*, a relative of cashews (*Anacardium occidentale*), mangos (*Mangifera indica*), and poison ivy (*Toxicodendron radicans*). Both words are cognate with Latin *mastichus* and Greek *mastikon*, to chew, from a use of the sap as chewing gum. Mastic sap also has other applications; it is used as glue, as a protective coating, for coloring, and in an alcoholic liqueur flavored with mastic sap and anise seed (Austin 1981).

Europeans found the people in the Caribbean using a similar sap. Today these American trees are *Sideroxylon foetidissimum,* a name applied by Nicolaus Jacquin in 1760. The species is *foetidissimum* (most offensive smelling) because of the foul odor of the flowers. Walk downwind of a tree in flower and you will think that something is dead and decaying.

However, by the 1760s the American plants had usurped the Old World names, and were "mastic" (Florida to Panama) and *almástiga* (Spanish). Bernard Romans ([1775] 1961), who surveyed coastal Florida in the 1760s, listed the trees near today's Vero Beach. Romans's use was a Middle English spelling "mastick," but the name was firmly entrenched.

Today people call these trees Barbados mastic (Barbados), bully-mastic [mastic-bully] (Bahamas, Jamaica, Puerto Rico, Virgin Islands), false mastic (Florida, Puerto Rico), mastic jungle-plum (Florida), mastic-wood (Jamaica, Virgin Islands), mastwood (Virgin Islands, Antigua, St. Vincent), and wild-mastic (Florida).

In the Caribbean, the Taino people called the "mastic" *jocuma* [*acoma, acomat, coma, jacuma, jocume, tocuma*] (Cuba, Haiti, Dominican Republic, Guadeloupe, Trinidad). Because that language died along with most of these people, spelling variants occur in literature from the 1500s and 1600s. Moreover, people speaking modern languages have their own variations. People now call the trees *jocuma amarilla* (yellow *jocuma,* Cuba), *jocuma blanca* (white *jocuma,* Cuba), *jocuma lechera* (milky *jocuma,* Cuba), or *jocuma prieta* (black *jocuma,* Cuba). Sometimes *jocuma amarilla* is rendered *tocuma amarillo* (yellow *jocuma*). In French, they are *acoma bâtard* (false *jocuma,* Guadeloupe), *acoma franc* (free *jocuma,* Guadeloupe), *acouma jaune* (yellow *jocuma,* French Antilles), or *caimite acoma* (*jocuma* resembling *Chrysophyllym cainito,* Dominica). The name in the Dominican Republic has even become *caya* [*calla*] *amarilla* (yellow *caya,* another Taino name for the tree), *caya blanca* (white *jocuma*), and *caya prieta* (black *jocuma*).

Mastic is one of the few plants in the southeastern United States with indigenous names having cognates in the Caribbean (see also discussions under *Ipomoea:* Moonvine and *Smilax*: Green-Brier). These trees are known to the Seminoles as *okó:ma* (Creek) and *okó: mî* (Mikasuki). Surely, those words are etymologically related to *jocuma.*

Hard wood from these trees was prized by historic and prehistoric people. Glades groups made domestic and war tools from the trees (Austin 1980, 1981, 1997). The Seminoles used the wood for mortars, pestles, and food paddles (Sturtevant 1955), and Caribbean people used the wood in cabinets and boats. Traits of the wood have led people to compare them with other prized materials. The mastic is called *tortuga* [*tortugo*] *amarillo* (yellow turtle, Puerto Rico, Panama), *tortugo colorado* (red turtle, Puerto Rico), *tortugo prieto* (black turtle, Puerto Rico), and *ebano amarillo* (yellow ebony, *Diospyros ebenum,* Cuba). Comparing the wood to turtle shell and ebony is high praise. The thick sapwood is yellow and the heartwood is yellow to orange. Wood is hard, heavy, with a specific gravity of 0.9, strong, and durable. Other names referring to the wood are ironwood (Florida), *sabina* (roan-colored one, Puerto Rico), *sanguinaria* (bloody one, Puerto Rico), sweetwood (applied by 1607, Barbuda), sourwood (applied to *Oxydendrum arboreum* by 1856, Bahamas), and *tabloncillo* (little plank, Puerto Rico).

Cubans use the Taino name *caguaní* [*coguani*], which may be linguistically related to a river in Puerto Rico called *Caguana. Ausubo* (Puerto Rico), the Taino name of *Manilkara bidentata,* has been dubiously applied to *S. foetidissimum.*

Fruits are edible in *S. foetidissimum* and the related and similar *S. salicifolium,* and the Seminoles ate both (Sturtevant 1955). Morton (1968) described *S. foetidissimum* as "gummy, acid and sometimes bitter." I never found them bitter, but the other two descriptors are accurate. In Florida, people also have called them jungleplum and wild olive because of their fruits. Another name mentioning fruits is *abricot des bois* (tree apricot, French Antilles). That designation seems to be comparable to almond, which is applied to *S. salicifolium.* Comparison to the apricot and almond are possibly a corruption of *almástiga,* as in *almendro* (almond, USA, West Indies), *almendrillo* (little almond, Cuba), *almendro silvestre* (wild almond, Cuba), and *almendrón* (false almond, Puerto Rico).

Medicinally, *jocuma* sap has been used to treat hernias (Roig 1945). The Miccosukees considered the *okó:mî* an important ingredient in a medicine for "washing the body" (Sturtevant 1955). Triterpenoid saponins have been isolated from *S. foetidissimum* (Nicolas et al. 1995), and similar chemicals are known from a related species (Jiang et al. 1994). Those compounds may be responsible for the anti-inflammatory action known in the genus (Desmarchelier et al. 1999).

The mainland subspecies (ssp. *gaumeri*) is called *caracolillo* (little snail, Campeche, Quintana Roo), cream tree (Belize), rosewood (Belize), *subul* (black bird, Maya, Yucatán, Belize), and white candlewood (Belize). *Tempixte* [*tempiote*] (from *tempixquiztli,* dribbling on the lips, Náhuatl, Guatemala) is applied to this tree and others in Mexico (Standley 1920–1926, Pennington 1990), and may be a reference to the sticky fruits. This subspecies is protected for its edible fruits (Pennington 1990).

"Bustic" is a better-known name for *S. salicifolium* (willow-leafed) used in English, Dutch, and French. No etymology has been found, but "bu-stic" may be a contraction of "bully-mastic." Other variants on "bustic" are willow bustic (USA, Puerto Rico) or willow-leaf bustic (Florida). Other English speakers in the Caribbean now call this tree bolletrie witte (white bully-tree, Guyana), bully-tree (Barbados), bullet-tree (Barbados), red bullet (Jamaica), white bull-tree (Jamaica), white bulletwood (Jamaica), and white bullywood (Jamaica). All those names were derived from "bully," which originally meant "friend" or "kinsman," a term of endearment since the time of Shakespeare in 1590. Part of that endearment is because the tree gave such a good material to chew.

The Taino name *jocuma* (Cuba) also has been applied to *S. salicifolium*, as in *jocuma blanca* (white *jocuma*, Cuba). Variants of that name are *acomât batârd* (false *acomât*, Guadeloupe), *acomât marron* (wild *acomât*, Haiti), *acomât petite feuille* (little-leaf *acomât*, Guadeloupe), *acomât rouge* (red *acomât*, Haiti), *caya Colorado* (red *caya*, Dominican Republic), and *cuyá* (Cuba). The name also seems to have cognates in one Seminole language with *toca:tî* (Creek), which may be akin to the Taino variant *tocuma.* There is no obvious linguistic relationship with *ahikitíscî* (red tree, Mikasuki). However, I have to wonder if that name is not related to Muskogee *vhecetv*, "to look ahead." Both may allude to scanning the horizon to see if there is a hammock (tree island) or some other item of interest. Another Taino name for the trees is *jaiquí,* which is also applied to *Manilkara jaimiqui* in Hispaniola.

Sap from *S. salicifolium* also was used as chewing gum. The Maya name for *S. salicifolium* still is *sak-chum* [*xac-chum*] (*sak*, white, *chum*, trunk, Yucatán, Belize). From this white-trunked tree, they extracted *tsits-ya* [*txitxya*] (slender zapote), which had many of the properties of the chicle zapote (see *Manilkara:* Wild Dilly). It is known that *chicleros*, harvesters of chicle, tapped the latex of *S. salicifolium*, and it was used separately and probably as an adulterant in "true" chicle (Standley and Record 1936).

Other names for *S. salicifolium* are varied. Some compare the species to others, such as black haw (Florida) and wild mespel (*mespel*, medlar, from Latin *mespilus*, *Mespilus germanica*, Puerto Rico). They are even called tough buckthorn (buck, as a male deer, plus thorn, Florida; see also *Rhamnus*). In Spanish, trees are *zapote faisán* or simply *faisán* (pheasant zapote, Campeche, Belize, West Indies) and *zapotillo* (little zapote, Mexico). Those names are comparable to *sapotille marron* (wild zapote tree, Haiti) and *sapotillier marron* (wild zapote tree, Haiti). *Weiden-blattrige-diphole* (willow-leaf dipholis, Dutch Antilles) is merely a translation of an old scientific name. Perhaps *m'panache* (*panaché*, plumed, Haiti) is a descriptive name for the branches, and *jubilla* (little jubilee, Cuba) may refer to the happiness at finding edible fruits.

The remaining names are of obscure meaning and derivation. An old name in Cuba is *carolina* (from Carolina or Carlos). Presumably, that refers to a person because the species does not grow in the Carolinas. Both *cháchiga* [*chachiga*] (Belize) and *mijico* (Belize) resemble indigenous words, but do not seem to be Mayan. They are perhaps remnants of the Miskito who formerly lived in the region. The name galimeta-wood was first used in 1756 in Jamaica. That has also been rendered *galemeta, gallamenta*, and even *galimetaholz* (German *Holz*, wood), but its origin is dubious. In Peten, Guatemala, and Belize, the trees are *silillon* or *silion de sabano* (savanna chair?, from *silla*?, Spanish), and in Belize they may be silly young. Finally, the Bahamian names *cassada*, cassada-wood, and wild cassada resemble the Taino *casabé*, whence *cassaba* (*Manihot esculenta*), although any connection is obscure.

Attempts to eat the fruits on either of these species will quickly make it clear why they were associated with mastic and chewing. The taste is pleasant, but there is a tendency to stick the lips together. Large quantities of the fruits also tend to burn the lips of some individuals. So, let the consumer beware.

Silphium

(From Greek *silphion*, the name of some resinous plant, transferred by Linnaeus to this plant)

Silphium compositum (compound, alluding to pinnatifid leaf blades) (=*S. terebinthinaceum*)
prairie-burdock
[kidney-leaf] rosin-weed [rosinweed]; rosin-plant

In 1776, William Bartram wrote about this rosinweed in West Florida before the species had a name. Bartram ([1791] 1958) gave the use as a chewing gum by Creeks and traders, and this is not recorded elsewhere. He wrote: "the flower stem, which is eight or ten feet in length when standing erect, terminates upwards with a long heavy spike of large golden yellow radiated flowers; the stem is usually seen bowing on one side or the other, occasioned by the weight of the flowers, and many of them are broken, just under the panicle or spike, by their own weight, after storms and heavy rains, which often crack or split the stem, from whence exudes a gummy or resinous substance, which the sun and air harden into semi-pellucid drops or tears of pale amber colour. This resin possesses a very agreeable fragrance and bitterish taste, somewhat like frankencense or turpentine; it is chewed by the Indians and traders, to cleanse their teeth and mouth." The species was not described until 1803 when French visitor and explorer of the eastern United States André Michaux published his *Flora Boreali-Americana* (Flora of North America).

Silphium is a genus of 23 species endemic to eastern North America (Mabberley 1997). Several of the species are indicators of prairie vegetation, while others may not be. *Silphium compositum* is similar to *S. laciniatum* in that they both have large basal and reduced cauline leaves. Both also often have pinnately divided leaves, but there are differences in their ranges. *Silphium compositum* grows from southeastern Virginia and eastern Tennessee to Georgia and Florida. While *S. laciniatum* is primarily a plant of the Great Plains, it grows east to Ohio and Alabama where it overlaps the range of *S. compositum* (Perry 1937, Settle and Fisher 1970, Sweeney 1970, Barkley 1986).

The only use for *S. compositum* other than the one recorded by Bartram was among the Cherokee. These people used rosin-weed as a stimulant for "weakly" females (Hamel and Chiltoskey 1975). Use was probably more widespread because Moerman (1998) found records of three other species among indigenous tribes. *Silphium integrifolium* was used by the Meskwaki; *S. laciniatum* was used by the Dakota, Meskwaki, Omaha, Pawnee, Ponca, and Winnebago; and *S. perfoliatum* was used by the Iroquois, Meskwaki, Ojibwa, Omaha, Ponca, and Winnebago. Moreover, Jones (1936) found "considerable amounts of pieces of chewed or beaten roots" of *S. laciniatum* at the Newt Kash Shelter in Mennifee County, Kentucky. He commented that it had been found at the Ozark Bluff-Dweller site, but it was not included in Gilmore's (1931) paper.

Names of *S. laciniatum* and information gathered by Gilmore (1919) in the Missouri River region show that it was still being used in the same way that Bartram reported for the Florida species in the 1770s. The Omaha called their species *zha-pa* (*zha*, weed, *pa*, bitter), and *makan-tanga* (big medicine or root). Winnebago said the plant was *shokanwa-hu* (*shokonwa*, gum, *hu*, plant). Especially children gathered the resin from broken stems for chewing gum. This is perhaps what the Chickasaw call *oktaak taakchili'* (*oktaak*, prairie, *taakchili'*, gum [wax]). Lewis and Elvin-Lewis (1977) suggested that all species of *Silphium* were used by indigenous people to clean their teeth and as chewing gum.

The Pawnee thought that *S. laciniatum* was a *khats-tawas* (*khats*, medicine, *tawas*, rough) and subsequent settlers agreed. Hocking (1997) considered it a strong stimulant, and the root was boiled with mullein (*Verbascum thapsus*) to treat swelling. It seems strange that Porcher (1863), Millspaugh (1892), Ellingwood and Lloyd (1898), Culbreth (1910), and Felter (1922) did not mention the genus.

Two species are included by Duke et al. (2002). They found that *S. perfoliatum* had been considered as alterative, analgesic, antimetic, antispasmodic, diaphoretic, emetic, emmenagogue, restorative, stimulant, and tonic. They consider it not as safe as a cup of coffee. More positive results were noted for *S. laciniatum,* which they gave two plusses. *Silphium laciniatum* has been used as an alterative, antibacterial, antipyretic, antispasmodic, bitter, diaphoretic, diuretic, emetic, expectorant, tonic, and vermifuge.

The only species of *Silphium* that has a variety of chemical studies is *S. perfoliatum.* That species produces two apiose-containing kaempferol triosides, nine flavonoids (El-Sayed et al. 2002), two carterochaetols (Pcolinski et al. 1994), triterpene glycosides (Davidyants et al. 1984, 1985a,b,c, 1986), and water-soluble polysaccharides (Dudkin et al. 1979). Other studies have found inhibitors of trypsin and chymotrypsin in seeds and vegetative parts of *S. perfoliatum* (Konarev et al. 2002), and monoterpenes (alpha-pinene, beta-pinene, and camphene) in *S. laciniatum* and *S. compositum* (Tooker et al. 2002).

Complex interactions take place between rosinweed and its herbivores. *Silphium* uses the enzyme inhibitors to reduce herbivory by bacteria, fungi, and insects, while male gall wasps *Antistrophus rufus* (Hymenoptera: Cnipidae) alters the chemistry of the plants to enable them to locate females (Tooker et al. 2002). If these interactions occur between plants and other organisms, why should the plants not also be used by humans?

Simarouba: Paradise Tree

(From Carib *simarouba* or Arawak *simaruppa*, or both)

When I first arrived in Florida, I was constantly misspelling two names. The two trees are in different families, but there is a single letter difference between them. The plants are the gumbo limbo (*Bursera simaruba*) and the paradise tree (*Simarouba glauca*). At one point, a reviewer chastized me severely for misspelling the species name of the gumbo limbo, so I learned. The similarity between the generic name *Simarouba* and *Bursera simaruba* is not accidental. The binomials are the result of confusion about plants found in the French Guiana region in the early 1700s.

Early explorers in French Guiana found the Caribs and Arawaks using a medicinal bark; they

***Simarouba glauca**. Drawn by P.N. Honychurch.*

wrote its name phonetically as *simarouba*. These medicinal plants first were taken back to Europe about 1713 under that name. Although Europeans did not know its true source, *simarouba* gained a reputation as a wondrous drug.

In the 1750s, Linnaeus had learned the word *simaruba*; yet, he still did not know which plant it was. Linnaeus's confusion caused him to create the name *Pistacia simaruba* for a Caribbean tree. Later, in 1762, he corrected that mistake and coined the name *Bursera simaruba*, when he realized it was not related to *Pistacia*. That same year he named a new genus *Quassia*, but as yet did not know the relationship between it and the true *simarouba*.

It was French explorer, linguist, pharmacist, and botanist Jean Baptiste Christophore Fuseé Aublet who finally sorted out the differences and created the genus *Simarouba* in 1775. Aublet called the trees *S. amara*, and pointed out that they were the medicine long used by indigenous people in French Guiana. The words those indigenous people used for the trees are still extant. In the Guiana region, the plants are *maruba* [*marupa, mboera*] (Brazil, Guyana, Peru) and *simarouba* [*siemaroepa, simaruba, simarupa, semarouba*] (French Guiana, Guyana, Suriname). Lawrence (1951) said the words were Carib, and the OED (1971) cited the *-ba* ending as evidence of that. However, Pittier (1926) was equally confident that the word was Arawakan, and that the correct spelling was *simaruppa*. They were both correct (Klooster et al. 2003).

The trees that Linnaeus named *Quassia* are relatives of *Simarouba*. Indeed, both are in the same family, the Simaroubaceae, although that name was not created until 1811. According to legend, a black slave in Suriname named Graman (=grand-man) Quassi [Quacy, Quasee, Kwasi] discovered the virtues of the roots of the tree about 1730. Edward S. Ayensu in Grimé (1976) suggested that *Quassi* was from Ghanan *Kwesi*, a name given to all males born on Sunday. According to the OED (1971), "Quassi communicated his discovery to C.B. Dahlberg, by whom it was made known to Linnaeus." Linnaeus named the genus *Quassia* after the African in 1762.

I consider that story a "legend," because Quassi was a newcomer to the New World. Like many other Africans, he had learned about the plants from indigenous people who had been using them for thousands of years. A professor of medicine and botany at Halle, Kurt Polykarp Joachim Sprengle, in his revision of Linnaeus's *Systema Vegetabilium* in 1825–1828, moved them all to the genus *Quassia* to create *Q. amara* and *Q. simarouba*. He may have known something that has subsequently been lost. On the other hand, maybe we have rediscovered with molecular data something Sprengle already knew. A recent idea is again to submerge *Simarouba* under *Quassia* (Fernando et al. 1995).

Either way, for now there are two genera distinguished in the Simaroubaceae, *Quassia* and *Simarouba*. Within *Simarouba*, the indigenous plants in the Guianas are *S. amara*, while Florida trees belong to *S. glauca*, a name given by Augustin-Pyramus de Candolle in 1811. The Florida species, *S. glauca*, ranges through the Bahamas, Cuba, Jamaica, Hispaniola, and from Mexico to Panama. Conversely, *S. amara* grows from the Guianas up through Central America and is sympatric with *S. glauca* in southern Mexico. Sorting out identifications and common names for these two trees is not always done by taxonomists, foresters, or people working in the field although they are often different (Franceschinelli et al. 1998). Maybe it does not matter, because both are used in essentially the same ways.

Paradise tree (Florida, Bahamas) is the most paradoxical name given to these trees. Which of the multiple meanings of the word "paradise" was intended by those who coined that name? The word was taken from Old Persian, *pairidaeza*, a park or enclosure, and became *paradeisos* in Greek, and *paradisus* in Latin. The original concept was a garden, a place where plants and animals were well-watered in contrast with the hostile, dry, desert outside. By A.D. 1000, a "paradise" was already considered the abode of God. By A.D. 1300, it was a place on earth being compared to God's Paradise, and by the 1370s it reverted to the original meaning and applied generally to a garden or pleasure garden. Some have even suggested that the name came into being because the seed kernel contains poisons (Morton 1968b), and eating them would send you to Paradise.

Many of the names compare the fruits to olives, and indeed, both are black when ripe and yield oils that can be used for cooking, manufacturing margarine, or making soap (Morton 1968b, Chavelas-P. 1981, Lognay et al. 1981a,b, Satpathi 1984, Jeyarani and Reddy 2001). Some names making this comparison are *aceitillo* (Puerto Rico), *aceituno* [*aceituna*] (olive tree, Chiapas, Honduras, Nicaragua, El Salvador, Belize, Costa Rica, Panama), *aceituno negrito* (black olive tree, Nicaragua), *manteco* (butter), and *olivo* (olive, Chiapas, Costa Rica).

Mostly, however, the names allude to the bitter taste that led to medical applications. Sometimes the tree is just *amargo* (bitter, Venezuela). More often it is bitter damson (Jamaica), mountain damson (English Antilles), bitter ash (Dominican Republic), bitterwood (English Antilles), *bois amer* (bitter wood, Haiti), *palo amargo* (bitter wood, Dominican Republic), *quassia amarga* (bitter quassia, Dominican Republic), *quinquina d'Europe* (European quinine, Haiti), or *simarouba* [*siemaroepa, simaruba, simarupa, semarouba*] (Costa Rica, Cuba). Other names alluding to medicinal use are *bois négresse* (black woman's tree, Haiti), dysentery bark (Belize), *Juan primero* (the first John), and *negrito* (little black one, Mexico, Belize, Honduras, Nicaragua). Perhaps some of the other names make similar allusions, but they are more obscure. These include *bois cayan* (tree from Cayenne, French Antilles), *chapascuapul* (maybe Náhuatl, Mexico), *gavilán* (hawk, Cuba), *jucunuco* (El Salvador), *lagartillo* (little alligator, Costa Rica), and *pujilté* [*pajulilte*] (Chiapas).

The undersides of the leaflets give a decided white glint when they are buffeted by winds and contrast pleasingly with the dark green upper surfaces. That difference may have led to names such as *bois blanc* (white wood, Haiti), *palo blanco* (white tree, Cuba), and *xpaxakil* [*pac, paque, pasa, pa-sac, pa-sak, pasa-ak, pa'sa'ak', pasac, pasaque, pazaque, pask', x-pak, x-pak'il, x-pasak'il, xpazakil*] (*sak*, white, *pa*, something thrown away, Maya, Yucatán, Belize, Guatemala). The Mayan name is applied because mature fruits give off a disagreeable odor. Incredibly, in spite of that odor, the pulp of the fruits is still eaten by people everywhere.

Several of the names compare the trees with others, often those in temperate regions. For example, it is *bois frêne* (ash tree, Haiti), *frêne* (ash, Haiti), and *roblecillo* (little oak, Cuba). Still, there are parallels within the tropics. There the tree is sometimes called *daguilla* [*daguillo*] (comparing with *Lagetta lagetto*, Thymeliaceae, Dominican Republic). However, that use is more complicated because the same name is given to *Philodendron* and *Yucca*, probably because each has useful fibers. In Guatemala, *Simarouba* is *jocote de mico* (monkey's plum, Guatemala), which highlights a similarity to *Spondias mombin* (Anacardiaceae). In Mexico, it is *rabo de largarto blanco* (white lizard's tail, Oaxaca), which notes that the leaves are similar to *rabo de largarto* (*Acacia paniculata*, Fabaceae).

One name, stavewood (Jamaica), notes the use of the wood, which is white-yellow or somewhat purplish, bitter, fine, smooth, has a specific gravity of 0.40, and easily worked but not durable. Apparently, the wood lasts long enough for staves in barrels. The high tannin content resulted in the name *zapatero* (shoemaker, Mexico), and that compound has been extracted for tanning leather in many areas (Hocking 1997).

A variety of triterpenes, quassinoid glucosides, and lipids in the bark and root are bitter and astringent, and have been used as an amebicide against *Entamoeba histolitica*, as an analgesic, to treat anemia, as an antibacterial, antimicrobial, antileukemic, antimalarial, antirheumatic agent, against chronic dysentery, diarrhea, diarrhea with colic, dyspepsia, as an emmenagogue, to control excessive menstruation, fevers, gonorrhea, hemorrhage, internal bleeding, sores, and as a stomachic, sudorific, and vermifuge (Standley 1920–1926, Spencer et al. 1947, Martínez 1969, Liogier 1974, O'Neill et al. 1988, Wright et al. 1988, Cáceres et al. 1990, Arvigo and Balick 1993, Franssen et al. 1997, rain-tree.com 2002). Fruits in El Salvador are made into a liqueur esteemed for stomach problems (Martínez 1969). However, in large doses extracts are purgative and vomative (Liogier 1974). In addition, compounds from *Simarouba amara* have been shown to be cytotoxic and antineoplastic (Polonsky et al. 1978, Bhatnagar et al. 1984).

The number of uses to which people have put these trees is staggering. Perhaps that is why they gave them the name Paradise tree—they are so useful, they must be from Paradise.

Sisyrinchium

(Theophrastus, 372–287 B.C., used the Greek *sisyrinchium*, based on *sys*, pig, *rhynchos*, snout, for what later authors called *Gynandriris sisyrinchium*, a plant related to *Iris*; Linnaeus reapplied the name to American herbs)

Binsenlilie (rush lily, German)

blue-eyed grass (the yellow center on the blue flower has been referred to as an "eye" since classical Roman times, when it was *ocellus* in

Latin; however, the name is doubly incorrect, as Diggs et al. 1999 and earlier authors have pointed out—the plants have a yellow eye, not blue, and are not grasses; perhaps we should write it "blue eyed-grass")
purole (Coahuila)
sisyrinchium (French)

Sisyrinchium angustifolium (narrow-leaved)
batchelor's-button (Massachusetts)
Bermuda-flag; *bermudienne* (from Bermuda, Quebec; this is based on a comparison with *S. bermudiana*)
blue-eyed lily [Mary]; blue-grass [bluegrass] (Massachusetts); pointed blue-eyed-grass
forget-me-not (typically applied to *Myosotis*, Boraginaceae, a plant that also has a blue corolla limb and yellow eye, Maine)
grass-flower (Massachusetts)
Grasschwertel schmalblättriges (small-leaved sword-grass, German)
pig-root [pigroot] (used in English by the 1890s, but surely older; perhaps simply a translation of *sisyrinchium*)
rush-lily
satin-flower [-lily]
Spanish nut (first in Gerarde's *Herball* [1597] 1975, for a different member of the Iridaceae)
star-eyed-grass [stargrass] (Massachusetts)

Sisyrinchium nashii (named for George V. Nash, 1864–1921, agrostologist and head gardener at the New York Botanical Garden)
cikto ika aha:ka: (*cikto*, snake, *ika*, head, *vhaketv*, resembling, Creek); *cintyó:sâ:bî* (snake's head replica, Mikasuki)
colô:ntkî (Mikasuki)

Back in the early 1970s, I ventured to study *Sisyrinchium* with my friend and colleague Royce L. Oliver (1929–1997). He had been working on the genus for years, and thought that we might solve one of the taxonomic problems for the Florida species. Eventually, we published our results, with which no other systematists in the world have agreed (Austin and Oliver 1974). In spite of other botanists' insistence on lumping *S. solstitiale* under the name *S. xerophyllum*, none of our colleagues has ever been able to explain how it is that a species that blooms in the spring (*S. xerophyllum*) is able to maintain a single gene pool with fall-flowering plants (*S. solstitiale*). If they are genetically isolated by different flowering seasons, how can they be the same? Dan Ward (personal communication, Oct. 2003) has finally agreed with our conclusion of two species, but tells me that we erred in using the name *S. solstitiale*. The fall-flowering plants have no species name.

I found many aspects of the morphology, and hence the taxonomy of the group, to be based on environmentally influenced aspects, especially fires. Given the disagreement on the limits of species, it is not surprising that *S. angustifolium* and *S. albidum* have been considered conspecific with each other and two more named species (Diggs et al. 1999). Different aspects of the genus, like pollination, habitat preference, and association with humans are fascinating. Records exist for all species, except the endemic *S. xerophyllum*, having been used by indigenous people. The restricted range of *S. xerophyllum* in peninsular Florida and the quick extirpation of native people in that area after European arrival probably explain its absence from the records.

There are perhaps 80 species in *Sisyrinchium*. Kartesz (1994) listed 38 in North America, and the rest are in Central and South America. Because the genus was entirely native in the New World, Europeans largely compared the herbs with plants they knew at home. Mabberley (1997) thought that *S. bermudiana* might be native to the Old World, because it occurred in Ireland. However, the only other place in the world that it occurs is Bermuda, and Dan Ward (personal communication 2003) tells me that that the Irish plants are actually *S. angustifolium*.

John Gerarde's ([1597] 1975) *Herball* includes the entry: "Spanish-nut hath smal grassie leaues." He further wrote, "The Spanish nut is eaten at the tables of rich and delicious, nay vicious persons, in sallads or otherwise, to procure lust and lecherie." Some have thought that was a *Sisyrinchium*, but it apparently refers to an African plant. John C. Loudon's *An Encyclopedia of Plants* published in 1829 attaches the name *Gynandriris sisyrinchium* (as *Moraea*) with the common name Spanish-nut. *Gynandriris* is an African genus.

Outside Florida, the Menomini of Wisconsin mixed *S. albidum* with horse food to make the animal sleek, vicious, and to give it a poisonous bite (Moerman 1998). They also kept a plant in the house or pocket to ward off snakes.

Sisyrinchium angustifolium was mixed with other greens and eaten by the Cherokee (Moerman 1998). Considering the other uses, that application blurs the distinction between medicine and food. The Cherokee also gave an infusion of the root to children to control diarrhea. The Iroquois of New England used it to relieve constipation and diarrhea (Moerman 1998). *Sisyrinchium nashii* was used by the Mikasuki as an analgesic, and for "Moving Sickness," with symptoms including pain moving in the region of the waist (Sturtevant 1955).

Several other species were used by different tribes in North America, and at least one species is used in Venezuela (Morton 1981, Moerman 1998). Most *Sisyrinchium* are considered laxative, and some, like *S. albidum*, are considered poisonous. Cronquist (1981) noted that many members of the family often have long, prismatic crystals of calcium oxalate, are sometimes poisonous, and have terpenoids, the glycoxanthone mangiferin, and naphthaquinones or anthraquinones, and are occasionally cyanogenic. The calcium oxalate alone would account for the reputation of being "poisonous" because of the burning sensation they cause on ingestion. Some of the other compounds are known laxatives, especially anthraquinones (see *Rhamnus*).

Sium

(From Greek *sion*, the name of some marsh plant)

berle (from Latin *berula*, of Celtic origin; akin to the Gallois *berwr*, fountain cress, or old Provençal, *berla*, or Spanish *berro*, cress; still in use in the countryside, but more often replaced by the modern *cresson*, French)
erba cannella (cinnamon herb, Italian)
Merk (German)

Sium suave (sweet)
berle douce (sweet cress, Quebec)
fo:similpá (*fo:si*, bird, *im*, its, *ilpá*, food, Koasati)
hemlock (of obscure etymology, but from Old English *hymlice*, feminine, and *hymlic* masculine, from about A.D. 700; no cognates in other languages; originally applied to *Cicuta*); hemlock-water-parsnip [hemlock waterparsnip]
hlue tla^h le^h (Slave, Canada)
queue de rat (rat's tail, Quebec)
siwaskatask (sweet carrot, Cree); *uskotask* (Plains Cree)
water-parsnip (comparison to the original "parsnip," from Middle English *passenep*, in turn from Latin *pastinaca*; the Latin is from *pastinare*, to dig and trench the ground, and *pastinum*, a two-pronged digging fork)

When the Spanish arrived in the New World, they were already familiar with *Sium*. One species, *S. latifolium*, is native to the Mediterranean from Spain eastward (*Flora Europaea* 2003). At least people in Italy cooked the leaves and ate them (Hedrick 1919). As the New World was beginning to be known, a second species called "skirret" was introduced into Europe. That herb, *S. siarum*, is native to Eurasia from Bulgaria, Hungary, and Romania eastward. From its region of nativity, the plants reached England some time after 1538 but before 1548 (Hedrick 1919). John Gerarde commented on skirret in his *Herball* in 1597, and by then the species was in demand as far north as England for its edible roots (Austin 2002).

By the 1750s, Linnaeus was familiar with both European species and some elsewhere (Linnaeus [1753] 1957). If the American plants had been discovered by that time, he did not know them. Indeed, it was 1788 when Thomas Walter described *S. suave* in his *Flora Caroliniana*. It seems likely that people familiar with European plants took this American native to be the Old World skirret. However, in the absence of preserved herbarium material, that will never be substantiated. It is known that the settlers were cultivating a remarkable number of European root crops by the 1700s. For example, in the 1760s, Bernard Romans found skirret in Mobile, Alabama (Romans [1775] 1961). Skirret was being grown there with other Old World root crops, including carrots (*Daucus carota*), turnips (*Brassica rapa*), radishes (*Raphanus sativus*), leeks (*Allium porrum*), and scallions (*Allium cepa*).

Sir John Richardson wrote in 1852 that, from the Ojibwa, "who were encamped on the Otter Lake, we procured a quantity of small white root, about the thickness of a goose quill, which had an agreeable nutty flavor. I ascertained that it was the root of the *Sium lineare* [=*S. suave*]. The poisonous roots of *Cicuta* ... are often mistaken for the edible one, and have proved fatal to several laborers in the [Hudson Bay] Company's service. The natives distinguish the proper kind by the last year's stem, which has the rays of its umbel ribbed or angled, while the *Cicutae* have round and smooth flower-stalks" (Richardson in Fernald et al. 1958).

Several indigenous tribes used *Sium* as food (Moerman 1998). The Algonquin, Cree, and Ojibwa are among the northeastern tribes that ate the roots. However, a number of more distant people in the western and northwestern United States also ate them. These tribes include the Bella Coola, Carrier, Klamath, Montana, Okanagan-Coville, Shuswap, Slave, and Thompson.

There are a few records of people using the herbs in medicine or magic (Moerman 1998). The Iroquois made a poultice of mashed roots to relieve pain in broken or dislocated limbs and sprained muscles, and made a decoction for treating people suffering from epilepsy. The Dakota used the roots to sooth stomachache. The Shuswap thought the flowers were poisonous, although they ate the roots. The Ojibwa smoked the seeds to drive away and blind the evil spirit that steals one's hunting luck.

Records vary about whether or not the plants are poisonous, even among tribes that used them as food. Some considered part of the plant poisonous, while the root was thought to be edible. Turner (1978) has found them eaten by indigenous people in British Columbia. Turner and Szczawinski (1991) and Diggs et al. (1999) found records of the herb poisoning livestock. Similarly, *S. latifolium* is considered poisonous to livestock in Europe (Polunin 1969), but eaten by people (Hedrick 1919). Maybe the different plant parts contain distinctive chemicals, or people and animals react to them differently. Indeed, Hocking (1997) says that the flowers are poisonous, while the young leaves, stems, and roots are edible when cooked.

Perhaps the problem is that the herbs are so similar to water hemlock (*Cicuta maculata*) that they have been mistaken for them. For the related *S. siarum,* Duke et al. (2002) suggest that the only danger in using the herb is confusion with poisonous members of the Apiaceae.

Smilax: Green-Brier

(Greek *smilax, smilakos,* a twining plant; the Latin *smilax* was used by Pliny, A.D. 23–79, for what is now *Smilax aspera,* but he also applied it to the yew and an evergreen oak)

Shortly after we moved into our Florida home, birds perched on the Spanish bayonet (*Yucca aloifolia*) and false sisal (*Agave decipiens*) in our front-yard xeric garden and "planted" a green-brier (*Smilax auriculata*). After we had torn out the climbing stems dozens of times, my wife, Sandra, decided that she was going to "eliminate" the pesky vine and began digging up the rhizomes. After a long work session, she had over a bushel of the woody horizontal stems. A few days later the stems were weaving themselves into their spiny supports, and it was as if she had done nothing. At least, I had taken a picture of the rhizomes to use in teaching classes, and I had an entertaining story to tell students.

Europeans already knew *Smilax* when they came to the New World. The species they knew best was *S. aspera,* which is a southern European plant native from Portugal to Yugoslavia. Plants were cultivated as far north as Germany during Leonard Fuchs's time in the mid-16th century (Meyer et al. 1999). Europeans called this plant sharpe smilax (English 1568), *grosz stechend Windt* (big spiny vine, German 1542), *sharpff Windt* (sharp spiny vine, German 1542), *groote stekende wranghe* (big spiny vine, Dutch 1549), *smilax picquant* (spiny smilax, French 1549), and *smilace aspra* (spiny smilax, Italian 1551). Speakers of Arabic called it *'ushba* (herb) or *'ulliq* or *'ullayk* (generally applied to *Rubus,* Rosaceae). People speaking Spanish called the climbers *zarzaparilla* (*zarza,* brier, *parilla,* small grape vine), and that became the source of the English word "sasparilla."

Cowboy movie stars Roy Rogers and Gene Autry were popular in the 1940s and 1950s. They usually played a scene in each movie where the indestructible hero walked into a saloon full of bad guys and ordered a bottle of "sasparilla." Sometimes a fight ensued where the single good guy outgunned a dozen or so criminals without bloodshed—not a bad feat with guns that held only six bullets.

At the time, I had no idea what "sasparilla" was, but it obviously was not a drink preferred by the toughs. Later I learned that the same basic ingredient in "sasparilla" was the basis of "root beer." At the time, "A&W Rootbeer" businesses were common in rural towns. Both "sarsaparilla" and "root beer" came from the genus *Smilax,* and had a long history as a tonic drink in Europe before the New World was discovered. Early American settlers made root beer from *Smilax* rhizomes by combining them with molasses and parched corn, and then allowing the brew to ferment (Morton 1968b). That formulation was essentially unchanged during the Civil War when Confederate soldiers made the same drink (Porcher 1863).

Not only was the native European *S. aspera* used, but from 1535 onward, people knew of the "China root" or simply "China," another species now called *S. china* (Linnaeus [1753] 1957). Some other early names were *Radix China* (China root, Latin), *tuber chine* (China root, French), and *chob-chini* (China root, Persian). In Portuguese, it was *"alguma raiz ou pão de China"* (some root or bread from China), cited by Garcia de Orta in 1563. These rhizomes are now known to contain steroidal sapogenins, and the glycosides of the steroids smilagenin and sarsasapogenin (Itoh et al. 1977, Lewis and Elvin-Lewis 1977, Foster and Duke 1990) that were the basis of the tonics.

The Asian species was considered an "infallible remedy for gout" (OED 1971). Gerarde ([1597] 1975) wrote that the "China" was used "to cure the French Pox." Culpeper (1653) prescribed medicinal preparations from *Smilax* to be used to neutralize poisons, and to serve as a diuretic, an eyewash, against colds, to stop flatulence, sores, and venereal disease, and to serve as a tonic. Hocking (1997) expanded that to say that the rhizomes contain tannins, gums, resins, smilax saponins, coloring matter, and starch, and were used as *Radix Chinae ponderosae* to treat gout, syphilis, frambesia (a venereal disease), as an antidote in mercury poisoning, and as an aphrodisiac.

When they arrived in the New World, Europeans were delighted to discover more kinds of *Smilax.* In

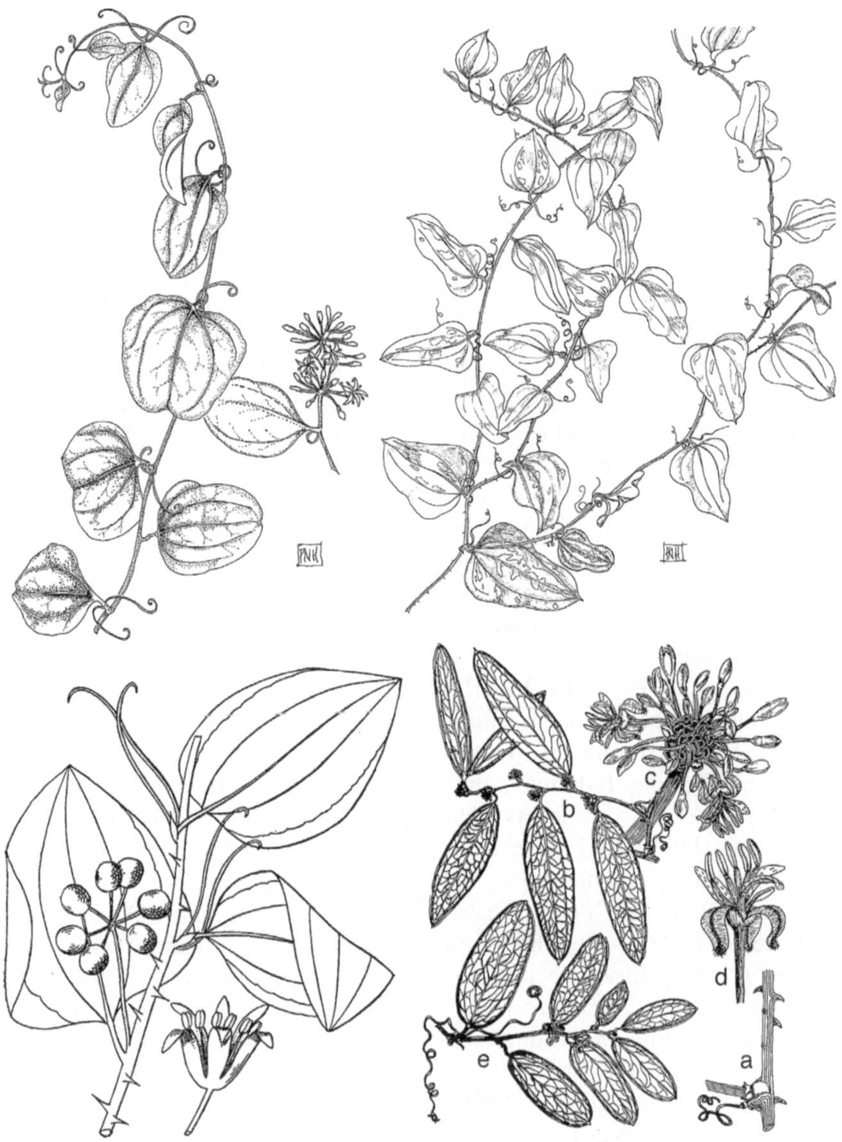

Smilax. *Smilax auriculata* (top left). *Drawn by P.N. Honychurch. Smilax bona-nox* (top right). *Drawn by P.N. Honychurch. Smilax glauca* (lower left). From Britton and Brown 1896. *Smilax laurifolia* (lower right). a. Segment of stem showing thorns and tendrils. b. Branch with clusters of flowers. c. Cluster of flowers. d. Staminate flower. e. Branch with fruits. *Drawn by Vivian Frazier*. From Correll and Correll 1972.

Frampton's ([1577] 1925) translation, the Spaniard Nicholas Monardes says that a merchant "shewed to mee a greate Roote, and other little rootes, and he asked mee what rootes they were, I aunswered that they were Rootes of the China." He added, "the Indians be healed of grevous diseases. And so they have it in great estimation, they doe heale al manner of large diseases therewith: and also the sharpe diseases: especially Agues with the Water of it, provokyng Swets, and by this waie they heale many." Near the same time, Thomas Harriot ([1590] 1972) wrote that plants in North Carolina were like "The China root brought from the East Indies."

The name "China root" became so entrenched in English that it was applied to most *Smilax* species. In the Bahamas *S. auriculata* is China-brier. *Smilax glauca, S. bona-nox, S. havanensis, S. laurifolia,* and *S. tamnoides* may be *bejuco chino* (China vine, Cuba), China-brier [chainey brier, chaney-brier] (Carolinas, Florida, Bahamas), China-root (USA), or *raiz de China* (Cuba). In more tropical areas, *S. havanensis* is China-briar, but it is also called chaney-winder, chaney-wine, and chaney-vine (USA, Bahamas), all variations of the same name. The same species is *liane siguine* (Haiti), where "*seguine*" is probably a corruption of "China." Sometimes it is just *siguine* [*tsiguine*]

(Haiti). Those seem to be variants of *bois d'eschine* (China root, French) from the 1500s. All the species of *Smilax* in Nicaragua are called chaney root (Creole), China root, *chiny* (Miskito), but they are also *cuculmeca* (*cuculí* is said to be Quechua for a dove, Spanish), *salnalai* (Sumu), and *wasalanka* (Sumu).

"Brier" in the names is an English word of unknown origin, already in use about A.D. 1000, and occurring as *bráer* (West Saxon), *brér* (Anglican), and *brere* (Middle English). In all of these, it was a prickly, thorny shrub or bush, often confined to the wild rose (*Rosa*) or blackberries (*Rubus*). Only later was "brier" applied to other thorny plants such as *Smilax*.

We know only the name in Timucua, which Crawford (1988) gave as *hassez*. He thought it referred to *Zamia*, but the fact that historical documents recorded that it was "roots from which they make their bread" argues against that. There were never enough *Zamia* in the northern part of the peninsula to support many people. Besides, it was *Smilax* that was the mainstay of southeastern tribes.

Florida has 12 species of *Smilax* (Wunderlin 1998), and all were probably used similarly. Some common names are applied indiscriminately. As cat-briers (USA) or saw-briers (Florida, Bahamas) they scratch; as squirrel-briars (USA) they climb; as bull-brier (USA) and horse-brier (USA) they have stout growth. The concept of prickly climbers is not confined to English, and they are *bakcî* (brier, Mikasuki) and *kvco* (Creek, Muskogee) to the Seminoles. The Alabama said *bakcho* (thorny brier), the Koasati *bakcó*, and the Choctaw *bisakchakina* (*bissa*, blackberry). *Smilax auriculata* is known to the Seminoles as *cintalô:símbakcî* (black-snake brier, Mikasuki) and *ciktolastocikinkacô* (*cetto*, snake, *lvste*, black, *-uce*, little, *em*, its, *kvco*, brier, Creek). Alabama speakers said *bakcho hollo* (*bakcho*, brier, *hollo*, evil) or *bakchokchako* (*bakcho*, brier, *okchakko*, blue). *Kvco láne* (*kvco*, brier, *láne*, yellow) is the name among the Muskogee speakers in Oklahoma. The Osage said *wa-xa'-ga tu-hu* (*wa-xa'-ga*, prickly, *to-ho*, green). The Koasati named three species—*kontí* (*S. bona-nox*), *bakcó ok:cakkó* (*bakcó*, briar, *okcakkó*, green, *S. rotundifolia*?), and *bakcoholló* (*bakcó*, briar, *holló*, taboo, *S. walteri*).

The same ideas are found in other American names and in Old World languages. To ancient people in the Middle East, *Smilax* was *melu* (ascent, as in climbing, Akkadian). However, the cognate word *mula* (Sanskrit) notes the edible rhizome. That view is related to Seminole *a:hî* (root, Mikasuki), and their relatives the Choctaw say *ahe*. Probably, the Biloxi *ato'* [*ado'*] is cognate. Usually, however, when the Seminoles refer to these rhizomes they say *kunti* [*kantik, kontik*]. The Seminoles call the rhizomes of several *Smilax* species *kunti chate* [*conti chatee, coontie chabi, coonti chacli, coonti chatee*] (*kunte*, root, *cate*, red, Creek), which is sometimes translated as "red flour" (Speck 1941, Sturtevant 1955). Related words are *kanta* (Alabama), *kanták* [*cantaque*] (Houma), *kantak* (Chickasaw, Choctaw), *kontí* (Koasati), and *kúnte* (Muskogee). Indeed, in the Muskogean languages, the names refer to a starch extracted from the rhizomes. The Choctaw even have the preparation *kantak paska* (smilax bread), making it from *ku<u>n</u>shak ahe* (reed potatoes). Koasati called the bread *kontí tabáhka*. Although their language was not related, the Biloxi called the plant *atuti'* (*ato'*, potato, *uti'*, stem or plant) and the flour from them *atudi' nupxi'* (root meal).

In an unrelated Algonquian language, rhizomes are *tsinaw* (North Carolina, Harriot [1590] 1972), which indicates their widespread importance as a starch source, although Geary (1955) took that word to be their pronunciation of English "China." Perhaps the Powhatan word *cawindgus*, "bramble or briar," belongs here. All that is certain is that the cognate word in Delaware was *kawinsi*, thorn (Siebert 1975). While Siebert (1975) did not mention the Carolina Algonquian name, it seems to be cognate with both Powhatan and Delaware. The Ontario Delaware *nzúkii-lehlookihláashak* (*nzúkeew*, black, *lehlookíh-laash*, raspberry) for the black raspberry does not seem to be cognate (O'Meara 1996). Choctaw *shinu-kyolulli* (*shinuk*, sand, *yolulli* =?) seems to be unique.

Smilax rhizomes were a source of starch for various indigenous groups of the South Atlantic States (Morton 1981). In addition to the Seminoles, indigenous people eating the starch or rhizomes included the Algonquians of coastal North Carolina (Harriot [1590] 1972), the Biloxi (Dorsey and Swanton 1912), Choctaw (Bushnell 1909), Cherokee (Hamel and Chiltoskey 1975), Houma (Speck 1941), and Glades and Timucua people (Austin 1980, Hahn 1986). That same notation of edible rhizomes is made with *bejuco* [*de*] *ñamé* (yam vine, Cuba) and *ñamé cimarrón* (wild yam) for *S. havanensis.* Both Cuban names compare the plants with the true yam, *Dioscorea.* Although he does not indicate the people using them in Alabama, Hocking (1997) noted that *S. pumila* and *S. smallii* rhizomes also were food plants there.

Preparation to make the rhizomes edible is said to be simple or complex depending on age. When rhizomes are young and tender, they can be cooked and eaten (Morton 1968b). Harriot ([1590] 1972) reported: "From these roots while they be new or fresh beeing chopt into small pieces & stampt, is strained with water a juice that maketh bread, & also being boiled, a very good spoonmeate in maner of gelly, and is much better in tast if it be temperd with oyle." Among other people, the mature rhizomes were chopped, pounded with a mortar and pestle, mixed with water, and strained. The sediment became a fine, reddish meal when dry; it took about 50 pounds of rhizomes to make 5 pounds of flour. A small amount of flour with warm water and honey became a reddish jelly. The meal, mixed with corn flour, was fried in bear grease to make hot cakes or fritters (Morton 1968b).

Williams ([1837] 1962) described the Seminole method of using *Smilax*, which he called "China briar." He wrote: "The Indians grate them, or bruise them in a large wooden mortar, then throw on water, strain the starch through baskets, dry and pulverize it; the color is reddish brown. They mix it with a fine homony, and make cakes; with honey and warm water, it becomes a fine jelly; toasted and mixed with sweet milk, it is a delicious food." The Seminoles called their preparation *sáfki* [*sofkee, sofkey*], which the Choctaw call *tanfula* and the Cherokees call *ganoheni* (Hudson 1976). We call the preparation "grits" (ground hominy) and make it from corn (*Zea mays*, which see), as they also did, without the *Smilax*.

Several different species are green-brier (Florida, Bahamas, Puerto Rico). Sometimes there are variants on the name to denote individual species such as laurel-leaved green-brier (*S. laurifolia*, Carolinas) and prickly greenbriar (*S. havanensis*, Florida, Bahamas). For obscure reasons, these climbers are also called wild bamboo (Florida) or bamboo [vine] (Florida, Carolinas).

One of the most widespread and commonly used species is *S. bona-nox*. The name was taken from *buenas noches* (good night) applied to the species in the Spanish West Indies and recorded by Charles de l'Escluse because of fragrant nocturnal blossoms. Due to its armament, that plant has been called blaspheme-vine (Florida, Carolinas), hellfetter (USA), and tramp's trouble (Florida, USA). The Natchez, distant relatives of the Seminoles, call the plant *we-bula*. At least as common and popular with the Seminoles was *S. laurifolia*, which they call *co inkontî: kacô* (*eco*, deer, *em*, its, *kunte*, root, *kvco*, briar, Creek) and *i:cinkantikî* [*iicinkantik*] (deer's root, Mikasuki), because deer eat rhizomes and sprouts (Sturtevant 1955).

Although there are no records of indigenous people eating the young stem tips, people of European ancestry eat them. These people compare them with other edible Old World plants by calling them wild asparagus (Carolinas) or wild spinach (Carolinas). The young stems are tender and succulent, and may be eaten raw or cooked. There is, however, considerable variation and some are bitter when raw, probably due to tannin concentration (Hocking 1997). Students always loved to see me react when I sampled them in the field and got a particularly bitter one.

Fruits of *Smilax* were eaten by the Meskwaki (Smith 1928), Omaha (Gilmore 1913), and probably many other people. One name used for several species of *Smilax* in Texas is "stretch-berry," ostensibly because of the rubbery texture of fruits (Tull 1999). Fruits of some species are considered inedible, such as of *S. bona-nox*, while others like *S. lasioneuron* (in literature as *S. herbacea*) are eaten.

Perhaps secondary to food, native Americans used the rhizomes to obtain dyes. A red-pink dye may be extracted by crushing the rhizomes in water and then soaking the skin, or simply by holding the skins over smoke from burning roots (Bennett 1997). This coloring source is favored because it helps the hides repel rain. Berries of most species supply dyes of blue, violet, and purple (Coon 1974, Tull 1999).

Like people in other parts of the world, indigenous groups in the southeastern United States made a variety of medicines with *Smilax*. Sturtevant (1955) reported that the Seminoles used *S. laurifolia* and *S. auriculata* to treat chronic sicknesses and in birth ceremonies, although Alice Micco Snow seems no longer to use either (Snow and Stans 2001). Sometimes *Smilax* medicine was used alone for chronic problems, and at other times, it was mixed with many other plants. The Cherokee made medicine from stem bark for burns and skin problems, and made rhizomes into a tonic (Hamel and Chiltoskey 1975). The Choctaw also made a tonic from the rhizomes (Bushnell 1909). Indigenous people in the Carolinas used a root infusion of *S. bona-nox* to "mundify and sweeten the blood" and to treat scurvy, fever, gonorrhea, and syphilis (Vogel 1970). The Houma used the root bark for urinary infections (Speck 1941). Hocking (1997) recorded people in Alabama using *S. pumila* and *S. smallii* to treat rheumatism, as an alterative, and as an aphrodisiac.

It is well known that numerous *Smilax* species contain steroidal sapogenins (Okanishi et al. 1965, Jia and Ju 1992, Kubo et al. 1992, Sashida et al. 1992, Ju and Jia 1993, Ju et al. 1993, 1994, Bernardo et al. 1996, Santos et al. 1997). Several other types of bioactive compounds also have been isolated, including flavanones (Chen et al. 1999), glycosides (Ju and Jia 1993, Li et al. 1996, Chen et al. 1996, 2000), isoflavones (Yi et al. 1998), and phenolics (Li et al. 2002).

In addition to being a tonic, *Smilax* has also been shown in laboratory studies to be antirheumatic (Ageel et al. 1989), antimutagenic and mutagenic (Lee and Lin 1988, de Sa Ferreira and Ferrão Vargas 1999), to treat gastrointestinal disorders (Cáceres 1990), and hypoglycemia (Fukunaga et al. 1997), and to heal skin infections (Cáceres et al. 1991). There is evidence that *Smilax* may be effective in free radical scavenging, antioxidant enzyme fortifying (Lee et al. 2001), and even in partial neutralization of the hemorrhagic effect of snakebite (Castro et al. 1999). Many of these experimental results support the historical use of the genus in medicines.

The name *Smilax* was originally recorded by Pliny, A.D. 23–79, in the 1st century A.D. He applied the classical Greek name (*smilakos*) to *S. aspera.* However, in Greek mythology Smilax was also a nymph who fell in love with mortal Krokos. It was a hopeless passion, so the gods changed her into the twining herb (Baumann 1993). Perhaps the analogy is good because the nymph lives on to help and bother humans. At least she bothered my wife and me for three decades in our front-yard garden.

Solanum: Nightshades

(From Latin *solamen,* a comfort, or *sol, solis,* the sun)

When Leonard Fuchs published his *Great Herbal* in 1542, he knew of eight plants we now place in the Solanaceae. Of those, five were native to Europe. The others had been imported by explorers in the Americas and Asia. Two of the native species in Fuchs's time were known as "nightshades" by speakers of English—*Atropa belladonna* (deadly nightshade, nyghte shad) and *Solanum nigrum* (black nightshade). Only the *Solanum* was a "nightshade" to speakers of other languages.

Beginning some time before A.D. 1000, speakers of English began calling certain European plants the "nightshades." Speakers of other Germanic languages called them the same with *nihtscada* (Old English), *nachtschade* (Dutch), *natskade* (Danish), *nattskatta* (Swedish), and *Nachtschatten* (German). All of those names include the word "night" and "shade," apparently alluding to the poisonous berries although no one is certain of the original reference. Perhaps it was a reference to death and ghosts.

Solanum. *Solanum bahamense* (left). *Drawn by P.N. Honychurch. Solanum erianthum* (right). a. Fruiting branch. b. Flowering branch. c. Trichome from the leaf. d. Flower, side view. e. Flower, longitudinally dissected. f. Stamens enlarged. g. Floral diagram. *Drawn by Priscilla Fawcett* From Correll and Correll 1982.

Maybe the name "nightshade" was originally applied to *Solanum* in England, and then also given to *Atropa* as a more potent representative. Either way, plants in the family were so famous as poisons that New World members were suspect when they were first introduced.

Solanum is an ancient name for a European plant, and many think it was originally *S. nigrum*. The origin of *Solanum* is uncertain. One school contends that it is derived from *solamen*, a comfort, an allusion to the soothing properties of the plant. The related word in Akkadian *sululu* means happy, glad. Others, however, think that it came from Latin *sol, solis*, sun. That interpretation would be more consistent with the 16th-century pharmaceutical name *solatrum* and the Italian *solatro hortolano*. Moreover, the plants are heliophiles, and their yellow stamens surrounded by white petals might be viewed as miniature suns. While we may never know the correct etymology, it was just like Linnaeus to continue usage of a name that has multiple interpretations.

Although Fuchs did not mention the second species, there were two *Solanum* species familiar to Europeans, *S. nigrum* (black nightshade) and *S. dulcamara* (bittersweet). Fruits of bittersweet were woven into the collarette of Egypt's Tutankahmun's third coffin, so both were long part of the pharmacopoeia of Eurasians (Manniche 1989).

Other Europeans made no reference to the poisonous nature of the *Solanum*, with the French saying *morelle* (little blackberry), a name evident in the 16th-century English names "petimorel" and "morel." The Spanish of the time said it was the *yerba mora de los huertos* (blackberry herb of gardens).

When Europeans arrived in the New World, they found an array of "nightshades" that were new to them. They also found that the people in the Americas used those new species in many ways. Some of the "nightshades" that were new to Europeans were tobacco (*Nicotiana tabacum*), potato (*S. tuberosum*), and tomatoes (*Lycopersicon esculentum*, also called *S. esculentum*). It would take centuries for some Europeans to adopt tomatoes, but they found smoking tobacco more appealing after their first shock of seeing people "swallow smoke" (see *Nicotiana*).

Florida shares several species with neighboring areas, but three are prominent in the flora. Those with the most records of use by people are *S. bahamense, S. erianthum*, and *S. verbascifolium*.

Solanum bahamense is now called the Bahama nightshade in Florida, but that is surely a recent translation of the scientific name. More likely names actually used by the people are Bahamian canker berry, sometimes rendered kanky berry. "Canker" in those names seem to indicate a use in treating "thrush," especially of the mouth (von Reis and Lipp 1982). Thrush is an infection of the fungus *Candida*. Morton (1981) recorded that it was also called *ajicón* (big pepper), but did not indicate who uses that name. Mashed fruits are used to treat thrush, and to help children with nocturnal incontinence. Women also use the berries mashed in water as a drink to "check" to see if they are pregnant or have missed a menstrual period for some other reason (Morton 1981).

The Seminoles used two other species that are not that similar, but their names have been confused (Sturtevant 1955). Both have been incorrectly called *S. verbascifolium*. Now the most commonly used plants are known to be *S. erianthum*, and those formerly called *S. blodgettii* or *S. verbascifolium* are actually *S. donianum*. *Solanum donianum* is rarely used, even though the Seminoles call it *akkónónô:wî* (Mikasuki), or *akkononô:wâ* (Creek). Sturtevant (1955) considered the Mikasuki a simple term loaned from Creek. Wood from its stems were formerly used for arrows. Mullen nightshade [mullein night-shade] has been recorded as a common name for *S. donianum*, but that was probably an error for *S. erianthum*. Leaves on *S. erianthum* are woolly, much as they are in mullein (*Verbascum thapsus*, Scrophulariaceae).

Solanum erianthum is called *acolaki i:hici* (*vculvke*, old people, *em*, their, *hece*, tobacco, Creek) and *honaknõ:sá:lî imakcomî* (Mikasuki), with both names meaning "old people's tobacco." Others who compare it with tobacco call it *tabac marron* (wild tobacco, Hispaniola), tobacco tree, *tobaco bobo* (crazy tobacco), *tobaco cimarrón* (wild tobacco, Cuba), *tobaco del monte* (wild tobacco), *tabacón* (big tobacco), *tabacón afelpado* (hairy tobacco, Puerto Rico), *tabacón peludo* (hairy tobacco, Puerto Rico), *tabaquillo* [*tobaquillo, tabacuelo*] (little tobacco, Dominican Republic), and wild tobacco (Puerto Rico). Apparently none of these people use the leaves as tobacco, but simply note the similarity. It is well that it is not smoked because it contains, among others, the alkaloids solanidine and solasodine (Morton 1981).

The names *amourette* (little love, Haiti), *zamorette* [*bâtard*] ([false] little love, Haiti), *zamorette male* (bad little love, Haiti), and *zamorette marron* (wild little love, Haiti) perhaps allude to the similarity to the tomato (*Lycopersicon esculentum*) that is sometimes called *pomme d'amour* (love apple). Mayan people also see that similarity, and say *tom-paap* [*toon-paap*] (*tom*, wild, *pak'*, tomato, Yucatán). The same relationship may include the Spanish name *galantea* (lover, Oaxaca). Other Spanish speakers sometimes say *guastomate* (*huaqui*, dry, *tomatl*, tomato, Náhuatl, Mexico).

Allusions to other members of the family Solanaceae are in the names *berengena* [*berenjena*] (eggplant, Veracruz), *berengena cimarrona* (wild eggplant, Puerto Rico), and *berengena de paloma* (dove's eggplant, Cuba, Puerto Rico). Some English-speaking people say it is the potato-tree, noting its similarity to *S. tuberosum*.

Several of the common names allude to some physical trait. Because the leaves have a foul odor when touched, *S. erianthum* has been called *hediondilla* (stinker), *pendejera hedionda* (stinking hanger, Cuba), and *pendejera macho* (male hanger, Cuba). Probably *zorillo* (little fox, skunk or opossum) makes the same reference because those are mammals with strong and unpleasant (at least to humans) body odors. To people in northwestern Mexico the flowers are *cornetón de monte* (wild horn, Sonora). The name *trompillo* (little funnel, Venezuela) also notes flower shape. *Hoja blanca* (white leaf) comments on the pubescent leaves. "Male *pito seco*" (Belize) is a mixture of English and Spanish meaning "male dry flower," unless it is a version of *malo pito seco* (evil dry flower). Also in Belize, they are *palo blanco* (white tree).

Several of the common names are untranslated and others make obtuse references. These include *allay muy* (Belize), *buey nini* (*buey*, ox in Spanish, Mayo), *lengua de buey* (ox tongue, Sonora), *mambia* (Mayo), *tai-mollaca*, *tapalayote* (maybe from *tapalacyoa*, referring to broken tile, but reference obscure, Náhuatl, El Salvador), and *xtuhuy* (said to be Mayan, but no significance known, Yucatán). In Florida and Puerto Rico, the fruits are sometimes called turkey berry. That may be related to *zoza* [*sosa*] (tasteless, insipid, Morelos, Veracruz, San Luis Potosí, Sonora), and mean that the fruits are good only for turkeys. *Hoja de manteca* (butter leaf, Oaxaca) and *saca manteca* (butter maker, Sinaloa) are from using plant parts (which vary locally) to curdle milk.

Wild susumber looks like it should be a Creole variant of French *succomber*, and Latin *succumbere*, related to our English "succumb," or to overpower. "Susumber" is a reference either to medical uses or to similarity with *S. torvum*, a popular seasoning for foods (Grigson 1986). Either way, people throughout their ranges have utilized both species.

In Cuba and Mexico, the heated leaves of *S. erianthum* are used to relieve headaches by placing them on the forehead (Morton 1981). Similarly, leaves are poulticed on ulcers, boils, and burns (Standley 1920–1926, Martínez 1969). Venezuelans crush the leaves in oil to treat skin irritations, tumors, or swelling (Morton 1981). This use as a healing topical salve is reflected in big saab (saab = salve, Bahamas), *salvadora* (health giver, Tamaulipas), and salve bush (Bahamas). Leaf decoctions are used in Cuba as a depurative drink (Roig 1945), while Mexicans use the liquid to reduce fever (Martínez 1969). Bahamians make a remedy of the leaves for colds and coughs (Eldridge 1975).

Medicinal application surely resulted in the name *gordo lobo* [*guardolobo*] (fat wolf, Nuevo León). Since the plants are such good remedies, they are also called *funga* (to supply, maybe from *fungir*, based on the old Latin *fungor*, Mexico), and *hierba de San Pedro* (St. Peter's herb, Nuevo León). Although no record of being laxative has been found in the Americas, the name *friega-plato* (plate breaker, Veracruz, Dominican Republic) suggests that usage. In Asia *S. erianthum* is used to treat dysentery (Hocking 1997).

Leaves and fruits are known to contain the alkaloids solaidine and solasodine (Morton 1981), among others (Zhou and Ding 2002). In spite of those potent compounds, they are considered only "slightly toxic" to cattle (Morton 1981).

My favorite names for these plants come from the Yucatán peninsula. There the plants are *yerba soler* (commonly used herb) in Spanish. That name succinctly condenses the many applications. The Mayas of Yucatán call them *xaxox* (*xux*, wasp, *ek*, star, or morning star). They do so because the flowers are open early in the morning and that is the same time that the wasps are most active. Many Mayan names are purely descriptive, but this one reveals their poetic soul.

Solidago

(Latin for *solido*, to make firm, whole, or strengthen; decoctions were applied to cure wounds and ulcers)

- *a'djidamo'wano* (squirrel tail, Ojibwa)
- *amotci' hayi'* (always dwells-in fields, Biloxi)
- *chamizo blanco* (white sage, New Mexico, fide Cobos 1983, but Curtin 1947 applied this name to *Chrysothamnus*)
- *eeche chatehche* [*itciçatistci, i:chicá:tíhcí*] (blind deer plant, Mikasuki); *éco-heceko* [*'coheceko, cohicikô*] (*eco*, deer, *heceko*, blind, Creek, Muskogee)
- flowers-of-gold
- *fuinseag-coille* [*fuinnseag-coille*] (*fuinseag*, setting sun, *coille*, woods, Gaelic)
- goldenrod (William Turner, in his herbal of 1568, wrote, "*Virga aurea*, may be called in English Golden-rod"); *Goldrute* (golden rod, German); *gullris* (golden rod, Norwegian); *vara de oiro* (golden rod, Portuguese); *vara de oro* (golden rod, Spanish); *verge d'or* (golden rod, French,

Solidago. *Solidago canadensis* (top left). a. Habit. b. Flower head. c. Disk flower. d. Ray flower. e. Achene. f. Leaves, upper (u) and lower (l) surfaces. *Drawn by Regina O. Hughes.* From Reed 1971. *Solidago fistulosa* (top middle). From Britton and Brown 1898. *Solidago gigantea* (top right). From Britton and Brown 1898. *Solidago odora* (bottom left). *Drawn by P.N. Honychurch.* *Solidago sempervirens* (bottom middle). From Britton and Brown 1898. *Solidago stricta* (bottom right). From Britton and Brown 1898.

Quebec); *virga aurea* (*virga*, rod, *aurea*, golden, Latin; a name in use since Roman times); *slat òir* (*slat*, twig or rod, *òir*, gold, Gaelic)
hathacipa (Alabama); *hissittiłopótli* (*híssi,* hair, *ittiłí*, eye, *pótli*, to feel, Koasati)
kowicikó (*kowí*, mountain lion, *cikó*, son, Koasati)
mariquilla (New Mexico)
okhinsh balali (*ikhinsh*, medicine, *balali*, creeper, Choctaw)
pluma de oro (golden plume, Spanish)
pukwachi (Plains Cree)
solidage (French)
yellow-tops [-weeds] (New Brunswick, Pennsylvania, Vermont)
zha-sage-zi (*zha,* plant, *sage,* hard, *zi,* yellow, Omaha-Ponca)

Solidago canadensis (of Canada) (fide Snow and Stans 2001)
chachamos kakew (it makes one sneeze, Cree)
[Canada, high] golden-rod; *verge d'or géante* (big goldenrod, Quebec)
mariquilla (New Mexico, Mexico)
owesa'wenokûk (yellow top, Potawatomi)
yellow weed

Solidago fistulosa (hollow and cylindric) (fide Snow and Stans 2001)
pine-barren golden-rod [pinebarren goldenrod] (Florida)

Solidago gigantea (gigantic) (=*Solidago leavenworthii*)
[giant, late] golden-rod [goldenrod]

Solidago nemoralis (growing in shady places) (fide Speck 1941)
[compass, dyers weed, gray, old-field] golden-rod [goldenrod]
l'herbe à St. Jean (St. John's herb, Houma, Louisiana)

Solidago odora (fragrant)
Blue-Mountain tea
Bohea-tea ("Bohea" was derived from Chinese *wu-i-(shan)* or *wu-i-cha*, referring to the *Wu-i* hills in northern Fuhkien; it was from that region that black tea was first brought to England; about 1701, the name "Bohea-tea" was used for the imported product; after the Boston Tea Party, the name began being used for *Solidago* tea exported to China)
[anise-scented, fragrant, sweet-[scented], true] golden-rod [goldenrod]
Liberty tea (a name popular during the Revolutionary War)
wound-wort [woundwort]

Solidago rugosa (wrinkled)
tall hairy golden-rod

Solidago sempervirens* var. *mexicana (*sempervirens*, evergreen; *mexicana*, of Mexico) (fide Bennett 1997)
[seaside, narrow-leaved] golden-rod
pluma de oro (golden-rod, Cuba)

Solidago stricta (narrow) (fide Sturtevant 1955)
willow-leaf goldenrod

Solidago ulmifolia (elm-leaved)
elm-leaved golden-rod

Linnaeus used a number of methods to gain the interest of his students, among them planting a "floral calendar" at his home. He was not the first to use plants as indicators of the seasons; non-industrial people had been doing that for thousands of years. Gilmore (1919) recorded that application among the Omaha of the Missouri River region. These people considered *Solidago* a part of the natural almanac. He wrote: "Goldenrod served the Omaha as a mark or sign in their floral calendar. They said that its time of blooming was synchronous with the ripening of the corn; so when they were on the summer buffalo hunts on the Platte River or the Republican River, far from their homes and fields, the sight of the goldenrod as it began to bloom caused them to say, 'Now our corn is beginning to ripen at home.'"

Before exploration of the New World began, Europeans knew a single species of goldenrod, *S. virgaurea.* That herb had been considered medicinal from the time of the classical Romans, probably earlier. Linnaeus ([1753] 1957) adopted the old Latin name for the species, *virga aurea* (goldenrod). Known in the pharmaceutical trade as *Radix et Herba Virgae aurea, Herba* [*Solidaginis*] *Virgaureae*, or *Herba Consolidae Saraceniae*, the species was considered carminative, diuretic, nervine, extitant, and digestive (Uphof 1968, Hocking 1997).

By the 16th century, the Old and New World species were famous for their medical properties. Typical of the time are the comments by John Gerarde ([1597] 1975) who wrote of the goldenrod: "It is extolled above all other herbes for the stopping of bloud in sanguinolent ulcers and bleeding wounds; and hath in times past beene had in greater estimation and regard than in these dayes: for in my rememberance I have knowne the dry herbe which came from beyond the sea sold in Bucklers Bury in London for halfe a crowne an ounce. But since it was found in Hampstead wood, even as it were at our townes end, no man will give halfe a crowne for an hundred weight of it; which plainly setteth forth our inconstancie and sudden mutabilitie, esteeming no longer of any thing, how pretious soever it be, than while it is strange and rare. This verifieth our English proverbe, Far fetcht and deare bought is best for ladies." He should have said "people."

Since Linnaeus ([1753] 1957) wrote about these plants, more species have been found. *Solidago* is now a genus of about 80 species with a few species in South America, Macaronesia, and Eurasia (Mabberley 1997, Diggs et al. 1999). Formerly there were more species in *Solidago* (see Fernald 1950, Steyermark 1963), but many have been moved to other genera.

Moerman (1998) listed 19 species used across North America by many tribes. Throughout the world, many other species have been used by different cultures (Gerarde [1597] 1975, Culpeper 1653, Uphof 1968, Hocking 1997, Mors et al. 2000).

The genus was formerly a well-known dye plant, flowers producing a yellow-green (Tull 1999). Seminoles still use the flowers and stems as the source of a dye (Bennett 1997). Porcher (1863) also recorded that *S. canadensis* was "[u]sed in Canada as a most valuable dye. ... The leaves and flowers of the English species are used for making a yellow dye; said to be as good as woad [*Isatis tinctoria*]."

Porcher (1863) listed, "Among our native substitutes for hemp ... *Solidago canadensis* ... Its stalks are numerous, straight, and grow almost five feet in height; they afford very strong fibres if treated in the same manner as hemp." The Keres also used *S. gigantea* to weave rough baskets (Moerman 1998).

Among the Florida species used by different people are *S. canadensis, S. gigantea, S. nemoralis, S. odora, S. rugosa, S. sempervirens* var. *mexicana*, and *S. ulmifolia.* Several reports are simply listed under the genus, and more species may have been employed.

The Iroquois used *Solidago canadensis* to treat side pains, as an emetic, to "kill" or counteract a love potion, as a "gambling medicine," and as a medicine to treat babies who awaken suddenly. The Meskwaki used Canada golden-rod in a wash for a child who does not talk or laugh. The Ojibwa applied it to burns, boils, and ulcers. The Potawatomi treated some fevers with it (Smith 1933). The species was also medicinal among the Okanagan-Coville, Shuswap, Thompson, Zuni, and the New Mexican Hispanics (Curtin 1947).

Subsequently, *S. canadensis* has been considered a carminative, diuretic, and stimulant in herbal stores (Hocking 1997). It is the source of Canadian goldenrod oil. Duke et al. (2002) retain the Latin name but note that the material available in the markets might come from any of a number of species. They list the plants as having been used against cancer, inflammation, spasms, and as both diuretic and emetic.

Although Uphof (1968) considered *S. canadensis* an emergency plant food, it was not listed as edible by Fernald et al. (1958). Still, *S. canadensis* and *S. nemorealis* were eaten by the Gosiute and Navajo (Yanovsky 1936), and Foster and Duke (1990) note that *S. canadensis* seeds were eaten as survival food.

Solidago gigantea was used by the Keres as a strong physic. The Menomini, Ojibwa, and Potawatomi treated fever with a flower decoction (Densmore 1928, Smith 1933, King 1984, Moerman 1998). Hocking (1997) noted that this species had been used as a gargle tea in treating pharyngitis, tonsillitis, and stomach problems.

Solidago nemoralis was used by the Houma to treat jaundice (Speck 1941). The Iroquois treat kidneys with a decoction of the roots (Moerman 1998). The Navajo still use it for incense (Moerman 1998). Hocking (1997) recorded that the species was used in domestic medicines.

Perhaps the most famous use of *S. odora* is as a tea substitute. This tea was especially popular during the Revolutionary War. Fernald et al. (1958) thought that Johann D. Schoepf, who was the chief surgeon of one of the bodies of German troops sent to America by George II during the Revolution, was one of the first to record this beverage. While visiting North America in 1783–1784, Schopf wrote from Bedford County, Pennsylvania: "Here we were introduced to still another domestic tea-plant, a variety of *Solidago.* The leaves were gathered and dried over a slow fire. It was said that around Fort Littleton many 100 pounds of this Bohea-tea, as they call it, had been made as long as the Chinese was scarcer. Our hostess praised its good taste, but this was not conspicuous in what she brewed."

When the Americans rebelled against British taxes by dumping a cargo of taxable tea into Boston Harbor, they decided to use a native substitute. This "Liberty Tea," as they called it, was made from the same leaves that Schopf decried. Not everyone was as negative about the tea from *S. odora*, and Frederick Pursh wrote in 1807, "The flowers, gathered when fully expanded, and carefully dried, give a most agreeable substitute for tea, which for some time has been an article of exportation to China, where it fetches a high price."

Rafinesque (1828–1830) also regarded the tea as highly as Pursh. He wrote: "This [species is] easily known by its sweet scent near to aniseed. Essential oil of it has the same sweet scent, much used for head ache, in frictions. Whole plant aromatic, stimulant, diaphoretic, carminative, useful in flatulence, nausea, spasms of the stomach; chiefly used as a grateful tea. Leaves prepared like tea have been sent to China, much used in some parts of our country, used in fevers by Cherokees."

During the Civil War, Porcher (1863) encouraged people in the South to use this drink. He wrote, "According to Pursh, the dried flowers are a pleasant and wholesome substitute for tea."

Later people describing the tea from *S. odora* thought that the air-dried leaves had an odor like licorice, and said that they were used to make "goldenrod tea" (Correll and Johnston 1970). Hocking (1997) thought that it had the odor of fennel or anise. He also considered it carminative, stimulant, and diaphoretic when used to make a beverage tea (as Blue Mountain tea), with a flavor similar to catnip tea.

There were also medical uses of *S. odora.* Because the plant played such an important part in the Revolutionary War and served as an export to China, it seems odd that only one tribe is known to have used it. The Cherokee treated colds, coughs, diarrhea, female problems, fevers, nerves, sore mouth, and tuberculosis with these plants as a diaphoretic, stimulant, and tonic (Hamel and Chiltoskey 1975).

Jacob Bigelow wrote in his *American Medical Flora* of 1817–1820: "The claims of the *Solidago* to stand as an article of the Materia Medica are of an humble, but not despicable kind. We import and consume many foreign drugs which possess no virtue beyond that of being aromatic, pleasant to taste, gently stimulant, diaphoretic and carminative. All these properties the Golden Rod seems to fully possess."

Porcher (1863) wrote of *S. odora*: "An aromatic, moderately stimulant, and carminative plant, and like other substances of the same class, diaphoretic in warm infusion. It is used to allay pain from flatulence, lessen nausea, and cover the taste or correct the operation of irritating or unpleasant medicines. Mérat states that the infusion is also employed as an astringent in dysentery, and in ulceration of the intestines. . . . When the leaves are subjected to distillation a very aromatic, volatile oil collects, and an essence may be made by dissolving this in proof spirits. This will also stop vomiting and correct the taste of medicines, even laudanum and castor oil; . . . it is valuable in allaying the pain from headache, externally applied. It is much used in the Eastern states, and Bigelow thinks it will entirely supplant more expensive articles."

Solidago rugosa is recorded as being used only among the Iroquois. This alliance of tribes treated liver disease, weakness, dizziness, and sunstroke with the species (Moerman 1998). Hocking (1997) recorded that it was used to treat fever by Mississippi blacks.

Solidago sempervirens has been used to treat wounds and warts (Hocking 1997). No records have been found of it being used by indigenous tribes other than the Seminoles. Porcher (1863) considered the species, "[v]ery efficacious in the cure of wounds."

Solidago ulmifolia is known to have been used by only two tribes. The Meskwaki and Ojibwa used smoke from the burning herb to revive unconscious persons (Densmore 1928, Vogel 1970, Hocking 1997).

There are records of several other tribes using *Solidago*. No species identifications were made, and they may have been other species, or they might be those listed above. The Alabama used some species to treat colds and toothaches (Swanton 1928a, Taylor 1940). The Algonquin treated heart disease; the Blackfoot chewed roots for sore throat and nasal congestion (Moerman 1998); the Cherokee used it to treat fever; the Iroquois used it to stop internal bleeding; and the Ojibwa applied roots externally to stop cramps (Moerman 1998).

Swanton (1928) found the Creek using *Solidago*, as did their relatives the Alabamas, Houmas, and Miccosukees. Miccosukees called the plant "blind deer plant," but Sturtevant's (1955) informant considered it useless. That lapse of information is typical of the spotty knowledge of individuals, because Alice Snow now uses goldenrod, applying the same name, as an ingredient for cough medicine (Snow and Stans 2001). In addition, other individuals gave information to Bennett (1997).

Medical use of *S. virgaurea* is founded on saponins, which are antifungal, and rutin and phenolic glycosides, which are anti-inflammatory (Bown 1995).

Solidago odora was the only species to be listed in the U.S. Pharmacopoeia, from 1820 to 1882. Its properties were defined as stimulant, carminative, and diaphoretic (Vogel 1970).

In addition to the chemicals known from the European species, the American plants contain a variety of compounds. Golden-rod oil (volatile oils with pinenes, limonene) has been extracted from *S. canadensis, S. gigantea, S. nemoralis, S. odora*, and *S. rugosa* (Uphof 1968). Sometimes, that from *S. canadensis* is distinguished as "Canadian Golden-rod oil." Perhaps it has a distinct name because the volatile oils in *S. canadensis* also contain cyclocolorenone, carophyllene, borneol, cadinene, and terpene hydrocarbons (pinene, phellandrene, diapentene) (Hocking 1997). The others vary somewhat: *S. nemoralis* has pinenes and phenols; *S. odora* has estragol, borneol, bornyl acetate, and limonene; and *S. rugosa* contains pinenes and limonene.

Solidago canadensis also contains quercitin and rutin; *S. gigantea* has saponins; and *S. sempervirens* has dihydromatricaria ester in the roots (Hocking 1997). Inventor Thomas A. Edison found that *S. sempervirens* produced caoutchouc, and for some years he explored it as an alternate rubber (Hocking 1997). There has recently been a renewed interest in *Solidago* as an alternate rubber source (Anonymous 1990a,b, Tanaka et al. 1995).

Sophora

(Linnaeus took the Arabic name *sophera* or *safayra*, yellow, and applied it to *S. alopecuroides*)

Sophora tomentosa (woolly)
coast [yellow] sophora (book names, Bahamas, Texas)
haricot bâtard (false kidney bean, Guadeloupe, Martinique)
micar (Cayman Islands)
necklace pod (Florida)
sea-coast laburnum (Australia)
silver-bush (USA)
tambalista [*tambalisa*] (Cuba)

Sophora tomentosa*. Drawn by P.N. Honychurch.*

Linnaeus coined *Sophora* in 1738 and formalized it in 1753. There are now 45 species in the genus, which is known from both tropical and temperate regions. Europe has 2 species, and there are 9 in North America (Kartesz 1994). Vestal and Schultes (1939) and then Merrill (1977) made the southwestern North American species *S. secundiflora* one of the most famous in the genus because of its use among the Kiowa.

Reputed to be diuretic, sudorific, and purgative, *Sophora* has been used in the West Indies against venereal disease (Standley 1920–1926). In the East Indies, the root bark and seeds were used as a remedy for cholera, diarrhea, and the consequences of food poisoning, especially ciguatera (Burkill 1966). Powdered seeds were used by Malays against diarrhea, and both roots and seeds were used to treat external wounds and other skin problems (Burkill 1966). People in the Philippines use the herbage and seeds as a remedy for stomach ailments, and in New South Wales in bilious sickness (Uphoff 1968).

Seeds are poisonous; they contain cytisine and related alkaloids with properties similar to nicotine (Cribb and Cribb 1981, Lampe and McCann 1985). All parts of plants contain poisonous alkaloids (Standley 1920–1926). The juice of the plant has been used as a fish poison in east Africa and as an insecticide in Fiji (Hocking 1997). Presumably, this species contains sophorine; it may have been used in the Americas as an intoxicant as was *S. secundiflora* (Merrill 1977). Beans of *S. tomentosa* are used in China for intoxication (Hocking 1997).

Sparganium

(Greek name used for the European species by Pliny, A.D. 23–79, and Dioscorides, fl. A.D. 40–80, from *sparganion*; derived from *sparganon*, a swaddling band, an allusion to the ribbon-like leaves)

bur-reed ("bur" or "burr" is of uncertain etymology, but it alludes to a rough flower head or seed head; cognate with Danish *borre*, bur, and Swedish *borre*, sea-urchin; John Gerarde wrote in 1597, "These plants of some are called Sparganium . . . I rather call them Burre Reede")
Igelkolben (hedgehog bloom, German)
piggknopp (spike bud, Norwegian)
sala (literally "twill," but plant material used "to stuff with straw" or "to cane a chair," Italian)
seisg madaidh (*seisg*, sedge, *madaidh*, dog, Gaelic)

Sparganium americanum (of America)
[American, three-square] bur-reed [burreed]
ho'-xthon̲ ta-xe (Osage; a sacred plant used in ceremonial rites)
rubanier (ribbon plant, Quebec)

Dioscorides used two related words in markedly different contexts. One of the words was *sparganion*, from which Linnaeus derived the generic name

Sparganium americanum. From Institute of Food and Agricultural Sciences.

Sparganium. Botanical sources say that Dioscorides and Linnaeus based the name on the ribbon-like leaves of these wetland plants (e.g., Fernald 1950, Munz 1973, Quattrocchi 1999). However, Dioscorides also wrote about *sparganosis* or *spargosis*. Those two words are based on Greek *spargar*, to swell, and refer to breasts swollen with milk (OED 1971). Could it be that both Dioscorides and Linnaeus were talking about the swollen fruit clusters in the plants? At least since the time of Gerarde, these cattail relatives have been called "bur-reed," in reference to those enlarged fruits. That duality of meaning is just the kind of complexity that Linnaeus loved to use in names.

Sparganium grows from the Mediterranean through most of Europe, and the species that Dioscorides knew still has the name he gave it. Linnaeus knew only two European species. Today there are 14 in the genus worldwide, with 7 in Europe and 10 in North America (Mabberley 1997). Biologists have long thought that reed-maces (cattails) and bur-reeds were closely related. Although Wunderlin (1998) continued using distinct families for them, others combine them into the single family, Typhaceae (Judd et al. 2002).

Indigenous people knew bur-reed as a source of food, although considerable energy had to be expended for a comparatively small return. The creeping roots bear tubers that have been gathered and eaten, although actual records have been found for only the Klamath of Oregon eating them (Fernald et al. 1958, Moerman 1998). While cattails are famous for their edible young staminate inflorescences, Judd et al. (2002) indicate that *Sparganium* may be similarly eaten. A practice related to food, but not actually eating these plants, is recorded for the Okanagan-Coville who lined cooking pits with the leaves (Moerman 1998).

In the East, only the Iroquois are known to have used bur-reed in medicine. They made an infusion that was drunk to relieve chills, and put the herbs into a more complex mixture for another purpose. The Iroquois believed that it was possible to cast a spell on another individual, and one symptom of the enchantment was that men were "sore all over." The complex herbal mixture was used to counteract that magic (Moerman 1998).

A side benefit to *Sparganium* that indigenous people surely utilized is that the fruits are a favored food for wildlife. As Martin et al. (1951) wrote, "They are more plentiful in the North, and it is here that they have their principal importance for wildlife." Their book *American Wildlife & Plants* listed 19 different species of waterfowl, marsh, and shorebirds that ate the seeds. Among those are many that were favored food for indigenous people such as 11 species of ducks, cranes, and snipe. Muskrats (*Ondatra zibethicus*), another animal favored by many tribes, "use the whole plant" (Martin et al. 1951). In Florida, that species is replaced by the round-tailed muskrat (*Neofiber alleni*), and it too may have favored the plants.

Spartina

(Johann Christian Daniel von Schreber, 1739–1810, named the genus with Greek *spartine*, *sparton*, a rope; akin to Latin *spartum*, a grass used for cordage, nets, and mats, Akkadian *sabaru* and Hebrew *safar*, to bind)

cord grass (USA, England)
marsh-grass (USA); *marskgras* (marsh-grass, Norwegian)
Reisgras (rice grass, German)

Schlickgrass (slime grass, German)
spartine (French)

Spartina alterniflora (alternate leaved)
[salt-marsh, saltmarsh, saltwater, smooth] cord-grass [cordgrass] (Florida, northeastern USA, Texas)
herbe salée (salt grass, Quebec)

Spartina bakeri (named for C.H. Baker, who collected the type specimen in Florida on 19 April 1898)
sand cord-grass [sand cordgrass] (Florida)

Spartina patens (spreading or diverging, from the spikes bending out from the axis)
[marshhay, salt-marsh, salt-meadow, saltmeadow] cord-grass [cordgrass] (east coast USA to Texas, Bahamas)
highwater grass (northeastern USA)
musotte (not French; perhaps an indigenous word, Quebec)
salt grass (Puerto Rico)
yerba de sal (salt grass, Puerto Rico)

Spartina spartinae (a cord)
Gulf cord-grass [Gulf cordgrass] (Florida, Texas, Bahamas)
sacahuista (*zacatl*, grass, *cuahuitl*, tree, Náhauatl, Texas)

Jonathan Dickinson's was one of the few historical documents commenting on the houses built by the Glades people (Andrews and Andrews 1945). He mentioned thatch only with palm leaves. The earlier account by Cabeza de Vaca, speaking of the people on the Gulf Coast, said, "Their huts were of mats erected on piles of oyster shells" (Cabeza de Vaca in Gilliland 1975). Mats could have been made of a number of plants. However, the Marco Island archaeological site provided more evidence of housing materials. Gilliland (1975) quoted from archaeologist John M. Goggin's manuscript that "these [houses] consisted of palmetto and marsh grass thatch." Unhappily, none of the fibers recovered was preserved well enough for Gilliland to make a positive identification.

However, Goggin's comments, and the record of other people using *Spartina* in thatching is compelling. Gilmore (1919) records that the people of the Missouri River region used slough grass (*S. pectinata*) to thatch their earth lodges. The Omaha-Ponca called it *sidu-hi*. The Iroquois are documented as using *S. alterniflora* in a different way, as "forage," presumably for livestock (Moerman 1998).

Spigelia

(With this name, Linnaeus commemorated the Dutch physician Adriaan van der Spiegel, 1578–1625, professor of anatomy and surgery in Padua)

Spigelia anthelmia (against worms)
arapabaca (cow killer, Minas Gerais, Brazil)
Brinvillière [*Brinvilliers, herbe-Brinvilliers*] (Guadeloupe, Martinique)
caricia (dear one, Colombia)
cuntv-heleswv (*cuntv*, worm, *heleswv*, medicine, Muskogee)
dronkegoeman [*droengoeman, droegoman*] (drugman, Sranan, Suriname)
espigelia (Hispaniola)
guambía (child's luck, Colombia)
herb-poison (poison herb, Guadeloupe, Martinique)
inanusu (Cuna)
kinchan (Zoque-Popoluca, Veracruz)
koroka (Carib, Suriname)
loggerhead-weed [loggerhead weed] (Barbados)
lombricera [*lombricero*] (wormer, Puerto Rico, Mexico, Panama, Colombia, Venezuela); *lombricera del pasto* (worm grass, Puerto Rico); *lombrigueiro* (wormer, Minas Gerais, Brazil), *pega pinto* (symptom stopper, Peru); worm grass [bush] (Jamaica, Virgin Islands, Nicaragua, Panama); *yerba* [*hierba*] *lombricera* (worm herb, Cuba); *yerba de lombrices* (worm herb, Dominican Republic); *zacate de lombriz* (worm grass, Nicaragua)
pink (Bahamas); pink weed (Jamaica); West Indian pinkroot (Florida)
poudre à vers [*poudre aux vers*] (powerful powder, Haiti, Guadeloupe, Martinique)
quiteria (expeller, Colombia)
tebeyuballi (*tebeyu*, a name for five other genera, *balli*, resembling, Arawak, Suriname)
water grass (this name and the following apparently allude to watery stools, Virgin Islands); water weed (Barbados)

Spigelia marilandica (of Maryland)
ala imokhinsh (*ala*, arrival, *im*, its, *ikhinsh*, medicine, Choctaw; used as a "tonic"); perhaps this is *chokfisholosh* (*chokfi*, rabbit, *im*, its, *sholosh*, shoe, Chickasaw)
mi-ka-k'e [*mekaa, mikaa*] (Osage; literally, "star"; called "starflower" by Hunter [1823] 1973; Millspaugh 1892 does not give his source for identifying the plants as *Spigelia*, but it was probably from Rafinesque's *Medical Flora* of

Spigelia. *Spigelia anthelmia* (right). a. Habit. b. Papillae on surface of leaves, greatly enlarged. c. Flower bud. d. Flower, side view. e. Flower, longitudinally dissected. f. Floral diagram. g. Fruit. h. Seed. *Drawn by Priscilla Fawcett.* From Correll and Correll 1982. *Spigelia marylandica* (left). From Britton and Brown 1897.

1828–1839); starbloom; starflower (from Hunter [1823] 1973)
[Carolina, Indian]-pink
[woodland] pinkroot (in use since at least the 1720s when Catesby was in South Carolina; the first record found by OED 1971 was 1763 in a record of the *Produce of South Carolina*—one cask was listed for export)
snakeroot
unsteetla [*unstilla*] (Cherokee)
wormroot

Marggraf brought the Europeans' attention to these medicinal plants in early 1648. He called one species *Arapabaca brasiliensibus dicta planta* (the plant Brazilians call *arapabaca*). Charles Plumier also used the Brazilian name in 1703. Linnaeus did not believe in other people using "foreign" (i.e., non-European) names for genera, and coined the genus with *Spigelia anthelmia.* There are now 50 species in the tropical and warm parts of the Americas (Mabberley 1997).

Vogel (1970) thought that the North American species *S. marilandica* became known in Europe when physician Alexander Garden of Charleston learned of its use from the Cherokee. By 1752, Garden had sent specimens to European colleagues, and in 1768 he sent a specimen of it to physician Cadwallader Colden with the comment that it was "an Excellent anthelmintic Given either in powder or Decoction" (Vogel 1970). In spite of Vogel's (1970) version of medical use, the actual European history is considerably older.

Spigelia marilandica was first collected by Hugh Jones in Maryland. After arriving in 1696 as a chaplain, Jones did not live up to the expectations of Royal Governor Francis Nicholson, and he was assigned to the Christ Church Parish in Calvert County where he spent much of his time collecting plants (Reveal 1992a).

John Ray in 1704 was one of the first to mention these American herbs in his *Historia plantarum.* He called them *Periclymeni virginiani flore coccinea planta marilandica* (Virginia honeysuckle with a red flower, plants from Maryland). Probably Ray's examples were cultivated and no herbarium samples made. Important specimens in Europe were sent there from Charleston by John Clayton in the 1730s. Jan Gronovius used the Clayton specimen, now preserved in the British Museum of Natural History in London. In his *Flora Virginica* of 1739–1743, Gronovius called the plants *Lonicera spicis terminalibus, foliis ovato-oblongis acu-*

minatis distinctis sessilibus (Honeysuckle with terminal spikes, and distinctly sessile ovate-oblong and acuminate leaves). Gronovius knew about the medical uses of the plants because he cited Catesby (1734). Catesby had ample opportunity to learn about them because he was in Williamsburg, Virginia, between 1712 and 1719. He returned to England, then went back to Charleston, South Carolina, in May of 1722. Somewhere between 1712 and his return to England in 1726 to publish his book, he learned about the vermifuge properties of "pinkroot" and wrote, "A Decoction made of this Plant is good against Worms" (Catesby 1734). Physician Benjamin Rush, in 1774, considered *S. marylandica* and *S. anthelmia* to be "two of the most powerful vermifuge medicines we are acquainted with."

The Cherokee, Choctaw, Creek, and Osage used *S. marilandica* to expel worms (Hunter [1823] 1973, Swanton 1931, Taylor 1940, King 1984, Moerman 1998), and others within its range (Maryland to Oklahoma and south to Texas and Florida) probably also did. Vogel (1970) wrote that, among settlers, "For almost two centuries this [*S. marilandica*] was the most celebrated vermifuge on the American continent."

From the earliest reports, the Creeks considered *S. anthelmia* a "well-known" remedy for expelling worms in children (Swanton 1928a). Since the herb has always been confined to what the Seminoles called *ekan-yuksa* (point of land) of southern Florida (Miami-Dade and Monroe Counties), they either learned about it from their predecessors, or through extrapolation from formerly having used the more widespread *S. marilandica.*

In the Caribbean and elsewhere within the range of *S. anthelmia,* most people have used the herb to expel intestinal worms. Roig (1945) found it being used in Cuba, and common names and records show use in the other islands (Petersen 1974, Proctor 1984). People from the Chocó and Cuna of Panama to French Guiana and Brazil also use *S. anthelmia* to eliminate parasites (Duke and Vásquez 1994, Mors et al. 2000).

In the Caymans, people know that cut and wilted pink weed "is extremely dangerous and may cause fatalities among livestock; the fresh plant is apparently less dangerous" (Proctor 1984). Small doses are effective vermifuges, but large doses can kill (Duke and Vásquez 1994). Caribbean people know that pink weed has been used in criminal poisonings (Uphof 1968). The plants contain isoquinoline and one iridoid, both of which are cardiotonically active (Wagner et al. 1986).

Other uses include the leaves to repel cockroaches and flies. In light doses it is effective against stomachaches; Chocó use it as laxative (Duke 1972). In addition, a root infusion is used to bathe children as a sedative and tranquilizer in northwestern Amazonia (Schultes and Raffof 1990).

Spiranthes

(Named from Greek *speira,* a spiral or coil, and *anthos,* flowers, in reference to the spiral or twist of the inflorescence)

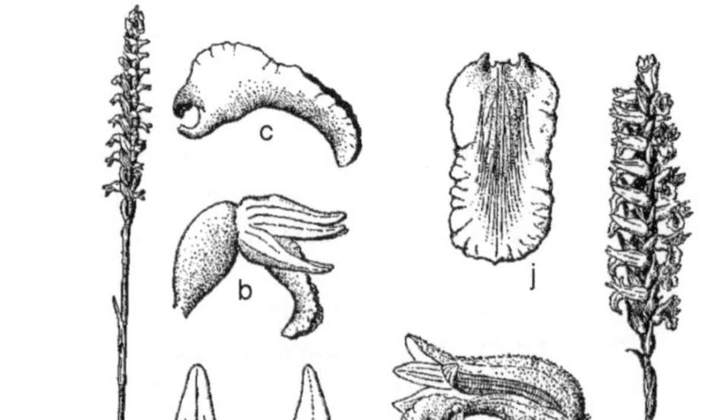

Spiranthes. a to g. *Spiranthes ovalis*; a. Habit of plant. b. Flower, side view. c. Lip, side view. d. Lip, spread out. e to g. Variation in leaf tip venation and shape. h to j. *Spiranthes cernua*; h. Habit of plant. i. Flower, side view. j. Lip, spread out. k. *Spiranthes odorata* with lip spread out. *Drawn by Gordon W. Dillon.* From Correll and Correll 1972.

bayonet-lily (Newfoundland)
Drehwurz (twisted herb, German)

ibis (allusion obscure, *ibis* the Greek and Latin name for a wading bird; for *S. torta*, Puerto Rico)
ladies'-dresses (a corruption or variation on either "ladies'-tresses" or ladies'-traces")
ladies'-traces (comparing the flower cluster to a braid; a name that has been used since at least the time of Turner's *Names of Herbes* of 1548, when he wrote, "Satyrion ... bryngeth furth whyte floures in the ende of harueste, and it is called Lady Traces"; Dodoens wrote in 1554, "The sweete Orchids, or Ladie Traces are moste commonly to be found in high, untilled, and dry places"; some spelling variant of ladies'-traces was used at least until 1794)
ladies'-tresses [lady's-tresses, ladiestresses orchid] (a spelling given to these orchids since 1842 when used in C.W. Johnson's *Farmer's Encyclopedia*; the name compares them to women's braided hair)
orphrys (from Middle English *orfreis*, ultimately from Latin *auriphrygium*, Phrygian or gold-embroidered garments; akin to Old French *orefris*, Portuguese *aurfres*; name for *S. tortilis* used by Williams [1837] 1962)
pearl-twist
Satyrion à trois couillons (literally, "Satyr with three testicles," but translated by John Cotgraves as "triple orchis, or Ladies traces" in his 1611 book *A Dictionary of the French and English Tongues*); *testicolo odoroso* (fragrant testicle, Italian)
skrueaks (*skru*, screw, Norwegian); spiral-orchid (New Hampshire, Texas); *spiranthe* (French)
Wendelhäre (spiral-hair, German)

These terrestrial orchids have been known as "ladies'-traces" or "ladies'-tresses" since at least the 1540s when they were listed by British herbalist William Turner. Indeed, Linnaeus knew some of them in 1753, but he included them within his genus *Ophrys* (eyebrow, because of hairy flower labellum or lip). It was not until after Louis Claude Marie Richard (1754–1821) had studied plants in South America and the Caribbean that the genus *Spiranthes* was created for them in 1817.

Europe has three or perhaps four *Spiranthes*, and most of the others are in the northern temperate zones (Mabberley 1997). Liogier and Martorell (1982) included eight species in Puerto Rico, although at least four of those are now placed in other genera. Kartesz (1994) recognized 35 species and a named hybrid in North America and Greenland, although Mabberley (1997) said there were only 30 species in the genus. Those differences are typical of individual interpretations of the genus.

As other names like *Satyrion* (the Greek *satyr*, who was lascivious), *Orchis* (testicle, Greek), orchid (testicle), and even the family name Orchidaceae, suggest, there has long been an association of these monocots and sexuality. Those names came into existence because some of the plants bear two rounded bulbs at their base with an uncanny resemblance to human anatomy. The beliefs in Europe may be illustrated by one example. In 1597, John Gerarde wrote of *Spiranthes*, "The full and sappy roots of Ladie traces eaten, or boyled in milke and drunke, provoke venery, nourish and strengthen the bodye, and be good for such as are fallen into a consumption and Fever Hectique." Many more examples exist for *Orchis* and others (e.g., Bedigian 2001, Gibson 2001).

Apparently, that same notion existed among indigenous Americans. The Gosiute, for example, used *S. romanzoffiana* against venereal disease (Moerman 1998). The same ideas may have been inherent in the Cherokee use of an *S. lucida* infusion as a wash for infants to ensure fast healthy growth, and mixing it with roots of tway-blade (*Liparis lilifolia*) for urinary trouble (Hamel and Chiltoskey 1975). At least, the power they believed resided in the plants is clear from the Ojibwa use of roots of *S. lacera* var. *gracilis* as an ingredient of the hunting charm to bring game to the hunter (Moerman 1998).

The only record found of Florida tribes using *Spiranthes* was made by Small (1933). Under the heading of the Ladies'-tresses he wrote, "A decoction made from the roots of some of the species is used by the Seminoles as a blood purifier." That may be a valid medical use, as other members of the genus are known to be medically active. For example, *S. diuretica* has been used to treat the urinary system (Lewis and Elvin-Lewis 1977).

While Gerarde recorded that Europeans ate orchid roots, and other genera are notably edible (Bedigan 2001, Gibson 2001), there is a single record of indigenous tribes eating *Spiranthes*. The Paiute ate roots of an unidentified *Spiranthes* (Moerman 1998).

Luer (1972) thought that the common name ladies'-traces came from "the traces or strings of a woman's bodice." Actually, the application is older than that women's clothing style. The word "trace" and its plural "traces" have meant a braid or plait of hair since at least A.D. 1380. By 1539, it was broadened to mean a braid of gold or silver used to trim the border of a robe, and by 1678, a string of maize (*Zea mays*) ears plaited together by the husks (OED 1971).

However, the word "tress" is just as old. It too appears in English in the 1300s. Chaucer himself wrote in 1386 about a woman that "Hir yelow heer was broyded in a tresse Bihynde hir bak, a yerde longe I gesse." Both "tress" and "trace" are cognates with French *tresse*, which came from Old French *tresce* (used in the 12th century for a braid of hair). Those words are also related to *tressa* or *treza* (Portuguese) and *treccia* (Italian). There are also infinitives *tresser* (French), *trecier* (Old French), and *trecciare* (Italian). The derivation of these words in all these languages may be from Greek *tricha*, "threefold, or a triple plait."

Sporobolus

(From Greek *spora*, seed, *bolos*, casting; the scientific and common name '"dropseed" both refer to the seeds being dispersed)

Sporobolus virginicus. *Drawn by Edna May Whitehorn*. From Hitchcock and Chase 1950.

Sporobolus pyramidatus (pyramid-shaped, alluding to the inflorescence)
brak grass [*brakgrass*] (break grass, Dutch Antilles); *brakgras doeshi* [*bragaadushi*] (sweet break grass, Dutch Antilles)
chiendent (dog tooth, Haiti)
match grass (Bahamas)
whorled dropseed (Bahamas, USA)
yerba fini (fine or slender herb, Dutch Antilles)

Sporobolus virginicus (from Virginia)
chiendient [*chien dient*] (dog tooth, Haiti)
crab grass (Barbados)
[seashore, coastal] dropseed ("dropseed" was first used for a *Muhlenbergia* in 1866 and Robert Brown created *Sporobolus* in 1810, northeastern USA, Carolinas, Florida, Texas, Bahamas)
grama (grass, Spanish)
herb [*zeb*] *bord de mer* [*lan-mer*] (seaside herb, Guadeloupe, Martinique)
matojo de burro [*de playa*] (donkey [beach] thicket, Puerto Rico)
[sea-shore] rush-grass [seashore rush grass] (northeastern USA, Bahamas, Puerto Rico)
saladilla (little salty one, Venezuela); salt grass (Florida)

Robert Brown described *Sporobolus*, based on grasses from India. Much earlier, John Clayton had sent Linnaeus grasses from Virginia, and among them was one that he called *Agrostis virginicus*. Lamarck made the same choice of genus in 1791 when he named *Agrostis pyramidatus*. Both were moved to *Sporobolus* much later, and other species added to them brought the total to 160, with different kinds growing in Africa, the Americas, and Asia (Mabberley 1997).

In the Caicos Islands, dry *S. pyramidatus* is used for kindling; it ignites quickly and makes a hot fire (Morton 1981). A decoction of the grass is diuretic. The mixture is used to treat severe fever and colds, as a purge in Curaçao, and as a laxative in Aruba (Morton 1981).

In Coró, Venezuela, decoctions of *S. virginicus* are taken to cure urinary irritation and kidney complaints. The grass is also made into a gargle (Morton 1981, Hocking 1997).

Stachys

(From Greek *stachys*, ear of wheat or spike, because of the terminal spike or raceme)

ande'gobûg (crow foot, for *S. palustris*, Ojibwa)
betony (from Latin *betonica*, which Pliny, A.D. 23–79, said was a Gaulish name from *vettonica*, derived from Vettones, people of Lusitania, *S. officinalis*); *betoine* (Dutch, Norwegian); *bétoine* (in use by 1550, French); *betonica* (in use by 1551, Italian); *betonica* (in use by 1557, Spanish); *Betonik* (in use by 1542, German)
brisgean nam caorach (brittle one of the sheep, for *S. palustris*, Gaelic)
épaire (from *epi,* spike, because of the form of the inflorescence, Quebec)

hedge nettle (an allusion to where the plants grew and the similarity of their leaves to nettle or *Urtica*; in use by 1678)
kestron [*cestron*] (Greek for a graving tool or hammer, for *S. officinalis*, used by Dioscorides, fl. A.D. 40–80)
lus nam sgor (plant of the sharp rocks, for *S. sylvatica*, Gaelic)
ortiga hedionda (stinking nettle, for *S. officinalis*, Spanish)
psychotrophon (*psycho*, cold, *trophon*, turning, because it grows in cold dark places, for *S. officinalis*, Greek)
svinerot (*svine*, hog, *rot*, root, Norwegian)
woundwort (dating from at least 1548 when Turner wrote in his book *Names of Herbes*, "wounde-wurte," but he applied it to *Solidago*; John Gerarde perhaps used it for *Stachys* in 1597, but he also used the name for *Solidago*; the name was applied to *Stachys* by the 1790s)
Ziest (German)

Stachys floridana (of Florida)
Florida betony
Florida hedge-nettle [hedgenettle]
knotroot

Stachys tenuifolia (thin-leaved)
épiaire (Quebec)
slender-leaf betony
smooth hedge-nettle

Pliny wrote of what we now call *Stachys officinalis*, "The Vettones in Spain discovered the plant called *vettonica* in Gaul, *serratula* in Italy, and *cestros* or *psychotrophon* by the Greeks, a plant more highly valued than any others" (Meyer et al. 1999). Along with that species, two others were famous among Europeans before they arrived in the New World—*S. palustris* and *S. sylvatica* (Fuchs 1542, Linnaeus [1753] 1957, Dobelis 1986). Two well-known English names of these were betony and woundwort, and all were renowned as treatments for sores and open wounds.

The plants were not simply known as medicines. Most, perhaps all, species have enlarged tubers that were used as food from Europe to Asia before the New World was discovered. John Lightfoot, in his *Flora Scotica* of 1777, wrote of *S. palustris* tubers that they were "sweet, and in times of necessity ... eaten by men, either boiled, or dry'd, and made into bread." Fernald et al. (1958) had not tried that species, but spoke highly of related *S. hyssopifolia* in New England. They found the nutty tubers "white and as good a nibble or salad as one could wish."

Although Hocking (1997) put *S. sieboldii* and *S. affinis* in synonymy with *S. floridana*, Wunderlin (1998) did not consider them the same species. Presumably, in spite of that, Hocking's (1997) comments still apply to the Florida plants; he considered *S. floridana* edible, and thought the tubers had a taste similar to cauliflower. In France, the mint (called either *S. sieboldii* or *S. affinis*) is named *crosnes de Japan*. That name was given because the plants were introduced in 1822, 1886, or 1882 (there are conflicting dates recorded), into the French town of *Crosnes* near Paris (Hedrick 1919, Larousse 1987). Paradoxically, they then became known for the French town and for Japan.

No record has been located of *S. tenuifolia* being eaten, but perhaps it too was used; *S. floridana* tubers are eaten in Georgia (Von Reis 1973). Among the Meskwaki of Iowa, *S. tenuifolia* was made into an infusion to treat colds, and acted as an emetic (Moerman 1998).

Hocking (1997) said that the tubers of *S. floridana* contain stachydrine and stachyose. Those and many other chemicals are known from *S. sieboldii* and related species, and presumably from both Florida species (Kato et al. 1979, Derkach et al. 1980, Kobzar 1986a,b, Kobzar' and Nikonov 1987, Greutert et al. 1993, Miyase et al. 1996). Other studies have indicated that chemicals from *Stachys* are useful as antioxidants and to treat jaundice (Voitenko et al. 1990, Couladis et al. 2003).

Stachytarpheta: Blue Porter-Weed

(From Greek *stachys*, a spike, and *tarphys*, thick, referring to the flower spikes)

When I arrived in Florida in 1970, one of the first new plants that I learned was called blue porter-weed (Small 1933, Baker 1938). Most of the people I met in Florida and Jamaica simply called it "porter-weed." Scientifically, it is *Stachytarpheta jamaicensis.*

Maybe the hard-core drinkers and connoisseurs among those I talked with knew what a "porter" was, but I did not. Indeed, I doubt many of them knew how we got the word "porter" in our language. Porter is a dark brown beer with a bitter taste that is brewed from malt partly charred or browned by drying at high temperatures. The term is a shortened form of "porter's ale" or "porter's beer," and dates in English literature from the early 1700s. Originally, the brew was a cheap alcoholic drink consumed largely by porters and other laborers in England. "Porter-house steak," and "Porter-house chop" were derived from the establishments who sold these cuts to be eaten along with the "brown, clear, bitter, and wholesome"

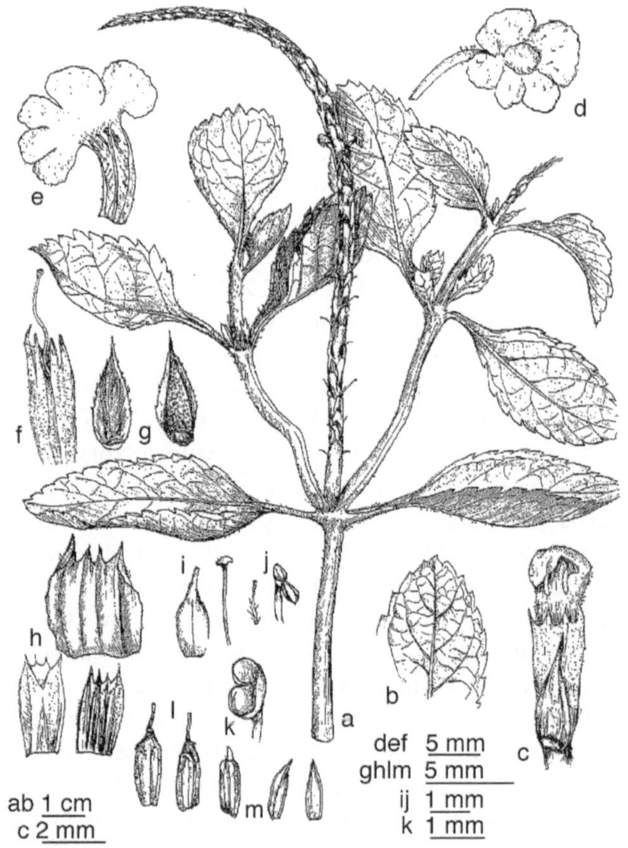

Stachytarpheta jamaicensis. *Drawn by Shih-Huei Chen.* From Chen and Wu 2002.

brew. The descriptive name "stout" is a strong form of Porter.

The connection to beer and these plants is not immediately obvious. However, *Stachytarpheta* came to be called blue porter-weed because it was used to make a mildly alcoholic beverage used for reasons other than just medicine (Morton 1968b). Still, porter was not the only beverage made of these herbs, and at one point the leaves were exported from Brazil to Europe as "Brazilian tea" (Standley 1920–1926). Not only was *Stachytarpheta* used as a tea substitute, but also as an adulterant for tea. The beverage was consumed simply for its taste and also as a medicine. Various people have claimed that the tea was anthelminthic, cathartic, emetic, an emmenagogue, used to treat baldness, bronchitis, chronic hepatitis, dropsy, erysipelas, fever, pneumonia, stomach problems, ulcers, venereal disease, worms (especially in children), and yellow fever. Blue porter is considered cathartic, diaphoretic, digestive, a drink for cooling the blood, a bath to reduce fever in children, emetic, an emmenagogue, a possible abortifacient, to relieve constipation, effective in ritual bathing, a sedative (in nervous disorders, or simply against insomnia), stimulant, and tonic (Standley 1920–1926, Martínez 1969, Duke 1972, Petersen 1974).

Just how much of this use should be attributed to *Stachytarpheta* and not the related *Verbena* is not clear. The two genera have been intermixed and confused since Europeans arrived in the New World. Linnaeus ([1753] 1957) even called blue porter-weed *Verbena jamaicensis* (Jamaican verbena). One must wonder if the common names Jamaica vervain (Bahamas, USA, Europe), *verbena de Jamaica* (Spanish), and *zapane de la Jamaique* (French) are indigenous or simply translations of Linnaeus's binomial. However, other common names reveal the extent of confusion between *Verbena* and *Stachytarpheta.*

When Europeans arrived in the New World, they brought with them an old tradition of using *V. officinalis.* That species and *S. jamaicensis* are both in the Verbenaceae, both are herbs, both have small blue flowers in spikes, and they generally look alike. Use of *V. officinalis* dates to at least the Celts, early Germans, and Romans (Dobelis 1986). The common name "verbena" and its variants appeared in English literature by the 4th century A.D. Some of the older spellings were *verueyn* (A.D. 300–500), *verveine* (A.D. 300 and 600), *warwayn* (A.D. 400), and *vervin* (A.D. 500–600). In 13th-century French it was *verveine,* and in Italian it was *vervena.* Those varied spellings spawned both "verbena" and "vervain" in English. Leonard Fuchs (Meyer et al. 1999) reported in 1542 that physicians were calling the European plants by the pharmaceutical names *verbena supina* (prostrate verbena) and *sacra herba* (holy herb). Other 16th-century names included sacra herbe, common vervaine (English), *Eisenhart* or *Eisenkraut wieble* (female iron heart or iron-herb, German), *iserkruyt wijfken* (female iron herb, Dutch), *veruaine femelle* (female verbena, French), and *verbenaca femina* (female vervena, Italian). This plant was part of the religious traditions of the pre-Christian people, and so the missionaries dubbed it "herb-of-the-cross," claiming that it had staunched Christ's wounds on Calvary.

When people arrived in the New World, they found another plant. They began comparing the new one with the one from the Old World. English speakers said bastard vervain (Jamaica) or vervain [also rendered as verryvine, *veng veng, vèven,* vervine, verbain] (Caymans, Jamaica, Montserrat, Barbados, Belize). In Portuguese, it became *gervão verdadeiro* (true vervain, originally *Verbena officinalis, gervão* from *urgebão,* both from *verbena,* Brazil). Those speaking Spanish said *verbena* (Veracruz, Hispaniola, Puerto Rico, Venezuela, Panama), *verbena azul* (blue verbena, Nicaragua), *verbena cimarrona* (wild verbena, Cuba), or *verbena mansa* (tame verbena, Hispaniola). Creole French rendered it as *vèven* [*gwo vèven,* big vervain*; vèven blanc,* white vervain; *gwan vèven,* old vervain; *vèven ble,* blue vervain; *vèven violèt,* blue vervain] (Haiti, Trinidad), *verveine* (Haiti, Dominica), *verveine* [*a*] *queue de rat* (rattail verbena, Hispaniola), and *verveine bleue* [*morada, violette*] (blue vervain, Hispaniola).

Several other names show European influence in a less pronounced fashion. The Dutch *ijzerkruid* (iron herb, Suriname) and *isri-wiwiri* (iron herb, Sranan, Suriname) hark back to the mid-16th century names. The same is true of *zèb sacré* (sacred herb, Haiti), although at first glance it might seem that this was just one of the many religious plants of Haitian voodoo (Beauvoir et al. 2001).

Some of the other common names show less European influence. To English-speaking people it is simply a blue flower (Bahamas, Jamaica, Montserrat) or blue flower vine (Jamaica, Montserrat). Several people note the similarity of the inflorescence to a tail and say *bretónica* (little tail, Puerto Rico) or *cola de millo* (millet tail, Panama). In much of the Americas, it is *chilillo* (Mexico, Central America). Literally *chilillo* is "little chile," but is widely applied to a variety of plants, usually denoting a small bud or flower.

Several of the indigenous names are either untranslated or obscure in meaning. These include *kuríbiu ákkuani* (Carib, Dominica), *kuiékuiéit* (Carib, Dominica), *lenga di baca* (cow tongue, Dutch Antilles), pound-cake bush (British Antilles), *rattestaart* (rat-tart or cake, Dutch Antilles), *rinchão* (based on *rincho,* the voice of a horse, or "big whinny," Brazil), rooster-comb (British Antilles), and *simbunugit* (Cuna).

However, a use rarely recorded in the literature is indicated by the names *ibinxiu* (*ibin',* placenta or umbellical cord, *xiu,* herb, Maya, Belize) and *talche* (*tal,* touch, *che',* tree, Maya, Mexico). Surely, those names developed because of the use of *Stachytarpheta* in childbirth to wash the hands and to bathe the newborn child as well as to groom the mother before and after giving birth (J.L. Tapia, personal communication 2002). A similar use extends into Veracruz (Vásquez and Jácome 1997). The names woe vine (British Antilles), eyebright (Montserrat), and wurra weed (wurra = worry, Virgin Islands) may all allude to that application.

The chemical constituents of *S. jamaicensis* include the glycoside stachytarphine, choline, three flavonic glucuronides, iridoid ipolamiide, and the phenylpropanoid glycoside verbascoside (Willaman and Li 1970, Duret et al. 1976, Wong 1976, Melita-R. and Castro 1996, Roengsumran et al. 2002). The related *S. cayennensis* contains less than 1% essential oils citral, geraniol, and salicylic acid. The more recent studies of the chemistry of blue porter-weed have concentrated on bioactivity. These studies found extracts of the plants to be anti-inflammatory, antidiarrhetic, antinociceptive, active against intestinal worms, microbes, and insects, to inhibit gastric acid secretion, and to promote sedation and analgesia (Robinson et al. 1990, Almeida et al. 1995, Melita-R. and Castro 1996, Rodriguez and Castro 1996, Vela et al. 1997, Schapoval et al. 1998, Chariandy et al. 1999). Not only were the plants useful for humans, but also for their dogs (Lans et al. 2000).

More recently, people have decided to cultivate *Stachytarpheta.* In the early 1980s, with the trend to cultivate Florida native species, wild plants from southern Florida were brought into nurseries. These herbs were propagated and spread across the state. Then, people began noticing that their cultivated plants were different from those in the wilder parts of southern Florida. When they looked more closely, they found that they had propagated and sold *S. urticifolia* from the Old World, and not our native species (cf. Moldenke and Moldenki 1983). Now, the Florida native, *S. jamaicensis,* has been discovered as an exotic plant in Taiwan (Chen and Wu 2002). Even when we humans try to do things right, we often mess them up. I cannot help but recall that old phrase from Robert Burns's (1785) poem, "The best-laid schemes o' mice an' men Gang aft a-gley."

Staphylea

(From Greek *staphyle,* a bunch of grapes, referring to the inflorescence)

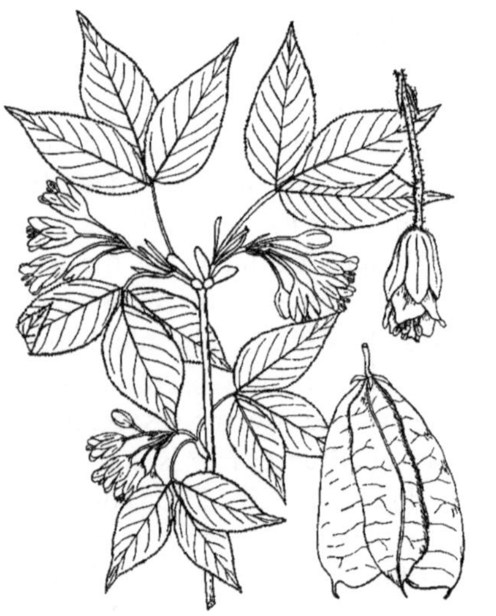

Staphyllea trifolia. From Britton and Brown 1897.

Klappernuss, Pimpernuss (*klappern* and *pimpern*, rattle, *nuss*, nut; *Pimpernuss* used for both *Staphylea* and pistachio, German)
pistacchio falso (false pistachio, Italian)
staphylier (French)

Staphylea trifolia (three-leaved)
bakokó (Koasati)
[American] bladder nut [bladdernut] (used for the plant by 1741, comparing the inflated fruits to the bladder; "bladder" from Old English *blaedre* and other Teutonic languages)

The first name appearing in the literature for the European species (*Staphylea pinnata*) was *Staphylodendron*, published by the French physician Jacques Daléchamps (1513–1588) in his *Historia generalis plantarum* (General History of Plants) appearing in 1586–1587. Daléchamps's name for these plants translates to "tree bearing a bunch of grapes." Gaspar Bauhin did somewhat better in 1650–1651 with *Pistacia sylvestris* (forest pistachio). At least Bauhin noted that the fruit was edible and that people considered it essentially a wild pistachio (*Pistacia vera*).

Not long after Bauhin's publication, a list of plants cultivated in Holland appeared. That book, written by Paul Hermann (1646–1695), was *Hortus academici Lugduno-Batavi catalogus* (Catalog of the Garden at the Leiden Academy) and published in 1687. Included was a plant he called *Staphylodendron virginianum trifoliatum* (grape tree from Virginia with three leaflets). Twelve years afterward Robert Morrison (1620–1683) wrote about a collection in northern France and called the same plants *Pistacia virginica sylvestris trifolia* (wild Virginia pistachio with three leaflets).

Linnaeus knew both species from cultivation. He had studied the central and southern European species at both the *Hortus Cliffortianus* and the *Hortus Upsaliensis.* American plants he knew only from the former garden. In 1753, Linnaeus gave binomials to both, with *S. pinnata* for the European plants and *S. trifolia* for the American plants. Subsequently, others were found and the genus contains 11 species in the north temperate zones (Mabberley 1997).

Seeds of the American species, like its European relative, contain a sweet oil and are eaten like pistachios (Hedrick 1919). No reference has been found to anyone who has actually tried them recently, but several groups in the "eastern States" ate them (Yanovsky 1936, King 1984).

Few records of people using these shrubs have been found. In the woodlands of New England, the Iroquois took an infusion of *S. trifolia* mixed with other plants for rheumatism, an infusion of bark for sores, as a wash for swollen breasts, and to keep children from crying (Moerman 1998).

In Iowa, the Meskwaki considered the seeds sacred and used them in rattles and in their medicine dance (King 1984). They also used the branches to make pipe stems.

Stenandrium

(Christian G.D. Nees named this with Greek *stenos*, narrow or tight, and *andros*, anther, an allusion to the stamens, which are pressed against each other in pairs)

Stenandrium dulce. *Drawn by P.N. Honychurch.*

Stenandrium dulce (sweet, although this reference is obscure)
shaawe loope [*sawlô:pî, sawloopi*] (raccoon's liver, from shape of leaf, Mikasuki); *wotkolopî* [*wotko impalko*] (raccoon's liver; *wotko,* raccoon, *em*, its, *lope*, liver, Creek)
[sweet] shaggyleaf (Florida)

The Miccosukees believe that babies who dream about a raccoon or opossum sometimes become ill. To calm the children, a bag containing *ayekche-hatke* (*Panax quinquefolius*), *to:lî* (*Persea borbonia*), *acinî* (*Juniperus virginiana*), and *shaawe loope* is placed around the child's neck. The last is used because of the relationship between its name and the cause of the ailment. The plant also is used to treat "Hog Sickness" (periodic unconsciousness with shallow and slow breathing) (Sturtevant 1955). Perhaps the Miccosukee use the plants in other ways because Bennett (1997) was told *shaawe loope* was used in an undisclosed medicine, and Snow and Stans (2001) listed the plant and its names but gave no use.

The common name "rattlesnake-flower" (Florida) was attributed to the Seminoles in 1919 by Sheehan, who claimed that they drank a hot decoction as a remedy for rattlesnake bite (von Reis and Lipp 1982). He said they called it *sin-ty-etsee* (may be *sentihyce*, little platform).

I must wonder about how long the Seminoles have been using these plants. Their malady involving raccoons and opossums must be old, but they have not been in contact with *Stenandrium* that long. The U.S. variety, *S. dulce* var. *floridanum*, is endemic to southern peninsular Florida and the Florida Keys. From there, the species is disjunct to southern South America, having first been collected in Chile. No other species occurs in mainland North America east of Texas, but there are several *Stenandrium* species endemic to the Bahamas and Antilles (Leon and Alain 1957–1963, Correll and Correll 1982, Liogier and Martorell 1982). At least the common names of *S. tuberosum* in Puerto Rico (*mata esperitista*, spirit bush; *yerba maravilla blanca*, white miracle herb) suggest use (Liogier and Martorell 1982).

Stillingia: Corkwood and Queen's Delight

(Linnaeus named the genus after the British naturalist Benjamin Stillingfleet, 1702–1771)

Stillingia sylvatica. From Britton and Brown 1897.

Almost as soon as I arrived in Florida, I began exploring areas that were still wild. In nearby ponds, I found a shrubby herb that had inconspicuous flowers. The stem was soft and corky, and I soon learned that it was corkwood. Old-timers in southeastern Florida told me that the stems floated nicely and made good corks for fishing. Alvin Wentworth Chapman had named the pond plants *Stillingia aquatica* (of wetlands) in 1860.

I soon learned that people of European descent were not the only ones who thought the soft, buoyant stems were useful. To the Seminoles, the plants were *ahisókpî* (*ahi*, tree, *sokpi*, spongy, Mikasuki). Sometimes they use its alternate equivalent name, *takɨsópkî* (*takɨî*, shrub, *sopki*, soft). The Creeks also said *Stillingia* was *itotahátka* (*etotale*, dead tree, *hvtke*, white, Creek).

It was not long before I found another species in the uplands of pine flatwoods. The pineland species was the one named by Linnaeus when he created the genus in 1767, *S. sylvatica* (of the woods). Since at least the late 1800s, that species has been known as "queen's delight" (Chapman 1897).

Although we are now taught that "queen" applies to a female of royal lineage, it was not always so. Originally, queen came from the Old English *cewn*, which meant simply "woman." Just when the Florida plants now called "queen's delight" came to be that is uncertain, but it always had an earthy connotation. That name alluded to an imagined similarity between the inflorescence and male genitalia. The name cock-up-hat [cocyshat], used in the Carolinas (Morton 1974), has a similar meaning.

Because the herbs are mostly confined to the southeastern United States, they have few names in languages other than English. Germans call them *Talgbaum* (tallow tree), although that has developed because of their relationships with the genus *Sapium*, which is usually called tallow wood. *Stillingia* is mostly known as "corkwood," whether it is *S. aquatica* or *S. sylvatica*. Sometimes *S. sylvatica* is called by the descriptive terms nettle potato or silver leaf. "Nettle potato" must refer to the caustic, vesicant nature of the raw sap, while "silver leaf" surely alludes to the sometimes white-striped leaves. That chemical trait may also be why it is called Indian flea root in Louisiana, where it was placed to repel those pests (von Reis 1973).

Two other common names in the southeastern states refer to medical uses—marcory, and yaw root. Although the first term resembles the English "marcor," from Latin *marcere*, to decay or become emaciated, it probably is not related. More likely, "marcory" is a mispronunciation of "mercury." Since the "standard" medical treatment for syphilis in the 1800s was to use the poisonous metallic element, the name was transferred to plants with similar functions.

Millspaugh (1892) recorded that *Stillingia* extracts were considered "an absolute specific in syphilis, entirely superceding the use of mercury."

The name yaw root notes that the plants were used to treat an infectious contagious tropical disease caused by the spirochete *Treponema pertenue.* "Yaws" is a Cariban word related to *yáya* in Calinago, an Arawakan language of the Lesser Antilles and Central America. The name was applied by 1679 to this disease, which, like syphilis, is characterized by ulcerating lesions with later bone involvement. The malady, thought to have come from sub-Saharan Africa, is closely related to the variant of syphilis called "pinta," which is caused by *Treponema carateum.* Causal organisms and symptoms are so similar that "pinta" was probably the original "yaws" in the New World (Desowitz 1997). Since about 1803, English-speaking people have also called the malady "frambesia" (from French *framboise,* raspberry) because of the appearance of the lesions.

Although Americans quickly adopted *Stillingia* as a remedy for venereal diseases, the native Americans did not use it as uniquely for that problem. Bartram ([1853] 1909) said that during the 1770s the Creek, Seminole, and Cherokee had "a high estimation of a species of either *Croton* or *Styllingia,* I am in doubt of which, as a cathartic." Indeed, later work among these people and others in the southern states showed that was but one of the uses for these plants. True, among the Cherokee a decoction or tincture of root was used for venereal disease (Hamel and Chiltoskey 1975). However, given the late recording of that information, it probably reflected a heavy non-Cherokee influence. More likely indicative of original uses were those recorded among other people.

The Creeks, for example, considered *Stillingia* abortifacient and cathartic. Creek women washed in root extracts after giving birth, and drank the liquid. A cold infusion was also taken by men to regain potency (Swanton 1928a, Taylor 1940, Howard 1984). They call the plants *yalang cati* (*yvlonkv,* root, *cate,* red, Creek) or *yelungadjadi hiliswa* (*yvlonkv,* root, *cate,* red, *heleswv,* medicine, Creek).

Their relatives, the Seminoles, considered the roots antidiarrheal. Because the leaves and stems have milky juice, they call it *owa:ctafânkî* (*owa:ct,* center, *efanki,* milk, Mikasuki). Those speaking Creek call them *pisi:ká:fkatî:ko* (*pese,* breast, *kafkv,* milking, *ʔtekoye,* pliable). Also, a decoction of the roots was taken by babies and adults for "Bird Sickness" (diarrhea, vomiting, and appetite loss). A similar decoction of roots was taken for "Menstruation Sickness," indicated by yellow eyes and skin, weakness, and a shaking head (Sturtevant 1955). Now, the Miccosukee healers usually admit only to using the remedy for stomachache (Bennett 1997). They are guarded in what they reveal, but they have told me that they still use the plants for women's problems.

Stillingia was perhaps less important among the unrelated Kiowa who considered *knoñ-klee-a* (turtle weed) a season indicator. They considered it a rain symbol because the dry land terrapins get under it for shade (Vestal and Schultes 1939).

Basically, the native Americans considered *Stillingia* root extracts a "blood purifier." That concept is rendered as an "alterative" in medical parlance, which means that it gradually restores the proper function of the body and increases health and vitality. However, Murphee (1965) inferred that it was also used to indicate anemia in the Panhandle where *S. sylvatica* was still in use. Queen's delight was official in the U.S. Pharmacopoeia between 1831 and 1926 and in the National Formulary between 1926 and 1947, where it was considered cathartic, alterative, and emetic.

Porcher (1863) considered the remedy "as an alterative in scrofula, in syphilis, in cutaneous diseases, in chronic hepatic affections, and in the composition of diet drinks; it adds to the efficacy of sarsaparilla." He went on to note: "A tincture is made with the root two ounces, of diluted alcohol a pint. Dose a fluid drachm. A decoction is made of the bruised root one ounce, water one and one-quarter pints. Boil to one pint. Dose, one or two fluidounces several times a day; an overdose is cathartic or emetic." During the same time, the plant in the Old South was "widely used to treat constipation; to induce vomiting, and as a remedy for syphilis, skin and pulmonary diseases, and liver ailments. A rubdown of boiled, mashed roots was used to treat women suffering from menstrual irregularity" (Krochmal and Krochmal 1973). In the Carolinas, a root decoction was used to cure boils and treat stomachaches, and it was an ingredient in "S.S.S. TONIC" (Morton 1974). Although it is not specifically stated in the literature, one of the properties of the remedy was as an antispasmodic. That trait would account for several of the historical uses.

The roots fell into disrepute after the 1940s, and were rarely used. However, some trade continued with them (Christensen 1946). Then, after a hiatus, they again became popular during the current herbal renaissance. There are dozens of companies on the Internet advertising them for a variety of ailments. For example, one company (abchomeopathy.com) recommends the plants for coughs, sore throat, genital problems, and syphilis. Herbal Alternatives (herbalalternatives.com) considers queen's delight alterative, depurative, expectorant, diaphoretic, sialagogue, astringent, and antispasmodic. Most of the modern companies note that *Stillingia* was one of the ingredients in the infamous Hoxsey anticancer compounds (cf. chatlink.com/~herbseed, herbalalternatives.com). That use has been discredited, although others have not. Moore (1989) recommended the preparation for skin problems and coughs, and to aid in recuperation from lung problems.

There are several bioactive chemicals known from *Stillingia*, including the alkaloid stillingine that was discovered about 1882 (Millspaugh 1892). The plants also contain apocinide, calcium oxylate, echiine, gnidilatin, hydrogen cyanide, phorbols, especially 5-beta-hydroxy 12-deoxy phorbol, prostatin, resin, silvacrol, tannin, and volatile oils (Felter 1922, Moore 1989, 2002). In spite of the thriving trade in the herbal market, no modern chemical studies seem to have been published on the genus.

Stillingia sylvatica has a wide range. It is known from southeastern Virginia to Florida, Texas, and New Mexico, north to Oklahoma, Kansas, and Arkansas. It was collected by C.W. Short in the "Barrens" of Kentucky in 1840, but apparently it has not been seen there since (Browne and Athey 1992). In spite of that range, *Stillingia* populations have steadily declined during the past few decades. Because of collection for the herbal market and habitat loss, *S. sylvatica* is now a rare plant. The species is so uncommon it was considered for listing as endangered in the late 1970s. Herbal doctors on the Big Cypress Seminole reservation lamented the disappearance of plants there during our research between 1998 and 2001. So, whenever we found new localities with plants, we made sure they were informed.

"Use it or lose it," is an admonishment that I heard often when growing up. Now, we learn that, by using our plant resources, we may well lose them. Sometimes, the brevity of old sayings underestimates the problems of the modern world.

Strumpfia

(Nicolaus Jacquin named this genus for Karl Strumpf, a professor in Halle during the middle of the 18th century)

Strumpfia maritima (of the sea, as it grows near coasts)
bai no bolbe [*bai no bolbera*] (go and do not return, Dutch Antilles)
faux rosmarin (false rosemary, Haiti); *romerin bord du mer* (seaside rosemary, Guadeloupe); *romero falso* (false rosemary, Cuba)
grankini (Bahamas)
lirio (lily, Puerto Rico)
pride-of-Big-Pine (Florida)
seaside strumpfia (a book name, Bahamas)

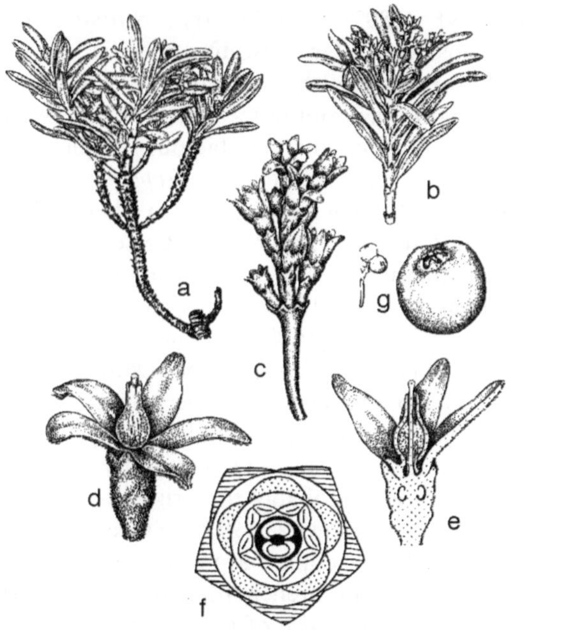

Strumpfia maritima. a. Old flowering branch. b. Young flowering branch. c. Inflorescence. d. Flower, front-side view. e. Flower, longitudinally dissected. f. Floral diagram. g. Fruit, smaller than life (left) and enlarged (right). *Drawn by Priscilla Fawcett.* From Correll and Correll 1982.

Jacquin named this species in 1760 from his collection in Jamaica. It is a monotypic genus, and the species is found in the Bahama Islands, Cuba, Jamaica, Puerto Rico, Yucatán, and the Lesser Antilles to the Netherland Antilles off South America (Boldingh 1909, 1914, Liogier 1962, Fournet 1978, Correll and Correll 1982).

The whole plant is stimulant and has been used against bites of poisonous animals. In severe fevers with delirium, *romero falso* is mixed with quinine (Roig 1945). In Curaçao, a weak decoction is used to treat colds, stomachache, and kidney stones. A stronger decoction permanently sterilizes women (Morton 1981).

The plant is burned in the Bahamas to repel mosquitoes (Von Reis 1973). In Curaçao, it was formerly burned outside homes to drive away bad luck (Morton 1981).

Stylosanthes

(Olaf Peter Swartz named this from Greek *stylos*, column, and *anthos*, a flower, because of the column-like calyx tube)

Stylosanthes hamata (hooked at the end)
Caribbean stylo
cheesy toes (Jamaica)

chichibe (something in the path that sticks, Maya, Yucatán)
comino sabanero (savanna dwarf, Cuba)
donkey weed (Jamaica, Virgin Islands)
lady's fingers (Jamaica)
Lucy Julia (Cayman Islands)
mother-sea-gel (Virgin Islands)
pencil flower (Florida, Bahamas, Puerto Rico, Jamaica, Cayman Islands)
petit trefle (little clover, Guadeloupe, Martinique); *trefle jaune* (yellow clover, Guadeloupe, Martinique)
pinda shimaron (wild *pinda*, Dutch Antilles); *wilde pinda* (wild *pinda*, Dutch Antilles)
tebenque (probably Taino, Cuba; also used for *Evolvulus* and *Pectis*, which see)
zarzabacoa enana (dwarf *zarzabacoa*, which is a variant of *guasabacoa*, a Taino word for *Desmodium*, Puerto Rico)

Stylosanthes was named by Swartz in 1788. Since then, the genus has grown to 25 species in the tropical and warm parts of the world (Mohlenbrock 1958, Mabberley 1997). The species are mycorrhizal and phosphate accumulators, and plants in Australia are famous for immobilizing and killing cattle ticks.

Jamaicans used "cheesy toes" as a cold remedy and for kidney trouble (Ayensu 1981, Morton 1981). Similarly, a decoction is drunk for kidney pain in Curaçao (Morton 1981). People in the Virgin Islands give it to teething children to relieve fever (Morton 1981).

There are several related species with other recorded uses. In Mississippi, *S. biflora* is called "afterbirth weed" or "squaw weed," and is used as a uterine sedative in cramps (Hocking 1997). A decoction is also used as an emmenogogue. In the West Indies, and from Mexico to Peru, *S. viscosa*, called *comino sabanero* or "poor man's friend," is used to treat colds, fever, headache, kidney disorders, and nervous problems (Hocking 1997). From Guatemala and Honduras to the Guianas, *S. guayanensis* is *yerba del campo* (countryside herb), and an infusion is used for "pains in the side" (Hocking 1997).

Suriana: Bay Cedar

(Named for Joseph Donat Surian)

In 1689, King Louis XIV sent two Frenchmen to the Caribbean where they explored for medicinal plants on Martinique, Guadeloupe, and Haiti. The best known of the two is the Franciscan priest Charles Plumier. Less known is Joseph Donat Surian, the physician and pharmacist from Marseilles who accompanied him. Although Surian returned to France in 1690 with a collection of plants now at Paris, he died in 1691. Perhaps because of Surian's death, Plumier named *Suriana* after him in 1703. *Suriana maritima* is now "bay cedar" (Bahamas, Florida).

Suriana maritima. *Drawn by P.N. Honychurch.*

Relationships of this beach shrub have been a mystery since its discovery. Plumier wrote of it, *Suriana foliis portulacacae angustis* (*Suriana* with narrow leaves like *Portulaca*). Hans Sloane said, *Thymelaeae facie frutex maritimis tetraspermos* (four-seeded marine shrub that resembles *Thymelia*). Leonard Plukenet penned, *Arbor americana, salicis foio* (American tree with leaves like willow). Linnaeus did not commit himself, simply using *Suriana maritima* (marine).

That "bay cedar" grows along ocean bays is obvious. Indeed, it grows along beaches around the tropics, in southern Florida, Bermuda, the Bahamas, the West Indies from Cuba to the Grenadines, Aruba, Bonaire, and Curaçao, and from Yucatán and Belize south to northern Venezuela. The shrubs also line beaches in the Caroline Islands, Tahiti, Guam, and other Pacific islands, and also in East Africa, Madagascar, islands in the Indian Ocean, Melasian islands, and northern Australia.

My problem with the Florida name was "cedar." I saw no obvious similarity to a gymnosperm of any kind. Others disagree, and say, "It does indeed resemble a refined conifer" (Anonymous 2002). To me the leaves are more like *Portulaca* or *Thymelia* or *Salix* as noted in the 1600s and 1700s. It took a long time before I finally discovered that the appellation "cedar" was a reference to the fragrance of the crushed leaves. At least to my nose, the odor is subtle and best enjoyed from bruised leaves in an almost-closed hand.

People long before me had known of that odor. To the Taino of Cuba, the plant was *cuabilla* (little *Amyris*) or *cuabilla de costa* [*playa*] (coastal [beach] little *Amyris*). Both names compare the fragrance with that of the much stronger *Citrus* relative (see *Amyris:* Torchwood). Other people simply say the plant is *incienso* (incense, Cuba), a compliment in Latin America. A similar compliment compares this New World plant with one from the Mediterranean. People say that *Suriana* is *romeín noir* (black rosemary, Guadeloupe, Martinique), *romero* (rosemary), or *romero de mar* (sea rosemary). All of those names compare the shrubs with *Rosmarinus officinalis,* the popular European medicinal spice.

Not all names refer to the odor. Moreover, several of bay cedar's other names note uses not found in the literature. Apparently indicating use as a flavoring for food are *crisse marine* [*christe marine*] (marine cress, Haiti), *oseiile bord de mer* (seaside sorrel, Guadeloupe, Martinique), and *salcedo* (salty, or a plant used in sauces). Allusions to medical uses are strangely few since many people use it. One apparent notation of medicine is *pantsil* [*pantzil*] (*pats',* smooth, *pats'il,* increase strength, Maya, Yucatán). Probably, *temporana* (early riser, Puerto Rico) has a similar implication. While any relationship is obscure, medicinal use may be inherent in *palo corra* (running or blowing [as wind] tree, Curaçao, Venezuela) and *perce-pierre* (stabs Peter, Haiti).

Elsewhere the stems have been made into a *cucharo* (spoon, Venezuela) and branches used as thatch-leaf (USA). Maybe tassel plant (Bahamas) indicates an ornamental or *incienso* use, but Workman (1980) thought it was simply because the leaves were clustered in "tassels" at the end of branches.

The heartwood is dark red or maroon, hard, heavy, polishes well, and is used for small articles (Morton 1981). Use of the wood perhaps is related to the common name *guazumilla* (little *guazuma,* Dominican Republic). The Taino name *guazuma* or *guásima,* refers to *Guazuma ulmifolia* (Malvaceae). That and several other Caribbean trees and shrubs bear variations on the old Taino name. It is tempting to conclude the same use for them as for the original. Of *guásima,* Bartolome Las Casas wrote, "*De este arbol solo sacaban fuego los indios: tomaban dos palos de él, muv secos, el uno tan gordo como dos dedos, e hacian en él con las unas o una piedra una mosquecita, e ponian este palo debajo de los pies, e el otro palo era más delgado, como un dedo, la punta redonda, iduesta en la muesca, con ambas palmas de las manos traianlo a manera de un taladro, ésti con mucha fuerza*" (From this tree the Indians only made fire. They took two branches of it, very dry, with one the size of two fingers, and they made in it with the fingernails or a rock a little groove, and put this stick between their feet. The other was narrower, like a finger, the point rounded, and of this in the notch, with both palms of the hands twirled it like a drill, with great force). Since *Suriana* is named after *Guazuma,* it too may have been used for making fire.

Another name for these shrubs is *jovero* [*jobero*] (Dominican Republic). That is an allusion to the *jobo,* a Taino name for *Spondias* (Coll y Toste 1972). Any relationship between the two is cryptic because they have no similarities in fruits, habit, or places where they grow. Another unexplained name is *guitarón* [*guitarán*] (presumably the same as *guitarrón,* rascal, good-for-nothing, Puerto Rico).

Not everyone agrees that *Suriana* is good-for-nothing. Bahamians treat mouth sores with it (von Reis and Lipp 1982). In Cuba, the leaves and branches are used in baths against rheumatism, a decoction of bark and leaves is used against sores, and powdered leaves are mixed with *atole* to treat bloody flux (Roig 1945). In Hispaniola, the bark and leaves are used against rheumatism, and to cure sores (Liogier 1974). In Inagua, people boil roots or twigs and inhale the vapor or hold the decoction in the mouth for toothache or to heal mouth sores (Morton 1981). In Mexico, Martínez (1969) says, "*es antídoto contra las morduras de cualquier pescado, llagas y heridas, fogazos de la boca y lengua; cámaguas de sangre, postemas e incordios*" (it is antidote for bites from all kinds of fish, sores and wounds, canker sores of the mouth and tongue, bloody fluxes, abscesses, and tumors). Specifically in Yucatán, a decoction of leaves and bark is used to clean sores, and dried powdered leaves are taken in cornmeal *atole* for rectal hemorrhages (Morton 1981). As far away as the Pacific Islands, crushed flowers are mixed with coconut milk to stop diarrhea and vomiting of blood (Weiner 1971, Whistler 1991).

Little seems to be known of the chemistry of the genus, but Watson and Dallwitz (2000) note that it has no cyanogenic compounds, no saponins, and no sapogenins. Hershenson and Quimby (1969) and Mitchell (1971) did find triterpenes and flavonol glycosides. Those classes of chemical would likely be responsible for many of the historical medical applications.

The mystery about relationships of this plant that began in the 1600s has barely abated. *Suriana* was included in the Simaroubaceae by Bentham and Hooker (1883) along with other families in the order Sapindales. Many floras since have followed that placement. Cronquist (1981) did not agree, and put *Suriana* with *Cadellia, Gulfoylia,* and *Stylobasium,* in the Surianaceae. There have been several convincing arguments for placement in a separate family. Some continued thinking *Suriana* was close to the Connaraceae, Sapindaceae, and Chrysobalanaceae. Others disagreed and put the Surianaceae in the Rosales, as did Cronquist.

Only within the past few years have people generally begun recognizing the family Surianaceae. Most include four genera, and now those are *Suriana, Cadellia, Gulfoylia,* and *Recchia. Stylobasium* now seems to be part of another lineage (Watson and Dallwitz 2000). Study of molecular genetics (Fernando et al. 1993, Kajita et al. 2001) and embryology (Heo and Tobe 1994) suggest monophyly of the family and kinship to the Polygalacaeae and Fabaceae. Art Cronquist would have been happy to know that he was right putting *Suriana* in the Rosales (Cronquist 1981). Or, maybe he would have just grinned and said, "Of course!"

Swietenia: Mahogany

(Named for Gerarde L.B. van Swieten)

There is a mahogany cross dated 1514 in the cathedral at Santo Domingo (Ciudad Trujillo), Hispaniola. In addition, the church contains other carved mahogany woodwork more than five centuries old (Little and Wadsworth 1964). Those items confirm the importance of these trees to indigenous people and to early European settlers. Given that antiquity, it seems odd that no common name for *Swietenia mahagoni* appeared in print until the 1600s, especially because names from three areas were in common usage.

In 1671, John Ogilby told about the Jamaican trees with "most curious and rich Woods, as Cedar, Mohogeney, Lignum-vitae." Languages other than English also use the Caribbean word for the trees. *Swietenia mahagoni* is also *Mahagoni* (German), *mahonie* (Dutch), *mahagoni* (French), *mogano* [*mogogane*] (Italian), and *mogno* (Portuguese). The Portuguese word *mogno* is said to be from an indigenous language (Ferreira ca. 1978), perhaps Taino. The Taino words *maga* or *magua* mean fertile lowland or plains. A related word *mahok,* which is akin to Taino *maho,* refers to trees with strong bast fibers (Coll y Toste 1972). In Suriname, *S. mahagoni* is still *mahoni.*

Mark Catesby gave the world the first illustration of mahogany. He was born and educated at Essex, England before he went to Virginia in 1712 at age 29. Wealthy and influential patrons back in Europe were so impressed with the plants he sent that, in May of 1722, Catesby was sent to Charles Town, South Carolina (Reveal 1992a). On a trip to the Bahama Islands Catesby drew "The Mahogony tree" (Catesby 1734–1747). Linnaeus traveled to Oxford in August of 1736 partly to examine Catesby's specimens and drawings. Subsequently, Linnaeus published *Cedrela mahagoni* in 1759 based on Catesby's drawing.

It was only the following year that Nikolaus Joseph Baron von Jacquin, director of the botanical garden in Vienna, coined the generic name *Swietenia* for this tree. Jacquin named the tree for Gerarde L.B. van Swieten (1700–1772), a Dutch physician for Empress Maria Teresa of Austria. He wrote the name as "Mahogoni," although Linnaeus had it *mahagoni.*

We are left with the question of why a tree with virtually no bast fibers got a name pinpointing them. Perhaps there was a confusion of indigenous names, with the *maho* having good bast fibers, and Europeans mistaking it for the name of this tree (see *Trema:* Nettle-Tree). Confusion seems likely because there is another Taino word that still means *S. mahagoni.* Cuban Tainos called the tree *caoba* [*caobo*] (Guerrero, Tabasco, Yucatán, Oaxaca, Cuba, Puerto Rico), and that became its name in Spanish. That Taino name remains in the Caribbean as *caoba americana* (American mahogany, Puerto Rico), *caoba de Cuba* (Cuban mahogany, Cuba), *caoba de Santo Domingo* (Dominican mahogany, Puerto Rico), *caoba dominicana* (Dominican mahogany, Puerto Rico), *caobilla* (little mahogany, Cuba), *palo de caoba* (mahogany tree, Puerto Rico), and *kuba* mahogany (variant of *caoba,* Cuba).

One of the reasons that names for mahogany are so intermingled is because people were trading two species. The original range of *S. mahagoni* was southern Florida, the Bahamas, Cuba, Jamaica, and Hispaniola. From there it was introduced into Puerto Rico and the Virgin Islands, Bermuda, the Lesser Antilles, Trinidad and Tobago, and Curaçao south through South America and elsewhere in the tropics (Little and Wadsworth 1964). Its names became mixed with those of *S. macrophylla,* which was native to much of the area where *S. mahagoni* was introduced.

Wood from the mahoganies was so important that even South American names became used in the Caribbean. A common Haitian appellation is *acajou* [*akajou, acaju, kajou, nacajou*]. That name came from

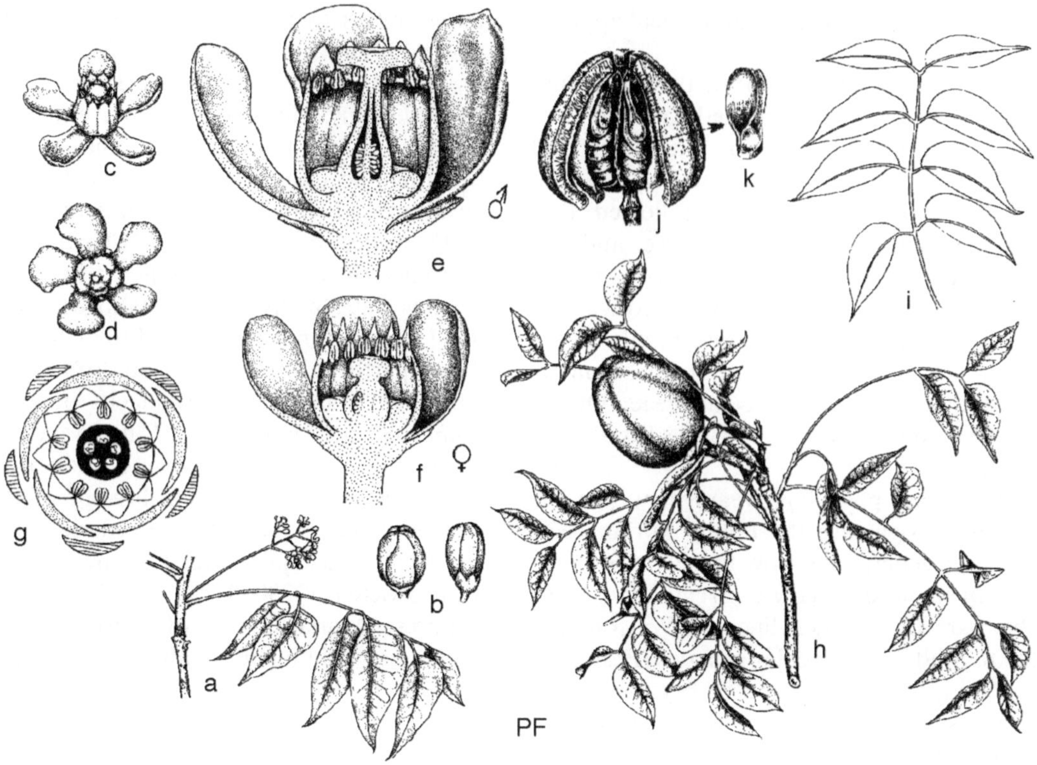

Swietenia mahagoni. (left). a. Flowering branch. b. Flower buds. c. Staminate flower, front-side view. d. Staminate flower, from below. e. Staminate flower longitudinally dissected. f. Pistillate flower, longitudinally dissected. g. Floral diagram. h. Fruiting branch. i. Leaf outline. j. Fruit, dehisced. k. Seed. *Drawn by Priscilla Fawcett.* From Correll and Correll 1982.

Tupí *aka 'vú*, originally applied to wood from the Brazilian *S. macrophylla*.

The exchange of the two species also was one of the reasons that so many geographic modifiers have been added to the word mahogany. Variations using *acajou* in the Caribbean include *acajou a meubles* (furniture mahogany, French Antilles), *acajou de Cuba* (Cuban mahogany, Cuba), *acajou de Haiti* (Haitian mahogany, Haiti), *acajou de St. Domingue* [*acajou San Domingo, kajou sen domeng*] (Dominican mahogany, Haiti, Guadeloupe), *acajou des Antilles* (Antillean mahogany, Haiti), *acajou mahogani* (mahogany mahogany, Guadeloupe), *acajou ronceux* (bramble mahogany, Haiti), *albero di acajou* (white mahogany, Dutch Antilles), *bwa kajou* (tree mahogany, Haiti), *kajou peyi* [*pyè kajou*] ([country] wild mahogany, Haiti), and *kajou ti fey* (little leaf mahogany, Haiti).

In English, the trees are known as American mahogany (West Indies), curlet mahogany (curly mahogany, Jamaica), medura redwood (medura meant for *madeira* from Spanish, USA), small-leaved mahogan (Virgin Islands), and true mahogany (West Indies). Depending on where the tree comes from, it may be Dominican mahogany (Puerto Rico), Florida mahogany (Florida), Honduras mahogany (Florida), Jamaica mahogany (Jamaica), Mexican mahogany (USA), San Domingo mahogany (West Indies), Spanish mahogany (Trinidad), or just West Indian mahogany (Trinidad).

Speakers of Spanish say *antillen* (from the Antilles, Cuba), *echtes kuba-mahagoni* (mahogany-mahogany sprout, Cuba), and *echtes mahagoni* (mahogany sprout, Cuba). Those speaking French say *mahogany du pays* ([country] wild mahogany, Guadeloupe) or *mahogany petites feuilles* (small-leaved mahogany, Windward Antilles). People in the Dutch Antilles say *Antillen-mahagoni* (Antillean mahogany), *Cuba mahonie* (Cuban mahogany), *mahonijboom* (mahogany tree), *mogano americano* (American mahogany), and *mogano di Cuba* (Cuban mahogany).

Some other names for *Swietenia* include *cedre des Antilles* (Antillean cedar, French Antilles), *madeira* (Spanish for wood, Bahamas, USA), *nieuxbladboom* (new leaf tree, Dutch Antilles), redwood (Bahamas), and sabica (Bahamas). The latter two names show comparison with wood from *Lysiloma* (see *Lysiloma:* Wild Tamarind).

Mexican trees are called *cóbano* (Guerrero, Michoacán, Colima), *zopilo-zontecoma-cuahuitl* (vulture-head-tree, in reference to the curved fruits that resemble buzzard heads from a distance, Náhuatl), *zopilote* [*tzopilotl, zopilocuahuitl, zopilotl*] (vulture

[tree], Guerrero, Tabasco, Yucatán), and *palo zopilote* (vulture tree, Oaxaca).

Plants are used as street and shade trees because of their spherical crown and dense shade. However, mahogany is best known for the wood used in furniture, cabinetmaking, interior finish, and veneers. The wood is easily worked and takes an attractive polish. Much older uses, from at least the time of Catesby and Romans ([1775] 1961), are in shipbuilding, construction, and for beams. Catesby (1734–1747) compared the wood of the mahogany with oak (*Quercus*), high praise indeed from a European of the time. Romans ([1775] 1961) complained that by the time of his visit to the Florida Keys most of the mahogany had been cut. In his instructions to navigators for the region between *Cayo Hueso* (bone key, now Key West) and Matacumbe Key, he says that, because of woodcutters from Providence in the Bahamas taking the mahogany trees, "little or none now remains here."

Less well known is that people use mahogany for medicines. Cubans use the gum exudates from the trunk as a bitter and astringent tonic and febrifuge (Roig 1945). Juices of young shoots are also used in Cuba to heal wounds and stop bleeding. Oil from *S. macrophylla* seeds (*pepitas de zopilote,* vulture seeds) was used by Aztecs as a cosmetic and in the 1920s in soaps (Standley 1920–1926). People in Hispaniola use the plants to treat anemia, bleeding, chest pain, diarrhea, dysentery, fever, loss of appetite, malaria, nerve disorders, toothache, and as an abortifacient, antiseptic, aphrodisiac, astringent, emmenagogue, and to provide iron and vitamins (Beauvoir 2001).

Mahogany bark contains catechin-like compounds, and the wood contains catechin (Ayensu 1981). These flavonoids and related chemicals are also in tea (*Camellia sinensis*) and red wines (*Vitis vinifera*), and have become famous for reducing risk of heart disease (e.g., Rosenkranz et al. 2002). Seeds contain swietenine, swietenolide, and related compounds (Ekimoto et al. 1991, Mootoo et al. 1999). Those chemicals are limonoids related to those found in neem (*Azadirachta indica*), chinaberry (*Melia azedarach*), and other members of the Meliaceae and Rutaceae (Bickii et al. 2000). Those compounds have become famous as insecticides (e.g., Mootoo et al. 1999, Bickii et al. 2000), have been shown to inhibit platelet aggregation (Ekimoto et al. 1991), to inhibit human cancer cell proliferation (Tian et al. 2001), and to exhibit antimalarial activity (Bickii et al. 2000).

Sometimes the common names themselves are a "generic" category. Anything that is not known becomes part of that umbrella. For example, in Dominica, no matter whom you ask if you wanted to know what a particular plant is, the answer was almost invariably *"p'tit feuille"* (little leaf, P. Honychurch, personal communication, 2002). That is always a good generic cover-all for almost any unknown bush. She used to have an overseer who taught her more than 25 species of *"p'tit feuilles."* So, never try to stretch the common names too far.

Symphoricarpos

(From Greek *symphorein,* to bear together, *carpos,* fruit, from the clustered fruits)

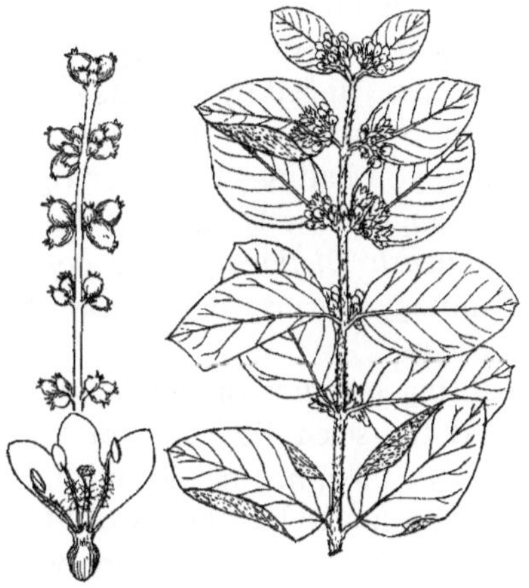

Symphoricarpos orbiculatus. From Britton and Brown 1898.

pianta della perle (pearl plant, Italian)
snow-berry [snowberry] (the European species has white fruits); *Schneebeere* (snow berry, German); *snøbaer* (snow berry, Norwegian)
symphorine (French)

Symphoricarpos orbiculatus (round)
buck-brush [buckbrush] (Kentucky)
cate fulotv (*cate,* red, *fulotv,* buck-brush, Muskogee)
coral-berry [coralberry] (name given by 1860, because of the red fruits)
gu-la-ko'-kee'a (Kiowa)
Indian-currant ("currant" came into English in A.D. 1234 as "raisins of Corauntz," from French *raisins de Corinthe,* a seedless grape from Greece; the name suggests that indigenous people ate the fruits, but no confirmation has been located; Core 1967 said of the fruits, "they are scarcely edible"; then maybe medicinal?)
inshtogahte-hi (*inshta,* eye, *gahte,* lotion, *hi,* plant, Omaha-Ponca)
snowberry (a form with white fruits)
tatepa'siki (twisted, Meskwaki)

[red] waxberry ("waxberry," a name originally given to *Myrica cerifera* by 1813, later transferred to these plants)
wolf berry (used since 1834; "wolf" is from Old English *wulf*, with similar words in German and Dutch, and related to Sanskrit *vrkas*)
zhon̲-çi hi zhin̲-ga (*zhon̲*, wood, *çi*, yellow, *hi*, stalk, *zhin̲-ga*, small, Osage)
zuzecha-ta-wote sapsapa (*zuzecha*, snake, *ta*, its, *wote*, food, *sapsapa*, black, Dakota)

I grew up calling these plants "buck-brush," a name my parents and grandparents used for them. As I floundered through the woods at night trying to keep up with my dad and the dogs as they chased raccoons, we were sometimes tripped by thickets of these shrubs. My impression was always that they were "buck-brush" because you had to "buck" or fight your way through their tangles. Surely, none of the compilers of the *Oxford English Dictionary* ever considered that possible etymology because it was too alien to their city dwelling lives.

It was during my first or second year in college that I learned the scientific name of "buck-brush." During that year, I skipped meals and did extra work to save enough money to buy a copy of M.L. Fernald's *Gray's Manual of Botany* (a 1632 pp. book for $15). Although that was in the early 1960s, I still have a vivid memory of my sense of accomplishment as I left the college bookstore guarding my prize. From that book of wonders, I learned that "buck-brush" was known to scientists as *Symphoricarpos orbiculatus*. At the time I had no clue what a history those words revealed. The events began when John Clayton sent a specimen to Europe where they were studied by Dillenius, Gronovius, and Linnaeus.

In 1732, Johann Dillenius called them *Symphoricarpos foliis alatis* (Cluster-fruit with winged leaves). Both Linnaeus and Jan Gronovius published on them within a year of each other (1738 and 1739), calling them *Lonicera pedunculis axilliaribus capitatis* (Honeysuckle with flowers in heads on axillary stalks). Later, Linnaeus ([1753] 1957) shortened the phrase name in *Species Plantarum*, where he coined the binomial *Lonicera symphoricarpos*.

It was only 1755 when French agronomist and forester Henri Louis Duhamel du Monceau (1700–1782) wrote on these plants in his *Traité des Arbres et Arbustes*. Duhamel adopted as a genus the one proposed by Dillenius, and created the name used today—*Symphoricarpos*.

This red-fruited shrub is native across much of southern North America. It is largely replaced in the north by white-fruited *S. albus*, which has an equally complicated history. These are 2 of the 17 species now known in *Symphoricarpos*. All but one of them is native to the Americas, and the other (*S. sinensis*) grows in China (Jones 1940, Mabberley 1997).

The only records found of people using *S. orbiculatus* are in the Missouri River region and southern Manitoba and Ontario. The Dakotas ate the berries (Yanovsky 1936). The Dakota, Ojibwa, Omaha, and Ponca steeped the leaves to make an infusion for weak or inflamed eyes (Gilmore 1919, Moerman 1998). The Meskwaki made a tea from the root which was used to expel the afterbirth, and was believed to "enable quicker convalescence" (King 1984, Kindscher 1992). The Missouri River tribes also utilized *S. occidentalis*, and their names and uses are identical for both species. The Dakota made charcoal from the wood and used it for making tattoos (Moerman 1998).

It seems likely that all of the eastern species were used because there are many more people recorded as using the related *S. albus*. Moerman (1998) lists 36 tribes using that species for medicine, fiber, food, and other purposes. Records also exist for three other species having been used by other tribes (Moerman 1998).

Hocking (1997) recorded that roots of *S. orbiculatus* roots were used as a diuretic and antipyretic, and that the roots were used as an astringent tonic. He did not say who used the plants, but presumably it was European and other settlers. Apparently, the only species that has been studied chemically is *S. albus*, and it contains saponins and tannins (Hocking 1997).

Symplocos

(From Greek *symplocos*, connected, from the union of the stamens)

Symplocos tinctoria (used as a dye)
horse-sugar [horse sugar] (in use by 1864)
[common] sweetleaf (so called because the leaves were relished by cattle and horses, and people, in use by 1829)
white root (South Carolina)
wild-laurel
yellow-wood [yellowwood] (named because the plants were in use to make yellow dyes before 1767; one of many trees called "yellow-wood")

In 1760, Nicolaus von Jacquin published his book *Enumeratio Systematica Plantarum* (Enumeration of Plant Systematics). In that volume, he described a new genus that he called *Symplocos martinicensis* for trees he had collected in Martinique. Then, 6 years later, Linnaeus published the name *Hopea tinctoria* for

Symplocos tinctoria. a. Flowering branch. b. Flower showing branched filaments. c. Flower showing calyx and bud scales. d. Fruiting branch. e. Drupe. *Drawn by Vivian Frazier.* From Correll and Correll 1972.

plants collected by Alexander Garden, a physician living in Charleston, South Carolina.

Subsequently, French magistrate Charles Louis L'Héritier de Brutelle realized that the trees discussed by Linnaeus were the same genus as those reported by Jacquin from Martinique. Therefore, in 1791, L'Héritier created the modern name *S. tinctoria.* There are now 250 species in the genus (Mabberley 1997). The family, Symplocaceae, which comprises this single genus, was created in 1820 by Réné Louiche Desfontaines (1750–1833), a French professor at the *Jardin des Plantes*, Paris.

Symplocos tinctoria is endemic to the southeastern United States where it grows from Florida to Texas and Arkansas, and north along the Coastal Plain and Blue Ridge to Delaware (Small 1933, Fernald 1950, Correll and Johnston 1970).

The first record of these trees in Florida was by William Bartram ([1791] 1958), who called them *Hopea tinctoria.* He found the plants along the St. Mary's River along the Georgia–Florida line in 1773, later on the St. Johns River above Lake Dexter, near Salt Springs in Marion County, and near the Alachua Savanna (Paynes Prairie of today) in Alachua County (Bartram [1791] 1958, Taylor and Norman 2002).

Linnaeus knew that the plants were used for dyes, as indicated by the species name, *tinctoria.* Small (1933) and others said that a yellow dye may be obtained from leaves and bark. Hocking (1997) said that the fruits also yield the dye.

The wood is close grained, and the heartwood is brown or red, but light and soft (Small 1933). Although never a commercially viable product, the wood has been used in turnery (Mabberley 1997). The sweet leaves are eaten by wildlife, cattle, and horses (Small 1933, Correll and Johnston 1970). Morton (1974) also found that woodsmen and campers chewed the leaves to relieve thirst.

The Choctaw used the roots to treat fever (Taylor 1940). Subsequently, settlers and blacks learned to use the plants. Standley (1920–1926) said that the root was used as a stomachic, to treat kidney ailments, and for syphilis and scrofula. Morton (1974) found people in the Low Country of South Carolina still using the roots for a "tea," which was drunk with sugar as a beverage during meals. This drink was so popular that workers at the Brookgreen Gardens had to be stopped from pulling up seedlings.

Hocking (1997) noted that the roots have been used as a tonic, stomachic, and digestive. He also noted that the plant has been used to treat asthma among the Gullah.

The genus contains tannins, ellagic and gallic acids, and some produce methyl salicylate or indole alkaloids of the carboline group (Cronquist 1981). Several of those chemicals are bioactive, and would produce some of the results desired in medical application. Not surprisingly, no chemical studies seem to have been done on the American species, but there are several on the Asian plants. In those, arabinogalactan, lignan glycosides, phenolic glycosides, triterpenoids, and sweetening agents have been found (Hussain et al. 1990, Ahmad et al. 2003, Li et al. 2003). Others found anti-AIDS agents, antimicrobial activity, antiplasmodial impacts, analgesic activity, and support for historical use against gastrointestinal problems and fevers (Inouye et al. 1973, Tiwari and Tripathi 1976, Grosvenor et al. 1995, Sampson et al. 2000, Horgen et al. 2001, Ishida et al. 2001, Khan et al. 2001).

T

Taxodium

(L.C.M. Richard named this from *taxus*, the classical Greek name of a European conifer, and *eidos*, resembling)

Taxodium distichum. From Sargent 1905.

Taxodium distichum (two-ranked, like a feather)
ahuehuete (from *ahuehuetl*, Náhuatl, Mexico)
amerikanische Zypresse [*amerikansk Cypress*] (German, North America)
asawî [*ashuit*] (Mikasuki); *vcenvho* [*acinahô, acinaho, achena, atchena, atcheanahoe, ashinahoo, uecheanatho, vcénvhó, vncénvhó*] (Creek, Muskogee; see also *Juniperus*); *a'tsina'* [*a'-tsi-na', a-tsi:-n'*] (Cherokee; loan-word from Muskogee); *waksiná* (maybe akin to Chitimacha *ho ʔaksus naʔa*, they are cypresses, Koasati)
[common] bald [pond] cypress [baldcypress, common-baldcypress] (USA); *calvo* (bald, Spanish, USA); *ciprés* [*cipresso*] *calvo* [*americano, de pantano, del sud*] ([American, swamp, southern] bald cypress, Spanish, Chiapas, USA); *cypres chauve* (bald cypress, French, USA); *taxodier chauve* (bald cypress, French, USA)
boscoyo à cypres (cypress knee, Houma, Louisiana)
ciprés de Montezuma (in the Gardens of Chapultepec, Mexico valley, was an old tree in the 1500s; estimated at 700 years old)
cipresso delle paludi (swamp cypress, Italian, USA)
cypres de la Louisiane [*cypres de Louisiane*] (Louisiana cypress, French, USA)
[black, buck, cow, deciduous, Gulf, Gulf red, knee, Louisiana black, Louisiana, Louisiana red, pond, red, river, southern, swamp, tidewater red, white, yellow] cypress (derived from Latin *cyparissus*, Greek *kyparissos*, the ancient name of an evergreen in the Mediterranean, now the genus *Cupressus*, USA)
gagüorí (Guarijío, Sonora)
moeras-cypres [*moerascypres*] (USA)
pecky cypress (from "peck" in the sense of "an impression or hole made by pecking," in English by 1591; the term was applied specifically to *Taxodium* with pockets of decay by 1848, USA)
pentamu [*pentamón*] (Tarascan, Tamaulipas)
pine barren cypress (for *T. ascendens,* Williams [1837] 1962)
pond *cyprés* (mixture of English and French, USA)
sabino (from the Latin for a *Juniperus*, which see, Sonora to Oaxaca)
satine faux (false satin, French, USA)
shan̲kolo [*shamgo'lo*] (Choctaw); *siwaakola* (Alabama); *shan̲kolo itibbi* (cypress knee; *itibbi*, battle, Choctaw); *siwaakola impisi* (cypress knee; *siwaakola*, cypress, *im*, its, *pisi*, breast, Alabama)
sump-cypress (swamp cypress, USA); *Sumpftaxodie* (swamp taxodium, German, USA), *Sumpfzypresse* [*Sumpcypress*] (swamp cypress, German, USA); *virginische Sumpfzedar* (Virginia swamp-cedar, German, USA)
tnuyucu or *yucu-ndatura* (Mixtec, Oaxaca)
txut (Atakapa)
yaga-chichicino or *yaga-guichi xiña* (*yaga*, tree, *guichi,* cedar, *xiña* from *xiñaa,* red, Zapotec, Oaxaca)

Depending on the taxonomy followed, the North American endemic genus *Taxodium* contains one, two, or three species. The latest view is that there is a single species (Watson 1993). Probably the first record of these trees was in the account of Hernándo Cortés weeping under one in the village of Popatela close to Mexico City. Supposedly, Cortés wept under the *Arbol de la Noche Triste* (tree of the sad night) on the night of 1 July 1520 after a disastrous battle (Stanley 1920–1926, Felger and Johnson 2001). The *Arbol de la Noche Triste* and the *Arbol del Tule* are now more than 500 years old. Although it appears to be several trunks fused, the *Arbol del Tule* at Santa Maria del Tule, Oaxaca, is considered to have the largest trunk circumference of any tree in the world (Dorado et al. 1996).

The trees are famous for providing timber, medicines, and resins, although there are other uses. Gilliland (1975) found that the Glades people used cypress wood before Columbus. From the Marco Island site dated at about A.D. 1300, trays, cups, bowls, and tubs were identified.

In historic times, Seminole relatives the Choctaw used the bark to make cords (Bushnell, 1909). The Houma used scrapings from the *boscoyo à cypres* (cypress knee) and mixed them with whiskey to purify the blood (Speck 1941). Houma also used the inner bark to cure jaundice. Sturtevant (1955) found the Miccosukee using cypress to build houses, canoes, dance posts, coffin logs, medicine bowls, spoons, food paddles, in tanning skins, to make arrowheads, drums, ox yokes and bows, heddles, mortars and pestles, and ball poles, spoon ball sticks, and dolls. They still use it. Their use likely reflects to a large extent that of their predecessors, the Glades people.

Standley (1920–1926) noted that Mexicans use the wood for planks, furniture, and general construction. In pre-Conquest times, the tree provided an acrid resin used to treat wounds, ulcers, skin diseases, toothache, and gout. The bark is used as an emmenagogue and diuretic, and the leaves relieve itch. Chips of wood are buried and fired, and yield a kind of pitch that is used to treat bronchitis and other chest problems. Martínez (1969) recorded the same basic uses in practice as in earlier records. Many of the same applications were made of the U.S. variations (Hocking 1997).

Taxus

(From Greek *taxos*, the yew-tree)

barlind (conifer-linden, Norwegian)
tasso (from Latin *taxus*, Italian); *teixo* (Portuguese); *tejo* (Spanish)
yew (from Old English *íw*, *éow*, dating from A.D. 725; akin to Old Saxon *íhm* Middle Dutch *íwe*, and Old Norse *yr*, bow); *eo* (Old Irish); *Eibe* (German); *hivin* (Cornish); *id* (Swedish); *if* (Quebec, France); *iff* [*ijf*, *ipe*] (Dutch); *iogh* (the ninth letter of the Gaelic alphabet); *iubhar* (Gaelic); *iuir* (Irish); *iva* (Spanish); *iven* (Breton); *ivos* (Old Celtic); *jevà* (Lithuanian); *ur* [*uir*] (Gaelic); *ywen* (Welsh); the French and Spanish may be derived from from Latin *ívus*, or all may be related to that source

Taxus floridana (of Florida)
Florida yew

Florida's endangered yew is confined to the margins of the Apalachicola River in Gadsden and Liberty Counties in the Panhandle (Coile 2000). The population of this and the related *Torreya taxifolia* are relicts of the last Pleistocene glaciation, and both are narrow endemics to that region. Like all *Taxus*, the Florida species is poisonous and would have been used by people living in its vicinity.

The yews are plants of ancient association with people in the Northern Hemisphere (de Cleene and Lejeune 2002). The similarity of names in European languages for the plants is one measure of its antiquity. These trees figure in classical Greek legends, and many of the views from that time persist in modern beliefs.

European yew (*Taxus baccata*) was dedicated to the Erinyes or Furies, goddesses of vengeance (Baumann 1993). These Furies punished human evil-doers in various ways, but one of their favorites was with poison from the yew. The needles contain an alkaloid so poisonous that 500 g is enough to kill a horse. Artemis, goddess of the chase, was commanded by her mother Leto to kill the children of Niobe. Those children had insulted Leto and she wanted revenge, so Artemis tipped her arrows with poison from yew. Artemis herself was worshiped in a temple that Theophrastus placed in a luxuriant forest of yews on the Artemision promontory of Arcadia.

Roman author Pliny (A.D. 23–79) wrote of the yew that the "berries are poisonous. Sextius states that the ... Arcadian variety has such an instantaneous effect that it is fatal to sleep under it or eat one's food beneath it. Some authorities claim that this is the reason why poisons were called *taxic* or *toxic* meaning 'that in which arrows are dipped.' " Editor and translator Healy (1991) put in a footnote regarding Pliny's text that *toxon* is Greek for bow, while *taxus* is Latin for yew. He could have added that "yew" meant "bow."

There is a remaining ambivalence toward the yew tree because of its history in Europe. The beneficent aspect of the yew lay in its wood, and it was highly prized in Europe because it was the best for making bows. Legend has it that Robin Hood's longbow was made of yew. Some also consider yew branches to make the best arrows (Vickery 1995).

Yew plants are associated with supernatural powers, and therefore considered capable of diverting evil. The same ideas were held by the Druids and other Celtic people, and the yew was more sacred to the Druids of Scotland than the oak (Dwelly 1933). Yews were planted by the Druids near their temples, and the Christians continued that practice. The shift may have been as simple as using an old Druid temple for Christian services and inheriting the trees with it, but it was probably more complicated than that.

Churches, whether Druidic or Christian, always had hallowed ground surrounding them. That holy land was used by communities as their graveyard. Because the yew has red heartwood and white sapwood, Christian leaders promoted the belief that those were the colors of the blood and body of Christ. Perhaps Druids used the same symbolism for their deities. Either way, the yew came to be considered a deterrent to evil spirits. Many cultures still plant a yew in a graveyard, believing it will disperse malevolent spirits. Except in cultures that have lost the old beliefs, yews are rarely planted near houses.

In Ireland, where palms occur only in greenhouses, the yew is a substitute on Palm Sunday—known as *Domhnach an iuir* (Yew Sunday) (Vickery 1995). The branches may be worn afterward and pieces are often placed in people's homes to bring good luck. Similarly, the yew substitutes for the ash (*Fraxinus*) on Ash Wednesday ceremonies. After the ceremony, some people put the branch in the cow byre, while others put it beside a religious picture in their kitchen.

Indigenous people in the Americas also knew and used *Taxus.* There are three species native to North America: *T. brevifolia* in the northwest, *T. canadensis* in the northeast, and *T. floridana* in Florida. Records exist that various tribes used the two most widespread species, *T. brevifolia* and *T. canadensis,* and adopted the European *T. baccata* and Japanese *T. cuspitada* into their pharmacopoeias (Moerman 1998).

Both native species have been used in a variety of medicines. Eastern tribes used *T. canadensis* in minute amounts to make tea to treat rheumatism, bowel ailments, fevers, colds, scurvy, to expel afterbirth, dispel clots, and as a diuretic (Foster and Duke 1990, Moerman 1998). The plant is so poisonous with the alkaloid taxine that as few as 50 needles ("leaves") have resulted in death (Foster and Duke 1990). Duke et al. (2002) consider all three species (*T. baccata, T. brevifolia, T. canadensis*) unsuitable for medicinal use.

Taxus is not only a plant, it is an animal. Since 1535, it has been the English derivation of Medieval Latin for the badger (OED 1971). The scientific name of the European badger is *Taxidea taxus.* How two such different organisms came to be called by the same name is a mystery. However, analysis of the animal's name reveals an unexpected connection. *Taxidea* is composed of Greek *taxos,* the yew, and the suffix *ide,* pertaining to trees. Thus, the badger is an animal that hides in the forests of yew.

Tephrosia

(From Greek *tephros,* ash-colored or hoary, a reference to the gray pubescence of the leaves of many species)

Tephrosia virginiana. From Britton and Brown 1897.

Tephrosia florida (of Florida)
Aschenwicke (ash-colored vetch, German)
Florida hoary-pea [hoarypea]

Tephrosia hispidula (with short, straight hairs)
sprawling hoary pea

Tephrosia virginiana (of Virginia)
a:tatiní:li iyyaksí (*a:tatiní:li,* bad man's [devil's], *iyyaksí,* claw, Koasati)
cat-gut [catgut] (applied by 1599 to violin strings, although no such use has been recorded beyond the name; Britton and Brown were using the plant name in the Americas by 1896)
distai'yí (the roots are tough, Cherokee)
goat's-rue (wild rue, a comparison to *Ruta graveolens,* perhaps because of the compound leaves; Britton and Brown were using the name in the Americas by 1896)
haashshachi' isholosh talakchi' (*haashshachi',* devil, *im,* his, *sholosh,* shoe, *talakchi',* tied up, Chickasaw); *siitan isholosh takachi'* (*siitan,* Satan, *im,* his, *isholosh,* shoe, *takachi',* tied up, Chickasaw); devil's shoestring (the long poisonous roots)
hó:łi:pî [*hó:rí:pî*] (Mikasuki); *hvloneske* [*aloniskâ:, aloníski, haloníski, hvnoleske, vlonéske*] (Creek); *łako* [*rako*] *niski* (by Howard 1984, Muskogee); *imallun̲sak* [*issimallun̲sak, imisin̲lusak*] (*isi,* deer, *in̲,* its, *allun̲sak,* baby, Choctaw)
loksimpolona (*loksi',* turtle, *im,* its, *polona,* string, Chickasaw)
rabbit's-pea (i.e., wild pea)
suckehihaw [*soo-ke-he-aw*] (young turkey's food; *çiú-ka,* turkey, *i-ha',* mouth, Osage); turkey pea

(translation of Osage name use by Hunter [1823] 1973)
Virginia tephrosia (a book name)
ya·mbara?' wi·ya'hewe'? (bad-man's [devil's] string, Catawba)

Tephrosia is a tropical genus with about 400 species (Mabberley 1997). The most famous species in the eastern United States is also the first that became known in Europe. Leonard Plukenet described these in 1696 as a *Cicer.* He mentioned them in another book published in 1700, but then changed his mind and called them *Orobus.* Finally, Linnaeus and Gronovius examined John Clayton's specimens and decided that they should be called *Clitoria foliis pinnatis* (butterfly-pea with pinnate leaves). Linnaeus realized that too was incorrect, and called the herbs *Cracca virginiana* in 1753. At the time *Cracca* was interpreted broadly, and included species now known to belong elsewhere. It was finally 1807 when C.H. Persoon described the genus *Tephrosia* to accommodate this and related species. The other two Florida species were first reported by André Michaux in 1803 under another genus.

Goat's rue or devil's shoestring (*T. virginiana*) became the most famous of these plants not only because of its early discovery and popularity among indigenous tribes, but also because chemists learned it contained rotenone. The species contains rotenoids and flavonoids, including rotenone, dehydrorotenone, tephrosin, and deguelin (Kinderscher 1992). Rotenone has shown anticancer activity, but is also described as carcinogenic. Duguelin, rotenone, and tephrosin can cause paralysis and death but are not as toxic to humans as they are to fish and insects. Roots and leaves are the parts used in medicines because the wood and seeds are toxic (Foster and Duke 1990, Kinderscher 1992).

Long before chemical studies were made, southeastern people used the plants to gather fish. The Miccosukee used it as fish poison even though it does not grow on the Big Cypress Seminole Reservation in southern Florida (Sturtevant 1955). The Miccosukee use was a tradition passed on from their ancestors and their Choctaw and Yuchi allies who all used the plants farther north (Byington et al. 1915, Sturtevant 1955). The retained name among the Oklahoma and Florida Seminoles indicates the antiquity and extent of use to obtain fish.

Most of the extant records for this species are for medicinal uses, as are all records for the other two species. The second most important application of *T. virginiana* was to eliminate intestinal worms. At least the Cherokee, Osage, Kansas, Koasati, and Pawnee used it for that purpose (Hunter [1823] 1973, Krochmal 1968, Hamel and Chiltoskey 1975, Kimball 1994). Settlers also adopted the remedy. Porcher (1863) wrote, "The roots were used by Indians, and are now employed in popular practice as a vermifuge; a decoction is said to act as powerfully and as efficiently as the pink root (*Spigelia*)." Millspaugh (1892) agreed.

The Creeks also mixed devil's shoestring with sassafras to treat bladder problems and coughs, including tuberculosis (Swanton 1928a). Creek women suffering from irregular periods bathed in liquid made from *Stillingia sylvatica* and *T. virginica* (Swanton 1928a). The same mixture was used to treat male impotence. The Chickasaw used the roots to "draw poison" from a hurting chest (Howard 1985). The Choctaw used it to treat fever and malaria (Byington et al. 1915). The Oklahoma Seminoles made a tea that is drunk as treatment for general malaise (Howard 1985), and the Cherokee did the same (Mooney 1885–1886).

The Cherokee had several other uses. Women washed their hair in a root decoction to prevent it from falling out. They believed the toughness of the roots would be transferred to their hair. Cherokee ball players rubbed the decoction on their limbs to toughen them (Mooney 1885–1886). Decoctions were given to Cherokee children to make them strong (Taylor 1940, Ajilvsgi 1984). The Catawba used the leaves to treat fever and rheumatism (Moerman 1998).

A decoction of mashed *T. florida* roots were applied to sores by the Choctaw (Taylor 1940). The Koasati applied an infusion of roots to snakebites (Taylor 1940). The Choctaw chewed roots of *T. hispidula* and swallowed the juice to stop coughs; too much would loosen the bowels (Campbell 1951).

The U.S. Dispensatory of 1885 considered the plant cathartic and the roots tonic and aperient (Vogel 1970). Duke et al. (2002) indicate that *T. virginiana* has been considered allergenic, analgesic, anticancer, anthelmintic, antileukemic, carcinogenic, diaphoretic, insecticidal, laxative, piscicideal, stimulant, and tonic. In the proper dose, they consider it as safe as drinking a cup of coffee.

Tetrazygia

(From Greek *tetras*, four, plus *zygon*, yoke, a reference to the four-parted flowers of *T. tetrandra*)

Tetrazygia bicolor (two-colored)
cordobancillo (probably based on *cordabán,* alluding to a use for tanning skins; ultimately from *Córdoba*, a city famous for leathers, Cuba)
Florida clover ash
tetrazygia (a book name)

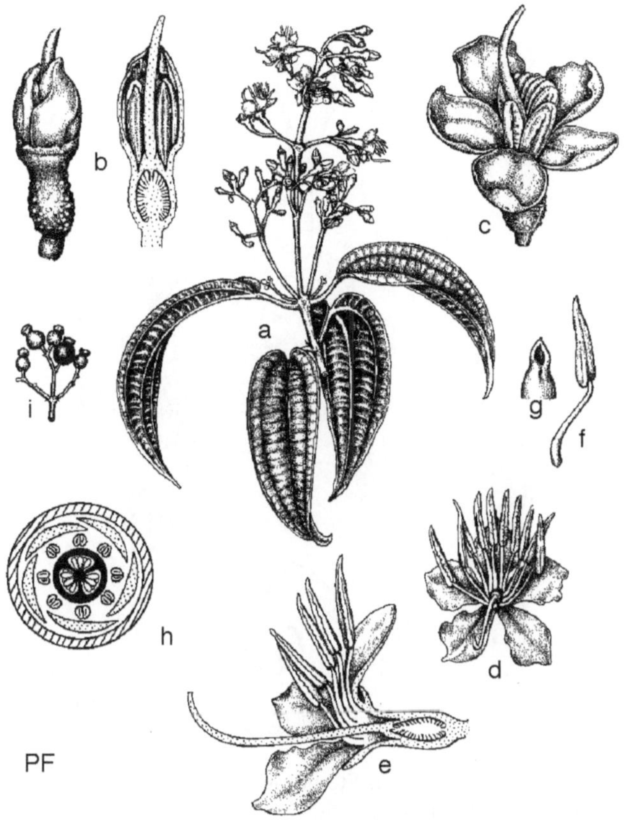

Tetrazygia bicolor. a. Branch with flowers. b. Flower, just before opening, in side view and in longitudinal section. c. Flower, just after opening, the stamens still folded. d. Open flower, from above. e. Open flower, longitudinally dissected. f. Stamen. g. Anther pore. h. Floral diagram. *Drawn by Priscilla Fawcett.* From Correll and Correll 1982.

In 1768, Philip Miller named plants growing in London *Melastoma bicolor*. That genus, created by Burmann in 1737 and adopted by Linnaeus ([1753] 1957), already had seven species, including four that Linnaeus knew grew in the Americas.

As more plants were found during exploration of the world, people realized that the Caribbean plants did not fit in *Melastoma*. Finally, in 1828, the genus *Tetrazygia* was created by L.C.M. Richard and A.-P. de Candolle, based on *T. tetrandra*. Belgian researcher Célestin Alfred Cogniaux (1841–1916) moved the Florida species to *Tetrazygia* in 1891. Now the genus contains 25 species, and it is known from Florida and the West Indies (Mabberley 1997).

Tetrazygia bicolor is found in Florida, the Bahamas, Cuba, and Hispaniola (Small 1933, Leon and Alain 1957–1963, Correll and Correll 1982). The species has been reported from Jamaica, but Adams (1972) separated their plants out as *T. pallens*. Jamaicans call *T. pallens* clover ash, which gave rise to the common name listed by Wunderlin (1998) for Florida plants. The other names of the Jamaican trees are ashes bush, ashes wood, and black ashes, so the name "clover ash" is not a comparison with *Fraxinus*.

Florida's plants are restricted to small parts of Miami-Dade and Monroe Counties. There the hammocks and pinelands are disappearing because of urbanization, and the species is considered threatened (Coile 2000).

The only record of use for *T. bicolor* is from the Bahamas. There a leaf decoction has been used to reduce night sweats (Higgs 1969). Unfortunately, Higgs gave little information, and Morton (1981) found no other mention of people using the plant.

The Cuban common name *cordobán* is applied to at least eight other members of the Melastomataceae. Similarly, *cordobancillo* is used for 16 other members of the family, including the medicinal *Miconia laevigata* (Morton 1981). Since a number of *Miconia* and other genera in the family are notable hemostats, it seems likely that *Tetrazygia* is also rich in tannins. Cronquist (1981) confirmed that several members of the family contain tannins.

Teucrium

(From *teukrion*, possibly for *Teucer* or *Teukros*, the founder of the town of Salamis in Cyprus)

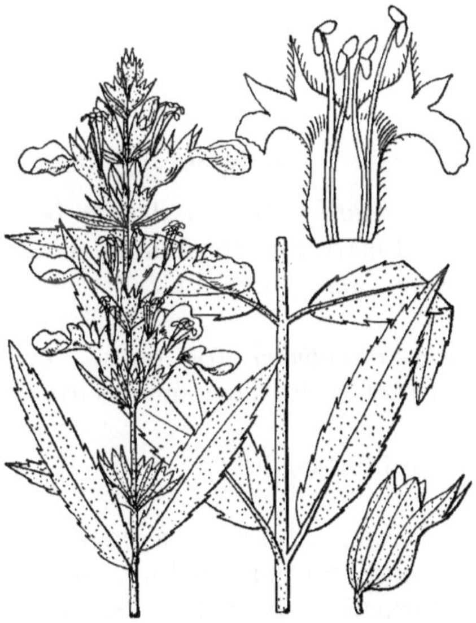

Teucrium canadense. From Britton and Brown 1898.

darag thalmhuinn (*darag*, oak, *thalmhuinn*, earth, Gaelic) germander (a corrupted form of Greek *chamaedryas*, ground oak; the original is commemorated in *Teucrium chamaedryas*, USA, England); *calamandrea* (Italian); *camadréo* (Spanish); *Gamander* (German); *germandrée* (French)

Teucrium canadense (of Canada)
agrimonia (Spanish; see also *Agrimonia*)
American germander
kofuckv [*kofucka, kofucke, kaputcka*] (Creek, Muskogee; also used for *Monarda*, which see)
snake-weed (Missouri)
wild basil [sage] (USA)

To Europeans, one of the best-known germanders was *Teucrium chamaedryas*, which has been used to make a medicinal tea since antiquity. Another well-known germander is *T. marum*, although we now call it "cat thyme" (Mabberley 1997).

The Seminoles in Oklahoma tie some of the leaves into a small sachet at the back of their neckerchief when attending the Stomp Dance (Howard 1984). They may have historically used the plant, or they may have imitated the people of the prairies and plains who used sachets of perfume in their War Dance bandoleers, or even the peyotists (Howard 1984).

A leaf tea among settlers was used to induce menstruation, urination, and sweating (Foster and Duke 1990). This species was also used like *Lycopus* for lung ailments, worms, and piles, and externally as a gargle and antiseptic dressing (Foster and Duke 1990).

Thalassia

(From Greek *thalassa, thalatta*, the sea, because these are submerged marine plants; akin to Akkadian *apsu*, sea, deep water, *sala'u*, besprinkle)

Thalassia testudinum (of turtles, because the plants grow where these animals live near shore)
seaweed
[Caribbean] turtle-grass

Thalassia was proposed as a genus by Sir Joseph Banks and Daniel Solander, but not published, and it was later formalized by C.D.E. Koening in 1805. There is one species in the Caribbean and another in the tropical Indian and Pacific Oceans (Mabberley 1997).

The leaves are used as fertilizer, at least in paddy fields and coconut plantations (Hocking 1997, Mabberley 1997).

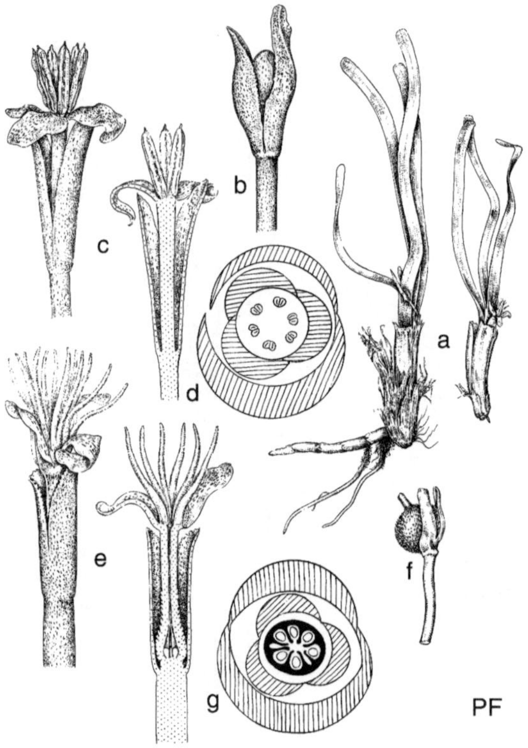

Thalassia testudinum. a. Sterile plant (left) and pistillate plant (right). b. Staminate flower bud. c. Staminate flower, side view (left) and longtitudinally dissected (right). d. Floral diagram of staminate flower. e. Pistillate flower, side view (left) and dissected (right). f. Fruit. g. Floral diagram of pistillate flower. *Drawn by Priscilla Fawcett.* From Correll and Correll 1982.

Thalia

(This is one of Linnaeus's double origin names; ostensibly he named it for Johannes Thal (1542?–1583), a German physician at Stolberg am Harz; however, he surely did not miss the allusion to the muse *Thalia*, who presided over comedy)

Thalia geniculata (with knees, meaning the jointed stems)
agutiguepe-obi [*agutigvepo-obi*] (first reported by Marcgraf and Piso in 1648, Brazil; according to Ferreira ca. 1978, *aguti* from *aku'ti*, the Tupí for the mammal *Dasyprocta araruta*; the *guepe* is unexplained; and *obi*, of African origin, akin to Twi *a'bi'a*; the implication is that the plant is eaten by the mammal, and that it is used in Macumba rituals, but no further information has been found)
alligator flag (presumably because it is one of the "flags" that grows in wetlands frequented by these large saurians, Florida)

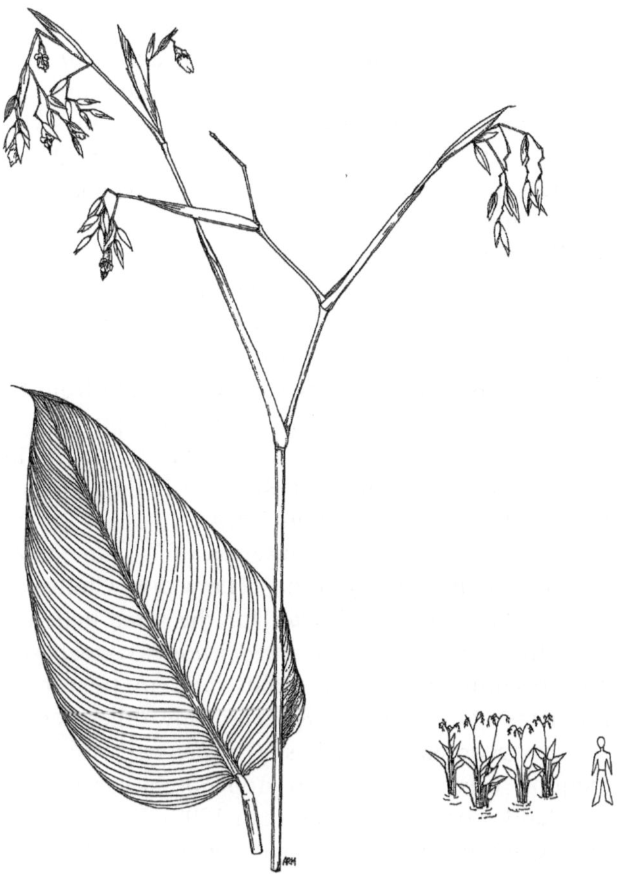

Thalia geniculata. From Institute of Food and Agricultural Sciences.

arumarana (*aruma*, from *aru'mã*, a name for *Ischnosiphon ovatus*, another member of the Marantaceae, *rana*, false, Tupí, Brazil)
co:kíta [*shogita, shongikee*] (to suck, Creek); *soki:kí* [*çonkigi, sonkiiki, shongiskehi*] (suck it, Mikasuki)
fire-flag [fireflag] (Florida)
glaïeul [*glade*] (gladiolus, Hispaniola); *gorgeau* (gullet, but probably from *glaïeu*, Haiti)
pámpano (probably from Quechua *pampa*, plain, prairie, Puerto Rico)
platanar (of bananas, Belize); *platanillo* [*de río*] (little [river] banana, Dominican Republic, Belize, Panama)
swamp lily (Panama)
waha (Miskito, Nicaragua)
zolupa (Hispaniola)

Bauhin knew these wetland herbs from cultivation in Europe in 1623. Marggraf and Piso reported them in eastern Brazil in 1648. Hans Sloane later recorded the plant in Jamaica in 1696. Thus, the plants had been in Europe a long time when Linnaeus finally gave them their name in 1753. Now, there are seven species in tropical America. In a range like that of *Annona glabra*, these herbs also reappear in western tropical Africa (Mabberley 1997).

Although Sturtevant (1955) recorded that the Miccosukee used *soki:kí* as food, he did not elaborate. Morton (1968) and Duke (1972) note that the inner part of the boiled root is edible. Morton's (1968) description of it as looking like coral-pink shrimp inside makes it sound palatable, but she says it has a "swampy" flavor.

Bennett (1997) was told that the fresh roots are eaten and that the flowers, rhizomes, and base of the stem were edible, and used especially during wartime. Leaves were also used to cook food. Corn bread dough was mixed with meat and rendered fat, and then boiled in the leaf, like a *tamale*, but the Miccosukee call it *paluee*. Beans sometimes were added. Corn meal and fat, wrapped in leaves, were boiled in four different pots at a time, as an unspoken "offering" to the gods of the four winds.

In western Africa, the stems were used for basketwork, and the leaves were used as a packing material (Dalziel 1937). Since there were no comparable leaves in Florida before the introduction of bananas, perhaps indigenous people also wove with them.

Grimé (1976) found an apparently unique application of the plants being used to neutralize poison in Jamaica in the 1790s. The Brazilians use the rhizomes as a medicine on sores and apply it as an antiseptic (central nervous system stimulant) (Mors et al. 2000).

Thalictrum

(From Greek *taliktron*, a name of some plant mentioned by Dioscorides, fl. A.D. 40–80, with coriander-like leaves; Pliny, A.D. 23–79, applied Latin *thalictrum* or *thalitruum* to the meadow-rue)

Thalictrum thalictroides. From Britton and Brown 1896.

frøstjerne (stone seed, Norwegian)
meadow-rhubarb; *rubarbio de los pobres* (poor people's rhubarb, Spanish); *ruibarba dos pobres* (poor people's rhubarb, Portuguese)
meadow-rue (in use by 1653 when Culpeper used the modifiers "meadow or wild" to distinguish between these plants and the cultivated rue, *Ruta graveolens*); *rù beag* (*rù*, rue, *beag*, little, Gaelic); *ruda cimarron* [*de la sierra*] (wild [mountain] rue, New Mexico); *Wiesenraute* (meadow rue, German)
nisude-hi (*nisude*, flute, *hi*, plant, Omaha-Ponca)
pigamo (from Greek *peganon*, the name of *Ruta*, Italian); *pigamon* (Quebec)
skadiks [*skariks*] (Pawnee)
talictro (from *Thalictrum*, Portuguese)
wazimma (*wazi*, pine, *mna*, to smell, Dakota)

Thalictrum cooleyi (named for George Ralph Cooley, 1896–1986, a wealthy benefactor of the botany programs at the University of North Carolina, Chapel Hill, and the University of South Florida, Tampa)
Cooley's meadow-rue [meadowrue]

Thalictrum macrostylum (long-styled)
piedmont meadow-rue [meadowrue]

Thalictrum revolutum (leaves with the edge rolled back)
maid-of-the-mist
[purple, skunk, skunk-leaved, wax-leaf, wax-leaved, waxy, waxy-leaf] meadow-rue [meadowrue]

Thalictrum thalictroides (resembling *Thalictrum*)
[rue] anemone (Ohio)
may [star, wind] flower (Massachusetts, Ohio)
wild-potato ("I know of mountainous districts in Pennsylvania where these roots are collected and eaten under the name of 'wild potato,' " Medsger 1939 in Coffey 1993)

The history of the European common names for *Thalictrum* is bound up with those for two other plants, rue (*Ruta graveolens*) and rhubarb (*Rheum rhaponicum*, from Greek *rheon*, root, and *Rha ponticum*, the root that grew on the *Rha*, Volga River). Because of Old World uses, and because of leaves similar to rue, *Thalictrum* was compared to both. Both *Rheum* and *Thalictrum* are purgative, and *Ruta* and *Thalictrum* have been used as spices.

Rue has as common names *Raute* (German), *ruda* (Portuguese, Spanish), *rue* (French), and *ruta* (Italian). *Ruta* is based on the Greek *rhyte* and Latin *ruta*, bitterness or a bitter herb. Related words are Akkadian *ratu*, to fortify, and Hebrew *ratam*, to bind fast. Rue was in use in English by ca. 1382. In Arabic, rue is *fegan*, which is cognate to Greek *peganon*.

Rhubarb is *rabarbaro* (Italian), *reubarba* (Portuguese), *Rhabarber* (German), *rhubarbe* (French), and *ruibarbo* (Spanish). These names were all derived from Latin *r(h)eumbarbum*, bearded root. These plants are not native to Europe. They were introduced from the Far East before 1400 when rhubarb appeared in English. For example, John Gerarde wrote in his *Herball* of 1597, "The best Rubarbe is that which is brought from China fresh and new."

The Cherokee took an infusion of *T. thalictroides* to stop diarrhea and vomiting (Hamel and Chiltoskey 1975). They used *T. dioicum* for the same maladies. The Iroquois also used *T. diocum* to treat sore eyes resulting from head colds, and to treat heart problems (Moerman 1998). The only other eastern species recorded as used is *T. pubescens*, which the Iroquois used in a bath to stop nosebleeds and for gallbladder problems (Moerman 1998). The Montagnais (Quebec) spice salmon with the leaves (Moerman 1998).

From the Great Plains west, Moerman (1998) documented six other species used as spices in foods, as deodorants, medicines, and toy flutes.

Thelypteris

(From Greek *thelys*, female, and *pteris*, fern; used by Theophrastus, 372–287 B.C., and Dioscorides, fl. A.D. 40–80, and resurrected by Casimir C. Schmidel in 1762)

Thelypteris kunthii (named for Carl S. Kunth, 1788–1850, professor of botany, Berlin) (= *T. normalis*)
fern (Bahamas)
[lady] maiden fern [palm] (apparently a comparison with maiden-hair fern, *Adiantum*, Florida, Bahamas, Belize)
southern shield-fern (Florida, Bahamas)
taapente (fern, Mikasuki)
tapínthískítilíkmî [*tapinthiçtilikmi*] (delicate foliage fern, Mikasuki)

Thelypteris palustris (of marshes)
marsh fern

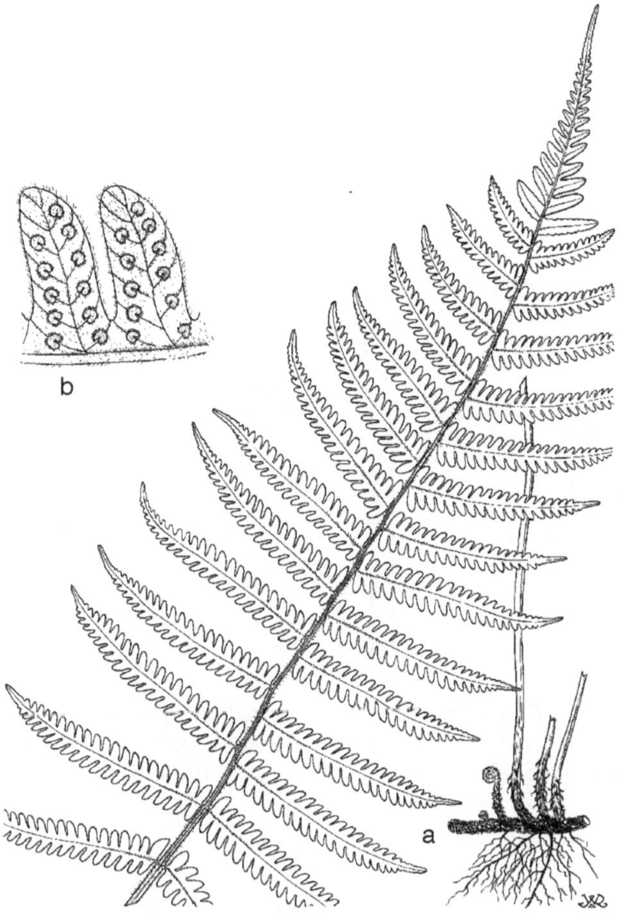

Thelypteris kunthii. a. Upper part of frond and rhizome. b. Two pinnules to show venation and sori. *Drawn by Jane W. Roller.* From Correll and Correll 1982.

Wunderlin (1998) included 14 species of *Thelypteris* native to Florida. Documentation of uses for only two was found, but it is likely that others were similarly applied.

Sturtevant (1955) was told that the leaves were used for "Old Paint Woman's Sickness" (weakness of limbs and neck). The Seminoles called this malady *hamocolo:pî* (paint old woman or paint witch, Mikasuki). There is no recorded connection to this use and the plant *hamockolo:píntapínti* (paint old-woman's fern, *Osmunda regalis,* which see).

In the Bahamas, they take leaves from the maiden fern, parch and powder them, and sprinkle the dust on dermatitis caused by poisonwood (Morton 1981). The Iroquois used the roots of *T. palustris* for "woman's troubles" (Moerman 1998).

Hocking (1997) recorded this species under both *T. dentata* and *T. kunthii,* apparently based on Sturtevant (1955) and Morton (1981).

Thrinax

(Olaf Swartz used the Greek for fan, in reference to the leaves)

Thrinax morrissii (named for D. Morris, assistant director of Kew Botanical Gardens, who first discovered this palm on Anguilla Island in 1890)
buffalo-top (Bahamas)
guano de sierra (mountain palm; *guano,* palm, Taino, Cuba)
miraguano (*mira* perhaps corruption of *lana,* ocean, *guano,* palm, Taino, Cuba)
palma de cojollo (little testicle palm, Puerto Rico); *palma de escoba* (broom palm, Puerto Rico); *palmito* (little palm, Cuba)
pandereta (panderer, Puerto Rico)
peaberry palm (Florida)
[brittle, broom, key, little, prickly, small-fruited] thatch [palm, thatch-palm] (Bahamas, Florida, Puerto Rico, West Indies)
yaray (presumably from *yara,* place, or *yerén,* fruit, Taino, Puerto Rico)

Thrinax radiata (rayed)
bayal (*baay,* net for carrying things, but usually a name given to *Desmoncus orthacanthos,* Maya, Yucatán)
ch'it (simple term, Maya, Yucatán, Belize)
fan palm (Belize)
guanillo (little palm, from *guano,* a palm, Taino, Dominican Republic); *guano Campeche* (Campeche palm, Mexico); *guano de costa* (coastal palm, Cuba); *guano de lana* (ocean palm; *guano,* palm, *lana,* ocean, Taino, Cuba)
kanalxan (*kanal,* tall, *xan,* feces, also the common name for *Sabal japa* or *S. mexicana,* Maya, Yucatán)
kultok (*kul,* short, *tok,* hard, resistant, Maya, Yucatán)
large-fruited thatch-palm (Bahamas)
latanier-de-la-mer [*latanier la mer, latanier lan mer*] (seaside *latania*; *latania* is an indigenous word from Mauritius for a palm there; apparently, this native plant became confused with the one from Africa, Haiti)
miraguano de lana (ocean palm, Taino, Cuba)
palma cana (cane palm, Cuba)
salt water palmetto [pimenta, pimento] (Belize)
sea thatch; [Florida] thatch palm (Florida)

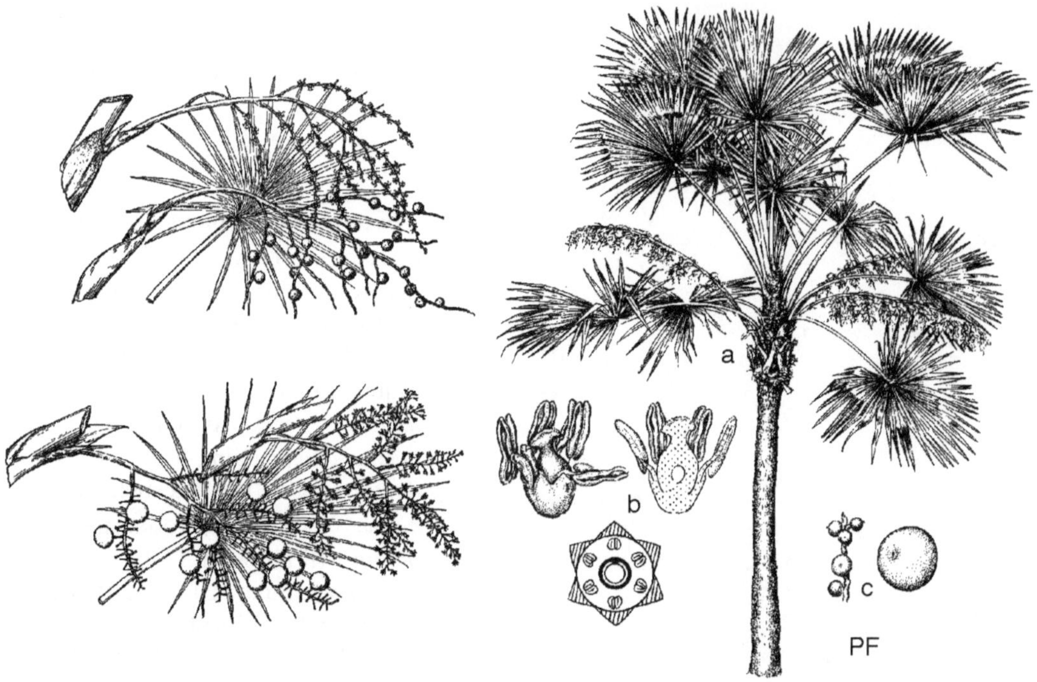

Thrinax. *Thrinax morrissii* (right). a. Habit. b. Complete flower on left; flower, longitudinally dissected on right; below, floral diagram. c. Twig with fruits and single fruit enlarged. *Drawn by Priscilla Fawcett*. From Correll and Correll 1982. Detail of flowering and fruiting branches and leaf (upper left). From Sargent 1905. *Thrinax radiata* (lower left). From Sargent 1905.

Gilliland (1975) recorded palm fiber cords and netting among the artifacts on the pre-Columbian site at Marco Island that had been used by Glades people. That particular site probably consisted of cords from *Sabal palmetto*, as that was the common palm in the region. In the Florida Keys and Caribbean, *Thrinax* was more common and the preferred species. As late as Vignoles ([1823] 1977) and Williams ([1837] 1962), people in the West Indies were still using palm fibers for ropes. As the common names suggest, the leaves of *Thrinax* were preferred for thatching where they existed. Probably, it was at least partly leaves of *Thrinax* that were used by the people on Matacumbe to make the women's shawls that Fontaneda described (see *Sabal*).

Roots and shoots are considered tonic and restorative on Hispaniola (Liogier 1974). In Yucatán, the palm has a reputation of being tonic, corroborant, vigorative, sedative, nutritive, and diuretic (Morton 1981). The palm has been used to treat tuberculosis, bronchitis, and other respiratory problems. *Thrinax* also has the reputation in Yucatán of developing the mammary glands and revitalizing the sexual organs (Morton 1981). Fruits are sweet and edible. The fiber has been used to stuff pillows and mattresses (Standley and Record 1936).

Tilia

(The classical Latin name of the tree, maybe based on Greek *ptilon*, wing, referring to the winglike bract of the flower clusters)

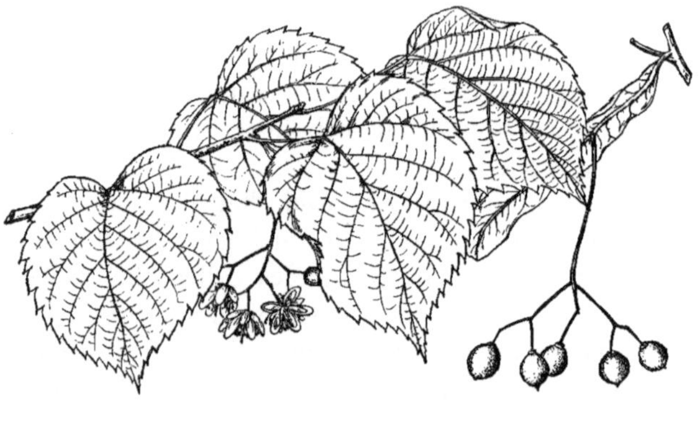

Tilia americana. From Sargent 1905.

craobh theile (*craobh*, tree, *theile*, lime or linden, Gaelic); *theile* [*teile*] (Gaelic); *tiglio* (Italian); *tilia* (Portuguese); *tilleul* (French); *tilo* (Spanish)
lenda (Old Teutonic; cognate with Greek *elate*, translated "silver tree" by the OED 1971, but *elate* is usually applied to fir or spruce); *lind*

(Danish, Norwegian, Swedish); *linde* (Dutch); *Linde* (German)

Tilia americana (of America)

[American, Carolina, Florida, Kendall, white, yellow] basswood ("bass" is a corruption by 1691 of "bast," fibers from the inner bark of trees; used with "wood" by 1824)

bast-tree ("bast" refers to fibers from the inner bark of trees, used in English since about A.D. 800; originally used uniquely for *Tilia*, but later applied to other plants with useful bark)

batahkó (Koasati); *panashuk* [*pishannuk*] (Choctaw); *ittobaksa* (*itto*, tree, *baksa*, thread, twine, Alabama; cf. also *Prunus*)

bee-tree

blacklime-tree

bois blanc (white tree, Quebec)

chesawk (perhaps this Virginia Powhatan name for the "rind of a tree like hemp" belongs here; recorded by Strachey in [1612] 1953)

daddy [monkey] nuts (an allusion to the fruits?)

hinde-hi (*hi*, plant, Omaha-Ponca); *hin'-dse* [*hin'-dse xtsi hi*] (Osage); *hinshke* (Winnebago); *hinta-chan* (*chan*, wood, Dakota)

i:-de-ha (Cherokee)

ittobitiima (*itto*, tree, *bitiima*, stink, Alabama)

lennikbi (basswood forest, Delaware)

[American, tree, limetree] lime ("Lime" as applied to *Tilia* is apparently derived from *lind*. From about A.D. 700, "lime" meant a glue put on limbs to catch birds; shortened from "birdlime." In the 1600s, "lime" was applied to the fruits of *Citrus aurantiifolia*, and about the same time to *Tilia*. The spelling "lime" came into English from distinct sources. For the glue, it came from Latin *limus*, mud, through Old Teutonic *limo*. The name of the fruit came from Arabic *limah*.)

linden [lin-tree, southern lind, white lind] (from Old English *lind* or *linde*, used since about A.D. 700; also rendered *linn* in English and Swedish)

ohòchsera (Onondaga)

spoonwood

tilleul d'Amerique (America linden, Quebec); *tilleul noir* (black linden, Quebec)

vhahwv [*uhawhu*] (Creek; according to OED 1971 the word meant *Tilia* or *Ulmus*; however, Martin and Mauldin 2000 equate the word with walnut, *Juglans*)

whistlewood (leaves used to make a whistling sound by holding them between the hands and blowing on the margin, Vickery 1995)

whitewood

wickyup (in English by 1852 as "wickiup." First recorded in [1612] 1953 by Strachey, who wrote *weihkippeis* for "hemp." Siebert 1975 has associated this with *wikepi*, bast or inner bark of linden or leatherwood. Derived from Algonquian languages; akin to Fox *wikiyap*, Potawatomi *wîgobi'mîc* (string tree), Ojibwa *wigub'imij* [*wigkub'imim*], Menomini *wikiop*, Cree *mekewap*, and Montagnais *mitshiuap*. Since the word means a dwelling, the association must be because the fibers were used to build the framework.)

The linden is eternally linked with Carolus Linnaeus (1707–1778), the egocentric genius who forever changed the way scientific names were written. At his home in Uppsala, Sweden, where he was professor at the university, he maintained a row of linden trees to emphasize his name.

It was in the poor Swedish town of Råshult that Carl [later Latinized to Carolus] was born to Christina and Nils Ingemarsson Linnaeus. Nils, a minister, as had been his father-in-law in the town, adopted the name "Linnaeus" during his years as a theological student. Tippo and Stern (1977) wrote that the name "was chosen from that of the linden tree ... which grew in the farmland of Nils' father." If Nils had not adopted the name, the "Prince of Botany" would have been known as Carl Ingemarsson. Of himself, Linnaeus once wrote, "*Potest e casa vir magnus exire*" (A great man can come forth from a cabin).

European common names for these trees are split between Germanic and Romance languages, with the former based on *lind* and the latter from *tilia*. Because of their utility in Europe for producing strong inner bark fibers, they were also known as the wood that produced "bast" or "bass." That later became "basswood."

When Europeans arrived in the New World, they found a new type of linden. Both Leonard Plukenet in 1696 and Jan Gronovius between 1739 and 1743 listed and discussed the New World species with phrase names. However, it was Linnaeus who gave it the binomial *Tilia americana*. Most of what Linnaeus knew about the American species was from his former student Peter Kalm.

In the village of Raccoon, New Jersey, on 17 April 1749, Kalm ([1753–1761] 1972) compared the New World tree to the one he knew in Sweden. Of the strength and toughness of *Tilia*, he wrote: "The *English* and the *Dutch* in many parts of *North America*, and the *French* in *Canada*, employ this bark in all cases where we make use of the *Lime-tree* bark in *Europe*. The tree itself is very tough, and you cannot easily separate its branches without the help of a knife: some people employ the twigs for rods."

Kalm probably did not mention the European medicinal uses because he expected his readers to be familiar with them. Tea made from the flowers is a "classical digestive end to a continental meal, and a treatment of insomnia, nervous tension, and overwrought children" (Bremness 1994). The tea also induces sweating, which helps in colds, headaches, flu, arteriosclerosis, and may lower blood pressure (Foster and Duke 1990). Some also consider the water a skin tonic and use it in bath preparations to soothe rheumatic aches (Duke et al. 2002).

Indigenous Americans also used the tree for fiber, medicine, and as food. Basswood fibers are known from both Hopewellian and rock-shelter sites in prehistoric context. Gilmore (1931) found the bast fibers used in twine and cordage among the Ozark Bluff-Dwellers of Arkansas. Jones (1936) identified *Tilia* at the Menifee County, Kentucky, site. Whitford (1941) noted linden fiber mixed with other kinds in fabric from both Ohio Hopewellian sites and rock-shelter areas.

Among historic people, Whitford (1941) examined Delaware wampum, Menomini, Sauk and Fox, Potawatomi, and Winnebago bags, and an Iroquois burden strap made from *Tilia.* He said that it was the most commonly and widely used of fiber sources among Eastern tribes, and he only lacked material of it from the "extreme south" (i.e., to him, Kentucky, Tennessee, and Louisiana).

Other people recorded *Tilia* bark use among historic Algonquin, Cherokee, Lakota, Malecite, Menomini, Meskwaki, Ojibwa, Omaha, Pawnee, Ponca, and Potawatomi (Gilmore 1919, Smith 1933, King 1984, Moerman 1998). These people used the bark to make bags, baskets, belts, fishnets, houses (*wickyup*), mats, netting in snowshoes, rope, sewing thread, shoes, and twine. They even used the thread for sewing up wounds (Vogel 1970).

Those same tribes used wood from the linden to make dugout canoes and lumber, but surely many more utilitarian items. Although not mentioned by Moerman (1998), the basswood was also the preferred tree for the carved masks used by the Iroquois Society of Faces (usually called False-Face Society) (Trigger 1978). One of the early records of that was given by John Bartram during his visit to the Onondaga in 1743 (Berkeley and Berkeley 1982). He wrote: "We were entertained by a comical fellow, disguised in as odd a dress as Indian folly could invent. He had on a clumsy visard of wood colour'd black, with a nose 4 or 5 inches long, a grinning mouth set awry, furnished with long teeth, round the eyes circles of bright brass, surrounded by a larger circle of white paint; from his forehead hung long tresses of buffaloes hair, and from the catch part of his head ropes made of plated husks of Indian corn." Bartram did not understand that entertainment was not the purpose of the display. Use of the mask was meant to dispel evil (Billard 1974, Trigger 1978). The seriousness of the event should have been conveyed to Bartram by Schikellamy, his Delaware guide, when he warned, "Lie still, John" at the beginning of the event. Before that moment, Bartram had never heard Schikellamy "speak so much plain English before."

The wood has been used for appliances, blinds, boxes and crates, carvings (e.g., decoys), caskets, casks, cheap furniture, cooperage, excelsior, fiber products, fixtures, furniture stock, handles, lumber, millwork, mobile homes, novelties, panels, paper pulp, plywood, shade rollers, shades, signs, sporting goods, toys, veneers, and woodenware (Vines 1977). The USDA technical sheet on *T. americana* indicates that it "works easily with tools, making it a premier carving wood. It is poor in holding nails and in bending, but moderate in gluing and good for holding paint or printing inks."

Only two records of use for food have been found, both among the Ojibwa. These people historically ate the young buds or cooked them as greens (Moerman 1998); they also used the sweet sap (Yanovsky 1936).

Numerous people used linden as medicine. Among the tribes recorded are the Algonquin, Cherokee, Iroquois, Malecite, Meskwaki, and Micmac (Vogel 1970, Moerman 1998). These people used the inner bark in a tea to treat burns and scalds, dysentery, eye problems, heartburn, intestinal worms, lung ailments, and weak stomach. They also made a poultice to draw out boils and help heal broken bones.

Inflorescences and flowers of *Tilia* contain polysaccharides (Yakovlev 1985), kaempferol, and quercetin (Karakaya and El 1999). Water extracts were antioxidant in laboratory studies (Yildirim et al. 2000). Duke et al. (2002) noted that ethanolic bark extracts were effective against *Candida, Escherichia,* and *Staphylococcus.* Although no chemicals were listed by Foster and Duke (1990) or by Duke et al. (2002), they issue the warning that frequent consumption of flower tea may cause heart damage.

Tillandsia

(Linnaeus commemorated Swedish physician Elias Erici Tillands, 1640–1693, professor at Åbo; according to legend, Tillands was so frightened of water that he would walk several miles around a lake rather than take a boat a few hundred feet across)

Tillandsia. *Tillandsia fasciculata* (left), with *Tillandsia usneoides* hanging from left side. *Tillandsia utriculata* (right), with *Tillandsia recurvata* as inset on left. *All drawn by P.N. Honychurch.*

Tillandsia fasciculata (drawn into a bundle)
ashome chobee [*asõmco:bî*] (*ashome,* Spanish moss, *chobee,* big, Mikasuki); *assonɬakko* [*assonrakko, ashonɬaku*] (*asónwa,* Spanish moss, *ɬakko,* big, Creek)
curujey de lámpara (lamp *curujey,* Spanish and Taino, Cuba)
dog-drink-water (Bahamas)

Tillandsia recurvata (curved backward, referring to the leaves)
agave de palo (agave on trees)
ball-moss [ballmoss] (Florida)
barba di cadushi [*kadoesji*] (*cadushi's* beard, Dutch Antilles)
curujey (Taino name of the plant, Cuba)
gallitos (little roosters)
guixi guii lace [*guijxi guij lace*] (*guixi,* herb, *guii,* medicine, *lace,* thin, Zapotec, Oaxaca)
heno (hay, Spanish)
huíribis cu'u (curve-billed thrasher's nest; *huíribis* is the name of the curve-billed thrasher, *Toxostoma curviostre,* Mayo, Sonora)
mari di palu (tree *mari,* Dutch Antilles)
mescalito (little mescal or *Agave,* Sonora); *mescalito de huithlcochi* (little mescal or *Agave* of the curve-billed thrasher, *Toxostoma curviostre,* Sonora)
nido de gundulén [*gungulén*] (*gundulén's* nest, Puerto Rico)
old man's beard (Jamaica); old man's hand (Virgin Islands)
pachtle [*paschtle, paixtli, pastel de mesquite*] (*pa:xtli,* moss, Náhuatl, San Luis Potosí)
thread-leaved wild pine [three-leaved wild pine] (Bahamas)
tiña
uchapoli (Guarijío, Sonora)

Tillandsia usneoides (resembling the lichen genus, *Usnea*)
asõ:mî [*ashome*] (Mikasuki); *asónwa* [*ashunwa, assúnwv*] (Creek, Muskogee); *somo* (Alabama); *hassolwá* (Koasati); *iti shumo* (*iti,* tree, *shumo,* maybe "moss" as cognate with Creek *asónwa,* Choctaw)
ayan̲' nanhi' (probably from *ayan̲',* tree, *nahin̲',* hair, Biloxi)
barba de palo (tree beard, Venezuela); *barba de viejo* (old man's beard, Costa Rica); *barba español* (Spanish beard, Cuba); old man's beard (recorded for *Tillandsia usneoides* by Patrick Brown for Jamaican plants in 1756; later applied

to a number of plants including *Clematis, Saxifraga,* and *Chionanthus,* Bahamas, Jamaica, Cayman Islands)
guajaca (probably Taino, resembles *Guajataca,* a Puerto Rican village, Cuba)
hichikoni (Guarijío)
ichokhiska hatka (*ichokhiska,* beard, *hatka,* white, Alabama)
[crape, Florida, long, Spanish] moss (southeast USA, Bahamas); *mousse* (moss, French, Houma, Louisiana); *musgo* (moss, Peru); *pastle* [*paixtle*] (*pa:xtli,* moss, Náhuatl, Veracruz)
pipisqui (Mayo)
salvaje (wild, Peru)
tcinasho'hi (Ofo)
wool-crape (southeastern USA)

Tillandsia utriculata (inflated or bag shaped in allusion to the leaf sheaths)
asõmco:bî (*ashome,* Spanish moss, *chobee,* big, Mikasuki); *assonłakko* [*assonrakkô*] (*asónwa,* Spanish moss, *łakko,* big, Creek)
[giant, swollen] wild pine ("pine" is abbreviation of "pineapple," Bahamas, Florida, Virgin Islands)

Hogan (1978) found pollen of *Tillandsia* in the Glades site at Fort Center on Lake Okeechobee. Since the flowers are pollinated by animals, the people must have been using them for something. Most have presumed they used *Tillandsia* for medicine, but it may not have been that simple. Whitford (1943) reported that "in the southern regions, the natives gathered the abundant Spanish moss from the trees." They then wove the vascular bundles into what he called "blankets." He had found the blankets among remains of the Koasati "and the people who built certain mounds in Florida used [Spanish moss] frequently" (Whitford 1941). Kimball (1994) recorded that moss is still used among the Koasati, where it is spun into strings, mattresses, and used in weaving. He quoted from one of his informants, "They looked for black Spanish moss in the forest, and when they brought it back they hung it up, scrubbed it with itself, and when they finished scrubbing it and finished lengthening it, they spun it." They even have the word *statihká:cin,* which is a heddle for weaving Spanish moss.

Bennett (1997) was told that the moss had been used as a lid for cooking pots. He was also told that the fibers had been used to weave skirts, especially after wars when cloth was not available, and that it made an emergency blanket. The Chitimacha of the southern side of the Mississippi Delta in Louisiana made cords of the fibers and mixed the stems with clay to plaster their wattle-and-daub houses (Bushnell 1917). A woman who may have been Chitimacha said that her people also used Spanish moss to make cords (Bushnell 1922). The technique she described was the same as the one Bushnell (1917) found earlier among that tribe. When Sturtevant (1955) worked with them, the Seminoles were using Spanish moss in tanning and to skim off the scum from the surface of cooking liquid.

The Houma used *T. usneoides* in a tea to relieve chills and fever (Speck 1941). The Houma also twisted the fibers into cords, which were used, among other purposes, to make mats for the floors of their homes (Read 1931).

Large water-tank bromeliads such as *T. fasciculata* and *T. utriculata* were used by the Seminoles to obtain water during the dry season (Sturtevant 1955). Modern Seminoles told Bennett (1997) that they still obtained water from the air plants.

Cubans have used mashed leaves of *T. usneoides* mixed with butter as a remedy for hemorrhoids (Roig 1945). The same use has been made of *T. recurvata* in Argentina (Morton 1981). In Curaçao, a leaf decoction of *T. recurvata* is considered an emmenagogue, a remedy for leukorrhea, and a treatment for gallbladder problems (Morton 1981). People in Veracruz treat digestive problems with *T. usneoides* (Vásquez and Jácome 1997). Costa Ricans consider the leaves caustic and vesicant, but they use the fibers for strings and ropes (Pittier 1957). The fiber is called *cabellos de ángel* (angel hair). In Venezuela, the fiber was used to make sacks (Pittier 1926).

Torreya

(Named for North American John Torrey, 1796–1873, physician and professor of chemistry, first at West Point and then at Columbia University in New York)

Torreya taxifolia (with leaves like *Taxus*)
fetid yew (yew with an offensive odor; from Latin *fetidus*)
gopher-wood [gopherwood] (from Hebrew *gopher,* a kind of wood used to build the ark; probably *Cupressus sempervirens* or a species of *Acacia*; applied to American trees by the 1850s)
Nusseibe (nut yew, German)
savin [sacin] (a comparison with the European *Juniperus sabina,* based on Latin *sabina*)
stinking cedar [sacin]
[Florida, yew-leaved] torreya [-tree]

Torreya is a genus of seven species. Florida has *T. taxifolia*; *T. californica* is in coastal California; and the remaining species are in eastern Asia. The genus is one of several that have a disjunction between the two regions. There are *Torreya* fossils from the mid-Jurassic, which indicates the antiquity of the gymnosperm lineage (Mabberley 1997). As ancient as these gymnosperms are, they were not discovered by the scientific community until the 1800s. Finally, the Florida plants were collected and then described by Scottish botanist George A.W. Arnott (1799–1868) in 1838. The first species described was *T. taxifolia*, and in 1854 *T. californica* was described by John Torrey.

Both U.S. species are relict populations, and the Florida plants are confined to the ravines and bluffs along the Apalachicola River valley in Florida and one adjacent county in Georgia. The Florida plants were found in three counties, but they have been extirpated from Jackson County, and they are now globally endangered by a fungus and habitat destruction (Little 1979, Coile 2000, Chafin 2001).

As several of the common names indicate, the foliage is fragrant when touched or bruised. Some considered it fetid (stinking) (Mabberley 1997).

Historically, wood from the trees was used for fencing, although the species is now so rare that is not an option (Mabberley 1997). The related species in California was used by several indigenous tribes. The Florida species were probably similarly used, but no information has been found, probably because of its rarity and comparatively late discovery. Nuts from the California species were used externally to treat headache and chills. Internally, they were made into a decoction to relieve indigestion and tuberculosis (Moerman 1998).

Tournefortia: Chiggery Grapes

(Named for Joseph Pitton de Tournefort, physician, naturalist, and professor of medicine and botany, 1656–1708)

In the late 1600s, Charles Plumier (1646–1704) sat on the Caribbean island of Hispaniola drawing a scrambling vine with plump white fruits. This French Franciscan monk must have been impressed with the plant because he devoted a new genus to it, and named it after a friend and colleague. That friend was the eminent Joseph Pitton de Tournefort. After returning to Europe, Plumier published his drawings and descriptions of plants he had studied in a book called *Nova plantarum americanarum genera* (New Genera of

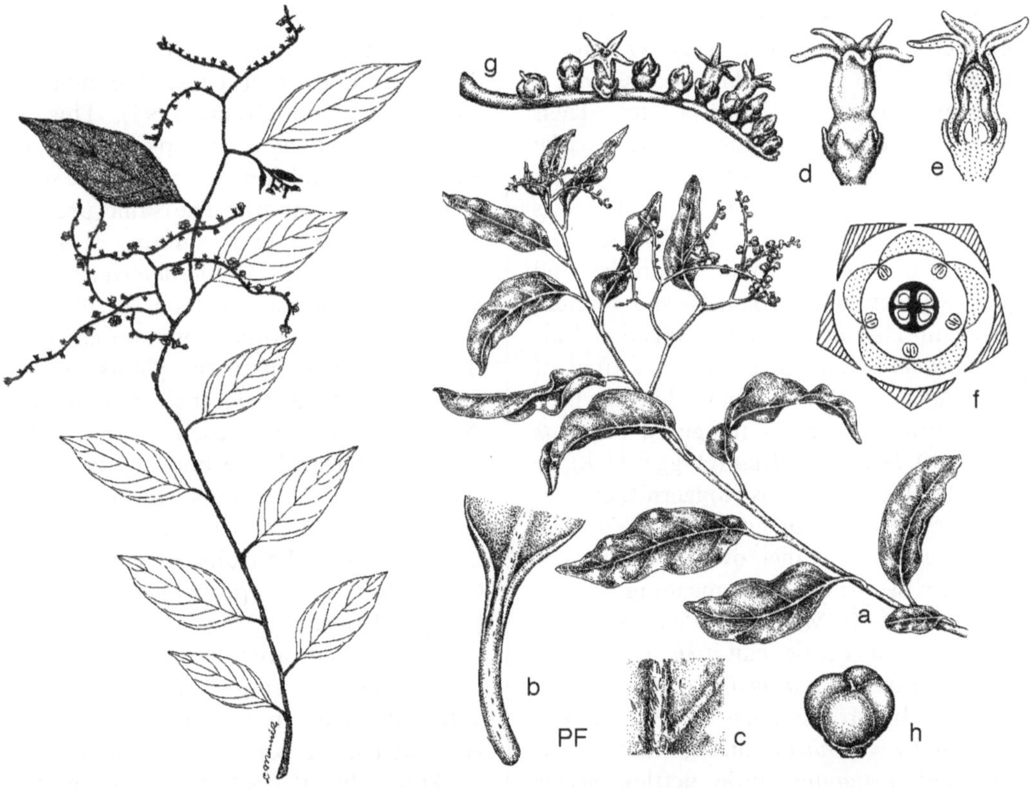

Tournefortia*. Tournefortia hirsutissima* (left). *Drawn by Leonardo Mourré.* From Nowicke 1969. *Tournefortia volubilis* (right). a. Branch with flowers and fruits. b. Base of leaf and petiole. c. Section of lower surface of leaf showing indument. d. Flower, side view. e. Flower, longitudinally dissected. f. Floral diagram. g. Branch of inflorescence with flowers and fruits. h. Mature fruit. *Drawn by Priscilla Fawcett.* From Correll and Correll 1982.

American Plants). He called the climbers *Pittonia hirsutissima & ramosissima, baccis albis* (pubescent and branched *Pittonia* with white berries). *Pittonia* is now a synonym of *Tournefortia.* His was the first European discovery of this tropical American relative of Old World borage (*Borago officinalis*), a medicinal plant of his homeland.

We now call the American plants *T. hirsutissima*, a name given to them by the Swedish botanist Carolus Linnaeus in 1753. Linnaeus often used names proposed for plants by people who preceded him, but he sometimes changed them to suit his own idiosyncratic tastes (see Austin 1993).

Since 1753, the plants have been found throughout the Greater Antilles, on scattered islands of the Lesser Antilles, and throughout the mainland from Mexico to Peru. *Tournefortia hirsutissima* also occurs in Florida, where it is one of the species on the Florida Department of Agriculture endangered list (Coile 2000). The species is so rare in Florida that few people know it. Those who do recognize the vine sometimes call it chiggery grapes, having imported a West Indian name. The vine is known from a half dozen or so scattered sites in Hendry, Collier, Dade, and Monroe Counties. In the Lesser Antilles it is also rare, known from Guadeloupe but not from nearby Dominica.

Even though this plant is apparently rare throughout its range, people living where it grows have a variety of names for it. To the Florida Miccosukee it is *cokashatkî* (*cokasî,* grapevine, *hatki,* white). Their relatives the Creek-speaking Seminoles call it *colo: fákâ* (*colóswv,* grapevine, *fvkv,* cord) or *colo:sfákaɬákko* [*colo:sfákarákko*] (*ɬakko,* big). In Cuba, it is *nigua* (chigger), [*bejuco*] *cayaya hembra* ([*bejuco,* vine], *cayaya,* shrub, Taino, *hembra,* female, Spanish), and *bejuco cachaza* (whiskey vine). Spanish- and French-speaking people in Hispaniola call it *nigua peluda* (hairy chigger), *chique-chique* (a *chique* is a quid of tobacco, or a chigger), or *liane chiques* (chigger vine). Puerto Ricans know it as *nigua* (chigger), *mata de nigua* (chigger bush), *bejuco de nigua* (chigger vine), or chiggery grapes. Jamaicans say it is chiggernit, chigger nut, cold withe, crocus bush, or hog hook. People in Belize just say crocus, or apply their own unique names. They say it is poison fish tie tie (tie tie meaning vine) or *robo de mico* (monkey tail). Mayans in Belize say *chik.*

People in the French Antilles call it *chique en fleur* (flower for chiggers) or *herbe á melingres* (herb for a weak constitution). In various parts of Mexico it is *amapa hasta* (*amapa* = ?, *hasta,* until), *mach'much'* (chew loudly vine), *ortiguilla* (little nettle), *perlas* (pearls), *tlachichinoa* (perhaps either based on *tlachichic,* referring to its bitter taste, *chiuchic* or *chichinoa,* for putting on "fire," reference to the itching, Náhuatl), *tsak ts'ohool* (red plant), *telpatli* (*patli,* medicine, Náhuatl), and *po'ko ka'hui'hui* (cool stomach, Totonac, Puebla). Nicaraguans call it *tiricia* (colloquial for *ictericia,* or jaundice) or *frutilla* (little fruit). It weeps the *lágrimas de San Pedro* (San Pedro's tears) from Veracruz to Colombia, and is *nigua* (chigger) or *nigüito* (little chigger) in Venezuela or *nigulilla* (little chigger) in Veracruz.

As several of the common names indicate, this plant is considered a remedy for chiggers (*Eutrombicula alfreddugesi*), those little red spider relatives that bite and leave itchy spots. In English, we also call them red-bugs, red mites, and harvest mites. The Creeks say *wasko.* In Latin America these tiny pests are called *tlalzahuatl, coloradillo* (little red one), and *bicho colorado* (red bug). Without some treatment, the itching will drive a person crazy enough to cause scratching. In a few cases, the bites may become infected. Some people even become so badly covered with bites that they go into shock. Under the best conditions the little red welts these mites cause are unpleasant. Even worse, they get in places that are not easily scratched in public without causing embarrassment. They particularly like to get in tight areas beneath underwear.

Many of the people living near chiggery grapes have learned that its mashed leaves applied to the skin alleviate this itching. Older literature says that the leaves "expel chiggers." We are now told that the animals never burrow but bite and inject us with digestive fluids that cause the itching through an allergic reaction (Nellis 1997). This application of mashed leaves has been expanded to other problems, and a leaf demulsion is used against skin diseases and ulcers on the mouth. The same preparation is used against diarrhea and vomiting (Roig 1945, Liogier 1974, Morton 1981). *Tournefortia volubilis* is used in much the same way (Mors et al. 2000).

A look at the other common names makes it clear that people have used the plants for other purposes. Some are obvious from the name; others are not. Leaf decoctions have been used as depuratives, a preparation that "cleans" the body by causing it to excrete wastes. Leaves and roots are diuretic and have been used against venereal diseases (Standley 1920–1926, Liogier 1974). The fruits are eaten in Hispaniola and Costa Rica, and probably throughout its range (Liogier 1974).

When plants are found that have widespread usage for the same or similar maladies, it usually may be assumed that they contain bioactive chemicals. Although not a lot is known about *Tournefortia,* we do know that it contains one or more pyrrolizidine alkaloids. Several unrelated plants contain variations on this group of alkaloids, and all that have been studied are poisonous. Plants containing the compounds include *Crotalaria* and *Senna* (Fabaceae),

Amsinckia, Argusia, Borago, Cynoglossum, Heliotropium, and *Tournefortia* (Boraginaceae), and *Senecio* (Asteraceae) (Lewis and Elvin-Lewis 1977). Many of those species known to contain pyrrolizidine alkaloids have been used as medicines.

Pyrrolizidine alkaloids may harm the human body by interfering with liver function, so ingestion of preparations using them is potentially more dangerous than the malady. Yet, people have used, and continue to use, chiggery grapes and other pyrrolizidine alkaloid-containing plants in medicines. A few people have done studies to see if groups using these preparations are suffering from higher incidences of liver problems than those not using them. The results are mixed, with some showing higher incidence of liver problems while others do not. Because the alkaloids cause liver damage in test rats in laboratories, it is thought that they cause the same or similar problems in people using the plants as medicines. That may or may not be true in all cases, because those using the plants are not extracting only the alkaloids. Instead, the people use an array of plant compounds. Could it be possible that those compounds work together to act as cures and not as liver poisons? The situation is not an either/or one, and is complicated by genetic variation in people and the plant chemicals mixed with those alkaloids in medical preparations.

Humans are actually the latest animals to discover the uses of pyrrolizidine alkaloids extracted from plants. Long before people discovered how to use the plants in medicines, insects had learned to use them in another way. Two groups of Lepidoptera are particularly well known for their use of these alkaloids—milkweed butterflies (Danaidae, *Danaus* spp.) like the monarch, and queens, and some less well-known Ctenuchid moths (Ctenuchidae). These butterflies and moths are not particularly closely related, and independently they have both learned to extract the alkaloids from plants that contain them. However, only the male insects gather the chemicals. The males then use the plant cocktails to "intoxicate" the females of their species before they mate with them (Plisk and Saltpeter 1971, Edgar et al. 1974, Pliske 1975). Without this chemical inducement, the females will have nothing to do with the males. So, these moths and butterflies owe their continued existence to the plants that support their reproduction. I cannot help being reminded of the Ogden Nash poem, "candy is dandy, but liquor is quicker."

Toxicodendron: Poison Ivy

(Named from Greek *toxikon*, poison for smearing on arrows, and *dendron*, tree)

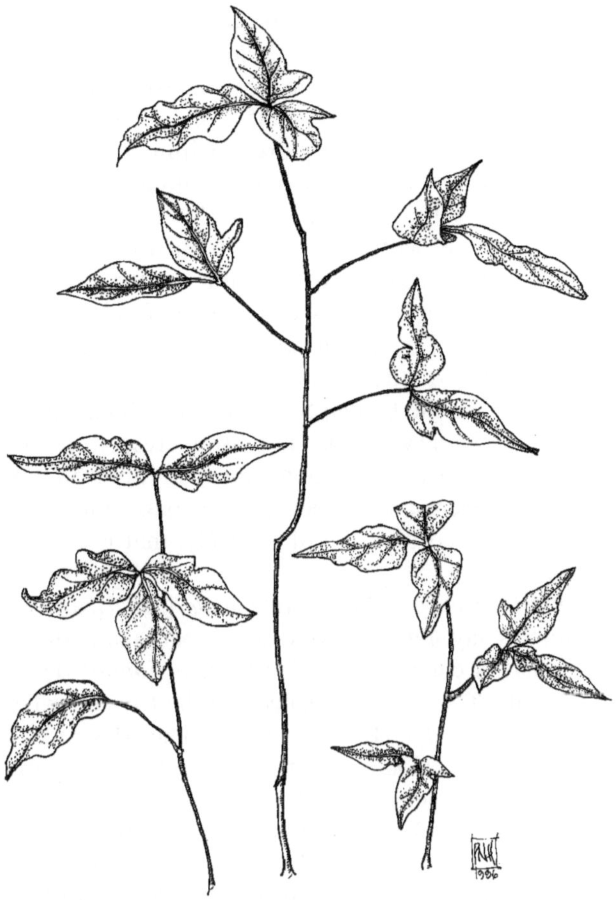

Toxicodendron radicans. *Drawn by P.N. Honychurch.*

Poison ivy and alligator fleas are two of the real "dangers" of visiting swamps. Wading into the swamp in places like Fakahatchee Strand State Preserve in southwestern Florida was always problematical. Often, we returned home with the beginnings of poison ivy (*Toxicodendron radicans*) rash on our legs. As the itchy spots were usually inside our trouser legs, oils floating on the water must have caused the problem. Tying the trouser legs around the tops of our boots always helped—with the added plus in warm weather of keeping out alligator fleas (Naucoridae). Those animals are true bugs (Hemiptera) about a half-inch long that make it a nasty habit to wait until they float up your pants legs to near the groin before biting. Their long, stiletto-like proboscis feels like a hot needle stabbing into the tender flesh of that region.

At least 50% (some say 75%) of us are susceptible to poisoning by *Toxicodendron*, relatives of cashews (*Anacardium giganteum*), pistachios (*Pistacia vera*), sumachs (*Rhus* spp.), pepper trees (*Schinus terebinthifolius*), mangos (*Mangifera indica*), and hog plums (*Spondias* spp.) (Lewis and Elvin-Lewis 1977, Nellis 1997). The active chemical, called urushiol but really a series of phenolic compounds, was discovered in the

1800s when it was called toxicodendric acid (Millspaugh 1892, Corbett and Billets 1975). This oily material, derived from catechols, is so potent that experiments with rats show that, from a few drops on the skin, it permeates every organ within 24 hours.

Toxicodendron radicans, and its multiple subspecies, ranges throughout most of the United States, north to Nova Scotia and west to British Columbia and Arizona, also Mexico south to the state of Yucatán. French speakers call it *bois de chien* (dog bush, French, Canada) or *herbe à la puce* (flea herb, an allusion to the itching, Canada, Houma in Louisiana). In the United States, it is climbing sumac, cow-itch, mercury [markry], poison creeper, poison ivy, or three-leaved ivy. People often call it poison oak, especially when it is a free-standing bush and not a twiner, but according to Gillis (1971), that is another species.

The Cherokee call the plants *uluda* [*ulvda*], and the Seminoles say *hahfa:lî* (Mikasuki), perhaps a simple term that is cognate with Creek *vfvlla* [*afala*]. However, caution is appropriate, because the Koasati called *Ipomoea* an *afalá* (Kimball 1994). The Mikasuki is also rendered *hili:háhlî.* Sturtevant's (1955) informants made the statement about the plants that *istilikwâycin* (*este,* person, *lekwicen,* gives sores, Creek) that he listed as a common name, but it is a statement, not a noun. The Alabama is different, with *baksakánko* (*baksa,* rope, *kano-ki-o,* evil). They also call it *ittokánko* (*itto,* tree, *kano-ki-o,* evil). The Chickasaws said *ofi' imittli'* [*ofimitti*] (*ofi',* dog, *imittli',* fight). The Koasati say it is *lambí pahí* (*lambí,* sting, *pahí,* grass). This is probably the *talun̲wa* of the Choctaw. In the Missouri River region the Omaha-Ponca call it *hthiwathe-hi* (*hthi,* sore, *wathe,* to make, *hi,* bush)—the plants that makes sores. Like people throughout its range, the Omaha knew and tried to avoid the poisonous plant (Gilmore 1919).

Some literature contends that the Glades people made an arrow poison from the leaves and stems. That seems to be an unsubstantiated extension of Caribbean use. In the Caribbean and Mexico, the plants are sometimes called *guau.* Originally, the Taino word *guau* applied to the Caribbean poisonous anacardiaceous shrub *Comocladia angulosa.* According to Bartolome Las Casas (Coll y Toste 1972): *la leche de este árbol es ponzoñosa e della e de otras cosas hacen los indios la yerba que ponen en las flechas con que matan* (the sap of this tree is poisonous and from it and other things the Indians make a mixture that they put on their arrow with which they kill). From that Caribbean application, it appears that Europeans assumed that both *Toxicodendron* and *Metopium* were used to poison arrows in Florida.

The scientific name did nothing to discourage that thinking. The genus *Toxicodendron* was created by Joseph Tournefort in 1700, used by Johann Jakob Dillenius in 1732, and resurrected by Philip Miller in 1754. Linnaeus ([1753] 1957) rejected their segregation of the poisonous species from those that are benign; he called the poisonous plants *Rhus toxicodendron* and *R. radicans.* However, Gillis (1971) argued that *Toxicodendron* should be retained. Perhaps there is something in the older literature that supports arrow poison views, but it has not been brought into more recent studies. Although the comparison with ivy (*Hedera*) goes back to at least 1635 with Jacques Philippe Cornut's book on the plants of Canada, "poison oak" appeared in English about 1743, while "poison ivy" was first published in 1784.

In Mexico, there are several names relating to the rash caused by these plants. To the Maya, it is *chechén* (*cheché,* unseasoned or raw), meaning the rash. Similar connotation is found in *lachi gobilla* [*lachi cobilla*] (*laa,* leaf, *na-chi,* fierce, *gobilla,* chaparro or oak thicket, Zapotec, Oaxaca), and *yaga beche topa* [*yaga peche topa, betz-tzaj*] (*yaga,* tree, *beche,* toast or burn, *topa,* short, Zapotec, Oaxaca). There is no translation of *bemberecua* [*bembiricua, bembericua, huembereua*] (Tarascan, Mexico), *mexie* [*mexi, meye*] (Otomí, Hidalgo), or *tumuraga* (Mountain Pima, Chihuahua), but they likely make a similar allusion.

Several names in Mexico refer to the problems caused by contact with *Toxicodendron. Fuego* (fire, Mexico) is obvious. Similarly, *yedra* or *hiedra* [*mala, venenosa*] ([bad, venomous] ivy, Sonora, Arizona, New Mexico) is analogous to "poison-ivy" in English. *Guadalagua* (perhaps from *guau del agua,* Mexico) may be use of a Caribbean word and notation of a preferred habitat. Several common names of irritating plants in Mexico are like *mala mujer* (bad woman) for *Toxicodendron.* Unless you are familiar with the allusion to testes, you might miss the meaning of *hincha huevos* (egg lifter, Mexico). Perhaps *dominguilla* (little Sunday, Mexico) is a notation of being embarrassed by the rash.

A loan-word is *sumaque* [*zumaque*] (sumac). This use was taken into English by the 1400s and earlier into Spanish from Arabic, *summaq* (see *Rhus*). The designation is used for members of both *Rhus* and *Toxicodendron.* While *sumaque* is applied in Mexico to poison ivy, there is no indication that people using the term confuse the poisonous with the nonpoisonous species.

Incredible as it is to those of us who are sensitive to these plants, poison ivy was used as medicine by native Americans, and it still figures in herbal remedies among some Europeans (Hocking 1997). The Houmas made a tonic and rejuvenating tea from it (Speck 1941).

Poison ivy was listed in the U.S. Pharmacopoeia from 1820 to 1905. Uses typical of these plants in that period are recorded by Porcher (1863). He wrote, "It is stimulant and narcotic, employed in paralysis and herpes; of the former disease, seventeen cases are reported by one physician to have been successfully treated with it." He continued with cases where different physicians had used poison ivy to treat "consumptive and anasarcous patients," "herpetic eruption," palsy, paralysis, "chronic ophthalmia of scrofulous infants," and ringworm. Teas from the leaves were considered antiarthritic, and leaves were once used to treat liver diseases (Foster and Duke 1990).

Porcher said, "The juice which exudes on plucking the stem makes a good indelible ink. It is dissolved by ether." Later, he continued, "The yellow, milky juice turns dark, and forms one of the best indelible inks for marking linen." However, producing such ink is risky because "[a]n acrimonious vapor ... exhales from a growing plant of the poison oak sumach during the night, can be collected in a jar, and is capable of inflaming and blistering the skin of persons of excitable constitution who plunge their arms into it."

All who have encountered the plant have tried remedies to cure the rash it produces. Porcher (1863) recommended, "Purging with neutral salts, the use of opium, blood-letting, and cold applications of acetate of lead are employed in case of poisoning from these plants." Thankfully, remedies have changed. More likely learned from native Americans were Porcher's use of the "bruised leaves of the *Collinsonia canadensis*" (Lamiaceae). Also, he commented, "*Verbena urticifolia* ... is likewise considered an antidote."

Now, there is the tendency to use corticosteroids, oral or otherwise, to treat the problem (Goodall 2002). Worse, there is a tendency among physicians to attribute any skin rash to poison ivy or poison oak (Guin et al. 1981, Stehlin 1996, McGovern et al. 2000). None of the dermatologists who treated me in Florida ever attributed my rashes to anything except poison ivy—I had to tell them the plant sources. In any event, the treatment is corticosteroids.

I learned early on to beware of the smoke of burning ivy (cf. Kollef 1995). The best solution to avoid dermatitis is to avoid the plants, but that is not easily done for swamp-stompers like me. A second-best solution to many is to control the plants. Chemicals are recommended (Paul 1993), but birds bring the seeds back rapidly, often the next season.

William T. Gillis (1933–1979) relished studying these plants because few others could. He delighted in quizzing visitors to where he lived at David Fairchild's home, the Kampong in Miami, about poisonous plants he had growing. Eventually, he fell victim to his own pranks. He was hospitalized with an extreme case of *Toxicodendron* poisoning. Apparently, he cried "wolf" once too often.

Tragia

(Linnaeus named this for Jerome Bock, 1498–1554; his name in Latin became *Hieronymus Tragus,* physician, teacher and herbalist who wrote the famous *New Kreütter Buch* in 1539)

Tragia saxicola (rock-loving)
noseburn (Florida)

Tragia saxicola is endemic to Florida, and there are no records that it was used. Its name "noseburn" is applied as a generic name for the genus. Although that name has not been tied to a date, the original "noseburn tree" was *Daphnopsis salicifolia,* Thymeliaceae. That is a notably irritant plant known to the Aztecs as *ahuejote* (maybe *ahue* from *ahuacatl,* testicle, *jote* from *xiuitl,* herb, Náhuatl). *Daphnopsis* leaves have been used like a mustard plaster, and presumably the irritating trichomes on *Tragia* led to it becoming "noseburn."

However, there is evidence that other members of *Tragia* have a long association with people. In Cuba, *T. volubilis* is *ortiga,* a name comparing it with the well-known European medicinal irritant *Urtica.* In Hispaniola, *T. volubilis* is *pingamoza* (child stinger), *ortiguilla* (little nettle), *picaculo* (tail sticker), and *pois gratté* (irritating pea). Leaves on *pingamoza* cause severe itching, yet the plant is used as a vermifuge. A tea of the leaves is also used to bathe sores, and a powder of the plant treats sores (Liogier 1974). The same species is called cow-itch [cowitch] or vine nettle in the Antilles (Hocking 1997). It is used in the Caribbean and Mexico as an analgesic, diuretic, and diaphoretic (Hocking 1997).

From the southwestern United States south to Central America, *T. nepetaefolia* is called *hoobox* or *popox* (black bark, Maya) and sometimes *mala mujer* (bad woman). *Popox* is used as a counterirritant in rheumatism, and a remedy for sores and cancer (Martínez 1969, Hocking 1997). To the Kiowa *T. ramosa* was *dan-sa-la* (stinging weed).

Trema: Nettle Tree

(Either from Greek *trema,* hole, aperture, because of the pitted stone, or from the Latin *trema,* to quiver)

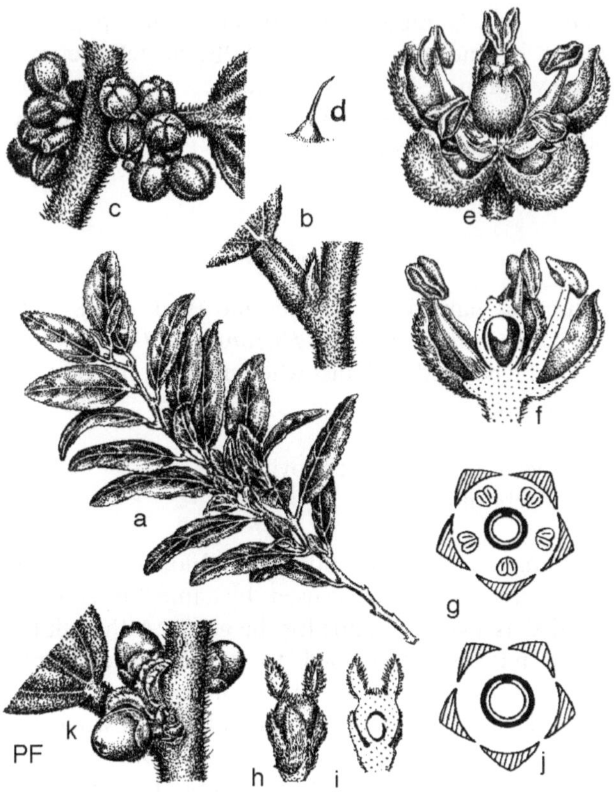

Trema lamarckianum. a. Branch. b. Node to show leaf base and stipule. c. Staminate flower buds. d. Pustulate trichome from leaf surface. e. Staminate flower, side view. f. Staminate flower, longitudinally dissected. g. Floral diagram of staminate flower. h. Pistillate flower, side view. i. Pistillate flower, longitudinally dissected. j. Floral diagram of pistillate flower. k. Fruits. *Drawn by Priscilla Fawcett.* From Correll and Correll 1982.

Before they built the Gumbo Limbo Nature Center in Boca Raton, there was no nettle tree (*Trema micrantha*) on the site. Shortly after the construction finished in 1984, I found a seedling nettle tree just north of the building.

Building the headquarters for the nature center had nothing to do with the appearance of nettle tree along the dunes in Palm Beach County. The first and last snow in that area occurred in the late 1970s. Since that time, fewer winters have been severe. In the first decade or so that I was in Florida, I watched *Trema* extend its range up into southern Palm Beach County during warm years, only to be killed back during a colder winter. Because the trees grow rapidly, it was possible to chart the changes from year to year. By the time Gumbo Limbo Nature Center was established, such cold winters had ceased. The last truly cold winter in the region was late 1989 and early 1990 when large figs (*Ficus* spp.) froze to the ground in Palm Beach. The frosts, however, did not kill the nettle tree at Gumbo Limbo. Not only was that first tree still there as of August 2001, but others have appeared. Global warming shows in small, almost unnoticed vegetation changes on the margin of the tropics.

Nettle trees are tropical members of the largely temperate elm family (Ulmaceae). Linnaeus and his colleagues knew only two genera—*Ulmus* and *Celtis.* Although Linnaeus studied two species of *Trema*, he put them in the wrong genera. What is now *T. micrantha*, Linnaeus called *Rhamnus micrantha.* The other species he knew was the African and Asian *T. orientalis,* which he called *Celtis orientalis.* It fell to German botanist Johann Friedrich Gmelin (1748–1804), editor of the 13th edition of *Species Plantarum* (1788–1793), to name the third North American member of the family, *Planera aquatica.* Later, Portuguese João de Loureiro created *Trema* in his *Flora Cochinchinensis* in 1790.

A.W. Chapman (1897) first recorded *Trema* in Florida. He noted that physician Abram Paschall Garber visited Florida, hoping to improve his tuberculosis, and collected *T. micrantha* on shell mounds in Estero Bay, near Fort Myers (Lee County). *Trema lamarckianum* was found later, maybe because it has such a restricted range in the southern parts of Miami-Dade County and the Keys. Now *Trema* contains 10 to 15 species, mostly in the tropics (Mabberley 1997).

None of the early accounts of *Trema* in the United States used common names, but the Seminoles knew the plants as *silî:tá:pî* [*shen lin aape*] (*shen lin,* slimy, *aape*, stem, Mikasuki). The Creeks use the cognate *lipá:pín* (*lepakfe*, slimy, *vpe*, stem). Sturtevant (1955) reported a use in birth medicines. However, Bennett's (1997) informants indicated only that a bark decoction was used to treat recurring indigestion but not everyday problems.

By the 1930s, *Trema* was being called the "nettle-tree" (Small 1933), although someone had invented the book names "Florida trema" (Florida, Puerto Rico) and "West Indian trema." "Nettle-tree" was originally applied to European *Celtis australis* and later extended to *Celtis* in North America. William Turner explained the usage in 1548 when he wrote, "It hathe a leaf lyke a Nettel, therefore it may be called in englishe Nettel tree." When the name for *Celtis* was changed to hackberry, people shifted nettle tree to *Trema.* Probably, the name was brought into the United States from the Caribbean because there *T. micrantha* is "Jamaica nettle tree."

Perhaps the first records of *Trema* were published by Bartolome de Las Casas and Gonzalo Fernández de Oviedo y Valdés in the early 1500s. Las Casas wrote the name as *guásima*, and Oviedo penned *guasuma* (Coll y Toste 1972). Technically, that Taino name refers to *Guazuma ulmifolia* (Malvaceae); however, even now most note the similarity of *Trema* to

Guazuma. Today, people say that *T. micrantha* is *guacimilla* [*guacimella, guazymillo*] [*boba, cimarróna*] ([crazy, wild] guazuma, Mexico, Cuba, Puerto Rico). To the Maya of Yucatán, *T. micrantha* is *sak pixoy* [*pixó*] (white *pixoy*), where *pixoy* is *Guazuma ulmifolia.* The Mayan names are applied because, although they are similar, the leaves in *Trema* are white on the underside. Similarly, *T. lamarckiana* is *guasimilla* (little guazuma, Cuba), and *Guazuma* is *guásima* (Cuba).

Guazuma and *Trema* are also compared because they produce fibers and edible fruits. Historically, the fibers were the most important product from the trees, but other sources overshadowed and almost obscured that application. *Trema micrantha* is *atadijo* [*atadyo*] (something used for tying things, Peru), *equipal* (a seat made of fiber, from Náhuatl *icpalli,* Michoacán), *mahua blanca* (*mahuat,* fiber tree, Taino, plus Spanish *blanca,* white, Belize), and *zaparote* (from Náhuatl *zapalotl,* originally an *Agave* fiber plant, Mexico). Probably, the name *kib* (candle or wax, Maya) alludes to the use of the fiber as a wick. *Ki'ik* (*ki,* henequen fiber, *'ik,* air or wind) is another Mayan version. *Kiim* (to die, Maya) is surely transcribed incorrectly. Perhaps *cuero de toro* (bull's skin), *pellejo de vieja* (old woman's skin, Mexico), *tortolero* (twisted one, Ecuador), and *yaco de cuero* (otter skin, Oaxaca) are all allusions to bark use.

Trema lamarckianum is called moho bush, *bois mahaut* (*mahaut* tree, Haiti), and *T. micrantha* is *mahaut piment* (pimento *mahaut,* Haiti). All are probably based on Taino *moho* (a tree with bast fibers). Other Taino names now applied to *Trema lamarckianum* are *majagua* (Dominican Republic) and *majagua de Cuba* (Dominican Republic). In Colombia, *T. micrantha* is rendered *majagua colorada* (red fiber tree). The original application is uncertain, but now *majagua* is also used for *Hibiscus tiliaceus* and *H. elatus* because of their useful bark. According to Coll y Toste (1972), *majagua* was a forest tree, *"que produce una corteza filamentosa textil, aprovechada en la industria de cordeleria criolla"* (that produces a filamentous bark textile, used in the Creole string industry). *Cabuya* was the Taino word for cords made from plants. It is clear from historical documents that *Guazuma ulmifolia,* both *Trema,* and probably the *Hibiscus* were involved in that industry and might have been subsumed under the name. Variants of *majagua* include *majaguillo, masaguilla, masaquila,* and *masauilla* (little *majagua,* Venezuela).

Guazuma and *Trema* produce edible fruits, and both are compared to *Prunus* (which see). In Cuba, *T. lamarckianum* is *capulí cimarrón* (wild cherry). Similar *T. micrantha* is *capulí* (cherry, Mexico, Guatemala), *capulín* [*capuleen, chapulin*] [*cimarrón, macho, montes, negro*] ([wild, male, country, black] cherry, Mexico, Belize, El Salvador, Honduras, Costa Rica), *capulincillo* (little cherry, El Salvador), and white capulin (Belize). Other allusions to the edible fruits are more circumspect. Because birds are fond of the drupes, the trees are called *fruta de paloma* (dove fruit, Argentina), *pie de paloma* (dove foot, Mexico), and *sapan* [*de paloma*] (Ecuador). Mayan people are more obtuse and say *ixpepe* [*is-pope, expepe*] (correctly *ix p'ep'el,* tree to hunt birds, Maya, Veracruz, Yucatán). Sometimes, people just say *guinda* (hanging, Veracruz), as fruit from a tree. Others say *T. micrantha* is a *periquiteira* (parrot tree, Brazil), or compare it with *puam* (variant of *puán,* an indigenous name for a fruit-bearing tree like *capulí,* Mexico).

Several names are shared by both *Trema.* Some applications are obvious while others are not. In Puerto Rico, *T. lamarckianum* is *cabrilla* (little goat) and *palo de cabrilla* (little goat tree), while *T. micrantha* is just *cabra* (goat) or *palo de cabra* (goat tree). These animals probably are associated with the trees because they browse the leaves (Morton 1981). Browsing is probably why *Trema* is called bo-hog (boar hog, English Antilles), *berraco* [*verraco*] (swine, Colombia), and *venaco* (for deer, Colombia). Several names include an apparently indigenous word *memiso,* but no etymology has been found. *Miso* is Spanish for "cat," but that seems an unlikely application. *Trema lamarckianum* is *memiso cimarrón* (wild *memiso,* Dominican Republic) or *memizo de majagua* (fiber *memiso,* Dominican Republic). *Trema micrantha* is *memiso* [*memizo, miso*] [*cimarrón, de palo*] ([wild, tree] *memiso,* Dominican Republic). French speakers compare these tropical trees with elms and say *orme petites feuilles* (little-leaf elm, Martinique, Dominique) for *T. lamarckianum.* They say either *petit orme* (little elm, French Antilles) or *bois de l'orme* (elm tree, French Antilles) for *T. micrantha.* Bahamians do not distinguish the two species by name, but note a medical application with pain-in-the-back.

To some the trees are simply a *vara blanca* (white stick, Mexico, Costa Rica). Others think of them as a *yana caspi* (*yana,* black, *caspi,* stick, Quechua, Peru). Comparisons to *Suriana maritima* made with names for *T. micrantha* such as bass cedar (Jamaica), bastard bay cedar (Belize), and wild bay-cedar (Belize) are obscure, unless perhaps they refer to the wood. Similarly, the Caribbean name "bit-root" is not clear.

The names *jaco* (Puerto Rico), *juco* (Costa Rica), *jucui* (Mexico), and *shuco* (Tabasco, El Salvador) have been applied to *T. micrantha.* Those are probably variants of what is mostly written *joco,* a word derived from Náhuatl *xococ,* meaning bitter. This designation is usually applied to fruits that have been fermented or dried like raisins. That meaning may be akin to *ceriúva* [*seriúva*] (wax grape, Brazil). Perhaps false *jaco-calalu*

(Puerto Rico) notes a similarity to an edible plant because *calalu* is a generic term for plants cooked as greens.

Brazilians say *quatiniaba* [*corindiúba, corindiba, candiúva, coatindiva, crindeúva, crindiúva, grandiúva*] (possibly Tupí). This is a name used for elm relatives, often the spiny types like *Celtis pallida*, but also applied to *T. micrantha*. The name *kurundi'y* (maybe Guaraní, Paraguay) is probably a cognate. *Quatiniaba* is perhaps akin to *raspador* (rough one, Colombia). Untranslated names are *afta colorado* (red *afta*, Argentina), *aisegerina* (Huitoto, Peru), *dushan* (Ecuador), *kabuya koro* (Arawak, Suriname), *moniki* (Guyana), *muchichilan* [*muchi chilón*] (Ecuador), *shalipu* (Ecuador), *sui siyi* (Ecuador), and *tokyéwé* (Carib, Suriname).

Churrusco (diarrhea tree, El Salvador) and *carraspero* [*carrasposo*] (astringent, Venezuela) suggest a remedy for bowel problems. Maybe the same is inherent in *palo-polvora* (powder tree, Argentina) and *pão de pólvora* (powder tree, Brazil). Perhaps the name *caca* [*kaka*] (feces, Ecuador) indicates the same. Decoctions of *T. lamarckianum* are used in the Bahamas to relieve colds and asthma (Morton 1981). Also in the Bahamas, a remedy of the bark together with other plants is used for menstrual problems and to increase female fertility (Morton 1981). Others in the Caribbean use decoctions of the leaves for skin eruptions.

Several bioactive compounds have been isolated from different species of *Trema*, including dihydrophenanthrene and phenyldihydroisocoumarin, vitexin, paprazine and terpenoids, a flavone glycoside, simiarenone, a triterpenoid alcohol, and xanthones (Ogunkoya et al. 1972, 1973, 1977, Rakotovao et al. 1988, Ernesto et al. 2000, Dijoux-Franca et al. 2001, Tchamo et al. 2001). These and associated chemicals are known to be poisonous to camels, horses, and rabbits (Hill et al. 1985, Trueman and Powell 1991, Traverso and Driemier 2000). Similar laboratory studies indicate that *Trema* extracts are antimalarial, work against hypertension and certain nerve problems, are analgesic and anti-inflammatory, and are used to treat gastrointestinal and urinary problems (Marquis 1974, Trovato et al. 1988, Barbera 1992, N'gouemo 1994, Grosvenor et al. 1995, Muñoz et al. 2000).

Humans may have valued the bark of *Trema* more than other parts, but birds prefer the fruit. Sit beside either species for a while when they are in fruit and you will see gray catbirds (*Dumatella carolinensis*), brown thrashers (*Toxostoma rufum*), mockingbirds (*Mimus polyglottos*), cardinals (*Cardinalis cardinalis*), and perhaps a dozen other kinds eating the orange drupes. They were using the fruits before humans were a part of the New World.

Triadenum

(From Greek *tri*, three, and *adena*, gland, referring to three large glands alternating with three sets of stamens)

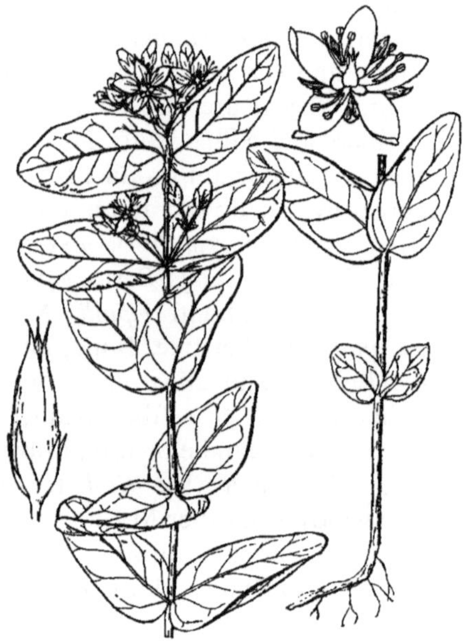

Triadenum virginicum. From Britton and Brown 1897.

Triadenum virginicum (of Virginia)

Big John root (a black slave whose life was an inspiration to slaves who wanted to rebel against their masters but could not do so openly)

High John the Conqueror ("John, said to be the son of an African king, was in captivity, but he never became subservient, and his cleverness at tricking his master supplied many a story with a pointed moral. If he was a real being, he soon acquired some of the characteristics of mythical trickster figures like the Native American Coyote, the African-American Br'er Rabbit, and the West African deity known variously as Eleggua, Legba, and Eshu"; from Yronwode 1999.)

Virginia marsh St. John's wort (named for St. John because his birthday was on 24 June, when the plant is supposed to bloom; the naming of the saint and the mystical figure are not accidental)

Hypericum and *Triadenum* have not always been considered distinct genera. For example, Small (1933) and Gleason and Cronquist (1963) separated the two, but Fernald (1950) and Steyermark (1963) put them under *Hypericum*. Now there is a consensus that they are distinct. They may be recognized by flower color—yellow in *Hypericum*, pink or reddish in *Triadenum*—although there are other differences

(Diggs et al. 1999). Habitat sometimes distinguishes them, with *Triadenum* confined to wetlands, as the common name "marsh St. John's wort" indicates. Part of the complication of genera dates from Linnaeus who described this species as *H. virginicum* in 1759. Rafinesque transferred the species to *Triadenum* in 1837 in his *Flora Telluriana*.

There are now six to ten species of *Triadenum* in eastern North America and eastern Asia (Mabberley 1997, Diggs et al. 1999). Kartesz (1994) recognized four species in North America.

Among the indigenous tribes of North America, records have been found of only one using these plants. The Potawatomi used an infusion of leaves to treat fevers (Smith 1933).

Porcher (1863), Millspaugh (1892), and Foster and Duke (1990) do not mention the plants. However, Hocking (1997) recorded that in the southeastern states these herbs were believed to be the most potent medicine of the forest. Many people carried a piece of the plant in a pocket as a charm. Among the blacks, a leaf tea was taken for chest diseases, the flowers were used to treat colic, and an ointment was used for wounds, sores, and swellings (Hocking 1997).

Nothing has been found published on the chemistry of the America species. However, *T. japonicum* is reasonably well studied, and it has only betulinic acid reported. That chemical is considered anti-HIV, anticarcinomic, anti-inflammatory, antimalarial, antiplasmodial, antitumor, antiviral, cytotoxic, and a prostaglandin-synthesis inhibitor (Duke 2003).

The name "High John the Conqueror" is not confined to this species, but also applied to *Ipomoea* (which see).

Trianthema

(From Greek *treis*, three, and *anthemon*, flower)

Trianthema portulacastrum (with flowers like *Portulaca*, purslane)

[desert] horse purslane [horsepurslane] (a notation that it is not the "true" purslane, *Portulaca*, Arizona, Florida, Texas, Sonora, Bahamas)

kasvañ [*kashviñ, kashvañ*] (Akimel O'odham [Pima], Tohono O'odham [Papago], Arizona)

pigweed (usually applied to *Amaranthus*, Arizona)

pourpier couran (running purslane, Martinique; also used for *Kallstroemia maxima*)

sea-purslane (a name usually given to *Sesuvium*; in both cases it compares the plants with purslane or *Portulaca*)

tostón (a name shared with *Boerhavia*, Cuba)

Trianthema portulacastrum. Flowering plant. a. Portion of a flowering branch. b. Flower enlarged. c. Fruit, two-crested with seeds. d. Seed. *Drawn by Lucretia Breazeale Hamilton.* From Parker 1972.

verdolaga [*blanca*] ([white] purslane, Spanish, Arizona, Dominican Republic, Río Mayo, Sonora, Texas); *verdolaga de hoja ancha* [*de mar*] (narrow leaf [seaside] purslane, Puerto Rico); *verdolagas de cochi* (Sonora)

verdolaga de sapo (toad purslane, Akimel O'odham [Pima], Sonora); *yerba de sapo* (toad herb)

Linnaeus ([1753] 1957) knew a single species of *Trianthema*, and described *T. portulacastrum* from specimens collected in Jamaica and Curaçao. Before him, Paul Hermann had it in cultivation in the Netherlands before 1689; Hans Sloane collected it in Jamaica after 1686; and John Ray cultivated it in England by 1704. The genus is now known to contain about 17 species in warm parts of the world, although most species are in Australia (12 species, 10 endemic) (Bogle 1970, Mabberley 1997).

These plants have been recorded as being eaten raw as a salad in New Mexico and served as greens in Arizona (Curtin 1949). Curtin's report of these plants may be correct, but it is probably through confusion with *Portulaca oleracea* (Rea 1997, Hodgson 2001).

Names throughout the range suggest possible confusion.

However, the plants were (and maybe are) eaten in Asia. Hedrick (1919) noted that they were a potherb in India in the 1800s, although in times of hunger. Hedrick (1919) further warned that they might produce diarrhea and paralysis if eaten in abundance. Watt ([1889] 1972) recorded edibility and medicinal uses of four other species in India. More recently, the plants have been recorded as diuretic (Vohora et al. 1983, Hocking 1997). Some contain alkaloids, and others are used for soap (Mabberley 1997).

Trianthema contains trainthemine, saponins, and tetraterpenoids (Bogle 1970, Hocking 1997, Nawaz et al. 2001). Studies in Asia have shown that the plants act as antifungal agents (Nawaz et al. 2001), and that they have potential as antihepatotoxics (Mandal et al. 1998), antiulcerogenics (Akhtar and Ahmad 1995), and to inhibit DNA damage and chromosomal aberrations (Sarkar et al. 1999), inhibit hepatocarcinogenesis (Bhattacharya and Chatterjee 1998a, 1999), and restore the antioxidant defense enzyme levels and hepatic biotransformation patterns (Bhattacharya and Chatterjee 1998b).

Trichostigma: Hoop Vine—The Plant That Wasn't There

(From Greek *trichos*, hair, *stigma*, the stigma, having hairlike projections from the stigma).

Back in the early 1970s the avid field botanist George N. Avery (1922–1983), of Fairchild Tropical Garden, told me that he had found *Trichostigma octandrum* on Chokoloskee Island just south of Everglades City. At the time, David McJunkin and I were studying the ethnoflora of that island on the northern fringe of the 10,000 Islands. It should have been simple enough to find the plants. Their stems grow to about 10 m long, scrambling over other plants to form a blanket of greenery. The stems are woody and pliant and grow larger in diameter than a person's thumb. We scoured every inch of the island but did not find this climber, now called "hoop vine" in Florida.

Later, in 1982, John Popenoe and Don Correll (1908–1983), also of Fairchild Tropical Garden, collected the vines on Chokoloskee. I went back and still could not find it. In March of 2000, Florida Atlantic University graduate student, Mike Anderson, brought in a sterile sample from Conservation Area 3 in Broward County. The sample looked to me like a woody "rougeberry" (*Rivina humilis*). Mike had found the elusive hoop vine.

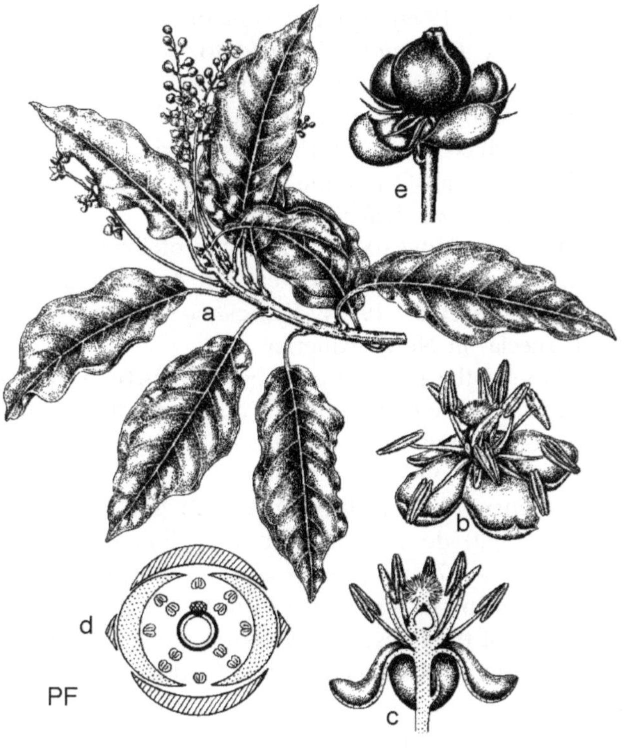

Trichostigma octandrum. a. Flowering branch. b. Flower, front-side view. c. Flower, longitudinally dissected. d. Floral diagram. e. Fruit. *Drawn by Priscilla Fawcett.* From Correll and Correll 1982.

It seems that Linnaeus thought the plant looked like rougeberry too back in 1756 when he named it *Rivina octandra* (*Rivina*, named for German botanist A.G. Rivinus, 1652–1723; *octandra*, with eight anthers). Subsequently, in 1909, another German biologist, Hans P.H. Walter (1882–?), realized that the species belonged in a separate genus. He discovered that the Frenchman Achille Richard (1794–1852) had created the proper home for the species in 1845, and so Walter moved Linnaeus's species name (*octandra*) to *Trichostigma.* So far as Mike and I could learn, we are the only biologists to see the live plants in Florida since 1982.

There has always been an unusual "hit-and-miss" story in the United States behind this relative of rougeberry and pokeweed (*Phytolacca americana*). Chapman's southeastern flora (1897) did not include the species. Later, when John K. Small (1933) wrote an updated southeastern flora, the species was there. Hoop vine has only been collected two or three times since, and it is now known from Collier, Monroe, and Broward Counties. Mike's discovery was a county record.

I cannot help but wonder, "Is this a rare native or an exotic species?" Indeed, the plants were so rampant when Mike found them, he was afraid he had discovered yet another alien species running amuck in the Everglades. After seeing the vigor with which hoop vine grows, I could even better understand his fear.

We were not the first to question this climber's nativity in this northern fringe of its range. Julia Morton (1981) wrote in her atlas on medicinal plants of the Americas that hoop vine was naturalized in the Bahamas. George Avery thought people might have introduced *Trichostigma* onto Chokoloskee Island (Austin and McJunkin 1978).

Why would someone introduce this plant? A search reveals that people have used the plant in several ways throughout its range from Cuba to Venezuela and Peru. The stems have been used to weave baskets and to make barrel hoops (Liogier 1974). In Cuba, it is *bejuco canesta* (basket vine) and *guaniquí* (Taino). On French-speaking islands it is *liane pannier* (basket vine) or *liane a barriques* (barrel or cask vine, Guadeloupe, Martinique). On English-speaking islands, reaching from Jamaica to Barbados, it is black basket wythe [wiss, with], cooper wythe, and hoop with. Some other names appear to allude to medicinal uses, like *bonbon codine* (codeine candy), *bois à terre* (wild bush), and *liane à terre* (wild vine, Guadeloupe, Martinique). For some reason, the plant was important enough to be given names like *sotacaballo* (substitute horse, in Costa Rica), *pabellón* (pavilion) or *pabellón del rey* (king's pavilion, in the Dominican Republic), *látigo* (lariat for catching horses, Guatemala), *entiil aaxux ts'aah* (like garlic vine, Huastec, Veracruz), and *zamurito* (little vulture).

Colombians use the leaves to help cure wounds (Morton 1981). Hispaniolans apply leaf decoctions to help someone overcome suffocation or choking and to treat asthma (Liogier 1974). Fruits and stems have the same uses. It is medicinal also among the Mayan speakers of Mexico. Most surprisingly, people grow it. Perhaps they grow it for medicine or weaving, but some apparently find it ornamental. This species was cultivated in the Bahamas in the 1970s, and the South American species *Trichostigma peruviana* is listed in *Hortus Third* (Hortorium 1976).

Nothing seems to be known of the bioactive compounds in this species, but several of its relatives are notorious chemical factories. Two genera, *Petiveria* and *Phytolacca*, are best known. Among their active compounds are several saponins, or soaplike compounds. If hoop vines are anything like those two, there is no wonder they have been used in medicines. Extracts of *Petiveria* have shown promise in treatment of protozoal infections, particularly those caused by *Trypanosoma cruzi,* which causes Chagas' disease in the tropics. That same plant may be useful in controlling tumors (mitogenic). Several species of *Phytolacca* have been studied and show promise against mitogenesis, in treating human immunodeficiency virus (HIV), in fighting vector-borne diseases in animals, and in killing disease-carrying snails (Berger et al. 1998, Rajamohan et al. 1999).

Still, the names *bonbon coq d'Inde* (Indian rooster candy), *bejuco de paloma* (dove vine), *pira de guacharaca* (chachalaca's nest; the chachalaca, genus *Ortalis,* is a bird somewhat resembling a chicken), *guacamaya* (macaw), and *murette* (little mulberry) show that people from Colombia to Hispaniola recognize that the fruits are spread by birds.

So, did the species come to Florida with people or in the stomachs of birds? Is it a rare and endangered native or an alien? We do not know. But, after a hiatus of 18 years, we know the species is still in Florida. It shows up about every couple of decades. Maybe this is a plant that appears only periodically, like the legendary Scottish village *Brigadoon.* Perhaps, like that village, the plants appear, thrive a short time, and then disappear back into the mists.

Triodanis

(From Greek *tri,* three, and *odontos,* teeth, possibly referring to the three calyx lobes of some flowers)

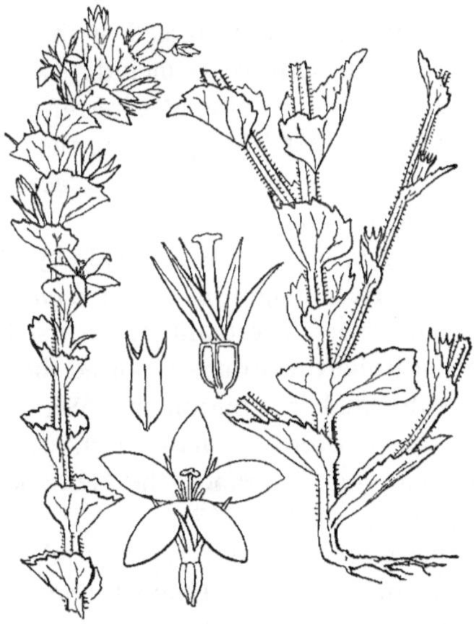

Triodanis perfoliata. From Britton and Brown 1898.

Triodanis perfoliata (perfoliate or having the stems appear to grow through the leaves)
clasping bell-flower [bellflower] ("bell-flower," a name originally given to *Campanula* in Europe, in use by 1578 when Henry Lyte translated Dodoens's *Cruydeboek* of 1554 into English)
clasping Venus's looking-glass ("looking-glass" was an alternate name for a mirror in use by

1526; "Lady's looking-glass" or "Venus' looking-glass" was in use by 1622, originally for *Campanula speculum*, and then later for other plants)
hens-and-chickens (in Europe, "hens-and-chickens" was the common name of the constellation Pleides by 1535; later it became the name of several plants, and by 1794 it was being applied to some daisies, Texas)

For decades the name *Triodanis* was largely ignored with preference given instead to *Specularia* (from Latin *specula*, looking-glass). *Specularia,* for example, was used by Fernald (1950). There was good precedent in using that name because of the European Venus's looking-glass, *Legousia speculum-veneris*, that has been put in synonymy with *Specularia.* Geoffrey Grigson wrote in 1955, "When the long capsule opens, the looking-glasses of the Virgin or Venus are revealed, the seeds of which are oval or elliptical, pale brown, exquisitely polished, and pellucid like a speculum" (Coffey 1993). That description really does not apply to the American plants, and it is odd that it was applied.

Rafinesque created *Triodanis* in 1894 in a book published after his death. McVaugh (1945, 1949) used that genus, but others disagreed with him. Now *Triodanis* is circumscribed to include six species in North America and one in the Mediterranean (Rosatti 1986, Mabberley 1997). The species in the Old World is *T. falcata*, and it too has been placed in both *Specularia* and *Legousia.*

Triodanis perfoliata was named by Linnaeus ([1753] 1957). He thought it was a *Campanula*, as had his predecessors. The plants were first known in Europe when Robert Morrison reported them in a list of plants cultivated in London in 1686. From that point, the herbs seem to have been continuously cultivated there and in France, the Netherlands, and Sweden (Linnaeus [1753] 1957).

The Cherokee used roots in a drink to relieve dyspepsia from overeating (Hamel and Chiltoskey 1975). The Meskwaki used it as an emetic and commented, "it will make one sick all day long." They also smoked it at ceremonies (Moerman 1998).

No comments have been found on poisons in the plants by Hardin and Arena (1974), Perkins and Payne (1978), Turner and Szczawinski (1991), or Foster and Caras (1994). Therefore, the plants must not be too poisonous. However, given the toxic chemicals known to occur in *Campanula* and *Lobelia* (which see), caution is advised.

Tripsacum

(Origin obscure, perhaps from Greek *tribein*, to rub, perhaps an allusion to the polished spike-like inflorescence; Quattrocchi 1999 suggested it might be Greek *tripsis*, rubbing, and *psakas*, a grain)

Tripsacum dactyloides. *Drawn by Mary Wright Gill and Agnes Chase.* From Hitchcock and Chase 1950.

Tripascum dactyloides (fingerlike, an allusion to the branches of the inflorescence)
Fakahatchee grass [Fakahatcheegrass] (a name applied to these plants by the horticulture industry in spite of the fact that the plants grow from New England and Michigan south)
[eastern] gama [gamma] grass [gamagrass]
sesame-grass

Tripsacum floridanum (of Florida)
Florida mock gamagrass

At the Ozark Bluff-Dweller site in Arkansas, Melvin Gilmore found gama-grass. He wrote: "A puzzling fact is the presence of great quantities of remains of *Tripsacum dactyloides* L. in the Bluff-Dweller habitations. The seeds would undoubtedly be good for food, but they are so deeply imbedded in the hard, tough rachis that one cannot see how the Indians could have separated them. The hulls might possibly have been cracked by pounding in a mortar and then winnowed out" (Gilmore 1931).

Mary Eubanks (2001) commented that it is not necessary to crack the hulls with a mortar to extracted the kernel from gama-grass. She tells me "how easy it actually is to extract the kernel from the fruitcase because the outer glume does not overlap the inner glume. You can quite literally take your fingernail and easily pop open the fruitcase of *Tripsacum*."

Kindscher (1987) found a report that the "popped kernels of *Tripascum* species are almost indistinguishable from those of open-pollinated strawberry popcorn." However, even that may have been the least common preparation of gama-grass. Galinat and Craighead (1964) reported that local Floridians carried "a pocket full of fruit cases to chew when out on hunting trips." The species they carried was the second species in southern Florida, *T. floridanum*. If it was used as food, certainly the same people used *T. dactyloides* because it is more common and produces more seeds. They should be eaten because the seeds contain, "10X the protein of corn." They are "delicious raw, roasted, or popped. I eat it. It's more flavorful than corn" (Mary Eubanks, personal communication Mar. 2003).

Typha

(Greek *typhe, tiphe* applied by Theophrastus, 372–287 B.C., Dioscorides, fl. A.D. 40–80, and Aristotle, 372–287 B.C., to a kind of aquatic plant used for stuffing beds and bolsters)

Typha. a. *Typha domingensis*. b. *Typha latifolia*. From Institute of Food and Agricultural Sciences.

Typha domingensis (from Santo Domingo)
Typha latifolia (broad-leaved)

aba'kweûck (shelter weed, Potawatomi); *apuk'we* (Ojibwa); *pakwe'ûck* [*pukyuk*] (Prairie Potawatomi)
aguapá (from Tewa *'aguap'a*, New Mexico, Mexico)
beecho [*peecho*] (mat or matting, or bundle of reeds, Zapotec, Oaxaca)
biwiê'skwînû (fruit for baby's bed, Potawatomi)
bodan-dubh (black penis, Gaelic)
bull-rush (USA, England)
cat-tail [cattail, cattes tayle] (from at least 1558, when Turner recorded the name as "cattis tayle," and "cats taile," USA and England to Panama)
cuigeal nam-ban-sithe (*cuigeal*, staff, *nam*, of the, *ban-sithe*, female fairy, Gaelic); *cuigeal nan-lo-gann* (*cuigeal*, staff, *nan*, of the, *lo*, day, *gann*, little, Gaelic); *cuile* (Gaelic)
down [cat-tail down] (referring to the fluff on the seeds which has been used like duck feathers in stuffing pillows, etc., Bahamas)
elephant grass (Belize)
énea [*éneas, anea, aneas*] (copper colored, Dominican Republic, Panama, Colombia); *yerba de éneas* (copper colored herb, Puerto Rico)
espadaxa (having a spike, the inflorescence, Colombia)
[*grand*] *gros jonc* [*natte*] (big [matting] rush, Haiti)
hashuk pancha (*hashuk*, herb, *pancha*, cattail; *panti* is the upper end of the stalk where the "down" grows, Choctaw); *hassikbacilká* (*hassí*, herb, Koasati)
hawahawa (Pawnee)
herbe a masses (mace herb, French); *macío* (mace, Spanish); *masse* (mace, from 1542, French); *massette* (little mace, Guadeloupe, Martinique); *mazzo-sorda* (mute or deaf [blunt?] mace, Italian); reed-mace [reedmace] (as "reide mace" by 1558 or before)

hoski (Chickasaw)
jonc (based on the Latin *iuncus*, for a rush, derived from *iugere*, to tie or bind, from the use of stems for tying, Haiti, Guadeloupe, Martinique); *junco* [*de estera, de pasion*] ([marsh] rush [for suffering], Colombia)
kalakalaru (Arawak, Suriname); *kara* (Carib, Suriname)
kirit-tacharush (*kirit*, eye, *tacharush*, itch, so named because flying down causes itching of eyes, Pawnee)
kowi' hasimbish (*kowi'*, cat, *hasimbish*, tail, Chickasaw)
ksho-hin (*ksho*, prairie chicken, *Tampanuchus cupido*, *hin*, feather; plucked down resembles finer feathers from the prairie chicken, Winnebago)
mi-ke'-the-stse-dse [*mi-ke'-the-stse-e*] (Osage; one of their life symbols, representing the water part of the earth)
nathita^{h} *tlho*^{h} (Slave, Canada); *tlh'ogh k'a* (fat grass, refers to the shoots, Chipewyan, Canada)
otawaskwa (water-edge plant, Cree)
pashini [*pasí:nî*] (Mikasuki); *passenv* [*pashinat*] (Creek)
puh (Maya)
roseau (reed, Houma, Louisiana, Guadeloupe, Martinique)
sacate ignea (burning grass, Belize)
slag (slippery with mud, English dialectic, Bahamas)
tule (from *tollin* or *tullin*, Náhuatl)
wahab' igaskonthe (*wahaba*, corn, *igaskonthe*, similar, referring to the appearance of the floral spikes synchronously with maturing of corn, Omaha-Ponca)
wihuta-hu (*wihuta*, bottom of tipi, *hu*, plant, Dakota)

The name "reed-mace" for *Typha latifolia* is the key to understanding the Romance language names for these herbs. According to William Turner in 1562, "It maye be called rede mace, because boys vse it in theyr handes in the stede of a mace." The mace (Old French, *masse*; Portuguese, *massa* or *maça*; Italian, *mazza*, Spanish, *maza*) was a weapon of war. Apparently, boys "playing war" used a cat-tail substitute and originated the "reed-mace."

People all over the world have used these plants—not just small boys. Names tend to be applied indiscriminately to either of the species, with *T. latifolia* the one from Europe, and *T. domingensis* dominant in southern North America. In Florida, the pollen of *Typha* was in the archaeological remains of Fort Center, formerly inhabited by Glades people (Hogan 1978). Because the plants are wind pollinated, that cannot be taken as firm evidence they were using the plants, but they probably were. Oddly, most people in North America except the Muskogean tribes, Houma excepted, left records of using *Typha* (Moerman 1998). Maybe the Muskogean absence is explained by the belief that handling the stems would "make the stomach big" (Bennett 1997). Still, there is a mat found in Tennessee caves that was within the geographic range of the more northern Muskogeans (Whitford 1941).

In the Missouri River region, the Omaha-Ponca used the down as dressings for burns and scalds; on infants to prevent chafing, like talcum; and fillings for pillows, padding for cradle boards, and quilting in baby wrappings. Stems were made into ceremonial objects called *niniba weawan* for the Wawan ceremony. *Typha* down or fluff was used to dry the mother and infant at birth. Pads of cat-tail down were used as diapers (Gilmore 1919). Even Sacagawea, the Shoshoni guide with the Lewis and Clark expedition, used the down as a diaper for her newborn in the Mandan village in North Dakota (De Voto 1981). The Omaha-Ponca uses essentially summarize those of other people. Seminole relatives the Houma boiled stalks into a tea to treat whooping cough (Speck 1941).

It is impossible to separate what Europeans brought from Old World uses of *Typha* and what they learned from the indigenous people. Both knew that the root could be peeled, boiled, and eaten as potatoes or macerated and boiled for sweet syrup. The root also yields flour, and the pithy portion where root joins sprouting new leaves is boiled and eaten. Young shoots may be eaten in the spring (Fernald et al. 1958, Morton 1968b, Liogier 1974). Also the pollen is used as food and may be substituted for *Lycopodiella* spores (which see).

Curtin (1947) recounted how the Paiute ate the seeds. These people spread the fruits on an open spot of ground and set fire to them. When the fluff and other material burned away, the toasted seeds remained, and could then be eaten.

In Central America and the Caribbean, the *lana* (wool) from the spikes has been used to stuff pillows and cushions (Liogier 1974, Ayensu 1981). Leaves were once used for thatching, the stem is diuretic, and some consider the root aphrodisiac (Liogier 1974).

Typha latifolia is sold in Mexican-American herbal stores as *aguapá*, and is used as a diuretic and for insect bites. Spikes were dipped in fat and burned as torches in colonial times (Hocking 1997). *Typha domingensis* is the source of *tabua* fibers, a substitute for kapok in Brazil (Hocking 1997).

U

Ulmus

(The classical Latin name for the elm tree)

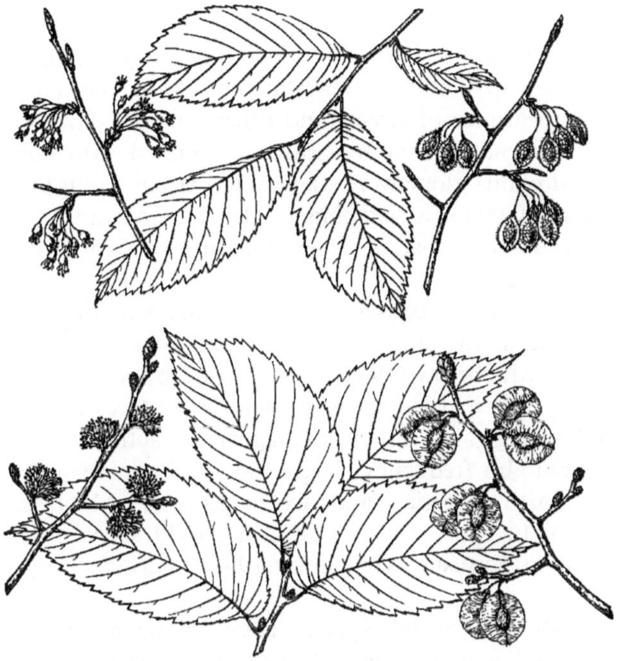

Ulmus. *Ulmus americana* (upper). *Ulmus rubra* (lower). Both from Sargent 1905.

é-hiu (Osage)
elm (derived from Latin *ulmus*; occurs about A.D. 1000 in Anglo-Saxon, with variations including *elme, elem, ellum, ulm, ulme*); *alim* (the first letter of the Gaelic alphabet); *alm* (Danish, Swedish, Norwegian); *álmr* (Old Norse); Elme (Turner [1548] 1965); *iep* or *olm* (Dutch); *olmo* (Italian, Spanish); *orme* (Quebec, France); *Ulme* (German); *u'ulum* (to bind, Akkadian)
lamequeiro (elm tree, probably based on Gaelic, Spanish)
leamhan (pronounced "leven," *U. glabra*, the Wych [supple, pliant] elm, Gaelic); *llwyfan* (pronounced "leven," *U. glabra*, Welsh)
ptelea (Greek)
ucha'pesk (Plains Cree)

Ulmus alata (winged)
[cork, corky, mountain, red, small-leaf, southern, water, wing, winged] elm
flygel-alm (winged elm, Norwegian); *vleugel iep* (winged elm, Dutch)
olmo corcho (cork elm, Spanish); *olmo sughero* (cork elm, Italian); *orme liege* (cork elm, French)
silhobá (Koasati); *úhawhu* (said to be Muskogee; Martin and Mauldin 2000 said the word *vhahwv* referred to the walnut, wood, nut, and tree); wahoo [whahoo, wahoo elm] (variant of Muskogee name, adapted into English by 1770); *whohao* (used by Williams [1837] 1962)
winged ash (an anomalous application)
witch elm (supple or pliant elm, from Old Teutonic *wik*, through Old English *wicca*, with the variant spelling of wych, the European name for *U. glabra*)

Ulmus americana (of America)
Amerikaanse treur-iep [*witte iep*] (American weeping [white] elm, Dutch)
amerikansk hang-alm (American hanging elm, Norwegian; could this be an indication of historic lynch parties?)
amerikansk vit-alm (American white elm, Norwegian, Canada, USA)
anib (Potawatomi)
baksá (thread, Koasati)
da-wa-tsi:-lai (Cherokee)
[American, American weeping, common, Florida, gray, gray hard, rock, soft, swamp, water, white] elm
ezhon zhon [*ezhon zhon ska*] (*ezhon,* elm, *zhon,* wood, *ska*, white, Omaha-Ponca)
ohóskere (Onondaga)
olmo bianco americano (American white elm, Italian, Canada, USA); *olmo blanco americano* (American white elm, Spanish, Canada, USA); *olmo lloron americano* (American weeping elm, Spanish); *olmo piangente americano* (American white elm, Italian); orhamwood (apparently a mixture of French, *orham* = *orme*, and English, Canada); *orme blanc* [*americain, d'Amerique*] ([American] white elm, French, Canada); *orme maigre* (meager elm, French); *orme pleureur americain* (American weeping elm, French)
pe chan (*pe,* elm, *chan,* wood, Dakota)
springwood (USA)
taitsako taka (*taitsako*, elm, *taka*, white, Pawnee)
úhawhu (said to be the Creek word for *Ulmus alata*; however, the Dakota *wahú*, arrowwood, was applied to *Euonymus*, and is not cognate)
vfosho [*afósho, tvfosho*] (maybe the source of "wahoo" in English, Muskogee)

Ulmus rubra (red)
achgükbi (Delaware); *cacî'bûb* (slippery bark, Potawatomi); *osasha'kûp* (Prairie Potawatomi)
afósho (Muskogee); *to-fósho* (*to* is an abbreviated form of *eto*, tree, Muskogee); *tohto* (Choctaw)
Amerikaanse rode iep (American red elm, Dutch, Canada, USA); *amerikansk rod-alm* (American red elm, Norwegian, Canada, USA)
atcinni pihi ayudi (*atcinni*, grease, *pihi*, smells good, *ayudi*, tree, Biloxi)
balop (Chicksaw); *balup* (Choctaw); *lopakv* [*lupákv, lupaka*] (Muskogee)
é-hiu hin'-dse gthi-gthi-e (*é-hiu*, elm, *hin'-dse*, linden, *gthi-gthi-e*, slippery; Hunter said it was *hon-kos-kao-ga-sha*, it won't go down, Osage); *ezhon zhide* (*ezhon*, elm, *zhide*, red, Omaha-Ponca); *ezhon zhide gthogthide* (*ezhon*, elm, *zhide*, red, *gthogthide*, slippery, Omaha-Ponca)
[gray, Indian, red, red-wooded, rock, slippery, slippery-bark, soft, sweet] elm (Canada, Europe, USA)
Fuchsbaum (fox elm, German)
gochráta (Onondaga)
it slips (English?, North America)
kaw-la [*ga'dal-a*] (buffalo tree, because buffalo congregate under its shade, Kiowa)
kee-awdl-ya-eep-eep (Kiowa)
moose elm (so called because moose, *Alces alces*, browse on young twigs and foliage, Canada)
olmo rojo americano (American red elm, Spanish, Canada, USA); *olmo rosso americano* (American red elm, Italian, Canada, USA)
oo-hoosk-ah (Onondaga)
orme fauve [*gras, rouge, rouge d'Amerique*] (fawn-colored [fat, red, American red] elm, French, Canada, USA)
pe tututupa [*pe tutuntu npa*] (*pe*, elm, Dakota, Teton)
taitsako pahat (*taitsako*, elm, *pahat*, red, Pawnee)
taw-a [*ta'ä, te-aw-ya-daw*] (saddle tree, because older saddles were made with a forked branch, Kiowa)
wakidikidik (Winnebago)
yap tcu'wi·kare (tree slipping [slimy], Catawba)

When I was growing up in western Kentucky, elm trees of several species were common. Larger trees were favorite places for adults and children. Streets were lined with elms, and where old specimens grew, people congregated in their shaded havens. As youngsters, we make bows from the branches, and twisted the inner bark to make bowstrings. We had no idea that in using the bark for string we were reenacting scenes as old as the Hopwellian cultures of the area's past.

Then, in the early 1960s, trees began dying. First one branch would turn brown, and soon the whole tree died. We learned that the malady was caused by the "Dutch elm disease" that had been accidentally introduced from Asia by way of infected timber from Holland. Beetles quickly spread the fungus, *Ophiostoma* (*Ceratocystis*) *ulmi*, from one tree to the next. The disease had almost annihilated the American elms by the early 1970s, and it began eliminating other species. The same thing happened in Europe, where in Great Britain 11 million trees were killed by the fungus between 1970 and 1978 (Mabberley 1997). The change in appearance of streets and other parts of cities was phenomenal. Economic losses associated with cultivated plants alone were staggering. At one point there were 400,000 diseased elms in Denver, Colorado, with an estimated cost of $500 million to remove them (Hocking 1997).

The change caused by loss of elms was the latest episode in the story of people's interaction with the genus. Mabberley (1997) suggests that by ca. 3000 B.C. Europeans had begun manipulating the trees through pollarding and other management activities that altered elm frequency. There are seven native species in North America and six in Europe (Mabberley 1997). In total, *Ulmus* contains 20–40 species, with most in Eurasia (Mabberley 1997, Sherman-Broyles 1997).

All of Florida's species have a long association with indigenous people. Although no tribes seem to have been recorded as using *U. alata*, its Creek name is *úhawhu*. Surely, they used at least the wood.

Wood was perhaps the most frequent item taken from elms. The wood from the different species is so similar that few can distinguish them. For example, Harrar and Harrar (1946) wrote that milled boards of *U. alata* "are mixed with those of other elms and thus lose their identify in the trade." The same is true of other species. Wood is close grained, heavy, hard, and light brown. Wood of all species is difficult to split (Uphof 1968).

Wood from elms was important to several tribes. More information has been found on Great Plains people than others, but it is likely southeastern tribes used it similarly (Gilmore 1919, Moerman 1998). Posts were used for building lodges by the Dakota, Lakota, Omaha, Pawnee, and Ponca. The same tribes used it to make mortars and pestles to grind corn, medicines, or perfumes, fuel, and saddle trees. Vines (1977) indicated that indigenous people (no tribes mentioned) used elm wood for canoes.

Settlers too used the wood in many ways. Elm was used for agriculture implements, baskets, boats and shipbuilding, boxes, cabinets, car construction, cooperage, crates, flooring, framework, fuel, furniture, handles, hubs of wheels, posts, saddle trees, sills, sporting goods, stock staves, ties, tool handles, toys, trunks, vehicles, veneer, and woodenware (Uphof 1968, Vines 1977, Hocking 1997).

Bark was probably second in importance to many tribes (Gilmore 1919, Yanovsky 1936, King 1984, Moerman 1998). Among the Ojibwa and Potawatomi, sheets of the bark were used to cover wigwams. Kalm ([1753–1761] 1972) described how his Mohegan guides made a canoe for them with the bark of *U. americana*, caulking the ends with *U. rubra* fibers. He wrote: "Of the [white elm] the boats made use of here are commonly made, it being tougher than the bark of any other tree. With the bark of the hiccory, which is employed as bast, they sew the elm-bark together, and with the bark of the red elm they join the ends of the boat so close as to keep the water out. ... That which we made was big enough to bear four persons, with our baggage, which weighed somewhat more than a man."

The Cherokee, Menomini, Omaha, Pawnee, Ponca, Potawatomi, and Winnebago used inner bark fibers to make baskets, dogsled harnesses, matting, nets, ropes and cords, and snowshoe laces. The Iroquois made dogsled harnesses from the bark. One of the Iroquois tribes, the Onondaga, used "the bast like hemp & oakam to stop leaks" (Zeisberger and Horsford 1887). When the outer bark had "weathered for several years till it glows with phosphorescence in the darkness," it was used to catch the spark in making fire among the Missouri River tribes (Gilmore 1919).

Those uses of the inner bark go back to the Hopwellian cultures, and Whitford (1941) was the first to note elm fibers in their woven fabrics in Ohio. Gilmore (1931) found elm fibers among the materials left by the Ozark Bluff-Dwellers in Arkansas, and Whitford (1941) also identified both *U. americana* and *U. rubra* in Iroquois burden straps.

Settlers learned from the indigenous uses. During the Civil War, the inner bark was touted as twine to fasten cotton bale covers. Porcher (1863) quoted a Montgomery County, Alabama, newspaper as saying, "Mr. [T. J.] Howard has used the wahoo rope with great success in bagging cotton on Col. Baldwin's place, and we can safely recommend his contrivance to the attention of planters. ... The manner of using the rope made of wahoo bark is altogether similar to that which has been in ordinary use." Later, the inner bark still was used locally for baling twine (Harrar and Harrar 1946).

Elms are also a source of food. John D. Hunter wrote in 1823 that he found the bark nutritive and had "subsisted for days on it" while living among the tribes in the Ozark region (Hunter [1823] 1973). Yanovsky (1936) found bark had been eaten by tribes in New York. The Cheyenne and Kiowa drank a "tea" of the inner bark like coffee and also stored it for winter use (Moerman 1998). They used the bark in much the same way as pioneers who used the bark to quench thirst (Vines 1977). The Omaha cooked bark with the fat of buffalo as a preservative to keep it from becoming rancid and to add flavor. Once the fat was rendered, children ate the bark pieces.

The most frequent medical use by indigenous people seems to have been to treat wounds, burns, inflammation, and other skin problems. While there are more records of *U. americana* and *U. rubra* than for *U. alata*, all three were used to treat medical problems. In 1714, John Lawson wrote: "The Indians take the Bark of [the elm] Root and beat it, whilst green, to a Pulp, and then dry it. ... This they use as a Sovereign Remedy to heal a Cut or green Wound, or any thing that is not corrupted. It is of a very glutinous Quality" (Lawson in Vogel 1970).

Southeastern uses include the Choctaw who made a decoction of *U. americana* inner bark to relieve menstrual cramps (Taylor 1940). The Houma mixed the bark with red oak (*Quercus falcata*) to treat dysentery (Speck 1941). The Koasati made a decoction of the bark to heal gunshot wounds (Taylor 1940). The Creeks used *Ulmus* to treat toothache (Swanton 1928a).

Ulmus rubra was used similarly. The Alabama and Creek made a decoction of the bark and believed that it induced labor (Swanton 1928a). The Cherokee made remedies of it to soothe the stomach, relieve dysentery, coughs, sore eyes, heartburn, catarrh, quinsies (inflamed uvula), consumption, and breast complaints. They also made a poultice to treat burns (Hamel and Chiltoskey 1975). Americans learned from indigenous people, and physicians used the bark as a poultice for gunshot wounds during the Revolutionary War and the Civil War (Hocking 1997). Sap of *U. rubra* is still used by Oklahoma Seminole women as a vaginal lubricant before sex and also in childbirth to ease delivery (Howard 1984).

Dried inner bark of *U. rubra* is mucilaginous and aromatic, and it contains an aromatic principle like fenugreek in odor (Hocking 1997). The bark was in the U.S. Pharmacopoeia from 1820 to 1936 and in the National Formulary between 1936 and 1960 (Vogel 1970). Those lists considered it protective, demulcent, and emollient. While the bark is no longer "official," various preparations of it are still marketed (Sherman-Broyles 1997).

American elm life spans were 150 to 175 years and their canopies created favored sites for special events. This use began with indigenous people and continued with settlers. One of these old trees was the Great Elm at Palmer, Massachusetts. There General George Washington stood in its shade to address his troops 3 days before he took command of the Continental Army. Another historic tree was the Penn Treaty Elm that blew down in 1810 (Harrar and Harrar 1946). Surely, one of the most notorious trees was the Daniel Boone Judgment Tree. After "settling down," Boone became a judge and held court under that venerable old tree. According to the legend, he selected that particular tree because it had served as a location for meetings by the indigenous people before him. When Boone deemed an individual guilty, the tree's branches allowed the hangman to do his job without moving. One of the offspring of that tree was reaching maturity at the Missouri Botanical Garden in the late 1960s when I watched as Dutch elm disease killed it. Our lives were greatly impoverished by the introduction of the disease that shortened the life span of elms.

Uvularia

(From Latin *uvula*, the flowers hanging like the uvula from the palate)

Uvularia perfoliata. From Britton and Brown 1896.

bellwort (a name usually applied to *Campanula*); merry-bells
wild-oats

Uvularia perfoliata (perfoliate or having the stems appear to grow through the leaves)
[cow, hay, straw] bells
[fragrant, mealy, perfoliate] bellwort
crow's foot (used by John Bartram in 1751)
Mohawk-weed ("Mohawk" one of the Five Nations of the Iroquois in upper New York)
strawflower

Uvularia sessiliflora (sessile-leaved, or leaf without stalks)
cornflower (Maine)
crow's foot (used by John Bartram in 1751)
sessile-leaf bellwort
straw-lillies (Connecticut)
tsu-hi-tsu-gi (Cherokee)
wild-oats

John Gerarde ([1597] 1975) called the European plants "throote wort or uvula woort, of the vertue it hath against the pain and swelling thereof." Linnaeus ([1753] 1957) accepted the analogy and named the herbs *Uvularia.* He described American *U. perfoliata* and *U. sessilifolia,* and European *U. amplexifolia.* Subsequently, the European species was moved to *Streptopus* so *Uvularia* is now endemic to eastern North America. As currently circumscribed this is a genus of five species (Mabberley 1997). Four of these (*U. grandiflora, U. perfoliata, U. pudica, U. sessiliflora*) grow in the northeastern states (Fernald 1950). *Uvularia floridana* grows from South Carolina to Georgia, Florida, and Alabama (Radford et al. 1968).

Uvularia perfoliata and *U. sessiliflora* were eaten by the Cherokee; the leaves were boiled or fried with fat as greens (Hamel and Chiltoskey 1975, Moerman 1998). Manasseh Cutler wrote of *U. sessiliflora* in 1785 that "the young shoots may be eaten as asparagus. The roots are nutritious and are used in diet drinks" (Fernald et al. 1958, Coffey 1993). Porcher (1863) agreed, and wrote of both *U. perfoliata* and *U. sessiliflora,* "The roots are ... edible when cooked, and the young shoots are a very good substitute for asparagus." Yanovsky (1936) said indigenous people throughout the eastern states ate roots and shoots.

Mostly, the plants are recorded as medicinal. The Cherokee made a medicine of *U. sessiliflora* roots to stop diarrhea, and as a poultice on boils (Hamel and Chiltoskey 1975). The Iroquois used *U. perfoliata* roots as cough medicine for children, to treat eye sores, and to help broken bones heal (Moerman 1998). They also used *U. sessiliflora* roots to purify the blood, and as a poultice to help heal broken bones. The Ojibwa used roots in hunting medicine (Moerman 1998).

In 1751, John Bartram called *U. perfoliata* "Uvulary," and said, "It was formerly taken for a species of Solomon's Seal, having smooth leaves like it; but the stalk grows through the leaf and the little yellowish flowers something resemble a Lilly; it grows about a foot high, the root is white, and spreads like a Crow's foot; some people call it by that name for that reason. It is a good root for gathering and breaking a boil, and makes a fine salve for healing wounds and ulcers; it makes a fine maturating poultice" (Bartram in Coffey 1993).

Porcher (1863) wrote of both *U. perfoliata* and *U. sessiliflora*, "The roots of the different species are subacid and mucilaginous when fresh; and a decoction of them has been employed as a domestic remedy in sore mouth and in affections of the throat; also considered as alexipharmic in snake bites."

V

Vaccinium

(Ancient name of bilberry, *V. myrtillis*; presumably, Linnaeus used the Latin *vaccinus*, of cows, as reflected in the Germanic folk-name *Kuhteke*; alternately, *vaccinus* may be a corruption of Greek *hyakinthos*, the color purple or dark red; related words may be Akkadian *bakkitu*, professional mourner; *bakku*, tearful, *bititi*, lamented)

Vaccinium arboreum (treelike)

cafaknv copo-peleksv (*cafaknv*, huckleberry, *copo*, anus, *peleksv*, puckered, Creek)

farkleberry (probably a corruption of sparkleberry)

gooseberry (properly a name for *Ribes*, USA)

hape'tka ha'yi (Biloxi)

[tree-, winter-] huckleberry (huckle- is presumed to have been derived from hurtle-, or whortle-)

hurt [hurtberry] (from ca. 1450 when it was applied to a European *Vaccinium*, but not combined with -berry until 1542; etymology uncertain)

osakohchi [*wasakohchi*] (Alabama); *owessv* [*owisa*] (Muskogee)

sparkleberry (appearing in literature by 1891, presumably because of the shiny fruits)

whortleberry [hurtleberry] (whortle- [hurtle-] derived from whort-, in a southwestern England

Vaccinium. *Vaccinium arboreum* (top left). From Sargent 1905. *Vaccinium corymbosum* (top right). From Britton and Brown 1897. *Vaccinium myrsinites* (lower left). *Drawn by P.N. Honychurch. Vaccinium stamineum* (lower right). From Britton and Brown 1897.

dialect, where the "w" is added, as in "whome" for "home" and "whole" for "hole," dating from 1570s; originally applied to a European species, then extended to American plants)
yun̲lo (Choctaw)

Vaccinium corymbosum (having flowers in a corymb or flat-topped cluster)
accoondews (this may be the species mentioned in [1612] 1953 by Strachey, who said the fruits were "blue berries of the bignes of grapes very pleasant"; Powhatan, Virginia)
bilberry [billberry] (from Danish *böllebaer*, where *bölle* means ball or ball-shaped, and *baer* is fruit. *Bölle* is sometimes used alone to designate the species. Usually applied to *V. myrtillus*, a native to Eurasia where its fruits are used to make pies and other foods. Williams [1837] 1962 identified this species by the Old World name.)
giant whortleberry
high-bush blueberry [highbush blueberry]
oksakohchi (*oksak*, hickory, *ohchi*, to draw water, Choctaw); *okciskó* (*okcí*, juice, *ísko*, drinker, Koasati)

Vaccinium darrowii (named for George McMillan Darrow, 1889–1983, who first recognized the species and brought it into cultivation at Beltsville, Maryland in 1940)
cafaknv sopahvtke (*cafaknv*, huckleberry, *sopahvtke*, gray, Creek)
Darrow's blueberry (a book name, Florida)

Vaccinium myrsinites (resembling *Myrsine*)
cafaknv [*câ:faknâ, tsafuckina, caafakna, chafakana*] (Creek); *cafaknv lvstuce* (little black huckleberry, Creek)
huckleberry (Florida)
olake [*olákî, oluckee, oláká:pî*] (*olaki*, huckleberry, *api*, tree, Mikasuki)
shiny blueberry (Florida)
whortleberry (Williams [1837] 1962)

Vaccinium stamineum (with prominent stamens)
[buck, deer, goose]-berry [berries]
bleuet [*bluet*] (blue one, Quebec)
[squaw]-huckleberry (USA)
whortleberry (Williams [1837] 1962)

Because *Vaccinium* has a circumboreal distribution, it has been used by people around the world (Mabberley 1997). Even the etymology of the generic name records long association with people. Moerman (1998) listed 16 species used across North America.

Five species are known to have been used in Florida, but surely another one, *V. tenellum*, was similarly valuable. At least the Algonquin and Iroquois used *V. corymbosum* (Yanovsky 1936, Moerman 1998). Maybe one of those is the unidentified species mentioned among the Alabama as medicinal (Swanton 1928a). Williams ([1837] 1962) called it "bilberry," and that implies that the fruits were edible.

Oddly, Moerman (1998) does not mention *V. arboreum* or *V. stamineum*. At least among the Seminoles of Florida and Oklahoma, the value of the former is well documented. In Oklahoma, the roots of *owisa* are boiled and the infusion is used on someone who is "out of his head" and talking and behaving erratically. Deer are fond of berries and hunters often look for animals near stands of them (Howard 1984). From New England west to Minnesota and Wisconsin, indigenous tribes ate fruits of both *V. stamineum* and *V. corymbosum*.

Among the Florida Seminoles, Alice Snow recognized a species she called *cafaknv copo-peleksv*. Her description of it being taller than *cafaknv lvstuce*, and growing on one stem with wider leaves certainly suggests *V. arboreum*, which extends into the Brighton Reservation region. She says the bigger berries are used for pies and jellies (Snow and Stans 2001), which is what people have used them for throughout its range. Farther north, decoctions of the leaves and bark of the roots were used to treat sore throat and diarrhea, and a drink made from fruits helped relieve chronic dysentery (Krochmal 1973). Surely the Seminoles, like the settlers, used the hard, brown, close-grained wood for tool handles (Small 1933).

Hogan (1978) listed *V. myrsinites* pollen from the Fort Center site on Lake Okeechobee, although it is likely that she used that species only because it was the common plant there. Regardless, *Vaccinium* has been associated with people in Florida since before the Europeans arrived. To the Seminoles on the Big Cypress and Brighton Reservations, this is the most important species. Sturtevant (1955) found them using it for "Sun Sickness" (eye disease, headache, high fever, diarrhea), "Hog Sickness" (unconsciousness), and during birth, death, and busk ceremonies (see *Pterocaulon*). Then and now the berries were eaten. Alice Snow makes medicine from *cafaknv*, which she considers "good for anything" (Snow and Stans 2001).

Farther north on the peninsula, *V. stamineum* is more common. South of Lake Okeechobee there are only a few restricted areas where it grows. To some people the fruits are unpalatable and supposed to be poisonous, but some individual plants are said to give delicious fruit (Fernald et al. 1958).

Vanilla

(Philip Miller named this from the Spanish *vaina* or *vainilla*, referring to the cylindrical, sheathlike "pod," actually fruit; the Spanish was based on Latin *vagina*, a sheath)

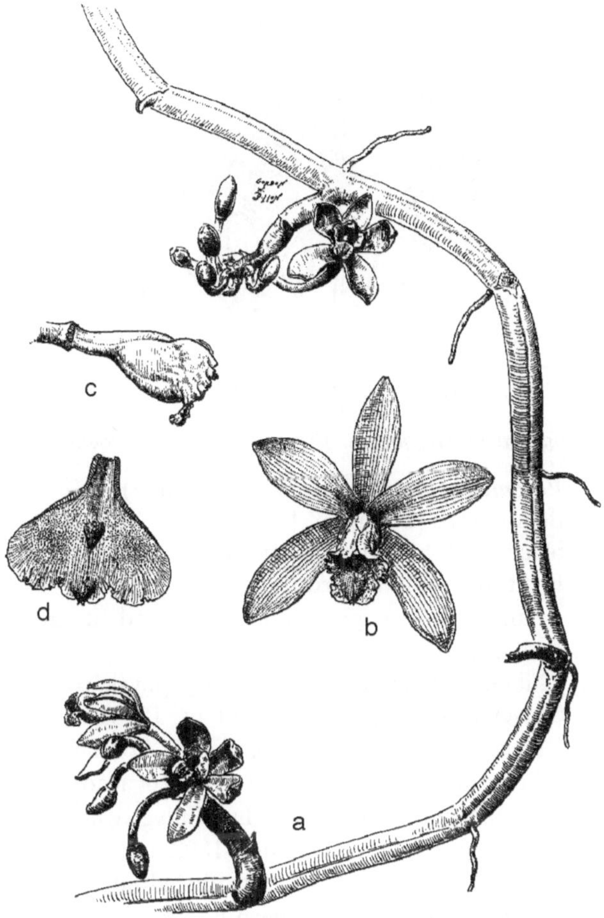

Vanilla barbellata. a. Upper part of plant. b. Flower, front view, lip in natural position, sepals and petals spread open. c. Lip, attached to column, in natural position, side view. d. Lip detached from column, spread open. *Drawn by Gordon W. Dillon.* From Correll and Correll 1982.

Vanilla barbellata (little beard, in reference to the "hairs" on the labellum)
bejuco de lombriz (worm vine, Cuba, Dominican Republic); *lombricera* (wormer, Puerto Rico); worm-vine (Florida, Bahamas)
liane couleuvre (snake vine, Haiti)
liane vers (*liane*, twiner, *vers*, about, Haiti)
link vine (Bahamas)

Vanilla phaeantha (gray flowers)
leafy vanilla

In both Cuba and Hispaniola, these and other species of *Vanilla* are used in remedies against worms (Liogier 1974). Pods of *V. phaeantha* are sold as a substitute for cultivated vanilla (*V. planifolia*) (Hocking 1997).

Verbesina

(Some say Linnaeus thought these had leaves resembling *Verbena*; others think the name was "metamorphosed" from that name; in either case, *Verbena* is an ancient Latin name given to a sacred plant)

Verbesina virginica. From Britton and Brown 1898.

Verbesina virginica (from Virginia)
crown-beard [crownbeard] ("crown" from Middle English *coroune* or *crowne*, from Old French *corone*, from Latin *corona*; first in English in the 12th century; "beard" refers to the pappus); white crown-beard (Missouri)
epte aape [*iptá:pî, itutapi, ip'tápi*] (frost stem, because the interiors of the stems freeze and the stems split open; *epte*, ice, *aapé*, stem or handle, Mikasuki); *hitotapî* (*hetute*, frost, *vpe*, stem, Creek)
frost-weed [frostweed]
holshish okwa stikbe ishkwo ("root water put in drink," maybe from *hakshih*, root, *oka*, water, *shikoli*, bundle, *ishko*, drink, Choctaw)
Indian-tobacco
tickweed [tickweed] (maybe a host for ticks?)

Frostweed must be an ancient name for these plants. Even Steyermark (1963) mentioned that the water frozen in the stems produces "frost flowers." He explained that these are "white, ribbonlike, fluted, and sometimes twisted ice formations ... [which] result from the rise of the cell sap and moisture from the still active root into the dead and dry stems." Apparently, the first use of "frost-weed" in English was in 1866, for *Helianthemum canadense.* However, by 1856 the name "frost-wort" was used for *Cistus canadensis.* Both of those were considered medicinal, having tonic and astringent properties.

Verbesina is associated with medicines throughout its range in the Americas. An infusion of *V. virginica* roots was taken by the Choctaw to relieve fever (Bushnell 1909, Taylor 1940). The Chickasaw used it as a "deobstructant," as a diuretic, for uterine problems, as a stimulant, and to treat venereal disease (Campbell 1951).

Sturtevant (1955) found the Seminoles using *epte aape* to treat "Bear Sickness" (fever, headache, thirst, constipation, blocked urination), "Fire Sickness" (fever, body aches), "Mist Sickness" (eye disease, fever, chills), and stomachache (root bark decoction). An emetic of root bark is also taken at funerals, at "doctor's school," and after the death of a patient.

Bennett (1997) found the Seminoles still using the herbs. The roots are scraped into water; the solution is taken orally and used to wash the patient to treat "night sweats," fever, or hot flashes. This is considered a "cold medicine." The plant also affords protection after lightning strikes, as people are vulnerable to "bad spirits" after lightning strikes.

Verbesina helianthoides (gravel-weed crownbeard, sunflower crownbeard) was called diabetes weed in the 1880s in Georgia (Hocking 1997). It was used as a diuretic to relieve edema and urinary problems (cystitis).

Morton (1981) also lists three other species. Those are used in Haiti and the Virgin Islands (*V. alata, V. gigantea*), Mexico (*V. gigantea*), and Costa Rica (*V. scabriuscula*). Both *V. alata* and *V. gigantea* are considered antispasmodics used to relieve stomachache and muscle strains, and are emetic, while *V. gigantea* is also used against toothache (Lewis and Elvin-Lewis 1977). *Verbesina scabriuscula* is used in healing baths. In the southwestern United States, *V. encelioides* (cowpen-daisy, *añil del muerto*) is used by the Hopi, Navajo, and Zuni; it too is emetic, and it is used to treat stomach and skin problems (Coffey 1993, Moerman 1998). Other species are used in Mexico, Venezuela, Guatemala, and Costa Rica (Hocking (1997).

Veronicastrum

("False Veronica"; based on *Veronica*, from *vera*, true, and *eicon*, image, from the early Christian legend picturing St. Veronica, wiping Christ's face on the way to Calvary, with a handkerchief that received a miraculous "true image" of his features, and the suffix–*astrum*, resembling)

Veronicastrum virginicum. From Britton and Brown 1897.

Veronicastrum virginicum (of Virginia)
Beaumont's [Bowman's] root
black-root [blackroot] (Missouri)
Brinton's root (according to Rafinesque in 1826, named for a physician who used it in his practice)
[Cubeun's] Cubeno-root [-physic]; physic root
Culver's physic (used in print by 1858); Culver's root [Gulver's root, an error by Porcher 1863] (recorded in Massachusetts in 1716; according to Rafinesque, named for a Dr. Culver who used it in his practice; Millspaugh 1892 and Core 1967 said that the plant was introduced into England in 1714; however, the herb had been mentioned by Plukenet in 1696)
high veronica; tall veronica; *veronique de Virginie* (Virginia veronica, French)
hini (from *hi-ni'-e*, toothache? Osage name according to homeopathic physician J.U. Lloyd, but the plant is not mentioned by Hunter [1823] 1973 or La Flesche 1932); Virginia-hini (maybe based on the Osage name)
Leptandra (from Greek *leptos*, slender, and *andros*, male, a generic name applied by Nuttall in 1818); *Leptandra-Wurzel* (Leptandra herb, German); purple-*Leptandra* (the purple form)
oxadaddy (English?)
quitel (Delaware name according to homeopathic physician J.U. Lloyd)
quitich (according to Britten and Holland's *A Dictionary of English Plant-Names* of 1886,

"quitich" is a "grass," but "in County Donegal [Scotland], every weed is so called")
tall speedwell ("speedwell," meaning "goodbye," is a generic common name for *Veronica*; it alludes to the rapid fall of the corollas if the spikes are picked; in use by 1578)
virginischer Ehrenpreis (Virginia honor-prize or speedwell, German)
whorly wort [whorlywort] (an allusion to the multiple leaves at each node along the stem)

Linnaeus ([1753] 1957) called these plants *Veronica virginica*, based on plants cultivated in Holland, Sweden, and herbarium specimens. Subsequently, Philipp Conrad Fabricius (1714–1774) created the generic name *Veronicastrum* in 1759. Nuttall did not realize that the herbs had a name and established the generic synonym *Leptandra* in 1818. Both the names *Veronica virginica* and *Leptandra virginica* were used in the older literature. It was not until 1917 that Oliver A. Farwell (1867–1944) created the name used today for this endemic North American species.

This herb grows from Massachusetts and Vermont to Manitoba and south to Florida, Mississippi, Louisiana, and eastern Texas (Fernald 1950). There is a sister species in eastern Asia (*V. sibirica*) that Millspaugh (1892) and other physicians considered the same as American plants. Although Czerepanov (1994) recognized other species in Asia, Mabberley (1997) thought there were only two.

The Cherokees used the root as a purgative and antiseptic. They applied the remedy to treat typhus and bilious fevers, induce sweating, as a tonic, and for inactive liver they mixed it with Indian physic (*Gillenia* spp.) and boneset (*Eupatorium perfoliatum*). The Cherokees also chewed the plant to relieve colic and drank a tea of it for backache (Hamel and Chiltoskey 1975). The Menominis used *V. virginicum* as an emetic and physic, and a "reviver," for purification when defiled by the touch of a bereaved person (Vogel 1970). The Meskwaki used the root to cure constipation and kidney stones, for ague, and as a medicine for women who were "weak" or in labor (Vogel 1970). The Ojibwa used a decoction of the crushed root as a cathartic, physic, and to clean scrofula sores caused by tuberculosis (Moerman 1998). The Senecas used the root as a cathartic (Vogel 1970). Other Iroquois considered it a treatment for chills and fever, diarrhea, rheumatism, as a physic, an emetic, a treatment for heart problems, and a "witch medicine" (Moerman 1998).

Probably because they considered the bowel to be the source of most maladies, European settlers in the New World were quick to adopt these plants as remedies. Cotton Mather wrote in 1716 to John Winthrop of New London, asking for some Culver's root to treat his eldest daughter who had consumption. She died the next month after the remedy arrived, and Oliver Wendell Holmes opined, "all the sooner, I have little doubt, for this uncertain and violent drug, with which the meddlesome pedant tormented her" (Vogel 1970). André Michaux wrote on 11 Dec. 1795 at Kaskaskia, Illinois, that he was "confirmed once more in my opinion that the root of *Veronica Virginiana* ... used as a decoction for a month, is effective for the cure of Venereal Disease" (Vogel 1970).

Porcher (1863) was more realistic that Michaux. He wrote: "The root is bitter and nauseous, yielding its active properties to boiling water. In the recent state [fresh] it is said to act violently, sometimes as a cathartic, and sometimes as an emetic. ... Under the name blackroot, Gulver's root, and the probably erroneous botanical name (*Leptandra alba*), the author of a work professing to describe the Indian mode of treating diseases, entitled the "Cherokee Physician," recommends the plant as an efficient purge: 'operating with mildness and certainty'; peculiarly adapted to typhoid and bilious fevers. Dose, a large teaspoonful of the root in a gill of boiling water, repeated in three hours. It is said to be also diaphoretic. The root may be given in any shape, and is thought to have a slow, alterative action. An extract is also used in making cathartic pills by concentrating the decoction, and using starch or liquorice root powder; or a syrup is made by adding molasses or sugar. It is laxative in tablespoonful doses. A principle called leptandrine, from the *Leptandra*, is much used in the Western states. An emetic decoction is made by the vegetable practitioners with the *Leptandra* root: half a pound American ipecacuanha, or the Indian physic one pound, put into a gallon of water and boiled down to a pint, of which the dose is an ounce every twenty minutes till vomiting is induced; or two teaspoonfuls of the powder may be given in an ounce of boiling water, to be repeated."

The drug was listed in the U.S. Pharmacopoeia under the name *Veronica* in 1820, and it remained there until 1840 (Vogel 1970). It was restored in 1860 under the name *Leptandra* and remained listed until 1916. It was in the National Formulary from 1916 to 1955. The dried roots and rhizomes were the parts used as a cathartic and emetic.

Hocking (1998) wrote that the herb had been used as a laxative, to treat rheumatism, as a cathartic, an emetic, a liver stimulant in hepatitis, a diaphoretic, a tonic, and applied to hemorrhoids. Foster and Duke (1990) warned that traditionally the dried root was the part used and that the fresh root was a violent laxative that was potentially toxic.

Roots and rhizomes of *V. virginicum* contain the bitter leptandrin (glycosidal amaroid), a glycoside resembling senegin, hormonal substances, resins, saponins, tannin, fatty oil sitosterol, and volatile oils (including esters of cinnamic acids) (Bown 1994, Hocking 1997). Duke et al. (2002) are guarded about the plants, noting that dried roots only are to be used. They give it a single plus—not as safe as coffee.

Viburnum

(The classical Latin name; some say of doubtful meaning, others affirm it means "wayfaring tree")

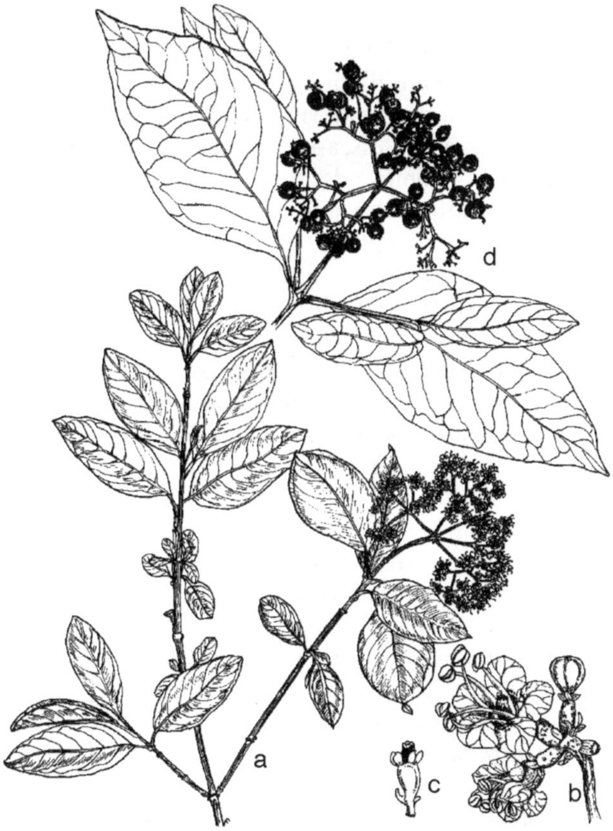

Viburnum nudum. a. Branch with flowers. b. Flowers. c. Calyx and ovary. d. Branch with fruit, showing enlargement of leaves by fruiting time. *Drawn by Vivian Frazier.* From Correll and Correll 1972.

arrow-wood; *łakí alyokcá* (*łakí*, arrow, *alyokcá*, makings, Koasati); *mon'-ci hi* (*mon'-ci*, arrow, *hi*, plant, Osage)
korsved (cross wood, Norwegian)
Schneeball (snow ball, German)
viburno (Italian, Spanish); viburnum (used in English by 1731); *viorne* (French)

Viburnum acerifolium (maple-leaved)
anib' (Ojibwa)
arrow-wood (applied to several *Viburnum* spp. by 1848)
dockmackie (obscure; maybe from Gaelic *dogha*, meaning *Rumex*, and *meacan*, root, an allusion to medical use of the roots; by analogy, the carrot, *Daucus carota*, is "mackenboy," which was derived from Gaelic *meacan buidhe*, yellow root)
Guelder-rose (originally applied to *V. opulus*; named from *Guelders*, a town in Prussia on the Holland border)
laurestinus (usually applied to *V. tinus*; originally two Latin words, *laurus*, laurel, *tinus*, a plant)
maple-leaf [maple-leaved] viburnum (a book name)
possum-haw [possumhaw] ("possum" is a colloquial form of "opossum," the American marsupial, *Didelphis virginiana*, alluding to the animal eating the fruits; "haw" is from Old English *haga*, hedge, akin to Dutch *haag*, German *hage*, Danish *have*, garden)
squash-berry
viorne (Quebec)

Viburnum dentatum (toothed)
conoská cobá (*conoská*, heart, *cobá*, big, Koasati)
mealy-wood (Peter Kalm called this "mealy-tree" when he found it at Lorette, Quebec in 1749; in doing so, he was using a European name for *V. opulus* that was in use by 1706; probably the name was given to note the scaly buds of the European plant)
southern arrow-wood [arrowwood]

Viburnum nudum (naked, from the definitely peduncled cyme)
[black, naked] alder
bilberry (of Norse origin, related to Danish *böllebaer*, from *bölle*, meaning unknown, *baer*, berry; originally applied to *Vaccinium myrtillus* ca. 1577, later to American plants)
[possum, Shawnee, swamp] haw [possumhaw] ("'possum" because of the insipid fruits)
nannyberry [nanny-tree] haw ("nanny" perhaps elliptical for nanny-goat, a female goat, first applied to the animal by 1788; however, we must wonder if the term "nanny" here is not adapted from indigenous languages like Dakota where *V. lentago* is known as *mna*, or *mna-ha* and *nan* in Omaha-Ponca)
southern wild-raisin
[smooth, naked] withe-rod [with-rod, with-wood] ("withe" or "wythe" is a piece of limber plant stem that is used for tying or binding, first used

about A.D. 1000; the word was derived from Old English *wippe*, related to Old Frisian *withethe*, with cognates in other Germanic languages, and often applied to willows)

Viburnum rufidulum (reddish)
łakakcó (*łakí*, arrow, Koasati)
rusby black-haw [blackhaw, black haul]
webasa'p (hauls bone, Catawba)

Southern Europeans knew about these tall shrubs at least by the time of Virgil (70–19 B.C.), who wrote about *lenta viburna* (*lenta*, pliant, *viburna*, maybe from *via*, path). Later, the Romans also called them *lentagenem*. That pliant nature led to them being called *lantana* by English speakers by the 1200s, and much earlier for people speaking Romance languages.

Still, there is no doubt what these names meant because there are no native European *Lantana* (in the modern sense of Verbenaceae). Therefore, it is believed that Virgil applied the name *lenta viburna* to one or more of the European *Viburnum* (Adoxaceae). One of the common names for the genus is "arrow-wood." Arrow shafts made of *V. lantana* accompanied the Neolithic "Iceman" found in the Alps (Mabberley 1997).

Perhaps John Gerarde in 1597 was simply translating the word *Viburnum* when he wrote that the shrubs were called "wayfarer's-tree." The OED (1971) noted that the name is an abbreviated form of "wayfaring man's tree." No further explanation is given, but that surely came into use because of the utility of the plants. Two of the best-known European species are *V. opulus* and *V. tinus*; the first is famous for its poisonous raw berries that become edible when cooked (Bown 1995). Both yield notable herbal medicines (Bown 1995, Hocking 1997).

By the time the New World was discovered, settlers were acquainted with European species. Exploration of the eastern coast revealed more species, and Linnaeus named the Florida plants in 1753. Three species were based on John Clayton's 1730s collections and originally discussed by Jan Gronovius in his *Flora Virginica*. Linnaeus knew 8 species, and there are now 150 recognized in temperate Eurasia and North America (Mabberley 1997). Kartesz (1994) accepted 21 species in North America, but that includes *V. lantana, V. sieboldii*, and *V. tinus*, which have become naturalized from the Old World.

Many of the species were recorded to make arrow shafts. *Viburnum acerifolium, V. dentatum,* and *V. rufidulum* are among the Florida species recorded (Vines 1977, Kimball 1994).

There are direct records of people eating the fruits from the northeastern states to the Great Plains (Gilmore 1919, Hedrick 1919, Moerman 1998). For example, fruits of *V. nudum* were eaten by Abenaki and Algonquin; Missouri River people ate *V. lentago*. However, drupes probably were more important in the past. *Viburnum acerifolium* was *anib'* to the Ojibwa. To the Missouri River people *V. lentago* was *mna hu* (Dakota), *nan shaman* (Omaha-Ponca), *wuwu* (Winnebago), and *akiwasas* (Pawnee). The North American form of *V. opulus* was known to the Cree as *nipinminan* (*nipin,* summer, *minan*, berry). That name was corrupted by settlers to *pembina*, which became the name of a river. The Ojibwa called the river *nepinminan sipi* (summer-berry river). Later, Pembina became the name of a mountain, town, and county in the northeastern corner of North Dakota. The Pembina River runs from south-central Manitoba into North Dakota, where it then flows across the northern part of Pembina County and empties into the Red River. Englishman John Lindley called fruits of *V. opulus* "a miserable food for savage nations" in 1846. Perhaps because of the thin-fleshed fruits, most tribes had abandoned them in favor of larger and more flavorful fruits from Europe by the time people began keeping records (Hedrick 1919).

The Delaware, Ojibwa, and Omaha mixed bark as part of a *kinnikinnick* (mixture, Algonquin) for smoking. Since the tribes were historically using *Nicotiana rusticum*, doubtless they were adding the *Viburnum* to dilute the strong tobacco taste.

By far the most important use among indigenous tribes was in medicine as an antispasmodic. Both *V. opulus* and *V. prunifolium* are known to contain scopoletin, a coumarin glycoside that has a sedative effect, especially on the uterus. Lewis and Elvin-Lewis (1977) indicated that coumarin glycosides are uncommon in plants, and it is suspected that most *Viburnum* have them. *Viburnum prunifolium* also contains salicin, and Bown (1995) recommended its use with *V. opulus*. The Cherokee, Iroquois, Menomini, and Ojibwa made an infusion of *V. acerifolium* to relieve cramps and colic (Hamel and Chiltoskey 1975, Moerman 1998). Iroquois women took a decoction of *V. dentatum* twigs as a contraceptive and made a poultice for swollen legs after childbirth. The Cherokee used *V. nudum* to prevent spasms.

Other medicines are related to these antispasmodic uses. The Iroquois took an infusion of *V. acerifolium* as a diuretic; they also drank an infusion and applied a poultice for pain they considered caused by witchcraft. Since *V. dentatum* is used as an adulterant for *V. prunifolium*, it too probably was used as a diuretic (Hocking 1997). Both *V. acerifolium* and *V. nudum* were made into tonics by the Cherokee (Hamel and Chiltoskey 1975).

The coumarin glycosides or chemicals called "viburnin" may be responsible for the other applications (Hocking 1997). The Cherokee used the root bark of *V. acerifolium* and *V. nudum* as a diaphoretic, febrifuge (including malaria), and treated sore tongues with it (Hamel and Chiltoskey 1975). The Menomini treated cramps and colic with a decoction of inner bark (King 1984); The Ojibwa made it into an emetic (Densmore 1928, Moerman 1998).

Vicia

(The classical Latin name, used by Varro, 116–27 B.C.; Quattrocchi (1999) thought it came from *vici*, I conquer, based on *vincere*, to be victorious; according to Saint Isidore, archbishop of Seville, "*herba vicia, id ist victorialis*," the herb vicia, it is victorious)

Vicia caroliniana. From Britton and Brown 1897.

Vicia sativa (the common European species; not in Florida)
arvejas (Spanish, 1557; from Latin *arvilia*, probably originally the same as what is now *V. ervilia*, European bitter vetch)
gall-pheasair (*gall*, stone, *pheasair*, peas, Gaelic)
ketsach (Hebrew, de Candolle [1886] 1959)
vetch (from Old French *vecce, vesse*, which was taken from Latin *vicia*); *bicio* (Fuchs 1542); *bikion* (Greek, Galen A.D. 129–?200); fetch (John Gerarde [1597] 1975); *fiatail* [*fiatghal*] (Gaelic); fiche (Turner [1548] 1965), *veccia* (Italian by 1551); *vesce* (French by 1575); *vesse* (French by 1550); *viccia* (modern Italian); *vitsen* (Dutch 1549); *Wicke* [*Wicken*] (German, Fuchs 1542)

Vicia caroliniana (of Carolina)
altsa'sti (a wreath for the head, Cherokee)
vesce (Quebec)
[Carolina, pale, woody] vetch

Not long ago my wife and I ate dinner with our friend Valeria Volin, who is from Padua in northern Italy. She was serving a "special" Italian dish that my wife and I had never eaten. This specialty consisted of raw fava beans (*Vicia faba*), sweet fresh figs (*Ficus carica*), and cheese. As I peeled the beans and munched them with the figs and cheese, I could not help thinking of the people in the Mediterranean who had to balance such a meal between too little and too much. Too little and they ran the risk of getting malaria; too much and they stood a good chance of dying.

Those people, largely of African black, Greek, and Italian heritage, have inherited a red blood cell enzyme deficiency called "favism" (glucose-6-phosphate dehydrogenase, often abbreviated G-6-DP) (Perkins and Payne 1978, Desowitz 1987, Foster and Caras 1994, Duke et al. 2002). When these people are exposed to certain substances, they develop intravascular hemolysis where the red blood cells disrupt. The sulfonamides and the antimalarial drug primaquine are two of the other chemicals causing that disruption. Yet, the most powerful hemolytic inducer is the fava bean. Those beans eaten by sensitive individuals can be fatal. Even inhaling the pollen may induce favism.

People long ago made the association between the dangers of fava beans and potential death (de Cleene and Lejeune 2002). They elected to continue eating the beans because they liked them, and another realization—eating favas lessens their chance of developing malaria. They did not know it, but the partial anemia is the reason for reduced risk. Favas remain so popular that they are sold fresh in some communities in the southern United States. Farther north, beans are sold canned and frozen.

Linnaeus gave us the scientific name *V. faba*, continuing the use of Gaspar Bauhin's *Pinax* from 1623. The plant producing the edible seed was known in classical Latin as *faba* and *cyamos* in Greek. Fuchs (1542) still called it *cyamo*, but included the synonym *faba*. William Turner's *Herball* of 1551 called it "beane," but John Gerarde said "great garden bean" in 1597. We now associate the word "bean" with the legume *Phaseolus*, a native of the New World. However, that is a recent application. The word "bean" comes from Old Teutonic *béan* and appeared in Anglo-Saxon about A.D. 1000, and two Germanic cognates are *boon* (Dutch) and *Bonhe* (German). The same root probably gave rise to the Gaelic *pònair.* The "bean" of pre-Columbian Europe was *V. fava.*

Other English names for these seeds are horse-bean and broad-bean. The former came into use with a belief (among the rich) that they were fit to eat only by horses. Poor people knew better, and the proverbial expression of "beans and bacon" meant *V. faba* well into the 20th century. Broad-bean came into use to distinguish *Vicia* from *Phaseolus* after the New World plant was brought to Europe. Although Turner called *Phaseolus* "kidney bean" in 1548, people confused broad-beans and kidney beans into the 1770s. Bernard Romans ([1775] 1961), for example, used the expression "beans and pease" in several places to indicate native American *Phaseolus.*

Herbalists in Europe usually mentioned only two *Vicia* (*V. faba* and *V. sativa*). Linnaeus ([1753] 1957) listed 17, but there are 54 native to Europe (*Flora Europaea* 2003). In total, there are 140 species in the genus, mostly in northern temperate regions (Mabberley 1997). There are 31 species in North America (Kartesz 1994), but several of them are naturalized from elsewhere in the world.

The Florida species that has a record of use by the Cherokee was not discovered until the late 1780s. Thomas Walter, a British-American planter living in Charleston, named *V. caroliniana* in his *Flora Caroliniana* published in 1788.

Comparatively few *Vicia* have been used by people, partly because they are so poisonous. Many contain both cyanogenic glycosides and lectinic compounds (*V. cracca* contains alpha-D-galpNAc; *V. graminea* has beta-D-Galp, anti-N [RBC site]), which render them dangerously toxic (Lewis and Elvin-Lewis 1977). Millspaugh (1892), for example, noted that seeds of *V. ervilia* (European bitter vetch) were a virulent poison to humans and horses. No reports are known of poisoning by *V. caroliniana* (Perkins and Payne 1978), but *V. sativa* elsewhere in North America has caused problems (Turner and Szczawinski 1991).

Europeans use some species as fodder for domestic stock (de Candolle [1886] 1959, Meyer et al. 1999). Both *V. ervilia* and V. *sativa* have long histories. De Candolle ([1886] 1959) noted that Greeks still were cultivating *V. ervilia* in his time. Seeds of *V. ervilia* are known from Turkey dating back to the 6th and 7th millennia B.C. Zohary (1982) also cited several Neolithic and Bronze Age sites with *V. sativa* in the Near East and Europe. This is still a minor crop in the Mediterranean basin and the Near East, although the seeds are poisonous. Once soaked in water, seeds are eaten, but only during famine (Zohary 1982).

Vicia sativa was said by Cato (234–149 B.C.) to have been cultivated for both seed and fodder by the Romans (de Candolle [1886] 1959). Zohary (1982) clarified that the seeds were used to propagate the plants, and not for human food.

Among indigenous American tribes, records have been found for use of only seven or eight species. Two of those, *V. faba* and *V. sativa,* are Old World introductions. Maybe they learned uses of those alien species from Europeans, because in England a decoction of vetches in water was given by nurses to "expel" smallpox and measles (Lightfoot 1777 in Vickery 1995). However, they are just as likely to have discovered uses on their own. In the eastern states, the Cherokee, Iroquois, and Rappahannock are known to use *Vicia* as medicine. The Iroquois used both *V. americana* and *V. sativa*; the Rappahannock used only *V. villosa* (Moerman 1998).

The Cherokee used *V. caroliniana* for several purposes, and they considered it one of their most valuable medicinal herbs. Mooney (1885–1886) was the first to record most of these applications, and he has largely been copied by Hamel and Chiltoskey (1975). Decoctions were drunk for dyspepsia and for pains in the back, rubbed on the stomach for cramps, and used to treat rheumatism. The same medicine was also important in treatment of the disease known as *unnagei,* in which one side of the body develops black spots and partial paralysis. To treat that problem, the affected area was scratched, and the liquid applied to the scarified region. The Cherokee also took an infusion as an emetic (Taylor 1940).

Ballgames were an important element in southeastern living, and the Cherokee were addicted to them like many other tribes. *Altsa'sti* was a crucial medicine for players. Their limbs were scratched and the infusion rubbed on to render their muscles tough (Mooney 1885–1886). Hamel and Chiltoskey (1975) found a variation where an infusion of *altsa'sti* was not used alone. A compound mixture was made with Virginia pipe (mistake for Indian pipe, *Monotropa uniflora*?) and sweet apple (*Malus pumila,* introduced) as a drink to improve ballplayers' wind during the game.

Vigna

(Dedicated to Dominico Vigna, Italian scientist of the 17th century)

Vigna luteola (yellow, the flowers)

bejuco [frijol] marullero (sea [bean] vine; *marullero* from *mar,* ocean, and *barullo,* confusion, Venezuela)

caraota de monte (wild *carao*; a *carao* is a tree legume)

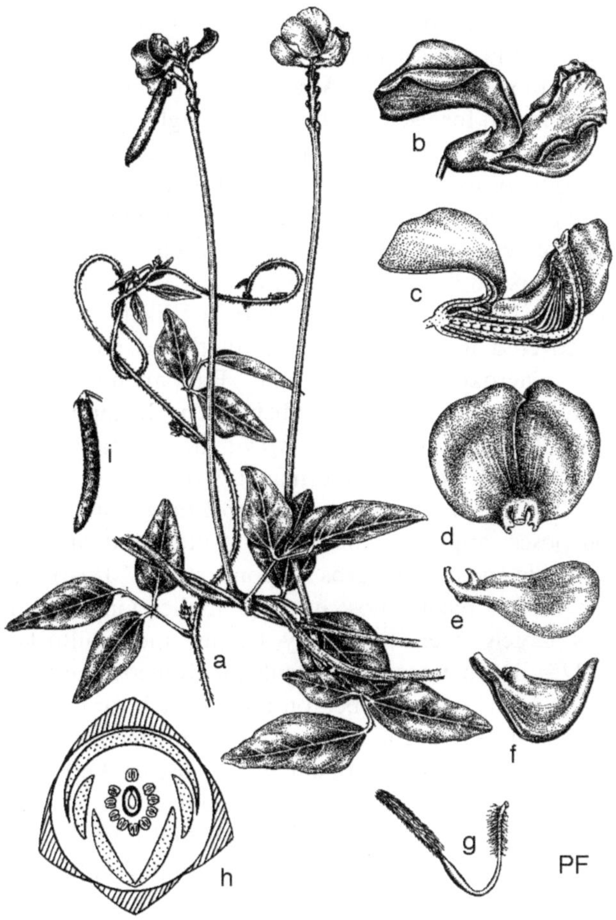

Vigna luteola. a. Flowering and fruiting branch. b. Flower, side view. c. Flower, longitudinally dissected. d. Standard. e. Wing. f. Keel. g. Pistil. h. Floral diagram. i. Fruit. *Drawn by Priscilla Fawcett*. From Correll and Correll 1982.

caupi de monte (wild cow pea, Netherland Antilles); hairy-pod cow-pea [hairypod cowpea]; *waakimbala* (*waaka'*, cow, *im*, its, *bala*, bean, Chickasaw; probably belongs here)
frijol cimarrón [*de arena, de cabra*] (wild [sand, goat] bean, Cuba); *frijol de monte* (wild bean, Mexico); *frijol de playa* (beach bean, Honduras); *frijol silvestre* (wild bean, Puerto Rico)
goat wiss (wiss = vine, Virgin Islands)
pois jaune [*pigeon, zombi*] (yellow [pigeon, zombi] bean, Guadeloupe, Martinique)
porotillo (based on *poroto*, bean, meaning little bean, Spanish)
sea bean (Belize)
wild pea (Virgin Islands)
yellow vigna (a book name, Bahamas)

Nicholas von Jacquin called this twining legume *Dolichos luteolus* in 1771, naming it from plants cultivated in Vienna. It was not until 1859 when George Bentham moved the species to *Vigna*, a genus created in 1824 by Italian botanist C. Gaëtano Savi (1769–1844). This species is thought to be native to the New World, but it has been overshadowed by its cultivated relative *V. unguiculata*, the cow-pea, better known as black-eyed pea. That plant was brought into cultivation in Ethiopia and subsequently spread around the world (Simpson and Conner-Ogorzaly 1995). The genus has about 150 species spread through the tropics, especially the Old World (Mabberley 1997).

John Small (1933) knew these twiners from Florida to Texas, and north to North Carolina, and from the Caribbean, Mexico, Central, and South America. For a long time, people thought the proper name was *V. repens*, and older literature uses that epithet.

In spite of the wide distribution, uses have been found only in two disjunct places. A decoction of the plant is used to treat coughs and colds in the Virgin Islands (Morton 1981). Balick et al. (2000) also found it being used in Belize, but gave no details.

Viola

(The classical Latin name for the violet; akin to Akkadian *asu*, and Hebrew *js*, to grow, Akkadian, *isum*, plant)

amor-perfeito (perfect-love, Portuguese)
bròg na cubhaig [*spòg na cubhaig*] (*bròg* or *spòg*, talon, *na*, of the, *cubhair*, cooper, Gaelic)
fail chuach [*dail chuach*] (*dail*, field, *fail*, wreath, *chuach*, plait or drinking cup, Gaelic); *sail chuach* (*sail*, willow, *chuach*, drinking cup, Gaelic)
goirmean-searradh (*goirmean*, indigo, *searradh*, cutting, Gaelic)
pansy ("pansy" was derived ca. 1500 from French *pensée*, thought; originally applied to European *V. tricolor*; perhaps the first suggestion of this use came in 1536 when Ruellius published *De natura stirpium* [On the Nature of Things], where he wrote "*Violae inodorae genus esse putaverim quam vulgus gallicus penseam vocat*" [This is the odorless violet genus that is said to be commonly called *thought* in France]); *pensée* (French)
Stiefmütterchen (little stepmother, German)
violet ("violet" diminutive of Latin *viola*, derived in English from French *violette*; in English ca. 1330, and applied to the rayless kinds according to Mabberley 1997); *fiol* (from Latin *viola*, Norwegian); *Veilchen* (*veil*, violet, *-chen*, little, German); *viola* (Italian, Portuguese); *violeta* (Spanish, Portuguese); *violette* (French); *viooltje*

Viola. *Viola conspersa* (left). *Viola sororia* (right). Both from Russell 1965.

[*violetje(s), violetten*] (Dutch); *vion* [*ion, io*] (Greek, cf. Russell 1965)
xtha-çka tsu-hu (*xtha*, flower, *çka*, white, *tu-hu*, blue?, Osage)

Viola bicolor (two-colored)
cupid's-delight (the same idea as the British name "love-in-idleness")
field daisy ("daisy" is usually applied to *Bellis perennis,* Asteraceae)
field pansy ("pansy" was derived ca. 1500 from French *pensée,* thought)
heart's-ease [heartsease] (first recorded in Chaucer's *Clerk's Tales* in the 1400s, where he wrote, "And wisly bringe hem alle in hartes eese"; the origin is uncertain, but the sense is tranquillity or peace of mind, England)
Johnny-jump-up ("Johnny" [Johnnie] is a contemptuous nickname given to Englishmen in the Mediterranean, and to Confederate soldiers during the Civil War)
kiss-me-at-the-garden-gate (England)
love-in-idleness (England)
three-faces-under-a-hood (from 1548, England)
wild pansy

Viola conspersa (sprinkled, from the dotted lower leaf surface)
American dog-violet (the "dog-violet" of Europe is *V. canina*; so named because it lacks scent; see Scott 1995)

Viola sororia (sisterly, resembling other species)
bayou violet ("bayou" is a variant of the Choctaw word *bayok*, meaning marshy branches of rivers and lakes; in use by 1763)
chicken-fights (Maryland); fighting cocks; *akankittibi'chi'* (*akanka'*, chicken, *ittibichi'*, to make fight, Chickasaw); *dinda'skwate'ski* (they pull each other's heads off, Cherokee)
Confederate violet (an allusion to the name Johnny-jump-up)
Johnny-jump-up[-and-kiss-me]
long-stemmed purple violet
rooster-hoods [roosters] (North Carolina)
sister violet (a "scholarly" translation without consulting people who use common names)
[blue prairie, common blue, downy blue, hooded blue, meadow] violet

What could the violet and the peacock moth possibly have in common? Both were dedicated to Io, the beautiful Greek priestess. The violet in Latin is *Viola,* which is related to the Greek *Vion,* a variant of *Io.* The moth is *Automeris io,* named by Philipp Conrad Fabricius (1714–1774) to honor that goddess—or perhaps to warn of her dangers as the larvae are covered with stinging hairs.

According to Greek legend, Zeus was enamored of Io and seduced her. That made his wife Hera furious, and to protect Io he transformed her into a white heifer. Io wept because of the coarse grass that she was forced to eat. One version of the story is that Zeus changed her tears into a sweet-smelling, dainty violet (*V. odorata*) as a special food (Lehner and Lehner 1960, Ward 1999).

Violets also figure prominently in the histories of Romans, Middle Eastern Muslims, and Napoleon Bonaparte. Romans drank violet-flavored wine and spent so much time cultivating the herb that Horace (65–8 B.C.) criticized them for spending more time on *Viola* than their olives (*Olea europaea*). Muslims extol the violet by saying, "The excellence of the violet is as the excellence of Islam above all other religions." Napoleon was nicknamed the *Caporal Violette*, and he used either that or *Père la Violette* as a *nom de plume*. He died wearing a locket of violets taken from his wife Josephine's grave (Bown 1995, Ward 1999).

There are 91 species of violet in Europe (Mabberley 1997), and the genus has been cultivated since 400 B.C. in Attica in eastern Greece (Mabberley 1997). *Viola odorata* has not only been used to flavor wine, but also in flavoring other materials, food (crystallized), and in scent making. Its fragrance still is used in creams, salves, soaps, oils, and shampoos (Lewis and Elvin-Lewis 1977). The main aromatic in *V. odorata* is ionone, and its synthesis in 1893 caused a dramatic decline in violet cultivation.

Newcomers to the New World were not surprised to find violets here. Strachey ([1612] 1953) included violets among the "many herbes in the Spring-tyme are Comonly dispersed throughout the woods good for broathes and sallets." Kartesz (1993) listed 119 species, including numerous hybrids. Most of these were discovered comparatively recently, and Linnaeus ([1753] 1957) knew only 19 species in the world—he knew none of Florida's species. Indeed, the Florida violets were described between 1809 and 1823.

Steyermark (1963) found the game called "Hens and Roosters" being played in the Ozarks. There they used *V. pedata*, and the "roosters" were the bicolored flowers, while the "hens" were uniformly colored. One person holds a rooster flower and another a hen. These are hooked together and the first person to pull off the other's flower is the winner.

Coffey (1993) was under the impression that this was an invention of southeastern settlers, probably because of the names "fight rooster," "fighting cocks," or "chicken fight." He found that people in the South used *V. sororia* instead of *V. pedata* (Coffey 1993). Moreover, ask anyone in the South who remembers the game, and they will tell you that it is a purely European-American game.

However, people from the Cherokee to the Omaha may have played the game before Europeans arrived in the New World (Mooney 1885–1886, Gilmore 1919). The Cherokee name recorded by *dinda'skwate'ski* (they pull each other's heads off) clearly describes the game (Mooney 1885–1886). Gilmore (1919) voices no opinion, but comments that the game was popular among Omaha boys. It seems likely that the indigenous tribes invented the game and taught it to settlers. No records were found of Old World people playing it before the 1500s.

All three of Florida's violets were used by indigenous people, and Moerman (1998) lists 14 other species used by different tribes. Although no records have been found of indigenous people eating the violet, that seems likely. Stephen Elliott (1771–1830), in his *A Sketch of the Botany of South Carolina and Georgia*, published between 1821 and 1824, treated all the palmately dissected leaf violets as "*V. esculenta*," and said, "This Violet is very mucilaginous and much used by negroes in their soup." Later, Porcher (1863) called it "wild okra" and said that the blue violets were used in making soup during wartime. Only the leaves are used in the soups because the roots are emetic. Fernald et al. (1958) wrote, "probably any of the 'blue' violets might be used" in soups. Kindscher (1987) added that some are more palatable than others. Zennie and Ogzewell (1977) found that the basal leaves of *V. papilonacea* have more vitamin A and C than any of 17 wild edible plants studied in Ohio and Kentucky. That report spawned the assumption that all violets have similar quantities of vitamins.

Most of the uses by indigenous people were in medicines. The Cherokee made a poultice of *V. bicolor* or *V. sororia* leaves for headache, took an infusion for dysentery, colds, and coughs, applied a poultice to boils, and considered it a spring tonic (for the blood) (Hamel and Chiltoskey 1975). An infusion of the roots was used as an insecticide to soak corn kernels before planting. The Seminoles in 1919 made a tea of some species to treat kidney trouble (von Reis and Lipp 1982). The Ojibwa made tea of *V. canadensis* for bladder pains and to treat heart pains (Vogel 1970). They used *V. conspersa* for heart trouble (King 1984). The Potawatomi also used *V. pubescens* to treat heart pains (Smith 1933, Vogel 1970). Those uses by northern tribes for heart remedies are reminiscent of the old European name "heart's ease," and Vickery (1995) found that application still practiced in England.

By far the most common utilization in Europe was for skin problems, and the Cherokee also used the plants for that. Porcher (1863) noted that violets were good for ulcers and other skin problems, and gave a formula for creating the remedy. Foster and Duke (1990) said that experiments with rats confirm efficacy for skin eruptions. Bown (1995) added that violets were being checked in England for use against skin cancer and childhood eczema.

The compounds that are used to treat skin problems are taken from the roots. Those known are saponins, and they may be toxic in large doses (Foster and Duke 1990). Moreover, rhizomes and seeds cause severe gastroenteritis, nervousness, and respiratory and circulatory depression in large doses (Lewis and Elvin-Lewis 1977).

Vitis: Grapes

(To bend, plat, or weave, Latin)

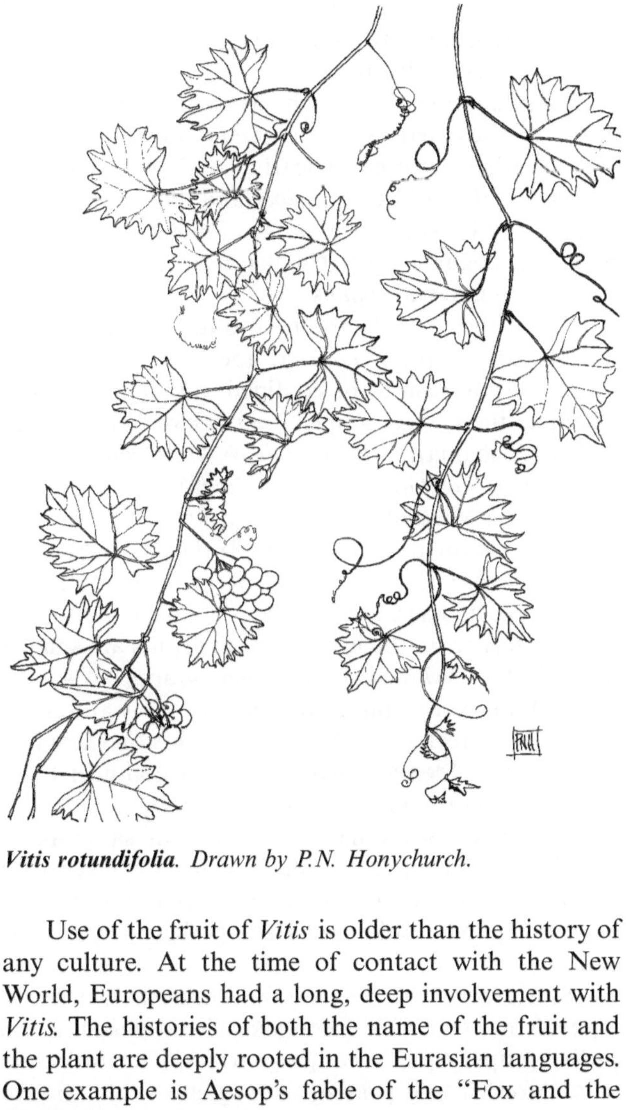

Vitis rotundifolia. *Drawn by P.N. Honychurch.*

Use of the fruit of *Vitis* is older than the history of any culture. At the time of contact with the New World, Europeans had a long, deep involvement with *Vitis.* The histories of both the name of the fruit and the plant are deeply rooted in the Eurasian languages. One example is Aesop's fable of the "Fox and the Grapes."

To the Greeks, the unripe fruit is *omphelos* or *omphax*, a dried fruit (raisin) is *astaphis*, and the plant *tamnus* (Brown 1954). In Hebrew, the plant is *gefen*, a vineyard is *kerem*, and *anavim* is grape (Zohary 1982). In Egyptian Arabic, the plant and grapes are *'inab*, and grapes are *zibib* (Manniche 1989).

Our English word "grape" came into the language about A.D. 1290. The *South English Legendary* or *Lives of Saints* of that year included the sentence, "A luytel foul . . . brochte a gret bouch Fol of grapus swyte rede" (a little fowl . . . brought a great bunch full of sweet red grapes). Before 1290, the word was *krappen* (Teutonic), *krappa* (Old High German), and *grap, grappe* (Old French), all meaning a hook. Presumably, the hook was the one used to harvest fruit from the vine. These words are related to *grappare* (Italian), to seize, and *grappa*, a hook, in Italian, Spanish, and Portuguese, but they use a different word for the fruit.

Uva (Italian, Spanish) and *uba* (Portuguese) are based on Latin *uva.* However, speakers of Latin distinguished between *uva* and *acinus, atrusca, bannanica, bumastus, irtiola, sircula, spionia, sticula*, and *talpona.* Those varietal names have not come into English, but *acinus* became *acino* (grape) in Italian. "Raisin" in French and English is based on Latin *racinus*, a bunch of grapes, although its meaning in English as dried fruit has deviated to the point that they hardly seem related. German has retained the idea of the bunch of grapes with *Traube*, which means either raceme or grape.

"Vine" in all English dialects except American specifically indicates *Vitis vinifera*—**the** vine (Austin 1994). Americans have expanded the word to mean all creeping, climbing, or twining plants. The Latin *vitis* was retained more nearly in original form in *vigne* (French) and *vid* (Spanish). A synonym of *vid* is *parra*, from a different source. Gaelic speakers had other names, including *fionan* and *muin* (the 11th letter of their alphabet).

Europeans found many grapes in the New World, all different from those they knew at home. Between 1570 and 1577 Francisco Hernández recorded that the Náhuatl name of grapes in Mexico was *xocomecatl* (sour rope) or *ceual-chilchiltic* (cool and red). The Zapotec say *bicholi* or *yaga-bicholi* (*yaga*, tree, *bicholi*, grape).

The most widespread species in southeastern North America is *V. cinerea*, known as graybark or pigeon grape. That plant also is known as the *hastanhanka* (Dakota), *chan wíkyape* (*chan*, tree, *wíyape*, twiner, Teton Dakota), *hazi-hi* (grapevine, Omaha-Ponca), *hapsintsh* (Winnebago), and *kisúts* (Pawnee). Probably, this is what the Cherokee called *uhnasuga*, and their relatives the *ojinquisere* (Onondaga). Delaware said *wisachgim*, and farther west, the Potawatomi called them *siwnwen.* To the Shawnee it is *teatepawtaquey* (Edgar 1891). Although their word is generic, the Yuchi called them *cá*.

To the Muskogee of Oklahoma *V. cinerea* is *pvłko* [*pv'rko*]. The Creeks of Florida say *wotko empvłko* (*wotko*, raccoon, *empvłko*, its grape), which is the same as *shaawe embałbi* (Mikasuki). *Sawimbałka* (Alabama), *sawá imbałká* (Koasati), and *shawimpakki* (Chickasaw) have the same meaning. Maybe the Tunica name *yi'sihe'sani* (*yi'si*, raccoon, *he'sani*, vine) belongs here.

The Miccosukee name for *V. shuttleworthii* is almost the same as for *V. cinerea*, but shortened. Spellings include *balbe, bałbi, bokbe, balbî*, and *cokasî:* (Sturtevant 1955, Snow and Stans 2001). The name *bałbe* is pronounced "puł-be," and is cognate with the Creek word *pałko* [*pvrko, párko*], which is pronounced "puł-go." This grape is also called *pałko łako* [*pvrko rako*] (big grape, Creek). Alabama speakers use the cognate *bałka*, and Koasati *bałká*. English speakers say summer grape or Calusa [Caloosa] grape.

The other prominent species in the southeastern United States is *V. rotundifolia*, called the muscadine, southern fox grape, bullace, and scuppernong grape. "Muscadine" may have come from Portuguese *muscat*, which is cognate to Italian *muscato*. Both words came from Latin *muscatus*, having the flavor of musk. The word was in English as "muskadyne" by 1541. Muskatel is a strong sweet wine made from the muscate grape, and is *moscadello* or *moscatello* in Italian. "Bullace" is from Old French *beloce*, and denotes a plum, *Prunus* ×*domestica* ssp. *insititia*. *Vitis rotundifolia* is the wild type from which people selected the cultivated scuppernong grape. That cultivar is said to be named for a river in North Carolina. Core (1967) says both fruit and locality are based on the indigenous word *askuponnong* (probably Algonquian). For those who have not tasted this giant-fruited cultivar, the fruits are more like plums than grapes, very different from the tart wild forms.

The Cherokee name for *V. rotundifolia* is *telu'lati* [*qualusi*] or *oonee tayluhn'dee*. Siebert (1975) thought the Powhatan name *wapapemins*, recorded by Strachey in [1612] 1953 as *wapapamindgas*, belonged here. The Powhatan word is composed of */*wa·p-/*, white, */*a·pye·-/*, vine, and */*-eminsy-/*, drupe-bearing bush (Siebert 1975). Cognate is Fox *kiwapyehi*, creeping vine. Perhaps the Powhatan words *marakimmins* (grapes) and *metucsmarakimins* (bunch of grapes) written by Strachey are also cognates. Siouan-speaking Biloxi said *maktcuhi'* for the fruit, and *maktcuhi' pan'him* (*maktcuhi'*, grape, *pan'him*, vine), Ofo said *o*<u>*n*</u>*tho'mofthu'* for the fruit and the plant was *anthu'hi fte'pi* (*anthu'hi*, creeper), while the Tunica called the fruit *so'su*, and the Atakapa *hila'ñ wol te'i* (*hila'ñ*, buzzard's, *wol*, eye, *te'i*, vine).

According to Snow and Stans (2001), the Mikasuki say *chokooche*. That is the same word that Sturtevant (1955) recorded as *cóko:cî* (little grape, Mikasuki), and others spelled *chokoche, cokooci, chuko aske, tcokotci*, and *tsuk-ko-chee*. The Creek cognates are *culoswuce* [*colo:skocî, chukochet, chalushwa, so-losh-ka, chu-los-sho-a-kee*] (little muscadine), and Muskogee *culoswv* [*colóswv*]. In Alabama, the modern name is *chokkapihchi* (*chokko*, grape, *apihchi*, body), and in Koasati *cokkó*. The Choctaw called the vines simply *pa*<u>*n*</u>*ki, p*<u>*a*</u>*ki vpi* (grape vine or stem), or *suko*. This also may be what they call *pa*<u>*n*</u>*kachu*<u>*n*</u>*si*.

At first, it does not seem that the Muskogean words for grapes are closely related. However, closer examination suggests that they are all cognates. The sound of the "p" in Creek is unaspirated, unlike the aspirated letter in English. Because of the unaspirated nature, the Creek "p" sounds to English speakers as though it were halfway between "p" and "b" (Martin and Mauldin 2000). Thus, the Muskogean words with either of these letters are closer in sound than they appear when written. Clearly, Alabama *bałka* is cognate with Mikasuki *bałbe*, and both words are cognate with Muskogee *pvłko*. Similarly, Alabama *bałka* is cognate with Chickasaw *pakki*, which in turn is cognate with *panhki*.

All of these words are used to denote the Florida or graybark grape, *V. cinerea*; however, their application does not seem to be confined to that species. For example, the Calusa grape, *V. shuttleworthii*, is called *bałka* (Alabama), *bałbi* [*bathbi, bokbe, balbi*] (Mikasuki), or *pvłko* [*páłko, pałko*] (Creek).

It is the Muscadine grape, *V. rotundifolia*, that has the most distinctive names in Muskogean. Probably that is because this is the most characteristic species in the southeast. The Calusa grape is unique in that it has no pubescence on the leaves, and the fruits are smaller and thicker skinned than other grapes. Also, as pointed out by Seminole Alice Snow (Snow and Stans 2001), the "green leaves have little points on them and ragged edges [dentate margins] and are almost heart shaped [subcordate]."

The Muskogean name for the Calusa grape is *culosw* [*coló:swv*] (Creek), but it is usually given with the diminutive ending as *culoswuce* (*culosw*, grape, *oce*, small). The Mikasuki *chokooche* [*cokasî:*] means the same.

Sturtevant (1955) said that Mikasuki word was equivalent to Creek *parkofakâ* [*pałkofaka*] (from *pvłko*, grape, *fvkv*, vine, grape vine). The Mikasuki stem word *cóko* [*choko*] is cognate with *chokko* (Alabama), and Choctaw [and Chickasaw?] *sokko*. Those words are also the source of the Mobilian Jargon name *sok(k)o*. Drechsel (1997:66) suggests that the names are probably derived from the word thick, *sok(k)o*, in reference to the skin of the fruit. The Choctaw apparently also use the word *paki* for fruits in this species, and that word is the same as *panhki*, their word for *V. aestivalis*.

Everyone who knows grapes, from the extinct Apalachee to Timucua and Calusa to the present, is aware of the three basic uses for them—young leaves and stems are cooked as greens; old stems give potable, safe drinking water; and fruits are edible (Gilmore 1919, Yanovsky 1936, Morton 1968, Austin 1980, Hahn 1986, Gilliland 1975, Hogan 1978). We do not know if native people cooked leaves and stems, but they probably did. American cultures did make full use of the water in the stems. Grapevines have notable vasculature such that no water escapes as long as cut stems are kept upright. When a stem segment about a foot long is inverted, up to a liter of water may be released.

Fruits were an important food source for native people of the South and early settlers (Medsger 1966). Swanton (1946) listed grapes as food among people from North Carolina to Texas, including the Cherokee, Creek, Chickasaw, and Caddo. Moerman (1998) added records of grapes being used as food among the Apache, Cahuilla, Cheyenne, Comanche, Crow, Dakota, Iroquois, Karok, Keres, Kiowa, Lakota, Mahuna, Meskwaki, Menomini, Ojibwa, Omaha, Pawnee, Ponca, Seminoles, Sioux, and Winnebago. It would be astounding if there was a group within the range of the genus who did not eat the fruits. Not only were fruits eaten fresh, but they were dried and stored for winter (Swanton 1946). George Catlin ([1844] 1973) was so taken with the process that he devoted a painting to grape gathering by the Caddo, a people who lived in Louisiana, Arkansas, and Texas.

The Lincecum manuscript also contained what appears to be a unique storage method used by the Choctaw (Campbell 1959). According to Lincecum's notes, the Choctaw gathered grapes near the end of the summer and put some of them in cane baskets that were then submerged in water. He says, "Grapes preserved in this way kept well through most of the winter." Campbell (1959) wrote at the time that this was the only record of storage in this way in the southeast. No other records of this practice have been found since his paper.

Europeans used the fruits for juice, syrup, jelly, jams, and wines, and they brought their Old World uses into the New. In their homeland, they also used fresh leaves as an astringent styptic. They made a medicine to combat edema, hemorrhoids, and easily broken capillary veins, and used the leaves in teas and in cooking—as stuffed grape leaf rolls. Juice from the clipped tendrils was used externally to treat rashes, and internally for intestinal bleeding, dysentery, and rheumatism. Unripe fruit juice (Syrup of Verjus) was used against obesity, and as an acidulant and a diuretic. Juice of stems was used to treat eyes. Raisins still are used as a laxative, diuretic, nutrient for hair, stomachic, for hoarseness, and in baked goods.

Southeastern people had no fermented drinks before the arrival of the Europeans, so they did not include that Old World use. Swanton (1928a) found Creeks and Sturtevant (1955) found Seminoles using grapevines for "Snake Disease" (itching spots like poison ivy, dreams of snakes), snakebites and in birth, death, and busk ceremonies (see *Pterocaulon*). Swanton (1928a) also found the Creeks using a decoction of tendrils and soft succulent ends of vines with ginseng for tonsillitis. The Choctaws used *V. aestivalis* to treat coughs (Campbell 1951). Bennett (1997) found Seminoles continuing some of those practices. Seminole Alice Snow uses *vtelokuce* (mixed medicine treatment), which includes *balbe*, with several other plants (Snow and Stans 2001). *Vtelokuce* is a treatment for a person who is weak and wants to sleep all the time. Snow instructs them to bathe in the decoction and drink some to be cured. According to Snow, *balbe* is used for many medicines, "whatever the doctor requires" including "death medicine" (Snow and Stans 2001).

Howard (1984) found the Oklahoma Seminoles using grape leaves and stems to make tea to treat diabetes. That tea is mixed with *Salix humilis* to make it more effective. Tendrils are chewed by children to prevent cavities, and used in sympathetic magic to "hold onto" a wife or husband. A cloth soaked in the tea is used to promote hair growth.

There is an ample literature on heart protection, cataract prevention, cancer protection, and slowing of aging through free radical scavenging by compounds in grapes and wines, particularly cyanidins (Maffei Facino et al. 1994, 1996, Costantini et al. 1999, Facino et al. 1999, Castillo et al. 2000, Kiesewetter et al. 2000, Wang et al. 2000, Carini et al. 2001, Ray et al. 2001, Sato et al. 2001, Singletary and Meline 2001, Bagchi et al. 2002, Cui et al. 2002, Fitzpatrick et al. 2002, Pataki et al. 2002, Preuss et al. 2002, Simonetti et al. 2002, Yamakoshi et al. 2002). Resveratrol and other stilbines have been shown to inhibit cancer, provide liver and kidney protection, act as an anti-inflammatory, and function as a natural profungicide (Oshima et al. 1995, Li et al. 1998, Bavaresco et al. 1999, Huang and Lin 1999, Paul et al. 1999, Huang et al. 2001, Soleas et al. 2001, Waffo-Teguo et al. 2001, Bertelli et al. 2002, Burns et al. 2002, Decendit et al. 2002, Gao et al. 2002, Juan et al. 2002, Schouten et al. 2002, Yoon et al. 2002).

As with everything, there are potential problems with using *Vitis* for food or medicine. Some people have developed allergies, even some that are life-threatening (Vaswani et al. 1999, Rodriguez et al. 2001, Senna et al. 2001). A few individuals have allergenic reactions resembling those from poison ivy, and even pollen can cause them (Feo Brito et al. 1999).

Still, grapes continue teaching those who are willing to learn. Each fall they showed me when hunting season opened on doves in southern Florida. During that week, the mourning doves (*Zenaida macroura*) flocked to the ecological preserve on the Boca Raton campus to feed on grapes. They were joined by mockingbirds (*Mimus polyglottos*), cardinals (*Cardinalis cardinalis*), opossums (*Didelphis virginiana*), raccoons (*Procyon lotor*), and gray foxes (*Urocyon cinereoargenteus*). True to its name, *Vitis* has woven its way into our food, drink, medicine, and arts. And that is none of Aesop's "sour grapes."

Vittaria

(Named by James E. Smith, 1759–1828, with Latin *vittat*, a stripe or band; akin to *vieere*, to bend, twist together; Akkadian *ebetu, ebiu*, to be tied, gird)

Vittaria lineata (marked with lines, meaning the marginal sori)
grass-fern (Bahamas)
ishta taapente (person fern, Mikasuki); *este hvlwat cokhesse* (*este*, person, *hvlwv*, high, *-at*, the one that is, *cokhesse*, beard, Creek); *yaat-chayhe echooshke* [*yá:tcáyhicô:skî, yaatcayhicooski, yaatchayhen echoonshke*] (tall person's whiskers, Mikasuki); *isti shopka chukisget* (*este*, person, *shopka*, giant, *chukisget*, beard, Creek); *istimá:ha cokhíssi* (*este*, person, *mahe*, tall, *cokhesse*, beard, Creek)
shoe-string fern (Florida, Bahamas)

Sturtevant (1955) found the Seminoles using shoestring fern to treat chronic sickness and insanity, and at birth ceremonies. Bennett (1997) learned that the plant is boiled and drunk or administered in a steambath to treat depression. Resurrection fern is usually added to the mixture, and it is also used to treat senility. Snow and Stans (2001) list the plant but give no remedy using it, so perhaps Snow, like Bennett's (1997) informants, uses it in a medicine they do not wish to divulge. The Seminoles also use the leaves to avert maladies caused by lightning striking near a person (Bennett 1997). The Guaymi of Panama use shoestring fern to treat headaches (Gupta 1995b).

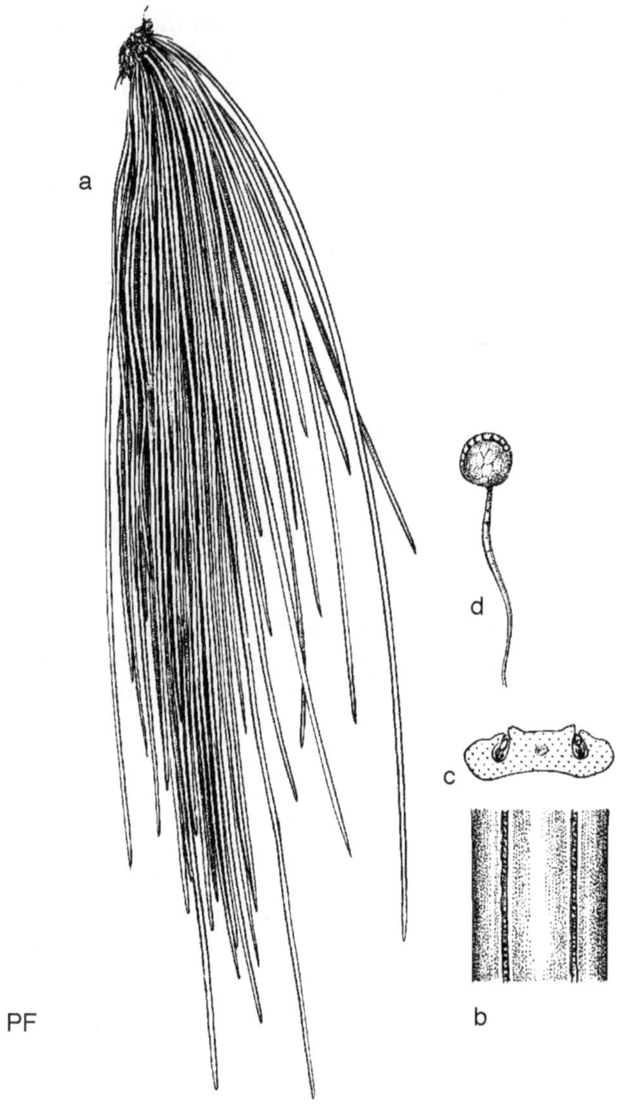

Vittaria lineata. a. Habit. b. Section through lower surface of fertile front showing linear sori. c. Cross section of fertile frond. d. Sporangium. *Drawn by Priscilla Fawcett*. From Correll and Correll 1982.

Voyria

(From a name used in French Guiana)

Voyria parasitica (parasitic, a misnomer; it is actually saprophitic) (=*Leiphiamos parasitica, Leiphiamos aphylla*)
geniciana de la tierra (wild gentian, Cuba)
[parasitic] ghost-plant (Florida)

Leiphiamos (from Greek *leiphaimeo, leiphaimos*, lacking blood, pallid, alluding to the absence of chlorophyll) was created by D.F.L. von Schelchtendal (1794–1866) and L.K.A. von Chamisso (1781–1838), and that generic name was used until recently. Paul

Maas determined recently that it should be a synonym of *Voyria* Aublet (e.g., Maas and Westra 1998). With either definition, it is a genus of 20 species growing in tropical America and West Africa (Mabberley 1997). It is one of the few mycotrophic members of the Gentianaceae.

The genus is used as a tonic, febrifuge, vermifuge, and against dyspepsia (Roig 1945).

X

Xanthorhiza

(From Greek *xanthos*, yellow, and *rhiza*, root)

Xanthorhiza simplissima. From Britton and Brown 1897.

Xanthorhiza simplissima (most-simple, unbranched)
shrub yellow-root [yellowroot]
wi̱·ti· nuse̱'? (*wi̱·ti*, root, *nuse̱'?*, yellow, Catawba)

Xanthorhiza is a deciduous shrub with pinnate leaves, brownish-purple fruits, and bitter yellow roots. The genus and species were described by the Pennsylvania botanist Humphry Marshall (1722–1801) in his *Arbustum Americanum* published in 1785. The species is endangered in Florida, having been found in only four counties in the Panhandle (Coile 2000). The shrubs are also rare in southeastern Texas, although they range from there to New York (Correll and Johnston 1970). This is the only species in a genus that is endemic to eastern North America (Mabberley 1997).

The Cherokee used the crushed plant to make a yellow dye (Hamel and Chiltoskey 1975).

The Catawba used the roots for colds, ulcerated stomach, and jaundice (Speck 1934, Moerman 1998). The Cherokee made a decoction of the root for cramps, to use as a blood tonic, to treat cancer, piles, sore eyes, sore mouth, and sore throat (Hamel and Chiltoskey 1975).

Ximenia: Hog-Plum

(Genus dedicated to the Spanish monk Francisco Ximénez, a native of Luna in the Kingdom of Aragon)

Hog-plum (*Ximenia americana*) can be identified without being seen. The flowers have a lilac-like fragrance. Even if you do not smell, you "hear" them because of all the visiting bees, flies, and wasps. If the shrubs are not in flower, people usually discover them by feel—when they run into the thorns.

Hog-plums are pantropical trees or shrubs that belong to a family of root parasites, the Olacaceae (DeFilipps 1969a,b). In 1703, Charles Plumier dedicated the genus to a Spanish monk. Ximénez came to New Spain in 1605, and joined the *Convento de Santo Domingo de México.* In 1615, he published the book *Quatro libros de la natureleza y virtudes de las plantas, animales que están receuidos en el uso de medicina de la Nueva España* (Four Books on the Nature and Virtues of Plants and Animals that are Known to be Used in the Medicine of New Spain). Ximénez said his book was a translation of Francisco Hernández's Latin publication, but it contains much original information (Standley 1920–1926). When Linnaeus studied the plants at *Hortus Cliffortianus* in 1738, he decided that Plumier's dedication was appropriate. He continued using *Ximenia* in *Species Plantarum* in 1753.

Other people who know these root parasites also note their armament. In Brazil, people call them *ambuy* (from *abo 'á*, with sharp hairs, Tupí). Elsewhere in northern South America, they are *espino [de] brujo* (wizard's spine, Colombia), *tigrito [ligrito]* (little tiger, Venezuela), or spiny mountain. To some, they are just a *croc* (hook, fang, Dominican Republic, Haiti), or they resemble a *macaby [macabi]* (a spiny fish, Taino, Haiti).

In many places, the shrubs are *pepenance [tepenance]* (Guatemala, El Salvador, Costa Rica, Brazil). That name is probably from *pepe* (from *pepenar*, to sieze), and *nance* (from *mantli*, wood, Náhuatl, now the fruit of *Malphigia* or *Byrsonima*). *Pepenance* is related to *nanchicacao* (chocolate wood, Mexico, Argentina). *Nanche[i]* is a variant of *nance*, plus *cacao* (*kaj*, bitter, *kab*, juice, Maya) for chocolate. Both names are probably an allusion to using *Ximenia* as "nurse trees" in chocolate plantations.

More names allude to the fruits than other features. The most frequent comparison is to the plum (*Prunus* ×*domestica*) or apricot (*P. armeniaca*). *Ximenia* is the American plum, blue sour plum, hog-plum (Florida, Belize), monkey plum, mountain plum (Puerto Rico), seaside plum (Bahamas), Spanish plum (Bahamas), tallow-plum (Florida), wild plum (Belize), or yellow plum. In the French Antilles they are *prune bord de mer* (coastal prune), *prune épice* or *prune epineau* (spiny prune). Brazilians think they are more

Ximenia americana. *Drawn by P.N. Honychurch.*

like the apricot and say *ameixa* or *ameixero* [*ameizero*] (from *a* = the + Latin *damascena*, Portuguese). More often they note its differences compared to *ameixeira do Brasil* (Brazilian apricot) or *ameixa de espinho* [*ameixeira de espinos, espinha de meicha, espinheiro da ameixa*] (spiny apricot). People in Argentina agree with *albaricoque* (apricot). Variations in Argentina include *albarillo de campo* [*alvarillo del campo, albarillo del campo*] (wild apricot), *albaricoquillo* (little apricot), *albarillo* (little apricot), *albericoquillo* (little apricot), and *albericoquillo del campo* (little wild apricot).

There are several other fruit comparisons. Some think *Ximenia* fruits are like cherries (*P. cerasus*). These people say *cerise de mer* (ocean cherry, Haiti), *ciruela cimarrón* (wild cherry, Cuba), *ciruelillo* (little cherry, Mexico, Cuba, Dominican Republic, Puerto Rico), and *ciruelo* (cherry, Colombia). To many, they are more like lemons or limes because of their sour taste. Names including this view are *citron de mer* (seaside lemon, Haiti), *limoncillo* [*de playa*] ([beach] little lemon, Dominican Republic, Puerto Rico, Colombia), muri lime (Guyana), *oranger de montagne* [*oranger des falaises*] (wild [false] orange, Guadeloupe), and wild lime (USA, Belize, Guyana).

Many compare the fruits with apples. Those names include *manzana* [*manzana del diable, manzana del diablo, manzana guayaba, manzana guayabo*] ([devil's, guava] apple, Venezuela), *manzanilla* (little apple, Puerto Rico, Guatemala, El Salvador, Honduras, Panama), and *pommiere de Cithère* (fiddle apple). Cubans mix Spanish and indigenous words with *jía manzanilla* (little apple like *achiote* or *Bixa*, probably Taino). Another comparison with an Old World fruit resembling an apple is *membrillo de monte* (wild quince, *Cydonia oblonga*, Guatemala).

Several comparisons are unique, as in *caimito de monte* (wild *caimito*, Taino for *Chrysophyllum*, Colombia), wild olive (USA, West Indies), and iguana-berry (Jamaica). *Hevmassoli* [*heymascoli*] (Guyana) and *poonish* (Mexico) have not been translated.

The Seminoles did not know the plants until they migrated south down the Florida peninsula. By that time, they had learned the virtues of the introduced Old World peach (*P. persica*). So, the Seminoles began calling the new wild fruit they found *i:cintohã:nî* (deer's peach, Mikasuki), and the cognate with the same meaning, *co impaká:na* (*eco*, deer, *em*, its, *pvkanv*, peach, Creek). From at least the 1700s, the Seminoles ate the fruits (Bartram 1943) and used the plant against "Cow Creek Sickness" (Sturtevant 1955).

Fruits are edible raw or cooked (Ledin 1951). The kernel can be roasted and eaten but should not be consumed in quantity as it is purgative (Standley 1920–1926, Roig 1945, Morton 1968). Some consider the seeds "rather dangerous," apparently because of their laxative nature (Duke 1972), or because they contain cyanide (Liogier 1974). Fruits are used against gastric upsets (Roig 1945) and made into beer in South America (Duke 1972). European-Panamanians make "sloe gin" from the fruits (Duke 1972).

Those laxative properties gave rise to the name purge-nut (Florida). Related names are tallow-nut, tallow-wood, and tallow-plum (Florida to Panama), because of the waxy texture of the flesh on the drupe. The laxative properties are included in the names *cagalera* [*cagalero, cagadera*] (diarrhea, Honduras, Belize) and *fransman moppe* [*fransmanmope*] (Frenchman's complaint, Sranan, Suriname). Probably, they are implied in *tocote de monte* (wild *tocote*, often a name for *Jatropha*, Guatemala). The name *unsicacá* [*uncincaca*] [*de montaña*] is said to be from the Zoque language of the Isthmus of Tehuantepec in Chiapas. Perhaps the Zoque word *caca* is related to Mayan *kaj-kab* (bitter). Still, the name looks suspiciously like Spanish *un sin caca* (one without feces), maybe from an advanced stage of purging by *Ximenia*.

The Arawak name *wiri* (herb, Arawak, Guyana, Suriname) could refer to the medicinal fruits or to the edible leaves cooked as a potherb (Morton 1968, Duke 1972, Freiberger et al. 1998). The concept of purging may also be related to the name *k'uk-ché* [*kic-che, k'uk-che, xkukche, xkuk-ché, xk'uk-ché*] (*k'uk*, renews, *che'*, tree, Maya, Yucatán, Belize). Purging and emesis are part of cleansing rituals throughout the world.

Wood of *Ximenia* is yellow-red, hard, heavy, and fragrant (Little et al. 1974). The odor prompted people to call it sandlewood (USA, West Indies) and false sandlewood (USA, West Indies). However, not everyone is that complimentary. To some, it is the *bois puant* (stinking tree, Windward Islands). Others think it is as hard as *ishiri* (*isri*, iron, Dutch Antilles). A compliment is to say it is *yaná* (originally buttonwood, *Conocarpus erecta*, Taino, Cuba), yellow sanders (USA, West Indies), or just *pata* [*pata de monte*] (mountain wood [literally foot, but colloquially for wood], Argentina). The allusion in *chocomico* (*choco mico*, black monkey, Honduras, Nicaragua) is not clear.

To people who appreciate the plant, it is *teukra* [*teu-kra*] (variant of *teocali*, from *teotl*, god, *calli*, house, Náhuatl, Costa Rica). However, *saaxnik* (*sak*, white, *nik*, small flower, Maya, Yucatán) is almost Japanese *haiku* poetry.

In addition to being a laxative, a syrup made from the fruit is used in the West Indies and Panama for dropsy, rheumatism, and other problems (Standley 1920–1926, Roig 1945, Duke 1972). Baths of the astringent bark have been used in Brazil against excessive menstruation (Roig 1945), and in many tropical countries to cure skin diseases (Liogier 1974). In Panama, the roots are chewed to stop toothache, against sleeping sickness, fevers, ringworm, venereal disease, and to stop vomiting and diarrhea (Duke 1972). Oil extracted from the seeds (Standley 1920–1926) is used for cooking, soap, and lubrication (Morton 1968b, Eromosele et al. 1994, Eromosele and Eromosele 2002).

Among the oils, the plants contain mandelonitrile lyase (Kuroki and Conn 1989), oleanene palmitate (Fatope et al. 2000b), and acetylenic fatty acids (Fatope et al. 2000a). Some of those compounds have potential as pesticides (Fatope et al. 2000a). Laboratory studies show that extracts from *Ximenia* also are antiviral (Asres et al. 2001), antibacterial (Fabry et al. 1996b, 1998), fungicidal (Fabry et al. 1996a), antimalarial (Benoit et al. 1996), insecticidal (Fatope 2000a), and molluscicidal (Kela et al. 1989), and may be useful in the treatment of sexually transmitted diseases (Ndubani and Hoger 1999).

After encountering them for weeks, some of my graduate students decided that hog-plum was incorrectly named. One of our tasks, while working for Palm Beach County, was to locate gopher tortoise (*Gopherus polyphemus*) dens. We usually found burrows when thorns on hog-plum growing beside them tore our skin and clothing. So, the students renamed the plants the "Tree from Hell." From hell maybe, but the fragrance of the flowers is heavenly.

Xyris

(Greek name for some plant with two-edged leaves, from *xyron*, a razor, reapplied by Linnaeus; some say *xyris* was applied by Pliny, A.D. 23–79, to a wild iris)

Degenbinse (sword rush, German)
yellow-eyed grass

Xyris ambigua (doubtful, alluding to the difficulty in distinguishing this species from others)
łakó:cfánî:kî [*rakó:cfánî:kî*] (*ła*, arrow, *ó:cfáni*, twisted, Mikasuki); *łi:opaykocín* [*ri:opaykocín*] (*łe*, arrow, *opiyetv*, twisted, *oce*, small, Creek)
coastal-plain yellow-eyed grass [coastalplain yelloweyed grass, yellow eyed-grass]
vare goldies

Xyris baldwiniana (named for its discoverer, William Baldwin, 1779–1818, a physician in Pennsylvania)
Baldwin's yellow-eyed grass [Baldwin's yelloweyed grass] (a book name)

Xyris brevifolia (short-leaved)
short-leaf yellow-eyed grass [shortleaf yelloweyed grass]

Xyris caroliniana (of Carolina)

Xyris chapmanii (named for Alvin Wentworth Chapman, 1809–1899, who wrote the first flora of the southeastern United States)

Xyris difformis. Drawn by P.N. Honychurch.

Carolina yellow-eyed grass [Carolina yelloweyed grass]

Xyris difformis (of two forms)
bog yellow-eyed grass [yelloweyed]

Xyris elliottii (for Stephen Elliott, 1771–1830, author of *A Sketch of the Botany of South Carolina and Georgia*, published between 1821 and 1824)
Elliott's yellow-eyed grass [yelloweyed] (a book name)

Jan Gronovius created the genus *Xyris* from an old Greek name of doubtful application in his *Flora Virginica.* He used the phrase-name *Xyris foliis gladiatis* (xyris with leaves like swords). At first, Linnaeus followed Gronovius in using the phrase name in the *Flora Zeylanica* of 1747 but then decided to call the plants *Xyris indica* in 1753. Linnaeus seemed not to be bothered with plants occurring on the opposite sides of the world, and wrote, "*Habitat in Indiis*" (Grows in the East and West Indies). In 1819, J.E. Smith recognized that the Linnaean concept was a mixture of New and Old World taxa. He restricted the Linnaean name to the Old World herb, describing *X. torta* for the American plant (Kral 1966, Steve Cafferty, personal communication 2003).

Thomas Walter described *X. caroliniana* (from South Carolinas) in 1788. André Michaux knew a single species (*X. jupicai*) in Florida, which he described (Taylor and Norman 2002); he also gave us *X. brevifolia* (from Georgia) in 1803. The other species in Florida were discovered as late as 1990 (cf. Kral 1983c, Bridges and Orzell 1990).

Since 1753, *Xyris* has grown to a genus of 200 to 400 species known in the tropics and warm regions of the world (20 Australia, 25 Africa) (Mabberley 1997). Kartesz (1994) listed 24 species for North America. It is understandable why early explorers did not take much notice of *Xyris*; they are grasslike plants intermixed with real grasses and sedges. Their flowers, with three yellow petals, all look similar, and it is only with more sophisticated magnification that people were able to examine the fine details that are used in distinguishing many species. Indigenous people, without magnifying lenses, perhaps recognized fewer species than Europeans.

All uses found among indigenous people were in the southeastern United States. Since the genus does not occur in the Great Plains (Barkley 1986) and is known from only two counties in Missouri (Steyermark 1963), it is not too surprising that Missouri River people did not utilize them (Gilmore 1919).

The Cherokee are the people farthest north known to use the plants. They made an infusion of *X. caroliniana* roots to treat diarrhea (Hamel and Chiltoskey 1975). While it is not clear, they may also have taught settlers other uses. According to Uphof (1968) and Lewis and Elvin-Lewis (1977) the same species has been used in home remedies for skin problems as well as colds and pulmonary ailments.

Xyris caroliniana was not the only species used for lung and chest problems. The Seminoles used "an infusion of [*X. ambigua*], and doubtless other, species of *Xyris* ... in cases of colds and pulmonary disorders" (Small 1933). *Xyris elliottii* was used identically, and an infusion of the herbage was rubbed on the chest to treat colds (von Reis and Lipp 1982).

Little is known about the chemistry of *Xyris.* Tannins, saponins, calcium oxalate crystals, isocoumarins, anthraquinones, and sometimes aluminum have been found in several species (Fournier et al. 1975, Cronquist 1981, Ruangrungsi et al. 1995). All of those are notoriously bioactive chemicals and may account for some of the medical applications.

Historically, there were magical applications of *Xyris* among the Seminoles. Josie Billie, one of

Sturtevant's (1955) informants, recounted an ancient use for *X. ambigua.* According to him, the *Xyris* flowering stem is twisted, as one of the supernatural thunders is also twisted. Therefore, the effectiveness of the magic depends on the thunders. To invoke this magic, the sorcerer fasts for 2 or 3 nights and 4 whole days, breaks off four *Xyris* flowering stems, and sends them "like arrows," with the flower at the butt end, for great distances ("thousand miles"). No one in the 1950s could do this any longer.

Y

Yucca: From Silk-Grass to Spanish Dagger

(From Taino *yuca*, originally the root of *Manihot esculenta*, while the plant was called *yucubia*; reapplied to this monocot because of confusion probably dating from Bauhin's *Pinax* of 1623)

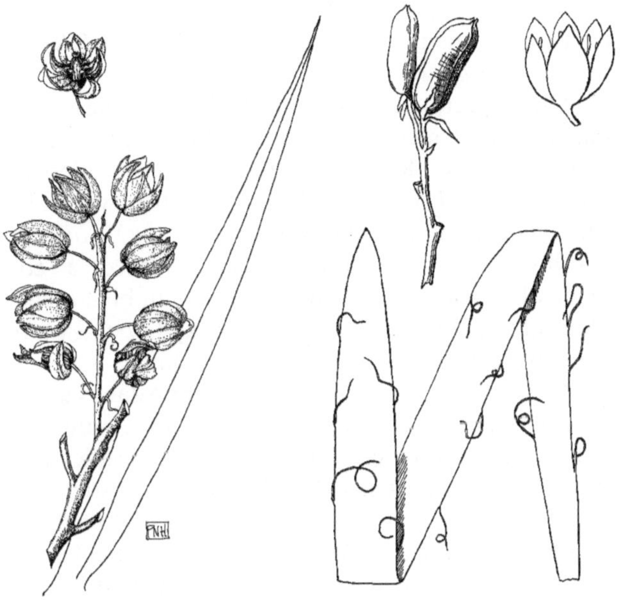

Yucca. *Yucca aloifolia* (left). *Drawn by P.N. Honychurch. Yucca filamentosa* (right), with fibers on leaf margins. From Britton and Brown 1896.

With classes on the beaches, I always tried to find a Spanish dagger (*Yucca aloifolia*). I would cut off a leaf while I explained how the plant was important to indigenous people. As we continued, I stripped the flesh from the fibers, always leaving the apical spine intact. When finished, I showed them the "needle and thread" and told them how its names changed from silk-grass to Spanish dagger.

When the Europeans arrived in the Caribbean, they found the Taino people using roots they called *yuca*. Somehow, the newcomers began calling two totally different kinds of plants by the single name.

The important food plant was *Manihot esculenta* (Hedrick 1919, Sauer 1969). Martyr (1555) wrote, "*iucca* is a roote, whereof the best and most delicate bread is made, both in the firme land [mainland] of these regions and also in ilandes." Gaspar Bauhin knew the same plant in 1623, when he called it *Manihot inodorum seu yucca foliis cannabinis* (odorless manihot or yucca with leaves like Cannabis). In the same book, Bauhin listed a second species (*Y. aloifolia*) that he called *Yucca foliis aloes* (yucca with leaves like aloe).

In Spanish, the original *yuca* of the Tainos was retained for *Manihot*. The *Agave* relative *Yucca* is usually *palmillo* [*palmito*] (little palm), or simply *palma* (palm), although most Caribbean people understand *yuca* for that plant too. Indeed, *yuca* [*yucca*] (Dominican Republic, Guadeloupe, Martinique) is a common name for both in several areas.

Why this confusion took place is not clear. All the original references were to *Manihot*. Of *Yucca*, Santamaria (1959) says, "*La raíz ni es comestible, ni es verdad que de ella se extraiga harina*" (The root is not edible, nor is it true that one may extract flour from it). In spite of this, Liogier (1974) says of *Y. aloifolia*, "*La raíz carnosa contiene almidón y es comestible al igual que la Yuca*" (The fleshy root contains starch, and is edible just like *Manihot*). The common name Indian bread plant (Hispaniola) makes me wonder if we have missed something.

A trait arguing against eating the starch from the roots of *Yucca* is their saponin content (Oser 1966, Dewidar and el-Munajjed 1970, Stohs et al. 1974, Sener and Turkoz 1987, Tanaka et al. 1996, Sen et al. 1998, Miyakoshi et al. 2000, Uematsu et al. 2000, Oleszek et al. 2001, Plock et al. 2001). Indeed, in Mexico, a generic term for the genus is *amole* (soap, Náhuatl), and the Mexicans use extracts from the roots for washing. Howard (1907) even recorded a Seminole woman at Lake Okeechobee in 1856 "digging some roots that had a saponaceous juice she powdered them and soaked them in water, washing her hair with the lathery substance till it was smooth and glossy." The lake is not a usual area for *Yucca* but, if Howard identified it correctly, the shrub would have been cultivated.

Regardless of its origins, by the 1600s *Manihot* and *Yucca* were confused. It took Linnaeus to straighten them out, and in 1753, he recognized four American species of *Yucca* — *Y. aloifolia, Y. gloriosa, Y. draconis,* and *Y. filamentosa. Yucca draconis* is now a variety of *Y. aloifolia.*

The oldest reference we have for *Y. aloifolia* in Florida was from H. d'Escalante Fontaneda in 1575 (Austin 1980). That former captive of the Calusa wrote of women's clothes on Matacumbe Key that they were "*vnas yerbas que nazen de unos arboles estos yerbas paresen lana aunque son differentes*" (some grasses that grows on trees, the herbs looks like wool although they are different). Of the women at Abalachi, he wrote that they covered themselves with "*vnas pajuelas que nazen de los arboles a manera de estope o lana y no es blanco sino pardo i con aquellas yerbas se cubren dellas a la redonda a la sinta*" (some straw that grows on trees like burlap or wool and it is not white but brown and with these herbs they cover themselves about the waist). The fibers were from *Y. aloifolia,* and not *Tillandsia usneoides* as Smith (1854) and Small (1933) contended. Farther north, Harriot ([1590] 1972) and Sir Thomas Gates, governor of Virginia in 1610, recorded indigenous skirts of *Y. filamentosa* much like those used in Florida (Coffey 1993). Indeed, Strachey ([1612] 1953) made numerous comments about "silk grass" use by the Virginian Algonquian tribes.

Dickinson (1699) first noted that the Hobé men near Jupiter Inlet wore loincloths of palm with a "horsetail of a bunch of silk-grass exactly resembling it, of a flaxen color." Bernard Romans ([1775] 1961) also called *Y. aloifolia* silk grass. Typically, William DeBrahm ([1772] 1974) could not make up his mind and said they were *iuca,* and *palmeta royal.* From the same period, Bartram ([1791] 1958) agreed with palmetto royal.

Small (1933) recorded that leaves of all southeastern species were used by pioneers to make ropes and string for hanging up cured meats. In the early 1970s, I found fishermen on Great Inagua near the Bahamas still using the cord made from *Yucca* to hang their bonefish (*Albula vulpes*) to dry. Fiber from the leaves is recorded in *maguey silvestre* (wild *maguey,* Taino, Cuba). People also say *pinguin* [*pingouin*] (Haiti), probably because the sharp leaves remind them of *Bromelia pinguin,* another fiber plant.

It was late when *Yucca* became Spanish-bayonet [bayonette] (by 1843, Florida, Dominican Republic) and Spanish-dagger (Florida), but those names reflect the Mexican name *izote* [*iczotli*] (from *ixtli, xotla,* and *ixotla,* stab someone in the face, Náhuatl). Related names are still used, as *bayoneta* (Cuba, Dominican Republic, Puerto Rico), *bayonette* (Haiti), *espino* (spine, Cuba, Dominican Republic), and sword plant (Bahamas). *Piñon de puñal* (dagger pine, Cuba) contains similar intent.

Aloe yucca (USA, Puerto Rico) harks back to the 1600s. Plukenet made the comparison in 1696 as *Aloe, yucae foliis* (Aloe with leaves like yuca), as did Kaspar Commelin in 1703 with *Aloe americana, yuccae folio* (American aloe with leaves like yucca).

Yucca to the Catawba was *nemẹ'? sara'k* [*surá·k*] (bear grass). The Choctaw said *nita i̱npisa* [*neta pisa*] (*nita,* bear, *in,* its, *pisa,* eyes). Choctaw relatives the Creeks say *sowenv* [*suwénv, sowi:nâ*], and to the other Seminoles the plant was *pasalátkî* (Mikasuki). The Koasati also say *passá.* Apparently, all are simple untranslatable names, although the Mikasuki and Koasati include *pasa* (*Eryngium yuccifolium*). The Alabama said *tosìina istatakka* (*tosìina,* from Spanish *tocino* for bacon, *ist-*, it is, *atákkaaka,* hanging). The name is derived from their use of the sharp point on the leaf and its fibers to hang meat for smoking (Sylestine et al. 1993).

There is no obvious reason why these spiny shrubs are *salesepareille* (sarsparilla, usually used for *Smilax,* Guadeloupe, Martinique). Similarly, it is not clear why the plants are called *flor de Jericó* (flower from Jerico, a city in Palestine, Dominican Republic) or just *Jericó* (Dominican Republic). Perhaps the reference is related to the old name "Lord's candlestick" (USA) for several *Yucca.*

One species called Our Lord's candles is *Y. filamentosa,* which grows from Maryland west to Tennessee and Kentucky, and south to Georgia, Florida, and Louisiana. That plant too was called silk-grass, and had the same uses as *Y. aloifolia* (Coffey 1993). Whitford (1941) recorded a cord made of *Y. filamentosa* among pre-Columbian archaeological remains from Ohio. Yanovsky (1936) said the fruits were eaten in the southeastern states.

Other names for *Y. filamentosa* are Adam's needle, bear-grass, bear's thread, Christmas bells, Confederate flax, curly hair [yucca], Eve's darning needle, Eve's thread, grass-cactus, needle palm, soaproot, soapweed, Spanish dagger, spoon-leaf yucca, and thread-and-needle. The Catawba, Cherokee, and Nanticoke used root extracts to treat skin diseases, rheumatism, and gonorrhea through the 1600s and 1700s (Speck 1937, Taylor 1940, Coffey 1993, Moerman 1998). They considered it sedative, among other properties. The Cherokee also used the roots to kill fish and wash blankets (Hamel and Chiltoskey 1975, Moerman 1998).

Much less common is *Y. gloriosa* (Brunswick, Dare, and Onslow Counties, North Carolina, Charleston, South Carolina, Georgia, and Leon and Franklin Counties, Florida), an endangered plant in Florida (Coile 2000). This species is famous for its names Lord's candlestick and Roman-candle (USA). It is also known as Spanish bayonet and palm lily, but it is usually called mound-lily yucca. The comparison of growth to a mound is reflected also in *sak tuk* (white mound, Maya, Yucatán) for *Y. aloifolia.*

Plants still are eaten in many places. People in most of the Caribbean region eat flowers in salads, and often cook them with eggs (Standley 1920–1926, Leon and Alain 1946–1954). Morton (1968b) contended that the buds are edible raw, but taste better when boiled or roasted as a vegetable. Taste varies, apparently from plant to plant and with the age of individual flowers. Students were often amused when I tasted flowers to show them they were edible and frowned at the bitter taste. Morton (1968) also considered the raw fruits edible, but most find them rubbery and bitter. Flower stalks are also peeled, boiled, and eaten.

According to Bushnell (1909), the Choctaw boiled the mashed root with grease or tallow and used the healing salve for various problems. Taylor (1940) also found the Koasati using *Yucca* medicinally. However, by the time Sturtevant (1955) worked with the Seminoles, they no longer used the plants. In Hispaniola, tea is made of the young leaves as a diuretic (Liogier 1974). *Yucca* has been used in many remedies probably because it contains saponins, glycosides, and steroids (Blunden et al. 1965, Lazur'evskii et al. 1975, Stohs et al. 1975, Nakano et al. 1991a,b, Kishor et al. 1992). Laboratory studies indicate that compounds from *Yucca* are antitumor (Ali et al. 1978), antiviral (Hayashi et al. 1992), antioxidant (VanderJagt et al. 2002), have antiplatelet effects (Olas et al. 2002), and are useful against giardiosis and *Leishmania* (McAllister et al. 2001, Plock et al. 2001). The resveratrol and other stilbines have been shown to inhibit cancer and give liver and kidney protection. They are anti-inflammatory and function as a natural profungicide (Oleszek et al. 2001, Uenobe et al. 1997). However, there are potential problems with allergenic reactions (Kanerva et al. 2001, Munno et al. 2001).

The most visible use in the past few decades for all *Yucca* in the Caribbean has been as living fences or as ornamentals in yards (Leon and Alain 1946–1954, Liogier 1974). Especially in rural areas, but also in places like Fort Lauderdale and Miami, *Yucca* is planted near houses and adorned with eggshells on the tips of the leaves. Some enterprising moderns substitute segments of egg cartons for the shells. Either way the daggers become more visible.

Z

Zamia: Coontie

(From *azaniea nueces,* dried-up nuts or Latin *zamia* or *samia,* to hurt or damage)

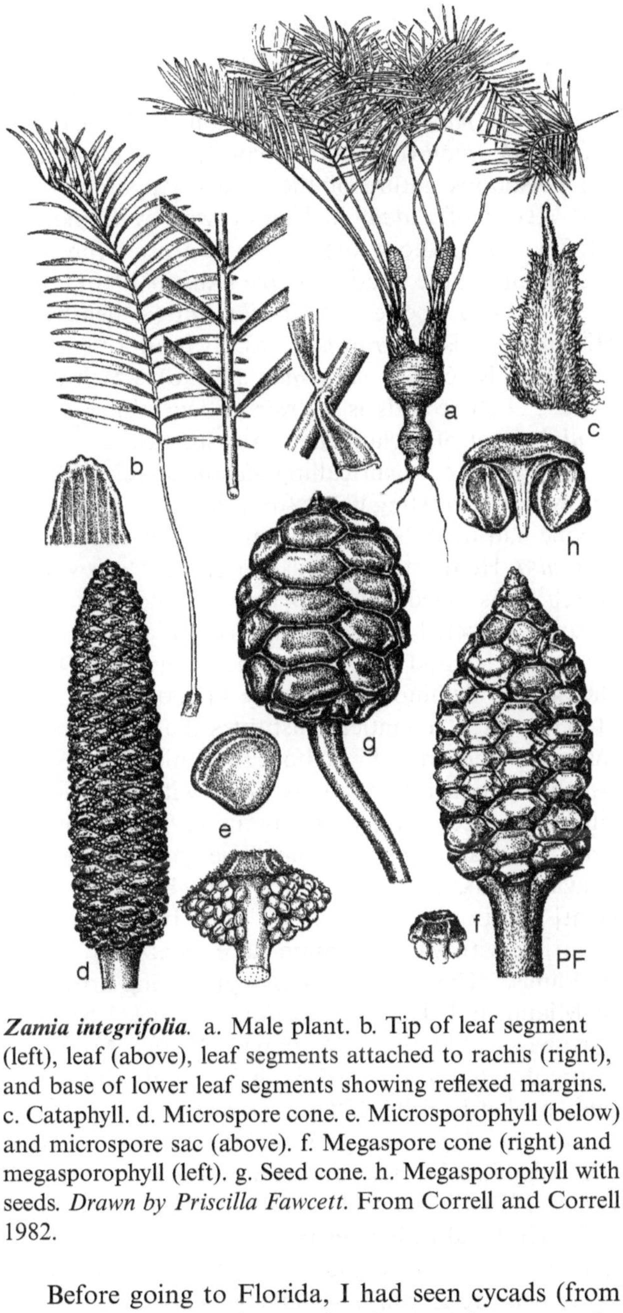

Zamia integrifolia. a. Male plant. b. Tip of leaf segment (left), leaf (above), leaf segments attached to rachis (right), and base of lower leaf segments showing reflexed margins. c. Cataphyll. d. Microspore cone. e. Microsporophyll (below) and microspore sac (above). f. Megaspore cone (right) and megasporophyll (left). g. Seed cone. h. Megasporophyll with seeds. *Drawn by Priscilla Fawcett.* From Correll and Correll 1982.

Before going to Florida, I had seen cycads (from *kykas,* a palm, Greek) growing in greenhouses at the Missouri Botanical Garden where I did my graduate work. These ancient conifer relatives have been called living fossils, and they were widespread and pervasive throughout the Mesozoic, 250–265 million years ago (Schneider et al. 2002). Partly because of that, I was anxious to see them in the wilds of southern Florida. In the Everglades National Park, I found coontie (*Zamia integrifolia*) in the pinelands. They were not very impressive, but they did look a little like palms.

I wondered why Linnaeus called them *Zamia,* and found that he followed Pliny the elder (A.D. 23–79). Pliny had called a pinecone *zamia* and for pine nuts wrote *azaniea nueces* (dried up nuts). In Linnaeus's usual delight for multiple meanings, the genus could either mean their cones looked dried up (they do), or come from Latin *zamia* or *samia* (to hurt or damage). As Linnaeus was a physician, he surely had learned of poisons in the plants.

The poisonous compounds in *Zamia* include the glucoside cycasin (Lampe and McCann 1985, Nellis 1997), flavones such as amentoflavone, bilobetin, sequoiaflavone, and ginkgetin (Meurer-Grimes and Stevenson 1994, Hocking 1997), vicilin-like seed proteins (Braun 1996), alkanes in the foliar waxes (Osborne et al. 1993), azoglucosides, and nonprotein amino acids (Schneider et al. 2002). It is no surprise that Morton (1968b) recorded, "Water used to leach the [*Zamia*] starch was red from poisonous compounds, and would kill cattle or other animals drinking it." Even turkeys eating the seeds were poisoned (Morton 1968b) in spite of *Zamia* dispersal by birds (Eckenwalder 1980a).

In another puzzle, I found that the word coontie [contee, conte, contie, comptie, coontah, coonta] came from Florida's Seminoles. William Bartram ([1791] 1958) first wrote "coontie" in 1791. He wrote of eating with Creeks on the Suwanee River a "cooling sort of jelly, which they call conte; this is prepared from the root of the China brier (*Smilax*)." He also wrote that the reddish flour, "mixed with warm water and sweetened with honey, when cool, becomes a beautiful, delicious jelly, very nourishing and wholesome; they also mix it with fine Corn flour, which being fried in fresh bear's oil makes very good hot cakes or fritters."

To the Creek people there are two words for edible roots. One Creek word for them is *vhv* [*áha*], while their relatives say *a:hí* (Mikasuki) or *ahe* (Choctaw). The other Creek word is *kunte* [*konte, kontî, kunti, kontik, kantik*], and its Muskogean cognates are *kantíkî* [*conteke*] (Mikasuki), *kanták* (Houma), and *kantak* (Choctaw). *Áha* is generic for roots, while *kunte* historically applied to the rhizomes of *Smilax.*

After the Seminoles were driven south down the Florida peninsula, they found a new plant with edible roots (technically stems), *Z. integrifolia* Since they now had two kinds of *konti,* they distinguished between them. The old familiar *Smilax* became *kunte cate* [*kontí chate, conti chatee, coontie chabi, coonti chacli, coonti chatee*] (red coontie, Creek), which is sometimes translated as "red flour" (Speck 1941, Sturtevant 1955). *Zamia* roots were dubbed *kunte hvteke* [*conti hatke*] (white [bread] root, Creek). Both kinds of *kunte* became the basis of the Seminole diet in the gruel or soup they call *sofkee*.

Because *Zamia* was poisonous, it required a more elaborate preparation than *Smilax*. Accounts of how the stems were made edible vary in details (Yanovsky 1936). According to Morton (1968b), stems were scraped or peeled, pounded, grated, ground, or boiled until soft and mashed with abundant water. The mixture was then drained and the resultant starch was allowed to dry in the sun. Bennett (1997) was told that the stems were grated (using corrugated roofing with nail holes) and collected in a vat. That material was then allowed to ferment for up to 7 days. The odor of the mixture indicated to those who knew the process when the coontie was ready. This mixture was spread on a cloth sheet, and squeezed to eliminate moisture. The squeezing process was repeated several times, and then the mixture was spread out and allowed to dry. A powdery flour is the final substance. To make *sofkee* Seminoles cooked the mash, instead of drying it, with meat and vegetables.

Coontie has an important place in Florida history although few residents in Fort Lauderdale know that the plants started their city. The Seminoles in the early 1800s sometimes called what is now Middle River the *Colohatchee* (*kala*, oak, or *colowa*, colored, *hatchee*, creek), but they also called it the "coonti grounds" (Simpson 1956). In the 1830s, Motte (1963) wrote that the river was "called by the Indians *Coonte Hatchee*, the country around it being the great coonte region, where this root abounds; which being prepared by the Indian women for their bread by a simple process of grating and washing." The product was so important it came to be called "Seminole bread" (Nellis 1997).

Into this important Seminole region came a white man named William Coolie. He, his family (wife, three children), the children's male tutor named Flinton, and their slaves settled beside the river in what is now downtown Fort Lauderdale. They proceeded to cut down the forest and created a coontie-processing business for their "plantation." On 6 Jan. 1836, several Seminoles attacked and killed everyone except Coolie himself, who was away, and the slaves. Between Dec. 1835 and Jan. 1836, nearly all the plantations between Cape Florida and St. Augustine, mostly on important coontie grounds, were destroyed (Motte 1963). Two years later Major William Lauderdale and his Tennessee Volunteers built the fort that later bore his name. While Patsy West and Betty Mae Jumper (1998, 2001) argued that the Coolie incident was coincidental to the other attacks, that incident and the others precipitated the Second Seminole War.

Although not widely known, coontie was familiar to Europeans long before the Seminole war. *Zamia* was recorded in Cuba by Bartolome de Las Casas in the 1550s when he wrote "*hay en los montes otras raices, que llamaban los indios guayaros*" (there are other wild roots that are called by the Indians *guayaros*) (Coll y Toste 1972). That word remains in Spanish with *guayará* (Taino, Cuba), the name for *Z. integrifolia*. Other Cuban names for *Z. integrifolia* are *yuquilla de paredón* (little execution wall yuca) and *yuquilla de ratón* (little mouse yuca), both allusions to the poisonous nature of the plants. Apparently, the first record of *Zamia* in Florida was by Fontaneda ([1575] 1944), who wrote of *pan de rraises* (bread of roots) among the Glades people (Austin 1980).

A related species, *Z. debilis*, grows nearby. In Hispaniola it is *guáyra* [*guáyiga*], a variant of *guayará*. In Puerto Rico, it is *marunquey* (Taino), which Coll y Toste (1972) records is a "*raiz de una planta que da fécula*" (root of a plant that gives starch).

People also began calling *Zamia* starch "Florida arrowroot" to distinguish it from the other kind. Hans Sloane had used "arrowroot" in 1696 for starch from *Maranta*. He did that because originally "arrowroot" starch was medicinal—to absorb poisons from wounds, especially those from poisoned arrows. Later, it came to be used for other medical purposes. During the Second Seminole War, army surgeon Jacob Mott (1963) wrote, "[coontie] constitutes the arrow-root of many Apothecaries throughout the United States, and which I frequently used as such with the sick in Florida." According to Culbreth (1910), arrowroot was "used as demulcent, nutritive food for infants, convalescents, bowel or urinary troubles; in 5 p[er] c[ent] solution with water or milk by boiling and flavoring with vanilla, lemon juice, etc.; also used in puddings." There was a factory producing arrowroot in Miami in 1911, and the flour was used to make "arrowroot biscuits" (Workman 1980). Arrowroot flour also has been used in baby food, chocolates, and spaghetti (Nellis 1997).

Zamia integrifolia grows in the Greater Antilles, the Bahamas, Florida, and southern coastal Georgia (Ward 1979, Eckenwalder 1980b, Correll and Correll 1982, Kral and Robinson 1983, Stevenson 1987, Jones and Coile 1988, Whitelock 2002). Its prominent synonyms are *Z. floridana* and *Z. pumila*. Oddly, the shrubs are called bay-rush (Bahamas) and bulrush [bull rush] (Cayman Islands) in the Islands. Typically, those common names are given to Cyperaceae or Typhaceae.

The same compounds that make raw coontie poisonous to humans are crucial elements in the development and maintenance of mutualism (pollination) and parasitism (herbivory) by cycad-linked insects (Schneider et al. 2002). The toxin cycasin now is implicated with retaining the relationship between the pollinator (the weevil, *Rhopalotria mollis*) and the specialist herbivore (the aposematic Atala butterfly, *Eumaeus atala*).

Coontie was almost extirpated from Florida along with its spectacularly colored butterfly herbivore by the 1950s (Klots 1951). Then, by cultivating the plant and moving the butterfly to the new populations, both seem to have stabilized (Kimball 1965, Ward 1979, Kral and Robinson 1983, Scott 1985). Organisms with such complicated and long histories, even with humans, should remain on the planet—they teach us a great deal.

Zanthoxylum: Prickly Ash, Hercules's Club, and Wild Lime

(From Greek *xanthos*, yellow, *xylon*, wood)

My introduction to *Zanthoxylum* was when I grabbed a tree to slow my descent down a hillside. The spines on the stems left what I thought would be permanent scars on the palm—and I fell anyhow. Later I learned that the plant was *Z. americanum,* prickly ash or *frêne épineux* (Quebec).

The tree that perforated my palm barely extends to Florida in Levy, Gadsden, and Liberty Counties. More common in the state is *Z. clava-herculis,* usually called Hercules's club. Linnaeus gave it that Latin name in 1753, and some say the English translation only appeared in 1847. The reverse seems likely, with the common name preceding the Latin.

Zanthoxylum clava-herculis reaches down the peninsula to Lake Worth in Palm Beach County, and to Tampa Bay on the west coast. Historically, the species reached Miami. Both it and *Z. americanum* figured prominently in native American medicine. The Houma called the trees *frêne piquant* (pungent ash). Otherwise, we have only English names for *Z. clava-herculis,* and Linnaeus ([1753] 1957) was obviously impressed with its armament. Many Americans, also impressed, say prickly yellow-wood, tear blanket, wait-a-bit, wild orange, sea ash, and yellow prickly-ash.

People who have tasted the bark call it pepper wood, rabbit gum, sting tongue, tongue-bush, and toothache [tree]. The flavor is much like the sedative the dentist swabs on your gums before giving a shot of novocaine. Maybe they are related compounds. Either way, using the plant to relieve toothache was a prominent application by native Americans and European immigrants. Millspaugh (1892) described his own experiment: "I found that, upon chewing the bark for relief of toothache, speedy mitigation of the pain followed, though the sensation of the acrid bark was nearly or fully as unpleasant as the ache, and so painful finally in itself that I abandoned its use." Formerly *Z. clava-herculis* was called "pillenterry" a corruption of "pellitory," and a misnomer by Catesby (1731–1732) and Barton (1810).

According to Moerman (1998) and others, uses of *Z. americana* and *Z. clava-herculis* cannot be separated. One or both were used by the Alabama, Algonquin, Atapaka, Biloxi, Cherokee, Choctaw, Comanche, Creeks, Delaware, Houma, Illinois-Miami, Iroquois, Koasati, Menomini, Meskwaki, Mohegan, Ojibwa, Omaha, Osage, Pawnee, Potawatomi, Seneca, and Winnebago.

To Alabama speakers *Zanthoxylum clava-herculis* was *ittohalokpa* (*itto*, tree, *halokpa*, sharp-pointed) and *natikasahlichi* [*innatikasahlichi, natikasaalichi*] (*nati*, tooth, *kasahlichi*, cooling). They made wedges from the hard wood and also applied the inner bark to gums to relieve toothache (Sylestine et al. 1993). The Choctaw said *nuti alikchi* (*nati*, tooth, *alikchi*, doctor) or *ikhi<u>n</u>sh patassa* (*ikhi<u>n</u>sh,* medicine, *patassa,* flat). The Creeks of Oklahoma call it *'tokvsvppe* [*tokvsap'pv*] (*eto*, tree, *kvsvppe*, cold), which is the same idea as *ittokasáhka* [*ittohalokpá*] (*ittó*, tree, *kasáhka*, lower fever, Koasati). The Biloxi said *anisni' hudi'* (*ani'*, water). The Atakapa called them *ci'wat te ne'c* (*ci'wat*, alligator, *te*, plural, *ne'c*, tree).

According to Rafinesque (1820), a common name for *Z. americanum* among the "Western tribes" was *hantola.* Hunter ([1823] 1973) confirmed that the Osage name was *han-to-la.* The Ojibwa call it *gawa'-komic.* The Pawnee simply called it *haknsits* (thorn). Omaha men used the fruits as perfume (Gilmore 1919).

Indigenous people also used these two plants to treat bronchitis, tuberculosis, for various infections (draw off pus, itch, oozing or ulcerated external conditions, quinsy, sore throat, sores, swellings, and ulcers), to treat hemorrhages and intermittent fever, for kidney problems (including gonorrhea), flatulence, dropsy, rheumatism, to purify the blood, and as a general tonic (Lawson [1714] 1937, Catesby 1731–1732, Carver 1779, Loskiel 1794, Schöpf [1798] 1903, Barton 1810, Hunter [1823] 1973, Millspaugh 1892, Michaux 1904–1907, Nuttall 1905, Lloyd 1921, Swanton 1928a, Speck 1941, Kinietz 1965, Murphee 1965). They were also used as a cathartic and an emetic (Vogel 1973).

Modern herbalists consider the bark and berries stimulants to the circulatory, digestive, and lymphatic systems. Like their native American teachers, they

Zanthoxylum. *Zanthoxylum americanum* (top left). From Britton and Brown 1897. *Zanthoxylum clava-herculis* (top right). Flowering and fruiting branches. From Sargent 1905. *Zanthoxylum fagara* (lower left). *Drawn by P.N. Honychurch. Zanthoxylum flavum* (lower right). Flowering and fruiting branches. From Sargent 1905.

prescribed both for rheumatism, skin disease, nervous headache, varicose veins, congestion, and as a convalescent tonic, which may have anticancer properties (Bremness 1994, Bown 1995, Duke et al. 2002). Bark was in the U.S. Pharmacopoeia from 1820 to 1926 and in the National Formulary between 1926 and 1947. Berries were in the National Formulary from 1916 to 1947.

The tropical *Z. fagara*, which grows in the southern two thirds of the Florida peninsula, was used at least as much as the others (Standley 1920–1926). The bark is sudorific and acts as an arterial and nervous stimulant. Leaves and bark have been used as a spice because of their pungent taste, and the bark gives yellow dye (Standley, 1920–1926). Some of *Z. fagara*'s uses are reflected in its common names, such as *cha'ach* (to chew, Maya, Yucatán; incorrectly recorded as *cha kah,* which is the name for *Bursera simaruba*), and *wi'puuy* [*huipuy*] (*puuy,* mouth, Huastec, San Luis Potosí). Surely *salitrero* (saltpeter) also is from the taste.

Many names allude to the tree's prickly nature. Some compare *Z. fagara* to spines as in *espino* (spine, Cuba), *espino amarillo* (yellow spine), and *espino rubial* [*pino rubial*] (blond spine, Puerto Rico). Others think *Z. fagara* more like claws and say *arañagato* [*aruña gato*] (cat catcher, Cuba, Venezuela), *o'ouse suctu* (lion's claws, Mayo, Sonora), and *uña de gato* [*uñagato*] (cat claw, Dominican Republic, Tamaulipas, Colombia). A few even compare *Z. fagara* to horns, with *chivo* (goat, Cuba), and *tarro de chivo* (goat horn, Cuba). To some, it is *amoroso* (lover, Isle of Pines), and others think it simply *chincho* (irritating tree, from *chinche*, Honduras). *Zanthoxylum fagara* is even considered an *alacrán* (scorpion, Chiapas).

People also compare *Z. fagara* with other prickly plants, as in *palo* [*de*] *zarza* (blackberry tree), *zarza de tomeguín* (hummingbird's blackberry, Cuba), or simply shorten the latter to *tomeguín* (a hummingbird, Cuba). The name *garabatillo* (little hook, Mexico) alludes to the similarly prickly *Mimosa* To others, it is *pinzanillo* (little pincers, related to *pinzan*, a common name for *Pithecellobium dulce*). Another comparison with *Pithecellobium* is *tenaza* (tenacious), a name used in the northern part of Mexico. The name *ceibilla malacapa* (*ceibilla,* little ceiba, Taino, and *malacapa* from *malacatl*, spindle, Náhuatl, but the name of spiny legume *Myroxylon horridum*) is a double allusion to the spiny trunks.

It is common to compare *Z. fagara* with its *Citrus* relatives. Some of those names are lime prickly-ash (USA), prickly lime (USA), *limoncillo* (little lemon, Sonora, Sinaloa, Cuba), *mata-la-sed* (thirst killer, Sonora), *naranjillo* (little orange, San Luis Potosí), and wild lime [tree] (Florida).

A widespread name in the Río Grande valley of northern Mexico and adjacent Texas is *colima* (from *colimaitl*, *colli*, man, technically uncle, *maitl*, hand or arm, Náhuatl, Nuevo León, Tamaulipas, Chihuahua). The reference to a man's hand is not clear, but the wide distribution of the name and its application to the Mexican state of Colima indicate its historical importance. The Aztec glyph for *colima* was a man's arm and hand (Santamaría 1959).

The importance of the wood among indigenous people and early Europeans is evident in the genus name originally applied by Mark Catesby in the 1730s and then formalized by Linnaeus in 1753—*Zanthoxylum*, or yellow wood. Not only is the wood colored, but a yellow dye is made from several species. The color of the wood is inherent in the names *palo mulato* (mulato tree, Jalisco), *pine jaune* (yellow pine, Haiti), *pino rubial* (blond pine, Dominican Republic), and satinwood (Bahamas). Other traits are noted in bastard ironwood (USA), and *correosa* [*corriosa*] (flexible, Texas).

To the Creeks *Z. fagara* was *tofvske* [*tofaskin*] (*eto*, tree, *fvske*, sharp), and to the Seminoles *caháhtî* (Mikasuki). The Mikasuki name was considered a simple term by Sturtevant. They used the wood for bows and arrows (Sturtevant 1955).

The people of Yucatán have several names for *Z. fagara*, but there is no clear relationship between uses and names. Their name *xik-ché* [*xuc-che, xuché*] (*xik*, wings, *che'*, tree, Maya) surely refers to the winged rachis. Two other names are related, *tanka-ché* [*tamcaz-che, tancas-che, tancaz-che*] (*tankas*, something round, *che'*, tree, Maya) and *uolé* [*wolé*] (*uole'*, something full, Maya). The apparent relationship to use is for preventing flatulence (Morton 1981). Elsewhere in Mexico and Guatamala a bark decoction is consided sudorific and a stimulant to both arteries and nerves (Morton 1981). *Zanthoxylum fagara* also has been used to treat syphilis.

The most restricted species in Florida is *Z. flavum*, known from the Keys. Elsewhere this species grows in Bermuda, the Bahamas, and from Cuba to St. Lucia; then it is disjunct to Bonaire. *Zanthoxylum flavum* shares names with several other species, including *espinillo* [*espinilla*] (little spiny one, Dominican Republic), Hercules's club, satinwood (USA, Bahamas, Puerto Rico), yellow prickly ash (USA), yellow sander (Jamaica, Puerto Rico), and yellow-wood [heart] (USA). Some names related to other species are *acetillo* (little oily one, Cuba, Puerto Rico), *bois a pians* (spiny tree, *pians* a variant of *épines*, French Antilles), and *kalabari* (maybe from *colibrí*, hummingbird, Bonaire).

This tree also has some unique names. Probably because of the pinnately compound leaves, it is *bois noyer* or simply *noyer* (walnut tree, Guadeloupe). However, the most unusual name is *misimieu* [*musimieu*] (Haiti). *Mise* means to put or to place an object in a chosen location. *Mieux* in French means "the best." Therefore, the name means "the best place to put this plant that will silence you if you trespass on my property." Boundary marking is not a use recorded elsewhere.

Chemicals in *Zanthoxylum* are resins, volatile oils, tannins, alkaloids of the benzophenanthridine-type (chelerythrine, nitidine) and aporphine-type (magnoflorine), with both considered subtypes of the benzyltetrahydroisoquinoline-type. *Zanthoxylum* also contains berberine, lauriflorine, mannoflorine, and lignans (sesamin and asarinin) (Fish and Waterman 1973, Fish et al. 1975, Morton 1981, Hocking 1998, HealthLink Online Resources 2000, E. Eich, personal communication, Sept. 2003). At least some of the alkaloids of the benzyltetrahydroisochinoline-type exhibit local anesthetic effects. Coumarins (dipetaline, alloxanthoxyletin, xanthoxyletin, and xanthyletin) are present in *Z. americanum* but absent from *Z. clava-herculis* (HealthLink Online Resources 2000, Ju et al. 2001).

Poisoning of livestock has resulted from animals eating *Z. clava-herculis* (Bowen et al. 1996). Both coumarins and lignans have shown inhibition of human leukemia cells (Ju et al. 2001). Chemicals from other species are amoebicidal and giardicidal (Arrieta et al. 2001), anthelmintic (Navarrete and Hong 1996), antifungal (Ngane et al. 2000), antimalarial (Gessler et al. 1994), antimicrobial and cytotoxic (Islam et al. 2001, Nissanka et al. 2001), antinociceptive and antidiarrheal (Rahman et al. 2002), have antiplatelet action (Chen et al. 1995, Sheen et al. 1996, Tsai et al. 2000), and have antispasmodic, leishmanicidal, and trypanocidal activity (Bastos et al. 1999, Ferreira et al. 2002, de Moura et al. 2002).

Temperate northeastern foresters Harrar and Harrar (1946) simply wrote "none" for "Importance" for *Z. fagara.* The many people around the Caribbean, who still use tropical species, were surely astonished to learn that their wood, spice, and medicine source had no value. Academics too often ignore other worldviews.

Zea

(Linnaeus reapplied the old Greek name *zeia* or *zea* for some cereal grass, now unknown, to this important New World food plant; akin to Akkadian, *se'u,* grain)

corn (an old Germanic word, meaning a grain or seed; cognate to Old Frisian *korn,* Old Saxon *corn,* Gothic *kourn*); Indian corn; *Turkishe corne* (Turkish grain, German, cf. Turner [1548] 1965); Turkish grain [Turkish millet] (cf. Turner [1548] 1965); *Welchkorn* (foreign grain, German)
maize (from the Taino *maisi* or *majisi*); *maïs* (French); *Mais* (German); *mais* (Italian); *maíz* (Spanish); *me* (Creole for *maïs,* Trinidad); *milho* (from Latin *milium,* millet, Portuguese)

Zea mays. From Fuchs 1542.

Zea mays (from the Taino name, *maisí*)
àʔù (Paya, Honduras); *af* (Guatuso, Costa Rica); *ai* (Rama, Nicaragua)
abati [*abaty, abatij, ubatim*] (contraction of *abatinga,* from *aba,* hair, *tinga,* white, because of the white silks, Tupí, Brazil ca. 1550, Paraguay)
abo [*holaba, tapolaba*] (Timucua)
Agwe'la (the "Old Woman," a name used in prayer formulas alluding to the origin legend of corn appearing from where the body of a woman was buried, Cherokee)
aspî [*aspee, aspeen*] (Mikasuki); *cassí* (Koasati); *chassi* (Alabama); *vce* [*v'ce, acî, atsche, atchee*] (Creek, Muskogee; cf. Simmons [1822] 1973)
atce'ki (Ofo)
awasi (Carib, Suriname); *awási* (Garífuna, Nicaragua)
aya (Miskito, Nicaragua)
carracony (Algonquin, Montreal in 1535); *chasquen* [*chasqueen, chassqueen*] (Delaware)
ha'-ba (Osage); *ha'hka* (Tunica); *wagmiza* (Lakota); *wahaba* (Omaha-Ponca)
ik (Cabécara, Bribrí, Costa Rica); *ip* (Térraba, Costa Rica)
ixim (Maya, Yucatán, Belize); *eem* [*em*] (Huastec, San Luis Potosí)
ku la (black maize, Tarahumara, Chihuahua); *ku mi now* (yellow maize, Tarahumara, Chihuahua); *ku Tow* (white maize, Tarahumara, Chihuahua)
ku's [*kus, kush*] (Catawba)
mandamîn (good berry, Potawatomi); *mahta'min* [*mahta'minak*] (Plains Cree); *manda'min* (cognate with *mano'min,* their name for wild rice, *Zizania palustris,* Ojibwa)
marishi (Arawak, Suriname)
mopj (Jicaque, Honduras)
necoum [*neshoum*] (Atakapa)

onèhha (Onondaga)
osapa (cited by some as "maize," but correctly the maize "field," Choctaw)
pe (Chocó, Panama)
poketawes [*pocuntavhca, pohcuwtoah, poketaws, pocuttawes*] (Powhatan, Virginia); *pagatowr* (Carolina Algonquians)
ru+krá (Brunka, Costa Rica)
sara (Quechua, Peru)
selu (Cherokee)
tanchi [*tonche*] (Choctaw)
tawmey (Shawnee, fide Edgar 1891)
tsotʔo' (Yuchi)
tuùpevu (Hopi)
yek' [*ye'ki*] (Biloxi)
yokotana (Arawak, Suriname)

I grew up in the land of the Chickasaws—western Kentucky. Bernard Romans ([1775] 1961) was among the first to record that Chickasaws claimed the territory north to where the Mississippi joins the Ohio River, and east to the confluence of the Tennessee and Ohio Rivers. Until late 1818, this region was west of the United States, and Kentucky ended on the east bank of the Tennessee River. Then, through promotion by Andrew Jackson, the land was obtained by a treaty between the United States and the Chickasaw Nation on October 19, 1818. After that, the area became known as the "Jackson Purchase." I recognized early that southeastern use of maize was a prominent feature of the region's culture.

My introduction to the cultivation of maize was helping my father plant garden plots. He "hilled" the planting site; small mounds of earth were raised with a hoe and the kernels planted in their loosened tops. Beans (*Phaseolus vulgaris*) were planted with maize seeds so their stems could grow up the cornstalks, and he pulled suckers (lateral stems) from the base of the stalks. All of those techniques were used by southeastern tribes (Hudson 1976).

When sweet corn was harvested, we ate it roasted on the cob, another indigenous method. The Alabama called them *chassapalyi* (*chassi*, maize, *apayli*, roast), the Choctaw *tanch nipasha* (*tanch*, maize, *nipasha*, roasting ear), or *tanchi hiloha* (*tanchi*, maize, *hiloha*, thunder), and to the Creeks they were *vce lowvcke* (*vce*, maize, *lowvcke*, soft). Capt. John Smith and Strachey wrote of this method among the Algonquians in Virginia in the 1600s (Swanton 1946). In the same region, Robert Beverley wrote in 1705, "They delight much to feed on Roasting-ears; that is, the *Indian* Corn, gathered green and milky, before it is grown to its full bigness, and roasted before the Fire, in the Ear."

Our daily bread when I was young was corn bread. It was only when I was in grade school that "light bread" (wheat bread) became a daily commodity. For special treats, my mother made what she called *tamales* (from *tamalli*, Náhuatl). Yet, her method was intermediate between the Chickasaw and Mexican preparation. She filled the cornmeal dough with cooked meat instead of the beans, sunflowers, or hickory nuts the Chickasaws would have used. However, she tied the ends of the corn husks covering the *tamale* before boiling it—a decidedly Muskogean method (Byington et al. 1915, Hudson 1976, Kimball 1994).

Perhaps because we lived in the borderlands between the "South" and the "North," we did not eat grits. My uncle had never seen grits in his life until, in his 60s, he visited us in Florida and was served the Creek preparation in a Georgia restaurant. Of course, the Creeks and their relatives put bear oil instead of butter on their grits, but otherwise the dish is unchanged (Hudson 1976).

No one in Europe had seen maize before Columbus found it being cultivated by the Taino people in the Caribbean in 1492 (Jane and Skelton 1960, Hawkes 1998). Benzoni's *History of the New World*, published in 1572, affirmed that Columbus brought seed of maize back to Spain in 1493 (Hedrick 1919).

Hedrick (1919) wrote that Columbus recorded *Zea* and called it *maiz* while he was in Cuba during November of 1492. My translation by Jane and Skelton (1960) does not have that passage, and Hawkes (1998) did not find it either. However, Columbus wrote on 16 October that the island he called Fernandina (Long Island, Bahamas) "is very green and flat and very fertile, and I have no doubt that all the year they sow and reap Indian corn." Unhappily, Jane and Skelton (1960) did not include the Spanish word used by Columbus.

Maize was cultivated in Seville in 1500 and near Avila in 1530 (Hawkes 1998). The grass was introduced from Cádiz into northwestern Portugal in 1520 (Hawkes 1998). Subsequently, it spread through Europe and was mentioned by Fuchs (1542), Monardes in 1569 (Monardes [1577] 1925), and Clusius in 1576 (Hawkes 1998).

By the time Fuchs (1542) wrote about maize, the knowledge that it came from the New World had been lost. That confusion is said to have begun in the 1530s (Hedrick 1919). Fuchs called it *turcicio frumento* (Turkish corn), but he also said it was *milium indicum* (Indian millet). Calling it "corn" in 1542 seems today like the obvious thing to do because that name is applied in most English-speaking countries. However, "corn" was an old word for wheat (*Triticum*), and Fuchs was really comparing the new grain with that old familiar food. Germans knew maize as *Türckish korn*, with the related Dutch *torckish coren* in use by 1549. The French said *bled sarrazin* in 1549, *blé turquois* in 1550, and *blé de Turquie* in 1551. Italians said *formento turckischo* by 1551, and Spanish *trigo de Turquia* by 1557. Those were really just different words for what the English speakers called it, like Henry Lyte's "Turkish or Indian wheate" in 1578 and John Gerarde's "Turky wheate" in 1597 (Meyer et al. 1999).

It fell to Pedro Mártir de Anghiera in 1493 to record, "the panicum of this country ... they call *maizim*" (Hedrick 1919). Oviedo in 1534–1537, and Bartolome Las Casas in the 1550s also record the Taino name for these grasses. Oviedo wrote, "*a esta clase de trigo le Llaman maiz*" (this type of wheat they call *maiz*). Las Casas (in the 1550s) said, "*Sembraban y cogian dos veces al trigo, el grano que llamaban mahiz, no para hacer pan del, sino para comer*" (Twice they planted and harvested the wheat, the grain that they call *mahiz*, not to make bread of it, but to eat). Coll y Toste (1972) suggested that the original word was probably *majisi*, since the "h" did not exist in Taino.

Bartolome Las Casas wrote, "*El indo-antillano comia el maiz tostado, el indio de Tierra Firme hacia pan de maiz, moliéndolo en un mortero de piedra y haciendo tortas llamadas arepas*" (The Antilles Indians eat maize toasted, the Indians of the mainland make maize into bread, grinding it in a pestle of stone and making cakes called *arepas*) (Coll y Toste 1972). *Arepas* is still the name of the corn cakes in Venezuela.

The first historical record of maize in Florida was from the 1528 Pánfilo de Narváez expedition; de Soto also found "*maes*" in 1540 (Hedrick 1919). Jean Ribaut found maize among the Timucua in northeastern Florida in 1562, but also used its Taino name *mahiz*. All of the historic Florida records are from Timucuan and Choctaw tribes. There was no maize grown by the Glades tribes in spite of Sears's (1982) claims to the contrary. I studied his slides made from the Fort Center site and showed Mary Eubanks photographs of the grains. We both conclude that Sears mistook *Tripascum* pollen for maize.

We retain perhaps more of the indigenous methods for eating maize than any of the other plants used by American tribes. In 1753, Dumont de Montigny wrote of the Natchez in what is now Mississippi. He first noted that there were three kinds of corn being grown, and then continued, "They can be prepared in 42 styles, each of which has its special name. ... It is sufficient to inform the reader that there is made of it bread, porridge [*bouillie*], cold meal [*farine froide*], ground corn [*farine grôlée*], smoked-dried meal or meal dried in the fire and smoke, which when cooked has the same taste as our small peas and is as sugary. That is also made which is called gruel [*grut*], that is to say that having beaten and pounded it for some time in a wooden mortar, along with a little water, the skin of envelope with which it is covered is removed. The grain thus beaten and dried is transported to great distances and keeps perfectly. The fines of that left behind is used in making hominy [*sagamité*, an Algonquian word], which is a kind of porridge cooked with oil or meat" (Swanton 1946). See Crawford (1978) and Drechsel (1997) for a discussion of the incorporation of the Algonquian word *sagamité* or *sacamité* into not only Louisiana French, but also into the Mobilian Jargon.

Smith (1992) noted that more than 90 dishes of maize were made in the southeast. So, it is an oversimplification of actual usage, but most of those dishes were variations on four basic preparations for ripe maize. The four preparations were hominy, cracked hominy, bread, and cornmeal.

Hominy was made by soaking the grains in lye from wood ashes until the aleurone, the outermost layer of the grain, was loosened. This covering was then separated from the "popped" starchy interior, the hominy. At least, that is the meaning of the word "hominy" now, but it was not always defined that way. Each tribe had a special term for this preparation—*chassoolbi* (*chassi*, corn, *oolbi*, soaked, Alabama), *vce sokv* (*vce*, maize, *sokv*, hominy, Muskogee), *olbí* (Koasati), and *newada* (Cherokee). To the Choctaw it was *tanlubo* [*tan hlabo*] or *tan-lubona* (*tanci*, corn, *lubo*, round). Siouxan speakers said *a'tcak-bi'ska* or *i̲ndaki'* [*in'taki*] (Ofo), *hauni'* (lye hominy, Biloxi), and *ko'tyuki* (Tunica).

Hominy was typically "cracked" before it was eaten. That grinding process was accomplished with a wooden mortar, usually of gum (*Nyssa*), hickory (*Carya*), oak (*Quercus*), or beech (*Fagus*). The pestle, often made of oak or hickory, was about 6 inches in diameter and 5 to 6 feet long. This was narrowed to about 2 inches for four fifths of its length and the remainder was left as a weight at the upper end.

Cracked hominy was made by placing dry corn kernels in a container with water and wood-ash lye. After soaking overnight, the corn was drained and the grains placed in the mortar where the pestle was used to crack them and loosen the hulls. A large flat basket with a shallow pocket on one side was used to separate the chaff. In the South, "cracked hominy" came to be called "hominy grits" or simply "grits." That name "grits" was an English term in use by 1579 for coarsely ground oatmeal. About 1886, the term "grits" was being applied to cracked hominy.

This product of "cracking" also had descriptive names—*chasbotáoka* (*chassi*, corn, *botooka*, ground up, Alabama), *tanchi' bota'* [*tabota*] (*tanchi'*, corn, *bota'*, ground up (Chickasaw), *tanchi bota* (*tanchi*, corn, *bota*, ground up, Choctaw), *vce-enfulotkv* (*vce*, corn, *enfulotkv*, ground up, Muskogee), and *selu isa* (*selu*, corn, *isa*, meal, Cherokee). Sometimes it was simply meal, as in *botooka* (Alabama), *bota* [*botah*] (Choctaw), *hockvte* (Muskogee), *isa* (Cherokee), and *tsukhá* (Yuchi). The Powhatans in Virginia and the Carolina Algonquians called cornmeal *rokohamin* [*äpärunmenan, rockahominy, rokehamen*] (Gerard 1907, Harrington 1955). Siebert (1975) wrote this phonetically as *rohkahamen*, and noted that the root was /**lo·xk-*/, soft, with cognates in Cree (*yo·sk-*), Fox (*no·kh-*), Menomini (*no·hk-*), Ojibwa (*no·kk-*), and Penobscot (*nohk-*). Penobscot speakers called it *nòh-khamen* (parched corn meal). Siouan people said *ye' nupxi'* (fine corn, Biloxi), *atcék anapha'si* (Ofo), and *ha'hkiri* (*ha'hka*, corn, *ki'ri*, ground, Tunica).

Breads were made by frying, baking, or boiling. The hominy was pounded into fine meal and mixed with water to form batter, in Alabama *olmo* (mixed with water) and in Chickasaw *paska toba'* (*paska*, bread, *toba'*, becoming). A thin batter was made into "fritters" by cooking in a flat-bottomed pot with bear grease. Thicker batter was made into loaves or "pones." The loaves were baked on thin, broad stones placed over or beside the fire, or in a flat-bottomed pot with another inverted over it and the two covered with hot coals. The Yuchi called those cakes *kánla*

This cooking method lives on in the eastern United States, where the food is called Johnny cake [Johnnycake, jonnycake], journey cake, and Shawnee cake. The name "Johnny cake" was in use by 1739, and might be from *jonakin* (American Heritage 2000). Another view is that early American settlers ate this on long trips, and that gave rise to "journey cake," which then produced "johnnycake." The third possibility is surely related to the indigenous northeastern "chief" named "Journeycake" (Mitchell 1895). Since the American Heritage dictionary gives no etymology of *jonakin*, the origin is unresolved. Johnny cake is now a New England specialty, especially in Rhode Island, where it is celebrated by the "Society for the Propagation of Johnny Cakes." The name "Shawnee cake," however, reveals that its origin was not just local. The Shawnees lived, among other places, in southern Illinois, Indiana, and north-central Kentucky (Grosvenor et al. 1972, Callendar 1978).

The Usquepaugh, Rhode Island, "Johnnycake Festival" features cakes made of white flint corn (because of the hard starch). Outside New England the name johnnycake is best known in the Upper Midwest, but the food itself is most popular in the South and South Midland states, where it is known as ashcake, batter bread, battercake, corn cake, cornpone (see "pone" below), or hoecake. The color of the cornmeal, the consistency of the batter, the size of the cake, and the cooking method vary from region to region. For example, an ashcake, according to one Georgia *aficionado*, is "made by wrapping cornbread batter in cabbage leaves and burying it gently at the back of the fireplace." As "hoecake" suggests, the bread was supposed to be cooked by field hands on the metal part of their work instruments. Surely, they used the hoes only as a spatula to insert and remove the bread from the fire—otherwise, how did they keep the handles from burning?

The Choctaw dish *paluska holbi* (*paluska*, bread, *holbi*, within) was made by wrapping the loaves in rehydrated dried corn husks, tying the ends and middle with a strip of husk, and baking them in hot coals. To the Alabama, this was *paspa hokfi*, the Muskogee said *taklike ofv*, and the Cherokee *gadu ayaʔa*.

The Creeks made *chuto-ahake* (*chuto*, bread, *ahake*, hard) by forming the batter into doughnut shapes, baking, and then drying the loaves in the sun until they were as hard as wood. These *chuto-ahake* were strung on cords for later use, but had to be stewed to make edible. This was *paska kallo'* (Chicksaw) and *tvklike takv'cwe* (Muskogee).

Sometimes the batter was made like *paluska holbi*, but boiled instead of baked. To the Alabama, this dumpling was *chassoochiika* (*chassi*, corn, *chiika*, made, Alabama), and their relatives said *hassabonká* (Koasati), *paska alhfola'* [*paskalhfola'*] (*paska*, bread, *aɬfola'*, dumpling, Chickasaw), *banaha* (Choctaw), and *tvklike-óme* (*tvklike*, bread, *óme*, like, Muskogee). The Cherokee knew them as *digalvnvhi*. Variants of this included the Choctaw *tanch hiloha palaska* (*tanch*, maize, *hiloha*, thunder, *palaska*, bread), *tanch hotokbi palaska* (*hotokbi*, moist), or *naksakawa* There was also the traditional boiled blue corn bread dumpling often eaten with salt pork grease. That blue dumpling was called *paspokchakko* (*paspa*, bread, *okchakko*, blue, Alabama), *paska okchamali* [*paskochamali'*] (*paska*, bread, *okachamali*, blue, Chickasaw), and *cvtvhakv* (Muskogee). A variant was *paspoochiika* (*papa*, bread, *chiika*, made), which was made from cornmeal mixed with a little flour and water, shaped into patties, boiled, and served with bacon grease (Sylestine et al. 1993).

These preparations are almost identical to the Mexican preparation of *tamales* (*tamalli*, Náhuatl). The cooked loaves could be eaten fresh or dried for later use. Instead of wrapping the batter in husks, they sometimes dropped balls or flat cakes of it into boiling water to make a corn dumpling. This dish remains as "hush-puppies," although they are deep-fried rather than boiled as the native Americans cooked them.

Southeastern people referred to cornmeal as "cold meal." This was what Frenchman Dumont de

Montigny called *farine froide.* Dry corn was soaked overnight in warm water. The kernels were parched in a pot of ashes until brown, then pounded in the mortar until they became fine meal. A fanner separated the hulls (*mayeta,* to fan corn, Muskogee; *talhpak* [*taɬpak*], winnowing basket, Chickasaw; *ufką* fanner, Choctaw). To store the meal for longer periods, it was dried further over a smoky fire. Originally, southeastern tribes had a variation of the cold meal (Romans [1775] 1961, Swanton 1946). Corn that was shelled when ripe but not dry was used to make *bota lanshpa* [*bota lanschpa*] (heated corn, Choctaw). This distinction was lost on Europeans and seems to have been abandoned later by the tribes (Campbell 1959).

To indigenous people, cold meal was *atschehotowaw* (*atsche,* maize, *hotowaw,* cold, Creek, cf. Simmons [1822] 1973), *botooka kasahka* (*botooka,* ground up, *kasahka,* cold, Alabama), *botah kapussa* or *tan̲sh fotoha* (*tan̲shi,* maize, *fotoha,* ground up, Choctaw), *hockvte kvsvp'pe* (Muskogee), *tashfoloha'* [*tashfolowa'*] (*tanchi',* corn, *foloha,* ground up, Chickasaw), or *gawi sida* (Cherokee). Virginia Algonquians said *yiketehamun.*

Botah kapussa was eaten dry or made into gruel by adding one portion of meal to two of water. Their gruel preparation was a variation of what the Mexicans call *atole* (from *atolli,* Náhuatl). This may be what the Alabama called *afolka,* the Chickasaw *tafola,* the Choctaw *tan̲sh lakchi* (maize dregs) and *tan̲fula,* and the Muskogee *nerkvce* (small seeds or grits).

In addition to the name maize, the grass has resulted in other contributions to the English language, including hominy, pone, succotash, and sofkee. "Hominy" is a Virginia Algonquian word first recorded by Capt. John Smith in 1629. He wrote, "Their servants commonly feed upon Milke Homini, which is bruized Indian corne pounded, and boiled thicke, and [hickory] milke for the sauce." Bartram ([1791] 1958) wrote that hickory milk "is an ingredient in most of their cookery, especially homony and corn cakes." Elsewhere Smith called it *Ustatahamen* (from *ûsekkutehemen,* crushed by pounding), William Strachey called it *Vsketehamum,* and Catesby (1734–1747) called it *roccahomony* (Mahr 1955b).

In 1634, another comment on Maryland tribes was that "Their ordinary diet is Poane and Omine, both made of corne." In 1772, John Josselyn wrote of the people of New England, "They beat the corn in a mortar and sift the flower [*sic*] out of it: the remainder they call Homminey." The first record of a dish called by the name in Florida was by Jonathan Dickinson in 1699, when he wrote, "Our chief dyet was Hommeney." As Swanton (1946) pointed out, all of these hominy preparations are equivalent to *pinole* (from Náhuatl *pinolli*) consumed by the Aztecs of Mexico.

The etymology of "hominy" is uncertain, but most sources compare it to Algonquian words like Natik *minne,* and Narraganset *min,* grain. Siebert (1975) glossed the Powhatan word as *apohomin,* "grain of parched corn." He based that on Strachey's *aphohomins,* and both have as their root /**apw-*/, heat, /*-*ahw-*/, instrument, and /*-*min-*/, grain. In other words, hominy was parched ground grain, just as he translated it. The prefix /**apw-*/ has been compared with Narraganset *appóon,* to bake. This *appóon* of the preparation "hominy" contributed the next loanword.

Pone (from *äpân,* baked, Powhatan) is a word from Virginia first recorded by Capt. John Smith in 1612. Speaking of their foods, he wrote of them "eating the broth with the bread which they call *Ponap.*" Strachey also recorded the word as *apones, appoans,* and *apetawh poan,* in 1612 (Harrington 1955). Siebert (1975) noted that phonetically this was *aponas,* "loaves of bread." The cognate in Menomini is *apuan* and in Penobscot *àpan* The word is related to Fox *aponewa* (he makes a roast for him) and Cree *apwan* (roast of meat). By 1716, Robert Beverley wrote, "The bread in Gentlemen's Houses, is generally made of Wheat, but some rather choose the Pone, which is the Bread made of Indian Meal ... so called from the Indian name *Oppone.*"

"Succotash" is another Algonquian contribution to English, and it was in use by 1751. The word is akin to Narraganset *msakwatas* or *msiquitash* and consists of beans and corn cooked together. James Fenimore Cooper, in his *The Last of the Mohicans* of 1826, spelled it "*suc-ca-tush.*"

John Bartram called one dish he ate "Indian corn soup, or thin homony" in 1751. Another reference in 1771 compared hominy to "rice-pudding." Those dishes were probably identical to what the Koasati called *cawáhka* [*cassapáya*], the Alabama *hoɬka,* the Creeks and Muskogees *sáfke* [*osafkv, safkv sofkee, sofki*], the Choctaw *tanfula* [*tan fula*], and the Cherokees *ganoheni.* This dish is cracked hominy cooked with water and enough wood-ash lye to turn it light yellow (Hudson 1976).

The Yuchi knew *sáfke* by what is probably the cognate word, *tso'ci* (sun fluid). According to Yuchi legend, a woman of the ancients cut a hole in the sky. From this hole there flowed a liquid. She tasted it and found it good. So, ever since, her people have called it "sun fluid" (Speck 1909).

This "sofkee" was southeastern hospitality food, and visitors were expected to help themselves to a bowl using a spoon kept alongside. Anyone who failed to eat a little was considered impolite (Hudson 1976).

Another dish was known as *sappaun* (Delaware), said to be a "hasty pudding" made from maize in 1654

(Hedrick 1919). The Powhatan in 1612 used the cognate *aspan* (Harrington 1955). Later, Catesby (1743–1747) also referred to a "*Mush* ... made of the meal, in the manner of hasty-pudding." This was probably the same dish that Dumont called *bouillie* (porridge) in 1753 among the Natchez. This became "samp" among the colonists.

Moerman (1998) wrote simply that the Seminoles used maize seeds as food, but Bartram ([1791] 1958) gave more details. During his visit to the Alachua Savanna (present-day Payne's Prairie near Gainesville), the Seminoles served Bartram "venison, stewed with bear's oil, fresh corn cakes, [hickory] milk and homony, and our drink honey and water, very cool and agreeable." When Bartram was at Fort Picoleta near Fort Augustine, he remembered that the Seminoles also used acorn oil (*Quercus virginiana*) in preparing hominy.

In addition to the food they prepared, Campbell (1959) recorded a unique storage method. Gideon Lincecum wrote that the Choctaw would "gather corn, dry it in the sun thoroughly, then lay it in small lots in a dry place, cover it with a coat of dry grass, then a thick coat of good clay mortar well mixed with grass. When it was completed, the numerous little piles of two or three baskets of corn, each covered and arranged side by side, looked like a big mud dauber's nest. In this manner they could keep their corn sound and sweet from year to year."

The Cherokee made corn husk dolls (Hamel and Chiltowsky 1975). If these had any religious significance, no reference has been found. However, among the Iroquois, corn husks were part of the Husk Face Society, which they called *katsíhsa?* (Cayuga, Seneca), *hatsíhsa?* (Onondaga), and *katíhsq* (Mohawk) (Trigger 1978). The tutelaries of the society were agricultural spirits. Individuals wearing these masks usually appeared in the Midwinter ceremony, and sometimes served as messengers for the False Faces. However, the Husk Face and False Face were two different societies, appeared separately, and performed cures. If anything like this existed among southeastern tribes, no records have been located.

If the southeastern people were at all like those in Mexico, they had multiple uses for corncobs. They called them *chassapi* (Alabama), *pahoha* (Choctaw), *talv'pe* (Muskogee), and *selu uda* (Cherokee), all meaning "corn stem." Powhatan said *okishec* [*okiic-shec*], which Siebert (1975) rendered as *ohkit*, or plural *ohkicis* He found cognates in Cree as *oskiciy* and Ojibwa *okkic*, both of which mean pipe stem. He concluded that the ancestral (Proto-Algonquian) word was /**wexkicyi*/ and that it meant spike, spadix, inflorescence.

Since many of the archaeological cobs are charred, it seems likely that they were sometimes used as fuel (Eubanks 1999, 2001). In Mexico, I often noticed corncobs stored on the roofs of houses. Finally, Bob Bye reminded me that there was another use for them—as substitute toilet paper. I had heard of that as a joke when growing up, but I never really believed anyone would use that rough item on such a tender spot. Given the frequency of the allusion to its use in the South, indigenous tribes and settlers had the same use as the Mexicans.

The only record of medical use found for the southeast was among the Cherokee. They used the "smut" from the plant (*Ustilago maydis*) as a salve, an infusion of the silks for kidney stones, and ate the parched grains for "long wind" (i.e., endurance) (Hamel and Chiltoskey 1975). However, the uses as salve and for kidney problems were widespread in tropical America (Morton 1981). Murphee (1965) found residents of the Panhandle using a tea of corn husks to treat measles and congestion and to lower fevers; they also smoked corn silk to cure sore throat.

Foster and Duke (1990) and Duke et al. (2002) confirm that corn silks (technically, the styles) have long been used as a diuretic, and as treatment for cystitis, gonorrhea, gout, and rheumatism. Experiments have established that maize has diuretic, hypoglycemic, and hypotensive activity. It is also known that the seeds contain allantoin, a cell-proliferant, wound-healing substance.

Zephyranthes: Rain Lilies

(Derived from *zephyros,* the Greek god Zephyr was the west wind that reawakened nature each spring, and *anthos,* flower)

Zephyranthes atamasco. (left) From Robert Morrison (1699), who called the herbs *Lilio narcissus indicus pumilus monanthos walbus*. (right) From Britton and Brown (1896), who used Linnaeus's name *Zephyranthes atamasco,* with a more technical drawing.

When the first Europeans arrived in the New World, they viewed all "plants" as potential "herbs." For example, in the late 1500s and early 1600s, a person could find 300 to 400 medicinal and culinary herbs in a single English garden. Virtually none was "just" decorative. All levels of society were permeated by this obsession because people's health and food depended on their knowledge of plants. Among the many species that Europeans considered medicinal are "Lilies," "Amaryllis," and "Crocus."

One of the early New World visitors discovered another "lily" growing in the coastal regions of what is now the State of Virginia. This lily became the first of the rain lilies known to the world (Kral 1983a, Meerow 1984, 1985).

This rain lily received its current name, *Zephyranthes atamasca*, in 1821 from William Herbert (1778–1847), a British politician and botanist who later became active in the Church of England and was dean of Manchester. However, European knowledge of these herbs began long before that. The first record of the plants in Europe seems to have been with John Parkinson who called them *Lilio Narcissus virginicus* (Daffodil lily from Virginia) in 1629. Then, in 1696, Leonard Plukenet called them *Lilio narcissus liliflorus carolinianus* (Daffodil lily from Carolina with lily flowers). Catesby (1731–1732) reverted to the Parkinson name.

Parkinson cited the common name in Virginia as *attamusca* Although Mish (1988) wrote that the word appeared in English in 1743, it actually appeared earlier. *Attamusco* was used by Parkinson in 1629, Plukenet in 1696, and Morrison's *History of Plants* published in 1699. Catesby (1731–1732) also wrote "The Attamusco Lily." Mish (1988) said that *attamusco* meant "it is red," but Gerard (1907) contended that it came from *ätamäsku* (*ätam*, under, *äsku*, grass. Virginia Algonquian).

Linnaeus ([1753] 1957) picked up the common name and called the herbs *Amaryllis atamasca* As more kinds of "lilies" became known to Europeans, they realized that several genera were involved. Now *Zephyranthes* contains 40 species in the warm-temperate and tropical parts of the Americas (Meerow 1984, 1985, Mabberley 1997). Florida has three species—*Z. atamasco, Z. simpsonii,* and *Z. treatiae* (Wunderlin 1998). Only information on *Z. atamasco* being used has been found, but perhaps all three were part of the ethnoflora.

When John K. Small wrote his *Manual of the Southeastern Flora* in 1933, he recorded common names for the four *Zephyranthes* he found. He and others called them amaryllis, atamasco-lilies, Easter-lilies, fairy-lilies, rain-lilies, stagger-grasses, and Zephyr-lilies. To Bahamians they are crocus or snowdrop. Colombians think they look like the *jacinto* (hyacinth). Cubans call them *brujita* (little witch); Puerto Ricans and Dominicans say *duende* (little devil) or *adelfa* (Puerto Rico). Haitians say they are a *tulipe* (tulip), while they are *petit lis* (little lily) or *lis-savane* (savanna lily) in Guadeloupe and Martinique.

These common names tell us several things about what people think of the plants. First, Europeans and their descendants in the Americas saw resemblances between Old World plants (lilies, crocus, amaryllis, snowdrops, hyacinth) and those found in the New World. Instead of incorporating new names for the American plants, sometimes they simply transplanted old ones for them. At other times, the newcomers adopted indigenous names. The specific name *atamasco* and the common name atamasco-lilies are derived from *attamusco* used for the plants by speakers of an Algonquian language in Virginia. *Atamasco* is a reference to where the plants grew.

The generic name of rain lilies speaks of plant biology. *Zephyranthes* is an allusion to their tendency to flower suddenly after a spring or summer downpour, appearing as if by magic from what had seemed to be grass leaves. The names *brujita*, *duende*, fairy-lilies, rain lilies, and Zephyr-lilies all refer to the abrupt emergence.

Stagger-grasses alludes to the symptoms shown by cattle and horses when they graze on *Zephyranthes* (Gerard 1907). Indeed, *Zephyranthes* contains a cocktail of alkaloids including poisons (e.g., lycorine, glycosides, pancratistatin) and antitumor agents (Krishna Rao 1969, Kobayashi et al. 1977, Nakayama et al. 1978, Krishna Rao and Krishna Rao 1979, Petitt et al. 1984, 1990) Ghosal et al. 1985, 1986, 1987, Kojima et al. 1998). Hedrick (1919) and Yanovsky (1936) say that the bulbs of Atamasco-lily were eaten by the Creek Indians in time of scarcity, but that would be risky. Perhaps stagger-grass was made acceptable by roasting, as is *Z. candida* in tropical America (Hocking 1997). Seminoles also used *Zephyranthes* as an analgesic against toothache (Small 1933). Other species have been used to treat diabetes, to reduce fevers, in poultices against abscesses, against colds, coughs, and tuberculosis, and to treat external infections (Eldridge 1975, Morton 1981, Ayensu 1981).

Zephyranthes species are under different controls on flowering, and there sometimes seems to be no rhyme or reason as to when blossoms appear (Broyles and Watt 1991, Mori and Imanishi 1997). Less is known about the introduced and cultivated species, but many are stimulated to flower by fires, mowing, and spring or summer rains. When fires burn throughout our remaining pine flatwoods, the rain lilies are among the species responding. Even though the timing of the rains varies, if the temperatures change enough the plants are stimulated into flowering. Mowing is an artificial "fire" from the plant's view. When these herbs are in lawns, roadsides, or pastures, mowing causes them to flower in the "wrong" season, especially if the soil is moist enough. Yet, Florida natives are especially sensitive to Zephyr when he starts the spring rains, which then bring the rain lilies to their peak.

Old World Relatives

Amaryllis—This word appeared in the Greek poetry of Theocritus (310–250 B.C.), and was used later by Latin poets Ovid (43 B.C.–A.D. 18) and Virgil (70–19 B.C.). The original name was of a Greek country girl or shepherdess. By 1637, herbalist-physicians had begun using "amaryllis" for liliaceous plants. Finally, Linnaeus started applying it to relatives of *Hippeastrum* (common name amaryllis) in 1738. As a genus, *Amaryllis* is restricted now to *A. belladonna*, the belladonna lily from South Africa. However, by 1753 this African species was already well known in Europe, the Caribbean, and South America.

Crocus—Another Greek classical name, *krokos*, has cognates in Hebrew (*karkom*), and Arabic (*kurkum*). Crocus was a young Greek shepherd who fell in love with Smilax, a shepherdess. She ignored him, and when he died of a broken heart, the gods changed him into a flower bearing his name. *Crocus* spread in Latin northward from the Mediterranean and became part of Old English (*croh*) and *kroch* in Irish and Gaelic.

"Crocus" turns out to be two botanical genera in different families. *Crocus sativus* (Iridaceae, iris family) is the source of saffron (from Arabic *za'faran*). Although originally used as a reddish dye, saffron eventually became the signature ingredient in certain Mediterranean foods. Then it was discovered that those who used the seasoning had fewer cardiovascular problems than their neighbors. Not only does the spice reduce heart and other circulatory problems, but it is also the richest known source of vitamin B_2 (riboflavin).

The other "crocus," *Colchicum autumnale* (Colchicaceae), is the source of colchicum (*tinctoria colcichi*), which is used as a pain killer for gout. The active ingredient, colchicine, is an alkaloid that also causes errors in cell division, resulting in doubled chromosomes. Colchicine was first used in plant breeding, and then in treating certain kinds of cancer.

See also *Lilium*.

Zizania

(Adapted from the Greek *zizanion*, a weed of wheat fields)

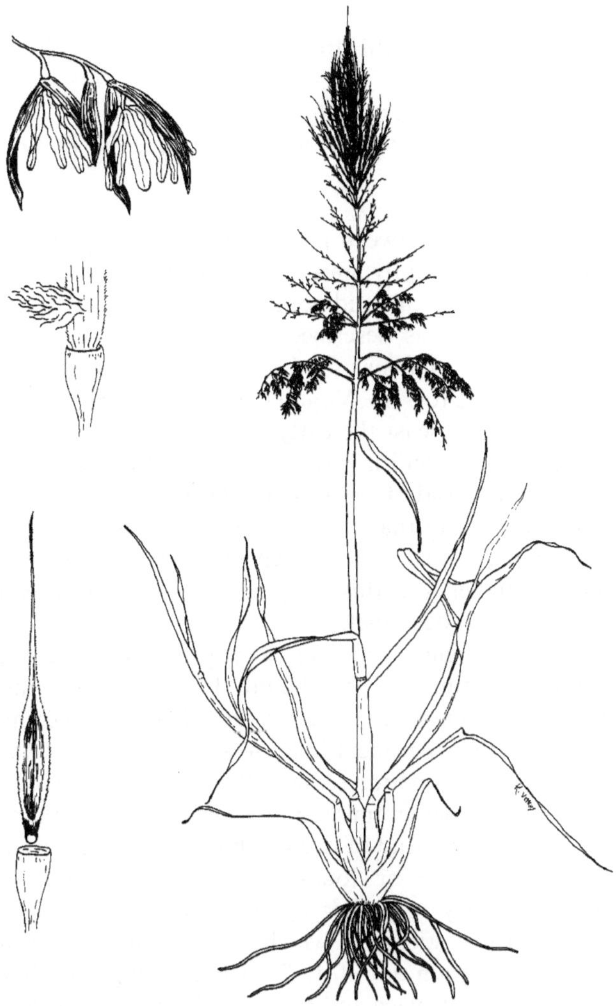

Zizania aquatica. From Institute of Food and Agricultural Sciences.

Zizania aquatica (of wetlands)

apetélewimin (laughing grains, Penobscot); *matomin* (Capt. John Smith and William Strachey wrote the Virginia Powhatan word *mattoume*); cognates include *mal8min* (Abenaki); *malo·mina* (Miami); *malomin* (Nipmuck-Pocumtuck); *mandamîn* (good berry, Potawatomi); *manomeh* (Menomini); *manomin* (Swampy Cree, a loanword from Ojibwa); *manomina* (Fox); *menomin* [*mano'min, minomina, menomen*] (Ojibwa); *omanomen* (the Menomini tribe, whose name was taken from *omanomen*, rice, and *inini*, people)

bo'-çu (Osage); *psin* (Dakota); *sin* (Winnebago); *sinwaninda* (Omaha-Ponca)

folle avoine (wild or foolish rice, Quebec)

honoshe (Choctaw; Bushnell 1909); *aloshommá* (*aló:so*, rice, *hommá*, red, Koasati)

[annual, Canadian, Indian, wild] rice [wildrice]

riz sauvage (wild rice, Quebec)

te'ue-u (Atakapa)

Tuscarora rice (this Iroquoian tribe originally lived in Virginia)
water oats [wateroats]
widlogouil (Natchez, fide Du Pratz 1758 in Gilmore 1931)

The chasm between people of the real world and "science" constantly amazes me. Europeans in the scientific community did not know wild rice until after 1743 when it was described by Jan Gronovius from Virginia. The plants as food, however, were known to European visitors in Virginia and the Great Lakes region from at least the early 1600s.

Capt. John Smith, writing in the first two decades of the 1600s, told of the Virginia Powhatans eating the grain. He wrote that it "groweth as our bents do in meadows ... seed is not much unlike to rye, though much smaller ... this they use for dainty bread buttered with deer suet."

Later, French explorer Pierre Esprit Radisson wrote of eating the rice before 1660 when he and Médard Chouart Groseliers returned from exploring the *Trois-Rivières*, opening up new territory in the far end of Lake Superior (Allen 1999). Radisson wrote: "Our songs being finished, we began our teeth to work. We had a kind of rice much like oats. It grows in the water three or four feet deep. They have a particular way to gather up that grain. Two take a boat and two sticks, by which they get the ear down and get the grain out of it. Their boat being full, they bring it to a fit place and dry it, and this is their food for the most part of the winter, and they do dress it thus: for each man a handful of that they put in a pot, that swells so much that it can suffice a man" (Radisson in Olson 1963).

Father Marquette in 1673 spoke of the tall grass growing in the rivers and swampy places, which "[t]he Savages Gather and prepare for food in the month of September." He then described how they shook the seeds from the stems into their canoes, dried it on a grating over a low fire, and trod the grain to separate the chaff.

The grass was so important to dwellers in the Great Lakes region, that early French explorers called one tribe the *Folles Avoines*, although its real name was Menomini. Even the State of Wisconsin may owe its name to *Zizania* (Olson 1963). Although that name is said to come from "an Algonquin river name" (Billard 1974), it probably came from *weese-co-seh*, meaning a good place to live. In other words, a place where wild rice grew.

Another early account of the grass was written by Father Hennepin who, in 1680, ate the grain on the upper Mississippi River and pronounced it better and more wholesome than rice (Hedrick 1919). Finally, Jan Gronovius named the plants *Zizania* Linnaeus ([1753] 1957) accepted his designation as *Zizania*, and added the species epithet *aquatica*.

Not long after receiving a formal Latin name, these grasses were discussed by other explorers. In 1784, Jonathan Carver wrote of wild rice as the most valuable of all the spontaneous productions of the "Northwest" [Great Lakes Region]. Regarding this edible grass, Porcher (1863) continued, "This plant was experimented with by Sir Joseph Banks, by removing it from Canada to England in 1791. At first it could scarcely endure the climate, but gradually improved and became thoroughly acclimated. It became 'in fourteen generations as strong and as vigorous as our indigenous plant.' 'It abounds in all the shallow streams of North America, feeds immense flocks of wild swans and other water-fowl, contributes largely to the support of the wandering tribes of Indians. ... This grain has become acclimated in Middlesex, producing bland, farinaceous seeds, which afford a very good meal.' ... It would perhaps reward the trouble to experiment with this plant at the South, in order by cultivation to procure a new cereal."

Zizania is a small genus, containing three or four species. While Hitchcock and Chase (1950), Fernald (1950), and Watson and Dallwitz (1992) recognized three species, Kartesz (1994) included four. The difference is whether or not the plants in the Great Lakes region are distinct species or varieties (Warwick and Aiken 1986, Duvall and Biesboer 1988, Reveal 1992b, Terrell et al. 1997). If they are considered distinct species, then North America has *Z. aquatica* (mostly coastal regions), *Z. texana* (Texas), and *Z. palustris* (Great Lakes vicinity). If combined, *Z. aquatica* has priority over *Z. palustris*

The other species, *Z. latifolia,* grows in Asia, and it too has been considered only a variety of *Z. aquatica* Young shoots of *Z. latifolia*, known as *kaw-sun*, are eaten in Chinese food to give texture; spores from a smut fungus growing on the plant are used to dye eyebrows and hair (Hedrick 1919). Mats are made from the leaves (Mabberley 1997).

Zizania texana has always been a restricted endemic in Texas, confined to the San Marcos River region of Hays County. When Correll and Johnston (1970) wrote the flora of the state, they considered it rare. Now the species is endangered (Poole and Bowles 1999).

Thomas Jefferys wrote in 1760 that the people of Louisiana gathered the seeds and made them into bread (Hedrick 1919). That seems to be only the second record (after Capt. John Smith) of early use by people in the southeast. However, the name "Tuscarora rice" suggests that the grain was also eaten by that tribe in Virginia and by the Powhatans there. Indeed, it was a statement written in the 1860s that probably explains the lack of information about these grasses in the South. Porcher (1863) wrote that, in the opinion of some, *Z. aquatica* "seems destined ... to become the bread-corn of the North." The "bread-corn" of the South was *Zea mays* Although *Zizania* grew in the southeast, reaching as far west as Texas, indigenous people and then settlers and slaves were accustomed to eating maize; any other grain was secondary. In Texas, *Z. texana* was eaten like *Z. aquatica* (Hocking 1997), but the rarity of this species has precluded its widespread use.

The grains of *Zizania* were important food to the Dakota (North and South Dakota), Meskwaki (Iowa), Menomini (upper peninsula Michigan), Ojibwa (Ontario), Omaha (Nebraska), Ponca (South Dakota and Nebraska), Potawatomi (lower peninsula Michigan), Thompson (British Columbia), and Winnebago (Minnesota, northern Illinois) (Gilmore 1919, Smith 1933, Olson 1963, Moerman 1998). Maybe the Iroquois also ate it because Yanovsky (1936) indicated that it had been eaten in New York. That is surely an underestimate of its importance because Hedrick (1919) noted that the grass was abundant on the Hudson and Delaware Rivers in New York. At the higher latitudes where there were abundant wetlands, this grass was an important element in the diet of all the tribes (Gilmore 1919, Hedrick 1919). Olson (1963) reminded us that "Bloody tribal wars were once fought for its possession, and those whose lands included strands of it were considered wealthy and insured against starvation and want."

Lorenz (1981) found that most of the modern commercial rice comes from Minnesota, with smaller quantities from Wisconsin and southern Canada. *Zizania* rice is relatively high in protein when compared with other cereals, a good source of the B vitamins, thiamin, riboflavin, and niacin, and contains common mineral elements in amounts comparable to those in oats, wheat, and corn. Hocking (1997) added that, apart from being a popular side dish in restaurants and groceries, with or without Old World rice (*Oryza sativa*), it is also used in hash, sauces, and seafoods. He noted that medicinally *Zizania* was made into a gruel and mixed with wild licorice tea to treat cholera (Hocking 1997), but he failed to tell by whom.

Zizia

(Named for Johann Baptist Ziz, 1779–1829, a botanist of the Rhine region of Germany)

Zizia aurea (golden)

golden Alexander(s) ("Alexanders" is a name given to *Smyrnium* in Europe; perhaps the name is a corruption of *olus atrum*, black oil, but the presence of *alexandre* about A.D. 940 seems to contradict that view; French *alexandre* is cognate; "golden" was added to distinguish the American from the Old World plants)

golden zizia (a book name)

[golden] meadow parsnip ("parsnip" is corrupted from Middle English *passenep*, ultimately from Latin *pastinaca*, for *Pastinaca sativa*; in use by A.D. 1389; cognate words are German *Pastinak(e)*, Dutch *pastinak*, Italian *pastinaca*, and French *panais*; meadow parsnip was originally in use by 1562 for both a European *Heracleum* and American *Thaspium;* later applied to *Zizia*); *goldene Pastinake* (German, fide Millspaugh 1892)

wild parsley ("parsley" is from Latin *petroselenium* and Greek *petroselinion*, rock parsley, now *Petroselinum crispum*; cognates are German *Petersilie*, Dutch *peterselie*, Italian *petrosellino*; the name "wild parsley" has been in use since A.D. 1265 when it was applied to Europcan Apiaceae other than cultivated parsley)

Linnaeus ([1753] 1957) called these plants *Smyrnium aureum,* following the name given to them by Leonard Plukenet in 1696. The name was particularly apt, because he was comparing them with myrrh (Greek *smyrna*) because of the fragrance, and gold (*aurea*) because of the yellow flowers. Linnaeus was using material studied by Jan Gronovius for *Flora Virginica*, Robert Morrison's list of plants cultivated in England, and specimens grown at the *Hortus Cliffortianus.* Subsequently, Wilhelm Daniel Joseph Koch (1771–1849), a German professor of botany at Erlangen, realized that these American and the European plants were not the same. One of the major differences between the Old and New World plants was that those in Europe were edible, being used as a substitute for celery and parsnips. The American *Smyrnium aureum* was inedible. There are seven Old World species of *Smyrnium* (Mabberley 1997).

In 1825, Koch segregated out the American plants and put them in *Zizia.* Golden Alexanders grows from Saskatchewan and Quebec to New Brunswick, New England, Georgia, Florida, Tennessee, Missouri, Oklahoma, and Texas (Fernald 1950). Later other species were found, and there are now four, all endemic to North America (Mabberley 1997).

Only one indigenous tribe was found to have used these herbs. The Meskwaki powdered Philadelphia fleabane (*Erigeron philadelphicus*) and golden Alexander, used it as a snuff to stop headache, and took the roots to reduce fevers (Gilmore 1919, Vogel 1970).

Among Europeans visiting the New World, this herb was apparently first spoken of by Rafinesque (1830) as vulnerary, antisyphilitic, and sudorific. Millspaugh (1892) knew of a young lady who had eaten a large root and subsequently vomited violently. He thought that, because she ejected the root, she was spared any other toxic action. Later, he experimented on himself with the plants and those experiences made him wonder about its safety. He wrote that his "tongue felt fuzzy and numb. This sensation was followed by a feeling as if the tongue had been scalded with hot tea; my eyes began to water and smart; I ceased writing, and threw myself upon my lounge (12 M); my face then began to feel suffused with blood and soon became hot, especially the cheeks and forehead; drowsiness followed, and I fell into a distressingly dreamy sleep, lasting an hour. When I awoke (1:30 P.M.) all my symptoms had passed away except the scalded sensation of the tongue, which lasted fully an hour longer."

Foster and Duke (1990) wrote, "Amateurs fooling with plants in the parsley family are playing herbal roulette." Their comment was written specifically for this species, but it applies to the whole family.

In spite of all these adverse reactions to golden Alexander, the root is still used as a diuretic in European homeopathy (Hocking 1997). Probably it is fortunate that Bremness (1994), Bown (1995), and Duke et al. (2002) do not list the species.

There are comparatively few reports of chemicals in *Zizia*, but *Z. aptera* has coumarins (Schramm 1978) and the glucoside apterin (Steck and Wetter 1974). Hocking (1997) recorded that *Z. aurea* contains aromatic volatile oils.

Zornia

(Named for German apothecary Johannes Zorn, 1739–1799)

Zornia bracteata (with bracts on the inflorescence)
bracted zornia (a book name)
cinquefoil (five-finger, from French; usually used for *Potentilla*); five-finger grass
viperina (little viper, Florida, Texas)

Plants now known as *Zornia* began in confusion. Except for the Florida species, most of us still find the names baffling.

Relatives of Florida's *viperina* appeared in the European literature at the same time from opposite sides of the world. Hedrik van Rheede used the Indian name *nelam-mari* (Malayalam) in the 1690s, and at the same time Hans Sloane described them from Jamaica under the Latin *Hedysarum minus diphyllum* (the smallest *Hedysarum* with two leaves). Later both N.L. Burmann and Linnaeus studied specimens for flora of Ceylon, and finally Linnaeus called them *Hedysarum diphyllum* in 1753. Oddly, Linnaeus ([1753] 1957) said that they were only found in India (*habitat in Indiis*). Unhappily, that lapse set the stage for continued confusion regarding what became the genus *Zornia*, because he had included two species under one name. Now the Indian plants are *Z. diphylla*, and the Jamaican plants are *Z. reticulata*

Zornia bracteata. From Britton and Brown 1897.

It was not until 1792 that the genus *Zornia* was named by J.F. Gmelin in the 13th edition of *Systema Naturae*. The species he used to establish the genus was *Z. bracteata*. That plant grows from Virginia through Florida and into eastern Texas (Fernald 1950, Mohlenbrock 1961, Correll and Johnston 1970). Small (1933) indicated that the species grew in Mexico, but neither Mohlenbrock (1961) nor floras indicate that. Now *Zornia* is considered a genus of 86 species in the warm parts of the world, with 19 species endemic to Australia (Mabberley 1997).

Zornia bracteata has been used as a diuretic and laxative (Hocking 1997). Those are the same uses for the widespread *Z. gemella* Given the incredible confusion in the scientific names and species limits, perhaps all American species are used interchangeably.

Most of the records of American plants are under the name *Z. diphylla,* but recent studies suggest that the correct name is *Z. gemella.* This herb grows in Texas, New Mexico, Arizona, and south through Mexico (at least Durango, Jalisco, Oaxaca, Queretaro, San Luis Potosí, Sinaloa, Sonora, Zacatecas) and Central America to Argentina and Paraguay. The legume also grows in Cuba (Leon and Alain 1946–1953) but not in the other islands of the Antilles. Adams (1972) referred Jamaican plants to

Z. reticulata. Other names for this species are *alverjilla* (sweet pea, Argentina), *barba de burro* (donkey beard, El Salvador), *cargadita* (little strong one, Colombia), *encarrujada* (curly one, Venezuela), *hierba de pujo* (puke herb, Venezuela), *urinaria* (urine maker, Brazil), and *zarzabacoa de dos hojas* (two-leaved trefoil, corrupted from *guasabacoa*, the Taino name, Puerto Rico).

Roig (1945) found the species called *trencilla* (adornment) in Cuba, where it was used to treat dysentery and fever, and as a diuretic. Martínez (1969) and Ford (1975) noted that *Z. diphylla* is called *hierba de la víbora* [*yerba de la víbora*] (viper herb) in Mexico, and that it was used to lower fever. Rojas et al. (1999) found the Otomí of Mexico using it against gastrointestinal disorders. Pérez-Arbelaez (1978) says that *Z. gemella* is used in Panama and Brazil to stop dysentery and lower fever, and that it is diuretic. Morton (1981) listed identical medical uses in Panama, El Salvador, Costa Rica, Guatemala, and elsewhere in Central America, and in Brazil.

Study by Rojas et al. (1999) suggested that the extracts from at least *Z. gemella* "tend to support ethnomedical use of the plants for the treatment of gastrointestinal disorders." The presence of other bioactive compounds in the genus is suggested by the name *maconha branca* (white marijuana) recorded for *Z. latifolia* in Brazil by Ghillean T. Prance et al. (Von Reis and Lipp 1982). Kloos et al. (1987) found that leaf extracts of *Z. setosa* killed 100% of the *Biomphalaria pfeifferi* snails in their study of African plants.

References

Aalbersberg, W.G.L., Hussein, S., Sotheeswaran, S., and Parkinson, S. 1993. Carotenoids in the leaves of *Morinda citrifolia. Journal of Herbs, Spices and Medicinal Plants* 2(1):51–54.

Abdel-Khalik, S.M., Miyase, T., Melek, F.R., el-Shabraway, O.A., Mahmoud, I.I., and Mina, S.A. 2002. New steroidal saponins from *Agave lophantha* Schiede and their pharmacological evaluation. *Pharmazie* 57(8): 562–566.

Abdel-Mogib, M., Basaif, S.A., Asiri, A.M., Sobahi, T.R., and Batterjee, S.M. 2001. New clerodane diterpenoid and flavonol-3-methyl ethers from *Dodonaea viscosa. Pharmazie* 56(10):830–831.

Abe, Y. 1993. Asparaginyl endopeptidase of jack bean seeds: purification, characterization, and high utility in protein sequence analysis. *The Journal of Biological Chemistry* 268(5):3525–3529.

Abeygunawardena, C., Kumar, V., Marshall, D.S., Thomson, R.H., and Wickramaratne, D.B.M. 1991. Furanonaphthoquinones from two *Lantana* species. *Phytochemistry* 30(3):941–945.

Abraham, E.A.V. 1912. Materia medica Guiana Britani Teheheri. *Journal of the Royal Agriculture and Commerce Society* (3rd ser.) 2(1):179–196.

Acevedo-Rodríguez, P. 2003. *Bejucos y Plantas Trepadoras de Puerto Rico e Islas Virgines.* Smithsonian Institution, Washington, D.C.

Acevedo-Rodríguez, P. and Collaborators. 1996. *Flora of St. John U.S. Virgin Islands.* New York Botanical Garden, Bronx, NY.

Acharya, R.N. and Chaubal, M.G. 1968. Essential oil of *Anemopsis californica. Journal of Pharmaceutical Sciences* 57(6):1020–1022.

Achenbach, H., Gross, J., Bauereiss, P., Dominguez, X.A., Vega, H.S., Star, J.V., and Rombold, C. 1989. Neolignans and nor-neolignans from *Krameria lanceolata. Phytochemistry* 28:1959–1962.

Achhireddy, N.R., Singh, M., Achhireddy, L.L., Nigg, H.N., and Nagy, S. 1985. Isolation and partial characterization of phytotoxic compounds from lantana (*Lantana camara* L.). *Journal of Chemical Ecology* 11(8):979–988.

Acuña, U.M., Atha, D.E., Ma, J., Nee, M.H., and Kennelly, E.J. 2002. Antioxidant capacities of ten edible North American plants. *Phytotherapy Research* 16(1):63–65.

Adair, M. 2003. Great Plains Paleoethnobotany. Pp. 258–346. In: Minnis, P.E., Ed. *Plants and People in Ancient Eastern North America.* Smithsonian Institution Press, Washington, D.C.

Adams, C.D. 1971. *The Blue Mahoe and Other Bush.* McGraw-Hill Far Eastern, Singapore.

Adams, C.D. 1972. *Flowering Plants of Jamaica.* University of the West Indies, Mona.

Adams, K.R. and Adams, R.K. 1998. How does our agave grow? Reproductive biology of a suspected ancient Arizona cultivar, *Agave murpheyi* Gibson. *Desert Plants* 14(2):11–14.

Adams, R.P. 1993. *Juniperus.* Pp. 412–420. In: Morin, N. and editorial committee, Eds. *Flora of North America.* Vol. 2. *Pteridophytes and Gymnosperms.* Oxford University Press, New York.

Adams, R.P., Zanoni, T.A., Beck, T.A. van, Posthumus, M. A., and Haar, C. van der. 1998. Essential leaf oil of *Amyris diatrypa* Sprengel from the Dominican Republic. *Journal of Essential Oil Research* 10(2): 175–178.

Adesanya, S.A., Nia, R., Martin, M.T., Boukamcha, N., Montagnac, A., and Pais, M. 1999. Stilbene derivatives from *Cissus quadrangularis. Journal of Natural Products* 62(12):1694–1695.

Adesogan, E.K. 1979. Oruwacin, a new iridoid ferulate from *Morinda lucida. Phytochemistry* 18(1):175–176.

Adewunmi, C.O. and Kayode Adesogan, E. 1984. Anthraquinones and oruwacin from *Morinda lucida* as possible agents in fascioliasis and schistosomiasis control. *Fitoterapia* 5:259–263.

Adicks, R., Ed. 1978. *LeConte's Report on East Florida.* Florida Technological University, University Presses of Florida, Orlando.

Adjibade, Y., Hue, B., Pelhate, M., and Anton, R. 1991. Action of calycanthine on nervous transmission in cockroach central nervous system. *Planta Medica* 57(2):99–101.

Ageel, A.M., Mossa, J.S., al-Yahya, M.A., al-Said, M.S., and Tariq, M. 1989. Experimental studies on antirheumatic crude drugs used in Saudi traditional medicine. *Drugs under Experimental Clinical Research* 15 (8):369–372.

Ahmad, A., Davies, J., Randall, S., and Skinner, G.R. 1996. Antiviral properties of extract of *Opuntia streptacantha. Antiviral Research* 30(2–3):75–85.

Ahmad, I., Mehmood, Z., and Mohammad, F. 1998. Screening of some Indian medicinal plants for their antimicrobial properties. *Journal of Ethnopharmacology* 62(2):183–193.

Ahmad, S.A., Catalano, S., Marsili, A., Morelli, I., and Scartoni, V. 1977. Chemical examination of the leaves of *Ardisia solanacea. Plant Medica* 32(2): 162–164.

Ahmad, S., Malik, A., Afza, N., and Yasmin, R. 1999. A new withanolide glycoside from *Physalis peruviana. Journal of Natural Products* 62(3):493–494.

Ahmad, V.U. and Fizza, K. 1988a. Arguticinin, a sesquiterpene from *Pluchea arguta. Phytochemistry* 27(6): 1861–1862.

Ahmad, V.U. and Fizza, K. 1988b. Argutin: a new cuauhtemone ester from *Pluchea arguta. Planta Medica* 54 (5):462–463.

Ahmad, V.U., Bano, N., and Bano, S. 1984a. Officigenin, a new sapogenin of *Guaiacum officinale. Journal of Natural Products* 47(6):977–982.

Ahmad, V.U., Bano, N., and Bano, S. 1984b. Sapogenins from *Guajacum officinale*. *Phytochemistry* 23(11): 2613–2616.

Ahmad, V.U., Bano, N., and Bano, S. 1985. A new prosapogenin from *Guaiacum officinale*. *Journal of Natural Products* 48(5):826–827.

Ahmad, V.U., Bano, N., and Bano, S. 1986a. A saponin from the stem bark of *Guaiacum officinale*. *Phytochemistry* 25(4):951–952.

Ahmad, V.U., Bano, N., Bano, S., and Fatima, A. 1986b. Guaianin, a new saponin from *Guaiacum officinale*. *Journal of Natural Products* 49(5):784–786.

Ahmad, V.U., Fizza, K., and Amber, A.U.R. 1989a. Odontin and odonticin, two new eudesmane sesquiterpenes from *Pluchea arguta*. *Journal of Natural Products* 52 (4):861–863.

Ahmad, V.U., Uddin, S., Bano, S., and Fatima, I. 1989b. Two saponins from fruits of *Guaiacum officinale*. *Phytochemistry* 28(8):2169–2171.

Ahmad, V.U., Perveen, S., and Bano, S. 1989c. Guaiacin A and B from the leaves of *Guaiacum officinale*. *Planta Medica* 55(3):307–308.

Ahmad, V.U., Perveen, S., and Bano, S. 1990a. Saponins from the leaves of *Guaiacum officinale*. *Phytochemistry* 29(10):3287–3290.

Ahmad, V.U., Uddin, S., and Bano, S. 1990b. Isolation and structure elucidation of saponins from the fruit of *Guaiacum officinale*. *Journal of Natural Products* 53 (5):1168–1175.

Ahmad, V.U., Fizza, K., Khan, M.A., and Farooqui, T.A. 1991. Sesquiterpenes from *Pluchea arguta*. *Phytochemistry* 30(2):689–690.

Ahmad, V.U., Farooqui, T.A., Fizza, K., Sultana, A., and Khatoon, R. 1992a. Three new eudesmane sesquiterpenes from *Pluchea arguta*. *Journal of Natural Products* 55(6):730–735.

Ahmad, V.U., Farooqui, T.A., Sultana, A., Fizza, K., and Khatoon, R. 1992b. Two sesquiterpenes from *Pluchea arguta*. *Phytochemistry* 31(8):2888–2890.

Ahmad, V.U., Abbasi, M.A., Hussain, H., Akhtar, M.N., Farooq, U., Fatima, N., and Choudhary, M.I. 2003. Phenolic glycosides from *Symplocos racemosa*: natural inhibitors of phosphodiesterase I. *Phytochemistry* 63(2):217–220.

Ahmed, A.A., El-Seedi, H.R., Mahmoud, A.A., El-Aziz, A., El-Douski, A.E.A.A., Zeid, I.F., and Bohlin, L. 1998. Eudesmane derivatives from *Laggera crispata* and *Pluchea carolinensis*. *Phytochemistry* 49(8): 2421–2424.

Ahmed, M., Shikha, H.A., Sadhu, S.K., Rahman, M.T., and Datta, B.K. 2001. Analgesic, diuretic, and antiinflammatory principle from *Scoparia dulcis*. *Pharmazie* 56(8):657–660.

Ahn, B.T., Lee, S., Lee, S.B., Lee, E.S., Kim, J.G., Bok, S.H., and Jeong, T.S. 2001. Low-density lipoprotein-antioxidant constituents of *Saururus chinensis*. *Journal of Natural Products* 64(12):1562–1564.

Ajilvsgi, G. 1984. *Wildflowers of Texas*. Shearer Publishers, Bryan, TX.

Akhondzadeh, S., Kashani, L., Mobaseri, M., Hosseini, S. H., Nikzad, S., and Khani, M. 2001. Passionflower in the treatment of opiates withdrawal: a double-blind randomized controlled trial. *Journal of Clinical Pharmacology Therapy* 26(5):369–373.

Akhtar, A.H. and Ahmad, K.U. 1995. Anti-ulcerogenic evaluation of the methanolic extracts of some indigenous medicinal plants of Pakistan in aspirin-ulcerated rats. *Journal of Ethnopharmacology* 46 (1):1–6.

Akpapunam, M.A. and Sefa-Dedeh, S. 1997. Some physicochemical properties and anti-nutritional factors of raw, cooked and germinated Jack bean (*Canavalia ensiformis*). *Food Chemistry* 59(1):121–125.

Alabama-Coushatta Tribe of Texas. 2003. United Houma Nation. URL: www.geocities.com/houmaindians/.

Alam, M.I., Auddy, B., and Gomes, A. 1996. Viper venom neutralization by Indian medicinal plant (*Hemidesmus indicus* and *Pluchea indica*) root extracts. *Phytotherapy Research* 10(1):58–61.

Alanis-Guzman, M.G., González-Quijada, M.R., and Mercado-Hernández, R. 1998. Effect of cooking on the chemical and nutritional value of the *Pithecellobium flexicaule* (Benth.) Coult. seed [in Spanish]. *Archivos Latinoamericanos de Nutricion* 48(4):328–333.

Alarcon-A., F.J., Roman-R., R., Pérez-G., S., Aguilar-C., A., Contreras-Weber, C.C., and Flores-S., J.L. 1998. Study of the anti-hyperglycemic effect of plants used as antidiabetics. *Journal of Ethnopharmacology* 61(2):101–110.

Alcock, J. 1985. *Sonoran Desert Spring*. University of Chicago Press, Chicago.

Alcorn, J.B. 1984. *Huastec Mayan Ethnobotany*. University of Texas Press, Austin.

Alden, H.A. 1994a. Technology Transfer Fact Sheet on *Gleditsia triacanthos*. United States Forest Service. Madison, WI.

Alden, H.A. 1994b. Technology Transfer Fact Sheet on *Maclura pomifera*. United States Forest Service. Madison, WI.

Alexander, I.C., Pascoe, K.O., Manchard, P., and Williams, L.A.D. 1991. An insecticidal diterpene from *Croton linearis*. *Phytochemistry* 30(6):1801–1803.

Ali, A.M., Ismail, N.H., Mackeen, M.M., Yazan, L.S., Mohamed, S.M., Ho, A.S.H., and Lajis, N.H. 2000. Antiviral, cytotoxic and antimicrobial activities of anthraquinones isolated from the roots of *Morinda elliptica*. *Pharmacien Biologiste* 38(4):298–301.

Ali, M.S., Sharma, G.C., Asplund, R.O., Nevins, M.P., and Garb. S. 1978. Isolation of antitumor polysaccharide fractions from *Yucca glauca* Nutt. (Lilliaceae). *Growth* 42(2):213–223.

Ali, M.S., Azhar, I., Ahmad, F., Ahmad, V.U., and Usmanghani, K. 2001. Antimicrobial screening of mimosaceous plants. *Pharmacien Biologiste* 39(1): 43–46.

Ali-Shtayeh, M.S. and Abu Ghdeib, S.I. 1999. Antifungal activity of plant extracts against dermatophytes. *Mycoses* 42(11–12):665–672.

Allen, C.M. 2003. *Paspilidium*, Pp. 558–560. In: Editorial Commitee. *Flora of North America North of Mexico.* Vol. 25. Commelinidae (in part): Poaceae, Part 2. Oxford University Press, New York.

Allen, J.S. 1999. Trois-Rivières to Trail's End, available at http://www.bikexprt.com/menard/menardc5.htm.

Almeida, C.E., Karnikowski, M.G., Foleto, R., and Baldisserotto, B. 1995. Analysis of antidiarrhoeic effect of plants used in popular medicine. *Revista de Saude Publica* 29(6):428–433.

Almeida-Alves, T.M. de, Nagem, T.J., Ribeiro, A., Mendes, N.M., Gomez, J.D., Zani, C.L., Hamburger, M., and Hostettmann, K. 1996. Molluscicidal saponins from *Guaiacum officinale* (Zygophyllaceae). *International Journal of Pharmacognosy* 34(2):81–86.

Al-Shamma, A., Drake, S.D., Guagliardi, L.E., Mitscher, L. A., and Swayze, J.K. 1982. Antimicrobial alkaloids from *Boehmeria cylindrica. Phytochemistry* 21(2): 485–487.

Alvarenga, N. and Ferro, E.A. 2000. A new lupane caffeoyl ester from *Hippocratea volubilis. Fitoterapia* 71 (6):719–721.

Amabeoku, G.J., Eagles, P., Scott, G., Mayeng, I., and Springfield, E. 2001. Analgesic and antipyretic effects of *Dodonaea angustifolia* and *Salvia africana-lutea. Journal of Ethnopharmacology* 75(ER2–3):117–124.

Amador, T.A., Verotta, L., Nunes, D.S., and Elisabetsky, E. 2000. Antinociceptive profile of hodgkinsine. *Planta Medica* 66(8):770–772.

Amador, T.A., Verotta, L., Nunes, D.S., and Elisabetsky, E. 2001. Involvement of NMDA receptors in the analgesic properties of psychotridine. *Phytomedicine* 8(3):202–206.

American Heritage. 2000. *American Heritage Dictionary of the English Language,* 4th ed. Houghton Mifflin, New York. URL: Bartlby.com.

Amladi, S.R. 1975. Danaid butterflies attracted to *Heliotropium indicum* (Boraginaceae), an alkaloid containing plant. *Journal of the Bombay Natural History Society* 72(2):585–587.

Anaya, A.L., Mata, R., Rivero-Cruz, F., Hernández-Bautista, B.E., Chavez-Velasco, D., and Gomez-Pompa, A. 1999. Allelochemical potential of *Metopium brownei. Journal of Chemical Ecology* 25(1):141–156.

Andallu, B., Suryakantham, V., Lakshmi Srikanthi, B., and Kesava Reddy, G. 2001. Effect of mulberry (*Morus indica* L.) therapy on plasma and erythrocyte membrane lipids in patients with type 2 diabetes. *Clinica Chimica Acta* 314(ER1–2):47–53.

Anderson, A. 1975. *Plains Cree Dictionary in the "y" Dialect.* Rev. ed. published by author, Edmonton, Canada.

Anderson, D.M.W., Weiping, W., and Lewis, G.P. 1990. The composition and properties of eight gum exudates (Leguminosae) of American origin. *Biochemical Systematics and Ecology* 18(1):39–42.

Anderson, E.F. 1986. Ethnobotany of hill tribes of northern Thailand. I. Medicinal plants of Akha. *Economic Botany* 40:48–51.

Ando, M., Arai, K., Kikuchi, K., and Isogai, K. 1994. Synthetic studies of sesquiterpenes with a *cis*-fused decalin system, 4. Synthesis of (+)-5 beta H-eudesma-3,11-diene, (–)-5 beta H-eudesmane-4 beta,11-diol, and (+)-5 beta H-eudesmane-4 alpha,11-diol, and structure revision of a natural eudesmane-4,11-diol isolated from *Pluchea arguta. Journal of Natural Products* 57(9):1189–1199.

Andrews, E.G. and Windham, M.D. 1993. *Pleopeltis.* In: Morin, N., Ed. *Flora of North America.* Vol. 2. *Pteridophytes and Gymnosperms.* Oxford University Press, New York.

Andrews, J. 1984. *Peppers. The Domesticated Capsicums.* University of Texas Press, Austin.

Andrews, E.W. and Andrews, C.M., Eds. 1945. *Jonathan Dickinson's Journal or God's Protecting Providence.* Yale University Press, New Haven, CT.

Anghiera, Peter Martyr de. [1555] 1895. The Decades of the New World, or West India. In: Arber, E. The first three English books on America. [?1511]–1555 A.D. Being chiefly translations, compilations, &c., by Richard Eden, from the writings, maps, &c., of Pietro Martire, of Anghiera (1455–1526) Sebastian Münster, the cosmographer (1489–1552) Sebastian Cabot, of Bristol (1474–1557) with extracts, &c., from the works of other Spanish, Italian, and German writers of the time. A. Constable, Westminster, U.K.

Angliclici hortulani (A Society of Gardeners). 1730. *Catalogus plantarum ... A catalogue of trees, shrubs, plants, and flowers.* London. 11. t. 4. cited in Linnaeus 1753.

Ankli, A., Heilmann, J., Heinrich, M., and Sticher, O. 2000. Cytotoxic cardenolides and antibacterial terpenoids from *Crossopetalum gaumeri. Phytochemistry* 54(5): 531–537.

Anonymous. 1990a. Weeds and yeasts may produce premium rubber. *Bioprocess Technology* 12(11):6–7.

Anonymous. 1990b. Weeds and yeast: tomorrow's rubber factories? *BioEngineering News* 11(44):6.

Anonymous. 2002. Wildflower nirvana, available at URL: www.wfnirvana.com/pr/surimari.html.

Anonymous. 2003. Beaufort, South Carolina History, available at URL: www.beaufort-sc.com/history/.

Antoun, M.D., Ramos, Z., Vazques, J., Oquendo, I., Proctor, G.R., Gerena, L., and Franzblau, S.G. 2001. Evaluation of the flora of Puerto Rico for *in vitro* antiplasmodial and antimycobacterial activities. *Phytotherapy Research* 15(7):638–642.

Apata, D.F. 1994. Biochemical evaluation of some Nigerian legume seeds. *Food Chemistry* 49(4):333–338.

Apori, S.O., Long, R.J., Castro, F.B., and Orskov, E.R. 2000. Chemical composition and nutritive value of leaves and stems of tropical weed *Chromolaena odorata. Grass and Forage Science* 55(1):77–81.

Aquino, R., Ciavatta, M.L., de Simone, F., and Pizza, C. 1990. A flavanone glycoside from *Hamelia patens. Phytochemistry* 29(7):2358–2360.

Aquino, R., de Simone, F., Pizza, C., Cerri, R., and Mello, J. F. de. 1988. Quinovic acid glycosides from *Guettarda platypoda. Phytochemistry* 27(9):2927–2930.

Aquino, R., de Simone, F., Pizza, C., Conti, C., and Stein, M.L. 1989a. Plant metabolites. Structure and *in vitro* antiviral activity of quinovic acid glycosides from *Uncaria tomentosa* and *Guettarda platypoda. Journal of Natural Products* 52(4):679–685.

Aquino, R., de Simone, F., Pizza, C., and de Mello, J.F. 1989b. Further quinovic acid glycosides from *Guettarda platypoda. Phytochemistry* 28(1):199–201.

Arber, A. 1986. *Herbals, Their Origin and Evolution: A Chapter in the History of Botany 1470–1670.* 3rd ed., with an introduction and annotations by William T. Stearn. Cambridge University Press, Cambridge, U.K.

Arenas, P. and Moreno-Axorero, R. 1977a. Plants of common use in Paraguayan folk medicine for regulating fertility. *Economic Botany* 31:298–301.

Arenas, P. and Moreno-Azorero, P. 1977b. Plants used as means of abortion, contraception, sterilization, and fecundation by Paraguayan indigenous people. *Economic Botany* 31:302–306.

Arizaga, S., Ezcurra, E., Peters, E., de Arellano, F.R., and Vega, E. 2000a. Pollination ecology of *Agave macroacantha* (Agavaceae) in a Mexican tropical desert. I. Floral biology and pollination mechanisms. *American Journal of Botany* 87(7):1004–1010.

Arizaga, S., Ezcurra, E., Peters, E., de Arellano, F.R., and Vega, E. 2000b. Pollination ecology of *Agave macroacantha* (Agavaceae) in a Mexican tropical desert. II. The role of pollinators. *American Journal of Botany* 87(7):1011–1017.

Arriaga, F.J. and Borges-del-Castillo, J. 1985. A new eudesmane derivative from *Pluchea odorata. Planta Medica* 51(3):290.

Arriaga-Giner, F.J., Borges-del-Castillo, J., Manresa-Ferrero, M.T., Vázquez-Bueno, P., and Rodriquez-Luis, F. 1983. Eudesmane derivatives from *Pluchea odorata. Phytochemistry* 22(8):1767–1769.

Arrieta, J., Reyes, B., Calzada, F., Cedillo-Rivera, R., and Navarrete, A. 2001. Amoebicidal and giardicidal compounds from the leaves of *Zanthoxylum liebmannianun. Fitoterapia* 72(3):295–297.

Aruna, K. and Sivaramakrishnan, V.M. 1992. Anticarcinogenic effects of some Indian plant products. *Food and Chemical Toxicology* 30(11):953–956.

Arvigo, R. and Balick, M. 1993. *Rainforest Remedies. One Hundred Healing Herbs of Belize.* Lotus Press, Twin Lakes, WI.

Asano, N., Kato, A., Oseki, K., Kizu, H., and Matsui, K. 1995. Calystegins of *Physalis alkekengi* var. *francheti* (Solanaceae). Structure determination and their glycosidase inhibitory activities. *European Journal of Biochemistry* 229(2):369–376.

Asano, N., Kato, A., Matsui, K., Watson, A.A., Nash, R.J., Molyneux, R.J., Hackett, L., Topping, J., and Winchester, B. 1997. The effects of calystegines isolated from edible fruits and vegetables on mammalian liver glycosidases. *Glycobiology* 7(8):1085–1088.

Asano, N., Yamashita, T., Yasuda, K., Ikeda, K., Kizu, H., Kameda, Y., Kato, A., Nash, R.J., Lee, H.S., and Ryu, K.S. 2001. Polyhydroxylated alkaloids isolated from mulberry trees (*Morus alba* L.) and silkworms (*Bombyx mori* L.). *Journal of Agriculture and Food Chemistry* 49(9):4208–4213.

Asano, S., Mizutani, M., Hayashi, T., Morita, N., and Takeguchi, N. 1990. Reversible inhibitions of gastric H+,K(+)-ATPase by scopadulcic acid B and diacetyl scopadol. New biochemical tools of H+,K(+)-ATPase. *Journal of Biological Chemistry* 265(36): 22167–22173.

Asch, D.L. and N.B. Asch. 1978. The economic potential of *Iva annua* and its prehistoric importance in the lower Illinois Valley. Pp. 301–341. In: Ford, R.I. et al., Eds. *The Nature and Status of Ethnobotany.* University of Michigan Museum of Anthropology, Ann Arbor.

Asch, D.L. and N.B. Asch. 1982. A chronology for the development of prehistoric horticulture in Westcentral Illinois. Center for American Archaeology, Archeobotanical Laboratory Report 46.

Asch, D.L. and N.B. Asch. 1985. Prehistoric plant cultivation in West-Central Illinois. Pp. 149–204. In: Ford, R.I., Ed. *Prehistoric Food Production in North America.* Anthropological Papers, Museum of Anthropology, University of Michigan, Ann Arbor.

Asibal, C.F., Gelbaum, L.T., and Zalkow, L.H. 1989. Pyrrolizidine alkaloids from *Heliotropium rotundifolium. Journal of Natural Products* 52(4):726–731.

Asres, K., Bucar, F., Kartnig, T., Witvrouw, M., Pannecouque, C., and De Clercq, E. 2001. Antiviral activity against human immunodeficiency virus type 1 (HIV-1) and type 2 (HIV-2) of ethnobotanically selected Ethiopian medicinal plants. *Phytotherapy Research* 15(1):62–69.

Asuzu, I.U. and Chineme, C.N. 1990. Effects of *Morinda lucida* leaf extract on *Trypanosoma brucei brucei* infection in mice. *Journal of Ethnopharmacology* 30 (3):307–313.

Aublet, J.F. 1775. *Histoire des Plantes de la Guiane Françoise.* Didot, Paris.

Austin, D.F. 1978. Gumbo Limbo. *The Florida Naturalist* 51 (3):6–7.

Austin, D.F. 1979. Plants without beginning or end. *Bulletin of the Fairchild Tropical Garden* 34(3):17–19.

Austin, D.F. 1980a. Vishnu's halo. *Bulletin of the Fairchild Tropical Garden* 35(2):26–29.

Austin, D.F. 1980b. Convolvulaceae. In: Dassenayake, M.D. and Fosberg, F.R., Eds. *A Revised Handbook of the Flora of Ceylon,* Vol. 1, 288–363. Amerind Publishing Company, New Delhi.

Austin, D.F. 1980c. Historically important plants of southeastern Florida. *Florida Anthropologist* 33:17–31.

Austin, D.F. 1981. Mastic. *The Florida Naturalist* 54(1):5.

Austin, D.F. 1984. Blue daze. *Bulletin of the Fairchild Tropical Garden* 39:18–21.

Austin, D.F. 1990. Vegetation of the Florida Atlantic University Ecolological Site. *Florida Scientist* 53 (1):11–27.

Austin, D.F. 1991. American bays. *The Palmetto* 11(2):12–13.

Austin, D.F. 1992. Black calabash in southern Florida. *The Palmetto* 12(1):6–7.
Austin, D.F. 1993. The nuance and wit of Carolus Linnaeus. *The Palmetto* 13(4):8–11.
Austin, D.F. 1994. Creepers, climbers, vines and wines. *The Palmetto* 14(3):6.
Austin, D.F. 1995. Black calabash bears fruit in Florida! *The Palmetto* 15(2):6–7.
Austin, D.F. 1997. Glades Indians and the plants they used. An ethnobotanical study of Florida's long-gone people. *The Palmetto* 17(2):7–10.
Austin, D.F. 1998a. Christmas botany or how reindeer learned to fly. *The Palmetto* 17(3):12–14, 23.
Austin, D.F. 1998b. Xixicamátic or wood rose (*Merremia tuberosa*, Convolvulaceae): origins and dispersal. *Economic Botany* 52(4):412–422.
Austin, D.F. 1998c. The indiscriminate vector: human distribution of *Dichondra micrantha* (Convolvulaceae). *Economic Botany* 52(1):88–106.
Austin, D.F. 2000a. Chiggery grape. *The Palmetto* 20(2). 6–9.
Austin, D.F. 2000b. Rain Lilies. *The Palmetto* 20(3):15–16.
Austin, D.F. 2000c. The search for "kaladana" (*Ipomoea*, Convolvulaceae). *Economic Botany* 54:114–118.
Austin, D.F. 2001a. Hoop vine: the plant that wasn't there. *The Palmetto* 20(4):10–12.
Austin, D.F. 2001b. Pond apple. *The Palmetto* 21(1):10–11.
Austin, D.F. 2002a. Chapter 57. Roots as a Source of Food, Pp. 1025–1043. In: Waisel, Y., Eshel, A., and Kafkaki, U., Eds. *Plant Roots. The Hidden Half*, 3rd ed., revised and expanded. Marcel Dekker, New York.
Austin, D.F. 2002b. Lizard's tail. *The Palmetto* 22(1):12–14.
Austin, D.F. 2002c. Milkworts. *The Palmetto* 21(2):10–11.
Austin, D.F. 2002d. Moonvine. *The Palmetto* 21(4):10–11.
Austin, D.F. 2002e. Sundews. *The Palmetto* 21(3):12–13.
Austin, D.F. 2003. Prickly pear. *The Palmetto* 22(2):10–12.
Austin, D.F. and Bourne, G.R. 1992. Notes on Guyana medical ethnobotany. *Economic Botany* 46(3):293–298.
Austin, D.F. and McJunkin, D.M. 1978. An ethnoflora of Chokoloskee Island, Collier County, Florida. *Journal of the Arnold Arboretum* 59(1):50–67.
Austin, D.F. and Oliver, R.L. 1974. *Sisyrinchium solstitale* (Iridaceae): a Florida endemic. *Journal of the Arnold Arboretum* 55:291–299.
Austin, D.F. and Smith, E. 1997. Pine Rockland Plant Guide. A Field Guide to the Plants of South Florida's Pine Rockland Community. 1997. Department of Environmental Resource Management, Environmentally Endangered Lands, Miami-Dade County, FL.
Austin, D.F. and Weise, J. 1972. An annotated checklist of the Boynton Beach hammock. *Quarterly Journal of the Florida Academy of Sciences* 35:145–154.
Austin, D.F., Jones, J.L., and Bennett, B.C. 1986. The Fakahatchee Strand. *The Palmetto* 6(2):3–6.
Austin, D.F., Cholewa, A.F., Lassiter, R.B., and Hansen, B.F. 1987a. *The Florida of John Kunkel Small.* Contributions from the New York Botanical Garden, Vol. 18. Bronx, New York.
Austin, D.F., Posin, F.R., and Burch, J.N. 1987b. Scrub species patterns on the Atlantic Coastal Ridge, Florida. *Journal of Coastal Research* 3(4):491–498.
Austin, D.F., Honychurch, P.N., and Bass, S. 1989. *Coastal Park Plant Guide. The Native Trees, Shrubs and Vines of Boca Raton's Hammock and Mangrove Parklands.* Gumbo Limbo Nature Center, Boca Raton, FL. Reprinted 1991, with corrections as Coastal Dune Plants. A Pocket Guide to the Common Plants of Southeast Florida's Ocean-side Communities. Published by the City of Boca Raton Department of Parks and Recreation, Boca Raton, FL.
Austin, D.F., Jones, J.L., and Bennett, B.C. 1990a. Vegetation of the Fakahatchee Strand State Preserve. *Florida Scientist* 53(2):89–117.
Austin, D.F., Jones, J.L., and Bennett, B.C. 1990b. Endangered plants of the Fakahatchee Strand State Preserve. *Rhodora* 92:27–35.
Austin, D.F., Honychurch, P.N., and Bass, S. 1991. *Coastal Dune Plants.* Gumbo Limbo Nature Center of South Palm Beach County, Boca Raton, FL.
Austin, D.F., Honychurch, P.N., and Bass, S. 1993. *Scrub Plant Guide. A Pocket Guide to the Common Plants of Southern Florida's Scrub Community.* Gumbo Limbo Nature Center, Boca Raton, FL.
Austin, D.F., Kitajima, K., Yoneda, Y., and Qian, L. 2001. A putative tropical American plant, *Ipomoea nil* (Convolvulaceae), in Pre-Columbian Japanese art. *Economic Botany* 55(4):515–527.
Austin, D.F., Honychurch, P.N., and Bass, S. 2003. *Flatwoods Plant Guide. A Pocket Guide to the Common Plants of Southern Florida's Pine Flatwoods Community.* Gumbo Limbo Nature Center of South Palm Beach County, Boca Raton, FL.
Axelius, B. 1996. The phylogenetic relationships of the physaloid genera (Solanaceae) based on morphological data. *American Journal of Botany* 83(1):118–124.
Ayensu, E.S. 1981. *Medicinal Plants of the West Indies.* Reference Publishers, Algonac, MI.
Azam, A. 1993. A triterpenoidal sapoginin from the seeds of *Dodonaea viscosa* Linn. *Indian Journal of Chemistry. Section B*, 32(4):513.
Badawi, M.M., Seida, A.A., Kinghorn, A.D., Cordell, G.A., and Farnsworth, N.R. 1981. Potential anticancer agents. XVIII. Constituents of *Amyris pinnata. Journal of Natural Products* 44(3):331–334.
Bagchi, D., Bagchi, M., Stohs, S., Ray, S.D., Sen, C.K., and Preuss, H.G. 2002. Cellular protection with proanthocyanidins derived from grape seeds. *Annals of the New York Academy of Sciences* 957:260–270.
Bailey, C. and Danin, A. 1981. Bedouin plant utilization in Sinai and the Negev. *Economic Botany* 35:145–162.
Bailey, H.E., Mooken, J.D., and Bailey, V.L. 1971. Studies on the geographical distribution of some alkaloids and coumarins in leaf extracts of *Ptelea* species. *Lloydia* 34(4):377–382.

Bailey, V.L. 1962. Revision of the genus *Ptelea* (Rutaceae). *Brittonia* 14:1–45.

Baker, M.F. 1938. *Florida Wild Flowers,* rev. ed. Macmillan, New York.

Baldini, V.L.S., Iaderoza, M., and Draetta, I.S. 1995. Anthocyanins from *Ardisia humilis. Tropical Science* 35(2):130–134.

Balick, M.J. and Cox, P.A. 1996. *Plants, People, and Culture. The Science of Ethnobotany.* Scientific American Library/W.H. Freeman, New York.

Balick, M.J., Nee, M.H., and Atha, D.E. 2000. *Checklist of the Vascular Plants of Belize. With Common Names and Uses.* New York Botanical Garden Press, Bronx, NY.

Ballot, D., Baynes, R.D., Bothwell, T.H., Gillooly, M., MacFarlane, B.J., MacPhail, A.P., Lyons, G., Derman, D.P., Bezwoda, W.R., Torrance, J.D. et al. 1987. The effects of fruit juices and fruits on the absorption of iron from a rice meal. *British Journal of Nutrition* 57(3):331–343.

Bambode, R.S. and Shukla, V.N. 1974. Studies on fungitoxic properties of *Plumbago zeylanica* L. extract. *Journal Punjabrao Krishi Vidyapeeth* 2(2):110–112.

Bankole, S.A. 1997. Effect of essential oils from two Nigerian medicinal plants (*Azadirachta indica* and *Morinda lucida*) on growth and aflatoxin B1 production in maize grain by a toxigenic *Aspergillus flavus. Letters in Applied Microbiology* 24(3):190–192.

Baranov, A.I. 1967. Wild vegetables of the Chinese in Manchuria. *Economic Botany* 21:140–155.

Barber, S.C. and Estes, J.R. 1978. Comparative pollination ecology of *Pyrrhopappus geiseri* and *P. carolinianus. American Journal of Botany* 65:562–566.

Barbera, G., Carimi, F., and Inglese, P. 1992. Past and present role of the Indian-fig (*Opuntia ficus-indica* (L.) Miller, Cactaceae) in the agriculture of Sicily. *Economic Botany* 46:10–20.

Barbera, R. 1992. Analgesic and antiinflammatory activity in acute and chronic conditions of *Trema guineense* (Schum. et Thonn.) Ficalho and *Trema micrantha* Blume extracts in rodents. *Phytotherapy Research* 6 (3):146–148.

Barbosa, P., Gross, P., Provan, G.J., Pacheco, D.Y., and Stermitz, F.R. 1990. Allelochemicals in foliage of unfavored tree hosts of the gypsy moth, *Lymantria dispar* L. 1. Alkaloids and other components of *Liriodendron tulipifera* L. (Magnoliaceae), *Acer rubrum* L. (Aceraceae), and *Cornus florida* L. (Cornaceae). *Journal of Chemical Ecology* 16(5): 1719–1730.

Barcia Carballido y Zuñiga, Andrés Gonález de. 1951. *Chronological History of the Continent of Florida.* Translation by Anthony Kerrigan from the original Spanish edition of 1723. University of Florida Press, Gainesville; reprinted by Greenwood Press, Westport, CT, 1970.

Barkley, T.M., Ed., 1986. *Flora of the Great Plains.* University Press of Kansas, Lawrence.

Barkley, T.M. and Cronquist, A. 1978. *Erechtites. North American Flora II.* 10:139–142.

Barkworth, M.E. 2003. *Ctenium.* Pp. 232–234. In: Editorial Committee. *Flora of North America North of Mexico.* Vol. 25. Commelinidae (in part): Poaceae, Part 2. Oxford University Press, New York.

Barneby, R.C. 1977. Daleae imagines. An illustrated revision of *Errazurizia* Philippi, *Psorothamnus* Rydberg, *Marina* Liebmann, and *Dalea* Lucanus emend. Barneby, including all species of Leguminosae tribe *Amorpheae* Borissova ever referred to *Dalea. Memoirs of the New York Botanical Garden* 27, 892 pp.

Barneby, R.C. and Grimes, J.W. 1996. Silk Tree, Guanacaste, Monkey's Earring. A Generic System for the Synandrous Mimosaceae of the Americas. Part I. Abarema, Albizia, and Allies. *Memoirs of the New York Botanical Garden* 74, Part 1. The New York Botanical Garden, Bronx, NY.

Barre, J.T., Bowden, B.F., Coll, J.C., DeJesus, J., De La Fuente, V.E., Janario, G.C., and Ragasa, C.Y. 1997. A bioactive triterpene from *Lantana camara. Phytochemistry* 45(2):321–324.

Barrett, B. 1994. Medicinal plants of Nicaragua's Atlantic coast. *Economic Botany* 48:8–20.

Barrientos, B. [1567] 1965. *Vida y Hechos de Pero. Menéndez de Aviles.* Translated by Anthony Kerrigan, and a facsimile reprinted as *Pedro Menéndez de Aviles, Founder of Florida.* University of Florida Press, Gainesville.

Barros, M.E., Schor, N., and Boim, M.A. 2003. Effects of an aqueous extract from *Phyllantus niruri* on calcium oxalate crystallization *in vitro. Urological Research* 30(6):374–379.

Barton, B.S. 1810. *Collections for an Essay towards a Materia Medica of the United States,* 3rd ed. Edward Earle, Philadelphia.

Bartram, W. [1791] 1958. Facsimile in Harper, F., Ed. *The Travels of William Bartram, Naturalist's Edition.* Yale University Press, New Haven, CT.

Bartram, W. [1853] 1909. Observations on the Creek and Cherokee Indians. With prefatory and supplementary notes by E.G. Squier. *Transactions of the American Ethnological Society* 3(1):1–81. Facsimile reprint.

Bartram, W. 1943. Travels in Georgia and Florida, 1773–1774. A report to Dr. John Fothergill. (Francis Harper ed.). *Transactions American Philosophical Society* 33, pt. 2.

Bartsch, G., Dreikom, K., and Schontoger, P.S. 1998. Combined *Sabal* and *Urtica* extract vs. finasteride in benign prostatic hyperplasia (Alken stages I to II). Comment on the contribution by J. Sokeland and J. Albrech. *Der Urologe. Ausg. A.* 1997. 36(4):327–333 [in German]. *Der Urologe. Ausg. A.* 37(1):83–85.

Baruah, R.N. and Leclercq, P.A. 1993. Constituents of the essential oil from the flowers of *Chromolaena odorata. Planta Medica* 59(3):283.

Baruah, R.N., Sharma, R.P., Thyagarajan, G., and Hertz, W. 1978. Flavonoids of *Chromolaena odorata. Phytochemistry* 17(10):1807–1808.

Bashir, A.K., Ross, M.S.F., and Turner, T.D. 1986. HPLC separation of harman alkaloids from *Grewia villosa* and of coumarins from *Randia nilotica. Fitoterapia* 3:190–192.

Basnet, P., Kadota, S., Terashima, S., Shimizu, M., and Namba. T. 1993. Two new 2-arylbenzofuran derivatives from hypoglycemic activity-bearing fractions of *Morus insignis. Chemical and Pharmaceutical Bulletin* (Tokyo) 41(7):1238–1243.

Bastos, J.K., Albuquerque, S., and Silva, M.L. 1999. Evaluation of the trypanocidal activity of lignans isolated from the leaves of *Zanthoxylum naranjillo. Planta Medica* 65(6):541–544.

Basu, N.G., Ghosal, P.K., and Thakur, S. 1986. Some structural features of an arabinoglucan from the fruits of *Cordia dichotoma* Forst. *Carbohydrate Research* 146(2):350–351.

Baumann, H. 1993. *The Greek Plant World in Myth, Art and Literature.* Timber Press, Portland, OR.

Bavaresco, L., Fregoni, C., Cantu, E., and Trevisan, M. 1999. Stilbene compounds: from the grapevine to wine. *Drugs under Experimental Clinical Research* 25 (2–3):57–63.

Beauvoir, M.G., DeFilipps, R.A., and Wolpert, B.J. with Crepin, J. 2001. Selected Medicinal Plants of Haitian Vodou. Department of Systematic Biology-Botany, National Museum of Natural History, Smithsonian Institution, Washington, D.C.

Beddows, C.G., Jagait, C., and Kelly, M.J. 2000. Preservation of alpha-tocopherol in sunflower oil by herbs and spices. *International Journal of Food Sciences and Nutrition* 51(5):327–339.

Bedi, S.J. 1978. Ethnobotany of the Ratan Mahal Hills, Guajarat, India. *Economic Botany* 32:278–284.

Bedigan, D. 2001. Review of *Seasoning Savvy. Economic Botany* 55:176.

Bedour, M.S., Elgamal, M.H., and El-Tawil, B.A. 1979. Steroid sapogenins, part XV. The constituents of *Agave utahensis* var. *nevadensis, A. lophantha* and *A. parasana. Planta Medica* 36(2):180–181.

Beek, T.A. van, Kleis, R., Posthumus, M.A., and Veldhuizen, A. van. 1989. Essential oil of *Amyris balsamifera. Phytochemistry* 28(7):1909–1911.

Beg, A.Z. and Ahmad, I. 2000. Effect of *Plumbago zeylanica* extract and certain curing agents on multidrug resistant bacteria of clinical origin. *World Journal of Microbiology and Biotechnology* 16(8/9):841–844.

Begum, S., Wahab, A., Siddiqui, B.S., and Qamar, F. 2000. Nematicidal constituents of the aerial parts of *Lantana camara. Journal of Natural Products* 63 (6):765–767.

Beith, M. 1995. *Healing Threads. Traditional Medicines of the Highlands and Islands.* Polygon, Edinburgh.

Belcher, R.O. 1956. A revision of the genus *Erechtites* (Compositae) with inquiries into *Senecio* and *Arrhenechites. Annals of the Missouri Botanical Garden* 43:1–85.

Benkrief, R., Ranarivelo, Y., Skaltsounis, A.L. Tillequin, F., Koch, M., Pusset, J., and Sevenet, T. 1998. Monoterpene alkaloids, iridoids and phenylpropanoid glycosides from *Osmanthus austrocaledonica. Phytochemistry* 47(5):825–832.

Bennati, E. 1971. Quantitative determination of harmane and harmine in the extract of *Passiflora incarnata* [in Italian]. *Bolletin Chimica Farmacologica* 110(11): 664–669.

Bennett, B.C. 1997. Seminole Ethnobotany. Notes from Big Cypress Seminole Reservation. Unpublished.

Bennett, B.C. and Hicklin, J.R. 1998. Uses of saw palmetto (*Serenoa repens,* Arecaceae) in Florida. *Economic Botany* 52(4):381–393.

Bennett, B.C. and Prance, G.T. 2000. Introduced plants in the indigenous pharmacopoeia of northern South America. *Economic Botany* 54 (1):90–102.

Bennett, B.C., Bell, C.R., and Boulware, R.T. 1990. Geographic variation in alkaloid content of *Sanguinaria canadensis* (Papaveraceae). *Rhodora* 92(870):57–69.

Bennett, R.S., Ed. 1982. *Lost Empires, Living Tribes.* National Geographic Society, Washington, D.C.

Benoit, F., Valentin, A., Pelissier, Y., Diafouka, F., Marion, C., Kone-Bamba, D., Kone, M., Mallie, M., Yapo, A., and Bastide, J.M. 1996. *In vitro* antimalarial activity of vegetal extracts used in West African traditional medicine. *American Journal of Tropical Medicine and Hygiene* 54(1):67–71.

Bentham, G. and Hooker, J.D. 1883. *Genera plantarum.* Reeve, Williams and Norgate, London.

Beretz, A., Roth-Georger, A., Corre, G., Kuballa, B., Anton, R., and Cazenave, J.P. 1985. Polyindolinic alkaloids from *Psychotria forsteriana.* Potent inhibitors of the aggregation of human platelets. *Planta Medica* 28 (4):300–330.

Bergen, F.D. 1892. Some bits of plant lore. *Journal of American Folk-lore* 5:19–22.

Bergen, F.D. 1892–1898. Popular American plant names. *Journal of American Folk-lore* 5:89–106. 1892; 6:135–142. 1893; 7:89–104. 1894; 9:179–193. 1896; 10:49–54. 1897; 11:221–230, 273–283. 1898.

Berger, I., Barrientos, A.C., Cáceres, A., Hernández, M., Rastrelli, L., Passreiter, C.M., and Kubelka, W. 1998. Plants used in Guatemala for the treatment of protozoal infections: II. Activity of extracts and fractions of five Guatemalan plants against *Trypanosoma cruzi. Journal of Ethnopharmacology* 62(2): 107–115.

Berkeley, E. and Berkeley, D.S. 1982. *The Life and Travels of John Bartram: From Lake Ontario to the River St. John.* University Presses of Florida, Tallahassee.

Bermejo-Benito, P., Díaz-Lanza, A.M., Silvan-Sen, A.M., De Santos-Galindez, J., Fernández-Matellano, L., Sanz-Gómez, A., and Abad-Martínez, M.J. 2000. Effects of some iridoids from plant origin on arachidonic acid metabolism in cellular systems. *Planta Medica* 66(4):324–328.

Bernardello, L.M. 1986. Revisión taxonómica de las especies sudamericanas de *Lycium* (Solanaceae). *Boletín de la Academia Nacional de Ciencias de Córdoba* 57:173–356.

Bernardo, R.R., Pinto, A.V., and Parente, J.P. 1996. Steroidal saponins from *Smilax officinalis. Phytochemistry* 43 (2):465–469.

Bertelli, A.A., Migliori, M., Panichi, V., Origlia, N., Filippi, C., Das, D.K., and Giovannini, L. 2002. Resveratrol, a component of wine and grapes, in the prevention of kidney disease. *Annals of the New York Academy of Sciences* 957:230–238.

Bhatnagar, S., Polonsky, J., Prange, T., and Pascard, C. 1984. New toxic quassinoid glucosides from *Simarouba glauca* (X-ray analysis). *Tetrahedron Letters* 25(3): 299–302.

Bhattacharya, S. and Chatterjee, M. 1998a. Protective role of *Trianthema portulacastrum* against diethylnitrosoamine-induced experimental hepatocarcinogenesis. *Cancer Letters* 129(1):7–13.

Bhattacharya, S. and Chatterjee, M. 1998b. *Trianthema portulacastrum* restores the antioxidant defense enzyme levels and hepatic biotransformation patterns in experimental rat hepatocarcinogenesis. *Italian Journal of Biochemistry* 47(4):225–232.

Bhattacharya, S. and Chatterjee, M. 1999. Inhibitory effect of *Trianthema portulacastrum* L. diethylnitrosoamine-induced phenobarbital promoted hepatocarcinogenesis. *Neoplasma* 46(2):105–111.

Bhattacharyya, J. and Carvalho, V.R. de. 1986. Epi-isoshinanolone from *Plumbago scandens. Phytochemistry* 25(3):764–765.

Bhattacharyya, J. and Cunha, E.V.L. 1992. A triterpenoid from the root-bark of *Chiococca alba. Phytochemistry* 31:2546–2547.

Bhattacharyya, J., Kokpol, U., and Miles, D.H. 1976. The isolation and synthesis of betulinaldehyde: a rare triterpene from insectivorous plant *Sarracenia flava. Phytochemistry* 15(3):432–433.

Bianchi, E. and Cole, J.R. 1969. Antitumor agents from *Agave schottii* (Amaryllidaceae). *Journal of Pharmaceutical Science* 58(5):589–591.

Bianchini, F. and Corbetta, F. 1975. *The Complete Book of Fruits and Vegetables.* Crown Publishers, New York.

Bickii, J., Njifutie, N., Foyere, J.A., Basco, L.K., and Ringwald, P. 2000. *In vitro* antimalarial activity of limonoids from *Khaya grandifoliola* C.D.C. (Meliaceae). *Journal of Ethnopharmacology* 69(1):27–33.

Bieber, L.W., Messana, I., Lins, S.C.N., Silva Filho, A.A. da, Chiappeta, A.A., and Mello, J.F. de. 1990. Meroterpenoid naphthoquinones from *Cordia corymbosa. Phytochemistry* 29(6):1955–1959.

Bieber, L.W., Krebs, H.C., and Schafer, W. 1994. Further meroterpenoid naphthoquinones from *Cordia corymbosa. Phytochemistry* 35(4):1027–1028.

Bilia, A.R., Morelli, I., and Méndez, J. 1996. New lupane derivatives from the leaves of *Licania pyrifolia. Journal of Natural Products* 59(3):297–300.

Billard, J.B.. Ed. 1974. *The World of the American Indian.* National Geographic Society, Washington, D.C.

Billing, J. and Sherman, P.W. 1998. Antimicrobial functions of spices: why some like it hot. *Quarterly Review of Biology* 73(1):3–49.

Binder, R.G. and French, R.C. 1994. Volatile components of Canada thistle. *Journal of Agricultural and Food Chemistry* 42(12):2942–2945.

Birecka, H., Frohlich, M.W., Hull, L., and Chaskes, M.J. 1980. Pyrrolizidine alkaloids of *Heliotropium* from Mexico and adjacent U.S.A. *Phytochemistry* 19(3): 421–426.

Birecka, H., Bitonti, A.J., and McCann, P.P. 1985. Assaying ornithine and arginine decarboxylases in some plant species. *Plant Physiology* 79(2):509–514.

Birecka, H., Birecki, M., and Frohlich, M.W. 1987. Evidence for arginine as the endogenous precursor of necines in *Heliotropium. Plant Physiology* 84(1):42–46.

Black, M. 1963. The distribution and archaeological significance of the marsh elder *Iva annua* L. *Papers of the Michigan Academy of Science, Arts and Letters* 48:541–547.

Blackmon, W.J. and Reynolds, B.D. 1986. The crop potential of *Apios americana* — preliminary evaluations. *Hortscience* 21:1334–1336.

Blackwell, W.H. 1990. *Poisonous and Medicinal Plants.* Prentice Hall, Englewood Cliffs, NJ.

Bloch, O. and von Wartburg, W. 1994. *Dictionnaire etymologique de la langue française.* Presses Universitaires de France, Paris.

Bloodworth, B.E. and Morris, A.C. 1978. *Places in the Sun. The History and Romance of Florida Place-Names.* University Press of Florida, Gainesville.

Blunden, G., Hardman, R., and Wensley, W.R. 1965. Effects of enzymes on *Yucca glauca* Nutt. and other steroid-yielding monocotyledons. *The Journal of Pharmacy and Pharmacology* 17(5):274–280.

Blunden, G., Yi, Y., and Jewers, K. 1974. A reinvestigation of the steroidal sapogenins of *Agave sisalana. Lloydia* 37(1):10–16.

Blunden, G., Carabot, A., Cripps, A.L., and Jewers, K. 1980. Ruizgenin, a new steroidal sapogenin diol from *Agave lecheguilla. Steroids* 35(5):503–510.

Blunden, G., Patel, A.V., and Crabb, T.A. 1986. Barbourgenin, a new steroidal sapogenin from *Agave sisalana* leaves. *Journal of Natural Products* 49(4):687–689.

Bock, M.A., Sanchez-Pilcher, J., McKee, L.J., and Ortiz, M. 1995. Selected nutritional and quality analyses of tomatillos (*Physalis ixocarpa*). *Plant Foods and Human Nutrition* 48(2):127–133.

Boeykens, M., Kimpe, N. de, Debenedetti, S.L., Nadinic, E. L., Gómez, M.A., Coussio, J.D., Abyshev, A.Z., and Gindin, V.A. 1994. Revision of the structure of obtusifol. *Phytochemistry* 36(6):1559–1560.

Bogle, A.L. 1970. The genera of Molluginaceae and Aizoaceae in the southeastern United States. *Journal of the Arnold Arboretum* 51:431–462.

Bogler, D.J. and Simpson, B.B. 1995. A chloroplast DNA study of the Agavaceae. *Systematic Botany* 20 (2):191–205.

Bogler, D.J. and Simpson, B.B. 1996. Phylogeny of Agavaceae based on ITS rDNA sequence variation. *American Journal of Botany* 83(9):1225–1235.

Bohlmann, F. and Zdero, C. 1976. Natürlich vorkommende Terpen Derivate. 66. Über neue Inhaltsstoffe aus *Pluchea odorata* Cass. *Chemische Berichte der Gesellschaft Deutscher Chemiker* 109(8):2653–2656.

Bohlmann, F. and Zdero, C. 1979a. Neue Cadinan-Derivate aus *Heterotheca subaxillaris. Phytochemistry* 18(7): 1185–1187.

Bohlmann, F. and Zdero, C. 1979b. Zwei neue Eudesman-Derivate aus *Iva annua. Phytochemistry* 18(12): 2034–2035.

Bohlmann, F., Zdero, C., Robinson, H., and King, R.M. 1979. Neue Cadinen- und Norcadinen-Derivate aus *Heterotheca grandiflora. Phytochemistry* 18(10): 1675–1680.

Bohlmann, F., Abraham, W.R., King, R.M., and Robinson, H. 1981a. Diterpenes from *Koanophyllon* species. *Phytochemistry* 20(8):1903–1906.

Bohlmann, F., Abraham, W.R., King, R.M., and Robinson, H. 1981b. Thiophene acetylenes and flavonols from *Pterocaulon virgatum. Phytochemistry* 20:825–827.

Bohlmann, F., Gupta, R.K., King, R.M., and Robinson, H. 1982. New cadinene derivatives from *Heterotheca latifolia* sesquiterpenes. *Phytochemistry* 21(12): 2982–2984.

Bohlmann, F., Scheidges, C., King, R.M., and Robinson, H. 1984. Five labdane derivatives from *Koanophyllon conglobatum. Phytochemistry* 23(5):1190–1192.

Bohlmann, F., Wolfrum, C., Jakupovic, J., King, R.M., and Robinson, H. 1985. Further cadinene derivatives from *Heterotheca latifolia. Phytochemistry* 24(5): 1101–1103.

Boisacq, E. 1923. *Dictionnaire Étymologique de la Langue Grecque.* Librarie C. Klincksieck, Paris.

Boldingh, I. 1909. *The Flora of the Dutch West Indian Islands, Vol. I. St. Eustatius, Saba and St. Martin.* E.J. Brill Publications, Leyden.

Boldingh, I. 1914. *The Flora of the Dutch West Indian Islands, Vol. II. Curaçao, Aruba and Bonaire.* E.J. Brill Publications, Leyden.

Bolli, R. 1994. Revision of the genus *Sambucus. Dissertationes Botanicae 22.* K.J. Cramer, Berlin.

Booker, K.M., Hudson, C.M., and Rankin, R.L. 1992. Place-name identification and multilingualism in the sixteenth-century Southeast. *Ethnohistory* 39(4): 399–451.

Borges, J., Manresa, M.T., Ramon, J.L.M., Pascual, C., and Rumbero, A. 1979. Two new oxindole alkaloids isolated from *Hamelia patens* Jacq. *Tetrahedron Letters* 34:3197–3204.

Borges Del Castillo, J., Manresa-Ferrero, M.T., Martin-Ramon, J.L., Pascual-Rigau, C., and Valles-Fernández, L. 1977. Alcaloides oxindolicos en las hojas de *Hamelia patens* Jacq. *Boletin de la Sociedad Quimica Peruana* 43(2):85–92.

Borland, H. 1965. *Our Natural World.* Doubleday, New York.

Borror, D.J. 1960. *Dictionary of Word Roots and Combining Forms.* Mayfield, Palo Alto, CA.

Borror, D.J. and DeLong, D.M. 1954. *An Introduction to the Study of Insects,* rev. ed. Holt, Rinehart & Winston, New York.

Bose, P. and Banerji, J. 1991. Synthesis of 4-phenylcoumarins from *Dalbergia volubilis* and *Exostema caribaeum. Phytochemistry* 30(7):2438–2439.

Bowen, J.M., Cole, R.J., Bedell, D., and Schabdach, D. 1996. Neuromuscular effects of toxins isolated from southern prickly ash (*Zanthoxylum clava-herculis*) bark. *American Journal of Veterinary Research* 57(8): 1239–1244.

Bown, D. 1995. *The Herb Society of America Encyclopedia of Herbs and Their Uses.* Dorling Kindersley, New York.

Boyd, C.E. 1968. Fresh-water plants: a potential source of protein. *Economic Botany* 22:359–368.

Boyd, C.E. and McGinty, P.S. 1981. Percentage digestible dry matter and crude protein in dried aquatic weeds. *Economic Botany* 35:296–299.

Braca, A., De Tommasi, N., Méndez, J., and Morelli, I. 1999a. Flavonoids and triterpenoids from *Licania heteromorpha* (Chrysobalanaceae). *Biochemical Systematics and Ecology* 27(5):527–530.

Braca, A., De Tommasi, N., Méndez, J., Morelli, I., and Pizza, C. 1999b. Three flavonoids from *Licania heteromorpha. Phytochemistry* 51(8):1121–1124.

Braca, A., Morelli, I., Méndez, J., Braghiroli, L., and Mazzanti, G. 2000. Antimicrobial triterpenoids from *Licania heteromorpha. Planta Medica* 66 (8):768–769.

Brako, L. and Zarucchi, J.L. 1993. *Catalogue of the Flowering Plants and Gymnosperms of Peru/Catálogo de las angiospermas y gimnospermas del Perú.* Missouri Botanical Garden, Monographs in Systematic Botany Vol. 45, St. Louis.

Brako, L., Rossman, A.Y., and Farr, D.F. 1995. *Scientific and Common Names of 7,000 Vascular Plants in the United States.* American Phytopathological Society, St. Paul, MN.

Braun, H. 1996. A vicilin-like seed protein of cycads: similarity to sucrose-binding proteins. *Plant Molecular Biology* 31(1):35–44.

Bremer, B. and Jansen, R.K. 1991. Comparative restriction site mapping of chloroplast DNA implies new phylogenetic relationships within Rubiaceae. *American Journal of Botany* 78(2):198–213.

Bremness, L. 1994. *Eyewitness Handbooks. Herbs.* Dorling Kindersley, London.

Bressani, R., Brenes, R.G., Garcia, A., and Elias, L.G. 1987. Chemical composition, amino acid content and protein quality of *Canavalia* spp. seeds. *Journal of the Science of Food and Agriculture* 40(1):17–23.

Breu, W., Hagenlocher, M., Redl, K., Tittel, G., Stadler, F., and Wagner. H. 1992. Antiinflammatory activity of sabal fruit extracts prepared with supercritical carbon dioxide. *In vitro* antagonists of cyclooxygenase and 5-lipoxygenase metabolism [in German]. *Arzneimittel-Forschung* 42(4):547–551.

Breuer, H., Rangel, M., and Medina, E. 1982. Pharmacological properties of melochinine, an alkaloid producing Central American cattle paralysis. *Toxicology* 25(2–3):223–242.

Brickell, C. and Zuk, J.D., Eds. 1997. *The American Horticultural Society A–Z Encyclopedia of Garden Plants.* DK Publishing, New York.

Bridges, E.L. and Orzell, S. L. 1990. *Xyris chapmanii*, a new species from the Gulf Coastal Plain of the southern United States. *Phytologia* 68(5):382–389.

Britten, J. and Holland, R. [1886] 1965. *A Dictionary of English Plant-Names.* Publications (English Dialect Society) Vol. 10. Reprinted by Kraus Reprint, Vaduz.

Britton, N.L. and Brown, H.A. 1896–1898. *An Illustrated Flora of the Northern United States, Canada and the British Possessions.* Charles Scribner's Sons, New York.

Britton, N.L. and Millspaugh, C.F. 1920. *The Bahama Flora.* Reprint 1962, Hafner, New York.

Brooklyn Online. 2002. URL: www.brooklynonline.com/history/hubbard.html.

Brown, C.H. 2001. Linguistic ethnobiology: Amerindian oak nomenclature. Pp. 111–147. In: Ford, R.I., Ed. 2001. *Ethnobiology at the Millennium. Past, Promise, and Future Prospects.* Anthropological Papers, Museum of Anthropology, University of Michigan, No. 91, Ann Arbor.

Brown, J.A., Kerber, R.A., and Winter, H.D. 1990. Trade and the evolution of exchange relations at the beginning of the Mississippian Period. Pp. 251–280. In: Smith, B.D., Ed. *The Mississippian Emergence.* Smithsonian Institution Press, Washington, D.C.

Brown, J.W. 1990. Urban biology of *Leptotes marina* (Reakirt) (Lycaenidae). *Journal of the Lepidopterists' Society* 44(3):200–201.

Brown, R.W. 1954. *Composition of Scientific Words.* Smithsonian Institution Press, Washington, D.C.

Browne, E.T.J. and Athey, R. 1992. *Vascular Plants of Kentucky. An Annotated Checklist.* University Press of Kentucky, Lexington.

Broyles, S.B. and Wyatt, R. 1991. The breeding system of *Zephyranthes atamasco* (Amaryllidaceae). *Bulletin of the Torrey Botanical Club* 118(2):137–140.

Brummitt, R.K., Ed. 1992. *Vascular Plant Families and Genera.* Royal Botanic Gardens, Kew, U.K.

Bruneau, A. and Anderson, G.J. 1988. Reproductive biology of diploid and triploid *Apios americana* (Leguminosae). *American Journal of Botany* 75:1876–1883.

Brussell, D.E. 1997. *Poisons, Potions, and Panaceas. An Ethnobotanical Study of Montserrat.* Southern Illinois University Press, Carbondale, IL.

Brutsch, M.O. and Zimmerman, H.G. 1993. The prickly pear (*Opuntia ficus-indica* [Cactaceae]) in South Africa: utilization of the naturalized weed, and the cultivated plants. *Economic Botany* 47:154–163.

Buchholtz, K.P., Ed. 1968. *Weeds of the North Central States.* University of Illinois Agricultural Experiment Station Circular 718, Urbana.

Budde, W.M. and Shriner, R.L. 1950. Exploratory studies on the constituents of *Scoparia dulcis* (Linn.). *Proceedings Iowa Academy of Sciences* 57:161–164.

Budinsky, A., Wolfram, R., Oguogho, A., Efthimiou, Y., Stamatopoulos, Y., and Sinzinger, H. 2001. Regular ingestion of *Opuntia robusta* lowers oxidation injury. *Prostaglandins Leukotrienes Essential Fatty Acids* 65 (1):45–50.

Buhrmester, R.A., Ebingerla, J.E., and Seigler, D.S. 2000. Sambunigrin and cyanogenic variability in populations of *Sambucus canadensis* L. (Caprifoliaceae). *Biochemical Systematics and Ecology* 28(7):689–695.

Burger, W.P., Naude, T.W., Van Rensburg, I.B., Botha, C.J., and Pienaar, A.C. 1994. Cardiomyopathy in ostriches (*Struthio camelus*) due to avocado (*Persea americana* var. *guatemalensis*) intoxication. *Journal of the South African Veterinary Association* 65(3): 113–118.

Burke, B.A. and Parkins, H. 1978. *Amyris* of Jamaica. Nicotinamides of *Amyris plumieri* D.C. (Rutaceae). *Tetrahedron Letters* 19(30):2763–2726.

Burke, B.A. and Parkins, H. 1979. Coumarins from *Amyris balsamifera. Phytochemistry* 18 (6):1073–1075.

Burke, H.R., Clark, W.E. Cate, J.R., and Fryxell, P.A. 1986. Origin and dispersal of the boll weevil. *Bulletin of the Entomological Society of America* 32:228–238.

Burkill, I.H. 1966. *A Dictionary of the Economic Products of the Malay Peninsula.* Ministry of Agriculture and Cooperatives, Kuala Lumpur.

Burns, J., Yokota, T., Ashihara, H., Lean, M.E., and Crozier, A. 2002. Plant foods and herbal sources of resveratrol. *Journal of agriculture and food chemistry* 50 (11):3337–3340.

Burns, R. 1785. To a mouse. Reprinted in Felleman, H., Ed. 1965. *Poems That Live Forever.* Doubleday, Garden City, NY.

Burwell, T. 1995. Bootlegging on a desert mountain: the political ecology of Agave (*Agave* spp.) Demographic change in the Sonora River Valley, Sonora, Mexico. *Human Ecology* 23(3):407–431.

Bushnell, D.I., Jr. 1909. The Choctaw of Bayou Lacomb, St. Tammany Parish, Louisiana. Smithsonian Institution, Bureau of American Ethnography Bulletin 48.

Bushnell, D.I., Jr. 1917. The Chitimacha of Bayou La Fourche, Louisiana. *Journal of the Washington Academy of Sciences* 7(10):301–307.

Bushnell, D.I., Jr. 1922. Some new ethnologic data from Louisiana. *Journal of the Washington Academy of Sciences* 12:303–307.

Butler, C.L. and Mullen, L.E. 1955. Investigation of *Piscidia erythrina* (Jamaica dogwood). *Acta Phytotherapeutica* 2(8):1.

Bwititi, P.T., Musabayane, C.T., and Nhachi, C.F. 2000. Effects of *Opuntia megacantha* on blood glucose and kidney function in streptozotocin diabetic rats. *Journal of Ethnopharmacology* 69(3):247–252.

Bwititi, P.T., Machakaire, T., Nhachi, C.B., and Musabayane, C.T. 2001. Effects of *Opuntia megacantha* leaves extract on renal electrolyte and fluid handling in streptozotocin (STZ)-diabetic rats. *Renal Failure* 23(2):149–158.

Bye, R.A., Jr. 1979. An 1878 ethnobotanical collection from San Luis Potosí: Dr. Edward Palmer's first major Mexican collection. *Economic Botany* 33:135–162.

Bye, R.A., Jr. 1986. Medicinal plants of the Sierra Madre: comparative study of Tarahumara and Mexican market plants. *Economic Botany* 40:103–124.

Bye, R.A., Jr. 1993. The role of humans in the diversification of plants in Mexico. Pp. 707–731. In: T.P. Ramamoorthy, R.A. Bye, A. Lot, and J. Fa, *Biological Diversity of Mexico: Origins and Distribution.* Oxford University Press, New York.

Byington, C., Swanton, J.R., and Halbert, H.S. 1915. A Dictionary of the Choctaw Language. Smithsonian Institution, Bureau of American Ethnology Bulletin 46. Washington, D.C.

Byrd, William. [1728] 1980. The history of the dividing line. Pp. 15–27. In: Bergon, F. *The Wilderness Reader.* A Mentor Book, New American Library, Penguin Books, New York.

Caballero-George, C., Vanderheyden, P.M., Solis, P.N., Pieters, L., Shahat, A.A., Gupta, M.P., Vauquelin, G., and Vlietinck, A.J. 2001. Biological screening of selected medicinal Panamanian plants by radioligand-binding techniques. *Phytomedicine* 8(1):59–70.

Cabrera, A.L. and Ragonese, A.M. 1978. Revision del genero *Pterocaulon* (Compositae). *Darwiniana* 21 (2/4):185–257.

Cáceres, A., Cano, O., Samayoa, B., and Aguilar, L. 1990. Plants used in Guatemala for the treatment of gastrointestinal disorders. 1. Screening of 84 plants against enterobacteria. *Journal of Ethnopharmacology* 30(1):55–73.

Cáceres, A., Alvarez, A.V., Ovando, A.E., and Samayoa, B. E. 1991a. Plants used in Guatemala for the treatment of respiratory diseases. 1. Screening of 68 plants against Gram-positive bacteria. *Journal of Ethnopharmacology* 31(2):193–208.

Cáceres, A., López, B.R., Giron, M.A., and Logemann, H. 1991b. Plants used in Guatemala for the treatment of dermatophytic infections. 1. Screening for antimycotic activity of 44 plant extracts. *Journal of Ethnopharmacology* 31(3):263–276.

Cáceres, A., López, B., Juarez, X., del Aguila, J., and Garcia, S. 1993. Plants used in Guatemala for the treatment of dermatophytic infections. 2. Evaluation of antifungal activity of seven American plants. *Journal of Ethnopharmacology* 40(3):207–213.

Cáceres, A., Menéndez, H., Méndez, E., Cohobon, E., Samayoa, B.E., Jauregui, E., Peralta, E., and Carrillo, G. 1995. Antigonorrhoeal activity of plants used in Guatemala for the treatment of sexually transmitted diseases. *Journal of Ethnopharmacology* 48(2):85–88.

Cai, Y., Sun, M., and Corke, H. 2001. Identification and distribution of simple and acylated betacyanins in the Amaranthaceae. *Journal of Agricultural and Food Chemistry* 49(4):1971–1978.

Calera, M.R., Mata, R., Lotina-Hennsen, B., and Anaya, A. L. 1996. Uncoupling behavior of the 4-phenylcoumarins in spinach chloroplasts: structure–activity relationships. *Journal of Agricultural and Food Chemistry* 44(10):2966–2969.

Callendar, C. 1978. Shawnee. Pp. 622–635. In: Trigger, B.G., Ed. 1978. *Northeast. Handbook of North American Indians.* Vol. 15. Smithsonian Institution, Washington, D.C.

Calzada, F. and Mata, R. 1995. Hippocrateine III, a sesquiterpene alkaloid from *Hippocratea excelsa. Phytochemistry* 40(2):583–585.

Calzada, F., Mata-R., R., Linares, E., Bye, R.A., Barreto, V. M., and Rio, F. de. 1991. Friedelanes and triterpenoid quinone methides from *Hippocratea excelsa. Planta Medica* 57(2):194–195.

Camp, B.J., Bridges, C.H., Hill, D.W., Patamalai, B., and Wilson, S. 1988. Isolation of a steroidal sapogenin from the bile of a sheep fed *Agave lecheguilla. Veterinary and Human Toxicology* 30(6):533–535.

Campbell, T.N. 1951. Medicinal plants used by Choctaw, Chickasaw, and Creek Indians in the early nineteenth century. *Journal of the Washington Academy of Sciences* 41(9):285–290.

Campbell, T.N. 1959. Choctaw subsistence: ethnographic notes from the Lincecum manuscript. *Florida Anthropologist* 12:9–24.

Cannon, J. and Cannon, M. 2003. *Dye Plants and Dyeing,* rev. ed. Timber Press, Portland, OR, in association with Royal Botanic Gardens, Kew, U.K.

Capasso, A., Balderrama, L., Sivila, S.C., De Tommasi, N., Sorrentino, L., and Pizza, C. 1998. Phytochemical and pharmacological studies of *Guettarda acreana. Planta Medica* 64(4):348–352.

Capinera, J.L. 2001. *Handbook of Vegetable Pests.* Academic Press, San Diego, CA.

Capron, L. 1953. The medicine bundle of the Florida Seminole and the Green Corn Dance. *Anthropological Papers 35. Bulletin of the Bureau of American Ethnology* 151:155–210.

Caras, R.A. 1967. *North American Mammals.* Galahad Books, New York.

Carbin, B.E., Larsson, B., and Lindahl, O. 1990. Treatment of benign prostatic hyperplasia with phytosterols. *British Journal of Urology* 66(6):639–641.

Cárdenas-Medellin, M.L., Serna-Saldivar, S.O., and Velazco de la Garza, J. 1998. Effect of raw and cooked nopal (*Opuntia ficus-indica*) ingestion on growth and profile of total cholesterol, lipoproteins, and blood glucose in rats [in Spanish]. *Archivos Latinoamericanos de Nutricion* 48(4):316–323.

Carini, M., Stefani, R., Aldini, G., Ozioli, M., and Facino, R.M. 2001. Procyanidins from *Vitis vinifera* seeds inhibit the respiratory burst of activated human neutrophils and lysosomal enzyme release. *Planta Medica* 67(8):714–717.

Carlini, C.R. and Udedibie, A.B. 1997. Comparative effects of processing methods on hemagglutinating and antitryptic activities of *Canavalia ensiformis* and *Canavalia braziliensis* seeds. *Journal of Agricultural and Food Chemistry* 45(11):4372–4377.

Carney, J.A. 2001. *Black Rice: The African Origins of Rice Cultivation in the Americas.* Harvard University Press, Cambridge, MA.

Carter, M. and Feeny, P. 1999. Host-plant chemistry influences oviposition choice of the spicebush swallowtail butterfly. *Journal of Chemical Ecology* 25 (9):1999–2009.

Carver, J. 1779. *Travels through the Interior Parts of North America in the Years 1766, 1767, and 1768.* S. Price et al., Dublin.

Casagrande, C., Canonica, L., and Severini-Ricca, G. 1975. Studies on proaporphine and aporphine alkaloids. VII. Stereochemistry of reduced proaporphines of *Croton sparsiflorus* and *Croton linearis. Journal of the Chemical Society, Perkin Transactions I* 17:1659–1663

Castetter, E.F. and Bell, W.H. 1942. *Pima and Papago Indian Agriculture.* Inter-Americana Studies 1. University of New Mexico Press, Albuquerque. Reprinted 1980, AMS Press, New York.

Castillo, J., Benavente-Garcia, O., Lorente, J., Alcaraz, M., Redondo, A., Ortuno, A., and Del Rio, J.A. 2000. Antioxidant activity and radioprotective effects against chromosomal damage induced *in vivo* by X-rays of flavan-3-ols (Procyanidins) from grape seeds (*Vitis vinifera*): comparative study versus other phenolic and organic compounds. *Journal of Agricultural and Food Chemistry* 48(5):1738–1745.

Castro, O., Gutierrez, J.M., Barrios, M., Castro, I., Romero, M., and Umana, E. 1999. Neutralization of the hemorrhagic effect induced by *Bothrops asper* (Serpentes: Viperidae) venom with tropical plant extracts [in Spanish]. *Revista de Biologia Tropical* 47(3):605–616.

Catesby, M. 1731–1732. *The Natural History of Carolina, Florida and the Bahama Islands: Containing the Figures of Birds, Beasts, Fishes, Serpents, Insects and Plants.* Vol. I. C. Marsh, London.

Catesby, M. 1734–1747. *The Natural History of Carolina, Florida and the Bahama Islands: Containing the Figures of Birds, Beasts, Fishes, Serpents, Insects and Plants.* Vol. 2. C. Marsh, London.

Catlin, G. [1844] 1973. *Letters and notes on the manners, customs, and conditions of the North American Indians. Written during eight year's travel (1832–1839) amongst the wildest tribes of Indians in North America.* 2 vols. Reprinted by Dover, New York.

Catling, P.M., Dumouchel, L., and Brownell, V.R. 1998. Pollination of the Miccosukee gooseberry (*Ribes echinellum*). *Castanea* 63(4):402–407.

Cava, M.P., Rao, K.V., Kota, V., Douglas, E., and Weisbach, J.A. 1968. Alkaloids of *Cassytha americana* Nees. *The Journal of Organic Chemistry* 33:2443–2446.

Cavalcante, P.B. 1974. *Frutas comestíveis da Amazônia. II.* Museu Paraense Emilio Goeldi (Publ. Avulsas 27), Belém.

Cavalcante, P.B. 1976. *Frutas comestíveis da Amazônia.* Museu Paraense Emilio Goeldi, Belém.

Cavalcante, P.B. 1979. *Frutas comestíveis da Amazônia.* III. Museu Paranse Emilio Goeldi (Publ. Avulsas 33), Belém.

Cechinel-Filho, V., Santos, A.R.S., Campos, O.G., and Calixto, J.B. 1996a. Chemical and pharmacological studies of *Phyllanthus caroliniensis* in mice. *The Journal of Pharmacy and Pharmacology* 48:1231–1236.

Cechinel-Filho, V., Vaz, Z.R., Nunes, R.J., Calixto, J.B., and Yunes, R.A. 1996b. Antinociceptive activity of phyllanthimide analogues in mice: structure–activity relationships. *Pharmaceutical Sciences* 2:199–201.

Cechinel-Filho, V., Santos, A.R., De Campos, R.O., Miguel, O.G., Yunes, R.A., Ferrari, F., Messana, I., and Calixto, J.B. 1996c. Chemical and pharmacological studies of *Phyllanthus caroliniensis* in mice. *The Journal of Pharmacy and Pharmacology* 48(12): 1231–1236.

Cervantes-Peredo, L. 1999. Description of the immature stages, adult morphology and biology of *Pantochlora vivida* Stal (Heteroptera: Pentatomidae: Edessinae). *Journal of the New York Entomological Society* 107 (4):372–385.

Chafin, L.G. 2001. *Field Guide to the Rare Plants of Florida.* Florida Natural Areas Inventory, Tallahassee.

Chang, F.R., Yang, P.Y., Lin, J.Y., Lee, K.H., and Wu, Y.C. 1998. Bioactive kaurane diterpenoids from *Annona glabra. Journal of Natural Products* 61(4):437–439.

Chang, P., Lee, K.H., Shingu, T., Hirayama, T., and Hall, I. H. 1982. Antitumor agents 50. Morindaparvin-A, a new antileukemic anthraquinone, and alizarin-1-methyl ether from *Morinda parvifolia,* and the antileukemic activity of the related derivatives. *Journal of Natural Products* 45(2):206–210.

Channel, R.B. 1957. A revisional study of the genus *Marshallia* (Compositae). *Contributions to the Gray Herbarium of Harvard University* 181:41–130.

Chapman, A.W. 1897. *Flora of the Southern United States.* Cambridge Botanical Supply Co., Cambridge, MA.

Chariandy, C.M., Seaforth, C.E., Phelps, R.H., Pollard, G. V., and Khambay, B.P. 1999. Screening of medicinal plants from Trinidad and Tobago for antimicrobial and insecticidal properties. *Journal of Ethnopharmacology* 64(3):265–270.

Chase, M.W., Soltis, D.E., Olmstead, R.G., Morgan, D. Hes, D.H., Mischler, D.B., Duvall, M.R., Price, R.A. Hills, H.G., Qiu, Y.-L., Kron, K.A., Rettig, J.H., Conti, E., Palmer, J.D., Manhart, J.R., Sytsma, K.J., Michaels, H.J., Kress, W.J., Karol, K.G., Clark, W. D., Hedrén, M., Gaut, B.S., Jansen, R.K., Kim, K.-J., Wimpee, C.F., Smith, J.F., Furnier, G.R., Strauss, S.H., Xiang, Q.-Y., Plunkett, G.M., Soltis, P.S., Swensen, S.M., Williams, S.E., Gadek, P.A., Quinn, C.J., Eguarte, L.E., Golenberg, E., Learn, G.H., Jr., Graham, S.W., Barrett, S.C.H., Dayanandan, S., and Albert, V.A. 1993. Phylogenetics of seed plants: an analysis of nucleotide sequences from the plastid gene *rbcL. Annals of the Missouri Botanical Garden* 80:528–580.

Chatterjee, G.K., Burman, T.K., Nagchaudhuri, A.K., and Pal, S.P. 1983. Antiinflammatory and antipyretic activities of *Morus indica. Planta Medica* 48 (2):116–119.

Chaudhuri, P.K. and Thakur, R.S. 1991. *Hamelia patens*: a new source of ephedrine. *Planta Medica* 57(2):199.

Chavelas-Polito, J. 1981. "El negrito" (*Simarouba glauca* D. C.), una especie nativa de uso multiple. *Ciencias forestales* 6(29):3–16.

Chávez, P.I., Sanchez, L.A., Gonzalez, F.A., Rodríguez, J.L., and Axelrod, F. 1997. Cytotoxicity correlations of Puerto Rican plants using a simplified brine shrimp lethality screening procedure. *International Journal of Pharmacognosy* 35(4):222–226.

Cheatham, S., Johnston, M.C., and Marshall, L. 1995. *The Useful Wild Plants of Texas, the Southeastern and Southwestern United States, the Southern Plains, and Northern Mexico.* Vol. 1. Useful Wild Plants, Inc., Austin, TX.

Chel-Guerrero, L., Pérez-Flores, V., Betancur-Ancona, D., and Davila-Ortiz, G. 2002. Functional properties of flours and protein isolates from *Phaseolus lunatus* and *Canavalia ensiformis* seeds. *Journal of Agricultural and Food Chemistry* 50(3):584–591.

Chen, C.-M. and Chen, M.-T. 1976. 6-Methoxybenzoxazolinone and triterpenoides from roots of *Scoparia dulcis. Phytochemistry* 15:1997–1999.

Chen, C.P., Lin, C.C., and Namba, T. 1989. Screening of Taiwanese crude drugs for antibacterial activity against *Streptococcus mutans. Journal of Ethnopharmacology* 27(3):285–295.

Chen, C.Y., Chang, F.R., Cho, C.P., and Wu, Y.C. 2000. ent-Kaurane diterpenoids from *Annona glabra. Journal of Natural Products* 63(7):1000–1003.

Chen, F., Nakashima, N., Kimura, I., Kimura, M., Asano, N., and Koya, S. 1995a. Potentiating effects on pilocarpine-induced saliva secretion, by extracts and N-containing sugars derived from mulberry leaves, in streptozocin-diabetic mice. *Biological and Pharmaceutical Bulletin* 18(12):1676–1680.

Chen, F., Nakashima, N., Kimura, I., and Kimura, M. 1995b. Hypoglycemic activity and mechanisms of extracts from mulberry leaves (*Folium mori*) and *Cortex mori radicis* in streptozotocin-induced diabetic mice [in Japanese]. *Yakugaku Zasshi* 115 (6):476–482.

Chen, G., Shen, L., and Jiang, P. 1996. Flavanonol glucosides of *Smilax glabra* Roxb. [Chinese]. *Zhongguo Zhong Yao Za Zhi* 21(6):355–357, 383.

Chen, H.C., Chang, M.D., and Chang, T.J. 1985. Antibacterial properties of some spice plants before and after heat treatment [in Chinese]. *Zhonghua Min Guo Wei Sheng Wu Ji Mian Yi Xue Za Zhi* 18(3):190–195.

Chen, I.S., Lin, Y.C., Tsai, I.L., Teng, C.M., Ko, F.N., Ishikawa, T., and Ishii, H. 1995. Coumarins and anti-platelet aggregation constituents from *Zanthoxylum schinifolium. Phytochemistry* 39(5):1091–1097.

Chen, S.-H. and Wu, M.-J. 2002. Remarks on the species of *Stachytarpheta* (Verbenaceae) of Taiwan. *Botanical Bulletin of Academia Sinica* 44:167–174.

Chen, T., Li, J., Cao, J., Xu, Q., Komatsu, K., and Namba, T. 1999. A new flavanone isolated from *Rhizoma Smilacis Glabrae* and the structural requirements of its derivatives for preventing immunological hepatocyte damage. *Planta Medica* 65(1):56–59.

Chen, T., Li, J.X., and Xu, Q. 2000. Phenylpropanoid glycosides from *Smilax glabra. Phytochemistry* 53 (8):1051–1055.

Chen, Y., Peumans, W.J., and Van Damme, E.J. 2002. The *Sambucus nigra* type-2 ribosome-inactivating protein SNA-I′ exhibits in planta antiviral activity in transgenic tobacco. *FEBS Letters* 516(1–3):27–30.

Chiang, H.C., Jaw, S.M., Chen, C.F., and Kan, W.S. 1992a. Antitumor agent, physalin F from *Physalis angulata* L. *Anticancer Research* 12(3):837–843.

Chiang, H.C., Jaw, S.M., and Chen, P.M. 1992b. Inhibitory effects of physalin B and physalin F on various human leukemia cells *in vitro. Anticancer Research* 12(4):1155–1162.

Chiang-Cabrera, F. 1981. A Taxonomic Study of the North American Species of *Lycium* (Solanaceae). Ph.D. dissertation. University of Texas at Austin.

Chiapella, C., Radovan, R.D., Moreno, J.A., Casares, L., Barbe, J., and Llagostera, M. 2000. Plant activation of aromatic amines mediated by cytochromes P450 and flavin-containing monooxygenases. *Mutation Research* 470(2):155–160.

Chichon, P.G. 2000. Herbs and the common cold. *Advance for Nurse Practitioners* 8(8):31–32.

Chien, M.M., Svoboda, G.H., Schiff, P.L., Jr., Slatkin, D.J., and Knapp, J.E. 1979. Chemical constituents of *Echites hirsuta* (Apocynaceae). *Journal of Pharmaceutical Sciences* 68(2):247–249.

Childs, R.F. and Cole, J.R. 1965. Phytochemical and pharmacological investigation of *Anemopsis californica. Journal of Pharmaceutical Sciences* 54(5):789–791.

Chitra, M., Shyamala Devi, C.S., and Sukumar, E. 2003. Antibacterial activity of embelin. *Fitoterapia* 74 (4):401–403.

Chong, Y.K., Galbraith, M.N., and Horn, D.H.S. 1970. Isolation of deoxycrustecdysone, deoxyecdysone, and alpha-ecdysone from the fern *Blechnum minus. Chemical Communications* (Cambridge, England) 18:1217–1218.

Christensen, B.V. 1946. *Collection and Cultivation of Medicinal Plants in Florida.* School of Pharmacy, University of Florida, Gainesville.

Christensen, J. and Jaroszewski, J.W. 2001. Natural glycosides containing allopyranose from the passion fruit plant and circular dichroism of benzaldehyde cyanohydrin glycosides. *Organic Letters* 3(14):2193–2195.

Cilliers, C.J. and Neser, S. 1991. Biological control of *Lantana camara* (Verbenaceae) in South Africa. *Agriculture, Ecosystems and Environment* 37(1/3):57–75.

Cimanga, K., Bruyne, T. de, Lasure, A., Li, Q., Pieters, L.L., Claeys, M., Berghe, D.V., Kambu, K., Tona, L., and Vlietinck, A. 1995a. Flavonoids *O*-glycosides from the leaves of *Morinda morindoides. Phytochemistry* 38(5):1301–1303.

Cimanga, K., Bruyne, T. de, Lasure, A., Poel, B. van, Pieters, L., Berghe, D. vanden, Vlietinck, A., Kambu, K., and Tona, L. 1995b. *In vitro* anticomplementary activity of constituents from *Morinda morindoides. Journal of Natural Products* 58(3):372–378.

Cimanga, K., Bruyne, T. de, Poel, B. van, Ma, Y.L., Glaeys, M., Pieters, L., Kambu, K., Tona, L., Bakana, P., and Berghe, D. vanden. 1997. Complement-modulating properties of a kaempferol 7-*O*-rhamnosylsophoroside from the leaves of *Morinda morindoides. Planta Medica* 63(3):220–223.

Clapham, A.R., Tutin, T.G., and Moore, D.M. 1987. *Flora of the British Isles.* Cambridge University Press, Cambridge, U.K.

Clement, E.J. and Foster, M.C. 1994. *Alien Plants of the British Isles.* Botanical Society of the British Isles, London.

Clute, W.N. 1923. *A Dictionary of American Plant Names.* W.N. Clute, Joliet, IL.

Cobos, R. 1983. *A Dictionary of New Mexico and Southern Colorado Spanish.* Museum of New Mexico Press, Santa Fe.

Coe, F.G. and Anderson, G.J. 1996. Ethnobotany of the Garífuna of eastern Nicaragua. *Economic Botany* 50:71–107.

Coe, S. 1994. *America's First Cuisines.* University of Texas, Austin.

Coffey, T. 1993. *The History and Folklore of North American Wildflowers.* Houghton Mifflin, Boston.

Coile, N.C. 2000. Notes on Florida's Endangered and Threatened Plants. Florida Department of Agriculture and Consumer Services, Contribution 38, 2nd ed. Gainesville, FL.

Colard, J.M., Compernolle, F., and Dumont, P.A. 1976. Contribution à l'étude des saponines de *Colubrina arborescens* Mill: isolement du colubrinoside et etablissement de la structure de sa chaine osidique. *Journal de Pharmacologie Belgique* 31(2):199–214.

Coleta, M., Campos, M.G., Cotrim, M.D., and Proenca da Cunha, A. 2001. Comparative evaluation of *Melissa officinalis* L., *Tilia europaea* L., *Passiflora edulis* Sims, and *Hypericum perforatum* L. in the elevated plus maze anxiety test. *Pharmacopsychiatry* 34 (Suppl. 1):520–521.

Coll y Toste, C. 1972. *Clásicos de Puerto Rico,* 2nd ed. Ediciones Latinoamericanas. S.A., Puerto Rico; basis for Web site — http://members.dandy.net/~orocobix/terms1.htm — by the Taino Language Project TITC! (sm), Millville, NJ, 1993, 1999.

Colunga García-Marin, P. and Fliogonio, M.-P. 1997. Morphological variation of henequen (*Agave fourcroydes,* Agavaceae) germplasm and its wild ancestor (*A. angustifolia*) under uniform growth conditions: diversity and domestication. *American Journal of Botany* 84(11):1449–1465.

Colunga García-Marin, P., Coello-Coello, J., Equiarde, L.E., and Pinero, D. 1999. Isozymatic variation and phylogenetic relationships between henequen (*Agave fourcroydes*) and its wild ancestor *A. angustifolia* (Agavaceae). *American Journal of Botany* 86 (1):115–123.

Colunga García-Marin, P., Estrada-Loera, E., and Filogonio, M.-P. 1996. Patterns of morphological variation, diversity, and domestication of wild and cultivated populations of *Agave* in Yucatan, Mexico. *American Journal of Botany* 83(8):1069–1082.

Colwell, R.K. 1995. Effects of nectar consumption by the hummingbird flower mite *Proctolaelaps kirmsei* on nectar availability in *Hamelia patens. Biotropica* 27 (2):206–217.

Comerford, S.C. 1996. Medicinal plants of two Mayan healers from San Andrés, Petén, Guatemala. *Economic Botany* 50:327–336.

Comin, J., Gonçalves de Lima, O., Grant, H.N., Jackson, L. M., Keller-Schierlein, W., and Prelog, V. 1963. Über die Konstitution des Biflorins, eines o-Chinons der Diterpen-Reihe. *Helvetica Chimica Acta* 46:409–415.

Constantinidis, T., Harvala, C., and Skaltsounis, A.L. 1993. Pyrrolizidine *N*-oxide alkaloids of *Heliotropium hirsutissimum. Phytochemistry* 32(5):1335–1337.

Contractor, Z., Hussain, F., and Jabbar, A. 1999. Postprandial glucose response to mango, banana and sapota. *The Journal of the Pakistan Medical Association* 49 (9):215–216.

Cooke, V. and Sampley, J. 1947. *Palmetto Braiding and Weaving.* Manual Arts Press, Peoria.

Coon, N. 1974. *The Dictionary of Useful Plants.* Rodale Press, Emmaus, PA.

Corbett, M.D. and Billets, S. 1975. Characterization of poison oak urushiol. *Journal of Pharmaceutical Science* 64(10):1715–1718.

Core, E.L. 1967. Ethnobotany of the southern Appalachian aborigines. *Economic Botany* 21:198–214.

Correa, M.D. and Wilbur, R.L. 1969. A revision of the genus *Carphephorus* (Compositae: Eupatorieae). *Journal of the Elisha Mitchell Society* 85:79–91.

Correll, D.S. 1950. *Native Orchids of North America North of Mexico.* Chronica Botanica Co., Waltham, MA.

Correll, D.S. and Correll, H.B. 1972. Aquatic and Wetland Plants of Southwestern United States. Environmental Protection Agency, Water Pollution Control Research Series, Washington, D.C.

Correll, D.S. and Correll, H.B. 1982. *Flora of the Bahama Archipelago.* J. Cramer, Vaduz.

Correll, D.S. and Johnston, M.C. 1970. *Manual of the Vascular Plants of Texas.* Texas Research Foundation, Renner.

Costantini, A., De Bernardi, T., and Gotti, A. 1999. Clinical and capillaroscopic evaluation of chronic uncomplicated venous insufficiency with procyanidins extracted from *Vitis vinifera* [in Italian]. *Minerva Cardioangiologia* 47(1–2):39–46.

Cotterill, F.P.D. and Fergusson, R.A. 1993. Floral essential oil of *Guettarda poasana* inhibits yeast growth. *Biotropica* 25(4):483–486.

Cotton, C.M. 1996. *Ethnobotany. Principles and Applications.* John Wiley & Sons, Chichester, West Sussex, U.K.

Couladis, M., Tzakou, O., Verykokidou, E., and Harvala, C. 2003. Screening of some Greek aromatic plants for antioxidant activity. *Phytotherapy Research* 17(2): 194–195.

Cox, P.A. and Leslie, P. 1991. *Texas Trees.* Corona, San Antonio, TX.

Coyote Creek. 2003. URL: www.coyote-creek.net/vines.html.

Craighead, F.C. 1971. *The Trees of South Florida.* University of Miami Press, Coral Gables.

Crawford, J.M. 1978. *The Mobilian Trade Language.* University of Tennessee Press, Knoxville.

Crawford, J.M. 1988. On the relationship of Timucua to Muskogean. Pp. 157–164. In: Shipley, W., Ed. *In Honor of Mary Haas.* Mouton de Gruyer, Berlin.

Cribb, A.B. and Cribb, J.W. 1981. *Wild Medicine in Australia.* Collins, Sydney.

Crins, W.J. 1991. The genera of Paniceae (Gramineae: Panicoideae) in the Southeastern United States. *Journal of the Arnold Arboretum Supplemental Series* 1:171–312.

Cripps, A.L. and Blunden G. 1978. A quantitative gas-liquid chromatographic method for the estimation of hecogenin and tigogenin in the leaves, juice and sapogenin concentrates of *Agave sisalana. Steroids* 31(5):661–669.

Crites, G.D. and Terry, R.D. 1984. Nutritive value of Maygrass, *Phalaris caroliniana. Economic Botany* 38:114–129.

Crombie, L. and Whiting, D.A. 1998. Biosynthesis in the rotenoid group of natural products: applications of isotope methodology. *Phytochemistry* 49(6):1479–1507.

Cronquist, A. 1981. *An Integrated System of Classification of Flowering Plants.* Columbia University Press, New York.

Crosby, A.W.J. 1972. *The Columbian Exchange. Biological and Cultural Consequences of 1492.* Greenwood Press, Westport, CT.

Cruttwell-McFadyen, R.E. 1989. Siam weed: a new threat to Australia's north. *Plant Protection Quarterly* 4(1):3–7.

Cui, J., Cordis, G.A., Tosaki, A., Maulik, N., and Das, D.K. 2002. Reduction of myocardial ischemia reperfusion injury with regular consumption of grapes. *Annals of the New York Academy of Sciences* 957:302–307.

Culbreth, D.M.R. 1910. *A Manual of Materia Medica and Pharmacology.* Reprinted by Eclectic Medical, Portland, OR, 1983.

Culpeper, N. 1653. *Complete Herbal: Consisting of a comprehensive description of nearly all herbs with their Medicinal Properties and directions for compounding the medicines extracted from them.* Reprinted, n.d., by W. Foulsham, London.

Cunningham, A.B. and Milton, S.J. 1987. Effects of basket-weaving industry on mokola palm and dye plants in northwestern Botswana. *Economic Botany* 41:386–402.

Curtin, L.S.M. 1947. *Healing Herbs of the Upper Río Grande.* Laboratory of Anthropology, Santa Fe, NM.

Curtin, L.S.M. 1949. *By the Prophet of the Earth. Ethnobotany of the Pima.* University of Arizona Press, Tucson. Reprinted 1984.

Cushing, F.H. 1897. Exploration of ancient key-dwelling remains on the Gulf Coast of Florida. *Proceedings of the American Philosophical Society* 25(153).

Cutler, H.C. 1975. Two kinds of gourds from Marco Island. Pp. 255, 256. In: Gilliland, M.S. 1975. *The Material Culture of Key Marco Florida.* University Press of Florida, Gainesville.

Cutler, H.C. and Whittaker, T.W. 1961. History and distribution of cultivated cucurbits in the Americas. *American Antiquities* 26(4):469–485.

Czerepanov, S.K. 1994. *Vascular Plants of Russia and Adjacent States (the former USSR).* Cambridge University Press, New York.

da Silva, B.P., de Sousa, A.C., Silva, G.M., Mendes, T.P., and Parente, J.P. 2002. A new bioactive steroidal saponin from *Agave attenuata. Zeitschrift für Naturforschung [C]* 57(5–6):423–428.

Dalziel, J.M. 1937. *The Useful Plants of West Tropical Africa.* Crown Agents for Overseas Governments and Administrations, London.

Daniels-Sakim, C.R. 1983. *Muskogee Words and Ways.* Moskogee Press, Wakulla Springs.

D'Aoust, J.Y. 1994. Salmonella and the international food trade. *International Journal of Food Microbiology* 24 (1–2):11–31.

Darlington, P.J.J. 1957. *Zoogeography: The Geographical Distribution of Animals.* Robert E. Krieger, Malabar, FL.

Datta, P.C. and Dasgupta, A. 1977. Comparison of vegetative anatomy of Piperales. I. Juvenile xylem of twigs. *Acta Biologica* 28(1):81–96.

Datta, S.K., Datta, K., and Chattopadhyay, S. 1983. Pharmacognosy of *Kallstroemia pubescens*, a potential source of diosgenin. *Tropical Plant Science Research* 1(4):299–302.

Dattagupta, S. and Datta, P.C. 1977. Pharmacognostic study of the leaf of *Heliotropium indicum* Linn. (Boraginaceae). *Quarterly Journal of Crude Drug Research* 15(3):141–151.

Daumeyer, R. 2002. Examining the Fort Ancient. *American Anthropologist* 6(2):27–32.

Davicino, J.G., Pestchanker, M.J., and Giordano, O.S. 1988. Pyrrolizidine alkaloids from *Heliotropicum curassavicum. Phytochemistry* 27(3):960–962.

Davidse, G. 1978. A systematic study of the genus *Lasiacis* (Gramineae: Paniceae). *Annals of the Missouri Botanical Garden* 65(4):1133–1254.

Davidson, A. 1999. *The Oxford Companion to Food.* Oxford University Press, Oxford.

Davidson, J.R. and Ortiz de Montellano, B.R. 1983. The antibacterial properties of an Aztec wound remedy. *Journal of Ethnopharmacology* 8(2):149–161.

Davidyants, E.S., Putieva, Zh.M., Bandyukova, V.A., and Abubakirov, N.K. 1984. Triterpene glycosides of *Silphium perfoliatum* [Isolated from the epigeal part]. *Chemistry of Natural Compounds* 20(1):121–122 (translated from *Khimiia prirodnykh soedinenii* 20:120–121).

Davidyants, E.S., Putieva, Zh.M., Bandyukova, V.A., and Abubakirov, N.K. 1985a. Triterpene glycosides of

Silphium perfoliatum. III. Structure of silphioside E. *Chemistry of Natural Compounds* 20(6):708–710 (translated from *Khimiia prirodnykh soedinenii* 20: 750–753).

Davidyants, E.S., Putieva, Zh.M., Bandyukova, V.A., and Abubakirov, N.K. 1985b. Triterpene glycosides of *Silphium perfoliatum.* II. *Chemistry of Natural Compounds* 20(5):636 (translated from *Khimiia prirodnykh soedinenii* 21(5):666–667).

Davidyants, E.S., Putieva, Zh.M., Bandyukova, V.A., and Abubakirov, N.K. 1985c. Triterpene glycosides of *Silphium perfoliatum.* IV. Structure of siliphioside C. *Chemistry of Natural Compounds* 21(4):486–489 (translated from *Khimiia prirodnykh soedinenii* 21 (4):519–522, 1985).

Davidyants, E.S., Putieva, Zh.M., Bandyukova, V.A., and Abubakirova, N.K. 1986. Triterpene glycosides of *Silphium perfoliatum.* V. The structure of silphioside A. *Chemistry of Natural Compounds* 22(1):58–60 (translated from *Khimiia prirodnykh soedinenii* 22 (1):63–66, 1986).

Davioud, E. and Bailleul, F. 1988. Iridoids of *Randia ruiziana. Planta Medica* 54(1):87.

Dawidar, A.M., Winternitz, F., and Johns, S.R. 1977. Structure of ipomine, a new alkaloid from *Ipomoea muricata* Jacq. *Tetrahedron* 33(14):1733–1734.

Dayton, W.A. 1937. *Range Plant Handbook.* Dover, New York.

De Barrios, V.B. 1971. *A Guide to Tequila, Mezcal and Pulque.* Minutiae Mexicana, Mexico City.

De Brahm, W.G. [1772] 1974. *The Atlantic Pilot.* Bicentennial Floridiana Facsimile Series, De Vorsey, L., Jr., Ed. University Press of Florida, Gainesville.

De Brahm, W.G. [1775] 1971. *DeBrahm's Report of the General Survey in the Southern District of North America.* De Vorsey, L., Jr., Ed. Tricentennial Edition, No. 3. University of South Carolina Press, Columbia.

de Candolle, A. [1886] 1959. *Origin of Cultivated Plants.* Reprinted by Hafner, New York.

de Cleene, M. and Lejeune, M.C. 2002. *Compendium of Symbolic and Ritual Plants in Europe.* Man and Culture Publishers, Ghent, Belgium.

De Feo, V., Capasso, A., De Simone, F., and Sorrentino, L. 1996. CNS pharmacological effects of aqueous extract from *Iresine herbstii. International Journal of Pharmacognosy* 34(3):184–188.

de la Cruz, A.A. 1978. The production of pulp from marsh grass. *Economic Botany* 32:46–50.

De La Vega, G. [1605] 1962. *The Florida of the Inca.* Translated by Varner, J.G. and J.I. Varner. University of Texas Press, Austin.

de Moura, N.F., Morel, A.F., Dessoy, E.C., Zanatta, N., Burger, M.M., Ahlert, N., Porto, G.P., and Baldisserotto, B. 2002. Alkaloids, amides and antispasmodic activity of *Zanthoxylum hyemale. Planta Medica* 68(6):534–538.

De Oliveira, D.B., De Almeida, A.P., De B A Simões, G.L., Auvin, C., Kaiser, C.R., and Costa, S.S. 2003. First isolation of a symmetrical glycosylated methylene bisflavonoid. *Planta Medica* 69(4):382–384.

de Sa Ferreira, I.C. and Ferrao Vargas, V.M. 1999. Mutagenicity of medicinal plant extracts in *Salmonella*/microsome assay. *Phytotherapy Research* 13 (5):397–400.

de Souza, M.M., Bittar, M., Cechinel-Filho, V., Yunes, R.A., Messana, I., Delle Monache, F., and Ferrari, F. 2000. Antinociceptive properties of morusin, a prenylflavonoid isolated from *Morus nigra* root bark. *Zeitschrift für Naturforschung [C]* 55(3–4):256–260.

De Tommasi, N., Piacente, S., De Simone, F., and Pizza, C. 1993. Characterization of three new triterpenoid saponins from *Ardisia japonica. Journal of Natural Products* 56(10):1669–1675.

De Voto, B., Ed. 1981. *The Journals of Lewis and Clark.* Houghton Mifflin, Boston.

De Vries, L. 1959. *German-English Science Dictionary.* McGraw-Hill, New York.

de Winter, W.P. and Amoroso, V.B., Eds. 2003. *Plant Resources of South-East Asia.* No. 15(2). *Cryptogams: Ferns and Fern Allies.* Prosea Foundation, Backhuys Publishers, Leiden, the Netherlands.

Debenedetti, S.L., Ferraro, G.E., and Coussio, J.D. 1981. Coumarins from *Pterocaulon virgatum. Planta Medica* 42(1):97–98.

Debenedetti, S.L., Nadinic, E.L., Coussio, J.D., de Kimpe, N., Feneau-Dupont, J., and Declarco, J.P. 1991. Purpurenol, a highly oxygenated coumarin from *Pterocaulon purpurascens. Phytochemistry* 30 (8):2757–2758.

Debenedetti, S.L., Nadinic, E.L., Gómez, M.A., Coussio, J. D., Kimpe, N. de, and Boeykens, M. 1992. Purpuraslo, a highly oxygenated coumarin from *Pterocaulon purpurascens. Phytochemistry* 31(9):3284–3285.

Debenedetti, S.L., de Kimpe, N., Boeykens, M., Coussio, J. D., and Kesteleyn, B. 1997. Structural revision of four coumarins from *Pterocaulon* species. *Phytochemistry* 45(7):1515–1517.

Debenedetti, S.L., Palacios, P.S., Nadinic, E.L., Coussio, J. D., Kimpe, N. de, Boeykens, M., Feneau-Dupont, J., and Declercq, J.P. 1994. 5-(3-Methyl-2-butenyloxy)-6,7-methylenedioxycoumarin, a 5,6,7-trioxygenated coumarin from *Pterocaulon virgatum. Journal of Natural Products* 57(11):1539–1542.

Debenedetti, S.L., Nadinic, E.L., Gómez, M.A., Coussio, J. D., Kimpe, N. de, and Boeykens, M. 1996. Purpurasolol, a highly oxygenated coumarin from *Pterocaulon purpurascens. Phytochemistry* 42(2):563–564.

Debenedetti, S.L., Nadinic, E.L., Coussio, J.D., de Kimpe, N., and Boeykens, M. 1998. Two 6,7-dioxygenated coumarins from *Pterocaulon virgatum. Phytochemistry* 48(4):707–710.

Debenedetti, S.L., Tehrani, K.A., Puyvelde, L. van, and Kimpe, N. de. 1999. Isopurpurasol, a coumarin from *Pterocaulon virgatum. Phytochemistry* 51(5):701–703.

Decendit, A., Waffo-Teguo, P., Richard, T., Krisa, S., Vercauteren, J., Monti, J.P., Deffieux, G., and Merillon, J.M. 2002. Galloylated catechins and stilbene diglucosides in *Vitis vinifera* cell suspension cultures. *Phytochemistry* 60(8):795–798.

Decker, D.S. 1988. Numerical analysis of archaeological *Cucurbita pepo* seeds from Hontoon Island, Florida. *Journal of Ethnobiology* 8(1):35–44.

Decker-Walters, D.S., Walters, T.W., Cowan, C.W. and Smith, B.D. 1993. Isozymic characterization of wild populations of *Cucurbita pepo. Journal of Ethnobiology* 13(1):55–72.

DeFilipps, R.A. 1969a. Parasitism in *Ximenia* (Olacaeae). *Rhodora* 71(787):439–443.

DeFilipps, R.A. 1969b. A revision of *Ximenia* (Plum.) L. (Olacaceae). *Dissertation Abstracts Section B* 29 (10):3634B.

Degener, O. 1973. *Plants of Hawaii National Park Illustrative of Plants and Customs of the South Seas.* Braun-Brumfeld, Ann Arbor, MI.

Delgado, G., Olivares, M. del S., Chávez, M.I., Ramírez-Apan, T., Linares, E., Bye, R., and Espinosa-García, F.J. 2001. Antiinflammatory constituents from *Heterotheca inuloides. Journal of Natural Products* 64 (7):861–864.

Della Loggia, R., Zilli, C., Del Negro, P., Redaelli, C., and Tubaro, A. 1988. Isoflavones as spasmolytic principles of *Piscidia erythrina. Progress in Clinical Biological Research* 280:365–368.

Dellacasa, A., Bailac, P., Duschatzky, C., Carrascull, A., and Ponzi, M. 1999. Antimicrobial activity of the essential oil of *Heterotheca latifolia* Buckley (Compositae). *Journal of Essential Oil Research* 11(2):262–263.

Delle Monache, F., Ferrari, F., and Menichini, F. 1984. Two isoflavones from *Piscidia erythrina. Phytochemistry* 23(12):2945–2947.

Dellinger, S.C. 1932. Baby cradles of Ozark Bluff Dwellers. *American Antiquity* 1:197–214.

Densmore, F. 1928. *Uses of Plants by the Chippewa Indians.* Reprinted 1974: *How Indians Use Wild Plants for Food, Medicine and Crafts.* Dover, New York.

De-Oliveira, M.M., Nakamura, I.T., Joussef, A.C., and Giannotti Filho, O. 1985. Intratumor chemotherapy in combination with a systemic antimetastatic drug in the treatment of Lewis-lung carcinoma. *Brazilian Journal of Medical Biological Research* 18(3):293–302.

Derkach, A.I., Komissarenko, N.F., Gordienko, V.G., Sheremet, I.P., Kovalev, I.P., and Pakaln, D.A. 1980. Flavonoids of *Stachys spectabilis. Chemistry of Natural Compounds* 16(2):128–130 (translated from *Khimiia prirodnykh soedinenii* 172–174).

Deshpande, V.H., Rama-Rao, A.V., Venkataraman, K., and Wakharkar, P.V. 1974. Wood phenolics of *Morus* species. III. Phenolic constituents of *Morus rubra* bark. *Indian Journal of Chemistry* 12(5):431–436.

Desmarchelier, C., Novoa Bermudez, M.J., Coussio, J., Ciccia, G., and Boveris, A. 1997. Antioxidant and prooxidant in aqueous extracts of Argentine plants. *Pharmaceutical Biology* (formerly *International Journal of Pharmacognosy*) 35(2):116–120.

Desmarchelier, C., Romao, R.L., Coussio, J., Ciccia, G. 1999. Antioxidant and free radical scavenging activities in extracts from medicinal trees used in the 'Caatinga' region in northeastern Brazil. *Journal of Ethnopharmacology* 67(1):69–77.

Desmyter, S., Vandenbussche, F., Van Damme, E.J., and Peumans, W.J. 2001. Preparation of monospecific polyclonal antibodies against *Sambucus nigra* lectin related protein, a glycosylated plant protein. *Preparative biochemistry and biotechnology* 31(3):209–216.

Desowitz, R.S. 1987. *New Guinea Tapeworms and Jewish Grandmothers.* W.W. Norton and Co., New York.

Desowitz, R.S. 1991. *The Malaria Capers.* W.W. Norton, New York.

Desowitz, R.S. 1997. *Who Gave Pinta to the Santa Maria? Torrid Diseases in a Temperate World.* W.W. Norton, New York.

Devi, P.U., Solomon, F.E., and Sharada, A.C. 1999. Plumbagin, a plant naphthoquinone with antitumor and radiomodifying properties. *Le Pharmacien biologiste* 37(3):231–236.

Devlin, R.M. and Deubert, K.H. 1982. Control of wild bean on cranberry bogs *Apios americana*, weed. *Proceedings of the Annual Meeting of the Northeastern Weed Science Society* 36:97–100.

Dewidar, A.M. and el-Munajjed, D. 1970. The steroid sapogenin constituents of *Agave americana, A. variegata* and *Yucca gloriosa. Planta Medica* 19(1): 87–91.

Dey, D. and Das, M.N. 1988. Pharmacognosy of antidysenteric drugs of Indian medicine. *Acta Botanica Indica* 16(2):216–226.

Deyo, J.A. and Kerkvliet, N.I. 1990. Immunotoxicity of the pyrrolizidine alkaloid monocrotaline following subchronic administration to C57B1/6. *Fundamental and Applied Toxicology* 14(4):842–849.

Dhar, R., Dawar, H., Garg, S., Basir, S.F., and Talwar, G.P. 1996. Effect of volatiles from neem and other natural products on gonotrophic cycle and oviposition of *Anopheles stephensi* and *An. culicifacies* (Diptera: Culicidae). *Journal of Medical Entomology* 33(2): 195–201.

Dhawan, K., Kumar, S., and Sharma, A. 2001a. Anti-anxiety studies on extracts of *Passiflora incarnata* Linnaeus. *Journal of Ethnopharmacology* 78(2–3): 165–170.

Dhawan, K., Kumar, S., and Sharma, A. 2001b. Anxiolytic activity of aerial and underground parts of *Passiflora incarnata. Fitoterapia* 72(8):922–926.

Dhawan, K., Kumar, S., and Sharma, A. 2001c. Comparative biological activity study on *Passiflora incarnata* and *P. edulis. Fitoterapia* 72(6):698–702.

Díaz del Castillo, B. 1956. *Historia verdadera de la conquista de la Nueva España* [The discovery and conquest of Mexico, 1517–1521]. García, Genaro, Ed. from the only exact copy of the original MS (and published in Mexico); translated with an introduction and notes by A.P. Maudslay; introduction to the American edition by I.A. Leonard. Farrar, Straus & Cudahy, New York.

Diggs, G.M.J., Lipscomb, B.L., and O'Kennon, R.J. 1999. *Shinners and Mahler's Illustrated Flora of North Central Texas.* Center for Environmental Studies and Department of Biology, Austin College, Sherman, TX and Botanical Research Institute of Texas (BRIT), Fort Worth.

Dijoux-Franca, M.G., Tchamo, D.N., Cherel, B., Cussac, M., Tsamo, E., and Mariotte, A.M. 2001. New dihydrophenanthrene and phenyldihydroisocoumarin constituents of *Trema orientalis. Journal of Natural Products* 64(6):832–835.

Dinda, B. 1994. A dihydroflavonol from *Plumbago indica. Phytochemistry* 35(4):1083–1084.

Dinda, B. and Chel, G. 1992. 6-Hydroxyplumbagin, a naphthoquinone from *Plumbago indica. Phytochemistry* 31(10):3652–3653.

Ding, Y., Tian, R.H., Yang, C.R., Chen, Y.Y., and Nohara, T. 1993. Two new steroidal saponins from dried fermented residues of leaf-juices of *Agave sisalana* forma Dong No. 1. *Chemical and Pharmaceutical Bulletin* (Tokyo) 41(3):557–560.

Ditmars, R.L. 1931. *Snakes of the World.* Macmillan, New York.

Dittmar, A. 1993. *Morinda citrifolia* L. — use in indigenous Samoan medicine. *Journal of Herbs, Spices and Medicinal Plants* 1(3):77–92.

Dixon, A.R., McMillen, H., and Etkin, N.L. 1999. Ferment this: the transformation of Noni, a traditional Polynesian medicine (*Morinda citrifolia*, Rubiaceae). *Economic Botany* 53(1):51–68.

Dobelis, I.N., Ed. 1986. *Magic and Medicine of Plants.* Reader's Digest Association, Pleasantville, NY.

Dobyns, H.F. 1988. Piman Indian historic *Agave* cultivation. *Desert Plants* 9(2):49–53.

Dollahite, J.W. 1975. Toxicity of *Kallstroemia parviflora* (warty caltrop) to sheep, goats and rabbits. *Southwestern Veterinarian* 28(2):135–139.

Domergue, F., Helms, G.L., Prusky, D., and Browse, J. 2000. Antifungal compounds from idioblast cells isolated from avocado fruits. *Phytochemistry* 54(2):183–189.

Domínguez-S., X.A., Cano, G., Luna, I., and Dieck, A. 1977. Two coumarins from the aereal parts of *Amyris madrensis. Phytochemistry* 16(7):1096.

Domínguez-S., X.A. and Alcorn, J.B. 1985. Screening of medicinal plants used by Huastec Mayans of Northeastern Mexico. *Journal of Ethnopharmacology* 13 (2):139–156.

Domínguez-S., X.A. and Zamudio, A. 1972. Beta-amyrin acetate and campesterol from *Pluchea odorata. Phytochemistry* 11(3):1179.

Domínguez-S., X.A., Franco, R., Cano, G., Villarreal, R., Bapuji, M., and Bohlmann, F. 1981. Three eudesmanolides from *Pluchea rosea. Phytochemistry* 20(9): 2297–2298.

Domínguez-S., X.A., Watson, W.H., García, S., and Martínez Davila, I.A. 1985. 25S-spirost-4-ene-3, 12-dione, a new sapogenin from *Kallstroemia maxima. Planta Medica* 51(6):534–535.

Dominique, R. and Altamirano, F. 1877. Gac. Medicina México. 12:77 cited in Erythrina Symposium 1974.

Dorado, O., Avila, G., Arias, D.M., Ramirez, R., Salinas, D., and Valladares, G. 1996. The Arbol del Tule (*Taxodium mucronatum* Ten.) is a single genetic individual. *Madroño* 43:445–452.

Doran, G., Dickel, D., and Newsom, L. 1990. A 7,290-year-old bottle gourd from the Windover Site, Florida. *American Antiquity* 55:354–360.

Dornenburg, H. and Knorr, D. 1994a. Effectiveness of plant-derived and microbial polysaccharides as elicitors for anthraquinone synthesis in *Morinda citrifolia* cultures. *Journal of Agricultural and Food Chemistry* 42 (4):1048–1052.

Dornenburg, H. and Knorr, D. 1994b. Elicitation of chitinases and anthraquinones in *Morinda citrifolia* cell cultures. *Food Biotechnology* 8(1):57–65.

Dorsey, J.O. and Swanton, J.R. 1912. A dictionary of the Biloxi and Ofo language: accompanied with thirty-one Biloxi texts and numerous Biloxi phrases. Smithsonian Institution, Bureau of American Ethnology Bulletin 47. Washington, D.C.

Douglas, M.S. 1947. *The Everglades: River of Grass.* Hurricane House, Coconut Grove, FL.

Downie, S.R. and Palmer, J.D. 1994. A chloroplast DNA phylogeny of the Caryophyllales based on structural and inverted repeat restriction site variation. *Systematic Botany* 19(2):236–252.

Drechsel, E.J. 1997. *Mobilian Jargon. Linguistic and Sociohistorical Aspects of a Native American Pidgin.* Clarendon Press, Oxford.

Dreikorn K. 2002. The role of phytotherapy in treating lower urinary tract symptoms and benign prostatic hyperplasia. *World Journal of Urology* 19(6):426–435.

Dressler, R.L. 1957. The genus *Pedilanthus* (Euphorbiaceae). *Contributions to the Gray Herbarium Harvard University* No. 182. 188 pp.

Dubois, M.A., Benze, S., and Wagner, H. 1990. New biologically active triterpene-saponins from *Randia dumetorum. Planta Medica* 56(5):451–455.

Dudkin, M.S., Cherno, N.K., and Skhantova, N.G. 1979. Characteristics of the water-soluble polysaccharides of the leaves of *Silphium perfoliatum. Chemistry of Natural Compounds* 15(6):680–682 (translation of *Khimiia prirodnykh soedinenii* pp. 771–774).

Duke, J.A. 1968. *Darien Ethnobotanical Dictionary.* Battelle Memorial Institute, Columbus, OH.

Duke, J.A. 1970. Ethnobotanical observations on the Chocó Indians. *Economic Botany* 24:344–366.

Duke, J.A. 1972. *Isthmian Ethnobotanical Dictionary.* Published by the author, Fulton, MD.

Duke, J.A. 1985. *CRC Handbook of Medicinal Herbs.* CRC Press, Boca Raton, FL.

Duke, J.A. 1997. *The Green Pharmacy.* Rodale Press, Emmanus, PA.

Duke, J.A. 2003. Dr. Duke's Phytochemical and Ethnobotanical Databases. URL: http://www.ars-grin.gov/cgi-bin/duke/ethnobot.pl.

Duke, J.A. and DuCellier, J.L. 1993. *CRC Handbook of Alternative Cash Crops.* CRC Press, Boca Raton, FL.

Duke, J.A. and Vásquez, R. 1994. *Amazonian Ethnobotanical Dictionary.* CRC Press, Boca Raton, FL.

Duke, J.A., with Bogenschutz-Godwin, M.J., duCellier, J., and Duke, P.-A. K. 2002. *Handbook of Medicinal Herbs,* 2nd ed. CRC Press, Boca Raton, FL.

Duncan, A.C., Jager, A.K., and van Staden, J. 1999. Screening of Zulu medicinal plants for angiotensin converting enzyme (ACE) inhibitors. *Journal of Ethnopharmacology* 68(1–3):63–70.

Dunhill, M.P. and Fowden, L. 1965. Amino acids of seeds of the Cucurbitaceae. *Phytochemistry* 4:933–944.

Dunmire, W.W. and Tierney, G.D. 1997. *Wild Plants and Native Peoples of the Four Corners.* Museum of New Mexico Press, Santa Fe.

Duret, S., Jacquemin, H., and Paris, R.R. 1976. Plantes malagaches XIX. Sur la composition chimique de *Stachytarpheta jamaicensis* (L.) Vahl, Verbenaceae. *Plantes Médicinales et Phytothérapie* 10:96–104.

Durga, R., Sridhar, P., and Polasa, H. 1990. Effects of plumbagin on antibiotic resistance in bacteria. *The Indian Journal of Medical Research* 91:18–20.

Durodola, J.I. 1974. Anti-neoplastic property of a crystalline compound extracted from *Morinda lucida. Planta Medica* 26(3):208–211.

Duschatzky, C., Carrascull, A., Bailac, P., Ponzia, M., Baren, C. van, Muschietti, L., and Coussio, J. 1998. Composition of the essential oil from flowers and leaves of *Heterotheca latifolia* Buckley (Compositae). *Journal of Essential Oil Research* 10(1):101–103.

Duvall, M.R. and Biesboer, D.D. 1988. Anatomical distinctions between the pistillate spikelets of the species of wild-rice (*Zizania*, Poaceae). *American Journal of Botany* 75(1):157–159.

Dwelly, E. 1933. *The Illustrated Gaelic-English Dictionary.* Birlinn, Edinburgh.

Dyas, L., Threlfall, D.R., and Goad, L.J. 1994. The sterol composition of five plant species grown as cell suspension cultures. *Phytochemistry* 35(3):655–660.

Ebeling, W. 1986. *Handbook of Indian Foods and Fibers of Arid America.* University of California Press, Berkeley.

Eccleston, V.S. and Harwood, J.L. 1995. Solubilisation, partial purification and properties of acetyl-CoA: glycerol-3-phosphate acetyltransferase from avocado (*Persea americana*) fruit mesocarp. *Biochimica Biophysica Acta* 1257(1):1–10.

Echeverri, F., Arango, V., Quinones, W., Torres, F., Escobar, G., Rosero, Y., and Archbold, R. 2001. Passifloricins, polyketides alpha-pyrones from *Passiflora foetida* resin. *Phytochemistry* 56(8):881–885.

Eckenwalder, J.E. 1980a. Dispersal of the West Indian cycad, *Zamia pumila* L. *Biotropica* 12(1):79–80.

Eckenwalder, J.E. 1980b. Taxonomy of the West Indian cycads. *Journal of the Arnold Arboretum* 61(4):701–722.

Eden, R. 1555. *The Decades of the New World or West India,* cited in OED 1971.

Edgar, J.A., Culvenor, C.C.J., and Pliske, T.E. 1974. Coevolution of Danaid butterflies with their host plants. *Nature* 250(5468):646–648.

Edgar, M. Ridout. 1891. *Ten Years of Upper Canada in Peace and War, 1805–1815; Being the Ridout Letters.* T.F. Unwin, London.

Editorial Committee. 2003. *Flora of North America North of Mexico. Vol. 26. Magnoliophyta: Liliidae: Liliales and Orchidales.* Oxford University Press, New York.

Edlin, H.L. and Mitchell, A.F. 1985. *Broadleaves.* Her Majesty's Stationery Office, Forestry Commission Booklet 20, London.

Ekimoto, H., Irie, Y., Araki, Y., Han, G.Q., Kadota, S., and Kikuchi, T. 1991. Platelet aggregation inhibitors from the seeds of *Swietenia mahagoni*: inhibition of *in vitro* and *in vivo* platelet-activating factor-induced effects of tetranortriterpenoids related to swietenine and swietenolide. *Planta Medica* 57(1):56–58.

El Abbadi, N., Weniger, B., Lobstein, A., Quirion, J.C. and Anton, R. 1989. New alkaloids of *Chiococca alba. Planta Medica* 55:603–604.

El Kossori, R.L., Villaume, C., El Boustani, E., Sauvaire, Y., and Mejean, L. 1998. Composition of pulp, skin and seeds of prickly pears fruit (*Opuntia ficus-indica* sp.). *Plant Foods and Human Nutrition* 52(3):263–270.

El-Dahmy, S., Sarg, T., Farrag, N.M., Ateya, A.M., Jakupovic, J., Bohlmann, F., and King, R.M. 1986. Further cadinene derivatives from *Heterotheca grandiflora. Phytochemistry* 25(6):1474–1475.

Eldridge, J. 1975. Bush medicine in the Exumas and Long Island, Bahamas: A field study. *Economic Botany* 29:307–332.

Elghamry, M.I. and Haensel, R. 1969. Activity and isolated phytoestrogen of shrub palmetto fruits (*Serenoa repens* Small), a new estrogenic plant. *Experientia* 25(8):828–829.

El-Hafiz, M.A.A., Weniger, B., Quirion, J.C., and Anton, R. 1991. Keto-alcohols, lignans and coumarins from *Chiococca alba. Phytochemistry* 39(6):2029–2031.

Ellingwood, F. and Lloyd, J. 1898. *American Materia Medica, Therapeutics and Pharmacognosy.* Eclectic Medical Publications, Portland, OR. Reprint, 1983.

Elliott, S. 1821. *A Sketch of the Botany of South-Carolina and Georgia.* Vol. 1, J.R. Schenck, Charleston, SC.

Elmore, F.H. 1944. Ethnobotany of the Navajo. *University of New Mexico Bulletin 392* 1(2):1–136.

El-Sayed, N.H., Miski, M., Whittemore, A.T., and Mabry, T. J. 1988. Sesquiterpene lactones from *Chromolaena opadoclinia. Phytochemistry* 27(10):3312–3314.

El-Sayed, N.H., Wojcinska, M., Drost-Karbowska, K., Matlawska, I., Williams, J., and Mabry, T.J. 2002. Kaempferol triosides from *Silphium perfoliatum. Phytochemistry* 60(8):835–838.

El-Sedawy, A.I., Hattori, M., Kobashi, K., and Namba, T. 1989a. Metabolism of gentiopicroside (gentiopicrin) by human intestinal bacteria. *Chemical and Pharmaceutical Bulletin* (Tokyo) 37(9):2435–2437.

El-Sedawy, A.I., Shu, Y.Z., Hattori, M., Kobashi, K., and Namba, T. 1989b. Metabolism of swertiamarin from

Swertia japonica by human intestinal bacteria. *Planta Medica* 55(2):147–150.

Encyclopedia Britannica. 2003. URL: www.britannica.com.

Erlund, I., Marniemi, J., Hakala, P., Alfthan, G., Meririnne, E., and Aro, A. 2003. Consumption of black currants, lingonberries and bilberries increases serum quercetin concentrations. *European Journal of Clinical Nutrition* 57(1):37–42.

Ernesto-Frimmel, A., Luna-Bilheiro Peixoto, J., Helena-Sarragiotto, M., and Vidotti, G.J. 2000. Vitexin, paprazine and terpenoids from *Trema micrantha*. *Biochemical Systematics and Ecology* 28(5):495–496.

Ernst, E. 2002. The risk–benefit profile of commonly used herbal therapies: ginkgo, St. John's wort, ginseng, echinacea, saw palmetto, and kava. *Annals of Internal Medicine* 136(1):42–53.

Eromosele, C.O. and Eromosele, I.C. 2002. Fatty acid compositions of seed oils of *Haematostaphis barteri* and *Ximenia americana*. *Bioresource Technology* 82 (3):303–304.

Eromosele, I.C., Eromosele, C.O., Akintoye, A.O., and Komolafe, T.O. 1994. Characterization of oils and chemical analyses of the seeds of wild plants. *Plant Foods and Human Nutrition* 46(4):361–365.

Erythrina Symposium. 1974. *Lloydia* 37(3):321–487, 37(4): 543–588.

Esposito-A., M., Brown, P., Tejeira, I., Bruitrago, R., Barrios, L., Sánchez, C., Gupta, M.P., and Cedéno, J. 1985. Pharmacological screening of Panamanian medicinal plants. Part 1. *International Journal of Crude Drug Research* 23(1):17–25.

Eubanks, M.W. 1999. *Corn in Clay. Maize Paleoethnobotany in Pre-Columbian Art*. University Press of Florida, Gainesville.

Eubanks, M.W. 2001. The mysterious origin of maize. *Economic Botany* 55(4):492–514.

Fabry, W., Okemo, P., and Ansorg, R. 1996a. Fungistatic and fungicidal activity of East African medicinal plants. *Mycoses* 39(1–2):67–70.

Fabry, W., Okemo, P., and Ansorg, R. 1996b. Activity of East African medicinal plants against *Helicobacter pylori*. *Chemotherapy* 42(5):315–317.

Fabry, W., Okemo, P., and Ansorg, R. 1998. Antibacterial activity of East African medicinal plants. *Journal of Ethnopharmacology* 60(1):79–84.

Facino, R.M., Carini, M., Aldini, G., Berti, F., Rossoni, G., Bombardelli, E., and Morazzoni, P. 1999. Diet enriched with procyanidins enhances antioxidant activity and reduces myocardial post-ischaemic damage in rats. *Life Science* 64(8):627–642.

Fairbanks, C.H. 1978. The ethno-archeology of the Florida Seminole. Pp. 141–162. In: Milanich, J.T. and Proctor, S., Eds. *Tacachale: Essays on the Indians of Florida and Southeastern Georgia during the Historic Period*. University Press of Florida, Gainesville.

Fairchild, D. 1947. *The World Grows Round My Door*. Charles Scribner's Sons, New York.

Fang, S., Li, L., Niu, C., and Tseng, K. 1961. Chemical studies of *Cucurbita moschata* Duch. I. Isolation and structural studies of cucurbitine, a new amino acid. *Science Sinica* (Peking) 10:845–851.

Fantz, P.R. 1996. Taxonomic notes on the *Centrosema pubescens* Bentham complex in Central America (Leguminosae: Phaseoleae: Clitoriinae). *Sida* 17 (2):321–332.

Fantz, P.R. and Predeep, S.V. 1992. Comments on four legumes (*Clitoria, Centrosema*) reported as occurring in India. *Sida* 15(1):1–7.

Farine, J.P., Legal, L., Moreteau, B., and Le Quere, J.L. 1996. Volatile components of ripe fruits of *Morinda citrifolia* and their effects on *Drosophila*. *Phytochemistry* 41(2):433–438.

Farnsworth, N.R., Hart, N.K., Johns, S.R., Lamberton, J.A., and Messmer, W. 1969. Alkaloids of *Boehmeria cylindrica* (family Urticaceae): identification of a cytotoxic agent, highly active against Eagle's 9KB carcinoma of the nasopharynx in cell culture, as crytopleurine. *Australian Journal of Chemistry* 22(8): 1805–1807.

Farsam, H., Yassa, N., Sarkhail, P., and Shafiee, A. 2000. New pyrrolizidine alkaloids from *Heliotropium crassifolium*. *Planta Medica* 66(4):389–391.

Fatope, M.O., Adoum, O.A., and Takeda, Y. 2000a. C(18) acetylenic fatty acids of *Ximenia americana* with potential pesticidal activity. *Journal of Agricultural and Food Chemistry* 48(5):1872–1874.

Fatope, M.O., Adoum, O.A., and Takeda, Y. 2000b. Oleanene palmitate from *Ximenia americana*. *Le Pharmacien biologiste* 38(5):391–393.

Feeling, D. 1975. *Cherokee-English Dictionary*. Cherokee Nation of Oklahoma, Tahlequah.

Felger, R.S. 2000. *Flora of the Gran Desierto and Río Colorado Delta*. University of Arizona Press, Tucson.

Felger, R.S. and Moser, M.B. 1985. *People of the Desert and Sea. Ethnobotany of the Seri Indians*. University of Arizona Press, Tucson.

Felger, R.S., Johnson, M.S., and Wilson, M.F. 2001. *The Trees of Sonora, Mexico*. Oxford University Press, New York.

Fellows, E.J. and Smith, C.S. 1938. Chemistry of *Passiflora incarnata*. *Journal of the American Pharmacists Association* 27:565–573.

Felter, H.W. 1922. *The Eclectic Materia Medica, Pharmacology and Therapeutics*. Eclectic Medical Publications, Portland, OR. Reprint, 1983.

Feng, Y.X. 1985. Pharmacognostical identification of four species of Saururaceae [in Chinese]. *Zhong Yao Tong Bao* 10(5):10–12.

Feo Brito, F., Martínez, A., Palacios, R., Mur, P., Gómez, E., Galindo, P.A., Borja, J., and Martínez, J. 1999. Rhinoconjunctivitis and asthma caused by vine pollen: a case report. *The Journal of Allergy and Clinical Immunology* 103(2 Pt. 1):262–266.

Fernald, M.L. 1950. *Gray's Manual of Botany*. American Book, New York.

Fernald, M.L., Kinsey, A.C., and Rollins, R.C. 1958. *Edible Wild Plants of Eastern North America*. Harper & Row, New York.

Fernández, M.L., Trejo, A., and McNamara, D.J. 1990. Pectin isolated from prickly pear (*Opuntia* sp.) modifies low density lipoprotein metabolism in cholesterol-fed guinea pigs. *The Journal of Nutrition* 120(11):1283–1290.

Fernández, M.L., Lin, E.C., Trejo, A., and McNamara, D.J. 1992. Prickly pear (*Opuntia* sp.) pectin reverses low density lipoprotein receptor suppression induced by a hypercholesterolemic diet in guinea pigs. *The Journal of Nutrition* 122(12):2330–2340.

Fernández, M.L., Lin, E.C., Trejo, A., and McNamara, D.J. 1994. Prickly pear (*Opuntia* sp.) pectin alters hepatic cholesterol metabolism without affecting cholesterol absorption in guinea pigs fed a hypercholesterolemic diet. *The Journal of Nutrition* 124(6):817–824.

Fernando, E.F. and C.J. Quinn. 1995. Picramniaceae, a new family, and a recircumscription of Simaroubaceae. *Taxon* 44:177–181.

Fernando, E.S., Gadek, P.A., Crayn, D.M., and Quinn, C.J. 1993. Rosid affinities of Surianaceae: molecular evidence. *Molecular and Phylogenetic Evolution* 2 (4):344–350.

Fernando, E.S., Gadek, P.A., and Quinn, C.J. 1995. Simaroubaceae, an artificial construct: evidence from rbcL sequence variation. *American Journal of Botany* 82 (1):92–103.

Ferreira, A.B. de Holanda. ca. 1978. *Novo Dicionário da Língua Portuguesa.* Editora Nova Fronteira, Rio de Janeiro.

Ferreira, E.O. and Dias, D.A. 2000. A methylenedioxyflavonol from aerial parts of *Blutaparon portulacoides.* *Phytochemistry* 53(1):145–147.

Ferreira, M.E., Rojas de Arias, A., Torres de Ortiz, S., Inchausti, A., Nakayama, H., Thouvenel, C., Hocquemiller, R., and Fournet, A. 2002. Leishmanicidal activity of two canthin-6-one alkaloids, two major constituents of *Zanthoxylum chiloperone* var. *angustifolium.* *Journal of Ethnopharmacology* 80(2–3): 199–202.

Fields, K.T. (Hutke). 2003. Natchez Indians: Historical Overview. URL: www.natcheznation.gq.nu.

Filho, J.M.B., Medeiros, D.F., Agra, M. de F., and Bhattacharyya, J. 1989. Spirostanes of *Kallstroemia tribuloides*: identification of C-25 epimers in mixture by 13C NMR spectroscopy. *Phytochemistry* 28(7): 1985–1986.

Fish, F. and Waterman, P.G. 1973. Proceedings: alkaloids in the bark of *Zanthoxylum clava-herculis.* *The Journal of Pharmacy and Pharmacology* 25(Suppl.): 115P–116.

Fish, F., Gray, A.I., Waterman, P.G., and Donachie, F. 1975. Alkaloids and coumarins from North American *Zanthoxylum* species. *Lloydia* 38(3):268–270.

Fisher, A.A., Purcell, P., and Le Couteur, D.G. 2000. Toxicity of *Passiflora incarnata* L. *Journal of Toxicology. Clinical Toxicology* 38(1):63–66.

Fitzpatrick, D.F., Bing, B., Maggi, D.A., Fleming, R.C., and O'Malley, R.M. 2002. Vasodilating procyanidins derived from grape seeds. *Annals of the New York Academy of Sciences* 957:78–89.

Flamm, J., Kiesswetter, H., and Englisch, M. 1979. An urodynamic study of patients with benign prostatic hypertrophy treated conservatively with phytotherapy or testosterone [in German]. *Wiener Klinische Wochenschrift* 91(18):622–627.

Flora Europaea. 2003. URL: http://rbg-web2.rbge.org.uk/FE/fe.html.

Folkers, K. and Koniuszy, F. 1940. Isolation of erythroidine, an alkaloid of curare action from *Erythrina americana.* *Journal of the American Chemical Society* 59:1580.

Folsom-Dickerson, W.E.S. 1965. *The White Path.* Naylor, San Antonio, TX.

Fontaneda, H.D. [1575] 1944. *Memoires of Do. d'Escalente Fontaneda Respecting Florida.* Translated by Buckingham Smith. 1854. Reprinted with revisions, Historical Association of South Florida, Misc. Publ. 1:1–77, Miami.

Forbes, J.G. [1821] 1964. *Sketches, Historical and Topographical, of the Floridas; most particularly of East Florida.* A facsimile reproduction, with an introduction by J.W. Covington. University of Florida, Gainesville.

Ford, K.C. 1975. *Las Yerbas de la Gente: A Study of Hispano-American Medicinal Plants.* Anthropological Papers, Museum of Anthropology, University of Michigan, No. 60, Ann Arbor.

Forestieri, A.M., Monforte, M.T., Ragusa, S., Trovato, A., and Iauk, L. 1996. Antiinflammatory, analgesic and antipyretic activity in rodents of plant extracts used in African medicine. *Phytotherapy Research* 10(2): 100–106.

Foster, S. and Caras, R. 1994. *A Field Guide to Venomous Animals and Poisonous Plants. North America North of Mexico.* Houghton Mifflin, Boston.

Foster, S. and Duke, J.A. 1990. *A Field Guide to Medicinal Plants.* Houghton Mifflin, Boston.

Foster, S. and Hobbs, C. 2002. *Western Medicinal Plants and Herbs.* Houghton Mifflin, Boston.

Fournet, J. 1978. *Flore illustrée des phanerogames de Guadelupe et de Martinique.* Institut National de la Recherche Agronomique, Paris.

Fournet, J. and Hammerton, J.L. 1991. *Weeds of the Lesser Antilles/Mauvaises herbes des petites antilles.* Institut National de la Recherche Agronomique, Paris.

Fournier, G., Bercht, C.A.L., and Paris, R.R. 1975. 3-Methoxychrysazin, a new anthraquinone from *Xyris semifuscata.* *Phytochemistry* 14(9):2099.

Fraga, B.M., Terrero, D., Gutierrez, C., and Gonzalez-Coloma, A. 2001. Minor diterpenes from *Persea indica*: their antifeedant activity. *Phytochemistry* 56 (4):315–320.

Frampton, J. [1577] 1925. *Joyfull Newes out of the Newe Founde Worlde.* Reprint by Alfred A. Knopf, New York.

Franceschinelli, E.V., Yamamoto, K., and Shepherd, G.J. 1998. Distinctions among three *Simarouba* species. *Systematic Botany* 23(4):479–488.

Frank, J., Kamal-Eldin, A., Lundh, T., Maatta, K., Torronen, R., and Vessby, B. 2002. Effects of dietary anthocyanins on tocopherols and lipids in rats. *Journal of Agricultural and Food Chemistry* 50(25): 7226–7230.

Franssen, F.F., Smeijsters, L.J., Berger, I., and Medinilla Aldana, B.E. 1997. *In vivo* and *in vitro* antiplasmodial activities of some plants traditionally used in Guatemala against malaria. *Antimicrobial Agents and Chemotherapy* 41(7):1500–1503.

Frati-Munari, A.C., Fernández-Harp, J.A., de la Riva, H., Ariza-Andraca, R., and del Carmen Torres, M. 1983. Effects of nopal (*Opuntia* sp.) on serum lipids, glycemia and body weight. *Archivos de Investigacion Medica* (Mexico) 14(2):117–125.

Frati-Munari, A.C., Altamirano-Bustamante, E., Rodriguez-Barcenas, N., Ariza-Andraca, R., and López-Ledesma, R. 1989a. Hypoglycemic action of *Opuntia streptacantha* Lemaire: study using raw extracts [in Spanish]. *Archivos de Investigacion Medica* (Mexico) 20(4):321–325.

Frati-Munari, A.C., de Leon, C., Ariza-Andraca, R., Banales-Ham, M.B., López-Ledesma, R., and Lozoya, X. 1989b. Effect of a dehydrated extract of nopal (*Opuntia ficus-indica* Mill.) on blood glucose [in Spanish]. *Archivos de Investigacion Medica* (Mexico) 20(3):211–216.

Frati-Munari, A.C., Del Valle-Martínez, L.M., Ariza-Andraca, C.R., Islas-Andrade, S,, and Chavez-Negrete, A. 1989c. Hypoglycemic action of different doses of nopal (*Opuntia streptacantha* Lemaire) in patients with type II diabetes mellitus [in Spanish]. *Archivos de Investigacion Medica* (Mexico) 20(2):197–201.

Frati-Munari, A.C., Rios-Gil, U., Ariza-Andraca, C.R., Islas-Andrade. S., and López-Ledesma, R. 1989d. Duration of the hypoglycemic action *of Opuntia streptacantha* Lem. [in Spanish]. *Archivos de Investigacion Medica* (Mexico) 20(4):297–300.

Frati-Munari, A.C., Licona-Quesada, R., Araiza-Andraca, C.R., López-Ledesma, R., and Chávez-Negrete, A. 1990. Activity of *Opuntia streptacantha* in healthy individuals with induced hyperglycemia [in Spanish]. *Archivos de Investigacion Medica* (Mexico) 21(2):99–102.

Frati-Murari, A.C., Xilotl-Díaz, N., Altamirano, P., Ariza, R., and López-Ledesma, R. 1991. The effect of two sequential doses of *Opuntia streptacantha* upon glycemia [in Spanish]. *Archivos de Investigacion Medica* (Mexico) 22(3–4):333–336.

Frati-Munari, A.C., Vera-Lastra, O., and Ariza-Andraca, C. R. 1992. Evaluation of nopal capsules in diabetes mellitus [in Spanish]. *Gaceta Medicina Mexicana* 128(4):431–436.

Freiberger, C.E., Vanderjagt, D.J., Pastuszyn, A., Glew, R.S., Mounkaila, G., Millson, M., and Glew, R.H. 1998. Nutrient content of the edible leaves of seven wild plants from Niger. *Plant Foods and Human Nutrition* 53(1):57–69.

Freiburghaus, F., Kaminsky, R., Nkunya, M.H., and Brun, R. 1996. Evaluation of African medicinal plants for their in vitro trypanocidal activity. *Journal of Ethnopharmacology* 55(1):1–11.

Freire, S.M., Torres, L.M., Roque, N.F., Souccar, C., and Lapa, A.J. 1991. Analgesic activity of a triterpene isolated from *Scoparia dulcis* L. (vassourinha). *Memorias do Instituto Oswaldo Cruz* 86(Suppl. 2): 149–151.

Freire, S.M., Torres, L.M., Souccar, C., and Lapa, A.J. 1996. Sympathomimetic effects of *Scoparia dulcis* L. and catecholamines isolated from plant extracts. *The Journal of Pharmacy and Pharmacology* 48(6):624–628.

Freitas, A.M., Schor, N., and Boim, M.A. 2002. The effect of *Phyllanthus niruri* on urinary inhibitors of calcium oxalate crystallization and other factors associated with renal stone formation. *Blackwell Journal Urological International* 89(9):829–834.

Freytag, G.F. and Debouck, D.G. 2002. *Taxonomy, Distribution, and Ecology of the Genus Phaseolus (Leguminosae — Papilionoideae) in North America, Mexico, and Central America.* Sida, Botanical Miscellany 23, Botanical Research Institute, Fort Worth, TX.

Fritz, G.J. 2000a. Native farming systems and ecosystems in the Mississippi River Valley. Pp. 225–249. In: Lentz, D.L., Ed. *Imperfect Balance. Landscape Transformations in the Pre-Colombian Americas.* Columbia University Press, New York.

Fritz, G.J. 2000b. Levels of native biodiversity in Eastern North America. Pp. 223–247. In: Minnis, P.E. and Elisens, W.J., Eds. *Biodiversity and Native America.* University of Oklahoma Press, Norman.

Frost, R. 1992. *Robert Frost. Selected Poems.* Gramercy Books, New York.

Frulow, J.J. 1990. The genera of Betulaceae in the southeastern United States. *Journal of the Arnold Arboretum* 71(1):1–67.

Fryxell, P.A. 1988. *Malvaceae of Mexico.* Systematic Botany Monographs Vol. 25. American Society of Plant Taxonomists, Laramie, WY.

Fuchs, L. 1542. *De historia stirpium commentarii insignes.* Basel.

Fuente, G. de la, Reina, M., and Timon, I. 1991. Chromene amides from *Amyris texana. Phytochemistry* 30(8): 2677–2684.

Fujimoto, M., Mihara, S., Nakajima, S., Ueda, M., Nakamura, M., and Sakurai, K. 1992. A novel nonpeptide endothelin antagonist isolated from bayberry, *Myrica cerifera. FEBS Letters* 305(1):41–44.

Fukunaga, T., Miura, T., Furuta, K., and Kato, A. 1997. Hypoglycemic effect of the rhizomes of *Smilax glabra* in normal and diabetic mice. *Biological and Pharmaceutical Bulletin* 20(1):44–46.

Fun, C.E. and Svendsen, A.B. 1990. The essential oil of *Cordia cylindrostachya* Roem. and Schult. grown on Aruba. *Journal of Essential Oil Research* 2(4):209–210.

Fuzikawa, F., Nakamura, J., and Asami, K. 1940. On the occurrence of dulcitol in *Cassytha filiformis* L. *Journal of the Pharmaceutical Society of Japan* 60:209.

Gadek, P.A., Fernando, E.S., Quinn, C.J., Hoot, S.B., Terrazas, T., Sheahan, M.C., and Chase, M.W.

1996. Sapindales: molecular delimitation and infraordinal groups. *American Journal of Botany* 83:802–811.

Gaiser, L.O. 1946. The genus *Liatris. Rhodora* 48:165–183, 216–263, 326, 331–382, 393–412.

Galati, E.M., Monforte, M.T., Tripodo, M.M., d'Aquino, A., and Mondello, M.R. 2001. Antiulcer activity of *Opuntia ficus-indica* (L.) Mill. (Cactaceae): ultrastructural study. *Journal of Ethnopharmacology* 76 (1):1–9.

Galati, E.M., Tripodo, M.M., Trovato, A., Miceli, N., and Monforte, M.T. 2002. Biological effect of *Opuntia ficus-indica* (L.) Mill. (Cactaceae) waste matter. Note I: diuretic activity. *Journal of Ethnopharmacology* 79(1):17–21.

Galinat, W.C. and Craighead, F.C. 1964. Some observations on the dissemination of *Tripsacum. Rhodora* 66: 371–374.

Gallardo, T., Aragon, R., Tormo, J.R., Blazquez, M.A., Zafra-Polo, M.C., and Cortes, D. 1998. Acetogenins from *Annona glabra* seeds. *Phytochemistry* 47(5): 811–816.

Gamboa-Leon, R. and Scott Chilton, W. 2000. Isobutylamide numbing agents of toothache grass, *Ctenium aromaticum. Biochemical Systematics and Ecology* 28 (10):1019–1021.

Gann, G.D., Bradley, K.A., and Woodmansee, S.W. 2002. *Rare Plants of South Florida: Their History, Conservation, and Restoration.* Institute for Regional Conservation, Miami, FL.

Gao, X., Xu, Y.X., Divine, G., Janakiraman, N., Chapman, R.A., and Gautam, S.C. 2002. Disparate *in vitro* and *in vivo* antileukemic effects of resveratrol, a natural polyphenolic compound found in grapes. *The Journal of Nutrition* 132(7):2076–2081.

Garca-Mateos, R., Soto-Hernández, M., and Kelly, D. 1998. Alkaloids from six *Erythrina* species endemic to Mexico — isolation and characterisation of alkaloids from seven *Erythrina* species. *Biochemical Systematics and Ecology* 26(5):545–551.

García, J.G.L., MacBryde, B., Molina, A.R., and Herrera-MacBryde, O. 1975. *Malezas prevalentes de America Central/Prevalent Weeds of Central America.* International Plant Protection Center, San Salvador, El Salvador.

García, M.D., Quilez, A.M., Saenz, M.T., Martínez-Dominguez, M.E., de la Puerta, R. 2000. Antiinflammatory activity of *Agave intermixta* Trel. and *Cissus sicyoides* L., species used in the Caribbean traditional medicine. *Journal of Ethnopharmacology* 71 (3):395–400.

García-Mateos, R., Soto-Hernández, M., and Vibrans, H. 2001. *Erythrina americana* Miller ("Colorin"; Fabaceae), a versatile resource from Mexico: a review. *Economic Botany* 55(3):391–400.

García-Pelayo y Gross, R. 1980. *Pequeño Larousse (Diccionario Español).* Ediciones Larousse, Paris.

Gardner, P.S. 1994. Plant remains. Pp. 189–212. In: Kelley, D.B., Ed. The McLelland and Joe Clark Sites: Protohistoric-Historic Caddoan Farmsteads in Southern Bossier Parish, Louisiana. Submitted to U.S. Army Corps of Engineers, Vicksburg District, Contract No. DACW38-91-D-0014 by Coastal Environments, Inc. Baton Rouge, LA.

Garín-Aguilar, M.E., Ramírez-Luna, J.E., Soto-Hernández, M., Valencia del Toro, G., and Martínez-Vázquez, M. 2000. Effect of crude extracts of *Erythrina americana* Mill. on aggressive behavior in rats. *Journal of Ethnopharmacology* 69(2):189–196.

Garrett, W.E., Darley, R.J., and Shupp, J.F. 1982. *[Map of the] Indians of South America.* National Geographic Society, Washington, D.C.

Gatschet, A.S. 1877. The Timucua language. *Proceedings of the American Philosophical Society* 99:626–642.

Gatschet, A.S. 1878. The Timucua language. *Proceedings of the American Philosophical Society* 101:490–504.

Gatschet, A.S. 1880. The Timucua language. *Proceedings of the American Philosophical Society* 105:465–502.

Gatschet, A.S. and Swanton, J.R. 1932. A Dictionary of the Atakapa Language. Smithsonian Institution, Bureau of American Ethnology Bulletin *108.* Washington, D.C.

Geary, J. 1955. List of Indian words found in the documents, with notes on their meanings. 2:884–900. In: Quinn, D.B. *The Roanoke Voyages, 1584–90. Documents to illustrate the English voyages to North America under the patent granted to Walter Raleigh in 1584.* Hakluyt Society, London.

Genders, R. 1988. *Edible Wild Plants.* van der Marck Editions, New York.

Gene, R.M., Segura, L., Adzet, T., Marin, E., and Iglesias, J. 1998. *Heterotheca inuloides*: antiinflammatory and analgesic effect. *Journal of Ethnopharmacology* 60 (2):157–162.

Gentry, H.S. 1942. *Rio Mayo Plants.* Carnegie Institution of Washington Publication 527. Washington, D.C.

Gentry, H.S. 1963. The Warihio Indians of Sonora-Chihuahua: An Ethnographic Survey. Smithsonian Institution Bureau of American Ethnological Bulletin 186, Anthropology Papers 65, U.S. Government Printing Office, Washington, D.C.

Gentry, H.S. 1972. The Agave Family in Sonora. Agricultural Research Service, U.S. Department of Agriculture, Agriculture Handbook 399, Washington, D.C.

Gentry, H.S. 1982. *Agaves of Continental North America.* University of Arizona Press, Tucson.

Gérard, J., Miller, R.B., and ter Welle, B.J.H. 1996. *Major Timber Trees of Guyana. Timber Characteristics and Utilization.* Tropenbos Foundation, Wageningen, the Netherlands.

Gerard, W.R. 1907. Virginia's Indian contributions to English. *American Anthropologist* 9(1):87–112.

Gerarde, John. [1597] 1975. *The herball, or general historie of plants* ... very much enlarged and amended by Thomas Johnson. 1633. Reprinted by Dover, New York.

Gerberg, E.J. and Arnett, R.H.J. 1989. *Florida Butterflies.* Natural Science, Baltimore, MD.

Gerth Van Wijk, H.L. 1909–1916. *A Dictionary of Plant Names.* 2 vols. originally pub. by M. Nyhoff, The

Hague. Reprinted by courtesy of the Dutch Academy in Haarlem. A. Asher, Amsterdam, 1962.

Gessler, M.C., Nkunya, M.H., Mwasumbi, L.B., Heinrich, M., and Tanner, M. 1994. Screening Tanzanian medicinal plants for antimalarial activity. *Acta Tropica* 56(1):65–77.

Ghisalberti, E.L. 1998. Ethnopharmacology and phytochemistry of *Dodonea* species. *Fitoterapia* 69 (2):99–113.

Ghosal, S., Singh, A.K., Sharma, P.V., and Chaudhuri, R.K. 1974. Chemical constituents of Gentianaceae. IX. Natural occurrence of erythrocentaurin in *Enicostema hyssopifolium* and *Swertia lawii. Journal of Pharmaceutical Science* 63(6):944–945.

Ghosal, S., Razdan, A., and Razdan, S. 1985. (+)-Epimaritidine, an alkaloid from *Zephyranthes rosea. Phytochemistry* 24(3):635–637.

Ghosal, S., Singh, S.K., and Srivastava, R.S. 1986. Alkaloids of *Zephyranthes flava. Phytochemistry* 25(8):1975–1978.

Ghosal, S., Singh, S.K., and Unnikrishnan, S.G. 1987. Phosphatidylpyrrolophenanthridine alkaloids from *Zephyranthes flava. Phytochemistry* 26(3):823–828.

Gibbs, R.D. 1974. *Chemotaxonomy of Flowering Plants.* 4 vols. McGill-Queen's University Press, Montreal.

Gibson, D.J. 2001. Review of *Orchid Fever. Economic Botany* 55:173, 174.

Gill, J.E. and Read, B.R., Eds. 1975. *Born of the Sun.* Florida Bicentennial Commemorative Journal, Hollywood.

Gilliland, M.S. 1975. *The Material Culture of Key Marco Florida.* University Press of Florida, Gainesville.

Gillis, W.T. 1971. The systematics and ecology of poison-ivy and the poison-oaks (*Toxicodendron*, Anacardiaceae) *Rhodora* 73(793):72–159; 73(794):161–237; 73(795):370–443.

Gillis, W.T. 1974. Name changes for the seed plants in the Bahama flora. *Rhodora* 76:67–138.

Gillis, W.T. 1977. *Pluchea* revisited. *Taxon* 26(5/6):587–591.

Gillott, C. 1980. *Entomology.* Plenum Press, New York.

Gilmore, M.R. 1919. *Uses of Plants by the Indians of the Missouri River Region.* Reprinted, University of Nebraska Press, Lincoln, 1977.

Gilmore, M.R. 1931. Vegetal remains of the Ozark bluff-dweller culture. *Papers of the Michigan Academy of Arts and Letters* 14:83–102.

Glassberg, J. 2001. *Butterflies through Binoculars. The West.* Oxford University Press, New York.

Gleason, H.A. and Cronquist, A. 1963. *Manual of Vascular Plants of Northeastern United States and Adjacent Canada.* D. Van Nostrand Company, Princeton, NJ.

Gledhill, D. 2002. *The Names of Plants,* 3rd ed. Cambridge University Press, New York.

Gnostic Garden. 2001. URL: www.gnosticgarden.com/seeds.html/.

Godfrey, R.K. and Wooton, J.W. 1979. *Aquatic and Wetlands Plants of Southeastern United States, Monocotyledons.* University of Georgia Press, Athens.

Goggin, J.M. 1940. The Tekesta Indians of southern Florida. *Florida Historical Quarterly* 18(4):274–284.

Goggin, J.M. and Sturtevant, W.C. 1964. The Calusa: A stratified, nonagricultural society (with notes on sibling marriage). Pp. 179–219, In: Goodenough, W.H., Ed. *Explorations in Cultural Anthropology.* McGraw-Hill, New York.

Goldblatt, P. 1991. In: Johnson, M. Cytology. Pp. 15–22 in Pennington, T.D. *The Genera of Sapotaceae.* Royal Botanic Gardens, Kew, U.K.

Gonçalves de Lima, O., d'Albuquerque, I.L., and Loureiro, P. 1953. Novas observações sobre a biflorina, antibiótico isolado de *Capraria biflora* L. (Scrophulariaceae). *Anais da Sociedad de Biologicade de Pernambuco* 11:3–9.

Gonçalves de Lima, O., Keller-Schierlein, W., and Prelog, V. 1958. Über das Biflorin. *Helvetica Chimica Acta* 41:1386–1390.

Gonçalves de Lima, O., d'Albuquerque, I.L., Maciel, G.M., and Maciel, M.C.N. 1968. Substâncias antimicrobianas de plantas superiores. XXVII. Isolamento de plumbagina de *Plumbago scandens* L. *Revista do Instituto de Antibióticos, Universidade Federal de Pernambuco* 8(1):95–97.

González, A., Ferreira, F., Vázquez, A., Moyna, P., and Paz, E.A. 1993. Biological screening of Uruguayan medicinal plants. *Journal of Ethnopharmacology* 39 (3):217–220.

González-Coloma, A., Gutierréz, C., Hubner, H., Achenbach, H., Terrero, D., and Fraga, B.M. 1999. Selective insect antifeedant and toxic action of ryanoid diterpenes. *Journal of Agriculture and Food Chemistry* 47(10):4419–4424.

Goodall, J. 2002. Oral corticosteroids for poison ivy dermatitis. *Canadian Medical Association Journal* 166(3):300–301.

Goodding, E.G.B. 1940–1942. Facts and beliefs about Barbadian plants. *Journal of the Barbados Museum and Historical Society* 7(4):170–174, 8(1):32–35; 8 (2):70–73; 8(3):103–106; 8(4):194–197; 9(1):17–19; 9(2):84–88; 9(3):126–129; 9(4):192–194; 10(1):3–6.

Goodding, E.G.B., Loveless, A.R., and Proctor, G.R. 1965. *Flora of Barbados.* Her Majesty's Stationery Office, London.

Goodspeed, T.H. 1954. *The Genus Nicotiana.* Chronica Botanica, Waltham, MA.

Gordon, A. and Keating, R.C. 2001. Light microscopy and determination of *Eryngium yuccifolium* Michaux leaf material in twined slippers from Salts Cave, Kentucky. *Journal of Archaeological Science* 28(1):55–60.

Gourley, J.M. 1969. The structure of ipalbine, a new hexahydroindolizine alkaloid, isolated from *Ipomoea alba* L. *Chemical Communications* (Cambridge, U.K.) 13:709–710.

Gramain, J.C., Remuson, R., Vallee-Goyet, D., Guilhem, J., and Lavaud, C. 1991. Total synthesis and determination of structure of the pyrrolizidine alkaloid curasanecine. *Journal of Natural Products* 54(4):1062–1067.

Gramain, J.C., Remuson, R., Vallee-Goyet, D., Guilhem, J., Lavaud, C., Gramosa, N.V., Lemos, T.L.G., and

Braz-Filho, R. 1997. Volatile constituents isolated from *Capparis flexuosa* of Brazil. *Journal of Essential Oil Research* 9(6):709–712.

Grant, R., Basson, P.A., Booker, H.H., Hofherr, J.B., and Anthonissen, M. 1991. Cardiomyopathy caused by avocado (*Persea americana* Mill) leaves. *South African Veterinary Association* 62(1):21–22.

Gray, A. 1875. *Gray's Lessons in Botany and Vegetable Physiology. Glossary, or Dictionary of Botanical Terms Included.* Ivison, Blakeman, Taylor, New York.

Greenlee, R.F. 1944. Medicine and curing practices of the modern Florida Seminoles. *American Anthropologist* 46(3):317–328.

Greutert, H. and Keller, F. 1993. Further evidence for stachyose and sucrose/H+ antiporters on the tonoplast of Japanese artichoke (*Stachys sieboldii*) tubers. *Plant Physiology* 101(4):1317–1322.

Griffin, J.B. 1967. Eastern North American archeology: a summary. *Science* 156:175–191.

Grigson, J. 1986. *Exotic Fruits and Vegetables.* Henry Holt and Co., New York.

Grimé, W.E. 1976. *Botany of the Black Americans.* Scholarly Press, St. Clair Shores, MI.

Grossman, M.L. and Hamlet, J. 1964. *Birds of Prey of the World.* Bonanza Books, New York.

Grosvenor, G.B., Peele, W.T., and Cook, D.W. 1972. *[Map of the] Indians of North America.* National Geographic Society, Washington, D.C.

Grosvenor, P.W., Gothard, P.K., McWilliam, N.C., Supriono, A., and Gray, D.O. 1995. Medicinal plants from Riau province, Sumatra, Indonesia. Part 1: Uses. *Journal of Ethnopharmacology* 45(2):75–95.

Gu, Z., Zhou, D., and Gordon, J. 1999. Quantitative evaluation of annonaceous acetogenins in monthly samples of paw paw (*Asimina triloba*) twigs by liquid chromatography/electrospray ionization/tandem mass spectrometry. *Phytochemical Analysis* 10:32.

Guevara, J.M., Chumpitaz, J., and Valencia, E. 1994. The *in vitro* action of plants on *Vibrio cholerae* [in Spanish]. *Revista de Gastroenterologia del Peru* 14(1):27–31.

Guilhon, G.M.S.P. and Muller, A.H. 1996. Eudesmane derivatives from *Pluchea quitoc. Phytochemistry* 43 (2):417–421.

Guilhon, G.M.S.P. and Muller, A.H. 1998. Eudesmane sesquiterpenoids from *Pluchea quitoc. Phytochemistry* 47(2):227–229.

Guin, J.D., Gillis, W.T., and Beaman, J.H. 1981. Recognizing the Toxicodendrons (poison ivy, poison oak, and poison sumac). *The Journal of the American Academy of Dermatologists* 4(1):99–114.

Guinaudeau, H., Leboeuf, M., Cave, A., Duret, S., and Paris, R.R. 1976. Alcaloides et composes polyphenoliques du *Colubrina faralaotra* ssp. *sinuata*; composes polyphenoliques du *Colubrina faralaotra* ssp. *faralaotra* (Rhamnacees). *Planta Medica* 30(3):201–210.

Gunn, C.R. and Dennis, J.V. 1976. *World Guide to Tropical Drift Seeds and Fruits.* Quadrangle/New York Times Book Co., New York.

Gupta, A., Gupta, A., and Singh, J. 1999. New naphthoquinones from *Plumbago zeylanica. Le Pharmacien biologiste* 37(4):321–323.

Gupta, M.P., Ed. 1995a. *270 Plantas Medicinales Iberoamericanas.* Convenio Andres Bello (CYTED, SECAB, UNESCO), Bogota, Colombia.

Gupta, M.P. 1995b. *Panamanian Flora: Source of Bioactive Compounds in Phytochemistry of Plants Used in Traditional Medicine.* Oxford University Press, New York, cited in Bennett 1997.

Gupta, S.K. and Banerjee, A.B. 1972. Screening of selected West Bengal plants for antifungal activity. *Economic Botany* 26:255–259.

Gutiérrez, M., García de Boto, M.J., Cantabrana, B., and Hidalgo, A. 1996a. Mechanisms involved in the spasmolytic effect of extracts from *Sabal serrulata* fruit on smooth muscle. *General Pharmacology* 27 (1):171–176.

Gutiérrez, M., Hidalgo, A., and Cantabrana, B. 1996b. Spasmolytic activity of a lipidic extract from *Sabal serrulata* fruits: further study of the mechanisms underlying this activity. *Planta Medica* 62(6):507–511.

Haag, M. and Willis, H. 2001. *Choctaw Language and Culture.* University of Oklahoma Press, Norman.

Haas, M.R. 1940. *Tunica.* Extracted from *Handbook of American Indian Languages.* J.J. Augustin, New York.

Habek, D.H. and Bennett, F.D. 1990. *Cactoblastis cactorum* Berg (Lepidoptera: Pyralidae), a Phycitine New to Florida. Florida Department of Agricultural Consumer Services, Division Plant Industry. Entomology Circular 333.

Hagendoorn, M.J.M., Plas, L.H.W., and van der Segers, G.J. 1994. Accumulation of anthraquinones in *Morinda citrifolia* cell suspensions. A model system for the study of the interaction between secondary and primary metabolism. *Plant Cell, Tissue and Organ Culture* 38(2/3):227–234.

Haggag, M.Y. and Taha, K.F. 1981. Isolation and determination of liriodenine from *Annona glabra* L. growing in Egypt. *Egyptian Journal of Pharmaceutical Sciences* 22(1/4):153–157.

Hale, E.M. 1891. *Ilex cassine*: the aboriginal North American tea — its history, distribution, and use among the native American Indians. *United States Department of Agriculture Division of Botany Bulletin* 14:1–22.

Hamel, P.B. and Chiltoskey, M.U. 1975. *Cherokee Plants and Their Uses — A 400 Year History.* Herald Publishers, Sylva, NC.

Hamilton, C.W. 1989a. A revision of Mesoamerican subgenus *Psychotria* (Rubiaceae). I. Introduction and species 1–16. *Annals of the Missouri Botanical Garden* 76(1):67–111.

Hamilton, C.W. 1989b. A revision of Mesoamerican *Psychotria* subgenus *Psychotria* (Rubiaceae), part II: species 17–47. *Annals of the Missouri Botanical Garden* 76(2):386–429.

Hamilton, C.W. 1989c. A revision of Mesoamerican subgenus *Psychotria* (Rubiaceae). III. Species 48–61 and appendices. *Annals of the Missouri Botanical Garden* 76(3):886–916.

Hammer, M.L. and Johns, E.A. 1993. Tapping an Amazonian plethora: four medicinal plants of Marajo Island, Para (Brazil). *Journal of Ethnopharmacology* 40(1):53–75.

Han, A.Y.T., Van Gorkom, L., and Barrow, K.D. 1995. Prosapogenins from *Dodonaea attenuata. Phytochemical Analysis* 6(3):157.

Han, Y.N., Choo, Y., Lee, Y.C., Moon, Y.I., Kim, S.D., and Choi, J.W. 2001. Monoamine oxidase B inhibitors from the fruits of *Opuntia ficus-indica* var. *saboten. Archives of Pharmaceutical Research* 24(1):51–54.

Hann, J.H. 1986. The use and processing of plants by Indians of Spanish Florida. *Southeastern Archaeology* 5(2):1–102.

Hann, J.H., Ed. 1991. *Missions to the Calusa.* University Press of Florida, Gainesville.

Hansen, B.F. and Wunderlin, R.P. 1986. *Pentalion* Voigt, an earlier name for *Urechites* Müell. Arg. (Apocynaceae). *Taxon* 35:166–168.

Hanson, A.D., Rivoal, J., Paquet, L., and Gage, D.A. 1994. Biosynthesis of 3-dimethylsulfoniopropionate in *Wollastonia biflora* (L.) D.C. Evidence that *S*-methylmethionine is an intermediate. *Plant Physiology* 105(1):103–110.

Hara, H. 1969. Remarkable examples of speciation in Asiatic plants. *American Journal of Botany* 56:732–737.

Haraguchi, H., Saito, T., Ishikawa, H., Sánchez, Y., Ogura, T., and Kubo, I. 1996. Inhibition of lipid peroxidation by sesquiterpenoid in *Heterotheca inuloides. The Journal of Pharmacy and Pharmacology* 48(4):441–443.

Haraguchi, H., Ishikawa, H., Sánchez, Y., Ogura, T., Kubo, Y., and Kubo, I. 1997. Antioxidative constituents in *Heterotheca inuloides. Bioorganic and Medicinal Chemistry* 5(5):865–871.

Harborne, J.B. 1967. Comparative biochemistry of the flavonoids. IV. Correlations between chemistry, pollen morphology and systematics in the family Plumbaginaceae. *Phytochemistry* 6:1415–1418.

Hardin, J.W. and Arena, J.M. 1974. *Human Poisoning from Native and Cultivated Plants.* Duke University Press, Durham, NC.

Hargis, A.M., Stauber, E., Casteel, S., and Eitner, D. 1989. Avocado (*Persea americana*) intoxication in caged birds. *Journal of the American Veterinary Medical Association* 194(1):64–66.

Hargreaves, R.T., Johnson, R.D., Millington, D.S., Mondal, M.H., Beavers, W., Becker, L., Young, C., and Rinehart, K.R. 1974. Alkaloids of the American species of *Erythrina. Lloydia* 37:569–580.

Harrar, E.S. and Harrar, J.G. 1946. *Guide to Southern Trees.* Reprinted 1962, Dover, New York.

Harrington, J.P. 1955. The original Strachey vocabulary of the Virginia Indian language. Bulletin of the Bureau of American Ethnology 157. Washington, D.C.

Harriot, Thomas. [1590] 1972. *A Briefe and True Report of the New Found Land of Virginia.* Reprinted with an introduction by P. Hulton, Dover, New York.

Harris, S.A., Robinson, J.P., and Juniper, B.E. 2002. Genetic clues to the origin of the apple. *Trends in Genetics* 18 (8):426–430.

Hartwell, J.L. 1971. Plants used against cancer. A survey. *Lloydia* 34:103–160, 204–255, 310–361, 386–438.

Hartwell, J.L. and Abbott, B.J. 1969. Antineoplastic principles in plants: recent developments in the field. *Advances in Pharmacology and Chemotherapy* 7: 117–209.

Hasbun, C. and Castro, O. 1988. A new benzamide from bark of *Amyris brenesii. Journal of Natural Products* 51(4):817–818.

Hasbun, C., Castro, O., and Andrade, R. 1989. Two new chromenilated dihydrostyrilamides from *Amyris sylvatica. Journal of Natural Products* 52(4):868–870.

Hashimura, H., Ueda, C., Kawabata, J., and Kasai, T. 2001. Acetyl-CoA carboxylase inhibitors from avocado (*Persea americana* Mill) fruits. *Bioscience, Biotechnology and Biochemistry* 65(7):1656–1658.

Hassanali, A. and Lwande, W. 1989. Antipest secondary metabolites from African plants. *American Chemical Society* 387:78–94.

Hatano, T., Yasuhara, T., Abe, R., and Okuda, T. 1990. A galloylated monoterpene glucoside and a dimeric hydrolysable tannin from *Cornus officinalis. Phytochemistry* 29(9):2975–2978.

Hawkes, J.G. 1998. The introduction of New World crops into Europe after 1492. Pp. 147–159. In: Prendergast, H.D.V., Etkin, N., Harris, D.R., and Houghton, P.J., Eds. *Plants for Food and Medicine.* Royal Botanic Garden, Kew, U.K.

Hay, C.L., Linton, R.L., Lothrop, S.K., Shapiro, H.L., and Vaillant, G.C., Eds. 1940. *The Maya and Their Neighbors.* Dover, New York.

Hayashi, K., Niwayama, S., Hayashi, T., Nago, R., Ochiai, H., and Morita, N. 1988a. *In vitro* and *in vivo* antiviral activity of scopadulcic acid B from *Scoparia dulcis*, Scrophulariaceae, against herpes simplex virus type 1. *Antiviral Research* 9(6):345–354.

Hayashi, T., Ushida, J., Hayashi, K., Niwayama, S., and Morita, N. 1988b. A cytotoxic flavone from *Scoparia dulcis* L. *Chemical and Pharmaceutical Bulletin* (Tokyo) 36:4849–4851.

Hayashi, T., Kishi, M., Kawasaki, M., Arisawa, M., and Morita, N. 1988c. The crystal structure of scopadulcic acid A from Paraguayan crude drug "typychá kuratu" (*Scoparia dulcis*). *Journal of Natural Products* 51:360–363.

Hayashi, T., Okamura, K., Kakemi, M., and Asano, S. 1990a. Scopadulcic acid B, a new tetracyclic diterpenoid from *Scoparia dulcis* L. Its structure, H+, K (+)-adenosine triphosphatase inhibitory activity and pharmacokinetic behaviour in rats. *Chemical and Pharmaceutical Bulletin* (Tokyo) 38(10):2740–2745.

Hayashi, T., Kawasaki, M., Miwa, Y., Taga, T., and Morita, N. 1990b. Antiviral agents of plant origin. III.

Scopadulin, a novel tetracyclic diterpene from *Scoparia dulcis* L. *Chemical and Pharmaceutical Bulletin* (Tokyo) 38(4):945–947.

Hayashi, K., Hayashi, T., Morita, N., and Niwayama, S. 1990c. Antiviral activity of an extract of *Cordia salicifolia* on herpes simplex virus type 1. *Planta Medica* 56(5):439–443.

Hayashi, T., Asano, S., Mizutani, M., Takeguchi, N., Kojima, T., Okamura, K., and Morita, N. 1991. Scopadulciol, an inhibitor of gastric H+, K(+)-ATPase from *Scoparia dulcis*, and its structure–activity relationships. *Journal of Natural Products* 54(3):802–809.

Hayashi, K., Nishino, H., Niwayama, S., Shiraki, K., and Hiramatsu, A. 1992a. Yucca leaf protein (YLP) stops the protein synthesis in HSV-infected cells and inhibits virus replication. *Antiviral Research* 17 (4):323–333.

Hayashi, T., Kawasaki, M., Okamura, K., Tamada, Y., Morita, N., Tezuka. Y., Kikuchi, T., Miwa, Y., and Taga, T. 1992b. Scoparic acid A, a beta-glucuronidase inhibitor from *Scoparia dulcis*. *Journal of Natural Products* 55(12):1748–1755.

Hayashi, K., Kamiya, M., and Hayashi, T. 1995. Virucidal effects of the steam distillate from *Houttuynia cordata* and its components on HSV-1, influenza virus, and HIV. *Planta Medica* 61(3):237–241.

He, K., Shi, G., and McLaughlin, J.L. 1996. Three new adjacent bis-tetrahydrofuran acetogenins with four hydroxyl groups from *Asimina triloba*. *Journal of Natural Products* 59(11):1029

HealthLink Online Resources. 2000. http://www.healthlink.com.au/nat_lib/htm-data/htm-herb/bhp777.htm.

Healy, J.F., Ed. 1991. *Pliny the Elder. Natural History. A Selection*. Penguin Books, New York.

Hearon, R. 1993. *Salsa. Music for Your Mouth*. Chronicle Books, San Francisco.

Hedrick, U.P., Ed. 1919. *Sturtevant's Notes on Edible Plants*. J.B. Lyon, Albany, NY.

Heinicke, R.M. 1985. The pharmacologically active ingredient of noni. *Bulletin of the Pacific Tropical Botanical Garden* 15(1):10–14.

Heinrich, M. and Rimpler, H. 1989. Harpagide and 8-*O*-benzoyl-harpagide from the Mixe medicinal plant *Capraria biflora*. *Planta Medica* 55:626.

Heiser, C.B., Jr. 1951. The sunflower among the North American Indians. *Proceedings of the American Philosophical Society* 95:432–448.

Heiser, C.B., Jr. 1969. *Nightshades. The Paradoxical Plants*. W.H. Freeman, San Francisco.

Heiser, C.B., Jr. 1976. *The Sunflower*. University of Oklahoma Press, Norman.

Heiser, C.B., Jr. 1978. The totora (*Scirpus californicus*) in Ecuador and Peru. *Economic Botany* 32:222–236.

Heiser, C.B., Jr. 1979. *The Gourd Book*. University of Oklahoma Press, Norman.

Heiser, C.B., Jr. 1985. *Of Plants and People*. University of Oklahoma Press, Norman.

Heiser, C.B., Jr. 1993. Ethnobotany and economic botany. In: Editorial Committee, Eds. *Flora of North America* 1:199–206. Oxford University Press, New York.

Heiser, C.B., Jr. 2001. Letter to editor: about sunflowers.... *Economic Botany* 55:470–471.

Heller, W. and Tamm, C. 1975. Fukiinsaure und 3′-*O*-Methyl-fukiinsaure, zwei phenolische Hydroxycarbonsauren aus *Piscidia erythrina*. *Helvetica Chimica Acta* 58(4):974–979.

Hemlata and Kalidhar, S.B. 1994. Chemical components of *Dodonaea viscosa* stems. *Journal of the Indian Chemical Society* 71(4):213.

Henshaw, H.W. 1890. Indian origin of maple sugar. *American Anthropologist* 3(4):341–352.

Heo, K. and Tobe, H. 1994. Embryology and relationships of *Suriana maritima* L. (Surianaceae). *Journal of Plant Research* 107(1085):29–37.

Herbert, J.M., Maffrand, J.P., Taoubi, K., Augereau, J.M., Fouraste, I., and Gleye, J. 1991. Verbascoside isolated from *Lantana camara*, an inhibitor of protein kinase C. *Journal of Natural Products* 54 (6):1595–1600.

Herklots, G.A.C. 1972. *Vegetables in South-east Asia*. George Allen & Unwin, Hong Kong.

Hernan, E., Cordova, B., and Garelli-M, L.N. 1974. A new coumarin in *Amyris simplicifolia*. *Phytochemistry* 13 (4):758–760.

Hernández, F. 1651. *Nova plantarum, animalium et mineralium mexicanorum historia....* Blassi Deversini et Zanobij Mazotti, Rome.

Herrin, A. 1999. *White Bear Newsletter*. Cherokee Nation, Oklahoma. URL: www.intellex.com/ ~herrin.

Hershenson, B.R. and Quimby, M.W. 1969. Preliminary phytochemical investigation of the stem wood of *Suriana maritima* Linn (Simarubaceae). *Journal of Pharmaceutical Science* 58(11):1411–1412.

Herz, W. and Sosa, V.E. 1988. Sesquiterpene lactones and other constituents of *Arnica acaulis*. *Phytochemistry* 27(1):155–159.

Herzog, F., Farah, Z., and Amado, R. 1994. Composition and consumption of gathered wild fruits in the V-Baoule, Cote d'Ivoire. *Ecology of Food and Nutrition* 32(3/4):181–196.

Hickman, J.C., Ed. 1993. *The Jepson Manual. Higher Plants of California*. University of California Press, Berkeley.

Hiermann, A. 1989. The contents of sabal fruits and testing of their antiinflammatory effect [in German]. *Archiv der Pharmazie* (Weinheim) 322(2):111–114.

Higgs, L. 1969. *Bush Medicine in the Bahamas*. Bahamas Nassau Guardian, Nassau.

Hill, B.D., Wills, L.D., and Dowling, R.M. 1985. Suspected poisoning of horses by *Trema aspera* (poison peach). *Australian Veterinary Journal* 62(3):107–108.

Hirazumi, A. and Furusawa, E. 1999. An immunomodulatory polysaccharide-rich substance from the fruit juice of *Morinda citrifolia* (Noni) with antitumour activity. *Phytotherapy Research* 13(5):380–387.

Hitchcock, A.S. and Chase, A. 1950. *Manual of the Grasses of the United States.* Reprinted in 1971 by Dover, New York.

Hitchcock, C L. 1932. A monographic study of the genus *Lycium* of the Western Hemisphere. *Annals of the Missouri Botanical Garden* 19:179–375.

Hocking, G.M. 1997. *A Dictionary of Natural Products. Terms in the Field of Pharmacognosy Relating to Natural Medicinal and Pharmaceutical Materials and the Plants, Animals, and Minerals from Which They Are Derived.* Plexus, Medford, NJ.

Hodge, W.H. and Taylor, D. 1957. The ethnobotany of the Island Caribs of Dominica. *Webbia* 12(2):513–644.

Hodgson, W. 2001. *Food Plants of the Sonoran Desert.* University of Arizona Press, Tucson.

Hogan, J. 1978. Palynology of Ft. Center. Environmental Interpretations and Cultural Implications for a Central Florida Hopewellian Ceremonial Center. M.A. thesis, Florida Atlantic University, Boca Raton.

Holm, L.G., Plucknett, D.L., Pancho, J.V., and Herberger, J. P. 1991. *The World's Worst Weeds. Distribution and Biology.* Krieger, Malabar, FL.

Honychurch, P.N. 1987. *Caribbean Wild Plants and Their Uses.* Macmillan, London.

Horgen, F.D., Guinaudeau, H., Pezzuto, J.M., Soejarto, D. D., Farnsworth, N.R., Agcaoili, F., Reyes, G. de los, and Edrada, R.A. 1997. Isolation and structure elucidation of ardisenone: a new, cytotoxic alkenylphenol from *Ardisia iwahigensis. Journal of Natural Products* 60(5):533–535.

Horgen, F.D., Edrada, R.A., de los Reyes, G., Agcaoili, F., Madulid, D.A., Wongpanich, V., Angerhofer, C.K., Pezzuto, J.M., Soejarto, D.D., and Farnsworth, N.R. 2001. Biological screening of rain forest plot trees from Palawan Island (Philippines). *Phytomedicine* 8 (1):71–81.

Hortorium, L.H. Bailey. 1976. *Hortus Third.* Macmillan, New York.

Hosamani, K.M. 1995. A minor source of vernolic, malvalic, and sterculic acids in *Pithecollobium dulce* (syn. *Inga dulcis*) seed oil. *Journal of the American Oil Chemists' Society* 72(4):489–492.

Hosler, D., Burkett, S.L. and Tarkanian, M.J. 1999. Prehistoric polymers: rubber processing in ancient Mesoamerica. *Science* 284(5422):1988–1991 (also pages 1898 and 1899).

Houle, G. and Bouchard, F. 1990. Hackberry (*Celtis occidentalis*) at the northeastern limit of its distribution in North America: population structure and radial growth patterns. *Canadian Journal of Botany* 68(12):2685–2692.

Howard, J.H. 1968. The southeastern ceremonial complex and its interpretation. Memoir 6, Missouri Archaeological Society, Columbia, MO.

Howard, J.H. 1984. *Oklahoma Seminoles Medicines, Magic, and Religion.* University of Oklahoma Press, Norman.

Howard, O.O. 1907. *My Life and Experiences among Our Hostile Indians.* Worthington, Hartford, CT (quoted in Sturtevant 1955).

Howard, R.A. 1996. *An Almanac of Botanical Trivia.* Published by the author, Acton, MA.

Howard, R.A. and Powell, D. 1963. The introduction of rubber-producing species in the West Indies. *Economic Botany* 17:337–349.

Howard, R.A. and Staples, G.W. 1983. The modern names for Catesby's plants. *Journal of the Arnold Arboretum* 64:511–546.

Hsu, H.-Y. 1986. *Oriental Materia Medica.* Keats, New Canaan, CT.

Hsu, H.-Y., Chen, P., and Kakisawa, H. 1971. Structure of hautriwaic acid. *Phytochemistry* 10:2813–2814.

Huang, H., Layne, D.R., and Peterson, R.N. 1997. Using isozyme polymorphisms for identifying and assessing genetic variation in cultivated pawpaw [*Asimina triloba* (L.) Dunal]. *Journal of the American Society for Horticulture* 122(4):504.

Huang, H., Layne, D.R., and Riemenschneider, D.E. 1998. Genetic diversity and geographic differentiation in pawpaw [*Asimina triloba* (L.) Dunal] populations from nine states as revealed by allozyme analysis. *Journal of the American Society for Horticulture* 123 (4):635.

Huang, H., Layne, D.R., and Kubisiak, T.L. 2000. RAPD inheritance and diversity in pawpaw (*Asimina triloba*). *American Society for Horticultural Science* 125(4):454.

Huang, J., Ogihara, Y., Zhang, H., Shimizu, N. and Takeda, T. 2000. Triterpenoid saponins from *Ardisia mamillata. Phytochemistry* 54(8):817–822.

Huang, K.S. and Lin, M. 1999. Oligostilbenes from the roots of *Vitis amurensis. Journal of Asian Natural Products Research* 2(1):21–28.

Huang, K.S., Lin, M., and Cheng, G.F. 2001. Antiinflammatory tetramers of resveratrol from the roots of *Vitis amurensis* and the conformations of the seven-membered ring in some oligostilbenes. *Phytochemistry* 58(2):357–362.

Hübel, W. and Nahrstedt, A. 1979. Cardiosperminsulfat, ein schwefelhaltiges cyanogenes Glukosid aus *Cardiospermum grandiflorum. Tetrahedron Letters* 4395–4397.

Hudson, C.M., Ed. 1975. *Four Centuries of Southern Indians.* University of Georgia Press, Athens.

Hudson, C.M. 1976. *The Southeastern Indians.* University of Tennessee Press, Knoxville.

Hudson, C.M., Ed. 1979. *Black Drink. A Native American Tea.* University of Georgia Press, Athens.

Hudson, C.M., DePratter, C.B., and Smith, M.T. 1989. Hernando de Soto's Expedition through the Southern United States. Pp. 77–98. In: Milanich, J.T. and Milbrath, S., Eds. *First Encounters, Spanish Exploration in the Caribbean and the United States, 1492–1570.* University Press of Florida, Gainesville.

Hultén, E. 1968. *Flora of Alaska and Neighboring Territories.* Stanford University Press, Stanford, CA.

Hunter, J.D. [1823] 1973. *Memoires of a Captive among the Indians of North America.* Edited by Richard Drinnon, Schocken Books, New York.

Hussain, R.A., Lin, Y.M., Poveda, L.J., Bordas, E., Chung, B.S., Pezzuto, J.M., Soejarto, D.D., and Kinghorn, A.D. 1990. Plant-derived sweetening agents: saccharide and polyol constituents of some sweet-tasting plants. *Journal of Ethnopharmacology* 28(1):103–115.

Hussey, J.S. 1974. Some useful plants of early New England. *Economic Botany* 28:311–337.

Hutchinson, J. and Dalziel, J.M. 1937. *The Useful Plants of West Tropical Africa.* Crown Agents for Overseas Governments and Administrations, London.

Hutson, L. and Marshall, J.I. 1995. *Tequila! Cooking with the Spirit of Mexico.* Ten Speed Press, Berkeley, CA.

Hwang, B.Y., Lee, J.H., Nam, J.B., Kim, H.S., Hong, Y.S., and Lee, J.J. 2002. Two new furanoditerpenes from *Saururus chinenesis* and their effects on the activation of peroxisome proliferator-activated receptor gamma. *Journal of Natural Products* 65(4):616–617.

Ikhiri, K., Koulodo, D.D.D., Garba, M., Mamane, S., Ahond, A., Poupat, C., and Potier, P. 1987. New indolizinic type alkaloids isolated from *Ipomoea alba. Journal of Natural Products* 50(2):152–156.

Illinois Natural History Survey. 2003. Based on Herkert, J. R., and J.E. Ebinger, Eds. 2002. *Endangered and Threatened Species of Illinois: Status and Distribution,* Vol. 1: *Plants.* Illinois Endangered Species Protection Board, Springfield, IL.

Imai, H., Osawa, K., Yasuda, H., Hamashima, H., Arai, T., and Sasatsu, M. 2001. Inhibition by the essential oils of peppermint and spearmint of the growth of pathogenic bacteria. *Microbios* 106(Suppl. 1):31–39.

Inagaki, N., Nishimura, H., Okada, M., and Mitsuhashi, H. 1991. Verbascoside production by plant cell cultures. *Plant Cell Reports* 9(9):484–487.

Inoshiri, S., Saiki, M., Kohda, H., Otsuka, H., and Yamasaki, K. 1988. Monoterpene glucosides from *Berchemia racemosa. Phytochemistry* 27(9):2869–2871.

Inoue, T., Nagashima, S.I., Ohata, S., Shinoda, M., and Sakurai, N. 1990. Two phenol compounds from *Berchemia racemosa. Planta Medica* 56(1):120–121.

Inouye, H., Takeda, Y., and Nishimura, H. 1973. On the lignan glucosides of *Symplocos lucida* [in Japanese]. *Yakugaku Zasshi* 93(1):44–46.

Inouye, H., Inoue, K., and Nishioka, T. 1975. Two new iridoid glucosides from *Osmanthus fragrans. Phytochemistry* 14(9):2029–2032.

Ionescu-Tirgoviste, C., Popa, E., Mirodon, Z., Simionescu, M., and Mincu, I. 1989. The effect of a plant mixture on the metabolic equilibrium in patients with type-2 diabetes mellitus [in Romanian]. *Revista de medinica interna, neurologe, psihiatrie, neurochirurgie, dermato-venerologie. Medicina interna* 41(2):185–192.

Ioset, J.R., Marston, A., Gupta, M.P., and Hostettmann, K. 1998. Antifungal and larvicidal meroterpenoid naphthoquinones and a naphthoxirene from the roots of *Cordia linnaei. Phytochemistry* 47(5):729–734.

Irobi, O.N. 1997. Antibiotic properties of ethanol extract of *Chromolaena odorata* (Asteriaceae). *International Journal of Pharmacognosy* 35(2):111–115.

Irvine, F.R. 1952. Supplementary and emergency food plants of West Africa. *Economic Botany* 6:23–40.

Ishida, J., Wang, H.K., Oyama, M., Cosentino, M.L., Hu, C. Q., and Lee, K.H. 2001. Anti-AIDS agents. 46. Anti-HIV activity of harman, an anti-HIV principle from *Symplocos setchuensis,* and its derivatives. *Journal of Natural Products* 64(7):958–960.

Islam, A., Sayeed, A., Bhuiyan, M.S., Mosaddik, M.A., Islam, M.A., and Astaq Mondal Khan, G.R. 2001. Antimicrobial activity and cytotoxicity of *Zanthoxylum budrunga. Fitoterapia* 72(4):428–430.

Ismail, N. and Alam, M. 2001. A novel cytotoxic flavonoid glycoside from *Physalis angulata. Fitoterapia* 72 (6):676–679.

Ismail, N.H., Ali, A.M., Aimi, N., Kitajima, M., Takayama, H., and Lajis, N.H. 1997. Anthraquinones from *Morinda elliptica. Phytochemistry* 45(8):1723–1725.

Ismail, S.A., Dea, T., Abd El-Rahman, H., Yassien, M.A., and Beuchat, L.R. 2001. Effectiveness of immersion treatments with acids, trisodium phosphate, and herb decoctions in reducing populations of *Yarrowia lipolytica* and naturally occurring aerobic microorganisms on raw chicken. *International Journal of Food Microbiology* 64(1–2):13–19.

Itoh, T., Tamura, T., Tsuhashi, T., and Matskumoto, T. 1977. Steroids of Liliaceae. *Phytochemistry* 16:140–141.

Iwashina, T., Kamenosono, K., and Ueno, T. 1999. Hispidulin and nepetin 4′ glucosides from *Circium oligophyllum. Phytochemistry* 51(8):1109–1111.

Iwu, M.M. 1993. *Handbook of African Medicinal Plants.* CRC Press, Boca Raton, FL.

Iwu, M.W., Unaeze, N.C., Okunji, C.O., Corley, D.G., Sanson, D.R., and Tempesta, M.S. 1991. Antibacterial aromatic isothiocyanates from the essential oil of *Hippocratea welwitschii* roots. *International Journal of Pharmacognosy* 29(2):154–158.

Jackson, B.J. 2003. *Yuchi Ceremonial Life. Performance, Meaning, and Tradition in a Contemporary American Indian Community.* University of Nebraska Press, Lincoln.

Jackson, R.C. 1960. A revision of the genus *Iva. Kansas University Science Bulletin* 41:793–874.

Jain, R., Singh, M., and Dezman, D.J. 1989. Qualitative and quantitative characterization of phenolic compounds from lantana (*Lantana camara*) leaves. *Weed Science* 37(3):302–307.

Jain, S.K. 1965. Medicinal plant lore of the tribals of Bastar. *Economic Botany* 19:236–250.

Jain, S.K. and Tarafder, C.R. 1970. Medicinal plant-lore of the Santals. *Economic Botany* 24:241–278.

Jakupovic, J., Boeker, R., Bohlmann, F., King, R.M., and Robinson, H. 1987a. Humulene derivatives from *Heterotheca villosa. Phytochemistry* 26(2):445–448.

Jakupovic, J., Castro, V., and Bohlmann, F. 1987b. Further sesquiterpene lactones from Costa Rican Eupatorieae. *Phytochemistry* 26(2):451–455.

James, A. 2003. *Pseudophoenix sargentii* in Dominica. *PALMS* 47:57–61.

James, C.W. 1956. A revision of *Rhexia* (Melastomataceae). *Brittonia* 8:201–230.

Jane, C. and Skelton, R.A., Eds. 1960. *Journal of Christopher Columbus.* Clarkson N. Potter, New York.

Jansakul, C., Baumann, H., Kenne, L., and Samuelsson, G. 1987. Ardisiacrispin A and B, two utero-contracting saponins from *Ardisia crispa. Planta Medica* 53 (5):405–409.

Jansakul, C., Intarit, K., Itharat, A., Phadungcharoen, T., Ruangrungsi, N., Merica, A., and Lange, G.L. 1999. Biological activity of crude extract and saponin pseudoginsenoside-RT1 derived from the fruit of *Randia siamensis. Le Pharmacien biologiste* 37 (1):42–45.

Janzen, D.H. 1982. Cenizero tree (Leguminosae: *Pithecellobium saman*) delayed fruit development in Costa Rican deciduous forests. *American Journal of Botany* 69(8):1269–1276.

Janzen, D.H., Ed. 1983. *Costa Rican Natural History.* University of Chicago Press, Chicago.

Jefferies, P.R. and Payne, T.G. 1967. Diterpenes of the cascarallin group from *Dodonea viscosa. Tetrahedron Letters* 4777–4782.

Jerga, C., Merfort, I., and Willuhn, G. 1990a. Flavonoid-aglyka aus den Bluten von *Heterotheca inuloides. Planta Medica* 56 (1):122–123.

Jerga, C., Merfort, I., and Willuhn, G. 1990b. Flavonoid-glykoside und andere hydrophile Inhaltsstoffe aus den Bluten von *Heterotheca inuloides. Planta Medica* 56(4):413–415.

Jeyarani, T. and Yella Reddy, S.Y. 2001. Cocoa butter extender from *Simarouba glauca* fat. *Journal of the American Oil Chemists' Society* 78(3):271–276.

Jia, Z., Koike, K., Ohmoto, T., and Ni, M. 1994. Triterpenoid saponins from *Ardisia crenata. Phytochemistry* 37(5):1389–1396.

Jia, Z.H. and Ju, Y. 1992. Steroidal saponins from *Smilax lebrunii. Phytochemistry* 31(9):3173–3175.

Jiang, Y., Ali, A.O., Guillaume, D., Weniger, B., and Anton, R. 1994a. Triterpenoid saponins from the root of *Sideroxylon cubense. Phytochemistry* 35(4):1013–1015.

Jiang, Y., Weniger, B., Quirion, J.C., Muller, C., and Anton, R. 1994b. 3a,5a-Tetrahydrodeoxycordifoline from *Guettarda ovalifolia. Planta Medica* 60(3):294.

Jimenez-E., M., Reyes-C., R., and Van Calsteren, M.R. 2000. Two novel diels — alder adducts from *Hippocratea celastroides* roots and their insecticidal activity. *Canadian Journal of Chemistry* 78(2):248.

Jiménez-M., C.A., Rojas-H., N.M., and López-A., A.M. 1979. Contribution to the biological evaluation of Cuban plants. 6. *Revista Cubana Medicina Tropical* 31:45–51.

Jiu, J. 1966. A survey of some medicinal plants of Mexico for selected biological activities. *Lloydia* 29(3):250–259.

Jizba, J. and Herout, V. 1974. Ponasterone A and 22-hopanol, characteristic components of the fern *Blechnum spicant* L. *Collection of Czechoslovak Chemical Communications* 39(12):3756–3759.

Johansen, O.P., Andersen, O.M., Nerdal, W., and Aksnes, D. W. 1991. Cyanidin 3-[6-(*p*-coumaroyl)-2-(xylosyl)-glucoside]-5-glucoside and other anthocyanins from fruits of *Sambucus canadensis. Phytochemistry* 30 (12):4137–4141.

Johnston, A. 1970. Blackfoot Indian utilization of the flora of the northwestern Great Plains. *Economic Botany* 24:301–324.

Johnston, J.R. 1909. Flora of the islands of Margarita and Choche, Venezuela. *Proceedings of the Boston Society of Natural History* 7:163–312+123–130.

Johnston, M.C. 1971. Revision of *Colubrina* (Rhamnaceae). *Brittonia* 23(1):2–53.

Jones, G.N. 1940. A monograph of the genus *Symphoricarpos. Journal of the Arnold Arboretum* 21:201–252.

Jones, S.B.J. and Coile, N.C. 1988. *The Distribution of the Vascular Flora of Georgia.* University of Georgia Press, Athens.

Jones, V. 1936. The vegetal remains of Newt Kash Hollow shelter. Pp. 147–165. In: Webb, W.S. and Funkhouser, W.D. 1936. *Rock Shelters in Menifee County, Kentucky.* Publications of the University of Kentucky Department of Anthropology and Archaeology 3(4):105–167.

Jorgensen, P.M. and León-Yánez, S., Eds. 1999. *Catalogue of the Vascular Plants of Ecuador/Catálogo de las Plantas Vasculares del Ecuador.* Missouri Botanical Garden Press, St. Louis.

Joyal, E. 1987. Ethnobotanical field notes from Ecuador: Camp Prieto. *Economic Botany* 41:163–189.

Ju, Y. and Jia, Z.J. 1992. Steroidal saponins from the rhizomes of *Smilax menispermoidea. Phytochemistry* 31(4):1349–1351.

Ju, Y. and Jia, Z.J. 1993. Minor steroidal glycosides from the roots of *Smilax lebrunii. Phytochemistry* 33(5):1193–1195.

Ju, Y., Du, M., Jia, Z.J., and Sun, X.J. 1993. Studies on chemical constituents of *Smilax menispermoidea* D.C. [in Chinese]. *Zhongguo Zhong Yao Za Zhi* 18 (10):611–613, 639.

Ju, Y., Jia, Z.J., and Sun, X. 1994. Steroidal saponins from *Smilax menispermoidea* and *S. lebrunii. Phytochemistry* 37(5):1433–1436.

Ju, Y., Still, C.C., Sacalis, J.N., Li, J., and Ho, C.T. 2001. Cytotoxic coumarins and lignans from extracts of the northern prickly ash (*Zanthoxylum americanum*). *Phytotherapy Research* 15(5):441–443.

Juan, M.E., Vinardell, M.P., and Planas, J.M. 2002. The daily oral administration of high doses of trans-resveratrol to rats for 28 days is not harmful. *The Journal of Nutrition* 132(2):257–260.

Juang, J.K., Huang, H.W., Chen, C.M., and Liu, H.J. 1989. A new compound, withangulatin A, promotes type II DNA topoisomerase-mediated DNA damage. *Biochemical and Biophysical Research Communications* 159(3):1128–1134.

Judd, W.S. 1981. A monograph of *Lyonia* (Ericaceae). *Journal of the Arnold Arboretum* 62:315–436.

Judd, W.S., Campbell, C.S., Kellogg, E.A., Stevens, P.F., and Donoghue, M.J. 2002. *Plant Systematics. A Phylogenetic Approach,* 2nd ed. Sinauer Associates, Sunderland, MA.

Julien, M. and Griffiths, J.W., Eds. 1998. *Biological Control of Weeds, A Catalogue of Agents and Their Target Weeds.* 4th ed. CABI Publishing, New York.

Jurney, D.H. 1994. The original distribution of the bois d'arc. *Caddoan Archaeology Newsletter* 5(2):6–13.

Kaaiakamanu, D.M. and Akina, J.K. 1976. *Hawaiian Herbs of Medicinal Value.* Pacific Book House, Honolulu.

Kaiser, R. and Lamparsky, D. 1978a. Inhaltsstoffe des *Osmanthus* absolues. 1. 2,5 Epoxy megastigma 6,8 diene. *Helvetica Chimica Acta* 61(1):373–382.

Kaiser, R. and Lamparsky, D. 1978b. Inhaltsstoffe des *Osmanthus* absolues. 2. 2,7 Epoxy megastigma 4,8 (E) dien. *Helvetica Chimica Acta* 61(1):383–386.

Kaiser, R. and Lamparsky, D. 1978c. Inhaltsstoffe des *Osmanthus* absolues. 4. Megastigma-5,7(E),9-trien-4-on und Megastigma-5,8(E)-dien-4-on. *Helvetica Chimica Acta* 61(7):2328–2335.

Kaiser, R., Kappeler, A., and Lamparsky, D. 1978. Inhaltsstoffe des *Osmanthus* absolues. 3. Derivate der Theaspirane. *Helvetica Chimica Acta* 61(1):387–400.

Kaith, B.S. 1996. Neolupenol and antiinflammatory activity. *International Journal of Pharmacognosy* 34(1):73–75.

Kajita, T., Ohashi, H., Tateishi, Y., Bailey, C.D., and Doyle, J.J. 2001. rbcL and legume phylogeny, with particular reference to Phaseoleae, Millettieae, and allies. *Systematic Botany* 26(3):515–536.

Kakuda, R., Iijima, T., Yaoita, Y., Machida, K., and Kikuchi, M. 2001. Secoiridoid glycosides from *Gentiana scabra. Journal of Natural Products* 64 (12):1574–1575.

Kakuda, R., Iijima, T., Yaoita, Y., Machida, K., and Kikuchi, M. 2002. Triterpenoids from *Gentiana scabra. Phytochemistry* 59(8):791–794.

Kalm, P. [1753–1761] 1972. *Travels into North America.* Reprint of 1772 translation by J.R. Forster. Imprint Society, Barre, MA.

Kamanyi, A., Njamen, D., and Nkeh, B. 1994. Hypoglycaemic properties of the aqueous root extract of *Morinda lucida* (Benth) (Rubiaceae). Studies in the mouse. *Phytotherapy Research* 8(6):369–371.

Kanchanapoom, T., Kasai, R., and Yamasaki, K. 2001. Acetylated triterpene saponins from the Thai medicinal plant, *Sapindus emarginatus. Chemical and Pharmaceutical Bulletin* (Tokyo) 49(9):1195–1197.

Kanerva, L., Estlander, T., Petman, L., and Makinen-Kiljunen, S. 2001. Occupational allergic contact urticaria to yucca (*Yucca aloifolia*), weeping fig (*Ficus benjamina*), and spathe flower (*Spathiphyllum wallisii*). *Allergy* 56(10):1008–1011.

Kan-Fan, C., Brillanceau, M.H., Pusset, J., Chauviere, G., and Husson, H.P. 1985. Three quinicine derived alkaloids from *Guettarda trimera. Phytochemistry* 24 (11):2773–2775.

Kanis, A. 1984. Arbitrary lectotypification of generic names: the case of *Philoxerus* (Amaranthaceae). *Taxon* 33 (1):97–99.

Kapadia, G.J., Paul, B.D., Silverton, J.V., Fales, H.M., and Sokoloski, E.A. 1975. Melochinone, a novel quinolinone from *Melochia tomentosa* L. *Journal of the American Chemical Society* 97(23):6814–6819.

Kapadia, G.J., Paul, B.D., Chung, E.B., Ghosh, B., and Pradhan, S.N. 1976. Carcinogenicity of *Camellia sinensis* (tea) and some tannin-containing folk medicinal herbs administered subcutaneously in rats. *Journal of the National Cancer Institute* 57 (1):207–209.

Kaplan, L., Ed. 2001. *Economic Botany. Index to Volumes 1–50.* New York Botanical Garden, New York.

Kaplan, L. and Lynch, T.F. 1999. *Phaseolus* (Fabaceae) in archaeology: AMS radiocarbon dates and their significance for Pre-Columbian agriculture. *Economic Botany* 53:261–272.

Karakaya, S. and El, S.N. 1999. Quercetin, luteolin, apigenin and kaempferol contents of some foods. *Food Chemistry* 66(3):289–292.

Kartesz, J.T. 1994. *A Synonymized Checklist of the Vascular Flora of the United States, Canada and Greenland. Vol. 1. Checklist.* Timber Press, Portland, OR.

Kassu, A., Dagne, E., Abate, D., Castro, A., and Van Wyk, B.E. 1999. Ethnomedical aspects of the commonly used toothbrush sticks in Ethiopia. *East African Medical Journal* 76(11):651–653.

Kato, K., Yokoi, S., Inagaki, H., and Ueno, Y. 1979. Methyl alpha-d-galactopyranoside from the tubers of *Stachys affinis. Agricultural and Biological Chemistry* 43 (1):187–188.

Kato-Noguchi, H. 2000. Assessment of the allelopathic potential of extracts of *Evolvulus alsinoides. Weed Research* 40(4):343.

Kawagishi, H., Fukumoto, Y., Hatakeyama, M., He, P., Arimoto, H., Matsuzawa, T., Arimoto, Y., Suganuma, H., Inakuma, T., and Sugiyama, K. 2001. Liver injury suppressing compounds from avocado (*Persea americana*). *Journal of Agriculture and Food Chemistry* 49(5):2215–2221.

Kawai, M., Yamamoto, T., Makino, B., Yamamura, H., Araki, S., Butsugan, Y., and Saito, K. 2001. The structure of physalin T from *Physalis alkekengi* var. *franchetti. Journal of Asian Natural Products Research* 3(3):199–205.

Kawasaki, M., Hayashi, T., Arisawa, M., Shimizu, M., Horie, S., Ueno, H., Syogawa, H., Suzuki, S., Yoshizaki, M., Morita N. et al. 1987. Structure of scoparic acid A, a new labdane-type diterpenoid from a Paraguayan crude drug "Typycha Kuratu" (*Scoparia dulcis* L.). *Chemical Pharmacological Bulletin* (Tokyo) 35(9):3963–3966.

Kay, M.A. 1996. *Healing with Plants in the American and Mexican West.* University of Arizona Press, Tucson.

Kearney, T.H. and Peebles, R.H. 1942. Flowering plants and ferns of Arizona. Miscellaneous Publications 423, U.S. Department of Agriculture. Washington, D.C.

Kearney, T.H., Peebles, R.H., and collaborators. 1951. *Arizona Flora.* University of California Press, Berkeley.

Kela, S.L., Ogunsusi, R.A., Ogbogu, V.C., and Nwude, N. 1989. Screening of some Nigerian plants for molluscicidal activity. *Revue Eleve de Medecine Veterinaire du Pays Tropical* 42(2):195–202.

Kemal, M., Khalil, S.K.W., Rao, N.G.S., and Woolsey, N.F. 1979. Isolation and identification of a cannabinoid-like compound from *Amorpha* species. *Journal of Natural Products* 42(5):463–468.

Kenk, V.C. 1963. The importance of plants in heraldry. *Economic Botany* 17:169–179.

Kennelly, E.J., Lewis, W.H., and Winter, R.E.K. 1993. Triterpenoid saponins from *Gouania lupuloides. Journal of Natural Products* 56(3):402.

Khan, I.A., Clark, A.M., and McChesney, J.D. 1997. Antifungal activity of a new triterpenoid glycoside from *Pithecellobium racemosum* (M.). *Pharmaceutical Research* 14(3):358–361.

Khan, M.R., Kihara, M., and Omoloso, A.D. 2001a. Antimicrobial activity of *Psychotria microlabastra. Fitoterapia* 72(7):818–821.

Khan, M.R., Kihara, M., and Omoloso, A.D. 2001b. Antimicrobial activity of *Symplocos cochinensis. Fitoterapia* 72(7):825–828.

Khan, R. and Jarvis, C.E. 1989. The correct name for the plant known as *Pluchea symphytifolia* (Miller) Gillis (Asteraceae). *Taxon* 38(4):659–662.

Khanna, A.K., Rizvi, F., and Chander, R. 2002. Lipid lowering activity of *Phyllanthus niruri* in hyperlipemic rats. *Journal of Ethnopharmacology* 82(1): 19–22.

Khanna, K.R., Srivastava, S.N., Dixit, B.S., and Singh, S.P. 1981. *Kallstroemia pubescens* as diosgenin yielding crop. *Indian Journal of Pharmaceutical Sciences* 43 (5):172–175.

Kiesewetter, H., Koscielny, J., Kalus, U., Vix, J.M., Peil, H., Petrini, O., van Toor, B.S., and de Mey, C. 2000. Efficacy of orally administered extract of red vine leaf AS 195 (*folia vitis viniferae*) in chronic venous insufficiency (stages I–II). A randomized, double-blind, placebo-controlled trial. *Arzneimittelforschung* 50(2):109–117.

Ki-Joong, K. and Turner, B.L. 1992. Systematic overview of *Krigia* (Asteraceae: Lactuceae). *Brittonia* 44:173–198.

Kim, E.-J., Tian, F., and Woo, M.-H. 2000. Asitrocin, (2,4)-*cis*- and *trans*-Asitrocinones: novel bioactive mono-tetrahydrofuran acetogenins from *Asimina triloba* seeds. *Journal of Natural Products* 63:1503–1506.

Kim, H.M., Han, S.B., Lee, K.H., Lee, C.W., Kim, C.Y., Lee, E.J., and Huh, H. 2000. Immunomodulating activity of a polysaccharide isolated from mori cortex radicis. *Archives of Pharmacal Research* 23(3):240–242.

Kim, K.-J. and Turner, B.L. 1990. An overview of the genus *Pyrrhopappus* (Asteraceae: Lactuceae) with emphasis on chloroplast DNA restriction site data. *American Journal of Botany* 77:845–850.

Kim, O.K., Murakami, A., Nakamura, Y., Takeda, N., Yoshizumi, H., and Ohigashi, H. 2000. Novel nitric oxide and superoxide generation inhibitors, persenone A and B, from avocado fruit. *Journal of Agriculture and Food Chemistry* 48(5):1557–1563.

Kim, S.Y., Gao, J.J., Lee, W.C., Ryu, K.S., Lee, K.R., and Kim, Y.C. 1999. Antioxidative flavonoids from the leaves of *Morus alba. Archives of Pharmacal Research* 22(1):81–85.

Kim, S.Y., Gao, J.J., and Kang, H.K. 2000. Two flavonoids from the leaves of *Morus alba* induce differentiation of the human promyelocytic leukemia (HL-60) cell line. *Biological and Pharmaceutical Bulletin* 23 (4):451–455.

Kim, Y.K., Yoon, S.K., and Ryu, S.Y. 2000. Cytotoxic triterpenes from stem bark of *Physocarpus intermedius. Planta Medica* 66(5):485–486.

Kimball, C.P. 1965. *The Lepidoptera of Florida — An Annotated Checklist.* Division of Plant Industry, Florida Department of Agriculture, Gainesville.

Kimball, G.D. 1994. *Koasati Dictionary.* University of Nebraska Press, Lincoln.

Kimura, M., Ikeda, T., Fukumoto, D., Yamasaki, N., and Yonekura, M. 1995. Primary structure of a cysteine proteinase inhibitor from the fruit of avocado (*Persea americana* Mill). *Bioscience, Biotechnology and Biochemistry* 59(12):2328–2329.

Kindscher, K. 1987. *Edible Wild Plants of the Prairie. An Ethnobotanical Guide.* University of Kansas Press, Lawrence.

Kindscher, K. 1992. *Medicinal Wild Plants of the Prairie. An Ethnobotanical Guide.* University of Kansas Press, Lawrence.

King, F.B. 1984. *Plants, People, and Paleoecology: Biotic Communities and Aboriginal Plant Usage in Illinois.* Scientific papers (Illinois State Museum) Vol. 20. Illinois State Museum, Springfield.

King, R.M. and Robinson, H. 1987. *Monographs in Systematic Botany 22 The Genera of the Eupatorieae (Asteraceae).* Missouri Botanical Garden, St. Louis.

Kinietz, W.V. 1965. *Indians of the Western Great Lakes.* Ann Arbor Paperbacks, Ann Arbor, MI.

Kintja, P.K., Bobejko, W.A., Krochmaljuk, W.W., and Tschirva, W.J. 1975. The saponins in the leaves of *Agave americana* L. [in German]. *Pharmazie* 30 (6):396–397.

Kioy, D.W. 1989. A Study of the Chemistry and Pharmacology of the Phytochemicals from Three Species of Canellaceae. Ph.D. dissertation, University of Strathclyde.

Kioy, D.W., Gray, A.I., and Waterman, P. 1989. Further drimane sesquiterpene from the stem bark of *Canella winterana. Journal of Natural Products* 52(1):174–177.

Kirkpatrick, Z.M. 1992. *Wildflowers of the Western Plains.* University of Texas Press, Austin.

Kishor, N., Sati, O.P., Sakakibara, J., and Kaiya, T. 1992. A spirostanol glycoside from *Yucca aloifolia. Phytochemistry* 31(2):706–707.

Kjaer, A. and Schuster, A. 1971. Glucosinolates in *Capparis flexuosa* of Jamaican origin. *Phytochemistry* 10 (12):3155–3160.

Klooster, C.I.E.A. van't, Lindeman, J.C., and Jansen-Jacobs, M.J. 2003. Index of vernacular plant names in Suriname. *Blumea Supplement* 15:1–322.

Kloss, P. 1966. Steam vaporizable constituents of pressed juice of *Sabal serrulatum* (Roem. et Schult.) [in German]. *Arzneimittelforschung* 16(1):95–96.

Klots, A.B. 1951. *A Field Guide to the Butterflies.* Houghton Mifflin, Boston.

Knox, Y.M., Suzutani, T., Yosida, I., and Azuma, M. 2003. Anti-influenza virus activity of crude extract of *Ribes nigrum* L. *Phytotherapy Research* 17(2):120–122.

Kobayashi, S., Ishikawa, H., Kihara, M., Shingu, T., and Hashimoto, T. 1977. Isolation of carinatine and pretazettine from the bulbs of *Zephyranthes carinata* herb (Amaryllidaceae). *Chemical and Pharmaceutical Bulletin* 25(9):2244–2248.

Kobzar', A.Ya. 1986a. Phytochemical study of *Betonica officinalis.* I. Isolation of biologically active substances from the epigeal part of the plant. *Chemistry of Natural Compounds* 22(2):223–224. (translated from *Khimiia prirodnykh soedinenii* 22(2):239).

Kobzar', A.Ya. 1986b. Phytochemical study of *Betonica officinalis.* II. Acids from the epigeal part of *Betonica. Chemistry of Natural Compounds* 22(2):224–225 (translated from *Khimiia prirodnykh soedinenii* 22(2):239–240).

Kobzar', A.Ya. and Nikonov, G.K. 1987. Flavonoids of the epigeal part of *Betonica officinalis. Chemistry of Natural Compounds* 22(5):600–601 (translated from *Khimiia prirodnykh soedinenii* 22(5):636–637).

Koch, E. 2001. Extracts from fruits of saw palmetto (*Sabal serrulata*) and roots of stinging nettle (*Urtica dioica*): viable alternatives in the medical treatment of benign prostatic hyperplasia and associated lower urinary tract symptoms. *Planta Medica* 67(6):489–500.

Kohler, I., Jenett-Siems, K., Mockenhaupt, F.P., Siems, K., Jakupovic, J., Gonzalez, J.C., Hernandez, M.A., Ibarra, R.A., Berendsohn, W.G., and Bienzle, U. 2001. *In vitro* antiplasmodial activity of 4-phenylcoumarins from *Exostema mexicanum. Planta Medica* 67(1):89–91.

Kojima, K., Mutsuga, M., Inoue, M., and Ogihara, Y. 1998. Two alkaloids from *Zephyranthes carinata. Phytochemistry* 48(7):1199–1202.

Kollef, M.H. 1995. Adult respiratory distress syndrome after smoke inhalation from burning poison ivy. *Journal of the American Medical Association* 274(4):358–359.

Konarev, A.V., Anisimova, I.N., Gavrilova, V.A., Vachrusheva, T.E., Konechnaya, G.Y., Lewis, M., and Shewry, P.R. 2002. Serine proteinase inhibitors in the Compositae: distribution, polymorphism and properties. *Phytochemistry* 59(3):279–291.

Kondas, J., Philipp, V., and Dioszeghy, G. 1996. *Sabal serrulata* extract (Strogen forte) in the treatment of symptomatic benign prostatic hyperplasia. *International Urology and Nephrology* 28(6):767–772.

Kondas, J., Philipp, V., and Dioszeghy, G. 1997. *Sabal serrulata* extract in the management of symptoms of prostatic hypertrophy [in Hungarian]. *Orvosi Hetilap* 138(7):419–421.

Konno, C., Oshima, Y., and Hikino, H. 1977. Morusinol, isoprenoid flavone from *Morus* root barks. *Planta Medica* 32(2):118–124.

Konoshima, T., Terada, H., Kokumai, M., Kozuka, M., Tokuda, H., and Estes, J.R. 1993. Studies on inhibitors of skin tumor promotion. XII. Rotenoids from *Amorpha fruticosa. Journal of Natural Products* 56(6):843–848.

Koua, H.K., Han, S.H., and d'Almeida, M.A. 1998. Histopathology of *Anopheles gambiae* s.l. Giles, 1902 (Diptera, Culicidae) subjected to the larvicidal activity of the aqueous extract of *Persea americana* Miller, 1768 (Lauraceae) [in French]. *Bulletin de la Societe de Pathologie Exotique* 91(3):252–256.

Koumaglo, K., Gbeassor, M., Nikabu, O., de Souza, C., and Werner, W. 1992. Effects of three compounds extracted from *Morinda lucida* on *Plasmodium falciparum. Planta Medica* 58(6):533–534.

Kowalchik, C. and Hylton, W.H., Eds. 1987. *Rodale's Illustrated Encyclopedia of Herbs.* Rodale Press, Emmanus, PA.

Kral, R. 1960. A Revision of *Asimina* and *Deeringothamnus* (Annonaceae). *Brittonia* 12:233–278.

Kral, R. 1966. *Xyris* (Xyridaceae) of the Continental United States and Canada. *Sida* 2:177–260.

Kral, R. 1983a. Amaryllidaceae: *Zephyranthes simpsonii* Chapm. Technical publication R8-TP. USDA Forest Service, Southern Region (2, pt.1):241–244.

Kral, R. 1983b. Saxifragaceae: *Ribes echinellum* (Coville) Rehder. Technical publication R8-TP. USDA Forest Service, Southern Region (2, pt.1):574–577.

Kral, R. 1983c. The Xyridaceae in the southeastern United States *Xyris* spp. description. *Journal of the Arnold Arboretum* 64(3):421–429.

Kral, R. and Bostick, P.E. 1969. The genus *Rhexia* (Melastomataceae). *Sida* 3(6):387–440.

Kral, R. and Robinson, A.1983. Cycadaceae: *Zamia integrifolia* Ait. Technical publication R8-TP. USDA Forest Service, Southern Region (2, pt.1):20–23.

Kratochvil, J.F., Burris, R.H., Seikel, M.K., and Harkin, J. M. 1971. Isolation and characterization of alpha-guaiaconic acid and the nature of guaiacum blue. *Phytochemistry* 10(10):2529–2531.

Kreiger, P.J. and Austin, D.F. 1975. *Liguus*: the Boynton Beach colony after forty years. *Nautilus* 89:97–98.

Krishna Rao, R.V. 1969. Alkaloidal components of *Zephyranthes robusta. The Indian Journal of Pharmacy* 31 (2):62–63.

Krishna Rao, R.V. and Seshagiri Rao, J.V.L.N. 1979. Occurrence of the rare alkaloid maritidine in *Zephyranthes robusta* and *Zephyranthes sulphurea. Current Science* 48(3):110–111.

Krishnan, H.B. 1998. Identification of genistein, an anticarcinogenic compound, in the edible tubers of the America groundnut (*Apios americana* Medikus). *Crop Science* 38:1052–1056.

Krmpotic, E., Farnsworth, N.R., and Messmer, W.M. 1972. Cryptopleurine, an active antiviral alkaloid from *Boehmeria cylindrica* (L.) Sw. (Urticaceae). *Journal of Pharmaceutical Science* 61(9):1508–1509.

Krochmal, A. 1968. Medicinal plants and Appalachia. *Economic Botany* 22:332–337.

Krochmal, A. and Krochmal, C. 1973. *A Guide to the Medicinal Plants of the United States.* New York Times Book Co., New York.

Kruthiventi, A.K. and Krishnaswamy, N.R. 2000. Constituents of the flowers of *Persea gratissima. Fitoterapia* 71(1):94–96.

Kubo, I., Taniguchi, M., Chapya, A., and Tsujimoto, K. 1980. An insect antifeedant and antimicrobial agent from *Plumbago capensis. Planta Medica Supplement* 1980:185–187.

Kubo, I., Uchida, M., and Klocke, J.A. 1983. An insect ecdysis inhibitor from the African medicinal plant, *Plumbago capensis* (Plumbaginaceae); a naturally occurring chitin synthetase inhibitor biological control. *Agricultural and Biological Chemistry* 47 (4):911–913.

Kubo, I., Muroi, H., Kubo, A., Chaudhuri, S.K., Sanchez, Y., and Ogura, T. 1994. Antimicrobial agents from *Heterotheca inuloides. Planta Medica* 60(3):218–221.

Kubo, I., Ishiguro, K., Chaudhuri, S.K., Kubo, Y., Sanchez, Y., and Ogura, T. 1995. A plant growth inhibitory sesquiterpenoid from *Heterotheca inuloides. Phytochemistry* 38(2):553–554.

Kubo, I., Chaudhuri, S.K., Kubo, Y., Sanchez, Y., Ogura, T., Saito, T., Ishikawa, H., and Haraguchi, H. 1996. Cytotoxic and antioxidative sesquiterpenoids from *Heterotheca inuloides. Planta Medica* 62(5):427–430.

Kubo, I., Kinst-Hori, I., Chaudhuri, S.K., Kubo, Y., Sanchez, Y., and Ogura, T. 2000. Flavonols from *Heterotheca inuloides*: tyrosinase inhibitory activity and structural criteria. *Bioorganic and Medicinal Chemistry* 8(7):1749–1755.

Kubo, M., Ido, Y., Matsuda, H., and Mikami, H. 2001. Anti-diabetic effect of protoplast preparation from fresh leaves of mulberry (*Morus alba*). *Natural Medicines* 55(4):181–186.

Kubo, S., Mimaki, Y., Sashida, Y., Nikaido, T., and Ohmoto, T. 1992. Steroidal saponins from the rhizomes of *Smilax sieboldii. Phytochemistry* 31(7):2445–2450.

Kugelman, M., Liu, W.C., Axelrod, M., McBride, T.J., and Rao, K.V. 1976. Indicine-*N*-oxide: the antitumor principle of *Heliotropium indicum. Lloydia* 39(2/3): 125–128.

Kuroki, G.W. and Conn, E.E. 1998. Mandelonitrile lyase from *Ximenia americana* L.: stereospecificity and lack of flavin prosthetic group. *Proceedings of the National Academy of Science* 86(18):6978–6981.

Kurz, H. and Godfrey, R.K. 1962. *Trees of Northern Florida.* Regency Press, Gainesville.

Kusano, G., Orihara, S., Tsukamoto, D., Shibano, M., Coskun, M., Guvenc, A., and Erdurak, C.S. 2002. Five new nortropane alkaloids and six new amino acids from the fruit of *Morus alba* LINNE growing in Turkey. *Chemical and Pharmaceutical Bulletin* 50 (2):185–192.

Kut-Lasserre, C., Miller, C.C., Ejeil, A.L., Gogly, B., Dridi, M., Piccardi, N., Guillou, B., Pellat, B., and Godeau, G. 2001. Effect of avocado and soybean unsaponifiables on gelatinase A (MMP-2), stromelysin 1 (MMP-3), and tissue inhibitors of matrix metalloproteinase (TIMP-1 and TIMP-2) secretion by human fibroblasts in culture. *The Journal of Periodontology* 72(12):1685–1694.

Kwaasi, A.A., Parhar, R.S., al-Mohanna, F.A., Harfi, H.A., Collison, K.S., and al-Sedairy, S.T. 1998. Aeroallergens and viable microbes in sandstorm dust. Potential triggers of allergic and nonallergic respiratory ailments. *Allergy* 53(3):255–265.

Kwo, V.T. and Craker, L.E. 1996. Screening Cameroon medicinal plant extracts for antimicrobial activity. *Acta Horticulturae* 426:147–155.

La Flesche, F. 1932. *A Dictionary of the Osage Language.* Smithsonian Institution, Bureau of American Ethnology Bulletin 109. Washington, D.C.

Labbiento, L., Menichini, F., and Delle Monache, F. 1986. Further flavonoids from *Piscidia erythrina. Phytochemistry* 25(6):1505–1506.

Laferrière, J.E. 1994. Vegetation and flora of the Mountain Pima village of Nobogame, Chihuahua, Mexico. *Phytologia* 77(2):102–140.

Laguna, A. 1984. Alkaloids and coumarins from the leaves of *Amyris diatripa. Planta Medica* 50(1):112.

Laguna, A. 1985. Coumarins from the leaves of *Amyris lineata. Planta Medica* (6):528–529.

Laguna, A., Fajardo, M., and Alvarez, E. 1987. Coumarins from the bark of *Amyris lineata. Planta Medica* 53 (4):391–392.

Lamarck, J.B. 1823. Recueil de Planches de Botanique de l'Encyclopédie [Méthodique]. Chez Agasse, Paris.

Lampe, K.F. and McCann, M.A. 1985. *AMA Handbook of Poisonous and Injurious Plants.* American Medical Association, Chicago.

Landrum, L.R. 1992. *Mosiera* (Myrtaceae) in Mexico and Mesoamerica. *Novon* 2:26–29.

Lang, J.M. and Isely, D. 1982. *Eisenhardita* (Leguminosae: Papilionoideae). *Iowa State Journal of Research* 56: 394–417.

Lankford, G.E., Ed. 1987. *Native American Legends. Southeastern Legends.* August House, Little Rock, AR.

Lans, C., Harper, T., Georges, K., and Bridgewater, E. 2000. Medicinal plants used for dogs in Trinidad and Tobago. *Preventive Veterinary Medicine* 45(3–4): 201–220.

Larousse. 1987. *Grand Larousse Universel.* Librairie Larousse, Paris.

Larson, L.H. 1980. *Aboriginal Subsistence Technology on the Southeastern Coastal Plain during the Late Prehistoric Period.* University Press of Florida, Gainesville.

Latorre, D.L. and Latorre, F.A. 1977. Plants used by the Mexican Kickapoo Indians. *Economic Botany* 31: 340–357.

Laudonnière, R.G. de. [1586] 1928. *L'histoire notable de la Floride située ès Indes Occidentales.* Translated by

Richard Hakluyt and reprinted as the "History of Florida." Pp. 13–134. In: *The Voyages, Traffiques and Discoveries of Foreign Voyagers.* E.P. Dutton, New York.

Lavaud, C., Beauviere, S., Massiot, G., Le Men-Olivier, L., and Bourdy, G. 1996. Saponins from *Pisonia umbellifera. Phytochemistry* 43(1):189–194.

Lawless, J. 1995. *The Illustrated Encyclopedia of Essential Oils. The Complete Guide to the Use of Oils in Aromatherapy and Herbalism.* Element Books, Rockport, MA.

Lawrence, G.H.M. 1951. *Taxonomy of Vascular Plants.* Macmillan, New York.

Lawson, J. [1714] 1937. *History of North Carolina.* Reprinted by Garrett and Massie, Richmond, VA.

Lazur'evskii, G.V., Kintia, P.K., and Dragalin, I.P. 1975. Structure of steroid glycosides from the roots of *Yucca filamentosa* L. [in Russian]. *Dokladi Akademii Nauk SSSR.* 221(2):481–483.

Le Moyne, J. [1564] 1968. *Narratives of Le Moyne, an artist who accompanied the French Expedition to Florida under Laudonnière.* Reprinted as "Narratives." Pp. 185–227. In: Bennett, C.E. *Settlement in Florida.* University Press of Florida, Gainesville.

Leal, M.B. and Elisabetsky, E. 1996. Absence of alkaloids in *Psychotria carthagenensis* Jacq. (Rubiaceae). *Journal of Ethnopharmacology* 54(1):37–40.

LeCointe, P. 1947. *Amazonia Brasileira III. Arbores e Plantas Uteis.* Companhia Editora Nacional, São Paulo.

Lederhouse, R.C., Ayres, M.P., Nitao, J.K., and Scriber, J.M. 1992. Differential use of lauraceous hosts by swallowtail butterflies, *Papilio troilus* and *P. palamedes* (Papilionidae). *Oikos* 63:244–252.

Ledin, R.B. 1957. Tropical and subtropical fruits in Florida (other than citrus). *Economic Botany* 11:349–375.

Lee, E.B., Hyun, J.E., da Li, W., and Moon, Y.I. 2002. Effects of *Opuntia ficus-indica* var. Saboten stem on gastric damages in rats. *Archives of Pharmacal Research* 25(1):67–70.

Lee, H. and Lin, J.Y. 1988. Antimutagenic activity of extracts from anticancer drugs in Chinese medicine. *Mutation Research* 204(2):229–234.

Lee, M.W., Morimoto, S., Nonaka, G., and Nishioka, I. 1992. Flavan-3-ol gallates and proanthocyanidins from *Pithecellobium lobatum. Phytochemistry* 31(6): 2117–2120.

Lee, S.E., Ju, E.M., and Kim, J.H. 2001. Free radical scavenging and antioxidant enzyme fortifying activities of extracts from *Smilax china* root. *Experimental and Molecular Medicine* 33(4):263–268.

Lee, W.C., Lin, K.Y., Chen, C.M., Chen, Z.T., Liu, H.J., and Lai, Y.K. 1991. Induction of heat-shock response and alterations of protein phosphorylation by a novel topoisomerase II inhibitor, withangulatin A, in 9L rat brain tumor cells. *Journal Cell Physiology* 149(1):66–76.

Legal, L., Chappe, B., and Jallon, J.M. 1994. Molecular basis of *Morinda citrifolia* (L.): toxicity on *Drosophila. Journal of Chemical Ecology* 20(8):1931–1943.

Lehner, E. and Lehner, J. 1960. *Folklore and Symbolism of Flowers, Plants and Trees.* Tudor, New York.

Leistner, E. 1995. *Morinda* species: biosynthesis of quinones in cell cultures. Medicinal and aromatic plants VIII. *Biotechnology in Agriculture and Forestry* 33:296–307.

Lemmich, E., Cornett, C., Furu, P., Jorstian, C.L., Knudsen, A.D., Olsen, C.E., Salih, A., and Thilborg, S.T. 1995. Molluscicidal saponins from *Catunaregam nilotica* [=*Randia nilotica*]. *Phytochemistry* 39(1):63–68.

Lemus, I., Garcia, R., Delvillar, E., and Knop, G. 1999. Hypoglycaemic activity of four plants used in Chilean popular medicine. *Phytotherapy Research* 13(2):91–94.

Lentz, D.L. 1986. Ethnobotany of the Jicaque of Honduras. *Economic Botany* 40:210–219.

Lentz, D.L. 1993. Medicinal and other economic plants of the Paya of Honduras. *Economic Botany* 47:358–370.

Lentz, D.L., Ed. 2000. *Imperfect Balance. Landscape Transformations in the Pre-Columbian Americas.* Columbia University Press, New York.

Lentz, D.L., Pohl, M.E.D., Pope, K.O., and Wyatt, A.R. 2001a. Prehistoric sunflower (*Helianthus annuus* L.) domestication in Mexico. *Economic Botany* 55:370–376.

Lentz, D.L., Pohl, M.E.D., and Pope, K.O. 2001b. Author's reply to letter to editor. *Economic Botany* 55:471–472.

Leon, H. and Alain, H. 1946–1953. *Flora de Cuba,* Vol. 1. Reprinted Otto Koeltz Science Publishers, Koenigstein, 1974.

Leon, H. and Alain, H. 1957–1963. *Flora de Cuba,* Vol. 2. Reprinted Otto Koeltz Science Publishers, Koenigstein, 1974.

Leon De Pinto, G., Martínez, M., Gutierrez De Gotera, O., Vera, A., Rivas, C., and Ocando, E. 1995. Comparison of two *Pithecellobium* gum exudates. *Biochemical Systematics and Ecology* 23(7/8):849–853.

Lequesne, M., Maheu, E., Cadet, C., and Dreiser, R.L. 2002. Structural effect of avocado/soybean unsaponifiables on joint space loss in osteoarthritis of the hip. *Arthritis and Rheumatism* 47(1):50–58.

Lewington, A. 1990. *Plants for People.* Oxford University Press, New York.

Lewis, W.H. and Elvin-Lewis, M.P.F. 1977. *Medical Botany. Plants Affecting Man's Health.* John Wiley & Sons, New York.

Li, C.M., Tan, N.H., Mu, Q., Zheng, H.L., Hao, X.J., Liang, H.L., and Zhou, J. 1998. Cyclopeptides from the seeds of *Annona glabra. Phytochemistry* 47(7):1293–1296.

Li, C.M., Tan, N.H., Zheng, H.L., Mu, Q., Hao, X.J., He, Y. N., and Zhou, J. 1999. Cyclopeptides from the seeds of *Annona glabra. Phytochemistry* 50(6):1047–1052.

Li, L., Wang, H.K., Chang, J.J., McPhail, A.T., McPhail, D. R., Terada, H., Konoshima, T., Kokumai, M., Kozuka, M., and Eetes, J.R. 1993. Antitumor agents. 138. Rotenoids and isoflavones as cytotoxic constituents from *Amorpha fruticosa. Journal of Natural Products* 56(5):690–698.

Li, S.Y., Fuchino, H., Kawahara, N., Sekita, S., and Satake, M. 2002. New phenolic constituents from *Smilax bracteata. Journal of Natural Products* 65(3):262–266.

Li, W., Li, B., and Chen, Y. 1998. Flexuosol A, a new tetrastilbene from *Vitis flexuosa. Journal of Natural Products* 61(5):646–647.

Li, X.H., Shen, D.D., Li, N., and Yu, S.S. 2003. Bioactive triterpenoids from *Symplocos chinensis. Journal of Asian Natural Products Research* 5(1):49–56.

Li, Y.Q., Yi, Y.H., Tang, H.F., and Xiao, K. 1996. Studies on the structure of isoastilbin [in Chinese]. *Yao Xue Xue Bao* 31(10):761–763.

Lin, S.C., Lin, Y.H., Shyuu, S.J., and Lin, C.C. 1994. Hepatoprotective effects of Taiwan folk medicine: *Alternanthera sessilis* on liver damage induced by various hepatotoxins. *Phytotherapy Research* 8(7): 391–398.

Lin, Y.S., Chiang, H.C., Kan, W.S., Hone, E., Shih, S.J., and Won, M.H. 1992. Immunomodulatory activity of various fractions derived from *Physalis angulata* L. extract. *The American Journal of Chinese Medicine* 20(3–4):233–243.

Linares, E. and Aguirre, J., Eds. 1992. *Los Quelites, un Tesoro Culinario.* Universidad Nacional Autónoma de México, Mexico City.

Lincoln, D.E. and Lawrence, B.M. 1984. The volatile constituents of camphorweed, *Heterotheca subaxillaris. Phytochemistry* 23(4):933–934.

Linnaeus, C. [1749] 1979. *Flora Oeconomica or Household Uses of Wild Plants in Sweden.* Reprinted by Rediviva Publishing House, Stockholm.

Linnaeus, C. [1753] 1957. *Species Plantarum.* Facsimile edition with introduction by W.T. Stearn, Ray Society, London.

Liogier, A.E. 1962. *Flora de Cuba,* Vol. 5. Editorial Universitaria, Universidad de Puerto Rico, Rio Piedras.

Liogier, A.H. 1974. *Diccionario Botanico de Nombres Vulgares de la Espaniola.* Impresora University Nacional Pedro Henriquez Ureña, Santo Domingo.

Liogier, H.A. and Martorell, L.F. 1982. *Flora of Puerto Rico and Adjacent Islands: A Systematic Synopsis.* Editorial de la Universidad, Río Piedras.

Lipp, F.J. 1971. Ethnobotany of the Chinantec Indians, Oaxaca, Mexico. *Economic Botany* 25:234–244.

Lis, A. and Gora, J. 2001. Essential oil of *Amorpha fruticosa* L. *Journal of Essential Oil Research* 13(5):340–342.

Little, E.L.J. 1971. Atlas of the United States Trees. Vol. 1. Conifers and Important Hardwoods. U.S. Department of Agriculture, Miscellaneous Publications 1146, Washington, D.C.

Little, E.L.J. 1978. Atlas of the United States Trees. Vol. 5. Florida. U.S. Department of Agriculture, Miscellaneous Publications 1361, Washington, D.C.

Little, E.L.J. 1979. Checklist of United States Trees (Native and Naturalized). U.S. Department of Agriculture, Washington, D.C.

Little, E.L.J. and Wadsworth, F.H. 1964. Common Trees of Puerto Rico and the Virgin Islands. U.S. Department of Agriculture, Forest Service, Agriculture Handbook 249, Washington, D.C.

Little, E.L.J., Woodbury, R.O., and Wadsworth, F.H. 1974. Trees of Puerto Rico and the Virgin Islands, Vol. 2. U.S. Department of Agriculture, Forest Service, Agriculture Handbook 449, Washington, D.C.

Liu, X.X., Alali, F.Q., Pilarinou, E., and McLaughlin, J.L. 1998. Glacins A and B: two novel bioactive mono-tetrahydrofuran acetogenins from *Annona glabra. Journal of Natural Products* 61(5):620–624.

Liu, X.X., Alali, F.Q., Pilarinou, E., and McLaughlin, J.L. 1999a. Two bioactive mono-tetrahydrofuran acetogenins, annoglacins A and B, from *Annona glabra. Phytochemistry* 50(5):815–821.

Liu, X.X., Pilarinou, E., and McLaughlin, J.L. 1999b. Pondaplin: a novel cyclic prenylated phenylpropanoid from *Annona glabra. Tetrahedron Letters* 40 (3):399–402.

Liu, X.X., Pilarinou, E., and McLaughlin, J.L. 1999c. Two novel acetogenins, annoglaxin and 27-hydroxybullatacin, from *Annona glabra. Journal of Natural Products* 62(6):848–852.

Ller, A., Boppre, M., Witte, L., and Hartmann, T. 1994. Pyrrolizidine alkaloids in *Chromolaena odorata.* Chemical and chemoecological aspects. *Phytochemistry* 35(3):615–619.

Lloyd, J.U. 1921. *Origin and History of All the Pharmacopeial Vegetable Drugs, Chemicals, and Preparations.* Caxton Press, Cincinnati, OH.

Loayza, I., Deslauriers, H., Jean, F.I., and Collin, G.J. 1992. Volatile constituents of the essential oil of *Pluchea fastigiata* Griseb. *Journal of Essential Oil Research* 4 (2):191–193.

Lodhi, M.A.K. and Rice, E.L. 1971. Allelopathic effects of *Celtis laevigata. Bulletin of the Torrey Botanical Club* 98(2):83–89.

Lognay, G., Wathelet, J.P., Severin, M., and Ergo, A. 1981a. Composition chimique de l'endocarpe de *Simarouba glauca* D.C. *Revue Française des Corps Gras* 28(7/8): 319–321.

Lognay, G., Wathelet, J.P., Severin, M., and Ergo, A. 1981b. Etude chimique de la graine de *Simarouba glauca* D.C. Determination de la structure des triglycerides. *Revue Française des Corps Gras* 28(2):67–70.

Long, J. 1996. *Conquista y Comida. Consecuencias del Encuentro de dos Mundos.* Universidad Nacional Autónoma de México, Mexico City.

Long, R.W. and Lakela, O. 1971. *A Flora of Tropical Florida.* University of Miami Press, Miami.

López-A., A.M., Rojas-H., N.M., and Jiménez-M., C.A. 1979. Plant extracts with cytostatic properties growing in Cuba. I. [in Spanish]. *Revista Cubana Medicina Tropical* 31(2):97–104.

López-A., A.M., Rojas-H., N.M., and Jiménez-M, C.A. 1981. Potential antineoplastic activity of Cuban plants. 4. *Revista Cubana Farmacologica* 15(1):71–77.

Lorence, D.H. 1999. A nomenclator of Mexican and Central American Rubiaceae. *Monographs in Systematic Botany from the Missouri Botanical Garden* 73:1–177.

Lorenz, K. 1981. Wild rice: the Indian's staple and the white man's delicacy. *Critical Reviews in Food Science and Nutrition* 15(3):281–319.

Loro, J.F., del Río, I., and Pérez-Santana, L. 1999. Preliminary studies of analgesic and antiinflammatory properties of *Opuntia dillenii* aqueous extract. *Journal of Ethnopharmacology* 67(2):213–218.

Loskiel, G.H. 1794. *History of the Mission of the United Brethren among the Indians of North America.* 3 pts. Translated from German by C.I. LaTrobe. Brethren's Society for the Furtherance of the Gospel, London.

Loughridge, R.M., and Hodge, D.M. 1914. *English and Muskokee Dictionary.* Collected from Various Sources and Revised. Westminster Press, Philadelphia.

Lowe, F.C. 2001. Phytotherapy in the management of benign prostatic hyperplasia. *Urology* 58(6 Suppl. 1):71–77.

Lowe, F.C. and Fagelman, E. 2002. Phytotherapy in the treatment of benign prostatic hyperplasia. *Current Opinion in Urology* 12(1):15–18.

Luckow, M. 1993. Monograph of *Desmanthus* (Leguminosae—Mimosoideae). *Systematic Botany Monographs* 38:1–166.

Luer, C.A. 1972. *The Native Orchids of Florida.* New York Botanical Garden, New York.

Lutomski, J. and Wrocinski, T. 1960. Pharmacodynamic properties of *Passiflora incarnata* preparations. The effect of alkaloid and flavenoid components on pharmacodynamic properties of the raw material. *Biuletyn Instytutu Roslin Leczniczych* 6:176–184.

Lym, R.G., and Christianson, K.M. 1996. The Thistles of North Dakota. URL: www.ext.nodak.edu/extpubs/plantsci/weeds/w1120/w1120w.htm.

Ma, M. and Ruan, J. 2001. Antiinflammatory effect of Saucernetilignan-8 [in Chinese]. *Zhong Yao Cai* 24 (1):42–43.

Ma, W.W., Anderson, J.E., Chang, C.J., Smith, D.L., and McLaughlin, J.L. 1989. Majorenolide and majorynolide: a new pair of cytotoxic and pesticidal alkene-alkyne delta-lactones from *Persea major. Natural Products* 52(6):1263–1266.

Maatta, K., Kamal-Eldin, A., and Torronen, R. 2001. Phenolic compounds in berries of black, red, green, and white currants (*Ribes* sp.). *Antioxidant Redox Signal* 3(6):981–993.

Mabberley, D.J. 1997. *The Plant-Book.* Cambridge University Press, Cambridge, U.K.

MacCauley, C. 1887. The Seminole Indians of Florida. Annual Report, Bureau of (American) Ethnology 5:469–531.

Macedo, A.F., Barbosa, N.C., Esquibel, M.A., Souza. M. M., and Cechinel-Filho, V. 1999. Pharmacological and phytochemical studies of callus culture extracts from *Alternanthera brasiliana. Pharmazie* 54(10): 776–777.

Macfadyen, J. 1837. *The flora of Jamaica; a description of the plants of that island, arranged according to the natural orders. With an appendix, containing an enumeration of the genera according to the Linnaean system, and an essay on the geographical distribution of the species.* Longman, Orme, Brown, Green, and Longman, London.

Mackeen, M.M., Ali, A.M., Abdullah, M.A., Nasir, R.M., Mat, N.B., Razak, A.R., and Kawazu, K. 1997. Antinematodal activity of some Malaysian plant extracts against the pine wood nematode, *Bursaphelenchus xylophilus. Pesticide Science* 51(2):165–170.

MacLeod, J.K. and Rasmussen, H.B. 1999. A hydroxy-beta-caryophyllene from *Pterocaulon serrulatum. Phytochemistry* 50(1):105–108.

Madsen, J.E., Mix, R.L., and Balslev, H. 2001. *Flora of Puná Island. Plant Resources on a Neotropical Island.* Aarhus University Press, Aarhus, Denmark.

Maffei Facino, R., Carini, M., Aldini, G., Bombardelli, E., Morazzoni, P., and Morelli, R. 1994. Free radicals scavenging action and anti-enzyme activities of procyanidines from *Vitis vinifera.* A mechanism for their capillary protective action. *Arzneimittelforschung* 44(5):592–601.

Maffei Facino, R., Carini, M., Aldini, G., Berti, F., Rossoni, G., Bombardelli, E., and Morazzoni, P. 1996. Procyanidines from *Vitis vinifera* seeds protect rabbit heart from ischemia/reperfusion injury: antioxidant intervention and/or iron and copper sequestering ability. *Planta Medica* 62(6):495–502.

Magalhães, A.F., Magalhães, E.G., Leitao Filho, H.F., Frighetto, R.T.S., and Barros, S.M.G. 1981. Coumarins from *Pterocaulon balansae* and *Pterocaulon lanatum. Phytochemistry* 20(6):1369–1371.

Magalhães, A.F., Magalhães, E.G., Junior, V.N., and Leitão Filho, H.F. 1989. Polyacetylenes from *Pterocaulon* species. *Phytochemistry* 28(9):2497–2499.

Mahato, S.B., Das, M.C., and Sahu, N.P. 1981. Triterpenoids of *Scoparia dulcis. Phytochemistry* 20:171–173.

Mahmoud, A.A. 1997. 7-epi-Eudesmanes, eudesmanoic acids, eudesmanolides and other sesquiterpenes from *Pluchea dioscoridis. Phytochemistry* 45(8): 1633–1638.

Mahr, A.C. 1955a. Semantic analysis of eighteenth-century Delaware Indian names for medicinal plants. *Ethnohistory* 2(3):11–28.

Mahr, A.C. 1955b. Eighteenth century terminology of Delaware Indian cultivation and use of maize: a semantic analysis. *Ethnohistory* 2(3):209–240.

Mahon, J.K. 1985. *History of the Second Seminole War, 1835–1842.* University Press of Florida, Gainesville.

Major, C. 1994. *Juba to Jive: A Dictionary of African-American Slang.* Penguin Books, New York.

Majumder, P. and Bhattacharyya, M. 1975. Furoguaiaoxidin — a new enedione lignan of *Guaiacum officinale* L.: a novel method of sequential introduction of alkoxy functions in the 3- and 4-methyl groups of 2,5-diaryl-3,4-dimethylfurans. *Chemical Communications* (Cambridge, U.K.) Sept 3, 17:702–703.

Majumder, P.L. and Bhattacharyya, M. 1974. Structure of furoguaiacidin: a new furanoid lignan of the heartwood of *Guaiacum officinale* L. *Chemistry and Industry London* 2:77–78.

Mallick, B., Mandal, S., and Chatterjee, A. 1998. Occurrence of triacontanoic acid esters and a coumarin in rootbark of *Randia dumetorum* Lam. Syn. *Xeromphis spinosa* Thunb. *Journal of the Indian Chemical Society* 75(2):111.

Manandhar, N.P. 1991. Medicinal plant-lore of Tamang Tribe of Kabhrepalanchok District, Nepal. *Economic Botany* 45:58–71.

Manandhar, N.P. 2002. *Plants and People in Nepal.* Timber Press, Portland, OR.

Manchand, P.S., Blount, J.F., McCabe, T., and Clardy, J. 1979. Structure of evillosin, a novel labdane diterpenoid lactone from *Eupatorium villosum* Sw. *Journal of Organic Chemistry* 44(8):1322–1324.

Mandal, A., Karmakar, R., Bandyopadhyay, S., and Chatterjee M. 1998. Antihepatotoxic potential of *Trianthema portulacastrum* in carbon tetrachloride-induced chronic hepatocellular injury in mice: reflection in haematological, histological and biochemical characteristics. *Archives of Pharmacal Research* 21 (3):223–230.

Mandrile, E.L. and Dongiorno de Pfirter, G. 1990. Principios activos naturalies con acción alucinógena. VIII. Alacaloides derivados de la ergolina. *Acta Farmacologica Bonaerense* 9(1):41–52.

Manganaris, A.G. and Alston, F.H. 1992. Genetics of esterase isoenzymes in *Malus. Theoretical and Applied Genetics* 83(4):467–475.

Manniche, L. 1989. *An Ancient Egyptian Herbal.* University of Texas Press, Austin.

Marafioti, R.L. 1970. Meanings of generic names of important economic plants. *Economic Botany* 24: 189–207.

Marcgrave, G. 1942. *História natural do Brasil.* Impreso do Estado, São Paulo.

Marles, R.J., Clavelle, C., Monteleone, L., Tays, N., and Burns, D. 2000. *Aboriginal Plant Use in Canada's Northwest Boreal Forest.* University of British Columbia Press and Natural Resources Canada, Canadian Forest Service, Vancouver, BC, Canada.

Marquis, V.O. 1974. Proceedings: Preliminary report on the aqueous alcoholic extract and fraction of *Trema orientalis* on hypertensive rats. *West African Journal of Pharmacology and Drug Research* 2(1):1–62.

Marston, A., Potterat, O., and Hostettmann, K. 1988. Isolation of biologically active plant constituents by liquid chromatography. *Journal of Chromatography* 450(1):3–11.

Martin, A.C., Zim, H.S., and Nelson, A.L. 1951. *American Wildlife and Plants. A Guide to Wildlife Food Habits.* Dover, New York.

Martin, J.B. and Mauldin, M.M. 2000. *A Dictionary of Creek/Muskogee.* University of Nebraska Press, Lincoln.

Martin, P.S., Yetman, D., Fishbein, M., Jenkins, P., Van Devender, T.R., and Wilson, R.K., Eds. 1998. *Gentry's Rio Mayo Plants. The Tropical Deciduous Forest and Environs of Northwest Mexico.* University of Arizona Press, Tucson.

Martínez, M. 1969. *Las Plantas Medicinales de Mexico.* Ediciones Botas, Mexico City.

Martínez, M. 1979. Católogo de nombres vulgares y científicas de plantas mexicanas. Fondo de Cultura Económica, Mexico City.

Martínez-Vazquez, M., Apan, T.O.R., Lastra, A.L., and Bye, R. 1998. A comparative study of the analgesic and antiinflammatory activities of pectolinarin isolated from *Cirsium subcoriaceum* and linarin isolated from *Buddleia cordata. Planta Medica* 64(2):134–137.

Martino, V.S., Debenedetti, S.L., and Cousso, J.D. 1979. Caffeolyquinic acids from *Pterocaulon virgatum* and *Pluchea sagittalis. Phytochemistry* 18:2052.

Mata, R. 1993. Chemical studies and biological aspects of some Mexican plants used in traditional medicine. *Recent Advances in Phytochemistry* 27:41–64.

Mata, R., Calzada, F., Garcia, M.R., and Reguero, M.T. 1987. Chemical studies on Mexican plants used in traditional medicine. III. New 4-phenylcoumarins from *Exostema caribaeum. Journal of Natural Products* 50(5):866–871.

Mata, R., Calzada, F., and Garcia, M. del Rosario. 1988. Chemical studies on Mexican plants used in traditional medicine. VI. Additional new 4-phenylcoumarins from *Exostema caribaeum. Journal of Natural Products* 51(5):851–856.

Mata, R., Calzada, F., Díaz, E., and Toscano, R.A. 1990a. Chemical studies on Mexican plants used in traditional medicine. XV. Sesquiterpene evoninoate alkaloids from *Hippocratea excelsa. Journal of Natural Products* 53(5):1212–1219.

Mata, R., Calzada, F., Díaz, E., and Toscano, R.A. 1990b. Sequiterpene evoninoate alkaloids from *Hippocratea excelsa. Planta Medica* 56(6):517.

Mata, R., Albor, C., Pereda-Miranda, R., and McLaughlin, J.L. 1990c. Cytotoxic constituents of *Exostema mexicanum. Planta Medica* 56(2):241.

Matos, M.E.O., Sousa, M.P., Machado, M.I.L., and Braz Filho, R. 1986. Quinovic acid glycosides from *Guettarda angelica. Phytochemistry* 25(6):1419–1422.

May, G. and Willuhn, G. 1978. Antiviral activity of aqueous extracts from medicinal plants in tissue cultures. *Arzneimittel-Forschung* 28(1):1–7.

Mayes, V.O. and Lacy, B.B. 1989. *Nanisé. A Navajo Herbal.* Navajo Community College Press, Tsaile, AZ.

Mazur, W.M., Duke, J.A., Wahala, K., Rasku, S., and Adlercreutz, H. 1998. Isoflavonoids and lignans in legumes: nutritional and health aspects in humans. *Journal of Nutritional Biochemistry* 9:193–200.

McAllister, T.A., Annett, C.B., Cockwill, C.L., Olson, M.E., Wang, Y., and Cheeke, P.R. 2001. Studies on the use of *Yucca schidigera* to control giardiosis. *Veterinary Parasitology* 97(2):85–99.

McChesney, J.D. and Adams, R.P. 1985. Co-evaluation of plant extracts as petrochemical substitutes and for biologically active compounds. *Economic Botany* 39:74–86.

McClure, F.A. 1973. *Genera of Bamboos Native to the New World (Gramineae: Bambusoideae).* Smithsonian Contributions to Botany 9. Smithsonian Institution, Washington, D.C.

McClure, F.A. 1993. *The Bamboos.* Smithsonian Institution Press, Washington, D.C.

McGovern, T.W., LaWarre, S.R., and Brunette, C. 2000. Is it, or isn't it? Poison ivy look-a-likes. *American Journal Contact Dermatitis* 11(2):104–110.

McJunkin, D.M. 1991. Logwood: An Inquiry into Histological Biogeography of *Haematoxylum campechianum* L. and Related Dyewoods of the Neotropics. Ph.D. dissertation, University of California, Berkeley.

McKenzie, R.A. and Brown, O.P. 1991. Avocado (*Persea americana*) poisoning of horses. *Australian Veterinary Journal* 68(2):77–78.

McLaughlin, J.L., Ye, Q., and He, K. 1996. Longimicins A–D: novel bioactive acetogenins from *Asimina longifolia* (Annonaceae) and structure–activity relationships of asimicin type of annonaceous acetogenins. *Journal of Medicinal Chemistry* 39(9):1790.

McVaugh, R. 1945. The genus *Triodanis* Rafinesque and its relationships to *Specularia* and *Campanula. Wrightia* 1:13–52.

McVaugh, R. 1948. Generic status of *Triodanis* and *Specularia. Rhodora* 50:38–49.

Meade, N.A., Staat, R.H., Langley, S.D., and Doyle, R.J. 1980. Lectin-like activity from *Persea americana. Carbohydrate Research* 78(2):349–363.

Mears, J.A. 1982a. A summary of *Blutaparon* Rafinesque including species earlier known as *Philoxerus* R. Brown (Amaranthaceae). *Taxon* 31(1):111–117.

Mears, J.A. 1982b. The typification of *Philoxerus* R. Brown, an Australasian genus of the Gomphrenoideae (Amaranthaceae). *Taxon* 31(1):105–110.

Meckes-Lozyoa, M. and Roman-Ramos, R. 1986. *Opuntia streptacantha*: a coadjutor in the treatment of diabetes mellitus. *The American Journal of Chinese Medicine* 14(3–4):116–118.

Medeiros, R.M.T., Barbosa, R.C., Lima, E.F., Simões, S.V. D., and Riet-Correa, F. 2001. Intoxication of goats by *Plumbago scandens* in Northeastern Brazil. *Veterinary and Human Toxicology* 43(3):167–169.

Medina, E., Lugo, A.E., and Novelo, A. 1995. Contenido mineral del tejido foliar de especies de manglar de la laguna de Sontecomapan (Veracruz, Mexico) y su relacion con la salinidad. [Plant tissue mineral content of mangrove swamp species in the Sontecomapan lagoon.] *Biotropica* 27(3):317–323.

Medina, J.H., Paladini, A.C., Wolfman, C., Levi de Stein, M., Calvo, D., Díaz, L.E., and Pena, C. 1990. Chrysin (5,7-di-OH-flavone), a naturally-occurring ligand for benzodiazepine receptors, with anticonvulsant properties. *Biochemical Pharmacology* 40 (10):2227–2231.

Medsger, O.P. 1966. *Edible Wild Plants.* Macmillan, Toronto, Canada.

Medvedev, A.A., Siniakova, L.A., and Zaitsev, A.V. 2000. Prostaplant treatment of benign prostatic hyperplasia [in Russian]. *Urologiia* 2000(4):13–15.

Meerow, A.W. 1984. From *Amaryllis* to *Zephyranthes. Horticulture* 63(9):40–46.

Meerow, A.W. 1985. Notes on Florida *Zephyranthes. Herbertia* 41:86–94.

Mehta, C.R. and Shah, N.B. 1958. Chemical examination of *Evolvulus alsinoides. Science and Culture* (Calcutta) 24:180–181.

Melchor, G, Armenteros, M, Fernandez, O, Linares, E., and Fragas, I. 2001. Antibacterial activity of *Rhizophora mangle* bark. *Fitoterapia* 72(6):689–691.

Meléndez, E.N. 1989. *Plantas Medicinales de Puerto Rico.* Editorial de la Universidad de Puerto Rico, Rio Piedras.

Melita-Rodriguez, S. and Castro, O. 1996. Pharmacological and chemical evaluation of *Stachytarpheta jamaicensis* (Verbenaceae) [in Spanish]. *Revista de Biologia Tropical* 44(2A):353–359.

Mello, A.C. and Afiatpour, P. 1985. Presence of acetylcholine in the fruit of *Physalis angulata* (Solanaceae). *Ciência e Cultura* (São Paulo) 37:799–805.

Melo, A.M., Jardim, M.L., De Santana, C.F., Lacet, Y., Lobo Filho, J. De Lima, and Ivan Leoncio, O.G. 1974. First observations on the topical use of Primin, Plumbagin and Maytenin in patients with skin cancer [in Portuguese]. *Revista do Instituto de Antibióticos, Universidade Federal de Pernambuco* 14(1–2):9–16.

Menkovic, N., Savikin-Fodulovic, K., Momcilovic, I., and Grubisic, D. 2000. Quantitative determination of secoiridoid and gamma-pyrone compounds in *Gentiana lutea* cultured *in vitro. Planta Medica* 66 (1):96–98.

Merchant, J.R. and Desai, H.K. 1973. Isolation of nantenine from *Cassytha filiformis* Linn. and its synthesis. *Indian Journal of Chemistry* 11:342–344.

Merfort, I. 1994. Flavonoid glycosides in flowers of *Arnica acaulis*, subgenus *Montana. Biochemical Systematics and Ecology* 22(2):229.

Merrill, W.L. 1977. *An Investigation of Ethnographic and Archaeological Specimens of Mescalbeans (Sophora secundiflora).* American Museum Technical Report 6, Museum of Anthropology, University of Michigan, Ann Arbor.

Messer, E. 1978. Zapotec plant knowledge: classification, uses, and communication about plants in Mitla, Oaxaca, Mexico. Prehistory and human ecology of the Valley of Oaxaca. In: K.V. Flannery and R.E. Blanton. *Memoirs of the Museum of Anthropology, University of Michigan No. 10,* 5:1–140.

Meurer-Grimes, B. and Stevenson, D.W. 1994. The biflavones of the Cycadales revisited: biflavones in *Stangeria eriopus, Chigua restrepoi* and 32 other species of Cycadales. *Biochemical Systematics and Ecology* 22(6):595–603.

Meyer, F.G. 1997. Hamamelidaceae. In: Editorial Committee, Eds. *Flora of North America* 3:362–367. Oxford University Press, New York.

Meyer, F.G., Trueblood, E.E., and Miller, J.L. 1999. *The Great Herbal of Leonhart Fuchs. De historia stirpium commentarii insignes, 1542.* Stanford University Press, Stanford, CA.

Meyer, J.E. and Meyer, C. 1986. *The Herbalist.* Meyerbooks, Glenwood, IL.

Meyer-A., A.L., Sarragiotto, M.H., Fujimura, A., and Bacchi, E.M. 2001. Cytotoxic-activity of *Sapindus saponaria* L. fruits on Ehrlich ascitic tumor cells. *Acta Farmaceutica Bonaerense* 20(3):169–172.

Michaux, A. 1904–1907. Journal of travels into Kentucky [in 1795], Vol. III, p. 79. In: Thwaites, R.G., Ed. *Early Western Travels, 1748–1846.* Arthur H. Clark, Cleveland, OH.

Michener, D.C. 1993. *Chamaecyparis.* In: Editorial Committee, Eds. *Flora of North America* 2:408–410. Oxford University Press, New York.

Mihara, S. and Fujimoto, M. 1993. The endothelin ETA receptor-specific effect of 50-235, a nonpeptide endothelin antagonist. *European Journal of Pharmacology* 246(1):33–38.

Mihara, S., Hara, M., Nakamura, M., Sakurawi, K., Tokura, K., Fujimoto, M., Fukai, T., and Nomura, T. 1995. Non-peptide bombesin receptor antagonists, kuwanon G and H, isolated from mulberry. *Biochemical and Biophysical Research Communications* 213(2):594–599.

Milanich, J.T. 1975. Where have all the Indians gone? Pp. 41–49. In: Gill, J.E. and Read, B.R., Eds. *Born of the Sun.* Florida Bicentennial Commemorative Journal, Hollywood.

Milanich, J.T. and Fairbanks, C.H. 1980. *Florida Archaeology.* Academic Press, New York.

Milanich, J.T. and Milbrath, S., Eds. 1989. *First Encounters, Spanish Exploration in the Caribbean and the United States, 1492–1570.* University Press of Florida, Gainesville.

Milanich, J.T. and Proctor, S., Eds. 1978. Tacachale: Essays on the Indians of Florida and Southeastern Georgia during the Historic Period. University Press of Florida, Gainesville.

Miles, D.H. and Kokpol, U. 1976. Tumor inhibitors II: constituents and antitumor activity of *Sarracenia flava. Journal of Pharmaceutical Science* 65:284.

Miles, D.H., Kokpol, U., Zalkow, L.H., Steindel, S.J., and Nabors, J.B. 1974. Tumor inhibitors. I. Preliminary investigation of antitumor activity of *Sarracenia flava. Journal of Pharmaceutical Science* 63(4): 613–615.

Miles, D.H., Kokpol, U., and Mody, N. 1975. Volatiles in *Sarracenia flava. Phytochemistry* 14(3):845–846.

Miles, D.H., Kokpol, U., Bryson, T.A., Wilson, C.A., Bhattacharyya, J.A., Atwood, J.L., and Stone, K. 1976. Structure of sarracenin. An unusual enol diacetal monoterpene from the insectivorous plant *Sarracenia flava. Journal of the American Chemical Society* 98:1569–1573.

Miller, P. 1730. *Catalogus plantarum officinalium quae in Horto Botanico Chelseyano.* Londini.

Miller, R. and Wiedenhoeft, A. 2000. Center for Wood Anatomy Research. URL: www2.fpl.fs.fed.us/.

Millspaugh, C.F. 1892. *American Medicinal Plants.* Reprint 1974, Dover, New York.

Millspaugh, C.F. 1907. Flora of the Sand Keys of Florida. *Field Collection Museum, Botanical Series* 2(5):191–245.

Millspaugh, C.F. 1913. The genera *Pedilanthus* and *Cubanthus* and other American Euphorbiaceae. *Field Museum Publications Botany* 2:353–373.

Minno, M.C. and Minno, M. 1999. *Florida Butterfly Gardening. A Complete Guide to Attracting, Identifying, and Enjoying Butterflies of the Lower South.* University Press of Florida, Gainesville.

Mione, T., Olmstead, R.C., Jansen, R.K., and Anderson, G. J. 1994. Systematic implications of chloroplast DNA variation in *Jaltomata* and selected physaloid genera (Solanaceae). *American Journal of Botany* 81(7): 912–918.

Miraglia, M.D.C.M., Padua, A.P. de., Mesquita, A.A.L., and Gottlieb, O.R. 1985. Flavonoids from the ferns *Blechnum regnellianum* and *Pityrogramma ebenea. Phytochemistry* 24(5):1120.

Mish, F.C., Ed. 1988. *Webster's Ninth New Collegiate Dictionary.* Merriam-Webster, Springfield, MA.

Misra, L. and Laatsch, H. 2000. Triterpenoids, essential oil and photo-oxidative 28 to 13-lactonization of oleanolic acid from *Lantana camara. Phytochemistry* 54 (8):969–974.

Mitchell, J.D. 1990. The poisonous Anacardiaceae genera of the world. *Advances Economic Botany* 8:103–129.

Mitchell, R.E. and Geissman, T.A. 1971. Constituents of *Suriana maritima*: a triterpene diol of novel structure and a new flavonol glycoside. *Phytochemistry* 10 (7):1559–1567.

Mitchell, S.H. 1895. *The Indian Chief, Journeycake.* American Baptist Publication Society, Philadelphia.

Mitich, L.W. 1998. Bull thistle, *Cirsium vulgare. Weed Technology* 12(4):761–763.

Mitscher, L.A., Bathala, M.S., Clark, G.W., and Beal, J.L. 1975a. Antimicrobial agents from higher plants. The quaternary alkaloids of *Ptelea trifoliata. Lloydia* 38 (2):109–116.

Mitscher, L.A., Bathala, M.S., Clark, G.W., and Beal, J.L. 1975b. Antimicrobial agents from higher plants. The antimicrobially inactive components of *Ptelea trifoliata* L. *Lloydia* 38(2):117–124.

Mitscher, L.A., Park, Y.H., Al-Shamma, A., Hudson, P.B., and Haas, T. 1981. Amorfrutin A and B, bibenzyl antimicrobial agents from *Amorpha fruticosa. Phytochemistry* 20(4):781–785.

Mitscher, L.A., Gollapudi, S.R., Drake, S., and Oburn, D.S. 1985. Amorphastilbol, an antimicrobial agent from *Amorpha nana. Phytochemistry* 24(7):1481–1483.

Miyakoshi, M., Tamura, Y., Masuda, H., Mizutani, K., Tanaka, O., Ikeda, T., Ohtani, K., Kasai, R., and Yamasaki, K. 2000. Antiyeast steroidal saponins from *Yucca schidigera* (Mohave yucca), a new antifood-deteriorating agent. *Journal of Natural Products* 63(3):332–338.

Miyase, T., Yamamoto, R., and Ueno, A. 1996. Phenylethanoid glycosides from *Stachys officinalis. Phytochemistry* 43(2):475–479.

Mody, N.V., Henson, R., Hedin, P.A., Kokpol, U., and Miles, D.H. 1976. Isolation of the insect paralyzing agent conline from *Sarracenia flava. Experientia* 32:829–830.

Moerman, D.E. 1991. The medicinal flora of native North America, an analysis. *Journal of Ethnopharmacology* 31:1–42.

Moerman, D.E. 1998. *Native American Ethnobotany.* Timber Press, Portland, OR.

Mohan, S.B. and Kekwick, R.G. 1980. Acetyl-coenzyme A carboxylase from avocado (*Persea americana*) plastids and spinach (*Spinacia oleracea*) chloroplasts. *The Biochemical Journal* 187(3):667–676.

Mohanraj, S. and Subramanian, P.S. 1978. Curassavine, an alkaloid from *Heliotropium curassavicum* Linn. with a C8 necic acid skeleton. *Chemical Communications* (Cambridge, U.K.) 10:423–424.

Mohlenbrock, R.H. 1958. A Revision of the Genus *Stylosanthes. Annals of the Missouri Botanical Garden* 44:299–355.

Mohlenbrock, R.H. 1961. A monograph of the leguminous genus *Zornia. Webbia* 16:1–141.

Mohlenbrock, R.H. and Voight, J.W. 1959. *A Flora of Southern Illinois.* Southern Illinois University Press, Carbondale.

Moldenke, H.N. 1958. Materials toward a monograph of the genus *Citharexylum.* II. *Phytologia* 6:262–352; III. 6:332–368; IV. 6:448–505.

Moldenke, H.N. and Moldenke, A.L. 1983. Verbenaceae. In: Dassanayake, M.D, Ed. *A Revised Handbook to the Flora of Ceylon* IV:196–487 (*Stachytarpheta* pp. 246–267). Amerind Publishing Co., New Delhi.

Monardes, N. [1577] 1925. *Joyfull Newes out of the Newe Founde Worlde.* Translated by John Frampton. Reprint by Alfred A. Knopf, New York.

Mongelli, E., Pampuro, S., Coussio, J., Salomon, H., and Ciccia, G. 2000. Cytotoxic and DNA interaction activities of extracts from medicinal plants used in Argentina. *Journal of Ethnopharmacology* 71(1–2): 145–151.

Montagnac, A., Litaudon, M., and Pais, M. 1997. Quinine- and quinicine-derived alkaloids from *Guettarda noumena. Phytochemistry* 46(5):973–975.

Moondove. 1997. URL: home.earthlink.net/~deanna1jc/moondoves_spiral_8.htm.

Mooney, J. 1885–86. *Myths of the Cherokee and Sacred Formulas of the Cherokees.* Reprinted 1982, Charles and Randy Elder-Booksellers, Nashville, TN.

Moore, M. 1989. *Medicinal Plants of the Desert and Canyon West.* Museum of New Mexico Press, Santa Fe.

Moore, M. 2002. URL: www.ibiblio.org/london/alternative-healthcare/Southwest-School-of-Botanical-Medicine/Constituents/Stillingia_sylvatica.txt.

Mootoo, B.S., Ali, A., Motilal, R., Pingal, R., Ramlal, A., Khan, A., Reynolds, W.F., and McLean, S. 1999. Limonoids from *Swietenia macrophylla* and *S. aubrevilleana. Journal of Natural Products* 62(11): 1514–1517.

Mori, G. and Imanishi, H. 1997. Effects of temperature on the growth and flowering of *Zephyranthes candida* Herb. *Engei Gakkai Zasshi* 66(1):133.

Moriyama, M., Tahara, S., Ingham, J.L., and Mizutani, J. 1992. Isoflavones from the root bark of *Piscidia erythrina. Phytochemistry* 31(2):683–687.

Moriyama, M., Tahara, S., Ingham, J.L., and Mizutani, J. 1993. Isoflavonoid alkaloids from *Piscidia erythrina. Phytochemistry* 32(5):1317–1325.

Morrison, R. 1699. *Hortus regius Blesensis auctus: cum notulis durationis & charactismis plantarum tam additarum, quam non scriptarum; item plantarum in eodem horto Regio Blesensi aucto contentarum, nemini hucusque scriptarum, brevis & succincta delineatio: quibus accessêre observationes generaliores (plantarum in eodem horto Regio Blesensi aucto contentarum) rei herbariae studiosis valdè necesariae, & cognitu perutiles; praeludiorum botanicorum pars prior.* London.

Mors, W.B., Rizzini, C.T., Pereira, N.A., and DeFilipps, R. A., Eds. 2000. *Medicinal Plants of Brazil.* Reference Publications, Algonac, MI.

Morton, J.F. 1968a. A survey of medicinal plants of Curaçao. *Economic Botany* 22:87–102.

Morton, J.F. 1968b. *Wild Plants for Survival in South Florida.* Hurricane House, Miami, FL.

Morton, J.F. 1970. Tentative correlations of plant usage and esophageal cancer zones. *Economic Botany* 24:217–226.

Morton, J.F. 1971. *Plants Poisonous to People.* Hurricane House, Miami.

Morton, J.F. 1974. *Folk Remedies of the Low Country.* E.A. Seemann, Miami.

Morton, J.F. 1977. Medicinal and other plants used by people in the North Caicos (Turks and Caicos Islands, West Indies). *Quarterly Journal of Crude Drugs Research* 15:1–24.

Morton, J.F. 1981. *Atlas of Medicinal Plants of Middle America, Bahamas to Yucatán.* Charles C Thomas, Springfield, IL.

Mossberg, B. and Stenberg, L. 1994. *Gyldendals Store Nordiske Flora.* Gyldendal Norsk Forlag, Oslo, Norway.

Motte, J.R. 1963. *Journey into Wilderness. An Army Surgeon's Account of Life in Camp and Field during the Creek and Seminole Wars 1836–1842.* Edited by J.F. Sunderman. University Press of Florida, Gainesville.

Moyer, R.A., Hummer, K.E, Finn, C.E, Frei, B., and Wrolstad, R.E. 2002. Anthocyanins, phenolics, and antioxidant capacity in diverse small fruits: vaccinium, rubus, and ribes. *Journal of Agriculture and Food Chemistry* 50(3):519–525.

Muhayimana, A., Chalchat, J.C., and Garry, R.P. 1998. Chemical composition of essential oils of *Lantana trifolia* L. from Rwanda. *Journal of Essential Oil Research* 10(5):547–549.

Munno, G., Giannoccaro, F., Riva, G., and Antonicelli, L. 2001. Allergy to yucca. *Allergy* 56(9):921–922.

Muñoz, V., Sauvain, M., Bourdy, G., Callapa, J., Rojas, I., Vargas, L., Tae, A., and Deharo, E. 2000. The search for natural bioactive compounds through a multidisciplinary approach in Bolivia. Part II. Antimalarial activity of some plants used by Mosetene Indians. *Journal of Ethnopharmacology* 69(2):139–155.

Munro, P. and Willmond, C. 1994. *Chickasaw. An Analytical Dictionary.* University of Oklahoma, Norman.

Munz, P.A. 1973. *A California Flora.* University of California Press, Berkeley.

Murcia, M.A., Jiménez, A.M., and Martínez-Tome, M. 2001. Evaluation of the antioxidant properties of Mediterranean and tropical fruits compared with common food additives. *Journal of Food Protection* 64(12):2037–2046.

Murphee, A.H. 1965. Folk medicine in Florida: remedies using plants. *Florida Anthropologist* 18(3, pt.1):175–185.

Murty, Y.L.N., Jairaj, M.A., and Sree, A. 1989. Triterpenoids from *Randia dumetorum. Phytochemistry* 28 (1):276–277.

Musselman, L.J. 1975. Parasitism and haustorial structure of *Krameria lanceolata* (Krameriaceae): a preliminary study. *Phytomorphology* 25:416–422.

Musselman, L.J. 1977. Seed germination and seedlings of *Krameria lanceolata* (Krameriaceae). *Sida* 7:224–225.

Myagmar, B.E. and Aniya, Y. 2000. Free radical scavenging action of medicinal herbs from Mongolia. *Phytomedicine* 7(3):221–229.

Nagai, M., Sakurai, N., Yumoto, N., Nagumo, S., and Seo, S. 2000. Oleanane acid from *Myrica cerifera. Chemical and Pharmaceutical Bulletin* (Tokyo) 48(10): 1427–1428.

Naidoo, K., Denysyk, D., and Brady, J. 1997. Molecular dynamics simulations of the N-linked oligosaccharide of the lectin from *Erythrina corallodendron. Protein Engineering* 10(11):1249–1261.

Nair, S.P. and Madhusudana Rao, J. 1993. Gouanic acid from the leaves of *Gouania microcarpa. Phytochemistry* 33(3):711.

Nakamura, N., Kojima, S., Lim, Y.A., Meselhy, M.R., Hattori, M., Gupta, M.P., and Correa, M. 1997. Dammarane-type triterpenes from *Cordia spinescens. Phytochemistry* 46(6):1139–1141.

Nakano, K., Midzuta, Y., Hara, Y., Murakami, K., Takaishi, Y., and Tomimatsu, T. 1991a. 12–Keto steroidal glycosides from the caudex of *Yucca gloriosa. Phytochemistry* 30(2):633–636.

Nakano, K., Hara, Y., Murakami, K., Takaishi, Y., and Tomimatsu, T. 1991b. The Constituents of *Yucca gloriosa*, Part 4. 12-Hydroxy steroidal glycosides from the caudex of *Yucca gloriosa. Phytochemistry* 30(6):1993–1995.

Nakatani, N., Kikuzaki, H., Hikida, J., Ohba, M., Inami, O., and Tamura, I. 1995. Acylated anthocyanins from fruits of *Sambucus canadensis. Phytochemistry* 38(3): 755–757.

Nakayama, M., Horie, T., Tsukayama, M., Masumura, M., and Hayashi, S. 1978. Isolation of kaempferol-3-*O*-rhamnoglucoside, a flavonoid glycoside from *Zephyranthes candida. Zeitschrift für Naturforschung [C]* 33(7/8):587–588.

Nambiar, G.R. and Mehta, A.R. 1981. Influence of sugars on ergot alkaloid production by cell suspensions of *Evolvulus alsinoides* L. *Indian Journal of Experimental Biology* 19(6):535–537.

Narayana, K. Raj, Reddy, M. Sripal, Chaluvadi, M.R., and Krishna, D.R. 2001. Bioflavonoids: classification, pharmacological, biochemical effects, and therapeutic potential. *The Indian Journal of Pharmacy* 33:2–16.

Nascimento, S.C., Chiappeta, A. de A., and Lima, R.M.O.C. 1990. Antimicrobial and cytotoxic activities in plants from Pernambuco, Brazil. *Fitoterapia* 61(4):353–355.

Nascimento, S.C., de Mello, J. F., and Chiapetta, A. de A. 1985. Cytotoxic agents: experiments with KB cells. *Revista do Instituto de Antibióticos, Universidade Federal de Pernambuco* 22(1/2):19–26.

Nash, O. 1962. *The Pocket Book of Ogden Nash.* Pocket Books, New York.

Nath, M.C. and Banerjee, S.R. 1943. The new antidiabetic principle (amellin) occurring in nature. II. Effects on glycosuria and hyperglycemia in cases of human diabetes. *Annals of Biochemistry and Experimental Medicine* (India) 3:55–62.

Nauman, C.E. 1993. *Phelbodium.* Pp. 323–324. In: Morin, N. and Editorial Committee, Eds. *Flora of North America.* Vol. 2. *Pteridophytes and Gymnosperms.* Oxford University Press, New York.

Navarrete, A and Hong, E. 1996. Anthelmintic properties of alpha-sanshool from *Zanthoxylum liebmannianum. Planta Medica* 62(3):250–251.

Nawaz, H.R., Malik, A., and Ali, M.S. 2001. Trianthenol: an antifungal tetraterpenoid from *Trianthema portulacastrum* (Aizoaceae). *Phytochemistry* 56(1):99–102.

Ndubani, P. and Hojer, B. 1999. Traditional healers and the treatment of sexually transmitted illnesses in rural Zambia. *Journal of Ethnopharmacology* 67(1):15–25.

Nellis, D.W. 1997. *Poisonous Plants and Animals of Florida and the Caribbean.* Pineapple Press, Sarasota, FL.

Nelson, E.C. 2000. *Sea Beans and Nickar Nuts. A Handbook of Exotic Seeds and Fruits Stranded on Beaches in North-Western Europe* [with illustrations by W. Walsh, A. Hathway, and B. Shine]. BSBI handbook 10. Botanical Society of the British Isles, London.

Nepokroeff, M., Bremer, B., and Sytsma, K.J. 1999. Reorganization of the genus *Psychotria* and tribe Psychotrieae (Rubiaceae) inferred from ITS and rbcL sequence data. *Systematic Botany* 24(1):5–27.

Nesom, G. 1994. Review of the taxonomy of *Aster sensu lato* (Asteraceae: Astereae), emphasizing the New World species. *Phytologia* 77:141–297.

Neves, A.N. 1975. *Diccionario de Americanismos,* 2nd ed. Sopena Argentina, Buenos Aires.

Newall, C.A., Anderson, L.A., and Phillipson, J.D. 1996. *Herbal Medicine — A Guide for Health-Care Professionals.* Pharmaceutical Press, London.

Newcomb, W.W., Jr. 1961. *The Indians of Texas.* University of Texas Press, Austin.

Newman, T., Ibrahim, S., Wheeler, J.W., McLaughlin, W.B., Petersen, R.L., and Duffield, R.M. 2000. Identification of sarracenin in four species of *Sarracenia* (Sarraceniaceae). *Biochemical Systematics and Ecology* 28:193–195.

Newsom, B. 1989. *Gentiana catesbei. Journal of the South Carolina Medical Association* 85(2):89.

Newsom, L.A., Webb, S.D., and Dunbar, J.S. 1993. History and geographic distribution of *Cucurbita pepo* gourds in Florida. *Journal of Ethnobiology* 13:75–97.

Ngane, A.N., Biyiti, L., Zollo, P.H., and Bouchet, P. 2000. Evaluation of antifungal activity of extracts of two Cameroonian rutaceae: *Zanthoxylum leprieurii* Guill. et Perr. and *Zanthoxylum xanthoxyloides* Waterm. *Journal of Ethnopharmacology* 70(3):335–342.

N'gouemo, P. 1994. Some pharmacological effects of an ethanolic extract of *Trema guineensis* on the central nervous system in rodents. *Planta Medica* 60(4): 305–307.

Nguyen, H.A., Ripperger, H., Schmidt, J., Porzel, A., Tran, V.S., and Adam, G. 1996. Resorcinol derivatives form two *Ardisia* species. *Planta Medica* 62(5):479–480.

Nguyen, X.D., Le, K.B., and Leclerq, P.A. 1992. The constituents of the leaf oil of *Chromolaena odorata* (L.) R.M. King and H. Robinson from Vietnam. *Journal of Essential Oil Research* 4(3):309–310.

Nicolas, G., Oulad-Ali, A., Guillaume, D., Lobstein, A., Weniger, B., and Anton, R. 1995. Triterpenoid saponins from the root of *Sideroxylon foetidissimum. Phytochemistry* 38(1):225–228.

Nicolls, J.M., Birner, J., and Forsell, P. 1973. Passicol, an antibacterial and antifungal agent produced by *Passiflora* plant species: qualitative and quantitative range of activity. *Antimicrobial Agents and Chemotherapy* 3(1):110–117.

Nicolson, D.H. 1991. *Flora of Dominica,* Part 2: *Dicotyledoneae.* Smithsonian Contributions to Botany 77: iii+274 pp.

Nicolson, D.H., Suresh, C.R., and Manilal, K.S. 1988. *Interpreting Rheede's Hortus Malabaricus.* Regnum Vegetabili, Utrecht 119:1–378.

Nierenstein, M. 1931. Rhatany catechin. *Journal of the Chemical Society* 2809.

Nigam, S.K., Gopal, M., Uddin, R., Yoshikawa, K., Kawamoto, M., and Arihara, S. 1997. Pithedulosides A–G, oleanane glycosides from *Pithecellobium dulce. Phytochemistry* 44(7):1329–1334.

Nikolov, H. 1996. *Dictionary of Plant Names in Latin, German, English and French.* J. Cramer, Stuttgart, Germany.

Nishino, H., Hayashi, T., Arisawa, M., Satomi, Y., and Iwashima, A. 1993. Antitumor-promoting activity of scopadulcic acid B, isolated from the medicinal plant *Scoparia dulcis* L. *Oncology* 50(2):100–103.

Nissanka, A.P., Karunaratne, V., Bandara, B.M., Kumar, V., Nakanishi, T., Nishi, M., Inada, A., Tillekeratne, L. M., Wijesundara, D.S., and Gunatilaka, A.A. 2001. Antimicrobial alkaloids from *Zanthoxylum tetraspermum* and *caudatum. Phytochemistry* 56(8): 857–861.

Niu, R.J. 1986. A study of the preventive and therapeutic effects of *Alternanthera philoxeroides* on influenza [in Chinese]. *Zhong Xi Yi Jie He Za Zhi* 6(1):29–30, 34.

Nixon, K.C. 1997. Fagaceae. Pp. 436–506. In: Editorial Committee, Eds. *Flora of North America,* Vol. 3. Oxford University Press, New York.

Nojima, H., Kimura, I., Chen, F.J., Sugihara, Y., Haruno, M., Kato, A., and Asano, N. 1998. Antihyperglycemic effects of N-containing sugars from *Xanthocercis zambesiaca, Morus bombycis, Aglaonema treubii,* and *Castanospermum australe* in streptozotocin-diabetic mice. *Journal of Natural Products* 61(3):397–400.

Nomura, T. 2001. Chemistry and biosynthesis of prenylflavonoids [in Japanese]. *Yakugaku Zasshi* 121(7): 535–556.

Nomura, T., Fukai, T., Hano, Y., Yoshizawa, S., Suganuma, M., and Fujiki, H. 1988. Chemistry and anti-tumor promoting activity of *Morus* flavonoids. *Progress in Clinical Biological Research* 280:267–281.

Noster, S. and Kraus, L. 1990. *In vitro* antimalarial activity of *Coutarea latiflora* and *Exostema caribaeum* extracts on *Plasmodium falciparum. Planta Medica* 56(1):63–65.

Nowicke, J.W. 1969. Flora of Panama. *Annals of the Missouri Botanical Garden* 56:41, fig. 3.

Nuttall, T. 1905. A journal of travels into the Arkansa Territory during the year 1819. Vol. XIII, In: Thwaites, R.G., Ed. *Early Western Travels, 1748–1846.* Arthur H. Clark, Cleveland, OH.

Oberlies, N.H., Rogers, L.L., Martin, J.M., and McLaughlin, J.L. 1998. Cytotoxic and insecticidal constituents of the unripe fruit of *Persea americana. Journal of Natural Products* 61(6):781–785.

OED. 1971. *The Compact Edition of the Oxford English Dictionary.* Oxford University Press, Oxford, U.K.

Oelrichs, P.B., Ng, J.C., Seawright, A.A., Ward, A., Schaffeler, L., and MacLeod, J.K. 1995. Isolation and identification of a compound from avocado (*Persea americana*) leaves which causes necrosis of the acinar epithelium of the lactating mammary gland and the myocardium. *Natural Toxins* 3(5):344–349.

Ognyanov, I. and Somleva, T. 1980. Rotenoids and 7,2′,4′,5′-tetramethoxyisoflavone in *Amorpha fruticosa* L. fruits. *Planta Medica* 38 (3):279–280.

Ogunkoya, L., Olubajo, O.O., and Sondha, D.S. 1972. Triterpenoid alcohols from *Trema orientalis. Phytochemistry* 11(10):3093–3094.

Ogunkoya, L., Olubajo, O.O., and Sondha, D.S. 1973. Simiarenone from *Trema orientalis. Phytochemistry* 12(3):732–733.

Ogunkoya, L., Olubajo, O.O., and Sondha, D.S. 1977. A new triterpenoid alcohol from *Trema orientalis. Phytochemistry* 16(10):1606–1608.

Oguntimein, B.O. and Elakovich, S.D. 1991. Allelopathic activity of the essential oils of Nigerian medicinal plants. *International Journal of Pharmacognosy* 29 (1):39–44.

Ohwi, J. 1965. *Flora of Japan (in English)*. Smithsonian Institution, Washington, D.C.

Ohyama, M., Tanaka, T., and Iinuma, M. 1998. A prenylated flavanone from roots of *Amorpha fruticosa. Phytochemistry* 48(5):907–909.

Okanishi T, Akahori A, and Yasuda F. 1965. Studies on the steroidal components of domestic plants. XLVII. Constituents of the stem of *Smilax sieboldi* Miq. (1). The structure of laxogenin. *Chemical and Pharmaceutical Bulletin* (Tokyo) 13(5):545–550.

Okolie, N.P. and Ugochukwu, E.N. 1989. Cyanide contents of some Nigerian legumes and the effect of simple processing. *Food Chemistry* 32(3):209–216.

Olajide, O.A., Awe, S.O., and Makinde, J.M. 1998. The effects of *Morinda lucida* Benth. (Rubiaceae) extract on the gastrointestinal tract of rodents. *Phytotherapy Research* 12(6):439–441.

Olas, B., Wachowicz, B., Stochmal, A., and Oleszek, W. 2002. Anti-platelet effects of different phenolic compounds from *Yucca schidigera* Roezl. bark. *Platelets* 13(3):167–173.

Oleszek, W., Sitek, M., Stochmal, A., Piacente, S., Pizza, C., and Cheeke, P. 2001a. Resveratrol and other phenolics from the bark of *Yucca schidigera* Roezl. *Journal of Agriculture and Food Chemistry* 49(2):747–752.

Oleszek, W., Sitek, M., Stochmal, A., Piacente, S., Pizza, C., and Cheeke, P. 2001b. Steroidal saponins of *Yucca schidigera* Roezl. *Journal of Agriculture and Food Chemistry* 49(9):4392–4396.

Oliverio, G.R.M. and Rankin, R.L. 2003. On the subgroupings of the Virginia Siouan languages. Pp. 165–180. In: Rudes, B.A. and Costa, D.J., Eds. *Essays in Algonquian, Catawban and Siouan Linguistics in Memory of Frank T. Siebert, Jr.* Memoir Algonquian and Iroquoian Linguistics 16. Algonquian and Iroquoian Linguistics, Winnipeg.

Olson, S.F. 1963. Wild rice. In: Borland, H. 1965. *Our Natural World.* Doubleday, New York.

Omura, H., Honda, K., and Hayashi, N. 2000. Floral scent of *Osmanthus fragrans* discourages foraging behavior of cabbage butterfly, *Pieris rapae. Journal of Chemical Ecology* 26(3):655–666.

Ondrizek, R.R., Chan, P.J., Patton, W.C., and King, A. 1999. Inhibition of human sperm motility by specific herbs used in alternative medicine. *Journal of Assisted Reproductive Genetics* 16(2):87–91.

O'Neill, M.J., Bray, D.H., Boardman, P., Wright, C.W., Phillipson, J.D., Warhurst, D.C., Gupta, M.P., Correya, M., and Solis, P. 1988. Plants as sources of antimalarial drugs, Part 6: Activities of *Simarouba amara* fruits. *Journal of Ethnopharmacology* 22 (2):183–190.

O'Neill, M.J., Lewis, J.A., Noble, H.M., Holland, S., Manst, C., Farthing, J.E., Foster, G., Noble, D., Lane, S.J., and Sidebottom, P.J. 1998. Isolation of translactone-containing triterpenes with thrombin inhibitory activities from the leaves of *Lantana camara. Journal of Natural Products* 61(11):1328–1331.

Ooststroom, S.J. van. 1934. A monograph of the genus Evolvulus. *Mededeelingen botanisch museum herbarium rijks universiteit Utrecht* 14:1–267.

Ortega, A., Garcia, P.E., Cardenas, J., Mancera, C., Marquina, S., Carmen Garduno, M.I., and Maldonado, E. 2001. Methyl dodonates, a new type of diterpenes with a modified clerodane skeleton from *Dodonaea viscosa. Tetrahedron* 57(ER15):2981–2989.

Ortega, M. 1990. Purification and characterization of a lectin from *Erythrina americana* by affinity chromatography. *Plant Science* 72(1):133.

Osborne, R., Salatino, A., Salatino, M.L.F., Sekiya, C.M., and Vázquez Torres, M. 1993. Alkanes of foliar epicuticular waxes from five Cycad genera in the Zamiaceae. *Phytochemistry* 33(3):607–609.

Oser, B.L. 1966. An evaluation of *Yucca mohavensis* as a source of food grade saponin. *Food and Cosmetics Toxicology* 4(1):57–61.

Oshima, Y., Namao, K., Kamijou, A., Matsuoka, S., Nakano, M., Terao, K., and Ohizumi, Y. 1995. Powerful hepatoprotective and hepatotoxic plant oligostilbenes, isolated from the Oriental medicinal plant *Vitis coignetiae* (Vitaceae). *Experientia* 51 (1):63–66.

Osorio, C., Duque, C., and Fujimoto, Y. 2000. Oxygenated monoterpenoids from badea (*Passiflora quadrangularis*) fruit pulp. *Phytochemistry* 53(1):97–101.

Otero, R., Nuñez, V., Barona, J., Fonnegra, R., Jiménez, S. L., Osorio, R.G., Saldarriaga, M., and Díaz, A. 2000. Snakebites and ethnobotany in the northwest region of Colombia. Part III: Neutralization of the haemorrhagic effect of *Bothrops atrox* venom. *Journal of Ethnopharmacology* 73(1–2):233–241.

Ott, J. 2001. *Shamanic Snuffs or Entheogenic Errhines.* Ethnobotanica, Solothum, Switzerland.

Oulad-Ali, A., Guillaume, D., Weniger, B., Jiang, Y., and Anton, R. 1994. Mabiosides C-E: triterpenoid saponins from the bark of *Colubrina elliptica. Phytochemistry* 36(2):445–448.

Oviedo y Valdés, G. Fernández de. [1526] 1969. *De la Natural Hystoria de las Indias.* University of North Carolina Press, Chapel Hill. Facsimile Edition reprinted in *Studies in the Romance Languages and Literature* 85.

Oyedapo, O.O. 1996. Studies on bioactivity of the root extract of *Plumbago zeylanica. International Journal of Pharmacognosy* 34(5):365–369.

Paciorek, C.J., Moyer, B.R., Levin, R.A., and Halpern, S.L. 1995. Pollen consumption by the hummingbird flower mite *Proctolaelaps kirmsei* and possible fitness effects on *Hamelia patens. Biotropica* 27(2): 258–262.

Padmaja, V., Thankamany, V., Hara, N., Fujimoto, Y., and Hisham, A. 1995. Biological activities of *Annona glabra. Journal of Ethnopharmacology* 48(1):21–24.

Pal, S. and Chaudhuri, A.K.N. 1989. Studies on the effects of *Pluchea indica* less root extract on gastroduodenal ulcer models in rats and guinea pigs. *Phytotherapy Research* 3(4):156–158.

Palacios, J., Mata, R., and López, R. 1989. *Hippocratea excelsa* (Hippocrateaceae), a new source of trans polisoprene. *Economic Botany* 43 (4):508–509.

Palacios, P.S., Rojo, A.A., Coussio, J.D., Kimpe, N. de, and Debenedetti, S.L. 1999. 6,7-Dioxygenated and 5,6,7-trioxygenated coumarins from *Pterocaulon polystachium. Planta Medica* 65(3):294–295.

Palmer, A.C. 1975. Derrengue, a paralysis of cattle in El Salvador ascribed to ingestion of *Melochia pyramidata. Veterinary Record* 96(25):547–548.

Panati, C. 1987. *Extraordinary Origins of Everyday Things.* Harper & Row, New York.

Paris, R.R. and Nothis, A. 1970. Plantes de Nouvelle Calédonie. II. Plantes contenand des derivés poplphénolques. *Plantes Médicinales Phytothérapeutiques* 4:64–74.

Park, E.H., Kahng, J.H., and Paek, E.A. 1998. Studies on the pharmacological action of cactus: identification of its antiinflammatory effect. *Archives of Pharmacal Research* 21(1):30–34.

Park, E.H., Kahng, J.H., Lee, S.H., and Shin, K.H. 2001. An antiinflammatory principle from cactus. *Fitoterapia* 72(3):288–290.

Parker, K.F. 1972. *An Illustrated Guide to Arizona Weeds.* University of Arizona Press, Tucson.

Parkinson, J. [1629] 1976. *Paradisi in Sole: Paradisius Terrestris [A Garden of Pleasant Flowers].* Reprinted by Dover, New York.

Parks, J.C. 1973. Revision of North American and Caribbean *Melanthera. Rhodora* 75:169–210.

Parrotta, J.A. 2001. *Healing Plants of Peninsular India.* CABI Publishing, New York.

Pascarella, J.B. 1996. The biology of *Periploca* sp. (Lepidoptera: Cosmopterigiae): a gall maker on *Ardisia escallonioides* (Myrsinaceae). *Florida Entomologist* 79(4):606–610.

Pascarella, J.B. 1997. Pollination ecology of *Ardisia escallonioides* (Myrsinaceae). *Castanea* 62(1):1–7.

Pascoe, K.O., Burke, B.A., and Chan, W.R. 1986. Caesalpin F, a new furanoditerpene from *Caesalpinia bonducella. Journal of Natural Products* 49:913–915.

Pataki, T., Bak, I., Kovacs, P., Bagchi, D., Das, D.K., and Tosaki, A. 2002. Grape seed proanthocyanidins improved cardiac recovery during reperfusion after ischemia in isolated rat hearts. *American Journal Clinical Nutrition* 75(5):894–899.

Patterson, G.W., Xu, S., and Salt, T.A. 1991. Sterols of Caryophyllales with emphasis on Amaranthaceae. *Phytochemistry* 30(2):523–526.

Paul, B., Masih, I., Deopujari, J., and Charpentier, C. 1999. Occurrence of resveratrol and pterostilbene in age-old darakchasava, an ayurvedic medicine from India. *Journal of Ethnopharmacology* 68(1–3):71–76.

Paul, B.D, Rao, G.S., and Kapadia, G.J. 1974. Isolation of myricadiol, myricitrin, taraxerol, and taraxerone from *Myrica cerifera* L. root bark. *Journal of Pharmaceutical Sciences* 63(6):958–959.

Paul, J.S. 1993. Goats control poison ivy, oak and sumac. *Small Farm Today* 10(3):44.

Payne, W.W. and Jones, V.H. 1961. The taxonomic status and archaeological significance of a giant ragweed from Prehistoric Bluff Shelters in the Ozark Plateau Region. *Michigan Academy of Science, Arts, and Letters Papers* 47:147–163.

Pcolinski, M.J., Doskotch, R.W., Lee, A.Y., and Clardy, J. 1994. Chlorosilphanol A and silphanepoxol, labdane diterpenes from *Silphium perfoliatum. Journal of Natural Products* 57(6):776–783.

Peana, A.T., Moretti, M.D., Manconi, V., Desole, G., and Pippia, P. 1997. Antiinflammatory activity of aqueous extracts and steroidal sapogenins of *Agave americana. Planta Medica* 63(3):199–202.

Peattie, D.C. 1953. *A Natural History of Western Trees.* Reprinted 1980, University of Nebraska Press, Lincoln.

Pelotto, J.P. and Martínez, M.A.P. del. 1998. Flavonoid aglycones from Argentinian *Capparis* species (Capparaceae). *Biochemical Systematics and Ecology* 26 (5):577–580.

Peña, D. 1949. Journal of an expedition to Apalachee and Apalachicola in 1716. Translated and interpreted by M.F. Boyd. *Florida Historical Quarterly* 28(1):1–27.

Penders, A. and Delaude, C. 1994. Triterpenoid saponins from *Melanthera scandens. Phytochemistry* 37 (3):821–825.

Peng, C.-I., Ed. 2000. *Revised Draft of Index to Codes of Vascular Plants of Taiwan.* Academia Sinitica, Taipei, Taiwan.

Peng, C.-I. and Yang, K.C. 1998. Unwelcome naturalization of *Chromolaena odorata* (Asteraceae) in Taiwan. *Taiwania* 43(4):289–294.

Pennington, C.W. 1958. Tarahumar fish stupification plants. *Economic Botany* 12:95–102.

Pennington, T.D. 1990. Sapotaceae. In: *Flora Neotropica* 52:1–771, New York Botanical Garden, Bronx, NY.

Peraza-Sánchez, S.R. and Peña-Rodríguez, L.M. 1992. Isolation of picropolygamain from the resin of *Bursera simaruba. Journal of Natural Products* 55 (12):1768–1771.

Pereira, N.A. 1949. Contribução ao estudo da tapixava (*Scoparia dulcis* L.). *Revista Flora Medica* (Rio de Janeiro) 16:363–381.

Pereira-Martins, S.R., Takahashi, C.S., Tavares, D.C., and Torres, L.M. 1998. *In vitro* and *in vivo* study of the clastogenicity of the flavone cirsitakaoside extracted from *Scoparia dulcis* L. (Scrophulariaceae). *Teratogenesis, Carcinogenesis, and Mutagenesis* 18(6):293–302.

Perera, L.M., Ruedas, D., and Gómez, B.C. 2001. Gastric antiulcer effect of *Rhizophora mangle* L. *Journal of Ethnopharmacology* 77(1):1–3.

Pérez, G., Pérez, C., Sousa-Cavada, B., Moreira, R., and Richardson, M. 1991. Comparison of the amino acid sequences of the lectins from seeds of *Dioclea lehmanni* and *Canavalia maritima. Phytochemistry* 30(8):2619–2621.

Pérez, R.M., Pérez, S., Zavala, M.A., and Salazar, M. 1995. Antiinflammatory activity of the bark of *Hippocra-*

tea excelsa. *Journal of Ethnopharmacology* 47(2):85–90.

Pérez-Arbelaez, E. 1978. *Plantas útiles de Colombia.* Litografia Arco, Bogota.

Pérez-G., F., Marin, E., and Adzet, T. 1995. The antiinflammatory effect of several Compositae from South America. Extracts in rats. *Phytotherapy Research* 9 (2):145–146.

Pérez-G., R.M., Ocegueda-A., A., Muñoz-L., J.L., Avila-A., J.G., and Morrow, W.W. 1984. A study of the hypoglucemic [*sic*] effect of some Mexican plants. *Journal of Ethnopharmacology* 12:253–262.

Pérez-G., R.M., Ramírez-L., E., and Vargas-S., R. 2001. Effect of *Cirsium pascuarense* on blood glucose levels of normoglycaemic and alloxan-diabetic mice. *Phytotherapy Research* 15(6):552–554.

Pérez-G., S., Pérez-G., R.M., Pérez-G., C., Zavala-S., M.A., and Vargas-S., R. 1993. Cicatrizing activity of *Randia echinocarpa* in gastric ulcers. *Phyton* 54 (2):157–162.

Pérez-G., S., Zavala-S., M.A., Vargas-S., R., Hernandez-Z., E., and Pérez-G., R.M. 1996. Antidiarrhoeal activity of *Hamelia patens* methanol extract in rats and mice. *Phytotherapy Research* 10(8):686–688.

Pérez-G., S., Zavala-S., M.A., Arias-G., L., Pérez-G., C., and Pérez-G., R.M. 2001. Antimicrobial study of bark from five tree species. *Phytotherapy Research* 15(4): 356–359.

Pérez-Garcia, F., Marin, E., Canigueral, S., and Adzet, T. 1996. Antiinflammatory action of *Pluchea sagittalis*: involvement of an antioxidant mechanism. *Life Science* 59(24):2033–2040.

Perkins, K.D. and Payne, W.W. 1978. *Guide to the Poisonous and Irritant Plants of Florida.* Institute of Food and Agricultural Sciences, Gainesville.

Pernet, R. 1972. Phytochemistry of the Burseraceae. *Lloydia* 35(3):280–287.

Perry, L.M. 1937. Notes on *Silphium. Rhodora* 39:281–297.

Petersen, A. 1974. *Herbs and Proverbs of the Virgin Islands.* St. Thomas Graphics, St. Thomas, Virgin Islands.

Petersen, K.A., Elisens, W.J., and Estes, J.R. 1990. Allozyme variation in *Pyrrhopappus multicaulis* and *P. carolinianus* (Asteraceae): relation to mating system and purported hybridization. *Systematic Botany* 15:534–543.

Petit-Paly, G., Montagu, M., Viel, C., Rideau, M., and Chenieux, J.C. 1987. Dihydrofuro[2,3-b]quinolinium alkaloids in cultured cells of *Ptelea trifoliata* L. — isolation of a new alkaloid, ptelecultinium. *Plant Cell Reports* 6(4):309–312.

Petit-Paly, G., Montagu, M., Merienne, C., Ambrose, J.D., Rideau, M., Viel, C., and Chenieux, J.C. 1989. New alkaloids from *Ptelea trifoliata. Planta Medica* 55 (2):209–210.

Petitt, G.R., Cragg, G.M., Singh, S.B., Duke, J.A., and Doubek, D.L. 1990. Antineoplastic Agents, 162, *Zephyranthes candida. Journal of Natural Products* 53(1):176.

Pettit, G.R., Gaddamidi, V., and Cragg, G.M. 1984. Antineoplastic agents. 105. *Zephyranthes grandiflora. Journal of Natural Products* 47(6):1018–1020.

Petkov, E., Uzunov, P., Kostova, I.T., and Ognyanov, I. 1983. Inhibition of rat heart phosphodiesterase by some rotenoids and isoflavonoids isolated from the petroleum ether extract of ripe *Amorpha fruticosa. Planta Medica* 47(4):237–239.

Petlevski, R., Hadzija, M., Slijepcevic, M., and Juretic, D. 2001. Effect of "antidiabetis" herbal preparation on serum glucose and fructosamine in NOD mice. *Journal of Ethnopharmacology* 75(2–3):181–184.

Phipps, J.B., Robertson, K.R., Smith, P.G., and Rohrer, J.R. 1990. A checklist of the subfamily Maloideae (Rosaceae). *Canadian Journal of Botany* 68:2209–2269.

Piacente, S., Pizza, C., De Tommasi, N., and Mahmood, N. 1996. Constituents of *Ardisia japonica* and their *in vitro* anti-HIV activity. *Journal of Natural Products* 59(6):565–569.

Picoche, J. 1994. *Dictionnaire Etymologique du Français.* Dictionnaires Le Robert, Paris.

Pietro, R.C., Kashima, S., Sato, D.N., Januario, A.H., and Franca, S.C. 2000. *In vitro* antimycobacterial activities of *Physalis angulata* L. *Phytomedicine* 7(4): 335–338.

Pietta, P. and Zio, C. 1983. Isolation of jamaicin and lisetin and their determination in *Piscidia erythrina* L. by high-performance liquid chromatography. Analgesic and insecticidal properties of the root-bark. *Journal of Chromatography* 260(2):497–501.

Pignatti, S. 1982. *Flora d'Italia.* Edagricole, Bologna.

Pilcher, J.D. 1916. The action of various "female" remedies on the excised uterus of the guinea pig. *Journal of the American Medical Association* 67:490.

Pinkley, H.V. 1969. Etymology of *Psychotria* in view of a new use of the genus. *Rhodora* 71(788):535–540.

Pino, J.A., Rosado, A., Bello, A., Urguiola, A., and Garcia, S. 2000. Essential oil of *Amyris elimifera* L. from Cuba. *Journal of Essential Oil Research* 12(1):39–40.

Pinto, G.L. de., Nava, M., Martínez, M., and Rivas, C. 1993. Gum polysaccharides of nine specimens of *Laguncularia racemosa. Biochemical Systematics and Ecology* 21(4):463–466.

Pittier, H. 1926. *Manual de las Plantas Usuales de Venezuela.* Lithografia del Comercio, Caracas.

Pittier, H. 1957. *Ensayo sobre Plantas Usuales de Costa Rica.* Universidad de Costa Rica, Serie Ciencias Naturales 2, San Jose, Costa Rica.

Pizarro, C.M. 1959. *Sinopsis de la Flora Chilena.* Editorial Universitaria, S.A., Santiago, Chile.

Platt, S.G. and Brantley, C.G. 1997. Canebrakes: an ecological and historical perspective. *Castanea* 62 (1):8–21.

Pliske, T.E. 1975. Courtship behavior and use of chemical communication by males of certain species of ithomiine butterflies (Nymphalidae: Lepidoptera) [Danainae]. *Annals of the Entomological Society of America* 68(6):935–942.

Pliske, T.E. and Salpeter, M.M. 1971. The structure and development of the hairpencil glands in males of the queen butterfly, *Danaus gilippus berenice. Journal of Morphology* 134(2):215–242.

Plock, A., Beyer, G., Hiller, K., Grundemann, E., Krause, E., Nimtz, M., and Wray, V. 2001a. Application of MS and NMR to the structure elucidation of complex sugar moieties of natural products: exemplified by the steroidal saponin from *Yucca filamentosa* L. *Phytochemistry* 57(3):489–496.

Plock, A., Sokolowska-Kohler, W., and Presber, W. 2001b. Application of flow cytometry and microscopical methods to characterize the effect of herbal drugs on *Leishmania* spp. *Experimental Parasitology* 97(3): 141–153.

Plotkin, M., Boom, B.M., and Allison, M. 1991. *The Ethnobotany of Aublet's Histoire des Plantes de la Guiane Françoise (1775).* Missouri Botanical Garden, St. Louis.

Plouvier, V. 1958. Sur le recherche des ethers methyliques des isnositols dans quelques groupes botaniques. *Compte-rendu l'Académie Sciences* (Paris) 247:2423–1426.

Plowman, T. 1969. Folk use of New World aroids. *Economic Botany* 23:97–122.

Podimuang, V., Mongkolsuk, S., Yoshihira, K., and Natori, S. 1971. Constituents of three Thai medicinal plants: *Ardisia polycephala* (Myrsinaceae), *Rhabdia lycioides* (Boraginaceae), and *Balanophora polyandra* (Balanophoraceae). *Chemical and Pharmaceutical Bulletin* 19(1):207–208.

Polonsky, J., Varon, Z., Jacquemin, H., and Pettit, G.R. 1978. The isolation and structure of 13,18-dehydroglaucarubinone, a new antineoplastic quassinoid from *Simarouba amara. Experientia* 34(9):1122–1123.

Polunin, O. 1969. *Flowers of Europe. A Field Guide.* Oxford University Press, New York.

Polunin, O. and Smythies, B.E. 1988. *Flowers of South-West Europe. A Field Guide.* Oxford University Press, Oxford.

Ponce-Macotela, M., Navarro-Alegria, I., Martínez-Gordillo, M.N., and Álvarez-Chacon, R. 1994. *In vitro* effect against *Giardia* of 14 plant extracts [in Spanish]. *Revista de Investigacion Clinica* 46 (5):343–347.

Pongprayoon, U. 1991. Pharmacognostic studies on the Thai medicinal plant *Ipomoea pes-caprae* (L.). *Acta Pharmaceutica Nordica* 3(3):186.

Pongprayoon, U., Baeckstrom, P., and Jacobsson, U. 1991. Compounds inhibiting prostaglandin synthesis isolated from *Ipomoea pes-caprae. Planta Medica* 57 (6):515.

Pongprayoon, U., Baeckstrom, P., and Jacobsson, U. 1992. Antispasmodic activity of beta-phenylpropanoids and E-phytol isolated from *Ipomoea pes-caprae. Planta Medica* 58(1):19.

Poole, J. and Bowles, D.E. 1999. Habitat characterization of Texas wild-rice (*Zizania texana* Hitchcock), and endangered aquatic macrophyte from the San Marcos River, TX, USA. *Aquatic Conservation* 9 (3):291–302.

Porcher, F.P. 1863. *Resources of the Southern Fields and Forests, Medical, Economical, and Agricultural. Being also a Medical Botany of the Confederate States; with Practical Information on the Useful Properties of the Trees, Plants, and Shrubs.* Steam-Power Press of Evans and Cogswell, Charleston, SC.

Porras-Reyes, Bh., Lewis, W.H., Roman, J. et al. 1993. Enhancement of wound healing by the alkaloid taspine defining mechanism of action. *Proceedings of the Society for Experimental Biology and Medicine* 203:18–25.

Porter, D.M. 1969. The genus *Kallstroemia* (Zygophyllaceae). *Contributions to the Gray Herbarium* 198:41–153.

Poulsen, C., Heijden, R. van der, and Verpoorte, R. 1991. Assay of isochorismate synthase from plant cell cultures by high-performance liquid chromatography. *Phytochemistry* 30(9):2873–2876.

Powell, M.A. 1978. Systematics of *Flaveria* (Flaveriinae-Asteraceae). *Annals of the Missouri Botanical Garden* 65:590–636.

Prager, N., Bickett, K., French, N., and Marcovici, G. 2002. A randomized, double-blind, placebo-controlled trial to determine the effectiveness of botanically derived inhibitors of 5-alpha-reductase in the treatment of androgenetic alopecia. *Journal of Alternative and Complementary Medicine* 8(2):143–152.

Prakash, A.O. 1985. Biological evaluation of some medicinal plant extracts for contraceptive efficacy in females [in French]. *Contraception, Fertilité, Sexualité* (Paris) 13(4):649–655.

Prakash, A.O., Saxena, V., Shukla, S., Tewari, R.K., Mathur, S., Gupta, A., Sharma, S., and Mathur, R. 1985. Anti-implantation activity of some indigenous plants in rats. *Acta Europa Fertilica* 16(6):441–448.

Prance, G.T. 1970. The genera of Chrysobalanaceae in the southeastern United States. *Journal of the Arnold Arboretum* 51:521–528.

Prance, G.T. 1972. *Chrysobalanus.* In: *Flora Neotropical* 9. New York Botanical Garden, Bronx, NY.

Prance, G.T. and White, F. 1988. The genera of Chrysobalanaceae: a study in practical and theoretical taxonomy and its relevance to evolutionary biology. *Philosophical Transactions of the Royal Society B* 320:1–184.

Prendergast, H.D.V. and Pearman, G. 2001. Comparing uses and collections — the example of *Dodonaea viscosa* Jacq. [Sapindaceae]. *Economic Botany* 55(2):184–186.

Prescott, W.H. 1892. *History of the Conquest of Peru, with a Preliminary View of the Civilization of the Incas.* David McKay, Philadelphia.

Preuss, H.G., Bagchi, D., and Bagchi, M. 2002. Protective effects of a novel niacin-bound chromium complex and a grape seed proanthocyanidin extract on advancing age and various aspects of syndrome X. *Annals of the New York Academy of Sciences* 957: 250–259.

Proctor, G.R. 1984. *Flora of the Cayman Islands.* HMSO, Kew Bulletin Additional Series XI, London.

PubMed. 2003. URL: http://www4.ncbi.nlm.nih.gov/entrez/query.fcgi.

Purseglove, J.W. 1968. *Tropical Crops. Dicotyledons,* Vols. 1 and 2. Longmans, Green, London.

Qian-Cutrone, J., Huang, S., Trimble, J., Li, H., Lin, P.F., Alam, M., Klohr, S.E., and Kadow, K.F. 1996. Niruriside, a new HIV REV/RRE binding inhibitor from *Phyllanthus niruri. Journal of Natural Products* 59(2):196–199.

Qu, C.F., Yang, Z.Q., and Xiang, J.M. 1993. *Alternanthera philoxeroides* (Mart.) Griseb. Protection against fetal epidemic hemorrhagic fever virus infection in suckling mice [in Chinese]. *Zhongguo Zhong Yao Za Zhi* 18(5):304–305, 320.

Quattrocchi, U. 1999. *CRC World Dictionary of Plant Names. Common Names, Scientific Names, Eponyms, Synonyms, and Etymology.* CRC Press, Boca Raton, FL.

Quershi, M.A. and Thakur, R.S. 1977. Chemical constituents of *Randia tetrasperma. Plant Medica* 32(3): 229–232.

Quinn, D.B. 1955. *The Roanoke Voyages, 1584–90. Documents to Illustrate the English Voyages to North America under the Patent Granted to Walter Raleigh in 1584.* Hakluyt Society, London.

Radford, A.E., Ahles, H.E., and Bell, C.R. 1968. *Manual of the Vascular Flora of the Carolinas.* University of North Carolina Press, Chapel Hill.

Raffaele, H.A. 1983. *A Guide to the Birds of Puerto Rico and the Virgin Islands.* Fondo Educativo Interamericano, San Juan.

Rafii, Z.A., Dodd, R.S., and Fromard, F. 1996. Biogeographic variation in foliar waxes of mangrove species. *Biochemical Systematics and Ecology* 24 (4):341–345.

Rafinesque, C.S. 1828–1830. *Medical Flora.* Atkinson and Alexander, Philadelphia.

Ragasa, C.Y., Tremor, N., and Rideout, J.A. 2002. Ionone derivatives from *Alternanthera sessilis. Journal of Asian Natural Products Research* 4(2):109–115.

Ragazzini, G. and Biagi, A. 1986. *Il Nouvo Concise Dizionario Inglese e Italiano/Italian and English Dictionary.* Zanichelli/Longman, Bologna, Italy.

Rahman, M., Alimuzzaman, M., Ahmad, S., and Chowdhury, A. 2002. Antinociceptive and antidiarrhoeal activity of *Zanthoxylum rhetsa. Fitoterapia* 73(4): 340.

Rajamohan, F., Venkatachalam, T.K., Irvin, J.D., and Uckun, F.M. 1999. Pokeweed antiviral protein isoforms PAP-I, PAP-II, and PAP-III depurinate RNA of human immunodeficiency virus (HIV)-1. *Biochemical and Biophysical Research Communications* 260(2):453–458.

Rajbhandari, I., Takamatsu, S., and Nagle, D.G. 2001. A new dehydrogeranylgeraniol antioxidant from *Saururus cernuus* that inhibits intracellular reactive oxygen species (ROS)-catalyzed oxidation within HL-60 cells. *Journal of Natural Products* 64(6):693–695.

Rakotovao, M., Voirin, B., Bayet, C., Favre-Bonvin, J., and Andriantsiferana, M. 1988. 3′-*O*-beta-xylosyltricetin, a novel flavone glycoside from *Trema humbertii. Phytochemistry* 27(8):2655–2656.

Raloff, J. 1992. Look what's hidden in the pawpaw. *Science News* 141:143.

Rankin, R.L. 1988. Quapaw: genetic and areal affiliations. Pp. 629–650. In: Shipley, W., Ed. *In Honor of Mary Haas.* From the Haas Festival Conference on Native American Linguistics. Mouton de Gruyter, Berlin.

Rao, C.V., Heidelberger, M., and Grosvenor, W.P. 1971. Immunochemical studies of mangle gum (*Rhizophora mangle* L.). *Immunochemistry* 8(7):657–663.

Rao, K.V. 1962. Chemical examination of the leaves of *Dodonea viscosa. Journal of the Indian Chemical Society* 39:561–562.

Rao, K.V. and Rao, N.S. 1990. Chemistry of *Saururus cernuus.* VI: Three new neolignans. *Journal of Natural Products* 53(1):212–215.

Rao, K.V. and Reddy, G.C. 1990. Chemistry of *Saururus cernuus.* V. Sauristolactam and other nitrogenous constituents. *Journal of Natural Products* 53(2):309–312.

Rao, K.V., Puri, V.N., Diwan, P.K., and Álvarez, F.M. 1987. Preliminary evaluation of manassantin A, a potential neuroleptic agent from *Saururus cernuus. Pharmacological Research Communications* 19(9):629–638.

Rao, K.V., Puri, V.N., and el-Sawaf, H.A. 1990. Further studies on the neuroleptic profile of manassantin A. *European Journal of Pharmacology* 179(3):367–376.

Rapisarda, A., Ragusa, S., and De Pasquale, A. 1993. Hepatotoxic effect of the leaves of some *Cordia* species. *Acta Horticulturae* 332:237–242.

Rastogi, N., Abaul, J., Goh, K.S., Devallois, A., Philogene, E., and Bourgeois, P. 1998. Antimycobacterial activity of chemically defined natural substances from the Caribbean flora in Guadeloupe. *FEMS Immunological and Medical Microbiology* 20(4):267–273.

Rath, G., Ndonzao, M., and Hostettmann, K. 1995. Antifungal anthraquinones from *Morinda lucida. International Journal of Pharmacognosy* 33(2):107–114.

Ray, S., Bagchi, D., Lim, P.M., Bagchi, M., Gross, S.M., Kothari, S.C., Preuss, H.G., and Stohs, S.J. 2001. Acute and long-term safety evaluation of a novel IH636 grape seed proanthocyanidin extract. *Research Communications in Molecular Pathology Pharmacology* 109(3–4):165–197.

Rea, A.M. 1997. *At the Desert's Green Edge. An Ethnobotany of the Gila River Pima.* University of Arizona Press, Tucson.

Reanmongkol, W., Matsumoto, K., Itharat, A., and Watanabe, H. 1994. Antinociceptive effects of *Randia siamensis* extract and its constituent, pseudoginsenoside-RP1 in experimental animals. *Biological and Pharmaceutical Bulletin* 17(9):1206–1209.

Record, S.J. 1926. The wood of *Krugiodendron ferreum* (Vahl) Urban. *Tropical Woods* 8:13–15.

Redaelli, C. and Santaniello, E. 1984. Major isoflavonoids of the Jamaican dogwood *Piscidia erythrina*. *Phytochemistry* 23(12):2976–2977.

Reddy, J.S., Rao, P.R., and Reddy, M.S. 2002. Wound healing effects of *Heliotropium indicum*, *Plumbago zeylanicum* and *Acalypha indica* in rats. *Journal of Ethnopharmacology* 79(2):249–251.

Reed, C.F. 1971. *Common Weeds of the United States*. Dover, New York.

Reichelt, J. and Strasnky, K. 1978. Steroidal sapogenins. I. Determination of hecagenins and tigogenins in *Agave sisalana* Perrine using gas chromatography [in Czech]. *Ceskoslovenska Farmacie* 7(3):136–139.

Reinders, M. 1993. Medicinal Plants and Their Uses and the Ideas about Illness and Healing among the Warao of Guyana. Thesis, Utrecht University, 261 pp.

Reko, B.P. 1945. *Mitobotánica zapoteca*. Tacubaya, Mexico.

Reveal, J.L. 1983. Proposal to conserve 9157. *Heliopsis* Persoon (1807) with the conserved type *Buphthalmum heliathoides* L. (Asteraceae). *Taxon* 32:353, 354.

Reveal, J.L. 1992a. *Gentle Conquest. The Botanical Discovery of North America with Illustrations from the Library of Congress*. Starwood Publishing, Washington, D.C.

Reveal, J.L. 1992b. Typification of the Linnaean species of *Zizania* (Poaceae). *Phytologia* 72(1):5–8.

Reynolds, B.D., Blackmon, W.J., Wickremesinhe, E., Wells, M.H., and Constantin, R.J. 1990. Domestication of *Apios americana*. Pp. 436–442. In: J. Janick, and J.E. Simon, Eds. *Advances in New Crops: Proceedings of the First National Symposium NEW CROPS*, Research, Development, Economics, Indianapolis, IN, Oct. 23–26, 1988. Timber Press, Portland, OR.

Rhodes, D.G. 1997. Menispermaceae. Pp. 295–299. In: Editorial Committee, Eds. *Flora of North America*. Vol. 3. Oxford University Press, New York.

Richards, J.H. and Koptur, S. 1993. Floral variation and distyly in *Guettarda scabra* (Rubiaceae). *American Journal of Botany* 80(1):31–40.

Rico-G., V., Garcia-F., J.G., Chemas, A., Puch, A., and Sima, P. 1990. Species composition, similarity, and structure of Mayan homegardens in Tixpeual and Tixcacaltuyub, Yucatán, México. *Economic Botany* 44:470–487.

Riel, M.A., Kyle, D.E., and Milhous, W.K. 2002. Efficacy of scopadulcic acid A against *Plasmodium falciparum in vitro*. *Journal of Natural Products* 65(4):614–615.

Rivera, M., Figueroa, V., Aguilo, R., Hernández, R., and Alarcon, M. 1980. Reporte de la toxicidad de la planta *Melanthera deltoidea* Mich (boton de plata). *Revista Cubana de Ciencias Veterinarias* 11(1):47–51.

Rivero-Cruz, J.F., Chávez, D., Hernandez Bautista, B., Anaya, A.L., and Mata, R. 1997. Separation and characterization of *Metopium brownei* urushiol components. *Phytochemistry* 45(5):1003–1008.

Robinson, J.P., Harris, S.A., and Juniper, B.E. 2001. Taxonomy of the genus *Malus* Mill. (Rosaceae) with emphasis on the cultivated apple, *Malus domestica* Borkh. *Plant Systematics and Evolution* 226(1–2):35–58.

Robinson, R.D., Williams, L.A., Lindo, J.F., Terry, S.I., and Mansingh, A. 1990. Inactivation of strongyloides stercoralis filariform larvae *in vitro* by six Jamaican plant extracts and three commercial anthelmintics. *West Indian Medical Journal* 39(4):213–217.

Robinson, R.W. and Decker-Walters, D.S. 1997. *Cucurbits*. CAB International, Oxon, U.K.

Rodríguez, A., Trujillo, M.J., Matheu, V., Baeza, M.L., Zapatero, L., and Martínez, M. Allergy to grape: a case report. 2001. *Pediatric Allergy and Immunology* 12(5):289–290.

Rodríguez, S.E. and Castro, O. 1996. Pharmacological and chemical evaluation of *Stachytarpheta jamaicensis* (Verbenaceae). *Revista de Biologia Tropical* 44 (2A):353.

Roeder, E., Breitmaier, E., Birecka, H., Frohlich, M.W., and Badzies-Crombach, A. 1991. Pyrrolizidine alkaloids of *Heliothropium spathulatum*. *Phytochemistry* 30 (5):1703–1706.

Roengsumran, S., Sookkongwaree, K., Jaiboon, N., Chaichit, N., and Petsom, A. 2002. Crystal structure of ipolamiide monohydrate from *Stachytarpheta indica*. *Analytical Sciences* 18(9):1063–1064.

Rogers, G.K. 1985. The genera of Phytolaccaceae in the southeastern United States. *Journal of the Arnold Arboretum* 66(1):1–37.

Rogers, G.K. 1987. The genera of Cinchonoideae (Rubiaceae) in the southeastern United States. *Journal of the Arnold Arboretum* 68(2):137–183.

Rohmer, M., Schwartz, A.C., and Anton, R. 1977. Sesquiterpenes from essential oil of *Amyris balsamifera*. *Phytochemistry* 16(6):773–774.

Rohwer, J.G. 1993. Lauraceae: *Nectandra*. In: *Flora Neotropica* 60:1–332. Botanical Garden, Bronx, NY.

Roig y Mesa, J.T. 1945. *Plantas Medicinales Aromáticas o Venenosas de Cuba*. Ministerio de Agricultura, Havana.

Rojas, A., Hernández, L., Pereda-Miranda, R., and Mata, R. 1992. Screening for antimicrobial activity of crude drug extracts and pure natural products from Mexican medicinal plants. *Journal of Ethnopharmacology* 35(3):275–283.

Rojas, A., Cruz, S., Ponce-Monter, H., and Mata, R. 1996. Smooth muscle relaxing compounds from *Dodonea viscosa*. *Planta Medica* 62(2):154–159.

Rojas, A., Bah, M., Rojas, J.I., Serrano, V., and Pacheco, S. 1999. Spasmolytic activity of some plants used by the Otomi Indians of Queretaro (Mexico) for the treatment of gastrointestinal disorders. *Phytomedicine* 6(5):367–371.

Rojas, A., Bah, M., Rojas, J.I., and Gutierrez, D.M. 2000. Smooth muscle relaxing activity of gentiopicroside isolated from *Gentiana spathacea*. *Planta Medica* 66 (8):765–767.

Rojas-H., N.M. and Coto-P., O. 1978. Antimicrobial properties of extracts from *Rhizophora mangle* L. [in Spanish]. *Revista Cubana Medicina Tropical* 30(3): 181–187.

Roman-Ramos, R., Flores-Saenz, J.L., Partida-Hernandez, G., Lara-Lemus, A., and Alarcon-Aguilar, F. 1991. Experimental study of the hypoglycemic effect of some antidiabetic plants. *Archivos de Investigacion Medica* (Mexico) 22(1):87–93.

Roman-Ramos, R., Alarcon-Aguilar, F., Lara-Lemus, A., and Flores-Saenz, J.L. 1992. Hypoglycemic effect of plants used in Mexico as antidiabetics. *Archives of Medical Research* 23(1):59–64.

Romans, B. [1775] 1961. *A Concise Natural History of East and West Florida.* Reprinted, with an introduction by L. Richardson, Pelican, New Orleans, LA.

Romics, I., Schmitz, H., and Frang, D. 1993. Experience in treating benign prostatic hypertrophy with *Sabal serrulata* for one year. *International Urology and Nephrology* 25(6):565–569.

Rosatti, T.J. 1986. The genera of Sphenocleaceae and Campanulaceae in the Southeastern United States. *Journal of the Arnold Arboretum* 67:1–64.

Rosatti, T.J. 1989. The genera of suborder Apocyninae (Apocynaceae and Asclepiadaceae) in the southeastern United States. *Journal of the Arnold Arboretum* 70(3):307–401.

Rosengarten, F.J. 1969. *The Book of Spices.* Livingston, Wynnewood, PA.

Rosengarten, F.J. 1984. *The Book of Edible Nuts.* Walker, New York.

Rosenkranz, S., Knirel, D., Dietrich, H., Flesch, M., Erdmann, E., and Bohm, M. 2002. Inhibition of the PDGF receptor by red wine flavonoids provides a molecular explanation for the "French paradox." *FASEB Journal* 16(14):1958–1960.

Rosenthal, G.A. and Berge, M.A. 1989. Catabolism of l-canavanine and l-canaline in the jack bean, *Canavalia ensiformis* (L.) D.C. (Leguminosae). *Journal of Agricultural and Food Chemistry* 37(3):591–595.

Ross, K., Ed. 1978. *Codex Mendoza, Aztec Manuscript.* Miller Graphics, Productions Liber, Fribourg.

Rouw, A. de. 1991. The invasion of *Chromolaena odorata* (L.) King and Robinson (ex *Eupatorium odoratum*), and competition with the native flora, in a rain forest zone, south-west Cote d'Ivoire. *Journal of Biogeography* 18(1):13–23.

Rozsa, Z., Hohmann, J., Reisch, J., Mester, I., and Szendrei, K. 1982. Amorinin, a prenylated chromenoflavanone from *Amorpha fruticosa. Phytochemistry* 21(7): 1827–1828.

Ruangrungsi, N., Sekine, T., Phadungcharoen, T., Suriyagan, S., and Murakoshi, I. 1995. Isocoumarins from *Xyris indica. Phytochemistry* 450(5622):481–483.

Rudd, V.E. 1969. A synopsis of the genus *Piscidia* (Leguminosae). *Phytologia* 18(8):473–499.

Rudes, B.A. and Costa, D.J., Eds. 2003. *Essays in Algonquian, Catawban and Siouan Linguistics in Memory of Frank T. Siebert, Jr.* Memoir Algonquian and Iroquoian Linguistics 16. Algonquian and Iroquoian Linguistics, Winnipeg.

Ruffa, M.J., Ferraro, G., Wagner, M.L., Calcagno, M.L., Campos, R.H., and Cavallaro, L. 2002. Cytotoxic effect of Argentine medicinal plant extracts on human hepatocellular carcinoma cell line. *Journal of Ethnopharmacology* 79(3):335–339.

Russell, A.B., Hardin, J.W., Grand, L., and Fraser, A. 1997. Poisonous Plants of North Carolina, North Carolina State University. Computer programming, M.A. Buendia; graphics, B. Capel. URL: http://www.ces.ncsu.edu/depts/hort/consumer/poison/indcoa_e.htm.

Russell, F. 1908. The Pima Indians. *Annual Report, Bureau of American Ethnology* 26:3–389. Reprinted 1975, University of Arizona Press, Tucson.

Russell, N.H. 1965. Violets (*Viola*) of central and eastern United States: an introductory survey. *Sida* 2(1):1–113.

Rutten, A.M. and Statius van Eps, L.W. 1998. Poisoning with toxic plants in Curacao in 1766 [in Dutch]. *Nederlandesch Tijdschrift Geneeskundige* 142(51): 2796–2798.

Rwangabo, P.C., Claeys, M., Pieters, L., Corthout, J., and Vanden Berghe, D.A. 1988. Umuhengerin, a new antimicrobially active flavonoid from *Lantana trifolia. Journal of Natural Products* 51(5):966–968.

Saadoun, I. and Hameed, K.M. 1999. Antibacterial activity of *Orobanche cernua* extract. *Journal of Basic Microbiology* 39(5–6):377–380.

Sachdev, K. and Kulschreshtha, D.K. 1983. Flavonoids from *Dodonea viscosa. Phytochemistry* 22:1253–1256.

Saenz, C., Estevez, A.M., Sepulveda, E., and Mecklenburg, P. 1998. Cactus pear fruit: a new source for a natural sweetener. *Plant Foods and Human Nutrition* 52(2): 141–149.

Sahagún, Bernardino de. [1590] 1964. *Códices matritenses de la Historia general de las cosas de la nueva España.* Reprint: Ediciones José Porrua Turanzas, Madrid.

Saharia, G.S. and Seshadri, V. 1980. Chemical investigation on *Randia* saponins, isolation and characterisation of randioside A. *Indian Journal of Forestry* 3(1):6–8.

Sahpaz, S., Gupta, M.P., and Hostettmann, K. 2000. Triterpene saponins from *Randia formosa. Phytochemistry* 54(1):77–84.

Sahu, N.P. and Mahato, S.B. 1994. Antiinflammatory triterpene saponins of *Pithecellobium dulce*: characterization of an echinocystic acid bisdesmoside. *Phytochemistry* 37(5):1425–1427.

Sainty, D., Delaveau, P., Bailleul, F., and Moretti, C. 1982. 10-cafeyl desacetyldaphylloside nouvel iridoide de *Randia formosa. Journal of Natural Products* 45(6): 676–678.

Sakurai, K., Ishii, H., Kobayashi, S., and Iwao, T. 1976. Isolation of 4 beta-hydroxywithanolide E, a new withanolide from *Physalis peruviana* L. *Chemical and Pharmaceutical Bulletin* (Tokyo) 24(6):1403–1405.

Sakurawi, K., Yasuda, F., Tozyo, T., Nakamura, M., Sato, T., Kikuchi, J., Terui, Y., Ikenishi, Y., Iwata, T., Takahashi, K., Konoike, T., Mihara, S., and Fujimoto, M. 1996. Endothelin receptor antagonist triterpenoid, myriceric acid A, isolated from *Myrica cerifera*, and structure activity relationships of its derivatives. *Chemical and Pharmaceutical Bulletin* (Tokyo) 44(2):343–351.

Saleh, M., Kamel, A., Li, X., and Swaray, J. 1999. Antibacterial triterpenoids isolated from *Lantana camara. Le Pharmacien biologiste* 37(1):63–66.

Salinas, M.L., Ogura, T., and Soffchi, L. 2001. Irritant contact dermatitis caused by needle-like calcium oxalate crystals, raphides, in *Agave tequilana* among workers in tequila distilleries and agave plantations. *Contact Dermatitis* 44(2):94–96.

Salvat, A., Antonnacci, L., Fortunato, R.H., Suarez, E.Y., and Godoy, H.M. 2001. Screening of some plants from Northern Argentina for their antimicrobial activity. *Letters in Applied Microbiology* 32(5):293–297.

Sampaio, T. 1970. *Vocabulário Geográfico Brasileiro. O Tupi na Geografia Nacional.* Editora da Universidad de São Paulo, São Paulo.

Sampson, J.H., Phillipson, J.D., Bowery, N.G., O'Neill, M.J., Houston, J.G., and Lewis, J.A. 2000. Ethnomedicinally selected plants as sources of potential analgesic compounds: indication of *in vitro* biological activity in receptor binding assays. *Phytotherapy Research* 14 (1):24–29.

Sánchez Palacios, A. 2001. Latex allergy. Diagnosis and therapeutic aspects [in Spanish]. *Allergolia Immunopathologia* (Madrid) 29(5):212–221.

Sandved, K.B., Prance, G.T., and Prance, A.E. 1993. *Bark. The Formation, Characteristics, and Uses of Bark around the World.* Timber Press, Portland, OR.

Sang, S.M., Cheng, X.F., Zhu, N.Q., Wang, M.F., Jhoo, J. W., Stark, R.E., Badmaev, V., Ghai, G., Rosen, R.T., and Ho, C.T. 2001. Iridoid glycosides from the leaves of *Morinda citrifolia. Journal of Natural Products* 64 (6):799–800.

Sankaram, A.V.B., Rao, A., and Shoolery, J.N. 1979. Zeylanone and isozeylanone, two novel quinones from *Plumbago zeylanica. Tetrahedron* 35(14):1777–1782.

Sanoko, R., Speranza, G., Pizza, C., and De Tommasi, N. 1999. Triterpene saponins from *Alternanthera repens. Phytochemistry* 51(8):1043–1047.

Santa-Cruz, L.H., Turner, C.E., Knapp, J.E., Schiff, P.L., Jr., and Slatkin, D.J. 1975. Moretenol and other constituents of *Celtis laevigata. Phytochemistry* 14(11): 2532–2533.

Santamaría, F.J. 1959. *Diccionario de Mejicanismos.* Editorial Porrua, Mexico City.

Santamaría, F.J. and García Icazbalceta, J. 1974. *Diccionario de mejicanismos: razonado, comprobado con citas de autoridades, comparado con el de americanismos y con los vocabularios provinciales de los más distinguidos diccionaristas hispanamericanos.* 2nd ed. Editorial Porrúa, Mexico.

Santos, A.R., De Campos, R.O., Miguel, O.G., Cechinel-Filho, V., Yunes, R.A., and Calixto, J.B. 1999. The involvement of K+ channels and Gi/o protein in the antinociceptive action of the gallic acid ethyl ester. *European Journal of Pharmacology* 379(1):7–17.

Santos, W.R., Bernardo, R.R., Pecanha, L.M., Palatnik, M., Parente, J.P., and de Sousa, C.B. 1997. Haemolytic activities of plant saponins and adjuvants. Effect of *Periandra mediterranea* saponin on the humoral response to the FML antigen of *Leishmania donovani. Vaccine* 15(9):1024–1029.

Sanvordeker, D.R. and Chaubal, M.G. 1969. Essential oil of *Anemopsis californica.* II. Minor constituents. *Journal of Pharmaceutical Science* 58(10):1213–1217.

Sargent, C.S. 1905. *Manual of the Trees of North America.* Houghton Mifflin, Boston.

Sarkar, A., Pradhan, S., Mukhopadhyay, I., Bose, S.K., Roy, S., and Chatterjee, M. 1999. Inhibition of early DNA-damage and chromosomal aberrations by *Trianthema portulacastrum* L. in carbon tetrachloride-induced mouse liver damage. *Cell Biology International* 23(10):703–708.

Sarker, S.D., Whiting, P., Sik, V., and Dinan, L. 1999. Ecdysteroid antagonists (cucurbitacins) from *Physocarpus opulifolius* (Rosaceae). *Phytochemistry* 50(7): 1123–1128.

Sashida, Y., Kubo, S., Mimaki, Y., Nikaido, T., and Ohmoto, T. 1992. Steroidal saponins from *Smilax riparia* and *S. china. Phytochemistry* 31(7):2439–2443.

Sastry, K.N.S. and Nayudamma, N. 1966. Leucocyanidin from *Dodonea viscosa* bark. *Leather Science* (India) 13:147.

Sati, O.P., Pant, G., Nohara, T., and Sato, A. 1985. Cytotoxic saponins from *Asparagus* and *Agave. Pharmazie* 40(8):586.

Sati, O.P., Chaukiyal, D.C., Nishi, M., Miyahara, K., and Kawasaki, T. 1986. An iridoid from *Randia dumetorum. Phytochemistry* 25(11):2658–2660.

Sati, O.P., Bahuguna, S., Uniyal, S., and Bhakuni, D.S. 1989. Triterpenoid saponins from *Randia uliginosa* fruits. *Phytochemistry* 28(2):575–577.

Sato, M., Bagchi, D., Tosaki, A., and Das, D.K. 2001. Grape seed proanthocyanidin reduces cardiomyocyte apoptosis by inhibiting ischemia/reperfusion-induced activation of JNK-1 and C-JUN. *Free Radical Biology and Medicine* 31(6):729–737.

Satpathi, B. 1984. *Simarouba glauca*: an oil-yielding tree. *Indian Farming* 34(6):22–23.

Sauer, C.O. 1969. *The Early Spanish Main.* University of California Press, Berkeley.

Sauer, J.D. 1964. Revision of *Cavanalia. Brittonia* 16:106–181.

Savolainen, V., Chase, M.W., Hoot, S.B., Morton, C.M. Soltis, D.E., Boyer, C., Fay, M.F., de Bruijn, A.Y., Sulivan, S., and Qui, Y.-L. 2000. Phylogenetics of flowering plants based on combined analysis of plastid *atpB* and *rbcL* sequences. *Systematic Botany* 49:306–362.

Saxena, M.K. 2000. Aqueous leachate of *Lantana camara* kills water hyacinth. *Journal of Chemical Ecology* 26 (10):2435–2447.

Sazima, M. and Sazima, I. 1989. Oil-gathering bees visit flowers of eglandular morphs of the oil-producing Malpighiaceae. *Botanica Acta* 102(1):106.

Schapoval, E.E., Vargas, M.R., Chaves, C.G., Bridi, R., Zuanazzi, J.A., and Henriques, A.T. 1998. Anti-inflammatory and antinocioceptive activities of extracts and isolated compounds from *Stachytarpheta cayennensis. Journal of Ethnopharmacology* 60 (1):53–59.

Schapoval, E.E.S., Schenekl, E.P., Chaves, C.M.G., Shenkel, I.L., Zanata, A., Mentz, L.A., and Irgang, B.E. 1983. Ensaios químicos e farmacológicos com a *Chiococca alba* (L.) Hitch. Nota Prévia. *Anais 2º Simpronan (João Pessoa)* 289–294.

Schilcher, H. 1998. Herbal drugs in the treatment of benign prostatic hyperplasia. Pp. 62–73. In: Lawson, L.D. and Bauer, R., Eds. *Phytomedicines of Europe. Chemistry and Biological Activity.* ACS Symposium Series 691, American Chemical Society, Washington, D.C.

Schilcher, H. 1999. Is there a rational therapy for symptomatic treatment of benign prostatic hyperplasia with phytogenic drugs? Illustrated with the example of the prostate agent from *Serenoa repens* (Sabal fructus) [in German]. *Wiener Klinische Wochenschrift* 149(8–10):236–240.

Schimming, T., Tofern, B., Mann, P., Richter, A., Jennett-Siems, K., Drager, B., Asano, N., Gupta, M.P., Correa, M.D., and Eich, E. 1998. Distribution and taxonomic significance of calystegines in the Convolvulaceae. *Phytochemistry* 49(7):1989–1995.

Schimming, T., Jenett-Siems, K., Siems, K., Witte, L., Gupta, M.P., and Eich, E. 1999. Animartinines, novel piperidine alkaloids from *Aniseia martinicensis* (Convolvulaceae). Joint meeting of the American Society of Pharmacognosy, Association Française pour l'Enseignement et la Recherche en Pharmacognosie, Gesellschaft für Arzneipflanzenforschung and the Phytochemical Society of Europe, July 28–30 Amsterdam, the Netherlands.

Schirarend, C. 1987. Zur Holz- und Blattanatomie der neotropischen Gattung *Krugiodendron* Urban (Rhamnaceae) [Wood and leaf anatomy of the Neotropical genus *Krugiodendron* Urban (Rhamnaceae)]. *Feddes Repertorium* 98(9/10):515–519.

Schlemper, S.R, Schlemper, V., da Silva, D., Cordeiro, F., Cruz, A.B., Oliveira, A.E., and Cechinel-Filho, V. 2001. Antibacterial activity of *Persea cordata* stem barks. *Fitoterapia* 72(1):73–75.

Schlesinger, V. 2002. *Animals and Plants of the Ancient Maya. A Guide.* University of Texas Press, Austin.

Schmutz, E.M. and Hamilton, L.B. 1979. *Plants That Poison. An Illustrated Guide to Plants Poisonous to Man.* Northland Press, Flagstaff, AZ.

Schneider, D., Wink, M., Sporer, F., and Lounibos, P. 2002. Cycads: their evolution, toxins, herbivores and insect pollinators. *Naturwissenschaften* 89(7):281–294.

Schneider, H.J., Honold, E., and Masuhr, T. 1995. Treatment of benign prostatic hyperplasia. Results of a treatment study with the phytogenic combination of *Sabal* extract WS 1473 and *Urtica* extract WS 1031 in urologic specialty practices [in German]. *Fortschritte in der Medizin* 113(3):37–40.

Schnell, D.E. 1976. *Carnivorous Plants of the United States and Canada.* John F. Blair, Winston-Salem, NC.

Scholz, E. and Rimpler, H. 1989. Proanthocyanins from *Krameria triandra* root. *Planta Medica* 55:379–384.

Scholz, E., Heinrich, M., and Hunkler, D. 1990. New caffeic acid esters from *Pluchea symphytifolia* and their antimicrobial activity. *Planta Medica* 56(6):648.

Scholz, E., Heinrich, M., and Hunkler, D. 1994. Caffeoylquinic acids and some biological activities of *Pluchea symphytifolia. Planta Medica* 60(4):360–364.

Schöpf, J.D. [1798] 1903. *Materia Medica Americana Potissimum Regni Vegetabilis.* Reprinted by Lloyd Library Bulletin 6, Reproduction Series 3, Cincinnati, OH.

Schouten, A., Wagemakers, L., Stefanato, F.L., van der Kaaij, R.M., and van Kan, J.A. 2002. Resveratrol acts as a natural profungicide and induces self-intoxication by a specific laccase. *Molecular Microbiology* 43(4):883–894.

Schramm, L.C. 1978. The coumarin constituents of *Zizia aptera. Lloydia* 41(4):381–382.

Schultes, R.E. and Hofmann, A. 1979. *Plants of the Gods.* McGraw-Hill, New York.

Schultes, R.E. and Raffauf, R.F. 1990. *The Healing Forest. Medicinal and Toxic Plants of the Northwest Amazonia.* Dioscorides Press, Portland, OR.

Schultes, R.E. and Raffauf, R.F. 1992. Vine of the Soul. Medicine Men, Their Plants and Rituals in Colombian Amazonia. Syngeretic Press, Oracle, AZ.

Schultes, R.E. and von Reis, S., Eds. 1995. *Ethnobotany. Evolution of Discipline.* Dioscorides Press, Portland, OR.

Schuster, A., Castro, V., Poveda, L., Papastergiou, F., and Jakupovic, J. 1992. Sesquiterpene lactones from *Koanophyllon albicaule. Phytochemistry* 31 (89):3145–3146.

Scott, J.A. 1986. The Butterflies of North America. A Natural History and Field Guide. Stanford University Press, Stanford, CA.

Scott, M. 1995. *Scottish Wild Flowers.* HarperCollins, London.

Seaforth, C.E., Mohammed, S., Maxwell, A., Tinto, W.F., and Reynolds, W.F. 1992. Mabioside A, a new saponin from *Colubrina elliptica. Tetrahedron Letters* 33(29):4111–4114.

Sears, W.H. 1982. Fort Center, An Archaeological Site in the Lake Okeechobee Basin. University Press of Florida, Gainesville.

Secoy, D.M. and Smith, A.E. 1983. Use of plants in control of agricultural and domestic pests. *Economic Botany* 34:28–57.

Seeram, N.P., Momin, R.A., Nair, M.G., and Bourquin, L. D. 2001. Cyclooxygenase inhibitory and antioxidant cyanidin glycosides in cherries and berries. *Phytomedicine* 8(5):362–369.

Segura, J.J., Zavala, A., Medinilla, G., Del-Valle, J., Puc-Dzul, M., Guerrero, A., and Calzado-Flores, C. 1991. Preliminary toxicological evaluation of *Agave* sp. "amole." *Proceedings of the Western Pharmacological Society* 34:69–70.

Segura, L., Freixa, B,. Ringbom, T., Vila, R., Perera, P., Adzet, T., Bohlin, L., and Canigueral, S. 2000. Antiinflammatory activity of dichloromethane extract of *Heterotheca inuloides in vivo* and *in vitro. Planta Medica* 66(6):553–555.

Semple, J.C. 1977. Cytotaxonomy of *Chrysopsis* and *Heterotheca* (Compositae-Astereae): a new interpretation of phylogeny. *Canadian Journal of Botany* 55:2503–2513.

Semple, J.C. 1978. A revision of the genus *Borrichia* Adans. (Compositae). *Annals of the Missouri Botanical Garden* 65:681–693.

Semple, J.C. 1981. A revision of the goldenaster genus *Chrysopsis* (Nutt.) Ell. nom. cons. (Compositae-Astereae). *Rhodora* 83:323–384.

Semple, J.C. 1992. The goldenrods of Ontario: *Solidago* and *Euthamia*. *University of Waterloo Biology Series* 36: 1–82.

Semple, J.C. 1993. *Euthamia*. In: Hickman, C., Ed. *The Jepson Manual of Higher Plants of California*. University of California Press, Berkeley.

Semple, J.C. 1996. A revision of *Heterotheca* sect. *Phyllotheca* (Nutt.) Harms (Compositae: Astereae): the prairie and montane goldenasters of North America. *University of Waterloo Biology Series* 37:1–1164.

Semple, J.C. and Bowers, F.D. 1985. A revision of the goldenaster genus *Pityopsis* Nutt. (Compositae: Astereae). *University of Waterloo Biology Series* 29:1–34.

Semple, J.C. and Bowers, F.D. 1987. Cytogeography of *Pityopsis* Nutt., the grass-leaved goldenasters (Compositae: Astereae). *Rhodora* 89 (860):381–389.

Semple, J.C., and K.S. Semple. 1977. *Borrichia* × *cubana* (*B. frutescens* × *arborescens*) interspecific hybridization in the Florida Keys. *Systematic Botany* 2:292–301.

Semple, J.C., Blok, V.C., and Heiman, P. 1980. Morphological, anatomical, habit, and habitat differences among the goldenaster genera, *Chrysopsis, Heterotheca*, and *Pityopsis* (Compositae-Astereae). *Canadian Journal of Botany* 58:147–163.

Semple, S.J., Reynolds, G.D., O'Leary, M.C., and Flower, R. L. 1998. Screening of Australian medicinal plants for antiviral activity. *Journal of Ethnopharmacology* 60 (2):163–172.

Semple, S.J., Nobbs, S.F., Pyke, S.M., Reynolds, G.D., and Flower, R.L.P. 1999. Antiviral flavonoid from *Pterocaulon sphacelatum*, an Australian Aboriginal medicine. *Journal of Ethnopharmacology* 68(1/3):283–288.

Sen, S., Makkar, H.P., Muetzel, S., and Becker, K. 1998. Effect of *Quillaja saponaria* saponins and *Yucca schidigera* plant extract on growth of *Escherichia coli*. *Letters in Applied Microbiology* 27(1):35–38.

Sen, T. and Nag Chaudhuri, A.K. 1990. Further studies on the antiinflammatory profile of *Pluchea indica* root extracts. *Planta Medica* 56(6):661–662.

Sen, T. and Nag Chaudhuri, A.K. 1991. Antiinflammatory evaluation of a *Pluchea indica* root extract. *Journal of Ethnopharmacology* 33(1–2):135–141.

Sen, T., Ghosh, T.K., and Chaudhuri, A.K. 1993a. Studies on the mechanism of antiinflammatory and antiulcer activity of *Pluchea indica* — probable involvement of 5-lipooxygenase pathway. *Life Science* 52 (8):737–743.

Sen, T., Basu, A., Ray, R.N., and Nag Chaudhuri, A.K. 1993b. Hepatoprotective effects of *Pluchea indica* (Less) extract in experimental acute liver damage in rodents. *Phytotherapy Research* 7(5):352–355.

Sener, B. and Turkoz, S. 1987. High-pressure liquid chromatographic separation of steroidal sapogenins isolated from *Yucca aloifolia* L. *Journal de Pharmacologie Belgique* 42(4):270–271.

Senna, G., Mistrello, G., Roncarolo, D., Crivellaro, M., Bonadonna, P., Schiappoli, M., and Passalacqua, G. 2001. Exercise-induced anaphylaxis to grape. *Allergy* 56(12):1235–1236.

Sertie, J.A.A., Basile, A.C., Panizza, S., Matida, A.K., and Zelnik, R. 1990a. Antiinflammatory activity and sub-acute toxicity of artemetin. *Planta Medica* 56 (1):36–40.

Sertie, J.A.A., Basile, A.C., Panizza, S., Oshiro, T.T., Azzolini, C.P., and Penna, S.C. 1990b. Pharmacological assay of *Cordia verbenacea*. III. Oral and topical antiinflammatory activity and gastrotoxicity of a crude leaf extract. *Journal of Ethnopharmacology* 31(2):239–247.

Settle, W.J. and Fisher, T.R. 1970. The varieties of *Silphium integrifolium*. *Rhodora* 72:536–543.

Shah, N.K. 1971. A study of an indigenous drug Maswin as an oral contraceptive. *Current Medical Practice* 15 (2):614–616.

Shapiro, K. and Gong, W.C. 2002. Natural products used for diabetes. *Journal of the American Pharmacists Association* (Washington) 42(2):217–226.

Sharma, H.C. 1984. Insect antifeedants and growth inhibitors from *Azadirachta indica* and *Plumbago zeylanica*. *Schriftenreihe der Gesellschaft für Technische Zusammenarbeit* 161:291–320.

Sharma, I., Gusain, D., and Dixit, V.P. 1991. Hypolipidaemic and antiatherosclerotic effects of plumbagin in rabbits. *Indian Journal of Physiology and Pharmacology* 35(1):10–14.

Sharma, O.P., Dawra, R.K., and Makkar, H.P.S. 1989. Toxicity of isolated Lantana (*Lantana camara* L.) constituents to male and female guinea pigs. *Veterinary and Human Toxicology* 31(1):10–13.

Sharma, O.P., Dawra, R.K., and Ramesh, D. 1990. A triterpenoid acid, lantadene D from *Lantana camara* var. *aculeata*. *Phytochemistry* 29(12):3961–3962.

Sharma, O.P., Vaid, J., and Sharma, P.D. 1991. Comparison of lantadenes content and toxicity of different taxa of the *Lantana* plant. *Journal of Chemical Ecology* 17(11):2283–2291.

Sharma, R.S., Rajalakshmi, M., Sharma, R.S., and Jeyaraj, D.A. 2001. Current status of fertility control methods in India. *Bioscience* 26(4 Suppl.):391–405.

Sheahan, M.C. and Chase, M.W. 1996. Phylogenetic relationships within Zygophyllaceae R. Br. based on morphological, anatomical, and *rbcL* DNA sequence data. *Botanical Journal of the Linnaean Society* 113:227–262.

Sheen, W.S., Tsai, I.L., Teng, C.M., Ko, F.N., and Chen, I.S. 1996. Indolopyridoquinazoline alkaloids with antiplatelet aggregation activity from *Zanthoxylum integrifoliolum. Planta Medica* 62(2):175–176.

Sheldon, P.S., Kekwick, R.G., Sidebottom, C., Smith, C.G., and Slabas, A.R. 1990. 3-Oxoacyl-(acyl-carrier protein) reductase from avocado (*Persea americana*) fruit mesocarp. *Biochemistry Journal* 271(3):713–720.

Sheridan, P.M. and Griesbach, R.J. 2001. Anthocyanidins of *Sarracenia* L. flowers and leaves. *HortScience* 36 (2):384.

Sherman-Broyles, S.L. 1997. *Ulmus.* Pp. 369–375. In: Editorial Committee, Eds. *Flora of North America.* Vol. 3. Oxford University Press, New York.

Shi, Y.-Q., Fukai, T., Chang, W.-J., Yang, P.-Q., Wang, F.-P., and Nomura, T. 2001a. Phenolic constituents of the root bark of Chinese *Morus australis. Natural Medicines* 55(3):143–146.

Shi, Y.-Q., Fukai, T., Sakagami, H., Chang, W.-J., Yang, P.-Q., Wang, F.-P., and Nomura, T. 2001b. Cytotoxic flavonoids with isoprenoid groups from *Morus mongolica. Journal of Natural Products* 64(2):181–188.

Shimoma, F., Kondo, H., Yuuya, S., Suzuki, T., Hagiwara, H., and Ando, M. 1998. Enantioselective total syntheses of (−)-7betaH-eudesmane-4alpha,11-diol and (+)-ent-7betaH-eudesmane-4alpha,11-diol. *Journal of Natural Products* 61(1):22–28.

Siddiqui, A.A. 1998. Chemical and pharmacological evaluation of *Dodonaea viscosa* Linn. *Asian Journal of Chemistry* 10(1):14.

Siebert, F.T., Jr. 1975. Resurrecting Virginia Algonquian from the dead: the reconstituted and historical phonology of Powhatan. Pp. 285–453. In: Crawford, J.M., Ed. *Studies in Southeastern Indian Languages.* University of Georgia Press, Athens.

Sieren, D.J. 1981. The taxonomy of the genus *Euthamia. Rhodora* 83:551–579.

Siméon, R. [1885] 1981. *Diccionario de la Lengua Náhuatl o Mexicana.* Reprinted by Siglo Veintiuno Editores, Mexico City.

Simmonds, P.L. 1858. *The Dictionary of Trade Products, Manufacturing, and Technical Terms with a Definition of the Moneys, Weights, and Measures of All Countries.* G. Routledge, London.

Simmons, W.H. [1822] 1973. *Notices of East Florida.* Bicentennial Floridiana Facsimile Series, Reprint with introduction and index by G.E. Buker. University Press of Florida, Gainesville.

Simonetti, G. 1990. *Guide to Herbs and Spices.* Simon & Schuster, New York.

Simonetti, P., Ciappellano, S., Gardana, C., Bramati, L., and Pietta, P. 2002. Procyanidins from *Vitis vinifera* seeds: *in vivo* effects on oxidative stress. *Journal of Agriculture and Food Chemistry* 50(21):6217–6221.

Simpson, B.B. 1989. *Krameriaceae. Flora Neotropica* 49. New York Botanical Garden, Bronx, New York.

Simpson, B.B. 1991. The past present uses of Rhatany (*Krameria*, Krameriaceae). *Economic Botany* 45 (3):397–409.

Simpson, B.B. and Conner-Ogorzaly, M. 1995. *Economic Botany. Plants in Our World.* McGraw-Hill, New York.

Simpson, B.B. and Salywon, A. 1999. Krameriaceae. In: A New Flora of Arizona. *Journal of the Arizona-Nevada Academy of Sciences* 32(1):57–61.

Simpson, B.B., Weeks, A., Megan Helfgott, D., and Larkin, L.L. 2004. Species relationships in Krameria (Krameriaceae) based on ITS sequences and morphology: implications for character utility and biogeography. *Systematic Botany* 29:97–108.

Simpson, B.J. 1988. *A Field Guide to Texas Trees.* Texas Monthly Press, Austin.

Simpson, J.C. 1956. *A Provisional Gazetteer of Florida Place-Names of Indian Derivation.* Florida Geological Survey, Tallahassee.

Singh, J. and Tiwari, R.D. 1976. Flavone glycosides from the flowers of *Morinda citrifolia. Journal of the Indian Chemical Society* 53(4):424.

Singh, P.D., Simon, W.R., and West, M.E. 1984. Acute toxicity of seeds of the sapodilla (*Achras sapota* L.). *Toxicon* 22(1):145–147.

Singh, S.K., Tripathi, V.J., and Singh, R.H. 1990. 3 beta,24-Dihydroxyolean-12-en-28-oic acid, a pentacyclic triterpene acid from *Lantana indica. Phytochemistry* 29 (10):3360–3362.

Singletary, K.W. and Meline, B. 2001. Effect of grape seed proanthocyanidins on colon aberrant crypts and breast tumors in a rat dual-organ tumor model. *Nutrition and Cancer* 39(2):252–258.

Slauson, L.A. 2000. Pollination biology of two chiropterophilous agaves in Arizona. *American Journal of Botany* 87(6):825–836.

Sloane, H. 1696. *Catalogus plantarum quae in insual Jamaica sponte proveniunt.* London.

Sloane, H. 1707–1725. *The Natural History of Jamaica.* 2 vols. London.

Sluis, W.G. van der, Nat, J.M. van der, Spek, A.L., Ikeshiro, Y., and Labadie, R.P. 1983. Gentiogenal, a conversion product of gentiopicrin (Gentiopicroside). *Planta Medica* 49(4):211–215.

Small, J.K. 1913. *Flora of Miami.* Published by the author, New York.

Small, J.K. 1922. Wild pumpkins. *Journal of the New York Botanical Garden* 23:19–23.

Small, J.K. 1929. *From Eden to Sahara. Florida's Tragedy.* Science Press, Lancaster, PA.

Small, J.K. 1933. *Manual of the Southeastern Flora.* University of North Carolina Press, Chapel Hill.

Smartt, J. and Simmonds, N.W. 1995. *Evolution of Crop Plants.* 2nd ed. Longman Scientific and Technical, Essex, U.K.

Smith, A.C. 1940. The American species of Hippocrateaceae. *Brittonia* 3(3):341–555.

Smith, A.C. 1985. *Flora Vitiensis nova. A New Flora of Fiji (Spermatophytes Only).* Vol. 3. Pacific Tropical Botanical Garden, Lawaii, HI.

Smith, B.D. 1984. *Chenopodium* as a prehistoric domesticate in Eastern North America: evidence from Russell Cave, Alabama. *Science* 226(4671):165–167.

Smith, B.D. 1985. The role of *Chenopodium* as a domesticate in premaize garden systems of the eastern United States. *Southeastern Archaeology* 4:51–72.

Smith, B.D. 1992. *Rivers of Change: Essays on Early Agriculture in Eastern North America.* Smithsonian Institution Press, Washington, D.C.

Smith, B.D. and Funk, V.A. 1985. A newly described subfossil cultivar of *Chenopodium* (Chenopodiaceae). *Phytologia* 57(7):445–448.

Smith, C.E., Jr. 1968. The New World centers of origin of cultivated plants and the archaeological evidence. *Economic Botany* 22:253–266.

Smith, E.B. 1994. *Keys to the Flora of Arkansas.* University of Arkansas Press, Fayetteville.

Smith, H.H. 1923. Ethnobotany of the Menomini Indians. *Bulletin of the Public Museum of the City of Milwaukee* 4(1):1–174

Smith, H.H. 1928. Ethnobotany of the Meskwaki Indians. *Bulletin of the Public Museum of the City of Milwaukee* 4:175–326.

Smith, H.H. 1933. Ethnobotany of the Forest Potawatomi Indians. *Bulletin of the Public Museum of the City of Milwaukee* 7(1):1–230.

Smith, L.B. and Downs, R.J. 1977. *Tillandsioideae (Bromeliaceae).* Organization for Flora Neotropica/Hafner Press, New York.

Smith, L.W. and Culvenor, C.C.J. 1981. Plant sources of hepatotoxic pyrrholizidine alkaloids. *Journal of Natural Products* 44:129–152.

Snow, A.M. and Stans, S.E. 2001. *Healing Plants. Medicine of the Florida Seminole Indians.* University Press of Florida, Gainesville.

Sokeland, J. 2000. Combined sabal and urtica extract compared with finasteride in men with benign prostatic hyperplasia: analysis of prostate volume and therapeutic outcome. *British Journal of Urology* 86(4):439–442.

Sokeland, J. and Albrecht, J. 1997. Combination of *Sabal* and *Urtica* extract vs. finasteride in benign prostatic hyperplasia (Aiken stages I to II). Comparison of therapeutic effectiveness in a one year double-blind study [in German]. *Der Urologe. Ausgabe A.* 36 (4):327–333.

Solbakken, A.M., Rorbakken, G., and Gundersen, T. 1997. Nature medicine as intoxicant [in Norwegian]. *Tidsskrift for den Norske Laegeforening* 117(8): 1140–1141.

Soleas, G.J., Diamandis, E.P., and Goldberg, D.M. 2001. The world of resveratrol. *Advances in Experimental Medicine and Biology* 492:159–182.

Solis, P.N., Lang'at, C., Gupta, M.P., Kirby, G.C., Warhusrst, D.C., and Phillipson, J.D. 1995. Bio-active compounds from *Psychotria camponutans. Planta Medica* 61(1):62–65.

Solomon, J. 2003. TROPICOS. URL: http://mobot.mobot.org/W3T/Search/vast.html.

Somleva, T. and Ognyanov, I. 1985. New rotenoids in *Amorpha fruticosa* fruits. *Planta Medica* 49(3): 219–221.

Sotheeswaran, S., Bokel, M., and Kraus, W. 1989. A hemolytic saponin, randianin, from *Randia dumetorum. Phytochemistry* 28(5):1544–1546.

Sousa, M.P., Matos, M.E.O., Machado, M.I.L., Braz Filho, R., Vencato, I., and Mascarenhas, Y.P. 1984. Triterpenoids from *Guettarda angelica. Phytochemistry* 23 (11):2589–2592.

Soustelle, J. 1962. The Daily Life of the Aztecs on the Eve of the Spanish Conquest. Macmillan, New York.

Souza-Novelo, N. [1940] 1975. *Plantas meliferas y poliniferas que viven en Yucatán.* Edicion facsimilar reacondicionada. J. Díaz-Bolio, Mérida, Yucatán, Mexico.

Sow, A.I., Koyalta, D., Boye, C.S., Diedhiou-Badiane, D., and Bassene, E. 1995. Antibacterial activity of essential oils from mint in *Senegal Dakar [French]. Medicine* 40(2):193–195.

Speck, F.G. 1909. *Ethnology of the Yuchi Indians.* Anthropological Publications of the University Museum, University of Pennsylvania, Vol. 1, No. 1. Philadelphia.

Speck, F.G. 1934. *Catawba Texts.* Columbia University Contributions to Anthropology. Vol. XXIV. Columbia University Press, New York.

Speck, F.G. 1937. Catabawa medicines and curing practice. *Publications of the Philadelphia Anthropological Society* 1:179–198 (25th Anniversary Studies, D.S. Davidson, Ed.).

Speck, F.G. 1941. A list of plant curatives obtained from the Houma Indians of Louisiana. *Primitive Man* 14:49–73.

Speck, F.G. 1944. Catabawa herbals and curing practice. *Journal of American Folk-lore* 57(223):37–50.

Spencer, C.F., Koniuszy, F.R., Rogers, E.F., Shavel, Jr., J., Easton, N.R., Kaczka, E.A., Kuehl, Jr., F.A., Phillips, R.F., Walti, A., Folkers, K., Malanga, C., and Seeler, A.O. 1947. Survey of plants for antimalarial activity. *Lloydia* 10:145–174.

Speroni, E. and Minghetti, A. 1988. Neuropharmacological activity of extracts from *Passiflora incarnata. Planta Medica* 54(6):488–491.

Spjut, R.W. 1994. *A Systematic Treatment of Fruit Types.* Memoirs of the New York Botanical Garden Vol. 70, New York.

Spoehr, A. 1939. Field notes on Florida Seminole ethnography. Quoted in Sturtevant 1955.

Srivastava, M. and Singh, J. 1993. A new anthraquinone glycoside from *Morinda citrifolia. International Journal of Pharmacognosy* 31(3):182–184.

Srivastava, S.K. and Srivastava, S.D. 1979. Taxifolin 3,5-dirhamnoside from the seeds of *Cordia obliqua. Phytochemistry* 18(12):2058–2059.

Srivastava, V., Varma, N., Tandon, J.S., and Srimal, R.C. 1990. Antiinflammatory activity of *Pluchea lanceolata*: isolation of an active principle. *International Journal of Crude Drug Research* 28(2):135–137.

Stadler, P., van Rensburg, I.B., and Naude, T.W. 1991. Suspected avocado (*Persea americana*) poisoning in goats. *Journal South African Veterinary Association* 62(4):186–188.

Standifer, M.S., Kuttruff, J., and Tucker, S.C. 2003. Plant padding in prehistoric footwear from Arnold Research Cave, Missouri. Talk given at 44th Annual Meeting, Society for Economic Botany, Tucson, AZ.

Standley, P.C. [1920–1926] 1971. *Trees and Shrubs of Mexico.* Reprinted from Contributions U.S. Herbarium Vol. 23, Smithsonian Press, Washington, D.C.

Standley, P.C. 1926. The wood of *Krugiodendron ferreum* (Vahl) Urban. *Tropical Woods* 8:131.

Standley, P.C. 1928. Flora of the Panama Canal Zone. *Contributions from the United States National Herbarium* 27:1–416.

Standley, P.C. 1931. *Flora of the Lancetilla Valley Honduras.* Field Museum of Natural History, Publication 283, Botany Series Vol. 10, Chicago.

Standley, P.C. 1937. *Flora of Costa Rica,* Part 1. Field Museum of Natural History, Publ. 391, Chicago.

Standley, P.C. and Record, S.J. 1936. *The Forests and Flora of British Honduras.* Field Museum of Natural History, Chicago.

Standley, P.C. and Steyermark, J.A. 1946. Flora of Guatemala, Part 4. *Fieldiana: Botany* 24(4):1–493.

Steck, W. and Wetter, L.R. 1974. Apterin, an unusual glucoside of *Zizia aptera. Phytochemistry* 13(9): 1925–1927.

Steggerda, M. 1943. Some ethnological data concerning one hundred Yucatan plants. *Anthropological Papers 29, Bulletin 29, Bureau of American Ethnology,* Smithsonian Institution, Washington, D.C, 193–226.

Stehlin, I.B. 1996. Outsmarting poison ivy and its cousins. *FDA Consumer* 30(7):25–28.

Steiner, K.E. and Whitehead, V.B. 1991. Oil flowers and oil bees: further evidence for pollinator adaptation. *Evolution* 45(6):1493.

Stermitz, F.R. 1998. Iridoid glycosides of *Cornus canadensis*: a comparison with some other *Cornus* species. *Biochemical Systematics and Ecology* 26 (8):845–849.

Stevens, W.D., Ulloa Ulloa, C., Pool, A., and Montriel, O. M., Eds. 2001. *Flora de Nicaragua. Introduccion, Gimnospermas y Angiospermas.* Vol. 1. Missouri Botanical Garden Press, St. Louis.

Stevenson, D.W. 1987. Again the West Indian Zamias. *Bulletin of the Fairchild Tropical Garden* 42(3):23–27.

Steyermark, J.A. 1963. *Flora of Missouri.* Iowa State University Press, Ames.

Steynberg, J.P. and Hemingway, R.W. 1994. Condensed tannins: must enzymes control the stereochemistry in profisetinidin biosynthesis? *Proceedings of the Third Annual Southern Station Chemical Sciences Meeting,* February 7–8, 1990, Alexandria, edited by T.G. Rials. Louisiana. General Technical Report SO/ Mar. 101:58–60.

Stirling, G.M. 1941. Preliminary report on the Mikisuki Division of the Seminole Indians of Florida. Part I. Myths and Dictionary of the Mikisuki Language. Mss. in Peabody Museum, Harvard University, Cambridge, MA.

Stohs, S.J., Sabatka, J.J., Obrist, J.J., and Rosenberg, H. 1974. Sapogenins of *Yucca glauca* tissue cultures. *Lloydia* 37(3):504–505.

Stohs, S.J., Rosenberg, H., and Billets, S. 1975. Sterols of *Yucca glauca* tissue cultures and seeds. *Planta Medica* 27(3):257–261.

Stormer, T. 1993. Penn Center, South Carolina. URL: www.angelfire.com/sc/jhstevens/penncenter.html.

Strachey, Wm. [1612] 1953. *The Historie of Travell into Virginia Britania.* London. (Wright, L.B. and Freund, V., Eds. Reprinted by Hakluyt Society, London.)

Struttmann, J.M. 2002. Moths of North America, URL: www.npwrc.usgs.gov/resource/distr/lepid/moths/fl/1091.htm.

Stuart, K.L. and Graham, L. 1973. Alkaloid biosynthesis in *Croton linearis. Phytochemistry* 12(8):1973–1975.

Stucker, M., Memmel, U., Hoffmann, M., Hartung, J., and Altmeyer, P. 2001. Vitamin B(12) cream containing avocado oil in the therapy of plaque psoriasis. *Dermatology* 203(2):141–147.

Sturtevant, W.C. 1953. Chakaika and the "Spanish Indians": documentary sources compared with Seminole traditions. *Tequesta* 13:35–73.

Sturtevant, W.C. 1954. The medicine bundles and busks of the Florida Seminoles. *Florida Anthropologist* 7:31–70.

Sturtevant, W.C. 1955. *The Mikasuki Seminole: Medical Beliefs and Practices.* University Microfilms, Ann Arbor, MI.

Sturtevant, W.C. 1978. The last of the South Florida aborigines. Pp. 141–162. In: Milanich, J.T. and Proctor, S., Eds. *Tacachale: Essays on the Indians of Florida and Southeastern Georgia During the Historic Period.* University Press of Florida, Gainesville.

Sugiyama, M. and Kikuchi, M. 1991. Phenolic glycosides from *Osmanthus asiaticus. Phytochemistry* 30(9): 3147–3149.

Sugiyama, M. and Kikuchi, M. 1993. Phenylethanoid glycosides from *Osmanthus asiaticus. Phytochemistry* 32(6):1553–1555.

Sugiyama, M., Machida, K., Matsuda, N., and Kikuchi, M. 1993a. A secoiridoid glycoside from *Osmanthus asiaticus. Phytochemistry* 34(4):1169–1170.

Sugiyama, M., Nagayama, E., and Kikuchi, M. 1993b. Lignan and phenylpropanoid glycosides from *Osmanthus asiaticus. Phytochemistry* 33(5):1215–1219.

Suksamrarn, A., Wilkie, J.S., and Horn, D.H.S. 1986. Blechnosides A and B: ecdysteroid glycosides from *Blechnum minus. Phytochemistry* 25(6):1301–1304.

Sullivan, J.R. 1985. Systematics of the *Physalis viscosa* complex (Solanaceae). *Systematic Botany* 10(4): 426–444.

Sung, S.H. and Kim, Y.C. 2000. Hepatoprotective diastereomeric lignans from *Saururus chinensis* herbs. *Journal of Natural Products* 63(7):1019–1021.

Sung, S.H., Lee, E.J., Cho, J.H., Kim, H.S., and Kim, Y.C. 2000. Sauchinone, a lignan from *Saururus chinensis,* attenuates CCl4-induced toxicity in primary cultures of rat hepatocytes. *Biological and Pharmaceutical Bulletin* 23(5):666–668.

Sung, S.H., Huh, M.S., and Kim, Y.C. 2001. New tetrahydrofuran-type sesquilignans of *Saururus chinensis* root. *Chemical and Pharmaceutical Bulletin* (Tokyo) 49(9):1192–1194.

Svanidze, N., Sanchez, A., Lanovenki, V., Soler, B., Rodríguez, P., Suarez, A., and Méndez, G. 1974. Resultados de la intraducción y estudio farmacognosico de la *Passiflora incarnata* L. en las condiciónes de Cuba. *Revista Cubana de Farmacia* 8:309–314.

Swanton, J.R. 1928a. Religious beliefs and medical practices of the Creek Indians. *Annual Report of the Bureau of American Ethnology* 42:473–672.

Swanton, J.R. 1928b. Social and religious beliefs and usage of the Chickasaw. *Annual Report Bureau of American Ethnology* 44:169–273.

Swanton, J.R. 1929. *Myths and Tales of the Southeastern Indians.* Smithsonian Institution Bureau of American Ethnology, Bulletin 88, Washington, D.C.

Swanton, J.R. 1939. *Final Report of the United States De Soto Expedition Commission.* Smithsonian Institution Press, Washington, D.C. Reprint edition of the original as U.S. House of Representatives Document 71, 76th Congress, 1st Session, 1985, Smithsonian Institution Press, Washington, D.C.

Swanton, J.R. 1946. *The Indians of the Southeastern United States.* Bureau of American Ethnology, Bulletin 145, Reprint, Smithsonian Institution Press, Washington, D.C., 1979.

Swanton, J.R. 1931 [2001]. *Source Material for the Social and Ceremonial Life of the Choctaw Indians.* Contemporary American Indian Studies. University of Alabama Press, Tuscaloosa. Originally published: Smithsonian Institution, *Bulletin of the Bureau of American Ethnology* 103. Washington, 1931.

Swanton, J.R. 1953. Uchee or Yuchi Indians. Pp. 104–120. In: *Indians Tribes of North America.* Bureau of American Ethnology, Bulletin 137, Washington, D.C.

Sweeney, C.R. 1970. Monograph of the genus *Silphium.* I. *Silphium compositum* Michaux (Compositae). *The Ohio Journal of Science* 70(4):226–233.

Swerdlow, J.L. 2000. *Nature's Medicine. Plants That Heal.* National Geographic Society, Washington, D.C.

Sylestine, C., Hardy, H., and Montler, T. 1993. *Dictionary of the Alabama Language.* University of Texas Press, Austin.

Szendrei, K., Novak, I., Petz, M., Reisch, J., Bailey, H.E., and Bailey, V.L. 1973. Coumarins and alkaloids from *Ptelea trifoliata* ssp. *pallida* var. *confinis. Lloydia* 36(3):333–337.

Taglioli, V., Bilia, A.R., Ghiara, C., Mazzi, G., Mercati, V., and Vincieri, F.F. 2001. Evaluation of the dissolution behaviour of some commercial herbal drugs and their preparations. *Pharmazie* 56(11):868–870.

Tahara, S., Moriyama, M., Ingham, J.L., and Mizutani, J. 1991. 5,7,3′4′,5′-Pentaoxygenated and 2′,6′-diprenylated isoflavones from *Piscidia erythrina. Phytochemistry* 30(8):2769–2775.

Tahara, S., Moriyama, M., Ingham, J.L., and Mizutani, J. 1992. Structure revision of piscidone, a major isoflavonoid in the root bark of *Piscidia erythrina. Phytochemistry* 31(2):679–682.

Tahara, S., Moriyama, M., Ingham, J.L., and Mizutani, J. 1993a. Isoflavone atropisomers from *Piscidia erythrina. Phytochemistry* 34(2):545–552.

Tahara, S., Moriyama, M., Ingham, J.L., and Mizutani, J. 1993b. Diverse oxygenated isoflavonoids from *Piscidia erythrina. Phytochemistry* 34(1):303–315.

Takagi, K., Park, E.H., and Kato, H. 1980. Antiinflammatory activities of hederagenin and crude saponin isolated from *Sapindus mukorossi* Gaertn. *Chemical and Pharmaceutical Bulletin* (Tokyo) 28(4):1183–1188.

Takaishi, Y., Okuyama, T., Masuda, A., Nakano, K., Murakami, K., and Tomimatsu, T. 1990. Acetylenes from *Cirsium japonicum. Phytochemistry* 29 (12):3849–3852.

Takano, S. and Kawaminami, S. 1988. Cyperenyl acetate and cyperenal from *Cirsium dipsacolepis. Phytochemistry* 27(4):1197–1199.

Takeda, Y., Takechi, A., Masuda, T., and Otsuka, H. 1998. An acyclic monoterpene glucosyl ester from *Lantana lilacia. Planta Medica* 64 (1):78–79.

Tamura, Y., Mizutani, K., Ikeda, T., Ohtani, K., Kasai, R., Yamasaki, K., and Tanaka, O. 2001. Antimicrobial activities of saponins of pericarps of *Sapindus mukurossi* on dermatophytes. *Natural Medicines* 55 (1):11–16.

Tanaka, N., Tanaka, T., Fujioka, T., Fujii, H., Mihashi, K., Shimomura, K., and Ishimaru, K. 2001. An ellagic compound and iridoids from *Cornus capitata* root cultures. *Phytochemistry* 57(8):1287–1291.

Tanaka, O., Tamura, Y., Masuda, H., and Mizutani, K. 1996. Application of saponins in foods and cosmetics: saponins of Mohave yucca and *Sapindus mukurossi. Advances in Experimental Medicine and Biology* 405:1–11.

Tanaka, Y., Kawahara, S., Aik-Hwee, E., Shiba, K., and Ohya, N. 1995. Initiation of biosynthesis in *cis* polyisoprenes. *Phytochemistry* 39(4):779–784.

Taoubi, K., Fauvel, M.T., Gleye, J., Moulis, C., and Fouraste, I. 1997. Phenylpropanoid glycosides from *Lantana camara* and *Lippia multiflora. Planta Medica* 63(2):192–193.

Tapia-M., J.L. and Contréras, F. In press. *Evolvulus.* In: *Flora Ilustrada de la Peninsula de Yucatán.* Universidad Autónoma de Yucatán, Mérida, Yucatán, Mexico.

Tarayre, J.P., Delhon, A., Lauressergues, H., Stenger, A., Barbara, M., Bru, M., Villanova, G., Caillol, V., and Aliaga, M. 1983. Anti-edematous action of a hexane extract of the stone fruit of *Serenoa repens* Bartr [in French]. *Annales de Pharmacologie Françaises* 41 (6):559–570.

Tassou, C.C., Drosinos, E.H., and Nychas, G.J. 1995. Effects of essential oil from mint (*Mentha piperita*) on *Salmonella enteritidis* and *Listeria monocytogenes* in model food systems at 4 degrees and 10 degrees C. *Journal of Applied Bacteriology* 78(6):593–600.

Taylor, D. 1977. *Languages of the West Indies.* Johns Hopkins University Press, Baltimore, MD.

Taylor, L.A. 1940. *Plants Used as Curatives by Certain Southeastern Tribes.* Botanical Museum of Harvard University, Cambridge, MA.

Taylor, W.K. and Norman, E.M. 2002. *André Michaux in Florida. An Eighteenth-Century Botanical Journey.* University Press of Florida, Gainesville.

Tchamo, D.N., Cartier, G., Dijoux-Franca, M.G., Tsamo, E., and Mariotte, A.M. 2001. Xanthones and other constituents of *Trema orientalis. Le Pharmacien biologiste* 39(3):202–205.

Tebeau, C.W. 1955. *The Story of the Chokoloskee Bay Country.* University of Miami Press, Coral Gables.

Tebeau, C.W. 1966. *Florida's Last Frontier. The History of Collier County.* University of Miami Press, Coral Gables.

Tebeau, C.W. 1968. *Man in the Everglades, 2000 Years of Human History in the Everglades National Park.* University of Miami Press, Coral Gables.

Tedlock, D. 1985. *Popul Vuh: The Mayan Book of the Dawn of Life.* Simon & Schuster, New York.

Terrell, E.E., Peterson, P.M., and Duvall, M.R. 1997. Taxonomy of North American species of *Zizania* (Poaceae). *Sida* 17(3):533–549.

Texas A&M. 2000. Uvalde Research and Extension Center. URL: uvalde.tamu.edu/herbarium/rihu.htm.

Thomas, J.L. 1960. A monographic study of the Cyrillaceae. *Contributions to the Gray Herbarium of Harvard University* 186:1–114.

Thomas, R.K. 1954. The Origin and Development of the Redbird Smith Movement. M.A. thesis, University of Arizona, Tucson.

Tian, Q., Miller, E.G., Ahmad, H., Tang, L., and Patil, B.S. 2001. Differential inhibition of human cancer cell proliferation by citrus limonoids. *Nutrition and Cancer* 40(2):180–184.

Tian, Z., Chang, M.N., Sandrino, M., Huang, L., Pan, J.X., Arison, B., Smith, J., and Lam, Y.K.T. 1987. Quinones from *Ardisia cornudentata. Phytochemistry* 26(8):2361–2362.

Timbrook, J. 1990. Ethnobotany of the Cumash Indians, California, based on collections by John P. Harrington. *Economic Botany* 44:236–253.

Tincusi, B.M., Jiménez, I.A., Ravelo, A.G., and Missico, R. 1998. New sesquiterpenes from *Crossopetalum tonduzii. Journal of Natural Products* 61(12):1520–1523.

Tippo, O. and Stern, W.L. 1977. *Humanistic Botany.* W.W. Norton, New York.

Tiwari, K.P. and Srivastava, S.S.D. 1979. Chemical investigation of the stem bark of *Cordia obliqua. Planta Medica* 36(2):191–192.

Tiwari, R.D. and Tripathi, H L. 1976. The structure of an arabinogalactan from the leaves of *Symplocos spicata. Planta Medica* 29(4):376–379.

Tokarnia, C.H. and Dobereiner, J. 1982. Intoxicacão experimental por *Plumbago scandens* (Plumbaginaceae) em bovinos Brazil. *Pesquisa Veterinaria Brasileira* 2(3): 105–112.

Toledo, M.C.F., Reyes, F.G.R., Iaderoza, M., Francis, F.J. and Draetta, I.S. 1983. Anthocyanins from anil trepador (*Cissus sicyoides,* Linn.). *Journal of Food Science* 48(4):1368–1369.

Tomlinson, P.B. 1980. *The Biology of Trees Native to Tropical Florida.* Published by the author, Allston, MA.

Tomlinson, P.B. 1986. *The Botany of Mangroves.* Cambridge University Press, Cambridge, U.K.

Tona, L., Mesia, K., Ngimbi, N.P., Chrimwami, B., Okond'Ahoka, Cimanga, K., de Bruyne, T., Apers, S., Hermans, N., Totte, J., Pieters, L., and Vlietinck, A.J. 2001. *In-vivo* antimalarial activity of *Cassia occidentalis, Morinda morindoides* and *Phyllanthus niruri. Annals of Tropical Medicine and Parasitology* 95 (1):47–57.

Tooker, J.F., Koenig, W.A., and Hanks, L.M. 2002. Altered host plant volatiles are proxies for sex pheromones in the gall wasp *Antistrophus rufus. Proceedings of the National Academy of Sciences of the United States of America* 99(24):15486–15491.

Touchstone, S.J. 1983. *Herbal and Folk Medicine in Louisiana and Adjacent States.* Folk-Life Books, Princeton, LA.

Traverso, S.D. and Driemeier, D. 2000. Experimental *Trema micrantha* (Ulmaceae) poisoning in rabbits. *Veterinary Human Toxicology* 42(5):301–302.

Trejo-González, A., Gabriel-Ortiz, G., Puebla-Pérez, A.M., Huizar-Contreras, M.D., Munguia-Mazariegos, M. R., Mejia-Arreguin, S., and Calva, E. 1996. A purified extract from prickly pear cactus (*Opuntia fuliginosa*) controls experimentally induced diabetes in rats. *Journal of Ethnopharmacology* 55(1):27–33.

Trigger, B.G., Ed. 1978. *Handbook of North American Indians,* Vol. 15. *Northeast.* Smithsonian Institution, Washington, D.C.

Trigo, J.R., Witte, L., Brown, K.S., Jr., Hartmann, T., and Barata, L.E.S. 1993. Pyrrolizidine alkaloids in the arctiid moth *Hyalurga syma. Journal of Chemical Ecology* 19(4):669–679.

Trovato, A., Forestieri, A.M., Barbera, R., and Galati, E.M. 1988. Effects of *Trema guineense* decoction on some metabolic activities in the rat. *Pharmacological Research Communications* 20(Suppl. 5):63–65.

Trueman, K.F. and Powell, M.W. 1991. Suspected poisoning of camels by *Trema tomentosa* (poison peach). *Australian Veterinary Journal* 68(6):213–214.

Tsai, I.L., Hsieh, C.F., and Duh, C.Y. 1998. Additional cytotoxic neolignans from *Persea obovatifolia. Phytochemistry* 48(8):1371–1375.

Tsai, I.L., Lin, W.Y., Teng, C.M., Ishikawa, T., Doong, S.L., Huang, M.W., Chen, Y.C., and Chen, I.S. 2000. Coumarins and antiplatelet constituents from the root bark of *Zanthoxylum schinifolium. Planta Medica* 66(7):618–623.

Tschesche, R. and Reutel, I. 1968. Alkaloids from Sterculiaceae. I. Peptide alkaloids from *Melochia corchorifolia* [in German]. *Tetrahedron Letters* 35:3817–3818.

Tucker, A.O., Maciarello, M.J., Wofford, B.E., and Dennis, W.M. 1997. Volatile leaf oils of *Persea borbonia* (L.) Spreng., *P. humilis* Nash, and *P. palustris* (Raf.) Sarg. (Lauraceae) of North America. *Journal of Essential Oil Research* 9(2):209–211.

Tucker, G.C. 1988. The genera of Bambusoideae (Gramineae) in the southeastern United States. *Journal of the Arnold Arboretum* 69:239–273.

Tull, D. 1999. *Edible and Useful Plants of Texas and the Southwest. A Practical Guide.* University of Texas Press, Austin.

Turner, B.L. 1958. Chromosome numbers in the genus *Krameria*: evidence for familial status. *Rhodora* 60:101–106.

Turner, N.C. 1978. *Food Plants of British Columbia Indians.* Part 2. *Interior Peoples.* British Columbia Provincial Museum Handbook 36. Victoria.

Turner, N.C. and Bell, M.A.M. 1971. The ethnobotany of the Coastal Salish Indians of Vancouver Island. *Economic Botany* 25:63–104; appendix 335–339.

Turner, N.J. and Szczawinski, A.F. 1991. *Common Poisonous Plants and Mushrooms of North America.* Timber Press, Portland, OR.

Turner, W. [1548] 1965. *The Names of Herbes (with an introduction, an index of English names and an identification of the plants enumerated by Turner).* Britten, J., Ed. Published for the English Dialect Society by N. Trübner, London, 1881. Reprint by Kraus Reprint, Vaudz.

Tussac, F.R. 1808–1827. *Flore des Antilles.* Vols. 1–4. Paris.

Tutupalli, L.V. and Chaubal, M.G. 1975. Saururaceae. V. Composition of essential oil from foliage of *Houttuynia cordata* and chemo systematics of Saururaceae. *Lloydia* 38(2):92–96.

Tutupalli, L.V., Chaubal, M.G., and Malone, M.H. 1975. Saururaceae. VI. Hippocractic screening of *Anemopsis californica. Lloydia* 38(4):352–354.

Twain, M. [1897] 1992. *Following the Equator.* Reprinted by Ecco Press, Hopewell, NJ.

Tyler, J.D. and Jenkins, G.L. 1979. Notes on some fall foods of mourning doves in southwestern Oklahoma. *Proceedings of the Oklahoma Academy of Science* 59:82–84.

Tyler, V. 1993. *The Honest Herbal.* Pharmaceutical Products Press, New York.

Uchiyama, T., Miyase, T., Ueno, A., and Usmanghani, K. 1991. Terpene and lignan glycosides from *Pluchea indica. Phytochemistry* 30(2):655–657.

Udem, S., Madubunyi, Asuzu, I.U., and Anika, S.M. 1993. Trypanocidal effects of some tropical plant extracts — a review. *Acta Horticulturae* 344:538–543.

Udino, L., Abaul, J., Bourgeois, P., and Mauléon, H. 1994. Algunos resultados sobre la actividad insecticidal de *Canella winterana.* Primer Congreso Latinoamericano y del Caribe sobre el NIM y otros insecticidas vegetales, marzo 7–11.

Uematsu, Y., Hirata, K., Saito, K., and Kudo, I. 2000. Spectrophotometric determination of saponin in *Yucca* extract used as food additive. *Journal of Association of Official Agricultural Chemists International* 83(6):1451–1454.

Uenobe, F., Nakamura, S., and Miyazawa, M. 1997. Antimutagenic effect of resveratrol against Trp-P-1. *Mutation Research* 373(2):197–200.

Uesato, S., Ali, E., Nishimura, H., Kawamura, I., and Inouye, H. 1982. Four iridoids from *Randia canthioides. Phytochemistry* 21(2):353–357.

Uhegbu, F.O. and Maduagwu, E.N. 1995. Occurrence of nitrosatable amines in some Nigerian medicinal plants. *Bulletin of Environmental Contamination and Toxicology* 55(5):643–649.

Uncini Manganelli, R.E., Chericoni, S., and Baragatti, B. 2000. Ethnopharmacobotany in Tuscany: plants used as antihypertensives. *Fitoterapia* 71(Suppl. 1):95–100.

Uniyal, G.C., Agrawal, P.K., Sati, O.P., and Thakur, R.S. 1991. A spirostane hexaglycoside from *Agave cantala* fruits. *Phytochemistry* 30(12):4187–4189.

Uphof, J.C.T. 1968. *Dictionary of Economic Plants.* J. Cramer, Lehre, Germany.

Urteaga, H.H. 1978. El arte de navaegar entre los antiguosperuanos. Pp. 659–676. In: Ravines, R., Ed. *Technologia Andina.* Instituto de Estudios Peruanos, Lima.

Utech, F.H. 2002. *Clintonia* and *Medeola*, Pp. 150–153. In: Morin, N. and Editorial Committee, Eds. *Flora of North America.* Vol. 26. *Magnoliophyta: Liliidae.* Oxford University Press, New York.

Vadivel, V. 2001. Diversity in nutritional composition of wild jack bean (*Canavalia ensiformis* L. D.C) seeds collected from south India. *Food Chemistry* 74 (4):507–511.

Vahlensieck, W., Jr., Volp, A., Lubos, W., and Kuntze, M. 1993. Benign prostatic hyperplasia — treatment with sabal fruit extract. A treatment study of 1,334 patients [in German]. *Fortschritte in der Medizin* 111 (18):323–326.

Vahlensieck, W., Jr., Fabricius, P.G., and Hell, U. 1996. Drug therapy of benign prostatic hyperplasia [in German]. *Fortschritte in der Medizin* 114(31):407–411.

Valdez, O.T., Cabrera-Cano, E.F., Linares-Mazari, E., and Bye, R. 1989. *Las Plantas de Cozumel.* Instituto de Biología, Universidad Nacional Autónoma de México, Mexico City.

Van Goor. 1979. *Woordenboek Engels-Nederlands en Nederlands-Engels.* van Goor Zonen, Amsterdam.

van Heerden, F.R., Viljoen, A.M., and van Wyk, B.-E. 2000. The major flavonoid of *Dodonaea angustifolia. Fitoterapia* 71(5):598–600.

Van Valkenburg, J.L.C.H. and Bunyapraphatsara, N., Eds. 2001. *Plant Resources in South-East Asia.* No. 12(2). *Medicinal and Poisonous Plants* 2. Prosea Foundation, Leiden, the Netherlands.

VanderJagt, T.J., Ghattas, R., VanderJagt, D.J., Crossey, M., and Glew, R.H. 2002. Comparison of the total antioxidant content of 30 widely used medicinal plants of New Mexico. *Life Science* 70(9):1035–1040.

Varshney, I.P., Pal, R., and Srivastava, H.C. 1978. Saponins from *Randia dumetorum* Lamk. fruit pulp. *Journal of the Indian Chemical Society* 55(4):397–400.

Varshney, L.P., Jain, D.C., Beg, M.F.A., and Srivastava, H. C. 1985. Samanin-C and Samanin-E new saponins from *Pithecolobium saman. Fitoterapia* 56(5):281–283.

Vásquez Torres, M. and Jácome Castillo, E. 1997. *Flora Medicinal de Veracruz. I. Inventario etnobotánico.* Universidad Veracruzana, Xalapa.

Vaswani, S.K., Chang, B.W., Carey, R.N., and Hamilton, R. G. 1999. Adult onset grape hypersensitivity causing life threatening anaphylaxis. *Annals of Allergy, Asthma, and Immunology* 83(1):25–26.

Vecera, R., Skottova, N., Vana, P., Kazdova, L., Chmela, Z., Svagera, Z., Waltera, D., Ulrichova, J., and Simanek, V. 2003. Antioxidant status, lipoprotein profile and liver lipids in rats fed on high-cholesterol diet containing currant oil rich in *n*-3 and *n*-6 polyunsaturated fatty acids. *Physiological Research* 52 (2):177–187.

Vela, S.M., Souccar, C., Lima-Landman, M.T., and Lapa, A. J. 1997. Inhibition of gastric acid secretion by the aqueous extract and purified extracts of *Stachytarpheta cayennensis. Planta Medica* 63(1):36–69.

Velde, V.V., Lavie, D., Zeinik, R., Matida, A.K., and Panizza, S. 1982. Cordialin A and B, two new triterpenes from *Cordia verbenacea* D.C. *Journal of the Chemical Society. Perkin Transactions I. Organic and Bio-organic Chemistry* 11:2697–2700.

Venkateshwar Rao, G. and Rao, P.S. 1983. Chemical examination of the leaves, stem bark and root bark of *Morinda tinctoria* var. *tomentosa. Journal of the Indian Chemical Society* 60(6):585–586.

Vera, N., Bardon, A., Catalan, C.A.N., Gedris, T.E., and Herz, W. 2001. New coumarins from *Pterocaulon polystachyum. Planta Medica* 67(7):674–677.

Verastegui, M.A., Sanchez, C.A., Heredia, N.L., and Garcia-Alvarado, J.S. 1996. Antimicrobial activity of extracts of three major plants from the Chihuahuan desert. *Journal of Ethnopharmacology* 52 (3):175–177.

Verhoek, S. 1975. A study of the tribe Poliantheae (including Manfreda) and revisions of *Manfreda* and *Prochyanthes* (Agavaceae). Doctoral dissertation, Cornell University, Ithaca, NY. *Dissertation Abstracts International* 36:356.

Verhoek, S. 1977. Additional location and county record for *Manfreda virginica* in North Carolina. *Castanea* 42 (3):255.

Verhoek, S. 1978. Huaco and amole: a survey of the uses of *Manfreda* and *Prochnyanthes. Economic Botany* 32 (2):124–130.

Vermes, B. and Wagner, H. 1980. Synthesis and structure proof of morindone 6-*O*-gentiobioside from *Morinda tinctoria. Phytochemistry* 19(11):2493–2494.

Verotta, L., Orsinim, F., Sbacchi, M., Scheildler, M.A., Amador, T.A., and Elisabetsky, E. 2002. Synthesis and antinociceptive activity of chimonanthines and pyrrolidinoindoline-type alkaloids. *Bioorganic and Medicinal Chemistry* 10(7):2133–2142.

Vessal, M., Rasti, M., and Kooshesh, F. 1996. Modulation of the pituitary and basomedial hypothalamic lysylaminopeptidase activities by beta-estradiol and/or an aqueous extract of *Physalis alkekengi* fruits. *Comparative Biochemistry and Physiology B. Biochemical Molecular Biology* 115(2):267–271.

Vestal, P.A. and Schultes, R.E. 1939. *The Economic Botany of the Kiowa Indians as It Relates to the History of the Tribe.* Botanical Museum Harvard University, Cambridge, MA.

Vickery, R. 1995. *A Dictionary of Plant-Lore.* Oxford University Press, Oxford.

Vignoles, C. [1823] 1977. *Observations upon the Floridas.* Bicentennial Floridiana Facsimile Series reprint. University Press of Florida, Gainesville.

Vijver, L.M. van der and Lotter, A.P. 1971. The constituents in the roots of *Plumbago auriculata* Lam. and *Plumbago zeylancia* Lam. responsible for antibacterial activity. *Planta Medica* 20(1):8–13.

Vilegas, W., Boralle, N., Cabrera, A., Bernardi, A.C., Pozetti, G.L., and Arantes, S.F. 1995. Coumarins and a flavonoid from *Pterocaulon alopecuroides. Phytochemistry* 38(4):1017–1019.

Villarroel, L., Torres, R., Urzua, A., Reina, M., Cabrera, R., and Gonzalez-Coloma, A. 2001. *Heliotropium huascoense* resin exudate: chemical constituents and defensive properties. *Journal of Natural Products* 64 (9):1123–1126.

Vines, R.A. 1977. *Trees of East Texas.* University of Texas Press, Austin.

Voeks, R.A. 1997. Sacred Leaves of Candomblé. African Magic, Medicine, and Religion in Brazil. University of Texas Press, Austin.

Vogel, V.J. 1970. *American Indian Medicine.* University of Oklahoma Press, Norman.

Vohora, S.B., Shah, S.A., Naqvi, S.A., Ahmad, S., and Khan, M.S. 1983. Studies on *Trianthema portulacastrum. Planta Medica* 47(2):106–108.

Voitenko, G.N., Umel', V.B., Tarapon, O., and Zosim, G.L. 1990. Use of stachyrene in combined treatment of patients with obstructive jaundice [in Russian]. *Klin Khir* (11):26–27.

Volz, H.P. 2001. Phytochemicals as means to induce sleep [in German]. *Zeitschrift für Ärztliche Fortbildung und Qualitätssicherung* 95(1):33–34.

von Hutten, U. 1540. *Of the Wood Called Guaiacum.* Translated by Thomas Paynel. Thomas Bertheletregii, London.

Von Reis-Altschul, S. 1973. *Drugs and Foods from Little-Known Plants.* Harvard University Press, Cambridge, MA.

Von Reis-Altschul, S. and Lipp, F.J.J. 1982. *New Plant Sources for Drugs and Foods from the New York Botanical Garden Herbarium.* Harvard University Press, Cambridge, MA.

Vora, R.S. and Labus, Z. 1988. An investigation of Texas ebony seed provenances in the lower Rio Grande

Valley, Texas. *The Texas Journal of Science* 40(4): 452–454.

Vuilleumier, B.S. 1969. The genera of Senecioneae in the southeastern United States. *Journal of the Arnold Arboretum* 50:104–121.

Waage, S.K. and Hedin, P.A. 1985. Biologically active flavonoids in *Heterotheca camporum*. *Journal of the Mississippi Academy of Sciences* 30:55–58.

Waffo-Teguo, P., Hawthorne, M.E., Cuendet, M., Merillon, J.M., Kinghorn, A.D., Pezzuto, J.M., and Mehta, R. G. 2001. Potential cancer-chemopreventive activities of wine stilbenoids and flavans extracted from grape (*Vitis vinifera*) cell cultures. *Nutrition and Cancer* 40 (2):173–179.

Wagner, D.H. 1993. *Polystichum*. Pp. 290–299. In: Editorial Committee, Eds. *Flora of North America 2. Pteridophytes and Gymnosperms.* Oxford University Press, New York.

Wagner, H. 1999. Phytomedicine research in Germany. *Environmental Health Perspective* 107(10):779–781.

Wagner, H. and Flachsbarth, H. 1981. A new antiphlogistic principle from *Sabal serrulata*, I [in German]. *Planta Medica* 41(3):244–251.

Wagncr, H., Iycngar, M.A., and Hörhammcr, L. 1972. Flavonol-3-glucosides in eight *Eupatorium* species. *Phytochemistry* 11:1504–1505.

Wagner, H., Seegert, K., Gupta, M.P., Esposito Avella, E., and Soltis, P. 1986. Cardiotone Wirkstoffe aus *Spigelia anthelmia*. *Planta Medica* 52:376–381.

Wagner, H., Ludwig, C., Grotjahn, L., and Kahn, M.S.Y. 1987. Biologically active saponins from *Dodonea viscosa*. *Phytochemistry* 26:697–701.

Wagner, W.H. and Beitl, J.M. 1993. Lycopodiaceae, Pp. 18–37. In: Morin, N. and Editorial Committee, Eds. *Flora of North America.* Vol. 2. *Pteridophytes and Gymnosperms.* Oxford University Press, New York.

Wagstaff, S.J. and Olmstead, R.G. 1997. Phylogeny of Labiatae and Verbenaceae inferred from rbcL sequences. *Systematic Botany* 22:165–179.

Waheenee Goodbird, E. and Wilson, G.L. 1987. *Agriculture of the Hidatsa Indians. Buffalo Bird Woman's Garden.* Minnesota Historical Society Press, St. Paul.

Walker, E.H. 1976. *Flora of Okinawa and the Southern Ryukyu Islands.* Smithsonian Institution Press, Washington, D.C.

Walker, W. 1981. Cherokee curing and conjuring, identity, and the Southeastern co-tradition. Pp. 86–105. In: Castille, G.P. and Kushner, G. *Persistent Peoples: Cultural Enclaves in Perspective.* University of Arizona Press, Tucson.

Walters, T.W. and Decker-Walters, D.S. 1993. Systematics of the endangered Okeechobee gourd (*Cucurbita okeechobeensis*: Cucurbitaceae). *Systematic Botany* 18 (2):175–187.

Wang, J.N., Hano, Y., Nomura, T., and Chen, Y.J. 2000. Procyanidins from the seeds of *Vitis amurensis*. *Phytochemistry* 53(8):1097–1102.

Wang, M., Kikuzaki, H., Jin, Y., Nakatani, N., Zhu, N., Csiszar, K.C., Rosen, R.T., Ghai, G., and Ho, C.T. 2000. Novel glycosides from noni (*Morinda citrifolia*). *Journal of Natural Products* 63(8):1182–1183.

Wang, M.T., Guan, X.T., Han, X.W., and Hong, S.H. 1992. A new triterpenoid saponin from *Ardisia crenata*. *Planta Medica* 58(2):205–207.

Wani, M.C., Taylor, H.L., and Wall, M.E. 1973. Plant antitumour agents: colubrinol acetate and colubrinol, antileukaemic ansa macrolides from *Colubrina texensis*. *Chemical Communications* (Cambridge, U.K.) 12:390.

Ward, B.J. 1999. A Contemplation upon Flowers. Garden Plants in Myth and Literature. Timber Press, Portland, OR.

Ward, D.B., Ed. 1979. *Rare and Endangered Biota of Florida,* Vol. 5, *Plants.* University Presses of Florida, Gainesville.

Warthen, D., Gooden, E.L., and Jacobson, M. 1969. Tumor inhibitors: liriodenine, a cytotoxic alkaloid from *Annona glabra*. *Journal of Pharmaceutical Sciences* 58(5):637–638.

Warwick, S.I. and Aiken, S.G. 1986. Electrophoretic evidence for the recognition of two species in annual wild rice (*Zizania*, Poaceae). *Systematic Botany* 11 (3):464–473.

Waterfall, U.T. 1967. *Physalis* in Mexico, Central America, and the West Indies. *Rhodora* 69:82–120, 203–239, 319–329.

Watson, F.D. 1993. *Taxodium.* Pp. 403–404. In: Morin, N. and Editorial Committee, Eds. *Flora of North America.* Vol. 2. *Pteridophytes and Gymnosperms.* Oxford University Press, New York.

Watson, L. and Dallwitz, M.J. 1992. *The Grass Genera of the World.* CAB International, London.

Watson, L. and Dallwitz, M.J. 2000. The Families of Flowering Plants: Descriptions, Illustrations, Identification, and Information Retrieval. Version: 14 December 2000. URL: biodiversity.uno.edu/delta/.

Watt, G. [1889] 1972. *A Dictionary of the Economic Products of India.* Reprint by Bishen Singh Mahendra Pal Singh, Dehra Dun, India.

Webb, W.S. and Funkhouser, W.D. 1936. Rock shelters in Menifee County, Kentucky. *Publications of the University of Kentucky Department of Anthropology and Archaeology* 3(4):105–167.

Weber, W.A. 1987. *Colorado Flora: Western Slope.* Colorado Associated University Press, Boulder.

Webster, G.L. 1955. Studies of the Euphorbiaceae, Phyllanthoideae I. Taxonomic notes on the West Indian species of *Phyllanthus. Contributions to the Gray Herbarium of Harvard University* 176:53–63.

Webster, G.L. 1956. Studies of the Euphorbiaceae, Phyllanthoideae II. The American species of *Pyllanthus* described by Linnaeus. *Journal of the Arnold Arboretum* 37:1–14.

Webster, G.L. 1970. A revision of *Phyllanthus* (Euphorbiaceae) in the continental United States. *Brittonia* 22:44–76.

Webster, G.L. 1983. A botanical Gordian knot: the case of *Ateramnus* and *Gymnanthes* (Euphorbiaceae). *Taxon* 32(2):304–305.

Webster, G.L. 2002. A synopsis of the Brazilian taxa of *Phyllanthus* section *Phyllanthus* (Euphorbiaceae). *Lundellia* 5:1–26.

Wedge, D.E., Camper, N.D. and Tainter, F.H. 1995. Establishment of a tissue culture protocol for pharmacologically important Cornaceae. In: Gustine, D.L. and H.E. Flores, Eds. Phytochemicals and Health: Proceedings, Tenth Annual Penn State Symposium in Plant Physiology, May 18–20. *Plant Physiology* 15:324–325.

Weekley, C. 2000. The natural history of *Tanaostigmodes pithecellobiae* (Hymenoptera: Tanaostigmatidae), a gall-maker on blackbead (*Pithecellobium keyense*). *Florida Entomologist* 83 (1):31–41.

Weiner, M.A. 1971. Ethnomedicine in Tonga. *Economic Botany* 25:423–450.

Weiss, P. 1999. Tropical fruits linked to parkinsonism. *Science News* 156:69.

Weisser, H., Tunn, S., Behnke, B., and Krieg, M. 1996. Effects of the *Sabal serrulata* extract IDS 89 and its subfractions on 5 alpha-reductase activity in human benign prostatic hyperplasia. *Prostate* 28(5):300–306.

Weisser, H., Behnke, B., Helpap, B., Bach, D., and Krieg, M. 1997. Enzyme activities in tissue of human benign prostatic hyperplasia after three months' treatment with the *Sabal serrulata* extract IDS 89 (Strogen) or placebo. *European Urology* 31(1):97–101.

Welsh, S.L., Atwood, N.D., Higgins, L.C., and Goodrich, D. 1987. *A Utah Flora.* Brigham Young University, Provo.

Wen, G.Y., Yu, F.L., Wang, H.T., Kang, Z.Q., Jiang, K., and Bian, Y.M. 1983. Chemical constituents of the essential oils from flower of *Osmanthus fragrans* (gui hua) [in Chinese]. *Acta Botanica Sinica* 25 (5):468–471.

Wen, G.Y., Mato, A., Wisniewski, H.M., Malik, M.N., Jenkins, E.C., Sheikh, A.M., and Kim, K.S. 1995. Light and electron microscopic immunocytochemical localization of two major proteins in garlic bulb. *Journal of Cellular Biochemistry* 58(4):481–489.

Wendel, J.F. and Albert, V.A. 1992. Phylogenetics of the cotton genus (*Gossypium*): Character-state weighted parsimony analysis of chloroplast-DNA restriction site data and its systematic and biogeographic implications. *Systematic Botany* 17:115–143.

Wendel, J.F., Brubaker, C.L., and Percival, A.E. 1992. Genetic diversity in *Gossypium hirsutum* and the origin of upland cotton. *American Journal of Botany* 79:1291–1310.

West, J.G. 1984. A revision of *Dodonea* Miller (Sapindaceae) in Australia. *Brunonia* 7:1–194.

West, P. 1998. *The Enduring Seminoles.* University Press of Florida, Gainesville.

West, P. and Jumper, B.M. 2001. *A Seminole Legend.* University Press of Florida, Gainesville.

Westerkamp, C. and Paul, H. 1993. *Apios americana*, a fly-pollinated papilionaceous flower. *Plant Systematics and Evolution* 187(1/4):135–144.

Wherry, E.T. 1964. *The Southern Fern Guide.* Doubleday, Garden City, NY.

Whistler, W.A. 1991. *The Ethnobotany of Tonga: The Plants, Their Tongan Names, and Their Uses.* Bishop Museum Press, Honolulu.

White, L.A. 1941. Further data on the cultivation of tobacco by Pueblo Indians. *Science* 96(2481):59, 60.

White, L.A. 1942. *Nicotiana rustica* cultivated by Pueblo Indians. *Science* 94(2429):64, 65.

Whitelock, L.M. 2002. *The Cycads.* Timber Press, Portland, OR.

Whitford, A.C. 1941. Textile fibers used in eastern aboriginal North America. *Anthropological Papers of the American Museum of Natural History* 38(1):5–22.

Whitford, A.C. 1943. Fiber plants of the North American aborigines. *Journal of the New York Botanical Garden* 44(518):25–34.

Whittaker, H. 1948. Amellin for diabetes. *British Medical Journal* 1:546–547.

Wickens, G. 2000. Defining economic botany and ethnobotany. *Economic Botany* 54(3):251.

Widrlechner, M.P. 1980. History and utilization of *Rosa damascena. Economic Botany* 35(1):42–58.

Wiersema, J.W. and Hellquist, C.B. 1994. Nomenclatural changes in Nymphaeaceae for the Flora of North America. *Rhodora* 96:170–178.

Wiersema, J.W. and Hellquist, C.B. 1997. Nymphaeaceae. Pp. 66–77. In: Editorial Committee, Eds. *Flora of North America.* Vol. 3. Oxford University Press, New York.

Wiersema, J.W. and León, B. 1999. *World Economic Plants. A Standard Reference.* CRC Press, Boca Raton, FL.

Wiggins, I.L. 1980. *Flora of Baja California.* Stanford University Press, Stanford, CA.

Wilbur, R.L. 1955. A revision of the North American genus *Sabatia* (Gentianaceae). *Rhodora* 57:1–33, 43–71, 78–104.

Wilbur, R.L. 1975. A revision of the North American genus *Amorpha* (Leguminosae: Psoraleae). *Rhodora* 77: 337–409.

Wilbur, R.L. 1990. Identification of the plants illustrated and described in Catesby's Natural History of the Carolinas, Florida, and the Bahamas. *Sida* 14 (1):29–48.

Wilde, M.I. and Goa, K.L. 1999. Finasteride: an update of its use in the management of symptomatic benign prostatic hyperplasia. *Drugs* 57(4):557–581.

Wildman, S.G., McMullan, A.V., and Griggs, R. 1943. Isolation of an active substance from *Calonyction aculeatum* capable of coagulating *Castilla* latex. *Science* 97(2525):471, 472.

Will, L.E. 1964a. *Cracker History of Okeechobee.* Great Outdoors Press, St. Petersburg.

Will, L.E. 1964b. Custard apple. Pp. 34–37. In: *Cracker History of Okeechobee.* Great Outdoors Press, St. Petersburg.

Willaman, J.J. and Li, H.L. 1970. Alkaloid-bearing plants and their contained alkaloids 1957–1968. *Lloydia* 33 (3A, Suppl.):1–286.

Willaman, J.J. and Schubert, B.G. 1961. Alkaloid-Bearing Plants and Their Contained Alkaloids. Technical Bulletin 1234, U.S. Department of Agriculture, Agricultural Research Service, Washington, D.C.

Williams, J.L. [1837] 1962. *The Territory of Florida.* Floridana Facsimile and Reprint Series. University Press of Florida, Gainesville.

Williams, J.L. 1827. *A View of West Florida, Embracing Its Geography, Topography, etc. ...* Tanner, Philadelphia.

Williams, J.J. 2002. Psychic phenomena of Jamaica. URL: www.sacred-texts.com/afr/ppj/.

Williams, L.A. 1999. *Rhizophora mangle* (Rhizophoraceae) triterpenoids with insecticidal activity. *Naturwissenschaften* 86(9):450–452.

Williams, M.C. and Molyneux, R.J. 1987. Occurrence, concentration, and toxicity of pyrrolizidine alkaloids in *Crotalaria* seeds. *Weed Science* 35(4):476–481.

Williams, M.L. 1993. *Toumeyella lignumvitae*, a new species of scale insect from the Florida Keys (Homoptera: Coccidae). *The Florida Entomologist* 76(4):566–572.

Williams, S.E., Albert, V.A., and Chase, M.W. 1994. Relationships of Droseraceae: a cladistic analysis of rbcL sequence and morphological data. *American Journal of Botany* 81:1027–1037.

Williamson, R.A. 1987. *Living the Sky. The Cosmos of the American Indian.* University of Oklahoma Press, Norman.

Willoughby, C.C. 1907. The Virginia Indians of the seventeenth century. *American Anthropologist* 9(1):57–86.

Wilson, H.D. 1981. Domesticated *Chenopodium* of the Ozark Bluff Dwellers. *Economic Botany* 35:233–239.

Wilson, H.D. and Heiser, C.B. 1979. The origin and evolutionary relationships of "Huauzontle" (*Chenopodium nuttalliae* Safford), domesticated chenopod of Mexico. *American Journal of Botany* 66:198–206.

Wilson, P.W., Pichardo, F.J., Liuzzo, J.A., Blackmon, W.J., and Reynolds, B.D. 1987. Amino acids in the American groundnut (*Apios americana*). *Journal of Food-Science* 52:224–225.

Wilson, R.D., Witzel, D.A., and Verlander, J.M. 1982. Somatosensory-evoked response of ataxic Angora goats in suspected haloxon-delayed neurotoxicity. *American Journal of Veterinary Research* 43(12): 2224–2226.

Winter, J.C. 2001. *Tobacco Use by Native North Americans: Sacred Smoke and Silent Killer.* Civilization of the American Indian series. Vol. 236. University of Oklahoma Press, Norman.

Witherup, K.M., Bogusky, M.J., Anderson, P.S., Ramjit, H., Ransom, R.W., Wood, T., and Sardana, M. 1994. Protein cyclopsychotride A, a biologically active, 31-residue cyclic peptide isolated from *Psychotria longipes. Journal of Natural Products* 57(12):1619–1625.

Wollenweber, E. and Roitman, J.N. 1996. A novel methyl ether of quercetagetin from *Chromolaena odorata* leaf exudate. *Biochemical Systematics and Ecology* 24(5):479–480.

Wollenweber, E., Schober, I., Clark, W.D., and Yatskievych, G. 1985. Flavonoid aglycones from leaf resins of two species of *Heterotheca* (Compositae). *Phytochemistry* 24(9):2129–2131.

Wollenweber, E., Dorr, M., and Muniappan, R. 1995. Exudate flavonoids in a tropical weed, *Chromolaena odorata* (L.) R.M. King et H. Robinson. *Biochemical Systematics and Ecology* 23(7/8):873–874.

Wolverton, B.C. 1982. Hybrid wastewater treatment system using anaerobic microorganisms and reed (*Phragmites communis*). *Economic Botany* 36:373–380.

Wong, W. 1976. Some folk medicinal plants from Trinidad. *Economic Botany* 30:103–142.

Woo, M.H., Zeng, L., and McLaughlin, J.L. 1995. 16,19-*cis*-Murisolin and murisolin A, two novel bioactive mono-tetrahydrofuran annonaceous acetogenins from *Asimina triloba* seeds. *Bioorganic and Medicinal Chemistry Letters* 5(11):1135.

Wood, R. and Peterson, S. 1999. Lipids of the pawpaw fruit: *Asimina triloba. Lipids* 34(10):1099.

Wooton, E.O. and Standley, P.C. 1915. *Flora of New Mexico.* Contributions from the United States National Herbarium, Vol. 19. Washington, D.C.

Workman, R.W. 1980. *Growing Native. Native Plants for Landscape Use in Coastal South Florida.* Sanibel-Captiva Conservation Foundation, Sanibel.

Wright, C.W., O'Neill, M.J., Phillipson, J.D., and Warhurst, D.C. 1988. Use of microdilution to assess *in vitro* antiamoebic activities of *Brucea javanica* fruits, *Simarouba amara* stem, and a number of quassinoids. *Antimicrobial Agents and Chemotherapy* 32 (11):1725–1729.

Wright, T. 1884. *Anglo-Saxon and Old English Vocabularies.* 2nd ed. Edited by R.P. Wülcker. Trübner, London.

Wu, Z.-y. and Raven, P.H., Eds. 1998. *Flora of China.* Vol. 18. *Scrophulariaceae through Gesneriaceae.* Science Press and Missouri Botanical Garden Press, St. Louis.

Wunderlin, R.P. 1998. *Guide to the Vascular Plants of Florida.* University Press of Florida, Gainesville.

Wunderlin, R.P. and Hansen. B.F. 2002. Atlas of Florida Vascular Plants (www.plantatlas.usf.edu/). [S.M. Landry and K.N. Campbell (application development), Florida Center for Community Design and Research.] Institute for Systematic Botany, University of South Florida, Tampa.

Wunderlin, R.P. and Hansen, B.F. 2003. *Guide to the Vascular Plants of Florida.* 2nd ed. University Press of Florida, Gainesville.

www.est.uniandes.edu.co/ ~in-sanch/Ingrid4.html.

www.humboldt.org.co/conservacionListas_Preliminares.htu.

Xu, L.X. and Xu, Y.Z. 1986. Coulometric titration of the total flavonoids and hyperin in *Saururus chinensis* (Lour.) Ball [in Chinese]. *Yao Xue Xue Bao* 21(4): 306–309.

Xu, L.X. and Zhou, T.H. 1982. Colorimetric method for the determination of hecogenin in *Agave americana* L. residue [in Chinese]. *Yao Xue Xue Bao* 17(8):609–614.

Yakovlev, A.I. 1985. Polysaccharides of the inflorescences of Tilia cordata. *Chemistry of Natural Compounds* 21

(1):114–115. Translated from *Khimiia prirodnykh soedinenii*, pp. 116–117.
Yamaguchi, H. and Nishimoto, K. 1965. Studies on the alkaloids of the root of *Physalis alkekengi.* (1). Isolation of 3-alpha-tigloyloxytropane. *Chemical and Pharmaceutical Bulletin* (Tokyo) 13(2):217–220.
Yamakoshi, J., Saito, M., Kataoka, S., and Tokutake, S. 2002. Procyanidin-rich extract from grape seeds prevents cataract formation in hereditary cataractous (ICR/f) rats. *Journal of Agriculture and Food Chemistry* 50(17):4983–4988.
Yang, R.Z. and Tang, C.S. 1988. Plants used for pest control in China: a literature review. *Economic Botany* 42:376–406.
Yang, T.H., Chen, C.M., and Kuan, S.S. 1971. Studies on the alkaloids of *Annona glabra* L. I. The isolation of (−)-*N*-methylactinodaphnine. *Chinese Chemical Society Journal* (ser. 2) 18(3):133–136.
Yang, Z.Q., Zhang, M.Y., Liu, J.J., Hu, Z.J., Zhu, B.L., Liu, Y.W., Wang, G.Z., Wan, M., and Wu, X.L. 1989. Extraction of effective parts of *Alternanthera philoxeroides* (Mart.) Griseb. and its antiviral effect [in Chinese]. *Zhongguo Zhong Yao Za Zhi* 14(8):488–90, 511–512.
Yanovsky, E. 1936. Food Plants of the North American Indians. U.S. Department of Agriculture Miscellaneous Publications 237. U.S. Government Printing Office, Washington, D.C.
Yarnell, R.A. 1964. *Aboriginal Relationships between Cultures and Plant Life in the Upper Great Lakes Region.* Anthropology Papers of the Museum of Anthropology, University of Michigan No. 23, Ann Arbor.
Yarnell, R.A. 1978. Domestication of sunflower and sumpweed in eastern North America. Pp. 289–299. In: Ford, R.I. et al., Eds. *The Nature and Status of Ethnobotany.* University of Michigan Museum of Anthropology, Ann Arbor.
Yarnell, R.A. 1986. A survey of prehistoric crop plants in eastern North America. *Missouri Archaeologist* 47: 47–59.
Yassin, W.I.E.-E.-N.A. 2001. Hypolipidemic effect of aqueous extract from dried leaves of *Morus alba. Journal of Egyptian German Society of Zoology* 36(A):143–154.
Yazaki, K. and Okuda, T. 1989. Gallotannin production in cell cultures of *Cornus officinalis* Sieb. et Zucc. *Plant Cell Reports* 8(6):346–349.
Yen, G.C., Wu, S.C., and Duh, P.D. 1996. Extraction and identification of antioxidant components from the leaves of mulberry (*Morus alba* L.). *Journal of Agricultural and Food Chemistry* 44(7):1687–1690.
Yetman, D. 2002. *The Guarjíos of the Sierra Madre. Hidden People of Northwestern Mexico.* University of New Mexico, Albuquerque.
Yetman, D. and Van Devender, T.R. 2001. *Mayo Ethnobotany. Land, History, and Traditional Knowledge in Northwest Mexico.* University of California Press, Berkeley.
Yi, Y., Cao, Z., Yang, D., Cao, Y., Wu, Y., and Zhao, S. 1998. Studies on the chemical constituents of *Smilax glabra* [in Chinese]. *Yao Xue Xue Bao* 33(11):873–875.
Yildirim, A., Mavi, A., Oktay, M., Kara, A.A., Algur, O.F., and Bilaloglu, V. 2000. Comparison of antioxidant and antimicrobial activities of tilia (*Tilia argentea* Desf ex D.C), sage (*Salvia triloba* L.), and black tea (*Camellia sinensis*) extracts. *Journal of Agricultural and Food Chemistry* 48(10):5030–5034.
Yokosuka, A., Mimaki, Y., Kuroda, M., and Sashida, Y. 2000. A new steroidal saponin from the leaves of *Agave americana. Planta Medica* 66(4):393–396.
Yoon, S.H., Kim, Y.S., Ghim, S.Y., Song, B.H., and Bae, Y.S. 2002. Inhibition of protein kinase CKII activity by resveratrol, a natural compound in red wine and grapes. *Life Science* 71(18):2145–2152.
Yoshikawa, K., Katsuta, S., Mizumori, J., and Arihara, S. 2000a. Four cycloartane triterpenoids and six related saponins from *Passiflora edulis. Journal of Natural Products* 63(9):1229–1234.
Yoshikawa, K., Katsuta, S., Mizumori, J., and Arihara, S. 2000b. New cycloartane triterpenoids from *Passiflora edulis. Journal of Natural Products* 63(10): 1377–1380.
Yoshikawa, K., Suzaki, Y., Tanaka, M., Arihara, S., and Nigam, S.K. 1997. Three acylated saponins and a related compound from *Pithecellobium dulce. Journal of Natural Products* 60(12):1269–1274.
Young, J.V. 1990. *Kokopelli. Casanova of the Cliff Dwellers.* Filter Press, Palmer Lake, CO.
Young, M. and Siebert, F.T., Jr. 2003. Catawba onomastics. Pp. 271–273. In: Rudes, B.A. and Costa, D.J., Eds. 2003. *Essays in Algonquian, Catawban and Siouan Linguistics in Memory of Frank T. Siebert, Jr.* Memoir Algonquian and Iroquoian Linguistics 16. Algonquian and Iroquoian Linguistics, Winnipeg.
Younos, C., Rolland, A., Fleurentin, J., Lanhers, M.C., Misslin, R., and Mortier, F. 1990. Analgesic and behavioural effects of *Morinda citrifolia. Planta Medica* 56(5):430–434.
Yronwode, C. 1999. Lucky Mojo Curio Co. http://www.luckymojo.com/luckymojocatalogue.html.
Yusuf, U.K. 1994. Flavonoid glycosides in the fern *Blechnum orientale* Linn. *American Fern Journal* 84(2):69–70.
Zani, C.L., Alves, T.M.A., Oliveira, A.B. de, Murta, S.M.F., Ceravolo, I.P., and Romanha, A.J. 1994. Trypanocidal components of *Pluchea quitoc* L. *Phytotherapy Research* 8(6):375–377.
Zanobini, A., Firenzuoli, A.M., and Bianchi, A. 1974. Isolation and determination of vitamin D in avocado (*Persea gratissima*) [in Italian]. *Bollettino Societá Biologico Sperimentale* 50(12):887–891.
Zardini, E. 1992. *Madia sativa* Mol. (Asteraceae-Heliantheae-Madiinae): an ethnobotanical and geographical disjunct. *Economic Botany* 46:34–44.
Zavala, M.A., Pérez, S., Pérez, C., Vargas, R., and Pérez, R. M. 1988. Antidiarrhoeal activity of *Waltheria americana, Commelina coelestis* and *Alternanthera repens. Journal of Ethnopharmacology* 61(1):41–47.

Zdero, C. and Bohlmann, F. 1989. Furoeudesmanes and other constituents from representatives of the *Pluchea* group. *Phytochemistry* 28(11):3097–3100.

Zeisberger, D. and Horsford, E.N. 1887. *Zeisberger's Indian Dictionary. English, German, Iroquois — the Onondaga, and Algonquin — the Delaware.* John Wilson & Sons, Cambridge, MA.

Zennie, T.M. and Ogzewalla, C.D. 1977. Ascorbic acid and vitamin A content of edible wild plants of Ohio and Kentucky. *Economic Botany* 31:76–79.

Zhang, S.M., He, Y.S., Tabba, H.D., and Smith, K.M. 1988. Inhibitor against the human immunodeficiency virus in aqueous extracts of *Alternanthera philoxeroides. Chinese Medical Journal (English)* 101(11):861–866.

Zhou, B.N., Blasko, G., and Cordell, G.A. 1988. Alternanthin, a C-glycosylated flavonoid from *Alternanthera philoxeroides. Phytochemistry* 27(11):3633–3636.

Zhou, L.X. and Ding, Y. 2002. A cinnamide derivative from *Solanum verbascifolium* L. *Journal of Asian Natural Products Research* 4(3):185–187.

Zhou, Z.Q. and Li, Y.N. 2000. The RAPD evidence for the phylogenetic relationship of the closely related species of cultivated apple. *Genetic Resources and Crop Evolution* 47(4):353–357.

Zohary, M. 1982. *Plants of the Bible.* Cambridge University Press, Cambridge, U.K.

Zona, S. 1990. A monograph of *Sabal* (Aracaceae: Coryphoideae). *Aliso* 12(4):583–666.

Zona, S. 1996. *Roystonea* (Arecaceae: Arecoideae). *Flora Neotropica* 71:1–36.

Zona, S. 2002. A revision of *Pseudophoenix. PALMS* 46:19–38.

Index

A

B

C

D

E

F

G

H

I

J

K

M

O

P

Q

R

S

T

W

X

Y

Z